Flora of North America

Flora of North America

North of Mexico

Edited by FLORA OF NORTH AMERICA EDITORIAL COMMITTEE

VOLUME 24

Magnoliophyta: *Commelinidae* (in part): *Poaceae*, part 1

Edited by Mary E. Barkworth, Kathleen M. Capels, Sandy Long, Laurel K. Anderton, and Michael B. Piep

Illustrated by Cindy Talbot Roché, Linda Ann Vorobik, Sandy Long, Annaliese Miller, Bee F. Gunn, and Christine Roberts

NEW YORK OXFORD · OXFORD UNIVERSITY PRESS · 2007

Oxford University Press, Inc., publishes works that further
Oxford University's objective of excellence
in research, scholarship, and education.

Oxford   New York
Auckland   Cape Town   Dar es Salaam   Hong Kong   Karachi
Kuala Lumpur   Madrid   Melbourne   Mexico City   Nairobi   New Delhi
Shanghai   Taipei   Toronto

Copyright ©2007 by Utah State University
The account of *Avena* is reproduced by permission of Bernard R. Baum for the
Department of Agriculture and Agri-Food, Government of Canada, ©Minister of Public Works and Government Services, Canada, 2007.
The accounts of *Arctophila, Dupontia, Schizachne, Vahlodea*, ×*Arctodupontia*, and ×*Dupoa* are reproduced
by permission of Jacques Cayouette and Stephen J. Darbyshire for the
Department of Agriculture and Agri-Food, Government of Canada, ©Minister of Public Works and Government Services, Canada, 2007.
The accounts of *Eremopoa, Leucopoa, Schedonorus*, and ×*Pucciphippsia* are reproduced by permission of Stephen J. Darbyshire for the
Department of Agriculture and Agri-Food, Government of Canada, ©Minister of Public Works and Government Services, Canada, 2007.

Published by Oxford University Press, Inc.
198 Madison Avenue, New York, New York 10016
www.oup.com

Oxford is a registered trademark of Oxford University Press

Library of Congress Cataloging-in-Publication Data
(Revised for volume 24)
Flora of North America north of Mexico
edited by Flora of North America Editorial Committee.
Includes bibliographical references and indices.
Contents: v.1. Introduction—v. 2. Pteridophytes and gymnosperms—
v. 3. Magnoliophyta: Magnoliidae and Hamamelidae—
v. 22. Magnoliophyta: Alismatidae, Arecidae, Commelinidae (in part), and Zingiberidae—
v. 26. Magnoliophyta: Liliidae: Liliales and Orchidales—
v. 23. Magnoliophyta: Commelinidae (in part): Cyperaceae—
v. 25. Magnoliophyta: Commelinidae (in part): Poaceae, part 2—
v. 4. Magnoliophyta: Caryophyllidae (in part): part 1—
v. 5. Magnoliophyta: Caryophyllidae (in part): part 2—
v. 19 Magnoliophyta: Asteridae (in part): Asteraceae, part 1—
v. 20 Magnoliophyta: Asteridae (in part): Asteraceae, part 2—
v. 21 Magnoliophyta: Asteridae (in part): Asteraceae, part 3—
v. 24. Magnoliophyta: Commelinidae (in part): Poaceae, part 1

ISBN 978-0-19-531071-9 (v. 24)
1. Botany —North America.
2. Botany—United States.
3. Botany—Canada.
4. Botany—Grasses.
1. Mary E. Barkworth, Kathleen M. Capels, Sandy Long, Laurel K. Anderton, and Michael B. Piep.
QK110.F55 2003 581.97 92-30459

9 8 7 6 5 4 3 2 1
Printed in the United States of America
on acid-free paper

Contents

Dedication vi

Historical Introduction vii

Acknowledgments ix

  Authors ix

  Data Contributors xi

    Individuals xii

    Herbaria xii

  Reviewers xiv

  Specimen Loans xvi

  Nomenclatural Consultants xvii

  Financial Supporters xviii

Introduction xix

  Taxonomic Treatment xix

  Content and Layout xx

  Nomenclature xxi

  Distribution Maps xxii

  Bibliographic Indices xxii

  Numerical Summary xxiii

    Native Species xxiii

    Introduced Species xxiii

  Copyright and Citation xxvi

  Volume Preparation xxvi

  Other xxvii

  Illustrators xxvii

Flora of North America Association Board of Directors xxviii

COMMELINIDAE

Poaceae (part 1) 3

  Pharoideae 11

    Phareae 11

  Bambusoideae 14

    Bambuseae 15

    Olyreae 29

  Ehrhartoideae 32

    Ehrharteae 33

    Oryzeae 36

  Poöideae 57

    Brachyelytreae 59

    Nardeae 62

    Diarrheneae 64

    Meliceae 67

    Stipeae 109

    Brachypodieae 187

    Bromeae 192

    Triticeae 238

    Poeae 378

Volume 25: Additions, Corrections, and Comments 790

Geographic Bibliography 794

General Bibliography 810

Names and Synonyms 831

Nomenclatural Index 876

Numerical Listing of the Family, Subfamilies, Tribes, and Genera, Volume 24 910

Numerical Listing of the Subfamilies, Tribes, and Genera, Volume 25 911

To the two giants on whose shoulders we stand

Albert Spear Hitchcock
1865–1935

Mary Agnes Chase
1869–1963

Historical Introduction

The Manual of Grasses of the United States first appeared as "U.S.D.A. Miscellaneous Publication 200" in May of 1935 and cost $1.75. Its first printing was exhausted within two months, and by 1938 it was in its eighth printing. The path that led to its appearance is straightforward. When Albert Spear Hitchcock arrived in Washington, D.C. in 1901 from Manhattan, Kansas, to work as assistant chief of the Division of Agrostology, U.S. Department of Agriculture, he found a magnificent herbarium, especially rich in grasses, that had been built up over the years, most notably by George Vasey and Frank Lamson-Scribner, Chief of the Division. Carl Piper curated the herbarium at the time. Hitchcock traveled throughout the southeastern states and from Colorado and Wyoming to the Pacific coast, focusing "officially" on farming conditions and forage crops and "unofficially" on taxonomically significant information on grasses. Part of his work on the control of sand dunes took him to the dune regions of Europe. Although economic agrostology occupied much of his time, Hitchcock managed to pursue grass taxonomy in both his spare moments and leisure hours. In 1905 he and Piper exchanged billets to their mutual satisfaction.

Two years before that, Mary Agnes Chase had joined Hitchcock as an illustrator with high recommendations from Charles Frederick Millspaugh of The Field Museum of Chicago. After recognizing Chase's keen eye for discriminating species, Hitchcock dropped the role of instructor and treated her as a collaborator in the Division's studies. During three decades of research, including exploration of most areas of the nation, both Hitchcock and Chase published monographs on North American species of *Panicum* (1910 and 1915), while Chase alone completed one on *Paspalum* (1929), all of which led to the completion of the *Manual*.

Hitchcock and Chase were children of the tallgrass prairie. Although born in Owosso, Michigan, in 1865, Hitchcock grew up in Nebraska and Kansas. At Iowa State Agricultural College, Professors Charles Bessey, Herbert Osborn, and B.D. Halstead taught him chemistry and botany. Hitchcock himself became an engaging instructor. Chase (*née* Merrill) was born on a farm in Iroquois County, Illinois, but grew up in Chicago, her widowed mother moving there shortly after the fire of 1871. She finished only grammar school, but learned plants in the fields in and around Chicago with Ellsworth Jerome Hill and later with Millspaugh at The Field Museum. Her career was typical of women scientists of her day—mentored by a male professor and advancing only as far as governmental or university regulations would allow, in her case, that of Senior Botanist. Despite the constraints, she succeeded in becoming the acknowledged "Dean of Agrostologists" before her death in 1963 at the age of 94.

Hitchcock should be remembered as both a botanical explorer and taxonomist, these two careers beginning while he worked at the Missouri Botanical Garden in 1890. He championed the now taken-for-granted typification process for publishing the description of new species, especially at the 1935 International Botanical Congress in Amsterdam. On the return voyage from that Congress, Hitchcock suffered a heart attack and died on 16 December 1935, aboard the *S.S. City of Norfolk*. Both he and Chase had spent many months determining and locating type specimens in European herbaria during the years preceding the production of the *Manual*. The documentation in that and other taxonomic works of this dymanic pair reflects painstaking hours sequestered in dark and sometimes dingy herbaria.

Neither should we forget the humanitarian aid they extended to European colleagues who had been devastated by the economic collapse in Europe that followed the Great War. They sent replacement sets of grass specimens to many in Europe, regardless of their political alliances, along with monographs and basic floristic works, mostly at their own expense. Joined by many colleagues in America, they also forwarded money for food and clothing to some of their indigent botanical friends. Chase continued this aid during and after World War II, relying on friends and colleagues to supplement her own meager resources as a retired Research Associate at the Smithsonian, where the grass herbarium had relocated in 1912.

Although Chase never botanized in Africa or Asia as Hitchcock had, she did make two extended trips to Brazil and one to Venezuela, enriching our knowledge of the grass species of South America and adding thousands of specimens to the grass herbarium in Washington. Anyone who has ever used the grass herbarium at the Smithsonian, especially for its type collection, owes much to Hitchcock and Chase.

By about 1939, sales had exhausted supplies of the *Manual*. The Department of Agriculture would not authorize a ninth printing, but it did sanction a revision, which Jason R. Swallen prepared in the next two years. However, World War II pre-empted any further work. After the war, the Botanical Society of America proposed that the Secretary of Agriculture have the revision brought up to date and published for the many agricultural colleges that relied upon it for a basic agrostology text. Swallen was then working for the U.S.D.A. research division in Beltsville, Maryland, and so it fell to Chase, acting Custodian of Grasses at the Smithsonian—without pay, as she stressed—to carry out the project. She sent her additional illustrations, with the revised manuscript, to the publications office in October 1947. After she learned that the engravings for the 1,696 illustrations in the 1935 *Manual* had been taken for war metal, Chase was thankful for her cache of the original drawings that she had wisely kept in the grass herbarium office. After sending these along, Chase felt that the new *Manual* would be out soon.

Not so. Congress slashed Agriculture's funds that year, so the work returned to the author. While awaiting a new Congressional allocation, Chase began adding new information on the distribution maps, until finally—after a few years—the new edition's proofs appeared on her desk. After many weeks proofreading with help from friends, including Egbert Walker, she returned the page proofs to the printing office. To keep expenses down, the publications officer did not send her the revised proofs, so many errors slipped into the 1951 *Manual*. During this time, Richard W. Pohl had encouraged Chase with the suggestion that, if the government refused to print the work, he might get Iowa State University to do so.

Frank Lamson-Scribner, Hitchcock's former chief, who had begun studying grasses in 1866, wrote to him on 15 May 1935, to thank him for the copy of the *Manual* that Hitchcock had sent him and congratulated the author on the work, calling it "a fitting monument commemorating your achievements in the field of Agrostology." It was one of many, both for Hitchcock and for Chase. What these two people accomplished in their lifetimes, with only typewriters, index cards, and precious little outside money, is astounding! It has taken a cadre of their successors to bring these two volumes to birth. It gives one pause to admire the accomplishments of these two pioneer agrostologists of the United States, Albert Spear Hitchcock and Mary Agnes Chase.

Acknowledgments

Like all volumes in the *Flora of North America* series, this volume has benefited from the assistance of numerous individuals and institutions. We extend our profound thanks to all who have helped us in completing this, the second of the two volumes on grasses in the series.

Authors

Preparing a taxonomic treatment is a demanding task, requiring synthesis of information from diverse sources, including descriptions written by early explorers, field observations, crossing experiments, studies of breeding relationships, and molecular studies of phylogenetic relationships and development. Tentative conclusions, best thought of as hypotheses concerning the taxa to recognize, and provisional keys are then tested in the field and herbarium.

The treatment of some species is controversial. Many of the problems are a consequence of trying to fit the products of the extraordinary reproductive versatility exhibited by grasses into the neat boxes preferred by humans writing reports. Others stem from lack of the resources—human, temporal, and financial—to perform the studies that might yield a definitive answer.

The descriptions in this volume are original. In some instances, existing descriptions were modified and amplified to meet the needs of the volume; other descriptions are completely new. Most of the identification keys are new, because this volume represents the first attempt to prepare a grass treatment covering all grasses from North America north of Mexico.

We thank the individuals listed below for their commitment and generosity in preparing the treatments presented in this volume, particularly those who submitted a treatment when the grass project began. They are listed below in alphabetical order, in addition to being identified with the treatment(s) they authored. The Grass Phylogeny Working Group comprised 13 individuals who asked that their contributions be attributed to the group as a whole, with the members being listed, if at all, alphabetically.

Susan G. Aiken
Canadian Museum of Nature
Ottawa, Ontario

Kelly W. Allred
New Mexico State University
Las Cruces, New Mexico

Laurel K. Anderton
Utah State University
Logan, Utah

Mirta O. Arriaga
Museo Argentino de Ciencias
  Naturales Bernardino Rivadavia
Buenos Aires, Argentina

Claus Baden†
The Royal Veterinary and Agricultural
  University
Frederiksberg, Denmark

Nigel P. Barker
Rhodes University
Grahamstown, South Africa

Mary E. Barkworth
Utah State University
Logan, Utah

Bernard R. Baum
Agriculture and Agri-Food Canada
Ottawa, Ontario

Roland von Bothmer
Swedish University of Agricultural
  Sciences
Alnarp, Sweden

David M. Brandenburg
The Dawes Arboretum
Newark, Ohio

Paul B.H. But
Chinese University of Hong Kong
Hong Kong, China

Julian J.N. Campbell
The Nature Conservancy
Lexington, Kentucky

Jack R. Carlson
Natural Resources Conservation
  Service
Washington, D.C.

Jacques Cayouette
Agriculture and Agri-Food Canada
Ottawa, Ontario

Lynn G. Clark
Iowa State University
Ames, Iowa

Hans J. Conert
Herbarium Senckenbergianum
Frankfurt/Main, Germany

Laurie L. Consaul
Canadian Museum of Nature
Ottawa, Ontario

William J. Crins
Ontario Ministry of Natural Resources
Peterborough, Ontario

Thomas F. Daniel
California Academy of Sciences
San Francisco, California

Stephen J. Darbyshire
Agriculture and Agri-Food Canada
Ottawa, Ontario

Jerrold I. Davis
Cornell University
Ithaca, New York

Melvin Duvall
Northern Illinois University
DeKalb, Illinois

Signe Frederiksen
Københavns Universitet
Copenhagen, Denmark

Grass Phylogeny Working Group
  Nigel P. Barker
  Lynn G. Clark
  Jerrold I. Davis
  Melvin Duvall
  Gerald F. Guala
  Catherine Hsiao
  Elizabeth A. Kellogg
  H. Peter Linder
  Roberta J. Mason-Gamer
  Sarah Y. Mathews
  Mark P. Simmons
  Robert J. Soreng
  Russell E. Spangler

Craig W. Greene†
College of the Atlantic
Bar Harbor, Maine

M.J. Harvey
Dalhousie University
Halifax, Nova Scotia

Stephan J. Hatch
Texas A&M University
College Station, Texas

Richard J. Hebda
Royal British Columbia Museum
Victoria, British Columbia

Catherine Hsiao
U.S. Department of Agriculture-
  Agricultural Research Service
Logan, Utah

Elizabeth A. Kellogg
University of Missouri-St. Louis
St. Louis, Missouri

Surrey W.L. Jacobs
Royal Botanic Gardens Sydney
Sydney, Australia

Niels H. Jacobsen
The Royal Veterinary and Agricultural
  University
Frederiksberg, Denmark

Emmet J. Judziewicz
University of Wisconsin-Stevens Point
Stevens Point, Wisconsin

H. Peter Linder
Universität Zürich
Zürich, Switzerland

Robert I. Lonard
University of Texas-Pan American
Edinburg, Texas

Sandy Long
Utah State University
Logan, Utah

Kendrick L. Marr
Royal British Columbia Museum
Victoria, British Columbia

Roberta J. Mason-Gamer
University of Illinois
Chicago, Illinois

Sarah Y. Mathews
Harvard University
Cambridge, Massachusetts

Laura A. Morrison
Oregon State University
Corvallis, Oregon

Leon E. Pavlick†
Royal British Columbia Museum
Victoria, British Columbia

Michael B. Piep
Utah State University
Logan, Utah

Grant L. Pyrah
Missouri State University
Springfield, Missouri

Nancy F. Refulio
Rancho Santa Ana Botanic Garden
Claremont, California

Zulma E. Rúgolo de Agrasar
Instituto de Botánica Darwinion
Buenos Aires, Argentina

John H. Rumely
Montana State University
Bozeman, Montana

Jeffery M. Saarela
University of British Columbia
Vancouver, British Columbia

Björn Salomon
Swedish University of Agricultural
* Sciences*
Alnarp, Sweden

Sandra Saufferer
Pullman, Washington

Mark P. Simmons
Ohio State University
Columbus, Ohio

James P. Smith, Jr.
Humboldt State University
Arcata, California

Neil W. Snow
University of Northern Colorado
Greeley, Colorado

Robert J. Soreng
Smithsonian Institution
Washington, D.C.

Russell E. Spangler
University of Minnesota
St. Paul, Minnesota

Lisa A. Standley
Vanasse Hangen Brustlin, Inc.
Watertown, Massachusetts

Christopher M.A. Stapleton
Royal Botanic Gardens, Kew
Surrey, England

Stephen N. Stephenson
Michigan State University
East Lansing, Michigan

Edward E. Terrell
Smithsonian Institution
Washington, D.C.

John W. Thieret†
Northern Kentucky University
Highland Heights, Kentucky

Jimmy K. Triplett
Iowa State University
Ames, Iowa

Gordon C. Tucker
Eastern Illinois University
Charleston, Illinois

Francisco M.Vázquez
Servicio de Investigación y Desarrollo
* Tecnológico*
Badajoz, Spain

J.K. Wipff
Barenbrug USA
Albany, Oregon

Thomas Worley
Lucerne, California

Data Contributors

Ideally, treatment authors would prepare the distribution maps for each species in their genus, based on a combination of fieldwork and examination of herbarium specimens. This is a daunting task, particularly for a region of approximately 21.5×10^6 km^2, with over 700 herbaria. To address this problem, we created a geographic database. Data in this database have come from a wide variety of sources, including authors, reviewers, herbarium databases, publications, and individuals. In recent years, to increase the verifiability of the maps, we have asked that information on relevant voucher specimens accompany the distribution data. A list of the published sources that we drew on is provided in the Geographic Bibliography at the back. We list below the individuals not otherwise connected with the volume who provided us with distribution data.

All the maps in this volume were submitted to the treatment authors for approval. The need to obtain such approval precluded creating maps by dynamic links to online resources. Larger, more informative versions of the maps are available on the Web at http://herbarium.usu.edu/webmanual/. They are updated as well-documented information is submitted.

Individuals

Ken Allison
Canadian Food Inspection Agency
Ottawa, Ontario

Brock Benson
Natural Resources Conservation
  Service
Tremonton, Utah

Curtis Randall Björk
University of Idaho
Moscow, Idaho

Pam Brunsfeld
University of Idaho
Moscow, Idaho

Charles A. Davis
Ecological Consultant
Lutherville, Maryland

Steven A. Dewey
Utah State University
Logan, Utah

Myrna Fleming
Flathead Valley Community College
Flathead, Montana

Ben Franklin
Utah Natural Heritage Program
Salt Lake City, Utah

Alton M. Harvill, Jr.
Longwood College
Farmville, Virginia

Stuart G. Hay
Université de Montréal
Montréal, Québec

Douglass Henderson†
University of Idaho
Moscow, Idaho

Harold Hinds†
University of New Brunswick
Fredericton, New Brunswick

Ron Hise
Idaho Department of Parks and
  Recreation
Plummer, Idaho

Judy Hoy
Stevensville, Montana

John T. Kartesz
Biota of North America Program
Chapel Hill, North Carolina

Wesley M. Knapp
Maryland Department of Natural
  Resources
Wye Mills, Maryland

Roger Q. Landers, Jr.
Extension Specialist
San Angelo, Texas

Steve W. Leonard
Wiggins, Mississippi

S. Frank Lomer
New Westminster, British Columbia

Hughes B. Massicotte
University of Northern British
  Columbia
Prince George, British Columbia

James D. Morefield
Nevada Natural Heritage Program
Carson City, Nevada

Jeanette C. Oliver
Flathead Valley Community College
Flathead, Montana

Elizabeth L. Painter
Affiliate, Jepson Herbarium
Santa Barbara, California

Steve J. Popovich
U.S. Department of Agriculture-Forest
  Service
Fort Collins, Colorado

Hans Roemer
University of British Columbia
Vancouver, British Columbia

Welby Smith
Minnesota Natural Heritage Program
St. Paul, Minnesota

Margriet Wetherwax
University of California-Berkeley
Berkeley, California

Theo Witsell
Arkansas Natural Heritage
  Commission
Little Rock, Arkansas

Vernon Yadon
Pacific Grove Museum of Natural
  History
Pacific Grove, California

Herbaria

Some herbaria provided copies of information in their databases for our use; others provided information on individual taxa; still others make their data available via the Web. The number of herbaria now providing specimen data over the Web has increased beyond our ability to ensure that the contributors were able to review the maps for their genera prior to publication. Consequently, the maps do not reflect all the records now available for download. They do, however, convey a reasonable picture of the distribution of the species in the *Flora* region, with some exceptions that are noted in the text. The maps on the Web will continue to be updated at irregular intervals.

The herbaria that provided data are listed below, alphabetized by country, state or province, and city. The letters in parentheses are internationally recognized codes for the herbaria concerned. A complete list of these codes and herbarium addresses is available from Index Herbariorum at http://sciweb.nybg.org/science2/IndexHerbariorum.asp. If a code is not given, it means that the herbarium concerned does not have an official code.

Royal Alberta Museum (PMAE)
Edmonton, Alberta

University of Alberta (ALTA)
Edmonton, Alberta

Ministry of Forests, British Columbia
  Herbarium
Prince George, British Columbia

Northern Interior Forest Region
  Herbarium
Prince George, British Columbia

Prince Rupert Forest Region (SMI)
Smithers, British Columbia

University of British Columbia (UBC)
Vancouver, British Columbia

Royal British Columbia Museum (V)
Victoria, British Columbia

University of Manitoba (WIN)
Winnipeg, Manitoba

Chignecto Herbarium—Coastal
  Wetlands Institute
Mount Allison University
Sackville, New Brunswick

National Herbarium of Canada (CAN)
Canadian Museum of Nature
Ottawa, Ontario

Vascular Plant Herbarium (DAO)
Agriculture and Agri-Food Canada
Ottawa, Ontario

Natural Heritage Information Centre
  (NHIC)
Ontario Ministry of Natural Resources
Peterborough, Ontario

Algoma University College Herbarium
Sault Sainte Marie, Ontario

Great Lakes Forestry Centre, Canadian
  Forest Service (SSMF)
Sault Sainte Marie, Ontario

Ontario Forest Research Institute
  Herbarium
Sault Sainte Marie, Ontario

Claude Garton Herbarium (LKHD)
Lakehead University
Thunder Bay, Ontario

Algonquin Provincial Park (APM)
Whitney, Ontario

McGill University, Macdonald Campus
  (MTMG)
Sainte-Anne-de-Bellevue, Québec

John D. Freeman Herbarium (AUA)
Auburn University
Auburn, Alabama

University of Alabama (UNA)
Tuscaloosa, Alabama

University of Alaska Museum (ALA)
Fairbanks, Alaska

U.S. Department of Agriculture-Forest
  Service, Alaska Region (TNFS)
Sitka, Alaska

Deaver Herbarium (ASC)
Northern Arizona University
Flagstaff, Arizona

Arizona State University (ASU)
Tempe, Arizona

University of Arizona (ARIZ)
Tucson, Arizona

Humboldt State University (HSC)
Arcata, California

Jepson Herbarium (JEPS)
University of California-Berkeley
Berkeley, California

University of California-Berkeley (UC)
Berkeley, California

California State University (CHSC)
Chico, California

University of California-Davis (DAV)
Davis, California

University of California-Irvine (IRVC)
Irvine, California

University of California-Riverside
  (UCR)
Riverside, California

Carl Sharsmith Herbarium (SJSU)
San Jose State University
San Jose, California

Santa Barbara Botanic Garden (SBBG)
Santa Barbara, California

University of California-Santa Barbara
  (UCSB)
Santa Barbara, California

University of California-Santa Cruz
  (UCSC)
Santa Cruz, California

Stanislaus Herbarium (SHTC)
California State University
Turlock, California

University of Colorado, Boulder
  (COLO)
Boulder, Colorado

Colorado State University (CS)
Fort Collins, Colorado

Colorado National Monument
  Herbarium
Fruita, Colorado

Mesa State College (MESA)
Grand Junction, Colorado

United States National Arboretum,
  USDA/ARS (NA)
Washington, D.C.

United States National Herbarium (US)
Smithsonian Institution
Washington, D.C.

Fairchild Tropical Botanic Garden
  (FTG)
Coral Gables, Florida

Florida Museum of Natural History
  (FLAS)
University of Florida
Gainesville, Florida

R.K. Godfrey Herbarium (FSU)
Florida State University
Tallahassee, Florida

University of South Florida (USF)
Tampa, Florida

University of Georgia (GA)
Athens, Georgia

Valdosta State University (VSC)
Valdosta, Georgia

Stillinger Herbarium (ID)
University of Idaho
Moscow, Idaho

Illinois Natural History Survey (ILLS)
Champaign, Illinois

Stover-Ebinger Herbarium (EIU)
Eastern Illinois University
Charleston, Illinois

Ada Hayden Herbarium (ISC)
Iowa State University
Ames, Iowa

University of Iowa (IA)
Iowa City, Iowa

R.L. McGregor Herbarium (KANU)
University of Kansas
Lawrence, Kansas

Louisiana State University (LSU)
Baton Rouge, Louisiana

Mississippi Museum of Natural Science
  (MMNS)
Jackson, Mississippi

Institute for Botanical Exploration
  (IBE)
Mississippi State, Mississippi

Mississippi State University (MISSA)
Mississippi State, Mississippi

Pullen Herbarium (MISS)
University of Mississippi
University, Mississippi

Missouri Botanical Garden (MO)
St. Louis, Missouri

Montana State University (MONT)
Bozeman, Montana

Rocky Mountain Research Station
(MRC)
Missoula, Montana

University of Montana (MONTU)
Missoula, Montana

University of Nebraska-Omaha (OMA)
Omaha, Nebraska

Nevada State Museum (NSMC)
Carson City, Nevada

Wesley E. Niles Herbarium (UNLV)
University of Nevada-Las Vegas
Las Vegas, Nevada

San Juan College (SJNM)
Farmington, New Mexico

New York State Museum (NYS)
Albany, New York

University of North Carolina (NCU)
Chapel Hill, North Carolina

Northern Prairie Research Center
(NPWRC)
Jamestown, North Dakota

Ohio University (BHO)
Athens, Ohio

Cleveland Museum of Natural History
(CLM)
Cleveland, Ohio

Museum of Biological Diversity (OS)
Ohio State University
Columbus, Ohio

Tom S. and Miwako K. Cooperrider
Herbarium (KE)
Kent State Universtiy
Kent, Ohio

W.S. Turrell Herbarium (MU)
Miami University
Oxford, Ohio

Robert Bebb Herbarium (OKL)
University of Oklahoma
Norman, Oklahoma

Oklahoma State University (OKLA)
Stillwater, Oklahoma

Malheur, Umatilla, and Wallowa-
Whitman National Forests
Herbarium
Baker City, Oregon

Morton E. Peck Herbarium (WILLU)
Oregon State University
Corvallis, Oregon

Oregon State University (OSC)
Corvallis, Oregon

Andrew Charles Moore Herbarium
(USCH)
University of South Carolina
Columbia, South Carolina

Austin Peay State University (APSC)
Clarksville, Tennessee

Lundell Herbarium (LL)
University of Texas at Austin
Austin, Texas

University of Texas at Austin (TEX)
Austin, Texas

S.M. Tracy Herbarium (TAES)
Texas A&M University
College Station, Texas

Texas A&M University (TAMU)
College Station, Texas

Botanical Research Institute of Texas
(BRIT)
Fort Worth, Texas

Robert A. Vines Environmental Science
Center (SBSC)
Houston, Texas

S.R. Warner Herbarium (SHST)
Sam Houston State University
Huntsville, Texas

Stephen F. Austin State University
(ASTC)
Nacogdoches, Texas

Rob & Bessie Welder Wildlife
Foundation (WWF)
Sinton, Texas

Intermountain Herbarium (UTC)
Utah State University
Logan, Utah

Massey Herbarium (VPI)
Virginia Polytechnic Institute and State
University
Blacksburg, Virginia

Western Washington University
(WWB)
Bellingham, Washington

Marion Ownbey Herbarium (WS)
Washington State University
Pullman, Washington

Burke Museum of Natural History and
Culture (WTU)
University of Washington
Seattle, Washington

Pacific Lutheran University
Tacoma, Washington

West Virginia University (WVA)
Morgantown, West Virginia

University of Wisconsin (WIS)
Madison, Wisconsin

Reviewers

Before being finalized, each treatment was reviewed by regional specialists, i.e., taxonomists with particular knowledge of the flora in some part of the *Flora* region. These individuals provided a wide range of advice, including the provision of additional distributional information, suggestions for improving the keys, and observations on the ecology of taxa in their region. We join the contributors in thanking them for their assistance in improving the quality of the treatments.

Charles M. Allen
*Center for the Environmental
  Management of Military Lands*
Fort Polk, Louisiana

Geraldine A. Allen
University of Victoria
Victoria, British Columbia

Kelly W. Allred
New Mexico State University
Las Cruces, New Mexico

Edward R. Alverson
The Nature Conservancy
Portland, Oregon

Loran C. Anderson
Florida State University
Tallahassee, Florida

Ray Angelo
New England Botanical Club
Cambridge, Massachusetts

Alan R. Batten
University of Alaska-Fairbanks
Fairbanks, Alaska

Bruce Bennett
Yukon Department of the Environment
Whitehorse, Yukon

David E. Boufford
Harvard University
Cambridge, Massachusetts

Edwin L. Bridges
Botanical and Ecological Consultant
Bremerton, Washington

Luc Brouillet
Université de Montréal
Montréal, Québec

Julian J.N. Campbell
The Nature Conservancy
Lexington, Kentucky

Jacques Cayouette
Agriculture and Agri-Food Canada
Ottawa, Ontario

Adolf Ceska
*British Columbia Ministry of the
  Environment*
Victoria, British Columbia

Kenton L. Chambers
Oregon State University
Corvallis, Oregon

Edward W. Chester
Austin Peay State University
Clarksville, Tennessee

Anita F. Cholewa
University of Minnesota
St. Paul, Minnesota

William J. Cody
Agriculture and Agri-Food Canada
Ottawa, Ontario

Tom S. Cooperrider
Kent State University
Kent, Ohio

William Crins
Ontario Ministry of Natural Resources
Peterborough, Ontario

H. R. DeSelm
University of Tennessee-Knoxville
Knoxville, Tennessee

George W. Douglas†
*British Columbia Ministry of
  Sustainable Resource Management*
Victoria, British Columbia

Barbara Ertter
University of California-Berkeley
Berkeley, California

Bruce A. Ford
University of Manitoba
Winnipeg, Manitoba

Donna Ford-Werntz
West Virginia University
Morgantown, West Virginia

Robert W. Freckmann
University of Wisconsin-Stevens Point
Stevens Point, Wisconsin

Craig C. Freeman
University of Kansas
Lawrence, Kansas

Joyce Gould
*Alberta Natural Heritage Information
  Centre*
Edmonton, Alberta

Arthur Haines
New England Wild Flower Society
Framingham, Massachusetts

Richard Halse
Oregon State University
Corvallis, Oregon

H. David Hammond
Northern Arizona University
Flagstaff, Arizona

Bruce F. Hansen
University of South Florida
Tampa, Florida

Vernon L Harms
University of Saskatchewan
Saskatoon, Saskatchewan

Ronald L. Hartman
University of Wyoming
Laramie, Wyoming

Stephan L. Hatch
Texas A&M University
College Station, Texas

Robert R. Haynes
University of Alabama
Tuscaloosa, Alabama

Richard J. Hebda
Royal British Columbia Museum
Victoria, British Columbia

G. Frederic Hrusa
*California Department of Food and
  Agriculture*
Courtland, California

Charles G. Johnson
*U.S. Department of Agriculture-Forest
  Service*
Baker, Oregon

Walter S. Judd
University of Florida
Gainesville, Florida

Gordon Leppig
*California Department of Fish and
  Game*
Eureka, California

Aaron Liston
Oregon State University
Corvallis, Oregon

Charles T. Mason, Jr.
University of Arizona
Tucson, Arizona

Michelle McMahon
Washington State University
Pullman, Washington

James D. Morefield
Nevada Natural Heritage Program
Carson City, Nevada

Nancy R. Morin
Point Arena, California

Marian Zinck Munro
*Nova Scotia Museum of Natural
  History*
Halifax, Nova Scotia

David F. Murray
University of Alaska-Fairbanks
Fairbanks, Alaska

Rachel Newton
Field Museum of Natural History
Chicago, Illinois

Francis E. Northam
Arizona Department of Agriculture
Phoenix, Arizona

Richard R. Old
XID Services, Inc.
Pullman, Washington

Michael J. Oldham
Ontario Natural Heritage Information
Centre
Peterborough, Ontario

Harold Ornes
Southern Utah University
Cedar City, Utah

John G. Packer
University of Alberta
Edmonton, Alberta

James J. Pojar
Canadian Parks and Wilderness
Society
Whitehorse, Yukon

Elizabeth Punter
University of Manitoba
Winnipeg, Manitoba

Cindy Roché
Consultant
Medford, Oregon

Bruce A. Sorrie
Longleaf Ecological
Whispering Pines, North Carolina

Russell Spangler
University of Minnesota
St. Paul, Minnesota

Mary C. Stensvold
U.S. Department of Agriculture-Forest
Service
Sitka, Alaska

Larry Strich
U.S. Department of Agriculture-Forest
Service
Washington, D.C.

John L. Strother
University of California-Berkeley
Berkeley, California

Scott Sundberg†
Oregon State University
Corvallis, Oregon

David M. Sutherland
University of Nebraska-Omaha
Omaha, Nebraska

John W. Thieret†
Northern Kentucky University
Highland Heights, Kentucky

Rahmona A. Thompson
East Central University
Ada, Oklahoma

Lowell E. Urbatsch
Louisiana State University
Baton Rouge, Louisiana

Edward G. Voss
University of Michigan
Ann Arbor, Michigan

Alan S. Weakley
University of North Carolina-Chapel
Hill
Chapel Hill, North Carolina

Thomas F. Wieboldt
Virginia Polytechnic Institute and State
University
Blacksburg, Virginia

Richard P. Wunderlin
University of South Florida
Tampa, Florida

James L. Zarucchi
Missouri Botanical Garden
St. Louis, Missouri

Peter F. Zika
University of Washington
Seattle, Washington

Specimen Loans

Loaning specimens creates considerable work and expense for both the loaning and the borrowing institutions, for each must keep track of the specimens as they are loaned, received, and returned, and each incurs shipping costs. We thank all herbaria that loaned specimens for the project. The herbaria listed below are those that loaned specimens to the Intermountain Herbarium. We particularly thank those who responded to our appeal for unreasonably expedited service as we sought the specimens we needed to complete the illustrations, and those who have allowed us to keep specimens beyond the usual time period.

Bailey Hortorium Herbarium (BH)
Cornell University
Ithaca, New York

Botanical Museum (LD)
Lund University
Lund, Sweden

Botanical Research Institute of Texas
(BRIT)
Fort Worth, Texas

Botanischer Garten und Botanisches
Museum Berlin-Dahlem (B)
Zentraleinrichtung der Freien
Universität Berlin
Berlin, Germany

Burke Museum of Natural History and
Culture (WTU)
University of Washington
Seattle, Washington

California Academy of Sciences (CAS)
San Francisco, California

Gray Herbarium (GH)
Harvard University
Cambridge, Massachusetts

Intermountain Herbarium (UTC)
Utah State University
Logan, Utah

Jepson Herbarium (JEPS)
University of California-Berkeley
Berkeley, California

Missouri Botanical Garden (MO)
St. Louis, Missouri

Museo Argentino de Ciencias
  Naturales Bernardino Rivadavia (BA)
Buenos Aires, Argentina

National Herbarium of New South
  Wales (NSW)
Sydney, Australia

Plant Pest Diagnostics Branch
  Herbarium (CDA)
*California Department of Food and
  Agriculture*
Sacramento, California

Range Science Herbarium (NMCR)
New Mexico State University
Las Cruces, New Mexico

Royal Botanic Gardens (K)
Kew, England

Royal British Columbia Museum (V)
Victoria, British Columbia

S.M. Tracy Herbarium (TAES)
Texas A&M University
College Station, Texas

Stillinger Herbarium (ID)
University of Idaho
Moscow, Idaho

Sul Ross State University (SRSC)
Alpine, Texas

United States Department of
  Agriculture-Forest Service, Alaska
  Region (TNFS)
Sitka, Alaska

United States National Herbarium (US)
Washington, D.C.

University of Alaska Museum (ALA)
Fairbanks, Alaska

University of Arizona (ARIZ)
Tucson, Arizona

University of British Columbia (UBC)
Vancouver, British Columbia

University of California-Berkeley (UC)
Berkeley, California

University of Manitoba (WIN)
Winnipeg, Manitoba

University of Michigan (MICH)
Ann Arbor, Michigan

University of Montana (MONTU)
Missoula, Montana

Vascular Plant Herbarium (DAO)
Agriculture and Agri-Food Canada
Ottawa, Ontario

West Virginia University (WVA)
Morgantown, West Virginia

In addition, Jerrold I. Davis, Laurie L. Consaul, and Robert J. Soreng loaned several specimens from their personal collections for use in preparing the illustrations. We thank them for their assistance.

Nomenclatural Consultants

One of the most difficult aspects of a flora for authors at institutions lacking extensive holdings of older taxonomic literature is nomenclature. Determining the correct scientific name for a taxon frequently requires consulting and interpreting literature that is available at few institutions. The task is particularly onerous for the grass volumes, as our nomenclatural file, which will be published separately, accounts for many more synonyms than are mentioned in other volumes of the *Flora of North America*. We sent the nomenclatural questions for this volume to Kanchi Gandhi (Harvard University Herbaria) and Dan H. Nicolson (U.S. National Herbarium). We thank both, and especially Kanchi, to whom most of the questions were sent, for their prompt and clear answers to our questions. We also thank members of the Committee for Spermatophyta of the International Association for Plant Taxonomy, whose assistance Kanchi sought on particularly difficult questions.

We also thank Steve Cafferty (The Natural History Museum, England) for answering numerous questions concerning the typification of Linnaean names, and J.F. Veldkamp (National Herbarium Nederland, Leiden branch, The Netherlands) for responding to a variety of questions, many relating to the nomenclature of southeast Asian species that are now established in North America, and for *Grass Literature*, his intermittently published summaries of taxonomic literature on grasses.

Answers to some of the nomenclatural questions that were raised while preparing the grass volumes have been posted to http://herbarium.usu.edu/webmanual/, where they can be located under the "notes" button. We hope that this will save others from reinvestigating questions that have already been addressed.

Financial Supporters

Producing a flora, particularly one which includes maps and illustrations and involves many contributors, requires substantial financial support. The major funding for completing this volume came from the Forage and Range Research Laboratory (FRRL) of the U.S. Department of Agriculture's Agricultural Research Service. We express our deep appreciation to Dr. N.J. Chatterton, the laboratory's research leader, for his consistent support of the grass project, including the financial support that enabled us to concentrate on completing this volume—and our delight in being able to tell him that it is done. We also received significant support from the National Science Foundation (NSF DEB 0425979) during this period. The combination of the two made it feasible to work more closely with some of the contributors and carry out more detailed studies of some controversial taxa than would otherwise have been the case.

Utah State University and the Utah Agricultural Experiment Station have supported the project since its inception. The university, through the Department of Biology, has provided most of the office supplies, mailing costs, and computer services. The experiment station has provided direct financial support for the project. Both the university and the experiment station support the Intermountain Herbarium, a resource that has been essential to the project.

Although it was the funding from the FRRL and NSF that made completion of this volume possible, several other donors also contributed to developing the resources that we drew on during this phase. For this reason, we express our thanks to all those listed below for their help in making the two grass volumes in the *Flora of North America* possible.

Anonymous

Thad and Jenny Box

Chanticleer Foundation

Idaho Botanical Center

National Fish and Wildlife Foundation

National Science Foundation

U.S. Department of Agriculture:
    Agricultural Research Service, Forage and Range Research Laboratory
    Animal and Plant Health Inspection Service
    Forest Service
    Natural Resources Conservation Service

U.S. Department of the Interior:
    Bureau of Land Management
    National Park Service

Utah Agricultural Experiment Station

Utah State University

Introduction

This volume is, both logically and numerically, the first of the two volumes in the *Flora of North America* devoted to the *Poaceae* (Grass Family). It contains the family description and a key to all the tribes recognized in the two volumes. In addition, it includes treatments for genera in four subfamilies: *Pharoideae*, *Bambusoideae*, *Ehrhartoideae*, and *Poöideae*.

Publication of the volume was postponed until FNA 25, the other grass volume, had been completed so that the keys to the tribes and genera could be checked for consistency with the generic treatments. There are some disadvantages to this approach. Recognition of a new subfamily, the *Micrairoideae*, is now recommended, transforming the PACCAD clade into the PACMCAD clade, and there is no tribe 15 because it was decided to include the *Hainardeae* in the *Poeae* after the tribes in volume 25 had been given numbers. We considered these problems to be minor compared to the value of having FNA 25 available at an earlier date than would otherwise have been possible.

Taxonomic Treatment

The subfamily classification, as in volume 25, is based on the recommendations of the Grass Phylogeny Working Group (GPWG 1986). The tribal classifications adopted in the *Pharoideae*, *Bambusoideae*, and *Ehrhartoideae* are widely supported; that adopted in the *Poöideae* is more controversial. The most controversial aspect is the decision to incorporate the *Aveneae*, together with the *Agrostideae*, *Hainardeae*, and *Phalarideae*, in the *Poeae*. Future work will probably support recognizing additional taxa within the expanded *Poeae*, possibly at the tribal level. It is not yet clear how such taxa should be delimited.

For the most part, the authors of the treatments determined the limits of the genera, species, and infraspecific taxa. The most controversial generic decisions are those in the *Triticeae*. At the species level, changes from existing treatments will be found in several genera, notably *Poa*, *Elymus*, *Calamagrostis*, *Deschampsia*, and *Puccinellia*. Further research on these and other genera is encouraged. To aid in such work, most accounts include a relatively long list of references. The general bibliography includes all the references cited in this volume. Combined citations for the two volumes are available at http:// herbarium.usu.edu/grassbib.htm.

The order of presentation of the tribes, genera, and species is intended to reflect their relationships. If authors were not comfortable suggesting such an ordering, the order is based on the identification key.

Both approaches tend to group morphologically similar taxa together. In practical terms, this means that if an identification does not seem quite right, it is worth looking at the treatments of the adjacent taxa. Infraspecific taxa are treated in alphabetical order.

Content and Layout

The material in the two FNA grass volumes was originally prepared for publication as a successor to Hitchcock's (1935, 1951) *Manual of Grasses of the United States*. This is most evident in the large number of illustrations provided, and the relegation of most nomenclatural information to the back of the book. It is also reflected in the decision to include, in addition to native and established species, several non-native species that are not known to be established. These additional species fall into three categories: cultivated species, waifs, and noxious weeds. Some of the cultivated species are grown for food, others as part of breeding programs, and some as ornamentals. Waifs are species that have been found in the region but have not become established. Some of the noxious weeds have not yet been found in the region. They are included because the U.S.D.A. has identified them as having the potential to seriously impact agriculture in North America. One species, *Oryza rufipogon*, that was initially included only because it was on the noxious weed list, has since become established in the *Flora* region.

The coverage of ornamental grasses is not complete. One reason is that the number of such species is increasing relatively rapidly, as is evident from the increasing number available through horticultural outlets. The other is that ornamental species are rather poorly represented in herbaria. The most poorly represented group of cultivated species are the woody bamboos, over 240 of which are available for sale in the United States according to the American Bamboo Society.

The format of the text material differs slightly from that of other volumes. For instance, a bipartite number is used to indicate the position of genera within the grass volumes. The number before the period indicates the tribal placement; the number after the period, its placement within the tribe. These numbers, rather than page numbers, are used to indicate the location of a treatment within the volume; they are provided in the running heads on each page. Alphabetic and numerical listings of the genera in both volumes, together with their page numbers, are located on the end pages.

Scientific names are either *italicized* or in **boldface**, including those above the generic level. This practice is recommended, but not required, by the International Code of Botanical Nomenclature (Greuter et al. 2000). **Boldface** is used to indicate that a name is the accepted name for the taxon under discussion. It is sometimes used for names of taxa not in the *Flora* region, for instance when comparing two taxa, one of which does not occur in the *Flora* region. Both taxa are accepted, and hence their names are in **boldface**, but only the taxon that is found in the region is described.

Measurements, unless stated otherwise, refer to lengths. They may be preceded by the number of structures present, e.g., anthers 3, 4–6 mm, but are usually placed before the comments on shape or other aspects of the structure concerned. The order in which structures are described follows the same general pattern in all descriptions, particularly within a genus, to facilitate comparison of different species. Nevertheless, changes were made if they resulted in a significantly clearer description.

Taxonomic descriptions, including those in this work, represent an attempt to provide a summary of the salient features of the taxa recognized. Grasses, however, in addition to being genetically variable, are highly susceptible to environmental modification. Moreover, they are often reproductively versatile. The keys and descriptions should fit the majority of specimens; users are expected to be aware that exceptional specimens exist. There are also some genera that are notorious for their ability to hybridize. Hybrids tend to be intermediate in their morphology, but if they backcross to their parents, the range of variation may be considerable.

In editing the treatments, it became evident that there is considerable inconsistency among individual authors in the use of various terms, particularly those used to describe vestiture. To aid users, an indication of the length of the hairs has frequently been provided. This should be taken only as a guide.

In the treatments, accent marks are shown only for people's names or the vernacular names of plants, not for names of places (e.g., "Mexico", not "México"). Similarly, the English spelling of place names is adopted (e.g., "Greenland", not "Grønland"). In citing references, the language of the original article is used unless a non-Roman language is involved; words in non-Roman alphabets are transliterated according to internationally recognized conventions. Journals cited in the references are abbreviated according to *Botanico-Periodico-Huntianum* (Lawrence et al. 1968) and its supplement (Bridson and Smith 1991).

Some of the information that, in other volumes of the *Flora*, is presented in a standardized format, is presented in the grass volumes as part of the comments on a taxon. For instance, a summary of the number of species in each genus, both globally and within the *Flora* region, is included in the comments following the generic description. Information on the rarity or horticultural value of individual species is provided, not by boxed codes but in the comment section. Flowering time is rarely provided, even for species with a narrow distribution, because authors were not asked to provide it.

Nomenclature

In general, the treatments include only the accepted name of a taxon and an abbreviated citation of the author(s) of the name. The abbreviations follow Brummitt and Powell (1992) or, for authors not in that work, the usage of the International Plant Names Index (IPNI). These abbreviated citations are appropriate for use in scientific journals and other publications. Full publication information for the names mentioned in this volume can be found at http://www.ipni.org/, the Web site of IPNI, or at http://mobot.mobot.org/W3T/Search/vast.html, the Missouri Botanical Garden's nomenclatural site. The full names of the authors of the names can be found at the IPNI Web site. Frequently encountered synonyms of the names used in the volume are presented in the list of Names and Synonyms (p. 831).

English names are provided for most of the species, with French names being given for most species that grow in eastern Canada. Spanish names are included only if they were provided by the author. Vernacular names in other languages are not provided.

The vernacular names provided were obtained from many sources, including the contributors. Sources for English names were (Hitchcock 1951) and the U.S. Department of Agriculture's PLANTS database (http://plants.usda.gov/). French names were obtained from the Centre ARICO's information for noxious weeds in Quebec (http://www.mapaq.gouv.qc.ca/dgpar/arico/herbierv/index2.htm), the federal Terminology Office, Montreal, and the Université de Montréal. If no English name was found, the editors followed the model of other species in the genus, giving preference to names that incorporated the generic name.

Many vernacular names are simply a translation of a scientific name, frequently one that reflects a taxonomic treatment that is no longer in use. In some controversial groups, the vernacular name may be less contentious than the scientific name. The disadvantage of vernacular names is that they vary both regionally and linguistically. Moreover, some species have more than one vernacular name and a given vernacular name may apply to more than one species.

Distribution Maps

The maps in the grass volumes are prepared from a geographic database maintained at the Intermountain Herbarium. The database was initially populated entirely from publications such as regional checklists and maps developed by others. Around 1990, many herbaria started databasing their holdings; several of these institutions have allowed us to incorporate their data into our databases (see Acknowledgments, p. xii).

Before generating the maps used in this volume, the contributors were sent copies and asked to approve them. The necessity of obtaining a contributor's approval of the maps precluded our making maps through dynamic links to existing databases. If there were records outside the area that the contributor considered part of the distribution area of a species, an attempt was made to examine the specimens documenting the questionable distribution. If this was not feasible, we deleted the records from the map.

The basic mapping unit for the contiguous United States is the county, because most herbarium specimens show the county from which they were obtained. In addition, there are county-level checklists and state atlases that provide information at this level. For Canada, Alaska, and Greenland, the basic mapping unit is the locality. If georeferenced data were available, we also plotted locality records from within the contiguous United States. The dots used for localities in Alaska, Canada, and Greenland have a diameter of 60 km, so they vary in size with the scale of the map. Dots used for locations in the contiguous United States do not vary in size with the scale of the map.

Specimen labels are increasingly likely to include the geographic coordinates of the collection site, but many of the records in the database were georeferenced retrospectively, using resources such as the Web sites of the Geographic Names Information Service for Canada (http://geonames.nrcan.gc.ca/) and the Geographic Names Information Service for the United States (http://geonames.usgs.gov/gnishome.html). Most specimens from Greenland included geographic coordinates. For those that did not, the foreign names section of the U.S. Geographic Names Information Service site was consulted.

Plant distributions are constantly changing. The maps in the grass volumes show where a species has been found. This is not the same as showing where it can now be seen. Most native species have been eliminated from some parts of their range; some have spread beyond their historical range. Records known to be based on cultivated specimens are not shown; some records may reflect an escape from cultivation that has not persisted in a particular area. We encourage those interested in the current distribution of a species in a region to consult herbaria and individuals in the region concerned.

We have stored information about the distribution of infraspecific taxa in the database but, because few of the resources we initially consulted provided such information, we have not provided maps for such taxa.

We shall update the maps at http://herbarium.usu.edu/webmanual/ when we receive additional information. Questions, errors, and corrections should be sent to Mary Barkworth (mary@biology.usu.edu). New records must be supported by a herbarium specimen deposited in a herbarium that is willing to make loans.

Bibliographic Indices

The *Geographic Bibliography* (p. 794) lists all the published sources that were consulted in preparing the distribution maps for the two grass volumes. The *General Bibliography* (p. 810) contains all the references cited under "Selected References" in this volume. An integrated general bibliography is available for the two volumes at http://herbarium.usu.edu/grassbib.htm. Names of journals are abbreviated according to *Botanico-Periodicum-Huntianum* (Lawrence et al. 1968) and its supplement (Bridson and Smith 1991), and names of books according to *Taxonomic Literature*, edition 2 and its supplements (Stafleu and Cowan 1976-1988; Stafleu and Mennega 1992+).

Numerical Summary

This volume contains the family description and a key to the 25 grass tribes represented by the 236 genera and 1373 species that are treated in the two grass volumes in the *Flora* series (Table 1), including the additions to volume 25 presented on p. 790 of this volume. Of these species, about 895 are native and about 290 are established introductions. These numbers are approximate because there are some species for which it is not clear whether they are native or introduced and others for which it is not known whether or not they have become established. The additional 180 or so species include some that have been found once but are not thought to be established, some that are of interest because of their relationship to such economically important grasses as wheat and corn, some that are grown as ornamentals, a few that have been reported from the region based on misidentifications, and some that have not yet been found in the region but, because they present a serious threat to agriculture, should be eliminated if found.

Native species

The three subfamilies with the largest number of native species are the *Poöideae* (394 species), *Chloridoideae* (240 species), and *Panicoideae* (211 species) (Table 2). Six of the tribes treated (*Olyreae*, *Ehrharteae*, *Nardeae*, *Brachypodieae*, *Gynerieae*, and *Thysanolaeneae*) are represented only by introduced species.

Of the native species, about 54% are also found outside the region, 22% are confined to the contiguous United States, 15% grow in at least two of the three countries within the region, usually Canada and the United States, 3 species (*Deschampsia mackenzieana*, *Festuca pseudovivipara*, and *Poa ammophila*) are known only from Canada, and two (*Festuca groenlandica* and *Puccinellia groenlandica*) only from Greenland.

Of the species that extend outside the region, most continue southward, some only into Mexico but many, particularly in the *Chloridoideae* and *Panicoideae*, reach further south. The tribes with the highest proportion of species that extend into temperate Asia, Eurasia, or Europe are the *Meliceae*, *Triticeae*, and *Poeae*.

Seventy-nine species are narrow endemics within the United States, that is, restricted to a small region of a large state or a small region at the boundary of two states. Of these endemics, 44 are in California (with 2 being shared with Oregon), 13 in Florida, 8 in Texas, and 6 in Oregon (2 of which are also in California and another 2 in Washington). No other state has more than 2 narrow endemics.

Introduced species

Of the non-native species treated, about 290 are thought to be established in the region. Both the *Ehrharteae* and *Brachypodieae* are represented only by established introductions; the *Arundineae* and *Danthonieae* have more established introductions than native species. Of the remaining tribes, the *Bromeae* has a much higher ratio of established to native species, (0.81), than any other tribe, the next highest being the *Triticeae* (0.43), and the *Paniceae* and *Andropogoneae*, (0.41 each).

In the context of the *Flora* series, "introduced" means introduced from outside the *Flora* region. There are, however, examples of introductions from one side of the continent to another, for instance in *Spartina*.

Table 1. Summary of the taxonomic diversity represented in the two grass volumes in the *Flora of North America* series.

Subfamily	Tribe	Genera	Species	Infraspecies	Hybrid species
Pharoideae	Phareae	1	1	0	0
Bambusoideae	Bambuseae	4	9	0	0
	Olyreae	1	1	0	0
	Total	5	10	0	0
Ehrhartoideae	Ehrharteae	1	3	0	0
	Oryzeae	6	18	4	0
	Total	7	21	4	0
Poöideae	Brachyelytreae	1	2	0	0
	Diarrheneae	1	2	0	0
	Nardeae	1	1	0	0
	Meliceae	4	42	14	1
	Stipeae	14	71	13	2
	Bromeae	1	52	10	0
	Brachypodieae	1	5	0	0
	Triticeae	14	99	45	33
	Poeae	63	344	97	9
	Total	100	618	179	45
Arundinoideae	Arundineae	4	4	2	0
Chloridoideae	Cynodonteae	53	304	46	5
	Pappophoreae	3	5	0	0
	Orcuttieae	3	8	0	0
	Total	59	317	46	5
Danthonioideae	Danthonieae	6	19	0	0
Aristidoideae	Aristideae	1	29	18	0
Centothecoideae	Centotheceae	1	5	0	0
	Thysanolaeneae	1	1	0	0
	Total	2	6	0	0
Panicoideae	Gynerieae	1	1	0	0
	Paniceae	26	265	101	0
	Andropogoneae	30	101	26	0
	Total	57	367	127	0
Volume 24		113	650	183	45
Volume 25		123	723	193	5
Both volumes		236	1373	376	50

Table 2. Summary of the taxonomic diversity of grass taxa native to the *Flora* region. The figures in the last column show the number of native species as a percentage of the number of species treated.

Subfamily	Tribe	Genera	Species	Infraspecies	Hybrid species	Native percent
Pharoideae	Phareae	1	1	0	0	100%
Bambusoideae	Bambuseae	1	3	0	0	33%
	Olyreae	0	0	0	0	0%
	Total	1	3	0	0	30%
Ehrhartoideae	Ehrharteae	0	0	0	0	0%
	Oryzeae	4	11	4	0	61%
	Total	4	11	4	0	52%
Poöideae	Brachyelytreae	1	2	0	0	100%
	Diarrheneae	1	2	0	0	100%
	Nardeae	0	0	0	0	0%
	Meliceae	4	35	14	1	83%
	Stipeae	7	51	11	2	72%
	Bromeae	1	27	4	0	52%
	Brachypodieae	0	0	0	0	0%
	Triticeae	5	50	37	32	51%
	Poeae	35	227	80	6	66%
	Total	54	394	146	41	64%
Arundinoideae	Arundineae	1	1	1	0	25%
Chloridoideae	Cynodonteae	37	228	34	3	75%
	Pappophoreae	3	4	0	0	80%
	Orcuttieae	3	8	0	0	100%
	Total	43	240	34	3	76%
Danthonioideae	Danthonieae	1	7	0	0	37%
Aristidoideae	Aristideae	1	29	18	0	100%
Centothecoideae	Centotheceae	1	5	0	0	100%
	Thysanolaeneae	0	0	0	0	0%
	Total	1	5	0	0	83%
Panicoideae	Gynerieae	0	0	0	0	0%
	Paniceae	19	160	90	0	60%
	Andropogoneae	10	51	19	0	50%
	Total	29	211	109	0	57%
Volume 24		61	409	150	42	63%
Volume 25		75	486	162	3	67%
Both volumes		136	895	312	45	65%

Other within-region introductions are harder to discern. In general, native species are more likely to expand their historical range, often along transportation routes, than to become established in a disjunct portion of the region.

Many of the introduced but not established species were collected in the late 1800s and early 1900s from ballast dumps. Most have not become established and, in many cases, the site involved has been through many phases of development. Nevertheless, some have persisted without spreading far beyond their original location.

At least one of the noxious weeds, *Oryza rufipogon*, has become established in Europe and the United States since work began on the grass volumes. It is thought to have arrived as a contaminant in seed of *O. sativa*, the cultivated species. Some species that were first brought in as ornamentals have since become established and even weedy species, e.g., *Cortaderia*. Others may do so in the future.

Copyright and Citation

With the exception of a few treatments, copyright for all the work in the grass volumes—text, illustrations, and maps—belongs to Utah State University. The exceptions are treatments prepared by employees of the U.S. and Canadian federal governments. United States law does not permit copyrighting work conducted by its employees so the text of treatments prepared by Jack R. Carlson, Paul M. Peterson, Robert J. Soreng, and Robert D. Webster is not copyrighted. Under Canadian law, the Crown retains the copyright on all material prepared by Canadian federal employees. Consequently, inquiries about using the text of treatments by Bernard R. Baum, Jacques Cayouette, and Stephen Darbyshire should be directed to the Minister of Public Works and Government Services, Canada.

If cited individually, this volume should be cited as:

Barkworth, M.E, K.M. Capels, S. Long, L.K. Anderton, and M.B. Piep, eds. 2007. *Magnoliophyta*: *Commelinidae* (in part): *Poaceae*, part 1. Flora of North America North of Mexico, volume 24. Oxford University Press, New York and Oxford.

If cited as one of several volumes in the *Flora* series, it should be included in a general citation:

Flora of North America Editorial Committee, eds. 1993+. Flora of North America North of Mexico 13+ volumes. Oxford University Press, New York and Oxford.

It should, of course, be catalogued as a volume in the *Flora of North America North of Mexico* series.

Volume Preparation

The editors for this volume were initially identical to those for volume 25, Barkworth, Capels, Long, and Piep, but shortly after publication of that volume, Michael B. Piep was appointed the Assistant Curator of the Intermountain Herbarium. Fortunately, we were able to bring in Laurel K. Anderton to replace him.

Kathleen M. Capels has been the general and bibliographic editor for both volumes, and has been responsible for maintaining the geographic database and the nomenclatural files. Sandy Long has been an assistant scientific editor, illustrations editor, and illustrator for this volume. In her role as scientific editor, she had primary responsibility for editing *Agrostis*, *Calamagrostis*, and *Poa*. Anderton, in addition to being an assistant scientific editor, undertook a substantial revision of *Bromus*, and has been responsible for page composition.

The illustrations have been prepared by Cindy Roché, Linda A. Vorobik, Sandy Long, Annaliese Miller, Bee F. Gunn, Hana Pazdírková, Karen Klitz, and Christine Roberts. In some instances, a single illustrator prepared the plates from initial design to completion; in others, a second illustrator completed the illustrations based on the primary illustrator's drawings. The initials of both the primary and, if applicable, the second illustrator, appear on all the plates. A key to these initials is given on the right. Sandy Long, as illustrations editor, reviewed all the plates for compatibility with the text, and was also responsible for labeling them after scanning.

Christine M. Garrard designed the geographic database and wrote the mapping program used to generate the maps in the two grass volumes. She also designed and implemented the project's primary Web site (http://herbarium.usu.edu/ webmanual/).

Kanchi Gandhi (Harvard University Herbaria), as nomenclatural editor of the Flora of North America Association, has answered innumerable questions swiftly, clearly, and courteously. The treatments were sent out for review to the Flora of North America Association's regional reviewers and to John L. Strother (University of California-Berkeley). After reviewing the comments, Long, Anderton, and Barkworth referred them back to the authors or addressed them by consulting specimens in the Intermountain Herbarium and/or descriptions in other publications. All treatment authors were sent a copy of their modified manuscript(s) and maps for approval, together with a letter explaining the changes that had been made and why. Emendations requested by the authors were made unless they concerned matters of style rather than substance.

The composed pages were reviewed by James L. Zarucchi, Editorial Director of the Flora of North America Association, whose prompt attention to our requests for review was greatly appreciated, and the page proofs were reviewed by Barkworth, Anderton, and Capels before being submitted for publication. Final decisions concerning the content and appearance of the volume were made by Barkworth. All comments and corrections should be sent to her.

BFG	Bee F. Gunn
K	Karen Klitz
SL	Sandy Long
AM	Annaliese Miller
HP	Hana Pazdírková
CR	Christine Roberts
CR	Cindy Roché
AV	Linda Ann Vorobik

ILLUSTRATORS

Other

As the acknowledgments and the last section make apparent, production of this volume, like the last volume, has required the dedicated commitment of many individuals. As lead editor I thank all those involved, particularly those who have been involved in the countless details of converting treatments, illustrations, and mapping data into this book: Kathleen Capels, Sandy Long, and Laurel Anderton. I also thank my sisters, Delia Rossi and Helen Vette, for their constant encouragement and support, and the inventors of the Internet, of which we have made abundant use. I am constantly amazed by how much Hitchcock and Chase accomplished without the assistance of digital technology.

Flora of North America Association Board of Directors

The Flora of North America Association is an organization dedicated to the goal of completing a synoptic floristic account of all the plants of North America north of Mexico. Its governing body is the Flora of North America Board of Directors, current members of which are listed below. Emeritus/a members of the committee are designated by an asterisk.

George W. Argus*
Merrickville, Ontario

Guy Baillargeon
Agriculture and Agri-Food Canada

Mary E. Barkworth*
Utah State University

David E. Boufford
Harvard University Herbaria

Luc Brouillet
Université de Montréal

Marshall R. Crosby
Missouri Botanical Garden

Claudio Delgadillo M.
Universidad Nacional Autónoma de México

Wayne J. Elisens
University of Oklahoma

Bruce A. Ford
University of Manitoba

Craig C. Freeman
University of Kansas

Kanchi Gandhi
Harvard University Herbaria

Lynn J. Gillespie
Canadian Museum of Nature

Ronald L. Hartman
University of Wyoming

Marie L. Hicks*
Moab, Utah

Marshall C. Johnston*
Austin, Texas

Robert W. Kiger
Carnegie Mellon University

Geoffrey A. Levin
Illinois Natural History Survey

Barney Lipscomb
Botanical Research Institute of Texas

Aaron Liston
Oregon State University

James Macklin
Harvard University Herbaria

John McNeill*
Royal Botanic Garden, Edinburgh

Nancy R. Morin
Point Arena, California

Barbara M. Murray*
University of Alaska Museum of the North

David F. Murray
University of Alaska Museum of the North

J. Scott Peterson
U.S. Department of Agriculture-Natural Resources Conservation Service

Jackie M. Poole
Texas Parks and Wildlife Department

Richard K. Rabeler
University of Michigan

Jay A. Raveill
Central Missouri State University

Leila M. Shultz
Utah State University

Alan R. Smith*
University of California-Berkeley

Richard W. Spellenberg*
New Mexico State University

Lloyd R. Stark
University of Nevada

John L. Strother
University of California-Berkeley

Barbara M. Thiers*
The New York Botanical Garden

Rahmona A. Thompson*
East Central University

Gordon C. Tucker
Eastern Illinois University

Frederick H. Utech
Carnegie Mellon University

Dale H. Vitt
Southern Illinois University

Michael Vincent
Miami University [Ohio]

Alan S. Weakley
University of North Carolina-Chapel Hill

Kay Yatskievych
Missouri Botanical Garden

Richard H. Zander
Missouri Botanical Garden

James L. Zarucchi
Missouri Botanical Garden

Flora of North America

211. POACEAE Barnhart
GRAMINEAE Adans., alternate name

• Grass Family

Lynn G. Clark

Elizabeth A. Kellogg

Plants annual or perennial; usually terrestrial, sometimes aquatic; tufted, mat-forming, cespitose, pluricespitose, or with solitary *culms* (flowering stems), rhizomes and stolons often well developed. **Culms** annual or perennial, herbaceous or woody, usually erect or ascending, sometimes prostrate or decumbent for much of their length, occasionally climbing, rarely floating; **nodes** prominent, sometimes concealed by the leaf sheaths; **internodes** hollow or solid, bases meristematic; **branching** from the basal nodes only or from the basal, middle, and upper nodes; **basal branching** extravaginal or intravaginal; **branching from the upper nodes** intravaginal, extravaginal, or infravaginal. **Leaves** alternate, 2-ranked, each composed of a sheath and blade encircling the culm or branch; **sheaths** usually open, sometimes closed, the margins fused for all or part of their length; **auricles** (lobes of tissue extending beyond the margins of the sheaths on either side) sometimes present; **ligules** usually present at the sheath-blade junction, particularly on the adaxial surface, abaxial ligules common in the *Bambusoideae*, membranous, sometimes ciliate, adaxial ligules usually present, of membranous to hyaline tissue, a line of hairs, or a ciliate membrane; **blades** usually linear to lanceolate, occasionally ovate to triangular, bases sometimes *pseudopetiolate* (having a petiole-like constriction), venation usually parallel, sometimes with evident cross veins, occasionally divergent. **Inflorescences** (*synflorescences*) usually compound, composed of simple or complex aggregations of primary inflorescences, aggregations paniculate, spicate, or racemose or of spikelike branches, often with an evident *rachis* (central axis), primary inflorescences *spikelets*, *pseudospikelets*, or spikelet equivalents; **inflorescence branches** usually without obvious bracts. **Spikelets** with (0–1)2(3–6) *glumes* (empty bracts) subtending 1–60 florets, glumes and florets distichously attached to a *rachilla* (central axis); **pseudospikelets** with bud-subtending bracts below the glumes. **Glumes** usually with an odd number of veins, sometimes awned. **Florets** bisexual, staminate, or pistillate, usually composed of a *lemma* (lower bract) and *palea* (upper bract), lodicules, and reproductive organs, often laterally or dorsally compressed, sometimes round in cross section; **lemmas** usually with an odd number of veins, often awned, bases frequently thick and hard, forming a callus, backs rounded or keeled over the midvein, awns usually 1(–3), arising basally to terminally; **paleas** usually with 2 major veins, with 0 to many additional veins between the major veins, sometimes also in the margins, often keeled over the major veins; **lodicules** (0)2–3, inconspicuous, usually without veins, bases swelling at anthesis; **stamens** usually 3, sometimes 1(2) or 6+, filaments capillary, anthers versatile, usually all alike within a floret, sometimes 1 or 2 evidently longer than the others; **ovaries** 1-loculed, with (1)2–3(4) styles or style branches, stigmatic region usually plumose. **Fruits** caryopses, pericarp usually dry and adhering to the seed, sometimes fleshy or dry and separating from the seed at maturity or when moistened; **embryos** $^1/_5$ as long as to almost equaling the caryopses, highly differentiated with a *scutellum* (absorptive organ), a shoot with leaf primordium covered by the *coleoptile* (shoot sheath), and a root covered by the *coleorhiza* (root sheath); **hila** punctate to linear. *x* = 5, 6, 7, 9, 10, 11, 12. The formulaic name, *Poaceae*, is based on *Poa*, the largest grass genus; the alternate name, *Gramineae*, comes from the Latin, '*gramen*', grass.

The *Poaceae* or grass family includes approximately 700 genera and 11,000 species (Chen et al. 2006). The two grass volumes in this series treat 10 subfamilies, 25 tribes, 236 genera, and 1373 species. Of these, all the subfamilies, 22 tribes, 136 genera, and 892 species are native to the *Flora* region; 2 tribes, 78 genera, and 290 species have become established in the region. The remaining taxa include ornamental species; species grown for research purposes; species that, if introduced to the region, would pose a threat to important agricultural species; and a few waifs, i.e., species that have been found in the region but have not become established. Most of the waifs are species that were found on ballast dumps near ports around the turn of the last century.

Grasses constitute the fourth largest plant family in terms of number of species. Nevertheless, the family is clearly more significant than any other plant family in terms of geographic, ecological, and economic importance. Grasses grow in almost all terrestrial environments, including dense forests, open deserts, and freshwater streams and lakes. There are no truly marine grasses, but some species grow within reach of the highest tides.

In addition to being widely distributed, grasses are often dominant or co-dominant over large areas. The importance of such areas to humans is reflected in the many words that exist for grasslands, words such as meadow, palouse, pampas, prairie, savanna(h), steppe, and veldt. Not surprisingly, given their abundance and prevalence, grasses are of great ecological importance as soil stabilizers and as providers of shelter and food for many different animals.

The economic importance of grasses to humans is almost impossible to overestimate. The wealth of individuals and countries is dependent on the availability of such sources of grain as *Triticum* (wheat), *Oryza* (rice), *Zea* (corn or maize), *Hordeum* (barley), *Avena* (oats), *Secale* (rye), *Eragrostis* (tef), and *Zizania* (wild rice). Most countries invest heavily in research programs designed to develop better strains of these grasses and the many other grasses that are used for livestock, soil stabilization, and revegetation. Developing improved grasses for recreation areas, such as playing fields, golf courses, and parks, is also a major industry in many parts of the world; increasing recognition of the aesthetic value of grasses is reflected in their prominence in horticultural catalogs.

There are, of course, grasses that are considered undesirable, at least in some parts of the world, but even the most obnoxious grasses may be well-regarded over a portion of their range. For instance, *Bromus tectorum* (cheatgrass) is a noxious, fire-prone invader of western North American ecosystems; it is also welcomed as a source of early spring feed in some parts of the *Flora* region. *Cynodon dactylon* (bermudagrass) is listed as a noxious weed in some jurisdictions; in others it is valued as a lawn grass.

Although grasses are widespread and often dominant in open areas, all evidence points to an origin of the family in forests, most likely in the Southern Hemisphere, at least 55–70 mya

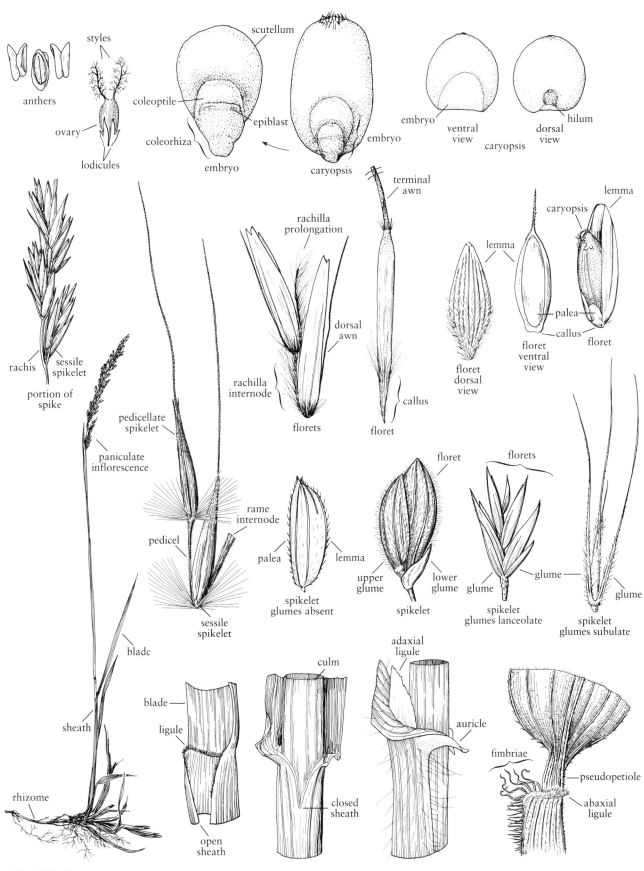

anthers

styles

ovary

lodicules

scutellum

coleoptile

epiblast

coleorhiza

embryo

caryopsis

embryo

embryo

ventral view

dorsal view

hilum

caryopsis

terminal awn

rachilla prolongation

dorsal awn

rachilla internode

florets

callus

floret

lemma

caryopsis

lemma

palea

callus

floret

floret ventral view

floret

floret dorsal view

rachis

sessile spikelet

portion of spike

pedicellate spikelet

paniculate inflorescence

pedicel

rame internode

sessile spikelet

palea

lemma

spikelet glumes absent

upper glume

lower glume

spikelet

floret

florets

glume

glume

spikelet glumes lanceolate

glume

glume

spikelet glumes subulate

blade

sheath

rhizome

blade

ligule

open sheath

culm

closed sheath

adaxial ligule

auricle

fimbriae

pseudopetiole

abaxial ligule

POACEAE

(Grass Phylogeny Working Group 2001). Recent evidence from phytoliths (isolated silica bodies commonly produced inside the epidermal cells of grasses and some other plants) embedded in fossil coprolites strongly suggests that grasses evolved earlier in the Cretaceous than previously thought (Prasad et al. 2005). Living representatives of the three earliest lineages of the grass family, together comprising about 30 species, are perennial, broad-leaved plants of relatively small stature, native to tropical or subtropical forests in South America, Africa, southeast Asia, some Pacific Islands, and northern Australia. The major diversification of the family probably occurred in the mid-Cenozoic, and was associated with climatic changes that produced more open habitats. All major lineages of the grass family were present by the middle of the Miocene (Jacobs et al. 1999), and C_4 photosynthesis in grasses had evolved by then, as well.

Molecular and morphological data unequivocally support a single origin for the *Poaceae* (Grass Phylogeny Working Group 2001). The caryopsis, a single-seeded, usually dry and indehiscent fruit with the pericarp usually strongly adherent to the seed, and the laterally positioned, highly differentiated embryo are unique to grasses. Beyond the three early-diverging lineages (*Anomochlooïdeae* Potztal, *Pharoideae*, and *Puelioideae* L.G. Clark et al.), the great diversity of grasses can be divided into two major lineages: the BEP clade (*Bambusoideae*, *Ehrhartoideae*, and *Poöideae*); and the PACMCAD clade (*Panicoideae*, *Arundinoideae*, *Chloridoideae*, *Micrairoideae* Pilger, *Centothecoideae*, *Aristidoideae*, and *Danthonioideae*, i.e., the PACCAD clade of volume 25 plus the *Micrairoideae*, support for recognition of which was obtained after publication of that volume). Relationships among the BEP grass lineages remain uncertain, and some evidence points to the *Poöideae* as being more closely related to the PACMCAD clade than to the *Bambusoideae* or *Ehrhartoideae*. The PACMCAD clade includes all known C_4 or warm-season grasses.

The closest relatives of the *Poaceae* lie within a group of six families, all native primarily to the Southern Hemisphere: *Joinvilleaceae* Toml. & A.C. Sm., *Ecdeiocoleaceae* D.F. Cutler & Airy Shaw, *Restionaceae* R. Br., *Centrolepidaceae* Endl., *Anarthriaceae* D.F. Cutler & Airy Shaw, and *Flagellariaceae* Dumort. (*Poales* Small, *sensu stricto*). *Joinvilleaceae*, *Ecdeiocoleaceae*, and *Poaceae* constitute a three family clade, with *Ecdeiocoleaceae* probably being closer than *Joinvilleaceae* to the *Poaceae* (Bremer 2000; Bremer 2002; Michelangeli et al. 2003). Rudall et al. (2005), based on a study of reproductive structures in the *Ecdeiocoleaceae*, suggest that the grass caryopsis may represent "one end of a transformation series embodied by the reduced gynoecial structure and indehiscent fruit of other *Poales* such as *Flagellaria* and *Ecdeiocolea*" (p. 1441).

SELECTED REFERENCES **Bremer, K.** 2000. Early Cretaceous lineages of monocot flowering plants. Proc. Natl. Acad. Sci. U.S.A. [PNAS] 97:4707–4711; **Bremer, K.** 2002. Gondwanan evolution of the grass alliance of families (Poales). Evolution 56:1374–1387; **Briggs, B.G., A.D. Marchant, S. Gilmore** and **C.L. Porter.** 2000. A molecular phylogeny of Restionaceae and allies. Pp. 661–671 *in* K.L. Wilson and D.A. Morrison (eds.). Monocots: Systematics and Evolution. CSIRO Publishing, Collingwood, Victoria, Australia. 738 pp.; **Chen, S.-L., B. Sun, L. Liu, Z. Wu, S. Lu, D. Li, Z. Wang, Z. Zhu, N. Xia, L. Jia, G. Zhu, Z. Guo, G. Yang, W. Chen, X. Chen, S.M. Phillips, C. Stapleton, R.J. Soreng, S.G. Aiken, N.N. Tzvelev** [Tsvelev], **P.M. Peterson, S.A. Renvoize, M.V. Olonova,** and **K.H. Ammann.** 2006. Poaceae (Gramineae). Pp. 1–2 *in* Z.-Y. Wu, P.H. Raven, and D.-Y. Hong (eds.). Flora of China, vol. 22 (Poaceae). Science Press, Beijing, Peoples Republic of China and Missouri Botanical Garden Press, St. Louis, Missouri, U.S.A. 653 pp. http://flora.huh.harvard.edu/china/mss/volume22/index.htm.; **Grass Phylogeny Working Group.** 2001. Phylogeny and subfamilial classification of the grasses (Poaceae). Ann. Missouri Bot. Gard. 88:373–457; **Jacobs, B.F., J.D. Kingston** and **L.L. Jacobs.** 1999. The origin of grass-dominated ecosystems. Ann. Missouri Bot. Gard. 86:590–643; **Michelangeli, F.A., J.I. Davis,** and **D.W. Stevenson.** 2003. Phylogenetic relationships among Poaceae and related families as inferred from morphology, inversion in the plastid genome, and sequence data from the mitochondrial and plastid genomes. Amer. J. Bot. 90:93–106; **Prasad, V., C.A.E. Strömberg, H. Alimohammadian,** and **A. Sahni.** 2005. Dinosaur coprolites and the early evolution of grasses and grazers. Science 310:1177–1180; **Rudall, P.J., W. Stuppy, J. Cunniff, E.A. Kellogg,** and **B.G. Briggs.** 2005. Evolution of reproductive structures in grasses (Poaceae) inferred by sister-group comparison with their putative closest living relatives, Ecdeiocoleaceae. Amer. J. Bot. 92:1432–1443.

Key to Tribes

Mary E. Barkworth

Many of the differences among the subfamilies of the *Poaceae* lie in their anatomical characters. For that reason, the primary morphological key presented is a key to tribes. Even some of the tribes are difficult to characterize morphologically, as is evident from the number of tribes that have been brought out more than once. The volume and page number for each tribe appear at the end of the terminal leads, e.g. [FNA 24:100].

1. Leaf blades with divergent veins; spikelets unisexual and dimorphic, the pistillate lemmas with uncinate hairs (*Pharoideae*; FNA 24:11) . . . . 1. *Phareae*
1. Leaf blades with parallel veins; spikelets bisexual, unisexual, or modified into plantlets, the pistillate lemmas never with uncinate hairs.
  2. Culms perennial, woody or herbaceous, often developing complex branching systems from the upper nodes; leaves on the upper portion of the culms, or distal on the branches, usually pseudopetiolate (*Bambusoideae*).
    3. Culms woody, to 30 m tall; leaves strongly dimorphic, those of the main culms (*culm leaves*) with expanded sheaths and often with reduced, non-photosynthetic blades, those of the branches (*foliage leaves*) with abaxial ligules; blades of the distal leaves not folding at night or under stress; florets bisexual; plants native or introduced, often cultivated (FNA 24:15) . . . . 2. *Bambuseae*
    3. Culms herbaceous, to 3.5 m tall or climbing; leaves not strongly dimorphic; blades of the distal leaves often folding at night or under stress; florets unisexual; plants known only in cultivation in the *Flora* region (FNA 24:29) . . . . 3. *Olyreae*
  2. Culms usually annual, sometimes facultatively perennial, rarely woody, sometimes branching from the upper nodes but the branching system not complex; leaves usually not pseudopetiolate.
    4. Spikelets almost always with 2 florets, the lower florets in the spikelets always sterile or staminate, frequently reduced to lemmas, occasionally missing, the upper florets bisexual, staminate, or sterile, unawned or awned from the lemma apices or, if the lemmas bilobed, from the sinuses; glumes membranous and the upper lemma stiffer than the lower lemma, or both florets reduced and concealed by the stiff to coriaceous glumes; rachilla not prolonged beyond the second floret (*Panicoideae*, in part).
      5. Glumes flexible, membranous, the lower glumes usually shorter than the upper glumes, sometimes missing, the upper glumes usually subequal to or exceeded by the upper floret; lower lemmas membranous; upper lemmas usually coriaceous to indurate, sometimes membranous; upper paleas similar in texture; spikelets usually single or in pairs, occasionally in triplets and all pedicellate, often shortly so) (FNA 25:353) . . . . 25. *Paniceae*
      5. Glumes stiff, coriaceous to indurate, often subequal, at least 1 and usually both exceeding the upper floret (excluding the awn); both lemmas hyaline; paleas hyaline or absent; most spikelets in pairs or triplets, at least 1 spikelet in each group usually sessile; pedicels shorter or only a little longer than the sessile spikelets (FNA 25:602) . . . . 26. *Andropogoneae*
    4. Spikelets either with other than 2 florets or, if with 2, the lower floret bisexual or the upper floret awned from the back or base of the lemma, or the spikelets bulbiferous; glumes usually membranous; lemmas scarious to indurate; rachilla sometimes prolonged beyond the distal floret.
      6. Spikelets with 1 floret; lemmas terminating in a 3-branched awn (the lateral branches sometimes greatly reduced); callus well developed; ligules usually of hairs, sometimes ciliate membranes, the cilia longer than the membranous base (*Aristidoideae*; FNA 25:314) . . . . 21. *Aristideae*
      6. Spikelets with more than 1 floret or, if only 1, the lemma not terminating in a 3-branched awn; callus development various; ligules various.

7. Spikelets with 1 sexual floret or the spikelets bulbiferous; glumes absent or less than ¼ as long as the adjacent floret; lower glumes, if present, without veins, upper glumes, if present, veinless or 1-veined.

    8. Upper glumes present, 1-veined; lower glumes absent or much shorter than the upper glumes and lacking veins (*Poöideae*, in part; FNA 24:57) . . . . 6. *Brachyelytreae*

    8. Both glumes absent or lacking veins.

        9. Inflorescences 1-sided spikes; triangular in cross section, (*Poöideae*, in part; FNA 24:62) . . . . 7. *Nardeae*

        9. Inflorescences panicles; spikelets laterally compressed or terete.

            10. Culms aerenchymatous, 20–500 cm long; plants of wet places, often emergent, sometimes floating; lemmas of the bisexual or pistillate florets 3–14-veined; paleas 3–10-veined (*Ehrhartoideae*, in part; FNA 24:36) . . . . 5. *Oryzeae*

            10. Culms not aerenchymatous, 2–300 cm tall; plants of wet or dry habitats but not emergent or floating; lemmas of the bisexual or pistillate florets 1–3-veined; paleas 2-veined.

                11. Culms 2–19 cm tall; plants of cold or damp habitats, not rhizomatous; sheaths of the flag leaves closed for at least ½ their length; caryopses exposed at maturity (*Poöideae*, in part; FNA 24:378) . . . . 14. *Poeae* (in part)

                11. Culms 5–300 cm tall; plants usually of warm or dry habitats, often rhizomatous; sheaths of the flag leaves open to the base; caryopses not exposed at maturity (*Chloridoideae*, in part; FNA 25:14) . . . . 17. *Cynodonteae* (in part)

7. Spikelets usually with more than 1 sexual floret; usually with 2 glumes, 1 or both glumes often longer than ¼ the length of the adjacent floret and/or with more than 1 vein, always longer in taxa with 1 sexual floret.

    12. Lemmas unawned, flabellate or with (5)7–15 awnlike teeth (*Chloridoideae*, in part).

        13. Plants not viscid, usually perennial; ligules present, composed of hairs (FNA 25:285) . . . . 18. *Pappophoreae*

        13. Plants viscid annuals; ligules absent (FNA 25:290) . . . . 19. *Orcuttieae*

    12. Lemmas awned or unawned, lanceolate, rectangular, or ovate, apices entire, mucronate, bilobed, or bifid, occasionally 4-lobed or 4–5-toothed, sometimes erose.

        14. Cauline leaf sheaths closed for ½ their length or more; glumes usually exceeded by the distal florets, sometimes greatly so (*Poöideae*, in part).

            15. Spikelets 5–80 mm long, not bulbiferous; lemmas usually awned, often bilobed or bifid, veins convergent distally; ovary apices hairy (FNA 24:193) . . . . 12. *Bromeae*

            15. Spikelets 0.7–60 mm long, sometimes bulbiferous; lemmas often unawned, not both bilobed/bifid and with convergent veins; ovary apices usually glabrous.

                16. Lemma veins (4)5–15, usually prominent, parallel distally; spikelets 2.5–60 mm long, not bulbiferous (FNA 24:67) . . . . 9. *Meliceae*

                16. Lemmas veins 1–9, often inconspicuous, usually convergent distally; spikelets 0.7–18(20) mm long, sometimes bulbiferous (FNA 24:378) . . . . 14. *Poeae* (in part)

        14. Cauline leaf sheaths open for at least ½ their length; glumes exceeding or exceeded by the distal florets.

            17. Spikelets with 1 floret; lemmas terminally or subterminally awned, the junction of the awn and lemma conspicuous; rachillas notprolonged beyond the base of the floret (*Poöideae*, in part; FNA 24:109) . . . . 10. *Stipeae* (in part)

17. Spikelets with 1–60 florets; lemmas unawned or awned, awns basal to terminal, if terminal or subterminal, the lemma-awn junction not conspicuous; rachillas often prolonged beyond the base of the distal floret.

  18. Ligules, at least of the flag leaves, of hairs, a ciliate ridge or membrane bearing cilia longer than the basal ridge or membrane; leaves usually hairy on either side of the ligule; auricles absent.

    19. Lemmas of the fertile florets with 3–11 inconspicuous veins, never glabrous, if with 3 veins, pilose throughout or with transverse rows of tufts of hair, if with 5–11 veins, the margins pilose proximally, the hairs not papillose-based; lemma apices usually bilobed or bifid and awned or mucronate from the sinus, if acute to acuminate, the lemmaspilose; awns twisted proximally (*Danthonioideae*; FNA 25:298) . . . . 20. *Danthonieae*

    19. Lemmas of the fertile florets usually with 1–3 conspicuous veins, sometimes with 3 inconspicuous veins or 5–11 veins, often glabrous, if with 3 veins, usually glabrous throughout or hairy over the veins, sometimes the margins with papillose-based hairs; lemma apices acute to obtuse, bilobed, or 4-lobed, often mucronate or awned from the sinuses; awns usually not twisted.

      20. Lemmas 1–11-veined, veins glabrous or hairy, margins without papillose-based hairs; rachillas and calluses not pilose, sometimes strigose or strigulose; basal internodes of the culms not persistent, not swollen and clavate (*Chloridoideae*, in part; FNA 25:14) . . . . 17. *Cynodonteae* (in part)

      20. Lemmas 3(5)-veined, veins glabrous, margins sometimes with papillose-based hairs; rachillas or calluses pilose or the basal internodes of the culms persistent, oftenswollen and clavate (*Arundinoideae*, in part; FNA 25:7) . . . . 16. *Arundineae* (in part)

  18. Ligules membranous, if ciliate, the cilia shorter than the membranous base; leaves usually glabrous on either side of the ligule; auricles present or absent.

    21. Inflorescences panicles or unilateral racemes, not spikelike, without spike-like branches; spikelets solitary, the lowest 0–4 florets in a spikelet sterile or staminate, the distal florets sexual.

      22. Spikelets with (1)2–25 bisexual florets; all lemmas similar in size and shape; glumes and lemmas membranous (*Centothecoideae*).

        23. Culms 35–150 cm tall; spikelets with (2)3–26 florets, including the lowest (0)1–4 sterile or staminate florets; lower glumes(1)2–9-veined (FNA 25:344) . . . . 22. *Centotheceae*

        23. Culms 150–400 cm tall; spikelets with 2–4 florets, including the lowest sterile floret; glumes 0–1-veined (FNA 25:349) . . . . 23. *Thysanolaeneae*

      22. Spikelets with 1 bisexual or unisexual floret; lemmas of the sterile florets usually differing in size and shape from those of the sexual floret; glumes membranous, lemmas of the sexual florets firmer.

24. Lemmas of the lower florets coriaceous, at least the upper exceeding the sexual floret (*Ehrhartoideae*, in part; FNA 24:33) . . . . 4. *Ehrharteae*

24. Lemmas of the lower florets membranous, often both much shorter than the sexual floret, sometimes subequal to it, sometimes only 1 sterile floret present (*Poöideae*, in part; FNA 24:378) . . . . 14. *Poeae* (in part)

21. Inflorescences panicles, racemes, or spikes; spikelets sometimes in pairs or triplets, sterile florets, if any, distal to the bisexual or pistillate florets.

[⇐revert to left, Ed.]

25. Lemmas with 1–3 or 9–11 conspicuous veins; sheaths open; blade cross sections with Kranz leaf anatomy (*Chloridoideae*, in part; FNA 25:14) . . . . 17. *Cynodonteae* (in part)

25. Lemmas with (1)3–15 often inconspicuous veins, if with 3 conspicuous veins, the sheaths closed; sheaths open or closed; blade cross sections without Kranz leaf anatomy.

26. Inflorescences spikes or spikelike; spikelets 1–5+ per node, at least 1 spikelet sessile or subsessile (*Poöideae*, in part).

27. Upper glumes 5–9-veined; spikelets subsessile and solitary at the nodes; auricles absent (FNA 24:187) . . . . 11. *Brachypodieae*

27. Upper glumes 1–5-veined; spikelets 1–5+ per node, usually at least 1 sessile at each node, sometimes highly reduced branches present; auricles present or absent.

28. Inflorescences with 1–5 spikelets at a node, if 3, usually with 1 sessile and 2 pedicellate spikelets, if 1, the spikelet tangential to or embedded in the rachis, with 2 glumes, the glumes facing each other; ovaries with hairy apices; auricles often present (FNA 24:238) . . . . 13. *Triticeae* (in part)

28. Inflorescences spikelike panicles with highly reduced branches, or spikes with spikelets radial to the rachises and all but the terminal spikelet with only 1 glume, or spikes with spikelets tangential to the rachises and having 2 glumes adjacent to each other; ovaries with glabrous apices; auricles usually absent (FNA 24:378) . . . . 14. *Poeae* (in part)

26. Inflorescences panicles, with no sessile spikelets.

29. Caryopses with a thick pericarp forming a distinct apical knob or beak at maturity; lemmas 3(5)-veined (*Poöideae*, in part; FNA 24:64) . . . . 8. *Diarrheneae*

29. Caryopses usually with a thin pericarp, never with a distinct apical beak or knob; lemmas 3–9-veined.

30. Glumes subulate, stiff; lemmas unawned or with awns to 4 mm long (*Poöideae*, in part; FNA 24:238) . . . . 13. *Triticeae* (in part)

30. Glumes lanceolate, membranous; lemmas awned or unawned, awn length varied.

31. Rachillas hairy, hairs 2–3 mm long; lemmas coriaceous; plants established in California, sometimes cultivated as ornamentals (*Poöideae*, in part; FNA 24:109) . . . . 10. *Stipeae* (in part)

31. Rachillas glabrous to hairy, hairs shorter than 2 mm; lemmas membranous to coriaceous; plants native, established, or cultivated.

32. Leaves to 2 cm wide, usually not conspicuously distichous; culms 0.01–2.75 m tall, usually less than 1 cm thick (*Poöideae*, in part; FNA 24:378) . . . . 14. *Poeae* (in part)

32. Leaves 2–10 cm wide, often conspicuously distichous; culms 2–10(15) m tall, often more than 1 cm thick.

33. Lower cauline blades disarticulating, upper cauline blades forming a flat, fan-shaped arrangement (*Panicoideae*, in part; FNA 25:352) . . . . 24. *Gynerieae*

33. Lower cauline blades persistent, upper cauline blades not forming a flat, fan-shaped arrangement (*Arundinoideae*, in part; FNA 25:7) . . . . 16. *Arundineae* (in part)

1. PHAROIDEAE L.G. Clark & Judz.

Grass Phylogeny Working Group

Plants perennial, rhizomatous; monoecious. **Culms** annual, not woody, solid or hollow. **Leaves** mostly cauline, evidently distichous; **sheaths** open; **auricles** absent; **abaxial ligules** absent; **adaxial ligules** membranous, ciliate; **pseudopetioles** prominent, twisted so the blades are resupinate; **blades** with the lateral veins evident, straight, diverging obliquely from the central vein; **mesophyll** nonradiate; **adaxial palisade layer** absent; **fusoid cells** well developed, large; **arm cells** weakly to moderately developed; **Kranz anatomy** not developed; **midribs** complex; **inflated interstomatal cells** present; **bulliform cells** poorly developed or absent; **stomates** with parallel-sided to dome-shaped subsidiary cells; **bicellular microhairs** absent; **papillae** absent. **Inflorescences** panicles, rachises and branches sometimes disarticulating, hairy, hairs uncinate. **Spikelets** unisexual, with 1 floret, mostly in staminate-pistillate pairs on short branchlets, some pistillate spikelets solitary. **Staminate spikelets** smaller than the pistillate spikelets, short- to long-pedicellate; **glumes** 1–2, membranous, shorter than the floret; **lodicules** absent or 3 and minute, elliptic, glabrous, veinless; **anthers** 6. **Pistillate spikelets** sessile or shortly pedicellate; **glumes** 2, shorter than the floret; **lemmas** tubular or inflated, covered wholly or in part with uncinate hairs, unawned; **paleas** well developed; **lodicules** absent; **ovaries** glabrous, without an apical appendage; **styles** 1, with 3 branches. **Caryopses: hila** subequal to the caryopses, linear; **endosperm** hard, without lipid; **embryos** small; **epiblasts** present; **scutellar cleft** present but short; **mesocotyl internode** absent; **embryonic leaf margins** overlapping. $x = 12$.

The *Pharoideae* include one tribe, the *Phareae*, three genera, and twelve species. It is pantropical, but in the Americas, it is represented by one genus, *Pharus*, that extends from Florida to Uruguay and Argentina. The *Pharoideae* is a basal lineage of the *Poaceae*, and the first subfamily in which an adaxial ligule and true spikelets are found.

SELECTED REFERENCES **Clark, L.G.** and **E.J. Judziewicz.** 1996. The grass subfamilies Anomochlooideae and Pharoideae (Poaceae). Taxon 45:641–645; **Clark, L.G., W. Zhang,** and **J.F. Wendel.** 1995. A phylogeny of the grass family (Poaceae) based on *ndh*F sequence data. Syst. Bot. 20:436–460; **Grass Phylogeny Working Group.** 2001. Phylogeny and subfamilial classification of the grasses (Poaceae). Ann. Missouri Bot. Gard. 88:373–457.

1. PHAREAE Stapf

Mary E. Barkworth

Plants perennial; rhizomatous, sometimes cespitose; monoecious. **Culms** annual, 10–300 cm, erect to decumbent; **internodes** usually solid. **Ligules** scarious, sometimes ciliolate; **pseudopetioles** present, twisted, placing the abaxial surface of the blades uppermost; **blades** linear to oblong, not folding or drooping at night, lateral veins diverging obliquely from the midveins, cross venation evident. **Inflorescences** panicles, usually espatheate; **ultimate branches** with 1–2 pistillate spikelets and 1 terminal, staminate spikelet; **disarticulation** beneath the pistillate florets and in the panicle branches. **Spikelets** unisexual, heteromorphic, usually in staminate-pistillate pairs on branchlets, with 1 floret; **rachillas** not prolonged beyond the florets. **Staminate spikelets** pedicellate, smaller than the pistillate spikelets, lanceolate to ovate, caducous; **glumes** unequal; **lower glumes** absent or much shorter than the upper glumes; **upper glumes** somewhat shorter than the floret; **lodicules** minute or absent; **anthers** 6. **Pistillate spikelets** sessile or shortly pedicellate, terete, sometimes inflated; **glumes** unequal to subequal, shorter than the florets, scarious, entire, sometimes persistent; **lemmas** chartaceous, becoming coriaceous, veins 5 or more, margins involute or utriculate, partly or wholly covered with uncinate hairs, not terminating in a branched awn; **paleas** 2-veined; **lodicules** absent; **styles** 1, 3-branched. **Caryopses** oblong to linear; **hila** as long as the caryopses. $x = 12$.

The *Phareae* include three genera, all of which grow in tropical and subtropical forests. The tribe is represented by one genus in the Western Hemisphere, *Pharus*.

SELECTED REFERENCES **Clark, L.G.** and **E.J. Judziewicz.** 1996. The grass subfamilies Anomochloöideae and Pharoideae (Poaceae). Taxon 45:641–645; **Clark, L.G., W. Zhang,** and **J.F. Wendel.** 1995. A phylogeny of the grass family (Poaceae) based on *ndh*F sequence data. Syst. Bot. 20:436–460; **Grass Phylogeny Working Group.** 2001. Phylogeny and subfamilial classification of the grasses (Poaceae). Ann. Missouri Bot. Gard. 88:373–457.

1.01 **PHARUS** P. Browne

Emmet J. Judziewicz

Gerald F. Guala

Plants perennial, some apparently monocarpic; rhizomatous, sometimes cespitose or stoloniferous; monoecious. **Culms** 10–130 cm, erect to decumbent; **internodes** solid, frequently with prominent prop roots at the lower nodes. **Sheaths** open, glabrous; **ligules** usually scarious, sometimes membranous and ciliate; **pseudopetioles** conspicuous, twisted 180° distally, inverting the blades; **blades** linear to ovate, usually broad, usually tessellate, lateral veins diverging obliquely from the midvein. **Inflorescences** terminal panicles, ovate, open; **rachises** terminating in a staminate spikelet or naked; **branches** with uncinate hairs, spikelets appressed. **Spikelets** unisexual, dimorphic, sexes paired or pistillate spikelets solitary, with 1 floret; **rachillas** not prolonged beyond the florets; **disarticulation** above the glumes and in the panicle branches. **Staminate spikelets** smaller than the pistillate spikelets, attached below the pistillate spikelets on appressed pedicels; **lower glumes** shorter than the upper glumes or absent; **lemmas** longer than the glumes, ovate, 3-veined; **lodicules** 3, minute; **anthers** 6. **Pistillate spikelets** larger than the staminate spikelets, subsessile, elongate; **glumes** subequal, lanceolate, (3)5–9(11)-veined, purple or green; **lemmas** cylindrical, longer than the glumes, indurate, involute, with uncinate hairs over at least a portion of the surface, 7-veined, margins inrolled, concealing the palea; **lodicules** absent; **staminodes** 6, minute; **styles** 1, 3-branched, stigmas hispid. $x = 12$. Name from the Greek *pharos*, 'cloak' or 'mantle'.

Pharus includes eight species. It extends from central Florida through Mexico to Argentina and Uruguay, and grows in moist to wet lowland forests. One species, *Pharus glaber*, is native to the *Flora* region.

The uncinate hairs and disarticulating panicle branches of *Pharus* promote dispersion by attaching to the coats of passing animals. The inverted, pseudopetiolate leaf blades and oblique venation make the genus easily distinguished, even in its vegetative state. Well-preserved female spikelets resembling those of *Pharus mezii* Prod. have been found in 30–45 million-year-old amber.

SELECTED REFERENCES **Clark, L.G.** and **E.J. Judziewicz.** 1996. The grass subfamilies Anomochloöideae and Pharoideae (Poaceae). Taxon 45:641–645; **Hitchcock, A.S.** 1951. Manual of the Grasses of the United States, ed. 2, rev. A. Chase. U.S.D.A. Miscellaneous Publication No. 200. U.S. Government Printing Office, Washington, D.C., U.S.A. 1051 pp.; **Judziewicz, E.J.** 1987. Taxonomy and morphology of the Tribe Phareae (Poaceae: Bambusoideae). Ph.D. dissertation, University of Wisconsin, Madison, Wisconsin, U.S.A. 557 pp.

1. **Pharus glaber** Kunth [p. 13]

UPSIDEDOWN GRASS

Plants cespitose. **Culms** 25–95 cm, generally decumbent and rooting at the nodes. **Sheaths** glabrous, extensively overlapping; **ligules** 1–2 mm; **pseudopetioles** 8–60 mm; **blades** 7–30 cm long, 2–6.5 cm wide, narrowly elliptic to obovate, often acuminate, lacking intercostal fibrous bands, sometimes whitened beneath, lateral veins diverging from the midvein at a 4–8° angle. **Panicles** 10–40 cm, sparsely flowered; **branches** solitary, with uncinate hairs, usually tipped with a staminate spikelet. **Staminate spikelets** 2.5–3.5 mm, on 4–11 mm pedicels, subtending the pistillate spikelets, purple; **lower glumes** 1–2 mm; **upper glumes** 1.5–3.2 mm, 1- or 3-veined; **lemmas** 2.5–3.5 mm; **paleas** about ³/₄ the length of the lemmas; **anthers** 0.9–1.1 mm. **Pistillate spikelets** 7.5–12 mm, diverging slightly from the branches; **glumes** brown; **lower glumes** 5–7 mm, 5–7-veined; **upper glumes** 6–8 mm, 3–5-veined; **lemmas** 7.5–12 mm, linear-oblong, abruptly short-beaked, with uncinate hairs nearly to the base; **paleas** equaling the lemmas. $2n = 24$.

Pharus glaber grows on limestone-influenced sand in the hammocks of central Florida. Only two remaining populations are known in the United States, but the species is still widely present elsewhere in the Neotropics. Hitchcock (1951) erroneously referred this species to *Pharus parvifolius* Nash, which differs primarily in the presence of intercostal fibrous bands on the adaxial surfaces of the leaf blades.

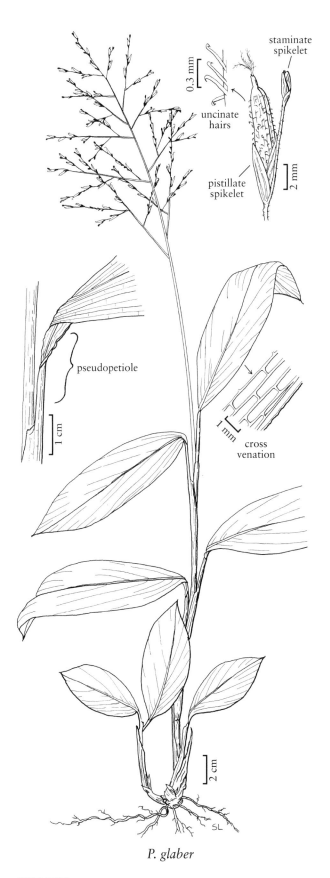

P. glaber

PHARUS

2. BAMBUSOIDEAE Luerss.

Grass Phylogeny Working Group

Plants usually perennial, rarely annual; rhizomatous. **Culms** woody or herbaceous, hollow or solid; often developing complex vegetative branching; **leaves** distichous, if complex vegetative branching present, leaves of the culms (*culm leaves*) differing from those of the vegetative branches (*foliage leaves*); **auricles** often present; **abaxial ligules** rarely present on the culm leaves, usually present on the foliage leaves; **adaxial ligules** membranous or chartaceous, ciliate or not; **pseudopetioles** sometimes present on the culm leaves, usually present on the foliage leaves; **blades** usually relatively broad, venation parallel, often with evident cross venation; **mesophyll** nonradiate; **adaxial palisade layer** usually absent; **fusoid cells** usually well developed, large; **arm cells** usually well developed and highly invaginated; **Kranz anatomy** not developed; **midribs** complex or simple; **stomates** with dome-shaped, triangular, or more rarely parallel-sided subsidiary cells; **adaxial bulliform cells** present; **bicellular microhairs** present, terminal cells tapered; **papillae** common and abundant. **Inflorescences** spicate, racemose, or paniculate, comprising spikelets or pseudospikelets, the spikelets lacking subtending bracts and prophylls, completing their development during 1 period of growth, the pseudospikelets having subtending bracts, prophylls, and basal bud-bearing bracts developing 2 or more orders of true spikelets with different phases of maturity. **Spikelets** bisexual or unisexual, with 1 to many florets. **Glumes** absent or 1–2+; **lemmas** without uncinate hairs, sometimes awned, awns single; **paleas** well developed; **lodicules** (0)3(6+), membranous, vascularized, often ciliate; **anthers** usually 2, 3, or 6, rarely 10–120; **ovaries** glabrous or hairy, sometimes with an apical appendage; **haustorial synergids** absent; **styles** or **style branches** 1–4. **Caryopses: hila** linear, usually as long as the caryopses; **endosperm** hard, without lipid, containing compound starch grains; **embryos** small relative to the caryopses; **epiblasts** present; **scutellar cleft** present; **mesocotyl internode** absent; **embryonic leaf margins** overlapping. $x = 7, 9, 10, 11, 12$.

The *Bambusoideae* includes two tribes, the woody *Bambuseae* and the herbaceous *Olyreae*. Their range includes tropical and temperate regions of Asia, Australia, and the Americas, primarily Central and South America. Three species of *Bambuseae* are native to the *Flora* region; there are no native species of *Olyreae*.

Members of the *Bambusoideae* grow in temperate and tropical forests, high montane grasslands, along riverbanks, and sometimes in savannahs. They are mainly forest understory or margin plants with a limited ability to reproduce, disperse, or survive outside their forest environment. Many have relatively small geographic ranges, and there is a high degree of endemism. The conservation status of most bamboos is not known; all are intrinsically vulnerable because of their breeding behavior and reliance upon a benign forest habitat. Only the C_3 photosynthetic pathway is found in the subfamily.

SELECTED REFERENCES **Grass Phylogeny Working Group**. 2001. Phylogeny and subfamilial classification of the grasses (Poaceae). Ann. Missouri Bot. Gard. 88:373–457; **Judziewicz, E.J., L.G. Clark, X. Londoño, and M.J. Stern**. 1999. American Bamboos. Smithsonian Institution Press, Washington, D.C., U.S.A. 392 pp.

1. Culms woody, usually taller than 1 m, developing complex vegetative branching from the upper nodes; abaxial ligules present on the foliage leaves, rarely present on the culm leaves . . . . 2. *Bambuseae*
1. Culms herbaceous, usually shorter than 1 m; complex vegetative branching not developed; abaxial ligules not present . . . . 3. *Olyreae*

2. BAMBUSEAE Nees

Christopher M.A. Stapleton

Plants perennial; rhizomatous, shrubby to arborescent, self-supporting to climbing; **rhizomes** well developed, pachymorphic or leptomorphic, rarely both. **Culms** perennial, woody, to 30 m tall, internodes usually hollow, initially unbranched and bearing thickened overlapping culm leaves, subsequently developing foliage leaves on the complex branch systems from buds at the internode bases. **Culm leaves** thickened, usually early deciduous; **auricles** and/or **fimbriae** often present; **abaxial ligules** usually lacking; **adaxial ligules** present; **blades** poorly to well developed, erect or reflexed, the base as wide as or narrower than the sheath apex, sometimes pseudopetiolate. **Foliage leaves: auricles** and **fimbriae** present or absent; **abaxial** and **adaxial ligules** present; **pseudopetioles** nearly always present; **blades** deciduous, venation parallel, cross venation often evident, particularly at the base. **Inflorescences** determinate or indeterminate, bracteate or ebracteate, racemose to paniculate, composed of pseudospikelets or spikelets. **Spikelets** or **pseudospikelets** with 1 to many florets, the lower floret(s) often sterile, the others bisexual; **glumes** often subtending the buds; **lemmas** often unawned, usually multiveined; **lodicules** usually 3, with vascular tissue; **anthers** usually 3 or 6, sometimes fewer or up to 7, very occasionally many; **ovaries** glabrous or pubescent; **styles** or **style branches** 1–4; **Caryopses** with or without a thickened fleshy pericarp.

The *Bambuseae* include about 80–90 genera and around 1400 species. They are most abundant in Asia and South America, but are also found in Africa, Australia, Central America and North America. One genus, *Arundinaria*, is native to the *Flora* region, where it is represented by 3 species. Many other genera and species are cultivated in the region for their ornamental value. These often persist for decades; some have become established beyond the original planting. Identification of these introduced bamboos is hampered by the lack of taxonomic studies in their countries of origin, particularly studies of vegetative features, and the large number of taxa that have not yet been described. Identification is further hindered by their infrequent flowering.

Woody bamboos differ from all other grasses in the division of vegetative growth into two phases. Growth commences with the production of unbranched new culms which bear protective leaves called *culm leaves* (or *culm sheaths*) at their nodes. Later in the same season, or in the following season, buds at the nodes of the new culms develop into branches, which produce *foliage leaves* with photosynthetic blades. The leaves on the culms and foliage branches are homologous, each consisting of a sheath, adaxial ligule, and blade, but the culm leaves have less well-developed blades and quickly become non-photosynthetic or fall away, while the foliage leaves have more persistent, photosynthetic, well-developed blades.

Another feature characteristic of some, but not all, bamboos is the 'pseudospikelet', which differs from a determinate spikelet in its indeterminate growth from lateral buds in the axils of the basal glumelike bracts. This can lead to extensive ramification of the pseudospikelet into a capitate cluster [see McClure (1966) for a more detailed discussion].

Woody bamboos are also distinguished by perennial lignified culms which often have complex branch complements at the nodes, and often by cyclical flowering at intervals of up to 150 years. In cyclical flowering, an entire population or even species will flower in a given year, after which the parents usually die, regeneration being through slow-growing and vulnerable seedlings.

Most woody bamboos are tropical or subtropical in their distribution, but about 25 genera are found in temperate regions, mainly in eastern Asia. Two genera are native to North America; *Arundinaria* is the only genus native to the *Flora* region. It is thought to have crossed the Bering Strait from eastern Asia, and may represent the sole remnant of a much larger preglacial bamboo population. The other genus, *Otatea* (McClure & E.W. Sm.) C.E. Calderón & Soderstr., is probably a relatively recent entrant into Mexico from South America.

The taxonomy of woody bamboos has developed slowly because of the scarcity of flowering material, and the distribution of the species in predominantly inaccessible or less-developed parts of the world. Concentration on floral characters in earlier classification systems has made taxonomy and identification even more difficult. Cytological information has been of little value beyond the separation of two major Asian groups, one tropical with $2n = 56–72$ and the other temperate with $2n = 48$. There appears to be more variation among American bamboos.

Molecular data have thrown doubt upon the phylogenetic validity of many earlier attempts to classify the woody bamboos by floral characters alone. Nevertheless, they are illuminating interesting patterns of evolution and dispersal. A wider range of morphological characters is increasingly being studied and applied. Because many species have only been described in recent years, while many cultivated bamboos have been grown under misapplied or speculative names, the taxonomy of bamboos will require continued study for many years to come. Fieldwork, herbarium study, and laboratory investigations will all be necessary, along with extensive international collaboration. It is highly likely that a substantial proportion of forest bamboos will become extinct before they have been properly documented.

The American Bamboo Society lists over 450 taxa of bamboo, representing 240 species in 40 genera, as being commercially available in the United States. Although bamboos are increasingly widely cultivated in the *Flora* region, they are most common in the coastal and southern states. Most of the cultivated species are Asian in origin, but in recent years numerous Central and South American taxa have been introduced. Most introduced taxa will persist indefinitely without cultivation. In favorable climates, many will spread beyond the original plantings to become naturalized, especially those with long rhizomes. Wider dispersal occurs when sections of the rhizome are transported from one location to another, as may happen during floods. Clump-forming bamboos may eventually spread through seed dispersal.

Bamboos are multipurpose plants of immense utility to mankind. Traditional uses, such as building, basketry, and fodder, have been supplemented by industrial-scale paper-pulp production, the canning of edible shoots, and production of advanced board products, laminates, and cloth.

This treatment is limited to a full treatment of the native genus, *Arundinaria*, and descriptions and representative illustrations for the three genera and seven species thought to have become established in the *Flora* region.

SELECTED REFERENCES **Bystriakova, N., V. Kapos, C.M.A. Stapleton,** and **J. Lysenko.** 2003. Bamboo Biodiversity: Information for Planning Conservation and Management in the Asia–Pacific Region. UNEP–WCMC Biodiversity Series, no. 14. United Nations Environment Programme–World Conservation Monitoring Centre and International Network for Bamboo and Rattan, Cambridge, England. 71 pp.; **Calderón, C.E.** and **T.R. Soderstrom.** 1980. The genera of Bambusoideae (Poaceae) of the American continent: Keys and comments. Smithsonian Contr. Bot. 44:1–27; **Judziewicz, E.J., L.G. Clark, X. Londoño** and **M.J. Stern.** 1999. American Bamboos. Smithsonian Institution Press, Washington, D.C. , U.S.A. 392 pp.; **McClure, F.A.** 1966. The Bamboos: A Fresh Perspective. Harvard University Press, Cambridge, Massachusetts, U.S.A.. 347 pp.; **McClure, F.A.** 1973. Genera of bamboos native to the New World (Gramineae: Bambusoideae). Smithsonian Contr. Bot. 9:1–148; **Soderstrom, T.R.** and **R. Ellis.** 1987. The position of bamboo genera and allies in a system of grass classification. Pp. 225–238 *in* T.R. Soderstrom, K.W. Hilu, C.S. Campbell, and M.E. Barkworth (eds.). Grass Systematics and Evolution. Smithsonian Institution Press, Washington, D.C., U.S.A. 473 pp.; **Stapleton, C.M.A.** 1997. Morphology of woody bamboos. Pp. 251–267 *in* G.P. Chapman (ed.). The Bamboos. Academic Press, San Diego, California, U.S.A. 370 pp.; **Stapleton, C.M.A.** 1998. Form and function in the bamboo rhizome. J. Amer. Bamboo Soc.12: 21–29; **Stapleton, C.M.A., G.N. Chonghaile,** and **T.R. Hodkinson.** 2004. *Sarocalamus,* a new Sino–Himalayan bamboo genus (Poaceae–Bambusoideae). Novon 14:345–349.

1. Rhizomes pachymorphic, short, thicker than the culms; culms forming separate, well-defined clusters . . . . .2.02 *Bambusa*
1. Rhizomes leptomorphic, long, thinner than the culms; culms solitary, loosely clumped, or both.
  2. Culm internodes grooved their whole length, often doubly sulcate above the branches; branches usually without compressed internodes at the base; spikelets sessile . . . . .2.03 *Phyllostachys*
  2. Culm internodes mostly terete, flattened or shallowly sulcate above the branches; branches usually with 1–5 compressed internodes at the base, sometimes without any compressed internodes; spikelets pedicellate.
    3. Culm leaves persistent or deciduous, usually with auricles; culms to 3 cm thick; culm buds open, margins not fused; plants native in the *Flora* region .2.01 *Arundinaria*
    3. Culm leaves persistent, usually without auricles; culms to 1.5 cm thick; culm buds closed, margins fused; plants cultivated in the *Flora* region, occasionally escaped . . . . . . .2.04 *Pseudosasa*

2.01 ARUNDINARIA Michx.

Lynn G. Clark

J.K. Triplett

Plants arborescent or subarborescent, spreading or loosely clumped; **rhizomes** leptomorphic. **Culms** 0.5–8 m tall, to 3 cm thick, erect; **nodes** not swollen; **supranodal ridges** not prominent; **internodes** terete to slightly flattened or shallowly sulcate above the branches. **Culm leaves: sheaths** persistent or deciduous, mostly glabrous, abaxial surfaces sparsely pilose towards the margins and apices, margins ciliate; **auricles** usually present; **blades** erect or becoming reflexed, narrowly triangular to strap-shaped, abaxial surfaces sparsely pilose; **leaves at tips of new shoots** crowded into distinctive fan-shaped clusters or *topknots*, blades expanded as on the foliage leaves. **Branch complements** of 1 primary branch and 0–2 subequal secondary branches on young culms, rebranching to produce to 40+ secondary branches on older culms. **Foliage leaves: sheaths** persistent on the lower branch nodes; **auricles** usually present; **fimbriae** to 10 mm; **blades** finely cross veined abaxially, acuminate, blades of the ultimate branchlets often smaller, crowded into flabellate clusters of 3–7 leaves. **Inflorescences** open racemes or panicles; **disarticulation** below and between the florets. **Spikelets** 3–7 cm, with 6–12 florets, basal floret occasionally sterile, laterally compressed. **Glumes** 1–2, shorter than the lowest lemmas; **lemmas** to 2 cm, sometimes awned, awns about 4 mm; **anthers** 3; **styles** 3; **paleas** 2-keeled, not exceeding the lemmas. $x = 12$. Name from the Latin *arundo*, 'reed'.

Arundinaria is a north-temperate genus with three native North American species. The most consistent differences among the North American species are seen in the vegetative characters, including the topknot leaf blades, foliage leaf blades, and features of the branch complements.

Arundinaria is sometimes taken as including several Asian species. Genera that used to be treated in *Arundinaria* include, for example, *Fargesia* Franch. and *Sasa* Makino & Shibata.

SELECTED REFERENCES McClure, F.A. 1973. Genera of bamboos native to the New World (Gramineae: Bambusoideae). Smithsonian Contr. Bot. 9:1–148; Judziewicz, E.J., L.G. Clark, X. Londoño, and M.J. Stern. 1999. American Bamboos. Smithsonian Institution Press: Washington, D.C., U.S.A. 392 pp.; Zhu, Z., C-D. Chu, and C. Stapleton. 2006. *Arundinaria*. Pp. 113–115 *in* Z.-Y. Wu, P.H. Raven, and D.-Y. Hong (eds.). Flora of China, vol. 22 (Poaceae). Science Press, Beijing, Peoples Republic of China and Missouri Botanical Garden Press, St. Louis, Missouri, U.S.A. 653 pp. http://flora.huh.harvard.edu/china/mss/volume22/index.htm.

1. Primary branches with 0–1 compressed basal internodes; culm internodes usually sulcate; culm leaves deciduous . . . . .1. *A. gigantea*
1. Primary branches with 2–5 compressed basal internodes; culm internodes usually terete; culm leaves persistent to tardily deciduous.

2. Foliage blades coriaceous, persistent, abaxial surfaces densely pubescent or glabrous, strongly cross veined; primary branches usually longer than 50 cm, basal nodes developing secondary branches; topknot blades 20–30 cm long . . . . .2. *A. tecta*
2. Foliage blades chartaceous, deciduous, abaxial surfaces pilose or glabrous, weakly cross veined; primary branches usually shorter than 35 cm, basal nodes not developing secondary branches; topknot blades 9–22.5 cm long . . . . .3. *A. appalachiana*

1. Arundinaria gigantea (Walter) Muhl. [p. 19]
RIVER CANE, GIANT CANE

Rhizomes normally remaining horizontal, sometimes hollow-centered, air canals absent. **Culms** 2–8 m tall, to 3 cm thick; **internodes** typically sulcate distal to the branches. **Culm leaves** deciduous; **sheaths** 9–15 cm; **fimbriae** 2.2–7 mm; **blades** 1.5–3.5 cm. **Topknots** of 6–8 leaves; **blades** 16–24 cm long, 2–3.2 cm wide, lanceolate to ovate-lanceolate. **Primary branches** to 25 cm, erect or nearly so, with 0–1 compressed basal internodes, lower elongated internodes flattened in cross section. **Foliage leaves: abaxial ligules** usually ciliate, sometimes glabrous; **blades** subcoriaceous, persistent, evergreen, 8–15 cm long, 0.8–1.3 cm wide, bases rounded, abaxial surfaces glabrous or pubescent, strongly cross veined, adaxial surfaces glabrous or almost so. **Spikelets** 4–7 cm, greenish or brownish, with 8–12 florets. **Glumes** unequal, glabrous or pubescent, lowest glumes obtuse to acuminate or absent; **lemmas** 1.2–2 cm, usually appressed-hirsute to canescent, sometimes pubescent only towards the base and margins. **Caryopses** oblong, beaked, without a style branch below the beaks. $2n = 48$.

Arundinaria gigantea forms extensive colonies in low woods, moist ground, and along river banks. It was once widespread in the southeastern United States, but cultivation, burning, and overgrazing have destroyed many stands.

2. Arundinaria tecta (Walter) Muhl. [p. 20]
SWITCH CANE

Rhizomes normally horizontal for only a short distance before turning up to form a culm, hollow-centered, air canals present. **Culms** usually shorter than 2.5 m tall, to 2 cm thick; **internodes** terete in the vegetative parts. **Culm leaves** persistent to tardily deciduous; **sheaths** 11–18 cm; **fimbriae** 1.5–8.5 mm; **blades** 2.5–4 cm. **Topknots** of 9–12 leaves; **blades** 20–30 cm long, 1.8–3.2 cm wide, lanceolate to ovate-lanceolate.

Primary branches usually 50+ cm, basally erect and distally arcuate, terete, with 3–4 compressed basal internodes, basal nodes developing secondary branches, lower elongated internodes terete in cross section. **Foliage leaves: abaxial ligules** fimbriate to lacerate, sometimes ciliate; **blades** 7–23 cm long, 1–2 cm wide, coriaceous, persistent, evergreen, bases rounded, abaxial surfaces densely pubescent or glabrous, strongly cross veined, adaxial surfaces pubescent. **Spikelets** 3–5 cm, with 6–12 florets, the first occasionally sterile. **Glumes** unequal, glabrous or pubescent; **lowest glume** obtuse to acuminate or absent; **lemmas** 1.2–2 cm, glabrous or nearly so. **Caryopses** oblong, beaked, a rudimentary hooked style branch present below the beak. $2n =$ unknown.

Arundinaria tecta grows in swampy woods, moist pine barrens, live oak woods, and along the sandy margins of streams, preferring moister sites than *A. gigantea*. It grows only on the coastal plain of the southeastern United States.

3. Arundinaria appalachiana Triplett, Weakley & L.G. Clark [p. 20]
HILL CANE

Rhizomes normally horizontal for only a short distance before turning up to form a culm, sometimes hollow-centered, air canals sometimes present. **Culms** 0.5–1 (1.8) m tall, 0.2–0.6 cm thick; **internodes** terete. **Culm leaves** persistent to tardily deciduous; **sheaths** 5.5–11 cm; **fimbriae** 1–4.6 mm; **blades** 0.8–1.4 cm. **Topknots** of 6–12 leaves; **blades** 9–22.5 cm long, 1.4–2.8 cm wide, linear, linear-lanceolate, or ovate-lanceolate. **Primary branches** usually shorter than 35 cm, erect, terete, with 2–5 compressed basal internodes, basal nodes not developing secondary branches. **Foliage leaves: abaxial ligules** glabrous or ciliate, fimbriate or lacerate; **blades** 5–20 cm long, 0.8–2 cm wide, chartaceous, deciduous, bases rounded, abaxial surfaces pilose or glabrous, weakly cross veined, adaxial surfaces pilose. **Spikelets** 3–5.5 cm, usually somewhat reddish purple, with 5–8 florets. $2n =$ unknown.

Arundinaria appalachiana grows on moist to dry slopes and in seeps. It is restricted to the southern Appalachians and upper piedmont.

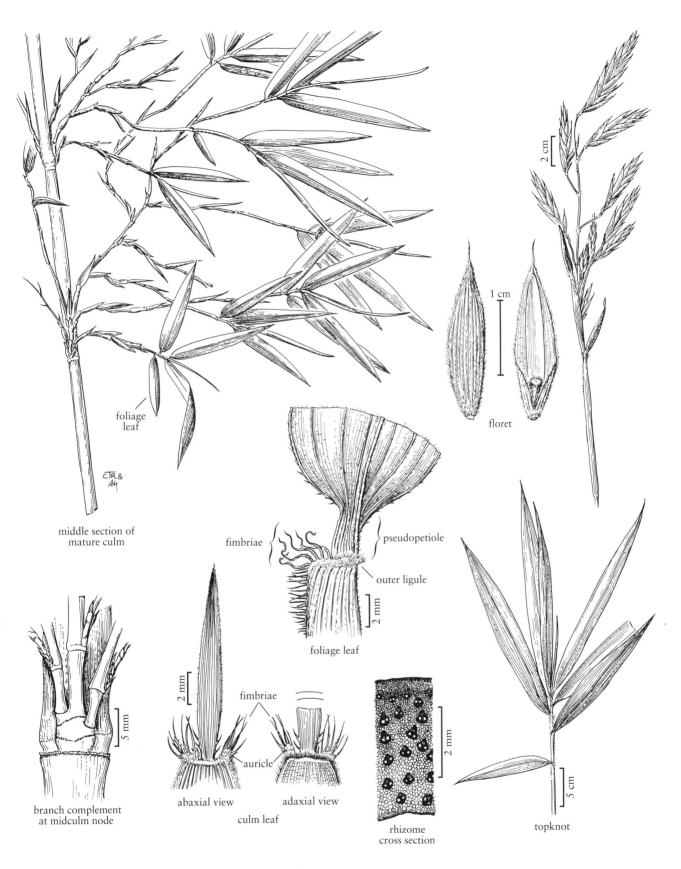

foliage
leaf

middle section of
mature culm

CTR &
AM

branch complement
at midculm node

fimbriae

auricle

abaxial view　　adaxial view

culm leaf

fimbriae

pseudopetiole

outer ligule

2 mm

foliage leaf

floret

1 cm

2 cm

rhizome
cross section

2 mm

topknot

5 cm

A. gigantea

ARUNDINARIA

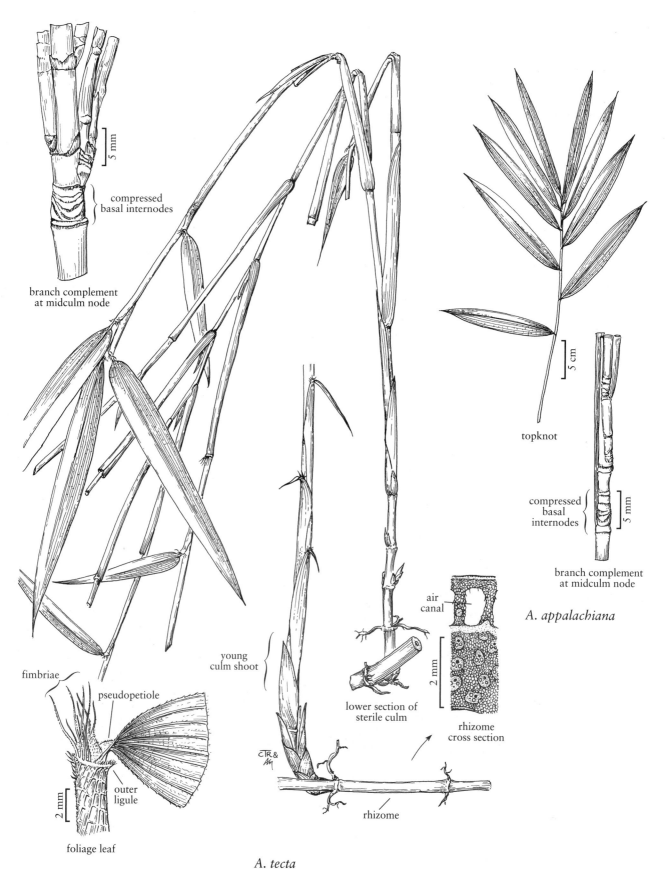

5 mm

compressed
basal internodes

branch complement
at midculm node

fimbriae

pseudopetiole

outer
ligule

2 mm

foliage leaf

young
culm shoot

lower section of
sterile culm

rhizome

air
canal

2 mm

rhizome
cross section

5 cm

topknot

compressed
basal
internodes

5 mm

branch complement
at midculm node

A. appalachiana

A. tecta

ARUNDINARIA

2.02 BAMBUSA Schreb.

Christopher M.A. Stapleton

Plants usually arborescent, in well defined or rather loose clumps; **rhizomes** pachymorphic, with short necks. **Culms** 0.5–30(35) m tall, 0.5–18(20) cm thick, woody, perennial, usually self-supporting; **nodes** not swollen; **supranodal ridges** obscure; **internodes** terete, usually thinly covered initially with light-colored wax. **Branch complements** usually with a dominant primary central branch and 2 smaller co-dominant lateral branches, usually similar at all nodes; **bud scales** 2-keeled, thickened, initially closed at the back and front; **branches** all subtended by bracts, higher order branchlets at the lower nodes sometimes thornlike. **Culm leaves** usually promptly deciduous, initially lightly waxy, sometimes with short, stiff hairs, subsequently losing the wax and becoming glabrous; **auricles** usually well developed; **fimbriae** usually present; **blades** triangular to broadly triangular, usually erect. **Foliage leaves: sheaths** usually deciduous from the lower nodes of the branches, persistent at the distal nodes; **blades** to 30 cm long, to 6 cm wide, not distinctly cross veined. **Inflorescences** usually spicate, rarely capitate, bracteate; **prophylls** 2-keeled, narrow. **Pseudospikelets** 1–5 cm, with 3–12 florets; **disarticulation** above the glumes and below the florets, rapid; **rachilla internodes** usually long. **Glumes** several, subtending the buds; **lemmas** narrowly ovate, acute, unawned; **paleas** not exceeding the lemmas, 2-keeled, not winged; **anthers** 6; **ovaries** usually suboblong; **styles** short, with (2)3–4 plumose branches. $2n$ = 56–72. Name a Latinized form of *bambu*, a local name of Malayan origin.

Bambusa is a tropical and subtropical genus of 75–100+ species. It is native to southern and southeastern Asia, but is widely cultivated and naturalized throughout the tropics. *Bambusa vulgaris* and *B. multiplex* grow widely in Florida and Texas, having spread to some extent after being planted as ornamentals. Other species are known only in cultivation. The American Bamboo Society lists over 40 species as being commercially available in North America in 2005. This treatment includes a few of the more commonly cultivated species.

SELECTED REFERENCES **But, P.P.-H., L.-C. Chia, H.-L.F., and S.-Y. Hu.** 1985. Hong Kong Bamboos. Urban Council, Hong Kong. 85 pp.; **Dransfield, S. and E.A. Widjaja** (eds.). 1995. Plant Resources of South-East Asia [PROSEA] No. 7: Bamboos. Backhuys, Leiden, The Netherlands. 189 pp.; **Edelman, D.K., T.R. Soderstrom, and G.F. Deitzer.** 1985. Bamboo introduction and research in Puerto Rico. J. Amer. Bamboo Soc. 6: 43–57; **McClure, F.A.** 1955. *Bambusa* Schreb. Fieldiana, Bot. 24²: 52–60; **Pohl, R.W.** 1994. *Bambusa* Schreber. Pp. 193–194 *in* G. Davidse, M. Sousa S., and A.O. Chater (eds.). Flora Mesoamericana, vol. 6: Alismataceae a Cyperaceae. Universidad Nacional Autónoma de México, Instituto de Biología, México, D.F., México. 543 pp.; **Soderstrom, T.R. and R.P. Ellis.** 1988. The woody bamboos (Poaceae: Bambuseae) of Sri Lanka: A morphological–anatomical study. Smithsonian Contr. Bot. 72:1–75; **Stapleton, C.M.A.** 1994. The bamboos of Nepal and Bhutan, Part I: *Bambusa, Dendrocalamus, Melocanna, Cephalostachyum, Teinostachyum,* and *Pseudostachyum* (Gramineae: Poaceae, Bambusoideae). Edinburgh J. Bot. 51:1–32; **Stapleton, C.M.A.** 2002. *Bambusa ventricosa* versus *Bambusa tuldoides*. Bamboo 23:17–18; **Widjaja, E.A.** 1997. New taxa in Indonesian bamboos. Reinwardtia 11:57 152; **Wong, K.M.** 1995. The Morphology, Anatomy, Biology, and Classification of Peninsular Malayan Bamboos. University of Malaya Botanical Monographs No. 1. University of Malaya, Kuala Lumpur, Malaysia. 189 pp.; **Xia, N.H.** and **C.M.A. Stapleton.** 1997. Typification of *Bambusa bambos* (Gramineae, Bambusoideae). Kew Bull. 52:693–698.

1. Branchlets of the lower branches recurved, hardened, thornlike . . . . 1. *B. bambos*
1. Branchlets of the lower branches not thornlike.
    2. Culm sheath auricles well developed, to 5 cm long . . . . 2. *B. vulgaris*
    2. Culm sheath auricles absent or poorly developed.
        3. Culm internodes antrorsely hispid; culms 0.5–7 m tall, broadly arched above . . . . 3. *B. multiplex*
        3. Culm internodes glabrous; culms 6–15 m tall, erect . . . . 4. *B. oldhamii*

1. Bambusa bambos (L.) Voss [not illustrated]
GIANT THORNY BAMBOO

Plants densely clumped, with intertwined thorny branches. **Culms** to 20(35) m tall, 12–18 cm thick, thick-walled, sometimes almost solid; **internodes** 20–40 cm, green, waxy at first, becoming dull. **Branches** forming at the basal and upper nodes, central branches slightly dominant, branchlets of the lower branches recurved, hardened and thornlike. **Culm leaves** dark green, initially sparsely hairy, sometimes more densely hairy on the margins and auricles, hairs dark brown, deciduous; **auricles** subequal, wrinkled, wide; **fimbriae** absent; **ligules** to 2 mm, ciliate; **blades** erect or reflexed, merging into the auricles, adaxial surfaces densely brown-velvety. **Foliage leaves: sheaths** glabrous; **ligules** short, entire; **auricles** small; **fimbriae** few, erect; **blades** 6–22 cm long, 1–3 cm wide, glabrous. **Inflorescences** initially spicate, becoming dense globular clusters. **Pseudospikelets** 10–30 mm, with 3–7 florets. **Lemmas** 7–8 mm, glabrous; **anthers** to 5 mm. $2n = 70$–72.

Bambusa bambos is native to India and Indochina, but is cultivated throughout the tropics. It was the first bamboo species to be given a scientific name, being described as treelike, thorny, and a source of tabashir, lumps of pure silica that form in the internodal cavities. *Bambusa arundinacea* (Retz.) Willd. is a synonym of *B. bambos* that still appears in some listings of bamboos.

2. Bambusa vulgaris Schrad. *ex* J.C. Wendl.
[not illustrated]
COMMON BAMBOO

Plants forming moderately loose clumps, without thorny branches. **Culms** 10–20 m tall, 4–10 cm thick, erect, sinuous or slightly flexuous; **nodes** slightly inflated, flaring at the pubescent sheath scar; **internodes** 20–45 cm, glossy green, yellow, yellow with green stripes, or green with yellowish green stripes, all similar or the basal internodes swollen and shorter than those above. **Branches** developing from the midculm nodes and above, occasionally also at the lower nodes, several to many branches per node, branchlets of the lower branches not thornlike. **Culm leaves** promptly deciduous, with dense, appressed, brown pubescence, lower sheaths broader than long, apices broader than the base of the blades; **auricles** well developed, to 5 cm long and 1.5 cm wide, equal, ovoid to falcate-spreading, dark; **fimbriae** to 15 mm, dense, wavy, light; **blades** 4–5 cm long, 5–6 cm wide, appressed to the culm, usually persistent, triangular, abaxial surfaces glabrous, adaxial surfaces densely dark pubescent towards the base, basal margins ciliate or with stiff hairs; **ligules** about 3 mm, shortly ciliate. **Foliage leaves: sheaths** glabrous to sparsely hispidulous; **ligules** 0.5–1.5 mm, glabrous, truncate, entire; **auricles** 0.5–1.5 mm, falcate, hardened,

persistent; **fimbriae** few, 0.5–1.5 mm, spreading; **blades** 6–30 cm long, 1–4 cm wide, glabrous, abruptly acuminate. **Pseudospikelets** 12–35 mm, with 5–10 florets, always strongly grooved along the center, appearing 2-cleft. $2n = 64$.

Bambusa vulgaris probably originated in tropical Asia. It is now the most widely cultivated tropical bamboo, largely because of the ease with which the branches and culm sections take root. Many different cultivars exist, including forms with variously green and yellow-striped culms which are sometimes placed in distinct varieties or even species. 'Wamin' is a cultivated form with ventricose to very short, concertina-like internodes. Like *B. tuldoides* 'Buddha's-Belly', plants of *B. vulgaris* 'Wamin' can develop abbreviated internodes when grown in pots or under extreme environmental conditions; they readily return to normal growth when these conditions are ameliorated.

3. Bambusa multiplex (Lour.) Raeusch. ex Schult. & Schult. f. [p. 23]
HEDGE BAMBOO

Plants densely clumping, without thorny branches. **Culms** 0.5–7 m tall, 1–2.5 cm thick, emerging at an angle, broadly arching above, usually thin-walled and hollow, solid in some cultivars; **nodes** not swollen; **internodes** all similar, 3–60 cm. **Branches** to 20 per node, erect to spreading, the central branch slightly dominant, often becoming densely congested and forming tangled clusters of rhizomes, aborted shoots, and stunted roots, branchlets of the lower branches not thornlike. **Culm leaves** 12–15 cm, narrowly triangular, tardily deciduous, initially light green, becoming reddish brown to stramineous, glabrous; **auricles** and **fimbriae** developed; **blades** 1–2 cm, initially appressed to the culm, initially antrorsely hispid on both surfaces, becoming glabrous. **Foliage leaves: sheaths** glabrous; **ligules** to 0.5 mm; **auricles** absent; **fimbriae** sometimes present; **blades** 7–15 cm long, 1–2 cm wide, abaxial surfaces glaucous and slightly pubescent, adaxial surfaces dark green and glabrous. **Pseudospikelets** 30–40 mm, with up to 10 florets. $2n = 72$.

Bambusa multiplex is native to southeast Asia. It is now widely planted around the world. The dense foliage with many leaves on each branchlet makes it well suited to hedging. A large number of cultivars are available, some with striped culms and leaves, others with greatly reduced stature and leaf size suitable for bonsai culture or hedging. The tangled branch clusters allow natural dispersal and easy propagation in hot, humid climates. Plants listed as *B. glaucescens* (Willd.) Sieb. *ex* Munro in North America probably belong to *B. multiplex*.

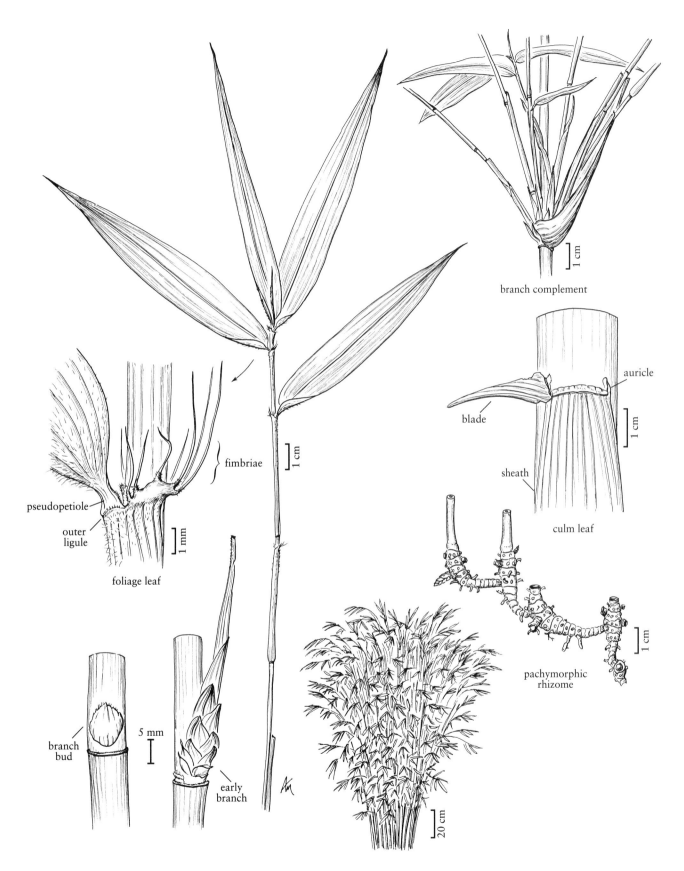

branch complement

auricle

blade

sheath

culm leaf

fimbriae

pseudopetiole

outer
ligule

foliage leaf

pachymorphic
rhizome

branch
bud

early
branch

B. multiplex

BAMBUSA

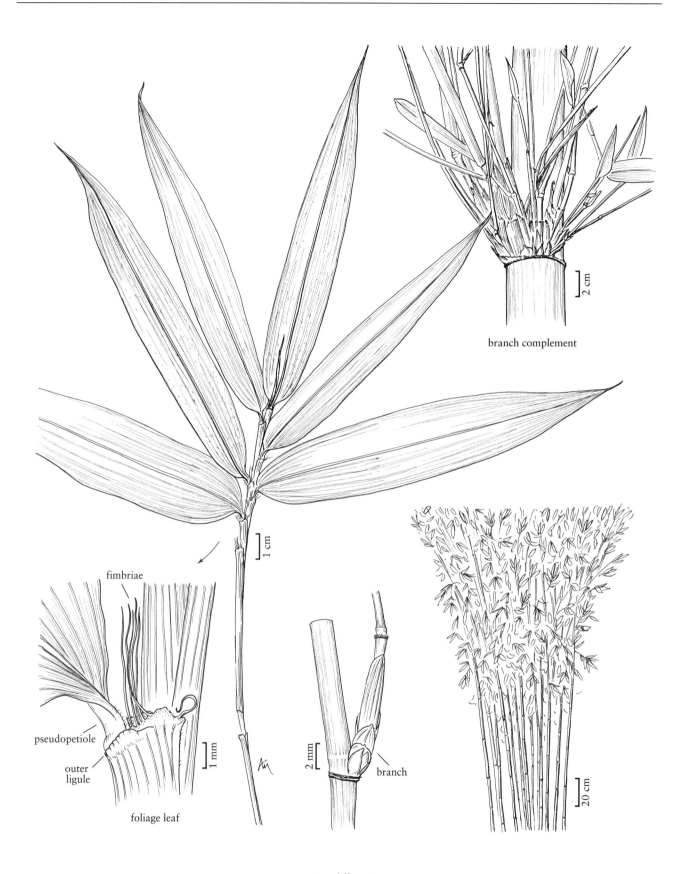

branch complement

fimbriae

pseudopetiole

outer
ligule

foliage leaf

branch

B. oldhamii

BAMBUSA

4. Bambusa oldhamii Munro [p. 24]

OLDHAM'S BAMBOO

Plants forming dense to moderately loose clumps, without thorny branches. **Culms** 6–15 m tall, 3–13 cm thick, erect; **internodes** all similar, hollow, walls about 1 cm thick, pale green, glabrous, glaucous below the nodes. **Branches** very short, not thorny, the central branch at each node often tardily developed, branches not developing from the lower nodes, branchlets of the lower branches not thornlike. **Culm leaves** promptly deciduous, oblong, initially brown-sericeous, becoming glabrous, rounded distally; **auricles** absent or very small and rounded; **fimbriae** few, to 3 mm, curved; **ligules** to 2 mm, entire or finely serrulate; **blades** broadly subtriangular, usually with concave margins, abaxial surfaces glabrous, adaxial surfaces antrorsely hispid, apices acuminate. **Foliage leaves: sheaths** striate, glabrous or sparsely hispidulous, margins very shortly ciliate; **auricles** very small, rounded; **fimbriae** many, to 5 mm, fine, wavy; **ligules** to 1 mm, truncate, glabrous, entire; **blades** 15–30 cm long, 3–6 cm wide, oblong-lanceolate, abruptly acuminate, abaxial surfaces pubescent initially, becoming glabrous, adaxial surfaces glabrous. **Pseudospikelets** with 6–10 florets. $2n =$ unknown.

Bambusa oldhamii is native to low-lying areas of eastern China and Taiwan. It is the most commonly grown large, clump-forming bamboo in the United States, where it is grown mostly in Florida and California. With its upright culms and short branches it makes an excellent tall hedge.

2.03 PHYLLOSTACHYS Siebold & Zucc.

Christopher M.A. Stapleton

Mary E. Barkworth

Plants shrublike to arborescent, in open or dense, spreading clumps or thickets; **rhizomes** leptomorphic. **Culms** 3–10(20) m tall, 3–10(15) cm thick, self-supporting, erect or nodding, diffuse or pluricespitose, rarely solitary; **nodes** slightly swollen; **supranodal ridge** prominent; **internodes** strongly flattened for their whole length, doubly sulcate above the branches, glabrous, smooth. **Branches** 2(3) per midculm node, unequal, initially erect, becoming deflexed, basal internodes not compressed. **Culm leaves** coriaceous, very quickly deciduous; **blades** usually strap-shaped and narrow, usually reflexed. **Foliage leaves: sheaths** deciduous; **blades** small to medium-sized, usually glossy and thickened, indistinctly cross veined. **Inflorescences** open or congested, sometimes spicate to subcapitate, fully bracteate, bracts usually bearing a small blade at the apex. **Spikelets** or **pseudospikelets** with 2 to several florets, the uppermost rudimentary. **Lemmas** lanceolate; **paleas** not exceeding the lemmas, strongly to very weakly 2-keeled, often bifid; **anthers** 3; **styles** or **style branches** 3. $x = 12$. Name from the Greek *phyllos*, 'leaf', and *stachys*, 'spike', referring to the reduced blades often seen on persistent bracts proximal to the spikelets.

Phyllostachys is a hardy, temperate, Asiatic genus of at least 50 species, native mainly to China, from Hainan to the Yellow River, and from Yunnan to Taiwan, but introduced to surrounding countries, especially Japan. Many species and a large number of cultivars have been introduced. The genus is characterized by the two unequal branches at most nodes, a result of a complete lack of internodal compression, along with the almost universal presence of buds at all nodes. *Phyllostachys* is the most distinct genus of hardy temperate bamboos, of enormous economic importance in eastern Asia, and increasingly valued in North America and Europe.

All species are ornamental, especially those having cultivars with colored culms. Almost all species are likely to be invasive. Rhizomes may extend as far as the height of the culms. Root barriers should generally be installed if uncontrolled spreading is not acceptable.

SELECTED REFERENCES **McClure. F.A.** 1957. Bamboos of the Genus *Phyllostachys* Under Cultivation in the United States. U.S. Department of Agriculture, Agricultural Research Service, Agriculture Handbook No. 114. U.S. Government Printing Office, Washington, D.C., U.S.A. 69 pp.; **Wang, C.-P., Z.-H. Yu,** and **G.-H. Ye.** 1980. A taxonomical study of *Phyllostachys* in China [parts 1 & 2]. Acta Phytotax. Sin. 18:15–19, 168–193.

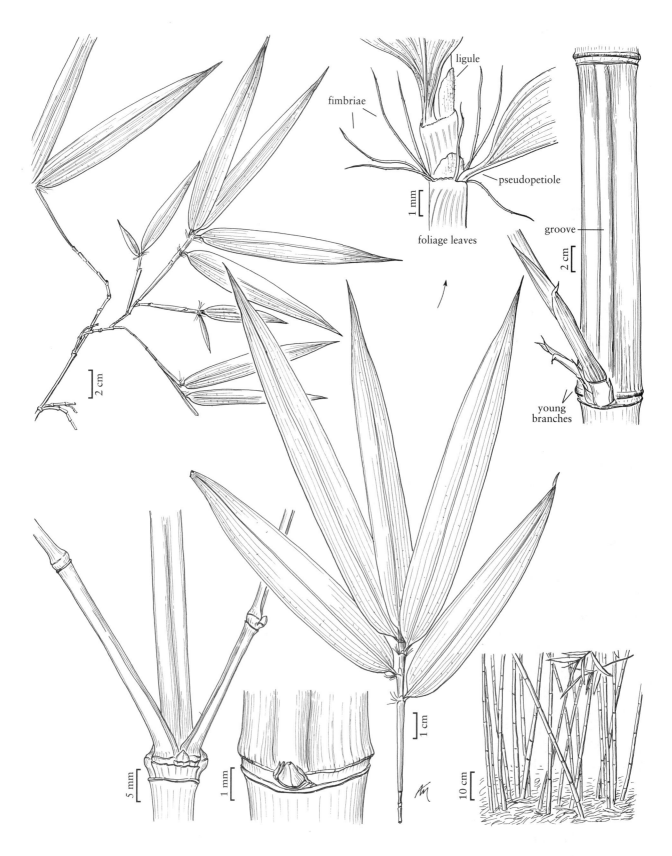

ligule

fimbriae

pseudopetiole

1 mm

foliage leaves

groove

2 cm

young
branches

2 cm

5 mm

1 mm

1 cm

10 cm

P. bambusoides

PHYLLOSTACHYS

1. Neither auricles nor fimbriae present on any culm leaves . . . . 1. *P. aurea*
1. Auricles present on the upper culm leaves; fimbriae present on all culm leaves . . . . 2. *P. bambusoides*

1. Phyllostachys aurea Carrière *ex* Rivière & C. Rivière [not illustrated]

FISHPOLE BAMBOO, GOLDEN BAMBOO

Culms to 10 m tall, 1–4 cm thick, straight; **internodes** glabrous, initially green, becoming gray, glaucous soon after sheath-fall, some culms in every clump with 1 to several short internodes; **nodal ridges** moderately prominent; **sheath scars** not flared, fringed with short, persistent, white hairs. **Culm leaves: sheaths** with a basal line of minute white hairs, otherwise glabrous, pale olive-green to rosy-buff, with a sparse scattering of small brown spots and wine-colored or pale green veins, not glaucous; **auricles** and **fimbriae** absent; **ligules** short, slightly rounded, ciliate; **blades** lanceolate, somewhat crinkled below, upper blades pendulous. **Foliage leaves: auricles** and **fimbriae** well developed or lacking; **ligules** very short, glabrous or sparsely ciliolate; **blades** 4–15 cm long, 5–23 mm wide. $2n = 48$.

Phyllostachys aurea is native to China, but it is widely cultivated in temperate and subtropical regions. In North America, it grows as far north as Vancouver, British Columbia, in the west and Buffalo, New York, in the east. The young shoots are very palatable, even when raw, but the mature culms are very hard when dried. They are sometimes used for fishpoles. This species differs from other species of *Phyllostachys*, including those with brighter yellow culms, in having a raised collar below the nodes and irregularly compressed basal culm nodes.

2. Phyllostachys bambusoides Siebold & Zucc. [p. 26]

GIANT TIMBER BAMBOO, MADAKE

Culms to 22 m tall and 15 cm thick, erect or leaning towards the light, base sinuous in some cultivars; **internodes** glabrous, usually green, in cultivars golden yellow, or with yellow and green stripes, lustrous; **nodal ridges** usually prominent (scarcely discernible in 'Crookstem' forms); **sheath scars** thin, not strongly flared, glabrous. **Culm leaves: sheaths** glabrous or pubescent, greenish to ruddy-buff, more or less densely dark-brown-spotted; **auricles** absent from the basal sheaths, narrow to broadly ovate or falcate on the upper sheaths; **fimbriae** greenish, crinkled; **ligules** rounded and ciliolate to truncate and ciliate with coarse hairs; **blades** short, lanceolate, reflexed and crinkled on the lower leaves, those above longer and recurved, green or variously striped. **Foliage leaves: auricles** and **fimbriae** usually well developed; **ligules** well developed; **blades** to 20 cm long and 3.2 cm wide, usually puberulent to subglabrous. $2n = 48$.

Phyllostachys bambusoides, a widely cultivated species, is hardy to −17°C. Several cultivars are available, differing in the color of their culms and leaves.

2.04 PSEUDOSASA Makino *ex* Nakai

Christopher M.A. Stapleton

Plants shrublike, spreading or loosely to densely clumped; **rhizomes** leptomorphic. **Culms** 0.5–13 m tall, to 4 cm thick, self-supporting, erect or nodding, pluricespitose; **nodes** not or slightly swollen; **supranodal ridge** not evident; **internodes** mainly terete, only slightly flattened immediately above the branches, glabrous, with light wax below the nodes. **Branches** initially 1–3, erect to arcuate, often short, central branch dominant, with compressed basal nodes, branches fully sheathed, lateral branches arising either from the basal nodes or from more distal nodes, sheaths and prophylls more or less glabrous, persistent, tough. **Culm leaves** coriaceous and very persistent; **blades** erect or reflexed, narrowly triangular to strap-shaped. **Foliage leaves: sheaths** persistent; **blades** cross veined, medium to large for the size of the culm, without marginal necrosis in winter, their arrangement random. **Inflorescences** racemose or paniculate; **branches** subtended by much reduced or quite substantial bracts. **Spikelets** 2–20 cm, with 3–30 florets; **rachillas** sinuous; **disarticulation** below the florets. **Glumes** 2, shorter than the first lemma; **lemmas** to 1 cm; **anthers** 3; **styles** 3; **paleas** 2-keeled. Named for the similar, but 6-anthered genus *Sasa* Makino & Shibata, in which it had been included.

Pseudosasa includes about 36 species, all of which are native to China, Japan, and Korea.

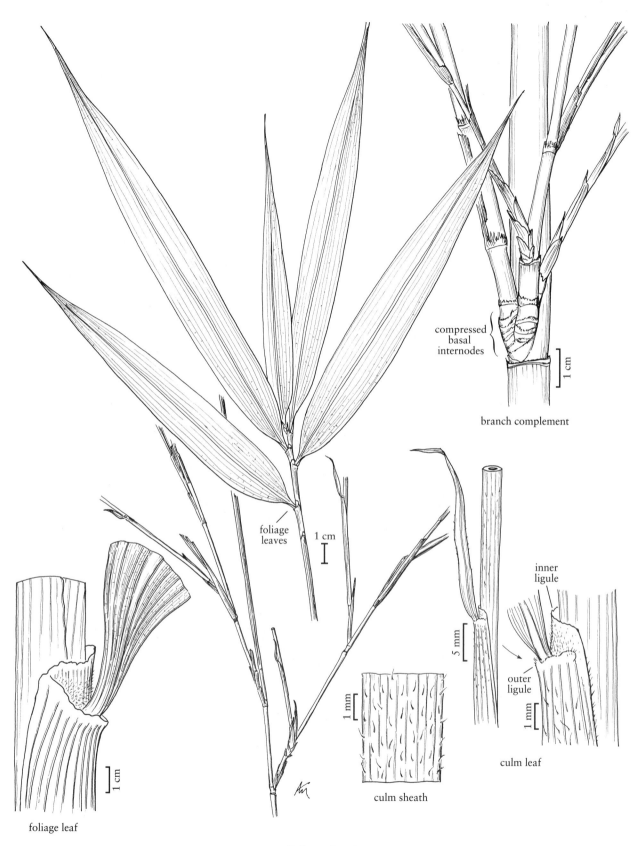

compressed
basal
internodes

1 cm

branch complement

foliage
leaves

1 cm

inner
ligule

outer
ligule

culm leaf

5 mm

1 mm

culm sheath

1 mm

1 cm

foliage leaf

P. japonica

PSEUDOSASA

1. **Pseudosasa japonica** (Siebold & Zucc. *ex* Steud.) Makino *ex* Nakai [p. 28]

JAPANESE ARROW BAMBOO, METAKE, YADAKE

Culms 1–3(5) m tall, to 1.5 cm thick, erect or nodding, finely ridged; **nodes** slightly raised; **sheath scar** large; **internodes** long, finely mottled, with a light ring of wax below the nodes. **Culm sheaths** to 25 cm, basally glabrous, distally appressed-hispid, persistent; **auricles** and **fimbriae** absent; **blades** 2–5 cm, erect, abaxial surfaces glabrous. **Branches** usually 1 per node, with no basal buds or branches, sometimes rebranching from more distal nodes. **Foliage leaves: sheaths** glabrous, edges membranous; **auricles** absent or small and erect; **fimbriae** absent or scarce, erect; **ligules** long, oblique, erose, slightly pubescent, abaxial ligules glabrous to finely ciliate; **blades** 15–35 cm long, 1.5–5 cm wide, glabrous or abaxial surfaces sporadically shortly red-brown tomentose, light green to glaucous, adaxial surfaces dark green, glossy, glabrous. **Spikelets** 3.5–10 cm, narrowly cylindrical, curved, with 5–20(25) florets. **Lemmas** 1.2–1.5 cm, glabrous, often mucronate, mucros about 2 mm; **paleas** nearly equaling the lemmas, glabrous, keels finely cilate.

Pseudosasa japonica is a widely cultivated ornamental species that used to be grown for arrows in Japan. There are no known wild populations. It forms a tough and effective screen, and has become naturalized in British Columbia and the eastern United States. A shorter cultivar with partially ventricose culms, 'Tsutsumiana', and cultivars with variegated leaves are also available.

3. OLYREAE Kunth

Mary E. Barkworth

Plants usually perennial; cespitose, stoloniferous. **Culms** perennial, 3–350 cm or climbing, not woody; **nodes** often with 2 thick circumferential ridges, with more elastic tissue between; **branches** usually not developed at the middle and upper nodes. **Leaves** often crowded towards the culm tips; **abaxial ligules** absent; **adaxial ligules** membranous; **pseudopetioles** 1–2 mm, not twisted; **blades** usually persistent, usually folding at night or when stressed, venation parallel, cross venation sometimes evident, particularly at the base, the bases and apices often asymmetric. **Inflorescences** spicate or paniculate, usually produced at the middle and upper nodes of the leafy culms; **disarticulation** above or below the glumes. **Spikelets** unisexual and dimorphic, usually mixed within an inflorescence, with 1 floret. **Pistillate spikelets** on clavate pedicels; **glumes** usually exceeding the floret, many-veined; **lemmas** usually coriaceous to indurate, pale when immature, mottled with dark spots or uniformly dark when mature, glabrous or with non-uncinate hairs, unawned; **paleas** usually shorter than and enclosed by the lemmas; **lodicules** 3; **style branches** 2(3). **Caryopses** dry; **embryos** small relative to the caryopses; **hila** usually linear. **Staminate spikelets** deciduous; **glumes** usually lacking; **lemmas** membranous or hyaline; **lodicules** 3 or absent; **anthers** 3, 2, or multiples of 6. $x = 7, 9, 10, 11$.

The *Olyreae* is primarily a New World tribe that extends from Mexico to Argentina and southern Brazil. It includes about 20 genera and 110 species. One species has been found in tropical Africa and Sri Lanka; it is not clear whether it is native or introduced to those regions. One species is sold as an ornamental in the United States. It is included here as a representative of this fascinating group of grasses.

SELECTED REFERENCE Judziewicz, E.J., L.G. Clark, X. Londoño, and M.J. Stern. 1999. American Bamboos. Smithsonian Institution Press, Washington, D.C., U.S.A. 392 pp.

3.01 LITHACHNE P. Beauv.

Mary E. Barkworth

Plants perennial; loosely cespitose. **Culms** 6–75 cm, not woody; **nodes** numerous, swollen. **Sheaths** open; **adaxial ligules** membranous; **pseudopetioles** 1–2 mm; **blades** lanceolate to ovate, folding downwards at night or when stressed, bases and apices asymmetric. **Inflorescences** racemes, partly enclosed by the leaf sheaths; **axillary racemes** unisexual or bisexual, if bisexual, the staminate spikelets below 1 to several pistillate spikelets, pedicels clavate; **terminal racemes** staminate, pedicels filiform. **Spikelets** with 1 floret. **Pistillate spikelets** much larger than the staminate spikelets, floret borne on a persistent peglike rachilla internode; **glumes** exceeding the floret, subequal, several-veined; **lemmas** helmet-shaped, indurate, laterally compressed, the margins from overlapping the palea margins to nearly concealing the paleas; **style branches** 2, plumose. **Staminate spikelets** lanceolate, early deciduous; **glumes** absent; **lemmas** translucent, 3-veined; **anthers** 3. *x* = 11. Name from the Greek *litho*, 'stony', and *achne*, 'scale' or 'chaff', a reference to the indurate lemma and palea of the pistillate floret.

Lithachne has four species. Its primary range extends south from Mexico and the Caribbean islands to Ecuador, but there are disjunct populations in Peru, southern Brazil, northern Argentina, and Paraguay. One species is sold as an ornamental in the United States.

SELECTED REFERENCES **Judziewicz, E.J.** 1990. Flora of the Guianas: 187. Poaceae (series ed. A.R.A. Görts-van Rijn). Koeltz Scientific Books, Koenigstein, Germany. 727 pp.; **Judziewicz, E.J., L.G. Clark, X. Londoño, and M.J. Stern.** 1999. American Bamboos. Smithsonian Institution Press, Washington, D.C., U.S.A. 392 pp.; **Paisooksantivatana, Y. and R.W. Pohl.** 1992. Morphology, anatomy and cytology of the genus *Lithachne* (Poaceae: Bambusoideae). Revista Biol. Trop. 40:47–72; **Pohl, R.W.** 1994. *Lithachne* P. Beauv. Pp. 215–216 *in* G. Davidse, M. Sousa S., and A.O. Chater (eds.). Flora Mesoamericana, vol. 6: Alismataceae a Cyperaceae. Universidad Nacional Autónoma de México, Instituto de Biología, México, D.F., México. 543 pp.; **Soderstrom, T.R.** 1980. A new species of *Lithachne* (Poaceae: Bambusoideae) and remarks on its sleep movements. Brittonia 32:495–501.

1. Lithachne humilis Soderstr. [p. 31]
SMALL LITHACHNE

Culms 15–27 cm, erect; **nodes** 5–6, glabrous, with prominent infra- and supranodal ridges. **Foliage leaves: sheaths** glabrous or puberulent, hairs to 0.2 mm; **pseudopetioles** 0.7–1 mm; **ligules** about 0.2 mm, petiolar region sparsely pilose abaxially, hirsute adaxially; **blades** 2–6 cm long, 0.4–1 cm wide, bases asymmetric, midveins thickened near the base, both surfaces hairy, abaxial surfaces more densely hairy, hairs about 0.2 mm. **Pistillate racemes** exserted from the lower leaf sheaths, with 1 spikelet; **rachis** 8–10 mm, with a thin, rudimentary flap of tissue below the pistillate spikelet. **Pistillate spikelets** 7.3–9.5 mm, triangular in side view, gibbous; **glumes** glabrous; **lower glumes** 5.9–9.5 mm, 7–9-veined; **upper glumes** 5–7.5 mm, 5–7-veined; **florets** subtended by a persistent rachilla internode of about 1 mm; **lemmas** 2.8–3.5 mm, white, glabrous, indurate; **paleas** slightly shorter than the lemmas, 2-veined, veins inconspicuous; **lodicules** 3;

ovaries with 1 style and 2 plumose style branches. **Staminate racemes** 1–4, exserted terminally, extending about 2 cm beyond the subtending leaf sheath, with 1–2 pairs of unequally pedicellate spikelets, an upper short-pedicellate spikelet and a terminal spikelet. **Staminate spikelets** 4.2–7.5 mm; **glumes** absent; **lemmas** membranous, 3-veined, glabrous; **paleas** 2-veined; **lodicules** 3; **anthers** 3, 3.5–5.4 mm, yellow, almost basifixed. **Caryopses** brown, gibbous; **hila** linear. 2*n* = unknown.

Lithachne humilis is endemic to Honduras. It is sold as an ornamental in the United States. Soderstrom (1980) postulated that the flap below the pistillate spikelet is a vestige of a staminate spikelet; the wide-ranging *L. pauciflora* (Sw.) P. Beauv. has 1–5 staminate spikelets below each of the pistillate spikelets, while the narrow endemics *L. pineti* (C. Wright *ex* Griseb.) Chase and *L. horizontalis* Chase (from eastern Brazil and eastern Cuba, respectively) have no staminate spikelets and no flap of tissue beneath the pistillate spikelets.

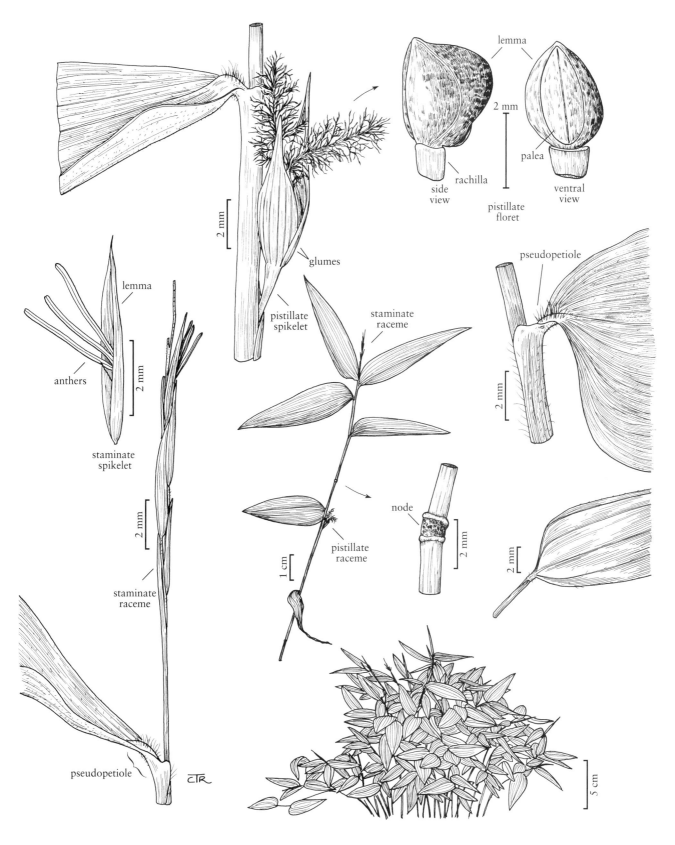

L. humilis

LITHACHNE

3. EHRHARTOIDEAE Link

Grass Phylogeny Working Group

Plants annual or perennial. **Culms** annual, sometimes woody, hollow or solid. **Leaves** distichous; **sheaths** open; **auricles** sometimes present; **abaxial ligules** absent; **adaxial ligules** membranous, scarious, or of hairs; **pseudopetioles** sometimes present; **blades** rarely cordate or sagittate at the base, venation parallel; **mesophyll** not radiate; **adaxial palisade layer** usually absent; **fusoid cells** sometimes present; **arm cells** absent or present; **Kranz anatomy** not developed; **midribs** simple or complex; **adaxial bulliform cells** present; **stomates** with dome-shaped or triangular subsidiary cells; **bicellular microhairs** present, terminal cells tapered; **papillae** sometimes present. **Inflorescences** panicles, racemes, or spikes, rarely with bracts other than those of the spikelets; **disarticulation** usually above the glumes, sometimes beneath the spikelets or at the base of the primary branches. **Spikelets** bisexual or unisexual, with 1 pistillate or bisexual floret, sometimes with 1–2 sterile florets below the functional floret. **Glumes** absent or 2; **lemmas** without uncinate hairs, sometimes terminally awned, awns single; **paleas** well-developed, lacking in sterile florets; **lodicules** 2, usually membranous, rarely fleshy, heavily vascularized; **anthers** (1)3–6(16); **ovaries** glabrous, without an apical appendage; **styles** 2, free to the base to fused throughout, 2-branched. **Fruits** caryopses or achenes; **hila** long-linear; **endosperm** without lipid, usually containing compound starch grains, rarely with simple starch grains; **embryos** to $^1/_3$ the length of the caryopses; **epiblasts** usually present; **scutellar cleft** usually present; **mesocotyl internode** absent or very short; **embryonic leaf margins** usually overlapping. $x = 12 \ (10, 15, 17)$.

The *Ehrhartoideae* encompasses three tribes, one of which, the *Oryzeae*, is native to the *Flora* region; the *Ehrharteae* is represented by introduced species. The third tribe, *Phyllorachideae* C.E. Hubb., is native to Africa and Madagascar. It was included in the subfamily on the basis of its morphological similarity to the other two tribes. There are approximately 120 species in the *Ehrhartoideae*. They grow in forests, open hillsides, and aquatic habitats.

Molecular data provide strong support for the close relationship of the *Oryzeae* and *Ehrharteae* (Grass Phylogeny Working Group 2001). Morphologically, they are characterized by spikelets that have a distal unisexual or bisexual floret with up to two proximal sterile florets, and the frequent presence of six stamens in the staminate or bisexual florets.

SELECTED REFERENCES **Gibbs Russell, G.E.,** L. Watson, M. Koekemoer, L. Smook, N.P. Barker, H.M. Anderson, and M.J. Dallwitz. 1991. Grasses of Southern Africa (ed. O.A. Leistner). National Botanic Gardens, Botanical Research Institute, Pretoria, Republic of South Africa. 437 pp.; **Grass Phylogeny Working Group.** 2001. Phylogeny and subfamilial classification of the grasses (Poaceae). Ann. Missouri Bot. Gard. 88:373–457.

1. Spikelets with 2 sterile florets below the functional floret, both well-developed, at least the upper sterile floret as long as or longer than the functional floret; glumes from $^1/_2$ as long as the spikelets to exceeding the florets; culms not aerenchymatous; plants of dry to damp habitats . . . . 4. *Ehrharteae*
1. Spikelets with 0–2 sterile florets below the functional floret, when present, sterile florets $^1/_8 - ^9/_{10}$ as long as the functional floret; glumes absent or highly reduced; culms aerenchymatous; plants of wet habitats . . . . 5. *Oryzeae*

4. EHRHARTEAE Nevski

Mary E. Barkworth

Plants annual or perennial. **Culms** annual, (1)6–200 cm, sometimes woody, not aerenchymatous, sometimes branched above the base. **Sheaths** open, usually rounded on the back, glabrous or not, sometimes scabrous; **collars** frequently with tuberculate hairs; **auricles** usually present, often ciliate; **ligules** usually membranous, sometimes merely a membranous rim or of hairs; **pseudopetioles** not present; **blades** linear, venation parallel, cross venation not evident , abaxial surfaces with microhairs and variously shaped silica bodies, cross sections non-Kranz; **first seedling leaves** with well-developed, erect blades. **Inflorescences** terminal, panicles or unilateral racemes; **disarticulation** above the glumes, florets falling as a cluster. **Spikelets** solitary, terete or laterally compressed, with 3 florets, lower 2 florets sterile, terminal floret bisexual, at least the upper sterile floret as long as or longer than the bisexual floret; **rachillas** sometimes shortly prolonged beyond the base of the bisexual floret. **Glumes** 2, from $^1/_2$ as long as to exceeding the florets, (3)5–7-veined. **Sterile florets: lemmas** coriaceous, 5–7-veined, awned or unawned; **paleas** lacking. **Bisexual florets: lemmas** lanceolate or rectangular, firmly cartilaginous to coriaceous, 5–7-veined, veins inconspicuous, apices entire, unawned; **paleas** 0–2(5)-veined; **lodicules** 2, free; **anthers** 1–6; **styles** 2, fused or free to the base, stigmas linear, plumose. **Caryopses** ellipsoid; **hila** linear, at least $^1/_2$ as long as the caryopses; **embryos** up to $^1/_3$ the length of the caryopses, waisted, without an epiblast, with a scutellar tail and a minute mesocotyl internode. $x = 12$.

The number of genera recognized in the *Ehrharteae* varies from one to four (Willemse 1982; Edgar and Connor 2000; Wheeler et al. 2002). The largest genus, *Ehrharta*, is native to Africa, the other three being Australasian. Only one genus, *Ehrharta*, has been found in the *Flora* region.

SELECTED REFERENCES **Edgar, E.** and **H.E. Connor.** 2000. Flora of New Zealand, vol. 5. Manaaki Whenua Press, Lincoln, New Zealand. 650 pp.; **Wheeler, D.J.B., S.W.L. Jacobs,** and **R.D.B. Whalley.** 2002. Grasses of New South Wales, ed. 3. University of New England, Armidale, New South Wales, Australia. 445 pp.; **Willemse, L.P.M.** 1982. A discussion of the Ehrharteae (Gramineae) with special reference to the Malesian taxa formerly included in *Microlaena*. Blumea 28:181–194.

4.01 EHRHARTA Thunb.

Mary E. Barkworth

Plants annual or perennial; synoecious. **Culms** 6–200 cm, sometimes woody, erect to decumbent, sometimes branched above the base, usually pubescent; **basal branching** intravaginal. **Leaves** basal or basal and cauline; **sheaths** terete, open; **auricles** present, often ciliate; **ligules** 0.5–3 mm, truncate, membranous or of hairs; **blades** linear to lanceolate, sometimes disarticulating from the sheaths. **Inflorescences** racemes or panicles; **primary branches** spreading to ascending; **disarticulation** above the glumes, not between the florets. **Spikelets** 2–17 mm, solitary, pedicellate, terete or laterally compressed, with 3 florets, lower 2 florets sterile, at least the upper equaling or exceeding the distal floret, distal floret bisexual. **Glumes** from about $^1/_2$ as long as to exceeding the florets, (3)5–7-veined. **Sterile florets** consisting only of lemmas; **sterile lemmas** firmer than the glumes, glabrous or pubescent, stipitate or non-stipitate, smooth to rugose, unawned or awned, lowest lemma often with lateral earlike appendages at the base, upper lemma subequal to or longer than the distal floret. **Bisexual florets: lemmas** often indurate at maturity, glabrous, 5–7-veined, keeled, unawned, sometimes mucronate; **paleas** thinner than the lemmas, 1–2(5)-veined; **anthers** (1–5)6, yellow; **styles** 2, fused or free to the base, white or brown. **Caryopses** laterally compressed. $x = 12$.

Named for Jakob Friedrich Ehrhart (1742–1795), a German botanist of Swiss origin who studied under Linnaeus.

Ehrharta is a genus of approximately 25 species, most of which are native to southern Africa. Three species, all from southern Africa, are established in California.

SELECTED REFERENCES **Gibbs Russell, G.E.** 1991. *Ehrharta* Thunb. Pp. 121–129 *in* G.E. Gibbs Russell, L. Watson, M. Koekemoer, L. Smook, N.P. Barker, H.M. Anderson, and M.J. Dallwitz. Grasses of Southern Africa (ed. O.A. Leistner). National Botanic Gardens, Botanical Research Institute, Pretoria, Republic of South Africa. 437 pp.; **Gibbs Russell, G.E.** and **R.P. Ellis.** 1987. Species groups in the genus *Ehrharta* (Poaceae) in southern Africa. Bothalia 17:51–65; **Jacobs, S.W.L.** and **S.M. Hastings.** 1993. *Ehrhata.* Pp. 652–654 *in* G.J. Harden (ed.). Flora of New South Wales, vol. 4. New South Wales University Press, Kensington, New South Wales, Australia. 775 pp.

1. Upper glumes $^3/_4$–$^9/_{10}$ the length of the spikelets; sterile lemmas hairy, not transversely rugose . . . . 1. *E. calycina*
1. Upper glumes less than $^3/_4$ the length of the spikelets; sterile lemmas glabrous or sparsely hispidulous, often transversely rugose distally.
   2. Lower sterile lemmas unawned . . . . 2. *E. erecta*
   2. Lower sterile lemmas awned, the awns 2–20 mm long . . . . 3. *E. longiflora*

1. Ehrharta calycina Sm. [p. 35]
PERENNIAL VELDTGRASS

Plants perennial; cespitose, often rhizomatous. **Culms** 30–75(180) cm, erect, glabrous. **Sheaths** finely striate, smooth, sometimes densely pubescent, with short hairs between the veins, usually purplish; **auricles** ciliate; **ligules** about 1 mm, lacerate, glabrous; **blades** 2–9 cm long, 2–7 mm wide, flat or involute, surfaces glabrous, sometimes scabridulous, margins hairy, wavy. **Panicles** 7–22 cm, sometimes partially enclosed in the upper leaf sheaths, sometimes nodding; **pedicels** curved or bent, sometimes straight. **Spikelets** 4–9 mm, U-shaped, purplish. **Glumes** subequal, 3–8 mm long, $^3/_4$–$^9/_{10}$ the length of the spikelets, 7-veined; **sterile lemmas** hairy, smooth, lower sterile lemmas from $^2/_3$ the length of to equaling the upper sterile lemmas, bases with earlike appendages, apices of both lemmas mucronate or shortly awned; **bisexual lemmas** slightly shorter than the upper sterile lemmas, 5–7-veined, glabrous or sparsely pubescent; **paleas** shorter than the lemmas, 2-veined; **anthers** 6, 2.5–3.5 mm. **Caryopses** about 3 mm. $2n = 24$–28, 30.

Ehrharta calycina is native to southern Africa. It was introduced to Davis, California, as a drought-resistant grass for rangelands, but it is unable to withstand heavy grazing. It is now common on the coastal sand dunes at San Luis Obispo and San Diego, California, and has been reported from Nevada and Texas. Jacobs and Hastings (1993) describe it as "moderately useful . . . on light soils of low fertility and rainfall between 330 and 760 mm" in New South Wales. Four varieties have been described; they are not treated here.

2. Ehrharta erecta Lam. [p. 35]
PANIC VELDTGRASS

Plants perennial; weakly cespitose. **Culms** (20)30–100(200) cm, erect or ascending from a decumbent base, sometimes rooting at the lower nodes. **Sheaths** finely striate, glabrous or shortly pubescent; **auricles** ciliate; **ligules** to 3 mm, lacerate, glabrous; **blades** 5–15 cm long, 2–15 mm wide, flat, lax, usually glabrous, smooth or minutely roughened, margins often wavy. **Panicles** 5–21 cm, erect or nodding, open to contracted; **pedicels** usually straight, sometimes curved. **Spikelets** 3–5 mm, oval, greenish. **Glumes** unequal, membranous to chartaceous; **lower glumes** 1–2 mm, $^1/_3$–$^2/_3$ the length of the spikelets, 3–5-veined; **upper glumes** 2–2.5 mm, to $^3/_4$ the length of the spikelets, wider than the lower glumes, 5-veined; **sterile lemmas** 2.5–4.5 mm, indurate, glabrous or sparsely hispidulous, unawned, lower sterile lemmas often with a basal appendage, upper sterile lemmas transversely rugose distally; **bisexual lemmas** 2.5–3.5 mm, firm, glabrous, obscurely 5–7-veined, often cross veined, unawned; **anthers** 6, 0.7–1.2 mm. **Caryopses** about 2 mm. $2n = 24$.

Ehrharta erecta was introduced to California from South Africa. It prefers shady, somewhat moist locations, and is best known from the eastern San Francisco Bay area, San Diego, and the campus of the University of California at Riverside. In Australia, it is considered to be a weed of moist, shady places (Jacobs and Hastings 1993). Three varieties have been described; they are not treated here.

sterile
fertile floret
sterile
florets
2 mm
spikelet
E. calycina

upper glume   lower glume
spikelet
sterile
fertile floret
sterile
florets
2 mm
2 cm
E. erecta

ligule
auricle
2 mm

sterile florets
2 mm
fertile floret
spikelet
1 cm
E. longiflora

1 cm

EHRHARTA

3. Ehrharta longiflora Sm. [p. 35]
LONGFLOWERED VELDTGRASS, ANNUAL VELDTGRASS

Plants annual. **Culms** 15–90 cm, erect or often geniculate basally, often with secondary inflorescences developing from the lower nodes. **Sheaths** obviously veined, keeled, submembranous, sparsely pubescent at the base; **auricles** ciliate; **ligules** 1–2.5 mm, lacerate; **blades** 6–20 cm long, (1)2.5–15 mm wide, flat, usually softly pubescent or glabrescent, sometimes scabridulous or smooth. **Primary panicles** 9–15 cm, erect, open, sometimes reduced to a raceme; **branches** usually ascending, sometimes spreading; **pedicels** usually curved or bent, sometimes flexuous. **Spikelets** 8–30 mm, including the awns. **Lower glumes** 3–3.5 mm long, 0.7–0.8 mm wide, 5-veined; **upper glumes** 4–4.5 mm long, to $^3/_4$ the length of the spikelets, 1–1.5 mm wide, 7-veined; **sterile lemmas** 6–13 mm, indurate, glabrous or sparsely hispidulous, sometimes faintly transversely rugose, scabrous distally, lower sterile lemmas 3–7-veined, awned, awns 2–20 mm, upper sterile lemmas with a short stalklike base and 2 inconspicuously bearded ridges, awned or unawned; **bisexual lemmas** 4–7 mm, 7-veined, unawned; **anthers** 3, about 1.2 mm. **Caryopses** about 3–4 mm. $2n = 24, 48$.

Ehrharta longiflora is a southern African species, well-established in Australia, that in the *Flora* region is established near Torrey Pines State Park in southern California. It is said to prefer shaded areas on hillsides and disturbed areas such as gardens and roadsides, usually growing in light sandy to loamy soils. Three varieties have been described; they are not treated here.

5. ORYZEAE Dumort.

Edward E. Terrell

Plants annual or perennial; synoecious or monoecious. **Culms** annual, 20–500 cm tall, aerenchymatous, sometimes floating. **Leaves** aerenchymatous; **auricles** present or absent; **ligules** membranous or scarious, sometimes absent; **pseudopetioles** sometimes present; **blades** with parallel veins, cross venation not evident; **abaxial blade epidermes** with microhairs and transversely dumbbell-shaped silica bodies; **first seedling leaf** without a blade. **Inflorescences** usually panicles, sometimes racemes or spikes; **disarticulation** below the spikelets, not occurring in cultivated taxa. **Spikelets** laterally compressed or terete, with 1 bisexual or unisexual floret, sometimes with 2 sterile florets below the sexual floret, these no more than $^1/_2(^9/_{10})$ the length of the fertile floret; **unisexual spikelets** in the same or different panicles; **rachillas** not prolonged. **Glumes** absent or highly reduced, forming an annular ring or lobes at the pedicel apices; **sterile florets** $^1/_8–^1/_2(^9/_{10})$ as long as the spikelets; **fertile lemmas** 3–14-veined, membranous or coriaceous, apices entire, unawned or with a terminal awn; **paleas** similar to the lemmas, 3–10-veined, 1-keeled; **lodicules** 2; **anthers** usually 6(1–16); **styles** 2, bases fused or free, stigmas linear, plumose. **Fruits** usually caryopses, sometimes achenes, ovoid, oblong, or cylindrical; **embryos** of the F+FP or F+PP type, small or elongate, with or without a scutellar tail; **hila** usually linear. $x = 12, 15, 17$.

The *Oryzeae* include about 10–12 genera and 70–100 species. Its members are native to temperate, subtropical, and tropical regions. *Oryza sativa* is one of the world's most important crop species. Four genera are native to the *Flora* region; two are introduced.

SELECTED REFERENCES **Clayton, W.D.** and **S.A. Renvoize.** 1986. Genera Graminum: Grasses of the World. Kew Bull., Addit. Ser. 13. Her Majesty's Stationery Office, London, England. 389 pp.; **Guo, Y.-L.** and **S. Ge.** 2005. Molecular phylogeny of Oryzeae (Poaceae) based on DNA sequences from chloroplast, mitochondrial, and nuclear genomes. Amer. J. Bot. 92:1548–1558; **Liu, L.** and **S.M. Phillips.** 2006. *Oryzeae.* P. 183 *in* Z.-Y. Wu, P.H. Raven, and D.-Y. Hong (eds.). Flora of China, vol. 22 (Poaceae). Science Press, Beijing, Peoples Republic of China and Missouri Botanical Garden Press, St. Louis, Missouri, U.S.A. 653 pp. http://flora.huh.harvard.edu/china/mss/volume22/index.htm/.

1. Lemma margins free; fruits achenes, ellipsoid, obovoid, ovoid or subglobose, beaked with a shell-like pericarp.
    2. Lemmas of the pistillate spikelets awned; plants emergent, more than 1 m tall . . . . 5.05 *Zizaniopsis*
    2. Lemmas of the pistillate spikelets unawned; plants emergent and less than 1 m tall or submerged aquatics . . . . 5.06 *Luziola*
1. Lemmas and paleas clasping along their margins; fruits caryopses, cylindrical or laterally compressed, not beaked.
    3. Spikelets unisexual; caryopses terete . . . . 5.04 *Zizania*
    3. Spikelets bisexual; caryopses laterally compressed or terete.
        4. Sterile florets present below the fertile floret, $^{1}/_{8}-^{1}/_{2}$ $(^{9}/_{10})$ as long as the spikelets . . . . 5.01 *Oryza*
        4. Glumes absent.
            5. Leaf blades aerial, not pseudopetiolate, linear to broadly lanceolate; spikelets pedicellate, without stipelike calluses; lemmas unawned; widespread native species . . . . 5.02 *Leersia*
            5. Leaf blades floating, pseudopetiolate, elliptic to ovate or ovate-lanceolate; spikelets on stipelike calluses (1)2–10 mm long; lemmas awned; aquatic ornamental species, not known to be established in the *Flora* region . . . . 5.03 *Hygroryza*

5.01 ORYZA L.

Mary E. Barkworth
Edward E. Terrell

Plants annual or perennial; usually aquatic, rooted and emergent or floating, sometimes terrestrial; rhizomatous and/or cespitose; synoecious. **Culms** to 3.3(5) m, erect, decumbent, or prostrate, sometimes rooting at the lower nodes, aerenchymatous, emergent or immersed, branched or unbranched. **Leaves** cauline and basal; **sheaths** open, lower sheaths often slightly inflated, upper sheaths not inflated; **auricles** usually present; **ligules** membranous, often veined; **pseudopetioles** absent; **blades** linear to narrowly lanceolate, flat, margins smooth or scabridulous. **Inflorescences** terminal panicles; **disarticulation** above the glumes, beneath the sterile florets in wild taxa, spikelets of cultivated taxa not disarticulating. **Spikelets** bisexual, laterally compressed, with 3 florets, lower 2 florets sterile, terminal floret functional. **Glumes** absent or reduced to lobes at the pedicel apices; **sterile florets** glumelike, 1.2–10 mm, $^{1}/_{8}-^{1}/_{2}(^{9}/_{10})$ as long as the spikelets, linear or subulate to narrowly ovate, coriaceous, 1-veined, acute to acuminate; **functional florets: calluses** usually inconspicuous and flat to rounded, sometimes conspicuous and stipelike, glabrous; **lemmas** coriaceous or indurate, with vertical rows of tubercles separated by longitudinal furrows, 5-veined, keeled, margins clasping the margins of the paleas, apices obtuse or acute to acuminate, awned or unawned; **paleas** with surfaces similar to the lemmas, 3-veined, unawned; **lodicules** 2; **anthers** 6; **styles** 2, bases fused or not, stigmas laterally exserted, plumose. **Caryopses** laterally compressed; **embryos** usually $^{1}/_{4}-^{1}/_{3}$ as long as the caryopses; **hila** linear. $x = 12$. Name from the Greek *oryza*, 'rice'.

Oryza is a tropical and subtropical genus of about 20 species that grow in shallow water, swamps, and marshes in seasonally inundated areas, or along streams, rivers, or lake edges. *Oryza sativa* (rice) is one of the three most economically valuable cereals, and constitutes a major portion of the diet for half of the world's population. In the Flora region, *O. sativa* is cultivated and several weedy forms have become established. These are thought to be derived from introgression between *O. sativa* and *O. rufipogon* and *O. punctata*. The latter two species and *O. longistaminata* are included here because of the threat they pose to cultivated rice.

Spikelets of *Oryza* have sometimes been interpreted as comprising one functional and two sterile florets with two highly reduced glumes (Duistermat 1987), sometimes as comprising a single floret, subtended by two glumes borne on a bilobed pedicel (Terrell et al. 2001). Molecular developmental studies (Komatsu et al. 2003) show that the former interpretation is correct.

SELECTED REFERENCES **Duistermaat, H.** 1987. A revision of *Oryza* (Gramineae) in Malesia and Australia. Blumea 32:157–193; **Komatsu, M., A. Chujo, Y. Nagato, K. Shimamoto,** and **J. Kyozuka.** 2003. *FRIZZY PANICLE* is required to prevent the formation of axillary meristems and to establish floral meristem identity in rice spikelets. Development 130:3841–3850; **Launert, E.** 1971. *Oryza* L. Pp. 31–36 *in* A. Fernande, E. Launert, and H. Wild (eds.). Flora Zambesiaca, vol. 10¹. Crown Agents for Oversea Governments and Administrations, London, England. 152 pp.; **Londo, J.P., Y.-C. Chiang, K.-H. Hung, T.-Y. Chiang,** and **B.A. Schaal.** 2006. Phylogeography of Asian wild rice, *Oryza rufipogon*, reveals multiple independent domestications of cultivated rice, *Oryza sativa*. Proc. Natl. Acad. Sci. U.S.A. [PNAS] 103(25):9578–9583; **Lu, B.-R., E.B. Naredo, A.B. Juliano,** and **M.T. Jackson.** 2000. Preliminary studies on taxonomy and biosystematics of the AA genome *Oryza* species (Poaceae). Pp. 51–58 *in* S.W.L. Jacobs and J. Everett (eds.). Grasses: Systematics and Evolution. CSIRO Publishing, Collingwood, Victoria, Australia. 406 pp.; **Terrell, E.E., P.M. Peterson** and **W.P. Wergin.** 2001. Epidermal features and spikelet micromorphology in *Oryza* and related genera (Poaceae: Oryzeae). Smithsonian Contr. Bot. 91. 50 pp.; **Vaughan, D.A.** 1989. The Genus *Oryza* L.–Current Status of Taxonomy. International Rice Research Institute Research Paper Series 138. International Rice Research Institute, Los Baños, Laguna, Philippines. 21 pp.

1. Ligules truncate to rounded, 1.5–10 mm long; sterile florets 1.2–2 mm long; disarticulation scar centric or slightly eccentric . . . . 3. *O. punctata*
1. Ligules acute, 4–45 mm long; sterile florets 1.3–10 mm long; disarticulation scar lateral.
    2. Anthers 1–2.5 mm long; spikelets persistent; lemmas usually unawned, plants not rhizomatous; auricles absent or to 5 mm long; blades 5–20 mm wide . . . . 4. *O. sativa*
    2. Anthers 3.5–7.4 mm long; spikelets deciduous; lemmas awned; plants usually rhizomatous; auricles absent or to 15 mm long; blades 7–50 mm wide.
        3. Caryopses 5–7 mm long; lemma-awn junctions purplish, pubescent; lemma awns 4–16 cm long; plants cespitose or rhizomatous; auricles absent or to 7 mm long . . . . 2. *O. rufipogon*
        3. Caryopses 7.5–8.5 mm long; lemma-awn junctions similar in color to the lemmas, glabrous; lemma awns 2.6–8 cm long; plants strongly rhizomatous; auricles present, to 15 mm long . . . . 1. *O. longistaminata*

1. **Oryza longistaminata** A. Chev. & Roehr. [p. 39]
LONGSTAMEN RICE

Plants perennial; extensively rhizomatous. **Culms** to 2+ m tall, to 25+ mm thick at the base, soft and spongy, erect or decumbent, sometimes floating, rooting at the lower nodes. **Sheaths** shorter or nearly as long as the internodes, smooth, glabrous; **auricles** to 15 mm; **ligules** 15–45 mm, acute, often splitting; **blades** to 45 cm long, 10–50 mm wide, linear-lanceolate to narrowly elliptic, glabrous, smooth or scabrous abaxially, scabrous adaxially, acute or 2-cleft. **Panicles** (16)20–30(40) cm long, 2.5–8 cm wide, exserted; **branches** 7.5–15 cm, ascending; **pedicels** 0.5–4(7) mm. **Spikelets** 7–9 mm long, 2–3 mm wide, narrowly oblong, deciduous, obliquely articulated with the pedicel, disarticulation scar lateral. **Sterile florets** 2–4 mm, ¹/₅–³/₅ as long as the spikelets, sometimes trilobed, lateral lobes shorter than the central lobe. **Functional florets: lemmas** 7–9 mm long, 1.8–2.5 mm wide, narrowly elliptic, scabrous to hispid over the veins, awned, lemma-awn junctions glabrous, colored as the lemmas, awns (2.6)3.5–8 cm; **paleas** usually slightly shorter than the lemmas, narrower, acute or tapering; **anthers** 5–7 mm. **Caryopses** 7.5–8.5 mm, oblong, light brown, glossy. **Haplome** A. 2*n* = 24.

Oryza longistaminata is native to Africa; it has not yet been found in North America. It is included here because its establishment in North America would seriously impact North American agriculture. The U.S.

Department of Agriculture considers it a noxious weed; plants found growing within the United States should be promptly reported to that agency.

In its native range, *Oryza longistaminata* grows in swampy areas, pond and lake edges, irrigation canals, and streams and rivers in areas that are permanently wet or seasonally dry.

2. **Oryza rufipogon** Griff. [p. 39]
RED RICE, BROWNBEARD RICE

Plants annual or perennial; cespitose, sometimes rhizomatous, rhizomes elongated. **Culms** 0.6–3.3(5) m tall, 6–15 mm thick, decumbent or prostrate, rooting and branching at both the lower and submerged upper nodes. **Sheaths** smooth, glabrous; **auricles** sometimes present, 1–7 mm, erect; **ligules** 7–45 mm, acute, finely splitting; **blades** 10–80 cm long, 7–24 mm wide, smooth or scabrous. **Panicles** 12–30 cm long, 1–7 cm wide; **branches** 2.5–12(20) cm, ascending or divergent; **pedicels** 1–3 mm. **Spikelets** 4.5–11 mm long, 1.6–3.5 mm wide, oblong or elliptic, deciduous, obliquely articulated with the pedicel, disarticulation scar lateral. **Sterile florets** 1.3–7 mm long, ¹/₄–¹/₂(³/₄) as long as the spikelets, 0.3–0.7 mm wide, acute. **Functional florets: lemmas** 6–11 mm long, 1.4–2.3 mm wide, hispid, apices

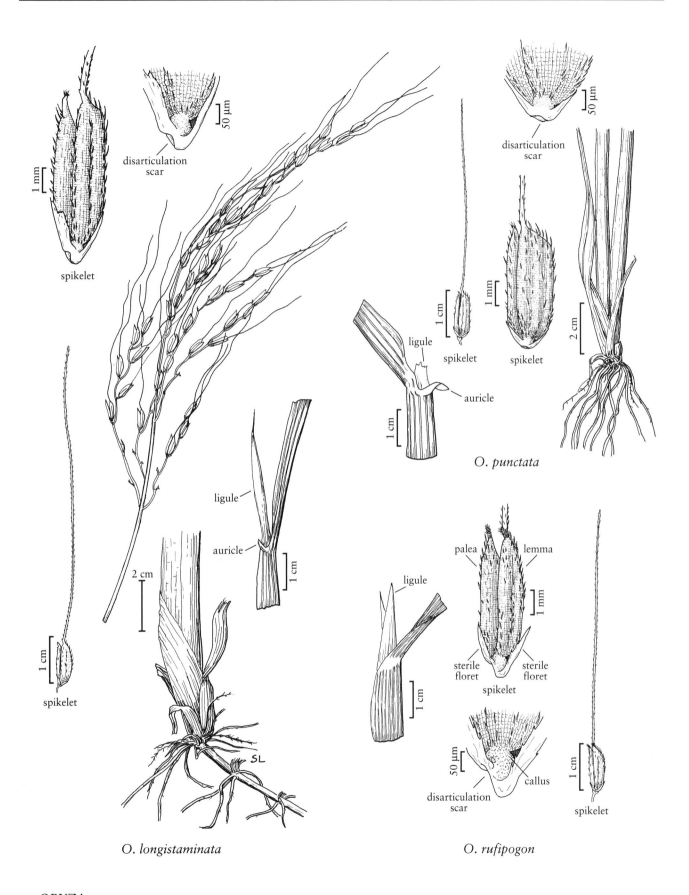

disarticulation
scar

50 μm

1 mm

spikelet

disarticulation
scar

50 μm

1 mm

2 cm

ligule

auricle

1 cm

spikelet

spikelet

O. punctata

ligule

auricle

2 cm

1 cm

1 cm

spikelet

palea   lemma

1 mm

sterile
floret

sterile
floret

spikelet

ligule

1 cm

50 μm

callus

disarticulation
scar

1 cm

spikelet

O. longistaminata   *O. rufipogon*

SL

ORYZA

beaked, beaks to 1 mm, straight or curved, lemma-awn junctions marked by a purplish, pubescent constriction, awns 4–11(16) cm; **paleas** 0.5–1.2 mm wide, acuminate to awned to 2.3 mm; **anthers** 3.5–7.4 mm, yellow or brown. **Caryopses** 5–7 mm long, 2.2–2.7 mm wide, broadly elliptic or oblong, reddish-brown to dark red; **embryos** 1–1.5 mm. **Haplome A.** $2n = 24$.

Oryza rufipogon is native to southeast Asia and Australia, where it grows in shallow, standing or slow-moving water, along irrigation canals, and as a weed in rice fields. It is the ancestor of *O. sativa* (Londo et al. 2006). The vernacular name 'Red Rice' is used to refer to *O. rufipogon*, *O. punctata*, and weedy forms of *O. sativa*, some of which are probably derivatives of introgression from the first two species.

Oryza rufipogon is a weedy taxon that hybridizes readily with *O. sativa*, forming partially fertile hybrids. This makes it a serious threat to rice growers. Some of the known populations have been eradicated, for instance those in Everglades National Park, Miami-Dade County, Florida, and in the Sacramento Valley, California. Nevertheless, weedy red rice has become an increasingly serious problem in rice fields throughout rice-growing regions of the world, including the contiguous United States, probably through the presence of fertile seed in commercial seed.

3. Oryza punctata Kotschy *ex* Steud. [p. 39]
RED RICE

Plants annual or perennial; cespitose. **Culms** 0.6–1.2 m tall, usually 4 mm or more thick, erect, branched, spongy. **Sheaths** smooth, glabrous; **auricles** present; **ligules** 1.5–10 mm, truncate to rounded, whitish, tending to split at maturity; **blades** 20–60 cm long, 10–30 mm wide, linear, usually scabrous, rarely glaucous. **Panicles** 20–35 cm long, 3–17 cm wide, loose; **branches** to 20 cm, divergent; **pedicels** 2–5 mm. **Spikelets** 4.6–9.2 mm long, 2–2.5 mm wide, asymmetrically elliptic-oblong to broadly oblong, deciduous, horizontally or slightly obliquely articulated with the pedicels, disarticulation scar centric or slightly eccentric. **Sterile florets** 1.2–2 mm, $^1/_5$–$^2/_5$ as long as the spikelets, acute. **Functional florets:** lemmas 4.6–9.2 mm long, 1.9–2.6 mm wide (about 2.5 times longer than wide), hispid, awned, awns (1)2–7.5 cm; **paleas** usually slightly shorter than the lemmas, narrower, acute or tapering; **anthers** 1.5–3.2 mm. **Caryopses** 4–4.8 mm long, 1.5–1.8 mm wide, oblong, light brown. **Haplomes B, BC.** $2n = 24, 48$.

Oryza punctata is an African grass that is not established in North America. It is considered a noxious weed by the U.S. Department of Agriculture; if any populations are found in the United States, that agency should be notified. It has two morphological types. The tetraploid race is perennial, with spikelets to 5.5 mm long and at least 2.3 mm wide. It grows in semi-open

or shaded forest habitats, in swampy areas, water holes, pools, and river areas that flood to 1 m in depth. The diploid race is annual, with spikelets 5.5 mm or more long and at most 2.3 mm wide. It grows in open to semi-open forest margins, grasslands, and thickets, scrub, and open bush, in swampy areas, water holes, pools, and river areas that flood to 1 m in depth.

The vernacular name 'Red Rice' is used to refer to *Oryza rufipogon*, *O. punctata*, and weedy forms of *O. sativa*, some of which may be the result of introgression from the first two species.

4. Oryza sativa L. [p. 41]
RICE

Plants usually annual, sometimes perennial; cespitose, not rhizomatous. **Culms** 0.3–2 m tall, 4–20 mm thick, erect or ascending, branching at the base, usually rooting at both the lower and submerged upper nodes. **Sheaths** smooth, glabrous, lowest sheaths usually longer than the internodes, upper sheaths shorter than the internodes; **auricles** often present, 1–5 mm; **ligules** (4)10–36 mm, acute; **blades** 20–70 cm long, 5–20 mm wide, glabrous, sometimes scabrous. **Panicles** 10–50 cm long, 1–8 cm wide, often nodding; **branches** 2–13 cm, ascending or divergent; **pedicels** 1–7 mm. **Spikelets** 6–11 mm long, 2.5–4 mm wide, broadly elliptic, sometimes with obvious rows of white papillae, persistent, obliquely articulated with the pedicels. **Sterile florets** 1.5–3(10) mm long, $^1/_4$–$^1/_2$($^9/_{10}$) as long as the spikelets, 0.5–1.5 mm wide. **Functional florets:** lemmas 6–11 mm long, 2–3 mm wide, glabrous or with stiff hairs to 1.5 mm, apices beaked, beaks 0.3–1(2) mm, rigid, usually unawned, sometimes awned, awns to 6(15) cm; **paleas** 1–1.7 mm wide, acute to acuminate or mucronate to 0.5 mm; **anthers** 1–2.5 mm, white or yellow; **styles** white, yellow, red, or blackish-purple. **Caryopses** 4.5–8 mm long, 2–3.5 mm wide, broadly elliptic or broadly oblong, brown, tan, or white; **embryos** 1.4–1.7 mm. **Haplome A.** $2n = 24$.

Oryza sativa is cultivated in California, Arkansas, Texas, Louisiana, Mississippi, and Florida and is sometimes found as an adventive in moist or wet places, particularly in the southeastern United States, but it is not established in the *Flora* region. It used to be extensively cultivated in the Carolinas and Georgia, but no rice plantations are currently known to be in operation in those states. Many cultivars have been developed; there is considerable morphological, as well as ecological, variability in the cultivated crop.

O. sativa

ORYZA

5.02 LEERSIA Sw.

Grant L. Pyrah

Plants usually perennial, rarely annual; terrestrial or aquatic; rhizomatous or cespitose; synoecious. **Culms** 20–150 cm (occasionally longer in floating mats), erect or decumbent, often rooting at the nodes, branched or unbranched. **Leaves** equitably distributed along the culm; **sheaths** open; **auricles** absent; **ligules** membranous; **pseudopetioles** absent; **blades** aerial, linear to broadly lanceolate, flat or folded, sometimes involute when dry. **Inflorescences** terminal panicles, usually exserted, axillary panicles sometimes also present; **disarticulation** beneath the spikelets. **Spikelets** bisexual, with 1 floret; **florets** laterally compressed, linear to suborbicular in sideview. **Glumes** absent; **calluses** not stipelike, glabrous; **lemmas** and **paleas** subequal, chartaceous to coriaceous, ciliate-hispid or glabrous, tightly clasping along the margins; **lemmas** 5-veined, obtuse or acute to acuminate, sometimes mucronate, usually unawned; **paleas** 3-veined, unawned; **lodicules** 2; **anthers** 1, 2, 3, or 6; **styles** 2, bases fused, stigmas laterally exserted, plumose. **Caryopses** laterally compressed; **embryos** about ⅓ as long as the caryopses; **hila** linear. $x = 12$. Named for Johann Daniel Leers (1727–1774), a German botanist and pharmacist.

Leersia is a genus of about 17 aquatic to mesophytic species, growing primarily in tropical and warm-temperate regions. Five species are native to the *Flora* region. *Leersia* is closely allied to *Oryza*. It is unusual in the variability in stamen numbers among its species.

SELECTED REFERENCE **Pyrah**, G.L. 1969. Taxonomic and distributional studies in *Leersia* (Gramineae). Iowa State Coll. J. Sci. 44:215–270.

1. Spikelets 1.5–2 mm long, glabrous; plants not rhizomatous . . . . 1. *L. monandra*
1. Spikelets 2.5–6.5 mm long, usually ciliate on the margins and keel, and glabrous or pubescent elsewhere; plants rhizomatous.
   2. Spikelets nearly as wide as long . . . . 2. *L. lenticularis*
   2. Spikelets not more than ½ as wide as long.
      3. Anthers 2; spikelets 2.5–3.6 mm long; panicle branches single at all nodes . . . . 3. *L. virginica*
      3. Anthers 3 or 6; spikelets 3.2–6.5 mm long; panicle branches 1–2 or more at the lower nodes, single at the upper nodes.
         4. Panicles exserted, 5–15 cm long; branches appressed to ascending, spikelet-bearing to near the base; anthers 6; spikelets 3.2–4.7(5) mm long . . . . 4. *L. hexandra*
         4. Panicles exserted or enclosed, 10–30 cm long; branches spreading on exserted panicles, naked on the lower ⅓; anthers 3; spikelets (4)4.2–6.5 mm long . . . . 5. *L. oryzoides*

1. Leersia monandra Sw. [p. 43]
BUNCH CUTGRASS, CANYONGRASS, CEDAR WHITEGRASS

Plants perennial; cespitose, without rhizomes. **Culms** 40–120 cm tall, 0.8–1.2 mm thick, erect, unbranched; **nodes** nearly glabrous to densely retrorsely hispidulose, adjacent portions of the internodes glabrous or moderately retrorsely hispidulose. **Sheaths** glabrous or moderately retrorsely hispid-scabrous; **ligules** 1.5–4 mm; **blades** 20–30 cm long, 3.5–6 mm wide, ascending, usually involute on drying, veins and margins smooth or scabridulous. **Panicles** 5–15 cm, with 1 branch per node; **branches** spreading to somewhat ascending, spikelets closely appressed, confined to the distal ⅓ of the branches. **Spikelets** 1.5–2 mm long, 1–1.3 mm wide, laterally compressed, broadly elliptic or ovate. **Lemmas** and **paleas** glabrous, acute or obtuse; **anthers** 2. **Caryopses** 1–1.3 mm, ovate, yellow. $2n = 48$.

Leersia monandra grows in rather dry, rocky, limestone soils in open woods, grasslands, and bluffs, from Texas and Florida south to the Yucatan Peninsula, Mexico, and the Antilles. It is also sold as an ornamental. There have been few collections from the *Flora* region in the last two decades. In areas with heavy grazing, *L. monandra* tends to disappear, surviving only in areas where shrubs provide some measure of protection. It flowers throughout the year.

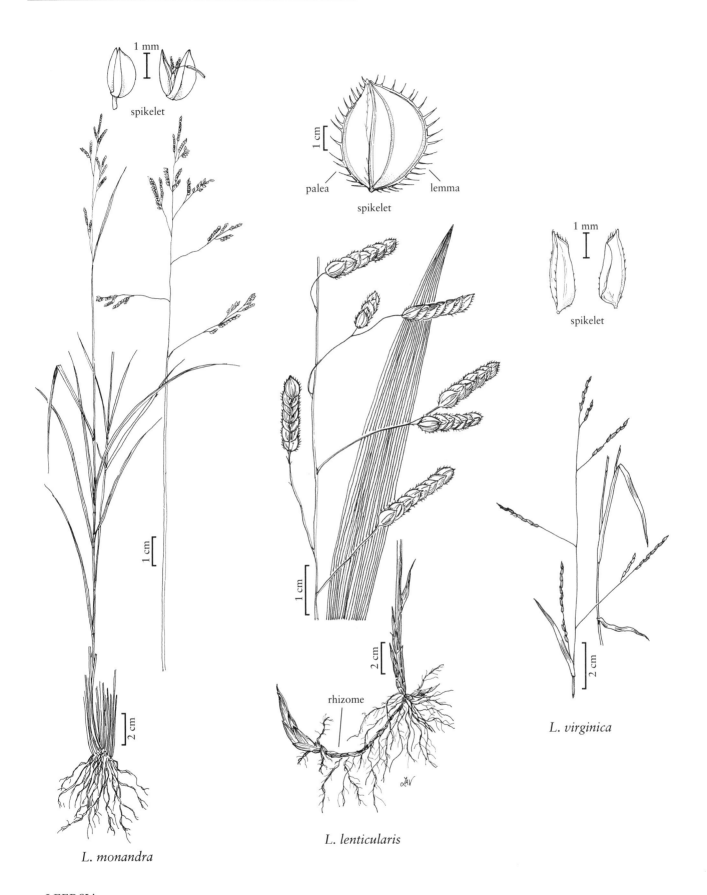

1 mm

spikelet

1 cm

palea　　lemma

spikelet

1 mm

spikelet

1 cm

1 cm

2 cm

rhizome

2 cm

L. virginica

L. monandra

L. lenticularis

LEERSIA

2. Leersia lenticularis Michx. [p. 43]
CATCHFLY GRASS, OATMEAL GRASS

Plants perennial; rhizomatous, rhizomes moderately elongate, scaly. **Culms** 50–150 cm tall, 1–3 mm thick, usually ascending, unbranched or branched; **nodes** retrorsely hispidulous, adjacent portion of the internodes glabrous. **Sheaths** glabrous or scabrous; **ligules** 0.5–1.5 mm; **blades** 4–35 cm long, 5–22 mm wide, spreading to somewhat ascending, abaxial surfaces glabrous or scabridulous, adaxial surfaces glabrous or pubescent, margins usually scabrous. **Panicles** 4–25 cm, exserted, with 1(2) branches per node; **branches** 8–15 cm, spreading, secund, lower branches naked on the lower ⅓, spikelets strongly imbricate. **Spikelets** 4–5.5 mm long, 3–4 mm wide, broadly elliptic to suborbicular. **Lemmas** coarsely ciliate on the keels, variously pubescent on the margins and body, mucronate; **paleas** ciliate on the keels; **anthers** 2. **Caryopses** 3.5–4 mm, reddish-brown. $2n = 48$.

Leersia lenticularis grows in river bottoms and moist woods of the midwestern and southeastern United States. It flowers from July to November. Ohio and Maryland list it as an endangered species.

3. Leersia virginica Willd. [p. 43]
WHITE CUTGRASS, WHITEGRASS, LÉERSIE DE VIRGINIE

Plants perennial; rhizomatous, rhizomes short, scaly, scales imbricate, giving the rhizomes a "braided" appearance. **Culms** 30–140 cm tall, 1–1.5 mm thick, branched, sometimes rooting at the nodes, more or less glabrous, pubescent near the nodes. **Sheaths** glabrous or slightly scabrous; **ligules** 1–3 mm; **blades** 4–20 cm long, (1)6–15 mm wide, flaccid, surfaces glabrous or puberulent or the abaxial surfaces sometimes densely pilose, margins hispid. **Panicles** 10–25 cm, long-exserted at maturity, with 1 branch per node; **branches** 4–8 cm, spreading, naked on the lower ⅓, spikelets more or less appressed, scarcely imbricate. **Spikelets** 2.5–3.6 mm long, 0.4–1.2 mm wide, oblong or ovate. **Lemmas** ciliate to nearly glabrous on the keels and margins, glabrous or short-pubescent on the body; **paleas** glabrous or slightly ciliate on the keels; **anthers** 2. **Caryopses** 2–2.4 mm, slightly compressed, reddish-brown. $2n = 48$.

Leersia virginica grows in moist places in woods and along stream courses east of the Rocky Mountains. The western Wyoming record may represent an introduction. *Leersia virginica* flowers from July to October.

4. Leersia hexandra Sw. [p. 45]
SOUTHERN CUTGRASS

Plants perennial; rhizomatous, rhizomes elongate, not scaly. **Culms** 25–150 cm tall, 1–1.5 mm thick, decumbent, rooting at the nodes, terminal portions erect, often floating, branched or unbranched; **nodes** pubescent, adjacent portions of the internodes glabrous or coarsely scabrous. **Sheaths** glabrous or coarsely scabrous, margins often ciliate; **ligules** 1–3 mm; **blades** 5–25 cm long, 3–15 mm wide, ascending, glabrous or pubescent. **Panicles** 5–15 cm, exserted at maturity, with 1(2) branches per node; **branches** 3–10 cm, appressed to ascending, spikelet-bearing to near the base, spikelets appressed to slightly divergent, slightly imbricate. **Spikelets** 3.2–4.7(5) mm long, 0.5–2 mm wide, ovate to elliptic. **Lemmas** ciliate on the keels and margins, short hispid or glabrous elsewhere, apices acute to acuminate; **paleas** ciliate on the keels; **anthers** 6, 2–3.2 mm. **Caryopses** about 2 mm, usually not developed. $2n = 48$.

Leersia hexandra is found in wet areas, usually in fresh water along streams and ponds, where it sometimes forms floating mats. It grows in the southeastern United States and throughout much of the neotropics; the California record probably represents a recent introduction.

Leersia hexandra is sometimes a weed in rice. It usually flowers in late fall, but may flower throughout the year. Very little seed is set.

5. Leersia oryzoides (L.) Sw. [p. 45]
RICE CUTGRASS, LÉERSIE FAUX-RIZ

Plants perennial; rhizomatous, rhizomes elongate, scaly, scales not imbricate. **Culms** 35–150 cm tall, 1–3 mm thick, branching, decumbent, sprawling, rooting at the nodes, terminal portions erect. **Sheaths** scabrous; **ligules** 0.5–1 mm; **blades** 7–30 cm long, 5–15 mm wide, spreading to slightly ascending, both surfaces usually scabrous. **Panicles** 10–30 cm, terminal, also axillary, exserted or enclosed at maturity, spreading on exserted panicles, usually 2 or more branches at the lowest nodes, 1 at the upper nodes; **branches** 4–10 cm, the lower ⅓ naked, spikelets imbricate. **Spikelets** (4)4.2–6.5 mm long, 1.3–1.7 mm wide, elliptic. **Lemmas** and **paleas** usually ciliate on the keels and margins, glabrous or puberulent elsewhere; **anthers** 3, 1.5–2(3) mm in chasmogamous spikelets, 0.4–0.7 mm in cleistogamous spikelets. **Caryopses** 2–3.5 mm, asymmetrically pyriform to obovoid, whitish to dark brown. $2n = 48$.

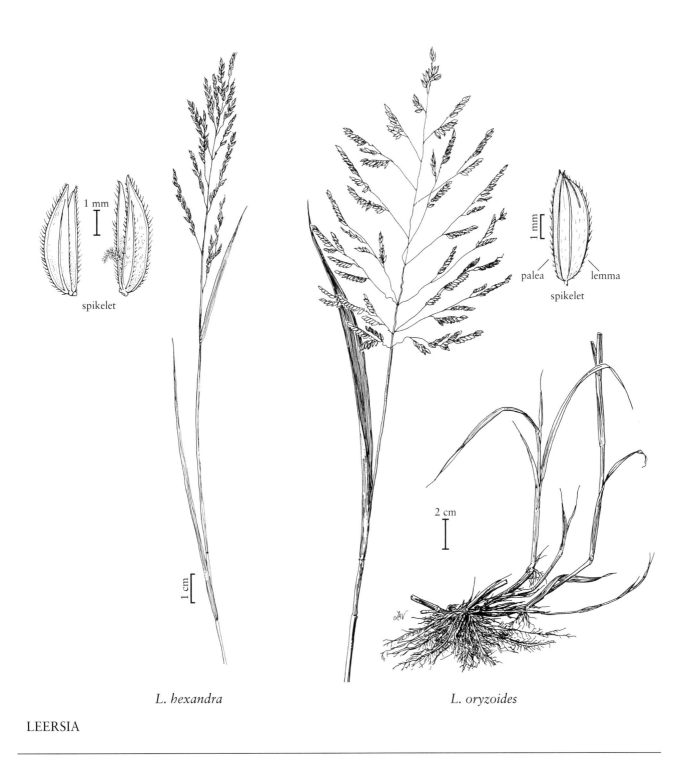

L. hexandra

L. oryzoides

spikelet

1 mm

spikelet

palea   lemma

1 mm

1 cm

2 cm

LEERSIA

Leersia oryzoides grows in wet, heavy, clay or sandy soils, and is often aquatic. It is found across most of southern Canada, extending south throughout the contiguous United States into northern Mexico, and flowers from July to October. It has also become established in Europe and Asia.

5.03 **HYGRORYZA** Nees

J.K. Wipff

Plants perennial; aquatic, producing long, floating culms; synoecious. **Culms** 50–150 cm, spongy, developing adventitious roots at the nodes, branched; **branches** erect, leafy. **Leaves** cauline, glabrous, veins tessellate; **sheaths** open, inflated, serving as floats; **ligules** absent or hyaline; **pseudopetioles** present; **blades** elliptic, ovate, ovate-lanceolate, or oblong. **Inflorescences** terminal panicles, aerial, lowermost branches whorled; **disarticulation** beneath the spikelet calluses. **Spikelets** bisexual, laterally compressed, with 1 floret. **Glumes** absent or an annular rim; **calluses** (1)2–10 mm, stipelike, glabrous, junction with the pedicels marked by a tan constriction; **lemmas** 5-veined, margins clasping the paleas, apices acuminate, awned, awns terminal, antrorsely scabridulous; **paleas** similar to the lemmas, 3-veined, 1-keeled, acute-acuminate, unawned; **lodicules** 2, glabrous; **anthers** 6; **styles** 2, bases not fused, stigmas laterally exserted, plumose. **Caryopses** terete, fusiform; **embryos** small; **hila** linear, almost as long as the embryo. $x = 12$. Name from the Greek *hygros*, 'wet' or 'moist', and *oryza*, 'rice', referring to its aquatic habit and similarity to rice.

Hygroryza is a monospecific Asian genus that grows in India, Ceylon, and throughout southeast Asia. It forms floating masses, often of considerable extent, in lakes and slow-moving streams, and is sometimes a weed in rice.

SELECTED REFERENCES **Bor, N.L.** 1960. The Grasses of Burma, Ceylon, India and Pakistan (Excluding Bambuseae). International Series of Monographs on Pure and Applied Biology, Division: Botany, vol. 1. Pergamon Press, New York, Oxford, London, and Paris. 767 pp.; **Koyama, T.** 1987. Grasses of Japan and Its Neighboring Regions: An Identification Manual. Kodansha, Ltd., Tokyo, Japan. 370 pp.; **Watson, L.** and **M.J. Dallwitz.** 1992. The Grass Genera of the World. C.A.B. International, Wallingford, England. 1038 pp.

1. **Hygroryza aristata** (Retz.) Nees [p. 46]
ASIAN WATERGRASS, WATER STARGRASS

Culms 50–150 cm, floating, branched, flexuous; **nodes** rooting, roots feathery, whorled. **Sheaths** glabrous, open, inflated; **ligules** absent or 0.5–0.8 mm, hyaline, truncate; **pseudopetioles** shorter than 1 mm; **blades** 2–8 cm long, 5–20 mm wide, flat, bases rounded to cordate, apices blunt to rounded. **Panicles** 3–8 cm, pyramidal, with 4–9 branches; **lower branches** 2–4 cm, whorled, spreading or deflexed, glabrous; **pedicels** 0.2–2 mm, sometimes absent. **Florets** (6)7–18 mm (including the stipelike callus), narrowly lanceolate; **calluses** (1)2–10 mm, stipelike; **lemmas** 5–8 mm, chartaceous, keeled, veins and margins with hairs, glabrous or pubescent between the veins, awns 5–14 mm; **paleas** as long as the lemmas, chartaceous, glabrous, keels ciliate or scabrous; **anthers** 6, about 3.5 mm. **Caryopses** about 3.5 mm. $2n = 24$.

Hygroryza aristata is native to tropical Asia, where it has occasionally been used as forage for cattle (Bor 1960). It is sold for ponds and aquaria, where its long, feathery, adventitious roots have a decorative effect, but it has the potential to become a significant weed problem in the southern United States.

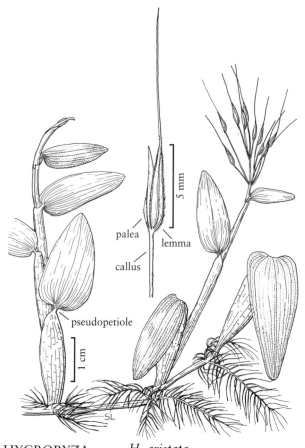

HYGRORYZA   *H. aristata*

5.04 ZIZANIA L.

Edward E. Terrell

Plants annual or perennial; aquatic, usually rooted in the substrate; sometimes rhizomatous or stoloniferous; monoecious. **Culms** to 5 m, erect and emergent or floating. **Leaves** concentrated on the lower portion of the stem or evenly distributed; **sheaths** open, not inflated; **ligules** membranous or scarious, glabrous; **pseudopetioles** absent; **blades** flat, aerial or floating, scabrous or smooth. **Inflorescences** terminal panicles; **branches** usually unisexual, lower branches staminate, upper branches pistillate, middle branches sometimes with staminate and pistillate spikelets intermixed; **pedicel apices** cupulate; **disarticulation** beneath the spikelets, in cultivated strains disarticulation delayed, the spikelets tending not to shatter until harvested. **Spikelets** unisexual, with 1 floret. **Glumes** absent; **calluses** inconspicuous; **lemmas** 5-veined; **paleas** 3-veined; **lodicules** 2, membranous. **Staminate spikelets** pendant, terete or appearing so; **lemmas** membranous; **paleas** membranous, loosely enclosing the stamens; **anthers** 6. **Pistillate spikelets** terete; **lemmas** chartaceous or coriaceous, margins involute and clasping the margins of the paleas, apices acute to acuminate, sometimes awned, awns terminal, slender, scabridulous; **styles** 2, bases not fused, stigmas laterally exserted, plumose. **Caryopses** cylindrical; **embryos** linear, often as long as the caryopses; **hila** linear. $x = 15$. Name from the Greek *zizanion*, a weed growing in grain.

Zizania includes three North American and one eastern Asian species. *Zizania aquatica* and *Z. palustris* are important constituents of aquatic plant communities in North America, providing food and shelter for numerous animal species. *Zizania palustris* is also an important food source for humans. *Zizania texana* is federally listed as an endangered species in the United States. *Zizania latifolia*, an Asian species, is available through horticultural outlets despite its potential for harboring a fungus that would devastate the native species (for additional information, see the comment following the species description).

SELECTED REFERENCES Aiken, S.G., P.F. Lee, D. Punter, and J.M. Stewart. 1988. Wild Rice in Canada. Agriculture Canada Publication 1830. NC Press, Toronto, Ontario, Canada. 130 pp.; **Dore, W.G.** 1969. Wild Rice. Canada Department of Agriculture Publication No. 1393. Information Canada, Ottawa, Ontario, Canada. 84 pp.; **Duvall, M.R.** and **D.D. Biesboer.** 1988. Nonreciprocal hybridization failure in crosses between annual wild-rice species (*Zizania palustris* × *Z. aquatica*: Poaceae). Syst. Bot. 13:229–234; **Environment Walkato.** 2002–2007. Regional Pest Management Strategy. Walkato Regional Council, Hamilton East, New Zealand. http://www.ew.govt.nz/policyandplans/rpmsintro/rpms2002/operative5.2.7.htm/; **Liu, L.** and **S.M. Phillips.** 2006. *Zizania*. Pp. 187–188 *in* Z.-Y. Wu, P.H. Raven, and D.-Y. Hong (eds.). Flora of China, vol. 22 (Poaceae). Science Press, Beijing, Peoples Republic of China and Missouri Botanical Garden Press, St. Louis, Missouri, U.S.A. 653 pp. http://flora.huh.harvard.edu/china/mss/volume22/index.htm/; **Terrell, E.E.** and **L.R. Batra.** 1982. *Zizania latifolia* and *Ustilago esculenta*, a grass-fungus association. Econ. Bot. 36:274–285; **Terrell, E.E., W.H.P. Emery,** and **H.E. Beaty.** 1978. Observations on *Zizania texana* (Texas wildrice), an endangered species. Bull. Torrey Bot. Club 105:50–57; **Terrell, E.E., P.M. Peterson, J.L. Reveal,** and **M.R. Duvall.** 1997. Taxonomy of North American species of *Zizania* (Poaceae). Sida 17:533–549; **Warwick, S.I.** and **S.G. Aiken.** 1986. Electrophoretic evidence for the recognition of two species in annual wild rice (*Zizania*, Poaceae). Syst. Bot. 11:464–473.

1. Culms decumbent, completely immersed or the upper parts of the culm emergent; known only from the San Marcos River in Hays County, Texas . . . . 3. *Z. texana*
1. Culms usually erect at maturity, rarely completely immersed; plants not known from Texas.
  2. Plants rhizomatous, perennial; middle branches of the panicles with both staminate and pistillate spikelets, other branches with either staminate or pistillate spikelets; plants cultivated as ornamentals . . . . 4. *Z. latifolia*
  2. Plants without rhizomes, annual; all panicle branches unisexual, with either staminate or pistillate spikelets; plants native and widespread, also cultivated for grain.
    3. Lemmas of the pistillate spikelets flexible and chartaceous, dull or sublustrous, bearing short, scattered hairs, these not or only slightly more dense towards the apices; aborted pistillate spikelets 0.4–1 mm wide; pistillate inflorescence branches usually divaricate at maturity . . . . 1. *Z. aquatica*

3. Lemmas of the pistillate spikelets stiff and coriaceous or indurate, lustrous, glabrous or with lines of short hairs, the apices more densely hairy; aborted pistillate spikelets 0.6–2.6 mm wide; pistillate inflorescence branches usually appressed at maturity, or with 1 to few, somewhat spreading branches . . . . 2. *Z. palustris*

1. Zizania aquatica L. [p. 49]

Plants annual. **Culms** 0.2–5 m, usually emergent. **Sheaths** glabrous or scabridulous; **ligules** 5–30 mm, upper ligules truncate to ovate or acuminate, often erose or irregularly lobed; **blades** to 1.5 m or longer, (3)10–55(75) mm wide, abaxial and adaxial surfaces scabrous or glabrate. **Panicles** 20–120 cm long, (5)10–50 cm wide; **branches** unisexual. **Staminate branches** ascending to reflexed; **pedicel apices** 0.2–0.6 mm wide. **Staminate spikelets** 5–12.5 mm, lanceolate, acuminate or awned, awns to 3 mm. **Pistillate branches** divaricate, sometimes appressed if immature or bearing only aborted spikelets; **pedicel apices** 0.5–1.2 mm wide. **Pistillate spikelets** 5–24 mm long, 1–2.5 mm wide, lanceolate or oblong, chartaceous and flexible, dull or sublustrous, with scattered short hairs, these not or scarcely denser at the apices, awned, awns to 10 cm; **lemmas** and **paleas** often partly separating at maturity; **aborted pistillate spikelets** 0.4–1 mm wide, linear, shriveled, often threadlike. **Caryopses** 6–22 mm long, 0.8–2 mm wide. 2*n* = 30.

Zizania aquatica is native from the central plains to the eastern seaboard. It is sometimes planted for wildfowl food. The records from western North America reflect such plantings. Most, possibly all, have since died out. The population in northern Arizona was discovered in 1967, and persisted until the early 1990s. It is presumed to have been extirpated because the area has since been developed into a golf course, and no plants have been found downstream.

1. Plants to 5 m tall; blades (5)10–75 mm wide; pistillate spikelets 7–24 mm long; awns to 10 cm long . . . . var. *aquatica*
1. Plants 0.2–1 m tall; blades 3–12(20) mm wide; pistillate spikelets 5–11 mm long; awns 1–8 mm long . . . . var. *brevis*

Zizania aquatica L. var. aquatica [p. 49]
SOUTHERN WILDRICE, ZIZANIE AQUATIQUE

Plants to 5 m. **Blades** (5)10–75 mm wide. **Pistillate spikelets** 7–24 mm; **awns** to 10 cm.

Zizania aquatica var. *aquatica* grows in fresh or somewhat brackish marshes, swamps, streams, and lakes. Its native range extends from southeastern Minnesota to southern Maine, and south to central Florida and southern Louisiana. Plants in the Wading River, New Jersey, grow mostly immersed; narrow-leaved populations occur locally in the New England states and near Ottawa, Ontario.

Zizania aquatica L. var. brevis Fassett [p. 49]
ESTUARINE WILDRICE

Plants 0.2–1 m. **Blades** 3–12(20) mm wide. **Pistillate spikelets** 5–11 mm; **awns** 1–8 mm.

Zizania aquatica var. *brevis* is known from tidal mud flats along the St. Lawrence River, about 80 km up- and downstream from Quebec City, and from a small delta along the northern shore of the northwest Miramichi River estuary in New Brunswick.

2. Zizania palustris L. [p. 49]

Plants annual. **Culms** to 3 m, erect, usually at least partly immersed. **Sheaths** glabrous or with scattered hairs; **ligules** 3–16 mm, upper ligules truncate, lanceolate or triangular, erose; **blades** 20–60 cm long, 3–21(40+) mm wide, glabrous, margins glabrate or scabrous. **Panicles** 24–60 cm long, 1–20(40) cm wide; **branches** unisexual. **Staminate branches** ascending or divergent; **pedicel apices** 0.2–0.4 mm wide. **Staminate spikelets** 6–17 mm, lanceolate, acuminate or awned, awns to 2 mm. **Pistillate branches** mostly appressed or ascending, a few sometimes divergent; **pedicel apices** 0.7–1.2 mm wide. **Pistillate spikelets** 8–33 mm long, 1–2.6 mm wide, lanceolate or oblong, coriaceous or indurate, lustrous, glabrous or with lines of short hairs, apices usually hirsute and abruptly narrowed, awned, awns to 10 cm; **lemmas** and **paleas** remaining clasped at maturity; **aborted pistillate spikelets** 0.6–2.6 mm wide. **Caryopses** 6–30 mm long, 0.6–2 mm wide. 2*n* = 30.

Zizania palustris grows mostly to the north of *Z. aquatica*, but the two species overlap in the Great Lakes region, eastern Canada, and New England. It is cultivated as a crop in some provinces and states, with California being the largest producer. All records from the western part of the *Flora* region reflect deliberate plantings; none are known to have persisted. In cultivated strains, the pistillate spikelets remain on the plant at maturity.

1. Lower pistillate branches with 9–30 spikelets; pistillate part of the inflorescence 10–40 cm or more wide, the branches ascending to widely

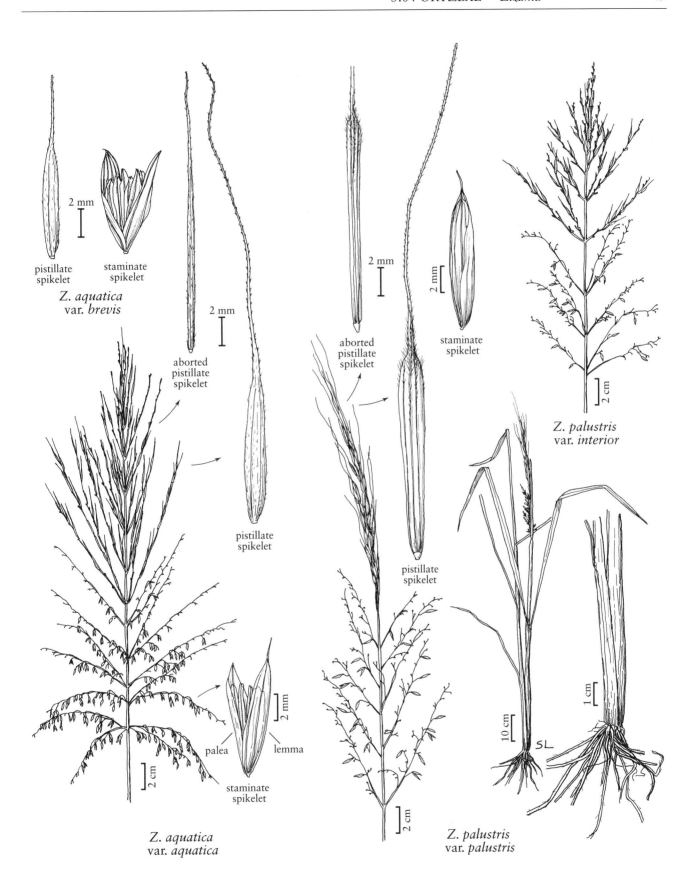

pistillate
spikelet

2 mm

staminate
spikelet

Z. aquatica
var. *brevis*

2 mm

aborted
pistillate
spikelet

2 mm

pistillate
spikelet

aborted
pistillate
spikelet

2 mm

staminate
spikelet

2 mm

Z. palustris
var. *interior*

2 cm

palea   lemma

2 mm

staminate
spikelet

Z. aquatica
var. *aquatica*

2 cm

pistillate
spikelet

2 cm

Z. palustris
var. *palustris*

10 cm   SL   1 cm

divergent; plants 1–3 m tall; blades 10–40+
mm wide . . . . var. *interior*
1. Lower pistillate branches with 2–8 spikelets;
pistillate part of the inflorescence 1–8(15) cm
wide, the branches appressed or ascending, or
a few branches somewhat divergent; plants to
2 m tall; blades 3–21 mm wide . . . . var. *palustris*

Zizania palustris L. var. interior (Fassett) Dore [p. 49]
INTERIOR WILDRICE

Plants 1–3 m. **Blades** 10–40+ mm wide. **Pistillate part of inflorescences** 10–40+ cm wide; **branches** ascending to widely divergent; **lower pistillate branches** with 9–30 spikelets.

Zizania palustris var. *interior* grows on muddy shores and in shallow water, mainly in the north central United States and adjacent Canada. It resembles *Z. aquatica* in its vegetative characters, and *Z. palustris* var. *palustris* in its pistillate spikelets. Crossing experiments (Duvall and Biesboer 1988) suggest that hybridization between the two may occur. Isozyme data (Warwick and Aiken 1986) do not support the hypothesis that *Z. palustris* var. *interior* is a hybrid.

Zizania palustris L. var. palustris [p. 49]
NORTHERN WILDRICE, ZIZANIE DES MARAIS,
FOLLE AVOINE, RIZ SAUVAGE

Plants to 2 m. **Blades** 3–21 mm wide. **Pistillate part of inflorescences** 1–8(15) cm wide; **branches** appressed or ascending, or with 1 to few branches somewhat divergent; **lower pistillate branches** with 2–8 spikelets.

Zizania palustris var. *palustris* grows in the shallow water of lakes and streams, often forming extensive stands in northern lakes. It has been introduced to British Columbia, Nova Scotia, Idaho, Arizona, and West Virginia for waterfowl food; some of the stands in the Canadian prairies may also have resulted from planting (Aiken et al. 1988).

3. Zizania texana Hitchc. [p. 51]
TEXAS WILDRICE

Plants perennial; stoloniferous. **Culms** 1–2(5) m, decumbent, geniculate, floating or the distal portions emergent. **Sheaths** glabrous; **ligules** 4–12 mm, upper ligules caudate or acuminate; **blades** to 1 m long and 13(25) mm wide, glabrous. **Panicles** 16–31 cm long, 1–10 cm wide; **branches** unisexual. **Staminate branches** ascending or somewhat divergent; **pedicel apices** about 3 mm wide. **Staminate spikelets** 6.5–11 mm, ovate or oblong, acute to acuminate. **Pistillate branches** appressed or ascending; **pedicel apices** 0.5–0.9 mm

wide. **Pistillate spikelets** 9–12.5 mm long, 1.2–1.8 mm wide, lanceolate, somewhat coriaceous, somewhat lustrous, with scattered short hairs, apices scabrous or hispidulous, awned, awns 9–35 mm; **aborted pistillate spikelets** 0.7–1.5 mm wide. **Caryopses** 4.3–7.6 mm long, 1–1.5 mm wide. $2n = 30$.

Zizania texana grows only in the headwaters of the San Marcos River, in San Marcos, Texas (Terrell et al. 1978). It is officially listed as an endangered species in the United States.

4. Zizania latifolia (Griseb.) Turcz. ex Stapf [p. 51]
ASIAN WILDRICE

Plants perennial; rhizomatous. **Culms** 1–2.5(4) m, erect, rooting at the lower nodes, emergent. **Sheaths** glabrous, lower sheaths tessellate; **ligules** 10–15 mm, triangular; **blades** (30)50–100 cm long, 1–3.5 cm wide, abaxial surfaces smooth, adaxial surfaces scabrous. **Panicles** 30–50 cm long, 10–15 cm wide; **branches** unisexual or bisexual, lower branches with staminate spikelets, middle branches with staminate and pistillate spikelets, upper branches with pistillate spikelets. **Staminate branches** spreading; **pedicel apices** 0.2–0.5 mm wide. **Staminate spikelets** 8–12 mm, elliptic-oblong, margins ciliate, awned, awns 2–8 mm, scabrous; **anthers** 5–8 mm. **Pistillate branches** erect; **pedicel apices** 0.3–0.6 mm wide. **Pistillate spikelets** 15–25 mm, linear, veins scabrous, awned, awns 15–30 mm, scabrous. **Caryopses** about 1 cm. $2n = 30, 34$.

Zizania latifolia is native to Asia, extending from northeast India and Russia through China and Myanmar to Korea and Japan. In its native range, it grows in the shallow waters of lakes and swamps, forming large patches.

The rhizomes and basal parts of the culms of *Z. latifolia* are edible, and become swollen when infected with the fungus *Ustilago esculenta* Henn. The infection also prevents the plants from flowering and fruiting. If infected plants were introduced into North America, the fungus might also infect the native species of *Zizania* and likewise prevent their flowering (Terrell and Batra 1982), a possibility that should be strenuously resisted. Plants of *Z. latifolia* should not be brought into North America. Many states do not permit importation of plants of *Z. latifolia* from another state without examination by a state-approved plant pathology laboratory.

New Zealand has designated *Z. latifolia* a prohibited plant because it "displaces all species by its dense growth, blocks drainage and access to water, and increases the chance of flooding. It forms dense colonies in swampy areas, thus affecting productive farm land" (Environment Walkato 2002–2007).

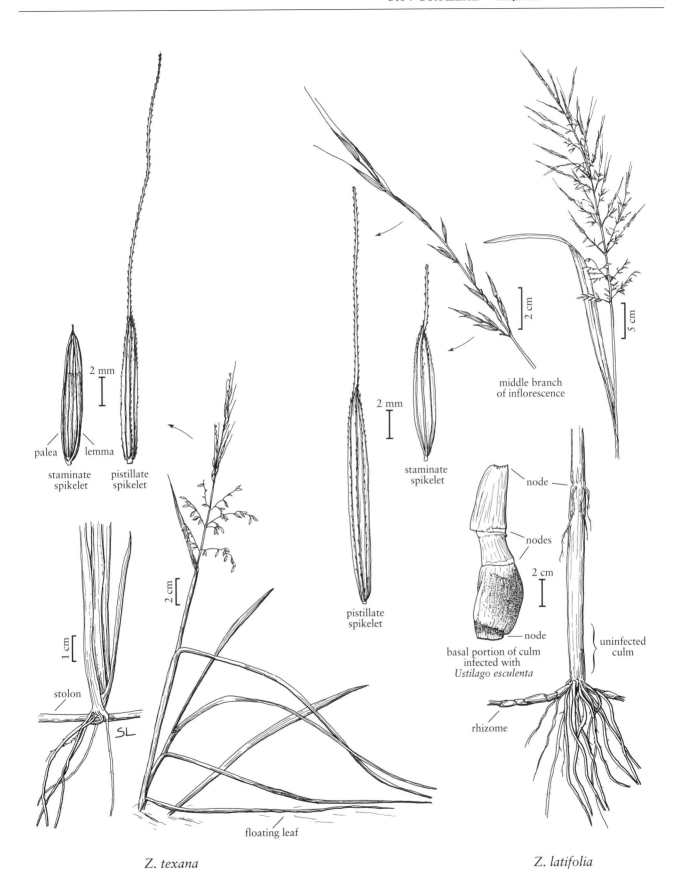

palea   lemma

staminate   pistillate
spikelet   spikelet

2 mm

stolon

SL

1 cm

2 cm

floating leaf

Z. texana

2 mm

staminate
spikelet

pistillate
spikelet

2 cm

middle branch
of inflorescence

5 cm

node

nodes

2 cm

node

basal portion of culm
infected with
Ustilago esculenta

rhizome

uninfected
culm

Z. latifolia

ZIZANIA

5.05 ZIZANIOPSIS Döll & Asch.

Edward E. Terrell

Plants perennial or annual; aquatic, rooted and emergent; rhizomatous; monoecious. **Culms** 1–4 m, erect or decumbent, sometimes rooting at the nodes. **Leaves** basal and cauline; **sheaths** open, somewhat laterally compressed; **ligules** scarious; **pseudopetioles** absent; **blades** flat or folded at the base, lanceolate. **Inflorescences** terminal panicles, staminate and pistillate spikelets on the same branches, staminate spikelets proximal, pistillate spikelets distal; **disarticulation** beneath the spikelets; **pedicel apices** cupulate. **Spikelets** unisexual, laterally compressed to subterete, lemma margins not clasping the paleas, with 1 floret. **Glumes** absent; **calluses** glabrous. **Staminate lemmas** membranous, 5–7-veined, acuminate or terminally awned; **paleas** similar to the lemmas, 3-veined; **lodicules** 2; **anthers** 6. **Pistillate lemmas** membranous, 7-veined, terminally awned; **paleas** similar to the lemmas, 3-veined, awned or unawned; **styles** 2, bases fused, stigmas terminally exserted, plumose. **Fruits** achenes, ellipsoid or obovoid, beaked by the persistent style base; **pericarps** shell-like, partially free from the seed, smooth, coriaceous or crustaceous; **seeds** oblong, subterete, or 2-angled; **embryos** basal; **hila** linear. $x = 12$. Name based on the generic name *Zizania* and the Greek *opsis*, 'appearance', alluding to the similarity to *Zizania*.

Zizaniopsis grows from the southern United States to Argentina. All of its five species grow in wet habitats. Only *Zizaniopsis miliacea* is native to and found in the *Flora* region.

SELECTED REFERENCES Fox, A.M. and W.T. Haller. 2000. Production and survivorship of the functional stolons of giant cutgrass, *Zizaniopsis miliacea* (Poaceae). Amer. J. Bot. 87:811–818; McVaugh, R. 1983. Flora Novo-Galiciana: A Descriptive Account of the Vascular Plants of Western Mexico, vol. 14; Gramineae (series ed. W.R. Anderson). University of Michigan Press, Ann Arbor, Michigan, U.S.A. 436 pp.; Terrell, E.E. and H. Robinson. 1974. Luziolinae, a new subtribe of oryzoid grasses. Bull. Torrey Bot. Club 101:235–245.

1. **Zizaniopsis miliacea** (Michx.) Döll & Asch. [p. 53]
GIANT CUTGRASS, WATER MILLET

Plants perennial; rhizomatous, rhizomes to 1.5 cm thick. **Culms** 1–4 m tall, to 3.5 cm thick, erect or decumbent, glabrous, readily rooting at the nodes when decumbent and producing leafy buds. **Sheaths** thick, glabrous; **ligules** to 2 cm, glabrous; **blades** to 1 m long, 6–30 mm wide, sometimes scabrous, bluish-green, margins scabrous. **Panicles** to 80+ cm long, usually 4–20 cm wide, open; **pedicels** to 10 mm long, apices 0.1–0.4 mm wide. **Staminate lemmas** 5–10 mm, lanceolate to elliptic, glabrous, acuminate or awned, awns to 2 mm; **paleas** acuminate or awned, awns to 1 mm; **anthers** 2.5–5 mm. **Pistillate lemmas** 4–8 mm, ovate or elliptic, awned, awns to 9 mm; **paleas** caudate-acuminate or awned, awns to 1 mm; **style bases** 1–3 mm, stigmas 2–6 mm, conspicuously exserted. **Achenes** 2.5–4 mm long, 1–2 mm wide, ellipsoid or obovoid, smooth, lustrous, beaked by the persistent style base. $2n = 24$.

Zizaniopsis miliacea grows in shallow, fresh- or brackish-water marshes, swamps, streams, lakes, and ditches. It is most common on the eastern coastal plain of the United States, extending south to Florida and west to Illinois, Oklahoma and Texas. It has also been reported growing as a disjunct in central Mexico (McVaugh 1983).

Fox and Haller (2000) found that decumbent flowering culms readily produce roots and axillary shoots at the nodes. The decumbent culms act as functional stolons, allowing for rapid colonization; thus plants become established up to 3–4 m away from the parent plant.

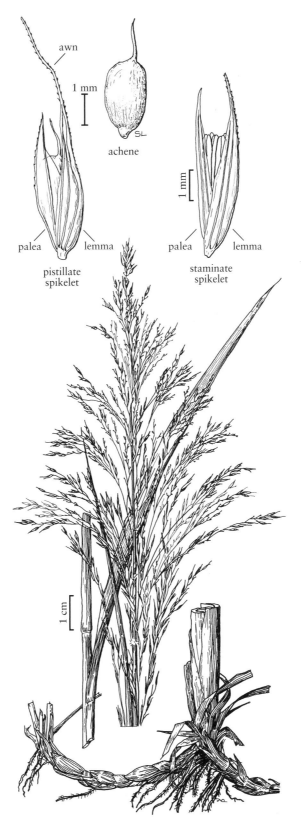

Z. miliacea

ZIZANIOPSIS

5.06 LUZIOLA Juss.

Edward E. Terrell

Plants perennial; aquatic, usually rooted, sometimes floating; stoloniferous, sometimes mat-forming; monoecious. **Culms** 10–100+ cm, erect or prostrate, sometimes rooting at the nodes, branched, emergent or immersed. **Leaves** cauline; **sheaths** open, not inflated or somewhat inflated; **ligules** hyaline; **pseudopetioles** present or absent; **blades** flat, linear to lanceolate or narrowly elliptic, glabrous, pubescent, or scabrous. **Inflorescences** panicles, racemes, or spikes, exserted or enclosed, staminate and pistillate spikelets usually in separate inflorescences, pistillate inflorescences at the lower or middle nodes, staminate inflorescences usually terminal; **disarticulation** below the spikelets. **Spikelets** unisexual, laterally compressed to subterete, with 1 floret. **Glumes** absent; **calluses** glabrous; **lemmas** and **paleas** subequal, ovate or lanceolate, membranous or hyaline, unawned; **lodicules** 2. **Staminate lemmas** and **paleas** obscurely few- to several-veined; **anthers** 6–16. **Pistillate lemmas** 5–14-veined, margins not clasping the margins of the paleas, unawned; **paleas** 3–10-veined; **styles** 2, bases fused, stigmas laterally or terminally exserted, plumose. **Fruits** achenes, ovoid, ellipsoid, or subglobose, beaked by the persistent style bases; **pericarps** shell-like, partially free from the seed, smooth or striate, crustaceous; **seeds** ovoid to subglobose; **embryos** basal; **hila** linear. $x = 12$. Name modified from *Luzula*, a genus of the *Juncaceae*.

Luziola is a genus of about 12 species that range from the southeastern United States to Argentina. Only *L. fluitans* is native to the *Flora* region; two other species have been introduced and are established. The species are emergent or immersed in shallow, fresh to brackish water.

SELECTED REFERENCES **Anderson, L.C.** and **D.W. Hall**. 1993. *Luziola bahiensis* (Poaceae): New to Florida. Sida 15:619–622; **Pohl, R.W.** and **G. Davidse**. 2001. *Luziola*. Pp. 2072–2073 *in* W.D. Stevens, C.U. Ulloa, A. Pool, and M. Montiel (eds.). Flora de Nicaragua, Vol. 3: Angiosperms (Pandanaceae–Zygophyllaceae). Missouri Botanical Garden Press, St. Louis, Missouri, U.S.A. 2666 pp. [for vols. 1–3]; **Swallen, J.R.** 1965. The grass genus *Luziola*. Ann. Missouri Bot. Gard. 52:472–475; **Terrell, E.E.** and **H. Robinson**. 1974. Luziolinae, a new subtribe of oryzoid grasses. Bull. Torrey Bot. Club 101:235–245.

1. Culms prostrate, usually immersed; leaves floating or streaming in currents, 1–5(8) cm long, usually more numerous towards the ends of the culms; pistillate inflorescences mostly included in the sheaths, only the stigmas visible . . . . 1. *L. fluitans*
1. Culms suberect to erect, from fully emergent to immersed; leaves not conspicuously floating or streaming, longer than 6 cm, basal or scattered along the culms; pistillate inflorescences all or mostly exserted, their branches and spikelets evident.
    2. Pistillate florets 3–5 mm long; achenes striate . . . . 2. *L. bahiensis*
    2. Pistillate florets 2–2.5 mm long; achenes smooth . . . . 3. *L. peruviana*

1. **Luziola fluitans** (Michx.) Terrell & H. Rob. [p. 56]
SILVERLEAF GRASS, WATERGRASS

Plants stoloniferous; mostly immersed. **Culms** to 1+ m, bases prostrate and rooting at the nodes. **Leaves** mainly cauline, usually crowded distally, streaming in the current or floating on the surface, distal leaves sometimes emergent; **sheaths** glabrous; **ligules** 0.5–2 mm, subtriangular to truncate, sometimes erose; **pseudopetioles** sometimes present; **blades** 1–5(8) cm long, 1–4 mm wide, glabrous, sometimes pilose basally, scabridulous. **Inflorescences** spikes or racemes, spikelets sessile or shortly pedicellate. **Staminate inflorescences** 1–2 cm, terminal, partially exserted from the sheaths, emergent at anthesis, with 2–4 spikelets; **staminate florets** 3–5 mm; **lemmas** narrowly ovate, usually 5-veined; **paleas** similar, usually 3-veined; **anthers** 6, 2–3.7 mm, linear. **Pistillate inflorescences** 0.5–2 cm, mostly included in the subtending sheaths, only the stigmas visible, with 2 to few spikelets; **pistillate florets** 1.4–3 mm, caducous; **lemmas** 5–9-veined, acuminate; **paleas** similar to the lemmas, 3–7-veined; **stigmas** 3–6 mm. **Achenes** 1.3–2 mm, ovoid, asymmetrical, minutely striate, lustrous. $2n = 24$.

Luziola fluitans grows in fresh to slightly saline lakes and streams in the southeastern United States and eastern Mexico. It is most common in the coastal plain, and also occurs in the Piedmont. Plants in the United States and eastern Mexico belong to **L. fluitans** (Michx.) Terrell & H. Rob var. **fluitans**, which differs from **L. fluitans** var. **oconnorii** (R. Guzmán) G. Tucker of the uplands of central Mexico in having shorter and narrower leaf blades, shorter spikelets, and shorter anthers.

2. Luziola bahiensis (Steud.) Hitchc. [p. 56]
BRAZILIAN WATERGRASS

Plants stoloniferous, emergent or immersed. **Culms** 10–50 cm, suberect to erect. **Leaves** mostly basal or scattered along the culms, not conspicuously floating or streaming in the current; **sheaths** glabrous or ciliate; **ligules** to about 6 mm, lanceolate, acuminate; **pseudopetioles** sometimes present; **blades** 3–30(39) cm long, 1–6 mm wide, glabrous or puberulent. **Inflorescences** panicles; **pedicels** capillary. **Staminate panicles** 4–6 cm, terminal, with 9–17 spikelets; **staminate florets** 4–8 mm; **lemmas** lanceolate, faintly 5–9-veined; **paleas** similar, 3-veined; **anthers** 6, 2–4 mm, linear. **Pistillate panicles** to about 12 cm, arising from the lower nodes of the culms, all or mostly exserted, with 9–17 spikelets; **branches** slender, divergent to reflexed; **pistillate florets** 3–5 mm, caducous; **lemmas** 8–13-veined, acuminate or caudate; **paleas** similar, 6–8-veined; **stigmas** 2–4 mm. **Achenes** 1.5–2.2 mm, ovoid or broadly ellipsoid, lightly to moderately striate, somewhat lustrous. 2*n* = 24.

Luziola bahiensis is native from the Caribbean south to Argentina. It has been found at scattered locations in southern Louisiana, Mississippi, Alabama, and northwestern Florida. It grows in wet places or shallow water along streams and lakes.

3. Luziola peruviana J.F. Gmel. [p. 56]
PERUVIAN WATERGRASS

Plants stoloniferous, fully or mostly emergent. **Culms** 10–40 cm, suberect, branching, fully or mostly emergent. **Leaves** mostly basal or scattered along the culms, not conspicuously floating or streaming in the current; **sheaths** glabrous or ciliate; **ligules** to 5 mm, lanceolate, acuminate; **blades** 3–30 cm long, 0.5–4 mm wide, glabrous, sometimes scabridulous. **Inflorescences** panicles. **Staminate panicles** 1–3 cm, terminal, with 10–25 spikelets; **staminate florets** 3.5–8 mm, caducous; **lemmas** lanceolate, 5-veined; **paleas** similar, 3-veined; **anthers** 6, 2–5 mm. **Pistillate panicles** to 7 cm, arising from the lower to middle nodes of the culms, partially exserted from the sheaths, with 10–25 spikelets; **branches** filiform, spreading to reflexed; **pistillate florets** 2–2.5 mm, caducous; **lemmas** 7–14-veined, erose to acuminate or caudate; **paleas** similar to the lemmas, 5–10-veined; **stigmas** 2–3 mm. **Achenes** 1–1.3 mm, broadly ellipsoid to subglobose, smooth, lustrous. 2*n* = 24.

Luziola peruviana has been found at scattered locations from Texas to Florida. It is native to the Caribbean, Central America, and South America, and grows in wet places and shallow water along streams and lakes.

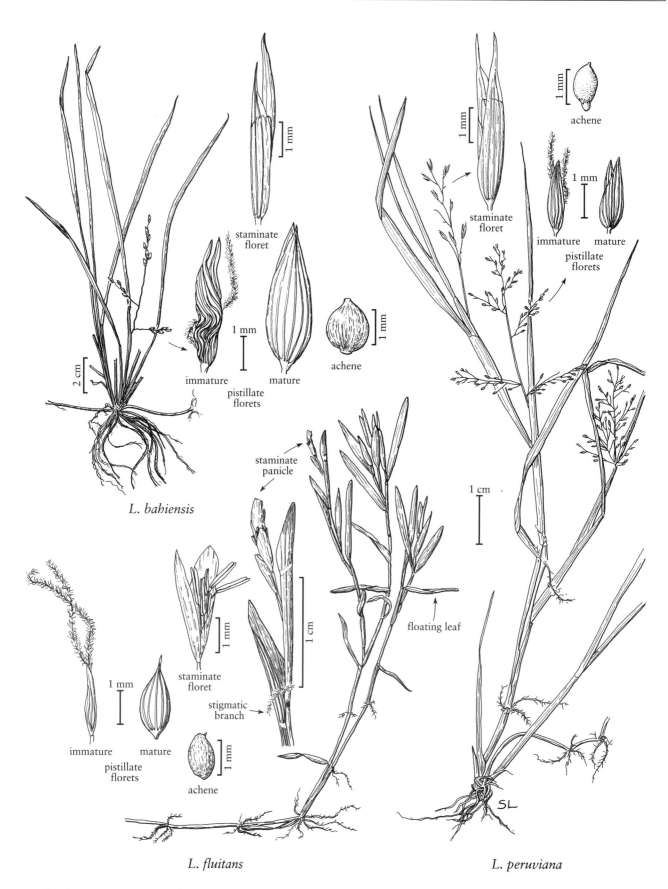

L. bahiensis

staminate floret

immature   mature
pistillate
florets

achene

staminate
floret

immature   mature
pistillate
florets

achene

staminate
panicle

floating leaf

staminate
floret

stigmatic
branch

immature   mature
pistillate
florets

achene

L. fluitans

L. peruviana

LUZIOLA

3. POÖIDEAE Benth.

Grass Phylogeny Working Group

Plants annual or perennial; sometimes matlike, sometimes cespitose, sometimes stoloniferous, sometimes rhizomatous. **Culms** usually hollow, sometimes solid. **Leaves** distichous; **sheaths** usually open to the base, varying to closed for nearly their full length; **auricles** present or absent; **abaxial ligules** absent; **adaxial ligules** scarious or membranous, sometimes puberulent or scabridulous, usually not ciliate, cilia sometimes shorter than the base; **pseudopetioles** rarely present; **blades** usually linear, sometimes broadly so, venation parallel; **cross sections** non-Kranz, mesophyll nonradiate, adaxial palisade layer absent, fusoid and arm cells usually absent; **midribs** usually simple; **adaxial bulliform cells** present; **stomates** with parallel-sided subsidiary cells; **epidermes** usually lacking bicellular microhairs, sometimes with unicellular microhairs, papillae usually absent, when present, rarely more than 1 per cell. **Inflorescences** usually terminal, panicles, spikes, or racemes, usually ebracteate; **disarticulation** usually below the florets, sometimes below the glumes, at the rachis nodes, or at the inflorescence bases. **Spikelets** usually bisexual, infrequently unisexual or mixed, usually laterally compressed or not compressed, occasionally dorsally compressed, with 1–30 sexual florets, distal floret(s) often reduced, infrequently spikelets with 1–2 reduced or staminate basal florets and a single terminal sexual floret. **Glumes** usually 2, upper or lower glumes sometimes absent, rarely both glumes absent; **lemmas** without uncinate hairs, awned or not, awns single, basal to apical; **paleas** usually well-developed, sometimes reduced or absent; **lodicules** 2(3), usually lanceolate and broadly membranous distally, rarely truncate and fleshy, usually not veined or obscurely veined, sometimes distinctly veined, sometimes ciliate; **anthers** (1, 2)3; **ovaries** glabrous or sometimes hairy distally, sometimes with an apical appendage; **haustorial synergids** absent; **styles** (1)2(–4), bases close together, sometimes fused. **Caryopses: hila** linear, elliptic, ovate, or punctate; **endosperm** usually hard, sometimes soft or liquid, with or without lipids, starch grains compound or simple; **embryos** less than $^1/_2$ the length of the caryopses; **epiblasts** usually present; **scutellar cleft** usually absent; **mesocotyl internode** usually absent; **embryonic leaf margins** overlapping. $x = 7, 10$.

The subfamily *Poöideae* includes approximately 3300 species, making it the largest subfamily in the *Poaceae*. It reaches its greatest diversity in cool temperate and boreal regions, extending across the tropics only in high mountains.

The circumscription and relationships of tribes within the *Poöideae* are unsettled (see, for example, Catalán et al. 1997, 2004; Soreng and Davis 1998). In this flora, some previously recognized tribes have been combined with the *Poeae*. Recognition of some of these as subtribes is well supported; among these is the *Hainardieae* Greuter (which, at the subtribal level, is called the *Parapholiinae* Caro). Members of other traditional tribal groupings, such as the *Aveneae* Dumort., appear to be widely dispersed within the *Poeae sensu lato*. Further work will probably support the division of the expanded *Poeae* into additional tribes; there is as yet no clear indication as to what the boundaries of such tribes should be.

SELECTED REFERENCES Catalán, P., E.A. Kellogg, and R.G. Olmstead. 1997. Phylogeny of Poaceae subfamily Poöideae based on chloroplast *ndh*F gene sequences. Molecular Phylogenetics and Evolution 8:150–166; Catalán, P., P. Torrecilla, J.A.L. Rodríguez, and R.G. Olmstead. 2004. Phylogeny of the festucoid grasses of subtribe Loliinae and allies (Poeae, Poöideae) inferred from ITS and *trn*L-F sequences. Molecular Phylogenetics and Evolution 31:517–541. **Grass Phylogeny Working Group.** 2001. Phylogeny and subfamilial classification of the grasses (Poaceae). Ann. Missouri Bot. Gard. 88:373–457; **Soreng, R.J. and J.I. Davis.** 1998. Phylogenetics and Character Evolution in the Grass Family (Poaceae): Simultaneous Analysis of Morphological and Chloroplast DNA Restriction Site Character Sets. Bot. Rev. 64: 1–85.

1. Inflorescences 1-sided spikes, the spikelets radial to and partially embedded in the rachises; spikelets with 1 floret each . . . . 7. *Nardeae*
1. Inflorescences panicles, racemes, or 2-sided spikes with spikelets radial or tangential to the rachises, sometimes embedded in the axes, never both radial and embedded; spikelets with 1–30 florets.
   2. Cauline leaf sheaths closed for at least $^3/_4$ their length; lemmas longer than (4.5)6.5 mm or awned or with prominent, parallel veins.
      3. Ovary apices glabrous; styles fused at the base, divergent, naked on the lower portion, plumose distally; lemmas often with a purplish band in the distal $^1/_2$, usually unawned; distal 1–3 florets often reduced to lemmas, the lower 1–2 lemmas often enclosing the terminal lemmas; lodicules about 0.2–0.5 mm long, truncate, fleshy, without a distal membranous portion . . . . 9. *Meliceae*
      3. Ovary apices hairy; styles separate and plumose to the base; lemmas usually without a purplish band, sometimes with purplish bases, usually awned; distal 1–2 florets sometimes reduced, each separate with lemma and palea; lodicules usually more than 1 mmlong, fleshy at the base, with a distal membranous portion . . . . 12. *Bromeae*
   2. Cauline leaf sheaths usually open for most or all of their length; if the sheaths closed, the lemmas shorter than 7 mm, unawned and with lemma veins inconspicuous and converging distally.
      4. Inflorescences usually spikes or spikelike racemes, sometimes panicles, lateral spikelets on pedicels less than 3 mm long; if inflorescences with 1 spikelet per node, the spikelets tangential to the rachises or pedicellate and the lemmas unawned or terminally awned; ovary apices hairy.
         5. Glumes unequal, exceeded by the lowest lemmas, lanceolate, apices obtuse to acuminate or mucronate, rarely awned; inflorescences spikelike racemes, all spikelets pedicellate; pedicels 0.5–2.5 mm long . . . . 11. *Brachypodieae*
         5. Glumes equal to unequal, sometimes absent, frequently exceeding the lowest lemmas, subulate to lanceolate, ovate, or obovate, apices truncate to acuminate, frequently awned; inflorescences usually spikes or spikelike, with 1 or more sessile spikelets per node, sometimes a panicle; pedicels absent or up to 4 mm long . . . . 13. *Triticeae*
      4. Inflorescences usually panicles, sometimes racemes with pedicels more than 2.5 mm long, or spikes with 1 spikelet per node and the spikelets radial or tangential to the rachises; if spikelets 1 per node and tangential, the lemmas awned from midlength to subapically, never terminally, if spikes with radial spikelets, the lemmas unawned or awned, awns basal to terminal; ovary apices usually glabrous, sometimes hairy.
         6. Lower glumes absent or highly reduced; inflorescences panicles . . . . 6. *Brachyelytreae*
         6. Lower glumes usually well-developed, sometimes present only on the terminal spikelets; inflorescences panicles, racemes, or spikes.
            7. Caryopses beaked; blades tapering both basally and apically, midveins usually eccentric . . . . 8. *Diarrheneae*
            7. Caryopses not beaked; blades usually tapering only apically, midveins usually centric.
               8. Spikelets with 1 floret; lemmas terminally awned, the junction of the lemma and awn abrupt, evident; glumes equal to or longer than the florets . . . . 10. *Stipeae* (in part)
               8. Spikelets with 1–22 florets; lemmas unawned or dorsally to terminally awned, if terminally awned, the transition from lemma to awn gradual, not evident; glumes absent or shorter than to longer than the adjacent florets.
                  9. Lemmas membranous, bidentate or bifid; both surfaces of the leaf blades deeply ribbed; ovary apices hairy; culms with solid internodes; plants cultivated or established at a few locations . . . . 10. *Stipeae* (in part)
                  9. Lemmas hyaline to membranous, entire or minutely bidentate; leaf blades rarely deeply ribbed on both sides; ovary apices usually glabrous; culms usually with hollow internodes; plants mostly native or established throughout the *Flora* region, sometimes cultivated . . . . 14. *Poeae*

6. BRACHYELYTREAE Ohwi

Stephen N. Stephenson

Plants perennial; with knotty rhizomes. **Culms** annual, not branching above the base; **internodes** solid. **Sheaths** open, margins not fused; **collars** glabrous, without tufts of hair at the sides; **auricles** absent; **ligules** scarious, not ciliate, those of the upper and lower cauline leaves usually similar; **pseudopetioles** absent; **blades** tapering basally and distally, venation parallel, cross venation not evident, secondary veins parallel to the midvein; **cross sections** non-Kranz, with arm and fusoid cells; **epidermes** without microhairs, cells not papillate. **Inflorescences** terminal panicles. **Spikelets** scarcely compressed, with 1 floret, floret bisexual; **rachillas** prolonged beyond floret base; **disarticulation** above the glumes, beneath the floret. **Glumes** unequal, lanceolate; **lower glumes** absent or highly reduced; **upper glumes** less than ¼ as long as the florets, 1-veined; **florets** 8–12 mm, dorsally compressed; **calluses** rounded, antrorsely hairy, hairs 0.2–0.5 mm; **lemmas** coriaceous, unawned, rounded dorsally, 5-veined, veins converging distally, apices entire; **paleas** subequal to the lemmas, 2-veined, ridged over the veins; **lodicules** 2, glabrous, veined; **anthers** 3; **ovaries** glabrous; **styles** 2, elongate, hairy, bases free. **Caryopses** grooved, styles persistent; **hila** linear; **embryos** less than ½ as long as the caryopses. $x = 11$.

There is only one genus, *Brachyelytrum*, in the *Brachyelytreae*. The subfamilial placement of the tribe has been disputed. Campbell et al. (1986), after examining a range of characters, concluded that it should be included in the *Bambusoideae*, with its probable closest relatives being the herbaceous bamboos. They acknowledged, however, that the *Brachyelytreae*, which are entirely north-temperate in distribution, are biogeographically distinct from the herbaceous *Bambusoideae*, all of which are native to tropical South America. The tribe is treated here as a member of the *Poöideae*, based on the findings of the Grass Phylogeny Working Group (2001). It is anomalous within the subfamily in having arm and fusoid cells and broad seedling leaves, features that are generally associated with the *Bambusoideae*.

SELECTED REFERENCES **Campbell, C.S., P.E. Garwood** and **L.P. Specht.** 1986. Bambusoid affinities of the north temperate genus *Brachyelytrum* (Gramineae). Bull. Torrey Bot. Club 113:135-141; **Grass Phylogeny Working Group.** 2001. Phylogeny and subfamilial classification of the grasses (Poaceae). Ann. Missouri Bot. Gard. 88:373–457; **Stephenson, S.N.** 1971. The biosystematics and ecology of the genus *Brachyelytrum* (Gramineae) in Michigan. Michigan Bot. 10:19-33.

6.01 BRACHYELYTRUM P. Beauv.

Stephen N. Stephenson

Jeffery M. Saarela

Plants perennial; rhizomatous, rhizomes knotty. **Culms** 28–102 cm, erect, not branched above the bases; **internodes** solid; **nodes** glabrous or retrorsely pubescent. **Leaves** mostly cauline; **sheaths** open; **auricles** absent; **ligules** membranous; **lower leaf blades** absent or reduced; **upper leaf blades** flat, tapering both basally and apically. **Inflorescences** terminal panicles, contracted; **branches** appressed, with 1–3(5) spikelets. **Spikelets** pedicellate, terete to dorsally compressed, with 1 floret; **rachillas** prolonged beyond the floret base, glabrous; **disarticulation** above the glumes, beneath the floret. **Glumes** 1 or 2; **lower glumes** 0.1–1.1 mm, sometimes absent; **upper glumes** 0.2–7 mm, clearly exceeded by the florets; **florets** 8–12 mm; **calluses** about 0.8 mm, blunt, with hairs; **lemmas** membranous to coriaceous, scabrous, enclosing the paleas, 5-veined, tapering, awned, awns terminal, lemma-awn transition gradual; **awns** 9.5–32.5 mm, longer than the lemma bodies, straight, scabrous; **paleas** subequal to the lemmas, 2-veined; **lodicules** 2, veined; **anthers** 3, yellow; **styles** 2, bases free, white. **Caryopses** linear, longitudinally grooved, apices beaked, pubescent; **hila** linear. $x = 11$. Name from the Greek *brachys*, 'short', and *elytron*, 'husk' or 'involucre', a reference to the short glumes.

Brachyelytrum includes three species, two native to eastern North American and one to eastern Asia (Saarela et al. 2003). The ranges of the two North American species overlap but, although they often grow closely together, neither mixed populations nor apparent hybrids have been found (Stephenson 1971; Saarela et al. 2003). Saarela et al. (2003) were unable to detect any differences in the ecological preferences of the two North American taxa.

SELECTED REFERENCES **Campbell, C.S., P.E. Garwood** and **L.P. Specht.** 1986. Bambusoid affinities of the north temperate genus *Brachyelytrum* (Gramineae). Bull. Torrey Bot. Club 113:135–141; **Koyama, T.** and **S. Kawano.** 1964. Critical taxa of grasses with North American and eastern Asiatic distribution. Canad. J. Bot. 42:859–884; **Saarela, J.M., P.M. Peterson, R.J. Soreng,** and **R.E. Chapman.** 2003. A taxonomic revision of the eastern North American and eastern Asian disjunct genus *Brachyelytrum* (Poaceae): Evidence from morphology, phytogeography and AFLPs. Syst. Bot. 28:674–692; **Stephenson, S.N.** 1971. The biosystematics and ecology of the genus *Brachyelytrum* (Gramineae) in Michigan. Michigan Bot. 10:19–33.

1. Lemmas hispid, hairs 0.2–0.9 mm long, visible at 10× magnification; anthers 3.5–6 mm long; awns 13–17(20) mm long . . . . 1. *B. erectum*
1. Lemmas scabrous, scabrules 0.08–0.14(0.2) mm long; anthers 2–3.5 mm long; awns (14)17–24(26) mm long . . . . 2. *B. aristosum*

1. **Brachyelytrum erectum** (Schreb.) P. Beauv. [p. 61]
SOUTHERN SHORTHUSK, BRACHYELYTRUM DRESSÉ

Culms 34–102 cm long, 0.7–1.4 mm thick, erect; **nodes** densely pilose; **lowest internodes** mostly glabrous, usually retrorsely pubescent near the nodes. **Sheaths** hispid; **ligules of middle and upper cauline blades** 2–3.5 mm, truncate to acute, lacerate or erose; **blades** 8.8–17.5 cm long, (9)11–17(20) mm wide, abaxial surfaces pilose on the veins and often between the veins, adaxial surfaces glabrous or slightly hispid, margins scabrous, with (2)5–11(14) prickles and 0–2(14) macrohairs per mm. **Panicles** (5.5)9.1–14.3(18.5) cm. **Spikelets,** including the awns, (25)29–36(42) mm. **Lower glumes** (0.1)0.3–0.7(1.1) mm, sometimes absent; **upper glumes** (0.2)0.9–3.5(7) mm, sometimes aristate; **calluses** hairy, hairs 0.2–0.5 mm; **lemmas** 9–13 mm long, 0.8–1.7 mm wide, veins hispid, hairs 0.2–0.9 mm, midveins more prominent than the lateral veins; **awns** 13–17(20) mm; **paleas** 7–12 mm; **anthers** 3.5–6 mm. **Caryopses** 5.5–7.5 mm. $2n = 22$.

Brachyelytrum erectum grows in woodlands, occasionally over limestone bedrock, and in moist woods and forests. It extends from Lake of the Woods, Ontario, east to Newfoundland, and in the United States from Minnesota to New England and south to the Gulf Coast and Florida.

Koyama and Kawano (1964), among others, treated *Brachyelytrum erectum* var. *glabratum* (Vasey) T. Koyama & Kawano as the northern taxon *B. aristosum*, but the holotype of *B. erectum* var. *glabratum* belongs to *B. erectum sensu stricto*. This means that, nomenclaturally, *B. erectum* var. *glabratum* is a synonym of *B. erectum*, although most of the specimens identified as var. *glabratum* belong to *B. aristosum*.

2. **Brachyelytrum aristosum** (Michx.) P. Beauv. *ex* Branner & Coville [p. 61]
NORTHERN SHORTHUSK

Culms (28)41–78(96) cm long, 0.6–1 mm thick; **nodes** densely pilose; **internodes** glabrous or hispid, occasionally retrorsely pubescent near the nodes. **Sheaths** pubescent; **ligules of middle and upper cauline blades** 1.8–2.5 mm, acute, erose; **blades** (6.9)8.6–13 (16.1) cm long, 8–16 mm wide, abaxial surfaces sparsely pilose, adaxial surfaces with some hairs usually restricted to the veins, margins scabrous, with (1)4–10(12) prickles and (1)1–9 macrohairs per mm. **Panicles** (6.6)9.5–17.5 cm. **Spikelets,** including the awns, 23–36 mm. **Lower glumes** 0.1–0.4(0.9) mm, sometimes absent; **upper glumes** 0.6–1.7(3) mm; **calluses** hairy, hairs 0.2–0.5 mm; **lemmas** 8–10(11) mm long, 0.7–1.4 mm wide, veins scabridulous, scabrules 0.08–0.14(0.2) mm, all veins equally prominent; **awns** (14)17–24(26) mm; **paleas** 7.7–11.5 mm; **anthers** 2–3.5 mm. **Caryopses** 5.5–7.5 mm. $2n = 22$.

Brachyelytrum aristosum, like *B. erectum*, grows in moist woods and forests, but its primary distribution is more northern, extending from Ontario to Newfoundland,

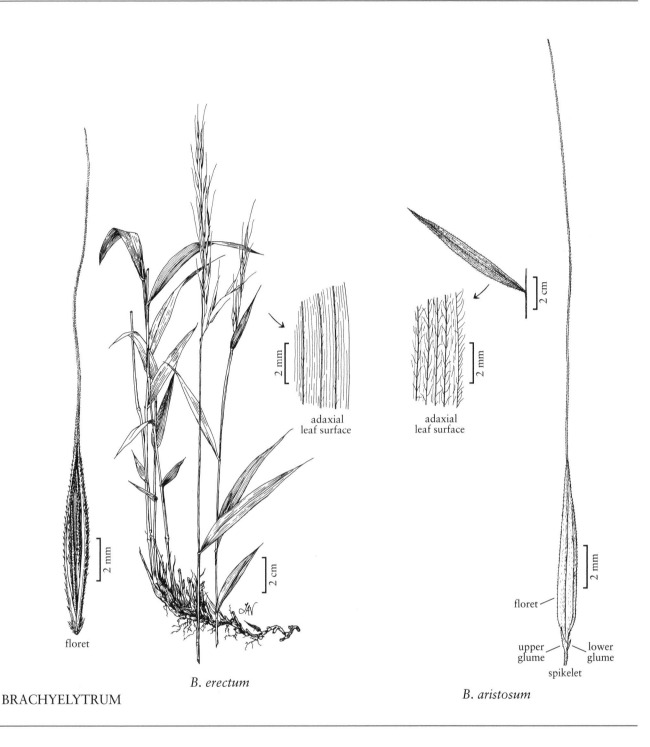

floret

adaxial
leaf surface

adaxial
leaf surface

floret

upper
glume   lower
                     glume

spikelet

BRACHYELYTRUM

B. erectum

B. aristosum

Minnesota, and Pennsylvania, and south through the Appalachian Mountains to the junction of Tennessee, North Carolina, and Georgia.

Some authors (e.g., Koyama and Kawano 1964) have called this taxon *Brachyelytrum erectum* var. *glabratum*.

As discussed under *B. erectum*, that name is a nomenclatural synonym of *B. erectum sensu stricto*. Nevertheless, most plants identified as *B. erectum* var. *glabratum* will be found to be *B. aristosum*.

7. NARDEAE W.D.J. Koch

Mary E. Barkworth

Plants perennial; cespitose. **Culms** annual, to 60 cm; **internodes** hollow. **Sheaths** open, margins not fused; **collars** glabrous, without tufts of hair at the sides; **auricles** absent; **ligules** scarious, not ciliate, those of the upper and lower cauline leaves usually similar; **pseudopetioles** not present; **blades** filiform, venation parallel, cross venation not evident, secondary veins parallel to the midvein; **cross sections** non-Kranz, without arm or fusoid cells, **adaxial epidermes** with bicellular microhairs, not papillate. **Inflorescences** terminal spikes, 1-sided, spikelets solitary, radial to the rachises; **rachises** with the spikelets partially embedded. **Spikelets** not compressed, triangular in cross section, with 1 floret, floret bisexual; **rachillas** not prolonged beyond the floret base; **disarticulation** above the glumes, beneath the floret. **Glumes** absent or vestigial; **lower glumes** a cupular rim; **upper glumes** absent or vestigial; **florets** 5–10 mm, not compressed; **calluses** poorly developed, glabrous; **lemmas** chartaceous, 3-veined, angled over the veins, most strongly so over the lateral veins, apices entire, awned, awns terminal, not branched, lemma-awn transition gradual, not evident; **paleas** subequal to the lemmas, hyaline, 2-keeled; **lodicules** absent; **anthers** 3; **ovaries** glabrous; **styles** 1. **Caryopses** fusiform, style bases not persistent; **hila** linear, more than $\frac{1}{2}$ as long as the caryopses; **embryos** about $\frac{1}{6}$ the length of the caryopses. $x = 13$.

There is only one genus, *Nardus*, in the *Nardeae*. Its relationships are obscure. Embryo characters, ligule texture, and DNA sequence data suggest it belongs in the *Poöideae*; the bicellular microhairs suggest a bambusoid or arundinoid affiliation. Its inclusion here in the *Poöideae* reflects the findings of the Grass Phylogeny Working Group (2001).

SELECTED REFERENCES **Davis, J.I.** and **R.J. Soreng.** 1993. Phylogenetic structure in the grass family (Poaceae) as inferred from chloroplast DNA restriction site variation. Amer. J. Bot. 80:1444–1454; **Grass Phylogeny Working Group.** 2001. Phylogeny and subfamilial classification of the grasses (Poaceae). Ann. Missouri Bot. Gard. 88:373–457.

7.01 NARDUS L.

Mary E. Barkworth

Plants perennial; cespitose. **Culms** 3–60 cm, erect; **basal branching** intravaginal. **Leaves** mostly basal; **sheaths** open; **auricles** absent; **ligules** membranous, entire, rounded; **blades** filiform, tightly convolute, **epidermes** with bicellular microhairs. **Inflorescences** terminal spikes, 1-sided, spikelets in 2 rows, loosely to closely imbricate; **rachises** terminating in a bristle; **disarticulation** below the floret. **Spikelets** triangular in cross section, with 1 floret, floret bisexual. **Lower glumes** a highly reduced, cupular rim; **upper glumes** absent or vestigial; **florets** 5–10 mm; **lemmas** linear-lanceolate to lanceolate-oblong, chartaceous, enveloping the paleas, 3-veined, awned; **paleas** hyaline, 2-veined, 2-keeled; **lodicules** absent; **anthers** 3; **styles** 1. $x = 13$. Name from the Greek *nardos*, referring to spikenard, an aromatic herb. It is not clear why the name was applied to this genus; its only species is not scented.

Nardus is a monospecific European genus. Its relationships to other genera are unclear.

SELECTED REFERENCES **Hubbard, C.E.** 1984. Grasses: A Guide to their Structure, Identification, Uses, and Distribution in the British Isles, ed. 3, rev. J.C.E. Hubbard. Penguin Books, Hammondsworth, Middlesex, England and New York, New York, U.S.A. 476 pp.; **Tutin, T.G.** 1980. *Nardus* L. P. 255 *in* T.G. Tutin, V.H. Heywood, N.A. Burges, D.M. Moore, D.H. Valentine, S.M. Walters, and D.A. Webb (eds.). Flora Europaea, vol. 5. Cambridge University Press, Cambridge, England. 452 pp.

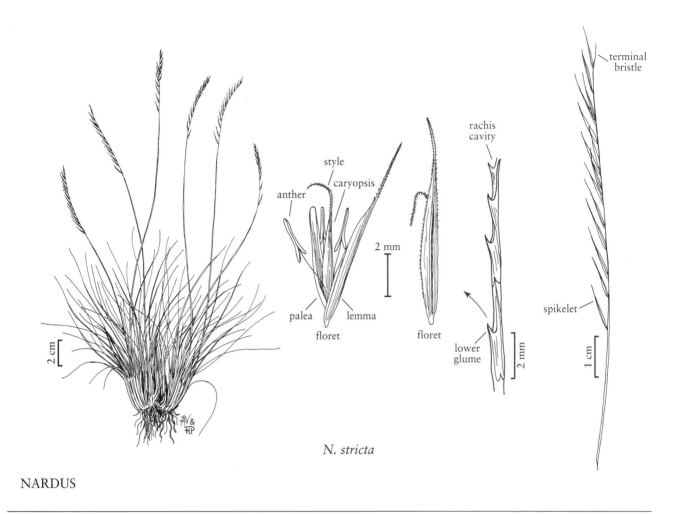

N. stricta

NARDUS

1. Nardus stricta L. [p. 63]

MATGRASS, NARDE RAIDE

Culms (3)10–40(60) cm, stiff, wiry, frequently gray-green; nodes 1(2) per culm, restricted to the lower portion of the culms, pubescent; internodes glabrous. Sheaths smooth, whitish, tough; ligules 0.5–1(2) mm, blunt; blades 4–30 cm long, 0.5–1 mm wide, stiff, tightly convolute, abaxial surfaces hispid, hairs about 0.3 mm, adaxial surfaces scabridulous, ribbed over the veins, apices sharply acute. Spikes (1)3–8 cm, terminating in a single bristle, bristle to 1 cm. Spikelets 5–10 mm, narrowly linear, triangular in cross section, bluish or purplish; lemmas 5–10 mm, 2–3-keeled, awned, awns 1–4.5 mm; paleas slightly shorter than the lemmas; anthers 1–4 mm. Caryopses 3–4.5 mm, tightly enclosed by the lemma and palea. 2*n* = 26.

Nardus stricta is a widespread xerophytic and glycophytic species in Europe, usually growing in open areas on sandy or peaty soils. In the *Flora* region, it is found in scattered locations from upper Michigan to Newfoundland and Greenland, and in Oregon and Idaho, where it is listed as a state noxious weed. The stiff, sharp leaves make it unpalatable; hence it tends to survive in areas of heavy grazing. This, combined with its broad ecological range, makes its potential for spreading in western rangelands a matter of concern.

8. DIARRHENEAE C.S. Campb.

Mary E. Barkworth

Plants perennial; rhizomatous. **Culms** annual, not branching above the base. **Sheaths** open, margins not fused; **collars** glabrous, without tufts of hair at the sides; **auricles** sometimes present; **ligules** stiff, scarious, ciliolate, those of the upper and lower cauline leaves usually similar; **pseudopetioles** absent; **blades** tapering both basally and apically, midveins usually eccentric, venation parallel, cross venation not evident; **cross sections** non-Kranz, without arm or fusoid cells; **epidermes** without microhairs or with unicellular microhairs, cells not papillate. **Inflorescences** terminal panicles. **Spikelets** laterally compressed, pedicellate, with (2)3–5(7) florets, distal floret(s) reduced and sterile, sometimes concealed by the subterminal florets; **rachillas** not prolonged beyond the terminal, sterile floret; **disarticulation** above the glumes and beneath the florets. **Glumes** 2, 1–5-veined, at least the upper glumes longer than $^1\!/_4$ the length of the adjacent floret; **florets** laterally compressed; **calluses** glabrous or with a few hairs, rounded; **lemmas** lanceolate, cartilaginous to thinly coriaceous, 3(5)-veined, veins inconspicuous, apices unawned, sometimes mucronate; **paleas** from $^1\!/_2$ as long as to subequal to the lemmas, 2-veined; **lodicules** 2, membranous, ciliate; **anthers** (1)2(3); **styles** 2, bases free. **Caryopses** obliquely ellipsoid, pericarp thick, easily peeled away at maturity, forming a conspicuous knob or beak, styles not persistent; **hila** linear; **embryos** $^1\!/_4$–$^1\!/_3$ as long as the fruits. $x = 10, 19$.

There are 1–2 genera in the *Diarrheneae*. The tribe is sometimes placed in the *Bambusoideae*, sometimes in the *Poöideae*. Its inclusion in the *Bambusoideae* is supported by embryo characteristics and not strongly opposed by others, but it lacks some of the diagnostic features of that subfamily, e.g., fusoid cells, leaf blades with evident cross venation, and a complex midvein. Its inclusion in the *Poöideae* reflects the findings of the Grass Phylogeny Working Group (2001).

SELECTED REFERENCES **Grass Phylogeny Working Group.** 2000. A phylogeny of the grass family (Poaceae), as inferred from eight character sets. Pp. 3–7 *in* S.W.L. Jacobs and J. Everett (eds.). Grasses: Systematics and Evolution. CSIRO Publishing, Collingwood, Victoria, Australia. 408 pp.; **Grass Phylogeny Working Group.** 2001. Phylogeny and subfamilial classification of the grasses (Poaceae). Ann. Missouri Bot. Gard. 88:373–457; **Koyama, T.** and **S. Kawano.** 1964. Critical taxa of grasses with North American and eastern Asiatic distribution. Canad. J. Bot. 42:859–864; **Tateoka, T.** 1960. Cytology in grass systematics: A critical review. Nucleus (Calcutta) 3:81–110.

8.01 DIARRHENA P. Beauv.

David M. Brandenburg

Plants perennial; rhizomatous, rhizomes 1.5–5 mm thick, scaly. **Culms** 48–131 cm tall, 1–3 mm thick, slender and arching, unbranched, usually clumped, rarely solitary. **Leaves** basally concentrated or proximal; **sheaths** open, longer than the internodes, margins narrowly hyaline, entire, sometimes ciliate; **collars** cartilaginous, thickened, light green or yellowish, somewhat flared marginally; **auricles** sometimes present; **ligules** stiffly membranous, rounded, ciliolate; **blades** flat, tapering basally, long-tapering apically, midveins usually eccentric. **Inflorescences** panicles, contracted, exserted, arching, racemose distally; **branches** 1 or 2 per node, ascending or appressed, terminating in a spikelet. **Spikelets** cylindrical when young, laterally compressed at maturity, with (2)3–5(7) florets, distal floret reduced and sterile, sometimes including an additional rudimentary floret; **disarticulation** above the glumes and beneath the florets. **Glumes** unequal, chartaceous, lanceolate, glabrous, keeled, sometimes scabridulous near the keels distally, margins entire or ciliolate, apices acute; **lower glumes** $^1/_3$–$^2/_3$ shorter than the upper glumes, less than $^1/_3$ as long as the adjacent lemmas, 1–3(5)-veined; **upper glumes** (3)5-veined; **calluses** glabrous or with a few hairs, hairs about 0.5 mm; **lemmas** mostly chartaceous, veins 3, prominent, convergent, margins hyaline, entire, sometimes ciliate, apices sharply cuspidate, cusps 1–2 mm; **paleas** from $^1/_2$ as long as to slightly shorter than the lemmas, chartaceous, keeled, sides narrowly hyaline; **lodicules** about 1.5 mm, lanceolate to elliptic, apices ciliolate; **anthers** 2, yellow. **Caryopses** prominently beaked, style bases usually persistent, pericarp loose, at least partially. $x = 10$. Name from the Greek *dias*, 'twice', and *arren*, 'male', alluding to the 2 anthers.

Diarrhena is an odd and distinctive genus whose relationships are not clear. Two of its approximately six species grow in the woodlands of eastern North America; the remainder, which are sometimes placed in the segregate genus *Neomolinia* Honda, occupy similar habitats in eastern Asia. The Asian species have $x = 19$. The above description pertains to the North American species.

Although *Diarrhena americana* and *D. obovata* grow in similar habitats and overlap in their ranges, no intermediates have been found. Earlier reports of intermediate specimens are based on the use of less reliable characters for distinguishing between the two species.

SELECTED REFERENCES **Brandenburg, D.M., J.R. Estes,** and **S.L. Collins.** 1991. A revision of *Diarrhena* (Poaceae) in the United States. Bull. Torrey Bot. Club 118:128–136; **Koyama, T.** and **S. Kawano.** 1964. Critical taxa of grasses with North American and eastern Asiatic distribution. Canad. J. Bot. 42:859–864; **Tateoka, T.** 1960. Cytology in grass systematics: A critical review. Nucleus (Calcutta) 3:81–110.

1. Calluses pubescent on all but the lowest mature lemma; lemma of the lowest floret in each spikelet (6)7.1–10.8 mm long, widest below the middle, tapering gradually to the apex; mature fruits 1.3–1.8 mm wide, gradually tapering to a blunt beak . . . . 1. *D. americana*
1. Calluses glabrous on all mature lemmas; lemma of the lowest floret in each spikelet 4.6–7.5 mm long, widest near or above the middle, abruptly contracted to the apex; mature fruits 1.8–2.5 mm wide, abruptly contracted to a bottlenose-shaped beak . . . . 2. *D. obovata*

1. **Diarrhena americana** P. Beauv. [p. 66]
AMERICAN BEAKGRAIN

Culms 60–131 cm, glabrous or pubescent. **Sheaths** often pubescent; **collars** usually pubescent; **auricles** pubescent; **ligules** 0.5–1.8 mm; **blades** 25–51 cm long, 7–20 mm wide, glabrous or scabridulous on both surfaces, margins scabridulous or ciliate. **Panicles** 9–30 cm, with 4–23 spikelets. **Spikelets** 10–20 mm, oblong to elliptic, with (2)4–5(7) florets. **Glumes** green to stramineous; **lower glumes** 1.7–4.2 mm long, 0.3–0.7 mm wide in profile, (1)3(5)-veined; **upper glumes** 2.8–6.4 mm long, 0.6–1.2 mm wide in profile, (3)5-veined; **calluses** pubescent on all but the lowest mature lemma; **lemmas** (3.8)5.3–10.8 mm, widest below the middle, lanceolate in profile, tapering gradually, apices sharply cuspidate; **paleas** glabrous or scabridulous, apices usually bifid, sinuses to 0.7 mm deep; **anthers** (1.7)2–2.9(3.5) mm. **Caryopses** 4.6–5.8 mm long, 1.3–1.8 mm wide, narrowly lanceolate in outline, gradually tapering to a blunt beak, wrinkled or smooth, usually blackish-brown to black, except for the straw-colored beak, rarely orange-brown. $2n$ = unknown.

Diarrhena americana is restricted to the United States, where it grows in rich, moist woods from Missouri to Maryland and south to Oklahoma and Alabama. Its range is primarily to the east of the range of *D. obovata*.

2. **Diarrhena obovata** (Gleason) Brandenburg [p. 66]
OBOVATE BEAKGRAIN

Culms 48–131 cm, glabrous. **Sheaths, collars,** and **auricles** glabrous or pubescent; **ligules** 0.2–1 mm; **blades** 24–72 cm long, 6–18 mm wide, abaxial surfaces glabrous or scabridulous, adaxial surfaces similar or pubescent, margins usually scabridulous, rarely smooth. **Panicles** 5–30 cm, with 4–33 spikelets. **Spikelets** 7–17 mm, oblong to ovate, with (2)3–5(7) florets. **Glumes** green to stramineous; **lower glumes** 1.7–3.7 mm long,

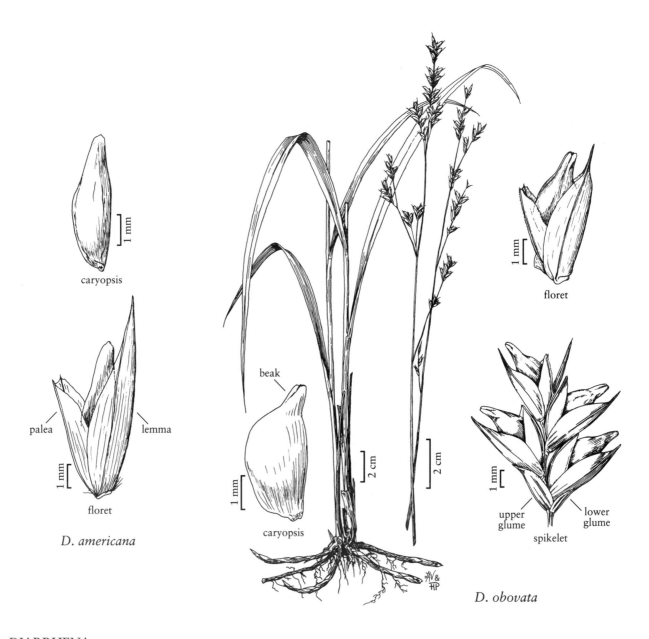

caryopsis

palea   lemma

floret

D. *americana*

beak

caryopsis

2 cm

2 cm

floret

upper
glume   lower
                               glume
            spikelet

D. *obovata*

DIARRHENA

0.3–0.6 mm wide in profile, 1–3-veined; **upper glumes** 2.2–5.2 mm long, 0.75–1.5 mm wide in profile, (3)5-veined; **calluses** glabrous; **lemmas** 3.7–7.5 mm, widest near or above the middle, obovate or elliptic in profile, abruptly tapering, apices sharply cuspidate; **paleas** glabrous, keeled, apices usually bifid, rarely truncate, sinuses 0.05–0.3(0.5) mm deep; **anthers** 1.4–2 mm. **Caryopses** 4.1–6 mm long, 1.8–2.5 mm wide, broadly elliptic to obovate in outline, narrowing abruptly to a bottlenose-shaped beak, usually smooth, occasionally wrinkled, shiny, usually straw-colored, with occasional brown areas basally. $2n = 60$.

Diarrhena obovata is restricted to the *Flora* region, growing in rich woodlands from South Dakota to Ontario and New York and south to Texas, Tennessee, and Virginia. It is most common in the prairie states.

9. MELICEAE Endl.

Mary E. Barkworth

Plants usually perennial, sometimes annual; cespitose, sometimes rhizomatous. **Culms** annual, not woody, not branching above the base; **internodes** hollow. **Sheaths** closed for their whole length or almost so; **collars** without tufts of hair on the sides; **auricles** sometimes present; **ligules** hyaline, glabrous, often lacerate, occasionally ciliate, those of the lower and upper cauline leaves usually similar; **pseudopetioles** absent; **blades** linear to narrowly lanceolate, venation parallel, cross venation sometimes evident; **cross sections** non-Kranz, without arm or fusoid cells; **epidermes** without microhairs, sometimes papillate. **Inflorescences** terminal panicles or racemes; **disarticulation** above the glumes and beneath the florets or below the glumes. **Spikelets** 2.5–60 mm, not viviparous, slightly to strongly laterally compressed, with 1–30 florets, proximal florets bisexual, distal 1–3 florets usually sterile, sometimes pistillate, sometimes reduced and amalgamated into a knob- or club-shaped *rudiment*; **rachillas** prolonged beyond the base of the distal floret. **Glumes** exceeded by the distal florets, shorter than to longer than the adjacent lemmas, mostly membranous, scarious distally, 1–11-veined, apices usually rounded to acute; **florets** laterally or dorsally compressed; **calluses** blunt, glabrous or with hairs; **lemmas of sexual florets** rectangular or ovate, mostly membranous, scarious distally, often with a purplish band adjacent to the scarious apices, (4)5–15-veined, veins not converging distally, often prominent, unawned or awned, awns not branched, apices entire to bilobed or bifid, awns straight, subterminal or from the sinuses; **paleas** from shorter than to longer than the lemmas, similar in texture, 2-veined, veins keeled, sometimes winged; **lodicules** 2, fleshy, usually connate into a single structure, without a membranous wing, truncate, not ciliate, not or scarcely veined; **anthers** 1, 2, or 3; **ovaries** glabrous; **styles** 2-branched, bases persistent, branches plumose distally. **Caryopses** ovoid to ellipsoid, longitudinally grooved or not; **hila** usually linear; **embryos** less than $^1/_3$ as long as the caryopses. $x = (8)9, 10$.

There are approximately 130 species and 8 or 9 genera in the *Meliceae*. Four of the genera are monotypic. *Melica* and *Glyceria*, the two largest genera, are well represented in North America. *Pleuropogon* and *Schizachne* are primarily North American, but extend into eastern Asia.

Molecular studies (e.g., Soreng and Davis 2000; Grass Phylogeny Working Group 2001) show the tribe to be monophyletic and somewhat basal within the *Poöideae*. Members of the tribe are most easily recognized by the combination of closed leaf sheaths, scarious lemma apices, and non-converging lemma veins. The tribe also differs from other tribes in the *Poöideae* in having 2 unwinged lodicules that are usually connate into a single structure, and a base chromosome number of 9 or 10. *Catabrosa* and *Briza*, whose inclusion in the tribe was suggested by the preliminary results of Mejia-Saulés and Bisby (2000), have more membranous lemma margins and free, winged lodicules. *Briza* also has open leaf sheaths and more convergent lemma veins. Their inclusion is not supported by the molecular data.

SELECTED REFERENCES **Catalán, P., E.A. Kellogg,** and **R.G. Olmstead.** 1997. Phylogeny of Poaceae subfamily Poöideae based on chloroplast *ndb*F sequences. Molec. Phylogenet. Evol. 8:150–166; **Grass Phylogeny Working Group.** 2001. Phylogeny and subfamilial classification of the grasses (Poaceae). Ann. Missouri Bot. Gard. 88:373–457; **Mejia-Saulés, T.** and **F.A. Bisby.** 2000. Preliminary views on the tribe Meliceae (Gramineae: Poöideae). Pp. 83–88 *in* S.W.L. Jacobs and J. Everett (eds.). Grasses: Systematics and Evolution. CSIRO Publishing, Collingwood, Victoria, Australia. 408 pp.; **Soreng, R.J.** and **J.I. Davis.** 2000. Phylogenetic structure in Poaceae subfamily Poöideae as inferred from molecular and morphological characters: Misclassification versus reticulation. Pp. 61–74 *in* S.W.L. Jacobs and J. Everett (eds.). Grasses: Systematics and Evolution. CSIRO Publishing, Collingwood, Victoria, Australia. 408 pp.

1. Calluses hairy; lemmas awned, awns 8–15 mm long, twisted, divergent to slightly geniculate . . . . *9.03 Schizachne*
1. Calluses glabrous; lemmas unawned or awned, awns to 12 mm long, straight.
   2. Inflorescences racemes; palea keels winged, the wings notched and awned . . . . *9.04 Pleuropogon*
   2. Inflorescences usually panicles, racemes in depauperate specimens; palea keels not winged or the wings entire and unawned.
      3. Lower glumes 1-veined, 0.3–4.5 mm long; disarticulation always above the glumes; lemmas unawned, never with hairs more than 1 mm long; culms never with cormous bases; distal florets in the spikelets sometimes reduced, not forming a morphologically distinct rudiment; plants of wet meadows and streamsides . . . . *9.01 Glyceria*
      3. Lower glumes 1–9-veined, 2–16 mm long; disarticulation above or below the glumes; lemmas sometimes awned, sometimes with hairs longer than 1 mm; culms sometimes with cormous bases; distal florets in the spikelets often forming a morphologically distinct rudiment; plants of drier or well drained habitats . . . . *9.02 Melica*

9.01 GLYCERIA R. Br.

Mary E. Barkworth

Laurel K. Anderton

Plants usually perennial, rarely annual; rhizomatous. **Culms** (10)20–250 cm, erect or decumbent, freely rooting at the lower nodes, not cormous based. **Sheaths** closed for at least ³/₄ their length, often almost entirely closed; **ligules** scarious, erose to lacerate; **blades** flat or folded. **Inflorescences** terminal, usually panicles, sometimes racemes in depauperate specimens, branches appressed to divergent or reflexed. **Spikelets** cylindrical and terete or oval and laterally compressed, with 2–16 florets, terminal floret in each spikelet sterile, reduced; **disarticulation** above the glumes, below the florets. **Glumes** much smaller than to equaling the adjacent lemmas, 1-veined, obtuse or acute, often erose; **lower glumes** 0.3–4.5 mm; **upper glumes** 0.6–7 mm; **calluses** glabrous; **lemmas** membranous to thinly coriaceous, rounded over the back, smooth or scabrous, glabrous or hairy, hairs to about 0.1 mm, 5–11-veined, veins usually evident, often prominent and ridged, not or scarcely converging distally, apical margins hyaline, sometimes with a purplish band below the hyaline portion, apices acute to rounded or truncate, entire, erose, or irregularly lobed, unawned; **paleas** from shorter than to longer than the lemmas, keeled, keels sometimes winged; **lodicules** thick, sometimes connate, not winged; **anthers** (1)2–3; **ovaries** glabrous; **styles** 2-branched, branches divergent to recurved, plumose distally. *x* = 10. Name from the Greek *glukeros*, 'sweet', the caryopses of the type species being sweet.

Glyceria includes approximately 35 species, all of which grow in wet areas. All but five species are native to the Northern Hemisphere. The genus is represented in the *Flora* region by 13 native and 3 introduced species, as well as 3 named hybrids. One additional European species, *G. notata*, is included in this treatment because it has been reported to be present in the region.

All native species of *Glyceria* are palatable to livestock. They are rarely sufficiently abundant to be important forage species. Some grow in areas that are soon degraded by grazing. *Glyceria maxima* can cause cyanide poisoning in cattle. Species in sects. *Striatae* and *Hydropoa* have potential as ornamentals.

Glyceria resembles *Puccinellia* in the structure of its spikelets and its preference for wet habitats; it differs in its inability to tolerate highly alkaline soils, and its usually more flexuous panicle branches, closed leaf sheaths, and single-veined upper glumes. Some species are apt to be confused with *Torreyochloa pallida*, another species associated with wet habitats but one that, like *Puccinellia*, has open leaf sheaths.

Glyceria includes several species that appear to intergrade. In some cases, the distinctions between such taxa are more evident in the field, particularly when they are sympatric. Recognition of such taxa at the specific level is merited unless it can be shown that all the distinctions between them are inherited as a group.

The three named North American hybrids are *Glyceria* ×*gatineauensis* Bowden, *G.* ×*ottawensis* Bowden, and *G.* ×*occidentalis* (Piper) J.C. Nelson. The first two were named as hybrids; they are not included in the key and are mentioned only briefly in the descriptions. *Glyceria* ×*occidentalis* has hitherto been treated as a species. Studies finished shortly before completion of this volume indicate that it, too, consists of hybrids (Whipple et al. [in press]). It is included in the key and provided with a full description.

Culm thickness is measured near midlength of the basal internode; it does not include leaf sheaths. Unless otherwise stated, ligule measurements reflect both the basal and upper leaves. Ligules of the basal leaves are usually shorter than, but similar in shape and texture to, those of the upper leaves. The number of spikelets on a branch is counted on the longest primary branches, and includes all the spikelets on the secondary (and higher order) branches of the primary branch. Pedicel lengths are measured for lateral spikelets on a branch, not the terminal spikelet. Lemma characteristics are based on the lowest lemmas of most spikelets in a panicle. There is often, unfortunately, considerable variation within a panicle.

SELECTED REFERENCES **Anderson, J.E.** and **A.A. Reznicek**. 1994. *Glyceria maxima* (Poaceae) in New England. Rhodora 96:97–101; **Borrill, M.** 1955. Breeding systems and compatibility in *Glyceria*. Nature 175:561–563; **Bowden, W.M.** 1960. Chromosome numbers and taxonomic notes on northern grasses: III. Festuceae. Canad. J. Bot. 38:117–131; **Chester, E.W., B.E. Wofford, H.R. DeSelm**, and **A.M. Evans**. 1993. Atlas of Tennessee Vascular Plants, vol. 1. Austin Peay State University Miscellaneous Publication No. 9. The Center for Field Biology, Austin Peay State University, Clarksville, Tennessee, U.S.A. 118 pp.; **Church, G.L.** 1949. Cytotaxonomic study of *Glyceria* and *Puccinellia*. Amer. J. Bot. 36:155–165; **Conert, H.J.** 1992. *Glyceria*. Pp. 440–457 *in* G. Hegi. Illustrierte Flora von Mitteleuropa, ed. 3. Band I, Teil 3, Lieferung 6 (pp. 401–480). Verlag Paul Parey, Berlin and Hamburg, Germany; **Dore, W.G.** and **J. McNeill**. 1980. Grasses of Ontario. Research Branch, Agriculture Canada Monograph No. 26. Canadian Government Publishing Centre, Hull, Québec, Canada. 568 pp.; **Hitchcock, C.L., A. Cronquist**, and **M. Ownbey**. 1969. Vascular Plants of the Pacific Northwest. Part 1: Vascular Cryptogams, Gymnosperms, and Monocotyledons. University of Washington Press, Seattle, Washington, U.S.A. 914 pp.; **Komarov, V.L.** 1963. Genus 176. *Glyceria* R. Br. Pp. 356–365 *in* R.Yu. Rozhevits [R.J. Roshevitz] and B.K. Shishkin [Schischkin] (eds.). Flora of the U.S.S.R., vol. 2, trans. N. Landau (series ed. V.L. Komarov). Published for the National Science Foundation and the Smithsonian Institution, Washington, D.C. by the Israel Program for Scientific Translations, Jerusalem, Israel. 622 pp. [English translation of Flora SSSR, vol. II. 1934. Botanicheskii Institut Im. V.L. Komarova, Akademiya Nauk, Leningrad, Russia. 778 pp.]; **Koyama, T.** 1987. Grasses of Japan and Its Neighboring Regions: An Identification Manual. Kodansha, Ltd., Tokyo, Japan. 370 pp.; **Scoggan, H.** 1978. Flora of Canada, part 2: Pteridophyta, Gymnosperms, Monocotyledoneae. National Museum of Natural Sciences Publications in Botany No. 7[2]. National Museums of Canada, Ottawa, Ontario, Canada. 545 pp.; **Voss, E.G.** 1972. Michigan Flora: A Guide to the Identification and Occurrence of the Native and Naturalized Seed-Plants of the State, part 1. University of Michigan, Ann Arbor, Michigan, U.S.A. 488 pp.; **Whipple, I.G., B.S. Bushman**, and **M.E. Barkworth**. [in press]. *Glyceria* in North America.

1. Spikelets laterally compressed, lengths 1–4 times widths, oval in side view; paleal keels not winged (sects. *Hydropoa* and *Striatae*).
    2. Upper glumes 2.5–5 mm long, longer than wide.
        3. Blades 3–7 mm wide; culms 2.5–4 mm thick, 60–90 cm tall; anthers 0.7–1.2 mm long . . . . 2. *G. alnasteretum*
        3. Blades 6–20 mm wide; culms 6–12 mm thick, 60–250 cm tall; anthers (1)1.2–2 mm long . . . . 3. *G. maxima*
    2. Upper glumes 0.6–3.7 mm long, if longer than 3 mm, then shorter than wide.
        4. Panicles ovoid to linear; panicle branches appressed to strongly ascending; ligules of the upper leaves 0.5–0.9 mm long.
            5. Panicles 5–15 cm long, 2.5–6 cm wide, ovoid, erect . . . . 4. *G. obtusa*
            5. Panicles 15–25 cm long, 0.8–1.5 cm wide, linear, nodding . . . . 5. *G. melicaria*
        4. Panicles pyramidal; panicle branches strongly divergent or drooping; ligules of the upper leaves 1–7 mm long.
            6. Lemma apices almost flat; anthers 3; veins of 1 or both glumes in each spikelet usually extending to the apices . . . . 1. *G. grandis*
            6. Lemma apices prow-shaped; anthers 2; veins of both glumes terminating below the apices.

7. Glumes tapering from below midlength to the narrowly acute (< 45°) apices; lemma lengths more than twice widths .... 6. *G. nubigena*

7. Glumes narrowing from midlength or above to the acute (≥ 45°) or rounded apices; lemma lengths less than twice widths.

    8. Spikelets (2.5)3–5 mm wide; lemma veins evident but not raised distally; palea lengths 1.5–1.8 times widths .... 10. *G. canadensis*

    8. Spikelets 1.2–2.9 mm wide; lemma veins distinctly raised throughout; palea lengths 1.5–3.5 times widths.

        9. Lemmas 2.5–3.5 mm long; glume lengths about 3 times widths, glume apices broadly acute; lower glumes 1.5–2 mm long; upper glumes 2–2.6 mm long .... 7. *G. pulchella*

        9. Lemmas 1.2–2.2 mm long; glume lengths up to twice widths, glume apices rounded or acute; lower glumes 0.5–1.5 mm long; upper glumes 0.6–1.5 mm long.

            10. Blades 2–6 mm wide; anthers 0.2–0.6 mm long; culms 1.5–3.5 mm thick .... 8. *G. striata*

            10. Blades 6–15 mm wide; anthers 0.5–0.8 mm long; culms 2.5–8 mm thick .... 9. *G. elata*

1. Spikelets cylindrical and terete, except at anthesis when slightly laterally compressed, lengths more than 5 times widths, rectangular in side view; paleal keels usually winged distally (sect. *Glyceria*).

    11. Lemmas tapering from near midlength to the acuminate or narrowly acute apices; paleas exceeding the lemmas by 0.7–3 mm; palea apices often appearing bifid, the teeth 0.4–1 mm long .... 13. *G. acutiflora*

    11. Lemmas not tapered or tapering only in the distal ¼, apices truncate, rounded, or acute; paleas shorter or to 1(1.5) mm longer than the lemmas; palea apices not or shortly bifid, the teeth to 0.5 mm long.

        12. Lemma apices with 1 strongly developed lobe on 1 or both sides, entire to crenulate between the lobes; blades 3–12 cm long; primary panicle branches 1.5–9.5 cm long ... 17. *G. declinata*

        12. Lemma apices not or more or less evenly lobed; blades 5–30 cm long; primary panicle branches 3–18 cm long.

            13. Lemmas 5–8 mm long.

                14. Anthers 0.6–1.6 mm long; lemma apices usually slightly lobed or irregularly crenate .... 15. *G. ×occidentalis* (in part)

                14. Anthers 1.5–3 mm long; lemma apices usually entire .... 16. *G. fluitans*

             13. Lemmas 2.4–5 mm long.

                15. Lemmas usually smooth between the veins, if scabridulous the prickles between the veins smaller than those over the veins.

                    16. Lemmas usually acute, sometimes obtuse, entire or almost so; adaxial surfaces of the midcauline blades usually densely papillose, glabrous ...... 11. *G. borealis*

                    16. Lemmas truncate to obtuse, crenate; adaxial surfaces of the midcauline blades rarely densely papillose, sometimes sparsely hairy.

                        17. Culms 73–182 cm tall; pedicels 0.7–1.7 mm .... 12. *G. septentrionalis* (in part)

                        17. Culms 25–80 cm tall; pedicels 1–6 mm .... 18. *G. notata*

                15. Lemmas scabridulous or hispidulous between the veins, the prickles between the veins similar in size to those over the veins.

                  18. Lemma apices acute.

                      19. Lemmas 2.4–4.8 mm long; pedicels 0.7–1.7 mm long; plants from east of the Rocky Mountains .... 12. *G. septentrionalis* (in part)

                      19. Lemmas 4.5–5.9 mm long; pedicels 1.5–8 mm long; plants from west of the Rocky Mountains .... 15. *G. ×occidentalis* (in part)

                  18. Lemma apices truncate to obtuse.

                    20. Pedicels 0.7–1.7 mm long; anthers 0.5–1.8 mm long; plants from east of the Rocky Mountains .... 12. *G. septentrionalis* (in part)

                    20. Pedicels 2–5 mm long; anthers 0.3–0.9 mm long; plants from British Columbia and the Pacific states .... 14. *G. leptostachya*

Glyceria sect. Hydropoa (Dumort.) Dumort.

Plants perennial. **Sheaths** not or only slightly compressed, midvein sometimes conspicuous distally. **Panicles** 12–30 cm wide, usually open; **branches** divergent to strongly divergent and drooping; **pedicels** 0.8–15 mm. **Spikelets** oval in side view, lengths 1–4 times widths, somewhat laterally compressed, usually not appressed to the panicle branches. **Lemmas** with apices acute to rounded or truncate, more or less flat to slightly prow-shaped; **paleas** keeled, keels not winged, truncate to notched between the keels; **lodicules** wholly to partially connate; **anthers** 1, 2, or 3. **Caryopses** ovoid-oblong; **hila** ovoid to oblong.

Glyceria sect. *Hydropoa* includes approximately five species. Three species grow in the *Flora* region; one is introduced. They grow along streams and at the edges of lakes and ponds.

1. **Glyceria grandis** S. Watson [p. 72]
AMERICAN GLYCERIA, AMERICAN MANNAGRASS

Plants perennial. **Culms** 50–150 (200) cm tall, 8–12 mm thick, erect or decumbent and rooting at the base. **Sheaths** smooth or scabridulous, keeled; **ligules** 1–5 (7) mm, truncate to rounded, ligules of the lower leaves stiff at the base, ligules of the upper leaves flexible throughout; **blades** 25–43 cm long, 4.5–15 mm wide. **Panicles** 16–42 cm long, 12–20 cm wide, open; **branches** (7)10–18 cm, lax, widely divergent to drooping, with 35–80+ spikelets; **pedicels** 1–15 mm. **Spikelets** 3.2–10 mm long, 2–3 mm wide, somewhat laterally compressed, oval to elliptic in side view, with 4–10 florets. **Glumes** mostly hyaline, usually the midvein of 1 or both glumes extending to the apices, apices acute; **lower glumes** 1–2.3 mm; **upper glumes** 1.5–2.7 mm; **rachilla internodes** 0.5–0.8 mm; **lemmas** 1.8–3 mm, prominently (5)7-veined, veins often scabridulous, intercostal regions smooth, apices rounded to truncate, sometimes erose, almost flat at maturity; **paleas** from shorter than to slightly longer than the lemmas, lengths more than 3 times widths, keels not winged, ciliolate, tips not strongly incurved, truncate to notched between the keels; **anthers** 3, 0.5–1.2 mm. **Caryopses** 1–1.5 mm. 2*n* = 20.

Glyceria grandis grows on banks and in the water of streams, ditches, ponds, and wet meadows, from Alaska to Newfoundland and south in the mountains to California, Arizona, and New Mexico in the western United States, and to Virginia and Tennessee in the eastern United States. It is similar to *G. maxima*, differing primarily in its shorter, flatter lemmas and shorter anthers. It is also confused with *G. elata* and *Torreyochloa pallida*. It differs from the former in having acute glumes with long veins, more evenly dark florets, flatter lemma apices, and paleal keel tips that do not point towards each other. It differs from

Torreyochloa pallida in its closed leaf sheaths and 1-veined glumes.

1. Spikelets 3.2–6.4 mm long, with 4–8 florets . . . . var. *grandis*
1. Spikelets 6–10 mm long, with 5–10 florets . . . . var. *komarovii*

Glyceria grandis S. Watson var. **grandis** [p. 72]
GIANT GLYCERIA, GIANT MANNAGRASS, GLYCÉRIE GÉANTE

Spikelets 3.2–6.4 mm, with 4–8 florets.

Glyceria grandis var. *grandis* is the more widespread of the two varieties, growing throughout the range of the species.

Glyceria grandis var. **komarovii** Kelso [p. 72]

Spikelets 6–10 mm, with 5–10 florets.

Glyceria grandis var. *komarovii* is restricted to Alaska and the Yukon Territory.

2. **Glyceria alnasteretum** Kom. [p. 72]
ALEUTIAN GLYCERIA

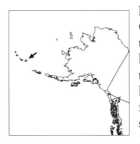

Plants perennial, rhizomatous. **Culms** 60–90 cm tall, 2.5–4 mm thick, erect. **Sheaths** smooth, not keeled; **ligules** 2–3 mm, rounded to truncate; **blades** 5–20 cm long, 3–7 mm wide, abaxial surfaces smooth, adaxial surfaces scabrous, apices acute. **Panicles** 15–22 cm long, 12–16 cm wide, open, pyramidal, erect to nodding; **branches** 8–10 cm, lower branches widely divergent to drooping. **Spikelets** 7–9 mm long, 3–4.5 mm wide, with 5–8 florets. **Glumes** unequal, lanceolate, acute; **lower glumes** 2–3.5 mm; **upper glumes** 2.5–3.5 mm, longer than wide; **lemmas** 3–5.5 mm, 7-veined, obtuse to acute; **paleas** shorter than or subequal to the lemmas, keels not winged, apices not strongly incurved, emarginate between the keels; **anthers** 3, 0.7–1.2 mm. **Caryopses** not seen. 2*n* = 20.

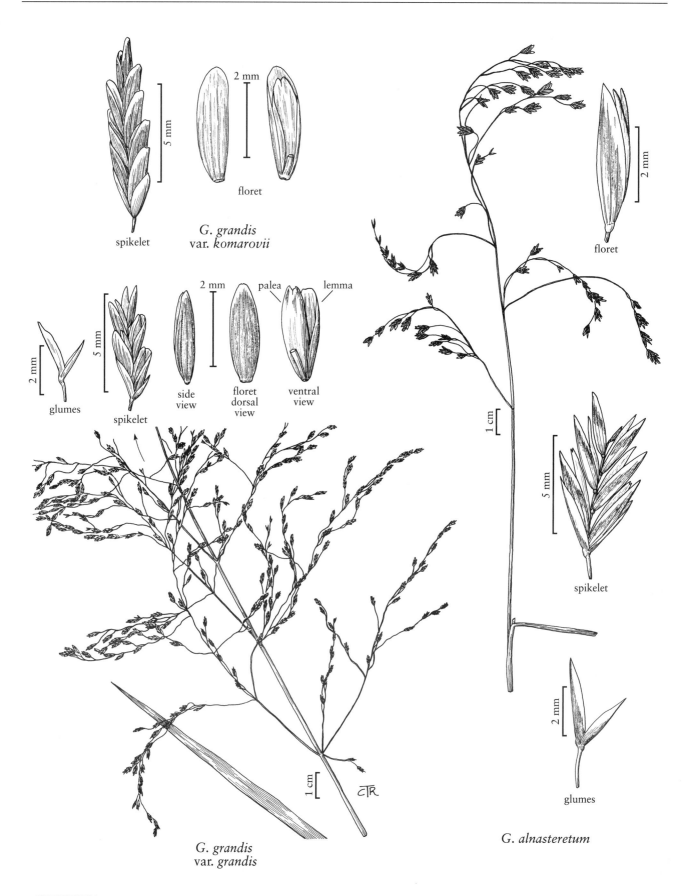

spikelet

G. grandis
var. *komarovii*

floret

2 mm

5 mm

glumes

2 mm

spikelet

5 mm

side
view

floret
dorsal
view

ventral
view

palea   lemma

2 mm

floret

2 mm

G. grandis
var. *grandis*

1 cm

spikelet

5 mm

glumes

2 mm

G. alnasteretum

1 cm

CTR

GLYCERIA

Glyceria alnasteretum is included in this treatment with some hesitation, based on *van Schaack 724* (WTU 152646) and *van Schaack 887* (MO 1710727), both collected at Signal Point, Attu Island, Alaska in 1945. The above description is based on Komarov (1963) and Koyama (1987), modified to reflect the wider panicles and longer glumes and lemmas of the van Schaack specimens. The difference in habitat is troubling. The van Schaack specimens were found "in a beachside meadow" and "near beach." Koyama describes the habitat of *G. alnasteretum* as "wet meadows and marshes at high altitudes as well as subarctic zone" (p. 114). Nevertheless, the van Schaack specimens fit the description of *G. alnasteretum* better than any other taxon in this treatment. Clearly, further investigation is called for; it should include plants from both sides of the Bering Strait.

3. Glyceria maxima (Hartm.) Holmb. [p. 74]

TALL GLYCERIA, ENGLISH WATERGRASS,
GLYCÉRIE AQUATIQUE

Plants perennial. **Culms** 60–250 cm tall, 6–12 mm thick, erect. **Sheaths** scabridulous, keeled; **ligules** 1.2–6 mm, rounded or with a central point, ligules of the lower leaves thick, stiff, and opaque, ligules of the upper leaves thinner and translucent; **blades** 30–60 cm long, 6–20 mm wide, both surfaces smooth or adaxial surfaces scabridulous. **Panicles** 15–45 cm long, to 30 cm wide, open; **branches** 8–20 cm, lax, strongly divergent or drooping at maturity, scabridulous, primary branches with 50+ spikelets; **pedicels** 0.8–10 mm. **Spikelets** 5–12 mm long, 2–3.5 mm wide, somewhat laterally compressed, oval in side view, with 4–10 florets. **Glumes** unequal, usually the midvein of 1 or both reaching to the apices; **lower glumes** 2–3 mm; **upper glumes** 3–4 mm, longer than wide; **rachilla internodes** 0.5–1 mm; **lemmas** 3–4 mm, 7-veined, veins scabridulous, apices broadly acute to rounded, slightly prow-shaped; **paleas** subequal to the lemmas, lengths more than 3 times widths, keels not winged, ciliate, tips not strongly incurved, curved to broadly notched between the keels; **anthers** 3, (1)1.2–2 mm. **Caryopses** 1.5–2 mm. $2n = 60$.

Glyceria maxima is native to Eurasia. It grows in wet areas, including shallow water, at scattered locations in the *Flora* region. It is an excellent fodder grass, and may have been planted deliberately at one time (Dore and McNeill 1980). At some sites, the species appears to be spreading, largely vegetatively. It is easily confused with large specimens of *G. grandis*, but differs in its firmer, more prow-tipped lemmas as well as its larger lemmas and usually larger anthers.

Glyceria sect. Striatae G.L. Church

Plants perennial. **Sheaths** not or weakly compressed, midvein often conspicuous distally. **Panicles** 0.8–30 cm wide; **branches** usually ascending to strongly divergent or drooping, sometimes appressed. **Spikelets** oval in side view, lengths 1–4 times widths, laterally compressed. **Lemmas** with apices acute or obtuse to rounded, prow-shaped; **paleas** slightly shorter to slightly longer than the lemmas, keels well-developed, tips strongly incurved, apices narrowly notched between the keels; **lodicules** free; **anthers** 2. **Caryopses** usually obovoid, sometimes ovoid; **hila** punctate or linear.

Members of *Glyceria* sect. *Striatae* grow along streams, in swamps, and in shallow, fresh water. The section includes seven species, all of which are native to the *Flora* region.

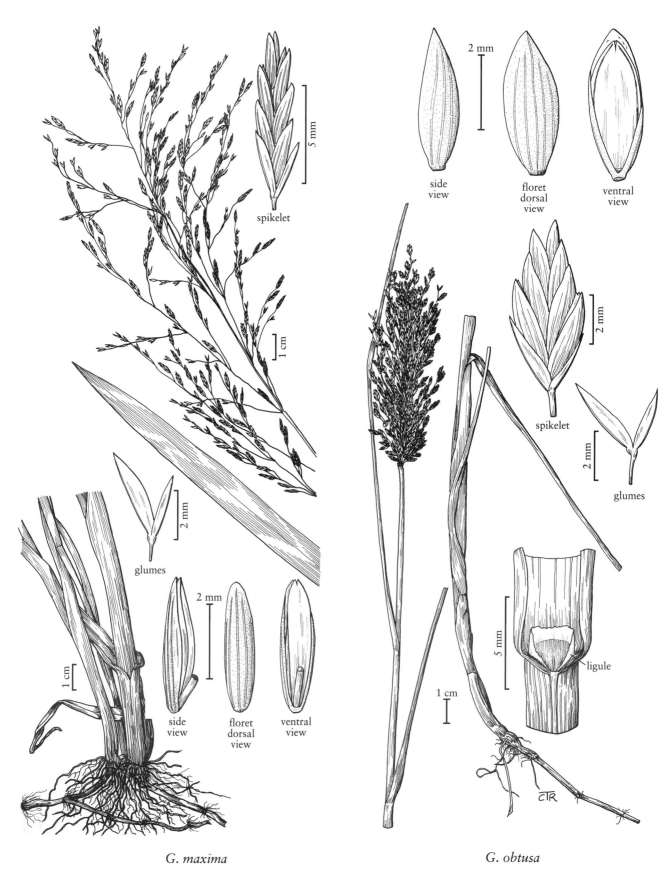

side
view

floret
dorsal
view

ventral
view

spikelet

spikelet

glumes

glumes

side
view

floret
dorsal
view

ventral
view

ligule

G. maxima

G. obtusa

GLYCERIA

4. Glyceria obtusa (Muhl.) Trin. [p. 74]
ATLANTIC MANNAGRASS

Plants perennial. **Culms** 60–100 cm tall, 2.5–5 mm thick, often decumbent at the base. **Sheaths** glabrous, smooth, not keeled, midvein prominent; **ligules** 0.5–0.8 mm, not translucent, truncate to slightly rounded; **blades** 15–40 cm long, 2–8 mm wide, abaxial surfaces smooth, adaxial surfaces scabridulous. **Panicles** 5–15 cm long, 2.5–6 cm wide, ovoid, erect, dense; **branches** 2.5–8 cm, strongly ascending, with 10–30 spikelets; **pedicels** 1–14 mm. **Spikelets** 4–7 mm long, 2.5–4 mm wide, somewhat laterally compressed, oval in side view, with 4–7 florets. **Glumes** keeled, 1–veined, veins not extending to the apical margins, apical margins hyaline, acute, entire or often splitting with age; **lower glumes** 1.6–2.5 mm, lanceolate to narrowly ovate or obovate; **upper glumes** 1.7–3.5 mm, ovate-elliptic to obovate, obtuse to rounded; **rachilla internodes** 0.2–0.4 mm; **lemmas** 3–3.9 mm, 5–9-veined, veins not raised, apices rounded, somewhat prow-shaped; **paleas** subequal to the lemmas, lengths 2–2.8 times widths, keels well-developed, not winged, tips pointing towards each other, narrowly notched between the keels; **anthers** 2, 0.6–0.8 mm. **Caryopses** 1.5–1.8 mm. $2n = 40$.

Glyceria obtusa is a distinctive species that grows in wet woods, swamps, and shallow waters, primarily on the eastern seaboard of North America, from Nova Scotia and New Brunswick to South Carolina.

5. Glyceria melicaria (Michx.) F.T. Hubb. [p. 76]
MELIC MANNAGRASS, GLYCÉRIE MÉLICAIRE

Plants perennial. **Culms** 50–100 cm tall, 3–5 mm thick, erect. **Sheaths** smooth, not or only weakly keeled; **ligules** 0.2–0.9 mm, translucent, rounded to truncate; **blades** 25–40 cm long, 2–7 mm wide, abaxial surfaces smooth, adaxial surfaces scabridulous. **Panicles** 15–25 cm long, 0.8–1.5 cm wide, linear, nodding; **branches** 5–9 cm, appressed, scarcely surpassing the node above, with 30–60+ spikelets; **pedicels** 0.5–2 mm. **Spikelets** 3.5–5 mm long, 1–2.5 mm wide, laterally compressed, oval in side view, with 3–4 florets. **Glumes** lanceolate, veins terminating below the apices, apices acute; **lower glumes** 1.3–2.4 mm; **upper glumes** 1.7–3 mm; **rachilla internodes** 0.4–0.6 mm; **lemmas** 1.9–2.8 mm, smooth or scabridulous, 5–7-veined, veins raised, apices acute, prow-shaped; **paleas** slightly shorter to slightly longer than the lemmas, lengths 2.5–4 times widths, keels well developed, not winged, tips incurved, narrowly notched between the keels; **anthers** 2, 0.3–0.5 mm. **Caryopses** 1–1.5 mm; **hila** subequal to the caryopses. $2n = 40$.

Glyceria melicaria grows in swamps and wet soils. Its range extends from southeastern Ontario east to Nova Scotia, south to Illinois and the northeastern United States and, in the Appalachian Mountains, to northern Georgia. **Glyceria ×gatineauensis** Bowden is a sterile hybrid between *G. melicaria* and *G. striata*. For further comments, see p. 77.

6. Glyceria nubigena W.A. Anderson [p. 76]
GREAT SMOKY MOUNTAIN MANNAGRASS,
GREAT SMOKY MOUNTAIN GLYCERIA

Plants perennial. **Culms** 100–200 cm tall, 3–5 mm thick, smooth. **Sheaths** smooth or scabridulous, weakly keeled; **ligules** 1–1.5 mm, truncate; **blades** to 45 cm long, 6–10 mm wide, abaxial surfaces smooth or scabrous, adaxial surfaces scabrous. **Panicles** 20–30 cm long, 7.5–14 cm wide, open, pyramidal; **branches** 7.5–14 cm, spreading or reflexed, lax, with 16–80 spikelets; **pedicels** 2–7 mm. **Spikelets** 3.5–5.5 mm long, 2–3(3.5) mm wide, laterally compressed, oval in side view, with 3–5 florets. **Glumes** tapering from below midlength to the narrowly (< 45°) acute apices, veins not extending to the apices; **lower glumes** 0.8–1.5 mm; **upper glumes** 1.8–2.2 mm; **rachilla internodes** about 0.5 mm; **lemmas** 2.2–2.7 mm, 0.9–1.1 mm wide in dorsal view, veins distinctly raised, usually smooth over and between the veins, sometimes scabridulous over the veins, apices acute, prow-shaped; **paleas** slightly shorter than the lemmas, lengths 2–2.7 times widths, keels not winged, tips incurved, apices narrowly notched between the keels; **anthers** 2, about 1.5 mm, dehiscent at maturity. **Caryopses** about 1.5 mm. $2n = 40$.

Glyceria nubigena is known only from moist areas of balds and high ridges in the Great Smoky Mountains of North Carolina and Tennessee.

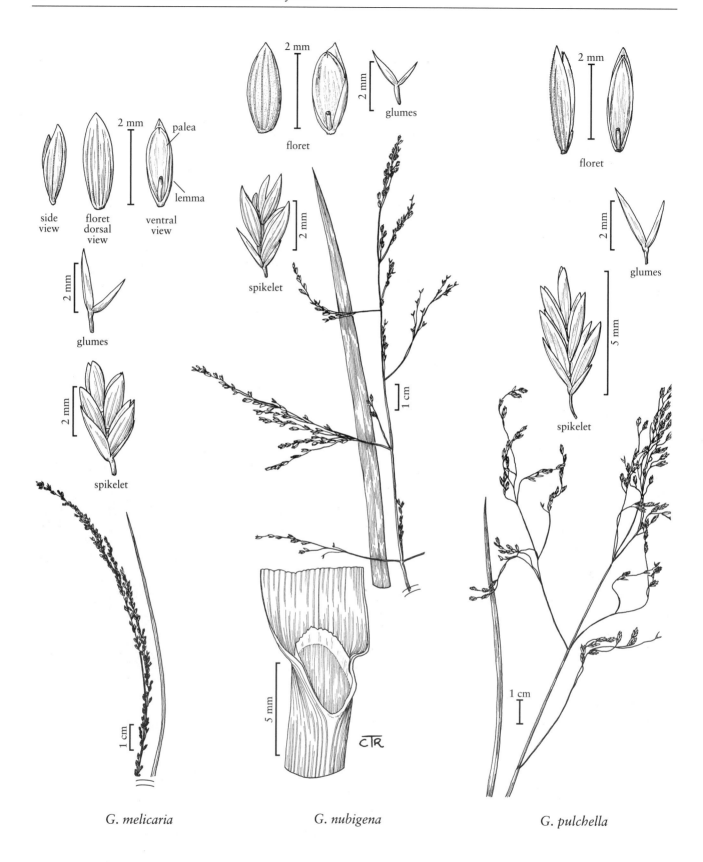

side
view

floret
dorsal
view

palea

lemma

ventral
view

glumes

spikelet

spikelet

floret

glumes

spikelet

floret

floret

glumes

spikelet

G. melicaria   *G. nubigena*   *G. pulchella*

GLYCERIA

7. Glyceria pulchella (Nash) K. Schum. [p. 76]

BEAUTIFUL GLYCERIA, MACKENZIE VALLEY MANNAGRASS

Plants perennial. **Culms** 40–60 cm tall, 1.5–5 mm thick, erect. **Sheaths** scabridulous, not or weakly keeled; **ligules** 1.5–4 mm; **blades** 9–29 cm long, 2–7.5 mm wide, scabrous. **Panicles** 15–25 cm long, 6–15(20) cm wide, open, pyramidal, nodding; **branches** 8–12 cm, ascending to divergent, flexuous, often smooth, with 30–40+ spikelets; **pedicels** 0.3–6 mm. **Spikelets** 3.5–6 mm long, 1.4–2.5 mm wide, about 2.5 times longer than wide, laterally compressed, oval in side view, with 3–6 florets. **Glumes** unequal, lengths about 3 times widths, narrowing beyond midlength, veins terminating below the apices, apices acute, forming an angle of about 45°; **lower glumes** 1.5–2 mm; **upper glumes** 2–2.6 mm, erose; **rachilla internodes** 0.4–0.6 mm; **lemmas** 2.5–3.5 mm, oval in dorsal view, 7-veined, veins raised, scabridulous, apices broadly acute to obtuse, prow-shaped; **paleas** from shorter than to equaling the lemmas, lengths 3–3.5 times widths, keels not winged, tips incurved, apices narrowly notched between the keels; **anthers** 2, 0.5–0.9 mm. **Caryopses** about 1 mm. $2n$ = unknown.

Glyceria pulchella grows in marshes, muskegs, ponds, and ditches, from central Alaska and the Northwest Territories to southern British Columbia and central Manitoba. In overall aspect, *G. pulchella* resembles *G. striata* and *G. elata*. It differs in having somewhat stiffer and straighter panicle branches, in addition to larger spikelets and florets.

8. Glyceria striata (Lam.) Hitchc. [p. 78]

RIDGED GLYCERIA, GLYCÉRIE STRIÉE

Plants perennial. **Culms** 20–80 (100) cm tall, (1.5)2–3.5 mm thick, not or only slightly spongy, sometimes rooting at the lower nodes. **Sheaths** smooth to scabridulous, keeled, sometimes weakly so; **ligules** 1–4 mm, usually rounded, sometimes acute to mucronate, erose-lacerate; **blades** 12–30 cm long, 2–6 mm wide, abaxial surfaces smooth or scabridulous, adaxial surfaces scabridulous to scabrous. **Panicles** 6–25 cm long, 2.5–21 cm wide, pyramidal, open, nodding; **branches** 5–13 cm, straight to lax, lower branches usually strongly divergent to drooping at maturity, sometimes ascending, with 15–50 spikelets, these often confined to the distal $^2/_3$; **pedicels** 0.5–7 mm. **Spikelets** 1.8–4 mm long, 1.2–2.9 mm wide, laterally compressed, oval in side view, with 3–7 florets. **Glumes**

ovate, 1–1.5 times longer than wide, narrowing from midlength or above, veins terminating below the apical margins, apices often splitting with age; **lower glumes** 0.5–1.2 mm, rounded to obtuse; **upper glumes** 0.6–1.2 mm, acute or rounded; **rachilla internodes** 0.1–0.6 mm; **lemmas** 1.2–2 mm, ovate in dorsal view, veins raised, scabridulous over and between the veins, apices acute, prow-shaped; **paleas** slightly shorter than to equaling the lemmas, lengths 1.5–3 times widths, keeled, keels not winged, tips pointing towards each other, apices narrowly notched between the keels; **anthers** 2, (0.2)0.4–0.6 mm, purple or yellow. **Caryopses** 0.5–2 mm. $2n$ = 20 [reports of 28 are questionable].

Glyceria striata grows in bogs, along lakes and streams, and in other wet places. Its range extends from Alaska to Newfoundland and south into Mexico. Plants from the eastern portion of the range have sometimes been treated as *G. striata* var. *striata*, and those from the west as *G. striata* var. *stricta* (Scribn.) Fernald. Eastern plants tend to have somewhat narrower leaves and thinner culms than western plants, but the variation appears continuous. In the west, larger specimens are easy to confuse with *G. elata*. The two species are sometimes found growing together without hybridizing; this and molecular data (Whipple et al. [in prep.]) support their recognition as separate species. The differences between the two in growth habit and stature are evident in the field; they are not always evident on herbarium specimens. In its overall aspect, *G. striata* also resembles *G. pulchella*, but it has somewhat more lax panicle branches in addition to smaller spikelets and florets.

Glyceria ×gatineauensis Bowden is a sterile hybrid between *G. striata* and *G. melicaria*. It resembles *G. melicaria* but has longer (up to 12 cm), less appressed panicle branches and is a triploid with $2n$ = 30. It was described from a population near Eardley, Quebec. An additional specimen, tentatively identified as *G. ×gatineauensis*, was collected in 1929 from French Creek in Upshur County, West Virginia.

Glyceria ×ottawensis Bowden is a sterile hybrid between *G. striata* and *G. canadensis*. It is intermediate between the two parents, and is known only from the original populations near Ottawa. It has sometimes been included in *G. ×laxa* (Scribn.) Scribn. [≡ *G. canadensis* var. *laxa*]; that taxon often produces viable seed, indicating that it is not a hybrid.

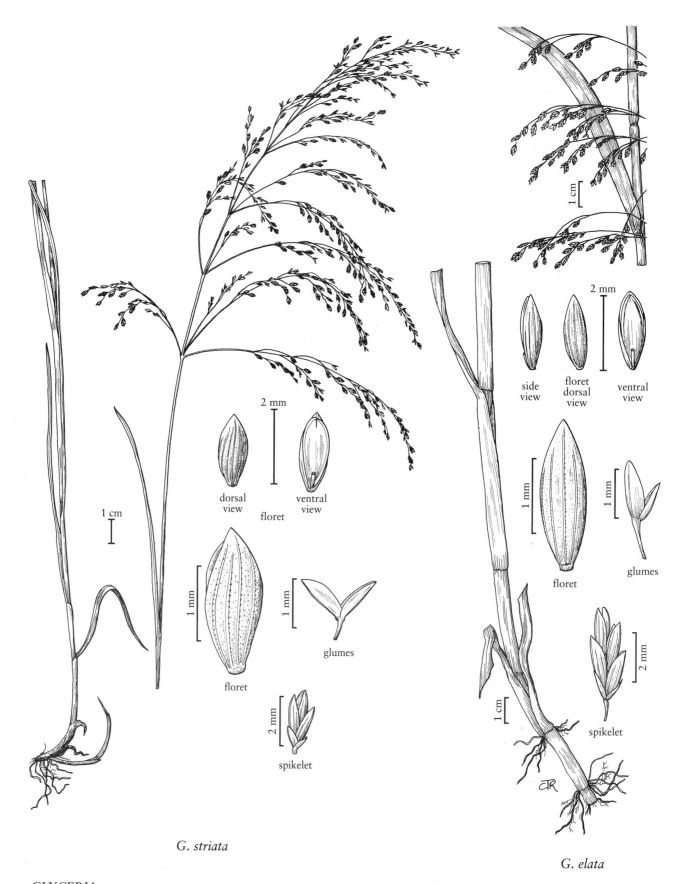

side
view

floret
dorsal
view

ventral
view

2 mm

floret

glumes

spikelet

dorsal
view

ventral
view

floret

2 mm

floret

glumes

spikelet

1 cm

G. striata

G. elata

GLYCERIA

9. Glyceria elata (Nash) M.E. Jones [p. 78]

TALL MANNAGRASS

Plants perennial. Culms 75–150 cm tall, 2.5–8 mm thick, spongy, decumbent and rooting at the lower nodes. Sheaths scabridulous or hirtellous, not or weakly keeled; ligules 2.5–4(6) mm, truncate to acute, erose, puberulent; blades 19–40+ mm long, 6–12(15) mm wide, abaxial surfaces smooth or scabridulous, adaxial surfaces usually scabrous, sometimes scabridulous. Panicles 15–30 cm long, 12–30 cm wide, pyramidal, open; branches 12–17 cm, divergent to drooping, lax, with 30–50+ spikelets; pedicels 0.3–5 mm. Spikelets 3–6 mm long, 1.5–2.8 mm wide, laterally compressed, oval in side view, with 3–4(6) florets. Glumes 1–1.5(2) times longer than wide, narrowing beyond midlength, veins terminating below the apical margins, apices obtuse to rounded; lower glumes 0.7–1.5 mm; upper glumes 1–1.5 mm; rachilla internodes 0.5–0.6 mm; lemmas 1.7–2.2 mm, oval in dorsal view, 5–7-veined, veins raised throughout, scabridulous, apices rounded, prow-shaped; paleas subequal to or often slightly longer than the lemmas, lengths 2.4–3 times widths, oval in dorsal view, keels not winged, tips pointing towards each other, apices narrowly notched between the keels; anthers 2, 0.5–0.8 mm. Caryopses 0.8–1.5 mm long, 0.5–0.7 mm wide; hila as long as the caryopses. 2*n* = 20.

Glyceria elata grows in wet meadows and shady moist woods, from British Columbia east to Alberta and south to California and New Mexico. It is not known from Mexico. The anomalous record from Georgia may represent an inadvertent introduction. It is very similar to, and sometimes confused with, *G. striata*, but the two sometimes grow together and show no evidence of hybridization. Their differences in growth habit and stature are evident in the field. Molecular data (Whipple et al. [in press]) confirm that *G. elata* and *G. striata* are distinct, closely related entities.

Glyceria elata is also sometimes confused with *G. grandis*. It differs in having rounded glumes with veins that terminate below the apices, more readily disarticulating florets, and greener lemmas with more prow-shaped apices, as well as in having paleal keel tips that point towards each other. In its overall aspect, it also resembles *G. pulchella*, but has somewhat more lax panicle branches than that species, in addition to smaller spikelets and florets.

10. Glyceria canadensis (Michx.) Trin. [p. 80]

Plants perennial. Culms 60–150 cm tall, 2.5–5 mm thick, erect or the bases decumbent. Sheaths retrorsely scabridulous to scabrous, keeled; ligules 2–6 mm; blades 8–36 cm long, 3–8 mm wide, abaxial surfaces smooth or scabrous, adaxial surfaces scabridulous to scabrous. Panicles 10–30 cm long, 10–20 cm wide, pyramidal, open, nodding; branches 7–20 cm, lax, divergent, often drooping, with 15–60+ spikelets; pedicels 2.5–9 mm. Spikelets 3–8 mm long, (2.5)3–5 mm wide, laterally compressed, oval in side view, with 2–10 florets. Glumes narrowing from midlength or above to the broadly (≥ 45°) acute or rounded apices, 1-veined, veins terminating below the apices; lower glumes 0.6–2.4 mm, ovate to rectangular; upper glumes 1.5–2.5 mm, lanceolate; rachilla internodes 0.2–0.5 mm; lemmas 1.8–4 mm, ovate in dorsal view, 5–7-veined, veins evident but not raised distally, smooth over and between the veins, apices acute, prow-shaped; paleas 0.1–0.8 mm shorter than lemmas, lengths 1.5–1.8 times widths, almost round in dorsal view, keels well developed, not winged, tips incurved, apices narrowly notched between the keels; anthers 2, 0.4–0.5 mm, dehiscent at maturity. Caryopses 1.5–2 mm.

Glyceria canadensis is an attractive native species that grows in swamps, bogs, lakeshore marshes, and wet woods throughout much of eastern North America, extending from eastern Saskatchewan to Newfoundland, Illinois, and northeastern Tennessee. It is now established in western North America, having been introduced as a weed in cranberry farms. It forms sterile hybrids with *G. striata*; the hybrids are called G. ×ottawensis Bowden. For further comments, see p. 77.

1. Lemmas 2.4–4 mm long; spikelets 5–8 mm long, with 4–10 florets; lower glumes 1.6–2.4 mm long; upper glumes acute . . . . var. *canadensis*
1. Lemmas 1.8–2.5 mm long; spikelets 3–5 mm long, with 2–5 florets; lower glumes 0.6–1.3 mm long; upper glumes usually rounded, sometimes acute . . . . var. *laxa*

Glyceria canadensis (Michx.) Trin. var. canadensis [p. 80]

CANADIAN GLYCERIA, CANADIAN MANNAGRASS, RATTLESNAKE MANNAGRASS, GLYCÉRIE DU CANADA

Spikelets 5–8 mm, with 4–10 florets. Lower glumes 1.6–2.4 mm; upper glumes 1.8–2.5 mm, acute; lemmas 2.4–4 mm. 2*n* = 60.

Glyceria canadensis var. *canadensis* grows throughout the range of the species. The spikelets bear some

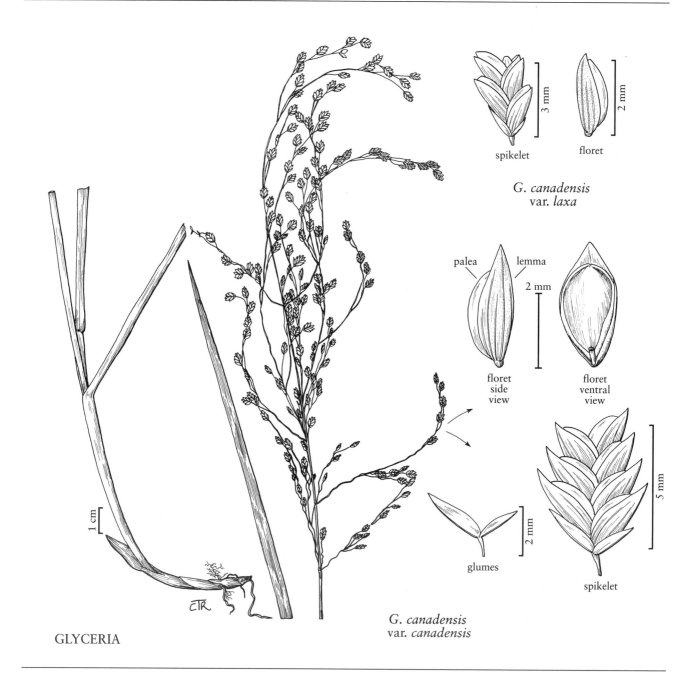

spikelet   floret

3 mm   2 mm

G. canadensis
var. *laxa*

palea   lemma

2 mm

floret
side
view

floret
ventral
view

glumes

2 mm

5 mm

spikelet

GLYCERIA

G. canadensis
var. *canadensis*

resemblance to those of *Bromus briziformis*, otherwise known as rattlesnake brome, hence the vernacular name "rattlesnake mannagrass".

Glyceria canadensis var. **laxa** (Scribn.) Hitchc. [p. 80]
LIMP MANNAGRASS

Spikelets 3–5 mm, with (2)3–5 florets. **Lower glumes** 0.6–1.3 mm; **upper glumes** 1.5–2.3 mm, usually rounded, sometimes acute; **lemmas** 1.8–2.5 mm. 2n = 60.

Glyceria canadensis var. *laxa* grows in swamps, bogs, and wet woods, primarily along the eastern seaboard of North America from Nova Scotia to northeastern Tennessee. It is sometimes treated as a hybrid, *G.* ×*laxa* (Scribn.) Scribn., but several specimens have dehiscent anthers and well-formed caryopses, indicating that they are not hybrids. The report of 2n = 30 is based on counts for *G.* ×*ottawaensis*.

Glyceria R. Br. sect. Glyceria

Plants perennial, rarely annual. **Sheaths** compressed, usually at least weakly keeled. **Inflorescences** 0.5–5 cm wide if the branches appressed, to 20 cm wide if divergent; **branches** usually appressed to ascending, divergent at anthesis; **pedicels** 0.5–5 mm. **Spikelets** cylindrical and terete, except at anthesis when slightly laterally compressed, rectangular in side view, appressed to the panicle branches, lengths 5–22 times widths. **Lemmas** rounded over the back, apices acute to rounded or truncate, entire to irregularly lobed; **paleas** keeled, keels usually winged distally, tips parallel or almost so, sometimes extending into teeth, truncate, rounded, or notched between the keels; **lodicules** connate; **anthers** 3. **Caryopses** ovoid-oblong; **hila** about as long as the caryopses, linear.

Glyceria sect. *Glyceria* includes about 15 species. Seven species grow in the *Flora* region, three of which are introduced. In addition, there is one named hybrid. They grow in and beside shallow, still or slowly moving fresh water, such as along the edges of lakes and ponds and in low areas in wet meadows.

11. **Glyceria borealis** (Nash) Batch. [p. 82]
    BOREAL GLYCERIA, BOREAL MANNAGRASS,
    GLYCÉRIE BORÉALE

Plants perennial. **Culms** 60–100 cm tall, 1.5–5 mm thick, often decumbent and rooting at the lower nodes. **Sheaths** glabrous, keeled; **ligules** 4–12 mm; **blades** 9–25 cm long, 2–7 mm wide, often floating, abaxial surfaces smooth, adaxial surfaces of the midcauline leaves densely papillose, glabrous. **Panicles** 18–40(50) cm long, 0.5–2(5) cm wide, arching, usually narrow, open at anthesis, bases often enclosed in the upper leaf sheath at maturity; **branches** 5–10(15) cm, usually 1–3(5) per node, usually appressed to strongly ascending, occasionally spreading, longer branches with 3–6 spikelets; **pedicels** 1.2–5 mm. **Spikelets** 9–22 mm long, 0.8–2.5 mm wide, cylindrical and terete, except at anthesis when slightly laterally compressed, rectangular in side view, with 8–12 florets. **Glumes** elliptic, apices rounded to obtuse, sometimes erose; **lower glumes** 1.2–2.2 mm; **upper glumes** 2–3.8 mm, rounded; **rachilla internodes** 0.6–3.5 mm; **lemmas** 2.7–5.4 mm, veins raised, scabridulous or smooth, intercostal regions usually smooth, sometimes scabridulous, midvein terminating about (0.1)0.2 mm short of the apical margin, apices usually acute, sometimes obtuse, entire or almost so; **paleas** usually shorter than to equaling the lemmas, sometimes exceeding them by up to 0.5 mm, keels narrowly winged, apices bifid, teeth to 0.2 mm, parallel to weakly incurved; **anthers** 3, 0.4–1.5 mm. **Caryopses** 1.2–2 mm. $2n = 20$.

Glyceria borealis is a widespread native species that grows in the northern portion of the *Flora* region, extending southward through the western mountains into northern Mexico. It grows along the edges and muddy shores of freshwater streams, lakes, and ponds. In the southern portion of its range, *G. borealis* is restricted to subalpine and alpine areas. The midcauline leaves of *G. borealis* almost always have densely papillose upper leaf surfaces. Voss (1972) stated that such surfaces are non-wettable and develop on the floating leaves.

Glyceria borealis differs from *G. notata* in having acute lemmas and, usually, densely papillose midcauline leaves.

12. **Glyceria septentrionalis** Hitchc. [p. 82]
    NORTHERN GLYCERIA, NORTHERN MANNAGRASS,
    GLYCÉRIE SEPTENTRIONALE

Plants perennial. **Culms** 73–182 cm tall, to 8 mm thick, often decumbent and rooting from the lower nodes. **Sheaths** smooth or scabridulous, keeled; **ligules** 5–16 mm; **blades** 18–32 cm long, 2–15 mm wide, abaxial surfaces scabrous, adaxial surfaces scabridulous, usually glabrous, midcauline leaves sometimes papillose. **Panicles** 15–60 cm long, 1–3.5 cm wide; **branches** 3–17 cm, usually erect to strongly ascending, sometimes spreading at anthesis, usually straight, sometimes lax, with 1–9 spikelets; **pedicels** 0.7–1.7 mm. **Spikelets** (6.5)10–23 mm long, 1–3 mm wide, cylindrical and terete, except at anthesis when slightly laterally compressed, rectangular in side view, with 8–16 florets. **Glumes** elliptic to obovate, apices rounded to acute; **lower glumes** (0.3)1.5–3.7

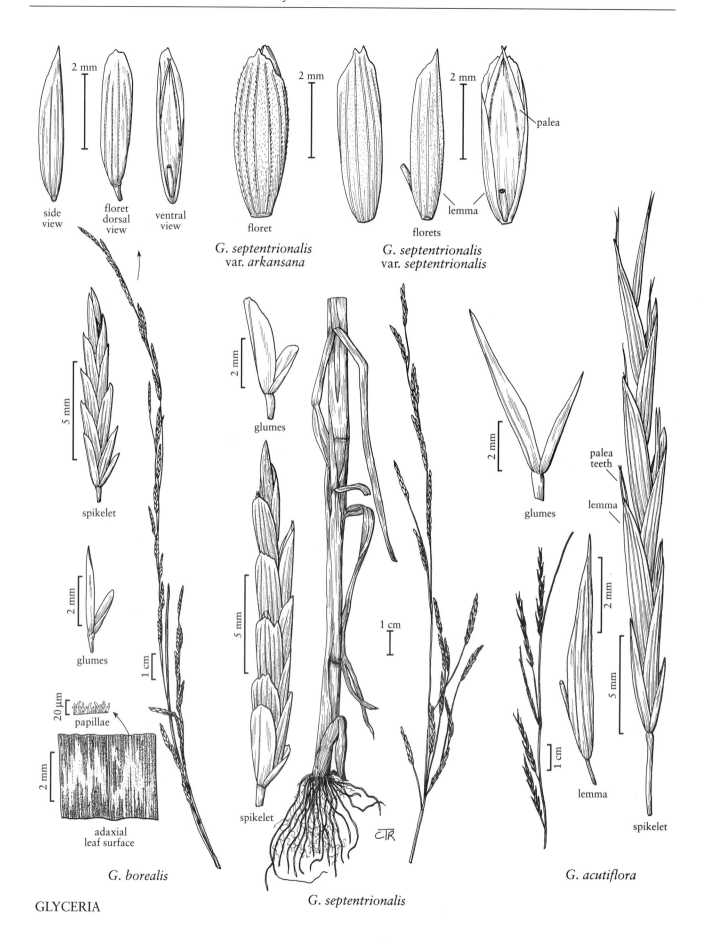

side
view

floret
dorsal
view

ventral
view

2 mm

floret

G. septentrionalis
var. *arkansana*

2 mm

palea

lemma

florets

G. septentrionalis
var. *septentrionalis*

2 mm

5 mm

spikelet

2 mm

glumes

1 cm

20 μm
papillae

2 mm

adaxial
leaf surface

G. borealis

GLYCERIA

2 mm

glumes

5 mm

spikelet

1 cm

G. septentrionalis

glumes

2 mm

1 cm

lemma

5 mm

palea
teeth

lemma

2 mm

spikelet

G. acutiflora

mm; **upper glumes** (1.9)2.3–5.2 mm; **rachilla internodes** 1.1–1.8 mm; **lemmas** 2.4–4.8 mm, veins scabrous or hispidulous, intercostal regions scabridulous, scabrous, or hispidulous, midveins extending to within 0.1 mm of the apical margins, apices truncate to obtuse or acute, apical margins crenate to entire; **paleas** from slightly shorter than to exceeding the lemmas, apices bifid, teeth to 0.2 mm; **anthers** 3, 0.5–1.8 mm. **Caryopses** 1.5–2 mm; **hila** about as long as the caryopses.

Glyceria septentrionalis is native and restricted to North America. It grows in shallow water or very wet soils, from southern Quebec to the east coast and south to eastern Texas and South Carolina. Voss (1972) stated that it is the floating leaves of *G. septentrionalis* that develop papillose, non-wettable adaxial surfaces. They seem to be developed less often than in *G. borealis*; whether this reflects a difference in habitat or growth habit is not known.

Glyceria septentrionalis resembles *G. notata* in its rather short, truncate to rounded lemmas, but it tends to have fewer spikelets on its branches. In addition, the veins of its leaf sheaths appear completely smooth, even under high magnification. That said, many specimens will be hard to identify if their provenance is not known.

1. Lemmas hispidulous over the veins, hairs about 0.1 mm long . . . . var. *arkansana*
1. Lemmas scabrous over the veins, prickles about 0.05 mm long . . . . var. *septentrionalis*

Glyceria septentrionalis var. **arkansana** (Fernald) Steyerm. & Kučera [p. 82]

Ligules 5–14 mm; **blades** 6–12 mm wide. **Panicle branches** 3–7 cm, with 3–5 spikelets. **Rachilla internodes** 0.5–1.5 mm. **Lemmas** hispidulous over the veins, hairs about 0.1 mm, hispidulous, scabrous, or scabridulous between the veins, apices rounded to acute, crenate. 2n = unknown.

Glyceria septentrionalis var. *arkansana* grows in roadside ditches and on the edges of swamps, lakes, and ponds in the flood plain of the Mississippi River, from southern Illinois and Indiana to the Gulf coast. There is also one record from central Tennessee (Chester et al. 1993). The size of its stomates suggests that var. *arkansana*, like var. *septentrionalis*, is a tetraploid.

Glyceria septentrionalis Hitchc. var. **septentrionalis** [p. 82]
Ligules 5–10 mm; **blades** 2–15 mm wide. **Panicle branches** 3–17 cm, with 1–9 spikelets. **Rachilla internodes** 1.1–1.8 mm. **Lemmas** scabrous over the veins, prickles about 0.05 mm, scabrous or scabridulous between the veins, apices almost truncate to obtuse or acute, apical margins crenate to entire. 2n = 40.

Glyceria septentrionalis var. *septentrionalis* grows throughout the range of the species, but is less common in the lower floodplain of the Mississippi River and Kentucky than var. *arkansana*. It is found in shallow water or wet soils. In reviewing specimens for this treatment, some were found to have acute lemmas that usually exceeded the paleas, and lemma midveins that were clearly longer than the other veins; others had truncate to obtuse lemmas that were usually shorter than or equaling the paleas, and lemma midveins that were barely longer than the lateral veins. Further study is needed to determine whether the two kinds merit separate recognition.

13. Glyceria acutiflora Torr. [p. 82]
CREEPING MANNAGRASS

Plants perennial. **Culms** 30–100 cm tall, 3–6 mm thick, spongy, usually decumbent and rooting at the lower nodes. **Sheaths** smooth, weakly keeled; **ligules** 5–9 mm; **blades** 10–15 cm long, 3–8 mm wide, abaxial surfaces smooth, adaxial surfaces of the midcauline leaves often papillose. **Inflorescences** often racemes, sometimes panicles, 15–35 cm long, 1–2 cm wide, open at anthesis, bases often enclosed in the flag leaf sheaths at maturity; **branches** 5.5–8 cm (absent in racemose plants), solitary or in pairs, appressed, most branches with 1–3 spikelets, the lower branches sometimes with more than 3; **pedicels** 1.5–2.5 mm. **Spikelets** 20–45 mm long, 2.5–3 mm wide, cylindrical and terete except slightly laterally compressed at anthesis, rectangular in side view, with 5–12 florets. **Glumes** unequal, acute; **lower glumes** 1.3–4.5 mm; **upper glumes** 3–7 mm; **rachilla internodes** 2–3 mm; **lemmas** 6–8.5 mm, scabridulous, 7-veined, gradually tapering from near midlength to the narrowly acute (< 45°) or acuminate apices; **paleas** 0.7–3 mm longer than the lemmas, keels winged, tips parallel, intercostal region truncate, often splitting, apices appearing bifid, with 0.4–1 mm teeth; **anthers** 3, 1–2 mm. **Caryopses** about 3 mm. 2*n* = 40.

Glyceria acutiflora grows in wet soils and shallow water of the northeastern United States, extending from Michigan and Missouri to the Atlantic coast between southwestern Maine and Delaware. Its long paleas make *G. acutiflora* the most distinctive North American species of sect. *Glyceria*.

side view

floret dorsal view

palea   lemma

2 mm

ventral view

2 mm

floret

floret

side view

floret dorsal view

2 mm

ventral view

spikelet

5 mm

2 mm

glumes

spikelet

5 mm

1 cm

G. leptostachya

5 mm

1 cm

spikelet

G. ×occidentalis

1 cm

CTR

spikelet

5 mm

2 mm

glumes

G. fluitans

GLYCERIA

14. Glyceria leptostachya Buckley [p. 84]
NARROW MANNAGRASS

Plants perennial. **Culms** 50–100 (150) cm tall, 3–8 mm thick, spongy, erect to decumbent and rooting at the lower nodes. **Sheaths** finely scabridulous, not or weakly keeled; **ligules** 4.5–12 mm, lacerate; **blades** 12–30 cm long, 3.5–11 mm wide, both surfaces sometimes scabridulous, adaxial surfaces sometimes sparsely papillose. **Panicles** 20–40 cm long, 2.5–8 cm wide; **branches** 4.2–14.7 cm, appressed to ascending, with 3–8(10) spikelets; **pedicels** 2–5 mm, scabrous. **Spikelets** 9–20 mm long, 0.4–3 mm wide, cylindrical and terete, except at anthesis when slightly laterally compressed, rectangular in side view, with 6–15 florets. **Glumes** broadly rounded to acute; **lower glumes** 0.6–2.1 mm; **upper glumes** 1.4–3.4 mm; **rachilla internodes** 1–1.5 mm; **lemmas** 2.6–4.5 mm, somewhat indented below the apical margins at maturity, veins raised, scabridulous to scabrous over and between the veins, prickles about 0.05 mm, midveins extending to within 0.1 mm of the apical margins, apices truncate to obtuse, crenulate; **paleas** shorter than or equaling the lemmas, keels winged, tips parallel, intercostal region truncate or rounded, sometimes exceeding the keel tips; **anthers** 3, 0.3–0.9 mm. 2*n* = 40.

Glyceria leptostachya grows in swamps and along the margins of streams and lakes, on the western side of the coastal mountains from southern Alaska to San Francisco Bay. It is similar to the European *Glyceria notata*, differing primarily in its tendency to have fewer spikelets [3–8(10) vs. 5–15(19)] on its branches.

15. Glyceria ×occidentalis (Piper) J.C. Nelson [p. 84]
WESTERN MANNAGRASS

Plants perennial. **Culms** 60–160 cm tall, 2.5–5 mm thick, erect or decumbent and rooting from the lower nodes. **Sheaths** smooth to scabridulous, keeled, sometimes weakly so; **ligules** 7–12 mm; **blades** 20–30 cm long, (2.5)4–12 mm wide, adaxial surfaces scabridulous, occasionally papillose. **Panicles** 20–50 cm long, 2–15 cm wide, usually narrow, open at anthesis; **branches** 4.5–18 cm, somewhat lax, usually ascending, strongly divergent at anthesis, with 2–8 spikelets, **pedicels** 1.5–8 mm. **Spikelets** 13–23 mm long, 1.5–3.5 mm wide, cylindrical and terete, except at anthesis when slightly laterally compressed, rectangular in side view, with 6–13 florets. **Glumes** acute to obtuse; **lower glumes** 1.1–2.8 mm; **upper glumes** 2.9–3.7 mm, about twice as long as the lower glumes; **rachilla internodes** 1–2.8 mm; **lemmas** 4.5–5.9 mm, scabridulous, midveins extending to within 0.1 mm of the apical margins, apices acute, usually slightly lobed or irregularly crenate; **paleas** usually shorter than or equaling the lemmas, sometimes slightly longer, keels winged, apices shallowly notched to slightly bifid, teeth to 0.2 mm, parallel; **anthers** 2, 0.6–1.6 mm. 2*n* = 40.

Glyceria ×*occidentalis* has hitherto been considered an uncommon native species that grows along lakes, ponds, and streams, and in marshy areas of western North America. It differs from other species in the region primarily in its longer lemmas and anthers. Studies of chloroplast DNA in western North American species of *Glyceria* demonstrated that, contrary to C.L. Hitchcock's (1969) conclusion, *G. fluitans* is present in western North America, and that all specimens being identified as *G.* ×*occidentalis* had cpDNA resembling that of *G. leptostachya* or *G. fluitans*; there was no distinctive *G.* ×*occidentalis* cpDNA (Whipple et al. [in press]). This strongly suggests that *G.* ×*occidentalis* is a series of reciprocal hybids, and probably backcrosses, between *G. fluitans* and *G. leptostachya*. As the key indicates, *G.* ×*occidentalis* is intermediate between its two putative parents. The cpDNA study also confirmed that *G. declinata* is distinct from *G.* ×*occidentalis* (see discussion under that species).

16. Glyceria fluitans (L.) R. Br. [p. 84]
WATER MANNAGRASS, GLYCÉRIE FLOTTANTE

Plants perennial. **Culms** 20–150 cm tall, 2–4 mm thick, erect or spreading, sometimes decumbent and rooting from the lower nodes, distal portion sometimes floating in shallow water. **Sheaths** glabrous, keeled; **ligules** 5–15 mm; **blades** 5–25 cm long, 3–10 mm wide, both surfaces smooth. **Panicles** 10–50 cm long, 2–3 cm wide; **branches** 3–5 cm, paired or solitary, usually appressed to ascending, divergent at anthesis, with 1–4 spikelets; **pedicels** 0.8–20 mm. **Spikelets** (15)18–39 mm long, 1.7–3.3 mm wide, cylindrical and terete, except slightly laterally compressed at anthesis, rectangular in side view, with 8–16 florets. **Lower glumes** 1.3–3.9 mm; **upper glumes** 2.7–5 mm; **rachilla internodes** 1.9–2.5 mm; **lemmas** 5.2–8 mm, midveins extending to within 0.1 mm of the apical margins, scabrous over and between the veins, prickles about 0.05 mm, apices acute, usually entire; **paleas** from shorter than to 0.6(1.5) mm longer than the lemmas, keels winged, apices bifid, teeth 0.1–0.4 mm, parallel to convergent, sometimes crossing when dry; **anthers** 1.5–3 mm, usually purple. **Caryopses** 2–3 mm. 2*n* = 40.

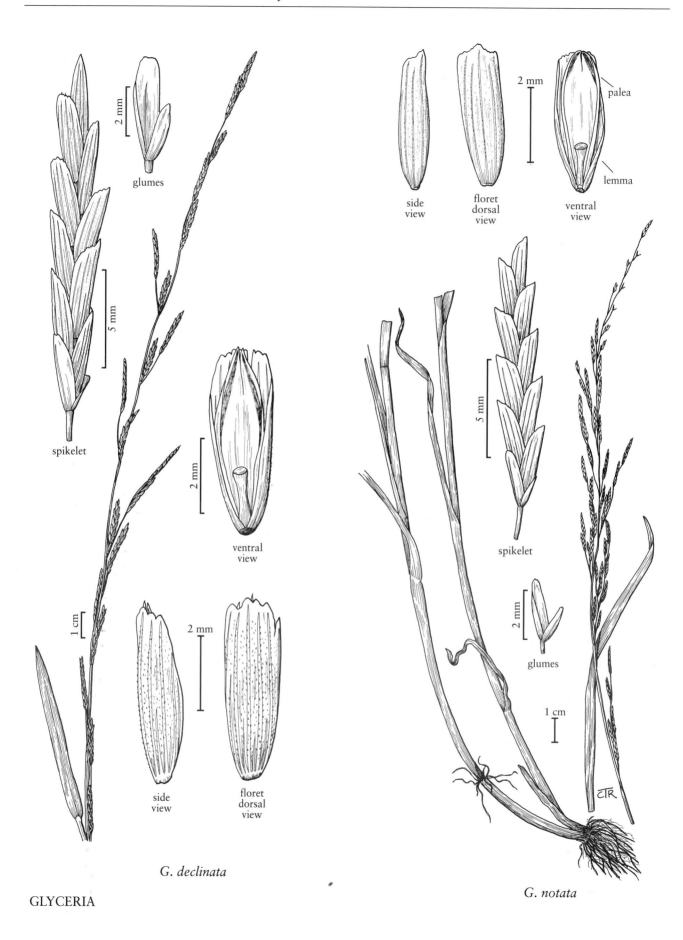

2 mm

glumes

spikelet

5 mm

1 cm

ventral
view

2 mm

side
view

floret
dorsal
view

2 mm

G. declinata

2 mm

side
view

floret
dorsal
view

palea

lemma

ventral
view

5 mm

spikelet

2 mm

glumes

1 cm

CTR

G. notata

GLYCERIA

Glyceria fluitans is a Eurasian species. In the Americas, it has been collected from British Columbia to California on the west coast, in South Dakota, and from Newfoundland to Pennsylvania on the eastern seaboard. In Europe, it grows in rich, organic, wet soils, often near *G. notata*, with which it hybridizes. It is less tolerant of trampling than *G. notata*. Many earlier reports from eastern Canada are based on *G. borealis* or *G. septentrionalis* (Dore and McNeill 1980; Scoggan 1978). In western North America, it has been confused with *G. ×occidentalis*. It tends to differ from all three in its longer lemmas and anthers. Nevertheless, identification of some specimens will prove troublesome. For further discussion, see under the species mentioned.

17. Glyceria declinata Bréb. [p. 86]
LOW GLYCERIA

Plants usually perennial, rarely annual. **Culms** (10)20–92 cm tall, 1.5–2.5 mm thick, ascending to erect from a decumbent, branching base. **Sheaths** glabrous, keeled; **ligules** 4–9 mm; **blades** (2)3–12 cm long, 4–8 mm wide, adaxial surfaces not papillose, apices abruptly acute. **Panicles** 6–30 cm long, 1–2.5 cm wide; **branches** 1.5–9.5 cm, ascending, with 1–5 spikelets; **pedicels** 1–2.5 mm. **Spikelets** 11–24 mm long, 1.3–3 mm wide, cylindrical and terete, except slightly laterally compressed at anthesis, rectangular in side view, with 8–15 florets. **Glumes** oval; **lower glumes** 1.4–3.5 mm; **upper glumes** 2.5–4.9 mm; **rachilla internodes** 1.2–1.8 mm; **lemmas** (3.5)4–6 mm, 7-veined, veins and intercostal regions scabridulous, prickles about 0.05 mm, midveins extending to within 0.1 mm of the apical margins, apices acute, with a well-developed lobe on one or both sides opposite the lateral veins, entire to crenulate between the lateral lobes; **paleas** exceeding the lemmas by 0.2–1(1.5) mm, keels winged, apices bifid, teeth 0.3–0.5 mm; **anthers** 0.5–1.4 mm, usually purple. **Caryopses** 1.8–2.5 mm. $2n = 20$.

Glyceria declinata is a European species that is established on the western seaboard of North America from southern British Columbia to southern California, and in northeastern Nevada, Arizona, the lower portion of the Mississippi valley, and on Long Island, New York. In Europe, it grows in low-calcium, acidic soils and tolerates drier conditions than other European species of *Glyceria* (Conert 1992). In Denmark, it tends to grow in areas that are highly trampled (Niels Jacobsen and Signe Frederiksen, pers. comm.). It is invading vernal pools in California.

In western North America, *G. declinata* has been confused with *G. ×occidentalis*. The most reliable distinguishing characteristics are the lateral lemma lobes of *G. declinata* and its rather short, straight panicle branches. The two species also differ in their ploidy level, *G. declinata* being diploid and *G. ×occidentalis* tetraploid (Church 1949). This is reflected in the length of their guard cells, those of *G. declinata* being 0.2–0.3 μm and those of *G. ×occidentalis* being 0.4–0.5 μm.

S.F. Hrusa found plants (Hrusa 13681, 15858, 16267; specimens in CDA) that have an annual growth habit. Apart from this, they fit within the circumscription of *G. declinata*, except that two of the three specimens have narrower (2–3 mm) leaves than normal; they were also collected relatively early in the season. For now, it seems best to include the plants in *G. declinata* pending a better understanding of their relationship to perennial members of the species.

18. Glyceria notata Chevall. [p. 86]
MARKED GLYCERIA

Plants perennial. **Culms** 25–80 cm, rooting at the nodes. **Sheaths** usually scabridulous or hirtellous; **ligules** 2–8 mm; **blades** 5–30 cm long, 3–11(14) mm wide, abaxial surfaces scabrous, adaxial surfaces sometimes scabridulous to scabrous, sometimes sparsely hairy, sometimes papillose. **Panicles** 10–45 cm; branches 2–5 per node, eventually widely spreading; **branches** to 12 cm, with 5–15(19) spikelets; **pedicels** 1–6 mm. **Spikelets** 10–25 mm long, 1.5–3 mm wide, cylindrical and terete except slightly laterally compressed at anthesis, rectangular in side view, with 7–16 florets. **Glumes** obtuse to rounded; **lower glumes** 1–2.5 mm; **upper glumes** 2.5–4.5 mm; **lemmas** 3.5–5 mm, the submarginal veins often longer than those adjacent to the midvein, veins scabridulous, smooth or scabridulous between the veins, apices truncate to rounded, crenulate; **paleas** from slightly shorter to slightly longer than the lemmas, keels winged distally, apices bifid, teeth about 0.2 mm; **anthers** 0.8–1.5 mm. **Caryopses** 1.5–2.5 mm. $2n = 40$.

Glyceria notata is a Eurasian species that has been reported from scattered locations in the *Flora* region; the reports have not been verified. In Europe, *G. notata* grows in rich, organic, wet soils, often near *G. fluitans*, with which it hybridizes. It is more tolerant of trampling than *G. fluitans*.

There is no single morphological characteristic that separates *Glyceria notata* from *G. septentrionalis* and *G. leptostachya*. It more frequently has lemmas with short veins adjacent to the midvein than the other two species, is more frequently smooth between the veins, more frequently has scabridulous leaf sheaths, and tends to have more spikelets on its branches. The limited cpDNA data indicate that the three are distinct taxa (Whipple et al. [in press]). An intensive examination of the three species is needed.

9.02 MELICA L.

Mary E. Barkworth

Plants perennial; cespitose or soboliferous, not or only shortly rhizomatous. **Culms** (4)9–250 cm, sometimes forming a basal corm; **nodes** and **internodes** usually glabrous. **Sheaths** closed almost to the top; **auricles** sometimes present; **ligules** thinly membranous, erose to lacerate, usually glabrous, those of the lower leaves shorter than those of the upper leaves; **blades** flat or folded, glabrous or hairy, particularly on the adaxial surfaces, sometimes scabrous. **Inflorescences** terminal panicles; **primary branches** often appressed; **secondary branches** appressed or divergent; **pedicels** either more or less straight or sharply bent below the spikelets, scabrous to strigose distally; **disarticulation** below the glumes in species with sharply bent pedicels, above the glumes in other species. **Spikelets** with 1–7 bisexual florets, terminating in a sterile structure, the *rudiment*, composed of 1–4 sterile florets; **rudiments** sometimes morphologically distinct from the bisexual florets, sometimes similar but smaller. **Glumes** membranous or chartaceous, distal margins wide, translucent; **lower glumes** 1–9-veined; **upper glumes** 1–11-veined; **calluses** glabrous; **lemmas** membranous basally, sometimes becoming coriaceous at maturity, glabrous or with hairs, (4)5–15-veined, usually unawned, sometimes awned, awns to 12 mm, straight; **paleas** from $^1/_2$ as long as to almost equaling the lemmas, keels usually ciliate; **lodicules** fused into a single, collarlike structure extending $^1/_2$–$^2/_3$ around the base of the ovaries; **anthers** (2)3. **Caryopses** usually 2–3 mm, smooth, glabrous, longitudinally furrowed, falling from the floret when mature. $x = 9$. From the Latin *mel*, 'honey', a classical name for an unknown, but presumably sweet, plant.

Melica includes approximately 80 species, which grow in all temperate regions of the world except Australia, usually in shady woodlands on dry stony slopes (Mejia-Saulés and Bisby 2003). The species are relatively nutritious, but are rarely sufficiently abundant to be important as forage.

Nineteen species of *Melica* grow in the *Flora* region. Two European species are grown as ornamentals in North America. Many of the seventeen native species merit such use.

Several proposals have been made for dividing *Melica* into smaller units. American taxonomists have tended to favor Thurber's (1880) recognition of two subgenera: *Melica* and *Bromelica*. In subg. *Melica*, the pedicels are straight and disarticulation is above the glumes; in subg. *Bromelica*, the pedicels are sharply bent and the spikelets disarticulate below the glumes. Hempel (1970) recognized three subgenera in *Melica*, but his groups do not correspond well to the pattern of morphological variation seen in North America. More recently, Mejia-Saulés and Bisby (2003) examined the variation in lemma silica bodies and hooked papillae within *Melica*. Their results are not consistent with either Thurber's or Hempel's treatment, but provide some support for Papp's (1928) recognition of two groups, based on the presence or absence of hairs on the lemmas and the compression of the spikelets.

In the following key and descriptions, unless otherwise stated, comments on the panicle branches apply to the longest branches within the panicle; glume widths are measured from side to side, at the widest portion; lemma descriptions are for the lowest floret in the spikelets; and rachilla internode comments apply to the lowest internode in the spikelets.

SELECTED REFERENCES **Boyle, W.S.** 1945. A cytotaxonomic study of the North American species of *Melica*. Madroño 8:1–26; **Farwell, O.A.** 1919. *Bromelica* (Thurber): A new genus of grasses. Rhodora 21:76–78; **Hempel, W.** 1970. Taxonomische und chorologische Untersuchungen an Arten von *Melica* L. subgen. *Melica*. Feddes Repert. 81:131–145; **Hitchcock, A.S.** 1951. Manual of the Grasses of the United States, ed. 2, rev. A. Chase. U.S.D.A. Miscellaneous Publication No. 200. U.S. Government Printing Office, Washington, D.C., U.S.A. 1051 pp.; **Mejia-Saulés, T.** and **F.A. Bisby.** 2003. Silica bodies and hooked papillae in lemmas of *Melica* species (Gramineae: Poöideae). Bot. J. Linn. Soc. 143:447–463; **Papp, C.** 1928. Monographie der Südamerikanischen Arten der Gattung *Melica* L. Repert. Spec. Nov. Regni Veg. 25:97–160; **Thurber, G.** 1880. *Melica* Linn. Pp. 302–305 *in* S. Watson. Geological Survey of California: Botany, vol. 2. Little, Brown, Boston, Masssachusetts, U.S.A. 559 pp.

1. Spikelets disarticulating below the glumes; pedicels sharply bent just below the spikelets.
   2. Lemmas with hairs.
      3. Lemmas with hairs on the lower portion of the lemmas, the hairs twisted . . . . 15. *M. montezumae*
      3. Lemmas with hairs on the marginal veins, the hairs not twisted . . . . 18. *M. ciliata*
   2. Lemmas glabrous, sometimes scabridulous to scabrous.
      4. Rudiments acute to acuminate, similar to but smaller than the bisexual florets.
         5. Spikelets broadly V-shaped when mature, 5–13 mm wide; upper glumes 6–18 mm long . . . . 13. *M. stricta*
         5. Spikelets parallel-sided when mature, 1.5–5 mm wide; upper glumes 5–8 mm long . . . . 14. *M. porteri*
      4. Rudiments clublike, not resembling the bisexual florets.
         6. Rudiments at an angle to the rachilla; panicle branches with 2–5 spikelets . . . . 16. *M. mutica*
         6. Rudiments in a straight line with the rachilla; panicle branches with 5–20 spikelets.
            7. Panicle branches often divergent to reflexed; glumes unequal, lower glumes shorter and more ovate than the upper glumes . . . . 17. *M. nitens*
            7. Panicle branches strongly ascending to appressed; glumes subequal in length and similar in shape . . . . 19. *M. altissima*
1. Spikelets disarticulating above the glumes; pedicels more or less straight.
   8. Rudiments truncate to acute, not resembling the lowest florets.
      9. Bisexual florets 1(2); paleas almost as long as the lemmas.
         10. Rudiments shorter than the terminal rachilla internode; bisexual lemmas scabridulous, sometimes hairy . . . . 1. *M. torreyana*
         10. Rudiments longer than the terminal rachilla internode; bisexual lemmas glabrous, sometimes scabrous . . . . 2. *M. imperfecta*
      9. Bisexual florets 2–7; paleas $^1/_2$–$^3/_4$ the length of the lemmas.
         11. Culm bases not forming distinct corms . . . . 6. *M. californica*
         11. Culm bases forming distinct corms.
            12. Glumes usually less than $^1/_2$ as long as the spikelets; ligules 0.1–2 mm long; corms connected to the rhizomes by a rootlike structure . . . . 3. *M. spectabilis* (in part)
            12. Glumes from ($^1/_2$)$^2/_3$ as long as to equaling the spikelets; ligules 2–6 mm long; corms almost sessile on the rhizomes . . . . 4. *M. bulbosa* (in part)
   8. Rudiments tapering, smaller than but otherwise similar to the lowest florets in shape.
      13. Lemmas awned.
         14. Awns shorter than 3 mm.
            15. Panicle branches appressed; lemmas usually with 0.7–1.3 mm hairs on the margins . . . . 7. *M. harfordii* (in part)
            15. Panicle branches widespread to reflexed; lemmas glabrous . . . . 8. *M. geyeri* (in part)
         14. Awns 3–12 mm long.
            16. Panicle branches 4–6 cm long, appressed or ascending; blades 2–6 mm wide . . . . . 9. *M. aristata*
            16. Panicle branches 7–11 cm long, spreading to reflexed; blades 5–12 mm wide . . . . 10. *M. smithii*
      13. Lemmas unawned.
         17. Lemmas strongly tapering and acuminate, the veins usually hairy . . . . 11. *M. subulata*
         17. Lemmas acute to obtuse, the veins hairy or not.
            18. Lemmas pubescent, the hairs on the marginal veins clearly longer than the hairs elsewhere . . . . 7. *M. harfordii* (in part)

18. Lemmas glabrous, scabrous, or pubescent, never with clearly longer hairs on
the marginal veins.
  19. Rachilla internodes swollen when fresh, wrinkled when dry . . . . 12. *M. fugax*
  19. Rachilla internodes not swollen when fresh, not wrinkled when dry.
    20. Panicle branches with 5–15 spikelets; paleas about ¹/₂ as long as the
    lemmas; culms not forming corms . . . . 5. *M. frutescens*
    20. Panicle branches with 1–6 spikelets; paleas from ²/₃ as long as to
    equaling the lemmas; culms forming corms.
      21. Panicle branches 3–11 cm long, divergent to reflexed, flexuous;
      lowest rachilla internodes 2–3 mm long . . . . 8. *M. geyeri* (in part)
      21. Panicle branches 2–6.5 cm long, usually appressed to ascending,
      straight, sometimes strongly divergent and flexuous; lowest
      rachilla internodes 1–2 mm long.
        22. Ligules 0.1–2 mm long; glumes usually less than ¹/₂ the
        length of the spikelets; corms not attached directly to the
        rhizomes . . . . 3. *M. spectabilis* (in part)
        22. Ligules 2–6 mm long; glumes from (¹/₂)²/₃ as long as to
        equaling the spikelets; corms almost sessile, directly attached
        to the rhizomes . . . . 4. *M. bulbosa* (in part)

1. Melica torreyana Scribn. [p. 92]
TORREY'S MELIC

Plants densely cespitose, not rhizomatous. **Culms** 30–100 cm, not forming corms; **internodes** smooth. **Sheaths** glabrous or sparsely pilose, sometimes pilose only at the throat, sometimes scabridulous; **ligules** 1–5 mm; **blades** 1–2.5 mm wide, sometimes pilose on both surfaces, sometimes scabridulous. **Panicles** 6–25 cm; **branches** 1–5 cm, usually appressed, occasionally divergent, with 5–37 spikelets; **pedicels** straight; **disarticulation** above the glumes. **Spikelets** 3.5–7 mm, with 1(2) bisexual florets. **Lower glumes** 3–5 mm long, about 1 mm wide, 1–5-veined; **upper glumes** 3.3–7 mm long, 1–2 mm wide, 3–5-veined; **lemmas** 3.5–6 mm, scabridulous, sometimes hairy, distal hairs longer than those below, 7-veined, veins inconspicuous, apices rounded to emarginate, unawned or awned, awns 1–2 mm; **paleas** slightly shorter than the lemmas; **anthers** 1.5–2.5 mm; **rudiments** 0.5–4 mm, clearly distinct from the bisexual florets, shorter than the terminal rachilla internode, truncate to acute. 2*n* = 18.

Melica torreyana grows from sea level to 1200 m, in thickets and woods in California. It is common throughout chaparral areas and coniferous forests but, on serpentine soils, grows only in shady locations. The shape and size of the rudiments make *M. torreyana* unique among the species found in North America. Boyle (1945) obtained vigorous, almost completely sterile hybrids between *M. imperfecta* and *M. torreyana*, but found no examples of natural hybrids.

2. Melica imperfecta Trin. [p. 92]
LITTLE CALIFORNIA MELIC

Plants densely cespitose, not rhizomatous. **Culms** 35–120 cm, not forming corms; **internodes** scabridulous immediately above the nodes. **Sheaths** glabrous or pilose; **ligules** 0.8–6.5 mm; **blades** 1–6 mm wide, abaxial surfaces glabrous or puberulent, adaxial surfaces with hairs. **Panicles** 5–36 cm; **branches** 2.5–9 cm, appressed to reflexed, straight or flexuous, with 5–30 spikelets; **pedicels** not sharply bent; **disarticulation** above the glumes. **Spikelets** 3–7 mm, with 1(2) bisexual florets; **rachilla internodes** 0.3–0.6 mm. **Lower glumes** 2–5 mm long, 1–2 mm wide, 1-veined; **upper glumes** 2.5–6 mm long, 1.5–2.5 mm wide, 1-veined; **lemmas** 3–7 mm, glabrous, sometimes scabrous, with 7+ veins, veins prominent, apices rounded to acute, unawned; **paleas** almost as long as the lemmas; **anthers** 1.5–2.5 mm; **rudiments** 1–4 mm, not resembling the lower florets, longer and thicker than the terminal rachilla internode, truncate to obtuse. 2*n* = 18.

Melica imperfecta grows from sea level to 1500 m, on stable coastal dunes, dry, rocky slopes, and in open woods, from California and southern Nevada south to Baja California, Mexico. Plants vary with respect to size, panicle shape, and pubescence, but no infraspecific taxa merit recognition. Boyle (1945) obtained vigorous, almost completely sterile hybrids between *M. imperfecta* and both *M. torreyana* and *M. californica*, but found no examples of natural hybrids.

3. Melica spectabilis Scribn. [p. 92]
PURPLE ONIONGRASS

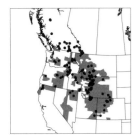

Plants loosely cespitose, rhizomatous. **Culms** 45–100 cm, forming corms, corms connected to the rhizomes by a rootlike, 10–30 mm structure, which usually remains attached to the corm; **internodes** smooth. **Sheaths** usually glabrous, often pilose at the throat and collar; **ligules** 0.1–2 mm; **blades** 2–5 mm wide, abaxial surfaces scabridulous over the veins, adaxial surfaces usually glabrous. **Panicles** 5–26 cm; **branches** 2–5 cm, usually appressed, sometimes divergent and flexuous, with 2–3 spikelets; **pedicels** not sharply bent; **disarticulation** above the glumes. **Spikelets** 7–19 mm, with 3–7 bisexual florets, base of the distal florets concealed at anthesis; **rachilla internodes** 1–2 mm, not swollen when fresh, not wrinkled when dry. **Glumes** usually less than $^1/_2$ the length of the spikelets; **lower glumes** 3.5–6.4 mm long, 1.5–3 mm wide, 1–3-veined; **upper glumes** 5–7 mm long, 2.3–3.5 mm wide, 5–7-veined; **lemmas** 6–9 mm, glabrous, scabridulous, 5–11-veined, veins inconspicuous, apices rounded to acute, unawned; **paleas** about $^2/_3$ the length of the lemmas; **anthers** 1.5–3 mm; **rudiments** 1.5–3.5 mm, acute, distinct from the bisexual florets, sometimes surrounded by a small sterile floret similar in shape to the bisexual florets. $2n = 18$.

Melica spectabilis grows in moist meadows, flats, and open woods, from 1200–2600 m, primarily in the Pacific Northwest and the Rocky Mountains. It is often confused with *M. bulbosa*, differing in its shorter glumes, "tailed" corm, and the more marked and evenly spaced purplish bands of its spikelets.

4. Melica bulbosa Geyer *ex* Porter & J.M. Coult. [p. 92]
ONIONGRASS

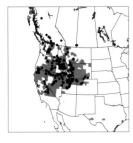

Plants loosely cespitose, rhizomatous. **Culms** 29–100 cm, forming corms, corms almost sessile on the connecting rhizomes; **internodes** scabridulous above the nodes. **Sheaths** usually scabridulous, sometimes sparsely pilose; **ligules** 2–6 mm; **blades** 1.5–5 mm wide, abaxial surfaces scabridulous, adaxial surfaces with hairs. **Panicles** 7–30 cm; **branches** 2–6.5 cm, appressed, usually straight, with 1–5 spikelets; **pedicels** straight; **disarticulation** above the glumes. **Spikelets** 6–24 mm, with 4–7 bisexual florets, base of the distal florets concealed at anthesis; **rachilla internodes** 1–2 mm, not swollen when fresh, not wrinkled when dry. **Glumes** from ($^1/_2$)$^2/_3$ as long as to equaling the spikelets; **lower glumes** 5.5–10.5 mm long, 2–3 mm wide, 3–5-veined; **upper glumes** 6–14 mm long, 2.3–3.5 mm wide, 5–7-veined; **lemmas** 6–12 mm, glabrous, smooth or scabrous, 7–11-veined, veins prominent, apices emarginate to acute, unawned; **paleas** about $^3/_4$ the length of the lemmas; **anthers** 3, 1.5–4 mm; **rudiments** 1.5–5 mm, truncate to tapering, sometimes resembling the bisexual florets in shape. $2n = 18$.

Melica bulbosa grows from 1370–3400 m, mostly in open woods on dry, well-drained slopes and along streams. It is restricted to the western half of the *Flora* region. Two records from Texas, in Jeff Davis and Sutton counties, have not been verified.

Melica bulbosa differs from *M. spectabilis* in its sessile corm and longer glumes. In addition, in *M. bulbosa* the spikelets have purplish bands which appear to be concentrated towards the apices; in *M. spectabilis* the bands appear more regularly spaced. It differs from *M. californica* in its more narrowly acute spikelets, more strongly colored lemmas, and lack of corms, and from *M. fugax* in not having swollen rachilla internodes.

5. Melica frutescens Scribn. [p. 94]
WOODY MELIC

Plants densely cespitose, not rhizomatous. **Culms** 60–200 cm, not forming corms, often branched from the lower nodes; **internodes** smooth. **Sheaths** glabrous, sometimes scabridulous, sometimes purplish; **ligules** 2.5–9 mm; **blades** 2–5 mm wide, abaxial sufaces scabridulous, adaxial surfaces puberulent. **Panicles** 12–40 cm; **branches** 3.5–9 cm, appressed, with 5–15 spikelets; **pedicels** straight; **disarticulation** above the glumes. **Spikelets** 9–18 mm, with 3–5 bisexual florets; **rachilla internodes** 1–1.3 mm, not swollen when fresh, not wrinkled when dry. **Lower glumes** 7–12 mm long, 2–3 mm wide, 5–7-veined; **upper glumes** 8–15 mm long, 2.5–3.5 mm wide, 5–7-veined; **lemmas** 8–11 mm, glabrous, chartaceous for the distal $^1/_3$ or more, 7–9-veined, sometimes purplish basally, veins inconspicuous, apices rounded to acute, unawned; **paleas** about $^1/_2$ the length of the lemmas; **anthers** 3, 1–2 mm; **rudiments** 2–6 mm, blunt, enclosed in empty lemmas resembling those of the bisexual florets. $2n = 18$.

Melica frutescens grows from 300–1500 m in the dry hills and canyons of southern California, Arizona, and adjacent Mexico. Boyle (1945) stated that its seeds remain viable longer than those of other North American species of *Melica*; he gave no information on how long.

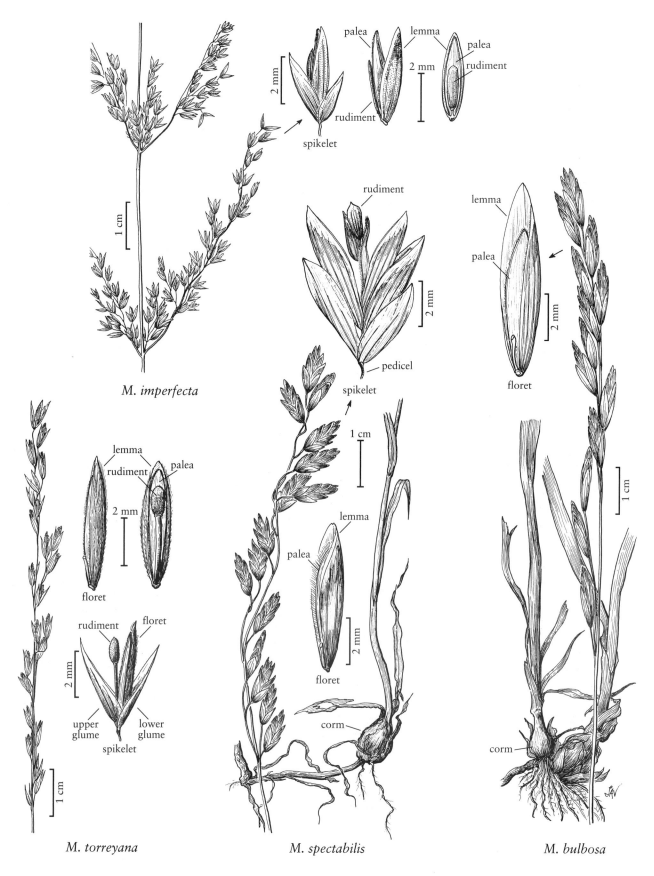

spikelet

palea   lemma

rudiment

palea

rudiment

2 mm

2 mm

1 cm

M. imperfecta

rudiment

spikelet

pedicel

2 mm

lemma

palea

floret

M. bulbosa

2 mm

corm

lemma

rudiment   palea

floret

1 cm

2 mm

rudiment   floret

upper
glume

lower
glume

spikelet

1 cm

lemma

palea

floret

2 mm

corm

M. torreyana   *M. spectabilis*   *M. bulbosa*

6. Melica californica Scribn. [p. 94]
CALIFORNIA MELIC

Plants densely cespitose, not rhizomatous. **Culms** 50–130 cm, not forming corms; **lower nodes** strigose; **internodes** usually smooth, sometimes puberulent below the nodes, lower 2–3 internodes usually swollen. **Sheaths** glabrous or pilose; **ligules** 1.5–4 mm; **blades** 1.5–5 mm wide, strigose on both surfaces. **Panicles** 4–30 cm; **branches** 3–6 cm, appressed, straight, with 4–15 spikelets; **pedicels** straight; **disarticulation** above the glumes. **Spikelets** 5–15 mm, with 2–5 bisexual florets; **rachilla internodes** 1.1–1.6 mm. **Lower glumes** 3.5–12 mm long, 2.5–3 mm wide, 3–5-veined; **upper glumes** 5–13 mm long, 2–2.5 mm wide, 5–7-veined; **lemmas** 5–9 mm, glabrous, smooth to scabrous, 7–9-veined, veins inconspicuous, apices rounded to broadly acute, unawned; **paleas** about ³/₄ the length of the lemmas; **anthers** 3, 1.8–3 mm; **rudiments** 1.4–3 mm, clublike, not resembling the bisexual florets, truncate to acute. $2n = 18$.

Melica californica grows from sea level to 2100 m, in a wide range of habitats, from dry, rocky, exposed hillsides to moist woods. Its range extends from Oregon to California. It differs from *M. bulbosa* in its more obtuse spikelets and less strongly colored lemmas, as well as in not having corms.

Melica californica var. *nevadensis* Boyle supposedly differs from var. *californica* in having shorter spikelets (averaging 8, rather than 10, mm), more acute glumes and lemmas, blunter rudiments, and in being restricted to the lower Sierra Nevada; the two varieties intergrade, both morphologically and geographically.

Boyle (1945) obtained vigorous sterile hybrids from crosses between *M. californica* and *M. imperfecta*, but found no natural hybrids.

7. Melica harfordii Bol. [p. 94]
HARFORD MELIC

Plants cespitose, not rhizomatous. **Culms** 35–120 cm, not forming corms; **internodes** smooth. **Sheaths** glabrous or pilose, often most pilose at the throat and collar; **ligules** 0.5–1.5 mm; **blades** 1.5–4.5 mm wide, abaxial surfaces smooth, adaxial surfaces scab-ridulous, glabrous or puberulent. **Panicles** 6–25 cm; **branches** 3–8 cm, appressed, with 2–6 spikelets; **pedicels** straight; **disarticulation** above the glumes. **Spikelets** 7–20 mm, with 2–6 bisexual florets; **rachilla internodes** 2–2.4 mm. **Glumes** obtuse to subacute; **lower glumes** 4–10 mm long, 1.5–2.5 mm wide, 3–5-veined; **upper glumes** 5–11 mm long, 1.8–2.5 mm wide, 5–7-veined; **lemmas** 6–16 mm, hairy, hairs to 0.75 mm on the back, 0.7–1.3 mm on the margins, 9–11-veined, veins inconspicuous, apices mucronate to rounded, usually awned, awns 0.5–3 mm, fragile; **paleas** about ³/₄ as long as to nearly equaling the length of the lemmas; **anthers** 3, 2.2–4 mm; **rudiments** 2.5–6 mm, tapering, resembling the bisexual florets. **Caryopses** about 5 mm. $2n = 18$.

Melica harfordii grows primarily in the Pacific coast ranges from Washington to California, as well as in the Sierra Nevada and a few other inland locations, usually on dry slopes or in dry, open woods. The awns in *M. harfordii* often escape attention because they do not always extend beyond the lemma.

8. Melica geyeri Munro Munro [p. 94]
GEYER'S ONIONGRASS

Plants cespitose, rhizomatous. **Culms** 65–200 cm, glabrous, forming corms, corms sessile on the rhizomes; **internodes** smooth. **Sheaths** scabridulous to scabrous, sometimes sparsely pilose, particularly at the throat and collar; **ligules** 0.8–5 mm; **blades** 2–8 mm wide, abaxial surfaces scabridulous, adaxial surfaces with hairs. **Panicles** 10–30 cm; **branches** 3–11 cm, divergent to reflexed, flexuous, with 1–6 spikelets; **pedicels** straight; **disarticulation** above the glumes. **Spikelets** 8–24 mm, with 4–7 bisexual florets, base of the distal florets exposed at anthesis; **rachilla internodes** 2–3 mm, not swollen when fresh, not wrinkled when dry. **Glumes** usually less than ¹/₂ the length of the spikelets; **lower glumes** 3.5–7 mm long, 1.5–2 mm wide, 5–9-veined; **upper glumes** 5–11 mm long, 2–2.5 mm wide, 5–11-veined; **lemmas** 7.5–12.5 mm, glabrous or scabrous, 7-veined, veins inconspicuous, apices rounded to acute, sometimes toothed, unawned or awned, awns to 2 mm; **paleas** about as long as the lemmas; **anthers** 3, 2.5–4 mm; **rudiments** 3–7 mm, tapering, resembling the bisexual florets. **Caryopses** 3–4 mm. $2n = 18$.

Melica geyeri grows to 2000 m, primarily in dry, open woods, in Oregon and California. Its large size and open panicle distinguish *M. geyeri* from most other North American species of *Melica*.

1. Lemma apices awned, awns 0.5–2 mm long
. . . . var. *aristulata*
1. Lemma apices unawned . . . . var. *geyeri*

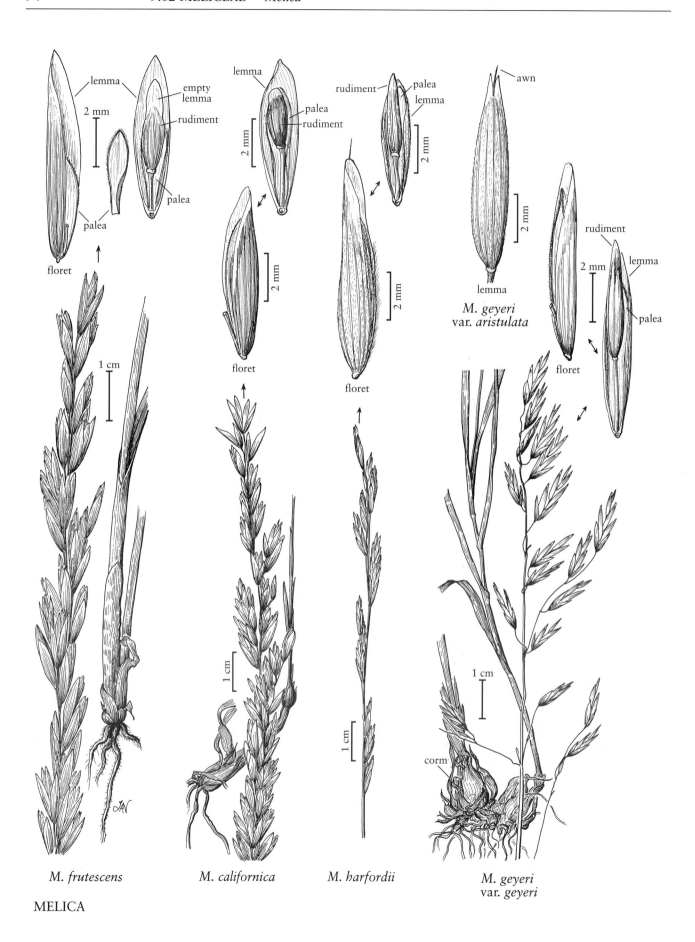

MELICA

Melica geyeri var. **aristulata** J.T. Howell [p. 94]

Lemmas toothed, awned from between the teeth, awns 0.5–2 mm.

Melica geyeri var. *aristulata* grows in Marin, and possibly Shasta, counties in California.

Melica geyeri Munro var. **geyeri** [p. 94]

Lemmas apices obtuse, unawned.

Melica geyeri var. *geyeri* grows throughout the range of the species.

9. **Melica aristata** Thurb. *ex* Bol. [p. 96]
AWNED MELIC

Plants cespitose, not rhizomatous. **Culms** 40–120 cm, not forming corms; **internodes** smooth. **Sheaths** glabrous, scabrous, sometimes sparsely pilose; **ligules** 2.5–5 mm; **blades** 5.5–15 cm long, 2–6 mm wide, often sparsely pilose on both surfaces. **Panicles** 10–26 cm; **branches** 4–6 cm, appressed or strongly ascending, with 1–4 spikelets per branch; **pedicels** not sharply bent; **disarticulation** above the glumes. **Spikelets** 11–21 mm, with (2)3–5 bisexual florets; **rachilla internodes** 3.4–3.8 mm. **Lower glumes** 9–11 mm long, 1.5–2.5 mm wide, 3–5-veined; **upper glumes** 11–12 mm long, 2–3 mm wide, 5–7-veined; **lemmas** 8–13 mm, with 0.3–0.6 mm hairs on the marginal veins, glabrous or with hairs to 0.1 mm elsewhere, 5–7-veined, veins prominent, apices bifid to emarginate, awned from the sinuses, awns 5–12 mm; **paleas** about ³/₄ the length of the lemmas; **anthers** 2, 2–3 mm; **rudiments** 2.5–6 mm, tapering, resembling the bisexual florets. **Caryopses** 5–6 mm. $2n = 18$.

Melica aristata grows from 1000–3000 m in open fir and pine woods. It is restricted to the *Flora* region, being native from Washington to southern California. It has also been found in Kentucky, possibly as an introduction from contaminated seed. *Melica aristata* is easily distinguished from most species of *Melica* by its conspicuous awns.

10. **Melica smithii** (Porter *ex* A. Gray) Vasey [p. 96]
SMITH'S MELIC

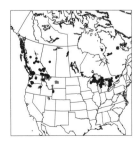

Plants loosely cespitose, not rhizomatous. **Culms** 60–160 cm, thickened basally, sometimes appearing cormous; **internodes** sometimes pubescent below the nodes. **Sheaths** usually glabrous, sometimes pilose or retrorsely scabrous, particularly at the throat, veins often prominent; **ligules** 2–4 mm; **blades** 15–25 cm long, 5–12 mm wide, both surfaces usually scabridulous, glabrous, sometimes the adaxial surfaces with hairs. **Panicles** 12–30 cm; **branches** 7–11 cm, spreading to reflexed, with 4–7 spikelets, spikelets restricted to the distal portion, axils frequently with brownish pulvini; **pedicels** straight; **disarticulation** above the glumes. **Spikelets** 12–18 mm, with 3–5 bisexual florets; **rachilla internodes** 2.5–3 mm. **Lower glumes** 4.5–7 mm long, 1–1.5 mm wide, 1–3-veined; **upper glumes** 6.5–9 mm long, 1.2–1.8 mm wide, 3–5-veined; **lemmas** 9.5–12 mm, glabrous or scabrous, 7-veined, apices bifid to emarginate, awned, awns 3–10 mm; **paleas** about ²/₃ the length of the lemmas; **anthers** 1.3–2.5 mm; **rudiments** 3.5–6 mm, tapering, resembling the bisexual florets. $2n$ = unknown.

Melica smithii grows in cool, moist woods from British Columbia and Alberta south to Oregon and Wyoming and, as a disjunct, from the Great Lakes region to western Quebec. It often forms colonies in the eastern portion of its range. Its disjunct distribution pattern is unusual among North America's grasses.

11. **Melica subulata** (Griseb.) Scribn. [p. 96]
ALASKAN ONIONGRASS, TAPERED ONIONGRASS

Plants cespitose, rhizomatous. **Culms** 55–125 cm, forming corms, corms attached to the rhizomes; **internodes** scabridulous basally. **Sheaths** usually scabridulous, sometimes glabrous or pilose; **ligules** 0.4–5 mm, to 1.5 mm on the lower leaves, to 5 mm on the upper leaves; **blades** 2–10 mm wide, abaxial surfaces smooth or scabridulous, adaxial surfaces scabridulous, glabrous or with hairs. **Panicles** 8–25 cm, lax; **branches** 1.7–9 cm, usually appressed to ascending, occasionally divergent, with 1–5 spikelets; **pedicels** not sharply bent; **disarticulation** above the glumes. **Spikelets** 10–28 mm, with 2–5 bisexual florets; **rachilla internodes** 1.8–2 mm. **Lower glumes** 4–8 mm long, 1.3–2.2 mm wide, 1–3-veined; **upper glumes** 5.5–11.5 mm long, 2–3 mm wide, 3–5-veined; **lemmas** 5.5–18 mm, usually strigose over the veins, hairs longest towards the base, 7–9-veined, veins prominent, apices strongly tapering and acuminate, unawned; **paleas** ¹/₂–³/₄ the length of the lemmas; **anthers** 1.5–2.5 mm; **rudiments** 4–9 mm, tapering, resembling the bisexual florets. **Caryopses** 4–5 mm. $2n = 18$.

Melica subulata grows from sea level to 2300 m in mesic, shady woods. Its range extends from the Aleutian Islands of Alaska through British Columbia to California, east to Lawrence County, South Dakota, and into Colorado.

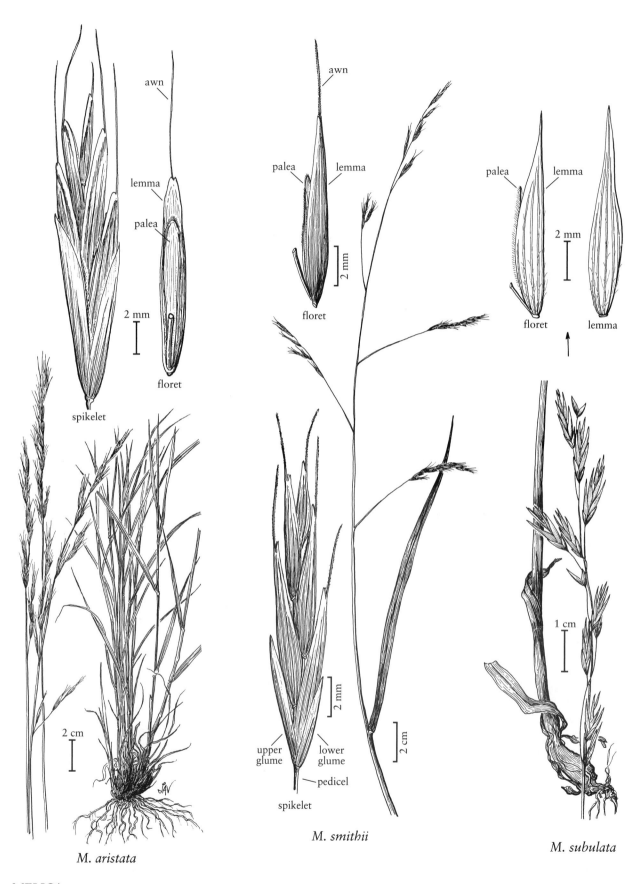

awn

lemma

palea

2 mm

floret

spikelet

M. aristata

awn

palea   lemma

2 mm

floret

upper   lower
glume   glume

pedicel

spikelet

M. smithii

palea   lemma

2 mm

floret   lemma

1 cm

M. subulata

2 cm

MELICA

12. Melica fugax Bol. [p. 97]
LITTLE MELIC

Plants cespitose, not rhizomatous. **Culms** 10–60 cm, forming corms; **internodes** smooth or scabridulous. **Sheaths** scabridulous to scabrous; **ligules** 0.5–2.6 mm; **blades** 1.2–5 mm wide, sometimes pilose on both surfaces. **Panicles** 4.5–18 cm; **branches** 0.8–4 cm, appressed to ascending, with 1–5 spikelets; **pedicels** straight. **Spikelets** 4–17 mm, with 2–5 bisexual florets; **rachilla internodes** 2.1–2.3 mm, swollen when fresh, wrinkled when dry; **disarticulation** above the glumes. **Lower glumes** 3–5 mm long, 1.5–2.5 mm wide, 1–3-veined; **upper glumes** 3.5–7 mm long, 2.5–3.5 mm wide, 5-veined; **lemmas** 4–7 mm, glabrous or scabrous, 4–11-veined, veins inconspicuous, apices rounded to acute, unawned; **paleas** almost as long as the lemmas; **anthers** 3, 1–2 mm; **rudiments** 1.5–3.5 mm, tapering, resembling the bisexual florets. $2n = 18$.

Melica fugax grows at elevations to 2200 m on dry, open flats, hillsides, and woods, from British Columbia to California and east to Idaho and Nevada. It is usually found on soils of volcanic origin, and rarely below 1300 m.

Melica fugax is often confused with *M. bulbosa*, but its rachilla internodes are unmistakable and unique among the species in the *Flora* region, being swollen when fresh and wrinkled when dry. One specimen, *C.L. Hitchcock 15521* [WTU 114265] from Elmore County, Idaho, appears to be a hybrid. It has shrunken caryopses and combines the rachilla of *M. fugax* with the lemma pubescence, size, and overall appearance of *M. subulata*, but lacks corms.

13. Melica stricta Bol. [p. 99]
ROCK MELIC

Plants densely cespitose, not rhizomatous. **Culms** 9–85 cm, not forming corms; **basal internodes** often thickened; **internodes** smooth. **Sheaths** scabridulous; **ligules** 2.5–5 mm; **blades** 1.5–5 mm wide, abaxial surfaces glabrous, scabridulous, adaxial surfaces sometimes strigose, sometimes glabrous or scabridulous. **Panicles** 3–30 cm; **branches** 0.5–10 cm, appressed, with 1–5 spikelets; **pedicels** sharply bent below the spikelets; **disarticulation** below the glumes. **Spikelets** 6–23 mm long, 5–13 mm wide, broadly V-shaped when mature, with 2–4 bisexual florets; **rachilla internodes** 1.8–2.1 mm. **Lower glumes** 6–16 mm long, 3.5–5 mm wide, 4–7-veined; **upper glumes** 6–18 mm long, 3–5 mm wide, 5–9-

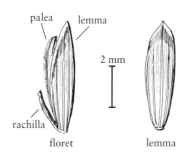

palea lemma

2 mm

rachilla

floret   lemma

1 cm

M. fugax

MELICA

veined; **lemmas** 6–16 mm, glabrous, scabridulous, 5–9-veined, veins inconspicuous, apices acute, unawned; **paleas** $^1/_2$–$^3/_4$ the length of the lemmas; **anthers** 1–3 mm; **rudiments** 2–7 mm, resembling the lower florets, acute to acuminate. **Caryopses** 4–5 mm. $2n = 18$.

Melica stricta grows from 1200–3350 m on rocky, often dry slopes, sometimes in alpine habitats. Its range extends from Oregon and California to Utah. Boyle (1945) recognized two varieties, more on their marked geographical separation than on their morphological divergence.

1. Paleas about $^3/_4$ the length of the lemmas; anthers 2–3 mm long .... var. *albicaulis*
1. Paleas about $^1/_2$ the length of the lemmas; anthers 1–2 mm long .... var. *stricta*

Melica stricta var. **albicaulis** Boyle [p. 99]

Sheaths of the lower portion pale. **Glumes** at least 4 mm wide; **paleas** about $^3/_4$ the length of the lemmas; **anthers** 2–3 mm.

Melica stricta var. *albicaulis* is restricted to the mountains of southern California.

Melica stricta Bol. var. **stricta** [p. 99]

Sheaths usually purplish, becoming dark brown. **Glumes** 3–4 mm wide; **paleas** about $^1/_2$ the length of the lemmas; **anthers** 1–2 mm.

Melica stricta var. *stricta* is the more widespread of the two varieties, growing throughout the range of the species except in the mountains of southern California.

14. Melica porteri Scribn. [p. 99]
PORTER'S MELIC

Plants not or loosely cespitose, shortly rhizomatous. **Culms** 55–100 cm, not forming corms; internodes smooth, basal internodes not thickened. **Sheaths** often scabrous on the keels, otherwise smooth; **ligules** 1–7 mm; **blades** 2–5 mm wide, both surfaces glabrous, scabridulous. **Panicles** 13–25 cm; **branches** 1–9 cm, straight and appressed or flexible and ascending to strongly divergent, with 1–12 spikelets; **pedicels** sharply bent below the spikelets; **disarticulation** below the glumes. **Spikelets** 8–16 mm long, 1.5–5 mm wide, parallel-sided when mature, with 2–5 bisexual florets; **rachilla internodes** 1.9–2.1 mm. **Glumes** green, pale, or purplish-tinged; **lower glumes** 3.5–6 mm long, 2–3 mm wide, 3–5-veined; **upper glumes** 5–8 mm long, 2–3 mm wide, 5-veined; **lemmas** 6–10 mm, glabrous, chartaceous on the distal $^1/_3$, 5–11-veined, veins conspicuous, apices rounded to acute, unawned; **paleas**

about $^2/_3$ the length of the lemmas; **anthers** 1–2.5 mm; **rudiments** 1.8–5 mm, acute to acuminate, resembling the bisexual florets. $2n = 18$.

Melica porteri grows on rocky slopes and in open woods, often near streams. It grows from Colorado and Arizona to central Texas and northern Mexico. Living plants are sometimes confused with *Bouteloua curtipendula*; the similarity is superficial.

1. Panicle branches flexible, ascending to strongly divergent; glumes purplish-tinged ..... var. *laxa*
1. Panicle branches straight, appressed; glumes green or pale .... var. *porteri*

Melica porteri var. **laxa** Boyle [p. 99]

Panicles open; **branches** flexible, ascending to patent. **Glumes** purplish-tinged.

Melica porteri var. *laxa* grows from southern Arizona east to the Chisos Mountains, Texas, and south into northern Mexico.

Melica porteri Scribn. var. **porteri** [p. 99]

Panicles narrow; **branches** straight, appressed. **Glumes** green or pale.

Melica porteri var. *porteri* grows from northern Colorado to Arizona and central Texas, and south to the Sierra Madre Occidental, Mexico.

15. Melica montezumae Piper [p. 101]
MONTEZUMA MELIC

Plants cespitose, not rhizomatous. **Culms** 14–100 cm, not forming corms; **internodes** smooth. **Sheaths** glabrous or scabrous; **ligules** 2.5–7 mm; **blades** 1.2–3 mm wide, abaxial surfaces glabrous, scabridulous, adaxial surfaces puberulent. **Panicles** 5–25 cm; **branches** 1–5 cm, appressed to reflexed, straight, with 2–9 spikelets; **pedicels** sharply bent below the spikelets; **disarticulation** below the glumes. **Spikelets** 6–8 mm, with 1 bisexual floret. **Lower glumes** 5.5–8 mm long, 1.8–3 mm wide, 5-veined; **upper glumes** 5–8 mm long, 0.7–1.5 mm wide, 3–5-veined; **lemmas** 4.5–8 mm, 9–15-veined, veins prominent, tuberculate, proximal portion with flat, twisted hairs, distal portion glabrous, chartaceous, apices emarginate to acute, unawned; **paleas** about $^3/_4$ the length of the lemmas; **anthers** 1.5–3 mm; **rudiments** 2–3 mm, obovoid or obconic, clublike, not resembling the bisexual florets. $2n = 18$.

Melica montezumae grows primarily in shady locations in the mountains of western Texas and adjacent Mexico.

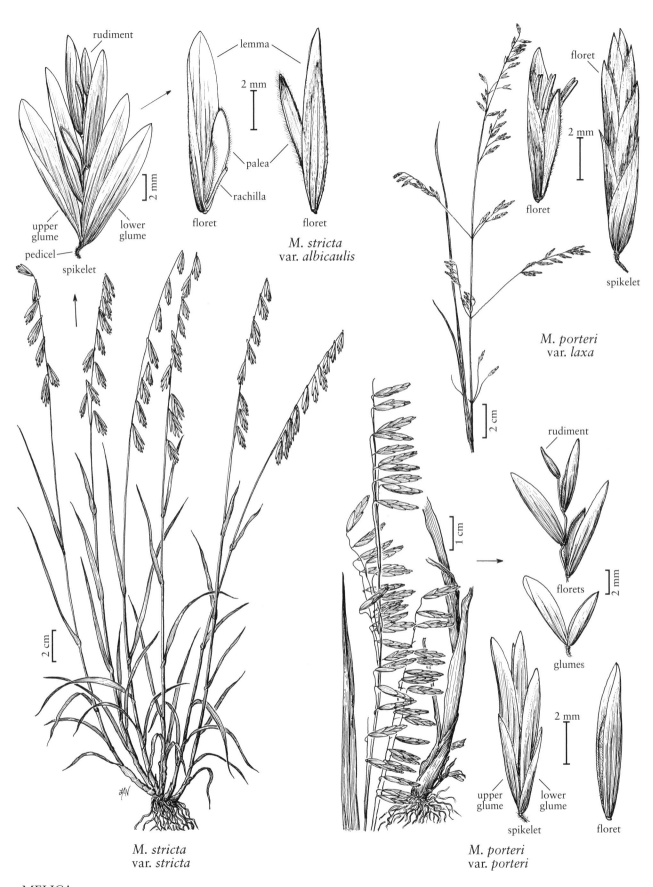

rudiment

lemma

2 mm

palea

rachilla

floret   floret

upper glume

lower glume

pedicel

spikelet

M. stricta
var. *albicaulis*

floret

2 mm

floret

spikelet

M. porteri
var. *laxa*

2 cm

2 mm

2 cm

1 cm

rudiment

florets

2 mm

glumes

upper glume

lower glume

2 mm

spikelet

floret

M. stricta
var. *stricta*

M. porteri
var. *porteri*

MELICA

16. Melica mutica Walter [p. 101]
TWO-FLOWER MELIC

Plants not or loosely cespitose, shortly rhizomatous. **Culms** 45–100 cm, not forming corms; **internodes** sometimes scabridulous above the nodes. **Sheaths** glabrous or pilose; **ligules** 0.5–1.5 mm; **blades** 1.8–6 mm wide, abaxial surfaces glabrous, scabridulous, adaxial surfaces with hairs. **Panicles** 4–25 cm; **branches** 3.5–6 cm, appressed to spreading, straight, with 2–5 spikelets; **pedicels** sharply bent below the spikelets; **disarticulation** below the glumes. **Spikelets** 6–11 mm, with (1)2(4) bisexual florets, floret apices at about the same level; **rachilla internodes** 1.5–1.7 mm. **Lower glumes** 4.5–8 mm long, 3–4 mm wide, 5–7-veined; **upper glumes** 5–9 mm long, 2.5–3.5 mm wide, 5–6-veined; **lemmas** 6–11 mm, glabrous or scabrous, indurate, 9–11-veined, veins prominent, apices rounded to acute, unawned; **paleas** about ³/₄ the length of the lemmas; **anthers** 1–3 mm; **rudiments** 2–3 mm, clublike, not resembling the bisexual florets, at a sharp angle to the rachilla. $2n = 18$.

Melica mutica grows in moist or dry areas in open woods and thickets, from Iowa and Texas east to Maryland and Florida. It is unique among the North American species in having a clublike rudiment at a sharp angle to the rachilla.

17. Melica nitens (Scribn.) Nutt. *ex* Piper [p. 101]
THREE-FLOWER MELIC

Plants not or loosely cespitose, shortly rhizomatous. **Culms** 55–130 cm, not forming corms; **internodes** smooth. **Sheaths** glabrous or scabridulous; **ligules** 1–6.5 mm; **blades** 3.5–11 mm wide, flat, abaxial surfaces smooth or scabridulous, adaxial surfaces scabridulous. **Panicles** 9–26 cm; **branches** 3.5–6 cm, often divergent to reflexed, straight, with 5–20 spikelets; **pedicels** sharply bent and hairy below the spikelets; **disarticulation** below the glumes. **Spikelets** 8–12 mm, with 2–3(4) bisexual florets, apices of the lowest 2 florets not at the same level; **rachilla internodes** 2.3–2.4 mm. **Glumes** unequal; **lower glumes** 5–9 mm long, 3.5–4.5 mm wide, more ovate than the upper glumes, 3–9-veined; **upper glumes** 6–11 mm long, 2.5–3.5 mm wide, 3–7-veined; **lemmas** 6.5–11.5 mm, glabrous or scabrous, somewhat indurate, with 9+ veins, veins prominent, apices rounded, unawned; **paleas** about ³/₄ the length of the lemmas; **anthers** 1.7–3.2 mm; **rudiments** 2–3 mm, clublike, not resembling the bisexual florets, in a straight line with the rachilla. $2n = 18$.

Melica nitens grows in dry to moist woodlands, often in rocky areas with rich soil. It grows primarily from Minnesota to Pennsylvania and southwest to Texas.

18. Melica ciliata L. [p. 102]
CILIATE MELIC, SILKY-SPIKE MELIC, HAIRY MELIC

Plants cespitose, sometimes shortly rhizomatous. **Culms** 20–60(100) cm, not forming corms. **Sheaths** glabrous or shortly and sparsely pubescent; **ligules** 1–4 mm; **blades** 7–15 cm long, 1–4 mm wide, usually involute. **Panicles** 4–8(25) cm, narrowly cylindrical, lax, pale; **branches** 1.5–4 cm, appressed to ascending, with 3–12(15) spikelets; **pedicels** sharply bent below the spikelets; **disarticulation** below the glumes. **Spikelets** 6–8 mm, with 1 bisexual floret, sometimes purple-tinged. **Lower glumes** 4–6 mm long, 1.5–2.5 mm wide, ovate, 1–5-veined, acute; **upper glumes** 6–8 mm long, about 1.5 mm wide, lanceolate, acute to acuminate; **lemmas** 4–6.5 mm, lanceolate, 7–9-veined, papillose, margins and marginal veins pubescent, hairs 3.5–5 mm, not twisted; **rudiments** 1–1.7 mm, ovoid, not resembling the bisexual florets. $2n = 18, 36$.

Melica ciliata is grown as an ornamental in North America and is not known to have escaped. It is native to Europe, northern Africa, and southwestern Asia, where it grows on damp to somewhat dry soils.

19. Melica altissima L. [p. 102]
TALL MELIC, SIBERIAN MELIC

Plants loosely cespitose. **Culms** 60–250 cm, not forming corms, scabrous below the panicles. **Sheaths** retrorsely scabridulous; **ligules** 3–5 mm; **blades** to 20 cm long, 5–15 mm wide, flat, lax. **Panicles** 10–20 cm long, 1–2(5) cm wide, cylindrical, pale or purplish; **branches** about 3 cm, strongly ascending to appressed, often with 15+ spikelets; **pedicels** sharply bent below the spikelets; **disarticulation** below the glumes. **Spikelets** 7–11 mm, with 1–2(3) bisexual florets. **Glumes** subequal in length and similar in shape, 7–10.5 mm long, 3–4 mm wide, glabrous, ovate-elliptic, obtuse to acute, ivory or purple, 7-veined; **lemmas** 7–11 mm, glabrous, scabridulous, 9–13-veined, scarious, apices acute; **paleas** about ²/₃ the length of the lemmas; **rudiments** 2.5–3 mm, pyriform. **Caryopses** about 3 mm. $2n = 18$.

Melica altissima is native to Eurasia. It is grown as an ornamental in North America and is reported to have escaped and become established in Oklahoma and Ontario. In its native region, it grows on the moist soils of shrubby thickets and forest edges, and on rocky slopes. Plants with dark purple glumes and lemmas can be called *M. altissima* var. *atropurpurea* Host.

M. nitens

upper glume   lower glume
spikelet

lemma
rudiment
2 mm
palea
floret

rudiment   lemma
2 mm
0.5 mm
floret

M. montezumae

1 cm

floret
2 mm

rudiment

pedicel   spikelet
2 mm

2 cm

M. mutica

MELICA

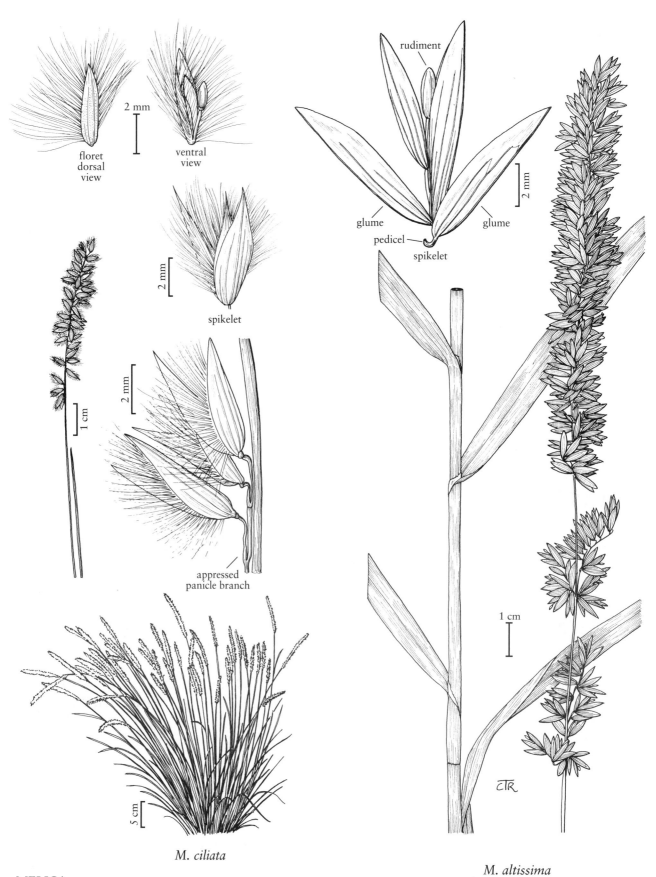

floret
dorsal
view

2 mm

ventral
view

2 mm

spikelet

2 mm

appressed
panicle branch

1 cm

5 cm

M. ciliata

rudiment

2 mm

glume

pedicel

glume

spikelet

1 cm

CTR

M. altissima

MELICA

9.03 SCHIZACHNE Hack.

Jacques Cayouette

Stephen J. Darbyshire

Plants perennial; loosely cespitose. **Culms** 30–110 cm, glabrous, often decumbent at the base; **nodes** glabrous, becoming dark. **Sheaths** closed almost to the top; **ligules** membranous, margins often united in front; **blades** folded or loosely involute, glabrous or pilose. **Inflorescences** panicles or racemes, with 4–20 spikelets; **branches** straight and appressed to lax and drooping. **Spikelets** slightly laterally compressed, with 3–6 florets; **disarticulation** above the glumes and beneath the florets. **Glumes** exceeded by the lowest lemma in each spikelet, chartaceous, often anthocyanic below, the upper $^1/_3$ hyaline; **calluses** rounded, with hairs; **lemmas** chartaceous, slightly scabrous, 7–9-veined, veins parallel, conspicuous, apices scarious, bifid, awned from below the teeth, awns 8–15 mm, divergent or slightly geniculate; **paleas** shorter than the lemmas, 2-veined, veins ciliate, keeled; **lodicules** truncate; **anthers** 3; **ovaries** glabrous. **Caryopses** 3.2–3.8 mm, smooth, shiny, falling free of the lemma and palea. $x = 10$. Name from the Greek *schizo*, 'split', and *achne*, 'chaff', alluding to the split (bifid) lemma.

Schizachne is a monospecific genus that extends across North America in boreal regions and southwards in the montane areas, as well as from the Ural Mountains of Russia to Kamchatka and Japan.

SELECTED REFERENCES **Koyama, T.** and **S. Kawano**. 1964. Critical taxa of grasses with North American and eastern Asiatic distribution. Canad. J. Bot. 42:859–884; **McNeill, J.** and **W.G. Dore**. 1976. Taxonomic and nomenclatural notes on Ontario grasses. Naturaliste Canad. 103:553–567; **Tsvelev, N.N.** 1976. Zlaki SSSR. Nauka, Leningrad [St. Petersburg], Russia. 788 pp.

1. Schizachne purpurascens (Torr.) Swallen [p. 104]
FALSE MELIC, SCHIZACHNÉ POURPRÉ

Culms (30)50–80(110) cm, sometimes slightly decumbent at the base, otherwise erect. **Ligules** 0.5–1.5 mm; **blades** 2–4(5) mm wide, glabrous or adaxial surfaces pilose. **Panicles** 7–13(17) cm, open or closed, often reduced to racemes in depauperate plants. **Spikelets** 11.5–17 mm. **Glumes** glabrous, acute; **lower glumes** 4.2–6.2 mm, faintly (1)3–5-veined; **upper glumes** 6–9 mm, faintly (3)5-veined; **lemmas** 8–10.5(12) mm; **awns** 8–15 mm, as long as or longer than the lemma bodies, somewhat twisted and divergent or slightly geniculate; **anthers** 1.4–2 mm. $2n = 20$.

In North America, *Schizachne purpurascens* grows in moist to mesic woods, from south of the tree line in Alaska and northern Canada through the Rocky Mountains to New Mexico in the west, and to Kentucky and Maryland in the east.

9.04 PLEUROPOGON R. Br.

Paul P.H. But

Plants annual or perennial; cespitose or rhizomatous. **Culms** 5–160 cm, erect or geniculate at the base, glabrous; **basal branching** extravaginal. **Sheaths** closed almost to the top; **ligules** membranous; **blades** flat to folded, adaxial surfaces with prominent midribs. **Inflorescences** terminal, racemes, rarely panicles. **Spikelets** laterally compressed, with 5–20(30) florets, upper florets reduced; **disarticulation** above the glumes and beneath the florets. **Glumes** unequal to subequal, shorter than the adjacent lemmas, membranous to subhyaline, margins scarious; **lower glumes** 1-veined; **upper glumes** 1–3-veined; **rachilla internodes** in some species swollen and glandular basally, the glandular portion turning whitish when dry; **calluses** rounded, glabrous; **lemmas** thick, herbaceous to membranous, 7(9)-veined, veins parallel, margins

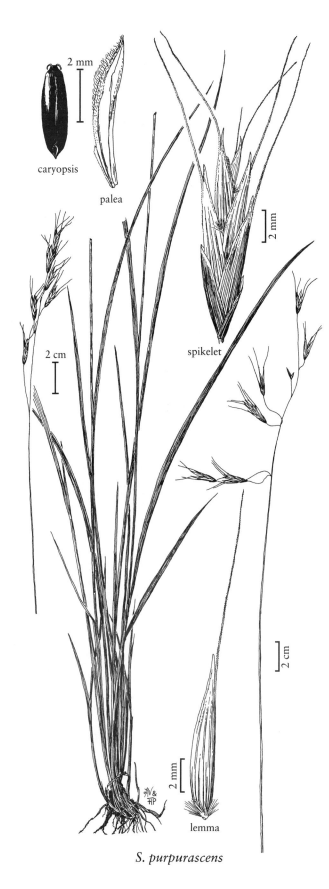

caryopsis

palea

2 mm

2 mm

spikelet

2 cm

2 cm

2 mm

lemma

S. purpurascens

SCHIZACHNE

scarious, apices scarious, entire or emarginate, midvein sometimes extended into an awn, awns straight; **paleas** subequal to the lemmas, 2-veined, keeled over each vein, keels winged, with 1 or 2 awns or a flat triangular appendage; **lodicules** 2, completely fused; **anthers** 3, opening by pores; **ovaries** glabrous. x = 8, 9, 10. Name from the Greek, *pleura*, 'side', and *pogon*, 'beard', a reference to the awns on the sides of the palea in some species.

Pleuropogon is a genus of five hydrophilous species, one circumboreal in the arctic, the other four restricted to the Pacific coast of North America, extending from southern British Columbia to central California. The Pacific coast species are sometimes treated as a separate genus, *Lophochlaena* Nees, but are here regarded as constituting a subgenus of *Pleuropogon*.

The flat, triangular paleal appendages differ from bristly or flattened awns in being wider at the base, and smooth rather than scabrous.

SELECTED REFERENCES **But, P.P.H.** 1977. Systematics of *Pleuropogon* R. Br. (Poaceae). Ph.D. dissertation, University of California, Berkeley, California, U.S.A. 229 pp.; **But, P.P.H., J. Kagan, V. Crosby,** and **J.S. Shelly.** 1985. Rediscovery and reproductive biology of *Pleuropogon oregonus* (Poaceae). Madroño 32:189–190.

1. Paleal keels each with 2 awns, the lower awn 1–3 mm long, the upper awn 0.3–1 mm long; lemmas 3.5–5 mm long; plants of the arctic (subg. *Pleuropogon*) . . . . 5. *P. sabinei*
1. Paleal keels each with 1 awn 3–9 mm long, or a triangular appendage; lemmas 4.5–10 mm long; plants of the Pacific Northwest and California (subg. *Lophochlaena*).
  2. Lowest lemma in each spikelet 4.5–7.5 mm long; culms 15–95 cm tall; caryopses 2.5–3.1 mm long.
    3. Paleal keels unawned, with a triangular appendage; rhizomes absent or poorly developed; rachilla internodes with a glandular swelling at the base . . . . 1. *P. californicus*
    3. Paleal keels with an awn 3–9 mm long, without a triangular appendage; rhizomes strongly developed; rachilla internodes without a glandular swelling at the base . . . . 3. *P. oregonus*
  2. Lowest lemma in each spikelet 8–10 mm long; culms mostly 100–160 cm tall; caryopses 3.5–6 mm long.
    4. Lemma awns 0.2–4 mm long; pedicels usually erect, rarely reflexed, the spikelets erect or ascending at maturity . . . . 2. *P. hooverianus*
    4. Lemma awns (5)9–20 mm long; pedicels reflexed, the spikelets pendent at maturity . . . . 4. *P. refractus*

1. **Pleuropogon californicus** (Nees) Benth. *ex* Vasey [p. 106]

CALIFORNIA SEMAPHOREGRASS

Plants annual or perennial; cespitose, with many innovations, not rhizomatous, or with poorly developed rhizomes. **Culms** 15–95 cm, erect or geniculate at the base, sometimes rooting at the lower nodes; **nodes** 3–6. **Sheaths** glabrous; **ligules** 2–6 mm; **blades** 3–29 cm long, 3–8 mm wide, adaxial surfaces slightly scabridulous. **Racemes** 8–35 cm, with 6–13 spikelets; **internodes** 1–6 cm; **pedicels** 1–5.5(9) mm, ascending to spreading. **Spikelets** 15–60 mm, ascending to spreading, with 7–20(30) florets, lower florets bisexual, upper florets pistillate, terminal florets sterile. **Glumes** lanceolate to broadly ovate; **lower glumes** 1–4.5 mm; **upper glumes** 2–6.5 mm, 1–3-veined; **rachilla internodes** 1–2.8 mm long, 0.2–0.5 mm thick, with a glandular swelling at the base; **lemmas** 4.5–7.5 mm, 7(9)-veined, veins distinct, often prominent, apices truncate or emarginate, unawned or awned, awns to 11 mm; **paleal keels** scabridulous, each with a 0.5–2.5 mm triangular, wing-like appendage, often denticulate beyond the appendage; **anthers** 2.5–4 mm. **Caryopses** 2.5–3.1 mm.

Pleuropogon californicus is a Californian endemic with two varieties.

1. Plants annual or facultative perennials; lemmas usually with awns 5–11 mm long, rarely unawned; paleal appendages 0.5–2.5 mm long; spikelets 15–30 mm long . . . . . var. *californicus*
1. Plants perennial; lemmas unawned, sometimes mucronate, mucros to 1.5 mm long; paleal appendages 0.5–1 mm long; spikelets 25–60 mm long . . . . var. *davyi*

upper glume

lower glume

spikelet

rachilla

floret

P. californicus
var. *californicus*

2 mm

rachilla

floret

P. californicus
var. *davyi*

palea

lemma

rachilla

floret

P. hooverianus

PLEUROPOGON

Pleuropogon californicus (Nees) Benth. *ex* Vasey var. **californicus** [p. 106]
ANNUAL SEMAPHOREGRASS

Plants annual or facultative perennials. **Spikelets** 15–30 mm. **Lemmas** usually awned, awns 5–11 mm, rarely unawned; **paleal appendages** 0.5–2.5 mm. $2n = 16$.

Pleuropogon californicus var. *californicus* grows in vernal pools, marshy grasslands, orchards, and roadside ditches in California, from southern Humboldt County south to San Luis Obispo County, and east to Amador County.

Pleuropogon californicus var. **davyi** (L.D. Benson) But [p. 106]
DAVY'S SEMAPHOREGRASS

Plants perennial. **Spikelets** 25–60 mm. **Lemmas** unawned, sometimes mucronate, mucros to 1.5 mm; **paleal appendages** 0.5–1 mm. $2n = 16$.

Pleuropogon californicus var. *davyi* is the more restricted of the two varieties, being known only from vernal pools, sloughs, and marshy grasslands in Mendocino and Lake counties, California.

2. Pleuropogon hooverianus (L.D. Benson) J.T. Howell [p. 106]
HOOVER'S SEMAPHOREGRASS

Plants perennial; not cespitose, rhizomatous. **Culms** 1–1.6 m, erect. **Sheaths** glabrous, retrorsely scabridulous; **ligules** 3–6.5 mm; **blades** 3–30 cm long, 4–10 mm wide, apices acute to acuminate, mucronate, flag leaves often with reduced spinose blades. **Racemes** 21–33 cm, with 7–10 spikelets; **internodes** 1.8–8 cm; **pedicels** 1.5–5 mm, erect or ascending, rarely reflexed. **Spikelets** 28–42 mm, erect or ascending, with 9–16 florets, usually all but the terminal floret bisexual. **Lower glumes** 3–5.6 mm; **upper glumes** 4.5–7.2 mm, 1–3-veined; **rachilla internodes** 2–3 mm long, about 0.4 mm thick, basal ¹/₂ developing into a glandular swelling; **lemmas** 8–9 mm, 7-veined, lateral veins strongly ribbed, apices toothed, usually rounded, rarely acute or erose, awned, awns 0.2–4 mm; **paleal keels** unawned, each with a 0.6–1.5 mm triangular appendage; **anthers** 4–4.8 mm. **Caryopses** 3.5–4 mm. $2n = 16, 36$.

Pleuropogon hooverianus grows in wet and marshy areas, usually in shady locations. Several of the populations are around redwood groves. It is known only from Mendocino, Sonoma, and Marin counties in California. It is listed as rare by the state of California.

3. Pleuropogon oregonus Chase [p. 108]
OREGON SEMAPHOREGRASS

Plants perennial; not cespitose, rhizomatous. **Culms** 40–95 cm tall, 2–3.5 mm thick, erect. **Sheaths** glabrous, smooth or scabridulous; **ligules** 5–10 mm, rounded or acute, often erose; **blades** 5–17 cm long, 4–9 mm wide, smooth or scabridulous over the veins, apices spinose. **Racemes** 13–20 cm, with 6–7 spikelets; **lower internodes** 3.5–7.2 cm; **upper internodes** shorter; **pedicels** 2–5(12) mm. **Spikelets** 20–40(50) mm, with 7–14 florets, lower florets bisexual, upper florets pistillate, terminal florets usually sterile. **Glumes** lanceolate to ovate, acute, erose; **lower glumes** 2–3 mm; **upper glumes** 2.5–4.5 mm; **rachilla internodes** 2–3 mm long, 0.1–0.2 mm thick, without a glandular swelling at the base; **lemmas** 5.5–7 mm, scabridulous, 7-veined, veins prominent, apices truncate, sometimes erose, awned, awns 5–12 mm; **paleal keels** awned, awns 3–9 mm, inserted ¹/₃–¹/₂ of the way from the base; **anthers** about 4 mm. **Caryopses** 2.5–3 mm. $2n$ = unknown.

Pleuropogon oregonus grows in swampy ground, wet meadows, and stream banks. It is known, even historically, from only a few locations in Union and Lake counties, Oregon. In 1975 it was thought to be extinct, but a population has since been discovered at one location in Lake County. The species is listed as threatened by the state of Oregon.

4. Pleuropogon refractus (A. Gray) Benth. *ex* Vasey [p. 108]
NODDING SEMAPHOREGRASS

Plants perennial; not cespitose, rhizomatous. **Culms** (85)100–150 cm, erect. **Sheaths** glabrous or pubescent, sometimes scabridulous; **ligules** 2–7 mm; **blades** 10–40 cm long, 5–14 mm wide, apices acute or acuminate and sharply mucronate, flag leaves often reduced to spinose tips. **Racemes** (10)20–35 cm, with 6–14 spikelets; **internodes** 2–5.6(8) cm; **pedicels** 2–3 mm long, reflexed at maturity. **Spikelets** (20)25–28 mm, pendent, with 7–14 florets, usually all but the terminal floret bisexual. **Glumes** frequently scabrous, apices obtuse, sometimes erosely notched; **lower glumes** 3–6 mm; **upper glumes** 4–7(8.3) mm, 3-veined; **rachilla internodes** 2–3 mm long, about 0.1 mm thick, bases sometimes weakly differentiated into a glandular swelling; **lemmas** 8–10 mm, faintly 7(9)-veined, lateral veins occasionally

P. oregonus

P. refractus

P. sabinei

PLEUROPOGON

prominent, apices truncate, sometimes toothed, awned, awns (5)9–20 mm; **paleal keels** each with a 0.2–0.6(1) mm triangular appendage; **anthers** 3.5–4 mm. Caryopses 4.5–6 mm. 2*n* = 32, 36.

Pleuropogon refractus grows in wet meadows, riverbanks, and shady places, from sea level to about 1000 m. Its range extends from British Columbia south to California.

5. Pleuropogon sabinei R. Br. [p. 108]
FALSE SEMAPHOREGRASS, PLEUROPOGON DE SABINE

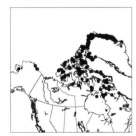

Plants perennial; not cespitose, rhizomatous. **Culms** 5–35 cm tall, 1–3 mm thick. **Sheaths** glabrous; **ligules** 1.5–3.5 mm; **blades** 2–35(50) cm long, 1.5–3 mm wide, often floating, sometimes scabridulous on the midribs and margins, apices acute. **Racemes** 2.8–10 cm, with 5–8 spikelets; internodes 4–30 mm; **pedicels** 1.5–3 mm long. **Spikelets** 10–19 mm, with 5–12 florets, lower florets bisexual, upper florets pistillate. **Lower glumes** 1–2.5 mm; **upper glumes** 2–3.5 mm; **rachilla internodes** 1–1.5(2) mm long, 0.1–0.2 mm thick, without glandular swellings; **lemmas** 3.5–5 mm, 7(9)-veined, densely scabridulous, apices truncate to rounded, entire, emarginate, unawned or awned, awns 0.2–1 mm; **paleal keels** scabridulous, winged, each with 2 awns, awns flattened, purple, scabridulous, lower awn 1–3 mm, upper awn 0.3–1 mm; **anthers** about 2 mm, occasionally to 2.6 mm. **Caryopses** 2.5–3 mm. 2*n* = 40, 42.

Pleuropogon sabinei grows in open, wet places, frequently partially submerged, around lakes, ponds, marshy areas, and riverbanks. Its range extends from eastern Siberia and the Altai Mountains to northern Alaska, Canada, and Greenland.

10. STIPEAE Dumort.

Mary E. Barkworth

Plants usually perennial; usually tightly to loosely cespitose, sometimes rhizomatous. **Culms** annual or perennial, not woody, branches 1 to many at the upper nodes. **Leaves** basally concentrated to evenly distributed; **sheaths** open, margins not fused, sometimes ciliate distally, basal sheaths sometimes concealing axillary panicles (*cleistogenes*), sometimes wider than the blade; **collars** sometimes with tufts of hair at the sides extending to the top of the sheaths; **auricles** absent; **ligules** scarious, often ciliate, cilia usually shorter than the base, ligules of the lower and upper cauline leaves sometimes differing in size and vestiture; **pseudopetioles** absent; **blades** linear to narrowly lanceolate, venation parallel, cross venation not evident, cross sections non-Kranz, without arm or fusoid cells; **epidermes** of adaxial surfaces sometimes with unicellular microhairs, cells not papillate. **Inflorescences** usually terminal panicles, occasionally reduced to racemes in depauperate plants, sometimes 2–3 panicles developing from the highest cauline node. **Spikelets** usually with 1 floret, sometimes with 2–6 florets, laterally compressed to terete; **rachillas** not prolonged beyond the base of the floret in spikelets with 1 floret, prolonged beyond the base of the distal floret in spikelets with 2–6 florets, prolongation hairy, hairs 2–3 mm; **disarticulation** above the glumes and beneath the florets. **Glumes** usually exceeding the floret(s), always longer than ¼ the length of the adjacent floret, 1–10-veined, narrowly lanceolate to ovate, hyaline or membranous, flexible; **florets** usually terete, sometimes laterally or dorsally compressed; **calluses** usually well-developed, rounded or blunt to sharply pointed, often antrorsely strigose; **lemmas** lanceolate, rectangular, or ovate, membranous to coriaceous or indurate, 3–5-veined, veins inconspicuous, apices entire, bilobed, or bifid, awned, lemma-awn junction usually conspicuous, awns 0.3–30 cm, not branched, usually terminal and centric or eccentric, sometimes subterminal, caducous to persistent, not or once- to twice-geniculate, if geniculate, proximal segment(s) twisted, distal segment straight, flexuous, or curled, not or scarcely twisted; **lodicules** 2 or 3; **anthers** 1 or 3, sometimes differing in length within a floret; **ovaries** glabrous throughout or pubescent distally; **styles** 2(3–4)-branched. **Caryopses** ovoid to fusiform, not beaked, pericarp thin; **hila** linear; **embryos** less than $^1/_3$ the length of the caryopses. x = 7, 8, 10, 11, 12.

The tribe *Stipeae* includes about 15 genera and approximately 500 species. It grows in Africa, Australia, South and North America, and Eurasia. In Australia, South America, and Asia, it is often the dominant grass tribe over substantial areas. It is not present in southern India, and is represented by only one native species in southern Africa. Most species grow in arid or seasonally arid, temperate regions.

Morphological considerations have led to the *Stipeae* being placed in three different subfamilies (*Poöideae*, *Bambusoideae*, and *Arundinoideae*) in the past, and even to recognition as a subfamily. Molecular data support its treatment as an early diverging lineage within the *Poöideae* (Soreng and Davis 1998; Grass Phylogeny Working Group 2001) that is more closely related to the *Meliceae* than the core poöid tribes.

Decker (1964) suggested including *Ampelodesmos* in the *Stipeae* on the basis of the cross sectional anatomy of its leaf blades. His suggestion is supported, not always strongly, by molecular studies (Soreng and Davis 1998; Grass Phylogeny Working Group 2001; Jacobs et al. 2006). The usual alternative is to treat *Ampelodesmos* as the only genus of a closely related, monospecific tribe, the *Ampelodesmeae* (Conert) Tutin, because it is so distinct from other members of the *Stipeae*, being, for example, the only member of the tribe with more than 1 floret in its spikelets and rachillas that are prolonged beyond the base of the terminal floret in a spikelet.

The lowest chromosome number known in the *Stipeae* is $2n$ = 18 (Prokudin et al. 1977), suggesting that all members of the tribe are ancient polyploids. The wide range of base numbers listed is based on numbers for the various genera. The primary basic chromosome number for the tribe is probably 5 or 6, with higher numbers reflecting ancient euploidy.

The hybrid genus ×*Achnella* is not included in the key; it is treated on p. 169.

SELECTED REFERENCES **Barkworth, M.E.** 1993. North American Stipeae (Gramineae): Taxonomic changes and other comments. Phytologia 74:1–25; **Barkworth, M.E. and J. Everett.** 1987. Evolution in the Stipeae: Identification and relationships of its monophyletic taxa. Pp. 251–264 *in* T.R. Soderstrom, K.W. Hilu, C.S. Campbell, and M.E. Barkworth (eds.). Grass Systematics and Evolution. Smithsonian Institution Press, Washington, D.C., U.S.A. 473 pp.; **Decker, H.F.** 1964. Affinities of the grass genus *Ampelodesmos*. Brittonia 16:76–79; **Grass Phylogeny Working Group.** 2001. Phylogeny and subfamilial classification of the grasses (Poaceae). Ann. Missouri Bot. Gard. 88:373–457; **Hsiao, C., S.W.L. Jacobs, N.J. Chatterton and K.H. Asay.** 1999. A molecular phylogeny of the grass family (Poaceae) based on the sequences of nuclear ribosomal DNA (ITS). Austral. Syst. Bot. 11:667–688; **Jacobs, S.W.L., R. Bayer, J. Everett, M.O. Arriaga, M.E. Barkworth, A. Sabin-Badereau, M.A. Torres, F. Vázquez, and N. Bagnall.** 2006. Systematics of the tribe Stipeae using molecular data. Aliso 23:349–361; **Jacobs, S.W.L. and J. Everett.** 1996. *Austrostipa*, a new genus, and new names for Australasian species formerly included in *Stipa* (Gramineae). Telopea 6:579–595; **Johnson, B.L.** 1945. Cytotaxonomic studies in *Oryzopsis*. Bot. Gaz. 107:1–32; **Prokudin, Y.N., A.G. Vovk, O.A. Petrova, E.D. Ermolenko, and Y.V. Vernichenklo.** 1977. Zlaki Ukrainy. Naukava Dumka, Kiev, Russia. 517 pp.; **Soreng, R.J. and J.I. Davis.** 1998. Phylogenetics and character evolution in the grass family (Poaceae): Simultaneous analysis of morphological and chloroplast DNA restriction site character sets. Bot. Rev. (Lancaster) 64:1–85.

1. Spikelets with 2–6 florets . . . . 10.01 *Ampelodesmos*
1. Spikelets with 1 floret.
  2. Paleas sulcate, longer than the lemmas; lemma margins involute, fitting into the paleal groove; lemma apices not lobed . . . . 10.09 *Piptochaetium*
  2. Paleas flat, from shorter than to longer than the lemmas; lemma margins convolute or not overlapping; lemma apices often lobed or bifid.
    3. Prophylls exceeding the leaf sheaths; plants cultivated as ornamentals.
      4. Panicles contracted; lemma awns once-geniculate . . . . 10.05 *Macrochloa*
      4. Panicles open; lemma awns twice-geniculate . . . . 10.06 *Celtica*
    3. Prophylls concealed by the leaf sheaths; plants native, introduced, sometimes cultivated as ornamentals.
      5. Flag leaf blades up to 12 mm long; basal leaves overwintering . . . . 10.10 *Oryzopsis*
      5. Flag leaf blades more than 10 mm long; basal leaves not overwintering.

6. Plants with multiple stiff branches from the upper nodes; pedicels sometimes plumose; species cultivated as ornamentals in the *Flora* region . . . . 10.15 *Austrostipa*
6. Plants not branching at the upper nodes, or with a few, flexible branches; pedicels never plumose; species native, established introductions, or cultivated as ornamentals.
    7. Apices of the leaf blades sharp and stiff; caryopses obovoid, often with 3 smooth ribs at maturity; cleistogenes usually present . . . . 10.14 *Amelichloa*
    7. Apices of the leaf blades acute to acuminate, never both sharp and stiff; caryopses fusiform, ovoid or obovoid, without ribs; cleistogenes sometimes present.
        8. Lemma margins strongly overlapping their whole length at maturity, lemma bodies usually rough throughout, apices not lobed; paleas $^1/_4$–$^1/_2$ the length of the lemmas, without veins, glabrous . . . . 10.12 *Nassella*
        8. Lemma margins usually not or only slightly overlapping for some or all of their length at maturity, strongly overlapping in some species with smooth lemmas, lemma bodies usually smooth on the lower portion, apices often 1–2-lobed; paleas from $^1/_3$ as long as to equaling or slightly exceeding the lemmas, 2-veined at least on the lower portion, usually with hairs or both lemmas and paleas glabrous.
            9. Calluses 1.5–6 mm long, sharply pointed; plants perennial or annual, if perennial, awns 65–500 mm long, if annual, awns 50–100 mm long; panicle branches straight.
                10. Lower ligules densely hairy, upper ligules less densely hairy or glabrous; plants perennial . . . . 10.13 *Jarava* (in part)
                10. Ligules glabrous or inconspicuously pubescent, lower and upper ligules alike in vestiture; plants perennial or annual.
                    11. Plants perennial; florets 7–25 mm long; awns scabrous or pilose on the first 2 segments, the terminal segment scabrous, or if pilose, the hairs 1–3 mm long . . . . 10.08 *Hesperostipa*
                    11. Plants annual or perennial, if annual, the florets 4–7 mm long and the awns not plumose, if perennial, the florets 18–27 mm long and the awns plumose on the terminal segment, the hairs 5–6 mm long . . . . 10.07 *Stipa*
            9. Calluses 0.1–2 mm long, blunt to sharply pointed; plants perennial; awns 1–70 mm; panicle branches straight or flexuous.
                12. Florets usually dorsally compressed at maturity, sometimes terete; paleas as long as or longer than the lemmas and similar in texture and pubescence; lemma margins separate for their whole length at maturity . . . . 10.04 *Piptatherum*
                12. Florets terete or laterally compressed at maturity; paleas often shorter than the lemmas, sometimes less pubescent, sometimes as long as the lemmas and similar in texture and pubescence; lemma margins often overlapping for part or all of their length at maturity.
                    13. Glumes without evident venation, glume apices rounded to acute; plants subalpine to alpine, sometimes growing in bogs . . . . 10.03 *Ptilagrostis*
                    13. Glumes with 1–3(5) evident veins or the glume apices attenuate; plants growing from near sea level to subalpine or alpine habitats, not growing in bogs.
                        14. Lemma bodies with evenly distributed hairs of similar length or completely glabrous, sometimes with longer hairs around the base of the awn; basal segment of the awns sometimes with hairs up to 2 mm long . . . . 10.02 *Achnatherum*
                        14. Lemma bodies with hairs to 1 mm long over most of their length, with strongly divergent hairs 3–8 mm long on the distal $^1/_4$, or the basal segment of the awns with hairs 3–8 mm long . . . . 10.13 *Jarava* (in part)

10.01 **AMPELODESMOS** Link

James P. Smith, Jr.

Plants perennial; cespitose, rhizomatous. **Culms** 60–350 cm, annual, internodes solid. **Leaves** mostly basal; **cleistogenes** not developed; **prophylls** shorter than the sheaths; **sheaths** open; **ligules** membranous, ciliate; **blades** initially flat, becoming involute, bases becoming indurate and curved. **Inflorescences** panicles, loosely contracted, somewhat 1-sided. **Spikelets** pedicellate, laterally compressed, with 2–6 florets; **rachillas** hairy, hairs 2–3 mm, prolonged beyond the distal florets; **disarticulation** above the glumes and beneath the florets. **Glumes** subequal, more than ¹/₂ as long as the adjacent lemmas, scarious or chartaceous, 3–5-veined, awn-tipped; **florets** 10–12 mm; **calluses** 0.2–0.5 mm, rounded, strigose; **lemmas** coriaceous, smooth, 5–7-veined, mostly glabrous, hairy over and adjacent to the basal ¹/₂ of the midvein, hairs 1–2 mm, apices bidentate or bilobed, mucronate or awned from the sinuses, lemma-awn junction not conspicuous; **paleas** subequal to the lemmas, 2-keeled, keels extending as teeth, flat between the keels; **lodicules** 3, lanceolate, membranous, ciliate; **anthers** 3, 6–8 mm; **ovaries** pubescent distally; **styles** 2, white. **Caryopses** fusiform, subterete, grooved adaxially, not ribbed; **hila** linear; **starch grains** simple. $x = 12$. Name from the Greek *ampelos*, 'vine', and *desmos*, 'a bond', referring to the use of the leaves to tie grape vines together.

Ampelodesmos is a monospecific, xerophytic genus that is native to the Mediterranean. It is now established in California. It is somewhat similar in overall shape to *Cortaderia*, but differs in its membranous ligules, drooping and somewhat one-sided panicles, and deeply ribbed leaves.

Ampelodesmos was initially included in the *Stipeae* by Decker (1964), who was struck by the similarity of its leaf cross sections to those of some members of the *Stipeae*. Other characteristics it shares with at least some members of the *Stipeae* are its 3 lodicules, relatively small chromosomes, pubescent ovaries, and deeply ribbed leaves. Molecular data also support its inclusion in the *Stipeae* (Soreng and Davis 1998; Hsiao et al. 1999; Jacobs et al. 2006). It is anomalous within the *Stipeae* in having more than one floret per spikelet and prolonged rachillas.

SELECTED REFERENCES **Decker, H.F.** 1964. Affinities of the grass genus *Ampelodesmos*. Brittonia 16:76–79; **Hsiao, C., S.W.L. Jacobs, N.J. Chatterton,** and **K.H. Asay.** 1999. A molecular phylogeny of the grass family (Poaceae) based on the sequences of nuclear ribosomal DNA (ITS). Austral. Syst. Bot. 11:667–688; **Jacobs, S.W.L., R. Bayer, J. Everett, M.O. Arriaga, M.E. Barkworth, A. Sabin-Badereau, M.A. Torres, F. Vázquez,** and **N. Bagnall.** 2006. Systematics of the tribe Stipeae using molecular data. Aliso 23:349–361; **Soreng, R.J.** and **J.I. Davis.** 1998. Phylogenetics and character evolution in the grass family (Poaceae): Simultaneous analysis of morphological and chloroplast DNA restriction site character sets. Bot. Rev. 64:1–88.

1. **Ampelodesmos mauritanicus** (Poir.) T. Durand & Schinz [p. 113]
MAURITANIAN GRASS

Plants rhizomatous, densely cespitose, clumps to 1 m in diameter. **Culms** 60–350 cm tall, 5–8 mm thick. **Sheaths** smooth, striate; **ligules** 8–15 mm; **blades** to 100 cm long, 3–9 mm wide, adaxial surfaces strongly ribbed, margins serrate, apices long-tapering. **Panicles** to 50 cm, lax; **branches** subsecund, drooping. **Spikelets** 10–15 mm, stramineous to purplish. **Lower glumes** 7–10 mm; **upper** glumes 9–12 mm; **calluses** 0.2–0.5 mm, rounded; **lemmas** 9–14.5 mm, distinctly keeled, bidentate, mucronate or awned, awns to 2 mm. **Caryopses** about 7 mm. $2n = 48$.

Ampelodesmos mauritanicus is sparingly established in California: in dry oak woodlands in Napa County, and beneath a mixed evergreen canopy on Mount St. Helena in Sonoma County. It is cultivated in other parts of the United States. The plants dry out rapidly in the summer, making them fire-prone. The amount of seed set varies substantially between years. In its native range, which lies along the drier portions of the Mediterranean coast, the leaves and culms are used for mats, vine ties, brooms, baskets, and thatching.

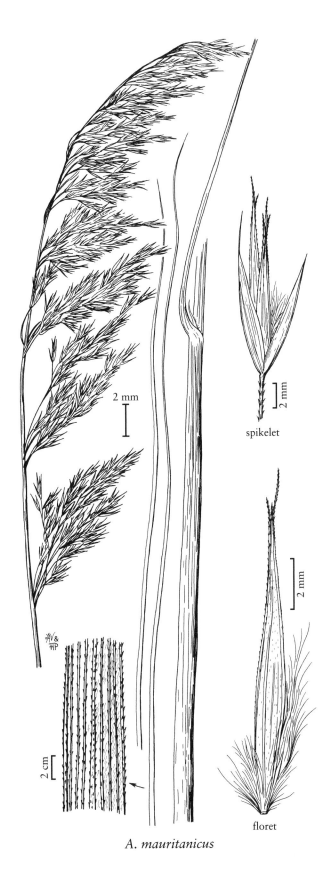

spikelet

2 mm

2 mm

floret

2 mm

2 cm

A. mauritanicus

AMPELODESMOS

10.02 ACHNATHERUM P. Beauv.

Mary E. Barkworth

Plants perennial; tightly to loosely cespitose, sometimes shortly rhizomatous. **Culms** 10–250 cm, erect, not branching at the upper nodes; **basal branching** extra- or intravaginal; **prophylls** shorter than the sheaths. **Leaves** sometimes concentrated at the base; **sheaths** open, margins often ciliate distally; **cleistogenes** not present in the basal leaf sheaths; **collars** sometimes with hairs on the sides; **auricles** absent; **ligules** hyaline to membranous, glabrous or pubescent, sometimes ciliate; **blades** flat, convolute, or involute, apices acute, flexible, basal blades not overwintering, flag leaf blades more than 10 mm long. **Inflorescences** terminal panicles, usually contracted, sometimes 2 forming at the terminal node; **branches** usually straight, sometimes flexuous. **Spikelets** usually appressed to the branches, with 1 floret; **rachillas** not prolonged beyond the floret; **disarticulation** above the glumes, beneath the floret. **Glumes** exceeding the floret, usually lanceolate, 1–7-veined, acute to acuminate, sometimes obtuse; **florets** usually terete, fusiform or globose, sometimes somewhat laterally compressed; **calluses** 0.1–4 mm, blunt to sharp, usually strigose; **lemmas** stiffly membranous to coriaceous, smooth, usually hairy, sometimes glabrous, hairs on the lemma body to 6 mm, usually evenly distributed, hairs on the upper $^{1}/_{4}$ sometimes somewhat longer than those below, not both markedly longer and more divergent, apical hairs to 7 mm, lemma margins usually not or only weakly overlapping, firmly overlapping in some species with glabrous lemmas, usually terminating in 0.05–3 mm lobes, sometimes unlobed, lobes usually membranous and flexible, sometimes thick, apices with a single, terminal, centric awn, awn-lemma junction evident; **awns** 3–80 mm, centric, readily deciduous to persistent, usually scabrous to scabridulous, sometimes hairy in whole or in part, if shorter than 12 mm, usually deciduous, not or once-geniculate and scarcely twisted, if longer than 12 mm, usually persistent, once- or twice-geniculate and twisted below, terminal segment usually straight, sometimes flexuous; **paleas** from $^{1}/_{3}$ as long as to slightly longer than the lemmas, usually pubescent, 2-veined, not keeled over the veins, flat between the veins, veins usually terminating below the apices, sometimes prolonged 1–3 mm, apices usually rounded; **lodicules** 2 or 3, membranous, not lobed; **anthers** 3, 1.5–6 mm, sometimes penicillate; **ovaries** with 2 style branches, branches fused at the base. **Caryopses** fusiform, not ribbed, style bases persistent; **hila** linear, almost as long as the caryopses; **embryos** $^{1}/_{5}$–$^{1}/_{3}$ the length of the caryopses. $x = 10$ or 11. Name from the Greek *achne*, 'scale', and *ather*, 'awn', a reference to the awned lemma.

As interpreted here, *Achnatherum* is one of the larger and more widely distributed genera in the *Stipeae*. It is difficult to estimate how many species it contains because its boundaries are still unclear. Of the 28 species in the *Flora* region, only *A. splendens*, from Europe, is introduced.

Most species of *Achnatherum* used to be included in *Stipa*, a genus that at one time included almost all *Stipeae* with an elongated floret. Keng (cited in Tsvelev 1977) transferred some Chinese species of *Stipa sensu lato* with blunt calluses and less indurate lemmas than *Stipa sensu stricto* to *Achnatherum*, a realignment that Tsvelev (1977) supported. Thomasson (1978) demonstrated that several North and South American species of *Stipa* had lemma epidermes similar to those of the Eurasian species of *Stipa* that had been transferred to *Achnatherum*. After considering various additional characters (Barkworth 1981, 1982), Barkworth (1993) transferred most North American species of *Stipa* and some of *Oryzopsis* into the expanded *Achnatherum*. In retrospect, her transfer of South American species to *Achnatherum* was ill-advised. Some have since been transferred to *Amelichloa* (Arriaga and Barkworth 2006), others to *Jarava* (Peñailillo 2002). With its current boundaries, *Achnatherum* is probably still polyphyletic (Jacobs et al. 2006), but the evidence does not support return of the North American species treated as *Achnatherum* to either *Stipa* or *Oryzopsis*.

In the key, glume widths are the distance between the midvein and the margin. Floret lengths include the callus, but not the apical lobes. Floret thickness refers to the thickest part of the floret.

SELECTED REFERENCES **Arriaga, M.O.** and **M.E. Barkworth.** 2006. *Amelichloa*: A new genus in the Stipeae (Poaceae). Sida 22:145–149; **Barkworth, M.E.** 1981. Foliar epidermes and the taxonomy of North American Stipeae (Gramineae). Syst. Bot. 6:136–152; **Barkworth, M.E.** 1982. Embryological characters and the taxonomy of the Stipeae (Gramineae). Taxon 31:233–243; **Barkworth, M.E.** 1993. North American Stipeae: Taxonomic changes and other comments. Phytologia 74:1–25; **Barkworth, M.E.** and **J. Linman.** 1984. *Stipa lemmonii* (Vasey) Scribner (Poaceae): A taxonomic and distributional study. Madroño 31:48–56; **Cheeke, P.R.** and **L.R. Shull.** 1985. Natural Toxicants in Feeds and Poisonous Plants. AVI Publishing Company, Westport, Connecticut, U.S.A. 492 pp.; **Epstein, W., K. Gerber,** and **R. Karler.** 1964. The hypnotic constituent of *Stipa vaseyi*, sleepy grass. Experientia (Basel) 20:390; **Freitag, H.** 1985. The genus *Stipa* in southwest and south Asia. Notes Roy. Bot. Gard. Edinburgh 42:355–487; **Hitchcock, A.S.** 1951. Manual of the Grasses of the United States, ed. 2, rev. A. Chase. U.S.D.A. Miscellaneous Publication No. 200. U.S. Government Printing Office, Washington, D.C., U.S.A. 1051 pp.; **Jacobs, S.W.L., R. Bayer, J. Everett, M.O. Arriaga, M.E. Barkworth, A. Sabin-Badereau, M.A. Torres, F. Vázquez,** and **N. Bagnall.** 2006. Systematics of the tribe Stipeae using molecular data. Aliso 23:349–361; **Johnson, B.L.** 1945. Natural hybrids between *Oryzopsis hymenoides* and several species of *Stipa*. Amer. J. Bot. 32:599–608; **Johnson, B.L.** 1960. Natural hybrids between *Oryzopsis* and *Stipa*: I. *Oryzopsis hymenoides* × *Stipa speciosa*. Amer. J. Bot. 47:736–742; **Johnson, B.L.** 1962. Amphiploidy and introgression in *Stipa*. Amer. J. Bot. 49:253–262; **Johnson, B.L.** 1962. Natural hybrids between *Oryzopsis* and *Stipa*: II. *Oryzopsis hymenoides* × *Stipa nevadensis*. Amer. J. Bot. 49:540–546; **Johnson, B.L.** 1963. Natural hybrids between *Oryzopsis* and *Stipa*: III. *Oryzopsis hymenoides* × *Stipa pinetorum*. Amer. J. Bot. 50:228–234; **Johnson, B.L.** 1972. Polyploidy as a factor in the evolution and distribution of grasses. Pp. 18–35 *in* V.B. Youngner and C.M. McKell (eds.). The Biology and Utilization of Grasses. Academic Press, New York, New York, U.S.A. 426 pp.; **Johnson, B.L.** and **G.A. Rogler.** 1943. A cytotaxonomic study of an intergeneric hybrid between *Oryzopsis hymenoides* and *Stipa viridula*. Amer. J. Bot. 30:49–56; **Matthei, O.** 1965. Estudio crítico de las gramíneas del género *Stipa* en Chile. Gayana, Bot. 13:1–137; **Maze, J.** 1962. A revision of the Stipas of the Pacific Northwest with special references to *S. occidentalis* Thurb. ex Wats. Master's thesis, University of Washington, Seattle, Washington, U.S.A. 95 pp.; **Maze, J.** 1965. Notes and key to some California species of *Stipa*. Leafl. W. Bot. 10:157–180; **Maze, J.** 1981. A preliminary study on the root of *Oryzopsis hendersonii* (Gramineae). Syesis 14:151–153; **Peñailillo, P.** 2002. El género *Jarava* Ruiz et Pavón (Stipeae–Poaceae): Delimitación y nuevas combinaciones. Gayana, Bot. 59:30; **Pohl, R.W.** 1954. The allopolyploid *Stipa latiglumis*. Madroño 12:145–150; **Scagel, R.K.** and **J. Maze.** 1984. A morphological analysis of local variation in *Stipa nelsonii* and *S. richardsonii* (Gramineae). Canad. J. Bot. 62:763–770; **Shechter, Y.** 1969. Electrophoretic investigation of the hybrid origin of *Oryzopsis contracta*. Pp. 19–25 *in* L. Chandra (ed.). Advancing Frontiers of Plant Sciences, vol. 23. Impex India, New Delhi, India. 201 pp.; **Shechter, Y.** and **B.L. Johnson.** 1968. The probable origin of *Oryzopsis contracta*. Amer. J. Bot. 55:611–618; **Thomasson, J.R.** 1978. Epidermal patterns of the lemma in some fossil and living grasses and their phylogenetic significance. Science 199:975–977; **Torres, M.A.** 1993. Revisión del Género *Stipa* (Poaceae) en la Provincia de Buenos Aires. Monografia 12. Comisión de Investigaciones Científicas, Provincia de Buenos Aires, La Plata, Argentina. 62 pp.; **Tsvelev, N.N.** 1976. Zlaki SSSR. Nauka, Leningrad [St. Petersburg], Russia. 788 pp.; **Tsvelev, N.N.** 1977. [On the origin and evolution of the feathergrasses (*Stipa* L.)]. Pp. 139–150 *in* Problemii Ekologii, Geobotaniki, and Botaniicheskoi Geografii i Floristickii. Nauka, Leningrad [St. Petersburg], Russia. 225 pp. [In Russian; translation of article by K. Gonzales, available from the Intermountain Herbarium, Utah State University, Logan, Utah 84322-5305, U.S.A.].

1. Awns persistent, basal segments pilose, at least some hairs 0.5–8 mm long.
  2. Flag leaves with ligules 3–8 mm long; lemmas with 1 apical lobe, the lobe to 0.1 mm long, thick, coriaceous . . . . 9. *A. thurberianum*
  2. Flag leaves with ligules 0.3–3 mm long; lemmas usually with 2 apical lobes, sometimes not lobed, lobes to 1 mm long, thin, membranous.
     3. Basal awn segments with hairs of mixed lengths, the longer hairs scattered among the shorter hairs; apical lemma hairs longer than most basal awn hairs.
        4. Florets 8–9 mm long; glumes 1.3–1.9 mm wide from midvein to margin . . . . . 7. *A. latiglume* (in part)
        4. Florets 5–7.5 mm long; glumes 0.6–1 mm wide from midvein to margin.
           5. Calluses 0.5–0.7 mm long; paleas $^{1}/_{2}$–$^{3}/_{4}$ as long as the lemmas; palea apices with hairs usually about 1 mm long . . . . 4. *A. nevadense*
           5. Calluses 0.8–1.2 mm long; paleas $^{2}/_{5}$–$^{3}/_{5}$ as long as the lemmas; palea apices with hairs usually less than 1 mm long . . . . 5. *A. occidentale* (in part)
     3. Basal awn segments with hairs that gradually and regularly decrease in length distally; apical lemma hairs usually similar in length to the longest basal awn hairs, sometimes longer on the adaxial side.
        6. Basal blades curling with age, forming circular arcs; paleas $^{1}/_{4}$–$^{1}/_{3}$ as long as the lemmas; panicles 7–11 cm long . . . . 18. *A. curvifolium*
        6. Basal blades straight to lax, not forming circular arcs; paleas $^{2}/_{5}$–$^{4}/_{5}$ as long as the lemmas; panicles 5–30 cm long.

7. Florets 5.5–7.5 mm long; paleas $^2/_5$–$^3/_5$ as long as the lemmas; glumes less than
   1 mm wide from midvein to margin ....5. *A. occidentale* (in part)
7. Florets 8–9 mm long; paleas $^3/_5$–$^4/_5$ as long as the lemmas; glumes 1.3–1.9 mm
   wide from midvein to margin ....7. *A. latiglume* (in part)
1. Awns deciduous or persistent, basal segments scabrous or with hairs shorter than 0.5 mm.
  8. Lemmas evenly hairy, hairs 1.2–6 mm long, hairs on the lemma body usually not evidently
    shorter than those at the apices.
    9. Awns persistent.
      10. Plants sterile, the anthers indehiscent, with few pollen grains (see discussion
        following *A. hymenoides*) .... hybrids of 26. *Achnatherum hymenoides* (in part)
      10. Plants fertile, the anthers dehiscent, with many pollen grains.
        11. Sheaths not becoming flat and ribbonlike with age; blades usually involute and
          0.2–0.4 mm in diameter, 0.5–1 mm wide when flat; awns twice-geniculate ..... 21. *A. pinetorum*
        11. Sheaths becoming flat and ribbonlike with age; blades 0.5–1.5 mm in diameter
          when convolute, to 7 mm wide when flat; awns once- or twice-geniculate.
          12. Awns twice-geniculate, culms 3–6 mm thick .... 10. *A. coronatum* (in part)
          12. Awns once-geniculate, culms 0.8–2 mm thick .... 11. *A. parishii* (in part)
    9. Awns rapidly deciduous.
      13. Florets at least 4.5 mm long, fusiform, anthers sometimes indehiscent.
        14. Anthers dehiscent, the pollen grains well formed ....22. *A. webberi*
        14. Anthers indehiscent, the pollen grains poorly formed.
          15. Anthers dimorphic, 1 longer than the other 2; lemmas with 7 veins .... see 10.11 ×*Achnella*
          15. Anthers all alike; lemmas with 5 veins (see discussion of hybrids on p. 142)
          .... hybrids of 26. *Achnatherum hymenoides* (in part)
      13. Florets 2.5–4.5 mm long, usually ovoid to obovoid, sometimes fusiform, anthers
        dehiscent.
        16. Panicle branches terminating in a pair of spikelets on conspicuously divaricate,
          unequal to subequal pedicels, most shorter pedicels at least $^1/_2$ as long as the
          longer pedicels ....26. *A. hymenoides*
        16. Panicle branches terminating in a pair of spikelets on loosely appressed,
          unequal pedicels, most shorter pedicels less than $^1/_2$ as long as the longer
          pedicels.
          17. Panicles 0.5–2.8 cm wide, branches 0.5–5 cm long, strongly ascending;
            spikelets evenly distributed over the branches ....27. *A. arnowiae*
          17. Panicles 7–15 cm wide, branches 5–8 cm long, ascending to strongly
            divergent; spikelets confined to the distal $^1/_2$ of the branches ....28. *A. contractum*
  8. Lemmas glabrous or with hairs 0.2–1.5(2) mm long at midlength, glabrous or with hairs
    distally, the hairs at midlength often evidently shorter than those at the lemma apices.
    18. Apical lemma hairs 2–7 mm long, usually 1+ mm longer than those at midlength.
      19. Calluses sharp; paleas 1/3–1/2 as long as the lemmas ....19. *A. scribneri*
      19. Calluses blunt to acute; paleas 1/2–9/10 as long as the lemmas.
        20. Awns twice-geniculate; culms 3–6 mm thick ....10. *A. coronatum* (in part)
        20. Awns once-geniculate; culms 0.8–2 mm thick ....11. *A. parishii* (in part)
    18. Apical lemma hairs absent or to 2.2 mm long, usually less than 1 mm longer than
      those at midlength.
      21. Awns 5–12 mm long, readily deciduous, not or only once-geniculate.
        22. Lemmas glabrous.
          23. Panicles lax, the branches flexuous, diverging ....24. *A. wallowaense*
          23. Panicles erect, the branches straight, ascending to appressed ....25. *A. hendersonii*
        22. Lemmas pubescent.
          24. Culms 30–250 cm long; plants cultivated ornamentals ....1. *A. splendens*
          24. Culms 15–25 cm long; plants native in the *Flora* region ....23. *A. swallenii*
      21. Awns 10–80 mm long, persistent, once- or twice-geniculate.

25. Terminal awn segment flexuous.
    26. Panicles contracted, all branches straight, appressed or strongly ascending; ligules on the flag leaves to 1.5 mm long . . . . 15. *A. aridum*
    26. Panicles open, the lower branches flexuous, ascending to widely divergent; ligules on the flag leaves to 4.5 mm long . . . . 16. *A. eminens*
25. Terminal awn segment straight or slightly arcuate.
    27. Panicle branches flexuous, ascending to strongly divergent; spikelets pendulous . . . . 17. *A. richardsonii*
    27. Panicle branches straight, usually appressed to ascending, sometimes divergent; spikelets appressed to the branches.
        28. Flag leaves with a densely pubescent collar, the hairs 0.5–2 mm long; paleas $^2/_3$–$^3/_4$ as long as the lemmas . . . . 12. *A. robustum*
        28. Flag leaves glabrous or sparsely pubescent on the collar, the hairs shorter than 0.5 mm; paleas from $^1/_3$ as long as to longer than the lemmas.
            29. Lemma apices 2-lobed, lobes 1–3 mm long; palea veins extending beyond the palea body, reaching to the tips of the lemma lobes . . . . 2. *A. stillmanii*
            29. Lemma apices unlobed or with lobes to 1.2 mm long; palea veins terminating before or at the palea apices.
                30. Apical lemma lobes thick, stiff, about 0.1 mm long; florets somewhat laterally compressed . . . . 8. *A. lemmonii*
                30. Apical lemma lobes membranous, 0.1–1.2 mm long; florets terete.
                    31. Lower cauline internodes densely pubescent for 3–9 mm below the nodes, more shortly and less densely pubescent elsewhere . . . . 13. *A. diegoense*
                    31. Lower cauline internodes glabrous or slightly pubescent to 5 mm below the nodes, usually glabrous elsewhere.
                        32. Glumes subequal, the lower glumes exceeding the upper glumes by less than 1 mm.
                            33. Paleas $^3/_5$–$^9/_{10}$ as long as the lemmas, the apical hairs exceeding the apices; blades 0.5–2 mm wide; awns 12–25 mm long . . . . 3. *A. lettermanii*
                            33. Paleas $^1/_3$–$^2/_3$ as long as the lemmas, the apical hairs usually not exceeding the apices; blades (0.5)1.2–5 mm wide; awns 19–45 mm long . . . . 6. *A. nelsonii*
                        32. Glumes unequal, the lower glumes exceeding the upper glumes by 1–4 mm.
                            34. Apical lemma hairs erect; lemma lobes 0.5–1.2 mm long . . . . 14. *A. lobatum*
                            34. Apical lemma hairs divergent to ascending; lemma lobes 0.2–0.5 mm long . . . . 20. *A. perplexum*

1. Achnatherum splendens (Trin.) Nevski [p. 119]
Jiji Grass

Plants cespitose, not rhizomatous. **Culms** 30–250 cm tall, 2–3 mm thick, glabrous, smooth; **nodes** 1–3. **Basal sheaths** glabrous or the margins ciliate, becoming fibrous with age; **collars** glabrous, including the sides; **basal ligules** 1–3 mm, membranous, glabrous, truncate to acute; **upper ligules** to 12 mm, acute; **blades** to 60 cm long, 2–5 mm wide, abaxial surfaces smooth, adaxial surfaces scabrous. **Panicles** 12–50 cm long, 4–15 cm wide; **branches** ascending, usually whorled, longest branches 2.5–15 cm, with 15+ spikelets. **Glumes** 4–8.5 mm, acute; **lower glumes** 0–1.7 mm shorter than the upper glumes; **upper glumes** 0.5–0.9 mm wide; **florets** 4–7.2 mm, fusiform; **calluses** 0.3–0.5 mm, blunt; **lemmas** evenly hairy, hairs at midlength to 1 mm, apical hairs to 1.5 mm, apical lobes 0.5–1 mm; **awns** 5–12 mm, readily deciduous, indistinctly once-geniculate or flexuous, scabrous; **paleas** slightly shorter than the lemmas, pubescent; **anthers** 3–4.5 mm, dehiscent, penicillate, yellow. **Caryopses** 2–4 mm, fusiform. $2n$ = 42, 48.

Achnatherum splendens is native from the Caspian Sea to eastern Siberia and south through central Asia to the inner ranges of the Himalayas. According to Freitag (1985), it is a common and typical plant of cold, semidesert regions, growing in groundwater-influenced habitats at elevations of 2100–3800 m. It is rarely eaten by grazing animals, so that it increases in abundance in overgrazed meadows. It is being considered as a potential soil binder for areas in Asia that are too cold for *Chrysopogon zizanioides* (L.) Roberty.

A.S. Hitchcock (1951) reported *A. splendens* to be "sparingly cultivated" in the United States. In view of Freitag's comments, its cultivation in North America should be discouraged.

2. Achnatherum stillmanii (Bol.) Barkworth [p. 119]
STILLMAN'S NEEDLEGRASS

Plants shortly rhizomatous, forming open clumps. **Culms** 60–150 cm tall, 2–5 mm thick, often geniculate at the lowest node, mostly glabrous; **nodes** 2–3, often puberulent. **Basal sheaths** mostly glabrous, sometimes ciliate distally, intact at maturity; **collars** glabrous or pubescent, often with hairs at the sides, hairs shorter than 0.5 mm; **basal ligules** 0.2–0.5 mm, membranous, truncate, ciliate, cilia 0.2–0.3 mm; **upper ligules** shorter than the basal ligules; **blades** 3–7 mm wide, lax. **Panicles** 10–24 cm long, 1.5–3 cm wide, contracted; **branches** straight, appressed to strongly ascending, lower branches 2.5–3.5 cm. **Spikelets** appressed to the branches. **Glumes** subequal, 14–16 mm; **lower glumes** 1–3-veined; **upper glumes** 0.6–1.5 mm wide, 3–5-veined; **florets** 8–10 mm, fusiform; **calluses** 0.5–1.2 mm, rounded; **lemmas** evenly hairy, hairs about 0.5 mm, apical hairs similar in length, apices 2-lobed, lobes 1–3 mm, narrow; **awns** 18–25 mm, persistent, scabrous, once- or twice-geniculate, terminal segment straight; **paleas** as long as or slightly longer than the lemmas, hairy, hairs about 0.5 mm, veins prolonged, reaching almost to the tips of the lemma lobes; **anthers** 4–6 mm, penicillate, dehiscent. **Caryopses** fusiform. $2n$ = unknown.

Achnatherum stillmanii grows at scattered locations in coniferous forests in northern California, at 900–1500 m, possibly being edaphically restricted. Its combination of large size, long, narrow lemma lobes, and paleal morphology distinguish *A. stillmanii* from all other North American species of *Achnatherum*.

3. Achnatherum lettermanii (Vasey) Barkworth [p. 119]
LETTERMAN'S NEEDLEGRASS

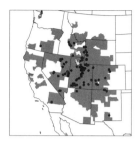

Plants tightly cespitose, not rhizomatous. **Culms** 15–90 cm tall, 0.5–0.8 mm thick, usually glabrous, sometimes puberulent to 5 mm below the lower nodes; **nodes** 2–3. **Basal sheaths** smooth, glabrous, margins not ciliate; **collars**, including the sides, glabrous or sparsely pubescent, collars of the flag leaves glabrous; **ligules** 0.2–1.5(2) mm, without tufts of hair on the sides, truncate to rounded; **blades** 0.5–2 mm wide, abaxial surfaces smooth to scabridulous, adaxial surfaces scabrous or puberulent. **Panicles** 7–19 cm long, 0.5–1 cm wide; **branches** straight, appressed to strongly ascending, longest branches 1.2–2.5 cm. **Spikelets** appressed to the branches. **Glumes** 6.5–9 mm, subequal; **lower glumes** 1(3)-veined; **upper glumes** to 0.5 mm shorter than the lower glumes, 0.6–1 mm wide, 1-veined; **florets** 4.5–6 mm long, 0.8–1 mm thick, fusiform, terete, widest below midlength; **calluses** 0.4–1 mm, blunt; **lemmas** evenly hairy, hairs at midlength about 0.5 mm, apical hairs 0.7–1.5(2) mm, apical lobes 0.3–0.8 mm, membranous, flexible; **awns** 12–25 mm, persistent, twice-geniculate, scabrous, terminal segment straight; **paleas** 3–4 mm, $^3/_4$–$^4/_5$($^9/_{10}$) as long as the lemmas, veins terminating at or before the apices, apices round, flat, apical hairs 0.5–1 mm, extending beyond the palea body; **anthers** 1.5–2 mm, dehiscent, not penicillate. **Caryopses** about 4 mm, fusiform. $2n$ = 32.

Achnatherum lettermanii grows in meadows and on dry slopes, from sagebrush to subalpine habitats, at 1700–3400 m. Its range extends from Oregon and Montana to southern California, Arizona, and New Mexico; it is not known from Mexico. When sympatric with *A. nelsonii*, *A. lettermanii* tends to grow in shallower or more disturbed soils. It can be distinguished from that species by its generally finer leaves and more tightly cespitose growth habit, as well as its blunter calluses and longer paleas. Its relatively long paleas also distinguish *A. lettermanii* from *A. perplexum*.

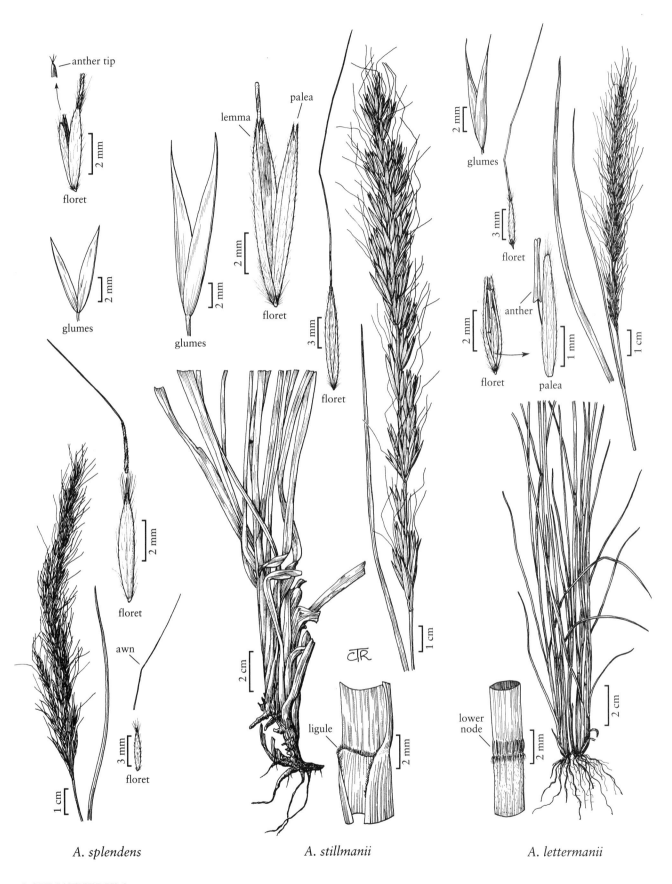

anther tip

floret

2 mm

glumes

2 mm

A. splendens

lemma   palea

glumes

2 mm

floret

2 mm

floret

2 mm

awn

floret

3 mm

A. stillmanii

floret

3 mm

2 cm

ligule

2 mm

CTR

1 cm

glumes

2 mm

floret

3 mm

anther

floret   palea

2 mm   1 mm

1 cm

lower
node

2 mm

2 cm

A. lettermanii

ACHNATHERUM

4. **Achnatherum nevadense** (B.L. Johnson) Barkworth [p. 120]

NEVADA NEEDLEGRASS

Plants cespitose, not rhizomatous. **Culms** 20–85 cm tall, 0.8–1.2 mm thick, usually retrorsely pubescent below the lower nodes, sometimes glabrous, sometimes pubescent over the whole of the internodes; **nodes** 3–4. **Basal sheaths** glabrous or pubescent, sometimes scabridulous, usually glabrous at the throat, becoming brown to gray-brown; **collars**, including the sides, glabrous; **basal ligules** 0.2–0.7 mm, truncate; **upper ligules** 0.3–1 mm, often wider than the blades; **blades** usually 10–25 cm long, 1–3 mm wide, usually involute, abaxial surfaces glabrous, adaxial surfaces more or less puberulent. **Panicles** 6–25 cm long, 0.4–1.5 cm wide; **branches** appressed, lower branches 2–7.5 cm. **Spikelets** appressed to the branches. **Glumes** subequal, 7–14 mm; **lower glumes** 0.6–0.9 mm wide; **florets** 5–6.5 mm long, 0.6–1 mm thick, fusiform, terete; **calluses** 0.5–0.7 mm, sharp, dorsal boundary of the glabrous tip with the callus hairs rounded to acute; **lemmas** evenly hairy, hairs 0.5–1 mm at midlength, apical hairs 0.8–2 mm, longer than those at midlength and those at the base of the awn, apical lobes membranous, about 0.2 mm; **awns** 20–35 mm, persistent, twice-geniculate, first 2 segments pilose, with hairs 0.5–1.5 mm and of mixed lengths, terminal segment scabridulous to smooth; **paleas** 2.8–4.2 mm, ¹/₂–³/₄ as long as the lemmas, pubescent, distal hairs usually about 1 mm, extending well beyond the apices, apices rounded; **anthers** about 2.5 mm, dehiscent, not penicillate. **Caryopses** 3–5.5 mm, fusiform. $2n = 68$.

Achnatherum nevadense grows in sagebrush and open woodlands, from Washington to south-central Wyoming and south to California and Utah. Johnson (1962) argued that it is an alloploid derivative of *A. occidentale* and *A. lettermanii*.

The apical lemma hairs of *Achnatherum nevadense* appear longer than the lowermost awn hairs. This difference is not reflected in the lengths shown because a few of the basal awn hairs may be as long as those at the top of the lemma, but the majority are shorter. This is the best character for distinguishing *A. nevadense* and *A. occidentale* subsp. *californicum* from *A. occidentale* subsp. *pubescens*. In addition, in *A. nevadense* and *A. occidentale* subsp. *californicum*, the hairs on the first awn segments tend to look untidy because of their varied lengths and the different angles they make with the awn; those of *A. occidentale* subsp. *occidentale* and subsp. *pubescens* have basal awn segments with tidier looking hairs.

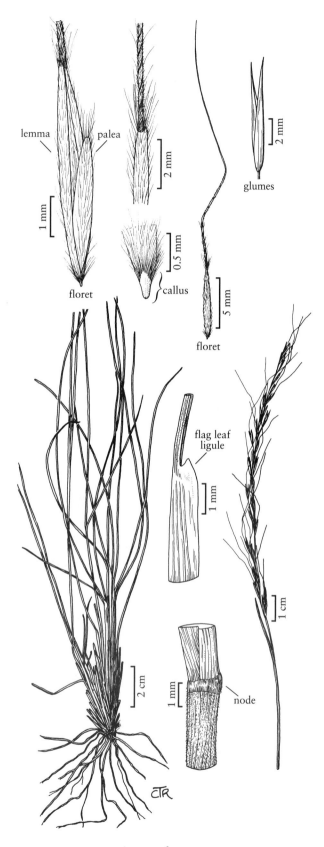

A. nevadense

ACHNATHERUM

Differentiating between *Achnatherum nevadense* and *A. occidentale* subsp. *californicum* can be difficult, but they differ in the shape of the boundary between the glabrous and strigose portions of the callus. In addition, *A. nevadense* is usually pubescent below the lower cauline nodes, and has paleas that are longer in relation to the lemmas. Plants of *A. nevadense* from Convict Creek Basin, California, are unusual in having culms that are glabrous below the lower nodes, but in other respects they correspond to *A. nevadense. Achnatherum nevadense* also resembles *A. latiglume*, but the latter species has blunter calluses and paleas that tend to be thicker and somewhat longer in comparison to the lemmas than those of *A. nevadense*.

5. Achnatherum occidentale (Thurb.) Barkworth [p. 122]

Plants tightly cespitose, not rhizomatous. **Culms** 14–120(180) cm tall, 0.3–2 mm thick, internodes glabrous or puberulent to densely pubescent; **nodes** 2–4, glabrous or pubescent. **Basal sheaths** glabrous or puberulent to densely pubescent, often ciliate at the throat; **collars** often with tufts of hair at the sides; **ligules** 0.2–1.5 mm, often ciliate; **blades** 0.5–3 mm wide and flat, or convolute and 0.1–0.8 mm in diameter, lax to straight. **Panicles** 5–30 cm long, 0.5–1.5 cm wide; **branches** appressed, straight, longest branches 1–7 cm. **Spikelets** appressed to the branches. **Glumes** subequal, 9–15 mm long, 0.6–0.9 mm wide; **florets** 5.5–7.5 mm long, 0.5–0.9 mm thick, fusiform, terete; **calluses** 0.8–1.2 mm, sharp, dorsal boundary of the glabrous tip with the callus hairs narrowly acute; **lemmas** evenly hairy, hairs 0.5–1.5 mm at midlength, apical hairs somewhat longer than those below, sometimes similar in length to those at the base of the awns, sometimes longer, apical lobes 0.3–0.5 mm, membranous; **awns** 15–55 mm, twice-geniculate, first 2 segments evidently hairy, terminal segment glabrous or partly to wholly pilose, sometimes scabrous; **paleas** 2.2–3.5 mm, $^2/_5$–$^3/_5$ as long as the lemmas, hairs at the tip usually shorter than 1 mm, frequently extending beyond the apices, apices rounded; **anthers** 2.5–3.5 mm, dehiscent, not penicillate. **Caryopses** 4–6 mm, fusiform. $2n = 36$.

Achnatherum occidentale, which extends from British Columbia to California, Utah, and Colorado, varies considerably in pubescence and size. The three subspecies recognized here occasionally occur together.

1. Terminal awn segment usually pilose; culms 0.3–1 mm thick, glabrous even on the basal internodes; glumes often purplish . . . .
. . . . subsp. *occidentale*
1. Terminal awn segment usually scabrous or glabrous, occasionally pilose at the base; culms 0.5–2 mm thick; glumes usually green.
 2. First 2 awn segments scabrous or pilose with hairs of mixed lengths; apical lemma hairs longer than the basal awn hairs . . . . subsp. *californicum*
 2. First 2 awn segments pilose, the hairs gradually and evenly becoming shorter towards the first geniculation; apical lemma hairs similar in length to the basal awn hairs . . . . subsp. *pubescens*

Achnatherum occidentale subsp. californicum (Merr. & Burtt Davy) Barkworth [p. 122]
CALIFORNIA NEEDLEGRASS

Culms 30–100(180) cm tall, 0.5–2 mm thick, internodes glabrous or pubescent, sometimes densely pubescent; **nodes** 2–4, glabrous or pubescent. **Sheaths** glabrous, pubescent, or pilose; **collars** usually with tufts of hair at the sides; **blades** 0.8–2 mm wide, usually erect or ascending, adaxial surfaces pilose. **Panicles** 8–30 cm; **branches** appressed, longest branches to 7 cm. **Glumes** usually green; **lower glumes** 0.6–0.8 mm wide; **lemmas** with hairs to 0.8 mm at midlength, apical hairs to 1.8 mm, longer than the basal awn hairs; **awns** 18–55 mm, scabrous or pilose on the first 2 segments, with hairs to 0.5 mm and of mixed lengths, terminal segment usually scabrous, occasionally pilose at the base. $2n = 36$.

Achnatherum occidentale subsp. *californicum* grows from Washington through Idaho to southwestern Montana and south to California and Nevada, with disjunct records from south-central Wyoming and southwestern Utah. Its elevation range is 2000–4000 m.

Johnson (1962) postulated that *Achnatherum occidentale* subsp. *californicum* is a hybrid derivative of *A. nelsonii* and *A. occidentale*; it intergrades with both. The scattering of longer hairs among shorter hairs on the basal awn segments, combined with the long apical lemma hairs, give florets of subsp. *californicum* a more untidy appearance than those of the other two subspecies. It resembles *A. nevadense* in this respect, but differs from that species in the shape of the boundary between the glabrous and strigose portions of the callus, in usually being glabrous below the lower cauline nodes, and in having paleas that are shorter in relation to the lemmas. Plants with scabrous awns are often confused with *A. nelsonii* subsp. *nelsonii*; they differ in having sharper calluses, a more elongated extension of the glabrous callus area into the strigose portion of the callus, and, usually, longer awns.

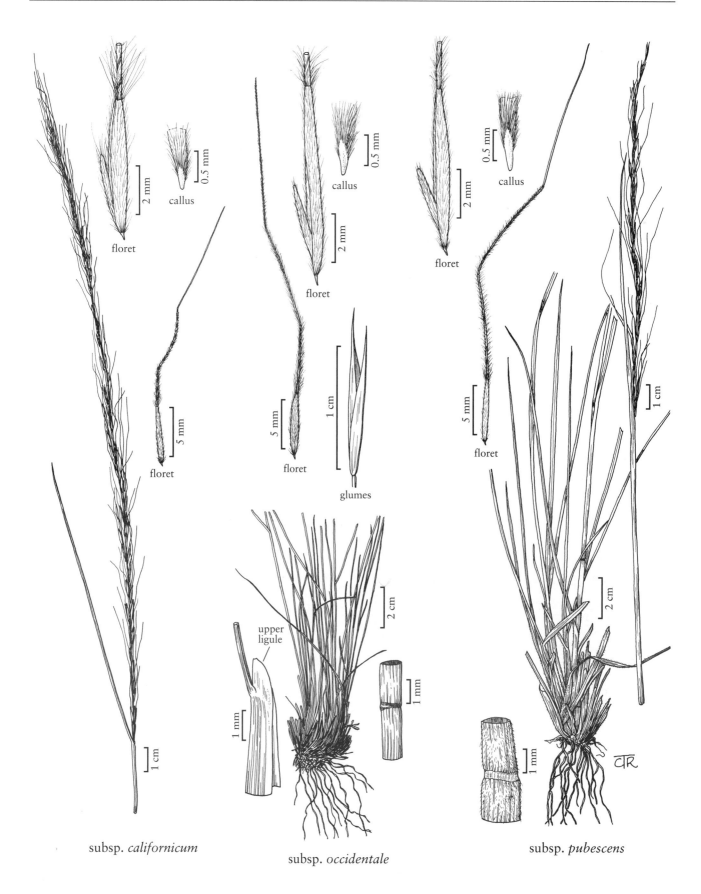

subsp. *californicum*

subsp. *occidentale*

subsp. *pubescens*

ACHNATHERUM OCCIDENTALE

Achnatherum occidentale (Thurb.) Barkworth subsp. **occidentale** [p. 122]
WESTERN NEEDLEGRASS

Culms 14–50 cm tall, 0.3–1 mm thick, internodes glabrous. **Basal sheaths** mostly glabrous, often ciliate or sparsely tomentose at the throat; **collars** usually without tufts of hair at the sides; **blades** 0.3–2 mm wide when flat, usually convolute and 0.1–0.2 mm in diameter. **Panicles** 5–20 cm; **longest branches** 1.5–4 cm. **Glumes** often purplish; **lemmas** with apical pubescence similar in length to the basal awn pubescence; **awns** 15–42 mm, first 2 segments always pilose, terminal segment usually pilose, hairs becoming shorter distally, occasionally scabridulous or smooth.

Achnatherum occidentale subsp. *occidentale* grows above 2400 m, primarily in California. It differs from *A. occidentale* subsp. *pubescens* in having culms that are glabrous throughout and awns with terminal segments that are usually pilose. One specimen from Idaho has been seen; it has the palea/lemma ratio of subsp. *californicum* but the pilose terminal awn segments, slender habit, and purplish coloration of subsp. *occidentale*.

Achnatherum occidentale subsp. **pubescens** (Vasey) Barkworth [p. 122]
COMMON WESTERN NEEDLEGRASS

Culms 32–120 cm tall, 0.8–1.3(2) mm thick, basal internodes puberulent to pubescent. **Basal sheaths** usually pubescent, sometimes densely pubescent, occasionally glabrous; **collars** often with tufts of hair at the sides; **blades** 1–3 mm wide, adaxial surfaces pubescent to pilose. **Panicles** 10–30 cm; **longest branches** 2–7 cm. **Glumes** usually green; **lemmas** with apical pubescence similar in length to the basal awn pubescence; **awns** 24–50 mm, pilose on the first 2 segments, with hairs gradually becoming shorter distally, terminal segment scabrous or glabrous. $2n = 36$.

Achnatherum occidentale subsp. *pubescens* grows from Washington to California and eastward to Wyoming, at 1300–4700 m. It is the most widespread and variable subspecies of *A. occidentale*, intergrading with subsp. *californicum*, *A. nelsonii*, and *A. lettermanii*. It differs from the latter two in its shorter paleas and its pilose awns.

6. **Achnatherum nelsonii** (Scribn.) Barkworth [p. 124]

Plants cespitose, not rhizomatous. **Culms** 40–175 cm tall, 0.7–2.4 mm thick, lower cauline internodes usually glabrous, sometimes slightly pubescent below the lower nodes; **nodes** 2–5. **Basal sheaths** glabrous or sparsely to densely pubescent, margins sometimes ciliate; **collars** glabrous or somewhat pubescent, without tufts of hair on the sides, collars of the flag leaves glabrous or sparsely pubescent; **basal ligules** 0.2–0.7 mm, membranous, truncate to rounded, usually not ciliate; **upper ligules** 1–1.5 mm, acute; **blades** (0.5)1.2–5 mm wide. **Panicles** 9–36 cm long, 0.8–2 cm wide; **branches** ascending to appressed, straight. **Spikelets** appressed to the branches. **Glumes** 6–12.5 mm long, 0.7–1.1 mm wide; **lower glumes** exceeding the upper glumes by 0.2–0.8 mm; **florets** 4.5–7 mm long, 0.6–0.9 mm thick, fusiform; **calluses** 0.2–1 mm, blunt to sharp, dorsal boundary of the glabrous tip with the callus hairs almost straight to acute; **lemmas** evenly hairy, hairs at midlength 0.5–1 mm, hairs at the apices to 2 mm, erect to ascending, apical lobes 0.1–0.4 mm, membranous, flexible; **awns** 19–45 mm, persistent, twice-geniculate, first 2 segments scabrous or with hairs shorter than 0.5 mm, terminal segment straight; **paleas** 2–4 mm, $^1/_3$–$^2/_3$ as long as the lemmas, pubescent, hairs usually not exceeding the apices, veins terminating before the apices, apices rounded; **anthers** 2–3.5 mm, dehiscent, not penicillate. **Caryopses** 3–4 mm, fusiform. $2n = 36, 44$.

Achnatherum nelsonii grows in meadows and openings, from sagebrush steppe and pinyon-juniper woodlands to subalpine forests, at 500–3500 m. It flowers in late spring to early summer, differing in this respect from *A. perplexum*. It is sometimes sympatric with *A. lettermanii*, from which it differs in its shorter paleas and wider leaves, and its tendency to grow in deeper or less disturbed soils. It differs from *A. lemmonii* in having wider leaf blades, shorter paleas, and membranous lemma lobes, and from *A. nevadense* and *A. occidentale* in its scabrous awns and the truncate to acute boundary of the glabrous tip of the callus with the callus hairs.

The two subspecies intergrade to some extent. There is also intergradation with *Achnatherum occidentale*, possibly as a result of hybridization and introgression.

1. Calluses blunt, dorsal boundary of the glabrous tip and the callus hairs almost straight to rounded; awns 19–31 mm long
. . . . subsp. *dorei*

1. Calluses sharp, dorsal boundary of the glabrous tip and the callus hairs acute; awns 19–45 mm long . . . . subsp. *nelsonii*

Achnatherum nelsonii subsp. dorei (Barkworth & J.R. Maze) Barkworth [p. 124]
DORE'S NEEDLEGRASS

Calluses blunt, glabrous tips 0.02–0.06 mm, dorsal boundary of the glabrous tip and the callus hairs almost straight to rounded; **awns** 19–31 mm.

   Achnatherum nelsonii subsp. *dorei* grows from the southern Yukon Territory to California and Wyoming. In regions where both subspecies grow, subsp. *dorei* is at higher elevations than subsp. *nelsonii*. It differs from *A. robustum* in the sparsely hairy collars of its flag leaves.

   Reports of *Achnatherum nelsonii* subsp. *dorei* (identified as *Stipa columbiana* Macoun by many authors) from New Mexico and Arizona are probably based on *A. perplexum*, which differs in having sparse, narrow inflorescences and slightly recurved glumes. The two also differ in flowering time, *A. nelsonii* subsp. *dorei* flowering in late spring to early summer and *A. perplexum* in the fall. *Achnatherum nelsonii* subsp. *nelsonii* is present in New Mexico. Scagel and Maze (1984) concluded that putative hybrids between *A. nelsonii* subsp. *dorei* and *A. richardsonii* were merely large plants of *A. nelsonii* subsp. *dorei* that varied in the direction of *A. richardsonii*.

Achnatherum nelsonii (Scribn.) Barkworth subsp. nelsonii [p. 124]
NELSON'S NEEDLEGRASS

Calluses sharp, acute to acuminate, glabrous tips 0.05–0.15 mm, dorsal boundary of the glabrous tip and the callus hairs acute; **awns** 19–45 mm.

   Achnatherum nelsonii subsp. *nelsonii* intergrades with subsp. *dorei* in Montana and Wyoming, and with *A. occidentale* subsp. *pubescens* in California. Its range extends from Idaho and Montana south to Nevada. It tends to grow at lower elevations than subsp. *dorei*.

7. Achnatherum latiglume (Swallen) Barkworth [p. 126]
WIDE-GLUMED NEEDLEGRASS

Plants tightly cespitose, not rhizomatous. **Culms** 50–110 cm tall, 0.7–1.2 mm thick, lower internodes retrorsely pilose, upper internodes glabrous; **nodes** 2–4. **Basal sheaths** usually retrorsely pubescent, brown to gray-brown, flat when mature; **collars** usually glabrous, sometimes with a few hairs at the sides; **basal ligules** 0.2–2.5 mm, truncate to rounded; **upper ligules** 1.2–3 mm, rounded to acute, ciliate; **blades** 0.7–3 mm wide,

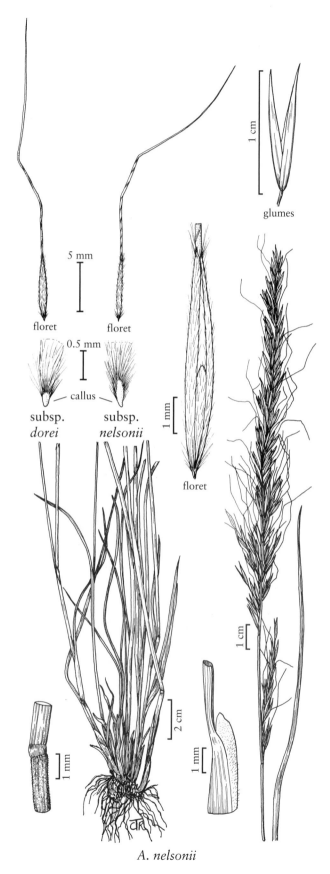

glumes

floret   floret

0.5 mm

callus

subsp.   subsp.
dorei   *nelsonii*

floret

A. nelsonii

ACHNATHERUM

straight to lax, abaxial surfaces smooth, glabrous, adaxial surfaces pubescent, scabrous. **Panicles** 15–30 cm long, 0.8–2 cm wide; **branches** appressed to strongly ascending, longest branches 2.5–6.5 cm. **Glumes** subequal, 12–15 mm long, 1.3–1.9 mm wide, 3-veined; **florets** 8–9 mm long, 0.9–1.4 mm thick, fusiform, terete; **calluses** 0.7–1 mm, blunt to sharp; **lemmas** evenly hairy, hairs 0.5–1.5 mm at midlength, apical hairs 1–2 mm, apical lobes to 1 mm, membranous; **awns** 33–45 mm, persistent, twice-geniculate, first 2 segments pilose, with hairs 0.5–2 mm, terminal segment mostly scabrous, straight; **paleas** 4–5 mm, $^3/_5$–$^4/_5$ as long as the lemmas, pubescent; **anthers** not seen. **Caryopses** not seen. $2n = 70$.

Achnatherum latiglume usually grows on dry slopes in yellow pine forests of southern California. Pohl (1954) demonstrated that it is an alloploid derivative of *A. nelsonii* and *A. lemmonii*. He reported being told that it was a fairly common species in the Yosemite Valley, and suggested that the isolated occurrences in Riverside and Fresno counties might represent separate origins of the species.

Achnatherum latiglume resembles *A. nevadense* and *A. occidentale*, but the latter two species have sharper calluses, and their paleas tend to be thinner and somewhat shorter relative to the lemmas than those of *A. latiglume*.

8. Achnatherum lemmonii (Vasey) Barkworth [p. 126]
LEMMON'S NEEDLEGRASS

Plants tightly cespitose, not rhizomatous. **Culms** 15–90 cm tall, 0.7–1 mm thick, glabrous, pubescent, or tomentose; **nodes** 3–4. **Basal sheaths** glabrous, pubescent, or tomentose; **collars**, including the sides, glabrous or sparsely pubescent, hairs shorter than 0.5 mm; **basal ligules** 0.5–1.2 mm, hyaline, glabrous, truncate to acute; **upper ligules** to 2.5 mm; **basal blades** 0.5–1.5 mm wide, folded to convolute, abaxial surfaces smooth, glabrous, adaxial surfaces prominently ribbed, often with 0.3–0.5 mm hairs, sometimes glabrous; **upper blades** to 2.5 mm wide, otherwise similar to the basal blades. **Panicles** 7–21 cm long, about 1 cm wide; **branches** straight, strongly ascending to appressed, longest branches 4–5 cm. **Spikelets** appressed to the branches. **Glumes** subequal, 7–11.5 mm; **lower glumes** 0.9–1.1 mm wide, 4–5-veined; **upper glumes** 3-veined; **florets** 5.5–7 mm long, 0.8–1.3 mm thick, fusiform, somewhat laterally compressed; **calluses** 0.4–1.2 mm, blunt; **lemmas** coriaceous, evenly pubescent, hairs 0.4–1 mm, apices 1-lobed, lobe about 0.1 mm long, thick,

stiff, apical lemma hairs 0.4–0.8 mm; **awns** 16–30 mm, persistent, (once)twice-geniculate, all segments scabrous, terminal segment straight; **paleas** 4.5–6.5 mm, from $^3/_4$ as long as to equaling the lemmas, sparsely to moderately pubescent, hairs not exceeding the apices, veins terminating below the apices, apices flat or pinched; **anthers** 2.3–3.5 mm, dehiscent, not penicillate. **Caryopses** 4–5 mm, fusiform. $2n = 34$.

Achnatherum lemmonii grows in sagebrush and yellow pine associations, from southern British Columbia to California and east to Utah. It has been confused in the past with *A. nelsonii*; it differs in having narrower leaves, laterally compressed florets with a thick apical lobe, and longer paleas.

1. Lower sheaths and culms glabrous or pubescent, not tomentose, the hairs to 0.2 mm long . . . . subsp. *lemmonii*
1. Lower sheaths and culms tomentose, the hairs 0.4–0.6 mm long . . . . subsp. *pubescens*

Achnatherum lemmonii (Vasey) Barkworth subsp. lemmonii [p. 126]

Sheaths and **culms** glabrous or pubescent, hairs to 0.2 mm, sometimes varying within a population.

Achnatherum lemmonii subsp. *lemmonii* grows throughout the range shown on the map, on both serpentine and non-serpentine soils.

Achnatherum lemmonii subsp. pubescens (Crampton) Barkworth [p. 126]

Lower sheaths and **culms** densely tomentose, hairs 0.4–0.6 mm.

Achnatherum lemmonii subsp. *pubescens* is restricted to serpentine soils in southern California. Barkworth and Linman (1984) rejected subsp. *pubescens*, not appreciating that it differed in having tomentose, rather than pubescent, culms.

9. Achnatherum thurberianum (Piper) Barkworth [p. 126]
THURBER'S NEEDLEGRASS

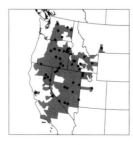

Plants cespitose, not rhizomatous. **Culms** 30–75 cm tall, 0.5–1.7 mm thick, internodes pubescent or glabrous, pubescence more common on the lower internodes, particularly just below the nodes; **nodes** 2–3, lower nodes retrorsely pubescent, upper nodes glabrous or pubescent. **Basal sheaths** glabrous, usually smooth, brown or gray-brown; **collars** glabrous, without tufts of hair at the sides; **basal ligules** 1.5–6

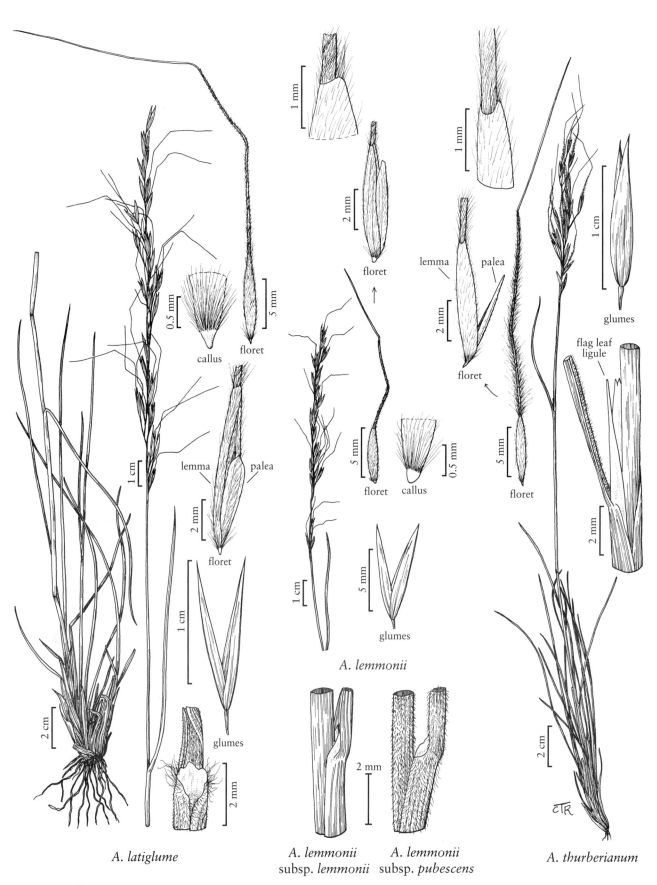

callus

floret

lemma   palea

floret

glumes

A. latiglume

floret

floret   callus

glumes

A. lemmonii

lemma   palea

floret

glumes

flag leaf
ligule

A. thurberianum

A. lemmonii
subsp. *lemmonii*

A. lemmonii
subsp. *pubescens*

ACHNATHERUM

mm, hyaline, rounded to acute, lacerate; **upper ligules** to 8 mm, hyaline, acute, glabrous; **blades** 0.5–2 mm wide, convolute, abaxial surfaces scabrous, adaxial surfaces scabrous or hairy, hairs about 0.3 mm. **Panicles** 7–15 cm long, 0.5–2.5 cm wide, often included in the upper leaf sheaths at the start of anthesis; **branches** 1.5–6 cm, appressed to strongly ascending, with 1–6 spikelets. **Glumes** often purplish; **lower glumes** 10–15 mm long, 1.2–2 mm wide; **upper glumes** to 2 mm shorter; **florets** 6–9 mm long, 0.7–1.2 mm thick, fusiform, terete; **calluses** 0.9–1.5 mm, sharp; **lemmas** coriaceous, evenly pubescent or the back glabrate distally, hairs 0.5–0.8 mm, apices lobed on 1 margin, lobe about 0.1 mm long, thick, apical lemma hairs 0.5–0.8 mm; **awns** 32–56 mm, twice-geniculate, first 2 segments pilose, hairs 0.8–2 mm, terminal segment glabrous, often scabrous; **paleas** 4.6–6.1 mm, $^3/_4$–$^9/_{10}$ as long as the lemmas, sparsely pubescent towards the base; **anthers** 2.5–3.5 mm, dehiscent, not penicillate. **Caryopses** 5–7 mm, fusiform. $2n = 34$.

Achnatherum thurberianum grows in canyons and foothills, primarily in sagebrush desert and juniper woodland associations, from Washington to southern Idaho and southwestern Montana and from California to Utah, at 900–3000 m. Its long ligules and pilose awns make it one of the easier North American species of *Achnatherum* to identify.

10. **Achnatherum coronatum** (Thurb.) Barkworth [p. 128]

CRESTED NEEDLEGRASS

Plants loosely cespitose, shortly rhizomatous, bases knotty. **Culms** 55–210 cm tall, 3–6 mm thick, internodes usually glabrous, lower internodes sometimes puberulent; **nodes** 1–2, glabrous. **Basal sheaths** mostly glabrous, often puberulent on the lower portion, flat, ribbon-like with age, margins hairy distally, hairs 1–2.5 mm; **collars** mostly glabrous; **ligules** 0.4–1.6(3) mm, truncate to slightly rounded, abaxial surfaces pubescent, ciliate, cilia about 0.5 mm; **blades** usually flat, 2.5–7 mm wide, both surfaces scabrous. **Panicles** 15–60 cm long, 2–4 cm wide; **branches** widely spreading to ascending, longest branches 4–13 cm. **Glumes** lanceolate, glabrous, tapering to awnlike apices; **lower glumes** 16–21 mm long, 1–1.3 mm wide, midveins scabrous; **upper glumes** 11–18 mm; **florets** 6.5–10 mm long, about 1 mm thick, fusiform, terete; **calluses** 0.5–2 mm, blunt to acute; **lemmas** densely hairy, hairs at midlength 1.5–4 mm, apical hairs 2–5 mm; **awns** 25–45 mm, persistent, twice-geniculate, all segments scabrous, terminal segment straight; **paleas** 3.5–5.5 mm, $^3/_5$–$^9/_{10}$ as long as the

lemmas, sparsely hairy between the veins, apices flat, rounded; **anthers** 3–4 mm, dehiscent, not penicillate. **Caryopses** 5–7 mm, fusiform. $2n = 40$.

Achnatherum coronatum grows on gravel and on rocky slopes, mostly in chaparral associations of the Coast Range from Monterey County, California, to Baja California, Mexico. It is similar in size to *A. diegoense*, but differs in its mostly glabrous internodes and longer paleas. It differs from *A. parishii*, an inland species, in its twice-geniculate awns, more robust habit, and more sparsely pubescent paleas. Occasional plants combine the characteristics of both species.

11. **Achnatherum parishii** (Vasey) Barkworth [p. 128]

Plants tightly cespitose, not rhizomatous. **Culms** 14–80 cm tall, 0.8–2 mm thick, internodes glabrous or pubescent below the nodes; **nodes** 3–5, glabrous. **Basal sheaths** mostly glabrous, sometimes pubescent at the base, flat and ribbonlike with age, margins sometimes hairy distally, hairs adjacent to the ligules 0.5–3 mm; **collars** glabrous; **ligules** truncate, abaxial surfaces pubescent, ciliate, cilia as long as or longer than the basal membrane, ligules of basal leaves 0.3–0.8 mm, of upper leaves 0.5–1.5 mm, asymmetric; **blades** 4–30+ cm long, 1–4.2 mm wide, usually flat and more or less straight, sometimes tightly convolute and arcuate. **Panicles** 7–15 cm long, 1.5–4 cm wide; **branches** strongly ascending at maturity, longest branches 1.5–4 cm. **Glumes** unequal to subequal, narrowly lanceolate, 3–5-veined; **lower glumes** 9–15 mm long, 0.9–1.2 mm wide; **upper glumes** 8–15 mm; **florets** 4.8–6.5 mm long, 0.8–1 mm thick, fusiform, terete; **calluses** 0.2–0.8 mm, acute; **lemmas** evenly and densely hairy, hairs 1.5–3.5 mm at midlength, apical hairs 2.5–5 mm; **awns** 10–35 mm, persistent, once-geniculate, first segment scabrous or strigose, hairs to 0.3 mm, terminal segment straight; **paleas** 2.5–4.5 mm, $^1/_2$–$^4/_5$ times the length of the lemmas, hairy between the veins, hairs often as long as those on the lemmas but not as dense, apices usually rounded, occasionally somewhat pinched; **anthers** 2.3–4.5 mm, dehiscent, not penicillate. **Caryopses** 3–6 mm, fusiform. $2n = $ unknown.

Achnatherum parishii grows from the coastal ranges of California to Nevada and Utah, south to Baja California, Mexico, and to the Grand Canyon in Arizona. It differs from *A. coronatum* in its once-geniculate awns, more densely pubescent paleas, and generally smaller stature; from *A. scribneri* in its shorter, blunter calluses and more abundant lemma hairs; and from *A. perplexum* in having longer hairs on its lemmas.

lemma

palea

2 mm

floret   callus

0.5 mm

5 mm

floret

2 mm

2 cm

A. coronatum

floret

2 mm

callus

0.5 mm

5 mm

floret

callus

0.5 mm

2 cm

2 cm

CTR

2 mm

subsp. *depauperatum*

floret

2 mm

5 mm

callus

0.5 mm

ligule

2 mm

subsp. *parishii*

A. parishii

ACHNATHERUM

1. Basal sheath margins glabrous or hairy distally, hairs to 0.5 mm long; culms 14–35 cm tall . . . . subsp. *depauperatum*
1. Basal sheath margins hairy distally, hairs 1–3.2 mm long; culms 20–80 cm tall . . . . subsp. *parishii*

Achnatherum parishii subsp. depauperatum (M.E. Jones) Barkworth [p. 128]

LOW NEEDLEGRASS

Culms 14–35 cm tall, 0.8–1.7 mm thick, glabrous. **Basal sheath margins** glabrous or hairy distally, hairs to 0.5 mm; **blades** 4–15 cm long, 1–2.5 mm wide, usually 0.5–1.5 mm in diameter and tightly convolute, conspicuously arcuate distally, abaxial surfaces smooth, glabrous, adaxial surfaces scabrous. **Panicles** 7–12 cm long, 1.5–2.5 cm wide. **Florets** 4.8–6 mm; **paleas** densely hairy between the veins, some hairs as long as those on the lemmas; **awns** 10–17 mm; **anthers** 2.3–2.8 mm. **Caryopses** 3–4 mm, fusiform. $2n$ = unknown.

Achnatherum parishii subsp. *depauperatum* grows in gravel and on rocky slopes, in juniper and mixed desert shrub associations, from central Nevada to western Utah. It differs from *A. webberi* in its persistent awns and thicker leaves that tend to curl when dry, and from *A. parishii* subsp. *parishii* in its smaller stature, glabrous or shortly hairy sheath margins, and densely hairy paleas.

Achnatherum parishii (Vasey) Barkworth subsp. parishii [p. 128]

PARISH'S NEEDLEGRASS

Culms 20–80 cm tall, 1.5–2 mm thick, mostly glabrous, pubescent below the nodes. **Basal sheath margins** hairy distally, hairs 1–3.2 mm; **blades** 11–30+ cm long, 2.5–4.2 mm wide, usually flat or only partly closed, sometimes completely convolute, straight to somewhat arcuate distally. **Panicles** 11–15 cm long, 2–4 cm wide. **Florets** 5.5–6.5 mm; **awns** 15–35 mm; **paleas** sparsely hairy between the veins, hairs about $^1/_2$ as long as the lemma hairs, apices usually rounded, occasionally somewhat pinched; **anthers** 3.5–4.5 mm, glabrous. **Caryopses** 5–6 mm, fusiform. $2n$ = unknown.

Achnatherum parishii subsp. *parishii* grows on dry, rocky slopes, in desert shrub and pinyon-juniper associations, from the coastal ranges of California to northeastern Nevada, eastern Utah, and the Grand Canyon in Arizona. Its range extends into Baja California, Mexico. It differs from *A. coronatum* in its shorter culms and once-geniculate awns, and from subsp. *depauperatum* in its longer culms, hairy sheath margins, and sparsely hairy paleas.

12. Achnatherum robustum (Vasey) Barkworth [p. 130]

SLEEPYGRASS

Plants cespitose, not rhizomatous. **Culms** 100–230 cm tall, 2–4.5 mm thick, mostly glabrous, often pubescent below the nodes, the pubescence antrorse or retrorse; **nodes** 4–5. **Basal sheaths** mostly glabrous, margins usually ciliate distally; **collars** hairy, those of the flag leaves densely hairy, hairs 0.5–2 mm, sides glabrous; **basal ligules** 1–2 mm; **upper ligules** to 4 mm, truncate, rounded, or obtuse, glabrous; **blades** 6–10 mm wide, glabrous, abaxial surfaces smooth, adaxial surfaces prominently ribbed, ribs scabrous. **Panicles** 15–30 cm long, 0.8–3.5 cm wide; **branches** straight, appressed to ascending, lower branches 3–9 cm. **Spikelets** appressed to the branches. **Glumes** subequal, 9–11.5 mm long, 1–1.4 mm wide; **florets** 5.9–8.5 mm long, 0.9–1.2 mm thick, fusiform, terete; **calluses** 0.3–1 mm, blunt; **lemmas** evenly hairy, hairs at midlength 0.3–0.8 mm, apical hairs to 1.5 mm; **awns** 20–32 mm, persistent, twice-geniculate, scabridulous to scabrous, scabrules to about 0.1 mm, longest on the middle segment, terminal segment straight; **paleas** 3.7–5.6 mm, $^2/_3$–$^3/_4$ as long as the lemmas, hairy, hairs about 0.5 mm, not exceeding the apices, veins terminating below the apices, apices rounded; **anthers** 4–5 mm, dehiscent, penicillate. **Caryopses** 5–6 mm. $2n$ = 64.

Achnatherum robustum grows on dry plains and hills, in open woods and forest clearings, and along roadsides, from Wyoming through Colorado to Arizona, New Mexico, and northern Mexico. Records from Kansas represent recent introductions; it is not clear whether the species has persisted there. *Achnatherum robustum* is sometimes confused with *A. nelsonii* subsp. *dorei* and *Nassella viridula*; it differs from both in the densely hairy collars of its flag leaves. Although not widely available, it has potential as an ornamental grass, particularly in arid regions with cold winters.

The English-language name refers to the effect some samples, particularly those from the Sacramento and Sierra Blanca mountains, New Mexico, have on livestock, especially horses and cattle. "Mildly poisoned animals are dejected, inactive, and withdrawn; severely poisoned animals lie on their sides in a profound slumber" (Cheeke and Shull 1985). The active ingredient is diacetone alcohol (Epstein et al. 1964).

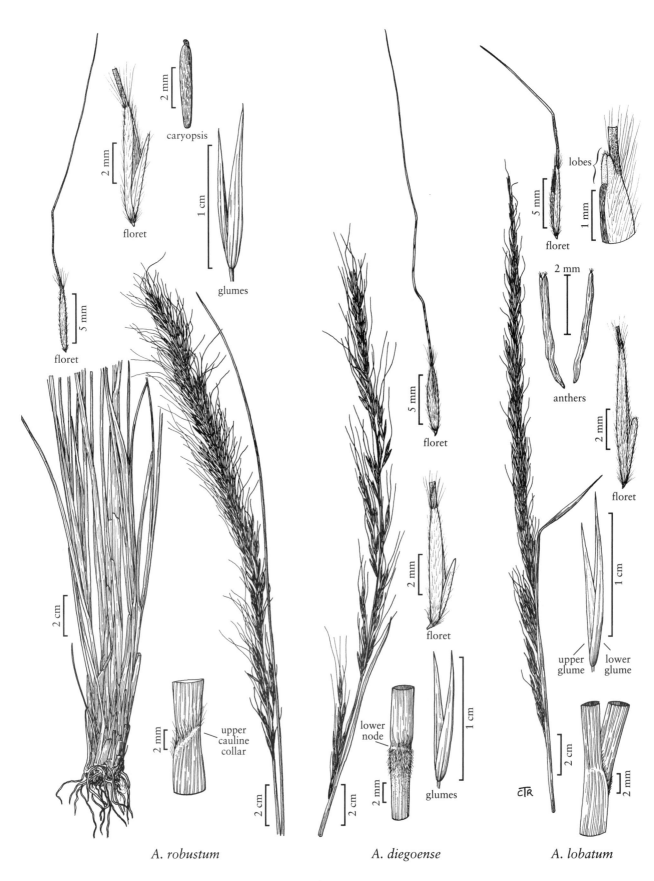

caryopsis

floret

glumes

2 mm

1 cm

floret

5 mm

2 cm

2 mm

upper
cauline
collar

2 cm

A. robustum

floret

5 mm

floret

2 mm

lower
node

2 mm

glumes

1 cm

2 cm

A. diegoense

lobes

floret

5 mm

1 mm

anthers

2 mm

floret

2 mm

upper
glume

lower
glume

1 cm

2 cm

2 mm

CTR

A. lobatum

ACHNATHERUM

13. **Achnatherum diegoense** (Swallen) Barkworth [p. 130]

SAN DIEGO NEEDLEGRASS

Plants cespitose, not rhizomatous. **Culms** 110–140 cm tall, 2.5–4 mm thick, internodes densely and retrorsely pubescent for 3–9 mm below the nodes, particularly the lower nodes, glabrous or retrorsely puberulent elsewhere; **nodes** 3, pubescent or glabrate. **Basal sheaths** mostly glabrous or puberulent, margins ciliate distally; **collars** glabrous or with hairs, hairs mostly to 0.5 mm, sides with tufts of 1.5–2 mm hairs; **ligules** 0.4–2 mm, rounded to acute, abaxial surfaces hairy, hairs to 0.5 mm; **upper ligules** 1–3 mm, similar in structure and pubescence; **blades** 1–3.5 mm wide, abaxial surfaces smooth or scabrous, adaxial surfaces prominently ribbed, hairy, hairs 2–3 mm. **Panicles** 21–25 cm long, (2)4–8 cm wide; **branches** strongly divergent to ascending, straight, lower branches 5–7 cm. **Spikelets** appressed to the branches. **Glumes** subequal, 8–11.5 mm; **lower glumes** 0.5–1 mm wide, 3–5-veined; **upper glumes** 3-veined; **florets** 5.5–7.5 mm long, 0.7–1 mm thick, fusiform, terete; **calluses** 0.25–1.2 mm, acute; **lemmas** evenly hairy, hairs at midlength and at the apices 0.5–1 mm, apical lobes 0.2–0.4 mm, membranous, flexible; **awns** 20–50 mm, persistent, twice-geniculate, all segments scabrous to scabridulous, terminal segment straight; **paleas** 2.6–3.8 mm, ¹/₂–³/₄ as long as the lemmas, pubescent, hairs not extending beyond the apices, veins terminating below the apices, apices rounded; **anthers** 2.5–4 mm, dehiscent, not penicillate. $2n$ = unknown.

Achnatherum diegoense grows in chaparral and coastal sage scrub, on rocky soil near streams or the coast, at 0–350 m, on the Channel Islands of Santa Barbara County, California, and, on the mainland, in Ventura and San Diego counties south into Baja California, Mexico.

14. **Achnatherum lobatum** (Swallen) Barkworth [p. 130]

LOBED NEEDLEGRASS

Plants cespitose, not rhizomatous. **Culms** 40–100 cm tall, 0.6–2.6 mm thick, glabrous or sparsely pubescent to 5 mm below the lower nodes; **nodes** 4. **Basal sheaths** becoming flat and papery in age, margins sometimes ciliate distally, cilia to 0.5 mm; **collars**, including the sides, glabrous or sparsely pubescent, collars of the flag leaves glabrous; **basal ligules** 0.2–1.3 mm, membranous, truncate, erose to ciliate, cilia about 0.05 mm; **upper ligules** 0.3–1 mm; **blades** 1–4 mm wide, abaxial surfaces smooth, adaxial surfaces scabrous. **Panicles** 12–28 cm long, 0.5–1.5 cm wide; **branches** ascending to appressed, straight, longest branches 3–6 cm. **Spikelets** appressed to the branches. **Glumes** unequal; **lower glumes** 9.5–12.5 mm long, 0.8–1.2 mm wide, 3(5, 7)-veined, apices straight to somewhat recurved; **upper glumes** 2–3.5 mm shorter, 3-veined; **florets** 5.5–7.5 mm long, 0.6–1.1 mm thick, terete, widest about midlength; **calluses** 0.3–0.5 mm, blunt; **lemmas** evenly hairy, hairs at midlength 0.7–1.2 mm, fusiform, terete, apical hairs 1.3–2.2 mm, erect, usually less than 1 mm longer than those at midlength, apical lobes 0.5–1.2 mm, membranous, flexible; **awns** 10–22 mm, persistent, once- or twice-geniculate, scabrous, terminal segments straight; **paleas** 3–4.3 mm, ³/₅–³/₄ as long as the lemmas, pubescent, hairs exceeding the apices, veins terminating below the apices, apices flat, rounded; **anthers** 3–4 mm, dehiscent, sparsely penicillate, hairs about 0.1 mm. $2n$ = unknown.

Achnatherum lobatum grows on rocky, open slopes in pinyon-pine and white fir associations of southern Arizona, New Mexico, Texas, and northern Mexico, at 2100–2800 m. It flowers from mid- to late summer.

Achnatherum lobatum is similar to *A. scribneri* and *A. perplexum*. It differs from *A. scribneri* in its shorter apical lemma hairs and blunt calluses, and from *A. perplexum* in having longer lemma lobes and erect apical hairs.

15. **Achnatherum aridum** (M.E. Jones) Barkworth [p. 132]

MORMON NEEDLEGRASS

Plants cespitose, not rhizomatous. **Culms** 35–85 cm tall, 0.9–2.5 mm thick, usually glabrous and smooth, sometimes scabridulous or puberulent; **nodes** 2–3. **Basal sheaths** glabrous, upper sheath margins hyaline distally; **collars** of the basal sheaths occasionally with a small tuft of 0.8 mm hair on the sides, collars of the upper leaves glabrous, scabridulous, or sparsely puberulent; **ligules** 0.2–1.5 mm, truncate to rounded, erose, sometimes ciliate, cilia about 0.05 mm; **blades** 0.9–3 mm wide, abaxial surfaces smooth or scabridulous, glabrous, adaxial surfaces hirtellous, hairs to 0.5 mm. **Panicles** 5–17 cm long, 1–1.5 cm wide, contracted, bases often enclosed at anthesis; **branches** appressed or strongly ascending, straight, lower

A. aridum

A. eminens

A. richardsonii

ACHNATHERUM

branches 1.5–4 cm. **Lower glumes** 8–15 mm long, 0.6–0.8 mm wide; **upper glumes** 1–5 mm shorter; **florets** 4–6.5 mm long, 0.6–1.1 mm thick, fusiform, terete; **calluses** 0.2–1 mm, sharp; **lemmas** evenly hairy on the lower portion, hairs 0.2–0.5 mm, the distal $^1/_5$–$^1/_4$ often glabrous, apical hairs absent or fewer than 5, to 1.5 mm; **awns** 40–80 mm, persistent, obscurely once-geniculate, scabridulous, terminal segment flexuous; **paleas** 2–3.2 mm, $^1/_2$–$^3/_4$ as long as the lemmas, pubescent, hairs exceeding the apices, apices rounded, flat; **anthers** 2–3.5 mm, dehiscent, not penicillate. $2n =$ unknown.

Achnatherum aridum grows on rocky outcrops, in shrub-steppe and pinyon-juniper associations, from southeastern California to Colorado and New Mexico, at 1200–2000 m. It has also been reported from Texas, but no specimens documenting these reports have been located. It has not been found in Mexico.

16. Achnatherum eminens (Cav.) Barkworth [p. 132]
SOUTHWESTERN NEEDLEGRASS

Plants cespitose, shortly rhizomatous, bases knotty. **Culms** 50–100 cm tall, 0.8–1.5 mm thick, glabrous; **nodes** 2–3. **Basal sheaths** mostly glabrous, ciliate on the margins; **collars** glabrous on the back, usually with tufts of hair on the sides, hairs about 0.8 mm; **basal ligules** 0.8–1.6 mm, membranous, glabrous, rounded to acute; **upper ligules** to 4.5 mm, acute; **blades** 0.7–3.5 mm wide, abaxial surfaces smooth to scabridulous, adaxial surfaces prominently ribbed, scabridulous or sparsely to densely pubescent, hairs about 0.1 mm. **Panicles** 20–55 cm long, 3–8 cm wide, open, often enclosed to midlength at anthesis; **lower branches** 5–8 cm, ascending to divergent, flexuous. **Lower glumes** 5–12 mm long, 0.5–0.7 mm wide, 3–5-veined; **upper glumes** 1–4 mm shorter, 3-veined; **florets** 4–7.5 mm long, 0.5–0.9 mm thick, fusiform, terete; **calluses** 1–2 mm, sharp; **lemmas** evenly hairy, hairs 0.4–0.8 mm throughout, apical lobes not present; **awns** 35–70 mm, persistent, twice-geniculate, first 2 segments scabrous, terminal segment flexuous; **paleas** 1–2 mm, $^1/_3$–$^1/_2$ as long as the lemmas, sparsely to moderately pubescent, apices rounded, flat; **anthers** 3–3.5 mm, dehiscent, a few penicillate, hairs about 0.3 mm. **Caryopses** about 4 mm, fusiform. $2n = 44, 46$.

Achnatherum eminens grows on dry, rocky slopes and valleys in the mountains of the southwestern United States, primarily in desert scrub, at 600–2600 m. Its range extends into Mexico. It is easy to recognize because of its open panicle, flexuous branches, and flexuous awns. It is superficially similar to *Nassella*

cernua, but differs in its longer, glabrous ligules, not or weakly overlapping lemma margins, pubescent paleas, and geographic distribution.

17. Achnatherum richardsonii (Link) Barkworth [p. 132]
RICHARDSON'S NEEDLEGRASS

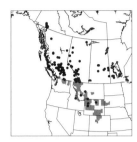

Plants tightly cespitose, not rhizomatous. **Culms** 30–100 cm tall, 1–1.5 mm thick, glabrous; **nodes** usually 3. **Basal sheaths** glabrous, margins ciliolate; **collars** glabrous, without tufts of hair on the sides; **ligules** 0.1–0.5 mm, truncate, ciliolate; **blades** 0.8–3 mm wide, convolute when dry, abaxial surfaces scabridulous, adaxial surfaces glabrous. **Panicles** 7–25 cm long, 7–15 cm wide; **branches** divergent, flexuous, longest branches 7–10 cm, with the spikelets confined to the distal $^1/_4$. **Spikelets** pendulous. **Lower glumes** 7.5–11 mm long, 0.9–1.2 mm wide; **upper glumes** 2–3 mm shorter; **florets** 5–6 mm long, 0.6–0.9 mm thick, fusiform, terete; **calluses** 0.4–0.7 mm, blunt; **lemmas** evenly hairy on the lower portion, often glabrate distally, body and apical hairs 0.2–0.5 mm, apical lobes not or scarcely developed, to 0.1 mm; **awns** 15–25 mm, persistent, twice-geniculate, first 2 segments strigulose, hairs about 0.1 mm, terminal segment straight; **paleas** 2.2–3.6 mm, $^1/_2$–$^3/_5$ as long as the lemmas, pubescent, hairs not exceeding the apices, apices rounded; **anthers** 2.5–3 mm, dehiscent, penicillate, hairs 0.1–0.5 mm. **Caryopses** 3–4 mm, fusiform. $2n = 44$.

Achnatherum richardsonii grows in open woodlands and grasslands, often on sand or gravel, from the Yukon Territory to Washington and Manitoba, and south in the Rocky Mountains through Montana and Wyoming to western South Dakota and northern Colorado. Its elevation range is 1000–3100 m. It is readily recognized by its combination of flexuous panicle branches, drooping spikelets, and straight distal awn segments. Scagel and Maze (1984) concluded that putative hybrids between *A. richardsonii* and *A. nelsonii* subsp. *dorei* were merely large plants of subsp. *dorei* that varied in the direction of *A. richardsonii*.

glumes

floret

callus

lemma   palea

floret

A. curvifolium

floret

floret

caryopsis

callus

A. scribneri

glumes

floret

floret

floret

A. perplexum

ACHNATHERUM

CTR

18. Achnatherum curvifolium (Swallen) Barkworth [p. 134]
CURLYLEAF NEEDLEGRASS

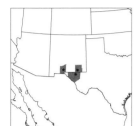

Plants tightly cespitose, not rhizomatous. **Culms** 25–55 cm tall, 0.7–1 mm thick, glabrous; **nodes** 3. **Basal sheaths** usually puberulent, hairs 0.1–0.2 mm, sometimes densely tomentose at the base, brown to gray-brown when old; **collars** glabrous, sometimes with tufts of hair on the sides, hairs to 0.5 mm; **ligules** truncate, pubescent, hairs about 0.1; **basal ligules** about 0.3 mm, **upper ligules** to 0.6 mm; **blades** normally valvate to involute, about 0.5 mm in diameter, strongly arcuate, abaxial surfaces pubescent near the base, glabrous and smooth distally, adaxial surfaces densely hairy, hairs to 0.2 mm. **Panicles** 7–11 cm long, 1–2 cm wide; **branches** appressed to strongly ascending, longest branches 3–4 cm. **Glumes** subequal, 10–14 mm long, 0.7–0.9 mm wide; **florets** 6–8 mm long, 0.4–0.8 mm thick, fusiform, terete; **calluses** 1–1.5 mm, sharp; **lemmas** evenly hairy, hairs at midlength 0.3–1 mm, apical hairs 1–1.5 mm, apical lobes not developed; **awns** 22–38 mm, once-geniculate, first segment pubescent, hairs 1–2 mm, gradually decreasing in length distally; **paleas** 2–2.3 mm, $^1/_4$–$^1/_3$ as long as the lemmas, glabrous; **anthers** about 3.5 mm, dehiscent, not penicillate. **Caryopses** about 4 mm, fusiform. $2n = 44$.

Achnatherum curvifolium grows on cliffs and in disturbed, rocky, limestone habitats. It is known from relatively few locations in the *Flora* region; it is more common in northern Mexico. It is most readily distinguished from other species of *Achnatherum* in the *Flora* region by its combination of curly leaves and hairy awns.

19. Achnatherum scribneri (Vasey) Barkworth [p. 134]
SCRIBNER'S NEEDLEGRASS

Plants cespitose, not rhizomatous. **Culms** 25–90 cm tall, 0.5–1.6 mm thick, glabrous; **nodes** 3. **Basal sheaths** becoming flat and papery, margins ciliate distally; **collars** glabrous, with tufts of hair on the sides, hairs on the basal leaves to 1.5 mm, hairs on the flag leaves 1–2.5 mm; **basal ligules** 0.3–0.8 mm, truncate, erose, ciliate, cilia 0.2–0.4 mm; **upper ligules** to 1.5 mm, asymmetric, obliquely truncate for most of their width, abruptly longer on 1 side; **blades** to 30 cm long, 2–5 mm wide, flat or involute, long-tapering. **Panicles** 7–21

cm long, 0.5–1 cm wide; **branches** appressed to ascending, straight. **Lower glumes** 10–17 mm long, 0.7–1.2 mm wide, exceeding the upper glumes by 2.5–4.5 mm, apices tapering, often slightly recurved; **florets** 6–9.5 mm long, 0.6–1.1 mm thick, fusiform, terete, widest at or below midlength; **calluses** 0.5–1.5 mm, sharp; **lemmas** evenly hairy, hairs at midlength to 1 mm, apical hairs 2–3 mm, ascending, apical lobes 0.3–0.5 mm; **awns** 13–25 mm, persistent, usually once-geniculate, first segment scabrous, terminal segment straight; **paleas** 2.5–3.5 mm, $^1/_3$–$^1/_2$ as long as the lemmas, pubescent, hairs not exceeding the apices, apices rounded; **anthers** 3–5 mm, dehiscent, not penicillate. **Caryopses** 5–6 mm, fusiform. $2n = 40$.

Achnatherum scribneri grows on rocky slopes, in pinyon-juniper and ponderosa pine associations at 1500–2700 m, from southeastern Wyoming through Colorado to Arizona, New Mexico, western Oklahoma, and Texas, and in Capital Reef National Park, Utah. At present, the Utah population appears to be disjunct from the species' primary range; this may reflect a lack of collecting. *Achnatherum scribneri* is similar to *A. parishii*, *A. robustum*, *A. perplexum*, and *A. lobatum*, differing from all of them in its sharp calluses.

20. Achnatherum perplexum Hoge & Barkworth [p. 134]
PERPLEXING NEEDLEGRASS

Plants cespitose, not rhizomatous. **Culms** 35–90 cm tall, 0.7–2.2 mm thick, lower internodes glabrous, puberulent to 5 mm below the nodes; **nodes** 2–3. **Basal sheaths** mostly glabrous, margins ciliolate distally; **collars** glabrous, including the sides; **basal ligules** 0.2–0.5 mm, truncate, ciliolate, cilia to 0.1 mm; **upper ligules** 0.2–3.5 mm, rounded to acute; **blades** to 30 cm long, 1–3 mm wide. **Panicles** 10–25 cm long, 0.5–1.5 cm wide; **branches** ascending to appressed, straight. **Spikelets** appressed to the branches. **Glumes** unequal; **lower glumes** 10–15 mm long, 0.5–1.1 mm wide, exceeding the upper glumes by 1–3(4) mm; **florets** 5.5–11 mm long, 0.7–1 mm thick, fusiform, terete, widest at or below midlength; **calluses** 0.4–0.6 mm, blunt; **lemmas** evenly hairy, hairs at midlength about 1 mm, apical hairs 1–2 mm, ascending to divergent, apical lobes 0.2–0.5 mm, membranous, flexible; **awns** 10–19 mm, persistent, once(twice)-geniculate, basal segments scabrous, terminal segments straight; **paleas** 2.8–5.6 mm, $^1/_2$–$^2/_3$ as long as the lemmas, hairy, hairs not or scarcely exceeding the apices, veins terminating at or before the apices, apices acute to rounded; **anthers**

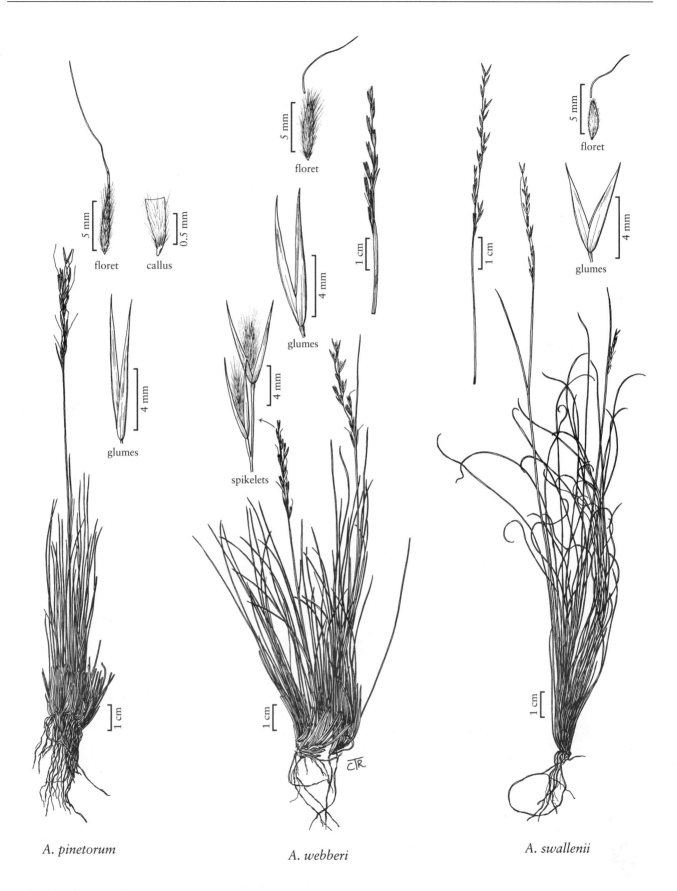

A. *pinetorum*   A. *webberi*   A. *swallenii*

ACHNATHERUM

2.5–4 mm, dehiscent, not penicillate. **Caryopses** 3–6 mm, fusiform. $2n$ = unknown.

Achnatherum perplexum grows on slopes in pinyon-pine associations of the southwestern United States and adjacent Mexico, at 1500–1700 m. It flowers in late summer to early fall. It has generally been confused with *A. scribneri*, *A. nelsonii*, and *A. lobatum*. It differs from *A. scribneri* in the glabrous collar margins of its basal leaves and its blunt calluses; from *A. nelsonii* and *A. lettermanii* in its unequal glumes; from *A. lettermanii* in its relatively short paleas; and from *A. lobatum* in its shorter lemma lobes and ascending to divergent apical lemma hairs.

21. Achnatherum pinetorum (M.E. Jones) Barkworth [p. 136]
PINEWOODS NEEDLEGRASS

Plants tightly cespitose, not rhizomatous. **Culms** 14–50(80) cm tall, 0.4–0.9 mm thick, mostly glabrous, lower internodes often puberulent or pubescent, particularly below the nodes; **nodes** 2–3. **Basal sheaths** not becoming flat and ribbonlike with age, usually glabrous, throat sometimes with a few hairs, hairs about 0.2 mm; **collars** glabrous, including the sides; **basal ligules** 0.2–0.8 mm, truncate to rounded, membranous, glabrous; **upper ligules** to 2 mm, rounded; **blades** usually involute and 0.2–0.4 mm in diameter, 0.5–1 mm wide if flat, often arcuate distally, abaxial surfaces scabridulous, adaxial surfaces hairy, hairs about 0.1 mm. **Panicles** 4.5–20 cm long, 0.5–1 cm wide, contracted; **branches** appressed, lower branches 1–5 cm, with 2–7 spikelets. **Glumes** subequal, 7–11 mm long, 0.6–0.9 mm wide, lanceolate, not saccate; **florets** 3.5–5.5 mm long, 0.6–0.8 mm thick, fusiform, terete; **calluses** 0.4–0.6 mm, sharp; **lemmas** densely and evenly pilose, hairs at midlength 1.5–3.5 mm, apical hairs to 5 mm, apical lobes 0.3–2 mm, thin; **awns** 13–25 mm, persistent, twice-geniculate, first 2 segments scabrous; **paleas** 2.5–4 mm, from $^2/_3$ as long as to equaling the lemmas, hairy; **anthers** 1.8–2.6 mm, dehiscent, not penicillate. **Caryopses** 2.5–4 mm, fusiform. $2n$ = 32.

Achnatherum pinetorum usually grows on rocky soil, in pinyon-juniper to subalpine associations, at 2100–3300 m. Its range extends from Oregon, Idaho, and Montana south to California, Nevada, and Colorado. It differs from *A. webberi* in its longer, persistent awns, and from *A. lettermanii* in its sharp calluses and longer lemma hairs.

22. Achnatherum webberi (Thurb.) Barkworth [p. 136]
WEBBER'S NEEDLEGRASS

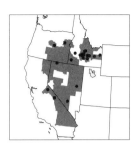

Plants tightly cespitose, not rhizomatous. **Culms** 12–35 cm tall, 0.4–0.7 mm thick, smooth or antrorsely scabridulous; **nodes** 2–3. **Basal sheaths** glabrous, smooth or scabridulous; **collars** glabrous, without tufts of hair on the sides; **basal ligules** 0.1–1 mm, truncate to rounded; **upper ligules** 1–2 mm, acute; **blades** 0.5–1.5 mm wide when flat, usually folded to involute and about 0.5 mm in diameter, stiff, abaxial surfaces smooth or scabrous, adaxial surfaces scabrous. **Panicles** 2.5–7 cm long, 0.5–2 cm wide, contracted; **branches** appressed, longest branches 1–2 cm. **Glumes** subequal, 6–10 mm long, 0.6–0.9 mm wide, lanceolate, not saccate; **florets** 4.5–6 mm long, 0.7–1 mm thick, fusiform, terete; **calluses** 0.3–0.8 mm, blunt; **lemmas** evenly and densely pilose, hairs 2.5–3.5 mm, apical lobes 0.6–1.9 mm, membranous; **awns** 4–11 mm, readily deciduous, straight to once-geniculate, scabrous; **paleas** 4–5.6 mm, from as long as to slightly longer than the lemmas; **anthers** 1.6–2 mm, dehiscent, not penicillate. **Caryopses** 3.5–4.5 mm, fusiform. $2n$ = 32.

Achnatherum webberi grows in dry, open flats and on rocky slopes, often with sagebrush, at 1500–2500 m. It grows at scattered locations from Oregon and Idaho to California and Nevada. It differs from *A. hymenoides* in its cylindrical floret and non-saccate glumes, and from *A. pinetorum* and *A. parishii* subsp. *parishii* in its shorter, deciduous awns. It also has narrower blades than *A. parishii* subsp. *depauperatum*.

23. Achnatherum swallenii (C.L. Hitchc. & Spellenb.) Barkworth [p. 136]
SWALLEN'S NEEDLEGRASS

Plants tightly cespitose, not rhizomatous. **Culms** 15–25 cm tall, 0.5–1 mm thick, glabrous; **nodes** 2–3. **Basal sheaths** mostly glabrous, pubescent at the base, throats glabrous; **collars**, including the sides, glabrous; **basal ligules** 0.2–0.3 mm, obtuse to rounded, glabrous; **upper ligules** to 0.5 mm, rounded to broadly acute; **blades** 0.4–0.7 mm wide, arcuate, abaxial surfaces smooth or scabridulous, adaxial surfaces with hairs shorter than 0.5 mm. **Panicles** 3–6.5 cm long, 0.3–0.7 cm wide; **branches** appressed, lower branches 1–5 cm, with 1–5 spikelets. **Glumes** subequal, 4–5.5 mm, not saccate, apices narrowly acute to acuminate, midveins often prolonged into an awnlike tip; **lower glumes** 0.6–1 mm wide, apices narrowly acute; **florets** 2.5–3.5 mm long, 0.6–0.8 mm thick, fusiform, terete; **calluses** 0.1–0.2 mm, blunt; **lemmas** evenly hairy, hairs 0.3–0.5 mm, all similar in length, apical lobes 0.3–0.5 mm, thickly membranous; **awns** 5–6 mm, once-geniculate, readily deciduous, basal segment scabridulous; **paleas** 2.2–2.5 mm, slightly shorter than the lemmas; **anthers** 1.5–2 mm, dehiscent, not penicillate. **Caryopses** 2–3 mm, ovoid. 2n = 34.

Achnatherum swallenii grows on open, rocky sites, frequently with low sagebrush, in Idaho and western Wyoming, at 1500–2200 m. It is a dominant species in parts of eastern Idaho, although it is poorly represented in collections.

24. Achnatherum wallowaense J.R. Maze & K.A. Robson [p. 138]

WALLOWA NEEDLEGRASS

Plants tightly cespitose, not rhizomatous. **Culms** (10)15–40 (45) cm tall, 0.5–0.7 mm thick, glabrous; **nodes** 1–2. **Basal sheaths** becoming flat with age, glabrous; **collars** glabrous, including the sides; **basal ligules** 0.8–1.3 mm, membranous, truncate to broadly acute, glabrous; **upper ligules** to 1.6 mm; **blades** tightly valvate to involute, 0.5–0.8 mm in diameter, abaxial surfaces scabridulous, adaxial surfaces hairy, sometimes densely hairy, hairs shorter than 0.05 mm. **Panicles** (4)5–13(15) cm long, to 10 cm wide, lax; **branches** divergent, flexuous, longest branches 2–10 cm, with spikelets confined to the distal portions, drooping. **Glumes** obtuse to acute; **lower glumes** 3.5–7 mm long, 0.8–1.3 mm wide, 5(7)-veined; **upper glumes** 3–6.5 mm, 3-veined; **florets** 3–5.5 mm long, 1–1.5 mm thick, fusiform, terete; **calluses** 0.1–0.2 mm, blunt; **lemmas** coriaceous, shiny, glabrous, black to dark brown at maturity, margins overlapping at maturity, apices thickened dorsally; **awns** 5–11 mm, readily deciduous, not or weakly geniculate, scabrous; **paleas** 2.8–4.5 mm, similar to the lemmas in texture, glabrous; **anthers** 1.6–1.8 mm, dehiscent, not penicillate. **Caryopses** 2–4 mm, ovoid. 2n = unknown.

Achnatherum wallowaense grows in shallow, rocky soil at scattered localities, from 1000–1600 m, in the Wallowa and Ochoco mountains, Oregon.

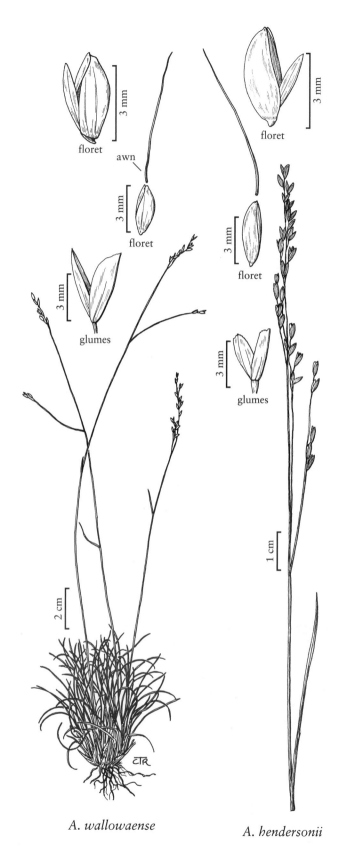

A. wallowaense   *A. hendersonii*

ACHNATHERUM

25. Achnatherum hendersonii (Vasey) Barkworth [p. 138]

HENDERSON'S NEEDLEGRASS

Plants tightly cespitose, not rhizomatous. **Culms** 10–35 cm tall, 0.3–0.9 mm thick, pubescent below the nodes, glabrous or sparsely puberulent elsewhere; **nodes** 1–2. **Basal sheaths** completely or mostly glabrous, margins sometimes ciliate distally; **collars** glabrous; **ligules** 0.4–1 mm, hyaline, glabrous or pubescent, rounded; **blades** tightly folded or convolute, to 1 mm wide or thick, abaxial surfaces scabrous, adaxial surfaces pubescent. **Panicles** 4–12 cm long, 2–5 cm wide, erect; **branches** and **pedicels** straight, appressed to strongly ascending, longest branches 2–7 cm. **Glumes** subequal, 3.5–5.5 mm long, 1–1.5 mm wide, 5-veined; **lower glumes** obtuse, apices rounded to acute; **upper glumes** rounded to obtuse, subequal or to 1 mm shorter than the lower glumes; **florets** 3.5–4.5 mm long, 1–1.5 mm wide, fusiform, laterally compressed; **calluses** 0.3–0.5 mm, blunt; **lemmas** coriaceous, glabrous, shiny, apical lobes about 0.2 mm long, thick; **awns** 6–10 mm, readily deciduous, not geniculate, scabrous; **paleas** about 3 mm, from $^3/_4$ as long as to equaling the lemmas, indurate, glabrous, apices rounded, flat; **anthers** about 2.5 mm, dehiscent, penicillate. **Caryopses** 2.5–4 mm. $2n = 34$.

Achnatherum hendersonii grows in dry, rocky, shallow soil, in sagebrush or ponderosa pine associations. It is known from only three counties: Yakima and Kittitas counties, Washington, and Crook County, Oregon. Maze (1981) noted that, at one site, *A. hendersonii* was restricted to areas subject to frost heaving, although under cultivation, it can grow without such disturbance. He hypothesized that its survival in such sites is attributable to a competitive advantage gained by the structure of its root system. Unlike *Poa secunda*, which grew in the surrounding, undisturbed areas, the outer cortex and epidermis of the roots of *A. hendersonii* form a sheath around the stele and inner cortex. When the roots are pulled, this sheath slips and breaks but the internal structures remain intact. In *Poa secunda*, the outer part of the root is attached to the central core and, when the roots are pulled, they break. *Achnatherum hendersonii* also differs from *P. secunda* in having relatively few (9–12), evenly distributed roots that extend to 30 cm.

26. Achnatherum hymenoides (Roem. & Schult.) Barkworth [p. 140]

INDIAN RICEGRASS

Plants tightly cespitose, not rhizomatous. **Culms** 25–70 cm tall, 0.7–1.3 mm thick, glabrous or partly scabridulous; **nodes** 3–4. **Sheaths** glabrous or scabridulous, sometimes puberulent on the distal margins, hairs to 0.8 mm; **collars** glabrous, sometimes with tufts of hair on the sides, hairs to 1 mm; **basal ligules** 1.5–4 mm, hyaline, glabrous, acute; **upper ligules** to 2 mm; **blades** usually convolute, 0.1–1 mm in diameter, abaxial surfaces smooth or scabridulous, adaxial surfaces pubescent. **Panicles** 9–20 cm long, 8–14 cm wide; **branches** ascending to strongly divergent, longest branches 3–15 cm; **pedicels** paired, conspicuously divaricate, shorter pedicels in each pair usually at least $^1/_2$ as long as the longer pedicels. **Glumes** subequal, 5–9 mm long, 0.8–2 mm wide, saccate below, puberulent, hairs about 0.1 mm, tapering above midlength, apices acuminate; **lower glumes** 5-veined at the base, 3-veined at midlength; **upper glumes** 5–7-veined at the base; **florets** 3–4.5 mm long, 1–2 mm thick, obovoid; **calluses** 0.4–1 mm, sharp; **lemmas** indurate, densely and evenly pilose, hairs 2.5–6 mm, easily rubbed off, apices not lobed; **awns** 3–6 mm, rapidly deciduous, not geniculate, scabrous; **paleas** subequal to the lemmas in length and texture, glabrous, apices pinched; **anthers** 1.5–2 mm, penicillate, dehiscent, well-filled. **Caryopses** 2–3 mm. $2n = 46, 48$.

Achnatherum hymenoides grows in dry, well-drained soils, primarily in the western part of the *Flora* region and northern Mexico. Specimens from further east may be introduced; it is unknown whether they have persisted. The roots of *A. hymenoides* are often surrounded by a rhizosheath formed by mucilaginous secretions to which soil particles attach. This rhizosheath harbors nitrogen-fixing organisms that probably contribute to the success of the species as a colonizer.

Native Americans used the seeds of *Achnatherum hymenoides* for food. It is also one of the most palatable native grasses for livestock. Several cultivars have been developed for use in restoration work, and it is becoming increasingly available for use as an ornamental.

Achnatherum hymenoides forms natural hybrids with other members of the *Stipeae*. See discussion on p. 142.

spikelets

glumes

spikelets

caryopsis

callus

lemma

palea

floret

floret

awn

floret

floret

callus

glumes

floret

glumes

glumes

A. hymenoides

A. arnowiae

A. contractum

CTR

ACHNATHERUM

27. **Achnatherum arnowiae** (S.L. Welsh & N.D Atwood) Barkworth [p. 140]
ARNOW'S RICEGRASS

Plants tightly cespitose, not rhizomatous. **Culms** 15–75 cm tall. **Sheaths** smooth, mostly glabrous, margins ciliate; **collars** glabrous, with or without tufts of hair at the sides, hairs 0.7–2 mm; **ligules** 1–4 mm, glabrous or sparsely hairy, acute, sometimes ciliate; **blades** 1–2 mm wide, involute, 0.5–1 mm in diameter, abaxial surfaces scabridulous or smooth, adaxial surfaces densely hairy, hairs about 0.2 mm. **Panicles** 5–20 cm long, 0.5–2.8 cm wide, loosely contracted; **branches** strongly ascending, longest branches 0.5–2.5(5) cm. **Spikelets** evenly distributed along the branches; **pedicels** loosely appressed to the branches, paired, unequal, shorter pedicels in each pair usually less than ¹/₂ as long as the longer pedicels. **Glumes** slightly unequal, saccate below, tapering from about midlength, veins and sometimes also the intercostal regions puberulent, hairs to 0.1 mm, apices acute to acuminate; **lower glumes** 5.1–6.1 mm; **upper glumes** 4.3–5.2 mm; **florets** 2.8–4.2 mm, ovoid; **calluses** 0.2–0.4 mm, acute; **lemmas** indurate, dark gray-brown, smooth, densely pilose, hairs at midlength and at the apices similar, 2–3 mm, easily rubbed off, apices not lobed; **awns** 3–4.4 mm, rapidly deciduous, not geniculate, scabrous; **paleas** similar to the lemmas in length and texture, glabrous, apices pinched; **anthers** about 1 mm, dehiscent, penicillate. **Caryopses** 1–1.7 mm long, 0.8–1 mm in diameter, globose to obovoid. 2*n* = unknown.

Achnatherum arnowiae grows in pinyon-juniper, sagebrush, and mixed desert shrub communities in Utah, at 1400–2000 m. Welsh and Atwood (2003) state that specimens belonging to *A. arnowiae* are often filed as *A. ×bloomeri*, and they suggest that this species may also be hybrid in origin, possibly with *Hesperostipa comata* as one of its parents. Another possibility is that it is a derivative of *A. hymenoides* that is adapted to particular soil types.

28. **Achnatherum contractum** (B.L. Johnson) Barkworth [p. 140]
CONTRACTED RICEGRASS

Plants tightly cespitose, not rhizomatous. **Culms** 30–50 cm tall, 1–1.3 mm thick, glabrous; **nodes** 3–4. **Sheaths** glabrous; **collars** puberulent, hairs about 0.2 mm, sometimes the margins with poorly developed tufts of hair to 0.5 mm; **ligules** 1.8–5 mm, broadly to narrowly acute, glabrous; **blades** 0.5–1.5 mm wide, flat or convolute, abaxial surfaces smooth, adaxial surfaces scabridulous. **Panicles** 6–25 cm long, 7–15 cm wide; **branches** ascending to strongly divergent, longest branches 5–8 cm; **pedicels** appressed to the branches, paired, unequal, shorter pedicels in each pair usually less than ¹/₂ as long as the longer pedicels. **Spikelets** confined to the distal ¹/₂ of the branches. **Glumes** saccate below, tapering at midlength, glabrous, midveins sometimes scabridulous, apices acuminate; **lower glumes** 5.5–7 mm long, 0.9–1.5 mm wide; **upper glumes** about 0.3 mm shorter; **florets** 2.5–3.5 mm long, 0.7–1.5 mm thick, fusiform to obovoid; **calluses** 0.3–0.5 mm, blunt; **lemmas** densely pilose, hairs at midlength and on the apices similar, 1.2–2 mm, apical lobes 0.5–0.6 mm; **awns** 6.5–9 mm, readily deciduous, scabrous; **paleas** similar to the lemmas in length, texture, and pubescence, distal hairs exceeding the paleal apices, apices rounded, flat; **anthers** about 1.5 mm, penicillate, dehiscent, well-filled. **Caryopses** 1.5–2.5 mm, globose to obovoid. 2*n* = 48.

Achnatherum contractum grows in rocky grasslands in eastern Idaho, southwestern Montana, and Wyoming. It is a fertile derivative of a *Piptatherum micranthum* × *Achnatherum hymenoides* hybrid (Shechter and Johnson 1968; Shechter 1969). Immature specimens of *A. hymenoides* are sometimes confused with *A. contractum* because they have contracted panicles with appressed branches and pedicels; they differ in having pedicel pairs in which the shorter pedicel is more than half as long as the longer pedicel.

Achnatherum ×bloomeri (Bol.) Barkworth and other hybrids involving *A. hymenoides* [p. 142]

Numerous natural hybrids exist between *Achnatherum hymenoides* and other members of the *Stipeae*. Johnson (1945, 1960, 1962, 1963, 1972) described several of these; all are sterile. Using the treatment adopted here, Johnson's hybrids have as the second parent *A. occidentale* (all subspecies), *A. thurberianum*, *A. scribneri*, *A. robustum*, *Jarava speciosa*, and *Nassella viridula*. Evidence from herbarium specimens suggests that *A. hymenoides* also forms sterile hybrids with other species of *Achnatherum*. The name **Achnatherum ×bloomeri** applies only to hybrids between *A. hymenoides* and *A. occidentale* subsp. *occidentale*, but plants keying here may include any of the other interspecific hybrids. They all differ from *A. hymenoides* in having more elongated florets and awns 10–20 mm long, and from their other parent, in most instances, in having longer lemma hairs and more saccate glumes. Identification of the second parent is best made in the field by noting which other species of *Stipeae* are present, bearing in mind that species that are not in anthesis at the same time in one year might have sufficient overlap for hybridization in other years.

Of the two intergeneric hybrids mentioned above, that with *Nassella viridula* is treated as ×*Achnella caduca* (see p. 169). It differs from *Achnatherum hymenoides* in its longer glumes and florets, and from other *A. hymenoides* hybrids in having a readily deciduous awn. No binomial has been proposed for the hybrid with *Jarava speciosa*. There is one fertile intergeneric hybrid involving *A. hymenoides*, *A. contractum*. It is included in *Achnatherum* and described below because it resembles other members of *Achnatherum* more than it does *Piptatherum*, the genus of the other parent.

Sterile hybrids have anthers that do not dehisce, and contain few, poorly formed pollen grains. They also fail to form good caryopses, but this is also true of some non-hybrid plants. In the case of non-hybrid plants, failure to form good caryopses can result from failure to capture pollen or from incompatibility between the pollen grain and the pistillate plant. It is not known which, if either, of these explanations accounts for the large number of empty caryopses found in *Achnatherum hymenoides*.

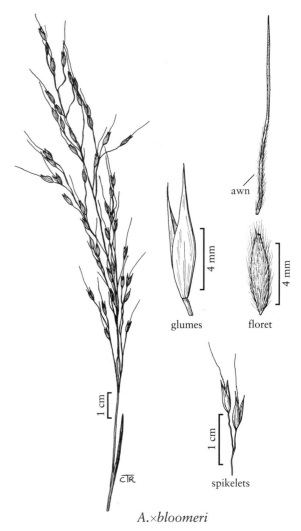

A.×bloomeri

ACHNATHERUM

10.03 PTILAGROSTIS Griseb.

Mary E. Barkworth

Plants perennial; tightly cespitose, not rhizomatous. **Culms** 10–60 cm tall, 0.4–1.2 mm thick, erect, glabrous, not branching at the upper nodes; **nodes** 1, exposed; **basal branching** intravaginal; **prophylls** shorter than the sheaths. **Leaves** mostly basal, not overwintering; **cleistogenes** not developed; **sheaths** open for most of their length, glabrous, smooth to somewhat scabrous; **auricles** absent; **ligules** 0.2–3 mm, membranous, rounded to acute, glabrous, not ciliate; **blades** convolute, 0.2–0.6 mm in diameter, apices stiff, flag leaf blades longer than 1 cm. **Inflorescences** terminal panicles; **branches** capillary, often flexuous, sometimes straight, glabrous or sparsely hirtellous; **disarticulation** above the glumes, beneath the floret. **Spikelets** 3.4–7 mm, with 1 floret; **rachillas** not prolonged beyond the floret. **Glumes** subequal, hyaline, mostly purplish, venation not evident, apices rounded to acute; **florets** slightly shorter than the glumes, terete; **calluses** 0.1–0.8 mm, blunt, hairy; **lemmas** thickly membranous, smooth, hairy over the basal portion or throughout, hairs 0.2–0.4 mm, margins flat, not overlapping at maturity, apices lobed, lobes 0.1–1 mm, membranous, apices with a single terminal awn, lemma-awn junction evident; **awns** 5–30 mm, centric, persistent, once- or twice-geniculate, sometimes weakly so, scabrous or hairy, hairs to 2 mm; **paleas** slightly shorter to slightly longer than the lemmas, hairy, hairs to 0.5 mm, 2-veined, not keeled over the veins, flat between the veins, veins ending before the apices, apices rounded; **lodicules** 3, free, membranous; **anthers** 3, 0.4–3.3 mm, sometimes penicillate; **ovaries** glabrous; **styles** 2, white, free to the base. **Caryopses** 2–5 mm, fusiform, not ribbed. *x* = unknown. Name from the Greek *ptilon*, 'feather', and *agrostis*, 'grass', a reference to the feathery awn of the type species.

Ptilagrostis is an alpine and subalpine genus of about 9 species. It grows in central Asia and the high mountains of western North America, sometimes in bogs. Its leaves differ from those of most other members of the *Stipeae* in lacking sclerenchyma pillars or girders. Two species are native to the *Flora* region.

SELECTED REFERENCES Barkworth, M.E. 1983. *Ptilagrostis* in North America and its relationship to other Stipeae (Gramineae). Syst. Bot. 8:395–419; Johnston, B.C. 2006. *Ptilagrostis porteri* (Rydberg) W.A. Weber (Porter's False Needlegrass): A technical conservation assessment. Species Conservation Project Report. U.S. Department of Agriculture-Forest Service, Rocky Mountain Region, Lakewood, Colorado, U.S.A. 62 pp. http://www.fs.fed.us/r2/projects/scp/assessments/ptilagrostisporteri.pdf.

1. Awns not hairy; panicles loosely contracted . . . . 1. *P. kingii*
1. Awns hairy, hairs on the lowest segment 1–2 mm long; panicles open or loosely contracted . . . . . . 2. *P. porteri*

1. Ptilagrostis kingii (Bol.) Barkworth [p. 144]
KING'S PTILAGROSTIS, SIERRA PTILAGROSTIS

Culms 15–38 cm tall, 0.4–0.8 mm thick. **Basal ligules** 1–2 mm, acute; **upper ligules** to 2.5 mm; **blades** filiform, about 0.3 mm in diameter. **Panicles** 6–10 cm, loosely contracted; **branches** ascending to appressed. **Glumes** 3–4.5 mm; **florets** 2.8–4.2 mm; **calluses** 0.3–0.7 mm; **lemmas** with 0.1–0.4 mm lobes; **awns** 10–14 mm, scabridulous, not hairy, weakly once- or twice-geniculate; **anthers** 0.5–1.5 mm, penicillate. **Caryopses** 1.5–2.3 mm. 2*n* = 22.

Ptliagrostis kingii grows along damp streambanks and wet meadows of the Sierra Nevada, at elevations from 2700–3500 m. It differs from most species in the genus in its scabridulous, rather than plumose, awns and short lemma lobes.

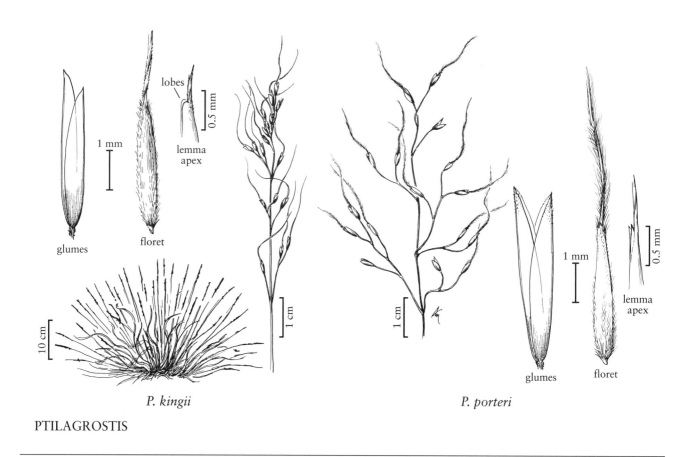

P. kingii   *P. porteri*

PTILAGROSTIS

---

2. Ptilagrostis porteri (Rydb.) W.A. Weber [p. 144]
PORTER'S PTILAGROSTIS,
ROCKY-MOUNTAIN PTILAGROSTIS

Culms 23–50 cm tall, 0.6–1.2 mm thick. **Lower ligules** 0.7–1.5 mm; **upper ligules** to 2.5 mm; **blades** filiform, 0.3–0.6 mm in diameter. **Panicles** 7.5–12 cm, open or loosely contracted; **branches** spreading. **Glumes** 4.5–6 mm; **florets** 2.5–4 mm; **calluses** 0.1–0.8 mm; **lemmas** with 0.2–0.8 mm lobes; **awns** 5–25 mm, hairy, hairs on the lowest segment 1–2 mm; **anthers** 1.2–3 mm, glabrous. 2*n* = unknown.

Ptilagrostis porteri grows on hummocks of poorly drained wetlands, at 2700–3600 m, in central Colorado. It is often associated with *Salix* spp. and *Deschampsia cespitosa*. There are 29 known populations.

The proportion of plants having closed and open panicles varies among the populations; in some populations all or almost all plants have open panicles, in others all or almost all plants have closed panicles (Johnston 2006). Barkworth (1983) treated *P. porteri* as a subspecies of *P. mongolica* (Turcz. *ex* Trin.) Griseb. because of their morphological similarity. They differ ecologically and geographically, *P. mongolica sensu stricto* growing in rocky alpine habitats of central Asia.

10.04 PIPTATHERUM P. Beauv.

Mary E. Barkworth

Plants perennial; cespitose or soboliferous, sometimes rhizomatous. **Culms** 10–140(150) cm, erect, usually glabrous, usually smooth; **nodes** 1–6; **branching** intra- or extravaginal at the base, not branching above the base; **prophylls** concealed by the leaf sheaths. **Leaves** sometimes basally concentrated; **cleistogenes** not present; **sheaths** open, glabrous, smooth to scabrous; **auricles**

absent; **ligules** 0.2–15 mm, membranous to hyaline; **blades** 0.5–16 mm wide, flat, involute, valvate, or folded, often tapering in the distal ⅓, apices acute to acuminate, not stiff, basal blades not overwintering, sometimes not developed, flag leaf blades well developed, longer than 1 cm. **Inflorescences** 3–40 cm, terminal panicles, open or contracted; **branches** straight or flexuous, usually scabrous, rarely smooth; **pedicels** often appressed to the branches. **Spikelets** 1.5–7.5 mm, with 1 floret; **rachillas** not prolonged beyond the floret; **disarticulation** above the glumes, beneath the floret. **Glumes** from 1 mm shorter than to exceeding the florets, subequal or the lower glumes longer than the upper glumes, membranous, 1–9-veined, veins evident, apices obtuse to acute or acuminate; **florets** 1.5–10 mm, usually dorsally compressed, sometimes terete; **calluses** 0.1–0.6 mm, glabrous or with hairs, blunt; **lemmas** 1.2–9 mm, smooth, coriaceous or stiffly membranous, tawny or light brown to black at maturity, 3–7-veined, margins flat, separated and parallel for their whole length at maturity, apices not lobed or lobed, glabrous or hairy, hairs about 0.5 mm, not spreading, awned, lemma-awn junction evident; **awns** 1–18(20) mm, centric, often caducous, almost straight to once- or twice-geniculate, scabrous; **paleas** as long as or slightly longer than the lemmas, similar in texture and pubescence, 2(3)-veined, not keeled over the veins, flat between the veins, veins terminating near the apices, apices often pinched; **anthers** 3, 0.6–5 mm, sometimes penicillate; **styles** 2 and free to their bases, or 1 with 2–3 branches. **Caryopses** glabrous, ovoid to obovoid; **hila** ½ as long as to equaling the length of the caryopses. $x = 11, 12$. Name from the Greek *pipto*, 'fall', and *ather*, 'awn'.

Piptatherum has approximately 30 species, most of which are Eurasian. They extend from lowland to alpine regions, and grow in habitats ranging from mesic forests to semideserts.

The pistils in *Piptatherum* exhibit variability in the development of the styles, a feature that can be seen only in florets shortly before or at anthesis. This variability is reported in the descriptions, but the number of specimens examined per species is low, sometimes only one.

SELECTED REFERENCES **Curto, M.L.** and **D.H. Henderson**. 1998. A new *Stipa* (Poaceae: Stipeae) from Idaho and Nevada. Madroño 45:57–63; **Freitag, H.** 1975. The genus *Piptatherum* (Gramineae) in southwest Asia. Notes Roy. Bot. Gard. Edinburgh 42:355–489; **Jacobs, S.W.L., R. Bayer, J. Everett, M.O. Arriaga, M.E. Barkworth, A. Sabin-Badereau, M.A. Torres, F. Vázquez,** and **N. Bagnall**. 2006. Systematics of the tribe Stipeae using molecular data. Aliso 23:349–361; **Johnson, B.L.** 1945. Cytotaxonomic studies in *Oryzopsis*. Bot. Gaz. 107:1–32.

1. Basal leaf blades 0–2 cm long; cauline leaf blades 8–16 mm wide; florets 4.5–7.5 mm long . . . 6. *P. racemosum*
1. Basal leaf blades 4–45 cm long; cauline leaf blades 0.5–10 mm wide; florets 1.5–6 mm long.
  2. Lemmas and calluses usually glabrous, occasionally sparsely pubescent; florets 1.5–2.5 mm long.
    3. Blades 0.5–2.5 mm wide, often involute; panicles 5–20 cm long, the lower nodes with 1–3 branches . . . . 4. *P. micranthum* (in part)
    3. Blades 2–10 mm wide, flat; panicles 10–40 cm long, the lower nodes usually with 3–7 branches, sometimes with 15–30+ branches . . . . 7. *P. miliaceum*
  2. Lemmas evenly pubescent; calluses hairy; florets 1.5–6 mm long.
    4. Awns 3.9–15 mm long, persistent, once- or twice-geniculate.
      5. Primary panicle branches straight, appressed; awns 3.9–7 mm long; florets 3–6 mm long . . . . 1. *P. exiguum*
      5. Primary panicle branches somewhat flexuous, often divergent; awns 5–15 mm long; florets 2.2–4.5 mm long . . . . 2. *P. canadense*
    4. Awns 1–8 mm long, caducous, often absent from herbarium specimens, straight or arcuate.
      6. Awns 4–8 mm long; florets 1.5–2.5 mm long . . . . 4. *P. micranthum* (in part)
      6. Awns 1–2.5 mm long; florets 2.2–4.5 mm long.
        7. Lower panicle branches straight; ligules 0.5–2.5 mm long . . . . 3. *P. pungens*
        7. Lower panicle branches flexuous; ligules 1.8–5.5 mm long . . . . 5. *P. shoshoneanum*

1. Piptatherum exiguum (Thurb.) Dorn [p. 147]
LITTLE PIPTATHERUM

Plants tightly cespitose, not rhizomatous. **Culms** 12–40 cm, sometimes not exceeding the basal leaves, scabridulous; **basal branching** mostly intravaginal. **Leaves** basally concentrated; **sheaths** mostly smooth, sometimes scabridulous distally; **ligules** 1.2–3.5 mm, acute; **basal blades** 9–30 cm long, 0.6–1.4 mm wide when open, usually valvate and 0.4–0.8 mm in diameter, both surfaces scabrous. **Panicles** 3.5–9 cm, lower nodes with 1–2 branches; **branches** straight, tightly appressed to the rachises, lower branches 1–2 cm, with 1–2 spikelets. **Glumes** subequal, 3.5–6 mm long, 1.8–2.2 mm wide, ovate, apices acute; **florets** 3–6 mm, terete; **calluses** 0.2–0.5 mm, hirsute, disarticulation scars circular; **lemmas** evenly pubescent, tan or gray-brown at maturity, margins not overlapping at maturity, apical lobes 2, 0.3–0.4 mm, thick; **awns** 3.9–7 mm, persistent, strongly once-geniculate, basal segment twisted; **paleas** not exceeding the lemma lobes, similar in texture and pubescence; **anthers** 1.5–3 mm; **ovaries** with a conelike extension bearing a 3-branched style. **Caryopses** about 2.5 mm long, 1 mm thick; **hila** linear, ⁹/₁₀ as long as to equaling the caryopses. 2*n* = 22.

Piptatherum exiguum grows on rocky slopes and outcrops in upper montane habitats, from central British Columbia to southwestern Alberta and south to northern California, Nevada, Utah, and northern Colorado. The limited DNA evidence available suggests that it is a basal species within *Piptatherum* (Jacobs et al. 2006).

2. Piptatherum canadense (Poir.) Dorn [p. 147]
CANADIAN PIPTATHERUM, ORYZOPSIS DU CANADA

Plants cespitose, not rhizomatous. **Culms** 30–90 cm, glabrous; **basal branching** mostly intravaginal. **Leaves** basally concentrated; **sheaths** smooth or scabridulous; **ligules** 1–4 mm, hyaline, truncate, rounded, or acute; **basal blades** 4–15 cm long, 1–1.5 mm wide when flat, 0.5–0.8 mm in diameter when folded or convolute. **Panicles** 9–15 cm, lower nodes with 1–2 branches; **branches** 1–6 cm, somewhat flexuous, ascending to divergent. **Glumes** subequal, 3–6 mm long, 1.3–2 mm wide, ovate, 1–3-veined, apices acute to mucronate; **florets** 2.2–4.5 mm, obovoid, dorsally compressed; **calluses** 0.2–0.5 mm, hairy, disarticulation scars elliptic; **lemmas** coriaceous, evenly pubescent, tan

at maturity, margins widely separated even when immature; **awns** 5–15 mm, persistent, once- or twice-geniculate, first segments strongly twisted; **paleas** similar to the lemmas in length, texture, and pubescence; **anthers** 1–2 mm; **ovaries** developing 2 conelike style bases, each bearing a single, unbranched style. **Caryopses** about 2.5 mm long, 0.5 mm thick; **hila** linear, almost equaling the caryopses. 2*n* = 22.

Piptatherum canadense grows in grasslands and open woods, from the British Columbia–Alberta border east to Newfoundland, extending south into the Great Lakes region and the northeastern United States. Its persistent, longer awns distinguish *P. canadense* from *P. pungens*.

3. Piptatherum pungens (Torr.) Dorn [p. 147]
SHARP PIPTATHERUM

Plants cespitose, not rhizomatous. **Culms** 10–90 cm, usually glabrous, occasionally puberulent beneath the nodes; **basal branching** intravaginal. **Leaves** basally concentrated; **sheaths** smooth or somewhat scabrous; **ligules** 0.5–2.5 mm, truncate to acute; **blades** (6)18–45 cm long, 0.5–1.8 mm wide, flat to convolute, abaxial surfaces scabridulous to scabrous, adaxial surfaces scabrous. **Panicles** 4–6 cm, lower nodes with 1–2 primary branches; **branches** 0.8–4 cm, straight, usually strongly ascending, ascending to divergent at anthesis. **Glumes** subequal, 3.5–4.5 mm long, 1.4–2 mm wide, from 1 mm shorter than to slightly exceeding the florets, ovate, usually 1-veined, sometimes 3–5-veined near the base, apices rounded or acute; **florets** 3–4 mm, dorsally compressed; **calluses** 0.2–0.3 mm, rounded, hairy, disarticulation scars circular; **lemmas** evenly pubescent, margins not overlapping at maturity; **awns** 1–2 mm, straight, slightly twisted, caducous, often absent even from immature florets; **paleas** equaling or almost equaling the lemma lobes, similar in texture and pubescence to the lemmas; **anthers** 0.8–1.8 mm, usually not penicillate; **ovaries** with a conelike extension bearing a 2-branched style. **Caryopses** about 1.8 mm long, about 0.9 mm wide; **hila** linear, ⁹/₁₀ as long as to equaling the caryopses. 2*n* = 22, 24.

Piptatherum pungens grows in sandy to rocky soils and open habitats, from southern Yukon Territory across Canada to the Great Lakes region and eastern Pennsylvania, and, as a disjunct, in the western Great Plains and the southern Rocky Mountains. Its apparent absence from Idaho and Montana, and almost complete absence from Wyoming, is puzzling. The awns of *P. pungens* fall off so rapidly that it is sometimes mistaken for *Milium* or *Agrostis*, but the only perennial species of

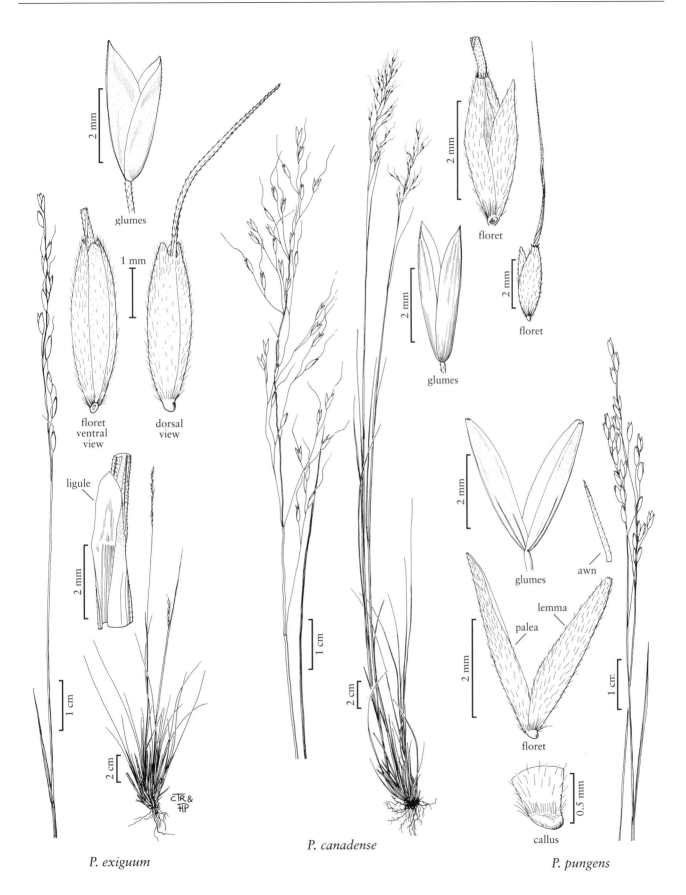

glumes

2 mm

1 mm

floret
ventral
view

dorsal
view

ligule

2 mm

1 cm

2 cm

CTR &
HP

P. exiguum

P. canadense

1 cm

2 cm

2 mm

glumes

floret

2 mm

floret

2 mm

floret

2 mm

glumes

awn

lemma

palea

2 mm

floret

callus

0.5 mm

1 cm

P. pungens

PIPTATHERUM

Milium in the Flora region has leaf blades 8–17 mm wide, and no species of *Agrostis* has such stiff lemmas and well-developed paleas. Its deciduous, shorter awns distinguish it from *P. canadense*.

4. Piptatherum micranthum (Trin. & Rupr.) Barkworth [p. 149]
SMALL-FLOWERED PIPTATHERUM

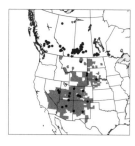

Plants loosely cespitose, not rhizomatous. **Culms** 20–85 cm, glabrous; **basal branching** extravaginal. **Leaves** basally concentrated; **sheaths** glabrous; **ligules** 0.4–1.5(2.5) mm, truncate; **blades** 5–16 cm long, 0.5–2.5 mm wide, usually involute. **Panicles** 5–20 cm, lower nodes with 1–3 branches; **branches** 2–6 cm, divergent to reflexed at maturity, with 3–10(15) spikelets, secondary branches appressed to the primary branches. **Glumes** 2.5–3.5 mm, acute; **lower glumes** 1(3)-veined; **upper glumes** 3-veined; **florets** 1.5–2.5 mm, dorsally compressed; **calluses** 0.1–0.2 mm, glabrous or sparsely hairy, disarticulation scars circular; **lemmas** usually glabrous, sometimes sparsely pubescent, brownish, shiny, 5-veined, margins not overlapping at maturity; **awns** 4–8 mm, straight or almost so, caducous; **anthers** 0.6–1.2 mm, not penicillate; **ovaries** truncate to rounded, bearing 2 separate styles. **Caryopses** about 1.2 mm long, about 0.8 mm wide; **hila** linear, $^{3}/_{4}$–$^{9}/_{10}$ as long as the caryopses. $2n = 22$.

Piptatherum micranthum grows on gravel benches, rocky slopes, and creek banks, from British Columbia to Manitoba and south to Arizona, New Mexico, and western Texas. The combination of small, dorsally compressed florets and appressed pedicels distinguishes this species from all other native North American *Stipeae*.

Achnatherum contractum is the fertile derivative of hybridization between *Piptatherum micranthum* and *A. hymenoides*. It is placed in *Achnatherum* because it resembles that genus more than *Piptatherum*.

5. Piptatherum shoshoneanum (Curto & Douglass M. Hend.) P.M. Peterson & Soreng [p. 149]
SHOSHONE PIPTATHERUM

Plants tightly cespitose, not rhizomatous. **Culms** 20–50 cm, internodes smooth to scabridulous; **nodes** glabrate; **basal branching** intravaginal. **Leaves** basally concentrated; **ligules** 1.8–5.5 mm, hyaline, acute, often lacerate; **blades** 4–16 cm long, 1–2.5 mm wide when flat, 0.6–1 mm in diameter when involute, abaxial surfaces smooth or scabridulous, adaxial surfaces scabridulous. **Panicles** 3.3–22 cm, lower nodes with 1–2(4) branches; **branches** 1.8–11.6 cm, flexuous, initially appressed, becoming strongly divergent to reflexed, secondary and tertiary branches appressed to the primary branches. **Glumes** subequal, 3.2–5.3 mm, exceeding the florets by 0.2–1.5 mm, ovate to broadly lanceolate, 1–9-veined, apices acute to acuminate; **florets** 2.4–4.1 mm, dorsally compressed; **calluses** about 0.3 mm, hairy, disarticulation scars round; **lemmas** coriaceous, evenly pubescent, hairs to 0.5 mm, becoming tawny with age, margins not overlapping at maturity; **awns** 1–2.5 mm, straight or slightly arcuate, caducous; **paleas** 2.1–3.6 mm, similar to the lemmas in texture and pubescence; **anthers** 1.7–2.2 mm, penicillate; **ovaries** truncate, bearing 2 styles. **Caryopses** 1.8–2 mm long, about 0.8 mm thick; **hila** linear, about $^{9}/_{10}$ the length of the caryopses. $2n = 20$.

Piptatherum shoshoneanum is known best from eastern Idaho, where it grows in the canyons of the Middle Fork of the Salmon River and its tributaries. It has also been found 750 km to the southwest in the Belted Range of southwestern Nevada. So far, fieldwork on the intervening mountain ranges has not revealed additional populations. It usually grows in moist crevices of igneous, metamorphic, or sedimentary cliffs and rock walls.

6. Piptatherum racemosum (Sm.) Eaton [p. 150]
MOUNTAIN RICEGRASS

Plants loosely cespitose to soboliferous, rhizomatous. **Culms** 48–80 cm, scabrous or pubescent adjacent to the nodes; **basal branching** extravaginal. **Leaves** not basally concentrated; **sheaths** usually smooth and glabrous, occasionally scabridulous and inconspicuously pubescent near the margins; **ligules of upper leaves** 0.3–0.7 mm, truncate; **blades of basal leaves** 0–2 cm; **blades of upper leaves** 10–27 cm long, 8–16 mm wide, abaxial surfaces evenly but sparsely pubescent, adaxial surfaces with straight hairs to 0.3 mm on the primary veins and flexuous hairs of 0.5–0.9 mm on the minor veins, tapering from near midlength to the apices. **Panicles** 12–25 cm, lower nodes with 1–2 branches; **branches** 3–9.5 cm, straight, strongly ascending to strongly divergent, with 2–5 spikelets. **Glumes** 6–8 mm, from subequal to the florets to exceeding the florets by 2 mm, ovate, 5–7-veined, acuminate; **florets** 4.5–7.5 mm; **calluses** 0.3–0.6 mm, disarticulation scars circular; **lemmas** coriaceous,

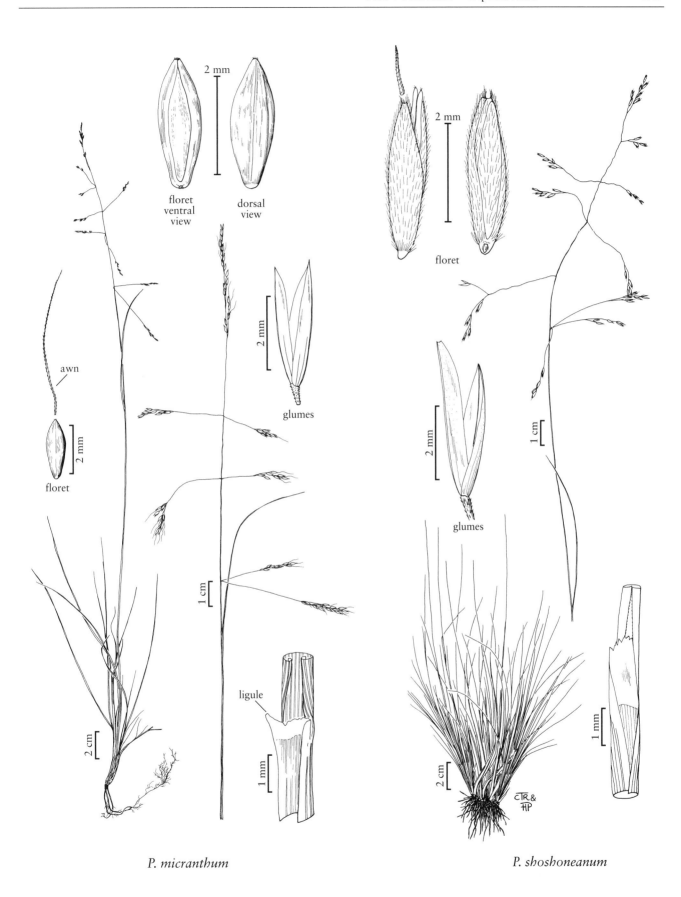

floret
ventral
view

dorsal
view

2 mm

floret

glumes

2 mm

awn

floret

2 mm

glumes

2 mm

ligule

1 mm

2 cm

1 cm

2 mm

1 cm

1 mm

2 cm

CTR&
HP

P. *micranthum*

P. *shoshoneanum*

PIPTATHERUM

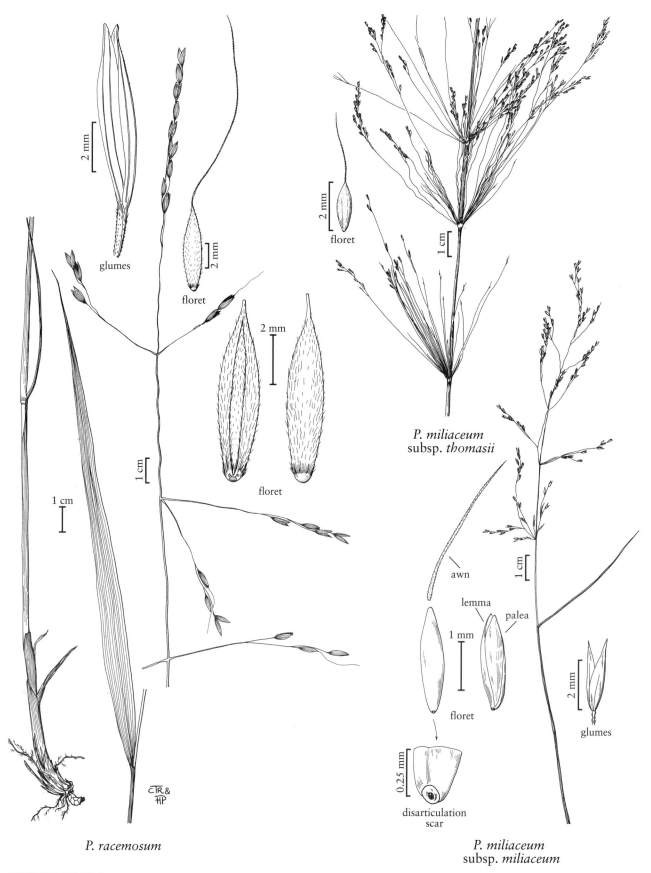

glumes

floret

2 mm

floret

2 mm

1 cm

1 cm

floret

2 mm

floret

P. miliaceum
subsp. *thomasii*

1 cm

awn

lemma

palea

floret

1 mm

2 mm

glumes

0.25 mm

disarticulation
scar

CTR &
HP

P. racemosum

P. miliaceum
subsp. *miliaceum*

PIPTATHERUM

sparsely pubescent to glabrate throughout, margins fused at the base, not overlapping, shiny dark brown to black at maturity; **awns** 10–25 mm, deciduous, slightly twisted, flexuous; **anthers** 3.5–5.5 mm, not penicillate; **ovaries** developing 2 conelike extensions, each terminating in a style. **Caryopses** 5–6 mm; **hila** linear, $^4/_5$–$^9/_{10}$ as long as the caryopses. $2n = 46, 48$.

Piptatherum racemosum usually grows in deciduous woods, and less often in open pine woods, in rocky, mountainous areas, from the St. Lawrence and Ottawa rivers south to the Missouri River, Tennessee, and Virginia, and east to Maine. The absence of basal blades and the dark, shiny lemmas distinguish it from all other North American *Stipeae*. It is highly palatable to livestock, but is never sufficiently abundant to be important as forage.

7. Piptatherum miliaceum (L.) Coss. [p. 150]
SMILO GRASS

Plants loosely cespitose, not rhizomatous. **Culms** 40–150 cm, glabrous, often branching at the lower cauline nodes; **basal branching** extravaginal. **Leaves** not basally concentrated; **sheaths** glabrous, persistent; **ligules of lower leaves** 0.5–1.5 mm, truncate; **ligules of upper leaves** 1.5–4 mm, rounded to sharply acute; **blades** 5–30 cm long, 2–10 mm wide, flat, smooth on both surfaces. **Panicles** 10–40 cm, lax, lower nodes either with 3–7 branches bearing 10–40 functional spikelets, or with 15–30+ branches with no functional spikelets; **primary branches** spreading to ascending; **lower branches** 3–8 cm; **secondary branches** diverging from the primary branches. **Glumes** 2.5–3.5 mm, acuminate, 3-veined; **florets** 1.5–2 mm, dorsally compressed; **calluses** about 0.3 mm, glabrous, disarticulation scars circular; **lemmas** stiffly membranous, glabrous, margins fused at the base, not overlapping,

light brown at maturity; **awns** 3–4 mm; **anthers** 2–2.5 mm, penicillate; **ovaries** rounded, bearing two styles. **Caryopses** 1.5–1.7 mm long, about 0.8 mm thick; **hila** linear, about $^1/_2$ as long as the caryopses. $2n = 24$.

Piptatherum miliaceum is a Eurasian introduction that is now established in several parts of the world. In its native range it grows, often as a common species, primarily in disturbed areas, wadis, and oases, penetrating into the semidesert regions of northern Africa and western Asia. It is used as a fodder plant in northern Africa. Within the *Flora* region, *P. miliaceum* is known from Arizona and California, growing in disturbed sites. It has also been found on a ballast dump in Maryland.

1. Panicle branches loosely whorled, lower nodes with 3–7 branches, all spikelet-bearing . . . . subsp. *miliaceum*
1. Panicle branches densely whorled, lower nodes with 15–30+ branches, some with highly reduced or no spikelets . . . . subsp. *thomasii*

Piptatherum miliaceum (L.) Coss. subsp. miliaceum [p. 150]

Branches loosely whorled; **lower nodes** with 3–7 spikelet-bearing branches.

Piptatherum mileaceum subsp. *miliaceum* is the most common of the two subspecies, and the only one known to be established in the *Flora* region.

Piptatherum miliaceum subsp. thomasii (Duby) Soják [p. 150]

Branches densely whorled; **lower nodes** with 15–30+ branches, some sterile, bearing highly reduced or no spikelets.

Piptatherum miliaceum subsp. *thomasii* has a native range similar to that of subsp. *miliaceum*, except that it does not grow in semidesert regions. In the *Flora* region, it is known only from cultivated specimens.

10.05 MACROCHLOA Kunth

Francisco M. Vázquez

Plants perennial; cespitose, not rhizomatous. **Culms** 60–170 cm, mostly smooth, scabrous beneath the panicles; **basal branching** intravaginal; **prophylls** longer than the sheaths, 2-awned, awns 2–3 cm, velutinous; **nodes** concealed by the sheaths. **Leaves** basally concentrated; **cleistogenes** not developed; **sheaths** open to the base, smooth, glabrous or hairy, hairs sometimes curly, margins extending into 2 awnlike extensions; **auricles** absent; **ligules** truncate, velutinous; **blades** conduplicate or convolute, about 1 mm in diameter. **Inflorescences** panicles, contracted, erect. **Spikelets** 26–30 mm, with 1 floret; **rachillas** not prolonged beyond the base of the floret; **disarticulation** above the glumes, beneath the floret. **Glumes** exceeding the floret, linear to lanceolate, gradually attenuate; **florets** 9–12 mm, terete; **calluses** well developed, sharp; **lemmas** thickly membranous to somewhat indurate, tan to brown, pubescent, margins flat, not overlapping at maturity, apices bifid, awned from between the teeth, teeth 1–6 mm, linear, scarious, awns once-geniculate, first segment twisted, terminal segment straight, scabrous; **paleas** equaling or exceeding the lemmas, pubescent, 2-veined, not keeled over the veins, flat between the veins, veins terminating before the apices, apices scarious, thinner than the palea body; **lodicules** 3, glabrous, ovate, acute, posterior lodicule smaller than the lateral lodicules; **anthers** 3, 10–15 mm, penicillate; **ovaries** glabrous; **style** 1, with 2 pilose branches. **Caryopses** fusiform; **hila** linear, about as long as the caryopses. $x = 8$. Name from the Greek *makros*, 'long', and *chloa*, 'grass'.

Macrochloa includes one to two species. It is native to the Mediterranean region, where it grows in basic and argillaceous soils. It has traditionally been included in *Stipa* because of its elongated florets and persistent, long awns. It differs from that genus in its strongly bifid lemmas, well-developed prophylls, lemma anatomy, and chromosome base number. The cross-sectional anatomy of its leaf blades is unique among the *Stipeae* (Vázquez and Barkworth 2004). One species is cultivated in the *Flora* region.

SELECTED REFERENCES **Darke, R.** 1999. The Color Encyclopedia of Ornamental Grasses: Sedges, Rushes, Restios, Cat-Tails, and Selected Bamboos. Timber Press, Portland, Oregon, U.S.A. 325 pp.; **Hitchcock, A.S.** 1951. Manual of the Grasses of the United States, ed. 2, rev. A. Chase. U.S.D.A. Miscellaneous Publication No. 200. U.S. Government Printing Office, Washington, D.C., U.S.A. 1051 pp.; **Vázquez, F.M.** and **M.E. Barkworth.** 2004. Resurrection and emendation of *Macrochloa* (Gramineae: Stipeae). Bot. J. Linn. Soc. 144:483–495.

1. Macrochloa tenacissima (L.) Kunth [p. 153]
ESPARTO GRASS

Plants initially forming dense, closed clumps, clumps becoming annular with age. **Culms** 60–170 cm tall, 4–6 mm thick; **prophyll awns** 2.5–3 cm, velutinous, projecting from the throats of the subtending leaves. **Sheaths** glabrous or hairy, marginal extensions 1–12 mm, velutinous; **collars** glabrous on the lower leaves, pubescent on the flag leaves; **ligules** to 0.8 mm in the center, densely tomentose, hairs completely concealing the membranous base; **blades** 30–120 cm, conduplicate to strongly convolute, about 1 mm in diameter when dry, abaxial surfaces smooth, adaxial surfaces strongly scabrous, apices sharply pointed. **Panicles** 25–35 cm, contracted, dense, with 3–4 branches per node; **branches** appressed, axils densely pubescent; **pedicels** shorter than the spikelets. **Glumes** lanceolate, membranous, glabrous, smooth; **lower glumes** 26–30 mm, 3-veined; **upper glumes** 24–27 mm, 3–5-veined; **florets** 9–12 mm; **calluses** 1–1.5 mm; **lemmas** pubescent, hairs about 1 mm, apices bidentate, teeth about 1 mm, membranous; **awns** 4.5–6.5 cm; **paleas** subequal to the lemmas, apices scarious; **anthers** 10–15 mm. $2n = 40, 64$.

Macrochloa tenacissima is native to the western Mediterranean regions of Europe and Africa. Hitchcock (1951) stated that it was "sparingly cultivated" in the United States; Darke (1999) described it as a coarse, rarely cultivated plant. It would be a striking feature in a xeric landscape.

Macrochloa tenacissima was used until the 1800s as a source of fiber for paper in Spain. It is currently used for erosion control in the Mediterranean basin, and for making hats and baskets. Its leaf cross sections are unique within the *Stipeae* in having midveins with multiple vascular bundles (Vásquez and Barkworth 2004).

10.06 CELTICA F.M. Vázquez & Barkworth

Francisco M. Vázquez

Plants perennial; cespitose, not rhizomatous. **Culms** 100–250 cm, erect, smooth, glabrous, uppermost node often exposed; **basal branching** intravaginal; **prophylls** exceeding the subtending leaf sheaths, awned, ciliate or glabrous. **Leaves** basally concentrated; **cleistogenes** not developed; **sheaths** open to the base, smooth, glabrous except at the throat; **auricles** absent; **ligules** membranous, rounded, abaxial surfaces densely pubescent, margins ciliate; **blades** flat to

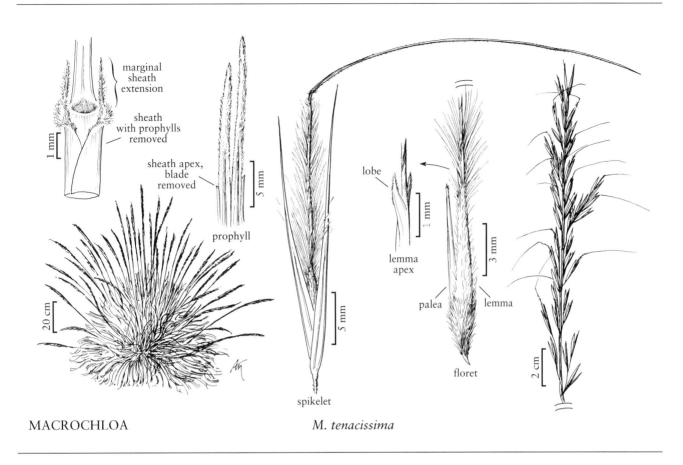

MACROCHLOA   *M. tenacissima*

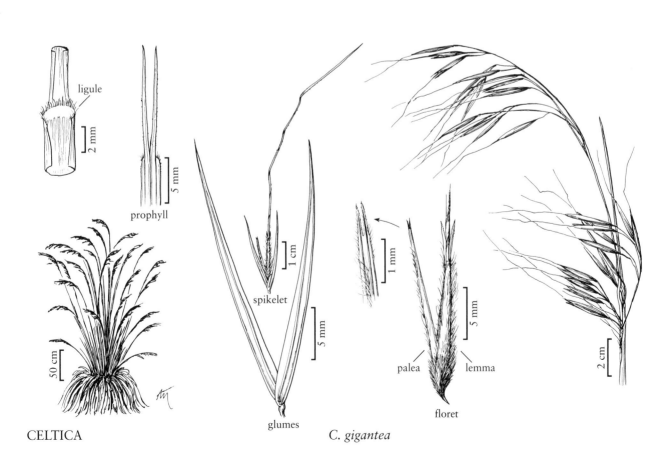

CELTICA   *C. gigantea*

involute, abaxial surfaces smooth, adaxial surfaces scabrous or hirtellous. **Inflorescences** panicles, nodding, open. **Spikelets** 25–32 mm, with 1 floret; **rachillas** not prolonged beyond the base of the floret; **disarticulation** above the glumes, beneath the floret. **Glumes** lanceolate, exceeding the floret, 3-veined; **florets** 14–16 mm, terete to laterally compressed; **calluses** sharp, strigose; **lemmas** coriaceous, evenly pubescent, hairs 1–2 mm, margins flat, overlapping at maturity, apices bifid, awned from between the teeth, teeth scarious; **awns** persistent, twice-geniculate, first segment twisted, terminal segment straight; **paleas** subequal to or longer than the lemmas, membranous, dorsally pubescent, veins forming 2 awnlike extensions; **lodicules** 3, glabrous, lanceolate, posterior lodicule larger than the lateral lodicules; **anthers** 3, penicillate; **ovaries** glabrous; **styles** 2. **Caryopses** fusiform; **hila** linear, about as long as the caryopses. $x = 12$. Named for the Celts, the genus being most abundant in the portion of the Iberian Peninsula to which the Celts were driven by the Romans.

Celtica is a monospecific genus that was formerly included in *Stipa* or *Macrochloa*.

SELECTED REFERENCES **Darke, R.** 1999. The Color Encyclopedia of Ornamental Grasses: Sedges, Rushes, Restios, Cat-Tails, and Selected Bamboos. Timber Press, Portland, Oregon, U.S.A. 325 pp.; **Vázquez, F.M. and M.E. Barkworth.** 2004. Resurrection and emendation of *Macrochloa* (Gramineae: Stipeae). Bot. J. Linn. Soc. 144:483–495.

1. **Celtica gigantea** (Link) F.M. Vázquez & Barkworth [p. 153]
GIANT FEATHERGRASS

Plants forming dense clumps. **Culms** 1–2.5 m tall, 5–7 mm thick. **Prophyll awns** to 1 cm, ciliate or glabrous, projecting from the throats of the subtending leaves. **Sheaths** mostly glabrous, throats ciliate; **ligules** about 0.5 mm, pubescent dorsally, ciliate; **blades** to 70 cm, involute, abaxial surfaces smooth, adaxial surfaces scabrous or hirtellous, hairs about 0.3 mm, apices acute. **Panicles** 30–50 cm, open, lower nodes with more than 1 primary branch, upper nodes with 1 branch, axils glabrous; **branches** strongly divergent; **pedicels** equaling or exceeding the spikelets. **Glumes** glabrous, smooth, 3-veined; **lower glumes** 25–32 mm; **upper glumes** slightly longer; **florets** 14–16 mm; **calluses** 1.5–2.5 mm; **lemmas** pubescent, hairs 1–2 mm, apical teeth to 6 mm; **awns** 6–9 cm; **lateral lodicules** about 1.5 mm; **anthers** 10–12 mm. $2n = 96$.

Celtica gigantea is native to the western and southern portions of the Iberian Peninsula and northern Africa. It is grown as an ornamental in the *Flora* region. Darke (1999) described it as "one of the most elegant and stately of the ornamental grasses." No attempt has been made to determine which of the infraspecific taxa is grown in the *Flora* region.

10.07 **STIPA** L.

Mary E. Barkworth

Plants annual or perennial; tufted or cespitose, not rhizomatous. **Culms** 10–200 cm, herbaceous, not branching at the upper nodes; **basal branching** usually intravaginal; **prophylls** shorter than the sheaths. **Leaves** mostly basal; **cleistogenes** usually not developed; **sheaths** open; **auricles** absent; **ligules** membranous, sometimes stiffly so, upper and lower ligules similar or upper ligules longer than those below; **blades** prominently ribbed, usually tightly convolute when dry. **Inflorescences** terminal panicles, usually contracted. **Spikelets** 12–90 mm, with 1 floret; **rachillas** not prolonged beyond the base of the floret; **disarticulation** above the glumes, beneath the floret. **Glumes** much longer than the floret, hyaline to membranous, usually acuminate, 1–3-veined; **florets** 3–27 mm, terete to slightly laterally compressed; **calluses** (1)1.5–6 mm, sharp or blunt, antrorsely hairy; **lemmas** coriaceous to indurate, tan to brown, smooth, glabrous or hairy, hairs sometimes uniformly distributed, sometimes in lines, margins flat, slightly overlapping at maturity, apices awned, lemma-awn junction evident; **awns** 50–500 mm, persistent, usually once- or twice-geniculate, sometimes plumose in whole or in part, basal segment often strongly twisted; **paleas** from shorter than to subequal to the lemmas, glabrous,

2-veined, not keeled, flat between the veins, apices sometimes scarious, sometimes similar in texture to the body; **lodicules** 2 or 3, glabrous or pilose; **anthers** 3; **styles** 2(3,4), free at the base, if 3 or 4, then 1 or 2 distinctly shorter. **Caryopses** fusiform, not ribbed. $x = 11$. Name from the Latin *stipa*, 'oakum' (a loose bunch of fibers), alluding both to the feathery inflorescences and the use of *Stipa tenacissima* L. [≡ *Macrochloa tenacissima*] as a source of cordage.

As treated here, *Stipa* is a genus of 150–200 species, all of which are native to Eurasia or northern Africa. Until recently, the genus was interpreted as including almost all species of *Stipeae* with cylindrical florets. In several parts of the world, this broader interpretation still prevails. Jacobs et al. (2006) found that even the European members of *Stipa* included in their study appeared to be polyphyletic. The most appropriate circumscription of the genus, and its size, is difficult to determine in the absence of a study that encompasses the Eurasian and North African members of the tribe.

Two species of *Stipa* grow in the *Flora* region. *Stipa pulcherrima* is cultivated as an ornamental; *Stipa capensis* has been introduced, probably accidentally.

SELECTED REFERENCES **Freitag, H.** 1985. The genus *Stipa* in southwest and south Asia. Notes Roy. Bot. Gard. Edinburgh 42:355–487; **Jacobs, S.W.L., R. Bayer, J. Everett, M.O. Arriaga, M.E. Barkworth, A. Sabin-Badereau, M.A. Torres, F. Vázquez,** and **N. Bagnall.** 2006. Systematics of the tribe Stipeae using molecular data. Aliso 23:349–361.

1. Plants perennial; glumes 60–90 mm long; awns plumose on the distal segment, hairs 5–6 mm long . . . . 1. *S. pulcherrima*
1. Plants annual; glumes 12–20 mm long; awns glabrous on the distal segment . . . . 2. *S. capensis*

1. Stipa pulcherrima K. Koch [p. 156]
BEAUTIFUL FEATHERGRASS

Plants perennial; cespitose, not rhizomatous. **Culms** 40–100 cm, glabrous or pubescent below the panicles. **Sheaths** longer than the internodes, smooth, mostly glabrous, margins sometimes ciliate; **ligules** 0.5–2 mm on the innovations, to 9 mm on the cauline leaves, stiffly membranous, pubescent, entire to erose; **blades** 2–5 mm wide when flat, 0.3–0.9(1.5) mm in diameter if involute, abaxial surfaces glabrous, smooth or scabrous, adaxial surfaces glabrous, scabrous, or hirsute, hairs to 0.6 mm. **Panicles** 10–15 cm, contracted, usually partially enclosed in the upper sheath; **branches** appressed to ascending, with 1–4 spikelets. **Glumes** 60–80(90) mm, long-attenuate, mostly hyaline; **florets** (18)20–27 mm; **calluses** 3–6 mm; **lemmas** with lines of hair over the veins and marginal veins, lines over the marginal veins longest, extending to the lemma apices; **awns** (250)300–500 mm, twice-geniculate, first 2 segments glabrous or hairy, hairs to 2 mm, terminal segment plumose, hairs 5–6 mm, spreading; **paleas** subequal to the lemmas, usually hairy on the keels, apices scarious; **anthers** 5–10 mm, not penicillate, yellowish or purplish; **styles** 2. **Caryopses** 10–18 mm. $2n = 44$.

Stipa pulcherrima is native from France and Germany to Armenia, Azerbaijan, Turkey, and Iran. Freitag (1985) regarded it as a subspecies of *S. pennata* L. Its long, plumose awns make it a striking ornamental.

2. Stipa capensis Thunb. [p. 156]

Plants annual, tufted. **Culms** 10–100 cm, erect or geniculate, glabrous, sometimes branching from the lowermost nodes. **Sheaths** glabrous or pilose; **collars** with tufts of hair at the sides; **ligules** 0.4–0.7 mm, membranous, ciliate; **blades** to 3 mm wide, flat or convolute, abaxial surfaces glabrous, sparsely pubescent, or pilose, adaxial surfaces scabrous or hairy, hairs about 0.8 mm. **Panicles** 3–15 cm, contracted, often partially enclosed in the upper sheath; **branches** scabrous; **pedicels** shorter than the spikelets. **Glumes** 12–20 mm, narrowly lanceolate to linear, 3-veined, tapering to the hairlike apices; **lower glumes** equal to or exceeding the upper glumes; **florets** 4–7 mm, terete; **calluses** 1.7–2.3 mm, sharp; **lemmas** indurate, with overlapping margins, dorsally constricted below the apices; **awns** 50–100 mm, twice-geniculate, first 2 segments twisted and pilose, hairs about 1 mm, terminal segment straight, glabrous; **paleas** 1.2–1.5 mm, 2-veined, glabrous; **lodicules** 2; **anthers** 3, 2–2.5 mm. $2n = 36$.

Stipa capensis is known from two locations in Riverside County, California: one in Palm Springs, and the other near the mouth of Chino Canyon. A.C. Sanders (University of California, Riverside) described the latter population as a "common annual on roadside and spreading into desert vegetation" (UTC 230476).

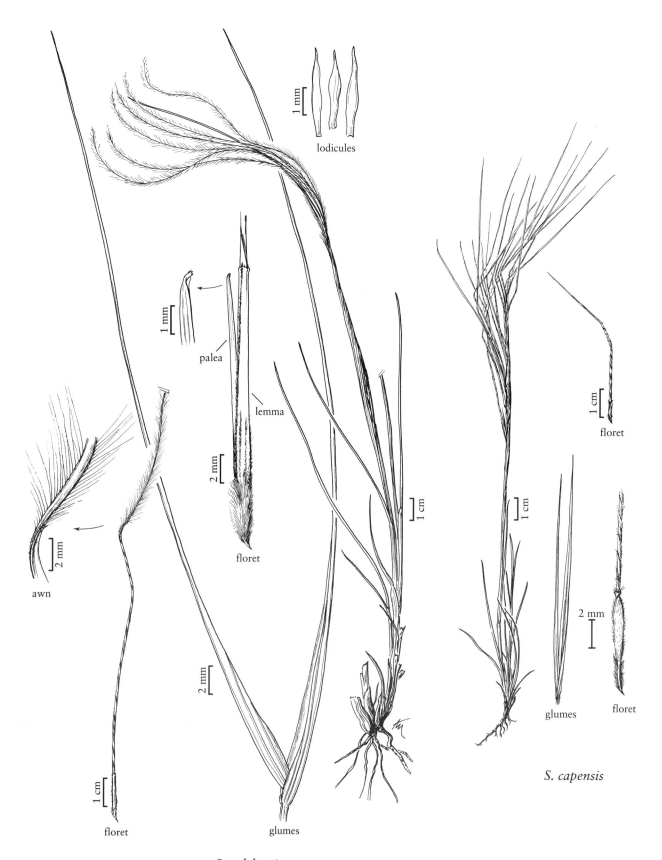

lodicules

palea

lemma

floret

awn

floret

glumes

S. pulcherrima

floret

glumes

floret

S. capensis

STIPA

10.08 **HESPEROSTIPA** (M.K. Elias) Barkworth

Mary E. Barkworth

Plants perennial; cespitose, not rhizomatous. **Culms** 12–110 cm, erect, not branching at the upper nodes; **prophylls** shorter than the sheaths. **Leaves** not overwintering, not basally concentrated; **cleistogenes** not developed; **sheaths** smooth; **auricles** absent; **ligules** membranous, frequently ciliate; **blades** 4–40 cm long, 0.5–4.5 mm wide, usually tightly involute, adaxial surfaces conspicuously ridged, apices narrowly acute, not sharp. **Inflorescences** terminal panicles, contracted or open. **Spikelets** 15–60 mm, with 1 floret; **rachillas** not prolonged beyond the base of the floret; **disarticulation** above the glumes and beneath the floret. **Glumes** 15–60 mm long, 2–4 mm wide, tapering from near the base to a hairlike tip; **florets** 7–25 mm, narrowly cylindrical; **calluses** 2–6 mm, sharp, densely strigose distally; **lemmas** indurate, smooth, margins flat, slightly overlapping at maturity, the upper portion fused into a papillose, ciliate crown, awned, lemma-awn junction distinct; **awns** 50–225 mm, persistent, twice-geniculate, often weakly so, lower segments twisted and scabrous to pilose, terminal segment not twisted, usually scabridulous or pilose; **paleas** equal to the lemmas, flat, pubescent, coriaceous, 2-veined, veins terminating at the apices, apices indurate, prow-tipped; **anthers** 3, 1.2–9 mm. **Caryopses** fusiform, not ribbed. $x = 11$. Name from the Greek *hesperos*, 'west', and the generic name *Stipa*.

Hesperostipa is a North American endemic that resembles the Eurasian *Stipa sensu stricto* in overall morphology, but is more closely related to the primarily South American genera *Piptochaetium* and *Nassella*. It differs from *Stipa sensu stricto* in its indurate palea apices and its lemma epidermal patterns. There are five species in the genus, four of which are found in the *Flora* region. The fifth species, from southern Mexico, is known only from the type specimen.

SELECTED REFERENCES **Barkworth, M.E.** 1977. A taxonomic study of the large-glumed species of *Stipa* (Gramineae) in Canada. Canad. J. Bot. 56:606–625; **Barkworth, M.E.** and **J. Everett.** 1987. Evolution in the Stipeae: Identification and relationships of its monophyletic taxa. Pp. 251–264 *in* T.R. Soderstrom, K.W. Hilu, C.S. Campbell, and M.E. Barkworth (eds.). Grass Systematics and Evolution. Smithsonian Institution Press, Washington, D.C., U.S.A. 473 pp.; **Elias, M.K.** 1942. Tertiary Prairie Grasses and Other Herbs from the High Plains. Geological Society of America Special Paper No. 41. The [Geological] Society [of America, New York, New York, U.S.A.]. 176 pp.; **Misra, K.C.** 1961. Geography, morphology, and environmental relationships of certain *Stipa* species in the northern Great Plains. Ph.D. dissertation, University of Saskatchewan, Saskatoon, Saskatchewan, Canada. 141 pp.; **Misra, K.C.** 1963. Phytogeography of the genus *Stipa* L. Trop. Ecol. 4:1–20 [reprint pagination]; **Thomasson, J.R.** 1979. Late Cenozoic Grasses and Other Angiosperms from Kansas, Nebraska, and Colorado: Biostratigraphy and Relationships to Living Taxa. Kansas Geological Survey, Bulletin No. 218. University of Kansas, Lawrence, Kansas, U.S.A. 68 pp.

1. Awns pilose on all segments, the terminal segment with hairs 1–3 mm long . . . . 2. *H. neomexicana*
1. Awns scabrous to strigose on the first 2 segments, the terminal segment scabridulous.
    2. Lemmas usually evenly white-pubescent, sometimes glabrous immediately above the callus; lower ligules often lacerate . . . . 1. *H. comata*
    2. Lemmas unevenly pubescent with brown to beige hairs; lower ligules not lacerate.
        3. Florets 8.5–14(17) mm long; awns 50–105 mm long; lower nodes usually glabrous, occasionally evenly pubescent . . . . 3. *H. curtiseta*
        3. Florets 15–25 mm long; awns 90–190 mm long; lower nodes usually with lines of pubescence . . . . 4. *H. spartea*

1. Hesperostipa comata (Trin. & Rupr.) Barkworth [p. 159]

NEEDLE-AND-THREAD

Culms 12–110 cm; **lower nodes** glabrous or pubescent. **Lower sheaths** glabrous or pubescent, not ciliate; **ligules of lower leaves** 1–6.5 mm, scarious, usually acute, sometimes truncate, often lacerate; **ligules of upper leaves** to 7 mm; **blades** 0.5–4 mm wide, usually involute. **Panicles** 10–32 cm, contracted. **Glumes** 16–35 mm, 3–5-veined; **lower glumes** 18–35 mm; **upper glumes** 1–3 mm shorter; **florets** 7–13 mm; **calluses** 2–4 mm; **lemmas** evenly pubescent, hairs about 1 mm, white, sometimes glabrous immediately above the callus; **awns** 65–225 mm, first 2 segments scabrous to strigose, hairs shorter than 1 mm, terminal segment scabridulous.

Hesperostipa comata is found primarily in the cool deserts, grasslands, and pinyon-juniper forests of western North America. The two subspecies overlap geographically, but are only occasionally sympatric. Both are primarily cleistogamous.

1. Terminal awn segment 40–120 mm long, sinuous to curled at maturity; lower cauline nodes usually concealed by the sheaths; panicles often partially enclosed in the uppermost sheath at maturity .... subsp. *comata*
1. Terminal awn segment 30–80 mm long, straight; lower cauline nodes usually exposed; panicles usually completely exserted at maturity .... subsp. *intermedia*

Hesperostipa comata (Trin. & Rupr.) Barkworth subsp. comata [p. 159]

Lower cauline nodes usually concealed by the sheaths. **Panicles** often partially included in the uppermost sheath at maturity. **Awns** 75–225 mm, terminal segment 40–120 mm, sinuous to curled. $2n = 38, 44, 46$.

Hesperostipa comata subsp. *comata* grows on well-drained soils of cool deserts, grasslands, and sagebrush associations, at elevations of 200–2500 m. It is widespread and often abundant in western and central North America, particularly in disturbed areas. It is similar to *H. neomexicana*, differing primarily in having awns that are either not hairy or have hairs that are no more than 0.5 mm long, and in having thinner, longer ligules. Intermediates to *H. neomexicana* exist but are not common.

Hesperostipa comata subsp. intermedia (Scribn. & Tweedy) Barkworth [p. 159]

Lower cauline nodes usually exposed. **Panicles** usually fully exserted at maturity. **Awns** 65–130 mm, terminal segment 30–80 mm, straight. $2n = 44–46$.

Hesperostipa comata subsp. *intermedia* is found in pinyon-juniper woodlands, at elevations of 2175–3075 m, in the Sierra Nevada and Rocky Mountains, from southern Canada to New Mexico. It resembles *H. curtiseta*, but differs in its evenly pubescent lemmas and its often lacerate ligules.

2. Hesperostipa neomexicana (Thurb.) Barkworth [p. 159]

NEW MEXICAN NEEDLEGRASS

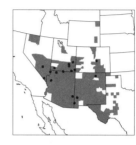

Culms 40–100 cm; **lower nodes** glabrous. **Lower sheaths** glabrous or puberulent, not ciliate; **ligules of lower leaves** 0.5–1 mm, thickly membranous, rounded; **ligules of upper leaves** to 3 mm, scarious, acute; **blades** 0.5–1 mm wide. **Panicles** 10–30 cm. **Glumes** subequal, 30–60 mm; **florets** 15–18 mm; **calluses** 4–5 mm; **lemmas** evenly pubescent, hairs shorter than 1 mm; **awns** 120–220 mm, first 2 segments hairy, hairs mostly 0.2–1 mm, terminal segment flexible, pilose, hairs 1–3 mm. $2n = 44$.

Hesperostipa neomexicana grows in grassland, oak, and pinyon pine associations, from 800–2400 m, usually in well-drained, rocky areas in the southwestern United States and adjacent Mexico. It is similar to *H. comata* subsp. *comata*, differing in its longer awn hairs and shorter ligules.

3. Hesperostipa curtiseta (Hitchc.) Barkworth [p. 160]

SMALL PORCUPINEGRASS

Culms 24–65 cm; **lower nodes** usually glabrous, sometimes evenly pubescent, often concealed by the lower sheaths. **Lower sheaths** usually glabrous; **ligules of lower leaves** 0.2–1 mm, truncate to rounded, often highest at the sides, stiff; **ligules of upper leaves** to 3.5 mm, usually acute to rounded, sometimes truncate; **blades** 1.3–3 mm wide. **Panicles** 6–24 cm. **Glumes** subequal, 15–30 mm; **florets** 8.5–14 mm; **calluses** 3–5 mm; **lemmas** unevenly pubescent, densely pubescent on the margins, more sparsely pubescent elsewhere on the lower portion of the lemmas, usually glabrous distally, hairs brown at maturity; **awns** 50–105 mm, scabrous to scabridulous throughout, terminal segment straight. $2n = 46$.

Hesperostipa curtiseta grows on light to clay loams in the prairies and northern portion of the central plains and northern intermontane grasslands, at elevations from 750–2050 m, extending primarily from British

glumes

floret

palea

floret

lower
ligule

lower
ligule

lemma

callus

H. comata
subsp. *intermedia*

H. comata
subsp. *comata*

H. neomexicana

HESPEROSTIPA

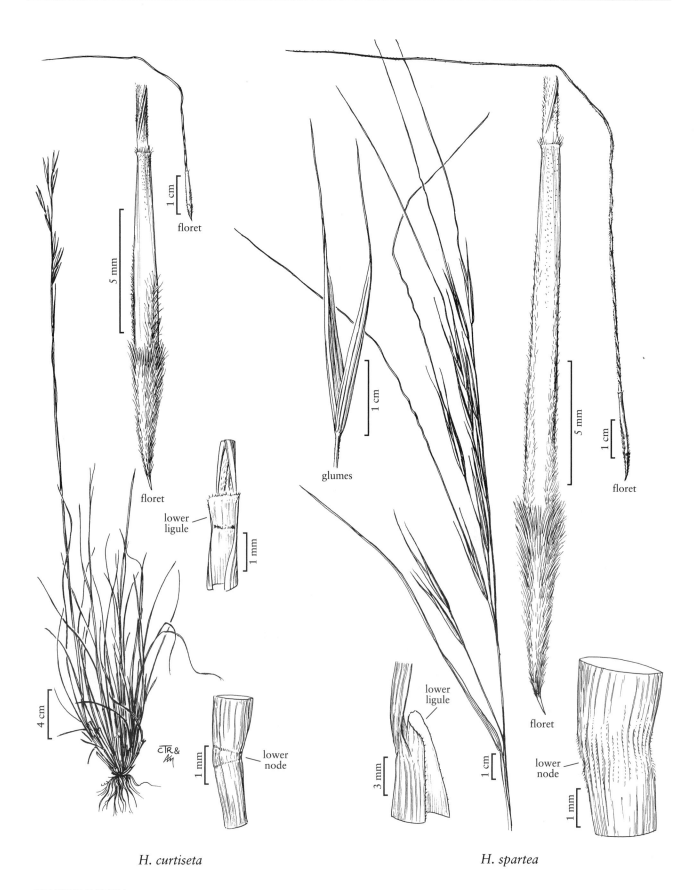

H. curtiseta   H. spartea

HESPEROSTIPA

Columbia to Manitoba and North Dakota. It resembles *H. comata* subsp. *intermedia*, but differs in having unevenly pubescent lemmas and non-lacerate ligules. It is also very similar to *H. spartea*, differing in its smaller size, usually glabrous or evenly pubescent culm nodes, usually glabrous sheaths, and shorter florets. Misra (1961) argued that its shorter florets and awns restrict *H. curtiseta* to more mesic sites than *H. spartea*.

4. Hesperostipa spartea (Trin.) Barkworth [p. 160]
PORCUPINEGRASS

Culms 45–90 (145) cm; **lower nodes** usually crossed by lines of pubescence, occasionally glabrous. **Lower sheaths** usually with ciliate margins; **ligules of lower leaves** 0.3–3 mm, stiff, truncate to rounded, usually entire; **ligules of upper leaves** 3–7.5 mm, thin, acute, often lacerate; **blades** 1.5–4.5 mm wide. **Panicles** 10–25 cm. **Glumes** 22–45 mm, subequal; **florets** 15–25 mm; **calluses** 3.5–6 mm; **lemmas** unevenly pubescent, densely pubescent on the margins and in lines on the lower portion of the lemmas, glabrous distally, hairs brown at maturity; **awns** 90–190 mm, terminal segment straight, scabridulous. $2n = 44, 46$.

Hesperostipa spartea grows at elevations of 200–2600 m, primarily in the grasslands of the central plains and southern prairies of the *Flora* region. In its more northern locations, it tends to grow on sandy soils. The southern and western specimens from outside its primary range may represent introductions. It was once a common species, but its habitat is now intensively cultivated. It differs from *H. curtiseta* in its larger size, unevenly pubescent culm nodes, usually ciliate lower sheaths, and longer florets. Misra (1961) argued that its longer florets and awns enabled *H. spartea* to survive in drier sites than *H. curtiseta*. Native Americans used bundles of the florets for combs.

10.09 PIPTOCHAETIUM J. Presl

Mary E. Barkworth

Plants perennial; cespitose, not rhizomatous. **Culms** 4–150 cm, usually erect, sometimes decumbent, glabrous, not branched above the base; **basal branching** intravaginal; **prophylls** shorter than the sheaths, mostly glabrous, keels usually with hairs, apices bifid, teeth 1–3 mm; **cleistogenes** not developed. **Sheaths** open to the base, margins glabrous; **ligules** membranous, decurrent, truncate to acute, sometimes highest at the sides, sometimes ciliate; **blades** convolute to flat, translucent between the veins, often sinuous distally. **Inflorescences** terminal panicles, open or contracted, spikelets usually confined to the distal ½ of each branch. **Spikelets** 4–22 mm, with 1 floret; **rachillas** not prolonged beyond the base of the floret; **disarticulation** above the glumes, beneath the floret. **Glumes** subequal, longer than the floret, lanceolate, 3–7(8)-veined; **florets** globose to fusiform, terete to laterally compressed; **calluses** well developed, sharp or blunt, glabrous or antrorsely strigose, hairs yellow to golden brown; **lemmas** coriaceous to indurate, glabrous or pubescent, striate, particularly near the base, smooth, papillose, or tuberculate, often smooth on the lower portion and papillate to tuberculate distally, margins involute, fitting into the grooved palea, apices fused into a crown, awned, lemmas often narrowed below the crown, crowns usually ciliate; **awns** caducous to persistent, usually twice-geniculate, first 2 segments usually twisted and hispid, terminal segment straight and scabridulous; **paleas** longer than the lemmas, similar in texture, glabrous, sulcate between the veins, apices prow-tipped; **lodicules** 2 or 3, membranous, glabrous, blunt or acute; **anthers** 3; **ovaries** glabrous; **styles** 2. **Caryopses** terete to globose or lens-shaped. $x = 11$. Name from the Greek *pipto*, 'fall', and *chaite*, 'long hair'.

Piptochaetium is primarily South American, being particularly abundant in Argentina. It has 27 species. Four species are native in the *Flora* region; two South American species are established at a single location in Marin County, California.

All stipoid species with elongate florets were included in *Stipa* by Hitchcock (1935, 1951) and other North American taxonomists. Parodi (1944) argued that *Piptochaetium* should be expanded to include those species of *Stipa* with elongated florets that shared with *Piptochaetium* its distinctive lemma and palea morphology, hair color, and leaf anatomy. His interpretation is now universally accepted.

The basal chromosome number of *Piptochaetium* is probably 11. This interpretation implies that counts of $2n = 42$ (Gould 1965 and Reeder 1968 for *P. fimbriatum*; Reeder 1977 for *P. pringlei*) represent an aneuploid reduction from $2n = 44$. It is also possible that the base number is 7, as is common in the *Poöideae*.

SELECTED REFERENCES **Cialdella, A.M.** and **M.O. Arriaga.** 1998. Revisión de las especies sudamericanas del género *Piptochaetium* (Poaceae, Poöideae, Stipeae). Darwinia 36:105–157; **Cialdella, A.M.** and **L.M. Giussani.** 2002. Phylogenetic relations of the genus *Piptochaetium* (Poaceae: Poöideae, Stipeae): Evidence from morphological data. Ann. Missouri Bot. Gard. 89:305–336; **Gould, F.W.** 1958. Chromosome numbers in southwestern grasses. Amer. J. Bot. 45:757–767; **Gould, F.W.** 1965. Chromosome numbers in some Mexican grasses. Bol. Soc. Bot. México 29:49–62; **Hitchcock, A.S.** 1935. Manual of the Grasses of the United States. U.S. Government Printing Office, Washington, D.C., U.S.A. 1040 pp.; **Hitchcock, A.S.** 1951. Manual of the Grasses of the United States, ed. 2, rev. A. Chase. U.S.D.A. Miscellaneous Publication No. 200. U.S. Government Printing Office, Washington, D.C., U.S.A. 1051 pp.; **Parodi, L.** 1944. Revisión de las gramíneas australes americanas del género *Piptochaetium*. Revista Mus. La Plata 6:213–310; **Reeder, J.R.** 1968. Notes on Mexican grasses VIII: Miscellaneous chromosome numbers–2. Bull. Torrey Bot. Club 95:69–86; **Reeder, J.R.** 1977. Chromosome numbers in western grasses. Amer J. Bot. 64:102–110; **Thomasson, J.R.** 1979. Late Cenozoic Grasses and Other Angiosperms from Kansas, Nebraska, and Colorado: Biostratigraphy and Relationships to Living Taxa. Kansas Geological Survey, Bulletin No. 218. University of Kansas Publications, Lawrence, Kansas, U.S.A. 68 pp.

1. Florets 6.5–22 mm long; culms 40–130 cm tall.
  2. Lemmas hairy; awns 19–35 mm long . . . . 1. *P. pringlei*
  2. Lemmas glabrous; awns 40–120 mm long.
    3. Florets 7–13 mm long; awns 40–75 mm long . . . . 2. *P. avenaceum*
    3. Florets 13.5–22 mm long; awns 62–120 mm long . . . . 3. *P. avenacioides*
1. Florets 2.3–5.5 mm long; culms 20–95 cm tall.
  4. Lemmas golden brown, hairy, the hairs easily rubbed off . . . . 4. *P. fimbriatum*
  4. Lemmas dark brown, glabrous.
    5. Awns 10–16 mm long; blades 0.8–1.5 mm wide; distal margin of the lemma crowns
     straight . . . . 5. *P. setosum*
    5. Awns 15–25 mm long; blades 0.2–0.4 mm wide; distal margin of the lemma crowns
     sometimes slightly to strongly revolute . . . . 6. *P. stipoides*

1. **Piptochaetium pringlei** (Beal) Parodi [p. 163]
PRINGLE'S SPEARGRASS

Culms 50–125 cm, mostly glabrous, pubescent below the nodes; **nodes** 2–3, dark, glabrous or slightly pubescent. **Sheaths** smooth to scabridulous; **ligules** of basal leaves 0.5–2.8 mm, truncate to rounded, of upper leaves 1–3.5 mm, rounded to acute; **blades** 10–30 cm long, 1–3.5 mm wide, 3–5-veined, abaxial surfaces glabrous, smooth, adaxial surfaces smooth or scabrous over the veins, margins smooth or scabrous. **Panicles** 6–20 cm, open, with 10–25 spikelets; **branches** ascending, flexuous; **pedicels** to 1 mm, flattened, hispid. **Glumes** subequal, 9–12 mm long, 2.5–3.5 mm wide; **lower glumes** 5–7-veined; **upper glumes** 7-veined; **florets** 6.5–10 mm long, 1.5–2.1 mm thick, terete to somewhat laterally compressed; **calluses** 0.6–1.9 mm, blunt to acute, strigose; **lemmas** golden brown to dark brown at maturity, shiny or not, smooth to spiny-tuberculate distally or for almost their entire length, pubescent, hairs tawny to golden brown, evenly distributed or somewhat more abundant on the basal $^1/_2$, apices tapering to the crown; **crowns** 0.5–0.6 mm, inconspicuous, straight, hairy, hairs 0.5–1 mm; **awns** 19–27(35) mm, persistent, twice-geniculate, sometimes inconspicuously so; **paleas** 6.3–9.5 mm; **lodicules** 2, 1–1.5 mm, acute; **anthers** 3.5–5.5 mm, sometimes penicillate. **Caryopses** about 7 mm, fusiform. $2n = 42$.

Piptochaetium pringlei grows in oak woodlands, often on rocky soils, in the southwestern United States and northwestern Mexico. It is often confused with *P. fimbriatum*; it differs from that species in having longer florets and sharper calluses.

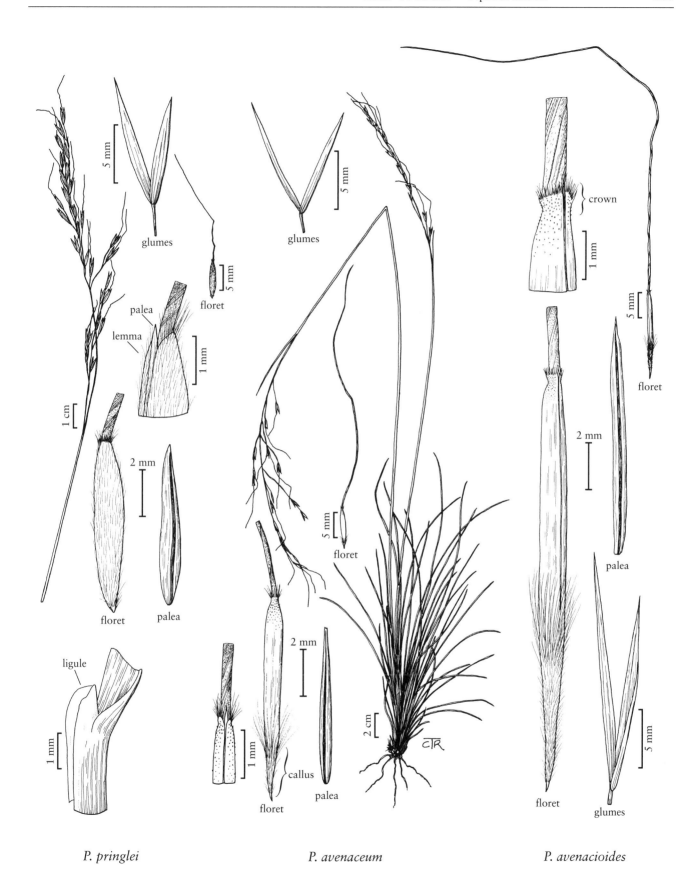

P. pringlei

P. avenaceum

P. avenacioides

PIPTOCHAETIUM

2. **Piptochaetium avenaceum** (L.) Parodi [p. 163]
BLACKSEED SPEARGRASS

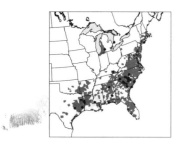

Culms (30)40–100 cm, glabrous; **nodes** 2–3, narrowed, yellowish to reddish. **Sheaths** glabrous; **ligules** rounded, sometimes highest at the sides, entire, of basal leaves 0.4–3.3 mm, of upper leaves 1.8–2.5 mm; **blades** 8–30 cm long, 0.6–3 mm wide, usually involute and about 0.5 mm in diameter, 3-veined, abaxial surfaces glabrous, smooth, adaxial surfaces scabrous over the veins, margins scabrous. **Panicles** 14–22 cm, open, with (10)15–25 spikelets; **branches** lax, divergent, spikelets confined to the distal ¹/₂; **pedicels** 15–50 mm, flattened, hispid. **Glumes** subequal, 9–15 mm long, 0.9–1.9 mm wide, acute; **lower glumes** 3(5)-veined; **upper glumes** 5-veined; **florets** 7–13 mm long, 1–1.2 mm thick, terete; **calluses** 2–3 mm, sharp, strigose, hairs golden brown at maturity; **lemmas** glabrous, tan to brown at maturity, smooth below, sharply tuberculate in the distal ¹/₃, constricted below the crown; **crowns** 0.5–0.6 mm wide, straight, not revolute, hairy, hairs 0.2–0.5 mm, golden brown; **awns** 40–75 mm, persistent, twice-geniculate, basal segment hispid, terminal segment scabrous; **paleas** 7–14 mm; **anthers** 0.3–0.5 mm or 3–4 mm, not penicillate. **Caryopses** 3.5–6 mm, terete. 2*n* = 22. Cialdella & Giussani (2002) mistakenly cited Gould (1958) as having reported 2*n* = 28.

Piptochaetium avenaceum grows in open oak and pine woods, often on sandy soils, throughout most of the coastal plain of the eastern United States, extending north up the Mississippi valley, and also on the east side of Lake Michigan. With the exception of one record from southern Ontario, Canada (collected in 1965 and not seen in Canada since, even though it has been searched for, *fide* Michael Oldham, pers. comm.), *P. avenaceum* is known only from the contiguous United States.

Piptochaetium avenaceum is very similar to *P. avenacioides*, differing only in its smaller size and more widespread distribution. It is also similar to *P. leianthum* (Hitchc.) Beetle, a species of northeastern Mexico, from which it differs in it larger size. The existence of two ranges of anther length suggests that the species is sometimes cleistogamous.

3. **Piptochaetium avenacioides** (Nash) Valencia & Costas [p. 163]
FLORIDA SPEARGRASS

Culms 70–130 cm, mostly glabrous, sometimes pubescent below the nodes; **nodes** yellowish, glabrous. **Sheaths** glabrous; **ligules** blunt to acute, of basal leaves 0.4–0.7 mm, of upper leaves to 3 mm; **blades** 15–30 cm long, 0.8–1.5 mm wide, usually involute and 0.5 mm in diameter, 3-veined, abaxial surfaces usually glabrous and smooth, sometimes scabrous, adaxial surfaces usually scabrous over the veins, sometimes smooth, sometimes hairy. **Panicles** 10–31 cm, open, with 10–50 spikelets; **pedicels** 15–20 mm, scabrous. **Glumes** 15–22 mm, (3)5-veined; **florets** 13.5–22 mm long, 1–2 mm thick, terete; **calluses** 3.5–8 mm, sharp, strigose, hairs golden brown at maturity; **lemmas** glabrous, tan to brown at maturity, mostly smooth, sharply tuberculate distally, contracted below the crown; **crowns** 0.6–0.7 mm wide, hairy, hairs 0.2–0.6 mm; **awns** 62–120 mm, persistent, twice-geniculate; **paleas** 9–12 mm; **lodicules** 2; **anthers** 4–7 mm. 2*n* = unknown.

Piptochaetium avenacioides grows in dry woods, generally on sandy ridges. It is endemic to Florida, growing primarily in the central peninsula. Morphologically, it is very similar to *P. avenaceum*, differing only in its larger size and more restricted distribution.

4. **Piptochaetium fimbriatum** (Kunth) Hitchc. [p. 165]
PINYON RICEGRASS

Culms 35–95 cm, usually glabrous, sometimes pubescent below the nodes; **nodes** 2–3, often dark, glabrous. **Sheaths** glabrous, smooth; **ligules** truncate to rounded, of basal leaves 0.4–1.8 mm, of upper leaves 1.5–2 mm; **blades** 6–26 cm long, usually involute and 0.3–5 mm in diameter, sometimes flat and 0.5–1(1.5) mm wide, 3-veined, both surfaces glabrous, veins often scabridulous, margins scabrous. **Panicles** 6.5–25 cm, open, often partially enclosed in the upper leaf sheath, with 20–60 spikelets; **branches** flexuous; **pedicels** 4–12 mm, flattened, hispid. **Glumes** subequal, 4–6.2 mm long, 1.8–3.1 mm wide, 5–7-veined, often partly purplish; **florets** 3–5.5 mm long, 0.6–1.9 mm thick, somewhat laterally compressed, rectangular to slightly obovate in side view; **calluses** 0.2–0.7 mm, blunt, strigose; **lemmas** tan to light chocolate brown, shiny, smooth, evenly pubescent when immature, hairs easily

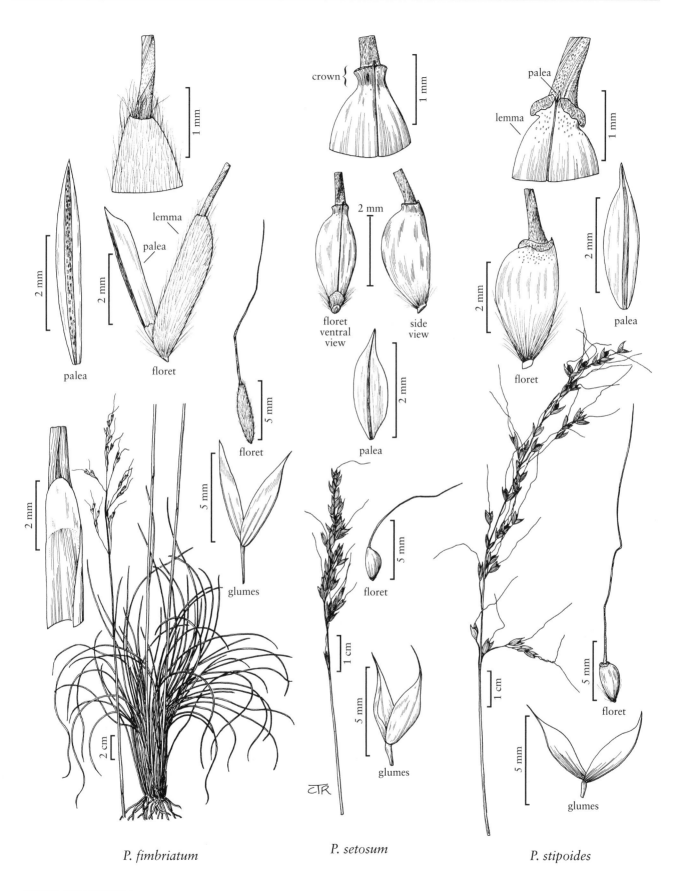

P. fimbriatum

P. setosum

P. stipoides

PIPTOCHAETIUM

rubbed off; **crowns** about 0.8 mm wide, inconspicuous, glabrous or glabrate; **awns** 11–20 mm, persistent, twice-geniculate; **paleas** about 3.5 mm; **lodicules** 2, about 1 mm; **anthers** 0.3–0.5 mm, not penicillate. **Caryopses** 2.5–3 mm long, about 0.6 mm thick, fusiform. $2n = 42$.

Piptochaetium fimbriatum is an attractive species that grows in oak and pinyon woods of the southwestern United States and adjacent Mexico, and merits consideration as an ornamental. It has also been reported from Guatemala; the report has not been verified.

Piptochaetium fimbriatum is not easily confused with other species in our range. Hitchcock (1951) treated it as including **P. seleri** (Pilg.) Henrard, a Mexican species with dull, rough, oblanceolate lemmas and persistent lemma hairs, an interpretation that is no longer accepted. It is occasionally confused with *P. pringlei*; it differs from that species in having shorter florets and blunt calluses.

5. Piptochaetium setosum (Trin.) Arechav. [p. 165]
BRISTLY RICEGRASS

Culms 20–40 cm, prostrate to ascending; **nodes** 2, dark, glabrous. **Sheaths** glabrous, smooth; **ligules** 0.5–2.5 mm, obtuse, membranous, glabrous; **blades** (3)5–12.5 cm long, 0.8–1.5 mm wide, glabrous or hispidulous, margins scabridulous. **Panicles** 3–15 cm long, 2–3 cm wide, with (5)10–30 spikelets; **branches** appressed to ascending, glabrous or hispid; **pedicels** 2–6 mm, hispidulous. **Glumes** subequal, 5–7 mm long, 1–2 mm wide, purplish at the base; **lower glumes** (3)5(7)-veined; **upper glumes** 5(7)-veined; **florets** 2.5–3 mm long, 1.2–1.8 mm thick, globose to slightly laterally compressed, gibbous; **calluses** 0.2–0.5 mm, obtuse, antrorsely strigose, hairs whitish to golden; **lemmas** glabrous, longitudinally striate, constricted below the crown, chestnut brown at maturity; **crowns** 0.5–0.8 mm wide, straight, not strongly differentiated, distal margins papillose; **awns** 10–16 mm, once- or twice-geniculate; **paleas** to 3.2 mm; **anthers** about 0.5 mm. **Caryopses** 2–2.5 mm, spherical to ellipsoid. $2n =$ unknown.

Piptochaetium setosum is native to central Chile. There is an established population in Marin County, California, that grows intermingled with *P. stipoides*, another South American species. The two species grow in the middle of a dirt track and in the adjacent meadow. The California plants of *P. setosum* are notable for their prostrate culms. This characteristic was not mentioned by Parodi (1944) or Cialdella and Arriaga (1998).

The origin of the California population is not known. It has been suggested that the seeds might have been brought in by birds, as the area was a bird refuge at one time.

6. Piptochaetium stipoides (Trin. & Rupr.) Hack.
[p. 165]
STIPOID RICEGRASS

Culms 20–60 cm, erect to ascending; **nodes** 2–4, dark, glabrous. **Sheaths** glabrous or hispidulous towards the collar; **ligules** 0.8–2 mm, glabrous, abaxial surfaces scabridulous, margins occasionally ciliate; **blades** (5)14–30 cm long, 0.2–0.4 mm wide, linear, glabrous or villous, margins scabridulous. **Panicles** 4–15 cm long, 1.5–3 cm wide, with 10–70 spikelets; **branches** ascending, scabridulous; **pedicels** 1–11 mm, hispid. **Glumes** subequal, 4–8.5 mm long, 1.5–2 mm wide, purple towards the base, glabrous, 5-veined, apices aristulate; **florets** 2.3–4(5) mm long, 0.8–2.3 mm thick, obovoid, globose to laterally compressed; **calluses** 0.5–0.6 mm, blunt, hairs white to golden tan; **lemmas** shiny, glabrous, striate, dark brown to black at maturity, wholly smooth to conspicuously verrucose or sharply papillose, at least distally, constricted below the crown; **crowns** well-developed, 0.6–1.6 mm wide, distal margins slightly to strongly revolute, inner surfaces densely covered with hooks and hairs; **awns** 15–25 mm, eccentric, twice-geniculate, tardily deciduous; **paleas** 2.5–5 mm; **lodicules** 2, linear; **anthers** about 0.5 mm. **Caryopses** 1.5–2.5 mm, spherical to ellipsoid. $2n =$ unknown.

Piptochaetium stipoides is native to South America. There is one known population in the *Flora* region, in Marin County, California, which grows with *P. setosum* in a meadow adjacent to an old dirt road. The origin of the population is not known; it has been suggested that the seeds might have been brought in by birds, as the area was a bird refuge at one time.

The Californian plants belong to **Piptochaetium stipoides** (Trin. & Rupr.) Hack. var. **stipoides**, which differs from the only other variety recognized by Cialdella and Arriaga (1998), **P. stipoides** var. **echinulatum** Parodi, in having lemmas that are mostly smooth as well as a less revolute crown.

10.10 **ORYZOPSIS** Michx.

Mary E. Barkworth

Plants perennial; cespitose, not rhizomatous. **Culms** 25–65 cm, erect or spreading, **basal branching** extravaginal; **prophylls** not visible; **nodes** glabrous. **Leaves** mostly basal; **cleistogenes** not developed; **sheaths** open, glabrous; **auricles** absent; **ligules** membranous, longest at the sides or rounded, ciliate; **blades** of basal leaves 30–90 cm, remaining green over winter, erect when young, recumbent in the fall, bases twisted, placing the abaxial surfaces uppermost, cauline leaf blades reduced, flag leaf blades 2–12 mm, conspicuously narrower than the top of the sheath. **Inflorescences** panicles, contracted. **Spikelets** 5–7.5 mm, with 1 floret; **rachillas** not prolonged beyond the base of the floret; **disarticulation** above the glumes, beneath the floret. **Glumes** subequal, 6–10-veined, apices mucronate; **florets** terete to laterally compressed; **calluses** usually less than ¹/₅ the length of the florets, blunt, distal portions pilose; **lemmas** coriaceous, pubescent at least basally, 3–5(9)-veined, margins strongly overlapping at maturity, awned, lemma-awn junction conspicuous, lobed, lobes 0.1–0.2 mm; **awns** more or less straight, deciduous; **paleas** similar to the lemmas in length, texture, and pubescence, concealed by the lemmas, 2-veined, flat between the veins; **lodicules** 2, free, membranous, 2-veined; **anthers** 3; **styles** 1, with 2 branches; **ovaries** glabrous. **Caryopses** falling with the lemma and palea. x = 11, 12. Name from the Greek *oryza*, 'rice', and *opsis*, 'appearance', in reference to a supposed resemblance to rice.

Oryzopsis is treated here as a monospecific genus that is restricted to North America. Hitchcock (1951) and Johnson (1945) treated it as including both Eurasian and North American taxa; Freitag (1975) and Tutin (1980) placed the Eurasian species in a separate genus, *Piptatherum*. Kam and Maze (1974) demonstrated that *O. asperifolia* differs from both North American and Eurasian species previously included in *Oryzopsis* in the development of its floret and callus, and in having 2-veined lodicules. Phylogenetic studies based on ITS sequence data have not yielded clear support for any particular treatment of *Oryzopsis*; they are consistent with the treatment presented here. The North American species previously included in *Oryzopsis* have been transferred to *Achnatherum* and *Piptatherum*.

SELECTED REFERENCES **Dore, W.G.** and **J. McNeill.** 1980. Grasses of Ontario. Research Branch, Agriculture Canada Monograph No. 26. Canadian Government Publishing Centre, Hull, Québec, Canada. 566 pp.; **Freitag, H.** 1975. The genus *Piptatherum* (Gramineae) in southwest Asia. Notes Roy. Bot. Gard. Edinburgh 33:341–408; **Hitchcock, A.S.** 1951. Manual of the Grasses of the United States, ed. 2, rev. A. Chase. U.S.D.A. Miscellaneous Publication No. 200. U.S. Government Printing Office, Washington, D.C., U.S.A. 1051 pp.; **Jacobs, S.W.L., R. Bayer, J. Everett, M.O. Arriaga, M.E. Barkworth, A. Sabin-Badereau, M.A. Torres, F. Vázquez,** and **N. Bagnall.** 2006. Systematics of the tribe Stipeae using molecular data. Aliso 23:349–361; **Johnson, B.L.** 1945. Cytotaxonomic studies in *Oryzopsis*. Bot. Gaz. 107:1–32; **Kam, Y.K.** and **J. Maze.** 1974. Studies on the relationships and evolution of supraspecific taxa utilizing developmental data: II. Relationships and evolution of *Oryzopsis hymenoides*, *O. virescens*, *O. kingii*, *O. micrantha*, and *O. asperifolia*. Bot. Gaz. 135:227–247; **Tutin, T.G.** 1980. *Piptatherum* Beauv. Pp. 246–247 *in* T.G. Tutin, V.H. Heywood, N.A. Burges, D.M. Moore, D.H. Valentine, S.M. Walters, and D.A. Webb (eds.). Flora Europaea, vol. 5. Cambridge University Press, Cambridge, England. 452 pp.

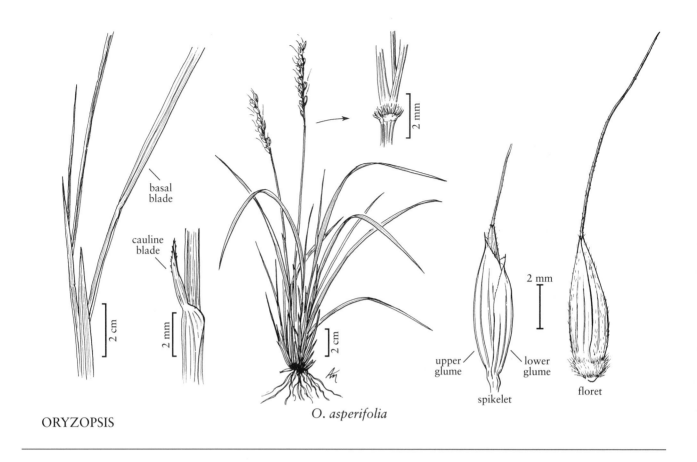

ORYZOPSIS   *O. asperifolia*

---

1. Oryzopsis asperifolia Michx. [p. 168]
ROUGHLEAF RICEGRASS, WINTER GRASS,
ORYZOPSIS À FEUILLES RUDES

Culms 25–65 cm. **Basal ligules** 0.2–0.7 mm, rounded, sometimes longest at the sides; **blades** of basal leaves 30–90 cm long, 4–9 mm wide, flag leaf blades 8–12 mm. **Panicles** 3.5–13 cm, contracted. **Glumes** subequal, 5–7.5 mm long, 2.5–4 mm wide, 6–10-veined, apices mucronate; **florets** 5–7 mm; **calluses** 0.8–2 mm, blunt, distal portions with a dense ruff of soft hairs; **lemmas** coriaceous or indurate at maturity, pale green, white, or yellowish, sometimes purple-tinged, glossy or dull, pubescent at least basally, margins overlapping, concealing the paleas; **awns** 7–15 mm; **paleas** similar to the lemmas; **anthers** 2–4 mm, usually penicillate. **Caryopses** 4–6.5 mm. $2n = 46$.

Oryzopsis asperifolia grows in both deciduous and coniferous woods, generally on open, rocky ground in areas with well-developed duff. It is found from the Yukon and Northwest Territories south to New Mexico along the Rocky Mountains, and from British Columbia east to Newfoundland and Maryland. It is listed as endangered or threatened in Indiana, Ohio, New Jersey, Maryland, and Virginia.

Leaf development in *Oryzopsis asperifolia* is unusual in that the leaves start to develop in midsummer, the blades growing upright. As the year progresses, they bend over, but stay alive and green through winter and the following spring. The part of the sheaths that remains below the level of the duff is usually bright purple (Dore and McNeill 1980).

10.11 ×ACHNELLA Barkworth

Mary E. Barkworth

Plants perennial; cespitose, not rhizomatous. **Culms** to 90 cm, erect, glabrous, not branching at the upper nodes; **prophylls** shorter than the sheaths. **Sheaths** mostly glabrous, margins sparsely ciliate, hairs longer distally; **cleistogenes** not present in the basal sheaths; **collars** without tufts of hair at the sides; **auricles** absent; **ligules** scarious, glabrous; **blades** of basal leaves to 40 cm long, 1–1.5 mm wide, convolute when dry, tapering to the narrowly acute apices, flag leaf blades longer than 10 mm, bases about as wide as the top of the sheaths. **Inflorescences** panicles. **Spikelets** 6–8.5 mm, with 1 floret; **rachillas** not prolonged beyond the base of the floret; **disarticulation** above the glumes, beneath the floret. **Glumes** saccate-lanceolate, tapering from above midlength into elongate apices, midveins extending to or nearly to the apices, apices narrowly acute to acuminate; **lower glumes** 6–8.5 mm; **upper glumes** 6.5–7.5 mm, slightly narrower than the lower glumes; **florets** 4–5 mm, lengths more than 3 times the widths; **calluses** about 0.7 mm, blunt; **lemmas** coriaceous, evenly hairy throughout, hairs 1–2 mm, apical hairs not longer than those below; **awns** twisted, not or once-geniculate, readily deciduous; **paleas** more than ²/₃ as long as the lemmas, veins ending at the apices; **anthers** poorly developed, indehiscent, sometimes with 1 or 2 penicillate hairs.

  ×*Achnella* comprises sterile hybrids between *Achnatherum* and *Nassella*. Only one such hybrid is known; it is restricted to the *Flora* region.

SELECTED REFERENCE **Johnson, B.L.** and **G.A. Rogler.** 1943. A cytotaxonomic study of an intergeneric hybrid between *Oryzopsis hymenoides* and *Stipa viridula*. Amer. J. Bot. 30:49–56.

1. ×**Achnella caduca** (Beal) Barkworth [p. 170]
  DROPAWN

Culms 50–90 cm, glabrous. **Basal sheaths** mostly glabrous, margins and throats usually ciliate; **ligules** 0.5–1.7 mm, truncate-rounded; **blades** 1–3.5 mm wide. **Panicles** 15–18 cm; **branches** ascending. **Lower glumes** 6–8.5 mm long, 1–1.6 mm wide; **florets** 4–5 mm; **calluses** about 0.7 mm; **lemmas** 7-veined, hairy, hairs 1–2 mm, apices awned; **awns** 9–16 mm, readily deciduous; **paleas** 2.5–3.3 mm, ²/₃–³/₄ the length of the lemmas, hairy; **anthers** 2, 1.7–2.3 mm, differing in length within the florets, indehiscent. 2n = 65.

  ×*Achnella caduca* is a sterile hybrid between *Achnatherum hymenoides* and *Nassella viridula* that occurs infrequently in Montana and Wyoming. It differs from *Achnatherum hymenoides* in its longer glumes and florets; from *Nassella viridula* in its more saccate glumes, longer lemma hairs, and well-developed palea; and from other hybrids involving *Achnatherum hymenoides* in its readily deciduous awn.

awn

2 mm

floret   palea

1 cm

2 mm

glumes

×ACHNELLA   ×*A. caduca*

10.12 NASSELLA (Trin.) E. Desv.

Mary E. Barkworth

Plants usually perennial, rarely annual; cespitose, occasionally rhizomatous. **Culms** 10–175(210) cm, sometimes branched at the upper nodes, branches flexible; **prophylls** not evident, shorter than the sheaths. **Leaves** mostly basal, not overwintering; **sheaths** open; **cleistogenes** sometimes present; **auricles** absent; **ligules** membranous, sometimes pubescent or ciliate; **blades** of basal leaves 3–60 cm long, 0.2–8 mm wide, apices narrowly acute to acute, not sharp, flag leaf blades 1–80 mm, bases about as wide as the top of the sheaths. **Inflorescences** terminal panicles, sometimes partially included at maturity. **Spikelets** 3–22 mm, with 1 floret; **rachillas** not prolonged beyond the base of the floret; **disarticulation** above the glumes, beneath the floret. **Glumes** longer than the floret, narrowly lanceolate or ovate, basal portion usually purplish at anthesis, color fading with age, (1)3–5-veined, sometimes awned; **florets** usually terete, sometimes slightly laterally compressed; **calluses** blunt or sharp, glabrous or antrorsely strigose; **lemmas** usually papillose or tuberculate, at least distally, sometimes smooth throughout, glabrous or variously hairy, strongly convolute, wrapping 1.2–1.5 times around the caryopses, apices not lobed, fused distally into *crowns*, these often evident by their pale color and constricted bases; **crowns** mostly glabrous, rims often bearing hairs with bulbous bases; **awns** terminal, centric or eccentric, deciduous or persistent, usually twice-geniculate, second geniculation often obscure; **paleas** up to $^1/_2$ as long as the lemmas, glabrous, without veins, flat; **lodicules** 2 or 3, if 3, the third somewhat shorter than the other 2; **anthers** 1 or 3, if 3, often of 2 lengths, penicillate; **ovaries** glabrous; **styles** 2, bases free. **Caryopses** glabrous, not ribbed; **hila** elongate; **embryos** to $^2/_5$ as long as the caryopses. $x = 7, 8$. Name not explained by Desvaux (1854), but possibly from the Latin *nassa*, a narrow-necked basket for catching fish.

Nassella used to be interpreted as a South American genus of approximately 14 species. It is now interpreted as including at least 116 species (Barkworth and Torres 2001), the majority of which are South American. The additional species were previously included in *Stipa*. There are eight species in the *Flora* region, one of which is introduced; two additional species treated here were found in the region at one time, but have not become established. The strongly convolute lemmas distinguish *Nassella* from all other genera of *Stipeae* in the Americas and, in combination with the reduced, ecostate, glabrous paleas, from all other genera in the tribe worldwide. Molecular data (Jacobs et al. 2006) support the expanded interpretation of *Nassella*. Relationships among the species have not been explored.

Many species of *Nassella* develop both cleistogamous and chasmogamous florets in the terminal panicle. The cleistogamous florets have 1–3 anthers that are less than 1 mm long; the chasmogamous florets have 3 anthers that are significantly longer. In addition, some species develop panicles in the axils of their basal sheaths. Spikelets of cleistogenes have reduced or no glumes, and florets with no or very short awns.

SELECTED REFERENCES **Barkworth, M.E.** 1990. *Nassella* (Gramineae: Stipeae): Revised interpretation and nomenclatural changes. Taxon 39:597–614; **Barkworth, M.E.** 1993. *Nassella*. Pp. 1274–1276 *in* J.C. Hickman (ed.). The Jepson Manual: Higher Plants of California. University of California Press, Berkeley and Los Angeles, California, U.S.A. 1400 pp.; **Barkworth, M.E.** and **M.A. Torres.** 2001. Distribution and diagnostic characters of *Nassella* (Poaceae: Stipeae). Taxon 50:439–468; **Brown, W.V.** 1952. The relation of soil moisture to cleistogamy in *Stipa leucotricha*. Bot. Gaz. 113:438–444; **Desvaux, E.** 1854. Gramineas. Pp. 233–469 *in* C. Gay. Flora Chilena [Historia Fisica y Politica de Chile], vol. 6. Museo Historia Natural, Santiago, Chile. 551 pp. [1853 on title page; printed March 1854]; **Dyksterhuis, E.J.** 1949. Axillary cleistogenes in *Stipa leucotricha* and their role in nature. Ecology 26:195–199; **Hamilton, J.G.** 1997. Changing perceptions of pre-European grasslands in California. Madroño 44:311–333; **Jacobs, S.W.L., R. Bayer, J. Everett, M.O. Arriaga, M.E. Barkworth, A. Sabin-Badereau, M.A. Torres, F. Vázquez,** and **N. Bagnall.** 2006. Systematics of the tribe Stipeae using molecular data. Aliso 23:349–361; **Jacobs, S.W.L., J. Everett,** and **M.E. Barkworth.** 1995. Clarification of morphological terms used in the Stipeae (Gramineae), and a reassessment of *Nassella* in Australia. Taxon 44:33–41; **Love, R.M.** 1946. Interspecific hybridization in *Stipa*: I. Natural hybrids. Amer. Naturalist 80:189–192; **Love, R.M.** 1954. Interspecific hybridization in *Stipa*: II. Hybrids of *S. cernua*, *S. lepida*, and *S. pulchra*. Amer. J. Bot. 41:107–110.

1. Florets 1.5–3 mm long; blades 0.2–1.5 mm wide, usually tightly convolute.
   2. Florets widest about midlength; awns 45–100 mm long, almost centric . . . . 7. *N. tenuissima*
   2. Florets widest near the top; awns 7–35 mm long, eccentric.
      3. Awns 15–35 mm long; lemmas strongly tuberculate, particularly distally . . . . 8. *N. trichotoma*
      3. Awns 7–10 mm long; lemmas smooth . . . . 9. *N. chilensis*
1. Florets 3.4–13 mm long; blades 0.4–8 mm wide, flat to convolute.
   4. Terminal segment of the awns cernuous.
      5. Awns 12–55 mm long, 0.1–0.2 mm thick at the base . . . . 4. *N. lepida*
      5. Awns 50–110 mm long, 0.2–0.3 mm thick at the base . . . . 6. *N. cernua*
   4. Terminal segment of the awns straight.
      6. Florets 3.4–5.5 mm long; lemmas not constricted below the crown; awns 19–32 mm long . . . . 10. *N. viridula*
      6. Florets 6–13 mm long; lemmas constricted below the crown; awns 30–120 mm long.

7. Lemmas hairy between the veins at maturity . . . . 5. *N. pulchra*
7. Lemmas glabrous between the veins at maturity.
   8. Crowns usually wider than long, the rims with hairs to 0.5 mm long; florets
     widest just below the crowns . . . . 1. *N. neesiana*
   8. Crowns usually longer than wide, the rims with hairs 1–2 mm long; florets widest
     near or slightly above midlength.
      9. Florets 6.5–13 mm long; crowns often flaring distally; plants native to Texas
       and adjacent states . . . . 2. *N. leucotricha*
      9. Florets 6–8 mm long; crowns more or less straight-sided; plants introduced,
       established in California . . . . 3. *N. manicata*

1. **Nassella neesiana** (Trin. & Rupr.) Barkworth [p. 173]
URUGUAYAN TUSSOCKGRASS

Plants perennial; cespitose, not rhizomatous. **Culms** 30–140 cm tall, 1–1.8 mm thick, erect or geniculate, internodes glabrous; **nodes** usually 2–4, exposed, sericeous, hairs to 1.2 mm. **Sheaths** glabrous or slightly hispid, basal leaf sheaths often with cleistogenes; **collars** glabrous, often brown or purple-tinged, with tufts of hair at the sides, hairs 0.5–1.2 mm; **ligules** 1–4 mm, glabrous or pubescent, truncate; **blades** to 30 cm long, 2–8 mm wide, flat to convolute, sometimes scabrous, sometimes with hairs. **Terminal panicles** 5–40 cm, exserted, erect to nodding; **branches** 2.5–8.5 cm, with 2–5 spikelets; **pedicels** 1–8 mm, angled, scabrous, pubescent. **Glumes** subequal, 10–22 mm long, 1.8–2.3 mm wide, narrowly lanceolate, glabrous, 3–5-veined; **florets** 6–13 mm long, 1–1.5 mm wide, terete, widest just below the crown; **calluses** 2–4.5 mm, sharp, strigose; **lemmas** often purple, finely rugose-papillose, particularly near the crown, constricted below the crown, midveins pilose proximally, glabrous between the veins at maturity; **crowns** 0.4–1.6 mm, usually wider than long, sides usually flaring somewhat distally, rims with hairs to 0.5 mm; **awns** 50–120 mm, clearly twice-geniculate, terminal segment straight; **anthers** 3–3.5 mm, penicillate. **Caryopses** 3–5 mm. 2*n* = 28.

Nassella neesiana is native to South America, growing from Ecuador to Argentina, primarily in steppe habitats. It was found on ballast dumps in Mobile, Alabama but has not persisted in the *Flora* region. It has become established in Australia, where it is considered a noxious weed.

2. **Nassella leucotricha** (Trin. & Rupr.) R.W. Pohl [p. 173]
TEXAN NASSELLA, TEXAN NEEDLEGRASS

Plants perennial; cespitose, not rhizomatous. **Culms** 33–120 cm tall, 1–1.2 mm thick, erect, internodes glabrous; **nodes** 3–5, usually pubescent. **Sheaths** often conspicuously hairy, sometimes glabrous, basal leaf sheaths often with cleistogenes; **collars** glabrous, often brown or purple-tinged, with tufts of hair at the sides, hairs 0.5–1 mm; **ligules** 0.2–1.2 mm, glabrous, truncate, sometimes longest at the sides; **blades** 10–35 cm long, 1–3.6 mm wide, flat to convolute, abaxial surfaces sparsely coarsely hairy, adaxial surfaces glabrous. **Terminal panicles** 5–55 cm, open; **branches** 3–7 cm, ascending or spreading, angled, scabrous, glabrous or hairy, with 1–4 spikelets; **pedicels** 4–9 mm. **Glumes** subequal, 10–21 mm long, 0.7–3.2 mm wide, narrowly lanceolate, glabrous, 3–5-veined; **florets** 6.5–13 mm long, 1.1–1.4 mm wide, terete, widest near or slightly above midlength; **calluses** 1–5.5 mm, sharp, strigose; **lemmas** straw-colored to light brown, papillose distally, constricted below the crown, veins pubescent proximally, particularly the midveins, glabrous between the veins; **crowns** 0.75–2 mm, usually longer than wide, often flaring distally, rims irregular, with 1–2 mm hairs; **awns** 40–90 mm, clearly twice-geniculate, terminal segment straight; **anthers** 3 and 3.5–5 mm in chasmogamous florets, 1–3 and shorter than 0.7 mm in cleistogamous florets. **Caryopses** about 4 mm. 2*n* = 28.

The range of *Nassella leucotricha* extends from the southern United States, where it was once one of the dominant species, into northern Mexico. It is now established in Australia (Jacobs et al. 1995). In North America, *N. leucotricha* grows mostly in open grasslands, but it is also found in woodlands. It provides good spring forage and increases in abundance with moderate grazing, primarily because of its cleistogenes. It resembles *N. manicata*, but has longer florets and less strongly developed crowns.

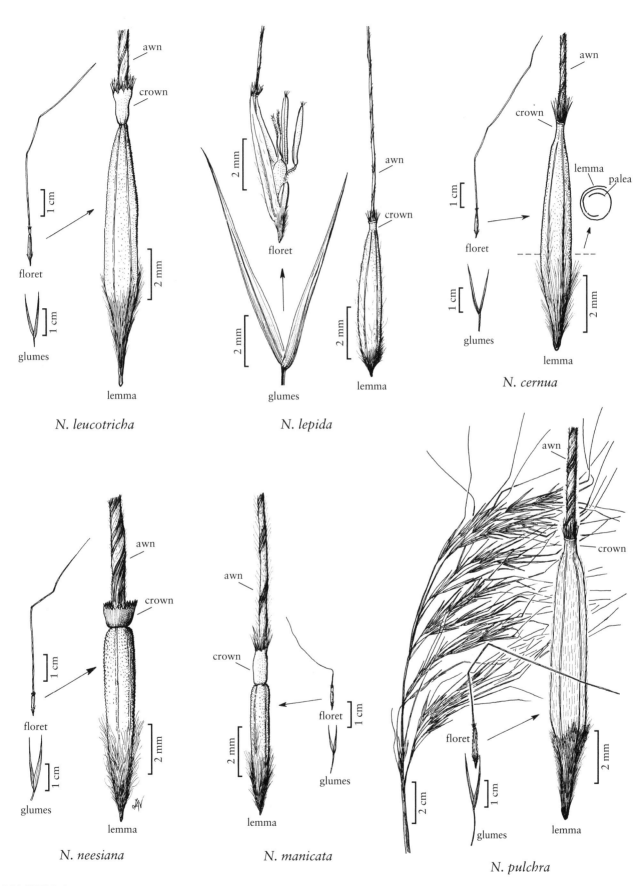

N. *leucotricha*

N. *lepida*

N. *cernua*

N. *neesiana*

N. *manicata*

N. *pulchra*

NASSELLA

The sharp callus easily sticks to skin and clothing, and can cause wounds, especially in the mouths of grazing animals. The wounds often retain the fruit, which may require surgical removal for proper healing.

Individual plants produce both chasmogamous and cleistogamous florets in their terminal panicles, with the terminal florets usually being cleistogamous. The proportion of cleistogamous florets is influenced by soil moisture, a higher proportion being produced if the soil moisture content is low (Brown 1952). Spikelets of the axillary panicles usually mature before those of the terminal panicles (Dyksterhuis 1949).

3. Nassella manicata (E. Desv.) Barkworth [p. 173]
ANDEAN TUSSOCKGRASS

Plants perennial; cespitose, not rhizomatous. **Culms** 40–80 cm tall, 1.5–2.5 mm thick, erect, internodes glabrous; **nodes** 2–3, pubescent. **Sheaths** glabrous; **collars** mostly glabrous, with tufts of hair at the sides, hairs 0.5–1 mm; **ligules** 0.1–0.3 mm, glabrous, truncate; **blades** 12–25 cm long, 1.5–2.5 mm wide, flat to convolute, pubescent. **Panicles** 10–20 cm, open; **branches** 1–3 cm, ascending to somewhat spreading, scabridulous, with 1–8 spikelets; **pedicels** 1–8 mm, pubescent. **Glumes** subequal, 10–15 mm long, 1.5–2 mm wide, narrowly lanceolate, 3-veined, keeled, keels scabrous; **florets** 6–8 mm long, 0.9–1.1 mm wide, terete, widest near or slightly above midlength; **calluses** 1.5–2.5 mm, sharp, strigose; **lemmas** papillose-tuberculate, constricted and purplish below the crown, midveins and exposed marginal veins pubescent over the proximal ²/₃, glabrous between the veins at maturity; **crowns** about 1 mm long, about 0.5 mm wide, conspicuous, more or less straight-sided, purple, rims with 1–1.5 mm hairs; **awns** 30–50 mm, clearly twice-geniculate, terminal segment straight; **anthers** 3–4 mm in putatively chasmogamous florets, 0.3–0.5 mm in cleistogamous florets, both ranges sometimes present within a panicle. **Caryopses** about 3 mm. 2*n* = unknown.

Nassella manicata is native to Ecuador, Chile, Argentina, and Uruguay, growing on the foothills of the Andes Mountains. It is established in three California counties, growing in disturbed sites, including grazed meadows and old gold tailings. It has also been recorded from Mississippi; it is not known whether the Mississippi population has persisted.

Nassella manicata resembles *N. leucotricha* and *N. pulchra*. It differs from both in its shorter florets and more strongly developed crowns. It was misidentified as *Nassella formicarum* (Delile) Barkworth in the Jepson Manual (Barkworth 1993).

4. Nassella lepida (Hitchc.) Barkworth [p. 173]
FOOTHILLS NASSELLA, FOOTHILLS NEEDLEGRASS

Plants perennial; cespitose, not rhizomatous. **Culms** 35–100 cm tall, 0.7–1.2 mm thick, erect, bases sometimes decumbent, internodes glabrous, or the lower internodes pubescent just below the nodes or throughout, varying within a plant; **nodes** 3–4, pubescent. **Sheaths** glabrous or coarsely hairy, sometimes scabridulous, margins glabrous; **collars** hairy, particularly towards the sides, hairs at the sides 0.2–0.5 mm; **ligules** 0.1–0.6 mm, glabrous, truncate to rounded; **blades** 12–23 cm long, 1–3.5 mm wide, flat to convolute, lax, abaxial surfaces scabridulous, adaxial surfaces coarsely hairy. **Panicles** 9–55 cm, open; **branches** 1–8 cm, ascending to spreading, with 1–6 spikelets; **pedicels** 1–5 mm. **Glumes** subequal, 5.5–15 mm long, 0.5–1 mm wide, narrowly lanceolate, glabrous, acuminate; **florets** 4–7 mm long, 0.5–0.7 mm wide, terete, widest near or slightly above midlength; **calluses** 0.4–1.6 mm, sharp, strigose; **lemmas** papillose, initially evenly pubescent, becoming glabrous between the veins at maturity, tapering to the crown; **crowns** 0.25–0.3 mm long, 0.15–0.2 mm wide, straight-sided, rims with 0.3–0.6 mm hairs; **awns** 12–55 mm long, 0.1–0.15 mm thick at the base, first geniculation distinct, second geniculation obscure, terminal segment cernuous; **anthers** 3, 2–2.5 mm, penicillate. **Caryopses** 3–4 mm. 2*n* = 34.

Nassella lepida usually grows on dry hillsides in chaparral habitats, from California into northern Mexico. It is most likely to be confused with *N. cernua*, but differs from that species in its shorter, thinner awns and more numerous spikelets. It occasionally hybridizes with *N. pulchra*.

5. Nassella pulchra (Hitchc.) Barkworth [p. 173]
PURPLE NASSELLA, PURPLE NEEDLEGRASS

Plants perennial; cespitose, not rhizomatous. **Culms** 35–100 cm tall, 1.8–3.1 mm thick, erect or geniculate at the lowest nodes, sometimes scabrous below the panicles, internodes mostly glabrous, lower internodes sometimes pubescent below the nodes; **nodes** 2–3, pubescent. **Sheaths** glabrous or hairy, sometimes mostly glabrous, sometimes the distal margins ciliate, varying within a plant; **collars** with tufts of hair at the sides, hairs 0.5–0.8 mm; **ligules** 0.3–1.2 mm, glabrous, truncate to rounded; **blades** 10–20 cm long, 0.8–3.5 mm wide, flat to convolute, abaxial surfaces glabrous or sparsely

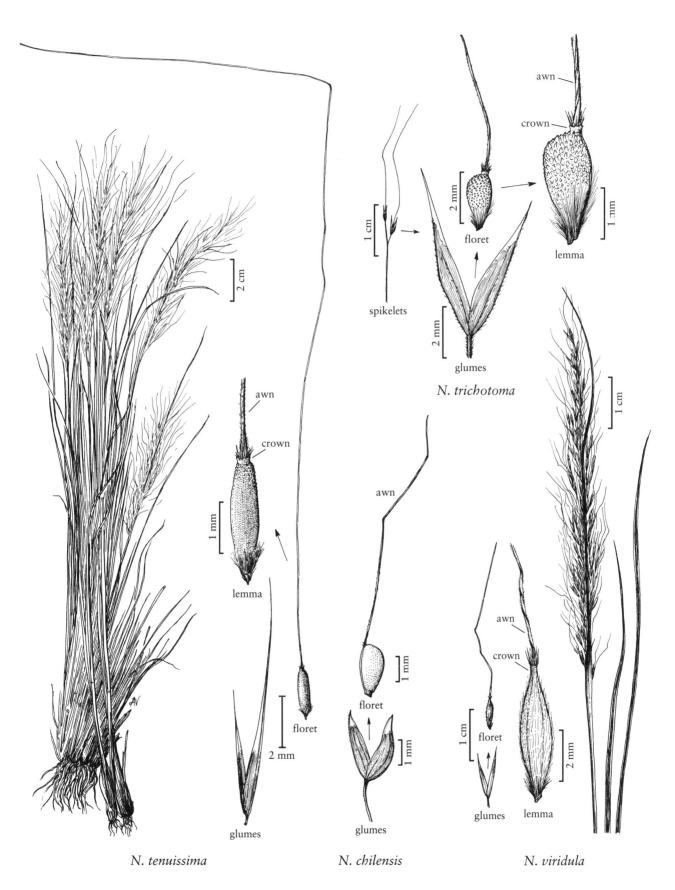

N. trichotoma

N. tenuissima

N. chilensis

N. viridula

NASSELLA

pilose. **Panicles** 18–60 cm, open; **branches** 3–9 cm, spreading, flexuous, often pilose at the axils, with 2–6 spikelets; **pedicels** 3–10 mm. **Glumes** subequal, 12–20 mm long, 1.1–2.2 mm wide, narrowly lanceolate, glabrous; **florets** 7.5–11.5 mm long, about 1.2 mm thick, terete; **calluses** 1.8–3.5 mm, sharp, strigose; **lemmas** papillose, evenly pubescent at maturity, constricted below the crown; **crowns** 0.6–1.1 mm long, 0.5–0.7 mm wide, straight-sided to slightly flared, rims with 0.8–0.9 mm hairs; **awns** 38–100 mm long, 0.3–0.45 mm thick at the base, strongly twice-geniculate, terminal segment straight; **anthers** 3.5–5.5 mm, penicillate. **Caryopses** 4.5–6 mm. $2n = 64$.

Nassella pulchra grows in oak chaparral and grassland communities of the coast ranges and Sierra foothills of California, extending south into Mexico. It probably never formed extensive grasslands (Hamilton 1997), flourishing primarily in moderately disturbed areas. It resembles *N. manicata*, but has longer florets and less strongly developed crowns. *Nassella pulchra* and *N. cernua* sometimes hybridize.

6. Nassella cernua (Stebbins & Love) Barkworth [p. 173]
CERNUOUS NASSELLA, NODDING NEEDLEGRASS

Plants perennial; cespitose, not rhizomatous. **Culms** 30–100 cm tall, 1–2.2 mm thick, erect or geniculate at the basal nodes, internodes pubescent below the nodes, lowest internodes sometimes pubescent throughout; **nodes** 2–3, glabrous. **Leaves** usually glaucous; **sheaths** mostly glabrous, throats ciliate; **collars** mostly glabrous, with sparse tufts of hair at the sides, hairs 1–1.6 mm; **ligules** 0.2–1.6 mm, glabrous, truncate to rounded; **blades** 3–26 cm long, 0.4–1.2 mm wide, flat to convolute, abaxial surfaces scabridulous, adaxial surfaces hairy. **Panicles** 15–80 cm, open, often partially enclosed at maturity; **branches** 1–6 cm, flexuous or cernuous, with 1–8 spikelets on the distal $^1/_2$; **pedicels** 3–9 mm. **Glumes** glabrous, narrowly lanceolate, acuminate; **lower glumes** 12–22 mm long, 0.9–1.7 mm wide; **upper glumes** 3–4 mm shorter that the lower glumes; **florets** 4–9 mm long, 0.6–0.8 mm wide, terete; **calluses** 1.4–3.6 mm, sharp, strigose; **lemmas** minutely papillose, tapering to the crown, the proximal $^1/_4$ evenly pubescent, the distal $^3/_4$ pubescent only over the veins; **crowns** 0.2–0.5 mm long, 0.3–0.35 mm wide, straight-sided, rims with 0.8–1.1 mm hairs; **awns** 50–110 mm long, 0.2–0.3 mm thick at the base, first geniculation evident, second geniculation obscure, terminal segment cernuous; **anthers** 3.5–5.5 mm, penicillate. **Caryopses** 4.5–5.5 mm. $2n = 70$.

Nassella cernua grows in grasslands, chaparral, and juniper associations of the inner coast ranges of California and Baja California, Mexico. Small specimens resemble *N. lepida*, but have longer and thicker awns and fewer florets. Large specimens resemble *N. pulchra*, but have thinner awns with cernuous, rather than straight, terminal segments. It is superficially similar to *Achnatherum eminens*, but differs in its shorter ligules, strongly overlapping lemma margins, glabrous paleas, and geographic distribution.

7. Nassella tenuissima (Trin.) Barkworth [p. 175]
FINELEAVED NASSELLA

Plants perennial; tightly cespitose, not rhizomatous. **Culms** 25–100 cm tall, 0.4–0.7(1.1) mm thick, usually erect, basal nodes sometimes geniculate, internodes mostly glabrous, pubescent just below the lower nodes; **nodes** 2–4, glabrous. **Sheaths** glabrous, even on the margins, sometimes scabridulous; **collars** glabrous, without tufts of hair at the sides; **ligules** 1–5 mm, glabrous, acute; **blades** 7–60 cm long, 0.2–1.5 mm wide, usually convolute, stiff, glabrous, scabridulous. **Panicles** 8–50 cm, loosely contracted, often partly enclosed at maturity; **branches** 2–8 cm, glabrous; **pedicels** 1–11 mm. **Glumes** subequal, 5–13 mm long, 0.5–1.2 mm wide, narrowly lanceolate, glabrous, aristate; **florets** (1.5)2.5–3 mm long, about 0.5 mm wide, widest at about midlength, somewhat laterally compressed; **calluses** 0.2–0.5 mm, blunt, strigose, hairs reaching to about $^1/_4$–$^1/_3$ the length of the lemmas; **lemmas** finely tuberculate, rounded to the crown, midveins pubescent on the proximal $^1/_2$; **crowns** 0.1–0.2 mm long, 0.2–0.25 mm wide, straight-sided, rims with hairs shorter than 0.5 mm; **awns** 45–100 mm, almost centric, cernuous throughout, twice-geniculate, usually both geniculations obscure; **anthers** 3, 1.2–1.5 mm. **Caryopses** about 2 mm, linear, dark brown. $2n = 40$.

Nassella tenuissima grows on rocky slopes, frequently in oak or pine associations but also in open, exposed grasslands. Its native range extends from the southwestern United States into northern Mexico. It is now also established in the San Francisco Bay area, having been introduced as a garden plant. It is an attractive species, available through some horticultural outlets, but it readily escapes from cultivation into nearby disturbed sites.

8. **Nassella trichotoma** (Nees) Hack. *ex* Arechav. [p. 175]

SERRATED TUSSOCKGRASS, YASS TUSSOCKGRASS

Plants perennial; cespitose, not rhizomatous. **Culms** 20–60 cm tall, about 1 mm thick, erect, internodes glabrous; **nodes** 2–4, pubescent. **Sheaths** glabrous, smooth; **collars** glabrous, without tufts of hair at the sides; **ligules** 0.5–2.5 mm, glabrous, obtuse; **blades** 15–45 cm long, 0.2–0.6 mm wide, convolute, stiff, scabridulous. **Panicles** 8–25 cm, open, lax, sparse; **branches** 2–6 cm, with 1–8 spikelets; **pedicels** 3–12 mm. **Glumes** subequal, 4–10 mm long, 0.9–1.2 mm wide, narrowly lanceolate, scabridulous, apices aristate; **florets** 1.5–2.5 mm long, 0.7–0.9 mm wide, terete, widest near the top; **calluses** 0.1–0.3 mm, acute, strigose, hairs reaching to midlength on the lemmas; **lemmas** strongly tuberculate, particularly distally, mostly glabrous, narrowing abruptly to the crown, midveins pubescent proximally; **crowns** about 0.5 mm long, about 0.2 mm wide, straight-sided, rims entire or irregularly lacerate, glabrous; **awns** 15–35 mm, eccentric, straight to twice-geniculate; **anthers** 3, 1–1.5 mm. **Caryopses** about 1.2 mm, oblong, dark brown. $2n = 36$.

Nassella trichotoma is a native of South America, and has been accidentally introduced into the United States. Because it is on the U.S. Department of Agriculture's noxious weed list, all known populations have been eliminated. New populations should be reported to the Department.

9. **Nassella chilensis** (Trin.) E. Desv. [p. 175]

CHILEAN TUSSOCKGRASS

Plants perennial; shortly rhizomatous, appearing cespitose, rhizomes slender, somewhat woody. **Culms** 30–100 cm tall, 0.4–0.7 mm thick, bases somewhat bulblike, erect, geniculate and often branching intravaginally at the lower cauline nodes, internodes glabrous; **nodes** 5–8+, glabrous. **Sheaths** mostly glabrous, throats sometimes ciliate; **collars** sparsely hairy, with tufts of hair at the sides, hairs 0.5–1.3 mm; **ligules** 0.2–0.3 mm, truncate, usually ciliate; **blades** 3–10 cm long, 1–1.5 mm wide, strongly convolute, stiff, abaxial surfaces glabrous, adaxial surfaces with coarse hairs. **Panicles** 2–20 cm; **branches** 0.4–1.2 cm, with 1–4 spikelets; **pedicels** 0.5–4 mm. **Glumes** subequal, 3–4.5 mm long, 1.1–1.6 mm wide, ovate, 3-veined, glabrous or puberulent, acuminate; **florets** 1.6–2.2 mm long, 0.6–0.9 mm wide, obovate to oblong, terete, widest near the top; **calluses** 0.2–0.3 mm, obtuse, glabrous; **lemmas** glabrous, smooth, lustrous, transition to the crown not evident; **crowns** about 0.1 mm long and wide, not differing in texture from the lemmas; **awns** 7–10 mm, eccentric, rapidly deciduous; **anthers** about 1 mm or 0.3–0.4 mm, florets with longer anthers presumably chasmogamous, those with shorter anthers presumably cleistogamous. **Caryopses** about 1 mm. $2n = 42$.

Nassella chilensis is an Andean species that was once collected from a ballast dump in Portland, Oregon. It is not established in the *Flora* region.

10. **Nassella viridula** (Trin.) Barkworth [p. 175]

GREEN NASSELLA, GREEN NEEDLEGRASS

Plants perennial; cespitose, not rhizomatous. **Culms** 35–120 cm, erect or geniculate basally, internodes mostly glabrous, pubescent below the lower nodes; **nodes** 2–3, glabrous. **Sheaths** mostly glabrous, margins usually ciliate; **collars** of basal leaves hispidulous, with tufts of hair at the sides, hairs 0.5–1.8 mm, collars of flag leaves glabrous or sparsely pubescent; **ligules** 0.2–1.2 mm, glabrous, truncate to rounded; **blades** 10–30 cm long; 1.5–3 mm wide, flat to convolute, abaxial surfaces scabridulous, glabrous, adaxial surfaces glabrous. **Panicles** 2.9–7.2 cm, loosely contracted; **branches** 1–4 cm, appressed or ascending, with 3–7 spikelets; **pedicels** 1–9 mm. **Glumes** subequal, 6.8–13 mm long, 1–2.1 mm wide, narrowly lanceolate, glabrous, apiculate; **florets** 3.4–5.5 mm long, 1–1.2 mm wide, terete; **calluses** 0.7–1.4 mm, moderately sharp, strigose; **lemmas** papillose, evenly pubescent, not constricted below the crown; **crowns** 0.4–0.5 mm long, 0.3–0.5 mm wide, not conspicuous, straight-sided, rims with 0.5–0.75 mm hairs; **awns** 19–32 mm, evidently twice-geniculate, terminal segment straight; **anthers** (0.8)2–3 mm, sometimes penicillate. **Caryopses** about 3.5 mm. $2n = 82, 88$.

Nassella viridula grows in grasslands and open woods, frequently on sandy soils. It is the most widespread species of *Nassella* in North America. Its morphology, distribution, and high chromosome number suggest that it may be an alloploid between *Nassella* and *Achnatherum*. It is included in *Nassella* because it resembles *Nassella* more than *Achnatherum* in the characters distinguishing the two genera. It differs from *Achnatherum robustum* in its tightly convolute lemmas and in having glabrous to sparsely pubescent collars on its flag leaves. It differs from the hybrid with *Achnatherum hymenoides*, ×*Achnella caduca*, in its less saccate glumes, shorter lemmas hairs, and shorter paleas.

10.13 JARAVA Ruiz & Pav.

Mirta O. Arriaga

Plants perennial; cespitose, sometimes rhizomatous, rhizomes forming knotted bases. **Culms** 15–200 cm, not branching at the upper nodes; **basal branching** intravaginal or extravaginal; **prophylls** not evident, shorter than the leaf sheaths. **Leaves** mostly basal, not overwintering; **sheaths** open to the base; **cleistogenes** not present; **collars** with tufts of hair on either side; **auricles** absent; **ligules** membranous, truncate or shortest in the center and rounded, edges usually ciliate, hairs at the outer edges often longer than the central membranous portion, ligules of the lower leaves glabrous or hairy, sometimes densely hairy, those of the upper leaves glabrous or sparsely hairy; **blades** usually convolute, apices narrowly pointed, flag leaves longer than 10 mm. **Inflorescences** panicles, often partially included in the upper leaf sheath; **branches** straight. **Spikelets** 5.5–24 mm, with 1 floret; **rachillas** not prolonged beyond the floret; **disarticulation** above the glumes, beneath the floret. **Glumes** unequal, usually longer than the floret, sometimes shorter, hyaline, 0–5-veined; **florets** narrowly lanceoloid, terete; **calluses** 0.2–1.6(3) mm, acute, less than or equaling the floret diameter, antrorsely strigose distally, hairs white; **lemmas** thickly membranous, basal $^2/_3$ scabrous or shortly pubescent, distal $^1/_3$ often bearing a pappus of ascending to strongly divergent 3–8 mm hairs, sometimes glabrous or with appressed hairs shorter than 1 mm, margins not or only slightly overlapping at maturity, apices not fused into a crown, lobes to 0.2 mm, with a single, terminal awn, awn-lemma junction conspicuous; **awns** 9–45(80) mm, persistent or deciduous, scabrous, weakly once- or twice-geniculate, first segment scabrous or pilose, terminal segment glabrous or pilose, smooth or scabrous; **paleas** $^1/_3$–$^1/_2$ as long as the lemmas, flat between the veins, membranous to hyaline, glabrous or sparsely pubescent, 2-veined, veins poorly developed, apices rounded to irregular; **lodicules** 2–3, the third, if present, reduced; **anthers** 3. **Caryopses** fusiform, not ribbed; **hila** linear. x = unknown.

Jarava is a South American genus that used to be included in *Stipa*. Its limits are currently under study. As treated here, it is a genus of approximately 50 species, all of which are native to South America. It includes three groups that have been recognized as subgenera within *Stipa sensu lato*: *Jarava* (Ruiz. & Pav.) Trin. & Rupr., *Pappostipa* Speg., and *Ptilostipa* Speg. Jacobs and Everett (1997) recommended including only the approximately 14 species of *Stipa* subg. *Jarava* in *Jarava*. This treatment adopts a somewhat broader interpretation and includes the members of the other two subgenera, pending more detailed study of relationships among the American *Stipeae*. In this respect, the treatment presented here conforms with the treatment by Peñailillo (2002); it differs in excluding *Achnatherum* and *Amelichloa*.

One species of the *Pappostipa* group, *Jarava speciosa*, grows as a disjunct in the southwestern United States, its broadest distribution being in South America. Two other species, *J. ichu* and *J. plumosa*, both members of *Jarava sensu stricto*, have been found as escapes from cultivation in California.

Many species of *Jarava* have conspicuous hairs on the distal portion of the lemma, termed a *pappus*, or on the first and/or second segment of the awn. These are an adaptation to wind dispersal. *Jarava sensu stricto* shows an even stronger adaptation to wind dispersal, usually combining a well-developed pappus with light florets less than 5 mm long and 1 mm wide.

SELECTED REFERENCES **Arriaga, M.O.** 1983. Anatomía foliar de las especies de *Stipa* del subgénero *Pappostipa* (Stipeae-Poaceae) de Argentina. Revista Mus. Argent. Ci. Nat., Bernardino Rivadavia Inst. Nac. Invest. Ci. Nat., Bot. 6:89–141; **Caro, J.A.** and **E. Sánchez.** 1973. Las especies de *Stipa* (Gramineae) del subgénero *Jarava*. Kurtziana 7:61–116; **Jacobs, S.W.L.** and **J. Everett.** 1997. *Jarava plumosa* (Gramineae), a new combination for the species formerly known as *Stipa papposa*. Telopea 7:301–302; **Matthei, O.** 1965. Estudio crítico de las gramíneas del género *Stipa* en Chile. Gayana, Bot. 13:1–137; **Peñailillo, P.** 2002. El género *Jarava* Ruiz et Pavon (Stipeae-Poaceae): Delimitacion y nuevas combinaciones. Gayana, Bot. 59:27–34.

1. Awns 35–80 cm long, the basal segment pilose; ligules of the basal leaves softly and densely hairy . . . . 3. *J. speciosa*
1. Awns 9–30 mm long, the basal segment scabridulous or smooth; ligules of the basal leaves glabrous or almost so.
    2. Glumes clearly exceeding the florets; pappus hairs 3–4 mm long . . . . 1. *J. ichu*
    2. Glumes from shorter than to subequal to the florets; pappus hairs 5–8 mm long . . . . 2. *J. plumosa*

1. Jarava ichu Ruiz & Pav. [p. 180]
PERUVIAN NEEDLEGRASS

Plants densely cespitose, not rhizomatous. **Culms** (15)30–100 cm, bases dull brown, glabrous; **nodes** 2–4; **branching** intravaginal. **Sheaths** mostly glabrous, scabridulous, basal sheaths dull brown; **ligules** 0.3–1 mm, truncate, erose, abaxial surfaces glabrous or almost so, ciliate, hairs longest (to 2 mm) towards the sides of the leaves, at the top of the sheaths; **blades** (3)10–40 cm long, 0.5–1 mm wide, all alike, straight, erect, convolute, apices sharp. **Panicles** (3)10–25(30) cm, narrow, cylindrical to lanceoloid, dense, from partially to wholly exserted at anthesis, erect or nodding distally; **branches** strongly ascending. **Spikelets** 5.5–11 mm. **Glumes** subequal, clearly exceeding the florets, linear-lanceolate, tapering to attenuate apices; **lower glumes** 5.5–11 mm, 1–3-veined; **upper glumes** 5–10.5 mm, 3-veined; **florets** 2.3–3 mm, cylindrical to fusiform; **calluses** 0.2–0.4 mm, acute to broadly acute, strigose; **lemmas** hairy throughout, hairs on the lower portion about 0.15 mm, sparse, appressed, pappus hairs 3–4 mm; **awns** 9–15 mm, twice-geniculate, first 2 segments twisted, scabridulous; **paleas** 1–1.5 mm, sparsely pubescent, 2-veined, apices rounded; **lodicules** 2, 0.6–1 mm; **anthers** about 0.8 mm. **Caryopses** 1.8–2.2 mm long, 0.6–0.7 mm thick, cylindrical.

Jarava ichu is native to Mexico, Costa Rica, Venezuela, Colombia, Ecuador, Peru, Bolivia, and Argentina. It is abundant in much of this range. In the *Flora* region, it is sold as an attractive ornamental. The species could become a problem, because it is self-compatible and produces a large quantity of wind-dispersed seeds. In parts of its native range, *J. ichu* is highly valued for its ability to prevent soil erosion, and for its use in thatch, mats, and basketry. "Ichu" is a term used to describe any bunchgrass in some parts of South America.

2. Jarava plumosa (Spreng.) S.W.L. Jacobs & J. Everett [p. 180]
PLUMOSE NEEDLEGRASS

Plants cespitose, shortly rhizomatous, rhizomes forming knotted bases. **Culms** 15–85 cm, glabrous, bases dull gray-brown; **nodes** 2–6; **basal branching** mostly extravaginal, lower nodes sometimes with intravaginal branches. **Sheaths** glabrous, basal sheaths dull gray-brown; **ligules** 0.1–0.2 mm, truncate, abaxial surfaces puberulent, ciliolate to ciliate, hairs longest (1.5–4 mm) towards the sides of the leaves, at the top of the sheaths; **blades** 1–9(25) cm long, those of the innovations the longest, 1–1.5 mm wide and flat or conduplicate, or to 0.5 mm in diameter and convolute, straight to almost falcate, abaxial surfaces of the innovation leaves glabrous or pubescent, adaxial surfaces usually glabrous, sometimes slightly scabrous. **Panicles** 3–20 cm, ovoid, lax, partially included in the upper leaf sheaths; **branches** ascending to divergent; **pedicels** 1–1.5 mm. **Spikelets** 5–8 mm. **Glumes** from shorter than to subequal to the florets, linear-lanceolate, hyaline, smooth, ecostate or with 1 inconspicuous vein, apices attenuate; **lower glumes** 2.5–5 mm; **upper glumes** 4.5–6.5 mm; **florets** (4)5–7.5 mm; **calluses** 1–1.5 mm, strigose, hairs white; **lemmas** about 0.3 mm thick, mostly scabrous, strigose over the midvein, tapering to the apices, pappus hairs 5–8 mm; **awns** 15–30 mm, scabrous, weakly geniculate; **paleas** 1–2.5 mm, from $^1/_3$–$^1/_2$ the length of the lemmas, hyaline, glabrous, weakly 2-veined; **lodicules** 2, 0.8–1 mm, linear. **Caryopses** 4–5 mm, narrowly lanceoloid. $2n = 40$.

A native of Argentina, Chile, and Uruguay, *Jarava plumosa* was collected in Berkeley, California in 1983. It is not known to be established in the *Flora* region. In its native range, it often grows on poor, unstable soils. Matthei (1965) stated that it is a valuable forage species when young, but that it should not be overgrazed because of its value in preventing soil degradation.

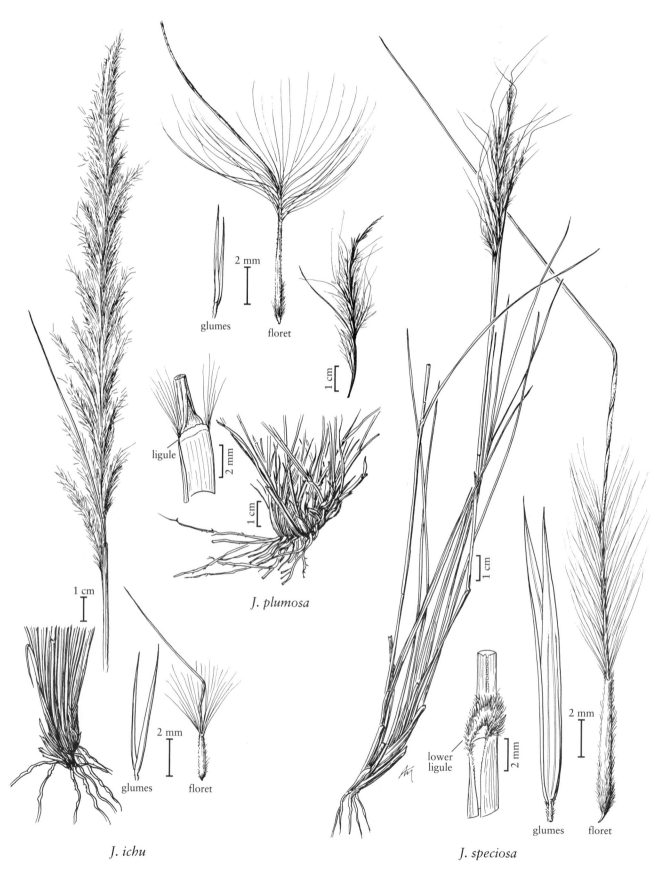

glumes

floret

2 mm

1 cm

ligule

2 mm

1 cm

J. plumosa

1 cm

glumes

floret

2 mm

J. ichu

1 cm

lower
ligule

2 mm

glumes

floret

2 mm

J. speciosa

JARAVA

3. Jarava speciosa (Trin. & Rupr.) Peñail. [p.180]
DESERT NEEDLEGRASS

Plants tightly cespitose, not rhizomatous. Culms 30–60 cm, bases orange-brown; nodes 3–6; basal branching intravaginal. Sheaths mostly glabrous, throats densely ciliate, basal sheaths reddish brown, flat and ribbonlike with age; ligules varying within a plant, lower ligules 0.3–1 mm, densely hairy and ciliate, hairs 0.2–1 mm, often longer than the basal membrane, upper ligules to 2.5 mm, hyaline to scarious, glabrous or hairy, usually less hairy than the lower ligules, sometimes ciliate; blades 10–30 cm long, 0.5–2 mm wide when flat, usually rolled, to 1 mm in diameter, abaxial surfaces glabrous, smooth, adaxial surfaces pilose. Panicles 10–15 cm, dense, frequently partially included in the upper leaf sheaths at maturity; branches ascending. Spikelets 16–24 mm. Glumes linear-lanceolate, glabrous, tapering from below midlength to the narrowly acute apices; lower glumes 16–24 mm, 1-veined; upper glumes 13–19 mm, 3–5-veined; florets (6)8–10 mm; calluses 0.8–1.6(3) mm, sharp; lemmas densely and evenly hairy, hairs about 0.5 mm, without a pappus; awns 35–45(80) mm, once-geniculate, first segment pilose, hairs 3–8 mm, terminal segment glabrous, smooth; paleas 3.2–5.1 mm, $^2/_5$–$^2/_3$($^4/_5$) the length of the lemmas, usually hairy, hairs about 0.5 mm. $2n = 66, 68$, about 74.

Jarava speciosa grows on rocky slopes in canyons of arid and semiarid regions of the southwestern United States and northern Mexico, and in Chile and northern to central Argentina. Several varieties are recognized in South America. It is not clear to which of these varieties, if any, the North American plants belong.

The reddish brown leaf bases, differing lower and upper ligules, and the pilose, once-geniculate awns make *Jarava speciosa* an easy species to recognize in North America. It is also an attractive species, well worth cultivating. It prefers open areas with well-drained soils. The growth of young shoots and flowering is stimulated by fire.

10.14 AMELICHLOA Arriaga & Barkworth

Mirta O. Arriaga

Plants perennial; cespitose. Culms erect, with 2–3 nodes, not branching at the upper nodes; basal branching intravaginal; prophylls concealed by the leaf sheaths, winged over the keels, apices bifid, teeth 0.5–3.5 mm. Leaves mostly basal; sheaths open, smooth, glabrous; cleistogenes often present, spikelets of cleistogenes 0.5–1 mm long, with thin glumes shorter than the florets, florets unawned or with reduced awns; auricles absent; ligules scarious, rounded to acute, ciliate; blades stiff, involute, apices stiff, brown, sharply pointed, blades of the flag leaves 5–13 cm long, bases similar in width to the top of the sheaths. Inflorescences panicles, the main panicle terminal, apparently wholly chasmogamous. Spikelets with 1 floret; disarticulation above the glumes, beneath the floret. Glumes exceeding the floret, acute to acuminate, 1–5-veined; florets fusiform, terete; calluses antrorsely strigose, blunt; lemmas pubescent, often more densely and/or more persistently so over the midvein and lateral veins, hairs on the proximal portion about 0.7–2 mm, hairs on the distal portion often longer; crowns not developed; awns once- or twice-geniculate, scabrous, persistent; paleas $^3/_4$ as long as to almost equaling the lemmas, flat, hairy, hairs 0.2–1 mm, veins terminating at or near the apices, apices similar in texture to the body; lodicules 3; anthers 3, anthers sometimes all of equal size and more than 2 mm, sometimes 1 longer than 2 mm and 2 much shorter, sometimes all shorter than 2 mm; ovaries glabrous; styles with 2 branches, united at the base, stigmas plumose. Caryopses obovoid, with 3 smooth, longitudinal ribs at maturity, stylar bases 1–2 mm, persistent, sometimes eccentric; hila linear, about as long as the caryopses. $x = $ unknown.

Amelichloa includes five species, four of which are South American. The fifth species, *A. clandestina*, grows in northern Mexico. Two species are established in the *Flora* region. A third species, *A. caudata*, was found on ballast dumps near Portland, Oregon, at the turn of the twentieth century; it is not established in the region.

Cattle avoid species of *Amelichloa* because of their sharply pointed leaves. This means that any of the species could become a serious problem in rangelands. Mowing favors their establishment and spread because it does not eliminate, and may disperse, the cleistogenes. The species are eaten by goats.

SELECTED REFERENCES **Caro, J.A.** and **E. Sánchez.** 1971. La identidad de *Stipa brachychaeta* Godron, *S. caudata* Trinius y *S. bertrandii* Philippi. Darwinia 16:637–653; **Torres, M.A.** 1993. Revisión del género *Stipa* (Poaceae) en la Provincia de Buenos Aires. Monografia 12. Comisión de Investigaciones Científicas, Provincia de Buenos Aires, La Plata, Argentina. 62 pp.

1. Mature caryopses with inclined, eccentric stylar bases; lemmas glabrous or hairy between the lateral and marginal veins, glabrous between the midvein and the lateral vein, even at the base . . . . .2. *A. caudata*
1. Mature caryopses with erect, usually centric stylar bases; proximal ¹/₂ of the lemmas pubescent between the lateral and marginal veins, at least initially, usually also between the midvein and lateral veins.
　　2. Florets 4–5.5 mm long; ligules 0.2–0.6 mm long; anthers 2–3 mm long . . . . .1. *A. brachychaeta*
　　2. Florets 5.5–8 mm long; ligules 0.5–1.5 mm long; anthers 3–4 mm long . . . . .3. *A. clandestina*

1. **Amelichloa brachychaeta** (Godr.) Arriaga & Barkworth [p.183]
PUNA NEEDLEGRASS

Plants with knotty, shortly rhizomatous bases. **Culms** 40–90 cm tall, 1–2(3) mm thick, erect, glabrous; **nodes** usually 3. **Basal sheaths** mostly glabrous, margins ciliate distally; **collars** glabrous, often with tufts of hair to 1.5 mm on the sides; **ligules** 0.2–0.6 mm, membranous, strigose, ciliate, cilia to 2 mm, slightly longer at the leaf margins; **blades** 8–35 cm long, usually convolute and 0.5–0.8 mm in diameter, 2–3 mm wide when flat, erect, abaxial surfaces smooth, adaxial surfaces usually glabrous. **Panicles** 10–25 cm long, 1–4 cm wide, bases sometimes included in the upper leaf sheaths; **branches** ascending to spreading, longest lower branches 4–12 cm. **Glumes** subequal, 6–8 mm, linear-lanceolate, 1–3-veined, midveins smooth, scabridulous, or with stiff hairs, varying within a panicle, apices acuminate; **florets** 4–5.5 mm long, about 0.8 mm thick, fusiform; **calluses** 0.4–0.5 mm, blunt, strigose, hairs 0.5–0.8 mm; **lemmas** pubescent over and between the veins on the proximal ¹/₂, at least initially, hairs 0.5–0.8 mm, sometimes glabrous at maturity between the midveins and lateral veins, distal portion glabrous, tapering to the apices, apices with 0.7–1 mm hairs around the base of the awn; **awns** 10–18 mm, glabrous or scabrous, usually once-geniculate; **paleas** ³/₄–⁹/₁₀ as long as the lemmas, pubescent over the central portion, apices involute; **lodicules** 3; **anthers** 2–2.4(3) mm, penicillate. **Caryopses** 2–3 mm long, 0.9–1 mm thick, fusiform; **style bases** straight, centric. $2n$ = unknown.

Amelichloa brachychaeta has been found at a few locations in California, where it is listed as a noxious weed. It is native to Uruguay and Argentina. It is avoided by cattle because of its sharply pointed leaves. The cleistogamous panicles, which may be at or below ground level, remain a source of seeds unless the plants are completely uprooted. *Amelichloa caudata* and *A. clandestina* (see below) are a greater problem in this regard, because they appear to produce such panicles more frequently.

2. **Amelichloa caudata** (Trin.) Arriaga & Barkworth [p. 183]
SOUTH AMERICAN NEEDLEGRASS

Plants cespitose, with knotty, rhizomatous bases. **Culms** 40–100 cm tall, 0.8–3(4.8) thick, erect, glabrous; **nodes** usually 3. **Basal sheaths** mostly glabrous, margins ciliate distally; **collars** glabrous, often with tufts of hair to 1.5 mm on the margins; **ligules** 0.2–0.6 mm, membranous, ciliate, cilia 1–2 mm, longest at the margins; **blades** 25–70 cm long, 2–7 mm wide when flat, usually convolute and 0.4–0.8 mm in diameter, erect, abaxial surfaces smooth, adaxial surfaces mostly glabrous. **Panicles** 10–45 cm long, 2–8 cm wide, bases sometimes included in the upper leaf sheaths; **branches** ascending to spreading, longest lower branches 4–12 cm. **Glumes** subequal, 5–11 mm, linear-lanceolate, 3-veined, midveins smooth, scabrous, or with stiff hairs, varying within a panicle, apices acuminate; **florets** 3.5–5(7) mm long, about 0.8 mm thick, fusiform; **calluses** 0.5–1 mm, blunt, strigose, hairs 0.5–0.8 mm;

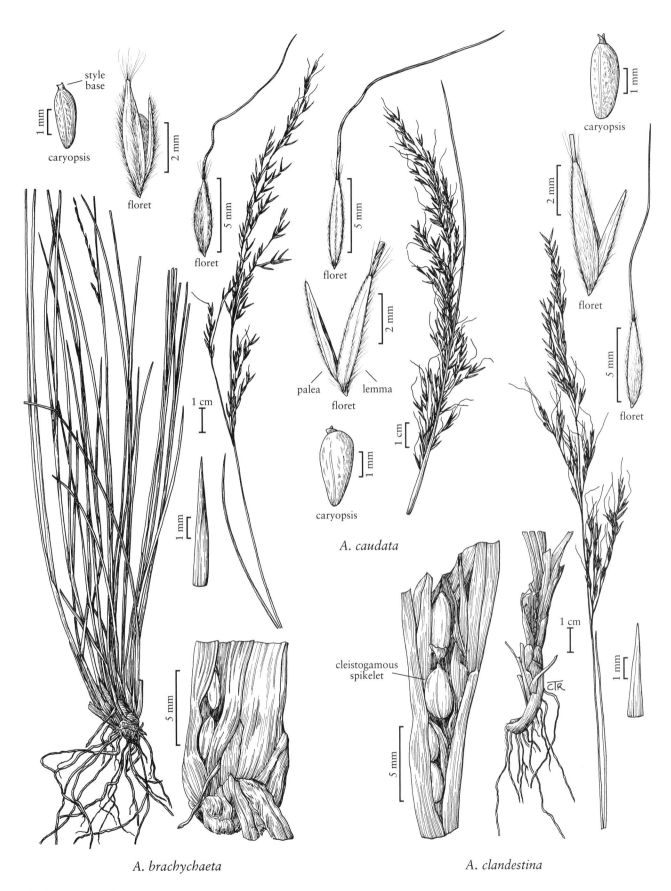

style base

caryopsis

floret

floret

floret

palea   lemma
floret

caryopsis

A. caudata

cleistogamous spikelet

caryopsis

floret

floret

A. brachychaeta

A. clandestina

AMELICHLOA

lemmas hairy over the veins, hairs 0.5–0.8 mm, glabrous between the mid- and lateral veins, glabrous or hairy between the lateral and marginal veins, distal portion glabrous, tapering to the apices, apices with 0.7–1 mm hairs around the base of the awn; **awns** 12–25 mm, glabrous or scabrous, once- or twice-geniculate; **paleas** 3.5–6.5 mm, $^3/_4$–$^9/_{10}$ as long as the lemmas, pubescent over the central portion, apices involute; **lodicules** 3; **anthers** 2–4 mm in chasmogamous florets, cleistogamous florets sometimes with anthers of 2 lengths, about 0.4 mm and 0.7 mm, or all anthers of the same length, the long anthers penicillate. **Caryopses** about 3 mm long, 1–1.4 mm thick; **style bases** inclined, eccentric. $2n$ = unknown.

Amelichloa caudata is native to South America, extending from central Chile to Uruguay and Argentina. It was collected, as *Stipa litoralis* Phil., on ballast dumps near Portland, Oregon, early in the twentieth century. Although it has not become established in the *Flora* region, it has done so in Australia. It is a potentially invasive weed. Species with anthers of three different lengths, and of two lengths within a floret, have been reported for *Nassella*; it appears to be the first report of this pattern in the species of *Amelichloa*.

3. Amelichloa clandestina (Hack.) Arriaga & Barkworth [p.183]
MEXICAN NEEDLEGRASS

Plants cespitose, with knotty, rhizomatous bases. **Culms** 50–90 cm tall, 1–2.9 thick, erect, glabrous; **nodes** usually 3. **Sheaths** mostly glabrous, margins ciliate distally; **collars** glabrous, often with tufts of hair to 1.5 mm on the margins; **ligules** 0.5–1.5 mm, membranous, ciliate, cilia 1–2 mm, slightly longer at the leaf margins; **blades** 10–50 cm long, usually tightly convolute and 0.4–1 mm in diameter, 2–4 mm wide when flat, erect, abaxial surfaces smooth, adaxial surfaces usually glabrous. **Panicles** 10–20 cm long, 1–5 cm wide, bases sometimes included in the leaf sheaths; **branches** ascending to spreading, longest lower branches 4–12 cm. **Glumes** subequal, 6.3–13 mm, linear-lanceolate, 3-veined, midveins smooth, scabrous, or with stiff hairs, varying within a panicle, apices acuminate; **florets** 5.5–8 mm long, about 0.8–1.3 mm thick, fusiform; **calluses** 0.3–0.6 mm, blunt, strigose, hairs 0.5–0.8 mm; **lemmas** pubescent on the proximal $^1/_2$, the pubescence over the midvein extending to the awn, hairs 0.5–0.8 mm, distal portion glabrous, tapering to the apices, apices with 0.7–1 mm hairs around the base of the awn; **awns** 11–23 mm, glabrous or scabrous, usually twice-geniculate; **paleas** 3.5–6.5 mm, $^3/_4$–$^9/_{10}$ as long as the lemmas, pubescent over the central portion, apices involute; **lodicules** 3; **anthers** 3–4 mm, penicillate. **Caryopses** about 3 mm long, 1–1.4 mm thick; **style bases** upright, centric. $2n$ = unknown.

Amelichloa clandestina is native from northern Mexico to Colombia. It has been accidentally introduced to pastures and roadsides in Texas, and is now established there. Reports from California need to be checked. They may reflect misidentifications of *A. brachychaeta*, which differs from *A. clandestina* in its shorter ligules, florets, and anthers.

10.15 AUSTROSTIPA S.W.L. Jacobs & J. Everett

Surrey W.L. Jacobs

Plants facultative perennials; to 2.5 m, sometimes shrublike, often with knotty bases. **Culms** often persistent, sometimes geniculate at the base, sometimes with stiff branches at the upper nodes; **basal branching** usually extravaginal; **prophylls** not evident. **Leaves** not overwintering; **sheaths** open to the base, margins sometimes extending beyond the base of the ligule, the extensions often conspicuously hairy, sometimes grading into the ligule; **auricles** not present; **ligules** membranous; **blades** flat, convolute, or terete, usually scabrous, sometimes pubescent, those of the flag leaves longer than 10 mm, bases usually as wide as the top of the sheaths, sometimes narrower. **Inflorescences** panicles, sometimes contracted, rachises persistent or disarticulating whole at maturity; **pedicels** smooth or scabrous, sometimes hairy, hairs to 3 mm; **disarticulation** above the glumes, sometimes also at the base of the panicle. **Spikelets** with 1 floret; **rachillas** not prolonged. **Glumes** usually longer than the florets, narrow, more or less keeled, hyaline to chartaceous, 1–5(7)-veined, apices usually acute or acuminate, rarely muticous or mucronate, remaining open after the floret falls; **florets** at least 5 times longer than wide; **calluses** strigose, usually sharp; **lemmas** usually coriaceous or indurate, 3–5(7)-veined, margins thin, usually convolute, rarely involute, apices glabrous or pilose, with 0–2 minute to small, membranous lobes, awned, lemma-awn junction evident; **awns** once- or twice-geniculate; **paleas** from shorter than to subequal to the lemmas, flat between the veins, 0–2-veined; **lodicules** 3 or 2; **anthers** 3, frequently penicillate; **ovaries** glabrous. **Caryopses** fusiform, terete; **hila** linear, nearly as long as the caryopses. x = unknown. Name from the Latin, *austra*, 'south', and *Stipa*, the name of the genus in which *Austrostipa* used to be included.

Austrostipa is a genus of 63 species, all of which are native to Australasia. Two species are cultivated in the *Flora* region.

SELECTED REFERENCE **Jacobs, S.W.L.** and **J. Everett.** 1996. *Austrostipa*, a new genus, and new names for Australasian species formerly included in *Stipa* (Gramineae). Telopea 6:579–595.

1. Plants shrubby; panicle branches and pedicels plumose, hairs 1.5 mm or longer . . . . 1. *A. elegantissima*
1. Plants bamboolike; panicle branches and pedicels not plumose, if hairy then the hairs to 0.3 mm long . . . . 2. *A. ramosissima*

1. Austrostipa elegantissima (Labill.) S.W.L. Jacobs & J. Everett [p.186]
AUSTRALIAN FEATHERGRASS

Plants perennial; to 2 m, shrubby; cespitose, shortly rhizomatous. **Culms** decumbent, branched, with 3–6(10) nodes, glabrous. **Leaves** not forming a basal cluster; **sheaths** becoming somewhat free, glabrous, often somewhat ribbed; **ligules** 2–3 mm, obtuse, erose, glabrous; **blades** (2)5–7.5 cm long, 1.5–2 mm wide, tightly rolled, glabrous, sometimes scabrous, margins glabrous, sometimes scabrous. **Panicles** 14–25 cm, open, with whorls of branches bearing few spikelets, detached at maturity; **branches** and **pedicels** plumose, hairs 1.5–3 mm. **Glumes** 7–12 mm, 3-veined basally, veins pilose; **florets** 4.5–10 mm, narrowly cylindrical; **calluses** 0.5–0.8 mm, nearly straight, sericeous; **lemmas** tuberculate, mostly glabrous, black when mature, margins with short strigose hairs proximally; **awns** 2–5 cm, once- or twice-geniculate, first segment scabrous; **paleas** about ½ the length of the lemmas; **anthers** 1.5–3 mm, penicillate. **Caryopses** 4–5 mm. $2n$ = unknown.

Austrostipa elegantissima is native to southern Australia. It is cultivated as an ornamental in the United States.

2. Austrostipa ramosissima (Trin.) S.W.L. Jacobs & J. Everett [p.186]
PILLAR-OF-SMOKE, AUSTRALIAN PLUMEGRASS, STOUT BAMBOOGRASS

Plants perennial; to 2.5 m, bamboolike; cespitose, shortly rhizomatous. **Culms** 0.5–7 mm thick, erect, glabrous, with (3)6–9 nodes, highly branched at the nodes. **Leaves** mostly cauline, rarely basal; **sheaths** becoming loose, glabrous; **ligules** 0.2–0.5 mm, membranous, erose; **blades** 35–40(80) cm long, 1–10 mm wide, linear, scabrous, readily deciduous, margins scabrous. **Panicles** 8–20(50) cm, exserted, diffuse; **branches** numerous, clustering at the nodes, scabrous, glabrous, or with hairs to 0.3 mm; **pedicels** glabrous, scabrous, or with hairs to 0.3 mm. **Spikelets** 2.3–5 mm. **Glumes** subequal, 2.5–3 mm, erose, inflated, scabridulous, 3-veined, apices blunt or acute; **florets** 1.8–2.5 mm, broadly cylindrical; **calluses** hairy, hairs white; **lemmas** 1.5–2.5 mm, tuberculate, glabrous or with a tuft of silky hair; **awns** (14)17–30 mm, strongly once-geniculate, scabrous; **paleas** about ⅓ the length of the lemmas, scabrous, acute, margins glabrous; **anthers** 1–1.3 mm, penicillate. **Caryopses** (1.2)1.5–1.6 mm. $2n$ = unknown.

Austrostipa ramosissima is native to eastern Australia. It is cultivated in the United States and southern British Columbia. In its native range *A. ramosissima* is drought tolerant, but prefers moist soils and well-drained gullies near forest or woodland margins.

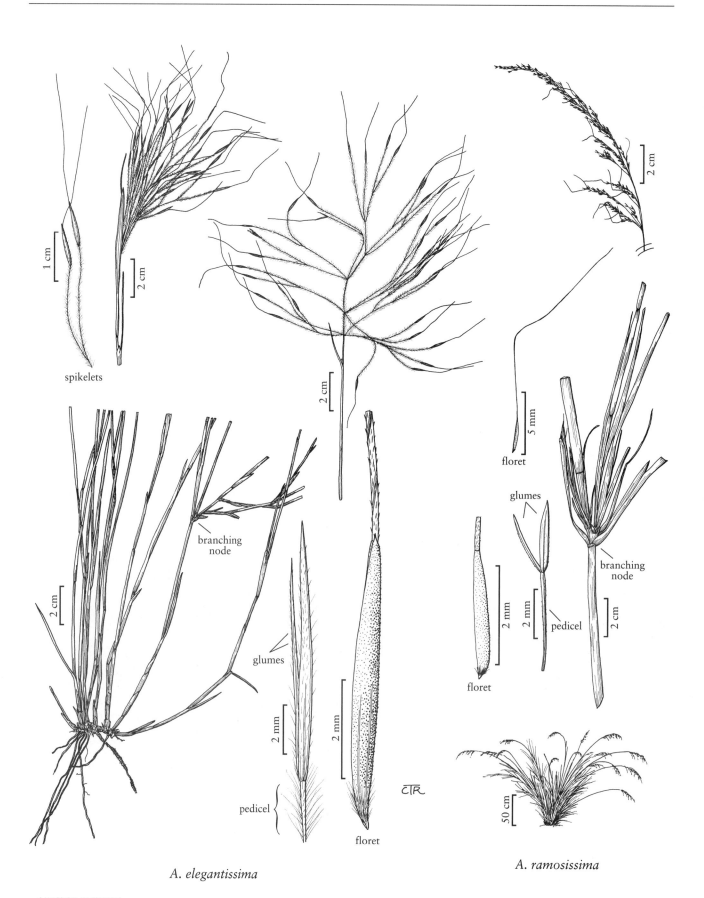

spikelets

branching node

glumes

pedicel

2 mm

2 mm

floret

floret

glumes

pedicel

branching node

floret

A. elegantissima

A. ramosissima

AUSTROSTIPA

11. BRACHYPODIEAE Harz

Mary E. Barkworth

Plants annual or perennial; rhizomatous or cespitose. **Culms** annual, not woody, ascending to erect or decumbent, sometimes branching above the base; **internodes** hollow. **Sheaths** open, margins overlapping for most of their length; **collars** without tufts of hair on the sides; **auricles** absent; **ligules** membranous, entire or toothed, sometimes shortly ciliate, those of the lower and upper cauline leaves usually similar; **pseudopetioles** absent; **blades** linear to narrowly lanceolate, venation parallel, cross venation not evident, without arm or fusoid cells, cross sections non-Kranz, epidermes without microhairs, not papillate. **Inflorescences** terminal, spikelike racemes, spikelets subsessile, solitary at all or most nodes; **pedicels** to 2.5 mm. **Spikelets** terete to slightly laterally compressed, with (3)5–24 florets, distal florets sometimes reduced, sterile; **disarticulation** above the glumes, beneath the florets; **rachillas** prolonged beyond the base of the distal floret. **Glumes** unequal, ½ as long as to equaling the adjacent lemmas, lanceolate, lower glumes 3–7-veined, upper glumes 5–9-veined; **florets** subterete to slightly laterally compressed; **calluses** glabrous, not well developed; **lemmas** lanceolate, usually membranous, rounded dorsally, (5)7–9-veined, veins not converging distally, inconspicuous, apices entire, obtuse or acute, unawned or terminally awned; **paleas** shorter than to slightly longer than the lemmas; **lodicules** 2, not veined, distal margins ciliate or the apices puberulent; **anthers** 3; **ovaries** with hairy apices; **styles** 2, bases free. **Caryopses** with hairy apices, longitudinally grooved; **hila** linear; **embryos** about ⅙ the length of the caryopses. $x = 5, 7, 9$.

The only genus of this tribe, *Brachypodium*, has sometimes been included in the *Bromeae* or *Triticeae* because of its large spikelets, apically hairy caryopses, and simple endosperm starch grains, but its smaller chromosomes and different base numbers cast doubt on its close relationship to either of these tribes, a doubt that is strongly supported by nucleic acid data. It now appears that *Brachypodium* is an isolated genus within the *Poöideae*, hence its treatment here as the sole genus in a distinct tribe.

SELECTED REFERENCE Soreng, R.J. and J.I. Davis. 1998. Phylogenetics and character evolution in the grass family (Poaceae): Simultaneous analysis of morphological and chloroplast DNA restriction site character sets. Bot. Rev. (Lancaster) 64:1–85.

11.01 BRACHYPODIUM P. Beauv.

Michael B. Piep

Plants perennial or annual; rhizomatous or cespitose, rhizomes often extensively branched. **Culms** 5–200 cm, erect or decumbent, often rooting at the lower nodes, sometimes branched above the base; **nodes** often pubescent. **Leaves** not basally concentrated; **sheaths** open, margins overlapping, not fused; **auricles** absent; **ligules** membranous, entire, toothed, or ciliate; **blades** flat or convolute, often attenuate. **Inflorescences** spikelike racemes, most or all nodes with 1 spikelet, sometimes some with 2–3, most or all spikelets appressed to strongly ascending; **disarticulation** above the glumes, beneath the florets. **Spikelets** 14–80 mm, terete to laterally compressed, with (3)5–24 florets. **Glumes** unequal, ½ as long as to equaling the adjacent lemmas, lanceolate, membranous, apices obtuse to acuminate, lower glumes 3–7-veined, upper glumes 5–9-veined; **lemmas** usually membranous, sometimes coriaceous at maturity, rounded on the back, (5)7–9-veined, apices obtuse or acute, unawned or terminally awned; **paleas** shorter than to slightly longer than the lemmas, with 2 well-developed veins, sometimes with minor veins in between, keeled over the well-developed veins, keels strongly ciliate; **lodicules** 2, oblong, attenuate distally, margins ciliate or apices puberulent; **anthers** 3; **styles** 2, free to the base, white. **Caryopses** oblong, flattened, apices pubescent; **hila** linear. $x = 5, 7, 9$. Name from the Greek *brachy*, 'short', and *podion*, 'foot', referring to the shortly pedicellate spikelets.

Brachypodium is a genus of about 18 species, with about 15 species in Eurasia, centered on the Mediterranean, and three in the Western Hemisphere, centered in Mexico. All five species in the *Flora* region are Eurasian. Four of the five species have been used in the western United States in seeding trials for mountain rangeland.

SELECTED REFERENCES **Catalán, P.** and **R.G. Olmstead**. 2000. Phylogenetic reconstruction of the genus *Brachypodium* P. Beauv. (Poaceae) from combined sequences of chloroplast *ndh*F gene and nuclear ITS. Pl. Syst. Evol. 200:1–19; **Catalán, P., Y. Shi, L. Armstrong,** and **C.A. Stace**. 1995. Molecular phylogeny of the grass genus *Brachypodium* P. Beauv. based on RFLP and RAPD analysis. Bot. J. Linn. Soc. 113:263–280; **False-Brome Working Group**. [viewed 2006]. Home page. http://www.appliedeco.org/FBWG.htm; **Hull, A.C., Jr.** 1974. Species for seeding mountain rangelands in southeastern Idaho, northeastern Utah, and western Wyoming. J. Range Managem. 27:150–153; **Khan, M.A.** and **C.A. Stace**. 1998. Breeding relationships in the genus *Brachypodium* (Poaceae: Poöideae). Nordic J. Bot. 19:257–269; **Lucchese, F.** 1990. Revision and distribution of *Brachypodium phoenicoides* (L.) Roemer et Schultes in Italy. Ann. Bot. (Rome) 48:163–177; **Nevski, S.A.** 1963. Genus 193. *Brachypodium* P.B. Pp. 472–474 *in* R.Yu. Rozhevits [R.J. Roshevitz] and B.K. Shishkin [Schischkin] (eds.). Flora of the U.S.S.R., vol. 2, trans. N. Landau (series ed. V.L. Komarov). Published for the National Science Foundation and the Smithsonian Institution, Washington, D.C. by the Israel Program for Scientific Translations, Jerusalem, Israel. 622 pp. [English translation of Flora SSSR, vol. II. 1934. Botanicheskii Institut Im. V.L. Komarova, Akademiya Nauk, Leningrad [St. Petersburg], Russia. 778 pp.]; **Rivas-Martinez, S., D. Sánchez-Mata,** and **M. Costa**. 1999. North American boreal and western temperate forest vegetation. Itinera Geobot. 12:3–331; **Schippmann, U.** 1991. Revision der europäischen Arten der Gattung *Brachypodium* Palisot de Beauvois (Poaceae). Boissiera 45:1–249.

1. Plants annual; spikelets laterally compressed; anthers 0.5–1.1 mm long . . . . 1. *B. distachyon*
1. Plants perennial; spikelets terete or subterete; anthers 2.8–6 mm long.
    2. Lemma awns 7–15 mm long, as long as or longer than the lemmas . . . . 2. *B. sylvaticum*
    2. Lemma awns absent or to 7 mm long, shorter than the lemmas.
        3. Blades with all veins more or less equally prominent on the adaxial surfaces . . . . 3. *B. phoenicoides*
        3. Blades with the primary veins separated by finer secondary veins on the adaxial surfaces.
            4. Leaf blades flat, dark green, abaxial surfaces scabrous, not shiny; lemmas usually hairy . . . . 4. *B. pinnatum*
            4. Leaf blades involute or flat, light green, abaxial surfaces smooth or almost so, conspicuously shiny; lemmas usually glabrous . . . . 5. *B. rupestre*

1. Brachypodium distachyon (L.) P. Beauv. [p. 189]

PURPLE FALSEBROME

Plants annual; loosely tufted; bright green or glaucous. **Culms** 5–35(45–50) cm, geniculate or stiffly erect, internodes glabrous; **nodes** conspicuously pubescent. **Leaves** cauline; **sheaths** usually glabrous; **ligules** 0.5–2 mm, pubescent; **blades** 10–40 cm long, 2–5 mm wide, flat, glaucous, sparsely hairy, hairs 0.5–1 mm, veins unequally prominent, margins thickened, sparsely hairy, hairs 0.8–1.3 mm. **Racemes** 2–7 cm, with 1–7 usually overlapping, appressed spikelets, the basal 1–2 spikelets sometimes diverging at maturity; **pedicels** 0.5–1 mm. **Spikelets** 15–40 mm, laterally compressed, with 7–15(17) florets. **Lower glumes** 5–6 mm, 5–7-veined; **upper glumes** 7–8 mm, 7–9-veined, veins prominent, apices acute; **lemmas** 7–10 mm, lanceolate, 7-veined, awned, awns 8–17 mm, usually straight, sometimes curved; **paleas** (6.5)7–9 mm, with 2–numerous veins, 2-keeled, keels stiffly ciliate; **lodicules** oblong, acute, sparsely hairy, ciliate; **anthers** 0.5–1.1 mm. **Caryopses** 5.7–7.8 mm, ellipsoid, apices hairy. $2n = 20, 28, 30$.

Brachypodium distachyon is native to dry, open habitats in southern Europe. It is now established in California and is known from scattered locations elsewhere in the *Flora* region. It is also established in Australia, where it grows in dry, disturbed areas on sandy or rocky soils.

Brachypodium distachyon is sometimes treated as the only member of *Trachynia* Link. It differs from other species of *Brachypodium* in being a cleistogamous annual with shorter pedicels and anthers, laterally compressed spikelets, and fewer spikelets per raceme. Molecular data (Catalán and Olmstead 2000) show it as the basal lineage within *Brachypodium*. It has been proposed as a model species for molecular work in the *Poaceae*.

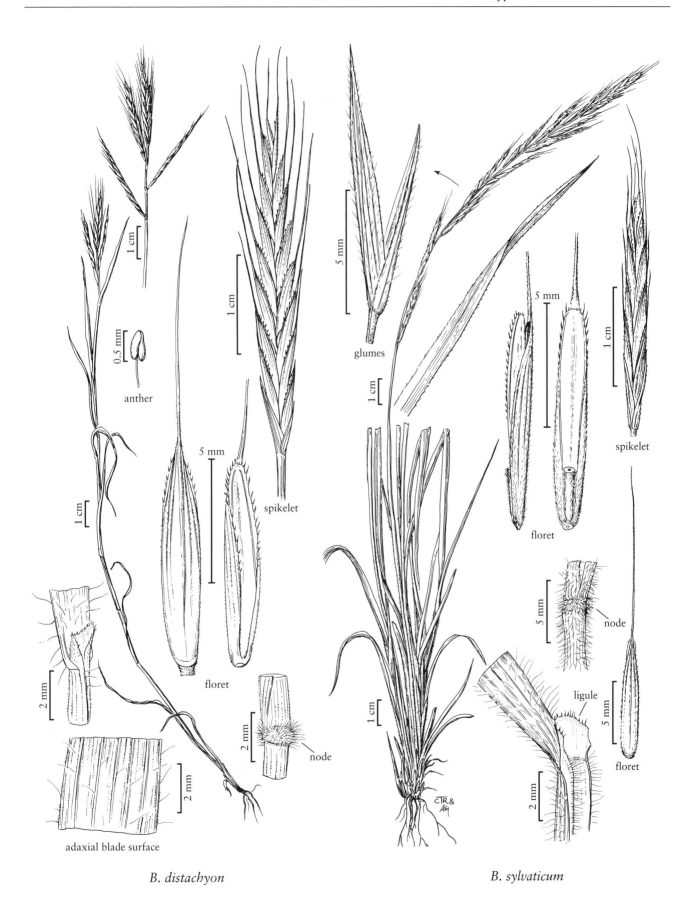

1 cm

0.5 mm

anther

5 mm

1 cm

spikelet

1 cm

5 mm

floret

2 mm

2 mm

node

adaxial blade surface

2 mm

B. distachyon

5 mm

glumes

5 mm

1 cm

5 mm

1 cm

spikelet

floret

5 mm

node

1 cm

ligule

5 mm

floret

2 mm

B. sylvaticum

BRACHYPODIUM

2. Brachypodium sylvaticum (Huds.) P. Beauv. [p. 189]
SLENDER FALSEBROME

Plants perennial; cespitose, sometimes shortly rhizomatous. **Culms** 30–120(200) cm, erect or ascending, usually unbranched, with (3)4–6(12) nodes, internodes smooth, pubescent adjacent to the nodes; **nodes** pubescent. **Sheaths** rounded or keeled distally, smooth, usually hairy, sometimes glabrous, hairs spreading to reflexed; **ligules** 1–3.5(5–6) mm, obtuse, pubescent, erose-ciliate; **blades** 8–35 cm long, (4)5–12(15) mm wide, flat, lax, scabrous, both surfaces sparsely pilose, hairs 0.8–1.8 mm, abaxial surfaces often with a tuft of hairs near the base, adaxial surfaces not prominently veined, veins pale, sparsely pilose. **Racemes** (2)4–20 cm, suberect to pendent, with (3)4–12 rather distant spikelets; **pedicels** 0.5–2 mm. **Spikelets** 17–30(55) mm, terete, with 6–16(22) florets. **Glumes** unequal, acute, usually pubescent; **lower glumes** 6–9 mm, 5–7-veined; **upper glumes** 8–11 mm, 7–9-veined; **lemmas** 6–12(13.5) mm, lanceolate, 7-veined, awned, awns 7–15 mm, straight or weakly flexuous; **paleas** (5)6–10(12) mm, narrowly oblong, 2-keeled; **anthers** 3–4(5.5) mm. **Caryopses** 5–8.7 mm. 2n = 16, 18.

Brachypodium sylvaticum is native to Eurasia and northern Africa, where it grows in woods and other shady places. In the *Flora* region, it is established in Oregon, where it is an aggressive weed that has spread to several western counties since its first discovery in Eugene, Lane County, in 1939 (False-Brome Working Group 2006). In December 2003 it was discovered in California. Many of the locations are in nature preserves of some kind, with humans being the primary dispersers; hikers are being urged to clean their shoes and clothing before and after hiking in affected locations. An additional report from Virginia has been confirmed. A report from Utah (Hull 1974) was not supported by voucher specimens; attempts to locate the species in 2004 and 2005 were unsuccessful.

Brachypodium sylvaticum has sometimes been cultivated as an ornamental, but because of its ability to become an aggressive weed, such use is discouraged in the *Flora* region.

Plants in the *Flora* region belong to **Brachypodium sylvaticum** (Huds.) P. Beauv. subsp. **sylvaticum**, which has retrorsely pubescent sheaths, green leaves, green hairy lemmas, and spikelet lengths within the lower portion of the range given. Specimens with glabrous sheaths belong to **B. sylvaticum** subsp. **glaucovirens** (St.-Yves) Murb.

3. Brachypodium phoenicoides (L.) Roem. & Schult. [p. 191]
THINLEAF FALSEBROME

Plants perennial; rhizomatous, rhizomes branched. **Culms** (30)40–90(110) cm, erect, with 3–5 nodes; **nodes** pubescent. **Sheaths** antrorsely scabrous, usually glabrous, sometimes pilose; **ligules** (0.4)0.7–1.5(1.8) mm, ciliate; **blades** 10–40 cm long, 3–5 mm wide, flat and more or less lax or convolute and somewhat rigid, particularly when dry, surfaces scabrous, usually glabrous, sometimes pilose, adaxial surfaces sparsely hairy, scabrous, with all veins more or less equally prominent. **Racemes** 10–20(30) cm, with 5–9(15) scarcely overlapping spikelets; **pedicels** (0.7)1–2(3.2) mm. **Spikelets** (20)30–60(80) mm, often slightly falcate, terete or subterete, with (7)9–18(31) florets. **Lower glumes** 4–7 mm, (4)5(7)-veined; **upper glumes** 5–8 mm, 7–8(9)-veined, mucronate or shortly awned; **lemmas** (7.5)8–10.3(11) mm, apices abruptly narrowed, mucronate or awned, awns 1–2.5 mm; **paleas** 6.5–10.5 mm, keeled, keels ciliate or hairy; **anthers** 4–6 mm. **Caryopses** 5–7 mm. 2n = 28.

Brachypodium phoenicoides is native to dry, usually open, and often sandy habitats of the northern Mediterranean region. In the *Flora* region, it is currently known only from Sonoma County, California, where it is established on coastal sand dunes.

4. Brachypodium pinnatum (L.) P. Beauv. [p. 191]
HEATH FALSEBROME, TOR GRASS

Plants perennial; rhizomatous, sometimes subcespitose, rhizomes scaly, sparingly branched. **Culms** 30–120 cm, erect, stiff, usually unbranched, with 2–3 nodes, not narrowed below the racemes, internodes smooth, usually glabrous; **nodes** pubescent, often scabrous. **Sheaths** rounded, smooth, glabrous or hairy, bases sometimes shortly pubescent; **ligules** 0.8–2.5 mm, truncate; **blades** 6–45 cm long, 2–7(10) mm wide, flat, lax, dark green, abaxial surfaces scabrous, not shiny, adaxial surfaces sparsely hairy, primary veins conspicuous, separated by finer secondary veins. **Racemes** 4–25 cm, erect, sometimes nodding, with (3)6–15 spikelets; **pedicels** 1–2 mm. **Spikelets** 15–50 mm, terete, with (3)7–24 florets. **Glumes** lanceolate to narrowly ovate, acute, firm, glabrous; **lower glumes** 3–6 mm, 3–6-veined; **upper glumes** 3–8 mm, 5–9-veined; **lemmas** 6–11 mm,

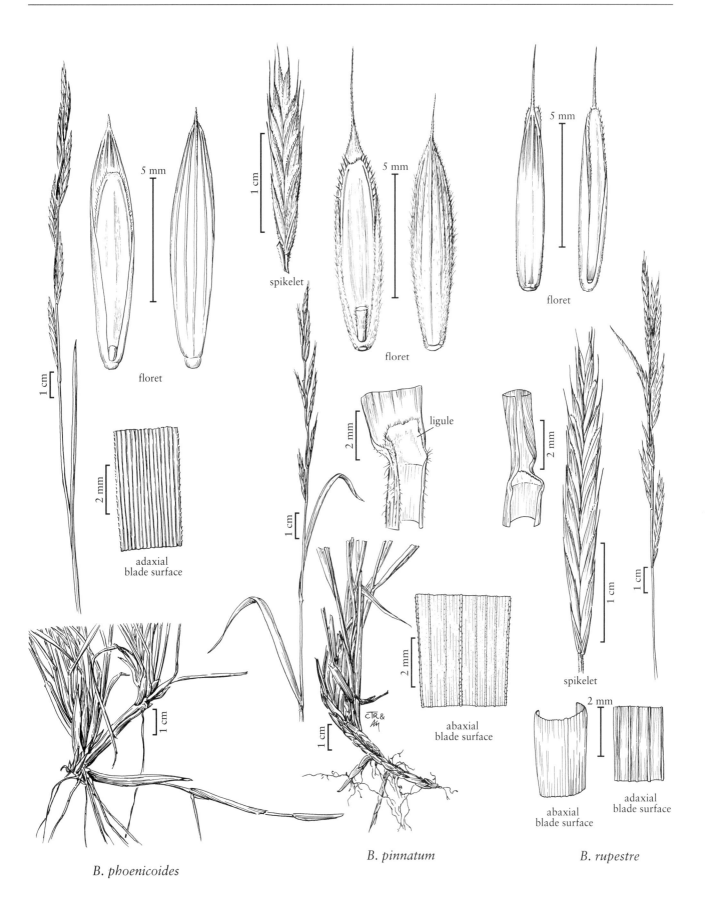

spikelet

floret

floret

floret

adaxial
blade surface

ligule

abaxial
blade surface

spikelet

abaxial
blade surface

adaxial
blade surface

B. phoenicoides

B. pinnatum

B. rupestre

BRACHYPODIUM

oblong to lanceolate, usually hairy, 7-veined, awned, awns 1–7 mm, straight; **paleas** (5)7–9(10) mm, narrowly oblong, 2-keeled, keels ciliate; **anthers** 3.5–5.5 mm. **Caryopses** 6–7.2 mm. $2n = 14, 16, 28$.

Brachypodium pinnatum is native to Eurasia. It is reportedly established in Sonoma County, California, and in Massachusetts. In its native range it prefers open woodlands, forest edges, and grassland habitats, and is fairly tolerant of hot, dry conditions.

5. Brachypodium rupestre (Host) Roem. & Schult.
[p. 191]
TUFTED FALSEBROME

Plants perennial; cespitose, shortly rhizomatous. **Culms** 30–95(125) cm tall, 0.5–0.8 mm thick, narrowed below the racemes; **nodes** pubescent. **Sheaths** glabrous or hairy; **ligules** 0.4–2.4 mm, truncate; **blades** (9)15–40(52) cm long, 2.5–6.5(9) mm wide, stiff, involute or flat, light green, abaxial surfaces smooth or almost so, glabrous, conspicuously shiny, adaxial surfaces somewhat scabridulous, primary veins conspicuous, separated by finer secondary veins. **Racemes** (4)10–30(40) cm, with 5–9(12) spikelets. **Spikelets** (14)20–33(46) mm, terete or subterete, usually glabrous, with (7)10–17(24) florets; **pedicels** 0.6–2(3) mm. **Glumes** unequal, less than ⅓ the length of the spikelets, lanceolate, mucronate; **lower glumes** 4–8.5 mm, 3–6-veined; **upper glumes** 5.5–10.5 mm, 6–7-veined; **lemmas** 7–10.5(12) mm, lanceolate, 5–7-veined, awned, awns (2)3–10 mm; **lodicules** lanceolate, apices hairy, hairs stiff; **anthers** about 6 mm. **Caryopses** 5–7 mm. $2n = 14, 18, 28, 36$.

Brachypodium rupestre is native to Europe and northern Turkey; it is not known to be established in the *Flora* region. Specimens of this species were grown by the Natural Resources Conservation Service in Pima County, Arizona and Bernalillo County, New Mexico; they were distributed either as *Brachypodium* sp. or as *B. cespitosum* (Host) Roem. & Schult. The description is based on Schippmann (1991).

12. BROMEAE Dumort.

Mary E. Barkworth

Plants annual or perennial; usually cespitose, sometimes rhizomatous. **Culms** annual, not woody, not branching above the base; **internodes** usually hollow, rarely solid. **Sheaths** closed, margins united for most of their length; **collars** without tufts of hair on the sides; **auricles** sometimes present; **ligules** membranous, sometimes shortly ciliate, those of the upper and lower cauline leaves usually similar; **pseudopetioles** absent; **blades** linear to narrowly lanceolate, venation parallel, cross venation not evident, without arm or fusoid cells, cross sections non-Kranz, epidermes without microhairs, not papillate. **Inflorescences** usually terminal panicles, sometimes reduced to racemes in depauperate plants; **disarticulation** above the glumes and beneath each floret. **Spikelets** 5–80 mm, not viviparous, terete to laterally compressed, with 3–30 bisexual florets, distal florets sometimes reduced; **rachillas** prolonged beyond the bases of the distal florets. **Glumes** usually unequal, rarely more or less equal, exceeded by the distal florets, usually longer than ¼ the length of the adjacent florets, lanceolate, 1–9(11)-veined; **florets** terete to laterally compressed; **calluses** glabrous, not well developed; **lemmas** lanceolate to ovate, rounded or keeled over the midvein, herbaceous to coriaceous, 5–13-veined, veins converging somewhat distally, apices usually minutely bilobed to bifid, rarely entire, usually awned, sometimes unawned, awns unbranched, terminal or subterminal, usually straight, sometimes geniculate; **paleas** usually shorter than the lemmas; **lodicules** 2, glabrous, not veined; **anthers** 3; **ovaries** with hairy apices; **styles** 2, bases free. **Caryopses** narrowly ellipsoid to linear, longitudinally grooved; **hila** linear; **embryos** about ⅙ the length of the caryopses. $x = 7$.

There are three genera in the *Bromeae*. One genus, *Bromus*, grows in the *Flora* region. The tribe was included in the *Festuceae* Dumort. [= *Poeae*] by earlier agrostologists (e.g., Hitchcock 1951) because it has paniculate inflorescences, spikelets with more than 1 floret, and glumes that are shorter than the lemmas. It is now considered to be most closely related to the *Triticeae*. This is indicated by the pubescent apices of the ovaries and simple endosperm starch grains. It is further supported by data from serology, nucleic acid sequences, and seedling development.

These data do not support a close relationship between the *Bromeae* and *Brachypodium*, a genus that has sometimes been included in the tribe.

SELECTED REFERENCES **Hitchcock, A.S.** 1951. Manual of the Grasses of the United States, ed. 2, rev. A. Chase. U.S.D.A. Miscellaneous Publication No. 200. U.S. Government Printing Office, Washington, D.C., U.S.A. 1051 pp.; **Kellogg, E.A.** 1992. Tools for studying the chloroplast genome in the Triticeae (Gramineae): An *Eco*RI map, a diagnostic deletion, and support for *Bromus* as an outgroup. Amer. J. Bot. 79:186–197.

12.01 BROMUS L.

Leon E. Pavlick†

Laurel K. Anderton

Plants perennial, annual, or biennial; usually cespitose, sometimes rhizomatous. **Culms** 5–190 cm. **Sheaths** closed to near the top, usually pubescent; **auricles** sometimes present; **ligules** membranous, to 6 mm, usually erose or lacerate; **blades** usually flat, rarely involute. **Inflorescences** panicles, sometimes racemes in depauperate specimens, erect to nodding, open to dense, occasionally 1-sided; **branches** usually ascending to spreading, sometimes reflexed or drooping. **Spikelets** 5–70 mm, terete to laterally compressed, with 3–30 florets; **disarticulation** above the glumes, beneath the florets. **Glumes** unequal, usually shorter than the adjacent lemmas, always shorter than the spikelets, glabrous or pubescent, usually acute, rarely mucronate; **lower glumes** 1–7(9)-veined; **upper glumes** 3–9(11)-veined; **lemmas** 5–13-veined, rounded to keeled, glabrous or pubescent, apices entire, emarginate, or toothed, usually terminally or subterminally awned, sometimes with 3 distinct awns or unawned; **paleas** usually shorter than the lemmas, ciliate on the keels, adnate to the caryopses; **anthers** (2)3. $x = 7$. Name from the Greek *bromos*, an ancient name for 'oats', which was based on *broma*, 'food'.

Bromus grows in temperate and cool regions. It is estimated to include 100–400 species, the number depending on how the species are circumscribed. Of the 52 species in the *Flora* region, 28 are native and 24 are introduced. The native perennial species provide considerable forage for grazing animals, with some species being cultivated for this purpose. The introduced species, all but three of which are annuals, range from sporadic introductions to well-established members of the region's flora. Many are weedy and occupy disturbed sites. Some are used for hay; others have sharp, pointed florets and long, rough awns that can injure grazing animals.

This treatment is based on one submitted by Pavlick, who died before it could be reviewed and edited. It has been substantially revised by Anderton to meet the requirements for publication in this volume. The majority of Pavlick's taxonomic concepts are retained, despite the necessity for overlap in many key leads; time constraints prevented a thorough investigation of problematic taxa. The treatment recognizes taxa at both the subspecies and varietal rank; this simply reflects the decisions of the original author. We thank Hildemar Scholz of the Botanic Garden and Botanical Museum Berlin-Dahlem, Free University Berlin, for providing accurately identified specimens of the weedy European species for use in preparing the illustrations, and for his helpful suggestions for the keys and descriptions.

In the keys and descriptions, the distances from the bases of the subterminal lemma awns to the lemma apices are measured on the most distal florets in a spikelet.

SELECTED REFERENCES **Ainouche, M.L., R.J. Bayer, J.-P. Gourret, A. Defontaine,** and **M.-T. Misset.** 1999. The allotetraploid invasive weed *Bromus hordeaceus* L. (Poaceae): Genetic diversity, origin and molecular evolution. Folia Geobot. 34:405–419; **Allred, K.W.** 1993. *Bromus*, section *Pnigma*, in New Mexico, with a key to the bromegrasses of the state. Phytologia 74:319–345; **Barkworth, M.E., L.K. Anderton, J. McGrew,** and **D.E. Giblin.** 2006. Geography and morphology of the *Bromus carinatus* (Poaceae: Bromeae) complex. Madroño. 53:235–245; **Bartlett, E., S.J. Novak,** and **R.N. Mack.** 2002. Genetic variation in *Bromus tectorum* (Poaceae): Differentiation in the eastern United States. Amer. J. Bot.

89:602–612; **Cope, T.A.** 1982. Flora of Pakistan, No. 143: Poaceae (E. Nasir and S.I. Ali, eds.). Pakistan Agricultural Research Council and University of Karachi, Islamabad and Karachi, Pakistan. 678 pp.; **Davis, P.H.** 1985. Flora of Turkey and the East Aegean Islands, vol. 9. Edinburgh University Press, Edinburgh, Scotland. 724 pp.; **Harlan, J.R.** 1945a. Cleistogamy and chasmogamy in *Bromus carinatus* Hook. & Arn. Amer. J. Bot. 32:66–72; **Harlan, J.R.** 1945b. Natural breeding structure in the *Bromus carinatus* complex as determined by population analyses. Amer. J. Bot. 32:142–147; **Hitchcock, A.S.** 1951. Manual of the Grasses of the United States, ed. 2, rev. A. Chase. U.S.D.A. Miscellaneous Publication No. 200. U.S. Government Printing Office, Washington, D.C., U.S.A. 1051 pp.; **Hitchcock, C.L.** 1969. Gramineae. Pp. 384–725 *in* C.L. Hitchcock, A. Cronquist, and M. Ownbey. Vascular Plants of the Pacific Northwest, Part 1: Vascular Cryptogams, Gymnosperms, and Monocotyledons. University of Washington Press, Seattle, Washington, U.S.A. 914 pp.; **Matthei, O.** 1986. El género *Bromus* L. (Poaceae) en Chile. Gayana, Bot. 43:47–110; **Mitchell, W.W.** and **A.C. Wilton.** 1966. A new tetraploid brome, section *Bromopsis*, of Alaska. Brittonia 18:162–166; **Pavlick, L.E.** 1995. *Bromus* L. of North America. Royal British Columbia Museum, Victoria, British Columbia, Canada. 160 pp.; **Peterson, P.M., J. Cayouette, Y.S.N. Ferdinandez, B. Coulman,** and **R.E. Chapman.** 2001. Recognition of *Bromus richardsonii* and *B. ciliatus*: Evidence from morphology, cytology and DNA fingerprinting. Aliso 20:21–36; **Saarela, J.M., P.M. Peterson,** and **J. Cayouette.** 2005. *Bromus hallii* (Poaceae), a new combination for California, U.S.A., and taxonomic notes on *Bromus orcuttianus* and *Bromus grandis*. Sida 21:1997–2013; **Sales, F.** 1993. Taxonomy and nomenclature of *Bromus* sect. *Genea*. Edinburgh J. Bot. 50:1–31; **Scholz, H.** 1970. Zur Systematik der Gattung *Bromus* (Gramineae) mit einer Abbildung. Willdenowia 6:139–159; **Scholz, H.** 2003. Die Ackersippe der Verwechselten Trespe (*Bromus commutatus*). Bot. Naturschutz in Hessen 16:17–22; **Spalton, L.M.** 2001. Brome-grasses with small lemmas. B.S.B.I. [Botanical Society of the British Isles] News 87:21–23; **Spalton, L.M.** 2002. An analysis of the characters of *Bromus racemosus* L., *B. commutatus* Schrad. and *B. secalinus* L. (Poaceae). Watsonia 24:193–202; **Stebbins, G.L., Jr.** 1947. The origin of the complex of *Bromus carinatus* and its phytogeographic implications. Contr. Gray Herb. 165:42–55; **Stebbins, G.L., Jr.** and **H.A. Tobgy.** 1944. The cytogenetics of hybrids in *Bromus*: 1. Hybrids within the section *Ceratochloa*. Amer. J. Bot. 31:1–11; **Veldkamp, J.F.** 1990. *Bromus luzonensis* is the correct name for *Bromus breviaristatus* Buckl. (Gramineae). Taxon 39:660; **Vogel, K.P., K.J. Moore,** and **L.E. Moser.** 1996. Bromegrasses. Pp. 535–567 *in* L.E. Moser, D.R. Buxton, and M.D. Casler (eds.). Cool-Season Forage Grasses. Agronomy Monograph No. 34. American Society of Agronomy, Crop Science Society of America, and Soil Science Society of America, Madison, Wisconsin, U.S.A. 841 pp.; **Wagnon, H.K.** 1952. A revision of the genus *Bromus*, section *Bromopsis*, of North America. Brittonia 7:415–480; **Yatskievych, G.** 1999. Steyermark's Flora of Missouri, vol. 1, rev. ed. Missouri Department of Conservation, Jefferson City, Missouri, U.S.A. 991 pp. http://biology.missouristate.edu/herbarium/.

1. Lemmas strongly keeled, at least distally; spikelets strongly laterally compressed; lower glumes 3–7(9)-veined . . . . sect. *Ceratochloa*
1. Lemmas rounded over the midvein; spikelets terete to moderately laterally compressed; lower glumes 1–5-veined.
   2. Awns, if present, arising less than 1.5 mm below the lemma apices; lemma apices entire, emarginate, or with teeth less than 1 mm long.
      3. Lower glumes 1–3-veined; upper glumes 3–5-veined; plants perennial or annual, if annual, the lower glumes 1-veined and the upper glumes 3-veined . . . . sect. *Bromopsis*
      3. Lower glumes 3–5-veined; upper glumes 5–9-veined; plants annual or biennial, if biennial, the upper glumes 7-veined and/or the lateral veins of the lemmas prominently ribbed . . . . sect. *Bromus* (in part)
   2. Awns arising 1.5 mm or more below the lemma apices, lemma apices entire, emarginate, or with teeth to 5 mm long.
      4. Awns usually geniculate, sometimes only divaricate, lemma teeth 2–3 mm long, usually aristate, sometimes only acuminate . . . . sect. *Neobromus*
      4. Awns straight, arcuate, or divaricate, not geniculate, sometimes absent; lemma teeth absent or to 5 mm long, acuminate.
         5. Lower glumes 1–3-veined; upper glumes 3–5-veined; spikelets with parallel or diverging sides in outline, often widening distally; lemma apices bifid, teeth (0.8)1–5 mm long . . . . sect. *Genea*
         5. Lower glumes 3–5-veined; upper glumes 5–9-veined; spikelets with parallel or converging sides in outline; lemma apices entire to bifid, teeth less than 1 mm long, apices sometimes split and teeth appearing longer . . . . sect. *Bromus* (in part)

Bromus sect. **Ceratochloa**

1. Lemmas unawned or with awns to 3.5 mm long; lemmas usually glabrous, sometimes pubescent distally, veins prominent for most of their length . . . . 1. *B. catharticus* (in part)
1. Lemmas awned, awns (2)4–17 mm long; lemmas pubescent or glabrous, veins obscure or prominent.
   2. Lower panicle branches shorter than 20 cm, with 1–3 spikelets on the distal ¹/₂, sometimes confined to the tips; culms 3–7 mm thick.

3. Lower panicle branches shorter than 20 cm, spreading to drooping . . . . 2. *B. sitchensis*
3. Lower panicle branches shorter than 10 cm, stiffly ascending . . . . 4. *B. aleutensis*
2. Lower panicle branches usually shorter than 10 cm, with 1–5 spikelets variously distributed; culms less than 4 mm thick.
    4. Upper glume about as long as the lowest lemma in each spikelet; lemmas glabrous or pubescent distally or throughout, the marginal hairs, if present, longer than those elsewhere . . . . 3. *B. arizonicus*
    4. Upper glume shorter than the lowest lemma in each spikelet; lemmas glabrous or pubescent only on the margins or throughout, if throughout, the marginal hairs similar in length to those elsewhere.
        5. Panicles dense; spikelets crowded, overlapping, usually longer than the pedicels and branches; culms 20–70 cm tall, sometimes geniculate at the base; blades glabrous; ligules 1–6 mm long . . . . 5. *B. maritimus*
        5. Panicles loose to compact; spikelets not crowded or overlapping, shorter than at least some pedicels and branches; culms 30–120(180) cm tall, erect or decumbent; blades glabrous or hairy; ligules 1–4 mm long.
            6. Lemmas and sheath throats glabrous . . . . 7. *B. polyanthus*
            6. Lemmas and/or sheath throats with hairs.
                7. Lemmas 9–13-veined, veins often raised and riblike distally or throughout . . . . 1. *B. catharticus* (in part)
                7. Lemmas 7–9-veined, veins usually not raised or riblike . . . . 6. *B. carinatus*

Bromus sect. Bromopsis

1. Plants rhizomatous.
    2. Culms 30–90 cm long, forming distinct clumps; rhizomes short . . . . 9. *B. riparius*
    2. Culms 50–135 cm long, single or few together; rhizomes short to long-creeping.
        3. Lemma backs sparsely to densely hairy throughout, or on the lower portion and margins, or along the marginal veins and keel; cauline nodes and leaf blades pubescent or glabrous; awns usually present, to 7.5 mm long, sometimes absent . . . . . . 10. *B. pumpellianus* (in part)
        3. Lemma backs usually glabrous, occasionally sparsely puberulent at the base and sometimes on the margins; cauline nodes and leaf blades usually glabrous, rarely hairy; awns absent or to 3 mm long . . . . 8. *B. inermis*
1. Plants not rhizomatous.
    4. Anthers (3.5)4–6(6.8) mm long; awns 2.5–7.5 mm long; plants of the Yukon River drainage of Alaska . . . . 10. *B. pumpellianus* (in part)
    4. Anthers 1–7 mm long; awns 1–12 mm long; plants of various locations in the *Flora* region, if in the Yukon River drainage of Alaska, anthers 1–1.4 mm long.
        5. Culms with 9–20 nodes; collars and throats densely pilose; auricles 1–2.5 mm long on most lower leaves . . . . 11. *B. latiglumis*
        5. Culms with (1)2–9 nodes; collars and throats pubescent or glabrous; auricles, if present, of various lengths.
            6. Most lower glumes within a panicle 3-veined, sometimes some 1-veined.
                7. Most upper glumes within a panicle 5-veined, sometimes some 3-veined.
                    8. Awns 1.5–3 mm long; anthers 1.5–2.5 mm long; ligules 0.5–1 mm long . . . . 13. *B. kalmii*
                    8. Awns 3–7 mm long; anthers 3–6 mm long; ligules to 4.2 mm long.
                        9. Glumes glabrous; ligules glabrous . . . . 12. *B. laevipes*
                        9. Glumes usually pubescent, rarely glabrous; ligules usually pubescent or pilose, sometimes glabrous.
                            10. Margins of the glumes and lemmas often bronze-tinged; ligules to 1.5 mm long; auricles usually present on the lower leaves, rarely absent . . . . 14. *B. pseudolaevipes*
                            10. Margins of the glumes and lemmas not bronze-tinged; ligules 1–3 mm long; auricles sometimes present . . . . 17. *B. grandis* (in part)

7. Most upper glumes within a panicle 3-veined, sometimes some 5-veined.
   11. Culms 70–180 cm tall; awns 3–8 mm long; anthers 3–6 mm long.
      12. Lower leaf sheaths pilose, hairs 2–4 mm long; blades glabrous or with pilose margins . . . . 16. *B. orcuttianus* (in part)
      12. Lower leaf sheaths densely pubescent, hairs to 1 mm long; blades densely pubescent.
         13. Blades 7.5–16.5 cm long; culm nodes 1–2(3) . . . . 15. *B. hallii* (in part)
         13. Blades (13)18–38 cm long; culm nodes 3–7 . . . . 17. *B. grandis* (in part)
   11. Culms 30–100 cm tall; awns 1–4 mm long; anthers (1)1.5–4 mm long.
      14. Leaf blades often glaucous; glumes usually glabrous, rarely slightly pubescent . . . . 18. *B. frondosus*
      14. Leaf blades not glaucous; glumes usually pubescent, rarely glabrous.
         15. Midrib of the culm leaves abruptly narrowed just below the collar; auricles frequently present on the lower leaves; plants of western Texas . . . . 19. *B. anomalus* (in part)
         15. Midrib of the culm leaves not abruptly narrowed just below the collar; auricles absent; plants of western North America, including Texas . . . . 20. *B. porteri* (in part)
6. Most lower glumes within a panicle 1-veined, sometimes some 3-veined.
   16. Upper glumes within a panicle consistently 5-veined; collars with a dense line of hairs; lower sheaths often sericeous; ligules 0.4–1 mm long . . . . 21. *B. nottowayanus*
   16. All or most upper glumes within a panicle 3-veined, sometimes some with 2 additional faint lateral veins; collars glabrous or hairy, hairs evenly distributed over the surface, not in a dense line; lower sheaths glabrous or hairy, not sericeous; ligules to 6 mm long.
      17. Plants annual; lemmas glabrous; ligules pubescent . . . . 22. *B. texensis*
      17. Plants perennial; lemmas usually pubescent on the backs and/or margins, sometimes glabrous; ligules usually glabrous, sometimes pubescent or pilose.
         18. Awns (4)6–12 mm long; ligules 2–6 mm long . . . . 23. *B. vulgaris*
         18. Awns 1–8 mm long; ligules to 4 mm long.
            19. Blades densely pubescent on both surfaces, 7.5–16.5 cm long; anthers 3–6 mm long; awns 3.5–7 mm long . . . . 15. *B. hallii* (in part)
            19. Blades glabrous or hairy on 1 or both surfaces, (3)5–60 cm long, if 7.5–16.5 cm long and densely pubescent on both surfaces, then anthers 1–4 mm long and/or awns 1–4 mm long.
               20. Panicle branches appressed to slightly spreading; culm nodes 1–4.
                  21. Awns 2–5 mm long; anthers 2–3.5 mm long; blades flat . . . . 26. *B. suksdorfii*
                  21. Awns (4)5–8 mm long; anthers 3–6.5 mm long; blades sometimes involute.
                     22. Culms 90–150 cm tall; ligules 1–3 mm long . . 16. *B. orcuttianus* (in part)
                     22. Culms 50–100 cm tall; ligules to 1.5 mm long . . . . 25. *B. erectus*
               20. Panicle branches ascending to drooping; culm nodes (1)2–8.
                  23. Midrib of the culm leaves abruptly narrowed just below the collar; auricles frequently present on the lower leaves; plants of western Texas . . . . 19. *B. anomalus* (in part)
                  23. Midrib of the culm leaves not abruptly narrowed just below the collar; auricles sometimes present; plants of various distribution, including Texas.
                     24. Glumes usually pubescent, rarely glabrous.
                        25. Upper glume mucronate . . . . 27. *B. mucroglumis*
                        25. Upper glume not mucronate.
                           26. Awns (1)2–3(3.5) mm long; blades 2–6 mm wide . . . . 20. *B. porteri* (in part)

26. Awns 3–7(8) mm long; blades 3–19 mm wide.
   27. Anthers 3–6 mm long; ligules densely pubescent to pilose . . . . 17. *B. grandis* (in part)
   27. Anthers 2–4(5) mm long; ligules glabrous.
      28. Ligules 2–4 mm long . . . . 24. *B. pacificus*
      28. Ligules 0.5–2 mm long . . . . 28. *B. pubescens*
24. Glumes usually glabrous, sometimes pubescent.
   29. Ligules 2–3.5 mm long; auricles present . . . . 30. *B. ramosus*
   29. Ligules 0.4–2 mm long; auricles sometimes present.
      30. Lemma margins and backs usually pubescent, sometimes nearly glabrous; awns 2–4 mm long; anthers 1.8–4 mm long . . . . 29. *B. lanatipes*
      30. Lemma margins conspicuously hirsute or densely pilose, at least along the lower $^1/_2$, the backs glabrous at least on the lower lemmas in a spikelet; awns 3–5 mm long; anthers 1–2.7 mm long.
         31. Backs of all lemmas glabrous; anthers 1–1.4 mm long; upper glumes 7.1–8.5 mm long . . . . 31. *B. ciliatus*
         31. Backs of the upper lemmas in a spikelet hairy; anthers 1.6–2.7 mm long; upper glumes 8.9–11.3 mm long . . . . 32. *B. richardsonii*

Bromus sect. Neobromus

This section includes one species, 33. *Bromus berteroanus*.

Bromus sect. Genea

1. Lemmas 20–35 mm long . . . . 34. *B. diandrus*
1. Lemmas 9–20 mm long.
   2. Spikelets usually shorter than the panicle branches; panicle branches ascending to spreading or drooping.
      3. Lemmas 14–20 mm long; panicles with spreading, ascending, or drooping branches, rarely with any branches with more than 3 spikelets . . . . 35. *B. sterilis*
      3. Lemmas 9–12 mm long; panicles with drooping branches, often with 1 or more branches with 4–8 spikelets . . . . 36. *B. tectorum*
   2. Spikelets longer than the panicle branches; panicle branches ascending to spreading, never drooping.
      4. Some panicle branches 1–3+ cm long, most branches visible . . . . 37. *B. madritensis*
      4. Panicle branches 0.1–1 cm long, usually not readily visible . . . . 38. *B. rubens*

Bromus sect. Bromus

1. Lemmas inflated, 6–8 mm wide; unawned or with awns up to 1 mm long; spikelets ovate . . . 39. *B. briziformis*
1. Lemmas not inflated, 1–7 mm wide; awns 2–25 mm long, rarely absent; spikelet shape various.
   2. Lemma margins inrolled at maturity; floret bases visible at maturity; rachilla internodes visible at maturity; caryopses sometimes thick, strongly inrolled.

3. Anthers 2.5–5 mm long; awns straight; spikelets often purple-tinged; lower leaf sheaths
   with soft, appressed hairs . . . . .40. *B. arvensis* (in part)
3. Anthers 0.7–2 mm long; awns straight or flexuous; spikelets not purple-tinged; lower
   leaf sheaths glabrous, loosely pubescent and glabrate, or evenly covered with stiff hairs.
   4. Lower leaf sheaths glabrous or loosely pubescent and glabrate; lemmas 6.5–8.5(10)
      mm long, margins evenly rounded; awns straight or flexuous . . . . .41. *B. secalinus*
   4. Lower leaf sheaths evenly covered with stiff hairs; lemmas 8–11.5 mm long, margins
      bluntly angled; awns straight . . . . .42. *B. commutatus* (in part)
2. Lemma margins not inrolled at maturity; floret bases concealed at maturity; rachilla
   internodes concealed at maturity; caryopses thin, weakly inrolled or flat.
   5. Lemmas 4.5–6.5 mm long, margins sharply angled; caryopses longer than the paleas . . . . . . 43. *B. lepidus*
   5. Lemmas 6.5–20 mm long, margins rounded or slightly to strongly angled; caryopses
      equaling or shorter than the paleas.
      6. Awns arising less than 1.5 mm below the lemma apices, erect or weakly divaricate,
         not twisted at the base.
         7. Panicle branches shorter than the spikelets; lemmas chartaceous, with prominent
            ribs over the veins, often concave between the veins; anthers 0.6–1.5 mm long
            . . . . .44. *B. hordeaceus* (in part)
         7. At least some panicle branches longer than the spikelets; lemmas coriaceous, veins
            obscure or distinct, not ribbed; anthers 0.7–5 mm long.
            8. Lower leaf sheaths with soft, appressed hairs; anthers 2.5–5 mm long; panicles
               11–30 cm long . . . . .40. *B. arvensis* (in part)
            8. Lower leaf sheaths with stiff hairs; anthers 0.7–3 mm long; panicles 4–16 cm
               long.
               9. Anthers 0.7–1.7 mm long; rachilla internodes 1.5–2 mm long; lemmas
                  8–11.5 mm long, margins bluntly angled . . . . .42. *B. commutatus* (in part)
               9. Anthers 1.5–3 mm long; rachilla internodes 1–1.5 mm long; lemmas 6.5–8
                  mm long, margins rounded . . . . .45. *B. racemosus*
      6. Awns arising 1.5 mm or more below the lemma apices, erect to strongly divaricate,
         often twisted at the base.
         10. Panicle branches shorter than the spikelets, slightly curved or straight, panicles
             erect.
             11. At least the upper lemmas in each spikelet with 3 awns . . . . .46. *B. danthoniae*
             11. All lemmas 1-awned.
                 12. Lemmas 11–20 mm long; spikelets 20–50 mm long.
                     13. Spikelets usually single at the nodes; glumes glabrous or
                         puberulent; panicles strongly contracted, even at maturity . . . . . . .47. *B. caroli-henrici*
                     13. Spikelets often 2 or more at each node; glumes pilose; panicles
                         contracted when immature, more open with age . . . . .48. *B. lanceolatus*
                 12. Lemmas 6.5–11 mm long; spikelets 11–25 mm long.
                     14. Lemmas 1.5–2 mm wide; panicles obovoid, branches sometimes
                         verticillate . . . . .49. *B. scoparius*
                     14. Lemmas 3–5 mm wide; panicles usually ovoid . . . . .44. *B. hordeaceus* (in part)
         10. At least some panicle branches as long as or longer than the spikelets, sometimes
             sinuous; panicles nodding.
             15. Lower glumes 7–10 mm long; upper glumes 8–12 mm long; panicle
                 branches conspicuously sinuous; awns erect to weakly spreading; lemma
                 margins rounded . . . . .50. *B. arenarius*
             15. Lower glumes 4–7 mm long; upper glumes 5–8 mm long; panicle branches
                 sometimes sinuous; awns erect to strongly divergent; lemma margins
                 slightly to strongly angled above the middle.
                 16. Anthers 2.5–5 mm long; spikelets often purple-tinged; culms 80–110
                     cm tall . . . . .40. *B. arvensis* (in part)
                 16. Anthers 1–1.5 mm long; spikelets not purple-tinged; culms 20–70 cm tall.

17. Lemmas with hyaline margins 0.3–0.6 mm wide, slightly angled above the middle; branches somewhat drooping, sometimes sinuous, often with more than 1 spikelet . . . . 51. *B. japonicus*
17. Lemmas with hyaline margins 0.6–0.9 mm wide, strongly angled above the middle; branches not drooping or sinuous, usually with 1 spikelet . . . . 52. *B. squarrosus*

Bromus sect. Ceratochloa (P. Beauv.) Griseb.

Plants annual, biennial, or perennial. **Spikelets** elliptic to lanceolate, strongly laterally compressed, with 3–12 florets. **Lower glumes** 3–7(9)-veined; **upper glumes** 5–9(11)-veined; **lemmas** lanceolate, laterally compressed, strongly keeled, at least distally, apices entire or with acute teeth, teeth shorter than 1 mm; **awns** straight, erect to slightly divaricate.

Bromus sect. *Ceratochloa* is native to North and South America, and contains about 25 species. It is marked by polyploid complexes; the major one in North America is the *Bromus carinatus* complex. This treatment recognizes six species in the complex: *B. aleutensis, B. arizonicus, B. carinatus, B. maritimus, B. polyanthus,* and *B. sitchensis.* The lowest chromosome number known for members of this complex is $2n = 28$, found in *B. carinatus*; the highest is $2n = 84$, found in *B. arizonicus.* The remaining species are hexaploids with $2n = 42$, or octoploids with $2n = 56$. One other species in the section, *B. catharticus,* has been introduced from South America and is also part of a polyploid complex.

There is morphological intergradation among the species recognized here, and some evidence that these intermediates are sometimes partially fertile (Harlan 1945a, 1945b; Stebbins and Tobgy 1944; Stebbins 1947). Stebbins and Tobgy (1944) commented that partial hybrid sterility between plants placed in different species on the basis of their morphology "supports the recognition of more than one species among the octoploid members of the complex," but later Stebbins (1981) stated that ". . . all the North American octoploids . . . should be united into a single species, in spite of the barriers of hybrid sterility that separate them."

1. Bromus catharticus Vahl [p. 200]
RESCUE GRASS

Plants annual, biennial, or perennial; loosely cespitose or tufted. **Culms** 30–120 cm tall, 2–4 mm thick, erect or decumbent. **Sheaths** usually densely, often retrorsely, hairy, hairs sometimes confined to the throat; **auricles** absent; **ligules** 1–4 mm, glabrous or pilose, obtuse, lacerate to erose; **blades** 4–30 cm long, 3–10 mm wide, flat, glabrous or hairy on both surfaces. **Panicles** 9–28 cm, usually open, erect or nodding; **lower branches** shorter than 10 cm, 1–4 per node, spreading or ascending, with up to 5 spikelets variously distributed. **Spikelets** (17)20–40 mm, shorter than at least some pedicels and branches, elliptic to lanceolate, strongly laterally compressed, not crowded or overlapping, with 4–12 florets. **Glumes** smooth or scabrous, glabrous or pubescent; **lower glumes** 7–12 mm, 5–7(9)-veined; **upper glumes** 9–17 mm, 7–9(11)-veined, shorter than the lowest lemma; **lemmas** 11–20 mm, lanceolate, laterally compressed, strongly keeled, usually glabrous, sometimes pubescent distally, smooth or scabrous, 9–13-veined, veins often raised and riblike, margins sometimes conspicuous, hyaline, whitish or partly purplish, apices entire or toothed, teeth acute, shorter than 1 mm; **awns** absent or to 10 mm; **anthers** 0.5–1 mm in cleistogamous florets, 2–5 mm in chasmogamous florets. $2n = 42$.

1. Awns absent or to 3.5 mm long . . . . var. *catharticus*
1. Awns (5)6–10 mm long . . . . var. *elatus*

Bromus catharticus Vahl var. catharticus [p. 200]

Plants annual or biennial; tufted. **Culms** 30–120 cm, erect or decumbent. **Sheaths** usually densely, often retrorsely, hairy, hairs sometimes confined to the throat; **ligules** 1–4 mm, glabrous or pilose, erose; **blades** 4–26 cm long, 3–10 mm wide, glabrous or hairy on both surfaces. **Panicles** 9–28 cm, open, erect or nodding; **lower branches** 1–4 per node, spreading or ascending,

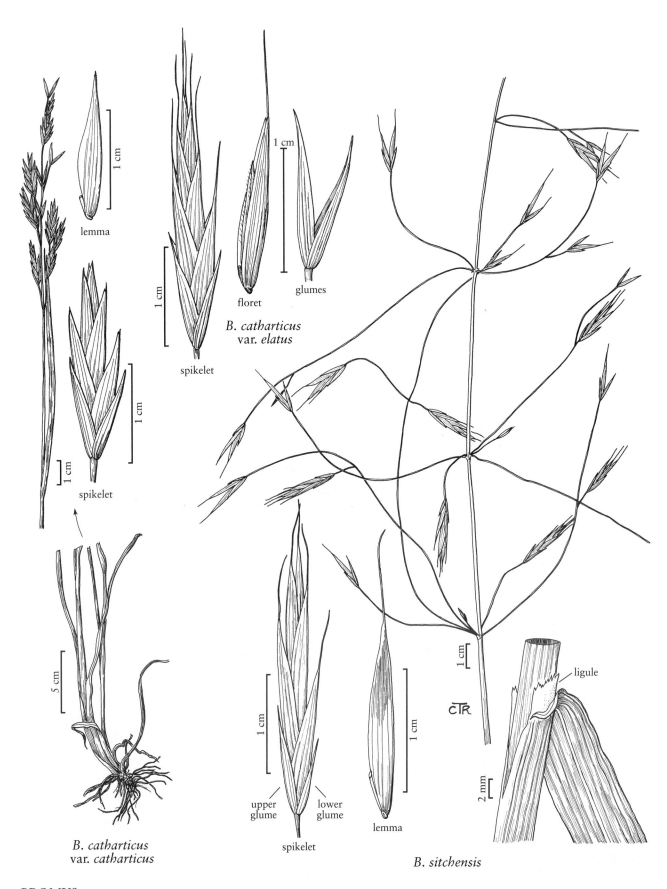

lemma

spikelet

spikelet

**B. *catharticus*
var. *catharticus***

floret

glumes

**B. *catharticus*
var. *elatus***

spikelet

upper
glume

lower
glume

spikelet

lemma

B. *sitchensis*

ligule

CTR

with 1–5 spikelets. **Spikelets** 20–30 mm, with 6–12 florets. **Lower glumes** 5–7-veined; **upper glumes** 9–13 mm, (7)9(11)-veined; **lemmas** 11–20 mm, glabrous or scabrous, sometimes pubescent distally, (9)11–13-veined; **awns** absent or to 3.5 mm; **anthers** about 0.5 mm in cleistogamous florets, 2–4 mm in chasmogamous florets. 2*n* = 42.

Bromus catharticus var. *catharticus* is native to South America. It has been widely introduced in the *Flora* region as a forage crop and is now established, particularly in the southern half of the United States. It usually grows on disturbed soils.

Bromus catharticus var. elatus (E. Desv.) Planchuelo [p. 200]

Plants perennial; loosely cespitose. **Culms** 50–110 cm, erect. **Sheaths** usually densely retrorsely pilose, sometimes densely villous generally or only on the throat; **ligules** 1–3.5 mm, glabrous, obtuse, lacerate to erose; **blades** 10–30 cm long, 3–5 mm wide, glabrous or hairy. **Panicles** 10–28 cm, lax, usually open; **lower branches** 2–3 per node, stiffly spreading to ascending, with 2–5 spikelets. **Spikelets** 20–40 mm, with 4–6(8) florets. **Lower glumes** 5–7(9)-veined; **upper glumes** 10–17 mm, 7–9-veined; **lemmas** 11–16 mm, scabrous, glabrous or pubescent, 9–11-veined; **awns** (5)6–10 mm; **anthers** 0.6–1 mm in cleistogamous florets, 3–5 mm in chasmogamous florets. 2*n* = 42.

Bromus catharticus var. *elatus*, a native of South America, now grows in disturbed soils in central California. It has also been reported from ballast dumps in Oregon. Although originally published as var. *elata*, the correct name is var. *elatus*.

2. Bromus sitchensis Trin. [p. 200]
SITKA BROME, ALASKA BROME

Plants perennial; loosely cespitose. **Culms** 120–180 cm tall, 3–5 mm thick, erect. **Sheaths** glabrous or sparsely pilose; **auricles** absent; **ligules** 3–4 mm, glabrous or hairy, obtuse, lacerate; **blades** 20–40 cm long, 2–9 mm wide, flat, sparsely pilose adaxially or on both surfaces. **Panicles** 25–35 cm, open; **lower branches** to 20 cm, 2–4(6) per node, spreading, often drooping, with 1–3 spikelets on the distal ¹/₂, sometimes confined to the tips. **Spikelets** 18–38 mm, elliptic to lanceolate, strongly laterally compressed, with (5)6–9 florets. **Glumes** glabrous, sometimes scabrous; **lower glumes** 6–10 mm, 3–5-veined; **upper glumes** 8–11 mm, 5–7-veined; **lemmas** 12–14(15) mm, lanceolate, laterally compressed, 7–11-veined, strongly keeled at least distally, usually glabrous, sometimes hirtellous, margins

sometimes sparsely pilose, apices entire or with acute teeth, teeth shorter than 1 mm; **awns** 5–10 mm; **anthers** to 6 mm. 2*n* = 42, 56.

Bromus sitchensis grows on exposed rock bluffs and cliffs, and in meadows, often in the partial shade of forests along the ocean edge, and on road verges and other disturbed sites. Its range extends from the Aleutian Islands and Alaska panhandle through British Columbia to southern California.

Bromus sitchensis resembles *B. aleutensis*, the two sometimes being treated as conspecific varieties. *Bromus sitchensis* is predominantly outcrossing, while *B. aleutensis* is predominantly self-fertilizing (C.L. Hitchcock 1969).

3. Bromus arizonicus (Shear) Stebbins [p.202]
ARIZONA BROME

Plants annual; tufted. **Culms** 30–90 cm tall, to 3 mm thick, erect. **Sheaths** retrorsely pilose, sometimes mostly glabrous, throats sometimes with hairs; **auricles** absent; **ligules** 1–4 mm, usually glabrous, obtuse, erose; **blades** 8–18 cm long, 3–9 mm wide, flat, sparsely pilose on both surfaces or the abaxial surfaces glabrous. **Panicles** 12–25 cm, somewhat contracted or open; **lower branches** shorter than 10 cm, 2–3(5) per node, initially erect to ascending, spreading at maturity, with 1–2 spikelets variously distributed. **Spikelets** 18–25 mm, elliptic to lanceolate, strongly laterally compressed, with 4–8 florets. **Glumes** subequal, smooth or scabrous; **lower glumes** 8–12.5 mm, 3-veined; **upper glumes** 9.5–14 mm, 7-veined, about as long as the lowest lemma; **lemmas** 9.5–14 mm, lanceolate, laterally compressed, prominently 7-veined, strongly keeled at least distally, glabrous or pubescent distally or throughout, marginal hairs, if present, longer than those elsewhere, apices entire or with acute teeth shorter than 1 mm; **awns** 6–13 mm, sometimes slightly geniculate; **anthers** 0.4–0.5 mm. 2*n* = 84.

Bromus arizonicus grows in dry, open areas and disturbed ground of the southwest, usually below 2000 m. Its range extends from California and southern Nevada into Arizona, New Mexico, and northern Mexico.

Stebbins et al. (1944) demonstrated that, like *Bromus carinatus* var. *carinatus*, *B. arizonicus* obtained three of its genomes from *B. catharticus* or a close relative, but the remaining three genomes are not homologous with those in *B. carinatus*, probably being derived from a species in a section other than *Ceratochloa*. The small anthers of *B. arizonicus* strongly suggest that most seed is produced by selfing.

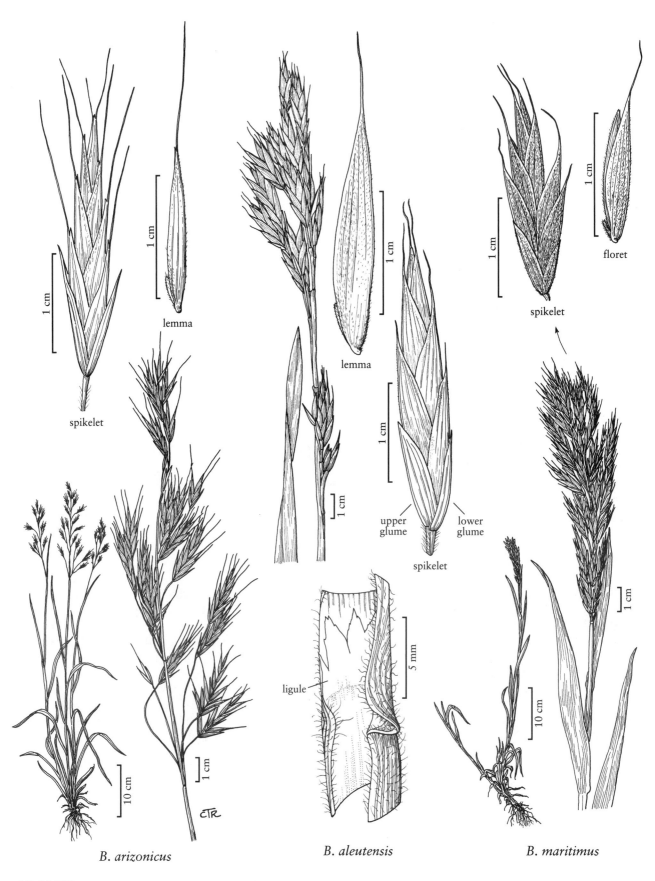

lemma

spikelet

B. arizonicus

lemma

ligule

5 mm

B. aleutensis

upper
glume

lower
glume

spikelet

1 cm

floret

spikelet

B. maritimus

CTR

BROMUS

4. Bromus aleutensis Trin. *ex* Griseb. [p. 202]
ALEUT BROME

Plants perennial; loosely cespitose. **Culms** 40–130 cm tall, 3–7 mm thick, often decumbent. **Sheaths** coarsely striate, pilose, hairs sparse to moderately dense, throats pilose; **auricles** rarely present; **ligules** 3.5–5 mm, usually glabrous, occasionally pubescent, lacerate; **blades** 13–35 cm long, 6–15 mm wide, flat, usually sparsely to moderately pilose on both surfaces, sometimes glabrous. **Panicles** 10–28 cm, erect, open or somewhat contracted; **lower branches** to 10 cm, 1–2 per node, stiffly ascending, with (1)2–3 spikelets on the distal ¹/₂, sometimes confined to the tips. **Spikelets** 25–40 mm, elliptic to lanceolate, strongly laterally compressed, with 3–6 florets. **Glumes** glabrous or pubescent; **lower glumes** 9–13 mm, 3–5-veined; **upper glumes** 10–15 mm, 7(9)-veined; **lemmas** 12–17 mm, lanceolate, laterally compressed, usually softly pubescent, sometimes glabrous, strongly keeled at least distally, 9(11)-veined, veins conspicuous distally, apices entire or with acute teeth shorter than 1 mm; **awns** (3)5–10 mm; **anthers** 2.2–4.2 mm. 2*n* = 56.

Bromus aleutensis grows in sand, gravel, and disturbed soil along the Pacific coast, from the Aleutian Islands of Alaska to western Washington, and on some lake shores of central British Columbia. It has also been found further east in Canada and in northern Idaho, always in disturbed sites, such as road edges.

Bromus aleutensis might represent a modified version of *B. sitchensis*, in which reproduction occurs at a relatively early developmental state in response to the climatic conditions of the Aleutian Islands (Hultén 1968). *B. aleutensis* is predominantly self-fertilizing, and *B. sitchensis* is predominantly outcrossing. Anther lengths close to 4.2 mm suggest that at least some plants of *B. aleutensis* are outcrossing (Hitchcock 1969). *Bromus aleutensis* intergrades with *B. carinatus* var. *marginatus* to the south.

5. Bromus maritimus (Piper) Hitchc. [p. 202]
MARITIME BROME

Plants perennial; loosely cespitose. **Culms** 20–70 cm tall, to 3 mm thick, sometimes geniculate at the base. **Sheaths** usually smooth or scabridulous, sometimes slightly pubescent distally, not pilose at the throat; **auricles** absent; **ligules** 1–6 mm, densely hairy to ciliolate, acute to obtuse, erose; **blades** 6–13 cm long, 6–8 mm wide, flat, both surfaces glabrous, sometimes scabrous. **Panicles** 9–20 cm long, 2–2.5 cm wide, dense; **lower branches** shorter than 10 cm, 2–4 per node, erect, with 1–2 spikelets variously distributed. **Spikelets** 20–40 mm, usually longer than the branches and pedicels, elliptic to lanceolate, strongly laterally compressed, crowded, overlapping, with 3–7 florets. **Glumes** pubescent; **lower glumes** 8–12 mm, (3)5(7)-veined; **upper glumes** 10–13 mm, 7(9)-veined, shorter than the lowest lemma; **lemmas** 12–14 mm, lanceolate, laterally compressed, distinctly 9–11-veined, strongly keeled at least distally, more or less uniformly hairy, often with bronze hyaline margins, apices entire or with acute teeth shorter than 1 mm; **awns** (2)4–7 mm; **anthers** 2–4 mm. 2*n* = 56.

Bromus maritimus grows in coastal sands from Lane County, Oregon, to Los Angeles County, California.

6. Bromus carinatus Hook. & Arn. [p. 204]

Plants annual, biennial, or perennial; loosely cespitose. **Culms** 45–120(180) cm tall, usually less than 3 mm thick, erect. **Sheaths** mostly glabrous or retrorsely soft pilose, throats usually hairy; **auricles** sometimes present on the lower leaves; **ligules** 1–3.5(4) mm, glabrous or sparsely hairy, acute to obtuse, lacerate or erose; **blades** 8–30 cm long, 1–12 mm wide, flat or becoming involute, glabrous or sparsely pilose to pubescent on 1 or both surfaces. **Panicles** 5–40 cm, lax, open or erect; **lower branches** usually shorter than 10 cm, 1–4 per node, ascending to strongly divergent or reflexed, with 1–4 spikelets variously distributed. **Spikelets** 20–40 mm, shorter than at least some pedicels and branches, elliptic to lanceolate, strongly laterally compressed, not crowded or overlapping, sometimes purplish, with 4–11 florets. **Glumes** glabrous or pubescent; **lower glumes** 7–11 mm, 3–7(9)-veined; **upper glumes** 9–13 mm, shorter than the lowest lemma, 5–9(11)-veined; **lemmas** 10–16(17) mm, lanceolate, laterally compressed, strongly keeled distally, usually more or less uniformly pubescent or pubescent on the margins only, sometimes glabrous or scabrous, 7–9-veined, veins usually not raised or riblike, apices entire or with acute teeth shorter than 1 mm; **awns** 4–17 mm, sometimes slightly geniculate; **anthers** 1–6 mm. 2*n* = 28, 42, 56.

Bromus carinatus is native from British Columbia to Saskatchewan and south to Mexico. It has been introduced to various more eastern locations and to the southern Yukon Territory. The two varieties recognized here are sometimes recognized as species.

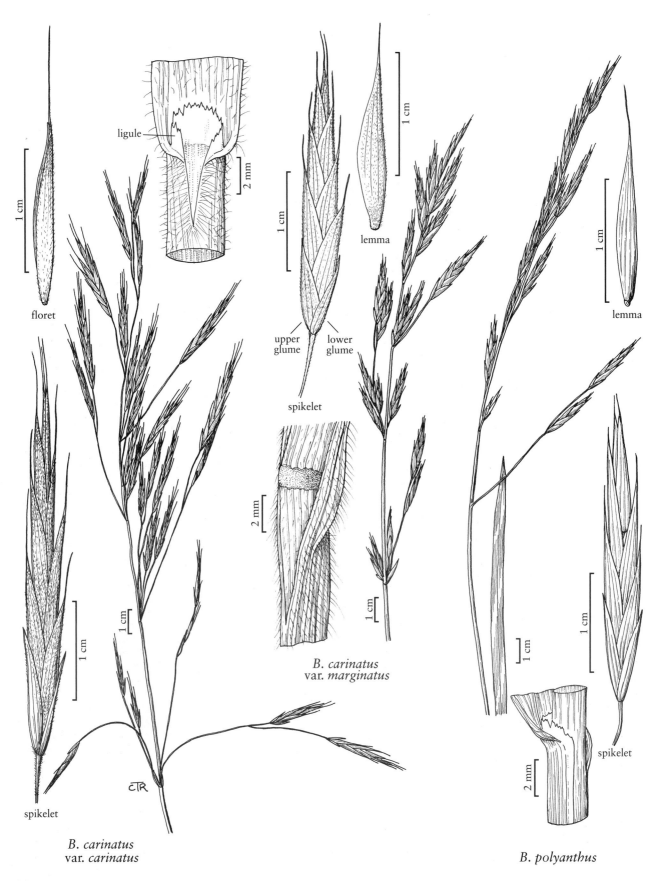

floret

ligule

2 mm

lemma

1 cm

upper
glume   lower
glume

spikelet

lemma

B. carinatus
var. *marginatus*

2 mm

1 cm

1 cm

CTR

spikelet

B. carinatus
var. *carinatus*

1 cm

2 mm

spikelet

B. polyanthus

BROMUS

1. Most awns 8–17 mm long . . . . var. *carinatus*
1. Most awns 4–7 mm long . . . . var. *marginatus*

Bromus carinatus Hook. & Arn. var. **carinatus** [p. 204]
CALIFORNIA BROME

Plants annual or biennial. **Culms** 50–100 cm tall. **Sheaths** mostly glabrous or retrorsely soft pilose, throats usually hairy; **auricles** absent; **ligules** 1–3(4) mm, usually glabrous, obtuse, lacerate or erose; **blades** 10–30 cm long, 3–6 mm wide, flat, usually sparsely pilose on both surfaces, sometimes glabrous. **Panicles** 15–40 cm, lax, open; **lower branches** 2–4 per node, ascending to strongly divergent or reflexed. **Spikelets** with 6–11 florets. **Lower glumes** 8–10 mm, 3(5)-veined; **upper glumes** 9.5–12 mm, 5(7)-veined; **lemmas** 12–16 mm, usually more or less uniformly pubescent, sometimes scabrous, 7-veined; **awns** 8–17 mm, sometimes slightly geniculate; **anthers** 1–5 mm. $2n = 56$.

Bromus carinatus var. *carinatus* is primarily coastal and grows in shrublands, grasslands, meadows, and openings in chaparral and oak and yellow pine woodlands. It ranges from southern British Columbia through Washington, Oregon, and California to Baja California, Mexico, and extends eastward through Arizona to New Mexico.

Bromus carinatus var. *carinatus* intergrades with var. *marginatus*, which tends to grow at higher elevations and extends further inland.

Bromus carinatus var. **marginatus** (Nees) Barkworth & Anderton [p. 204]
MOUNTAIN BROME

Plants perennial. **Culms** 45–120(180) cm tall. **Sheaths** usually sparsely retrorsely pilose throughout, ranging from densely pilose to glabrous, except at the throat, throats always pilose; **auricles** sometimes present on the lower leaves; **ligules** 1–3.5 mm, sparsely hairy, acute to obtuse, erose or lacerate; **blades** 8–25 cm long, 1–12 mm wide, flat or involute, glabrous or sparsely pilose to pubescent on 1 or both surfaces. **Panicles** 5–20(30) cm, erect; **lower branches** 1–4 per node, erect or ascending. **Spikelets** with 4–9 florets. **Lower glumes** 7–11 mm, 3–7(9)-veined; **upper glumes** 9–13 mm, 5–9(11)-veined; **lemmas** 10–14(17) mm, pubescent on the backs and margins, on the margins only, or glabrous, 7–9-veined; **awns** 4–7 mm; **anthers** 1–6 mm. $2n = 42$.

Bromus carinatus var. *marginatus* is primarily an inland species and grows on open slopes, grass balds, shrublands, meadows, and open forests, in montane and subalpine zones. It grows from British Columbia to Saskatchewan, south throughout the western United States, and also extends into northern Mexico. Its elevational range is 350–2200 m in the northern part of its distribution, and 1500–3300 m in the south.

Bromus carinatus var. *marginatus* is variable and intergrades with *B. carinatus* var. *carinatus* to the west, *B. aleutensis* to the north, and *B. polyanthus* to the southeast. As treated here, *B. carinatus* var. *marginatus* includes *B. luzonensis* J. Presl, which has been recognized mainly on the basis of its canescent sheaths and blades; this trait is highly variable and may be environmentally determined. Although the name *Bromus carinatus* var. *marginatus* was attributed to Hitchcock by Scoggan, there is no evidence that either A.S. or C.L. Hitchcock actually made the combination.

7. Bromus polyanthus Scribn. [p. 204]
COLORADO BROME, GREAT-BASIN BROME

Plants perennial; loosely cespitose. **Culms** 60–120 cm tall, to 3 mm thick, erect, glabrous or puberulent. **Sheaths** usually smooth or scabrous, sometimes hairy except at the throat; **auricles** absent; **ligules** (1)2–2.5 mm, glabrous, obtuse, erose; **blades** 10–31 cm long, 2–9 mm wide, flat, sometimes scabrous, usually glabrous, rarely puberulent to pubescent near the collar. **Panicles** 10–20 cm, open to somewhat contracted; **lower branches** shorter than 10 cm, (1)2–3 per node, erect, ascending or spreading, with 1–2 spikelets variously distributed. **Spikelets** 20–35 mm, shorter than at least some pedicels and branches, elliptic to lanceolate, strongly laterally compressed, not crowded or overlapping, with 6–11 florets. **Glumes** smooth or scabrous; **lower glumes** (5.5)7–10(11.5) mm, 3-veined; **upper glumes** (7.5) 9–11(12.5) mm, 5–7-veined, shorter than the lowest lemma; **lemmas** 12–15 mm, lanceolate, laterally compressed, strongly keeled at least distally, glabrous, sometimes scabrous, 7–9-veined, veins usually not raised or riblike, apices entire or with acute teeth, teeth shorter than 1 mm; **awns** 4–7 mm; **anthers** 1–5 mm. $2n = 56$.

Bromus polyanthus grows on open slopes and in meadows. It is found primarily in the central Rocky Mountains, but the limits of its range include British Columbia in the north, California in the west, and Arizona, New Mexico, and western Texas in the south. It is not known from Mexico. It intergrades with *B. carinatus* var. *marginatus*. Plants with an erect, contracted panicle and awns 4–6 mm long can be called *B. polyanthus* var. *polyanthus*; those with with an open, nodding panicle and awns up to 8 mm long can be called *B. polyanthus* var. *paniculatus* Shear. Because the variation in both characters is continuous, the varieties are not recognized here.

Bromus sect. Bromopsis Dumort.

Plants usually perennial, sometimes annual. **Spikelets** elliptic to lanceolate, more or less terete initially, sometimes becoming laterally compressed at anthesis, with (3)4–14(16) florets. **Lower glumes** 1–3-veined; **upper glumes** 3–5-veined; **lemmas** elliptic to lanceolate, rounded over the midvein, apices subulate, acute, obtuse or rounded, entire or slightly emarginate; **awns** straight, arising less than 1.5 mm below the lemma apices.

Bromus sect. *Bromopsis* is sometimes incorrectly called sect. *Pnigma* Dumort. It is native to Eurasia as well as to North and South America, and has about 90 species.

8. Bromus inermis Leyss. [p. 208]
SMOOTH BROME, HUNGARIAN BROME, BROME INERME

Plants perennial; rhizomatous, rhizomes short to long-creeping. **Culms** 50–130 cm, erect, single or a few together; **nodes** (2)3–5(6), usually glabrous, rarely pubescent; **internodes** usually glabrous, rarely pubescent. **Sheaths** usually glabrous, rarely pubescent or pilose; **auricles** sometimes present; **ligules** to 3 mm, glabrous, truncate, erose; **blades** 11–35(42) cm long, 5–15 mm wide, flat, usually glabrous, rarely pubescent or pilose. **Panicles** 10–20 cm, open, erect; **branches** ascending or spreading. **Spikelets** 20–40 mm, elliptic to lanceolate, terete to moderately laterally compressed, sometimes purplish, with (5)8–10 florets. **Glumes** glabrous; **lower glumes** (4)6–8(9) mm, 1(3)-veined; **upper glumes** (5)7–10 mm, 3-veined; **lemmas** 9–13 mm, elliptic to lanceolate, rounded over the midvein, usually glabrous and smooth, sometimes scabrous, margins sometimes sparsely puberulent, the basal part of the backs less frequently so, apices acute to obtuse, entire; **awns** absent or to 3 mm, straight, arising less than 1.5 mm below the lemma apices; **anthers** 3.5–6 mm. 2*n* = 28, 56.

Bromus inermis is native to Eurasia, and is now found in disturbed sites in Alaska, Greenland, and most of Canada as well as south throughout most of the contiguous United States except the southeast. It has also been used for rehabilitation, and is planted extensively for forage in pastures and rangelands from Alaska and the Yukon Territory to Texas.

Bromus inermis is similar to *B. pumpellianus*, differing mainly in having glabrous lemmas, nodes, and leaf blades, but lack of pubescence is not a consistently reliable distinguishing character. *Bromus inermis* also resembles a recently introduced species, *B. riparius*, from which it differs primarily in its shorter or nonexistent awns.

9. Bromus riparius Rehmann [p. 208]
MEADOW BROME

Plants perennial; cespitose, shortly rhizomatous. **Culms** 30–90 cm, erect or decumbent, forming distinct clumps; **nodes** 2–3, glabrous or puberulent; **internodes** glabrous or puberulent. **Sheaths** glabrous or with hairs; **auricles** to 1 mm on the lower leaves; **ligules** 0.4–1.0 mm, glabrous or ciliate, truncate, erose; **blades** 10–20 cm long, 2–3 mm wide, scabridulous, glabrous or sparsely pilose, margins sometimes ciliate. **Panicles** 8–20 cm long, lax; **branches** scabridulous, with 1–2 spikelets. **Spikelets** 20–32 mm, lanceolate, becoming cuneate, with 5–8 florets. **Glumes** glabrous, sometimes scabridulous on the veins; **lower glumes** 6.5–10 mm, 1(3)-veined; **upper glumes** 7.5–12 mm, 3–5-veined; **lemmas** 10–13 mm, oblong to lanceolate, rounded over the midvein, 7-veined, glabrous or appressed-hairy, sometimes scabridulous, apices acute, entire or minutely bifid; **awns** 4–8 mm, straight or slightly spreading, arising less than 1.5 mm below the lemma apices; **anthers** 2.5–5.2 mm. 2*n* = 70.

Bromus riparius is an Asian species that was introduced to the United States in the late 1950s for cultivation as a pasture grass. Various cultivars are now grown, mainly in Canada and the northwestern United States. The description given here is derived in part from cultivated specimens. North American plants have sometimes been referred to, incorrectly, as *Bromus biebersteinii* Roem. & Schult. (Vogel et al. 1996). *Bromus riparius* differs from that species in having acute lemma apices and, usually, more pubescent leaf blades, sheaths, and lemmas.

The existence of *Bromus riparius* in the *Flora* region was not realized until shortly before this treatment was submitted for publication, making it impossible to fully investigate its similarities to *B. inermis* and *B. pumpellianus*, particularly subsp. *dicksonii*. It appears to differ from both species in having shorter culms on average, longer awns than *B. inermis*, and shorter rhizomes than *B. pumpellianus* subsp. *pumpellianus*. Its distribution in the *Flora* region is not known.

10. **Bromus pumpellianus** Scribn. [p. 208]

ARCTIC BROME

Plants perennial; usually rhizomatous, sometimes cespitose, rhizomes short to long-creeping. **Culms** 50–135 cm, erect or ascending, sometimes geniculate, usually single or few together, sometimes clumped; **nodes** 2–7, pubescent or glabrous; **internodes** glabrous or pubescent. **Sheaths** pilose, villous, or glabrous; **auricles** sometimes present on the lower leaves; **ligules** to 4 mm, glabrous, truncate or obtuse, erose; **blades** 7–30 cm long, 2.5–8.5(9) mm wide, flat, pubescent or glabrous on both surfaces, sometimes only the adaxial surface pubescent. **Panicles** 10–24 cm, open or contracted, erect or nodding; **branches** erect to spreading. **Spikelets** 16–32(45) mm, elliptic to lanceolate, terete to moderately laterally compressed, sometimes purplish, with 4–14 florets. **Glumes** glabrous or hairy; **lower glumes** (4)5–10 mm, 1(3)-veined; **upper glumes** (5)7.5–13 mm, 3-veined; **lemmas** 9–16 mm, lanceolate, rounded over the midvein, sparsely to densely hairy throughout, or on the margins and lower portion of the back, or along the marginal veins and keel, apices subulate to acute, entire or slightly emarginate, lobes shorter than 1 mm; **awns** usually present, sometimes absent, to 7.5 mm, straight, arising less than 1.5 mm below the lemma apices; **anthers** 3.5–7 mm. $2n = 28, 56$.

The range of *Bromus pumpellianus* extends from Asia to North America, where it includes Alaska, the western half of Canada, the western United States as far south as New Mexico, and a few other locations eastward. It is sometimes treated as a subspecies of *B. inermis*. It differs from that species primarily in its tendency to have pubescent lemmas, nodes, and leaf blades.

Two subspecies that differ in morphology and distribution are described below. Both strongly resemble the recently introduced *Bromus riparius*, differing in the case of *B. pumpellianus* subsp. *pumpellianus* in having longer rhizomes, or, in the case of *B. pumpellianus* subsp. *dicksonii*, in having a more restricted distribution. It is possible that the description and distribution of *B. pumpellianus* may be based in part on misidentification of *B. riparius*, as many taxonomists may have been unaware of the introduction of the latter species to North America.

1. Panicles usually open; plants cespitose, sometimes shortly rhizomatous; culms ascending, often geniculate; nodes glabrous or pubescent; plants of the Yukon River drainage
. . . . subsp. *dicksonii*

1. Panicles contracted to open; plants rhizomatous; culms erect; nodes usually pubescent; plants of the range of the species
. . . . subsp. *pumpellianus*

Bromus pumpellianus subsp. **dicksonii** W.W. Mitch. & Wilton [p. 208]

Plants cespitose, sometimes shortly rhizomatous. **Culms** 65–135 cm, ascending, often geniculate; **nodes** 3–7, glabrous or pubescent; **internodes** glabrous. **Sheaths** glabrous or hairy; **auricles** sometimes present; **ligules** 0.5–4 mm; **blades** 7–30 cm long, 2.5–8.5 mm wide, glabrous or hairy on both surfaces, sometimes only the adaxial surface hairy. **Panicles** 10–24 cm, usually open, nodding to partially erect; **branches** spreading to drooping. **Spikelets** 16–32(45) mm. **Glumes** glabrous or hairy; **lower glumes** 5–10 mm, 1(3)-veined; **upper glumes** 7.5–13 mm, 3-veined; **lemmas** 9–16 mm, sparsely to densely hairy throughout or along the marginal vein and keel; **awns** 2.5–7.5 mm; **anthers** (3.5)4–6(6.8) mm. $2n = 28$.

Bromus pumpellianus subsp. *dicksonii* grows in shallow, rocky soils of river banks and bluffs in the Yukon River drainage of Alaska. Apart from the more restricted distribution, it is not clear how this subspecies differs from the introduced *B. riparius*.

Bromus pumpellianus Scribn. subsp. **pumpellianus** [p. 208]

Plants rhizomatous. **Culms** 50–120 cm, erect; **nodes** 2–3(4), usually pubescent, sometimes glabrous; **internodes** glabrous or pubescent. **Sheaths** pilose, villous, or glabrous; **auricles** present on the lower leaves or absent; **ligules** to 3 mm; **blades** 9–17(25) cm long, (3)4–8(9) mm wide, abaxial surfaces glabrous or pilose, adaxial surfaces usually pilose, rarely glabrous. **Panicles** 10–20 cm, open or contracted, erect or nodding; **branches** erect to spreading. **Spikelets** 20–30 mm. **Glumes** glabrous, pubescent, or hirsute; **lower glumes** (4)5–9 mm, 1-veined; **upper glumes** (5)8–11 mm, 3-veined; **lemmas** 9–14 mm, pubescent on the lower portion of the back and along the margins; **awns** usually present, to 6 mm, sometimes absent; **anthers** 3.5–7 mm. $2n = 56$.

Bromus pumpellianus subsp. *pumpellianus* grows on sandy and gravelly stream banks and lake shores, sand dunes, meadows, dry grassy slopes, and road verges.

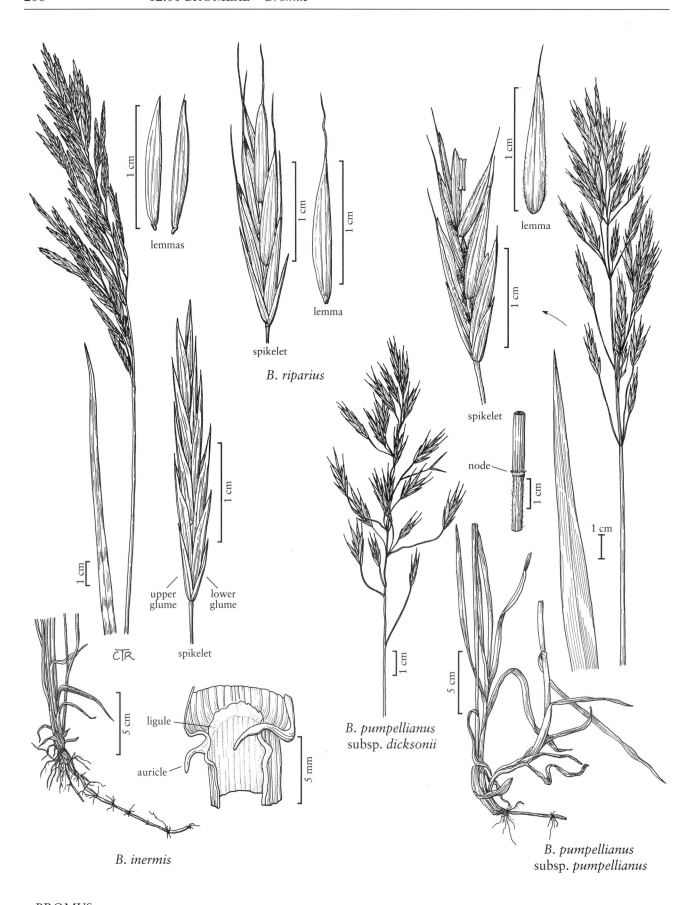

lemmas

B. riparius

spikelet

lemma

spikelet

upper glume

lower glume

spikelet

CTR

lemma

node

spikelet

ligule

auricle

B. pumpellianus
subsp. *dicksonii*

B. inermis

B. pumpellianus
subsp. *pumpellianus*

BROMUS

11. **Bromus latiglumis** (Scribn. *ex* Shear) Hitchc. [p. 210]

HAIRY WOODBROME, FLANGED BROME,
BROME À LARGES GLUMES

Plants perennial; not rhizomatous. **Culms** 80–150 cm, erect; **nodes** 9–20, glabrous, usually concealed by the leaf sheaths; **internodes** usually glabrous, sometimes hairy just below the nodes. **Sheaths** overlapping, densely to moderately retrorsely pilose or glabrous over most of their surface, throats and collars densely pilose; **auricles** 1–2.5 mm on most lower leaves; **ligules** 0.8–1.4 mm, hirsute, ciliate, truncate, erose; **blades** 20–30 cm long, 5–15 mm wide, flat, usually glabrous, rarely pilose, with 2 prominent flanges at the collar. **Panicles** 10–22 cm, open, nodding; **branches** spreading to ascending. **Spikelets** 15–30 mm, elliptic to lanceolate, terete to moderately laterally compressed, with 4–9 florets. **Glumes** pubescent or glabrous; **lower glumes** 4–7.5 mm, 1(3)-veined; **upper glumes** 6–9 mm, 3-veined, sometimes mucronate; **lemmas** 8–14 mm, elliptic to lanceolate, rounded over the midvein, backs glabrous or pilose to pubescent, margins long-pilose, apices obtuse to acute, entire; **awns** 3–4.5(7) mm, straight, arising less than 1.5 mm below the lemma apices; **anthers** 2–3 mm. $2n = 14$.

Bromus latiglumis grows in shaded or open woods, along stream banks, and on alluvial plains and slopes. Its range is mainly in the north-central and northeastern United States and adjacent Canadian provinces. Specimens with decumbent, weak, sprawling culms, densely hairy sheaths, and heavy panicles can be called *Bromus latiglumis* f. *incanus* (Shear) Fernald.

12. **Bromus laevipes** Shear [p. 210]

CHINOOK BROME

Plants perennial; not rhizomatous. **Culms** 50–150 cm, erect or basally decumbent, often rooting from the lower nodes; **nodes** 3–5(6), pubescent; **internodes** usually glabrous, often puberulent-pubescent just below the nodes, rarely puberulent throughout. **Sheaths** glabrous, sometimes slightly pubescent near the throat, sometimes with hairs in the auricular position; **auricles** absent or vestigial on the basal leaves; **ligules** 2–4.2 mm, glabrous, obtuse, lacerate; **blades** 13–26 cm long, 4–10 mm wide, light green or glaucous, flat, glabrous, sometimes scabrous on both surfaces. **Panicles** 10–20 cm, open, nodding; **branches** ascending to spreading, often drooping. **Spikelets** 23–35 mm, elliptic to lanceolate, terete to moderately laterally compressed, with 5–11 florets. **Glumes** glabrous, sometimes scabrous, margins often bronze-tinged; **lower glumes** 6–9 mm, 3-veined; **upper glumes** 8–12 mm, 5-veined; **lemmas** 12–16 mm, elliptic to lanceolate, rounded over the midvein, backs sparsely pilose, pubescent, or scabrous, margins densely pilose, at least on the lower $^1/_2$, often bronze-tinged, apices acute to obtuse, entire, rarely slightly emarginate, lobes shorter than 1 mm; **awns** 4–6 mm, straight, arising less than 1.5 mm below the lemma apices; **anthers** 3.5–5 mm. $2n = 14$.

Bromus laevipes grows from northern Oregon to southern California. It grows in shaded woodlands and on exposed brushy slopes, at 300–1500 m.

13. **Bromus kalmii** A. Gray [p. 210]

KALM'S BROME, BROME DE KALM

Plants perennial; not rhizomatous. **Culms** 50–100(110) cm, usually erect, sometimes decumbent at the base; **nodes** 3–5, pubescent, puberulent, or glabrous; **internodes** puberulent or glabrous. **Sheaths** and **throats** pilose or glabrous; **auricles** absent; **ligules** 0.5–1 mm, glabrous, truncate, erose; **blades** 10–17 cm long, 5–10 mm wide, flat, with prow-shaped tips, both surfaces glabrous or pilose or only the adaxial surfaces pilose. **Panicles** 8–13 cm, open, drooping; **branches** ascending to spreading, flexuous. **Spikelets** 15–25 mm, elliptic to lanceolate, terete to moderately laterally compressed, with 7–11 florets. **Glumes** pubescent, margins often hyaline; **lower glumes** 5–7.5 mm, 3-veined; **upper glumes** 6.5–8.5 mm, 5-veined; **lemmas** 7–11 mm, elliptic to lanceolate, rounded over the midvein, backs more or less uniformly pilose or pubescent, margins densely long-pilose, apices acute to obtuse, entire; **awns** 1.5–3 mm, straight, arising less than 1.5 mm below the lemma apices; **anthers** 1.5–2.5 mm. $2n = 14$.

Bromus kalmii grows in sandy, gravelly, or limestone soils in open woods and calcareous fens. Its range centers in the north-central and northeastern United States and adjacent Canadian provinces.

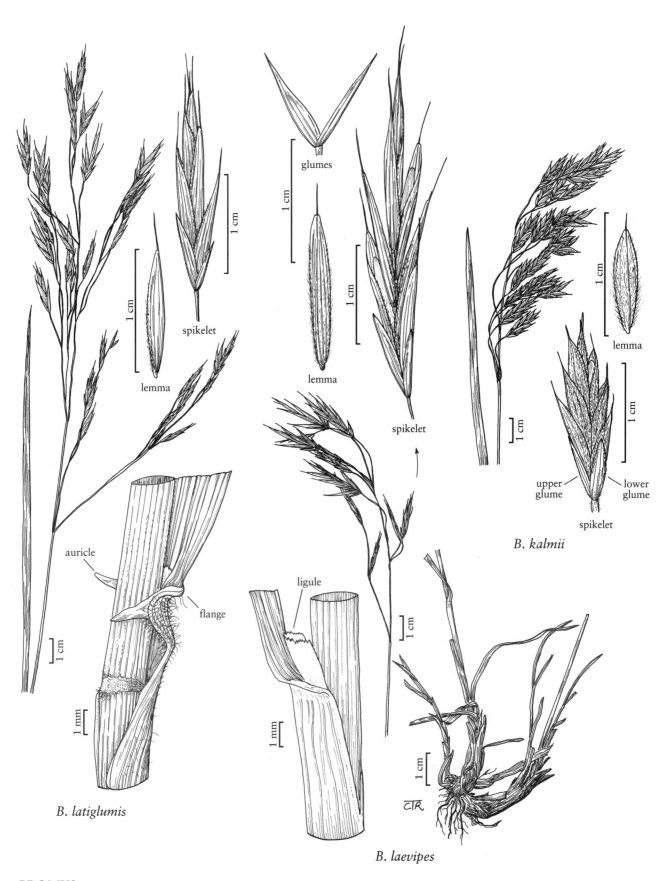

glumes

1 cm

1 cm

lemma

spikelet

lemma

spikelet

1 cm

lemma

spikelet

1 cm

1 cm

upper
glume

lower
glume

spikelet

B. kalmii

auricle

flange

1 cm

1 mm

B. latiglumis

ligule

1 mm

1 cm

1 cm

CTR

B. laevipes

BROMUS

14. Bromus pseudolaevipes Wagnon [p. 212]
WOODLAND BROME

Plants perennial; not rhizomatous. **Culms** 60–120 cm, erect or spreading; **nodes** 4–6, pubescent or puberulent; **internodes** mostly glabrous, sometimes pubescent to puberulent just below the nodes. **Sheaths** glabrous or pilose, often pilose near the auricles; **auricles** usually present on the lower leaves, rarely absent; **ligules** to 1.5 mm, usually pubescent, sometimes glabrous, truncate to obtuse, laciniate, ciliolate; **blades** 10–25 cm long, 3–9 mm wide, flat, glabrous, pilose on the margins or throughout. **Panicles** 10–20 cm, open, usually nodding; **branches** ascending to spreading or reflexed. **Spikelets** 15–35 mm, elliptic to lanceolate, terete to moderately laterally compressed, with 4–10 florets. **Glumes** usually pubescent, rarely glabrous, sometimes scabrous, margins often bronze-tinged; **lower glumes** 4–7 mm, 3-veined; **upper glumes** 6.5–9 mm, (3)5-veined; **lemmas** 10–13 mm, elliptic to lanceolate, rounded over the midvein, backs usually pubescent, sometimes glabrous distally, margins often bronze-tinged, pubescent nearly throughout, apices acute to obtuse, entire, rarely slightly emarginate, lobes shorter than 1 mm; **awns** 3–5 mm, straight, arising less than 1.5 mm below the lemma apices; **anthers** 3.5–5 mm. $2n = 14$.

Bromus pseudolaevipes grows in dry, shaded or semishaded sites in chaparral, coastal sage scrub, and woodland-savannah zones, from near sea level to about 900 m, in central and southern California. It is not known from Mexico.

15. Bromus hallii (Hitchc.) Saarela & P.M. Peterson [p. 212]
HALL'S BROME

Plants perennial; not rhizomatous. **Culms** 90–150 cm, erect; **nodes** 1–2(3), pubescent or puberulent; **internodes** usually puberulent, occasionally glabrous, pilose to densely pubescent below the nodes. **Sheaths** densely pubescent to pilose, hairs to 1 mm, collars pilose, hairs to 2 mm; **auricles** absent; **ligules** 0.5–2.5 mm long, sparsely to densely pubescent, obtuse, erose; **blades** 7.5–16.5 cm long, 3–12 mm wide, flat, densely pubescent on both surfaces. **Panicles** 5–16 cm, open; **branches** erect, ascending and appressed to slightly spreading. **Spikelets** 25–35(45) mm, terete to moderately laterally compressed, with 3–7 florets.

Glumes sparsely to densely pubescent; **lower glumes** 5–8(9) mm, 1(3)-veined; **upper glumes** (7)8–9 mm, 3-veined, sometimes mucronate; **lemmas** 10–14 mm, elliptic, rounded over the midvein, backs sparsely to densely pubescent, margins pubescent, apices entire; **awns** 3.5–7 mm, straight, arising less than 1.5 mm below the lemma apices; **anthers** 3–6 mm. $2n =$ unknown.

Bromus hallii grows in southern California on dry, open or shaded hillsides, rocky slopes, and in montane pine woods, from 1500–2700 m.

16. Bromus orcuttianus Vasey [p. 212]
ORCUTT'S BROME

Plants perennial; not rhizomatous. **Culms** 90–150 cm, erect; **nodes** 2–4, pubescent or puberulent; **internodes** glabrous to pubescent, pilose to densely pubescent below the nodes. **Basal sheaths** sparingly to densely pilose, hairs 2–4 mm, occasionally glabrous; **upper sheaths** hairy, hairs to 1 mm, collars glabrous or pilose, hairs to 4 mm; **auricles** absent; **ligules** 1–3 mm, usually glabrous, occasionally pilose, obtuse, erose; **blades** 7–24 cm long, 3–12 mm wide, flat, usually glabrous, sometimes hairy. **Panicles** 7–13.5 cm, open; **branches** erect, ascending and appressed to slightly spreading. **Spikelets** 20–40 mm, elliptic to lanceolate, terete to moderately laterally compressed, with 3–9(11) florets. **Glumes** usually glabrous, occasionally scabrous or pubescent; **lower glumes** 5–9 mm, 1(3)-veined; **upper glumes** 7–11 mm, 3(5)-veined, sometimes mucronate; **lemmas** 9–16 mm, elliptic, rounded over the midvein, backs usually pubescent, sometimes glabrous or scabrous, margins pubescent or scabrous, apices obtuse, entire; **awns** (4)5.5–8 mm, straight, arising less than 1.5 mm below the lemma apices; **anthers** 3–5 mm. $2n = 14$.

Bromus orcuttianus grows on dry hillsides and rocky slopes, and in open pine woods and meadows in the mountains, from 500–3500 m. It is found in the western United States, including Washington, Oregon, California, Nevada, and Arizona. It is not known from Mexico.

spikelet

1 cm

lemma

lemma

spikelet

lemma

spikelet

spikelet

1 cm

auricle

ligule

1 mm

lower leaf

upper leaf

B. pseudolaevipes

B. hallii

B. orcuttianus

BROMUS

17. Bromus grandis (Shear) Hitchc. [p. 214]
TALL BROME

Plants perennial; not rhizomatous. **Culms** 70–180 cm, erect; **nodes** 3–7, pubescent or puberulent; **internodes** pubescent, puberulent, or glabrous. **Sheaths** densely pubescent, hairs to 1 mm, collars with hairs to 2 mm, midrib of the culm leaves not abruptly narrowed just below the collar; **auricles** sometimes present; **ligules** 1–3 mm, densely pubescent to pilose, obtuse, lacerate; **blades** (13)18–38 cm long, 3–12 mm wide, flat, sparsely to densely pubescent on both surfaces. **Panicles** 14–27 cm long, very open, nodding; **branches** flexuous, usually widely spreading, with spikelets near the tips. **Spikelets** 20–37(45) mm, elliptic to lanceolate, terete to moderately laterally compressed, with 4–9(10) florets. **Glumes** pubescent, with prominent veins, margins not bronze-tinged; **lower glumes** 4–8.5 mm, (1)3-veined; **upper glumes** 7–10 mm, 3(5)-veined, not mucronate; **lemmas** 9.5–14 mm, lanceolate, rounded over the midvein, backs pilose, pubescent, or glabrous, margins pilose, not bronze-tinged, apices subulate to acute, entire; **awns** 3–7 mm, straight, arising less than 1.5 mm below the lemma apices; **anthers** 3–6 mm. $2n = 14$.

Bromus grandis grows on dry, wooded or open slopes, at elevations of 350–2500 m. Its range extends from central California into Baja California, Mexico.

18. Bromus frondosus (Shear) Wooton & Standl. [p. 214]
WEEPING BROME

Plants perennial; not rhizomatous. **Culms** 50–100 cm, erect to spreading; **nodes** 3–5, usually glabrous, rarely pubescent; **internodes** glabrous. **Sheaths** usually glabrous, sometimes pubescent or pilose, especially the lower sheaths, midrib of the culm leaves usually narrowed just below the collar; **auricles** absent; **ligules** 1–3 mm, glabrous, truncate to obtuse, laciniate; **blades** 10–20 cm long, 3–6 mm wide, flat, often glaucous, usually glabrous, sometimes scabrous, basal blades often pubescent. **Panicles** 10–20 cm, open; **branches** ascending and spreading or declining and drooping. **Spikelets** 15–30 mm, elliptic to lanceolate, terete to moderately laterally compressed, with (4)5–10 florets. **Glumes** usually glabrous, rarely slightly pubescent, 3-veined; **lower glumes** 5.5–8 mm; **upper glumes** 6.5–9 mm, often mucronate; **lemmas** 8–12 mm, elliptic to lanceolate, rounded over the midvein, backs pubescent or glabrous, margins usually with longer hairs, apices subulate to obtuse, entire; **awns** 1.5–4 mm, straight, arising less than 1.5 mm below the lemma apices; **anthers** 1.5–3.5 mm. $2n = 14$.

Bromus frondosus grows in open woods and on rocky slopes, at 1500–2500 m. Its range extends from Colorado, Arizona, and New Mexico into Mexico.

19. Bromus anomalus Rupr. *ex* E. Fourn. [p. 215]
MEXICAN BROME

Plants perennial; not rhizomatous. **Culms** 40–90 cm, erect; **nodes** (3)4–7(8), these and the **internodes** pubescent or glabrous. **Sheaths** glabrous or pilose, midrib of the culm leaves abruptly narrowed just below the collar; **auricles** often present on the lower leaves; **ligules** to 1 mm, truncate; **blades** 14–22(26) cm long, to 6 mm wide, flat, glabrous or pilose, not glaucous. **Panicles** 10–20 cm, open, nodding; **branches** ascending or spreading. **Spikelets** (14)15–30 mm, elliptic to lanceolate, terete to moderately laterally compressed, with 4–12 florets. **Glumes** usually pubescent, rarely glabrous; **lower glumes** 5–6 mm, 1–3-veined; **upper glumes** 6–8 mm, 3-veined, sometimes mucronate; **lemmas** 7–10 mm, elliptic to lanceolate, rounded over the midvein, backs and margins pubescent, apices acute to obtuse, entire; **awns** 1–3(5) mm, straight, arising less than 1.5 mm below the lemma apices; **anthers** 2–4 mm. $2n = 14$.

Bromus anomalus grows on rocky slopes in western Texas and adjacent Mexico. Many records of this species in the *Flora* region are here treated as *B. porteri*, a closely related species that has sometimes been included in *B. anomalus*. The main difference is that *B. anomalus* has auricles, and culm leaves with midribs that are narrowed just below the collar.

20. Bromus porteri (J.M. Coult.) Nash [p. 215]
NODDING BROME

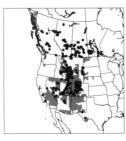

Plants perennial; not rhizomatous. **Culms** 30–100 cm, erect; **nodes** (2)3–4(5), glabrous or pubescent; **internodes** mostly glabrous, puberulent near the nodes. **Sheaths** glabrous or pilose, midrib of the culm leaves not abruptly narrowed just below the collar; **auricles** absent; **ligules** to 2.5 mm, glabrous, truncate or obtuse, erose or lacerate; **blades** (3)10–25(35) cm long, 2–5(6) mm wide, flat, not glaucous, both surfaces

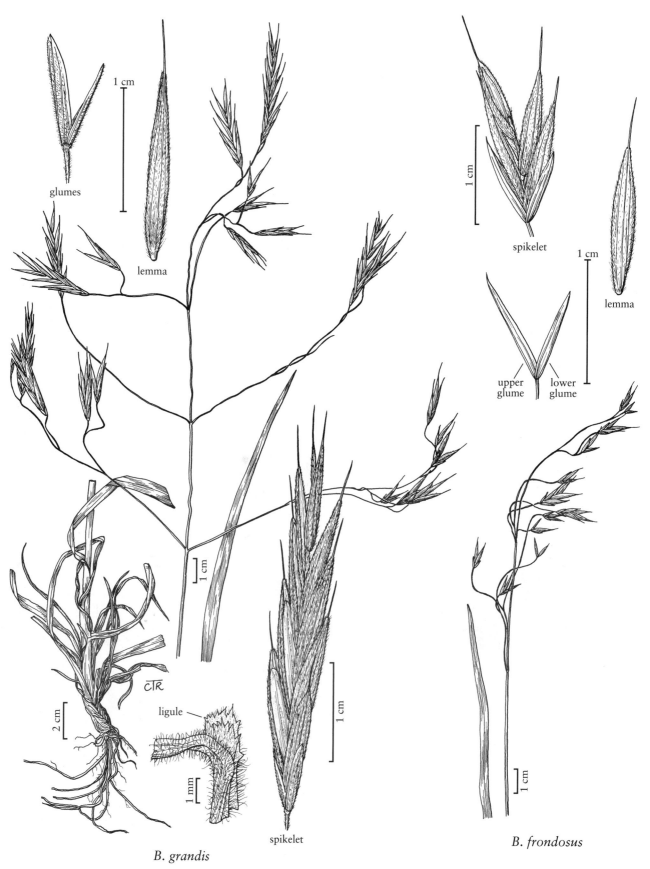

glumes

lemma

1 cm

spikelet

lemma

upper
glume

lower
glume

1 cm

spikelet

1 cm

ligule

1 mm

2 cm

C̄T̄R

B. grandis

B. frondosus

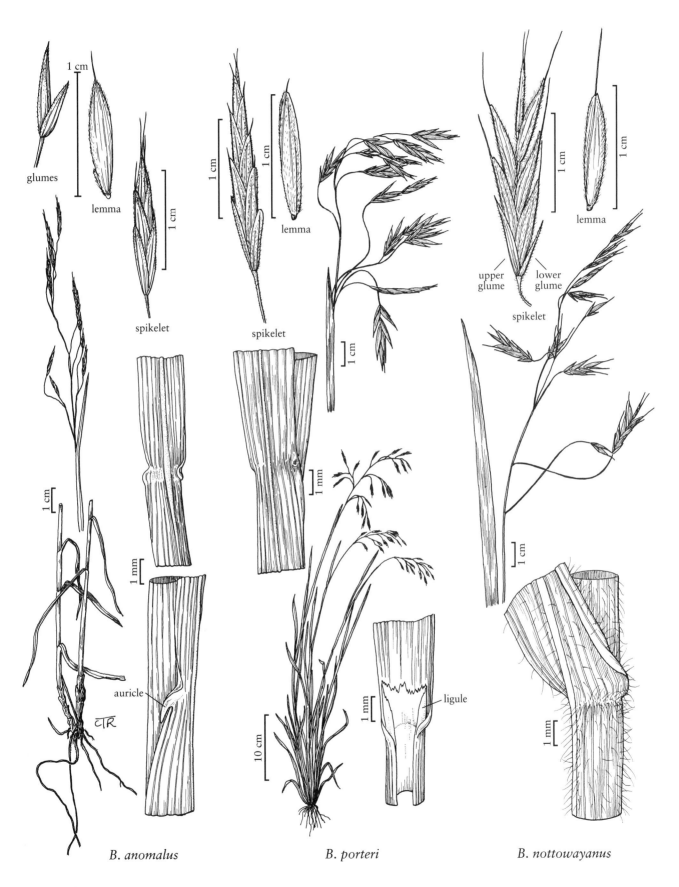

glumes

lemma

spikelet

spikelet

lemma

upper glume

lower glume

spikelet

lemma

auricle

ligule

B. anomalus

B. porteri

B. nottowayanus

BROMUS

usually glabrous, sometimes the adaxial surface pilose. **Panicles** 7–20 cm, open, nodding, often 1-sided; **branches** slender, ascending to spreading, often recurved and flexuous. **Spikelets** 12–38 mm, elliptic to lanceolate, terete to moderately laterally compressed, with (3)5–11(13) florets. **Glumes** usually pubescent, rarely glabrous; **lower glumes** 5–7(9) mm, usually 3-veined, sometimes 1-veined; **upper glumes** 6–10 mm, 3-veined, not mucronate; **lemmas** 8–14 mm, elliptic, rounded over the midvein, usually pubescent or pilose, margins often with longer hairs, backs and margins rarely glabrous, apices acute or obtuse to truncate, entire; **awns** (1)2–3(3.5) mm, straight, arising less than 1.5 mm below the lemma apices; **anthers** (1)2–3 mm. 2*n* = 14.

Bromus porteri grows in montane meadows, grassy slopes, mesic steppes, forest edges, and open forest habitats, at 500–3500 m. It is found from British Columbia to Manitoba, and south to California, western Texas, and Mexico. It is closely related to *B. anomalus*, and has often been included in that species. It differs chiefly in its lack of auricles, and in having culm leaves with midribs that are not narrowed just below the collar.

21. Bromus nottowayanus Fernald [p. 215]
VIRGINIA BROME

Plants perennial; not rhizomatous. **Culms** (60)70–140 cm, erect or spreading; **nodes** 5–9, pubescent or glabrous, often concealed by the sheaths; **internodes** usually glabrous. **Sheaths** usually retrorsely pilose, sometimes glabrous, with a dense line of hairs at the collar, lower sheaths often sericeous; **auricles** absent; **ligules** 0.4–1 mm, often hairy, truncate, erose, ciliolate; **blades** 15–30 cm long, 5–12 mm wide, often shiny yellow-green, flat, abaxial surfaces pilose, adaxial surfaces glabrous or pilose over the veins. **Panicles** 9–25 cm, open, nodding; **branches** ascending or spreading, often recurved. **Spikelets** 18–30 mm, elliptic to lanceolate, terete to moderately laterally compressed, often purplish, with 6–12 florets. **Glumes** usually pubescent; **lower glumes** 5.5–8 mm, 1(3)-veined; **upper glumes** 7–10 mm, 5-veined, often mucronate; **lemmas** 8–13 mm, elliptic to lanceolate, rounded over the midvein, usually uniformly densely hairy, or the backs less densely so, apices acute to obtuse, entire; **awns** 5–8 mm, straight, arising less than 1.5 mm below the lemma apices; **anthers** 2.8–3.5(5) mm. 2*n* = 14.

Bromus nottowayanus is native to the east-central and eastern United States from Iowa to New York, south to Oklahoma, northern Alabama, and Virginia. It

grows in damp, shaded woods, often in ravines and along streams.

22. Bromus texensis (Shear) Hitchc. [p. 217]
TEXAS BROME

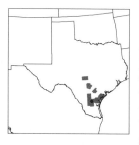

Plants annual. **Culms** 30–70 cm, erect or spreading; **nodes** 3–5, pubescent. **Sheaths** densely pubescent to pilose; **auricles** absent; **ligules** 2–3 mm, lanceolate, pubescent, obtuse, erose; **blades** 7–20 cm long, 3–7 mm wide, flat, usually pubescent to pilose, rarely glabrous. **Panicles** 8–15 cm, open, drooping; **branches** ascending to spreading. **Spikelets** 20–30 mm, elliptic to lanceolate, terete to moderately laterally compressed, with 4–7 florets. **Glumes** glabrous or hispidulous; **lower glumes** 6–9 mm, 1-veined; **upper glumes** 8–10.5 mm, 3-veined, usually acute, rarely mucronate; **lemmas** 9–15 mm, lanceolate, rounded over the midvein, glabrous, sometimes scabrous, apices subulate to acute, entire; **awns** 4–8 mm, straight, arising less than 1.5 mm below the lemma apices; **anthers** 3–5 mm. 2*n* = 28.

Bromus texensis grows in openings in brushy areas on rocky ground. It is rare, found only southern Texas and northern Mexico.

23. Bromus vulgaris (Hook.) Shear [p. 217]
COMMON BROME

Plants perennial; not rhizomatous. **Culms** 60–120 cm, erect or spreading; **nodes** (3)4–6(7), usually pilose; **internodes** glabrous. **Sheaths** pilose or glabrous; **auricles** absent; **ligules** 2–6 mm, glabrous, obtuse or truncate, erose or lacerate; **blades** 13–25(33) cm long, to 14 mm wide, flat, abaxial surfaces usually glabrous, sometimes pilose, adaxial surfaces usually pilose, sometimes glabrous. **Panicles** 10–15 cm, open; **branches** ascending to drooping. **Spikelets** 15–30 mm, elliptic to lanceolate, terete to moderately laterally compressed, with (3)4–9 florets. **Glumes** glabrous or pilose; **lower glumes** 5–8 mm, 1(3)-veined; **upper glumes** 8–12 mm, 3-veined; **lemmas** 8–15 mm, lanceolate, rounded over the midvein, backs sparsely hairy or glabrous, margins usually coarsely pubescent, sometimes glabrous, apices subulate to acute, entire; **awns** (4)6–12 mm, straight, arising less than 1.5 mm below the lemma apices; **anthers** 2–4 mm. 2*n* = 14.

Bromus vulgaris grows in shaded or partially shaded, often damp, coniferous forests along the coast, and inland in montane pine, spruce, fir, and aspen forests, from sea level to about 2000 m. Its range extends from

spikelet

lemma

1 cm

1 cm

upper glume

lower glume

spikelet

B. vulgaris

glumes

lemma

1 cm

1 cm

lemma

spikelet

1 cm

1 cm

ligule

1 mm

ligule

1 mm

1 cm

B. texensis

CTR

B. pacificus

BROMUS

coastal British Columbia eastward to southwestern Alberta and southward to central California, northern Utah, and western Wyoming.

Varieties have been described within *Bromus vulgaris*; because their variation is overlapping, none are recognized here.

24. Bromus pacificus Shear [p. 217]
PACIFIC BROME

Plants perennial; not rhizomatous. **Culms** 60–170 cm, erect; **nodes** (5)6–8, pubescent; **internodes** usually glabrous, sometimes pubescent near the nodes. **Sheaths** pilose, midrib of the culm leaves not abruptly narrowed just below the collar; **auricles** absent; **ligules** 2–4 mm, glabrous, truncate, erose or lacerate; **blades** 20–35(37) mm long, 6–16 mm wide, flat, abaxial surfaces glabrous, adaxial surfaces pilose. **Panicles** 10–25 cm, open, nodding; **branches** ascending, spreading, or drooping. **Spikelets** 20–30 mm, elliptic to lanceolate, terete to moderately laterally compressed, with (4)6–10 florets. **Glumes** pubescent; **lower glumes** 6–8.5 mm, 1(3)-veined; **upper glumes** 8–11.5 mm, 3-veined, not mucronate; **lemmas** 10–12 mm, lanceolate, rounded over the midvein, backs pubescent, margins more densely so, apices acute, entire; **awns** 3.5–7 mm, straight, arising less than 1.5 mm below the lemma apices; **anthers** 2–4 mm. $2n = 28$.

Bromus pacificus grows in moist thickets, openings, and ravines along the Pacific coast from southeastern Alaska to northern California, with a few occurrences further inland.

25. Bromus erectus Huds. [p. 219]
MEADOW BROME, UPRIGHT BROME, BROME DRESSÉ

Plants perennial; not rhizomatous. **Culms** 50–100 cm, erect; **nodes** (1)2–3, usually glabrous, rarely pubescent; **internodes** usually glabrous, rarely pubescent. **Sheaths** glabrous or pilose; **auricles** absent; **ligules** to 1.5 mm, glabrous, truncate, lacerate; **blades** 10–20 cm long, 2–6 mm wide, often involute or folded, glabrous or sparingly hairy. **Panicles** 10–20 cm, erect, contracted; **branches** erect or ascending. **Spikelets** 15–30 mm, elliptic to lanceolate, terete to moderately laterally compressed, with 5–8(12) florets. **Glumes** glabrous; **lower glumes** 7–9 mm, 1-veined; **upper glumes** 9–11 mm, 3-veined; **lemmas** 10–13(15) mm, lanceolate, rounded over the midvein, backs and margins glabrous

or sparsely pubescent, apices subulate to acute, entire; **awns** 5–7 mm, straight, arising less than 1.5 mm below the lemma apices; **anthers** 4–6.5 mm. $2n = 56$.

Bromus erectus is native to Europe. In the *Flora* region, it grows on disturbed soils, often over limestone. It is established in the eastern United States and Canada, and has been reported from other locations where it has not persisted.

26. Bromus suksdorfii Vasey [p. 219]
SUKSDORF'S BROME

Plants perennial; not rhizomatous. **Culms** 50–100 cm, erect; **nodes** 2–3(4), glabrous, **internodes** glabrous or puberulent just below the nodes. **Sheaths** glabrous; **auricles** absent; **ligules** to 1 mm, glabrous, truncate; **blades** (8)12–19(24) cm long, 4–8(14) mm wide, flat, glabrous, margins scabrous. **Panicles** 6–14 cm, erect, contracted; **branches** erect or ascending. **Spikelets** 15–30 mm, elliptic to lanceolate, terete to moderately laterally compressed, with (3)5–7 florets. **Glumes** glabrous or sparsely pubescent; **lower glumes** 7–11 mm, 1(3)-veined; **upper glumes** 9–12 mm, 3-veined; **lemmas** 12–15 mm, elliptic, rounded over the midvein, backs and margins pubescent or nearly glabrous, apices obtuse, entire; **awns** 2–5 mm, straight, arising less than 1.5 mm below the lemma apices; **anthers** 2–3.5 mm. $2n = 14$.

Bromus suksdorfii grows on open slopes and in open subalpine forests, at about 1300–3300 m, from southern Washington to southern California.

27. Bromus mucroglumis Wagnon [p. 219]
SHARPGLUME BROME

Plants perennial; not rhizomatous. **Culms** 50–100 cm, erect or spreading; **nodes** 5–7, pilose or pubescent; **internodes** glabrous. **Basal sheaths** pubescent or pilose, throats pilose; **upper sheaths** pubescent or glabrous, midrib of the culm leaves not abruptly narrowed just below the collar; **auricles** absent; **ligules** 1–2 mm, glabrous, truncate or obtuse; **blades** 20–30 cm long, (4)7–11 mm wide, flat, both surfaces pilose or the abaxial surface glabrous. **Panicles** 10–20 cm, open, nodding; **branches** ascending or spreading. **Spikelets** 20–30 mm, elliptic to lanceolate, terete to moderately laterally compressed, with 5–10 florets. **Glumes** usually pilose or pubescent, rarely glabrous; **lower glumes** 6–8 mm, 1-veined; **upper glumes** 8–8.5 mm, 3-veined, mucronate; **lemmas** 10–12 mm, elliptic to lanceolate,

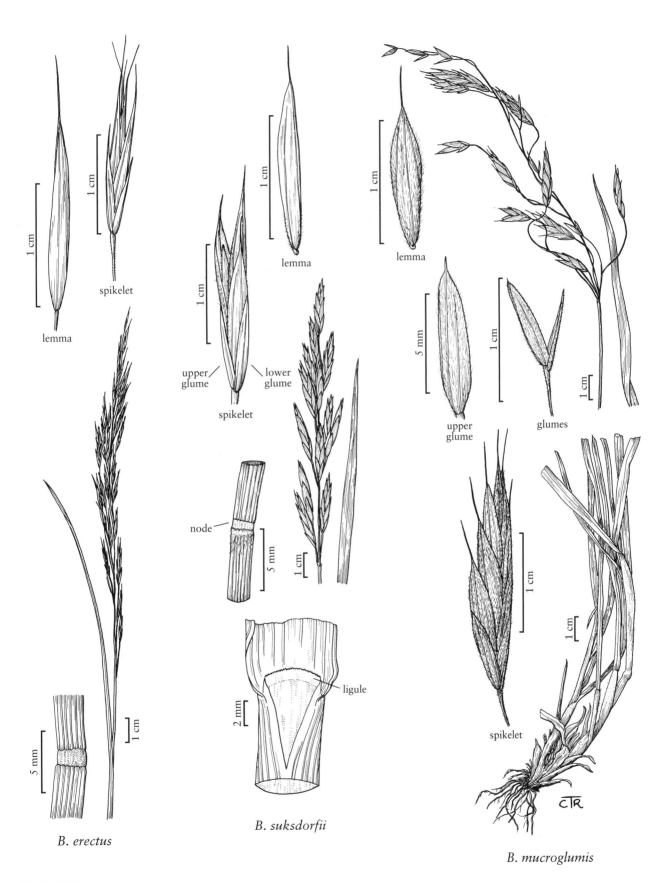

B. erectus

B. suksdorfii

B. mucroglumis

BROMUS

rounded over the midvein, backs and margins pilose or pubescent, apices acute to obtuse, entire; **awns** 3–5 mm, straight, arising less than 1.5 mm below the lemma apices; **anthers** 1.5–3 mm. $2n = 28$.

Bromus mucroglumis grows at 1500–3000 m in the southwestern United States and northern Mexico.

28. Bromus pubescens Muhl. *ex* Willd. [p. 221]
CANADA BROME

Plants perennial; not rhizomatous. **Culms** 65–120(150) cm, erect; **nodes** (3)5–7(9), usually pubescent, sometimes glabrous; **internodes** pubescent or glabrous. **Sheaths** retrorsely pilose, midrib of the culm leaves not abruptly narrowed just below the collar, collars hairy or glabrous; **auricles** absent; **ligules** 0.5–2 mm, glabrous, obtuse to truncate, erose; **blades** 12–32 cm long, 6–15(19) mm wide, flat, 1 or both surfaces glabrous or hairy. **Panicles** 10–25 cm, open, usually nodding; **branches** usually spreading, sometimes ascending, often drooping. **Spikelets** (13)15–30 mm, elliptic to lanceolate, terete to moderately laterally compressed, with (4)5–10(13) florets. **Glumes** usually pubescent, rarely glabrous; **lower glumes** 4–8 mm, 1-veined; **upper glumes** 5–10 mm, 3(5)-veined, not mucronate; **lemmas** 8–12 mm, lanceolate, rounded over the midvein, backs and margins usually hairy, sometimes glabrous or scabrous, apices subulate to acute, entire; **awns** (3)4–7(8) mm, straight, arising less than 1.5 mm below the lemma apices; **anthers** 2–4(5) mm. $2n = 14$.

Bromus pubescens grows in shaded, moist, often upland deciduous woods. Its range is centered in the eastern half of the United States, and extends northward to southern Manitoba, Ontario, and Quebec, westward in scattered locations to Arizona, and southward to eastern Texas and western Florida.

29. Bromus lanatipes (Shear) Rydb. [p. 221]
WOOLY BROME

Plants perennial; not rhizomatous. **Culms** 40–90 cm, erect; **nodes** 3–5(7), mainly pubescent; **internodes** mostly glabrous, puberulent near the nodes. **Basal sheaths** densely pilose or glabrous; **upper sheaths** glabrous or almost so, midrib of the culm leaves not abruptly narrowed just below the collar; **auricles** absent; **ligules** 1–2 mm, glabrous, truncate or obtuse, sometimes lacerate; **blades** 5–20 cm long, to 7 mm wide, flat, both surfaces glabrous, sometimes scabrous. **Panicles** 10–25

cm, open, nodding; **branches** ascending to spreading. **Spikelets** 10–30 mm, elliptic to lanceolate, terete to moderately laterally compressed, with 7–12(16) florets. **Glumes** usually glabrous, sometimes pubescent; **lower glumes** 5–6.5(7) mm, 1(3)-veined; **upper glumes** (6)7–9 mm, 3-veined, not mucronate; **lemmas** 8–11 mm, elliptic, rounded over the midvein, backs and margins pubescent, sometimes nearly glabrous, apices truncate or obtuse, entire, rarely emarginate, lobes shorter than 1 mm; **awns** 2–4 mm, straight, arising less than 1.5 mm below the lemma apices; **anthers** 1.8–4 mm. $2n = 28$.

Bromus lanatipes grows in a wide range of habitats at 800–2500 m, from Wyoming through the southwestern United States to northern Mexico.

30. Bromus ramosus Huds. [p. 221]
HAIRY BROME

Plants perennial; not rhizomatous. **Culms** 40–190 cm, erect; **nodes** 2–4, usually pubescent; **internodes** usually pubescent. **Sheaths** with long, stiff, retrorse hairs, at least on the lower portion, midrib of the culm leaves not abruptly narrowed just below the collar; **auricles** present; **ligules** 2–3.5 mm, glabrous or sparsely pilose, rounded to truncate, erose; **blades** 10–60 cm long, 6–15 mm wide, flat, drooping, glabrous or sparsely hairy. **Panicles** 15–40 cm long, open, lax, drooping; **branches** spreading or drooping. **Spikelets** 20–40 mm, elliptic to lanceolate, terete to moderately laterally compressed, with 3–10 florets. **Glumes** glabrous, scabridulous over the veins; **lower glumes** 5–8 mm, 1-veined; **upper glumes** 8–11 mm, 3-veined, mucronate; **lemmas** 10–14 mm, lanceolate, rounded over the midvein, margins and at least the lower ½ of the backs pubescent, apices acute, entire or emarginate, lobes shorter than 1 mm; **awns** 4–7 mm, straight, arising less than 1.5 mm below the lemma apices; **anthers** 2.5–4 mm. $2n = 14, 28, 42$.

Bromus ramosus is native to Asia, Europe, and northern Africa. It is included here based on Pavlick's (1995) statement that it is found sporadically in the southern and eastern United States; specimens to substantiate his statement have not been located.

spikelet

lemma

B. pubescens

glumes

lemma

spikelet

lemma

upper glume

lower glume

spikelet

B. lanatipes

ligule

auricle

B. ramosus

BROMUS

31. Bromus ciliatus L. [p. 223]
FRINGED BROME, BROME CILIÉ

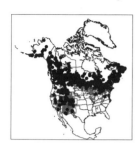

Plants perennial; not rhizomatous. **Culms** 45–120(150) cm, erect; **nodes** (3)4–7(8), all pubescent or the lower nodes sometimes glabrous; **internodes** glabrous. **Basal sheaths** usually retrorsely pilose, sometimes glabrous; **upper sheaths** glabrous, throats glabrous or pilose, midrib of the culm leaves not abruptly narrowed just below the collar; **auricles** sometimes present; **ligules** 0.4–1.4 mm, usually glabrous, rarely pilose, truncate, erose; **blades** 13–25 cm long, 4–10 mm wide, flat, abaxial surfaces usually glabrous, sometimes pilose, adaxial surfaces usually pilose, sometimes glabrous. **Panicles** 10–20 cm, open, nodding; **branches** ascending, spreading, or drooping. **Spikelets** 15–25 mm, elliptic to lanceolate, terete to moderately laterally compressed, with 4–9 florets. **Glumes** glabrous; **lower glumes** 5.5–7.5 mm, 1(3)-veined; **upper glumes** 7.1–8.5 mm, 3-veined, not mucronate; **lemmas** 9.5–14 mm, elliptic to lanceolate, rounded over the midvein, backs glabrous, sometimes scabrous, margins conspicuously hirsute on the lower $^1/_2$–$^2/_3$, apices obtuse to acute, entire; **awns** 3–5 mm, straight, arising less than 1.5 mm below the lemma apices; **anthers** 1–1.4 mm. $2n = 14$.

Bromus ciliatus grows in damp meadows, thickets, woods, and stream banks across almost all of northern North America except the high arctic, extending further south mainly through the western United States to Mexico. Some taxonomists have named plants with different degrees of sheath pubescence as different forms. Because the variation is continuous, such differences are not formally recognized in this treatment.

32. Bromus richardsonii Link [p. 223]
RICHARDSON'S BROME

Plants perennial; not rhizomatous. **Culms** 50–110(145) cm, erect to spreading; **nodes** (3)4–5(6), usually glabrous, sometimes pubescent; **internodes** usually glabrous. **Basal sheaths** often retrorsely pilose; **culm sheaths** glabrous, often tufted-pilose near the auricle position, midrib of the culm leaves not abruptly narrowed just below the collar; **auricles** absent; **ligules** 0.4–2 mm, glabrous, rounded, erose, ciliolate; **blades** 10–35 cm long, 3–12 mm wide, flat, glabrous. **Panicles** 10–20(25) cm, open, nodding; **branches** ascending to spreading or drooping, filiform. **Spikelets** 15–25(40) mm, elliptic to lanceolate, terete to moderately laterally compressed, with (4)6–10(15) florets. **Glumes** usually glabrous, sometimes pubescent; **lower glumes** 7.5–12.5 mm, 1(3)-veined; **upper glumes** 8.9–11.3 mm, 3-veined, often mucronate; **lemmas** 9–14(16) mm, elliptic, rounded over the midvein, margins more or less densely pilose on the lower $^1/_2$ or $^3/_4$, lower lemmas in a spikelet glabrous across the backs, uppermost lemmas with appressed hairs on the backs, apices obtuse, entire; **awns** (2)3–5 mm, straight, arising less than 1.5 mm below the lemma apices; **anthers** 1.6–2.7 mm. $2n = 28$.

Bromus richardsonii grows in meadows and open woods in the upper montane and subalpine zones, at 2000–4000 m in the southern Rocky Mountains, and at lower elevations northwards. Its range extends from southern Alaska to southern California and northern Baja California, Mexico; it is found as far east as Saskatchewan, South Dakota, and western Texas. Specimens with pubescent nodes and glumes are apparently confined to the southwestern United States.

Bromus sect. **Neobromus** (Shear) Hitchc.

Plants annual. **Spikelets** elliptic to lanceolate, more or less terete, with 3–9 florets. **Lower glumes** 1(3)-veined; **upper glumes** 3(5)-veined; **lemmas** lanceolate to linear-lanceolate, rounded over the midvein, apices acuminate, bifid, teeth aristate or acuminate; **awns** geniculate, divaricate, arising 1.5 mm or more below the lemma apices.

Bromus sect. *Neobromus* has two species, both of which are native to South America. *Bromus berteroanus* has become established in the *Flora* region.

spikelet

lemma

glumes

upper
lemma

lower
lemma

glumes

spikelet

lemma

spikelet

upper
glume

lower
glume

B. ciliatus

B. richardsonii

B. berteroanus

CTR

BROMUS

33. Bromus berteroanus Colla [p. 223]
CHILEAN CHESS

Plants annual; often tufted. Culms 30–60 cm, slender. Sheaths pilose-pubescent to nearly glabrous; blades 7–28 cm long, 2–9 mm wide, pilose or glabrous. Panicles 10–20 cm long, 3–9 cm wide, erect, dense; branches appressed to spreading, sometimes flexuous. Spikelets 15–20 mm, elliptic to lanceolate, more or less terete, with 3–9 florets. Glumes glabrous, acuminate; lower glumes 8–10 mm, 1-veined; upper glumes 12–16 mm, 3(5)-veined; lemmas 11–14 mm, lanceolate to linear-lanceolate, sparsely pubescent, 5-veined, rounded over the midvein, apices acuminate, bifid, teeth 2–3 mm, usually aristate, sometimes acuminate; awns 13–20 mm, geniculate, strongly to moderately twisted in the basal portion, arising 1.5 mm or more below the lemma apices; anthers 2–2.5 mm. 2*n* = unknown.

Bromus berteroanus is from Chile, and can now be found in dry areas in western North America, including British Columbia, Montana, California, Nevada, Arizona, southwestern Utah, and Baja California, Mexico.

Bromus sect. Genea Dumort.

Plants annual. Spikelets with parallel or diverging sides in outline, terete to moderately laterally compressed, with 4–11 florets. Lower glumes 1–3-veined; upper glumes 3–5-veined; lemmas lanceolate to linear-lanceolate, rounded over the midvein, apices acuminate, teeth 0.8–5 mm; awns straight or arcuate, arising 1.5 mm or more below the lemma apices.

Bromus sect. *Genea* is native to Europe and northern Africa; five of its six species are established in the *Flora* region.

34. Bromus diandrus Roth [p. 225]
GREAT BROME, RIPGUT GRASS

Plants annual. Culms 20–90 cm, erect or decumbent, puberulent below the panicle. Sheaths softly pilose, hairs often retrorse or spreading; auricles absent; ligules 2–3 mm, glabrous, obtuse, lacerate or erose; blades 3.5–27 cm long, 1–9 mm wide, both surfaces pilose. Panicles 13–25 cm long, 2–12 cm wide, erect to spreading; branches 1–7 cm, stiffly erect to ascending or spreading, with 1 or 2 spikelets. Spikelets 25–70 mm, sides parallel or diverging distally, moderately laterally compressed, with 4–11 florets. Glumes smooth or scabrous, margins hyaline; lower glumes 15–25 mm, 1–3-veined; upper glumes 20–35 mm, 3–5-veined; lemmas 20–35 mm, linear-lanceolate, scabrous, 7-veined, rounded over the midvein, margins hyaline, apices bifid, acuminate, teeth 3–5 mm; awns 30–65 mm, straight, arising 1.5 mm or more below the lemma apices; anthers 0.5–1 mm. 2*n* = 42, 56.

Bromus diandrus is native to southern and western Europe. It is now established in North America, where it grows in disturbed ground, waste places, fields, sand dunes, and limestone areas. It occurs from southwestern British Columbia to Baja California, Mexico, and eastward to Montana, Colorado, Texas, and scattered locations in the eastern United States. The common name 'ripgut grass' indicates the effect it has on animals if they consume the sharp, long-awned florets of this species.

Bromus diandrus, as treated here, includes *B. rigidus* Roth. Sales (1993) reduced these two taxa to varietal rank, pointing out that the differences between them in panicle morphology and callus and scar shape are subtle enough that identification of many specimens beyond *B. diandrus sensu lato* is often impossible.

35. Bromus sterilis L. [p. 225]
BARREN BROME

Plants annual. Culms 35–100 cm, erect or geniculate near the base, glabrous. Sheaths densely pubescent; auricles absent; ligules 2–2.5 mm, glabrous, acute, lacerate; blades 4–20 cm long, 1–6 mm wide, pubescent on both surfaces. Panicles 10–20 cm long, 5–12 cm wide, open; branches 2–10 cm, spreading, ascending or drooping, rarely with more than 3 spikelets. Spikelets 20–35 mm, usually shorter than the panicle branches, sides parallel or diverging distally, moderately laterally compressed, with 5–9 florets. Glumes smooth or scabrous, margins hyaline; lower glumes 6–14 mm, 1(3)-veined; upper glumes 10–20 mm, 3(5)-veined;

glumes

1 cm

glumes

1 cm

1 cm

spikelet

1 cm

lemma

1 cm

ligule

2 mm

CTR

B. diandrus

glumes

1 cm

lemma

upper glume

lower glume

spikelet

5 cm

B. sterilis

1 cm

1 cm

1 cm

lemma

spikelet

2 mm

B. tectorum

lemmas 14–20 mm, narrowly lanceolate, pubescent or puberulent, 7(9)-veined, rounded over the midvein, margins hyaline, apices acuminate, bifid, teeth 1–3 mm; **awns** 15–30 mm, straight, arising 1.5 mm or more below the lemma apices; **anthers** 1–1.4 mm. $2n = 14, 28$.

Bromus sterilis is native to Europe, growing from Sweden southward. In the *Flora* region, it grows in road verges, waste places, fields, and overgrazed rangeland. It is widespread in western and eastern North America, but is mostly absent from the Great Plains and the southeastern states.

36. Bromus tectorum L. [p. 225]
CHEATGRASS, DOWNY CHESS

Plants annual. **Culms** 5–90 cm, erect, slender, puberulent below the panicle. **Sheaths** usually densely and softly retrorsely pubescent to pilose, upper sheaths sometimes glabrous; **auricles** absent; **ligules** 2–3 mm, glabrous, obtuse, lacerate; **blades** to 16 cm long, 1–6 mm wide, both surfaces softly hairy. **Panicles** 5–20 cm long, 3–8 cm wide, open, lax, drooping distally, usually 1-sided; **branches** 1–4 cm, drooping, usually 1-sided and longer than the spikelets, usually at least 1 branch with 4–8 spikelets. **Spikelets** 10–20 mm, usually shorter than the panicle branches, sides parallel or diverging distally, moderately laterally compressed, often purplish-tinged, not densely crowded, with 4–8 florets. **Glumes** villous, pubescent, or glabrous, margins hyaline; **lower glumes** 4–9 mm, 1-veined; **upper glumes** 7–13 mm, 3–5-veined; **lemmas** 9–12 mm, lanceolate, glabrous or pubescent to pilose, 5–7-veined, rounded over the midvein, margins hyaline, often with some hairs longer than those on the backs, apices acuminate, hyaline, bifid, teeth 0.8–2(3) mm; **awns** 10–18 mm, straight, arising 1.5 mm or more below the lemma apices; **anthers** 0.5–1 mm. $2n = 14$.

Bromus tectorum is a European species that is well established in the *Flora* region and other parts of the world. It grows in disturbed sites, such as overgrazed rangelands, fields, sand dunes, road verges, and waste places. In the southwestern United States, *Bromus tectorum* is considered a good source of spring feed for cattle, at least until the awns mature. It is highly competitive and dominates rapidly after fire, especially in sagebrush areas. The resulting dense, fine fuels permanently shorten the fire-return interval, further hindering reestablishment of native species. It now dominates large areas of the sagebrush ecosystem of the western *Flora* region.

Specimens with glabrous spikelets have been called *Bromus tectorum* f. *nudus* (Klett & Richt.) H. St. John.

They occur throughout the range of the species, and are not known to have any other distinguishing characteristics. For this reason, they are not given formal recognition in this treatment.

37. Bromus madritensis L. [p. 227]
COMPACT BROME

Plants annual. **Culms** 34–70 cm, erect or ascending, glabrous or puberulent below the panicle. **Sheaths** densely short-pubescent or glabrous; **auricles** absent; **ligules** 1.5–2 mm, glabrous, obtuse, erose; **blades** 4–20 cm long, 1–5 mm wide, flat, both surfaces pubescent or glabrous. **Panicles** 3–15 cm long, 2–6 cm wide, open, erect; **branches** (at least some) 1–3+ cm, ascending to spreading, never drooping, usually visible, with 1 or 2 spikelets. **Spikelets** 30–50 mm, longer than the panicle branches, not densely crowded, with parallel sides or widening distally, moderately laterally compressed, with 6–10 florets. **Glumes** pilose, margins hyaline; **lower glumes** 5–10 mm, 1-veined; **upper glumes** 10–15 mm, 3-veined; **lemmas** 12–20 mm, linear-lanceolate, often arcuate, pubescent, with longer hairs near the margins, 5–7-veined, rounded over the midvein, margins hyaline, apices acuminate, teeth 1.5–3 mm; **awns** 12–23 mm, straight or arcuate, arising 1.5 mm or more below the lemma apices; **anthers** 0.5–1 mm. $2n = 28$.

Bromus madritensis is native to southern and western Europe. It is now established in North America, and grows in disturbed soil, waste places, banks, and road verges in southern Oregon, California, and Arizona.

38. Bromus rubens L. [p. 227]
FOXTAIL CHESS, RED BROME

Plants annual. **Culms** 10–40 cm, erect or ascending, often puberulent below the panicle. **Sheaths** softly pubescent to pilose; **auricles** absent; **ligules** 1–3(4) mm, pubescent, obtuse, lacerate; **blades** to 15 cm long, 1–5 mm wide, flat, pubescent on both surfaces. **Panicles** 2–10 cm long, 2–5 cm wide, erect, dense, often reddish brown; **branches** 0.1–1 cm, ascending, never drooping, not readily visible, with 1 or 2 spikelets. **Spikelets** 18–25 mm, much longer than the panicle branches, densely crowded, subsessile, with parallel sides or widening distally, moderately laterally compressed, with 4–8 florets. **Glumes** pilose, margins hyaline; **lower glumes**

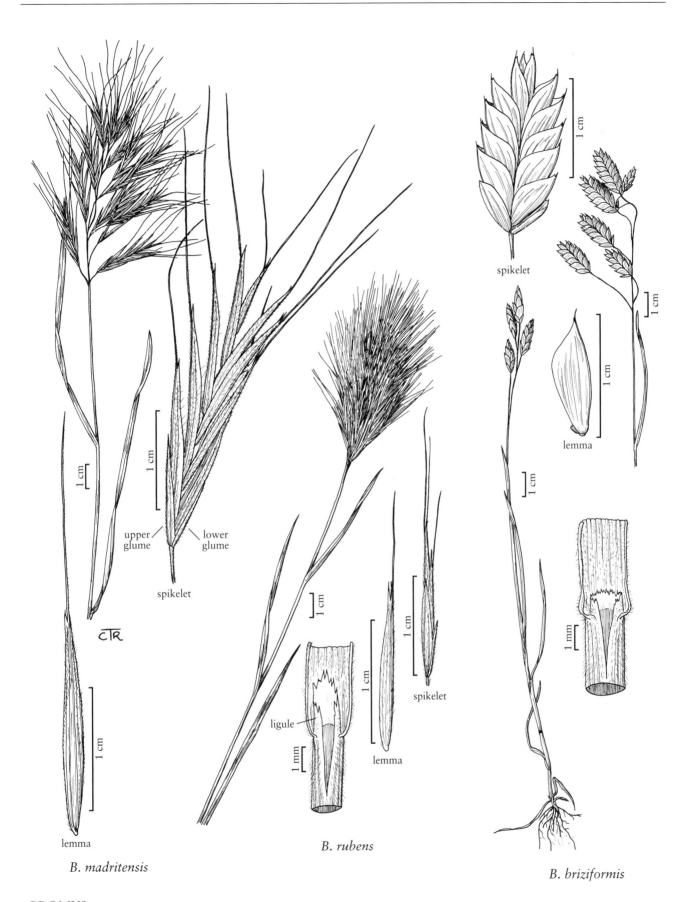

upper
glume
lower
glume

spikelet

CTR

lemma

B. madritensis

ligule

1 mm

lemma

spikelet

B. rubens

spikelet

lemma

1 mm

B. briziformis

BROMUS

5–8 mm, 1(3)-veined; **upper glumes** 8–12 mm, 3–5-veined; **lemmas** 10–15 mm, linear-lanceolate, pubescent to pilose, 7-veined, rounded over the midvein, margins hyaline, apices acuminate, teeth 1–3 mm; **awns** 8–20 mm, straight, reddish, arising 1.5 mm or more below the lemma apices; **anthers** 0.5–1 mm. $2n = 14, 28$.

Bromus rubens is native to southern and southwestern Europe. It now grows in North America in disturbed ground, waste places, fields, and rocky slopes, from southern Washington to southern California, eastward to Idaho, New Mexico, and western Texas. It was found in Massachusetts before 1900 in wool waste used on a crop field; it is not established there. The record from New York represents a rare introduction; it is not known whether it is established.

Bromus L. sect. Bromus

Plants usually annual, sometimes biennial. **Spikelets** with parallel or converging sides in outline, terete to moderately laterally compressed, with 4–30 florets. **Lower glumes** 3–5-veined; **upper glumes** 5–9-veined; **lemmas** elliptic, lanceolate, obovate, or rhombic, rounded over the midvein, apices subulate, acute, acuminate, or obtuse, notched, minutely bifid, or toothed, teeth shorter than 1 mm, apices sometimes split and the teeth appearing longer; **awns** (0)1(3), straight or flexuous, recurved or divaricate.

Bromus sect. *Bromus* has about 40 species that are native to Eurasia, northern Africa, and Australia; 14 species have been introduced to the *Flora* region.

39. Bromus briziformis Fisch. & C.A. Mey. [p. 227]
RATTLESNAKE BROME

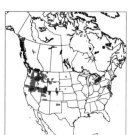

Plants annual. **Culms** 20–62 cm, erect or ascending. **Sheaths** densely pilose; **ligules** 0.5–2 mm, hairy, obtuse, erose; **blades** 3–13 cm long, 2–4 mm wide, pilose to pubescent on both surfaces. **Panicles** 5–15 cm long, 3–7 cm wide, open, secund, nodding; **branches** sometimes longer than the spikelets, curved to reflexed. **Spikelets** 15–27 mm long, 8–12 mm wide, ovate, laterally compressed; **florets** 7–15, bases concealed at maturity; **rachilla internodes** concealed at maturity. **Glumes** smooth or scabridulous; **lower glumes** 5–6 mm, 3–5-veined; **upper glumes** 6–8 mm, 7–9-veined; **lemmas** 9–10 mm long, 6–8 mm wide, inflated, obovate or rhombic, coriaceous, smooth or scabridulous, obscurely 9-veined, rounded over the midvein, margins hyaline, 1–1.3 mm wide, abruptly angled, not inrolled at maturity, apices acute to obtuse, bifid, teeth shorter than 1 mm; **awns** usually absent, sometimes to 1 mm, arising less than 1.5 mm below the lemma apices; **anthers** 0.7–1 mm. **Caryopses** equaling or shorter than the paleas, thin, weakly inrolled or flat. $2n = 14$.

Bromus briziformis grows in waste places, road verges, and overgrazed areas. It is native to southwest Asia and Europe, and is adventive in the *Flora* region, occurring from southern British Columbia to as far south as New Mexico, and in scattered locations eastward. The unique shape of its spikelets has led to its use in dried flower arrangements and as a garden ornamental. The common name may refer to the similarity of the spikelets to a rattlesnake's tail.

40. Bromus arvensis L. [p. 229]
FIELD BROME

Plants annual. **Culms** 80–110 cm, erect. **Lower sheaths** with dense, soft, appressed hairs; **ligules** 1–1.5 mm, hairy, obtuse, erose; **blades** 10–20 cm long, 2–6 mm wide, coarsely pilose on both surfaces. **Panicles** 11–30 cm long, 4–20 cm wide, open, erect or nodding; **branches** usually longer than the spikelets, ascending to widely spreading, slender, slightly curved or straight. **Spikelets** 10–25 mm, lanceolate, terete to moderately laterally compressed, often purple-tinged; **florets** 4–10, bases concealed or visible at maturity; **rachilla internodes** concealed or visible at maturity. **Glumes** glabrous; **lower glumes** 4–6 mm, 3-veined; **upper glumes** 5–8 mm, 5-veined; **lemmas** 7–9 mm long, 1.1–1.5 mm wide, lanceolate, obscurely 7-veined, rounded over the midvein, glabrous, coriaceous, margins slightly angled, inrolled or not at maturity, apices acute, bifid, teeth shorter than 1 mm; **awns** 6–11 mm, straight, arising at varying distances below the lemma apices; **anthers** 2.5–5 mm. **Caryopses** shorter than the paleas, weakly to strongly inrolled. $2n = 14$.

Bromus arvensis grows along roadsides and in fields and waste places at scattered locations in the *Flora* region. It is native to southern and south-central Europe.

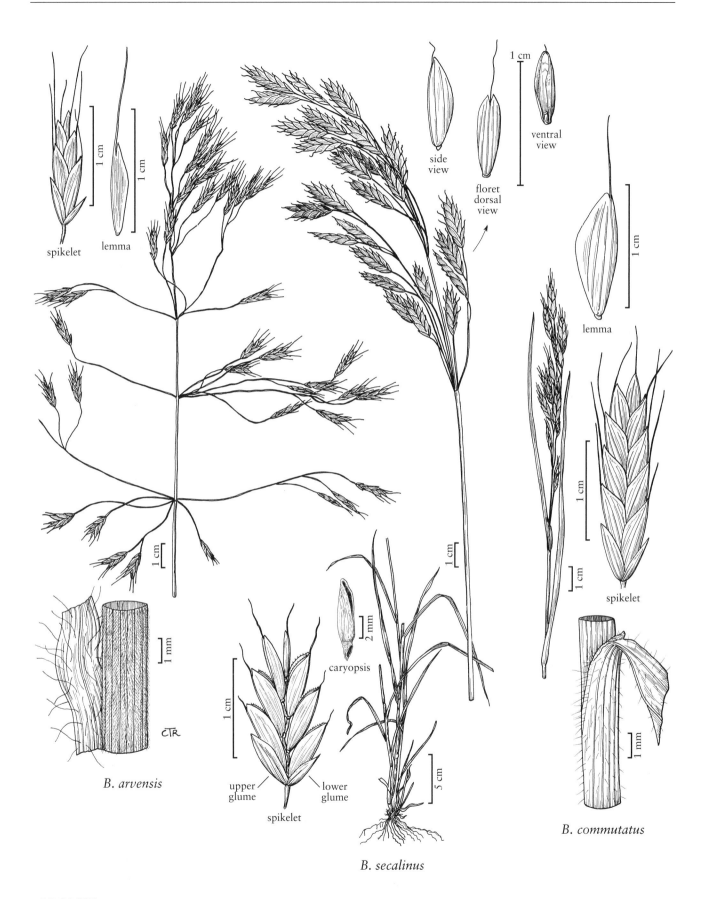

side view

floret dorsal view

ventral view

1 cm

lemma

spikelet

lemma

spikelet

B. commutatus

1 cm

1 mm

CTR

B. arvensis

caryopsis

2 mm

upper glume

lower glume

spikelet

5 cm

B. secalinus

1 mm

BROMUS

41. **Bromus secalinus** L. [p. 229]
RYEBROME, BROME DES SEIGLES

Plants annual. Culms 20–80 (120) cm, erect. **Lower sheaths** glabrous or loosely pubescent and glabrate; **ligules** 2–3 mm, glabrous, obtuse; **blades** 15–30 cm long, 2–4 mm wide, abaxial surfaces pilose or glabrous, adaxial surfaces pilose. **Panicles** 5–23 cm long, 2.5–12 cm wide, open, nodding; **branches** spreading to ascending; **lower branches** slightly drooping, often secund after anthesis, not sinuous. **Spikelets** 10–20 mm, shorter than at least some panicle branches, ovoid-lanceolate or ovate, laterally compressed, not purple-tinged; **florets** 4–9(10), ascending-spreading after flowering, bases visible at maturity; **rachilla internodes** visible at maturity. **Glumes** scabrous or glabrous; **lower glumes** 4–6 mm, 3–5-veined; **upper glumes** 6–7 mm, 7-veined; **lemmas** 6.5–8.5(10) mm long, 1.7–2.5 mm wide, elliptic, coriaceous, obscurely 7-veined, rounded over the midvein, backs usually glabrous, sometimes pubescent, scabrous to puberulent on the margins and near the apices, margins evenly rounded, inrolled at maturity, apices acute to obtuse, bifid, teeth shorter than 1 mm; **awns** (0)3–6(9.5) mm, straight or flexuous, arising less than 1.5 mm below the lemma apices; **anthers** 1–2 mm. **Caryopses** equaling the paleas, thick, strongly inrolled at maturity. $2n = 28$.

Bromus secalinus is native to Europe. It is widespread in the *Flora* region, where it grows in fields, on waste ground, and along roadsides. Specimens with pubescent spikelets may be called *B. secalinus* var. *velutinus* (Schrad.) W.D.J. Koch.

42. **Bromus commutatus** Schrad. [p. 229]
MEADOW BROME, HAIRY CHESS

Plants annual. Culms 40–120 cm, erect or ascending. **Lower sheaths** densely hairy, hairs stiff, often retrorse; **upper sheaths** pubescent or glabrous; **ligules** 1–2.5 mm, glabrous or pilose, obtuse, ciliolate; **blades** 9–18 cm long, 2–4 mm wide, pilose on both surfaces. **Panicles** 7–16 cm long, 3–6 cm wide, open, erect to ascending; **branches** sometimes longer than the spikelets, slender, ascending to spreading. **Spikelets** 14–18(30) mm, oblong-lanceolate, terete to moderately laterally compressed, not purple-tinged; **florets** 4–9(11), bases concealed or visible at maturity; **rachilla internodes** 1.5–2 mm, concealed or visible at maturity. **Glumes** usually glabrous, sometimes scabrous or pubescent; **lower glumes** 5–7 mm, 5-veined; **upper glumes** 6–9 mm, 7(9)-veined; **lemmas** 8–11.5 mm long, 1.7–2.6 mm wide, elliptic to lanceolate, coriaceous, backs usually glabrous, distinctly 7(9)-veined, not ribbed, rounded over the midvein, margins scabrous or pubescent, bluntly angled, inrolled or not at maturity, apices acute to obtuse, bifid, teeth shorter than 1 mm; **awns** 3–10 mm, straight, arising less than 1.5 mm below the lemma apices, awn of the lowest lemma shorter than the others; **anthers** 0.7–1.7 mm. **Caryopses** equaling or shorter than the paleas, weakly to strongly inrolled. $2n = 14, 28, 56$.

Bromus commutatus grows in fields, waste places, and road verges. It is native to Europe and the Baltic region; in the *Flora* region, it is found mainly in the United States and southern Canada. Hildemar Scholz (pers. comm.) recognizes three subspecies of *B. commutatus* in Europe; no attempt has been made to determine which subspecies are present in the *Flora* region.

43. **Bromus lepidus** Holmb. [p. 231]
SCALY BROME

Plants annual, rarely biennial. Culms 5–60 cm, erect. **Sheaths** pilose; **ligules** 0.5–1 mm, hairy, obtuse; **blades** 3–13 cm long, 2–4 mm wide. **Panicles** 2–10 cm long, 1.5–3 cm wide, erect, contracted or loose; **branches** shorter than the spikelets, ascending, slightly curved or straight. **Spikelets** 6–15 mm, lanceolate, shiny, terete to moderately laterally compressed; **florets** 5–12, bases concealed at maturity; **rachilla internodes** concealed at maturity. **Glumes** glabrous; **lower glumes** 4–4.6 mm, 3–5-veined; **upper glumes** 5.2–5.4 mm, 7-veined; **lemmas** 4.5–6.5 mm long, 1.5–1.7 mm wide, elliptic, glabrous, distinctly 7-veined, rounded over the midvein, margins broadly hyaline, sharply angled, not inrolled at maturity, apices notched, notch at least 0.6 mm deep; **awns** 2–6 mm, straight, arising from the base of the apical notch but less than 1.5 mm below the lemma apices; **anthers** 0.5–2 mm. **Caryopses** longer than the paleas, thin, weakly inrolled or flat. $2n = 28$.

Bromus lepidus grows in fields and waste places. It is native to Europe, and is reported from New York and Massachusetts; it probably also occurs elsewhere in the *Flora* region.

Bromus lepidus often resembles *B. hordeaceus* subsp. *pseudothominei* in lemma characteristics (e.g., length, smoothness, and margin angle), so that either may be misinterpreted. *Bromus lepidus* differs in the wide apical notch of its lemmas, and the length of its caryopses relative to the paleas.

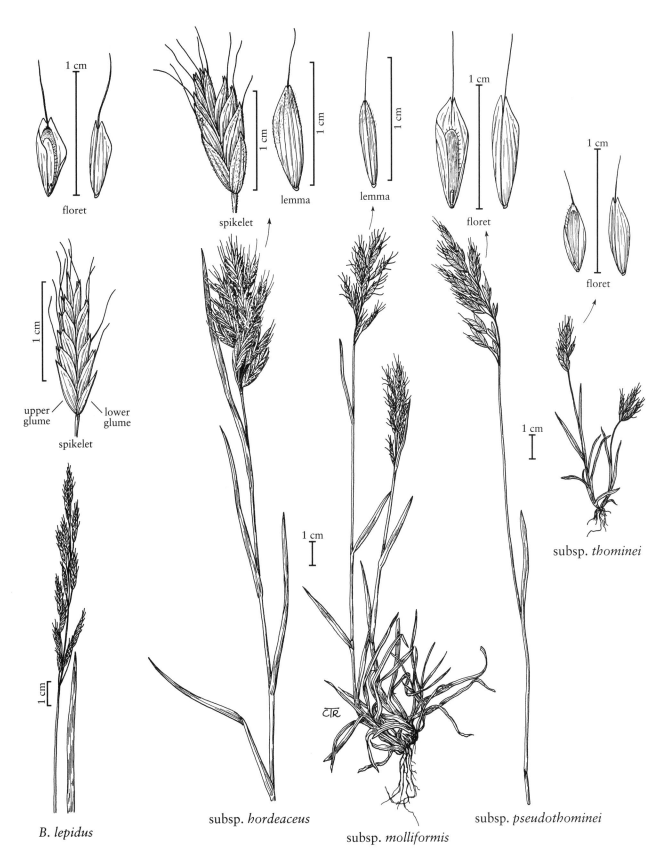

floret

spikelet

lemma

upper glume

lower glume

spikelet

lemma

floret

floret

B. lepidus

subsp. *hordeaceus*

subsp. *molliformis*

subsp. *pseudothominei*

subsp. *thominei*

B. hordeaceus

BROMUS

44. Bromus hordeaceus L. [p. 231]

LOPGRASS, BROME MOU

Plants annual or biennial. **Culms** 2–70 cm, erect or ascending. **Lower sheaths** densely, often retrorsely, pilose; **upper sheaths** pubescent or glabrous; **ligules** 1–1.5 mm, hairy, obtuse, erose; **blades** 2–19 cm long, 1–4 mm wide, abaxial surfaces glabrous or hairy, adaxial surfaces hairy. **Panicles** 1–13 cm long, 1–4 cm wide, erect, usually ovoid, open, becoming dense, occasionally reduced to 1 or 2 spikelets; **branches** shorter than the spikelets, ascending to erect, straight or almost so. **Spikelets** (11)14–20(23) mm, lanceolate, terete to moderately laterally compressed; **florets** 5–10, bases concealed at maturity; **rachilla internodes** concealed at maturity. **Glumes** pilose or glabrous; **lower glumes** 5–7 mm, 3–5-veined; **upper glumes** 6.5–8 mm, 5–7-veined; **lemmas** 6.5–11 mm long, 3–5 mm wide, lanceolate, chartaceous, antrorsely pilose to pubescent, or glabrous proximally or throughout, 7–9-veined, lateral veins prominently ribbed, rounded over the midvein, hyaline margins abruptly or bluntly angled, not inrolled at maturity, apices acute, bifid, teeth shorter than 1 mm; **awns** 6–8 mm, usually arising less than 1.5 mm below the lemma apices, straight to recurved at maturity; **anthers** 0.6–1.5 mm. **Caryopses** equaling or shorter than the paleas, thin, weakly inrolled to flat. 2*n* = 28.

Bromus hordeaceus is native to southern Europe and northern Africa. It is weedy, growing in disturbed areas such as roadsides, fields, sandy beaches, and waste places, and can be found in many locations in the *Flora* region, with the exception of the central Canadian provinces and most of the southeastern United States. Its origin is obscure. Ainouche et al. (1999) reviewed various suggestions, and concluded that at least one of its diploid ancestors may have been an extinct or undiscovered species related to *B. caroli-henrici*, a diploid species.

The four subspecies are usually morphologically distinct. Ainouche et al. (1999), however, found no evidence of genetic differentiation among them.

1. Lemmas (7)8–11 mm long, usually pubescent or pilose.
   2. Awns more than 0.1 mm wide at the base, straight, erect; culms (3)10–70 cm long . . . . subsp. *hordeaceus*
   2. Awns less than 0.1 mm wide at the base, often divaricate or recurved at maturity; culms 15–25(60) cm long . . . . subsp. *molliformis*
1. Lemmas 6.5–8(9) mm long, glabrous or pubescent.
   3. Culms (3)10–70 cm long; panicles up to 10 cm long, usually with more than 1 spikelet;

lemmas usually glabrous; caryopses usually as long as the paleas; habitat various . . . . subsp. *pseudothomineii*
   3. Culms 2–16 cm long; panicles 1–3 cm long, often reduced to 1 spikelet; lemmas pubescent or glabrous; caryopses shorter than the paleas; plants of maritime or lacustrine sands . . . . subsp. *thominei*

Bromus hordeaceus L. subsp. hordeaceus [p. 231]

Culms (3)10–70 cm. **Panicles** (3)5–10 cm, usually with more than 1 spikelet. **Lemmas** (7)8–11 mm, usually pilose or pubescent, margins bluntly angled; **awns** more than 0.1 mm wide at the base, straight, erect. **Caryopses** shorter than the paleas.

Bromus hordeaceus subsp. *hordeaceus* grows throughout the range of the species, being most prevalent in southwestern British Columbia, the western United States, and the northeastern coast.

Bromus hordeaceus subsp. molliformis (J. Lloyd *ex* Billot) Maire & Weiller [p. 231]

Culms 15–25(60) cm. **Panicles** to 10 cm, usually with more than 1 spikelet. **Lemmas** (7)8–11 mm, pubescent, margins rounded; **awns** less than 0.1 mm wide at the base, often divaricate or recurved at maturity. **Caryopses** shorter than the paleas.

Bromus hordeaceus subsp. *molliformis* grows in California and other scattered locations, including Idaho, New Mexico, and southern Michigan.

Bromus hordeaceus subsp. pseudothominei (P. M. Sm.) H. Scholz [p. 231]

Culms (3)10–70 cm. **Panicles** to 10 cm, usually with more than 1 spikelet. **Lemmas** 6.5–8(9) mm, usually glabrous, margins often abruptly angled; **awns** straight, erect. **Caryopses** usually as long as the paleas.

Bromus hordeaceus subsp. *pseudothominei* grows sporadically throughout the range of the species in the *Flora* region. Hitchcock (1951) included it in *B. racemosus*. *Bromus hordeaceus* subsp. *pseudothominei* often resembles *B. lepidus* in lemma characteristics (e.g., length, smoothness, and margin angle), so that either may be misinterpreted.

Bromus hordeaceus subsp. thominei (Hardouin) Braun-Blanq. [p. 231]

Culms 2–16 cm. **Panicles** 1–3 cm, often reduced to 1 spikelet. **Lemmas** 6.5–7.5 mm, pubescent or glabrous, margins bluntly angled; **awns** sometimes divaricate at maturity. **Caryopses** shorter than the paleas.

Bromus hordeaceus subsp. *thominei* grows along the Pacific coast of Canada, from the Queen Charlotte Islands to Vancouver Island, as well as at inland locations in British Columbia; it has also been recorded from California, Massachusetts, and Rhode Island.

45. Bromus racemosus L. [p. 234]
SMOOTH BROME, BROME À GRAPPES

Plants annual. Culms 20–110 cm, erect or ascending. **Lower sheaths** densely hairy, hairs stiff, often retrorse; **upper sheaths** glabrous or pubescent; **ligules** 1–2 mm, glabrous or hairy, erose; **blades** 7–18 cm long, 1–4 mm wide, pilose on both surfaces. **Panicles** 4–16 cm long, 2–3 cm wide, erect, open; **branches** sometimes longer than the spikelets, slender, usually ascending, slightly curved or straight. **Spikelets** 12–20 mm, lanceolate, terete to moderately laterally compressed; **florets** 5–6, bases concealed at maturity; **rachilla internodes** 1–1.5 mm, concealed at maturity. **Glumes** smooth to scabrous; **lower glumes** 4–6 mm, (3)5-veined; **upper glumes** 4–7 mm, 7-veined; **lemmas** 6.5–8 mm long, 3–4.5 mm wide, elliptic to lanceolate, coriaceous, backs smooth, distinctly 7(9)-veined, not ribbed, rounded over the midvein, margins scabrous, rounded, not inrolled at maturity, apices acute to obtuse, bifid, teeth shorter than 1 mm; **awns** 5–9 mm, all more or less equal in length, straight, arising less than 1.5 mm below the lemma apices; **anthers** 1.5–3 mm. **Caryopses** shorter than the paleas, thin, weakly inrolled or flat. 2*n* = 28.

Bromus racemosus grows in fields, waste places, and road verges. It is native to western Europe and the Baltic region, and occurs throughout much of southern Canada and the United States. Hitchcock (1951) included *B. hordeaceus* subsp. *pseudothominei* in *B. racemosus*.

46. Bromus danthoniae Trin. *ex* C.A. Mey. [p. 234]
THREE-AWNED BROME

Plants annual. Culms 5–40 cm, erect or ascending. **Sheaths** glabrous or pubescent; **ligules** 1.2–2.6 mm, puberulent, obtuse, laciniate; **blades** 2–15 cm long, 2–5 mm wide, pubescent on both surfaces. **Panicles** 2–12 cm long, 1–5 cm wide, dense, ovoid, stiffly erect, sometimes racemose; **branches** shorter than the spikelets, ascending, slightly curved or straight. **Spikelets** 10–40(45) mm long, 4–10 mm wide, lanceolate to elliptic or oblong, laterally compressed; **florets** 5–8(10), bases concealed at maturity; **rachilla internodes** concealed at maturity. **Glumes** glabrous or pubescent; **lower glumes** 5–8.5 mm, 3–5-veined, lanceolate; **upper glumes** 6.5–9.5 mm, 7–9(11)-veined, elliptic; **lemmas** 8–12(13.5) mm long, 6–7 mm wide, oblanceolate, veins glabrous, scabridulous, or ciliolate, glabrous or pubescent elsewhere, 9–11-veined, rounded over the midvein, margins broadly hyaline, bluntly angled above the middle, not inrolled at maturity, apices subulate to acute or obtuse, toothed, teeth shorter than 1 mm; **awns** usually 3 on the upper lemmas in each spikelet, arising 2–4 mm below the lemma apices, purple or deep red, central awn 5–25 mm, flattened at the base, divaricate and sometimes twisted at maturity, lateral awns 4–10 mm, erect or reflexed, sometimes absent or much reduced on the lower lemmas; **anthers** 1–1.8 mm. **Caryopses** equaling or shorter than the paleas, thin, weakly inrolled or flat. 2*n* = 14.

Bromus danthoniae is native from the western Asia to southern Russia and Tibet. It was collected in 1904 in Ontario; no other North American collections are known.

47. Bromus caroli-henrici Greuter [p. 234]

Plants annual. Culms 10–40 cm, erect or ascending. **Lower sheaths** densely retrorsely villous-pubescent, **upper sheaths** glabrous; **ligules** 0.5–1 mm, glabrous, truncate, dentate or lacerate; **blades** 5–20 cm long, 1.5–4 mm wide, pilose, sparingly pubescent, or subglabrous. **Panicles** 6–15 cm long, 1–2 cm wide, racemose, dense, strongly contracted, stiffly erect; **branches** shorter than the spikelets, stiff, erect, straight. **Spikelets** 17–45 mm long, 4–7 mm wide, narrowly oblong or lanceolate, terete to moderately laterally compressed, usually 1 per node; **florets** (4)8–12, bases concealed at maturity; **rachilla internodes** concealed at maturity. **Glumes** glabrous or puberulent; **lower glumes** 7–9 mm, 3-veined, **upper glumes** 8–11 mm, 5–7-veined; **lemmas** 11–15(17) mm long, 3–5 mm wide, lanceolate, glabrous or pubescent, 9-veined, rounded over the midvein, margins bluntly angled, not inrolled at maturity, apices and teeth acuminate, teeth shorter than 1 mm; **awns** 12–20 mm, strongly divaricate at maturity, flattened and often basally twisted, arising 1.5 mm or more below the lemma apices; **anthers** 0.75–1.5 mm. **Caryopses** equaling or shorter than the paleas, thin, weakly inrolled or flat. 2*n* = 14, 28.

Bromus caroli-henrici is native to Mediterranean Europe. In the *Flora* region, it grows in open, disturbed areas in Butte and Yolo counties, California. It has been misidentified as *B. alopecuros* Poir. It differs in having 1, rather than 2–3, spikelets at the rachis nodes, and acuminate, rather than broadly triangular, lemma teeth.

lemma

florets

spikelet

B. racemosus

upper glume

lower glume

spikelet

upper lemma

lower lemma

B. danthoniae

CTR

spikelet

lemma

B. caroli-henrici

BROMUS

48. Bromus lanceolatus Roth [p. 236]
LANCEOLATE BROME

Plants annual. Culms 30–70 cm, erect or ascending. Sheaths often densely hairy with soft, white hairs; ligules 1–2 mm, hairy, obtuse, erose; blades 10–30 cm long, 3–5 mm wide, glabrous or pubescent. Panicles 5–15 cm long, 2–9 cm wide, erect, densely contracted when immature, more open with age; branches usually shorter than the spikelets, rigid, ascending to slightly spreading, slightly curved or straight. Spikelets 20–50 mm, lanceolate, terete to moderately laterally compressed, often 2+ per node; florets 7–20, bases concealed at maturity; rachilla internodes concealed at maturity. Glumes pilose; lower glumes 5–9 mm, 3–5-veined; upper glumes 8–12 mm, 5–7-veined; lemmas 11–20 mm long, 1.8–2.5 mm wide, lanceolate, pilose, obscurely 7-veined, rounded over the midvein, margins rounded, not inrolled at maturity, apices acute, bifid, teeth shorter than 1 mm; awns 6–12 mm, to 20 mm on some distal lemmas, divaricate when mature, arising 1.5 mm or more below the lemma apices; anthers 1–1.5 mm. Caryopses equaling or slightly shorter than the paleas, thin, weakly inrolled or flat. 2n = 28, 42.

Bromus lanceolatus grows in waste places, and is also cultivated as an ornamental. It has been introduced to the *Flora* region from southern Europe, and is reported from scattered sites, e.g., Yonkers, New York (wool waste); College Station, Texas; and Pima County, Arizona.

49. Bromus scoparius L. [p. 236]
BROOM BROME

Plants annual. Culms (9)20–40 cm, erect or ascending. Sheaths sparsely pubescent or glabrous; ligules 0.8–1.5 mm, glabrous or hairy, obtuse; blades 5–20 cm long, 2–5 mm wide, abaxial surfaces glabrous, adaxial surfaces pilose. Panicles 2–7 cm long, 1–4 cm wide, erect, dense, obovoid, wedge-shaped at the base, sometimes interrupted; branches shorter than the spikelets, erect, straight or almost so, sometimes verticillate. Spikelets 12–25 mm, lanceolate, crowded, terete to moderately laterally compressed; florets 5–10, bases concealed at maturity; rachilla internodes concealed at maturity. Glumes scabrous to pubescent; lower glumes 3–4 mm, 3–5-veined; upper glumes 5–7 mm, 5–7-veined; lemmas 7–10 mm long, 1.5–2 mm wide, lanceolate, glabrous, obscurely 7-veined, rounded over the midvein, margins rounded, not inrolled at maturity, apices sharply acute, bifid, teeth shorter than 1 mm; awns 7–10 mm, flattened at the base, divaricate or recurved when mature, arising 1.5 mm or more below the lemma apices; anthers 0.3–0.5 mm. Caryopses shorter than the paleas, thin, weakly inrolled or flat. 2n = 14.

Bromus scoparius is native to southern Europe. It grows in waste places. In the *Flora* region, it has been recorded from California and New York.

50. Bromus arenarius Labill. [p. 236]
AUSTRALIAN BROME

Plants annual. Culms 20–40 cm, erect to ascending. Sheaths densely retrorsely pilose; ligules 1.5–2.5 mm, glabrous or pilose, obtuse, lacerate; blades 7–8 cm long, 3–6 mm wide, pilose on both surfaces. Panicles (4)10–15 cm long, 4–7 cm wide, open, nodding; branches sometimes longer than the spikelets, spreading or ascending, sinuous. Spikelets 10–20 mm, lanceolate, terete to moderately laterally compressed; florets 5–9(11), bases concealed at maturity; rachilla internodes concealed at maturity. Glumes densely pilose; lower glumes 7–10 mm, 3-veined; upper glumes 8–12 mm, (5)7-veined; lemmas 9–11(13) mm long, 1–1.8 mm wide, lanceolate, densely pilose, distinctly 7-veined, rounded over the midvein, margins rounded, not inrolled at maturity, apices acute, bifid, teeth shorter than 1 mm; awns 10–16 mm, straight to weakly spreading, arising 1.5 mm or more below the lemma apices; anthers 0.7–1 mm. Caryopses equaling or shorter than the paleas, thin, weakly inrolled. 2n = unknown.

Bromus arenarius grows in dry, often sandy slopes, fields, and waste places. Native to Australia, it is now widely scattered throughout California, and is also recorded from Oregon, eastern Nevada, Arizona, New Mexico, Texas, and Pennsylvania.

51. Bromus japonicus Thunb. [p. 237]
JAPANESE BROME

Plants annual. Culms (22)30–70 cm, erect or ascending. Sheaths usually densely pilose; upper sheaths sometimes pubescent or glabrous; ligules 1–2.2 mm, pilose, obtuse, lacerate; blades 10–20 cm long, 2–4 mm wide, usually pilose on both surfaces. Panicles 10–22 cm long, 4–13 cm wide, open, nodding; branches usually longer than the spikelets, ascending to spreading or somewhat drooping, slender, flexuous, sometimes sinuous, often

lemma

spikelet

1 cm

B. lanceolatus

ligule

1 mm

1 cm

spikelet

CTR

1 cm

upper glume

lower glume

spikelet

glumes

1 cm

lemma

spikelet

lemma

1 cm

B. scoparius

B. arenarius

BROMUS

with more than 1 spikelet. **Spikelets** 20–40 mm, lanceolate, terete to moderately laterally compressed; **florets** 6–12, bases concealed at maturity; **rachilla internodes** concealed at maturity. **Glumes** smooth or scabrous; **lower glumes** 4.5–7 mm, (3)5-veined; **upper glumes** 5–8 mm, 7-veined; **lemmas** 7–9 mm long, 1.2–2.2 mm wide, lanceolate, coriaceous, smooth proximally, scabrous on the distal ½, obscurely (7)9-veined, rounded over the midvein, margins hyaline, 0.3–0.6 mm wide, obtusely angled above the middle, not inrolled at maturity, apices acute, bifid, teeth shorter than 1 mm; **awns** 8–13 mm, strongly divergent at maturity, sometimes erect, twisted, flattened at the base, arising 1.5 mm or more below the lemma apices; **anthers** 1–1.5 mm. **Caryopses** equaling or shorter than the paleas, thin, weakly inrolled or flat. 2*n* = 14.

Bromus japonicus grows in fields, waste places, and road verges. It is native to central and southeastern Europe and Asia, and is distributed throughout much of the United States and southern Canada, with one record from the Yukon Territory.

52. Bromus squarrosus L. [p. 237]
SQUARROSE BROME

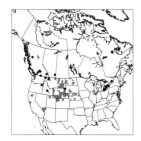

Plants annual. **Culms** 20–60 cm, erect or geniculately ascending. **Lower sheaths** densely pilose; **ligules** 1–1.5 mm, hairy, obtuse, erose, ciliolate; **blades** 5–15 cm long, 4–6 mm wide, densely pilose on both surfaces. **Panicles** 7–20 cm long, 4–8 cm wide, racemose, open, nodding, often with few spikelets, usually secund; **branches** sometimes longer than the spikelets, ascending-spreading, flexuous, slightly curved, usually with 1 spikelet. **Spikelets** 15–70 mm, broadly oblong or ovate-lanceolate, terete to moderately laterally compressed; **florets** 8–30, bases concealed at maturity; **rachilla internodes** concealed at maturity. **Glumes** smooth or scabrous; **lower glumes** 4.5–7 mm, 3–5(7)-veined; **upper glumes** 6–8 mm, 7-veined; **lemmas** 8–11 mm long, 2–2.4 mm wide, lanceolate, chartaceous, smooth or scabridulous, 7–9-veined, rounded over the midvein, margins hyaline, 0.6–0.9 mm wide, strongly angled above the middle, not inrolled at maturity, apices acute, bifid, teeth shorter than 1 mm; **awns** 8–10 mm, flattened and sometimes twisted at the base, divaricate at maturity, arising 1.5 mm or more below the lemma apices; **anthers** 1–1.3 mm. **Caryopses** equaling the paleas, thin, weakly inrolled or flat. 2*n* = 14.

Bromus squarrosus grows in overgrazed pastures, fields, waste places, and road verges. Native to central Russia and southern Europe, it can be found mainly in southern Canada and the northern half of the United States.

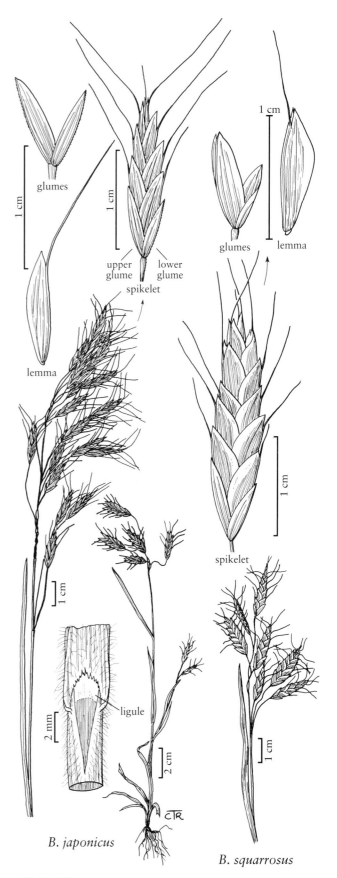

glumes

upper glume | lower glume
spikelet

lemma

glumes

lemma

spikelet

ligule

B. *japonicus*

B. *squarrosus*

BROMUS

13. **TRITICEAE** Dumort.

Mary E. Barkworth

Plants annual or perennial; sometimes cespitose, sometimes rhizomatous. **Culms** annual, not woody, usually erect, not branching above the base; **internodes** hollow or solid. **Sheaths** usually open, those of the basal leaves sometimes closed; **collars** without tufts of hair on the sides; **auricles** usually present; **ligules** membranous or scarious, sometimes ciliolate, those of the upper and lower cauline leaves usually similar; **pseudopetioles** absent; **blades** linear to narrowly lanceolate, venation parallel, cross venation not evident, without arm or fusoid cells, surfaces without microhairs, not papillate, cross sections non-Kranz. **Inflorescences** usually spikes or spikelike racemes, with 1–5 sessile or subsessile spikelets per node, occasionally panicles, sometimes with morphologically distinct sterile and bisexual spikelets within an inflorescence; **pedicels** absent or to 4 mm; **disarticulation** usually above the glumes and beneath the florets, sometimes in the rachises, sometimes at the inflorescence bases. **Spikelets** usually laterally compressed, sometimes terete, with 1–16 bisexual florets, the distal (or only) floret sometimes sterile; **rachillas** sometimes prolonged beyond the base of the distal floret. **Glumes** unequal to equal, shorter than to longer than the adjacent florets, subulate, lanceolate, rectangular, ovate, or obovate, 1–5-veined, absent or vestigial in some species; **florets** laterally compressed to terete; **calluses** glabrous or hairy; **lemmas** lanceolate to rectangular, stiffly membranous to coriaceous, sometimes keeled, 5(7)-veined, veins not converging distally, inconspicuous, apices entire, lobed, or toothed, unawned or awned, awns terminal, unbranched, lemma-awn junction not evident; **paleas** usually subequal to the lemmas, sometimes considerably shorter or slightly longer than the lemmas; **lodicules** 2, without venation, usually ciliate; **anthers** 3; **ovaries** with hairy apices; **styles** 2, bases free. **Caryopses** ovoid to fusiform, longitudinally grooved, not beaked, pericarp thin; **hila** linear; **embryos** about $\frac{1}{3}$ as long as the caryopses. $x = 7$.

The *Triticeae* are primarily north-temperate in distribution. The tribe includes 400–500 species, among which are several important cereal, forage, and range species. Its generic treatment is contentious. Linnaeus (1753) recognized five genera among the species now included in the tribe; *Hordeum* and *Secale* are the only two that still have his circumscription. *Hordeum* is also the only genus to include both annual and perennial species. The lack of agreement concerning the generic treatment of the tribe reflects the prevalence of natural hybridization, introgression, polyploidy, and reticulate relationships among its species. These factors preclude the circumscription of monophyletic groups, and mitigate against the delineation of morphologically coherent groups. Tzvelev (1975) argued that these same factors contribute to the tribe's success by maintaining a "generalist" genome.

The major disagreement in the treatment of the annual genera concerns *Triticum* and *Aegilops*. Some (e.g., Kimber and Feldman 1987) advocate treating them as a single genus in recognition of their close genetic similarity; others argue for maintaining them as separate genera (e.g., van Slageren 1994). Löve (1984) divided them among 14 genera. They are accepted here in their traditional senses, despite the strong argument for their combination, largely in deference to the wealth of literature, reports, and genetic resources that have been accumulated under these two names. Spontaneous hybridization and introgression between the two are common, and most species of *Triticum* are derived from hybrids between the two genera. Nevertheless, they differ in their ecology and, to some extent, in their morphology.

Treatment of the perennial species is more contentious. Restriction of *Agropyron* to what are known in English as the crested wheatgrasses is universally accepted; most taxonomists also accept the placement of alkaline-tolerant species that are strongly rhizomatous or have short, subulate glumes in *Leymus*. *Pseudoroegneria, Pascopyrum,* and *Thinopyrum* are less accepted. They are widely accepted by those working in genetic resources, but less so by those involved in floristics who prefer to include them in *Elymus*; all were traditionally included in *Agropyron*. Another area of disagreement is the treatment of *Elytrigia* Desv., *Roegneria* K. Koch, and *Hystrix* Moench. Species sometimes placed in *Elytrigia* are here included in *Elymus*, *Thinopyrum*, or *Pseudoroegneria*; species of *Roegneria* in *Elymus*; and species of *Hystrix* in *Elymus* or *Leymus*. Wide acceptance of a single treatment is hampered by the existence of differing taxonomic traditions, and by the lack of a coordinated international examination of morphological variation among the tribe's species.

The treatment followed here is strongly influenced by the treatments of Löve (1984) and Dewey (1984), particularly with respect to the perennial genera. Both advocated using genomic constitution as the basis for generic delimitation in the tribe. The genomic constitution of individual species is determined by observing meiotic chromosome pairing in hybrids. The base chromosome number in the tribe is seven. If a hybrid between two tetraploids forms 7 quadrivalents and 14 bivalents at meiosis, its parents are considered to have one similar set of chromosomes or *haplome,* and one dissimilar haplome. The three haplomes can then be assigned codes. For example, one parent might be said to have the **E** and **F** haplomes, or an **EF** genomic constitution, and the other the **E** and **L** haplomes, or an **EL** genomic constitution. The prevalence of polyploids and the ease of forming hybrids in the *Triticeae* has enabled cytogeneticists to build up a rather complete picture of the genomic constitution of its members. This led to the discovery that there is a strong, but not perfect, correlation between morphology and genomic constitution. The haplome codes used in this volume are those endorsed by the International Triticeae Consortium (http://herbarium.usu.edu/Triticeae/genmsymb.htm). Molecular tools reveal a pattern that is, in general, consistent with the cytogenetic data, but they often reveal an underlying complexity that cannot be discerned using only classical cytogenetic techniques.

The following key does not include intergeneric hybrids; they are treated in the text on the following pages: ×*Triticosecale* (p. 260), ×*Pseudelymus* (p. 282), ×*Elyhordeum* (p. 283), ×*Elyleymus* (p. 343), ×*Pascoleymus* (p. 351), and ×*Leydeum* (p. 369). In the field, they can usually be detected by their intermediate morphology and sterility. In sterile plants, the anthers are indehiscent, somewhat pointed, and tend to remain on the plants. Measurements of rachis internodes and spikelets should be made at midspike.

SELECTED REFERENCES **Barkworth, M.E.** and **D.R. Dewey**. 1985. Genomically based genera in the perennial Triticeae of North America: Identification and membership. Amer. J. Bot. 72:767–776; **Baum, B.R., C. Yen,** and **J.-L. Yang**. 1991. *Roegneria*: Its generic limits and justification for its recognition. Canad. J. Bot. 69:282–294; **Dewey, D.R.** 1984. The genomic system of classification as a guide to intergeneric hybridization in the perennial Triticeae. Pp. 209–279 *in* J.P. Gustafson (ed.). Gene Manipulation in Plant Improvement. Plenum Press, New York, New York, U.S.A. 668 pp.; **Kellogg, E.A.** 1989. Comments on genomic genera in the Triticeae. Amer. J. Bot. 76:796–805; **Linnaeus, C.** 1753. Species Plantarum. Impensis Laurentii Salvii, Stockholm, Sweden. 1200 pp.; **Löve, A.** 1984. Conspectus of the Triticeae. Feddes Repert. 95:425–521; **Svitashev, S., B. Salomon, T. Bryngelsson,** and **R. von Bothmer**. 1996. A study of 28 *Elymus* species using repetitive DNA sequences. Genome 39:1093–1101; **Tsvelev, N.N.** 1975. [On the possibility of despecialization by hybridogenesis for explaining the evolution of the Triticeae (Poaceae)]. Zhurn. Obshchei Biol. 36:90–99. [In Russian; translation of article by K. Gonzales, available at the Intermountain Herbarium, Utah State University, Logan, Utah 84322–5305, U.S.A.].

1. Spikelets 2–7 at all or most nodes.
    2. Spikelets 3 at each node, the central spikelets sessile, the lateral spikelets usually pedicellate, sometimes all 3 spikelets sessile in cultivated plants; spikelets with 1 floret, usually only the central spikelet with a functional floret, the florets of the lateral spikelets usually sterile and reduced, in cultivated plants all florets functional or those of the lateral spikelets functional and those of the central spikelet reduced . . . . .13.01 *Hordeum*

2. Spikelets usually other than 3 at each node, if 3, all three sessile; spikelets with 1–11 florets, if 1 floret, additional reduced or sterile florets present distal to the functional floret in at least 1 spikelet per node.
  3. Lemmas strongly keeled, keels conspicuously scabrous distally, scabrules 0.5–0.8 mm long; lemma awns straight . . . . 13.05 *Secale*
  3. Lemmas rounded proximally, sometimes keeled distally, keels not or inconspicuously scabrous distally; lemma awns straight, flexuous, or variously curved.
    4. Plants annual, weedy; spikelets with only 1 bisexual floret . . . . 13.04 *Taeniatherum*
    4. Plants perennial, usually not weedy; spikelets usually with more than 1 bisexual floret.
      5. Lemma awns (0)1–120 mm long; anthers 0.9–6 mm long; blades with well-spaced, unequally prominent veins on the adaxial surfaces . . . . 13.13 *Elymus* (in part)
      5. Lemmas usually unawned or with awns up to 7 mm long, if awns 16–35 mm long, anthers 6–8 mm; blades usually with closely spaced, equally prominent veins on the adaxial surfaces.
        6. Disarticulation in the spikelets, beneath the florets; plants sometimes cespitose, often rhizomatous . . . . 13.17 *Leymus* (in part)
        6. Disarticulation tardy, in the rachises; plants cespitose, not rhizomatous . . . 13.19 *Psathyrostachys*
1. Spikelets 1 at all or most nodes.
  7. Spikelets usually more than 3 times the length of the middle rachis internodes, usually divergent, sometimes ascending; rachis internodes 0.2–5.5 mm long.
    8. Glumes with 2 prominent keels, keels with tufts of hair . . . . 13.03 *Dasypyrum*
    8. Glumes initially with 1 keel, sometimes 2-keeled at maturity, keels glabrous or hairy, hairs never in tufts.
      9. Plants annual; anthers 0.4–1.4 mm long; spikes 0.8–4.5 cm long . . . . 13.02 *Eremopyrum*
      9. Plants perennial; anthers 3–5 mm long; spikes 1.3–15 cm long . . . . 13.09 *Agropyron*
  7. Spikelets ¹/₂–3 times the length of the middle rachis internodes, appressed or ascending; rachis internodes 3–28 mm long.
    10. Glumes subulate to narrowly lanceolate, tapering from below midlength, 1(3)-veined at midlength.
      11. Glumes lanceolate, tapering to acuminate apices from near midlength or below, keels curving to the side distally; plants always rhizomatous . . . . 13.15 *Pascopyrum*
      11. Glumes subulate to lanceolate, tapering from below midlength, keels straight or almost so; plants often rhizomatous . . . . 13.17 *Leymus* (in part)
    10. Glumes lanceolate, rectangular, ovate, or obovate, narrowing beyond midlength, often the in distal ¹/₄, (1)3–5(7)-veined at midlength.
      12. Plants annual; glumes often with lateral teeth or awns, midveins smooth throughout.
        13. Glumes rounded over the midveins; plants weedy . . . . 13.07 *Aegilops*
        13. Glumes keeled over the midveins; plants cultivated, sometimes escaping . . . . 13.08 *Triticum*
      12. Plants perennial; glumes without lateral teeth or awns, midveins sometimes scabrous.
        14. Glumes stiff, truncate, obtuse, or acute, unawned; glume keels smooth proximally, usually scabrous distally . . . . 13.20 *Thinopyrum*
        14. Glumes flexible, acute to acuminate, sometimes awn-tipped; glume keels usually uniformly smooth or scabrous their whole length, sometimes smooth proximally and scabrous distally.
          15. Spikelets distant, not or scarcely reaching the base of the spikelet above on the same side of the rachis; anthers 4–8 mm long . . . . 13.10 *Pseudoroegneria*
          15. Spikelets usually more closely spaced, reaching midlength of the spikelet above on the same side of the rachis; anthers 0.7–7 mm long . . . . . . 13.13 *Elymus* (in part)

13.01 HORDEUM L.

Roland von Bothmer

Claus Baden†

Niels H. Jacobsen

Plants summer or winter annuals or perennials; cespitose, sometimes shortly rhizomatous. **Culms** to 135(150) cm, erect, geniculate, or decumbent; **nodes** glabrous or pubescent. **Sheaths** open, pubescent or glabrous; **auricles** present or absent; **ligules** hyaline, truncate, erose; **blades** flat to more or less involute, more or less pubescent on both sides. **Inflorescences** usually spikelike racemes, sometimes spikes, all customarily called *spikes*, with 3 spikelets at each node, central spikelets usually sessile, sometimes pedicellate, pedicels to 2 mm, lateral spikelets usually pedicellate, pedicels curved or straight, sometimes all 3 spikelets sessile in cultivated plants; **disarticulation** usually in the rachises, the spikelets falling in triplets, cultivated forms generally not disarticulating. **Spikelets** with 1 floret; **glumes** awnlike, usually exceeding the floret. **Lateral spikelets** usually sterile or staminate, often bisexual in cultivated forms; **florets** pedicellate, usually reduced; **lemmas** awned or unawned. **Central spikelets** bisexual; **florets** sessile; **rachillas** prolonged beyond the floret; **lemmas** ovate, glabrous to pubescent, 5-veined, usually awned, rarely unawned; **paleas** almost equal to the lemmas, narrowly ovate, keeled; **lodicules** 2, broadly lanceolate, margins ciliate; **anthers** 3, usually yellowish. **Caryopses** usually tightly enclosed in the lemma and palea at maturity. $2n = 14, 28, 42$. Name from the old Latin name for barley.

Hordeum is a genus of 32 species that grow in temperate and adjacent subtropical areas, at elevations from 0–4500 m. The genus is native to Eurasia, the Americas, and Africa, and has been introduced to Australasia. The species are confined to rather moist habitats, even on saline soils. The annual species occupy seasonally moist habitats that cannot sustain a continuous grass cover.

Some species of *Hordeum*, such as *H. marinum* and *H. murinum*, are cosmopolitan weeds. *Hordeum vulgare* is widely cultivated for feed, malt, and flour. Archeological records suggest that *Hordeum* and *Triticum* were two of the earliest domesticated crops.

Eleven species of *Hordeum* grow in the *Flora* region: six are native, three are established weeds, and two are cultivated and occasionally persist as weeds. *Hordeum secalinum* has been reported from the *Flora* region, but the reports are based on misidentifications.

Four different haplomes are present in *Hordeum*. *Hordeum vulgare* and *H. bulbosum* have the **I** genome (often called the **H** genome by plant breeders), North American diploid species are based on the **H** genome, diploid *H. marinum* on the **X** genome, and diploids in the *H. murinum* group on the **Y** genome. Relationships among the polyploid taxa are complex (Jakob and Blattner 2006).

Spike measurements and lemma lengths, unless stated otherwise, do not include the awns.

SELECTED REFERENCES **Baum, B.R.** 1978. The status of *Hordeum brachyantherum* in eastern Canada, with related discussions. Canad. J. Bot. 56:107–109; **Baum, B.R.** and **L.G. Bailey.** 1990. Key and synopsis of North American *Hordeum* species. Canad. J. Bot. 68:2433–2442; **Blattner, F.R.** 2006. Multiple intercontinental dispersals shaped the distribution area of *Hordeum* (Poaceae). New Phytol. 169:603–614; **Bothmer, R. von, N. Jacobsen, C. Baden, R.B. Jørgensen,** and **I. Linde-Laursen.** 1995. An Ecogeographical Study of the Genus *Hordeum*, ed. 2. Systematic and Ecogeographic Studies on Crop Genepools No. 7. International Board of Plant Genetic Resources, Rome, Italy. 129 pp.; **Bothmer, R. von, N. Jacobsen,** and **R.O. Seberg.** 1993. Variation and taxonomy in *Hordeum depressum* and in the *H. brachyantherum* complex (Poaceae). Nordic J. Bot. 13:3–17; **Jakob, S.S.** and **F.R. Blattner.** 2006. A chloroplast genealogy of *Hordeum* (Poaeae): Long-term persisting haplotypes, incomplete lineage sorting, regional extinction, and the consequences for phylogenetic inference. Mol. Biol. Evol. 23:1602–1612; **Moyer, J.R.** and **A.L. Boswall.** 2002. Tall fescue or creeping foxtail suppresses foxtail barley. Canad. J. Pl. Sci. 82:89–92; **Petersen, G.** and **O. Seberg.** 2003. Phylogenetic analyses of the diploid species of *Hordeum* (Poaceae) and a revised classification of the genus. Syst. Bot. 28:293–306.

1. Plants perennial.
  2. Culms usually with a bulbous swelling at the base; auricles to 5.5 mm long, well developed. . . . . 10. *H. bulbosum*
  2. Culms not bulbous-based; auricles absent or no more than 1 mm long.
    3. Glumes of the central spikelet flattened near the base . . . . 6. *H. arizonicum* (in part)
    3. Glumes of the central spikelet usually setaceous throughout, rarely flattened near the base.
      4. Glumes 15–85 mm long, divergent to strongly divergent at maturity . . . . 5. *H. jubatum* (in part)
      4. Glumes 7–19 mm long, divergent or not at maturity.
        5. Anthers of the central spikelet 0.8–4 mm long; auricles absent . . . . 4. *H. brachyantherum*
        5. Anthers of the central spikelet 3.5–5 mm long; auricles present on the basal leaves
        . . . . 8. *H. secalinum*
1. Plants annual.
  6. Auricles to 8 mm long, well developed even on the upper leaves; lemmas of the lateral florets 6–15 mm long.
    7. Rachises disarticulating at maturity; glumes of the central spikelets ciliate; lemmas of the central florets to 2 mm wide, with awns 20–40 mm long; lateral spikelets staminate . . . . 9. *H. murinum*
    7. Rachises usually not disarticulating at maturity; glumes of the central spikelets pubescent; lemmas of the central florets at least 3 mm wide, unawned or with awns 30–180 mm long; usually 1 or both lateral spikelets at a node seed-forming . . . . 11. *H. vulgare*
  6. Auricles usually absent or to 0.3 mm long; lemmas of the lateral florets 1.7–8.5 mm long.
    8. Glumes bent, strongly divergent at maturity.
      9. Glumes of the central spikelets not flattened, (15)35–85 mm long . . . . 5. *H. jubatum* (in part)
      9. Glumes of the central spikelets slightly flattened towards the base, 11–28 mm long
      . . . . 6. *H. arizonicum* (in part)
    8. Glumes straight, ascending to slightly divergent at maturity.
      10. Lemmas of the lateral spikelets with awns 3–8 mm long . . . . 7. *H. marinum*
      10. Lemmas of the lateral spikelets unawned or with awns no more than 3 mm long.
        11. Glumes of the central spikelets setaceous to slightly flattened near the base.
          12. Spikes 4–8 mm wide; lemmas of the central spikelets with awns 3–12 mm long; ligules 0.3–0.8 mm long . . . . 3. *H. depressum*
          12. Spikes 6–20 mm wide; lemmas of the central spikelets with awns 10–22 mm long; ligules 0.6–1.8 mm long . . . . 6. *H. arizonicum* (in part)
        11. Glumes of the central spikelets distinctly flattened near the base.
          13. Lemmas of the lateral spikelets 1.7–4.4 mm long, usually unawned, rarely with awns to 1.2 mm long; sheaths with stripes of hairs . . . . 1. *H. intercedens*
          13. Lemmas of the lateral spikelets 2.5–5.7 mm long, usually awned, with awns to 1.8 mm long; sheaths glabrous . . . . 2. *H. pusillum*

1. Hordeum intercedens Nevski [p. 244]
BOBTAIL BARLEY

Plants annual; loosely tufted. **Culms** 5–40 cm, erect to geniculate; **nodes** usually pubescent. **Sheaths** with stripes of hairs; **ligules** 0.3–0.8 mm; **auricles** usually absent, shorter than 2 mm if present; **blades** to 9 cm long, to 4 mm wide, both surfaces sparsely to densely hairy, hairs spreading. **Spikes** 2.5–6.2 cm long, 4–6 mm wide, often partially enclosed at maturity, pale green. **Glumes** straight, usually slightly divergent at maturity.

Central spikelets: glumes to 17 mm long, to 0.8 mm wide basally, distinctly flattened near the base; **lemmas** 4.5–7.5 mm, usually sparsely pubescent towards the base, glabrous distally, awned, awns 5.6–9.8 mm, often slightly divergent at maturity; **anthers** 0.6–1.2 mm. **Lateral spikelets** usually sterile; **glumes** to 17.5 mm, distinctly flattened near the base; **lemmas** 1.7–4.4 mm, blunt to acute, usually unawned, rarely awned, awns to 1.2 mm. $2n = 14$.

Hordeum intercedens grows in vernal pools and flooded, often saline river beds and alkaline flats. It is restricted to southwestern California, including some of the coastal islands, and northwestern Baja California, Mexico.

2. Hordeum pusillum Nutt. [p. 244]
LITTLE BARLEY

Plants annual; loosely tufted. **Culms** 10–60 cm, erect, geniculate, or ascending; **nodes** glabrous. **Sheaths** glabrous or slightly pubescent; **ligules** 0.2–0.8 mm; **auricles** absent; **blades** to 10.5 cm long, to 4.5 mm wide, sparsely to densely pubescent on both sides. **Spikes** 2–9 cm long, 3–7 mm wide, erect, often partially enclosed at maturity, pale green. **Glumes** straight, not divergent at maturity. **Central spikelets: glumes** 8–17 mm long, 0.5–1.5 mm wide, distinctly flattened near the base; **lemmas** 5–8.5 mm, usually glabrous, sometimes sparsely to densely pubescent, awned, awns 3.5–9.5 mm; **anthers** 0.7–1.8 mm. **Lateral spikelets** usually sterile; **glumes** to 18 mm; **lower glumes** distinctly flattened, more or less winged basally; **lemmas** 2.5–5.7 mm, usually awned, awns to 1.8 mm, rarely unawned; **anthers** 0.6–1.2 mm. 2*n* = 14.

Hordeum pusillum grows in open grasslands, pastures, and the borders of marshes, and in disturbed places such as roadsides and waste places, often in alkaline soil. It is native, widespread, and often common in much of the *Flora* region. Its range extends into northern Mexico, but it is not common there.

3. Hordeum depressum (Scribn. & J.G. Sm.) Rydb. [p. 244]
LOW BARLEY

Plants annual; loosely tufted. **Culms** 10–55 cm, erect; **nodes** glabrous. **Basal sheaths** pubescent; **ligules** 0.3–0.8 mm; **auricles** absent; **blades** to 7.5(13.5) cm long, to 4.5 mm wide, sparsely to densely pubescent on both sides. **Spikes** 2.2–7 cm long, 4–8 mm wide, often partially enclosed at maturity, pale green or with a reddish tinge to the glumes and awns. **Glumes** straight, ascending to slightly divergent at maturity. **Central spikelets: glumes** 5.5–20.5 mm long, to 0.5 mm wide, setaceous to slightly flattened near the base; **lemmas** 5–9 mm, glabrous, awned, awns 3–12 mm; **anthers** 0.5–1.5 mm. **Lateral spikelets** sterile or staminate, occasionally bisexual; **glumes** 5–20 mm; **lower glumes** slightly flattened near the base; **upper glumes** setaceous throughout; **lemmas** 1.8–8.5 mm, unawned or awned, awns to 1 mm. 2*n* = 28.

Hordeum depressum grows in vernal pools and ephemeral habitats, often in alkaline soil. It is restricted to the western United States.

4. Hordeum brachyantherum Nevski [p. 246]

Plants perennial; loosely to densely cespitose. **Culms** to 90 cm, erect to geniculate, not bulbous; **nodes** glabrous. **Sheaths** glabrous or densely pubescent; **auricles** absent; **blades** to 19 cm long, to 8 mm wide, glabrous or with hairs on both surfaces, hairs sometimes of mixed lengths. **Spikes** 3–8.5 cm, green to somewhat purple. **Glumes** 7–19 mm, ascending to slightly divergent at maturity. **Central spikelets: glumes** 9–19 mm long, about 0.2 mm wide, setaceous throughout, rarely flattened near the base; **lemmas** 5.5–10 mm, usually glabrous, rarely pubescent, awned, awns 3.5–14 mm; **anthers** 0.8–4 mm. **Lateral spikelets** staminate; **glumes** 7–19 mm, setaceous; **lower glumes** sometimes flattened near the base; **lemmas** rudimentary to well developed, awns to 7.5 mm, rarely absent; **anthers** 0.8–4 mm. 2*n* = 14, 28, 42.

Hordeum brachyantherum is native to the Kamchatka Peninsula and western North America, and has been introduced to a few locations in the eastern United States. There is also a small disjunct population in Newfoundland and Labrador that Baum (1978) identified as *H. secalinum*. *Hordeum brachyantherum* grows in salt marshes, pastures, woodlands, subarctic woodland meadows, and subalpine meadows.

Two subspecies are recognized here, but there is so much overlap in their morphological variation that unambiguous determination of many specimens is impossible in the absence of a chromosome count. They are sometimes treated as two species.

1. Basal sheaths usually glabrous, sometimes sparsely pubescent; anthers 0.8–3.5 mm long; culms often robust, sometimes slender . . . . subsp. *brachyantherum*
1. Basal sheaths usually densely pubescent; anthers 1.1–4 mm long; culms usually slender . . . . subsp. *californicum*

Hordeum brachyantherum Nevski subsp. brachyantherum [p. 246]
MEADOW BARLEY, NORTHERN BARLEY, ORGE À ANTHÈRES COURTES, ORGE DES PRÉS

Plants densely cespitose. **Culms** 30–95 cm, often robust, sometimes slender. **Basal sheaths** usually glabrous, sometimes sparsely pubescent; **blades** to 19 cm long, to 8 mm wide, both sides usually glabrous, sometimes with hairs to 0.5 mm on both surfaces. **Glumes** 7–17 mm, usually straight at maturity; **lemmas** usually awned, awns to 6.5 mm, usually straight at maturity; **anthers** 0.8–3.5 mm. 2*n* = 28, 42.

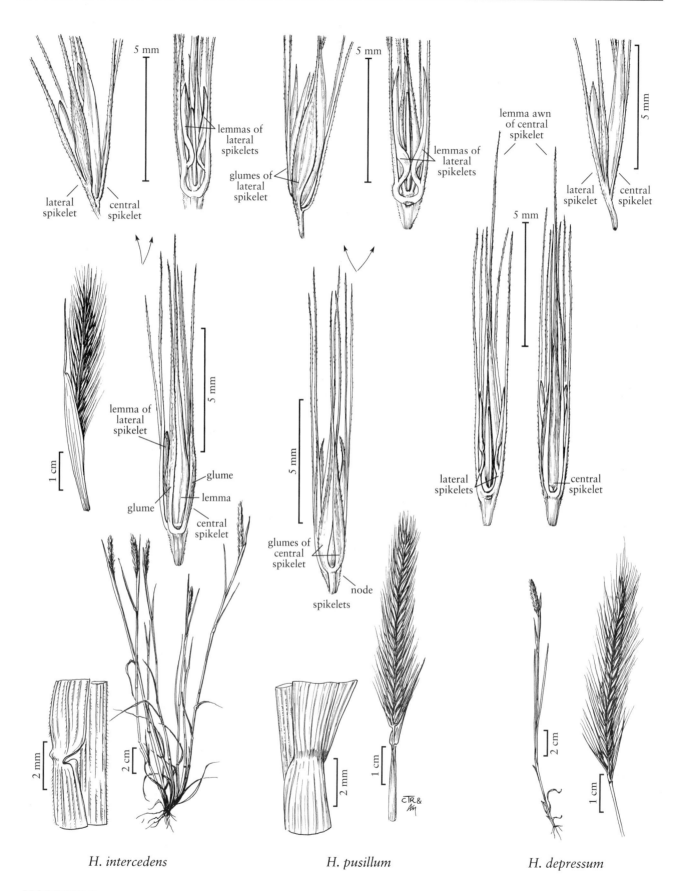

H. intercedens   *H. pusillum*   *H. depressum*

HORDEUM

Hordeum brachyantherum subsp. *brachyantherum* grows in pastures and along streams and lake shores, from sea level to 4000 m. Its range extends from Kamchatka through western North America to Baja California, Mexico. It is also known from disjunct locations in Newfoundland and Labrador and the eastern United Sates. The latter are probably recent introductions; the Newfoundland populations are harder to explain. One population from California is known to be hexaploid.

Hordeum brachyantherum subsp. **californicum** (Covas & Stebbins) Bothmcr, N. Jacobsen & Seberg [p. 246]
CALIFORNIA BARLEY

Plants loosely cespitose. **Culms** 20–65 cm, usually slender. **Basal sheaths** usually densely pubescent; **blades** to 11.5 cm long, to 3.5(5.5) mm wide, usually hairy with spreading hairs of mixed lengths on both sides, rarely glabrous or almost glabrous. **Glumes** 9–19 mm, usually spreading at maturity; **lemmas** usually awned, awns to 7.5 mm, usually divergent at maturity; **anthers** 1.1–4 mm. $2n = 14$.

Hordeum brachyantherum subsp. *californicum* is restricted to California. It grows on dry and moist grass slopes, in meadows and rocky stream beds, along stream margins, and around vernal pools, in oak woodlands and disturbed ground, and in serpentine, alkaline, and granitic soils, up to 2300 m. Records from outside California, and many from inside California, are based on misidentified specimens, usually of *H. brachyantherum* subsp. *brachyantherum*.

5. **Hordeum jubatum** L. [p. 247]

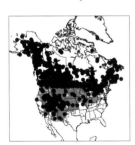

Plants perennial, sometimes appearing annual; cespitose. **Culms** 20–80 cm, geniculate to straight, not bulbous based; **nodes** glabrous. **Sheaths** glabrous or pubescent; **ligules** to 0.8 mm; **auricles** absent; **blades** to 15 cm long, to 5 mm wide, scabrous, sometimes hairy. **Spikes** 3–15 cm, usually nodding, whitish green to light purplish. **Glumes** 15–85 mm long, conspicuous, bent, divergent to strongly divergent at maturity. **Central spikelets: glumes** (15)35–85 mm, setaceous throughout, strongly spreading at maturity; **lemmas** 4–8.5 mm, glabrous, awned, awns 11–90 mm, straight to ascending; **paleas** 5.5–8 mm; **anthers** 0.6–1.2 mm. **Lateral spikelets** staminate or sterile; **glumes** 17–83 mm, setaceous; **lemmas** 4–6.5 mm, awned; **awns** 2–15 mm, divergent; **anthers** 1–1.5 mm. $2n = 28$.

Hordeum jubatum grows in meadows and prairies around riverbeds and seasonal lakes, often in saline habitats, and along roadsides and in other disturbed

sites. It is native from eastern Siberia through most of North America to Mexico, growing at elevations of 0–3000 m. It has been introduced to South America, Europe, and central Asia. It is grown in Russia and other areas outside its native range as an ornamental. In its native range, it is a weedy species.

Hordeum jubatum shows a wide range of variation in almost all characters; most such variation is not taxonomically significant. *Hordeum jubatum* subsp. *intermedium* is considered to be a subspecies of *H. jubatum* because no clear-cut discontinuities exist in the characters used to distinguish it from *H. jubatum* subsp. *jubatum*. These plants are fertile.

1. Glumes of the central spikelet 15–35 mm long; lemma awns of the central spikelets 11–35 mm long . . . . subsp. *intermedium*
1. Glumes of the central spikelet 35–85 mm long; lemma awns of the central spikelets 35–90 mm long . . . . subsp. *jubatum*

Hordeum jubatum subsp. **intermedium** Bowden [p. 247]
INTERMEDIATE BARLEY

Central spikelets: glumes 15–35 mm, spreading at maturity; **lemmas** awned, awns 11–35 mm.

Hordeum jubatum subsp. *intermedium* is most abundant in the dry prairies of the northern Rocky Mountains and northern plains, growing at 0–3000 m. It also grows, as a disjunct, in southern Mexico. It is sometimes treated as a species, either as *H. intermedium* Hausskn. or *H.* ×*intermedium* Hausskn., the latter reflecting a suspected hybrid origin involving *H. jubatum* and *H. brachyantherum*.

Hordeum jubatum L. subsp. **jubatum** [p. 247]
FOXTAIL BARLEY, SQUIRRELTAIL BARLEY, SQUIRRELTAIL GRASS, ORGE AGRÉABLE, QUEUE D'ÉCUREUIL, ORGE QUEUE D'ÉCUREUIL

Central spikelets: glumes 35–85 mm, strongly spreading at maturity; **lemmas** awned, awns 35–90 mm.

Hordeum jubatum subsp. *jubatum* is the more widespread of the two subspecies, extending from eastern Siberia through most of North America to northern Mexico. Native in western and northern portions of the *Flora* region, it is considered to be adventive in the eastern and southeastern portion of its range. It grows in moist soil along roadsides and other disturbed areas, as well as in meadows, the edges of sloughs and salt marshes, and on grassy slopes.

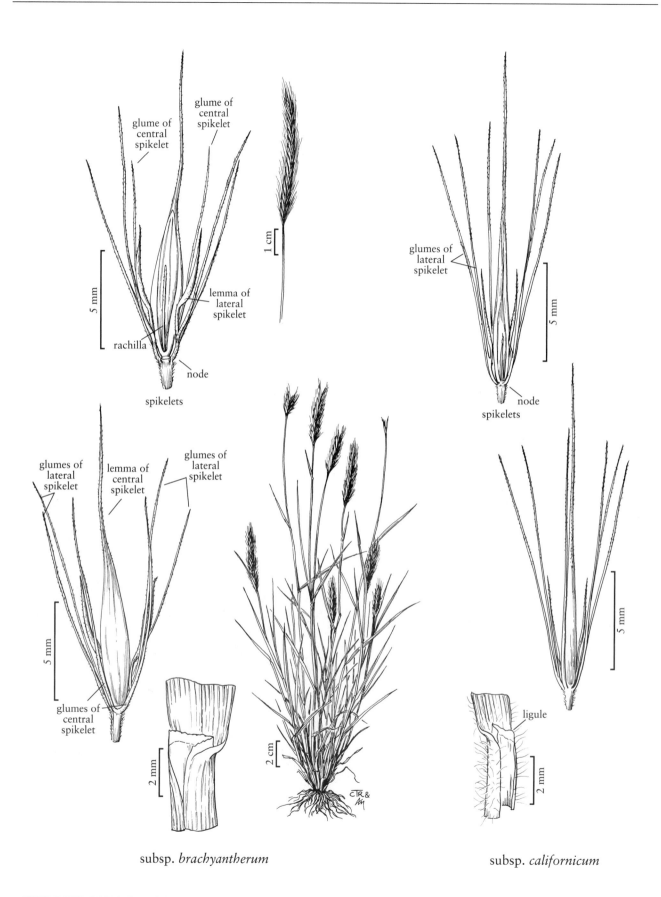

glume of central spikelet

glume of central spikelet

lemma of lateral spikelet

1 cm

5 mm

rachilla

node

spikelets

glumes of lateral spikelet

5 mm

node

spikelets

glumes of lateral spikelet

lemma of central spikelet

glumes of lateral spikelet

5 mm

glumes of central spikelet

2 mm

2 cm

5 mm

ligule

2 mm

subsp. *brachyantherum*

subsp. *californicum*

HORDEUM BRACHYANTHERUM

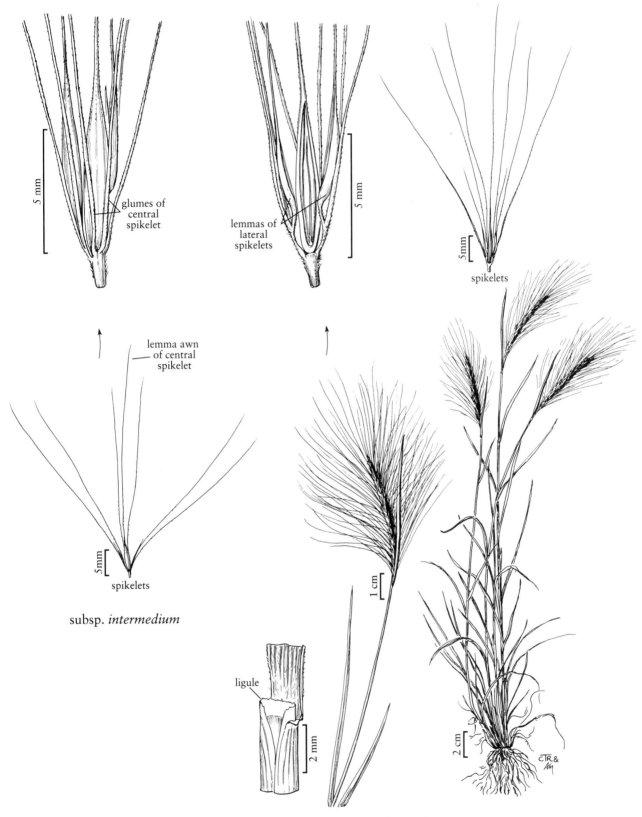

glumes of
central
spikelet

lemmas of
lateral
spikelets

spikelets

lemma awn
of central
spikelet

spikelets

subsp. *intermedium*

ligule

subsp. *jubatum*

HORDEUM JUBATUM

6. Hordeum arizonicum Covas [p. 249]
ARIZONA BARLEY

Plants annual to biennial, perennial under favorable conditions; forming small tufts. **Culms** 21–75 cm, rarely geniculate, not bulbous; **nodes** glabrous. **Lower sheaths** pubescent; **upper sheaths** glabrous; **ligules** 0.6–1.8 mm; **auricles** absent; **blades** to 13 cm long, to 4 mm wide, flat, glaucous, both surfaces scabrous, hairy, sometimes only sparsely hairy. **Spikes** 5–12 cm long, 6–10 mm wide, erect, often partially enclosed at maturity, pale green. **Glumes** bent, strongly divergent at maturity. **Central spikelets: glumes** 11–28 mm long, 0.2–0.4 mm wide, flattened near the base, setaceous above the middle; **lemmas** 5–9 mm, glabrous, awned, awns 10–22 mm; **anthers** 1–2.2 mm. **Lateral spikelets** sterile; **glumes** to 27 mm long, 0.2–0.4 mm wide, flattened near the base; **lemmas** 2.5–5 mm, unawned or awned, awns to 3 mm, lemma and awn together to 7 mm, the transition from the lemma to the awn gradual. $2n = 42$.

Hordeum arizonicum grows in saline habitats, along irrigation ditches, canals, and ponds in the southwestern United States and northern Mexico. It is a segmental allopolyploid between *H. jubatum* and either *H. pusillum* or *H. intercedens*.

7. Hordeum marinum Huds. [p. 249]

Plants annual; loosely tufted. **Culms** to 50 cm, straight to geniculate; **nodes** glabrous. **Basal sheaths** somewhat hairy; **ligules** 0.2–0.5 mm; **auricles** usually absent or to 0.3 mm; **blades** to 8 cm long, 1–6 mm wide, sparsely to densely hairy on both sides. **Spikes** 1.5–7 cm long, 5–10(20) mm wide, dense, green or more or less purplish on the awns and glumes; **rachises** disarticulating at maturity, triplets not breaking up immediately after maturity. **Glumes** straight, ascending to slightly divergent at maturity. **Central spikelets: glumes** 14–26 mm long, 0.2–1.1 mm wide basally, setaceous distally; **lemmas** 5–8 mm, usually smooth or somewhat scabrous, sometimes pubescent, awned, awns 6–18 mm; **anthers** 0.8–1.3 mm, yellowish. **Lateral spikelets** sterile; **lower glumes** setaceous or winged, wings 0.5–2.3 mm wide; **upper glumes** setaceous; **lemmas** 4–6 mm, awned, awns 3–8 mm. $2n = 14, 28$.

Hordeum marinum is native to Eurasia, where it grows in disturbed habitats. It has become established

in similar habitats in western North America, and in scattered locations elsewhere. Two subspecies are recognized; both are found in the *Flora* region.

1. Lower glumes of the lateral spikelets usually setaceous, not winged . . . . subsp. *gussoneanum*
1. Lower glumes of the lateral spikelets with a flattened wing, the wings 0.5–2.3 mm wide
. . . . subsp. *marinum*

Hordeum marinum subsp. gussoneanum (Parl.) Thell. [p. 249]
MEDITERRANEAN BARLEY, GENICULATE BARLEY

Plants summer annuals. **Leaves** 1.5–4 mm wide, flat or involute. **Lateral spikelets: lower glumes** usually setaceous, occasionally a little widened, but not winged. $2n = 14, 28$.

Hordeum marinum subsp. *gussoneanum* grows in grassy fields, waste places, and open ground. It was introduced to North America from the Mediterranean area, and it is now an established weed, especially in western North America.

Hordeum marinum Huds. subsp. marinum [p. 249]
SEA BARLEY

Plants summer or winter annuals. **Leaves** 1.5–8 mm wide, usually flat. **Lateral spikelets: lower glumes** winged on the side towards the central spikelets, wings 0.5–2.3 mm wide. $2n = 14$.

Hordeum marinum subsp. *marinum* is native to Eurasia, where it grows in disturbed habitats. Although it has been reported occasionally from the *Flora* region, it has not become established.

8. Hordeum secalinum Schreb. [p. 251]
FALSE-RYE BARLEY, MEADOW BARLEY

Plants perennial; loosely cespitose. **Culms** to 85 cm, erect, usually slender, not bulbous; **nodes** glabrous. **Sheaths**, at least the basal sheaths, densely pubescent; **auricles** to 1 mm, present on the basal leaves; **blades** flat or involute, surfaces scabrous, densely pilose. **Spikes** 3–7 cm. **Glumes** 8–15 mm, setaceous throughout, straight, slightly divergent at maturity. **Central spikelets** sessile; **lemmas** 5.7–9.9 mm, glabrous, awned, awns 5–15.4 mm, scabrous; **anthers** 3.5–5 mm, yellow. **Lateral spikelets** usually staminate, occasionally bisexual; **lemmas** 3.7–6.9 mm, usually glabrous, sometimes sparsely pilose, awns 1.4–5.8 mm; **anthers** 1.7–3.3 mm. $2n = 28$.

Hordeum secalinum is native to Europe, where it grows in moist, saline areas, often in coastal meadows. It does not grow in the *Flora* region; reports from North America (Baum 1978) were based on specimens of *Hordeum brachyantherum*, from which *H. secalinum* differs in having auricles on the basal leaves and longer

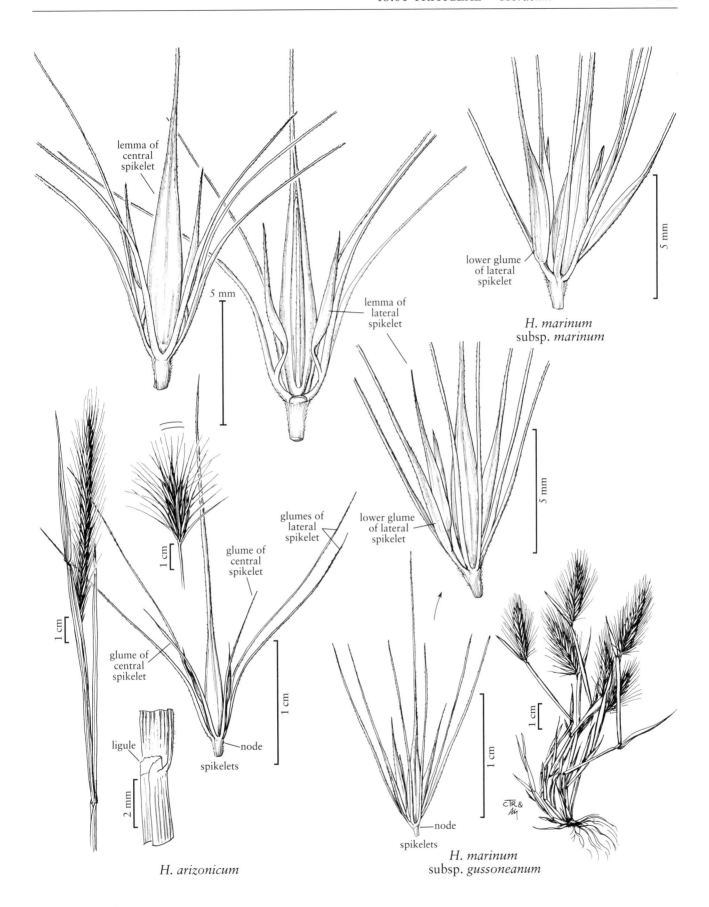

lemma of
central
spikelet

lemma of
lateral
spikelet

lower glume
of lateral
spikelet

5 mm

5 mm

H. marinum
subsp. *marinum*

glumes of
lateral
spikelet

lower glume
of lateral
spikelet

glume of
central
spikelet

glume of
central
spikelet

node

spikelets

ligule

2 mm

1 cm

1 cm

1 cm

node

spikelets

5 mm

1 cm

1 cm

H. arizonicum

H. marinum
subsp. *gussoneanum*

HORDEUM

anthers. It is treated here to aid those interested in distinguishing between the two species.

9. Hordeum murinum L. [p. 251]

Plants annual; loosely tufted. **Culms** to 110 cm, usually erect, sometimes almost prostrate; **nodes** glabrous. **Lower sheaths** often completely surrounding the culms, glabrous or somewhat pilose; **ligules** 1–4 mm; **auricles** to 8 mm, well developed even on the upper leaves; **blades** to 28 cm, usually flat, occasionally with involute margins, glabrous or sparsely pilose, sometimes scabrous. **Spikes** 3–8 cm long, 7–16 mm wide, pale green to distinctly reddish, especially the awns; **rachises** disarticulating at maturity. **Central spikelets** sessile, florets sessile or pedicellate, pedicels to 2 mm; **glumes** 11–25 mm long, 0.8–1.8 mm wide, flattened, margins usually distinctly ciliate; **lemmas** 8–14 mm long, to 2 mm wide, more or less smooth, awned, awns 20–40 mm; **lodicules** glabrous or with 1+ cilia; **anthers** 0.2–3.2 mm, gray to yellow, sometimes with purple spots. **Lateral spikelets** staminate, floret sessile; **glumes** flattened, margins ciliate; **lemmas** 8–15 mm, awned, awns 20–50 mm; **paleas** 8–15 mm; **rachillas** 2.5–6.5 mm, slender or gibbous, yellow. $2n = 14, 28, 42.$

Hordeum murinum is native to Eurasia, where it is a common weed in areas of human disturbance. It is thought to have originated around seasides, sandy riverbanks, and animal watering holes. It is now an established weed in the southwestern *Flora* region and other scattered locations. The records in Alaska are from the Anchorage area. Prostrate plants are associated with grazing. Three subspecies are recognized.

1. Central spikelets sessile to subsessile; lemmas of the central florets subequal to those of the lateral florets, the awns longer than those of the lateral florets; paleas of the lateral florets almost glabrous . . . . subsp. *murinum*
1. Central spikelets pedicellate; lemmas of the central florets from subequal to shorter than those of the lateral florets, the awns from shorter to longer than those of the lateral florets; paleas of the lateral florets scabrous to hairy.
    2. Lemmas of the central florets much shorter than those of the lateral florets; paleas of the lateral florets scabrous on the lower $^1\!/_2$; anthers of the central and lateral florets similar in size . . . . subsp. *leporinum*
    2. Lemmas of the central florets about equal to those of the lateral florets; paleas of the

lateral florets distinctly pilose on the lower $^1\!/_2$; anthers of the central florets 0.2–0.6 mm long, those of the lateral florets 1.2–1.8 mm long . . . . subsp. *glaucum*

Hordeum murinum subsp. glaucum (Steudel) Tzvelev [p. 251]
SMOOTH BARLEY

Plants summer annuals. **Culms** 15–40 cm. **Leaves** glaucous. **Spikes** sometimes glaucous, often brownish when mature. **Central floret** pedicellate; **lemmas** subequal to those of the lateral florets, awns as long as or longer than those of the lateral florets; **anthers** 0.2–0.6 mm, much shorter than those of the lateral florets, more or less covered with purple spots. **Lateral spikelets: paleas** more or less densely pilose, especially on the lower $^1\!/_2$; **anthers** 1.2–1.8 mm; **rachillas** about 0.3 mm, yellow. $2n = 14.$

Hordeum murinum subsp. *glaucum* grows in grasslands, fields, and waste places. It is native to the eastern Mediterranean area. It is now common in arid areas of the western United States, and is also known from scattered locations elsewhere in the *Flora* region.

Hordeum murinum subsp. leporinum (Link) Arcang. [p. 251]
MOUSE BARLEY

Plants winter annuals. **Culms** 30–110 cm. **Leaves** green. **Spikes** green at anthesis, often more or less purple just before maturity. **Central floret** pedicellate; **lemmas** and awns shorter than those of the lateral florets; **anthers** 0.9–3 mm. **Lateral spikelets: paleas** scabrous on the lower $^1\!/_2$; **anthers** 1.2–3.2 mm; **rachillas** about 0.25 mm, pale. $2n = 14, 28, 42.$

Hordeum murinum subsp. *leporinum* grows in waste places, roadsides, and disturbed areas in arid regions. It is native to the Mediterranean region. It is now established in the *Flora* region, being most common in the western United States. A hexaploid cytotype has been found in Turkey, Armenia, Turkmenistan, and Iran. It has been named *H. leporinum* var. *simulans* Bowden. It is treated here as part of *H. murinum* subsp. *leporinum.*

Hordeum murinum L. subsp. murinum [p. 251]
WALL BARLEY, FARMER'S FOXTAIL, WAY BARLEY

Plants winter annuals. **Culms** 30–60 cm. **Leaves** green. **Spikes** green. **Central floret** sessile to subsessile; **lemmas** subequal to those of the lateral florets, awns longer than those of the lateral florets; **anthers** 0.8–1.4 mm. **Lateral spikelets: paleas** almost glabrous; **anthers** 0.8–1.4 mm; **rachillas** about 0.15 mm, pale. $2n = 28.$

Hordeum murinum subsp. *murinum* grows in waste places that are somewhat moist. It is native to Europe. Within the *Flora* region, it has the most restricted

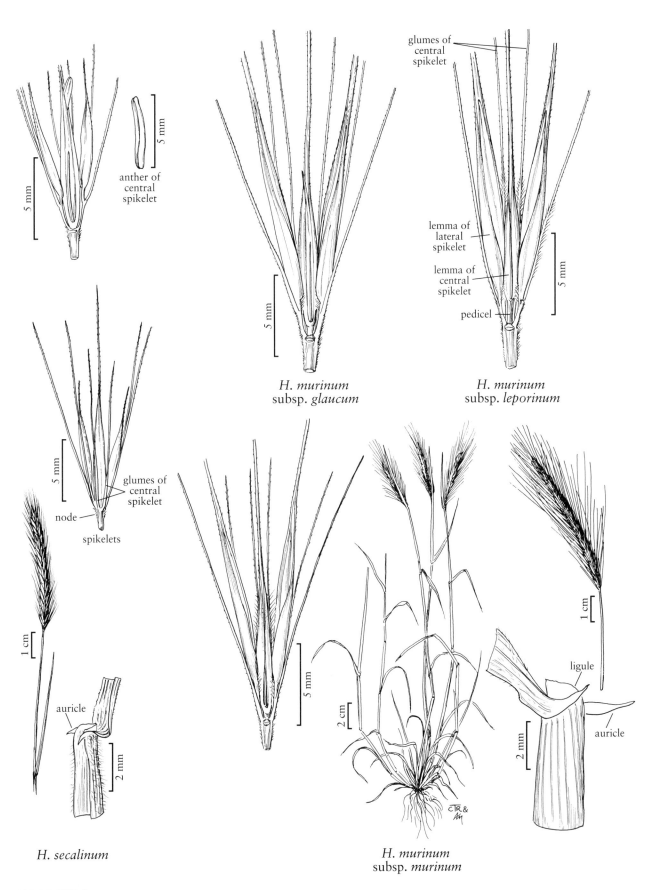

anther of
central
spikelet

5 mm

H. murinum
subsp. *glaucum*

glumes of
central
spikelet

lemma of
lateral
spikelet

lemma of
central
spikelet

pedicel

H. murinum
subsp. *leporinum*

glumes of
central
spikelet

node

spikelets

auricle

H. secalinum

ligule

auricle

H. murinum
subsp. *murinum*

HORDEUM

distribution of the three subspecies, being found from Washington to Arizona, and in scattered locations from Maine to Virginia.

10. Hordeum bulbosum L. [p. 253]
BULBOUS BARLEY

Plants perennial. **Culms** to 135 cm, erect or somewhat geniculate; **basal internodes** usually bulbous, with 1–4 ellipsoid to pyriform bulbs per culm; **nodes** glabrous. **Sheaths** glabrous or pubescent; **auricles** well developed, to 5.5 mm; **blades** to 6 mm wide, flat, scabrous, often also pubescent. **Spikes** 4.5–16.5 cm. **Central spikelets** subsessile; **glumes** 10–25 mm, flattened near the base, margins more or less ciliate; **lemmas** glabrous, awns 12–50 mm; **lodicules** densely pilose; **anthers** 4.5–10 mm, yellow to violet. **Lateral spikelets: lower glumes** flattened near the base; **upper glumes** setaceous; **lemmas** usually unawned, sometimes awned, awns to 14 mm; **anthers** to 9 mm. $2n = 14, 28$.

Hordeum bulbosum is native to the eastern Mediterranean and western Asia. In the *Flora* region, it is known as an occasional escape from breeding programs. In its native range it is found in a wide range of habitats, from wet meadows to dry hillsides, roadsides, and abandoned fields. It is one of two obligate outcrossers in the genus, *H. brevisubulatum* (Trin.) Link being the other.

11. Hordeum vulgare L. [p. 253]
BARLEY, ORGE, ORGE VULGAIRE

Plants summer or winter annuals; loosely tufted. **Culms** to 100(150) cm, usually erect; **nodes** glabrous. **Lower sheaths** pilose; **upper sheaths** glabrous; **auricles** to 6 mm, well developed even on the upper leaves; **blades** to 30 cm long, 5–15 mm wide, flat, scabrous or glabrous. **Spikes** 5–10 cm long, 0.8–2 cm wide, green to purplish or blackish; **nodes** 10–30, with 3 spikelets per node, 0–2 lateral spikelets, in addition to the central spikelets, forming seed at maturity (resulting in 2-, 4-, and 6-rowed barley); **rachises** usually not disarticulating at maturity. **Central spikelets** sessile; **glumes** 10–30 mm, pubescent, flattened near the base; **lemmas** 6–12 mm long, 3+ mm wide, glabrous, sometimes scabrous, particularly distally, unawned or awned, awns 30–180 mm, usually scabrous; **anthers** 6–10 mm, yellowish. **Lateral spikelets** usually sessile if seed-forming, pedicellate if sterile; **pedicels** to 3 mm; **lemmas** usually 6–15 mm, awned when fertile, obtuse to acute when sterile. $2n = 14 (28)$.

Hordeum vulgare is native to Eurasia. Plants in the *Flora* region belong to the cultivated subspecies, **H. vulgare** L. subsp. **vulgare**. The progenitor of cultivated barley, **H. vulgare** subsp. **spontaneum** (K. Koch) Thell., has a brittle rachis, tough awn, and, often, shrunken seeds. It does not grow in the *Flora* region.

Hordeum vulgare subsp. *vulgare* was first domesticated in western Asia. It is now grown in most temperate parts of the world. In the *Flora* region, it occurs as a cultivated species that is often found as an adventive in fields, roadsides, and waste places throughout the region, not just at the locations shown on the map. There are many distinctive, but interfertile, forms. Bothmer et al. (1995) presented an artificial classification of such forms.

13.02 EREMOPYRUM (Ledeb.) Jaub. & Spach

Signe Frederiksen

Plants annual. **Culms** 3–40 cm, geniculate. **Sheaths** open for most of their length; **auricles** present, often inconspicuous; **ligules** 0.4–2 mm, membranous, truncate; **blades** 1–6 mm wide, flat, linear. **Inflorescences** distichous spikes, 0.8–4.5 cm, with 1 spikelet per node, usually erect when mature; **rachis internodes** flat, margins glabrous or with hairs, hairs white; **middle internodes** 0.5–3 mm; **disarticulation** in the rachises, at the nodes beneath each spikelet, or at the base of each floret. **Spikelets** 6–25 mm, including the awns, more than 3 times the length of the internodes, divergent, laterally compressed, with 2–5 bisexual florets, sterile florets distal or absent. **Glumes** equal, 4–19 mm, including the awns, coriaceous, becoming indurate, 1-keeled

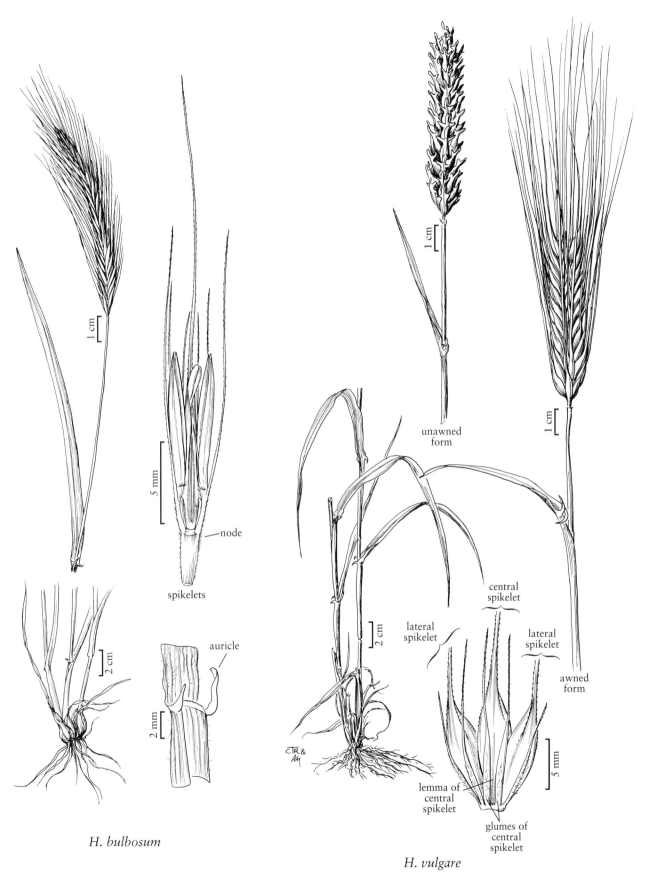

spikelets

node

auricle

2 mm

2 cm

H. bulbosum

unawned
form

1 cm

central
spikelet

lateral
spikelet

lateral
spikelet

awned
form

2 cm

5 mm

lemma of
central
spikelet

glumes of
central
spikelet

H. vulgare

HORDEUM

initially, sometimes 2-keeled at maturity, keels glabrous or hairy, never with tufts of hair, bases slightly connate, apices tapering to a sharp point or straight awn; **lemmas** 5–24 mm, coriaceous, rounded basally, keeled distally, 5-veined, unawned or shortly awned; **paleas** usually shorter and thinner than the lemmas, 2-keeled, ciliate or scabrous distally, keels sometimes prolonged into 2 toothlike appendages; **anthers** 3, 0.4–1.3 mm, yellow. **Ovaries** pubescent; **styles** 2, free to the base. $x = 7$. Haplomes **F, Xe**. Name from the Greek *eremia*, 'desert', and *pyros*, 'wheat'.

Eremopyrum includes 5–10 species that grow in steppes and semidesert regions from Turkey to central Asia and Pakistan. Three species have been found in North America; only *E. triticeum* is widepsread. In their native ranges, species of *Eremopyrum* are valuable fodder on ephemeral spring pastures.

SELECTED REFERENCES Frederiksen, S. 1991. Taxonomic studies in *Eremopyrum* (Poaceae). Nordic J. Bot. 11:271–285; **Melderis, A.** 1985. *Eremopyrum* (Ledeb.) Jaub. & Spach. Pp. 227–231 *in* P.H. Davis (ed.). Flora of Turkey and the East Aegean Islands, vol. 9. University Press, Edinburgh, Scotland. 724 pp.; **Sakamoto, S.** 1979. Genetic relationships among four species of the genus *Eremopyrum* in the tribe Triticeae, Gramineae. Mem. Coll. Agric. Kyoto Univ. 114:1–27.

1. Glumes 1-veined and 2-keeled at maturity; lemmas of the first floret in each spikelet pubescent on the lower $^1/_2$, glabrous distally, the other lemmas glabrous; disarticulation beneath the florets, sometimes at the base of the spikes . . . . .1. *E. triticeum*
1. Glumes usually 3–5-veined and 1-keeled; lemmas of all florets alike in their pubescence or lack thereof; disarticulation at the rachis nodes.
  2. Glume bases straight; spikes 1.3–2.8 cm wide . . . . .2. *E. bonaepartis*
  2. Glume bases arcuately curved; spikes 0.9–1.8 cm wide . . . . .3. *E. orientale*

1. Eremopyrum triticeum (Gaertn.) Nevski [p. 255]
ANNUAL WHEATGRASS

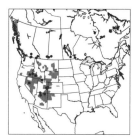

Culms to 30 cm, mostly glabrous, puberulent below the spikes. **Sheaths of upper leaves** inflated; **blades** 1–3(6) mm wide, scabrous or shortly pilose distally. **Spikes** 1.3–2.4 cm long, 0.8–2 cm wide, elliptic, ovate, or nearly circular in outline; **disarticulation** beneath each floret, sometimes at the base of the spikes, not in the rachises. **Spikelets** 6–12 mm, with 2–3 florets. **Glumes** 4–7.5 mm, glabrous, 1-veined and -keeled, becoming 2-keeled by the development of a ridge adjacent to the vein, bases prominently inflated and curved; **lemmas** 5–7.5 mm, prominently keeled towards the subulate apices, lowest lemma in each spikelet pubescent on the proximal $^1/_2$, hairs 0.1–0.15 mm, glabrous distally, the other lemmas glabrous; **palea keels** not prolonged. $2n = 14$.

Eremopyrum triticeum is known primarily from scattered disturbed sites in western North America, from southern Canada to Arizona and New Mexico. Like most weeds, it is probably more widely distributed than herbarium records indicate. It is tolerant of alkaline soils, and is summer-dormant.

2. Eremopyrum bonaepartis (Spreng.) Nevski [p. 255]

Culms to 30 cm, smooth, mostly glabrous, puberulent below the spikes. **Blades** 3–5 mm wide, scabrous distally. **Spikes** 1.4–4.5 cm long, 1.3–2.8 cm wide, oblong, obtuse, or truncate; **disarticulation** at the rachis nodes. **Spikelets** 10–25 mm, with 3–5 florets. **Glumes** 4–19 mm, scabrous or hairy, 3–5-veined, 1-keeled, lateral veins obscure, bases straight; **lemmas** 6–24 mm, glabrous, scabrous, or hirsute, all alike in their pubescence, apices subacute to shortly awned; **palea keels** prolonged into 2 toothlike appendages. $2n = 14, 28$.

In the *Flora* region, *Eremopyrum bonaepartis* is known only from a few collections in Arizona. Several infraspecific taxa have been recognized in Eurasia: subsp. *bonaepartis* has glabrous lemmas and glumes; specimens with pilose, hirsute, or fairly scabrous lemmas and glumes have been referred to as subsp. *hirsutum* (Bertol.) Melderis or subsp. *sublanuginosum* (Drobow) Á. Löve; and those with awned lemmas have been called subsp. *turkestanicum* (Gand.) Tzvelev. No attempt has been made to determine which are present in the *Flora* region.

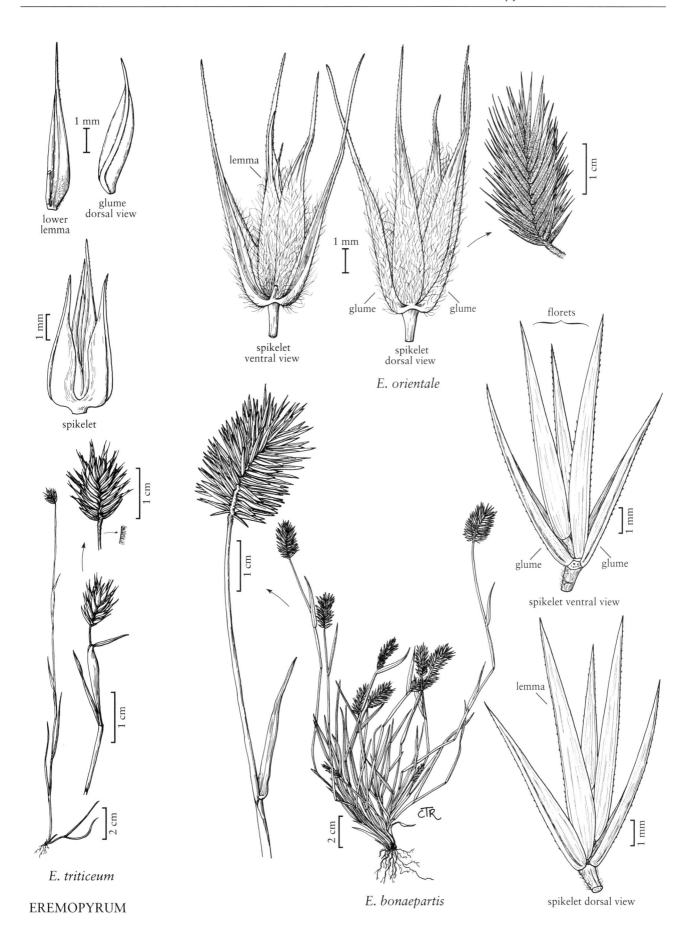

lower
lemma

glume
dorsal view

spikelet

lemma

spikelet
ventral view

glume   glume

spikelet
dorsal view

E. orientale

florets

glume   glume

spikelet ventral view

lemma

spikelet dorsal view

E. triticeum

EREMOPYRUM

E. bonaepartis

3. Eremopyrum orientale (L.) Jaub. & Spach [p. 255]

Culms to 15 cm, smooth, mostly glabrous, puberulent below the spikes. **Sheaths of upper leaves** often somewhat inflated; **blades** 2–3 mm wide, more or less scabrous on both surfaces. **Spikes** 1.3–3.5 cm long, 0.9–1.8 cm wide, usually ovate-elliptic; **disarticulation** at the rachis nodes. **Spikelets** 7–12 mm long, with 2–3 florets. **Glumes** 5–12 mm, lanceolate, 3–5-veined, 1-keeled, lateral 2–3 veins prominent, hispid, bases curved, apices gradually tapering to a 0.5–3 mm awn; **lemmas** 5–12 mm, hispid, hairs 0.5–2 mm, prominently keeled, awned, awns 0.5–4 mm; **paleas** pubescent between the keels, keels prolonged into 2 toothlike appendages. $2n = 28$.

Eremopyrum orientale has been collected from southern Manitoba, growing with *E. triticeum*, and has been reported from southeastern British Columbia and New York. It is not known to be established in the *Flora* region.

13.03 DASYPYRUM (Coss. & Durieu) T. Durand

Signe Frederiksen

Plants annual or perennial; shortly rhizomatous if perennial. **Culms** 20–100 cm. **Sheaths** open; **auricles** present, often inconspicuous; **ligules** 0.3–1 mm, membranous, truncate; **blades** 1–5 mm wide, flat, linear, central vein distinct on the abaxial side. **Inflorescences** terminal spikes, 4–12 cm long including the awns, 0.6–2 cm wide excluding the awns, compressed, dense, with 1 spikelet per node; **rachis internodes** flat, margins ciliate, hairs white; **middle internodes** 1–3 mm; **disarticulation** in the rachises, at the nodes beneath each spikelet. **Spikelets** 25–75 mm including the awns, 7–22 mm excluding the awns, more than 3 times the length of the rachis internodes, usually divergent, sometimes ascending, laterally compressed, with 2–4 florets, the lower 2 florets usually bisexual, the terminal florets sterile; **rachilla internodes** below the lower florets shorter than those below the terminal florets. **Glumes** equal, to 40 mm including the awns, to 8 mm excluding the awns, coriaceous, usually 5-veined, strongly 2-keeled, keels with 1–3 mm hairs, margins unequal, stiff, translucent, apices tapering into scabrous awns; **bisexual lemmas** 9–13 mm excluding the awns, lanceolate, keeled, usually 5-veined, apices acuminate, awned, awns to 60 mm; **sterile lemmas** smaller, awns to 10 mm; **paleas** narrowly lanceolate, membranous, 2-veined, 2-keeled; **lodicules** 2, free, membranous, ciliate or glabrous; **anthers** 3, 4–7 mm, yellow; **ovaries** pubescent; **styles** 2, free to the base. $x = 7$. Haplome **V**. Name from the Greek *dasys*, 'shaggy' or 'hairy', and *pyros*, 'wheat', an allusion to the hairy keels of the glumes.

Dasypyrum is a Mediterranean genus of two species; only one has been collected in the *Flora* region. The hairy, 2-keeled glumes make the genus easily distinguishable from other genera in the *Triticeae*.

SELECTED REFERENCE Frederiksen, S. 1991. Taxonomic studies in *Dasypyrum* (Poaceae). Nordic J. Bot. 11:135–142.

1. Dasypyrum villosum (L.) P. Candargy [p. 257]
MOSQUITOGRASS

Plants annual. **Culms** 20–100 cm, decumbent to geniculate, glabrous. **Blades** 1–5 mm wide, supple, light green. **Spikes** 4–12 cm long, 0.6–2 cm wide. **Spikelets** 7–22 mm excluding the awns. **Glumes** 9–40 mm including the awns, with tufts of hair on the keels, hairs 1.5–3 mm, white; **bisexual lemmas** 10–13 mm, glabrous proximally, sparsely pilose distally, with a tuft of stiff hairs below the awns, awns 15–60 mm, straight; **sterile lemmas** smaller, awns 7–10 mm; **paleas** to 14 mm; **anthers** 3, 4–7 mm. $2n = 14$.

Dasypyrum villosum is native from southern Europe to Turkey, the Crimea, and the Caucasus. The only known North American record is a collection made in Philadelphia County, Pennsylvania, in 1877.

Dasypyrum breviaristatum (H. Lindb.) Fred. differs from *D. villosum* in being perennial and shortly rhizomatous, and in having stiff, dark green leaves, glume keels with hairs that are not in tufts, and awns to 15 mm long.

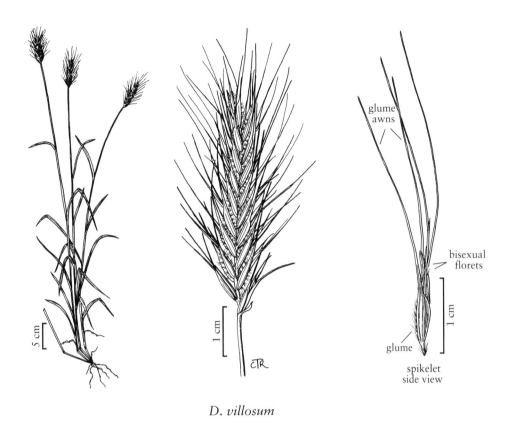

D. villosum

DASYPYRUM

13.04 **TAENIATHERUM** Nevski

J.K. Wipff

Plants annual. **Culms** (5)10–55(70) cm, erect, glabrous; **nodes** 3–6. **Leaves** evenly distributed; **sheaths** open, usually glabrous; **auricles** 0.1–0.5 mm, rarely absent; **ligules** membranous, truncate; **blades** flat to involute. **Inflorescences** spikes, erect; **nodes** 4–24(28), each with 2(3, 4) spikelets; **internodes** 0.5–3.5 mm. **Spikelets** with 2(3) florets, the lowest floret in each spikelet bisexual, the distal floret(s) highly reduced, sterile; **disarticulation** above the glumes. **Glumes** 5–80 mm, equal, awnlike, erect to spreading or reflexed, bases connate. **Bisexual florets: lemmas** 5-veined, glabrous or scabrous, margins flat, scabrous, apices terminally awned, awns 20–110 mm, longer than the lemmas, divergent, often cernuous; **paleas** as long as the lemmas, keels antrorsely ciliate, apices truncate; **lodicules** 2, lobed, ciliate. **Reduced florets: lemmas** 3-veined, awned; **paleas** absent; **anthers** 3, yellow to purple. **Caryopses** narrowly elliptic, with an adaxial groove, apices pubescent. *x* = 7. Haplome **Ta.** Name from the Greek *taenia*, 'ribbon', and *ather*, 'awn'.

Taeniatherum includes only one species. It is native to Eurasia.

SELECTED REFERENCES **Frederiksen, S.** 1986. Revision of *Taeniatherum* (Poaceae). Nordic J. Bot. 6:389–397; **Humphries, C.J.** 1978. Variation in *Taeniatherum caput-medusae* (L.) Nevski. Bot. J. Linn. Soc. 76:340–344.

1. Taeniatherum caput-medusae (L.) Nevski [p. 258]

MEDUSAHEAD

Culms (5)10–55(70) cm. **Auricles** 0.1–0.5 mm, rarely absent; **ligules** 0.2–0.6 mm; **blades** (0.2) 0.7–2.5 mm wide, flat to involute. **Spikes** 1.2–6 cm. **Spikelets** 6–45 mm; **glumes** (5)7–80 mm, awnlike, erect to reflexed. **Bisexual florets: lemmas** 5.5–8 mm, awns (20) 30–110 mm, divergent; **anthers** 0.8–1 mm. **Caryopses** 4–5.2 mm. $2n = 14$.

Taeniatherum caput-medusae is native from Portugal and Morocco east to Kyrgyzstan. It usually grows on stony soils, and flowers from May–June (July). It is an aggressive invader of disturbed sites in the western United States, where it has become a serious problem on rangelands. It has been found as a rare introduction at several sites in the eastern United States, but may not persist there. It is listed as a noxious weed by the U.S. Department of Agriculture.

Frederiksen (1986) recognized three subspecies within *Taeniatherum caput-medusae*, distinguishing among them on the basis of morphology and geography. Plants in the *Flora* region belong to **Taeniatherum caput-medusae** (L.) Nevski subsp. **caput-medusae**. It differs from the other two subspecies in its longer glumes and shorter lemmas.

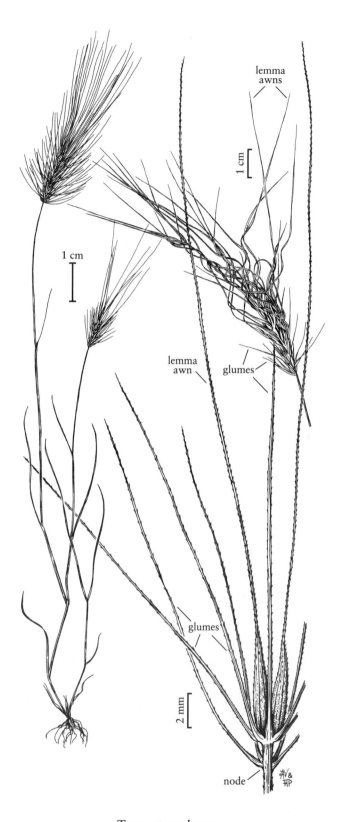

T. caput-medusae

TAENIATHERUM

13.05 SECALE L.

Mary E. Barkworth

Plants annual, biennial, or short-lived perennials; cespitose when perennial. **Culms** 25–120(300) cm. **Sheaths** open; **auricles** usually present, 0.5–1 mm; **ligules** membranous, truncate, often lacerate; **blades** flat or involute. **Inflorescences** laterally compressed, distichous spikes; **middle internodes** 2–4 mm, with 1 spikelet per node, spikelets strongly ascending; **disarticulation** in the rachises, below the spikelets, rachises not or tardily disarticulating in cultivated strains. **Spikelets** 10–18 mm, with 2(3) florets; **florets** bisexual. **Glumes** 8–20 mm, shorter than the adjacent lemmas, linear to subulate, scabrous, margins hyaline, 1-veined, keeled, keels terminating in an awn, awns to 35 mm; **lemmas** 8–19 mm, strongly laterally compressed, strongly keeled, keels conspicuously scabrous distally, scabrules 0.6–1.3 mm, apices tapering to a scabrous awn, awns 2–50 mm; **anthers** 3, 2.3–12 mm, yellow. $x = 7$. Haplome **R.** *Secale* is the classical Latin name for rye.

Secale is a genus of three species. All are native to the Mediterranean region and western Asia; two species have been collected in the *Flora* region. *Secale cereale* is cultivated as a crop and used along roadsides to prevent soil erosion, and is established in the *Flora* region. *Secale strictum* has been cultivated experimentally; it is not established in the *Flora* region.

Unlike other cereal grasses such as *Triticum*, *Hordeum*, and *Avena*, species of *Secale* are outcrossing, although *S. sylvestre* Host is reported to be self-compatible. All three species are diploids. Remains of cultivated rye dating to 6000 B.C. have been found in Turkey.

×*Triticosecale* is an artificially derived hybrid between *Triticum* and *Secale* that is now widely cultivated (see p. 261).

SELECTED REFERENCES Frederiksen, S. and G. Petersen. 1998. A taxonomic revision of *Secale* L. (Triticeae, Poaceae). Nordic J. Bot. 18:399–420; Hitchcock, A.S. 1951. Manual of the Grasses of the United States, ed. 2, rev. A. Chase. U.S.D.A. Miscellaneous Publication No. 200. U.S. Government Printing Office, Washington, D.C., U.S.A. 1051 pp.

1. Plants annual or biennial; rachises not or tardily disarticulating; lemmas 14–18 mm long . . . . 1. *S. cereale*
1. Plants perennial; rachises readily disarticulating; lemmas 8–16 mm long . . . . 2. *S. strictum*

1. Secale cereale L. [p. 260]
RYE, SEIGLE, SEIGLE CULTIVÉ

Plants annual or biennial. **Culms** (35)50–120(300) cm. **Blades** (3)4–12 mm wide, usually glabrous. **Spikes** (2) 4.5–12(19) cm, often nodding when mature; **disarticulation** tardy, in the rachises, at the nodes, or not occurring. **Glumes** 8–20 mm, keels scabrous, terminating in awns, awns 1–3 mm; **lemmas** 14–18 mm, awns 7–50 mm; **anthers** about 7 mm. $2n = 14, 21, 28$.

Secale cereale is one of the world's most important cereal grasses; it is also widely used in North America for soil stabilization and, particularly in Canada, for whisky. When dry, the spike is often distinctly nodding.

Frederiksen and Petersen (1998) placed cultivated plants with a non-disarticulating rachis into **Secale cereale** L. subsp. **cereale**, and wild or weedy plants with more fragile rachises into **S. cereale** subsp. **ancestrale** Zhuk.

2. Secale strictum (C. Presl) C. Presl [p. 260]

Plants perennial, short-lived; cespitose. **Culms** (35) 60–100(150) cm. **Blades** 2–8 mm wide, glabrous or scabridulous. **Spikes** (3.5)5–8(11) cm; **disarticulation** occurring readily, in the rachises, at the nodes. **Glumes** 8–11 mm, densely scabrous on the keels, acuminate or awned, awns 3–4 mm; **lemmas** 8–16 mm, awns 2–25 mm. $2n = 14$.

Secale strictum is native to Eurasia and, as a disjunct, to South Africa. It grows on dry, stony or sandy soils, often in mountainous areas. So far as is known, it is not established in the *Flora* region.

Hitchcock (1951) reported that *Secale strictum* had become established around the Agricultural Experiment Station in Pullman, Washington, but it is no longer present there. Prior to 1931, the station worked on development of a *S. cereale* × *S. strictum* strain that

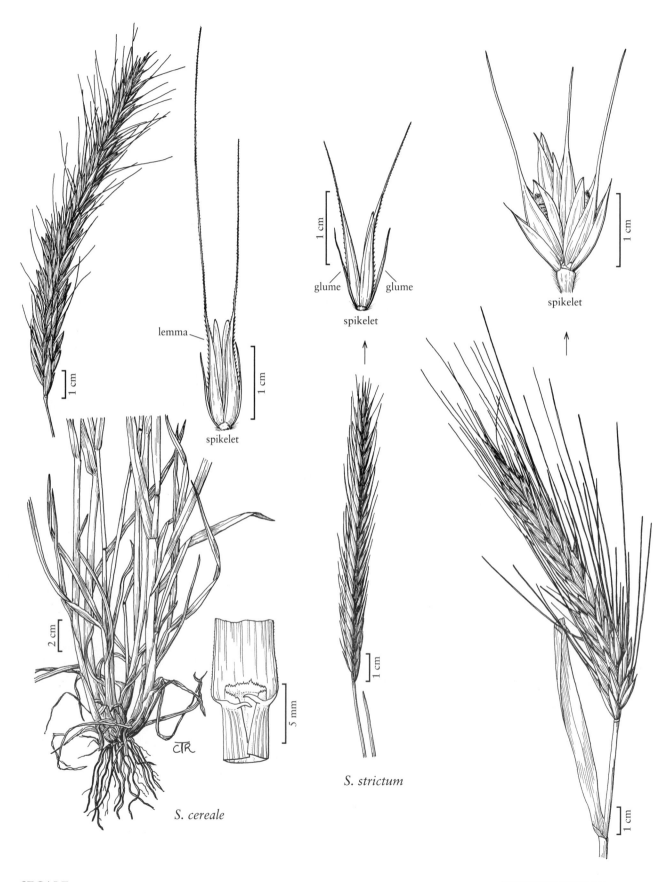

lemma

spikelet

glume   glume

spikelet

spikelet

S. strictum

S. cereale

CTR

would combine the perennial habit with good seed production. The attempt had been abandoned by 1931, but hybrid seed had been distributed as 'Michael's Grass'. The seed was originally thought to be derived from a *Triticum aestivum* × *Leymus racemosus* cross; subsequent studies, both morphological and cytological, revealed that it was *S. cereale* × *S. strictum*.

13.06 ×TRITICOSECALE Wittm. *ex* A. Camus [p. 260]

TRITICALE

Mary E. Barkworth

Plants annual. **Culms** to 130 cm, erect, straight or geniculate at the lowest node. **Leaves** mainly cauline; **ligules** 2–4 mm, membranous, truncate to rounded. **Inflorescences** terminal, distichous spikes, with solitary spikelets; **internodes** 3–5 mm, densely pilose, at least on the edges. **Spikelets** 10–17 mm, with 2–4 florets, distal floret usually reduced. **Glumes** 9–12 mm, asymmetrically keeled, keels stronger and sometimes conspicuously ciliate distally, apices retuse to acute, awned, awns 3–4 mm; **lemmas** 10–15 mm, laterally compressed, keeled, keels sometimes ciliate distally, terminally awned, awns 3–50 mm; **anthers** 3, yellow. $x = 7$.

×*Triticosecale* comprises hybrids between *Secale* and *Triticum*. Natural hybrids between the two genera are rare, but Triticale, which consists of cultivars derived from artificial hybrids between *S. cereale* and *T. aestivum*, is becoming an increasingly important cereal crop. The existing names in ×*Triticosecale* do not apply to Triticale because they involve different species of *Triticum*.

Triticale cultivars often have a complex ancestry, involving multiple hybridizations, backcrossings, and artificially induced chromosome doubling. Their genetic material varies from being derived almost entirely from *Triticum* to being derived almost entirely from *Secale*. For this reason, they are best identified as such, e.g., ×*Triticosecale* 'Newton' or ×*Triticosecale* 'Bokolo'. A hybrid formula would misrepresent their genetic and morphological diversity.

SELECTED REFERENCE Stace, C.A. 1987. Triticale: A case of nomenclatural mistreatment. Taxon 36:445–452.

13.07 AEGILOPS L.

Sandra M. Saufferer

Plants annual. **Culms** 14–80 cm, usually glabrous, erect or geniculate at the base, with (1)2–4(5) nodes. **Sheaths** open; **auricles** ciliate; **ligules** 0.2–0.8 mm, membranous, truncate; **blades** 1.5–10 mm wide, linear to linear-lanceolate, flat, spreading. **Inflorescences** terminal spikes, with 2–13 spikelets, usually with 1–3 additional rudimentary spikelets at the base; **internodes** 6–12 mm; **disarticulation** either at the base of the spikes or in the rachises, the spikelets falling attached to the internodes above or below. **Spikelets** solitary at each node, $^1/_2$–2(3) times the length of the internodes, tangential to the rachis, appressed or ascending, the upper spikelet(s) sometimes sterile; **fertile spikelets** 5–15 mm, with 2–7 florets, the distal florets often sterile. **Glumes** ovate to rectangular, rounded on the back, scabrous or pubescent, with several prominent veins, midveins smooth throughout, apices truncate, toothed, or awned, sometimes indurate at maturity; **lemmas** rounded on the back, apices toothed, frequently awned; **paleas** chartaceous, 2-keeled, keels ciliate; **anthers** 3, 1.5–4 mm, not penicillate; **ovaries** with pubescent apices. **Caryopses** lanceolate to lanceolate-ovate. $x = 7$. Haplomes **B, C, D, S, T, U, M, N**. The name is derived from from the Greek *aigilops*, a word which has multiple etymological interpretations (Slageren 1994, pp. 19–20, 118–119), including 'wild oats' and several variants related to the Greek *aigos*, 'goat'.

Aegilops has about 23 species, and is native to the Canary Islands, as well as from the Mediterranean region to central Asia. It is sometimes included in *Triticum* because the two form natural hybrids and both are involved in the evolution of the cultivated wheats, including *T. aestivum*. They are treated as distinct genera here, in keeping both with past practice and with their differing ecological attributes, *Aegilops* being a weedy genus.

Four species are established in the *Flora* region; only *Aegilops cylindrica* is widespread. The introductions occurred at the end of the nineteenth or beginning of the twentieth century. Three other species have been collected in the region; they are not known to have persisted.

In the key and descriptions, spike and spikelet lengths exclude the rudimentary spikelets and awns.

SELECTED REFERENCES **Kimber, G.** and **M. Feldman.** 1987. Wild Wheat: An Introduction. Special Report No. 353. College of Agriculture, University of Missouri-Columbia, Columbia, Missouri, U.S.A. 142 pp.; **Slageren, M.W. van.** 1994. Wild Wheats: A Monograph of *Aegilops* L. and *Amblyopyrum* (Jaub. & Spach) Eig. Wageningen Agricultural University Papers 94–7. Wageningen Agricultural University and International Center for Agricultural Research in the Dry Areas (ICARDA), Wageningen, The Netherlands and Aleppo, Syria. 512 pp.

1. Glumes unawned, or with a single awn to 2 cm long; spikes narrowly cylindrical to moniliform, not ovoid; disarticulation in the rachises, the spikelets falling attached to the internodes above.
    2. Spikes cylindrical to slightly moniliform . . . . .1. *A. tauschii*
    2. Spikes distinctly moniliform.
        3. Glumes mostly glabrous, the veins setulose; lemmas of the apical spikelets with awns to 4 cm long; spikelets with 2–5 florets, the distal 1 or 2 sterile . . . . .2. *A. ventricosa*
        3. Glumes appressed-velutinous; lemmas of the apical spikelets with awns 3–8.5 cm long; spikelets with 4–7 florets, the distal 2 sterile . . . . .3. *A. crassa*
1. Some glumes with awns 2–8 cm long; spikes narrowly cylindrical to ovoid, not moniliform; disarticulation near the base of the spikes, at least initially.
    4. Spikes narrowly cylindrical, about 0.3 cm wide . . . . .4. *A. cylindrica*
    4. Spikes subcylindrical to ovoid, widest at the base, 0.4–1.3 cm wide.
        5. Upper spikelets 7–9 mm long; lemmas of the lower fertile florets with 2–3 teeth, 1 tooth sometimes extending into an awn up to 10 mm long . . . . .5. *A. triuncialis*
        5. Upper spikelets 4–5 mm long; lemmas of the fertile florets 2–3-awned, awns 5–40 mm long.
            6. Rudimentary spikelet(s) usually 1, occasionally 2; spikes gradually tapering distally . . . 6. *A. geniculata*
            6. Rudimentary spikelets 3, occasionally 2; spikes abruptly contracted distally to a narrow cylinder . . . . .7. *A. neglecta*

1. **Aegilops tauschii** Coss. [p. 264]
TAUSCH'S GOATGRASS, ROUGH-SPIKED HARDGRASS

Culms 20–45 cm, geniculate at the base, usually forming many tillers. **Sheaths** with hyaline margins, margins of the lower cauline sheaths usually ciliate; **blades** 6–20 cm long, 3–6 mm wide. **Spikes** 5.5–8.2 cm long, 0.3–0.6 cm wide, narrowly cylindrical to slightly moniliform, with 6–13 spikelets; **rudimentary spikelets** absent; **disarticulation** in the rachises, the spikelets falling attached to the internodes above. **Spikelets** 7–8 mm long, 3–4 mm wide, cylindrical, scabrous, with 2–5 florets, the distal 2 florets sterile. **Glumes** 5–7 mm, coriaceous, scabrous, rims thickened, apices obtuse to truncate, minutely denticulate, unawned; **lemmas** 6–8 mm, mucronate or awned, awns usually solitary; **lemmas of lower spikelets** with shorter awns than those of the upper spikelets, the lower lemmas in a spikelet with shorter awns than those of the distal florets. **Caryopses** 5–6 mm, adhering to the lemmas and paleas. **Haplome D.** 2*n* = 14.

Aegilops tauschii is a weed of disturbed areas. In the *Flora* region, it is known only from Riverside County,

California; Cochise County, Arizona; and an old collection from Westchester County, New York. It is native from the Caucasus and southern shores of the Caspian Sea, eastward to Kazakhstan and western China, and southward to Iraq and northwestern India.

2. Aegilops ventricosa Tausch [p. 264]
SWOLLEN GOATGRASS, BELLY-SHAPED HARDGRASS

Culms 25–70 cm, erect to slightly geniculate at the base, forming few to many tillers. Sheaths with hyaline margins, ciliate; blades 7–15 cm long, 3–6 mm wide. Spikes 5–12 cm long, 0.3–0.6 cm wide, distinctly moniliform, with 3–11 fertile spikelets; rudimentary spikelets usually absent, sometimes 1–2; disarticulation in the rachises, the spikelets falling attached to the internodes above. Fertile spikelets 7–11 mm, urceolate, with 2–5 florets, the distal 1–2 florets sterile. Glumes of fertile spikelets mostly glabrous, veins setulose, apices truncate, with a sharply acute tooth to 3 mm; glumes of apical spikelets 7–8 mm, with a central tooth or awn flanked by 2 short teeth, central awns to 0.9 cm on the lower glumes, to 2 cm on the upper glumes; lemmas of fertile spikelets 8–10 mm, adaxial surfaces velutinous, apices awned, awns 0.3–3 cm, solitary; lemmas of apical spikelets awned, awns to 4 cm. Caryopses 5–7 mm, adhering to the lemmas and paleas. Haplomes DN. 2*n* = 28.

In the *Flora* region, *Aegilops ventricosa* was collected once in New Castle County, Delaware. It is native to the Mediterranean area. The Arabic name, *Oum el guemah*, translates as 'mother of [durum] wheat'. It occasionally forms hybrids with *Triticum durum*, although the two species have no haplomes in common.

3. Aegilops crassa Boiss. [p. 264]
PERSIAN GOATGRASS

Culms 30–50 cm, erect, slightly geniculate, with few to many tillers. Sheaths ciliate near the throat; blades 8–25 cm long, 3–4 mm wide. Spikes (4.5) 6–10(13) cm long, 0.5–0.7 cm wide, moniliform, with (4)6–12 fertile spikelets; rudimentary spikelets usually absent, occasionally 1–2; disarticulation in the rachises, the spikelets falling attached to the internodes above. Spikelets 7–14 mm long, 4–7 mm wide, ovate, pubescent, with 4–7 florets, the distal 2 florets sterile. Glumes 7–10 mm, coriaceous, appressed-velutinous, 2–3-toothed; lemmas of lower spikelets 8–10 mm,

toothed, usually awned; lemmas of apical spikelets awned, awns 3–8.5 cm, usually diverging. Caryopses about 7 mm, adhering to the lemmas and paleas. Haplomes DM, DDM. 2*n* = 28, 42.

The single record of *Aegilops crassa* for North America is a specimen collected "from about the Yonkers Wood Mill [in Yonkers, New York]" in 1898. The species is native from Egypt to central Asia.

4. Aegilops cylindrica Host [p. 266]
JOINTED GOATGRASS

Culms 14–50 cm, erect to decumbent at the base, usually with many tillers. Sheaths with hyaline margins, sometimes ciliate; blades 3–15 cm long, 2–5 mm wide. Spikes 2.2–12 cm long, about 0.3 cm wide, narrowly cylindrical, 10–45 times longer than wide, with (2)3–8(12) fertile spikelets; rudimentary spikelets absent or 1–2; disarticulation initially at the base of the spikes and secondarily in the rachises, the spikelets remaining attached to the internodes above. Fertile spikelets 9–12 mm, narrowly cylindrical, scabrous or pubescent, with 3–5 florets, the lower (1)2–3 florets fertile. Glumes of fertile spikelets awned, awns of the lower spikelets 2–5 mm; glumes of apical spikelets 7–9 mm, scabrid, awned, awns 3–6 cm, usually flanked by 2 lateral teeth; lemmas of lower fertile spikelets 9–10 mm, adaxial surfaces velutinous distally, apices mucronate or awned, awns 0.1–0.5 cm; lemmas of apical spikelets 1-awned, awns (2)4–8 cm, flanked by 2 teeth. Caryopses 6–7 mm, adhering to the lemmas and paleas. Haplomes DC. 2*n* = 28.

Aegilops cylindrica is a widespread weed in North America, being particularly troublesome in winter wheat. It usually grows in disturbed sites, such as roadsides, fields, and along railroad tracks. It is native to the Mediterranean region and central Asia, and is adventive in other temperate countries. Its apparent absence from Canada is somewhat remarkable.

Hybrids between *Aegilops cylindrica* and *Triticum aestivum*, called ×**Aegilotriticum sancti-andreae** (Degen) Soó [p. 266], have been found in various parts of North America. They often have a few functional seeds which, on maturing into reproductive plants, can backcross to either parent. For this reason, *A. cylindrica* is considered a serious weed in many wheat-growing areas within the *Flora* region.

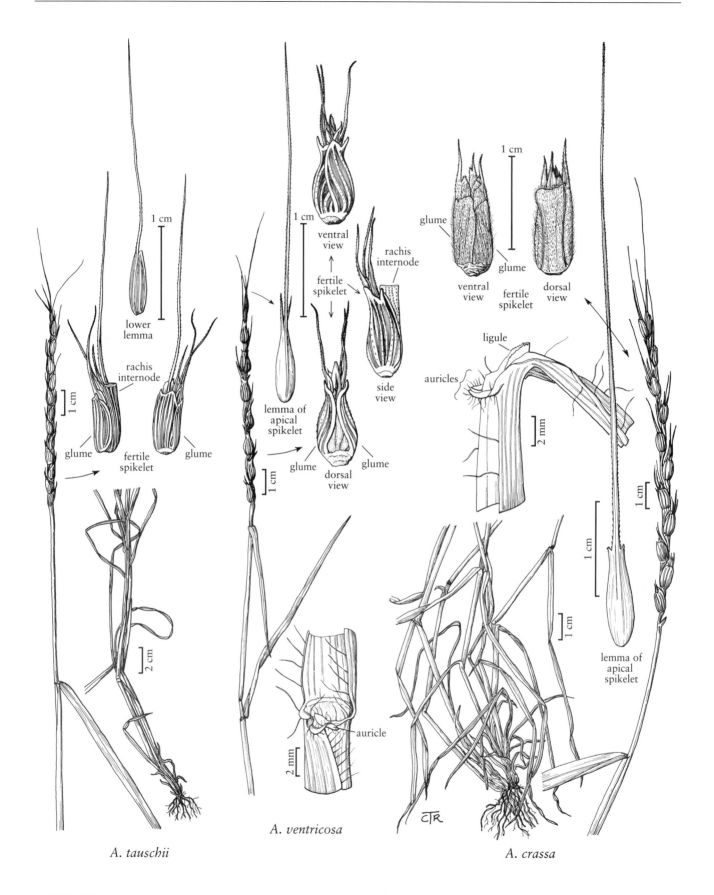

A. tauschii

A. ventricosa

A. crassa

AEGILOPS

5. Aegilops triuncialis L. [p. 266]
BARBED GOATGRASS

Culms 17–60 cm, geniculate to semiprostrate at the base, usually with several tillers. Sheaths with hyaline margins, lower cauline sheath margins usually ciliate; blades 1.5–7 cm long, 2–3 mm wide. Spikes 2.2–6 cm long, bases 0.4–0.5 cm wide, narrowly ellipsoid, becoming subcylindrical distally, with 2–7 fertile spikelets; rudimentary spikelets (2)3; disarticulation at the base of the spikes. Lower fertile spikelets 7–13 mm, lanceolate-ovate, with 3–5 florets, the first 1–2 florets fertile; upper spikelets 7–9 mm, reduced. Glumes of lower fertile spikelets 6–10 mm, 2–3-awned, awns 1.5–6 cm, glabrous, scabrous, or velutinous, if 3-awned, the central awn often shorter than the lateral awns, sometimes reduced to a tooth; glumes of apical spikelets 6–8 mm, 3-awned or with 1 awn and 2 lateral teeth, awns 2.5–8 cm, if 3-awned, the central awn the longest; lemmas of lower fertile spikelets 7–11 mm, with 2–3 teeth, if 3-toothed, the central tooth the longest, sometimes extending into a 10 mm awn. Caryopses 5–8 mm, falling free from the lemmas and paleas. Haplomes UC. $2n = 28$.

North American collections of *Aegilops triuncialis* are from disturbed sites, mostly roadsides and railroads. The native range of the species extends from the Mediterranean area east to central Asia and south to Saudi Arabia. Specimens from the *Flora* region belong to **Aegilops triuncialis** L. var. **triuncialis**, in which the glumes of the apical spikelets have a 5–8 cm central awn flanked by shorter (1–3 cm) lateral awns, and the glumes of the lower fertile spikelets have 2–3 awns of 1.5–6 cm. In *A. triuncialis* var. *persica* (Boiss.) Eig, the glumes of the apical spikelets have a 2.5–5.5 cm central awn, and 2 lateral awns of 0.6–2.5 cm that are sometimes reduced to teeth; the lower fertile spikelets have 2 teeth, or 1 tooth and 1 awn to 1 cm long.

6. Aegilops geniculata Roth [p. 267]
OVATE GOATGRASS

Culms 20–40 cm, geniculate at the base, usually with many tillers. Sheaths with hyaline margins, the distal portion of the lower cauline sheaths ciliate; blades 2–7.5 cm long, 2–5 mm wide. Spikes 1–3 cm long, bases 0.4–0.7 cm wide, narrowly ovoid to ellipsoid, gradually tapering distally, with (2)3–4 spikelets, the distal spikelet sterile; rudimentary spikelets 1(2);

disarticulation at the base of the spikes, above the rudimentary spikelets. Fertile spikelets 7–10 mm long, 3–4 mm wide, urceolate; lower spikelet with 3–4 florets, the lower 1–2 florets fertile; apical spikelets 4–5 mm long, 1–2 mm wide, narrowly obovoid, with 1 floret, floret reduced, sterile. Glumes of fertile spikelets 6–10 mm, ovate, smooth, scabrous, appressed-velutinous, (3)4(5)-awned, awns 2–4.5 cm; glumes of apical spikelets about 3 mm, 4-awned, awns usually 1–3.5 cm; lemmas of fertile spikelets 6–8 mm, adaxial surfaces often velutinous distally, apices 2–3-awned, awns 1–2.5 cm. Caryopses 4–6 mm, falling free from the lemmas and paleas. Haplomes MU. $2n = 28$.

In the *Flora* region, *Aegilops geniculata* is known only from Mendocino County, California, where it usually occurs along roadsides. It is native from the Mediterranean area to central Asia. In California, it grows in silty clay.

7. Aegilops neglecta Req. *ex* Bertol. [p. 267]
THREE-AWNED GOATGRASS

Culms 20–55 cm, geniculate and semiprostrate at the base, usually with many tillers. Sheaths with hyaline margins, ciliate; blades 5–8 cm long, 3–4 mm wide. Spikes 2–4.5 cm long, 0.4–1.3 cm wide, ovoid-ellipsoid proximally, abruptly contracted distally to a narrow cylinder, with 3–6 spikelets, the distal 1–3 spikelets sterile; rudimentary spikelets (2)3; disarticulation at the base of the spikes. Fertile spikelets 8–12 mm, subventricose, urceolate, with 2 fertile florets and 2–3 sterile florets; apical spikelets 4–5 mm, usually with 1 sterile floret, cultivated forms often with 1 fertile floret and 1–2 sterile florets. Glumes of fertile spikelets 7–10 mm, usually velutinous, sometimes scabrous, apices with (2)3 awns, awns 2–5.5 cm; lemmas of fertile spikelets 2-awned, awns 0.5–4 cm, sometimes with 1 lateral tooth. Caryopses 5–7 mm, falling free of the lemmas and paleas. Haplomes UM, UMN. $2n = 28, 48$.

Aegilops neglecta is native around the Mediterranean and in western Asia. It has been collected in Arlington County, Virginia, and near Corvallis, Oregon; the Oregon record indicates it is persisting from previous cultivation and becoming weedy.

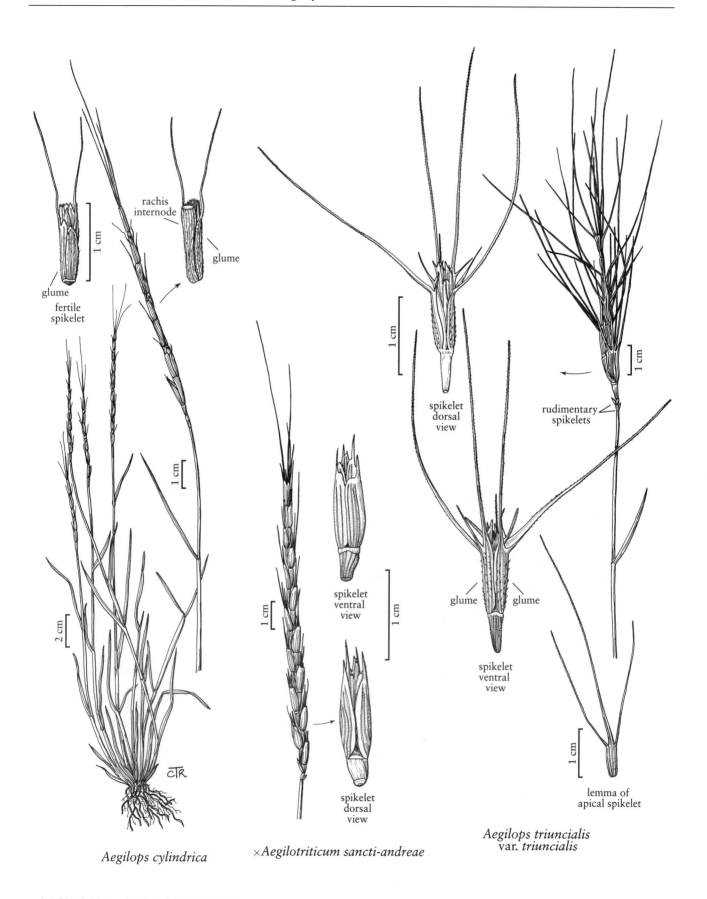

rachis internode

glume

glume

fertile spikelet

1 cm

1 cm

2 cm

CTR

Aegilops cylindrica

spikelet ventral view

1 cm

1 cm

spikelet dorsal view

×*Aegilotriticum sancti-andreae*

spikelet dorsal view

1 cm

rudimentary spikelets

1 cm

glume   glume

spikelet ventral view

1 cm

lemma of apical spikelet

Aegilops triuncialis var. *triuncialis*

AEGILOPS · ×AEGILOTRITICUM

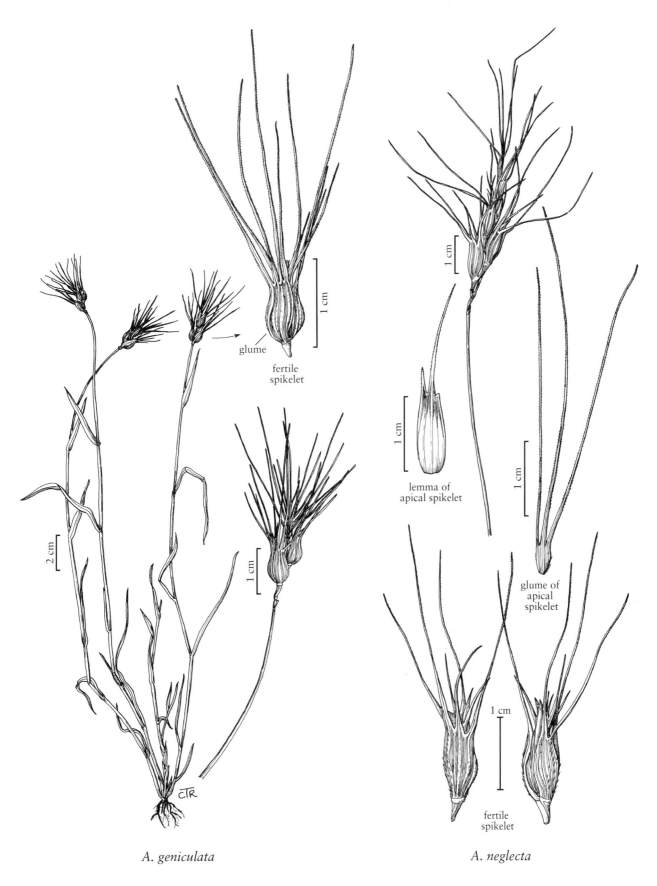

glume

fertile
spikelet

lemma of
apical spikelet

glume of
apical
spikelet

fertile
spikelet

A. geniculata

A. neglecta

AEGILOPS

13.08 TRITICUM L.

Laura A. Morrison

Plants annual. **Culms** 14–180 cm, solitary or branched at the base; **internodes** usually hollow throughout in hexaploids, usually solid for about 1 cm below the spike in diploids and tetraploids, even if hollow below. **Sheaths** open; **auricles** present, often deciduous at maturity; **ligules** membranous; **blades** flat, glabrous or pubescent. **Inflorescences** usually terminal spikes, distichous, with 1 spikelet per node, occasionally branched; **internodes** (0.5)1.4–8 mm; **disarticulation** in the rachises, the spikelets usually falling with the internode below to form a wedge-shaped diaspore, sometimes falling with the adjacent internode to form a barrel-shaped diaspore, domesticated taxa usually non-disarticulating, or disarticulating only under pressure. **Spikelets** 10–25(40) mm, usually 1–3 times the length of the internodes, appressed to ascending, with 2–9 florets, the distal florets often sterile. **Glumes** subequal, ovate, rectangular, or lanceolate, chartaceous to coriaceous, usually stiff, tightly to loosely appressed to the lower florets, with 1 prominent keel, at least distally, keels often winged and ending in a tooth or awn, a second keel or prominent lateral vein present in some taxa; **lemmas** keeled, chartaceous to coriaceous, 2 lowest lemmas usually awned, awns 3–23 cm, scabrous, distal lemmas unawned or awned, awns to 2 cm; **paleas** hyaline-membranous, splitting at maturity in diploid taxa; **anthers** 3. **Caryopses** tightly (hulled wheats) or loosely (naked wheats) enclosed by the glumes and lemmas, lemmas and paleas not adherent; **endosperm** flinty or mealy. $x = 7$. Haplomes **A**, **B**, **D**, and **G**. *Triticum* is the classical Latin name for wheat.

Triticum is a genus of approximately 25 wild and domesticated species. It was first cultivated in western Asia at least 9000 years ago and is now the world's most important crop, being planted more widely than any other genus.

Triticum is native to western and central Asia. It includes diploids (**A** haplome), tetraploids (**AB** or **AG** haplomes), and hexaploids (**ABD** or **AAG** haplomes). The world's only reserve designed to protect native populations of wild cereals, the Erebuni Reserve, is located just outside Yerevan, Armenia. It is home to three wild species of *Triticum*: *T. araraticum* Jakubz., *T. boeoticum*, and *T. urartu*.

Only *Triticum aestivum*, *T. durum*, and *T. spelta* are grown commercially in North America, *T. aestivum* being by far the most important. The remaining species in this treatment are those most frequently grown by North American plant breeders and wheat researchers. None of the species has become an established part of the North American flora, but they may be encountered as escapes near agricultural fields and research stations, or along transportation routes.

Triticum is sometimes treated as including *Aegilops*, but the taxa in these genera differ morphologically and ecologically. In addition, species of *Triticum sensu stricto* are unique in possessing the **A** haplome, of which there are two forms, **A**^{u} and **A**^{b}. Of the other three haplomes present in *Triticum*, the **D** haplome is derived from *Aegilops tauschii* (Dvorak et al. 1998), and the **B** and **G** haplomes are derived from *A. speltoides* Tausch (Giorgi et al. 2003).

The treatment presented here is based on Dorofeev and Migushova's (1979) monograph of *Triticum sensu stricto*, with some modification. The assignment of species status to domesticated forms is controversial. It is done here for convenience, and to aid in distinguishing between taxa with distinct morphological, ecological, and evolutionary traits. For a genomically based treatment, see Kimber and Sears (1987).

"Spring wheat" and "winter wheat" refer to the growing season. Spring wheat is planted in the spring and harvested in the summer of the same year; winter wheat is planted in the fall and harvested the following summer. "Hard wheat" and "soft wheat" are terms used to describe wheats with flinty or mealy endosperm, respectively. Flinty endosperm has a higher protein content and is harder than mealy endosperm. At the species level, soft wheat refers to *Triticum aestivum*; hard wheat refers to *T. durum*. Within *T. aestivum*, endosperm type also is graded as either soft or hard; it is never as hard (flinty) as in *T. durum*.

The width of a spike is the distance from one spikelet edge to the other across the two-rowed side of the spike; its thickness is the distance across the frontal face of spikelet, from one edge to the other. The spike and spikelet measurements do not include the awns. The glumes are measured from the base to the shoulder, and do not include any toothed tip.

SELECTED REFERENCES **Dorofeev, V.F.** and **E.F. Migushova.** 1979. Pshenitsa [Wheat]. (Vol. 1 *in* V.F. Dorofeev and O.N. Korovina (eds.). Kul'turnaia Flora SSSR [Flora of Cultivated Plants of the USSR]). Kolos, Leningrad [St. Petersburg], Russia. 346 pp. [In Russian; English translation (in prep.) used]; **Dvorak, J., Z.-C. Luo, Z.-L. Yang,** and **H.-B. Zhang.** 1998. The structure of the *Aegilops tauschii* genepool and the evolution of hexaploid wheat. Theor. Appl. Genet. 97:657–670; **Giorgi, D., R. D'Ovidio, O.A. Tanzarella, C. Ceoloni,** and **E. Porceddu.** 2003. Isolation and characterization of S genome specific sequences from *Aegilops* sect. *Sitopsis* species. Genome 46:478–489; **Kimber, G.** and **E.R. Sears.** 1987. Evolution in the genus *Triticum* and the origin of cultivated wheat. Pp. 154–164 *in* E.G. Heyne (ed.). Wheat and Wheat Improvement, ed. 2. American Society of Agronomy, Madison, Wisconsin, U.S.A. 765 pp.; **Morrison, L.A.** 2001. The Percival Herbarium and wheat taxonomy: Yesterday, today, and tomorrow Pp. 65–80 *in* P.D.S. Caligari and P.E. Brandham (eds.) Wheat Taxonomy: The Legacy of John Percival. Linnean Special Issue No. 3. Academic Press, London, England. 190 pp.; **Percival, J.** 1921. The Wheat Plant. Duckworth, London, England. 463 pp.

1. Culms usually hollow to the base of the spikes; glumes with only 1 keel, this often developed only in the upper $^1/_2$ of the glumes.
    2. Glumes loosely appressed to the lower florets; rachises not disarticulating, even under pressure . . . . 11. *T. aestivum*
    2. Glumes tightly appressed to the lower florets; rachises disarticulating under pressure.
        3. Spikes strongly flattened; glumes acute . . . . 5. *T. timopheevii* (in part)
        3. Spikes almost cylindrical; glumes truncate . . . . 12. *T. spelta*
1. Culms partially to completely solid 1 cm below the spikes; glumes with 1 fully developed keel, sometimes with a second keel.
    4. Rachises not disarticulating, even under pressure; glumes loosely appressed to the lower florets.
        5. Glumes chartaceous.
            6. Glumes 6–13 mm long; rachises not enlarged at the base of the glumes; spikelets usually producing 1 caryopsis . . . . 4. *T. monococcum* (in part)
            6. Glumes 20–40 mm long; rachises enlarged at the base of the glumes; spikelets producing 2–3 caryopses . . . . 9. *T. polonicum*
        5. Glumes coriaceous.
            7. Glumes awned, awns 1–6 cm long; spikes thicker than wide, never branched at the base . . . . 10. *T. carthlicum*
            7. Glumes toothed, teeth to 0.3 cm long; spikes about as wide as thick, sometimes branched at the base.
                8. Spikes 4–11 cm long, never branched at the base; plants 60–160 cm tall; blades usually glabrous; endosperm usually flinty . . . . 7. *T. durum*
                8. Spikes 7–14 cm long, sometimes branched at the base; plants 120–180 cm tall; blades hairy; endosperm mealy . . . . 8. *T. turgidum*
    4. Rachises disarticulating spontaneously or with pressure; glumes usually tightly appressed to the lower florets.
        9. Paleas splitting at maturity; spikelets 10–17 mm long.
            10. Spikelets elliptical to ovate; rachis internodes 1.4–2.5 mm long; rachises disarticulating with pressure . . . . 4. *T. monococcum* (in part)
            10. Spikelets rectangular; rachis internodes 3–5 mm long; rachises disarticulating spontaneously.
                11. Caryopses blue or amber or red mottled with blue; third lemma in each spikelet, if present, usually unawned; blades blue-green, hairs stiff, those on the veins longer than those between; anthers 3–6 mm long . . . . 1. *T. boeoticum*

11. Caryopses red; third lemma in each spikelet, if present, awned, the awns up to 10 mm long; blades yellow-green, hairs soft, of uniform length; anthers 2–4 mm long . . . . 2. *T. urartu*
9. Paleas not splitting at maturity; spikelets 10–25 mm long.
   12. Spikelets oblong to rectangular; rachises disarticulating spontaneously; glumes unequally 2-keeled, the more prominent keel winged to the base . . . . 3. *T. dicoccoides*
   12. Spikelets elliptical to ovate; rachises disarticulating only with pressure; glumes with 1 prominent keel, the keel not winged to the base.
      13. Spikelets 16–18 mm long; rachis internodes 1.5–2.5 mm long; spikes always wider than thick; culms partially solid to hollow for 1 cm below the spikes . . . . 5. *T. timopheevii* (in part)
      13. Spikelets 10–16 mm long; rachis internodes (0.5)2–5 mm long; spikes variously shaped, from cylindrical to wider than thick; culms usually solid for 1 cm below the spike . . . . 6. *T. dicoccum*

1. Triticum boeoticum Boiss. [p. 271]
WILD EINKORN

Culms to 160 cm, decumbent at the base; **nodes** pubescent; **internodes** mostly hollow, solid for 1 cm below the spike. **Blades** 5–15 mm wide, blue-green, hirsute, with long hairs over the veins, shorter hairs between the veins, hairs stiff. **Spikes** 5–14 cm, wider than thick; **rachises** densely ciliate at the nodes and margins; **internodes** 3–5 mm; **disarticulation** spontaneous, dispersal units wedge-shaped. **Spikelets** 12–17 mm, rectangular, with 2–3 florets, 1–2 seed-forming. **Glumes** 6–11 mm, coriaceous, tightly appressed to the lower florets, 2-keeled, with 2 prominent teeth; **lemmas** 10–14 mm, first (and sometimes the second) lemma awned, awns to 11 cm, third lemma usually unawned; **paleas** splitting at maturity; **anthers** 3–6 mm. **Caryopses** of the lowest floret in each spikelet usually blue, that of the second amber or red with blue mottling; **endosperm** flinty. Haplome A[b]. $2n = 14$.

Triticum boeoticum is a wild diploid wheat that is native from the Balkans through the Caucasus to Iran and Afghanistan and south to Iraq. It is morphologically similar to and, in its native range, sometimes sympatric with *T. urartu*, another wild diploid wheat. *Triticum monococcum* is the domesticated derivative of *T. boeoticum*.

Boissier published the combination for this species both as "*Triticum baeoticum*" and "*T. boeoticum*". Because the type specimen is from Boeotia [Greece], "*boeoticum*" is the correct spelling of the epithet.

2. Triticum urartu Thumanjan *ex* Gandilyan [p. 271]
RED WILD EINKORN

Culms to 145 cm, decumbent at the base; **nodes** pubescent; **internodes** mostly hollow, solid for 1 cm below the spikes. **Blades** 7–10 mm wide, yellow-green, puberulent, hairs uniform in length, soft. **Spikes** 6–12 cm, wider than thick; **rachises** densely ciliate at the nodes and margins; **internodes** 3–5 mm; **disarticulation** spontaneous, dispersal units wedge-shaped. **Spikelets** 12–16 mm, rectangular, with 2–3 florets, 1–2 seed-forming. **Glumes** 8–11 mm, coriaceous, tightly appressed to the lower florets, 2-keeled, 2-toothed, second tooth not well developed; **lemmas** 10–13 mm, awned, awns on the lower 2 lemmas to 7 cm, on the third lemma to 1 cm; **paleas** splitting at maturity; **anthers** 2–4 mm. **Caryopses** red, that of the lowest floret in each spikelet darker than the second; **endosperm** flinty. Haplome A[u]. $2n = 14$.

Triticum urartu is the wild diploid wheat that contributed the **A** haplome to the durum and bread wheat evolutionary lines. It does not have a diploid domesticated form. Because of its close morphological similarity to *T. boeoticum*, *T. urartu* was included in *T. boeoticum* until genetic analysis showed it to be a separate species. It has a more limited distribution than *T. boeoticum*, being known from disjunct regions in Turkey, Lebanon, Armenia, western Iran, and eastern Iraq.

3. Triticum dicoccoides (Körn.) Körn. *ex* Schweinf. [p. 271]
WILD EMMER

Culms to 100 cm, decumbent; **nodes** glabrous or puberulent; **internodes** mostly hollow, solid for 1 cm below the spikes. **Blades** 4–6 mm wide, pubescent. **Spikes** to 10 cm, wider than thick; **rachises** densely hairy at the nodes and margins; **internodes** 3–8 mm; **disarticulation** spontaneous, dispersal units wedge-shaped. **Spikelets** 15–25 mm, oblong to rectangular, with 3 florets, usually the lower 2 seed-forming. **Glumes** 10–15 mm, coriaceous, tightly appressed to the lower florets, 2-keeled, prominent keel winged to the base, 2-toothed, second tooth poorly developed; **lemmas** 10–15 mm, awned, awns on the lower 2 lemmas to 15 cm, on the third lemma to 2 cm; **paleas** not splitting at maturity. **Endosperm** flinty. Haplomes A[u]B. $2n = 28$.

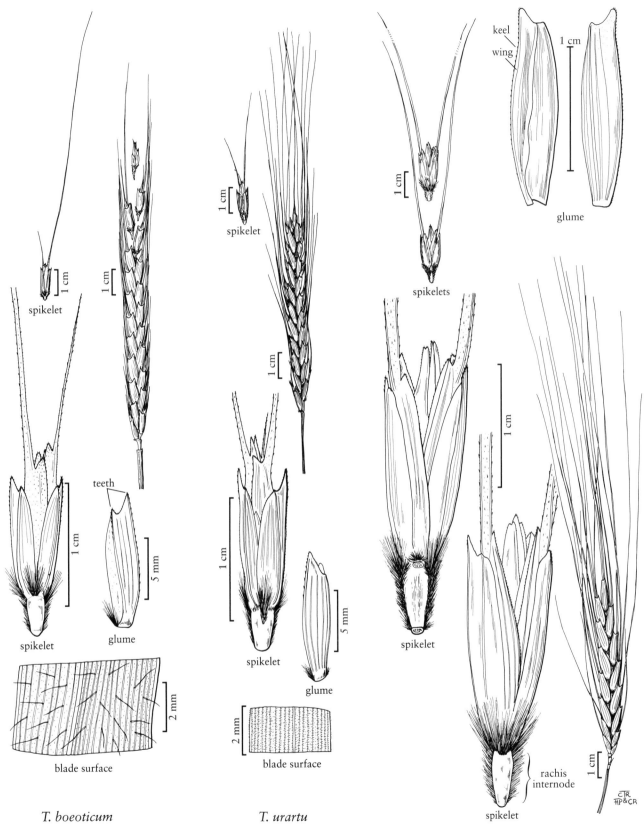

keel

wing

1 cm

glume

spikelet

spikelets

1 cm

spikelet

teeth

1 cm

spikelet

glume

5 mm

blade surface

T. boeoticum

spikelet

1 cm

spikelet

glume

5 mm

blade surface

T. urartu

spikelet

1 cm

rachis internode

spikelet

1 cm

T. dicoccoides

TRITICUM

Triticum dicoccoides is the wild counterpart of *T. dicoccum*, and is an ancestor of both *T. durum* and *T. aestivum*. Morphologically, it is almost indistinguishable from *T. araraticum* Jakubz., a wild tetraploid that differs from *T. dicoccoides* in combining the A^b and G haplomes. *Triticum dicoccoides* is native to the Fertile Crescent. Its distribution overlaps that of *T. araraticum* in the northern and eastern portions of the region.

4. Triticum monococcum L. [p. 273]
EINKORN, SMALL SPELT, PETIT ÉPEAUTRE

Culms to 120 cm; nodes pilose; internodes mostly hollow, solid for 1 cm below the spikes. Blades 6–7 mm wide, pubescent. Spikes 4–9 cm, strongly flattened, wider than thick; rachises glabrous or sparsely ciliate at the nodes and margins, not enlarged at the base of the glumes; internodes 1.4–2.5 mm, not disarticulating or disarticulating only with pressure, dispersal units wedge-shaped. Spikelets 10–12 mm, elliptical to ovate, with 2–3 florets, usually only 1 seed-forming. Glumes 6–8(13) mm, usually coriaceous and tightly appressed to the lower florets, sometimes chartaceous and only loosely appressed to the florets, 2-keeled, 2-toothed; lemmas 8–11 mm, lower 2 lemmas awned, awns 3–8 cm; paleas splitting at maturity. Caryopses amber; endosperm flinty. Haplome A^b. $2n = 14$.

Triticum monococcum is the domesticated derivative of *T. boeoticum*. Its primary range extends from the Balkans and Romania through the Crimea and Caucasus to northern Iraq and western Iran, and south to northern Africa. It was originally introduced to the *Flora* region as a food crop, but is now used primarily for plant breeding. It is still grown as a crop plant in some parts of the Balkans and in Romania.

Plants that originated from a spontaneous mutation and have tough rachises and chartaceous glumes that loosely enclose, but do not conceal, the florets have been named *Triticum sinskajae* A.A. Filat. & Kurkiev.

5. Triticum timopheevii (Zhuk.) Zhuk. [p. 273]
TIMOPHEEV'S WHEAT

Culms to 140 cm; nodes glabrous or pubescent; internodes mostly hollow, partially solid to hollow for 1 cm below the spikes. Blades to 10 mm wide, densely hairy, hairs 1.5–4 mm. Spikes 5–7 cm, strongly flattened, wider than thick; rachises ciliate at the nodes and margins; internodes 1.5–2.5 mm, disarticulating with pressure, dispersal units wedge-shaped. Spikelets 16–18 mm, elliptical to ovate, strongly flattened, with 3 florets, usually the lower 2 seed-forming. Glumes 7–10 mm, usually coriaceous and tightly appressed to the lower florets, sometimes chartaceous, acute, with 1 prominent keel, keel winged only in the distal $^2/_3$, terminating in a tooth; lemmas 10–12 mm, lower 2 lemmas awned, awns to 9 cm; paleas not splitting at

maturity. Endosperm flinty. Haplomes A^bG. $2n = 28$.

Triticum timopheevii is the domesticated derivative of *T. araraticum* Jakubz. It is established in Georgia, Armenia, and northeastern Turkey. It differs from other species of *Triticum* in its long leaf hairs and their relatively higher density. Plants with tough rachises and chartaceous glumes have been named *T. militinae* Zhuk. & Migush.

6. Triticum dicoccum Schrank ex Schübl. [p. 273]
EMMER, FARRO, FAR

Culms 80–150 cm, decumbent; nodes glabrous or pubescent; internodes mostly hollow, solid for 1 cm below the spikes. Blades to 20 mm wide, pubescent. Spikes 5–10 cm, about as wide as thick to wider than thick, cylindrical to strongly flattened; rachises glabrous or shortly ciliate at the nodes and margins; internodes (0.5)2–5 mm, disarticulating with pressure, dispersal units wedge-shaped. Spikelets 10–16 mm, elliptical to ovate, with 3–4 florets, usually only the lower 2 seed-forming. Glumes 6–10 mm, coriaceous, tightly appressed to the lower florets, with 1 prominent keel, keel winged only in the distal $^2/_3$, terminating in a tooth; lemmas 9–12 mm, awned, lower 2 lemmas awned, awns to 17 cm, upper lemmas unawned or shortly awned; paleas not splitting at maturity. Endosperm flinty. Haplomes A^uB. $2n = 28$.

Triticum dicoccum is the domesticated derivative of *T. dicoccoides*. It was once grown fairly extensively in central and southern Europe, southern Russia, northern Africa, and Arabia, because it can withstand poor, waterlogged soils. It is rarely grown now. It was introduced to the *Flora* region as a feed grain and forage for livestock. Currently, its primary use in the region is for plant breeding; it is also sold for human consumption as farro in specialty food markets.

7. Triticum durum Desf. [p. 275]
DURUM WHEAT, MACARONI WHEAT, HARD WHEAT, BLÉ DUR

Culms 60–160 cm; nodes glabrous; internodes mostly hollow, solid for 1 cm below the spikes. Blades 7–16 mm wide, usually glabrous. Spikes 4–11 cm, about as wide as thick, never branched; rachises ciliate to partially ciliate at the nodes and margins, not disarticulating; internodes 3–6 mm. Spikelets 10–15 mm, with 5–7 florets, 2–4 seed-forming. Glumes 8–12 mm, coriaceous, loosely appressed to the lower florets, with 1 prominent keel, terminating in a tooth, tooth to 0.3 cm; lemmas 10–12 mm, lower 2 lemmas awned, awns to 23 cm; paleas not splitting at maturity. Endosperm usually flinty, sometimes mealy. Haplomes A^uB. $2n = 28$.

Triticum durum is a domesticated spring wheat that is grown in temperate climates throughout the world. In

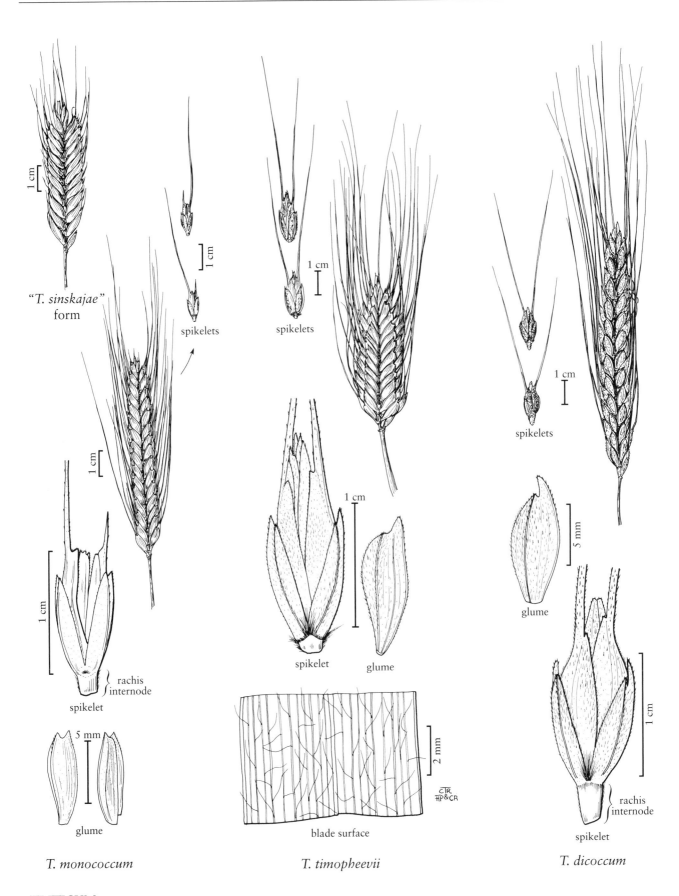

"*T. sinskajae*" form

spikelets

spikelets

spikelets

1 cm

1 cm

1 cm

1 cm

1 cm

1 cm

5 mm

5 mm

2 mm

rachis internode

spikelet

glume

spikelet

glume

blade surface

glume

rachis internode

spikelet

T. monococcum

T. timopheevii

T. dicoccum

TRITICUM

the *Flora* region, it is grown in the Canadian prairies and northern Great Plains as a spring wheat, and in the southwestern United States and Mexico as a winter wheat. *Triticum durum* is typically used for macaroni-type pastas, semolina, and bulghur. Durum imparts a yellowish color to bread, and is the traditional wheat for flat breads and pita. Cultivars grown in the *Flora* region represent a minor sampling of the overall diversity in the species.

The commercial cultivar Kamut® is durum wheat. Grown in the *Flora* region and worldwide, it encompasses a variable collection of forms. Kamut® has also been identified as *Triticum turanicum* Jakubz. (a durum-like wheat from Iran) or *T. polonicum*, although its presumed Egyptian origin and spike morphology do not agree with the original concept of these species.

8. Triticum turgidum L. [p. 275]
RIVET WHEAT, CONE WHEAT, BLÉ POULARD

Culms 120–180 cm; **nodes** glabrous; **internodes** mostly hollow, solid for 1 cm below the spikes. **Blades** to 18 mm wide, shortly pubescent to villous. **Spikes** 7–14 cm, about as wide as thick, except when branched below; **rachises** hairy at the nodes and margins, not disarticulating. **Spikelets** 10–16 mm, with 5–7 florets, 2–5 seed-forming. **Glumes** 8–11 mm, coriaceous, loosely appressed to the lower florets, with 1 prominent keel, terminating in a tooth, tooth to 0.3 cm; **lemmas** 10–13 mm, lowest 2 lemmas awned, awns to 20 cm; **paleas** not splitting at maturity. **Endosperm** mealy. **Haplomes** A^uB. $2n = 28$.

Triticum turgidum is the tallest of the wheats, and differs from other species of domesticated wheat in having branched-spike forms. It is grown primarily in southern Europe, northern Iraq, southern Iran, and western Pakistan. As treated here, *T. turgidum* is a narrowly distributed taxon of minor importance in plant breeding. Under genomic classifications, however, the name is applied to all A^uB taxa, e.g., to *T. polonicum*, *T. durum*, and *T. carthlicum*, as well as to *T. turgidum sensu stricto*.

9. Triticum polonicum L. [p. 275]
POLISH WHEAT

Culms 100–160 cm; **nodes** glabrous; **internodes** mostly hollow, solid for 1 cm below the spikes. **Blades** to 20 mm wide, glabrous or pubescent. **Spikes** 7–16 cm, wider than thick or about as wide as thick; **rachises** enlarged at the base of the glumes, sparsely hairy at the nodes and margins, not disarticulating. **Spikelets** 25–40 mm, with 4–5 florets, 2–3 seed-forming. **Glumes** 20–40 mm, often concealing the florets, lanceolate, chartaceous, loosely appressed to the lower florets, with 1 prominent keel, apices acute, terminating in a tooth; **lemmas** to 30 mm, chartaceous, toothed or awned, awns on the lower

2 lemmas to 15 cm; **paleas** not splitting at maturity. **Endosperm** flinty. **Haplomes** A^uB. $2n = 28$.

Triticum polonicum is a minor, durum-like, spring wheat species. It is grown in the Mediterranean basin and central Asia on a small scale. In the *Flora* region, it is grown principally for plant breeding. It differs from other domesticated wheats in its unusually long, chartaceous glumes and lemmas. The epithet "polonicum" reflects an early European botanical bias; the species did not originate in Poland.

10. Triticum carthlicum Nevski [p. 276]
PERSIAN WHEAT

Culms 60–100 cm; **nodes** glabrous or pubescent; **internodes** mostly hollow, solid for 1 cm below the spikes. **Blades** to 10 mm wide, puberulent. **Spikes** 8–16 cm, thicker than wide to about as thick as wide, not branched at the base; **rachises** glabrous or shortly ciliate at the nodes and margins, not disarticulating. **Spikelets** 10–15 mm, with 3–5 florets, 2–4 seed-forming. **Glumes** 7–9 mm, coriaceous, loosely appressed to the lower florets, with 1 prominent keel, terminating in an awn, awns 1–6 cm; **lemmas** 8.5–12.5 mm, lower 2 lemmas awned, awns to 13 cm; **paleas** not splitting at maturity. **Caryopses** red; **endosperm** flinty. **Haplomes** A^uB. $2n = 28$.

Triticum carthlicum is of evolutionary interest because, morphologically, its spikes resemble those of *T. aestivum* rather than those of free-threshing tetraploid wheats such as *T. durum*, *T. turgidum*, and *T. polonicum*. It is still occasionally cultivated in Georgia, Armenia, Azerbaijan, northern Iraq, and Iran because of its resistance to drought, frost, and ergot infection. A morphologically similar form of *T. aestivum* with awned glumes, known as 'carthlicoides', is often found intermixed with *T. carthlicum*.

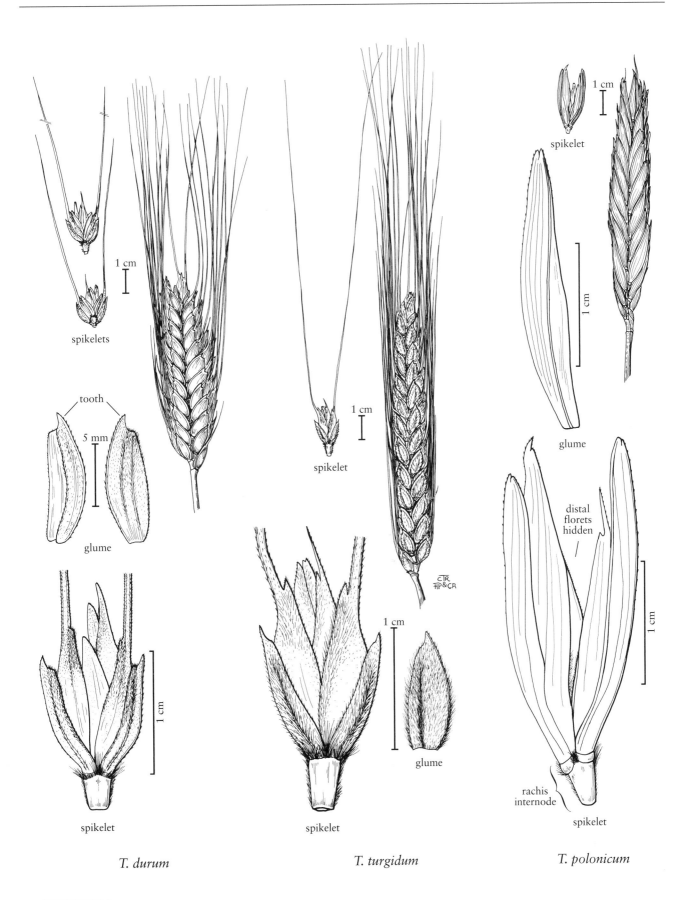

spikelets

tooth

5 mm

glume

spikelet

T. durum

spikelet

1 cm

glume

spikelet

T. turgidum

1 cm

spikelet

glume

distal
florets
hidden

1 cm

rachis
internode

spikelet

T. polonicum

TRITICUM

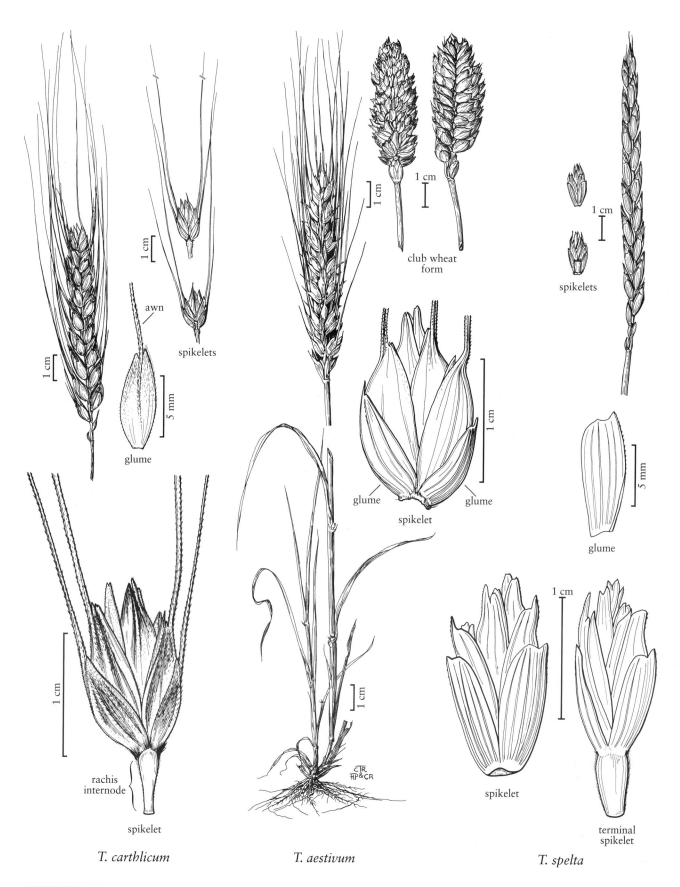

awn

glume

spikelets

1 cm

5 mm

rachis
internode

spikelet

T. carthlicum

club wheat
form

1 cm

1 cm

glume   glume

spikelet

1 cm

T. aestivum

spikelets

1 cm

glume

5 mm

spikelet

1 cm

terminal
spikelet

T. spelta

11. Triticum aestivum L. [p. 276]
WHEAT, BREAD WHEAT, COMMON WHEAT, SOFT WHEAT, BLÉ CULTIVÉ, BLÉ COMMUN

Culms 14–150 cm; **nodes** glabrous or pubescent; **internodes** usually hollow, even immediately below the spikes. **Blades** 6–15(20) mm wide, glabrous or pubescent. **Spikes** (3.5)6–18 cm, usually thicker than wide to about as thick as wide, wider than thick in compact forms; **rachises** shortly ciliate at the nodes and margins, not disarticulating. **Spikelets** 10–15 mm, appressed or ascending, with 3–9 florets, 2–5 seed-forming. **Glumes** 6–12 mm, coriaceous, loosely appressed to the lower florets, usually keeled in the distal $^1/_2$, sometimes prominently keeled to the base, terminating in a tooth or awn, awns to 4 cm; **lemmas** 10–15 mm, toothed or awned, awns to 12 cm; **paleas** not splitting at maturity. **Endosperm** mealy to flinty. Haplomes A^{u}BD. $2n = 42$.

Triticum aestivum is the most widely cultivated wheat. Both winter and spring types are grown in the *Flora* region. In addition to being grown for bread flour, *T. aestivum* cultivars are used for pastry-grade flour, Oriental-style soft noodles, and cereals.

Club wheats, sometimes called *Triticum compactum* Host, are cultivated in the Pacific Northwest for export to Asian markets. They have short (3.5–6 cm), compressed spikes, with up to 25 spikelets having 2–6 florets. Their spike shape varies from oblong or oval with uniformly distributed spikelets to club-shaped with spikelets crowded towards the apex.

No wild hexaploid progenitors of *Triticum aestivum* are known, but the two distinguishing characteristics of wild *Tritcum* species, fragile rachises breaking into wedge-shaped units and closely appressed glumes, are found in plants cultivated in Tibet and named *T. aestivum* subsp. *tibetanum* J.Z. Shao.

12. Triticum spelta L. [p. 276]
SPELT, DINKEL, ÉPEAUTRE, GRAND ÉPEAUTRE

Culms 80–120 cm; **nodes** glabrous or pubescent; **internodes** hollow, even immediately below the spikes. **Blades** 12–20 mm wide, sparsely pubescent. **Spikes** 6–20 cm, about as wide as thick, slender, almost cylindrical, narrowing distally; **rachises** glabrous or sparsely hairy at the nodes and margins, disarticulating with pressure, disarticulation units barrel-shaped or wedge-shaped. **Spikelets** 12–16 mm, with 3–5 florets, 1–3 seed-forming. **Glumes** 5–10 mm, coriaceous, tightly appressed to the lower florets, truncate, with 1 prominent keel, keel winged to the base, terminating in a tooth; **lemmas** 8–12 mm, toothed or awned, awns on the lower 2 lemmas to 10 cm, the third lemma sometimes awned, awns to 2 cm; **paleas** not splitting at maturity. **Endosperm** usually flinty. Haplomes A^{u}BD. $2n = 42$.

In the *Flora* region, *Triticum spelta* is grown for the specialty food and feed grain markets. It is known for yielding a pastry-grade flour not suitable for bread making unless mixed with *T. aestivum*, the bread-quality flour. Modern plant breeding programs are improving its gluten profile to upgrade its bread-making quality. Consequently, claims that *T. spelta* is a safe option for consumers with gluten intolerance should be treated with caution.

The ability of *Triticum spelta* to break under pressure into barrel-shaped units similar to those found in *Aegilops cylindrica* distinguishes it from all other members of *Triticum*.

13.09 AGROPYRON Gaertn.

Mary E. Barkworth

Plants perennial; densely to loosely cespitose, sometimes rhizomatous. **Culms** 25–110 cm, geniculate or erect. **Sheaths** open; **auricles** usually present; **ligules** membranous, often erose. **Inflorescences** spikes, usually pectinate; **middle internodes** 0.2–3(5.5) mm, basal internodes often somewhat longer. **Spikelets** solitary, usually more than 3 times as long as the internodes, usually divergent or spreading from the rachis, with 3–16 florets; **disarticulation** above the glumes and beneath the florets. **Glumes** shorter than the adjacent lemmas, lance-ovate to lanceolate, 1–5-veined, asymmetrically keeled, a secondary keel sometimes present on the wider side, keels glabrous or with hairs, hairs not tufted, apices acute and entire, sometimes awned, awns to 6 mm; **lemmas** 5–7-veined, asymmetrically keeled, acute to awned, awns to 4.5 mm; **paleas** from slightly shorter than to exceeding the lemmas, bifid; **anthers** 3, 3–5 mm, yellow. **Caryopses** usually falling with the lemmas and paleas attached. $x = 7$. Haplome P. Name from the Greek *agrios*, 'wild', and *pyros*, 'wheat'.

Agropyron, it is now agreed, should be restricted to perennial species of *Triticeae* with keeled glumes, i.e., *A. cristatum* and its allies, or the "crested wheatgrasses". The excluded species are distributed among *Pseudoroegneria*, *Thinopyrum*, *Elymus*, *Eremopyrum*, and *Pascopyrum*. This leaves *Agropyron* as a Eurasian genus that includes diploid, tetraploid, and hexaploid plants, all of which contain a single genome, designated the **P** genome by the International Triticeae Consortium. The genus is now widespread in western North America, frequently being used for soil stabilization on degraded rangeland and abandoned cropland, because it is highly tolerant of grazing and provides good spring forage.

Prior to the 1930s, most Soviet agrostologists recognized two species in the genus: *Agropyron cristatum*, with broad spikes; and *A. desertorum* (Fischer *ex* Link) Schult., with narrow spikes. Kosarev (1949) recognized four species, the two additional species being *A. pectiniforme* Roem. & Schult. and *A. sibiricum* (Willd.) P. Beauv. Tsvelev (1976) recognized 10 species, within one of which, *A. cristatum*, he recognized nine subspecies. He considered the widely distributed taxon introduced to many different countries to be *A. cristatum* subsp. *pectinatum* (M. Bieb.) Tzvelev. Chen and Zhu (2006) suggested that there are 15 species in the world, five of which are present in China.

Estimating the number of species present in the *Flora* region is difficult, because many seed samples were brought into the region, planted out in experimental plots, and subsequently developed for various agricultural uses. In reviewing the history of crested wheatgrass in North America, Dillman (1946) stated that, based on the identifications provided with some of the early seed accessions, two species of crested wheatgrass had been introduced into North America, *Agropyron cristatum* and *A. desertorum*. He described them as "quite distinct, both in seed and plant characters" (p. 248). According to Dewey (1986), a third species, now known as *A. fragile*, was introduced at about the same time; it apparently escaped Dillman's attention.

The problem is that "taxa introduced into North America soon lose their taxonomic identity and genetic integrity because of extensive intercrossing that occurs in nursery situations" (Dewey 1986, p. 34). Despite his observations, Dewey recognized three species of *Agropyron* in North America, and admitted that identifying individual plants "will often be difficult and unsatisfying. Variation is continuous between the morphological extremes of the unawned, linear-spiked *A. fragile* to the broad, pectinate-spiked *A. cristatum*" (p. 38).

This treatment recognizes two species within the *Flora* region, a very broadly interpreted *Agropyron cristatum*, which includes Dewey's *A. cristatum* and *A. desertorum*, and a traditionally interpreted *A. fragile*. *Agropyron cristatum* in North America reflects a process that might be called de-speciation.

SELECTED REFERENCES **Asay, K.H.** and **D.R. Dewey.** 1979. Bridging ploidy differences in crested wheatgrass with hexaploid × diploid hybrids. Crop Science 19:519–523; **Asay, K.H., K.B. Jensen, C. Hsiao,** and **D.R. Dewey.** 1992. Probable origin of standard crested wheatgrass, *Agropyron desertorum* (Fisch. *ex* Link) Schultes. Canad. J. Pl. Sci. 72:763–772; **Chen, S.-L.** and **G.-H. Zhu.** 2006. *Agropyron*. Pp. 439–441 *in* Z.-Y. Wu, P.H. Raven, and D.-Y. Hong (eds.). Flora of China, vol. 22 (Poaceae). Science Press, Beijing, Peoples Republic of China and Missouri Botanical Garden Press, St. Louis, Missouri, U.S.A. 653 pp. http://flora.huh.harvard.edu/china/mss/volume22/index.htm; **Dewey, D.R.** 1986. Taxonomy of the crested wheatgrasses (*Agropyron*). Pp. 31–42 *in* K.L. Johnson (ed.). Crested Wheatgrass: Its Values, Problems and Myths; Symposium Proceedings. Utah State University, Logan, Utah, U.S.A. 348 pp.; **Dillman, A.C.** 1946. The beginnings of crested wheatgrass in North America. J. Amer. Soc. Agron. 38:237–250; **Kosarev, M.G.** 1949. The variability of characters of crested wheatgrass. Selekts. & Semenov. 4:41–43; **Lesica, P.** and **T.H. DeLuca.** 1996. Long-term harmful effects of crested wheatgrass on Great Plains grassland ecosystems. J. Soil Water Conservation 51:408–409; **Tsvelev, N.N.** 1976. Zlaki SSSR. Nauka, Leningrad [St. Petersburg], Russia. 788 pp.

1. Lemmas usually awned, awns 1–6 mm long; spikelets diverging from the rachises at angles
   of 30–95°; spikes narrowly to broadly lanceolate, rectangular, or ovate in outline . . . . 1. *A. cristatum*
1. Lemmas unawned, sometimes mucronate; spikelets diverging from the rachises at an angle
   of less than 30(35)°; spikes linear to narrowly lanceolate in outline . . . . 2. *A. fragile*

1. **Agropyron cristatum** (L.) Gaertn. [p. 280]

CRESTED WHEATGRASS, AGROPYRON ACCRÊTÉ,
AGROPYRON À CRÊTE

Plants occasionally rhizomatous. **Culms** 25–110 cm, sometimes geniculate. **Ligules** to 1.5 mm; **blades** 1.5–6 mm wide, glabrous or pubescent. **Spikes** 1.3–10.5(15) cm long, 5–25 mm wide, narrowly to broadly lanceolate, rectangular, or ovate, sometimes tapering distally; **internodes** (0.2)0.7–5(8) mm, glabrous or pilose, sometimes all more or less equal, sometimes short and long internodes alternating within a spike, basal internodes often longer than those at midlength. **Spikelets** 7–16 mm, diverging at angles of 30–95° at maturity, with 3–6(8) florets. **Glumes** 3–6 mm, glabrous or with coarse hairs on the keels, acute, usually awned, awns 1.5–3 mm; **lemmas** 5–9 mm, glabrous or with hairs, keeled, keels sometimes scabrous distally, apices acute, usually awned, awns 1–6 mm; **anthers** 3–5 mm. $2n = 14, 28, 42$.

Agropyron cristatum is native from central Europe and the eastern Mediterranean to Mongolia and China. According to Tsvelev (1976), the most widely distributed taxon outside the Soviet Union is *A. cristatum* subsp. *pectinatum*. Within the *Flora* region, the reticulate genetic history of crested wheatgrass and the absence of any native populations argue against attempting recognition of subspecies.

Among the more commonly encountered variants of *Agropyron cristatum* in the *Flora* region are the cultivar 'Fairway', which was considered by Dillman (1946) and Dewey (1986) to belong to *A. cristatum* rather than *A. desertorum*, and its derivatives 'Parkway' and 'Ruff'. The name "Fairway" is also widely used in agricultural circles to refer to any crested wheatgrass that looks like the cultivar 'Fairway'. "Standard" crested wheatgrass, which Dewey (1986) and others placed in *A. desertorum*, originally referred to a particular seed lot (S.P.I. 19537) that the Montana Wheatgrowers' Association decided to use as a standard against which to compare the performance of other crested wheatgrass strains. The term is now applied by agronomists to all crested wheatgrasses that are less leafy and have more lanceolate spikes than "Fairway" crested wheatgrasses. There are numerous cultivars of crested wheatgrass available.

Because it is easy to establish, *Agropyron cristatum* has often been used to restore productivity to areas that have been overgrazed, burned, or otherwise disturbed. This ability, combined with its high seed production, tends to prevent establishment of most other species, both native and introduced.

2. **Agropyron fragile** (Roth) P. Candargy [p. 280]

SIBERIAN WHEATGRASS

Plants not rhizomatous. **Culms** 30–100 cm, rarely geniculate. **Ligules** to 1 mm; **blades** 1.5–6 mm wide. **Spikes** (5)8–15 cm long, 5–13 mm wide, linear to narrowly lanceolate; **internodes** 1.5–5 mm. **Spikelets** 7–16 mm, appressed or diverging up to 30(35)° from the rachises. **Glumes** 3–5 mm, glabrous or hairy, often awned, awns 1–3 mm; **lemmas** 5–9 mm, keels scabrous distally, apices unawned, sometimes mucronate, mucros up to 0.5 mm; **anthers** 4–5 mm. $2n = (14), 28, (42)$.

Agropyron fragile is native from the southern Volga basin through the Caucasus to Turkmenistan and Mongolia. It is more drought-tolerant than *A. cristatum*. Within the *Flora* region, *A. fragile* appears to be uncommon outside of experimental plantings. This may change as more cultivars become available.

13.10 **PSEUDOROEGNERIA** (Nevski) Á. Löve

Jack R. Carlson

Plants perennial; usually cespitose, sometimes rhizomatous. **Culms** 30–100 cm, usually erect, sometimes decumbent or geniculate. **Leaves** evenly distributed; **sheaths** open; **auricles** well developed; **ligules** membranous; **blades** flat to loosely involute. **Inflorescences** terminal spikes, erect, with 1 spikelet per node; **internodes** (7)10–20(28) mm at midlength, lower internodes often longer than those at midlength. **Spikelets** (8)12–25 mm, 1.1–1.5(2) times the length of the internodes, usually appressed, sometimes slightly divergent, with 4–9 florets; **disarticulation** above the glumes and beneath the florets. **Glumes** unequal, from shorter than to slightly longer than the lowest lemma in the spikelets, lanceolate to oblanceolate, (3)4–5(7)-veined, usually acute to obtuse, occasionally truncate, narrowing beyond midlength, veins prominent; **lemmas** inconspicuously 5-veined, unawned or terminally awned, awns straight to strongly bent and

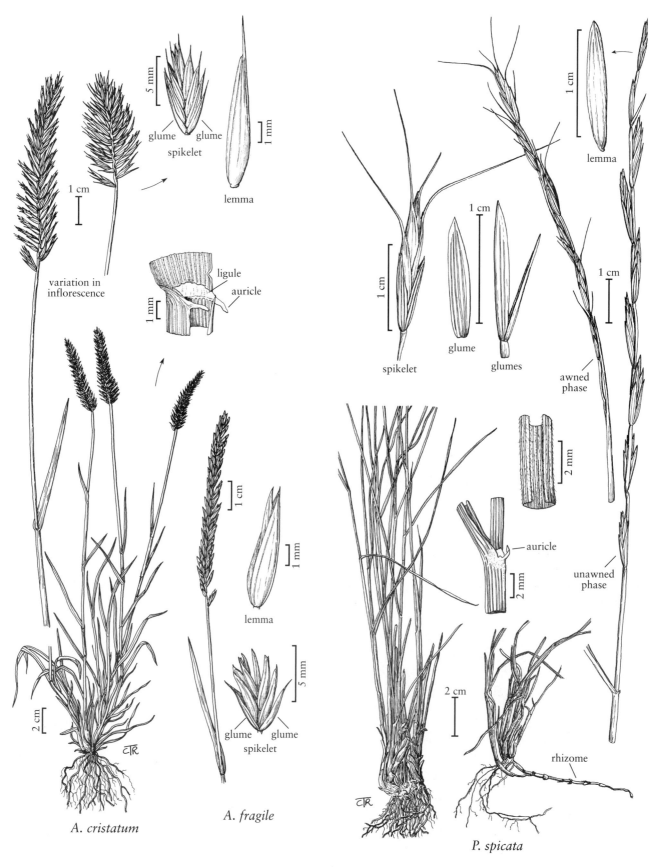

variation in
inflorescence

glume   glume
spikelet

lemma

ligule

auricle

A. cristatum

A. fragile

lemma

glume   glume
spikelet

spikelet

glume

glumes

lemma

awned
phase

auricle

unawned
phase

rhizome

P. spicata

AGROPYRON

PSEUDOROEGNERIA

divergent; **anthers** 4–8 mm. $x = 7$. Haplome **St**. Name from the Greek *pseudo*, 'false', and the genus *Roegneria*, an Asian taxon often included in *Elymus*.

Pseudoroegneria includes 15–20 species, one of which is North American and the remainder either Eurasian or Asian. All species currently included in the genus are obligate outcrossers, and almost all are diploids or autotetraploids (Jensen et al. 1992) containing the **St** haplome (designation by the International Triticeae Consortium). This genome is the most widely distributed in the *Triticeae*, being found in all species of *Elymus sensu lato* as well as some species of *Thinopyrum*.

The limits of *Pseudoroegneria* are not well established, whether it is treated as a genus, as here, or included in *Elytrigia* (Tsvelev 1976) or *Elymus* (Melderis 1980). Two species that were originally included have been transferred to *Douglasdeweya* C. Yen, J.L. Yang & B.R. Baum (Chen et al. 2005), because specimens grown at agricultural experiment stations were found to be **StP** allotetraploids (Jensen et al. 1992; Wang et al. 1986; Chen et al. 2005).

There are also questions concerning the delimitation of species in *Pseudoroegneria*. For instance, Jensen et al. (1995) suggested that *P. spicata*, *P. strigosa* (M.-Bieb.) Á. Löve, *P. geniculata* (Trin.) Á. Löve, *P. elytrigioides* (C. Yen & J.L. Yang) Bao-Rong Lu, and *Roegneria glaberrima* Keng & S.L. Chen [*Elymus glaberrimus* (Keng & S.L. Chen) S.L. Chen] should be considered members of a single transberingian species complex. They presented cytological, but not morphological, analyses supporting this conclusion.

SELECTED REFERENCES **Assadi, M.** and **H. Runemark**. 1995. Hybridization, genomic constitution and generic delimitation in *Elymus s.l.* (Poaceae, Triticeae). Pl. Syst. Evol. 194:189–205; **Dewey, D.R.** 1984. The genomic system of classification as a guide to intergeneric hybridization in the perennial Triticeae. Pp. 209–279 *in* J.P. Gustafson (ed.). Gene Manipulation in Plant Improvement. Plenum Press, New York, New York, U.S.A. 668 pp.; **Daubenmire, R.F.** 1939. The taxonomy and ecology of *Agropyron spicatum* and *A. inerme*. Bull. Torrey Bot. Club 66:327–329; **Daubenmire, R.F.** 1960. An experimental study of variation in the *Agropyron spicatum–A. inerme* complex. Bot. Gaz. 122:104–108; **Jensen, K.B,** **S.L. Hatch**, and **J. Wipff**. 1992. Cytogenetics and morphology of *Pseudoroegneria deweyi* (Poaceae: Triticeae), a new species from the Soviet Union. Canad. J. Bot. 70:900–909; **Jensen, K.B., M. Curto**, and **K.H. Asay**. 1995. Cytogenetics of Eurasian bluebunch wheatgrass and their relationship to North American bluebunch and thickspike wheatgrasses. Crop Sci. 35:1157–1162; **Melderis, A.** 1980. *Elymus*. Pp. 192–198 *in* T.G. Tutin, V.H. Heywood, N.A. Burges, D.M. Moore, D.H. Valentine, S.M. Walters. and D.A. Webb (eds.). Flora Europaea, vol. 5. Cambridge University Press, Cambridge, England. 452 pp.; **Tsvelev, N.N.** 1976. Zlaki SSSR. Nauka, Leningrad [St. Petersburg], Russia. 788 pp.; **Voss, E.G.** 1972. Michigan Flora: A Guide to the Identification and Occurrence of the Native and Naturalized Seed-Plants of the State, part 1. University of Michigan, Ann Arbor, Michigan, U.S.A. 488 pp.; **Wang, R.R-C., D.R. Dewey**, and **C. Hsiao**. 1986. Genome analysis of the tetraploid *Pseudoroegneria tauri*. Crop Sci. 26:723–727; **Yen, C., J.-L. Yang**, and **B.R. Baum**. 2005. *Douglasdeweya*: A new genus, with a new species and a new combination (Triticeae: Poaceae). Canad. J. Bot. 83:413–419.

1. **Pseudoroegneria spicata** (Pursh) Á. Löve [p. 280]
Bluebunch Wheatgrass

Plants loosely cespitose, sometimes rhizomatous. **Culms** 30–100 cm tall, 0.5–2 mm thick, sometimes glaucous. **Ligules** truncate, 0.1–0.4 mm on the lower leaves, 0.2–0.4 mm on the upper leaves; **blades** 2–6 mm wide, involute when dry, flag leaf blades strongly divergent when dry, abaxial surfaces smooth, glabrous, adaxial surfaces scabrous or hirsute. **Spikes** 8–15 cm long, 3–8(10) mm wide excluding the awns; **middle internodes** 7–20(25) mm, glabrous, scabrous on the angles. **Spikelets** 8–22(25) mm, with 4–9 florets. **Glumes** 6–13 mm long, 0.9–2.2 mm wide, about ¹/₂ the length of the spikelets, glabrous, sometimes scabrous over the veins, acute; **lemmas** 9–14 mm, unawned or with a terminal, strongly divergent awn, awns to 25 mm. $2n = 14, 28$.

Pseudoroegneria spicata is primarily a western North American species, extending from the east side of the coastal mountains to the western edge of the Great Plains, and from the Yukon Territory to northern Mexico. It was also collected by Farwell in Keenewaw County, Michigan in 1895 (Voss 1972). It grows on medium-textured soils in arid and semiarid steppe, shrub-steppe, and open woodland communities, and was one of the dominant species in grassland communities of the Columbia and Snake river plains (Daubenmire 1939, 1960). It is still an important forage plant in the northern portion of the Intermountain region. Several cultivars have been developed.

Rhizomatous plants are favored in relatively moist habitats, and cespitose plants in dry habitats (Daubenmire 1960). Daubenmire noted that rhizomatous plants produce few inflorescences and, possibly for this reason, are collected less frequently than cespitose plants. Daubenmire also found that awn length varies continuously within plants grown from seed. He concluded that the ability to produce rhizomes and unawned plants is heritable, that the two characters are not linked, and that the form which becomes dominant at a local site is determined by environmental conditions.

The unawned phase tends to be more restricted in its distribution than the awned phase, being dominant in the native grasslands of southern British Columbia, eastern Washington, northern Idaho, and northern and eastern Oregon; the awned phase is found throughout the range of the species. Many populations include awned and unawned plants, as well as some that have poorly developed awns on some lemmas. Awned autotetraploid populations grow in mesic grassland and woodland communities of the hills and mountains of southern British Columbia and eastern Washington.

Based on informal observations, plant breeders working with *Pseudoroegneria spicata* consider that awn presence is determined by a single major gene, and modified by some minor genes. The unawned condition is apparently dominant, as seed from crosses of heterozygotic, diploid, unawned parents gives rise to around 50% awned offspring.

The above observations make it clear that the awned and unawned phases of *Pseudoroegneria spicata* are of little taxonomic significance, despite their evident morphological difference. If it is considered necessary to distinguish between them, the awned phase can be called **Pseudoroegneria spicata** (Pursh) Á. Löve f.

spicata and the unawned phase **P. spicata** f. **inermis** (Scribn. & J.G. Sm.) Barkworth.

Plants with densely pubescent leaves are known from the east slope of the Cascade Mountains in Washington. Plants with nearly as densely pubescent leaves are found elsewhere in southern Washington and northeastern Oregon. Such pubescent plants may be called **Pseudoroegneria spicata** f. **pubescens** (Elmer) Barkworth.

Pseudoroegneria spicata used to be confused with *Elymus wawawaiensis*, from which it differs in its more widely spaced spikelets and wider, less stiff glumes. The two species are geographically sympatric, but *P. spicata* grows in medium- to fine-textured loess soils, and *E. wawawaiensis* in shallow, rocky soils. *Pseudoroegneria spicata* may also be confused with *E. arizonicus*, particularly with immature specimens of that species or specimens mounted so that they appear to have erect, rather than drooping, spikes. It differs in having shorter, truncate ligules and generally thicker culms than *E. arizonicus*, and in having a distribution that extends much further north.

Pseudoroegneria spicata has been suggested as one of the parents in numerous natural hybrids with species of *Elymus* in the *Flora* region. These hybrids are usually mostly sterile, but development of even a few viable seeds permits introgression to occur, as well as the formation of distinctive populations. It is often difficult to detect such hybrids, particularly if they involve the unawned form of *Pseudoroegneria*. The named hybrids are treated under ×*Pseudelymus* (see below). Others are discussed under the *Elymus* parent.

13.11 ×PSEUDELYMUS Barkworth & D.R. Dewey

Mary E. Barkworth

Plants perennial; sometimes rhizomatous. **Culms** 50–80 cm, erect, glabrous. **Leaves** not basally concentrated; **sheaths** glabrous or puberulent; **auricles** present; **ligules** truncate. **Inflorescences** distichous spikes, with 1(2) spikelet(s) per node. **Spikelets** appressed, with 3–5 florets. **Glumes** unequal, linear-lanceolate to lanceolate; **lemmas** awned or unawned; **paleas** slightly shorter than to slightly longer than the lemmas; **anthers** indehiscent.

×*Pseudelymus* comprises hybrids between *Pseudoroegneria* and *Elymus*. Only one species is treated here. Another species, *E. albicans*, is thought to be a similar hybrid, but it is treated as a species because it is frequently fertile.

SELECTED REFERENCE **Dewey, D.R.** 1964. Natural and synthetic hybrids of *Agropyron spicatum* × *Sitanion hystrix*. Bull. Torrey Bot. Club 91:396–405.

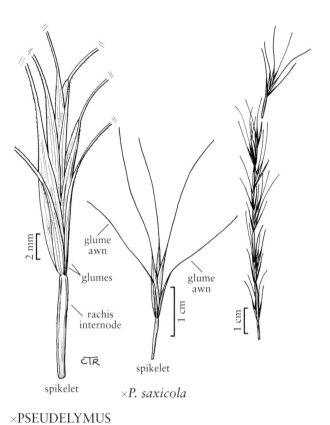

2 mm

glume
awn

glumes

rachis
internode

CTR

spikelet

×PSEUDELYMUS

glume
awn

spikelet

1 cm

×*P. saxicola*

1 cm

1. ×Pseudelymus saxicola (Scribn. & J.G. Sm.)
Barkworth & D.R. Dewey [p. 283]

Plants not rhizomatous. **Culms** 0.8–1 mm thick; **ligules**
0.2–0.4 mm; **blades** 1.5–2 mm wide. **Spikes** 7–14 cm,
with 1 spikelet per node; **internodes** 7–9 mm. **Spikelets**
12–15 mm excluding the awns; **disarticulation** beneath
the florets, sometimes also in the rachises. **Glumes**
15–21 mm including the awns, glume bodies 6–8 mm
long, 0.7–1 mm wide, (1)3–4-veined; **lemmas** 10–11
mm excluding the awns, (14)18–37 mm including the
awns, apices often bifid; **anthers** about 2.5 mm.

×*Pseudelymus saxicola* consists of hybrids between
Pseudoroegneria spicata and *Elymus elymoides*. It is a
rather common hybrid in western North America. It
differs from *E. albicans*, which is thought to be derived
from hybrids between *P. spicata* and *E. lanceolatus*, in
lacking rhizomes, having longer awns on its glumes and
lemmas, and having disarticulating rachises. It is more
likely to be confused with *E. ×saundersii*, but differs in
its longer glume and lemma awns.

13.12 ×ELYHORDEUM Mansf. *ex* Tsitsin & K.A. Petrova

Mary E. Barkworth

Plants perennial; usually cespitose, occasionally shortly rhizomatous. **Inflorescences** terminal,
spikes or spikelike, with 1–3(7) spikelets per node, lateral spikelets usually shortly pedicellate,
central spikelets sessile or nearly so; **disarticulation** tardy, at the rachis nodes and beneath the
florets. **Spikelets** with 1–4 florets. **Glumes** subulate to narrowly lanceolate, usually awned;
lemmas usually awned; **anthers** sterile. **Caryopses** rarely formed.

×*Elyhordeum* is the name given to hybrids between *Elymus* and *Hordeum*. These hybrids are
fairly common. All appear to be sterile, i.e., they do not produce good pollen or set seed. The
descriptions should be treated with reservation because, in some instances, only type specimens
have been examined. For that reason, no key is provided. Only named hybrids are described and
illustrated.

Interspecific hybrids between *Elymus elymoides* or *E. multisetus* and other species of *Elymus*
resemble the ×*Elyhordeum* hybrids in having tardily disarticulating, spikelike inflorescences and
awned glumes and lemmas, but are more likely to have solitary spikelets, even at the lowest
node. Distinguishing between them and ×*Elyhordeum* hybrids, without knowledge of other
species of *Triticeae* at a site, is challenging.

Inflorescence measurements, unless stated otherwise, do not include the awns.

SELECTED REFERENCES **Bowden, W.M.** 1958. Natural and artificial ×*Elymordeum* hybrids. Canad. J. Bot. 36:101–123; **Bowden, W.M.** 1960. Typification of *Elymus macounii* Vasey. Bull. Torrey Bot. Club. 87:205–208; **Jordal, L.K.** 1951. A floristic and phytogeographic survey of the southern slopes of the Brooks Range, Alaska. Ph.D. dissertation, University of Michigan, Ann Arbor, Michigan, U.S.A. 411 pp.; **Mitchell, W.W.** and **H.J. Hodgson.** 1965. A new ×*Agrohordeum* from Alaska. Bull. Torrey Bot. Club 92:403–407.

1. ×**Elyhordeum dakotense** (Bowden) Bowden [p. 285]

Culms 55–80 cm. **Ligules** about 0.2 mm, truncate; **blades** 1.5–2 mm wide. **Spikes** 7–10 cm long, 5–8 mm wide, nodding, with 2(3) spikelets per node. **Spikelets** with 2–4 florets. **Glumes** 20–50 mm including the awns, bases indurate; **lemmas** pubescent, awned, awns similar in length to those of the glumes; **anthers** not developed. $2n$ = unknown.

×*Elyhordeum dakotense* refers to hybrids between *Elymus canadensis* and *Hordeum jubatum*. They are known only from Brookings, South Dakota.

2. ×**Elyhordeum macounii** (Vasey) Barkworth & D.R. Dewey [p. 285]

Culms 50–100 cm. **Sheaths** usually glabrous; **ligules** truncate; **blades** 9–16 cm long, 2–5 mm wide, stiff, ascending, scabrous. **Spikes** 4–13 cm long, about 5 mm wide, erect, lower nodes with 1–2 spikelets, upper nodes with 1 spikelet, the spikelets imbricate. **Spikelets** with 1–3 florets, those at the lower nodes frequently with 3 glumes. **Glumes** 6–9 mm, not indurate at the base, awned, awns as long as or longer than the glume bodies; **lemmas** 6–11 mm, oblong-lanceolate, glabrous or sometimes scabrous distally, awned, awns 10–20 mm. $2n$ = 28.

×*Elyhordeum macounii* consists of hybrids between *Elymus trachycaulus* and *Hordeum jubatum*. It is quite common in western and central North America. Backcrosses to *E. trachycaulus* may have non-disarticulating rachises; they are likely to be identified as *E. trachycaulus*, falling between subsp. *trachycaulus* and subsp. *subsecundus*. Artificial, partially fertile octoploids were distributed to natural and experimental areas in several western states prior to 1960 (Bowden 1960); it is not known whether they have persisted.

3. ×**Elyhordeum pilosilemma** (W.W. Mitch. & H.J. Hodgs.) Barkworth [p. 285]

Culms 40–75 cm, erect, glabrous. **Sheaths** glabrous or pubescent; **blades** 6–15 cm long, 2–4 mm wide, flat, adaxial surfaces glabrous or hairy. **Spikes** 6–12 cm long, 5–7 mm wide including the awns, erect to arching, with 1–2 spikelets per node; **internodes** 2–5 mm, concealed by the spikelets. **Spikelets** 7–13 mm, with 2–3 florets and 2–3 glumes. **Glumes** 10–18 mm including the awns, not indurate at the base, linear to linear-lanceolate, hispid, 1–3-veined; **lemmas** 7–10 mm, evenly pilose, hairs about 0.2 mm, awns as long as to slightly longer than the lemma bodies; **anthers** 1–1.5 mm, indehiscent. $2n$ = 28.

×*Elyhordeum pilosilemma* is a hybrid between *Elymus macrourus* and *Hordeum jubatum* that occurs in many locations where the two parental species co-occur. It is very similar to ×*E. jordalii*, a hybrid between *E. macrourus* and *H. brachyantherum.*

4. ×**Elyhordeum jordalii** (Melderis) Tzvelev [p. 285]

Culms 25–30 cm. **Lower sheaths** puberulent; **blades** 2.5–4.5 mm wide, adaxial surfaces slightly scabrous. **Spikes** 4–12 cm, erect to slightly nodding, usually with 2 spikelets per node. **Spikelets** 7–12 mm excluding the awns, with 2–3 florets. **Glumes** 7–13 mm long, bases about 1 mm wide, not indurate, with a prominent, scabridulous midrib, apices gradually tapering into an awn shorter than 20 mm; **lemmas** 6–8 mm, shortly pubescent, the pubescence decreasing in density towards the apices, apices awned, awns 5–8 mm; **anthers** about 1.5 mm, indehiscent. $2n$ = unknown.

×*Elyhordeum jordalii* consists of hybrids between *Elymus macrourus* and *Hordeum brachyantherum*. It grows in the Brooks Range, Alaska. According to Jordal (1951), it was "only observed near settlements on the lowlands south of the range where it has a weedy habit. Dense tufts near cabins." It resembles ×*E. pilosilemma*, which differs in having *H. jubatum* as the *Hordeum* parent.

5. ×**Elyhordeum schaackianum** (Bowden) Bowden [p. 285]

Culms about 100 cm. **Sheaths** glabrous, smooth; **ligules** about 0.7 mm, truncate; **blades** 6–10 mm wide, flat, adaxial surfaces sparsely hairy, hairs about 0.8 mm. **Inflorescences** spikelike, 9–11 cm long, 10–12 mm wide, lax, with (2)3 spikelets per node. **Spikelets** with 1–2 florets, the second floret, if present, rudimentary, the lateral spikelets shortly pedicellate, the central spikelet sessile or nearly so. **Glumes** narrow, coarse, unawned or awned, awns to 15 mm; **lemmas** mostly glabrous, margins shortly hairy distally, apices unawned or awned, awns to 15(20) mm. $2n$ = 28.

×*Elyhordeum schaackianum* consists of hybrids between *Elymus hirsutus* and *Hordeum brachyantherum*. It is known only from Attu Island, Alaska, and the Queen Charlotte Islands, British Columbia.

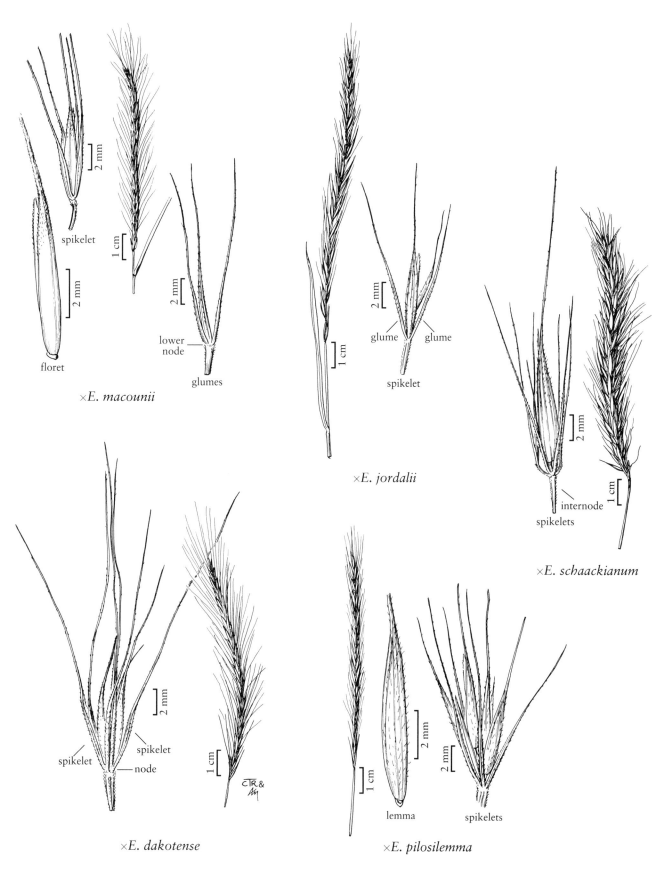

×*E. macounii*

×*E. jordalii*

×*E. schaackianum*

×*E. dakotense*

×*E. pilosilemma*

×ELYHORDEUM

6. ×**Elyhordeum stebbinsianum** (Bowden) Bowden [p. 287]

Culms 60–80 cm. **Blades** 3–5 mm wide, abaxial surfaces smooth, adaxial surfaces scabridulous. **Inflorescences** spikelike, 10–16 cm, with 3 spikelets per node; **internodes** averaging 3.4 mm. **Spikelets** with 1–3 florets, the lateral spikelets sessile or pedicellate, pedicels to 3 mm. **Glumes** 0.5–1 mm wide, terminating in a 4–7 mm awn; **lemmas** glabrous to scabrous, awned, awns 5–11 mm; **anthers** indehiscent. $2n$ = unknown.

×*Elyhordeum stebbinsianum* consists of hybrids between *Elymus glaucus* and *Hordeum brachyantherum*. Bowden (1958) reported that they appear to be completely sterile. They have been found at scattered locations in western North America.

7. ×**Elyhordeum iowense** R.W. Pohl [p. 287]

Culms 40–80 cm, erect. **Sheaths** hirsute; **ligules** about 0.3 mm; **auricles** short, acute; **blades** 6–8 mm wide, adaxial surfaces shortly velutinous. **Inflorescences** spikelike, 9–12 cm long, 5–10 mm wide, flexuous, with 3 spikelets per node. **Spikelets** with 2 florets, the lateral spikelets usually shortly pedicellate, the central spikelet sessile. **Glumes** about 2.5 cm, not indurate at the base; **lemmas** 6–8 mm, puberulent, awns 25–35 mm; **anthers** 1.6–1.8 mm, indehiscent in the lowest florets of each spikelet. $2n$ = 28.

×*Elyhordeum iowense* is a hybrid between *Elymus villosus* and *Hordeum jubatum* that has been found at scattered locations in the central plains. It probably occurs elsewhere, but is unlikely to be common, because *E. villosus* usually grows in more shady locations than *H. jubatum*.

8. ×**Elyhordeum arcuatum** W.W. Mitch. & H.J. Hodgs. [not illustrated]

Culms 40–85 cm. **Blades** 7–22 cm long, 3–10 mm wide, abaxial surfaces scabrous, adaxial surfaces pilose. **Inflorescences** spikelike, 6–15 cm, arcuate to nodding, with 3 spikelets per node; **central spikelets** sessile, with 1–3 florets; **lateral spikelets** shortly pedicellate, with 0–2 florets. **Glumes** to 25 mm, awnlike, scabrous; **lemmas** 9–11 mm, scabrous, awned, awns to 25 mm, arcuate and divergent at maturity; **anthers** 1–1.4 mm. $2n$ = 28.

×*Elyhordeum arcuatum* is probably a hybrid between *Elymus sibiricus* and *Hordeum jubatum* (Mitchell and Hodgson 1965). It was described from disturbed sites around Palmer, Alaska, from which it has since been eliminated. No additional reports are known. There is no illustration because the type specimens could not be located.

9. ×**Elyhordeum montanense** (Scribn. *ex* Beal) Bowden [p. 287]

Culms 60–100 cm. **Sheaths** glabrous; **blades** 5–8 mm wide, flat, lax. **Inflorescences** spikelike, 8–17 cm, nodding, with 3 spikelets per node. **Spikelets** with 1–4 florets, the lateral spikelets usually shortly pedicellate, with 1(2) sterile florets, the central spikelet sessile, with 2(3–4) sterile florets, the third and fourth florets rudimentary. **Glumes** 10–35 mm including the awns, slightly widened above the base; **lowest lemma of central florets** about 8 mm, glabrous towards the base, sparsely scabrous distally, awned, awns 15–25 mm, straight to slightly divergent at maturity. $2n$ = 28.

×*Elyhordeum montanense* applies to hybrids between *Elymus virginicus* and *Hordeum jubatum*. It is often found in disturbed areas where both parental taxa grow. Short-awned specimens may reflect the involvement of *E. submuticus* rather than *E. virginicus*, two taxa that have sometimes been treated as conspecific.

10. ×**Elyhordeum californicum** (Bowden) Barkworth [p. 287]

Culms 30–40 cm, glabrous throughout or hairy on the lower portion. **Sheaths** glabrous; **auricles** to 0.5 mm; **ligules** 0.4–1.5 mm, rounded; **blades** to about 15 cm long, 1–2(3) mm wide, abaxial surfaces glabrous or sparsely hairy, adaxial surfaces scabridulous, all veins equally prominent. **Inflorescences** spikelike, 2.5–5 cm long, 7–10 mm wide excluding the awns, 20–30 mm wide including the awns, straight, nodes with 3 spikelets; **internodes** 2.2–4 mm, bases about $^1\!/_2$ as wide as the apices. **Spikelets** 8–10 mm excluding the awns, 25–28 mm including the awns, with 1–2 florets, the lateral spikelets shortly pedicellate, the central spikelet sessile. **Glumes** 15–35 mm including the awns, setaceous, straight; **lemmas** 7–8 mm, smooth throughout or scabrous distally, awned, awns 10–20 mm, straight or slightly divergent at maturity; **anthers** about 1.5 mm.

×*Elyhordeum californicum* consists of hybrids between *Elymus elymoides* or *E. multisetus* and *Hordeum brachyantherum* subsp. *brachyantherum*. It was described by Bowden (1958) on the basis of specimens collected in California. It seems probable that it will be found at many locations where the two parents grow together.

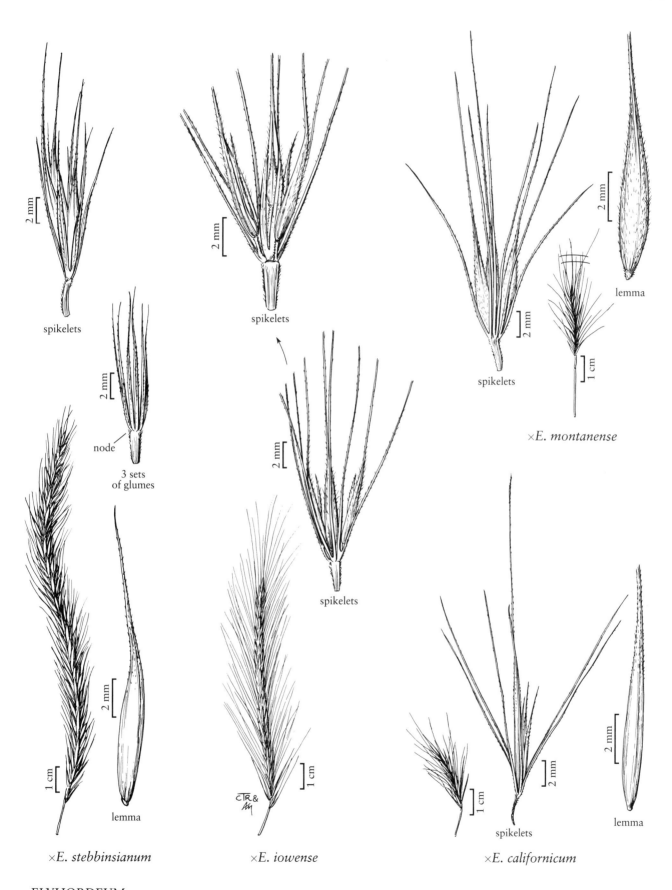

spikelets

spikelets

node

3 sets
of glumes

spikelets

spikelets

×*E. montanense*

lemma

lemma

lemma

spikelets

×*E. stebbinsianum*

×*E. iowense*

×*E. californicum*

×ELYHORDEUM

13.13 ELYMUS L.

Mary E. Barkworth

Julian J.N. Campbell

Björn Salomon

Plants perennial; sometimes cespitose, sometimes rhizomatous, sometimes stoloniferous. **Culms** 8–180(220) cm, usually erect to ascending, sometimes strongly decumbent to prostrate, usually glabrous. **Leaves** usually evenly distributed, sometimes somewhat basally concentrated; **sheaths** open for most of their length; **auricles** often present; **ligules** membranous, usually truncate or rounded, sometimes acute, entire or erose, often ciliolate; **blades** 1–24(25) mm wide, abaxial surfaces usually smooth or scabrous, sometimes with hairs, adaxial surfaces scabrous or with hairs, particularly over the veins, usually with unequal, not strongly ribbed, widely spaced veins, sometimes with equal, strongly ribbed, closely spaced veins. **Inflorescences** spikes, usually exserted, with 1–3(5) spikelets per node, **internodes** (1.5)2–26 mm; **rachises** with scabridulous, scabrous, or ciliate edges. **Spikelets** usually appressed to ascending, sometimes strongly divergent or patent, with 1–11 florets, the lowest florets usually functional, sterile and glumelike in some species, the distal florets often reduced; **disarticulation** usually above the glumes and beneath each floret, sometimes also below the glumes or in the rachises. **Glumes** usually 2, absent or highly reduced in some species, usually equal to subequal, sometimes unequal, usually linear-lanceolate to linear, setaceous, or subulate, sometimes oblanceolate to obovate, (0)1–7-veined, sometimes keeled over 1 vein, not necessarily the central vein, keel vein sometimes extending into an awn; **lemmas** linear-lanceolate, obscurely 5(7)-veined, apices acute, often awned, sometimes bidentate, teeth to 0.2 mm, sometimes with bristles, bristles to 10 mm, awns terminal or from the sinus, straight or arcuately divergent, not geniculate; **paleas** from shorter than to slightly longer than the lemmas, keels scabrous or ciliate, at least in part; **anthers** 3, 0.7–7 mm. **Caryopses** with hairy apices. $x = 7$. Haplomes **St, H, Y, P**. Name from the Greek *elyo*, 'rolled up', the caryopses being tightly embraced by the lemma and palea.

As interpreted here, *Elymus* is a widespread, north-temperate genus of about 150 species. It includes *Sitanion* Raf. and *Roegneria* K. Koch, but moves some taxa that others include in *Elymus* to *Leymus*, *Pascopyrum*, *Pseudoroegneria*, and *Thinopyrum*. Thirty-two species of *Elymus* are native to the *Flora* region. Of the seven non-native species treated, one is established (*E. repens*), two are distributed as forage (*E. dahuricus* and *E. hoffmannii*), two are known from ballast dumps and are not established (*E. tsukushiensis* and *E. ciliaris*), and two (*E. caninus* and *E. semicostatus*) have been attributed to the *Flora* region but specimens documenting the reports have not been located. Eight named, naturally occurring, intrageneric hybrids are described at the end of the treatment. They are not included in the key. As mentioned in the descriptions of the non-hybrid species, other interspecific hybrids undoubtedly exist. Because many of the hybrids are partially fertile, backcrossing and introgression occurs. Intergeneric hybrids are treated under ×*Elyhordeum* (p. 283), ×*Elyleymus* (p. 343), and ×*Pseudelymus* (p. 282); most are sterile.

The complex patterns of morphological diversity within *Elymus* in North America probably reflect a combination of multiple origins involving different progenitors, introgression, hybridization both within the genus and with other members of the tribe, and morphological plasticity. Little is known concerning the relative importance of these factors. Two infraspecific ranks have been used to aid in circumscribing the known variation. In general, infraspecies taxa that show great morphological and ecological distinction are treated as subspecies; others, as varieties.

All species of *Elymus* are alloploids that combine one copy of the **St** haplome present in *Pseudoroegneria* with at least one other haplome. So far as is known, all species that are native to North America, as well as many species native to northern Eurasia, are tetraploids with one additional haplome, the **H** genome from *Hordeum* sect. *Critesion*. Many Asian species combine the **St** haplome with the **Y** haplome, for which there are no known diploids; such species are sometimes placed in the segregate genus *Roegneria*. This treatment includes two such species, *E. ciliaris* and *E. semicostatus*. In addition, the treatment includes two hexaploid species, *E. tsukushiensis* and *E. dahuricus*, that combine all three haplomes. *Elymus repens* and *E. hoffmannii*, the other two hexaploid species in this treatment, basically combine two copies of the **St** haplome with one of the **H** haplome, but the molecular data for *E. repens* point to a more complex situation (Mason-Gamer 2001). For further discussion of generic delimitation in the *Triticeae*, see Barkworth (2000), Yen et al. (2005), and Barkworth and von Bothmer (2005).

Elymus is sometimes divided into multiple sections (see, for example, Tsvelev 1976; Löve 1984). There are, however, no detailed morphological descriptions of the sections, making it difficult to determine how to treat the North American species. It is notable that the species with solitary spikelets are concentrated in western and northern North America, whereas the species with multiple spikelets at a node are most prevalent east of the Rocky Mountains, from southern Canada to the Gulf Coast. There are exceptions to this statement. For instance, species with multiple spikelets and disarticulating rachises are primarily western in their distribution, yet *E. glaucus*, a species with multiple spikelets and non-disarticulating rachises, is western. Like the western species with solitary spikelets, and unlike most eastern species, its glumes have a hyaline margin.

In the key and descriptions, unless otherwise stated, the following conventions are observed: the number of culm nodes refers to the number of nodes above the base; measurements of spikes include the awns, while measurements of spikelets, glumes, and lemmas do not; rachis internodes are measured in the middle of the spike; glume widths of lanceolate to linear glumes are measured at the widest point, and those of linear to setaceous glumes about 5 mm above the base of the glumes; the number of florets in a spikelet includes the distal reduced, sterile florets; dates of anthesis, when provided, are for the central range of each species.

The curvature of the lemma awns is often important in identifying individual species. The curvature increases with maturity, and may vary within a spike. If a plant appears to have at least some strongly curved lemma awns, it should be taken through the "strongly curved" side of the key.

SELECTED REFERENCES **Barkworth, M.E.** 1997. Taxonomic and nomenclatural comments on the Triticeae in North America. Phytologia 83:302–311; **Barkworth, M.E.** 2000. Changing perceptions in the Triticeae. Pp. 110–120 *in* S.W.L. Jacobs and J. Everett (eds.). Grasses: Systematics and Evolution. CSIRO Publishing, Collingwood, Victoria, Australia. 408 pp.; **Barkworth, M.E.** and **R. von Bothmer.** 2005. Twenty-one years later: Löve and Dewey's genomic classification proposal. Czech J. Genet. Pl. Breed. 41 (Special Issue)3–9; **Bennett, B.A.** 2006. Siberian Wild Rye (*Elymus sibiricus* L., Poaceae) in western North America: Native or introduced? BEN [Botanical Electronic News] #366. http://www.ou.edu/cas/botany-micro/ben/; **Bödvarsdóttir, S.K.** and **K. Anamthawat-Jónsson.** 2003. Isolation, characterization, and analysis of *Leymus*-specific DNA sequences. Genome 46:673–682; **Booher, L.E.** and **R.M. Tryon.** 1948. A study of *Elymus* in Minnesota. Rhodora 50:80–91; **Bowden, W.M.** 1958. Natural and artificial ×*Elymordeum* hybrids. Canad. J. Bot. 36:101–123; **Bowden, W.M.** 1964. Cytotaxonomy of the species and interspecific hybrids of the genus *Elymus* in Canada and neighboring areas. Canad. J. Bot. 42:547–601; **Bowden, W.M.** 1967. Taxonomy of intergeneric hybrids of the tribe Triticeae from North America. Canad. J. Bot. 45:711–724; **Brooks, R.E.** 1974. Intraspecific variation in *Elymus virginicus* (Gramineae) in the central United States. Master's thesis, University of Kansas, Lawrence, Kansas, U.S.A. 112 pp.; **Brown, W.V.** and **G.A. Pratt.** 1960. Hybridization and introgression in the genus *Elymus*. Amer. J. Bot. 47:669–676; **Bush, B.F.** 1926. The Missouri species of *Elymus*. Amer. Midl. Naturalist 10:49–88; **Campbell, J.J.N.** 2000. Notes on North American *Elymus* species (Poaceae) with paired spikelets: I. *E. macgregorii sp. nov.* and *E. glaucus* ssp. *mackenzii comb. nov.* J. Kentucky Acad. Sci. 61:88–98; **Campbell, J.J.N.** 2002. Notes on North American *Elymus* species (Poaceae) with paired spikelets: II. The *interruptus* group. J. Kentucky Acad. Sci. 62:19–38; **Campbell, J.J.N.** and **A. Haines.** 2002. Corrections and additions to: "Campbell, J.J.N. 2000. Notes on North American *Elymus* species (Poaceae) with paired spikelets: I. *E. macgregorii sp. nov.* and *E. glaucus* ssp. *mackenzii comb. nov.* J. Ky. Acad. Sci. 61:88–98." J. Kentucky Acad. Sci. 62:65; **Church, G.L.** 1954. Interspecific hybridization in eastern *Elymus*. Rhodora 56:185–197; **Church, G.L.** 1958. Artificial hybrids of *Elymus virginicus* with *E. canadensis*, *E. interruptus*, *E. riparius*, and *E. wiegandii*. Amer. J. Bot. 45:410–417; **Church, G.L.** 1967a. Taxonomic and genetic relationships of eastern North American species of *Elymus* with setaceous glumes. Rhodora 69:121–162; **Church, G.L.** 1967b. Pine Hills *Elymus*. Rhodora 69:330–345; **Davies, R.S.** 1980. Introgression between *Elymus canadensis* and *E. virginicus* L. (Triticeae, Poaceae) in south central United States. Ph.D. dissertation, Texas A&M University, College Station, Texas, U.S.A. 232 pp.; **Dewey, D.R.** 1963. Natural hybrids of *Agropyron trachycaulum* and *Agropyron scribneri*. Bull. Torrey Bot. Club 90:111–120; **Dewey, D.R.** 1965. Morphology, cytology, and fertility of synthetic hybrids of *Agropyron spicatum* × *Agropyron dasystachyum–riparium*. Bot. Gaz. 126:269–275; **Dewey, D.R.** 1967a. Genome relations between *Agropyron scribneri* and *Sitanion hystrix*. Bull. Torrey Bot. Club 94:395–404; **Dewey, D.R.** 1967b. Synthetic *Agropyron–Elymus* hybrids: II. *Elymus canadensis* × *Agropyron dasystachum*. Amer. J. Bot. 54:1084–1089; **Dewey, D.R.** 1968. Synthetic hybrids of *Agropyron dasystachyum* × *Elymus glaucus* and *Sitanion hystrix*. Bot. Gaz. 129:316–322; **Dewey, D.R.** 1970. The origin of *Agropyron albicans*. Amer. J. Bot. 57:12–18; **Dewey, D.R.** 1974. Cytogenetics of *Elymus sibiricus* and its hybrids with *Agropyron tauri*, *Elymus canadensis*, and *Agropyron caninum*. Bot. Gaz. 135:80–87; **Dewey, D.R.** 1975. Introgression between *Agropyron dasystachyum* and *A. trachycaulum*. Bot. Gaz. 136:122–128; **Dewey, D.R.** 1976. Cytogenetics of *Agropyron pringlei* and its hybrids with *A. spicatum*, *A. scribneri*, *A. violaceum*, and *A. dasystachyum*. Bot. Gaz. 137:179–185; **Dewey, D.R.** 1982. Genomic and phylogenetic relationships among North American perennial Triticeae. Pp. 51–58 *in* J.R. Estes, R.J. Tyrl, and J.N. Brunken (eds.). Grasses and Grasslands. University of Oklahoma Press, Norman, Oklahoma, U.S.A. 312 pp.; **Gabel, M.L.** 1984. A biosystematic study of the genus *Elymus* (Gramineae: Triticeae). Proc. Iowa Acad. Sci. 91:140–146; **Gillett, J.M. and H.A. Senn.** 1960. Cytotaxonomy and infraspecific variation of *Agropyron smithii* Rydb. Canad. J. Bot. 38:747–760; **Gillett, J.M. and H.A. Senn.** 1961. A new species of *Agropyron* from the Great Lakes. Canad. J. Bot. 39:1169–1175; **Godley, E.J.** 1947. The variation and cytology of the British species of *Agropyron* and their natural hybrids. Master's thesis, Trinity College, Cambridge, England. 152 pp.; **Hitchcock, A.S.** 1935. Manual of the Grasses of the United States. U.S. Government Printing Office, Washington, D.C., U.S.A. 1040 pp.; **Hitchcock, A.S.** 1951. Manual of the Grasses of the United States, ed. 2, rev. A. Chase. U.S.D.A. Miscellaneous Publication No. 200. U.S. Government Printing Office, Washington, D.C., U.S.A. 1051 pp.; **Hitchcock, C.L., A. Cronquist, and M. Ownbey.** 1969. Vascular Plants of the Pacific Northwest, Vol. 1: Vascular Cryptogams, Gymnosperms, and Monocotyledons. University of Washington Press, Seattle, Washington, U.S.A. 914 pp.; **Hultén, E.** 1968. Flora of Alaska and Neighboring Territories. Stanford University Press, Stanford, California, U.S.A. 1008 pp.; **Jensen, K.B.** 1993. *Elymus magellanicus* and its intra- and inter-generic hybrids with *Pseudoroegneria spicata*, *Hordeum violaceum*, *E. trachycaulus*, *E. lanceolatus*, and *E. glaucus* (Poaceae: Triticeae). Genome 36:72–76; **Jensen, K.B. and K.H. Asay.** Cytology and morphology of *Elymus hoffmannii* (Poaceae: Triticeae): A new species from the Erzurum Province of Turkey. Int. J. Pl. Sci. 157:750–758; **Jones, S.B., Jr. and N.C. Coile.** 1988. The Distribution of the Vascular Flora of Georgia. University of Georgia, Athens, Georgia, U.S.A. 230 pp.; **Jozwik, F.X.** 1966. A biosystematic analysis of the slender wheatgrass complex. Ph.D. dissertation, University of Wyoming, Laramie, Wyoming, U.S.A. 112 pp.; **Lepage, E.** 1952. Études sur quelques plantes américaines: II. Hybrides intergénériques; *Agrohordeum* et *Agroelymus*. Naturaliste Canad. 79:241–266; **Lepage, E.** 1965. Revision généalogique de quelques ×*Agroelymus*. Naturaliste Canad. 92:205–216; **Löve, A.** 1984. Conspectus of the Triticeae. Feddes Repert. 95:425–521; **Mason-Gamer, R.J.** 2001. Origin of North American *Elymus* (Poaceae: Triticeae) allotetraploids based on granule-bound starch synthase gene sequences. Syst. Bot. 26:757–768; **Nelson, E.N. and R.J. Tyrl.** 1978. Hybridization and introgression between *Elymus canadensis* and *Elymus virginicus*. Proc. Oklahoma Acad. Sci. 58:32–34; **Pohl, R.W.** 1959. Morphology and cytology of some hybrids between *Elymus canadensis* and *E. virginicus*. Proc. Iowa Acad. Sci. 66:155–159; **Pohl, R.W.** 1966. ×*Elyhordeum iowense*, a new intergeneric hybrid in the Triticeae. Brittonia 18:250–255; **Porsild, A.E. and W.J. Cody.** 1980. Vascular Plants of the Continental Northwest Territories, Canada. National Museum of Natural Sciences, National Museums of Canada, Ottawa, Ontario, Canada. 667 pp.; **Pyrah, G.L.** 1983. *Agropyron arizonicum* (Gramineae: Triticeae) and a natural hybrid from Arizona. Great Basin Naturalist 43:131–135; **Salomon, B., R. von Bothmer, and N. Jacobsen.** 1991. Intergeneric crosses between *Hordeum* and North American *Elymus* (Poaceae: Triticeae). Hereditas 114:117–122; **Sanders, T.S., J.L. Hamrick, and L.R. Holden.** 1979. Allozyme variation in *Elymus canadensis* from the tallgrass prairie region: Geographic variation. Amer. Midl. Naturalist 101:1–12; **Smith, E.B.** (ed.). 1991. An Atlas and Annotated List of the Vascular Plants of Arkansas, ed. 2. Edwin B. Smith, Fayetteville, Arkansas, U.S.A. 489 pp.; **Snyder, L.A.** 1950. Morphological variability and hybrid development in *Elymus glaucus*. Amer. J. Bot. 37:628–636; **Snyder, L.A.** 1951. Cytology of inter-strain hybrids and the probable origin of variability in *Elymus glaucus*. Amer. J. Bot. 38:195–202; **Stebbins, G.L., Jr.** 1957. The hybrid origin of microspecies in the *Elymus glaucus* complex. Pp. 336–340 *in* International Union of Biological Sciences (eds.). Proceedings of the International Genetics Symposia, 1956: Tokyo & Kyoto, September 1956. Organizing Committee, International Genetics Symposia, Science Council of Japan, Tokyo, Japan. 680 pp.; **Stebbins, G.L., Jr. and L.A. Snyder.** 1956. Artificial and natural hybrids in the Gramineae, tribe Hordeae: IX. Hybrids between the western and eastern species. Amer. J. Bot. 43:305–312; **Steyermark, J.A.** 1963. Flora of Missouri. Iowa State University Press, Ames, Iowa, U.S.A. 1725 pp.; **Sun, G.-L. and B. Salomon.** 2003. Microsatellite variability and heterozygote deficiency in the arctic-alpine Alaskan wheatgrass (*Elymus alaskanus*) complex. Genome 46:729–737; **Sun, G.-L., B. Salomon, and R. von Bothmer.** 1998. Characterization of microsatellite loci from *Elymus alaskanus* and length polymorphism in several *Elymus* species (Triticeae: Poaceae). Genome 41:455–463; **Sun, G.L., H. Tang, and B. Salomon.** 2006. Molecular diversity and relationships of North American *Elymus trachycaulus* and the Eurasian *E. caninus* species. Genetica 127:55–64; **Svitashev, S., T. Bryngelsson, X. Li, and R.R.-C. Wang.** 1998. Genome-specific repetitive DNA and RAPD markers for genome identification in *Elymus* and *Hordelymus*. Genome 41:120–128; **Tsvelev, N.N.** 1976. Zlaki SSSR. Nauka, Leningrad [St. Petersburg], Russia. 788 pp.; **Vilkomerson, H.** 1950. The unusual meiotic behavior of *Elymus wiegandii*. Exp. Cell Res. 1:534–542; **Wilson, B.L., J. Kitzmiller, W. Rolle, and V.D. Hipkins.** 2001. Isozyme variation and its environmental correlates in *Elymus glaucus* from the California floristic province. Canad. J. Bot. 79:139–153; **Wilson, F.D.** 1963. Revision of *Sitanion* (Triticeae, Gramineae). Brittonia 15:303–323; **Yen, C., J.-L. Yang, and Y. Yen.** 2005. Hitoshi Kihara, Åskell Löve and the modern genetic concept of the genera in the tribe Triticeae (Poaceae). Acta Phytotax. Sin. 43:82–93; **Zhang, X-Q., B. Salomon, and R. von Bothmer.** 2002. Application of random amplified polymorphic DNA markers to evaluate intraspecific genetic variation in the *Elymus alaskanus* complex (Poaceae). Genet. Resources & Crop Evol. 49:397–407.

1. Spikelets 1 at all or most nodes; glumes with flat, non-indurate bases, glume bodies linear-
   lanceolate to obovate, margins hyaline, scarious, or chartaceous; lemmas awned or unawned
   [for opposite lead, see p. 292].
   2. Anthers 3–7 mm long; plants often strongly rhizomatous, sometimes not or only weakly
      rhizomatous.

3. At least some lemmas with strongly divergent, outcurving, or recurved awns.
    4. Culms prostrate to decumbent and geniculate, 20–50 cm tall; plants of subalpine
        and alpine habitats . . . . 32. *E. sierrae* (in part)
    4. Culms erect or decumbent only at the base, (15)40–130 cm tall; plants of valley and
        montane, but not subalpine or alpine, habitats.
        5. Plants strongly rhizomatous; blades 1–3 mm wide . . . . 34. *E. albicans*
        5. Plants cespitose or weakly rhizomatous; blades 1.5–6 mm wide.
            6. Spikes often drooping to pendent at maturity; rachis internodes 11–17 mm
                long; plants of the southwestern United States . . . . 29. *E. arizonicus*
            6. Spikes erect to slightly nodding at maturity; rachis internodes 5–12 mm long;
                plants of the northwestern contiguous United States . . . . 33. *E. wawawaiensis*
3. Lemmas unawned or with straight to flexuous awns.
    7. Lemmas 12–14 mm long; plants not or weakly rhizomatous.
        8. Palea keels straight or slightly outwardly curved below the apices, apices about
            0.2 mm wide between the vein ends . . . . 10. *E. glaucus* (in part)
        8. Palea keels distinctly outwardly curved below the apices; apices 0.3–0.7 mm wide
            between the vein ends . . . . 39. *E. semicostatus*
    7. Lemmas 7–12 mm long; plants not, weakly, or strongly rhizomatous.
        9. Glumes keeled distally, keels smooth and inconspicuous proximally, scabrous and
            conspicuous distally; lemmas glabrous.
            10. Adaxial surfaces of the blades usually sparsely pilose, sometimes glabrous,
                veins smooth, the primary veins separated by secondary veins; plants strongly
                rhizomatous . . . . 35. *E. repens*
            10. Adaxial surfaces of the blades glabrous, veins smooth or scabrous, all veins
                more or less equally prominent; plants slightly to moderately rhizomatous . . . 36. *E. hoffmannii*
        9. Glumes not keeled or keeled throughout their length, keels smooth or scabrous
            throughout, sometimes hairy, conspicuous or not; lemmas glabrous or hairy.
            11. Plants strongly rhizomatous; glumes 5–9 mm long; lemmas densely to
                sparsely hairy or glabrous . . . . 27. *E. lanceolatus* (in part)
            11. Plants cespitose or weakly rhizomatous; glumes 6–19 mm long; lemmas
              glabrous or pubescent, never densely hairy.
              12. Spikelets usually at least twice as long as the internodes; internodes 4–12
                mm long; glumes often awned, sometimes unawned; blades usually lax
                . . . . 10. *E. glaucus* (in part)
             12. Spikelets from shorter than to almost twice as long as the internodes;
                internodes 9–27 mm long; glumes unawned; blades usually straight . . . . . . 28. *E. stebbinsii*
2. Anthers 0.7–3 mm long; plants usually not or weakly rhizomatous, sometimes strongly
  rhizomatous.
    13. Culms prostrate or strongly decumbent at the base; disarticulation in the rachises or
        beneath the florets; plants of subalpine, alpine, and arctic habitats.
        14. Glumes unawned or with awns to 1 mm long; plants of arctic habitats . . . 26. *E. alaskanus* (in part)
        14. Glumes awned, awns 3–30 mm long; plants of subalpine and alpine habitats.
            15. Anthers 1–1.6 mm long; internodes 2.5–5(7) mm long; disarticulation
                initially in the rachises; spikelets appressed to ascending . . . . 31. *E. scribneri*
            15. Anthers 2–3.5 mm long; internodes 5–15 mm long; rachises not
                disarticulating; spikelets ascending to divergent . . . . 32. *E. sierrae* (in part)
    13. Culms usually ascending to erect, sometimes geniculate or weakly decumbent at the
        base; disarticulation beneath the florets; plants of sea level to subalpine habitats.
        16. Lemmas with coarse, stiff, marginal hairs up to 1 mm long; paleas $^2/_3$–$^4/_5$ as long
            as the lemmas, with wide, rounded apices . . . . 38. *E. ciliaris*
        16. Lemmas with the marginal hairs, if present, similar to those elsewhere on the
            lemma; paleas $^3/_4$ as long as to slightly longer than the lemmas, tapering to the
            apices.
            17. Lemmas awned, awns 7–40 mm long.
                18. Lemma awns strongly arcuate to outcurving or recurved.

19. Spikes 8–12 cm long, straight, erect or inclined; blades 2–4 mm wide .... 30. *E. bakeri*
19. Spikes 7–30 cm long, flexuous, nodding to pendent; blades 5–14 mm
    wide ....13. *E. sibiricus* (in part)
18. Lemma awns usually straight or flexuous, or, if shorter than 10 mm,
    sometimes weakly curving.
    20. Glumes with hairs on the adaxial (inner) surface, these often
        inconspicuous .... 23. *E. caninus*
    20. Glumes glabrous on the adaxial (inner) surface.
        21. Palea keels distinctly outwardly curved below the apices,
            winged, not or scarcely extending beyond the intercostal region;
            apices 0.3–0.5 mm wide .... 37. *E. tsukushiensis*
        21. Palea keels straight or slightly outwardly curved below the
            apices, not winged, often extending beyond the intercostal
            region, sometimes forming teeth; apices 0.1–0.3 mm wide.
            22. Glumes 1.8–2.3 mm wide, margins 0.2–0.3 mm wide
                ....22. *E. trachycaulus* (in part)
            22. Glumes 0.4–1.5(2) mm wide, margins 0.1–0.2 mm wide.
                23. Spikes erect or almost so, 0.5–2 cm wide .... 10. *E. glaucus* (in part)
                23. Spikes nodding to pendent, 2–5 cm wide .... 13. *E. sibiricus* (in part)
17. Lemmas unawned or with awns up to 7 mm long.
    24. Plants strongly rhizomatous .... 27. *E. lanceolatus* (in part)
    24. Plants not or only shortly rhizomatous.
        25. Glumes $^1\!/_3$–$^2\!/_3$ as long as the adjacent lemmas.
            26. Glumes 0.8–1.8 mm wide, lanceolate, margins subequal; lemmas
                evenly hairy or glabrous distally .... 25. *E. macrourus*
            26. Glumes 1.5–2 mm wide, oblanceolate to obovate, margins
                unequal; lemmas glabrous, evenly hairy, or more densely hairy
                distally .... 26. *E. alaskanus* (in part)
        25. Glumes $^3\!/_4$ as long as to slightly longer than the adjacent lemmas.
            27. Glumes 3(5)-veined; glume margins unequal, the wider margins
                0.3–1 mm wide, usually widest in the distal $^1\!/_3$; lemma awns
                0.5–3 mm long .... 24. *E. violaceus*
            27. Glumes 3–7-veined, glume margins equal, 0.1–0.5 mm wide,
                widest at or slightly beyond midlength; lemmas unawned or with
                awns to 40 mm long.
                28. Glumes 1.8–2.3 mm wide, margins 0.2–0.3 mm wide
                    ....22. *E. trachycaulus* (in part)
                28. Glumes 0.4–1.5(2) mm wide, margins 0.1–0.2 mm wide
                    ....10. *E. glaucus* (in part)
1. Spikelets 2–3(5) at all or most nodes; glumes often with subterete to terete, indurate bases,
    sometimes with flat, non-indurate bases, glume bodies linear-lanceolate to setaceous or
    subulate, margins usually firm, sometimes hyaline or scarious; lemmas usually awned, awns
    up to 120 mm long [for opposite lead, see p. 290].
    29. Rachises disarticulating at maturity; glumes 10–135 mm long including the awns,
        sometimes split longitudinally, flexuous to outcurving from near the base; lowest floret
        in each spikelet sometimes sterile; blades 1–6 mm wide.
        30. Glume awns split into 3–9 divisions; lemma awns about 0.2 mm wide at the base;
            rachis internodes 3–5 mm long .... 20. *E. multisetus*
        30. Glume awns entire or split into 2–3 divisions; lemma awns about 0.4 mm wide at
            the base; rachis internodes 3–10(15) mm long .... 21. *E. elymoides*
    29. Rachises not disarticulating at maturity; glumes 0–43 mm long including the awns,
        entire, straight or outcurving from well above the base; lowest floret in each spikelet
        functional; blades 2–25 mm wide.
        31. Glume bodies with 0–1(2) veins, linear or tapering from the base, 0.1–0.6 mm wide,
            0–24 mm long including the awns, often differing in length by more than 5 mm,

persistent after the florets disarticulate; rachis internodes 0.1–0.3(0.4) mm thick at the thinnest sections, often with green lateral bands.

32. Spikelets widely divergent to patent at maturity; lemma awns usually straight, rarely slightly curving; glumes vestigial or 1–3 mm long, occasionally some unequal glumes up to 10(20) mm long and 0.1–0.2 mm wide but with no distinct vein; spikes more or less erect .... 19. *E. hystrix*
32. Spikelets usually appressed, never widely divergent; lemma awns straight or curving; glumes sometimes vestigial, usually 1–24 mm long, 0.1–0.6 mm wide, often with 1(2) distinct veins; spikes erect, nodding, or pendent.
　　33. Glumes 12–30 mm long including the awns, subequal; lemma awns straight to moderately curving; spikes erect to slightly nodding.
　　　　34. Spikelets (6)9–15(22) mm long excluding the awns, each with 2–5 florets; lemma awns moderately outcurving at maturity; glumes (0.2)0.3–0.5(0.7) mm wide .... 9. *E. interruptus* (in part)
　　　　34. Spikelets 18–40 mm long excluding the awns, each with 3–8 florets; lemma awns straight to slightly curving at maturity; glumes 0.1–0.3(0.6) mm wide.
　　　　　　35. Anthers 2.5–4 mm long; lemmas scabrous-hispid to thinly strigose, at least distally; spikes 4–12 cm long; internodes 3–6 mm long, without green lateral bands, with hispid dorsal angles .... 14. *E. pringlei*
　　　　　　35. Anthers 4.5–6 mm long; lemmas smooth, glabrous; spikes 9–20 cm long; internodes (5)7–15(22) mm long, with green lateral bands, glabrous except for the ciliolate margins .... 15. *E. texensis*
　　33. Glumes 0–15(30) mm long including the awns, usually differing in length by at least 4 mm, 1 or both shorter than 12 mm, sometimes both essentially absent; lemma awns outcurving at maturity; spikes more or less nodding.
　　　　36. Rachis internodes 4–6(9) mm long; lemmas hirsute to strigose, at least near the margins, awns 20–35 mm long; sheaths glabrous; plants not glaucous or moderately glaucous .... 18. *E. diversiglumis*
　　　　36. Rachis internodes (4)6–13(18) mm long; lemmas glabrous or pubescent, awns (8)10–30(35) mm long; sheaths glabrous or villous; plants usually glaucous, sometimes strongly so.
　　　　　　37. Lemmas usually glabrous, veins occasionally hispidulous near the apices, awns (8)10–20(25) mm long; spikelets with (3)4–5 florets; rachis internodes (4)6–10(12) mm long, without green lateral bands, glabrous; adaxial surfaces of the blades usually villous; plants strongly glaucous .... 16. *E. svensonii*
　　　　　　37. Lemmas usually hairy, awns (10)20–30(35) mm long; spikelets with 3(5) florets; rachis internodes (5)7–13(18) mm long, with green lateral bands and hispid dorsal angles; adaxial surfaces of the blades glabrous or short-pilose; plants somewhat glaucous .... 17. *E. churchii*
31. Glume bodies with 2–5(8) veins, widening or parallel-sided above the base, (0.2)0.3–2.3 mm wide, 4–43 mm long including the awns, equal or subequal, persistent or disarticulating; rachis internodes 0.1–0.8 mm thick at the thinnest sections, usually lacking green lateral bands.
　　38. Glumes bases more or less terete, indurate, and without veins for 0.5–4 mm; glume bodies exceeding the adjacent lemmas by 1–5 mm or indistinguishable from the glume awns; lemma awns usually straight, occasionally contorted on the lower spikelets; rachis internodes (1.5)2–5(8) mm long.
　　　　39. Glumes persistent, glume bodies (0.2)0.3–0.8(1) mm wide, with 2–4 veins, the basal 0.5–2 mm straight or slightly curving; lemmas with hairs or scabrous; spikelets with 1–3(4) florets; spikes nodding, exserted from the sheath.
　　　　　　40. Adaxial surfaces of the blades densely villous with fine whitish hairs, rarely just pilose on the veins, dark glossy green; spikes 4–12 cm long;

internodes (1.5)2–3(4) mm long; spikelets with 1–2(3) florets; lemmas usually villous, sometimes glabrous, sometimes scabrous, 5.5–9 mm long, 0.5–1.5 mm longer than the paleas; anthesis usually in early June to early July . . . . 5. *E. villosus*

    40. Adaxial surfaces of the blades glabrous or scabrous, dull green; spikes 7–25 cm long; internodes 3–5(8) mm long; spikelets with 2–3(4) florets; lemmas hispidulous or scabrous, 7–14 mm long, 1–5 mm longer than the paleas; anthesis usually in late June to late July . . . . 6. *E. riparius*

39. Glumes disarticulating, glume bodies (0.5)0.7–2.3 mm wide, with (2)3–5(8) veins, the basal 1–4 mm clearly bowed out; lemmas often glabrous, sometimes scabrous; spikelets with 2–5(6) florets; spikes erect, exserted or sheathed.

  41. Spikes (0.5)0.7–2.2(2.5) cm wide including the awns, exserted or sheathed; glume awns 0–10(15) mm long; spikelets appressed to slightly spreading; blades usually glabrous or scabridulous.

    42. Lemma awns 5–15(20) mm long at midspike; blades of all leaves usually spreading or lax and flat, those of the lower leaves not markedly larger or more persistent than those of the upper leaves; anthesis in mid-June to mid-August, usually 1–2 weeks earlier than sympatric *E. curvatus* . . . . 3. *E. virginicus*

    42. Lemma awns 0.5–3(4) mm long at midspike; upper blades usually ascending and somewhat involute, blades of the lower leaves relatively short, narrow, and senescing earlier than those of the upper leaves; anthesis usually in late June to early August, 1–2 weeks later than sympatric *E. virginicus* . . . . 4. *E. curvatus*

  41. Spikes (1.7)2.2–4.5(5.5) cm wide including the awns, exserted; glume awns (10)15–30 mm long; spikelets spreading; blades glabrous or villous.

    43. Spikes with (6)9–16(20) nodes; internodes 4–7 mm long, about 0.3 mm thick at the thinnest portion; blades lax, dark glossy green under the glaucous bloom; auricles 2–3 mm long, often purplish black, at least in the central range of the species; anthesis usually in mid-May to mid-June . . . . 1. *E. macgregorii*

    43. Spikes with (10)18–30(36) nodes; internodes 3–5 mm long, 0.3–0.8 mm thick at the thinnest portion; blades lax, or ascending and involute, usually dull green, with or without a glaucous bloom; auricles 0–2 mm long, usually purplish brown; anthesis usually in mid-June to late July . . . . 2. *E. glabriflorus*

38. Glume bases flat and veined or, if subterete to terete, indurate and without veins for less than 1 mm; glume bodies shorter than or subequal to the lowest lemmas; lemma awns usually flexuous to curving, sometimes straight; rachis internodes (2)3–14 mm long.

  44. Glumes with more or less terete bases, without hyaline or scarious margins, always awned, awns (5)8–25(27) mm long or the glume bodies indistinguishable from the awns; spikelets (1)2–3(5) per node, spreading, not or rarely purplish; cauline nodes usually concealed by the sheaths.

    45. Spikes erect to slightly nodding, internodes (5)8–14 mm long; glumes 0.2–0.5(0.7) mm wide; lemmas 7–10 mm long, usually smooth or scabrous, occasionally hirtellous, especially near the margins, awns 15–22 mm long, straight to moderately outcurving; blades 3–9 mm wide; culms (40)60–100(120) cm tall, nodes usually exposed . . . 9. *E. interruptus* (in part)

    45. Spikes usually nodding to pendent, sometimes erect, internodes (2)3–8(12) mm long; glume bodies (0.2)0.4–1.6 mm wide; lemmas 8–15 mm long, glabrous or uniformly hairy, awns (10)15–40(50) mm long, moderately to strongly outcurving; blades 3–24 mm wide; culms

(40)60–180(220) cm tall, nodes usually concealed by the leaf sheaths.

 46. Rachis internodes (2)3–5(7) mm long; spikelets 2(3) at most nodes, occasionally 1 or up to 5 at some nodes; paleas acute; blades (3)4–15(20) mm wide, usually firm and somewhat involute, dull green, drying grayish . . . . 7. *E. canadensis*

 46. Rachis internodes 5–12 mm long; spikelets 2 per node; paleas narrowly truncate; blades (8)10–20(24) mm wide, flat, lax, dark green . . . . 8. *E. wiegandii*

44. Glumes with flat bases and hyaline or scarious margins, usually awned, awns 1–10 mm long, sometimes unawned; spikelets (1)2(3) per node, appressed to divergent, sometimes purplish; cauline nodes mostly exposed.

 47. Anthers 0.9–1.7 mm long; glumes 3–8 mm long; lowest lemmas 3–6 mm longer than the glumes, densely scabridulous to scabrous, awns usually outcurving; spikelets with (3)4–5(7) florets; spikes 2–5 cm wide, nodding to pendent; cauline nodes glabrous . . . . 13. *E. sibiricus* (in part)

 47. Anthers 1.5–4.5 mm long; glumes (4.5)6–14(19) mm long; lowest lemmas from shorter than to 2.5 mm longer than the glumes, smooth, sometimes hairy, awns straight, flexuous, or outcurving; spikelets with 2–4(7) florets; spikes (0.2)0.5–2.5 cm wide, erect, nodding, or pendent; cauline nodes occasionally with short hairs.

  48. Glume bodies (6)9–14(19) mm long; lemmas 8–16 mm long, awns usually straight to flexuous; auricles usually present, to 2.5 mm long . . . . 10. *E. glaucus* (in part)

  48. Glume bodies (4.5)6–10(11) mm long; lemmas 5–14 mm long, awns flexuous to moderately outcurving; auricles often absent, or to 1.5 mm long.

   49. Lemmas with hairs, the marginal hairs markedly longer than those elsewhere; paleas acute; spikes nodding to pendent; rachis internodes 3–12 mm long; leaves usually deep green; plants native to the Pacific coastal mountains . . . . 11. *E. hirsutus*

   49. Lemmas smooth, scabrous, or hispid, the marginal hairs, if present, not markedly longer than those elsewhere; paleas obtuse or truncate; spikes erect to slightly nodding; rachis internodes 3–6 mm long; leaves usually pale green, sometimes glaucous; plants introduced . . . . 12. *E. dahuricus*

1. Elymus macgregorii R. Brooks & J.J.N. Campb. [p. 297]
EARLY WILDRYE

Plants cespitose, not rhizomatous, usually glaucous. **Culms** 40–120 cm, erect or slightly decumbent; **nodes** 4–8, mostly exposed, glabrous. **Leaves** evenly distributed; **sheaths** usually glabrous, rarely villous; **auricles** 2–3 mm, usually purplish black when fresh, sometimes light brown; **ligules** shorter than 1 mm; **blades** 7–15 mm wide, lax, dark glossy green under the glaucous bloom, adaxial surfaces usually glabrous, occasionally villous. **Spikes** 4–12 cm long, (1.7)2.2–3(4)4 cm wide, erect, exserted, with (6)9–16(20) nodes and 2 spikelets at all or most nodes, sometimes with 3 at some nodes; **internodes** 4–7 mm long, about 0.3 mm thick and 2-angled at the thinnest sections, usually glabrous or scabridulous beneath the spikelets. **Spikelets** 10–15 mm, strongly divergent, glaucous, maturing to pale yellowish brown, with (2)3–4 florets, lowest florets functional; **disarticulation** below the glumes and each floret, the lowest floret often falling with the glumes. **Glumes** subequal, entire, the basal 1–3 mm terete or subterete, indurate, without evident venation, moderately bowed out, glume bodies 8–16 mm long, 1–1.8 mm wide, linear-lanceolate, widening or parallel-sided above the base, (2)4–5(8)-veined, usually glabrous, occasionally hirsute, sometimes scabrous, margins firm, awns (10)15–20(25) mm, straight except the awns of the lowest spikelets occasionally contorted; **lemmas** 6–12 mm, usually glabrous, sometimes scabrous, occasionally villous, awns (15)20–30 mm, straight; **paleas** 6–10 mm, apices obtuse; **anthers** 2–4 mm. **Anthesis** usually mid-May to mid-June. $2n = 28$.

Elymus macgregorii grows in moist, deep, alluvial or residual, calcareous or other base-rich soils in woods and thickets, mostly east of the 100th Meridian in the contiguous United States. It used to be confused with *E. glabriflorus* (p. 296) or *E. virginicus* (p. 298), but it reaches anthesis about a month earlier than sympatric populations of these species. In most of its range, *E. macgregorii* has purplish black auricles; light brown auricles may be locally abundant, particularly in populations at the limits of its range.

Elymus macregorii hybridizes with several species, but especially *E. virginicus* and *E. hystrix* (p. 316) (Campbell 2000). Western plants often have smaller, more condensed spikes and distinctly villous leaves, suggesting a transition to *E. virginicus* var. *jejunus* (p. 300). Transitions to *E. virginicus* var. *jejunus* can also be recognized to the north, where the dates of anthesis are delayed, but even in Maine, *E. macgregorii* reaches anthesis about 10 days earlier than *E. virginicus* (Campbell and Haines 2002). Plants with villous lemmas grow at scattered locations; they have not been reported in distinct habitats, nor in large enough populations to warrant taxonomic recognition.

2. Elymus glabriflorus (Vasey *ex* L.H. Dewey) Scribn. & C.R. Ball [p. 297]

SOUTHEASTERN WILDRYE

Plants cespitose, not rhizomatous, often glaucous. **Culms** 60–140 cm, erect; **nodes** 6–9, mostly concealed, glabrous. **Leaves** evenly distributed; **sheaths** glabrous or pubescent, often reddish brown; **auricles** absent or to 2 mm, usually purplish brown; **ligules** shorter than 1 mm; **blades** 7–15 mm wide, lax or somewhat involute and ascending, usually dull green, sometimes with a glaucous bloom, adaxial surfaces glabrous or densely short-villous. **Spikes** 6–20 cm long, (2) 2.5–4(5.5) cm wide, erect, exserted, with (10)18–30(36) nodes, usually with 2(3) spikelets per node, occasionally with up to 5 at some nodes; **internodes** 3–5 mm long, 0.3–0.8 mm thick and usually 4-angled at the thinnest sections, glabrous or pubescent beneath the spikelets. **Spikelets** 10–20 mm, strongly divergent, often reddish brown at maturity, with (2)3–5(6) florets, lowest florets functional; **disarticulation** below the glumes and each floret, or the lowest floret often falling with the glumes. **Glumes** equal or subequal, entire, the basal 1–3 mm terete, indurate, moderately bowed out, without evident

venation, glume bodies 7–18 mm long, (0.7)0.9–1.7 mm wide, linear-lanceolate, widening above the base, (3)4–5(7)-veined, smooth or scabrous, sometimes hirsute, margins firm, awns (10)15–25(30) mm, straight except the awns of the lowest spikelets frequently contorted; **lemmas** 6–13 mm, smooth, scabrous, or hirsute, awns (15)25–35(40) mm, straight except the awns of the lowest spikelets occasionally contorted; **paleas** 6–12 mm, obtuse; **anthers** 2–4 mm. **Anthesis** usually mid-June to late July. $2n = 28$.

Elymus glabriflorus grows on moist, damp, or dry soil in open woods, thickets, and tall grasslands, sometimes spreading into old fields and roadsides. It is found in most of the southeastern United States, extending north to Iowa, Illinois, Indiana, West Virginia, and along the Atlantic coast to Maine; it is rare north of Maryland. Anthesis is usually 2–4 weeks later than in *E. virginicus* (see next) and other sympatric taxa, even in Texas, where it occurs up to a month earlier than the dates given (Davies 1980).

Elymus glabriflorus varies greatly in its pubescence, but without clear taxonomic relevance. Plants that combine pubescent spikelets and, usually, pubescent leaves with somewhat shorter spikes (6–12 cm versus 9–20 cm) and lemmas (6–10 mm versus 7–13 mm) are typical on relatively dry, infertile soils, especially in hilly interior regions, and are less frequent on the southeastern coastal plain. They have been named *E. glabriflorus* var. *australis* (Scribn. & C.R. Ball) J.J.N. Campb. In contrast, glabrous to scabrous plants that are often more robust usually grow on relatively moist or damp soils of bottomlands and upland depressions.

Elymus glabriflorus is most closely related to *E. macgregorii* (see previous) and *E. virginicus*, forming occasional hybrids with both (Campbell 2000). It is sometimes confused with *E. villosus* (p. 302), from which it differs in having erect spikes, and glumes that are bowed out and disarticulate at maturity. It has also been confused with *E. canadensis*, especially *E. canadensis* var. *robustus* (p. 305), which may be derived from introgressants between the two species (Davies 1980). Hybrids with *E. hystrix* (p. 316) are also known, with apparent introgression at some range margins. Artificial crosses with other species failed in several cases (Church 1967a, 1967b).

E. macgregorii   *E. glabriflorus*

ELYMUS

3. Elymus virginicus L. [p. 299]
VIRGINIA WILDRYE, ÉLYME DE VIRGINIE

Plants cespitose, not rhizomatous, sometimes glaucous, especially in the spikes. **Culms** 30–130 cm, erect to slightly decumbent; **nodes** 4–9, concealed or exposed, usually glabrous, rarely pubescent. **Leaves** evenly distributed; **sheaths** usually glabrous, rarely hirsute, occasionally reddish or purplish; **auricles** absent or to 1.8 mm, pale brown; **ligules** shorter than 1 mm; **blades** 2–14(18) mm wide, usually spreading or lax, sometimes becoming involute, basal blades similar to the upper blades, adaxial surfaces usually smooth, sometimes scabridulous, usually glabrous, occasionally pubescent. **Spikes** (3)4–16(22) cm long, 1–2.2(2.5) cm wide, erect, the bases often sheathed, with 2 spikelets per node, rarely with 3 at some nodes; **internodes** 3–5 mm long, 0.25–0.5 thick at the thinnest sections, smooth and glabrous, or scabrous, or with hairs beneath the spikelets. **Spikelets** 10–15 mm, appressed to slightly divergent, with (2)3–4(6) florets, lowest florets functional; **disarticulation** below the glumes and each floret, or the lowest floret falling with the glumes. **Glumes** subequal or equal, the basal 1–4 mm terete, indurate, without evident venation, bowed out, yellowish, glume bodies 7–15 mm long, (0.5)0.7–2.3 mm wide, linear-lanceolate, widening above the base, 3–5(8)-veined, usually smooth or scabridulous, margins firm, awns 3–10(15) mm, straight; **lemmas** 6–10 mm, scabridulous, glabrous or villous-hirsute, awns (5)8–20(25) mm, straight; **paleas** 5–9 mm, obtuse; **anthers** 2–3.5(4) mm. **Anthesis** usually mid-June to late July (mid-August). 2*n* = 28.

Elymus virginicus is widespread in temperate North America, growing as far west as British Columbia and Arizona. It is infrequent to rare in the Rocky Mountains, western Great Plains, and southeastern coastal plain. It is a complex species, divided here into four intergrading varieties.

1. Spikelets hispidulous to villous-hirsute, usually glaucous; anthesis usually in early July to mid-August . . . . var. *intermedius*
1. Spikelets usually glabrous or scabrous, glaucous or not; anthesis usually in mid-June to late July.
  2. Spikes partly sheathed; glumes 1–2.3 mm wide, strongly indurate and bowed out in the basal 2–4 mm; plants not glaucous, becoming yellowish brown or occasionally somewhat purplish at maturity . . . . var. *virginicus*
  2. Spikes exserted; glumes (0.5)0.7–1.5(1.8) mm wide, moderately indurate and bowed

out in the basal 1–2 mm; plants usually glaucous, becoming yellowish or reddish brown at maturity.
    3. Culms usually 70–100 cm tall, with 6–8 nodes; blades 3–15 mm wide, flat; spikes 4–20 cm long, not strongly glaucous; glumes indurate only in the basal 1 mm . . . . var. *jejunus*
    3. Culms usually 30–80 cm tall, with 4–6 nodes; blades 2–9 mm wide, often becoming involute; spikes 3.5–11 cm long, often strongly glaucous; glumes usually indurate in the basal 1–2 mm . . . . var. *halophilus*

Elymus virginicus var. halophilus (E.P. Bicknell) Wiegand [p. 299]

Plants glaucous, often strongly so, becoming reddish brown at maturity. **Culms** usually 30–80 cm; **nodes** usually 4–6, often exposed; **ligules** and **auricles** often pronounced. **Blades** 2–9 mm wide, often becoming involute and ascending, glabrous or slightly scabrous. **Spikes** 3.5–11 cm, exserted; **spikelets** usually glabrous to scabrous, strongly glaucous; **glumes** 0.7–1.5 mm wide, usually somewhat indurate and bowed out in the basal 1–2 mm. **Anthesis** late June to late July.

Elymus virginicus var. *halophilus* grows in the moist to damp soil of dunes and brackish marsh edges along the northern Atlantic coast, from Nova Scotia to North Carolina. It could be considered a relatively small, disjunct form of the largely midwestern *E. virginicus* var. *jejunus*, but its glumes are more like those of var. *virginicus*. Transitions to *E. glabriflorus* (p. 296) may be found in southern regions. Rare northern plants with hirsute spikelets (sometimes called *E. virginicus* f. *lasiolepis* Fernald) differ from var. *intermedius* in having grayish green, involute blades.

Elymus virginicus var. intermedius (Vasey *ex* A. Gray) Bush [p. 299]

Plants often partly glaucous, becoming yellowish, reddish, or slightly purplish brown at maturity. **Culms** usually 60–120 cm; **nodes** usually 4–8, concealed or exposed; **auricles** and **ligules** usually poorly developed. **Blades** 4–18 mm wide, lax or involute, glabrous, scabridulous, or occasionally pubescent. **Spikes** 6–22 cm long, sheathed or exserted; **spikelets** hispidulous to villous-hirsute, usually glaucous; **glumes** 0.8–1.2(2) mm wide, indurate and bowed out in the basal 2–4 mm. **Anthesis** early July to mid-August.

Elymus virginicus var. *intermedius* grows in moist, base-rich soil in open forests and thickets, especially on rocky, gravelly, or sandy banks of larger streams. It grows from the central and southern Great Plains, through the central Mississippi and Ohio valleys, to the

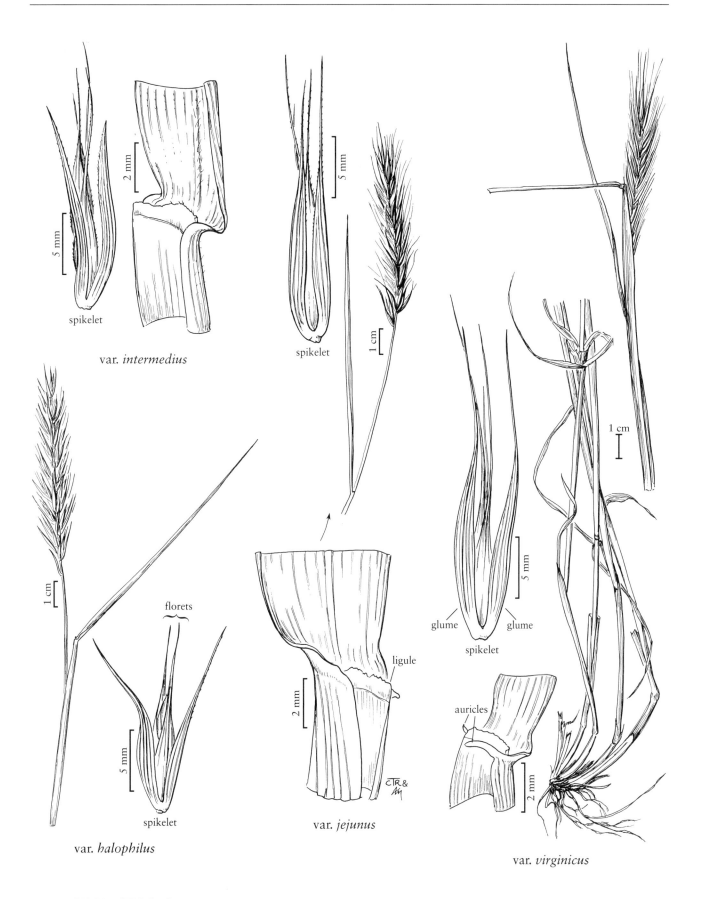

var. *intermedius*

spikelet

var. *halophilus*

spikelet

florets

spikelet

var. *jejunus*

ligule

glume   glume

spikelet

auricles

var. *virginicus*

ELYMUS VIRGINICUS

northeastern United States and adjacent Canada, but is rare to absent south of Oklahoma to Tennessee and to Maryland.

Elymus virginicus var. jejunus (Ramaley) Bush [p. 299]

Plants usually glaucous, but not strongly so, becoming yellowish or reddish brown at maturity. **Culms** (50)70–100(130) cm; **nodes** usually 6–8, often exposed; **auricles** and **ligules** often pronounced. **Blades** 3–15 mm wide, flat, usually scabridulous, rarely pubescent. **Spikes** 4–20 cm, exserted; **spikelets** usually glabrous to scabrous, sometimes glaucous; **glumes** (0.5)0.7–1.2(1.8) mm wide, indurate and bowed out only in the basal 1 mm. **Anthesis** mid-June to late July.

Elymus virginicus var. *jejunus* grows in moist to dry, sometimes alkaline or saline soil, in open, rocky, or alluvial woods, grasslands, glades, and disturbed places. It occupies the western range of the species, except for the Intermountain region. It is uncommon in the northeast, from Virginia to Newfoundland, and rare or absent in the southeast, beyond Texas and Indiana. It intergrades with var. *virginicus* (with spikes up to 22 cm long in intermediate plants) in the Great Plains, and with *E. glabriflorus* (p. 296) or *E. macgregorii* (p. 295) in some southern or eastern regions. Spike exsertion, and thus the ease of distinction from var. *virginicus*, can be influenced by the environment (Brooks 1974). Some northern pop-ulations, from Alberta and North Dakota to Michigan and Quebec, especially on damper soils, have blades that are villous-hirsute and often darker green (e.g., Booher and Tryon 1948). Blades are often only 4–8 mm wide in the Great Plains, and some exceptionally small northern plants (*E. virginicus* var. *micromeris* Schmoll) have blades (1)2–4(5.5) mm wide, spikes 3.2–7 cm long, glume awns 1–1.5 mm long, and lemma awns 3.5–13 mm long, suggesting a transition to *E. curvatus* (see discussion under the next species).

Elymus virginicus L. var. **virginicus** [p. 299]

Plants not glaucous, usually becoming yellowish brown, occasionally somewhat purplish at maturity. **Culms** (30)40–90(130) cm; **nodes** usually 4–8, concealed or exposed; **auricles** and **ligules** sometimes pronounced. **Blades** 3–18 mm wide, lax or involute, usually scabridulous, rarely pubescent. **Spikes** 4–16(22) cm, partly sheathed; **spikelets** smooth or scabridulous, not glaucous; **glumes** 1–2(2.3) mm wide, indurate and bowed out in the basal 2–4 mm. **Anthesis** mid-June to late July.

Elymus virginicus var. *virginicus* grows in moist to damp or rather dry soil, mostly on bottomland or fertile uplands, in open woods, thickets, tall forbs, or weedy sites. It is widespread and abundant in the eastern range of the species, but also overlaps with var. *jejunus* in the

Great Plains, east to Texas and Manitoba. Its dimensions have much genetic and phenotypic variation (Brooks 1974). It occasionally hybridizes with sympatric *Elymus* species, including *E. riparius* (p. 302), and even with *Hordeum* (Bowden 1958; Church 1958; Pohl 1959; Nelson and Tyrl 1978). In its eastern range, most plants are distinctively short, reaching only 30–90 cm, with sheathed spikes 6–10 cm long. In more open or drier environments, especially in midwestern regions, plants are often more glaucous, robust, and exserted, grading into var. *jejunus*. Awn length increases towards the south, suggesting introgression with *E. glabriflorus* (p. 296) (Davies 1980). Pubescent blades are generally absent, but appear more frequently in Wisconsin and perhaps other northern areas.

4. Elymus curvatus Piper [p. 301]
AWNLESS WILDRYE

Plants cespitose, not rhizom-atous, often glaucous. **Culms** 60–110 cm, stiffly erect, or the base sometimes geniculate; **nodes** 6–9, concealed or exposed, glabrous. **Leaves** evenly distributed; **sheaths** glabrous, often reddish brown; **auricles** to 1 mm, sometimes absent; **ligules** shorter than 1 mm, ciliolate; **blades** 5–15 mm wide, the lower blades usually lax, shorter, narrower, and senescing earlier, the upper blades usually ascending and somewhat involute, adaxial surfaces smooth or scabridulous, occasionally scabrous. **Spikes** 9–15 cm long, (0.5)0.7–1.3 cm wide, erect, exserted or the bases slightly sheathed, with 2 spikelets per node; **internodes** 2.5–4.5 mm long, about 0.25–5 mm thick at the thinnest sections, smooth or scabrous beneath the spikelets. **Spikelets** 10–15 mm, appressed, often reddish brown at maturity, with (2)3–4(5) florets, lowest florets functional; **disarticulation** below the glumes and beneath the florets, or the lowest floret falling with the glumes. **Glumes** equal or subequal, the basal 2–3 mm terete, indurate, strongly bowed out, without evident venation, glume bodies 7–15 mm long, 1.2–2.1 mm wide, linear-lanceolate, widening above the base, 3–5-veined, usually glabrous or scabrous, occasionally hispidulous, rarely hirsute on the veins, margins firm, awns 0–3(5) mm; **lemmas** 6–10 mm, glabrous or scabrous, rarely hirsute, awns (0.5)1–3(4) mm, rarely 5–10 mm on the lemmas of the distal spikelets, straight; **paleas** 6–10 mm, obtuse, often emarginate; **anthers** 1.5–3 mm. **Anthesis** late June to mid-August. $2n = 28, 42$.

Elymus curvatus grows in moist or damp soils of open forests, thickets, grasslands, ditches, and disturbed ground, especially on bottomland. It is widespread from

2 mm

glumes
glumes
node

spikelet

5 mm

1 cm

spikelet

5 mm

floret

2 mm

1 cm

E. curvatus

ELYMUS

1 cm

spikelet

floret

CTR&
AM

E. villosus

1 cm

spikelet

1 cm

2 mm

floret

5 mm

glume   glume

spikelet

E. riparius

British Columbia and Washington, through the Intermountain region and northern Rockies, to the northern Great Plains. It is infrequent or rare in the midwest, the Great Lakes region, and the northeast, and is virtually unknown in the southeast. It is similar to *E. virginicus* (p. 298), and has sometimes been included in that species as *E. virginicus* var. *submuticus* Hook., but is more distinct than the varieties of *E. virginicus* treated above. Although *E. virginicus* and *E. curvatus* overlap greatly in range, *E. curvatus* usually has a distinct growth form, and its anthesis is 1–2 weeks later (Brooks 1974). Its spikes range from being completely exserted, especially west of the Great Plains, to largely sheathed, especially east of the Mississippi River and in more stressed environments. This geographic trend parallels that within *E. virginicus*, but sheathed plants of *E. curvatus* can usually be distinguished by their short awns. Clear transitions to *E. virginicus*, usually var. *jejunus*, are rare, but, especially from Missouri to Wisconsin, there are occasional plants with 5–10 mm awns on a few lemmas, especially at the spike tips. Rarely, plants from Missouri and Iowa to Quebec have hispid to hirsute spikelets, suggesting introgression with *E. virginicus* var. *intermedius*. There are a few records of apparent hybrids with other species.

5. Elymus villosus Muhl. *ex* Willd. [p. 301]

DOWNY WILDRYE

Plants cespitose, not rhizomatous, often persistently deep green. **Culms** 40–130 cm, erect; **nodes** 4–8, concealed or exposed, glabrous. **Leaves** evenly distributed; **sheaths** villoushirsute, pilose, or occasionally glabrate, occasionally reddish brown; **auricles** 1–3 mm, brownish; **ligules** shorter than 1 mm, entire or erose; **blades** 4–12 mm wide, lax, dark glossy green, adaxial surfaces usually densely velutinous-villous with fine whitish hairs, rarely pilose only on the veins. **Spikes** 4–12 cm long, 1.5–3.5 cm wide, slightly or strongly nodding, exserted, usually with 2 spikelets per node, rarely with 1 or 3 at a few nodes; **internodes** (1.5)2–3(4) mm long, 0.15–0.25 mm thick at the thinnest sections, usually hairy below the spikelets, rarely glabrous. **Spikelets** 7–12 mm, moderately divergent, with 1–2(3) florets, lowest florets functional; **disarticulation** above the glumes and beneath each floret. **Glumes** equal, 12–25 mm including the often undifferentiated awns, the basal 0.5–2 mm terete, slightly indurate, straight or nearly so, without evident venation, glume bodies 7–10 mm long, (0.2)0.3–0.8 mm wide, linear-setiform, widening or parallel-sided above the base, 2–3(4)-veined, usually hirsute to hispid, occasionally scabrous

to scabridulous, margins firm, awns 5–15 mm, straight; **lemmas** 5.5–9 mm, usually villous with fine, whitish, spreading hairs, especially near the margins and apices, sometimes glabrous or with coarser hairs, sometimes scabrous, awns 9–33 mm, straight; **paleas** 5–7.5 mm, obtuse, occasionally emarginate; **anthers** (1.6)2–3(4) mm. **Anthesis** early June to early July. 2*n* = 28.

Elymus villosus grows in moist to moderately dry, often rocky soils in woods and thickets, especially in calcareous or other base-rich soils, but it is also frequent on drier, sandy soils or damper, alluvial soils in glaciated regions. It extends from the Great Plains east to southern Quebec, northern New York, and Vermont south to Texas, Georgia, and South Carolina. It is absent from the southern portion of the coastal plain.

Elymus villosus is relatively uniform and distinct, although it has sometimes been confused with hairy plants of *E. canadensis* (p. 303) and *E. glabriflorus* (p. 296). The hairs of *E. villosus* are fine, whitish, and consistently dense on the leaf blades, typically spreading in the spikelets; the hairs of the other species are typically stouter and more appressed in the spikelets. Plants called *E. villosus* var. *arkansanas* (Scribn. & C.R. Ball) J.J.N. Campb. are scabrous to glabrous in the spikes, except for the ciliate rachis margins, and often more robust. These are scattered over much of the species' range, except in the north (from Wisconsin to New England), and are locally more frequent than typical plants in the Ozark Mountains and other midwestern hills. Some other western plants (including those called *E. striatus* var. *ballii* Pammel) have unusually large, almost erect spikes, suggesting introgression from *E. virginicus* (p. 298). There are rare apparent hybrids with species in the *E. virginicus* group, but the only proven natural hybrid is with *Hordeum jubatum* (p. 245) (see ×*Elyhordeum*, p. 283). Artificial crosses with several species failed to produce healthy F$_1$ plants (Church 1958).

6. Elymus riparius Wiegand [p. 301]

EASTERN RIVERBANK WILDRYE, ÉLYME DES RIVAGES

Plants cespitose, not rhizomatous, often somewhat glaucous. **Culms** 70–160 cm, erect, sometimes rooting at the lower nodes; **nodes** 5–10, mostly concealed, glabrous. **Leaves** evenly distributed; **sheaths** usually glabrous or scabridulous, often reddish brown; **auricles** absent or to 2 mm, brown; **ligules** shorter than 1 mm; **blades** (5)8–15(25) mm wide, flat, lax, dull green, drying to grayish, adaxial surfaces glabrous or scabrous. **Spikes** 7–25 cm long, 2–4 cm wide, nodding, exserted, usually with 2 spikelets per node, rarely with 3 at some nodes; **internodes** 3–5(8) mm long, 0.2–0.35 thick at the thinnest sections, usually glabrous below the spikelets. **Spikelets** 10–20 mm, strongly divergent, with 2–3(4) florets, lowest florets functional; **disarticulation** above the glumes and beneath each floret. **Glumes** equal or subequal, 14–30 mm including the sometimes undifferentiated awn, the basal 0.5–2 mm terete, indurate, straight or nearly so, veins not evident, glume bodies 9–17 mm long, (0.3)0.5–0.8(1) mm wide, linear-setiform, entire, widening or parallel-sided above the base, 2–3(4)-veined, usually hispidulous or scabrous, rarely glabrous, margins firm, awns (5)8–18 mm, straight; **lemmas** 7–14 mm, usually hispidulous, sometimes scabrous, awns 15–35 mm, usually straight, those of the basal spikelets occasionally contorted; **paleas** 6–9 mm, usually acute, sometimes obtuse to truncate, bidentate; **anthers** 2–2.7 mm. **Anthesis** late June to late July. 2n = 28.

Elymus riparius grows in moist, usually alluvial and often sandy soils in woods and thickets, usually along larger streams and occasionally along upland ditches. It is widespread in most of temperate east-central North America. It is rare in southern Ontario and Quebec, and the eastern Great Plains. It is virtually absent from the southeastern coastal plain.

Elymus riparius is relatively uniform and distinct. It is sometimes confused with *E. canadensis* (see next), but that species has curving awns. It hybridizes occasionally with several other taxa, especially *E. virginicus* var. *virginicus* (p. 300) and *E. hystrix* (p. 316), but the hybrids produce only late, depauperate spikes or none at all (e.g., Church 1958).

7. Elymus canadensis L. [p. 304]
GREAT PLAINS WILDRYE, ÉLYME DU CANADA

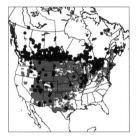

Plants loosely cespitose, rarely with rhizomes to 4 cm long and 1–2 mm thick, often glaucous. **Culms** (40)60–150(180) cm, erect or decumbent; **nodes** 4–10, mostly concealed by the leaf sheaths, glabrous. **Leaves** evenly distributed; **sheaths** smooth or scabridulous, glabrous or hirsute, often reddish brown; **auricles** 1.5–4 mm, brown or purplish black; **ligules** to 1(2) mm, truncate, ciliolate; **blades** (3)4–15(20) mm wide, usually firm, often ascending and somewhat involute, usually dull green, drying to grayish, adaxial surfaces usually smooth or scabridulous and glabrous, rarely sparsely hispid to villous. **Spikes** 6–30 cm long, 3–7 cm wide, usually nodding, sometimes pendent or almost erect, usually with 2(3) spikelets per node, occasionally to 5 at some nodes, rarely with 1 at some nodes but never throughout; **internodes** (2)3–5(7) mm long, or 5–10 mm long towards the base, 0.2–0.35 mm thick at the

thinnest sections, glabrous or with a few hairs below the spikelets. **Spikelets** 12–20 mm excluding the awns, more or less divergent, with (2)3–5(7) florets, lowest florets functional; **disarticulation** usually above the glumes and beneath each floret, rarely also below the glumes. **Glumes** usually equal, occasionally subequal, 11–40 mm including the awns, the basal 0–1 mm subterete and slightly indurate, glume bodies 6–13 mm long, 0.5–1.6 mm wide, linear-lanceolate to subsetaceous, entire, widening or parallel-sided above the base, 3–5-veined, glabrous to scabrous-ciliate, rarely villous on the veins, margins firm, awns (5)10–25(27) mm, straight to outcurving; **lemmas** 8–15 mm, glabrous, scabrous, hispid, or uniformly villous with the hairs generally appressed, awns (10)15–40(50) mm, moderately to strongly outcurving, often contorted at the spike bases; **paleas** 7–13 mm, acute, usually bidentate; **anthers** 2–3.5 mm. **Anthesis** May to July. 2n = 28, rarely 42.

Elymus canadensis grows on dry to moist or damp, often sandy or gravelly soil on prairies, dunes, stream banks, ditches, roadsides, and disturbed ground, or, especially to the south, in thickets and open woods near streams. It is widespread in most of temperate North America, extending from the southwestern Northwest Territories to Coahuila, Mexico, being especially common in the Great Plains. Reports from California and the southeastern states appear to be based on misidentifications. *E. canadensis* is considered a good forage species.

Elymus canadensis is sometimes confused with *E. riparius* (see previous), from which it differs in having curved rather than straight awns; and with *E. wiegandii* (p. 305), from which it differs in its less robust habit and narrower leaves. It can hybridize with *E. glabriflorus* (p. 296), *E. virginicus* (p. 298), *E. hystrix* (p. 316) and allies, *E. glaucus* (p. 306), *E. trachycaulus* (p. 321), *Pseudoroegneria spicata* (p. 281), and other species. Subsequent introgression may have contributed to much of the diversity within the genus (Pohl 1959; Brown and Pratt 1960; Nelson and Tyrl 1978; Davies 1980; Campbell 2002). The three varieties recognized here show clear differences in their typical expression and evidence some geographic separation, but they may prove to be artificial reference points within a more or less continuous variation (Sanders et al. 1979). Nevertheless, crossing barriers sometimes exist between the varieties, and even between some sympatric strains (Church 1954, 1958, 1967a).

1. Lemmas usually villous or hispid; spikes nodding to almost pendent; internodes 4–7 mm long, often strongly glaucous . . . . var. *canadensis*
1. Lemmas usually smooth or scabridulous, occasionally hirsute; spikes usually nodding, occasionally almost erect; internodes 3–4 mm long, not strongly glaucous.

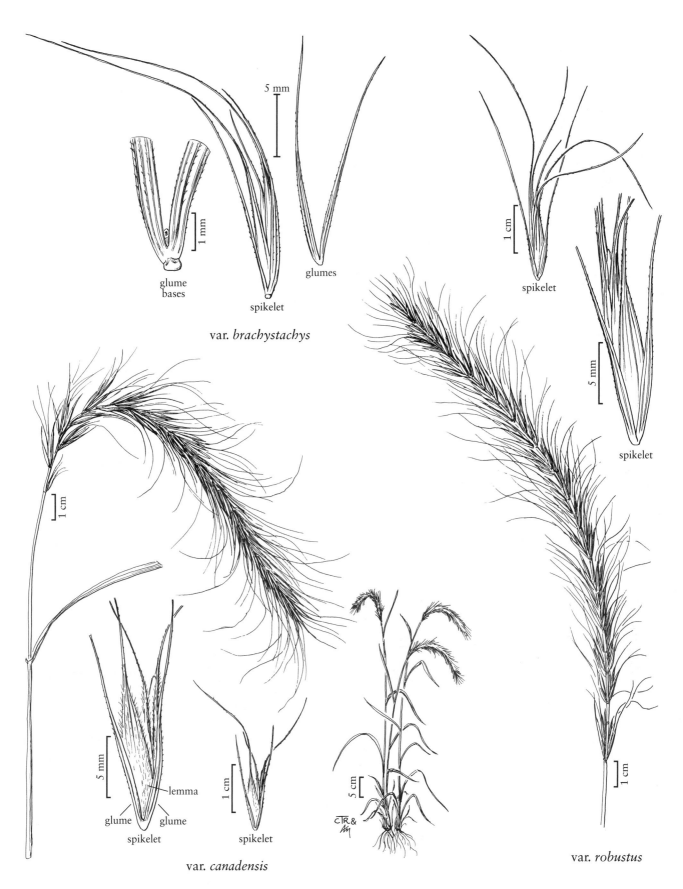

var. *brachystachys*

glume
bases

spikelet

glumes

spikelet

spikelet

1 cm

var. *canadensis*

lemma

glume   glume
spikelet

spikelet

var. *robustus*

ELYMUS CANADENSIS

2. Glumes not clearly indurate or bowed out at the base, awns 10–20 mm long; lemmas smooth or scabridulous, awns usually 20–30 mm long, moderately outcurving; spikes 6–20 cm long . . . . var. *brachystachys*
2. Glumes often slightly indurate and bowed out at the base, awns 15–25 mm long; lemmas occasionally hirsute, awns 30–40 mm long, often strongly outcurving; spikes 15–25(30) cm long . . . . var. *robustus*

Elymus canadensis var. **brachystachys** (Scribn. & C.R. Ball) Farw. [p. 304]

Spikes 6–20 cm, nodding, not strongly glaucous, often becoming yellowish or pale reddish brown, rarely with 3 spikelets per node; **internodes** mostly 3–4 mm. **Glumes** not clearly indurate or bowed out at the base, awns 10–20 mm; **lemmas** smooth or scabridulous, awns usually 20–30 mm, moderately outcurving.

Elymus canadensis var. *brachystachys* is widespread in the southern Great Plains from Nebraska to Mexico, where anthesis is from March to early June. It also occurs sporadically as far north as southern Canada, from British Columbia to Quebec. Plants of this variety occasionally appear introgressed from *E. virginicus* (p. 298), particularly *E. virginicus* var. *jejunus*, but not to the same extent as *E. canadensis* var. *robustus*.

Elymus canadensis L. var. **canadensis** [p. 304]

Spikes (6)10–25(30) cm, nodding to almost pendent, often strongly glaucous, often with 3 spikelets per node; **internodes** 4–7 mm. **Glumes** not clearly indurate or bowed out at the base, awns 10–25 mm; **lemmas** villous or hispid, awns 15–40 mm, moderately to strongly outcurving.

Elymus canadensis var. *canadensis* is widespread across the northern range of the species, where anthesis is from late June to August, but it is also frequent as far south as Arizona, New Mexico, and Oklahoma. Tentatively included here are *E. canadensis* var. *glaucifolius* (Muhl.) Torr., which is strongly glaucous, with scabrous blades and hirsute or scabrous lemmas, and *E. canadensis* var. *villosus* Bates, which has villous leaves and occurs rarely in the northern Great Plains.

Elymus canadensis var. **robustus** (Scribn. & J.G. Sm.) Mack. & Bush [p. 304]

Spikes 15–25(30) cm, moderately nodding, occasionally almost erect, not strongly glaucous, often becoming yellowish or pale reddish brown; **internodes** 3–4 mm. **Glumes** often slightly indurate and bowed out at the base, awns 15–25 mm; **lemmas** smooth to scabridulous, or occasionally hirsute, awns 30–40 mm, moderately or often strongly outcurving.

Elymus canadensis var. *robustus* occurs mostly in the east-central range of the species, from Illinois and Ohio to Oklahoma and Nebraska, locally becoming the most common variety. Anthesis can be earlier than in other sympatric *E. canadensis* varieties (Bush 1926). These rather heterogeneous plants tend to be large in most dimensions, and may have resulted from introgression with *E. virginicus* (p. 298) or *E. glabriflorus* (p. 296), with which they have often been confused (Davies 1980). F$_1$ hybrids of the other varieties with these species are similar to var. *robustus*, sometimes with erect spikes that are longer than those of either parent, but they are usually sterile. Spike pubescence may vary considerably, perhaps reflecting different hybrid origins.

8. **Elymus wiegandii** Fernald [p. 307]
NORTHERN RIVERBANK WILDRYE, ÉLYME DE WIEGAND

Plants cespitose, not rhizomatous, somewhat glaucous. **Culms** 100–180(220) cm, erect; **nodes** 9–16, mostly concealed by the leaf sheaths, glabrous. **Leaves** evenly distributed; **sheaths** usually glabrous, occasionally villous, often reddish brown; **auricles** 1–3 mm, brown; **ligules** to 1 mm; **blades** (8)10–20(24) mm wide, flat, lax, dark green, adaxial surfaces usually thinly pilose, with weakly spreading hairs on the veins at least near the margins, sometimes villous or glabrous. **Spikes** 10–30 cm long, 3–5 cm wide, pendent, the bases often barely exserted, with 2 spikelets per node; **internodes** 5–8(12) mm long, 0.2–0.3 mm thick at the thinnest sections, usually pubescent beneath the spikelets. **Spikelets** 12–20 mm, divergent, with (3)4–6(7) florets, lowest florets functional; **disarticulation** above the glumes and beneath each floret. **Glumes** equal or subequal, 12–30 mm including the often undifferentiated awns, the basal 0.5–1 mm subterete and slightly indurate, glume bodies 7–12 mm long, (0.2)0.4–0.9(1.1) mm wide, linear-setiform, entire, widening or parallel-sided above the base, 1–3(5)-veined, glabrous, hispidulous or villous, especially near the margins, margins firm, awns (5)8–15(18) mm, straight or flexuous; **lemmas** 10–15 mm, usually uniformly appressed-villous, rarely scabrous-hirtellous or glabrous, awns 15–25(30) mm, moderately to strongly outcurving; **paleas** 9–14 mm, narrowly truncate, minutely bidentate; **anthers** 2–3.5 mm. **Anthesis** from mid-July to early August. $2n = 28$.

Elymus wiegandii grows in moist or damp, rich, alluvial soil, especially on sandy river terraces and in woods and thickets, primarily from Saskatchewan through much of the Great Lakes region to Nova Scotia and Connecticut. It has abnormal neocentric chromosomes with meiotic irregularities that appear to limit the fertility of its hybrids, and even some crosses within the species (Vilkomerson 1950). It may be derived from hybrids between *E. canadensis* (p. 303) and perhaps *E. riparius* (p. 302). The latter species is similar to *E. wiegandii* and overlaps with it in range and habitat within the Great Lakes region, where there are a few plants that appear to be hybrids between the two. Plants with scabrous-hirtellous or glabrous lemmas (*E. wiegandii* f. *calvescens* Fernald) are known from Maine and New Hampshire.

Elymus wiegandii is often confused with sympatric *E. canadensis* and *E. diversiglumis* (p. 316), but it has a distinctive robust, broad-leaved habit. It is intermediate between the two in spike density and glume development. Occasional plants with glabrous leaves and less pendent spikes suggest introgression from *E. canadensis*, but artificial crosses produced no fertile F₁ plants (Church 1958).

9. Elymus interruptus Buckley [p. 307]
SOUTHWESTERN WILDRYE

Plants cespitose, not rhizomatous, usually glaucous. **Culms** usually (40)60–100(120) cm, erect or the bases somewhat decumbent; **nodes** 4–8, usually exposed, glabrous. **Leaves** evenly distributed; **sheaths** usually glabrous, occasionally hirsute; **auricles** 0–2 mm, pale or reddish brown; **ligules** to 1 mm; **blades** 3–9 mm wide, lax, pale green, adaxial surfaces densely short-pilose, hispidulous, or scabridulous, especially on the veins. **Spikes** 5–20 cm long, 2–5 cm wide, erect to slightly nodding, with 2 spikelets per node; **internodes** (5)8–14 mm long, 0.2–0.3 mm thick at the thinnest sections, without pronounced dorsal angles, often with green lateral bands, glabrous beneath the spikelets. **Spikelets** (6)9–15(22) mm, somewhat to strongly divergent, with 2–5 florets, lowest florets functional; **disarticulation** above the glumes and beneath each floret. **Glumes** subequal, 15–30 mm including the weakly differentiated awns, the basal 0–1 mm subterete and indurate, glume bodies, when distinguishable, about 6–10 mm long, (0.2)0.3–0.5(0.7) mm wide, linear-setiform to setaceous, entire, widening or parallel-sided above the base, 1–3-veined, glabrous or scabridulous, margins firm, awns straight or flexuous; **lemmas** 7–10 mm, smooth or scabridulous, occasionally hirtellous especially near the margins, awns 15–22 mm, straight to moderately outcurving; **paleas** 6–9 mm, obtuse or narrowly truncate, sometimes emarginate; **anthers** 2–4.5 mm. Anthesis May to July. 2*n* = 28.

Elymus interruptus grows in dry to moist, rocky soil, often in canyons, open woods, and thickets, in the southwestern United States and northern Mexico. Apparent intermediates between *E. interruptus* and *E. canadensis* (p. 303) have been collected north of the documented range of typical *E. interruptus* in Arizona, New Mexico, and Iowa. Plants in the Ozark and Ouachita mountains, especially in Arkansas, that were previously referred to *E. interruptus* are now included in *E. churchii* (p. 314).

Elymus interruptus is a poorly understood southern species that, at one extreme, used to be included in *E. canadensis* or, at the other extreme, used to include *E. churchii*, *E. svensonii* (p. 314), and *E. diversiglumis* (p. 316), three species that seem more closely allied to *E. hystrix* (p. 316). Campbell (2002) suggested *E. interruptus* may have arisen from the introgression of *E. hystrix* or a related species into *E. canadensis* var. *brachystachys*. Artificial crosses between *E. hystrix* and *E. canadensis* were generally unsuccessful, but yielded some plants resembling *E. interruptus* (Church 1954). *Elymus interruptus* has been crossed with *E. canadensis*, *E. hystrix*, *E. svensonii*, *E. virginicus* (p. 298), *E. glabriflorus* (p. 296) and *E. diversiglumis*; only the hybrids with *E. diversiglumis* were completely sterile (Church 1954, 1967a; Brown and Pratt 1960).

10. Elymus glaucus Buckley [p. 309]
COMMON WESTERN WILDRYE, BLUE WILDRYE

Plants densely to loosely cespitose, sometimes weakly rhizomatous, often glaucous. **Culms** 30–140 cm, erect or slightly decumbent; **nodes** 4–7, mostly exposed, usually glabrous, sometimes puberulent. **Leaves** evenly distributed; **sheaths** scabrous or smooth, glabrous or, particularly those of the lower leaves, retrorsely puberulent to hirsute, often purplish; **auricles** usually present, to 2.5 mm, often purplish; **ligules** to 1 mm, truncate, erose-ciliolate or entire; **blades** 2–13(17) mm wide, usually lax, sometimes slightly involute, adaxial surfaces glabrous, scabrous, or strigose on the veins, sometimes pilose to villous. **Spikes** 5–21 cm long, (0.2)0.5–2 cm wide, erect to slightly nodding, rarely somewhat pendent, usually with 2 spikelets per node, sometimes with 1 at all or most nodes, rarely with 3 at some nodes; **internodes** 4–8(12) mm long, 0.15–0.5 mm thick at the thinnest sections, angles scabrous, glabrous below the spikelets. **Spikelets** 8–25 mm, sometimes purplish at higher latitudes and elevations, appressed to slightly divergent, with (1)2–4(6) florets, lowest florets functional; **disarticulation** above the glumes and beneath each floret. **Glumes** subequal, ³/₄ as long as or

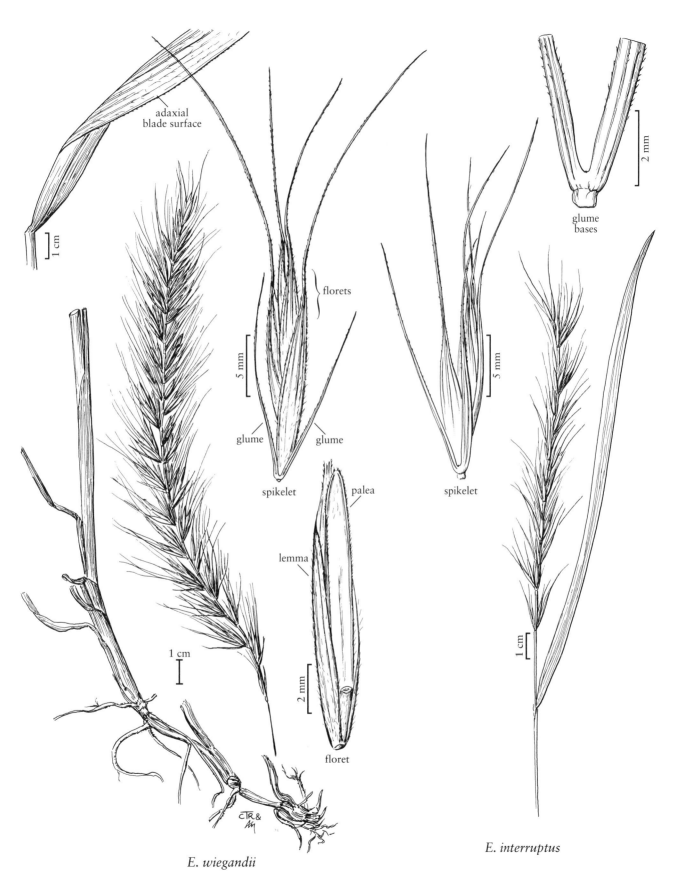

adaxial
blade surface

1 cm

florets

5 mm

glume   glume

spikelet

palea

lemma

2 mm

floret

glume
bases

2 mm

5 mm

spikelet

1 cm

E. wiegandii

E. interruptus

ELYMUS

equaling the adjacent lemmas, bases often overlapping, usually flat and thin with evident venation, glume bodies (6)9–14(19) mm long, 0.6–1.5(2) mm wide, linear-lanceolate, entire, widening above the base, (1)3–5(7)-veined, 2–3 veins extending to the apices, glabrous, veins smooth or evenly scabrous, margins 0.1–0.2 mm wide, whitish hyaline, tapering towards the apices, unawned or awned, awns to 5(9) mm, straight; **lemmas** (8)9–14(16) mm, glabrous, scabrous, or short-hirsute, awns (0)1–30(35) mm, usually straight to flexuous, sometimes slightly curving; **paleas** 7–13 mm, keels straight or slightly concave, usually scabrous to ciliate, apices often bidentate; **anthers** 1.5–3.5 mm. **Anthesis** from May to July. $2n = 28$.

Elymus glaucus grows in moist to dry soil in meadows, thickets, and open woods. It is widespread in western North America, from Alaska to Saskatchewan, and south to Baja California and New Mexico. It is also sporadic, sometimes appearing transitional to *E. trachycaulus* (p. 321), from the northern Great Plains to southern Ontario and New York and, as a disjunct, on rocky sites in the Ozark and Ouachita mountains.

Populations can differ greatly in morphology, especially in rhizome development, leaf width, pubescence, and the prevalence of solitary spikelets; their crossing relationships are partly correlated with such variation (Snyder 1950, 1951; Stebbins 1957, Wilson et al. 2001). Rhizome development and the production of solitary spikelets may also be environmental responses. Rhizomatous plants are more common on unstable slopes or sandy soils. Plants with solitary spikelets are more common on poor soil or in shade. They are often confused, particularly in the herbarium, with *E. stebbinsii* (p. 329) or *E. trachycaulus*. They differ from *E. stebbinsii* in their shorter anthers and awned glumes. Distinction from *E. trachycaulus* can be difficult with herbarium specimens, but is generally easy in the field, *E. glaucus* having more evenly leafy culms, laxer and wider blades, more tapered glumes that are almost always awned, and shorter anthers than the sympatric *E. trachycaulus*.

There are reports of natural hybrids with several other species of *Elymus*, including *E. elymoides* (p. 318), *E. multisetus* (p. 318) (see *E. ×hansenii*, p. 340), *E. trachycaulus*, and *E. stebbinsii*. These hybrids often appear at least partially fertile. *Elymus glaucus* can also form intergeneric hybrids with *Leymus* and *Hordeum* (see ×*Elyleymus*, p. 343, and ×*Leydeum*, p. 368).

The following three subspecies appear to be morphologically, ecologically, and geographically distinct. Plants found at elevations of up to 2200 m along the Pacific coast, with hairy leaf blades and lemma awns usually shorter than 20 mm, have been called subsp. *jepsonii* (Burtt Davy) Gould, but Wilson et al. (2001) demonstrated that such plants are neither genetically nor ecologically distinct from those with glabrous leaf blades; they are included here in subsp. *glaucus*.

1. Lemma awns (0)1–5(7) mm long; glume awns 0–2 mm long .... subsp. *virescens*
1. Lemma awns (5)10–30(35) mm long; glume awns (0.5)1–9 mm long.
    2. Blades 4–17 mm wide, adaxial surfaces glabrous or strigose, occasionally pilose to hirsute with hairs of fairly uniform length; glume awns (0.5)1–5(9) mm long ..... subsp. *glaucus*
    2. Blades 3–8 mm wide, densely short-pilose with scattered longer hairs; glume awns 3–8 mm long .... subsp. *mackenziei*

Elymus glaucus Buckley subsp. **glaucus** [p. 309]

Sheaths glabrous, scabrous or pubescent; **blades** 4–17 mm wide, adaxial surfaces glabrous or strigose, occasionally pilose to hirsute with hairs of fairly uniform length. **Glume awns** (0.5)1–5(9) mm; **lemma awns** (5)10–25(35) mm.

Elymus glaucus subsp. *glaucus* grows throughout the range of the species, from sea level to 2500 m. It is absent from the area where *E. glaucus* subsp. *mackenziei* grows. It resembles *E. hirsutus* (see next), differing in its erect spikes and in the pattern of its lemma pubescence. It also resembles the introduced *E. dahuricus* (p. 310), from which it differs in its palea shape.

Elymus glaucus subsp. **mackenziei** (Bush) J.J.N. Campb. [p. 309]

Sheaths usually puberulent; **blades** 3–8 mm wide, densely short-pilose, with scattered longer hairs. **Glume awns** 3–8 mm; **lemma awns** 20–30 mm.

Elymus glaucus subsp. *mackenziei* grows on limestone clifftops, rocky ledges, and glades, in open woods and thickets. It is known only from Arkansas, Missouri, and Oklahoma, at scattered sites in the Ozark Mountains and at Rich Mountain in the Ouachita Mountains. This subspecies is remarkably disjunct, at least 500 miles from the nearest known *E. glaucus* to the west and north.

Elymus glaucus subsp. **virescens** (Piper) Gould [p. 309]

Sheaths glabrous or scabrous; **blades** 2–10 mm wide, smooth or scabridulous to pubescent. **Glume awns** 0–2 mm; **lemma awns** (0)1–5(7) mm.

Elymus glaucus subsp. *virescens* generally grows in relatively dry or rocky soils along cliffs, bluffs, slopes, shores, and river banks, and in coniferous forests, chaparral, and other woodlands along the coast from Alaska to central California, at elevations from sea level to 1200 m.

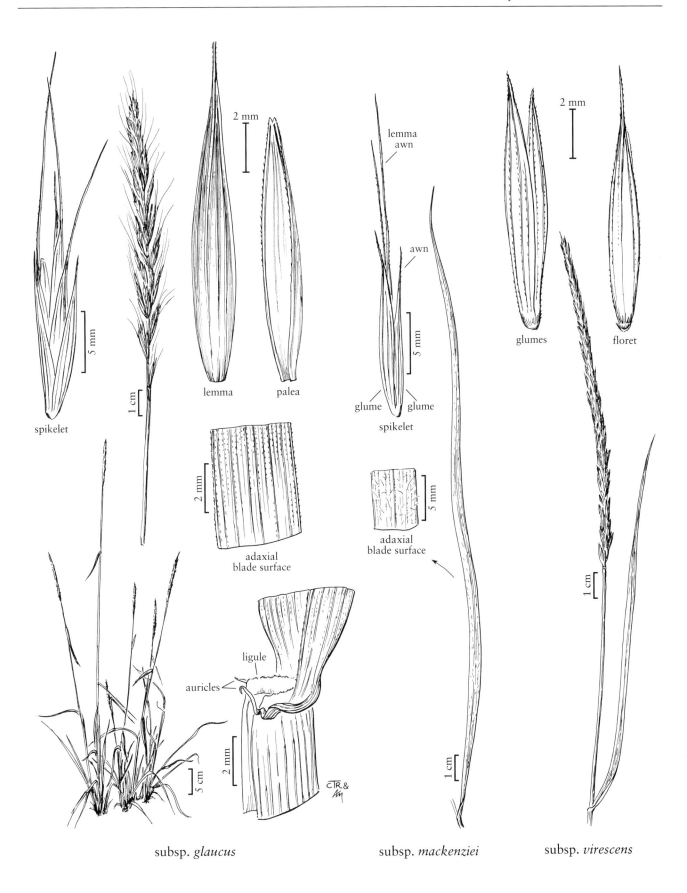

spikelet

lemma   palea

adaxial
blade surface

ligule

auricles

subsp. *glaucus*

lemma
awn

awn

glume   glume
spikelet

adaxial
blade surface

subsp. *mackenziei*

glumes   floret

subsp. *virescens*

ELYMUS GLAUCUS

11. Elymus hirsutus J. Presl [p. 311]
NORTHWESTERN WILDRYE

Plants cespitose, sometimes shortly rhizomatous. **Culms** 40–140 cm, usually somewhat decumbent; **nodes** 4–7, mostly exposed, usually glabrous, occasionally puberulent. **Leaves** evenly distributed; **sheaths** usually glabrous and smooth, occasionally scabridulous or retrorsely hairy, sometimes purplish; **auricles** to 1.5 mm, often absent; **ligules** to 1 mm; **blades** 4–12 mm wide, lax, usually deep green, adaxial surfaces usually pilose or villous, occasionally puberulent or scabridulous. **Spikes** 6–20 cm long, 0.5–2 cm wide, nodding to pendent, with 2 spikelets per node, rarely with 3 at some nodes; **internodes** 3–10(12) mm long, 0.2–0.7 mm thick at the thinnest sections, usually glabrous, sometimes sparsely hairy. **Spikelets** 12–20 mm, appressed to divergent, sometimes purplish at higher latitudes, with 2–4(7) florets, lowest florets functional; **disarticulation** above the glumes and beneath each floret. **Glumes** equal or subequal, the bases flat, occasionally indurate for 0.5 mm, veins usually evident, glume bodies (4.5)7–10(11) mm long, 0.7–1.5 mm wide, linear-lanceolate, entire, widening or parallel-sided above the base, 3–5-veined, usually scabridulous to scabrous, veins occasionally hirsute beyond midlength, margins hyaline or scarious, awns 1–10 mm, straight; **lemmas** 7–14 mm, smooth or scabridulous, lateral veins hairy, margins hairy beyond midlength, marginal hairs 0.5–1 mm, longer than those elsewhere, awns (2)8–30 mm, flexuous to moderately outcurving; **paleas** 6–13 mm, with hairs of varying lengths on the keels and apices, acute, bidentate; **anthers** 2–3.5 mm. Anthesis from May to July. 2*n* = 28.

Elymus hirsutus grows in moist to damp or dry soils in woods, thickets, and grasslands. Its range extends along the coastal mountains from the Aleutian Islands to northern Oregon, and inland to eastern British Columbia. Plants in the southern part of the range tend to have villous leaves and more erect spikes with shorter, straighter awns.

Elymus hirsutus is similar to *E. glaucus* (see previous), but its more pendent spikes, lemma pubescence pattern, and shorter glumes enable most specimens to be readily identified. Intermediates do exist; it is not known whether they reflect introgression or extremes of variation. *Elymus hirsutus* occasionally hybridizes with *Leymus mollis* (p. 356) and *Hordeum brachyantherum* (p. 243).

12. Elymus dahuricus Turcz. *ex* Griseb. [p. 311]

Plants cespitose, not rhizomatous, often glaucous. **Culms** 30–130 cm, erect; **nodes** 4–7, mostly exposed, usually glabrous, occasionally short-hairy. **Leaves** evenly distributed; **sheaths** glabrous; **auricles** minute or absent; **ligules** 0.5–1 mm; **blades** 3–18 mm wide, lax, usually pale green, sometimes glaucous, adaxial surfaces usually smooth or scabrous on the veins, sometimes sparsely pilose. **Spikes** 7–23 cm long, 1–2.5 cm wide, usually slightly nodding, sometimes erect, usually with 2 spikelets per node, occasionally with 1 spikelet at some nodes; **internodes** 3–6 mm long, 0.2–0.8 mm thick at the thinnest sections, angles usually with scattered hairs. **Spikelets** 10–15 mm, appressed to divergent, often purplish, with (2)3–4(5) florets, lowest florets functional; **disarticulation** above the glumes, beneath each floret. **Glumes** equal, the bases flat, not indurate, veins evident, glume bodies 6–9 mm long, 1–1.5 mm wide, linear-lanceolate, entire, widening or parallel-sided above the base, (1)3–5(7)-veined, veins scabrous, margins hyaline or scarious, awns (0)1–5 mm, straight or outcurving; **lemmas** (5)7–11 mm, usually glabrous and smooth throughout, sometimes scabrous to hispid distally and on the margins, marginal hairs not markedly longer than those elsewhere, awns (3)6–17(20) mm, usually somewhat outcurving from near the base; **paleas** 7–11 mm, keels spinose-ciliate, apices obtuse or truncate; **anthers** 1.5–3.5 mm. Anthesis from May to July. 2*n* = 42.

Elymus dahuricus is widespread in temperate central and eastern Asia. Like *E. tsukushiensis* (p. 336), it is a hexaploid with an **StYH** genome constitution. It has been introduced for reclamation in some parts of western North America. It is most likely to be confused with *E. glaucus* (p. 306), from which it differs in its palea shape. Because its presence in the region became known shortly before completion of this volume, its distribution in the region is not known. Several varieties have been described in Asia; only **Elymus dahuricus** Turcz. *ex* Griseb. var. **dahuricus** has been introduced to North America.

13. Elymus sibiricus L. [p. 311]
SIBERIAN WILDRYE

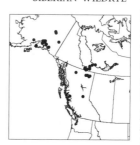

Plants usually cespitose, sometimes weakly rhizomatous, usually glaucous, occasionally strongly so. **Culms** 40–150 cm, erect or slightly geniculate at the base; **nodes** 6–9, usually exposed, glabrous. **Leaves** evenly distributed; **sheaths** glabrous or hirsute, often purplish; **auricles** to 1 mm, often absent; **ligules** to 1 mm; **blades** (3)5–14(16) mm wide, lax, adaxial surfaces usually

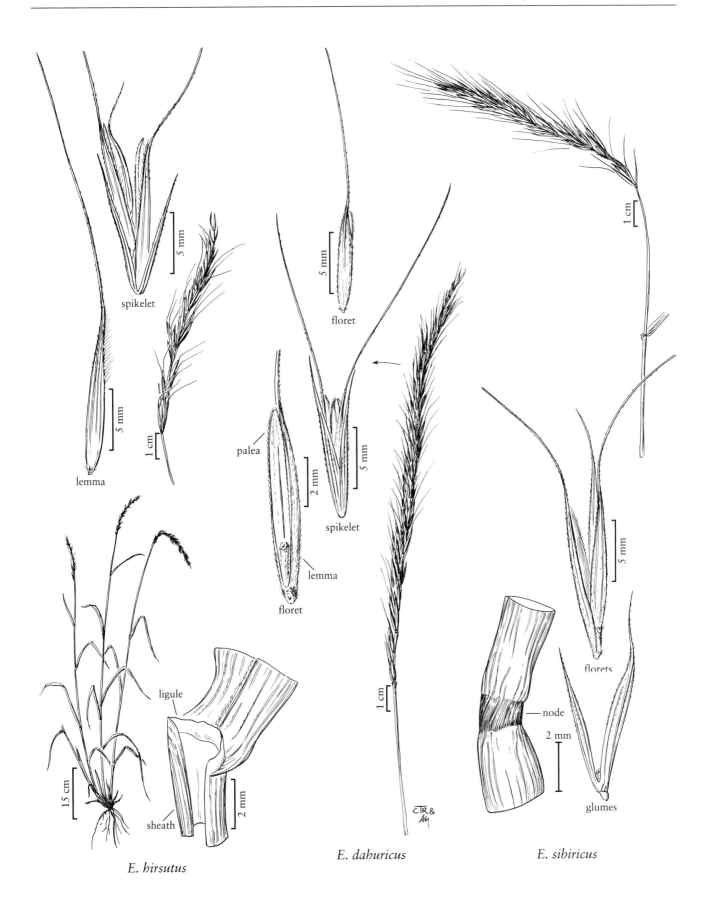

spikelet

5 mm

5 mm

lemma

1 cm

ligule

sheath

2 mm

15 cm

E. hirsutus

floret

5 mm

palea

2 mm

lemma

floret

spikelet

5 mm

1 cm

C̄T̄R̄ &
Aŋ

E. dahuricus

node

2 mm

glumes

florets

5 mm

1 cm

E. sibiricus

ELYMUS

pilose to hirsute on the veins, sometimes scabrous or smooth. **Spikes** 7–30 cm long, 2–5 cm wide, flexuous, nodding to pendent, with (1)2(3–4) spikelets per node, solitary spikelets usually basal or distal, rarely occurring throughout; **internodes** 5–10 mm long, 0.2–0.7 mm thick at the thinnest sections, mostly glabrous, sometimes scabrous below the spikelets, angles ciliate. **Spikelets** 10–18 mm, appressed to divergent, usually becoming purplish, with (3)4–5(7) florets, lowest florets functional; **disarticulation** above the glumes, beneath each floret. **Glumes** equal or subequal, the bases flat, evidently veined, not indurate, glume bodies 3–8 mm long, 0.4–1(1.2) mm wide, linear-lanceolate to subsetaceous, entire, widening or parallel-sided above the base, 3(5)-veined, veins smooth or scabrous, margins hyaline or scarious, awns 1–6 mm, straight; **lemmas** 8–13 mm, densely scabridulous to scabrous, at least along the outer veins, awns 10–25 mm, usually somewhat outcurving from near the base; **paleas** 8–12 mm, keels spinose-ciliate, bidentate, apices acute, 0.15–0.3 mm wide between the veins; **anthers** 0.9–1.7 mm. **Anthesis** from June to July. $2n = 28$.

Elymus sibiricus grows in dry to damp grasslands and thickets, on slopes, eroding river banks, mud flats, coastal benches, dunes, clearings, and other disturbed areas, in southern Alaska, the southern Yukon Territory, the southwestern MacKenzie District in the Northwest Territories, and central British Columbia. Porsild and Cody (1980) suggested that at least some of the populations are native to North America. In a more extensive analysis, Bennett (2006) concluded that all North American populations are the result of recent introductions. The species is widespread in cool temperate regions of central and eastern Asia. In China, it is considered an excellent forage grass, having a high protein content.

North American plants differ from Asian plants in several respects: they are up to 150 cm tall, versus 90 cm in Asia; their leaves are usually pubescent, rather than glabrous to scabrous; and their lemmas are scabridulous to scabrous, rather than glabrous to strigulose or pilose.

14. Elymus pringlei Scribn. & Merr. [p. 313]
MEXICAN WILDRYE

Plants cespitose, not rhizomatous, usually somewhat glaucous. **Culms** 50–110 cm, erect or somewhat geniculate at the base; **nodes** 6–9, mostly exposed, glabrous. **Leaves** evenly distributed; **sheaths** usually glabrous, occasionally pilose, hairs somewhat retrorse; **auricles** about 1 mm, pale or brownish; **ligules** about 1 mm, erose; **blades** 3–12 mm wide, lax, adaxial surfaces sparsely scabridulous, sometimes hispidulous to pilose on the veins, usually glaucous. **Spikes** 4–12 cm long, 2–3 cm wide, erect, the bases sometimes sheathed, with 2

spikelets per node; **internodes** 3–6 mm, about 0.2 mm thick at the thinnest sections, with 2 hispid dorsal angles, without green lateral bands. **Spikelets** 10–15 mm excluding the awns, 18–30 mm including the awns, appressed, with 3–5(6) florets, lowest florets functional; **disarticulation** above the glumes, beneath each floret. **Glumes** subequal, 12–22 mm long including the undifferentiated awns, 0.2–0.3(0.6) mm wide, setaceous, entire, 0–1(2)-veined, tapering from the base, glabrous, margins firm, awns more or less straight; **lemmas** 8–10 mm, usually scabrous-hispid or thinly strigose, at least distally, awns 8–22 mm, straight or flexuous; **paleas** 7–8 mm, obtuse, often emarginate; **anthers** 2.5–4 mm. **Anthesis** May to June. $2n$ = unknown.

Elymus pringlei grows on moist slopes and canyons, in pine and deciduous tree woods, at 1500–2300 m in the Sierra Madre Orientale of eastern Mexico. This poorly known species is similar to *E. texensis* (see next) and *E. interruptus* (p. 306). It is included here because it seems likely that it also grows in southern Texas, having been collected in Coahuila, Mexico, 54 miles from the border, near Big Bend National Park (Campbell 2002).

15. Elymus texensis J.J.N. Campb. [p. 313]
TEXAN WILDRYE

Plants cespitose, not rhizomatous, glaucous. **Culms** 70–110 cm, erect; **nodes** 4–6, mostly exposed, glabrous. **Leaves** evenly distributed; **sheaths** glabrous or ciliate; **auricles** to about 2 mm, pale to purplish brown; **ligules** 1–2 mm, erose; **blades** 2–9 mm wide, lax or somewhat involute, adaxial surfaces thinly scabrous to hirsute or densely pilose. **Spikes** 9–20 cm long, 2–2.5 cm wide, erect to slightly nodding, with 2 spikelets per node; **internodes** (5)7–15(22) mm long, 0.1–0.3 mm thick at the thinnest sections, glabrous except for the ciliolate margins, with slight dorsal angles and green lateral bands along the concave sides. **Spikelets** 13–20 mm excluding the awns, 20–40 mm including the awns, appressed, with 4–6(8) florets, lowest florets functional; **disarticulation** above the glumes, beneath each floret. **Glumes** subequal, 14–24 mm long including the undifferentiated awns, 0.1–0.3 mm wide, setaceous, entire, 0–1-veined, tapering from the base, glabrous, margins firm, awns more or less straight; **lemmas** 8–12 mm, smooth, glabrous, awns 8–25 mm, straight, flexuous or slightly curving; **paleas** 7–11 mm, obtuse or truncate; **anthers** 4.5–6 mm. **Anthesis** in May. $2n$ = unknown.

Elymus texensis is known only from calcareous bluffs and hills in juniper woods and grassy areas on the

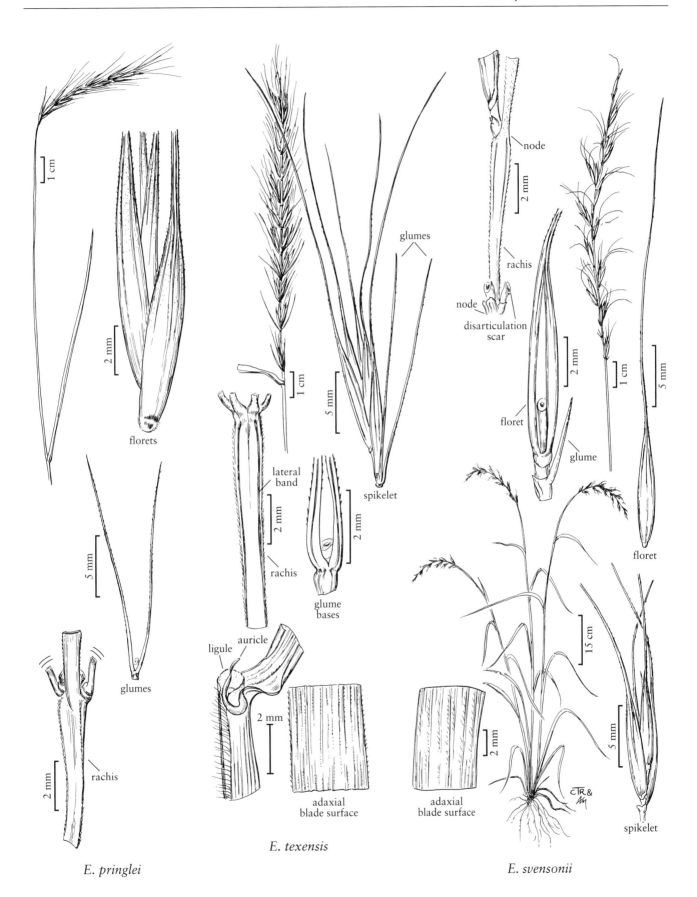

E. pringlei

E. texensis

E. svensonii

ELYMUS

Edwards Plateau of southwest Texas. It is known from only three collections and needs further study (Campbell 2002). It is similar to the Mexican species *E. pringlei* (see previous), but differs in its larger anthers, larger, less pubescent spikelets, and in its longer, glabrous rachis internodes with green lateral bands.

16. Elymus svensonii G.L. Church [p. 313]
SVENSON'S WILDRYE

Plants cespitose, not rhizomatous, strongly glaucous. **Culms** 50–110 cm, erect; **nodes** 6–8, mostly exposed, often reddish brown, glabrous. **Leaves** evenly distributed; **sheaths** glabrous or villous, often somewhat purplish; **auricles** 1–2 mm, purplish or reddish brown; **ligules** to 1 mm, often reddish brown; **blades** 4–8(10) mm wide, lax, usually pale green, adaxial surfaces usually villous. **Spikes** 10–16 cm long, 3–5 cm wide, nodding, with 2 spikelets per node; **internodes** (4)6–10(12) mm long, about 0.2 mm thick at the thinnest sections, flexuous, glabrous, without green lateral bands. **Spikelets** 10–16 mm, usually appressed, with (3)4–5 florets, lowest florets functional; **disarticulation** above the glumes, beneath each floret. **Glumes** usually differing in length by more than 5 mm, sometimes vestigial to absent from the upper spikelets or throughout, (0)1–15(18) mm long including the undifferentiated awns, indurate at the base, 0.1–0.3 mm wide, setaceous to subulate, entire, 0–1-veined, tapering from the base, glabrous, margins firm, awns often curving outward; **lemmas** 8–10 mm, glabrous, veins occasionally hispidulous near the lemma apices, awns (8)10–20(25) mm, moderately to strongly outcurving at maturity; **paleas** 7–9 mm, obtuse or truncate, occasionally emarginate; **anthers** 3–5 mm. Anthesis from mid-June to early July. 2*n* = unknown.

Elymus svensonii grows in dry, rocky soils in open woods of the interior low plateaus, mostly along bluffs of the Kentucky River and its tributaries in the bluegrass region of Kentucky, and along bluffs of the Cumberland River and its Caney Fork in the central basin of Tennessee. Most sites are on Ordovician limestone, but its discovery by Natural Heritage programs in Kentucky along the Green River on Mississippian limestone, and in Tennessee along the Piney River on Silurian limestone, suggest that it may be more widespread. It has been a candidate for federal protection in the United States.

Elymus svensonii, like *E. diversiglumis* (p. 316) and *E. churchii* (see next), may be derived from hybrids between *E. hystrix* (p. 316) and *E. canadensis* (p. 303) (Church 1967a), even though *E. canadensis* currently has its eastern limit 50–100 miles west of most *E. svensonii*.

Elymus svensonii hybridizes naturally with *E. hystrix*, *E. virginicus* (p. 298) and other species of *Elymus*. Plants with little glume development are frequent; they appear to be introgressed by *E. hystrix*. Artificial crosses with *E. interruptus* (p. 306) have been successful, but those with *E. diversiglumis* have not (Church 1967a). *Elymus svensonii* resembles *E. churchii*; it differs in having less open spikes, shorter awns, more florets per spikelet, and more pubescent, glaucous foliage.

17. Elymus churchii J.J.N. Campb. [p. 315]
CHURCH'S WILDRYE

Plants cespitose, not rhizomatous, often somewhat glaucous. **Culms** 50–120 cm, erect; **nodes** 4–8, exposed or concealed, often reddish brown or blackish, glabrous. **Leaves** evenly distributed; **sheaths** usually glabrous, sometimes pubescent at the summit; **auricles** 1–2 mm, often reddish brown or blackish; **ligules** to 1 mm, often reddish brown; **blades** 3–11 mm wide, lax, adaxial surfaces glabrous or short-pilose. **Spikes** 10–18 cm long, 3–5 cm wide, slightly nodding, with 2 spikelets per node; **internodes** (5)7–13(18) mm long, about 0.2 mm thick at the thinnest sections, flexuous, with green lateral bands, glabrous except the dorsal angles hispid. **Spikelets** 10–15 mm, usually appressed, with 3(5) florets, lowest florets functional; **disarticulation** above the glumes, beneath each floret. **Glumes** often differing in length by more than 5 mm, sometimes vestigial to absent from the upper spikelets or throughout, 0–15(20) mm long including the undifferentiated awns, indurate at the base, 0.1–0.3 mm wide, setaceous to subulate, entire, 0–1-veined, glabrous, margins firm, awns often outcurving; **lemmas** 8–10 mm, usually hairy, occasionally glabrous, awns (10)20–30(35) mm, slightly to strongly outcurving at maturity; **paleas** 7–9 mm, obtuse to truncate, sometimes emarginate; **anthers** 2.5–3 mm, evident in June. 2*n* = unknown.

Elymus churchii grows in dry, rocky, often relatively base-rich soils, in open woods on ridges, and on bluffs and river banks. Its range includes the central Ouachita Mountains and the western Ozark Mountains in Arkansas, Oklahoma, and Missouri.

Elymus churchii used to be included in *E. interruptus* (p. 306) (Steyermark 1963; Smith 1991). It is similar to the more eastern, disjunct *E. svensonii* (see previous), from which it differs in its more open spikes, longer awns, fewer florets per spikelet, and less pubescent, less glaucous foliage. Like *E. svensonii*, *E. churchii* may have originated from hybridization between *E. canadensis* (p. 303) and *E. hystrix* (p. 316); occasional intermediates with both species exist (Campbell 2002).

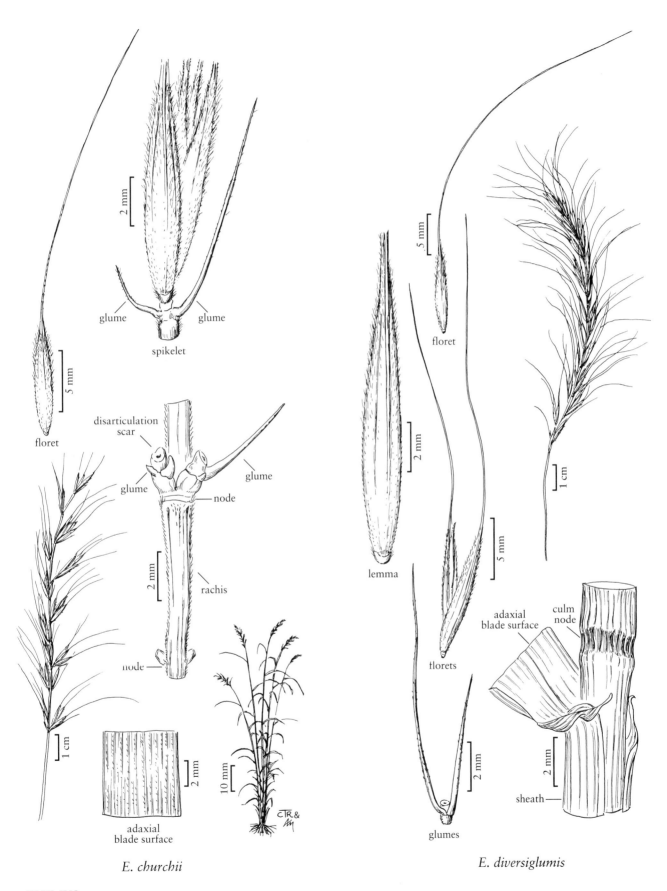

glume   glume

spikelet

floret

5 mm

disarticulation
scar

glume   glume

node

2 mm

rachis

node

1 cm

10 mm

adaxial
blade surface

E. churchii

2 mm

.5 mm

floret

lemma

2 mm

florets

5 mm

glumes

2 mm

1 cm

adaxial
blade surface

culm
node

sheath

2 mm

E. diversiglumis

ELYMUS

18. **Elymus diversiglumis** Scribn. & C.R. Ball [p. 315]
UNEQUAL-GLUMED WILDRYE

Plants cespitose, not rhizomatous, sometimes moderately glaucous. **Culms** 70–160 cm, erect; **nodes** 4–9, mostly exposed, glabrous. **Leaves** evenly distributed; **sheaths** glabrous, often purplish; **auricles** 1–2 mm, purplish or brownish black; **ligules** usually 1–2 mm; blades 5–17 mm wide, lax, adaxial surfaces usually pilose, at least on the veins, occasionally scabrous. **Spikes** 8–28 cm long, 3–5 cm wide, nodding to pendent, with 2 spikelets per node, rarely with 1 or 3 at a few nodes; **internodes** 4–6(9) mm long, 0.2–0.3 mm thick at the thinnest sections, margins and summits often pubescent. **Spikelets** 10–16 mm, appressed, with 2–4(5) florets, lowest florets functional; **disarticulation** above the glumes, beneath each floret. **Glumes** usually differing in length by at least (3)4 mm, occasionally obsolete, (1)2–15(20) mm long including the undifferentiated awns, indurate at the base, (0.1)0.2–0.4(0.6) mm wide, setaceous, 0–1-veined, tapering from the base, scabrous or hispidulous at least towards the apices, margins firm, awns often outcurving; **lemmas** 7–12 mm, usually silvery-hirsute to sericeous, occasionally hirtellous or strigose, at least near the margins, backs sometimes scabrous, awns 18–35 mm, moderately to strongly outcurving at maturity; **paleas** 7–10 mm, obtuse, occasionally emarginate; **anthers** 2–4 mm. **Anthesis** from early June to late July. 2*n* = 28.

Elymus diversiglumis grows in moist to dry, often base-rich and alluvial soils, in open woods, woodland margins, and thickets in the northern Great Plains, from Saskatchewan and Manitoba to Wyoming, Wisconsin, and Iowa.

Elymus diversiglumis is a variable species that, like *E. svensonii* (p. 314) and *E. churchii* (p. 314), may have originated from hybrids between *E. canadensis* var. *canadensis* (p. 305) and *E. hystrix* (see next), although part of its range extends further west than the current distribution of the latter species. *Elymus diversiglumis* usually reaches anthesis 2–4 weeks earlier than sympatric populations of *E. canadensis*. Church (1954, 1958, 1967a) found that most artificial *canadensis–hystrix* hybrids, as well as some plants of *E. diversiglumis* itself, are sterile. Those that were not sterile could occasionally form fertile backcrosses with *E. canadensis* and, to a lesser extent, with *E. hystrix*. Introgressant populations involving all three species are known. Artificial crosses with other species have not been successful.

19. **Elymus hystrix** L. [p. 317]
BOTTLEBRUSH GRASS, GLUMELESS WILDRYE

Plants cespitose, not rhizomatous, occasionally glaucous, particularly the spikes. **Culms** 50–140 cm, usually erect, occasionally geniculate below; **nodes** 4–8, exposed or concealed, glabrous. **Leaves** evenly distributed; **sheaths** usually glabrous, occasionally pilose, often purplish; **auricles** usually present, 0.5–3 mm, brown to black; **ligules** 1–2(3) mm; **blades** 4–16 mm wide, lax, usually deep glossy green, adaxial surfaces pilose or scabridulous. **Spikes** 7–20 cm long, 4–7 cm wide, more or less erect, usually with 2 spikelets per node, rarely with 3 at some nodes; **internodes** (3)4–8(10) mm long, (0.1)0.2–0.3(0.4) mm thick at the thinnest sections, flexuous, usually glabrous, sometimes scabrous or hirsute, usually with green lateral bands. **Spikelets** 10–18 mm, strongly divergent to patent at maturity, with (1)2–4(6) florets, lowest florets functional; **disarticulation** above the glumes, beneath each floret. **Glumes** usually vestigial, sometimes 1–3 mm long, about 0.1 mm wide, subulate, entire, with no evident veins, occasionally to 10(20) mm long including the undifferentiated awns and differing in length by more than 5 mm, 0.1–0.2 mm wide, setaceous, tapering from the base, usually glabrous, occasionally appressed-puberulent to strigose, sometimes scabrous, usually straight, rarely somewhat curving, margins firm; **lemmas** 8–11 mm, usually glabrous, occasionally appressed-puberulent to strigose, especially near the margins and apices, awns (12)20–40(47) mm, usually straight, rarely somewhat curving; **paleas** 7–11 mm, obtuse or truncate, occasionally emarginate; **anthers** 2.5–5 mm. **Anthesis** mid-June to early July. 2*n* = 28.

Elymus hystrix grows in dry to moist soils in open woods and thickets, especially on base-rich slopes and small stream terraces. It grows throughout most of temperate eastern North America, extending west to Manitoba and Oklahoma, but is absent from the southern portion of the coastal plain.

Plants with pubescent lemmas have been recognized as *Elymus hystrix* var. *bigelovianus* (Fernald) Bowden. These occur infrequently north of a line from South Dakota through Kentucky to New Jersey, and are often mixed with the typical variety; uniform populations are known in the northeastern United States. Plants with pubescent blades are also more prevalent to the north.

Elymus hystrix hybridizes with most eastern species of *Elymus*. Introgression may account for the considerable variation in glume development and spikelet appression among these species. Lack of glumes may be a recessive character, with even slight glume development indicating

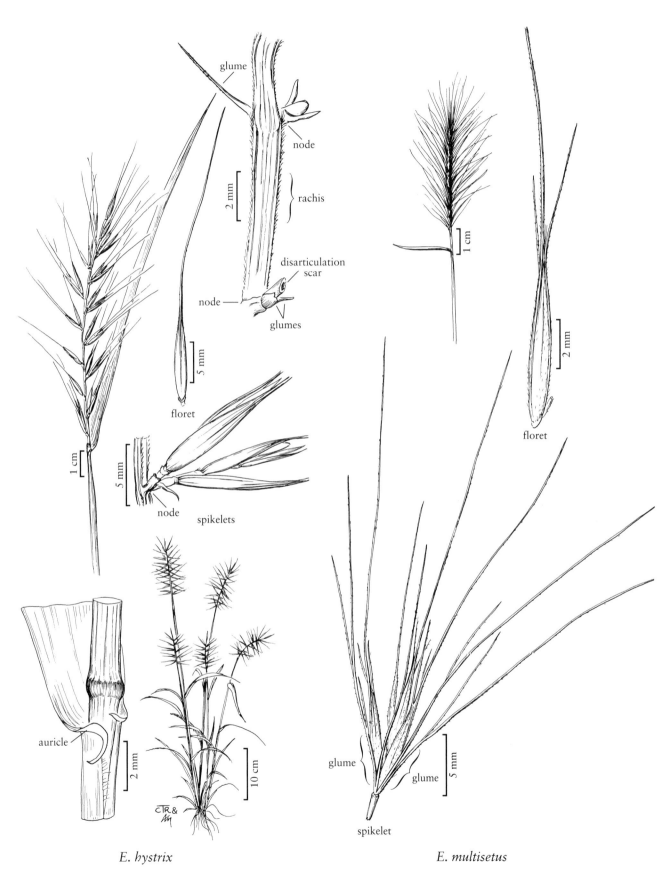

glume

node

rachis

2 mm

disarticulation scar

node

glumes

5 mm

floret

5 mm

node

spikelets

1 cm

auricle

2 mm

10 cm

1 cm

2 mm

floret

glume

glume

5 mm

spikelet

E. hystrix

E. multisetus

ELYMUS

introgression (Church 1967b). Plants with relatively well-developed, subequal glumes are presumed to be of hybrid origin. Such plants include most material from the Carolina piedmont region, where *E. glabriflorus* (p. 296) is the most likely source of introgression. The relatively frequent hybrids with *E. virginicus* (p. 298) are usually sterile, but Church (1967b) made crosses through three segregating generations. Within the ranges of *E. diversiglumis* (p. 316), *E. svensonii* (p. 314), and *E. churchii* (p. 314), there appear to be frequent introgressants between these species and *E. hystrix*. Further east, especially in the Appalachian regions of North Carolina, Virginia, West Virginia, and Maryland (including the shale barrens and nearby), there are scattered plants of *E. hystrix* with curving awns and, in a few cases, appressed spikelets (Campbell 2002). Whether these represent occasional variation within the *E. hystrix* gene pool, or whether they are outlying remnants of introgression with *E. canadensis* (p. 303) during a past eastward extension, is unknown.

20. Elymus multisetus (J.G. Sm.) Burtt Davy [p. 317]
BIG SQUIRRELTAIL

Plants cespitose, not rhizomatous. **Culms** 15–65 cm, erect to ascending, usually puberulent; **nodes** 4–6, mostly concealed, glabrous. **Leaves** evenly distributed; **sheaths** glabrous or white-villous; **auricles** usually present, 0.5–1.5 mm; **ligules** to 1 mm, truncate, entire or lacerate; **blades** 1.5–4(5) mm wide, often ascending and involute, adaxial surfaces scabrous, pilose, or villous. **Spikes** 5–20 cm long, 5–15 cm wide, erect, sometimes partially enclosed at the base, with 2 spikelets per node, rarely with 3–4 at some nodes; **internodes** 3–5(8) mm long, 0.1–0.3 mm thick at the thinnest sections, glabrous beneath the spikelets. **Spikelets** 10–15 mm, divergent, with 2–4 florets, lowest florets sterile and glumelike in 1 or both spikelets at each node; **disarticulation** initially at the rachis nodes, subsequently beneath each floret. **Glumes** subequal, (10)30–100 mm including the awns, the bases indurate and glabrous, glume bodies (2)5–10 mm long, 1–2 mm wide, setaceous, 2–3-veined, margins firm, awns (8)25–90 mm, each split into 3–9 unequal divisions, scabrous, flexuous to outcurving from near the glume bases at maturity; **fertile lemmas** 8–10 mm, smooth or scabrous near the apices, 2 lateral veins extending into bristles to 10 mm, awns (10)20–110 mm long, about 0.2 mm wide at the base, divergent to arcuate; **paleas** 7–9 mm, veins usually extending into about 1 mm bristles, apices acute to truncate; **anthers** 1–2 mm. **Anthesis** from late May to June. 2*n* = 28.

Elymus multisetus grows in dry, often rocky, open woods and thickets on slopes and plains, from central Washington and Idaho to southern California, Colorado, and northwestern Arizona, and from sea level to 2000 m. It has also been reported from Baja California, Mexico. It usually grows in less arid habitats than *E. elymoides* subsp. *elymoides* (p. 319), but the two taxa are sometimes sympatric.

Wilson (1963) reported a wide belt of introgression between *Elymus multisetus* and *E. elymoides* subsp. *elymoides* from southeastern California to southern Nevada, but not in other areas where they are sympatric. There are also probable hybrids with *E. glaucus* (p. 306) and *Pseudoroegneria spicata* (p. 281).

21. Elymus elymoides (Raf.) Swezey [p. 320]

Plants cespitose, often glaucous, not rhizomatous. **Culms** 8–65 (77) cm, erect or geniculate to slightly decumbent, sometimes puberulent; **nodes** 4–6, mostly concealed, usually glabrous, sometimes pubescent. **Leaves** evenly distributed; **sheaths** glabrous, scabrous, puberulent, or densely white-villous; **auricles** usually present, to about 1 mm, often purplish; **ligules** shorter than 1 mm, truncate, entire or lacerate; **blades** (1)2–4(6) mm wide, spreading or ascending, often involute, sometimes folded, abaxial surfaces glabrous to puberulent, adaxial surfaces scabrous, puberulent, hirsute, or white-villous. **Spikes** 3–20 cm long, 5–15 cm wide, erect to subflexuous, with 2–3 spikelets per node, rarely with 1 at some nodes; **internodes** 3–10(15) mm long, 0.1–0.4 mm thick at the thinnest sections, usually glabrous, sometimes puberulent beneath the spikelets. **Spikelets** 10–20 mm, divergent, sometimes glaucous, at least 1 spikelet at a node with 2–4(5) florets, 1–4(5) florets fertile, sometimes all florets sterile in the lateral spikelets; **disarticulation** initially at the rachis nodes, subsequently beneath each floret. **Glumes** subequal, 20–135 mm including the often undifferentiated awns, the bases indurate and glabrous, glume bodies 5–10 mm long, 1–3 mm wide, linear to setaceous, 1–3-veined, margins firm, awns 15–125 mm, scabrous, sometimes split into 2–3 unequal divisions, flexuous to outcurving from near the base at maturity; **fertile lemmas** 6–12 mm, glabrous, scabrous, or appressed-pubescent, 2 lateral veins extending into bristles to 10 mm, awns 15–120 mm long, about 0.4 mm wide at the base, often reddish or purplish, scabrous, flexuous to curved near the base; **paleas** 6–11 mm, veins often extending into bristles to 2(5) mm, apices acute to truncate; **anthers** 0.9–2.2 mm. **Anthesis** from late May to July. 2*n* = 28.

Elymus elymoides grows in dry, often rocky, open woods, thickets, grasslands, and disturbed areas, from sagebrush deserts to alpine tundra. It is widespread in western North America, from British Columbia to northern Mexico and the western Great Plains, and introduced in western Missouri, Illinois, and Kentucky. It is often dominant in overgrazed pinyon-juniper woodlands. Although palatable early in the season, the disarticulating, long-awned spikes irritate grazing animals later in the year.

Elymus elymoides intergrades with *E. multisetus* (see previous) in parts of its southern range (Wilson 1963). It is sometimes confused with *E. scribneri* (p. 330), but differs in having more than one spikelet per node, narrower glumes, and less tardily disarticulating rachises. Hybrids with several other species in the *Triticeae* are known; they can often be recognized by their tardily disarticulating rachises. Named interspecific hybrids (pp. 338–343) (and the other parent) are *E. ×saundersii* (*E. trachycaulus*), *E. ×pinalenoensis* (*E. arizonicus*), and possibly *E. ×hansenii* (*E. elymoides* or *E. multisetus* × *E. glaucus*). Hybrids with *E. sierrae* have not been named; they are common where the two species are sympatric. They have broader glume bases, shorter glume awns, and longer anthers than *E. elymoides*.

1. Rachis nodes with 3 spikelets, the central spikelet usually with 2 fertile florets, the florets of the lateral spikelets rudimentary to awnlike; lemma awns 15–30 mm long . . . . subsp. *hordeoides*
1. Rachis nodes usually with 2 spikelets, each spikelet usually with (1)2–4(5) fertile florets; lemma awns 15–120 mm long.
   2. No spikelets appearing to have 3 glumes, the lowermost floret in each spikelet well developed; paleas rarely with the veins extended as bristles . . . . subsp. *brevifolius*
   2. One or more of the spikelets at most nodes appearing to have 3 glumes, the lowest 1–2 florets sterile and glumelike; paleas usually with the veins extended as bristles.
      3. Glumes with awns 15–70 mm long, all glumes entire . . . . subsp. *californicus*
      3. Glumes with awns 35–85 mm long, one of the glumes at most nodes with the awn split into 2 or 3 divisions . . . subsp. *elymoides*

Elymus elymoides subsp. **brevifolius** (J.G. Sm.) Barkworth [p. 320]
LONGLEAF SQUIRRELTAIL

Culms 25–65(77) cm, erect. **Blades** usually puberulent abaxially, sometimes glabrous. **Spikes** 7–20 cm, usually exserted, usually with 2 spikelets per node. **Spikelets** with (1)2–4(5) florets, lowermost floret functional. **Glume awns** 50–125 mm, entire; **lemma awns** 50–120 mm; **paleas** rarely with the veins extended as bristles.

Elymus elymoides subsp. *brevifolius* has a wide ecological and elevation range, extending from the arid Sonoran Desert to subalpine habitats, from 600–3500 m. It extends further south than the other subspecies, into northern Mexico; it is rare in Canada.

Elymus elymoides subsp. **californicus** (J.G. Sm.) Barkworth [p. 320]
CALIFORNIA SQUIRRELTAIL

Culms 8–40 cm, erect or decumbent. **Blades** usually glabrous abaxially, sometimes puberulent. **Spikes** 3–10 cm, often partly included, usually with 2 spikelets per node. **Spikelets** with (1)2–3 fertile florets, lowest 1–2 florets sterile and glumelike. **Glume awns** 15–40(70) mm, entire; **lemma awns** 25–70 mm, usually exceeding those of the glumes; **paleas** often with the veins extended as 1–2 mm bristles.

Elymus elymoides subsp. *californicus* grows in midmontane to arctic-alpine habitats in western North America, at elevations of 1500–4200 m. Plants transitional to subsp. *elymoides* occur where the two are sympatric.

Elymus elymoides (Raf.) Swezey subsp. **elymoides** [p. 320]
COMMON SQUIRRELTAIL

Culms 15–45 cm, erect to decumbent. **Blades** usually puberulent abaxially, sometimes glabrous. **Spikes** 4–15 cm, exserted or partly included, usually with 2 spikelets per node. **Spikelets** with (1)2–3(4) fertile florets, lowest 1–2 florets sterile and glumelike. **Glume awns** 35–85 mm, often split into 2, sometimes 3, unequal divisions; **lemma awns** 25–75 mm, usually exceeded by those of the glumes; **paleas** with the veins extended as bristles.

Elymus elymoides subsp. *elymoides* grows in desert and shrub-steppe areas of western North America, extending to the western edge of the Great Plains and, as an adventive, occasionally further east. It is frequently associated with disturbed sites.

Elymus elymoides subsp. **hordeoides** (Suksd.) Barkworth [p. 320]
COMMON SQUIRRELTAIL

Culms 10–20 cm, erect. **Blades** glabrous or puberulent abaxially. **Spikes** 3–6 cm, exserted or partly included, with 3 spikelets per node. **Spikelets** in the central position usually with 2 fertile florets, the lateral spikelets usually with rudimentary to awnlike florets. **Glume awns** 15–50 mm, usually entire; **lemma awns** of fertile florets 15–30 mm; **paleas** with or without distinct bristles.

Elymus elymoides subsp. *hordeoides* grows in dry, rocky, often shallow soils, particularly in *Artemisia rigida–Poa secunda* communities, from eastern Washington and Idaho to northern California and Nevada. It resembles some *Elymus–Hordeum* hybrids.

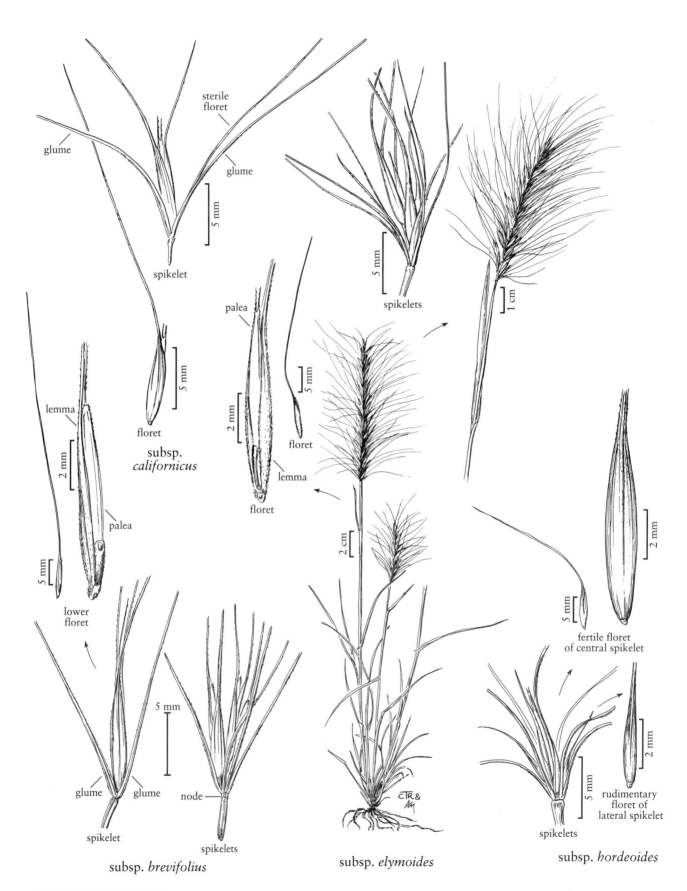

subsp. *californicus*

subsp. *brevifolius*

subsp. *elymoides*

subsp. *hordeoides*

ELYMUS ELYMOIDES

22. Elymus trachycaulus (Link) Gould [p. 323]

Plants usually cespitose, sometimes weakly rhizomatous. **Culms** 30–150 cm, ascending to erect; **nodes** usually glabrous. **Leaves** somewhat basally concentrated; **sheaths** usually glabrous, sometimes markedly retrorsely hirsute or villous; **auricles** absent or to 1 mm; **ligules** 0.2–0.8 mm, truncate; **blades** 2–5(8) mm wide, flat to involute, usually straight and ascending, abaxial surfaces usually smooth and glabrous, sometimes hairy, adaxial surfaces usually glabrous, sometimes conspicuously hairy. **Spikes** 4–25 cm long, 0.4–1 cm wide, erect, with 1 spikelet at all or most nodes; **internodes** (4)7–9(12) mm, edges scabrous, both surfaces smooth and glabrous. **Spikelets** 9–17(20) mm long, usually at least twice as long as the internodes, 3–6 mm wide, appressed, with 3–9 florets, lowest florets functional; **rachillas** glabrous or hairy, hairs to 0.3 mm; **disarticulation** above the glumes, beneath each floret. **Glumes** subequal, 5–17 mm long, from ³/₄ as long as to longer than the adjacent lemmas, 1.8–2.3 mm wide, lanceolate to narrowly ovate, widest about midlength, usually green, purple at higher latitudes and elevations, flat or asymmetrically keeled for their full length, 3–7-veined, the keel vein usually scabrous, the others smooth or scabrous, only 1 vein extending to the apex, adaxial surfaces glabrous, margins hyaline or scarious, usually more or less equal, 0.2–0.5 mm wide, widest at or slightly beyond midlength, apices acute to awned, awns to 11 mm; **lemmas** 6–13 mm, glabrous, usually smooth proximally, often scabridulous distally over the veins, apices acute, usually awned, awns to 40 mm, usually straight, sometimes weakly curved if shorter than 10 mm; **paleas** subequal to the lemmas, keels straight or slightly outwardly curved below the apices, tapering to the apices, apices truncate, 0.15–0.3 mm wide, keel veins often extending beyond the intercostal region, sometimes forming teeth; **anthers** (0.8)1.2–2.5 (3) mm. 2*n* = 28.

Elymus trachycaulus grows from sea level to 3300 m, usually in open or moderately open areas, but sometimes in forests. Its range extends from the boreal forests of North America east through Canada to Greenland and south into Mexico. It also grows, as an introduction, in Asia and Europe. It exhibits considerable variability in the presence or absence of rhizomes, the length and density of the spike, awn development on the glumes and lemmas, and glume venation. The variability in these features has often been used to circumscribe infraspecific taxa, but most such taxa, even though locally distinctive, appear to intergrade. Some of the features appear to be strongly influenced by environmental factors. For instance, plants growing in forested areas of northwestern North America tend to be slightly rhizomatous, more gracile, and later-flowering that those in adjacent, more exposed areas; whether they constitute a distinct taxon or merely a forest ecotype is not clear. Plants growing at higher elevations tend to have glumes with more widely spaced veins and broader, often unequal margins, resembling *E. violaceus* in these respects. Whether this reflects ecotypic differentiation, hybridization with *E. violaceus* (p. 324), or greater genetic continuity than is suggested by their placement in different species is not clear.

Jozwik (1966) recognized four groups within *Elymus trachycaulus*. Group I comprised unawned or shortly awned specimens; group II a polymorphic assemblage of awned specimens; group III a rather homomorphic group of specimens with secund spikes and relatively long awns; and group IV a relatively homomorphic group of unawned, high-elevation specimens. He concluded that group II consists of hybrids and backcrosses between *E. trachycaulus* and other species of *Triticeae*. He based this conclusion on consideration of field observations, artificial hybrids, the polymorphism of the specimens, and the geographic distribution of the group. This last was similar to that of unawned specimens of *E. trachycaulus*, but the populations were highly scattered within the area concerned. Jozwik's group III is treated here as *E. trachycaulus* subsp. *subsecundus*. His group IV is treated here as *E. violaceus*.

Elymus trachycaulus is often confused with *E. stebbinsii* (p. 329). It differs in having shorter anthers, shorter internodes, and glumes that are sometimes awned. It may also be confused, particularly in the herbarium, with specimens of *E. glaucus* having solitary spikelets at all the spike nodes; it usually differs in having shorter anthers and less acuminate glumes. When, as is sometimes the case, the two species grow together, *E. trachycaulus* can be distinguished by its stiffer leaves. *Elymus trachycaulus* also resembles *E. macrourus* (p. 324) and *E. alaskanus* (p. 326), but its glumes are longer relative to the lemmas. It also has less hairy rachillas than most plants of those species.

C.L. Hitchcock et al. (1969) treated *Elymus trachycaulus* as a subspecies of *E. caninus* (see next); it differs consistently from the latter species in glumes that are glabrous on the adaxial (inner) surface, in a chromosome interchange, and in its molecular characteristics (Sun et al. 1998). It also tends to have a more erect spike.

Elymus trachycaulus has been implicated in several interspecific and intergeneric hybrids. Named interspecific hybrids (pp. 338–343) (and the other parent) are *E. ×cayouetteorum* (*E. canadensis*), *E. ×palmerensis* (*E. sibiricus*), *E. ×pseudorepens* (*E.*

lanceolatus), and *E.* ×*saundersii* (*E. elymoides*). Hybrids with *E. hystrix* have been named ×*Agroelymus dorei* Bowden; the appropriate combination has not been made in *Elymus*. Named intergeneric hybrids are ×*Elyhordeum macounii* (p. 284) (*Hordeum jubatum*), ×*Elyleymus jamesensis* (p. 348) (*Leymus mollis*), and ×*Elyleymus ontariensis* (p. 346) (*Leymus innovatus*). Hybrids with *Elymus elymoides*, *E. multisetus*, and *Hordeum jubatum* have brittle rachises and tend to be awned. Others are harder to recognize.

1. Lemma awns 17–40 mm long, longer than the lemma body, straight; spikes somewhat 1-sided . . . . subsp. *subsecundus*
1. Lemmas unawned or with awns to 24 mm long, shorter or longer than the lemma body, straight or curved; spikes 2-sided.
  2. Lemma awns 9–24 mm long . . . . . see p. 321, group II
  2. Lemmas unawned or with awns to 9 mm long, the awns sometimes curved.
    3. Spike internodes 8–15 mm long; spikes 8–25 cm long; glumes unawned or with straight awns to 2 mm long; spikelet bases usually visible; lemmas unawned or with straight awns to 40 mm long . . . . subsp. *trachycaulus*
    3. Spike internodes 4–5 mm long; spikes 5–10 cm long; glumes awned, awns 1.8–4 mm long; spikelet bases usually concealed; lemmas awned, awns 2–3 mm long, slightly curved . . . . subsp. *virescens*

Elymus trachycaulus subsp. **subsecundus** (Link) Á. Löve & D. Löve [p. 323]
ONE-SIDED WHEATGRASS

Culms 40–110 cm. **Spikes** 7–25 cm, somewhat 1-sided. **Spikelets** with 3–7 florets, the bases usually visible. **Glumes** 11–17 mm, long-acuminate or awned, awns to 11 mm; **lemmas** awned, awns 17–40 mm, longer than the lemma body, straight.

Elymus trachycaulus subsp. *subsecundus* grows primarily in the Great Plains. It differs from plants of *E. glaucus* (p. 306) with solitary spikelets, in its 1-sided spike and stiffer, more basally concentrated leaves. It may comprise derivatives of *E. trachycaulus* subsp. *trachycaulus* × *Hordeum jubatum* (p. 245) hybrids that are adapted to moist prairies. The unilateral spike is particularly characteristic of artificial hybrids between the two species, and is uncommon in other hybrids (Jozwik 1966).

Elymus trachycaulus (Link) Gould subsp. **trachycaulus** [p. 323]
SLENDER WHEATGRASS, ÉLYME À CHAUMES RUDES, AGROPYRE À CHAUMES RUDES

Culms 30–150 cm. **Spikes** (4)8–30 cm long, 0.5–0.8 cm wide, 2-sided; **internodes** 8–15 mm. **Spikelets** with 3–9 florets, the bases usually visible. **Glumes** 5–17 mm, at least 1 vein scabrous to near the base, sometimes all veins scabrous, unawned or with straight awns shorter than 2 mm; **lemmas** unawned or awned, awns to 5 mm, straight.

Elymus trachycaulus subsp. *trachycaulus* grows throughout the habitat and range of the species, and exhibits considerably more variation than subsp. *subsecundus*. Two aspects of the variation that seem particularly worthy of further study are the glume venation and the spacing of spikelets in the spikes. Plants with glumes having 5–7 well-developed, narrowly spaced veins are restricted to lower elevations and the southern portion of the subspecies range; northern plants and plants at higher elevations generally have 3–5 weakly developed and widely spaced veins. The former glumes resemble those of *E. glaucus*, with which *E. trachycaulus* subsp. *trachycaulus* is often sympatric; the latter, those of *E. violaceus* (p. 324). Spikelet spacing also varies considerably. In at least some instances, plants with widely spaced spikelets appear to be associated with more shady habitats.

Elymus trachycaulus subsp. **virescens** (Lange) Á. Löve & D. Löve [p. 323]

Culms 20–80 cm. **Spikes** 5–10 cm long, 0.5–0.8(1) cm wide, 2-sided; **internodes** 5–10 mm. **Spikelets** usually with the bases concealed. **Glumes** 9.5–13.5 mm, 1 vein scabrous over most of its length, the basal ¼ usually smooth, the remaining veins scabrous or smooth, awned, awns 1.5–2 mm; **lemmas** awned, awns 2.5–10 mm, often curved.

Elymus trachycaulus subsp. *virescens* is restricted to Greenland. It is very consistent in its morphology.

23. **Elymus caninus** (L.) L. [p. 323]
BEARDED WHEATGRASS

Plants cespitose, not strongly rhizomatous. **Culms** 30–130 cm, erect or geniculate, usually hairy on or below the nodes. **Leaves** evenly distributed; **sheaths** glabrous; **auricles** to 1.5 mm; **ligules** 0.2–1.5 mm; **blades** 10–30 cm long, 4–10 mm wide, flat, both surfaces scabrous, adaxial surfaces sometimes with hairs over the veins, hairs to 0.5 mm, veins not prominent, widely spaced. **Spikes** 5–20 cm long, 0.5–1.5 cm wide including the awns, 5–8 mm wide excluding the awns, erect or arching, with 1 spikelet per node; **internodes** 4.5–7 mm, edges scabrous or ciliate, both surfaces hairy below the spikelets. **Spikelets** 10–15(20) mm long, 2–5(7) mm wide, appressed to slightly divergent, with 2–6 florets; **rachillas** scabridulous or pubescent, often more densely pubescent distally; **disarticulation** above the glumes,

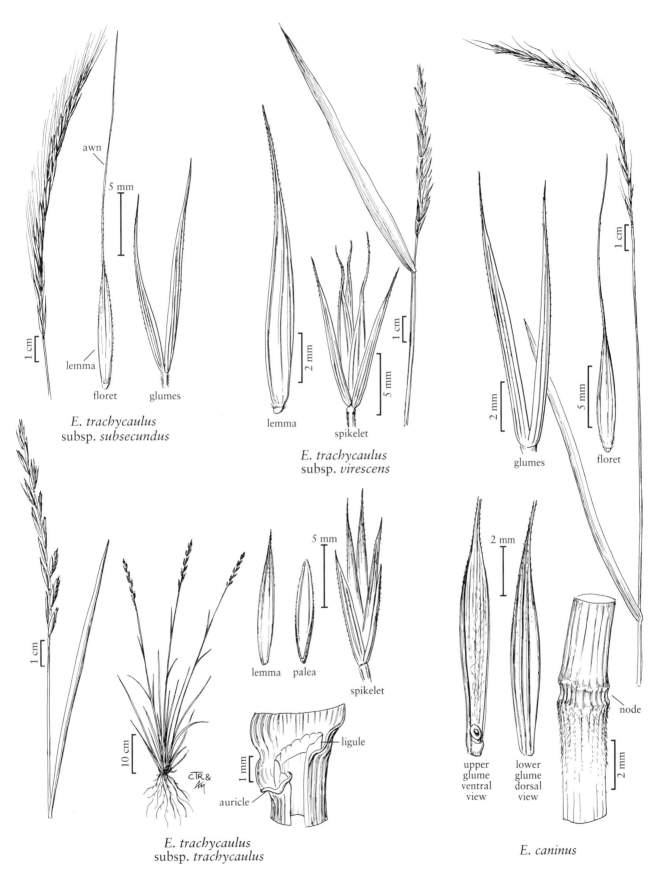

E. trachycaulus
subsp. *subsecundus*

E. trachycaulus
subsp. *virescens*

E. trachycaulus
subsp. *trachycaulus*

E. caninus

ELYMUS

beneath each floret. **Glumes** equal to unequal, 0.6–1 mm wide, lanceolate to narrowly ovate, usually green, flat or weakly keeled, keels eccentric, adaxial surfaces hairy, hairs often inconspicuous, hyaline margins sometimes widest distally, narrowing abruptly to the acute to acuminate apices; **lower glumes** 8–11 mm, 3-veined, usually awned, awns to 3 mm; **upper glumes** 10–13 mm, 3–5-veined, sometimes awn-tipped, awns to 0.3 mm; **lemmas** 9–13 mm, glabrous, smooth to somewhat scabridulous distally, rounded on the back proximally, awned, awns 7–20 mm, straight or flexuous; **paleas** subequal to the lemmas, keels finely and densely ciliate over most of their length, straight or slightly outwardly curved, tapering to the apices, apices about 0.2 mm wide; **anthers** 2–3 mm. $2n = 28$. Haplomes **StH**.

Elymus caninus is native to Eurasia; it is not known to be established in the *Flora* region. A.S. Hitchcock (1935, 1951) reported that it had been collected on ballast dumps in Portland, Oregon, but the specimens concerned belong to *E. ciliaris* (p. 336) and *E. tsukushiensis* (p. 336). *Elymus caninus* differs from *E. ciliaris* and *E. tsukushiensis* in having flatter glumes that are longer in relation to the lemmas, and palea keels that are straight or almost straight below the apices. Recent reports of its occurrence in the region reflect C.L. Hitchcock et al.'s (1969) treatment, in which *E. caninus* and *E. trachycaulus* were treated as conspecific subspecies. Because *E. caninus* is the older name, it is the correct name to use at the specific rank under such a treatment.

The hairs on the inside of the glumes are difficult to see. Nevertheless, this is the single most reliable morphological character for distinguishing *Elymus caninus* from all other species of *Elymus* in this treatment. *Elymus caninus* is most likely to be confused with awned plants of *E. trachycaulus* (p. 321). The two species also differ in their molecular characteristics, and in at least one chromosome interchange (Sun et al. 1998).

24. Elymus violaceus (Hornem.) Feilberg [p. 325]
ARCTIC WHEATGRASS, ÉLYME LATIGLUME

Plants cespitose, not rhizomatous. **Culms** 18–75 cm, often decumbent or geniculate; **nodes** usually glabrous. **Sheaths** glabrous; **auricles** about 0.5 mm; **ligules** 0.5–1 mm, truncate; **blades** 3–4 mm wide, flat, glabrous or hairy, abaxial surfaces less densely hairy and with shorter hairs than the adaxial surfaces, apices acute. **Spikes** 5–12 cm long, 0.4–0.7 cm wide excluding the awns, erect, with 1 spikelet per node; **internodes** 4–5.5 mm, edges ciliate. **Spikelets** 11–19 mm, appressed, with (3)4–5 florets; **rachillas** hairy, hairs about 0.4 mm;

disarticulation above the glumes, beneath each floret. **Glumes** 8–12 mm long, 1.2–2 mm wide, about $^3/_4$ as long as to equaling the adjacent lemmas, narrowly ovate to obovate, often purplish, glabrous, sometimes scabrous, flat or equally keeled the full length, keels and other veins usually smooth, sometimes scabrous, 3(5)-veined, adaxial surfaces glabrous, margins usually unequal, the wider margin 0.3–1 mm wide, usually widest in the distal $^1/_3$, apices acute to rounded, often awned, awns to 2 mm; **lemmas** glabrous or pubescent, hairs flexible, all similar, apices usually awned, awns 0.5–3 mm, straight; **paleas** subequal to the lemmas, tapering to the apices, apices about 0.4 mm wide; **anthers** 0.7–1.3 mm. $2n = 28$. Haplomes **StH**.

Elymus violaceus grows in arctic, subalpine, and alpine habitats, on calcareous or dolomitic rocks, from Alaska through arctic Canada to Greenland, and south in the Rocky Mountains to southern New Mexico. In western North America, it forms intermediates with *E. scribneri* (p. 330), *E. trachycaulus* (p. 321), and *E. alaskanus* (p. 326). It is treated here as including *E. alaskanus* subsp. *latiglumis* [≡ *Agropyron latiglume*], *E. alaskanus* being restricted to plants with relatively short glumes that are often found in valleys and at lower elevations than *E. violaceus*. Western plants of *E. violaceous* tend to be more glaucous, have shorter spikes and spikelets, and more obovate glumes than plants from Greenland but, until more is known about the extent and genetic basis of the variation in and among *E. violaceus*, *E. alaskanus*, and *E. trachycaulus*, formal taxonomic recognition seems inappropriate.

25. Elymus macrourus (Turcz. *ex* Steud.) Tzvelev [p. 325]
NORTHERN WHEATGRASS

Plants cespitose, sometimes appearing weakly rhizomatous. **Culms** 35–100 cm, ascending to erect; **nodes** sometimes pubescent. **Sheaths** glabrous; **auricles** absent; **ligules** 0.5–1 mm, truncate to rounded; **blades** 3–10 mm wide, flat, usually glabrous, abaxial surfaces smooth or scabridulous, adaxial surfaces scabrous. **Spikes** 5–20 cm long, 0.4–0.8 cm wide, erect, with 1 spikelet per node; **internodes** 7–8 mm long, about 0.5 mm wide, glabrous below the spikelets. **Spikelets** 12–20 mm, appressed, with 4–7 florets; **rachillas** hairy, hairs 0.3–0.5 mm; **disarticulation** above the glumes, beneath each floret. **Glumes** 6–10 mm long, $^1/_3$–$^2/_3$ the length of the spikelets and to about $^1/_2$ the length of the adjacent lemmas, 0.8–1.8 mm wide, widest at about midlength, lanceolate, flat, rounded, or symmetrically keeled, usually green or green tinged with purple, 3–4-veined,

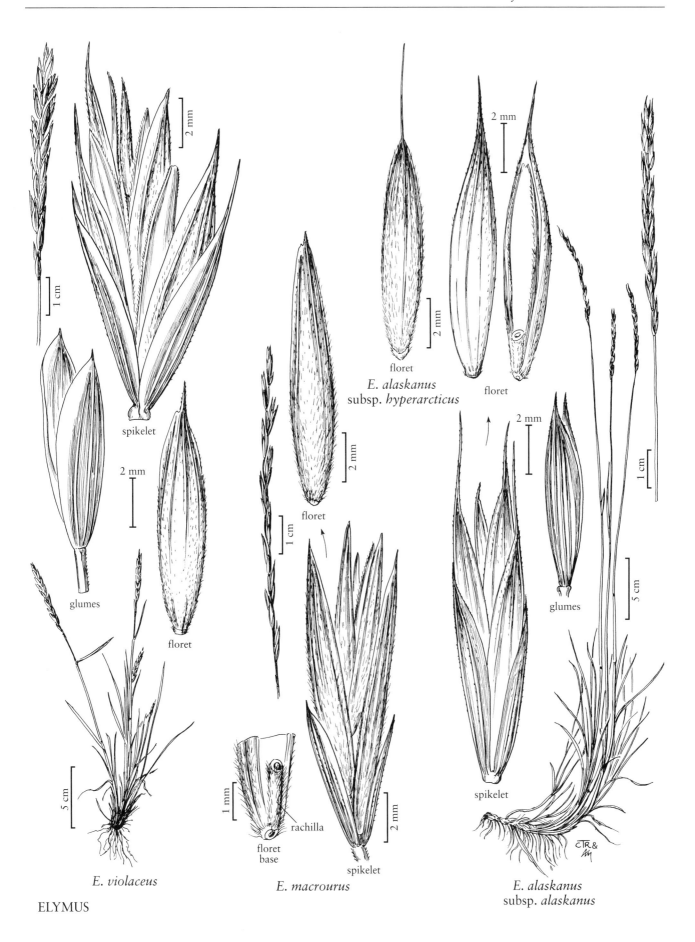

ELYMUS

spikelet

glumes

2 mm

floret

E. violaceus

1 cm

2 mm

5 cm

floret

1 cm

rachilla

floret
base

spikelet

2 mm

E. macrourus

floret

E. alaskanus
subsp. *hyperarcticus*

2 mm

floret

2 mm

glumes

2 mm

1 cm

5 cm

spikelet

E. alaskanus
subsp. *alaskanus*

veins scabridulous, scabrous, or with hairs to 0.3 mm, usually glabrous elsewhere, margins subequal, about 0.3 mm wide, widest near midlength, apices acute, unawned or awned, awns to 1 mm; **lemmas** 8–12 mm, hairy throughout or glabrous distally, hairs all alike, 0.2–0.3 mm, apices unawned or awned, awns to 7 mm, straight; **paleas** subequal to the lemmas, tapering to the apices, apices about 0.8 mm wide; **anthers** 1–2 mm. $2n = 28$.

Elymus macrourus grows on river banks and bars, lake shores, and hillsides in northwestern North America. Outside of North America, it grows across the Russian arctic, and extends south into the boreal forest. Plants growing on shifting river banks and bars often appear rhizomatous, as the lower internodes elongate in response to the disturbed substrate. Plants of *E. macrourus* differ from *E. alaskanus* (see next) in the shape of their glumes and their narrower glume margins, and from *E. trachycaulus* (p. 321) in their relatively short glumes and evidently hairy rachilla segments.

Three varieties of *Elymus macrourus* are recognized in Russian treatments. It is not clear to which, if any, of the Russian varieties North American plants belong. A circumboreal study is needed, using plants grown from seeds collected in the wild. Seeds available as *E. macrourus* through germplasm resources appear to be misidentified.

Elymus macrourus is one of the parents in both *E. ×palmerensis* (p. 340) and ×*Elyhordeum pilosilemma* (p. 284).

26. Elymus alaskanus (Scribn. & Merr.) Á.Löve [p. 325]

Plants cespitose or weakly rhizomatous. **Culms** 20–90 cm, sometimes decumbent at the base, ascending to erect above; **nodes** usually pubescent, sometimes glabrous. **Leaves** sometimes basally concentrated; **sheaths** smooth or scabrous, glabrous or pilose; **auricles** absent or to 0.5 mm; **ligules** 0.2–1 mm, erose, ciliolate; **blades** 3–7 mm wide, flat, both surfaces smooth, scabrous, or pubescent. **Spikes** 3.5–14 cm long, 0.5–0.8 cm wide, erect or nodding distally, usually with 1 spikelet per node, occasionally with 2 at the lower nodes; **internodes** 3–10 mm long, 0.5–0.8 mm wide, mostly glabrous and smooth, edges scabrous or ciliate. **Spikelets** 9–15(20) mm, 2–5 times longer than the internodes, appressed, with 3–6 florets, **rachillas** hispidulous; **disarticulation** above the glumes, beneath each floret. **Glumes** 4–8 mm long, (1.2)1.5–2 mm wide, $1/3$–$2/3$ as long as the adjacent lemmas, oblanceolate to obovate, flat, usually purplish, glabrous or hairy, hairs 0.3–0.5 mm, margins unequal, the widest margin 0.4–1

mm wide, both margins widest above the middle, apices unawned or awned, awns to 1 mm; **lemmas** 7–11 mm, glabrous or hairy, sometimes scabridulous, sometimes more densely hairy distally, hairs 0.2–0.6 mm, all alike, apices unawned or awned, awns to 7 mm, straight; **paleas** subequal to the lemmas, keels straight below the apices; **anthers** 1–2 mm. $2n = 28$.

Elymus alaskanus extends across the high arctic of North America to extreme eastern Russia. This treatment interprets *E. alaskanus* as having relatively short glumes, in accordance with its treatment by Hultén (1968). Large specimens resemble *E. macrourus* (see previous), but differ in the shape of their glumes and in their wider glume margins. *Elymus alaskanus* differs from *E. trachycaulus* (p. 321) in its greater cold tolerance and the distal widening of its glume margins. There is some intergradation, particularly with *E. violaceus* (p. 324) and *E. trachycaulus*, but these species have longer glumes. Moreover, in western North America, *E. violaceus* is restricted to rocky habitats at or above treeline, whereas *E. alaskanus* is often associated with valleys and flat areas. Reports of its extending to New Mexico are based on the inclusion of high-elevation forms of *E. trachycaulus*.

1. Glumes glabrous, scabrous or sparsely hairy, hairs to about 0.2 mm long; lemmas glabrous or with hairs to about 0.2 mm long . . . . subsp. *alaskanus*
1. Glumes and lemmas densely hairy, hairs 0.2–0.5 mm long . . . . subsp. *hyperarcticus*

Elymus alaskanus (Scribn. & Merr.) Á. Löve subsp. alaskanus [p. 325]
ALASKAN WHEATGRASS

Plants cespitose or shortly rhizomatous. **Culms** 25–90 cm, sometimes decumbent; **nodes** often exposed, glabrous or pubescent. **Ligules** 0.2–0.6 mm; **blades** 3–7 mm wide. **Spikes** 3.5–14 cm; **internodes** scabrous on the edges. **Glumes** glabrous, scabrous or sparsely hairy, hairs to about 0.2 mm; **lemmas** glabrous or hairy, sometimes more densely hairy distally, hairs about 0.2 mm, apices unawned or awned, awns to 7 mm, straight.

Elymus alaskanus subsp. *alaskanus* grows on river banks and hillsides, primarily north of 50° N latitude.

Elymus alaskanus subsp. hyperarcticus (Polunin) Á. Löve & D. Löve [p. 325]
HIGH-ARCTIC WHEATGRASS

Plants cespitose, not rhizomatous. **Culms** 20–35(50) cm, ascending to erect; **nodes** concealed. **Ligules** 0.5–1 mm, erose to ciliate; **blades** 2.5–5 mm wide. **Spikes** 4.5–7 cm; **internodes** 4–6 mm, ciliate, distal cilia longest. **Glumes** densely hairy, hairs 0.2–0.5 mm, apices mucronate or awned, awns to 1 mm; **lemmas** densely hairy, hairs 0.2–0.5 mm, apices awned, awns 1–5 mm, straight. $2n = 28$.

Elymus alaskanus subsp. *hyperarcticus* grows on river banks and hillsides. It extends from the Lake Taymyr basin in arctic Russia across northern North America to Greenland.

27. Elymus lanceolatus (Scribn. & J.G. Sm.) Gould [p. 328]

Plants strongly rhizomatous, sometimes glaucous. **Culms** 22–130 cm, erect; **nodes** glabrous. **Leaves** often mostly basal, sometimes more evenly distributed; **sheaths** glabrous or pubescent; **auricles** usually present on the lower leaves, 0.5–1.5 mm; **ligules** 0.1–0.5 mm, erose, sometimes ciliolate; **blades** 1.5–6 mm wide, generally involute, abaxial surfaces usually glabrous, adaxial surfaces strigose, ribs subequal in size and spacing. **Spikes** 3.5–26 cm long, 0.5–1 cm wide, erect to slightly nodding, usually with 1 spikelet per node, sometimes with 2 at a few nodes; **internodes** 3.5–15 mm long, 0.1–0.8 mm wide, glabrous or hairy. **Spikelets** 8–31 mm, 1.5–3 times longer than the internodes, appressed, with 3–11 florets; **rachillas** glabrous or hairy, hairs to 1 mm; **disarticulation** above the glumes, beneath each floret. **Glumes** subequal, 5–14 mm long, $^{1}/_{2}$–$^{3}/_{4}$ the length of the adjacent lemmas, 0.7–1.3 mm wide, lanceolate, glabrous or hairy, smooth or scabrous, 3–5-veined, flat or weakly, often asymmetrically keeled, keels straight, margins narrow, tapering from the base or from beyond midlength, apices acute to acuminate, sometimes mucronate or shortly awned; **lemmas** 7–12 mm, glabrous or hairy, hairs all alike, sometimes scabrous, acute to awn-tipped, awns to 2 mm, straight; **paleas** about equal to the lemmas, keels straight below the apices, smooth or scabrous proximally, sometimes hairy, scabrous distally, intercostal region glabrous or with hairs, apices 0.2–0.3 mm wide; **anthers** (2.5)3–6 mm. $2n = 28$.

Elymus lanceolatus grows in sand and clay soils and dry to mesic habitats. It is found primarily in the western half of the *Flora* region, between the coastal mountains and 95° W longitude, with the exception of *E. lanceolatus* subsp. *psammophilus*, which extends around the Great Lakes. Three subspecies are recognized, primarily on the basis of their lemma and palea pubescence.

Elymus lanceolatus is primarily outcrossing, and hybridizes with several species of *Triticeae*. *Elymus albicans* (p. 334) is thought to be derived from hybridization with the awned phase of *Pseudoroegneria spicata* (p. 281). Judging from specimens of controlled hybrids, hybridization with *E. trachycaulus* (p. 321) and unawned plants of *P. spicata* probably occur, but

would be almost impossible to detect without careful observation in the field. Experimental hybrids are partially fertile, and capable of backcrossing to either parent (Dewey 1965, 1967, 1968, 1975, 1976).

1. Lemmas densely hairy, hairs flexible, some 1 mm long or longer . . . . subsp. *psammophilus*
1. Lemmas glabrous or with stiff hairs shorter than 1 mm.
   2. Lemmas with hairs, not scabrous . . . . subsp. *lanceolatus*
   2. Lemmas smooth, sometimes scabrous distally, mostly glabrous, sometimes the lemma margins hairy proximally . . . . . subsp. *riparius*

Elymus lanceolatus (Scribn. & J.G. Sm.) Gould subsp. lanceolatus [p. 328]
THICKSPIKE WHEATGRASS

Culms 60–130 cm. **Spikes** 10–22 cm; **internodes** 7–15 mm, smooth, scabrous, or hairy distally. **Spikelets** 10–28 mm. **Lemmas** not scabrous, moderately hairy, hairs stiff, shorter than 1 mm.

Elymus lanceolatus subsp. *lanceolatus* grows in clay, sand, loam, and rocky soils, and is widely distributed in the western *Flora* region. It is most likely to be confused with the octoploid *Pascopyrum smithii* (p. 351); it differs morphologically from that species in having more evenly distributed leaves and acute glumes that tend to taper from midlength or higher, rather than acuminate glumes that tend to taper from below midlength. In addition, the midvein of the glumes of *E. lanceolatus* is straight, whereas that of *P. smithii* "leans" to the side distally.

Elymus lanceolatus subsp. psammophilus (J.M. Gillett & H. Senn) Á. Löve [p. 328]
SAND-DUNE WHEATGRASS

Culms 20–95 cm. **Spikes** 4–26 cm; **internodes** 3.5–13 mm, hairy at least distally. **Spikelets** 9–31 mm. **Lemmas** densely hairy, hairs flexible, usually many longer than 1 mm; **paleas** hairy between the keels, keels hairy proximally.

Elymus lanceolatus subsp. *psammophilus* tends to grow in sandy soils. It was originally described from around the Great Lakes, but plants with similar vesture have been found scattered throughout the western range of the species, almost always in association with sandy soils. Those from the Yukon and northern British Columbia tend to be shorter and have smaller spikelets and spikelet parts than those from Washington and Saskatchewan, but there is considerable overlap in these characters. Plants from around the Great Lakes (Gillett and Senn 1960) were almost completely pollen sterile. Despite this, Gillett and Senn rejected the notion that they were hybrids.

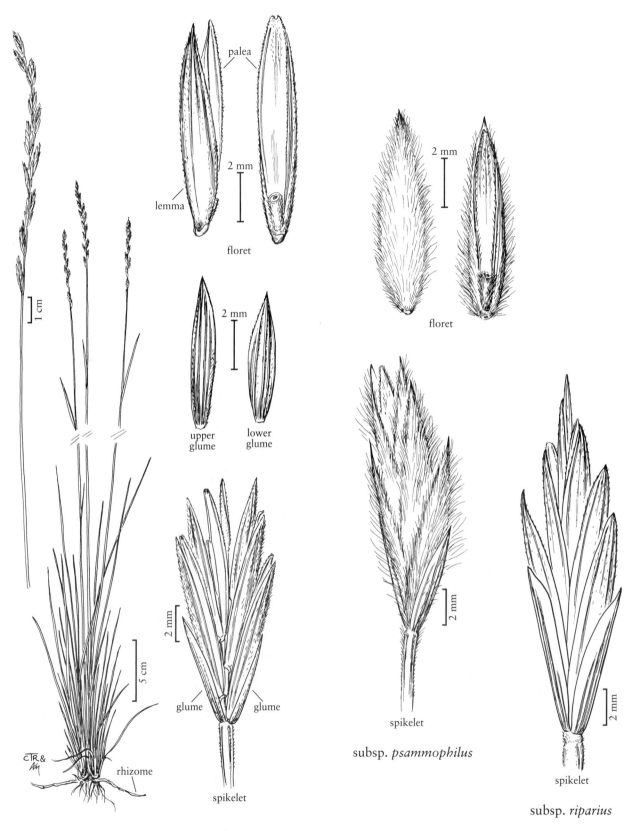

palea

lemma

floret

2 mm

upper
glume

lower
glume

2 mm

glume   glume

spikelet

2 mm

subsp. *lanceolatus*

floret

2 mm

spikelet

subsp. *psammophilus*

2 mm

spikelet

subsp. *riparius*

2 mm

rhizome

1 cm

5 cm

ELYMUS LANCEOLATUS

Elymus lanceolatus subsp. **riparius** (Scribn. & J.G. Sm.) Barkworth [p. 328]
STREAMBANK WHEATGRASS

Culms usually 22–60 cm. Spikes 6–10 cm; **internodes** 3.5–10 mm, glabrous, sometimes scabrous. **Spikelets** 10–17 mm. **Lemmas** smooth, sometimes scabrous distally, mostly glabrous, margins sometimes hairy proximally.

Elymus lanceolatus subsp. *riparius* grows throughout most of the western part of the range of *E. lanceolatus*, being more common in mesic habitats and clay soils than the other two subspecies.

28. **Elymus stebbinsii** Gould [p. 331]

Plants cespitose or shortly rhizomatous. **Culms** 60–140 cm; **nodes** glabrous or retrorsely pubescent. **Leaves** evenly distributed; **sheaths** glabrous or pubescent; **auricles** usually present, 0.5–2 mm; **ligules** 0.3–3.5 mm, truncate to acute, sometimes long-ciliate; **blades** 4–6.5 mm wide, flat or the margins involute, straight. **Spikes** 15–31 cm long, 0.4–1.5 cm wide including the awns, 0.4–0.8 cm wide excluding the awns, erect, with 1 spikelet per node; **internodes** 9–27 mm long, 1–1.3 mm wide, glabrous, smooth. **Spikelets** 13–29 mm long, from shorter than to almost twice as long as the internodes, 2.5–5 mm wide, appressed, with 5–7 florets; **rachillas** glabrous; **disarticulation** above the glumes and beneath each floret. **Glumes** subequal, 7.5–12 mm long, 1.2–1.5 mm wide, lanceolate, widest at about midlength, flat or rounded on the back, 5-veined, veins smooth, scabrous or just the midvein scabridulous, margins widest at about midlength, apices acute, unawned; **lemmas** 9–12 mm, glabrous, sometimes scabrous, acute, unawned or awned, awns to 28 mm, straight; **paleas** subequal to the lemmas, tapering, apices 0.2–0.3 mm wide; **anthers** (3.5)4–7 mm. $2n = 28$.

Elymus stebbinsii is restricted to California, where it grows on dry slopes, chaparral, and wooded areas, at elevations below 1600 m. It differs from other *Elymus* species primarily in its combination of long anthers and solitary spikelets. It is often confused with *E. glaucus* (p. 306) and *E. trachycaulus* (p. 321) with solitary spikelets. It differs from both in its longer anthers, and from most representatives of *E. glaucus* in its acute, but unawned, glumes.

1. Lemmas awned, awns 8–28 mm long; lower leaf sheaths rarely pubescent; spikelets 13–22 mm long . . . . subsp. *septentrionalis*
1. Lemmas unawned or with awns to 8(12) mm long; lower leaf sheaths pubescent or glabrous; spikelets 17–29 mm long . . . . subsp. *stebbinsii*

Elymus stebbinsii subsp. **septentrionalis** Barkworth [p. 331]
NORTHERN STEBBINS' WHEATGRASS

Lowest visible cauline node usually glabrous, rarely pubescent. **Lower leaf sheaths** usually glabrous, rarely pubescent. **Spike internodes** 9–21 mm. **Spikelets** 13–22 mm. **Lemmas** awned, awns 8–28 mm.

Elymus stebbinsii subsp. *septentrionalis* grows primarily in the Sierra Nevada. Its range extends from near the Oregon border to Tulare County, and includes the coastal mountains north of San Francisco Bay.

Elymus stebbinsii Gould subsp. **stebbinsii** [p. 331]
STEBBINS' WHEATGRASS

Lowest visible cauline node often pubescent. **Lower leaf sheaths** pubescent or glabrous. **Spike internodes** 16.3–27 mm. **Spikelets** 17–29 mm. **Lemmas** unawned or awned, awns to 8(12) mm.

Elymus stebbinsii subsp. *stebbinsii* is best known from the coastal mountains south of San José. It also grows at scattered locations from the central Sierra Nevada south to the Transverse Mountains.

29. **Elymus arizonicus** (Scribn. & J.G. Sm.) Gould [p. 331]
ARIZONA WHEATGRASS

Plants cespitose, not rhizomatous. **Culms** 45–100 cm, erect or decumbent at the base; **nodes** glabrous or almost so. **Leaves** evenly distributed over the lower $^1/_2$ of the culms; **sheaths** glabrous; **auricles** usually present, to 1 mm; **ligules** to 1 mm on the basal leaves, 1–3 mm on the flag leaves; **blades** 2.5–6 mm wide, lax, abaxial surfaces smooth and glabrous, adaxial surfaces scabrous, with scattered 0.5–1 mm hairs, veins close together. **Spikes** 12–25 cm long, 2.5–6 cm wide including the awns, 10–15 mm wide excluding the awns, flexuous, usually nodding or pendent at maturity, with 1 spikelet per node; **internodes** 11–17 mm long, 0.4–1 mm wide, glabrous, mostly smooth, scabrous on the edges. **Spikelets** 14–26 mm long, 6–8 mm wide, appressed to divergent, 1.5–2 times as long as the internodes, with 4–6 florets; **rachillas** glabrous; **disarticulation** above the glumes and beneath each floret. **Glumes** narrowly lanceolate, margins about 0.2 mm wide, 3(5)-veined, the bases flat, evidently veined, margins hyaline, widest at about midlength, acute or acuminate, unawned or awned, awns to 4 mm, straight; **lower glumes** 5–9 mm; **upper glumes** 8–10 mm; **lemmas** 8–15 mm, scabrous, rounded on the back, awns 10–25 mm, arcuately diverging; **paleas** as long as or longer than the lemmas, tapering, apices truncate, about 0.3 mm wide; **anthers** 3–5 mm. 2*n* = 28.

Elymus arizonicus grows in moist, rocky soil in mountain canyons of the southwestern United States and northern Mexico. When mature, the drooping spike and solitary spikelets make *E. arizonicus* easy to identify. Immature specimens, or those mounted so that the spike appears erect, are easily mistaken for *Pseudoroegneria spicata* (p. 281), but they have thicker culms and longer ligules, more basal leaves, and wider leaf blades.

30. Elymus bakeri (E.E. Nelson) Á. Löve [p. 333]
BAKER'S WHEATGRASS

Plants cespitose, not rhizomatous. **Culms** 30–50 cm tall, 1–2 mm thick, ascending to erect; **nodes** glabrous. **Leaves** not basally concentrated; **sheaths** glabrous; **auricles** 0.3–0.6 mm; **ligules** 0.5–1 mm; **blades** 12–20 cm long, 2–4 mm wide, stiff, abaxial surfaces smooth, glabrous, adaxial surfaces smooth or scabridulous, veins prominent, closely spaced. **Spikes** 8–12 cm long, 4–6 cm wide including the awns, about 1 cm wide excluding the awns, straight, erect or inclined, with 1 spikelet per node; **internodes** 5–9 mm long, about 0.8 mm wide, both surfaces glabrous, edges ciliate. **Spikelets** 10–19 mm long, about twice as long as the adjacent internodes, 4–10 mm wide, appressed, with 4–5 florets; **rachillas** scabrous or hirtellous; **disarticulation** above the glumes, beneath each floret. **Glumes** 7–12 mm long, 1.4–2 mm wide, narrowly oblong, usually green or green tinged with purple, the bases evidently veined or indurate for less than 0.5 mm, 5-veined, veins scabrous, margins narrow, widest distally, apices acute, sometimes bifid, awned, awns 2–8 mm, straight or divergent; **lemmas** scabrous or smooth, apices often shortly bidentate, awns 10–35 mm, arcuate to recurved; **paleas** equaling or slightly longer than the lemmas, tapering to the 0.2–0.4 mm wide apices; **anthers** 0.8–1.5 mm. 2*n* = 28.

Elymus bakeri grows in high, but not alpine, mountain meadows of Colorado and northern New Mexico. It resembles the awned phase of *Pseudoroegneria spicata* (p. 281), but differs in having rather thicker culms and spikes, and stouter lemma awns. W.A. Weber (University of Colorado, pers. comm., ca.

1999) stated that it often forms large stands in Colorado.

Reports of *Elymus bakeri* from Idaho appear to be based on fertile hybrids of *E. trachycaulus* (p. 321) or *E. violaceus* (p. 324) with *Pseudoroegneria spicata*; that for Wallowa County, Oregon, on a specimen of *E. glaucus* (p. 306).

31. Elymus scribneri (Vasey) M.E. Jones [p. 333]
SCRIBNER'S WHEATGRASS

Plants cespitose, not rhizomatous. **Culms** 15–35(55) cm, prostrate to strongly decumbent, at least at the base; **nodes** glabrous. **Sheaths** glabrous or shortly pilose; **auricles** usually present, 0.5–1 mm; **ligules** 0.2–0.4(0.7) mm, usually truncate, occasionally acute, entire to erose; **blades** 1.5–4 mm wide, usually involute, adaxial surfaces prominently ribbed. **Spikes** 3.5–10 cm long, 0.8–1.2 cm wide excluding the awns, 3–6 cm wide including the awns, usually with 1 spikelet per node, occasionally with 2 spikelets at the lower nodes; **internodes** 2.5–5(7) mm long, 0.5–1 mm wide, glabrous, mostly smooth, edges scabrous. **Spikelets** 9–15 mm long, 6–12 mm wide, appressed to ascending, with 3–6 florets; **rachilla internodes** 0.8–1.3 mm, scabridulous; **disarticulation** initially at the rachis nodes, subsequently beneath each floret. **Glumes** 4–9 mm long, 0.5–1 mm wide, mostly glabrous, midveins scabrous, 3–5-veined, entire, tapering into a divergent, 12–30 mm awn; **lemmas** 7–10 mm, usually glabrous, occasionally scabridulous, awned, awns 15–30 mm, divergent, scabridulous; **paleas** usually longer than the lemmas, apices ciliate, truncate or the veins extending into teeth, teeth about 0.5 mm; **anthers** 1–1.6 mm. 2*n* = 28.

Elymus scribneri grows in rocky areas in open subalpine and alpine regions, at 2500–3200 m, often in windswept locations, in southwestern Alberta and the western United States. It is often confused with *E. elymoides* (p. 318), but differs from that species in having only one spikelet per node, wider glumes, and more tardily disarticulating rachises. It also resembles *E. sierrae* (see next), from which it differs in its disarticulating rachises, denser spikes, and shorter anthers.

Dewey (1963) concluded that *Elymus trachycaulus* subsp. *andinus* consists of hybrids between *E. scribneri* and *E. trachycaulus* (p. 321). In addition, several taxonomists have suggested that *E. scribneri* consists of fertile hybrids between *E. violaceus* (p. 324) and *E. elymoides*. This suggestion is supported by the frequency with which the three taxa are sympatric, the morphological variation exhibited by *E. scribneri*, and cytogenetic data (Dewey 1967).

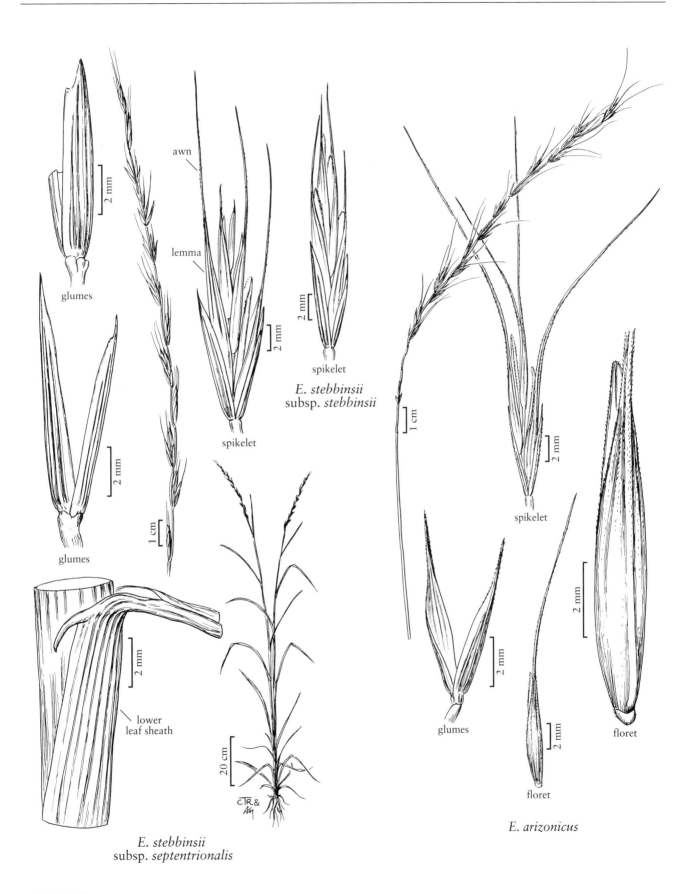

glumes

awn

lemma

glumes

spikelet

E. stebbinsii
subsp. *stebbinsii*

spikelet

lower
leaf sheath

E. stebbinsii
subsp. *septentrionalis*

spikelet

glumes

floret

floret

E. arizonicus

CTR&
AM

ELYMUS

32. Elymus sierrae Gould [p. 335]

SIERRA WHEATGRASS

Plants cespitose, not rhizomatous. **Culms** 20–50 cm, prostrate or decumbent and geniculate; **nodes** 1–2, exposed, glabrous. **Leaves** basally concentrated; **sheaths** glabrous; **auricles** usually present, to 1 mm on the lower leaves; **ligules** 0.2–0.5 mm, erose; **blades** 1–5 mm wide, flat, abaxial surfaces smooth, glabrous, adaxial surfaces prominently ridged over the veins, with scattered hairs, hairs to 0.2 mm, veins closely spaced. **Spikes** 5–15 cm long, 1.5–2.5 cm wide including the awns, 0.7–1.2 cm wide excluding the awns, flexuous, erect to nodding distally, with 1 spikelet at most nodes, occasionally some of the lower nodes with 2 spikelets; **internodes** 5–15 mm long, 0.2–0.5 mm wide, both surfaces glabrous, edges ciliate, not scabrous. **Spikelets** 15–20 mm, ascending to divergent, with 3–7 florets; **rachillas** glabrous; **disarticulation** above the glumes, beneath each floret. **Glumes** subequal, 6–9 mm long, 0.7–1 mm wide, lanceolate, glabrous, the bases evidently veined, apices entire, tapering into a 3–10 mm awn; **lemmas** 12–16 mm, glabrous, sometimes scabridulous, apices bidentate, awned, awns 15–30 mm, arcuately diverging to strongly recurved; **paleas** subequal to the lemmas, apices about 0.4 mm wide; **anthers** 2–3.5 mm. $2n = 28$.

Elymus sierrae is best known from rocky slopes and ridgetops in the Sierra Nevada, at 2100–3400 m, and is also found in Washington and Oregon. It resembles *E. scribneri* (see previous), differing in its non-disarticulating rachises, longer rachis internodes, and longer anthers. Hybrids with *E. elymoides* (p. 318) have glumes with awns 15+ mm long, and some spikelets with narrower glume bases and shorter anthers. Specimens with wide-margined glumes suggest hybridization with *E. violaceus* (p. 324).

33. Elymus wawawaiensis J.R. Carlson & Barkworth [p. 335]

SNAKERIVER WHEATGRASS

Plants cespitose, sometimes weakly rhizomatous. **Culms** (15)50–130 cm, erect, mostly glabrous; **nodes** usually glabrous, sometimes slightly pubescent. **Leaves** more or less evenly distributed; **basal sheaths** glabrate, margins not evidently ciliate; **auricles** absent or to 1.2 mm; **ligules** 0.1–1.1 mm; **blades** to 28 cm long, 1.7–5 mm wide, involute when dry, adaxial surfaces usually densely pubescent, rarely sparsely pubescent. **Spikes** 5–20 cm long, 2.5–3 cm wide including the awns, erect to slightly nodding, with 1 spikelet per node; **internodes** 5–12 mm long, about 0.2 mm thick, about 0.3 mm wide, glabrous beneath the spikelets. **Spikelets** 10–22 mm long, about twice as long as the internodes, 2–8.5 mm wide, appressed, with 4–10 florets; **rachillas** glabrous; **disarticulation** above the glumes, beneath each floret. **Glumes** 4–10 mm long, 0.5–1.3 mm wide, narrowly lanceolate, widest at or below midlength, glabrous, often glaucous, 1–3-veined, flat or weakly keeled, margins 0.1–0.2 mm wide, widest near midlength, apices usually acuminate, awned or unawned, awns to 6 mm; **lemmas** 6–12 mm, smooth or slightly scabrous, margins often sparsely pubescent proximally, apices awned, longest awns in the spikelets 9–28 mm, strongly divergent; **paleas** 7.2–10.5 mm, keels scabrous distally, tapering to the 0.2–0.3 mm wide apices; **anthers** 3.5–6 mm. $2n = 28$.

Elymus wawawaiensis grows primarily in shallow, rocky soils of slopes in coulees and reaches of the Salmon, Snake, and Yakima rivers of Washington, northern Oregon, and Idaho. There are also a few records from localities at some distance from the Snake River and its tributaries. These probably reflect deliberate introductions. C.V. Piper, who worked for the U.S. Department of Agriculture in southeastern Washington from 1892–1902, frequently distributed seed to farmers in the region from populations that he considered superior; he considered *E. wawawaiensis* to be a superior form of what is here called *Pseudoroegneria spicata* (p. 281). Another source of introduced populations is 'Secar', a cultivar of *E. wawawaiensis* that is recommended as a forage grass for arid areas of the northwestern United States.

Elymus wawawaiensis resembles a vigorous version of *Pseudoroegneria spicata*, and was long confused with that species. It differs in its more imbricate spikelets and narrower, stiff glumes. In its primary range, *E. wawawaiensis* is often sympatric with *P. spicata*, but the two tend to grow in different habitats, *E. wawawaiensis* growing in shallow, rocky soils and *P. spicata* in medium- to fine-textured loess soil. The two species also differ cytologically, *E. wawawaiensis* being an allotetraploid, and *P. spicata* consisting of diploids and autotetraploids.

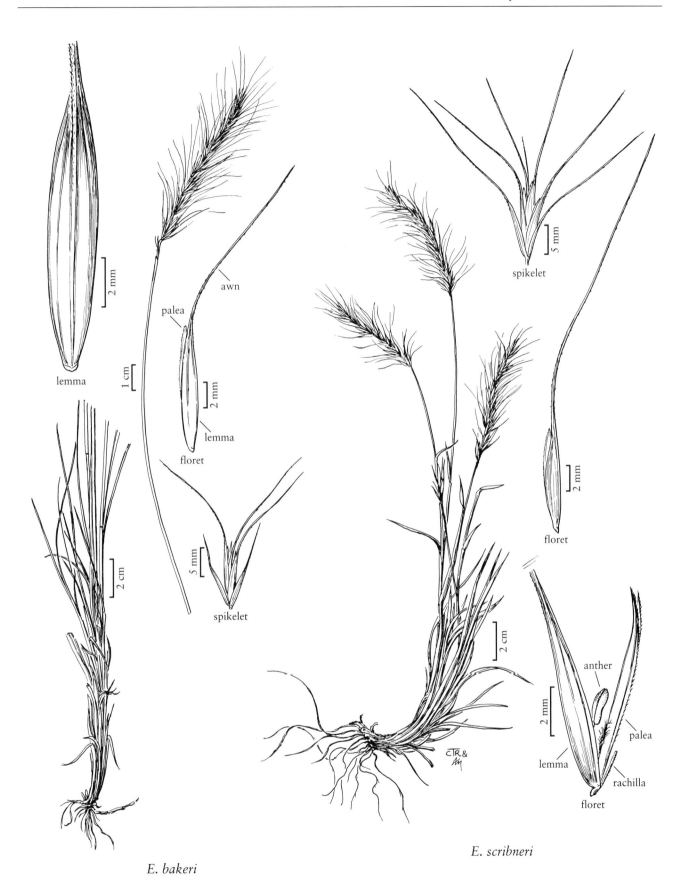

lemma

2 mm

awn

palea

lemma

floret

1 cm

2 mm

2 cm

spikelet

5 mm

E. bakeri

spikelet

5 mm

floret

2 mm

anther

palea

lemma

rachilla

floret

2 cm

2 mm

E. scribneri

ELYMUS

34. Elymus albicans (Scribn. & J.G. Sm.) Á. Löve [p. 335]

MONTANA WHEATGRASS

Plants strongly rhizomatous. **Culms** 40–100 cm, erect or decumbent only at the base, glabrous. **Leaves** somewhat basally concentrated; **sheaths** glabrous; **auricles** usually present, to 0.8 mm; **ligules** 0.2–0.5 mm, ciliolate; **blades** 1–3 mm wide, usually involute, adaxial surfaces scabrous to strigose. **Spikes** 4–14 cm long, 1.5–2.5 cm wide including the awns, 0.3–0.8 cm wide excluding the awns, erect, with 1 spikelet per node; **internodes** 6–14 mm long, 0.2–0.4 mm wide, glabrous or pubescent beneath the spikelets. **Spikelets** 10–18 mm, 1.5–2 times longer than the internodes, appressed to ascending, with 3–7 florets; **rachillas** strigillose; **disarticulation** above the glumes, beneath each floret. **Glumes** subequal, ¹/₂ as long as to almost equaling the adjacent lemmas, glabrous or hairy, weakly keeled, keels and adjacent veins smooth to evenly and strongly scabrous from the base to the apices, margins 0.2–0.3 mm wide, apices acute, acuminate, or shortly awned; **lower glumes** 4–8 mm; **upper glumes** 4.5–8 mm; **lemmas** 7.5–9.5 mm, glabrous or densely hairy, awns 4–12 mm, at least some strongly divergent; **paleas** subequal to the lemmas, tapering to the 0.1–0.3 mm wide apices; **anthers** 3–5 mm. 2*n* = 28.

Elymus albicans grows primarily in the central Rocky Mountains and the western portion of the Great Plains. It tends to grow in shallow, rocky soils on wooded or sagebrush-covered slopes, rather than in deep loams. It is derived from hybrids between *Pseudoroegneria spicata* (p. 281) and *E. lanceolatus* (p. 327). In practice, it is probably restricted to hybrids involving the awned variant of *Pseudoroegneria spicata*, because the hybrid origin of those involving the unawned variant would probably not be recognized.

Populations of *Elymus albicans* differ in their reproductive abilities (Dewey 1970). In some, most plants yield good seed; in others, most plants are sterile. Some fertile populations appear to be self-perpetuating; others appear to consist of recent hybrids and some backcrosses. Although treated here as a species, *E. albicans* could equally well be treated as a hybrid in ×*Pseudelymus* (p. 282), but the combination has not been published. Plants with glabrous lemmas, presumed to be derived from crosses with glabrous individuals of *E. lanceolatus*, have sometimes been treated as a distinct taxon, e.g., *Agropyron albicans* var. *griffithsii* (Scribn. & J.G. Sm.) Beetle or *A. griffithsii* Scribn. & J.G. Sm.; they are not formally recognized here.

35. Elymus repens (L.) Gould [p. 337]

QUACKGRASS, COUCHGRASS, CHIENDENT, CHIENDENT RAMPANT

Plants strongly rhizomatous, sometimes glaucous. **Culms** 50–100 cm. **Leaves** sometimes somewhat basally concentrated; **sheaths** pilose or glabrous proximally; **auricles** 0.3–1 mm; **ligules** 0.25–1.5 mm; **blades** 6–10 mm wide, usually flat, abaxial surfaces glabrous or sparsely pilose, adaxial surfaces usually sparsely pilose over the veins, sometimes glabrous, veins smooth, widely spaced, primary veins prominent, separated by the secondary veins. **Spikes** 5–15 cm long, 0.5–1.5 cm wide, erect, usually with 1 spikelet per node, occasionally with 2 at a few nodes; **internodes** 4–6(9.5) mm long, 0.5–1.2 mm wide, smooth or scabrous, glabrous, evenly puberulent, or sparsely pilose, hairs to 0.3 mm. **Spikelets** 10–27 mm, appressed to ascending, with 4–7 florets; **disarticulation** above the glumes, beneath each floret. **Glumes** oblong, glabrous, keeled distally, keels inconspicuous and smooth proximally, scabrous and conspicuous distally, lateral veins inconspicuous, hyaline margins present in the distal ¹/₂, apices acute, unawned or awned, awns to 3 mm; **lower glumes** 8.8–11.4 mm, 3–6-veined; **upper glumes** 7–12 mm, 5–7-veined; **lemmas** 8–12 mm, glabrous, mostly smooth, sometimes scabridulous distally, unawned or with a 0.2–4(10) mm awn, awns straight; **paleas** 7–9.5 mm, keels ciliate from ¹/₂ to almost the entire length, apices emarginate, truncate, or rounded; **anthers** 4–7 mm. 2*n* = 22, 42.

Elymus repens is native to Eurasia; it is now established through much of the *Flora* region, extending from Alaska to Greenland and south to California, Texas, and North Carolina. It grows well in disturbed sites, spreading rapidly via its long rhizomes, as well as by seed. It is also drought tolerant. Although it is listed a noxious weed in several states, it provides good forage. It differs from *E. hoffmannii* (see next) in having widely spaced, unequally prominent leaf veins and, usually, shorter awns.

Godley (1947) demonstrated that lemma awn development, glaucousness, and the pubescence of the rachises are each effectively controlled by single genes. Long-awned plants are homozygous recessive, and awn-tipped plants homozygous dominant; glaucousness is dominant over non-glaucousness, and glabrous rachises over pubescent rachises. Awned plants appear to be established along the coasts of Newfoundland and Nova Scotia. They have generally been identified as *Agropyron pungens* (Pers.) Roem. & Schult., a species that has obtuse, mucronate lemmas.

awn

floret

2 mm

1 cm

floret

glumes

2 mm

E. sierrae

1 cm

2 mm

floret

floret

2 mm

glume   glume

spikelet

2 mm

E. wawawaiensis

1 cm

2 mm

floret

glumes

2 mm

5 cm

rhizome

E. albicans

ELYMUS

Elymus repens is almost always a hexaploid. Most studies indicate that its genomic constitution is **StStH**, but Mason-Gamer (2001) demonstrated that it is genetically more complex than is implied by such a simple formula.

36. Elymus hoffmannii K.B. Jensen & Asay [p. 337]
HOFFMANN'S WHEATGRASS

Plants slightly to moderately rhizomatous. **Culms** 54–135 cm, glabrous. **Leaves** evenly distributed; **sheaths** glabrous; **auricles** absent or to 1 mm; **ligules** 0.6–1 mm, truncate, erose; **blades** 5–13 mm wide, flat to involute, abaxial surfaces smooth, glabrous, adaxial surfaces glabrous, veins closely spaced, all more or less equally prominent, smooth or scabrous. **Spikes** 10–50 cm long, 0.8–1.8 cm wide, with 1 spikelet per node, glabrous below the spikelets; **internodes** 5–8 mm long, about 0.2 mm thick, about 0.3 mm wide, both surfaces hairy, hairs 0.2–0.4 mm. **Spikelets** 15–27 mm, appressed to ascending, with 5–7 florets; **rachillas** scabridulous; **disarticulation** above the glumes, beneath the florets. **Glumes** equal, 5–11 mm long, 1.3–1.8 mm wide, stiff, lanceolate to linear-lanceolate, strongly rounded to keeled distally, keels inconspicuous and smooth on the proximal $^1/_3$–$^1/_2$, conspicuous and with a few teeth distally, lateral veins inconspicuous, hyaline margins 0.1–0.2 mm wide, apices acuminate to awned, awns to 8 mm; **lemmas** 7–12 mm, glabrous, smooth, apices unawned or awned, awns to 12 mm, straight; **paleas** ciliate on the keels, apices about 0.6 mm wide; **anthers** 4–7 mm. $2n = 42$.

Elymus hoffmannii was described from a breeding line of plants developed from seeds collected in Erzurum Province, Turkey by J.A. Hoffmann and R.J. Metzger (Jensen & Asay 1996). There is no information available about its native distribution. As indicated in the key, *E. hoffmannii* differs from *E. repens* (see previous) primarily in its evenly prominent, closely spaced leaf veins and, usually, in having longer awns.

The description of *Elymus hoffmannii* was explicitly written to encompass the cultivar 'NewHy' that is derived from an artificial cross between *E. repens* and *Pseudoroegneria spicata* (p. 281). Because of its morphological similarity to plants obtained from the Turkish seed, Jensen and Asay suggested that *E. hoffmannii* had a similar parentage. 'NewHy' was released as a cultivar in the 1980s. Its distribution within the *Flora* region is not known.

37. Elymus tsukushiensis Honda [p. 337]

Plants loosely cespitose, without conspicuous rhizomes. **Culms** 25–100 cm tall, 1.3–3.5 mm thick, erect; **nodes** 4–6, glabrous. **Leaves** basal and cauline; **sheaths** glaucous, glabrous or with hairs, margins glabrous or ciliate distally; **auricles** 1–2 mm; **ligules** 0.2–0.7 mm, truncate; **blades** 3–10 mm wide, flattish, often glaucous. **Spikes** (6.5)10–25 cm long, 1.4–4 cm wide including the awns, 0.7–20 cm wide excluding the awns, flexuous, nodding; **rachises** densely to sparsely hirsute on the edges, hairs about 0.2 mm, glabrous elsewhere, glaucous; **internodes** (5)8–20 mm. **Spikelets** 15–25 mm, loosely appressed or ascending, with 5–10 florets; **rachillas** hairy, hairs about 0.1 mm; **disarticulation** above the glumes, beneath the florets. **Glumes** lanceolate, tapering from about midlength, adaxial surfaces glabrous, hyaline margins about 0.1 mm wide, strongly keeled distally, midvein scabrous distally, other veins smooth or scabrous, apices acute to acuminate, sometimes awned, awns 2–5 mm; **lower glumes** 4–7 mm, 3–5-veined; **upper glumes** 5–8 mm, 5-veined; **calluses** glabrous; **lemmas** 8–12 mm, lanceolate, glabrous or pilose, apices acute, awned, awns 20–40 mm, straight or flexuous; **paleas** from slightly shorter than to longer than the lemmas, keels narrowly winged distally, not or scarcely extending beyond the intercostal region, distinctly outwardly curved below the apices, apices 0.3–0.5 mm wide; **anthers** 1.5–2.5 mm. $2n = 42$.

Elymus tsuskushiensis is native to northeastern China, Japan, and Korea. It was collected from ballast dumps in Portland, Oregon, but is not established in the *Flora* region. Hitchcock (1951) identified it and *E. ciliaris* as *Agropyron caninum* (L.) P. Beauv. [≡ *Elymus caninus*, p. 322], but that species has flatter glumes that are longer in relation to the lemmas than those of *E. tsuskushiensis*, and paleas with straight or slightly outwardly curved keels.

38. Elymus ciliaris (Trin.) Tzvelev [p. 339]

Plants loosely cespitose, without conspicuous rhizomes. **Culms** 30–130 cm tall, 1–5 mm thick, erect or weakly decumbent; **nodes** 3–4, glabrous, glaucous. **Leaves** basal and cauline; **sheaths** glaucous, glabrous or with hairs, lower sheaths sometimes hairy, upper sheaths glabrous, margins sometimes ciliate; **auricles** 1.5–2.5 mm; **ligules** about 0.3 mm; **blades** 10–25 cm long, 3–10 mm

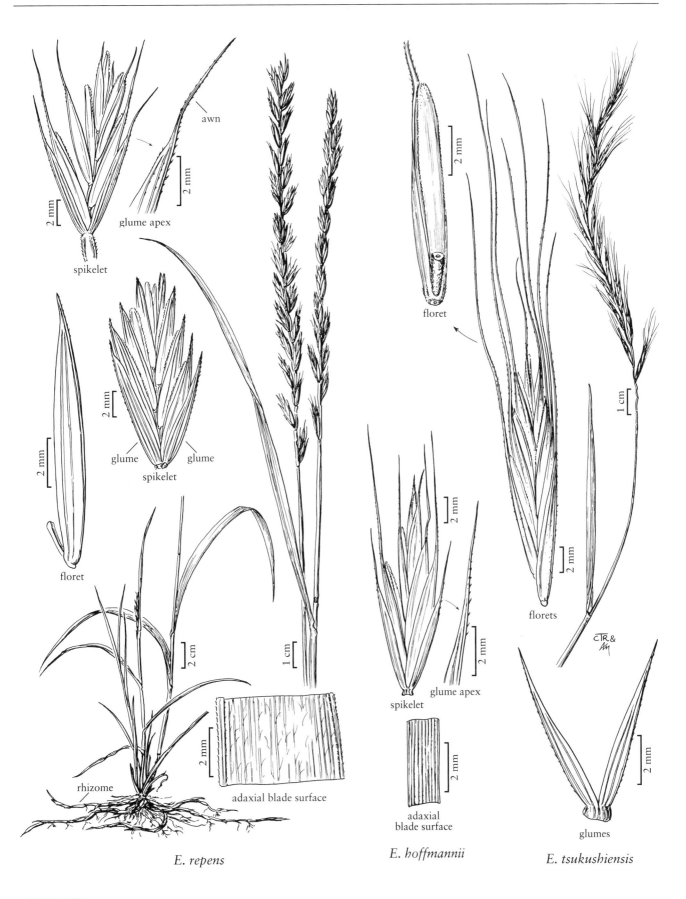

awn

2 mm

glume apex

2 mm

spikelet

2 mm

floret

2 mm

glume   glume

spikelet

2 mm

floret

floret

1 cm

2 mm

florets

1 cm

2 mm

spikelet

glume apex

adaxial blade surface

2 mm

adaxial
blade surface

2 mm

glumes

2 cm

1 cm

2 mm

rhizome

2 mm

adaxial blade surface

CTR &
AM

E. repens   *E. hoffmannii*   *E. tsukushiensis*

ELYMUS

wide, glabrous or pilose. **Spikes** 10–22 cm long, 1.5–2.5 cm wide including the awns, 0.8–1 cm wide excluding the awns, inclined to nodding, with 1 spikelet at all or most nodes; **rachises** scabrous on the edges, glabrous below the spikelets; **internodes** 10–25 mm. **Spikelets** 5–22 mm long, about 5 mm wide, appressed, with 4–12 florets; **disarticulation** above the glumes, beneath the florets. **Glumes** narrowly elliptic to lance-oblong, apices acute to acuminate; **lower glumes** 5–11 mm; **upper glumes** 7–13 mm; **lemmas** 7–12 mm, mostly glabrous, glabrate, or sparsely hairy, margins with coarse stiff hairs, hairs to 1 mm, apices abruptly narrowed, awned, awns 10–20 mm, scabrous, strongly outcurved to recurved; **paleas** $^2/_3$–$^4/_5$ the length of the lemmas, keels winged distally, distinctly outwardly curved below the apices, apices 0.5–0.6 mm wide, truncate to rounded; **anthers** about 2 mm. $2n = 28$. Genome StY.

Elymus ciliaris is native to northern China and Japan. It was collected from ballast dumps in Portland, Oregon, in 1899 and 1902; it is not established in the *Flora* region. A.S. Hitchcock identified both specimens on the sheet (US 1017954) as *Agropyron caninum* (L.) P. Beauv. [≡ *Elymus caninus*, p. 322], from which *E. ciliaris* differs in its short, rounded paleas and relatively short glumes with distinctly outwardly curving keels. The other specimen on that sheet is *E. tsuskushiensis* (p. 336).

39. Elymus semicostatus (Nees *ex* Steud.) Melderis [p. 339]

Plants cespitose, not rhizomatous. **Culms** 45–135 cm, erect or geniculately ascending, glabrous. **Sheaths** glabrous or villous; **auricles** to 1.5 mm; **ligules** 0.5–1.5 mm, truncate; **blades** 15–30 cm long, 4–12 mm wide, sometimes villous, adaxial surfaces smooth or scabrous, primary and secondary veins alternating. **Spikes** 8–30 cm long, 1–2 cm wide including the awns, 0.5–1 cm wide excluding the awns, erect or nodding, usually with 1 spikelet per node, sometimes with 2 spikelets at the lower nodes; **internodes** 10–20 mm long, about 0.8 mm wide, scabrous on the margins and on the surfaces, marginal prickles larger than those on the surfaces, hirtellous just below the spikelets. **Spikelets** 16–30 mm, loosely appressed, with 6–8 florets; **rachilla internodes** about 0.8 mm, strigose, hairs to about 0.3 mm; **disarticulation** above the glumes, beneath each floret. **Glumes** subequal, 10–18 mm long, 1.1–2 mm wide, elliptic-lanceolate, green, not keeled, 5–7-veined, veins more or less equally prominent, scabrous, apices acute to acuminate; **lemmas** 10–14 mm, scabrous or puberulent dorsally, awned, awns (4)12–18 mm, straight; **paleas** $^3/_4$ as long as to slightly shorter than the lemmas, keels outwardly curved below the apices, apices 0.3–0.7 mm wide, truncate; **anthers** 3–6 mm. $2n = 28$.

Elymus semicostatus is native to central Asia, from Afghanistan through Pakistan to northeastern India (Sikkim). Reports of its presence in the *Flora* region appear to be based on misidentifications.

Named hybrids

Elymus is notorious for its ability to hybridize. Most of its interspecific hybrids are partially fertile, permitting introgression between the parents. The descriptions provided below are restricted to the named interspecific hybrids. They should be treated with caution and some skepticism; some are based solely on the type specimen, because little other reliably identified material was available. Moreover, as the descriptions of the non-hybrid species indicate, many other interspecific hybrids exist.

The parentage of all hybrids is best determined in the field. Perennial hybrids, such as those in *Elymus*, can persist in an area after one or both parents have died out, but the simplest assumption is that both are present. Interspecific hybrids of *Elymus* that have disarticulating rachises presumably have *E. elymoides* or *E. multisetus* as one of their parents.

40. Elymus ×cayouetteorum (B. Boivin) Barkworth [p. 341]

Plants probably cespitose, not rhizomatous. **Culms** to 1 m tall, about 4 mm thick. **Leaves** not basally concentrated; **sheaths** smooth; **ligules** about 0.5 mm, glabrous; **blades** 20–30 cm long, about 10 mm wide, both surfaces scabrous. **Spikes** about 25 cm, lower nodes with 1 spikelet, most middle to upper nodes with 2; **internodes** about 18 mm. **Spikelets** about 40 mm including the awns, about 20 mm excluding the awns, appressed, with 5–8 florets. **Glumes** 12–15 mm, not or scarcely indurate, mostly smooth, veins scabrous, awns 3–5 mm; **lemmas** about 14 mm, glabrous, awns 18–22 mm, not to moderately divergent; **anthers** 1.8–2 mm, indehiscent.

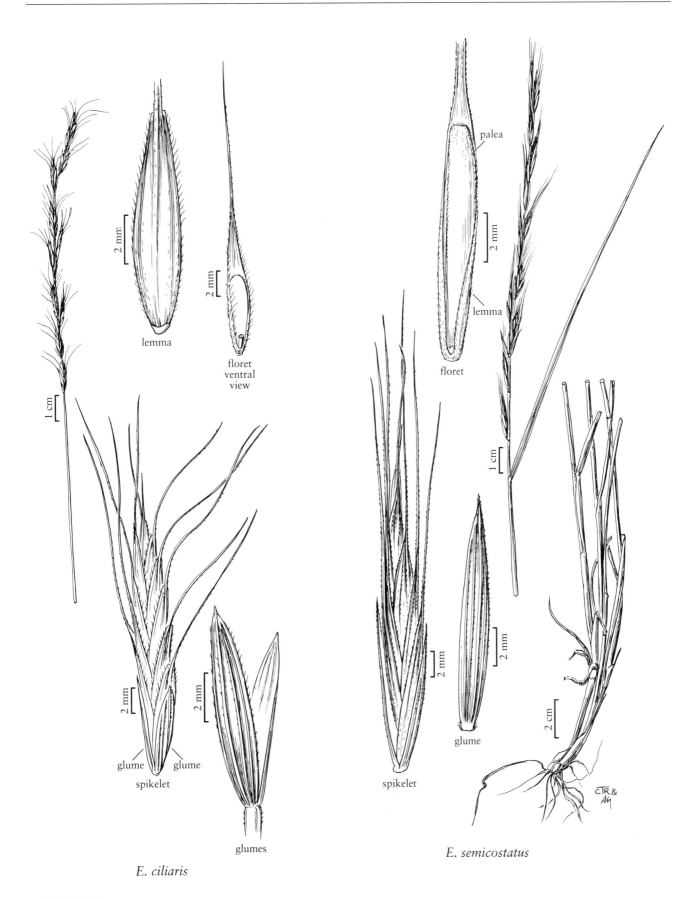

palea

lemma

floret

2 mm

lemma

floret
ventral
view

2 mm

2 cm

2 mm

spikelet

glume   glume

spikelet

glume

glumes

1 cm

E. ciliaris

E. semicostatus

CTR&
AM

ELYMUS

Elymus ×cayouetteorum consists of hybrids between *E. trachycaulus* (p. 321) and *E. canadensis* (p. 303). The above description is based on the type specimen, which was collected on the Îlets Jérémie, Quebec. It is not known how widespread such hybrids are.

41. Elymus ×palmerensis (Lepage) Barkworth & D.R. Dewey [p. 341]

Plants densely cespitose, shortly rhizomatous. **Blades** (2)4–8(15) mm wide, abaxial surfaces scabrous, adaxial surfaces pilose. **Spikes** 14–30 cm long, 1–2.5 cm wide including the awns, 0.3–0.8(1.5) cm wide excluding the awns, drooping, with 1–2 spikelets per node; **internodes** 5–20 mm, scabrous on the angles. **Spikelets** with 3–9 florets. **Glumes** 4–10.5 mm long, 0.8–1.5 mm wide, oblong to lanceolate, scabrous, gradually or abruptly narrowing in the distal $^{1}/_{3}$–$^{1}/_{4}$, 3(4)-veined, margins scarious, apices unawned or awned, awns to 15 mm; **rachillas** hairy; **lemmas** 8.5–15 mm, hairy, conspicuously keeled distally, awned, awns 3–10 mm; **paleas** 8.5–15 mm, retuse or truncate; **anthers** 1–2 mm.

Elymus ×palmerensis is the name for hybrids between *E. macrourus* (p. 324) and *E. sibiricus* (p. 310). It is known from disturbed sites around Palmer, Alaska, and in south-central Alaska. Bowden (1967) also reported it from Fort Liard, in the MacKenzie District, Northwest Territories. Lepage (1952) originally identified the parents as *Agropyron sericeum* [= *E. macrourus*] and *E. canadensis* (p. 303). Later, Lepage (1965) stated that the second parent was *E. sibiricus*. The above description includes *×Agroelymus hodgsonii* Lepage, which, according to Bowden (1967), is a synonym.

42. Elymus ×pseudorepens (Scribn. & J.G. Sm.) Barkworth & D.R. Dewey [p. 341]
FALSE QUACKGRASS

Plants rhizomatous. **Culms** 30–100 cm, ascending or erect, glabrous, mostly smooth, sometimes scabrous below the nodes. **Leaves** evenly distributed; **sheaths** glabrous; **ligules** to 0.5 mm; **blades** 10–25 cm long, 2–7 mm wide, involute when dry, both surfaces scabrous, adaxial surfaces sparsely pilose, hairs 0.7–1 mm. **Spikes** 5.5–13 cm long, 0.4–0.6 cm wide, with 1 spikelet per node; **internodes** 3.5–5 mm. **Glumes** 8–18 mm, equaling or exceeded by the adjacent lemmas, more or less flat, 5–9-veined, margins unequal, the wider margins to about 0.3 mm, narrowly acute, sometimes awned, awns to 1 mm; **lemmas** 7.5–15 mm, smooth and glabrous proximally, scabrous distally, mucronate or awned, awns to 3 mm; **anthers** 1.5–2 mm.

Elymus ×pseudorepens consists of hybrids between *E. lanceolatus* (p. 327) and *E. trachycaulus* (p. 321). It appears to be fairly common, having been reported from Alberta to Michigan and south to Arizona, New Mexico, and Arkansas.

43. Elymus ×yukonensis (Scribn. & Merr.) Á. Löve [p. 341]

Plants rhizomatous. **Culms** about 60 cm, erect. **Leaves** somewhat basally concentrated; **blades** 3.5–6 mm wide. **Spikes** 6–10 cm long, 0.7–1.7 cm wide, with 1 spikelet per node; **internodes** 6–11 mm. **Glumes** 4–5.5 mm, about $^{1}/_{2}$ the length of the adjacent lemmas, lanceolate, flat, hairy, apices acute or awn-tipped, awns shorter than 1 mm; **rachillas** densely hairy; **lemmas** 6–9 mm, densely villous, unawned; **anthers** 3.3–3.6 mm.

The parents of *Elymus ×yukonensis* have not been identified. Morphological and geographic considerations suggest that they may be *E. lanceolatus* subsp. *psammophilus* (p. 327) and *E. alaskanus* (p. 326).

44. Elymus ×saundersii Vasey [p. 342]

Plants cespitose, not rhizomatous. **Culms** 50–80 cm, erect, glabrous or pilose. **Leaves** somewhat basally concentrated; **sheaths** retrorsely hairy; **auricles** to 1 mm; **ligules** 0.5–1 mm; **blades** 10–15 cm long, 4–5 mm wide, flat, becoming involute when dry, tapering to the apices. **Spikes** 10–25 cm long, 1–2.5 cm wide, with 1 spikelet per node; **internodes** 4–5 mm; **disarticulation** at the rachis nodes. **Spikelets** 10–25 mm excluding the awns, with 3–6 florets. **Glumes** 6–8 mm, linear-lanceolate to lanceolate, 3–5-veined, veins scabrous, apices sometimes toothed, awned, awns 8–13 mm; **lemmas** 10–13 mm, glabrous, smooth proximally, scabrous distally, awned, awns 20–45 mm, outcurving; **anthers** 1.5–2 mm, usually indehiscent.

Elymus ×saundersii comprises hybrids between *E. trachycaulus* (p. 321) and *E. elymoides* (p. 318). Such hybrids are found throughout much of the western portion of the contiguous United States, mostly in disturbed areas. The hybrids are generally sterile and, as in all hybrids involving *E. elymoides* or *E. multisetus* (p. 318), the rachises disarticulate at maturity.

45. Elymus ×hansenii Scribn. [p. 342]

Plants cespitose, not rhizomatous. **Culms** 60–120 cm. **Leaves** evenly distributed; **sheaths** smooth; **ligules** to 1 mm; **blades** 10–30 cm long, 2–8 mm wide, flat or the margins involute. **Spikes** 5–20 cm, straight or nodding, with 2+ spikelets per node; **internodes** about 10 mm; **disarticulation** in the rachises. **Spikelets** about 15 mm, with 3–5 florets. **Glumes** narrowly lanceolate, 2–3-veined, awned, awns 25–35 mm; **lemmas** 10–12 mm, awned, awns 40–50 mm, outcurving; **paleas** subequal to the lemmas, truncate or bidentate.

Elymus ×hansenii refers to hybrids between *E. glaucus* (p. 306) and either *E. elymoides* or *E. multisetus* (p. 318). It is not clear which of the latter two species is involved. It is a fairly common hybrid in those parts of western North America where both parents grow. The

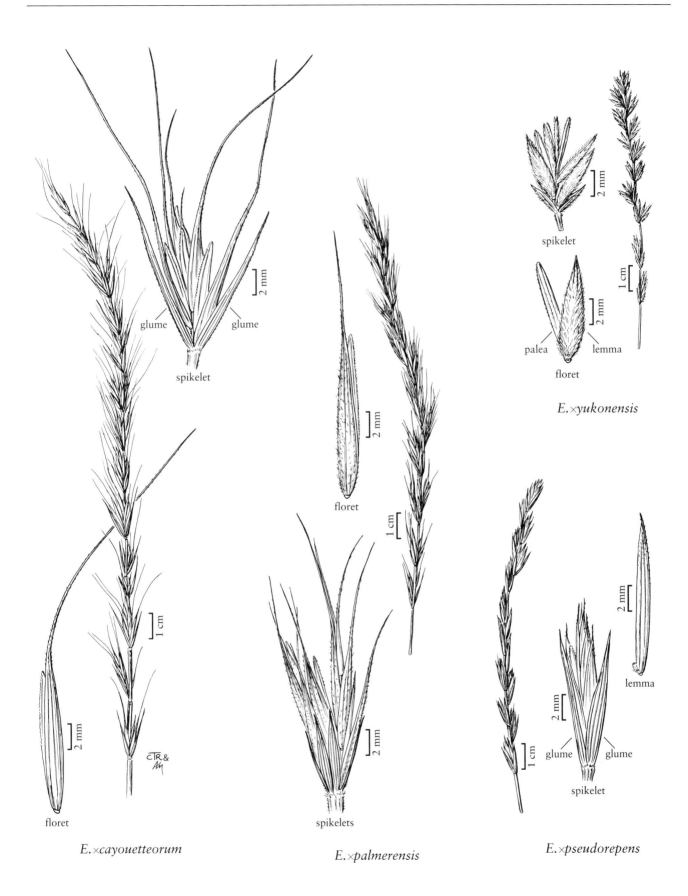

spikelet

glume   glume

spikelet

floret

floret

E.×cayouetteorum

floret

spikelets

E.×palmerensis

spikelet

palea   lemma

floret

E.×yukonensis

lemma

glume   glume

spikelet

E.×pseudorepens

ELYMUS NAMED HYBRIDS

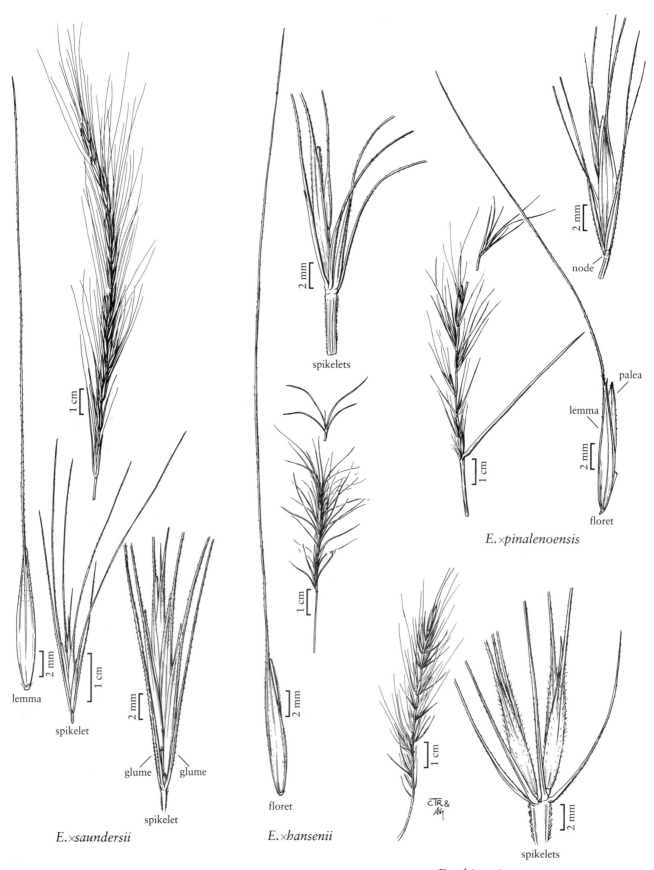

E.×*pinalenoensis*

palea

lemma

floret

lemma

spikelet

glume   glume

spikelet

E.×*saundersii*

ELYMUS NAMED HYBRIDS

spikelets

floret

E.×*hansenii*

node

spikelets

E.×*ebingeri*

glumes of the type specimen are as wide as those in *E. glaucus*, and some are divided longitudinally, as in *E. elymoides* and *E. multisetus*. As in other hybrids involving *E. elymoides* and *E. multisetus*, the rachis of *E.* ×*hansenii* disarticulates at maturity.

46. Elymus ×pinalenoensis (Pyrah) Barkworth & D.R. Dewey [p. 342]

Plants cespitose, not rhizomatous. **Culms** 60–90 cm tall, about 2 mm thick. **Leaves** somewhat basally concentrated; **sheaths** smooth or scabridulous; **auricles** to 1.5 mm; **ligules** 0.5–0.7 mm, rounded; **blades** 2–2.5 mm wide, those of the basal leaves 7–15 cm long, those of the flag leaves 2–7 cm long, abaxial surfaces smooth or scabridulous, glabrous or sparsely hairy basally, veins not evident, adaxial surfaces scabrous, sometimes with scattered hairs, veins prominently and equally ribbed. **Spikes** 8–14 cm long, about 5 cm wide including the awns, about 1 cm wide excluding the awns, nodding, lower nodes with 1–2 spikelets, upper nodes with 1 spikelet; **internodes** about 10 mm; **disarticulation** in the rachises and beneath each floret. **Glumes** 22–31 mm including the awns, glume bodies 5–10 mm, midvein evident, scabrous, awns 17–26 mm, divergent, scabrous; **lemmas** 11–13 mm, smooth, awns 20–40 mm, divergent, sometimes strongly divergent; **anthers** about 2 mm long, 1–1.5 mm thick.

Elymus ×*pinalenoensis* consists of hybrids between *E. elymoides* subsp. *brevifolius* (p. 319) and *E.*

arizonicus (p. 329) (Pyrah 1983). It has been found in the Pinaleno and Santa Catalina mountains of Graham County, Arizona, in areas disturbed by logging, road building, summer home development, and recreation.

47. Elymus ×ebingeri G.C. Tucker [p. 342]

Plants cespitose, not rhizomatous. **Culms** 50–135 cm, usually smooth, glabrous, sometimes retrorsely hairy; **nodes** glabrous or hairy. **Leaves** evenly distributed; **sheaths** smooth or scabridulous, hairy; **ligules** to 1 mm; **blades** 14–26 cm long, to 12 mm wide, flat, both surfaces scabridulous, adaxial surfaces hairy. **Spikes** 8–17 cm, with (1)2+ spikelets per node; **internodes** 4–8 mm. **Spikelets** diverging at about 45° from the rachises, not patent. **Glumes** usually subequal, 15–20 mm, bases indurate, bodies lanceolate to subulate, scabrous, sometimes the lower glumes reduced to a stub; **lemmas** 8–10 mm, glabrous or strigose, sometimes scabrous, awns 23–29 mm, straight, scabrous; **anthers** 2–2.9 mm. **Caryopses** seldom formed.

Elymus ×*ebingeri* is the name for hybrids between *E. virginicus* (p. 298) and *E. hystrix* (p. 316). It is frequently found where the two parental species grow together, often with later hybrid generations and introgressants to the two parents. It has been reported from southern Ontario, and from Wisconsin to New York and Illinois. Most published reports simply refer to the existence of these hybrids, the name itself not having been published until 1996.

13.14 ×ELYLEYMUS B.R. Baum

Mary E. Barkworth

Plants perennial; sometimes rhizomatous. **Culms** 40–235 cm, erect. **Inflorescences** usually spikes, sometimes spikelike racemes, 5–35 cm, erect, with 1–3 spikelets per node, pedicels, when present, to 3 mm. **Spikelets** with 2–8 florets; **disarticulation** usually above the glumes and beneath the florets, sometimes below the glumes, sometimes in the rachises, usually tardy. **Glumes** linear to lanceolate, often awn-tipped; **lemmas** 6–25 mm, glabrous or hairy, usually awned, awns to 15 mm; **anthers** 1.5–5 mm.

×*Elyleymus* consists of hybrids between *Elymus* and *Leymus*. So far as is known, they are completely sterile, having thin anthers (usually less than 0.5 mm thick) and failing to develop mature caryopses. Only the named hybrids are accounted for below. Each of the entities appears to be distinct, but identification of the parents is, in some instances, tentative. The descriptions are offered with considerable reservation, some being based solely on the type material. All the illustrations are based on type specimens.

The hybrids fall into two groups. Those with *Leymus mollis* as the *Leymus* parent (species 7–11) tend to have wider and flatter glumes than those with one of the inland species of *Leymus* as the *Leymus* parent.

Unless stated otherwise, measurements of the spikes include the awns; measurements of the spikelets, glumes, and lemmas do not. No attempt has been made to develop distribution maps.

SELECTED REFERENCES **Barkworth, M.E.** 2006. A new hybrid genus and 12 new combinations in North American grasses. Sida 22:495–501; **Bowden, W.M.** 1959. Chromosome numbers and taxonomic notes on northern grasses: I. Tribe Triticeae. Canad. J. Bot. 37:1143–1151; **Bowden, W.M.** 1967. Cytotaxonomy of intergeneric hybrids of the tribe Triticeae from North America. Canad. J. Bot. 45:711–724; **Dewey, D.R.** and **A.H. Holmgren.** 1962. Natural hybrids of *Elymus cinereus* and *Sitanion hystrix*. Bull. Torrey Bot. Club 89:217–228; **Lepage, E.** 1952. Études sur quelques plantes américaines: II. Hybrides intergénériques; *Agrohordeum* et *Agroelymus*. Naturaliste Canad. 79:241–266.

1. ×**Elyleymus turneri** (Lepage) Barkworth & D.R. Dewey [p. 345]

Plants not cespitose, rhizomatous. **Culms** 110–130 cm, pubescent below the spikes. **Leaves** glaucous; **auricles** present, at least on the innovations; **blades** 3–5 mm wide, abaxial surfaces scabrous or smooth, adaxial surfaces scabrous. **Inflorescences** spikelike racemes, 11–20 cm long, lax, with (1)2 spikelets per node, spikelets at the lower nodes unequally pedicellate, longer pedicels 1.5–3 mm; **internodes** 6–15(30) mm, those at the base of the spike longest, angles hispid, convex surfaces pubescent or pilose. **Spikelets** 15–25 mm, with 5–8 florets; **disarticulation** above or below the glumes. **Glumes** unequal, 2–17 mm long, 0.5–1.2 mm wide, subulate to narrowly lanceolate, scabrous or pilose, (0)1–3-veined; **rachilla internodes** pubescent; **lemmas** 6–16 mm, pubescent, pilose, or villous, backs often glabrate, apices awned, awns 0.7–8 mm; **paleas** 6.5–10 mm, puberulent or pubescent between the veins; **anthers** (2.8)3.5–5 mm. 2*n* = 28.

×*Elyleymus turneri* is treated here as the name for hybrids between *Elymus lanceolatus* and *Leymus innovatus*, in agreement with Bowden (1952). The type material was found on the banks of the Saskatchewan River, 2 miles below Fort Saskatchewan, Alberta. Lepage (1952) noted that Dr. Turner, the collector, reported that both *Agropyon smithii* [≡ *Pascopyrum smithii*] and *A. dasystachyum* [= *Elymus lanceolatus*] grew in the vicinity, and argued for *A. smithii* as the *Elymus* parent. Bowden (1959) reported 2*n* = 28 in specimens from the type locality, and used this to argue for the *Elymus* parent being either *A. dasystachyum* [= *Elymus lanceolatus*] or *A. trachycaulum* [≡ *Elymus trachycaulus*].

2. ×**Elyleymus aristatus** (Merr.) Barkworth & D.R. Dewey [p. 345]

Plants not or shortly rhizomatous. **Culms** 60–130 cm, glabrous. **Leaves** evenly distributed on the culms; **sheaths** smooth, glabrous; **auricles** poorly developed, to 0.5 mm; **ligules** 1–2.5 mm, scarious, rounded; **blades** about 5.5 mm wide, abaxial surfaces glabrous, mostly smooth, scabrous near the margins, adaxial surfaces scabridulous, primary veins separated by about 3 secondary veins. **Inflorescences** spikes, 6–15 cm long, 10–15 mm wide including the awns, 7–10 mm wide excluding the awns, erect, with 2–3 sessile or subsessile spikelets per node; **internodes** 4–7 mm, concealed by the spikelets; **disarticulation** tardy, in the rachises and beneath the florets. **Spikelets** 10–20 mm excluding the awns, to 18 mm including the awns, with 3–4 florets. **Glumes** 8–15 mm long, 0.3–0.5 mm wide, subequal to unequal, scabrous; **lemmas** 7.5–12 mm, glabrous, smooth or scabrous, sometimes only scabrous distally, midveins prominent and scabrous distally, awns 4–5 mm; **anthers** 2.2–2.4 mm.

Dewey and Holmgren (1962) argued that ×*Elyleymus aristatus* comprises hybrids between *Elymus elymoides* and *Leymus cinereus* or *L. triticoides*. It has been found at many locations where the parents are sympatric.

3. ×**Elyleymus colvillensis** (Lepage) Barkworth [p. 345]

Plants loosely cespitose, shortly rhizomatous. **Culms** about 60 cm, glabrous. **Auricles** often present; **ligules** 0.3–1.2 mm, truncate; **blades** 2–4 mm wide, scabrous on the adaxial surfaces. **Inflorescences** spikes, 6.5–12 cm, somewhat lax, with 1(2) sessile or subsessile spikelets per node; **internodes** 4–9 mm, angles pilose or hispid, convex surfaces sparsely villous distally. **Spikelets** 10–15 mm, with 3–5 florets; **disarticulation** above the glumes, beneath the florets. **Glumes** 6.5–11.5 mm long, 1.1–1.5 mm wide, narrowly lanceolate to subulate, 1–3(4)-veined, margins narrow, scarious; **lemmas** 8–11 mm, villous, slightly to strongly keeled, awned, awns 2–10 mm; **paleas** glabrous or puberulent between the veins, apices slightly dentate or retuse; **anthers** 1.7–2.2 mm.

×*Elyleymus colvillensis* consists of hybrids between *Leymus innovatus* and, probably, *Elymus alaskanus*. The original collections were made on the banks of the Colville River at Umiat, Alaska. It is not known how widely it is distributed.

4. ×**Elyleymus hirtiflorus** (Hitchc.) Barkworth [p. 347]

Plants rhizomatous. **Culms** 40–90 cm tall, 2–2.5 mm thick, glabrous. **Leaves** somewhat basally concentrated; **sheaths** smooth, glabrous; **auricles** 0.3–0.5 mm; **ligules** 0.2–0.5 mm, truncate; **blades** 5–20 mm long, 2–4 mm wide, usually involute, sometimes flat, abaxial surfaces smooth, glabrous or sparsely hairy, particularly towards the base, veins not prominent, adaxial surfaces scabridulous or scabrous, varying within a plant, all veins equally prominent, apices narrowly acute. **Inflorescences** spikes, 5–18 cm long, 8–10 mm wide, with 1–2 sessile or subsessile spikelets per node; **internodes** 4–5 mm, partially exposed on the sides, hairy. **Spikelets** 11–14 mm, to 23 mm including the awns, with 3–6 florets; **disarticulation** above the

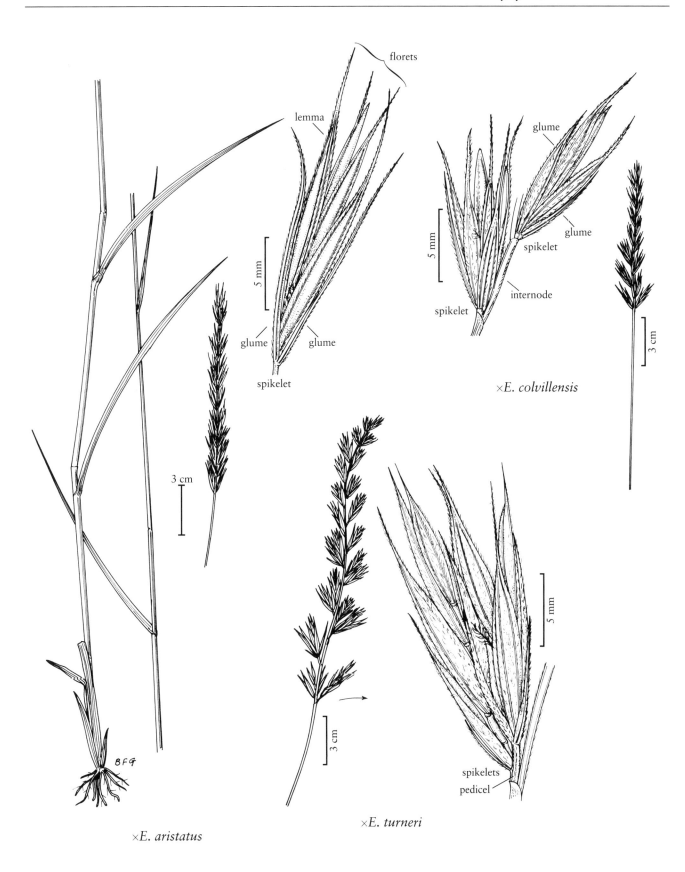

florets

lemma

glume

glume

5 mm

spikelet

glume

glume

spikelet

internode

spikelet

3 cm

×*E. colvillensis*

3 cm

BFG

×*E. aristatus*

3 cm

5 mm

spikelets

pedicel

×*E. turneri*

×ELYLEYMUS

glumes, beneath the florets. **Glumes** 12–17 mm long including the awnlike apices, 1–3 mm wide, widest at about ¼ length, 1(3)-veined, keeled, sparsely to densely hairy, hairs 0.3–0.5 mm; **lemmas** 8.5–10 mm, sparsely to densely hairy, hairs 0.3–0.5 mm, apices awned, awns 5–10 mm, straight; **anthers** 1.8–2 mm long.

Bowden (1967) suggested that ×*Elyleymus hirtiflorus* consisted of hybrids between *Elymus trachycaulus* and *Leymus innovatus*, and included in it plants from British Columbia. The name, however, is based on collections from the banks of the Green River, Wyoming, where neither putative parent grows. The more likely parents are *E. lanceolatus* and *L. simplex*. Admittedly, the short anthers argue for *E. trachycaulus* rather than *E. lanceolatus* as the *Elymus* parent. The Canadian specimens are here treated as belonging to ×*Elyleymus ontariensis*, a name that Bowden treated as a synonym of ×*Elyleymus hirtiflorus*.

5. ×Elyleymus mossii (Lepage) Barkworth [p. 347]

Plants cespitose, sometimes rhizomatous. **Culms** to 80 cm, glabrous. **Sheaths** smooth, glabrous; **auricles** 0.5–1.5 mm; **ligules** 0.5–1.2 mm; **blades** 3–7 mm wide, flat, adaxial surfaces and margins scabrous. **Inflorescences** spikes, 8–15 cm long, 10–25 mm wide including the awns, 8–15 mm wide excluding the awns, with (1)2 sessile or subsessile spikelets per node; **internodes** 4–6 mm, scabrous or hispid on the angles. **Spikelets** 12–17 mm, usually with 5 florets. **Glumes** 9–15 mm long including the awns, 0.8–1 mm wide, 1–3-veined, hairy; **lemmas** 8–13 mm, villous, awned, awns 4–15 mm; **paleas** 8–10 mm, glabrous between the keels, margins shortly ciliate; **anthers** 2–3 mm.

Lepage (1965) stated that it was obvious that *Elymus canadensis* was one parent of this hybrid, but that it would be necessary to discover which species of *Agropyron* [in the traditional sense] grew in the neighborhood to determine the other parent. He gave "*Agropyron* (?) *trachycaulum*" as a possibility. *Elymus canadensis*, however, is generally absent from the region around Lake Louise, Alberta (Moss 1983), where the holotype was collected. Barkworth (2006) argued that the parents are probably *E. glaucus* and *Leymus innovatus*, both species that are common in the holotype area.

6. ×Elyleymus ontariensis (Lepage) Barkworth [p. 347]

Plants rhizomatous. **Culms** to 75 cm tall, 2–2.5 mm thick, clustered, glabrous. **Leaves** more or less evenly distributed; **sheaths** glabrous, smooth; **auricles** absent or to 0.3 mm; **ligules** about 0.5 mm, truncate; **blades** 10–25 cm long, 2–3 mm wide, involute, abaxial surfaces scabridulous, glabrous, veins not evident, adaxial surfaces smooth or scabridulous, glabrous, veins all equally prominent, apices narrowly acuminate.

Inflorescences spikes, 8–12 cm long, 6–12 mm wide, with 1(2) sessile or subsessile spikelets per node; **internodes** 6–9 mm, partly exposed, angles hairy, hairs 0.4–0.6 mm; **disarticulation** above the glumes, beneath the florets. **Spikelets** 15–20 mm, with 3–5 florets. **Glumes** 9.5–16 mm long including the awns, 0.7–1.3 mm wide, lanceolate, widest about midlength, 3-veined, hairy or scabrous over the veins, awns 1–1.5 mm; **lemmas** 10–15 mm, hairy, hairs 0.3–0.4 mm, awns 3–7 mm; **anthers** about 2.8 mm.

×*Elyleymus ontariensis*, according to Bowden (1967), comprises hybrids between *Elymus trachycaulus* and *Leymus innovatus*. It differs from ×*Elyleymus hirtiflorus* in having wider, more parallel-sided glumes and longer rachis internodes.

7. ×Elyleymus uclueletensis (Bowden) B.R. Baum [p. 349]

Plants rhizomatous. **Culms** 170–235 cm tall, 4–5 mm thick, smooth, mostly glabrous, hairy for 12–15 cm below the spikes, hairs 0.1–0.2 mm. **Leaves** evenly distributed on the culms; **sheaths** smooth, glabrous: **auricles** absent; **ligules** 0.5–0.7 mm, truncate; **blades** 6–10 mm wide, smooth, glabrous, abaxial surfaces with the primary veins evident, adaxial surfaces with the primary and secondary veins evident. **Inflorescences** spikes, 25–35 cm long, about 15 mm thick, with 2 sessile or subsessile spikelets per node; **internodes** 7–14 mm. **Spikelets** 14–20 mm excluding the awns, to 25 mm including the awns, with 2–3 florets; **disarticulation** above the glumes, beneath the florets. **Glumes** 12–25 mm long including the awns, 0.8–2 mm wide, widest near midlength, flexible, hairy, hairs about 0.4 mm, bases indurate for about 0.3 mm, 3–5 veins evident at midlength, margins about 0.5 mm wide, apices tapering into awns, awns 1–3 mm; **lemmas** 20–25 mm, hairy, hairs about 0.4 mm, apices awned, awns 2.5–9(11) mm; **anthers** 2.5–3.5 mm.

×*Elyleymus uclueletensis* comprises hybrids between *Leymus mollis* and *Elymus glaucus*. It is known from two locations, near Ucluelet and along Gold River, both on the west coast of Vancouver Island, British Columbia.

8. ×Elyleymus aleuticus (Hultén) B.R. Baum [p. 349]

Plants rhizomatous. **Culms** to 75 cm tall, mostly glabrous, hairy for about 1 cm below the spikes, hairs about 0.1 mm. **Blades** to 15 mm wide, adaxial surfaces with the primary and secondary veins evident, subequal. **Inflorescences** spikes, to 15 cm long, 20 mm wide, erect, with (1)2 subsessile spikelets per node, pedicels to 1 mm; **internodes** about 5 mm. **Spikelets** with 3–5 florets; **disarticulation** above the glumes, beneath the florets. **Glumes** 10.5–13 mm long, 1.3–2(2.3) mm wide, linear-lanceolate to linear, flexible, hairy, 3–4(5)-veined, veins

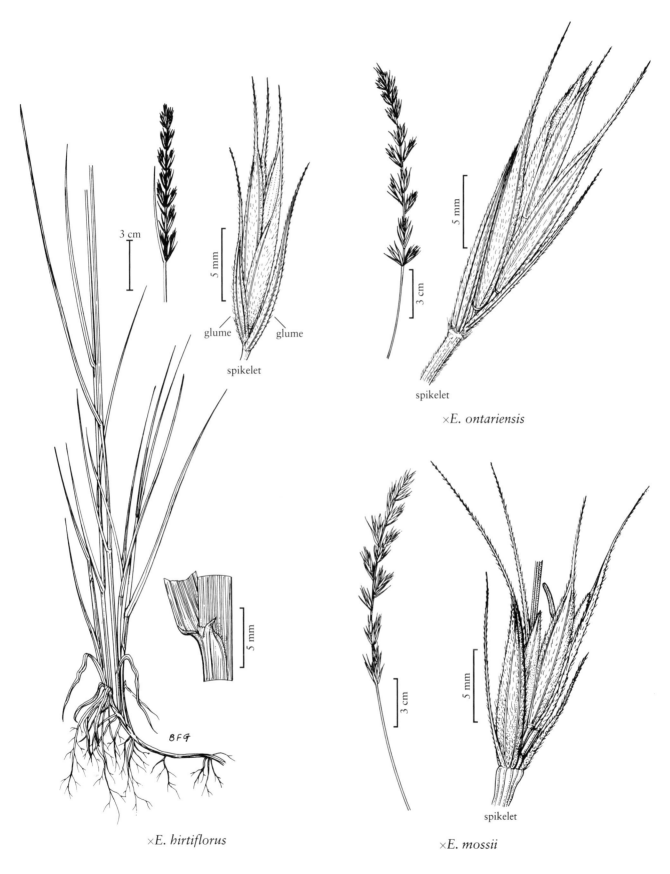

glume   glume

spikelet

×*E. hirtiflorus*

×*E. ontariensis*

spikelet

×*E. mossii*

spikelet

BFG

×ELYLEYMUS

prominent, apices unawned or awned, awns to 1.2 mm; **lemmas** 15–18 mm, lemmas awned, awns 1–7 mm, more or less straight; **anthers** 2.3–3.9 mm.

×*Elyleums aleuticus* comprises hybrids between *Elymus hirsutus* and *Leymus mollis*. It is known only from the type locality, Atka, Alaska. It probably occurs at other locations where the two parents are sympatric.

9. ×**Elyleymus hultenii** (Melderis) Barkworth [p. 349]

Plants cespitose, shortly rhizomatous. **Culms** to 60 cm, glabrous, hairy below the spikes, nodes shortly hairy. **Sheaths** smooth, glabrous; **auricles** absent or to 0.5 mm; **ligules** 0.2–0.8 mm, truncate; **blades** 3–4 mm wide, adaxial surfaces densely strigulose over the veins, all veins equally prominent. **Inflorescences** spikes, 9–17 cm long, 8–10 mm wide, erect, with solitary, sessile or subsessile spikelets. **Spikelets** 15–20 mm, with 4–6 florets. **Glumes** 9–11 mm long, 1.5–3 mm wide, with 4–5 veins, hairy, acute, unawned; **lemmas** 10–12 mm, hairy, unawned or awned, awns to about 1 mm; **paleas** with short and long hairs on the keels; **anthers** 3–4 mm.

×*Elyleymus hultenii* consists of hybrids between *Elymus alaskanus* subsp. *alaskanus* and *Leymus mollis*. The original collection is from Deering, Alaska.

10. ×**Elyleymus jamesensis** (Lepage) Barkworth [p. 350]

Plants rhizomatous. **Culms** 90–130 cm tall, about 5 mm thick. **Leaves** somewhat basally concentrated; **sheaths** smooth, glabrous; **auricles** well developed; **ligules** about 0.3 mm on the lower leaves, to 3 mm on the upper leaves; **blades** 5–6 mm wide, veins equally prominent on the adaxial surfaces. **Inflorescences** spikes, 15–25 cm long, 6–12 mm wide, with 1–2 sessile or subsessile spikelets per node; **internodes** 8–15 mm, angles scabrous or hispid. **Spikelets** 16–28 mm, with 3–4(5) florets. **Glumes** 12–20 mm long, 2–3 mm wide, 3–5-veined, scabrous and sparsely hairy, tapering gradually from about midlength, margins scarious, apices subulate or awned, awns to 4 mm; **rachilla internodes** pubescent or villous; **lemmas** 8.7–16 mm, with appressed hairs at the base, glabrous elsewhere, awned, awns 1–3 mm; **paleas** with ciliate keels and retuse to bidentate apices; **anthers** 2–3 mm.

×*Elyleymus jamesensis* comprises hybrids between *Elymus trachycaulus* and *Leymus mollis*. Lepage (1952) recognized three different infraspecific taxa, depending on the variety of *Agropyron trachycaulum* [≡ *Elymus trachycaulus*] involved. Because all the varieties concerned are treated here as part of *E. trachycaulus* subsp. *trachycaulus*, no attempt has been made to distinguish Lepage's infraspecific taxa. The above description includes ×*Agroelymus adamsii* J. Rousseau, which, according to Bowden (1967), is a synonym of ×*E. jamesensis*.

11. ×**Elyleymus ungavensis** (Louis-Marie) Barkworth [p. 350]

Plants rhizomatous. **Culms** to about 100 cm tall, 3–5 mm thick, glabrous, lower internodes glaucous. **Leaves** somewhat basally concentrated; **sheaths** smooth, glabrous; **auricles** present on some basal leaves, to 0.4 mm; **ligules** about 0.5 mm, truncate; **blades** of the culm leaves 5–7 mm wide, abaxial surfaces smooth, glabrous, primary and secondary veins evident, adaxial surfaces scabrous, glabrous, all veins more or less equally prominent, blades of the innovations to 3 mm wide. **Inflorescences** spikes, to 18 cm long, about 10 mm wide, with 1 sessile or subsessile spikelet per node; **internodes** 7–9 mm, mostly glabrous, angles hairy, hairs about 0.3 mm. **Spikelets** 15–25 mm, with 5–7 florets; **disarticulation** above the glumes, beneath the florets. **Glumes** 15–20 mm long including the awnlike tip, 2.5–4 mm wide, widest at or just beyond midlength, purplish, sparsely scabrous to hairy over the veins, margins about 1 mm wide, apices acuminate, unawned; **lemmas** 15–17 mm, evenly hairy, hairs 0.4–0.5 mm, apices mucronate, mucros about 0.5 mm; **anthers** about 1.8 mm.

×*Elyleymus ungavensis* is a northern hybrid, collected along the sandy banks of the Koksoak River, near Fort Chimo [= Kuujjuaq], at the southern end of Ungava Bay, Quebec. It consists of hybrids between *Elymus violaceus* and *Leymus mollis* subsp. *mollis*. The involvement of *E. violaceus* is suggested by the wide glume margins; that of *L. mollis* subsp. *mollis* by the relatively thick culms.

13.15 PASCOPYRUM Á. Löve

Mary E. Barkworth

Plants perennial; rhizomatous. **Culms** 20–100 cm. **Leaves** basally concentrated; **sheaths** striate when dry, smooth, usually glabrous, rarely pilose; **auricles** present; **ligules** membranous. **Inflorescences** terminal, distichous spikes, spikelets usually 1 per node, occasionally in pairs at the lower nodes, spikelets at the lower 4–6 nodes often sterile; **lowest internodes** to 26 mm,

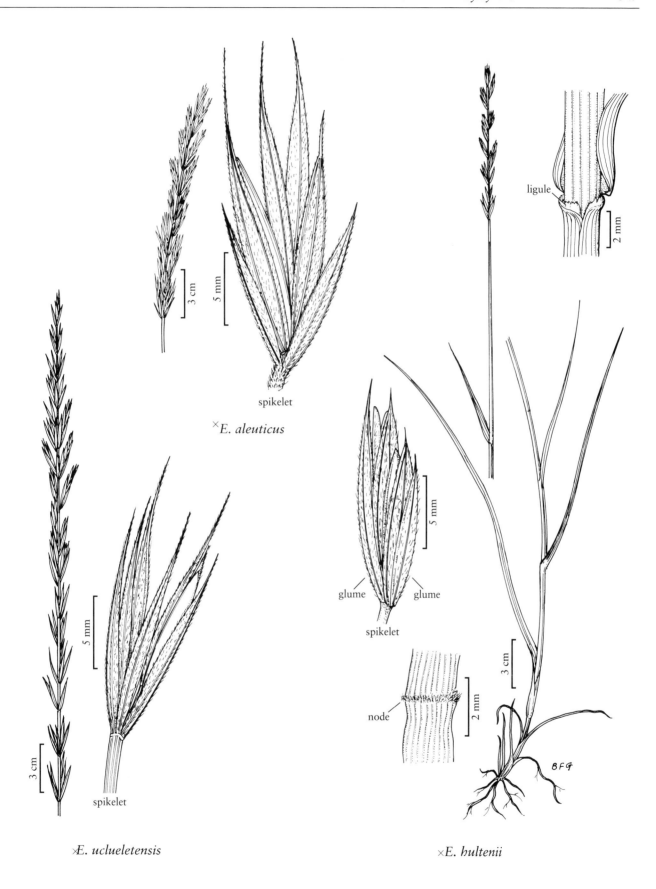

×*E. aleuticus*

×*E. uclueletensis*

×*E. hultenii*

×ELYLEYMUS

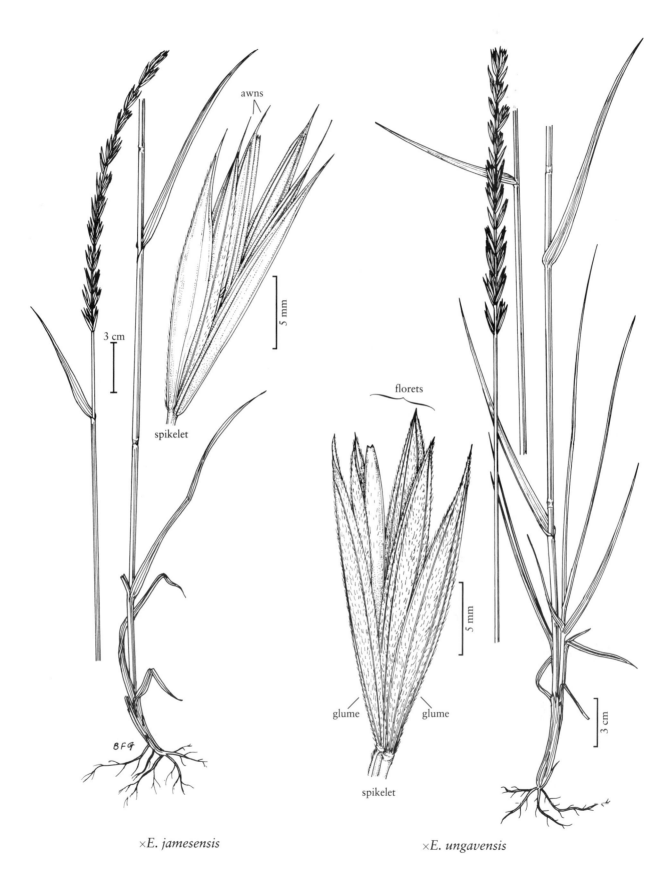

×*E. jamesensis*   ×*E. ungavensis*

×ELYLEYMUS

2 times as long as the middle internodes. **Spikelets** 12–26(30) mm, 1–3 times the length of the internodes, straight, usually ascending, not appressed, with 2–12 florets; **disarticulation** above the glumes, beneath the florets. **Glumes** 5–15 mm, $^1/_2$–$^2/_3$ the length of the spikelets, usually narrowly lanceolate, stiff, tapering from midlength or below, slightly curving to the side distally, not keeled, 3–5-veined basally, 1-veined distally, apices acuminate; **lemmas** lanceolate, rounded on the back, acute, mucronate to awned, awns to 5 mm, straight; **paleas** slightly shorter than the lemmas; **anthers** 3, 2.5–6 mm. **Caryopses** 4–5 mm, falling with the lemmas and paleas. $2n$ = 56. Haplomes **St**, **H**, **Ns**, and **Xm**. Name from the Latin *pascuum*, 'pasture', and the Greek *pyros*, 'wheat'.

Pascopyrum is a North American allooctoploid genus with one species. It combines the genomes of *Leymus* with those of *Elymus*. There are no other species that combine these two tetraploid genomes, although there are many species of both *Elymus* and *Leymus* in Eurasia and North America.

SELECTED REFERENCES **Dewey, D.R.** 1975. The origin of *Agropyron smithii*. Amer. J. Bot. 62:524–530; **Gillett, J.M.** and **H.A. Senn.** 1960. Cytotaxonomy and infraspecific variation of *Agropyron smithii* Rydb. Canad. J. Bot. 38:747–760.

1. **Pascopyrum smithii** (Rydb.) Barkworth & D.R. Dewey [p. 352]
WESTERN WHEATGRASS

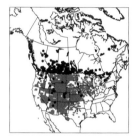

Culms 20–100 cm, glabrous. **Auricles** 0.2–1 mm, often purple; **ligules** about 0.1 mm; **blades** 2–26 cm long, 1–4.5 mm wide, decreasing in length upwards, spreading, rigid, adaxial surfaces with prominent veins. **Spikes** 5–17 cm; **middle internodes** 4.5–11 mm. **Spikelets** 12–26(30) mm, with 2–12 florets; **lowest rachilla internodes in each spikelet** 0.8–2 mm long, 0.5–0.9 mm wide at the top. **Glumes** 5–15 mm, lower glumes usually exceeded by the upper glumes; **lower glumes** 0.15–0.8 mm wide at $^3/_4$ length; **lemmas** 6–14 mm, unawned or awned, awns 0.5–5 mm. $2n$ = 56.

Pascopyrum smithii is native to sagebrush deserts and mesic alkaline meadows, growing in both clay and sandy soils. *Pascopyrum smithii* is probably derived from a *Leymus triticoides*–*Elymus lanceolatus* cross (Dewey 1975); it is frequently confused with both. *Leymus triticoides* differs in usually having 2 spikelets per node and glumes that are narrower at the base. In *E. lanceolatus*, the leaves tend to be more evenly distributed and the glumes have straight midveins, become narrow beyond midlength, and tend to be wider at $^3/_4$ length (0.35–1.6 mm). In addition, the first rachilla internodes of *E. lanceolatus* are often longer and narrower (the length/width ratio averaging 2.6, versus 1.8 in *P. smithii*). No infraspecific taxa of *P. smithii* are recognized here.

13.16 ×PASCOLEYMUS (B. Boivin) Barkworth

Mary E. Barkworth

Plants perennial; rhizomatous. **Culms** to 120 cm, glabrous. **Leaves** evenly distributed; **sheaths** open; **auricles** present; **ligules** 0.5–1 mm, ciliate; **blades** 3–10 mm wide, slightly glaucous, long-tapering to the narrowly acute apices, abaxial surfaces scabridulous, glabrous, adaxial surfaces scabrous. **Inflorescences** distichous spikes, with 2 spikelets per node. **Spikelets** sessile, with 8–9 florets; **disarticulation** above the glumes, beneath the florets. **Glumes** unequal, linear-lanceolate to subulate; **lemmas** awned, awns shorter than the lemma body; **paleas** subequal to the lemmas, apices about 0.2 mm wide; **anthers** about 3.5 mm, indehiscent.

×*Pascoleymus* consists, as its name indicates, of hybrids between *Pascopyrum* and *Leymus*. Only one such hybrid has been recognized.

SELECTED REFERENCE **Boivin, B.** 1967. Énumérations des plantes du Canada VI—Monopsides (2ème partie). Naturaliste Canad. 94: 471–528.

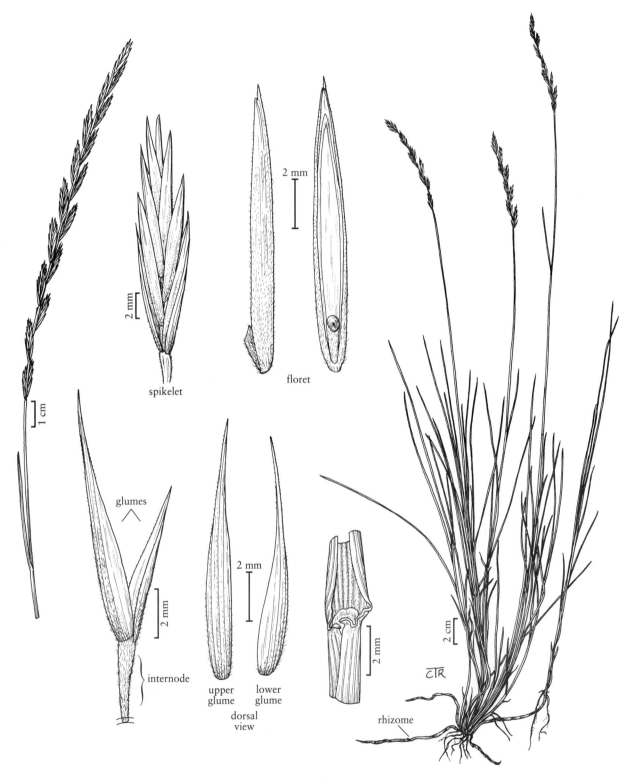

spikelet

floret

glumes

internode

upper
glume

lower
glume

dorsal
view

rhizome

CTR

P. smithii

PASCOPYRUM

1. ×**Pascoleymus bowdenii** (B. Boivin) Barkworth
   [p. 353]

Culms 2–3 mm thick. **Auricles** to 1 mm; **blades** 5–10 mm wide. **Rachis internodes** 7–12 mm. **Spikelets** 12–22 mm. **Glumes** 1.2–9 mm; **lemmas** about 10 mm, hairy, hairs about 1 mm, awns 1.5-4 mm.

The holotype of ×*Pascoleymus bowdenii* was collected at Beaverlodge, Alberta. Boivin (1967) considered it a hybrid between *Agropyon smithii* [≡ *Pascopyrum smithii*] and *Elymus innovatus* [≡ *Leymus innovatus*]. The tapering leaf blades and hairy lemmas support the involvement of *L. innovatus*; support for the involvement of *P. smithii* is less evident. ×*Pascoleymus bowdenii* differs it its wide blades from ×*Elyleymus turneri*, another hybrid having *L. innovatus* as one of its parents, but these do not provide any particular support for identifying *P. smithii* as the other parent. Some of the additional specimens that Boivin placed in ×*Pascoleymus bowdenii* belong to *L. triticoides*.

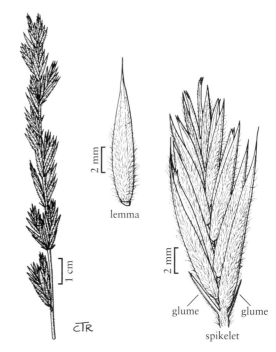

×*P. bowdenii*

×PASCOLEYMUS

13.17 **LEYMUS** Hochst.

Mary E. Barkworth

Plants perennial; sometimes cespitose, often rhizomatous. **Culms** 10–350 cm, erect, with extravaginal branching. **Leaves** basal or evenly distributed; **sheaths** open; **auricles** usually present; **ligules** membranous, truncate to rounded; **blades** often stiff, adaxial surfaces usually with subequal, closely spaced, prominently ribbed veins, sometimes with unequal, widely spaced, not prominently ribbed veins. **Inflorescences** usually distichous spikes with 1–8 spikelets per node, sometimes panicles with (2)3–35 spikelets associated with each rachis node; **rachises** with scabrous or ciliate edges; **internodes** 3.5–12(15) mm. **Spikelets** $^{1}/_{2}$–$3^{3}/_{4}$ times the length of the rachis internodes, usually sessile, sometimes pedicellate, pedicels to 5 mm, appressed to ascending, with 2–12 florets, the terminal floret usually reduced; **disarticulation** above the glumes, beneath the florets. **Glumes** usually 2, usually equal to subequal, the lower or both glumes sometimes reduced or absent, lanceolate and narrowing in the distal $^{1}/_{4}$, or lanceolate to subulate and tapering from below midlength, pilose or glabrous, sometimes scabrous, 0–3(7)-veined, veins evident at least at midlength, sometimes keeled, keels straight or almost so, apices acute, acuminate, or tapering to an awnlike tip, if distinctly awned, awns to 4 mm; **lemmas** glabrous or with hairs, sometimes scabrous distally, inconspicuously 5–7-veined, rounded over the back proximally, sometimes keeled distally, keels not conspicuously scabrous distally, apices acute, unawned or awned, awns usually to 7 mm, sometimes 16–33 mm, straight; **paleas** slightly shorter than to slightly longer than the lemmas, keels usually scabrous or ciliate on the distal portion, sometimes throughout; **lodicules** 2, shortly hairy, lobed; **anthers** 3, 2.5–10 mm. **Caryopses** with hairy apices. $x = 7$. Haplomes **NsNs** or **NsXm** (see below). Name an anagram of *Elymus*.

Leymus is a genus of approximately 50 species; all are native to temperate regions in the Northern Hemisphere. They are most abundant in eastern Asia, with North America being a secondary center. Of the 17 species treated, 11 are native to the *Flora* region, 4 are introduced, and 2 are naturally occurring hybrids.

Most species of *Leymus*, including most North American species, grow well in alkaline soils. They are used for soil stabilization and forage. All the species are self-incompatible, outcrossing polyploids. One of the haplomes present is the **Ns** genome; this genome is also found in *Psathyrostachys*, most species of which are diploids. There is disagreement concerning the second haplome. Wang and Jensen (1994) argued that there are two different haplomes present, the origin of the second one being unknown and designated **Xm**. Bödvarsdóttir and Anamthawat-Jónsson (2003) found no molecular probes that would distinguish between the two genera, from which they argued that *Leymus* is a segmental allopolyploid with only one basic haplome, **Ns**. Morphologically, *Psathyrostachys* and *Leymus* are very similar, the major differences being that *Psathyrostachys* is never rhizomatous, has disarticulating rachises, and, usually, distinctly awned lemmas.

Leymus arenarius and *L. mollis* are sometimes mistaken for *Ammophila*, which grows in the same habitats and has a similar habit. *Ammophila* differs from *Leymus*, however, in having only one floret per spikelet.

In most species of *Leymus*, at least some of the spikelets are on pedicels up to 2 mm long. Despite this, it is customary to identify the inflorescence of such species as a spike rather than a raceme, as is done in this treatment. Culm thicknesses are measured on the lower internodes. Descriptions of rachis nodes, unless stated otherwise, apply to the internodes at midspike.

SELECTED REFERENCES Anamthawat-Jónsson, K. 2005. The *Leymus* Ns-genome. Czech J. Genet. Pl. Breed. 41(Special Issue):13–20; Barkworth, M.E. and R.J. Atkins. 1984. *Leymus* Hochst. (Gramineae: Triticeae) in North America: Taxonomy and distribution. Amer. J. Bot. 71:609–625; Bödvarsdóttir, S.K. and K. Anamthawat-Jónsson. 2003. Isolation, characterization, and analysis of *Leymus*-specific DNA sequences. Genome 46:673–682; Bowden, W.M. 1957. Cytotaxonomy of section *Psammelymus* of the genus *Elymus*. Canad. J. Bot. 35:951–993; Bowden, W.M. 1959. Chromosome numbers and taxonomic notes on northern grasses: I. Tribe Triticeae. Canad. J. Bot. 37:1143–1151; Chen, S.-L. and G.-H. Zhu. 2006. *Leymus*. Pp. 386–394 *in* Z.-Y. Wu, P.H. Raven, and D.-Y. Hong (eds.). Flora of China, vol. 22 (Poaceae). Science Press, Beijing, Peoples Republic of China and Missouri Botanical Garden Press, St. Louis, Missouri, U.S.A. 653 pp. http://flora.huh.harvard.edu/china/mss/volume22/index.htm; Hole, D.J., K.B. Jensen, R.R.-C. Wang, and S.M. Clawson. 1999. Molecular analysis of *Leymus flavescens* and chromosome pairing in *Leymus flavescens* hybrids (Poaceae: Triticeae). Int. J. Plant Sci. 160:371–376; Jensen, K.B. and R.R.-C. Wang. 1997. Cytological and molecular evidence for transferring *Elymus coreanus* from the genus *Elymus* to *Leymus* and molecular evidence for *Elymus californicus* (Poaceae: Triticeae). Int. J. Pl. Sci. 158:872–877; Mason-Gamer, R.J. 2001. Origin of North American *Elymus* (Poaceae:Triticeae) allotetraploids based on granule-bound starch synthase gene sequences. Syst. Bot. 26:757–768; Tsvelev, N.N. 1976. Zlaki SSSR. Nauka, Leningrad [St. Petersburg], Russia. 788 pp.; Tsvelev, N.N. 1995. *Leymus*. Pp. 300–306 *in* J.G. Packer (ed., English edition). Flora of the Russian Arctic, vol. 1, trans. G.C.D. Griffiths. University of Alberta Press, Edmonton, Alberta, Canada. 330 pp. [English translation of A.I. Tolmachev (ed.). 1964. Arkticheskaya Flora SSSR, vol. 2. Nauka, Leningrad [St. Petersburg], Russia. 272 pp.]; Wang, R.R.-C. and K.B. Jensen. 1994. Absence of J genome in *Leymus* species (Poaceae: Triticeae): Evidence from DNA hybridization and meiotic pairing. Genome 37:231–235; Wang, R.R.-C., J.-Y. Zhang, B.S. Lee, K.B. Jensen, M. Kishi, and H. Tsuijimoto. 2006. Variations in abundance of 2 repetitive sequences in *Leymus* and *Psathyrostachys* species. Genome 49:511–519; Zhang, H.B. and J. Dvorák. 1991. The genome origin of tetraploid species of *Leymus* (Poaceae: Triticeae) inferred from variation in repeated nucleotide sequences. Amer. J. Bot. 78:871–884.

1. Glumes absent or shorter than 1 mm; lemmas awned, awns 16–33 mm long . . . . 17. *L. californicus*
1. Glumes developed, 3+ mm long, at least 1 on each spikelet; lemmas unawned or awned, awns to 7 mm long.
   2. Glumes flat or rounded on the back, tapering from midlength or above, flexible, the central portion scarcely thicker than the margins . . . . 3. *L. mollis*
   2. Glumes keeled, at least distally, tapering from below midlength, stiff, the central portion thicker than the margins.

3. Anthers usually indehiscent; plants rhizomatous, restricted to coastal regions from British Columbia to California.

    4. Glumes pubescent distally; lemmas awned, awns to 4 mm long; inflorescences spikes, not branched . . . . 4. *L. ×vancouverensis*

    4. Glumes glabrous; lemmas acute to awned, awns to 1.8 mm long; inflorescences sometimes with strongly ascending branches . . . . 11. *L. ×multiflorus*

3. Anthers dehiscent; plants rhizomatous or cespitose, widespread, including coastal regions from British Columbia to California.

    5. Inflorescences with 2–4 branches to 6 cm long at the proximal nodes; culms 115–350 cm tall . . . . 10. *L. condensatus*

    5. Inflorescences without branches; culms 10–270 cm tall.

        6. Lemmas densely hairy, hairs 0.7–3 mm long, occasionally glabrate.

            7. Lemmas awned, awns 2–4 mm long; lemma hairs 0.7–2.5 mm long . . . . 15. *L. innovatus*

            7. Lemmas unawned or the awns to 2 mm long; lemma hairs 2–3 mm long . . . . . . 16. *L. flavescens*

        6. Lemmas usually wholly or partly glabrous, or if hairy, the hairs shorter than 0.5(0.8) mm.

            8. Leaves equaling or exceeding the spikes; culms 10–30(60) cm tall; spikes 2–8 cm long, with 1–2 spikelets per node; plants of California coastal bluffs . . . . 5. *L. pacificus*

            8. Leaves exceeded by the spikes; culms 35–270 cm tall; spikes 3–35 cm long, with 1–8 spikelets per node; plants widespread in the western part of the *Flora* region, including the coastal bluffs of California.

                9. Plants cespitose, not or weakly rhizomatous, culms several to many together.

                    10. Spikes with 2–7 spikelets per node; blades 3–12 mm wide; culms (70)100–270 cm tall . . . . 12. *L. cinereus*

                    10. Spikes with 1 spikelet at the distal nodes, often at all nodes, sometimes with 2(3) at the lower nodes; blades 1–6 mm wide; culms 35–140 cm tall.

                        11. Blades with 5–9 adaxial veins; lemma awns to 2.5 mm long . . . . 13. *L salina*

                        11. Blades with (9)11–17 adaxial veins; lemma awns 1.3–7 mm long . . . . 14. *L. ambiguus*

                9. Plants rhizomatous, culms solitary or few together.

                    12. Culms 1–3 mm thick; glumes 4–16 mm long.

                        13. Spikes with 1 spikelet at all or most nodes, sometimes with 2 at a few nodes; lemma awns 2.3–6.5 mm long; culms 35–55 cm tall . . . . 6. *L. simplex*

                        13. Spikes with 2+ spikelets at most nodes; lemma awns to 3 mm long; culms 45–125 cm tall.

                            14. Adaxial surfaces of the blades usually with closely spaced, prominently ribbed, subequal veins; calluses usually glabrous, occasionally with a few hairs about 0.1 mm long . . . . 7. *L. triticoides*

                            14. Adaxial surfaces of the blades usually with widely spaced, not prominently ribbed veins, the primary veins evidently larger than the intervening secondary veins; calluses with hairs about 0.2 mm long . . . . 8. *L. multicaulis*

                  12. Culms 2.5–12 mm thick; glumes 10–30 mm long.

                      15. Spikelets 3–8 per node; lemmas hairy proximally, glabrous distally . . . . 1. *L. racemosus*

                      15. Spikelets 2–3 per node; lemmas glabrous or hairy their whole length.

                            16. Anthers 6–9 mm long; blades 3–11 mm wide; glumes with hairs to 1.3 mm long; plants established around the Great Lakes and the coast of Greenland, also found at a few other scattered locations, including western North America, sometimes cultivated . . . . 2. *L. arenarius*

                            16. Anthers 3–5 mm long; blades 5–7 mm wide; glumes glabrous, sometimes scabrous; plants cultivated . . . . 9. *L. angustus*

1. Leymus racemosus (Lam.) Tzvelev [p. 357]
MAMMOTH WILDRYE

Plants not or only weakly cespitose, strongly rhizomatous. often glaucous. **Culms** 50–100 cm tall, 8–12 mm thick, solitary or few together, mostly smooth and glabrous, scabridulous or pubescent below the spikes, hairs to 0.5 mm. **Leaves** exceeded by the spikes; **ligules** 1.5–2.5 mm; **blades** 20–40 cm long, 8–20 mm wide. **Spikes** 15–35 cm long, 10–20 mm wide, dense, with 3–8 spikelets per node; **internodes** 8–11 mm, surfaces hairy, hairs to 1 mm, on the edges to 1.5 mm. **Spikelets** 12–25 mm, sessile, with 4–6 florets. **Glumes** 12–25 mm long, to 2 mm wide, usually exceeding the lemmas, linear-lanceolate at the base, tapering from below midlength, stiff, glabrous at least at the base, the central portion thicker than the margins, keeled and subulate distally, 1-veined, veins inconspicuous at midlength; **lemmas** 15–20 mm, pubescent proximally, glabrous distally, tapering to an awn, awns 1.5–2.5 mm; **palea keels** usually glabrous, sometimes ciliate distally; **anthers** about 5 mm, dehiscent. 2n = 28.

Leymus racemosus is native to Europe and central Asia, where it grows on dry, sandy soils. It has been introduced into the *Flora* region, and collected at various locations, particularly in the northwestern contiguous United States; it is not clear how many of the populations represented by these specimens are still extant. Tsvelev (1976) recognized 4 subspecies. Because there are few North American specimens, and these are incomplete, no attempt has been made to determine to which subspecies the North American plants belong.

2. Leymus arenarius (L.) Hochst. [p. 357]
EUROPEAN DUNEGRASS, LYMEGRASS,
ÉLYME DES SABLES D'EUROPE

Plants weakly cespitose, rhizomatous, strongly glaucous. **Culms** 50–150 cm tall, (2)3–6 mm thick, usually glabrous throughout, occasionally pubescent distally to 5 mm below the spike. **Leaves** exceeded by the spikes; **ligules** 0.3–2.5 mm; **blades** 3–11 mm wide, with 15–40 adaxial veins. **Spikes** 12–35 cm long, 15–25 mm wide, usually with 2 spikelets per node; **internodes** 8–12 mm, surfaces glabrous, edges ciliate. **Spikelets** 12–30 mm, with 2–5 florets. **Glumes** 12–30 mm long, 2–3.5 mm wide, lanceolate, tapering from below midlength, stiff, glabrous towards the base and usually distally, sometimes pubescent distally, the central portion thicker than the margins, 3(5)-veined at midlength, keeled or rounded over the midvein, midveins and sometimes the margins with hairs to about 1.3 mm, apices acuminate; **lemmas** 12–25 mm, densely villous, hairs 0.3–0.7 mm, 5–7-veined, acute, occasionally awned, awns to 3 mm; **anthers** 6–9 mm, dehiscent. 2n = 56.

Leymus arenarius is native to Europe. It has become established in sandy habitats around the Great Lakes and the coast of Greenland. It has also been found at a few other widely scattered locations. It is sometimes cultivated, forming large, attractive, blue-green clumps, but its tendency to spread may be undesirable.

3. Leymus mollis (Trin.) Pilg. [p. 357]
AMERICAN DUNEGRASS, SEA LYMEGRASS,
ÉLYME DES SABLES D'AMERIQUE, SEIGLE DE MER

Plants not cespitose, strongly rhizomatous, occasionally slightly glaucous. **Culms** 12–170 cm tall, 3–6 mm thick, usually densely pubescent below the spikes for 10–40+ mm. **Leaves** exceeded by the spikes; **auricles** to 0.7 mm; **ligules** 0.2–2.5 mm; **blades** 10–94 cm long, 3–15 mm wide, adaxial surfaces scabridulous to scabrous, 20–40-veined, veins subequal, prominently ribbed, closely spaced. **Spikes** 5–34 cm long, 10–20 mm wide, with 3–33 nodes, usually with 2 spikelets per node; **internodes** 4.5–9.5 mm, surfaces and edges similar, hairs on the surfaces 0.1–0.5 mm, on the edges to 0.7 mm. **Spikelets** 15–34 mm, with 3–6 florets. **Glumes** 9–34 mm long, 1.5–4 mm wide, lanceolate, tapering from midlength or above, flat or rounded on the back, flexible, usually strigillose to pilose or villous, rarely almost glabrous, the central portion scarcely thicker than the margins, 3(5)-veined at midlength, apices acute; **lemmas** 11–20 mm, densely hairy, hairs 0.5–1 mm, soft, apices acute, unawned; **anthers** 4–9 mm, dehiscent.

Leymus mollis is native to Asia and North America. It is treated here as having two very similar subspecies that have somewhat different ranges. The subspecies are sometimes treated as separate species, but they may be little more than environmentally induced variants. Both subspecies grow primarily on coastal beaches, close to the high tide line, and along some inland waterways, particularly in the arctic. Reports of *L. ajanensis* (V.N. Vassil.) Tzvelev from North America are based on specimens of *L. mollis* (D. Murray, University of Alaska, pers. comm. 2006).

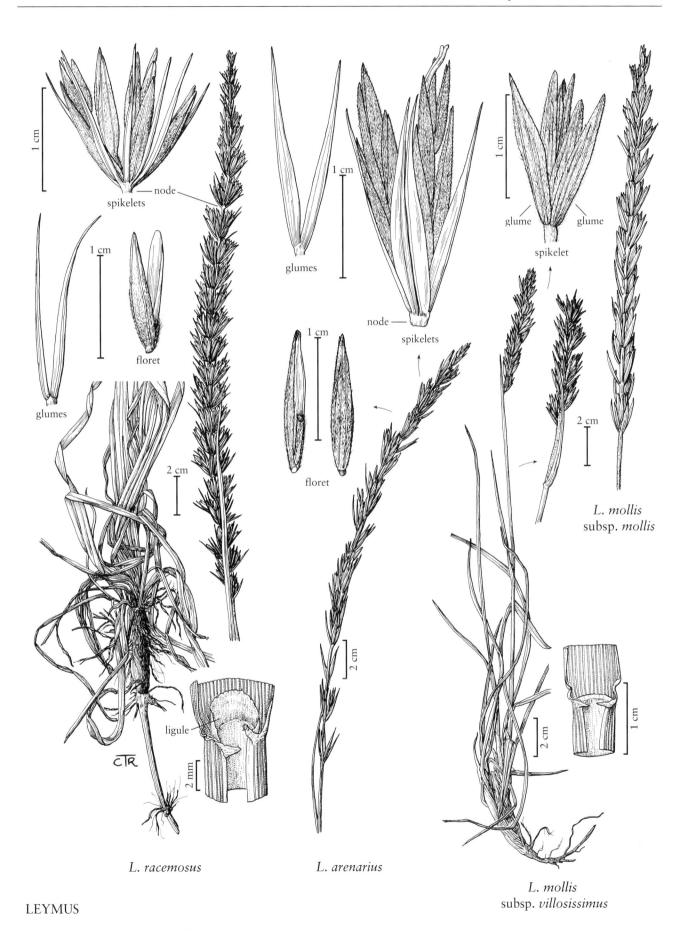

spikelets

node

glumes

floret

glumes

node

spikelets

floret

glume   glume

spikelet

L. mollis
subsp. *mollis*

ligule

CTR

L. racemosus

L. arenarius

L. mollis
subsp. *villosissimus*

LEYMUS

1. Spikes 12–34 cm long, with 12–33 nodes; basal blades 5–15 mm wide; culms 50–170 cm tall . . . . subsp. *mollis*
1. Spikes 5–13(16) cm long, with 3–14 nodes; basal blades 3–8 mm wide; culms 12–70 cm tall . . . . subsp. *villosissimus*

Leymus mollis (Trin.) Pilg. subsp. **mollis** [p. 357]

Rhizomes 4–6 mm thick. **Culms** (50)70–170 cm. **Basal blades** 22–94 cm long, 5–15 mm wide. **Spikes** 12–34 cm, with 12–33 nodes. **Glumes** 13–34 mm long, 3–4 mm wide, often longer than the lemmas, usually pilose, sometimes almost glabrous. 2*n* = 28.

In the *Flora* region, *Leymus mollis* subsp. *mollis* grows primarily on the west coast; on the east coast, it grows in New Brunswick and Nova Scotia, particularly along the St. Lawrence River, and on the coast of Greenland. It does not grow along the arctic coast. Outside the *Flora* region, it is native in the coastal region of eastern Asia, growing primarily along the coast and in the mouths of larger rivers, and on the shores of large lakes near the coast from the Korean Peninsula to the Kamchatka Peninsula. It was introduced to Iceland, but is now rare there.

Leymus ×vancouverensis is thought to be a hybrid between *L. mollis* subsp. *mollis* and *L. triticoides*, although its range extends beyond the current range of *L. triticoides*.

Leymus mollis subsp. **villosissimus** (Scribn.) Á. Löve [p. 357]

Rhizomes about 2 mm thick. **Culms** 12–70 cm. **Basal blades** 10–31 cm long, 3–8 mm wide. **Spikes** 5–13(16) cm, with 3–14 nodes. **Glumes** 9–14(21) mm long, 1.5–2.5 mm wide, often longer than the lemmas, villous. 2*n* = 28.

Leymus mollis subsp. *villosissimus* is an arctic taxon found primarily in eastern Siberia, Alaska, and northwestern Canada. It grows mostly on arctic coasts, but is also known from a few inland locations.

4. **Leymus ×vancouverensis** (Vasey) Pilg. [p. 359]
VANCOUVER WILDRYE

Plants not cespitose, rhizomatous, green or slightly glaucous. **Culms** 60–122 cm tall, 2–3.5 mm thick, sparsely to densely pubescent below the spike. **Leaves** exceeded by the spikes; **auricles** to 1 mm; **ligules** 0.4–1.2 mm; **blades** to 9 mm wide, veins prominently ribbed, subequal, closely spaced. **Spikes** 7–32 cm long, 7–11 mm wide, sometimes glaucous, often purplish or green with traces of purple, with 1–2 spikelets per node;

internodes 8–12 mm, surfaces hairy distally, hairs 0.1–0.3 mm, edges ciliate, hairs to 1 mm, coarser than the surface hairs. **Spikelets** 15–20 mm, with 2–6 florets. **Glumes** 9–28 mm long, 1.5–4 mm wide, lanceolate, glabrous towards the base, pubescent distally, midveins glabrous or with hairs to about 1.3 mm, stiff, keeled or flat proximally, keeled distally, the central portion thicker than the margins, 3(5)-veined at midlength, tapering from below midlength to an awn, awns to 4 mm; **lemmas** 8–10 mm, usually completely glabrous, margins and apices sometimes pubescent, apices tapering to an awn, awns to 4 mm; **anthers** 3.3–7.3 mm, indehiscent. 2*n* = 28, 42.

Leymus ×vancouverensis grows at scattered locations on beaches along the Pacific coast, from southern British Columbia to California. It is a sterile hybrid, probably between *L. mollis* and *L. triticoides* (Bowden 1957). The northern populations are outside the current range of *L. triticoides*.

5. **Leymus pacificus** (Gould) D.R. Dewey [p. 359]
PACIFIC WILDRYE

Plants not cespitose, strongly rhizomatous. **Culms** 10–30(60) cm tall, 1–2 mm thick, solitary or few together, glabrous or sparsely pubescent near the nodes. **Leaves** equaling or exceeding the spikes; **sheaths** glabrous; **auricles** to 1.4 mm; **ligules** 0.2–0.3 mm, truncate, erose; **blades** 10–30 cm long, 2–4 mm wide, abaxial surfaces glabrous, adaxial surfaces scabrous, veins about 15, subequal, prominently ribbed. **Spikes** 2–8 cm long, 7–12 mm wide, with 1–2 spikelets per node; **internodes** 3.5–4 mm, surfaces glabrous and smooth, edges weakly scabrous distally. **Spikelets** 12–15 mm, with 4–6 florets. **Glumes** subequal, (5)7–15 mm long, 0.5–2.5 mm wide, narrowly lanceolate, tapering from near the base, stiff, keeled, the central portion thicker than the margins, bases mostly glabrous, margins ciliate, 1–3(5)-veined, veins inconspicuous at midlength; **calluses** scarcely developed, glabrous; **lemmas** 7–11 mm, glabrous, smooth, apices acute to awn-tipped, awns to 0.8 mm; **anthers** 3–4 mm, dehiscent. 2*n* = 28.

Leymus pacificus is found on coastal bluffs from Mendocino to Santa Barbara counties, California. It is poorly represented in herbaria. In some years it grows almost entirely vegetatively, often being represented by scattered innovations with somewhat curved leaves.

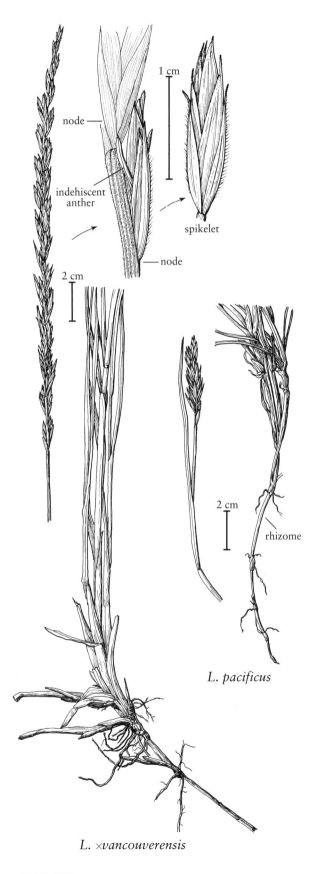

node

indehiscent
anther

2 cm

1 cm

spikelet

node

L. pacificus

2 cm

rhizome

L. ×vancouverensis

LEYMUS

6. Leymus simplex (Scribn. & T.A. Williams) D.R. Dewey [p. 361]

ALKALI WILDRYE

Plants not cespitose, strongly rhizomatous, often glaucous. **Culms** 35–75 cm tall, 1–2.5 mm thick, solitary or few together, glabrous or sparsely pubescent near the nodes. **Leaves** exceeded by the spikes; **sheaths** glabrous, smooth; **auricles** infrequently present, to 0.8 mm, the auricular location often with hairs to 2 mm; **ligules** 0.3–0.5 mm, truncate, erose; **blades** 4–29 cm long, 1–2(5) mm wide, flat, becoming involute when dry, stiff, adaxial surfaces scabrous, with scattered hairs to 2 mm, veins 7–11, subequal, prominently ribbed. **Spikes** 1.5–27 cm long, 4–15 mm wide, with 1 spikelet per node at midspike, sometimes with 2 at the lower nodes; **internodes** 7–20 mm, surfaces glabrous or strigillose, edges ciliate, cilia to 1 mm. **Spikelets** 16–25 mm, pedicellate, pedicels 1–2(5) mm, with 3–12 florets. **Glumes** subequal, 8–12 mm long, 0.5–1.5 mm wide, subulate, tapering from about ¼ of their length, stiff, glabrous at least at the base, the central portion thicker than the margins, keeled, 0–1(3)-veined, veins inconspicuous at midlength; **lemmas** 7–12 mm, glabrous, awned, awns 2.3–6.5(12) mm; **anthers** 3.7–4.5 mm, dehiscent. $2n = 28$.

Leymus simplex is found in meadows and drifting sand in southern Wyoming, and along the Green River in northeastern Utah.

1. Culms 55–75 cm tall; spikes 10–27 cm long; internodes 10–20 mm . . . . var. *luxurians*
1. Culms 35–55 cm tall; spikes 1.5–13 cm long; internodes 7–9 mm . . . . var. *simplex*

Leymus simplex var. **luxurians** (Scribn. & T.A. Williams) Beetle [p. 361]

Culms 55–75 cm tall, 2–2.5 mm thick. **Spikes** 10–27 cm long; **internodes** 10–20 mm. **Spikelets** 20–25 mm, with 6–12 florets.

Leymus simplex var. *luxurians* grows at a few locations in Wyoming. It sometimes grows close to var. *simplex*. It may represent clones that have access to more water and/or more nutrients, but the absence of intermediate plants suggests a genetic distinction.

Leymus simplex (Scribn. & T.A. Williams) D.R. Dewey var. **simplex** [p. 361]

Culms 35–55 cm tall, 1–2 mm thick. **Spikes** 1.5–13 cm long; **internodes** 7–9 mm. **Spikelets** 16–22 mm, with 3–6 florets.

Leymus simplex var. *simplex* is found throughout the range of the species, sometimes in close proximity to var. *luxurians*. The two may be environmentally induced variants, but the lack of intermediates suggests a genetic distinction.

7. Leymus triticoides (Buckley) Pilg. [p. 361]

BEARDLESS WILDRYE

Plants not cespitose, strongly rhizomatous. **Culms** 45–125 cm tall, 1.8–3 mm thick, solitary or few together. **Leaves** exceeded by the spikes, often basally concentrated; **sheaths** glabrous or hairy, hairs 0.5–1 mm; **auricles** to 1 mm; **ligules** 0.2–1.3 mm, truncate, erose; **blades** 10–35 cm long, 3.5–10 mm wide, flat to involute, usually stiffly ascending, adaxial surfaces usually scabrous, often also sparsely hairy, hairs to 0.8 mm, most abundant proximally, veins 11–27, closely spaced, subequal, prominently ribbed. **Spikes** 5–20 cm long, 5–15 mm wide, with 2 spikelets at midspike, sometimes 1 or 3 at other nodes; **internodes** 5–11.5 mm, usually mostly smooth and glabrous, sometimes strigillose distally, edges ciliate, cilia to 0.4 mm. **Spikelets** 10–22 mm, with 3–7 florets. **Glumes** 5–16 mm long, 0.5–1.2 mm wide, bases not overlapping, glabrous and smooth proximally, scabrous distally, tapering from below midlength to the subulate apices, stiff, keeled, the central portion thicker than the margins, 1(3)-veined, veins inconspicuous at midlength; **calluses** usually glabrous, occasionally with a few hairs, hairs about 0.1 mm; **lemmas** 5–12 mm, usually glabrous, occasionally sparsely hairy, hairs to 0.3 mm, apices acute, usually awned, awns to 3 mm; **anthers** 3–6 mm, dehiscent. 2*n* = 28.

Leymus triticoides grows in dry to moist, often saline meadows. Its range extends from southern British Columbia to Montana, south to California, Arizona, and New Mexico, but its populations are widely scattered. It is not known from Mexico. There is considerable variation within the species, but no pattern of variation suggesting the existence of infraspecific taxa is known. It is very similar to *L. multicaulis*, strains of which were initially released as *L. triticoides* by the U.S. Department of Agriculture. The most consistent differences between them appear to be in the venation of the leaf blades and the vestiture of the calluses. *Leymus triticoides* is also very similar to *L. simplex*, differing from it in the number of spikelets at the midspike nodes.

Leymus triticoides hybridizes with other species of *Leymus*; hybrids with *L. mollis* are called *L.* ×*vancouverensis* (see p. 358), those with *L. condensatus*

are called *L.* ×*multiflorus* (see p. 362). Hybrids with *L. cinereus* are known, but have not been formally named. Plants identified as *Elymus arenicolus* Scribn. & J.G. Sm. are here included in *L. flavescens*, but may represent hybrids between *L. triticoides* and *L. flavescens*.

8. Leymus multicaulis (Kar. & Kir.) Tzvelev [p. 361]

MANY-STEM WILDRYE

Plants somewhat cespitose, rhizomatous. **Culms** 50–80 cm tall, 1.5–3 mm thick, usually few together, glabrous, mostly smooth, scabrous beneath the spikes. **Leaves** exceeded by the spikes; **sheaths** glabrous, smooth; **auricles** to 1 mm; **ligules** 1–2 mm; **blades** 3–8 mm wide, flat or the margins slightly involute, grayish green, sometimes glaucous, abaxial surfaces smooth, adaxial surfaces glabrous, with both primary and secondary veins, primary veins 5–7, not prominently ribbed. **Spikes** 5–14 cm long, 6–13 mm wide, with 2–4(6) spikelets per node; **internodes** 4–6 mm, glabrous or strigillose, hairs about 0.1 mm, edges ciliate, cilia to 0.4 mm. **Spikelets** 8–15 mm, with 2–6 florets. **Glumes** 4–10 mm long, to 1 mm wide, stiff, keeled, glabrous, scabrous, the central portion thicker than the margins, bases not overlapping, tapering from below midlength to the subulate apices, inconspicuously 1-veined at midlength; **calluses** usually with at least some hairs, hairs about 0.2 mm; **lemmas** 5–9 mm, mostly glabrous and smooth, scabrous distally, apices tapering to an awn, awns 2–3 mm, scabrous; **anthers** 3–4 mm, dehiscent. 2*n* = 42.

Leymus multicaulis is native to Eurasia, extending from the Volga River delta in Russia to Xinjiang, China. In its native range, it grows in alkaline meadows and saline soils, and as a weed in fields, near roads, and around human habitations. It is very similar to *L. triticoides*, and hybrids with that species are highly fertile. A cultivar of *L. multicaulis*, 'Shoshone', that was originally thought to be a productive strain of *L. triticoides*, has been widely distributed for forage. *Leymus multicaulis* differs from *L. triticoides* primarily in having both primary and secondary veins in its blades, and small hairs on its calluses. Because it has only recently been realized that *L. multicaulis* has been introduced to North America, its distribution in North America is unknown.

9. Leymus angustus (Trin.) Pilg. [p. 363]

ALTAI WILDRYE

Plants somewhat cespitose, rhizomatous. **Culms** 60–120 cm tall, 2.5–7 mm thick, solitary or few together, glabrous or pubescent below the nodes. **Leaves** exceeded by the spikes, basally concentrated; **sheaths** smooth, scabridulous, or hairy; **auricles** to 1 mm; **ligules** 0.5–1 mm, rounded to obtuse, sometimes erose; **blades**

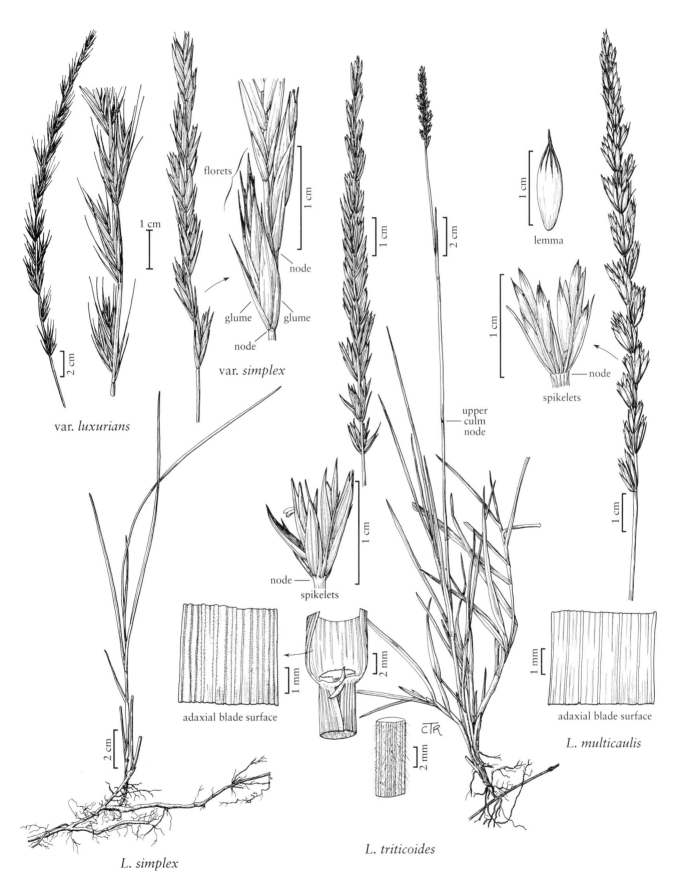

var. *luxurians*

florets

1 cm

node

glume   glume

node

var. *simplex*

lemma

node

spikelets

upper
culm
node

node
spikelets

adaxial blade surface

L. simplex

node
spikelets

adaxial blade surface

L. multicaulis

L. triticoides

LEYMUS

15–20 cm long, 5–7 mm wide, glaucous, stiff, involute, abaxial surfaces glabrous or hairy, sometimes scabridulous, adaxial surfaces scabrous, with 7–17 closely spaced subequal veins. **Spikes** 10–25 cm long, 7–10 mm wide, with 2(3) spikelets per node; **internodes** 8–10 mm, surfaces strigillose, hairs to 0.3 mm, edges ciliate, cilia to 1 mm. **Spikelets** 10–19 mm, with 2–3 florets. **Glumes** 10–13 mm long, 0.5–2.5 mm wide, exceeded by the florets, narrowly lanceolate, tapering from the base, stiff, keeled, the central portion thicker than the margins, (0)1(3)-veined at midlength, bases expanded, overlapping, concealing the base of the lowest floret, scabrous; **lemmas** 8–13 mm, densely hairy and not glaucous, hairs to 0.4 mm, or glabrous and glaucous, apices unawned or awned, awns to 2.5 mm; **anthers** 3–4 mm, dehiscent. $2n = 84$.

Leymus angustus is a Eurasian species that, in its native range, grows in alkaline meadows, and on sand and gravel in river and lake valleys. Several cultivars of *L. angustus* have been developed for use as forage, particularly in Canada. Some of the better known are 'Prairieland', 'Eejay', and 'Pearl'. The distribution of *L. angustus* in the *Flora* region is not known.

Chen and Zhu (2006) describe *Leymus angustus* as always being puberulent. Some accessions cultivated under this name by the U.S. Department of Agriculture (Plant Introduction Numbers 110,079; 406,461), have glabrous, glaucous lemmas and glumes that tend to exceed the lemmas, suggesting that they belong to another taxon, possibly *L. karelinii* (Turcz.) Tzvelev, a species for which $2n = 56$.

10. **Leymus condensatus** (J. Presl) Á. Löve [p. 363]
    GIANT WILDRYE

Plants cespitose, weakly rhizomatous. **Culms** 115–350 cm tall, 6–10 mm thick, usually several to many together. **Leaves** exceeded by the inflorescences; **auricles** absent; **ligules** 0.7–6 mm on the basal leaves, 4–7.5 mm on the flag leaves; **blades** 10–28 mm wide, abaxial surfaces glabrous, smooth, adaxial surfaces scabridulous, veins numerous, subequal or unequal. **Inflorescences** panicles, 17–44 cm long, 20–60 mm wide, lower nodes with 2–6 branches, branches to 8 cm, ascending, with 5–35 spikelets, upper nodes with pedicellate and sessile spikelets; **internodes** 3.5–10 mm, glabrous. **Spikelets** 9–25 mm, usually pedicellate, pedicels 0.8–2 mm, with 3–7 florets. **Glumes** 6–16 mm long, 0.5–2.5 mm wide, narrowly lanceolate, stiff, keeled, the central portion thicker than the margins, glabrous, smooth proximally, scabrous distally, 0–1(3)-veined, veins inconspicuous at midlength, apices tapering almost imperceptibly into an awn, awns subequal to the glume body; **lemmas** 7–14 mm, usually glabrous, apices acute, sometimes awned, awns to 4 mm; **anthers** 3.5–7 mm, dehiscent. $2n = 28, 56$.

Leymus condensatus is found primarily on dry slopes and in open woodlands of the coastal mountains and offshore islands of California, at elevations of 0–1500 m. Both its large size and paniculate inflorescence tend to make it a distinctive species in the *Triticeae*. Hybrids between *L. condensatus* and *L. triticoides*, known as *Leymus ×multiflorus*, are relatively common where the parents are sympatric.

11. **Leymus ×multiflorus** (Gould) Barkworth & R.J. Atkins [p. 363]
    MANY-FLOWERED WILDRYE

Plants cespitose, rhizomatous. **Culms** 65–210 cm tall, 3–5 mm thick. **Leaves** exceeded by the inflorescences; **auricles** absent; **ligules** 0.5–2 mm, truncate, erose; **blades** 6–15 mm wide, both surfaces glabrous, adaxial surfaces with numerous closely spaced, unequal veins. **Inflorescences** 15–35 cm long, 9–25 mm wide, usually spikes, with 2–6 spikelets per node, occasionally some nodes with 1–2 branches, branches to 60 mm, strongly ascending; **internodes** 8–10 mm, usually glabrous, sometimes scabrous. **Spikelets** 17–25 mm, pedicellate, pedicels 0.5–2(5) mm, with 6–9 florets. **Glumes** 9–25 mm long, differing in length by 1–4 mm, usually exceeding the lowest lemmas, 0.5–2.5 mm wide, subulate to narrowly lanceolate, stiff, keeled, the central portion thicker than the margins, tapering from the bases, glabrous, smooth proximally, scabrous distally, 1(3)-veined, veins inconspicuous at midlength; **lemmas** 8–12 mm, glabrous, acute to awned, awns to 1.8 mm; **anthers** about 6 mm, indehiscent. $2n = 42$.

Leymus ×multiflorus is a sterile hybrid between *Leymus condensatus* and *L. triticoides* that occurs near the coast of central and southern California.

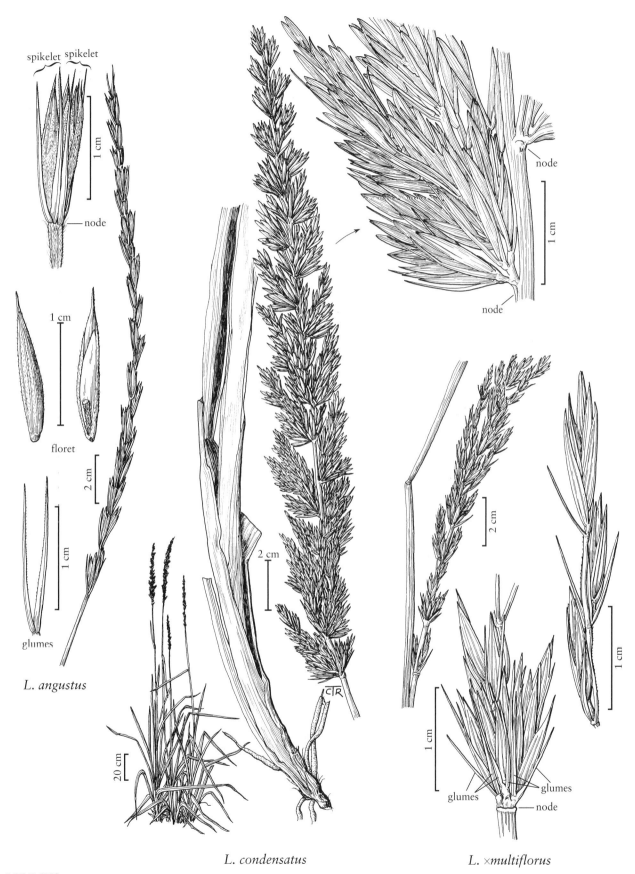

spikelet spikelet

1 cm

node

1 cm

floret

2 cm

1 cm

glumes

L. angustus

20 cm

node

1 cm

node

CTR

L. condensatus

2 cm

1 cm

1 cm

glumes

glumes

node

L. ×multiflorus

LEYMUS

12. Leymus cinereus (Scribn. & Merr.) Á. Löve [p. 365]
GREAT BASIN WILDRYE

Plants strongly cespitose, weakly rhizomatous, usually bright green, not glaucous. **Culms** 70–270 cm tall, 2–5 mm thick, many together, lowest nodes often pubescent, sometimes pubescent up to 1.5 cm below the inflorescence. **Leaves** exceeded by the spikes; **sheaths** glabrous or hairy; **auricles** to 1.5 mm; **ligules** 1.5–8 mm; **blades** 15–45 cm long, 3–12 mm wide, strongly involute to flat, abaxial surfaces glabrous, adaxial surfaces scabrous, 11–25-veined, veins subequal, prominently ribbed. **Spikes** 10–29 cm long, 8–17 mm wide, with 14–28 nodes and 2–7 spikelets per node; **internodes** 4–9 mm. **Spikelets** 9–25 mm, with 3–7 florets. **Glumes** 8–18 mm long, 0.5–2.5 mm wide, subulate distally, stiff, keeled, the central portion thicker than the margins, tapering from below midlength, smooth or scabrous, 0–1(3)-veined, veins inconspicuous at midlength; **lemmas** 6.5–12 mm, glabrous or hairy, hairs 0.1–0.3 mm, apices acute or awned, awns to 3 mm; **anthers** 4–7 mm, dehiscent. $2n = 28, 56$.

Leymus cinereus grows along streams, gullies, and roadsides, and in gravelly to sandy areas in sagebrush and open woodlands. It is widespread and common in western North America. *Leymus cinereus* resembles *Psathyrostachys juncea*, differing in its non-disarticulating rachises, larger spikelets with more florets, and longer ligules. Spontaneous hybridization between *L. cinereus* and *L. triticoides* is known; the hybrids do not have a scientific name. The rhizomes found in some specimens may reflect introgression from *L. triticoides* through such hybrids.

13. Leymus salina (M.E. Jones) Á. Löve [p. 365]

Plants cespitose, sometimes weakly rhizomatous. **Culms** 35–140 cm tall, 1.5–3 mm thick, several together. **Leaves** exceeded by the spikes; **auricles** to 1 mm; **ligules** 0.1–1 mm, truncate; **blades** 1–5 mm wide, flat to strongly involute, adaxial surfaces glabrous or sparsely to densely hirsute, with 5–9 prominently ribbed, subequal veins. **Spikes** 4–14 cm long, 4–11 mm wide, nodes below midspike with 1–2(3) spikelets, distal nodes with 1 spikelet; **internodes** 3.5–9 mm, surfaces glabrous, edges scabrous or strigillose. **Spikelets** 9–21 mm, pedicellate, pedicels to 1 mm, with 3–6 florets. **Glumes**

unequal to subequal, to 12.5 mm long, 0.5–3.2 mm wide, subulate, stiff, keeled, the central portion thicker than the margins, tapering from below midlength, 0–1(3)-veined, veins inconspicuous at midlength; **lower glumes** 0–12 mm; **upper glumes** 3.5–12.5 mm; **lemmas** 7–12.5 mm, usually glabrous, sometimes sparsely strigillose, unawned or awned, awns to 2.5 mm; **anthers** 2.5–7.5 mm, dehiscent.

The three subspecies of *Leymus salina* differ in their pubescence and geographic distribution, with subsp. *salina* being the most common of the three. The specific epithet comes from the locality of the type collection: Salina Pass, Utah.

1. Basal sheaths and blades conspicuously hairy on the abaxial surfaces . . . . subsp. *salmonis*
1. Basal sheaths glabrous; blades usually glabrous on the abaxial surfaces.
  2. Blades strongly involute, usually densely hairy just above the ligules . . . . subsp. *salina*
  2. Blades flat or almost flat, not densely hairy above the ligules . . . . subsp. *mojavensis*

Leymus salina subsp. mojavensis Barkworth & R.J. Atkins [p. 365]
MOJAVE WILDRYE

Culms 35–90 cm. **Basal sheaths** glabrous; **blades** flat or almost flat, abaxial surfaces nearly always glabrous, adaxial surfaces scabridulous, not densely hairy above the ligules, occasionally sparsely hairy. **Spikes** with 1 spikelet at most nodes, 2–3 central nodes with 2 spikelets. $2n =$ unknown.

Leymus salina subsp. *mojavensis* grows at scattered locations on steep, north-facing slopes of the New York, Providence, and Clark mountains in California, and of the south rim of the Grand Canyon in Arizona.

Leymus salina (M.E. Jones) Barkworth subsp. salina [p. 365]
SALINA WILDRYE

Culms 39–102 cm. **Basal sheaths** glabrous; **blades** strongly involute, abaxial surfaces glabrous, adaxial surfaces pubescent, usually densely hairy above the ligules. **Spikes** with 1 spikelet at most nodes, including those at midspike. $2n = 28$.

Leymus salina subsp. *salina* grows on rocky hillsides, primarily in eastern Utah and western Colorado, extending into southern Wyoming and northern Arizona and New Mexico.

Leymus salina subsp. salmonis (C.L. Hitchc.) R.J. Atkins [p. 365]
SALMON WILDRYE

Culms 60–140 cm. **Basal sheaths** conspicuously hairy; **blades** open to involute, abaxial surfaces conspicuously

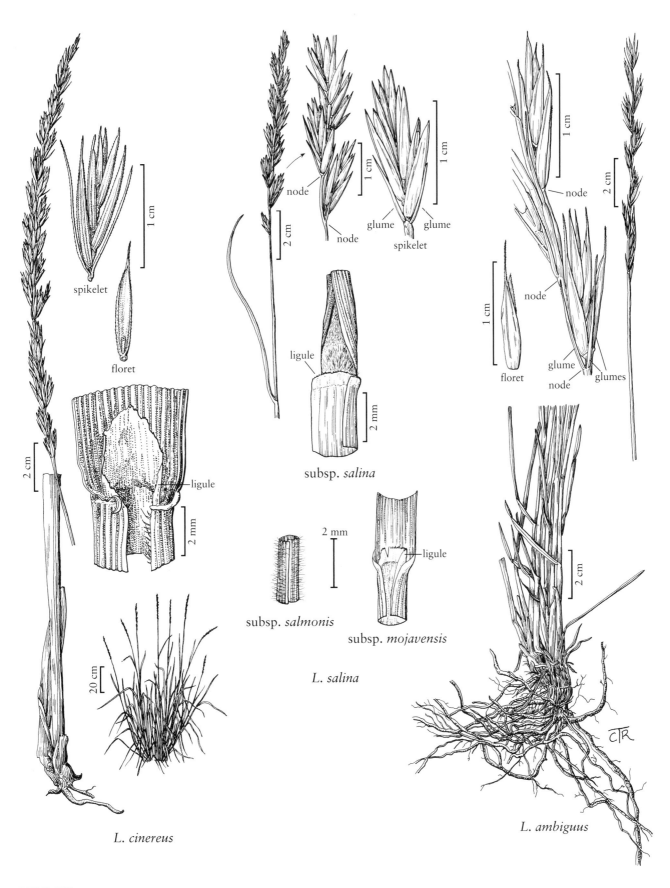

spikelet

floret

ligule

2 cm

20 cm

L. cinereus

node

node

2 cm

glume   glume

spikelet

ligule

2 mm

subsp. *salina*

2 mm

subsp. *salmonis*

ligule

subsp. *mojavensis*

L. salina

1 cm

node

node

2 cm

floret

node

glume

node   glumes

2 cm

L. ambiguus

CTR

LEYMUS

hairy, adaxial surfaces evenly strigillose to strigose. **Spikes** with 1–2 spikelets at the central nodes, usually 1 at the distal nodes. $2n = 28$.

Leymus salina subsp. *salmonis* grows at scattered locations on rocky hillsides in the mountains of southern Idaho, Nevada, and western Utah.

14. Leymus ambiguus (Vasey & Scribn.) D.R. Dewey [p. 365]
COLORADO WILDRYE

Plants loosely cespitose, occasionally rhizomatous. **Culms** 60–110 cm tall, 1–1.5 mm thick, many together. **Leaves** exceeded by the spikes; **sheaths** glabrous or sparsely pubescent; **auricles** to 1.1 mm; **ligules** 0.2–1.2 mm, truncate; **blades** 2.5–6 mm wide, flat, adaxial surfaces scabridulous, glabrous, veins (9)11–17, unequal, not crowded. **Spikes** 8–17 cm long, 5–10 mm wide, erect, with 2 spikelets at most nodes; **internodes** 4–11 mm, surfaces hairy distally, hairs 0.2–0.8 mm, edges ciliate. **Spikelets** 12–23 mm, with 2–7 florets. **Glumes** unequal, 0.5–2.5 mm wide, tapering from below midlength to the subulate apices, stiff, keeled, the central portion thicker than the margins, glabrous, scabrous, particularly distally, bases not overlapping, 0–1(3)-veined, veins inconspicuous at midlength; **lower glumes** 2–9.5 mm; **upper glumes** 6–14 mm; **calluses** with hairs, hairs about 0.2 mm; **lemmas** 8–14.5 mm, glabrous or sparsely strigose, awned, awns 1.3–7 mm; **anthers** 3.5–7 mm, dehiscent. $2n = 28$.

Leymus ambiguus grows on steep, often boulder-strewn hillsides at scattered locations in Colorado and New Mexico.

15. Leymus innovatus (Beal) Pilg. [p. 367]
DOWNY RYEGRASS, BOREAL WILDRYE

Plants sometimes cespitose, strongly rhizomatous. **Culms** 18–105 cm tall, 2–3 mm thick. **Leaves** exceeded by the spikes; **sheaths** glabrous or hairy, often most densely hairy in the collar region; **auricles** to 1.4 mm; **ligules** 0.1–0.5 mm; **blades** 2–6 mm wide, involute, abaxial surfaces scabridulous or smooth, adaxial surfaces scabrous, occasionally with scattered hairs to 1.5 mm, veins unequal, not crowded. **Spikes** 3–16 cm long, 8–20 mm thick, erect, usually well exserted, with 2–3 spikelets per node; **internodes** 4–6 mm, hairy throughout, edges with hairs to 2.5 mm. **Spikelets** 10–18 mm, with 3–7 florets. **Glumes** often unequal,

sometimes absent, 2.5–12 mm long, 0.5–1 mm wide, hairy, stiff, keeled, the central portion thicker than the margins, bases not overlapping, tapering from near the base to the subulate apices, 0–1(3)-veined, veins inconspicuous at mid-length; **calluses** hairy; **lemmas** 7–12 mm, usually conspicuously villous or velutinous, occasionally glab-rate, hairs 0.7–2.5 mm, awned, awns 2–4 mm; **anthers** 3.5–10 mm, dehiscent. $2n = 28, 56$.

Leymus innovatus is a North American species that grows in open woods and forests, riverbanks, open prairies, and rocky soils, and often in sandy, gravelly, or silty soils, primarily from northern Alaska to Hudson Bay, and south into the Black Hills region of Wyoming and South Dakota. Morphologically, the two subspecies show some overlap. Bowden recognized them in part because of their difference in ploidy level.

1. Spikes 8–16 cm long, 8–15 mm wide; lemma hairs 0.7–2.5 mm long . . . . subsp. *innovatus*
1. Spikes 3–8 cm long, 15–20 mm wide; lemma hairs 1.5–2.5 mm long . . . . subsp. *velutinus*

Leymus innovatus (Beal) Pilg. subsp. innovatus [p. 367]

Spikes 8–16 cm long, 8–15 mm wide. **Glumes** present; **lemma hairs** 0.7–2.5 mm. $2n = 28$.

Leymus innovatus subsp. *innovatus* is the more widespread of the two subspecies, extending across North America from the southern Yukon Territory to Ontario, south in the Rocky Mountains to northern Montana, and, as a disjunct, to the Black Hills region of Wyoming and South Dakota. Closer study is needed to determine its range more exactly.

Leymus innovatus subsp. velutinus (Bowden) Tzvelev [p. 367]

Spikes 3–8 cm long, 15–20 mm wide. **Glumes** sometimes absent. **Lemma hairs** 1.5–2.5 mm. $2n = 56$.

Leymus innovatus subsp. *velutinus* is the more northern of the two subspecies, growing in Alaska, the Yukon Territory, and the western Northwest Territories.

16. Leymus flavescens (Scribn. & J.G. Sm.) Pilg. [p. 367]
YELLOW WILDRYE

Plants sometimes cespitose, strongly rhizomatous. **Culms** 40–120 cm tall, 2–4 mm thick, pubescent beneath the nodes. **Leaves** exceeded by the spikes; **sheaths** glabrous; **auricles** absent, sometimes with a few hairs in the auricular position; **ligules** 0.3–1.5 mm; **blades** 3–4 mm wide, usually involute, adaxial surfaces scabrous, sometimes with scattered hairs, hairs to 1 mm, with

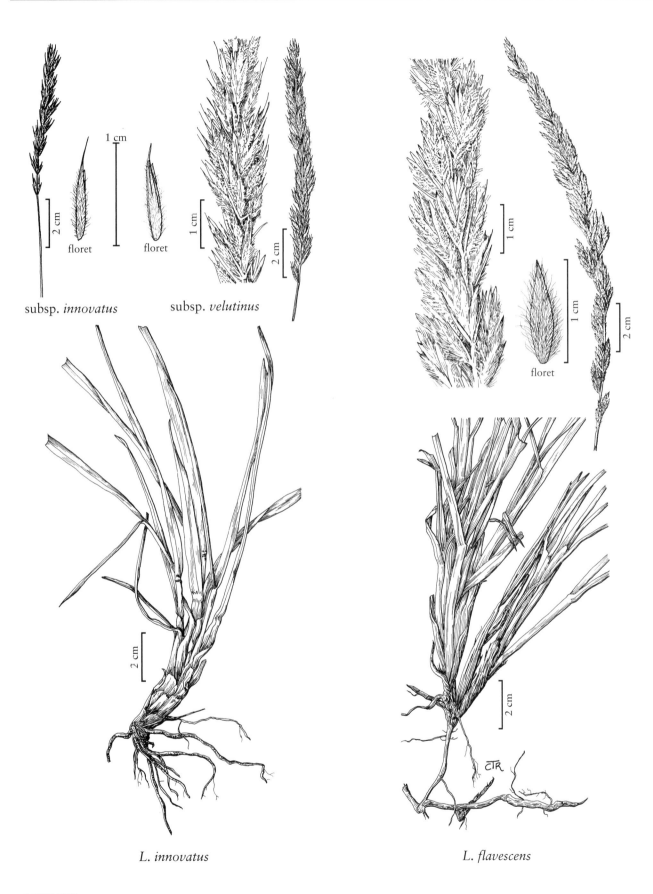

subsp. *innovatus*   subsp. *velutinus*

floret   floret

floret

L. innovatus   *L. flavescens*

LEYMUS

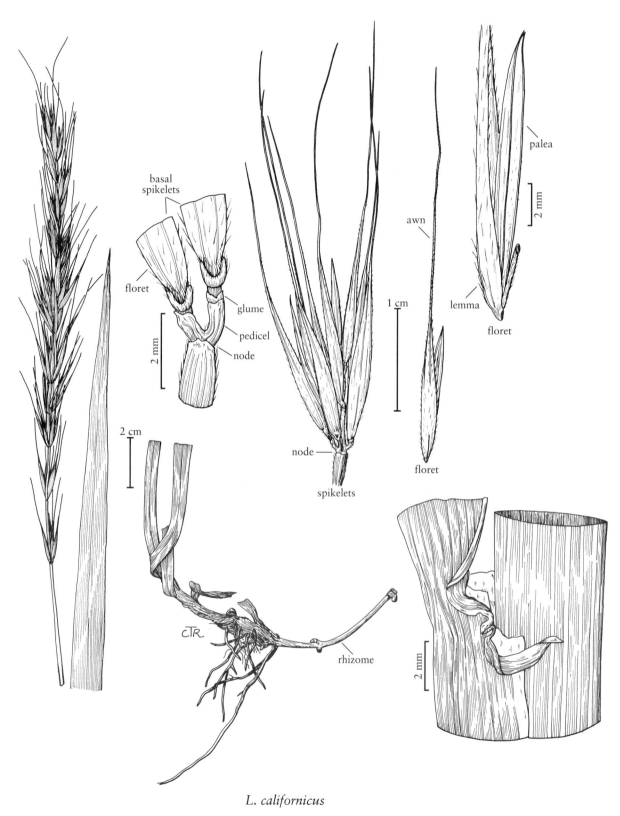

basal
spikelets

floret

glume
pedicel
node

2 mm

2 cm

node

spikelets

1 cm

awn

floret

palea

2 mm

lemma

floret

CTR

rhizome

2 mm

L. californicus

LEYMUS

about 15 closely spaced, subequal, mostly prominently ribbed veins. **Spikes** 10–20 cm long, 12–20 mm thick, with 12–20 nodes and 2 spikelets per node; **internodes** 7–10 mm, densely hairy. **Spikelets** 13.5–25 mm, with 4–9 florets. **Glumes** 8.5–16 mm long, 0.5–2.5 mm wide, stiff, keeled distally, the central portion thicker than the margins, tapering from below midlength to the subulate apices, hairy, 0–1(3)-veined, veins inconspicuous at midlength; **lower glumes** 8.5–13.5 mm; **upper glumes** 10–16 mm; **calluses** poorly developed; **lemmas** 10.5–15 mm, densely villous, hairs 2–3 mm, apices unawned or awned, awns to 2 mm; **anthers** 4.5–7 mm, dehiscent. $2n = 28$.

Leymus flavescens grows on sand dunes and open sandy flats, and ditch- and roadbanks, of the Snake and Columbia river valleys. The central Washington population is growing on a road cut; it seems to be well established there.

Plants identified as *Elymus arenicolus* Scribn. & J.G. Sm. are included here, but they may represent hybrids between *Leymus flavescens* and *L. triticoides*. Leckenby, the collector of the type specimen, noted that they grew on sand or sand drifts along the Columbia River, but could not withstand flooding. He could find no seed.

17. Leymus californicus (Bol. *ex* Thurb.) Barkworth [p. 368]

CALIFORNIA BOTTLEBRUSH

Plants loosely cespitose, rhizomatous. **Culms** 70–200 cm, erect, solitary. **Leaves** exceeded by the spikes; **sheaths** with stiff, 1–3 mm hairs; **auricles** 1–6 mm; **ligules** 1–5 mm; **blades** 6–28 mm wide, lax, flat, scabrous, glabrous, or thinly pilose above. **Spikes** 10–30 cm long, 2–5 cm wide, erect to nodding, lax, nodes with 2–4(5) spikelets, the basal spikelets often on 1–3 mm pedicels; **internodes** 7–11(15) mm. **Spikelets** 12–17 mm, appressed to spreading, with 2–5 florets. **Glumes** absent or shorter than 1 mm; **lemmas** 10–15 mm, sparsely scabrous to appressed hispid, awned, awns 16–33 mm, straight; **anthers** 6–8 mm, dehiscent. $2n = 56$.

Leymus californicus is endemic to coniferous forests near the coast in western California, from Sonoma to Santa Cruz counties, at elevations from near sea level to 300 m. It used to be included in *Hystrix* Moench, a genus that was described as lacking glumes. The type species of *Hystrix* has since been shown to be more closely related to species of *Elymus* than to other species placed in *Hystrix* which, with the exception of *L. californicus*, are native to eastern Asia. Transfer of *L. californicus*, and some of the other species formerly placed in *Hystrix*, to *Leymus* is supported by molecular data (Jensen and Wang 1997; Mason-Gamer 2001). The situation with respect to *L. californicus* illustrates the danger of circumscribing a taxon by its lack of a character. In this case, it appears that reduction in the glumes has taken place within both *Elymus* and *Leymus*.

Leymus californicus is unusual among the other species of *Leymus* in the contiguous United States in growing in a forested habitat, but *L. innovatus* also grows in forests, and some of the Chinese species that have traditionally been placed in *Hystrix* are also reported to grow in forest habitats. Indeed, these similarities, plus its restriction to the vicinity of San Francisco, suggest the possibility that *L. californicus* is an introduced Chinese species.

13.18 ×LEYDEUM Barkworth

Mary E. Barkworth

Plants perennial; rhizomatous, sometimes shortly so. **Culms** to 140 cm tall, 1–3 mm thick. **Spikes** 10–15 cm long, 5–12 mm wide excluding the awns, erect, sometimes lax, nodes with 2–3 spikelets; **internodes** 3–5 mm; **disarticulation** in the rachises, sometimes delayed. **Spikelets** appressed, with 1–3 florets. **Glumes** equal or unequal, 10–25 mm long, 0.2–1.5 mm wide, tapering from below midlength or subulate from the base; **lemmas** glabrous or hairy, awned, awns 1–10 mm; **anthers** 1.8–3 mm long, 0.1–0.3 mm thick. **Caryopses** not developed.

×*Leydeum* consists of hybrids between *Hordeum* and *Leymus*. The number of named ×*Leydeum* hybrids is substantially lower than that for hybrids between *Hordeum* and *Elymus*.

This probably reflects the lower likelihood of such hybrids being formed, because *Leymus* does not incorporate the **H** genome of *Hordeum*, whereas *Elymus* does. It is also possible that such hybrids are less likely to be recognized, because inland species of *Leymus*, like *Hordeum*, have narrow glumes. Of the three species recognized, two involve the coastal species *L. mollis*, which has flat glumes; these hybrids have intermediate glumes and rachises that tend to disarticulate. ×*Leydeum piperi*, the only hybrid involving one of the inland species of *Leymus*, differs from *Leymus* in its disarticulating rachises, and from *Hordeum* in having 2 spikelets per node and 2–3 florets in the larger spikelets.

SELECTED REFERENCES **Bowden, W.M.** 1967. Taxonomy of intergeneric hybrids of the tribe Triticeae from North America. Canad. J. Bot. 45:711–724; **Covas, G.** 1949. Taxonomic observations on the North American species of *Hordeum*. Madroño 9:233–264; **Hodgson, H.J.** and **W.W. Mitchell.** 1965. A new *Elymordeum* hybrid from Alaska. Canad. J. Bot 43:1355–1358; **Hultén, E.** 1968. Comments on the flora of Alaska and Yukon. Ark. Bot., n.s., 7:1–147.

1. ×**Leydeum piperi** (Bowden) Barkworth [p. 371]

Plants shortly rhizomatous. **Culms** to 80 cm tall, about 1.5 mm thick. **Leaves** evenly distributed; **sheaths** smooth; **auricles** to 1.5 mm; **ligules** 0.5–0.8 mm, truncate to rounded; **blades** 10–20 cm long, 1.5–2.5 mm wide, tapering from near the base, abaxial surfaces scabridulous, adaxial surfaces scabrous, with about 12 usually more or less equally prominent veins, apices narrowly acute. **Spikes** 10–15 cm long, 10–15 mm wide including the awns, 5–7 mm wide excluding the awns, lax, nodes with 2(3) spikelets, 1 (central if 3) spikelet sessile, 1(2) spikelet(s) pedicellate, pedicels about 0.2 mm; **internodes** 3–5 mm, completely or mostly concealed by the spikelets; **disarticulation** tardy, in the rachises. **Spikelets** about 15 mm including the awns, about 10 mm excluding the awns, with 1–3 florets, the larger or central spikelets with 2–3 florets, the lateral or smaller spikelets with 1(2) floret(s). **Glumes** 10–15 mm long, about 0.2 mm wide, subulate, scabrous; **lemmas** glabrous, mostly smooth, scabrous distally, the largest lemmas of the larger spikelets 6–7 mm, with awns about 9 mm, the largest lemmas of the smaller or lateral spikelets about 5 mm, with awns about 5.5 mm; **anthers** 1.8–2.2 mm long, about 0.3 mm thick.

Covas (1949) and Bowden (1967) agreed that that the parents of ×*Leydeum piperi* are *Hordeum jubatum* and *Leymus triticoides*. It is not known how common or widespread it is. It differs from ×*Elyhordeum macounii* in its subulate, rather than narrowly linear, glumes.

2. ×**Leydeum dutillyanum** (Lepage) Barkworth [p. 371]

Plants cespitose, rhizomatous. **Culms** about 60 cm tall, 2–2.5 mm thick. **Leaves** evenly distributed on the culms; **sheaths** smooth, glabrous; **auricles** absent or to 0.3 mm; **ligules** 0.2–0.5 mm, truncate, entire; **blades** 20–30 cm long, 5.5–7 mm wide, glabrous, tapering from below midlength, abaxial surfaces with prominent midveins, all veins more or less equally prominent or a few less prominent, apices narrowly acute. **Spikes** about 15 cm long, about 25 mm wide including the awns, 10–12 mm wide excluding the awns, nodes with 2 spikelets; **internodes** 3–5 mm, concealed by the spikelets; **disarticulation** in the rachises. **Spikelets** 18–25 mm, with 2–3 florets, the distal floret reduced. **Glumes** unequal, subulate from the base, to 0.5 mm wide, hairy, hairs about 0.8 mm; **lower glumes** 13–17 mm; **upper glumes** 20–25 mm; **lemmas** 10–12 mm, strigose, hairs about 0.5 mm, awned, awns 8–10 mm; **anthers** 2–2.4 mm.

×*Leydeum dutillyanum* consists of hybrids between *Hordeum jubatum* and *Leymus mollis* (Bowden 1967). It has been reported only from Vieux-Comptoir, Quebec. It appears to disarticulate more readily than ×*L. littorale*.

3. ×**Leydeum littorale** (H.J. Hodgs. & W.W. Mitch.) Barkworth [p. 371]

Plants rhizomatous. **Culms** to 140 cm tall, about 3 mm thick. **Leaves** evenly distributed; **sheaths** smooth, glabrous; **auricles** 0.8–1 mm; **ligules** 0.2–0.5 mm, truncate, entire; **blades** to 20 cm long, 6–9 mm wide, glabrous, tapering from near the base, abaxial surfaces smooth, with 20+ veins, veins more or less equally prominent or the primary veins slightly more prominent than the secondary veins, adaxial surfaces scabridulous, apices narrowly acute. **Spikes** 6.5–15 cm long, 12–15 mm wide including the awns, 8–12 mm wide excluding the awns, nodes with 2–3 spikelets; **internodes** 3.5–5 mm, concealed by the spikelets; **disarticulation** in the rachises, possibly also in the spikelets. **Spikelets** 12–15 mm, with 1–3 florets, the distal florets reduced. **Glumes** equal, 11–13 mm long, 0.8–1.5 mm wide, 3-veined at midlength, hairy, hairs about 0.5 mm, somewhat divergent, tapering from below midlength, apices awned, awns 3–4 mm long; **lemmas** 13–15 mm, hairy, awns 1–3 mm; **anthers** about 2.8 mm long, 0.1–0.3 mm thick.

×*Leydeum littorale* consists of hybrids between *Hordeum brachyantherum* and *Leymus mollis*. It has been collected in the Matanuska Valley, Alaska, and on the coast of Vancouver Island, British Columbia; it may be more widespread.

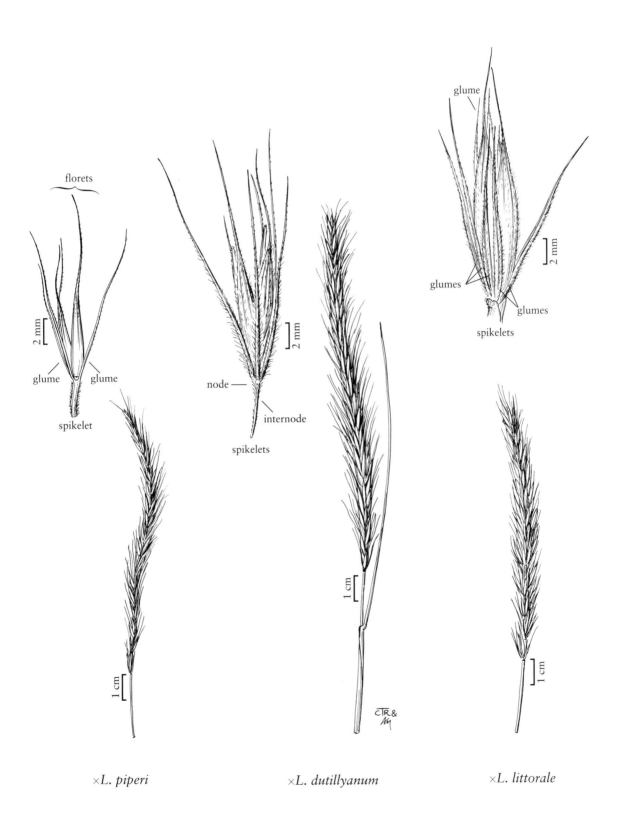

florets

2 mm

glume   glume

spikelet

node —

internode

spikelets

2 mm

glume

glumes

glumes

spikelets

2 mm

1 cm

1 cm

1 cm

×*L. piperi*   ×*L. dutillyanum*   ×*L. littorale*

×LEYDEUM

13.19 **PSATHYROSTACHYS** Nevski

Claus Baden†

Plants perennial; cespitose, forming dense to loose clumps, sometimes stoloniferous, sometimes rhizomatous. **Culms** 15–120 cm, erect or decumbent. **Sheaths** of the basal leaves closed, becoming fibrillose, of the upper cauline leaves open; **auricles** sometimes present; **ligules** 0.2–0.3 mm, membranous; **blades** with prominently ribbed veins on the adaxial surfaces. **Inflorescences** spikes, with 2–3 spikelets per node; **disarticulation** in the rachises. **Spikelets** appressed to ascending, with 1–2(3) florets, often with additional reduced florets distally. **Glumes** equal to unequal, (3.5)4.2–48.5(65) mm including the awns, subulate, stiff, scabrous to pubescent, obscurely 1-veined, not united at the base; **lemmas** 5.5–14.3 mm, narrowly elliptic, rounded, glabrous or pubescent, 5–7-veined, veins often prominent distally, apices sharply acute to awned, sometimes with a minute tooth on either side of the awn base, awns 0.8–34 mm, straight, ascending to slightly divergent, sometimes violet-tinged; **paleas** equaling or slightly longer than the lemmas, membranous, scabrous or pilose on and sometimes also between the keels, bifid; **anthers** 3, 2.5–6.8(7) mm, yellow or violet; **lodicules** 2, acute, entire, ciliate. **Caryopses** pubescent distally, tightly enclosed by the lemmas and paleas at maturity. $x = 7$. Name from the Greek *psathyros*, 'fragile', and *stachys*, 'spike'.

Psathyrostachys has eight species, all of which are native to arid regions of central Asia, from eastern Turkey to eastern Siberia, Russia, and Xinjiang Province, China. One species, *P. juncea*, was introduced to North America as a potential forage species, and is now established in the *Flora* region.

Psathyrostachys is very similar to *Leymus*, particularly the cespitose species of *Leymus*. The major differences are that *Psathyrostachys* has disarticulating rachises and, usually, distinctly awned lemmas.

SELECTED REFERENCES Anamthawat-Jónsson, K. 2005. The Leymus Ns-genome. Czech J. Genet. Pl. Breed. 41(Special Issue):13–20; **Baden, C.** 1991. A taxonomic revision of *Psathyrostachys* (Poaceae). Nordic J. Bot 11:3–26; **Bödvarsdóttir, S.K.** and **K. Anamthawat-Jónsson.** 2003. Isolation, characterization, and analysis of *Leymus*-specific DNA sequences. Genome 46:673–682.

1. **Psathyrostachys juncea** (Fisch.) Nevski [p. 373]
RUSSIAN WILDRYE

Plants densely cespitose. **Culms** (20)30–80(120) cm, erect or decumbent at the base, mostly glabrous, pubescent below the spikes. **Basal sheaths** glabrous, grayish-brown, old sheaths more or less persistent; **auricles** 0.2–1.5 mm; **blades** (1)2.5–18 (30) cm long, (1)5–20 mm wide, flat or involute, abaxial surfaces smooth or scabridulous, often glaucous. **Spikes** (3)6–11(16) cm long, 5–17 mm wide, erect, with (2)3 spikelets per node; **rachises** hirsute on the margins, puberulent elsewhere; **internodes** 3.5–6 mm. **Spikelets** 7–10(12) mm excluding the awns, strongly overlapping, lateral spikelets slightly larger than the central spikelets. **Glumes** (3.5)4.2–9.4 mm, subulate, scabrous or with 0.3–0.8 mm hairs; **lemmas** 5.5–7.5 mm, lanceolate, glabrous or with 0.3–0.8 mm hairs, sharply acute or awned, awns 0.8–3.5 mm; **paleas** 5.8–7.6 mm, scabrous, acute; **anthers** 2.5–5.1 mm; **lodicules** 1.3–1.5 mm. **Caryopses** 4.3–5 mm. $2n = 14$, rarely 28.

Psathyrostachys juncea is native to central Asia, primarily to the Russian and Mongolian steppes. It has become established at various locations, from Alaska to Arizona and New Mexico. It is drought-resistant and tolerant of saline soils. In its native range, it grows on stony slopes and roadsides, at elevations to 5500 m.

Psathyrostachys juncea closely resembles *Leymus cinereus*, differing primarily in having shorter ligules and a rachis that breaks up at maturity. Immature plants can be identified by the more uniform appearance of the spikelets. *Psathyrostachys juncea* also tends to have smaller spikelets with fewer florets than *L. cinereus*. Plants with pilose florets have been treated as a distinct taxon; such recognition is not merited.

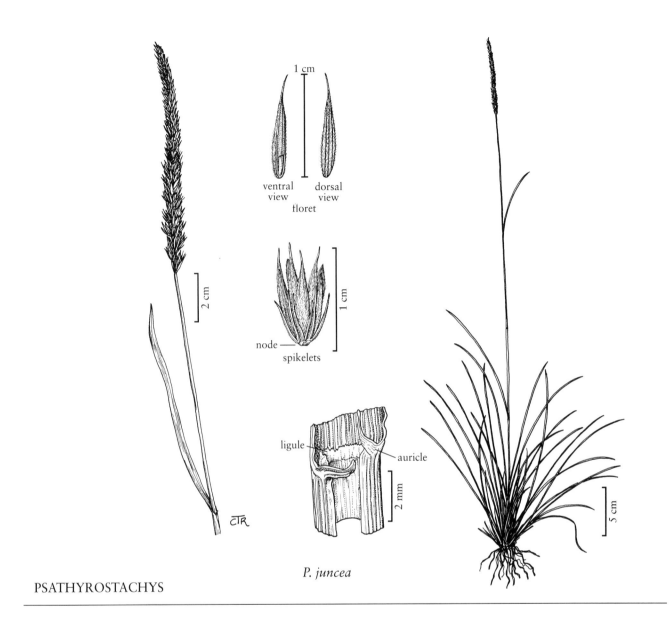

ventral view · dorsal view
floret

node — spikelets

ligule — auricle

P. juncea

PSATHYROSTACHYS

13.20 THINOPYRUM Á. Löve

Mary E. Barkworth

Plants perennial; cespitose or not, sometimes rhizomatous. **Culms** 10–250 cm, usually erect. **Sheaths** open, glabrous or ciliate; **auricles** 0.2–1.8 mm or absent; **ligules** membranous; **blades** convolute or flat. **Inflorescences** terminal, distichous spikes, usually not disarticulating at maturity, with 1 spikelet at all or most nodes; **internodes** 5–30 mm. **Spikelets** 1–3 times the length of the middle internodes, solitary, appressed to ascending, often diamond-shaped in outline and arching outwards at maturity; **disarticulation** tardy, usually beneath the florets, sometimes in the rachises. **Glumes** rectangular to lanceolate, narrowing beyond midlength, stiff, indurate to coriaceous, glabrous or with hairs, keeled or rounded at the base, usually more strongly keeled distally than proximally, 4–9-veined, midveins usually scabrous distally, margins often hyaline, apices truncate to acute, sometimes mucronate, unawned, without lateral teeth; **lemmas** 5-veined, coriaceous, glabrous or with hairs, truncate, obtuse, or acute, sometimes mucronate or awned, awns to 3 cm; **anthers** 3, 2.5–12 mm. $x = 7$. Name from the Greek *thino*, a shore weed, and *pyros*, 'wheat'.

Thinopyrum includes approximately ten species, most of which are alkaline tolerant. It is native from the Mediterranean region to western Asia. Four species are established in the *Flora* region; only *T. intermedium* and *T. ponticum* are common. The genus is sometimes included in *Elytrigia* Desv. or *Elymus*.

Thinopyrum differs from the other *Triticeae* in its thick, stiff glumes and lemmas. These are several cells thick, even between the veins (Jarvie and Barkworth 1992b).

In the key and descriptions, measurements and comments about rachis internodes refer to the internodes at midspike. The lowest internodes of a spike are usually 2–4 times as long as those at midspike.

SELECTED REFERENCES **Assasi, M.** 1994. The genus *Elymus* L. (Poaceae) in Iran: Biosystematic studies and generic delimitation. Ph.D. dissertation, Lund University, Lund, Sweden. 104 pp.; **Barkworth, M.E.** 1997. Taxonomic and nomenclatural comments on the Triticeae in North America. Phytologia 83:302–311; **Darbyshire, S.J.** 1997. Tall Wheatgrass, *Elymus elongatus* subsp. *ponticus*, in Nova Scotia. Rhodora 99:161–165; **Jarvie, J.K.** 1992. Taxonomy of *Elytrigia* sect. *Caespitosae* and sect. *Junceae* (Gramineae: Triticeae). Nordic J. Bot. 12:155–169; **Jarvie, J.K. and M.E. Barkworth.** 1992a. Anatomical variation in some perennial Triticeae. Bot. J. Linn. Soc. 108:287–301; **Jarvie, J.K. and M.E. Barkworth.** 1992b. Morphological variation and genome constitution in some perennial Triticeae. Bot. J. Linn. Soc. 108:167–189; **Liu, Z.W. and R.R.-C. Wang.** 1993. Genome constitutions of *Thinopyrum curvifolium*, *T. scirpeum*, *T. distichum*, and *T. junceum* (Triticeae: Gramineae). Genome 36:641–651; **Ogle, D.** 2001. Intermediate wheatgrass, *Thinopyrum intermedium* (Host) Barkworth & D.R. Dewey. Plant Fact Sheet, U.S. Department of Agriculture, Natural Resource Conservation Service Plant Materials Program. http://plants.usda.gov.

1. Plants not rhizomatous; glumes truncate, midveins about equal in length and prominence to the lateral veins . . . . .4. *T. ponticum*
1. Plants rhizomatous; glumes obliquely truncate or obtuse to acute, midveins usually slightly longer and more prominent than the lateral veins.
  2. Glumes 9–18 mm long . . . . .3. *T. pycnanthum*
  2. Glumes 4.5–8.5 mm long.
    3. Lemmas 7.5–10 mm long, glabrous or hairy; rachis internodes 7–12 mm long; plants widespread in the *Flora* region, particularly in the western United States . . . . .1. *T. intermedium*
    3. Lemmas 10–17 mm long, glabrous; rachis internodes 12–28 mm long; plants known only from a few coastal locations in the *Flora* region . . . . .2. *T. junceum*

1. **Thinopyrum intermedium** (Host) Barkworth & D.R. Dewey [p. 375]

Plants rhizomatous, often glaucous. **Culms** 50–115 cm, glabrous or hairy, sometimes hairy only on the nodes; **lowest internode plus sheath** 3–5 mm thick. **Sheaths** mostly glabrous, often ciliate on the margins; **auricles** 0.5–1.8 mm; **ligules** 0.1–0.8 mm; **blades** 2–8 mm wide, flat, abaxial surfaces glabrous, adaxial surfaces usually sparsely strigose, sometimes with hairs of mixed lengths, with 7–30 ribs, ribs not prominent, margins whitish, thicker than the veins. **Spikes** 8–21 cm, erect or lax; **internodes** 7–12 mm; **rachises** glabrous or with hairs, scabrous on the edges, particularly distally, not disarticulating at maturity. **Spikelets** 11–18 mm, with 3–10 florets; **disarticulation** beneath the florets. **Glumes** oblong, glabrous and mostly smooth, or strigose with 1–1.5 mm hairs, hairs usually evenly distributed, weakly keeled distally, keels scabrous, at least distally, midvein usually more prominent and longer than the lateral veins, margins not hyaline or hyaline near the apices, apices obliquely truncate or obtuse to acute, sometimes mucronate; **lower glumes** 4.5–7.5 mm long, 1.5–2.5 mm wide, 5–6-veined; **upper glumes** 5.5–8.5 mm long, 2–3 mm wide, 5–7-veined; **lemmas** 7.5–10 mm, glabrous or with 1–1.5 mm hairs, hairs usually evenly distributed, sometimes only on the outer portion of the lemmas, apices occasionally awned, awns to 5 mm; **paleas** 7–9.5 mm, keels usually scabrous for $^1/_2$ their length; **anthers** 5–7 mm. $2n = 42, 43$.

Thinopyrum intermedium is native to Europe and western Asia. It is widely established in western North America, having been introduced for erosion control, revegetation, forage, and hay. It also occurs in scattered locations further east. One of its advantages for erosion

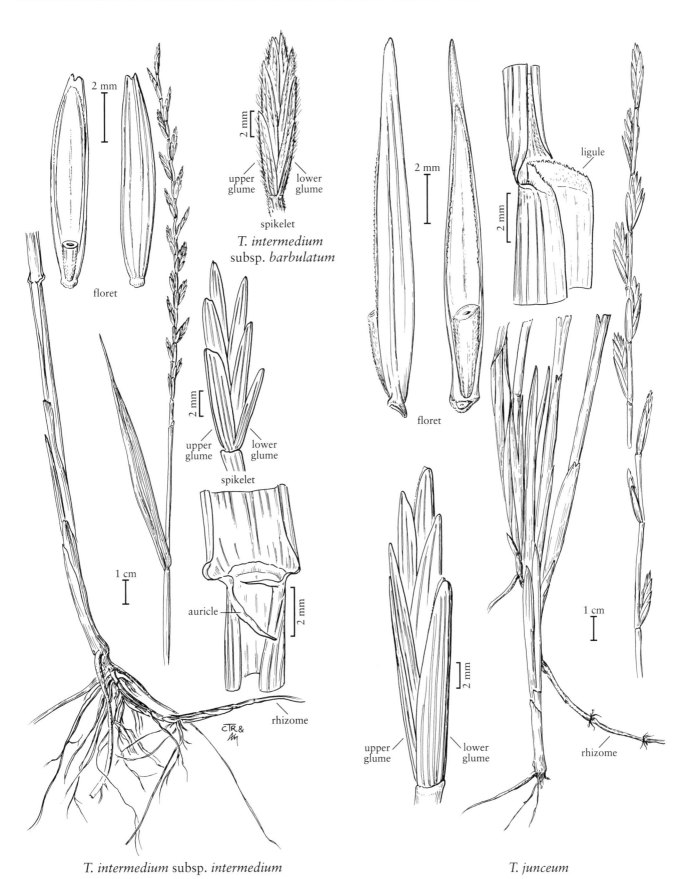

2 mm

floret

2 mm

upper
glume

lower
glume

spikelet

T. *intermedium*
subsp. *barbulatum*

2 mm

upper
glume

lower
glume

spikelet

auricle

2 mm

rhizome

1 cm

T. intermedium subsp. *intermedium*

2 mm

floret

2 mm

ligule

2 mm

upper
glume

lower
glume

1 cm

rhizome

T. junceum

THINOPYRUM

control and revegetation is that it establishes rapidly in many different habitats. In its native range, it grows in dry areas with sandy or stony soils. In Europe, it forms sterile hybrids with *Elymus repens*; no such hybrids are known from North America.

Several subspecies have been recognized within *Thinopyrum intermedium*, usually based on differences in the vestiture of the glumes and lemmas, the presence or absence of lemma awns, and the color of the plants. Assadi (1994) commented that there was little correlation between the different character states. He grew seeds from several wild plants and, even when most of the offspring resembled the parent plant, there was often segregation of some variants. Crossing experiments showed that hybrids between the morphological variants were fertile, and usually had regular meiosis. He noted, however, that the plants with glabrous spikelets tended to grow in mesophytic habitats, those with hairy glumes and lemmas on dry slopes, and those with ciliate glumes and lemmas at the edges of fields and in wet places. This difference in habitat preference was reiterated by Ogle (2001). Because of this ecological distinction, they are formally recognized here as subspecies. Plants with hairs only near the lemma margins are included under *T. intemedium* subsp. *intermedium*. They may be derived from crosses between the hairy and glabrous plants, a possibility that has not been experimentally evaluated. There seems to be little correlation between spikelet vestiture and that of the leaves and stems.

1. Lemmas and glumes glabrous ...... subsp. *intermedium*
1. Lemmas with hairs, sometimes only on the margins, hairs 1–1.5 mm long; glumes usually hairy throughout, sometimes glabrous but scabrous over the veins .... subsp. *barbulatum*

Thinopyrum intermedium subsp. **barbulatum** (Schur) Barkworth & D.R. Dewey [p. 375]
HAIRY WHEATGRASS

Glumes usually hairy throughout, sometimes merely scabrous on the veins; **lemmas** hirsute throughout, sometimes only on the margins, hairs 1–1.5 mm.

There is no known difference in geographic distribution between subsp. *barbulatum* and subsp. *intermedium* in the *Flora* region. Ogle (2001) states that *T. intermedium* subsp. *barbulatum* is adapted to areas with 11–12 inches of rainfall per year.

Thinopyrum intermedium (Host) Barkworth & D.R. Dewey subsp. **intermedium** [p. 375]
INTERMEDIATE WHEATGRASS

Glumes usually glabrous, smooth or scabrous over 1 or more veins; **lemmas** usually glabrous, sometimes with hairs on the margins.

There is no known difference in geographic distribution between subsp. *intermedium* and subsp. *barbulatum* in the *Flora* region. Ogle (2001) states that *T. intermedium* subsp. *intermedium* is adapted to areas with 12–13 inches of rainfall per year.

2. **Thinopyrum junceum** (L.) Á. Löve [p. 375]
RUSSIAN WHEATGRASS

Plants not cespitose, rhizomatous. **Culms** 27–49 cm, glabrous; **lowest internode plus sheath** 3–8 mm thick. **Sheaths** glabrous; **auricles** absent; **ligules** 0.5–1.5 mm; **blades** 2–4 mm wide, convolute, abaxial surfaces glabrous, adaxial surfaces scabrous to densely pubescent, with 3–8 ribs, ribs narrow, prominent, margins not conspicuously thickened. **Spikes** 4–55 cm, erect; **internodes** 12–28 mm; **rachises** glabrous; **disarticulation** in the rachises. **Spikelets** 14–30 mm, appressed to the rachises, with 4–8 florets. **Glumes** lanceolate, glabrous, midveins slightly longer and more prominent than the lateral veins, apices obtuse to acute, often mucronate, margins not hyaline; **lower glumes** 10–18 mm, keeled, keels prominent; **upper glumes** 9–16 mm; **lemmas** 10–17 mm long, 2–4.5 mm wide, glabrous; **paleas** 9–14 mm, keels ciliate for almost their entire length; **anthers** 6–12 mm. $2n = 42, 56$.

Thinopyrum junceum is native to the coast of Portugal, the Mediterranean, and the Black Sea. In the *Flora* region, it has been found on the coasts of southern California and Nova Scotia. In its native range, it grows on maritime rocky coasts, shifting beach sands, and, occasionally, by brackish water near river mouths.

3. **Thinopyrum pycnanthum** (Godr.) Barkworth [p. 377]
TICK QUACKGRASS

Plants rhizomatous. **Culms** 10–120 cm, glabrous; **lowest internode plus sheath** 2–3 mm thick. **Lower sheaths** ciliate; **auricles** absent or to 0.5 mm; **ligules** 0.3–0.6 mm, truncate, ciliate; **blades** to 35 cm long, 2–6 mm wide, flat or convolute, glaucous, adaxial surfaces with 3–20 ribs, ribs prominent, crowded, flattened, scabrous. **Spikes** 4–20 cm; **rachises** glabrous; **internodes** 5–9 mm. **Spikelets** 10–20 mm, with 3–10 florets; **disarticulation** beneath the florets. **Glumes** 4.5–8 mm, glabrous, oblong-lanceolate, 4–7-veined, weakly keeled, keels asymmetric, scabridulous, midveins slightly longer and

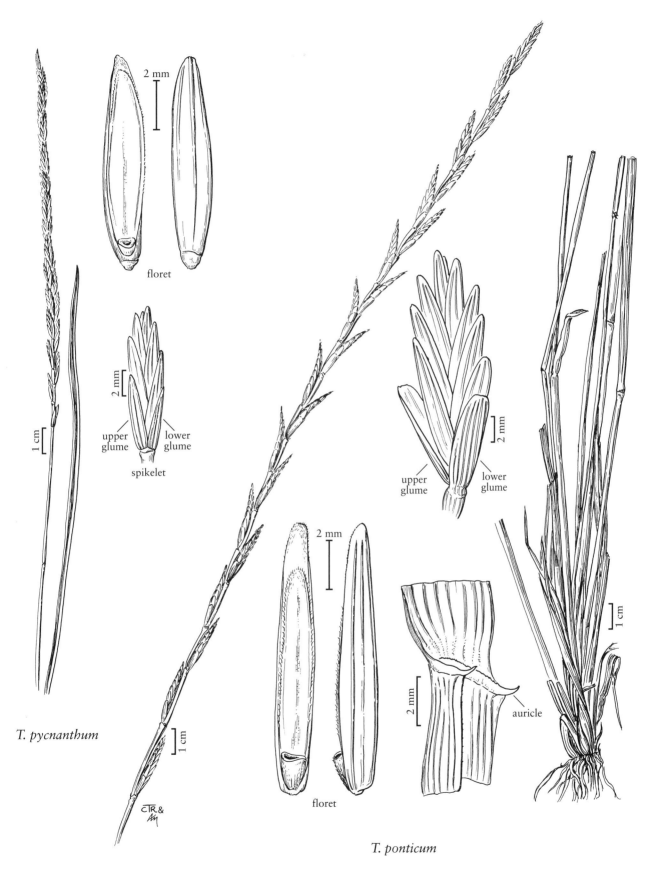

floret

2 mm

upper
glume   lower
                         glume

spikelet

2 mm

1 cm

T. pycnanthum

1 cm

upper
glume   lower
               glume

2 mm

2 mm

floret

2 mm

auricle

1 cm

T. ponticum

CTR&
AM

THINOPYRUM

more prominent than the lateral veins, apices acute; **lemmas** 7–11 mm, glabrous, subobtuse, unawned, mucronate, or awned, awns to 10 mm; **paleas** keeled, keels ciliate; **anthers** 5–7 mm. $2n = 42$.

Thinopyrum pycnanthum is native to the coasts of western and southern Europe. It is reported from scattered locations in the western United States, and from Nova Scotia to Pennsylvania in eastern North America. In its native range, it grows in maritime sands and gravels, or river gravels.

4. Thinopyrum ponticum Barkworth & D.R. Dewey [p. 377]

TALL WHEATGRASS, RUSH WHEATGRASS

Plants cespitose, not rhizomatous. **Culms** 50–200 cm, glabrous; **lowest internode plus sheath** about 3.5 mm thick. **Sheaths** ciliate on the lower margins; **auricles** 0.2–1.5 mm; **ligules** 0.3–1.5 mm; **blades** 2–6.5 mm wide, generally convolute, adaxial surfaces with 1–8 ribs, ribs rounded, prominent, spinulose, margins usually thinner than the ribs. **Spikes** 10–42 cm, erect; **internodes** 9–19 mm; **rachises** glabrous, not disarticulating at maturity. **Spikelets** 13–30 mm, with 6–12 florets; **disarticulation** beneath the florets. **Glumes** oblong, glabrous, 5–9-veined, midveins about equal in length and prominence to the lateral veins, margins about 0.5 mm wide, hyaline, apices truncate; **lower glumes** 6.5–10 mm, midveins occasionally scabrous distally; **upper glumes** 7–10 mm; **lemmas** 9–12 mm, glabrous; **paleas** 7.5–11 mm, keeled, keels ciliate; **anthers** 4–6 mm. $2n = 69, 70$.

Thinopyrum ponticum is native to southern Europe and western Asia. In its native range, it grows in dry and/or saline soils. In the *Flora* region, *T. ponticum* is planted along roadsides for soil stabilization, and is spreading naturally in cooler areas because of its tolerance of the saline conditions caused by salting roads in winter. It is sometimes treated as a subspecies of *T. elongatum* (Host) D.R. Dewey, a diploid species that grows in maritime regions of western Europe.

14. POEAE R. Br.

Mary E. Barkworth

Plants annual or perennial; cespitose, rhizomatous, or stoloniferous. **Culms** annual, not woody, not branching above the base; **internodes** usually hollow. **Sheaths** usually open for most of their length, sometimes closed; **collars** without tufts of hair on the sides; **auricles** usually absent; **ligules** membranous to hyaline, sometimes ciliate, those of the upper and lower cauline leaves usually similar; **pseudopetioles** not developed; **blades** linear to narrowly lanceolate, venation parallel, cross venation not evident, without arm or fusoid cells, epidermes without microhairs, not papillate, cross sections non-Kranz. **Inflorescences** terminal, usually panicles, sometimes spikes, panicles sometimes spikelike or reduced to racemes in depauperate specimens; **disarticulation** usually above the glumes and beneath the florets, sometimes below the glumes. **Spikelets** 0.7–50 mm, laterally compressed, sometimes weakly so, sometimes viviparous, usually with 2–22 florets, sometimes with 1, sterile florets usually distal to the reproductively functional florets, sometimes with 1 or 2 staminate or sterile florets below a bisexual floret, sterile florets often reduced in size; **rachillas** sometimes prolonged beyond the base of the distal florets. **Glumes** (0,1)2, equal or unequal, shorter or longer than the adjacent florets, sometimes exceeding the distal florets; **florets** laterally compressed; **calluses** glabrous or hairy, not well developed; **lemmas** lanceolate to ovate, 1–7(9)-veined, unawned or awned, veins usually converging distally, sometimes parallel, awns from basal to terminal on the lemmas, straight or bent; **paleas** 2-keeled, from shorter than to longer than the lemmas, sometimes absent or minute; **lodicules** 2, membranous, not or weakly veined; **anthers** 3; **ovaries** usually glabrous, sometimes hairy distally; **styles** 2, bases free. **Caryopses** longitudinally grooved or not, not beaked, pericarp thin; **hila** punctate to linear; **embryos** from $^{1}/_{4}$–$^{1}/_{3}$ as long as the caryopses. $x = 7$.

The *Poeae* constitute the largest tribe of grasses, encompassing around 115 genera and 2500 species. The species are primarily cool-temperate to arctic in their distribution. In the *Flora* region, there are 63 non-hybrid genera with 344 species, and 4 hybrid genera, each of which has one species. Many of the tribe's species are well known as lawn and pasture grasses, for example, *Poa pratensis* (Kentucky bluegrass), *Dactylis glomerata* (orchard grass), and *Phleum pratense* (timothy).

The tribe's circumscription and its infratribal taxonomy are unclear. It is interpreted here as including generic groups that are, or have been, treated in other works as tribes (e.g., *Agrostideae* Dumort., *Aveneae* Dumort., *Hainardeae* Greut., and *Phalarideae* Dumort.). Some of these are sometimes recognized as subtribes, often with modified circumscriptions. Recent studies (e.g., Catalán et al. 1997, 2004; Soreng and Davis 1998) indicate that there are some infratribal groupings that, based on chloroplast DNA data, appear stable; other groupings do not. In addition, there is little support for the monophyly of some genera, notably *Festuca* and its allies.

The following key does not include these four hybrid genera: ×*Agropogon* (*Agrostis* × *Polypogon*, p. 668), ×*Arctodupontia* (*Arctophila* × *Dupontia*, p. 604), ×*Dupoa* (*Dupontia* × *Poa*, p. 601), and ×*Pucciphippsia* (*Puccinellia* × *Phippsia*, p. 477). They are described on the pages indicated. In the key that follows, branch measurements include spikelets, but not awns.

SELECTED REFERENCES **Catalán, P., E.A. Kellogg**, and **R.G. Olmstead**. 1997. Phylogeny of Poaceae subfamily Poöideae based on chloroplast *ndh*F gene sequences. Molec. Phylogenet. Evol. 8:150–166; **Catalán, P., P. Torrecilla, J.Á. López Rodríguez**, and **R.G. Olmsted**. 2004. Phylogeny of the festucoid grasses of subtribe Loliinae and allies (Poeae, Poöideae) inferred from ITS and *trn*L–F sequences. Molec. Phylogenet. Evol. 31:517–541; **Macfarlane, T.D.** 1987. Poaceae subfamily Poöideae. Pp. 265–276 *in* T.R. Soderstrom, K.W. Hilu, C.S. Campbell and M.E. Barkworth (eds.). Grass Systematics and Evolution. Smithsonian Institution Press, Washington, D.C., U.S.A. 473 pp.; **Soreng, R.J.** and **J.I. Davis**. 1998. Phylogenetics and character evolution in the grass family (Poaceae): Simultaneous analysis of morphological and chloroplast DNA restriction site character sets. Bot. Rev. (Lancaster) 64:1–85.; **Soreng, R.J., J.I. Davis**, and **J.J. Doyle**. 1990. A phylogenetic analysis of chloroplast DNA restriction site variation in Poaceae subfam. Poöideae. Pl. Syst. Evol. 172:83–97.

1. All or almost all spikelets viviparous, the spikelets producing plantlets [if sexual spikelets are common, take the alternate lead].
  2. Panicle branches smooth or slightly scabrous, the scabrules widely spaced; blades with a translucent line on either side of the midvein, apices usually prowlike . . . . 14.13 *Poa* (in part)
  2. Panicle branches scabrous; blades without a translucent line on either side of the midvein, apices usually not prowlike.
    3. Sheaths closed for ¹/₂ or more of their length; ligules 0.1–0.6 mm long . . . . 14.01 *Festuca* (in part)
    3. Sheaths open; ligules 1.5–13 mm long . . . . 14.26 *Deschampsia* (in part)
1. Some, usually all, spikelets sexually functional, with 1–25 bisexual or unisexual florets, sometimes with sterile and sexual spikelets mixed within an inflorescence.
  4. Inflorescences with 2 morphologically distinct forms of spikelets.
    5. Spikelets in pairs, the pedicels not fused at the base, smooth or slightly scabrous; disarticulation above the glumes and beneath the florets . . . . 14.37 *Cynosurus*
    5. Spikelets in fascicles, the pedicels fused at the base, glabrous, hispid or strigose; disarticulation at the base of the fused pedicels.
      6. Secondary panicle branches sharply bent below the pedicels; glumes not winged . . . . 14.11 *Lamarckia*
      6. Secondary panicle branches straight below the pedicels; glumes winged . . . . . . 14.61 *Phalaris* (in part)
  4. Inflorescences with all spikelets morphologically alike.
    7. Glumes with pilose awns . . . . 14.30 *Lagurus*
    7. Glumes, if present, unawned or with glabrous awns.
      8. Inflorescences spikes with 1–2(4) spikelets per node, or spikelike racemes with 1 spikelet at all or most nodes.
        9. Spikelets with 1 functional floret, sometimes a reduced, sterile floret also present.
          10. Glumes membranous, flexible; all spikelets pedicellate, pedicels 0.5–1 mm long, 0.1–0.2 mm thick . . . . 14.59 *Mibora*

10. Glumes coriaceous, stiff; lower spikelets sessile, upper spikelets sometimes pedicellate.
 11. Spikelets radial to the rachises, most spikelets with 1 glume, the terminal spikelets with 2 glumes . . . . 14.40 *Hainardia*
 11. Spikelets tangential to the rachises, all with 2 glumes.
  12. Lemmas unawned . . . . 14.38 *Parapholis*
  12. Lemmas awned, awns 2–4 mm . . . . 14.39 *Scribneria*
9. Spikelets with 2–25 functional florets.
 13. Lemmas awned from about midlength, awns 8–26 mm long, twisted proximally.
  14. Adaxial surfaces of the leaves ribbed; rachillas pilose on all sides; ligules truncate to rounded, 0.5–1.5 mm long . . . . 14.47 *Helictotrichon* (in part)
  14. Adaxial surfaces of the leaves unribbed; rachillas glabrous on the side adjacent to the paleas, hairy elsewhere; ligules acute to truncate, 0.5–7 mm long . . . . 14.45 *Avenula* (in part)
 13. Lemmas unawned or apically awned, awns straight.
  15. Spikelets sessile; lemmas 2–12 mm long.
   16. Spikelets radial to the rachises, most spikelets with 1 glume, only the terminal spikelet with 2 glumes . . . . 14.05 *Lolium*
   16. Spikelets tangential to the rachises, all spikelets with 2 glumes . . . . 14.50 *Gaudinia*
  15. Spikelets subsessile to pedicellate, pedicels 0.5–3 mm long.
   17. Plants perennial . . . . 14.01 *Festuca* (in part)
   17. Plants annual.
    18. Inflorescences usually exceeded by the leaves; spikelets with (2)3–4(7) florets; lemmas (5)7–9-veined, apices round to emarginate, not bifid; culms usually prostrate or procumbent . . . . 14.09 *Sclerochloa* (in part)
    18. Inflorescences usually exceeding the leaves; spikelets with 4–25 florets; lemmas 5-veined, apices acute to obtuse, sometimes bifid; culms procumbent to erect . . . . 14.35 *Desmazeria* (in part)
8. Inflorescences panicles or racemes, with more than 1 spikelet associated with each node.
 19. Inflorescences racemes or spikelike panicles, with all branches shorter than 1 cm [for opposite lead, see p. 382].
  20. Leaves usually exceeding the inflorescences; culms usually prostrate to procumbent; lemmas indurate at maturity . . . . 14.09 *Sclerochloa* (in part)
  20. Leaves usually exceeded by the inflorescences; culms usually erect or decumbent at the base; lemmas usually membranous or papery, sometimes coriaceous, not indurate.
   21. Spikelets disarticulating below the glumes or, if the spikelets are attached to stipes, at the base of the stipes; glume bases sometimes fused.
    22. Spikelets weakly laterally compressed, with stipes that fall with the spikelets; glume bases not fused; glumes usually awned . . . . 14.28 *Polypogon* (in part)
    22. Spikelets strongly laterally compressed, without stipes; glume bases sometimes fused; glumes unawned or awned.
     23. Lemmas dorsally awned; spikelets oval in outline; glumes often connate at the base, often winged distally, keels sometimes ciliate, apices never abruptly truncate . . . . 14.66 *Alopecurus* (in part)
     23. Lemmas usually unawned, occasionally subterminally awned; spikelets often U-shaped in outline, sometimes oval; glumes not connate at the base, not winged, often strongly ciliate on the keels and abruptly truncate to an awnlike apex . . . . 14.31 *Phleum* (in part)
   21. Spikelets disarticulating above the glumes; glume bases not fused.
    24. Spikelets with 2–25 bisexual florets, the sterile or staminate florets, if present, distal to the bisexual florets.

25. Sheaths closed for at least ¹/₂ their length.
   26. Lemma midveins sometimes excurrent up to 2.2 mm, other
      veins not excurrent; plants native, arctic . . . . 14.15 *Dupontia* (in part)
   26. Lemmas with 3–5 veins excurrent, forming awnlike teeth;
      plants cultivated . . . . 14.34 *Sesleria*
25. Sheaths open for all or almost all of their length.
   27. Distal lemmas, sometimes all lemmas, awned from below
      midlength . . . . 14.22 *Aira* (in part)
   27. All lemmas unawned or apically awned.
      28. Lemmas coriaceous at maturity, unawned, sometimes
         mucronate . . . . 14.35 *Desmazeria* (in part)
      28. Lemmas membranous, apically awned, awns 0.3–22
         mm long.
         29. Lemma margins involute, not scarious . . . . 14.04 *Vulpia* (in part)
         29. Lemma margins flat, scarious . . . . 14.58 *Rostraria* (in part)
24. Spikelets with 1 bisexual floret, sometimes with 1–2 sterile florets
   below the bisexual floret, the sterile florets sometimes reduced to
   lemmas, sometimes resembling tufts of callus hair.
   30. Spikelets with 1–2 sterile or staminate florets below the
      bisexual florets, these from larger than to much smaller than
      the bisexual florets, sometimes resembling tufts of hair; glumes
      sometimes winged distally.
      31. Fresh leaves not sweet-smelling when crushed; sterile
         lemmas unawned; bisexual lemmas usually hairy,
         sometimes sparsely so; glumes subequal, sometimes winged
         distally . . . . 14.61 *Phalaris* (in part)
      31. Fresh leaves sweet-smelling when crushed; sterile lemmas
         awned; bisexual lemmas glabrous; glumes unequal, not
         winged . . . . 14.60 *Anthoxanthum* (in part)
   30. Spikelets without sterile or staminate florets below the bisexual
      floret; glumes not winged distally.
      32. Lemmas dorsally awned, awns geniculate; lateral lemma
         veins excurrent, forming 4 teeth, teeth sometimes awnlike
         . . . . 14.44 *Bromidium* (in part)
      32. Lemmas unawned or with only 1 awn, awns not strongly
         geniculate; lateral lemma veins not excurrent.
         33. Spikelets 8–15 mm long; lemmas more than ³/₄ as long
            as the glumes; plants strongly rhizomatous . . . 14.64 *Ammophila* (in part)
         33. Spikelets 1.2–7 mm long; lemmas less than ³/₄ as long
            as the glumes; plants rhizomatous or not.
            34. Sheaths closed for at least ¹/₂ their length.
               35. Calluses glabrous; exposed at maturity;
                  lemmas 1–3-veined, unawned; [other genera
                  may develop long caryopses when infected by
                  nematodes or fungi; such caryopses are
                  usually deformed and filled with eggs, larvae,
                  or spores] . . . . 14.08 *Phippsia* (in part)
               35. Calluses with a ring of stiff hairs, hairs to
                  about 1 mm long; lemmas 3–11-veined
                  . . . . 14.15 *Dupontia* (in part)
            34. Sheaths open for most of their length.
               36. Spikelet bases usually U-shaped, sometimes
                  cuneate; glumes equal, midveins usually
                  strongly ciliate . . . . 14.31 *Phleum* (in part)
               36. Spikelet bases cuneate; glumes unequal,
                  midveins not strongly ciliate.

37. Both glumes twice as long as the lemmas; lemmas pubescent . . . . 14.32 *Gastridium* (in part)
37. Glumes from slightly shorter than to slightly longer than the lemmas; lemmas glabrous, sometimes scabridulous or scabrous.
    38. Lemma awns 4–16 mm long; plants annual; paleas from ³/₄ as long as to slightly longer than the lemmas . . . . 14.67 *Apera*
    38. Lemma awns to 10 mm, if longer than 4 mm, plants perennial and/or paleas less than ¹/₂ as long as the lemmas . . . . 14.27 *Agrostis* (in part)
19. Inflorescences panicles, dense to open, sometimes compact, usually at least some branches longer than 1 cm [for opposite lead, see p. 380].
    39. Caryopses usually as long as or longer than the lemmas, exposed at maturity; lemmas 1–3-veined, unawned; spikelets with 1 floret; sheaths of the flag leaves closed for at least ¹/₂ their length; calluses glabrous [other genera may develop long caryopses when infected by nematodes or fungi; such caryopses are usually deformed or filled with eggs, larvae, or spores].
        40. Lemmas 1-veined, narrowed to awnlike apices; sheaths strongly inflated; glumes absent; plants of temperate habitats . . . . 14.23 *Coleanthus*
        40. Lemmas 1–3-veined, apices acute to rounded; sheaths not inflated; glumes developed, caducous or persistent; plants of arctic or alpine habitats . . . . 14.08 *Phippsia* (in part)
    39. Caryopses shorter than the lemmas, concealed at maturity; lemmas 3–11-veined; spikelets with 1 or more florets; leaf sheaths open or closed; calluses glabrous or with hairs.
        41. Panicle branches secund, appearing 1-sided; spikelets strongly imbricate, subsessile.
            42. Culms usually prostrate or procumbent; glumes obtuse to emarginate . . . . 14.09 *Sclerochloa* (in part)
            42. Culms erect or ascending; glumes apiculate to awn-tipped.
                43. Lemmas awned, awns of the lowest lemmas 0.3–22 mm long . . . . 14.04 *Vulpia* (in part)
                43. Lemmas unawned, sometimes awn-tipped.
                    44. Spikelets circular to ovate or obovate in outline, with 1–2 florets; glumes almost entirely concealing the sides of the florets; disarticulation below the glumes . . . . 14.12 *Beckmannia*
                    44. Spikelets oval in outline, longer than wide, with 2–6 florets; glumes partially exposing the sides of the florets; disarticulation above the glumes . . . . 14.10 *Dactylis*
        41. Panicle branches not secund; spikelets usually widely spaced to somewhat imbricate, usually clearly pedicellate, sometimes subsessile, sometimes on stipes.
            45. All or most spikelets in an inflorescence with 1 bisexual floret, sometimes with 1–2 sterile or staminate florets below the bisexual floret, the sterile florets sometimes resembling tufts of hair . . . . *Poeae* Subkey I
            45. All or most spikelets in an inflorescence with 2–25 sexual florets, usually all florets bisexual or the distal florets sterile or unisexual, sometimes all florets unisexual, sometimes the plants unisexual . . . . . . *Poeae* Subkey II

POEAE SUBKEY I

Synoecious or monoecious grasses with panicles having at least some branches longer than 1 cm and spikelets with 1 bisexual floret, sometimes with 1–2 sterile or staminate florets below the bisexual floret.

1. Spikelets with 1–2 staminate or sterile florets below the bisexual floret, sterile florets sometimes knoblike or resembling tufts of hair.
  2. Spikelets with 2 florets of similar size, the lower floret staminate; lower lemmas awned, the lemmas of the terminal floret unawned or awned . . . . 14.54 *Arrhenatherum*
  2. Spikelets with 2–3(4) florets, the lower 1–2 florets staminate or sterile, sometimes knoblike or resembling tufts of hair, sometimes larger than the bisexual floret; lemmas of the lower florets awned or unawned, the lemmas of the terminal floret unawned.
    3. Lower sterile florets 2, from shorter than to exceeding the bisexual floret; fresh leaves sweet-smelling when crushed . . . . 14.60 *Anthoxanthum* (in part)
    3. Lower sterile florets 1–2, varying from knoblike projections on the callus of the bisexual floret to linear or lanceolate lemmas up to ³/₄ as long as the bisexual floret; fresh leaves not sweet-smelling when crushed . . . . 14.61 *Phalaris* (in part)
1. Spikelets without staminate or sterile florets below the bisexual florets.
  4. Spikelets 15–50 mm long; lemmas usually dorsally awned, awns 20–90 mm long, sometimes unawned . . . . 14.52 *Avena* (in part)
  4. Spikelets 1–15 mm long; lemmas unawned or awned, awns to 18 mm long, basal, dorsal, subterminal, or terminal.
    5. Glume bases gibbous and subcoriaceous; disarticulation above the glumes . . . . 14.32 *Gastridium* (in part)
    5. Glumes bases not gibbous, usually membranous; disarticulation above or below the glumes.
      6. Lemmas awned, awns longer than 2 mm.
        7. Glumes coriaceous, rigid, hispid or scabrous; lemmas awned, awns 5–14.5 mm long, subterminal . . . . 14.63 *Limnodea*
        7. Glumes membranous, flexible, glabrous or with soft hairs, usually smooth; lemmas awned, awns 0.5–18 mm long, sometimes subterminal.
          8. Disarticulation below the glumes.
            9. Spikelets borne on stipes; disarticulation at the base of the stipes; lemmas 0.5–2 mm long; glumes usually awned, sometimes unawned . . . . 14.28 *Polypogon* (in part)
            9. Spikelets borne on pedicels; disarticulation immediately below the glumes; lemmas 1.5–7.5 mm long; glumes usually unawned.
              10. Paleas absent or greatly reduced; lemma awns attached at midlength or below; glume bases often fused; rachillas not prolonged beyond the floret base . . . . 14.66 *Alopecurus* (in part)
              10. Paleas from ³/₄ to nearly as long as the lemmas; lemma awns subterminal; glume bases not fused; rachillas usually prolonged beyond the base of the distal floret as a minute stub or slender bristle . . . . 14.62 *Cinna*
          8. Disarticulation above the glumes.
            11. Rachillas not prolonged beyond the base of the distal floret; paleas absent, minute, or subequal to the lemmas; lemmas 0.5–4 mm long.
              12. Lemmas usually glabrous, sometimes pubescent, unawned or awned, if awned, the awns usually shorter than 4.5 mm, sometimes to 10 mm long, basal, dorsal, subterminal, or terminal; veins usually not excurrent, if excurrent, not forming awnlike teeth; panicles often open, sometimes contracted and cylindrical . . . . 14.27 *Agrostis* (in part)
              12. Lemmas pilose and dorsally awned, awns 4.5–6 mm long; lateral lemma veins excurrent, forming 4 teeth, teeth sometimes awnlike; panicles dense . . . . 14.44 *Bromidium* (in part)

11. Rachillas prolonged beyond the base of the distal floret; paleas at least ¹/₂ as long as the lemmas; lemmas 1–8 mm long.

    13. Plants annual; calluses glabrous or sparsely hairy; lemma apices entire; marginal veins not excurrent; awns subterminal . . . . 14.67 *Apera*

    13. Plants perennial; calluses usually abundantly, sometimes sparsely hairy, hairs 0.2–6.5 mm long; lemma apices denticulate or the marginal veins excurrent; awn attachment from nearly basal to subterminal.

        14. Lemma surfaces mostly glabrous; lemma apices denticulate; marginal lemma veins not excurrent . . . . 14.49 *Calamagrostis* (in part)

        14. Lemma surfaces hairy; lemma apices erose or toothed; marginal lemma veins excurrent . . . . 14.43 *Lachnagrostis*

6. Lemmas unawned or, if awned, awns shorter than 2 mm.

  15. Disarticulation below the glumes.

    16. Glumes attached to stipes, disarticulation at the base of the stipes; glumes usually awned, awns flexuous . . . . 14.28 *Polypogon* (in part)

    16. Glumes attached to pedicels, disarticulation immediately beneath the glumes; glumes unawned or with stiff awns.

        17. Lemma awns subterminal; glume bases not fused; paleas from ³/₄ to nearly as long as the lemmas; rachillas prolonged beyond the base of the distal floret for 0.1–1.3 mm . . . . 14.62 *Cinna*

        17. Lemma awns attached at midlength or below; glume bases often fused; paleas absent or greatly reduced; rachillas not prolonged beyond the base of the distal floret . . . . 14.66 *Alopecurus* (in part)

  15. Disarticulation above the glumes.

    18. Glumes 8–15 mm long; plants strongly rhizomatous . . . . 14.64 *Ammophila* (in part)

    18. Glumes 1–10 mm long; plants rhizomatous or not.

        19. Spikelets dorsally compressed; lemmas dark, coriaceous, lustrous, and glabrous . . . . 14.65 *Milium*

        19. Spikelets laterally compressed, sometimes weakly so; lemmas not simultaneously dark, coriaceous, lustrous, and glabrous.

            20. Lower glumes exceeded by the florets, upper glumes exceeded by to exceeding the florets; sheaths usually closed for up to ¹/₅ their length . . . . 14.33 *Arctagrostis*

            20. Both glumes subequal to or exceeding the florets; sheaths open to the base.

                21. Paleas absent or minute to subequal to the lemmas, not veined; rachillas not prolonged beyond the base of the distal florets; lemmas often unawned, sometimes awned, awn attachment basal to terminal . . . . 14.27 *Agrostis* (in part)

                21. Paleas more than ¹/₂ as long as the lemmas, 2-veined; rachillas prolonged beyond the base of the floret by at least 0.1 mm; lemmas often awned, awn attachment usually on the proximal ¹/₂ of the lemmas.

                    22. Calluses hairy, hairs 0.5–4.5 mm long; lemmas usually awned, awns usually attached to the proximal ¹/₂, if the attachment higher, the callus hairs longer than 2 mm and/or the awns geniculate . . . . 14.49 *Calamagrostis* (in part)

                    22. Calluses glabrous or with hairs to about 1 mm long; lemmas unawned or terminally awned, awns to 1(2.2) mm long.

                        23. Glumes 4–9 mm long; sheaths closed for ¹/₂–²/₃ their length; plants of arctic and subarctic regions . . . . 14.15 *Dupontia* (in part)

                        23. Glumes 1.6–4.3 mm long; sheaths open; plants of western North America, from Alaska to California . . . . 14.42 *Podagrostis*

POEAE SUBKEY II

Synoecious, monoecious, or dioecious grasses with spikelets having 2–22 sexual florets, the lower florets sexual, the distal florets sometimes sterile.

1. One or both glumes exceeding the adjacent lemmas, sometimes exceeding the distal floret.
  2. All lemmas within a spikelet unawned or with awns shorter than 2 mm.
    3. Spikelets usually with 2 florets, lemmas of the lower florets unawned, lemmas of the upper florets awned, the awns strongly curved or hooked . . . . 14.53 *Holcus* (in part)
    3. Spikelets with 2–22 florets, all lemmas unawned or if awned, the awns straight.
      4. Leaf sheaths closed for at least ¹/₂ their length; caryopses falling free of the lemma and palea; plants of arctic or subarctic regions.
        5. Lemma apices obtuse; paleas subequal to the lemmas . . . . 14.17 *Arctophila* (in part)
        5. Lemma apices acute to acuminate; paleas shorter than the lemmas . . . . 14.15 *Dupontia* (in part)
      4. Leaf sheaths open for most of their length; caryopses usually falling with the lemma and palea attached; plants of temperate, arctic, or subarctic regions.
        6. Glumes 15–50 mm long; plants annual . . . . 14.52 *Avena* (in part)
        6. Glumes 0.4–9 mm long; plants annual or perennial.
          7. Lemmas inflated, about as wide as long, with broadly rounded backs; calluses glabrous; spikelets pendulous . . . . 14.21 *Briza* (in part)
          7. Lemmas not inflated, longer than wide, keeled to rounded over the midvein; calluses usually with hairs, sometimes glabrous; spikelets not pendulous.
            8. Plants annual; spikelets with 2 florets; lemmas evenly hairy, 3-veined . . . . 14.46 *Dissanthelium*
            8. Plants usually perennial, sometimes annual; spikelets with 2–10 florets; lemmas usually glabrous or with unevenly distributed hairs, never both annual and with evenly distributed hairs, 3–9-veined.
              9. Rachilla internodes hairy, hairs at least 1 mm long.
                10. Lemma apices truncate, erose to 2–4-toothed . . . . 14.26 *Deschampsia* (in part)
                10. Lemmas apices acute, bifid . . . . 14.56 *Trisetum* (in part)
              9. Rachilla internodes glabrous or with hairs shorter than 1 mm on the distal portion.
                11. Plants strongly rhizomatous; glumes 5–9 mm long . . . . 14.51 *Scolochloa* (in part)
                11. Plants not or weakly rhizomatous; glumes 0.4–9 mm long.
                  12. Panicle branches densely pubescent, hairs 0.1–0.2 mm long; lemma apices entire, sometimes mucronate; lemma veins converging distally . . . . 14.57 *Koeleria* (in part)
                  12. Panicles branches glabrous, sometimes scabrous; lemma apices entire or serrate to erose, not mucronate; lemma veins more or less parallel distally . . . . 14.06 *Puccinellia* (in part)
  2. One or all lemmas within a spikelet awned, the awns at least 2 mm long.
    13. Lemmas 14–40 mm long; glumes 7–11-veined . . . . 14.52 *Avena*
    13. Lemmas 1.3–16 mm long; glumes 1–9-veined.
      14. Lemmas 7–16 mm long.
        15. Adaxial surfaces of the leaves ribbed; rachillas pilose on all sides; ligules truncate to rounded, 0.5–1.5 mm long . . . . 14.47 *Helictotrichon* (in part)
        15. Adaxial surfaces of the leaves unribbed; rachillas glabrous on the side adjacent to the paleas, hairy elsewhere; ligules acute to truncate, 0.5–7 mm long . . . . 14.45 *Avenula* (in part)
      14. Lemmas 1.3–7 mm long.
        16. Lemmas 1-veined, awned, awns articulated near the middle, the proximal segment yellow-brown to dark brown, the distal segment pale green to whitish, the junction marked by a ring of minute, conical protuberances . . . . 14.55 *Corynephorus*
        16. Lemmas 3–7-veined, at least some lemmas awned, awns not articulated.

17. Disarticulation below the glumes.
　18. Spikelets usually with 2 florets, the lower florets bisexual with unawned lemmas, the upper florets staminate or sterile with awned lemmas . . . . 14.53 *Holcus* (in part)
　18. Spikelets with 2–5 florets, all florets bisexual or sometimes the distal florets sterile; all lemmas awned . . . . 14.56 *Trisetum* (in part)
17. Disarticulation above the glumes.
　19. Lowest lemma within a spikelet unawned or with a straight awn up to 4 mm long, the distal lemmas within a spikelet always awned, awns 10–16 mm long, geniculate . . . . 14.36 *Ventenata* (in part)
　19. All lemmas within a spikelet similarly awned or the awns of the lower lemmas longer than those of the upper lemmas, or the upper lemmas with awns shorter than 10 mm.
　　20. Callus hairs about $^1\!/_2$ as long as the lemmas; rachillas not prolonged or prolonged about 0.5 mm or less beyond the base of the distal floret; plants loosely cespitose . . . . 14.41 *Vahlodea*
　　20. Calluses usually glabrous or the hairs much shorter than $^1\!/_2$ the length of the lemmas, if about $^1\!/_2$ as long, the rachillas prolonged more than 0.5 mm beyond the base of the distal floret and the plants usually densely cespitose.
　　　21. Plants annual; culms 1–60 cm tall; rachillas not prolonged beyond the base of the distal florets . . . . 14.22 *Aira* (in part)
　　　21. Plants perennial or annual; culms 5–150 cm tall; rachillas prolonged beyond the base of the distal florets, the prolongations hairy.
　　　　22. Rachilla internodes glabrous or with hairs shorter than 1 mm on the distal portion; panicle branches densely pubescent, not scabrous . . . . 14.57 *Koeleria* (in part)
　　　　22. Rachilla internodes hairy, hairs at least 1 mm long; panicles branches usually glabrous, sometimes scabrous.
　　　　　23. Lemma apices truncate, erose or 2–4-toothed . . . . 14.26 *Deschampsia* (in part)
　　　　　23. Lemmas apices acute, bifid . . . . 14.56 *Trisetum* (in part)
1. Both glumes shorter than or subequal to the adjacent lemmas.
　24. Upper lemma(s) in a spikelet with hooked or geniculate awns, awns 2–16 mm long; lowest lemmas unawned or terminally awned, awns straight, to 4 mm long.
　　25. Spikelets 9–15 mm long, with 2–20 florets; awns of the distal florets 10–16 mm long . . . . 14.36 *Ventenata* (in part)
　　25. Spikelets 3–7 mm long, with 2 florets; awns of the distal floret 2–5 mm long . . . 14.53 *Holcus* (in part)
　24. Lemmas all similarly awned or unawned.
　　26. Lower lemmas with awns longer than 2 mm.
　　　27. Calluses hairy; rachillas prolonged beyond the base of the distal florets.
　　　　28. Glumes shorter than the adjacent lemmas; ligules 4.5–20 mm long . . . . . 14.48 *Amphibromus*
　　　　28. Glumes subequal to the adjacent lemmas; ligules 0.5–6 mm long . . . . 14.56 *Trisetum* (in part)
　　　27. Calluses glabrous or sparsely hairy; rachillas sometimes prolonged beyond the base of the distal florets.
　　　　29. Panicles dense, spikelike; plants annual . . . . 14.59 *Rostraria* (in part)
　　　　29. Panicles not both dense and spikelike; plants perennial or annual.
　　　　　30. Anthers 1; plants annual . . . . 14.04 *Vulpia* (in part)
　　　　　30. Anthers 3; plants perennial.
　　　　　　31. Leaves without auricles; blades flat, conduplicate, involute, or convolute . . . . 14.01 *Festuca* (in part)
　　　　　　31. Lower leaves with auricles; blades flat . . . . 14.03 *Schedonorus* (in part)
　　26. Lower lemmas unawned, mucronate, or with awns up to 2 mm long.
　　　32. Lemmas inflated, about as wide as long; spikelets pendulous . . . . 14.21 *Briza* (in part)

32. Lemmas not inflated, longer than wide; spikelets appressed to divergent, not
pendulous.
   33. Lemmas apices rounded, truncate, obtuse, or emarginate.
      34. Lemmas conspicuously 3-veined; lower glumes 0–3-veined.
         35. Lower glumes without veins; lemmas not keeled over the lateral
         veins . . . . 14.19 *Catabrosa*
         35. Lower glumes 1–3-veined; lemmas keeled over each vein . . . . 14.20 *Cutandia* (in part)
      34. Lemmas (3)5–9-veined, the veins often inconspicuous; lower glumes
      1–5-veined.
         36. Inflorescences usually exceeded by the leaves; lemmas indurate at
         maturity; pedicels 0.5–0.8 mm thick; culms usually prostrate to
         procumbent, sometimes ascending; upper glumes 2.6–6.2 mm
         long . . . . 14.09 *Sclerochloa* (in part)
         36. Inflorescences exceeding the leaves at maturity; lemmas usually
         membranous at maturity, sometimes coriaceous; pedicels less than
         0.5 mm thick; culms usually erect; upper glumes 0.7–4.5(9) mm
         long.
            37. Lower glumes about as long as the upper glumes but no more
            than $^1/_2$ as wide; disarticulation below the glumes . . . .
            . . . . 14.25 *Sphenopholis* (in part)
            37. Lower glumes shorter than the upper glumes or subequal and
            more than $^1/_2$ as wide; disarticulation above the glumes.
               38. Sheaths closed for more than $^1/_2$ their length . . . . . . 14.17 *Arctophila* (in part)
               38. Sheaths open their entire length.
                  39. Panicle branches stiff; lemmas coriaceous; plants
                  annual; culms to 60 cm tall . . . . 14.35 *Desmazeria* (in part)
                  39. Panicle branches flexible; lemmas usually mem-
                  branous, sometimes coriaceous; plants usually
                  perennial, sometimes annual or biennial; culms 2–145
                  cm tall.
                     40. Lemma veins excurrent, lemma apices indistinctly
                     3-lobed or toothed; plants strongly rhizomatous,
                     rhizomes succulent . . . . 14.51 *Scolochloa* (in part)
                     40. Lemma veins not excurrent, lemma apices entire,
                     serrate, or erose; plants sometimes rhizomatous,
                     rhizomes not succulent.
                        41. Lemma veins (5)7–9, prominent; plants of
                        non-saline and non-alkaline habitats . . . .
                        . . . . 14.18 *Torreyochloa* (in part)
                        41. Lemma veins (3)5(7), inconspicuous or
                        prominent; plants of saline and alkaline
                        habitats . . . . 14.06 *Puccinellia* (in part)
   33. Lemma apices acute to acuminate, sometimes mucronate or shortly awn-
   tipped.
      42. Lemmas (3)5–9-veined, veins more or less parallel distally, conspicuous.
         43. Lemma veins (5)7–9; plants rhizomatous, growing in non-saline
         and non-alkaline habitats . . . . 14.18 *Torreyochloa* (in part)
         43. Lemma veins (3)5(7); plants not truly rhizomatous, sometimes the
         culms rooting at buried lower nodes, growing in saline and alkaline
         habitats . . . . 14.06 *Puccinellia* (in part)
      42. Lemmas 3–9-veined, veins converging distally, usually inconspicuous,
      sometimes conspicuous.
         44. Lemmas conspicuously 3-veined, keeled over each vein; panicle
         branches divaricate; plants annual . . . . 14.20 *Cutandia* (in part)

44. Lemmas inconspicuously (3)5–9-veined, sometimes keeled over the midvein, not over the other veins; panicle branches divaricate or not; plants annual or perennial.
    45. Disarticulation below the glumes; lower glumes subequal to the upper glumes but no more than $^1/_2$ as wide . . . . 14.25 *Sphenopholis* (in part)
    45. Disarticulation above the glumes, sometimes above the basal floret; lower glumes shorter than the upper glumes or, if subequal, more than $^1/_2$ as wide.
        46. Panicle branches smooth, hairy, hairs soft . . . . 14.57 *Koeleria* (in part)
        46. Panicle branches smooth or scabrous, glabrous or strigose, never covered with soft hairs.
            47. Rachillas pilose, hairs at least 2 mm long (see *Stipeae*, p. 110) . . . . 10.01 *Ampelodesmos*
            47. Rachillas glabrous or with hairs shorter than 1 mm.
                48. Basal leaves with auricles . . . . 14.03 *Schedonorus* (in part)
                48. No leaves with auricles.
                    49. Lemma veins parallel distally; plants of saline and alkaline habitats . . . . 14.06 *Puccinellia* (in part)
                    49. Lemma veins converging distally; plants of many habitats, including saline habitats.
                        50. Leaf blades with translucent lines on either side of the midvein, apices often prow-tipped; lemmas often with a tuft of hair at the base of the midvein; hila round to oval . . . . 14.13 *Poa* (in part)
                        50. Leaf blades without translucent lines on either side of the midvein, apices not prow-tipped, often flat; lemmas without a tuft of hair below the midvein hila usually linear, always linear in perennial species.
                            51. Plants perennial.
                                52. Plants and florets bisexual; glumes not translucent; caryopses obovoid-oblong . . . . 14.01 *Festuca* (in part)
                                52. Plants unisexual; glumes translucent; caryopses fusiform . . . . 14.02 *Leucopoa*
                            51. Plants annual.
                                53. Ligules up to 1 mm long; lemma apices mucronate or awned . . . . 14.04 *Vulpia* (in part)
                                53. Ligules 1–4 mm long; lemma apices never awned, sometimes mucronate.
                                    54. Panicle branches up to 2 cm long, stiff, spikelet-bearing to the base; culms procumbent to erect . . . . 14.35 *Desmazeria* (in part)
                                    54. Panicles branches 2–10 cm long, flexible, spikelets confined to the distal portion; culms erect . . . . 14.24 *Eremopoa*

14.01 FESTUCA L.

Stephen J. Darbyshire

Leon E. Pavlick†

Plants perennial; bisexual; usually densely to loosely cespitose, with or without rhizomes, occasionally stoloniferous. **Culms** 5–150(275) cm, usually glabrous and smooth throughout, sometimes scabrous or densely pubescent below the inflorescences. **Sheaths** from open to the base to closed almost to the top, in some species sheaths of previous years persisting and the blades usually deciduous, in other species the senescent sheaths rapidly shredding into fibers and decaying between the veins and the blades not deciduous; **collars** inconspicuous, usually glabrous; **auricles** absent; **ligules** 0.1–2(8) mm, membranous, sometimes longest at the margins, usually truncate, sometimes acute, usually ciliate, sometimes erose; **blades** flat, conduplicate, involute, or convolute, sometimes glaucous or pruinose, abaxial surfaces usually glabrous or scabrous, sometimes puberulent or pubescent, rarely pilose, adaxial surfaces usually scabrous, sometimes hirsute or puberulent, with or without ribs over the major veins; **abaxial sclerenchyma tissue** varying from longitudinal strands at the margins and opposite the midvein to adjacent to some or all of the lateral veins, longitudinal strands sometimes laterally confluent with other strands into an interrupted or continuous band, sometimes reaching to the veins and forming *pillars*; **adaxial sclerenchyma tissue** sometimes present in strands opposite the veins at the epidermis, the strands sometimes extending to the veins and, in combination with the abaxial sclerenchyma, forming *girders* of sclerenchyma tissue extending from one epidermis to the other at some or all of the veins. **Inflorescences** usually open or contracted panicles, sometimes reduced to racemes, usually with 1–2(3) branches at the lower nodes; **branches** usually erect, spreading to widely spreading at anthesis, sometimes the lower branches reflexed; **Spikelets** with (1)2–10 mostly bisexual florets, distal florets reduced or abortive; **rachillas** usually scabrous or pubescent, sometimes smooth and glabrous; **disarticulation** above the glumes, beneath the florets. **Glumes** subequal or unequal, usually exceeded by the florets, ovate to lanceolate, acute to acuminate; **lower glumes** from shorter than to about equal to the adjacent lemmas, 1(3)-veined; **upper glumes** 3(5)-veined; **calluses** usually wider than long, usually glabrous and smooth, sometimes scabrous, occasionally pubescent; **lemmas** usually chartaceous, sometimes coriaceous, bases more or less rounded dorsally, slightly or distinctly keeled distally, veins 5(7), prominent or obscure, apices acute to attenuate, sometimes minutely bidentate, usually terminally or subterminally awned or mucronate; **paleas** from shorter than to slightly longer than the lemmas, veins sparsely to densely scabrous-ciliate, intercostal region usually smooth and glabrous at the base, usually scabrous and/or puberulent distally, bidentate; **anthers** 3; **ovaries** glabrous or with hispidulous apices, hairs persisting on the mature caryopses. **Caryopses** obovoid-oblong, adaxially grooved, usually free of the lemmas and paleas, sometimes adhering along the groove, sometimes adhering more broadly; **hila** linear, from ½ as long as to almost as long as the caryopses. $x = 7$. Name from the Latin *festuca*, 'stalk', 'stem', or 'straw'—a name used by Pliny for a weed.

Festuca is a widespread genus, probably having more than 500 species. The species grow in alpine, temperate, and polar regions of all continents except Antarctica. There are 37 species native to the *Flora* region, 2 introduced species that have become established, and 5 introduced species that are known only as ornamentals or waifs. One species, *F. rubra*, is represented by both native and introduced subspecies.

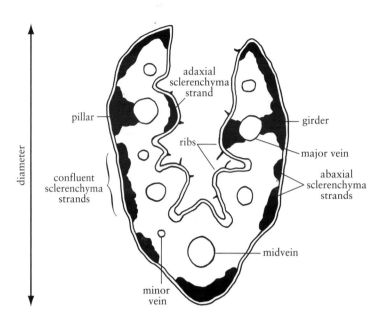

Figure 1. *Festuca* **blade cross section.**

Many native species provide good forage in western North American grasslands and montane forests. Important cultivated species include *Festuca rubra*, grown for forage and as a turf grass, and *F. trachyphylla*, used as a turf grass and for erosion control. Both these species have been widely introduced to many parts of the world. A number of species are cultivated as ornamentals—including *F. amethystina* L., *F. cinerea* Vill., *F. drymeia* Mert. & W.D.J. Koch, *F. elegans* Boiss., *F. gauthieri* (Hack.) K. Richt., *F. glauca* Vill., *F. kasmiriana* Stapf, *F. mairei* St.-Yves, *F. muelleri* Vickery, *F. pallens* Host, *F. pseudoeskia* Boiss., *F. rupicaprina* (Hack.) A. Kern., *F. spectabilis* Jan, and *F. varia* Haenke. These species have not become established in the *Flora* region; only *F. amethystina* and *F. glauca* are included in this account.

The distribution of some taxa that are grown for turf, revegetation, and, to a lesser extent, horticulture—such as *Festuca rubra* subsp. *rubra*, *F. trachyphylla*, *F. filiformis*, and *F. valesiaca*—is continually expanding because of their wide commercial availability. The occurrence of these in the *Flora* region is no doubt much more extensive than current herbarium collections indicate.

The taxonomy of the genus is problematic and contentious, and this treatment is far from definitive. Keying the species ultimately relies on characters that are sometimes difficult to detect on herbarium specimens, such as ovary pubescence and leaf blade sclerenchyma patterns. Because of the intraspecific variability in many characters, combinations of overlapping characters must be employed for identification.

The distribution of sclerenchyma tissue within the vegetative shoot leaves is often an important diagnostic character in *Festuca*. Taxa in a small region can often be identified reliably without resorting to consideration of these patterns but, for the *Flora* region as a whole, their use is essential. These patterns should be observed in cross sections made from mature, but not senescent, leaves of vegetative shoots, ¹⁄₄ to halfway up the blades; they can be made freehand,

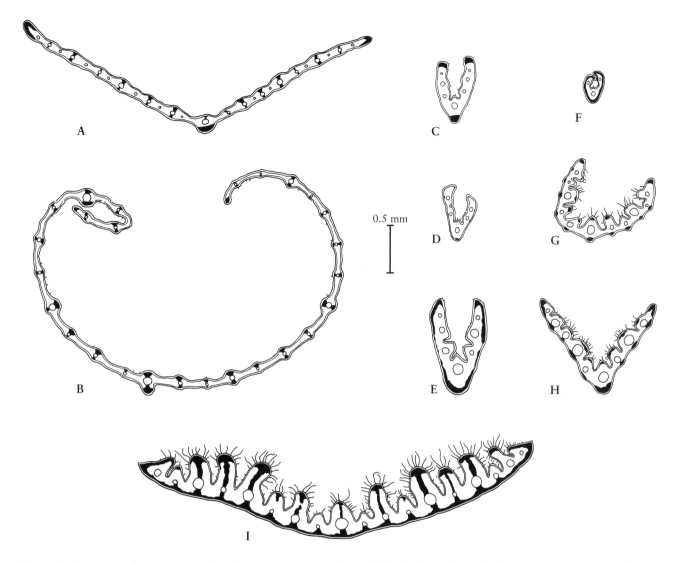

Figure 2. *Festuca* **sclerenchyma distribution patterns.** A–Leaf blade flat, ribs indistinct, sclerenchyma girders at most veins (*F. subverticillata*); B–Leaf blade loosely convolute, ribs indistinct, sclerenchyma girders at most veins (*F. subverticillata*); C–Leaf blade conduplicate, ribs indistinct, sclerenchyma in broad strands at margins and midvein (*F. lenensis*); D–Leaf blade conduplicate, ribs indistinct to distinct, sclerenchyma in narrow abaxial strands opposite veins (*F. hyperborea*); E–Leaf blade conduplicate, ribs distinct, sclerenchyma in broad, mostly confluent strands (*F. trachyphylla*); F–Leaf blade conduplicate, ribs indistinct, sclerenchyma in a continuous abaxial cylinder or band (*F. filiformis*); G–Leaf blade loosely rolled, ribs distinct, sclerenchyma in narrow abaxial and adaxial strands (*F. dasyclada*); H–Leaf blade loosely folded; ribs distinct, sclerenchyma in broad abaxial and adaxial strands forming pillars (*F. viridula*); I–Leaf blade flat, ribs distinct, sclerenchyma in a continuous abaxial cylinder or band, forming girders at most veins (*F. californica*).

with a single-edged razor blade. Sections are best viewed at 40× or greater magnification, and with transmitted light (polarized if possible). Important features seen in the blade cross section are identified in Fig. 1, p. 390.

There are five main sclerenchyma distribution patterns in *Festuca* (Fig. 2, above). Almost all species have a strand of sclerenchyma tissue along the margins and opposite the midvein against the abaxial epidermis (Fig. 2C). Strands may be narrow (about as wide as the adjacent veins or narrower, Fig. 2D) to broad (wider than the adjacent veins, Fig. 2E). Additional strands are often present at the abaxial surface opposite the veins; these strands may be confluent (Figs. 1, 2E), sometimes combining to form a cylinder around the leaf and appearing as a continuous ring or band in the cross sections (Figs. 2F, I). Some species have additional strands on the adaxial surface opposite some or all of the veins (Figs. 2G, H). Another variant is for the abaxial sclerenchyma strands to extend inwards to some or all of the vascular bundles (veins), forming *pillars* in the cross sections (Figs. 2H, I). If both the abaxial and adaxial strands extend inward to the vascular bundles, they are said to form *girders* (Figs. 2A, B, I).

Some of the patterns described may co-occur within a leaf. For instance, some veins may be associated with pillars, others with girders; some sclerenchyma strands within a leaf may be confluent, whereas others are not. Although there may be considerable variation in the extent of sclerenchyma development, the general pattern within a species is usually constant. It is this that makes such patterns useful diagnostic characters, particularly for those needing to identify plants in vegetative condition. They have not, however, been examined for all species.

Descriptions of leaf blades are based on the leaves of the basal vegetative shoots, where present. For those without basal tufts of vegetative shoots, the cauline leaves are described. Width measurements are provided for leaves that are usually flat, or almost so, when encountered in the field or herbarium. "Diameter" is given for leaves that are usually folded or conduplicate when encountered; for leaves that are oval in cross section when folded, it is the largest diameter (or width).

Closure of the leaf sheaths should be checked on young leaves, because the sheaths often split with age, leading to underestimations of the extent of their closure. The fraction of the leaf sheath that is closed varies within and between species of *Festuca*, but the species can be divided into three categories in this regard: those such as *F. rubra*, in which the leaves are closed for at least ³/₄ their length; those such as *F. saximontana*, in which they are closed from ¹/₃ to slightly more than ¹/₂ their length; and those such as *F. trachyphylla*, in which they are not closed or closed for less than ¹/₄ their length. The descriptions indicate to which of these categories each species belongs. Lemma awns tend to be longer, and should be measured, on the distal florets within a spikelet.

Under adverse conditions, many species may *proliferate vegetatively*, where leafy bulbils or shoots form in place of some or all spikelets. Some populations of *Festuca* are largely (or completely) sterile, reproducing almost entirely through such bulbils, a process termed *pseudovivipary*. Pseudoviviparous plants may be common or even abundant in certain areas and habitats. Since these stabilized forms are largely reproductively isolated, often of unusual ploidy, and largely morphologically distinct, they are treated as separate species. Although the lower bracts in pseudoviviparous spikelets are usually more or less normal in form, they are sometimes elongated or distorted, as are the upper bracts.

SELECTED REFERENCES **Aiken, S.G.** and **L.L. Consaul**. 1995. Leaf cross sections and phytogeography: A potent combination for identifying members of *Festuca* subgg. *Festuca* and *Leucopoa* (Poaceae), occurring in North America. Amer. J. Bot. 82:1287–1299; **Aiken, S.G., L.L. Consaul, J.I. Davis** and **P.S. Manos**. 1993. Systematic inferences from variation in isozyme profiles of arctic and alpine cespitose *Festuca* (Poaceae). Amer. J. Bot. 80:76–82; **Aiken, S.G., L.L. Consaul** and **L.P. Lefkovitch**. 1995. *Festuca edlundiae* (Poaceae), a high arctic, new species compared enzymatically and morphologically with similar *Festuca* species. Syst. Bot. 20:374–392; **Aiken, S.G.** and **S.J. Darbyshire**. 1990. Fescue Grasses of Canada. Agriculture Canada Publ. 1844/E. Canadian Government Publishing Centre, Ottawa, Ontario, Canada. 113 pp.; **Aiken, S.G., M.J. Dallwitz, C.L. McJannet** and **L.L. Consaul**. 1997. Biodiversity among *Festuca* (Poaceae) in North America: Diagnostic evidence from DELTA and clustering programs, and an INTKEY package for interactive, illustrated identification and information retrieval. Canad. J. Bot. 75:1527–1555; **Aiken, S.G., S.J. Darbyshire** and **L.P. Lefkovitch**. 1985. Restricted taxonomic value of leaf sections in Canadian narrow-leaved *Festuca* (Poaceae). Canad. J. Bot. 63:995–1007; **Aiken, S.G.** and **G. Fedak**. 1991. Cytotaxonomic observations on North American *Festuca* (Poaceae). Canad. J. Bot. 70:1940–1944; **Aiken, S.G.** and **S.E. Gardiner**. 1991. SDS-PAGE of seed proteins in *Festuca* (Poaceae): Taxonomic implications. Canad. J. Bot. 69:1425–1432; **Aiken, S.G., S.E. Gardiner**, and **M.B. Forde**. 1992. Taxonomic implications of SDS-PAGE analysis of seed proteins in North

American taxa of *Festuca* subgenus *Festuca* (Poaceae). Biochem. Syst. & Ecol. 20:615–629; **Aiken, S.G.** and **L.P. Lefkovitch**. 1984. The taxonomic value of using epidermal characteristics in the Canadian rough fescue complex (*Festuca altaica*, *F. campestris*, *F. hallii*, "*F. scabrella*"). Canad. J. Bot. 62:1864–1870; **Aiken, S.G.** and **L.P. Lefkovitch**. 1993. On the separation of two species within *Festuca* subg. *Obtusae* (Poaceae). Taxon 42:323–337; **Alexeev, E.B.** 1977. To the systematics of Asian fescues (*Festuca*): I. Subgenera *Drymanthele*, *Subulatae*, *Schedonorus*, *Leucopoa*. Byull. Moskovsk. Obshch. Isp. Prir., Otd. Biol., n.s., 82(3):95–102. [In Russian]; **Alexeev, E.B.** 1980. *Festuca* L.: Subgenera et sectiones novae ex America boreali et Mexico. Novosti Sist. Vyssh. Rast. 17:42–53. [In Russian]; **Alexeev, E.B.** 1982. New and little known fescues (*Festuca* L.) of North America. Byull. Moskovsk. Obshch. Isp. Prir., Otd. Biol., n.s., 87(2):109–118. [In Russian]; **Alexeev, E.B.** 1985. *Festuca* L. (Poaceae) in Alaska et Canada. Novosti Sist. Vyssh. Rast. 22:5–35. [In Russian]; **Auquier, P.** 1971. *Festuca rubra* subsp. *pruinosa* (Hack.) Piper: Morphologie, écologie, taxonomie. Lejeunia, n.s., 56:1–16; **Barker, C.M.** and **C.A. Stace**. 1982. Hybridization in the genera *Vulpia* and *Festuca*: The production of artificial F$_1$ plants. Nordic J. Bot. 2:435–444; **Beal, W.J.** 1896. Grasses of North America, vol. 2. Henry Holt & Company, New York, New York, U.S.A. 706 pp.; **Consaul, L.L.** and **S.G. Aiken**. 1993. Limited taxonomic value of palea intercostal characteristics in North American *Festuca* (Poaceae). Canad. J. Bot. 71:1651–1659; **Darbyshire, S.J.** and **S.G. Warwick**. 1992. Phylogeny of North American *Festuca* (Poaceae) and related genera using chloroplast DNA restriction site variation. Canad. J. Bot. 70:2415–2429; **Dore, W.G.** and **J. McNeill**. 1980. Grasses of Ontario. Research Branch, Agriculture Canada Monograph No. 26. Canadian Government Publishing Centre, Hull, Québec, Canada. 568 pp.; **Dubé, M.** 1983. Addition de *Festuca gigantea* (L.) Vill. (Poaceae) à la flore du Canada. Naturaliste Canad. 110:213–215; **Dubé, M.** and **P. Morisset**. 1987. Morphological and leaf anatomical variation in *Festuca rubra sensu lato* (Poaceae) from eastern Quebec. Canad. J. Bot. 65:1065–1077; **Dubé, M.** and **P. Morisset**. 1995. La variation des caractères épidermiques foliaires chez le *Festuca rubra sensu lato* (Poaceae) dans l'est du Canada. Canad. J. Bot. 74:1425–1438; **Dubé, M.** and **P. Morisset**. 1996. La plasticité phénotipique des caractères anatomiques foliaires chez le *Festuca rubra* L. (Poaceae). Canad. J. Bot. 74:1708–1718; **Dubé, M.**, **P. Morisset** and **J. Murdock**. 1985. Races chromosomiques chez le *Festuca rubra sensu lato* (Poaceae) dans l'est du Québec. Canad. J. Bot. 63:227–231; **Fernald, M.L.** 1933. Recent discoveries in the Newfoundland flora. Rhodora 35:120–140; **Frederiksen, S.** 1978. *Festuca brevissima* Jurtz. in Alaska. Bot. Not. 131:409–410; **Frederiksen, S.** 1979. *Festuca minutiflora* Rydb., a neglected species. Bot. Not. 132:315–318; **Frederiksen, S.** 1981. *Festuca vivipara* (Poaceae) in the North Atlantic area. Nordic J. Bot. 1:277–292; **Frederiksen, S.** 1982. *Festuca brachyphylla*, *F. saximontana* and related species in North America. Nordic J. Bot. 2:525–536; **Frederiksen, S.** 1983. *Festuca auriculata* in North America. Nordic J. Bot. 3:629–632; **Harms, V.L.** 1985. A reconsideration of the nomenclature and taxonomy of the *Festuca altaica* complex (Poaceae) in North America. Madroño 32:1–10; **Kerguélen, M.** and **F. Plonka**. 1989. Les *Festuca* de la flore de France (Corse comprise). Bull. Soc. Bot. Centre-Ouest, Numéro Spécial 10:1–368; **Kerguélen, M.**, **F. Plonka** and **É. Chas**. 1993. Nouvelle contribution aux *Festuca* (Poaceae) de France. Lejeunia, n.s., 142:1–42; **Markgraf-Dannenberg, I.** 1980. *Festuca* L. Pp. 125–153 *in* T.G. Tutin, V.H. Heywood, N.A. Burges, D.M. Moore, D.H. Valentine, S.M. Walters and D.A. Webb (eds.). Flora Europaea, vol. 5. Cambridge University Press, Cambridge, England. 439 pp.; **Pavlick, L.E.** 1983a. *Festuca viridula* Vasey (Poaceae): Re-establishment of its original lectotype. Taxon 32:117–120; **Pavlick, L.E.** 1983b. Notes on the taxonomy and nomenclature of *Festuca occidentalis* and *F. idahoensis*. Canad. J. Bot. 61:337–344; **Pavlick, L.E.** 1983c. The taxonomy and distribution of *Festuca idahoensis* in British Columbia and northwestern Washington. Canad. J. Bot. 61:345–353; **Pavlick, L.E.** 1984. Studies of the *Festuca ovina* complex in the Canadian Cordillera. Canad. J. Bot. 62:2448–2462; **Pavlick, L.E.** 1985. A new taxonomic survey of the *Festuca rubra* complex in northwestern North America, with emphasis on British Columbia. Phytologia 57:1–17; **Pavlick, L.E.** and **J. Looman**. 1984. Taxonomy and nomenclature of rough fescues, *Festuca altaica*, *F. campestris* (*F. scabrella* var. *major*), and *F. hallii*, in Canada and the adjacent United States. Canad. J. Bot. 62:1739–1749; **Piper, C.V.** 1906. North American species of *Festuca*. Contr. U.S. Natl. Herb. 10^1:1–48, vi–ix; **Ramesar-Fortner, N.S.**, **S.G. Aiken**, and **N.G. Dengler**. 1995. Phenotypic plasticity in leaves of four species of arctic *Festuca* (Poaceae). Canad. J. Bot. 73:1810–1823; **Saint-Yves, A.** 1925. Contribution à l'étude des *Festuca* (subgen. *Eufestuca*) de l'Amérique du Nord et du Mexique. Candollea 2:229–316; **Scholander, P.F.** 1934. Vascular plants from northern Svalbard with remarks on the vegetation in North-East Land. Skr. Svalbard Nordishavet 62:1–155; **Yatskievych, G.** 1999. Steyermark's Flora of Missouri, vol. 1, rev. ed. Missouri Department of Conservation, Jefferson City, Missouri, U.S.A. 991 pp.

1. Blades usually flat, sometimes loosely conduplicate or convolute, 1.8–10 mm wide; sheaths closed to about ¹/₂ their length, never about ³/₄ their length; sclerenchyma girders or pillars associated with at least some of the major veins; ovary apices usually pubescent, rarely glabrous; plants rarely pseudoviviparous [for opposite lead, see p. 395].

  2. Primary and secondary inflorescence branches stiffly and strongly divaricate at maturity, angles densely scabrous or ciliate; spikelets with 2(3) florets . . . . 44. *F. dasyclada* (in part)

  2. Primary inflorescence branches lax or stiff, erect to ascending or spreading at maturity, sometimes lax, secondary branches not stiffly divaricate, angles smooth or scabrous; spikelets with 2–6(10) florets.

    3. Lemma awns (1.3)1.5–20 mm long, occasionally absent [for opposite lead, see p. 394].

      4. Lemma awns usually less than ¹/₃ the lemma length, rarely absent; anthers (3)4–8.5 mm long; plants densely cespitose . . . . 13. *F. californica* (in part)

      4. Lemma awns usually more than ¹/₃ the lemma length; anthers 1.5–5.7 mm long; plants usually loosely cespitose, rarely densely cespitose.

        5. Lemma calluses longer than wide, pubescent, at least basally; awns flexuous or kinked . . . . 7. *F. subuliflora*

        5. Lemma calluses wider than long, glabrous, sometimes slightly scabrous; awns usually straight or slightly curved, bent or kinked.

          6. Anthers 1.5–3 mm long; lemmas entire, glabrous, sometimes sparsely scabrous; awns terminal, (2.5)5–15(20) mm long; leaf blades 3–10 mm wide . . . . . . 4. *F. subulata*

6. Anthers (3)3.4–5.7 mm long; lemmas bidentate or entire, puberulent or scabrous; awns subterminal or terminal, 1–5(8) mm long, occasionally absent; leaf blades usually less than 6 mm wide.

  7. Upper glumes 3–4.6 mm long; lemma awns (1.5)2–5(8) mm long; blades 1.8–6 mm wide, blades of the vegetative shoots narrower than the cauline blades, usually flat or convolute or loosely conduplicate; plants of forests, usually below 500 m . . . . 6. *F. elmeri*

  7. Upper glumes (4)5.5–7(8) mm long; lemma awns 1–3(3.5) mm long; blades 1.5–3 mm wide, the vegetative shoot and cauline blades similar in width, usually loosely conduplicate, or sometimes flat; plants of subalpine and low alpine habitats . . . . 43. *F. washingtonica* (in part)

3. Lemmas unawned, mucronate, or with awns shorter than 2 mm [for opposite lead, see p. 393].

  8. Ligules 2–9 mm long; lemmas unawned, sometimes mucronate, mucros to 0.2 mm long.

    9. Rhizomes present; blades with (5)7–9 veins; lemmas 4–6.5 mm long, unawned; anthers 1.5–2.6 mm long . . . . 11. *F. ligulata* (in part)

    9. Rhizomes absent; blades with 9–15 veins; lemmas 6–10 mm long, unawned, sometimes with a mucro to 0.2 mm long; anthers 3–4.5 mm long . . . . 12. *F. thurberi* (in part)

  8. Ligules 0.1–1.5(2) mm long; lemmas awned, mucronate, or unawned.

    10. Lemmas 3–5(5.2) mm long, unawned; anthers (0.7)1–2(2.5) mm long.

      11. Inflorescence branches usually reflexed at maturity; spikelets not or only slightly imbricate, elliptic to ovate; upper glumes 3–4(4.7) mm . . . . 2. *F. subverticillata*

      11. Inflorescence branches ascending to spreading at maturity; spikelets closely imbricate, elliptic to obovate; upper glumes (3.5)4–5(5.5) mm . . . . 3. *F. paradoxa*

    10. Lemmas (4.8)5–12 mm long, unawned or with awns to 2 mm long; anthers 1.6–6 mm long.

      12. Senescent sheaths not persistent, rapidly shredding into fibers; plants loosely cespitose; cauline nodes usually exposed.

        13. Lower glumes 4–7 mm long; lemmas smooth and glabrous, apices sometimes sparsely scabrous, unawned, sometimes mucronate . . . . 1. *F. versuta*

        13. Lower glumes 1.5–4.5 mm long; lemmas scabrous or puberulent, unawned or awned, awns to 2 mm long . . . . 5. *F. sororia*

      12. Senescent sheaths persistent or only slowly shredding into fibers; plants densely cespitose; cauline nodes usually not exposed.

        14. Panicle branches more or less erect, stiff; abaxial sclerenchyma forming continuous or interrupted bands; lower glumes from shorter than to about equal to the adjacent lemmas.

          15. Spikelets with 2–3(4) florets; glumes about equaling or slightly exceeding the upper florets; lemmas 5.5–8(9) mm long . . . . 9. *F. hallii* (in part)

          15. Spikelets with (3)4–5(7) florets; glumes exceeded by the upper florets; lemmas (6.2)7–8.5(10) mm long . . . . 10. *F. campestris* (in part)

        14. Panicle branches lax, loosely erect, spreading, recurved, or reflexed; abaxial sclerenchyma in strands about the same width as the adjacent veins, not forming a continuous or interrupted band; lower glumes distinctly shorter than the adjacent lemmas.

          16. Senescent sheaths persistent, not shredding into fibers; blades deciduous; spikelets lustrous; ovary apices usually sparsely pubescent, rarely glabrous; plants densely cespitose . . . . 8. *F. altaica* (in part)

          16. Senescent sheaths persistent or slowly shredding into fibers; blades not deciduous; spikelets not lustrous; ovary apices usually densely pubescent, sometimes sparsely pubescent; plants loosely to densely cespitose.

            17. Blades of the lower cauline leaves much shorter and stiffer than those of the upper cauline leaves; lemmas smooth or slightly scabrous, unawned or the awns to 1.5(2) mm long . . . . 42. *F. viridula* (in part)

17. Blades of the lower cauline leaves similar in length and stiffness to those of the upper cauline leaves; lemmas scabrous or puberulent distally, the awns usually longer than 1 mm, rarely absent . . . . 43. *F. washingtonica* (in part)

1. Blades usually conduplicate or folded and less than 2.5 mm in diameter, sometimes convolute or flat, sometimes the leaves of the vegetative shoots conduplicate and the cauline leaves more or less flat, up to 6(7) mm wide when flat; sheath closure varied, from completely open to closed for about ³/₄ their length, if the blades 2+ mm wide then the sheaths closed for about ³/₄ their length; sclerenchyma girders usually absent, present if the blades 3+ mm wide, pillars sometimes present; ovary apices pubescent or glabrous; plants sometimes pseudoviviparous [for opposite lead, see p. 393].

   18. Collars usually pubescent, at least at the margins, sometimes glabrous; lemmas (7)7.5–11 mm long, scabrous or pubescent; spikelets not pseudoviviparous; ovary apices densely pubescent . . . . .13. *F. californica* (in part)

   18. Collars glabrous; lemmas 2–10(11) mm long, smooth or scabrous, glabrous or with hairs; spikelets sometimes pseudoviviparous; ovary apices glabrous or pubescent or ovaries not developed.

      19. Ligules 2–9 mm long.

         20. Rhizomes present; blades with (5)7–9 veins; lemmas 4–6.5 mm long, unawned; anthers 1.5–2.6 mm long . . . . 11. *F. ligulata* (in part)

         20. Rhizomes absent; blades with 9–15 veins; lemmas 6–10 mm long, unawned, sometimes with a mucro to 0.2 mm long; anthers 3–4.5 mm long . . . . 12. *F. thurberi* (in part)

      19. Ligules to 1.5(2) mm long.

        21. Most or all spikelets pseudoviviparous; anthers and ovaries usually absent or abortive.

          22. Rhizomes present; sheaths closed for about ³/₄ their length or more, senescent sheaths rapidly shredding into fibers.

            23. Cauline blades 0.3–1 mm wide, conduplicate or folded; inflorescences sometimes racemose or subracemose, with 1–3 spikelets on the lower branches; plants of boreal and alpine eastern North America . . . . 15. *F. prolifera*

            23. Cauline blades 1.4–2.5 mm wide, flat; inflorescences paniculate, with 2–5 spikelets on the lower branches; plants known only from the Queen Charlotte Islands . . . . 16. *F. pseudovivipara*

          22. Rhizomes absent; sheaths closed for less than ³/₄ their length, senescent sheaths sometimes persistent, sometimes shredding into fibers.

            24. Abaxial sclerenchyma in broad, sometimes confluent strands that together cover ¹/₂ or more of the abaxial surface; glumes densely pubescent throughout; inflorescences 1.5–10 cm long . . . . 35. *F. frederikseniae*

            24. Abaxial sclerenchyma in narrow strands that together cover less than ¹/₂ the abaxial surface; glumes glabrous or pubescent; inflorescences 1–4.8 cm long . . . . 36. *F. viviparoidea*

        21. No spikelets pseudovivaparous; anthers and ovaries well developed.

          25. Glumes about equaling or slightly exceeding the upper florets; lemma awns absent or 0.5–1.3 mm long; anthers 4–6 mm long . . . . 9. *F. hallii* (in part)

          25. Glumes distinctly exceeded by the upper florets; lemma awns various; anthers 0.3–6 mm long.

            26. Rhizomes usually present; sheaths of the vegetative shoots closed for about ³/₄ their length, glabrous or pubescent, hairs retrorse or antrorse, senescent sheaths rapidly shredding into fibers.

              27. Anthers 1.8–4.5 mm long; ovary apices glabrous . . . . 14. *F. rubra*

              27. Anthers 0.6–1.4 mm long; ovary apices densely pubescent . . . . . . 17. *F. earlei* (in part)

            26. Rhizomes absent; sheaths of the vegetative shoots usually closed for less than ²/₃ their length, sometimes closed for about ³/₄ their length, usually glabrous and smooth, sometimes scabrous or puberulent, hairs rarely retrorse, senescent sheaths usually persistent for several years, sometimes slowly shredding into fibers.

28. Primary and secondary inflorescence branches stiffly and strongly divaricate at maturity; spikelets with 2(3) florets . . . . 44. *F. dasyclada* (in part)

28. Primary inflorescence branches stiffly erect or laxly spreading at maturity, secondary branches not divaricate; spikelets with (1)2–10 florets.

   29. Culms densely scabrous or densely pubescent below the inflorescences; ligules 0.5–1.5(2) mm long; anthers (2)3–4(4.2) mm long . . . . 39. *F. arizonica*

   29. Culms usually smooth and glabrous below the inflorescences, sometimes scabrous or sparsely pubescent, if scabrous or pubescent then the ligules to 0.6 mm long; ligules 0.1–0.8(1) mm long; anthers 0.3–4.5(5) mm long.

      30. Lemmas unawned, mucronate, or with awns to 3.5 mm long, if awned then the leaf blades with adaxial sclerenchyma strands present.

         31. Blades 0.2–1.2(1.5) mm in diameter; lemmas usually unawned, sometimes mucronate, mucros to 0.4 mm long; ovary apices glabrous or sparsely pubescent; adaxial sclerenchyma absent, pillars and girders not formed; plants introduced, usually of disturbed habitats.

            32. Blades with (5)7–9 veins and 5–9 indistinct or distinct ribs; sclerenchyma in (5)7–9 sometimes partly confluent abaxial strands; inflorescences (3)8–18(25) cm long; spikelets (5)6–8.5(10) mm long; lemmas (3.5)4–5.6(6.6) mm long; anthers (2)3–4 mm long . . . . 20. *F. amethystina*

            32. Blades with 5(7) veins and 1 distinct rib; sclerenchyma in a continuous or almost continuous abaxial band; inflorescences 1–6(14) cm long; spikelets 3–6(6.5) mm long; lemmas 2.3–4(4.4) mm long; anthers (1)1.5–2.2 mm long . . . . 23. *F. filiformis*

         31. Blades 0.3–3 mm in diameter; lemmas unawned or with awns to 3.5 mm long; ovary apices pubescent, sometimes sparsely so; usually at least some veins associated with adaxial sclerenchyma and pillars or girders; plants native, of western alpine, subalpine, and montane habitats.

            33. Panicle branches erect to stiffly spreading; abaxial sclerenchyma forming continuous or interrupted bands; lower glumes from shorter than to about equal to the adjacent lemmas . . . . 10. *F. campestris* (in part)

            33. Panicle branches lax, loosely erect, spreading, recurved, or reflexed; abaxial sclerenchyma in strands about the same width as the adjacent veins, not forming a continuous or interrupted band; lower glumes distinctly shorter than the adjacent lemmas.

               34. Senescent sheaths persistent, not shredding into fibers; blades deciduous; spikelets lustrous; ovary apices usually sparsely pubescent, rarely glabrous; plants densely cespitose; plants of alpine or arctic habitats from central British Columbia northward . . . . 8. *F. altaica* (in part)

34. Senescent sheaths persistent or slowly shredding into fibers; blades not deciduous; spikelets not lustrous; ovary apices usually densely pubescent, sometimes sparsely pubescent; plants loosely to densely cespitose; plants of alpine and subalpine habitats from southern British Columbia southward.

    35. Blades of the lower cauline leaves shorter and stiffer than those of the upper cauline leaves; lemmas glabrous, smooth or slightly scabrous . . . . 42. *F. viridula* (in part)

    35. Blades of the lower cauline leaves similar in length and stiffness to those of the upper cauline leaves; lemmas scabrous or puberulent distally . . . . 43. *F. washingtonica* (in part)

30. Lemmas usually awned, occasionally unawned, awns 0.3–12 mm long; leaf blades without adaxial sclerenchyma strands.

    36. Anthers (1.8)2–4.5 mm long; plants mostly not of arctic, subarctic, or alpine habitats (except *F. auriculata* and *F. lenensis*) . . . . Subkey I

    36. Anthers 0.3–1.8(2) mm long; plants mostly of arctic, subarctic, or alpine habitats (except *F. occidentalis*, *F. ovina*, and *F. saximontana*) . . . . Subkey II

Festuca Subkey I

1. Ovary apices pubescent.

  2. Sheaths closed for about ³/₄ their length or more; vegetative shoot leaf blades narrow and conduplicate, the cauline blades broader and flat; anthers 2.5–4.5 mm long . . . . 18. *F. heterophylla*

  2. Sheaths closed for no more than ¹/₂ their length; vegetative shoot and cauline leaf blades similar, conduplicate; anthers 1–3.5 mm long.

    3. Lower inflorescence branches usually reflexed at maturity; lemma awns 3–12 mm long; ovary apices densely pubescent . . . . 37. *F. occidentalis*

    3. Lower inflorescence branches erect at maturity; lemma awns 1–2.5 mm long; ovary apices sparsely pubescent . . . . 38. *F. calligera* (in part)

1. Ovary apices glabrous, or with up to 5 hairs in *F. calligera*.

  4. Abaxial sclerenchyma forming continuous or interrupted bands; lemmas (2.6)3–6(6.2) mm long; blades with 1–3(5) indistinct ribs; anthers 1.4–3 mm long; species used as ornamentals or for turf or soil stabilization, rarely spreading from cultivation.

    5. Plants usually not glaucous or pruinose; sheaths glabrous; plants used for turf . . . . . . 21. *F. ovina* (in part)

    5. Plants usually glaucous or pruinose; sheaths pubescent or glabrous; plants grown as ornamentals . . . . 22. *F. glauca*

  4. Abaxial sclerenchyma usually in 3–7 discrete or somewhat confluent strands, if in a continuous band, the lemmas 3.8–7(8.2) mm long; blades with 1–9 distinct ribs; anthers 0.4–4.5 mm long; most species native, a few introduced for turf.

    6. Abaxial sclerenchyma in 5–7 strands or in interrupted to continuous bands; blades with (1)3–9 well-defined ribs.

      7. Lemmas 3.8–5(6.5) mm long, usually scabrous or pubescent distally, especially on the margins, rarely entirely pubescent; lemma awns usually less than ¹/₂ the length of the lemma bodies . . . . 24. *F. trachyphylla*

      7. Lemmas 5–10 mm long, scabrous distally; lemma awns usually more than ¹/₂ the length of the lemma bodies.

        8. Blades with (1)3–5 ribs; adaxial surfaces of the blades pubescent or scabrous; inflorescence branches usually somewhat spreading at maturity . . . . 40. *F. idahoensis*

8. Blades with 5–9 ribs; adaxial surfaces of the blades glabrous or pubescent, sometimes scabrous; inflorescence branches erect to slightly spreading at maturity . . . . 41. *F. roemeri*

6. Abaxial sclerenchyma strands usually restricted to the margins and midvein, rarely with additional strands in between; blades with 1–5 well-defined ribs.

    9. Sheaths closed distinctly less than ¹/₂ their length; culms 15–65 cm tall; inflorescences 5–15 cm long; plants native to the southwestern United States . . . . .38. *F. calligera* (in part)

    9. Sheaths closed for about ¹/₂ their length; culms 8–50(60) cm tall; inflorescences 1.5–10 cm long; plants introduced or native to the extreme northwest of the *Flora* region.

        10. Inflorescences (3)5–10 cm long, panicles; abaxial blade surfaces glabrous or pubescent, not pilose; lower glumes 2–3 mm long; anthers 2.2–2.6 mm long; plants introduced . . . . 19. *F. valesiaca*

        10. Inflorescences 1.5–5(5.5) cm long, panicles or racemes; abaxial blade surfaces glabrous, pubescent, or pilose, varying within individual plants; lower glumes 2.5–3.4 mm long; anthers (2)2.4–3.5 mm long; plants native to the extreme northwest of the *Flora* region.

            11. Abaxial sclerenchyma strands distinctly narrower than the veins; spikelets 5–6.5(8) mm long . . . . .25. *F. auriculata*

            11. Abaxial sclerenchyma strands about the same width as or wider than the veins; spikelets (5)7–9(11) mm long . . . . .26. *F. lenensis* (in part)

Festuca Subkey II

1. Ovary apices pubescent, sometimes with only a few hairs.

    2. Ovary apices densely and conspicuously pubescent.

        3. Lemma awns 0.3–1.5 mm long; lemmas 3–4.5 mm long . . . . 17. *F. earlei* (in part)

        3. Lemma awns 3–12 mm long; lemmas (4)4.5–6.5(8) mm long . . . . 37. *F. occidentalis*

    2. Ovary apices sparsely and inconspicuously pubescent.

        4. Culms densely pubescent or shortly pilose below the inflorescence; lemmas 3.5–6 mm long; anthers 0.3–0.7(1.1) mm long . . . . .32. *F. baffinensis* (in part)

        4. Culms glabrous below the inflorescence; lemmas (2)2.2–3.5(4) mm long; anthers (0.4)0.6–1.2 mm long . . . . .34. *F. minutiflora* (in part)

1. Ovary apices glabrous.

    5. Abaxial sclerenchyma in a continuous band; anthers longer than 1.4 mm; plants persisting from historical use for turf and soil stabilization . . . . .21. *F. ovina* (in part)

    5. Abaxial sclerenchyma in 3+ discrete or confluent strands, rarely in a continuous band; anthers various; plants native in the *Flora* region.

        6. Abaxial sclerenchyma strands at least twice as wide as high, varying to more or less confluent, rarely forming a continuous band; plants of various habitats.

            7. Anthers (0.8)1.2–1.7(2) mm long; inflorescences 2–10(13) cm long; lemmas (3)3.4–4(5.6) mm long; spikelets (3)4.5–8.8(10) mm long; plants widespread in continental North America . . . . .29. *F. saximontana*

            7. Anthers 0.8–1.3 mm long; inflorescences 1.5–5 cm long; lemmas (2.5)3–3.5(4) mm long; spikelets 4–5.6(6) mm long; plants known only from Greenland . . . . 33. *F. groenlandica*

        6. Abaxial sclerenchyma strands usually less than twice as wide as high, not confluent, never forming a continuous band; plants restricted to arctic or alpine habitats.

            8. Blades usually with 3 abaxial sclerenchyma strands, 2 in the margins and 1 opposite the midvein, occasionally with 2 additional abaxial strands opposite the lateral veins.

                9. Abaxial sclerenchyma strands 3(5), narrower than the veins; spikelets 5–6.5(8) mm long . . . . .25. *F. auriculata* (in part)

                9. Abaxial sclerenchyma strands 3, about the same width as or wider than the veins; spikelets (5)7–9(11) mm long . . . . .26. *F. lenensis* (in part)

        8. Blades with 3–7(9) abaxial sclerenchyma strands.

10. Blades (0.2)0.3–0.4(0.6) mm in diameter; spikelets 2.5–5 mm long, with (1)2–3(5) florets; lemmas (2)2.2–3.5(4) mm long; lemma awns 0.5–1.5(1.7) mm long; plants of alpine habitats. . . . . 34. *F. minutiflora* (in part)

10. Blades (0.3)0.4–1.2 mm in diameter; spikelets (3)3.5–8.5 mm long, with 2–6 florets; lemmas 2.5–7 mm long; lemma awns (0.2)0.5–3.5 mm long; plants of arctic or alpine habitats.

  11. Culms densely pubescent or pilose below the inflorescences; anthers 0.3–0.7(1.1) mm long . . . . 32. *F. baffinensis* (in part)

  11. Culms usually glabrous and smooth below the inflorescences, occasionally slightly scabrous or sparsely puberulent; anthers (0.3)0.4–1.3 mm long.

    12. Inflorescences usually panicles; lower branches with 2+ spikelets; flag leaf sheaths not inflated; flag leaf blades (0.3)1–3 cm long . . . . 28. *F. brachyphylla*

    12. Inflorescences often racemes; lower branches with 1–2(3+) spikelets; flag leaf sheaths usually slightly to distinctly inflated; flag leaf blades 0.2–5(8) cm long.

      13. Culms erect, more than twice as tall as the basal tuft of leaves; glumes ovate-lanceolate to lanceolate; lemmas (3)4–5.5(7) mm long; plants of Alaska to the western Northwest Territories . . . . 27. *F. brevissima*

      13. Culms erect to prostrate, to twice as tall as the basal tuft of leaves; glumes ovate to ovate-lanceolate; lemmas 2.9–5.2 mm long; plants of Alaska to the eastern arctic.

        14. Culms usually erect, sometimes semi-prostrate; flag leaf blades 0.5–5(8) mm long; blades often curved or somewhat falcate; spikelets (3)4–5.5(7) mm long; upper glumes 2.2–3.2 mm long; lemma apices usually minutely bidentate; awns usually slightly subterminal . . . . 30. *F. hyperborea*

        14. Culms usually geniculate to prostrate, becoming erect at anthesis; flag leaf blades (0.3)0.5–2 cm long; most blades straight; spikelets 4.5–8.5 mm long; upper glumes 2.9–4.3 mm long; lemma apices entire; awns usually terminal, sometimes slightly subterminal . . . . 31. *F. edlundiae*

Festuca subg. **Montanae** (Hack.) Nyman

Plants loosely cespitose, without rhizomes. **Innovations** extravaginal, with or without basal cataphylls. **Blades** flat or convolute; **ribs** shallow, indistinct; **sclerenchyma girders** present at the major veins. **Calluses** wider than long; **lemmas** chartaceous, lanceolate to attenuate, entire, unawned; **ovary apices** densely pubescent.

Festuca subg. *Montanae* contains approximately 20–25 species. It has been divided into seven sections that are widely distributed on all continents except Africa and Antarctica. One monospecific section occurs in North America.

Festuca sect. **Texanae** E.B. Alexeev

Plants loosely cespitose, without rhizomes. **Innovations** extravaginal, without basal cataphylls. **Blades** lax, flat or convolute; **ribs** shallow, indistinct; **sclerenchyma girders** present at the major veins. **Calluses** wider than long, scabrous at the margins; **lemmas** chartaceous, unawned; **ovary apices** densely pubescent.

Three species have been placed in this section; one occurs in the *Flora* region, and the other two are Central American.

1. Festuca versuta Beal [p. 401]
TEXAS FESCUE

Plants loosely cespitose, without rhizomes. **Culms** 50–100 cm, glabrous, somewhat glaucous; **nodes** usually exposed. **Sheaths** closed for less than ¹/₃ their length, glabrous or sparsely pubescent, shredding into fibers; **ligules** 0.5–1 mm; **blades** 2–10 mm wide, flat, loosely conduplicate, or involute, abaxial surfaces smooth or scabrous, glabrous or puberulent, adaxial surfaces smooth or scabrous, veins 13–35, ribs obscure; **sclerenchyma** in abaxial and adaxial strands; **girders** formed at most major veins. **Inflorescences** (8)10–30(40) cm, open, with 1–2 branches per node; **branches** lax, spreading, spikelets borne towards the ends of branches. **Spikelets** 6–11 mm, sometimes glaucous, with (2)3–5 florets. **Glumes** lanceolate, smooth or scabrous, acuminate; **lower glumes** 4–7 mm; **upper glumes** 5–7.5 mm; **lemmas** 5–8 mm, chartaceous, lanceolate, glabrous, usually smooth, sometimes scabrous towards the apices, apices acute to acuminate, unawned, sometimes mucronate; **paleas** as long as or slightly shorter than the lemmas, intercostal region puberulent distally; **anthers** 2–3 mm; **ovary apices** densely pubescent. $2n$ = unknown.

Festuca versuta grows in moist, shaded sites on rocky slopes in open woods, from Oklahoma and Arkansas to Texas. It is an uncommon species.

Festuca subg. Obtusae E.B. Alexeev

Plants loosely cespitose, without rhizomes. **Innovations** extravaginal. **Blades** lax, flat or convolute; **ribs** shallow and indistinct; **sclerenchyma girders** present at the major veins. **Calluses** wider than long, scabrous at the margins; **lemmas** stiffly chartaceous, entire, unawned; **ovary apices** densely pubescent.

Festuca subg. *Obtusae* has been divided into two sections. The two species which occur in eastern North America belong to **Festuca** sect. **Obtusae** E.B. Alexeev. A third species, *F. japonica* Makino, belongs in **Festuca** sect. **Fauria** E.B. Alexeev.

2. Festuca subverticillata (Pers.) E.B. Alexeev [p. 401]
NODDING FESCUE

Plants loosely cespitose, or culms solitary to few in a tuft, without rhizomes. **Culms** (40) 50–100(150) cm, glabrous, erect or decumbent at the base. **Sheaths** closed for less than ¹/₃ their length, glabrous or sparsely pilose, shredding into fibers; **ligules** (0.2)0.5–1(2) mm; **blades** (3)5–10 mm wide, flat or loosely convolute, glabrous or sparsely pilose, smooth or scabrous, veins (11)15–39, ribs obscure; **abaxial sclerenchyma** in narrow strands; **adaxial sclerenchyma** developed; **girders** or **pillars** usually associated with the major veins. **Inflorescences** 13–25 cm, open, with 1–2(3) branches per node; **branches** lax, usually reflexed, sometimes spreading, spikelets borne towards the ends of the branches, not or only slightly imbricate. **Spikelets** 4–5(7) mm, elliptic to ovate, with 2–4(6) florets. **Glumes** ovate-lanceolate, scabrous on the veins and distal margins; **lower glumes** 2.5–3.5 mm, usually distinctly shorter than the adjacent lemmas; **upper** glumes 3–4(4.7) mm; **lemmas** 3–4.5 mm, ovate-lanceolate to ovate, stiffly chartaceous, glabrous, obtuse or somewhat acute, unawned; **paleas** as long as or slightly shorter than the lemmas, intercostal region smooth or scabridulous distally; **anthers** (0.8)1–1.7(2.2) mm; **ovary apices** pubescent. $2n$ = 42.

Festuca subverticillata grows in moist to dry, deciduous or mixed forests with organic rocky soils, from Manitoba to Nova Scotia, south to eastern Texas, Florida, and north-eastern Mexico. Plants that are sparsely pilose over the sheaths and blades have been named *F. subverticillata* f. *pilosifolia* (Dore) Darbysh. They frequently grow in mixed populations with *F. subverticillata* (Pers.) E.B. Alexeev f. *subverticillata*.

Festuca subverticillata resembles *F. paradoxa* (see next), but its spikelets are less crowded on the branches.

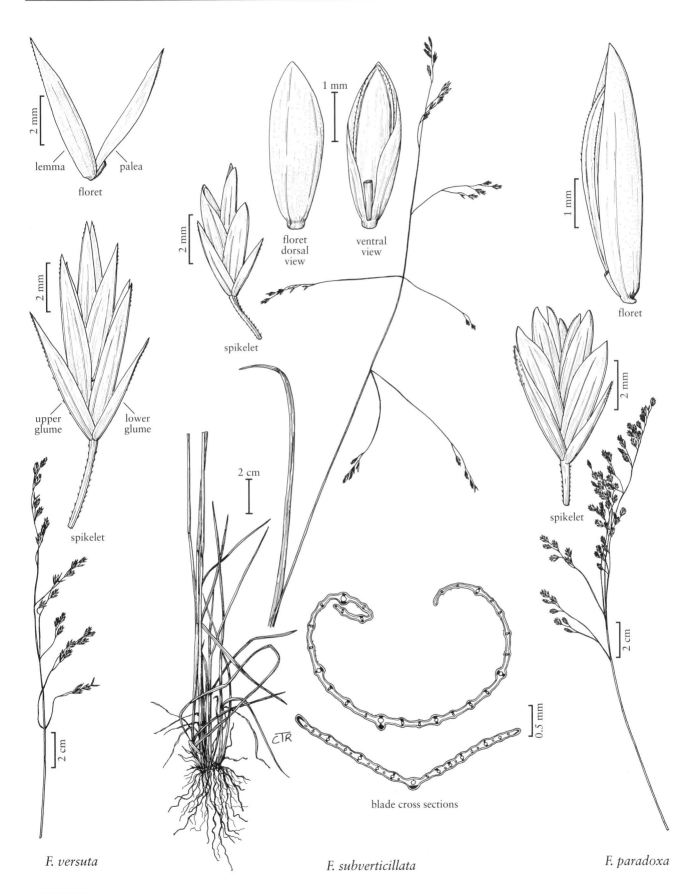

2 mm

lemma   palea

floret

1 mm

floret
dorsal
view

ventral
view

spikelet

1 mm

floret

2 mm

2 mm

upper
glume   lower
glume

spikelet

2 mm

2 cm

spikelet

2 mm

2 cm

CTR

0.5 mm

blade cross sections

F. versuta

F. subverticillata

F. paradoxa

FESTUCA

3. Festuca paradoxa Desv. [p. 401]
CLUSTER FESCUE

Plants loosely cespitose, without rhizomes. **Culms** 50–120 cm, glabrous. **Sheaths** closed for less than ¹/₃ their length, glabrous, shredding into fibers; **ligules** (0.2)0.5–1.5 mm; **blades** 2–8 mm wide, flat or loosely convolute, smooth or scabrous, veins 9–35, ribs obscure; **abaxial sclerenchyma** in narrow strands; **adaxial sclerenchyma** developed; **girders** or **pillars** usually associated with the major veins. **Inflorescences** (5)10–20 cm, open, with 1–2 branches per node; **branches** lax, ascending to spreading, spikelets clustered towards the ends of the branches, closely imbricate. **Spikelets** 4–7(7.5) mm, elliptic to obovate, with 3–5(8) florets. **Glumes** lanceolate to ovate-lanceolate, scabrous at least on the veins; **lower glumes** (2.5)3–4(5) mm, usually almost as long as the adjacent lemmas; **upper glumes** (3.5)4–5(5.5) mm; **lemmas** 4–5(5.2) mm, stiffly chartaceous, ovate to obovate, glabrous, somewhat acute, unawned; **paleas** as long as or slightly shorter than the lemmas, intercostal region smooth or scabrous distally; **anthers** (0.7)1–2(2.5) mm; **ovary apices** pubescent. $2n$ = unknown.

Festuca paradoxa grows in prairies, open woods, thickets, and low open ground, from Wisconsin to Pennsylvania, south to northeastern Texas and northern Georgia. It resembles *F. subverticillata* (see previous), but its spikelets are more crowded on the branches.

Festuca subg. Subulatae (Tzvelev) E.B. Alexeev

Plants loosely cespitose, without rhizomes. **Innovations** extravaginal. **Blades** lax, flat or convolute; **ribs** shallow and indistinct; **sclerenchyma** forming **pillars** or **girders** at the major veins. **Calluses** sometimes wider than long, sometimes narrower, scabrous at the margins; **lemmas** chartaceous, sometimes minutely bidentate, usually awned, sometimes unawned; **ovary apices** densely pubescent.

Festuca subg. *Subulatae* contains about 30–35 species. It is known from eastern Asia and western North America, as well as Central and South America. Three of the five sections that have been described occur in North America.

Festuca sect. Subulatae (Tzvelev)

Plants loosely cespitose. **Innovations** extravaginal. **Blades** lax, flat or convolute; **ribs** shallow and indistinct; **sclerenchyma** forming **pillars** or **girders** at the major veins. **Calluses** sometimes wider than long, sometimes narrow, scabrous at the margins; **lemmas** chartaceous, entire, awned or unawned; **ovary apices** densely pubescent.

Festuca sect. *Subulatae* is the largest section in this subgenus and contains about 20–25 species. Its range includes eastern Asia, western North America, and Central and South America.

4. Festuca subulata Trin. [p. 403]
BEARDED FESCUE

Plants loosely cespitose, without rhizomes, with short extravaginal tillers. **Culms** (35)50–100(120) cm, erect or decumbent at the base, scabrous. **Sheaths** closed for less than ¹/₃ their length, glabrous or sparsely pubescent, shredding into fibers; **collars** glabrous; **ligules** 0.2–0.6(1) mm; **blades** 3–10 mm wide, flat or loosely convolute, abaxial and adaxial surfaces scabrous or puberulent, veins 13–29, ribs obscure; **sclerenchyma** in narrow abaxial and adaxial strands; **pillars** or **girders** formed at the major veins. **Inflorescences** 10–40 cm, open, with 1–2(5) branches per node; **branches** lax, usually spreading, sometimes reflexed. **Spikelets** 6–12 mm, with (2)3–5(6) florets. **Glumes** sparsely scabrous towards the apices, acuminate to subulate; **lower glumes** (1.8)2–3(4)

callus

spikelet

floret

ligule

collar

2 mm

2 cm

1 mm

0.5 mm

F. subulata

floret

sheath

spikelet

CTR

F. sororia

FESTUCA

mm; **upper glumes** (2)3–6 mm; **calluses** wider than long, glabrous, smooth or slightly scabrous; **lemmas** 5–9 mm, glabrous, sometimes sparsely scabrous, lanceolate, apices entire, acute to acuminate, awned, awns (2.5) 5–15(20) mm, terminal, straight, sometimes curved or kinked; **paleas** about as long as or slightly longer than the lemmas, intercostal region puberulent distally; **anthers** 1.5–2.5(3) mm; **ovary apices** pubescent. $2n = 28$.

Festuca subulata grows on stream banks and in open woods, meadows, shady forests, and thickets, to about 2800 m. Its range extends from the southern Alaska panhandle eastward to southwestern Alberta and western South Dakota, and southward to central California and Colorado.

Festuca subulata differs from *F. subuliflora* (p. 406) in having blunter, glabrous calluses and glabrous, often scabrous or puberulent leaf blades that are obscurely ribbed.

5. Festuca sororia Piper [p. 403]
RAVINE FESCUE

Plants loosely cespitose, without rhizomes, with short extravaginal tillers. **Culms** 60–100 (150) cm, erect, glabrous; **nodes** usually exposed. **Sheaths** closed for less than ⅓ their length, glabrous or scabrous, shredding into fibers; **ligules** 0.3–1.5 mm; **blades** 3–6(10) mm wide, flat, lax, margins scabrous, abaxial and adaxial surfaces glabrous, adaxial surfaces sometimes scabrous, veins 13–25, ribs obscure; **abaxial sclerenchyma** in narrow strands; **adaxial sclerenchyma** developed; **girders** or **pillars** formed at the major veins. **Inflorescences** 10–20(40) cm, open or somewhat contracted, with 1–2(3) branches per node; **branches** lax, more or less spreading, spikelets borne towards the ends of the branches. **Spikelets** 7–12 mm, with (2)3–5 florets. **Glumes** lanceolate, scabrous at least on the midvein, acute to acuminate; **lower glumes** (1.5) 2.5–4(4.5) mm; **upper glumes** (3)4–6.5 mm; **calluses** wider than long, glabrous, smooth or slightly scabrous; **lemmas** (5)6–8(9) mm, lanceolate, scabrous or puberulent, acuminate, unawned or awned, awns to 2 mm; **paleas** about as long as or shorter than the lemmas, intercostal region puberulent distally; **anthers** 1.6–2.5 mm; **ovary apices** pubescent. $2n =$ unknown.

Festuca sororia grows in open woods and on shaded slopes and stream banks, at 2000–3000 m. It is restricted to the United States, growing from central Utah and Colorado to Arizona and New Mexico. A single puzzling specimen is the basis for the reported occurrence of this species in Missouri (Yatskievych 1999).

Festuca sect. Elmera E.B. Alexeev

Plants loosely cespitose. **Innovations** extravaginal. **Blades** lax, flat or convolute; **ribs** shallow and indistinct; **sclerenchyma girders** present at the major veins. **Calluses** wider than long, scabrous at the margins; **lemmas** chartaceous, minutely bidentate, awned; **ovary apices** densely pubescent.

One species has been placed in this section.

6. Festuca elmeri Scribn. & Merr. [p. 405]
COAST FESCUE, ELMER'S FESCUE

Plants loosely cespitose. **Culms** 40–100(120) cm, glabrous, erect or slightly decumbent at the base. **Sheaths** closed for less than ⅓ their length, glabrous, smooth or slightly scabrous, shredding into fibers; **collars** glabrous, smooth or slightly scabrous; **ligules** 0.1–0.5(0.7) mm; **blades** 1.8–6 mm wide, vegetative shoot blades narrower than the cauline blades, flat or loosely conduplicate or convolute, abaxial surfaces glabrous, adaxial surfaces slightly scabrous or pubescent, veins 7–19, ribs obscure to prominent; **abaxial sclerenchyma** in narrow strands; **adaxial sclerenchyma** developed; **pillars** or **girders** present at the major veins. **Inflorescences** 10–20 cm, open, with 1–2 branches per node; **branches** lax, more or less spreading, spikelets borne towards the ends of the branches. **Spikelets** (7)7.5–11 mm, with 2–6(7) florets. **Glumes** lanceolate, glabrous, smooth or the apices slightly scabrous, acuminate; **lower glumes** 2–4 mm; **upper glumes** 3–4.6 mm; **calluses** wider than long, smooth or slightly scabrous, glabrous; **lemmas** 5.5–7 mm, lanceolate, scabrous or puberulent, minutely bidentate, awned, awns (1.5)2–5(8) mm, subterminal, straight to slightly curved or kinked; **paleas** as long as or longer than the

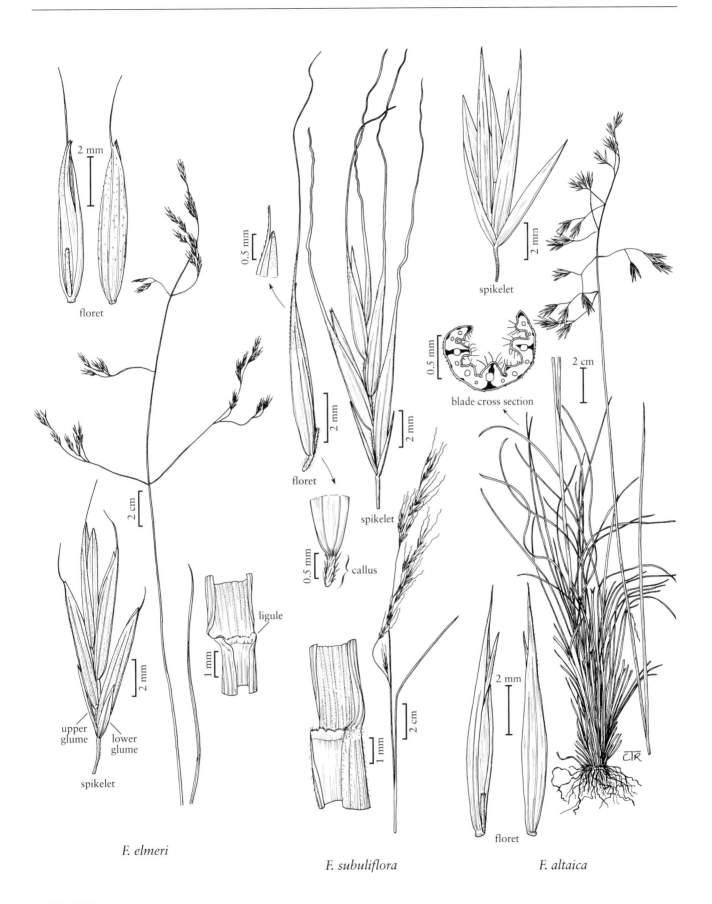

floret

spikelet

upper
glume

lower
glume

spikelet

F. elmeri

0.5 mm

floret

spikelet

callus

ligule

F. subuliflora

spikelet

blade cross section

floret

F. altaica

FESTUCA

lemmas, intercostal region puberulent distally; **anthers** (3)3.4–4 mm; **ovary apices** pubescent. 2*n* = 28.

Festuca elmeri grows on moist wooded slopes, usually below 300(500) m, from Oregon to south-central California. The more southerly populations, which have larger spikelets with 5–6, rather than 3–4, florets and a more compact inflorescence with more or less erect panicle branches, have been named *F. elmeri* subsp. *luxurians* Piper.

Festuca sect. **Subuliflorae** (E.B. Alexeev) Darbysh.

Plants loosely cespitose; **innovations** extravaginal; **blades** lax, flat or convolute; **ribs** obscure to prominent; **sclerenchyma girders** present at the major veins. **Calluses** longer than wide, bearded with a ring of hairs; **lemmas** chartaceous, minutely bidentate, awned; **ovary apices** densely pubescent.

Only one species has been placed in this section.

7. Festuca subuliflora Scribn. [p. 405]
CRINKLE-AWN FESCUE, COAST RANGE FESCUE

Plants loosely cespitose, without rhizomes. **Culms** 40–125 cm, glabrous. **Sheaths** closed for less than ⅓ their length, glabrous or pubescent, shredding into fibers; **collars** glabrous or pubescent; **ligules** 0.1–0.5 mm; **blades** 2–6(8) mm wide, flat or loosely convolute, abaxial surfaces glabrous or sparsely pubescent, adaxial surfaces pubescent, veins 13–29, ribs obscure to prominent; **abaxial sclerenchyma** in strands; **adaxial sclerenchyma** developed; **pillars** or **girders** present at the major veins. **Inflorescences** 7–20 cm, open, with 1(2) branches per node; **branches** lax, spreading. **Spikelets** 8–12.5 mm, with (2)3–5 florets. **Glumes** glabrous, lanceolate to subulate; **lower glumes** (2)2.5–4 mm; **upper glumes** 3.5–5.5(6) mm; **calluses** much longer than wide, pubescent at least basally; **lemmas** 6–9 mm, lanceolate, puberulent, particularly towards the bases, sometimes slightly scabrous, particularly towards the apices, apices minutely bidentate, awned, awns 10–15 mm, slightly subterminal, flexuous, or kinked; **paleas** about as long as the lemmas, intercostal region puberulent distally; **anthers** (2)2.5–4 mm; **ovary apices** pubescent. 2*n* = 28.

Festuca subuliflora grows in shady sites in dry to moist forests, usually below 700 m. Its range extends from southwestern British Columbia to central California. Superficially, it resembles *F. subulata* (p. 402); it differs in having more elongated and distinctly hairy calluses, and often in having softly pubescent foliage and more strongly ribbed blades.

Festuca L. subg. **Festuca**

Plants loosely or densely cespitose, with or without rhizomes. **Innovations** intravaginal or extravaginal. **Blades** usually more or less stiff, setaceous if lax, usually conduplicate, sometimes convolute or flat; **ribs** usually distinct; **sclerenchyma girders** sometimes present at the major veins. **Calluses** wider than long, scabrous on the margins; **lemmas** usually membranous or chartaceous, rarely somewhat coriaceous, usually entire, sometimes minutely bidentate, usually awned, sometimes unawned; **ovary apices** glabrous or sparsely to densely pubescent.

Festuca subg. *Festuca* is most abundant in the Northern Hemisphere, but it is distributed on all continents except Antarctica. Estimating the number of species in this subgenus is difficult in the absence of adequate treatments for many parts of the world, but it probably exceeds 400.

Festuca sect. **Breviaristatae** Krivot.

Plants usually densely cespitose, sometimes loosely cespitose, rhizomes short or absent. **Innovations** mostly intravaginal. **Blades** usually more or less stiff, conduplicate, sometimes convolute or flat; **ribs** usually distinct; **sclerenchyma girders** usually present at the major veins, rarely absent. **Calluses** wider than long, scabrous on the margins; **lemmas** chartaceous, sometimes somewhat coriaceous, apices entire, awned or unawned; **ovary apices** usually pubescent, sometimes sparsely pubescent, rarely glabrous.

Festuca sect. *Breviaristatae* is distributed in Asia and North America. It contains about 15 species.

8. Festuca altaica Trin. [p. 405]
NORTHERN ROUGH FESCUE, ALTAI FESCUE, FÉTUQUE D'ALTAI

Plants densely cespitose, rarely with short rhizomes. **Culms** (25)30–90(120) cm, glabrous or slightly scabrous; **nodes** usually not exposed. **Sheaths** closed for less than $^1/_3$ their length, glabrous or scabrous, persistent, not shredding into fibers; **collars** glabrous; **ligules** 0.2–0.6 (1) mm; **blades** deciduous, 2–4 mm wide, convolute, conduplicate, sometimes flat, 1–2.5 mm in diameter when conduplicate, yellow-green to dark green, abaxial surfaces scabrous, adaxial surfaces glabrous or pubescent, smooth or scabrous, veins 7–15(17), ribs 5–9; **abaxial sclerenchyma** in strands about as wide as the adjacent veins; **adaxial sclerenchyma** present; **girders** associated with the major veins. **Inflorescences** 5–16 cm, open, often secund, with 1–2(3) branches per node; **branches** lax, spreading, lower branches usually recurved or reflexed, spikelets borne towards the ends of the branches. **Spikelets** 8–14 mm, usually purple, lustrous, with 3–4(6) florets. **Glumes** glabrous or slightly scabrous, distinctly shorter than the adjacent lemmas; **lower glumes** 4–6.8(8.5) mm; **upper glumes** (4.5)5.3–7.5(10) mm; **lemmas** (6.5)7.5–9(12) mm, chartaceous, scabrous, at least on the veins, keeled on the lower $^1/_2$, veins 5, prominent, apices attenuate or short-awned, awns 0.2–0.7 mm; **paleas** about as long as or a little shorter than the lemmas, intercostal region puberulent distally; **anthers** 2.6–4.5(5) mm; **ovary apices** usually sparsely pubescent, rarely glabrous. $2n = 28$.

Festuca altaica is a plant of rocky alpine habitats, arctic tundra, and open boreal or subalpine forests. Its primary distribution extends from Alaska eastward to the western Northwest Territories, and south in the alpine regions of British Columbia and west-central Alberta. Disjunct populations occur in Quebec, western Labrador and Newfoundland, and in Michigan, where it may be introduced. From the Bering Sea it extends westward to the Altai Mountains of central Asia.

The spikelets of *Festuca altaica* are lustrous and usually intensely purplish; plants with greenish spikelets have been named *F. altaica* f. *pallida* Jordal. A form producing pseudoviviparous spikelets, *F. altaica* f. *vivipara* Jordal, has been described from Alaska.

9. Festuca hallii (Vasey) Piper [p. 409]
PLAINS ROUGH FESCUE

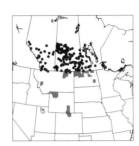

Plants densely cespitose, usually with short rhizomes. **Culms** (16)25–65(85) cm, glabrous, smooth or scabrous near the inflorescence; **nodes** usually not exposed. **Sheaths** closed for less than $^1/_3$ their length, glabrous, smooth or scabrous, persistent; **collars** glabrous; **ligules** 0.3–0.6 mm; **blades** usually conduplicate and 0.5–1.2 mm in diameter, rarely flat and 1–2.5 mm wide, gray-green, deciduous, abaxial surfaces scabrous, adaxial surfaces scabrous or puberulent, veins (5)7–9, ribs 5–7, conspicuous; **abaxial sclerenchyma** usually forming continuous or interrupted bands; **adaxial sclerenchyma** present; **girders** developed at the 3(5) major veins; **pillars** developed at most other veins. **Inflorescences** 6–16 cm, usually more or less contracted, open at anthesis, with 1–2(3) branches per node; **branches** erect or stiffly spreading, spikelets borne towards the ends of the branches. **Spikelets** (6.5)7–9.5 mm, with 2–3(4) florets. **Glumes** about equaling or slightly exceeding the upper florets; **lower glumes** 5–8(9.5) mm, about equaling or slightly longer than the adjacent lemmas; **upper glumes** 6.2–8.5(9.5) mm; **lemmas** 5.5–8(9) mm, chartaceous to somewhat coriaceous, scabrous, rounded below midlength, veins somewhat obscure, apices unawned or awned, awns 0.5–1.3 mm; **paleas** somewhat shorter than the lemmas, intercostal region puberulent distally; **anthers** 4–6 mm; **ovary apices** sparsely pubescent. $2n = 28$.

Festuca hallii is a major component of grasslands in the northern Great Plains and the grassland-boreal forest transition zone, where it is an important source of forage. Its range extends from the Rocky Mountains of Canada east to western Ontario and south to Colorado. At the southern end of its range in Colorado, it grows in alpine meadows.

Festuca hallii differs from *F. campestris* (see next) in usually having short rhizomes, stiffly erect panicles, and smaller spikelets. Where the two species are sympatric, as in the foothills of the Rocky Mountains, *F. hallii* is usually found at lower elevations.

10. **Festuca campestris** Rydb. [p. 409]
MOUNTAIN ROUGH FESCUE

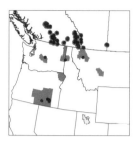

Plants densely cespitose, usually without rhizomes, occasionally with short rhizomes. **Culms** (30)40–90(140) cm, scabrous near the inflorescence; **nodes** usually not exposed. **Sheaths** closed for less than ¹/₃ their length, glabrous or scabrous, persistent; **collars** glabrous; **ligules** 0.1–0.5 mm; **blades** 0.8–2 mm in diameter, usually conduplicate, rarely convolute, gray-green, deciduous, abaxial surfaces scabrous, adaxial surfaces scabrous or puberulent, veins (8)11–15(17), ribs (6)7–11; **abaxial sclerenchyma** usually forming a more or less continuous band; **adaxial sclerenchyma** developed; **girders** at the 5–7 major veins; **pillars** at some of the other veins. **Inflorescences** (5)9–18(25) cm, open or loosely contracted, with(1)2(3) branches per node; **branches** erect to stiffly spreading. **Spikelets** 8–13(16) mm, with (3)4–5(7) florets. **Glumes** exceeded by the distal florets; **lower glumes** 4.5–7.5(8.5) mm, shorter than or about equaling the adjacent lemmas; **upper glumes** 5.3–8.2(9) mm; **lemmas** (6.2)7–8.5(10) mm, chartaceous to somewhat coriaceous, scabrous, backs rounded below the middle, veins more or less obscure, apices mucronate or shortly awned, awns to 1.5 mm; **paleas** somewhat shorter than the lemmas, intercostal region puberulent distally; **anthers** (3.3) 4.5–6 mm; **ovary apices** pubescent. 2*n* = 56.

Festuca campestris is a common species in prairies and montane and subalpine grasslands, at elevations to about 2000 m. Its range extends from southern British Columbia, Alberta, and southwestern Saskatchewan south through Washington, Oregon, Idaho, and Montana. It is highly palatable and provides nutritious forage.

Festuca campestris differs from *F. hallii* (see previous) in having larger spikelets, less stiffly erect panicles and, usually, in lacking rhizomes. Where the two are sympatric, *F. campestris* tends to grow at higher elevations.

11. **Festuca ligulata** Swallen [p. 409]
GUADALUPE FESCUE

Plants loosely to densely cespitose, with short rhizomes. **Culms** 45–80 cm, erect or the bases decumbent, scabrous near the inflorescence. **Sheaths** closed for less than ¹/₃ their length, glabrous or finely scabrous; **collars** glabrous; **ligules** 3–5(8) mm; **blades** 1–3 mm wide when flat, 0.6–1.2 mm in diameter when conduplicate, persistent, abaxial surfaces glabrous, smooth to sparsely scabrous, adaxial surfaces scabrous, veins (5)7–9, ribs 5–9; **abaxial sclerenchyma** in strands opposite the veins, rarely a discontinuous band; **adaxial sclerenchyma** sometimes present; **girders** sometimes present at the major veins; **pillars** usually present if the girders not developed. **Inflorescences** 6–10(16) cm, contracted or loosely open, with 1–2(3) branches per node; **branches** erect or spreading, lower branches sometimes reflexed, spikelets borne towards the ends of the branches. **Spikelets** 6–8.5 mm, with 2–3(4) florets. **Glumes** scabrous, acute; **lower glumes** 3–4(5.5) mm; **upper glumes** 3.5–5.5(6.5) mm; **lemmas** 4–6.5 mm, ovate-lanceolate, glabrous, smooth or sparsely scabrous towards the apices, unawned; **paleas** as long as to slightly longer than the lemmas, intercostal region puberulent distally; **anthers** 1.5–2.6 mm; **ovary apices** pubescent. 2*n* = unknown.

Festuca ligulata grows on moist, shady slopes in the mountains of western Texas and north-central Mexico. It is listed as an endangered species under the Endangered Species Act of the United States.

12. **Festuca thurberi** Vasey [p. 411]
THURBER'S FESCUE

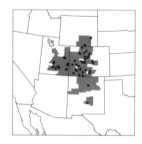

Plants densely cespitose, without rhizomes. **Culms** (45) 60–100(120) cm, glabrous, smooth or scabrous below the inflorescence. **Sheaths** closed for less than ¹/₃ their length, smooth or scabrous, persistent; **collars** glabrous; **ligules** 2–5(9) mm, entire or lacerate, not ciliate; **blades** 1.5–3 mm wide, 0.8–1.8 mm in diameter when conduplicate, deciduous, abaxial surfaces scabrous, adaxial surfaces scabrous or pubescent, veins 9–15, ribs 7–13; **abaxial sclerenchyma** a more or less continuous band; **adaxial sclerenchyma** present; **girders** usually formed at the major veins, sometimes only **pillars** present. **Inflorescences** (7)10–15(17) cm, open, with 1–2(3) branches per node; **branches** 4.5–9 cm, lax, erect or spreading, spikelets borne towards the ends of the branches. **Spikelets** (8)10–14 mm, with (3)4–5(6) florets. **Glumes** unequal to subequal, ovate-lanceolate, scabrous or smooth, acute; **lower glumes** (2)3.5–5.5 mm; **upper glumes** (2.5)4.5–6.5(7) mm; **lemmas** 6–10 mm, lanceolate to ovate-lanceolate, scabrous or smooth, unawned, sometimes mucronate, mucros to 0.2 mm; **paleas** shorter than to as long as the lemmas, intercostal region puberulent distally; **anthers** 3–4.5 mm; **ovary apices** densely pubescent. 2*n* = 28, 42.

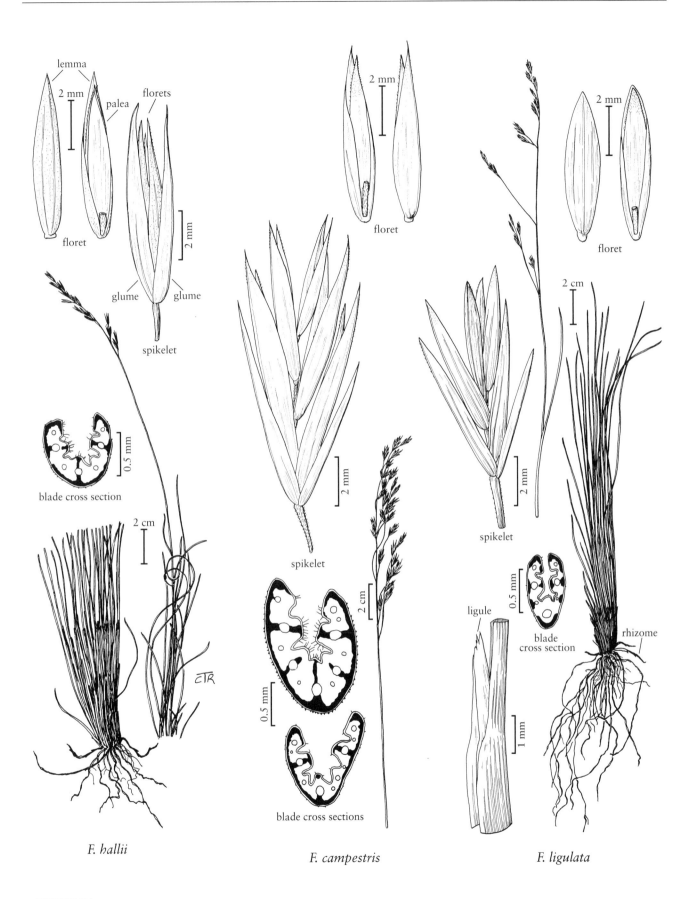

F. *hallii*

F. *campestris*

F. *ligulata*

FESTUCA

Festuca thurberi is a large bunchgrass of dry, rocky slopes and hills, open forests, and meadows in montane and subalpine regions, at (1000)2000–3500 m. Its range extends from southern Wyoming south through Utah and Colorado to New Mexico.

13. Festuca californica Vasey [p. 411]
CALIFORNIA FESCUE

Plants densely cespitose, without rhizomes. **Culms** 30–150 (200) cm, glabrous or pubescent, sometimes scabrous. **Sheaths** closed for less than ⅓ their length, persistent, glabrous or pilose, smooth or scabrous, sometimes scabrous or pilose only distally or on the distal margins; **collars** usually densely pubescent or with a few hairs at the margins, sometimes glabrous; **ligules** 0.2–5 mm, usually ciliate, abaxial surfaces puberulent; **blades** 1–6.5 mm wide, conduplicate, convolute, or flat, 0.5–2(2.5) mm in diameter when convolute, deciduous, abaxial surfaces scabrous or smooth, glabrous or the bases pubescent, adaxial surfaces puberulent to pubescent, veins 9–15(17), ribs (3)5–15(17); **abaxial sclerenchyma** forming more or less continuous bands, sometimes reduced to small strands; **adaxial sclerenchyma** sometimes present; **girders** or **pillars** present at most veins. **Inflorescences** 10–25(30) cm, open, with (1)2(4) branches per node; **branches** spreading and lax. **Spikelets** 8–18(20) mm, borne towards the ends of the branches, with 3–6(8) florets. **Glumes** lanceolate, glabrous or sparsely scabrous at the apices; **lower glumes** (4)4.5–6.7(8) mm; **upper glumes** (5)6–10 mm; **lemmas** (7)7.5–11 mm, lanceolate, scabrous, puberulent, sometimes minutely bidentate, acute, usually awned, rarely unawned, awns (1)2–3(4) mm; **paleas** shorter than to longer than the lemmas, pubescent or glabrous on the margins, intercostal region usually puberulent distally; **anthers** (3)4–7.5(8.5) mm; **ovary apices** densely pubescent. $2n = 56$.

Festuca californica grows on dry, open slopes and moist streambanks in thickets and open woods, from sea level to 2000 m. Its range extends from Clackamas County, Oregon, to the Sierra Nevada and southern California; it is not known to extend into Mexico. It is the largest species of *Festuca* in the *Flora* region.

1. Culms 30–80(100) cm tall, usually pubescent for more than 5 mm below the nodes; lower sheaths densely retrorsely pubescent; vegetative shoot blades with (3)5–9 ribs, the ribs to about ½ as deep as the blade thickness; abaxial sclerenchyma in small strands or forming continuous bands; adaxial sclerenchyma strands present or absent; sclerenchyma pillars rarely formed; girders not developed; spikelets with 3–4(5) florets . . . subsp. *parishii*

1. Culms 60–150(200) cm tall, glabrous or pubescent for less than 5 mm below the nodes; lower sheaths glabrous or pubescent, if pubescent then usually not densely retrorsely hairy; vegetative shoot blades with 7–15(17) ribs, the ribs usually more than ½ as deep as the blade thickness; abaxial sclerenchyma forming a continuous band; adaxial sclerenchyma in strands; sclerenchyma pillars or girders usually associated with most of the veins; spikelets with (3)4–6(8) florets.

  2. Ligules 0.2–1(1.2) mm long, ciliate; spikelets (8)13–18(20) mm long . . . . subsp. *californica*

  2. Ligules (1)1.5–5 mm long, ciliate or not; spikelets 8–12(17) mm long . . . . subsp. *hitchcockiana*

Festuca californica Vasey subsp. californica [p. 411]

Culms 60–150(200) cm, densely pubescent for less than 5 mm below the nodes, particularly the upper nodes. **Sheaths** glabrous or pubescent; **collars** usually densely pubescent, at least at the margins, sometimes glabrous; **ligules** 0.2–1(1.2) mm, ciliate; **blades** 30+ cm long, 3–6.5 mm wide, conduplicate or flat, ribs 7–15(17), usually more than ½ as deep as the blade thickness, abaxial surfaces glabrous or sparsely pubescent proximally, adaxial surfaces pubescent; **abaxial sclerenchyma** forming a continuous band; **adaxial sclerenchyma** present; **girders** usually formed at most veins. **Inflorescences** 15–25(30) cm. **Spikelets** (8)13–18(20) mm, with (3)4–6(8) florets. **Lemmas** usually entire, sometimes minutely bidentate. $2n = 56$.

Festuca californica subsp. *californica* is the most widespread variety, growing from west-central Oregon to central California. The lower leaf sheaths are typically glabrous and scabrous, but sometimes have spreading hairs. This subspecies differs from subsp. *parishii* in having wider and longer leaf blades and more extensively developed sclerenchyma.

Festuca californica subsp hitchcockiana (E.B. Alexeev) Darbysh. [p. 411]

Culms 60–120 cm, glabrous or pubescent for less than 5 mm below the nodes. **Sheaths** usually glabrous or scabrous, sometimes pubescent; **collars** glabrous or indistinctly pubescent; **ligules** (1)1.5–5 mm, sometimes ciliate, abaxial surfaces pubescent; **blades** 3–6.5 mm wide, conduplicate or flat, ribs 7–15(17), usually more than ½ as deep as the blade thickness, abaxial surfaces glabrous or sparsely pubescent proximally, adaxial surfaces pubescent; **abaxial sclerenchyma** forming continuous bands; **adaxial sclerenchyma** present; **girders** usually formed at most veins. **Inflorescences** 15–25 cm. **Spikelets** 8–12(17) mm, with (4)5–6(8)

spikelet

floret

F. thurberi

spikelet

floret

2 mm

lemma

anther

ovary

blade
cross section

spikelet

ligule

node

2 mm

upper
sheath

lower
sheath

F. californica
subsp. *parishii*

node

ligule

blade cross section

collar

ligule

F. californica subsp. *californica*

F. californica
subsp. *hitchcockiana*

CTR

FESTUCA

florets. **Lemmas** usually entire, sometimes minutely bidentate. $2n$ = unknown.

Festuca californica subsp. *hitchcockiana* is distinguished by its relatively long ligules. It is known only from Santa Clara and San Luis Obispo counties, California.

Festuca californica subsp. **parishii** (Piper) Darbysh. [p. 411]

Culms 30–80(100) cm, densely pubescent for more than 5 mm below the nodes, particularly the upper nodes. **Sheaths** densely retrorsely pubescent; **collars** usually densely pubescent, at least at the margins; **ligules** (0.2)0.5–1.5(2) mm, ciliate; **blades** 10–30 cm long, 1–3 mm wide, 0.5–1.2(1.5) mm in diameter when conduplicate, abaxial surfaces glabrous or pubescent proximally, adaxial surfaces pubescent, ribs (3)5–9, to about $^{1}/_{2}$ as deep as the blade thickness; **abaxial sclerenchyma** in small strands or forming continuous bands; **adaxial sclerenchyma** sometimes present; **girders** not developed; **pillars** rarely formed. **Inflorescences** 10–20 cm. **Spikelets** 11–16 mm, with 3–4(5) florets. **Lemmas** usually minutely bidentate, sometimes entire. $2n$ = unknown.

Festuca californica subsp. *parishii* grows in southern California, in the San Bernardino, San Gabriel, and Palomar mountains. Its leaf blades tend to be narrower and shorter than in subsp. *californica* (10–30 cm long versus more than 30 cm long), and the sclerenchyma is less developed, with pillars only sometimes present and girders absent. The lower leaf sheaths are densely retrorsely pubescent.

Festuca L. sect. Festuca

Plants loosely or densely cespitose, with short rhizomes or without rhizomes. **Innovations** mostly intravaginal. **Blades** more or less stiff, setaceous if lax, usually conduplicate, sometimes convolute or flat; **ribs** usually distinct; **sclerenchyma** usually only developed on the adaxial surface, sometimes forming **pillars** or **girders** at the major veins. **Calluses** wider than long, scabrous on the margins; **lemmas** chartaceous, apices usually entire, rarely minutely bidentate, usually awned, sometimes unawned; **ovary apices** usually pubescent, sometimes sparsely pubescent, rarely glabrous.

Festuca sect. *Festuca* is most abundant in the Northern Hemisphere. Its species are native to all continents except Antarctica. There are perhaps 400 or more species in this section, with new ones constantly described.

14. Festuca rubra L. [pp. 416, 417]
RED FESCUE, FÉTUQUE ROUGE

Plants usually rhizomatous, usually loosely to densely cespitose, culms sometimes single and widely spaced, sometimes stoloniferous. **Culms** (8)10–120 (130) cm, erect or decumbent, glabrous and smooth. **Sheaths** closed for about $^{3}/_{4}$ their length when young, readily splitting with age, usually pubescent, at least distally, hairs retrorse or antrorse, sometimes glabrous, not persistent, older vegetative shoot sheaths shredding into fibers; **collars** glabrous; **ligules** 0.1–0.5 mm; **blades** usually conduplicate or convolute and 0.3–2.5 mm in diameter, sometimes flat and 1.5–7 mm wide, abaxial surfaces glabrous, smooth or scabrous, adaxial surfaces scabrous or pubescent, veins 5–9(13), ribs (3)5–7(9), usually conspicuous; **abaxial sclerenchyma** in 5–9(13) discrete or partly confluent strands, rarely forming a complete band; **adaxial sclerenchyma** sometimes present in fascicles opposite the veins; **girders** and **pillars** not developed. **Inflorescences** (2)3.5–25(30) cm, usually open or loosely contracted panicles, occasionally racemes, with 1–3 branches per node, lower branches with 2+ spikelets; **branches** erect or spreading, stiff or lax, glabrous, scabrous, or pubescent. **Spikelets** (6)7–17 mm, with 3–10 florets. **Glumes** ovate-lanceolate to lanceolate, exceeded by the distal florets; **lower glumes** (1.5)2–6(7) mm; **upper glumes** (3)3.5–8.5 mm; **lemmas** 4–9.5 mm, usually glabrous and smooth, sometimes scabrous towards the apices, sometimes densely pubescent throughout, attenuate or acuminate in side view, awned, awns (0.1)0.4–4.5 mm; **paleas** slightly shorter than to about equaling the lemmas, intercostal region puberulent distally; **anthers** 1.8–4.5 mm; **ovary apices** glabrous. $2n$ = 28, 42, 56, 70.

Festuca rubra is interpreted here as a morphologically diverse polyploid complex that is widely distributed in the arctic and temperate zones of Europe, Asia, and North America. Its treatment is complicated by the fact that Eurasian material has been introduced in other parts of the world. In addition, hundreds of forage and turf cultivars have been developed, many of which have also been widely distributed.

Within the complex, morphologically, ecologically, geographically, and/or cytogenetically distinct taxa have been described, named, and given various taxonomic ranks. In some cases these taxa represent extremes, and in other cases they are morphologically intermediate between other taxa. Moreover, hybridization and/or introgression between native taxa, and between native and non-native taxa, may be occurring. In Iceland and southern Greenland, putative hybrids between *Festuca frederikseniae* and *F. rubra* have been reported, and named *F. villosa-vivipara* (Rosenv.) E.B. Alexeev (see under *F. frederikseniae*, p. 436).

Overlap in morphological characters between most taxa in the complex has led some taxonomists to ignore the variation within the complex, calling all its members *Festuca rubra* without qualification. This obscures what is known about the complex, and presents an extremely heterogeneous assemblage of plants as a single "species"—or a mega-species. The following account attempts to reflect the genetic diversity of the *F. rubra*

complex in the *Flora* region. All the taxa are recognized as subspecies, but they are not necessarily equivalent in terms of their distinction and genetic isolation. Much more work on the taxonomy of the *F. rubra* complex is needed before the boundaries of individual taxa can be firmly established. Some variants that need attention are (1) plants growing on the sandy shores of the Great Lakes that have glaucous leaves and spikelets, sometimes treated as *F. rubra* var. *juncea* (Hack.) K. Richt., (2) native plants along the James Bay and Hudson Bay shore that are ecologically distinct from *F. rubra* subsp. *rubra*, (3) native plants growing in marshes, sometimes called *F. rubra* var. *megastachys* (Gaudin) Hegi (Dore and McNeill 1980), (4) seashore variants along the Atlantic coast of North America, (5) plants with glaucescent leaves and spikelets which are widely distributed in the *Flora* region and have been called *F. rubra* subsp. *glaucodea* Piper, (6) the widespread variant with pubescent to villous lemmas, sometimes called *F. rubra* f. *squarrosa* (Hartm.) Holmb.

Festuca earlei (p. 420) is sometimes confused with *F. rubra*. It differs in having pubescent ovary apices.

1. Plants not rhizomatous, densely cespitose.
  2. Anthers 2.3–3.2 mm long; lemma awns 0.1–3 mm long; plants of natural habitats in coastal areas . . . . . subsp. *pruinosa*
  2. Anthers 1.8–2.2(3) mm long; lemma awns 1–3.3 mm long; plants of lawns, road verges, and other
    disturbed areas . . . . subsp. *commutata*
1. Plants rhizomatous, usually loosely to densely cespitose, sometimes with solitary culms.
  3. Vegetative shoot blades usually flat or loosely conduplicate; plants strongly rhizomatous; adaxial
    sclerenchyma strands always present . . . . subsp. *fallax*
  3. Vegetative shoot blades usually conduplicate, sometimes flat; plants strongly or weakly rhizomatous; adaxial sclerenchyma strands sometimes present.
    4. Plants not or only loosely cespitose, the culms usually single and widely spaced; plants of moist meadows in montane and subalpine regions of the western cordillera, usually above 1000 m . . . . subsp. *vallicola*
    4. Plants loosely to densely cespitose, with several culms arising from the same tuft; plants of various habitats and elevations.
      5. Inflorescence branches scabrous or pubescent; lemmas usually moderately to densely pilose, sometimes only partially pilose, occasionally glabrous; lemma awns (0.2)0.5–1.6 (2.5) mm long; plants of subalpine, alpine, boreal, and arctic regions, both littoral and inland . . . . subsp. *arctica*
      5. Inflorescence branches scabrous; lemmas usually glabrous, the lemmas of littoral plants sometimes hairy; lemma awns (0.1)0.4–5 mm long; plants of various habitats.
        6. Plants widely distributed, sometimes coastal.
          7. Lower glumes 3–4.5 mm long; inflorescences 7–12 cm long, lanceolate; plants of disturbed habitats throughout temperate and mesic regions . . . . subsp. *rubra*
          7. Lower glumes 2.2–3.2(4.5) mm long; inflorescences 3–10 (20) cm long, linear to lanceolate; plants of natural habitats in coastal areas . . . . subsp. *pruinosa*
        6. Plants of the Pacific coast, often growing close to the littoral zone.
          8. Cauline leaf sheaths tightly enclosing the culms; mature inflorescences usually completely exserted from the sheaths.
            9. Lemmas 4.5–6.5 mm long; sheaths glabrous or pubescent; plants of coastal rocks, cliffs, and sands . . . . subsp. *pruinosa*
            9. Lemmas 6–9.5 mm long; sheaths pubescent; plants of maritime sands and gravels . . . . subsp. *arenaria*
          8. Cauline leaf sheaths loosely or tightly enclosing the culms; mature inflorescences usually partly included in the uppermost sheaths.

10. Lemmas 4.5–6 mm long, acuminate in side view . . . . subsp. *mediana*
10. Lemmas 5.8–9 mm long, attenuate in side view.
　　11. Inflorescences 10–25 cm long; cauline leaf blades 2–4 mm wide, usually flat
　　　　or loosely conduplicate, not glaucous; lemmas 6–9 mm long, usually glabrous
　　　　. . . . . subsp. *aucta*
　　11. Inflorescences 7.5–12 cm long; cauline leaf blades to 2.5 mm wide when flat,
　　　　usually loosely to tightly conduplicate, sometimes glaucous; lemmas 5.8–6.6
　　　　mm long, glabrous or hairy . . . . subsp. *secunda*

Festuca rubra subsp. arctica (Hack.) Govor. [p. 416]
ARCTIC RED FESCUE, FÉTUQUE DE RICHARDSON

Plants rhizomatous, loosely to densely cespitose, with several culms arising from the same tuft. **Culms** (8)10–40(60) cm. **Sheaths** pubescent, slowly shredding into fibers; **vegetative shoot blades** (0.7)1–2 mm in diameter, usually conduplicate, veins 5–7(9), ribs 5(7), abaxial surfaces usually green, smooth or slightly scabrous, adaxial surfaces scabrous or pilose on the ribs; **cauline blades** usually conduplicate, sometimes flat; **flag leaf blades** usually (1.5)2–6 cm; **abaxial sclerenchyma** in 5–7 small strands; **adaxial sclerenchyma** rarely developed. **Inflorescences** (2)3.5–7 cm, sparsely branched panicles or racemes, well exserted, usually congested, sometimes open; **branches** usually stiff and erect, scabrous or pubescent. **Spikelets** (6)7–13 mm, mostly reddish or purplish, with (3)5–7 florets. **Glumes** ovate to ovate-lanceolate, acute to acuminate, often pilose near the apices; **lower glumes** (1.5)2.5–3.5(4) mm; **upper glumes** (3)3.5–5 mm; **lemmas** (4)4.5–6(6.5) mm, ovate to lanceolate, usually densely to moderately pilose, sometimes only partially pilose, rarely glabrous throughout, awned, awns (0.2)0.5–1.6(2.5) mm; **anthers** (2.3)2.5–3(3.7) mm. 2*n* = 42.

Festuca rubra subsp. *arctica* grows in sands, gravels, silts, and stony soils of river banks, bars, and flats; in periglacial outwashes, beaches, sand dunes, muskegs, solifluction slopes, and scree slopes in tundra, subarctic forest, and barren regions; and subalpine areas in the mountains. It extends from Alaska, the southern part of the Canadian arctic archipelago, and Greenland to northwestern British Columbia, the coast of Hudson Bay and James Bay, and Quebec and Labrador, extending farthest south in the Rocky Mountains of Alberta. It also grows in arctic and subarctic Europe and Asia, and in the Ural Mountains.

Festuca rubra subsp. arenaria (Osbeck) F. Aresch. [p. 416]
FÉTUQUE ROUGE DES SABLES

Plants strongly rhizomatous, usually loosely cespitose, with several culms arising from the same tuft. **Culms** (10)30–60 cm. **Sheaths** pubescent, shredding into fibers, cauline leaf sheaths tightly enclosing the culms; **vegetative shoot blades** 0.8–1(1.5) mm in diameter, conduplicate, veins (5)7–9(13), ribs (3)5–7(9), abaxial surfaces smooth or scabrous, green or glaucous, adaxial surfaces scabrous or pubescent; **cauline blades** (1)5–15(20) cm, conduplicate or flat; **abaxial sclerenchyma** in 5–7(9) broad strands, often confluent, sometimes forming a continuous band; **adaxial sclerenchyma strands** usually present. **Inflorescences** (5.5)7–15(16) cm, open, lanceolate, usually completely exserted from the uppermost leaf sheaths; **branches** scabrous. **Spikelets** (7)9–10(13) mm, with (4)6–8(9) florets. **Glumes** ovate-lanceolate, acute; **lower glumes** 3–6(6.5) mm; **upper glumes** (3.5)4–6.5(8.5) mm; **lemmas** 6–8(9.5) mm, lanceolate, green, sometimes glaucous, glabrous or villous, apices acute to acuminate, awned, awns (0.5)1–2(3) mm; **anthers** 2.6–4.5 mm. 2*n* = 56.

Festuca rubra subsp. *arenaria* is a European taxon that grows in maritime sands and gravels. It is known in the *Flora* region only from one specimen collected on Vancouver Island; it is not known to have persisted. The description is based on the range of variation seen in Europe. In the *Flora* region, the name has long been misapplied to *F. richardsonii* Hook. [= *F. rubra* subsp. *arctica*], which also has hairy lemmas.

Festuca rubra subsp. aucta (V.I. Krecz. & Bobrov) Hultén [p. 416]
ALEUT FESCUE

Plants rhizomatous, loosely cespitose, with several culms arising from the same tuft. **Culms** 30–120 cm. **Sheaths** pubescent, shredding into fibers, cauline leaf sheaths loosely enclosing the culms; **vegetative shoot blades** usually loosely conduplicate, sometimes flat, deep green, not glaucous, abaxial surfaces more or less uniformly scabridulous or roughened, ribs hispid or pilose; **cauline blades** 2–4 mm wide, usually flat or loosely conduplicate; **abaxial sclerenchyma** in small strands; **adaxial sclerenchyma** sometimes present. **Inflorescences** 10–25 cm, often open, lax, secund or partially secund, often partially included in the uppermost leaf sheaths at maturity; **branches** scabrous. **Spikelets** 9–14 mm, oblong, with (4)5–8(10) florets, mostly deep green or glaucous. **Glumes** ovate-lanceolate to lanceolate, acute to acuminate; **lower glumes** 3.5–6 mm; **upper glumes** 5.5–8.5 mm; **lemmas** 6–9 mm, attenuate in side view, usually glabrous, sometimes pubescent, usually mostly smooth, margins and apices scabrous, sometimes almost completely scabrous, deep green or the margins violet, sometimes glaucous, occasionally pubescent, apices awned, awns 2.5–4.5 mm; **anthers** 2.5–3.5 mm. 2*n* = unknown.

Festuca rubra subsp. *aucta* is a coastal taxon, growing above the high tide line in the sand of stabilized sand dunes, beaches, etc., or in silt deposits. Its range extends along the Pacific coast from the Kamchatka Peninsula through the Aleutian Islands, Queen Charlotte Islands, and Vancouver Island and the adjacent continental coastline. *Festuca pseudovivipara* (p. 419) has been described as a form of *F. rubra* subsp. *aucta*, but differs from that taxon in having pseudoviviparous spikelets. It is also ecologically, altitudinally, and probably reproductively isolated from *F. rubra* subsp. *aucta*.

Festuca rubra subsp. **commutata** Gaudin [p. 416]
CHEWING'S FESCUE

Plants without rhizomes, usually densely cespitose. **Culms** 25–90 cm. **Sheaths** red-brown, scarious near the base, puberulent or pubescent, slowly shredding into fibers; **blades** 0.3–0.7(1) in diameter, conduplicate, sometimes glaucous, abaxial surfaces scabrous or smooth; **abaxial sclerenchyma** in narrow to broad strands; **adaxial sclerenchyma** rarely present. **Inflorescences** 4–13(30) cm, more or less contracted, often secund; **branches** spreading at anthesis, scabrous on the angles. **Spikelets** 7–11 mm, with 3–9 florets. **Glumes** ovate-lanceolate, acute; **lower glumes** 2.5–4 mm; **upper glumes** 3.5–5 mm; **lemmas** 4.5–6 mm, green or reddish violet distally, glabrous, smooth, awned, awns 1–3.3 mm; **anthers** 1.8–2.2(3) mm. 2*n* = 28, 42.

Festuca rubra subsp. *commutata* is extensively used for lawns and road verges. It is native to Europe, growing from southern Sweden southward, but is widely introduced elsewhere in the world. In the *Flora* region, it is common south of Alaska, Yukon Territory, and the Northwest Territories.

Festuca rubra subsp. **fallax** (Thuill.) Nyman [p. 416]
FLATLEAF RED FESCUE, FÉTUQUE TROMPEUSE

Plants cespitose, strongly rhizomatous, stoloniferous, sometimes with solitary culms. **Culms** 50–90(130) cm. **Sheaths of innovations** red-brown at the base, scarious, pubescent on the upper parts, shredding into fibers; **blades** usually flat and 2–7 mm wide, or loosely conduplicate and 0.8–2.5 mm in diameter; **abaxial sclerenchyma** in 5–9 broad strands; **adaxial sclerenchyma** always present. **Inflorescences** 9–15 cm, open, lax or erect. **Spikelets** 8–14(17) mm, with (4)6–10 florets. **Glumes** ovate-lanceolate to lanceolate, acute to acuminate; **lower glumes** 3.5–4.5(7) mm; **upper glumes** 5–6.5(8.5) mm; **lemmas** 5–7.5(9.5) mm, green or glaucous, often red-violet along the margins, glabrous, smooth, awned, awns 0.8–3.2 mm; **anthers** 3.5–4.5 mm. 2*n* = 42, 56, 70.

Festuca rubra subsp. *fallax* is a robust taxon that grows in damp, often disturbed places. It is native to northern and central Europe, but has been introduced widely in the *Flora* region, occurring from British Columbia to eastern Quebec and south to California. It is now common in some areas, occasional in others.

Festuca rubra var. *fraterculae* Rasm., an unusual form described from the nesting colonies of Atlantic puffins (*Fratercula arctica*) on the Faeroe Islands and reported as introduced to southern Greenland, is included here in *F. rubra* subsp. *fallax*. Its luxuriant growth (flat leaves 4–7 mm wide, long stolons, and rhizomes) appears to be a phenotypic response to the soil conditions created by the puffins.

Festuca rubra subsp. **mediana** (Pavlick) Pavlick [p. 417]
DUNE RED FESCUE

Plants rhizomatous, densely cespitose. **Culms** 20–30 cm. **Sheaths** reddish brown, glabrous or pubescent, shredding into fibers, cauline leaf sheaths loosely or tightly enclosing the culms; **blades** 0.7–1.5 mm in diameter, conduplicate; **abaxial sclerenchyma** in narrow strands; **adaxial sclerenchyma** sometimes present. **Inflorescences** mostly congested, lanceolate, often partially included in the uppermost leaf sheaths at maturity; **branches** scabrous. **Spikelets** to about 12 mm, with 4–8 florets. **Lower glumes** 2.5–3.2 mm; **upper glumes** 3.5–4.6(5) mm; **lemmas** 4.5–6 mm, glabrous or pilose, acuminate in side view, apices awned, awns 0.5–2 mm; **anthers** 2.5–3 mm long. 2*n* = 42 [for *Festuca rubra* var. *littoralis* Vasey *ex* Beal].

Festuca rubra subsp. *mediana* grows in sand beaches and dunes along exposed coasts, from Vancouver Island to Oregon.

Festuca rubra subsp. **pruinosa** (Hack.) Piper [p. 417]
ROCK FESCUE, FÉTUQUE PRUINEUSE

Plants usually densely, sometimes loosely, cespitose, usually with obscure rhizomes and numerous vegetative shoots, sometimes with conspicuous rhizomes. **Culms** (15)20–40(70) cm. **Sheaths** reddish brown, glabrous to pubescent, shredding into fibers, cauline leaf sheaths tightly enclosing the culms; **blades** conduplicate, often glaucous, abaxial surfaces smooth or scabridulous,

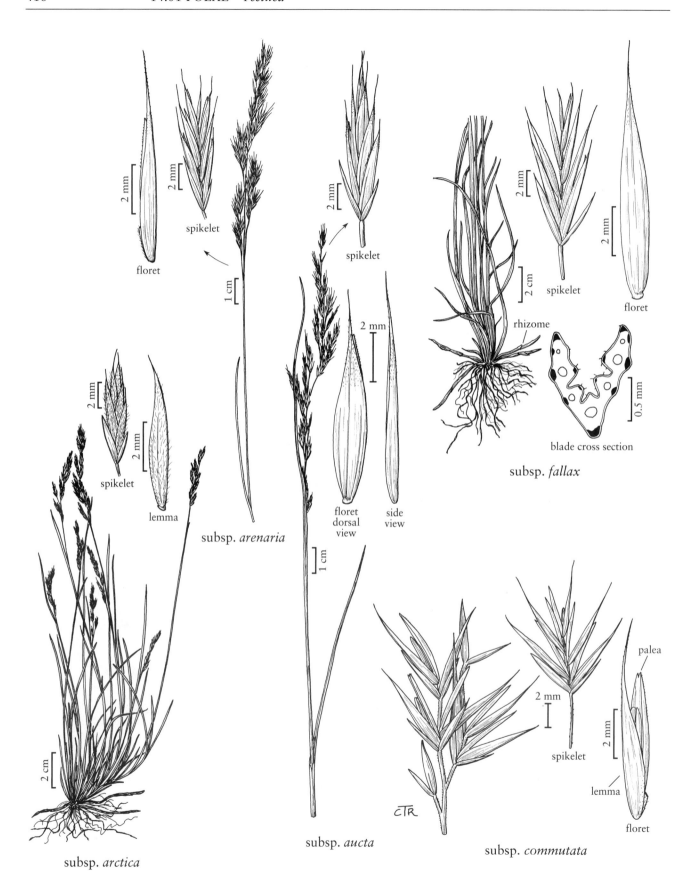

floret

spikelet

subsp. *arctica*

spikelet

lemma

subsp. *arenaria*

floret
dorsal
view

side
view

subsp. *aucta*

spikelet

rhizome

floret

blade cross section

subsp. *fallax*

spikelet

palea

lemma

floret

subsp. *commutata*

FESTUCA RUBRA (1 of 2)

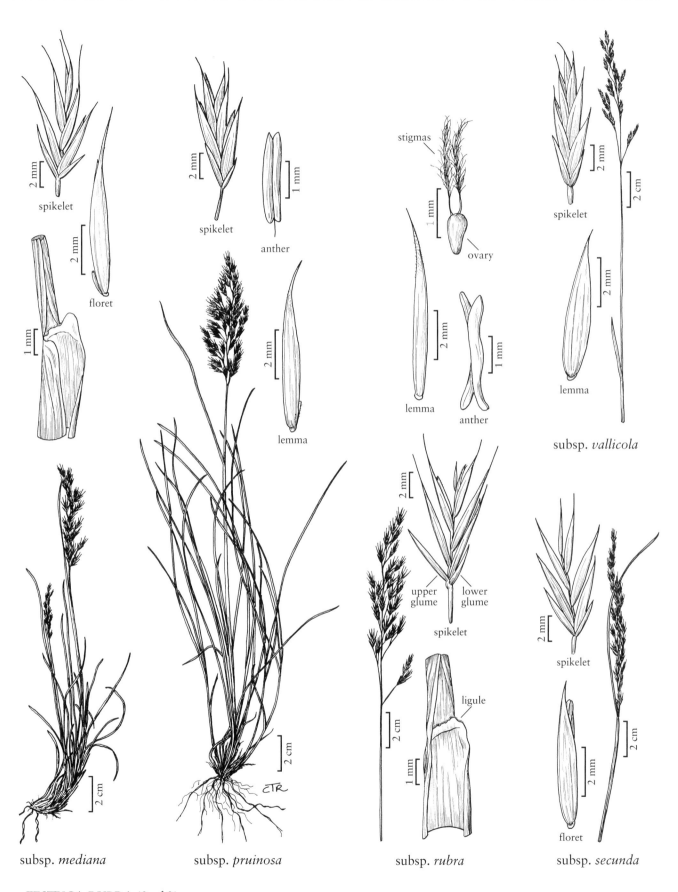

spikelet

2 mm

floret

1 mm

spikelet

anther

1 mm

lemma

2 mm

stigmas

ovary

1 mm

lemma

anther

2 mm

1 mm

spikelet

2 mm

2 cm

lemma

2 mm

subsp. *vallicola*

2 mm

upper glume

lower glume

spikelet

ligule

2 cm

1 mm

spikelet

2 mm

floret

2 mm

2 cm

subsp. *mediana*

2 cm

subsp. *pruinosa*

2 cm

CTR

subsp. *rubra*

subsp. *secunda*

FESTUCA RUBRA (2 of 2)

adaxial surfaces scabrous on the ribs, veins 5–9, ribs 5–7; **abaxial sclerenchyma** in narrow or wide strands; **adaxial sclerenchyma** rarely present. **Inflorescences** (3)4–10(20) cm, congested to more or less open, linear to lanceolate, usually completely exserted from the uppermost leaf sheaths; **branches** stiff, erect, scabrous. **Spikelets** 7.5–14 mm, with 4–7(9) florets. **Lower glumes** 2.2–3.2(4.5) mm; **upper glumes** 3.5–4.5(6.5) mm, margins and apices scabrous, apices acute or mucronate; **lemmas** 4.5–6(6.5) mm, lanceolate, green to violet, sometimes glaucous, scabrous near the apices, apices acute to acuminate, awned, awns (0.1)0.4–3 mm; **anthers** 2.3–3.2 mm. $2n = 42$.

Festuca rubra subsp. *pruinosa* grows in the crevices of rocks, in pilings, and occasionally on pebble or sand beaches, extending upward from the upper littoral zone of the Pacific and Atlantic coasts of North America and Europe. Plants growing on coastal sands from California to Vancouver Island that are loosely cespitose and have abaxial sclerenchyma in large strands are sometimes distinguished as *F. rubra* subsp. *arenicola* E.B. Alexeev [= *F. ammobia* Pavlick]. The rhizomes are rarely present on herbarium specimens.

Festuca rubra L. subsp. rubra [p. 417]
RED FESCUE, FÉTUQUE ROUGE TRAÇANTE

Plants rhizomatous, usually loosely cespitose, with several culms arising from the same tuft, vegetative shoots 8–22(30) cm. **Culms** (20)40–90 cm. **Sheaths** reddish brown, scarious, pubescent, shredding into fibers; **blades** 0.5–2 mm in diameter, usually conduplicate, sometimes flat, abaxial surfaces smooth or scabrous, green or glaucous, adaxial surfaces scabrous or pubescent on the ribs; **abaxial sclerenchyma** in 5–7(9) small strands; **adaxial sclerenchyma** rarely present. **Inflorescences** 7–12 cm, open, lanceolate; **branches** scabrous. **Spikelets** 9–14.5 mm, with 5–8 florets. **Lower glumes** 3–4.5 mm; **upper glumes** 4–6.4 mm; **lemmas** (4)6–7.5(8) mm, lanceolate, usually green with red-violet borders, sometimes mostly red-violet, margins sometimes scabrous, apices scabrous, acute to acuminate, awned, awns 0.6–3.2(4) mm; **anthers** 2.4–3.5 mm. $2n = 42$.

Festuca rubra subsp. *rubra* grows in disturbed soil. It is often planted as a soil binder, or as turf or forage grass, in mesic temperate parts of the *Flora* region. Originally from Eurasia, it has been widely introduced elsewhere in the world, including most of the *Flora* region, from southern Alaska east to Newfoundland and Greenland and south to California and Georgia. It also grows in Mexico. Because *F. rubra* subsp. *rubra* has often been misunderstood, confounded, and lumped with other taxa of the *F. rubra* complex, statements about its distribution, including that given here, should be treated with caution. It is to be expected throughout the *Flora* region, in all but the coldest and driest habitats.

Festuca rubra subsp. secunda (J. Presl) Pavlick [p. 417]
SECUND RED FESCUE

Plants rhizomatous, usually loosely cespitose, with several culms arising from the same tuft. **Culms** (20)30–70(80) cm. **Sheaths** glabrous or pubescent, sometimes glaucous, shredding into fibers, cauline sheaths usually wide, loosely enclosing the culms; **vegetative shoot blades** 0.7–1.5 mm in diameter, conduplicate, sometimes glaucous, abaxial surfaces more or less evenly scabrous, adaxial surfaces pilose on the ribs; **cauline blades** sometimes to 2.5 mm wide, sometimes flat; **abaxial sclerenchyma** in small strands; **abaxial sclerenchyma** rarely present. **Inflorescences** 7.5–12 cm, more or less open, usually secund, often partially included in the uppermost leaf sheaths at maturity; **branches** lax, scabrous. **Spikelets** 9.5–13 mm, green to violet, sometimes glaucous, with 4–7 florets. **Lower glumes** 3.1–4.5 mm; **upper glumes** (4.5)5–5.6 mm; **lemmas** 5.8–6.6 mm, attenuate in side view, mostly green, sometimes with violet borders, glabrous or more or less uniformly hairy, margins and apices scabrous, apices awned, awns 1–5 mm; **anthers** 2.8–3.9 mm. $2n = 42$.

Festuca rubra subsp. *secunda* grows on pebble beaches and in soil pockets on rocks, meadows, cliffs, banks, and stabilized sand dunes along seashores with high annual rainfall, on the Pacific coast of North America from Alaska south to Oregon.

Festuca rubra subsp. vallicola (Rydb.) Pavlick [p. 417]
MOUNTAIN RED FESCUE

Plants rhizomatous, often with widely spaced single culms and only a few vegetative shoots, varying to loosely cespitose. **Culms** (20)25–70(100) cm. **Sheaths** tightly enclosing the culms; **vegetative shoot blades** 0.5–1.5 mm in diameter, conduplicate, deep green, abaxial surfaces smooth to slightly scabrous, adaxial surfaces scabrous on the ribs; **cauline blades** conduplicate or flat; **abaxial sclerenchyma** in narrow strands; **adaxial sclerenchyma** absent. **Inflorescences** 5–8 cm, closed or open. **Spikelets** 8–11 mm, with 4–7 florets. **Lower glumes** 2–3 mm; **upper glumes** 4–5 mm, acute to acuminate; **lemmas** 5–6 mm, pale green with violet borders, smooth or scabrous near the apices, apices awned, awns 1–2.5(4) mm; **anthers** 2–2.6 mm. $2n$ = unknown.

Festuca rubra subsp. *vallicola* grows in moist meadows, lake margins, and disturbed soil, at 1000–2000 m, in montane and subalpine habitats from the Yukon Territory/British Columbia border area south to Wyoming.

15. Festuca prolifera (Piper) Fernald [p. 421]
PROLIFEROUS FESCUE

Plants usually loosely cespitose, often mat-forming, sometimes with solitary culms, rhizomatous. **Culms** (10)20–41 cm, smooth, glabrous throughout or pubescent near the inflorescence, bases often geniculate. **Sheaths** closed for about ³/₄ their length, often splitting with age, coarsely ribbed, shredding into fibers, bases reddish brown, scarious; **collars** glabrous; **ligules** 0.1–0.4(0.6) mm; **blades** 0.3–0.8(1) mm in diameter, conduplicate, green or glaucous, abaxial surfaces glabrous, smooth or scabrous, adaxial surfaces scabrous or puberulent; **abaxial sclerenchyma** in 5–7(9) small strands; **adaxial sclerenchyma** absent. **Inflorescences** (3)5–12 cm, usually paniculate, sometimes racemose or subracemose, compact or open, with 1–2 branches per node; **branches** stiff or somewhat lax, lower branches with 1–3 spikelets. **Spikelets** pseudoviviparous, varying in length with the stage of vegetative proliferation, the glumes and often 1 or 2 adjacent florets more or less normally developed or only slightly elongated, glabrous or pubescent, the distal florets vegetative. **Glumes** more or less normally developed, ovate to lanceolate; **lower glumes** (2.5)3–3.5(5.5); **upper glumes** 3.5–4.5(6.5); **lowest lemma in each spikelet** usually normally developed, acute, unawned, usually without reproductive structures or the structures abortive; **subsequent lemmas** modified into leafy bracts; **paleas** usually absent, shorter than the lemmas if present, intercostal region puberulent distally; **anthers** usually aborted, when present 1.5–2.3(3) mm; **ovaries** rarely present, apices glabrous. 2*n* = 49, 50, 63.

Festuca prolifera is often abundant, and may be a dominant component in some habitats. The leafy bulbils or plantlets sometimes root when the top-heavy inflorescence is bent to the ground.

Festuca prolifera has two varieties: **Festuca prolifera** (Piper) Fernald var. **prolifera**, with glabrous lemmas; and **Festuca prolifera** var. **lasiolepis** Fernald, with pubescent lemmas. *Festuca prolifera* var. *prolifera* grows in arctic, alpine, or boreal rocky areas, in calcareous, basic or neutral soils, and is found in the James Bay area, Ungava Bay, western Newfoundland, Cape Breton, the Gaspé Peninsula, the White Mountains (New Hampshire), and Katahdin (Maine). *Festuca prolifera* var. *lasiolepis* is found in moist, sandy riverbanks, lake shores, rocky areas, and cliffs, often on limestone, from the southeastern Northwest Territories to northern Quebec, Anticosti Island, and western Newfoundland. Proliferous plants from southern Greenland with extravaginal shoots, named *F. villosa-vivipara* (Rosenv.) E.B. Alexeev, are similar to *F. prolifera*, but appear to be hybrids between *F. rubra* and *F. frederikseniae* (see under *F. frederikseniae*, p. 436).

16. Festuca pseudovivipara (Pavlick) Pavlick [p. 421]
PSEUDOVIVIPAROUS FESCUE

Plants loosely cespitose, rhizomatous. **Culms** 30–60 cm. **Sheaths** closed for about ³/₄ their length, glabrous or scabrous-pubescent, shredding into fibers, bases red-scarious; **collars** glabrous; **ligules** 0.5–1 mm long; **vegetative shoot blades** to about 2 mm wide when flat, 0.5–1 mm in diameter when loosely conduplicate, deep green, abaxial surfaces more or less uniformly scabrous, adaxial surfaces hispid or pilose on the ribs; **abaxial sclerenchyma** in 5–9 small strands; **adaxial sclerenchyma** absent; **cauline blades** 1.4–2.5 mm wide, flat. **Inflorescences** (4)7–12(15) cm, open, lax, secund or partially secund, with 1–2 branches per node; **branches** somewhat stiff or lax, lower branches with 2–5 spikelets. **Spikelets** pseudoviviparous, varying in length with the stage of vegetative proliferation, most florets replaced by bracts, the glumes and sometimes the lowest floret more or less normally developed or only slightly elongated, mostly deep green or reddish tinged. **Glumes** more or less normally developed, lanceolate, apices scabrous; **lower glumes** (2.5)3.5–6 mm; **upper glumes** 4.5–6.5(8) mm; **lemmas** and **bracts** glabrous or pubescent, smooth or scabrous, sometimes mucronate, mucros to 0.5 mm; **paleas**, if present, about as long as or shorter than the lemmas, intercostal region puberulent distally; **anthers** not developed or abortive, to 2 mm; **ovaries** not developed. 2*n* = ca. 70.

Festuca pseudovivipara grows on coastal mountainsides, scree slopes, and other rocky areas, at 300–800 m. It is known only from the Queen Charlotte Islands, British Columbia.

Festuca pseudovivipara has been described as a form of *F. rubra* subsp. *aucta* (p. 414), but differs from that taxon in having pseudoviviparous spikelets. It is also eco-logically, altitudinally, and probably reproductively isolated from *F. rubra* subsp. *aucta*.

17. **Festuca earlei** Rydb. [p. 421]
EARLE'S FESCUE

Plants loosely cespitose, often with short rhizomes. **Culms** (15)20–40(45) cm, glabrous, smooth. **Sheaths** closed for about ¹/₂ their length, glabrous, shredding into fibers, sometimes slowly; **collars** glabrous; **ligules** 0.1–0.5(1) mm; **blades** to 3 mm wide when flat, 0.5–1 (1.5) mm in diameter when conduplicate, veins (3)5, ribs (1)3(5), abaxial surfaces smooth or slightly scabrous, adaxial surfaces sparsely scabrous; **abaxial sclerenchyma** in 3–5 narrow strands less than twice as wide as high; **adaxial sclerenchyma** absent. **Inflorescences** 3–5(8) cm, contracted, with 1–3 branches per node; **branches** stiff, erect, scabrous, lower branches with 2+ spikelets. **Spikelets** (4.5)5–6.5(7) mm, with 2–5 florets. **Glumes** exceeded by the upper florets, lanceolate to ovate-lanceolate, mostly smooth, sometimes scabrous distally; **lower glumes** 1.5–3 mm; **upper glumes** 2.5–3.8 mm; **lemmas** 3–4.5 mm, glabrous or puberulent near the apices, awns (0.3)1–1.5 mm, terminal; **paleas** as long as or slightly shorter than the lemmas, intercostal region puberulent distally; **anthers** 0.6–0.9(1.4) mm; **ovary apices** densely and conspicuously pubescent. 2*n* = unknown.

Festuca earlei grows in rich subalpine and alpine meadows, at 2800–3800 m, in Utah, Colorado, Arizona, and New Mexico. It often grows with the nonrhizomatous species *F. brachyphylla* subsp. *coloradensis* (p. 428) and *F. minutiflora* (p. 434). It can be distinguished from the former by its pubescent ovary apices, and from the latter by its larger spikelets and lemmas. Because of its short rhizomes (which are often missing from herbarium specimens), *F. earlei* is sometimes confused with members of the *F. rubra* (p. 412) complex. It differs from them in having pubescent ovary apices and shorter anthers.

18. **Festuca heterophylla** Lam. [p. 423]
VARIOUS-LEAVED FESCUE, FÉTUQUE HÉTÉROPHYLLE

Plants densely to loosely cespitose, without rhizomes. **Culms** 60–120(150) cm, glabrous, smooth. **Sheaths** closed for about ³/₄ their length, slowly shredding into fibers; **collars** glabrous; **ligules** 0.1–0.3 mm; **blades** varying within a plant, blades of the vegetative shoots to 60 cm long, (0.2)0.3–0.6 mm in diameter, conduplicate, veins 3–5(7), ribs 1(3), abaxial surfaces smooth or sparsely scabrous, adaxial surfaces scabrous, cauline blades to 25 cm long, 2–4 mm wide, flat; **abaxial sclerenchyma** of the vegetative shoot blades in 3–5 small strands less than twice as high as high, of the upper cauline blades in 7–11 small strands; **adaxial sclerenchyma** absent. **Inflorescences** 6–18 cm, open or contracted, somewhat secund, with 1–2 branches per node; **branches** more or less erect, scabrous, lower branches with 2+ spikelets. **Spikelets** 7–14 mm, with (2)3–6(9) florets. **Glumes** exceeded by the upper florets; ovate-lanceolate to lanceolate, mostly smooth or scabrous on the upper midvein; **lower glumes** 3–5.5 mm; **upper glumes** 4–6.5(7) mm; **lemmas** (4.7)5–8.5 mm, lanceolate, mostly smooth, sometimes scabrous near the apices, awns 1.5–6 mm; **paleas** as long as the lemmas, intercostal region smooth or scabrous distally; **anthers** 2.5–4.5 mm; **ovary apices** pubescent. 2*n* = 28.

Festuca heterophylla is native to open forests and forest edges in Europe and western Asia. In the *Flora* region, it used to be planted as a turf grass for shady areas, and sometimes persists in old lawns.

19. **Festuca valesiaca** Schleich. *ex* Gaudin [p. 423]
VALAIS FESCUE, FÉTUQUE DU VALAIS

Plants densely cespitose, without rhizomes. **Culms** 20–50(60) cm, erect, glabrous, smooth. **Sheaths** closed for about ¹/₂ their length, usually glabrous, sometimes pubescent distally, persistent; **collars** glabrous; **ligules** 0.1–0.5 mm; **blades** (0.3) 0.5–0.8(1.2) mm in diameter, conduplicate, veins 5–7, ribs (1)3–5, abaxial surfaces glabrous or pubescent, not pilose, adaxial surfaces smooth or scabrous; **abaxial sclerenchyma** in 3 broad strands, sometimes with additional narrow strands between the midrib and margins; **adaxial sclerenchyma** absent. **Inflorescences** (3)5–10 cm, panicles, contracted, with 1–2 branches per node; **branches** erect or somewhat spreading, at least at anthesis, lower branches with 2+ spikelets. **Spikelets** (4.8)5.5–6.5(8.5) mm, with 3–5(8) florets. **Glumes** exceeded by the upper florets, ovate-lanceolate to lanceolate, mostly smooth or sometimes scabrous on the upper midvein; **lower glumes** 2–3 mm; **upper glumes** (2.3)2.5–4(4.3) mm; **lemmas** (3.2)3.5–4.5(5.3) mm, ovate-lanceolate to lanceolate, smooth throughout or scabrous distally, awns (0.5)1–2 mm, terminal; **paleas** as long as the lemmas, intercostal region puberulent distally; **anthers** 2.2–2.6 mm; **ovary apices** glabrous. 2*n* = 14.

Festuca valesiaca is widely distributed through central Europe and northern Asia, where it grows in steppes, dry meadows, and open rocky or sandy areas. It is sold in the North American seed trade as *F. pseudovina* Hack. *ex* Wiesb., and has been collected at a few scattered localities in the *Flora* region, apparently having become established from deliberate seeding.

stigmas

1 mm

ovary

palea

1 mm

lemma

floret
ventral
view

2 mm

glume   glume

spikelet

2 mm

floret   side
dorsal   view
view

2 mm

spikelet

immature
inflorescence

1 cm

2 mm

spikelet   spikelet

2 mm

spikelet

0.5 mm

blade cross section

cauline
blade

1 mm

2 cm

rhizome

F. prolifera

2 cm

cauline
blade

ligule

1 mm

F. pseudovivipara

2 cm

CTR

F. earlei

FESTUCA

The taxonomy of the *Festuca valesiaca* complex is controversial, with different authors naming morphological variants and polyploid populations within it. No attempt has been made to determine which are present in the *Flora* region.

20. Festuca amethystina L. [p. 423]
TUFTED FESCUE, FÉTUQUE À COULEUR D'AMÉTHYSTE

Plants densely cespitose, without rhizomes. **Culms** (25)50–80(105) cm, erect, glabrous and smooth throughout. **Sheaths** closed for about ¹/₂ their length, glabrous, smooth, usually reddish blue towards the bases, persistent, flag leaf sheaths tightly enclosing the culms; **collars** glabrous; **ligules** 0.1–0.5 mm; **blades** 0.6–1.2(1.5) mm in diameter, conduplicate, abaxial surfaces glabrous, smooth or scabrous, adaxial surfaces scabrous, veins (5)7–9, ribs 5–9; **abaxial sclerenchyma** in (5)7–9 broad strands, rarely some strands confluent; **adaxial sclerenchyma** absent. **Inflorescences** (3)8–18 (25) cm, open or loosely contracted, with 1–2(3) branches per node; **branches** somewhat lax, erect to spreading, lower branches with 2+ spikelets. **Spikelets** (5)6–8.5(10) mm, with (3)4–6(8) florets. **Glumes** exceeded by the upper florets, lanceolate, mostly smooth and glabrous, sometimes scabrous distally; **lower glumes** 2.5–4 mm; **upper glumes** 3–5(5.5) mm; **lemmas** (3.5)4–5.6(6.6) mm, usually smooth, sometimes scabrous near the apices, lanceolate, usually acute, sometimes obtuse, unawned, occasionally mucronate; **paleas** about as long as the lemmas, intercostal region smooth or slightly scabrous distally; **anthers** (2)3–4 mm; **ovary apices** glabrous or sparsely pubescent. $2n = 14$.

Festuca amethystina is sometimes cultivated as an ornamental species; it may occasionally escape.

21. Festuca ovina L. [p. 425]
SHEEP FESCUE, FÉTUQUE DES OVINS

Plants densely cespitose, without rhizomes; usually not glaucous. **Culms** (10)30–50(70) cm, glabrous, smooth. **Sheaths** closed for about ¹/₂ their length, glabrous, smooth or scabrous distally, persistent; **collars** glabrous; **ligules** shorter than 0.3 mm; **blades** 0.3–0.7(1.2) mm in diameter, conduplicate, abaxial surfaces smooth or scabrous, adaxial surfaces scabrous, veins 5–7(9), ribs 1–3, indistinct; **abaxial sclerenchyma** usually a continuous band; **adaxial sclerenchyma** absent. **Inflorescences** (2)5–10(12) cm, contracted, with 1–2(3) branches per node; **branches** usually erect, sometimes spreading at anthesis, lower branches with 2+ spikelets. **Spikelets** 4–6(7.3) mm, with 3–6(8) florets. **Glumes** exceeded by the upper florets, lanceolate to ovate-lanceolate, mostly smooth and glabrous, sometimes scabrous distally; **lower glumes** 1–2(3) mm; **upper glumes** (2.2)2.6–4(4.6) mm; **lemmas** (2.6)3–4(5) mm,

ovate-lanceolate, mostly smooth, sometimes scabrous or hispid near the apices, awns 0.5–2 mm, terminal, sometimes absent; **paleas** about equal to the lemmas, intercostal region puberulent distally; **anthers** (1.4)2–2.6 mm; **ovary apices** glabrous. $2n = 14, 28$.

Festuca ovina was introduced from Europe as a turf grass. It is not presently used in the North American seed trade. The sporadic occurrences are mostly from old lawns and cemeteries, or sites seeded for soil stabilization.

Festuca ovina used to be interpreted very broadly in North America, including almost any fine-leaved fescue that lacked rhizomes. Consequently, much of the information reported for *F. ovina*, and many of the specimens identified as such, belong to other species. The only confirmed recent reports are from Ontario (Dore & McNeill 1980); Piatt County, Illinois; and Okanogan County, Washington. Species in this treatment that have frequently been included in *F. ovina* are *F. arizonica* (p. 438), *F. auriculata* (p. 424), *F. baffinensis* (p. 432), *F. brachyphylla* (p. 428), *F. brevissima* (p. 426), *F. calligera* (p. 437), *F. edlundiae* (p. 432), *F. frederikseniae* (p. 436), *F. hyperborea* (p. 432), *F. idahoensis* (p. 438), *F. lenensis* (p. 426), *F. minutiflora* (p. 434), *F. saximontana* (p. 430), *F. trachyphylla* (p. 424), and *F. viviparoidea* (p. 436).

22. Festuca glauca Vill. [p. 425]
BLUE FESCUE, GRAY FESCUE, FÉTUQUE GLAUQUE

Plants densely cespitose, without rhizomes; usually glaucous or pruinose. **Culms** (15)22–35(50) cm, erect, glabrous, smooth. **Sheaths** closed for about ¹/₂ their length, pubescent or glabrous, persistent; **collars** glabrous; **ligules** 0.1–0.4 mm; **blades** 0.6–1 mm in diameter, conduplicate, veins 5–7, ribs 1–3(5), indistinct; **abaxial sclerenchyma** forming continuous or interrupted bands; **adaxial sclerenchyma** absent. **Inflorescences** 2.5–9(11) cm, compact, erect, with 1–2 branches per node; **branches** stiff, erect, smooth or scabrous, lower branches with 2+ spikelets. **Spikelets** 5.5–9(11) mm, with 3–6(7) florets. **Glumes** exceeded by the upper florets, ovate-lanceolate to ovate, glabrous or pubescent distally; **lower glumes** (1.8)2–3(4) mm; **upper glumes** 2.8–4(5.1) mm; **lemmas** (3.5)4–6(6.2) mm, lanceolate to ovate-lanceolate, smooth or scabrous near the apices, sometimes pubescent distally, awns (0.6)1–1.5(2) mm, terminal; **paleas** about equal to the lemmas, intercostal region puberulent distally; **anthers** (1.8)2–3 mm; **ovary apices** glabrous. $2n = 42$ [in European literature for the horticultural forms].

Festuca glauca is widely grown as an ornamental in the *Flora* region because of its attractive dense tufts of glaucous foliage. It is not known to have escaped cultivation. Several other Eurasian species of fescue with white or bluish foliage are also sold in the

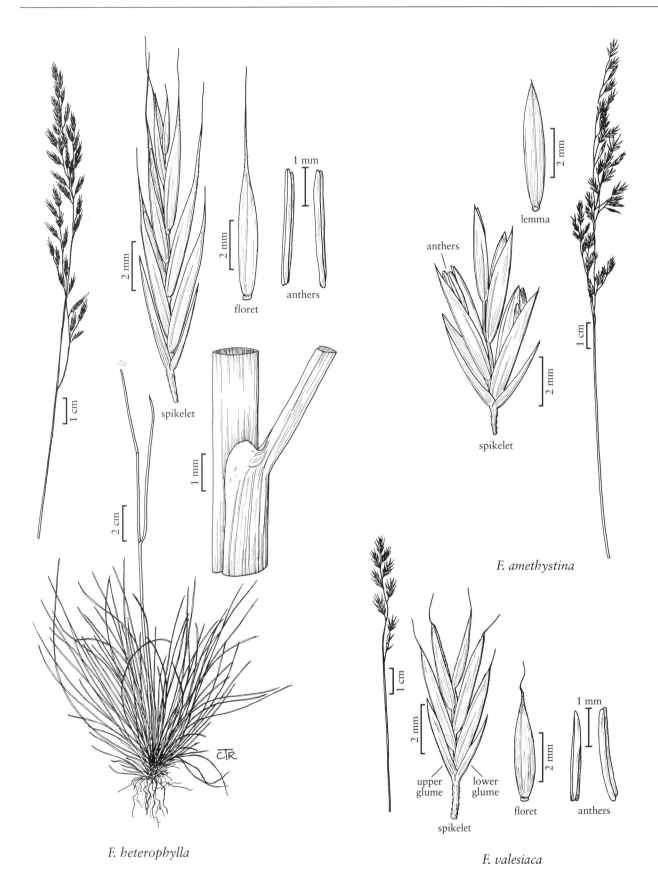

floret

anthers

spikelet

F. heterophylla

lemma

anthers

spikelet

F. amethystina

upper
glume

lower
glume

spikelet

floret

anthers

F. valesiaca

CTR

FESTUCA

horticultural trade as "*Festuca glauca*". Determining the species involved is beyond the scope of this treatment.

23. Festuca filiformis Pourr. [p. 425]
HAIR FESCUE, FINE-LEAVED SHEEP FESCUE, FÉTUQUE CHEVELUE

Plants densely cespitose, without rhizomes. **Culms** 18–40(60) cm, mostly scabrous or puberulent below the inflorescence. **Sheaths** closed for less than ¹/₃ their length, smooth or scabrous, glabrous or finely puberulent, persistent; **collars** glabrous; **ligules** 0.1–0.4 mm; **blades** 0.2–0.4(0.6) mm in diameter, conduplicate, abaxial surfaces smooth, adaxial surfaces scabrous, veins 5(7), ribs 1, distinct; **abaxial sclerenchyma** forming a continuous or almost continuous band; **adaxial sclerenchyma** absent. **Inflorescences** 1–6(14) cm, usually contracted, with 1–2 branches per node; **branches** usually erect, lower branches with 2+ spikelets. **Spikelets** 3–6(6.5) mm, with 2–6(8) florets. **Glumes** exceeded by the upper florets, ovate-lanceolate to lanceolate, glabrous; **lower glumes** 1–2.5 mm; **upper glumes** (1.7)2–3(3.9) mm; **lemmas** 2.3–4(4.4) mm, obtuse to acute, mostly smooth and glabrous, sometimes scabrous or pubescent distally, unawned, sometimes mucronate, mucros to 0.4 mm; **paleas** about as long as the lemmas, intercostal region smooth or scabrous distally; **anthers** (1)1.5–2.2 mm; **ovary apices** glabrous. 2*n* = 14 (28).

Festuca filiformis is a European species that has been introduced to the *Flora* region as a turf grass. It grows well on poor, dry soils and is becoming a ruderal weed in some areas. It is particularly common in the northeastern United States and southeastern Canada, but has been reported from scattered locations elsewhere.

24. Festuca trachyphylla (Hack.) Krajina [p. 427]
HARD FESCUE, SHEEP FESCUE, FÉTUQUE DRESSÉE À FEUILLES SCABRES

Plants densely cespitose, without rhizomes. **Culms** (15)20–75(80) cm, smooth, glabrous or with sparse hairs. **Sheaths** closed for less than ¹/₃ their length, usually glabrous, rarely pubescent, persistent; **collars** glabrous; **ligules** 0.1–0.5 mm; **blades** (0.5)0.8–1.2 mm in diameter, usually conduplicate, rarely flat, abaxial surfaces glabrous, puberulent, or scabrous, adaxial surfaces scabrous or puberulent to pubescent, veins 5–7(9), ribs 3–7, usually distinct; **abaxial sclerenchyma** usually in an irregular, interrupted or continuous band, rarely in 5–7 small strands, usually more than twice as wide as high; **adaxial sclerenchyma** absent. **Inflorescences** (2.5)3–13(16) cm, contracted, with 1–2 branches per node; **branches** erect or stiffly spreading, secondary branches not divaricate, lower branches with 2+ spikelets. **Spikelets** 5–9(10.8) mm, with 3–7(8) florets. **Glumes** exceeded by the upper florets, ovate-lanceolate to lanceolate, mostly smooth and glabrous, sometimes scabrous and/or pubescent distally; **lower glumes** (1.8)2–3.5(4) mm; **upper glumes** 3–5(5.5) mm; **lemmas** 3.8–5(6.5) mm, lanceolate, usually smooth and glabrous on the lower portion and scabrous or pubescent distally, especially on the margins, rarely entirely pubescent, awns 0.5–2.5(3) mm, usually less than ¹/₂ as long as the lemma body; **paleas** about as long as the lemmas, intercostal region puberulent distally; **anthers** (1.8)2.3–3.4 mm; **ovary apices** glabrous. 2*n* = 42.

Festuca trachyphylla is native to open forests and forest edge habitats of Europe. It has been introduced and has become naturalized in many temperate regions. In the *Flora* region, *F. trachyphylla* is generally sold under the name 'Hard Fescue', and is popular as a durable turf grass and soil stabilizer. It is particularly common in the eastern United States and southeastern Canada, but is probably grown throughout the temperate parts of the region. Its naturalized distribution can be expected to expand.

For many years, *Festuca trachyphylla* was known, inappropriately, under other names, e.g., *F. duriuscula* L., *F. ovina* var. *duriuscula* (L.) W.D.J. Koch, and *F. longifolia* Thuill. Some European authors treat it as *F. stricta* subsp. *trachyphylla* (Hack.) Patzke. It has frequently been included in *F. ovina* (p. 422).

25. Festuca auriculata Drobow [p. 427]
LOBED FESCUE

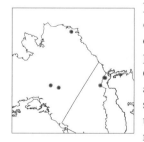

Plants densely cespitose, without rhizomes. **Culms** 8–25(40) cm, erect, smooth, glabrous or pubescent below the inflorescence. **Sheaths** closed for about ¹/₂ their length, smooth or scabrous, glabrous or puberulent, persistent; **collars** glabrous; **ligules** 0.1–0.5 mm; **blades** 0.4–0.8(1) mm in diameter, conduplicate, abaxial surfaces glabrous, pubescent, or pilose, all conditions often present on the same plant, scabrous at the apices, adaxial surfaces scabrous or pubescent on the ribs, veins 5–7, ribs 3–5, 1 distinct and 2–4 indistinct; **abaxial sclerenchyma** in 3 strands, 2 marginal and 1

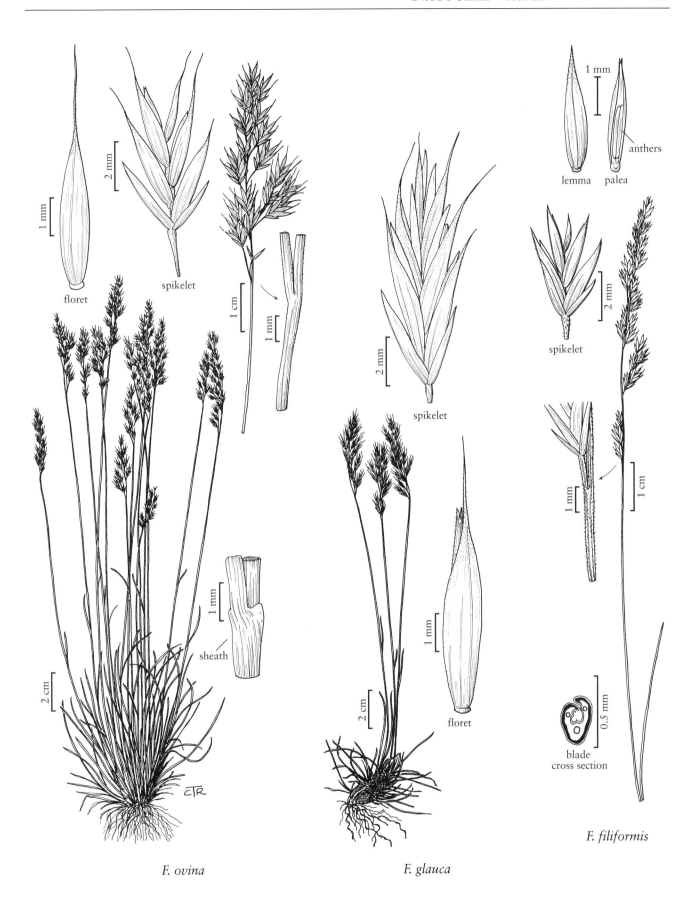

floret

spikelet

sheath

F. ovina

lemma palea

anthers

spikelet

blade
cross section

spikelet

floret

F. glauca

F. filiformis

FESTUCA

opposite the midvein, rarely with 2 additional strands, the strands narrower than the veins and usually less than twice as wide as high; **adaxial sclerenchyma strands** absent; **flag leaf blades** 0.5–2 cm. **Inflorescences** 2–3.5 cm, racemes or panicles, contracted, with 1(2) branches per node; **branches** erect, lower branches usually with 1–2 spikelets. **Spikelets** 5–6.5(8) mm, with 3–5(6) florets. **Glumes** exceeded by the upper florets, ovate to lanceolate, glabrous, sometimes scabrous distally; **lower glumes** about 3 mm; **upper glumes** about 4 mm; **lemmas** 4–5 mm, ovate to ovate-lanceolate, glabrous, sometimes scabrous towards the apices, apices acute, awns 0.8–2(2.5) mm, terminal; **paleas** about as long as the lemmas, intercostal region smooth or scabrous distally; **anthers** 2–3.5 mm; **ovary apices** glabrous. $2n = 14$.

Festuca auriculata is an amphiberingian species that extends from the Ural Mountains of Russia through Alaska to the western continental Northwest Territories. It grows on dry, rocky cliffs and slopes, in low arctic and alpine regions. In the *Flora* region, this species seems to intergrade with, and is sometimes included in, *F. lenensis* (see next). The two species tend to differ in their leaf surfaces as well as in the width of their sclerenchyma strands. *Festuca auriculata* has also frequently been included in *F. ovina* (p. 422).

26. Festuca lenensis Drobow [p. 427]
LENA FESCUE

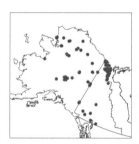

Plants densely cespitose, without rhizomes. **Culms** (8)10–35 (50) cm, erect, smooth, glabrous or sparsely pubescent below the inflorescence. **Sheaths** closed for about ¹/₂ their length, glabrous or pilose, persistent; **collars** glabrous; **ligules** 0.1–0.5 mm; **blades** 0.4–0.8(1) mm in diameter, conduplicate, sometimes glaucous, abaxial surfaces glabrous, pubescent, or pilose, all conditions often present on the same plant, adaxial surfaces scabrous or pubescent, veins 5–7, ribs 3, 1 distinct and 2 indistinct; **abaxial sclerenchyma** in 3 strands, 2 at the margins and 1 opposite the midvein, strands as wide or wider than the veins; **adaxial sclerenchyma** absent; **flag leaf blades** 0.3–2 cm. **Inflorescences** 1.5–4(5.5) cm, panicles or racemes, contracted, with 1(2) branches per node; **branches** erect, lower branches with 1–2 spikelets. **Spikelets** (5)7–9(11) mm, with (2)3–5(6) florets. **Glumes** exceeded by the upper florets, ovate to lanceolate, glabrous, sometimes scabrous distally; **lower glumes** 2.5–3.4 mm; **upper glumes** 3–4.3 mm; **lemmas** 4–5.5 mm, ovate to ovate-lanceolate, glabrous, sometimes scabrous towards the apices, awns 0.8–2.6(3) mm, terminal; **paleas** about as long as the

lemmas, intercostal region smooth or scabrous distally; **anthers** (2)2.4–3.5 mm; **ovary apices** glabrous. $2n = 14$.

Festuca lenensis is an amphiberingean species of dry, eroding, rocky slopes in alpine and low arctic habitats. Its range extends from Siberia, Russia, and Mongolia to Alaska and the Yukon Territory. In North America, this species seems to intergrade with, and is sometimes treated as including, *F. auriculata* (see previous). The two species usually differ in their leaf surfaces as well as in the width of their sclerenchyma strands. *Festuca lenensis* has been frequently included in *F. ovina* (p. 422).

27. Festuca brevissima Jurtsev [p. 429]
SHORT FESCUE

Plants densely cespitose, without rhizomes. **Culms** (3)10–15 (18) cm, more than twice as tall as the vegetative shoot leaves, erect, glabrous, sometimes slightly scabrous below the inflorescences. **Sheaths** closed for about ¹/₂ their length, glabrous, persistent; **collars** glabrous; **ligules** 0.3–0.5 mm; **blades** (0.3)0.4–0.8(1) mm in diameter, conduplicate, abaxial surfaces glabrous, adaxial surfaces scabrous or pubescent, veins 5–7(9), ribs (3)5(7); **abaxial sclerenchyma** in 5–7 strands, usually less than twice as wide as high; **adaxial sclerenchyma** absent; **flag leaf sheaths** often somewhat inflated; **flag leaf blades** 0.2–1 cm. **Inflorescences** (0.7)1–5 cm, usually racemes; **branches** erect, lower branches with 1(2) spikelets. **Spikelets** (4)5–7(8) mm, with 2–4(5) florets. **Glumes** exceeded by the upper florets, ovate-lanceolate to lanceolate; **lower glumes** (1.2)2.5–3.2 mm; **upper glumes** (2.4)3.2–4.8 mm; **lemmas** (3)4–5.5(7) mm, ovate-lanceolate to lanceolate, glabrous, scabrous distally, awns (0.2)0.5–2.5 mm, terminal; **paleas** about as long as the lemmas, intercostal region smooth or scabrous distally; **anthers** (0.6)0.9–1.2 mm; **ovary apices** glabrous. $2n = 14$.

Festuca brevissima is an amphiberingian diploid species that grows in rocky tundra habitats from the Russian Far East to Alaska and the western part of the Northwest Territories. It has frequently been included in *F. ovina* (p. 422).

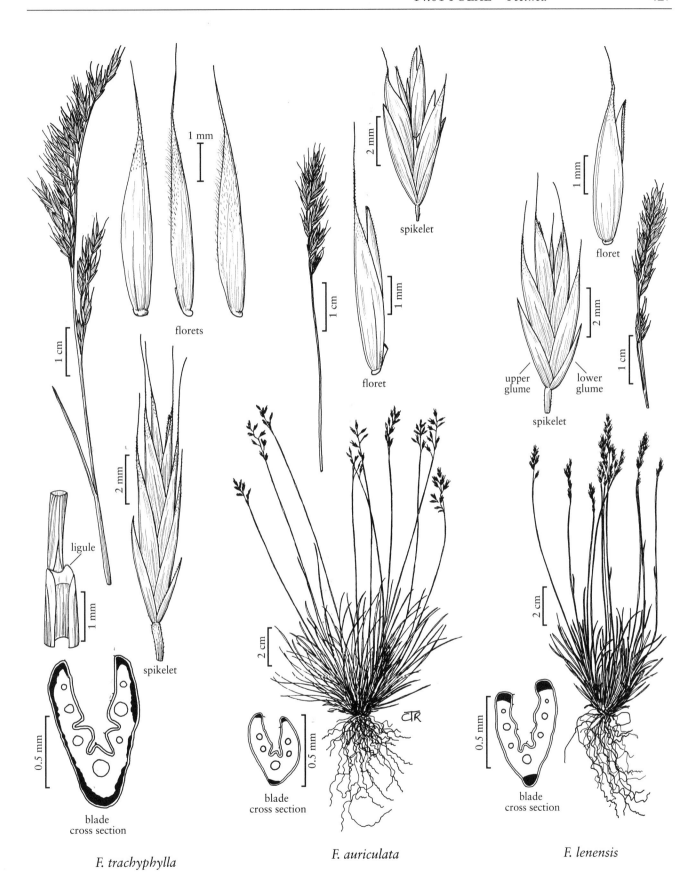

F. trachyphylla

F. auriculata

F. lenensis

FESTUCA

28. Festuca brachyphylla Schult. & Schult. f. [p. 429]
ALPINE FESCUE, FÉTUQUE À FEUILLES COURTES

Plants densely or loosely cespitose, without rhizomes. **Culms** (5)8–35(55) cm, erect, usually smooth and glabrous, sometimes sparsely scabrous or puberulent near the inflorescence. **Sheaths** closed for about $^1/_2$ their length, smooth or scabrous, persistent or slowly shredding into fibers; **collars** glabrous; **ligules** 0.1–0.4 mm; **blades** (0.3)0.5–1(1.2) mm in diameter, conduplicate, abaxial surfaces smooth or sparsely scabrous, adaxial surfaces scabrous, veins (3)5–7, ribs 3–5; **abaxial sclerenchyma** in 3–7(9) narrow strands, usually less than twice as wide as high; **adaxial sclerenchyma** absent; **flag leaf sheaths** not inflated, more or less tightly enclosing the culms; **flag leaf blades** (0.3)1–2.5(3) cm. **Inflorescences** 1.5–4(5.5) cm, contracted, usually panicles, very rarely racemes, with 1–2 branches per node; **branches** usually erect, sometimes spreading at anthesis, lower branches with 2+ spikelets. **Spikelets** 3.5–7(8.5) mm, with 2–4(6) florets. **Glumes** exceeded by the upper florets, ovate-lanceolate, usually glabrous and smooth, sometimes scabrous distally; **lower glumes** (1.2)1.8–3(3.5) mm; **upper glumes** (2.4)2.6–4(4.6) mm; **lemmas** 2.5–4.5(6) mm, ovate-lanceolate to lanceolate, scabrous towards the apices, awns (0.8)1–3(3.5) mm, terminal; **paleas** about as long as the lemmas, intercostal region scabrous or puberulent distally; **anthers** (0.5)0.7–1.1(1.3) mm; **ovary apices** glabrous. 2n = 28, 42, 44.

Festuca brachyphylla is a variable, circumpolar, arctic, alpine, and boreal species of open, rocky places. It is palatable to livestock, and is important in some areas as forage for wildlife. The spikelets are usually tinged red to purple by anthocyanin pigments; plants which lack anthocyanins in the spikelets have been named *F. brachyphylla* f. *flavida* Polunin. *Festuca brachyphylla* has frequently been included in *F. ovina* (p. 422), and it is closely related to *F. saximontana* (p. 430), *F. hyberborea* (p. 432), *F. edlundiae* (p. 432), *F. groenlandica* (p. 434), and *F. minutiflora* (p. 434). It may hybridize with *F. baffinensis* and/or other species to form *F. viviparoidea* (p. 436).

Three subspecies have been recognized in North America. *Festuca brachyphylla* subsp. *brachyphylla* is circumpolar and primarily arctic, subarctic, and boreal, extending southward in the northern Rocky Mountains. The other two subspecies are restricted to alpine regions in the western mountains.

1. Culms usually more than twice as long as the vegetative shoot leaves; spikelets 4.4–7(8.5) mm long; lemmas (3)3.5–4.5(6) mm long; plants boreal, arctic, and alpine in the northern cordillera . . . . subsp. *brachyphylla*
1. Culms up to twice as long as the vegetative shoot leaves; spikelets 3.5–5.5 mm long; lemmas (2.5)3.5–4 mm long; plants alpine in the southern cordillera.
 2. Culms usually twice as long as the vegetative shoot leaves; awns 2–3(3.2) mm long; spikelets 4.4–5.6(7) mm long; lemmas 3–4(4.5) mm long . . . . subsp. *coloradensis*
 2. Culms usually less than twice as long as the vegetative shoot leaves; awns 1–2(2.2) mm long; spikelets 3.5–5(5.5) mm long; lemmas 2.5–4 mm long . . . . subsp. *breviculmis*

Festuca brachyphylla Schult. & Schult. f. subsp. **brachyphylla** [p. 429]

Culms usually more than twice as long as the vegetative shoot leaves. **Spikelets** 4.4–7(8.5) mm. **Lemmas** (3)3.5–4.5(6) mm, awns (0.8)2–3(3.5) mm. 2n = 42, 44.

Festuca brachyphylla subsp. *brachyphylla* is circumpolar in its distribution. In the *Flora* region, it extends from Alaska to Newfoundland, south in the mountains to Washington in the west and in the high peaks of the Appalachian Mountains of eastern Quebec and New England in the east.

Festuca brachyphylla subsp. **breviculmis** Fred. [p. 429]

Culms usually less than twice as long as the vegetative shoot leaves. **Spikelets** 3.5–5(5.5) mm. **Lemmas** 2.5–4 mm; awns 1–2(2.2) mm. 2n = 28(?) [chromosome count is unknown, but has been inferred].

Festuca brachyphylla subsp. *breviculmis* is endemic to California, where it grows in alpine habitats in the Sierra Nevada and White Mountains.

Festuca brachyphylla subsp. **coloradensis** Fred. [p. 429]

Culms usually about twice as long as the vegetative shoot leaves. **Spikelets** 4.4–5.6(7) mm. **Lemmas** 3–4(4.5) mm, awns 2–3(3.2) mm. 2n = 28.

Festuca brachyphylla subsp. *coloradensis* is a common species in alpine areas of Colorado, Utah, Wyoming, Arizona, and New Mexico. It often grows with *F. earlei* (p. 420), from which it can be distinguished by its lack of rhizomes, and smaller spikelets and lemmas.

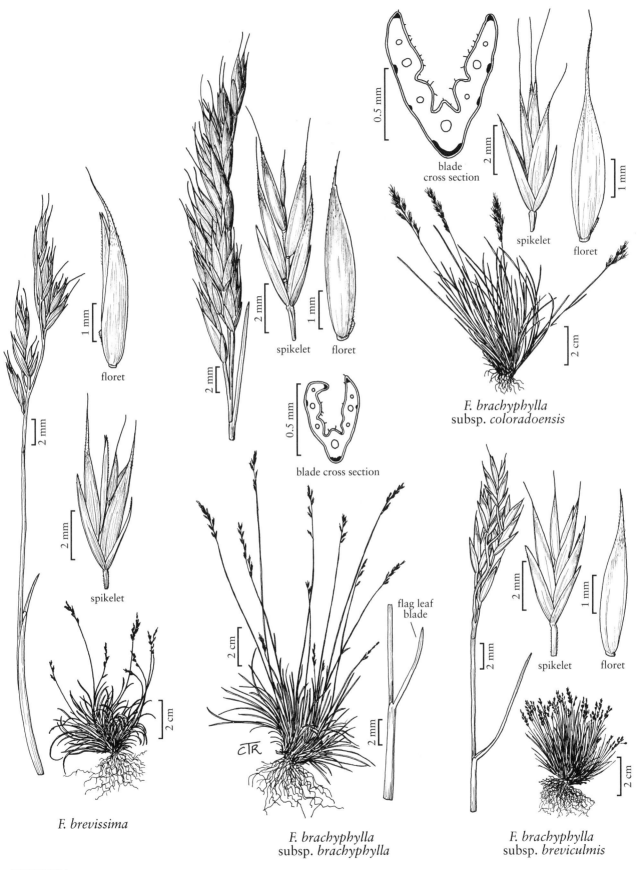

blade
cross section

spikelet

floret

F. brachyphylla
subsp. *coloradoensis*

floret

spikelet

2 mm

0.5 mm

blade cross section

spikelet

floret

F. brevissima

F. brachyphylla
subsp. *brachyphylla*

flag leaf
blade

CTR

F. brachyphylla
subsp. *breviculmis*

spikelet

floret

FESTUCA

29. Festuca saximontana Rydb. [p. 431]

ROCKY MOUNTAIN FESCUE, MOUNTAIN FESCUE,
FÉTUQUE DES MONTAGNES ROCHEUSES,
FÉTUQUE DES ROCHEUSES

Plants usually densely, sometimes loosely, cespitose, without rhizomes. **Culms** (5)8–50(60) cm, usually smooth and glabrous, occasionally sparsely scabrous or puberulent below the inflorescence. **Sheaths** closed for about ¹/₂ their length, glabrous, smooth or scabrous, usually persistent, rarely slowly shredding into fibers; **collars** glabrous; **ligules** 0.1–0.5 mm; **blades** 0.5–1.2 mm in diameter, conduplicate, abaxial surfaces glabrous or sparsely puberulent, adaxial surfaces scabrous or puberulent, veins 5–7(9), ribs 1–5; **abaxial sclerenchyma** in 3–7 strands, sometimes partly confluent or forming a continuous band, usually more than twice as wide as high; **adaxial sclerenchyma** absent; **flag leaf blades** 0.5–4 cm. **Inflorescences** (2)3–10(13) cm, contracted, with 1–2 branches per node; **branches** usually erect, spreading at anthesis, lower branches with 2+ spikelets. **Spikelets** (3)4.5–8.8(10) mm, with (2)3–5(7) florets. **Glumes** exceeded by the upper florets, ovate-lanceolate to lanceolate, scabrous distally; **lower glumes** 1.5–3.5 mm; **upper glumes** 2.5–4.8 mm; **lemmas** (3)3.4–4(5.6) mm, mostly smooth, often scabrous distally, awns (0.4) 1–2(2.5) mm; **paleas** as long as or slightly shorter than the lemmas, intercostal region puberulent distally; **anthers** (0.8)1.2–1.7(2) mm; **ovary apices** glabrous. $2n = 42$.

Festuca saximontana grows in grasslands, meadows, open forests, and sand dune complexes of the northern plains and boreal, montane, and subalpine regions in the *Flora* region, extending from Alaska to Greenland, south to southern California, northern Arizona, and New Mexico in the west and to the Great Lakes region in the east. It is also reported from the Russian Far East. *Festuca saximontana* provides good forage for livestock and wildlife. It is closely related to *F. brachyphylla* (see previous), and is sometimes included in that species as *F. brachyphylla* subsp. *saximontana* (Rydb.) Hultén. It has also frequently been included in *F. ovina* (p. 422).

The populations which grow in sandy areas around the upper Great Lakes have been named *Festuca canadensis* E.B. Alexeev; given the great variation in the species, there seems to be little justification for this. Three weakly differentiated taxa have been recognized at the varietal level in North America.

1. Culms 25–50(60) cm tall, usually 3–5 times the height of the vegetative shoot leaves; abaxial surfaces of the blades usually scabrous; abaxial sclerenchyma in 3–5 strands, sometimes partly confluent or forming a continuous band; plants of lowland, montane, or boreal habitats . . . . var. *saximontana*
1. Culms (5)8–37 cm tall, usually 2–3 times the height of the vegetative shoot leaves; abaxial surfaces of the blades smooth or scabrous; abaxial sclerenchyma in 5–7 narrow strands; plants of subalpine or lower alpine habitats.
   2. Culms (5)8–20(25) cm tall, usually glabrous below the inflorescence; outer vegetative shoot sheaths mostly stramineous; blades with hairs shorter than 0.06 mm on the ribs; lemmas usually scabrous towards the apices and often along the margins . . . . var. *purpusiana*
   2. Culms 16–37 cm tall, usually sparsely scabrous or pubescent below the inflorescence; outer vegetative shoot sheaths brownish on the lower ¹/₂; blades with hairs to 0.1 mm on the ribs; lemmas often scabrous on the distal ¹/₂ . . . . . . var. *robertsiana*

Festuca saximontana var. purpusiana (St.-Yves) Fred. & Pavlick [p. 431]

Culms (5)8–20(25) cm, usually 2–3 times the height of the vegetative shoot leaves, usually glabrous below the inflorescence. **Outer vegetative shoot sheaths** mostly stramineous; **blades** smooth or scabrous on the abaxial surfaces, ribs on the adaxial surfaces with hairs shorter than 0.06 mm; **abaxial sclerenchyma** in 5–7 narrow abaxial strands. **Lemmas** usually scabrous towards the apices and often along the margins.

Festuca saximontana var. *purpusiana* grows in subalpine or lower alpine habitats. The distribution of this taxon is poorly known; it probably extends from Alaska south to northern California. It is also reported from the Chukchi Peninsula in eastern Russia (Tzvelev 1976).

Festuca saximontana var. robertsiana Pavlick [p. 431]

Culms 16–37 cm, usually 2–3 times the height of the vegetative shoot leaves, usually sparsely scabrous or pubescent below the inflorescence. **Outer vegetative shoot sheaths** brownish on the lower ¹/₂; **blades** smooth or scabrous on the abaxial surfaces, ribs on the adaxial surfaces with hairs to 0.1 mm; **abaxial sclerenchyma** in 5–7 narrow strands. **Lemmas** often scabrous on the distal ¹/₂.

Festuca saximontana var. *robertsiana* grows in subalpine or lower alpine habitats. It has only been reported from British Columbia.

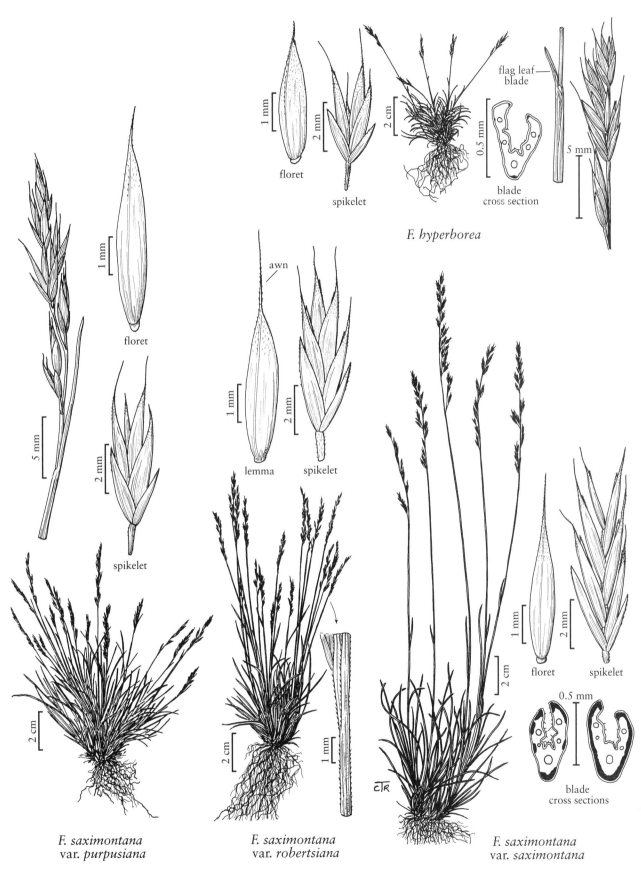

flag leaf
blade

floret

spikelet

blade
cross section

5 mm

F. hyperborea

floret

spikelet

awn

lemma   spikelet

floret   spikelet

blade
cross sections

F. saximontana
var. *purpusiana*

F. saximontana
var. *robertsiana*

F. saximontana
var. *saximontana*

FESTUCA

Festuca saximontana Rydb. var. **saximontana** [p. 431]

Culms 25–50(60) cm, usually 3–5 times the height of the vegetative shoot leaves, usually glabrous below the inflorescences, sometimes sparsely scabrous or pubescent; **blades** usually scabrous on the abaxial surfaces, scabrules to 0.1 mm; **abaxial sclerenchyma** in 3–5 strands, sometimes partly confluent or forming a continuous band. **Lemmas** smooth or scabrous distally.

Festuca saximontana var. *saximontana* grows throughout the range of the species.

30. **Festuca hyperborea** Holmen *ex* Fred. [p. 431]
    NORTHERN FESCUE

Plants densely cespitose, without rhizomes. **Culms** 5–15(20) cm, up to twice as tall as the vegetative shoot leaves, usually erect, sometimes semi-prostrate, glabrous, smooth. **Sheaths** closed for about ¹/₂ their length, glabrous, persistent; **collars** glabrous; **ligules** 0.1–0.5 mm; **blades** 0.5–1 mm in diameter, conduplicate, often curved or somewhat falcate, abaxial surfaces smooth or sparsely scabrous, adaxial surfaces scabrous, veins 5–7, ribs 3–5; **abaxial sclerenchyma** in 3–7 strands, usually less than twice as wide as high; **adaxial sclerenchyma** absent. **Flag leaf sheaths** usually somewhat inflated; **flag leaf blades** 0.5–5(8) mm. **Inflorescences** 1–2(2.5) cm, contracted, usually panicles, sometimes racemes, with 1–2 branches per node; **branches** erect, lower branches usually with 1–2 spikelets, sometimes more. **Spikelets** (3)4–5.5(7) mm, with 3–4(6) florets. **Glumes** exceeded by the upper florets, ovate to ovate-lanceolate, mostly glabrous and smooth, sometimes scabrous distally; **lower glumes** 1–3.5 mm; **upper glumes** 2.2–3.2 mm; **lemmas** 2.9–3.5(4.4) mm, ovate, apices scabrous and minutely bidentate, awns (0.5)1.4–2(3) mm, usually slightly subterminal, curved or slightly twisted; **paleas** about as long as or slightly longer than the lemmas, intercostal region smooth or scabrous distally; **anthers** 0.4–0.8(1.1) mm; **ovary apices** glabrous. 2n = 28.

Festuca hyperborea is a high arctic species that grows from Banks Island in the Canadian Arctic east to Greenland and south to Quebec. It differs from *F. brachyphylla* (p. 428) in its semi-prostrate habit, the loose sheaths and short blades of its flag leaves, the more pronounced ribs in its lower leaf blades, and its subterminal awn. It differs from *F. edlundiae* (see next) in having flag leaf blades shorter than 5 mm and smaller spikelets. It has frequently been included in *F. ovina* (p. 422).

31. **Festuca edlundiae** S. Aiken, Consaul & Lefk.
    [p. 433]
    EDLUND'S FESCUE

Plants densely cespitose, without rhizomes. **Culms** 2.5–10 (14) cm, up to twice as tall as the vegetative shoot leaves, usually geniculate to prostrate, erect at anthesis, glabrous, smooth. **Sheaths** closed for about ¹/₂ their length, smooth or slightly scabrous, persistent; **collars** glabrous; **ligules** 0.1–0.5 mm; **blades** (0.5)0.8–1.1 mm in diameter, conduplicate, usually straight, veins 5–7, ribs 3–5, abaxial surfaces smooth or sparsely scabrous, adaxial surfaces scabrous; **abaxial sclerenchyma** in 5–7(9) narrow strands, usually less than twice as wide as high; **adaxial sclerenchyma** absent. **Flag leaf sheaths** somewhat inflated; **flag leaf blades** (0.3)0.5–2 cm. **Inflorescences** 1.5–3.5 cm, often racemes, with 1–2 branches per node; **branches** erect, lower branches with 1–2(3+) spikelets. **Spikelets** 4.5–8.5 mm, with (2)3–6 florets. **Glumes** exceeded by the upper florets, ovate to ovate-lanceolate, mostly glabrous and smooth, sometimes scabrous distally; **lower glumes** 1.8–3.5 mm; **upper glumes** 2.9–4.3 mm; **lemmas** 3.6–5.2 mm, scabrous distally, apices entire, awns 1.1–2.9 mm, usually terminal, sometimes slightly subterminal; **paleas** about as long as or slightly longer than the lemmas, intercostal region scabrous distally; **anthers** 0.6–1.1 mm; **ovary apices** glabrous. 2n = 28.

Festuca edlundiae is a high arctic species that is closely related to *F. brachyphylla* (p. 428). It grows primarily on fine-grained and calcareous substrates in arctic regions of the Russian Far East, Alaska, the arctic islands of Canada, northern Greenland, and Svalbard. It resembles *F. hyperborea* (see previous), differing from it in having flag leaf blades that are usually at least 5 mm long and larger spikelets. *Festuca edlunieae* has frequently been included in *F. ovina* (p. 422).

32. **Festuca baffinensis** Polunin [p. 433]
    BAFFIN ISLAND FESCUE

Plants densely cespitose, without rhizomes. **Culms** 5–25(30) cm, densely pubescent or shortly pilose near the inflorescence. **Sheaths** closed for about ¹/₂ their length, glabrous, persistent or slowly shredding into fibers; **collars** glabrous; **ligules** 0.1–0.3 mm; **blades** (0.4)0.6–1(1.2) mm in diameter, conduplicate, abaxial surfaces smooth or sparsely scabrous, adaxial surfaces

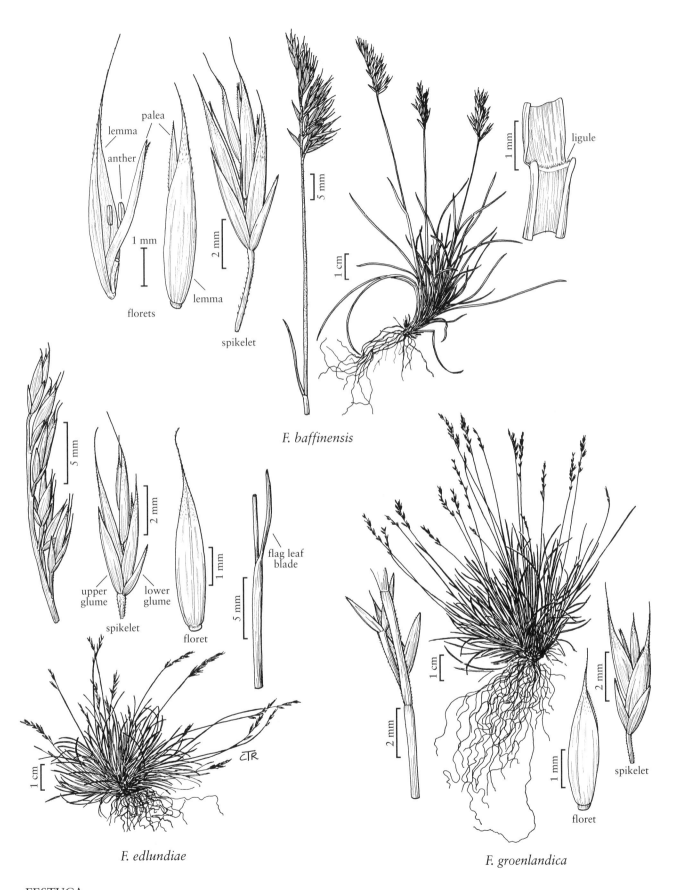

florets

lemma

palea

anther

lemma

spikelet

F. baffinensis

ligule

upper glume

lower glume

spikelet

floret

flag leaf blade

F. edlundiae

CTR

floret

spikelet

F. groenlandica

FESTUCA

scabrous or puberulent, veins (3)5–7, ribs 3–5; **abaxial sclerenchyma** in 3–7 small strands, usually less than twice as wide as high; **adaxial sclerenchyma** absent. **Flag leaf sheaths** usually somewhat loosely enclosing the culms; **flag leaf blades** 0.5–4 cm. **Inflorescences** 1.5–4(5) cm, contracted, usually panicles, rarely racemes, usually somewhat secund, with 1–2 branches per node; **branches** erect, lower branches with 2+ spikelets. **Spikelets** (4.5)5–7.5(8.5) mm, with (2)3–5(6) florets. **Glumes** exceeded by the upper florets, ovate-lanceolate, scabrous distally; **lower glumes** 2.2–3.7(4) mm; **upper glumes** 3–5 mm; **lemmas** (3.5)4–6 mm, scabridulous near the apices, awns 0.8–2.6(3.3) mm, terminal; **paleas** slightly shorter than to as long as the lemmas, intercostal region scabrous distally; **anthers** 0.3–0.7 (1.1) mm; **ovary apices** usually with a few hairs, rarely glabrous. $2n = 28$.

Festuca baffinensis grows chiefly in damp, exposed, gravelly areas in calcareous and volcanic regions. It is circumpolar in distribution, growing in arctic and alpine habitats and extending southward in the Rocky Mountains to Colorado. It has frequently been included in *F. ovina* (p. 422). It may hybridize with *F. brachyphylla* (p. 428) and/or other species to form *F. viviparoidea* (p. 436).

33. Festuca groenlandica (Schol.) Fred. [p. 433]
GREENLAND FESCUE

Plants densely cespitose, without rhizomes. **Culms** 10–37 cm, erect, glabrous, smooth. **Sheaths** closed for about ¹/₂ their length, glabrous, persistent; **collars** glabrous; **ligules** 0.2–0.5 mm; **blades** 0.4–0.7(1) mm in diameter, conduplicate, veins 5–7, ribs 1–5, abaxial surfaces smooth, adaxial surfaces scabrous or puberulent; **abaxial sclerenchyma** in 5–7 strands, at least twice as wide as high, often some strands confluent; **adaxial sclerenchyma** absent. **Flag leaf sheaths** tight or somewhat loose; **flag leaf blades** 1–5 cm. **Inflorescences** (1.5)2–5 cm, contracted, with 1–2 branches per node; **branches** erect, lower branches with 2+ spikelets. **Spikelets** 4–5.6(6) mm, with 3–4(5) florets. **Glumes** exceeded by the upper florets, ovate-lanceolate, mostly glabrous and smooth, sometimes scabrous distally; **lower glumes** 1.5–2 mm, lanceolate to ovate-lanceolate; **upper glumes** (2)2.2–2.7 mm, ovate; **lemmas** (2.5)3–3.5(4) mm, ovate-lanceolate to lanceolate, usually scabrous distally, sometimes smooth, usually awned, occasionally unawned, awns 0.7–2.1 mm, terminal; **paleas** about as long as the lemmas, intercostal region smooth or scabrous distally; **anthers** 0.8–1.3 mm; **ovary apices** glabrous. $2n = 42$.

Festuca groenlandica is endemic to Greenland. Scholander (1934) initially described it as a variety of *F. brachyphylla* (p. 428), but it differs from that species in having more extensive blade sclerenchyma, usually 7 broad abaxial strands rather than 5 narrow strands.

34. Festuca minutiflora Rydb. [p. 435]
LITTLE FESCUE, SMALL-FLOWERED FESCUE

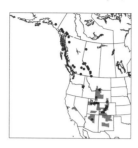

Plants loosely or densely cespitose, without rhizomes. **Culms** 4–30 cm, usually erect, sometimes semi-prostrate, glabrous, smooth. **Sheaths** closed for about ¹/₂ their length, glabrous, persistent; **collars** glabrous; **ligules** 0.1–0.3 mm; **blades** (0.2)0.3–0.4(0.6) mm in diameter, conduplicate, lax, abaxial surfaces glabrous, adaxial surfaces sparsely scabrous to puberulent, veins 3–5, ribs 1–3; **abaxial sclerenchyma** in 3–5 small strands, less than twice as wide as high; **adaxial sclerenchyma** absent. **Flag leaf blades** 0.7–3.5 cm. **Inflorescences** 1–4(5) cm, contracted, with 1–2 branches per node; **branches** erect, lower branches with 2+ spikelets. **Spikelets** (2.5)3–5 mm, with (1)2–3(5) florets. **Glumes** exceeded by the upper florets, ovate to ovate-lanceolate, sparsely scabrous distally; **lower glumes** 1.3–2.5 mm; **upper glumes** 2–3.5 mm; **lemmas** (2)2.2–3.5(4) mm, ovate-lanceolate, sparsely scabrous near the apices, apices abruptly acuminate, awns 0.5–1.5(1.7) mm; **paleas** about as long as or slightly shorter than the lemmas, intercostal region scabrous distally; **anthers** (0.4)0.6–1.2 mm; **ovary apices** usually with a few hairs, rarely glabrous. $2n = 28$.

Festuca minutiflora grows in alpine regions of the western mountains, from southeastern Alaska and the southwestern Yukon Territory to Arizona, New Mexico, and the Sierra Nevada of California. It has often been overlooked or included with *F. brachyphylla* (p. 428), from which it differs in its laxer and narrower leaves, looser panicles, smaller spikelets, more pointed lemmas, shorter awns, and scattered hairs on the ovary. In the southern Rocky Mountains, it may grow with *F. earlei* (p. 420), which has short rhizomes and larger spikelets and lemmas. *Festuca minutiflora* has frequently been included in *F. ovina* (p. 422).

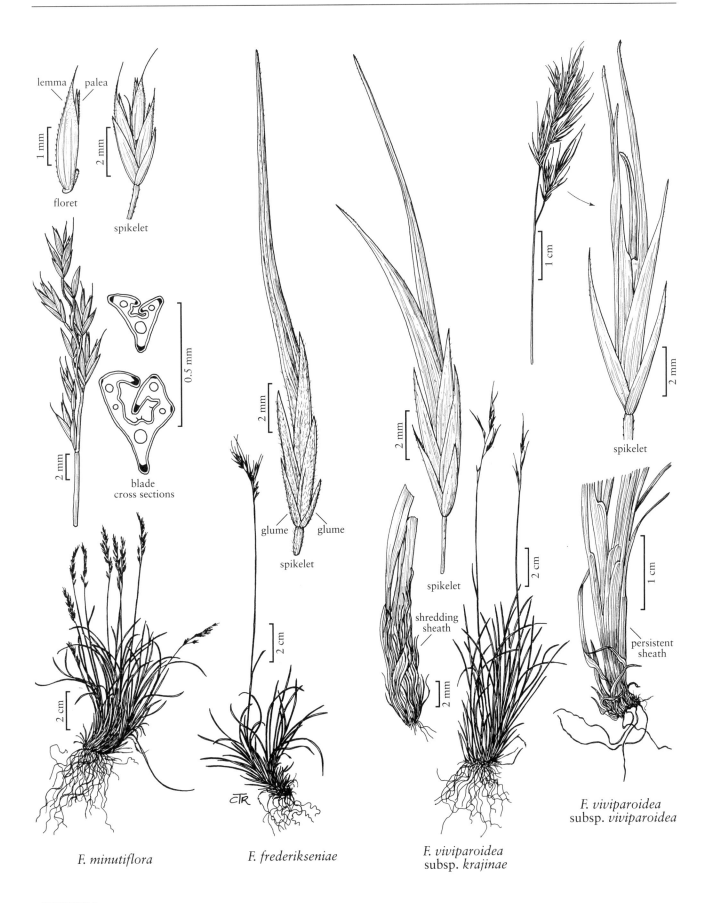

lemma palea

1 mm

floret

2 mm

spikelet

0.5 mm

2 mm

blade
cross sections

2 mm

2 cm

F. minutiflora

2 mm

glume glume

spikelet

2 cm

F. frederikseniae

2 mm

spikelet

shredding
sheath

2 mm

2 cm

F. viviparoidea
subsp. *krajinae*

1 cm

2 mm

spikelet

1 cm

persistent
sheath

F. viviparoidea
subsp. *viviparoidea*

FESTUCA

35. Festuca frederikseniae E.B. Alexeev [p. 435]
FREDERIKSEN'S FESCUE, FÉTUQUE DE FREDERIKSEN

Plants densely cespitose, without rhizomes. **Culms** (5)10–35 (45) cm, pubescent near the inflorescence. **Sheaths** closed for about ¹/₂ their length, glabrous or puberulent, persistent; **collars** glabrous; **ligules** 0.2–0.5 mm; **blades** 0.5–0.8 mm in diameter, conduplicate, abaxial surfaces glabrous, smooth or scabrous, adaxial surfaces scabrous or hirsute, veins (3)5–7, ribs 3–5, 1 distinct and 2–4 indistinct; **abaxial sclerenchyma** in 3–7 broad, sometimes confluent strands, covering ¹/₂ or more of the surface. **Inflorescences** (1.5)2–10 cm, contracted, with 1(2) branches per node; **branches** erect, stiff, lower branches with 2+ spikelets. **Spikelets** pseudoviviparous, varying in length with the stage of vegetative proliferation, the glumes and often 1 or 2 adjacent florets more or less normally developed or only slightly elongated, the distal florets replaced by leafy bracts. **Glumes** ovate-lanceolate, densely puberulent to pubescent throughout; **lower glumes** 2–4.5 mm; **upper glumes** (2.7)3.8–5.2 mm; **normal lemmas** 3.5–5 mm, densely hairy to pubescent, sometimes awned, awns to 0.2 mm; **vegetative bracts** unawned, leaflike, sometimes with ligules; **paleas** usually reduced or absent, well-formed paleas about as long as the lemmas; **anthers** usually poorly developed and the pollen sterile, well-formed anthers to about 2.5 mm; **ovary apices** glabrous. $2n = 28$.

Festuca frederikseniae grows on cliffs, rocky or sandy barrens, and alpine regions in southern Quebec (Mingan and Anticosti islands), Newfoundland, southern Labrador, and southern Greenland. It differs from *F. vivipara* (L.) Sm. of northern Europe and Asia in having densely pubescent spikelet bracts and fascicles, and an interrupted rather than continuous band of blade sclerenchyma. Frederiksen (1981) reported that *F. vivipara* occurs in southeastern Greenland, overlapping the range of *F. frederikseniae* and extending as far north as the southerly occurrences of *F. viviparoidea* subsp. *viviparoidea*; her paper should be consulted when trying to distinguish the complex pseudoviviparous fescues of Greenland.

In Iceland and southern Greenland, putative hybrids between *Festuca frederikseniae* or *F. vivipara* and *F. rubra* (p. 412) have been reported, and named *F. villosa-vivipara* (Rosenv.) E.B. Alexeev. These plants are highly variable but, unlike *F. frederikseniae*, produce extravaginal shoots, have closed sheaths, and have blades about 1 mm wide, with 7–9 small strands of abaxial sclerenchyma. Such hybrids can be expected within the range of *F. frederikseniae* in North America.

Festuca frederikseniae has frequently been included in *F. ovina* (p. 422).

36. Festuca viviparoidea Krajina *ex* Pavlick [p. 435]
VIVIPAROUS FESCUE

Plants loosely or densely cespitose, without rhizomes. **Culms** (11)13.5–25(28) cm, smooth and glabrous throughout or sparsely to densely scabrous or puberulent below the inflorescence. **Sheaths** closed for about ¹/₂ their length, glabrous or scabrous, stramineous or brownish, persistent or slowly shredding into fibers; **collars** glabrous; **ligules** 0.1–0.5 mm; **blades** 0.5–1 mm in diameter, conduplicate, abaxial surfaces glabrous, smooth or scabrous, adaxial surfaces scabrous, veins 5–7, ribs 3–5, 1 distinct and 2–4 indistinct; **abaxial sclerenchyma** in 3–7 small strands, covering less than ¹/₂ the abaxial surface and usually less than twice as wide as high. **Inflorescences** (1)3–4.8 cm, contracted, usually panicles, sometimes racemes, erect, with 1–2 branches per node; **branches** erect, lower branches with (1)2+ spikelets. **Spikelets** pseudoviviparous, their length varying with the stage of vegetative proliferation, the glumes and often 1 or 2 adjacent florets more or less normally developed, or only slightly elongated, the distal florets replaced by bracts. **Glumes** lanceolate, glabrous and smooth, sometimes scabrous towards the apices, or puberulent throughout or only towards the apices; **lower glumes** (2)3–6 mm; **upper glumes** (2.7)3–7 mm; **normal lemmas** 3.3–6 mm, mostly smooth or scabrous distally, glabrous or puberulent, awned or unawned, sometimes varying within a panicle, awns to 1 mm; **vegetative bracts** unawned, leaflike, sometimes with ligules; **paleas** usually reduced or absent, well-formed paleas about as long as the lemmas, intercostal region scabrous or puberulent distally; **anthers** usually not developed, well-formed anthers to about 2 mm; **ovaries** sometimes not developed; **ovary apices**, when present, glabrous. $2n = 49, 56$.

Festuca viviparoidea is circumboreal in distribution. It may consist of hybrids between *Festuca baffinensis* (p. 432) and *F. brachyphylla* (p. 428) and/or other species (see under *F. frederikseniae*, above). It has frequently been included in *F. ovina* (p. 422).

1. Plants loosely cespitose; culms usually glabrous and smooth throughout, rarely sparsely puberulent near the inflorescence; sheaths brownish, slowly shredding into fibers; abaxial sclerenchyma strands less than 2 times as wide as high; glumes and lemmas puberulent throughout or only near the apices . . . . subsp. *krajinae*

1. Plants densely cespitose; culms densely to sparsely puberulent below the inflorescence; sheaths stramineous, persistent; abaxial sclerenchyma strands 2–3 times wider than high; glumes and lemmas smooth or scabrous near the apices . . . . subsp. *viviparoidea*

Festuca viviparoidea subsp. krajinae Pavlick [p. 435]

Plants loosely cespitose. **Culms** usually glabrous and smooth throughout, rarely sparsely puberulent near the inflorescence. **Sheaths** brownish, slowly shredding into fibers; **abaxial sclerenchyma strands** about as wide as the adjacent veins, less than 2 times as wide as high. **Glumes** and **lemmas** puberulent throughout or only near the apices. $2n = 56$.

Festuca viviparoidea subsp. *krajinae* grows in alpine sites of the western cordillera, from southern Alaska and the Yukon Territory through British Columbia to southwestern Alberta.

Festuca viviparoidea Krajina *ex* Pavlick subsp. viviparoidea [p. 435]

Plants densely cespitose. **Culms** densely to sparsely puberulent below the inflorescence. **Sheaths** stramineous, persistent; **abaxial sclerenchyma strands** about twice as wide as the adjacent veins, 2–3 times as wide as high. **Glumes** and **lemmas** smooth or scabrous near the apices. $2n = 49, 56$.

Festuca viviparoidea subsp. *viviparoidea* is circumpolar and found in the high arctic, including Alaska, Yukon Territory, Nunavut, Greenland, Svalbard, and Russia.

37. Festuca occidentalis Hook. [p. 439]
WESTERN FESCUE

Plants densely to loosely cespitose, without rhizomes. **Culms** (25)40–80(110) cm, glabrous, smooth. **Sheaths** closed for much less than $^1/_2$ their length, glabrous, somewhat persistent or slowly shredding into fibers; **collars** glabrous; **ligules** 0.1–0.4 mm, usually longer at the sides; **blades** all alike, 0.3–0.7 mm in diameter, conduplicate, abaxial surfaces smooth or scabridulous, veins (3)5, ribs 1–5; **abaxial sclerenchyma** in 5–7 narrow strands, about as wide as the adjacent veins; **adaxial sclerenchyma** absent. **Inflorescences** (5)10–20 cm, open, with 1–2 branches per node; **branches** 1–15 cm, lax, widely spreading to reflexed, lower branches usually reflexed at maturity, with 2+ spikelets. **Spikelets** 6–12 mm, with 3–6(7) florets. **Glumes** exceeded by the upper florets, ovate to ovate-lanceolate, glabrous and smooth or slightly scabrous; **lower glumes** 2–5 mm; **upper glumes** 3–6 mm; **lemmas** (4)4.5–6.5(8) mm, ovate-lanceolate to attenuate, glabrous or finely puberulent, awns 3–12 mm, usually longer than the lemma bodies; **paleas** slightly shorter than the lemmas, intercostal region scabrous or puberulent distally; **anthers** (1)1.5–2(3) mm; **ovary apices** densely pubescent. $2n = 28$ [other numbers have been reported for this species, but are probably based on misidentifications].

Festuca occidentalis grows in dry to moist, open woodlands, forest openings, and rocky slopes, up to 3100 m. It extends from southern Alaska and northern British Columbia to southwestern Alberta, south to southern California and eastward to Wyoming, and, as a disjunct, around the upper Great Lakes in Ontario, eastern Wisconsin, and Michigan. It is sometimes important as a forage grass, but is usually not sufficiently abundant.

38. Festuca calligera (Piper) Rydb. [p. 439]
CALLUSED FESCUE

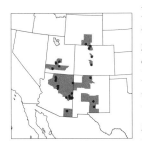

Plants densely cespitose, without rhizomes. **Culms** 15–65 cm, glabrous, smooth. **Sheaths** closed for less than $^1/_2$ their length, lower sheaths glabrous or retrorsely hirsute, persistent, upper sheaths glabrous; **collars** glabrous; **ligules** (0.2)0.3–0.5(1) mm; **blades** all alike, 0.4–0.8 mm in diameter, conduplicate, abaxial surfaces sparsely scabrous, adaxial surfaces scabrous to pubescent, veins 5–7, ribs (1)3–5; **abaxial sclerenchyma** in (3)5–7 narrow to broad strands, usually wider than the adjacent veins; **adaxial sclerenchyma** absent. **Inflorescences** 5–15 cm, loosely contracted, with 1–2(3) branches per node; **branches** erect, lower branches with 2+ spikelets. **Spikelets** (6)7–9(11) mm, with (2)4–6 florets. **Glumes** exceeded by the upper florets, lanceolate, scabrous distally; **lower glumes** 2.5–4 mm; **upper glumes** (2.8)3–5 mm; **lemmas** (3.8)4–6 mm, glabrous, smooth or scabrous distally, awns 1–2.5 mm; **paleas** slightly shorter than the lemmas, intercostal region scabrous or puberulent distally; **anthers** 2.2–3.5 mm; **ovary apices** sparsely pubescent. $2n = 28$.

Festuca calligera is a poorly known, often overlooked species. It grows in grasslands and open montane forests, at 2500–3400 m, from southern Utah to south-central Wyoming and central Colorado, south to Arizona and New Mexico. It is often found with *F. arizonica*. *Festuca calligera* has frequently been included in *F. ovina* (p. 422).

39. Festuca arizonica Vasey [p. 439]
ARIZONA FESCUE, PINEGRASS

Plants densely cespitose, without rhizomes. **Culms** 35–80 (100) cm, usually densely scabrous or densely pubescent below the inflorescences. **Sheaths** closed for less than $^1/_2$ their length, glabrous, smooth or scabrous, persistent; **collars** glabrous, smooth or scabrous; **ligules** 0.5–1.5(2) mm; **blades** 0.3–0.8 mm in diameter, conduplicate, abaxial surfaces scabrous or puberulent, adaxial surfaces scabrous to pubescent, veins 5–7, ribs (1)3–5(7), distinct; **abaxial sclerenchyma** in 5–7 broad strands, rarely forming a complete band, forming **pillars** with some veins; **adaxial sclerenchyma** not developed. **Inflorescences** (4)6–15(20) cm, loosely contracted or open, with 1–2 branches per node; **branches** erect or spreading, lower branches with 2+ spikelets. **Spikelets** (6)8–16 mm, with (3)4–6(8) florets. **Glumes** exceeded by the upper florets, lanceolate, glabrous, smooth or scabrous distally; **lower glumes** (3)3.3–5.5 mm; **upper glumes** 4.5–6.6(7) mm; **lemmas** 5.5–9 mm, glabrous, smooth or scabrous towards the apices, unawned or awned, awns 0.4–2(3) mm; **paleas** slightly shorter than the lemmas, intercostal region scabrous or puberulent distally; **anthers** (2)3–4(4.2) mm; **ovary apices** densely pubescent. $2n = 42$.

Festuca arizonica grows in dry meadows and openings of montane forests, in gravelly, rocky soil, at 2100–3400 m. Its range extends from southern Nevada and southern Utah east to Colorado and south to Arizona, western Texas, and northern Mexico. It is abundant and valuable forage in some parts of its range. It is often found with *F. calligera* (see previous).

Festuca arizonica differs from *F. idahoensis* (see next), with which it is sometimes confused, in its prominently ribbed blades and pubescent ovary apices. It has frequently been included in *F. ovina* (p. 422).

40. Festuca idahoensis Elmer [p. 441]
IDAHO FESCUE, BLUE BUNCHGRASS, BLUEBUNCH FESCUE

Plants densely cespitose, without rhizomes. **Culms** 25–85 (100) cm, usually smooth, glabrous, occasionally scabrous below the inflorescences. **Sheaths** closed for less than $^1/_2$ their length, smooth or scabrous, rarely pilose, persistent; **collars** glabrous; **ligules** 0.2–0.6 mm; **blades** (0.3)0.5–0.9(1.5) mm in diameter, conduplicate, abaxial surfaces smooth or scabrous, adaxial surfaces scabrous or pubescent, rarely pilose, often glaucous or bluish, veins (3)5(7), ribs (1)3–5, well defined; **abaxial sclerenchyma** in 5–7 wide, irregular strands; **adaxial sclerenchyma** absent. **Inflorescences** (5)7–15(20) cm, loosely contracted or open, with 1–2 branches per node; **branches** usually somewhat spreading at maturity, sometimes erect, rarely reflexed, lower branches with 2+ spikelets. **Spikelets** (5.8) 7.5–13.5(19) mm, with (2)4–7(9) florets. **Glumes** exceeded by the upper florets, ovate-lanceolate to lanceolate, mostly smooth, sometimes scabrous distally; **lower glumes** 2.4–5(6) mm; **upper glumes** 3–6(8) mm; **lemmas** 5–8.5(10) mm, scabrous at the apices, awns (1.5)3–6(7) mm, usually more than $^1/_2$ as long as the lemma bodies; **paleas** shorter than to about as long as the lemmas, intercostal region scabrous or puberulent distally; **anthers** 2.4–4.5 mm; **ovary apices** glabrous. $2n = 28$.

Festuca idahoensis grows in grasslands, open forests, and sagebrush meadow communities, mostly east of the Cascade Mountains, from southern British Columbia eastward to southwestern Saskatchewan and southward to central California and New Mexico. It extends up to 3000 m in the southern part of its range. It is often a dominant plant, and provides good forage. The young foliage is particularly palatable.

Festuca idahoensis differs from *F. arizonica* (see previous), with which it is sometimes confused, in its less prominently ribbed blades and glabrous ovary apices. It has frequently been included in *F. ovina* (p. 422).

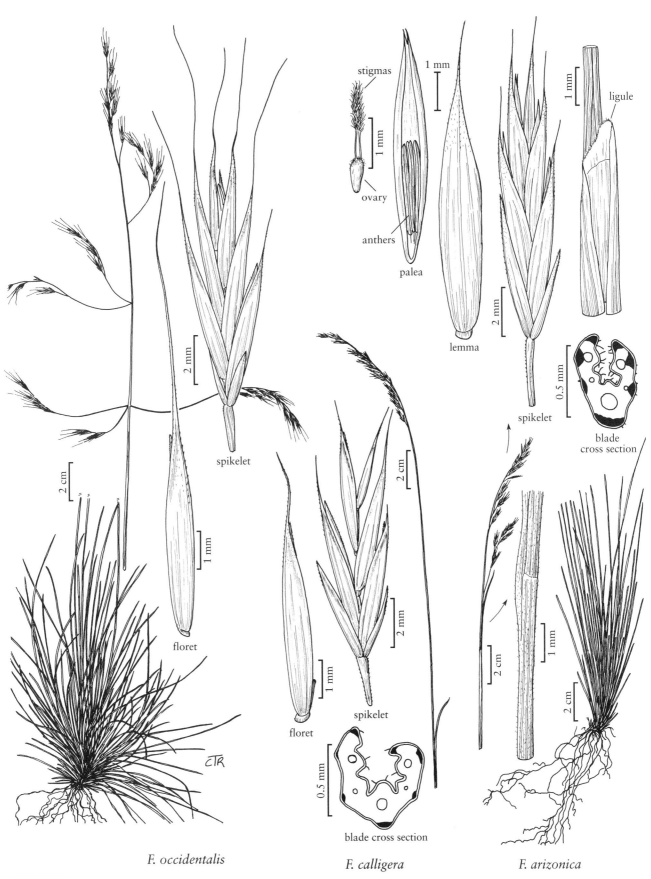

stigmas

ovary

anthers

palea

1 mm

1 mm

lemma

ligule

1 mm

spikelet

blade cross section

0.5 mm

spikelet

2 mm

spikelet

2 mm

floret

1 mm

floret

1 mm

2 mm

2 mm

floret

2 mm

1 mm

blade cross section

0.5 mm

spikelet

2 cm

2 cm

CTR

2 cm

2 cm

F. occidentalis

F. calligera

F. arizonica

FESTUCA

41. Festuca roemeri (Pavlick) E.B. Alexeev [p. 441]
OREGON FESCUE, ROEMER'S FESCUE

Plants densely cespitose, without rhizomes. **Culms** (35)50–90 (100) cm, erect, glabrous, smooth. **Sheaths** closed for less than ¹/₂ their length, glabrous, hirsute, or scabrous, persistent; **collars** glabrous; **ligules** 0.1–0.5 mm; **blades** 0.5–1(1.2) mm in diameter, conduplicate, abaxial surfaces glabrous or puberulent, adaxial surfaces sometimes scabrous, glabrous or pubescent, veins (5)7–9, ribs 5–9, well defined; **abaxial sclerenchyma** in 5–7 wide strands, sometimes confluent into a single band; **adaxial sclerenchyma** absent. **Inflorescences** (7)8–20(25) cm, loosely to densely contracted, with 1–2 branches per node; **branches** erect to slightly spreading, lower branches with 2+ spikelets. **Spikelets** 9–13.5 mm, with 4–6 florets. **Glumes** exceeded by the upper florets, ovate-lanceolate, smooth or scabrous distally; **lower glumes** (2)2.5–5 mm; **upper glumes** 4–6.2 mm; **lemmas** 5–7(8.2) mm, scabrous near the apices, awns (2)3–5 mm, terminal, usually more than ¹/₂ as long as the lemma bodies; **paleas** about as long as the lemmas, intercostal region scabrous or puberulent distally; **anthers** (2.6)2.8–3.6(4) mm; **ovary apices** glabrous. 2*n* = unknown.

Festuca roemeri grows in grasslands and open forests, primarily west of the Cascade Mountains, from southeastern Vancouver Island southward to northwestern California.

42. Festuca viridula Vasey [p. 441]
MOUNTAIN BUNCHGRASS, GREENLEAF FESCUE, GREEN FESCUE

Plants loosely or densely cespitose, without rhizomes. **Culms** 35–80(100) cm, smooth, glabrous throughout; **nodes** usually not exposed. **Sheaths** closed for less than ¹/₂ their length, usually glabrous, sometimes pubescent, strongly veined, persistent or slowly shredding into fibers; **collars** glabrous; **ligules** (0.2)0.3–0.8(1) mm; **blades** 0.5–1.3 mm in diameter when conduplicate, to 2.5 mm wide when flat, persistent, abaxial surfaces glabrous and smooth, adaxial surfaces scabrous or pubescent, veins 5–9(12), ribs 5–9, blades of the lower cauline leaves usually reduced to stiff horny points, blades of the upper cauline leaves longer and more flexuous; **abaxial sclerenchyma** in strands about as wide as the adjacent veins; **adaxial sclerenchyma** developed; **pillars** and girders often present. **Inflorescences** (4)8–15 cm, open or somewhat contracted, with 1–2 branches per node; **branches** lax, spreading or loosely erect, lower branches with 2+ spikelets. **Spikelets** 9–15 mm, with (2)3–6(7) florets. **Glumes** exceeded by the upper florets, ovate-lanceolate to lanceolate, glabrous, smooth or scabridulous distally; **lower glumes** (2.4)2.8–5 mm, distinctly shorter than the adjacent lemmas; **upper glumes** 4.5–7(8.5) mm; **lemmas** (4.8)6–8.5 mm, lanceolate to ovate-lanceolate, glabrous, smooth or slightly scabrous, apices acute, unawned or awned, awns 0.2–1.5(2) mm; **paleas** about as long as the lemmas, intercostal region scabrous or puberulent distally; **anthers** (2)2.5–4(5) mm; **ovary apices** densely pubescent. 2*n* = 28.

Festuca viridula grows in low alpine and subalpine meadows, forest openings, and open forests, at (900)1500–3000 m, from southern British Columbia east to Montana and south to central California and Nevada. It is highly palatable to livestock, and is an important forage species in some areas.

43. Festuca washingtonica E.B. Alexeev [p. 442]
WASHINGTON FESCUE, HOWELL'S FESCUE

Plants loosely or densely cespitose, without rhizomes. **Culms** 40–70(100) cm, smooth, glabrous throughout; **nodes** usually not exposed. **Sheaths** closed for less than ¹/₂ their length, glabrous or scabrous, persistent or slowly shredding into fibers; **collars** glabrous; **ligules** (0.2)0.3–0.5 mm; **blades** 1.5–3 mm in diameter, loosely conduplicate to flat, persistent, abaxial surfaces glabrous and smooth, adaxial surfaces scabrous or pubescent on the ribs, veins 7–13, ribs 7–10(13), blades of the lower and upper cauline leaves similar in length and stiffness; **abaxial sclerenchyma** in strands opposite and about as wide as the major veins; **adaxial sclerenchyma** often present opposite the major veins; **pillars** or **girders** often developed. **Inflorescences** 8–12(15) cm, loosely contracted, with 1–2 branches per node; **branches** lax, spreading or loosely erect, lower branches with 2+ spikelets. **Spikelets** 8–15(18) mm, with (3)4–6(10) florets. **Glumes** exceeded by the upper florets, lanceolate to ovate-lanceolate, scabrous distally; **lower glumes** (2)3.5–5.5 mm; **upper glumes** (4)5.5–7(8) mm; **calluses** wider than long, glabrous, sometimes slightly scabrous; **lemmas** (5.5)8–10(11) mm, lanceolate, scabrous or puberulent at least distally, attenuate, sometimes minutely bidentate, awns 1–3(3.5) mm, terminal or subterminal, straight, occasionally absent; **paleas** about as long as the lemmas, intercostal region scabrous or puberulent distally; **anthers** (3)3.7–5.7 mm;

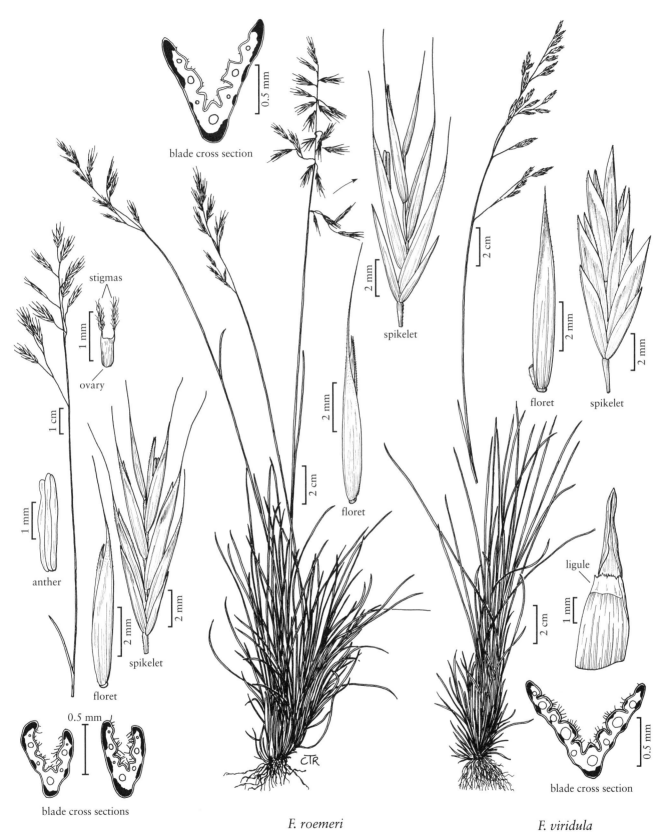

blade cross section

stigmas

ovary

0.5 mm

1 mm

1 cm

anther

spikelet

floret

0.5 mm

blade cross sections

F. idahoensis

spikelet

floret

CTR

F. roemeri

spikelet

floret

ligule

blade cross section

F. viridula

FESTUCA

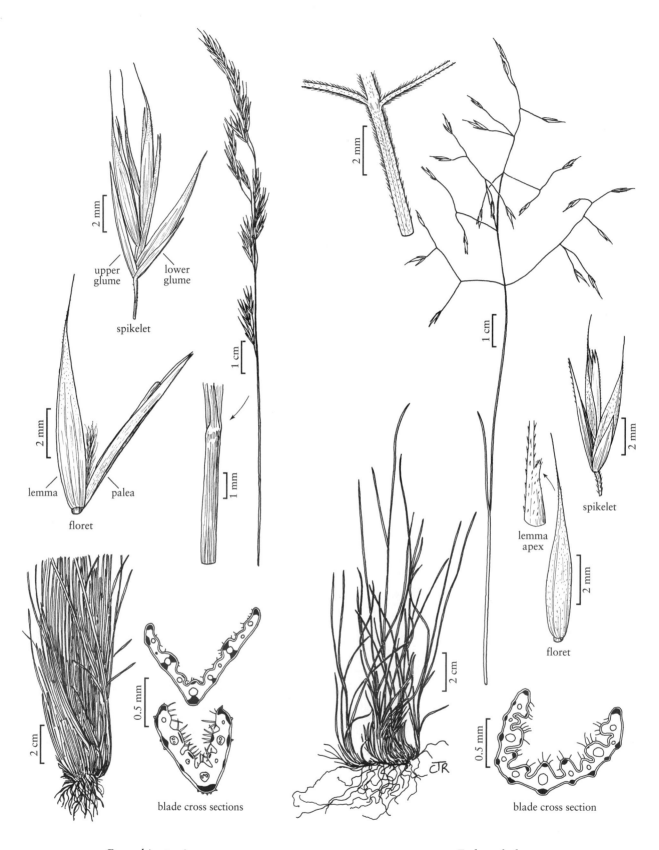

F. *washingtonica*   F. *dasyclada*

FESTUCA

ovary apices sparsely or densely pubescent. $2n =$ unknown.

Festuca washingtonica grows in subalpine to low alpine regions of British Columbia and Washington. It has also been reported from Oregon and northern California; these records have not been verified.

44. Festuca dasyclada Hack. *ex* Beal [p. 442]
OPEN FESCUE, INTERMOUNTAIN FESCUE

Plants loosely or densely cespitose, without rhizomes. Culms 20–40(50) cm, erect or somewhat geniculate at the base, densely scabrous or pubescent below the inflorescence; nodes usually not exposed, culms often breaking at the upper nodes at maturity. Sheaths closed for less than ¹/₂ their length, glabrous, persistent or slowly shredding into fibers; collars glabrous; ligules 0.2–0.5 mm; blades (1)1.2–2.5(3) mm wide, persistent, loosely conduplicate, convolute, or flat, abaxial surfaces glabrous, adaxial surfaces with stiff hairs, veins 7–13, ribs (6)7–13; abaxial sclerenchyma in strands opposite most of the veins, about as wide as the veins; adaxial sclerenchyma often present; pillars or girders sometimes present at the major veins. Inflorescences 6–12 cm, open, with 2–4 branches per node; branches stiffly divaricate, densely scabrous-ciliate on the angles, lower branches with 2+ spikelets; pedicels stiffly hairy. Spikelets 5.5–8 mm, with 2(3) florets. Glumes exceeded by the upper florets, lanceolate-acuminate, sparsely scabrous to puberulent; lower glumes 3.5–5 mm, distinctly shorter than the adjacent lemmas; upper glumes 5–7 mm; lemmas 5–7 mm, chartaceous, scabrous or puberulent, minutely bidentate, awned, awns 1.5–3 mm, subterminal; paleas about as long as or slightly longer than the lemmas, intercostal region scabrous or puberulent distally; anthers 1.5–2.5 mm; ovary apices pubescent. $2n = 28$.

Festuca dasyclada grows on rocky slopes in open forests and shrublands of western Colorado and central and southern Utah. For many years it was known only from the type collection. When the seeds are mature, the panicles break off the culms and are blown over the ground like a tumbleweed, shedding seeds as they travel. This and other unusual features, such as the divaricate branching pattern and hairy pedicels, prompted W.A. Weber to place it in the monotypic genus *Argillochloa* W.A. Weber.

14.02 LEUCOPOA Griseb.

Stephen J. Darbyshire

Plants perennial; unisexual. Culms 30–120 cm. Sheaths closed only at the base; auricles absent; ligules membranous; blades with sclerenchyma girders extending from the abaxial to adaxial surfaces. Inflorescences open or contracted panicles, usually erect to strongly ascending, not spikelike; branches glabrous, smooth or somewhat scabrous, at least some branches longer than 1 cm; pedicels sometimes longer than 3 mm, thinner than 1 mm. Spikelets pedicellate, somewhat dimorphic in unisexual plants, laterally compressed, with (2)3–5(6) florets; disarticulation above the glumes and beneath the florets. Glumes subequal to unequal, shorter than the adjacent lemmas, more or less equally wide, glabrous, sometimes scabrous, mostly hyaline and thinner than the lemmas, membranous adjacent to the midvein, unawned; lower glumes 1-veined; upper glumes 1–3-veined; calluses glabrous; lemmas membranous to chartaceous, smooth or scabrous, sometimes hirsute, 5-veined, veins converging distally, usually extending almost to the apices, apices entire, acute, usually unawned, sometimes awned, awns to 2 mm; paleas about equaling the lemmas, scabrous on the veins, scarious or membranous distally, veins terminating at the apex; lodicules 2, membranous; anthers 3; ovaries with glabrous or pubescent apices. Caryopses shorter than the lemmas, concealed at maturity, fusiform, usually adhering at least to the paleas; hila linear. $x = 7$. Name from the Greek *leuco*, 'white', and *poa*, 'grass'.

Leucopoa is a genus of about 10 species, most of which are Asian. One species is native to the *Flora* region. It is sometimes included in *Festuca*, but species of *Leucopoa* differ from those of *Festuca* in their dioecious habit, the hyaline glumes that are much thinner than the lemmas,

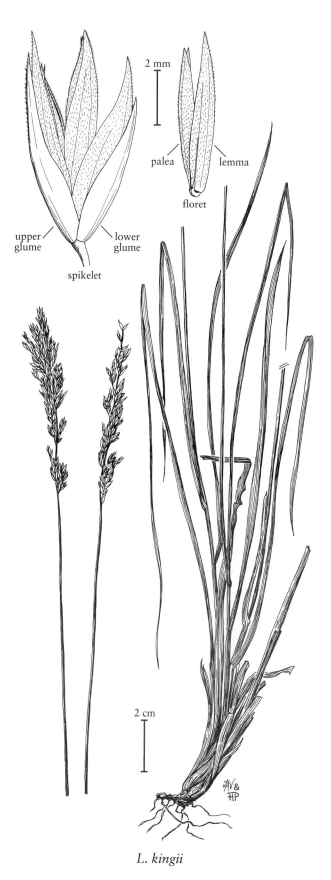

2 mm

palea   lemma

floret

upper   lower
glume   glume

spikelet

2 cm

L. kingii

LEUCOPOA

and their differing ovary and caryopsis morphology. Phylogenetic studies indicate that *Leucopoa* is more closely related to *Lolium* and *Schedonorus* than to *Festuca sensu stricto* (Darbyshire and Warwick 1992; Soreng and Davis 2000; Catalán et al. 2004).

SELECTED REFERENCES **Catalán, P., P. Torrecilla, J.Á. López Rodríguez,** and **R.G. Olmsted.** 2004. Phylogeny of the festucoid grasses of subtribe Loliinae and allies (Poeae, Poöideae) inferred from ITS and *trn*L–F sequences. Molec. Phylogenet. Evol. 31:517–541; **Darbyshire, S.J.** and **S.I. Warwick.** 1992. Phylogeny of North American *Festuca* (Poaceae) and related genera using chloroplast DNA restriction site variation. Canad. J. Bot. 70:2415–2429; **Soreng, R.J.** and **J.I. Davis.** 2000. Phylogenetic structure in Poaceae subfamily Poöideae as inferred from molecular and morphological characters: Misclassification versus reticulation. Pp. 61–74 *in* S.W.L. Jacobs and J. Everett (eds.). Grasses: Systematics and Evolution. CSIRO Publishing, Collingwood, Victoria, Australia. 408 pp.; **Swallen, J.R.** 1941. New species, names, and combinations of grasses. Proc. Biol. Soc. Wash. 54:43–46.

1. Leucopoa kingii (S. Watson) W.A. Weber [p. 444]
SPIKE FESCUE, SPIKEGRASS, WATSON'S FESCUE-GRASS, KING'S FESCUE

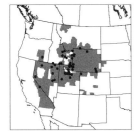

Plants unisexual; cespitose, rhizomatous (rhizomes usually absent from herbarium specimens). **Culms** 30–100(120) cm. **Sheaths** persisting, closed only at the base, glabrous, smooth or the lowest sometimes retrorsely scabrous; **ligules** 0.8–2(4) mm, truncate, erose-ciliate; **blades** 14–40(50) cm long, 1.5–7(10) mm wide, blades of the flag leaves often shorter, erect, somewhat stiff, flat or loosely convolute, glabrous, sometimes glaucous, abaxial surfaces usually smooth, rarely scabrous, adaxial surfaces sometimes somewhat scabrous or hirsute. **Panicles** 7–22 cm; **branches** erect or sometimes spreading, smooth or somewhat scabrous, spikelet-bearing to the base. **Spikelets** 6–12 mm long, 2.5–4.5 mm wide, staminate spikelets tending to be larger than pistillate spikelets, with (2)3–4(6) florets. **Glumes** usually unequal, sometimes subequal (especially in staminate spikelets) ovate or ovate-lanceolate (especially in pistillate spikelets) to lanceolate (especially in staminate spikelets), smooth or somewhat scabrous; **lower glumes** 3–5.5(6.5) mm, 1-veined; **upper glumes** 4–6.5(7.5) mm, shorter than or, particularly in staminate spikelets, equal to the lowest lemma, 1–3-veined; **calluses** wider than long, blunt; **lemmas** 4.5–8(10) mm, scabrous or hirsute, apices entire, acute, unawned or subterminally mucronate, mucro shorter than 1 mm; **paleas** 4.4–7(9) mm, scabrous or hirsute between the veins and on the margins; **lodicules** 1–2.6 mm; **anthers** (2.5)3.5–5(6) mm in staminate plants, vestigial and not functional in pistillate plants; **ovaries** in pistillate plants with pubescent apices and adjacent style bases, ovaries in staminate plants vestigial and non-functional. **Caryopses** fusiform, 3.5–5 mm long, 1.3–1.7 mm wide. $2n = 56$.

Leucopoa kingii grows from Oregon and Montana to Nebraska, south to southern California and northern New Mexico. It occurs in habitats from dry sagebrush plains to subalpine meadows, at 1700–3600 m. Although palatable to livestock in the early part of the season, *L. kingii* is only occasionally abundant enough to be an important forage species.

14.03 SCHEDONORUS P. Beauv.

Stephen J. Darbyshire

Plants perennial; cespitose, sometimes rhizomatous. **Culms** to 2 m, slender to stout, erect to decumbent. **Sheaths** open, rounded, smooth or scabrous; **auricles** present, usually falcate and clasping, sometimes an undulating flange; **ligules** membranous, glabrous; **blades** flat, linear. **Inflorescences** terminal panicles, erect, not spikelike; **branches** glabrous, smooth or scabrous, most branches longer than 1 cm; **pedicels** sometimes longer than 3 mm, thinner than 1 mm. **Spikelets** pedicellate, laterally compressed, with 2–22 florets; **disarticulation** above the glumes and between the florets. **Glumes** 2, shorter than the adjacent lemmas, more or less equally wide, lanceolate to oblong, rounded on the back, membranous, 3–9-veined, apices acute, unawned; **calluses** glabrous or sparsely hairy; **lemmas** lanceolate, ovate or oblong, rounded on the back, membranous, chartaceous, 3–7-veined, apices acute, sometimes hyaline, unawned or awned, awns to 18 mm, terminal or subterminal, straight; **paleas** narrower than the lemmas, membranous, usually smooth, keels ciliolate, veins terminating at or beyond midlength; **lodicules** 2, lanceolate to ovate; **anthers** 3; **ovaries** glabrous. **Caryopses** shorter than the lemmas, concealed at maturity, dorsally compressed, oblong, broadly elliptic, or ovate, longitudinally sulcate, adherent to the paleas; **hila** linear; **embryos** $1/5$–$1/3$ as long as the caryopses. $x = 7$. Name from the Greek *schedon*, 'near' or 'almost', and *oros*, 'mountain' or 'summit'.

Three species of the Eurasian genus *Schedonorus* are established in North America, having been widely introduced as forage and ornamental grasses.

Schedonorus has traditionally been included in *Festuca*, despite all the evidence pointing to its close relationship to *Lolium*. This evidence includes morphological features, such as the falcate leaf auricles, flat, relatively wide leaf blades, glabrous ovaries, subterminal stylar attachment, and adhesion of the mature caryopses to the paleas, none of which are found in *Festuca sensu stricto*. Fertile, natural hybrids between species of *Schedonorus* and those of *Lolium* are common in Europe, and several artificial hybrids have been registered for commercial use, primarily as forage grasses. *Schedonorus* and *Lolium* could appropriately be treated as congeneric subgenera (e.g., Darbyshire 1993). The two are treated as separate genera here for consistency with the treatments by Soreng and Terrell (1997), Holub (1998), and Edgar and Connor (2000).

SELECTED REFERENCES Aiken, S.G., M.J. Dallwitz, C.L. McJannet, and L.L. Consaul. 1997. Biodiversity among *Festuca* (Poaceae) in North America: Diagnostic evidence from DELTA and clustering programs, and an INTKEY package for interactive, illustrated identification and information retrieval. Canad. J. Bot. 75:1527–1555; Charmet, G., C. Ravel, and F. Balfourier. 1997. Phylogenetic analysis in the *Festuca–Lolium* complex using molecular markers and ITS rDNA. Theor. Appl. Genet. 94:1038–1046; Darbyshire, S.J. 1993. Realignment of *Festuca* subgenus *Schedonorus* with the genus *Lolium*. Novon 3:239–243; Dubé, M. 1983. Addition de *Festuca gigantea* (L.) Vill. (Poaceae) à la flore du Canada. Naturaliste Canad. 110:213–215; Edgar, E. and H.E. Connor. 2000. Flora of New Zealand, vol. 5. Manaaki Whenua Press, Lincoln, New Zealand. 650 pp.; Holub, J. 1998. Reclassifications and new names in vascular plants 1. Preslia 70:97–122; Jauhar, P.P. 1993. Cytogenetics of the *Festuca–Lolium* Complex. Monographs on Theoretical and Applied Genetics No. 18. Springer-Verlag, Berlin, Germany. 255 pp.; Kiang, A.-S., V. Connolly, D.J. McConnell, and T.A. Kavanagh. 1994. Paternal inheritance of mitochondria and chloroplasts in *Festuca pratensis–Lolium perenne* intergeneric hybrids. Theor. Appl. Genet. 87:681–688; Nihsen, M.E., E.L. Piper, C.P. West, R.J. Crawford, Jr., T.M. Denard, Z.B. Johnson, C.A. Roberts, D.A. Spiers, and C.F. Rosenkrans, Jr. 2004. Growth rate and physiology of steers grazing tall fescue inoculated with novel endophytes. J. Animal Sci. 82:878–883; Soreng, R.J. and E.E. Terrell. 1997 [publication date 1998]. Taxonomic notes on *Schedonorus*, a segregate genus from *Festuca* or *Lolium*, with a new nothogenus, ×*Schedololium*, and new combinations. Phytologia 83:85–88; Soreng, R.J., E.E. Terrell, J. Wiersema, and S.J. Darbyshire. 2001. Proposal to conserve the name *Schedonorus arundinaceus* (Schreb.) Dumort. against *Schedonorus arundinaceus* Roem. & Schult. (Poaceae: Poeae). Taxon 50:915–917.

1. Lemma awns 10–18 mm long, longer than the lemmas . . . . 2. *S. giganteus*
1. Lemmas unawned or the awns shorter than 4 mm, shorter than the lemmas.
  2. Auricles glabrous; panicle branches at the lowest node 1 or 2, if paired the shorter with 1–2(3) spikelets, the longer with 2–6(9) spikelets; lemmas usually smooth, sometimes slightly scabrous distally, unawned or with a mucro to 0.2 mm long . . . . 1. *S. pratensis*
  2. Auricles ciliate, having at least 1 or 2 hairs along the margins (check several leaves); panicle branches at the lowest node usually paired, the shorter with 1–13 spikelets, the longer with 3–19 spikelets; lemmas usually scabrous or hispidulous, at least distally, rarely smooth, unawned or with an awn up to 4 mm long . . . . 3. *S. arundinaceus*

1. **Schedonorus pratensis** (Huds.) P. Beauv. [p. 447]
MEADOW FESCUE, FÉTUQUE DES PRÉS

Plants perennial. **Culms** to 1.3 m. **Leaves** folded or convolute in young shoots; **auricles** glabrous; **ligules** to 0.5 mm; **blades** 10–25 cm long, 2–7 mm wide. **Panicles** (6)10–25 cm; **branches** at the lowest node 1 or 2, shorter branch with 1–2(3) spikelets, longer branch with 2–6(9) spikelets. **Spikelets** (8.5) 12–15.5(17) mm long, 2–5 mm wide, with (2)4–10(12) florets. **Lower glumes** (2)2.6–4.5 mm; **upper glumes** 3–5 mm; **lemmas** 5–8 mm, usually smooth, sometimes slightly scabrous distally, apices unawned, sometimes mucronate, mucros to 0.2 mm; **paleas** slightly shorter than the lemmas; **anthers** (1.5)2–4.6 mm. **Caryopses** 3–4 mm long, 1–1.5 mm wide. $2n = 14$.

Schedonorus pratensis is a Eurasian species that is now widely established in the *Flora* region. It used to be a popular forage grass in the contiguous United States and southern Canada, but is now rarely planted.

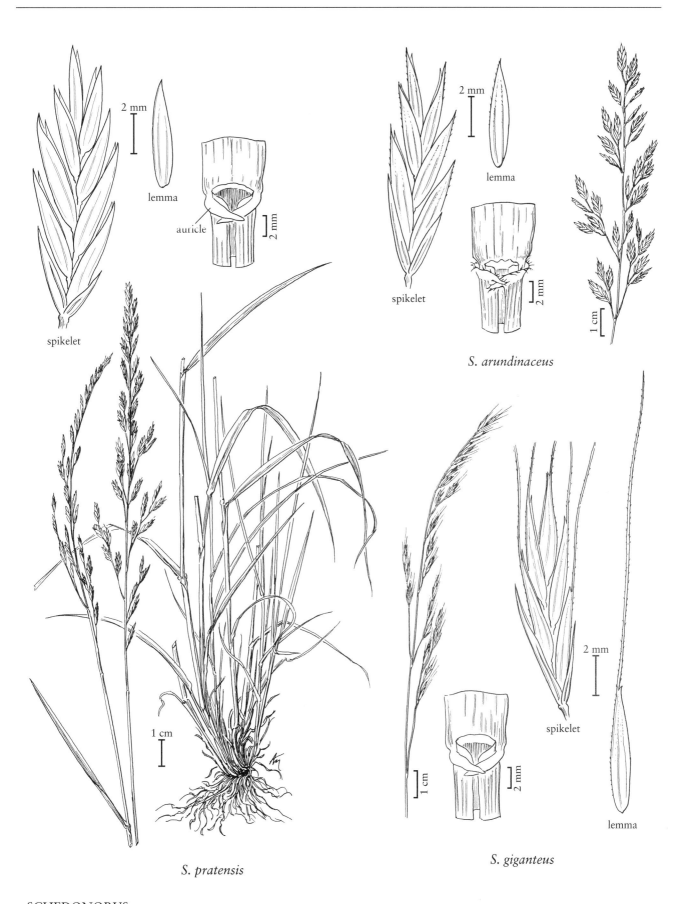

spikelet

lemma

auricle

2 mm

2 mm

spikelet

lemma

2 mm

2 mm

1 cm

S. arundinaceus

1 cm

S. pratensis

2 mm

spikelet

lemma

1 cm

2 mm

S. giganteus

SCHEDONORUS

2. **Schedonorus giganteus** (L.) Holub [p. 447]
GIANT FESCUE

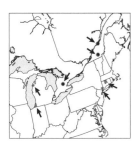

Plants perennial. **Culms** to 1.5 m. **Leaves** convolute in young shoots; **auricles** glabrous; **ligules** 0.5–2.5 mm; **blades** (10)20–40 cm long, 4–18 mm wide. **Panicles** 8–50 cm; **branches** usually 2 per node. **Spikelets** 8–13(20) mm long, 1.5–2.5 mm wide, with 3–10 florets. **Lower glumes** 4–7 mm; **upper glumes** 5–8 mm; **lemmas** 6–9 mm, usually scabrous or hispidulous, rarely smooth, awns 10–18 mm; **paleas** as long as to slightly longer than the lemmas; **anthers** 2.5–3 mm. **Caryopses** 3–4.6 mm long, 1–1.5 mm wide. 2*n* = 42.

Schedonorus giganteus is adventive from Europe. It is cultivated as an ornamental, and has escaped to woodland openings and edges and to shaded ravines, at isolated localities in Quebec, Ontario, Michigan, New York, and Connecticut.

3. **Schedonorus arundinaceus** (Schreb.) Dumort. [p. 447]
TALL FESCUE, FÉTUQUE ÉLEVÉE

Plants perennial, sometimes rhizomatous. **Culms** to 1.5(2) m. **Leaves** convolute in young shoots; **auricles** ciliate, having at least 1 or 2 hairs along the margins; **ligules** 1(2) mm; **blades** 11–30 cm long, 4–12 mm wide. **Panicles** 10–35 cm; **branches** at the lowest node usually 2, shorter branch with (1)2–9(13) spikelets, longer branch with (3)4–13(19) spikelets. **Spikelets** 8–15.5 mm long, 2–3.5 mm wide, with 3–6(9) florets. **Lower glumes** 3–6 mm; **upper glumes** 4.5–7(9) mm; **lemmas** (4)5–9(11.5) mm, usually scabrous or hispidulous, at least distally, rarely smooth, awns absent or to 4 mm, terminal or attached up to 0.4 mm below the apices; **paleas** slightly shorter than to slightly longer than the lemmas; **anthers** 2.5–4 mm. **Caryopses** 2–4 mm long, 0.9–1.6 mm wide. 2*n* = 28, 42, 56, 63, 70.

Schedonorus arundinaceus is a Eurasian species that has been introduced to the *Flora* region. It is grown for forage, soil stabilization, and coarse turf. It is now cultivated in all but the coldest and most arid parts of North America, and often escapes. It is frequently infected with the endophytic fungi *Neotyphodium coenophialum*, which confers insect and drought resistance to the plant, among other benefits; it also produces ergot alkaloids that are toxic to livestock. Varieties with endophyte strains that do not produce toxic ergot alkaloids have been developed (Nihsen et al. 2004).

14.04 VULPIA C.C. Gmel.

Robert I. Lonard

Plants usually annual, rarely perennial. **Culms** 5–90 cm, erect or ascending from a decumbent base, usually glabrous. **Sheaths** open, usually glabrous; **auricles** absent; **ligules** usually shorter than 1 mm, membranous, usually truncate, ciliate; **blades** flat or rolled, glabrous or pubescent. **Inflorescences** panicles or racemes, sometimes spikelike, usually with more than 1 spikelet associated with each node; **branches** 1–3 per node, appressed or spreading, usually glabrous, scabrous. **Spikelets** pedicellate, laterally compressed, with 1–11(17) florets, distal florets reduced; **disarticulation** above the glumes and beneath the florets, occasionally also at the base of the pedicels. **Glumes** shorter than the adjacent lemmas, subulate to lanceolate, apices acute to acuminate, unawned or awn-tipped; **lower glumes** much shorter than the upper glumes, 1-veined; **upper glumes** 3-veined; **rachillas** terminating in a reduced floret; **calluses** blunt, glabrous; **lemmas** membranous, lanceolate, 3–5-veined, veins converging distally, margins involute over the edges of the caryopses, apices entire, acute to acuminate, mucronate or awned;

paleas usually slightly shorter than to equaling the lemmas, sometimes longer; **anthers** usually 1, rarely 3 in chasmogamous specimens. **Caryopses** shorter than the lemmas, concealed at maturity, elongate, dorsally compressed, curved in cross section, falling with the lemma and palea. $x = 7$. Named for J.S. Vulpius, who studied the flora of Baden, Germany.

Vulpia, a genus of 30 species, is most abundant in Europe and the Mediterranean region (Cotton and Stace 1967). The *Flora* region has three native and three introduced species. Most species, including ours, are weedy, cleistogamous annuals, usually having one anther per floret. *Festuca*, in which *Vulpia* is sometimes included, consists of chasmogamous species having three anthers per floret. The two genera are closely related to each other. Sterile hybrids between *Vulpia* and *Festuca*, and *Vulpia* and *Lolium*, are known.

In the key and descriptions, the spikelet and lemma measurements exclude the awns.

SELECTED REFERENCES **Cotton, R.** and **C.A. Stace**. 1967. Taxonomy of the genus *Vulpia* (Gramineae): I. Chromosome numbers and geographical distribution of the Old World species. Genetica 46:235–255; **Lonard, R.I.** and **F.W. Gould**. 1974. The North American species of *Vulpia* (Gramineae). Madroño 22:217–230; **Stace, C.A.** 1975. Wild hybrids in the British flora. Pp. 111–125 *in* S.M. Walters (ed.). European Floristic and Taxonomic Studies. E.W. Classey, Faringdon, England. 144 pp.

1. Lower glumes less than $^1/_2$ the length of the upper glumes.
   2. Lemmas 5-veined, glabrous except the margins sometimes ciliate; rachilla internodes 0.75–1.9 mm long . . . . 1. *V. myuros*
   2. Lemmas 3(5)-veined, pubescent or glabrous, the margins ciliate; rachilla internodes 0.4–0.9 mm long . . . . 6. *V. ciliata*
1. Lower glumes $^1/_2$ or more the length of the upper glumes.
   3. Lemmas 2.5–3.5 mm long, the apices more pubescent than the bases; caryopses 1.5–2.5 mm long . . . . 2. *V. sciurea*
   3. Lemmas 2.7–9.5 mm long, if pubescent, the apices no more so than the bases but occasionally ciliate; caryopses 1.7–6.5 mm long.
      4. Panicle branches 1–2 per node; spikelets with 4–17 florets; rachilla internodes 0.5–0.7 mm long; awn of the lowermost lemma in each spikelet 0.3–9 mm long; caryopses 1.7–3.7 mm long . . . . 3. *V. octoflora*
      4. Panicle branches solitary; spikelets with 1–8 florets; rachilla internodes 0.6–1.2 mm long; awn of the lowermost lemma in each spikelet 2–20 mm long; caryopses 3.5–6.5 mm long.
         5. Panicle branches appressed to erect at maturity, without axillary pulvini; paleas equal to or shorter than the lemmas . . . . 4. *V. bromoides*
         5. Panicle branches spreading to reflexed at maturity, with axillary pulvini; paleas usually slightly longer than the lemmas . . . . 5. *V. microstachys*

1. **Vulpia myuros** (L.) C.C. Gmel. [p. 451]
FOXTAIL FESCUE, RATTAIL FESCUE

Culms 10–75(90) cm, solitary or loosely tufted, branched or unbranched distally. **Sheaths** usually glabrous; **ligules** 0.3–0.5 mm; **blades** 2.4–10.5(17) cm long, 0.4–3 mm wide, usually rolled, occasionally flat, usually glabrous. **Inflorescences** 3–25 cm long, 0.5–1.5(2) cm wide, dense panicles or spikelike racemes, with 1 branch per node, often partially enclosed in the uppermost sheaths at maturity, pulvini absent; **branches** spreading or appressed to erect. **Spikelets** 5–12 mm, with 3–7 florets; **rachilla internodes** 0.75–1(1.9) mm. **Glumes** glabrous; **lower glumes** 0.5–2 mm, $^1/_5$–$^1/_2$ the length of the upper glumes; **upper glumes** 2.5–5.5 mm; **lemmas** 4.5–7 mm, 5-veined, usually scabrous distally, glabrous except the margins sometimes ciliate, apices entire, awns 5–15(22) mm; **paleas** 4.7–6.4 mm, minutely bifid; **anthers** 0.5–1(2) mm. **Caryopses** 3–5 mm, fusiform, glabrous. $2n = 14$ [f. *myuros*], 42 [f. *myuros* and f. *megalura*].

Vulpia myuros grows in well-drained, sandy soils and disturbed sites. It is native to Europe and North Africa. **Vulpia myuros** f. **megalura** (Nutt.) Stace & R. Cotton differs from **Vulpia myuros** (L.) C.C. Gmel. f. **myuros** in having ciliate lemma margins. It was once thought to be native to North America, but it occurs throughout the European and North African range of f. *myuros*, even in undisturbed areas.

2. Vulpia sciurea (Nutt.) Henrard [p. 451]
SQUIRRELTAIL FESCUE

Culms 15–50(60) cm, solitary or tufted, erect or drooping at maturity. **Sheaths** glabrous; **ligules** 0.5–1 mm; **blades** usually shorter than 10 cm, 0.5–1 mm wide, flat or rolled, glabrous. **Panicles** 5–20 cm long, 0.5–1 cm wide, with 1–2 branches per node; **branches** appressed to erect. **Spikelets** 3.5–5.2 mm, with 3–6 florets; **rachilla internodes** 0.25–0.9 mm. **Lower glumes** 1.5–2.5 mm, about ²/₃ the length of the upper glumes; **upper glumes** 2.5–4 mm; **lemmas** 2.5–3.5 mm, 3-veined, evidently pubescent distally, glabrous or sparsely pubescent proximally, awns 4.5–10 mm; **paleas** subequal or equal to the lemmas; **anthers** about 0.5 mm. **Caryopses** 1.5–2.5 mm. $2n = 42$.

Vulpia sciurea, our most distinctive native species, is restricted to the *Flora* region. It can be recognized by its small spikelets and apically pubescent lemmas, and grows mostly in deep, sandy soils of open woodlands, old fields, roadside ditches, and sand hills in the southeastern *Flora* region. It is listed as endangered in New Jersey.

3. Vulpia octoflora (Walter) Rydb. [p. 451]
SIXWEEKS FESCUE

Culms 5–60 cm, solitary or loosely tufted, glabrous or pubescent. **Sheaths** glabrous or pubescent; **ligules** 0.3–1 mm; **blades** to 10 cm long, 0.5–1 mm wide, flat or rolled, glabrous or pubescent. **Panicles** 1–7(20) cm long, 0.5–1.5 cm wide, with 1–2 branches per node; **branches** appressed to spreading. **Spikelets** 4–10(13) mm, with (4)5–11(17) florets; **rachilla internodes** 0.5–0.7 mm. **Lower glumes** 1.7–4.5 mm, ¹/₂–²/₃ the length of the upper glumes; **upper glumes** 2.5–7.2 mm; **lemmas** 2.7–6.5 mm, 5-veined, smooth, scabrous, or pubescent, apices entire, no more pubescent than the bases, awns of the lowermost lemma in each spikelet 0.3–9 mm; **paleas** slightly shorter than the lemmas, apices entire or minutely bifid, teeth shorter than 0.2 mm; **anthers** 0.3–1.5 mm. **Caryopses** 1.7–3.7 mm. $2n = 14$.

Vulpia octoflora, a widespread native species, tends to be displaced by the introduced *Bromus tectorum* in the Pacific Northwest. It grows in grasslands, sagebrush, and open woodlands, as well as in disturbed habitats and areas of secondary succession, such as old fields, roadsides, and ditches. Three varieties are recognized here, but their characterization is not completely satisfactory, e.g., plants of the southwestern United States with spikelets in the size range of var. *glauca* often have densely pubescent lemmas, the distinguishing characteristic of var. *hirtella*.

1. Spikelets usually 4–6.5 mm long; awn of the lowermost lemma in each spikelet 0.3–3 mm long . . . . var. *glauca*
1. Spikelets usually 5.5–13 mm long; awn of the lowermost lemma in each spikelet 2.5–9 mm long.
　　2. Lemmas scabrous to pubescent . . . . var. *hirtella*
　　2. Lemmas usually smooth, sometimes scabridulous distally and on the margins . . . . var. *octoflora*

Vulpia octoflora var. glauca (Nutt.) Fernald [p. 451]

Panicle branches usually appressed, infrequently spreading distally, spikelets closely arranged. **Spikelets** usually 4–6.5 mm, subsessile or short-pedicellate. **Lemmas** glabrous or scabrous; **awns** of the lowermost lemma in each spikelet 0.3–3 mm.

Vulpia octoflora var. *glauca* is most frequent in southern Canada and the northern half of the United States, and is the most common representative of *V. octoflora* from North Dakota to western Kansas, and east to Maine and Virginia.

Vulpia octoflora var. hirtella (Piper) Henrard [p. 451]

Panicle branches appressed, spikelets closely imbricate. **Spikelets** usually 5.5–10 mm. **Lemmas** prominently scabrous to densely pubescent; **awns** of the lowermost lemma in each spikelet 2.5–6.5 mm.

Vulpia octoflora var. *hirtella* is most frequent from British Columbia south through the western United States and into Mexico. It is the most common variey of *V. octoflora* in the southwest.

Vulpia octoflora (Walter) Rydb. var. octoflora [p. 451]

Panicle branches erect to ascending, lower branches sometimes spreading distally. **Spikelets** usually 5.5–10(13) mm, usually not or only slightly overlapping. **Lemmas** usually smooth, sometimes scabridulous distally and on the margins; **awns** of the lowermost lemma in each spikelet 3–9 mm.

Vulpia octoflora var. *octoflora* is widespread throughout southern Canada, the United States, and Mexico, and has been introduced into temperate regions of South America, Europe, and Asia. It is most common from northern Oklahoma to Virginia, south to the Texas Gulf prairie and Florida.

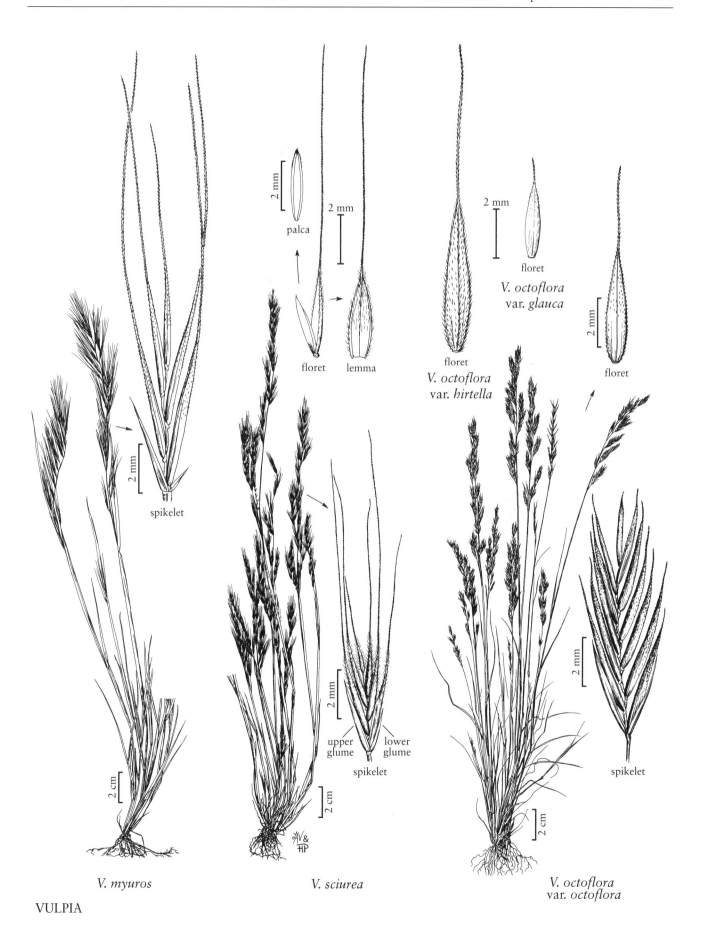

2 mm

palca

2 mm

floret   lemma

spikelet

2 mm

spikelet

upper
glume

lower
glume

spikelet

2 mm

floret

V. octoflora
var. *glauca*

floret

V. octoflora
var. *hirtella*

2 mm

floret

2 mm

spikelet

2 cm

2 cm

2 cm

V. myuros

V. sciurea

V. octoflora
var. *octoflora*

VULPIA

4. **Vulpia bromoides** (L.) Gray [p. 453]
BROME FESCUE

Culms 5–50 cm, solitary or loosely tufted, erect or decumbent, smooth, scabridulous, or puberulent, unbranched distally. **Sheaths** glabrous or puberulent; **ligules** to 0.5(1) mm; **blades** usually 2–10 cm long, 0.5–2.5 mm wide, rolled or flat, glabrous or puberulent. **Panicles** 1.5–15 cm long, 0.5–3 cm wide, conspicuously exserted, with 1 branch per node; **branches** usually appressed to erect at maturity, without axillary pulvini; **pedicels** flattened, sometimes clavate distally. **Spikelets** 5–10 mm, with 4–8 florets, not closely imbricate; **rachilla internodes** 0.6–1.1 mm. **Lower glumes** 3.5–5 mm, $^1/_2$–$^4/_5$ the length of the upper glumes; **upper glumes** 4.5–9.5 mm, midveins scabrous distally; **lemmas** 4–8 mm, 5-veined, scabrous distally, apices entire, awns of the lowermost lemma in each spikelet 2–13 mm; **paleas** 4–6.3 mm, equaling or shorter than the lemmas, minutely bifid; **anthers** 0.4–0.6(1.5) mm. **Caryopses** 3.5–5 mm. $2n = 14$.

Vulpia bromoides is a common European species that grows in wet to dry, open habitats. It is adventive and naturalized in North and South America. In North America, it is most common on the west coast, where it grows from British Columbia to northern Baja California; it occurs sparingly in other regions.

5. **Vulpia microstachys** (Nutt.) Munro [p. 453]
SMALL FESCUE

Culms 15–75 cm, solitary or loosely tufted, usually glabrous, occasionally puberulent. **Sheaths** glabrous or pubescent; **ligules** 0.5–1 mm; **blades** usually shorter than 10 cm, 0.5–1 mm wide, usually rolled, occasionally flat, glabrous or pubescent. **Inflorescences** 2–24 cm long, 0.8–8 cm wide, usually panicles, sometimes spikelike racemes; **branches** solitary, with axillary pulvini, appressed to erect when immature, spreading to reflexed at maturity. **Spikelets** 4–10 mm, with 1–6 florets, often purple-tinged; **rachilla internodes** 0.6–1.2 mm. **Glumes** smooth, scabrous, or pubescent; **lower glumes** 1.7–5.5 mm, $^1/_2$–$^3/_4$ the length of the upper glumes; **upper glumes** 3.5–7.5 mm; **lemmas** 3.5–9.5 mm, smooth, scabrous, or evenly pubescent, 5-veined, awns of the lowermost lemma in each spikelet (3)6–20 mm; **paleas** usually slightly longer than the lemmas, apices minutely bifid, teeth 0.2–0.5 mm; **anthers** 0.7–3 mm. **Caryopses** 3.5–6.5 mm. $2n = 42$.

Vulpia microstachys is native to western North America, growing from British Columbia south through the western United States into Baja California. Four varieties are recognized here on the basis of spikelet indumentum, but they frequently occur together, and intergrading forms are known. No difference in their geographic or ecological distribution is known.

1. Glumes and lemmas smooth or scabrous
. . . . var. *pauciflora*
1. Glumes and/or lemmas pubescent.
  2. Glumes and lemmas pubescent . . . . var. *ciliata*
  2. Glumes or lemmas, but not both, pubescent.
    3. Glumes pubescent; lemmas glabrous
. . . . var. *confusa*
    3. Glumes glabrous; lemmas pubescent
. . . . var. *microstachys*

Vulpia microstachys var. **ciliata** (A. Gray) Lonard & Gould [p. 453]
EASTWOOD FESCUE

Spikelets usually with 2–4 florets. **Glumes** and **lemmas** sparsely or densely pubescent.

Vulpia microstachys var. *ciliata* grows in loose, sandy soils.

Vulpia microstachys var. **confusa** (Piper) Lonard & Gould [p. 453]
CONFUSING FESCUE

Spikelets usually with 1–3 florets. **Glumes** pubescent; **lemmas** glabrous.

Vulpia microstachys var. *confusa* grows in sandy, open sites.

Vulpia microstachys (Nutt.) Munro var. **microstachys** [p. 453]
DESERT FESCUE

Spikelets with (1)2–5 florets. **Glumes** glabrous; **lemmas** sparsely to densely pubescent.

Vulpia microstachys var. *microstachys* grows most commonly in loose soil on open slopes and roadsides.

Vulpia microstachys var. **pauciflora** (Scribn. *ex* Beal) Lonard & Gould [p. 453]
PACIFIC FESCUE

Spikelets with 1–6 florets. **Glumes** and **lemmas** smooth or scabrous.

Vulpia microstachys var. *pauciflora* grows in sandy, often disturbed sites, and is the most common and widespread variety of the complex. It is often intermingled with plants of the other varieties.

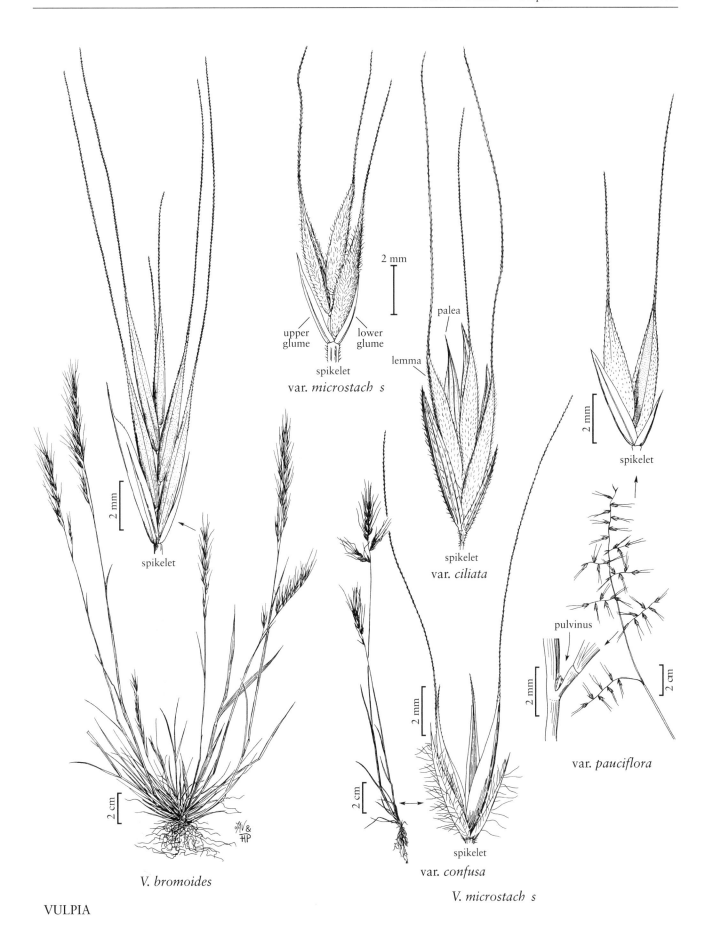

upper glume
lower glume
spikelet
var. *microstach s*

palea
lemma
spikelet
var. *ciliata*

2 mm

spikelet

spikelet

pulvinus

var. *pauciflora*

spikelet
var. *confusa*
V. *microstach s*

V. *bromoides*

6. **Vulpia ciliata** Dumort. [p. 454]

FRINGED FESCUE

Culms 6–45 cm, loosely tufted. **Sheaths** smooth, glabrous; **ligules** 0.2–0.5 mm; **blades** 3.5–10 cm long, about 0.4 mm wide, folded to involute, abaxial surfaces glabrous, adaxial surfaces puberulent. **Inflorescences** 3–20 cm long, 0.3–1.5 cm wide, panicles or spicate racemes, usually partially enclosed in the uppermost sheaths at maturity, with 1 branch per node, axillary pulvini absent. **Spikelets** 5–10.5 mm, with 4–10 florets; **rachilla internodes** 0.4–0.9 mm. **Glumes** glabrous; **lower glumes** 0.1–1.3 mm, less than ⅓ the length of the upper glumes; **upper glumes** 1.5–4 mm; **lemmas** 4–7.7 mm, 3(5)-veined, usually pubescent on the midvein, sometimes also on the body, rarely glabrous on both, margins ciliate, hairs to 1 mm, awns 6–15.3 mm; **paleas** slightly shorter than to equaling the lemmas, apices entire; **anthers** 0.4–0.6(1.6) mm. **Caryopses** 3.4–6.5 mm. $2n = 42$.

Vulpia ciliata is native to Europe, the Mediterranean area, and southwest and central Asia. It grows in open, dry habitats. It is easily distinguished from other members of the genus because of its upper glumes with broadly membranous tips that break off, making the glumes appear truncate or blunt. In the *Flora* region, it was known until recently only from an old ballast dump record from Philadelphia. In May 2004, it was collected immediately north of the Odgen Bay Waterfowl Management Area, Weber County, Utah, in an upland area of the site. The source of the seeds is not known.

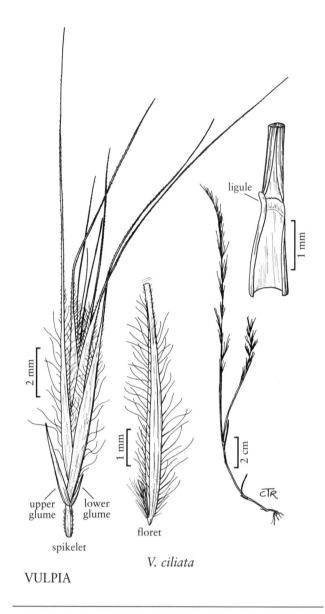

VULPIA   *V. ciliata*

14.05 LOLIUM L.

Edward E. Terrell

Plants annual or perennial; cespitose, sometimes shortly rhizomatous. **Culms** 10–150 cm, slender to stout, erect to decumbent, rarely prostrate. **Sheaths** open, rounded, glabrous, sometimes scabrous; **ligules** to 4 mm, membranous, glabrous; **auricles** sometimes present; **blades** flat, linear. **Inflorescences** distichous spikes, with solitary spikelets oriented radial to the rachises, perpendicular to the rachis concavities. **Spikelets** laterally compressed, with 2–22 florets, distal florets reduced; **rachillas** glabrous; **disarticulation** above the glumes, beneath the florets. **Glumes** usually 1, 2 in the terminal spikelets, lanceolate to oblong, rounded over the midvein, membranous to indurate, 3–9-veined, unawned; **lower glumes** absent from all but the terminal spikelet; **upper glumes** from shorter than to exceeding the distal florets; **calluses** short, blunt, glabrous; **lemmas** lanceolate, ovate or oblong, rounded over the midvein, membranous, chartaceous, 3–7-veined, apices sometimes hyaline, unawned or awned, awns subterminal, more or less straight; **paleas** membranous, usually smooth, keels ciliolate; **lodicules** 2, free, lanceolate to ovate; **anthers** 3; **ovaries** glabrous. **Caryopses** dorsally compressed, oblong, broadly elliptic or ovate, longitudinally sulcate; **hila** linear, in the furrow; **embryos** $^1/_5$–$^1/_3$ as long as the caryopses. $x = 7$. *Lolium*, first mentioned in Virgil's *Georgics*, is an old Latin name for darnel, *Lolium temulentum*.

As interpreted here, *Lolium* comprises five species that are native to Europe, temperate Asia, and northern Africa. All have been introduced to the *Flora* region, often as forage grasses; most have become established.

Lolium used to be included in the *Triticeae*, but evidence from genetics, morphology, and other studies shows its closest relationship to be to the species included here in *Schedonorus*. Artificial hybrids have been produced among *L. perenne*, *L. multiflorum*, *Schedonorus pratensis*, and *S. arundinaceus*. Cultivars of these crosses have been registered for commercial use and are sometimes used for forage. Natural hybrids are not uncommon in Europe.

SELECTED REFERENCES **Aiken, S.G., M.J. Dallwitz, C.L. McJannet**, and **L.L. Consaul**. 1997. Fescue Grasses of North America: Interactive Identification and Information Retrieval. DELTA, CSIRO Division of Entomology, Canberra, Australia. CD-ROM; **Dannhardt, G.** and **L. Steindl**. 1985. Alkaloids of *Lolium temulentum*: Isolation, identification and pharmacological activity. Pl. Med. (Stuttgart) 1985:212–214; **Dore, W.G.** 1950. Persian darnel in Canada. Sci. Agric. (Ottawa) 30:157–164; **Soreng, R.J.** and **E.E. Terrell**. 1997 [publication date 1998]. Taxonomic notes on *Schedonorus*, a segregate genus from *Festuca* or *Lolium*, with a new nothogenus, ×*Schedololium*, and new combinations. Phytologia 83:85–88; **Terrell, E.E.** 1968. A Taxonomic Revision of the Genus *Lolium*. Technical Bulletin, United States Department of Agriculture No. 1392. U.S. Government Printing Office, Washington, D.C., U.S.A. 65 pp.

1. Plants either long-lived perennials with 2–10 florets per spikelet, or annuals or short-lived perennials with 10–22 florets per spikelet.
  2. Plants long-lived perennials, with 2–10 florets per spikelet; lemmas unawned or awned, awns to about 8 mm long . . . . 1. *L. perenne*
  2. Plants annuals or short-lived perennials, with 10–22 florets per spikelet; lemmas usually awned, awns to 15 mm long, rarely unawned . . . . 2. *L. multiflorum*
1. Plants annuals, with 2–10(11) florets per spikelet.
  3. Spikelets somewhat sunken in the rachises and partly concealed by the glumes . . . . 3. *L. rigidum*
  3. Spikelets not sunken in the rachises and not concealed by the glumes.
    4. Lemmas 3.5–8.5 mm long; paleas from 1.2 mm shorter than to 0.8 mm longer than the lemmas; mature florets and caryopses 2–3 times longer than wide . . . . 4. *L. temulentum*
    4. Lemmas (5.2)7–12 mm long; paleas usually 0.5–1.8 mm longer than the lemmas; mature florets and caryopses 3.7–5 times longer than wide . . . . 5. *L. persicum*

1. Lolium perenne L. [p. 457]

PERENNIAL RYEGRASS, ENGLISH RYEGRASS, IVRAIE VIVACE, RAY-GRASS ANGLAIS

Plants long-lived perennials. **Culms** to 100 cm. **Leaves** folded in the bud; **blades** usually 10–30 cm long, (1)2–4(6) mm wide. **Spikes** 3–30 cm, with 5–37 spikelets; **rachises** 0.5–2.5 mm thick at the nodes, often flexuous. **Spikelets** 5–22 mm long, 1–7 mm wide, with (2)5–9(10) florets. **Glumes** 3.5–15 mm, ($^1/_3$)$^1/_2$–$^3/_4$ as long as to slightly exceeding the distal florets, membranous to indurate; **lemmas** 3.5–9 mm long, 0.8–2 mm wide, unawned or awned, awns to about 8 mm, attached 0.2–0.7 mm below the apices; **paleas** shorter than to slightly longer than the lemmas; **anthers** 2–4.2 mm. **Caryopses** 3–5.5 mm long, 0.7–1.5 mm wide, 3 or more times longer than wide. $2n = 14$.

Lolium perenne, a Eurasian species, is now established in disturbed areas throughout much of the *Flora* region. It is commercially important, being included in lawn seed mixtures as well as being used for forage and erosion prevention.

Lolium perenne intergrades and is interfertile with *L. multiflorum*; it also intergrades with *L. rigidum*. Typical *L. perenne* differs from *L. multiflorum* in being a shorter, longer-lived perennial with narrower leaves that are folded, rather than rolled, in the bud. Hybrids between the two species are called **Lolium ×hybridum** Hausskn.

2. Lolium multiflorum Lam. [p. 457]

ANNUAL RYEGRASS, ITALIAN RYEGRASS,
IVRAIE MULTIFLORE, RAY-GRASS D'ITALIE

Plants annuals or short-lived perennials. **Culms** to 150 cm. **Leaves** rolled in the bud; **blades** usually 10–30 cm long, (2)3–8 (13) mm wide. **Spikes** 15–45 cm, with 5–38 spikelets; **rachises** 0.8–2 mm thick at the nodes, not flexuous. **Spikelets** 8–31 mm long, 2–10 mm wide, with (10)11–22 florets. **Glumes** 5–18 mm, $^{1}/_{4}$–$^{1}/_{2}$ as long as the florets, membranous to indurate; **lemmas** 4–8.2 mm long, 1–2 mm wide, usually awned, awns to 15 mm, attached 0.2–0.7 mm below the apices, rarely unawned; **paleas** shorter than to slightly longer than the lemmas; **anthers** (2.5)3–4.5(5) mm. **Caryopses** 2.5–4 mm long, 0.7–1.5 mm wide, 3 or more times longer than wide. $2n = 14$.

Lolium multiflorum, a European species, now grows in most of the *Flora* region. It is planted as a cover crop, as a temporary lawn grass, for roadside restoration, and for soil or forage enrichment; it often escapes from cultivation, becoming established in disturbed sites.

Lolium multiflorum and *L. perenne* are interfertile and intergrade. *Lolium multiflorum* differs from *L. perenne* in being a taller, shorter-lived perennial or annual with wider leaves that are rolled, rather than folded, in the bud. Hybrids between the two species are called **Lolium ×hybridum** Hausskn. *Lolium multiflorum* also hybridizes with *L. rigidum*; those hybrids are called **Lolium ×hubbardii** Jansen & Wacht. *ex* B.K. Simon.

3. Lolium rigidum Gaudin [p. 457]

STIFF RYEGRASS

Plants annual. **Culms** to 70 cm. **Blades** to 17 cm long, 0.5–5(8) mm wide. **Spikes** 3–30 cm, with 2–20 spikelets; **rachises** 0.5–3.5 mm thick at the nodes, with the spikelets somewhat sunken in the rachises and partly concealed by the glumes. **Spikelets** 5–18 mm long, 1–3(7) mm wide, with 2–8(11) florets. **Glumes** 4–20(30) mm, usually from $^{3}/_{4}$ as long as to slightly exceeding the distal florets, rather indurate; **lemmas** 3–8.5(10.5) mm long, 0.9–2 mm wide; **paleas** slightly shorter than to slightly longer than the lemmas, usually unawned, sometimes awned, awns to 10 mm; **anthers** 1.2–3.2 mm. **Caryopses** 2.7–5.5 mm long, 1–1.5 mm wide, 3 or more times longer than wide. $2n = 14$.

Lolium rigidum is native to Europe, North Africa, and western Asia. It has been found as a weed of roadsides and waste places at scattered locations in the contiguous United States and Canada.

Lolium rigidum intergrades with *L. perenne, L. multiflorum*, and, occasionally, *L. temulentum*. Hybrids with *L. multiflorum* are called **Lolium ×hubbardii** Jansen & Wacht. *ex* B.K. Simon.

4. Lolium temulentum L. [p. 458]

Plants annual. **Culms** to 120 cm. **Blades** to 27 cm long, 1–12 mm wide. **Spikes** 2–40 cm, with 3–26 spikelets; **rachises** 0.5–3.5 mm thick at the nodes, spikelets not sunken in the rachises, not concealed by the glumes. **Spikelets** 5–28 mm long, 1–8 mm wide, with 2–10 florets. **Glumes** 5–28 mm, membranous to indurate; **lemmas** 3.5–8.5 mm long, 1.2–3 mm wide, unawned or awned, awns to 23 mm, attached 0.2–2 mm below the apices; **paleas** 1.2 mm shorter than to 0.8 mm longer than the lemmas, often wrinkled; **anthers** 1.5–4 mm. **Caryopses** 3.2–7 mm long, 1–3 mm wide, 2–3 times longer than wide, turgid. $2n = 14$.

Lolium temulentum is said to be the tares of the Bible. Its two subspecies differ mainly in quantitative characters.

1. Lemmas 3.5–5.5 mm long, 1.2–1.8 mm wide; glumes 5–16 mm long; caryopses 3.2–4.5 mm long, 1.2–1.8 mm wide; rachises slender . . . . subsp. *remotum*
1. Lemmas 4.5–8.5 mm long, 1.5–3 mm wide; glumes (5.5)7–28 mm long; caryopses (3.8)4–7 mm long, (1)1.5–3 mm wide; rachises rather stout . . . . subsp. *temulentum*

Lolium temulentum subsp. remotum (Schrank) Á. Löve & D. Löve [p. 458]

FLAX DARNEL, IVRAIE DU LIN

Blades 1–6.5 mm wide. **Spikes** 2–23 cm, with 3–20 spikelets; **rachises** slender. **Spikelets** 5–16 mm long, 1–5 mm wide. **Glumes** 5–16 mm, $(^{1}/_{2})^{2}/_{3}$ as long as to somewhat exceeding the distal florets; **lemmas** 3.5–5.5 mm long, 1.2–1.8 mm wide, usually unawned, rarely awned, awns to 10 mm, attached 0.2–1 mm below the apices. **Caryopses** 3.2–4.5 mm long, 1.2–1.8 mm wide.

Lolium temulentum subsp. *remotum* is native to Europe, Asia, and northern Africa. It originated as a weed in flax fields, through unintentional selection for seeds that could not be separated from flax seed using early harvesting techniques. It is a rare weed in the *Flora* region, being reported only from southern Ontario and California, where it grows in waste places and fields.

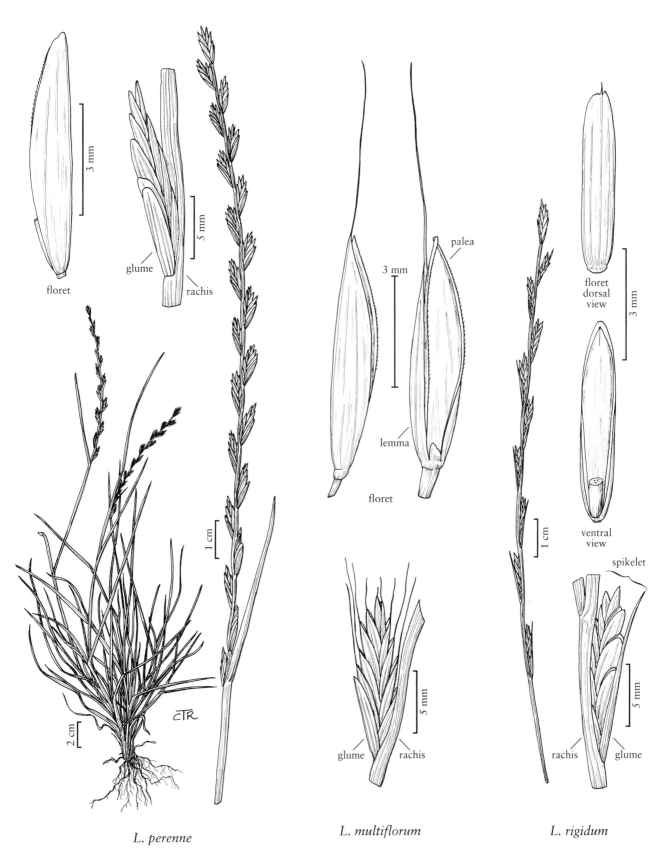

floret

glume
rachis

floret

palea

3 mm

lemma

floret

glume   rachis

L. perenne

L. multiflorum

floret
dorsal
view

3 mm

ventral
view

spikelet

rachis   glume

L. rigidum

LOLIUM

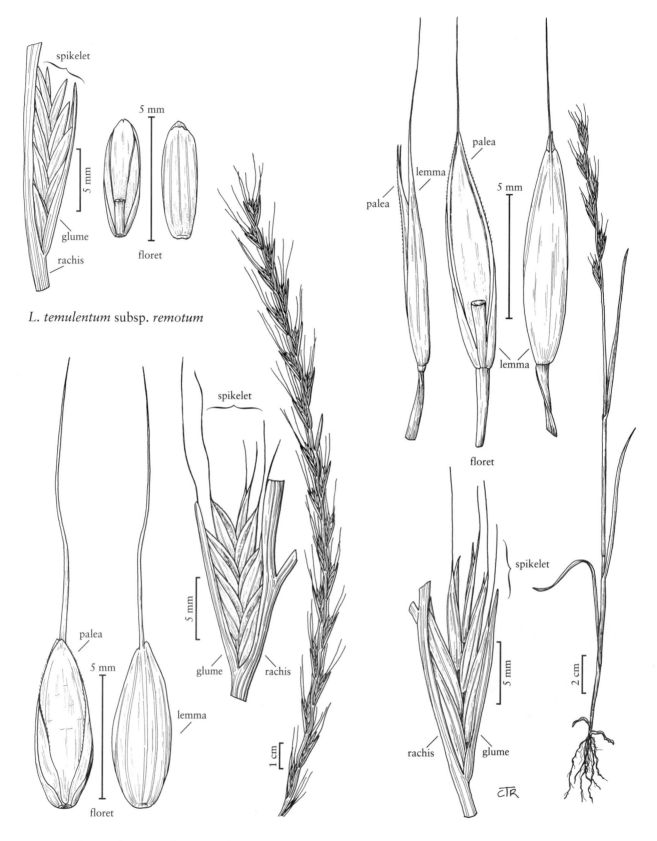

L. temulentum subsp. *remotum*

L. temulentum subsp. *temulentum*

L. persicum

Lolium temulentum L. subsp. **temulentum** L. [p. 458]
DARNEL, IVRAIE ENIVRANTE

Blades (1.5)3–10(12) mm wide. **Spikes** 5–40 cm, with 5–26 spikelets; **rachises** rather stout. **Spikelets** 8–28 mm long, 3–8 mm wide. **Glumes** (5.5)7–28 mm, from $^3/_4$ as long as to longer than the florets, somewhat indurate; **lemmas** 4.5–8.5 mm long, 1.5–3 mm wide, unawned or awned, awns to 23 mm, attached 0.5–2 mm below the apices. **Caryopses** (3.8)4–7 mm long, (1)1.5–3 mm wide.

Lolium temulentum subsp. *temulentum* is found occasionally in disturbed sites throughout much of the *Flora* region. It is native to the Eastern Hemisphere, where it is known only as a weed, especially of grain fields. Awn presence or absence and length vary, and have no taxonomic significance.

The seeds sometimes become infected with an endophytic fungus, assumed to be the source of the toxic pyrrolizidine alkaloids loline, 6-methyl loline, and lolinine, but not temuline, which is now considered an artifact of isolation (Dannhardt and Steindl 1985). Because primitive agricultural practices could not separate seeds of *Lolium temulentum* from those of wheat, infected seeds often resulted in poisonous flour.

5. Lolium persicum Boiss. & Hohen. [p. 458]
PERSIAN DARNEL

Plants annual. **Culms** 14–45(60) cm. **Blades** 3.5–20 cm long, 1.5–7 mm wide. **Spikes** 3–21 cm, with 3–12 spikelets; **rachises** 0.5–2 mm thick at the nodes, often flexuous, spikelets not sunken in the rachises, not concealed by the glumes. **Spikelets** 9–27 mm long, 1.5–7 mm wide, with 4–9 florets. **Glumes** (4.7)7.5–23 mm, from $^2/_3$ as long as to equaling the distal florets, somewhat indurate; **lemmas** (5.2)7–12 mm long, 1.5–2.7 mm wide, awns (1.5)5–18 mm, attached 0.2–1 mm below the apices; **paleas** usually 0.5–1.8 mm longer than the lemmas; **anthers** 1.5–3 mm. **Caryopses** 4.8–7 mm long, 1.2–2 mm wide, 3.7–5 times longer than wide. $2n = 14$.

Lolium persicum, a native of southwest Asia, has been found as a weed in grain fields and waste places in southern Canada, Montana, North Dakota, and Wyoming and, as an adventive, in New York and Missouri. It is now one of the top ten weeds of western Canadian cereal crops. It first became established in North America in Cavalier County, North Dakota, prior to 1910 (Dore 1950).

14.06 **PUCCINELLIA** Parl.

Jerrold I. Davis

Laurie L. Consaul

Plants annual, biennial, or perennial; usually cespitose, sometimes weakly or strongly stoloniferous and mat-forming. **Culms** 2–100 cm, erect or decumbent, sometimes geniculate; **internodes** hollow. **Sheaths** open to the base or nearly so; **auricles** absent; **ligules** membranous, acute to truncate, entire or erose; **blades** flat, folded, or involute. **Inflorescences** terminal panicles, open to contracted; **branches** smooth or scabrous, some branches longer than 1 cm; **pedicels** usually longer than 3 mm, thinner than 0.5 mm. **Spikelets** pedicellate, subterete to weakly laterally compressed, with 2–10 florets; **disarticulation** above the glumes, beneath the florets. **Glumes** usually unequal, sometimes subequal to equal, usually distinctly shorter than the lowest lemma in the spikelets, sometimes only slightly shorter, rarely longer, membranous, rounded or weakly keeled, veins obscure or prominent, apices unawned; **lower glumes** 1(3)-veined; **upper glumes** (1)3(5)-veined; **calluses** blunt, glabrous or pubescent; **lemmas** membranous to slightly or distinctly coriaceous, glabrous or pubescent, pubescence sometimes restricted to the bases of the veins, rounded or weakly keeled, at least distally, (3)5(7)-veined, veins obscure to prominent, more or less parallel distally, usually not extending to the apices, lateral veins sometimes reduced, apical margins with or without scabrules, apices usually acute to truncate, sometimes acuminate, entire or serrate to erose, unawned; **paleas** subequal to the lemmas, scarious or membranous distally, 2-veined, veins terminating at or beyond midlength; **lodicules** 2, free, glabrous; **anthers** 3; **ovaries** glabrous. **Caryopses** shorter than the lemmas, concealed at maturity, oblong, terete to dorsally flattened, falling free or with the palea or both the lemma and palea attached; **hila** oblong, about ⅓ or less the length of the caryopses. $x = 7$. Named for Benedetto Puccinelli (1808–1850), an Italian botanist.

Puccinellia, a genus of approximately 120 species, is most abundant in the middle and high latitudes of the Northern Hemisphere. There are 21 species in the *Flora* region, of which 3 are introduced. Ten are confined to the arctic, four are circumarctic and two are transberingian. Most species of *Puccinellia* are halophytes, either in coastal habitats or in saline or otherwise mineralized soils of interior habitats. Polyploidy, selfing, and hybridization are widespread in the genus, and many of the species boundaries are controversial. Several of the species with Arctic distributions have received different taxonomic treatments in North America and Eurasia.

The angle of the panicle branches (whether erect, ascending, etc.) refers to their position when the caryopses are mature. Lemma measurements should be made on the lowest lemma in the spikelets. Principal features of the lemmas, as used in the key and descriptions, are as follows. Scabrules (short, pointed hairs, similar in form to those that occur on the pedicels and inflorescence branches of many species of *Puccinellia*, and generally requiring magnification to observe) often occur along the distal margins of the lemmas. When present, they may be few and irregularly scattered, with gaps between them that are either wider than the individual scabrules (e.g., in some *P. pumila*, p. 472), or arranged in a continuous palisade-like row that lacks gaps (e.g., in *P. distans*, p. 474). Independent of the presence or absence of scabrules, the lemma margins may be entire (e.g., in *P. pumila* and *P. distans*) or serrate to erose (e.g., in *P. andersonii*, p. 474, and *P. vahliana*, p. 468).

SELECTED REFERENCES **Argus, G.W.** and **K.M. Pryer.** 1990. Rare Vascular Plants in Canada: Our Natural Heritage. Canadian Museum of Nature, Ottawa, Ontario, Canada. 191 pp.; **Consaul, L.L.** and **L.J. Gillespie.** 2001. A re-evaluation of species limits in Canadian Arctic island *Puccinellia* (Poaceae): Resolving key characters. Canad. J. Bot. 79:927–956; **Consaul, L.L., L.J. Gillespie,** and **K.I. MacInnes.** [in press]. Addition to the flora of Canada? A specimen from the Arctic Archipelago, Northwest Territories links two allopatric species of alkali grass; **Davis, J.I.** 1983. Phenotypic plasticity and the selection of taxonomic characters in *Puccinellia* (Poaceae). Syst. Bot. 8:341–353; **Fernald, M.L.** and **G.A. Weatherby.** 1916. The genus *Puccinellia* in eastern North America. Rhodora 18:1–23; **Porsild, A.E.** 1964. Illustrated Flora of the Canadian Arctic Archipelago, ed. 2, rev. Bulletin of the National Museum of Canada No. 146 [Biological Series No. 50]. R. Duhamel, Queen's Printer, Ottawa, Ontario, Canada. 218 pp.; **Scribner, F.L.** and **E.D. Merrill.** 1910. The grasses of Alaska. Contr. U.S. Natl. Herb. 13³:47–92; **Sørensen, T.J.** 1953. A revision of the Greenland species of *Puccinellia* Parl. with contributions to our knowledge of the arctic *Puccinellia* flora in general. Meddel. Grønl. 136:1–169; **Sørensen, T.J.** 1955. *Puccinellia agrostidea, Puccinellia bruggemannii, Puccinellia poacea.* Pp. 78–82 in A.E. Porsild. The Vascular Plants of the Western Canadian Arctic Archipelago. Bulletin of the National Museum of Canada No. 135 [Biological Series No. 45]. E. Cloutier, Queen's Printer, Ottawa, Ontario, Canada. 226 pp.; **Swallen, J.R.** 1944. The Alaskan species of *Puccinellia.* J. Wash. Acad. Sci. 34:16–32; **Tsvelev, N.N.** 1995. *Puccinellia.* Pp. 237–263 in J.G. Packer (ed., English edition). Flora of the Russian Arctic, vol. 1, trans. G.C.D. Griffiths. University of Alberta Press, Edmonton, Alberta, Canada. 330 pp. [English translation of A.I. Tolmachev (ed.). 1964. Arkticheskaya Flora SSSR, vol. 2. Nauka, Leningrad [St. Petersburg], Russia. 272 pp.]

1. Plants stoloniferous perennials, forming low, often extensive mats; most plants lacking inflorescences, the spikelets, when present, usually not producing mature pollen or caryopses. . . . . .1. *P. phryganodes*
1. Plants annual, biennial, or cespitose perennials, sometimes stoloniferous but not mat-forming; plants reproducing sexually, forming mature pollen and caryopses.
   2. Lemmas slightly to markedly coriaceous for most or all of their length; plants of temperate regions.
      3. Lemmas with hyaline apical margins; lemma midveins prominent .2. *P. rupestris*
      3. Lemmas with coriaceous apical margins; lemma midveins obscure.
         4. Lemmas 1.8–3 mm long; lower branches of the panicles ascending to erect, spikelet-bearing nearly to the base; anthers 0.6–1 mm long . . . . .3. *P. fasciculata*
         4. Lemmas 3–5 mm long; lower branches of the panicles erect to descending, spikelet-bearing from about midlength; anthers 1.5–2.6 mm long . . . . .4. *P. maritima*

2. Lemmas mostly membranous or herbaceous, apical margins sometimes hyaline; plants of temperate and arctic regions.
  5. Plants annual, of temperate regions.
    6. Lemma apices acute; lemmas 2.5–4 mm long, veins glabrous or hairy, particularly on the basal $^1/_2$, short (about 0.1 mm) hairs sparsely and evenly distributed between the veins . . . . 5. *P. simplex*
    6. Lemma apices obtuse to truncate; lemmas 1.8–2.2 mm long, veins densely hairy on the basal $^1/_2$–$^3/_4$, glabrous between the veins . . . . 6. *P. parishii*
  5. Plants perennial, of temperate and arctic regions.
    7. Palea veins with curly, intertwined hairs proximally, scabrous distally; plants of arctic regions.
      8. Pedicels smooth; apical margins of the lemmas smooth, veins obscure or distinct.
        9. Lower glumes $^2/_3$ to nearly as long as the adjacent lemmas; culms 5–15 cm; panicles 2–4 cm, anthers 0.8–1.5 mm . . . . 7. *P. vahliana*
        9. Lower glumes usually less than $^2/_3$ as long as the adjacent lemmas, culms 15–40 cm; panicles 5–8 cm, anthers 1.5–2.5 mm . . . . 8. *P. wrightii*
      8. Pedicels scabrous; apical margins of the lemmas scabrous, sometimes minutely so, veins obscure.
        10. Culms 50–65 cm tall; panicles 15–30 cm long . . . . 9. *P. groenlandica*
        10. Culms 5–35 cm tall; panicles 1–13 cm long.
          11. Lemmas 3.5–5.2 mm long; panicles (4)5–13 cm long . . . . 10. *P. angustata*
          11. Lemmas 2.8–3.8 mm long; panicles 1–4 cm long . . . . 11. *P. bruggemannii*
    7. Palea veins glabrous, shortly ciliate, or with fewer than 5 longer hairs proximally, never with curly intertwined hairs, scabrous or smooth distally; plants of temperate and arctic regions.
      12. Lemma margins smooth or with a few scabrules at and near the apices.
        13. Lemmas 2–2.5 mm long, usually purple with whitish margins, veins distinct, apices obtuse to truncate; lemmas and palea veins smooth and glabrous; pedicels smooth . . . . 12. *P. tenella*
        13. Lemmas 2.4–4.6 mm long, variously colored, margins not white, veins obscure to distinct, apices acute to truncate; lemmas and palea veins glabrous or hairy on the lower portion, often scabrous distally; pedicels smooth or scabrous.
          14. Lemmas glabrous or with a few hairs on the lower portion of the veins; lemma apices entire; plants of temperate regions or the low arctic, but not of the high arctic.
            15. Pedicels scabrous; palea veins scabrous distally; anthers 1–2 mm long; plants not littoral . . . . 13. *P. lemmonii* (in part)
            15. Pedicels smooth or with a few scattered scabrules; palea veins smooth or with a few scabrules distally; anthers 0.5–1.2 mm long; plants littoral . . . . 14. *P. pumila*
          14. Lemmas usually sparsely to moderately hairy, particularly on the vein bases, sometimes glabrous; lemma apices entire, irregularly serrate, or erose; plants of the low and high arctic.
            16. Panicles with (2)3–5 branches at the lowest node; lemmas 2.5–3.7 mm long, veins obscure to distinct, apices entire or slightly erose; anthers 1.2–2.2 mm long . . . . 15. *P. arctica* (in part)
            16. Panicles usually with 2 branches at the lowest node; lemmas 3–4.5 mm long, veins obscure, apices irregularly serrate or erose; anthers 0.8–1.2 mm long . . . . 16. *P. andersonii* (in part)
      12. Lemma margins densely scabrous at and near the apices.
        17. Lemmas 1.5–2.2 mm long, apices widely obtuse to truncate; anthers 0.4–0.8 mm long; lower panicle branches horizontal to descending . . . . 17. *P. distans*

17. Lemmas 2–5 mm long, apices usually acute to obtuse, occasionally acuminate or rounded; anthers 0.5–2.2 mm long; lower panicle branches erect to descending.

  18. Lemma apices irregularly serrate or erose; lemmas 3–4.5 mm long; anthers 0.8–1.2 mm long . . . . 16. *P. andersonii* (in part)

  18. Lemma apices entire or slightly erose; lemmas 2–4.5(5) mm long; anthers 0.5–2.2 mm long.

    19. Pedicels smooth or with a few scattered scabrules; lemmas glabrous or with a few hairs on the lower ¹/₂, principally along the veins; anthers 1.5–2 mm long; plants restricted to mineralized springs in California . . . . 18. *P. howellii*

    19. Pedicels smooth to uniformly scabrous; lemmas glabrous or sparsely to moderately hairy on the lower ¹/₂; anthers 0.5–2.2 mm long; plants of varied habitats, including hot springs.

      20. Lemma midveins often extending to the apical margins; lemma apices acute; lemmas mostly smooth, midveins often slightly scabrous distally; leaf blades involute, 1.2–1.9 mm wide when flattened; leaves concentrated at the base of plant; plants of inland, temperate habitats . . . . 13. *P. lemmonii* (in part)

      20. Lemma midveins usually not extending to the margins; lemma apices usually acute to obtuse, occasionally acuminate; lemmas scabrous or smooth distally; leaf blades involute or flat, 0.5–6 mm wide when flat, leaves ranging from nearly all basal to evenly distributed along the culms; plants of coastal and inland habitats in temperate and arctic regions.

        21. Culms 10–100 cm tall; lower glumes 0.5–1.6 mm long; plants usually growing south of 65° N latitude.

          22. Pedicel epidermal cells not tumid, pedicels uniformly scabrous; lower branches of the panicles erect to descending; lemmas (2)2.2–3(3.5) mm long; plants usually of interior habitats, occasionally of coastal habitats . . . . 19. *P. nuttalliana* (in part)

          22. Pedicel epidermal cells often tumid, pedicels sparsely to densely scabrous; lower branches of the panicles usually erect to ascending, occasionally spreading to descending; lemmas (2.2)3–4.5(5) mm long; plants of coastal habitats . . . . 20. *P. nutkaensis*

        21. Culms 6–30(40) cm tall; lower glumes 0.8–2.5 mm long; plants usually growing north of 65° N latitude.

          23. Anthers 1.2–2.2 mm long; lateral margins of the lemmas often inrolled . . . . 15. *P. arctica* (in part)

          23. Anthers 0.6–1.2 mm long; lateral margins of the lemmas usually not inrolled.

            24. Lemmas 2.8–4 mm long; panicles usually barely exserted from the sheaths . . . . 21. *P. vaginata*

            24. Lemmas 2–2.8 mm long; panicles usually distinctly exserted from the sheaths . . . . 19. *P. nuttalliana* (in part)

1. **Puccinellia phryganodes** (Trin.) Scribn. & Merr. [p. 464]

GOOSE GRASS, PUCCINELLIE RAMPANTE, PUCCINELLIE TROMPEUSE

Plants perennial; stoloniferous, often forming extensive low mats. **Culms** 2–15 cm, erect or decumbent. **Ligules** 0.4–1.5 mm, acute, obtuse, or truncate, entire; **blades** 0.4–2.2 mm wide when flat, 0.2–1 mm in diameter when involute. **Panicles** usually not developed, if developed, 1–7 cm, diffuse, lower branches ascending, spikelets usually confined to the distal $^1/_3$; **pedicels** scabrous, usually with tumid epidermal cells. **Spikelets** 6–9 mm, with 3–6(7) florets. **Glumes** rounded or slightly keeled over the back, veins obscure or distinct, apices acute or obtuse; **lower glumes** 1.5–2.2 mm; **upper glumes** 2.3–2.8 mm; **calluses** glabrous or almost so; **lemmas** 3.2–3.8(4.5) mm, membranous, usually glabrous, occasionally sparsely hairy, backs rounded, 5-veined, veins obscure, not extending to the margins, apices acute or rounded, apical margins whitish, hyaline, smooth, entire; **palea veins** glabrous; **anthers** (1.5)2–2.5 mm, almost always indehiscent, mature pollen rarely produced. **Caryopses** rarely developed. $2n = 21, 28$.

Puccinellia phryganodes is a widespread and common circumpolar arctic species that grows on seashores at or near the high tide line, in wet saline meadows, and in saline or brackish marshes.

2. **Puccinellia rupestris** (With.) Fernald & Weath. [p. 464]

STIFF SALTMARSH GRASS

Plants annual or biennial; tufted, not mat-forming. **Culms** 5–50 cm, erect to decumbent. **Ligules** 1–2 mm, obtuse to truncate, entire; **blades** 2–6 cm wide, flat or folded. **Panicles** 2–10 cm, contracted at maturity, lower branches ascending, spikelet-bearing nearly to the base; **pedicels** scabrous, sometimes only slightly so, sometimes with tumid epidermal cells. **Spikelets** 4–9 mm, with 2–6 florets. **Glumes** rounded to keeled over the back, midveins obscure or prominent, extending to the apices, often scabrous distally, lateral veins obscure, apices acute to obtuse; **lower glumes** 1–2 mm; **upper glumes** 2–3.2 mm; **calluses** with a few hairs; **lemmas** 2.5–4 mm, somewhat coriaceous for most of their length, glabrous or sparsely hairy on the lower $^1/_2$, particularly along the veins, backs rounded to keeled,

5-veined, midveins and sometimes all veins prominent, midveins extending to the margin, often excurrent, apical margins hyaline, smooth to densely scabrous, apices acute to obtuse, entire; **palea veins** shortly ciliate proximally, shortly ciliate to scabrous distally; **anthers** 0.8–1.2 mm. $2n = 42$.

Puccinellia rupestris grows in coastal and noncoastal habitats in Eurasia; North American collections were apparently introduced in ballast.

3. **Puccinellia fasciculata** (Torr.) E.P. Bicknell [p. 464]

BORRER'S SALTMARSH GRASS

Plants short-lived perennials; cespitose, not mat-forming. **Culms** 10–65 cm, usually decumbent and geniculate, sometimes erect. **Ligules** 1–2 mm, obtuse to truncate, entire; **blades** 2–7 mm wide, flat, folded, or involute. **Panicles** 4–16 cm, compact to diffuse at maturity, linear to pyramidal, lower branches ascending to erect, spikelet-bearing nearly to the base; **pedicels** slightly to densely scabrous, often with tumid epidermal cells. **Spikelets** 3–6 mm, with 2–6 florets. **Glumes** rounded over the back, veins prominent to obscure, apices acute to obtuse; **lower glumes** 0.8–1.6 mm; **upper glumes** 1.2–2.3 mm; **calluses** with a few hairs; **lemmas** 1.8–3 mm, slightly to markedly coriaceous throughout, glabrous or with a few hairs near the base, backs rounded, 5-veined, veins obscure, midveins usually excurrent, sometimes ending at the margins, apices acute to obtuse, entire, apical margins smooth or with a few scattered scabrules; **palea veins** glabrous proximally, scabrous to shortly hispid near midlength, scabrous distally; **anthers** 0.6–1 mm. $2n = 28$.

Puccinellia fasciculata is native to Europe. In the *Flora* region, it is found principally along the east coast, but it is also established at a few sites in Arizona and Utah, and has been reported from Nevada. All occurrences in the *Flora* region are probably the result of human introductions.

4. **Puccinellia maritima** (Huds.) Parl. [p. 466]

COMMON SALTMARSH GRASS, PUCCINELLIE MARITIME

Plants perennial; cespitose, often stoloniferous, not mat-forming. **Culms** 20–100 cm, erect to decumbent. **Ligules** 1–3.5 mm, obtuse to truncate, entire; **blades** 2–4.4 mm wide, flat to involute. **Panicles** 3–30 cm, compact to diffuse at maturity, lower branches erect

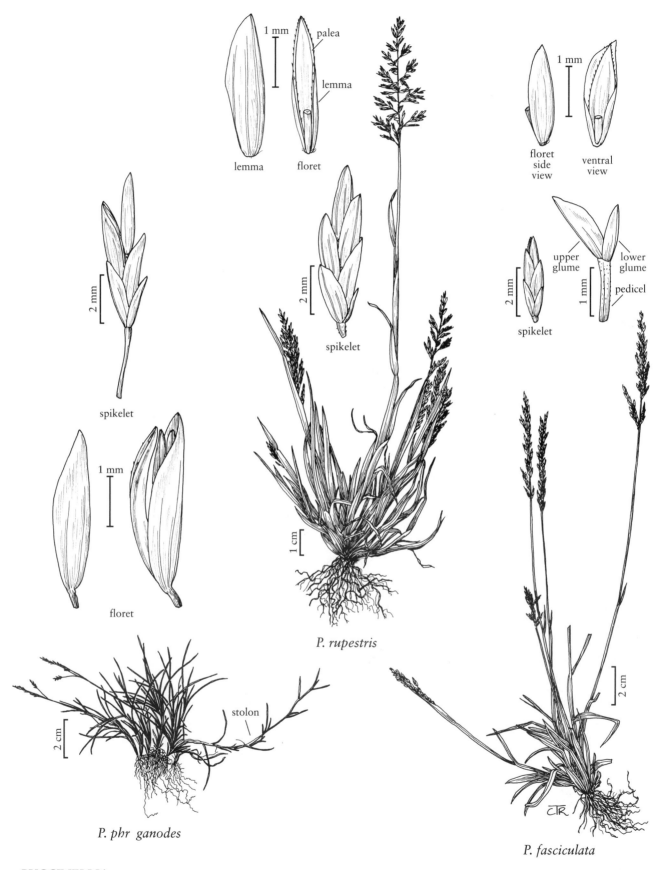

lemma

palea

lemma

floret

floret
side
view

ventral
view

spikelet

upper
glume

lower
glume

pedicel

spikelet

spikelet

floret

P. rupestris

P. phr ganodes

stolon

P. fasciculata

PUCCINELLIA

to descending, spikelet-bearing from about midlength; **pedicels** usually densely scabrous, occasionally with only a few scattered scabrules, often with tumid epidermal cells. **Spikelets** 5.5–13 mm, with 4–9 florets. **Glumes** rounded over the back, veins obscure, apices acute to obtuse; **lower glumes** 2–3.4 mm; **upper glumes** 3–4.5 mm; **calluses** with few to many hairs; **lemmas** 3–5 mm, slightly to markedly coriaceous throughout, sparsely to densely hairy in the lower ¹/₂, principally along the veins, backs rounded, 5-veined, veins obscure, usually not extending to the margins, sometimes the midvein doing so, apical margins usually smooth, sometimes with a few scattered scabrules, occasionally densely scabrous, apices acute to obtuse, entire; **palea veins** ciliate proximally, scabrous to short-ciliate distally; **anthers** 1.5–2.6 mm. $2n = 14$–77, usually 56.

Puccinellia maritima grows in coastal environments in North America and Greenland. It is native to Europe; most or all occurrences in the *Flora* region are probably the result of human introduction.

5. Puccinellia simplex Scribn. [p. 466]
WESTERN ALKALI GRASS

Plants annual; not mat-forming. **Culms** usually erect, 2–25 cm. **Ligules** 1–3 mm, acute to obtuse, entire; **blades** 0.7–2 mm wide, flat to involute. **Panicles** 1–18 cm, compact, mostly linear at maturity, primary branches usually spikelet-bearing to the base, lower branches erect; **pedicels** densely scabrous or with only a few scattered scabrules, often also with a few hairs, often with tumid epidermal cells. **Spikelets** 3.5–8 mm, with 2–7 florets. **Glumes** rounded to weakly keeled over the back, veins obscure to prominent, apices acute; **lower glumes** 1.3–2 mm; **upper glumes** 2.3–3 mm; **calluses** hairy; **lemmas** 2.5–4 mm, mostly herbaceous, usually with about 0.1 mm hairs distributed sparsely and evenly between the veins, longer hairs also usually present along the veins and near the base, basal hairs often longest, twisted and tangled, backs rounded, sometimes keeled distally, 5-veined, veins usually obscure, sometimes prominent, midveins sometimes reaching the margins, other veins usually not doing so, apical margins often hyaline, smooth or with a few scattered scabrules, apices acute, entire; **palea veins** with hairs, hairs on the proximal portion longer, twisted and somewhat tangled, hairs on the distal ²/₃ usually short and straight; **anthers** 0.2–0.5 mm. $2n = 56$.

Puccinellia simplex is widespread in, and mostly confined to, saline soils of central California. The records from Utah probably reflect introductions.

6. Puccinellia parishii Hitchc. [p. 466]
PARISH'S ALKALI GRASS

Plants annual; not mat-forming. **Culms** 3–22 cm, erect. **Leaves** basally concentrated; **ligules** 1–2 mm, obtuse to truncate, entire; **blades** 0.2–1.2 mm wide, flat to involute. **Panicles** 1–8.5 cm, compact to diffuse at maturity, lower branches erect to descending, usually spikelet-bearing to the base; **pedicels** densely scabrous or with a few scattered scabrules, often with tumid epidermal cells. **Spikelets** 3.5–5 mm, with 2–7 florets. **Glumes** rounded over the back, veins obscure to prominent, apices acute to obtuse; **lower glumes** 1–2 mm; **upper glumes** 1.8–2.2 mm; **calluses** hairy; **lemmas** 1.8–2.2 mm, mostly herbaceous, densely hairy over the proximal ¹/₂–³/₄ of the veins, glabrous between the veins, backs rounded, 5-veined, veins obscure to prominent, not extending to the margins, apical margins hyaline, smooth or with a few scattered scabrules, apices obtuse to truncate, entire; **palea veins** glabrous proximally, hairy at midlength, glabrous or scabrous-ciliate distally; **anthers** 0.4–0.5 mm. $2n = 14$.

Puccinellia parishii grows in saline seepage areas in California, Arizona, and New Mexico.

7. Puccinellia vahliana (Liebm.) Scribn. & Merr. [p. 468]
VAHL'S ALKALI GRASS, PUCCINELLIE DE VAHL

Plants perennial; cespitose, not mat-forming. **Culms** 5–15 cm, erect. **Ligules** 1–2.5 mm, acute to obtuse, entire; **blades** 2–8 mm wide, flat or folded. **Panicles** 2–4 cm, usually contracted and dense, sometimes slightly diffuse at maturity, lowest node usually with long and short branches, lower branches erect to ascending, spikelets usually confined to the distal ²/₃; **pedicels** glabrous and smooth, lacking tumid epidermal cells or with very small tumid epidermal cells. **Spikelets** 3.8–6.5 mm, with 2–4(5) florets. **Glumes** broadly ovate, enfolding the bases of the lower lemmas, rounded over the back, veins obscure or distinct, apices acute to obtuse; **lower glumes** 2–3.5 mm, at least ²/₃ as long as the adjacent lemmas; **upper glumes** 2.4–4 mm; **rachilla internodes** abruptly broadened at the point of attachment to the lemmas, less than 0.09 mm thick; **calluses** with a few hairs; **lemmas** 3–5.2 mm, usually herbaceous and mostly purplish, sometimes membranous with purple veins, basal ¹/₂ hairy over and

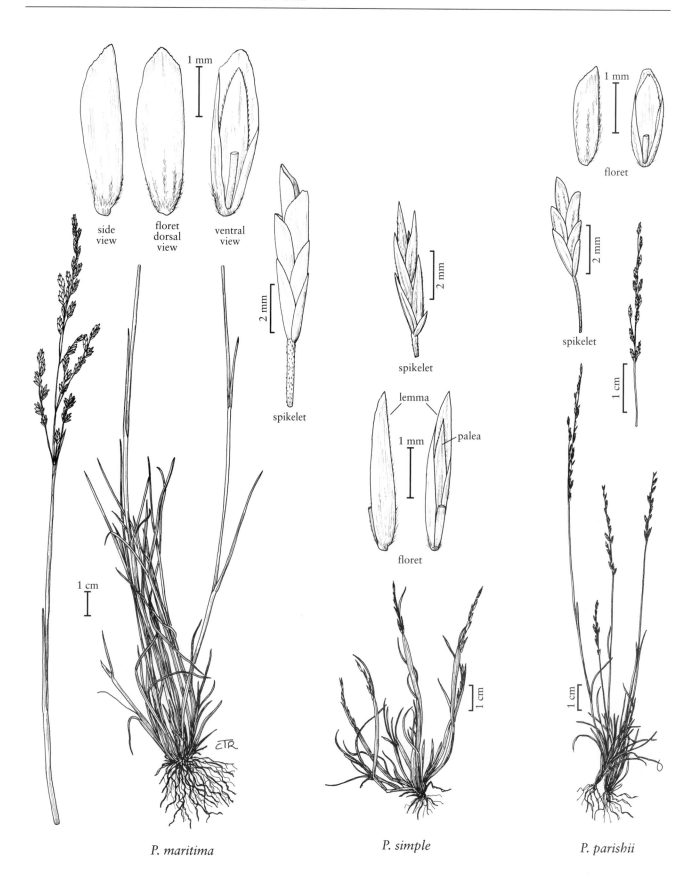

side
view

floret
dorsal
view

ventral
view

1 mm

spikelet

2 mm

spikelet

2 mm

lemma

palea

1 mm

floret

floret

1 mm

spikelet

2 mm

1 cm

1 cm

1 cm

1 cm

P. maritima

P. simple

P. parishii

PUCCINELLIA

between the veins, backs rounded, 5-veined, veins obscure or distinct, often dark purple, not extending to the margins, apical margins hyaline, often yellowish, smooth, apices acute, entire, becoming erose with age; **palea veins** with curly, intertwined hairs on the proximal portion, scabrous distally; **anthers** 0.8–1.5 mm. $2n = 14$.

Puccinellia vahliana is an arctic species that is circumpolar, except in the Beringian region. In the *Flora* region, it extends from Alaska through northern Canada to Greenland. It is generally non-halophytic, growing in calcareous gravel, sand, clay, or moss of imperfectly drained moist areas, and on seepage slopes from near sea level to 700 m, or, rarely, in seasonally dry, turfy sites. It is often a pioneering species in moist clay and silt by alpine brooks, ephemeral lakes, glacial runoff streams, and on snowbeds. The roots of this species and *P. wrightii* are characteristically thicker and more tightly curled than those of other *Puccinellia* species. It sometimes hybridizes with *Phippsia algida*.

8. Puccinellia wrightii (Scribn. & Merr.) Tzvelev [p. 468]
WRIGHT'S ALKALI GRASS

Plants perennial; cespitose, not mat-forming. **Culms** 15–40 cm, erect. **Ligules** 1.5–3 mm, acute to obtuse, entire; **blades** 2–8 mm wide, flat or folded. **Panicles** 5–8 cm, open and diffuse at maturity, lowest node usually with long branches, lower branches ascending, horizontal, or descending, spikelets usually confined to the distal $^1/_3$; **pedicels** glabrous and smooth, lacking tumid epidermal cells. **Spikelets** 4–7 mm, with 4–5 florets. **Glumes** narrow, not enfolding the base of the lower lemmas, rounded over the back, veins obscure or distinct, apices acute to obtuse; **lower glumes** 1.7–3 mm, usually less than $^2/_3$ as long as the adjacent lemmas; **upper glumes** 2.5–4 mm; **rachilla internodes** abruptly or gradually broadened to the point of attachment with the lemmas, at least the lowest internode usually more than 0.09 mm thick; **calluses** with a few hairs; **lemmas** 4–5 mm, mostly herbaceous, hairy along and between the veins on the proximal $^1/_2$, backs rounded, 5-veined, veins obscure or distinct, not extending to the margins, apical margins hyaline, smooth, apices acute, entire, becoming erose with age; **palea veins** with curly, intertwined hairs proximally, these hairs rarely lacking, scabrous distally; **anthers** 1.5–2.5 mm. $2n = 14$.

Puccinellia wrightii is an uncommon arctic species. Its range extends from the Chukotka Peninsula in the Russian Far East to western Alaska. Like *P. vahliana*, its roots are characteristically thicker and more tightly curled than those of other *Puccinellia* species.

9. Puccinellia groenlandica T.J. Sørensen [p. 468]
GREENLAND ALKALI GRASS

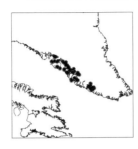

Plants perennial; cespitose, not mat-forming. **Culms** 50–65 cm, erect to decumbent. **Ligules** 2.5–3.3 mm, obtuse to truncate, entire; **blades** 1.8–3.2 mm wide, flat. **Panicles** 15–30 cm, diffuse at maturity, lower branches ascending to horizontal, spikelets usually confined to the distal $^2/_3$; **pedicels** uniformly scabrous, lacking tumid epidermal cells. **Spikelets** 7–10 mm, with 4–7 florets. **Glumes** rounded over the back, veins obscure, apices acute to obtuse; **lower glumes** 1.8–2.5 mm, up to approximately $^1/_2$ as long as the adjacent lemmas; **upper glumes** 2.5–3 mm; **rachilla internodes** slightly and gradually broadened to the point of attachment with the lemmas, at least the lowest internodes usually more than 0.09 mm thick; **calluses** with a few hairs; **lemmas** 3–4.4 mm, mostly herbaceous, hairy on the proximal $^1/_2$, principally along the veins, backs rounded or keeled distally, 5-veined, veins obscure, not extending to the margins, apical margins hyaline, uniformly scabrous, apices obtuse to acute, entire; **palea veins** with curly, intertwined hairs proximally, scabrous distally; **anthers** 1.1–1.5 mm. $2n = 56$.

Puccinellia groenlandica grows in littoral and nonlittoral environments. It is endemic to Greenland.

10. Puccinellia angustata (R. Br.) E.L. Rand & Redfield [p. 470]
TALL ALKALI GRASS, PUCCINELLIE ÉTROITE

Plants perennial; cespitose, not mat-forming. **Culms** 10–35 cm, erect to decumbent. **Ligules** (0.8)1–3(4) mm, acute, obtuse, or truncate, entire; **blades** usually involute and 0.5–1.2 mm in diameter, sometimes flat and 0.5–3 mm wide. **Panicles** (4)5–10(13)cm, usually contracted, occasionally diffuse at maturity, lower nodes with 2–3(4) branches, lower branches erect to ascending, spikelets usually confined to the distal $^1/_2$; **pedicels** scabrous, lacking tumid epidermal cells. **Spikelets** 4–10 mm, with 3–5(6) florets. **Glumes** rounded over the back, veins obscure, apices acute; **lower glumes** 1.5–2.8 mm, usually less than $^1/_2$ as long as the adjacent lemmas; **upper glumes** 2.2–4.2 mm; **rachilla internodes** slightly and gradually broadened to

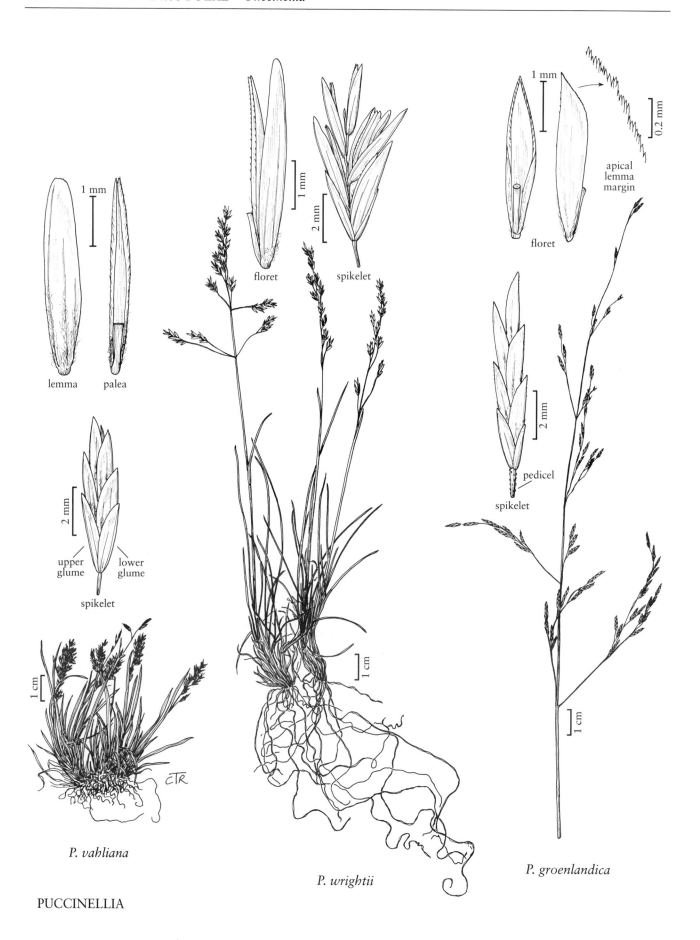

lemma   palea

floret   spikelet

1 mm

2 mm

1 mm

floret

apical
lemma
margin

0.2 mm

2 mm

upper
glume   lower
glume

spikelet

2 mm

pedicel

spikelet

1 cm

1 cm

1 cm

P. vahliana

P. wrightii

P. groenlandica

PUCCINELLIA

the point of attachment with the lemmas, at least the lowest internode usually more than 0.09 mm thick; **calluses** with a few hairs; **lemmas** 3.5–5.2 mm, herbaceous, often purplish, hairy on the proximal ¹/₂ along the veins, sometimes also between the veins, backs rounded, 5-veined, veins obscure, not extending to the margins, apical margins hyaline, scabrous, apices acute, entire or erose; **palea veins** with curly, intertwined hairs proximally, scabrous distally; **anthers** (0.6) 0.8–1.1(1.5) mm. $2n = 42$.

Puccinellia angustata is a common and widespread arctic species that grows in disturbed silty or sandy sediments. It is usually non-littoral, but when in coastal areas it grows above the influence of high tide.

11. Puccinellia bruggemannii T.J. Sørensen [p. 470]
BRUGGEMANN'S ALKALI GRASS

Plants perennial; cespitose, not mat-forming. **Culms** 5–12 cm, erect to decumbent. **Ligules** 0.8–2 mm, acute or obtuse, entire; **blades** usually involute and 0.5–1.1 mm in diameter, sometimes flat and 0.7–2.5 mm wide. **Panicles** 1–4 cm, contracted at maturity, lowest node with 2(3) branches, lower branches erect to ascending, spikelets usually confined to the distal ¹/₃; **pedicels** scabrous, lacking tumid epidermal cells. **Spikelets** 3.5–6.5 mm, with 2–4 florets. **Glumes** rounded over the back, veins obscure, apices acute to obtuse; **lower glumes** 0.7–2 mm, less than ¹/₂(²/₃) as long as the adjacent lemmas; **upper glumes** 1.5–2.8 mm, broadly elliptic; **rachilla internodes** slightly and gradually broadened to the point of attachment with the lemmas, at least the lowest internode usually more than 0.09 mm thick; **calluses** with a few hairs; **lemmas** 2.8–3.8 mm, herbaceous, hairy on the lower ¹/₂ along and between the veins, backs rounded, 5-veined, veins obscure, not extending to the margins, apical margins herbaceous, scabrous or scabridulous, apices acute or somewhat obtuse, entire or erose, slightly incurved; **palea veins** with curly, intertwined hairs proximally, scabrous distally; **anthers** 0.7–1.1 mm. $2n = 28$.

Puccinellia bruggemannii is restricted to arctic islands of Canada and northern Greenland. In Canada, it is a widespread yet local northern, western, and central arctic island species. It is found in calcareous, barren, gravelly, sandy, or silty sites, and is sometimes coastal. Although *P. bruggemannii* has been reported as non-littoral (Porsild 1964) and probably non-halophilous (Sørensen 1955), specimens keying to this species have been found near the sea coast: for example, paratype specimens collected on Beechey Island and in the vicinity of salt springs on Axel Heiberg Island.

Puccinellia bruggemannii sometimes superficially resembles *Poa abbreviata* or small *Poa glauca* in the field, because of its small, dense inflorescence.

12. Puccinellia tenella (Lange) Holmb. *ex* Porsild [p. 470]
TUNDRA ALKALI GRASS

Plants perennial; cespitose, not mat-forming. **Culms** 3–16 cm, erect. **Ligules** 0.5–1.7 mm, acute, obtuse, or truncate, entire; **blades** usually involute and 0.4–0.7 mm in diameter, occasionally flat and 0.5–1.5 mm wide. **Panicles** 1.6–5.5 cm, usually contracted, sometimes diffuse, lower branches erect or ascending, spikelets usually confined to the distal ¹/₂; **pedicels** smooth, with tumid epidermal cells. **Spikelets** 3.5–7 mm, with 3–6 florets. **Glumes** slightly keeled over the back, veins distinct, apices acuminate to acute; **lower glumes** 0.7–1.3 mm; **upper glumes** 1.3–1.8 mm; **calluses** glabrous or with 5 or fewer hairs shorter than 0.1 mm; **lemmas** 2–2.5 mm, herbaceous, usually purple with whitish margins, smooth, glabrous or with a few hairs on the bases of the lateral veins, backs rounded, 5-veined, veins distinct, not extending to the margins, apical margins smooth, apices obtuse to truncate, entire or slightly erose; **palea veins** smooth, glabrous; **anthers** 0.6–0.9 mm. $2n = 14$.

Puccinellia tenella is a halophytic, circumpolar subarctic and low arctic species. It is found above the high tide zone on sandy spits, in salt marshes, in silty soils, and among granitic rocks to 30 m above sea level. The above description applies to *P. tenella* subsp. *langeana* (Berlin) Tzevlev, the only subspecies in the *Flora* region. *Puccinellia alaskana* Scribn. & Merr., considered a subspecies of *P. langeana* (Berlin) T.J. Sørensen *ex* Hultén [= *P. tenella*] by Sørensen (1953), has acute lemmas and hairs on the palea. It is discussed under *P. pumila* (see p. 471).

13. Puccinellia lemmonii (Vasey) Scribn. [p. 472]
LEMMON'S ALKALI GRASS

Plants perennial; cespitose, not mat-forming. **Culms** 5–40 cm, usually erect. **Leaves** basally concentrated; **ligules** 0.8–2.2 mm, obtuse to acute, mostly entire, sometimes slightly erose; **blades** involute, 1.2–1.9 mm wide when flattened. **Panicles**

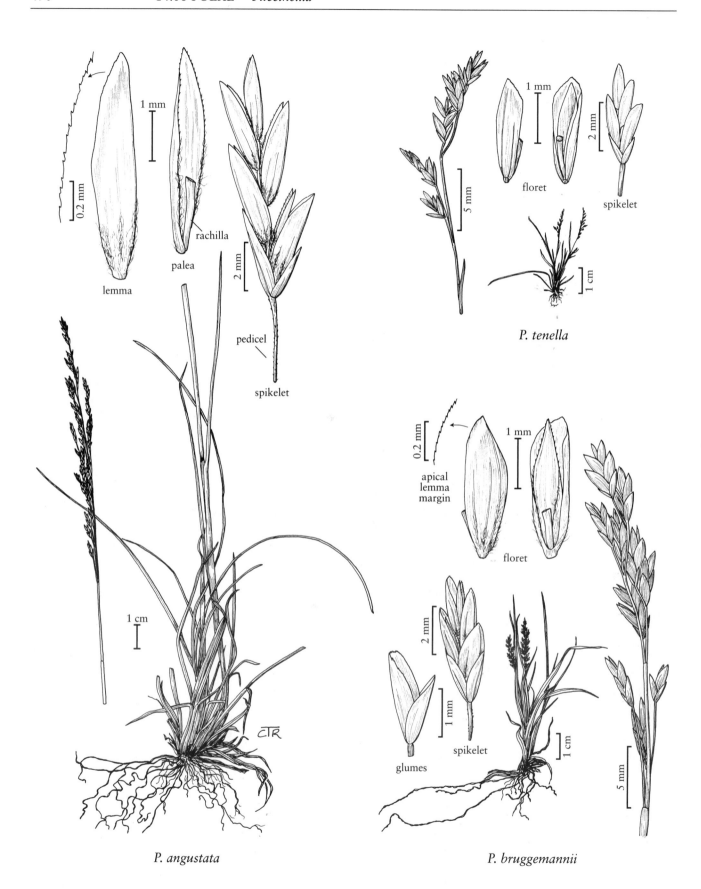

0.2 mm

1 mm

lemma

rachilla

palea

2 mm

pedicel

spikelet

1 cm

P. angustata

CTR

1 mm

floret

2 mm

spikelet

5 mm

1 cm

P. tenella

0.2 mm

apical
lemma
margin

1 mm

floret

2 mm

spikelet

glumes

1 mm

1 cm

5 mm

P. bruggemannii

PUCCINELLIA

2–18 cm, compact to diffuse at maturity, lower branches ascending to descending, usually spikelet-bearing to the base; **pedicels** scabrous, lacking tumid epidermal cells. **Spikelets** 3.5–8 mm, with 2–6 florets. **Glumes** rounded over the back, veins obscure, apices acute to obtuse; **lower glumes** 0.7–1.5 mm; **upper glumes** 1.4–3 mm; **calluses** with a few hairs; **lemmas** 2.4–4 mm, herbaceous, mostly smooth, usually glabrous, sometimes with a few hairs near the base, principally along the veins, backs usually rounded, sometimes weakly keeled distally, 5-veined, veins obscure, midveins often slightly scabrous and prominent in the distal $^1/_2$, often extending to the apical margins, lateral veins not extending to the margins, apical margins ranging from smooth to scabrous, entire, not white, apices acute, entire; **palea veins** glabrous or shortly ciliate proximally, uniformly scabrous distally; **anthers** 1–2 mm. $2n = 14$.

Puccinellia lemmonii grows in non-littoral saline environments in the western portion of the contiguous United States. Reports from Saskatchewan are probably based on depauperate specimens of *P. nuttalliana*.

14. **Puccinellia pumila** (Vasey) Hitchc. [p. 472]
Smooth Alkali Grass, Puccinellie Naine

Plants perennial; usually cespitose, occasionally appearing rhizomatous or stoloniferous after rooting at the nodes of buried stems, infrequently stoloniferous, not mat-forming. **Culms** 8–40 cm, erect to decumbent. **Ligules** 0.8–2.5 mm, obtuse to truncate, entire; **blades** 1–3 mm wide, flat to involute. **Panicles** 3–20 cm, dense to diffuse at maturity, lower branches ascending to descending, spikelets borne from near the bases or confined to the distal $^2/_3$; **pedicels** smooth or with a few scattered scabrules, often with tumid epidermal cells distally. **Spikelets** 4–9 mm, with 3–7 florets. **Glumes** rounded over the back, veins obscure to distinct, apices acute to obtuse; **lower glumes** 1.4–2(4) mm; **upper glumes** 2–3(9) mm; **calluses** glabrous or with a few hairs; **lemmas** 2.5–4.6 mm, herbaceous, glabrous or with a few hairs on the vein bases, backs rounded, 5-veined, veins obscure, not extending to the margins, apical margins usually smooth, occasionally with a few scattered scabrules, entire, not white, apices acute to obtuse, entire; **palea veins** glabrous, smooth or with a few scabrules distally; **anthers** 0.5–1.2 mm. $2n = 14$ [for *Puccinellia alaskana*], 42, 56.

Puccinellia pumila is primarily North American, growing on the Pacific, Arctic, and Atlantic coasts. It also grows in Kamchatka, Russia (Tsvelev 1995). It generally grows in sand and among stones in protected intertidal environments. A few specimens with exceptionally long glumes and lemmas were treated by Fernald and Weatherby (1916) as *P. paupercula* var. *longiglumis* Fernald & Weath.; they are regarded here as representing extremes of *P. pumila*.

Puccinellia alaskana Scribn. & Merr., here included in *P. pumila*, was considered a subspecies of *P. langeana* (Berlin) T.J. Sørensen *ex* Hultén [= *P. tenella*] by Sørensen (1953), but more closely resembles *P. pumila*. It differs morphologically from *P. pumila* mainly in its relatively distinct lemma veins. It also differs from most specimens of *P. pumila* in having smaller lemmas (2.5–3 mm) and anthers (0.5–0.9 mm), and in being diploid. It represents the Aleutian Islands component of the geographic distribution given for *P. pumila*. Its status is currently under investigation. Molecular data obtained as this volume went to press (Consaul et al. [in prep.]) tend to support recognition of *P. alaskana* as a distinct species.

15. **Puccinellia arctica** (Hook.) Fernald & Weath. [p. 472]
Arctic Alkali Grass

Plants perennial; cespitose. **Culms** 10–30(40) cm, erect. **Leaves** basally concentrated; **ligules** 0.9–3 mm, acute, obtuse, or truncate, entire, margins decurrent; **blades** usually flat and 0.5–2.2 mm wide, sometimes involute and 0.2–1.6 mm in diameter. **Panicles** 3–11 cm, diffuse or contracted at maturity, lowest nodes with (2)3–5 branches, lower branches ascending to horizontal, spikelets usually confined to the distal $^2/_3$; **pedicels** scabrous, without tumid epidermal cells. **Spikelets** 4.5–7(9.5) mm, with (2)3–6(8) florets. **Glumes** rounded over the back, veins distinct or obscure, lateral margins often inrolled, apices acute to obtuse; **lower glumes** 0.8–2.1(2.5) mm; **upper glumes** 1.8–3 mm; **calluses** with a few hairs; **lemmas** 2.5–3.7 mm, herbaceous or membranous, often translucent, often purplish, hairy, particularly on the bases of the veins, backs rounded, 5-veined, veins obscure to distinct, midveins scabrous or smooth distally, sometimes extending to the apical margin, sometimes excurrent, lateral veins not extending to the margins, lateral margins often inrolled, apical margins often hyaline and yellowish, scabrous, entire or slightly erose, apices acute to obtuse; **palea veins** glabrous, smooth proximally, scabrous from midlength or just below midlength to the apices; **anthers** 1.2–2.2 mm. $2n = 14$.

Puccinellia arctica is restricted to the North American arctic, where it grows in silt, clay, and sandy substrates near the coast, and on alkaline, sparsely

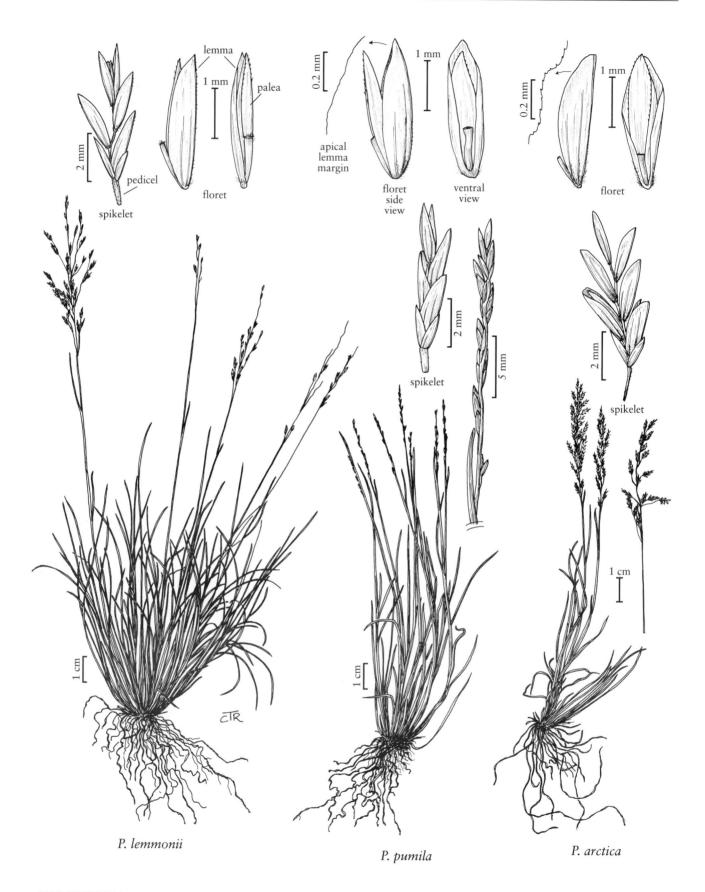

P. lemmonii

P. pumila

P. arctica

PUCCINELLIA

vegetated soils further inland. As treated here, it includes three entities that are sometimes treated as distinct species: *P. arctica sensu stricto*, *P. poacea* T.J. Sørensen, and *P. agrostidea* T.J. Sørensen. *Puccinellia arctica sensu stricto* is restricted to the southwestern arctic, *P. poacea* to the high arctic (Ellesmere and Axel Heiberg islands), and *P. agrostidea* to the southwestern arctic and possibly also Ellesmere Island. There are no morphological characters known for distinguishing these three entities. The first two may differ from the latter by the relatively frequent presence of small scabrules along the upper lemma midvein, slightly more distinct veins, and frequent yellowish margins to the lemma. The taxonomic validity of these characters was not completely understood at the time this treatment was written, but molecular analyses being conducted as this volume went to press (Consaul et al. [in prep.]) suggest that this group is best represented as a single species, *P. arctica*. Argus and Pryer (1990) stated that all three entities are rare in Canada.

16. Puccinellia andersonii Swallen [p. 474]
ANDERSON'S ALKALI GRASS

Plants perennial; cespitose, not mat-forming. **Culms** 10–25 cm, usually decumbent or geniculate. **Ligules** 1–2.8(3.3) mm, acute, obtuse, or truncate, entire or slightly erose; **blades** usually involute and 0.5–1 mm in diameter, sometimes flat and 0.8–2 mm wide. **Panicles** (3.5) 5–8 cm, diffuse or contracted at maturity, lowest node almost always with 2 branches, lower branches ascending to horizontal in fruit, spikelets usually confined to the distal $^1/_3$; **pedicels** smooth or scabrous, with tumid epidermal cells. **Spikelets** 5–7(9.5) mm, with (2)4–5(7) florets. **Glumes** rounded over the back, veins obscure, apices acute; **lower glumes** 1–2 mm; **upper glumes** 2–3 mm, often borne distinctly above the lower glumes; **calluses** glabrous or hairy; **lemmas** (3) 3.2–4(4.5) mm, herbaceous or membranous, glabrous or sparsely hairy on the bases of the veins, backs rounded, 5-veined, veins obscure, not extending to the margins, apical margins smooth or scabrous, not white, apices usually acute, occasionally rounded, irregularly serrate or erose; **palea veins** glabrous, smooth proximally, scabrous distally; **anthers** 0.8–1.2 mm. 2*n* = 56.

Puccinellia andersonii is a widespread, coastal arctic species. It grows near the tideline and on otherwise barren, reworked marine sediments of eroded flood plains. Its decumbent growth form often gives it an unhealthy appearance. It is unique among *Puccinellia* species in the *Flora* region in having blunt, rather than pointed, scabrules in the apical region of its lemmas.

17. Puccinellia distans (Jacq.) Parl. [p. 474]
EUROPEAN ALKALI GRASS,
PUCCINELLIE À FLEURS DISTANTES

Plants perennial; cespitose, not mat-forming. **Culms** 5–60 cm, erect to decumbent. **Ligules** 0.8–1.2 mm, obtuse to truncate, usually entire; **blades** 1–7 mm wide, flat to involute. **Panicles** 2.5–20 cm, diffuse at maturity, lower branches horizontal to descending, spikelets usually confined to the distal $^2/_3$; **pedicels** scabrous, lacking tumid epidermal cells. **Spikelets** 2.5–7 mm, with 2–7 florets. **Glumes** rounded over the back, veins obscure, apices acute to truncate; **lower glumes** 0.4–1.3 mm; **upper glumes** 0.9–1.8 mm; **calluses** with a few hairs; **lemmas** 1.5–2(2.2) mm, mostly herbaceous, glabrous or sparsely hairy on the lower $^1/_2$, principally along the veins, backs rounded, 5-veined, veins obscure, not extending to the margins, apical margins hyaline and often yellowish, uniformly and densely scabrous, apices widely obtuse to truncate, entire; **palea veins** shortly ciliate proximally, glabrous, sometimes scabrous distally; **anthers** 0.4–0.8 mm. 2*n* = 14, 28, 42.

Puccinellia distans is a Eurasian native, reportedly introduced in North America, where it is widespread, particularly as a weed in non-littoral environments, including the margins of salted roads. It is also found occasionally in coastal environments.

A specimen in the Smithsonian Institution attributed to *Puccinellia tenuiflora* (Griseb.) Scribn. & Merr. by Scribner and Merrill (1910) is a robust example of *P. distans*; others of this taxon in North America remain to be investigated. *Puccinellia hauptiana* (Trin. *ex* V.I. Krecz.) Kitag. has been reduced to *P. distans* subsp. *hauptiana* (Trin. *ex* V.I. Krecz.) W.E. Hughes. This taxon may represent a tetraploid component of *P. distans*; several specimens identified as *P. hauptiana* in Eurasia have a tetraploid chromosome count. Specimens from Alaska, the Yukon, and Saskatchewan, identified as *P. hauptiana* on the basis of their relatively narrow leaves (1–2 mm wide) and small anthers (0.5–0.6 mm), appear to be native in these regions and require further study with regard to their relationship with *P. distans*.

palea   rachilla   lemma   spikelet

lemma   palea   spikelet

floret   anthers   pedicel   spikelet

P. andersonii   *P. distans*   *P. howellii*

CTR

PUCCINELLIA

18. Puccinellia howellii J.I. Davis [p. 474]
HOWELL'S ALKALI GRASS

Plants perennial; cespitose, not mat-forming. **Culms** 7–40 cm, erect to ascending. **Ligules** 1.5–2.7 mm, obtuse, entire or minutely and irregularly serrate; **blades** involute, 1.4–2.2 mm wide when flattened. **Panicles** 2–13 cm, compact to diffuse at maturity, lower branches erect to descending, spikelet-bearing from near the base or confined to the distal 1/2; **pedicels** smooth or with a few scattered scabrules, sometimes with tumid epidermal cells. **Spikelets** 3–8 mm, with (1)2–5 florets. **Glumes** rounded or weakly keeled over the back, veins obscure, apices acute to obtuse; **lower glumes** 0.8–1.9 mm; **upper glumes** 1.7–2.5 mm; **calluses** with a few hairs; **lemmas** 2.4–3.3 mm, herbaceous, glabrous or with a few hairs near the base, principally along the veins, backs rounded or weakly keeled distally, 5-veined, veins obscure, not extending to the margins, apical margins uniformly and densely scabrous, apices acute to obtuse, entire; **palea veins** smooth and glabrous proximally, smooth or scabrous distally; **anthers** 1.5–2 mm. 2n = unknown.

Puccinellia howellii is known only from the type locality in Shasta County, California, where it is a dominant element of the vegetation associated with a group of three mineralized seeps. Isozyme profiles suggest that it is a polyploid.

19. Puccinellia nuttalliana (Schult.) Hitchc. [p. 476]
NUTTALL'S ALKALI GRASS

Plants perennial; cespitose, not mat-forming. **Culms** 10–100 cm, usually erect. **Leaves** either concentrated at the base or distributed along the culms; **ligules** 1–3 mm, obtuse, usually entire, sometimes slightly erose; **blades** 1–4 mm wide, flat to involute. **Panicles** 5–30 cm, compact to diffuse at maturity, usually distinctly exserted from the sheaths, lower branches usually erect to diverging, occasionally descending, spikelet-bearing from the base or on the distal 2/3; **pedicels** scabrous, lacking tumid epidermal cells. **Spikelets** 3.5–9 mm, with 2–7 florets. **Glumes** rounded over the back, veins obscure, apices acute to obtuse; **lower glumes** 0.5–1.5 mm, usually less than 1/2 as long as the adjacent lemmas; **upper glumes** 1–2.8 mm; **rachilla internodes** slightly and gradually broadened to the point of attachment with the lemmas; **calluses** with a few hairs; **lemmas** (2)2.2–3(3.5) mm, herbaceous, glabrous or sparsely hairy on the proximal 1/2, principally along the veins, backs rounded, 5-veined, veins obscure, not extending to the margins, smooth distally, lateral margins inrolled or not, apical margins uniformly and densely scabrous, apices acute to obtuse, entire; **palea veins** glabrous, short-ciliate, or with a few long hairs proximally, smooth or scabrous distally; **anthers** 0.6–2 mm. 2n = 28, 42, 56.

Puccinellia nuttalliana is a widespread and variable species, restricted to the *Flora* region. It grows principally in the interior, but is also found in coastal settings, where it is difficult to distinguish from *P. nutkaensis*. Northern, primarily boreal or southern arctic populations with relatively short lemmas and anthers (2–2.8 mm and 0.6–0.9 mm, respectively), and with a few long hairs on the lower palea veins, have sometimes been recognized as *P. borealis* Swallen.

20. Puccinellia nutkaensis (J. Presl) Fernald & Weath. [p. 476]
ALASKA ALKALI GRASS, PACIFIC ALKALI GRASS, PUCCINELLIE BRILLANTE

Plants perennial; cespitose, occasionally appearing rhizomatous or stoloniferous after rooting at the nodes of buried stems, not mat-forming. **Culms** 10–90 cm, usually erect, sometimes decumbent. **Leaves** usually distributed evenly along the culms; **ligules** 1–3 mm, obtuse to truncate, entire; **blades** 1.5–6 mm wide when flat, flat to involute. **Panicles** 5–30 cm, compact to diffuse at maturity, lower branches usually erect to ascending, occasionally spreading to descending, spikelet-bearing from near the base or the spikelets confined to the distal 1/2; **pedicels** from sparsely to densely scabrous, epidermal cells often tumid. **Spikelets** 3.5–12 mm, with 3–7 florets. **Glumes** rounded over the back, veins obscure, apices acute to truncate; **lower glumes** 1–1.6 mm; **upper glumes** 2–3 mm; **calluses** with a few hairs; **lemmas** (2.2)3–4.5(5) mm, herbaceous, glabrous or sparsely hairy on the proximal 1/2, principally along the veins, backs rounded, 5-veined, veins obscure, not extending to the margins, midveins smooth distally, apical margins uniformly and densely scabrous, apices usually acute to obtuse, sometimes acuminate, entire; **palea veins** glabrous or with short hairs proximally, scabrous distally; **anthers** 0.5–1.4 mm. 2n = 42, 56.

Puccinellia nutkaensis grows in coastal habitats of continental North America and Greenland, generally in sand and stones in protected intertidal environments. It is variable in form, ranging from diminutive plants that

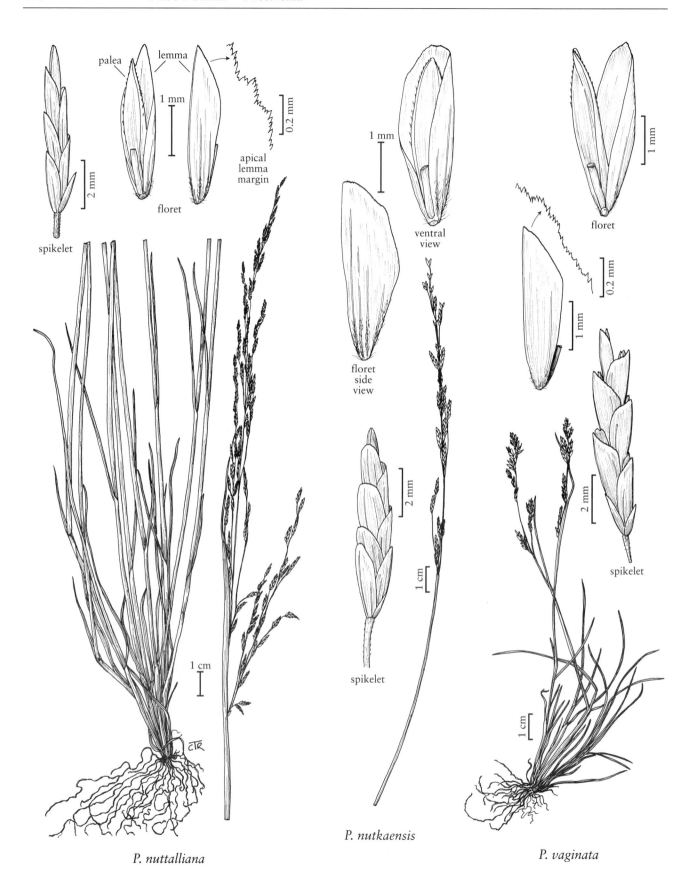

palea   lemma

1 mm

0.2 mm

apical
lemma
margin

floret

2 mm

spikelet

1 mm

ventral
view

floret
side
view

2 mm

spikelet

1 cm

P. nutkaensis

P. nuttalliana

1 cm

1 mm

floret

0.2 mm

1 mm

2 mm

spikelet

P. vaginata

1 cm

PUCCINELLIA

resemble *P. pumila* to tall, erect plants, often with dense or open inflorescences, resembling *P. nuttalliana*. Larger plants on the Pacific coast have been called *P. grandis* Swallen, and those on the Atlantic coast *P. lucida* Fernald & Weath., but there are many plants of intermediate stature.

21. Puccinellia vaginata (Lange) Fernald & Weath. [p. 476]

SHEATHED ALKALI GRASS, PUCCINELLIE ENGAINÉE

Plants perennial; cespitose, not mat-forming. **Culms** (6)9–20 (35) cm, erect or decumbent. **Leaves** basally concentrated on small plants, distributed along the culms in larger plants; **ligules** 1–3 mm, acute, obtuse, or truncate, entire; **blades** 1–2 mm wide when flat, 0.5–1.6 mm in diameter when involute. **Panicles** (3)6–12(14) cm, diffuse or contracted and usually barely exserted from the sheaths at maturity, lower branches ascending to horizontal, spikelets usually confined to the distal ²/₃; **pedicels** slightly scabrous, epidermal cells tumid. **Spikelets** 4–8 mm, with (2)4–5(6) florets. **Glumes** rounded over the back, veins distinct or obscure, margins not inrolled, apices broadly acute or obtuse; **lower glumes** 1.3–2.1 mm; **upper glumes** 1.3–2.6(3.4) mm; **calluses** with a few hairs; **lemmas** 2.8–4 mm, herbaceous, thin, often translucent, bases hairy, particularly on the veins, backs rounded, 5-veined, veins obscure, not extending to the margins, lateral margins usually not inrolled, apical margins uniformly and densely scabrous, apices broadly acute or obtuse, entire or slightly erose; **palea veins** glabrous proximally, scabrous distally; **anthers** 0.7–1.1(1.2) mm. $2n = 56$.

Puccinellia vaginata is a widespread North American arctic species. It is locally common in Greenland to about 78° N latitude; it is uncommon in Canada, where it is found primarily in the low arctic, rarely in the high arctic. One specimen is reported from eastern Siberia. It grows on coastal marine sediments and on eroding, raised marine sediments inland, and forms large tussocks near bird cliffs.

14.07 ×PUCCIPHIPPSIA Tzvelev

Stephen J. Darbyshire

Plants perennial; cespitose. **Culms** to 23 cm. **Sheaths** closed for ¹/₃–²/₃ their length, glabrous; **auricles** absent; **ligules** membranous; **blades** flat or folded, glabrous. **Inflorescences** compact panicles. **Spikelets** laterally compressed, with (1)2–3(4) florets; **rachillas** prolonged beyond the base of the distal floret; **disarticulation** above the glumes. **Glumes** usually 2, occasionally 1, unequal, membranous, hyaline, acute to obtuse; **lemmas** unawned; **paleas** 2-veined, veins spiculose; **anthers** indehiscent. **Caryopses** absent.

×*Pucciphippsia* consists of hybrids between *Puccinellia* and *Phippsia*. Two different hybrids are known. Only one, ×*Pucciphippsia vacillans*, grows in the *Flora* region. ×*Pucciphippsia czukczorum* Tzvelev is currently known only from the Chukotsk Peninsula in northeastern Russia, although its parents, *Puccinellia wrightii* and *Phippsia algida*, have both been found in Alaska on the Seward Peninsula, and in Greenland on the Hayes Peninsula and in the Ammassalik region. Both hybrids are completely sterile, as is indicated by their indehiscent anthers and lack of caryopses. They differ from *Puccinellia* in having small, compact panicles, and from *Phippsia* in usually having more than one floret per spikelet.

SELECTED REFERENCE **Hedberg, O.** 1962. The genesis of *Puccinellia vacillans*. Bot. Tidsskr. 58:157–167.

1. ×Pucciphippsia vacillans (Th. Fr.) Tzvelev [p. 479]

Culms 7–15(23) cm, erect, glabrous. **Sheaths** closed for ⅓–⅔ their length; **ligules** 0.5–2.7 mm, apices entire, truncate, rounded or acuminate; **blades** 1.5–6(8) cm long, 1.5–3 mm wide. **Panicles** 1.5–4.5 cm long, 1–2 cm wide; **branches** smooth; **primary branches** 1.3–14 mm, erect to ascending. **Spikelets** 2–3.5 mm, pedicellate; **rachillas** glabrous or pubescent, internodes 0.2–2.2 mm, prolonged beyond the uppermost floret. **Glumes** glabrous; **lower glumes** 0.4–0.8 mm, (0)1-veined, occasionally absent; **upper glumes** 0.5–1.4 mm, 1–3-veined, obtuse; **lemmas** 1.8–2.6 mm, (1)3-veined, pilose basally, particularly on the veins, apices entire or erose, truncate to acute; **paleas** 1.5–2.5 mm; **anthers** 1–3, 0.5–1 mm. 2*n* = 21.

×*Pucciphippsia vacillans* is a sterile hybrid between *Puccinellia vahliana* and *Phippsia algida*. It grows in damp, open habitats such as wet meadows, wet tundra, pond margins, and imperfectly drained areas in silt, till, and moss. It is currently known in North America from Axel Hedberg, Bathurst, Cornwallis, Ellesmere, Devon, and Baffin islands in arctic Nunavut, and from Midgaardsormen, Greenland. Outside North America, it is found on Svalbard and Novaya Zemlya islands, where both parents are also present.

14.08 PHIPPSIA (Trin.) R. Br.

Laurie L. Consaul

Susan G. Aiken

Plants usually perennial, annual in some environments that have a sufficient growing season for seed production in a single year; cespitose or matlike. **Culms** 2–19 cm, erect or procumbent, not rooting at the lower nodes, glabrous; **prophylls** 4–10 mm. **Sheaths** not inflated, those of the basal leaves usually fused only near the base, those of the flag leaves closed for at least ½ their length; **auricles** absent; **ligules** membranous, glabrous, acute; **blades** flat or conduplicate. **Inflorescences** panicles, dense or diffuse. **Spikelets** pedicellate, with 1 floret, laterally compressed to nearly terete; **rachillas** not prolonged beyond the florets; **disarticulation** above the glumes and beneath the florets. **Glumes** to ⅓ the length of the florets, ovate, without veins, caducous or persistent, unawned; **lower glumes** highly reduced; **calluses** short, glabrous; **lemmas** 1–3-veined, not strongly keeled, apices acute to rounded, unawned; **paleas** subequal to the lemmas; **lodicules** 2, free, glabrous; **anthers** 1 or 2; **ovaries** glabrous. **Caryopses** exceeding the lemmas and exposed at maturity. *x* = 7. Named for Constantine John Phipps (1744–1792), a captain in Britain's Royal Navy and an arctic explorer.

Phippsia has two species, one of which is found in arctic Eurasia, Greenland, and the Canadian arctic islands; the other is circumpolar in the arctic, and is also known from the Rocky Mountains of North America and the high Andes of South America.

SELECTED REFERENCES Aares, E., M. Nurminiemi, and C. Brochmann. 2000. Incongruent phylogeographies in spite of similar morphology, ecology, and distribution: *Phippsia algida* and *P. concinna* (Poaceae) in the North Atlantic region. Pl. Syst. Evol. 220:241–261; Aiken, S.G., L.L. Consaul, and M.J. Dallwitz. 1995 on. Grasses of the Canadian arctic archipelago: Descriptions, illustrations, identification, and information retrieval. http://www.mun.ca/biology/delta/arcticf/poa/index.htm; Bay, C. 1992. A phytogeographical study of the vascular plants of northern Greenland—north of 74° northern latitude. Meddel. Grønland, Biosci. 36:1–102; Steen, N.W., R. Elven, and I. Nordal. 2004. Hybrid origin of the arctic ×*Pucciphippsia vacillans* (Poaceae): Evidence from Svalbard plants. Pl. Syst. Evol. 245:215–238.

1. Panicles (0.5)1–2(3) cm long, 3–7 mm wide; spikelet length less than twice the width; glumes caducous, often colorless when present; lemmas often yellow-green, purple coloration, when present, not reaching the apices, glabrous or with a few soft hairs on the lower ⅓; caryopses ellipsoid, widest at or just above the middle . . . . 1. *P. algida*

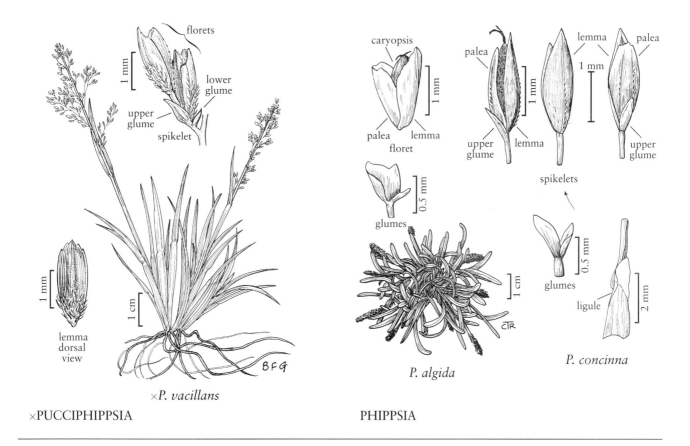

×*P. vacillans*

×PUCCIPHIPPSIA

PHIPPSIA

P. algida

P. concinna

1. Panicles (1)3–9 cm long, (4)5–15 mm wide; spikelet length 2–3 times the width; glumes, particularly the upper glumes, not caducous, often still present on the previous season's growth, often with some deep purple coloration; lemmas usually purplish-red with a strong coloration over the veins, the color over the midvein extending to the apex, veins with stiff hairs, soft or stiff hairs elsewhere over the lower ¹/₂–²/₃ of the surface; caryopses ovoid, widest below the middle . . . . 2. *P. concinna*

1. Phippsia algida (Sol.) R. Br. [p. 479]
ICEGRASS, PHIPPSIE FROIDE

Culms (2)3.5–15 cm. **Leaves** cauline or mostly basal; **sheaths** glabrous; **ligules** 0.3–1(1.6) mm, entire, acute; **blades** 0.6–2.8 cm long, 1.2–3 mm wide, glabrous. **Panicles** (0.5) 1–2(3) cm long, 3–7 mm wide, dense or slightly diffuse; **rachises** smooth; **primary branches** (0.9)3–8(9.5) mm, smooth, usually appressed, sometimes spreading, particularly at anthesis; **secondary branches** appressed. **Spikelets** (1)1.4–1.8 mm long, (0.5)0.7–0.9 mm wide, with 1 floret. **Glumes** caducous, often colorless when present; **lower glumes** 0.05–0.3 mm, obtuse; **upper glumes** 0.3–0.6 mm; **lemmas** 1.3–1.8 mm, broadly ovate, rounded on the back, often yellow-green, never

with purple coloration extending to the apices, glabrous or with a few soft hairs on the lower ¹/₃, sometimes sparsely scabridulous below, apices acute to rounded, entire or lacerate; **paleas** 1.1–1.3 mm, glabrous; **anthers** 0.3–0.7 mm. **Caryopses** 1.2–1.6 mm long, 0.3–0.5 mm wide, ellipsoid, widest at or just above the middle. $2n = 28$.

Phippsia algida is a circumpolar species that also grows at high elevations in the Rocky Mountains and the Andes. It is one of the first grasses to flower in the high arctic, which may contribute to its success as an early colonizer of disturbed areas. Although highly nitrophilous, it can tolerate a wide range of soils, from highly alkaline to peat and imperfectly drained mud flats. It sometimes hybridizes with *Puccinellia vahliana*. Plants of *P. algida* with slightly hairy lemmas have been recognized as *P. algida* f. *vestita* Holmb.

The original circumscription of *Phippsia algida* subsp. *algidiformis* (Harry Sm.) Á. Löve & D. Löve is unclear (Steen et al. 2004). Harry (K.A.H.) Smith subsequently included his specimens in *P. concinna*, indicating that he no longer recognized the taxon he had described. Some of his specimens, and those of Bay (1992) collected in Greenland, that are deposited at C and O have been reidentified as *P. algida* or *P. concinna* (some as depauperate *P. concinna*) by Reidar Elven (University of Oslo).

2. **Phippsia concinna** (Th. Fr.) Lindeb. [p. 479]
SNOWGRASS

Culms (3)6–19 cm. **Leaves** cauline or mostly basal; **sheaths** glabrous; **ligules** 0.3–1(1.6) mm, entire, acute; **blades** 1–9.5 cm long, 1–3 mm wide, glabrous. **Panicles** (1)3–9 cm long, (4)5–15 mm wide, dense or diffuse; **rachises** smooth; **primary branches** 3–10(34) mm, smooth, appressed or spreading, particularly at anthesis; **secondary branches** appressed. **Spikelets** (1.2)1.5–2.1 mm long, 0.4–1 mm wide, with 1 floret. **Glumes**, at least the upper glumes, not caducous, often still present on the previous season's growth, often with some deep purple coloration; **lower glumes** 0.3–0.6 mm, obtuse, sometimes absent; **upper glumes** 0.5–0.8 mm; **lemmas** 1.4–2 mm, broadly ovate, usually purplish-red, particularly over the veins, the color over the midveins extending to the apices, rounded on the back, veins with stiff hairs, soft or stiff hairs elsewhere over the lower ¹/₂–²/₃ of the surface; **apices** acute to rounded, entire or lacerate; **paleas** 1.2–1.8 mm, glabrous; **anthers** 0.3–0.7 mm. **Caryopses** 1.2–1.9 mm long, 0.4–0.7 mm wide, ovoid, widest below the middle, tapering to the apices. $2n = 28$.

The distribution of *Phippsia concinna* has previously been restricted to arctic Eurasia and Greenland. Specimens of *P. concinna* from latitudes near 60° N in Eurasia are relatively large plants, usually near 20 cm tall, with open inflorescences from spreading panicle branches. Specimens from northern Greenland latitudes near 80° N are usually smaller plants, with compact inflorescences and erect panicle branches. Greenland specimens from C, many of them annotated as *P. algida* subsp. *algidiformis* by Bay, were compared by Reidar Elven and Susan Aiken (May 2005) with similar specimens collected from Ellesmere Island at O. It was concluded that *P. concinna* is present in the Canadian arctic archipelago. As much of the Canadian arctic archipelago is colder than the area of Ellesmere Island near 80° N, candidate specimens for *P. concinna* are small plants, and are superficially like those of *P. algida* collected in the same areas. There are many specimens at CAN that have some of the characteristics of *P. concinna*, particularly the upper glume characteristics. The shape of the fruit character is often not available for consideration on pressed specimens. The number and characteristics of the hairs on the lemma are variable, and many specimens may be interpreted as intermediate in this character. A more thorough investigation is needed to determine the extent of *P. concinna* in North America.

14.09 SCLEROCHLOA P. Beauv.

David M. Brandenburg

Plants annual. **Culms** 2–30 cm. **Sheaths** open to closed; **auricles** absent; **ligules** membranous; **blades** flat or folded. **Inflorescences** terminal, usually racemes, sometimes reduced panicles, 1-sided, usually exceeded by the leaves. **Spikelets** laterally compressed, subsessile to pedicellate; **pedicels** 0.5–1 mm long, 0.5–0.8 mm thick, stout, with 2–7 florets; **rachillas** glabrous, lowest internodes thicker than those above; **disarticulation** tardy, not strongly localized. **Glumes** unequal, shorter than the lowest lemmas, glabrous, with wide hyaline margins, apices obtuse to emarginate, unawned; **lower glumes** (1)3–5-veined; **upper glumes** (3)5–9-veined; **calluses** blunt, glabrous; **lemmas** membranous, with hyaline margins, indurate at maturity, (5)7–9-veined, veins prominent, apices rounded to emarginate, entire, unawned; **paleas** shorter than to equaling the lemmas, dorsally compressed; **lodicules** 2, free, glabrous, entire to lacerate; **anthers** 3. **Caryopses** shorter than the lemmas, concealed at maturity, beaked from the persistent style base, falling free; **hila** round. $x = 7$. Name from the Greek *skleros*, 'hard', and *chloa*, 'grass', alluding to the leathery glumes and lemmas.

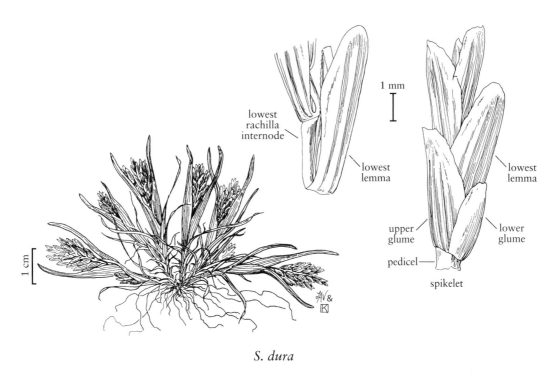

S. dura

SCLEROCHLOA

Sclerochloa is a genus of two species, both of which are native to southern Europe and western Asia. The species found in the *Flora* region, *S. dura*, is now a cosmopolitan weed.

SELECTED REFERENCES **Brandenburg, D.M., J.R. Estes,** and **J.W. Thieret.** 1991. Hard grass (*Sclerochloa dura*, Poaceae) in the United States. Sida 14:369–376; **Brandenburg, D.M.** and **J.W. Thieret.** 1996. *Sclerochloa dura* (Poaceae) in Kentucky. Trans. Kentucky Acad. Sci. 57:47–48; **Cusick, A.W., R.K. Rabeler,** and **M.J. Oldham.** 2002. Hard or fairgrounds grass (*Sclerochloa dura*, Poaceae) in the Great Lakes region. Michigan Bot. 41:125–135.

1. **Sclerochloa dura** (L.) P. Beauv. [p. 481]
HARDGRASS, FAIRGROUND GRASS

Plants often matted, occasionally with solitary culms. **Culms** 2–15(30) cm, usually prostrate to procumbent, sometimes ascending, glabrous. **Leaves** strongly overlapping, generally exceeding the inflorescences; **sheaths** completely open or closed to ½ their length; **ligules** (0.3)0.7–2(3.5) mm, glabrous, acute; **blades** 0.5–5(7) cm long, 1–4 mm wide, glabrous, apices prow-tipped. **Racemes** 1–4(5) cm, often partially enclosed in the upper leaf sheaths; **pedicels** 0.5–0.8 mm thick. **Spikelets** (3.4)5–12 mm, with (2)3–4(7) florets. **Lower glumes** 1.4–3(3.7) mm; **upper glumes** 2.6–5.4(6.2) mm; **lowest lemmas** (3.4)4.5–6(7) mm, midveins scabridulous distally; **distal lemmas** successively smaller than the lower lemmas; **anthers** 0.8–1.5 mm. **Caryopses** weakly trigonous, rugulose. $2n = 14$.

First collected in the United States in 1895, *Sclerochloa dura* is probably more widespread than indicated, because it is easily overlooked. It grows in lawns, campsites, roadsides, athletic fields, fairgrounds, and other disturbed sites. It is frequently found in severely compacted soils, because it can withstand heavy traffic by vehicles and pedestrians.

Sclerochloa dura is sometimes confused with *Poa annua*. The two species are superficially similar, occupy similar habitats, and have a similar phenology, but *S. dura* has blunt, glabrous lemmas and racemose inflorescences, whereas *P. annua* has obtuse to acute lemmas that are smooth and usually sericeous or crisply puberulent over the veins, and paniculate inflorescences. Plants of *S. dura* become stramineous in age,

making them easy to locate because areas dominated by this species change color.

14.10 DACTYLIS L.

Kelly W. Allred

Plants perennial; cespitose, sometimes with short rhizomes. **Culms** to 2.1+ m, bases laterally compressed; **internodes** hollow; **nodes** glabrous. **Leaves** mostly basal, glabrous; **sheaths** closed for at least ½ their length; **auricles** absent; **ligules** membranous; **blades** flat to folded. **Inflorescences** panicles; **primary branches** 1-sided, naked proximally, with dense clusters of subsessile spikelets distally, at least some branches longer than 1 cm. **Spikelets** oval to elliptic in outline, laterally compressed, with 2–6 florets; **rachillas** glabrous, not prolonged beyond the distal floret; **disarticulation** above the glumes and beneath the florets. **Glumes** shorter than the florets, lanceolate, 1–3-veined, ciliate-keeled, awn-tipped; **calluses** short, blunt; **lemmas** 5-veined, scabrous to ciliate-keeled, tapering to a short awn; **paleas** 2-keeled, tightly clasped by the lemmas, unawned, apices notched; **lodicules** 2, glabrous, toothed; **anthers** 3; **ovaries** glabrous. **Caryopses** shorter than the lemmas, concealed at maturity, oblong to ellipsoid, falling free or adhering to the lemma and/or palea; **hila** round. $x = 7$. Name from the Greek *daktylos*, 'finger'.

Dactylis is interpreted here as a variable monotypic genus, although five species are recognized by Russian taxonomists. Numerous infraspecific taxa have been recognized in Eurasia, where *Dactylis* is native, but it does not seem feasible to identify subspecies and varieties in North America.

SELECTED REFERENCE Stebbins, G.L., Jr. and D. Zohary. 1959. Cytogenetic and evolutionary studies in the genus *Dactylis*. Univ. Calif. Publ. Bot. 31:1–40.

1. **Dactylis glomerata** L. [p. 483]
ORCHARDGRASS, DACTYLE PELOTONNÉ

Culms to 2.1+ m, erect. **Leaves** dark green; **sheaths** longer than the internodes, glabrous, usually keeled; **ligules** 3–11 mm, truncate to acuminate; **blades** (2)4–8(10) mm wide, elongate, lax, with a conspicuous midrib and white, scabridulous to scabrous margins. **Panicles** 4–20 cm, typically pyramidal, lower branches spreading, upper branches appressed. **Spikelets** 5–8 mm, subsessile. **Glumes** 3–5 mm; **lemmas** 4–8 mm, scabridulous; **paleas** slightly shorter than the lemmas; **anthers** 2–3.5 mm. $2n = 14, 21, 27–31, 42$.

Dactylis glomerata grows in pastures, meadows, fence rows, roadsides, and similar habitats throughout North America. Native to Eurasia and Africa, it has been introduced throughout most of the cool-temperate regions of the world as a forage grass. It provides nutritious forage that is relished by all livestock, as well as by deer, geese, and rabbits. When abundant, the pollen can be a major contributor to hay fever.

The species includes both diploid and tetraploid populations. Although several infraspecific taxa have been described, based generally on the size of the stomata and pollen, variation in pubescence, and panicle features, formal taxonomic recognition does not seem warranted. Numerous cultivars have been developed for agricultural use.

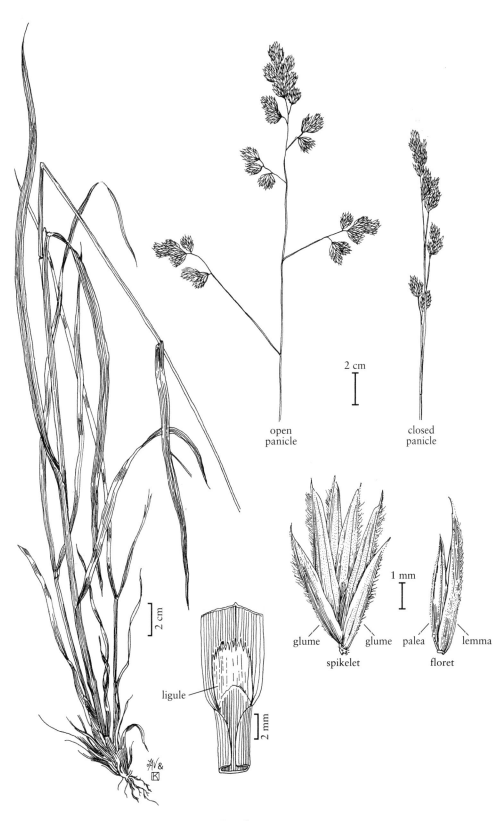

open
panicle

2 cm

closed
panicle

2 cm

ligule

2 mm

1 mm

glume　　　glume
spikelet

palea　　　lemma
floret

D. glomerata

DACTYLIS

14.11 LAMARCKIA Moench

Lynn G. Clark

Plants annual; tufted. **Culms** 5–40 cm, erect or decumbent at the base, glabrous. **Sheaths** open for at least ⅓ their length, glabrous, margins membranous and continuous with the ligules; **auricles** absent; **ligules** membranous, acute, glabrous, apices somewhat erose; **blades** flat, glabrous. **Inflorescences** panicles, dense, secund, golden-yellow to purplish; **primary branches** appressed to the rachis; **secondary branches** capillary, smooth, glabrous, flexuous, terminating in 1–4 fascicles of pedicellate spikelets, strongly bent below the junction with the fascicle base; **pedicels of spikelets** in each fascicle fused at the base, strigose. **Spikelets** dimorphic, the terminal spikelet of each fascicle fertile, the others sterile; **disarticulation** at the base of the fused pedicels. **Fertile spikelets** terete to somewhat laterally compressed, with 2 florets, 1 bisexual, the other a rudiment, borne on long rachilla internodes; **calluses** short, blunt, glabrous; **glumes** narrow, acuminate or short-awned, 1-veined; **bisexual lemmas** scarcely veined, each with a delicate, straight, subapical awn; **paleas** equal or subequal to the lemmas; **lodicules** 2, glabrous, toothed, **anthers** 3; **ovaries** glabrous; **rudimentary florets** highly reduced lemmas, each with a delicate, straight awn. **Sterile spikelets** linear, laterally compressed; **florets** 5–10; **glumes** similar to those of the fertile spikelets; **lemmas** imbricate, obtuse, unawned; **paleas** absent. **Caryopses** ovoid or ellipsoid, adhering to the lemmas and/or paleas. *x* = 7. Named for Jean Baptiste Antoine Pierre Monnet de Lamarck (1744–1829), a French biologist best known for his ideas about evolution by the inheritance of acquired characteristics.

Lamarckia is a monotypic genus that has been introduced into North America from the Mediterranean and the Middle East.

1. Lamarckia aurea (L.) Moench [p. 485]
GOLDENTOP

Culms (5)10–40 cm. **Ligules** 3–10 mm; **blades** 2.5–11(20) cm long, 2.5–7.5 mm wide. **Panicles** 2–8 cm long, 0.9–3.2 cm wide; **branches** 1–2 cm, with a few scattered, whitish hairs. **Fertile spikelets** 2.6–4.2 mm; **lower rachilla internodes** about 1 mm, glabrous; **upper rachilla internodes** 1.4–2.2(2.6) mm, puberulent or glabrous, sometimes scabrous; **glumes** 2.6–4.2 mm, bisexual florets 2.4–3 mm, minutely bifid, teeth shorter than 0.2 mm, apices awned, awns 5–7.2 mm; **anthers** 0.4–0.7 mm; **rudimentary florets** 0.4–0.6 mm, awns 4–5(7) mm. **Sterile spikelets** 6–9 mm; **lemmas** 1–2 mm. **Caryopses** 1–2 mm. 2*n* = 14.

Lamarckia aurea grows on open ground, rocky hillsides, and in sandy soil, at elevations from sea level to 700 m. Within the *Flora* region, it is known only from the southwest. It is an attractive, but rather weedy species.

14.12 BECKMANNIA Host

Stephan L. Hatch

Plants annual and tufted, or perennial and rhizomatous. **Culms** 20–150 cm, sometimes tuberous at the base, erect. **Leaves** mostly cauline; **sheaths** open, glabrous, ribbed; **auricles** absent; **ligules** membranous, acute; **blades** flat, glabrous. **Inflorescences** dense, spikelike panicles; **branches** 1-sided, racemosely arranged, secondary branches few, at least some branches longer than 1 cm, with closely imbricate spikelets; **disarticulation** below the glumes, the spikelets falling entire.

reduced
floret

bisexual
floret

reduced
floret

bisexual
floret

1 mm

florets

1 mm

upper
glume

lower
glume

fertile
spikelet

2 cm

L. aurea

2 cm

1 mm

floret

1 mm

glume   glume

spikelet

B. syzigachne

LAMARCKIA   BECKMANNIA

Spikelets laterally compressed, circular, ovate or obovate in side view, subsessile, with 1–2 florets; **rachillas** not prolonged beyond the base of the distal floret. **Glumes** subequal, slightly shorter than the lemmas, inflated, keeled, D-shaped in side view, unawned; **calluses** blunt, glabrous; **lemmas** lanceolate, inconspicuously 5-veined, unawned; **paleas** subequal to the lemmas; **lodicules** 2, free; **anthers** 3; **ovaries** glabrous. **Caryopses** shorter than the lemmas, concealed at maturity. $x = 7$. Named for Johann Beckmann (1739–1811), a German botanist and author of one of the first botanical dictionaries.

Beckmannia is a genus of two species: an annual species usually with one fertile floret per spikelet that is native to North America and Asia, and a perennial species with two fertile florets per spikelet that is restricted to Eurasia.

SELECTED REFERENCE Reeder, J.R. 1953. Affinities of the grass genus *Beckmannia* Host. Bull. Torrey Bot. Club 80:187–196.

1. Beckmannia syzigachne (Steud.) Fernald [p. 485]
AMERICAN SLOUGHGRASS, BECKMANNIE À ÉCAILLES UNIES

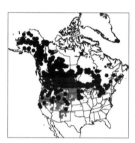

Plants annual; tufted. **Culms** 20–120 cm. **Ligules** 5–11 mm, pubescent, entire or lacerate, usually folded back; **blades** 4–10(20) mm wide, flat, scabrous. **Panicles** 7–30 cm; **branches** spikelike, usually 1–2 cm. **Spikelets** 2–3 mm, round to ovate in side view, with 1 floret, a second undeveloped or well-developed floret occasionally present. **Glumes** appearing inflated, strongly keeled, 3-veined, apiculate; **lemmas** 2.4–3.5 mm, unawned, sometimes mucronate; **paleas** subequal to the lemmas, acute; **anthers** 0.5–1(1.5) mm, pale yellow. **Caryopses** shorter than 2 mm, light to medium brown. $2n = 14$.

Beckmannia syzigachne grows in damp habitats such as marshes, floodplains, the edges of ponds, lakes, streams, and ditches, and in standing water. It is a good forage grass, but frequently grows in easily damaged habitats.

14.13 POA L.

Robert J. Soreng

Plants annual or perennial; usually synoecious, sometimes monoecious, gynodioecious, dioecious, and/or asexual; with or without rhizomes or stolons, densely to loosely tufted or the culms solitary. **Basal branching** intravaginal, pseudointravaginal, or extravaginal; **prophylls** of intravaginal shoots 2-keeled and open, of pseudointravaginal shoots not keeled and tubular, of extravaginal shoots scalelike. **Culms** 1–150 cm, hollow, usually unbranched above the base. **Sheaths** from almost completely open to almost completely closed, terete or weakly to strongly compressed; **auricles** absent; **ligules** membranous, truncate to acuminate; **blades** 0.4–12 mm wide, flat, folded, or involute, adaxial surfaces with a groove on each side of the midvein, other intercostal depressions shallow, indistinct, apices often prow-shaped. **Inflorescences** usually terminal panicles, rarely reduced and racemelike. **Spikelets** 2–12 mm, usually laterally compressed, infrequently terete to subterete, usually lanceolate, sometimes ovate; **florets** (1)2–6(13), usually sexual, sometimes bulb-forming; **rachillas** usually terete, sometimes prolonged beyond the base of the distal floret; **disarticulation** above the glumes and beneath the florets. **Glumes** usually shorter than the lowest lemma in the spikelet, usually keeled, 1–3(5)-veined, unawned; **calluses** blunt, usually terete or slightly laterally compressed, sometimes slightly dorsally compressed, glabrous or hairy, hairs often concentrated in 1(3) tufts or *webs*, sometimes distributed around the calluses below the lemmas as a *crown of hairs*; **lemmas**

usually keeled, infrequently weakly keeled or rounded, similar in texture to the glumes, 5(7–11)-veined, lateral veins sometimes faint, margins scarious-hyaline distally, apices scarious-hyaline, truncate or obtuse to acuminate, unawned; **paleas** from ²⁄₃ as long as to subequal to the lemmas, distinctly 2-keeled, margins and intercostal regions milky white to slightly greenish; **lodicules** 2, broadly lanceolate, glabrous, lobed; **functional anthers** (1–2)3, 0.1–5 mm; **ovaries** glabrous. **Caryopses** 1–4 mm, ellipsoidal, often shallowly ventrally grooved, solid, with lipid; **hila** subbasal, round or oval, to ¹⁄₆ the length of the caryopses. $x = 7$. Name from the Greek *poa*, 'grass'.

Poa includes about 500 species. It grows throughout the world, principally in temperate and boreal regions. Sixty-one species and five hybrid species are native to the *Flora* region; nine species are introduced.

Poa is taxonomically difficult because most species are polyploid, many are apomictic, and hybridization is common. A variety of sexual reproductive systems are represented within the genus, but individual species are usually uniform in this regard. Apomicts derived from bisexual species usually have functional anthers; they require fertilization to stimulate endosperm (and hence seed) development. Apomicts derived from dioecious species do not require fertilization; they are normally pistillate with vestigial anthers 0.1–0.2 mm long.

Herbivores find most species of *Poa* both palatable and nutritious. *Poa fendleriana*, *P. secunda*, and *P. wheeleri* are important native forage species in western North America; *P. alpina*, *P. arctica*, and *P. glauca* are common components of alpine and arctic vegetation. Species of *Poa* sect. *Abbreviatae* are found near the limits of vegetation in both arctic and alpine regions.

Several introduced species of *Poa* are economically important. *Poa pratensis* is commonly cultivated for lawns and pasture, and is a major forage species in cooler regions of North America; *P. compressa* and *P. trivialis* are widely planted for soil stabilization and forage; *P. annua* is one of the world's most widespread weeds. *Poa bulbosa* has been cultivated; it is now widely established in the *Flora* region.

Vegetative characteristics that may be useful for distinguishing *Poa* from other morphologically similar genera are: the more or less straight, rather than curly, roots; two-grooved, prow-shaped blades; partially or wholly closed flag leaf sheaths; and isomorphic collar margins. Useful spikelet characteristics include: terete rachillas; multiple, relatively small, unawned florets; webbed calluses; well-developed palea keels; and the greenish or milky white intercostal regions of the paleas.

There is a strong correlation between the type of basal branching, prophyll structure, and blade development of the initial leaves. Extravaginal shoots have scalelike prophylls 0.5–3 mm long and initial leaves that are bladeless; intravaginal shoots have prominently keeled prophylls 10–50 mm long that are open on the abaxial side and initial leaves with well-developed blades; pseudointravaginal shoots develop intravaginally but have tubular, indistinctly keeled prophylls, and initial leaves with rudimentary blades.

In bulbiferous spikelets, the upper florets form a single tardily disarticulating offset or bulb, each lemma being thickened at the base and leaflike distally. The bulb falls as a unit, with or without the basal floret. The basal floret(s) may have pistils and stamens, and occasionally sets seed. Generally, there is a progression within an inflorescence, the earlier spikelets being bulbiferous and the later spikelets normal.

Callus hairs in *Poa* follow one of three patterns. In the most common pattern, there is an isolated dorsal tuft of crinkled or pleated hairs, the *web*, below the lemma keel. In a few species, additional webs may be present below the marginal veins. In the second pattern, crinkled hairs are distributed around the lemma base, but are somewhat concentrated and longer towards the back; this pattern is called a *diffuse web*. Webbed calluses are found only in *Poa*. In the third pattern, the hairs are straight to slightly sinuous, and more or less evenly distributed around the lemmas bases; calluses with such a pattern are described as having a *crown of hairs*.

Three named infrasectional hybrids are included in this treatment. One, *Poa arida*, is accounted for in the key. The other two are not. *Poa ×limosa* is too variable, and *P. ×gaspensis* is known from too few specimens to make their inclusion in the key helpful. All three are described at the end of this treatment, with comments on the probable parental taxa.

Unless stated otherwise, sheath closure is measured on the flag leaf, and ligule length on the upper 1–2 culm leaves; spikelet, floret, callus, lemma, and palea measurements are on non-bulb-forming florets; floret pubescence is evaluated on the lower florets within several spikelets; length of the callus hairs refers to their length when stretched out; anther measurements are based on functional anthers, i.e., those that produce pollen, as indicated by their being plump or, after the pollen is shed, by their open sacs. For hair lengths, puberulent is to about 0.15 mm long, short-villous to about 0.3 mm long, and long-villous from 0.3–0.4+ mm long, but these are only guidelines, not discrete categories; some species are only on one end of the range, and ranges have not been confirmed for every species. Many species key more than once, due in part to infraspecific variation.

SELECTED REFERENCES **Bowden, W.M.** 1961. Chromosome numbers and taxonomic notes on northern grasses: IV. Tribe *Festuceae*; *Poa* and *Puccinellia*. Canad. J. Bot. 39:123–138; **Duckert-Henroid, M.M.** and **C. Favarger.** 1987. Contribution à la cytotaxonomie et à la cytogéographie des *Poa* (Poaceae = Gramineae) de la Suisse. Mémoires de la Société Helvétique des Sciences Naturelles No. 100. Birkhäuser, Basel, Switzerland. 130 pp.; **Gillespie, L.J.** and **R.J. Soreng.** 2005. A phylogenetic analysis of the bluegrass genus *Poa* based on cpDNA restriction site data. Syst. Bot. 30:84–105; **Gillespie, L.J., A. Archambault,** and **R.J. Soreng.** 2006. Phylogeny of *Poa* (Poaceae) based on *trn*T-*trn*F sequence data: Major clades and basal relationships. Aliso 23:420–434; **Hiesey, W.M.** and **M.A. Nobs.** 1982. Interspecific hybrid derivatives between facultatively apomictic species of bluegrass and their responses to contrasting environments. Experimental Studies on the Nature of Species 6, Publication No. 636. Carnegie Institution of Washington, Washington, D.C., U.S.A. 119 pp.; **Hitchcock, A.S.** 1951. Manual of the Grasses of the United States, ed. 2, rev. A. Chase. U.S.D.A. Miscellaneous Publication No. 200. U.S. Government Printing Office, Washington, D.C., U.S.A. 1051 pp.; **Hultén, E.** 1942. Flora of Alaska and Yukon [in part]. Acta Univ. Lund, n.s., 38:1–281; **Kellogg, E.A.** 1985. A biosystematic study of the *Poa secunda* complex. J. Arnold Arbor. 66:201–242; **Marsh, V.L.** 1952. A taxonomic revision of the genus *Poa* of the United States and southern Canada. Amer. Midl. Naturalist 47:202–250; **Munz, P.A.** 1959. A California Flora. University of California Press, Berkeley. California, U.S.A. 1681 pp.; **Soreng, R.J.** 1985. *Poa* L. in New Mexico with a key to middle and southern Rocky Mountain species. Great Basin Naturalist 45:395–422; **Soreng, R.J.** 1991a. Notes on new infraspecific taxa and hybrids in North American *Poa* (Poaceae). Phytologia 71:340–413; **Soreng, R.J.** 1991b. Systematics of the "Epiles" group of *Poa* (Poaceae). Syst. Bot. 16:507–528; **Soreng, R.J.** 1993. *Poa* L. Pp. 1284, 1286–1291 *in* J.C. Hickman (ed.). The Jepson Manual of Higher Plants of California. University of California Press, Berkeley, California, U.S.A. 1400 pp.; **Soreng, R.J.** 1998. An infrageneric classification for *Poa* in North America, and other notes on sections, species, and subspecies of *Poa, Puccinellia,* and *Dissanthelium* (Poaceae). Novon 8:187–202. **Soreng, R.J.** 2005. Miscellaneous chromosome number reports for *Poa* L. (Poaceae) in North America. Sida 21:2195–2203; **Soreng, R.J.** and **D. Keil.** 2003. Sequentially adjusted sex-ratios in gynomonoecism, and *Poa diaboli* (Poaceae), a new species from California. Madroño 50: 300–306; **Tsvelev, N.N.** 1976. Zlaki SSSR. Nauka Publishers, Leningrad [St. Petersburg], Russia. 788 pp.; **Tutin, T.G.** 1952. Origin of *P. annua*. Nature 169:160.

1. Culms with bulbous bases; spikelets often bulbiferous (sect. *Arenariae*) . . . . 8. *P. bulbosa*
1. Culms with non-bulbous bases; spikelets sometimes bulbiferous.
    2. Some or all spikelets bulbiferous . . . . Subkey I
    2. Spikelets not bulbiferous, florets developing normally.
        3. Anthers 0.1–1(1.2) mm long in all florets and well developed, or only the upper 1–2 florets with rudimentary anthers; plants annual or perennial . . . . Subkey II
        3. Some anthers (1.2)1.3–4 mm long, or the florets pistillate and all anthers vestigial and 0.1–0.2 mm long, or longer and poorly developed; plants perennial.
            4. Plants rhizomatous or stoloniferous, rhizomes or stolons usually longer than 5 mm; basal leaves of the erect shoots with well-developed blades; plants densely to loosely cespitose or the culms solitary . . . . Subkey III
            4. Plants neither rhizomatous nor stoloniferous; basal leaves of the erect shoots sometimes without blades; plants densely cespitose . . . . Subkey IV

Poa Subkey I

Culms without bulbous bases. **Leaf sheaths** not swollen at the base. **Spikelets** mostly bulbiferous, lower spikelets in each panicle frequently normal or subnormal, basal florets of the bulbiferous spikelets frequently normal or subnormal.

1. Sheaths closed for $^1/_{10}$–$^1/_5$ their length; plants without rhizomes; calluses glabrous or with a crown of hairs, rarely webbed.
   2. Panicles 2–4 cm long, open; plants delicate, 15–20 cm tall; blades 0.7–1.5 mm wide; lemmas 2–3 mm long, keels crisply puberulent; calluses glabrous; plants known only from the Brooks Range in Alaska . . . . 60. *P. pseudoabbreviata* (in part)
   2. Panicles 1–15 cm long, open or contracted; plants usually coarser, 10–60 cm tall; blades 1–7 mm wide; lemmas 3–8 mm long, keels villous; calluses often hairy; plants widespread.
      3. Basal branching all or nearly all extravaginal; calluses glabrous or webbed; plants strongly glaucous, rare, alpine and arctic . . . . 57. *P. glauca* (in part)
      3. Basal branching both extravaginal and intravaginal; calluses glabrous or with a crown of hairs; plants glaucous or not, not rare, arctic to subarctic or coastal and boreal.
         4. Basal branching strictly intravaginal; calluses glabrous; plants densely tufted . . . . . . 9. *P. alpina* (in part)
         4. Basal branching all or partly extravaginal; calluses glabrous or hairy; plants loosely to densely tufted.
            5. Panicle branches smooth or sparsely scabrous; blades folded and inrolled; lemmas loosely villous between the veins; glumes somewhat shiny, not glaucous; plants of the high arctic . . . . 67. *P. hartzii* (in part)
            5. Panicle branches densely scabrous, at least distally; blades flat; lemmas glabrous between the veins; glumes not shiny, often glaucous; plants subarctic or boreal and coastal . . . . 69. *P. stenantha* (in part)
1. Sheaths closed for ($^1/_5$)$^1/_4$–$^3/_4$ their length; plants with or without rhizomes; calluses usually webbed, sometimes glabrous.
   6. Basal branching intravaginal; plants densely tufted, neither rhizomatous nor stoloniferous; calluses glabrous . . . . 9. *P. alpina* (in part)
   6. Basal branching completely or partly extravaginal; plants loosely or densely tufted, rhizomatous or stoloniferous; calluses usually webbed, sometimes glabrous.
      7. Panicles open, longest branches (5)8–15 cm long; sheaths closed for $^1/_2$–$^3/_4$ their length; plants green, not anthocyanic; plants of coastal regions of British Columbia . . . . . 24. *P. laxiflora* (in part)
      7. Panicles contracted to open, longest branches 1–8 cm long; sheaths closed for ($^1/_5$)$^1/_4$–$^1/_2$ their length; plants strongly anthocyanic; plants widespread.
         8. Paleas glabrous intercostally; spikelets 4–5.5 mm long, lemmas 2.5–3.5 mm long, intercostal regions glabrous; panicles contracted, sometimes opening eventually; plants slender, mostly high arctic . . . . 13. *P. pratensis* (in part)
         8. Paleas puberulent or hispidulous intercostally; spikelets 5.5–12 mm long; lemmas 4.5–8 mm long, intercostal regions hairy; panicles open or loosely contracted; plants slender to stout, subarctic and coastal to high arctic.
            9. Paleas, if recognizable, hispidulous intercostally; glumes distinctly keeled, keels scabrous; plants stout, of subarctic coastal regions . . . . 15. *P. macrocalyx* (in part)
            9. Paleas softly puberulent intercostally; glumes indistinctly keeled, keels smooth or nearly so; plants slender to stout, mainly in alpine and arctic habitats, rarely coastal . . . . 16. *P. arctica* (in part)

Poa Subkey II

Plants annual or perennial. **Culms** not bulbous-based. **Basal leaf sheaths** not swollen at the base. **Spikelets** not bulbiferous, florets developing normally. **Anthers** 0.1–1(1.2) mm long in all florets, anther sacs usually well developed, the distal 1–2 fertile florets sometimes with rudimentary stamens.

1. Calluses glabrous; lemmas usually softly puberulent to long-villous on the keel and marginal veins, often also on the lateral veins, glabrous between the veins, non-alpine plants rarely glabrous throughout; palea keels smooth, usually short- to long-villous near the apices, rarely glabrous; panicle branches and glume keels smooth; plants annual, sometimes surviving for a second season, introduced, weedy species (sect. *Micrantherae*).

  2. Anthers 0.6–1(1.1) mm long, oblong prior to dehiscence; spikelets crowded or sparsely arranged on the branches; plants widespread . . . . 10. *P. annua*

  2. Anthers 0.1–0.5 mm long, round to elliptical prior to dehiscence; spikelets crowded on the branches; plants uncommon outside of California . . . . 11. *P. infirma*

1. Calluses webbed or glabrous, if glabrous, the lemma pubescence not as above or the palea keels at least slightly scabrous near the apices; panicle branches and glume keels smooth or scabrous; plants annual or perennial, native, sometimes growing in disturbed habitats.

  3. Calluses webbed; lemma keels glabrous throughout or, if hairy on the proximal $^1/_2$, the marginal veins glabrous.

    4. Culms 5–15(20) cm tall; plants alpine . . . . 61. *P. abbreviata* (in part)

    4. Culms 20–126 cm tall; plants of shady forests and forest openings.

      5. Lemmas hairy only on the keels; branches in whorls of (2)3–5(7) . . . . 2. *P. alsodes*

      5. Lemmas usually glabrous, marginal veins rarely sparsely hairy at the base, hairs to 0.15 mm long; branches 1–3 per node.

        6. Sheaths closed for at least $^9/_{10}$ their length . . . . 4. *P. marcida*

        6. Sheaths closed for $^1/_3$–$^3/_4$ their length.

          7. Plants perennial; panicles lax, less than $^1/_4$ the height of the plants; second rachilla internodes shorter than 1 mm; lemma apices obtuse to sharply acute or acuminate . . . . 1. *P. saltuensis* (in part)

          7. Plants usually annual, rarely longer-lived; panicles erect, $^1/_4$–$^1/_2$ the height of the plants; second rachilla internodes longer than (1)1.2 mm; lemma apices sharply acute . . . . 17. *P. bolanderi*

  3. Calluses webbed or glabrous, if webbed, the lemmas hairy on the keel and marginal veins.

    8. Plants annual, rarely persisting for a second season; calluses webbed.

      9. Anthers 1, 0.1–0.2(0.3) mm long; palea keels softly puberulent to long-villous at midlength; panicles eventually open; plants from east of the 100th meridian . . . . . . 19. *P. chapmaniana*

      9. Anthers (1–2)3 per floret, 0.2–1(1.2) mm long; palea keels scabrous or softly puberulent to short-villous at midlength; panicles contracted or eventually open; plants from west of the 100th meridian or Texas.

        10. Panicles eventually open; blade apices narrowly prow-shaped; lemmas with hairs of similar length over and between the veins; palea keels scabrous or softly puberulent at midlength . . . . 18. *P. howellii*

        10. Panicles contracted; blade apices broadly prow-shaped; lemmas long- to short-villous over the keels and marginal veins, glabrous or pilose between the veins; palea keels softly puberulent to short-villous at midlength . . . . 20. *P. bigelovii*

    8. Plants perennial; calluses glabrous or webbed.

      11. Panicles open, broadly rhomboidal to pyramidal, branches divaricately ascending to spreading, longest branches 1.5–5 cm long, pedicels often longer than the 3–5 mm long spikelets; calluses glabrous; lemmas crisply puberulent on the keel and marginal veins, glabrous elsewhere, rarely glabrous throughout . . . . 60. *P. pseudoabbreviata* (in part)

      11. Panicles contracted or open, if open and broadly rhomboidal to pyramidal, the branches not as above or, if approximately so, the calluses webbed or the lemma keels and marginal veins short- to long-villous.

        12. Sheaths closed from $^1/_{10}$–$^1/_4$($^1/_3$) their length; panicles 1–7(10) cm long, contracted, branches shorter than 1.5 cm; plants densely tufted, basal branching all or mainly intravaginal; lower glumes usually 3-veined, lanceolate to broadly lanceolate, upper glumes subequal to or longer than the lowest lemmas; culms 1–25(30) cm tall; plants of high alpine and tundra habitats (sect. *Abbreviatae*).

13. Lemmas short- to long-villous on the marginal veins and distal ³/₄ of the keel, glabrous or softly puberulent between the veins; calluses webbed or glabrous . . . . 61. *P. abbreviata* (in part)

13. Lemmas glabrous or the keel and marginal veins sparsely puberulent proximally, glabrous between the veins; calluses glabrous.

    14. Anthers 0.2–0.8 mm long; spikelets 3–4 mm long; lower glumes usually exceeding the lower lemmas; upper florets frequently exceeded by or only slightly exceeding the glumes; blades thin, flat, folded, or slightly inrolled . . . . 59. *P. lettermanii*

    14. Anthers 0.8–1.2 mm long; spikelets 3.5–7 mm long; lower glumes shorter than to equaling the lower lemmas; upper florets exceeding both glumes; blades moderately thick, often folded and inrolled on the margins.

        15. Adaxial surfaces of the innovation blades smooth or sparsely scabrous, long cells papillate (at 100×); upper culm blades with 7–15 closely spaced ribs on the adaxial surface; plants of California . . . . 62. *P. keckii* (in part)

        15. Adaxial surfaces of the innovation blades densely and minutely hispidulous, puberulent, or scabrous, rarely smooth, long cells not papillate (at 100×); upper culm blades with 5–9 well-spaced ribs on the adaxial suface; plants of British Columbia, Washington, and Oregon . . . . 63. *P. suksdorfii* (in part)

12. Sheaths closed for ¹/₁₀–⁷/₈ their length; panicles 1–40 cm long, loosely contracted to open, branches 0.5–20 cm long, or the panicles contracted to loosely contracted with the branches 0.5–2 cm long and the plants loosely tufted; basal branching mainly extravaginal or pseudointravaginal; lower glumes 1–3-veined, subulate to broadly lanceolate; upper glumes shorter than to subequal to the lowest lemmas; culms 5–150 cm tall; plants of various habitats.

    16. Sheaths closed for ¹/₁₀–¹/₅ their length; basal branching mainly extravaginal; lower 1–3 leaves of the culms and innovations bladeless; anthers 0.8–1.2 mm long, sometimes poorly developed.

        17. Flag leaf nodes at or above midculm length . . . . 55. *P. nemoralis* (in part)

        17. Flag leaf nodes usually in the basal ¹/₃ of the culms.

            18. Anthers poorly developed, mature anther sacs about 0.1 mm wide and indehiscent; panicles dense to moderately dense, ovoid, 1.5–3.5 cm long; panicle branches not glaucous, angles smooth or sparsely scabrous; glumes broadly lanceolate, equal; upper glumes 3.7–4.7 mm long, the length 3.6–4.1 times the width; lemmas 3.7–4.5 mm long, glabrous between the veins, lateral veins usually glabrous, infrequently softly puberulent . . . . 51. *P. laxa* × *glauca*

            18. Anthers well developed, mature anther sacs usually about 0.2 mm wide and dehiscent, rarely aborted; panicles dense to loose, ovoid to lanceoloid, 1–10 cm long; panicle branches glaucous, the angles scabrous, at least below the spikelets; glumes narrowly to broadly lanceolate, subequal; upper glumes 2–3.8(5.2) mm long, the length usually more than 4.1 times the width; lemmas 2.5–4 mm long, glabrous or softly puberulent between the veins, lateral veins usually sparsely softly puberulent to short-villous . . . . 57. *P. glauca* (in part)

    16. Sheaths closed for ¹/₅–⁷/₈ their length; basal branching extravaginal, mixed extra- and intravaginal, or pseudointravaginal; culms and innovations with or without bladeless leaves; anthers 0.2–1.2 mm long, well developed.

        19. Calluses usually glabrous, rarely sparsely and shortly webbed; palea keels softly puberulent to short-villous for much of their length; lemmas puberulent between the veins; panicles (5)8–20 cm long,

broadly pyramidal; panicle branches moderately to densely scabrous on the angles, longest branches 5–12 cm long, with 3–8 spikelets . . . . 6. *P. autumnalis* (in part)

19. Calluses webbed or glabrous, if glabrous then the palea keels glabrous, lemmas glabrous between the veins, panicles 2–8 cm long, usually loosely contracted, infrequently contracted, panicle branches smooth or sparsely scabrous, longest branches 1–3(4) cm long, with 1–8 spikelets.

20. Panicles 2–8 cm long, usually loosely contracted, infrequently contracted; panicle branches usually ascending or weakly divergent, infrequently erect, smooth or sparsely scabrous, sulcate, longest branches 1–3(4) cm long; calluses webbed or glabrous; anthers (0.6)0.8–1.1(1.3) mm long; sheaths closed for $^1/_5$–$^1/_3$ their length; plants alpine . . . . 50. *P. laxa* (in part)

20. Panicles 2.5–40 cm long, open; panicle branches loosely ascending, spreading, or reflexed, smooth or scabrous, angled, sulcate, or terete, longest branches usually longer than 3 cm; calluses webbed; anthers 0.2–1.2 mm long; sheaths closed for $^1/_4$–$^3/_4$ their length; plants alpine or not.

21. Panicle branches smooth or sparsely scabrous, usually terete or slightly sulcate; lower glumes subulate to broadly lanceolate; lemmas glabrous between the veins.

22. Lower glumes subulate to narrowly lanceolate, keels usually scabrous; lemma keels short- to long-villous for $^1/_3$–$^2/_3$ their length; lateral veins glabrous; lemma apices narrowly acute; spikelets lanceolate to narrowly lanceolate, green to purple; panicle branches sparsely scabrous, longest branches with (3)4–15 spikelets; palea keels evenly pectinate-ciliate or scabrous at midlength . . . . 53. *P. leptocoma* (in part)

22. Lower glumes narrowly to broadly lanceolate, keels usually smooth; lemma keels long-villous for $^1/_2$–$^4/_5$ their length; lateral veins glabrous or hairy; lemma apices acute; spikelets lanceolate to ovate; panicle branches smooth or sparsely scabrous, if sparsely scabrous, with 1–3 ovate, dark purple spikelets on the lax to drooping capillary branches; longest branches with 1–18 spikelets; palea keels softly puberulent or scabrous at midlength.

23. Longest panicle branches with (3)6–18 spikelets; palea keels scabrous or sparsely puberulent at midlength; lemmas usually sparsely puberulent on the lateral veins; lower branches usually reflexed . . . . 22. *P. reflexa*

23. Longest panicle branches with 1–3(5) spikelets; palea keels sparsely scabrous at midlength; lemmas glabrous on the lateral veins; lower branches usually laxly ascending to spreading . . . . 23. *P. paucispicula*

21. Panicle branches sparsely to densely scabrous, terete or angled; lower glumes subulate or broader; lemmas glabrous or puberulent between the veins.

24. Panicles conical, the lower nodes with (2)3–10 branches; branches eventually reflexed; upper sheaths often ciliate on the overlapping margins near the point of fusion; intercostal regions of the lemmas usually sparsely puberulent, lateral veins at least sparsely puberulent; palea keels puberulent . . . . 3. *P. sylvestris* (in part)

24. Panicles not conical, the lower nodes with 1–7 branches; branches usually ascending to spreading, sometimes drooping or reflexed; upper sheaths not ciliate on the margins; lemmas glabrous or puberulent between the keel and marginal veins; palea keels puberulent or glabrous.

  25. Lemmas usually puberulent on the lateral veins and between the veins; lower cauline sheaths and ligules densely retrorsely scabrous; panicles (6)12–40 cm long, with 2–7 branches at the lower nodes; lower glumes 1-veined; plants densely tufted . . . . 21. *P. occidentalis*

  25. Lemmas glabrous on the keel, lateral veins, and between the veins, rarely puberulent on the lateral veins and between the veins; sheaths and ligules smooth or sparsely to moderately densely retrorsely scabrous; panicles 3–30 cm long, branches 1–3(5) per node; lower glumes 1–3-veined; plants densely to loosely tufted.

    26. Plants loosely tufted or with solitary culms, long-rhizomatous; lower glumes lanceolate; palea keels scabrous; panicles 14–30 cm long . . . . 24. *P. laxiflora* (in part)

    26. Plants densely to loosely tufted, sometimes shortly rhizomatous; lower glumes subulate to lanceolate; palea keels scabrous or puberulent; panicles 3–15(18) cm long.

      27. Ligules 1.5–4(6) mm long, obtuse to acute; lemmas often purple, keels pubescent for $^1/_3$–$^2/_3$ their length, apices usually bronze-colored, sharply acute to acuminate; palea keels evenly pectinate-ciliate to scabrous; lower glumes 1-veined . . . . 53. *P. leptocoma* (in part)

      27. Ligules 0.3–2.1 mm long, truncate; lemmas green, keels pubescent for $^2/_3$ their length or more, apices white or faintly bronze, acute to obtuse; palea keels scabrous or puberulent; lower glumes 1–3-veined.

        28. Palea keels puberulent; anthers (0.5) 0.8–1.2 mm long; lemmas (2.5)3.2–4.7 mm long, lateral veins distinct . . . . 7. *P. wolfii*

        28. Palea keels scabrous; anthers 0.2–0.8 mm long; lemmas 2.5–4 mm long, lateral veins faint . . . . 52. *P. paludigena*

Poa Subkey III

Plants rhizomatous or stoloniferous, densely to loosely tufted or the culms solitary. **Anthers** longer than 1.2 mm, or the florets pistillate and all anthers vestigial and 0.1–0.2 mm long, or longer and poorly developed.

1. Calluses usually with a crown of hairs, hairs 1–2 mm long, sinuous; lemmas 4.5–7 mm long, 5–7-veined, outer margins usually with hairs to 0.2 mm long, marginal veins usually glabrous, sometimes long-villous; bases of the basal sheaths densely retrorsely strigose, hairs 0.1–0.2 mm long, thick; plants of subsaline boreal to low arctic coastal beaches and meadows (*Poa* subg. *Arctopoa*) . . . . 72. *P. eminens*

1. Calluses usually glabrous or webbed, sometimes with a crown of hairs; lemmas 2–8 mm long, 5(7)-veined, outer margins glabrous, marginal veins glabrous or not; bases of the basal sheaths glabrous; plants of various habitats.

  2. Culms and nodes strongly compressed; culms usually geniculate; lower culm nodes usually exserted; panicle branches angled, scabrous on the angles; sheaths closed for $^1/_{10}$–$^1/_5$ their length . . . . .58. *P. compressa*

  2. Culms terete to somewhat compressed, nodes not or only weakly compressed; culms geniculate or not; lower culm nodes exserted or not; panicle branches angled or terete, smooth or scabrous; sheath closure varied.

    3. Panicles 3–12(18) cm long, narrowly cylindrical or lobed, congested, usually with over 100 spikelets; plants unisexual; spikelets sexually dimorphic; pistillate plants: calluses webbed dorsally and below the marginal veins, lemmas 4.2–6.4 mm long, keels and marginal veins densely long-villous, panicle branches usually moderately to densely coarsely scabrous; staminate plants: calluses glabrous or sparsely webbed dorsally, rarely also webbed below the marginal veins, lemmas 3.5–5 mm long, keels and marginal veins glabrous or moderately densely and shortly pubescent, panicle branches sparsely to moderately scabrous; all plants: blades flat or folded, adaxial surfaces glabrous; plants native to the southern Great Plains, infrequently introduced elsewhere . . . . .48. *P. arachnifera*

    3. Panicles 1–30 cm long, contracted to open, infrequently narrowly cylindrical or lobed and congested with over 100 spikelets; plants unisexual or bisexual; spikelets not sexually dimorphic; calluses glabrous, webbed, or with a crown of hairs, rarely with 3 webs; lemma keels and marginal veins glabrous or hairy; panicle branches smooth or scabrous; blades flat, folded, or involute, adaxial surfaces sometimes hairy in plants with contracted or loosely contracted panicles and unisexual spikelets; plants widespread.

      4. Basal branching extravaginal, branches initiated as pinkish to purplish, fleshy-scaled buds, the scales becoming brownish and flabelliform after shoot development; sheaths closed for at least $^9/_{10}$ their length; florets unisexual . . . . .34. *P. sierrae*

      4. Basal branching extra- or intravaginal or both, branches not initiated as persistent pinkish to purplish, fleshy-scaled buds; sheaths closed for at least $^1/_{15}$ their length; florets bisexual or unisexual.

        5. Lemmas totally glabrous, often scabrous; calluses webbed or diffusely webbed, hairs at least (1)2 mm long.

          6. Panicles 10–20 cm long, open, pyramidal, sparse; sheaths closed for $^1/_{15}$–$^1/_5$ their length; florets bisexual; plants of coastal redwood forests in northern California . . . . .5. *P. kelloggii*

          6. Panicles 1–10.5 cm long, loosely contracted to open, lanceoloid to pyramidal, congested to sparse; sheaths closed for $^1/_3$–$^9/_{10}$ their length; florets unisexual or bisexual; plants of the Pacific coast states and provinces.

            7. Lemmas 4–7 mm long, smooth or sparsely scabrous between the veins.

              8. Blades flat or folded, margins smooth, adaxial surfaces smooth or sparsely scabrous; blades of culm leaves gradually reduced in length upwards; collars smooth or sparsely scabrous . . . . .33. *P. chambersii* (in part)

              8. Blades involute, margins scabrous, adaxial surfaces moderately to densely scabrous or pubescent, especially those of the innovations; blades of the culm leaves steeply reduced in length upwards, some collars usually sparsely hispidulous . . . . .39. *P. piperi*

            7. Lemmas 2.5–5 mm long, moderately to densely scabrous between the veins.

              9. Lemmas 2.5–4(4.5) mm long; rachilla internodes 0.8–1.1 mm long; panicles 1–5(7) cm long, ovoid, loosely contracted, congested or moderately congested, branches erect to ascending, longest branches 0.5–3 cm; blades 0.5–1(1.5) mm wide . . . . .37. *P. confinis* (in part)

              9. Lemmas (3.2)4.2–5 mm long; rachilla internodes 1–1.3 mm long; panicles (4) 5.5–10.5(12.5) cm long, ovoid to broadly pyramidal, open,

or eventually loosely contracted, sparse, branches laxly ascending, longest branches 2.1–4.5 (7) cm; blades 1.5–2.4 mm wide . . . . 38. *P. diaboli*

5. Lemmas variously pubescent or glabrous; calluses glabrous or not, webbed or not, hairs long or short; florets never with both glabrous lemmas and long-webbed calluses.

  10. Plants 8–12(20) cm tall; panicles 2.5–5 cm long, erect, with 10–25(30) spikelets, branches smooth or sparsely scabrous; calluses glabrous; palea keels smooth, glabrous or softly puberulent to short-villous; glume keels smooth; leaf blades thin, flat, soft; plants stoloniferous . . . . 12. *P. supina*

  10. Plants (5)10–150 cm tall; panicles 1–30(41) cm long, erect or lax, with 10–100+ spikelets, branches smooth or scabrous; calluses glabrous or with hairs; palea keels sometimes partially scabrous; glume keels smooth or scabrous; leaf blades various; plants stoloniferous or not.

    11. Lemma keels softly puberulent for ³/₅ their length, hairs usually sparse, marginal veins glabrous or puberulent to ¹/₄ their length, intercostal regions smooth and glabrous; lateral veins prominent; calluses webbed; palea keels smooth, muriculate, tuberculate, or scabridulous; lower glumes 1-veined, usually arched to sickle-shaped; ligules 3–10 mm long, acute to acuminate; panicle branches angled, angles densely scabrous; plants usually weakly stoloniferous . . . . 49. *P. trivialis* (in part)

    11. Lemmas glabrous or variously pubescent, if as above, the lateral veins faint or moderately prominent or the calluses glabrous or the palea keels distinctly scabrous or hairy or the lower glumes 3-veined; calluses glabrous or hairy; palea keels scabrous at least near the apices; lower glumes 1–3-veined, not arched, not sickle-shaped; ligules 0.5–18 mm long, truncate to acuminate; panicle branches terete or angled, smooth or scabrous; plants stoloniferous or not.

      12. Culm leaf blades steeply reduced in length upward, flag leaf blades absent or to 1(3) cm long, less than ¹/₉(¹/₅) the length of the sheath; calluses glabrous; lemmas usually villous on the keel and marginal veins, glabrous elsewhere, sometimes glabrous throughout; sheaths closed for about ¹/₃ their length; blades usually all involute and moderately firm, adaxial surfaces, at least those of the innovations, usually densely scabrous to puberulent; panicles contracted; spikelets laterally compressed; florets unisexual; plants of mountain slopes, never of low, wet ground . . . . 41. *P. fendleriana* (in part)

      12. Culm leaf blades gradually reduced in length upward or the midculm blades longer than those below; flag leaf blades usually over (0.5)1 cm long, usually more than ¹/₇ the length of the sheath or, if as above, most or all blades flat and the panicles open or the sheaths closed for ¹/₁₀–¹/₅(¹/₄) their length and the blades folded and firm and the adaxial surfaces smooth or nearly so and the florets bisexual or the spikelet lengths 4–5 times the widths; calluses glabrous or with hairs; lemmas glabrous or variously pubescent; sheaths closed for ¹/₁₀–⁹/₁₀ their length; blades as above or not; panicles contracted or open; spikelets laterally compressed or subterete; florets bisexual or unisexual; plants of various habitats, sometimes of low, wet ground.

        13. Calluses glabrous, diffusely webbed with hairs to ¹/₂ the lemma length, or with a crown of hairs, or sparsely and dorsally webbed with hairs to ¹/₄ the lemma length; lemmas glabrous or pubescent [for opposite lead, see p. 499].

          14. Spikelets subterete to weakly laterally compressed, the lengths (3.8)4–5 times the widths; panicles usually contracted, sometimes open at anthesis; sheaths closed for ¹/₁₀–¹/₄ their length; calluses glabrous or with a crown of

hairs; adaxial surfaces of the innovation blades glabrous, smooth or scabrous, not densely scabrous between the veins, flat and soon withering or folded and somewhat firm; florets bisexual . . . . 64. *P. secunda* (in part)

14. Spikelets laterally compressed, the lengths 2–3.5(3.8) times the widths; panicles contracted or open; sheaths closed for $^1/_{10}$–$^9/_{10}$ their length; calluses glabrous, diffusely webbed, with a crown of hairs, or dorsally webbed with hairs to $^1/_4$ the lemma length; adaxial surfaces of the innovation blades glabrous or with hairs, smooth or densely scabrous between the veins, flat and late withering, folded, soft and firm, or involute; florets bisexual or unisexual.

    15. Sheaths closed for $^1/_{10}$–$^1/_5$($^1/_4$) their length, smooth or sparsely scabrous, glabrous; panicles usually contracted, sometimes loosely contracted or open; paleas usually glabrous between the keels, if hairy, the panicles contracted; lemma keels, marginal veins, and, often, lateral veins short- to long-villous, intercostal regions usually glabrous, if hairy, the panicles contracted; lemma apices often blunt; calluses usually glabrous, occasionally dorsally webbed, hairs to $^1/_4$ the lemma length; innovation blades usually folded and firm, infrequently flat and somewhat soft, adaxial surfaces glabrous, smooth or moderately scabrous, mainly over the veins; florets bisexual; plants usually of low, wet, somewhat alkaline or subsaline soils, from the valleys of the eastern foothills of the Rocky Mountains to the Great Plains, sometimes extending to timberline, rarely on slopes west of the continental divide . . . . 73. *P. arida*

    15. Sheaths closed for $^1/_6$–$^9/_{10}$ their length, smooth or retrorsely scabrous, glabrous or with hairs; panicles contracted or open; paleas glabrous or hairy between the keels; lemmas glabrous or variously hairy, apices blunt or pointed; calluses glabrous, shortly webbed, or diffusely webbed; innovation blades flat, folded, or involute, soft or firm, adaxial surfaces glabrous or hairy, smooth or densely scabrous between the veins; florets bisexual or unisexual; plants widespread but not of subalkaline or subsaline soils from the eastern slope of the Rocky Mountains to the Great Plains.

        16. Sheaths closed for $^1/_3$–$^9/_{10}$ their length, sheaths of some leaves densely retrorsely scabrous or short-pubescent, at least on or near the collar margins; ligules of the lower culm leaves and innovations truncate, abaxial surfaces densely scabrous or softly puberulent; upper ligules 0.5–2 mm long; lemmas glabrous, or the keel and marginal veins softly puberulent to short-villous, intercostal region glabrous or hispidulous, infrequently softly puberulent; calluses usually glabrous, rarely shortly webbed.

            17. Sheaths hairy, hairs usually concentrated on and about the collars, collar margin hairs distinctly longer than those below the collar; sheaths closed for $^2/_3$–$^9/_{10}$ their length; blades flat or a

few folded, adaxial surfaces smooth or sparsely
scabrous, particularly over the veins; florets
bisexual and unisexual; plants from west of the
Cascade divide . . . . 30. *P. nervosa*

17. Sheaths retrorsely scabrous or pubescent for $^1/_4$
or more of the length below the collars, collar
and sheath vestiture not differing in length;
sheaths closed for $^1/_3$–$^3/_4$ their length; blades of
the innovations usually involute, adaxial
surfaces usually densely scabrous to hispidulous
on and between the veins; florets usually all
pistillate, rarely bisexual or staminate; plants
primarily from between the 100th meridian and
the Cascade and Sierra Nevada mountains of
western North America, rarely further west . . . . 31. *P. wheeleri*

16. Sheaths closed for $(^1/_6)^1/_5$–$^9/_{10}$ their length, glabrous,
collars glabrous, smooth or infrequently moderately
scabrous; ligules of the lower culm leaves and lateral
shoots truncate to acuminate, smooth or scabrous
abaxially, glabrous or softly puberulent; upper
ligules 0.5–7 mm long; lemmas glabrous or
variously pubescent; calluses glabrous, shortly
webbed, or with a crown of hairs.

18. Paleas pubescent between the keels; sheaths
closed for $(^1/_6)^1/_5$–$^2/_5$ their length, smooth or
slightly scabrous; lemma keels and marginal
veins long-villous, intercostal regions usually
short-villous, sometimes slightly softly
puberulent on the lower back; panicles loosely
contracted to open, branches smooth or
sparsely scabrous; calluses glabrous or dorsally
webbed; florets usually bisexual, anthers
aborted late in development or 1.4–2.5 mm
long; plants of subalpine to alpine and arctic
habitats . . . . 16. *P. arctica* (in part)

18. Paleas glabrous between the keels; sheaths
closed for $^1/_3$–$^9/_{10}$ their length, smooth or
scabrous; lemma keels and marginal veins
glabrous or pubescent, intercostal regions
usually glabrous, infrequently softly
puberulent; panicles contracted to open,
branches smooth or sparsely to densely
scabrous or hispidulous; calluses glabrous,
diffusely webbed or with a crown of hairs;
florets bisexual or unisexual; pistillate florets
with anthers 0.1–0.2 mm long; plants coastal to
subalpine.

19. Lemmas pubescent; calluses glabrous,
diffusely webbed, or with a crown of hairs;
blades involute, adaxial surfaces scabrous
or pubescent, frequently densely so between
the veins, or smooth and glabrous and the
blades 0.5–1(1.5) mm wide; sheaths closed
for $^1/_3$–$^2/_3$ their length; plants of sand dunes
and sandy soils along the Pacific coast.

20. Lemmas 2.5–4(4.5) mm long; panicles fairly tightly to loosely contracted; culms 0.4–0.9 mm thick; blades 0.5–1(1.5) mm wide, thin to moderately thick, soft, mostly filiform, adaxial surfaces sparsely scabrous; calluses diffusely webbed . . . . 37. *P. confinis* (in part)

20. Lemmas 5–11 mm long; panicles tightly contracted; culms 1–2 mm thick; blades 1–4 mm wide, thick, moderately firm to firm; adaxial surfaces densely scabrous or hispidulous; calluses glabrous, diffusely webbed, or with a crown of hairs.

   21. Panicle rachises and culms beneath the panicles densely hispidulous; lemmas 5–7.5 mm long . . . . 35. *P. douglasii*

   21. Panicle rachises and culms beneath the panicles glabrous, smooth or sparsely to moderately scabrous; lemmas (6)7.5–11 mm long . . . . 36. *P. macrantha*

19. Lemmas and calluses totally glabrous or, if the lemmas pubescent or the calluses dorsally webbed with hairs to ¼ the lemma length, then the blades flat or folded, 2–5 mm wide, smooth or sparsely scabrous adaxially; sheaths closed for ¼–⁹/₁₀ their length; plants of inland regions, not growing in sand.

   22. Panicles 3–7 cm long, densely contracted, branches smooth or sparsely scabrous distally; spikelets 3.5–5.5 mm long, compact, rachilla internodes about 0.5 mm long; lemmas and calluses smooth, glabrous . . . . 40. *P. atropurpurea*

   22. Panicles 2–22 cm long, densely contracted or open, if densely contracted, the branches sparsely to densely scabrous or spikelets 5.5–12 mm long; spikelets 3–12 mm long, looser, rachilla internodes 0.5–1.5 mm long; lemmas and calluses smooth or scabrous, glabrous or hairy.

      23. Basal branching nearly all intravaginal or mixed intra- and extravaginal; at least some innovation leaves with involute blades 0.5–2 mm wide and scabrous or pubescent on the adaxial surfaces; plants rarely rhizomatous, usually densely tufted; lemmas sparsely to densely scabrous, glabrous or sparsely

softly puberulent near the base of the keels and/or marginal veins . . . . 42. *P. cusickii* (in part)

23. Basal branching all or mainly extravaginal; blades flat or folded, 2–5 mm wide, adaxial surfaces smooth or sparsely scabrous; plants shortly rhizomatous, loosely tufted or the culms solitary; lemmas smooth or sparsely scabrous, glabrous or the keel and marginal veins hairy, intercostal regions rarely pubescent.

  24. Panicles (5)12–22 cm, open, branches spreading to eventually reflexed; calluses glabrous . . . . 28. *P. arnowiae*

  24. Panicles 2–9 cm long, tightly to loosely contracted, branches erect to ascending or scarcely spreading; calluses of some lemmas usually shortly webbed . . . . 33. *P. chambersii* (in part)

13. Calluses dorsally webbed, hairs over $(^1/_3)^1/_2$ the length of the lemmas, sometimes with additional webs below the marginal veins; lemma short- to long-villous on the keels and marginal veins [for opposite lead, see p. 495].

25. Sheaths closed for $^1/_{10}$–$^1/_5$ their length; spikelets 3–5 mm long; lemmas 2–3 mm long, glabrous between the keels and marginal veins; panicle branches angled, angles densely scabrous; plants sometimes stoloniferous, sometimes branching above the culm bases; florets bisexual . . . . . . 54. *P. palustris* (in part)

25. Sheaths closed for $(^1/_6)^1/_5$–$^9/_{10}$ their length; spikelets 3.5–12 mm long; lemmas 2–8 mm long, glabrous or hairy between the keels and marginal veins; panicle branches terete or angled, smooth or scabrous; plants rarely stoloniferous, usually rhizomatous, never branching above the culm bases; florets bisexual or unisexual.

  26. Sheaths closed for $(^2/_5)^1/_2$–$^9/_{10}$ their length, weakly to distinctly compressed, keels distinct, sometimes winged, wing to 0.5 mm wide, glabrous or the sides, collars, or throats pubescent; plants loosely tufted, shortly rhizomatous, never forming dense turf; culm blades flat or slightly folded, infrequently folded; innovations all or almost all extravaginal or a few intravaginal, with the intravaginal blades not involute and distinctly narrower than the culm blades; florets bisexual or unisexual, commonly some florets pistillate; anthers 0.1–0.2 mm or (1.3)2–4 mm long; plants mostly of forest openings and mountain thickets.

    27. Blades steeply reduced in length up the culms, flag leaf blades 0.2–3(6) long; panicles broadly pyramidal; usually at least some upper lemmas within the spikelets pubescent between the veins . . . . . 29. *P. cuspidata*

27. Blades not steeply reduced in length up the culm, midculm blades sometimes longer than those below, flag leaf blades (1.4)3–20 cm long; panicles loosely contracted to narrowly pyramidal; lemmas glabrous or sparsely pubescent between the veins.

    28. Panicles erect, usually narrowly pyramidal, (8)13–29 cm long, proximal internodes usually longer than 4 cm; usually some lemmas within the spikelets pubescent between the veins . . . . 27. *P. tracyi*

    28. Panicles nodding, ovoid, (2)4–10 cm long, proximal internodes 1.8–3 cm long; lemmas glabrous between the veins . . . . 32. *P. rhizomata*

26. Sheaths closed for ($^1/_6$)$^1/_5$–$^1/_2$($^3/_5$) their length, terete to compressed, with or without distinct keels, usually glabrous, the sides infrequently retrorsely scabrous or pubescent; plants densely to loosely tufted or with solitary culms, sometimes forming dense turf; culm blades flat or folded; innovations all extravaginal or some intravaginal, blades of the intravaginal shoots sometimes involute and distinctly narrower than the culm blades; florets bisexual; anthers usually 1.2–2.5 mm long, sometimes some anthers aborting late in development and 1–1.5 mm long; plants widespread, sometimes of coastal belts and alpine and arctic habitats.

    29. Glumes subequal in length and width, usually nearly equaling the adjacent lemmas, distinctly keeled, keels scabrous; lower glumes (4)4.5–7 mm long; lemmas (4)5–8 mm long, the intercostal regions usually moderately to densely scabrous or hispidulous, infrequently softly puberulent to short-villous near the base and moderately to densely scabrous to hispidulous in the middle $^1/_3$, rarely nearly smooth near the base and sparsely scabrous distally; intercostal regions of the paleas usually hispidulous, infrequently puberulent; blades (2)3–7 mm wide; culms usually stout, (20)30–120 cm tall; plants of coastal shores and low elevation wet meadows in Alaska and the low arctic . . . . . 15. *P. macrocalyx* (in part)

    29. Glumes unequal to subequal in length and width, distinctly shorter than to subequal to the adjacent lemmas, distinctly or weakly keeled, keels smooth or scabrous; lower glumes 1.5–5(6) mm long; lemmas 2–6(7) mm long, the intercostal regions smooth, glabrous or pilose to long-villous, and smooth or sparsely scabrous distally; intercostal regions of the paleas glabrous or pilose to short-villous; blades 0.4–6 mm wide; culms slender to stout, 10–70(100) cm tall; plants widespread.

       30. Palea keels usually pubescent, rarely nearly glabrous, intercostal regions usually at least sparsely and softly puberulent near the base, sometimes glabrous; glumes weakly to distinctly keeled, the keels smooth or sparsely to moderately scabrous; upper glumes usually

subequal to the lower lemmas or slightly shorter; lemma intercostal regions and lateral veins pubescent near the base; ligules smooth or sparsely scabrous, usually rounded or obtuse to acute, infrequently truncate, entire or lacerate, not ciliolate; panicle branches (1)2–5 per node, usually smooth or sparsely scabrous, infrequently moderately scabrous .... 16. *P. arctica* (in part)

30. Palea keels glabrous or pubescent, intercostal regions glabrous, rarely sparsely hispidulous; glumes distinctly keeled, the keels usually sparsely to densely scabrous distally, infrequently smooth; upper glumes usually distinctly shorter than the lower lemmas; lemma intercostal regions glabrous, lateral veins glabrous or pubescent; ligules smooth or scabrous, usually truncate or rounded, infrequently obtuse to acute, entire, glabrous, or ciliolate; panicle branches (1)2–7(9) per node, smooth or sparsely to fairly densely scabrous.

   31. Intercostal surfaces of the lemmas visible, not or only partly concealed by hairs; lemma keels and marginal veins moderately to densely long-villous, more or less straight, lateral veins glabrous or softly puberulent, infrequently short-villous; panicle branches and ligules smooth or sparsely to fairly densely scabrous, longest branches 1–9 cm; plants widespread .... 13. *P. pratensis* (in part)

   31. Intercostal surfaces of the lemmas concealed by the hairs over the keels and veins; lemma keels, marginal veins, and lateral veins copiously hairy, hairs of the keels and marginal veins cottony, those of the lateral veins somewhat shorter and sparser; panicle branches and ligules smooth or nearly so, longest branches 1–3 cm; plants of high arctic sands .... 14. *P. sublanata*

Poa Subkey IV

Plants perennial, not rhizomatous, not stoloniferous, loosely to densely tufted. **Culms** not bulbous-based. **Basal sheaths** not swollen. **Spikelets** not bulbiferous, florets developing normally. **Anthers** (1.2)1.3–4 mm long and dehiscent, or all rudimentary, having no or poorly formed pollen.

1. Calluses usually dorsally webbed, webs sometimes with 1 to few minute hairs, rarely the hairs somewhat diffuse [for opposite lead, see p. 503].
  2. Lemma lateral veins pronounced, keels pubescent, marginal veins glabrous or softly puberulent at the base, lemmas glabrous elsewhere; lower glumes 1-veined, subulate to narrowly lanceolate, usually arched to sickle-shaped; callus web well developed ..... 49. *P. trivialis* (in part)
  2. Lemma lateral veins obscure to pronounced, keels glabrous throughout or, if pubescent, the marginal veins distinctly pubescent for more than ¼ their length, lemma lateral veins

and intercostal regions glabrous or pubescent, or, if pubescent as in *P. trivialis*, then the callus web short, scant, poorly developed and the lower glumes 3-veined and lanceolate or broader.

3. Panicles open, conical, with whorls of (2)3–10, spreading to eventually reflexed, scabrous-angled branches at the lower nodes; lemmas hairy on the keel and veins, sometimes the intercostal regions also hairy; callus webs well developed . . . . 3. *P. sylvestris* (in part)

3. Panicles contracted to open, if open then not conical and without whorls of (2)3–10, eventually reflexed, scabrous-angled branches at the lower nodes; branches smooth or scabrous-angled; lemmas glabrous or hairy; calluses glabrous, with diffuse hairs, or with a scanty or well-developed web.

    4. Sheaths closed for ($^{1}/_{5}$)$^{1}/_{3}$–$^{3}/_{4}$ their length.

        5. Culms 8–35 cm tall, 0.5–0.8 mm thick; panicle branches smooth or sparsely scabrous; anthers to 1.3 mm long; plants alpine . . . . 50. *P. laxa* (in part)

        5. Culms 23–120 cm tall, 0.5–2 mm thick; panicle branches smooth or scabrous; anthers to 1.8 mm long; plants of many habitats, including alpine habitats.

            6. Lemmas usually hairy on the keel and marginal veins, usually also on the intercostal regions; palea keels softly puberulent to short-villous at midlength; panicles open and erect, broadly pyramidal at maturity; callus webs sparse, poorly developed . . . . 6. *P. autumnalis* (in part)

            6. Lemmas glabrous or with a few hairs at the base of the keel or marginal veins; palea keels scabrous, glabrous; panicles contracted to loosely contracted or open and lax; callus webs scant and short or well developed.

                7. Callus webs well developed; lemma keels glabrous; plants of eastern North America . . . . 1. *P. saltuensis* (in part)

                7. Callus webs minute, sometimes somewhat diffuse; lemma keels glabrous or sparsely softly puberulent near the base; plants of western North America . . . . 42. *P. cusickii* (in part)

    4. Sheaths closed for $^{1}/_{20}$–$^{1}/_{4}$($^{1}/_{3}$) their length.

        8. Basal branching all or mostly intravaginal; plants not stoloniferous.

            9. Culms 30–90 cm tall; panicles open; plants of the mountains in and around the Chihuahuan Desert . . . . 26 *P. strictiramea* (in part)

            9. Culms 5–15(20) cm tall; panicles contracted; alpine plants of the Rocky Mountains . . . . 61. *P. abbreviata* (in part)

         8. Basal branching all or mostly extravaginal, or extra- and intravaginal and the plants stoloniferous.

            10. Flag leaf nodes usually in the lower $^{1}/_{10}$–$^{1}/_{3}$ of the culms; flag leaf blades usually distinctly shorter than their sheaths; lemmas sometimes softly puberulent between the veins, lateral veins usually with at least a few minute hairs; ligules 1–4(5) mm long . . . . 57. *P. glauca* (in part)

            10. Flag leaf nodes usually in the upper $^{2}/_{3}$ of the culms; flag leaf blades shorter or longer than their sheaths; lemmas glabrous between the veins, lateral veins usually glabrous, rarely with 1 to several minute hairs; ligules 0.2–6 mm long.

                11. Spikelets narrowly lanceolate to lanceolate; glumes subulate to narrowly lanceolate, gradually tapering to narrowly acuminate apices; lower glume lengths 6.4–11 times the widths; ligules 0.2–0.5(1) mm long, truncate; flag leaf nodes at or above the middle of the culms; flag leaf blades usually longer than their sheaths; rachillas usually hairy, hairs to 0.15 mm long; webs usually short, scanty . . . . 55. *P. nemoralis* (in part)

                11. Spikelets and glumes not as above or, if so, the ligules 1.5–6 mm long, truncate to acute, and the rachillas glabrous; flag leaf nodes at or above the lower $^{1}/_{3}$ of the culm; flag leaf blades longer or shorter than their sheaths; webs short or long, scanty or not.

                    12. Panicles (9)13–30(41) cm long, branches 4–15 cm long; culms closely spaced to isolated at the base; lower glumes tapering to the apices, lengths 6.4–10 times the widths; lemma keels abruptly inwardly arched beneath the scarious apices; lemma margins

distinctly inrolled; rachillas usually muriculate, rarely sparsely hispidulous; web hairs usually longer than $^2/_3$ the length of the lemmas . . . . 54. *P. palustris* (in part)

  12. Panicles (1.5)3–15(17) cm long, branches 0.4–8(9) cm long; culms closely spaced at the base; lower glumes abruptly narrowing to the apices, lengths 4.5–6.3 times the widths; lemma keels not abruptly inwardly arched beneath the scarious apices; lemma margins not or slightly inrolled; rachillas usually muriculate or softly puberulent; web hairs shorter than $^1/_2$($^2/_3$) the length of the lemmas . . . . 56. *P. interior* (in part)

1. Callus glabrous or with a crown of hairs, hairs 0.1–2 mm long [for opposite lead, see p. 501].

  13. Lemmas and calluses glabrous [for opposite lead, see p. 505].

    14. Blades (4) 6–15 mm wide, flat or folded; sheaths closed for $^1/_2$–$^3/_4$ their length, strongly compressed, keeled, keels winged . . . . 25. *P. chaixii*

    14. Blades 0.5–5 mm wide, flat, folded, or involute; sheaths closed for $^1/_{20}$–$^4/_5$ their length, not strongly compressed, if keeled, keels not winged.

      15. Sheaths closed for $^2/_5$–$^4/_5$ their length; panicles 1–5(8) cm long, with (1)6–17(22) spikelets, nodes with 1–2 branches; branches appressed to spreading, smooth or sparsely scabrous; spikelets strongly compressed, lanceolate to broadly ovate; ligules hyaline, smooth, (1)2–4 mm long; blades 0.5–1 mm wide, thin, lax, filiform, soon withering; plants from the Columbia Plateau to southwestern Idaho and northwestern Nevada . . . . 45. *P. leibergii*

      15. Sheaths closed for $^1/_{20}$–$^3/_4$ their length, if for $^2/_5$–$^3/_4$, then the panicles longer than 8 cm or with more than 20 spikelets or the ligules of the innovations (and sometimes also the culms) 0.5–2.5 mm long and scabrous and often milky white; blades (0.5)1–5 mm wide, sometimes moderately thick and firm and holding their form; plants of many regions, including the range of *P. leibergii*.

        16. Panicles (7)10–30 cm long, open, pyramidal, nodes with 2–5 moderately to densely scabrous branches; sheaths closed for $^1/_{20}$–$^1/_{10}$ their length; basal branching intravaginal; plants of the Chisos Mountains of Texas to Mexico . . . . 26. *P. strictiramea* (in part)

        16. Panicles 1–25 cm long, contracted to loosely contracted or, if open, nodes with 1–3(5) smooth or scabrous branches; sheaths closed for $^1/_{20}$–$^3/_4$ their length; basal branching intravaginal, extravaginal, or both; plants of many regions, including the range of *P. strictiramea*.

          17. Sheaths closed for ($^1/_4$)$^1/_3$–$^3/_4$ their length; florets often unisexual, anthers 2–3.5 mm long or nonfunctional and to 1.8 mm long; uppermost ligules of the innovation leaves 0.2–0.5(2.5) mm long, scabrous, usually truncate; innovation blades usually involute; panicles contracted, loosely contracted, or open; lower glumes distinctly shorter than the lowest lemmas.

            18. Flag leaf blades usually absent or to 1 cm long; blades of the culm leaves sharply reduced in length upwards, similar in thickness and form to those of the innovations, moderately firm, usually involute; plants of southeastern Arizona and southwestern New Mexico . . . . 41. *P. fendleriana* (in part)

            18. Flag leaf blades usually present and 1+ cm long; blades of the culm leaves not sharply reduced in length upwards, sometimes differing in thickness or form from those of the innovations, soft, narrow and withering or broader and flat; plants from other parts of the *Flora* region.

              19. Panicles contracted or loosely contracted, branches smooth or sparsely to densely scabrous; innovation blades 0.5–2 mm wide, abaxial surfaces smooth or scabrous, adaxial surfaces usually densely scabrous or hispidulous; plants from southern Yukon Territory to California and Colorado . . . . 42. *P. cusickii* (in part)

19. Panicles open or slightly contracted, branches smooth or sparsely scabrous; innovation blades 1–2 mm wide, abaxial surfaces smooth, adaxial surfaces usually smooth or sparsely scabrous; plants of Alaska, Yukon Territory, and Northwest Territories . . . . .44. *P. porsildii*

17. Sheaths closed for $^1\!/_{20}$–$^2\!/_5$ their length; if sheaths closed for $^1\!/_4$–$^2\!/_5$ their length then all florets bisexual, or the functional anthers 1.2–1.8 mm long, or the ligules of the uppermost innovation leaves 2+ mm long and smooth or scabrous, or the lower glumes subequal to the lowest lemmas; blades of the innovation leaves involute or not; panicles contracted; lower glumes shorter than to equaling the lowest lemmas.

20. Culms 5–40 cm tall; panicles 3–7 cm long, densely contracted, nearly cylindrical; culm blades to 5 mm wide, often a bit fleshy and broader than those of the innovations, those of the innovations usually thin and soon withering, infrequently all blades flat and a bit fleshy; florets bisexual; plants of the Pacific coast . . . . . . 71. *P. unilateralis* (in part)

20. Culms 2–120 cm tall; panicles 1–25 cm long, densely to loosely contracted, not cylindrical; culm blades 0.5–5 mm wide and soft, culm and innovation blades not much differentiated or, if differentiated, then the basal blades moderately firm and involute; florets unisexual or bisexual; plants of non-coastal regions.

21. Culms 30–120 cm tall; panicles (4)5–25 cm long; spikelet lengths 3–5 times the widths; plants of saline or non-saline habitats, often below the subalpine zone, if of non-saline habitats, the spikelet lengths (3.8)4–5 times the widths and the panicles usually over 10 cm long.

22. Spikelets (4)7–10 mm long, subterete, narrowly lanceolate, lengths usually (3.8)4–5 times the widths; plants of many habitats, widespread . . . . .64. *P. secunda* (in part)

22. Spikelets 4.5–7 mm long, compressed, lengths 3–3.5 times the widths; plants of mineralized soils around hot springs in Napa County, California . . . . .70. *P. napensis* (in part)

21. Culms 2–40 cm tall; panicles 1–8 cm long; spikelet lengths 2–4 times the widths; plants of non-saline, subalpine or alpine habitats.

23. Ligules of the innovations 2.5–6 mm long, hyaline, smooth; panicles loosely contracted or contracted; basal branching extravaginal; lower glumes distinctly shorter than the lowest lemmas; florets often unisexual . . . . .46. *P. stebbinsii*

23. Ligules of the innovations 0.5–2.5 mm long, usually milky, often scabrous; panicles contracted; some or all basal branching intravaginal; lower glumes distinctly shorter than or subequal to the lowest lemmas; florets bisexual or unisexual.

24. Florets unisexual; anthers 2–4 mm long; blades involute . . . . .47. *P. pringlei*

24. Florets bisexual; anthers 0.6–3.5 long; blades flat, folded, or involute.

25. Anthers 2.2–3.5 mm long; culms 15–40 cm tall; longest culm blades 1–3 cm long and fairly firm, with thick white margins and broadly prow-tipped apices, basal blades similar; plants of serpentine soils in Washington .66. *P. curtifolia* (in part)

25. Anthers 0.6–1.2(2) mm long; culms 2–25 cm tall; basal and upper culm leaves not always similar,

culm leaves without the above combination of characteristics; plants of non-serpentine soils from British Columbia to California.

    26. Abaxial surfaces of the innovation blades smooth or sparsely scabrous, epidermes with papillae on the long cells (at 100×); abaxial surfaces of the flag leaf blades with 7–15 closely spaced ribs; culms 2–6(10) cm tall; plants of California . . . . 62. *P. keckii* (in part)

    26. Abaxial surfaces of the innovation blades densely hispidulous, scabrous, or softly puberulent, rarely smooth and glabrous, lacking papillae on the long cells (at 100×); abaxial surfaces of the flag leaf blades with 5–9 well-spaced ribs; culms 7–25 cm tall; plants of British Columbia, Washington, and Oregon . . . . 63. *P. suksdorfii* (in part)

13. Lemmas with hairs; calluses glabrous or hairy [for opposite lead, see p. 503].

    27. Sheaths closed for $^1/_2$–$^3/_4$ their length; lemmas mostly glabrous, lower lemmas of some spikelets usually sparsely softly puberulent near the base of the keels and/or marginal veins, lemmas glabrous elsewhere; panicles 4–7 cm long, with 13–50 spikelets, branches smooth or sparsely scabrous; spikelets 7–10 mm long, strongly laterally compressed; florets pistillate; plants of subalpine to alpine habitats, from southern British Columbia to California . . . . 42. *P. cusickii* (in part)

    27. Sheaths closed for $^1/_{20}$–$^3/_4$ their length, if closed for $^1/_2$–$^3/_4$ their length, the lemmas pilose between the veins or the panicle branches moderately to densely scabrous; lemmas variously pubescent, frequently pilose between the veins; panicles 1–30 cm long, with 9–100+ spikelets; branches smooth or sparsely to densely scabrous; spikelets 3–12 mm long, subterete to strongly laterally compressed; florets bisexual or unisexual; plants of various habitats, including subalpine to alpine habitats, widely distributed, including from British Columbia to California.

        28. Sheaths closed for $^1/_3$–$^1/_2$ their length; panicles (5)8–20 cm, erect or lax, broadly pyramidal at maturity, open, lower axils sometimes sparsely hairy; panicle branches spreading to reflexed, angled, longest branches 5–12 cm long, with 3–8 spikelets in the distal $^1/_3$–$^1/_4$; paleas pilose; florets bisexual; plants of eastern North American woods . . . . 6. *P. autumnalis* (in part)

        28. Sheaths closed for $^1/_{20}$–$^3/_4$ their length, if closed for $^1/_3$–$^3/_4$ their length, then the panicles contracted or loosely contracted, or the branches smooth, or the longest branches shorter than 5 cm, or the paleas glabrous, or the spikelets unisexual; panicles 1–40 cm long, lower axils glabrous; panicle branches erect, ascending, or widely divergent, terete or angled; plants widely distributed, including eastern North American woods.

            29. Basal branching mainly extravaginal, usually occurring late in the season; sheaths closed for $^1/_{10}$–$^1/_5$ their length; blades usually flat, sometimes folded, thin, soft; panicle branches usually scabrous-angled; lemmas distinctly keeled; spikelets laterally compressed, lengths 2–3 times the widths.

                30. Lemmas glabrous on the lateral veins and intercostal regions; culms usually with 1–2(3) nodes exserted, uppermost node usually at or above the lower $^1/_3$ of the culm . . . . 56. *P. interior* (in part)

                30. Lemmas usually at least sparsely softly puberulent on the lateral veins, intercostal regions with similar hairs or glabrous; culms with 0–1 nodes exserted, uppermost node usually in the lower $^1/_{10}$–$^1/_3$ of the culms . . . . 57. *P. glauca* (in part)

            29. Basal branching intra- or extravaginal or mixed, if mostly extravaginal, branching often occurring early in the season; sheaths closed for $^1/_{20}$–$^3/_4$

their length; blades involute, flat, or folded, thin and soft to thick and firm; panicle branches terete or angled, smooth or scabrous; lemmas weakly to distinctly keeled; spikelets subterete or laterally compressed, lengths 1.5–5 times the widths.

31. Spikelets ovate, rachilla internodes 0.5–0.8 mm; panicles 2–6(8) cm long, open or loosely contracted; branches terete, usually smooth or sparsely scabrous, rarely moderately densely scabrous, longest branches 1–3(4) cm; leaves mostly basal, blades 1–6(12) cm long, 2–4.5 mm wide, flat, soft; calluses glabrous; lemmas distinctly keeled, keels and marginal veins long- to short-villous, intercostal regions short-villous; paleas softly puberulent to short-villous at midlength; florets bisexual . . . . 9. *P. alpina* (in part)

31. Spikelets lanceolate to narrowly ovate, rachilla internodes 0.5–2 mm long; panicles 1–40 cm long, open or contracted; branches terete or angled, smooth or variously scabrous, longest branches 0.5–15 cm; leaves not as above; calluses glabrous or with a crown of hairs; lemmas weakly to distinctly keeled, variously hairy; paleas glabrous or softly puberulent; florets bisexual or unisexual.

    32. Panicles contracted; florets usually unisexual, rarely bisexual, commonly pistillate; blades usually involute.

        33. Sheaths closed for $^1/_7$–$^1/_3$ their length; lemmas weakly keeled; calluses usually with a crown of hairs around the base of the lemma; adaxial surfaces of the innovation blades smooth or somewhat scabrous; anthers late-aborted, 0.8–1.8 mm long; plants of the high arctic . . . . 67. *P. hartzii* (in part)

        33. Sheaths closed for $^1/_4$–$^3/_4$ their length; lemmas strongly keeled; calluses glabrous; adaxial surfaces of the innovation blades usually hispidulous to softly puberulent on and between the veins; anthers of pistillate plants rudimentary, 0.1–0.2 mm long; plants not arctic.

            34. Sheaths closed for about $^1/_3$ their length; culm leaf blades sharply reduced in length upwards, the flag leaf blades absent or vestigial, commonly less than 1 cm long, always less than $^1/_5$ the sheath length, when present usually firm, not withering; innovation blades usually 1–3 mm wide; lemmas short- to long-villous on the keel and marginal veins . . . . 41. *P. fendleriana* (in part)

            34. Sheaths closed for $^1/_4$–$^3/_4$ their length; culm leaf blades gradually reduced in length upward along the culm or some midculm blades longer than the lower culm blades, culm blades narrow, thin and withering; innovation blades usually 0.5–1(2) mm wide; lemmas usually softly puberulent, sometimes short-villous on the keel and marginal veins . . . . 43. *P. ×nematophylla*

    32. Panicles contracted or open; florets usually bisexual, if unisexual, the panicles open; blades flat, folded, or involute.

        35. Panicles open or loosely contracted at maturity, 5–30 cm long, spikelets not crowded.

            36. Spikelets laterally compressed, lengths usually 3–3.8 times the widths; lemmas distinctly keeled, intercostal regions glabrous or with hairs distinctly shorter than those over the keel and marginal veins.

                37. Lemmas 2.5–3.5 mm long, usually glabrous throughout, infrequently with keels and marginal veins softly puberulent to short- or long-villous and/or

intercostal regions sparsely softly puberulent; blades usually involute, rarely flat, scabrous; calluses usually glabrous, rarely sparsely and shortly webbed; panicles (7)10–30 cm long; plants of the Chisos Mountains in Texas and northern Mexico . . . . 26. *P. strictiramea* (in part)

37. Lemmas 4–6 mm long, keels and marginal veins, sometimes also the lateral veins, short- to long-villous, intercostal regions glabrous or sparsely pilose or hispidulous near the bases; blades flat or folded, smooth or sparsely scabrous; calluses glabrous or with a crown of hairs, hairs 0.2–2 mm; panicles 5–18(25) cm long; plants of coastal Alaska, the Pacific Northwest, and Rocky Mountains . . . . 69. *P. stenantha* (in part)

36. Spikelets subterete, lengths usually (3.8)4–5 times the widths; lemmas weakly keeled, usually at least sparsely softly puberulent, infrequently short-villous, between the veins, the hairs usually about the same length as those of the keel and marginal veins.

38. Ligules of the culm leaves usually 2–6 mm long, smooth or scabrous, truncate to acuminate; basal tuft of leaves narrow or loosely clumped; basal leaves reaching 2–20+ cm, blades filiform or to 3 mm wide; panicle branches capillary or stouter, smooth or scabrous; plants widespread, sometimes on serpentine soils, often in wet habitats . . . . 64. *P. secunda* (in part)

38. Ligules of the culm leaves 0.5–1.5(2.5) mm long, scabrous, apices truncate to obtuse (acute); basal tuft of leaves narrow, tightly clumped; basal leaves reaching 2–8(13) cm, basal blades filiform; panicle branches capillary, distinctly scabrous; plants of thin, early drying serpentine soils in the Sierra Nevada foothills of California . . . . 65. *P. tenerrima*

35. Panicles contracted at maturity, sometimes open during anthesis, 1–30 cm long, spikelets crowded or not.

39. Plants 2–6(10) cm tall; panicles 1–4(6) cm long; cauline blades soft, folded, 1–3.5(4.5) cm long, upper cauline blades 0.9–1.8 mm wide, abaxial surfaces with 7–15 ribs; spikelets 3.5–6 mm long; calluses glabrous; lemmas glabrous or the keel and marginal veins sparsely softly puberulent near the base; plants of high alpine habitats in the Sierra Nevada and adjacent ranges . . . . 62. *P. keckii* (in part)

39. Plants 5–120 cm tall; panicles 2–30 cm long; leaves not as above in all respects; spikelets 3–10 mm long; lemmas nearly glabrous to copiously pubescent, if hairy, only so near the base on the keel and marginal veins, the hairs softly to crisply puberulent; calluses glabrous or with a crown of hairs; plants widespread.

40. Plants 5–40 cm tall; panicles 3–7 cm long, usually densely contracted, rarely loosely contracted, nearly cylindrical; culm blades 2–5 mm wide, flat or folded; innovation blades usually 1–1.5 mm wide, involute, infrequently similar to the culm blades; anthers fully developed; plants of the Pacific coast . . . . 71. *P. unilateralis* (in part)

40. Plants 10–120 cm tall; panicles 2–30 cm long, densely to loosely contracted, not nearly cylindrical; culm and

innovation blades similar, 1–5 mm wide; anthers sometimes aborted late in development; plants of the high arctic or interior habitats of western North America.

41. Anthers usually sterile and to 1.5 mm long, infrequently well developed and 2–2.8 mm long; plants 10–33(45) cm tall; blades folded to involute, 1.5–3 mm wide, abaxial surfaces smooth or sparsely scabrous; spikelets lustrous; lemmas usually weakly keeled, more or less evenly and loosely short- to long-villous on the lower $^{1}/_{3}$–$^{1}/_{2}$, hairs mostly longer than 0.5 mm; calluses usually with a crown of hairs to 2 mm long; panicle branches smooth or sparsely to moderately scabrous; plants of the high arctic . . . . 67. *P. hartzii* (in part)

41. Anthers well developed, 1.2–3.5 mm long; plants 10–120 cm tall; blades flat, folded, or involute, 0.5–5 mm wide, abaxial surfaces smooth or scabrous; spikelets lustrous or not; lemmas weakly keeled or not, if the intercostal regions hairy, the hairs distinctly shorter than those on the keels or, if the lemmas more or less evenly hairy, then the hairs usually shorter than 0.5 mm; calluses usually glabrous, infrequently with a crown of hairs to 2 mm long; panicle branches smooth or sparsely to densely scabrous; plants of the high arctic or western North America, if of the high arctic, the lemma hairs shorter than 0.3 mm.

42. Lemmas usually evenly strigulose across the lower $^{1}/_{3}$–$^{1}/_{2}$, hairs 0.1–0.2 mm long, rarely to 0.3 mm on the keel and marginal veins; blades soft, involute; culms 10–30 cm tall; panicles 3–6 cm long, longest branches 1–3(4) cm long, smooth or sparsely scabrous . . . . 68. *P. ammophila*

42. Lemmas variously hairy, if as above, the panicle branches usually scabrous and/or the blades flat and soon withering; culms (10)20–120 cm tall; panicles 2–30 cm long; branches 1–15 cm long, sparsely to densely scabrous.

43. Culm blades 1–3 cm long, flat, (1)1.5–3 mm wide, with thick, white margins and broadly prow-shaped apices; panicles 4–8 cm long, narrowly lanceoloid; spikelets 7–9 mm long; plants of serpentine slopes in the Wenatchee Mountains of Washington . . . . 66. *P. curtifolia* (in part)

43. Culm blades longer or narrower than above, margins not thick and white, apices narrowly prow-shaped; panicles 2–30 cm long, narrowly lanceoloid to ovoid; spikelets (4)5–10 mm long; plants widespread.

44. Spikelets subterete to weakly laterally compressed, (4)5–10 mm long, lengths 3.5–5 times the widths; rachilla internodes usually 1–2 mm long; lemmas usually weakly keeled, 3.5–6 mm long, nearly glabrous or hairy all over the basal $^{2}/_{3}$; culms (10)15–120 cm tall; panicles 2–25(30) cm long . . . . 64. *P. secunda* (in part)

44. Spikelets laterally compressed, (4)4.5–7 mm long, lengths 3–3.5 times the widths; rachilla internodes usually shorter than 1 mm; lemmas distinctly keeled, 3–4 mm long, usually glabrous, keels and marginal veins rarely sparsely puberulent proximally; culms 30–100 cm tall; panicles 5–15 cm long . . . . 70. *P. napensis* (in part)

Poa L. subg. Poa

Plants annual or perennial; sometimes unisexual; with or without rhizomes or stolons, densely to loosely tufted or the culms solitary. **Basal branching** intra- and/or extravaginal or pseudointravaginal. **Culms** spindly to stout, terete or weakly to strongly compressed; **nodes** 0–5, exserted. **Sheaths** terete or weakly to strongly compressed, closed only at the base or up to full length, fusion of the margins not extended by a hyaline membrane, basal sheaths usually glabrous, rarely sparsely retrorsely strigose, hairs about 0.1 mm; **ligules** 0.1–18 mm, thinly membranous and white to milky white or hyaline, truncate to acuminate, entire or erose to lacerate, smooth or ciliolate; **blades** flat, folded, or involute, thin to thick, smooth or sparsely to densely scabrous, adaxial surfaces glabrous or hairy, hispidulous or puberulent, apices narrowly to broadly prow-shaped. **Panicles** 1–41 cm, erect to nodding or lax, tightly contracted to open, with 1–100+ spikelets; **branches** 0.5–20 cm, erect to reflexed, terete or angled, smooth or sparsely to densely scabrous, usually glabrous, rarely hispidulous, with 1 to many spikelets. **Spikelets** 2–12 mm, subterete to strongly laterally compressed, sometimes bulbiferous; **florets** (1)2–8(13); **rachilla internodes** smooth or scabrous, glabrous or pubescent. **Glumes** shorter than to slightly exceeding the adjacent lemmas, weakly to distinctly keeled, smooth or scabrous; **calluses** blunt, usually terete or slightly laterally compressed, sometimes slightly dorsally compressed, glabrous, dorsally webbed, diffusely webbed, or with a crown of hairs; **lemmas** 1.7–11 mm, rounded to weakly or distinctly keeled, thinly membranous to chartaceous, glabrous or hairy on the keel and veins, sometimes the intercostal regions also hairy, 5–7(11)-veined, margins smooth or scabrous, glabrous, apices obtuse to acuminate; **palea keels** usually scabrous, infrequently smooth, glabrous or with hairs; **anthers** (1–2)3, 0.1–4.5(5) mm.

Poa subg. *Poa* is the largest subgenus of *Poa*. Its distribution is essentially the same as that of the genus. It includes all but one of the 70 species of *Poa* in the *Flora* region; *P. eminens* is included in subg. *Arctopoa*.

Poa sect. Sylvestres V.L. Marsh *ex* Soreng

Plants perennial; usually non-rhizomatous and non-stoloniferous, sometimes shortly rhizomatous, usually loosely tufted, infrequently densely tufted. **Basal branching** usually mainly pseudointravaginal, sometimes mainly extravaginal. **Culms** 20–126 cm, terete or weakly compressed. **Sheaths** closed for ($^1/_{20}$)$^1/_3$ to about their full length, terete or weakly keeled, basal sheaths readily deteriorating; **ligules** 0.1–3(4) mm, smooth or sparsely scabrous, truncate to obtuse, entire or lacerate, smooth or ciliolate; **blades** smooth or scabrous, glabrous, apices narrowly prow-shaped. **Panicles** 4–36 cm, erect or lax, pyramidal or lanccoloid, usually sparse, lower rachis internodes usually longer than (2)3 cm; **nodes** with 1–10 branches; **branches** ascending to spreading or eventually reflexed, lax or straight, angled, angles scabrous, with spikelets confined to the distal $^1/_5$–$^1/_3$($^1/_2$). **Spikelets** 2.5–8.2 mm, laterally compressed; **florets** (1)2–5(6), normal, bisexual; **rachilla internodes** smooth, usually glabrous, sometimes puberulent. **Glumes** distinctly keeled, scabrous; **calluses** terete or slightly laterally compressed, usually dorsally webbed, sometimes glabrous; **lemmas** 2.1–5 mm, lanceolate to broadly lanceolate, distinctly keeled, glabrous or hairy, apices with narrow, clear or white margins; **palea keels** scabrous, glabrous or with hairs over the keels; **anthers** 3, 0.4–2(2.6) mm.

Poa sect. *Sylvestres* includes seven species, all of which are endemic to the *Flora* region. Chloroplast DNA shows it to be an early diverging lineage of *Poa* (Gillespie and Soreng 2005).

1. Poa saltuensis Fernald & Wiegand [p. 511]
OLDPASTURE BLUEGRASS

Plants perennial; not rhizomatous, not stoloniferous, loosely tufted. **Basal branching** mainly pseudointravaginal. **Culms** 20–95 cm tall, 0.8–1.5 mm thick. **Sheaths** closed for $^1/_3$–$^2/_3$ their length; **ligules** 0.2–3(4) mm, smooth or sparsely scabrous, truncate to obtuse; **blades** 1–3.6 (6) mm wide, flat, thin, lax, veins prominent. **Panicles** 4–20(24) cm long, less than $^1/_4$ the plant height, lax; **nodes** with 1–3 branches; **branches** ascending to spreading, lax, angled, angles prominent, scabrous. **Spikelets** 3–5.6 mm, laterally compressed; **florets** 2–5; **rachilla internodes** glabrous, usually shorter than 1 mm. **Glumes** $^2/_3$–$^3/_4$ as long as the adjacent lemmas, distinctly keeled; **lower glumes** 1(3)-veined; **upper glumes** shorter than or subequal to the lowest lemmas; **calluses** webbed; **lemmas** 2.4–4 mm, lanceolate to broadly lanceolate, distinctly keeled, usually glabrous, bases of marginal veins rarely sparsely softly puberulent, lateral veins prominent, intercostal regions smooth, minutely bumpy, apices obtuse to sharply acute or acuminate; **palea keels** scabrous; **anthers** 0.4–1.5 mm.

Poa saltuensis grows in woodlands of the north-central and northeastern United States and adjacent Canada, extending south to Tennessee. The two subspecies are sometimes treated as species. The variation between the two overlaps and is correlated to some extent with ecology and geography. *Poa marcida*

(p. 512), a western species once included in *P. saltuensis*, differs in having closed sheaths and attenuate lemmas.

1. Anthers 0.4–1 mm long; lemma apices obtuse to acute, firm or scarious for up to 0.25 mm . . . . subsp. *languida*
1. Anthers 0.9–1.5 mm long; lemma apices acute to acuminate, scarious for 0.25–0.5 mm . . . . subsp. *saltuensis*

Poa saltuensis subsp. languida (Hitchc.) A. Haines [p. 511]

Lemmas 2.4–3 mm, broadly lanceolate, apices obtuse to broadly acute and pointed, firm or scarious to 0.25 mm; **anthers** 0.4–0.9(1) mm. 2*n* = unknown.

Poa saltuensis subsp. *languida* grows in rich open woodlands and thickets with dry to mesic soils of moderate pH, and, where soils are thin, over limestone and marble substrates. It is most prevalent in the southern portion of the species' range. It is absent from Newfoundland, New Brunswick, Nova Scotia, Prince Edward Island, New Hampshire, and Maine.

Poa saltuensis Fernald & Wiegand subsp. saltuensis [p. 511]

Lemmas 2.4–4 mm, lanceolate, apices acute to acuminate and pointed, scarious for 0.25–0.5 mm; **anthers** 0.9–1.5 mm. 2*n* = 28.

Poa saltuensis subsp. *saltuensis* grows throughout the range of the species, in open forests and woodlands with low to moderate pH soils.

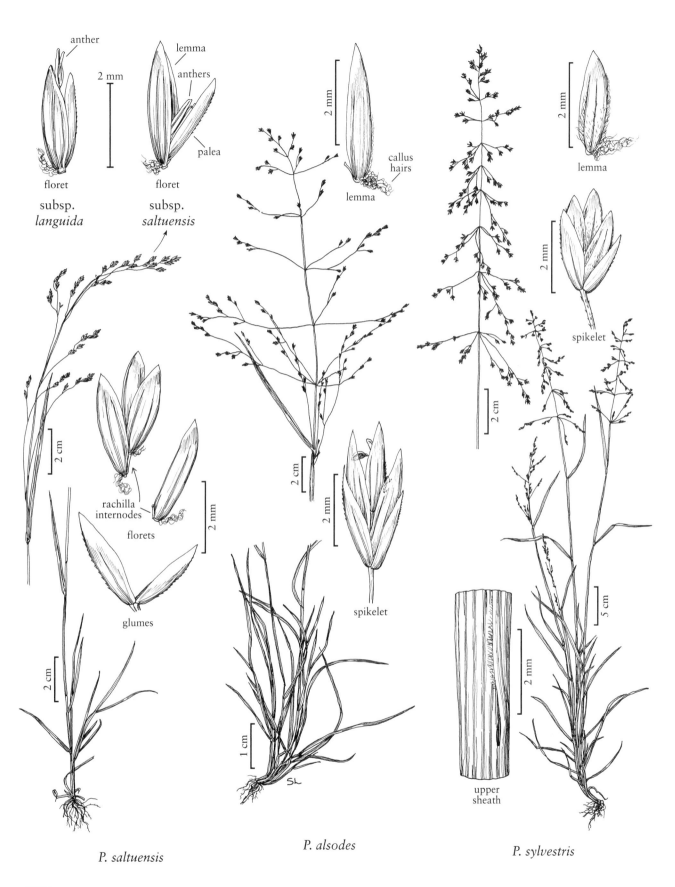

anther

lemma

anthers

2 mm

palea

floret

floret

subsp.
languida

subsp.
saltuensis

2 mm

lemma

callus
hairs

lemma

2 mm

2 mm

lemma

2 mm

spikelet

rachilla
internodes

florets

2 mm

glumes

2 cm

2 cm

2 mm

spikelet

2 cm

2 cm

5 cm

2 mm

1 cm

SL

upper
sheath

P. saltuensis

P. alsodes

P. sylvestris

2. Poa alsodes A. Gray [p. 511]
GROVE BLUEGRASS

Plants perennial; not rhizomatous, not stoloniferous, loosely tufted. **Basal branching** mainly pseudointravaginal. **Culms** 31–126 cm. **Sheaths** closed for $^1/_2$–$^7/_8$ their length; **ligules** 0.1–1.7(2.1) mm, smooth or sparsely scabrous, truncate to obtuse; **blades** 0.8–4.1 mm wide, flat, lax. **Panicles** 11.4–36 cm, erect or lax, narrowly pyramidal, usually open, infrequently contracted; **nodes** with (2)3–5(7) branches; **branches** spreading, straight, angled, angles sparsely to moderately scabrous. **Spikelets** 3.5–6.7 mm, laterally compressed; **florets** 2–4; **rachilla internodes** glabrous. **Glumes** ovate, distinctly keeled, keels scabrous; **lower glumes** 1-veined; **upper glumes** shorter than or subequal to the lowest lemmas; **calluses** webbed; **lemmas** 2.7–4.2(5) mm, lanceolate, distinctly keeled, keels short-villous to about midlength, marginal and lateral veins glabrous, lateral veins obscure or moderately prominent, intercostal regions glabrous, smooth, apices acute; **paleas** glabrous or ciliolate over the keels, apices finely scabrous; **anthers** 0.4–0.8 mm. $2n$ = unknown.

Poa alsodes grows in mesic woodlands of eastern Canada and the northeastern United States, extending south to Illinois, Tennessee, and North Carolina, particularly in the Appalachian Mountains.

3. Poa sylvestris A. Gray [p. 511]
WOODLAND BLUEGRASS

Plants perennial; not rhizomatous, not stoloniferous, loosely tufted, sometimes appearing shortly rhizomatous, loosely to densely tufted. **Basal branching** mainly pseudointravaginal. **Culms** 30–120 cm, bases often decumbent. **Sheaths** closed for ($^1/_{20}$) $^1/_2$–$^7/_8$ their length, terete, throats frequently ciliate near the point of fusion; **ligules** 0.5–2.7 mm, smooth or sparsely scabrous, truncate to obtuse; **blades** 0.7–5 mm wide, flat, thin, lax. **Panicles** (6.7)9–20 cm, open, narrowly conical at maturity; **nodes** with (2)3–10 branches per node; **branches** (2)3–7 cm, spreading to eventually reflexed, straight, angled, angles several, densely scabrous, with 1–11 spikelets. **Spikelets** 2.5–4.4 mm, laterally compressed; **florets** 2–3(4); **rachilla internodes** longer than (1)1.2 mm, smooth, glabrous. **Glumes** distinctly keeled, keels scabrous; **lower glumes** 1(3)-veined; **upper glumes** shorter than or subequal to the lowest lemmas; **calluses**

webbed; **lemmas** 2.1–3.1 mm, broadly lanceolate, distinctly keeled, keels and marginal veins short-villous, extending to near the apices on the keels, lateral veins prominent, softly puberulent to short-villous, intercostal regions usually sparsely softly puberulent, smooth, apices obtuse to acute; **palea keels** softly puberulent at midlength, apices finely scabrous; **anthers** 1–1.8 mm. $2n$ = 28.

Poa sylvestris grows in southeastern Canada and throughout much of the eastern United States, mainly at low elevations in woodlands, especially in riparian zones. It is easily distinguished from *P. wolfii* (p. 514) by its smaller, more numerous spikelets and lemmas that are usually sparsely hairy between the veins. Plants from the middle Appalachian Mountains have been confused with *P. paludigena* (p. 572); *P. sylvestris* is usually larger, has more than 2 branches per panicle node, is pubescent between the lemma veins and palea keels, and has larger anthers.

4. Poa marcida Hitchc. [p. 513]
WEEPING BLUEGRASS

Plants perennial; not rhizomatous, not stoloniferous, sometimes shortly rhizomatous, loosely to densely tufted. **Basal branching** mainly pseudointravaginal. **Culms** 20–80 cm. **Sheaths** closed for at least $^9/_{10}$ their length; **ligules** 0.5–2 mm, smooth, truncate; **blades** 1.5–5 mm wide, flat, lax. **Panicles** 6–22 cm, lax, narrowly lanceoloid, sparse; **nodes** with 1–3 branches; **branches** ascending, lax, angled, angles scabrous. **Spikelets** 3.5–7 mm, laterally compressed; **florets** (1)2(4); **rachilla internodes** about 1 mm, smooth, glabrous. **Glumes** distinctly keeled, keels scabrous; **lower glumes** 1-veined; **upper glumes** shorter than or subequal to the lowest lemmas; **calluses** webbed, webs sparse; **lemmas** 3.2–5 mm, narrowly lanceolate, distinctly keeled, glabrous, smooth, lateral veins moderately prominent, apices acuminate; **palea keels** scabrous, sometimes sparsely so; **anthers** 0.5–1.2 mm. $2n$ = unknown.

Poa marcida is an uncommon endemic of breaks in rich, mesic, generally old growth forests of the Pacific coast, from Vancouver Island through the western foothills of the northern Cascade Mountains to central Oregon. It differs from *P. saltuensis* (p. 510) in its closed sheaths and attenuate lemmas.

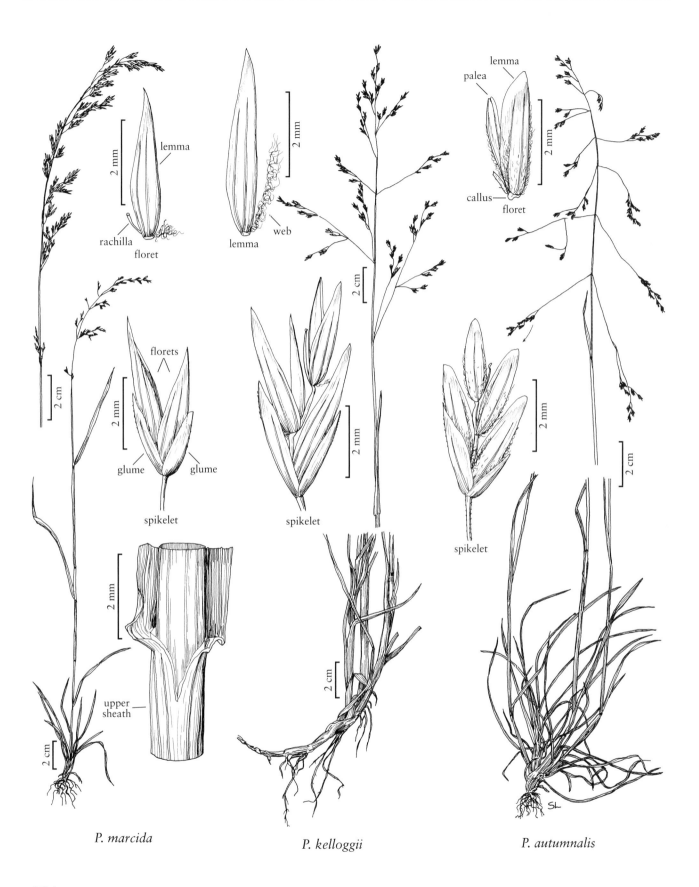

lemma

rachilla

floret

lemma

web

palea

lemma

callus

floret

florets

glume   glume

spikelet

spikelet

2 cm

2 mm

2 mm

spikelet

2 cm

upper
sheath

2 cm

2 cm

SL

P. marcida

P. kelloggii

P. autumnalis

POA

5. **Poa kelloggii** Vasey [p. 513]
KELLOGG'S BLUEGRASS

Plants perennial; rhizomatous, loosely tufted, culms solitary to several. **Basal branching** mainly extravaginal. **Culms** 28–85 cm, erect or the bases decumbent, terete or weakly compressed; **nodes** terete, 1–2 exserted. **Sheaths** closed for about $^1/_{15}$–$^1/_5$ their length, sometimes fused for a longer distance by a narrow hyaline membrane, terete, bases of basal sheaths glabrous, distal sheath lengths 0.8–1.5 times blade lengths; **ligules** 0.5–3 mm, scabrous, usually lacerate; **blades** scarcely reduced in length distally, 2–5 mm wide, flat, lax, apices narrowly prow-shaped, flag leaf blades 5–15 cm long. **Panicles** 10–20 cm, erect, pyramidal, open, sparse, with 25–70 spikelets; **nodes** with 1–3(5) branches; **branches** 5–15 cm, ascending, straight, spreading to eventually reflexed, angled, angles mostly moderately to densely scabrous, with 5–15 spikelets. **Spikelets** 4.5–6 mm, lengths 3.5 times widths, laterally compressed; **florets** 2–3; **rachilla internodes**, at least some, longer than 1 mm, smooth, glabrous. **Glumes** distinctly keeled; **lower glumes** 1–3-veined; **upper glumes** shorter than or subequal to the lowest lemmas; **calluses** webbed, hairs over $^1/_2$ the lemma length; **lemmas** 3.5–5 mm, narrowly lanceolate, distinctly keeled, smooth, glabrous throughout, lateral veins moderately prominent, apices acute to acuminate; **paleas** smooth to scabrous over the keels; **anthers** about 2 mm. $2n = 56$.

Poa kelloggii grows in rich coastal forests, especially redwood forests, in California. It is not common. Reports from Oregon and British Columbia are based on misidentifications of *P. laxiflora* (p. 538) and *P. howellii* (p. 534), respectively.

6. **Poa autumnalis** Muhl. *ex* Elliott [p. 513]
AUTUMN BLUEGRASS

Plants perennial; not rhizomatous, not stoloniferous, loosely tufted. **Basal branching** mainly pseudointravaginal. **Culms** 23–86 cm tall, 0.8–1.8 mm thick, bases often decumbent. **Sheaths** closed for $^1/_3$–$^1/_2$ their length; **ligules** 0.2–1.9(2.5) mm, smooth or sparsely scabrous, truncate to obtuse; **blades** (0.5)1–4 mm wide, flat or folded, thin. **Panicles** (5)8–20 cm, erect or lax, broadly pyramidal at maturity, open, sparse, lower axils sometimes sparsely pubescent; **nodes** with 1–2(4) branches; **branches** 5–12 cm, spreading to reflexed, straight, angled, angles scabrous, with 3–8 spikelets in the distal $^1/_4$–$^1/_3$.

Spikelets 3–8.2 mm, laterally compressed; **florets** 2–4(6); **rachilla internodes** smooth, sparsely softly puberulent. **Glumes** distinctly shorter than the adjacent lemmas, distinctly keeled, keels scabrous; **lower glumes** subulate to lanceolate, (1)3-veined; **upper glumes** lanceolate to broadly lanceolate; **calluses** usually glabrous, rarely sparsely and shortly webbed; **lemmas** (2.8)3–4.6 mm, lanceolate, distinctly keeled, keels and marginal veins short- to long-villous, hairs extending up $^3/_4$ of the keel, lateral veins prominent, intercostal regions softly puberulent, smooth, apices obtuse, blunt; **palea keels** softly puberulent to short-villous for much of their length, apices scabrous; **anthers** 1–1.4(2.6) mm. $2n = 28$.

Poa autumnalis grows primarily in the southeastern United States, being found in forests of the eastern and western Appalachian piedmont and coastal plain. It is readily distinguished from other perennial species of the eastern United States by its combination of glabrous calluses and pubescent palea keels.

7. **Poa wolfii** Scribn. [p. 515]
WOLF'S BLUEGRASS

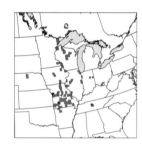

Plants perennial; not rhizomatous, not stoloniferous, loosely tufted. **Basal branching** mainly pseudointravaginal. **Culms** 25–90 cm. **Sheaths** closed for $^1/_2$–$^3/_4$ their length, smooth or sparsely scabrous, margins not ciliate; **ligules** 0.3–2.1 mm, smooth or sparsely scabrous, truncate to obtuse, ciliolate; **blades** 0.6–3.5 mm wide, flat. **Panicles** 7.5–15(18) cm, lax, pyramidal, open, sparse; **nodes** with 1–3(5) branches; **branches** 3–8 cm, ascending, straight to spreading, angled, angles prominent, scabrous. **Spikelets** 4–6.5 mm, laterally compressed; **florets** 2–5; **rachilla internodes** to 1 mm, smooth, glabrous. **Glumes** $^1/_2$–$^2/_3$ the length of the adjacent lemmas, distinctly keeled, keels scabrous; **lower glumes** subulate to narrowly lanceolate, (1)3-veined; **upper glumes** shorter than or subequal to the lowest lemmas; **calluses** webbed; **lemmas** (2.5)3.2–4.7 mm, lanceolate, green, distinctly keeled, keels and marginal veins long-villous, hairs extending up almost the whole keel length, lateral veins prominent, intercostal regions smooth, minutely bumpy, usually glabrous, rarely sparsely softly puberulent, apices acute, blunt, or pointed, white, not bronze; **palea keels** softly puberulent at midlength, apices scabrous; **anthers** (0.5)0.8–1.2(1.5) mm. $2n = 28$.

Poa wolfii is an uncommon species that grows in boggy areas of eastern deciduous forests, primarily west of the Appalachian divide. It differs from *P. sylvestris* (p. 512) in having fewer branches, larger spikelets, and lemmas that are usually glabrous between the veins.

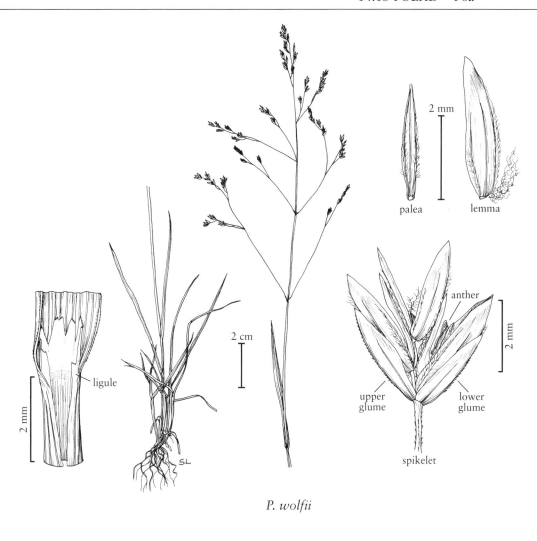

P. wolfii

POA

Poa sect. Arenariae (Hegetschw.) Stapf

Plants perennial; not rhizomatous, not stoloniferous, densely tufted. **Basal branching** intravaginal. **Culms** 2–60 cm, terete, bases bulbous. **Sheaths** closed for about ¹⁄₄ their length, lowest sheaths with swollen bases; **ligules** 1–6 mm, smooth or scabrous, obtuse to acute; **blades** (0.5)1–2.5 mm wide, flat, thin, lax, soon withering. **Panicles** (0.8)2–10 cm, ovoid, loosely contracted; **nodes** with 2–5 branches; **branches** usually ascending, infrequently spreading, terete, usually smooth or sparsely scabrous, rarely moderately scabrous. **Spikelets** 3–7 mm, laterally compressed, some or all bulbiferous; **florets** (2)3–7, forming a bulblet, sometimes the basal 1–2 florets normal. **Glumes** shorter than the adjacent lemmas, distinctly keeled, keels scabrous; **lower glumes** 3-veined; **calluses** terete or slightly laterally compressed, glabrous or dorsally webbed, hairs wrinkled; **lemmas** normal or leaflike, normal lemmas 2–4 mm, distinctly keeled, glabrous throughout or the keels and marginal veins villous, intercostal regions glabrous or puberulent, leaflike lemmas thickened at the base, bladelike distally; **paleas** scabrous, keels often softly puberulent at midlength; **anthers** 3, (0.6)1.2–2 mm, sometimes aborted late in development, sometimes not developed.

Poa sect. *Arenariae* is native to Eurasia and North Africa. It includes 14 species. These are easily recognized as members of the section by the bulbous bases of their new shoots. One species is established in the *Flora* region.

8. Poa bulbosa L. [p. 517]

BULBOUS BLUEGRASS

Plants perennial; densely tufted, not rhizomatous, not stoloniferous. **Basal branching** intravaginal. **Culms** 15–60 cm, erect or spreading, bases bulbous. **Sheaths** closed for about ¹/₄ their length, terete, lowest sheaths with swollen bases; **ligules** 1–3 mm, smooth or scabrous, apices obtuse to acute; **blades** 1–2.5 mm wide, flat, thin, lax, soon withering. **Panicles** 3–12 cm, ovoid; **nodes** with 2–5 branches; **branches** ascending to spreading, terete, usually smooth or sparsely scabrous, infrequently moderately scabrous. **Spikelets** 3–5 mm, laterally compressed, usually bulbiferous; **florets** 3–7, the basal floret, and sometimes additional florets, normal; **rachilla internodes** smooth, glabrous. **Glumes** keeled, keels scabrous; **lower glumes** 3-veined; **upper glumes** shorter than or subequal to the lowest lemmas; **calluses** webbed or glabrous; **lemmas** 3–4 mm, lanceolate, keeled, glabrous or the keels and marginal veins short- to long-villous, intercostal regions glabrous or softly puberulent, apices acute; **paleas** scabrous, keels often softly puberulent at midlength; **anthers** 1.2–1.5 mm and functional, sometimes aborted late in development, sometimes not developed. $2n = 14, 21, 28, 39, 42, 45$.

Poa bulbosa is a European species that is now established in the *Flora* region. In southern Europe and the Middle East, it is considered an important early spring forage.

1. Spikelets not bulbiferous . . . . subsp. *bulbosa*
1. All or some spikelets bulbiferous . . . . subsp. *vivipara*

Poa bulbosa L. subsp. bulbosa [p. 517]

Culms 15–25 cm. **Spikelets** not bulbiferous; **florets** all normal. **Calluses** webbed; **lemmas** short-villous on the keels and marginal veins, intercostal regions sparsely softly puberulent; **anthers** 1.2–1.5 mm. $2n =$ unknown.

Poa bulbosa subsp. *bulbosa* is common in its native Europe. It is uncommon in the *Flora* region, with the only known collections being from Drake, Butler, and Preble counties, Ohio. Whether these collections represent independent introductions or reversion to reproduction by seed is not known.

Poa bulbosa subsp. vivipara (Koel.) Arcang. [p. 517]

Culms 15–60 cm. **Spikelets** bulbiferous; **florets** modified into leafy bracts, sometimes the basal florets within a spikelet more or less normal. **Calluses** usually sparsely webbed, sometimes glabrous; **lemmas** glabrous or softly puberulent over the keel and lateral veins, sometimes between the veins; **anthers** in the least deformed florets 1.2–1.5 mm or aborted late in development, absent from modified florets. $2n = 21, 28, 31, 32, 33, 34, 35, 37, 39, 42+I, 44, 46, 48, 49$.

Poa bulbosa subsp. *vivipara* was introduced from Europe into the Pacific Northwest as a forage grass; it has since spread across temperate areas of the *Flora* region, particularly in the Pacific Northwest and northern Great Basin. It is highly tolerant of grazing and disturbance.

Poa sect. Alpinae (Hegetschw. *ex* Nyman) Stapf

Plants perennial; not rhizomatous, not stoloniferous. **Basal branching** intravaginal. **Culms** 10–40 cm, terete. **Leaves** mostly basal; **sheaths** closed for ¹/₂–²/₇ their length, terete, basal sheaths persistent, bases usually not swollen; **blades** flat, moderately thick, soft, straight, apices prow-shaped. **Panicles** 2–6(8) cm, erect, ovoid to pyramidal, open or loosely contracted at maturity; **nodes** with 1–2 branches; **branches** 1–3(4) cm, ascending to spreading, straight, terete, smooth or very sparsely scabrous, rarely moderately scabrous. **Spikelets** ovate, laterally compressed, occasionally bulbiferous; **florets** usually normal, bisexual. **Glumes** broadly lanceolate to narrowly ovate, shorter than to subequal to the adjacent lemmas, keeled, keels sparsely scabrous; **lower glumes** 3-veined; **calluses** terete, glabrous; **lemmas** broadly lanceolate, keeled, keels and marginal veins short- to long-villous, intercostal regions glabrous or sparsely to moderately short-villous; **palea keels** mostly softly puberulent to short-villous, scabrous distally; **anthers** 3, 1.3–2.3 mm.

Poa sect. *Alpinae* includes seven species. They are all cespitose perennials with intravaginal branching and broad leaves. One species is circumboreal; the other six are native to Europe.

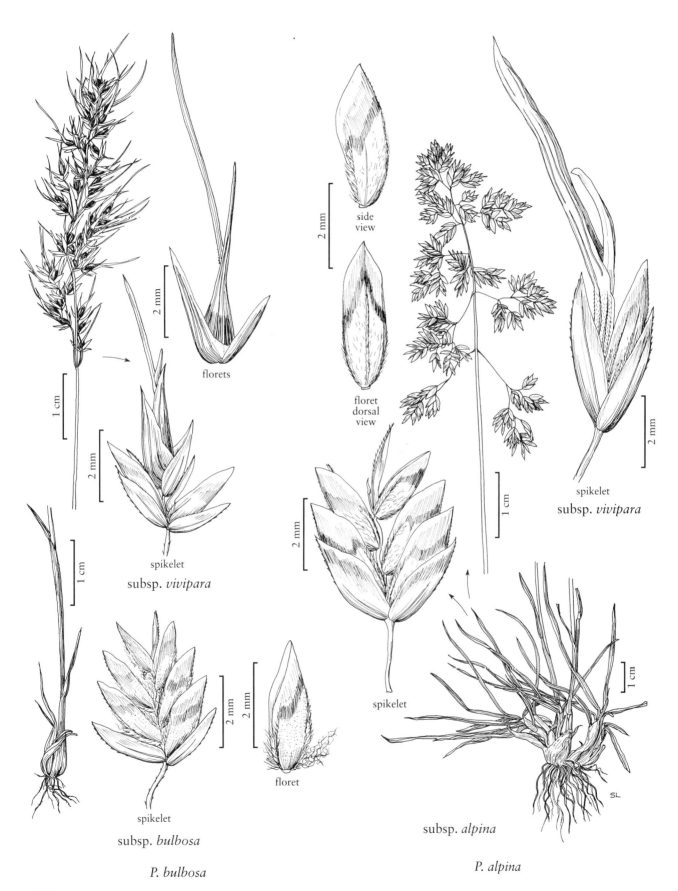

florets

spikelet

subsp. *vivipara*

spikelet

subsp. *bulbosa*

P. bulbosa

side
view

floret
dorsal
view

spikelet

floret

spikelet

subsp. *vivipara*

subsp. *alpina*

P. alpina

POA

9. Poa alpina L. [p. 517]

Plants perennial; not glaucous; densely cespitose, not rhizomatous, not stoloniferous. **Basal branching** intravaginal. **Culms** 10–40 cm. **Leaves** mostly basal; **sheaths** closed for $^1/_8$–$^2/_7$ their length, terete, basal sheaths persistent, overlapping, bases usually not swollen; **ligules** of innovations 1–2(3) mm, those of the upper cauline leaves to 4(5) mm, milky white, smooth, glabrous, obtuse; **blades** of innovations widely spreading, persisting through the season, blades of cauline leaves 1–5(12) cm long, 2–4.5 mm wide, flat, moderately thick, soft, straight, smooth or the margins sparsely scabrous, apices broadly prow-shaped, blades of upper cauline leaves much reduced in length. **Panicles** 2–6(8) cm, erect, ovoid to pyramidal, open or loosely contracted at maturity, fairly congested; **nodes** with 1–2 branches, lowest internodes 0.6–1(1.5) cm; **branches** 1–3(4) cm, ascending to spreading, straight, terete, usually smooth or sparsely scabrous, rarely moderately densely scabrous; **pedicels** divaricate, shorter than the spikelets. **Spikelets** 3.9–6.2 mm, ovate, lengths 1.5–2.5 times widths, laterally compressed, plump, sometimes bulbiferous; **florets** 3–7, usually normal; **rachilla internodes** 0.5–0.8 mm, smooth, glabrous or sparsely softly puberulent to short-villous. **Glumes** broadly lanceolate to narrowly ovate, keeled, keels sparsely scabrous; **lower glumes** 3-veined; **upper glumes** shorter than or subequal to the lowest lemmas; **calluses** glabrous; **lemmas** 3–5 mm, broadly lanceolate, keeled, keels and marginal veins short- to long-villous, lateral veins moderately prominent, intercostal regions sparsely to moderately short-villous, apices acute; **palea keels** softly puberulent to short-villous over most

of their length, apices scabrous; **anthers** 1.3–2.3 mm. $2n = 22, 23, 24, 25, 26, 27, 28, 28+II, 30, 31, 32, 32+I, 33, 34, 35, 36, 37, 39, 40+I, 41, 42,$ ca. $43, 44, 46,$ ca. $48, 56.$

Poa alpina is a fairly common circumboreal forest species of subalpine to arctic habitats, extending south in the Rocky Mountains to Utah and Colorado in the west, and to the northern Great Lakes region in the east. It often grows in disturbed ground and is calciphilic. *Poa ×gaspensis* (p. 601) is a natural hybrid which seems to be between *P. alpina* and *P. pratensis* subsp. *alpigena* (p. 525); it differs from *P. alpina* in its extravaginal branching, rhizomatous habit, and webbed calluses. The range of chromosome numbers suggests that *P. alpina* is predominantly apomictic.

1. Spikelets not bulbiferous . . . . subsp. *alpina*
1. Some or all spikelets bulbiferous . . . . subsp. *vivipara*

Poa alpina L. subsp. alpina [p. 517]
ALPINE BLUEGRASS

Spikelets not bulbiferous. **Anthers** 1.3–2.3 mm, well formed. $2n = 22, 23, 26, 27, 28, 28+I, 30, 31, 32, 32+I, 33, 34, 35, 36, 37, 39, 40+I, 41, 42,$ ca. $43, 44, 46,$ ca. $48, 56.$

Poa alpina subsp. *alpina* is the more common of the two subspecies. In the *Flora* region, it grows throughout the range of the species.

Poa alpina subsp. vivipara (L.) Arcang. [p. 517]

Spikelets all bulbiferous, or some normal and some bulbiferous. **Anthers** aborted late in development or not developed. $2n = 22, 24, 25, 26, 27, 28, 31, 32, 33.$

Poa alpina subsp. *vivipara* grows at scattered locations in Greenland, and has been reported for Alaska. It is common in alpine regions of northern and central Europe.

Poa sect. Micrantherae Stapf

Plants annual or perennial; green; usually neither rhizomatous nor stoloniferous, sometimes stoloniferous, densely to loosely tufted. **Basal branching** intravaginal. **Culms** 2–20(45) cm, terete or weakly compressed; **nodes** terete. **Sheaths** closed for $^1/_4$–$^1/_3$ their length, terete or weakly compressed, smooth, glabrous; **collars** smooth, glabrous; **ligules** 0.5–3(5) mm, smooth, glabrous, truncate to obtuse, entire; **blades** 1–3(6) mm wide, flat or weakly folded, thin, soft, smooth, margins usually slightly scabrous, apices broadly prow-shaped. **Panicles** 1–7(10) cm, erect, loosely contracted or open, ovoid to pyramidal; **nodes** with 1–2(5) branches; **branches** ascending to reflexed, straight, terete, smooth or sparsely scabrous. **Spikelets** 3–6 mm, lanceolate to narrowly ovoid, laterally compressed, not bulbiferous; **florets** 2–7, normal, upper 1–2 florets pistillate in some spikelets; **rachilla internodes** smooth, glabrous. **Glumes** distinctly keeled, smooth; **lower glumes** distinctly shorter than the lowest lemmas, 1-veined; **upper glumes** shorter than to subequal to the lowest lemmas; **calluses** terete, glabrous; **lemmas** 1.7–4 mm, distinctly keeled, smooth and glabrous or the keels, marginal veins, and, usually, lateral veins hairy, lateral veins moderately prominent to prominent, intercostal regions glabrous, margins smooth, glabrous, apices whitish, obtuse to acute; **palea keels** smooth, usually softly puberulent to long-villous, sometimes glabrous; **anthers** 3, 0.1–2.5 mm, sometimes vestigial in the upper 1–2 florets.

Poa sect. *Micrantherae* includes eight species, all of which are native to Eurasia and North Africa. They are gynomonoecious, with smooth or sparsely scabrous panicle branches. The calluses are glabrous in most species; the palea keels are usually hairy.

10. **Poa annua** L. [p. 520]
ANNUAL BLUEGRASS

Plants usually annual, rarely surviving for a second season; not rhizomatous, sometimes stoloniferous, densely tufted. **Basal branching** intravaginal, innovations common, similar to the culms. **Culms** 2–20(45) cm, prostrate to erect, slender; **nodes** terete, usually 1 exserted. **Sheaths** closed for about ⅓ their length, terete or weakly compressed, smooth; **ligules** 0.5–3(5) mm, smooth, glabrous, decurrent, obtuse to truncate; **blades** 1–10 cm long, 1–3(6) mm wide, flat or weakly folded, thin, soft, smooth, margins usually slightly scabrous, apices broadly prow-shaped. **Panicles** 1–7(10) cm, lengths 1.2–1.6 times widths, erect; **nodes** with 1–2(3) branches; **branches** ascending to spreading or reflexed, straight, terete, smooth, with crowded or loosely arranged spikelets. **Spikelets** 3–5 mm, laterally compressed; **florets** 2–6; **rachilla internodes** smooth, glabrous, concealed or exposed, distal internodes less than ½(¾) the length of the distal lemma. **Glumes** smooth, distinctly keeled, keels smooth; **lower glumes** 1-veined; **upper glumes** shorter than or subequal to the lowest lemma; **calluses** glabrous; **lemmas** 2.5–4 mm, lanceolate, distinctly keeled, smooth throughout, the keels, marginal veins, and, usually, lateral veins crisply puberulent to long-villous, rarely glabrous throughout, lateral veins prominent, intercostal regions glabrous, margins smooth, glabrous, apices obtuse to acute; **palea keels** smooth, usually short- to long-villous, rarely glabrous; **anthers** 0.6–1.1 mm, oblong prior to dehiscence, those of the upper 1–2 florets usually vestigial. 2*n* = 28.

Poa annua is one of the world's most widespread weeds. It thrives in anthropomorphic habitats outside of the arctic. A native of Eurasia, it is now well established throughout most of the *Flora* region.

Poa annua is a gynomonoecious tetraploid (possibly rarely polyhaploid), and is thought to have arisen from hybridization between *P. infirma* (see next) and *P. supina* (p. 521) (Tutin 1952). It is similar to *P. infirma*,

differing in having larger anthers. It differs from *P. chapmaniana* (p. 534) in having glabrous calluses and three larger anthers, rather than one. Forms with glabrous lemmas occur sporadically within populations.

11. **Poa infirma** Kunth [p. 520]
WEAK BLUEGRASS

Plants annual; neither rhizomatous nor stoloniferous, densely tufted. **Basal branching** intravaginal, sterile shoots common, similar to the culms. **Culms** 2–15 cm, prostrate to erect, slender; **nodes** terete, usually 1 exserted. **Sheaths** closed for about ⅓ their length, terete or weakly compressed, smooth; **ligules** 0.5–3 mm, smooth, glabrous, decurrent, obtuse to truncate; **blades** 1–3(4) mm wide, flat, thin, soft, smooth, margins usually slightly scabrous, apices broadly prow-shaped. **Panicles** 1–6 cm, lengths 1.5–3 times widths, erect; **nodes** with 1–2(5) branches; **branches** ascending, straight, terete, smooth, with crowded spikelets. **Spikelets** 3–5 mm, laterally compressed; **florets** 2–6; **rachilla internodes** smooth, glabrous, usually exposed in side view, distal internodes ½–¾ the length of tbe distal lemma. **Glumes** smooth, distinctly keeled, keels smooth; **lower glumes** 1-veined; **upper glumes** shorter than or subequal to the lowest lemmas; **calluses** glabrous; **lemmas** 2–2.5 mm, lanceolate, distinctly keeled, smooth throughout, the keels, marginal and lateral veins crisply puberulent to long-villous, lateral veins prominent, intercostal regions glabrous, margins smooth, glabrous, apices obtuse to acute; **palea keels** smooth, short- to long-villous; **anthers** 0.1–0.6 mm, more or less spherical to short-elliptical prior to dehiscence, those of the upper 1–2 florets commonly vestigial. 2*n* = 14.

Poa infirma was introduced from Europe to the Americas, and was first described from Colombia. It is sporadically established along the Pacific coast and in the central valleys of California, and has been collected in Charleston, South Carolina. It is rare elsewhere in the *Flora* region. *Poa annua* often resembles *P. infirma* (see

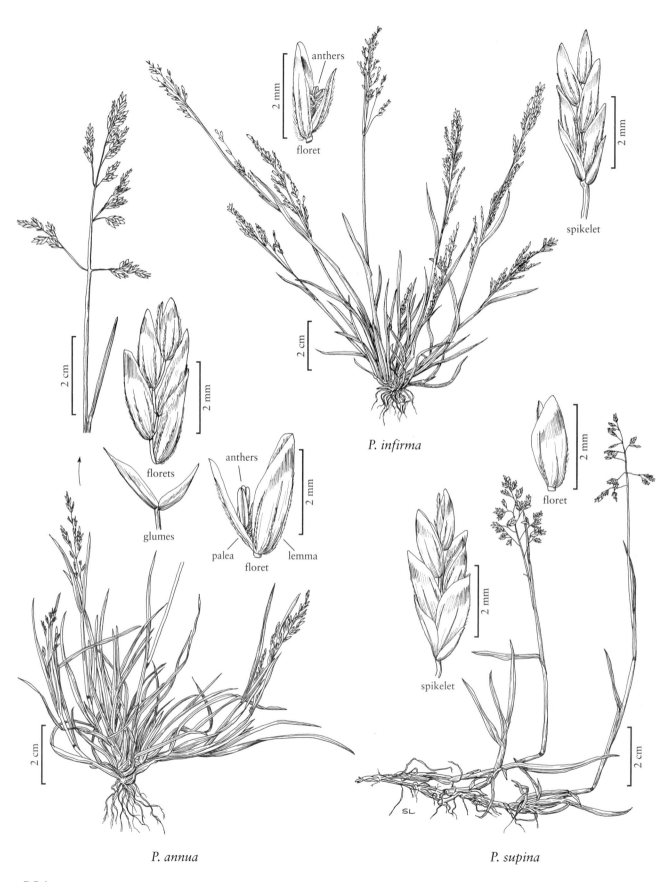

anthers

2 mm

floret

spikelet

2 mm

P. infirma

2 cm

2 mm

florets

anthers

2 mm

glumes

palea   lemma

floret

spikelet

2 mm

floret

2 mm

2 cm

2 cm

SL

P. annua

P. supina

POA

previous), which is thought to be one of its parents, but *P. annua* is tetraploid and has anthers 0.6–1.1 mm long. Both species are gynomonoecious.

12. Poa supina Schrad. [p. 520]
SUPINE BLUEGRASS

Plants perennial; stoloniferous, loosely tufted. **Basal branching** intravaginal. **Culms** 8–12(20) cm, slender, bases decumbent, terete or weakly compressed; **nodes** terete, 1 exserted. **Sheaths** closed for ¹/₄–¹/₃ their length, terete, smooth, glabrous, bases of basal sheaths glabrous, distal sheath lengths 2–4 times blade lengths; **collars** smooth, glabrous; **ligules** 0.6–1 mm, smooth, glabrous, truncate; **blades** 2–3 mm wide, flat, thin, soft, smooth, apices broadly prow-shaped, cauline blades subequal. **Panicles** 2.5–5 cm, lengths 1–2 times widths, erect, loosely contracted or open, ovoid to pyramidal, sparse, with 10–25(30) spikelets and 1–2 branches per node; **branches** 1–3 cm, spreading to reflexed, straight, terete, smooth or sparsely scabrous, with 2–5(8) spikelets. **Spikelets** 4–6 mm, laterally compressed; **florets** 3–7; **rachilla internodes** smooth, glabrous, more or less concealed, distal internode less than ¹/₂ the length of the distal lemma. **Glumes** distinctly keeled, keels smooth; **lower glumes** 1-veined; **calluses** glabrous;

lemmas 1.7–4 mm, lanceolate, distinctly keeled, smooth throughout, proximal lemmas glabrous throughout or the keels and marginal veins sparsely short-villous, distal lemmas glabrous or the keels and marginal veins short-villous to near the apices, lateral veins moderately prominent, intercostal regions glabrous, margins smooth, glabrous, apices obtuse to acute; **palea keels** smooth, sometimes sparsely softly puberulent to short-villous; **anthers** (1.25)1.5–2.5 mm, cylindrical prior to dehiscence, those of the upper 1–2 florets commonly vestigial. 2*n* = 14.

Poa supina is native to boreal to alpine regions of Eurasia. Beginning in the 1990s, the cultivar 'Supernova' has been introduced for seeding in wet to moist, cool, shady areas subject to heavy traffic. It has been tested in both Canada and the United States, and is expected to gradually escape cultivation, probably becoming established throughout the cool-temperate portion of the *Flora* region. Its current distribution is not known. *Poa supina* differs from *P. annua* (p. 519), of which is thought to be one of the parents, in having longer anthers and a more stoloniferous habit, as well as in being diploid. It is gynomonoecious.

Poa L. sect. Poa

Plants perennial; rhizomatous, rhizomes usually well developed and extensive, sometimes poorly developed, densely to loosely tufted or the shoots solitary. **Basal branching** mainly extravaginal or equally extra- and intravaginal. **Culms** 5–120 cm, terete or weakly compressed; **nodes** terete or weakly compressed. **Sheaths** closed for (¹/₆)¹/₄–³/₅ their length, terete to slightly compressed, smooth or sparsely scabrous, usually glabrous, infrequently sparsely to moderately hairy, distal sheaths usually longer than their blades; **collars** smooth, glabrous; **ligules** 0.9–7 mm, smooth or scabrous, truncate to acute, glabrous or ciliolate; **innovation blades** of intravaginal shoots involute and narrower or similar to the cauline blades and blades of extravaginal shoots; **cauline blades** subequal or the middle blades longest, flat, folded, or weakly involute, abaxial surfaces smooth, glabrous, adaxial surfaces smooth or sparsely scabrous, frequently sparsely hairy, hairs 0.2–0.8 mm, apices prow-shaped, sometimes narrowly prow-shaped, flag leaf blades 1.5–10 cm. **Panicles** 2–18(20) cm, loosely contracted to open, often slightly lax to nodding, sparsely to moderately congested, with 1–7(9) branches per node; **branches** 1–9 cm, ascending to widely spreading or somewhat reflexed, flexuous to straight, terete or angled, usually smooth or sparsely to moderately scabrous, infrequently densely scabrous. **Spikelets** 3.5–9(12) mm, lengths to 3.5 times widths, lanceolate to broadly lanceolate, laterally compressed, sometimes bulbiferous; **florets** 2–5(6), usually normal, bisexual; **rachilla internodes** smooth, glabrous or pubescent. **Glumes** unequal to subequal, distinctly shorter than to subequal to the adjacent lemmas, keels weak or distinct, smooth or scabrous; **lower glumes** 1- or 3-veined; **calluses** terete or slightly laterally compressed, usually dorsally webbed, sometimes with additional webs below the marginal veins, infrequently glabrous; **lemmas** 2–8 mm, lanceolate to broadly lanceolate, distinctly keeled, keels and marginal veins, and sometimes also the lateral veins, hairy, all veins prominent, intercostal regions glabrous or hairy; **palea keels** sometimes with hairs at midlength, intercostal regions glabrous or hairy; **anthers** 3, 1.2–2.5 mm, infrequently aborted late in development.

Poa section *Poa* includes 32 species. All the species are synoecious perennials; most are strongly rhizomatous.

13. Poa pratensis L. [pp. 524, 525]

KENTUCKY BLUEGRASS

Plants perennial; green or anthocyanic, sometimes glaucous; extensively rhizomatous, densely to loosely tufted or the shoots solitary. **Basal branching** mainly extravaginal or evenly extra- and intravaginal. **Culms** 5–70(100) cm, erect or the bases decumbent, not branching above the base, terete or weakly compressed; **nodes** terete or weakly compressed, 1–2(3) exposed, proximal node(s) usually not exserted. **Sheaths** closed for ¼–½ their length, terete to slightly compressed, glabrous or infrequently sparsely to moderately hairy, bases of basal sheaths glabrous, not swollen, distal sheath lengths 1.2–5(6.2) times blade lengths; **collars** smooth, glabrous; **ligules** 0.9–2(3.1) mm, smooth or scabrous, truncate to rounded, infrequently obtuse, ciliolate or glabrous; **blades** of extravaginal innovations like those of the culms, those of the intravaginal shoots sometimes distinctly narrower, 0.4–1 mm wide, flat to involute; **cauline blades** 0.4–4.5 mm wide, flat, folded, or involute, soft and lax to moderately firm, abaxial surfaces smooth, glabrous, adaxial surfaces smooth or sparsely scabrous, frequently sparsely hairy, hairs 0.2–0.8 mm, erect to appressed, slender, curving, sinuous or straight, apices usually broadly prow-shaped, sometimes narrowly prow-shaped, blades subequal, the middle blades longest, the flag leaf blades 1.5–10 cm. **Panicles** 2–15(20) cm, narrowly ovoid to narrowly or broadly pyramidal, loosely contracted to open, sparse to moderately congested, with (25) 30–100+ spikelets and (1)2–7(9) branches per node; **branches** (1)2–9 cm, spreading early or late, terete or angled, smooth or sparsely to moderately densely scabrous, with 4–30(50) spikelets usually fairly crowded in the distal ½. **Spikelets** 3.5–6(7) mm, lengths 3.5 times widths, laterally compressed, sometimes bulbiferous; **florets** 2–5, usually normal, sometimes bulb-forming; **rachilla internodes** usually shorter than 1 mm, smooth, glabrous. **Glumes** unequal to subequal, usually distinctly shorter than the adjacent lemmas, narrowly lanceolate to lanceolate, infrequently broadly lanceolate, distinctly keeled, keels usually sparsely to densely scabrous, infrequently smooth; **lower glumes** 1.5–4(4.5) mm, usually narrowly lanceolate to lanceolate, occasionally sickle-shaped, 1–3-veined; **upper glumes** 2–4.5(5) mm, distinctly shorter than to nearly equaling the lowest lemmas; **calluses** dorsally webbed, sometimes with additional webs below the marginal veins, hairs at least ½ as long as the lemmas, crimped; **lemmas** 2–4.3(6) mm, lanceolate, green or strongly purple-tinged, distinctly keeled, keels and marginal veins long-villous, lateral veins usually glabrous, infrequently short-villous to softly puberulent, lateral veins prominent, intercostal regions glabrous, lower portion smooth or finely muriculate, upper portion smooth or sparsely scabrous, margins narrowly to broadly hyaline, glabrous, apices acute; **paleas** scabrous, keels sometimes softly puberulent, intercostal regions narrow, usually glabrous, rarely sparsely hispidulous; **anthers** usually 1.2–2 mm, infrequently aborted late in development. 2n = 27, 28, 32, 35, 37, 41–46, 48–147.

Poa pratensis is common, widespread, and well established in many natural and anthropogenic habitats of the *Flora* region. The only taxa that are clearly native to the region are the arctic and subarctic subspp. *alpigena* and *colpodea*. Outside the *Flora* region, *P. pratensis* is native in temperate and arctic Eurasia. It is now established in temperate regions around the world.

Poa pratensis is a highly polymorphic, facultatively apomictic species, having what is probably the most extensive series of polyploid chromosome numbers of any species in the world. *Poa pratensis* is a hybridogenic species, i.e., it comprises numerous lineages with the same basic maternal genome, but different paternal genomes. The lineages are perpetuated by agamospermic and vegetative reproduction. Some major forms are recognized as microspecies or subspecies. These have some correlated ecological and morphological differences, but the morphological boundaries between them are completely bridged, and in some cases the taxa (e.g., subspp. *agassizensis* and *colpodea*) may represent environmentally induced plasticity.

Natural hybrids have been identified between *Poa pratensis* and *P. alpina*, *P. arctica*, *P. wheeleri*, and *P. secunda*. Many other artificial hybrids have been made; these involve many different, often distantly related, species. In addition, there are many cultivated forms of the species; these have been seeded widely throughout the *Flora* region for lawns, soil stabilization, and forage. Most cultivated forms favor subsp. *irrigata* morphologically; others tend towards subspp. *pratensis* and *angustifolia*, the latter occurring most commonly in xeric sites.

Poa rhizomata (p. 546) resembles *P. pratensis*, but has acute ligules and sparse inflorescences, florets that are usually unisexual, and generally larger spikelets; *Poa macrocalyx* (p. 527) looks like a robust *P. pratensis* with large spikelets, and lemmas and paleas that are generally hispidulous between the veins and palea keels. *Poa confinis* (p. 552) also resembles *P. pratensis*, but differs in having glabrous or sparsely hairy lemmas, and diffusely webbed calluses.

1. At least some spikelets bulbiferous; plants of the high arctic tundra . . . . subsp. *colpodea*
1. Spikelets not bulbiferous; plants widely distributed.
    2. Panicle branches smooth or almost smooth.
        3. Basal branching primarily extravaginal; blades flat or folded, soft, adaxial surfaces usually glabrous, sometimes sparsely hairy; plants of alpine and tundra regions . . . . subsp. *alpigena*
        3. Basal branching both intra- and extravaginal; blades folded or involute, somewhat firm, adaxial surfaces often sparsely hairy; plants widely distributed, but not in alpine or tundra regions . . . . subsp. *agassizensis*
    2. Panicles branches more or less scabrous.
        4. Intravaginal innovation shoots present, intra- and extravaginal blades alike, 0.4–1 mm wide, folded to involute, somewhat firm, adaxial surfaces often sparsely and softly hairy; plants of dry meadows and forests . . . . subsp. *angustifolia*
        4. Intravaginal innovation shoots present or absent, if present then differentiated or alike, at least some with blades 1.5–4.5 mm wide, flat or folded, adaxial surfaces rarely hairy; plants widespread, often of more mesic sites.
            5. Culms 8–30(50) cm tall, often somewhat glaucous, particularly the glumes; blades flat; intravaginal shoots absent or present and with blades similar to those of the extravaginal shoots; panicles with few spikelets per branch and 1–2(5) branches per node; plants of low, wet, often sandy ground . . . . subsp. *irrigata*
            5. Culms to 100 cm tall, not glaucous; blades flat or folded; intravaginal shoots present, with blades similar to those of the extravaginal shoots or distinctly narrower; panicles with several to many spikelets per branch and 3–5(7) branches per node; plants of various habitats, including those of subsp. *irrigata* . . . . subsp. *pratensis*

Cultivars of **Poa pratensis** L.

Plants densely to loosely tufted, often forming turf, shoots clustered. **Basal branching** intra- and extravaginal or mainly extravaginal. **Culms** 8–50 cm. **Innovation shoot blades** usually shorter than 45 cm, (0.4)1–4 mm wide, usually flat, sometimes some involute, usually soft, sometimes somewhat firm, adaxial surfaces usually glabrous; **cauline blades** flat or folded. **Panicles** 3–15 cm, broadly pyramidal, open or somewhat contracted, with 2–7(9) branches per node; **branches** ascending or widely spreading, sparsely to densely scabrous, with few to many spikelets per branch. **Spikelets** lanceolate to broadly lanceolate, not bulbiferous; **florets** normal. **Glume keels** strongly compressed, sparsely to moderately scabrous; **upper glumes** shorter than to nearly equaling the lowest lemmas; **lemmas** 2.8–4.3(6) mm, finely muriculate, lateral veins glabrous; **palea keels** scabrous, glabrous, intercostal regions glabrous. $2n = 41$–45, 48–59, 62, 64–74, 76, 78, 80, 81, 84–90, 95.

More than 60 cultivars of *Poa pratensis* have been released in the *Flora* region. Plants grown from commercially distributed seed have generally been placed in subsp. *pratensis* by North American authors, but they appear to include genetic contributions from at least three major subspecies, e.g., subspp. *angustifolia*, *pratensis*, and *irrigata*. These and intermediate forms, especially those favoring subspp. *irrigata* and *pratensis*, are best simply referred to as *Poa pratensis sensu lato* or labeled as cultivated material. The chromosome counts listed here are numbers reported for the species that are probably not subspp. *alpigena*, *angustifolia*, or *colpodea*; they may represent subspp. *irrigata* or *pratensis*.

Poa pratensis subsp. **agassizensis** (B. Boivin & D. Löve) Roy L. Taylor & MacBryde [p. 524]

Plants moderately to densely tufted, shoots clustered. **Basal branching** intra- and extravaginal. **Culms** 20–40(50) cm. **Innovation shoot blades** usually shorter than 15 cm, 0.8–2 mm wide, all involute or folded, adaxial surfaces sparsely pubescent. **Panicles** 4–8 cm, ovoid, narrowly pyramidal or loosely contracted, with 2–5 branches per node; **branches** ascending, smooth or sparsely to moderately densely scabrous, with several spikelets per branch. **Spikelets** lanceolate, not bulbiferous; **florets** normal. **Glume keels** strongly compressed, sparsely to moderately scabrous; **upper glumes** shorter than to nearly equaling the lowest lemmas; **lemmas** 2–4 mm, finely muriculate, lateral veins glabrous; **palea keels** scabrous, glabrous, intercostal regions glabrous. $2n = 56$.

Poa pratensis subsp. *agassizensis* was described as native to prairies and mountain grasslands of North

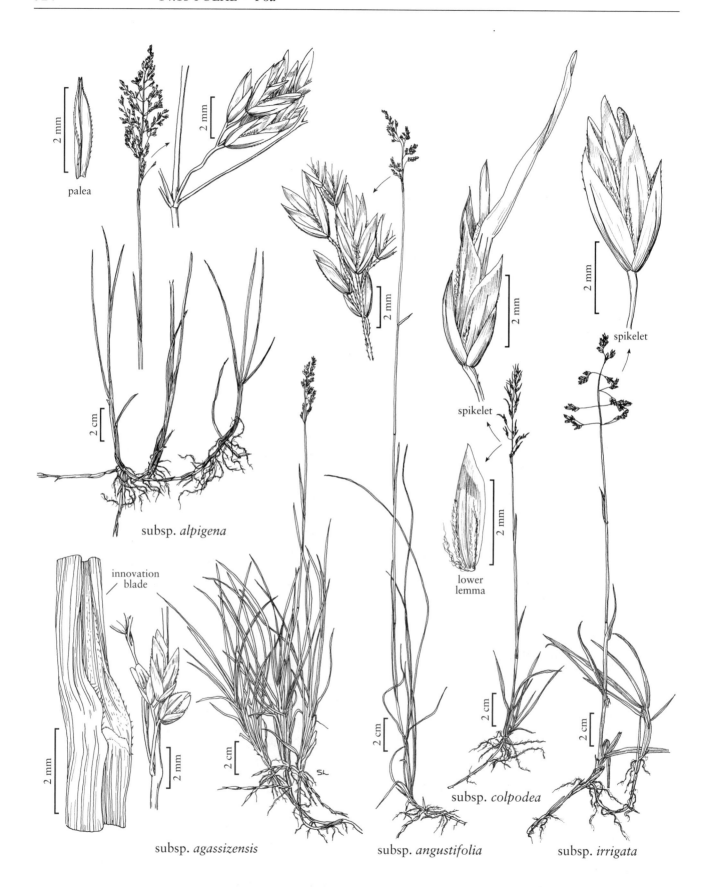

palea

subsp. *alpigena*

spikelet

spikelet

lower lemma

innovation blade

subsp. *agassizensis*

subsp. *angustifolia*

subsp. *colpodea*

subsp. *irrigata*

POA PRATENSIS (1 of 2)

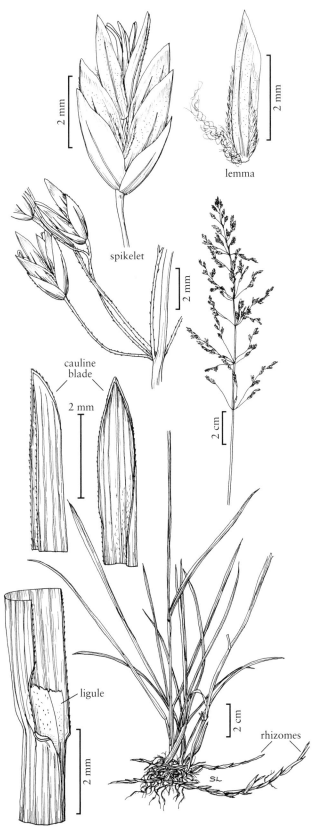

lemma

spikelet

cauline
blade

2 mm

ligule

2 mm

2 cm

rhizomes

SL

subsp. *pratensis*

POA PRATENSIS (2 of 2)

America; this has not been confirmed. It grows throughout the drier, cool-temperate range of the species in North America. It may consist of ecotypes derived from cultivated material, or be a native form that has adapted to xeric conditions. The least distinctive of the subspecies treated here, it closely approaches subsp. *angustifolia* in having involute leaves and small spikelets, but has shorter and broader leaves, and more condensed panicles.

Poa pratensis subsp. **alpigena** (Lindm.) Hiitonen [p. 524]
<small>ALPIGENE BLUEGRASS</small>

Plants strongly anthocyanic; moderately to loosely tufted, shoots usually solitary. **Basal branching** mainly extravaginal. **Culms** 15–70 cm. **Innovation shoot blades** shorter than 15 cm, 1–3.6 mm wide, flat or folded, soft, adaxial surfaces usually glabrous, sometimes sparsely pubescent; **cauline blades** flat or folded. **Panicles** 3–13(20) cm, narrowly pyramidal or contracted, expanding well after emergence from the sheath, with (1)2–5(7) branches per node; **branches** 1–6 cm, steeply ascending to eventually spreading or somewhat reflexed, smooth or sparsely scabrous, with 5–15 spikelets. **Spikelets** 4–5.5 mm, narrowly lanceolate, not bulbiferous; **florets** normal. **Glume keels** distinct, smooth or sparsely scabrous near the apices; **upper glumes** nearly equaling the lowest lemmas; **lemmas** 2.5–3.5 mm, smooth or finely muriculate, lateral veins frequently short-villous to softly puberulent; **palea keels** scabrous, often softly puberulent at midlength, intercostal regions usually glabrous, rarely sparsely hispidulous. $2n = 28, 32, 35, 42, 48, 50, 53, 56, 60, 63, 64, 65, 67, ca. 68, 69, 70, 72, 73, 74, 76, 77, 78, 79, 82, 84, 86, 88, 89, 92, 94.$

Poa pratensis subsp. *alpigena* is a circumpolar, mesophytic to subhydrophytic, arctic and alpine subspecies that extends into boreal forests in northern parts of the *Flora* region. It is infrequent south of Canada, with isolated collections being known from as far south as New Mexico in the Rocky Mountains, and New Hampshire and Maine in the east. It also grows in southern Patagonia.

Poa pratensis subsp. *alpigena* approaches *P. arctica* (p. 529). It differs in being glabrous between the lemma and palea veins, having somewhat more dense, later-opening panicles, and, usually, having smaller spikelets and more closely spaced palea keels. It differs from a likely hybrid with *P. alpina*, *P. ×gaspensis* (p. 601), in its truncate to rounded ligules, lemmas that are glabrous between the veins, and the lack of a basal tuft of leaves. In this treatment, bulbiferous plants are placed in subsp. *colpodea*, a subspecies which is more common than subsp. *alpigena* in the high arctic. The two sometimes grow together; there is some evidence of a shift in dominance from year to year.

Poa pratensis subsp. **angustifolia** (L.) Lej. [p. 524]
Plants moderately densely to densely tufted. **Basal branching** intra- and extravaginal, intravaginal shoots clustered. **Culms** 25–80 cm. **Innovation shoot blades** 10–45 cm long, 0.4–1 mm wide, all involute, sometimes narrower than the cauline blades, adaxial surfaces sparsely pubescent; **cauline blades** involute or folded, somewhat firm, adaxial surfaces sparsely pubescent. **Panicles** 8–18 cm, narrowly pyramidal or loosely contracted, branches ascending to spreading, smooth or sparsely to densely scabrous, with several to many spikelets per branch. **Spikelets** narrowly lanceolate, not bulbiferous; **florets** normal. **Glume keels** strongly compressed, sparsely to moderately scabrous; **upper glumes** shorter than to nearly equaling the lowest lemmas; **lemmas** 2.5–3.5 mm, finely muriculate, lateral veins glabrous; **palea keels** scabrous, glabrous, intercostal regions glabrous. $2n = 28, 46, 48–54, 56, 57, 59, 60, 61, 62, 63, 64, 65, 66, 68, 70, 72, 83$.

Poa pratensis subsp. *angustifolia* is a western Eurasian subspecies that is known from scattered locations throughout the temperate North American distribution of the species. It is characterized by the predominance of fascicles of elongate, narrow, involute blades on the intravaginal vegetative shoots, and slender panicles with small spikelets. Recent research has shown that it is primarily a low polyploid.

Poa pratensis subsp. **colpodea** (Th. Fr.) Tzvelev [p. 524]
Plants strongly anthocyanic; moderately densely to loosely tufted, shoots solitary. **Basal branching** mainly extravaginal. **Culms** 15–30 cm. **Innovation shoot blades** shorter than 15 cm, 1–3.6 mm wide, flat or folded, soft, adaxial surfaces usually glabrous, sometimes sparsely pubescent; **cauline blades** flat or folded. **Panicles** 4–8 cm, narrowly pyramidal or contracted, with (1)2–5 branches per node; **branches** 1–3(5) cm, ascending or eventually spreading, smooth or sparsely scabrous, with several spikelets per branch. **Spikelets** narrowly lanceolate, bulbiferous, the least deformed spikelets 4–5.5 mm; **florets** mostly bulb-forming. **Glume keels** distinct, smooth or sparsely scabrous distally; **upper glumes** 2–2.5(3) mm, nearly equaling the lowest lemmas; **lemmas** 2.5–3.5 mm, finely muriculate, lateral veins glabrous; **palea keels** scabrous, frequently softly puberulent at midlength, intercostal regions usually glabrous, rarely sparsely hispidulous; **anthers** usually aborted, sometimes a few fairly well developed. $2n =$ ca. 27, 35, 35, 37, 38, 42, 51, 52, 56, 60, 66, 68, 69, 72, 73, 74, 75, 76, 77, 78, 79, 80, 81.

Poa pratensis subsp. *colpodea* is circumpolar. In the *Flora* region, its range extends from Alaska and British Columbia to Greenland. It is more common than *P. pratensis* subsp. *alpigena* in the high arctic. The two sometimes grow together, and there is some evidence of a shift in dominance from year to year.

Poa pratensis subsp. **irrigata** (Lindm.) H. Lindb. [p. 524]
Plants moderately densely to loosely tufted, sometimes forming turf, culms solitary. **Basal branching** mainly extravaginal. **Culms** 8–30(50) cm. **Innovation shoot blades** usually shorter than 15 cm, 2–4.5 mm wide, adaxial surfaces usually glabrous; **cauline blades** flat. **Panicles** 2–10 cm, broadly pyramidal, open, with 1–2(5) branches per node; **branches** 1.5–6 cm, widely spreading, smooth or sparsely to moderately scabrous, with 4–8 spikelets. **Spikelets** lanceolate to broadly lanceolate, not bulbiferous; **florets** normal. **Glume keels** distinct, sparsely to moderately scabrous; **lower** and **upper glumes** often nearly equaling the lowest lemmas, often noticeably glaucous; **lemmas** 3–5(6) mm, finely muriculate, lateral veins glabrous; **palea keels** scabrous, glabrous, intercostal regions glabrous. $2n = 54, 56, 65, 80, 82–147$.

Poa pratensis subsp. *irrigata* is poorly understood in the *Flora* region. As interpreted here, it includes subarctic to boreal, coastal and lowland plants that differ from those of subsp. *pratensis* in their primarily extravaginal branching, shorter stature, wide, flat blades, short, stout panicles with fewer branches, larger spikelets, and relatively longer glumes that are often pruinose.

Poa pratensis L. subsp. **pratensis** [p. 525]
Plants densely to loosely tufted, often forming turf, culms clustered. **Basal branching** intra- and extravaginal. **Culms** 8–100 cm. **Innovation shoot blades** 10–45 cm long, 0.4–4 mm wide, some distinctly narrower than the cauline blades, all flat or some involute, usually soft, adaxial surfaces sparsely pubescent; **cauline blades** flat or folded. **Panicles** 5–18 cm, broadly pyramidal, open or somewhat contracted, with 3–5(7) branches per node; **branches** spreading to

somewhat reflexed, smooth or sparsely to fairly densely scabrous, with several to many spikelets per branch. **Spikelets** lanceolate to broadly lanceolate, not bulbiferous; **florets** normal. **Glume keels** strongly compressed, sparsely to moderately scabrous; **upper glumes** shorter than or nearly equaling the lowest lemmas; **lemmas** 2.8–4.3 mm, finely muriculate, lateral veins glabrous; **palea keels** scabrous, glabrous, intercostal regions glabrous. $2n = 43, 44, 48, 49, 50, 51, 52, 54, 56, 58, 59, 62, 65, 66, 67, 74, ca. 85, ca. 86, 88, 89, 95.$

Poa pratensis subsp. *pratensis* grows throughout most of the range of the species, but is absent from the high arctic, and only sporadic in the low arctic. It usually has a few narrow, flat or involute, intravaginal shoot leaves, in addition to some broader, extravaginal shoot leaves, and is intermediate between subspp. *angustifolia* and *irrigata*. For a comparison, see the descriptions of those subspecies.

14. Poa sublanata Reverd. [p. 528]
COTTONBALL BLUEGRASS

Plants perennial; usually anthocyanic; extensively rhizomatous, loosely tufted, culms solitary or a few together. **Basal branching** mainly extravaginal. **Culms** 20–40(75) cm, erect or the bases decumbent, not branching above the base, terete or weakly compressed; **nodes** terete, proximal nodes usually not exserted. **Sheaths** closed for $^2/_5$–$^1/_2$ their length, terete, glabrous, smooth or slightly scabrous, bases of basal sheaths glabrous, distal sheath lengths (1)1.2–3.5 times blade lengths; **collars** smooth, glabrous; **ligules** 1.5–4(6) mm, smooth, apices obtuse to acute, not ciliate; **blades** 1–3.5(5) mm wide, folded or flat, innovation shoot blades involute, soft, abaxial surfaces glabrous, adaxial surfaces smooth or sparsely scabrous, frequently sparsely hairy with 0.2–0.8 mm hairs, slender, erect to appressed, curving, sinuous or straight, apices narrowly prow-shaped, cauline blades subequal, flag leaf blades 1.5–8 cm. **Panicles** 5–9(17) cm, narrowly lanceoloid to narrowly pyramidal, loosely contacted to open, sparse, with 25–60 spikelets and 2–5 branches per node; **branches** 1–3 cm, ascending to spreading, slightly flexuous to fairly straight, terete, smooth or sparsely scabrous, usually with 1–5 spikelets per branch, the spikelets moderately crowded in the distal $^1/_2$. **Spikelets** 4–6.5(8) mm, lengths to 3 times widths, laterally compressed; **florets** 2–4(6); **rachilla internodes** shorter than 1 mm, smooth, usually glabrous, rarely with a few hairs. **Glumes** unequal to subequal, usually distinctly shorter

than the adjacent lemmas, lanceolate, distinctly keeled, keels smooth to sparsely scabrous; **lower glumes** 2.8–3.5 mm, narrowly lanceolate, (1)3-veined; **upper glumes** 4–4.5(5) mm, distinctly shorter than to nearly equaling the lowest lemmas; **calluses** dorsally webbed, webs copious, hairs at least $^1/_2$ times the lemma length, sometimes secondary webs present under the marginal veins; **lemmas** 3.7–4.5(5) mm, lanceolate, usually strongly purple, distinctly keeled, keels, lateral veins, and marginal veins copiously hairy, hairs cottony, lateral veins prominent, less hairy, intercostal regions glabrous, usually smooth or finely muriculate, sometimes sparsely scabrous distally, margins narrowly hyaline, glabrous, apices acute; **palea keels** sparsely scabrous, long-villous at midlength, intercostal regions narrow, glabrous; **anthers** 1.8–2.5 mm. $2n = 56, ca. 70.$

Poa sublanata grows in the high arctic of Alaska and Russia, usually on sandy ground. Bulbiferous plants are known from Russia; none have been found in the *Flora* region.

15. Poa macrocalyx Trautv. & C.A. Mey. [p. 528]
LARGE-GLUME BLUEGRASS

Plants perennial; green or strongly purplish; rhizomatous, rhizomes sometimes poorly developed, densely to loosely tufted. **Basal branching** all or mainly extravaginal. **Culms** (20)30–120 cm, usually stout, erect or decumbent, not branching above the base, terete or weakly compressed; **nodes** terete, 1–2 exserted. **Sheaths** closed for $^1/_4$–$^3/_5$ their length, terete to compressed, usually distinctly keeled, glabrous, sparsely to moderately scabrous, bases of basal sheaths glabrous, distal sheath lengths (0.8)1–2(2.5) times blade lengths; **collars** smooth, glabrous; **ligules** 2–5(6) mm, usually smooth, ligules of proximal leaves sometimes scabrous, obtuse, those of the distal leaves acute; **blades** 4–12(18) cm long, (2)3–7 mm wide, flat or loosely folded, glabrous, both surfaces smooth or the adaxial surfaces sparsely scabrous, apices broadly prow-shaped, cauline blades subequal, flag leaf blades 3–10 cm. **Panicles** (4)7–15(20) cm, lax, loosely contracted or open, moderately congested, with (15)20–100 spikelets and 2–4(5) branches per node; **branches** 3–8 cm, ascending or eventually spreading, flexuous, terete or weakly angled, usually sparsely scabridulous or scabrous, infrequently moderately densely scabrous, with (3)5–15(30) spikelets moderately crowded in the distal $^1/_2$. **Spikelets** (5)6–9(12) mm, lengths to 3 times widths, laterally compressed, sometimes bulbiferous; **florets** 2–5, infrequently bulb-forming; **rachilla internodes**

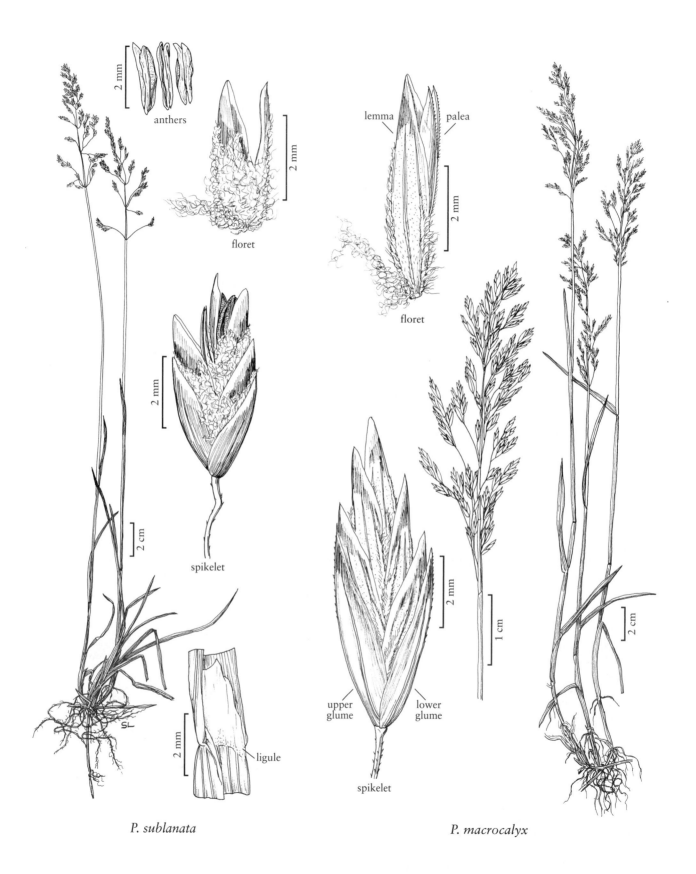

anthers

floret

lemma   palea

floret

2 mm

spikelet

ligule

upper
glume

lower
glume

spikelet

P. sublanata

P. macrocalyx

smooth or sparsely hispidulous. **Glumes** subequal in length and width, usually nearly equaling the adjacent lemmas, lanceolate to broadly lanceolate, distinctly keeled, keels sparsely to moderately scabrous distally, lateral veins prominent; **lower glumes** (4)4.5–7 mm, 3-veined; **upper glumes** (4)4.5–7(8) mm, 3(5)-veined; **calluses** dorsally webbed, hairs ⅓–⅔ times the lemma length, copious; **lemmas** (4)5–8 mm, lanceolate to broadly lanceolate, green or purple, distinctly keeled, keels and marginal veins long-villous, lateral veins prominent, usually softly puberulent to short-villous, intercostal regions sometimes hispidulous, infrequently softly puberulent or short-villous proximally and hispidulous in the central ⅓, sometimes merely muriculate proximally and scabrous distally, sometimes densely scabrous throughout, rarely glabrous and nearly smooth proximally and sparsely scabrous distally, margins broadly hyaline, glabrous, apices acute; **palea keels** scabrous, sometimes softly puberulent at midlength, intercostal regions broad, distinctly hispidulous or softly puberulent; **anthers** 1.5–2.5 mm. 2*n* = 42, 43, 44, 45, ca. 46, 49, 56, ca. 58, 63, ca. 64, ca. 65, 66, 67, 68, 69, 70, ca. 71, ca. 72, ca. 73, 74, 75, 76, 77, 78, 80, ca. 82, ca. 84, ca. 87, ca. 100.

Poa macrocalyx grows mainly in coastal areas of boreal Alaska, and from the eastern coast of Russia to northern Japan. Bulbiferous plants are occasionally found. *Poa macrocalyx* resembles an exaggeratedly robust *P. pratensis* (p. 522), with large spikelets and lemmas, proportionally longer glumes, and paleas that are generally hispidulous between the veins and palea keels. It is cytologically and morphologically complex, and is sometimes difficult to distinguish from *P. arctica* subsp. *lanata* (see below). *Poa norbergii* Hultén may belong to this species; alternatively, it may be a hybrid between *P. glauca* (p. 576) and *P. arctica* (p. 576) or *P. macrocalyx*.

16. Poa arctica R. Br. [p. 531]
ARCTIC BLUEGRASS

Plants perennial; usually strongly anthocyanic; rhizomatous, rhizomes usually well developed, sometimes poorly developed, shoots usually solitary. **Basal branching** mainly extravaginal. **Culms** 7.5–60 cm, slender to stout, terete or weakly compressed, bases usually decumbent, not branching above the bases; **nodes** terete, proximal nodes usually not exserted, 0–2 exserted above. **Sheaths** closed for (⅙)⅕–⅖ their length, terete, glabrous, smooth or sparsely scabrous, bases of basal sheaths glabrous, distal sheath lengths 1.4–4(5.3) times blade lengths; **collars** smooth, glabrous; **ligules** (1)2–7 mm, glabrous, smooth or sparsely to infrequently moderately scabrous, apices usually rounded to obtuse or acute, rarely truncate, entire or lacerate; **blades** 1–6 mm wide, flat or folded, somewhat involute, smooth, glabrous, apices broadly prow-shaped, cauline blades subequal or gradually reduced distally, flag leaf blades 0.7–9 cm. **Panicles** (2)3.5–15 cm, ovoid to broadly pyramidal, usually open, sparse, with 10–40(60) spikelets, proximal internodes shorter than 1.5(3) cm, with (1)2–5 branches per node; **branches** 1.5–6 cm, spreading soon after emergence from the sheath, thin, sinuous, and flexuous to fairly stout and straight, terete, smooth or sparsely to infrequently moderately scabrous, with (1)2–5 spikelets, the spikelets not crowded. **Spikelets** (3.5) 4.5–8 mm, lengths to 3.5 times widths, laterally compressed, sometimes bulbiferous; **florets** (2)3–6, infrequently bulb-forming; **rachilla internodes** smooth or muriculate, proximal internodes glabrous or sparsely softly puberulent to long-villous. **Glumes** lanceolate to broadly lanceolate, distinctly or weakly keeled, keels usually smooth, sometimes sparsely scabrous distally, lateral veins usually moderately pronounced; **lower glumes** (3)3.5–5(6) mm, 3-veined; **upper glumes** 3.5–5.5(6.5) mm, nearly equaling to slightly exceeding the lowest lemmas, or distinctly shorter; **calluses** glabrous or webbed, hairs sparse and short to over ⅓–⅔ the lemma length; **lemmas** (2.7)3–6(7) mm, lanceolate to broadly lanceolate, usually strongly purple, distinctly keeled, keels, marginal veins, and lateral veins long-villous, hairs on the lateral veins sometimes shorter, lateral veins prominent, intercostal regions short-villous to softly puberulent at least near the base, glabrous elsewhere, smooth to weakly muriculate and/or usually sparsely scabrous, infrequently moderately scabrous, margins broadly hyaline, glabrous, apices acute; **palea keels** usually short- to long-villous for most of their length, rarely nearly glabrous and scabrous, intercostal regions broad, usually at least sparsely softly puberulent, rarely glabrous, apices scabrous; **anthers** 1.4–2.5 mm, sometimes aborted late in development. 2*n* = 36, 42, 56, 60, 62–68, 70, ca. 72, 74–76, 78–80, 82–84, 86, 88, 99, 106.

Poa arctica is a common circumboreal species of arctic and alpine regions, growing mainly in mesic to subhydric, acidic tundra and alpine meadows, and on rocky slopes. It extends south in the Rocky Mountains to New Mexico. The frequency of sterile anthers in plants of the high arctic suggests that *P. arctica* is sometimes apomictic in that region. Over most of the rest of its range, *P. arctica* usually develops normal anthers. This and isozyme data for populations from alpine and low arctic regions suggest sexual reproduction is common in these habitats.

The most reliable way to distinguish *Poa arctica* from *P. pratensis* (p. 522), particularly subsp. *alpigena*, is by the wider paleas and the presence of hairs between the palea keels. Bulbiferous forms of *P. arctica* differ from *P. stenantha* var. *vivipara* (p. 594) in not being glaucous, and in having rhizomes and terete, smooth panicle branches. *Poa ×gaspensis* (p. 601) also resembles *P. arctica*, but it has sharply keeled, more scabrous glumes and a spikelet shape that is intermediate between *P. pratensis* and *P. alpina* (p. 518). *Poa arctica* forms natural hybrids with both *P. pratensis* and *P. secunda* (p. 586).

1. Plants lacking well-developed rhizomes; anthers aborted late in development; plants of the high arctic . . . . subsp. *caespitans*
1. Plants usually with well-developed rhizomes; anthers normal or plants not of the high arctic.
  2. Panicles erect, the branches relatively stout, fairly straight; longest branches of the lowest panicle nodes $^1/_4$–$^1/_2$ the length of the panicles; culms wiry, usually several together; calluses glabrous or shortly webbed; paleas sometimes glabrous; plants glaucous, growing in the southern Rocky Mountains and adjacent portions of the Intermountain region . . . . subsp. *aperta*
  2. Panicles lax to erect, the branches slender, flexuous to fairly stout and straight; longest branches of the lowest panicle nodes $^2/_5$–$^3/_5$ the length of the panicles; culms slender to stout, varying from solitary to several together; calluses glabrous or webbed, the hairs usually more than $^1/_2$ as long as the lemmas; paleas pubescent; plants sometimes glaucous, widespread in distribution.
    3. Calluses glabrous; spikelets not bulbiferous . . . . subsp. *grayana*
    3. Calluses webbed, often copiously so, sometimes glabrous in bulbiferous spikelets; spikelets sometimes bulbiferous.
      4. Spikelets (5)6–8 mm long; lemmas 4–6 mm long; blades 2–6 mm wide; rachillas usually hairy; plants primarily of the western arctic, extending to northwestern British Columbia . . . . subsp. *lanata*
      4. Spikelets (3.5)4–7 mm long; lemmas (2.7)3–4.5 mm long; blades 1.5–3 mm wide; rachillas commonly glabrous; plants widespread . . . . . subsp. *arctica*

Poa arctica subsp. **aperta** (Scribn. & Merr.) Soreng [p. 531]

Plants pale green, often glaucous; usually densely tufted, rhizomes usually short, usually well developed. **Culms** 20–60 cm, several together, wiry, bases decumbent.

Sheaths closed for ($^1/_6$)$^1/_5$–$^1/_3$ their length; **ligules** 3–7 mm, sparsely to moderately scabrous, acute; **blades** 1.5–2.5 mm wide, flat, folded, or somewhat involute. **Panicles** 4–15 cm, erect, loosely contracted or open, with 1–3 branches per node; **branches** ascending or widely spreading, fairly stout, fairly straight, smooth to very sparsely scabrous, proximal branches $^1/_4$–$^1/_2$ the panicle length. **Spikelets** narrowly lanceolate to lanceolate, not bulbiferous; **florets** 2–3(4), normal; **rachilla internodes** usually glabrous, infrequently sparsely softly puberulent; **calluses** glabrous or webbed, hairs to $^1/_4$ the lemma length; **lemmas** 3–4.5(6) mm; **palea keels** usually softly puberulent to long-villous at midlength, infrequently glabrous, intercostal regions usually softly puberulent; **anthers** aborted late in development or fully developed. $2n = 98+$I.

Poa arctica subsp. *aperta* is restricted to subalpine and low alpine habitats on the Wasatch Escarpment and high mountains of the Colorado Plateau in southern Utah, and the Rocky Mountains of southern Colorado and northern New Mexico. Many reports of *P. arida* (p. 599) growing west of the Rocky Mountains are based on misidentification of this subspecies. *Poa arctica* subsp. *aperta* may reflect introgression of genes from *P. secunda* (p. 586) into *P. arctica*. It has softer leaves, and is more densely hairy between the lemma veins and the palea keels, than subsp. *arctica*. It can be distinguished from subsp. *grayana* by its more wiry culms, and less contracted panicles with straighter branches.

Poa arctica R. Br. subsp. **arctica** [p. 531]

Plants usually loosely, sometimes densely, tufted, rhizomatous, rhizomes short or long, well developed. **Ligules** (1)2–4 mm, obtuse to acute; **blades** 1.5–2.5(3) mm wide, flat or folded, thin and soon withering, flag leaf blades 0.7–5.5 cm. **Panicles** lax to erect, open; **branches** ascending or widely spreading, sinuous and flexuous to fairly straight, smooth or sparsely scabrous, proximal branches $^2/_5$–$^3/_5$ the panicle length. **Spikelets** (3.5)4.5–6(7) mm, infrequently bulbiferous; **rachilla internodes** usually glabrous, infrequently sparsely softly puberulent to long-villous; **calluses** sparsely to copiously webbed; **lemmas** (2.7)3–4.5 mm; **palea keels** puberulent to long-villous at midlength, intercostal regions usually hairy, sometimes glabrous; **anthers** usually fully developed, except in the high arctic. $2n = 56, 60, 62, 63, 64, 65, 68, 70, 72, 74, 75, 76, 77, 78, 79, 80, 82, ca. 83, 84, 85, 88, 106.$

Poa arctica subsp. *arctica* is polymorphic and circumpolar. It grows in alpine and tundra habitats as far south as Wheeler Peak, New Mexico. Bulbiferous plants are known from alpine habitats in Alaska and British Columbia.

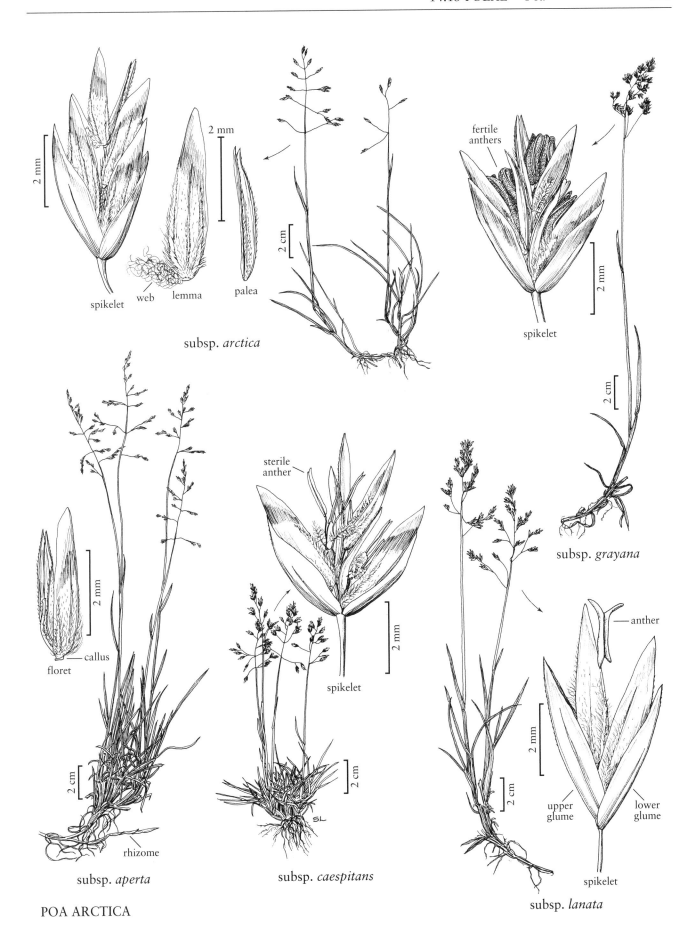

2 mm

spikelet   web   lemma   palea

2 mm

subsp. *arctica*

2 cm

fertile
anthers

2 mm

spikelet

subsp. *grayana*

2 cm

callus
floret

2 mm

sterile
anther

2 mm

spikelet

anther

2 mm

upper
glume

lower
glume

2 cm

2 cm

rhizome

2 cm

spikelet

subsp. *aperta*

subsp. *caespitans*

subsp. *lanata*

POA ARCTICA

Poa arctica subsp. *arctica* has tougher leaves, and is less densely hairy between the lemma veins and palea keels, than subsp. *aperta*. It often grows with subsp. *lanata*, but can be distinguished by its smaller and, usually, more numerous spikelets and narrower leaves. Paleas that are glabrous between the keels are frequent in plants from the Rocky Mountains. Such plants have been called *P. longipila* Nash, but do not merit recognition. Hultén (1942) recognized several variants within subsp. *arctica*; they are of ecotypic significance at best.

Poa arctica subsp. caespitans Simmons *ex* Nannf. [p. 531]

Plants densely to moderately densely tufted, rhizomatous, rhizomes very short, poorly developed. **Sheaths** closed for ($^1/_5$)$^1/_3$ their length, terete; **ligules** 2–4 mm, obtuse to acute; **blades** 1–2.5 mm wide, flat or folded, moderately thick. **Panicles** lax to erect, open; **branches** ascending or widely spreading, sinuous and flexuous to fairly straight, smooth or the angles moderately or densely scabrous, proximal branches $^1/_4$–$^1/_2$ the panicle length. **Spikelets** 4.5–8 mm, laterally compressed, infrequently bulbiferous; **rachilla internodes** usually glabrous, infrequently sparsely softly puberulent; **calluses** webbed; **lemmas** 3–6 mm; **palea keels** glabrous or long-villous; **anthers** usually aborted late in development, infrequently fully developed. 2n = 56, ca. 66, ca. 80.

Poa arctica subsp. *caespitans* grows in moist tundra of the high arctic, or infrequently in the low arctic of northeastern Canada and Greenland. It also grows in Norway and in the Russian high arctic; it is rare in both regions.

Many plants included here tend towards *Poa glauca* (p. 576, e.g., scabrous branches, and intermediate leaf and panicle forms), while others are distinguished from other *P. arctica* subspecies only by their more tufted habit and sterile anthers. As interpreted here, subsp. *caespitans* includes *P. trichopoda* Lange [= *P. tolmatchewii* Roshev.].

Poa arctica subsp. grayana (Vasey) Á. Löve, D. Löve & B.M. Kapoor [p. 531]

Plants sometimes glaucous; densely to loosely tufted, rhizomatous, rhizomes short or long, usually well developed, culms solitary or a few together. **Culms** 20–60 cm, bases decumbent, not wiry. **Sheaths** closed for $^1/_4$–$^2/_5$ their length; **ligules** (2)3–7 mm, smooth, obtuse to acute; **blades** 1–3 mm wide, flat or folded. **Panicles** lax to erect, open; **branches** ascending or widely spreading, somewhat sinuous and flexuous to

fairly straight, smooth to sparsely scabrous, proximal branches $^2/_5$–$^1/_2$ the panicle length. **Spikelets** (4)4.5–7 mm, not bulbiferous; **rachilla internodes** usually glabrous, infrequently sparsely softly puberulent; **calluses** glabrous; **lemmas** (2.7)3–5 mm; **palea keels** puberulent to long-villous at midlength; **anthers** usually fully developed. 2n = 36?

Poa arctica subsp. *grayana* grows only in the alpine regions of the middle and southern Rocky Mountains of Utah, Wyoming, Colorado, and New Mexico. It is characterized by its glabrous calluses, densely hairy lemmas, and paleas that are densely hairy between the keels. It has less wiry culms, and panicles with more flexuous branches, than subsp. *aperta* and, like that subspecies, can be difficult to distinguish from *P. arida* (p. 599).

Poa arctica subsp. lanata (Scribn. & Merr.) Soreng [p. 531]

Plants usually densely, infrequently loosely, tufted, rhizomes short, fairly stout, sometimes poorly developed. **Basal branching** frequently intravaginal. **Sheaths** persistent; **ligules** 2–4 mm, obtuse to acute; **blades** 2–6 mm wide, usually flat, firm and fairly persistent, flag leaf blades (1.5)2.5–9 cm. **Panicles** lax to erect, open; **branches** ascending or widely spreading, somewhat sinuous and flexuous to fairly stout and straight, smooth to sparsely or moderately scabrous, with (1)2–3(5) spikelets, proximal branches $^2/_5$–$^3/_5$ the panicle length. **Spikelets** (5)6–8 mm, infrequently bulbiferous; **rachilla internodes** sometimes muriculate, sparsely to moderately long-villous or glabrous. **Glumes** keeled, keels compressed, smooth or moderately scabrous; **calluses** webbed in normal plants, web copious, sometimes glabrous in bulbiferous plants; **lemmas** 4.5–6(7) mm; **palea keels** short- to long-villous, sometimes scabrous in bulbiferous plants; **anthers** usually fully developed in sexual plants, poorly developed in bulbiferous plants. 2n = 42, 56, 62, 70, 75, 78, 80.

Poa arctica subsp. *lanata* is amphiberingian in distribution, extending to northwestern British Columbia. It often grows with subsp. *arctica*, from which it differs in having larger and, usually, fewer spikelets and broader leaves. It intergrades with *P. macrocalyx* (p. 527) in one direction and *P. arctica* subsp. *arctica* in another.

Poa malacantha Kom. is included here in *P. arctica* subsp. *lanata*. It supposedly differs in having soft hairs on the rachilla internodes and smaller spikelets, but neither feature excludes it from subsp. *lanata*. Bulbiferous forms of *Poa arctica* subsp. *lanata* are known, primarily from uplands in the Alaska Range and Kenai Peninsula. The type of *P. lanata* var. *vivipara* Hultén is a robust plant that appears to belong to *P.*

macrocalyx; consequently the name cannot be applied to bulbiferous plants of subsp. *lanata*.

Poa sect. **Homalopoa** Dumort.

Plants annual or perennial; densely to loosely tufted or with solitary culms, shoots usually neither rhizomatous nor stoloniferous, infrequently rhizomatous. **Basal branching** both intra- and extravaginal or mainly extravaginal. **Culms** 2–120 cm, terete or somewhat compressed; **nodes** terete or weakly compressed. **Sheaths** usually closed for $^1/_2$–$^7/_8$ their length, sometimes only $^1/_{20}$–$^1/_{10}$ their length, terete to distinctly compressed, smooth or scabrous; **ligules** 0.7–12 mm, milky white, smooth or scabrous, truncate to acuminate; **innovation shoot blades** similar to the cauline blades; **cauline blades** 0.6–15 mm wide, flat or folded, thin or moderately thick, lax or moderately straight, abaxial surfaces usually smooth, sometimes scabrous over the midvein, adaxial surfaces smooth or scabrous over the veins, margins scabrous, apices narrowly to broadly prow-shaped. **Panicles** (1)2–40 cm, erect or nodding to lax, contracted or open, sparse or congested, with 1–7 branches per node; **branches** erect to reflexed, terete or angled, angles smooth or scabrous, smooth or sparsely scabrous between angles. **Spikelets** (2)2.4–9 mm, laterally compressed, rarely bulbiferous; **florets** (1)2–7, usually normal, sometimes the anthers aborting, rarely bulb-forming. **Glumes** unequal to subequal, distinctly shorter than the adjacent lemmas, usually bisexual, distinctly keeled; **lower glumes** 1–3-veined; **calluses** terete or slightly laterally compressed, usually dorsally webbed, sometimes glabrous; **lemmas** 2–6 mm, narrowly to broadly lanceolate, distinctly keeled, glabrous or hairy, lateral veins obscure to prominent, margins milky white, apices obtuse to narrowly acute; **palea keels** scabrous, glabrous or hairy at midlength; **anthers** (1, 2) 3, usually 0.1–1.1(1.8) mm, sometimes 1.5–3 mm and then sometimes aborting late in development.

Poa sect. *Homalopoa* is the largest and most heterogeneous section of the genus, having at least 170 species, including many annuals and short-lived perennials. Most species are cespitose, have sheaths closed for $^1/_4$–$^3/_4$ their length and anthers up to 1 mm long. The section is widespread in its distribution, growing almost everywhere the genus is native.

Poa chaixii is the type species of *Poa* sect. *Homalopoa*. It and other Eurasian species of the section have chloroplast genome markers like those of morphologically similar North American species, especially *P. occidentalis* (p. 536). For this reason, the sectional circumscription has been enlarged to include these and other species that are not readily placed elsewhere.

17. **Poa bolander**i Vasey [p. 535]
BOLANDER'S BLUEGRASS

Plants usually annual, rarely longer-lived; often glaucous; densely tufted, tuft bases narrow, sterile shoots few, not stoloniferous, not rhizomatous. **Basal branching** both intra- and extravaginal. **Culms** 20–60(70) cm, erect or geniculate at the base; **nodes** terete, usually 1–3 exserted. **Sheaths** closed for $^1/_2$–$^3/_4$ their length, usually compressed and keeled, usually smooth, infrequently scabrous; **ligules** 2.5–7 mm, smooth or scabrous, usually decurrent, obtuse to acute; **blades** 1.5–5 mm wide, usually flat, rarely folded, lax, soft, smooth or sparsely scabrous, margins scabrous, apices broadly prow-shaped, cauline blades 3–15 cm, flag leaf blades 1–4 cm. **Panicles** (5)10–15(25) cm long, $^1/_4$–$^1/_2$ the plant height, usually erect, infrequently slightly nodding, usually eventually open, sometimes interrupted, sparse, with 1–3(5) branches per node; **branches** initially erect and straight, usually some eventually spreading or reflexed, smooth or sparsely to moderately scabrous. **Spikelets** (3)4–7 mm, laterally compressed; **florets** 2–3(4); **rachilla internodes** usually 1–1.2+ mm, smooth or sparsely scabrous, glabrous. **Glumes** unequal, distinctly shorter than the adjacent lemmas, distinctly keeled, keels smooth or sparsely scabrous; **lower glumes** 1–3-veined, $^2/_3$ the length of the upper glume, $^1/_2$–$^2/_3$ the length of the lowest lemmas; **upper glumes** shorter than or subequal to the lowest lemmas; **calluses** of some or all florets sparsely webbed; **lemmas** 2.5–4 mm, lanceolate to narrowly lanceolate, distinctly keeled, smooth or scabrous throughout, glabrous, lateral veins obscure to moderately prominent, apices narrowly acute, usually anthocyanic near the tip; **palea keels** sparsely scabrous; **anthers** 3, 0.5–1(1.8) mm. $2n = 28$.

Poa bolanderi grows mainly in pine to fir forest openings of mountain slopes in the western United States, from Washington to California and Utah. It differs from *P. howellii* (see below) in having smooth to scabrous, rather than puberulent, lemmas; it also grows at higher elevations, mostly at 1500–3000 m.

18. Poa howellii Vasey & Scribn. [p. 535]
HOWELL'S BLUEGRASS

Plants usually annual, rarely longer-lived; densely tufted, tuft bases narrow; not stoloniferous, not rhizomatous. **Basal branching** intravaginal. **Culms** (10) 25–80(120) cm tall, 0.4–1.75 mm thick, usually erect; **nodes** terete, usually 1–2 exserted. **Sheaths** closed for $^1/_2$–$^7/_8$ their length, usually weakly compressed and keeled, usually scabrous, rarely smooth; **ligules** 1.5–5(10) mm, smooth or scabrous, acute; **blades** 1–7(10) mm wide, flat, lax, soft, finely scabrous, apices narrowly prow-shaped, cauline blades 2–10 cm. **Panicles** 10–25(30) cm, erect, eventually open, with (1)3–5(7) branches per node; **branches** eventually spreading or reflexed, fairly straight, angled, angles usually moderately to densely scabrous, rarely sparsely scabrous. **Spikelets** (2)4–6 mm, laterally compressed, with 2–5 florets; **rachilla internodes** about 1 mm, smooth, usually softly puberulent, infrequently glabrous. **Glumes** slightly unequal, lanceolate, distinctly keeled, keels and sometimes the lateral veins sparsely to moderately scabrous; **lower glumes** 1–3-veined; **upper glumes** shorter than or subequal to the lowest lemmas; **calluses** of some or all florets sparsely webbed; **lemmas** 2.5–3.5 mm, lanceolate to narrowly lanceolate, distinctly keeled, crisply puberulent proximally, hairs evenly distributed, finely scabrous distally, lateral veins obscure to prominent, margins narrowly hyaline, glabrous, apices narrowly acute, infrequently anthocyanic; **palea keels** sparsely scabrous, glabrous or softly puberulent at midlength, intercostal regions usually softly puberulent; **anthers** 3, 0.2–1 mm. $2n =$ unknown.

Poa howellii grows primarily on rocky banks and wooded slopes, from the coastal ranges of southern British Columbia to southern California. It differs from *P. bolanderi* (see above) in having puberulent, rather than smooth or scabrous, lemmas, and in growing at lower elevations, mostly from near sea level to 1000 m.

19. Poa chapmaniana Scribn. [p. 535]
CHAPMAN'S BLUEGRASS

Plants annual; densely tufted, tuft bases narrow, not stoloniferous, not rhizomatous. **Culms** 5–30(40) cm tall, 0.3–0.7 mm thick, erect or the bases geniculate; **nodes** terete, usually 1 exserted. **Sheaths** closed for $^1/_{10}$–$^1/_2$ their length, terete or weakly compressed, smooth; **ligules** 0.7–5 mm, decurrent, truncate to acute; **blades** of innovations and culms similar, 2–5(8) cm long, 0.6–2.8 mm wide, flat or folded, thin, soft, smooth, margins scabrous, apices narrowly prow-shaped. **Panicles** 2–9.6 cm, erect, eventually open, moderately to densely congested, with 1–4(7) branches per node; **branches** eventually ascending to spreading, rarely reflexed, terete, smooth or sparsely scabrous, spikelet-bearing to near the base or middle. **Spikelets** (2)2.4–4.5 mm, laterally compressed; **florets** (1)2–6; **rachilla internodes** usually shorter than 0.7 mm, smooth or scabrous, glabrous. **Glumes** subequal, about $^3/_4$ as long as to subequal to the adjacent lemmas, lanceolate, thin, distinctly keeled, keels scabrous; **lower glumes** 1–3-veined; **calluses** webbed; **lemmas** 1.9–3 mm, broadly lanceolate, distinctly keeled, smooth, keels and marginal veins short- to long-villous, hairs on the keels extending to near the apices, lateral veins obscure, usually softly puberulent, intercostal regions usually sparsely softly puberulent, apices obtuse to acute; **palea keels** softly puberulent to long-villous at midlength, scabrous near the apices; **anthers** 1, 0.1–0.2(0.3) mm. $2n =$ unknown.

Poa chapmaniana is native from the central part of the Great Plains east and southward to the coast. It grows in dry to mesic forests, forest openings, and the margins of bottomlands, often in disturbed ground and on acidic substrates. Records from New York probably represent introductions. Its web and single short anther distinguish *P. chapmaniana* from *P. annua* (p. 520) and most plants of *P. bigelovii*. It also differs from *P. bigelovii* (see below), probably its closest relative, in having narrower leaf blades, and panicle branches that are eventually spreading.

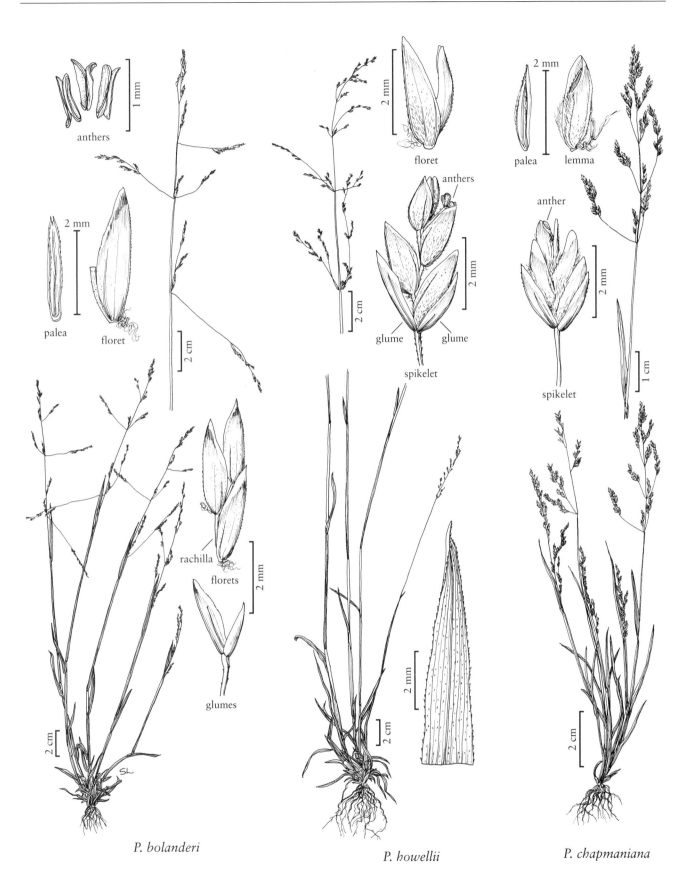

anthers

palea floret

P. bolanderi

rachilla

florets

glumes

floret

anthers

glume glume

spikelet

P. howellii

palea lemma

anther

spikelet

P. chapmaniana

20. **Poa bigelovii** Vasey & Scribn. [p. 537]
BIGELOW'S BLUEGRASS

Plants usually annual, rarely longer-lived; densely tufted, tuft bases narrow, usually without sterile shoots, not stoloniferous, not rhizomatous. **Basal branching** intravaginal. **Culms** (2)5–60 (70) cm tall, 0.3–1 mm thick, usually erect, bases rarely geniculate; **nodes** terete, usually 1 exserted. **Sheaths** closed for $^1/_4$–$^1/_2$ their length, usually compressed and keeled, smooth or the keels scabrous; **ligules** 2–6 mm, smooth or scabrous, usually decurrent, obtuse to acute; **blades** 1.5–5 mm wide, flat, thin, soft, finely scabrous, apices broadly prow-shaped, cauline blades (1)4–15 cm, flag leaf blades usually 1–4 cm. **Panicles** (1)5–15 cm, erect, cylindrical, contracted, sometimes interrupted, congested, with 2–3(5) branches per node; **branches** erect or steeply ascending, smooth or sparsely to densely scabrous. **Spikelets** 4–7 mm, laterally compressed; **florets** 3–7; **rachilla internodes** to 1 mm, smooth, glabrous. **Glumes** subequal, distinctly keeled, keels and sometimes the lateral veins scabrous; **lower glumes** 1(3)-veined; **upper glumes** shorter than or subequal to the lowest lemmas; **calluses** webbed; **lemmas** 2.6–4.2 mm, lanceolate, distinctly keeled, smooth, keels, marginal veins, and sometimes the lateral veins short- to long-villous, keels hairy to near the apices, marginal veins to $^2/_3$ their length, lateral veins obscure to moderately prominent, intercostal regions glabrous or softly puberulent, upper margins white, apices acute; **palea keels** softly puberulent to short-villous at midlength, scabrous near the apices, intercostal regions usually softly puberulent; **anthers** 1–3, 0.2–1 mm. $2n = 28, 28+I$.

Poa bigelovii grows in arid upland regions, particularly on shady, rocky slopes of the southwestern United States and northern Mexico. Plants from southeastern Arizona eastwards are usually glabrous between the lemma veins, whereas more western plants are usually puberulent between the lemma veins. Plants with 1 or 2 small anthers are found in the eastern portion of the species' range; they differ from *P. chapmaniana* (p. 534) in their persistently contracted panicles and broader leaf blades.

21. **Poa occidentalis** Vasey [p. 537]
NEW MEXICAN BLUEGRASS

Plants perennial, short-lived; densely tufted, tuft bases narrow or not, not rhizomatous, not stoloniferous. **Basal branching** mixed intra- and extravaginal. **Culms** 20–110 cm. **Sheaths** closed for ($^1/_5$)$^1/_4$–$^1/_2$ ($^3/_5$) their length, distinctly compressed and keeled, usually densely retrorsely scabrous, rarely sparsely scabrous, margins not ciliate; **ligules** 3–12 mm, densely scabrous, acute to acuminate; **blades** (1.2)1.5–6(10) mm wide, flat, lax, apices broadly prow-shaped. **Panicles** (6)12–40 cm, lax, eventually open, spikelets numerous, with 2–7 branches per node; **branches** (3)5–18(23) cm, eventually spreading or drooping, angled, angles densely scabrous, with (5)8–40(120) spikelets. **Spikelets** (3)4–7(8) mm, laterally compressed, with 3–7 florets; **rachilla internodes** shorter than 1 mm, smooth. **Glumes** distinctly keeled, keels scabrous; **lower glumes** 2–3.5 mm, 1-veined; **upper glumes** 2.5–4.2 mm, shorter than or subequal to the lowest lemmas; **calluses** webbed; **lemmas** 2.6–4.2 mm, narrowly lanceolate, distinctly keeled, scabrous distally, keels and marginal veins short- to long-villous, keel hairs extending to midlength, marginal vein hairs to $^1/_3$ the lemma length, lateral veins and intercostal regions usually sparsely softly puberulent, lateral veins prominent, apices narrowly acute; **palea keels** scabrous, glabrous; **anthers** 0.3–1 mm. $2n = 14, 28$.

Poa occidentalis grows in natural openings and disturbed sites in mixed coniferous forests of the southwestern United States. It is one of the three diploid species of *Poa* known to be native to North America. The tetraploid count was obtained from a single giant individual. *Poa occidentalis* has been confused with *P. tracyi* (p. 543), but *P. occidentalis* consistently has shorter, well-developed anthers and lacks rhizomes. It also usually has longer ligules relative to the blade width, and is shorter-lived. A few plants are intermediate in some characteristics. Small plants of *P. occidentalis* sometimes resemble *P. reflexa* (see next).

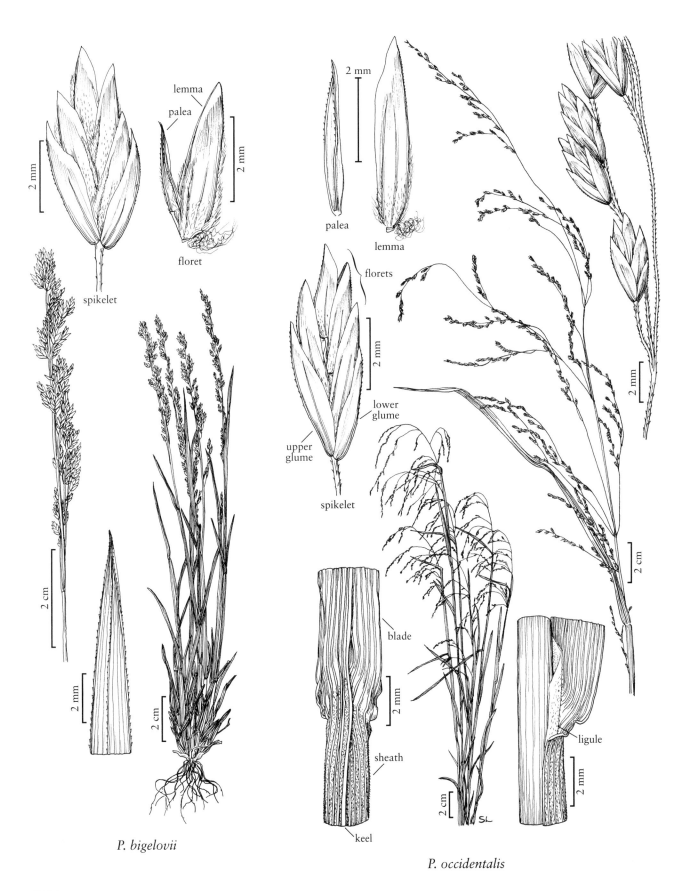

P. bigelovii

P. occidentalis

POA

22. Poa reflexa Vasey & Scribn. [p. 539]
NODDING BLUEGRASS

Plants perennial, short-lived; densely tufted, tuft bases narrow or not, not stoloniferous, not rhizomatous. **Basal branching** mixed intra- and extravaginal. **Culms** 10–60 cm. **Sheaths** closed for $^1/_3$–$^2/_3$ their length, terete, smooth; **ligules** 1.5–3.5 mm, smooth or sparsely scabrous; **blades** 1.5–4 mm wide, flat, thin, soft, apices broadly prow-shaped. **Panicles** 4–15 cm, nodding, open, with numerous spikelets and 1–2 branches per node; **branches** (2)3–7 cm, spreading to reflexed, lower branches usually reflexed, flexuous, usually terete, smooth or sparsely scabrous, with (3)6–18 spikelets. **Spikelets** 4–6 mm, lanceolate to broadly lanceolate, usually partly to wholly purplish, with 3–5 florets; **rachilla internodes** shorter than 1 mm, smooth. **Glumes** narrowly to broadly lanceolate, distinctly keeled, keels smooth or nearly so; **lower glumes** 1-veined; **upper glumes** shorter than or subequal to the lowest lemmas; **calluses** webbed; **lemmas** 2–3.5 mm, lanceolate, partly purple to fairly strongly purple, distinctly keeled, keels and marginal veins short- to long-villous, keels hairy for $^2/_3$–$^4/_5$ their length, lateral veins usually sparsely softly puberulent at least on 1 side, lateral veins obscure to moderately prominent, intercostal regions smooth, minutely bumpy, glabrous, apices acute, slightly bronze-colored or not; **palea keels** scabrous, usually softly puberulent at midlength; **anthers** 0.6–1 mm. 2*n* = 28.

Poa reflexa grows in subalpine forests, meadows, and low alpine habitats, primarily in the central and southern Rocky Mountains. It usually grows on drier and more disturbed sites, and appears shorter-lived, than the frequently sympatric or parapatric *P. leptocoma* (p. 573), from which it differs in usually having hairs on the palea keels and lateral veins of the lemmas, and smooth panicle branches. In addition, *P. reflexa* is tetraploid, whereas *P. leptocoma* is hexaploid. *Poa reflexa* may resemble small plants of *P. occidentalis* (see previous) in habit.

23. Poa paucispicula Scribn. & Merr. [p. 539]
FEW-FLOWER BLUEGRASS

Plants perennial; slightly or loosely tufted, not stoloniferous, not rhizomatous. **Basal branching** mainly extravaginal. **Culms** 10–30 cm. **Sheaths** closed for $^1/_4$–$^3/_5$ their length, terete; **ligules** 1–2 mm, smooth or sparsely scabrous, truncate to obtuse; **blades** 1–3 mm wide, flat, thin, soft, apices broadly prow-shaped. **Panicles** 2.5–10 cm, lax to nearly erect, open, sparse, with 1–2 branches per node; **branches** (2)3–6 cm, ascending to spreading, lower branches infrequently reflexed, lax or drooping, capillary, terete to slightly sulcate, usually smooth, rarely some branches within a panicle sparsely scabrous, with 1–3(5) spikelets. **Spikelets** 4–6 mm, laterally compressed, broadly lanceolate to ovate, usually dark purple, with 3–5 florets; **rachilla internodes** smooth, glabrous. **Glumes** lanceolate to broadly lanceolate, thin, distinctly keeled, keels smooth or nearly so; **lower glumes** 1-veined; **upper glumes** shorter than or subequal to the lowest lemmas; **calluses** sparsely webbed; **lemmas** 3–4 mm, broadly lanceolate, usually strongly purple, distinctly keeled, thin, keels and marginal veins short- to long-villous, keels hairy for $^1/_2$–$^2/_3$ their length, lateral veins glabrous, intercostal regions smooth, glabrous, margins glabrous, not infolded, apices acute, sometimes slightly bronze-colored; **palea keels** sparsely scabrous at midlength; **anthers** 0.4–1 mm. 2*n* = 28, 42.

Poa paucispicula grows in arctic and alpine regions, from the north coast of Alaska and the western Northwest Territories south to Washington, Idaho, and Wyoming; it also grows in arctic far east Russia. It is a delicate species that prefers open, mesic, rocky slopes. It has sometimes been included in *P. leptocoma* (p. 573), a member of *Poa* sect. *Oreinos*. It differs from *P. leptocoma* in having smoother branches, fewer spikelets, and broader glumes. Chloroplast DNA studies confirm that it is not closely related to species of sect. *Oreinos*; ITS data support its relationship to *P. leptocoma*.

24. Poa laxiflora Buckley [p. 539]
LAX-FLOWER BLUEGRASS

Plants perennial; green throughout; loosely tufted or with solitary shoots, long-rhizomatous. **Basal branching** extravaginal. **Culms** 50–120 cm. **Sheaths** closed for $^1/_2$–$^3/_4$ their length, usually sparsely to moderately retrorsely scabrous, margins not ciliate; **ligules** 2–3.5 mm, smooth or sparsely scabrous, obtuse to acute; **blades** 3–8 mm wide, flat, lax, apices narrowly prow-shaped. **Panicles** 14–30 cm, open, sparse, with 1–3(4) branches per node; **branches** (5.5)8–12(15) cm, widely spreading, fairly straight, angled, angles sparsely to moderately scabrous, with 3–13 spikelets. **Spikelets** 4–8 mm, laterally compressed, rarely bulbiferous; **florets** 2–4, usually normal, rarely bulb-forming;

spikelet

2 mm

floret
side views

palea

2 mm

palea   lemma

2 mm

anthers

1 mm

spikelet

2 mm

lemma

palea

2 mm

floret

2 cm

2 mm

upper
glume

lower
glume

spikelet

rhizome

2 cm

2 cm

2 cm

1 cm

2 cm

2 mm

SL

P. reflexa   *P. paucispicula*   *P. laxiflora*

POA

rachilla internodes about 1 mm, smooth, glabrous. **Glumes** distinctly keeled, keels scabrous; **lower glumes** lanceolate, 1–3-veined; **upper glumes** shorter than or subequal to the lowest lemmas; **calluses** webbed; **lemmas** 3.2–6 mm, lanceolate, distinctly keeled, smooth or sparsely finely scabrous, keels and marginal veins long-villous, keels hairy to $^2/_3$–$^3/_4$ their length, marginal veins sparsely hairy, lateral veins moderately prominent, usually glabrous, rarely sparsely softly puberulent, intercostal regions glabrous, apices acute; **paleas** scabrous, glabrous over the keels; **anthers** 0.5–1.1 mm. $2n$ = ca. 98.

Poa laxiflora is restricted to mesic, old growth, mixed conifer forests of the Pacific coast, from Alaska south through the western foothills of the northern Cascades to Oregon. It is not a common species. A bulbiferous specimen was collected in the Queen Charlotte Islands.

Inclusion of *Poa laxiflora* in *Poa* sect. *Homalopoa* is tentative; it may belong to sect. *Sylvestres*.

25. Poa chaixii Vill. [p. 541]
CHAIX'S BLUEGRASS

Plants perennial; densely tufted, not stoloniferous, not rhizomatous. **Basal branching** extravaginal. **Culms** 50–120 cm, stout. **Sheaths** closed for $^1/_2$–$^3/_4$ their length, distinctly compressed, keels winged, bases of basal sheaths glabrous; **ligules** 1–2 mm, smooth or sparsely scabrous, apices truncate; **blades** (4)6–15 mm wide, flat or folded, apices broadly and abruptly prow-shaped. **Panicles** 10–20 cm, ovoid to pyramidal, open, spikelets numerous, with 3–5 branches per node; **branches** ascending to spreading, angled, angles densely scabrous. **Spikelets** 4–9 mm, laterally compressed; **florets** 3–5; **rachilla internodes** about 1 mm, scabrous, glabrous. **Glumes** distinctly keeled, keels scabrous; **lower glumes** 1–3-veined; **calluses** glabrous; **lemmas** 3.5–4.5 mm, narrowly lanceolate, distinctly keeled, scabrous, glabrous throughout, lateral veins prominent, apices acute; **palea keels** scabrous, glabrous; **anthers** 1.5–3 mm. $2n$ = 14.

Poa chaixii was introduced from Europe as an attractive ornamental, and has occasionally escaped. A population in southwestern Quebec has been extirpated.

26. Poa strictiramea Hitchc. [p. 541]
BIG BEND BLUEGRASS

Plants perennial; densely tufted, not stoloniferous, not rhizomatous. **Basal branching** intravaginal. **Culms** 30–90 cm, slender to coarse. **Sheaths** closed for $^1/_{20}$–$^1/_{10}$ their length, terete, scabrous, glabrous; **collars** smooth to scabrous; **ligules** 0.5–4(6) mm, scabrous, apices truncate to acute, entire or lacerate; **innovation blades** 15–30 cm; **cauline blades** 1–4 mm wide, involute or rarely flat, moderately thick and firm, both surfaces sparsely to densely antrorsely scabrous, apices narrowly prow-shaped, flag leaf blades usually longer than their sheaths. **Panicles** (7)10–30 cm, erect, pyramidal, open, with 2–5 branches per node; **branches** 2–8(15) cm, spreading, straight, angled, angles moderately to densely scabrous, sometimes densely scabrous all over, with 10–30 spikelets. **Spikelets** 4–7 mm, lanceolate, laterally compressed; **florets** 2–5; **rachilla internodes** 0.8–1.5 mm, smooth or scabrous, sometimes sparsely hispidulous. **Glumes** sparsely to rarely densely scabrous; **lower glumes** 1–3-veined; **calluses** usually glabrous, rarely sparsely short-webbed; **lemmas** 2.5–3.5 mm, lanceolate, distinctly keeled, smooth or sparsely to densely scabrous, keels and marginal veins glabrous or softly puberulent or short- to long-villous, lateral veins moderately prominent to prominent, intercostal regions usually glabrous, infrequently sparsely softly puberulent, apices acute; **palea keels** scabrous; **anthers** aborted late in development, or 2.2–2.5 mm. $2n$ = 28+I, 28–29+II.

Poa strictiramea grows on shady, upland mountain slopes, usually below north-facing cliffs, in and around the Chihuahuan Desert. In the United States, it is known only from the Chisos Mountains, Texas. It used to be treated as *P. involuta* Hitchc. Plants from the eastern part of its range, including the Chisos Mountains, commonly have short, truncate ligules, whereas westward in Mexico, plants with long, acute ligules are more common.

anthers

palea

floret

lemma

spikelet

rachilla

floret

2 mm

spikelet

2 mm

2 cm

keel

2 mm

2 cm

P. chaixii

ligule

2 cm

1 mm

P. strictiramea

Poa sect. **Madropoa** Soreng

Plants perennial; densely to loosely tufted or with solitary shoots, sometimes stoloniferous, sometimes rhizomatous. **Basal branching** intra- and/or extravaginal. **Culms** (5)10–125 cm, terete or weakly compressed; **nodes** terete or slightly compressed. **Sheaths** closed from ¹/₇ their length to their entire length, terete to compressed, smooth or scabrous, glabrous or pubescent; **ligules** 0.2–18 mm, milky white or colorless, usually translucent, truncate to acuminate, glabrous or ciliolate; **innovation blades** with the adaxial surfaces usually moderately to densely scabrous or hispidulous on and between the veins, sometimes smooth and glabrous; **cauline blades** flat, folded, or involute, thin or thick, lax or straight, smooth or scabrous, adaxial surfaces sometimes hairy, apices narrowly to broadly prow-shaped. **Panicles** 1–29 cm, contracted to open, usually with fewer than 100 spikelets; **nodes** with 1–5 branches; **branches** 0.5–18 cm, terete or angled, smooth or scabrous, glabrous or hispidulous. **Spikelets** 3–17 mm, lengths 3.5 times widths, laterally compressed, not sexually dimorphic, not bulbiferous; **florets** 2–10(13) mm, normal; **rachilla internodes** smooth or scabrous, glabrous or hairy. **Glumes** distinctly keeled, keels smooth or scabrous; **lower glumes** 1, 3(or 5)-veined; **upper glumes** 3- or 5-veined; **calluses** terete or slightly laterally compressed, glabrous, webbed, or with a crown of hairs; **lemmas** 2.6–11 mm, lanceolate, distinctly keeled, keels, veins, and intercostal regions glabrous or hairy, 5–7(11)-veined; **palea keels** scabrous, glabrous or with hairs at midlength; **anthers** 3, vestigial (0.1–0.2 mm) or 1.3–4.5(5) mm.

Poa sect. *Madropoa* is confined to North America. Its 20 species exhibit breeding systems ranging from sequential gynomonoecy to gynodioecy and dioecy. The gynomonoecious species usually grow in forests and have broad, flat leaves. The gynomonoecious and dioecious species grow mainly in more open habitats. They have normally developed anthers that are 1.3–4 mm long, and involute innovation blades that, in several species, are densely scabrous or hairy on the adaxial surfaces.

There are two subsections in the *Flora* region: subsects. *Madropoa* and *Epiles*.

Poa nervosa Complex

Plants perennial; densely to loosely tufted or with solitary shoots, shortly rhizomatous. **Basal branching** all or mostly extravaginal. **Culms** 15–125 cm. **Sheaths** closed from (¹/₃)¹/₂ their length to their entire length, terete to compressed, distal sheaths shorter or longer than their blades; **ligules** 0.5–8 mm, smooth or scabrous, glabrous or softly puberulent, apices glabrous; **innovation blades** with the adaxial surfaces smooth or densely scabrous, glabrous or densely hispidulous; **cauline blades** 1–5.5 mm wide, flat or folded, lax or moderately firm, thin or moderately thick, smooth or sparsely scabrous, apices narrowly to broadly prow-shaped. **Panicles** 2–29 cm, erect or lax, sometimes nodding, contracted to open, with 1–5 branches per node; **branches** 0.9–18 cm, erect to reflexed, terete or angled, smooth or sparsely to moderately scabrous. **Spikelets** 3–12 mm, lengths to 3.5 times widths, not bulbiferous; **florets** 2–8. **Glumes** narrowly lanceolate to lanceolate; **upper glumes** 3-veined; **calluses** terete or slightly laterally compressed, glabrous or dorsally webbed; **lemmas** 2.6–7 mm, lanceolate, apices acute; **anthers** 3, vestigial (0.1–0.2 mm) or (1.3)1.8–4 mm.

The seven species of the *Poa nervosa* complex are typically forest species, with broad, flat leaf blades and short rhizomes. They exhibit breeding systems ranging from sequential gynomonoecy to dioecy. In populations of species with sequential gynomonoecy, plants with only bisexual florets exist in roughly the same number as plants that produce pistillate florets.

In most of those producing pistillate florets, the number of pistillate florets increases as the growing season progresses. The pistillate florets are initially concentrated in the lower spikelets of the panicles and the upper florets within these spikelets; they later develop throughout the panicle.

27. Poa tracyi Vasey [p. 544]
TRACY'S BLUEGRASS

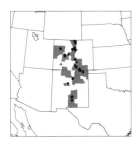

Plants perennial; loosely tufted, shortly rhizomatous. **Basal branching** mainly extravaginal. **Culms** (25)32–125 cm, erect or the bases decumbent, not branching above the base, terete or weakly compressed; **nodes** terete or slightly compressed, 1–2(3) exserted. **Sheaths** closed for $(^2/_5)^1/_2$–$^9/_{10}$ their length, compressed, distinctly keeled, keels winged, wing to 0.5 mm wide, smooth or sparsely to infrequently densely scabrous, glabrous or infrequently retrorsely pubescent, bases of basal sheaths glabrous, distal sheath lengths 0.7–1.6 times blade lengths; **collars** with vestiture similar to the sheaths; **ligules** 2–4.5 mm, smooth or scabrous, glabrous or softly puberulent, obtuse to acute; **innovation blades** similar to the cauline blades; **cauline blades** (1.5)2–5.5 mm wide, flat, lax, smooth or sparsely scabrous mainly over the veins, apices broadly prow-shaped, flag leaf blades 6–20 cm. **Panicles** (8)13–29 cm, erect, usually narrowly pyramidal, open, sparse, with 30–100 spikelets, proximal internodes usually 4+ cm, with (1)2–4(5) branches per node; **branches** 2.5–18 cm, spreading to eventually reflexed, fairly flexuous, terete to weakly angled, sparsely to moderately scabrous, with 3–34 spikelets. **Spikelets** 3–8 mm, lengths 3.5 times widths, laterally compressed, not sexually dimorphic; **florets** 2–8; **rachilla internodes** 1+ mm, smooth, glabrous. **Glumes** narrowly lanceolate, distinctly keeled; **lower glumes** 1.6–3.5 mm, 1(3)-veined, $^1/_2$–$^2/_3$ as long as the adjacent lemmas; **upper glumes** 2.2–4.9 mm; **calluses** webbed, hairs over $^1/_2$ the lemma length; **lemmas** 2.6–5 mm, lanceolate, distinctly keeled, keels and marginal veins long-villous, extending $^1/_2$–$^2/_3$ the keel length, $^1/_3$–$^1/_2$ the marginal vein length, lateral veins sometimes short-villous, the lateral veins obscure to moderately prominent, intercostal regions usually sparsely softly puberulent, margins glabrous, apices acute; **palea keels** scabrous, rarely softly puberulent at midlength; **anthers** vestigial (0.1–0.2 mm) or (1.3)2–3 mm. 2*n* = 28, 28+I.

Poa tracyi grows primarily in coniferous forest openings, sometimes with gambel oak, and in subalpine mesic meadows. It is restricted to the front ranges of the southern Rocky Mountains; it is not common. It differs from *P. occidentalis* (p. 536) in having longer and/or rudimentary anthers, shorter ligules relative to the leaf blade width, and a loose, shortly rhizomatous habit. Retrorsely pubescent sheaths are common in the more southern plants. It is sequentially gynomonoecious.

28. Poa arnowiae Soreng [p. 544]
WASATCH BLUEGRASS

Plants perennial; loosely tufted or with solitary shoots, short-rhizomatous. **Basal branching** all or mostly extravaginal. **Culms** (15)30–80 cm, erect or the bases decumbent, terete or weakly compressed; **nodes** terete, 1–3 exserted. **Sheaths** closed for $^1/_2$–$^9/_{10}$ their length, compressed, smooth, glabrous, bases of basal sheaths glabrous, distal sheath lengths 1–3 times blade lengths; **collars** smooth, glabrous; **ligules** 0.5–4 mm, smooth or sparsely scabrous, truncate to obtuse; **innovation blades** similar to the cauline blades; **cauline blades** 2–5 mm wide, flat, thin, smooth or sparsely scabrous mainly over the veins, apices broadly prow-shaped, middle and upper cauline blades subequal in length, flag leaf blades (2.5)4–7(11) cm long. **Panicles** (5)12–22 cm, usually narrowly pyramidal, open, sparse, with 20–70 spikelets, proximal internodes usually (3.5)4+ cm, with 2–3(4) branches per node; **branches** 3–8 cm, spreading to eventually reflexed, terete or weakly angled, sparsely to moderately scabrous, with 3–12 spikelets. **Spikelets** 5–9 mm, lengths to 3.5 times widths, laterally compressed, not sexually dimorphic; **florets** 2–6; **rachilla internodes** smooth, glabrous, distal internodes 1+ mm. **Glumes** lanceolate, distinctly keeled; **lower glumes** 1–3-veined; **calluses** glabrous; **lemmas** 3–6.5 mm, lanceolate, distinctly keeled, keels and marginal veins glabrous or short-villous to softly puberulent to $^1/_3$ their length, lateral veins obscure, intercostal regions glabrous or sparsely hispidulous, rarely softly puberulent, smooth or sparsely finely scabrous, margins glabrous, apices acute; **palea keels** scabrous, glabrous, intercostal regions glabrous; **anthers** vestigial (0.1–0.2 mm) or (1.3)2–3.6 mm. 2*n* = unknown.

Poa arnowiae grows in openings within the coniferous forests of the mountain ranges in southeastern Idaho, northern Utah, and adjacent Wyoming. It is sequentially gynomonoecious.

Poa arnowiae used to be called *Poa curta* Rydb., but the type of *P. curta* belongs in *P. wheeleri*.

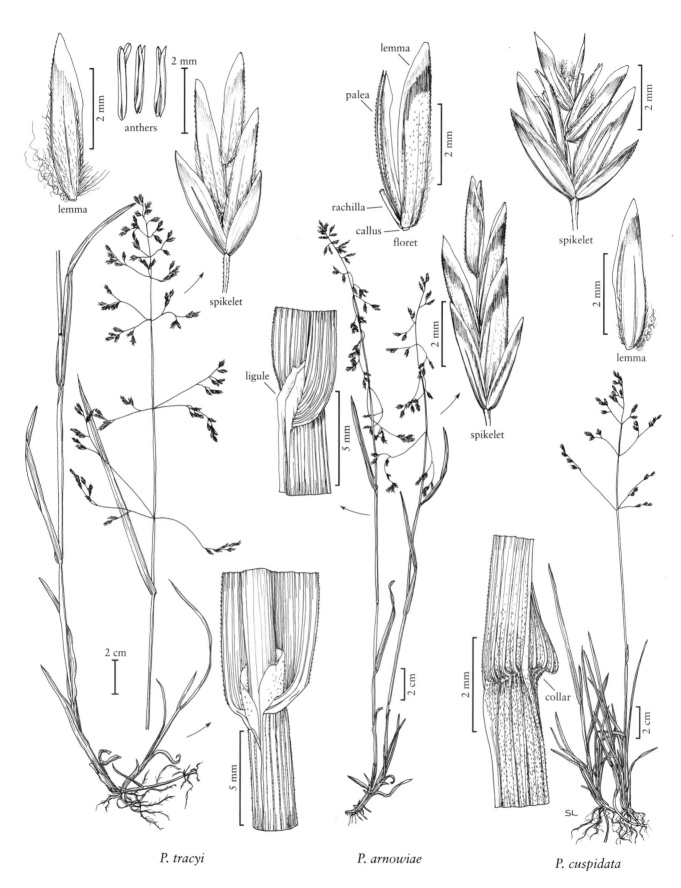

anthers

2 mm

lemma

spikelet

lemma

palea

rachilla

callus

floret

2 mm

spikelet

spikelet

2 mm

lemma

ligule

5 mm

2 cm

5 mm

2 cm

2 mm

collar

2 cm

SL

P. tracyi   P. arnowiae   P. cuspidata

29. Poa cuspidata Nutt. [p. 544]

EARLY BLUEGRASS

Plants perennial; loosely tufted or with solitary shoots, shortly rhizomatous. **Basal branching** mainly extravaginal. **Culms** 15–60 cm, erect or the bases decumbent, not branching above the base, terete or weakly compressed; **nodes** terete, 0–1 exserted. **Sheaths** closed for about ¹/₂ their length, slightly compressed, distinctly keeled, glabrous, bases of basal sheaths glabrous, distal sheath lengths 4–60 times blade lengths; **collars** of proximal leaves usually retrorsely scabrous or pubescent distally and about the throat; **ligules** 0.5–4 mm, smooth or scabrous, apices truncate to acute; **innovation blades** similar to the cauline blades; **cauline blades** 1–4 mm wide, usually flat, sometimes slightly folded, smooth or sparsely scabrous, primarily over the veins, apices broadly prow-shaped, blades steeply reduced in length distally, flag leaf blades 0.2–3(6) cm. **Panicles** 5–15 cm, erect or lax, pyramidal, open, sparse, with 20–80 spikelets, proximal internodes usually 3+ cm; **nodes** usually with 2 branches; **branches** (2)3–7(10) cm, spreading to reflexed, straight, angled, angles scabrous, with 2–8(10) spikelets. **Spikelets** 5–8 mm, lengths to 3.5 times widths, laterally compressed, not sexually dimorphic; **florets** 2–5; **rachilla internodes** smooth. **Glumes** narrowly lanceolate to lanceolate, distinctly keeled; **lower glumes** 1–3-veined; **calluses** webbed, hairs over ¹/₃ the lemma length; **lemmas** 3–6 mm, lanceolate, distinctly keeled, keels and marginal veins sparsely short- to long-villous, lateral veins moderately prominent, intercostal regions glabrous or the upper florets in the spikelets softly puberulent, margins glabrous, apices acute; **palea keels** scabrous, softly puberulent at midlength; **anthers** vestigial (0.1–0.2 mm) or 2–3.5 mm. 2*n* = 28.

Poa cuspidata is a common species of forest openings in the Appalachian Mountains. It is an eastern counterpart of *P. arnowiae* (see previous), *P. tracyi* (p. 543), and *P. nervosa* (see next). Like those species, it is sequentially gynomonoecious.

30. Poa nervosa (Hook.) Vasey [p. 547]

VEINY BLUEGRASS

Plants perennial; loosely tufted or with solitary shoots, short-rhizomatous. **Basal branching** mainly extravaginal. **Culms** 20–65 cm, erect or the bases decumbent, terete or weakly compressed; **nodes** terete, 1–2 exserted. **Sheaths** closed for

²/₃–⁹/₁₀ their length, terete to slightly compressed, smooth or sparsely scabrous, sometimes hairy, hairs about 0.15 mm, bases of basal sheaths glabrous, distal sheath lengths (0.7)1–2.2(2.8) times blade lengths; **collars** of proximal leaves usually hairy on and near the margins, marginal hairs longer than those of the sheaths; **ligules** 0.5–1.5 mm, smooth or scabrous, hairy, hairs about 0.15 mm, truncate to obtuse, those of the lower culm and innovation leaves 0.5–1 mm, scabrous or softly puberulent, truncate; **innovation blades** similar to or longer than the cauline blades; **cauline blades** 2–4.5 mm wide, usually flat, lax, adaxial surfaces smooth or sparsely scabrous, particularly over the veins, apices broadly prow-shaped, blades gradually reduced in length distally or the middle blades longest, flag leaf blades 3–8 cm long. **Panicles** 8–15 cm, erect or lax, ovoid to pyramidal, open or loosely contracted, sparse, with 25–80 spikelets, proximal internodes 1.8–3.5 cm; **nodes** with 3–5 branches; **branches** 2.5–8 cm, ascending to spreading, lax, terete to weakly angled, moderately scabrous, with 2–8 spikelets. **Spikelets** 4–7 mm, lengths to 3.5 times widths, laterally compressed, not sexually dimorphic; **florets** 3–8; **rachilla internodes** smooth or scabrous, glabrous or sparsely hispidulous. **Glumes** ²/₃–⁴/₅ as long as the adjacent lemmas, lanceolate, distinctly keeled; **lower glumes** 1–3(5)-veined; **calluses** usually glabrous, rarely minutely webbed; **lemmas** 3–4.5 mm, lanceolate, distinctly keeled, keels and marginal veins usually glabrous, infrequently sparsely softly puberulent to short-villous, intercostal regions glabrous or hispidulous, smooth or finely scabrous, margins glabrous, apices acute; **paleas** scabrous over the keels, intercostal regions glabrous; **anthers** usually 2.5–4 mm, sometimes vestigial (0.1–0.2 mm). 2*n* = 28, 28+I.

Poa nervosa occurs infrequently at low elevations in the western foothills of the northern Cascade Mountains and adjacent coast ranges, extending eastward up the Columbia Gorge as far as Multnomah Falls. It usually grows in wet habitats, such as mossy cliffs with seeps and around waterfalls, but it is also found in rich, old growth, mixed deciduous and conifer forests. It appears to be sexually reproducing and sequentially gynomonoecious.

Poa nervosa differs from *P. wheeleri* (see next) in having densely pubescent leaf collar margins, and glabrous or more sparsely and shortly pubescent sheaths. It also differs in usually having well-developed anthers, and in being tetraploid. The two species are geographically isolated and ecologically distinct. Plants from the Columbia River Gorge in Oregon, including **P. ×multnomae** Piper, that approach *P. tenerrima* (p. 588) are presumed to be derived from hybridization between *P. nervosa* (see previous) and *P. secunda* (p. 586).

31. Poa wheeleri Vasey [p. 547]
WHEELER'S BLUEGRASS

Plants perennial; densely to loosely tufted or with solitary shoots, shortly rhizomatous. **Basal branching** mainly extravaginal. **Culms** 35–80 cm, erect or the bases decumbent, terete or weakly compressed; **nodes** terete, 1–2 exserted. **Sheaths** closed for ¹/₃–³/₄ their length, terete to slightly compressed, at least some proximal sheaths densely retrorsely scabrous, hispidulous, or softly puberulent for the upper ¹/₄ of their length, bases of basal sheaths glabrous, distal sheath lengths (1.4)1.7–4.6(6.2) times blade lengths; **collars** of proximal leaves glabrous or with hairs the same length as those of their sheaths; **ligules** 0.5–2 mm, smooth or scabrous, sometimes puberulent, truncate, those of the lower culm and innovation leaves 0.5–1.5 mm, abaxial surfaces scabrous to softly puberulent, truncate; **innovation blades** folded or involute, infrequently flat, moderately thick, soft, adaxial surfaces usually densely scabrous to hispidulous; **cauline blades** 2–3.5 mm wide, flat or folded, smooth or sparsely scabrous, glabrous or hispidulous, apices narrowly to broadly prow-shaped, blades gradually reduced distally or the middle blades longest, flag leaf blades 1–10 cm long. **Panicles** 5–12(18) cm, erect or nodding, ovoid to pyramidal, loosely contracted to open, with 20–70 spikelets, proximal internodes usually shorter than 3.5 cm; **nodes** with 2–5 branches; **branches** (1)1.7–6.5 cm, ascending to spreading or reflexed, lax, terete or weakly angled, sparsely to moderately scabrous, with 2–8(12) spikelets. **Spikelets** 5.5–10 mm, lengths to 3.5 times widths, laterally compressed, not sexually dimorphic; **florets** 2–7; **rachilla internodes** smooth or scabrous, glabrous or sparsely to densely hispidulous. **Glumes** ¹/₄–²/₃(³/₄) as long as the adjacent lemmas, lanceolate, distinctly keeled; **lower glumes** 1–3-veined, ¹/₄–¹/₂ as long as the adjacent lemmas; **calluses** glabrous; **lemmas** 3–6 mm, lanceolate, distinctly keeled, keels and marginal veins glabrous or softly puberulent to short-villous, intercostal regions glabrous or hispidulous, infrequently puberulent, smooth or finely scabrous, lateral veins obscure to moderately prominent, margins glabrous, apices acute; **palea keels** scabrous, intercostal regions glabrous; **anthers** usually vestigial (0.1–0.2 mm) or aborted late in development and up to 2 mm, rarely normal. 2*n* = 56, 61, 62, 63, 64, 66, 67, 70, ca. 74, 75, 79, 80, 81, 87, 89, 90, 91.

Poa wheeleri is common at mid- to high elevations, generally on the east side of the coastal mountains from British Columbia to California, and from Manitoba to New Mexico. It generally grows in submesic coniferous forests to subalpine habitats. Most plants have densely retrorsely pubescent or scabrous sheaths, involute innovation blades that are pubescent adaxially, and pistillate florets.

Poa wheeleri, a high polyploid apomictic species, probably arose from hybridization between *P. cusickii* (p. 559) and another member of the *Poa nervosa* complex. It resembles *P. rhizomata* (see next) and *P. chambersii* (p. 548) more than *P. nervosa sensu stricto* (see previous). It differs from *P. chambersii* in having at least some proximal sheaths that are densely retrorsely scabrous or pubescent (sometimes obscurely so), and folded or involute innovation blades that are scabrous to hispidulous on the adaxial surfaces. For a comparison with *P. nervosa*, see p. 545. Natural hybrids have been found between *P. wheeleri* and *P. pratensis* (p. 522).

32. Poa rhizomata Hitchc. [p. 549]
RHIZOME BLUEGRASS

Plants perennial; usually unisexual; loosely tufted or with solitary shoots, shortly rhizomatous. **Basal branching** all or mainly extravaginal. **Culms** 20–65 cm, erect or the bases decumbent, not branching above the base, terete or weakly compressed; **nodes** terete, 1–2 exserted. **Sheaths** closed for ¹/₂–²/₃ their length, slightly compressed, keels moderately distinct, smooth or sparsely to moderately scabrous, glabrous, bases of basal sheaths glabrous, distal sheath lengths 1.5–4.4(5.7) times blade lengths; **collars** smooth, glabrous; **ligules** of cauline leaves 2–8 mm, smooth or scabrous, acute to acuminate, innovation ligules 2–5 mm; **innovation blades** to 20 cm, otherwise similar to the cauline blades; **cauline blades** gradually reduced in length distally, 1–3.5 mm wide, usually flat or folded, soft, thin, somewhat lax, smooth or sparsely scabrous, primarily over the veins and margins, distinctly keeled, apices narrowly to broadly prow-shaped, flag leaf blades (1.4)3–6(8) cm. **Panicles** (2)4–10 cm, nodding, ovoid, sparse, with 20–50 spikelets, proximal internodes usually 1.8–3 cm; **nodes** with 1–2(4) branches; **branches** 1.5–4.5 cm, ascending to spreading, lax, terete to weakly angled, angles sparsely to moderately scabrous, with 2–7 spikelets. **Spikelets** (4)6–9(12) mm, lengths to 3.5 times widths, laterally compressed, not sexually dimorphic; **florets** 3–8, usually unisexual; **rachilla internodes** smooth or sparsely scabrous, usually glabrous, infrequently sparsely puberulent. **Glumes** ³/₅–⁴/₅ as long as the adjacent lemmas, narrowly lanceolate to lanceolate,

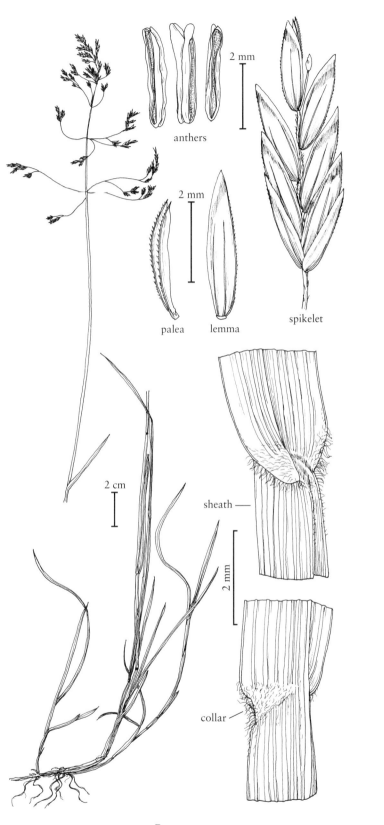

anthers

palea lemma spikelet

sheath

collar

2 cm

2 mm

P. nervosa

spikelet floret

2 mm

collar

sheath

2 mm

2 cm

SL

P. wheeleri

POA

distinctly keeled, keels scabrous; **lower glumes** 1–3(5)-veined; **calluses** webbed, hairs over ¹/₂ the lemma length; **lemmas** 4–6.5 mm, lanceolate, 5–7-veined, distinctly keeled, keels and marginal veins sparsely short- to long-villous, lateral veins moderately prominent, intercostal regions sparsely scabrous, glabrous, margins glabrous, apices acute; **palea keels** scabrous; **anthers** vestigial (0.1–0.2 mm) or 2.5–4 mm. 2*n* = 28.

Poa rhizomata is a rare species that grows in upper elevation, mixed coniferous forests on ultramafic (gabro or peridotite) rocks of the Klamath–Siskiyou region. It is subdioecious.

Poa rhizomata resembles *P. pratensis* (p. 522), differing in having acute ligules, sparse inflorescences, florets that are usually unisexual florets, and generally larger spikelets. It also resembles *P. chambersii* (see next), but has more open sheaths, longer ligules, more pubescent lemmas, and a more well-developed web. It used to include *P. piperi* (p. 554), which differs in having involute, adaxially hairy leaves and glabrous lemmas.

33. Poa chambersii Soreng [p. 549]

CHAMBERS' BLUEGRASS

Plants perennial; loosely tufted or with solitary shoots, short-rhizomatous. **Basal branching** all or mainly extravaginal. **Culms** 10–50 cm, erect or the bases decumbent, terete or weakly compressed; **nodes** terete, 0–1 exserted. **Sheaths** closed for ¹/₃–⁷/₈ their length, terete to slightly compressed, smooth, glabrous, bases of basal sheaths glabrous, distal sheath lengths (1.15)1.5–4.6(6.6) times blade lengths; **collars** smooth, glabrous; **ligules** 0.5–2(2.5) mm, smooth, truncate to obtuse; **innovation blades** similar to the cauline blades; **cauline blades** gradually reduced in length distally, 2–5 mm wide, flat or folded, smooth or the adaxial surfaces sparsely scabrous, primarily over the veins, apices broadly prow-shaped, flag leaf blades 0.7–6 cm. **Panicles** 2–9 cm, erect, lanceoloid to ovoid, tightly to loosely contracted, with 15–35 spikelets, proximal internodes shorter than 2 cm; **nodes** with 1–2 branches; **branches** 0.9–3.2 cm, erect to ascending or slightly spreading, terete, smooth or sparsely scabrous, with 1–4 spikelets. **Spikelets** 6–12 mm, lengths to 3 times widths, laterally compressed, not sexually dimorphic; **florets** 2–7; **rachilla internodes** 0.8–1.5 mm, smooth or sparsely scabrous, glabrous. **Glumes** ³/₅–⁴/₅ as long as the adjacent lemmas, distinctly keeled; **lower glumes** 3-veined; **calluses** of at least some proximal florets sparsely webbed, with 1–2 mm hairs, others glabrous,

rarely all glabrous; **lemmas** 5–7 mm, lanceolate, 5–7-veined, distinctly keeled, smooth or sparsely finely scabrous, glabrous throughout or the keels and marginal veins sparsely softly puberulent over the proximal ¹/₄, lateral veins moderately prominent, intercostal regions glabrous, margins glabrous, apices acute; **palea keels** sparsely scabrous, intercostal regions glabrous; **anthers** vestigial (0.1–0.2 mm), aborted late in development, or 1.8–3.7 mm. 2*n* = unknown.

Poa chambersii is known only from upland forest openings in the Cascades of western Oregon, where it is dioecious, and from high elevations on Steens Mountain in southeastern Oregon, where it is gynodioecious. It resembles *P. rhizomata* (see previous), but has more closed sheaths, shorter ligules, less pubescent or glabrous lemmas, and lacks a well-developed web. It approaches *P. cusickii* subsp. *purpurascens* (p. 562), but is rhizomatous and sexually reproducing. It differs from *P. wheeleri* (p. 546) in having glabrous sheaths and flat or folded, glabrous innovation blades.

34. Poa sierrae J.T. Howell [p. 550]

SIERRA BLUEGRASS

Plants perennial; loosely tufted or with solitary shoots, short-rhizomatous. **Basal branching** extravaginal, initiated as pinkish to purplish fleshy buds that persist as sets of short scales at the nodes of rhizomes and the proximal culm nodes, drying brownish and flabelliform after the shoots develop. **Culms** 20–60 cm, slender, erect or the bases decumbent, terete or weakly compressed; **nodes** terete, 1–2 exserted. **Sheaths** closed from ⁹/₁₀ their length to their entire length, terete, smooth or sparsely scabrous, glabrous, bases of basal sheaths glabrous, distal sheath lengths 0.18–0.8 times blade lengths; **collars** smooth, glabrous; **ligules** 3–6 mm, scabrous, acute to acuminate; **innovation blades** similar to the cauline blades; **cauline blades** gradually reduced in length distally, 1.5–2.5 mm wide, flat, thin, soft, smooth or sparsely scabrous, primarily over the veins, apices narrowly to broadly prow-shaped, flag leaf blades 8–12 cm. **Panicles** 4–15 cm, erect, ovoid, sparse, with fewer than 15(20) spikelets; **nodes** with 1–2 branches; **branches** 1–4.5 cm, spreading to reflexed, slender, terete, sparsely to moderately scabrous, with 1–3 spikelets. **Spikelets** 5–9 mm, lengths to 3.5 times widths, laterally compressed, not sexually dimorphic; **florets** 2–6; **rachilla internodes** smooth, sparsely hairy, hairs to 0.3 mm. **Glumes** ¹/₃–³/₄(⁴/₅) as long as the adjacent lemmas, keels sparsely scabrous; **lower glumes** 3-veined; **calluses** glabrous or webbed, hairs at least 1–2 mm;

floret

spikelet

ligule

spikelet

rhizome

P. rhizomata

spikelet

lemma

palea

collar

P. chambersii

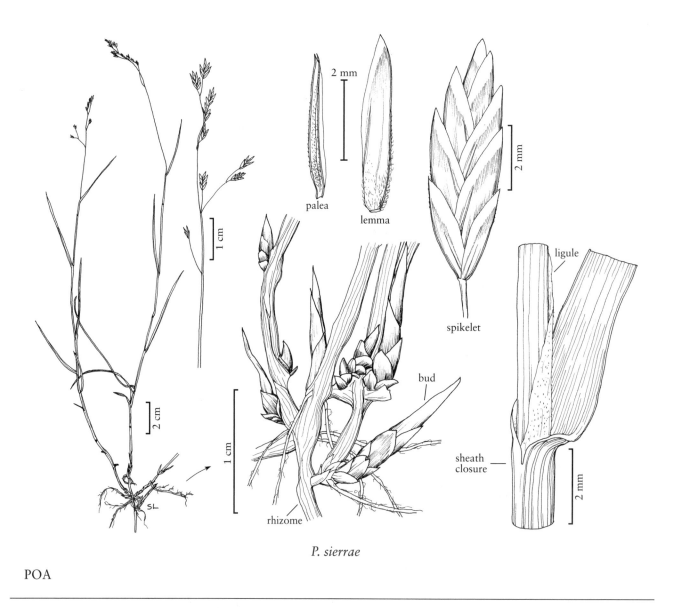

palea

lemma

spikelet

bud

ligule

sheath
closure

rhizome

SL

P. sierrae

POA

lemmas 4–7 mm, lanceolate, distinctly keeled, keels and marginal veins glabrous or short-villous, lateral veins obscure, glabrous, finely scabrous, intercostal regions glabrous or sparsely softly puberulent, margins glabrous, apices acute; **palea keels** scabrous, sometimes softly puberulent at midlength; **anthers** vestigial (0.1–0.2 mm) or 2–4 mm. $2n$ = ca. 58.

Poa sierrae, a distinctive dioecious species, is a narrow endemic of mid-elevation canyon slopes on the west side of the Sierra Nevada, California. It can be distinguished from all other *Poa* species by the scaly, pink to purplish buds on the rhizomes, and by the entirely or almost entirely closed upper culm sheaths that are shorter than their blades.

Poa subsect. **Madropoa** Soreng

Plants perennial; densely to loosely tufted, short- to long-rhizomatous and/or stoloniferous. **Basal branching** intra- and extravaginal. **Culms** (5)7–70 cm. **Sheaths** closed for $^1/_3$–$^7/_{10}$ their length, terete or weakly keeled, distal sheaths usually longer than their blades; **ligules** 0.2–18 mm, usually scabrous, sometimes ciliolate; **innovation blades** adaxially smooth or scabrous, glabrous or hispidulous; **cauline blades** 0.5–4 mm wide, usually involute, sometimes flat or folded, thin to thick, soft to firm, apices narrowly prow-shaped. **Panicles** 1–15(30) cm, erect, usually contracted to loosely contracted and lanceolate to ovoid, infrequently open and broadly pyramidal, congested to sparse, **nodes** with 1–2 branches; **branches** 0.5–8 cm, terete or angled, smooth or scabrous, glabrous or densely hispidulous, with 1–17(25) spikelets. **Spikelets** 3–17 mm, lengths to 3 times widths, not bulbiferous; **florets** 2–7(13). **Glumes** lanceolate to broadly lanceolate; **calluses** glabrous, webbed, or with a crown of hairs; **lemmas** 2.5–11 mm, lanceolate, sometimes narrowly lanceolate, glabrous or hairy, 5–7(11)-veined, lateral veins moderately prominent to prominent; **anthers** 3, vestigial (0.1–0.2 mm) or 1.5–4(5) mm.

The seven species of *Poa* subsect. *Madropoa* are strongly dioecious, usually rhizomatous, and usually have involute blades.

35. **Poa douglasii** Nees [p. 553]
DOUGLAS' BLUEGRASS

Plants perennial; loosely tufted, rhizomatous and stoloniferous, rhizomes and stolons to 1 m. **Basal branching** mainly intravaginal, some extravaginal. **Culms** (5)10–30 cm tall, 1.2–1.5 mm thick, bases decumbent, terete or weakly compressed, hispidulous beneath the panicles; **nodes** terete, 0(1) exserted. **Sheaths** closed for about ¹/₂ their length, terete, smooth or sparsely to moderately retrorsely scabrous near the collars, glabrous, bases of basal sheaths glabrous, distal sheath lengths 0.9–3.5 times blade lengths; **collars** sparsely to moderately retrorsely scabrous, glabrous; **ligules** 1–2 mm, scabrous, truncate to obtuse, ciliolate; **innovation blades** to 30 cm long, adaxial surfaces moderately to densely scabrous or hispidulous on and between the veins; **cauline blades** subequal in length, 1–2 mm wide, involute, moderately thick, moderately firm, arcuate, abaxial surfaces smooth or sparsely scabrous, adaxial surfaces moderately to densely scabrous or hispidulous on and between the veins, apices narrowly prow-shaped, flag leaf blades 1–9 cm. **Panicles** 1.5–6 cm, erect, compact, ovoid, contracted, infrequently interrupted, congested, with 15–50 spikelets; **nodes** with 1–2 branches, internodes densely hispidulous; **branches** 0.5–2 cm, erect, stiff, terete to weakly angled, densely hispidulous, with 1–5 spikelets. **Spikelets** 7–12 mm, lengths to 3 times widths, laterally compressed, not sexually dimorphic; **florets** 3–6; **rachilla internodes** usually shorter than 0.5 mm, smooth, glabrous. **Glumes** broadly lanceolate, ¹/₂ as long as to subequal to the adjacent lemmas, distinctly keeled; **lower glumes** 3-veined; **upper glumes** 4–4.5(7+) mm, 3-veined; **calluses** usually with a crown of hairs, sometimes glabrous or diffusely webbed; **lemmas** 5–7.5 mm, lanceolate, 5-veined, distinctly keeled, keels, marginal veins, and sometimes the lateral veins short- to long-villous or

softly puberulent, rarely glabrous, lateral veins moderately prominent, intercostal regions smooth, glabrous, margins glabrous, apices acute; **palea keels** scabrous to pectinate-ciliate, intercostal regions glabrous; **anthers** vestigial (0.1–0.2 mm) or (2)2.5–3.5 (4) mm. 2*n* = 28.

Poa douglasii is a dioecious endemic that grows on coastal sand dunes in California, a habitat that is being invaded by exotic species. It is rare north of Mendocino. Its hairy rachises distinguish *P. douglasii* from all other species of *Poa* in the *Flora* region. It differs from *P. macrantha* (see next), which occupies similar habitats, in this and in its usually longer glumes and lemmas.

36. **Poa macrantha** Vasey [p. 553]
DUNE BLUEGRASS

Plants perennial; loosely tufted, rhizomatous and stoloniferous, rhizomes and stolons to 4 m, stout, robust. **Basal branching** mostly intravaginal, some extravaginal. **Culms** (7)15–60 cm tall, 1.5–2 mm thick, bases decumbent, terete or weakly compressed, smooth or moderately scabrous below the panicles; **nodes** terete, 0(1) exserted. **Sheaths** closed for about ¹/₂ their length, terete, glabrous or sparsely retrorsely scabrous, bases of basal sheaths glabrous, distal sheath lengths 1.7–4(6) times blade lengths; **collars** smooth, glabrous; **ligules** 1–5 mm, scabrous, truncate to acute, ciliolate; **innovation blades** to 30 cm, moderately to densely scabrous or hispidulous on and between the veins; **cauline blades** subequal in length, 2–4 mm wide, involute, thick, somewhat arcuate, firm, abaxial surfaces smooth or moderately to densely scabrous or hispidulous on and between the veins, apices narrowly prow-shaped, flag leaf blades 1–10 cm. **Panicles** 3–15 cm, erect, ovoid to lanceolate, contracted, often interrupted, congested, with 15–80 spikelets, rachises glabrous, smooth to moderately scabrous; **nodes** with 1–2 branches; **branches** 1–6 cm, erect, stiff, terete to weakly angled, smooth or sparsely to moderately scabrous, with 3–17 spikelets. **Spikelets** 9–17 mm, lengths to 3 times widths, laterally compressed, not sexually dimorphic; **florets** 3–6(10); **rachilla internodes** smooth, usually hairy, hairs 0.3–0.4+ mm, rarely glabrous. **Glumes** broadly lanceolate, subequal to the adjacent lemmas, distinctly keeled, keels sparsely scabrous near the apices; **lower glumes** 3-veined; **upper glumes** usually 7+ mm, 3–5-veined; **calluses** usually with a crown of hairs, sometimes glabrous or diffusely webbed; **lemmas** (6)7.5–11 mm, lanceolate, 5–7(11)-veined, distinctly keeled, keels and marginal veins, and sometimes the lateral veins, short-villous to softly puberulent, intercostal regions smooth or scabrous, glabrous or softly puberulent, margins glabrous, apices acute; **palea keels** scabrous, intercostal regions glabrous; **anthers** vestigial (0.1–0.2 mm) or (2)3–4(5) mm. $2n = 28$.

Poa macrantha is a dioecious coastal sand dune species that grows from southern Alaska to northern California. It competes better than *P. douglasii* (see previous) with the invasion of its habitat by *Ammophila* and other exotic species. It used to be treated as a subspecies of *P. douglasii*; a few intermediates with that species have been found around the mouth of Little River, California. Although clearly related, the two species are reasonably divergent in a number of characters. *Poa macrantha* is readily distinguished from *P. douglasii* by its glabrous rachises and usually longer glumes and lemmas.

37. Poa confinis Vasey [p. 555]
COASTAL BLUEGRASS

Plants perennial; densely to loosely tufted, rhizomatous and stoloniferous, rhizomes and stolons to 1 m, slender. **Basal branching** mainly intravaginal, some extravaginal. **Culms** 7–30 (35) cm tall, 0.4–0.9 mm thick, slender, erect or the bases decumbent, terete or weakly compressed; **nodes** terete, 0–1 exserted. **Sheaths** closed for $^1/_3$–$^2/_3$ their length, terete, smooth, glabrous, bases of basal sheaths glabrous, distal sheath lengths (1)1.4–4.5 times blade lengths; **collars** smooth, glabrous; **ligules** 0.5–1.5(2.2) mm, scabrous, truncate to acute; **innovation blades** adaxially moderately to densely scabrous or hispidulous on and between the veins; **cauline blades** slightly reduced in length distally, 0.5–1(1.5) mm wide, involute, thin to moderately thick, usually filiform, soft, abaxial surfaces smooth, adaxial surfaces sparsely scabrous on and between the veins, apices narrowly prow-shaped, flag leaf blades (0.5)1–5 cm. **Panicles** 1–5(7) cm, erect, ovoid, fairly tightly to loosely contracted, congested or moderately congested, with fewer than 50 spikelets; **nodes** with 1–2 branches; **branches** 0.5–3 cm, erect to ascending, slightly lax, terete or angled, angles sparsely to densely scabrous, with 2–12 spikelets. **Spikelets** 3–6(8) mm, lengths to 3 times widths, laterally compressed, compact, not sexually dimorphic; **florets** 2–5; **rachilla internodes** 0.8–1.1 mm, usually not readily visible from the sides, glabrous or sparsely puberulent. **Glumes** slightly unequal, distinctly keeled, keels smooth or scabrous; **lower glumes** 2–4 mm, 1–3-veined, about $^2/_3$ the length of the adjacent lemmas; **upper glumes** 2.9–5 mm; **calluses** usually diffusely webbed, hairs 1–2 mm, infrequently glabrous; **lemmas** 2.5–4(4.5) mm, lanceolate, distinctly keeled, moderately to densely finely scabrous, glabrous throughout or the keels and sometimes the marginal veins sparsely puberulent proximally, margins narrowly scarious, glabrous, apices acute; **paleas** subequal to the lemmas, keels scabrous, intercostal regions glabrous; **anthers** vestigial (0.1–0.2 mm) or 1.5–2 mm. $2n = 42$.

Poa confinis grows on sandy beaches and forest margins of the west coast, a habitat that is being lost to invasion by exotic species and development. It is closely related to *P. diaboli* (see next), from which it differs by a suite of characters. The two species are ecologically and geographically distinct. *Poa confinis* differs from *P. pratensis* (p. 522) in having glabrous or sparsely hairy lemmas and diffusely webbed calluses. It is gynodioecious.

38. Poa diaboli Soreng & D.J. Keil [p. 555]
DIABLO BLUEGRASS

Plants perennial; loosely tufted, forming airy mounds to 30 cm across, shortly rhizomatous and stoloniferous. **Basal branching** extra-, pseudo-, and intravaginal. **Culms** 26–50 cm tall, 0.5–0.9 mm thick, bases decumbent or nearly erect, frequently branching above the base, terete or weakly compressed; **nodes** terete, 1–2 exserted. **Sheaths** closed for $^2/_5$–$^7/_{10}$ their length, weakly keeled, sparsely scabrous, glabrous, bases of basal sheaths glabrous, distal sheath lengths 0.6–2.4 times blade lengths; **collars** scabrous or pubescent on the margins; **ligules** (1)2–3 mm, moderately densely scabrous, truncate, obtuse, or acute, lacerate to entire; **innovation blades** to 20 cm, adaxial surfaces sparsely scabrous, glabrous or hispidulous on and between the veins; **cauline blades** 1.5–2.4 mm wide, folded or flat, thin, soft, abaxial surfaces smooth, veins prominent, keel and margins scabrous, adaxial surfaces moderately scabrous over the veins, sparsely scabrous between the veins,

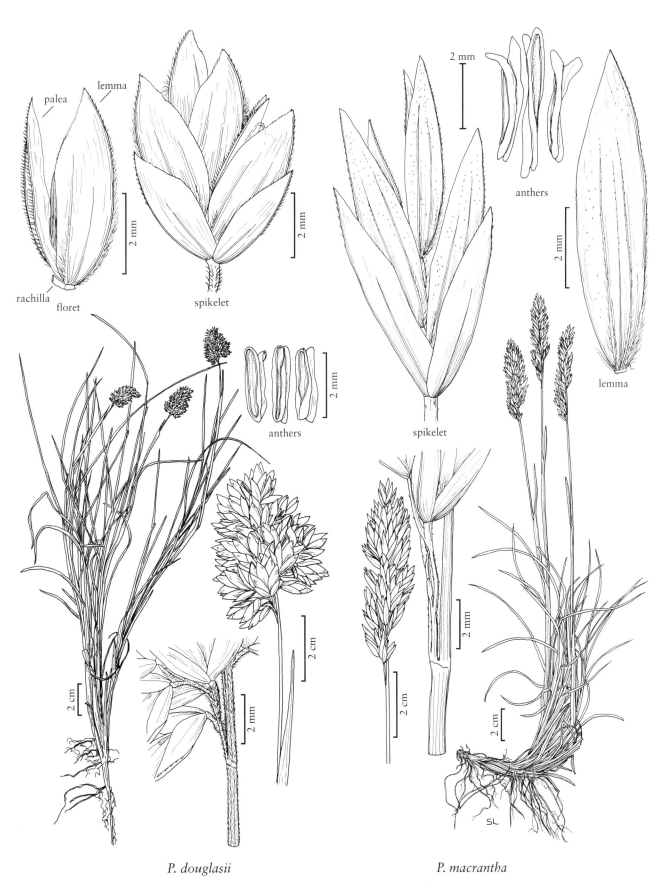

palea

lemma

rachilla

floret

spikelet

2 mm

2 mm

anthers

anthers

spikelet

2 mm

lemma

P. douglasii

P. macrantha

SL

POA

apices narrowly prow-shaped, flag leaf blades 2.9–8.6(11) cm. **Panicles** (4)5.5–10.5(12.5) cm, erect, ovoid to broadly pyramidal, open, or eventually loosely contracted, sparse, with 10–40 spikelets; **nodes** with 1–2 branches; **branches** 2.1–4.5(7) cm, ascending, lax, angled, angles moderately to densely scabrous, less scabrous between the angles, with 1–9 spikelets. **Spikelets** 5.3–9 mm, lengths to 3 times widths, laterally compressed, not sexually dimorphic; **florets** (2)3–6(7); **rachilla** internodes 1–1.3 mm, visible from the sides, usually sparsely to densely, coarsely scabrous, infrequently smooth. **Glumes** distinctly keeled; **lower glumes** (2)2.7–3.8 mm, 3-veined, **upper glumes** (2.3)2.9–3.9 mm; **calluses** diffusely webbed, hairs $^1/_3$–$^1/_2$ the lemma length; **lemmas** (3.2)4.25–5 mm, lanceolate to narrowly lanceolate, distinctly keeled, glabrous, moderately to densely, infrequently sparsely, scabrous, lateral veins prominent, margins narrowly scarious, glabrous, apices acute to narrowly acute; **paleas** $^3/_4$ as long as to subequal to the lemmas, keels scabrous, intercostal regions scabrous; **anthers** (1.4)1.75–2.6 mm, or vestigial (0.1–0.2 mm). $2n =$ unknown.

Poa diaboli, which is sequentially gynomonoecious, is endemic to upper shaly slopes, in soft coastal scrub and openings in Bishop Pine stands, in the coastal mountains of San Luis Obispo County, California. It is closely related to *P. confinis* (see previous), from which it differs by a suite of characters. The two species are also ecologically and geographically distinct.

39. Poa piperi Hitchc. [p. 557]
PIPER'S BLUEGRASS

Plants perennial; loosely tufted, rhizomatous. **Basal branching** extra- and intravaginal. **Culms** 20–55 cm, erect or the bases decumbent, terete or weakly compressed; **nodes** terete, 0–1 exserted. **Sheaths** closed for $^1/_3$–$^2/_3$ their length, terete, sparsely to moderately scabrous, glabrous or retrorsely hispidulous, bases of basal sheaths glabrous, distal sheath lengths 2.7–6.5(9.7) times blade lengths; **collars** of at least some leaves usually sparsely hispidulous; **ligules** 1–2 mm, scabrous, truncate to obtuse; **innovation blades** to 40 cm, adaxial surfaces moderately to densely scabrous or hispidulous on and between the veins; **cauline blades** steeply reduced in length distally, 1–3 mm wide, involute, moderately thick, soft, abaxial surfaces smooth, margins scabrous, apices narrowly prow-shaped, flag leaf blades 1–4.5 cm long. **Panicles** 4–8 cm, erect to nodding, lanceoloid to ovoid, loosely contracted, sparse, with 18–60 spikelets; **nodes** with 1–2 branches; **branches** 3–8 cm, ascending, lax, terete or weakly angled, moderately and sometimes coarsely scabrous, with 3–8 spikelets. **Spikelets** 6–9(11) mm, lengths to 3 times widths, laterally compressed, not sexually dimorphic; **florets** 2–5(7); **rachilla** internodes 1–2 mm, glabrous, scabrous, or sparsely to densely puberulent. **Glumes** subequal, distinctly keeled; **lower glumes** 3-veined; **calluses** diffusely webbed, hairs about $^1/_2$ the lemma length; **lemmas** 4–6(7) mm, lanceolate, distinctly keeled, glabrous, smooth or sparsely to moderately finely scabrous, keels scabrous, lateral veins moderately prominent, margins glabrous, apices acute; **palea keels** scabrous, sometimes softly puberulent at midlength; **anthers** vestigial (0.1–0.2 mm) or 2–3 mm. $2n = 28$.

Poa piperi grows in forests openings on serpentine rocks in the Coast Ranges of southwestern Oregon and northwestern California. It used to be included in *P. rhizomata* (p. 546), from which it differs in its involute leaves and glabrous lemmas. It is dioecious.

40. Poa atropurpurea Scribn. [p. 557]
SAN BERNARDINO BLUEGRASS

Plants perennial; loosely tufted, rhizomatous. **Basal branching** extra- and intravaginal. **Culms** 10–55 cm, erect or the bases decumbent, terete or weakly compressed; **nodes** terete, not exserted. **Sheaths** closed for about $^1/_3$ their length, terete, smooth, glabrous, bases of basal sheaths glabrous, distal sheath lengths 1.5–7.5 times blade lengths; **collars** smooth, glabrous; **ligules** 1–2 mm, smooth or sparsely scabrous, apices truncate to obtuse; **innovation blades** similar to the cauline blades, adaxial surfaces nearly smooth, glabrous on and between the veins; **cauline blades** fairly strongly reduced in length distally, 1–3 mm wide, folded to involute, moderately thick, moderately firm, abaxial surfaces smooth, apices narrowly prow-shaped, flag leaf blades 1–5.5 cm. **Panicles** 3–7 cm, erect, lanceoloid to ovoid, congested, with 20–70 spikelets; **nodes** with 1–2 branches; **branches** 0.5–3 cm, erect, terete, usually smooth, infrequently sparsely scabrous distally, with 3–12 spikelets. **Spikelets** 3.5–5.5 mm, lengths to 3 times widths, laterally compressed, very compact, not sexually dimorphic; **florets** 2–5; **rachilla** internodes about 0.5 mm, smooth, glabrous. **Glumes** broadly lanceolate, distinctly shorter than the adjacent lemmas, distinctly keeled, keels smooth or sparsely scabrous; **lower glumes** 3-veined; **calluses** glabrous; **lemmas** 2.5–3.5 mm, lanceolate, usually purplish, distinctly keeled, glabrous, smooth, margins glabrous, apices acute; **palea keels** scabrous, intercostal regions glabrous; **anthers** vestigial (0.1–0.2 mm) or 1.5–2 mm. $2n = 28$.

lemma

palea

anthers

spikelet

flag leaf
ligule

P. confinis

2 mm

1 cm

2 mm

2 cm

1 mm

SL

side
view

dorsal
view

floret

palea

anthers

spikelet

flag blade
apex

P. diaboli

2 mm

2 mm

2 cm

2 mm

2 cm

1 mm

Poa atropurpurea is a rare dioecious endemic of mesic upland meadows in southern California. It is federally listed as endangered.

41. Poa fendleriana (Steud.) Vasey [p. 558]
VASEY'S MUTTONGRASS

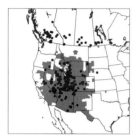

Plants perennial; densely to loosely tufted, rhizomatous, often weakly so, rhizomes usually short and inconspicuous. **Basal branching** mainly intravaginal, usually some extravaginal. **Culms** 15–70 cm, sometimes stout, erect or the bases decumbent, terete or weakly compressed; **nodes** terete, 0–1 exserted. **Sheaths** closed for about ¹/₃ their length, terete, smooth or scabrous, glabrous or occasionally retrorsely pubescent, bases of basal sheaths glabrous, distal sheath lengths usually (5)9+ times blade lengths; **collars** smooth or scabrous, glabrous or hispidulous; **ligules** 0.2–18 mm, smooth or scabrous, decurrent or not, apices truncate to acuminate, ciliolate or glabrous; **innovation blades** usually moderately to densely scabrous or hispidulous on and between the veins, infrequently nearly smooth and glabrous; **cauline blades** strongly reduced in length distally, (0.5)1–3(4) mm wide, usually involute, moderately thick and firm, infrequently moderately thin, abaxial surfaces usually smooth, infrequently scabrous, apices narrowly prow-shaped, steeply reduced in length distally along the culm, flag leaf blades often absent or very reduced, sometimes to 1(3) cm. **Panicles** 2–12(30) cm, erect, contracted, narrowly lanceoloid to ovoid, congested, frequently with 100+ spikelets; **nodes** with 1–2 branches; **branches** 1–8 cm, erect, terete to weakly angled, smooth or scabrous, with 3–15(25) spikelets. **Spikelets** (3)4–8(12) mm, lengths to 3 times widths, broadly lanceolate to ovate, laterally compressed, not sexually dimorphic; **florets** 2–7(13); **rachilla internodes** 0.8–1.3 mm, smooth, glabrous or hairy, hairs to 0.3 mm. **Glumes** lanceolate, distinctly keeled; **lower glumes** 1–3-veined, distinctly shorter than the lowest lemmas; **calluses** glabrous; **lemmas** 3–6 mm, lanceolate, distinctly keeled, keels, marginal veins, and lateral veins glabrous or short- to long-villous or softly puberulent, lateral veins moderately prominent, intercostal regions softly puberulent or glabrous, smooth or sparsely scabrous, margins glabrous, apices acute; **palea keels** scabrous, sometimes softly puberulent or long-villous at midlength, hairs to 0.4+ mm; **anthers** vestigial (0.1–0.2 mm) or 2–3 mm. 2*n* = 28+II, 56, 56–58, 58–64.

Poa fendleriana grows on rocky to rich slopes in sagebrush-scrub, interior chaparral, and southern (rarely northern) high plains grasslands to forests, and from desert hills to low alpine habitats. Its range extends from British Columbia to Manitoba and south to Mexico. It is one of the best spring fodder grasses in the eastern Great Basin, Colorado plateaus, and southern Rocky Mountains. It is dioecious. Each of the subspecies has regions of sexual reproduction in which staminate plants are common within populations, and extensive regions where only apomictic, pistillate plants are found. The sexual populations set little seed; the apomictic populations are highly fecund.

Poa fendleriana hybridizes with *Poa cusickii* subsp. *pallida* (p. 560). The hybrids are called *P. ×nematophylla* (p. 562).

1. Lemma keels and marginal veins glabrous or almost so . . . . subsp. *albescens*
1. Lemma keels and marginal veins conspicuously hairy.
  2. Ligules of the middle cauline leaves 0.2–1.2 (1.5) mm long, not decurrent, usually scabrous, apices truncate to rounded, upper margins ciliolate or scabrous . . . . subsp. *fendleriana*
  2. Ligules of the middle cauline leaves (1.5)1.8–18 mm long, decurrent, usually smooth to sparsely scabrous, apices obtuse to acuminate, upper margins usually smooth, glabrous . . . . subsp. *longiligula*

Poa fendleriana subsp. albescens (Hitchc.) Soreng [p. 558]

Collars often scabrous or hispidulous near the throat; **ligules of middle cauline leaves** 0.2–1.5 mm, smooth or scabrous, margins not decurrent, apices truncate, scabrous, ciliolate, or glabrous; **innovation blades** frequently glabrous adaxially. **Rachilla internodes** smooth, glabrous. **Lemmas** glabrous or the keels and marginal veins sparsely short-villous to softly puberulent. 2*n* = 28+II, 56.

Poa fendleriana subsp. *albescens* is endemic to the northern Sierra Madre Occidental, extending from the southwestern United States to Chihuahua and Sonora, Mexico. It grows mainly in upland forest openings. It intergrades with subsp. *fendleriana* where sexual populations have come into contact. Intermediate, pistillate populations with sparsely hairy lemmas are common in southeastern Arizona, and infrequent in southwestern New Mexico.

Poa fendleriana (Steud.) Vasey subsp. fendleriana [p. 558]

Collars often scabrous or hispidulous near the throat; **ligules of middle cauline leaves** 0.2–1.2(1.5) mm, scabrous, margins not decurrent, apices truncate to rounded, usually scabrous or ciliolate; **innovation blades** usually scabrous or puberulent adaxially.

floret

spikelet

anthers

ligule

collar

spikelet

lemma

palea

P. piperi

P. atropurpurea

POA

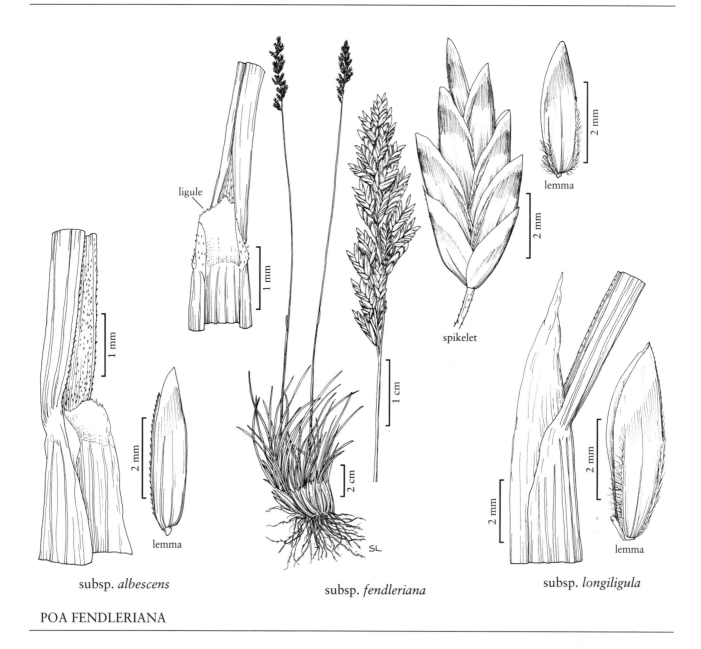

subsp. *albescens*   subsp. *fendleriana*   subsp. *longiligula*

POA FENDLERIANA

Rachilla internodes usually smooth and glabrous. **Lemmas** long-villous on the keels and marginal veins, intercostal regions usually glabrous, infrequently softly puberulent. $2n = 56, 58–60, 59, 58–64$.

Poa fendleriana subsp. *fendleriana* grows chiefly in the southern and middle Rocky Mountains, and in the mountains surrounding the Colorado plateaus. Sexually reproducing populations are mainly confined to Arizona, New Mexico, and Texas, are rare in California, and infrequent in Colorado and Utah. Pistillate populations are common from southern British Columbia to Manitoba and south to northern Mexico, but infrequent in the Great Basin. *Poa fendleriana* subsp. *fendleriana* intergrades with subspp. *albescens* and *longiligula* where sexual or partially sexual populations have come into contact.

Poa fendleriana subsp. **longiligula** (Scribn. & T. A. Williams) Soreng [p. 558]
LONGTONGUE MUTTONGRASS

Collars smooth to scabrous near the throat; **ligules of middle cauline leaves** (1.5)1.8–18 mm, smooth or sparsely scabrous, margins decurrent, apices obtuse to acuminate, usually smooth, glabrous; **innovation blades** usually scabrous, sometimes puberulent adaxially. **Rachilla internodes** usually sparsely hispidulous or sparsely softly puberulent. **Lemmas** long-villous on the keels and marginal veins, intercostal regions usually glabrous, infrequently softly puberulent. $2n = 56, 56–58$.

Poa fendleriana subsp. *longiligula* tends to grow to the west of the other two subspecies, in areas where winter precipitation is more consistent and summer precipitation less consistent. Apomixis is far more common and widespread than sexual reproduction in this subspecies. Apomictic populations range from southwestern British Columbia to Baja California, Mexico, throughout the Great Basin and Colorado

plateaus, and eastward across the Rocky Mountains. Sexual populations are mainly confined to northern Arizona, California, Nevada, and Utah.

Poa subsect. **Epiles** Hitchc. *ex* Soreng

Plants perennial; densely or rarely moderately densely tufted, usually neither stoloniferous nor rhizomatous, rarely shortly rhizomatous. **Basal branching** intra- or extravaginal, or both. **Culms** 5–60(70) cm. **Sheaths** closed for $^1/_7$–$^4/_5$ their length, terete, distal sheaths longer than their blades; **innovation blades** sometimes involute and firmer than the cauline blades, adaxial surfaces smooth or moderately to densely scabrous, glabrous or hispidulous on and between the veins; **cauline blades** 0.5–3 mm wide, flat, folded, or involute, thin to thick, soft to firm, apices usually narrowly prow-shaped, cauline blades sometimes broadly prow-shaped. **Panicles** 1–12 cm, erect or slightly nodding, contracted or open, usually narrowly lanceolate to ovate, sometimes pyramidal; **nodes** with 1–3(5) branches; **branches** 0.5–4(5) cm, terete to angled, smooth or sparsely to densely scabrous. **Spikelets** (3)4–10(12) mm, lengths to 3.5 times widths, lanceolate to broadly ovate, not bulbiferous; **florets** 2–8. **Glumes** lanceolate to broadly lanceolate; **lower glumes** 3-veined; **calluses** usually glabrous, sometimes shortly and sparsely webbed; **lemmas** (3)3.5–8 mm, lanceolate to broadly lanceolate, usually glabrous, sometimes sub-puberulent, or short-villous in hybrids; **palea keels** scabrous, glabrous or ciliate; **anthers** 3, vestigial (0.1–0.2 mm), aborted late in development, or 1.3–4.5 mm.

The five species of *Poa* subsect. *Epiles* are cespitose, gynodioecious or dioecious, and have involute or folded leaf blades.

42. **Poa cusickii** Vasey [p. 561]

CUSICK'S BLUEGRASS

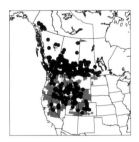

Plants perennial; usually densely tufted, rarely moderately densely tufted, usually neither rhizomatous nor stoloniferous, infrequently short-rhizomatous or stoloniferous, rarely with distinct rhizomes. **Basal branching** intravaginal or intra- and extravaginal. **Culms** 10–60(70) cm tall, 0.5–1.8 mm thick, erect or the bases decumbent, terete or weakly compressed; **nodes** terete, 0–2 exserted. **Sheaths** closed for $^1/_4$–$^3/_4$ their length, terete, smooth or scabrous, glabrous, bases of basal sheaths glabrous, distal sheath lengths 1.6–10 times blade lengths; **collars** smooth or scabrous, glabrous; **ligules** of cauline leaves 1–3(6) mm, smooth or scabrous, truncate to acute, ligules of the innovation leaves 0.2–0.5(2.5) mm, scabrous, usually truncate; **innovation blades** sometimes distinctly different from the cauline blades, 0.5–2 mm wide, involute, moderately thick, moderately firm, adaxial surfaces usually densely scabrous or hispidulous to softly

puberulent, infrequently nearly smooth and glabrous; **cauline blades** subequal or the midcauline blades longest or the blades gradually reduced in length distally, 0.5–3 mm wide, flat, folded, or involute, usually thin, usually withering, abaxial surfaces smooth or scabrous, apices narrowly to broadly prow-shaped, flag leaf blades 0.5–5(6) cm. **Panicles** 2–10(12) cm, usually erect, contracted or loosely contracted, narrowly lanceoloid to ovoid, congested or moderately congested, with 10–100 spikelets and 1–3(5) branches per node; **branches** 0.5–4(5) cm, erect or steeply ascending, fairly straight, slender to stout, terete to angled, smooth or scabrous, with 1–15 spikelets. **Spikelets** (3)4–10 mm, lengths to 3 times widths, broadly lanceolate to narrowly ovate, laterally compressed, not sexually dimorphic; **florets** 2–6; **rachilla internodes** 0.5–1.2 mm, smooth or scabrous. **Glumes** lanceolate, distinctly keeled; **lower glumes** 3-veined, distinctly shorter than the lowest lemmas; **calluses** glabrous or diffusely webbed, hairs less than $^1/_4$ the lemma length; **lemmas** (3)4–7 mm, lanceolate to broadly lanceolate, distinctly keeled, membranous to thinly membranous, smooth or sparsely to densely scabrous, glabrous or the keels and/or marginal veins puberulent proximally, lateral veins obscure to prominent, margins glabrous, apices acute; **palea keels** scabrous, intercostal regions glabrous; **anthers** vestigial (0.1–0.2 mm), aborted late in development, or 2–3.5 mm. $2n = 28, 28+II, 56, 56+II, 59$, ca. 70.

Poa cusickii grows in rich meadows in sagebrush scrub to rocky alpine slopes, from the southwestern Yukon Territory to Manitoba and North Dakota, south to central California and eastern Colorado. It is gynodioecious or dioecious.

Sexually reproducing plants of *Poa cusickii* subspp. *cusickii* and *pallida* grow in different geographic areas, but pistillate plants of these two subspecies have overlapping ranges. Only pistillate plants are known in *Poa cusickii* subspp. *epilis* and *purpurascens*. All the alpine plants studied were pistillate.

1. Panicle branches smooth or slightly scabrous, or the basal blades more than 1.5 mm wide and flat or folded; cauline blades more than 1.5 mm wide, often flat; some basal branching extravaginal; lemmas and calluses sometimes sparsely puberulent.
   2. Lemmas usually glabrous, rarely plants from the Rocky Mountains with puberulent keels and marginal veins; calluses glabrous; panicles erect, usually contracted; branches smooth to slightly scabrous .... subsp. *epilis*
   2. Lemmas rarely completely glabrous, at least some florets with sparsely puberulent keels, the marginal veins glabrous or puberulent; calluses frequently with a sparse, short web; panicles somewhat lax and loosely contracted; branches smooth or sparsely to moderately scabrous .... subsp. *purpurascens*
1. Panicle branches moderately to strongly scabrous; basal and cauline blades usually less than 1.5 mm wide, involute, rarely flat or folded; basal branching intravaginal; lemmas and calluses glabrous.
   3. Panicle branches longer than 1.7 cm in at least some panicles; panicles open or contracted .... subsp. *cusickii*
   3. Panicle branches up to 1.7 cm long, stout; panicles contracted .... subsp. *pallida*

Poa cusickii Vasey subsp. **cusickii** [p. 561]

Plants densely tufted. **Basal branching** intravaginal. **Culms** 10–60(70) cm, mostly erect, with 0–1 well-exserted nodes. **Sheaths** closed for $^1/_4$–$^2/_3$ their length, distal sheath lengths 3–10 times blade lengths; **innovation blades** 0.5–1 mm wide; **cauline blades** less than 1.5 mm wide, flat, folded, or involute, apices narrowly prow-shaped, flag leaf blades (0.5)1.5–5 cm. **Panicles** usually 5–10(12) cm, contracted or loosely contracted, with 20–100 spikelets; **nodes** with 1–5 branches; **branches** 1.7–4(5) cm, slender to stout,

moderately to densely scabrous, with 2–15 spikelets. **Spikelets** 4–10 mm. **Calluses** glabrous; **lemmas** 4–7 mm, glabrous; **anthers** vestigial (0.1–0.2 mm) or 2–3.5 mm. $2n = 28$.

Poa cusickii subsp. *cusickii* grows mainly in mesic desert upland and mountain meadows, on and around the Columbia plateaus of northern California, Oregon, southern Washington, and adjacent Idaho and Nevada. It is highly variable, with fairly open- to contracted-panicle populations, and from gynodioecious to dioecious populations. The modal and mean longest branch lengths of the narrower-panicled populations of subsp. *cusickii* serve to distinguish it from subsp. *pallida* in most cases. It appears to have hybridized with *P. pringlei* (p. 564) around Mount Shasta, California, and Mount Rose, Nevada. *Poa stebbinsii* (p. 564), an endemic in the high Sierra Nevada, is easily distinguished from *P. cusickii* subsp. *cusickii* by its long hyaline ligules.

Poa cusickii subsp. **epilis** (Scribn.) W.A. Weber [p. 561]
SKYLINE BLUEGRASS

Plants densely tufted. **Basal branching** intra- and extravaginal. **Culms** 20–45 cm, mostly erect, with 1–2 well-exserted nodes. **Sheaths** closed for $^1/_3$–$^3/_4$ their length, distal sheath lengths 2–5 times blade lengths; **innovation blades** 0.7–1 mm wide; **cauline blades** more than 1.5 mm wide, flat or folded, apices narrowly to broadly prow-shaped, flag leaf blades 1.5–5 cm, apices broadly prow-shaped. **Panicles** usually 2–7 cm, usually contracted, with 20–70 spikelets; **nodes** with 2–5 branches; **branches** 1–3 cm, moderately stout, smooth to sparsely scabrous, with 1–8 spikelets. **Spikelets** (3)4–8 mm. **Calluses** glabrous; **lemmas** 3–6 mm, glabrous or, rarely, the keels and marginal veins sparsely puberulent proximally; **anthers** usually aborted late in development. $2n = 56$, ca. 70.

Poa cusickii subsp. *epilis* tends to grow around timberline. It is strictly pistillate. It is usually quite distinct from subspp. *cusickii* and *pallida*, and differs from subsp. *purpurascens* in having on average more and shorter spikelets, lemmas that are shorter and rarely pubescent, and both intra- and extravaginal branching. It occurs throughout most of the range of the species, but is absent from the Yukon Territory, and uncommon in the Cascade Mountains. It is fairly uniform even though widespread.

Poa cusickii subsp. **pallida** Soreng [p. 561]

Plants densely tufted. **Basal branching** intravaginal. **Culms** 10–40(55) cm, mostly erect, with 0(1) scarcely exserted nodes. **Sheaths** closed for $^1/_4$–$^2/_3$ their length, distal sheath lengths 3.6–10 times blade lengths; **innovation blades** 0.5–1 mm wide, apices usually narrowly prow-shaped; **cauline blades** usually less than

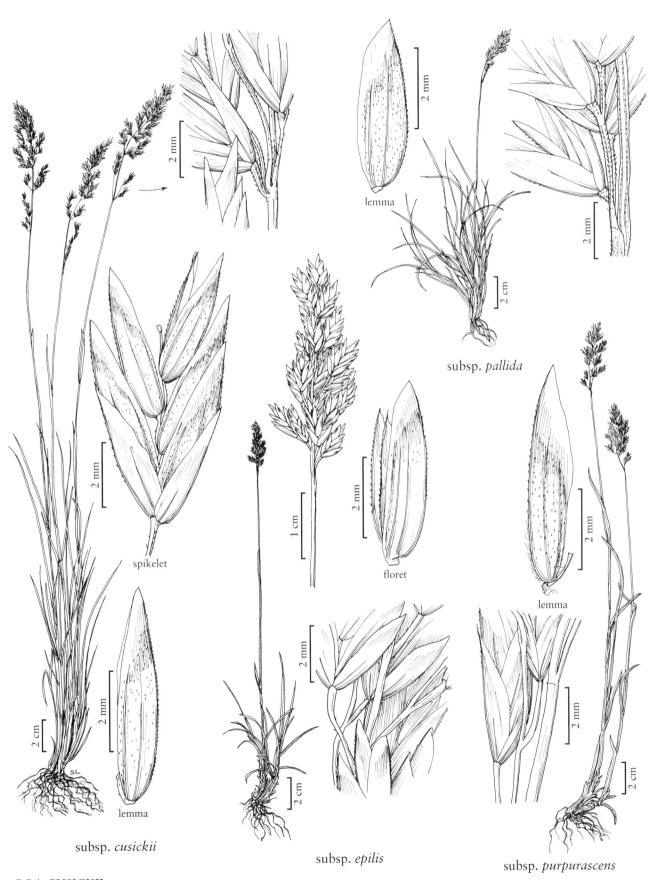

spikelet

lemma

subsp. *cusickii*

POA CUSICKII

floret

subsp. *epilis*

lemma

subsp. *pallida*

lemma

subsp. *purpurascens*

1.5 mm wide, flat, folded, or involute, usually narrowly prow-shaped, infrequently broadly prow-shaped, flag leaf blades 0.5–2(3) cm. **Panicles** 2–6 cm, contracted, with 10–40 spikelets; **nodes** with 1–3 branches; **branches** 0.5–1.7 cm, stout, moderately to densely scabrous, with 2–5 spikelets. **Spikelets** 4–10 mm. **Calluses** glabrous; **lemmas** 4–7 mm, glabrous; **anthers** vestigial (0.1–0.2 mm) or 2–3.5 mm. 2*n* = 56, 56+II, 59.

Poa cusickii subsp. *pallida* grows in forb-rich mountain grasslands to alpine habitats, from the southern Yukon Territory to California, across the Great Basin and through the Rocky Mountains to central Colorado. It is found mainly east and north of subsp. *cusickii*, but pistillate plants extend into the range of that subspecies in the eastern alpine peaks of California, Nevada, and Oregon. The shorter branch length serves to distinguish it from the narrow-panicled subsp. *cusickii* forms in most cases. It hybridizes with *P. fendleriana* (p. 556), forming *P.* ×*nematophylla* (see next). The hybrids may have hairy lemmas or, less often, broader leaf blades and glabrous lemmas. *Poa cusickii* subsp. *pallida* was included in Hitchcock's (1951) circumscription of *Poa pringlei*, along with *P. keckii* and *P. suksdorfii*.

Poa cusickii subsp. purpurascens (Vasey) Soreng [p. 561]

Plants densely to moderately densely tufted. **Basal branching** mostly extravaginal. **Culms** 25–50 cm, bases decumbent, with 1–2 well-exserted nodes. **Sheaths** closed for ¹⁄₂–³⁄₄ their length, distal sheath lengths 1.6–5 times blade lengths; **innovation blades** mostly 1–2 mm wide; **cauline blades** more than 1.5 mm wide, flat or folded, apices usually broadly prow-shaped, flag leaf blades 3–6 cm. **Panicles** usually 4–7 cm, slightly lax, ovate, loosely contracted, with 13–50 spikelets; **nodes** with 1–3 branches; **branches** 1–3(4) cm, moderately stout, smooth to moderately scabrous, with 1–8 spikelets. **Spikelets** 7–10 mm. **Calluses** of proximal lemmas usually sparsely and shortly webbed, hairs less than ¹⁄₄ the lemma length, sometimes glabrous, those of the distal lemmas glabrous; **lemmas** 4–7 mm, usually the keels and marginal veins of some proximal lemmas sparsely puberulent near the base, sometimes glabrous, distal lemmas glabrous; **anthers** usually aborted late in development. 2*n* = 28+II, 56.

Poa cusickii subsp. *purpurascens* grows in subalpine habitats in the coastal mountains from southern British Columbia to southern Oregon, with sporadic occurrences eastward in British Columbia to the Rocky Mountains and south to the central Sierra Nevada. It tends to differ from subsp. *epilis* in having predominantly extravaginal branching, fewer and longer spikelets, and longer lemmas that are usually sparsely hairy on the keel and marginal veins. It differs

from *P. chambersii* (p. 548) in lacking rhizomes and in being strictly pistillate; and from *P. porsildii* (p. 563) in its longer spikelets and in tending to have longer panicles with more spikelets.

43. Poa ×nematophylla Rydb. [p. 565]

Plants perennial; densely tufted, not stoloniferous, not rhizomatous. **Basal branching** intravaginal. **Culms** 10–35 cm, erect or the bases decumbent; **nodes** terete, 0–1 exserted. **Sheaths** closed for ¹⁄₄–³⁄₄ their length, terete, apices acuminate; **innovation blades** 0.5–1(2) mm wide, involute, moderately thick, moderately firm, abaxial surfaces smooth or scabrous, adaxial surfaces usually densely scabrous or hispidulous; **cauline blades** usually gradually reduced distally, 0.5–1(2) mm wide, flat, folded, or involute, thin, sometimes withering, abaxial surfaces smooth or scabrous, apices narrowly prow-shaped, sometimes the flag leaf blades vestigial. **Panicles** 2–8 cm, erect, narrowly lanceoloid to ovoid, contracted, congested; **nodes** with 1–2 branches; **branches** 0.5–3 cm, erect, terete to angled, scabrous. **Spikelets** 4–8 mm, lengths to 3 times widths, broadly lanceolate to narrowly ovate, laterally compressed, not sexually dimorphic; **florets** 2–5; **rachilla internodes** 0.5–1.2 mm, smooth or scabrous. **Glumes** lanceolate, distinctly keeled; **lower glumes** 3-veined, distinctly shorter than the lowest lemmas; **calluses** glabrous; **lemmas** 4–7 mm, lanceolate, distinctly keeled, membranous, keels and marginal veins usually softly puberulent, sometimes short-villous, intercostal regions usually glabrous, infrequently softly puberulent proximally, lateral veins moderately prominent, margins glabrous, apices acute; **palea keels** scabrous; **anthers** mostly vestigial (0.1–0.2 mm), rarely 2–3 mm. 2*n* = unknown.

Poa ×*nematophylla* is believed to consists of hybrids between *P. cusickii* subsp. *pallida* (see previous) and *P. fendleriana* (p. 556). It is mostly pistillate and apomictic; few staminate plants have been found. It usually resembles *P. cusickii* most, but grades towards *P. fendleriana*. It tends to grow on drier slopes than either parent, mainly in and around sagebrush desert/forest interfaces.

44. Poa porsildii Gjaerev. [p. 565]
PORSILD'S BLUEGRASS

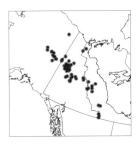

Plants perennial; densely tufted, not stoloniferous, not rhizomatous. **Basal branching** intravaginal. **Culms** (12)17–30(40) cm, erect or the bases decumbent, with (0)1(2) exserted nodes. **Sheaths** closed for $^1/_3$–$^2/_3$ their length, terete, bases of basal sheaths glabrous; **collars** smooth, glabrous; **ligules** of cauline leaves 1–2 mm, smooth, apices truncate to obtuse, ligules of the innovation leaves shorter than 0.5 mm, scabrous, truncate; **innovation blades** not or indistinctly differentiated from the cauline blades, flat or weakly involute, adaxial surfaces sparsely to fairly densely finely scabrous between the veins; **cauline blades** 1–3 mm wide, folded, fairly thin to moderately thick, soft, abaxial surfaces smooth, apices narrowly prow-shaped. **Panicles** 2–5(6) cm, erect or nodding, ovoid to pyramidal, slightly contracted or open, sparse, with fewer than 20 spikelets; **nodes** with 1–2 branches; **branches** 2–4 cm, ascending to widely spreading, occasionally reflexed, flexuous, lax, slender, terete, smooth or sparsely scabrous, with 1–3(4) spikelets. **Spikelets** 4–7 mm, lengths to 3 times widths, broadly lanceolate to narrowly ovate, laterally compressed, not sexually dimorphic, often strongly anthocyanic; **florets** 3–4; **rachilla internodes** smooth, glabrous or sparsely puberulent. **Glumes** thin, somewhat lustrous, distinctly keeled, keels smooth; **lower glumes** 3-veined, distinctly shorter than the lowest lemmas; **calluses** glabrous; **lemmas** 4–6 mm, lanceolate, distinctly keeled, thinly membranous, sparsely to moderately densely and finely scabrous, usually glabrous, rarely sparsely softly puberulent proximally, lateral veins moderately prominent, margins glabrous, apices acute; **palea keels** sparsely to moderately densely scabrous; **anthers** vestigial (0.1–0.2 mm) or 2.5–3 mm. $2n$ = unknown.

Poa porsildii is an alpine, calciphilic, mesophilic, dioecious species that grows from eastern Alaska to the western Northwest Territories. It differs from *P. cusickii* subsp. *purpurascens* (p. 562) in having panicles with laxer, smooth, and more slender branches, lemmas that are usually glabrous, and in having staminate plants.

45. Poa leibergii Scribn. [p. 565]
LEIBERG'S BLUEGRASS

Plants perennial; densely tufted, tufts slender, not stoloniferous, not rhizomatous. **Basal branching** intravaginal. **Culms** 5–35 cm tall, 0.5–0.7 mm thick, erect or the bases decumbent, with 0–1 exserted nodes. **Sheaths** closed for $^2/_5$–$^4/_5$ their length, terete, smooth and glabrous, bases of basal sheaths glabrous; **collars** smooth, glabrous; **ligules** (1)2–4 mm, colorless, transparent, smooth, margins decurrent or not, apices truncate to acute, ligules of innovation and cauline leaves alike; **innovation blades** smooth or sparsely scabrous abaxially; **cauline blades** 0.5–1 mm wide, flat, folded, or involute, thin, lax, filiform, usually soon withering, both surfaces smooth or sparsely scabrous, apices narrowly prow-shaped. **Panicles** 1–5(8) cm, erect to lax, lanceoloid to ovoid or pyramidal, contracted to open, sparse, with (1)6–17(22) spikelets; **nodes** with 1–2 branches; **branches** 1–4 cm, erect to spreading, slender, terete, smooth or sparsely to rarely moderately densely scabrous, with 1–2(3) spikelets. **Spikelets** 4–8 mm, lengths to 3 times widths, broadly lanceolate to broadly ovate, laterally compressed, not sexually dimorphic; **florets** 2–8; **rachilla internodes** glabrous. **Glumes** thin, somewhat lustrous, distinctly keeled; **lower glumes** 3-veined, distinctly shorter than the lowest lemmas; **calluses** glabrous; **lemmas** 3.5–7 mm, lanceolate, distinctly keeled, thinly membranous, smooth or scabrous, glabrous, lateral veins moderately prominent to prominent, margins glabrous, apices acute to truncate and erose; **palea keels** smooth or scabrous, glabrous or pectinately ciliate; **anthers** vestigial (0.1–0.2 mm) or 1.3–3 mm. $2n$ = unknown.

Poa leibergii grows on mossy ledges and around vernal pools and the outer margins of *Camassia* swales, in sagebrush desert to low alpine habitats, especially where snow persists. It is found primarily on and around the basaltic Columbia plateaus, and is gynodioecious. All reports of *P. leibergii* from California, and most of those from Nevada, are based on misidentified specimens of *P. cusickii* subsp. *cusickii* (p. 560) and *P. stebbinsii* (p. 564).

46. Poa stebbinsii Soreng [p. 567]
STEBBINS' BLUEGRASS

Plants perennial; densely tufted, not stoloniferous, not rhizomatous. **Basal branching** strictly extravaginal. **Culms** 10–30(40) cm, mostly erect, with 0–1 slightly exposed nodes. **Sheaths** closed for ¹/₅–²/₅ their length, terete, smooth and glabrous, bases of basal sheaths glabrous, distal sheath lengths 1.4–3.6 times blade lengths; **collars** smooth, glabrous; **ligules** of cauline leaves 3–8 mm, colorless, transparent, smooth, margins decurrent, apices obtuse to acuminate, ligules of the innovation leaves 2.5–6 mm; **innovation blades** similar to the cauline blades, 1–2 mm wide, involute, moderately thick, abaxial surfaces smooth, adaxial surfaces smooth or sparsely scabrous, sometimes sparsely hispidulous; **cauline blades** gradually reduced in length distally, 1–2 mm wide, folded or involute, moderately thick, soft, abaxial surfaces smooth, apices narrowly prow-shaped. **Panicles** 3–7 cm, erect or slightly nodding, narrowly lanceoloid to narrowly ovoid, often interrupted, contracted to loosely contracted, with 9–38(60) spikelets; **nodes** with 1–2 branches; **branches** 0.5–1.5(2.5) cm, erect at maturity, slender, terete to sulcate or weakly angled, sparsely to moderately scabrous, with 1–5 spikelets. **Spikelets** 4–6.5 mm, lengths to 3.5 times widths, lanceolate, laterally compressed, not sexually dimorphic, usually strongly anthocyanic, less so in pistillate plants; **florets** 2–4; **rachilla internodes** smooth, glabrous or sparsely hispidulous. **Glumes** unequal, lanceolate, thin, lustrous, distinctly keeled, keels and distal surface smooth or sparsely finely scabrous; **lower glumes** 3-veined, distinctly shorter than the lowest lemmas; **calluses** glabrous; **lemmas** 3.5–5.5 mm, lanceolate, distinctly keeled, thinly membranous, smooth or sparsely scabrous, glabrous, lateral veins moderately prominent, margins glabrous, apices acute; **palea keels** finely scabrous; **anthers** vestigial (0.1–0.2 mm) or 2–4.5 mm. $2n = 42, 81$ (both counts of uncertain application).

Poa stebbinsii is endemic to the high Sierra Nevada. It grows primarily in the outer margins of subalpine wet meadows, and is gynodioecious. It is easily recognized by its long hyaline ligules, thin glabrous lemmas, and the absence of intravaginal shoots. It was confused with *P. hansenii* Scribn. [= *P. cusickii* subsp. *cusickii*] by Keck in Munz (1959), and with *P. leibergii* by Hitchcock (1951).

47. Poa pringlei Scribn. [p. 567]
PRINGLE'S BLUEGRASS

Plants perennial; densely tufted, not stoloniferous, not rhizomatous. **Basal branching** intravaginal. **Culms** 5–35 cm tall, 0.5–0.9 mm thick, erect or the bases decumbent, with 0(1) exserted nodes. **Sheaths** closed for ¹/₇–¹/₃ their length, terete, smooth or sparsely scabrous, glabrous, bases of basal sheaths glabrous, distal sheath lengths 2–4 times blade lengths; **collar** margins smooth or scabrous to hispidulous; **ligules** of cauline leaves 1–6 mm, colorless, translucent, smooth or scabrous, truncate to acute, ligules of the innovations 1–2.5 mm; **innovation blades** similar to the cauline blades, 1.5–3 mm wide, involute, thick, frequently somewhat arcuate, abaxial surfaces smooth, adaxial surfaces densely scabrous or hispidulous; **cauline blades** becoming only slightly shorter distally, 1.5–3 mm wide, involute, moderately thick, soft to moderately firm, abaxial surfaces smooth, apices narrowly prow-shaped. **Panicles** 1–6 cm, erect, narrowly lanceoloid to ovoid, moderately congested, with 6–20(25) spikelets; **nodes** with 1–2 branches; **branches** 0.5–1.5(2) cm, erect, moderately stout, terete or weakly angled, angles smooth to fairly densely scabrous, with 1–3 spikelets. **Spikelets** 6–8(12) mm, lengths to 3.5 times widths, broadly lanceolate, laterally compressed, not sexually dimorphic, lustrous; **florets** 2–5; **rachilla internodes** smooth. **Glumes** subequal, isomorphic, lanceolate to broadly lanceolate, thin, lustrous, distinctly keeled, keels smooth or sparsely scabrous; **lower glumes** shorter than the adjacent lemmas, 3-veined; **calluses** glabrous; **lemmas** 5–8 mm, lanceolate, distinctly keeled, thinly membranous, smooth or sparsely finely scabrous, glabrous, lateral veins moderately prominent, margins glabrous, apices acute; **palea keels** coarsely scabrous; **anthers** vestigial (0.1–0.2 mm) or 2–4 mm. $2n =$ unknown.

Poa pringlei grows on rocky subalpine and alpine slopes in Oregon and California. Sexual populations, with approximately equal numbers of pistillate and staminate plants, are confined to the Klamath–Siskiyou region; Sierra Nevada populations are pistillate and apomictic. Hitchcock (1951) included *P. cusickii* subsp. *pallida* (p. 560), *P. keckii* (p. 584), and *P. suksdorfii* (p. 584) in *P. pringlei*; the illustration (Fig. 171) is of *P. cusickii* subsp. *pallida*.

Hybrids of *Poa pringlei* with *P. cusickii* (p. 560) have been found on Mount Shasta, California, Mount Rose, Nevada, and near Crater Lake, Oregon. *Poa pringlei* differs from *P. curtifolia* (p. 589) in being dioecious and in having blades that are involute, soft to moderately firm, and abaxially pubescent.

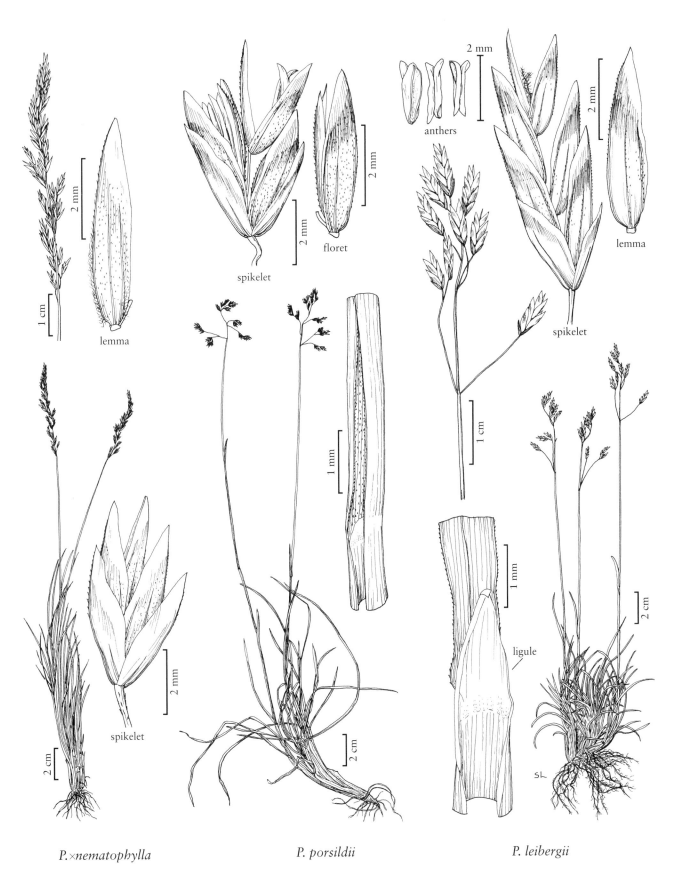

2 mm

anthers

2 mm

lemma

spikelet

floret

2 mm

lemma

spikelet

2 mm

lemma

1 cm

1 mm

1 cm

1 mm

ligule

spikelet

2 cm

2 cm

SL

2 cm

P. ×nematophylla

P. porsildii

P. leibergii

Poa sect. Dioicopoa E. Desv.

Plants perennial; usually rhizomatous, sometimes tufted or with solitary shoots. **Basal branching** intra- and extravaginal. **Culms** 20–85 cm, terete or weakly compressed. **Sheaths** closed firmly for $^1/_7$–$^1/_3$ their length, sometimes for a longer distance by a hyaline membrane, terete; **ligules** 1–4 mm (North America); **innovation blades** sparsely to densely scabrous, mainly over the veins; **cauline blades** flat or folded, rarely involute, abaxial and adaxial surfaces smooth or sparsely finely scabrous, glabrous, apices narrowly to broadly prow-shaped. **Panicles** 3–12(18) cm (North America), contracted or infrequently open, congested, branches erect to slightly ascending. **Spikelets** laterally compressed, sexually dimorphic, not bulbiferous (North America), of different sexes, slightly differentiated in floret number and lemma length, and sharply differentiated in vesture development; **florets** normal. **Glumes** shorter than the adjacent lemmas, distinctly keeled, keels scabrous; **lower glumes** 1–3-veined; **calluses** terete or slightly laterally compressed, those of staminate plants often glabrous, those of pistillate plants usually copiously pubescent, hairs arising as a single dorsal tuft and as single tufts from below the marginal veins, long-plicate or rarely closely crimped hairs or a crown of long hairs (in South America); **lemmas** distinctly keeled, those of pistillate plants pubescent (North America), usually the keels and marginal veins long-villous; **paleas** scabrous, glabrous, or medially softly puberulent to long-villous over the keels; **anthers** 3, vestigial (0.1–0.2 mm) or 1.6–2.7 mm.

Poa sect. *Dioicopoa* includes 29 species; all except the North American *P. arachnifera* are native to South America. The above description applies to the North American species. They are strictly dioecious. All appear to reproduce sexually; a few species are also bulbiferous. Many, including *P. arachnifera*, are characterized by having 3 well-developed webs on the calluses of the pistillate florets.

48. Poa arachnifera Torr. [p. 569]

TEXAS BLUEGRASS

Plants perennial; loosely tufted, rhizomatous, rhizomes slender. **Basal branching** intra- and extravaginal. **Culms** 20–85 cm, erect, terete or weakly compressed; **nodes** terete, 0–1 exserted. **Sheaths** closed firmly for $^1/_7$–$^1/_3$ their length, sometimes for a longer distance by a hyaline membrane, terete, smooth, glabrous, bases of basal sheaths glabrous; **ligules** 1–4 mm, smooth or scabrous; **innovation blades** 10–35 cm long, 1–3.5 mm wide; **cauline blades** 2–25 cm long, 1.5–4.5 mm wide, flat or folded, lax, both surfaces smooth or sparsely finely scabrous, glabrous, apices narrowly to broadly prow-shaped. **Panicles** 3–12(18) cm, erect, narrowly cylindrical, often interrupted or lobed, congested, with (70)100–200 spikelets; **nodes** with (2)3–7(9) branches; **branches** 1–3(5) cm, erect to slightly ascending, terete or weakly angled, sparsely to densely coarsely scabrous, with 8–30 spikelets. **Spikelets** 4–8(10) mm, sexually dimorphic, laterally compressed, pistillate spikelets larger, with fewer florets and more pubescence than the staminate spikelets; **florets** 2–10; **rachilla internodes** smooth. **Glumes** unequal, distinctly keeled, keels and lateral veins scabrous; **lower glumes** 1–3-veined.

Staminate florets: calluses glabrous or sparsely dorsally webbed, hairs plicate, rarely with additional webs under the marginal veins; **lemmas** 3.5–5 mm, lanceolate, distinctly keeled, keels and marginal veins sparsely short- to long-villous, margins glabrous, apices acute; **palea keels** scabrous, glabrous or softly puberulent to long-villous at midlength; **anthers** vestigial (0.1–0.2 mm) or 1.6–2.7 mm. **Pistillate florets: calluses** copiously 3-webbed, hairs 4–10 mm, mostly silky, plicate; **lemmas** 4.2–6.4 mm, lanceolate, 5–7 veined, distinctly keeled, glabrous, or the keels and marginal veins, sometimes also the lateral veins, densely long-villous, margins glabrous, apices acute; **palea keels** scabrous, glabrous or sometimes softly puberulent to long-villous at midlength. 2*n* = 42, ca. 54, 56, ca. 63, 84.

Poa arachnifera grows on moist, sandy to rich, black bottomlands of the southern Great Plains. At one time it was cultivated for winter pasture in the southeastern United States. It is strictly dioecious, with a 1:1 ratio of staminate to pistillate plants among herbarium samples. The variable and high chromosome numbers suggest it may be apomictic, but the occurrence of equal numbers of staminate and pistillate individuals in populations seems to suggest that reproduction is primarily sexual. It is the only non-South American species in the section. Its closest relatives appear to be *P. bonariensis* (Lam.) Kunth and *P. lanuginosa* Poir.

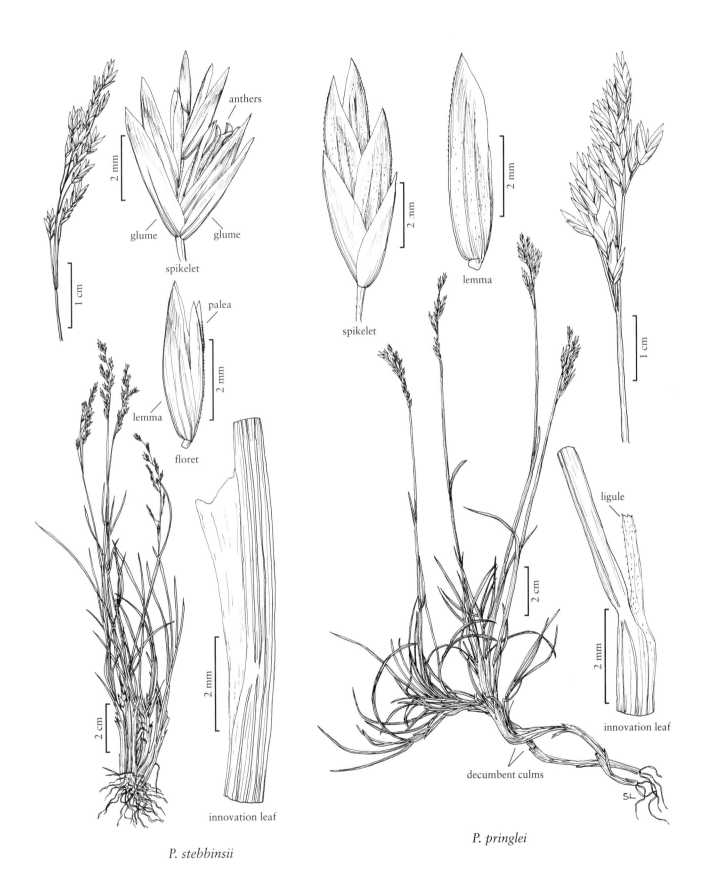

anthers

2 mm

glume   glume

spikelet

palea

lemma

floret

2 mm

lemma

2 mm

spikelet

2 mm

innovation leaf

1 cm

2 cm

2 mm

innovation leaf

P. stebbinsii

ligule

2 cm

2 mm

innovation leaf

decumbent culms

P. pringlei

SL

Poa sect. Pandemos Asch. & Graebn.

Plants perennial; sometimes stoloniferous, sometimes rhizomatous. **Basal branching** intra- and extravaginal. **Culms** 25–120 cm, terete or weakly compressed; **nodes** terete or slightly compressed. **Sheaths** closed for about ¼–½ their length, compressed, distal sheath lengths 0.5–4 times blade lengths; **ligules** 3–10 mm, scabrous, acute to acuminate; **blades** 1–5 mm wide, flat, lax, soft, veins and margins scabrous, apices narrowly prow-shaped. **Panicles** 8–25 cm, erect or lax, pyramidal, open; **nodes** with 3–7 branches; **branches** 2–8(10) cm, ascending to spreading, flexuous to fairly straight, angled, angles densely scabrous, crowded. **Spikelets** 2.3–3.5 mm, lengths to 3 times widths, laterally compressed, not bulbiferous; **florets** 2–4, bisexual. **Glumes** distinctly keeled, keels scabrous; **lower glumes** subulate to narrowly lanceolate, usually arched to sickle-shaped, 1-veined, distinctly shorter than the lowest lemmas; **calluses** terete or slightly laterally compressed, glabrous or dorsally webbed; **lemmas** 2.3–3.5 mm, lanceolate, distinctly keeled, keels hairy, glabrous elsewhere or the marginal veins pubescent, lateral veins prominent; **palea keels** smooth, muriculate, tuberculate, or minutely scabrous; **anthers** 3, 1.3–2 mm.

Poa sect. *Pandemos* includes two diploid species of European origin. One, *P. trivialis*, is now widespread around the world. Its chloroplast genome is related to the chloroplast genomes of sects. *Secundae* and *Stenopoa* (Gillespie and Soreng 2005); its nuclear ribosomal DNA markers suggest a relationship to sect. *Micrantherae* (Gillespie et al. 2006).

49. Poa trivialis L. [p. 569]
ROUGH BLUEGRASS

Plants perennial, short-lived; somewhat loosely to densely tufted, usually weakly stoloniferous. **Basal branching** intravaginal. **Culms** 25–120 cm, decumbent to erect, sometimes trailing and rooting at the nodes, terete or weakly compressed; **nodes** terete or slightly compressed, (0)1–3 exserted. **Sheaths** closed for about ⅓–½ their length, compressed, usually densely scabrous, bases of basal sheaths glabrous, distal sheath lengths 0.5–4 times blade lengths; **collars** smooth or scabrous, glabrous; **ligules** 3–10 mm, scabrous, acute to acuminate; **blades** 1–5 mm wide, flat, lax, soft, sparsely scabrous over the veins, margins scabrous, apices narrowly prow-shaped. **Panicles** 8–25 cm, erect or lax, pyramidal, open, with 35–100+ spikelets; **nodes** with 3–7 branches; **branches** 2–8(10) cm, ascending to spreading, flexuous to fairly straight, angled, angles densely scabrous, crowded, with 5–35 spikelets in the distal ½–¾. **Spikelets** 2.3–3.5 mm, lengths to 3 times widths, laterally compressed; **florets** 2–4, bisexual; **rachilla internodes** smooth or muriculate. **Glumes** distinctly keeled, keels scabrous; **lower glumes** subulate to narrowly lanceolate, usually arched to sickle-shaped, 1-veined, distinctly shorter than the lowest lemmas; **calluses** webbed, hairs over ⅔ the lemma length; **lemmas** 2.3–3.5 mm, lanceolate, distinctly keeled, keels usually sparsely puberulent to ⅗ their length, marginal veins usually glabrous, infrequently the proximal ¼ softly puberulent, intercostal regions smooth, glabrous, upper lemmas sometimes glabrous, lateral veins prominent, margins glabrous, apices acute; **palea keels** smooth, muriculate, tuberculate, or minutely scabrous; **anthers** 1.3–2 mm. 2*n* = 14.

Poa trivialis is an introduced European species. Only **Poa trivialis** subsp. **trivialis** is present in the *Flora* region. Several cultivars have been planted for pastures and lawns, and have often escaped cultivation. *Poa trivialis* sometimes grows with *P. paludigena* (p. 572), but has distinctly longer ligules and anthers. It is easily recognized by its flat blades, long ligules, sickle-shaped lower glumes, prominent callus webs, and lemmas with pubescent keels and pronounced lateral veins.

Poa sect. Oreinos Asch. & Graebn.

Plants perennial; densely to loosely tufted, sometimes shortly rhizomatous or stoloniferous. **Basal branching** mostly extravaginal or mixed intra- and extravaginal. **Culms** 5–100 cm tall, 0.5–1.5 mm thick, slender, sometimes weak, terete; **nodes** terete. **Sheaths** usually closed for

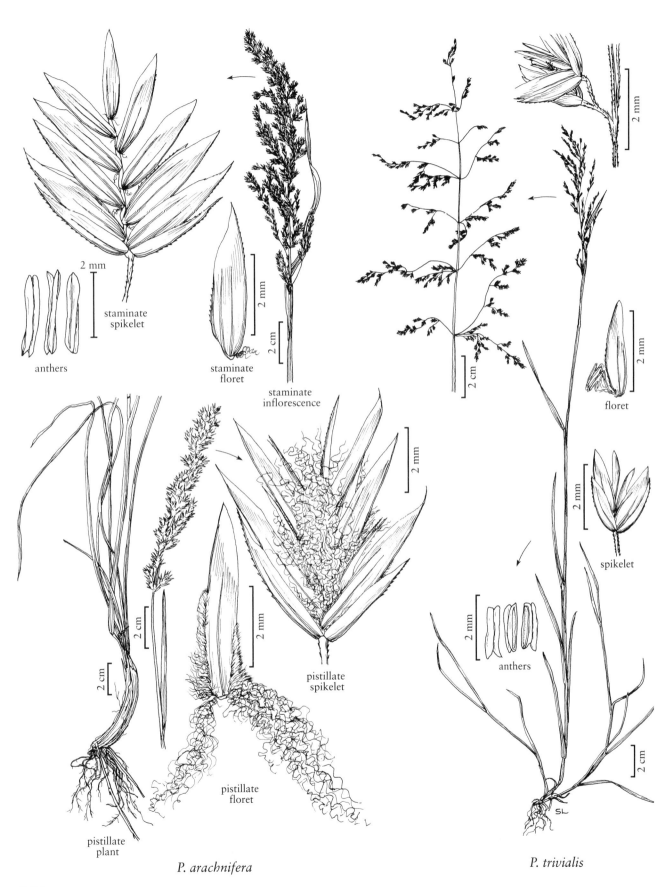

2 mm

anthers

staminate spikelet

staminate floret

staminate inflorescence

2 mm

2 cm

2 mm

2 cm

floret

2 mm

spikelet

2 mm

anthers

pistillate spikelet

2 cm

2 cm

2 mm

pistillate floret

pistillate plant

P. arachnifera

P. trivialis

$^1/_5$–$^3/_5$ their length, hybrids sometimes closed for $^1/_{10}$–$^1/_5$ their length, terete, smooth or sparsely scabrous; **ligules** 0.5–4(6) mm, smooth or sparsely scabrous, truncate to acute, sometimes lacerate; **innovation blades** similar to the cauline blades; **cauline blades** 0.8–4 mm wide, flat, thin, lax, soft, adaxial surfaces smooth or sparsely scabrous, narrowly prow-tipped. **Panicles** 1.5–15 cm, lax or slightly lax, loosely contracted to open; **nodes** with 1–3(5) branches; **branches** 1–8 cm, steeply ascending to reflexed, capillary to slender, drooping to fairly straight, sulcate or angled, smooth or the angles scabrous, with 1–15 spikelets. **Spikelets** 3.2–8 mm, lengths to 3.5 times widths, narrowly lanceolate to ovate, laterally compressed, not bulbiferous; **florets** 2–5, bisexual; **rachilla internodes** smooth, glabrous. **Glumes** subulate to broadly lanceolate, thin, distinctly keeled, keels smooth or scabrous; **lower glumes** 1–3-veined; **calluses** terete or slightly laterally compressed, glabrous or dorsally webbed; **lemmas** 2.5–4.6 mm, lanceolate to broadly lanceolate, distinctly keeled, thin, keels and marginal veins short- to long-villous, lateral veins usually glabrous, infrequently sparsely softly puberulent, lateral veins obscure or moderately prominent, intercostal regions glabrous; **palea keels** scabrous, usually glabrous, infrequently pectinately ciliate; **anthers** 3, 0.2–1.1(1.3) mm.

Poa sect. *Oreinos* is circumboreal. It includes seven species: four strictly Eurasian, one amphiatlantic, one primarily western North American with isolated occurrences in the Russian Far East, and one restricted to North America. The species are boreal, alpine to low arctic, and grow in bogs and on alpine slopes. They are primarily slender perennials with extravaginal tillering.

50. Poa laxa Haenke [p. 571]
LAX BLUEGRASS

Plants perennial; not or only slightly glaucous; densely tufted, not stoloniferous, not rhizomatous. **Basal branching** mixed, mainly extravaginal or mainly pseudointravaginal, sometimes intravaginal. **Culms** 8–35 cm tall, 0.5–0.9 mm thick, ascending to erect, slender; **nodes** terete, 0(1) exserted. **Sheaths** closed for $^1/_5$–$^1/_3$ their length, terete, smooth, glabrous, bases of basal sheaths glabrous; **collars** smooth or scabrous, glabrous; **ligules** 2–4 mm, smooth, apices acute, often lacerate; **innovation blades** similar to the cauline blades; **cauline blades** 1–2(3) mm wide, flat, thin, soft, smooth, narrowly prow-tipped, blades not strongly graduated or reduced upwards. **Panicles** 2–8 cm, slightly lax, usually loosely contracted and sparse, infrequently contracted and dense; **nodes** with 1–3(5) branches; **branches** 1–3(4) cm, usually ascending or weakly spreading, infrequently erect, fairly straight or flexuous, slender, sulcate or angled, smooth or the angles sparsely scabrous, with 1–8 spikelets. **Spikelets** 4–6 mm, lengths to 3 times widths, laterally compressed; **florets** 2–5; **rachilla internodes** shorter than 1 mm, smooth, glabrous. **Glumes** nearly equaling or slightly longer than the adjacent lemmas, lanceolate to broadly lanceolate, thin, distinctly keeled, keels smooth or sparsely scabrous; **lower glumes** 1–3-veined; **upper glumes** shorter than or subequal to the lowest lemmas; **calluses** glabrous or webbed, hairs usually shorter than $^1/_4$ the lemma length, sparse; **lemmas** 3–4.6 mm, lanceolate to broadly lanceolate, thin, distinctly keeled, keels and marginal veins short- to long-villous, lateral veins glabrous or sparsely softly puberulent, lateral veins obscure, intercostal regions glabrous, margins glabrous, apices acute; **paleas** sparsely scabrous over the keels; **anthers** (0.6)0.8–1.1(1.3) mm. 2*n* = 28, 42, 84.

Poa laxa is a low arctic to high alpine amphiatlantic species. It has been treated as a series of separate species, but the differences seem relatively minor and incomplete. Its short anthers and smoother branches usually distinguish it from *P. glauca* (p. 576), with which it can hybridize to form *P. laxa* × *glauca* (p. 572).

Poa laxa has four subspecies, two of which are native to the *Flora* region; subsp. *laxa* grows in central Europe; and subsp. *flexuosa* (Sm.) Hyl. in northwestern Europe.

1. Innovations primarily extravaginal; panicle branches fairly straight; calluses glabrous . . . . subsp. *banffiana*
1. Innovations primarily intravaginal; panicle branches flexuous, usually at least some florets having a webbed callus . . . . subsp. *fernaldiana*

Poa laxa subsp. banffiana Soreng [p. 571]

Basal branching mainly extravaginal. **Blades** thin. **Panicles** 2–8 cm, lax, loosely contracted, sparse, with 2–3(5) branches per node; **branches** steeply ascending,

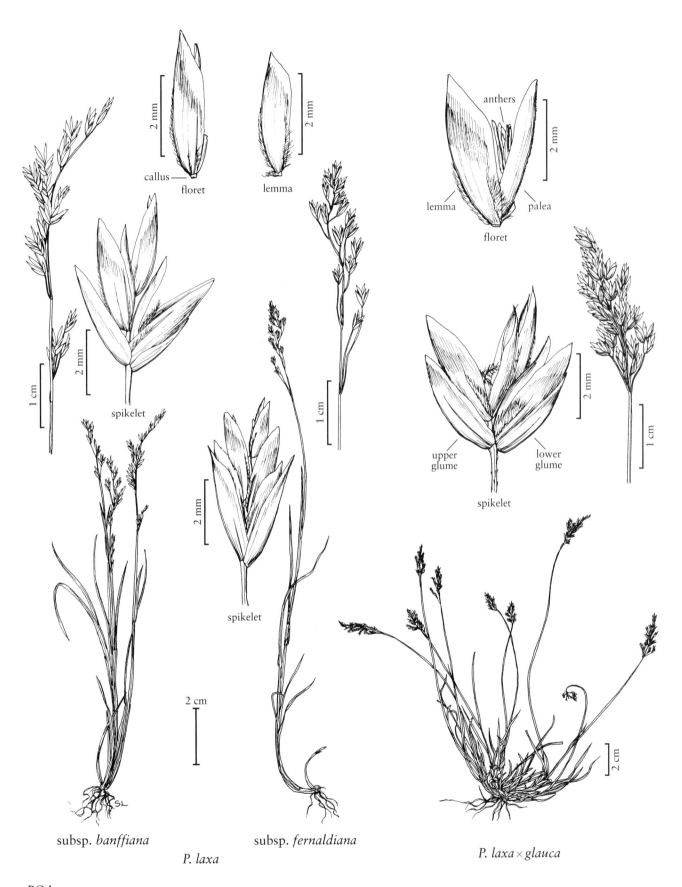

callus — floret

lemma

anthers

lemma — palea

floret

spikelet

2 mm

2 mm

2 mm

2 mm

2 mm

1 cm

1 cm

1 cm

upper glume   lower glume

spikelet

spikelet

2 cm

2 cm

subsp. *banffiana*

subsp. *fernaldiana*

P. laxa

P. laxa × *glauca*

fairly straight, usually sparsely scabrous, infrequently smooth. **Glumes** lanceolate to broadly lanceolate; **lower glumes** 3-veined; **calluses** glabrous; **lemmas** with keels short-villous for at least ½ their length, usually the lateral veins on at least 1 side of some florets sparsely softly puberulent, infrequently all the lateral veins glabrous; **anthers** 0.8–1.1 mm. $2n = 84$.

Poa laxa subsp. *banffiana* grows primarily in mesic alpine locations of the Rocky Mountains in Canada and the United States. It is sometimes difficult to distinguish from *P. glauca* (p. 576).

Poa laxa subsp. **fernaldiana** (Nannf.) Hyl. [p. 571]
MOUNT WASHINGTON BLUEGRASS

Basal branching mainly intravaginal. **Blades** very thin. **Panicles** 2–8 cm, lax, loosely contracted, sparse, with 1–3 branches per node; **branches** flexuous, smooth or infrequently very sparsely scabrous. **Glumes** lanceolate; **calluses** of at least some proximal florets sparsely webbed, rarely all glabrous; **lemmas** with keels short-villous for at least ½ their length, lateral veins glabrous; **anthers** (0.6)0.8–1.1(1.3) mm. $2n = 42$.

Poa laxa subsp. *fernaldiana* is native from Newfoundland south to New England.

51. Poa laxa × glauca [p. 571]

Plants perennial; not or only slightly glaucous; densely tufted, not stoloniferous, not rhizomatous. **Basal branching** mixed intra- and extravaginal. **Culms** 6–18 cm, with 0(1) exserted nodes, upper node in the lower ⅓ of the culms. **Sheaths** closed for ¹⁄₁₀–⅕ their length, terete, smooth or very sparsely scabrous, distal sheath lengths 0.8–1.8 times blade lengths; **ligules** 1.25–2.5 mm, smooth, apices obtuse, often lacerate; **blades** thin, sparsely scabrous adaxially, flag leaf blades 1.6–3.8 cm. **Panicles** 1.5–3.5 cm, slightly lax, ovoid, contracted to loosely contracted, dense to moderately dense, with 2–6 branches per node; **branches** steeply ascending, fairly straight, sulcate or angled, smooth or infrequently the angles sparsely scabrous, not glaucous. **Spikelets** laterally compressed; **florets** 2–5; **rachilla internodes** smooth, glabrous, lower internodes 0.8–1 mm. **Glumes** equal, broadly lanceolate, thin; **lower glumes** 0.75–1.05 mm wide, 3-veined; **upper glumes** 3.7–4.7 mm long, 0.9–1.3 mm wide, lengths 3.7–4.1 times widths; **calluses** all glabrous, or some proximal florets within a spikelet sparsely webbed; **lemmas** 3.7–4.5 mm, broadly lanceolate, distinctly keeled, thin, keels and marginal veins short- to long-villous, hairs extending ⅓–½ the keel length, lateral veins usually glabrous, or infrequently sparsely softly puberulent, intercostal regions glabrous; **palea keels** finely scabrous; **anthers** 0.8–1.2 mm, poorly formed, sacs not fully maturing, not dehiscing, about 0.1 mm in diameter. $2n = $ ca. 65, 70.

Poa laxa × *glauca* is an eastern low arctic entity which has passed under the name *P. flexuosa* Sm., *P. laxa* subsp. *flexuosa* (Sm.) Hyl., and, more recently, *P. laxiuscula* (Blytt) Lange. It has also been confused with *P. glauca* (p. 576). It can be distinguished from *P. laxa* (see previous) by its more open sheaths and poorly developed, indehiscent anthers. It differs from *P. glauca* in its broad, thin glumes and lemmas; compact panicles; smooth or nearly smooth, non-glaucous branches; and poorly developed, indehiscent anthers. It also grows in wetter habitats than *P. glauca*, often around seeps. Its chloroplast DNA is more like that of the American *P. laxa* subsp. *fernaldiana* than that of the European subspp. *flexuosa* and *laxa* or of *P. glauca*.

52. Poa paludigena Fernald & Wiegand [p. 575]
EASTERN BOG BLUEGRASS

Plants perennial; usually pale green; loosely tufted, slender, usually neither stoloniferous nor rhizomatous, occasionally with short, slender rhizomes. **Basal branching** mostly extravaginal. **Culms** 10–55 cm, very slender, weak. **Sheaths** closed for ¼–⅗ their length, terete, smooth or sparsely scabrous, margins not ciliate; **ligules** 0.5–2 mm, smooth or sparsely scabrous, truncate; **blades** 0.8–2 mm wide, flat, thin, soft, apices narrowly prow-shaped. **Panicles** 3–8(12) cm, lax, open, sparse; **nodes** with 1–2(3) branches; **branches** (2)3–7 cm, spreading to reflexed, capillary, angled, angles scabrous. **Spikelets** 3.2–5.2 mm, laterally compressed, broadly lanceolate to ovate; **florets** 2–3(5); **rachilla internodes** smooth, glabrous. **Glumes** narrowly lanceolate to lanceolate, thin, distinctly keeled, keels scabrous; **lower glumes** 1–3-veined; **upper glumes** shorter than or subequal to the adjacent lemmas; **calluses** sparsely webbed; **lemmas** 2.5–4 mm, lanceolate, green, distinctly keeled, keels and marginal veins short-villous, extending ⅔–⅘ the keel length, lateral veins fairly prominent, intercostal regions glabrous, apices obtuse to broadly acute, white, faintly bronze-colored or not; **palea keels** scabrous; **anthers** 0.2–0.8 mm. $2n = $ unknown.

Poa paludigena is an inconspicuous species restricted to the northeastern United States. It grows in shady bogs and fens, often underneath other plants. *Poa trivialis* (p. 568) sometimes grows with *P. paludigena*; the former has distinctly longer ligules and anthers. Plants from the middle Appalachian Mountains are sometimes confused with *P. sylvestris* (p. 512). *Poa paludigena* is generally shorter and more slender, has shorter panicles with only 1–2 branches per node, is glabrous between the lemma veins and on the palea keels, has shorter anthers, and grows in colder habitats.

53. Poa leptocoma Trin. [p. 575]

WESTERN BOG BLUEGRASS

Plants perennial; dark to light green, often anthocyanic in part; loosely tufted, usually neither stoloniferous nor rhizomatous, occasionally with short, slender rhizomes. **Basal branching** mostly extravaginal. **Culms** 15–100 cm, slender to middling. **Sheaths** closed for ¹/₄–³/₅ their length, terete, smooth or sparsely scabrous, margins not ciliate; **ligules** 1.5–4(6) mm, smooth to sparsely scabrous, obtuse to acute; **blades** 1–4 mm wide, flat, thin, lax, soft, apices narrowly prow-shaped. **Panicles** 5–15 cm, lax, open, sparse; **nodes** with 1–3(5) branches; **branches** (2)3–8 cm, spreading to reflexed, capillary, usually angled, infrequently only sulcate or subterete, angles usually moderately densely scabrous, sometimes only sparsely so, with (3)4–15 spikelets. **Spikelets** 4–8 mm, lanceolate or narrowly lanceolate, green or partly purple to dark purple; **florets** 2–5; **rachilla internodes** smooth, glabrous. **Glumes** subulate to lanceolate, thin, distinctly keeled, keels usually scabrous; **lower glumes** subulate to narrowly lanceolate, 1-veined; **upper glumes** distinctly shorter than to nearly equaling the lowest lemmas; **calluses** sparsely webbed; **lemmas** 3–4 mm, lanceolate, often partly purple, distinctly keeled, thin, smooth, or with sparse hooks apically, keels and marginal veins softly puberulent to long-villous, hairs extending ¹/₄–²/₃ the keel length, sometimes sparse, lateral veins and intercostal regions glabrous, margins glabrous, infolded, apices sharply acute to acuminate, usually bronze-colored; **palea keels** nearly smooth, scabrous, or pectinately ciliate; **anthers** 0.2–1.1 mm. 2n = 42.

Poa leptocoma grows around lakes and ponds and along streams, in subalpine and alpine to low arctic habitats, in western North America from Alaska to California and New Mexico, and on the Kamchatka Peninsula, Russia. It often grows with or near *P. reflexa* (p. 538), from which it differs in its more scabrous panicle branches, shorter anthers, glabrous or pectinately ciliate palea keels, and preference for wet sites. The two also differ in their ploidy level, *P. leptocoma* being hexaploid, and *P. reflexa* tetraploid. It differs from *P. paucispicula* (p. 538) in its more scabrous panicle branches, narrower glumes and lemmas, and its more sparsely hairy calluses and lemmas. Although its chloroplast haplotype is similar to that of species in sect. *Oreinos*, its ITS sequence is distinct and resembles that of *P. paucispicula*.

Poa sect. Stenopoa Dumort.

Plants perennial; densely to loosely tufted, not rhizomatous, infrequently stoloniferous. **Basal branching** all or mostly extravaginal. **Culms** 5–120 cm, terete or slightly compressed; **nodes** terete or slightly compressed. **Sheaths** closed for ¹/₁₀–¹/₅ their length, terete or slightly compressed, smooth or sparsely scabrous, distal sheaths shorter or longer than their blades; **ligules** 0.2–6 mm, usually scabrous, sometimes smooth, apices truncate or obtuse and usually ciliolate, or acute and not ciliolate; **blades** 0.8–8 mm wide, mostly flat, sometimes folded, moderately thin, abruptly ascending to spreading, lax or straight, margins scabrous, adaxial surfaces usually scabrous over the veins, apices narrowly prow-shaped. **Panicles** 1–30(41) cm, erect or lax, open, narrowly lanceoloid to ovoid, sparse to moderately congested; **nodes** with 2–9 branches; **branches** 0.4–15 cm, erect to reflexed, angled, angles scabrous. **Spikelets** 3–8(9) mm, lengths 2–3.5 times widths, narrowly lanceolate to narrowly ovate, laterally compressed, rarely bulbiferous; **florets** (1)2–5, bisexual, rarely bulb-forming; **rachilla internodes** mostly shorter than 1 mm, frequently muriculate or scabrous or pubescent. **Glumes** subulate to broadly lanceolate, distinctly keeled, keels smooth or sparsely scabrous; **lower glumes** 3-veined; **calluses** terete or slightly laterally compressed, glabrous or dorsally webbed; **lemmas** 2–4 mm, narrowly to broadly lanceolate, distinctly keeled, coriaceous-membranous, usually finely muriculate, keels and marginal veins long- to short-villous, intercostal regions glabrous or softly puberulent to short-villous, lateral veins obscure, apices usually partially bronze-colored; **palea keels** scabrous, sometimes softly puberulent at midlength, intercostal regions glabrous or puberulent; **anthers** 3, 0.8–2.5 mm.

Poa sect. *Stenopoa* includes 30 species. Most are Eurasian; three are native in, and one is restricted to, the *Flora* region. The North American species are cespitose or weakly stoloniferous, and have sheaths open for much of their length, scabrous panicle branches, and faint lateral lemma veins. The new shoots for the following year are initiated late in the growing season, after flowering and fruiting; vegetative and flowering shoots are usually not present at the same time.

54. Poa palustris L. [p. 575]
FOWL BLUEGRASS

Plants perennial; usually loosely, sometimes densely, tufted, frequently stoloniferous. **Basal branching** extravaginal or mixed extra- and intravaginal. **Culms** 25–120 cm, erect or the bases decumbent, sometimes branching above the base, terete or weakly compressed, scabrous below the panicle; **nodes** terete or slightly compressed, proximal nodes often slightly swollen, uppermost node at or above (¹/₃)¹/₂ the culm length. **Sheaths** closed for ¹/₁₀–¹/₅ their length, slightly compressed, glabrous or sparsely retrorsely scabrous, bases of basal sheaths glabrous, distal sheath lengths 0.7–2.2 times blade lengths; **ligules** (1)1.5–6 mm, smooth or sparsely to moderately scabrous, apices obtuse to acute, frequently lacerate, usually minutely ciliolate; **blades** 1.5–8 mm wide, flat, usually several per culm, steeply ascending or spreading to 80°, often lax distally, apices narrowly prow-shaped. **Panicles** (9)13–30(41) cm, lengths ¹/₃–¹/₂ times widths at maturity, lax, eventually open, sparsely to moderately congested, with 25–100+ spikelets; **nodes** with 2–9 branches; **branches** 4–15 cm, ³/₁₀–¹/₂ the panicle length, initially erect, eventually widely spreading to slightly reflexed, fairly straight, slender, angles densely scabrous. **Spikelets** 3–5 mm, lengths 3–3.5 times widths, narrowly to broadly lanceolate, laterally compressed; **florets** (1)2–5; **rachilla internodes** mostly shorter than 1 mm, usually muriculate, sometimes smooth, rarely sparsely hispidulous. **Glumes** subulate to lanceolate, distinctly keeled, keels smooth or sparsely scabrous; **lower glumes** with lengths 6.4–10 times widths, 3-veined, long-tapered to a slender point; **calluses** sparsely to moderately densely webbed, hairs (¹/₂)²/₃+ the lemma length; **lemmas** 2–3 mm, narrowly lanceolate to lanceolate, distinctly keeled, keels straight or gradually arched, usually abruptly inwardly arched at the junction of the scarious apices, keels and marginal veins short-villous, lateral veins obscure, intercostal regions muriculate, glabrous, margins distinctly inrolled, glabrous, apices obtuse or acute, usually partially bronze-colored, frequently incurved and blunt with a short, hyaline margin; **palea keels** scabrous, intercostal regions glabrous; **anthers** 1.3–1.8 mm. 2*n* = 28, 30, 32, 35, 42, 56, 84.

Poa palustris is native to boreal regions of northern Eurasia and North America, and is widespread in cool-temperate and boreal riparian and upland areas. European plants have also been introduced to other parts of North America. Plants in the Pacific Northwest and the southern United States are usually regarded as introduced, but some populations may be native. *Poa palustris* is used for soil stabilization and waterfowl feed.

Poa palustris from drier woods and meadows tends to resemble *P. interior* (p. 576). The best features for recognizing it include its loose growth habit, more steeply ascending leaf blades, well-developed callus webs, narrowly hyaline lemma margins, and incurving lemma keels. It also has a tendency to branch at the nodes above the base.

55. Poa nemoralis L. [p. 577]
WOODLAND BLUEGRASS

Plants perennial; green or glaucous; densely tufted, not stoloniferous, not rhizomatous. **Basal branching** all or mostly extravaginal. **Culms** 30–80 cm, mostly erect, smooth below the panicles; **nodes** slightly compressed, 2–5 exserted, top node at ¹/₂–³/₄ the culm length. **Sheaths** closed for ¹/₁₀–¹/₅ their length, terete, bases of basal sheaths glabrous, distal sheath lengths 0.45–1 (1.1) times blade lengths; **ligules** 0.2–0.8(1) mm, sparsely to densely scabrous, apices truncate, minutely ciliolate; **blades** 0.8–3 mm wide, mostly flat, appressed, abruptly ascending to spreading, straight or somewhat lax, apices narrowly prow-shaped. **Panicles** 7–16(20) cm, lengths usually 2.5–4 times widths at maturity, usually erect, lax in shade forms, narrowly lanceoloid to ovoid, slightly to moderately congested; **nodes** with 2–5 branches; **branches** ascending to widely spreading, fairly straight, slender to moderately stout, angled, angles moderately to densely scabrous. **Spikelets** 3–8 mm, lengths 2.5–3.5 times widths, narrowly lanceolate to lanceolate, laterally compressed, usually not glaucous; **florets** (1)2–5; **rachilla internodes** usually

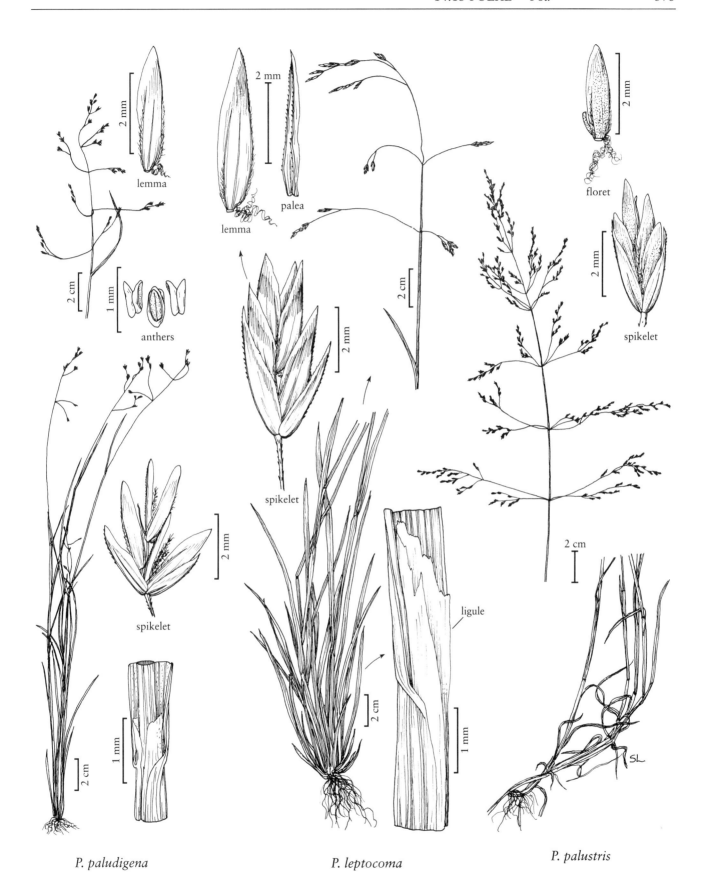

lemma

lemma

palea

floret

spikelet

2 cm

2 cm

anthers

spikelet

spikelet

spikelet

ligule

P. paludigena

P. leptocoma

P. palustris

SL

shorter than 1 mm, smooth, muriculate, or scabrous, usually puberulent, infrequently hispidulous or glabrous. **Glumes** subulate to narrowly lanceolate, distinctly keeled, keels smooth or sparsely scabrous, apices sharply acute to acuminate; **lower glumes** 3-veined, long-tapered to a slender point, lengths 6.4–11 times widths; **upper glumes** shorter than or subequal to the lowest lemmas; **calluses** webbed, hairs sparse, often short; **lemmas** 2.4–4 mm, proximal lemma widths less than $^1/_5$ times lengths, narrowly lanceolate to lanceolate, distinctly keeled, keels and marginal veins short-villous, lateral veins glabrous, obscure, intercostal regions smooth or muriculate, glabrous, margins glabrous, apices acute, usually partially bronze-colored; **palea keels** scabrous, intercostal regions glabrous; **anthers** 0.8–1.9 mm. $2n = 28, 35, 42, 48, 50, 56$.

Introduced from northern Eurasia, *Poa nemoralis* is established primarily at low elevations in deciduous and mixed conifer/deciduous forests. It is now common in southeastern Canada and the northeastern United States, and is spreading in the west. It can be distinguished from *P. glauca* (p. 576) and *P. interior* (see next) by its consistently short ligules, high top culm node, relatively long flag leaf blades, and narrow glumes and lemmas. It is usually hexaploid.

56. Poa interior Rydb. [p. 577]
INTERIOR BLUEGRASS

Plants perennial; green or less often glaucous; densely tufted, not stoloniferous, not rhizomatous. **Basal branching** all or mostly extravaginal. **Culms** 5–80 cm, usually slender, mostly erect or ascending, several to many arising together; **nodes** terete or slightly compressed, (0)1–2(3) exserted, top node usually at $^1/_3$–$^3/_5$ the culm length. **Sheaths** closed for $^1/_{10}$–$^1/_5$ their length, terete, bases of basal sheaths glabrous, distal sheath lengths (0.6)0.88–1.64 times blade lengths; **ligules** 0.5–1.5(3) mm, sparsely to densely scabrous, apices truncate to obtuse, ciliolate; **blades** 0.8–3 mm wide, mostly flat, thin, soft, appressed or abruptly ascending to spreading, straight or somewhat lax, apices narrowly prow-shaped. **Panicles** (1.5)3–15(17) cm, lengths generally 2.5–4 times widths at maturity, usually erect, lax in shade forms, narrowly lanceoloid to ovoid, sparsely to moderately congested; **nodes** with 2–5 branches; **branches** 0.4–8(9) cm long, $^1/_4$–$^1/_2$ the panicle length, ascending to widely spreading, fairly straight, slender to moderately stout, angled, angles moderately to densely scabrous. **Spikelets** 3–6 mm, lengths 2–3 times widths, lanceolate to narrowly ovate, laterally compressed, usually not glaucous; **florets** (1)2–3(5); **rachilla**

internodes usually shorter than 1 mm, smooth, muriculate, or scabrous, glabrous, hispidulous, or sparsely to densely puberulent, proximal internodes frequently curved. **Glumes** lanceolate to broadly lanceolate, distinctly keeled, keels smooth or sparsely scabrous; **lower glumes** 3-veined, long- or abruptly tapered to a slender point, lengths 4.5–6.3 times widths; **calluses** usually webbed, infrequently glabrous in depauperate alpine specimens, webs usually scant, less than $^1/_2$($^2/_3$) the lemma length, frequently minute; **lemmas** 2.4–4 mm, lanceolate, distinctly keeled, straight or gradually arched, not abruptly inwardly arched at the junction with the scarious apices, keels and marginal veins short-villous, hairs extending $^2/_3$–$^3/_4$ the keel length, lateral veins usually glabrous, rarely sparsely puberulent, obscure, intercostal regions smooth, sometimes weakly muriculate, glabrous, margins not or slightly inrolled, glabrous, apices acute, usually partially bronze-colored; **palea keels** scabrous, intercostal regions glabrous; **anthers** (1.1)1.3–2.5 mm. $2n = 28, 42, 56$.

Poa interior, a native species, grows from Alaska to western Quebec and New York, south to Arizona and New Mexico. It is restricted to the *Flora* region. It is fairly common from boreal forests to low alpine habitats of the Rocky Mountains. It grows in subxeric to mesic habitats, such as mossy rocks and scree, usually in forests. It is usually tetraploid.

In alpine habitats, *Poa interior* is often quite short, and often sympatric with *P. glauca* (see next). It is most reliably distinguished from *P. glauca* by lemmas that are glabrous between the marginal veins and keels or, rarely, sparsely puberulent on the lateral veins. It usually also differs from *P. glauca* subsp. *rupicola* in having at least a few hairs on its calluses. It can be distinguished from *P. nemoralis* (see previous) by its longer ligules, lower top culm node, and wider glumes and lemmas. It is sometimes difficult to distinguish from *P. palustris* (p. 574), but differs in having lemmas with wider hyaline margins and straight or gradually arched keels, a densely tufted habit, and scantly webbed calluses.

57. Poa glauca Vahl [p. 577]

Plants perennial; usually glaucous; densely tufted, not stoloniferous, not rhizomatous. **Basal branching** all or mostly extravaginal. **Culms** 5–40(80) cm, erect to spreading, straight, wiry, bases straight or slightly decumbent; **nodes** terete or slightly compressed, usually 0–1 exserted, top node at $^1/_{10}$–$^1/_3$ the culm length. **Sheaths** closed for $^1/_{10}$–$^1/_5$ their length, terete, bases of basal sheaths glabrous or sparsely minutely hairy, hairs

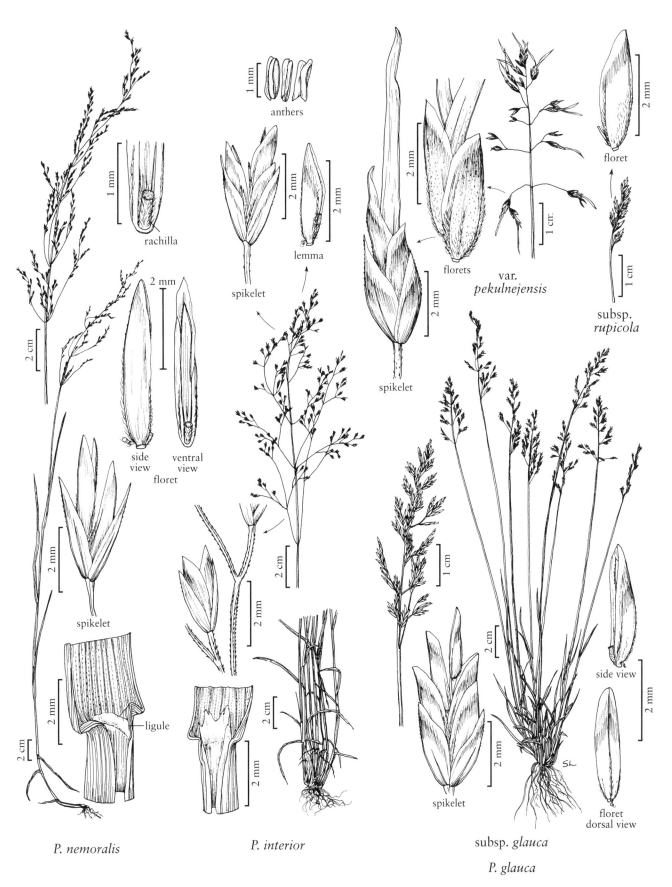

anthers

rachilla

lemma

spikelet

side view / ventral view floret

P. nemoralis

spikelet

ligule

spikelet

P. interior

florets

var. *pekulnejensis*

floret

subsp. *rupicola*

spikelet

spikelet

side view

floret dorsal view

subsp. *glauca*

P. glauca

POA

0.1–0.2 mm, distal sheath lengths 1.1–4 times blade lengths; **ligules** 1–4(5) mm, sparsely to densely scabrous, apices obtuse to acute, minutely ciliolate; **blades** 0.8–2.5 mm wide, flat or folded, thin, soft, appressed or abruptly ascending to spreading, straight, apices narrowly prow-shaped. **Panicles** 1–10(20) cm, lengths 3–5 times widths at maturity, rarely racemelike with branches of irregular length, erect, narrowly lanceoloid to ovoid, contracted to somewhat open, sparse, proximal internodes shorter than 1.5(4) cm; **nodes** with 2–3(5) branches; **branches** erect, ascending or weakly spreading, fairly straight, short, stout, angled, angles moderately to densely scabrous, rarely only scabrous distally, glaucous; **pedicels** usually shorter than the spikelets. **Spikelets** 3–7(9) mm, lengths 2–3 times widths, laterally compressed, rarely bulbiferous, usually glaucous; **florets** 2–5, rarely bulb-forming; **rachilla internodes** to 1.2 mm, smooth, muriculate, or scabrous, glabrous or sparsely to densely hispidulous or puberulent. **Glumes** subequal, narrowly to broadly lanceolate, distinctly keeled, keels smooth or sparsely scabrous, apices acute; **lower glumes** 3-veined; **upper glumes** 2–3.8(5.2) mm, lengths usually more than 4.1 times widths, distinctly shorter to subequal to the lowest lemmas; **calluses** glabrous or webbed, webs from minute to more than ¹⁄₂ the lemma length; **lemmas** 2.5–4 mm, lanceolate to broadly lanceolate, distinctly keeled, keels and marginal veins short-villous, lateral veins obscure, usually sparsely softly puberulent to short-villous, intercostal regions smooth, sometimes weakly muriculate, glabrous or puberulent, margins glabrous, apices usually partially bronze-colored, obtuse or acute; **palea keels** scabrous, glabrous or softly puberulent at midlength, intercostal regions glabrous or softly puberulent; **anthers** (1)1.2–2.5 mm, mature sacs 0.2 mm wide, rarely aborted late in development. $2n = 34$, 42, 44, 47, 48, 49, 50, 56, 56, 57, 58, 60, 63, 64, 65, 70, 75, 78, ca. 100.

Poa glauca is a common, highly variable, circumboreal, boreal forest to alpine and high arctic species. It grows from Alaska to Greenland, south to California and New Mexico in the west, and through Canada and the northeastern United States in the east. It also grows at scattered locations in Patagonia. It generally favors dry habitats and tolerates disturbance well. It can be distinguished from *P. nemoralis* (p. 574) and *P. interior* (see previous) by its longer ligules, lower top culm node, and wider glumes and lemmas. It can be difficult to distinguish from *P. laxa* subsp. *banffiana* (p. 570). *Poa glauca* is often confused in herbaria with *P. abbreviata* subsp. *pattersonii* (p. 582). It differs in having primarily extravaginal branching and, usually, longer anthers. It hybridizes with *P. laxa*, forming *P. laxa* × *glauca* (p. 572). It is also known to hybridize with *P. hartzii* (p. 589), and is suspected to hybridize

with *P. arctica* (p. 529) and *P. secunda* (p. 586). It is highly polyploid, and presumed to be highly apomictic.

1. All or some spikelets bulbiferous . . . . . var. *pekulnejensis*
1. Spikelets not bulbiferous.
   2. Calluses usually webbed, sometimes glabrous; lemmas glabrous or hairy between the veins . . . . subsp. *glauca*
   2. Calluses glabrous; lemmas hairy between the veins . . . . subsp. *rupicola*

Poa glauca Vahl subsp. **glauca** [p. 577]
GLAUCOUS BLUEGRASS

Culms 10–40(80) cm. **Panicles** 3.5–10(20) cm. **Spikelets** not bulbiferous; **florets** normal. **Calluses** webbed or glabrous; **lemmas** usually with lateral veins short-villous to softly puberulent, intercostal regions glabrous or short-villous to softly puberulent. $2n = 34, 42, 44, 47,$ 48, ca. 49, 50, 56, 60, 63, 64, 65, 70, 75, 78.

Poa glauca subsp. *glauca* is the widespread and common subspecies in the Northern Hemisphere. It is also disjunct in South America. Plants with glabrous calluses are found only in the arctic. In the Rocky Mountains, *P. glauca* subsp. *glauca* often grows with subsp. *rupicola* and *P. interior* (see previous). It does not grow in California, and is uncommon in the Great Basin and southern Rocky Mountains. It is highly variable, especially in the Great Lakes region. It is often confused in herbaria with subsp. *rupicola*, but can sometimes be distinguished by its webbed calluses and lemmas that are glabrous between the veins. The name *P. glaucantha* Gaud. has sometimes been applied to plants of *P. glauca* from southeastern Canada and the northeastern United States.

Poa glauca var. **pekulnejensis** (Jurtzev & Tzvelev) Prob. [p. 577]
PEKULNEI BLUEGRASS

Plants glaucous; densely tufted. **Culms** 5–20 cm. **Sheaths** closed for ¹⁄₁₀–¹⁄₅ their length. **Spikelets** bulbiferous; **florets** bulb-forming; **anthers** aborted, about 0.8 mm or undeveloped. $2n = $ unknown.

Poa glauca var. *pekulnejensis* is known only from sporadic records in Alaska and the Russian Far East. It can be difficult to distinguish from *P. hartzii* subsp. *vrangelica* (p. 589), but differs in its tighter habit and thicker glumes.

Poa glauca subsp. **rupicola** (Nash) W.A. Weber [p. 577]
TIMBERLINE BLUEGRASS

Culms to 5–15 cm. **Panicles** 1–5 cm, usually narrowly lanceoloid. **Spikelets** not bulbiferous; **florets** normal. **Calluses** glabrous; **lemmas** at least sparsely puberulent on the intercostal regions. $2n = 48, 48–50, 56, 56–58,$ ca. 100.

Poa glauca subsp. *rupicola* is endemic to dry alpine areas of western North America. It is often confused in herbaria with subsp. *glauca* and *P. interior* (p. 576), but its calluses lack even a vestige of a web, and its lemmas have at least a few hairs between the lemma veins. It is often sympatric with both taxa outside of California. It is not common in the northern Rocky Mountains.

Poa sect. Tichopoa Asch. & Graebn.

Plants perennial; rhizomatous, usually with solitary shoots, sometimes loosely tufted. **Basal branching** all or mostly extravaginal. **Culms** 15–60 cm, distinctly compressed; **nodes** distinctly compressed. **Sheaths** closed for $^1/_{10}$–$^1/_5$ their length, distinctly compressed; **ligules** 1–3 mm, moderately to densely scabrous, margins ciliolate, apices obtuse; **blades** 1.5–4 mm wide, flat, margins scabrous, adaxial surfaces smooth or scabrous mainly over the veins, apices narrowly prow-shaped. **Panicles** 2–10 cm, lengths usually 3–6 times widths, erect, linear to ovoid, mostly with 1–3 branches per node; **branches** 0.5–3 cm, angled, angles scabrous. **Spikelets** (2.3)3.5–7 mm, laterally compressed, bisexual, not bulbiferous; **florets** 3–7. **Calluses** terete or slightly laterally compressed, glabrous or pubescent, webbed; **lemmas** coriaceous-membranous, usually finely muriculate, lateral veins obscure, apices usually partially bronze-colored; **anthers** 3, 1.3–1.8 mm.

Poa sect. *Tichopoa* has two species, both of which are native to Europe. They are similar to species of *Poa* sect. *Stenopoa*, differing in having strongly compressed culms and nodes, and in being rhizomatous.

58. Poa compressa L. [p. 581]

CANADA BLUEGRASS

Plants perennial; usually with solitary shoots, sometimes loosely tufted, extensively rhizomatous. **Culms** 15–60 cm, wiry, bases usually geniculate, strongly compressed; **nodes** strongly compressed, some proximal nodes usually exserted. **Sheaths** closed for $^1/_{10}$–$^1/_5$ their length, distinctly compressed, bases of basal sheaths glabrous; **ligules** 1–3 mm, moderately to densely scabrous, ciliolate, apices obtuse; **blades** 1.5–4 mm wide, flat, cauline blades subequal. **Panicles** 2–10 cm, generally $^1/_6$–$^1/_3$ as wide as long, erect, linear, lanceoloid to ovoid, often interrupted, sparse to congested, with 15–80 spikelets and mostly with 1–3 branches per node; **branches** 0.5–3 cm, erect to ascending, or infrequently spreading, angles densely scabrous, at least in part, with 1–15 spikelets. **Spikelets** (2.3)3.5–7 mm, laterally compressed; **florets** 3–7; **rachilla internodes** usually shorter than 1 mm, smooth to muriculate. **Glumes** distinctly keeled; **lower glumes** 3-veined; **calluses** usually webbed, sometimes glabrous; **lemmas** 2.3–3.5 mm, lanceolate, distinctly keeled, keels and marginal veins short-villous, intercostal regions glabrous, lateral veins obscure, margins glabrous, apices acute; **paleas** scabrous over the keels; **anthers** 1.3–1.8 mm. $2n = 35, 42, 49, 50, 56, 84$.

Poa compressa is common in much of the *Flora* region. It is sometimes considered to be native, but this seems doubtful. It is rare and thought to be introduced in Siberia and only local in the Russian Far East, but is common in Europe. In the *Flora* region, it is often seeded for soil stabilization, and has frequently escaped. It grows mainly in riparian areas, wet meadows, and disturbed ground. Its distinctly compressed nodes and culms, exserted lower culm nodes, rhizomatous growth habit, and scabrous panicle branches make it easily identifiable.

Poa sect. Abbreviatae Nannf. *ex* Tzvelev

Plants perennial; densely tufted, not stoloniferous, not rhizomatous. **Basal branching** mainly intravaginal. **Culms** usually shorter than 25(30) cm, slender, terete; **nodes** terete. **Leaves** mostly basal; **sheaths** closed for $^1/_{10}$–$^1/_4$($^1/_3$) their length, terete; **ligules** 0.4–5.5 mm, milky white to hyaline, smooth or scabrous, apices truncate to acute, glabrous; **blades** 0.5–2 mm wide, flat, folded, or involute, thin to moderately thick, soft or moderately firm, apices narrowly prow-shaped. **Panicles** 1–7 cm, erect, usually contracted, sometimes open; **nodes** with 1–3 branches; **branches** 0.5–1.5(5) cm, usually erect to steeply ascending, sometimes ascending to spreading, sulcate to angled, smooth or the angles sparsely to densely scabrous. **Spikelets** 3–7 mm, laterally compressed, rarely bulbiferous; **florets** 2–5, usually bisexual, sometimes with vestigial anthers or anthers that abort late in the growing season, rarely bulb-forming; **rachilla internodes** usually glabrous, infrequently sparsely hispidulous. **Glumes** usually subequal to or slightly longer than the adjacent lemmas, distinctly keeled, keels smooth or sparsely scabrous; **lower glumes** (1)3-veined; **calluses** terete or slightly laterally compressed, glabrous or dorsally webbed; **lemmas** 2–5.8 mm, lanceolate to broadly lanceolate, distinctly keeled, thin, glabrous or the keels and marginal veins softly puberulent to long-villous, intercostal regions glabrous or softly puberulent to short-villous, obscurely 5-veined; **palea keels** scabrous, glabrous or softly puberulent to short-villous at midlength; **anthers** 3, 0.2–1.3(1.8) mm, rarely vestigial (0.1–0.2 mm) or aborted late in development.

Poa sect. *Abbreviatae* includes five North American species, two of which also grow in arctic regions of the Eastern Hemisphere. The species are principally high alpine to high arctic. Two of the species are known or reputed to be diploid.

59. Poa lettermanii Vasey [p. 581]
LETTERMAN'S BLUEGRASS

Plants perennial; not glaucous; densely tufted, not stoloniferous, not rhizomatous. **Basal branching** all or mainly intravaginal. **Culms** 1–12 cm, slender. **Sheaths** closed for $^1/_6$–$^1/_4$ their length, terete; **ligules** 1–3 mm, milky white to hyaline, smooth; **blades** 0.5–2 mm wide, flat or folded, or slightly inrolled, thin, without papillae (at 100×), apices narrowly prow-shaped. **Panicles** 1–3 cm, erect, contracted, usually exserted from the sheaths; **branches** to 1.5 cm, erect to steeply ascending, slender, sulcate or angled, smooth or the angles sparsely scabrous; **pedicels** shorter than the spikelets. **Spikelets** 3–4 mm, laterally compressed, green or anthocyanic; **florets** 2–3; **rachilla internodes** shorter than 1 mm, smooth. **Glumes** usually equaling or exceeding the lowest lemmas, sometimes also equaling or exceeding the upper florets, lanceolate to broadly lanceolate, distinctly keeled, keels smooth; **lower glumes** 3-veined; **calluses** glabrous; **lemmas** 2.5–3 mm, lanceolate, distinctly keeled, thin, usually glabrous, keels and marginal veins rarely sparsely puberulent proximally, apices acute; **palea keels** scabrous; **anthers** 0.2–0.8 mm. $2n = 14$.

Poa lettermanii grows on rocky slopes of the highest peaks and ridges in the alpine zone, from northern British Columbia to western Alberta and south to California and Colorado, usually in the shelter of rocks or on mesic to wet, frost-scarred slopes. It is one of only three known diploid *Poa* species native to the Western Hemisphere. Its glabrous calluses and lemmas usually distinguish it from *P. abbreviata* (p. 582); it also differs in having flat or folded leaf blades, and shorter spikelets with glumes that are longer than the adjacent florets. *Poa montevansii* E.H. Kelso is tentatively included here, although its slightly longer lemmas that slightly exceed the glumes suggest that it may represent rare, glabrous forms of *P. abbreviata*.

60. Poa pseudoabbreviata Roshev. [p. 581]
SHORT-FLOWERED BLUEGRASS

Plants perennial; glaucous; densely tufted, delicate, not stoloniferous, not rhizomatous. **Basal branching** all or mainly intravaginal. **Culms** 4–20(30) cm, to 18 cm in bulbiferous plants, slender. **Sheaths** closed for $^1/_{10}$–$^1/_6$ their length, terete; **ligules** 1–4 mm, smooth; **blades** 0.5–1.5(2) mm wide, flat or folded, thin, soft, apices narrowly prow-shaped. **Panicles** 2–7 cm, 2–4 cm in bulbiferous plants, widths equal to lengths, erect, broadly rhomboidal to pyramidal, open, exserted from the sheaths, sparse; **branches** 1.5–5 cm, ascending to spreading, divaricate, slender, sulcate or angled, angles sparsely to moderately scabrous; **pedicels** often longer than the spikelets. **Spikelets** 3–5 mm, laterally compressed, rarely bulbiferous, usually strongly anthocyanic, glaucous or not; **florets** 2–4, rarely bulb-forming; **rachilla internodes** shorter than 1 mm, smooth to scabrous. **Glumes** distinctly keeled; **lower glumes**

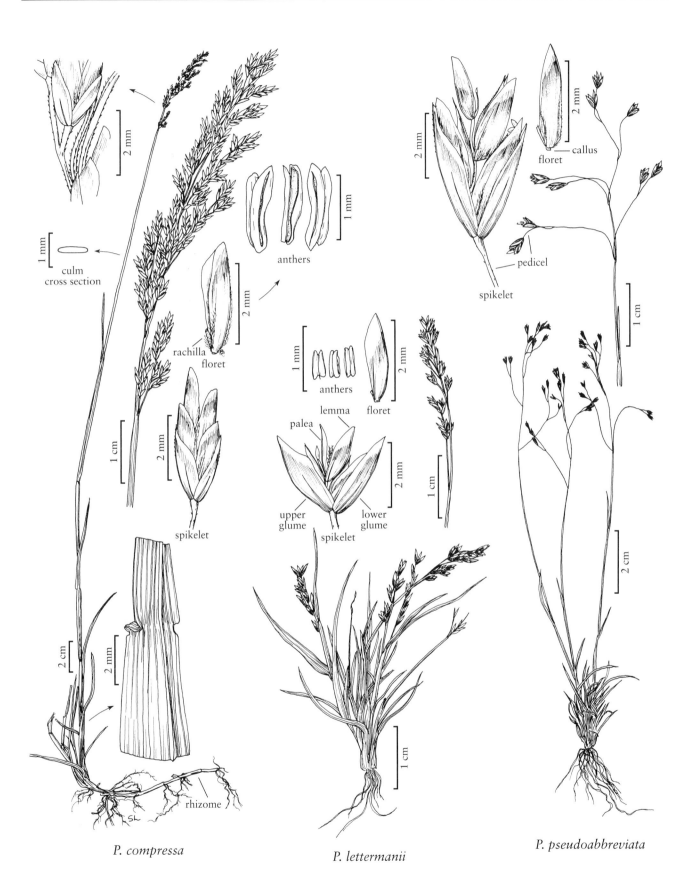

culm
cross section

anthers

rachilla
floret

spikelet

2 cm

2 mm

rhizome

P. compressa

anthers

lemma

palea

floret

upper
glume

lower
glume

spikelet

P. lettermanii

callus

floret

spikelet

pedicel

P. pseudoabbreviata

POA

subequal to equaling the lowest lemmas, 3-veined, **upper glumes** frequently longer than the lowest lemmas; **calluses** glabrous; **lemmas** 2–3 mm, lanceolate, distinctly keeled, thin, keels and marginal veins crisply puberulent, rarely glabrous, intercostal regions glabrous, apices acute; **paleas** scabrous over the keels; **anthers** 0.2–0.7 mm. $2n = 14$.

Poa pseudoabbreviata is a low arctic to subarctic and alpine species of Alaska, northwestern Canada, Siberia, and the Russian Far East. It grows mainly on frost scars, rocky slopes, and ridges, often on open ground. It is one of only three diploid species native to the Western Hemisphere.

Poa pseudoabbreviata is easily distinguished from all other alpine and arctic species of *Poa* by its spreading, capillary branches, long pedicels, short stature, small spikelets, and glabrous calluses. Bulbiferous plants are known only from the Brooks Range, Alaska.

61. Poa abbreviata R. Br. [p. 583]

Plants perennial; not or scarcely glaucous; densely tufted, not stoloniferous, not rhizomatous. **Basal branching** all or mainly intravaginal. **Culms** 5–15(20) cm, slender, leafless above the basal tuft. **Sheaths** closed for $^1/_{10}$–$^1/_4$ their length, terete; **ligules** 0.4–5.5 mm, milky white to hyaline, smooth or scabrous, apices truncate to acute; **blades** 0.8–1.5(2) mm wide, involute, moderately thick, soft, apices narrowly prow-shaped. **Panicles** 1.5–5 cm, erect, lanceoloid to ovoid, contracted, congested; **nodes** with 1–3 branches; **branches** to 1.5 cm, erect, slender, terete, sulcate or angled, smooth or the angles sparsely scabrous; **pedicels** usually shorter than the spikelets. **Spikelets** 4–6.5 mm, laterally compressed, rarely bulbiferous, frequently strongly anthocyanic; **florets** 2–5, rarely bulb-forming; **rachilla internodes** usually shorter than 1 mm, smooth or scabrous. **Glumes** subequal to slightly longer than the adjacent lemmas, lanceolate to broadly lanceolate, distinctly keeled, keels smooth; **lower glumes** (1)3-veined, lateral veins often faint and short; **upper glumes** exceeding or exceeded by the upper florets; **calluses** glabrous or webbed; **lemmas** 3–4.6 mm, lanceolate to broadly lanceolate, distinctly keeled, thin, keels and marginal veins usually short- to long-villous, hairs extending along $^3/_4$–$^5/_6$ of the keel, infrequently glabrous, intercostal regions glabrous or softly puberulent to short-villous, apices acute; **palea keels** scabrous, often short-villous to softly puberulent at midlength, sometimes glabrous; **anthers** 0.2–1.2(1.8) mm. $2n = 42$.

Poa abbreviata is an alpine and circumarctic species which has two subspecies in the western cordilleras, and

one in the high arctic. It grows mainly on frost scars and mesic rocky slopes, usually on open ground. In rare cases where the lemmas and calluses of *P. abbreviata* are glabrous, it can be confused with *P. lettermanii* (p. 580), but that species has shorter spikelets and glumes that are longer than the adjacent florets.

1. Lemmas glabrous; calluses webbed ...... subsp. *marshii*
1. Lemmas usually with hairs over the veins; calluses glabrous or webbed, rarely both the lemmas and calluses glabrous.
  2. Anthers 0.2–0.8 mm long; lemma intercostal regions hairy .... subsp. *abbreviata*
  2. Anthers 0.6–1.2(1.8) mm long; lemma intercostal regions glabrous or hairy .... subsp. *pattersonii*

Poa abbreviata R. Br. subsp. abbreviata [p. 583]
DWARF BLUEGRASS

Ligules 0.4–1.7(3) mm, apices truncate to acute. **Spikelets** not bulbiferous; **florets** normal. **Calluses** glabrous, rarely webbed; **lemmas** with keels and marginal veins short- to long-villous, hairs extending along $^5/_6$ of the keel, intercostal regions softly puberulent to short-villous; **anthers** 0.2–0.8 mm. $2n = 42$.

Poa abbreviata subsp. *abbreviata* is a common high arctic subspecies that also grows at scattered locations in the Brooks Range, Alaska, and in the Rocky Mountains of the Northwest Territories. It has a circumpolar distribution, but is not common in the Eurasian continental arctic.

Poa abbreviata subsp. marshii Soreng [p. 583]
MARSH'S BLUEGRASS

Ligules 1–3 mm, smooth, apices obtuse to acute. **Spikelets** not bulbiferous; **florets** normal. **Calluses** webbed; **lemmas** lanceolate, glabrous; **palea keels** glabrous; **anthers** 0.6–1.2 mm. $2n = $ unknown.

Poa abbreviata subsp. *marshii* is rather uncommon. It is known from scattered alpine peaks across the interior western United States: from the White Mountains of California, the Schell Creek Range of Nevada, the southern Rockies of Idaho, the Little Belt Mountains of Montana, and the Big Horn Mountains of Wyoming, mostly where the other subspecies are absent.

Poa abbreviata subsp. pattersonii (Vasey) Á. Löve, D. Löve & B.M. Kapoor [p. 583]
PATTERSON'S BLUEGRASS

Ligules 0.8–5.5 mm, smooth or scabrous, apices obtuse to acute. **Spikelets** rarely bulbiferous; **florets** rarely bulb-forming. **Calluses** usually webbed, rarely glabrous; **lemmas** long-villous along $^3/_4$ of the keel and the marginal veins, rarely glabrous, but then the calluses

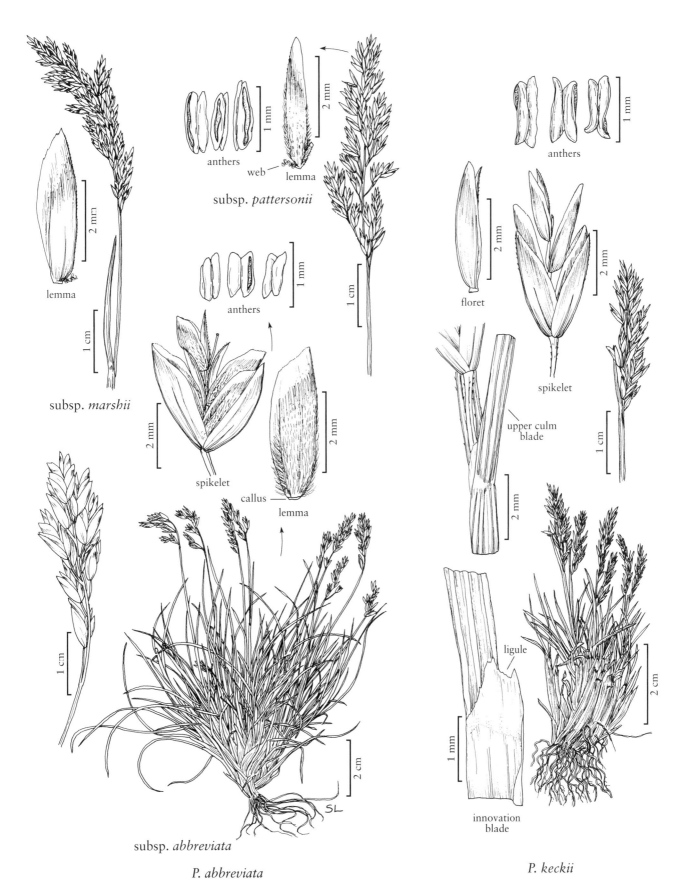

subsp. *pattersonii*

anthers

web   lemma

lemma

subsp. *marshii*

anthers

spikelet

callus   lemma

anthers

floret

spikelet

upper culm blade

ligule

innovation blade

subsp. *abbreviata*

P. abbreviata

P. keckii

also glabrous, intercostal regions glabrous or softly puberulent; **anthers** 0.6–1.2(1.8) mm, rarely vestigial. $2n = 42$.

Poa abbreviata subsp. *pattersonii* is an alpine taxon that extends from the Brooks Range, Alaska, to the Sierra Nevada, California, where it is rare, and through the Rocky Mountains to southern Colorado. It also grows in the Russian Far East. It is often confused in herbaria with *P. glauca* (p. 576), but differs in having predominantly intravaginal branching, an abundance of vegetative shoots, and usually shorter anthers. Plants from northern British Columbia to Alaska and Russia have been called *P. abbreviata* subsp. *jordalii* (A.E. Porsild) Hultén. They have webbed calluses, very short (occasionally nonexistent) lemma hairs, panicles often exserted well above the basal tuft of leaves, and particularly slender culms.

62. Poa keckii Soreng [p. 583]
KECK'S BLUEGRASS

Plants perennial; not glaucous; densely tufted, not stoloniferous, not rhizomatous. **Basal branching** all or mainly intravaginal. **Culms** 2–10(18) cm, erect to spreading; **nodes** terete, none exserted. **Sheaths** closed for $^1/_{10}$–$^1/_5$ their length, terete, smooth, glabrous, bases of basal sheaths glabrous, distal sheath lengths 1.5–7 times blade lengths; **collars** smooth, glabrous; **ligules** 1–3 mm, milky white, smooth or sparsely scabrous, apices obtuse to acute, ligules of upper innovation leaves shorter than 3 mm; **innovation blades** similar to the cauline blades; **cauline blades** 1–3.5(4.5) cm long, 0.9–1.8 mm wide, folded, moderately thick, soft, smooth, glabrous, adaxial surfaces infrequently sparsely scabrous, usually with papillae on the long cells (at 100×), apices narrowly prow-shaped, flag leaf blades folded, 1–1.8 mm wide, abaxial surfaces with 7–15 closely spaced, slightly protruding ribs. **Panicles** 1–4(6) cm, erect, ovoid to lanceoloid, contracted, congested, with 9–40 spikelets; **nodes** with 1–3 branches; **branches** 0.5–1.5 cm, erect, fairly straight, sulcate or angled, angles sparsely to densely scabrous, with 1–7 spikelets; **pedicels** shorter than the spikelets. **Spikelets** 3.5–6 mm long, lengths to 3.5(3.8) times widths, lanceolate, laterally compressed, fairly strongly anthocyanic, not glaucous; **florets** 2–3; **rachilla internodes** terete, to 1.5 mm, smooth, sometimes sparsely hispidulous. **Glumes** lanceolate, smooth, distinctly keeled, keels sparsely scabrous; **lower glumes** shorter than to equaling the lowest lemmas, 3-veined; **upper glumes** frequently exceeding the lowest lemmas, exceeded by the upper lemmas; **calluses** glabrous; **lemmas** 3–4.9 mm,

lanceolate, distinctly keeled, thin, smooth or finely scabrous, glabrous or the keels and marginal veins sparsely puberulent proximally, lateral veins obscure, margins glabrous, apices acute; **palea keels** scabrous; **anthers** 0.6–1.3(1.8) mm. $2n =$ unknown.

Poa keckii is endemic to high alpine frost scars and ledges, usually on open ground, in the Sierra Nevada and adjacent Sweetwater and White mountains of California. It is very similar to *Poa suksdorfii* (see next), but is consistently distinct in its details.

63. Poa suksdorfii (Beal) Vasey *ex* Piper [p. 587]
SUKSDORF'S BLUEGRASS

Plants perennial; not glaucous; densely tufted, not stoloniferous, not rhizomatous. **Basal branching** all or mainly intravaginal. **Culms** 7–25 cm. **Sheaths** closed for $^1/_7$–$^1/_4$($^1/_3$) their length, terete; **ligules** of cauline leaves 1–3 mm, milky white, usually densely scabrous, sometimes smooth, ligules of the upper innovation leaves 0.5–2.5 mm; **innovation blades** adaxially scabrous, hispidulous, or puberulent on and between the veins, lacking papillae on the long cells (at 100×); **cauline blades** folded to involute, moderately thick, soft or moderately firm, apices narrowly prow-shaped, flag leaf blades 1–2 mm wide, adaxial surfaces with 5–9 well-spaced ribs. **Panicles** 3–6 cm, erect, narrowly lanceoloid, contracted, moderately congested; **nodes** with 1–2 branches; **branches** to 1.5 cm, erect, slender, terete, sulcate or angled, smooth or the angles moderately scabrous; **pedicels** shorter than the spikelets. **Spikelets** 4.2–7 mm, laterally compressed, often strongly anthocyanic; **florets** 2–4; **rachilla internodes** 1–1.5 mm, smooth, sometimes sparsely hispidulous. **Glumes** lanceolate, distinctly keeled, keels smooth; **lower glumes** shorter than to equaling the lowest lemmas, 3-veined; **upper glumes** frequently exceeding the lowest lemmas, 3–5-veined, exceeded by the upper lemmas; **calluses** glabrous; **lemmas** 4.1–5.8 mm, narrowly lanceolate, distinctly keeled, thin, glabrous, apices acute; **palea keels** scabrous; **anthers** 0.8–1.2(1.7) mm, infrequently aborted late in development. $2n =$ unknown.

Poa suksdorfii is a high alpine species of open rocky ground in the Pacific Northwest. It used to be interpreted (Hitchcock 1951) as including California populations that are now placed in *Poa pringlei* (p. 564) or *P. keckii* (see previous). *Poa suksdorfii* has narrow panicles like *P. pringlei* and *P. curtifolia* (p. 589).

Poa sect. Secundae V.L. Marsh *ex* Soreng

Plants perennial; usually densely, infrequently loosely, tufted, rarely weakly rhizomatous or stoloniferous. **Basal branching** mixed intra- and extravaginal to completely intravaginal. **Culms** 10–120 cm, capillary to stout, terete or weakly compressed; **nodes** terete. **Sheaths** closed for $^{1}/_{10}$–$^{1}/_{3}$ their length, terete, smooth or scabrous, distal sheaths usually longer than their blades; **ligules** 0.5–7(10) mm, smooth or scabrous, apices truncate to acuminate; **blades** 0.4–3(5) mm wide, flat, folded, or involute, thin to moderately thick, soft and soon withering or moderately firm and persisting, smooth or scabrous mainly over the veins and margins, apices narrowly prow-shaped. **Panicles** 2–25(30) cm, erect or somewhat lax, narrowly lanceoloid to ovoid, usually contracted, sometimes open and pyramidal, sparse to congested, with 7–100(120) spikelets; **nodes** with 1–4(7) branches; **branches** 0.5–15 cm, erect to spreading, terete, sulcate or angled, smooth or the angles sparsely to densely scabrous, sometimes scabrous between the angles. **Spikelets** (4)4.5–10 mm, lengths 3–5 times widths, terete to weakly laterally compressed or distinctly compressed, sometimes bulbiferous; **florets** (2)3–5(10), usually normal and bisexual, sometimes bulb-forming. **Glumes** lanceolate to broadly lanceolate, shorter than to subequal to the adjacent lemmas, keels indistinct to distinct, smooth or scabrous; **lower glumes** 3-veined; **calluses** terete or slightly dorsally compressed, glabrous or with a crown of hairs, hairs to 2 mm; **lemmas** 3–7 mm, narrowly lanceolate to lanceolate or slightly oblanceolate, weakly to distinctly keeled, glabrous or the keels and marginal veins and sometimes the lateral veins with hairs, obscure, intercostal regions glabrous or with hairs; **anthers** 3, 1.2–3.5 mm, sometimes aborted late in development.

Poa sect. *Secundae* includes nine North American species. Two of the species also grow as disjuncts in South America. One species grows on high arctic islands in the Eastern Hemisphere. All the species tend to grow in arid areas, sometimes on wetlands within such areas. One species is confined to dry bluffs along the Pacific coast. All the species are primarily cespitose, but hybridization with members of *Poa* sect. *Poa* results in the formation of rhizomatous plants. Typically, members of sect. *Secundae* have sheaths that are closed for $^{1}/_{10}$–$^{1}/_{4}$ their length, contracted panicles, and anthers that are 1.2–3.5 mm long.

There are two subsections in the *Flora* region: subsects. *Secundae* and *Halophytae*.

Poa subsect. Secundae Soreng

Plants perennial; densely tufted, rarely weakly rhizomatous or stoloniferous. **Sheaths** closed for $^{1}/_{10}$–$^{1}/_{3}$ their length. **Panicles** 2–25(30) cm. **Spikelets** more or less terete, lengths 3.5–5 times widths, narrowly lanceolate to lanceolate, subterete to fairly compressed, somtimes bulbiferous; **florets** 2–5(6), normal or bulb-forming. **Glumes** lanceolate to broadly lanceolate, keels indistinct, smooth or sparsely scabrous; **lower glumes** 3-veined; **calluses** terete or slightly dorsally compressed; **lemmas** 3–7 mm, narrowly lanceolate to lanceolate or slightly oblanceolate, weakly keeled or somewhat weakly keeled; **anthers** 3, 1.5–3.5 mm, or aborted late in development and 0.8–1.8 mm.

Species of *Poa* subsect. *Secundae* usually have elongated, weakly compressed spikelets.

64. Poa secunda J. Presl [p. 587]

SECUND BLUEGRASS

Plants perennial; frequently anthocyanic, sometimes glaucous; densely tufted, basal leaf tufts 2–20+ cm, usually narrowly based, rarely with rhizomes. **Basal branching** intra- and extravaginal. **Culms** (10)15–120 cm, slender to stout, erect or the bases slightly decumbent, terete or weakly compressed; **nodes** terete, 0–2 exserted. **Sheaths** closed for $^{1}/_{10}$–$^{1}/_{4}$ their length, terete, smooth or scabrous, glabrous, bases of basal sheaths glabrous, distal sheath lengths (0.95)1.5–7(15) times blade lengths; **collars** smooth or scabrous, glabrous; **ligules** 0.5–6(10) mm, smooth or scabrous, truncate to acuminate, ligules of innovation leaves similar to those of the cauline leaves or shorter and truncate; **innovation blades** similar to the cauline blades; **cauline blades** gradually reduced in length upwards or the middle blades longest, 0.4–3(5) mm wide, flat, folded, or involute, thin, soft, and soon withering to thick, firm, and persisting, smooth or scabrous mainly over the veins, glabrous, apices narrowly prow-shaped, flag leaf blades 0.8–10(17) cm. **Panicles** 2–25(30) cm, erect or somewhat lax, narrowly lanceoloid to ovoid, usually contracted, more or less open at anthesis, infrequently remaining open at maturity, green or anthocyanic, sometimes glaucous, usually moderately congested, with 10–100+ spikelets; **nodes** usually with 1–3 branches; **branches** (0.5)1–8(10) cm, usually erect or ascending, infrequently spreading at maturity, terete to weakly angled, usually sparsely to densely scabrous on and between the angles, with (1)2–20(60+) spikelets in the distal $^{1}/_{2}$–$^{2}/_{3}$. **Spikelets** (4)5–10 mm, lengths (3.8)4–5 times widths, usually narrowly lanceolate, subterete to weakly laterally compressed, drab, green or strongly anthocyanic, sometimes glaucous; **florets** (2)3–5(10); **rachilla internodes** usually 1–2 mm, terete or slightly dorsally compressed, smooth or muriculate to scabrous. **Glumes** broadly lanceolate, keels indistinct; **lower glumes** 3-veined; **calluses** glabrous or with a crown of hairs, hairs 0.1–0.5(2) mm, crisp or slightly sinuous; **lemmas** 3.5–6 mm, lanceolate to narrowly lanceolate or slightly oblanceolate, usually weakly keeled, glabrous or the keels and marginal veins softly puberulent to short-villous, intercostal regions smooth or scabrous, glabrous, short-villous, crisply puberulent or softly puberulent over the basal $^{2}/_{3}$, hairs usually 0.1–0.5 mm, hairs of the keels and veins frequently similar in length to those between the veins, usually not or only slightly denser and extending further towards the apices, lateral veins obscure, margins strongly inrolled below, broadly scarious above, glabrous, apices obtuse to broadly acute, blunt, or pointed; **palea keels** scabrous, glabrous or softly puberulent to short-villous at midlength; **anthers** 1.5–3 mm. 2n = 42, 44+f, ca. 48, 56, ca. 62, 63, ca. 66, ca. 68, 70, ca.72, ca. 74, 78, ca. 80, 81, 82, ca. 83, 84–86, ca. 87, ca. 88, ca. 90, ca. 91, 93, ca. 94, ca. 97, ca. 98, ca. 99, 100, 104, 105–106.

Poa secunda is one of the major spring forage species of temperate western North America. It is very common in high deserts, mountain grasslands, saline wetlands, meadows, dry forests, and on lower alpine slopes, primarily from the Yukon Territory east to Manitoba and south to Baja California, Mexico. It also extends sporadically eastward across the Great Plains to the Gaspé Peninsula, Quebec. Both subspecies are present, as disjuncts, in Patagonia.

Poa secunda is highly variable. Hitchcock (1951) divided it into two groups, with a total of seven species. The two groups are recognized here as subspecies. They overlap almost completely in terms of morphology, but differ ecologically and cytologically.

Poa secunda is known or suspected to hybridize with several other species, including *P. arctica* (p. 529), *P. arida* (p. 599), *P. glauca* (p. 576), and *P. pratensis* (p. 522). Plants from the Columbia River Gorge in Oregon, including the type of *P. multnomae* Piper, that approach *P. tenerrima* (p. 588) are presumed to be derived from hybridization between *P. secunda* and *P. nervosa* (p. 545). *Poa secunda* differs from *P. curtifolia* (p. 589), with which it is sometimes confused, in having longer leaf blades that are sometimes folded or involute, and more spikelets per branch. Apomixis is common and facultative.

1. Lemmas usually glabrous, the keels and marginal veins infrequently sparsely puberulent at the base; basal branching mainly extravaginal; leaves slightly lax to firm, remaining intact through the growing season; ligules of the innovations to 2 mm long . . . . subsp. *juncifolia*
1. Lemmas sparsely to densely puberulent or short-villous on the basal $^{2}/_{3}$; basal branching mixed intra- or extravaginal or mainly intravaginal; leaves usually lax, withering with age; ligules of the innovations usually longer than 2 mm . . . . subsp. *secunda*

Poa secunda subsp. **juncifolia** (Scribn.) Soreng [p. 587]

ALKALI BLUEGRASS, BIG BLUEGRASS, NEVADA BLUEGRASS

Basal leaf tufts usually medium to robust, infrequently tiny. **Basal branching** mainly extravaginal. **Culms** 30–120 cm. **Ligules** of culm leaves 0.5–6 mm, those of the innovations 0.5–2 mm, scabrous, apices truncate to obtuse; **blades** 1–3(5) mm, moderately thick to thick, slightly lax to firm, tending to hold their form and persist. **Panicles** (4)10–25(30) cm, narrowly lanceoloid, contracted, congested; **branches** erect, scabrous.

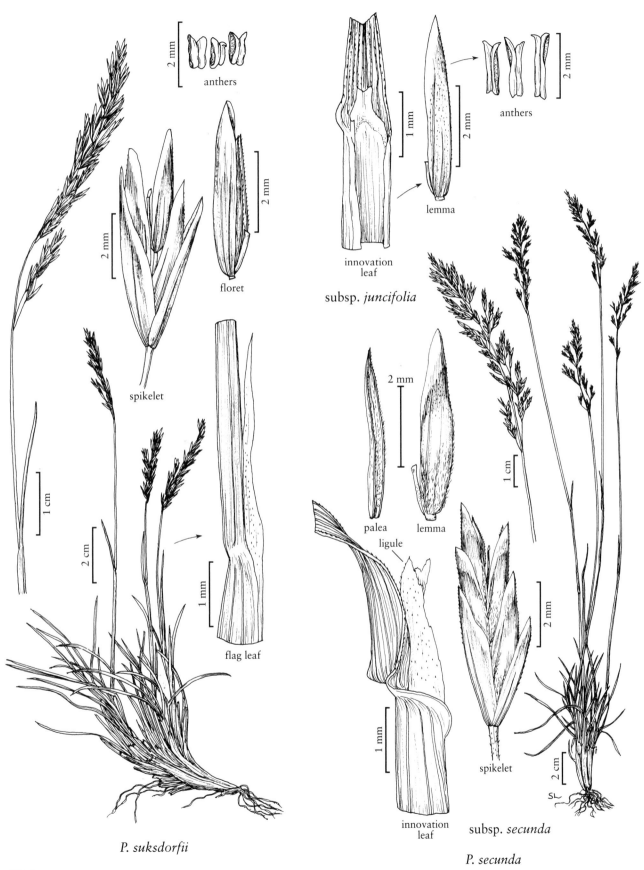

anthers

floret

spikelet

flag leaf

1 cm

2 cm

1 mm

P. suksdorfii

innovation leaf

subsp. *juncifolia*

anthers

lemma

1 mm

2 mm

palea

ligule

lemma

2 mm

spikelet

innovation leaf

1 mm

2 mm

1 cm

2 cm

subsp. *secunda*

P. secunda

POA

Spikelets (4)7–10 mm, lengths 4–5 times widths, narrowly lanceolate, subterete; **calluses** glabrous; **lemmas** sparsely to moderately scabridulous to scabrous, usually glabrous, keels and marginal veins infrequently crisply puberulent on the basal ¹/₄, hairs usually shorter than 0.2 mm; **paleas** glabrous. $2n$ = 42, 56, 60, 61, 62, 63, 64, ca. 65, ca. 66, 70, 78, 84, ca. 97.

Poa secunda subsp. *juncifolia* is usually more robust than subsp. *secunda*, and generally inhabits moister and sometimes saline habitats. It comprises two fairly distinct variants: a robust upland variant that is frequently used for revegetation (*P. ampla* Merr., Big Bluegrass) that grows in deep, rich, montane soils; and a riparian and wet meadow variant (*P. juncifolia* Scribn., Alkali Bluegrass). Apart from generally having glabrous lemmas, short ligules on the vegetative shoots, and leaf blades that hold their form better, *P. secunda* subsp. *juncifolia* differs anatomically in the predominance of sinuous-walled, rectangular long cells in the blade epidermis; smooth-walled, fusiform long cells are predominant in *P. secunda* subsp. *secunda*. Plants with glabrous lemmas and long ligules on the vegetative shoots have been called *P. nevadensis* Vasey *ex* Scribn.; they are intermediate between the subspecies. Chromosome numbers for *P. secunda* subsp. *juncifolia* center on $2n$ = 63, indicating a high degree of apomixis.

Poa secunda J. Presl subsp. **secunda** [p. 587]
PACIFIC BLUEGRASS, PINE BLUEGRASS, SANDBERG BLUEGRASS, CANBY BLUEGRASS

Basal leaf tufts usually tiny to medium, less often robust. **Basal branching** mixed intra- and extravaginal or mainly intravaginal. **Culms** (10) 15–100 cm, slender to middling. **Ligules** of culm leaves 2–6(10) mm, those of the innovations mostly 2–6 mm, smooth or scabrous, obtuse to acuminate; **blades** 0.4–3 mm, usually thin, lax, and soon withering, sometimes moderately thick, moderately firm, and somewhat persistent. **Panicles** 2–15(20) cm, usually narrowly lanceoloid to ovoid, contracted at maturity and congested, or occasionally pyramidal, open at maturity, and sparse; **branches** erect or ascending, infrequently widely spreading at maturity. **Spikelets** (4)5–8 mm; **calluses** glabrous or pubescent; **lemmas** with keels and marginal veins long-villous, crisply puberulent, or softly puberulent over the basal ²/₃, intercostal regions usually at least sparsely crisply or softly puberulent, hairs usually shorter than 0.5 mm; **palea keels** short-villous to softly puberulent at midlength, intercostal regions often softly puberulent. $2n$ = 42, 44+f, ca. 48, 56, ca. 62, 63, ca. 66, ca. 68, 70, ca. 72, ca. 74, ca. 78, ca. 80, 81, 82, ca. 83, 84, ca. 86, ca. 87, ca. 88, ca. 90, ca. 91, 93, ca. 94, ca. 98, ca. 99, 100, 104, 105–106.

Poa secunda subsp. *secunda* comprises several forms or ecotypes which intergrade morphologically and overlap geographically. Its chromosome numbers are centered on $2n$ = 84. It generally grows in more xeric habitats than subsp. *juncifolia*; it is also common in alpine habitats. Some of the major variants, and the names that have been applied to them, are: scabrous plants, primarily from west of the Cascade/Sierra Nevada axis (*P. scabrella* (Thurb.) Benth. *ex* Vasey, Pine Bluegrass); smoother, large plants extending eastward (*P. canbyi* (Scribn.) Howell, Canby Bluegrass); tiny, early-spring-flowering plants of stony and mossy ground (*P. sandbergii* Vasey, Sandberg Bluegrass); and slender, sparse plants, generally of mesic shady habitats, with panicles that remain open (*P. gracillima* Vasey, Pacific Bluegrass). Alpine plants have been called *P. incurva* Scribn. & T.A. Williams.

Poa secunda subsp. *secunda* can be difficult to separate from *P. stenantha* var. *stenantha* (p. 574). It differs in having more rounded lemma keels, hairs between the veins of the lemmas, and calluses that are glabrous or have hairs shorter than 0.2 mm. It also resembles *P. tenerrima* (see next), but lacks that species' combination of persistently wide, open panicles, very scabrous branches, short-truncate ligules, and very fine foliage.

65. **Poa tenerrima** Scribn. [p. 590]
DELICATE BLUEGRASS

Plants perennial; densely tufted, basal leaf tufts 2–8(13) cm, small, narrowly based, not stoloniferous, not rhizomatous. **Basal branching** intravaginal. **Culms** 15–50 cm tall, slender, 0.8–0.9 mm thick; **nodes** terete, 0–1 exserted. **Sheaths** closed for ¹/₆–¹/₄ their length, terete, scabrous, glabrous, bases of basal sheaths glabrous, distal sheath lengths 2.3–7.7 times blade lengths; **collars** sparsely scabrous, glabrous; **ligules** 0.5–1.5(2.5) mm, scabrous, apices usually truncate to obtuse, sometimes acute, ligules of innovations to 0.5 mm; **innovation blades** filiform; **cauline blades** gradually reduced in length upwards, 0.4–1.5 mm wide, mostly folded, thin, soft, soon withering, scabrous, apices narrowly prow-shaped. **Panicles** 5–15 cm, 1.3–2.2 times the branch lengths, erect, broadly rhomboidal to pyramidal, open, purple, sparse, proximal internodes 1.5–4.2 cm; **nodes** with 1–2(5) branches; **branches** 3–8.5 cm, widely spreading, capillary, straight, terete to weakly angled, moderately to mostly densely scabrous, with 3–9 spikelets in the distal ¹/₃. **Spikelets** 5–8 mm, lengths (3.8)4–5 times widths, usually narrowly lanceolate to lanceolate, subterete to weakly laterally compressed, drab, usually strongly anthocyanic; **florets** 3–5; **rachilla internodes** 1–1.5+ mm, terete or slightly dorsally compressed, muriculate or scabrous. **Glumes** lanceolate, distinctly shorter than the adjacent lemmas, scabrous distally, keels indistinct, scabrous, obtuse to acute; **lower glumes** 3-veined; **calluses** glabrous or with a crown of hairs, hairs to 0.3 mm; **lemmas** 3–4.2 mm, lanceolate, weakly keeled, keels, veins, and proximal $^2/_3$ of the intercostal regions puberulent, lateral veins obscure, margins strongly inrolled below, broadly scarious above, glabrous, apices obtuse to acute; **palea keels** scabrous, sometimes hairy at midlength; **anthers** 1.6–2.1 mm. 2*n* = 42.

Poa tenerrima is a rare species, endemic to serpentine barrens along the western base of the Sierra Nevada. It differs from *P. secunda* subsp. *secunda* (see previous) in combining consistently wide, open panicles, very scabrous branches, short-truncate ligules, and very fine foliage. A series of small, delicate, open-panicled plants from the California Coast Ranges, formerly included in *P. tenerrima* by Soreng (1993), differ in having smooth branches and longer ligules, and are better referred to *P. secunda* subsp. *secunda*. No intergradation is evident where the two taxa grow together. Plants from the Columbia River Gorge in Oregon, including the type of *P. multnomae* Piper, approach *P. tenerrima*, but are presumed to be derived from hybridization between *P. nervosa* (p. 545) and *P. secunda* (p. 586).

66. Poa curtifolia Scribn. [p. 590]
WENATCHEE BLUEGRASS

Plants perennial; densely tufted, not stoloniferous, not rhizomatous. **Basal branching** mainly intravaginal. **Culms** (15)20–40 cm, erect or decumbent, with 1–2 exserted nodes. **Sheaths** closed for $^1/_5$–$^1/_3$ their length, terete, smooth, bases of basal sheaths glabrous, distal sheath lengths 4–33 times blade lengths, smooth, glabrous; **ligules** (1.5)2–5 mm, smooth or sparsely scabrous, margins distinctly decurrent, apices obtuse to acute, ligules of innovations prominent, milky white; **innovation blades** similar to the cauline blades; **cauline blades** gradually reduced in length upwards, (1)1.5–3 mm wide, flat, thick, fairly firm, smooth or sparsely scabrous, margins white, apices broadly prow-shaped, flag leaf blades 0.2–1.8 cm, infrequently absent. **Panicles** 4–8 cm, erect, linear to narrowly lanceoloid, contracted, moderately congested, with 9–35 spikelets; **nodes** with 1–2 branches; **branches** 1–2.5 cm, erect to steeply ascending, straight, sulcate or angled, angles sparsely to moderately scabrous, with 1–4 spikelets in

the distal $^1/_2$. **Spikelets** 7–9 mm, lengths 3.5–4 times widths, lanceolate, fairly compressed, pale, slightly lustrous; **florets** (2)3–4; **rachilla internodes** 1–2 mm. **Glumes** lanceolate to broadly lanceolate, margins broadly scarious, keels indistinct, smooth or sparsely scabrous; **lower glumes** 3-veined, slightly to distinctly shorter than the lowest lemma; **calluses** glabrous; **lemmas** 4.5–6 mm, lanceolate, somewhat weakly keeled, glabrous or the keels, marginal veins, and intercostal regions very sparsely puberulent over the proximal $^1/_3$, lateral veins obscure, margins strongly inrolled proximally, broadly scarious distally, glabrous, apices acute; **palea keels** scabrous, glabrous or puberulent at midlength; **anthers** 2.2–3.5 mm. 2*n* = 42.

Poa curtifolia is endemic to upper serpentine slopes in the Wenatchee Mountains, Kittitas and Chelan counties, Washington. It has narrow panicles like *P. pringlei* (p. 564) and *P. suksdorfii* (p. 584). It differs from *P. secunda* (p. 586), with which it is sometimes confused, in having all blades short, flat, and firm, and few spikelets per branch.

67. Poa hartzii Gand. [p. 593]
HARTZ'S BLUEGRASS

Plants perennial; not glaucous; densely to loosely tufted, not rhizomatous, occasionally weakly stoloniferous. **Basal branching** extra- and intravaginal. **Culms** 10–33(45) cm, usually decumbent, terete; **nodes** terete, 0(1) exserted. **Sheaths** closed for $^1/_7$–$^1/_5$($^1/_3$) their length, terete, usually lustrous, bases of basal sheaths glabrous; **ligules** (1.5)2–7 mm, smooth or sparsely scabrous, margins usually decurrent, apices obtuse to acuminate; **innovation blades** similar in texture and shape to those of the culms; **cauline blades** 2–9 cm, gradually increasing or decreasing in length upwards, 1.5–3 mm wide, folded to involute, moderately thick, soft, abaxial surfaces smooth, adaxial surfaces smooth or somewhat scabrous, usually glabrous, infrequently sparsely hispidulous, apices narrowly prow-shaped. **Panicles** 2.5–6(12) cm, erect, narrowly lanceolate, contracted or narrowly ovate in some bulbiferous plants, moderately congested, with 7–40 spikelets; **nodes** with (1)2(4) branches; **branches** 1–3 cm, erect to ascending, straight, sulcate, smooth or sparsely to moderately scabrous, with 1–10 spikelets in the distal $^1/_3$–$^2/_3$. **Spikelets** 4.8–7.4 mm, lengths 3.5–4 times widths, lanceolate, weakly laterally compressed, sometimes bulbiferous, lustrous; **florets** (2)3–5(6), normal or bulb-forming; **rachilla internodes** 0.8–2 mm, smooth, sometimes sparsely hispidulous. **Glumes** mostly broadly scarious, somewhat lustrous, keels indistinct, smooth or sparsely

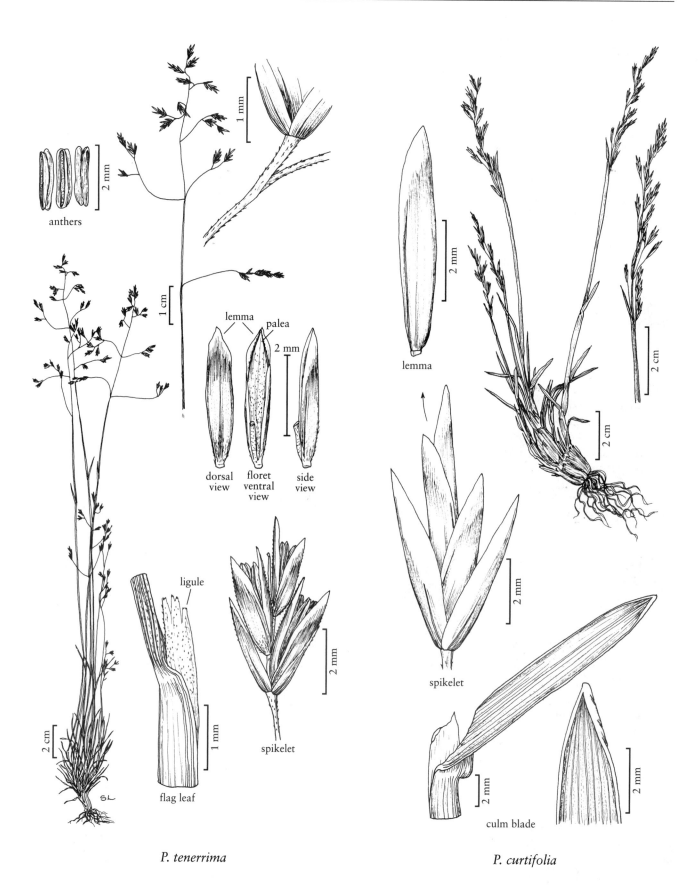

anthers

1 mm

lemma   palea

2 mm

dorsal
view

floret
ventral
view

side
view

ligule

flag leaf

1 mm

spikelet

2 mm

P. tenerrima

lemma

2 mm

2 cm

2 cm

spikelet

2 mm

culm blade

2 mm

2 mm

P. curtifolia

scabrous distally; **lower glumes** 3-veined; **upper glumes** frequently exceeding the lowest lemmas; **calluses** glabrous or with a crown of hairs, hairs to 2 mm; **lemmas** (3.3)3.5–7 mm, lanceolate, usually weakly keeled, more or less evenly and somewhat loosely to densely hairy over the proximal $^1/_3$–$^1/_2$, hairs usually longer than 0.5 mm, sparsely scabrous in the middle $^1/_3$, smooth distally, lateral veins obscure, margins weakly inrolled, broadly scarious, glabrous, apices long-scarious, acute to shortly obtuse, often erose, often bronze-colored below the apices; **palea keels** sparsely scabrous, softly puberulent at midlength, intercostal regions softly puberulent; **anthers** usually all aborted late in development and 0.8–1.8 mm, infrequently well developed and 2–2.8 mm. $2n = 63, 70$.

Poa hartzii grows only in the high arctic. It generally grows on open ground, on sandy or clayey soils, or on slumping slopes of old marine terraces. It carries two chloroplast genomes within its populations; one of these links it to *P. secunda* (p. 586) and *P. ammophila* (see next), the other to *P. glauca* (p. 576). Morphologically, it is closest to *P. secunda* and *P. ammophila*.

1. Spikelets bulbiferous . . . . subsp. *vrangelica*
1. Spikelets not bulbiferous.
    2. Lemmas 5.5–7 mm long; anthers well
       developed, 2–2.8 mm long . . . . subsp. *alaskana*
    2. Lemmas 3.3–5.4 mm long; anthers usually
       aborted and shorter than 1.5 mm . . . . . . subsp. *hartzii*

Poa hartzii subsp. alaskana Soreng [p. 593]

Plants loosely tufted. **Culms** 20–45 cm. **Ligules** 5–7 mm. **Panicles** 7–12 cm; **branches** smooth or sparsely to moderately scabrous. **Spikelets** 5–7 mm, not bulbiferous, lustrous; **florets** normal; **rachilla internodes** 1.5–2 mm. **Upper glumes** 5–6 mm; **calluses** usually with a crown of 1–2 mm hairs; **lemmas** 5.5–7 mm, sparsely hairy, hairs mostly longer than 0.5 mm, apices acute; **anthers** 2–2.8 mm. $2n$ = unknown.

Poa hartzii subsp. *alaskana* grows on the North Slope of Alaska, mainly in sandy places. It generally resembles robust plants of *P. hartzii* subsp. *hartzii*, but has a looser habit, longer lemmas, and well-developed anthers.

Poa hartzii Gand. subsp. hartzii [p. 593]

Culms 10–33 cm. **Ligules** (1.5)2–7 mm. **Panicles** 2.5–6 cm, with (1)2(3) branches per node; **branches** smooth or scabrous. **Spikelets** 4.8–7.4 mm, lustrous; **florets** normal; **rachilla internodes** 0.8–1.5 mm. **Upper glumes** 3.5–4.5 mm; **calluses** usually with a crown of hairs, longer hairs 1–2 mm long; **lemmas** (3.3)3.9–5.4 mm, sparsely hairy, hairs mostly longer than 0.5 mm, apices acute; **anthers** usually aborted, 0.5–1.5 mm. $2n = 63, 70$.

Poa hartzii subsp. *hartzii* is the common subspecies on high arctic islands and in Greenland. It grows at scattered locations on the continental margin, from the Mackenzie River Delta to the Ungava Peninsula, Canada. Outside the *Flora* region, it is known from Wrangel Island in Russia and from Svalbard, Norway. It is apomictic, setting seed despite rarely forming anthers. Curiously, this subspecies has two different chloroplast genome types within its northern populations, one like *P. glauca* and one like *P. secunda*. Robust plants of subsp. *hartzii* resemble those of subsp. *alaskana*, but have a tighter habit and poorly developed anthers that are usually aborted.

Poa hartzii subsp. vrangelica (Tzvelev) Soreng & L.J. Gillespie [p. 593]

Culms 30–40 cm, highest culm node in the proximal $^1/_5$–$^1/_4$. **Leaves** mostly basal. **Sheaths** closed for about $^3/_{10}$ their length, smooth; **ligules** 2–3 mm, smooth, margins decurrent, apices obtuse to acute; **blades** 1.5–2 mm wide, involute, moderately thick, erect or steeply ascending, cauline blades mostly shorter than 3 cm. **Panicles** 5–9 cm, erect, open, narrowly pyramidal, sparse, with 2–4 branches per node; **branches** 1–2.5 cm, widely spreading, slender, more or less terete, sparsely to moderately densely scabrous, each branch with 1–2 spikelets. **Spikelets** bulbiferous, about 5–6 mm excluding the bladelets, distinctly laterally compressed, strongly anthocyanic, lustrous; **florets** 2, bulb-forming; **rachilla internodes** not distinguishable. **Glumes** 4.5–5.5 mm, about equal in length, lanceolate, distinctly keeled, keels smooth or sparsely scabrous distally; **lower glumes** 3-veined; **calluses** poorly differentiated, or with a sparse crown of hairs; **lemmas** 4.5–5.5 mm, the few more or less normal basal florets lanceolate, keels and sometimes the marginal veins sparsely puberulent proximally, margins narrowly hyaline, apices acute; **palea keels** sparsely scabrous distally, intercostal regions smooth, sometimes puberulent; **anthers** aborted, about 0.8 mm.

Poa hartzii subsp. *vrangelica* can be difficult to distinguish from *P. glauca*, but it has a looser habit and thin glumes. It grows along the Sagavanirktok River, from the Franklin Hills to Prudhoe Bay, Alaska, as well as at scattered locations along the coast of the Beaufort Sea in Alaska, the Queen Elizabeth Islands in Nunavut, and Wrangell Island, Russia. It includes two varieties: **Poa hartzii** var. **vivipara** Polunin, which grows in the Canadian portion of its range; and **Poa hartzii** var. **vrangelica** (Tzvelev) Prob., which is more western and favors *P. glauca*.

68. Poa ammophila A.E. Porsild [p. 593]
SAND BLUEGRASS

Plants perennial; glaucous; densely tufted, not stoloniferous, not rhizomatous. **Basal branching** intravaginal. **Culms** 10–30 cm. **Sheaths** closed for ¹/₇–¹/₆ their length, terete, bases of basal sheaths glabrous; **ligules** 1.5–3 mm; **innovation blades** similar to the cauline blades; **cauline blades** 1–3 mm wide, involute, moderately thick, soft, abaxial surfaces smooth, adaxial surfaces smooth or sparsely scabrous, apices narrowly prow-shaped. **Panicles** 3–6 cm, congested or moderately congested; **nodes** with (1)2(3) branches; **branches** 1–3(4) cm, erect, terete, smooth or sparsely scabrous.

Spikelets 5–7 mm, lengths to 3.5 times widths, broadly lanceolate, weakly laterally compressed, fairly drab; **florets** 2–4; **rachilla internodes** usually 1–1.3 mm, smooth. **Glumes** lanceolate, distinctly keeled, keels smooth; **lower glumes** 3-veined; **calluses** glabrous; **lemmas** 3–4.6 mm, lanceolate, distinctly to weakly keeled, evenly and densely strigulose over the proximal ¹/₃–¹/₂, hairs mostly about 0.1 mm, some keel hairs to 0.2(0.3) mm, lateral veins obscure, margins broadly scarious, glabrous, apices acute; **palea keels** scabrous, softly puberulent at midlength, intercostal regions softly puberulent; **anthers** 1.5–1.8 mm. 2*n* = unknown.

Poa ammophila is endemic to the Mackenzie River Delta region, Northwest Territories. It grows primarily north of treeline and, as its name indicates, usually on sandy soils. Its close relative, *P. hartzii* (see previous), also reaches the continental coastline in this region.

Poa subsect. Halophytae V.L. Marsh *ex* Soreng

Plants perennial; not stoloniferous, not rhizomatous. **Culms** 5–100 cm. **Sheaths** closed for ¹/₁₀–¹/₅(¹/₄) their length, terete; **ligules** 2–6 mm, apices acute to acuminate; **blades** 1–5 mm wide. **Panicles** 3–18(25) cm; **nodes** with 2–7 branches. **Spikelets** (4)4.5–10 mm, lengths 3–3.6 times widths, lanceolate to narrowly ovate, laterally compressed, sometimes bulbiferous; **florets** 3–5(7), normal or sometimes bulb-forming. **Calluses** terete or slightly dorsally compressed; **lemmas** 3–6 mm, lanceolate, distinctly keeled, glabrous or the keels and marginal veins, and sometimes the lateral veins, hairy, intercostal regions glabrous or hairy, apices acute; **anthers** 3, 1.2–3 mm, aborted or undeveloped in bulbiferous spikelets.

Members of *Poa* subsect. *Halophytae* have spikelets that resemble those in other sections of *Poa* more closely than those of subsect. *Secundae*, being shorter and more compressed than those of that subsection.

69. Poa stenantha Trin. [p. 595]
NARROW-FLOWER BLUEGRASS

Plants perennial; glaucous or not; densely to loosely tufted, not stoloniferous, not rhizomatous. **Basal branching** mostly extravaginal, some intravaginal. **Culms** 20–60(100) cm, bases decumbent or sometimes erect, terete, with 1–2 exserted nodes. **Sheaths** closed for ¹/₁₀–¹/₅(¹/₄) their length, terete, bases of basal sheaths glabrous; **ligules** 2–5 mm, milky white, smooth or sparsely scabrous, acute to acuminate; **innovation blades** similar in texture and shape to the cauline blades; **cauline blades** not greatly reduced upwards, 1.5–4(5) mm wide, flat or folded, thin, lax, smooth or sparsely scabrous, apices narrowly prow-shaped. **Panicles** 5–18(25) cm, lax, loosely contracted to open, sparse, with 20–65

spikelets and usually 2(7) branches per node; **branches** 3–15 cm, ascending to spreading, angled, angles finely to coarsely, sparsely to fairly densely scabrous, infrequently smooth, with 3–10(15) spikelets in the distal ¹/₂. **Spikelets** 6–10 mm, lengths 3–3.6 times widths, lanceolate to narrowly ovate, laterally compressed, sometimes bulbiferous, drab, often slightly glaucous; **florets** 3–4(7), normal or bulb-forming; **rachilla internodes** 1.2–2 mm, slightly dorsally compressed, smooth or sparsely muriculate. **Glumes** subequal, lanceolate to broadly lanceolate, dull, frequently glaucous, obtuse to acute; **lower glumes** 3-veined; **upper glumes** (3.7)4.1–6.5 mm; **calluses** usually crowned with 0.2–2 mm hairs, sometimes glabrous; **lemmas** 4–6 mm, lanceolate, distinctly compressed, distinctly keeled, keels, marginal veins, and sometimes the lateral veins short- to long-villous, hairs extending for ³/₄ of the keel, intercostal regions glabrous, sparsely puberulent or hispidulous proximally, usually sparsely

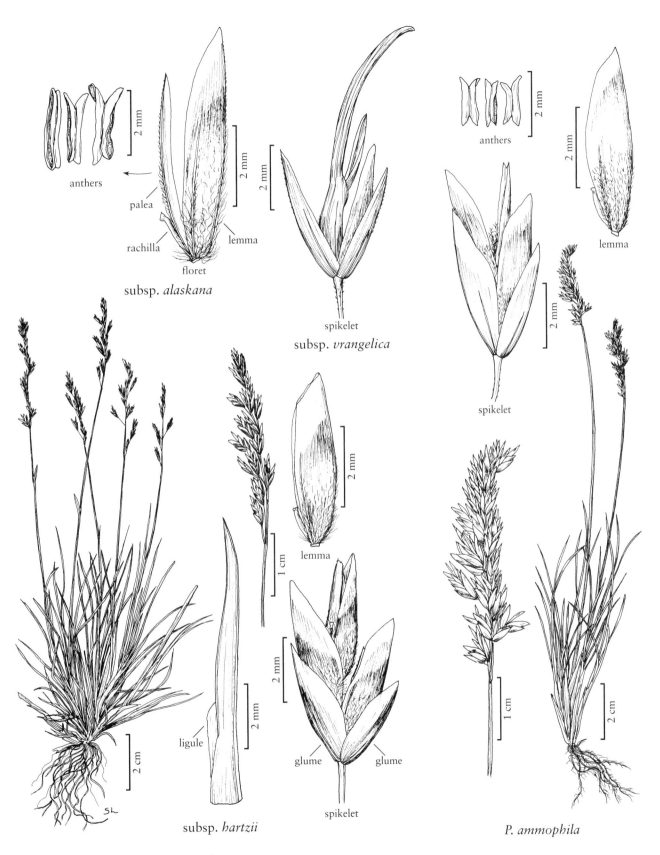

anthers

palea

rachilla

lemma

floret

subsp. *alaskana*

spikelet

subsp. *vrangelica*

lemma

spikelet

lemma

ligule

subsp. *hartzii*

glume   glume

spikelet

P. hartzii

anthers

P. ammophila

to moderately densely scabrous distally, hairs distinctly shorter than those of the keel and veins, margins weakly inrolled, broadly scarious, glabrous, apices acute; **palea keels** scabrous, often softly puberulent at midlength, intercostal regions glabrous or puberulent; **anthers** 1.2–2 mm, sometimes aborted late in development or undeveloped. $2n = 42$, [81, 84, 86?].

Poa stenantha grows in coastal meadows and on cliffs in subarctic and boreal forests; it is less common in moist, more southern subalpine and low alpine meadows and thickets. Its range extends from western Alaska to the northern Cascades and Rocky Mountains and, as a disjunct, to Patagonia. *Poa stenantha* was originally described as growing in Kamchatka, Russia, but the Russian plants have since been referred to other species.

1. Spikelets not bulbiferous .... var. *stenantha*
1. Spikelets bulbiferous .... var. *vivipara*

Poa stenantha Trin. var. stenantha [p. 595]

Spikelets not bulbiferous; **florets** normal. **Anthers** 1.2–2 mm.

Poa stenantha var. *stenantha* can be difficult to separate from *P. secunda* subsp. *secunda* (p. 588). Its main distinguishing features are its strongly keeled lemmas with glabrous intercostal regions, and, when present, callus hairs longer than 0.2 mm. Plants with large panicles and glabrous calluses have been called *P. macroclada* Rydb. Such plants grow infrequently in the U.S. Rocky Mountain portion of the species' range. They intergrade with the more compact typical form.

Poa stenantha var. vivipara Trin. [p. 595]

Spikelets bulbiferous; **florets** bulb-forming. **Glumes** usually glaucous, not lustrous; **anthers** usually aborted late in development or not developed.

Poa stenantha var. *vivipara* is the common form of the species in the Aleutian Islands; it extends eastward to Sitka, Alaska. It differs from bulbiferous forms of *P. arctica* (p. 529) in its lack of rhizomes, more open sheaths, and usually glaucous and scabrous panicle branches.

70. Poa napensis Beetle [p. 597]
NAPA BLUEGRASS

Plants perennial; fairly glaucous; densely tufted, not stoloniferous, not rhizomatous. **Basal branching** intravaginal. **Culms** 30–100 cm, erect, terete, with 0(1) exserted nodes. **Sheaths** closed for $^1/_{10}(^1/_8)$ their length, terete, bases of basal sheaths glabrous, distal sheath lengths

1.5–5 times blade lengths; **ligules** 4–6 mm, scabrous, obtuse to acute; **innovation blades** similar to the cauline blades; **cauline blades** 1–3 mm wide, folded to involute, thick, fairly firm, pale green, abaxial surfaces scabrous, apices narrowly prow-shaped. **Panicles** 5–18(21) cm, erect, narrowly to broadly lanceoloid, loosely contracted, congested, with 40–100+ spikelets; **nodes** with 2–3(5) branches; **branches** 3–10 cm, erect to ascending, straight, angles densely scabrous, with 5–27 spikelets in the distal $^1/_2$. **Spikelets** (4)4.5–7 mm, lengths 3–3.5 times widths, lanceolate, laterally compressed, drab; **florets** 3–5; **rachilla internodes** usually shorter than 1 mm, smooth. **Glumes** lanceolate, slightly unequal, pale, distinctly keeled, keels sparsely scabrous; **lower glumes** 3-veined; **calluses** glabrous, rarely with a crown of hairs, hairs to 0.1 mm; **lemmas** 3–4 mm, lanceolate, distinctly keeled, finely scabrous, usually glabrous, keels and marginal veins rarely sparsely puberulent proximally, lateral veins obscure to moderately prominent, intercostal regions muriculate, margins glabrous, apices acute; **paleas** scabrous over the keels; **anthers** 1.2–1.8 mm. $2n = 42$.

Poa napensis is endemic to mineralized ground around hot springs in Napa County, California. It is listed as an endangered species by the United States Fish and Wildlife Service. The sectional placement of the species is suggested by the rare occurrence of a minute crown of hairs around the callus and its possession of a chloroplast genome like that of *P. secunda* (p. 588).

71. Poa unilateralis Scribn. [p. 597]
SEA-BLUFF BLUEGRASS

Plants perennial; frequently glaucous; densely tufted, not stoloniferous, not rhizomatous. **Basal branching** all or mainly intravaginal. **Culms** 5–40 cm, erect or ascending, frequently decumbent, terete, with 0–2 exserted nodes. **Sheaths** closed for $^1/_{10}(^1/_5)$ their length, terete, smooth, glabrous, bases of basal sheaths glabrous, distal sheath lengths 1–4 times widths; **ligules** 2–6 mm, smooth or sparsely scabrous, obtuse to acute; **innovation blades** usually 1–1.5 mm wide, thin, soon withering, and distinctly narrower than the cauline blades, infrequently wider, flat, and a bit fleshy as in the cauline blades, or involute; **cauline blades** gradually reduced in length distally, 2–5 mm wide, flat or folded, soft, thin and soon withering or moderately thick and somewhat fleshy and retaining their form, smooth, apices narrowly to broadly prow-shaped. **Panicles** 3–7 cm, erect, nearly cylindrical, contracted, congested, with (20)30–80(120) spikelets; **nodes** with 3–7

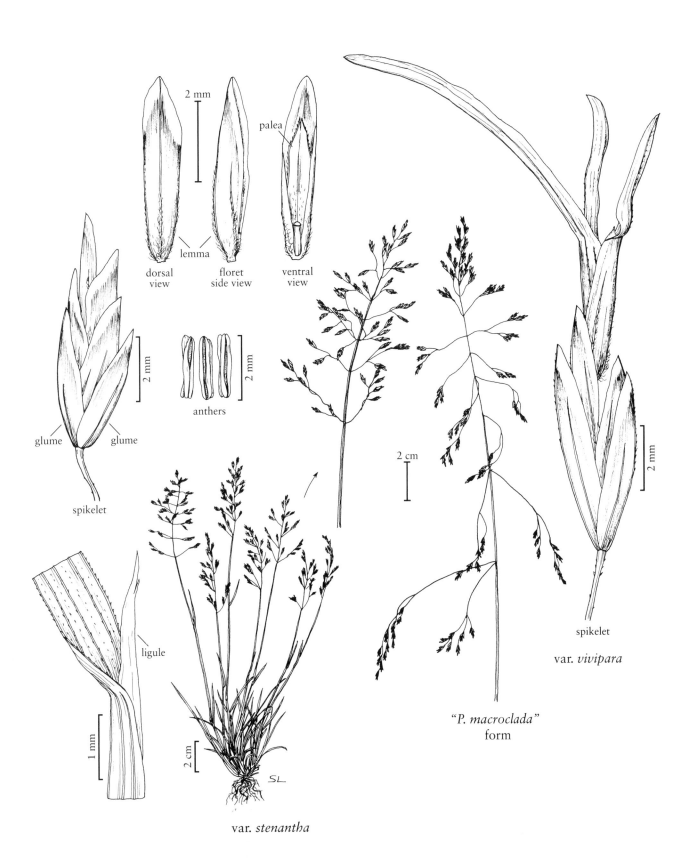

2 mm

palea

lemma

dorsal
view

floret
side view

ventral
view

2 mm

glume   glume

anthers

spikelet

2 cm

2 mm

spikelet

var. *vivipara*

ligule

1 mm

2 cm

var. *stenantha*

SL

"*P. macroclada*"
form

POA STENANTHA

branches; **branches** 0.5–1.5(4.5) cm, erect, usually angled, infrequently terete or sulcate, angles usually moderately to densely scabrous, infrequently smooth with dense papillae and weak-angled, with 2–10 spikelets in the proximal $^2/_3$; **pedicels** shorter than the spikelets. **Spikelets** 4.5–7 mm, lengths to 3.5 times widths, lanceolate to narrowly ovate, drab; **florets** 3–5; **rachilla internodes** usually shorter than 1 mm, smooth. **Glumes** lanceolate, slightly unequal, distinctly keeled, keels papillate or scabrous; **lower glumes** 3-veined; **calluses** glabrous or with a crown of hairs, hairs 0.1–0.2 mm; **lemmas** 3–4.5 mm, lanceolate, distinctly keeled, glabrous or the keels and marginal veins short-villous to midlength, intercostal regions sparsely puberulent near the base, margins glabrous, apices acute; **palea keels** scabrous, sometimes softly puberulent at midlength; **anthers** 1.5–3 mm. 2*n* = 42, 84.

Poa unilateralis grows on grassy bluffs and cliffs near the Pacific coast of North America, from Washington to California.

1. Lemmas villous on the keels and marginal veins for more than $^1/_3$ the length of the lemmas; blades usually involute, the cauline and innovation blades similar . . . . subsp. *pachypholis*
1. Lemmas glabrous or the keels and marginal veins villous or puberulent for less than $^1/_5$ the length of the lemmas; blades flat or folded, the cauline blades sometimes wider and thicker than the innovation blades . . . . subsp. *unilateralis*

Poa unilateralis subsp. **pachypholis** (Piper) D. D. Keck ex Soreng [p. 597]

Innovation blades similar to the cauline blades; **cauline blades** usually involute, slightly thick, not fleshy. **Calluses** with a crown of hairs, hairs 0.1–0.2 mm; **lemmas** villous on the keels and marginal veins for $^1/_3$–$^1/_2$ the lemma length, hairs about 0.13 mm. 2*n* = 42.

Poa unilateralis subsp. *pachypholis* is known from populations in Lincoln County, Oregon, and Pacific County, Washington.

Poa unilateralis Scribn. subsp. **unilateralis** [p. 597]

Innovation blades sometimes narrower and thinner than the cauline blades; **cauline blades** flat or folded, thin or thick and a bit fleshy. **Calluses** glabrous or with a crown of hairs, hairs to 0.1 mm; **lemmas** glabrous or the keels and marginal veins villous or puberulent for less than $^1/_5$ the lemma length. 2*n* = 42, 84.

The range of *Poa unilateralis* subsp. *unilateralis* extends from northern Oregon to central California.

Poa subg. Arctopoa (Griseb.) Prob.

Plants perennial; rhizomatous, rhizomes stout, 1.5–3 mm thick, culms mostly solitary. **Basal branching** extravaginal. **Culms** (15)20–115 cm tall, 2–3 mm thick, terete or weakly compressed. **Sheaths** closed for $^1/_5$–$^1/_3$ their length, sometimes fused by a hyaline membrane to $^3/_4$ their length, terete, bases of some basal sheaths densely retrorsely strigose, hairs 0.1–0.2 mm, thick; **ligules** 1–4(5.5) mm, white to off-white or yellow-cream to brown, truncate to obtuse, ciliolate; **blades** usually (2)4–11+ mm wide, flat, folded, or involute, thick, abaxial surfaces smooth, margins and sometimes the adaxial veins scabrous, apices narrowly to broadly prow-shaped. **Panicles** 5–35 cm, erect, contracted to wide open, with (1)2–5 branches per node; **branches** 1–20 cm, erect to widely spreading, straight or somewhat lax, terete or angled, smooth or scabrous. **Spikelets** 5–12 mm, laterally compressed, not bulbiferous; **florets** 2–6(8), bisexual; **rachilla internodes** smooth or scabrous, glabrous or hairy. **Glumes** subequal or the lower glumes to 2 mm longer than the upper glumes, outer margins smooth or scabrous, frequently ciliate proximally, distinctly keeled, keels smooth or scabrous; **lower glumes** 1–3(5) veined; **upper glumes** (1)3(5) veined; **calluses** obtusely angled, blunt or weakly pointed, terete, glabrous or with a crown of hairs; **lemmas** 3.8–7 mm, 5–7-veined, distinctly keeled, membranous to subcoriaceous, glabrous or the keels and marginal veins sparsely hairy on the proximal $^1/_2$, lateral veins faint, intercostal regions smooth or scabrous, margins usually with hairs to 0.2 mm proximally; **palea keels** scabrous, glabrous or with hairs at midlength, intercostal regions smooth or scabrous, usually glabrous, infrequently puberulent proximally; **anthers** 3, 1.6–3.2 mm.

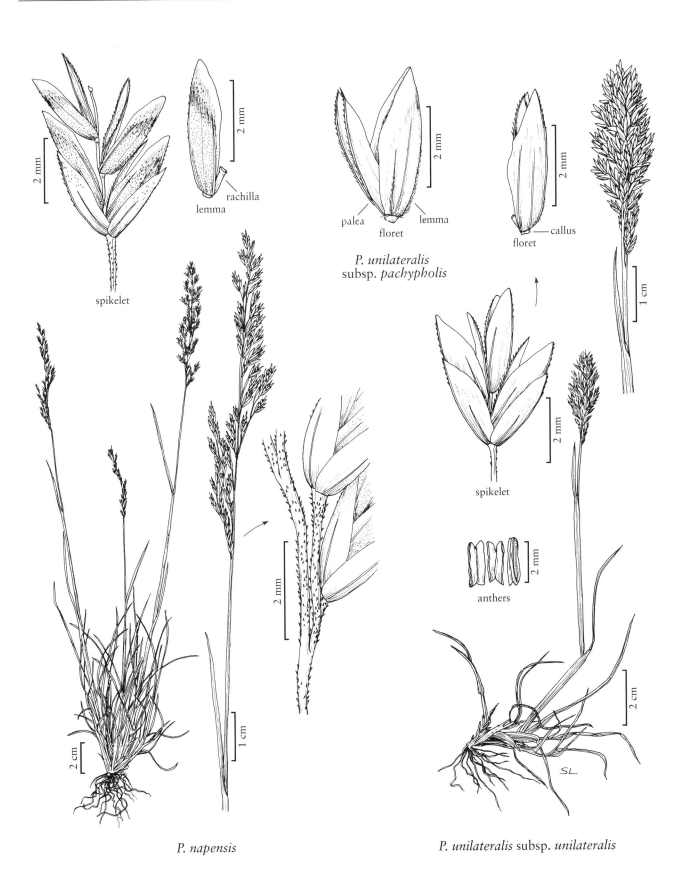

rachilla

lemma

spikelet

palea lemma

floret

P. unilateralis
subsp. *pachypholis*

callus

floret

spikelet

anthers

P. napensis

P. unilateralis subsp. *unilateralis*

Poa subg. *Arctopoa* includes five species in two sections; only one species grows in the *Flora* region. The species are generally robust, rhizomatous perennials of subsaline or subalkaline soils of wetlands. They have cilia along the lemma and sometimes the glume margins, and the bases of the basal sheaths have short, thick hairs. The subgenus is sometimes treated as a genus. The chloroplast genomes of species in the subgenus are similar to those in *Poa* sect. *Sylvestres*; the nuclear ribosomal DNA suggests a relationship with genera outside of *Poa*, such as *Beckmannia* (p. 484), *Dupontia* (p. 602), and *Arctophila* (p. 605).

Poa sect. Arctopoa (Griseb.) Tzvelev

Plants perennial. **Ligules** 1–3.5 mm, yellow-cream to brown; **blades** flat, thick, apices broadly prow-shaped. **Panicles** 8–30 cm, erect, loosely contracted; **branches** 3–10 cm, steeply ascending, terete. **Spikelets** not bulbiferous. **Glumes** distinctly keeled, keels smooth; **calluses of proximal lemmas** usually with a crown of hairs, hairs 1–2 mm; **lemmas** 4.5–7 mm, thinly membranous, glabrous or the keels and marginal veins long-villous, intercostal regions glabrous or hispidulous, moderately to densely scabrous; **palea keels** scabrous; **anthers** 3, 1.7–3.2 mm.

Poa sect. *Arctopoa* has one species. It grows along boreal and low arctic coasts in North America, and from the Russian Far East to northern Japan. **Poa** sect. **Aphydris** (Griseb.) Tzvelev, the other section in the subgenus, is restricted to central and east Asia.

72. Poa eminens J. Presl [p. 599]
EMINENT BLUEGRASS

Plants perennial; often glaucous; rhizomatous, rhizomes stout, about 2 mm thick, culms solitary. **Basal branching** extravaginal. **Culms** 20–100 cm tall, about 2 mm thick, terete or weakly compressed; **nodes** terete, 0–1 exserted. **Sheaths** closed for ¹/₆–¹/₃ their length, sometimes fused by a hyaline membrane to ³/₄ their length, terete, bases of some basal sheaths densely retrorsely hairy, hairs 0.1–0.2 mm, thick; **ligules** 1–3.5 mm, yellow-cream to brown, truncate, erose, ciliolate; **blades** (2)4–11 mm wide, flat, thick, smooth or sparsely scabrous, apices broadly prow-shaped. **Panicles** 8–30 cm, erect, loosely contracted, fairly congested, with 40–100+ spikelets; **branches** 3–10 cm, steeply ascending, terete, smooth or sparsely scabrous, sometimes with tufts of hair at the nodes, with 5–20 spikelets. **Spikelets** 5–12 mm, laterally compressed; **florets** 2–6; **rachilla internodes** smooth, infrequently sparsely puberulent. **Glumes** lanceolate, subequal or the upper glumes to 2 mm longer than the lower glumes, sometimes exceeding the lowest lemmas, distinctly keeled, smooth, often glaucous, acute to acuminate; **lower glumes** 4–9.5 mm, 1–3(5)-veined; **upper glumes** 5.5–10 mm, (1)3(5)-veined; **calluses of proximal lemmas** usually with a crown of hairs, hairs 1–2 mm; **lemmas** 4.5–7 mm, lanceolate, 5–7-veined, distinctly keeled, thinly membranous, glabrous or the keels and marginal veins long-villous, intercostal regions glabrous or hispidulous, moderately to densely scabrous, margins usually with hairs to 0.2 mm proximally, apices acute; **palea keels** scabrous; **anthers** 1.7–3.2 mm. $2n = 28$, 29+-, 42, 62.

Poa eminens grows along low arctic and boreal coasts and estuaries, in subsaline meadows and beaches. It also grows along the Asian coast from Hokkaido Island, Japan, to the Chukchi Peninsula, Russia. It hybridizes with *Dupontia* (see ×*Dupoa*, p. 601). Its nuclear ribosomal DNA appears to be related to an ancestor of *Dupontia* (p. 602) and *Arctophila* (p. 605); and its chloroplast DNA to *P. tibetica* Munro *ex* Stapf, an Asian member of **Poa** sect. **Aphydris** (Griseb.) Tzvelev (Gillespie & Soreng [in prep.]).

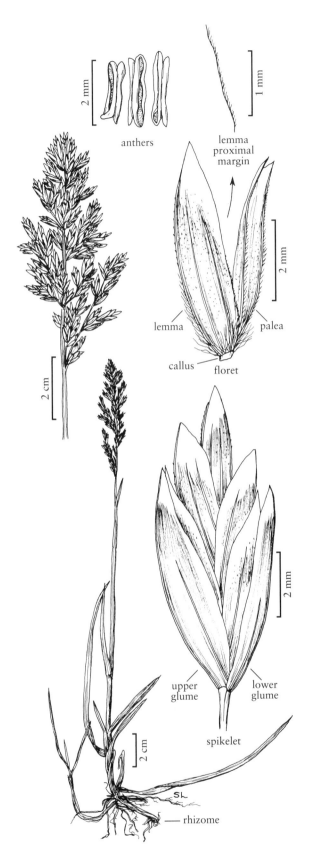

anthers

lemma proximal margin

lemma   palea

callus   floret

upper glume   lower glume

spikelet

rhizome

P. eminens

Named intersectional hybrids

73. **Poa arida** Vasey [p. 600]
PLAINS BLUEGRASS

Plants perennial; glaucous or not; densely to loosely tufted or the culms solitary, rhizomatous. **Basal branching** intra- and extravaginal. **Culms** 15–80 cm, erect or the bases decumbent, terete or weakly compressed; **nodes** terete, 0–1 exserted. **Sheaths** closed for $^1/_{10}$–$^1/_5$($^1/_4$) their length, terete, smooth or sparsely scabrous, glabrous, bases of basal sheaths glabrous, distal sheath lengths (1.2)1.5–9(20) times blade lengths; **ligules** (1)1.5–4(5) mm, smooth or sparsely to moderately scabrous, apices obtuse to acute; **blades** strongly to gradually reduced in length distally, 1.5–5 mm wide, flat and moderately thin to folded and moderately thick and firm, abaxial surfaces smooth, adaxial surfaces smooth or sparsely to moderately scabrous, primarily over the veins, apices narrowly prow-shaped, flag leaf blades (0.4)1–7(10) cm. **Panicles** (2.5)4–12(18) cm, erect, usually narrowly lanceoloid, contracted, sometimes interrupted, infrequently loosely contracted, usually congested, with 25–100 spikelets; **nodes** with 1–5 branches; **branches** 1–9 cm, erect to infrequently ascending, rarely spreading, terete to weakly angled, smooth or the angles sparsely to moderately scabrous, with 3–24 spikelets. **Spikelets** 3.2–7 mm, lengths to 3.5(3.8) times widths, laterally compressed; **florets** 2–7; **rachilla internodes** smooth, sometimes sparsely puberulent. **Glumes** lanceolate, distinctly keeled, smooth or sparsely scabrous; **lower glumes** 3-veined; **calluses** usually glabrous, infrequently webbed, hairs to $^1/_4$ the lemma length; **lemmas** 2.5–4.5 mm, lanceolate to narrowly lanceolate, distinctly to weakly keeled, keels and marginal veins short- to long-villous, lateral veins moderately prominent, glabrous or puberulent, intercostal regions usually glabrous, infrequently hairy, hairs to 0.3 mm, margins glabrous, apices acute or blunt; **palea keels** scabrous, glabrous or short-villous at midlength, intercostal regions usually glabrous, sometimes puberulent to short-villous; **anthers** 1.3–2.2 mm. 2n = 56, 56+I, 56–58, 63, 64, 70, 76, 84, ca. 90, 95+-5, 100, 103.

Poa arida grows mainly on the eastern slope of the Rocky Mountains and in the northern Great Plains, primarily in riparian habitats of varying salinity or alkalinity. It is spreading eastward along heavily salted highway corridors. Reports of its occurrence west of the Continental Divide and in southwestern Texas are mostly attributable to misidentifications of *P. arctica*

2 mm

palea

lemma

2 mm

dorsal
view

side
view

floret

2 cm

2 mm

2 cm

spikelet

P. arida

SL

2 mm

dorsal
view

side
view

floret

1 cm

2 mm

2 cm

spikelet

P.×limosa

floret

2 mm

2 cm

2 mm

spikelet

2 cm

P.×gaspensis

subsp. *aperta* (p. 530), *P. arctica* subsp. *grayana* (p. 532), and rhizomatous specimens of *P. fendleriana* (p. 556).

Poa arida may reflect past hybridization between *P. secunda* (p. 586) and a species of *Poa* sect. *Poa*. *Poa glaucifolia* Scribn. & T.A. Williams refers to specimens of the northern Great Plains that have a more lax growth form with broader leaves and occasionally somewhat open panicles, florets with a small web, and sometimes lacking hairs between the keel and marginal veins of the lemma. Plants with these characteristics have chromosome counts of $2n = 56$ and 70, whereas *P. arida sensu stricto* usually has $2n = 63$, 64, or greater than 70. It is suspected that some of the variability reflects introgression from *P. secunda*.

74. Poa ×limosa Scribn. & T.A. Williams [p. 600]

Plants perennial; densely to loosely tufted or the culms solitary, shortly rhizomatous. **Culms** 20–80 cm, erect or the bases decumbent. **Sheaths** usually closed for about $^1/_6$ their length; **ligules** 1–4 mm, smooth or sparsely scabrous, apices obtuse to acute; **innovation blades** 0.5–2 mm wide; **cauline blades** 0.5–5 mm wide, flat, folded, abaxial surfaces smooth or scabrous, apices narrowly prow-shaped. **Panicles** 5–15 cm, erect, usually contracted, sometimes interrupted; **branches** shorter than 4 cm, erect, angles somewhat scabrous. **Spikelets** 4–7 mm, weakly laterally compressed; **florets** 2–5; **rachilla internodes** smooth. **Lower glumes** 3-veined; **calluses** glabrous or webbed, hairs to $^1/_4$ the lemma length; **lemmas** 2.5–4.5 mm, narrowly lanceolate, distinctly to weakly keeled, glabrous throughout or the keels and marginal veins sparsely long-villous, apices acute; **palea keels** scabrous; **anthers** aborted late in development or 1.3–2.2 mm. $2n = 64$.

Poa ×limosa grows at scattered locations in western North America. It prefers wet to moist, often saline or alkaline meadows, primarily in the sagebrush zone. It is probably a hybrid between *P. pratensis* (p. 522) and *P. secunda* subsp. *juncifolia* (p. 586). Vigorous artificial hybrids of this parentage have been produced; they resemble *P. ×limosa*.

75. Poa ×gaspensis Fernald [p. 600]

Plants perennial; densely to loosely tufted, rhizomatous. **Basal branching** intra- and extravaginal. **Culms** 15–50 cm, erect or the bases decumbent. **Sheaths** closed for $^1/_4$–$^1/_2$ their length, terete; **ligules** acute; **blades** flat, thin, apices broadly prow-shaped. **Panicles** erect, narrowly lanceoloid to ovoid, contracted, with 2–4 branches per node; **branches** ascending to spreading, sparsely scabrous. **Spikelets** 3.5–6 mm, laterally compressed, with 3–4 florets. **Glumes** broadly lanceolate, distinctly keeled, distinctly scabrous on the distal $^1/_3$; **lower glumes** 3-veined; **calluses** shortly webbed; **lemmas** 2.5–4.5 mm, broadly lanceolate, keeled, keels and marginal veins long-villous, intercostal regions softly puberulent; **palea keels** scabrous, long-villous at midlength; **anthers** 1.2–1.4 mm. $2n$ = unknown.

Poa ×gaspensis is found in the coastal mountains of the Gaspé Pennisula. There are few plants that fit the description. It seems to consist of hybrids between *P. pratensis* subsp. *alpigena* (p. 525) and *P. alpina* (p. 518). *Poa ×gaspensis* differs from *P. alpina* in its extravaginal branching, rhizomatous habit, and webbed calluses; from *P. pratensis* in its acute ligules and more pubescent lemmas; and from *P. arctica* (p. 529) in its sharply keeled, more scabrous glumes and its spikelet shape, which approaches those of *P. alpina* and *P. pratensis*.

14.14 ×DUPOA J. Cay. & Darbysh.

Jacques Cayouette

Stephen J. Darbyshire

Plants perennial; rhizomatous. **Culms** 25–80 cm, glabrous, often glaucous, with 2–4 nodes. **Sheaths** closed for $^1/_3$–$^2/_3$ their length, glabrous, not conspicuously glaucous; **auricles** absent; **ligules** membranous, whitish- or yellowish-brown, acute to obtuse, lacerate to erose, ciliate; **blades** glabrous. **Inflorescences** panicles. **Spikelets** with (1)2–3(4) florets. **Glumes** usually not exceeding the florets, glabrous, often glaucous, acute to acuminate; **calluses** bearded, hairs often crinkled below the keel and/or absent from the sides; **lemmas** usually scabrous, sometimes smooth towards the base, acute; **paleas** smooth or slightly scabrous; **anthers** 3, indehiscent. **Caryopses** absent.

×*Dupoa* is a sterile hybrid between *Dupontia* and *Poa*. Only one species is known.

SELECTED REFERENCES Aiken, S.G., L.L. Consaul, and M.J. Dallwitz. 1995 on. Grasses of the Canadian Arctic Archipelago: Descriptions, illustrations, identification and information retrieval. http://www.mun.ca/biology/delta/arcticf/poa/index.htm; Cayouette, J. and S.J. Darbyshire. 1993. The intergeneric hybrid grass "*Poa labradorica*". Nordic J. Bot. 13:615–629; Darbyshire, S.J., J. Cayouette, and S.I. Warwick. 1992. The intergeneric hybrid origin of *Poa labradorica* Steudel (Poaceae). Pl. Syst. Evol. 181:57–76.

1. ×**Dupoa labradorica** (Steudel) J. Cay. & Darbysh. [p. 603]

LABRADOR BLUEGRASS, PÂTURIN DU LABRADOR

Rhizomes to 1.5 mm thick, nodes with few roots. **Culms** 25–80 cm. **Ligules** 1–4 mm; **blades** 5–20 cm long, 2–6 mm wide. **Panicles** 8–20 cm, open to somewhat dense; **branches** glabrous, spikelet-bearing to the base. **Spikelets** 5–10 mm. **Glumes** subequal to unequal; **lower glumes** 4–8.5 mm, 1–3-veined; **upper glumes** 6.5–10 mm, (1)3-veined; **lemmas** 4–7 mm, 3–5-veined; **paleas** 4–6 mm; **anthers** 1.8–2.5 mm. $2n = 43$–46 (43+0–3).

×*Dupoa labradorica* is known from southeastern Hudson Bay, eastern James Bay, and northern Labrador. Its distribution is completely within the sympatric range of the parental species. Additional records are to be expected from regions where both parental species are found, such as Alaska and the Russian Far East. It grows on seashores, lagoon margins, and salt marshes, usually in relatively deep organic matter over clay layers sometimes mixed with sandy or gravelly deposits. It sometimes forms dense and almost pure stands. The occurrence and spread of ×*Dupoa labradorica* are related primarily to geological processes associated with glacio-eustatic rebound and seasonal ice action, affecting its seashore habitats.

×*Dupoa labradorica* is a sterile hybrid between *Dupontia fisheri* (see next) and *Poa eminens* (p. 598). It is similar to *Poa eminens*, but has narrower leaves, stems, and rhizomes, and more open panicles. It differs from *Dupontia fisheri* in having glumes that usually do not exceed the florets, and some woolly or crinkly callus hairs. It differs from both parental species in having indehiscent anthers and no caryopses.

14.15 DUPONTIA R. Br.

Jacques Cayouette

Stephen J. Darbyshire

Plants perennial; rhizomatous. **Culms** 5–80 cm. **Sheaths** closed for $\frac{1}{2}$–$\frac{2}{3}$ their length, glabrous; **auricles** absent; **ligules** membranous, glabrous, truncate; **blades** flat or folded, usually glabrous, adaxial surfaces sometimes scabrous or shortly pubescent. **Inflorescences** panicles, diffuse to dense and spikelike, with few spikelets; **branches** 0.7–3.5(8.5) cm, smooth, stiff, erect to reflexed, secondary branches usually appressed. **Spikelets** pedicellate, slightly laterally compressed, with 1–4(5) florets, distal florets sterile; **rachillas** prolonged beyond the uppermost floret or terminating in a vestigial floret, glabrous; **disarticulation** above the glumes and beneath the fertile florets. **Glumes** subequal, equaling or usually exceeding the distal florets, ovate and obtuse to lanceolate-attenuate, membranous, glabrous, 1–3-veined, unawned; **calluses** short, blunt, with a ring of stiff hairs, hairs to about 1 mm; **lemmas** ovate to ovate-lanceolate, membranous to coriaceous, glabrous or pubescent, with 3(5) fine veins, apices acute to acuminate, midveins sometimes excurrent as a mucro or awn to 1(2.2) mm; **paleas** shorter than the lemmas, glabrous, sometimes scabrous on the veins; **lodicules** 2, membranous, glabrous; **anthers** 3; **ovaries** glabrous. **Caryopses** shorter than the lemmas, concealed at maturity, falling free of the lemma and palea; **hila** $\frac{1}{6}$–$\frac{1}{3}$ the length of the caryopses, ovate. x = not clear, 7 or 11. Named after the early nineteenth-century French botanist J.D. Dupont.

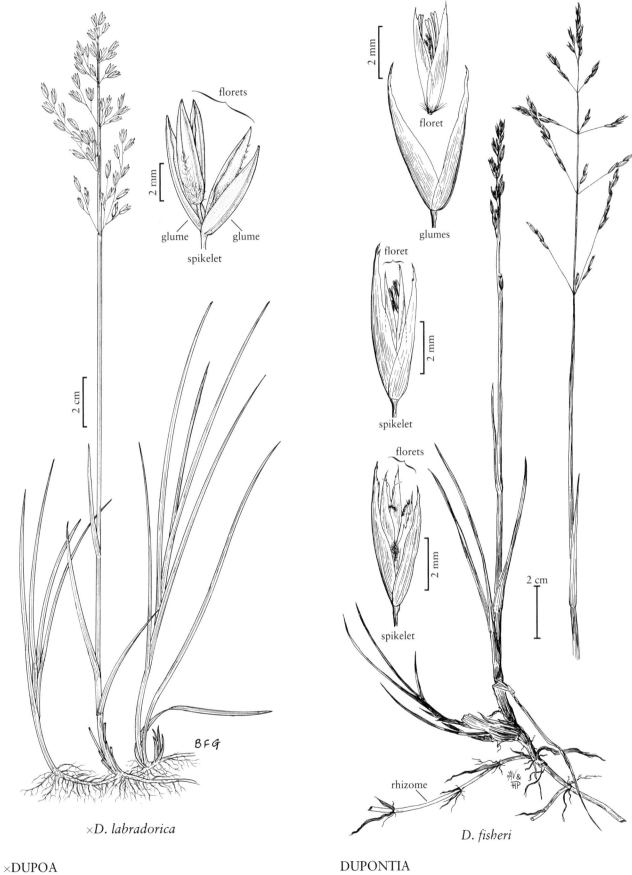

florets

2 mm

glume　　glume

spikelet

2 cm

2 mm

floret

glumes

floret

2 mm

spikelet

florets

2 mm

spikelet

2 cm

BFG

rhizome

AV &
HP

×*D. labradorica*

D. fisheri

×DUPOA　　　　　　　　　　　DUPONTIA

Dupontia is a monotypic genus of arctic and subarctic wetlands, found throughout the holarctic region except in Scandinavia. Hybrids with *Arctophila* are referred to ×*Arctodupontia* (see next); hybrids with *Poa eminens* are referred to ×*Dupoa* (see previous).

SELECTED REFERENCES **Brysting, A.K., S.G. Aiken, L.P. Lefkovitch,** and **R.L. Boles.** 2003. *Dupontia* (Poaceae) in North America. Canad. J. Bot. 81:769–779; **Brysting, A.K., M.F. Fay, I.J. Leitch,** and **S.G. Aiken.** 2004. One or more species in the arctic grass genus *Dupontia?*—A contribution to the Panarctic Flora project. Taxon 53:365–382; **Jurtsev, B.A.** 1995. *Dupontia*. Pp. 224–229 *in* J.G. Packer (ed., English edition). Flora of the Russian Arctic, vol. 1, trans. G.C.D. Griffiths. University of Alberta Press, Edmonton, Alberta, Canada. 330 pp. [English translation of A.I. Tolmachev (ed.). 1964. Arkticheskaya Flora SSSR, vol. 2. Nauka, Leningrad [St. Petersburg], Russia. 272 pp.].

1. Dupontia fisheri R. Br. [p. 603]

FISHER'S TUNDRAGRASS, DUPONTIE DE FISHER

Rhizomes 1–3 mm thick. **Culms** 5–80 cm, erect, glabrous. **Ligules** of lower leaves 0.4–3 mm; **ligules** of flag leaves 1–4(5.5) mm, usually lacerate; **blades** 1–13 cm long, 1–4 mm wide. **Panicles** 2.5–18 cm long, 1–6 cm wide. **Spikelets** 4–8.5(9) mm; **rachilla internodes** 1–1.5 mm. **Glumes** 4–8.5(9) mm; **lemmas** 3–6.5 mm; **paleas** 2.8–6 mm; **anthers** 1.5–3.5 mm. **Caryopses** 1.5–3 mm. $2n$ = 42, 44, 66, 84, 88, about 105, about 126, 132.

Dupontia fisheri grows in wet meadows, wet tundra, marshes, and along streams and the edges of lagoons, ponds, and lake shores, in sand, silt, clay, moss, and rarely in bogs.

Two subspecies of *Dupontia* are sometimes recognized in North America. *Dupontia fisheri* R. Br. subsp. *fisheri* supposedly differs from subsp. *psilosantha* (Rupr.) Hultén in being shorter than 40 cm, having erect panicle branches, 2–4 florets per spikelet, pubescent, obtuse lemmas, and $2n$ = 84, 88, or 132.

Dupontia fisheri subsp. *psilosantha* is taller, has reflexed panicle branches, 1–2 florets per spikelet, more or less glabrous, acute lemmas, and $2n$ = 42 or 44. Plants referable to subsp. *psilosantha* are restricted to coastal marshes, rarely penetrating inland along riparian habitats, from James Bay to the lower arctic archipelago. Plants referable to subsp. *fisheri* are less halophytic and more northerly in their distribution, being found in a variety of inland marshes and wet tundra habitats from northern Alaska to Ellesmere Island. Intermediates are readily found (e.g., hexaploid, $2n$ = 66 "hybrids" from Alaska) and the correlations among chromosome number, morphology, ecology, and distribution are relatively weak in North America and Greenland. For these reasons, no subspecific taxa are recognized in this treatment.

Dupontia fisheri hybridizes with *Poa eminens* to form ×*Dupoa labradorica* (p. 602). The hybrids differ from *D. fisheri* in having glumes that usually do not exceed the florets, and in having some woolly or crinkly callus hairs. The hybrid genus ×*Arctodupontia* is formed from *D. fisheri* and *Arctophila fulva*, and differs from *D. fisheri* in having lemmas with truncate, lacerate to dentate apices, rather than acute to acuminate apices.

14.16 ×ARCTODUPONTIA Tzvelev

Stephen J. Darbyshire

Jacques Cayouette

Plants perennial; rhizomatous. **Culms** to 38 cm. **Sheaths** closed for $^1/_3$–$^3/_4$ their length; **auricles** absent; **ligules** membranous, acute to truncate, lacerate; **blades** flat or folded, glabrous. **Inflorescences** partly open panicles; **branches** stiff, glabrous, lower branches sometimes reflexed. **Spikelets** somewhat compressed, with 2–4 florets. **Glumes** subequal, ovate, membranous, acute; **calluses** with scant, stiff hairs; **lemmas** ovate, membranous to subcoriaceous, 1(3)-veined, apices truncate and lacerate-dentate; **paleas** 2-keeled, glabrous; **anthers** indehiscent; **ovaries** glabrous; **lodicules** 2, membranous. **Caryopses** apparently absent.

×*Arctodupontia* is a sterile hybrid between the two monotypic genera *Dupontia* and *Arctophila*. It is intermediate between the two parents.

SELECTED REFERENCE **Brysting, A.K., S.G. Aiken, L.P. Lefkovitch,** and **R.L. Boles.** 2003. *Dupontia* (Poaceae) in North America. Canad. J. Bot. 81:769–779.

1. ×Arctodupontia scleroclada (Rupr.) Tzvelev [p. 606]

Culms 13–40 cm. Ligules 2–4 mm; blades 2–6 cm long, 1.5–4 mm wide. Panicles 4–12 cm, partly or completely exserted. Spikelets 3.5–7 mm. Glumes 3–6 mm; lower glumes 1-veined; upper glumes 1(3)-veined; lemmas 3–5 mm, paleas about as long as the lemmas; anthers to about 2 mm. 2*n* = unknown.

×*Arctodupontia scleroclada* is known from Mansfield Island in northern Hudson Bay, and from Malaya Zemlya and the Chukchi Peninsula in Russia. Plants seemingly intermediate between *Dupontia fisheri* and *Arctophila fulva* are common, but caution must be exercised in assigning them to ×*Arctodupontia*. The hybrids differ from their parents in being sterile (as indicated by their indehiscent anthers), and in having lemmas with truncate, lacerate to dentate apices. *Dupontia* has lemmas with acute to acuminate apices; *Arctophila* has lemmas with obtuse, entire apices.

14.17 ARCTOPHILA (Rupr.) Andersson

Jacques Cayouette

Stephen J. Darbyshire

Plants perennial; rhizomatous, sometimes producing aquatic leaves when submerged. **Culms** (5)10–100 cm, erect, glabrous, rooting at the lower nodes; **nodes** glabrous. **Sheaths of aquatic leaves** closed to near the apices, translucent, pale pinkish brown; **collars** inconspicuous; **ligules** acute; **blades** pinkish brown. **Sheaths of aerial leaves** usually closed for over ¹/₂ their length, sometimes open to the base, opaque, green to olive-green, glabrous; **collars** conspicuous as a zone of contrasting color; **auricles** absent; **ligules** membranous, truncate, lacerate; **blades** usually flat, glabrous, upper blades conspicuously longer than the lower blades. **Inflorescences** open panicles; **branches** stiff and ascending to pendulous, glabrous, some branches longer than 1 cm. **Spikelets** pedicellate, somewhat laterally compressed, with (1)2–7(9) florets; **rachillas** prolonged beyond the base of the distal floret, glabrous; **disarticulation** above the glumes, beneath the florets. **Glumes** subequal, broadly lanceolate to ovate, membranous to subcoriaceous, glabrous, 1–3(5)-veined, acute to obtuse, unawned; **calluses** short, blunt, shortly pubescent to almost glabrous; **lemmas** ovate, glabrous, membranous to subcoriaceous, with 3(5) obscure veins, lateral veins usually not reaching the lemma apices, apices entire or somewhat erose, obtuse, unawned or rarely the central vein extended as a short mucro to about 0.2 mm; **paleas** subequal to the lemmas; **lodicules** 2, free, glabrous, toothed or entire; **anthers** 3; **ovaries** glabrous. **Caryopses** shorter than the lemmas, concealed at maturity, falling free; **hila** broadly ovate, ¹/₆–¹/₅ the length of the caryopses. *x* = 7. From the Greek *arktos*, 'north', and *philia*, 'loving'.

Arctophila is a monospecific, but highly polymorphic, holarctic genus closely related to *Dupontia* (p. 602).

SELECTED REFERENCES **Aiken, S.G.** and **R.A. Buck**. 2002. Aquatic leaves and regeneration of last year's straw in the arctic grass *Arctophila fulva*. Canad. Field-Naturalist 116:81–86; **Brysting, A.K., M.F. Fay, I.J. Leitch,** and **S.G. Aiken**. 2004. One or more species in the arctic grass genus *Dupontia?*—A contribution to the Panarctic Flora project. Taxon 53:365–382.

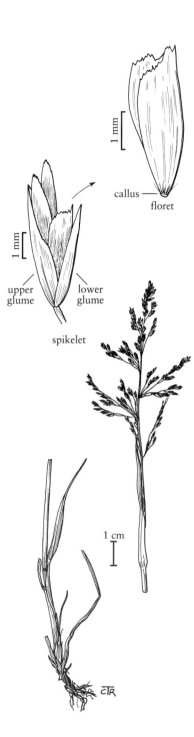

×*A. scleroclada*

A. fulva

×ARCTODUPONTIA

ARCTOPHILA

1. Arctophila fulva (Trin.) Andersson [p. 606]
PENDANT GRASS, ARCTOPHILE FAUVE

Culms (5)10–80(100) cm. **Ligules** (1)2–6(8) mm; **blades** 2–23 cm long, 1–5(10) mm wide. **Panicles** 3–20 cm long, (1.5)3–11 cm wide. **Spikelets** 2.5–7(8) mm. **Glumes** 1.5–4(5) mm; **lower glumes** exceeded by the lowest floret; **upper glumes** shorter than to longer than the lowest floret; **lemmas** 2.5–4 mm; **paleas** (1)1.8–4 mm; **anthers** 1.2–3 mm. **Caryopses** 1.5–2.2 mm. $2n = 42, 63$.

Arctophila fulva grows as an emergent species in shallow, standing water, or along slow-moving streams, wet meadows, marshes, and saturated soils of low arctic and subarctic regions, where it often forms pure stands. It is one of the few grasses that develop aquatic leaves. Field observations indicate that under some environmental conditions, *A. fulva* can propagate vegetatively from detached stems that have over-wintered (Aiken and Buck 2002). In the *Flora* region, it grows from Alaska through the Yukon, Northwest Territories, Nunavut, Ontario, Quebec, and Labrador to Greenland. Its range extends across Eurasia to arctic Scandinavia. It forms a sterile hybrid, ×*Arctodupontia scleroclada* (p. 605), with *Dupontia fisheri*. The hybrid differs from *Arctophila fulva* in having lemmas with truncate, lacerate to dentate apices, rather than obtuse, entire apices.

14.18 TORREYOCHLOA G.L. Church

Jerrold I. Davis

Plants perennial; rhizomatous. **Culms** 18–145 cm, usually erect, sometimes decumbent and rooting at the lower nodes; **internodes** hollow. **Sheaths** open to the base; **auricles** absent; **ligules** membranous; **blades** flat. **Inflorescences** terminal panicles; **branches** scabrous, usually densely scabrid distally; **pedicels** less than 0.5 mm thick. **Spikelets** pedicellate, laterally compressed to terete, with 2–8 florets; **disarticulation** above the glumes and beneath the florets. **Glumes** unequal, shorter than the lowest lemma, rounded to weakly keeled, membranous, veins obscure to prominent, unawned; **lower glumes** 1(3)-veined; **upper glumes** (1)3(5)-veined; **calluses** blunt, glabrous; **lemmas** membranous, rounded to weakly keeled, sometimes pubescent, particularly proximally, prominently (5)7–9-veined, veins more or less parallel, veins and interveins usually scabridulous, particularly distally, lateral veins usually reduced or absent, apices scabridulous and entire to serrate-erose, unawned; **paleas** subequal to the lemmas, 2-veined; **lodicules** 2, free, glabrous, entire or toothed; **anthers** usually 3; **ovaries** usually hairy, sometimes glabrous. **Caryopses** shorter than the lemmas, concealed at maturity, oblong, flattened dorsally, falling free; **hila** oblong, about ⅓ the length of the caryopses. $x = 7$. Named for John Torrey (1796–1873), an American botanist.

Torreyochloa grows in cold, wet, non-saline environments. It includes the two North American species treated below, plus two additional taxa in northeastern Asia (Koyama & Kawano 1964). Although similar to *Glyceria* and *Puccinellia*, *Torreyochloa* is not closely related to either (Church 1952; Soreng et al. 1990). It is distinguished from *Glyceria* by its open leaf sheaths, and from *Puccinellia* by the 7–9 (occasionally 5) prominent, rather than faint, lemma veins.

SELECTED REFERENCES **Church, G.L.** 1952. The genus *Torreyochloa*. Rhodora 54:197–200; **Koyama, T. and S. Kawano.** 1964. Critical taxa of grasses with North American and eastern Asiatic distribution. Canad. J. Bot. 42:859–884; **Soreng, R.J., J.I. Davis, and J.J. Doyle.** 1990. A phylogenetic analysis of chloroplast DNA restriction site variation in the Poaceae subfam. Poöideae. Pl. Syst. Evol. 172:83–97.

1. Mature inflorescences linear to narrowly elliptic, 0.3–1 cm wide, 5.5–19 times as long as wide; widest cauline blades 3.4–7.2 mm wide . . . . .1. *T. erecta*
1. Mature inflorescences conic, ovoid, or obovoid, 1–16 cm wide, 1–7.5 times as long as wide; widest cauline blades 1.5–18 mm wide . . . . .2. *T. pallida*

1. **Torreyochloa erecta** (Hitchc.) G.L. Church [p. 609]
SPIKED FALSE MANNAGRASS

Culms 20–62 cm tall, 1–1.4 mm thick, usually erect. **Ligules** of larger cauline leaves 2.6–6.5 mm, truncate or rounded to acute; **widest cauline blades** 3.4–6(7.2) mm wide. **Panicles** 5.5–11 cm long, 0.3–1 cm wide, (5.5)6–19 times as long as wide, linear to narrowly elliptic at maturity; **lowermost branches** stiff to flexuous, reflexed to erect at maturity. **Spikelets** 4.2–6.4 mm, with 4–7 florets. **Lower glumes** 0.8–1.6 mm; **upper glumes** 1–2.1 mm; **lemmas** 2.3–3.1 mm, obtuse to acute; **anthers** 0.6–0.8 mm. $2n = 14$.

Torreyochloa erecta grows at elevations above 2000 m, in the margins of subalpine and alpine lakes and streams of the Sierra Nevada and Cascade ranges.

2. **Torreyochloa pallida** (Torr.) G.L. Church [p. 609]

Culms 18–145 cm tall, 0.6–4.8 mm thick, erect to decumbent, sometimes matted. **Ligules** of larger cauline leaves 2–9 mm, truncate or acute to attenuate; **widest cauline blades** 1.5–18 mm wide. **Panicles** (3)5–25 cm long, (1)1.8–16 cm wide, 1–5.75(7.5) times as long as wide, narrowly to widely conic, ovoid, or obovoid at maturity; **lowermost branches** stiff to flexuous, reflexed to erect at maturity. **Spikelets** 3.6–6.9 mm, with 2–8 florets. **Lower glumes** 0.7–2.1 mm; **upper glumes** 0.9–2.7 mm; **lemmas** 2–3.6 mm, truncate to acute; **anthers** 0.3–1.5 mm. $2n = 14$.

All three varieties of *Torreyochloa pallida* grow in swamps, marshes, bogs, and the margins of lakes and streams. They are usually morphologically distinct, and tend to have different geographic ranges.

1. Widest cauline blades 1.5–3 mm wide; anthers of the lowest floret in each spikelet 0.3–0.6 mm long . . . . var. *fernaldii*
1. Widest cauline blades 2.8–18 mm wide; anthers of the lowest floret in each spikelet 0.5–1.5 mm long.
  2. Anthers of the lowest floret in each spikelet 0.7–1.5 mm long; basal diameter of the culms 1.2–3 mm; upper glumes 1.4–2.7 mm long; plants generally of eastern North America . . . . var. *pallida*
  2. Anthers of the lowest floret in each spikelet 0.5–0.7 mm long; basal diameter of the culms (1.3)1.8–4.8 mm; upper glumes 0.9–1.8 mm long; plants of western North America . . . . var. *pauciflora*

Torreyochloa pallida var. **fernaldii** (Hitchc.) Dore [p. 609]
FERNALD'S FALSE MANNAGRASS, GLYCÉRIE DE FERNALD

Culms 18–47 cm tall, 0.6–1.2 mm thick, erect to decumbent, matted. **Ligules** of larger cauline leaves 2–4 mm; **widest cauline blades** 1.5–3 mm wide. **Panicles** 5–8(9) cm long, 1.8–3(5) cm wide, 2.5–3.8 times as long as wide, conic to ovoid. **Spikelets** 4–4.8 mm, with 2–5(7) florets. **Lower glumes** 0.8–1.4 mm; **upper glumes** 1.3–1.6 mm; **lemmas** 2–2.4 mm, obtuse; **anthers** of lowest floret 0.3–0.6 mm.

Torreyochloa pallida var. *fernaldii* grows from Saskatchewan through southern Ontario and the Great Lakes area to New Brunswick and the New England states. *Torreyochloa pallida* var. *fernaldii* is often difficult to distinguish from *T. pallida* var. *pallida* where their ranges overlap.

Torreyochloa pallida (Torr.) G.L. Church var. **pallida** [p. 609]
PALE FALSE MANNAGRASS

Culms 47–110 cm tall, 1.2–3 mm thick, erect to decumbent. **Ligules** of larger cauline leaves 5–8 mm; **widest cauline blades** 2.8–9(11.4) mm wide. **Panicles** (7.5)11–19 cm long, 3–16 cm wide, 1–4.1 times as long as wide, ovoid to ellipsoidal. **Spikelets** 3.6–6.5 mm, with 2–6 florets. **Lower glumes** 1–2.1 mm; **upper glumes** 1.4–2.7 mm; **lemmas** 2.4–3.6 mm, obtuse to acute; **anthers** of lowest floret 0.7–1.5 mm.

The range of *Torreyochloa pallida* var. *pallida* extends from southeastern Manitoba to north of Lake Superior and Maine, and south to Missouri and Georgia. *Torreyochloa pallida* var. *pallida* can be difficult to distinguish from *T. pallida* var. *fernaldii*.

Torreyochloa pallida var. **pauciflora** (J. Presl) J.I. Davis [p. 609]
WEAK MANNAGRASS

Culms 20–145 cm tall, (1.3)1.8–4.8 mm thick, usually erect. **Ligules** of larger cauline leaves 3–9 mm; **widest cauline blades** 3.6–18 mm wide. **Panicles** (3)5–25 cm long, (1)2–14 cm wide, 1.2–5.75(7.5) times as long as wide, narrowly to widely conic, ovoid, or obovoid. **Spikelets** 3.6–6.9 mm, with (3)4–8 florets. **Lower glumes** 0.7–1.6 mm; **upper glumes** 0.9–1.8 mm; **lemmas** 2.2–3.3 mm, truncate or obtuse to acute; **anthers** of lowest floret 0.5–0.7 mm.

Torreyochloa pallida var. *pauciflora* grows in western North America, from sea level to 3500 m. Robust plants from Pacific lowland forests are often taller than 1 m; plants from farther east and higher elevations tend to be shorter.

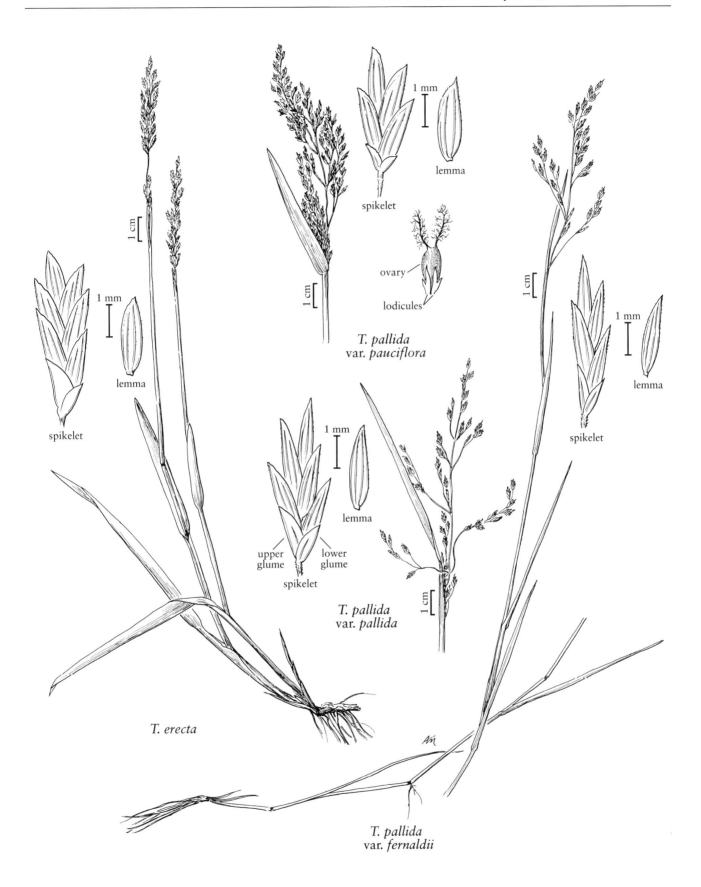

spikelet

lemma

T. pallida
var. pauciflora

ovary

lodicules

spikelet

lemma

spikelet

lemma

upper
glume

lower
glume

spikelet

T. pallida
var. pallida

spikelet

lemma

T. erecta

T. pallida
var. fernaldii

TORREYOCHLOA

14.19 CATABROSA P. Beauv.

Mary E. Barkworth

Plants perennial; not rhizomatous, sometimes stoloniferous. **Culms** 5–70 cm, usually decumbent and rooting at the lower nodes; **nodes** glabrous. **Sheaths** closed; **auricles** absent; **ligules** membranous; **blades** flat. **Inflorescences** open panicles, with at least some branches longer than 1 cm. **Spikelets** pedicellate, laterally compressed to terete, with (1)2(3) florets; **rachillas** glabrous, prolonged beyond the base of the distal fertile floret, empty or with reduced florets; **disarticulation** above the glumes and beneath the florets. **Glumes** unequal, much shorter than the lemmas, scarious, veinless or the upper glumes with 1 vein at the base, apices rounded to truncate, unawned; **calluses** short, blunt, glabrous; **lemmas** glabrous, conspicuously 3-veined, veins raised, rounded over the midvein, apices rounded to truncate, erose and scarious, unawned; **paleas** subequal to the lemmas, 2-veined; **lodicules** 2, truncate, irregularly lobed; **anthers** 3; **ovaries** glabrous. **Caryopses** shorter than the lemmas, concealed at maturity, fusiform; **hila** ovoid. $x = 5$. Name from the Greek *katabrosis*, 'eating up' or 'corrosion', a reference to the appearance of the lemma apices.

Catabrosa, a genus of two species, grows in marshes and shallow waters of the Northern Hemisphere and South America. It resembles members of the *Meliceae* in its closed leaf sheaths, truncate, scarious lemma apices, and chromosome base number, but lacks the distinctive lodicule morphology of that tribe. Some features support its inclusion in the *Poeae*. The lodicules are similar to those found elsewhere in the *Poeae*, and the closed leaf sheaths are not uncommon there. Chloroplast DNA data also support its placement in the *Poeae* (Soreng et al. 1990). The scarious glumes and prominently 3-veined lemmas are unusual in both the *Poeae* and *Meliceae*, but are also found in *Cutandia* of the *Poeae*. One species is native to the *Flora* region.

SELECTED REFERENCE Soreng, R.J., J.I. Davis, and J.J. Doyle. 1990. A phylogenetic analysis of chloroplast DNA restriction site variation in Poaceae subfam. Poöideae. Pl. Syst. Evol. 172:83–97.

1. Catabrosa aquatica (L.) P. Beauv. [p. 611]
BROOKGRASS, WATER WHORLGRASS,
CATABROSA AQUATIQUE

Plants often stoloniferous. **Culms** 10–60 cm, glabrous. **Sheaths** glabrous; **ligules** 1–8 mm, acute to truncate, erose to subentire; **blades** (1)3–15(20) cm long, 2–13 mm wide. **Panicles** 3–35 cm long, (1)2–10(12) cm wide; **nodes** distant, with 3 to many, often very unequal branches. **Spikelets** 1.5–3.5(4) mm, terete to somewhat dorsiventrally compressed, lowest floret sessile, second floret on an elongate internode; **rachilla internodes** 0.75–1.5 mm. **Lower glumes** 0.7–1.3 mm, often not reaching the base of the distal floret; **upper glumes** 1.2–2.2 mm; **lemmas** 2–3 mm; **anthers** 2–3 mm. $2n = 20, 30$.

Catabrosa aquatica grows in wet meadows and the margins of streams, ponds, and lakes in the *Flora* region, Argentina, Chile, Europe, and Asia. It is listed as endangered in Wisconsin. Although palatable, it is never sufficiently abundant to be important as a forage species. The species is regarded here as being variable, but having no infraspecific taxa.

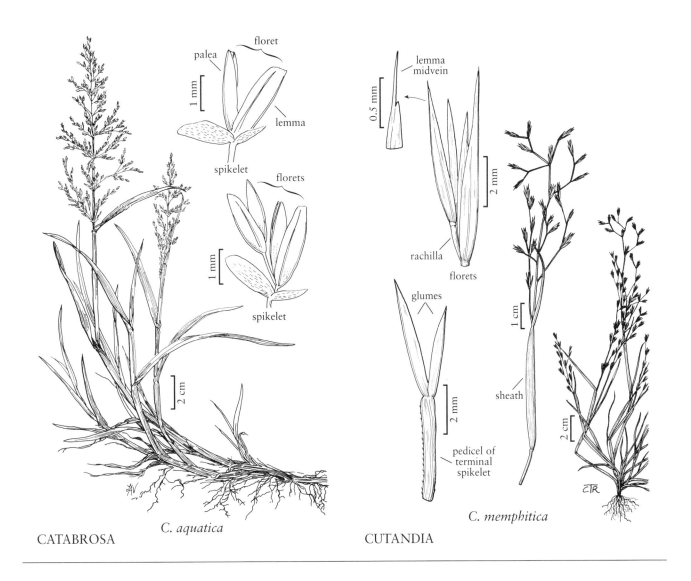

CATABROSA   *C. aquatica*   CUTANDIA   *C. memphitica*

14.20 CUTANDIA Willk.

Mary E. Barkworth

Plants annual. **Culms** (4)10–42 cm. **Sheaths** open to the base; **auricles** absent; **ligules** membranous; **blades** flat to convolute. **Inflorescences** usually panicles, rarely racemes; **branches** divaricate, some branches longer than 1 cm; **pedicels** of lateral spikelets 0.3–2 mm, pedicels of terminal spikelets on a branch longer. **Spikelets** laterally compressed, with 2–9(12) florets, distal florets often reduced; **disarticulation** occurring variously at the base of the florets, the pedicels, or the branches. **Glumes** equal or unequal, usually shorter than the adjacent lemmas, membranous, acuminate, rounded, or emarginate, unawned or shortly awned, awns glabrous; **lower glumes** 1–3-veined; **upper glumes** 1–5-veined; **calluses** short, glabrous; **lemmas** membranous, 3-veined, keeled on each vein, apices usually emarginate or bifid, varying to rounded or acuminate, unawned, mucronate, or awned from the sinus or slightly below; **paleas** subequal to the lemmas; **lodicules** 2, free, membranous, ciliate; **anthers** 3; **ovaries** glabrous.

Caryopses shorter than the lemmas, concealed at maturity, narrow, longitudinally grooved; **hila** oval. *x* = 7. Named for Vincente Cutanda (1804–1866), a Spanish botanist.

Cutandia is a Mediterranean and western Asian genus of six species, one of which has been collected in California.

SELECTED REFERENCE **Stace, C.A.** 1978. Notes on *Cutandia* and related genera. Bot. J. Linn. Soc. 76:350–352.

1. Cutandia memphitica (Spreng.) K. Richt. [p. 611]
MEMPHIS GRASS

Culms 4–35(42) cm, prostrate, geniculate, or erect, somewhat stiff. **Sheaths** inflated, glabrous; **ligules** 2–5 mm, truncate to acute, entire to erose; **blades** mostly 2–10 cm long, 1–4 mm wide, glabrous. **Panicles** 3–18 cm; **branches** usually 1(2) per node, branches and pedicels strongly divaricate at maturity; **pedicels of lateral spikelets** 0.3–0.6(1.5) mm, sharply angular. **Spikelets** 7–10.5 mm, oblanceolate, with 2–3(4) florets; **rachilla internodes** 1.5–3 mm. **Glumes** acute to acuminate, 1-veined, veins excurrent to 0.4 mm; **lower glumes** 3.5–5 mm; **upper glumes** 4.5–6 mm; **lemmas** 5.8–7.5 mm, smooth or scabridulous on the midvein, emarginate to shortly bifid, midveins excurrent to 1.2 mm; **anthers** 1–2 mm. 2*n* = 14.

Cutandia memphitica is native to maritime sands, inland dunes, sandy gravel, and gypsum soils in the Mediterranean region and Middle East. In the *Flora* region, it was collected in sandy soil at a nursery at Devil's Canyon in the San Bernardino Mountains, California, in 1933. How it got there and whether it persists are not known.

14.21 BRIZA L.

Neil Snow

Plants annual or perennial; cespitose. **Culms** 5–100 cm, usually erect, unbranched; **internodes** hollow; **nodes** glabrous. **Sheaths** sometimes less than ¹/₂ as long as the internodes, open; **auricles** absent; **ligules** hyaline; **blades** flat, usually erect. **Inflorescences** open panicles; **branches** sparsely strigose, capillary, spikelets usually pendulous, some branches longer than 1 cm. **Spikelets** pedicellate, pendulous, oval to triangular in side view, becoming light brown at maturity, laterally compressed but the glumes and lemmas with broadly rounded backs, glumes and florets strongly divergent from the rachillas, with 4–12(15) chartaceous florets, distal florets rudimentary; **rachillas** glabrous, not prolonged beyond base of the distal floret; **disarticulation** above the glumes and beneath the florets. **Glumes** subequal, shorter than to longer than the adjacent lemmas, naviculate, faintly 3–7-veined, margins more or less membranous, apices obtuse, unawned; **calluses** short, glabrous; **lemmas** inflated, about as wide as long, with broadly rounded backs, similar in shape to the glumes but somewhat cordate, margins becoming hyaline, frequently splitting perpendicular to the midveins, unawned; **paleas** shorter than the lemmas, scarious or chartaceous; **lodicules** 2, joined or free, usually entire, sometimes toothed; **anthers** 3; **ovaries** glabrous. **Caryopses** shorter than the lemmas, concealed at maturity, usually falling with the lemma and palea, ovoid to obovoid; **hila** round to elliptic. *x* = 5, 7. From the Greek *brizo*, 'to nod', in reference to the spikelets.

Briza, a genus of about 20 species, is native to Eurasia and South America. Most species have little to no fodder value because of the scant foliage. The ornamental value of the genus is more significant; the species are often grown for use in dried floral arrangements. Three European species are now scattered in the more temperate parts of southern Canada and the United States, and will undoubtedly be collected in areas not indicated here. *Briza* species can become weedy where established.

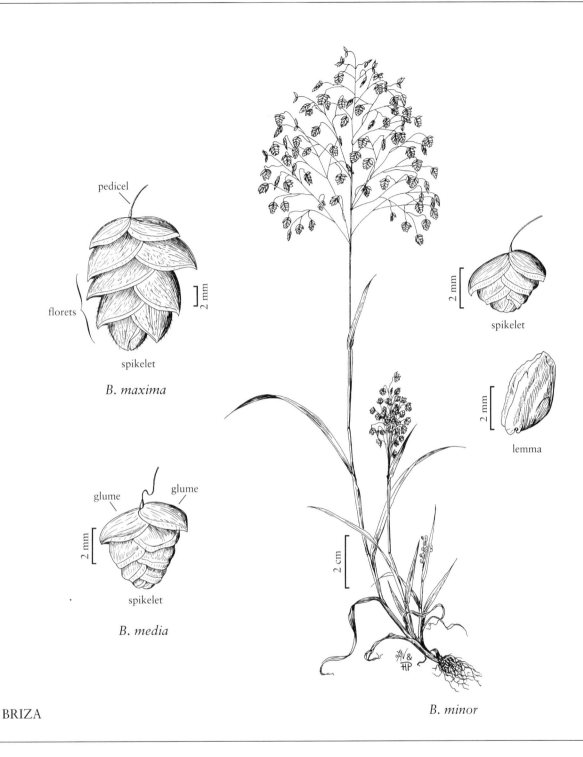

B. maxima

B. media

B. minor

BRIZA

SELECTED REFERENCES **Gould, F.W.** 1975. The Grasses of Texas. Texas A&M University Press, College Station, Texas, U.S.A. 653 pp.; **Murray, B.G.** and **N.R.N. Barker.** 1988. Pollen/stigma interactions and hybridization in the genus *Briza* L. (Gramineae). Evol. Trends Pl. 2:107–110.

1. Plants perennial; ligules about 0.5 mm long; sheaths open for about $^1/_2$ their length . . . . 1. *B. media*
1. Plants annual; ligules 3–13 mm long; sheaths open to near the base.
    2. Spikelets 10–20 mm long . . . . 2. *B. maxima*
    2. Spikelets 2–7 mm long . . . . 3. *B. minor*

1. Briza media L. [p. 613]
PERENNIAL QUAKINGGRASS, AMOURETTE COMMUNE, AMOUR DU VENT

Plants perennial, shortly rhizomatous. **Culms** 15–75 cm. **Leaves** mostly basal; **sheaths** about ½ the length of the internodes, open about ½ their length; **ligules** about 0.5 mm, usually not decurrent, sometimes erose at the apices, truncate; **blades** 4–16 cm long, blades of the upper leaves shorter than those below, 1.9–3.2 mm wide, glabrous or scabridulous, margins strigose. **Panicles** 8–20 cm long, to almost as wide; **pedicels** 5–20 mm. **Spikelets** 4–5.5 mm, mostly oval, with 3–6(10) florets. **Lower glumes** 2.5–3.2 mm; **upper glumes** 2.5–4 mm; **lowermost lemmas** 3–4 mm, indistinctly 9- or 10-veined, apices broadly obtuse; **paleas** about 3 mm, V-shaped in cross section, scarious, margins hyaline and ciliolate; **anthers** 1.3–2 mm. **Caryopses** 1.2–1.5 mm, distinctly flattened on 1 side. $2n = 10, 14, 28$.

Briza media is native to chalk and clay grasslands of Europe. It grows in acid to calcareous soils in moist to somewhat dry, sunny conditions, in meadow floodplains, forest clearings, old meadows, and pastures. It is often grown as an ornamental, and can colonize artificial habitats such as roadsides, but does not appear to invade recently disturbed locations. In the *Flora* region, it is most abundant in eastern North America, and is found in a few widely scattered locations elsewhere.

2. Briza maxima L. [p. 613]
BIG QUAKINGGRASS

Plants annual. **Culms** 20–80 cm. **Leaves** evenly distributed; **sheaths** frequently less than ½ as long as the internodes, open to near the base, margins overlapping; **ligules** 3–7 mm, sides sometimes decurrent, margins entire to erose, acute; **blades** 2.5–20 cm long, 2–8 mm wide, margins strigose or glabrous. **Panicles** 3.5–10 cm long, mostly 1–5 cm wide; **pedicels** 5–20 mm. **Spikelets** 10–20 mm, oval to elliptic, with 4–12(15) florets. **Lower glumes** 5–5.5 mm, 5-veined; **upper glumes** 6–6.5 mm, 7-veined; **lowermost lemmas** 7–9 mm, 7–9-veined, surfaces usually glabrous proximally, becoming villous distally, apices obtuse; **paleas** about 4 mm, more or less ciliolate along the margins; **anthers** 1.2–1.5 mm. **Caryopses** 2–3 mm, obovoid. $2n = 10, 14$.

Briza maxima is native to the Mediterranean region. Cultivated as an ornamental, it is possibly one of the earliest grasses grown for other than edible purposes. It occasionally becomes naturalized in dry to somewhat moist but well-drained, fine or sandy soil on banks, rocky places, open woodlands, and cultivated areas such as roadsides and pastures. In the *Flora* region, it is known from scattered locations, mostly in Oregon and California, where it is an invader of coastal dune habitat.

3. Briza minor L. [p. 613]
LITTLE QUAKINGGRASS

Plants annual. **Culms** 7.5–80 cm. **Leaves** evenly distributed; **sheaths** ½–¾ the length of the internodes, open to near the base, margins hyaline distally; **ligules** 4–13 mm, sides sometimes decurrent, margins at the base sometimes encasing the culms, truncate to acute; **blades** 5.5–12 cm long, 1–8(10) mm wide, slightly scabrous. **Panicles** (2)4–14(18) cm long, to 11 cm wide; **pedicels** 4–12 mm. **Spikelets** (2)3–4(7) mm, triangular to oval, with 4–7(13) florets. **Lower glumes** 2–2.5 mm; **upper glumes** 2–3.5 mm; **lowermost lemmas** 1.6–2 mm, frequently irregular in shape, becoming hyaline distally, glabrous, sometimes minutely scurfy, veins indistinct; **paleas** about 1.5 mm, often minutely scurfy; **anthers** 0.4–0.5 mm. **Caryopses** 0.8–1 mm, ovoid. $2n = 10, 14$.

Briza minor is native to the Mediterranean region. It is the most widespread species of *Briza* in the *Flora* region, growing in many habitats: swamp margins, seasonal wetlands and around vernal pools, open woodlands, sandhills, roadsides, and pastures. It appears to be established from southern British Columbia south through western Oregon to California and Arizona, and in the east from the Atlantic states to the Gulf Coast states, inland to Oklahoma and Arkansas.

14.22 AIRA L.

J.K. Wipff

Plants annual; tufted. **Culms** 1–55 cm, erect to decumbent. **Leaves** cauline; **sheaths** open for most of their length, glabrous, usually scabridulous, occasionally smooth; **auricles** absent; **ligules** membranous; **blades** of the uppermost leaves greatly reduced. **Inflorescences** open or contracted panicles, sometimes spikelike, with more than 1 spikelet associated with most nodes; **branches** longer than 5 mm, capillary, appressed to strongly divergent; **pedicels** capillary, appressed to divergent. **Spikelets** 1.5–3.8 mm, laterally compressed, with 2 bisexual florets, both usually awned, the lower floret sometimes unawned, occasionally both unawned; **rachillas** glabrous, lowest segments about 0.2 mm, florets appearing opposite, usually not or scarcely extended beyond the base of the distal floret; **disarticulation** above the glumes and beneath the florets. **Glumes** equal to subequal, longer than the florets, membranous, 1–3-veined, unawned; **calluses** puberulent; **lemmas** subcoriaceous, glabrous, scabridulous, 5-veined, apices bifid, awned or unawned, awns attached below midlength, usually geniculate, sometimes straight; **paleas** membranous, 2-veined; **lodicules** 2, free; **anthers** 3; **ovaries** glabrous. **Caryopses** shorter than the lemmas, concealed at maturity, adhering to the lemmas and/or paleas, longitudinally grooved, dorsally compressed. x = 7. Name from the Greek *aira*, the name for a weed.

Aira is a genus of eight species; two have been introduced into the *Flora* region. All members of the genus are native to temperate Europe and the Mediterranean region, North Africa, and western Asia. Frequently adventive, they are now widespread outside of their native range as weeds, although they are not considered particularly troublesome. They have little forage value because most are delicate, with extremely small leaves. All the species grow in open, disturbed places on usually dry, occasionally mesic, sandy to rocky soils.

1. Panicle branches ascending to divergent; panicles 1.2–13.5 cm long, 1.5–10 cm wide . . . . . . 1. *A. caryophyllea*
1. Panicle branches appressed to the rachises; panicles 0.5–4.1 cm long, 0.3–0.7 cm wide . . . . 2. *A. praecox*

1. Aira caryophyllea L. [p. 617]

Plants annual; tufted. **Culms** 4.5–55 cm, erect. **Sheaths** scabridulous, occasionally smooth; **ligules** 1.2–8 mm, abaxial surfaces scabridulous, acute to subobtuse, becoming lacerate; **blades** 0.3–13.5 cm long, 0.3–2.5 mm wide, antrorsely scabridulous, glabrous, apices prow-tipped. **Panicles** 1.2–13.5 cm long, 1.5–10 cm wide, open; **primary branches** to 7.3 cm, ascending to divergent, antrorsely scabridulous, occasionally smooth; **pedicels** 0.9–11.3 mm, apices enlarged. **Spikelets** 1.7–3.3(3.5) mm, silvery-green to stramineous or purplish; **rachillas** usually not prolonged beyond the base of the distal floret, sometimes prolonged, vestigial. **Glumes** subequal to equal, 1.3–3.3(3.5) mm, scabridulous on the upper ½; **callus hairs** 0.2–0.4 mm; **lemmas** 1.3–2.6 mm, apices bifid, sometimes only the upper lemma awned, awns 2.1–3.9 mm, straight or geniculate; **paleas** 0.9–1.7 mm; **anthers** 0.2–0.5 mm.

Caryopses 0.9–1.5 mm long, 0.3–0.5 mm wide, abaxial surfaces grooved in the distal ½, adaxial surfaces grooved the entire length. $2n = 14$.

Aira caryophyllea is native to Eurasia and Africa; it has become established in the *Flora* region, primarily on the Pacific, Gulf, and Atlantic coasts, and through much of the southeastern United States. It is usually found in disturbed areas, in vernally moist to dry, sandy to rocky, open sites, from sea level to subalpine elevations. It sometimes invades lawns or rock gardens.

1. Glumes subobtuse, usually denticulate, often mucronate; pedicels with abruptly thickened apices . . . . var. *cupaniana*
1. Glumes acute; pedicels gradually thickening to the apices.
  2. Pedicels usually 2–8 times as long as the spikelets; spikelets 1.7–2.5 mm long . . . . var. *capillaris*
  2. Pedicels usually 1–2 times as long as the spikelets; spikelets 2–3.5 mm long . . . var. *caryophyllea*

Aira caryophyllea var. **capillaris** (Mert. & W.D.J. Koch) Mutel [p. 617]
DELICATE HAIRGRASS

Pedicels 2–11.3 mm, usually 2–8 times as long as the spikelets, gradually thickening to the apices. **Spikelets** 1.7–2.5 mm, spreading, divergent from the secondary branches, often purplish to reddish-purple-tinged; **rachillas** not prolonged or vestigial. **Glumes** subequal, 1.7–2.5 mm, 1-veined, acute; **lower lemmas** 1.3–1.8 mm, apices bifid, teeth to 0.1 mm, awns absent or to 2.6 mm, straight or geniculate; **lower paleas** 1–1.3 mm; **upper lemmas** 1.7–2.1 mm, apices bifid, teeth 0.2–0.4 mm, awned, awns 2.1–3 mm, geniculate; **upper paleas** 1–1.3 mm; **anthers** 0.2–0.4 mm, yellow-orange or purple. **Caryopses** 0.9–1.2 mm long, about 0.3 mm wide, glabrous.

Aira caryophyllea var. *capillaris* is native to Europe, northern Africa, and western Asia. It usually grows in dry to somewhat moist, sandy loam soils of grassy banks, woodland openings, and disturbed sites such as pastures and roadsides.

Aira caryophyllea var. *capillaris* is the correct name for this taxon at the varietal level. If treated at the species level, its correct name is *Aira elegans* Willd. *ex* Roem. & Schult.

Aira caryophyllea L. var. **caryophyllea** [p. 617]
SILVER HAIRGRASS

Pedicels 0.9–7 mm, usually 1–2 times as long as the spikelets, gradually thickening to the apices. **Spikelets** 2–3.3(3.5) mm, usually appressed to the secondary branches, silvery-green to stramineous; **rachillas** not prolonged. **Glumes** 2–3.3(3.5) mm, subequal, 1-veined or the upper glumes with 2 lateral veins, these sometimes obscure and usually less than ¹/₂ the length of the glumes, acute; **lower lemmas** 2–2.4 mm, apices bifid, teeth 0.2–0.4 mm, awned, awns 2.4–3.5 mm, geniculate; **lower paleas** 1.5–1.7 mm; **upper lemmas** 2–2.6 mm, apices bifid, teeth 0.2–0.3 mm, awned, awns 2.5–3.9 mm; **upper paleas** 1.4–1.7 mm; **anthers** 0.2–0.5 mm, yellow to orange. **Caryopses** 1.4–1.5 mm long, 0.4–0.5 mm wide, glabrous.

Aira caryophyllea var. *caryophyllea* is native to the Mediterranean region. It usually grows in dry, sandy to rocky soil and on rock outcrops, in open and disturbed sites in woods, grassy flats, pastures, paths, and roadsides; it is occasionally found in damp ground at swamp or lagoon margins.

Aira caryophyllea var. **cupaniana** (Guss.) Fiori [p. 617]
SILVERY HAIRGRASS

Pedicels 2.8–6.5 mm, usually 1–3 times as long as the spikelets, abruptly thickened at the apices. **Spikelets** 1.3–2.6 mm, spreading, divergent from the secondary branches, silvery-green. **Glumes** equal to subequal, 1.3–2.6 mm, 1–3-veined, subobtuse, usually denticulate, often mucronate; **lower lemmas** 1.3–1.9 mm, usually unawned; **upper lemmas** 1.3–2 mm, apices bifid, awned, awns 1.8–2.5 mm, geniculate; **paleas** 0.9–1.4 mm; **anthers** 0.2–0.4 mm, yellow or purple. **Caryopses** about 1 mm long, 0.3–0.4 mm wide, glabrous.

Aira caryophyllea var. *cupaniana* is native to southern Europe and northern Africa, growing in mesic, open habitats in disturbed areas or open woodland. It was discovered in a prescribed burn area of Mount Diablo State Park in Contra Costa County, California, in 1995, but was not relocated in 1999.

2. **Aira praecox** L. [p. 617]
SPIKE HAIRGRASS

Plants annual; tufted. **Culms** 1–36 cm, erect. **Sheaths** usually scabridulous, occasionally smooth; **ligules** 1.4–5.3 mm, abaxial surfaces scabridulous, acute, becoming lacerate; **blades** (0.1) 0.25–5 cm long, 0.3–2.0 mm wide, antrorsely scabridulous, glabrous, prow-tipped. **Panicles** (0.5)1–4.1 cm long, 0.3–0.7 cm wide, sometimes reduced to a single spikelet in depauperate specimens; **primary branches** to 1.5 cm, appressed to the rachises, antrorsely scabridulous; **pedicels** 0.7–3.2 mm, enlarged at the apices. **Spikelets** 2.8–3.8 mm, green and purple-tinged; **rachillas** prolonged beyond the base of the distal floret. **Glumes** subequal, antrorsely scabridulous, especially on the midveins; **lower glumes** 2.8–3.6 mm, 1–3-veined; **upper glumes** 2.7–3.8 mm, 3-veined; **callus hairs** 0.3–0.5 mm; **lower lemmas** 2.4–3.3 mm, apices bifid, teeth 0.3–0.5 mm, awned, awns 3–4.5 mm; **paleas** 1.7–2.1 mm; **upper lemmas** similar; **anthers** 0.2–0.4 mm, yellow. **Caryopses** 1.3–1.7 mm long, 0.4–0.5 mm wide, glabrous, abaxial surfaces grooved 0.8–1 mm distally, adaxial surfaces grooved the entire length. $2n = 14$.

Aira praecox is native to Europe. In the *Flora* region, it grows mainly along or near the Pacific and Atlantic coasts, in dry to vernally moist sand dunes or in sandy to rocky soils, on rock faces and ledges, and in disturbed areas such as the edges of roads, railways, and airports. It is usually found in lowland areas, though it occasionally grows at montane to subalpine elevations.

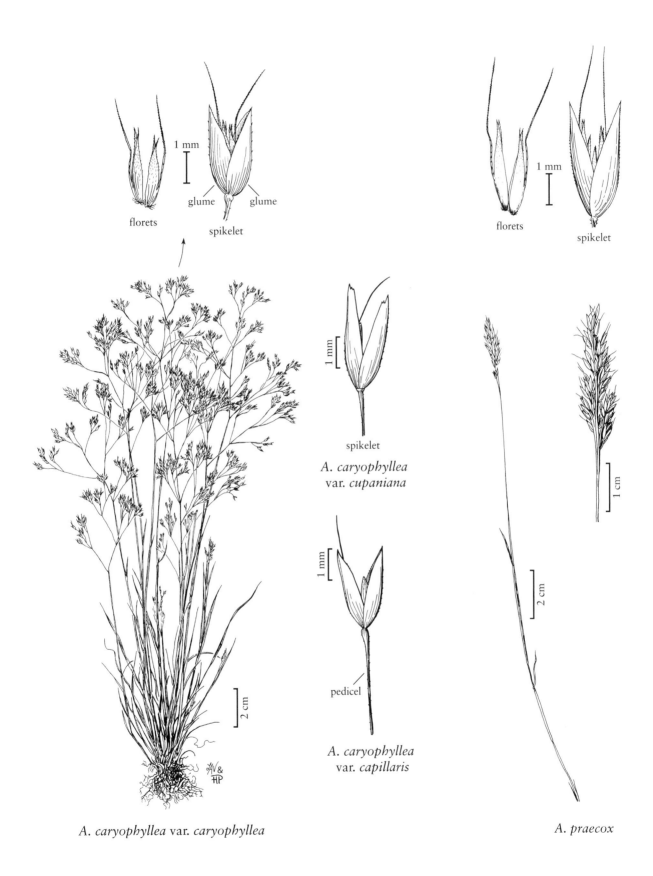

florets

1 mm

glume   glume

spikelet

florets

1 mm

spikelet

1 mm

spikelet

A. caryophyllea
var. *cupaniana*

1 mm

pedicel

A. caryophyllea
var. *capillaris*

2 cm

1 cm

2 cm

A. caryophyllea var. *caryophyllea*

A. praecox

AIRA

14.23 COLEANTHUS Seidl

Sandy Long

Plants annual, tufted. **Culms** 2–7(10) cm, ascending or decumbent. **Sheaths** closed, inflated; **auricles** absent; **ligules** membranous, ovate; **blades** flat. **Inflorescences** panicles; **branches** distant, verticillate. **Spikelets** slightly laterally compressed, pedicellate, pedicels hairy, with 1 floret; **rachillas** not prolonged beyond the base of the floret; **disarticulation** below the floret. **Glumes** absent; **calluses** short, glabrous; **lemmas** 1-veined, narrowed to awnlike apices; **paleas** 2–4-toothed, 2-keeled; **lodicules** absent; **anthers** 2; **ovaries** glabrous. **Caryopses** elliptic, exceeding the lemmas and paleas, falling free; **hila** oval. $x = 7$. Name from the Greek *koleos*, 'sheath', and *anthos*, 'flower', because the inflorescence grows out of an inflated leaf sheath.

Coleanthus is a monospecific genus that is native to Eurasia and North America.

SELECTED REFERENCES **Bernhardt, K.G., M, Koch, E. Ulbel,** and **J. Webhofer.** 2004. The soil seed bank as a resource for in situ and ex situ conservation of extinct species. [Pagination unknown] *in* E. Robbrecht and A. Bogaerts (eds.). EuroGard III. Scripta Botanica Belgica Series, No. 29. National Botanic Garden, Meise, Belgium. 177 pp.; **Conert, H.J.** 1992. *Coleanthus.* Pp. 434–437 *in* G. Hegi. Illustrierte Flora von Mitteleuropa, ed. 3. Band I, Teil 3, Lieferung 6 (pp. 401–480). Verlag Paul Parey, Berlin and Hamburg, Germany; **Hejný, S.** 1969. *Coleanthus subtilis* (Tratt.) Seidl in der Tschechoslowakei. Folia Geobot. Phytotax. 4:345–399; **Nečajev, A.P.** and **A.A. Nečajev.** 1972. *Coleanthus subtilis* (Tratt.) Seidl in the Amur Basin. Folia Geobot. Phytotax. 7:339–347; **Woike, S.** 1969. Beitrag zum Vorkommen von *Coleanthus subtilis* (Tratt.) Seidl (Feines Scheidenblütgras) in Europa. Folia Geobot. Phytotax. 4:401–413.

1. Coleanthus subtilis (Tratt.) Seidl [p. 619]

MOSS GRASS, MUD GRASS

Plants annual. **Culms** ascending or decumbent. **Sheaths** strongly inflated, uppermost sheath enclosing the base of the panicles; **ligules** 1–1.5 mm; **blades** 1–2 cm long, 0.5–1.5 mm wide. **Panicles** 1–5 cm, with 3–6 verticils. **Lemmas** 0.75–1.5 mm, ciliate on the keels, awnlike apices about equaling the lower portion; **paleas** about 0.5 mm, keels slightly prolonged; **anthers** about 0.3 mm. **Caryopses** about 1 mm. $2n = 14$.

Coleanthus subtilis is an ephemeral pioneer species of wet, open habitats. It grows on wet, muddy to sandy, calcium-deficient soils on the shores of lakes, sandbars, and islands. In the Flora region, it is known from the Columbia River, and around Hatzic, Arrow and Shuswap lakes in British Columbia. It also grows in Europe, Russia, and China. Throughout its range, *C. subtilis* is known from relatively few, scattered locations. It is easily overlooked because of its diminutive size, and because it flowers in early spring or late fall. It is not clear whether it is native or introduced in the *Flora* region.

14.24 EREMOPOA Roshev.

Stephen J. Darbyshire

Plants annual. **Culms** 5–60 cm. **Sheaths** open for most of their length; **auricles** absent; **ligules** glabrous; **blades** flat or convolute. **Inflorescences** panicles; **branches** glabrous, scabrous, some branches longer than 1 cm, some pedicels longer than 3 mm, usually less than 1 mm thick, flexible. **Spikelets** pedicellate, terete to slightly laterally compressed, with (1)2 to many florets, fertile florets (1)2–6, distal florets vestigial; **rachillas** straight or slightly bowed, scabrous or puberulent; **disarticulation** above the glumes and beneath the florets. **Glumes** unequal in length, subequal in width, usually shorter than the adjacent lemmas, unawned; **lower glumes** 1-veined; **upper glumes** 3-veined; **calluses** short, blunt, glabrous; **lemmas** lanceolate, glabrous or the bases slightly pilose, 5-veined, veins converging distally, apices not mucronate, unawned; **paleas** shorter than the lemmas; **lodicules** 2, free, glabrous; **anthers** 3; **ovaries**

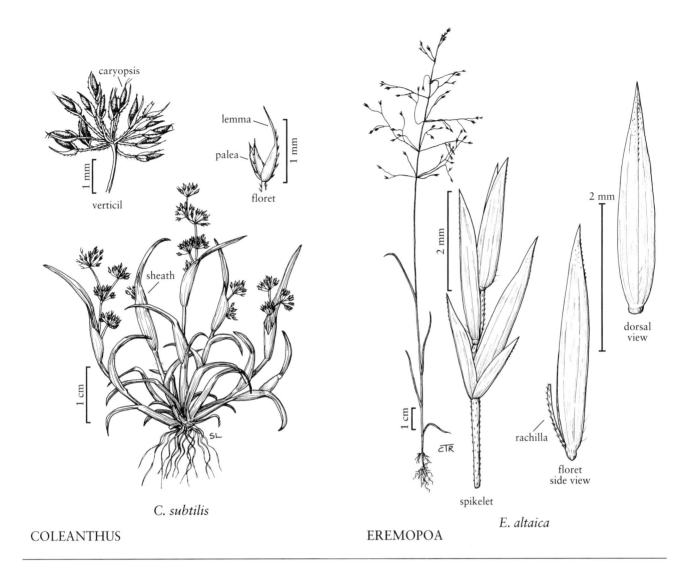

C. *subtilis*

COLEANTHUS

E. altaica

EREMOPOA

glabrous. **Caryopses** shorter than the lemmas, concealed at maturity, linear, adhering to the lemmas and/or paleas. $x = 7$. Name from the Greek *eremos*, a desert or uninhabited area, and *Poa*, the name of a grass genus.

Eremopoa is a genus of two to six species that are native from the eastern Mediterranean to western China.

SELECTED REFERENCE **Stevenson, G.A.** 1965. Notes on the more recently adventive flora of the Brandon area, Manitoba. Canad. Field-Naturalist 79:174–177.

1. Eremopoa altaica (Trin.) Roshev. [p. 619]

ALTAI GRASS

Culms (5)10–45 cm, solitary or tufted, erect, glabrous. **Leaves** cauline; **sheaths** closed for about $^{1}/_{10}$ their length, glabrous; **ligules** 1–4 mm, glabrous; **blades** 1–10 cm long, 0.5–4 mm wide, flat, usually glabrous, sometimes scabrous. **Panicles** (3)6–16(20) cm long, 1–6 cm wide, diffuse; **branches** 2-10 cm, whorled and strongly divergent, spikelets confined to the distal portion. **Spikelets** 2.8–6.2 mm, pedicellate, with (1)2–3(4) florets. **Glumes** lanceolate, glabrous, acute to acuminate; **lower glumes** 0.9–1.5(2) mm; **upper glumes** 1.5–2.5 mm; **lemmas** 2.5–4 mm, narrowly lanceolate, glabrous or with a few silky hairs on the lower keel, 5-veined, apices entire, acute to acuminate; **paleas** 2–3 mm, 2-keeled, glabrous; **anthers** 0.3–1 mm. **Caryopses** 1.9–2.6 mm. $2n = 14, 28, 42.$

Eremopoa altaica is native to the high deserts and mountain steppes of Asia, from Turkey through Afghanistan and Pakistan to the Altai region, the Himalayas, and western China. It was once found in railway yards and roadsides at Brandon, Manitoba (Stevenson 1965), but misidentified as *Eremopoa persica* (Trin.) Roshev. Although the population persisted for several years, recent efforts to relocate it have failed.

14.25 SPHENOPHOLIS Scribn.

Thomas F. Daniel

Plants usually perennial, rarely winter annuals; usually cespitose, sometimes the culms solitary. **Culms** (5)20–130 cm, leaves evenly distributed. **Sheaths** open; **auricles** absent; **ligules** membranous, erose; **blades** flat or involute, glabrous or pubescent. **Inflorescences** panicles, open or contracted, nodding to erect; **disarticulation** below the glumes, the distal floret sometimes disarticulating first. **Spikelets** pedicellate, 2.1–9.5 mm, laterally compressed, with 2–3 florets; **rachillas** glabrous or pubescent, prolonged beyond the base of the distal floret as a slender bristle. **Glumes** almost equaling the lowest floret, dissimilar in width, membranous to subcoriaceous, margins scarious, apices unawned; **lower glumes** narrower than the upper glumes, 1(3)-veined, strongly keeled, apices acute; **upper glumes** elliptical to oblanceolate, obovate, or subcucullate, 3(5)-veined, strongly to slightly keeled, apices acuminate, acute, rounded, or truncate; **calluses** glabrous; **lemmas** herbaceous, not indurate, rounded on the lower back, smooth or partly or wholly scabrous, usually keeled near the apices, 3(5)-veined, veins usually not visible, unawned or awned from just below the apices, awns straight or geniculate; **paleas** hyaline, shorter than the lemmas; **lodicules** 2, free, membranous, glabrous, toothed or entire; **anthers** 3; **ovaries** glabrous. **Caryopses** shorter than the lemmas, concealed at maturity, linear-ellipsoid, glabrous; **endosperm** liquid. $x = 7$. Name from the Greek *sphen*, 'wedge', and *pholis*, 'scale', in reference to the upper glumes.

Sphenopholis includes six species, all of which are native to the *Flora* region. Its greatest diversity is in the southeastern United States. One species, *Sphenopholis obtusata*, extends outside the region to southern Mexico and the Caribbean. It has also been collected in Hawaii, but is not known to be established there. Interspecific hybridization is known in the genus, but intermediate plants are not frequently encountered. *Trisetum interruptum* Buckl. is sometimes treated in *Sphenopholis* (e.g., Finot et al. 2004) because its spikelets disarticulate below the glumes.

Glume widths are measured in side view, from the lateral margin to the midvein.

SELECTED REFERENCES Erdman, K.S. 1965. Taxonomy of the genus *Sphenopholis* (Gramineae). Iowa State Coll. J. Sci. 39:289–336; Finot, V.L., P.M. Peterson, R.J. Soreng, and F.O. Zuloaga. 2004. A revision of *Trisetum*, *Peyritscia*, and *Sphenopholis* (Poaceae: Pooideae: Aveninae) in Mexico and Central America. Ann. Missouri Bot. Gard. 91:1–30; Smith, E.B. (ed.). 1991. An Atlas and Annotated List of the Vascular Plants of Arkansas, ed. 2. Edwin B. Smith, Fayetteville, Arkansas, U.S.A. 489 pp.

1. Distal lemmas awned, awns 3–9 mm long; spikelets 4.5–9.5 mm long . . . . 1. *S. pensylvanica*
1. Distal lemmas unawned or awned, awns 0.1–3 mm long; spikelets 2.1–5.2 mm long.
  2. Distal lemmas scabrous on the sides, unawned or awned; anthers (0.5)1–2 mm long.
    3. Lower glumes at least ⅓ as wide as the upper glumes; blades (1)2–5(7) mm wide, flat to slightly involute; awns absent or shorter than 1 mm . . . . 2. *S. nitida*
    3. Lower glumes less than ⅓ as wide as the upper glumes, rarely slightly wider; blades 0.3–1.5(2) mm wide, involute to filiform; awns absent or (0.1)1–3 mm long . . . . 3. *S. filiformis*
  2. Distal lemmas usually smooth on the sides, sometimes scabrous or scabridulous distally, unawned; anthers 0.2–1 mm long.
    4. Upper glumes subcucullate, the width/length ratio 0.3–0.5; panicles usually erect, often spikelike, the spikelets usually densely arranged . . . . 4. *S. obtusata*

4. Upper glumes not subcucullate, the width/length ratio 0.17–0.35; panicles usually nodding, not spikelike, the spikelets usually loosely arranged.

    5. Spikelets 2.1–4 mm long; lowest lemmas 2.1–3 mm long; upper glumes 1.9–2.9 mm long . . . . 5. *S. intermedia*

    5. Spikelets 4–5.2 mm long; lowest lemmas 3.1–4.2 mm long; upper glumes 3–3.9 mm long . . . . 6. *S. longiflora*

1. Sphenopholis pensylvanica (L.) Hitchc. [p. 622]
SWAMP OATS

Culms 30–120 cm, glabrous. Sheaths glabrous or pubescent; ligules 0.2–1 mm; blades 4–10(20) cm long, (1)2–8 mm wide, flat to slightly involute, smooth or scabridulous, sometimes pubescent. Panicles 7–35 cm long, 2–10 cm wide, erect to nodding, with relatively few, loosely arranged spikelets. Spikelets 4.5–9.5 mm. Lower glumes about ¹/₂ as wide as the upper glumes; upper glumes 3.6–6.2 mm, elliptical to oblanceolate, width/length ratio 0.15–0.28, apices acuminate to acute; lowest lemmas 3.5–6 mm, mostly smooth, apices scabridulous, unawned or awned, awns to 2.5 mm; distal lemmas scabrous, awned, awns 3–9 mm, slightly to evidently bent; anthers 0.5–1.8 mm. 2n = 14.

Sphenopholis pensylvanica grows in springheads, seepage areas, swamps, marshes, and other moist to wet places, at 0–1100 m, in the eastern and southeastern United States. It hybridizes with *S. obtusata* (see below). The hybrids, which are called **Sphenopholis ×pallens** (Biehler) Scribn., differ from *S. pensylvanica* in having generally shorter (3–5 mm) spikelets and shorter (0.1–4 mm), straight or bent awns on the distal lemmas. They differ from *S. obtusata* in having awns on the distal lemmas.

2. Sphenopholis nitida (Biehler) Scribn. [p. 622]
SHINY WEDGEGRASS

Culms 30–80 cm. Sheaths pubescent or the upper sheaths glabrous, sometimes scabridulous; ligules 1–2 mm; blades 2–15 cm long, (1)2–5(7) mm wide, flat to slightly involute, sometimes pubescent, flag leaf blades scabridulous. Panicles 6–20 cm long, 1–5 cm wide, nodding to erect, spikelets loosely to densely arranged. Spikelets 2.5–5 mm. Lower glumes at least ¹/₃ as wide as the upper glumes; upper glumes 2–3.2 mm, obovate, rarely subcucullate, width/length ratio 0.22–0.38; lowest lemmas 2.1–3.8 mm, smooth on the basal ¹/₃–¹/₂, scabrous distally, unawned; distal lemmas scabrous on the sides, usually unawned, rarely awned, awns 0.1–1 mm; anthers (0.5)1–2 mm. 2n = 14.

Sphenopholis nitida grows in moist to dry, deciduous and coniferous forests, on clay or silt banks and slopes, at 0–1200 m, in southern Ontario and the eastern United States. It can be confused with occasional forms of *S. obtusata* that have somewhat scabrous distal lemmas, but *S. nitida* has broader lower glumes.

3. Sphenopholis filiformis (Chapm.) Scribn. [p. 622]
SOUTHERN WEDGEGRASS

Culms 20–100 cm. Sheaths smooth, usually glabrous, sometimes pubescent; ligules 0.4–0.75 mm, erose-ciliate; blades 2–45 cm long, 0.3–1.5(2) mm wide, involute to filiform. Panicles 5–15 cm long, 0.5–1(2) cm wide, sometimes nodding, spikelets loosely to densely arranged. Spikelets 2.3–5 mm. Lower glumes less than ¹/₃ as wide as the upper glumes, rarely slightly wider; upper glumes 1.8–2.9 mm, obovate to oblanceolate, width/length ratio 0.16–0.45, apices rounded to truncate; lowest lemmas 2–3 mm, scabridulous distally; distal lemmas scabrous on the sides, unawned or infrequently awned, awns (0.1)1–3 mm; anthers (0.5)1–1.9 mm. 2n = 14.

Sphenopholis filiformis grows in sandy soils of pine and mixed pine forests, at 0–500 m, in the southeastern United States. It is found primarily in the coastal plain, but extends to the piedmont. Smith (1991) reported it for northern Arkansas (*Nielsen 4946*, identification not verified). *Sphenopholis filiformis* differs from occasional forms of *S. obtusata* with somewhat scabrous distal lemmas in having narrower leaves.

4. Sphenopholis obtusata (Michx.) Scribn. [p. 623]
PRAIRIE WEDGEGRASS, SPHENOPHOLIS OBTUS

Culms (9)20–130 cm. Sheaths glabrous or hairy, sometimes scabridulous; ligules (1)1.5–2.5 mm, erose-ciliate, more or less lacerate; blades 5–14 cm long, (1)2–8 mm wide, usually flat, rarely slightly involute, scabrous or pubescent. Panicles (2)5–15(25) cm long, 0.5–2 cm

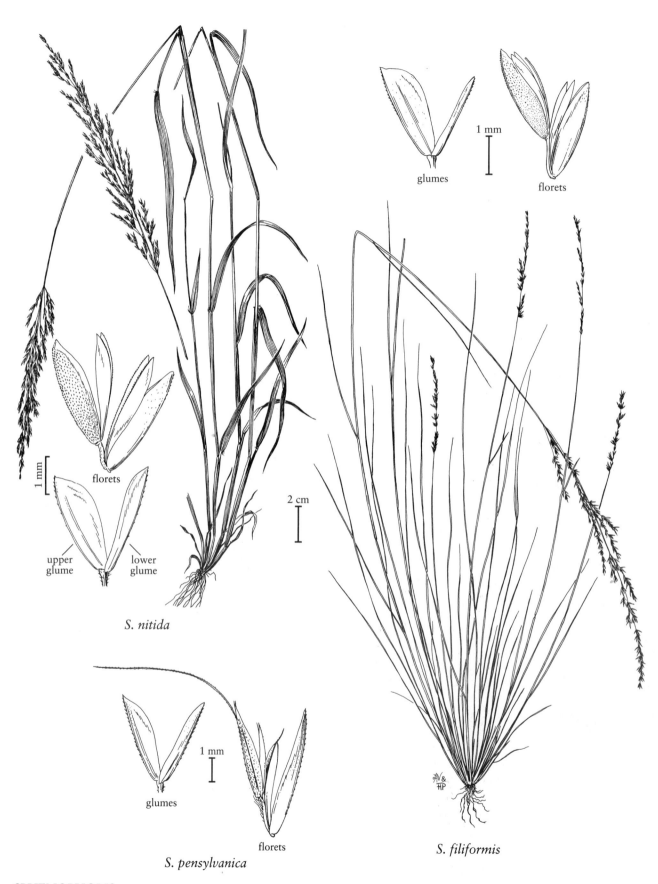

glumes

florets

florets

upper glume

lower glume

1 mm

S. nitida

2 cm

glumes

florets

S. pensylvanica

S. filiformis

SPHENOPHOLIS

wide, usually erect, often spikelike, spikelets usually densely arranged. **Spikelets** 2.2–3.6 mm. **Lower glumes** less than ¹/₃ as wide as the upper glumes; **upper glumes** 1.5–2.5 mm, subcucullate, width/length ratio 0.3–0.5, apices rounded to truncate; **lowest lemmas** 1.9–2.8 mm, usually scabridulous distally; **distal lemmas** usually smooth on the sides, occasionally scabrous, unawned; **anthers** 0.2–1 mm. $2n = 14$.

Sphenopholis obtusata grows in prairies, marshes, dunes, forests, and waste places, at 0–2500 m. Its range extends from British Columbia to New Brunswick, through most of the United States, to southern Mexico and the Caribbean. The distal lemmas of *S. obtusata* are occasionally somewhat scabrous. Such plants can be distinguished from *S. nitida* (p. 621) by their narrower lower glumes, from *S. filiformis* (p. 621) by their wider leaves, and from *S. pensylvanica* (p. 621) by their shorter, unawned spikelets. Hybrids with *S. pensylvanica*, called **Sphenopholis ×pallens**, have short (0.1–4 mm) awns on the distal lemmas.

5. Sphenopholis intermedia (Rydb.) Rydb. [p. 623]

SLENDER WEDGEGRASS, SPHENOPHOLIS INTERMÉDIAIRE

Culms (5)30–120 cm. **Sheaths** smooth or scabridulous, sometimes pubescent; **ligules** 1.5–2.5 mm, erose-ciliate, often lacerate; **blades** 8–15 cm long, (1)2–6 mm wide, flat to slightly involute. **Panicles** (2)7–20 cm long, (0.5)1–3 cm wide, usually nodding, not spikelike, spikelets usually loosely arranged. **Spikelets** 2.1–4 mm. **Lower glumes** less than ¹/₃ as wide as the upper glumes; **upper glumes** 1.9–2.9 mm, oblanceolate to obovate, not subcucullate, width/length ratio 0.23–0.35, apices acute, rounded, or subtruncate; **lowest lemmas** 2.1–3 mm, smooth or scabridulous; **distal lemmas** usually smooth on the sides, rarely scabridulous near the apices, unawned; **anthers** 0.2–0.8 mm. $2n = 14$.

Sphenopholis intermedia grows at 0–2500 m in wet to damp sites, sites that dry out after the growing season, and sites with clay soils that retain moisture. Restricted to the *Flora* region, it is found in forests, meadows, and waste places throughout most of the region other than the high arctic. It differs from *Koeleria macrantha* (p. 754), with which it is sometimes confused, in its more open panicles and in having spikelets that disarticulate below the glumes.

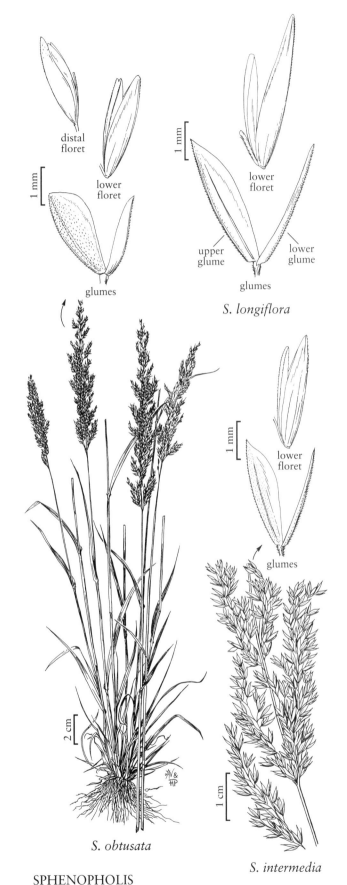

distal floret

lower floret

glumes

lower floret

upper glume

lower glume

glumes

S. longiflora

lower floret

glumes

S. obtusata

S. intermedia

SPHENOPHOLIS

6. Sphenopholis longiflora (Vasey *ex* L.H. Dewey) Hitchc. [p. 623]

BAYOU WEDGEGRASS

Culms to 70 cm. **Sheaths** smooth or retrorsely scabridulous, usually glabrous, sometimes pubescent; **ligules** 0.7–2.6 mm, glabrous, erose; **blades** 8–18 cm long, 2–10 mm wide, flat. **Panicles** 15–20 cm long, 1.5–2.5 cm wide, nodding, not spikelike, spikelets loosely arranged. **Spikelets** 4–5.2 mm. **Lower glumes** about ¹/₃ as wide as the upper glumes; **upper glumes** 3–3.9 mm, elliptical to oblanceolate, not subcucullate, width/length ratio 0.17–0.26, apices acute; **lowest lemmas** 3.1–4.2 mm, glabrous; **distal lemmas** usually smooth on the sides, rarely scabridulous distally, unawned; **anthers** 0.2–0.7 mm. 2*n* = 14.

Sphenopholis longiflora grows in forest bottoms along bayous and streams, from 0–50 m, in Texas, Arkansas, and Louisiana. Erdman (1965) treated both *S. longiflora* and *S. intermedia* as *S. obtusata* var. *major* (Torr.) Erdman. They are morphologically distinct from each other, and from *S. obtusata*.

14.26 DESCHAMPSIA P. Beauv.

Mary E. Barkworth

Plants usually perennial, sometimes annual; cespitose or tufted. **Culms** 5–140 cm, hollow, erect. **Leaves** usually mainly basal, often forming a dense tuft; **sheaths** open; **auricles** absent; **ligules** membranous, decurrent, rounded to acuminate; **blades** often all or almost all tightly rolled or folded and some flat, sometimes most flat, others rolled or folded. **Inflorescences** terminal panicles, open or contracted; **disarticulation** above the glumes, beneath the florets. **Spikelets** 3–9 mm, with 2(3) florets in all or almost all spikelets, florets usually bisexual, sometimes viviparous; **rachillas** hairy, usually prolonged more than 0.5 mm beyond the base of the distal floret, sometimes terminating in a highly reduced floret. **Glumes** subequal to unequal, usually exceeding the adjacent florets, often exceeding all florets, 1- or 3-veined, acute to acuminate; **calluses** antrorsely strigose; **lemmas** obscurely (3)5–7-veined, rounded over the back, apices truncate-erose to 2–4-toothed, awned, awns usually attached on the lower ¹/₂ of the lemmas, occasionally subapical, straight to strongly geniculate, slightly to strongly twisted proximally, straight distally; **paleas** shorter than the lemmas, 2-keeled, keels often scabrous; **lodicules** 2, lanceolate to ovate-lanceolate, usually entire; **anthers** 3; **ovaries** glabrous; **styles** 2. **Caryopses** oblong; **embryos** about ¹/₄ the length of the caryopses. *x* = 7. Named for Louise Auguste Deschamps (1765–1842), a French naturalist.

Deschampsia includes 20–40 species. It is best represented in the Americas and Eurasia, but it grows in cool, damp habitats throughout the world. Seven species are native to the *Flora* region; none of the remaining species have been introduced.

Deschampsia differs from *Vahlodea* (p. 691), which it used to include, in having primarily basal, rather than primarily cauline, leaves, and hairy rachillas that extend more than 0.5 mm beyond the base of the distal floret in a spikelet. *Trisetum* (p. 744) differs from *Deschampsia* primarily in its more acute, bifid lemmas, and in having awns that are inserted at or above the midpoint of the lemmas. In *Deschampsia*, the awns are usually inserted near the base.

Because the treatments of *Deschampsia brevifolia* and *D. sukatschewii* were revised shortly before going to press, the maps are preliminary, particularly with respect to the Canadian distribution of these two species.

Lemma length, awn attachment, and awn length should be examined on the lower florets within the spikelets. The upper florets often have shorter lemmas, and shorter awns that are attached higher on the back than those of the lower florets.

SELECTED REFERENCES **Aiken, S.G., L.L. Consaul,** and **M.J. Dallwitz.** 1995 on. Grasses of the Canadian Arctic Archipelago: Descriptions, illustrations, identification and information retrieval. http://www.mun.ca/biology/delta/arcticf/poa/index.htm; **Chiapella, J.** 2000. The *Deschampsia cespitosa* complex in central and northern Europe: A morphological analysis. Bot. J. Linn. Soc. 134:495–512; **Chiapella, J.** and **N.S. Probatova.** 2003. The *Deschampsia cespitosa* complex (Poaceae: Aveneae) with special reference to Russia. Bot. J. Linn. Soc. 142:213–228; **Clarke, G.C.S.** 1980. *Deschampsia* Beauv. Pp. 225–227 *in* T.G. Tutin, V.H. Heywood, N.A. Burges, D.M. Moore, D.H. Valentine, S.M. Walters, and D.A. Webb (eds.). Flora Europaea, vol. 5. Cambridge University Press, Cambridge, England. 452 pp.; **Hultén, E.** 1960. Flora of the Aleutian Islands and Westernmost Alaska Peninsula with Notes on the Flora of Commander Islands. J. Cramer, Weinheim, Germany. 376 pp.; **Kawano, S.** 1966. Biosystematic studies of the *Deschampsia caespitosa* complex with special reference to the karyology of Icelandic populations. Bot. Mag. (Tokyo) 79:293–307; **Lawrence, W.E.** 1945. Some ecotypic relations of *Deschampsia caespitosa*. Amer. J. Bot. 32:298–314; **McLachlan, K.I., S.G. Aiken, I.P. Lefkovitch,** and **S.A. Edlund.** 1989. Grasses of the Queen Elizabeth Islands. Canad. J. Bot. 67:2088–2105; **Tsvelev, N.N.** 1995. *Deschampsia.* Pp. 150–163 *in* J.G. Packer (ed., English edition). Flora of the Russian Arctic, vol. 1, trans. G.C.D. Griffiths. University of Alberta Press, Edmonton, Alberta, Canada [English translation of A.I. Tolmachev (ed.). 1964. Arkticheskaya Flora SSSR, vol. 2. Nauka, Leningrad [St. Petersburg], Russia].

1. All or most spikelets viviparous; panicle branches smooth .... 5. *D. alpina*
1. All or most spikelets bisexual or, if viviparous, the panicle branches scabrous.
  2. Plants annual; awns strongly geniculate .... 7. *D. danthonioides*
  2. Plants perennial; awns straight to strongly geniculate.
    3. Lemmas scabridulous or puberulent, dull .... 8. *D. flexuosa*
    3. Lemmas glabrous, shiny.
      4. Glumes mostly green, apices purple; panicles narrowly elongate, 0.5–1.5(2) cm wide, appearing greenish .... 6. *D. elongata*
      4. Glumes purplish proximally, sometimes over more than $^1/_2$ their surface, whitish to golden distally; panicles usually pyramidal or ovate, sometimes narrowly elongate, 0.5–30 cm wide, appearing bronze to dark purple (*D. cespitosa* complex).
      5. Spikelets 6–7.5 mm long; culms sometimes decumbent and rooting at the lower nodes; plants of sandy areas around lakes in the Northwest Territories and northern Saskatchewan .... 2. *D. mackenzieana*
      5. Spikelets 2–7.6 mm; culms erect, not rooting at the lower nodes; plants of gravels, wet meadows, and bogs, widely distributed in cooler regions of North America.
        6. Spikelets strongly imbricate, often rather densely clustered on the ends of the branches, sometimes evenly distributed on the branches; glumes and lemmas dark purple proximally for over more than $^1/_2$ their surface; lemmas 2.2–4 mm long .... 4. *D. brevifolia*
        6. Spikelets usually not or only moderately imbricate, not in dense clusters at the ends of the branches; glumes usually purple over less than $^1/_2$ their surface, often with a green base, a distal purple band, and pale apices; lemmas 2–5(7)mm long.
          7. Basal blades with 5–11 ribs, usually most or all ribs scabridulous or scabrous, outer ribs often more strongly so, sometimes the ribs only papillose or puberulent, usually at least some blades flat and 1–4 mm wide, the majority folded or rolled and 0.5–1 mm in diameter; lower glumes often scabridulous distally over the midvein; lower panicle branches often scabridulous or scabrous, sometimes smooth .... 1. *D. cespitosa*
          7. Basal blades with 3–5 ribs, ribs usually smooth or papillose, sometimes puberulent or the outer ribs scabridulous, all blades of the current year usually strongly involute and hairlike, 0.3–0.5(0.8) in diameter; lower glumes smooth over the midvein; lower panicle branches usually smooth, sometimes sparsely scabridulous; .... 3. *D. sukatschewii*

1. Deschampsia cespitosa (L.) P. Beauv. [p. 627]

Plants perennial; loosely to tightly cespitose. **Culms** (7) 35–150 cm, erect, not rooting at the lower nodes. **Leaves** mostly basal, sometimes forming a dense 10–35 cm tuft; **sheaths** glabrous; **ligules** 2–13 mm, scarious, decurrent, obtuse to acute; **blades** 5–30 cm long, usually at least some flat and 1–4 mm wide, the remainder folded or rolled and 0.5–1 mm in diameter, adaxial surfaces with 5–11 prominent ribs, ribs usually all papillose, scabridulous, or scabrous, sometimes puberulent, outer ribs sometimes more strongly so than the inner ribs. **Panicles** 8–30(40) cm, 4–30 cm wide, usually open and pyramidal, sometimes contracted and ovate; **branches** straight to slightly flexuous, usually strongly divergent, sometimes strongly ascending, lower branches often scabridulous or scabrous, particularly distally, with not or only moderately imbricate spikelets. **Spikelets** 2.5–7.6 mm, ovate to V-shaped, laterally compressed, usually bisexual, sometimes viviparous, bisexual spikelets usually with 2(3) florets, rarely with 1. **Glumes** lanceolate, acute; **lower glumes** 2.7–7 mm, entire, 1–3-veined, midvein sometimes scabridulous, at least distally; **upper glumes** 2–7.5 mm, 1–3-veined, lanceolate, midvein smooth or wholly or partly scabridulous; **callus hairs** 0.2–2.3 mm; **lemmas** 2–5(7) mm, smooth, shiny, glabrous, usually purple over less than ¹/₂ their surface, purple or green proximally, if green, often with a purple band about midlength, usually green or pale distally, usually awned, awns (0.5)1–8 mm, attached from near the base to about midlength, straight or geniculate, sometimes exceeding the glumes; **anthers** 1.5–3 mm. **Caryopses** 0.5–1 mm. $2n = 18, 24, 25, 26–28$, about 39, 52. The voucher specimens for these counts have not been examined.

Deschampsia cespitosa is circumboreal in the Northern Hemisphere, and also grows in New Zealand and Australia. It is an attractive taxon that grows in wet meadows and bogs, and along streams and lakes, from sea level to over 3000 m in cool-temperate, but not arctic, habitats.

There are widely varying opinions concerning the taxonomic treatment of *Deschampsia cespitosa*. Tsvelev, Aiken, Murray, and Elven (per Murray, pers. com. 2005) recommend a narrow circumscription, and consider *D. cespitosa* to be introduced and mostly ruderal in regions other than Europe and western Siberia. Chiapella and Probatova (2003) adopted a much broader interpretation of *D. cespitosa*, treating many of the species recognized in, for example, Tsvelev

(1995) as subspecies. There have been no interdisplinary, global studies of the complex. The circumscription adopted here is narrower than has been customary in North America. Some of the distribution records shown, particularly those from the northern part of the region, may reflect the broad interpretation of the species.

The name *Deschampsia cespitosa* is based on *Aira cespitosa* L. Linnaeus chose not to spell the epithet of the basionym "caespitosa". Consequently, the correct spelling of the epithet when combined with *Deschampsia* is "cespitosa".

Lawrence (1945) demonstrated that, in western North America, *Deschampsia cespitosa* exhibits both ecotypic differentiation and a high degree of plasticity. The following three subspecies intergrade.

1. Panicles contracted at anthesis, the branches appressed to ascending; glumes 4.5–5.8 mm long, midvein of the lower glumes scabrous distally . . . . subsp. *holciformis*
1. Panicles open at anthesis, the branches strongly divergent to drooping; glumes 2–7.5 mm long; midvein of the lower glumes smooth or scabridulous distally.
  2. Plants often glaucous; glumes 4.4–7.5 mm long; awns usually exceeding the lemmas; plants of the northwest coast of North America . . . . subsp. *beringensis*
  2. Plants not glaucous; glumes 2–6 mm long; awns exceeded by or exceeding the lemmas; plants widespread in North America . . . . subsp. *cespitosa*

Deschampsia cespitosa subsp. beringensis (Hultén) W.E. Lawr. [p. 627]

BERINGIAN HAIRGRASS

Plants loosely cespitose, often glaucous. **Culms** (15)70–140 cm. **Ligules** 4.5–13 mm; **blades** 5–12 cm long, 2–4 mm wide. **Panicles** 9–40 cm long, 8–30 cm wide, open, pyramidal; **branches** divergent, scabridulous to scabrous. **Spikelets** 4.5–8 mm, greenish, not to somewhat imbricate. **Glumes** from exceeding to exceeded by the distal floret, lengths usually 5+ times widths; **lower glumes** 4.3–7 mm, midveins smooth or scabridulous distally; **upper glumes** 4.4–7.5 mm; **callus hairs** 0.7–1.6 mm; **lemmas** 3–5(7) mm, apices 4-toothed or bifid, usually mostly green, awns 3.3–6.3 mm, straight to weakly geniculate, attached within the proximal ¹/₃ of the lemma; **anthers** (1.5)1.9–2.5 mm. $2n = 26$.

Deschampsia cespitosa subsp. *beringensis* is primarily a coastal species, growing up to 800 m along the Aleutian chain and the southern coast of Alaska south to Sonoma County, California, and west to the Kamchatka Peninsula, Russia. Typical plants are tall,

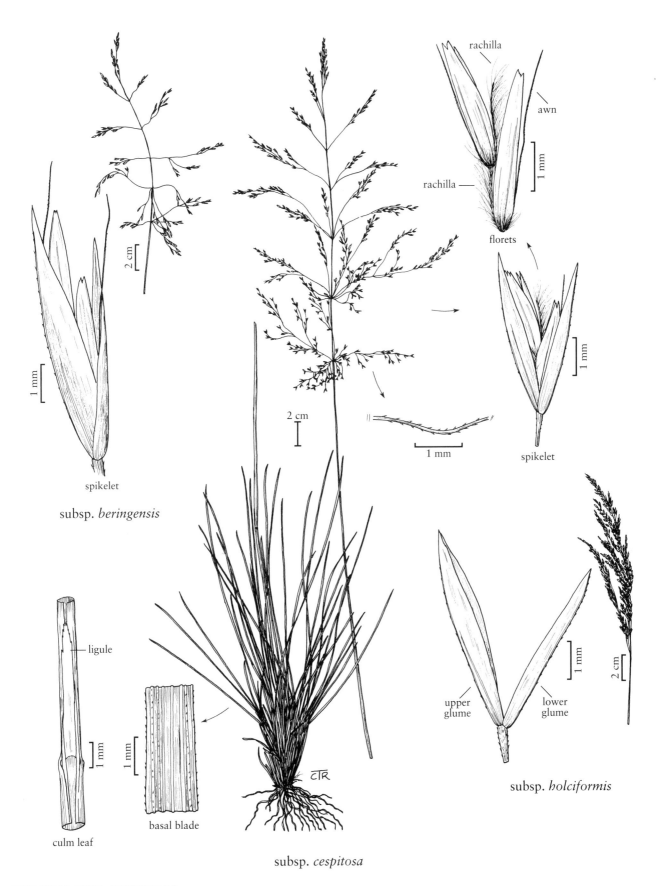

rachilla

awn

rachilla

1 mm

florets

1 mm

spikelet

subsp. *beringensis*

spikelet

1 mm

2 cm

2 cm

1 mm

ligule

1 mm

1 mm

basal blade

culm leaf

upper glume

lower glume

1 mm

2 cm

subsp. *holciformis*

subsp. *cespitosa*

DESCHAMPSIA CESPITOSA

glaucous, have long ligules and spikelets, and long, narrow glumes, but in the Pribiloff Islands and at scattered locations elsewhere, they intergrade with plants that are only 15–25 cm tall and also have smaller spikelet parts (Lawrence 1945). *Deschampsia cespitosa* subsp. *beringensis* differs from *D. mackenzieana* primarily in its coastal distribution and lower chromosome number. It supposedly differs from *D. cespitosa* subsp. *cespitosa* in having long glumes but, as the descriptions indicate, there is considerable overlap in this and other characters. The morphological, geographic, and ecological boundaries between the two subspecies need further study.

Deschampsia cespitosa (L.) P. Beauv. subsp. **cespitosa** [p. 627]

TUFTED HAIRGRASS, DESCHAMPSIE CESPITEUSE

Plants densely cespitose, not glaucous. **Culms** (7)35–150 cm. **Ligules** 2–8 mm; **blades** 5–25 cm long, 1.5–3.5 mm wide when flat. **Panicles** 8–30 cm long, 4–30 cm wide, open, nodding, pyramidal; **branches**, both primary and secondary, usually divergent, usually sparsely to moderately scabridulous or scabrous, sometimes smooth. **Spikelets** 2.5–7 mm, not to slightly imbricate. **Glumes** subequal to the distal floret, lengths often less than 5 times widths; **lower glumes** 2.5–5 mm, midveins smooth or scabridulous distally; **upper glumes** 2–6 mm; **lemmas** 2–4 mm, purple and/or green proximally, green to gold distally, the purple portion usually less than 1/2 the surface area, awns 1–8 mm, usually attached near the base, sometimes attached near midlength, straight or geniculate, exceeded by or exceeding the distal floret; **anthers** 1.5–2 mm.

Deschampsia cespitosa subsp. *cespitosa* is treated here as a circumboreal taxon that is most prevalent in boreal and temperate North America, growing at 0–3000 m; many reports from arctic and alpine North America refer to what are treated here as *D. sukatschewii* or *D. brevifolia*. Even with this narrower interpretation, *D. cespitosa* is highly polymorphic. Plants with long awns are more prevalent in western North America but, within that region, do not appear to show any geographic or ecological preference (Lawrence 1945). Larger plants are difficult to distinguish from *D. cespitosa* subsp. *beringensis*. The morphological, geographic, and ecological boundaries between *D. cespitosa* subsp. *cespitosa* and subsp. *beringensis* need further study.

Many cultivars of *Deschampsia cespitosa* subsp. *cespitosa* have been developed. At one time, the most frequently cultivated plants were distinguished by their combination of large (20–40 cm) panicles and small (2.5–4 mm) spikelets, and were called *D. cespitosa* var. *parviflora* (Thuill.) Coss. & Germ. or *D. cespitosa* subsp. *parviflora* (Thuill.) K. Richt. Such plants are treated here as one part of the spectrum of variation in subsp. *cespitosa*.

The name *Deschampsia cespitosa* var. *glauca* (Hartm.) Lindm. has been applied in eastern North America to glaucous plants less than 75 cm tall, with spikelets only 3–4.5 mm long. Unfortunately, the name is illegitimate; there is no legitimate name available for such plants.

Deschampsia cespitosa subsp. **holciformis** (J. Presl) W.E. Lawr. [p. 627]

Plants cespitose, sometimes glaucous. **Culms** 50–125 cm. **Ligules** 3–4.3 mm; **blades** 15–30 cm long, 1–4 mm wide when flat. **Panicles** 10–25 cm long, 3–8 cm wide, dense; **primary** and **secondary branches** appressed to ascending, scabridulous to densely scabrous. **Spikelets** 5.5–8 mm, usually strongly imbricate. **Glumes** usually exceeded by the distal floret, often purplish over more than 1/2 their area; **lower glumes** 4.6–5.8 mm, midveins scabrous distally; **upper glumes** 4.5–5.6 mm; **callus hairs** 1–2.3 mm; **lemmas** 3.8–4.5 mm, often purplish over more than 1/2 their area, awns 2–3 mm, straight to slightly geniculate, attached near or slightly above the middle of the lemma; **anthers** 2.5–3 mm. $2n = 26$.

Deschampsia cespitosa subsp. *holciformis* grows in coastal marshes and sandy soils, from the Queen Charlotte Islands, British Columbia, to central California. It intergrades and is interfertile with *D. cespitosa* subsp. *beringensis*, differing in its closed panicles, scabrous veins on the lower glumes, and more strongly imbricate spikelets. There are relatively few collections; it is not clear whether this reflects lack of collecting or rarity.

2. **Deschampsia mackenzieana** Raup [p. 630]

MACKENZIE HAIRGRASS

Plants loosely cespitose. **Culms** 30–80 cm, smooth, glabrous, sometimes decumbent at the base and rooting at the lower nodes. **Basal leaves** not forming a tuft; **sheaths** smooth; **ligules** 4–7.5 mm, acute; **blades** 1–3 mm wide, convolute to involute, abaxial surfaces smooth, adaxial surfaces scabrous. **Panicles** 10–20 cm long, 8–14 cm wide; **branches** ascending to laxly diverging or reflexed, somewhat scabrous, longest branches at the lower nodes usually undivided for 1/3–1/2 their length. **Spikelets** 6–7.5 mm, bisexual. **Glumes** acuminate, equaling or slightly longer than the distal floret; **callus hairs** 1.5–2 mm; **lemmas** 4.5–5.5 mm, awns attached on the lower 1/4–2/3, inconspicuous, weakly geniculate, from shorter than to exceeding the lemma by approximately 2 mm; **anthers** 1.5–2.7 mm. $2n = 52$.

Deschampsia mackenzieana grows on the sandy shores and dunes around Great Slave Lake, Northwest Territories, and Lake Athabasca, Saskatchewan. The decumbent culms of some plants may be a response to shifting substrate.

3. Deschampsia sukatschewii (Popl.) Roshev. [p. 630]

DESCHAMPSIE NAINE

Plants perennial; usually densely cespitose. **Culms** 5–70 cm, erect or strongly geniculate at the first node, glabrous. **Leaves** mostly basal, sometimes forming a dense, moss-like tuft 5–20 cm in diameter; **sheaths** smooth, glabrous; **ligules** 1.5–8 mm, acute; **blades** 0.5–8 cm long, usually strongly rolled and 0.5–1.3 mm in diameter, rarely flat and to 1.5(2) mm wide, abaxial surfaces smooth, adaxial surfaces with 3–5(6) ribs, ribs smooth, the outer ribs sometimes scabrous. **Panicles** 3.5–17 cm long, 1.5–9 cm wide, usually open and pyramidal, sometimes closed and ovate; **branches** 0.5–6 cm, spreading to reflexed, flexuous, smooth. **Spikelets** 3.5–5.2 mm, shiny, purplish, with 2(3) florets. **Glumes** lanceolate, sometimes purplish over the proximal $1/2$, acute to acuminate; **lower glumes** 2.7–4.8 mm, 0.8–0.9 times the length of the spikelets, 1–3-veined, veins smooth; **upper glumes** 3–5 mm, equaling or exceeding the lowest floret, 1–5-veined; **callus hairs** 0.3–1 mm; **lemmas** 2–4 mm, smooth, shiny, glabrous, sometimes purplish distally, apices rounded or truncate, erose, awns 0.8–2.5 mm, arising at or below midlength, straight, slender, only slightly or not exserted; **anthers** 0.7–2.5 mm. $2n = 26, 28, 36, ca. 39$.

Deschampsia sukatschewii is a circumboreal species that extends from northern Russia through Alaska, northern Canada, and Greenland to Svalbard, and southward in the Rocky Mountains to Nevada and Utah. It ranges from short plants that form dense, mossy tufts on the Arctic coast to larger plants in subalpine and alpine habitats of the Rocky Mountains that have frequently been included in *D. cespitosa* .

Arctic taxonomists recognize two subspecies of *Deschampsia sukatschewii* in arctic North America: the amphiberingian subsp. *orientalis* (Hultén) Tzvelev that extends to the northern coast of Alaska and the western Northwest Territories; and subsp. *borealis* (Trautv.) Tzvelev, which is circumpolar in the arctic. Chiapella and Probatova (2003) treated these two subspecies, and subsp. *sukatschewii*, as subspecies of *D. cespitosa*. Efforts to circumscribe infraspecific taxa of *D. sukatschewii* for this treatment failed.

4. Deschampsia brevifolia R. Br. [p. 630]

Plants perennial; cespitose, not glaucous. **Culms** 5–55 cm, erect, glabrous. **Leaves** often forming a basal tuft; **sheaths** glabrous; **ligules** 1–4.5 mm, acute or acuminate, entire; **blades** 2–12 (16) cm long, usually 0.3–0.8 mm in diameter, folded or con-volute, 0.5–2 mm wide when flat, abaxial surfaces glabrous, adaxial surfaces glabrous or sparsely hirtellous, sometimes scabrous, blades of flag leaves 0.8–3 cm. **Panicles** 1.5–10(12) cm long, 0.5–2(11) cm wide, usually dense, oblong-ovate to narrowly cylindrical; **branches** 1–3.6(6) cm, straight, usually stiff, erect to ascending, usually smooth or almost so, scabrules separated by 0.2+ mm, spikelet-bearing to near the base. **Spikelets** 2.3–6 mm, ovate to obovate, with 2(3) florets. **Glumes** subequal to equal, 2.5–5.6 mm, purplish over more than $1/2$ their surface, lanceolate, smooth, acuminate or acute; **lower glumes** 1-veined, smooth; **upper glumes** exceeding to exceeded by the lowest floret, 3-veined; **callus hairs** 0.2–2 mm; **lemmas** 2.2–4 mm, oblong or lanceolate, smooth, shiny, glabrous, awns (0.2)0.7–4 mm, usually equaling or exceeding the lemmas, straight or weakly geniculate, usually attached from near the base to midlength, occasionally connate almost their full length; **anthers** 1.2–2.5 mm. $2n = 26, 27, 28, about 50, 52$.

Deschampsia brevifolia is a circumboreal taxon that grows in wet places in the tundra, often in disturbed soils associated with riverbanks, frost-heaving, etc. It is interpreted here as extending southward through the Rocky Mountains to Colorado, where it grows at elevations up to 4300 m. It is to be expected from high elevations in British Columbia and Alberta; specimens currently identified as *D. cespitosa*, in which *D. brevifolia* is often included as a subspecies, need to be examined.

In its typical appearance, *Deschampsia brevifolia* is quite distinctive because of its dark, narrow panicles. Culm height can vary substantially from year to year, probably in response to the environment. Aiken et al. (1995 on) reported that plants transplanted from Eureka Sound, Ellesmere Island (80° 9' N 86° 0' W) to Iqaluit, Baffin Island (64° 44' N 68° 28' W) became smaller and more stunted; most of those transplanted to Ottawa, Ontario (45° 18' N 75° 50' W) died, but some grew larger than at the original site, and developed more diffuse panicles.

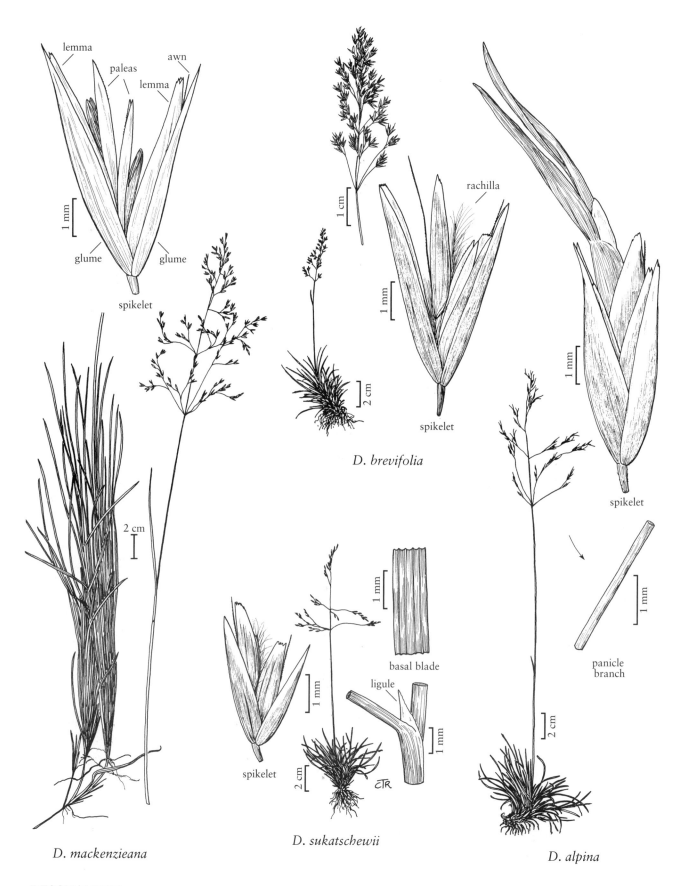

lemma

paleas

lemma

awn

1 mm

glume

glume

spikelet

rachilla

1 mm

spikelet

D. brevifolia

1 cm

1 mm

spikelet

1 mm

panicle
branch

2 cm

D. mackenzieana

spikelet

1 mm

basal blade

ligule

1 mm

1 mm

2 cm

CTR

D. sukatschewii

2 cm

D. alpina

DESCHAMPSIA

5. Deschampsia alpina (L.) Roem. & Schult. [p. 630]

ALPINE HAIRGRASS, DESCHAMPSIE ALPINE

Plants perennial; densely cespitose. **Culms** 8–45(65) cm, smooth, glabrous. **Leaves** forming a basal tuft; **sheaths** smooth, glabrous; **ligules** 1.5–7.5 mm, glabrous, acute to acuminate, entire; **blades** 2–8 cm long, 0.5–2 mm wide, usually folded or flat, sometimes some loosely involute, both surfaces glabrous, smooth. **Panicles** (4)8–16 cm; **branches** 2–8 cm long (excluding the blades of bulbous florets), straight, ascending, smooth. **Spikelets** usually viviparous, their length varying with age, rarely bisexual and 4–6.3 mm. **Glumes** subequal, exceeding the lowest floret in sexual spikelets, keels smooth, apices acuminate; **callus hairs** about 0.8 mm; **lemmas** 5–7 mm, smooth, shiny, glabrous, unawned or awned, awns to 4 mm, straight, attached from below midlength to near the apices; **paleas** vestigial or absent. $2n = 52, 56$.

Deschampsia alpina grows in damp, rocky places, on calcareous substrates with low organic content, in Greenland and northeastern Canada and, outside the *Flora* region, in the mountains of Scandinavia and Russia in the Kola Peninsula and Novaya Zemlya. Plants of *D. alpina* differ from viviparous plants of *D. cespitosa* in having smooth, rather than scabrous, panicle branches (Murray, pers com. 2005).

6. Deschampsia elongata (Hook.) Munro [p. 632]

SLENDER HAIRGRASS

Plants perennial; densely cespitose. **Culms** (10)30–120 cm. **Leaves** sometimes forming a basal tuft; **sheaths** glabrous; **ligules** 2.5–8(9) mm, acute to acuminate; **blades** 7–30 cm long, 0.2–2 mm wide, usually involute. **Panicles** 5–30(35) cm long, 0.5–1.5(2) cm wide, erect or nodding; **branches** erect to ascending. **Spikelets** 3–6.7 mm, bisexual, narrowly V-shaped, appressed to the branches. **Glumes** equaling or exceeding the florets, narrowly lanceolate, usually pale green, sometimes purple-tipped, 3-veined, acuminate; **lower glumes** (3)3.2–5.5(6.7) mm; **upper glumes** (3)3.1–5.4(6) mm; **callus hairs** 0.3–1.15 mm; **lemmas** 1.7–4.3 mm, smooth, shiny, glabrous, apices weakly toothed or erose, awns 1.5–5.5(6) mm, straight to slightly geniculate, attached from slightly below to slightly above the middle of the lemma, exceeding the florets by 1–2.5 mm; **anthers** 0.3–0.5(0.7) mm. $2n = 26$.

Deschampsia elongata grows in moist to wet habitats, from near sea level to alpine elevations, from Alaska and the Yukon south to northern Mexico and east to Montana, Wyoming, and Arizona. It also grows, as a disjunct, in Chile. The records from Maine and Colorado probably represent introductions.

7. Deschampsia danthonioides (Trin.) Munro [p. 632]

ANNUAL HAIRGRASS

Plants annual; tufted. **Culms** 10–40(70) cm, erect. **Leaves** not forming a basal tuft; **ligules** (0.5)2–3(4.7) mm, acute to acuminate, entire; **blades** 0.3–1.5 mm wide, involute or flat. **Panicles** 5–15(25) cm long, 2–8 cm wide, contracted to open, erect; **branches** with the spikelets confined to the distal portion. **Spikelets** 4–9 mm, bisexual, narrowly V-shaped, usually pale green. **Glumes** exceeding the distal florets, glabrous to scabridulous, 3-veined; **lower glumes** 4–9 mm; **upper glumes** 3.5–8.5 mm; **callus hairs** 0.4–1.6 mm; **lemmas** 1.5–3 mm, smooth, shiny, glabrous, pale green or purplish, apices blunt, erose to 4–toothed, ciliate, awns 4–9 mm, attached from near the base to about the middle of the lemmas, strongly geniculate, geniculation above the lemma apices, distal segment 1.5–5 mm; **anthers** 0.3–0.5 mm. $2n = 26$.

Deschampsia danthonioides grows in temperate and cool-temperate regions, usually in open, wet to dry habitats and often in disturbed ground. Its primary range extends from southern British Columbia, through Washington and Idaho, to Baja California, Mexico. It also grows, as a disjunct, in Chile and Argentina.

Records from the Yukon Territory date from the late 1800s and early 1900s; it has not been seen since in the region. Records from east of the primary region are also probably introductions; it is not known whether the species has persisted at these locations.

8. Deschampsia flexuosa (L.) Trin. [p. 632]

CRINKLED HAIRGRASS, WAVY HAIRGRASS,
DESCHAMPSIE FLEXUEUSE

Plants perennial; densely cespitose. **Culms** 30–80 cm, erect or geniculate at the base, usually with 2 nodes. **Leaves** mostly basal, sometimes forming a basal tuft; **sheaths** smooth, glabrous; **ligules** 1.5–3.6 mm, rounded to acute; **blades** 12–25 cm long, strongly rolled, 0.3–0.5

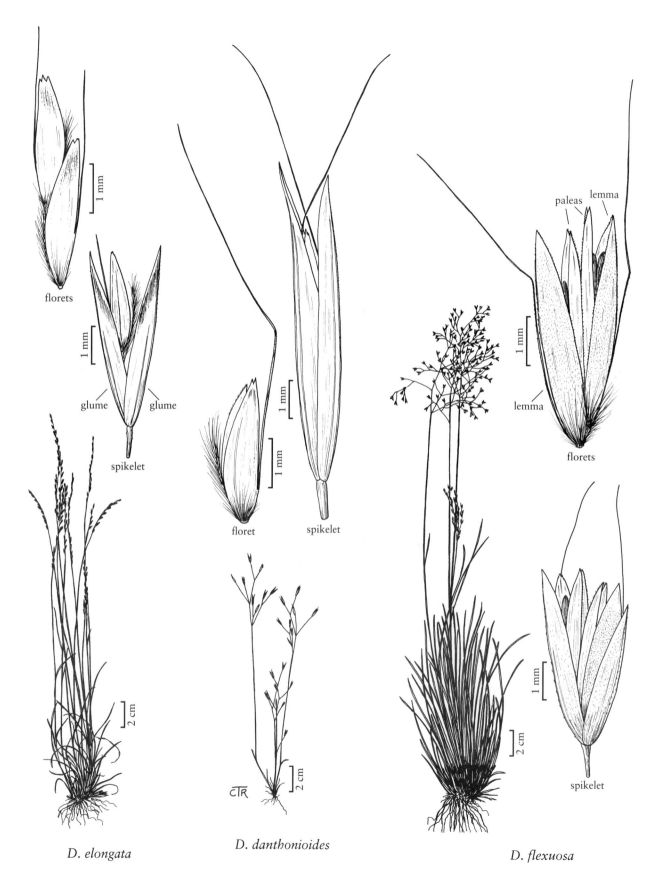

florets

1 mm

glume   glume

spikelet

florets

D. elongata

floret

spikelet

1 mm

1 mm

CTR

2 cm

2 cm

D. danthonioides

paleas   lemma

1 mm

lemma

florets

1 mm

spikelet

2 cm

D. flexuosa

DESCHAMPSIA

mm in diameter, abaxial surfaces smooth or scabridulous, glabrous or hairy, often scabridulous or hairy proximally and essentially smooth and glabrous distally, adaxial surfaces scabrous, flag leaf blades 5–8 cm. **Panicles** 5–15 cm long, (2)4–12 cm wide, narrow to open, often nodding; **branches** ascending to spreading, flexuous, smooth or scabridulous. **Spikelets** 4–7 mm, ovate or U-shaped. **Glumes** exceeded by or subequal to the adjacent florets, 1–veined, acute; **lower glumes** 2.7–4.5 mm; **upper glumes** 3.5–5 mm; **callus hairs** to 1 mm; **lemmas** 3.3–5 mm, scabridulous or puberulent, hairs to 0.1 mm, apices acute, erose to 4-toothed, awns 3.7–7 mm, attached near the base of the lemma, strongly geniculate, geniculation below the lemma apices, distal segment 2.5–4.5 mm, pale; **anthers** 2–3 mm. $2n = 14, 26, 28, 32, 42$.

Deschampsia flexuosa grows on dry, often rocky slopes, and in woods and thickets, often in disturbed sites. In the *Flora* region, it is primarily eastern in distribution, with records from west of the Great Lakes and Appalachians probably being introductions. It is also known from Mexico, Central America, South America, Borneo, the Philippines, and New Zealand.

14.27 AGROSTIS L.

M.J. Harvey

Plants usually perennial; usually cespitose, sometimes rhizomatous or stoloniferous. **Culms** (3)5–120 cm, usually erect. **Sheaths** open, usually smooth and glabrous, sometimes scabrous to scabridulous, rarely hairy; **collars** not strongly developed; **auricles** absent; **ligules** membranous, smooth or scabridulous dorsally, apices truncate, obtuse, rounded, or acute, usually erose to lacerate, the lacerations sometimes obscuring the shape, or entire; **blades** flat, folded, or involute, usually smooth and glabrous, sometimes scabridulous, adaxial surfaces somewhat ridged. **Inflorescences** terminal panicles, narrowly cylindrical and dense to open and diffuse; **branches** usually in whorls, usually more or less scabrous, rarely smooth, some branches longer than 1 cm; **secondary panicles** sometimes present in the leaf axils. **Spikelets** 1.2–7 mm, pedicellate, laterally compressed, lanceolate to narrowly oblong or ovate, with 1(2) florets; **rachillas** not prolonged beyond the base of the floret(s); **disarticulation** above the glumes, beneath the florets, sometimes initially at the panicle base. **Glumes** (1)1.3–2(4) times longer than the lemmas, 1(3)-veined, glabrous, usually mostly smooth, vein(s) often scabrous to scabridulous, backs keeled or rounded, apices acute to acuminate or awn-tipped; **lower glumes** usually 0.1–0.3 mm longer than the upper glumes, rarely equal; **calluses** poorly developed, blunt, glabrous or hairy, hairs to about ½ as long as the lemmas; **lemmas** thinly membranous to hyaline, usually smooth and glabrous, sometimes scabridulous, occasionally pubescent, rarely warty-tuberculate, 3–5-veined, veins not convergent, sometimes excurrent as 2–5 teeth, apices acute to obtuse or truncate, sometimes erose, unawned or awned, sometimes varying within an inflorescence, awns arising from near the lemma bases to near the apices, usually geniculate, sometimes straight; **paleas** absent, or minute to subequal to the lemmas, usually thin, veins not or only weakly developed; **lodicules** 2, free; **anthers** (1)3, 0.1–2 mm, not penicillate; **styles** 2, free to the base, white; **ovaries** glabrous. **Caryopses** with a hard, soft, or liquid endosperm, the latter resulting from the substitution of lipids for starch. $x = 7$. Name from the Greek *agros*, 'pasture' or 'green fodder'.

Agrostis in the older, broad sense is a genus comprised of species with the spikelets reduced to single florets. As such, it is found in all inhabited continents, is presumably of ancient origins, and many of the 150–200 species may be only distantly related. The shortage of clear-cut morphological features has hindered its subdivision into more natural units. This treatment follows Edgar (1995), Edgar and Connor (2000), and Jacobs (2001) in placing *A. avenacea* J.F. Gmel. in the Australasian genus *Lachnagrostis*, as *L. filiformis*; Rúgolo de Agrasar (1982) in treating *A. tandilensis* (Kuntze) Parodi as *Bromidium tandilense*; and Soreng (2003) in placing *A. aequivalvis* (Trin.) Trin. and *A. humilis* —together with several Central and South American species, including *A. sesquiflora* E. Desv.—in the genus *Podagrostis*.

Agrostis usually differs from both *Podagrostis* and *Lachnagrostis* in having no, or very reduced, paleas, and in rachillas that are not prolonged beyond the base of the floret. Some of the Eurasian species of *Agrostis* are exceptional in having paleas at least $^2/_5$ as long as the lemmas. *Agrostis* also differs from *Lachnagrostis* in certain features of the lemma epidermes (Jacobs 2001).

Agrostis is sometimes confused with *Apera* (p. 788), *Calamagrostis* (p. 706), or *Polypogon* (p. 662). It differs from *Apera* in having lemmas that are less firm than the glumes, paleas that are often absent or minute, and in lacking a rachilla prolongation. There is no single character that distinguishes all species of *Agrostis* from those of *Calamagrostis*. In general, *Agrostis* has smaller plants with smaller, less substantial lemmas and paleas than *Calamagrostis*, and tends to occupy drier habitats. It differs from *Polypogon* in having spikelets that disarticulate above the glumes.

Some taxonomists used the presence of a *trichodium net* for circumscribing *Trichodium* Michx. This net is formed by a series of transverse thickening bars developed on the inner wall of the dorsal epidermal cells of the lemma, and is found in several different genera, usually in species with a reduced palea.

Species of *Agrostis* growing in the temperate regions of the Northern Hemisphere and on tropical mountains are mostly perennials, with the annual species predominantly in warmer climates, such as the Mediterranean and the Southern Hemisphere. Of the 26 species known from the *Flora* region, 21 are native and 5 are introductions. Two additional species, *A. tolucensis* and *A. anadyrensis*, have been reported; the reports are dubious.

Some species of *Agrostis* make a modest contribution to forage, a few are agricultural weeds, and some are excellent lawn grasses in cool climates. Most North American native species are narrow habitat specialists, with many being western endemics. The introduced species are all widely distributed in temperate regions of the world.

Unusual specimens of *Agrostis* with elongate or leafy spikelets are caused by infection with the nematode *Anguillina agrostis*. Other pathogens may cause stunting.

Species with awns on the lemmas frequently exhibit a developmental gradient within the inflorescence. Upper florets may possess a well-developed geniculate awn inserted at the base or on the lower half of the lemma; mid-inflorescence spikelets may have a shorter, possibly non-geniculate awn inserted high on the lemma, while basal spikelets may possess only a terminal bristle on the lemma. This phenomenon is particularly sharply shown in *Agrostis castellana*, where a single side branch of only a dozen or so spikelets can show the whole sequence. When using the key, it is advised to examine spikelets from the upper parts of an inflorescence. Many species key more than once, due to the potential for awns to be either present or absent.

SELECTED REFERENCES **Björkman, S.O.** 1960. Studies in *Agrostis* and related genera. Symb. Bot. Upsal. 17:1–112; **Carlbom, C.G.** 1967. A biosystematic study of some North American species of *Agrostis* L. and *Podagrostis* (Griesb.) Scribn. & Merr. Ph.D. dissertation, Oregon State University, Corvallis, Oregon, U.S.A. 232 pp.; **Edgar, E.** 1995. New Zealand species of *Deyeuxia* P. Beauv. and *Lachnagrostis* Trin. (Gramineae: Aveneae). New Zealand J. Bot. 33:1–33; **Edgar, E. and H.E. Connor.** 2000. Flora of New Zealand, vol. 5. Manaaki Whenua Press, Lincoln, New Zealand. 650 pp.; **Hitchcock, A.S.** 1951. Manual of the Grasses of the United States, ed. 2, rev. A. Chase. U.S.D.A. Miscellaneous Publication No. 200. U.S. Government Printing Office, Washington, D.C., U.S.A. 1051 pp.; **Jacobs, S.W.L.** 2001. The genus *Lachnagrostis* (Gramineae) in Australia. Telopea 9:439–448; **Romero García, A.T., G. Blanca Lopez, and C. Morales Torres.** 1988. Revisión del género *Agrostis* L. (Poaceae) en la Península Ibérica. Ruizia 7:1–160; **Rúgolo de Agrasar, Z.E.** 1982. Revalidación del género *Bromidium* Nees et Meyen emend. Pilger (Gramineae). Darwiniana 24:187–216; **Rúgolo de Agrasar, Z.E. and A.M. Molina.** 1997. Las especies del género *Agrostis* L. (Gramineae: Agrostideae) de Chile. Gayana, Bot. 54:91–156; **Soreng, R.J.** 2003. *Podagrostis*. Contr. U.S. Natl. Herb. 48:581; **Tercek, M.T., D.P. Hauber, and S.P. Darwin.** 2003. Genetic and historical relationships among thermally adapted *Agrostis* (Bentgrass) of North America and Kamchatka: Evidence for a previously unrecognized, thermally adapted taxon. Amer. J. Bot. 90:1306–1312; **Tsvelev, N.N.** 1976. Zlaki SSSR. Nauka, Leningrad [St. Petersburg], Russia. 788 pp.

1. Paleas at least ²/₅ as long as the lemmas.
  2. Lemmas 0.5–0.8 mm long, transparent; paleas similar to the lemmas and almost as long; panicles usually over ¹/₂ the length of the culms, extremely diffuse . . . . .28. *A. nebulosa*
  2. Lemmas 1.2–2.5 mm long, opaque to translucent; paleas shorter than the lemmas; panicles less than ¹/₂ the length of the culms, diffuse or not.
    3. Panicles narrowly contracted, sometimes open at anthesis, 0.5–4(6) cm wide; branches ascending to appressed.
      4. Stolons present, plants mat-forming, rhizomes absent; paleas 0.7–1.4 mm long; anthers 0.9–1.4 mm long . . . . .4. *A. stolonifera* (in part)
      4. Stolons absent, plants usually cespitose, rhizomes sometimes present; paleas to 0.5 mm long; anthers 0.3–0.6 mm long . . . . 15. *A. exarata* (in part)
    3. Panicles open at maturity, sometimes somewhat contracted after anthesis, (1)2–15 cm wide; branches spreading to ascending.
      5. Ligules of the upper leaves longer than wide, 2–7.5 mm long; usually at least some lower panicle branches with spikelets to the base.
        6. Stolons absent, rhizomes present; culms 20–120 cm tall; panicles 8–30 cm long; longest lower panicle branches 4–9 cm long . . . . .3. *A. gigantea*
        6. Stolons present, rhizomes absent; culms 8–60 cm tall; panicles 3–20 cm long; longest lower panicle branches 2–6 cm long . . . . .4. *A. stolonifera* (in part)
      5. Ligules of the upper leaves usually shorter than wide, 0.3–3 mm long; lower panicle branches with spikelets confined to the distal ¹/₃–¹/₂.
        7. Calluses glabrous or with a few hairs to 0.1 mm long; adjacent pedicels divergent, giving well-separated spikelets; panicles stiffly erect, 3–20 cm long; awns rarely present, to 2 mm; lemmas glabrous . . . . 1. *A. capillaris*
        7. Calluses abundantly hairy, hairs to 0.6 mm long; adjacent pedicels not divergent, spikelets appearing clustered; panicles somewhat lax, 10–30 cm long; awns, if present, to 5 mm long on the terminal spikelet of a cluster; lemmas occasionally with hairs on the lower ¹/₂ . . . . .2. *A. castellana*
1. Paleas absent or less than ²/₅ as long as the lemmas.
  8. Lemmas awned [for opposite lead, see p. 637].
    9. Panicles dense, often spikelike, 0.2–4 cm wide; lower branches usually shorter than 2 cm, appressed to ascending, usually hidden by the spikelets.
      10. Lemma awns 3.5–10 mm long; calluses with hairs to 1 mm long; plants annual.
        11. Lemmas 2.5–4 mm long, teeth to 1.5 mm long; awns (5)8–10 mm long; blades 1–4.5 cm long . . . . .25. *A. hendersonii*
        11. Lemmas 1.5–2.3 mm long, teeth to 0.5 mm long; awns 3.5–8 mm long; blades 3–15 cm long . . . . 26. *A. microphylla*
      10. Lemma awns to 3.5 mm long; calluses glabrous or with hairs to 0.3 mm long; plants perennial.
        12. Blades less than 2 mm wide, usually involute or folded.
          13. Calluses glabrous; panicles often partly enclosed by the upper sheaths at maturity; lemma awns to 0.7 mm long . . . . .22. *A. blasdalei* (in part)
          13. Calluses hairy; panicles exserted from the upper sheaths at maturity; lemma awns to 3.5 mm long.
            14. Lemma apices acute, entire; lemma awns to 2.8 mm long, usually not exserted from the spikelets . . . . .21. *A. variabilis* (in part)
            14. Lemma apices truncate, denticulate; lemma awns 2–3.5 mm long, exserted from the spikelets . . . . .23. *A. tolucensis* (in part)
        12. Blades 2–10 mm wide, usually flat, sometimes folded.
          15. Lemma apices truncate to acute; blades to 4 mm wide, flat or involute; ligules 2–6.2 mm long; panicles 0.5–1.5 cm wide . . . . .23. *A. tolucensis* (in part)
          15. Lemma apices acute to obtuse; blades to 10 mm wide, flat; ligules 1–11.2 mm long; panicles 0.5–4 cm wide.

16. Anthers 0.3–0.6 mm long; paleas usually absent, rarely to 0.5 mm long and about $^1/_5$ the length of the lemmas; lemmas entire or with teeth to 0.12 mm long . . . . 15. *A. exarata* (in part)

16. Anthers 0.5–2 mm long; paleas 0.3–0.7 mm long, to about $^1/_3$ the length of the lemmas; lemmas usually with teeth to 0.3 mm long . . . . 16. *A. densiflora* (in part)

9. Panicles open or diffuse, or somewhat contracted but not spikelike, 0.4–20 cm wide; lower branches 1.5–12 cm long, erect to spreading, readily visible.

17. Leaves usually involute or becoming so, sometimes only the basal leaves involute, less than 1 mm in diameter when involute, 0.5–2 mm wide when flat; plants without rhizomes or stolons.

18. Anthers 1, 0.1–0.2 mm long, usually persistent at the apices of the caryopses; awns attached just below the apices of the lemmas, flexuous but not geniculate, deciduous . . . . 27. *A. elliottiana* (in part)

18. Anthers 3, 0.4–1.5 mm long, usually shed at anthesis; awns attached below midlength on the lemmas, usually geniculate, persistent.

19. Basal leaves usually withered at anthesis; lower sheaths finely tomentose; callus hairs abundant; plants endemic to coastal California . . . . 19. *A. hooveri*

19. Basal leaves persistent; lower sheaths smooth or scabrous; callus hairs sparse; plants widespread, especially in northern and montane parts of the *Flora* region, including California.

20. Panicles (2)3–10 cm long; branches not capillary, fairly stiff, smooth or sparsely scabridulous; callus hairs to 0.4 mm long; caryopses 1.4–2 mm long, endosperm solid . . . . 7. *A. mertensii* (in part)

20. Panicles (4)8–25(50) cm long; branches capillary, flexible, scabrous; callus hairs to 0.2 mm long; caryopses 0.9–1.4 mm long; endosperm liquid . . . . 10. *A. scabra* (in part)

17. Leaves usually remaining flat, 0.5–6 mm wide; plants with or without rhizomes or stolons.

21. Lemmas with 4 teeth up to 0.5 mm long, lemmas 2.5–3 mm long; awns 4–6 mm long . . . . 20. *A. howellii*

21. Lemmas usually entire, sometimes minutely toothed or erose, teeth to 0.4 mm long, lemmas 1–3 mm long; awns to 5 mm long.

22. Rhizomes and stolons absent; blades to 30 cm long; anthers 0.4–1.2 mm long.

23. Panicle branches widely divergent, the whole panicle often detaching at the base at maturity, forming a tumbleweed; cauline nodes usually 1–3; blades 1–2 mm wide; glume apices acuminate . . . . 10. *A. scabra* (in part)

23. Panicle branches usually erect to ascending, if widely divergent then the panicle not forming a tumbleweed; cauline nodes 2–10; blades 0.5–5 mm wide; glume apices acute to acuminate.

24. Lemma awns 1–4.4 mm long, geniculate, exserted; blades to 13 cm long.

25. Leaf blades 0.5–3 mm wide, flat to involute; panicles 2–10 cm long, usually open; awns 2–4.4 mm long, inserted just below midlength on the lemmas . . . . 7. *A. mertensii* (in part)

25. Leaf blades 3–4 mm wide, flat; panicles 6–20 cm long, somewhat contracted; awns 1–1.5 mm long, inserted just above midlength on the lemmas . . . . 8. *A. anadyrensis*

24. Lemma awns minute or to 2 mm long, straight, usually not exserted; blades 6–30 cm long.

26. Basal leaves usually withered by anthesis; culm leaves 3–10, as broad and substantial as the lower leaves; callus hairs dense; plants primarily from east of the 100th Meridian . . . . 12. *A. perennans* (in part)

26. Basal leaves persisting; culm leaves 5 or fewer, usually less substantial than the lower leaves; callus hairs sparse; plants primarily western . . . . 14. *A. oregonensis* (in part)

22. Rhizomes or stolons present; blades 1–10 cm long; anthers 0.7–1.8 mm long.

27. Rhizomes absent; stolons present, to about 25 cm long, producing tufts of shoots at the nodes; glumes 1.7–3 mm long; panicles open, branches erect to spreading . . . . 5. *A. canina* (in part)

27. Rhizomes present, to about 10 cm long; stolons absent; glumes 2–4 mm long; panicles open to constricted, branches more or less erect to ascending.

28. Lemma apices blunt, entire; lemmas usually awned from near the base, rarely unawned, awns 2–4.5 mm long, geniculate . . . . . . 6. *A. vinealis* (in part)

28. Lemma apices acute, entire or toothed; lemmas usually unawned, rarely awned from below the apices, awns to 0.5(2.7) mm long, straight . . . . 17. *A. pallens* (in part)

8. Lemmas unawned [for opposite lead, see p. 635].

29. Mature panicles dense; lower panicle branches to 3(4) cm long, often hidden by the spikelets; spikelets crowded.

30. Blades 0.5–2 mm wide, in dense basal tufts; panicles 0.2–2 cm wide; culms 5–30 cm tall.

31. Lemma veins not excurrent; anthers 0.4–1 mm long; plants of western alpine and subalpine zones . . . . 21. *A. variabilis* (in part)

31. Lemma veins excurrent to 0.2 mm; anthers 0.7–2 mm long; plants of western coastal cliffs, dunes, and shrublands . . . . 22. *A. blasdalei* (in part)

30. Blades 2–10 mm wide, not basally concentrated; panicles 0.5–4 cm wide; culms 8–100 cm tall.

32. Anthers 0.3–0.6 mm long; paleas usually absent, rarely to 0.5 mm long and about $^1/_5$ the length of the lemmas; lemmas entire or with teeth to 0.12 mm long . . . . 15. *A. exarata* (in part)

32. Anthers 0.5–2 mm long; paleas 0.3–0.7 mm long, to about $^1/_3$ the length of the lemmas; lemmas usually toothed, teeth to 0.3 mm long . . . . 16. *A. densiflora* (in part)

29. Mature panicles open to diffuse; lower panicle branches often longer than 3 cm, usually not hidden by the spikelets; spikelets crowded or not.

33. Blades 0.5–14 cm long, 0.5–2 mm wide, usually involute or becoming so; anthers 0.1–0.9 mm long.

34. Anthers 1, 0.1–0.2 mm long; callus hairs dense, to 0.6 mm long; plants annual . . . . 27. *A. elliottiana* (in part)

34. Anthers 3, 0.2–0.9 mm long; callus hairs sparse, to 0.3 mm long; plants perennial or annual.

35. Panicles (4)8–50 cm long; lower panicle branches 4–15 cm long.

36. Lemmas 1.4–2 mm long, exceeding the ripe caryopses by 0.3+ mm; anthers 0.4–0.8 mm long; pedicels to 9.6 mm long, spikelets not appearing clustered . . . . 10. *A. scabra* (in part)

36. Lemmas 0.8–1.2 mm long, exceeding the ripe caryopses by no more than 0.2 mm; anthers 0.2–0.5 mm long; pedicels to 3.5 mm long, spikelets appearing clustered . . . . 11. *A. hyemalis*

35. Panicles 1.5–13 cm long; lower panicle branches 1–4 cm long.

37. Anthers 0.3–0.6 mm long; upper culm sheaths not inflated; plants perennial, of western seepage areas and bogs . . . . 13. *A. idahoensis* (in part)

37. Anthers 0.5–0.9 mm long; upper culm sheaths inflated; plants annual, near hot springs . . . . 24. *A. rossiae*

33. Blades 1–30 cm long, 1–7.5 mm wide, usually flat; anthers 0.3–2.3 mm long.

38. Rhizomes present, to 50 cm long, stolons absent; panicle branches branching from midlength or to near the base; lower panicle branches 1–5 cm long; anthers 0.7–2.3 mm long.

    39. Lemma apices blunt, entire; callus hairs sparse, to 0.1 mm long; panicles 2–15 cm long; pedicels 0.5–2 mm long; blades 1–3 mm wide . . . . 6. *A. vinealis* (in part)

    39. Lemma apices usually acute, entire or toothed, teeth to about 0.2 mm long; callus hairs sparse or abundant, to 2 mm long; panicles 5–22 cm long; pedicels 0.5–7 mm long; blades 1–6 mm wide.

        40. Anthers 0.7–1.8 mm long; callus hairs to 0.3(1) mm long, sparse; leaf blades 1.5–11.5 cm long; caryopses 1–1.5 mm long . . . . 17. *A. pallens* (in part)

        40. Anthers 1.5–2.3 mm long; callus hairs 0.8–2 mm long, abundant; leaf blades 6–20 cm long; caryopses 1.5–2 mm long . . . . 18. *A. hallii*

38. Rhizomes absent, stolons sometimes present, to 25 cm long; panicle branches mostly branching at or beyond midlength; lower panicle branches 1–12 cm long; anthers 0.3–1.5 mm long.

    41. Stolons present, to about 25 cm long, producing tufts of shoots at the nodes; anthers 1–1.5 mm long; blades 1–3 mm wide; glume apices acute . . . . 5. *A. canina* (in part)

    41. Stolons absent; anthers 0.3–1.2 mm long; blades 0.5–7 mm wide; glume apices acute to acuminate.

        42. Blades to 2 mm wide, 1–14 cm long, flat or involute; leaves mostly basal.

            43. Lower panicle branches 4–12 cm long; whole panicle often detaching at the base at maturity, forming a tumbleweed; blades 4–14 cm long . . . . 10. *A. scabra* (in part)

            43. Lower panicle branches 1–4 cm long; panicle not detaching at maturity and forming a tumbleweed; blades 1–7 cm long . . . . 13. *A. idahoensis* (in part)

        42. Blades to 7 mm wide, usually at least some wider than 2 mm, 5–30 cm long, flat; leaves mostly cauline to mostly basal.

            44. Plants annual or short-lived perennials; glumes 1.5–2.8 mm long, subequal; lemmas smooth and glabrous; anthers 0.3–0.6 mm long; caryopses 0.9–1.3 long . . . . 9. *A. clavata*

            44. Plants perennial; glumes 1.8–3.6 mm long, unequal; lemmas smooth or scabridulous, sometimes pubescent; anthers 0.4–1.2 mm long; caryopses 1–1.9 mm long.

                45. Basal leaves usually withered by anthesis; cauline nodes 3–10; blades of the upper leaves as broad and substantial as those of the lower leaves; callus hairs abundant; plants primarily from east of the 100th Meridian . . . . 12. *A. perennans* (in part)

                45. Basal leaves persisting to anthesis; cauline nodes 5 or fewer; blades of the upper leaves usually less substantial than those of the lower leaves; callus hairs sparse; plants primarily western . . . . 14. *A. oregonensis* (in part)

1. Agrostis capillaris L. [p. 640]
BROWNTOP, RHODE ISLAND BENT, COLONIAL BENT, AGROSTIDE FINE

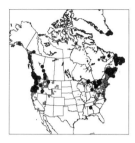

Plants perennial; rhizomatous or stoloniferous, rhizomes or stolons to 5 cm. **Culms** 10–75 cm, erect or geniculate, with 2–5 nodes. **Leaves** basal and cauline; **sheaths** smooth; **ligules** 0.3–2 mm, shorter than wide, dorsal surfaces usually scabridulous, sometimes smooth, apices truncate to rounded, erose-ciliolate, sometimes lacerate; **blades** 3–10 cm long, 1–5 mm wide, flat. **Panicles** 3–20 cm long, less than ¹/₂ the length of the culm, (1)2–12 cm wide, stiffly erect, widely ovate, open, exserted from the upper sheaths at maturity, lowest node with (2)3–9(13) branches; **branches** smooth or scabridulous, spreading during and after anthesis, spikelets usually confined to the distal ¹/₂, lower branches 1.5–7 cm; **pedicels** 0.4–3.3 mm, adjacent pedicels divergent. **Spikelets** lanceolate or oblong, purplish brown to greenish. **Glumes** subequal, 1.7–3 mm, 1-veined, acute; **lower glumes** scabridulous over the midvein towards the apices; **upper glumes** scabridulous or smooth over the midvein; **calluses** glabrous, or with a few hairs to 0.1 mm; **lemmas** 1.2–2.5 mm, smooth, glabrous, opaque to translucent, 3(5)-veined, veins typically prominent, apices obtuse to acute, usually entire, sometimes the veins excurrent to 0.5 mm, usually unawned, rarely awned, sometimes varying within a panicle, awns to 2 mm, mid-dorsal, straight or geniculate; **paleas** 0.6–1.2(1.4) mm, typically at least ¹/₂ the length of the lemmas, veins visible; **anthers** 3, 0.8–1.3 mm. **Caryopses** 0.8–1.5 mm; **endosperm** solid. 2n = 28.

Agrostis capillaris grows along roadsides and in disturbed areas. It was introduced from Europe, and is now well established in western and eastern North America. It is often used for fine-leaved lawns; commercial seed sold as *Agrostis tenuis* 'Highland' usually contains *A. capillaris*.

Agrostis capillaris differs from *A. gigantea* (p. 641) in its short ligules, especially on the vegetative shoots, and the open panicles that lack spikelets near the base of the branches. It differs from *A. castellana* (see next) in having diffuse rather than clustered spikelets, fewer rhizomes, divaricate panicle branches after anthesis, calluses that are glabrous or with hairs up to 0.1 mm long, and glabrous lemmas. It also tends to flower somewhat earlier than *A. castellana*. *Agrostis capillaris* readily hybridizes with *A. vinealis* (p. 643), the hybrids being somewhat intermediate between the two parents.

2. Agrostis castellana Boiss. & Reut. [p. 640]
HIGHLAND BENT, DRYLAND BROWNTOP

Plants perennial; loosely cespitose, rhizomatous, rhizomes to 10(40) cm, covered with inflated scales. **Culms** 30–80 cm, erect or geniculate, with up to 10 nodes, basal internodes devoid of leaf bases after anthesis in dry habitats, shoots proliferating from the upper nodes, especially of the innovations, later in the season. **Leaves** mostly cauline; **sheaths** smooth; **ligules** 0.5–3 mm, shorter than wide, dorsal surfaces smooth or scabridulous, apices truncate to rounded, ciliolate-erose; **blades** 4–10 cm long, 1–3 mm wide, flat. **Panicles** 10–30 cm long, less than ¹/₂ the length of the culm, 3–8 cm wide, loosely ovate, somewhat lax, somewhat contracted and linear-lanceolate after anthesis, lowest node with (1)2–7 branches; **branches** spreading to ascending, sparsely scabridulous, branching above the midpoint, spikelets usually confined to the distal ¹/₃ and in discrete clusters at the branch tips, lower branches 3–9 cm; **pedicels** 0.6–2.3(3) mm, usually shorter than the spikelets, adjacent pedicels not divergent. **Spikelets** lanceolate, yellowish green to stramineous or brownish, slightly to strongly suffused with purple. **Glumes** subequal, 2–3 mm, lanceolate, 1-veined, acute or acuminate; **lower glumes** scabrous over the midvein, at least distally; **upper glumes** smooth or scabrous to scabridulous over the midvein distally; **calluses** abundantly hairy, hairs to 0.3(0.6) mm; **lemmas** 1.3–1.9 mm, occasionally with hairs on the lower ¹/₂, translucent, (3)5-veined, veins prominent distally, apices usually truncate to obtuse, sometimes acute, entire or the lateral veins excurrent to 0.6 mm, awned or unawned, usually mixed in the inflorescence, terminal spikelets usually awned, in some plants all unawned, awns to 5 mm, awns arising from the lower ¹/₃ or occasionally as a minute bristle from above the center; **paleas** 0.6–1.1 mm, ¹/₂–²/₃ as long as the lemma; **anthers** 3, 1–1.5 mm. **Caryopses** about 1 mm. 2n = 28, 42.

Agrostis castellana is native to southern Europe. It was introduced to North America in the 1930s for use in lawns and golf greens, under the name *Agrostis tenuis* 'Highland'; commercial samples of 'Highland' often contain *A. capillaris* (p. 639). Escaped plants were collected at least as early as the 1950s, but were not recognized as belonging to *A. castellana* until the 1990s, when several collections were identified as such in Oregon. Recorded habitats have ranged from sunny gravel roadsides to moist ground alongside cranberry bogs, at elevations from near sea level to over 600 m.

lemma   palea
unawned
floret

awned
floret

anthers

spikelet

ligule

awn

spikelet

florets

rhizome

A. capillaris   *A. castellana*

AGROSTIS

In view of its extensive commercial use for over 70 years and its drought tolerance, it is likely that it is more widespread than shown.

Agrostis castellana belongs to a Eurasian group that includes *A. gigantea* (p. 642), *A. stolonifera* (p. 641), and *A. capillaris* (635). It differs from *A. gigantea* and *A. stolonifera* in having shorter, truncate ligules about as short as wide, and in not possessing extensive rhizomes and stolons. It differs from *A. capillaris* in having clustered rather than diffuse spikelets, more abundant rhizomes, somewhat constricted panicle branches after anthesis, abundantly hairy calluses with hairs up to 0.3(0.6) mm long, and lemmas that are sometimes dorsally pubescent. It also tends to flower somewhat later than *A. capillaris*.

3. Agrostis gigantea Roth [p. 642]
REDTOP, AGROSTIDE BLANCHE

Plants perennial; rhizomatous, rhizomes to 25 cm, not stoloniferous. **Culms** 20–120 cm, erect, sometimes geniculate at the base, sometimes rooting at the lower nodes, with 4–7 nodes. **Leaves** mostly cauline, **sheaths** smooth or sparsely scabridulous; **ligules** longer than wide, dorsal surfaces usually scabrous, sometimes smooth, apices rounded to truncate, erose to lacerate, basal ligules 1–4.5 mm, upper ligules 2–7 mm; **blades** 4–10 cm long, 3–8 mm wide, flat. **Panicles** 8–25(30) cm long, less than ¹/₂ the length of the culm, (1.5)3–15 cm wide, erect, open, broadly ovate, exserted from the upper sheaths at maturity, lowest node with (1)3–8 branches; **branches** scabrous, spreading during and after anthesis, usually some branches spikelet-bearing to the base, lower branches 4–9 cm, usually with many shorter secondary branches resulting in crowding of the spikelets, spikelets restricted to the distal ¹/₂ of the branches and not crowded in shade plants; **pedicels** 0.3–3.4(4.2) mm. **Spikelets** narrowly ovate to lanceolate, green and slightly to strongly suffused with purple. **Glumes** subequal, 1.7–3.2 mm, lanceolate, 1-veined, acute to apiculate; **lower glumes** scabrous on the distal ¹/₂ of the midvein; **upper glumes** scabridulous on the distal ¹/₂ of the midvein; **callus hairs** to 0.5 mm, sparse; **lemmas** 1.5–2.2 mm, opaque to translucent, smooth, 3–5-veined, veins usually obscure, sometimes prominent throughout or distally, often excurrent to 0.2 mm, apices usually acute, sometimes obtuse or truncate, usually unawned, rarely with a 0.4–1.5(3) mm straight awn arising from near the apices to near the base; **paleas** 0.7–1.4 mm, about ¹/₂ the length of the lemmas, veins visible; **anthers** 3, 1–1.4 mm. **Caryopses** 1–1.5 mm; **endosperm** solid. $2n = 42$.

Agrostis gigantea grows in fields, roadsides, ditches, and other disturbed habitats, mostly at lower elevations. It is a serious agricultural weed, as well as a valuable soil stabilizer. In the *Flora* region, its range extends from the subarctic to Mexico; it is considered to be native to Eurasia. It is more heat tolerant than most species of *Agrostis*.

Agrostis gigantea has been confused with *A. stolonifera* (see next), from which it differs in having rhizomes and a more open panicle. *Agrostis stolonifera* has elongated leafy stolons, mainly all above the surface, that root at the nodes, and the panicles are condensed and often less strongly pigmented than in *A. gigantea*. Its distribution tends to be more northern and coastal where ditches and pond margins are common habitats, and its stolons enable it to form loose mats. *Agrostis gigantea* is ecologically adapted to a more extreme climate—hot summers/cold winters and drought—than *A. stolonifera*. It is also similar to *A. capillaris* (p. 639) and *A. castellana* (p. 639); it differs from both in its longer ligules, from *A. capillaris* in its less open panicles with spikelets near the base of the branches, and from *A. castellana* in being more extensively rhizomatous.

When *Agrostis gigantea* grows in damp hollows under trees it becomes more like *A. stolonifera*, particularly when the inflorescence is young, not expanded, and pale. If the rootstock is not collected, identification is a major problem.

4. Agrostis stolonifera L. [p. 642]
CREEPING BENT, AGROSTIDE STOLONIFÈRE

Plants perennial; stoloniferous, stolons 5–100+ cm, rooting at the nodes, often forming a dense mat, without rhizomes. **Culms** (8)15–60 cm, erect from a geniculate base, sometimes rooting at the lower nodes, with (2)4–7 nodes. **Leaves** mostly cauline; **sheaths** smooth; **ligules** longer than wide, dorsal surfaces usually scabrous, rarely smooth, apices usually rounded, acute to truncate, erose to lacerate, basal ligules 0.7–4 mm, upper ligules 3–7.5 mm; **blades** 2–10 cm long, 2–6 mm wide, flat. **Panicles** (3)4–20 cm long, less than ¹/₂ the length of the culm, 0.5–3(6) cm wide, narrowly contracted, dense, oblong to lanceolate, exserted from the sheaths at maturity, lowest node with 1–7 branches; **branches** scabrous, ascending to appressed, except briefly spreading during anthesis, usually some branches at each node spikelet-bearing to the base, lower branches 2–6 cm; **pedicels** 0.3–3.3 mm. **Spikelets** lanceolate, green and slightly to strongly suffused with purple. **Glumes** subequal to unequal, 1.6–3 mm,

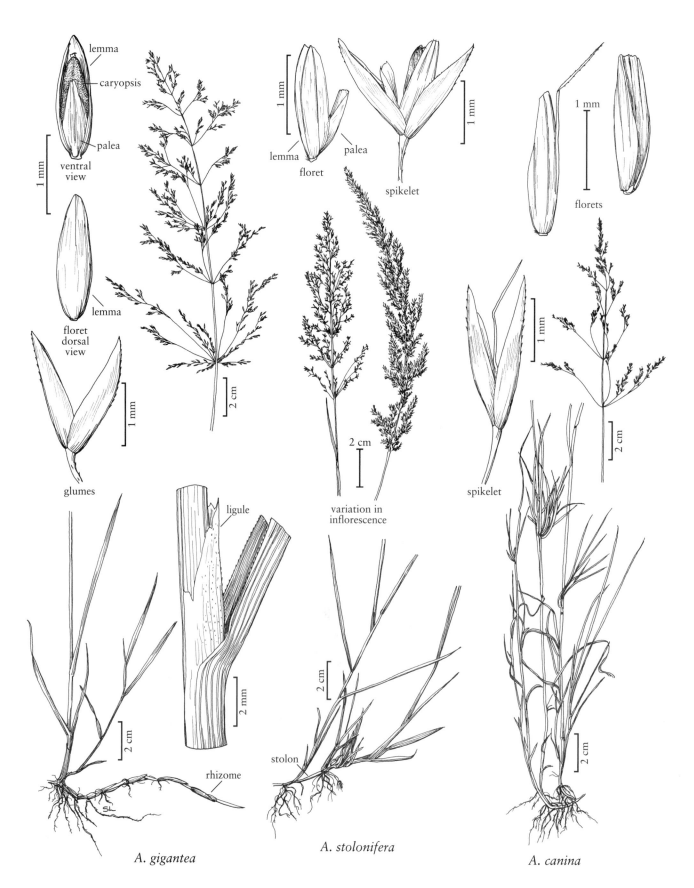

AGROSTIS

lanceolate, 1-veined, sometimes scabridulous distally, at least on the midvein, acute to acuminate or apiculate; **callus hairs** to 0.5 mm, sparse; **lemmas** 1.4–2 mm, opaque to translucent, smooth, 5-veined, veins obscure or prominent distally, apices acute to obtuse, entire or the veins excurrent to about 0.1 mm, usually unawned, rarely with a subapical straight awn to about 1 mm; **paleas** 0.7–1.4 mm, veins visible; **anthers** 3, 0.9–1.4 mm. **Caryopses** 0.9–1.3 mm; **endosperm** solid. $2n = 28$, 35, 42.

Agrostis stolonifera grows in areas that are often temporarily flooded, such as lakesides, marshes, salt marshes, lawns, and damp fields, as well as moist meadows, forest openings, and along streams. It will also colonize disturbed sites such as ditches, clearcuts, and overgrazed pastures. Its North American range extends from the subarctic into Mexico, mostly at low to middle elevations.

Agrostis stolonifera has been confused with *A. gigantea* (see previous). It is considered to be Eurasian, but some northern salt marsh and lakeside populations may be native. *Agrostis stolonifera* is also similar to *A. castellana* (p. 639); it differs in having longer, acute to truncate ligules that are longer than wide, and in possessing extensive stolons. The names *A. palustris* Huds. and *A. maritima* Lam. have been applied to plants with longer stolons; all forms intergrade. A hybrid between *A. stolonifera* and *Polypogon monspeliensis*, ×*Agropogon lutosus* (p. 668), has been found in the *Flora* region. It differs from *A. stolonifera* in having awned glumes and lemmas. *Agrostis stolonifera* readily hybridizes with *A. vinealis* (see below), the hybrids being somewhat intermediate between the two parents.

5. Agrostis canina L. [p. 642]
VELVET BENT, AGROSTIDE DES CHIENS

Plants perennial; appearing loosely cespitose and forming a rather closed turf, stoloniferous, stolons to 25 cm, slender, leafy, weakly rooting and eventually producing tufts of shoots at the nodes, without rhizomes. **Culms** 15–75 cm, erect, frequently geniculate at the base, sometimes weakly rooting at the lower nodes, with 2–4(6) nodes. **Leaves** basal and cauline; **sheaths** usually smooth, sometimes scabrous distally; **ligules** 1–4 mm, dorsal surfaces scabridulous, apices truncate to obtuse or acute, erose-lacerate; **blades** 1–10 cm long, 1–3 mm wide, usually flat, sometimes involute, usually scabrous, apices acute. **Panicles** 3–10 cm long, 1–7 cm wide, open, often lax, lanceolate to broadly ovate, lowest node with 3–7 branches; **branches** more or less scabrous, erect to

spreading, usually branched above midlength, spikelet-bearing in the distal $^1\!/_3$, patent at anthesis, lower branches 3–5 cm; **pedicels** (0.4)1–3 mm. **Spikelets** lanceolate or narrowly oblong, brownish yellow to purplish or rarely greenish. **Glumes** subequal, 1.7–3 mm, 1-veined, scabrous only on the distal part of the midveins, acute; **callus hairs** to 0.1 mm; **lemmas** 1–2 mm, about $^2\!/_3$ the length of the glumes, bases minutely pubescent, otherwise glabrous, translucent to opaque, 5-veined, veins prominent or obscure, apices acute to obtuse, entire, usually awned from near the base, awns to 5 mm, geniculate, rarely unawned, awned and unawned lemmas sometimes mixed within the panicle; **paleas** absent, or to about 0.2 mm and thin; **anthers** 3, 1–1.5 mm, at least $^1\!/_2$ as long as the lemmas. **Caryopses** 0.8–1.2 mm; **endosperm** solid. $2n = 14$.

Agrostis canina is a Eurasian species that is now established in both western and eastern North America, where it grows on roadsides and open ground in summer-cool climates. It is used for fine-textured lawns and golf greens. Similar to *A. vinealis* (see next), it may be differentiated from that species by its creeping, leafy stolons that form a dense carpet, and the finer, softer texture of its leaves. Unawned plants have been called *A. canina* var. *mutica* G. Sinclair.

6. Agrostis vinealis Schreb. [p. 645]
BROWN BENT

Plants perennial; densely cespitose, rhizomatous, rhizomes to about 10 cm, slender, scaly, not stoloniferous. **Culms** 10–60 cm, erect or geniculate at the base, slender, smooth, with 1–2(4) nodes. **Sheaths** smooth; **ligules** 0.6–5 mm, dorsal surfaces scabridulous, apices acute to obtuse, entire or lacerate to erose; **blades** 2–10 cm long, 1–3 mm wide, usually flat, sometimes involute, sometimes bristlelike, adaxial surfaces scabrous, abaxial surfaces sometimes scabrous. **Panicles** 2–15 cm long, (0.8)1–5.5(8) cm wide, lanceolate to oblong, somewhat open, often contracted after anthesis, lowest node with (1)3–8 branches; **branches** scabrous, readily visible, more or less erect, branched mostly at or below midlength, spikelets closely clustered, lower branches 3–5 cm; **pedicels** 0.5–2 mm. **Spikelets** lanceolate to narrowly oblong, greenish, purplish, or brownish. **Glumes** equal to subequal, 2–4 mm, membranous, acute to acuminate; **lower glumes** 1-veined, scabrous to scabridulous over the midvein; **upper glumes** usually shorter than the lower glumes, 1(3)-veined, almost smooth; **callus hairs** to 0.1 mm, sparse; **lemmas** 1.5–2.4 mm, about $^3\!/_4$ the length of the glumes, bases minutely pubescent, glabrous and smooth elsewhere, translucent to opaque, 5-veined, veins usually prominent distally, apices blunt, entire, usually awned from near the base, awns 2–4.5 mm, geniculate, rarely unawned; **paleas** to about 0.2 mm; **anthers** 3, 1–1.8 mm. **Caryopses** 0.8–1.3 mm; **endosperm** solid. 2*n* = 28.

Agrostis vinealis is native to Eurasia; it is not clear if populations in Greenland and Alaska represent a circumboreal distribution, or are introductions. It forms a fine, compact turf. It is similar to *A. canina* (see previous) in its habitat, except that it appears to be more heat tolerant and drought resistant. It used to be included in *A. canina*, but differs from that species in its subterranean rhizomes and lack of leafy stolons. *Agrostis vinealis* readily hybridizes with *A. capillaris* (p. 639) and *A. stolonifera* (p. 641), the hybrids being somewhat intermediate between the two parents.

7. Agrostis mertensii Trin. [p. 645]
NORTHERN BENT, AGROSTIDE DE MERTENS

Plants perennial; cespitose, not rhizomatous or stoloniferous. **Culms** (5)10–40 cm, erect, with 2–4 nodes. **Leaves** mostly basal or basal and cauline, basal leaves persistent; **sheaths** smooth or scabrous; **ligules** 0.7–3.3 mm, scabridulous or smooth, usually rounded, sometimes acute or truncate, erose, sometimes lacerate; **blades** 2.5–13 cm long, 0.5–3 mm wide, usually flat, occasionally involute or folded. **Panicles** (2)3–10 cm long, (0.5)1.5–5 cm wide, widely ovate to lanceolate, usually open, exserted from the upper sheaths at maturity, lowest node with (1)2–5 branches; **branches** erect, not capillary, readily visible, smooth or sparsely scabridulous, branched above midlength, spikelets in the distal ¹/₂ or beyond, lower branches (1.5)2–4 cm; **pedicels** 0.4–6.4 mm. **Spikelets** lanceolate to narrowly ovate, dark brown or purplish. **Glumes** subequal, 2–4 mm, elliptical to lanceolate, midveins scabrous to scabridulous, at least distally, 1-veined, acute; **callus hairs** to 0.4 mm, sparse; **lemmas** 1.6–2.6 mm, smooth or scabridulous, translucent to opaque, 5-veined, veins prominent to obscure, apices acute, entire or erose, awned from just below midlength, awns (2)3–4.4 mm, geniculate, exserted, persistent; **paleas** absent, or to 0.1 mm and thin; **anthers** 3, 0.5–0.8 mm, usually shed at anthesis. **Caryopses** 1.4–2 mm; **endosperm** solid. 2*n* = 56 [reports of 2*n* = 42 are for *Agrostis scabra*].

Agrostis mertensii grows on banks and gravel bars in river and lake valleys, and on open grasslands and rocky slopes of mountains and cliffs. It has a circumboreal distribution. In the *Flora* region, it extends from Alaska across Canada to Newfoundland and Greenland, south in the mountains to Wyoming and Colorado in the west, and West Virginia, Tennessee, and North Carolina in the east. It also grows in arctic Europe, Scandinavia, the mountainous regions of Mexico, and northwestern South America, where some unusually robust specimens have been somewhat dubiously referred to this species.

Agrostis mertensii is frequently confused with dwarf, awned forms of *A. scabra* (p. 646), but has larger spikelets, more culm nodes, larger anthers, slightly wider, flatter leaves, and panicles that are less expanded and less than ¹/₃ the culm length. *Agrostis mertensii* is also often confused with *A. idahoensis* (p. 649), but *A. mertensii* tends to grow in better-drained habitats. *Agrostis mertensii* differs from *A. anadyrensis* (see next) in being less robust, having narrower, less abundant basal leaves, smaller panicles, and minor differences in the insertion of the awns on the lemmas. In addition, the panicle branches are smooth to weakly scabrous, contrasting with the branches of *A. anadyrensis*, which are strongly scabrous, with long acicules throughout their length.

8. Agrostis anadyrensis Soczava [p. 645]
ANADYR BENT

Plants perennial; loosely cespitose, not rhizomatous or stoloniferous. **Culms** 20–50 cm, erect. **Basal leaves** usually numerous; **ligules** 0.6–3.5 mm, scabridulous, rounded, erose, sometimes lacerate, ciliate; **blades** to 10 cm long, 3–4 mm wide, flat. **Panicles** 6–20 cm long, 1–4(8) cm wide, lanceolate to ovate, somewhat contracted but not spikelike; **branches** erect, slender, scabrous; **pedicels** 0.5–8 mm. **Spikelets** lanceolate, greenish purple. **Glumes** subequal, 2–3.5 mm, 1-veined, ciliate on the keels, otherwise smooth, acute; **callus hairs** less than ¹/₆ the length of the lemma; **lemmas** 1.6–3 mm, translucent, 5-veined, veins obscure or prominent distally, apices acute, entire, awned from above midlength, awns 1–1.5 mm, geniculate, exserted; **paleas** absent; **anthers** 3, 0.5–0.8 mm. 2*n* = 56.

Agrostis anadyrensis grows in sand and gravel shores of rivers and lakes, and in meadows and shrubby valleys in eastern Siberia. It has been reported by Tsvelev (1976) from arctic Russia and southern Alaska. Specimens from Alaska purporting to be *A. anadyrensis* so far have proven to be *A. mertensii* (see previous). *Agrostis anadyrensis* differs from *A. mertensii* in being more robust, with wider, more abundant basal leaves, larger panicles, and minor differences in the insertion of the awns on the lemmas. In addition, the panicle branches are strongly scabrous, with long acicules throughout their length, contrasting with the smooth to weakly scabrous branches of *A. mertensii*.

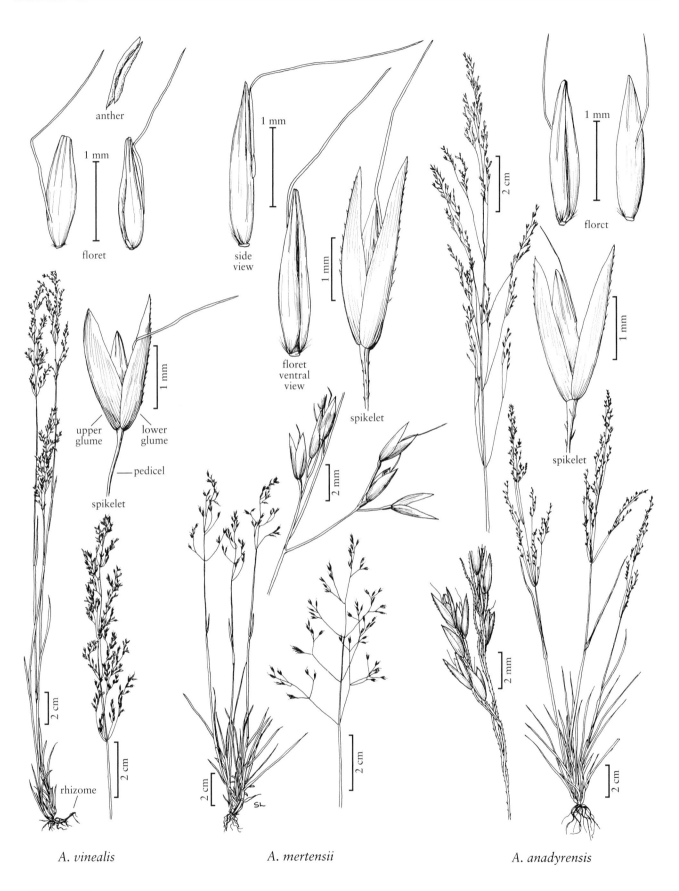

anther

1 mm

floret

1 mm

side
view

floret
ventral
view

spikelet

1 mm

florct

1 mm

2 cm

1 mm

upper
glume

lower
glume

pedicel

spikelet

2 mm

spikelet

2 mm

2 cm

2 cm

2 cm

2 cm

2 cm

rhizome

A. vinealis   *A. mertensii*   *A. anadyrensis*

AGROSTIS

9. **Agrostis clavata** Trin. [p. 646]
CLAVATE BENT

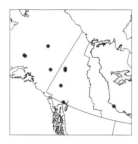

Plants annuals or short-lived perennials; densely tufted, not rhizomatous or stoloniferous. **Culms** 30–70 cm, erect, smooth, with 3–6 nodes. **Leaves** mostly basal or basal and cauline; **sheaths** smooth; **ligules** (0.5) 1.5–4.2 mm, dorsal surfaces scabrous, apices truncate to rounded, erose-lacerate, often ciliolate; **blades** 10–20 cm long, 1–5(7) mm wide, flat, scabrous at least along the margins and/or veins. **Panicles** 8–35 cm long, 3–10 cm wide, widely ovate, becoming lax, open, usually exserted, bases sometimes enclosed in the upper sheaths at maturity, lowest node with (1)2–8 branches; **branches** scabrous, spreading, branched above midlength, spikelet-bearing in the distal $^1/_2$–$^2/_3$, lower branches 3–12 cm; **pedicels** 0.5–5.5 mm, clavate; **secondary panicles** often present in the leaf axils, smaller than the primary panicles. **Spikelets** narrowly ovate to lanceolate, greenish or light brownish purple. **Glumes** subequal, 1.5–2.8 mm, lanceolate, 1-veined, keels somewhat aculeolate, apices usually acute, sometimes acuminate; **calluses** usually sparsely hairy, hairs to 0.2 mm, sometimes glabrous; **lemmas** 1.2–2 mm, smooth, glabrous, translucent to opaque, 5-veined, veins prominent distally or obscure, apices acute, entire, unawned; **paleas** absent, or to 0.2 mm and thin; **anthers** 3, 0.3–0.6 mm, to $^1/_2$ as long as the lemmas. **Caryopses** 0.9–1.3 mm. 2n = [28], 42.

Agrostis clavata grows in disturbed ground on sandbars and gravelbars, and in wet meadows and coniferous forests, from Sweden across northern Asia to Kamchatka. It was recently found in Alaska, the Yukon, and the Northwest Territories, and appears to be native there. It differs from the similarly large-panicled *A. scabra* (see next) in its much broader, flat leaves.

10. **Agrostis scabra** Willd. [p. 648]
TICKLEGRASS, ROUGH BENT, FOIN FOU, AGROSTIDE SCABRE

Plants perennial or annual; cespitose, not rhizomatous or stoloniferous. **Culms** (7.5) 15–90 cm, erect, nodes usually 1–3. **Leaves** mostly basal, basal leaves usually persistent; **sheaths** usually smooth, sometimes scabridulous; **ligules** 0.7–5 mm, dorsal surfaces scabrous, apices usually rounded, sometimes truncate or acute, erose-ciliolate, sometimes lacerate; **blades** 4–14 cm long, 1–2 mm wide, basal blades mostly involute, cauline blades

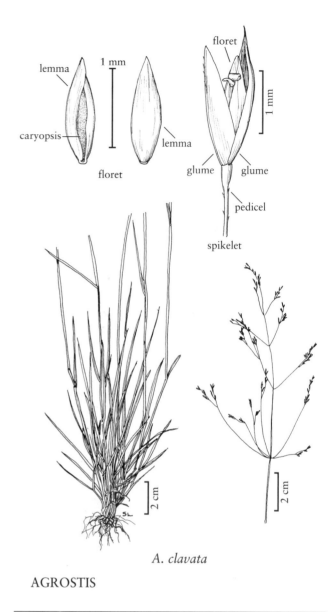

A. clavata

AGROSTIS

mostly flat. **Panicles** (4)8–25(50) cm long, 0.5–20 cm wide, broadly ovate, often nearly as wide as long, diffuse, the whole panicle often detaching at the base at maturity, forming a tumbleweed, exserted from the upper sheaths, lowest node with (1)2–7(12) branches; **branches** scabrous, capillary, flexible, wide-spreading, readily visible, branching beyond midlength, spikelets somewhat distant, not crowded, lower branches 4–12 cm; **pedicels** 0.4–9.6 mm. **Spikelets** lanceolate, greenish purple, frequently purple at maturity. **Glumes** unequal, 1.8–3.4 mm, lanceolate, 1-veined, keels scabrous at least towards the apices, apices acuminate; **callus hairs** to 0.2 mm, sparse; **lemmas** 1.4–2 mm, scabrous to scabridulous or smooth, translucent to opaque, 5-veined, veins prominent, apices acute to obtuse, usually entire, sometimes minutely toothed, unawned or awned from below midlength, awns 0.2–3 mm, exceeding the lemma apices by up to 2.5 mm, geniculate or straight, persistent; **paleas** absent or to 0.2 mm; **anthers** 3, 0.4–0.8 mm, usually shed at anthesis. **Caryopses** 0.9–1.4 mm; **endosperm** liquid. $2n = 42$.

Agrostis scabra grows in a wide variety of habitats, including grasslands, meadows, shrublands, woodlands, marshes, and stream and lake margins, as well as disturbed sites such as roadsides, ditches, and abandoned pastures. It occurs throughout much of the *Flora* region, but is not common in the Canadian high arctic or the southeastern United States. It extends south into Mexico; it is also native to the Pacific coast from Kamchatka to Japan and Korea, and has been introduced elsewhere.

Plants in the *Agrostis scabra* aggregate are variable. Awned and unawned plants often occur together, the difference presumably being caused by a single gene. At least three groups may be distinguished within the species as treated here: widespread, lowland, rather weedy plants capable of producing very large panicles that have been introduced into the southern United States; smaller, short-leaved, slow-growing plants of rocks and screes, which are widespread in the Rockies, the Appalachians, and much of Alaska, Canada, and Greenland; and luxuriant, broad-leaved plants that are characteristically found in sheltered, frost-free canyons of the southwestern United States. The second group has sometimes been called *A. scabra* var. *geminata* (Trin.) Swallen or *A. geminata* Trin.

Tercek et al. (2003) found that annual forms of *Agrostis scabra* with inflated upper sheaths and open panicles that were collected around hot springs in western North America were molecularly, and in some respects morphologically, more similar to plants identified as hot spring endemics such as *A. rossiae* and *A. pauzhetica* Prob., than they were to neighboring perennial plants of *A. scabra* that did not have inflated leaf sheaths. They differed, however, in having open, rather than contracted, panicles.

Agrostis scabra is often confused with a number of other species; for comparisons, see under the appropriate species description: *A. mertensii* (p. 644), *A. clavata* (see previous), *A. hyemalis* (see next), *A. perennans* (see this page), and *A. idahoensis* (p. 649).

11. Agrostis hyemalis (Walter) Britton, Sterns & Poggenb. [p. 648]
Winter Bent, Agrostide d'Hiver

Plants perennials or facultative annuals; cespitose, not rhizomatous or stoloniferous. **Culms** 15–82 cm, erect, with (3)4–7 nodes. **Leaves** cauline and basal; sheaths smooth; ligules (0.7) 1.2–4 mm, dorsal surfaces scabrous, apices usually rounded to truncate, sometimes acute, lacerate-erose; **blades** 3–10 cm long, 1–2 mm wide, flat, becoming involute, or folded. **Panicles** (5)10–25(36) cm long, (3)4–24 cm wide, broadly ovate, often nearly as wide as long, diffuse, the whole panicle often detaching at the base at maturity, forming a tumbleweed, bases often enclosed by the upper sheaths, lowest node with (3)5–11 branches; **branches** scabrous, capillary, flexible, wide-spreading, branching in the distal $^1/_4$, spikelets strongly clustered at the branch tips, lower branches 5–15 cm; **pedicels** 0.1–2.5(3.5) mm. **Spikelets** ovate to narrowly ovate, greenish or purplish. **Glumes** subequal, 1–2 mm, 1-veined, keeled, keels scabrous, sometimes the body also scabrous towards the apices, acute to acuminate; **callus hairs** to 0.2 mm, sparse; **lemmas** 0.8–1.2 mm, scabridulous, translucent to opaque, 5-veined, veins obscure or prominent distally, apices usually obtuse, sometimes acute, entire, unawned; **paleas** absent, or to 0.2 mm and thin; **anthers** 3, 0.2–0.5 mm. **Caryopses** 0.7–1 mm. $2n = 28$.

Agrostis hyemalis is most abundant along roadsides and in open pastures, scrub, and rocky areas. It is centered in the southeastern United States; historically it extended north to coastal Maine, where it may be extinct, west to Wisconsin and Texas, and south into the Caribbean, Mexico, and Ecuador. Records from further north and west in North America are confused; many reflect the former inclusion of the generally more northern *A. scabra* in *A. hyemalis*. *Agrostis hyemalis* differs from *A. scabra* in its smaller spikelets and anthers, more conspicuous culm leaves, and more clustered spikelets.

12. Agrostis perennans (Walter) Tuck. [p. 650]
Autumn Bent, Upland Bent, Agrostide Perennant

Plants perennial; cespitose, not rhizomatous or stoloniferous. **Culms** 20–80 cm, erect, sometimes rooting at the lower nodes, with 3–10 nodes. **Leaves** usually mostly cauline, basal leaves withering at anthesis; **sheaths** usually smooth, some-

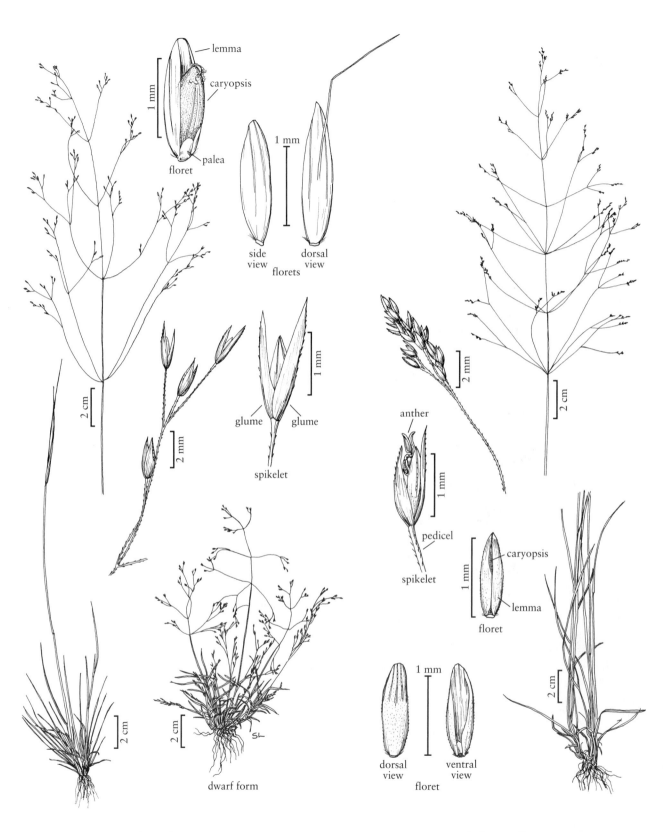

lemma

caryopsis

1 mm

palea

floret

1 mm

side
view

dorsal
view

florets

2 cm

2 mm

glume   glume

1 mm

spikelet

anther

1 mm

pedicel

spikelet

2 mm

2 cm

2 cm

dwarf form

caryopsis

1 mm

lemma

floret

1 mm

dorsal
view

ventral
view

floret

2 cm

A. scabra

A. hyemalis

AGROSTIS

times scabridulous, **ligules** (0.7)1.5–7.3 mm, dorsal surfaces scabrous, apices acute to truncate, erose to lacerate, often ciliolate; **blades** 6–20 cm long, 2–5 mm wide, flat, lax to stiff, cauline blades as substantial as the basal blades. **Panicles** 10–25 cm long, 2.5–11 cm wide, broadly ovate, open, bases usually exserted, sometimes enclosed in the upper sheaths at maturity, lowest node with (1)3–11(13) branches; **branches** scabridulous, capillary, wide-spreading, branching above midlength, spikelets somewhat aggregated towards the ends of the branches, lower branches 3–7 cm; **pedicels** 1–7.3 mm, spreading; **secondary panicles** sometimes present in the leaf axils. **Spikelets** lanceolate to narrowly ovate, green to tawny. **Glumes** unequal, 1.8–3.2 mm, lower glumes longer than the upper glumes, 1-veined, veins scabrous, acuminate to acute; **callus hairs** to 0.3 mm, abundant; **lemmas** 1.3–2.2 mm, smooth or scabridulous, translucent, 5-veined, veins prominent to obscure, apices acute to more or less truncate, entire or minutely denticulate, usually unawned, rarely awned from near midlength, awns to 2 mm, straight, not exserted; **paleas** absent, or to 0.1 mm and thin; **anthers** 3, 0.4–0.9 mm. **Caryopses** 1.1–1.9 mm; **endosperm** liquid. $2n = 42$.

Agrostis perennans grows along roadsides and in fields, fens, woodlands, and periodically inundated stream banks. It is widespread and common in eastern North America; it also grows from central Mexico to central South America. There are old records from Oregon and Washington, but *A. perennans* does not appear to be established in western North America. It is more tolerant of shade and moisture than *Agrostis scabra* (p. 646), from which it differs in its later flowering, leafier culms, and its basal leaves that usually wither by anthesis.

13. **Agrostis idahoensis** Nash [p. 650]
IDAHO REDTOP

Plants perennial; cespitose, not rhizomatous or stoloniferous. **Culms** 8–40 cm, slender, erect, with 2–5 nodes. **Leaves** mostly basal; **sheaths** usually smooth, sometimes scabridulous, not inflated; **ligules** (0.7)1–3.8 mm, dorsal surfaces scabridulous, apices rounded to truncate, rarely acute, erose to lacerate; **blades** 1–7 cm long, 0.5–2 mm wide, flat, becoming involute. **Panicles** 3–13 cm long, 1–6(8) cm wide, lanceolate to ovate, diffuse, exserted from the upper sheaths at maturity, lowest node with 1–6(10) branches; **branches** scabridulous, fairly stiff, more or less ascending, branching at or above midlength, spikelets not crowded, frequently solitary, lower branches 1–4 cm; **pedicels** 0.5–6.4 mm.

Spikelets lanceolate to narrowly ovate, purplish. **Glumes** subequal, 1.5–2.5 mm, 1-veined, usually scabrous to scabridulous, upper glumes sometimes smooth, apices acute to acuminate; **callus hairs** to 0.3 mm, sparse; **lemmas** 1.2–2.2 mm, usually smooth, sometimes scabridulous, translucent to opaque, 5-veined, veins usually prominent at least distally, sometimes obscure, apices acute to obtuse, entire, unawned; **paleas** absent, or to 0.2 mm and thin; **anthers** 3, 0.3–0.6 mm. **Caryopses** 1–1.3 mm. $2n = 28$.

Agrostis idahoensis grows in western North America, from British Columbia to California and New Mexico, in alpine and subalpine meadows along wet seepage areas and bogs, and in wet openings with *Sphagnum* in coniferous forests. It was recently discovered in Chile and Argentina; it is not known whether it is native or introduced there (Rúgolo de Agrasar and Molina 1997). *Agrostis idahoensis* is often confused with *A. mertensii* (p. 644) and dwarf forms of *A. scabra* (p. 646), both of which tend to grow in better-drained habitats.

14. **Agrostis oregonensis** Vasey [p. 650]
OREGON REDTOP

Plants perennial; cespitose, not rhizomatous or stoloniferous. **Culms** 12–75 cm, erect, with up to 5 nodes. **Leaves** mostly basal, basal leaves usually persistent; **sheaths** smooth or scabridulous, **ligules** 1.2–6.3 mm, usually scabridulous, sometimes smooth, truncate to rounded, lacerate-erose; **blades** 10–30 cm long, (1)2–4 mm wide, flat, culm blades usually less substantial than the basal blades. **Panicles** 8–35(60) cm long, (1.5)2.5–14 cm wide, lanceolate-ovate, usually 3 times longer than wide, open, bases exserted from the upper sheaths at maturity, lowest node with 1–15 branches; **branches** scabrous to scabridulous, ascending, mostly branching above midlength, lower branches occasionally branching to near the base, spikelets somewhat clustered towards the tips, lower branches 2–10 cm, mostly longer than 3 cm; **pedicels** (0.5)1–3.5(6) mm; **secondary panicles** not present. **Spikelets** lanceolate to narrowly ovate, yellowish green to purple. **Glumes** unequal, 2–3.6 mm, 1(3)-veined, scabrous on the midvein, occasionally also sparsely scabridulous over the body, acute to acuminate; **callus hairs** to 0.2 mm, sparse; **lemmas** 1.5–2.5 mm, usually smooth, sometimes scabridulous or pubescent, translucent to opaque, 5-veined, veins faint or prominent distally, apices acute to obtuse, usually entire, sometimes erose or toothed, teeth to about 0.4 mm, usually unawned, sometimes awned from midlength, awns to 2 mm, straight, not exserted,

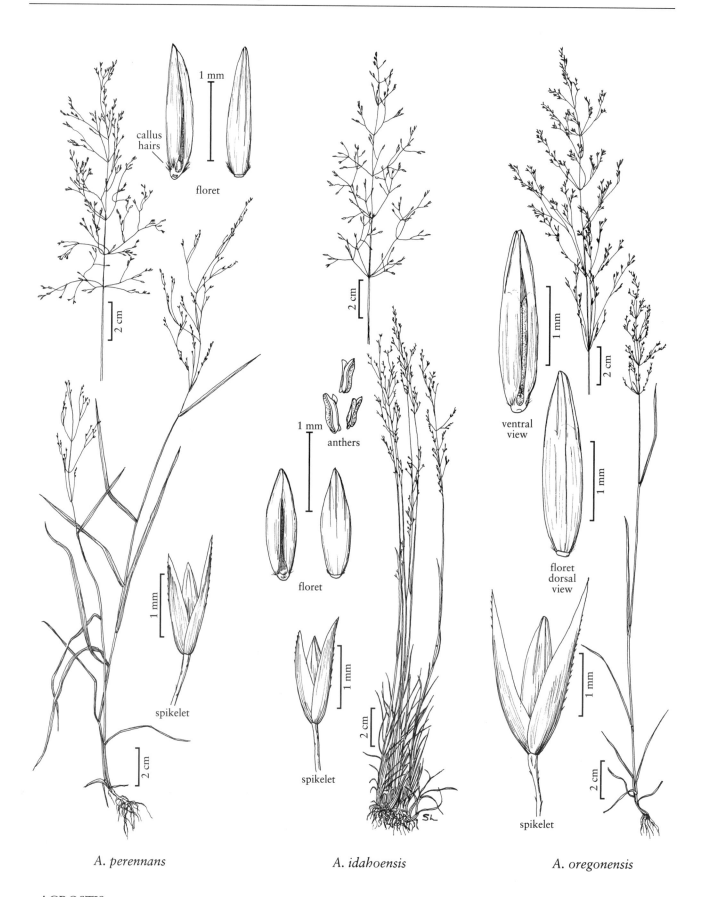

1 mm

callus
hairs

floret

anthers

1 mm

floret

2 cm

1 mm

spikelet

2 cm

ventral
view

1 mm

floret
dorsal
view

1 mm

2 cm

spikelet

2 cm

1 mm

spikelet

2 cm

A. perennans

A. idahoensis

A. oregonensis

AGROSTIS

sometimes awned and unawned spikelets present on the same plant; **paleas** absent, or to 0.2 mm and thin; **anthers** 3, 0.5–1.2 mm. **Caryopses** 1–1.6 mm; **endosperm** semisoft. 2*n* = 42.

Agrostis oregonensis grows in wet habitats, such as stream and lake margins, damp woods, and meadows, in western North America, primarily in the Pacific Northwest from British Columbia to California and Wyoming. It has not been found in Mexico.

15. Agrostis exarata Trin. [p. 652]
SPIKE BENT

Plants perennial; usually cespitose, sometimes rhizomatous, not stoloniferous. **Culms** 8–100 cm, erect or decumbent at the base, sometimes rooting at the lower nodes, with (2)3–6(8) nodes. **Leaves** mostly cauline; **sheaths** smooth or slightly scabrous; **ligules** (1)1.7–8(11.2) mm, dorsal surfaces scabrous, apices truncate to obtuse, lacerate to erose; **blades** 4–15 cm long, 2–7 mm wide, flat. **Panicles** (3)5–30 cm long, 0.5–4 cm wide, contracted, spikelike, oblong, or lanceolate, usually dense, rarely more open, sometimes interrupted near the base, bases usually exserted, rarely enclosed by the upper sheaths at maturity, lowest node with 1–5 branches; **branches** scabrous, ascending to appressed, spikelet-bearing to or near the base, usually hidden by the spikelets, spikelets crowded, lower branches 1–2(4) cm; **pedicels** 0.2–4.3 mm. **Spikelets** lanceolate to narrowly ovate, greenish to purplish. **Glumes** subequal to equal, 1.5–3.5 mm, scabrous on the midvein and sometimes on the back, 1(3)-veined, acute, elongate-acuminate, with an awnlike tip to 1 mm; **callus hairs** to 0.3 mm, sparse to abundant; **lemmas** 1.2–2.2 mm, smooth, translucent to opaque, 5-veined, veins prominent distally or obscure throughout, apices acute, entire or toothed, teeth no more than 0.12 mm, unawned or awned from above midlength, awns to 3.5 mm, straight or geniculate; **paleas** absent or to 0.5 mm; **anthers** 3, 0.3–0.6 mm. **Caryopses** 0.9–1.2 mm; **endosperm** solid or soft. 2*n* = 28, 42, 56.

Agrostis exarata is common and widely distributed in western North America, usually growing in moist ground in open woodlands, river valleys, tidal marshes, and swamp and lake margins; it also grows in dry habitats such as grasslands and shrublands. It extends from Alaska into Mexico, and is also found in Kamchatka and the Kuril Islands. Eastern North American records probably reflect introductions. It readily colonizes roadsides and bare soil, and exhibits ecological and developmental flexibility. *Agrostis exarata* is recognized here as a single, variable species

that includes what others have treated as distinct species or varieties. Cytotaxonomic study might clarify the basis of the observed variation. *Agrostis exarata* appears to be related to *A. densiflora* (see below).

16. Agrostis densiflora Vasey [p. 652]
CALIFORNIA BENT

Plants perennial; not rhizomatous or stoloniferous. **Culms** 9–85 cm, erect, sometimes decumbent at the base, sometimes rooting at the lower nodes, usually with 4–7 nodes. **Leaves** basal and cauline; **sheaths** smooth or scabrous; **ligules** 1–4.8(7.5) mm, dorsal surfaces scabrous, apices truncate to obtuse, erose-lacerate, sometimes ciliolate; **blades** 2–12 cm long, 2–10 mm wide, upper blades broader than those below, flat. **Panicles** 2–10 cm long, 0.5–2 cm wide, narrow and spikelike to dense, lobed-lanceolate, bases exserted or enclosed by the upper sheaths at maturity; **branches** to 1.5 cm, scabrous, appressed, branching at the base and mostly hidden by the spikelets; **pedicels** about 0.4–3 mm. **Spikelets** lanceolate to narrowly ovate, yellowish, sometimes tinged with purple, or greenish purple. **Glumes** generally equal, 2–3.3 mm, 1-veined, densely scabrous, aculeolate on the veins, narrowly acute to acuminate or somewhat mucronate, mucros to about 0.5 mm; **callus hairs** to 0.3 mm, usually dense; **lemmas** 1.5–2.1 mm, smooth or scabridulous, translucent, (3)5-veined, veins prominent to obscure, sometimes prominent only distally, apices acute to obtuse, veins usually extended as teeth up to 0.3 mm, unawned or awned from above midlength, awns to 3.5 mm, straight, readily deciduous; **paleas** 0.3–0.7 mm, thin; **anthers** 3, 0.5–2 mm. **Caryopses** 1–1.5 mm; **endosperm** solid. 2*n* = 42.

Agrostis densiflora is endemic to coastal Oregon and California. It grows in sandy soils, on cliffs, and in scrublands. It appears to be related to *A. exarata* (see previous), and hybridizes with *A. blasdalei* (p. 656).

17. Agrostis pallens Trin. [p. 652]
DUNE BENT

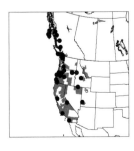

Plants perennial; rhizomatous, rhizomes to 10 cm, not stoloniferous. **Culms** 10–70 cm, erect, sometimes decumbent at the base, sometimes rooting at the lower nodes, with 3–7 nodes. **Leaves** usually cauline; **sheaths** usually smooth, sometimes scabridulous; **ligules** 1–6 mm,

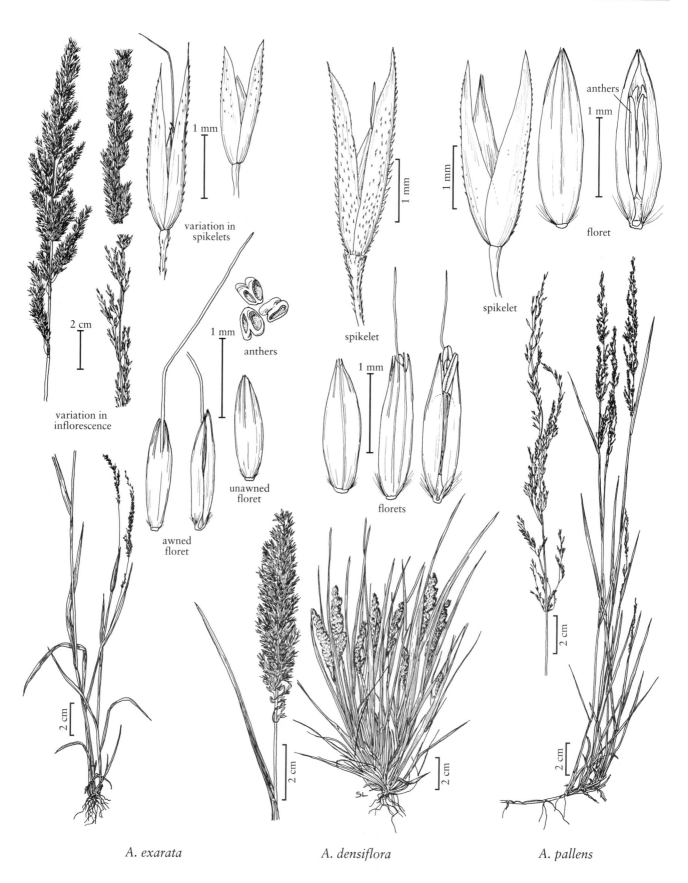

variation in
spikelets

anthers

awned
floret

unawned
floret

variation in
inflorescence

spikelet

spikelet

floret

florets

A. exarata

A. densiflora

A. pallens

AGROSTIS

dorsal surfaces scabrous, apices truncate to rounded or acute, often lacerate to erose; **blades** 1.5–11.5 cm long, 1–6 mm wide, flat, becoming involute. **Panicles** 5–20 cm long, 0.4–6(8) cm wide, lanceolate to narrowly ovate, open to contracted, exserted from the upper sheaths at maturity, lowest node with 1–8 branches; **branches** scabrous to scabridulous, usually ascending, branching below midlength, the majority spikelet-bearing to the base, lower branches 2–5 cm; **pedicels** 0.5–7 mm. **Spikelets** lanceolate to narrowly ovate, green to yellowish green or yellow, tinged with purple. **Glumes** equal to subequal, 2–3.5 mm, scabrous over the midvein and sometimes also sparsely over the body, 1(3)-veined, acute to acuminate; **callus hairs** to 0.3(1) mm, sparse; **lemmas** 1.5–2.5 mm, smooth or scabridulous or warty, 5-veined, veins prominent throughout or only distally, apices acute, entire or the veins excurrent to about 0.2 mm, usually unawned, rarely awned from below the apices, awns to 0.5(2.7) mm, straight; **paleas** absent, or to 0.2 mm and thin; **anthers** 3, 0.7–1.8 mm. **Caryopses** 1–1.5 mm; endosperm solid. $2n = 42, 56$.

Agrostis pallens grows on coastal sands and cliffs, in meadows, and in open, xeric woodlands to subalpine woodlands at 3500 m. It extends from British Columbia south into Baja California, Mexico, and east to western Montana and Utah. The relationship of the higher-elevation, more open-panicled plants to those of lower elevations merits further study.

18. Agrostis hallii Vasey [p. 653]
HALL'S BENT

Plants perennial; rhizomatous, rhizomes to 50 cm, not stoloniferous. **Culms** 17–100 cm, erect. **Leaves** mostly cauline or somewhat basally concentrated; **sheaths** smooth; **ligules** 2.3–7 mm, dorsal surfaces scabrous, apices acute, usually lacerate; **blades** 6–20 cm long, 2–5 mm wide, flat. **Panicles** 7–22 cm long, 1.5–5.5(7) cm wide, lanceolate to narrowly ovate, more or less open to dense, lowest node with 2–15 branches; **branches** scabridulous, ascending to more or less appressed, mostly branching at or above midlength, some branching near the base, lower branches 1–5 cm; **pedicels** 0.5–6 mm. **Spikelets** lanceolate, yellow-green, often tinged with purple. **Glumes** equal to subequal, 2.5–4 mm, 1-veined, scabrous to scabridulous on the midvein, at least distally, sometimes also sparsely scabridulous over the back, acute to acuminate; **callus hairs** (0.8)1–2 mm, abundant, conspicuous; **lemmas** 2–3 mm, smooth, translucent to opaque, 5-veined, veins prominent at least distally, apices acute, entire or erose,

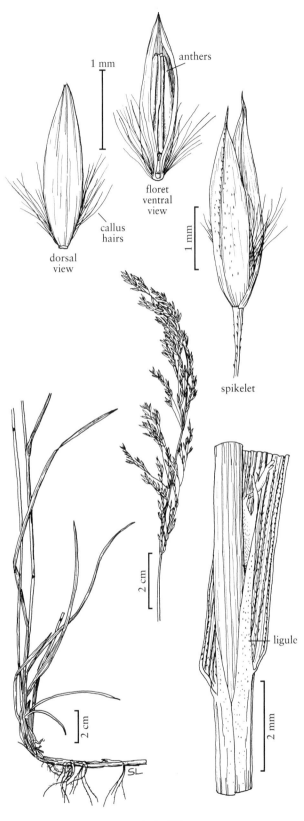

AGROSTIS

A. hallii

sometimes toothed, teeth to about 0.2 mm, unawned; **paleas** absent, or to 0.2 mm and thin; **anthers** 3, 1.5–2.3 mm. **Caryopses** 1.5–2 mm. $2n = 42$.

Agrostis hallii is primarily coastal, growing in open areas of oak and coniferous forests in Oregon and California.

19. Agrostis hooveri Swallen [p. 655]
HOOVER'S BENT

Plants perennial; cespitose, not rhizomatous, not stoloniferous. **Culms** 30–80 cm, erect, usually with more than 3 nodes. **Leaves** mostly on the lower $^1/_2$ of the culm; **basal leaves** withered by anthesis; **lower leaf sheaths** finely tomentose, the tomentum extending to below ground; **upper leaf sheaths** smooth; **ligules** 2.5–6 mm, dorsal surfaces scabridulous, apices acute to truncate, lacerate; **blades** 10–16 cm long, 0.5–1(2) mm wide, flat, becoming involute. **Panicles** (4)10–17 cm long, 2–5 cm wide, broadly lanceolate, usually open, exserted from the upper sheaths at maturity, lowest node with 1–8 branches; **branches** scabrous, generally ascending, mostly branching at about midlength, sometimes to near the base, lower branches 1.5–5 cm; **pedicels** 0.4–5 mm. **Spikelets** lanceolate, slightly purplish. **Glumes** equal to subequal, 1.8–3 mm, 1-veined, scabridulous on the veins, sometimes also on the body, acute; **callus hairs** to 0.3 mm, abundant; **lemmas** 1.5–2 mm, scabridulous to warty throughout or only on the veins, translucent to opaque, 5-veined, veins prominent distally, apices truncate, minutely toothed to about 0.2 mm, awned on the lower $^1/_3$, awns to 2.5 mm, geniculate, persistent; **paleas** absent or minute; **anthers** 3, 1–1.5 mm, usually shed at anthesis. **Caryopses** 1–1.5 mm; **endosperm** liquid. $2n = $ unknown.

Agrostis hooveri is an uncommon species, endemic to dry, sandy soils, open chaparral, and oak woodlands of San Luis Obispo and Santa Barbara counties, California.

20. Agrostis howellii Scribn. [p. 655]
HOWELL'S BENT

Plants perennial; cespitose, not rhizomatous, not stoloniferous. **Culms** 40–80 cm, geniculate and decumbent at the base, sometimes rooting at the lower nodes, with 3–6 nodes. **Leaves** mostly cauline; **basal leaves** withered by anthesis; **cauline leaves** persisting; **sheaths** smooth or scabridulous; **ligules** (0.9)2.7–5 mm, dorsal surfaces scabrous, apices truncate to acute, erose to lacerate; **blades** 15–20 cm long, 3–5 mm wide, flat. **Panicles** 10–25 cm long, 3.5–11 cm wide, ovate, open and diffuse, bases usually exserted, rarely enclosed in the upper sheaths, lowest node with (1)2–6 branches; **branches** scabridulous, flexuous, spreading, spikelets usually only on the distal $^1/_2$, lower branches 3–10 cm; **pedicels** 1.2–9.4 mm. **Spikelets** ovate to lanceolate, green, not or slightly tinged with purple. **Glumes** unequal, 2.3–3.5 mm, 1(3)-veined, veins scabridulous, acute to acuminate; **callus hairs** to 0.3 mm, abundant; **lemmas** 2.5–3 mm, usually smooth, sometimes scabridulous, translucent to opaque, 5-veined, veins prominent distally or obscure, apices acute to obtuse, lateral veins extending as 4 teeth to 0.5 mm, awned from the lower $^1/_3$, awns 4–6 mm, geniculate; **paleas** absent or minute; **anthers** 3, 1–1.3 mm. **Caryopses** 1.2–1.6 mm; **endosperm** liquid. $2n = 28$.

Agrostis howellii is a rare Washington and Oregon endemic, growing in shady woodlands and at the base of cliffs.

21. Agrostis variabilis Rydb. [p. 657]
MOUNTAIN BENT

Plants perennial; cespitose, rarely rhizomatous, rhizomes to 2 cm. **Culms** 5–30 cm, erect, sometimes geniculate at the base, with 2–5(7) nodes. **Leaves** mostly basal, forming dense tufts; **sheaths** smooth; **ligules** (0.7)1–2.8 mm, dorsal surfaces usually scabridulous, sometimes smooth, apices rounded to truncate, lacerate to erose; **blades** 3–7 cm long, 0.5–2 mm wide, flat, becoming folded or involute. **Panicles** (1)2.5–6 cm long, 0.3–1.2(2) cm wide, cylindric to lanceolate, usually dense, exserted from the upper sheaths at maturity, lowest node with 1–5 branches; **branches** usually scabridulous, sometimes smooth, ascending to erect, branching at or near the base and spikelet-bearing to the base, to branching in the distal $^2/_3$, lower branches 0.5–1.5 cm; **pedicels** 0.4–2.8(4.3) mm. **Spikelets** ovate to lanceolate, greenish purple. **Glumes** subequal to equal, 1.8–2.5 mm, smooth, or scabrous on the keel and sometimes elsewhere, 1-veined, acute to acuminate; **callus hairs** to 0.2 mm, sparse to abundant; **lemmas** 1.5–2 mm, smooth, translucent, (3)5-veined, veins usually prominent distally, sometimes obscure throughout, apices acute, entire, usually unawned, rarely awned, awns to 1(2.8) mm, arising beyond the midpoint, usually not reaching the lemma apices; **paleas** to 0.2 mm, thin; **anthers** 3, 0.4–0.7(1) mm. **Caryopses** 1–1.3 mm; **endosperm** soft. $2n = 28$.

Agrostis variabilis grows in alpine and subalpine meadows and forests and on talus slopes, at elevations

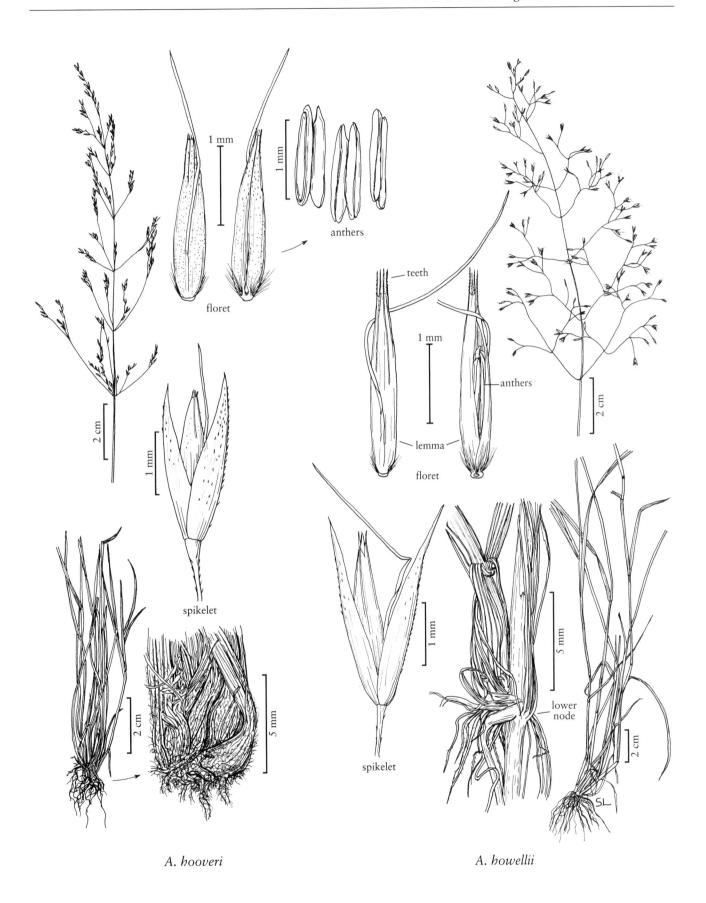

anthers

floret

teeth

1 mm

anthers

lemma

floret

1 mm

spikelet

2 cm

2 cm

1 mm

1 mm

2 cm

spikelet

5 mm

5 mm

lower node

2 cm

SL

A. hooveri   *A. howellii*

AGROSTIS

up to 4000 m, from British Columbia and Alberta south to California and New Mexico. It can appear similar to dwarf forms of *Podagrostis humilis* (p. 694), but differs from that species in not having paleas.

22. Agrostis blasdalei Hitchc. [p. 657]
BLASDALE'S BENT

Plants perennial; forming dense, stiff clumps, not rhizomatous or stoloniferous. **Culms** 6–30 cm, decumbent to erect. **Leaves** forming a dense, bristly basal tuft; **ligules** 0.7–2.3 mm, dorsal surfaces scabridulous, apices truncate to obtuse, often erose, sometimes lacerate or ciliolate; **blades** 2–5 cm long, less than 1 mm wide, soon becoming tightly inrolled and rigid. **Panicles** 2–8 cm long, 0.2–0.6 cm wide, narrowly cylindric, spikelike, dense, occasionally interrupted near the base, the base often enclosed by the upper sheaths; **branches** to 2 cm, scabrous, strongly appressed, hidden by the spikelets; **pedicels** 0.5–7 mm. **Spikelets** lanceolate to narrowly ovate, greenish to purplish. **Glumes** 1.8–4 mm, often 3-veined, midveins scabrous to smooth, acute to acuminate; **calluses** glabrous; **lemmas** 1.5–2.5 mm, 5-veined, veins obscure or prominent distally, extending as teeth to 0.2 mm, unawned or awned from above midlength, awns to 1.2 mm, usually scarcely exceeding the lemma apices, straight; **paleas** to 0.3 mm, thin; **anthers** 3, 0.7–2 mm. **Caryopses** 1–1.5 mm; **endosperm** liquid. 2*n* = 42.

Agrostis blasdalei is a xerophytic species that is known only from Mendocino to Santa Cruz counties, California, where it grows on coastal cliffs and dunes and in shrublands. It hybridizes with *A. densiflora* (p. 651).

23. Agrostis tolucensis Kunth [p. 657]

Plants perennial; cespitose. **Culms** (3.3)5.5–60 cm, erect, glabrous, with (2)3–4 nodes. **Leaves** mostly basal or evenly distributed; **sheaths** glabrous, scabridulous; **ligules** 2–6.2 mm, membranous, scabridulous dorsally, apices acute to more or less truncate, erose to lacerate; **blades** 4–19 cm long, 0.5–4 mm wide, involute or flat, scabrous or smooth over the veins. **Panicles** 1.5–14 cm long, 0.5–1.5 cm wide, lanceoloid, somewhat spikelike, dense to somewhat open, exserted at anthesis; **branches** appressed, shorter than 2 cm; **pedicels** 0.7–3(4.5) mm, scabrous. **Spikelets** purple to green, shiny. **Glumes** subequal, 2–3.5 mm, 1-veined, keeled, keels and back usually smooth, occasionally scabrous, apices acute, muticous; **lower glumes** wider than the upper glumes; **calluses** with 2 tufts of hair to 0.3 mm; **lemmas** 1.4–1.9 mm, glabrous, 5-veined, veins evident, apices truncate to acute, microdenticulate, teeth to about 0.1 mm,

dorsally awned from midlength or below, awns 2–3.5 mm, exserted, twisted, geniculate, scabridulous; **paleas** 0.1–0.2 mm, hyaline, linear; **anthers** 3, 0.5–1 mm. **Caryopses** 0.7–1.2 mm; **endosperm** soft. 2*n* = 28.

Agrostis tolucensis grows in alpine meadows, usually in damp areas by lakes or streams. It is native from Mexico to Chile, Bolivia, and Argentina, growing in the Andes at 1800–4900 m. Its presence in the *Flora* region is dubious; two specimens in the S.M. Tracy herbarium, from Brewster and Brown counties in Texas, are listed in the Flora of Texas online database (http://www.csdl.tamu.edu/FLORA/tracy/main1.html). Attempts to locate the specimens in 2005 were unsuccessful, suggesting the records may have been based upon misidentifications which have since been rectified.

24. Agrostis rossiae Vasey [p. 658]
ROSS' BENT

Plants annual. **Culms** 4–20 cm. **Sheaths** smooth, upper sheaths inflated; **ligules** (0.6)1–1.5 mm, dorsal surfaces scabrous, apices truncate to rounded, lacerate to erose-ciliolate; **blades** 1–2.5 cm long, 1–2 mm wide, flat or folded. **Panicles** 1.5–6 cm long, 0.4–2 cm wide, initially lanceolate, becoming ovate and diffuse, bases sometimes enclosed in the upper sheaths; **branches** 1–3 cm, erect to spreading, scabrous; **pedicels** 0.5–6.3 mm. **Spikelets** ovate, green, slightly to strongly tinged with purple. **Glumes** equal, 2–2.5 mm, 1-veined, acuminate; **lower glumes** scabrous on the midvein; **upper glumes** smooth or scabrous on the midvein; **callus hairs** to 0.1 mm, sparse; **lemmas** 1.3–1.7 mm, scabrous, translucent to opaque, 5-veined, veins mostly obscure, apices truncate, entire, erose, or the veins excurrent to about 0.12 mm, unawned; **paleas** to 0.2 mm, thin; **anthers** 3, 0.5–0.9 mm, often retained at the apices of the caryopses. **Caryopses** 1.2–1.5 mm; **endosperm** semisoft. 2*n* = unknown.

Agrostis rossiae is a rare species, originally known only from alkaline soils near hot springs in Yellowstone National Park, Wyoming. Tercek et al. (2003) showed that somewhat morphologically similar plants of *A. scabra* found around hot springs in Yellowstone and Lassen Volcanic national parks, and of *A. pauzhetica* Prob. found around hot springs on the Kamchatka Peninsula, Russia, are closely related to *A. rossiae*. They recommended, however, that the three be treated as separate species until more information has been obtained.

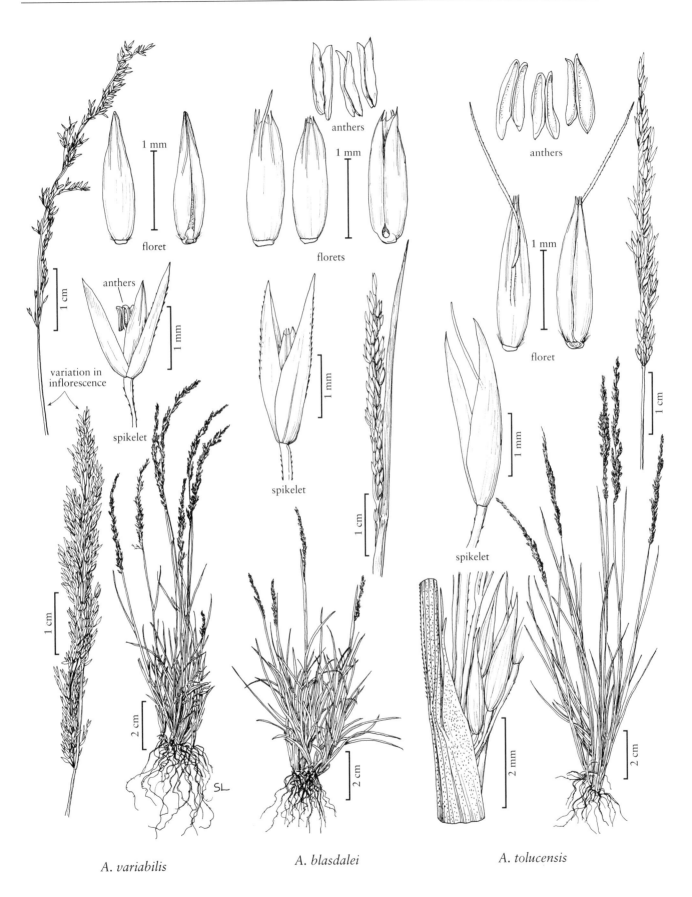

A. variabilis   *A. blasdalei*   *A. tolucensis*

AGROSTIS

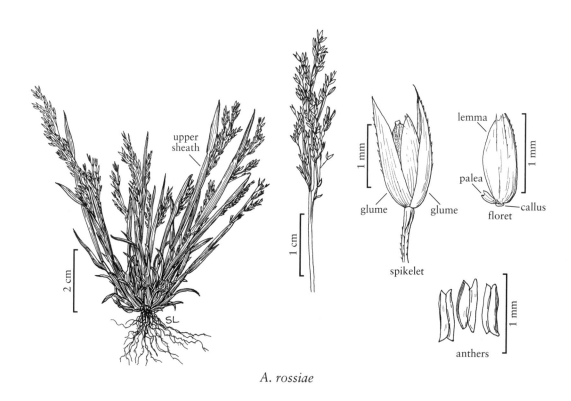

A. rossiae

AGROSTIS

25. Agrostis hendersonii Hitchc. [p. 659]
HENDERSON'S BENT

Plants annual. **Culms** 6–70 cm, erect, with 2–5 nodes. **Sheaths** smooth; **ligules** 0.5–5 mm, usually scabridulous, dorsal surfaces sometimes smooth, apices acute to obtuse, erose-lacerate, sometimes ciliolate; **blades** 1–4.5 cm long, 0.5–1(2) mm wide, flat or weakly involute. **Panicles** 1–5 cm long, 0.5–1.5 cm wide, cylindrical, spikelike, dense, sometimes interrupted near the base, usually well exserted from the upper sheaths; **branches** scabridulous, ascending to appressed, mostly hidden by the spikelets, lower branches 0.5–2.5 cm; **pedicels** 0.5–4(6.3) mm. **Spikelets** lanceolate, greenish to yellowish, tinged with purple. **Glumes** subequal, 5–7 mm, scabrous over the midvein, often sparsely scabridulous over the body, 1(3)-veined, apices narrowly acuminate to awn-tipped to 2 mm; **callus hairs** to 0.7 mm, abundant; **lemmas** 2.5–4 mm, scabridulous over the veins and sometimes over the body, opaque to translucent, 5-veined, veins prominent distally, apices acute, veins extended into 2 teeth, teeth 0.2–1.5 mm, awned from about midlength, awns (5)8–10 mm, more or less geniculate; **paleas** absent or to 0.9 mm; **anthers** 3, 0.4–0.7 mm. **Caryopses** 1.6–1.8 mm. 2*n* = 42.

Agrostis hendersonii is a rare species that grows below 600 m in clay or adobe, sometimes rocky, soils around the edges of vernal pools in Oregon and California.

Agrostis aristiglumis Swallen, known only from its type locality on the Point Reyes Peninsula, Marin County, California, is included here in *A. hendersonii*. It differs from *A. hendersonii* in having a palea up to 0.9 mm long, and a barely exserted panicle. The relationship of these to each other and to *A. microphylla* merits investigation. Their differences may be the result of the founder effect on inbreeding annuals.

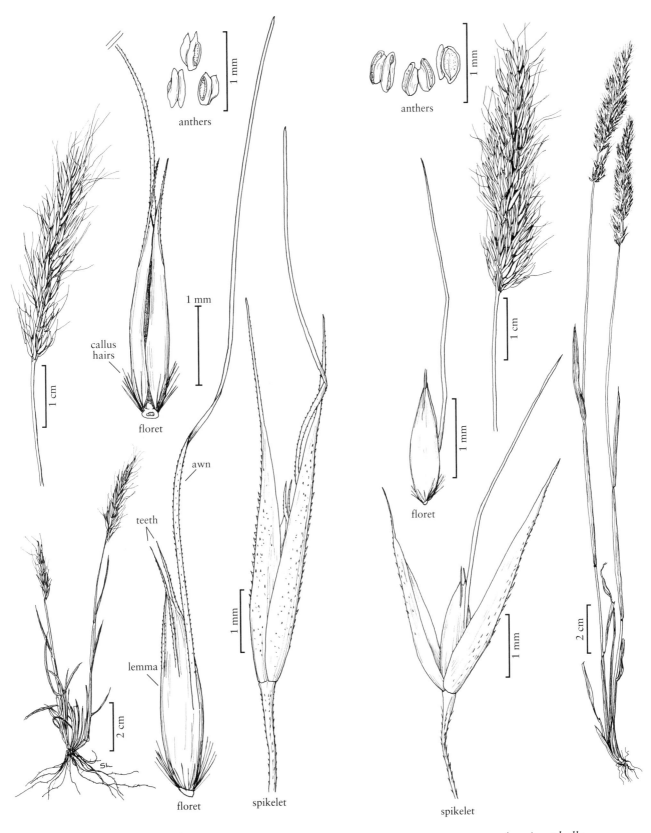

anthers

callus
hairs

floret

1 mm

awn

teeth

lemma

floret

spikelet

A. hendersonii

AGROSTIS

anthers

1 cm

floret

spikelet

A. microphylla

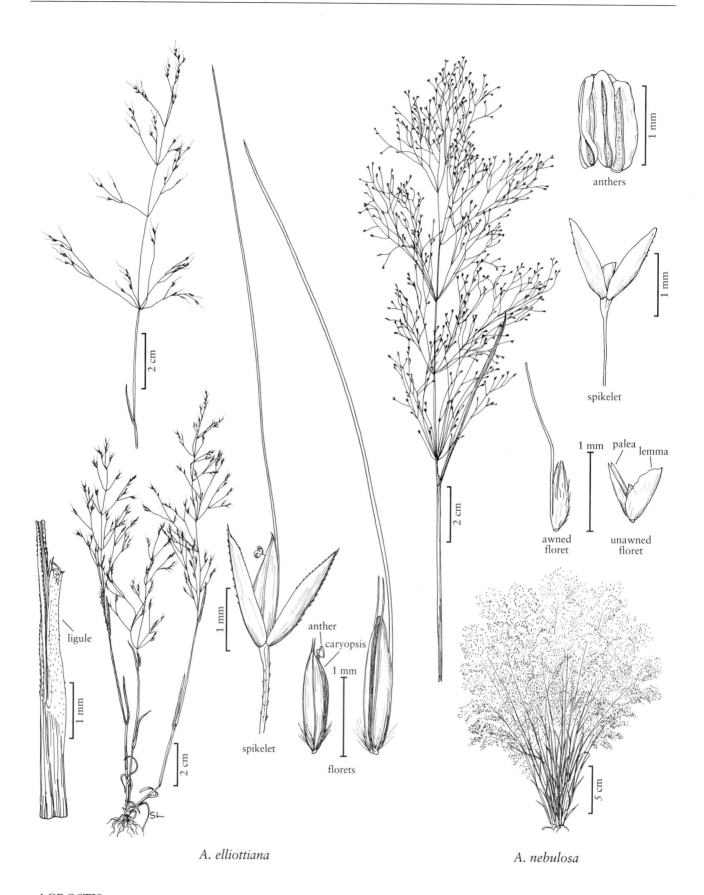

anthers

1 mm

spikelet

1 mm

palea lemma

1 mm

awned
floret

unawned
floret

ligule

1 mm

anther

caryopsis

1 mm

spikelet

florets

2 cm

2 cm

2 cm

5 cm

A. elliottiana

A. nebulosa

AGROSTIS

26. Agrostis microphylla Steud. [p. 659]
SMALL-LEAF BENT

Plants annual. Culms 8–45 cm, erect. Leaves usually mostly cauline, sometimes mostly basal; sheaths glabrous or pubescent; upper sheaths usually inflated; ligules 1.4–4.5 mm, dorsal surfaces scabridulous, apices truncate to acute, lacerate to erose; blades 3–15 cm long, 0.7–2.5 mm wide, flat, becoming involute, finely scabrous. Panicles 2–12 cm long, 0.4–2 cm wide, cylindrical, spikelike, dense, occasionally lobed or interrupted near the base, appearing bristly from the awns, usually exserted from the upper sheaths at maturity; branches 0.3–1.5 cm, scabrous to scabridulous, ascending to appressed, spikelet-bearing to the base and hidden by the spikelets; pedicels 0.3–3 mm. Spikelets lanceolate, greenish to yellowish, tinged with purple. Glumes equal to subequal, 2.5–5 mm, scabrous over the midvein and often scabridulous on the body, 1-veined, apices narrowly acuminate to awn-tipped, awns to 1.5 mm; callus hairs to 0.5(1) mm, usually dense; lemmas 1.5–2.3 mm, scabrous, translucent to opaque, 5-veined, veins prominent distally, apices acute, veins extended into 2(4) teeth, teeth 0.1–0.5 mm, awned from about midlength or above, awns 3.5–8 mm, geniculate; paleas absent, or to 0.2 mm and thin; anthers 3, 0.4–0.6 mm, often retained at the apices of the caryopses. Caryopses 0.9–1.3 mm; endosperm soft. 2*n* = 56.

Agrostis microphylla grows in thin, rocky soils, sandy areas, cliffs, vernal pools, and serpentine areas. It is a winter annual, flowering in late winter to spring, adapted to low-competition habitats with summer drought. It may be related to, or conspecific with, *Agrostis hendersonii* (see previous).

Agrostis microphylla grows mostly along the Pacific coast from British Columbia to northern Baja California, Mexico. Reports of *A. microphylla* from the Humboldt Mountains, Nevada, reflect Vasey's treatment of a specimen of *A. exarata* as the type of a new variety, *A. microphylla* var. *major* Vasey.

27. Agrostis elliottiana Schult. [p. 660]
ELLIOTT'S BENT

Plants annual. Culms 5–45 cm, erect, sometimes geniculate at the base, with (3)4–9 nodes. Leaves mostly basal or cauline; basal leaves withered at anthesis; sheaths smooth or scabridulous; ligules (0.7) 1.5–3.5 mm, dorsal surfaces scabrous, apices acute, rounded, or truncate, lacerate; blades 0.5–4 cm long, 0.5–1 mm wide, flat, becoming involute. Panicles 3–20 cm long, (0.5)2–12 cm wide, widely ovate, ultimately open and diffuse, the whole panicle detaching after maturity, blowing about as a tumbleweed, bases usually exserted, sometimes enclosed by the upper sheaths at maturity, lowest node with 1–6 branches; branches scabridulous, capillary, branching beyond midlength, initially ascending, becoming laxly spreading, spikelets clustered near the tips, lower branches 2–8 cm; pedicels 0.3–7.5 mm; secondary panicles sometimes present in the leaf axils. Spikelets narrowly elliptic to lanceolate, yellowish purple to greenish purple. Glumes equal, 1.5–2.2 mm, 1-veined, scabrous on the midvein, margins scabrous distally, acute; callus hairs to 0.6 mm, dense; lemmas 1–2 mm, smooth or scabrous to warty, translucent, 5-veined, veins prominent, apices acute, entire or 2–5-toothed, teeth minute, to 0.8 mm, usually awned from just below the apices, sometimes unawned, awns 3–10 mm, flexuous, not geniculate, deciduous; paleas absent or minute; anthers 1, 0.1–0.2 mm, lobes widely separated by the connective, usually retained at the apices of the caryopses. Caryopses 1–1.4 mm; endosperm liquid. 2*n* = 28.

Agrostis elliottiana grows in fields and scrublands and along roadsides. It has a disjunct distribution, occurring in western North America in northern California and southern Arizona and New Mexico; in eastern North America from Kansas and Texas east to Pennsylvania and northern Florida; and in Yucatan, Mexico. Although it has been introduced elsewhere, notably in Maine, it is not known to have become established at those locations.

Agrostis elliottiana resembles *A. scabra* (p. 646) and *A. hyemalis* (p. 647) in its diffuse panicle, but differs in its flexible awn and single anther. Small Californian plants have sometimes been called *A. exigua* Thurb.; they are otherwise identical to *A. elliottiana*.

28. Agrostis nebulosa Boiss. & Reut. [p. 660]
CLOUD GRASS

Plants annual. Culms 10–75 cm, erect or geniculate, with 2–7 nodes. Leaves mostly cauline; sheaths scabrous; ligules 1–6 mm, dorsal surfaces scabridulous, apices acute to rounded, erose to lacerate; blades 5–15 cm long, 1–4 mm wide, flat. Panicles 3–30 cm long, (2)5–20 cm wide, oblong to ovate, diffuse, usually over ¹/₂ the length of the culm, lowest node with (2)3–18 branches; branches scabrous, erect to spreading, lower branches

4–15 cm; **pedicels** (2.5)4–15 mm, much longer than the spikelets. **Spikelets** lanceolate, usually purplish, sometimes green to yellowish green. **Glumes** subequal, 1.3–2.1 mm, 1-veined, sparsely aculeolate on the veins, obtuse; **calluses** glabrous; **lemmas** 0.5–0.8 mm, transparent, thin, veins scarcely visible, smooth and glabrous or rarely hairy, apices truncate to rounded or acute, toothed or erose, usually unawned, rarely awned from near the base; **paleas** 0.4–0.7 mm, about as long as the lemmas and of similar texture; **anthers** 3, 1–1.4 mm. **Caryopses** 0.6–0.8 mm. $2n = 14$.

Agrostis nebulosa is native to Spain and Portugal. It is cultivated as an ornamental and for dried flower arrangements, but occasional escapes have been found on roadsides, ditches, and in fields in widely scattered locations in the *Flora* region.

14.28 POLYPOGON Desf.

Mary E. Barkworth

Plants annual or perennial; not rhizomatous. **Culms** 4–120 cm, erect to decumbent, rooting at the lower nodes, sparingly branched near the base. **Leaves** usually no more than 5 per culm, basal and cauline; **sheaths** open, smooth or scabridulous; **auricles** absent; **ligules** membranous or hyaline, acute to broadly rounded, erose, ciliate; **blades** flat to convolute. **Inflorescences** terminal panicles, dense, continuous or interrupted below; **branches** flexible, usually some longer than 1 cm; **pedicels** absent and the spikelets borne on a stipe, or present and terminating in a stipe; **stipes** scabrous, flaring distally; **disarticulation** at the base of the stipes. **Spikelets** 1–5 mm, weakly laterally compressed, with 1 bisexual floret; **rachillas** not prolonged beyond the base of the floret. **Glumes** exceeding the floret, lanceolate, bases not fused, apices entire to emarginate or bilobed, usually awned from the sinuses or apices, awns flexuous, glabrous, sometimes unawned; **lemmas** 1–3(5)-veined, often awned, awns usually terminal or subterminal, sometimes arising from just above midlength; **paleas** from ⅓ as long as to equaling the lemmas; **lodicules** 2, oblong-lanceolate to lanceolate; **anthers** 3; **ovaries** glabrous; **styles** separate. **Caryopses** slightly flattened, broadly ellipsoid to oblong-ellipsoid; **hila** ⅙–¼ as long as the caryopses, ovate. $x = 7$. Name from the Greek *poly*, 'many' or 'much', and *pogon*, 'beard', an allusion to the bristly appearance of the inflorescence.

Polypogon is a pantropical and warm-temperate genus of about 18 species. There are eight species in the *Flora* region; one species, *P. interruptus*, is native.

Polypogon is similar to *Agrostis*, and occasionally hybridizes with it. It differs from *Agrostis* in having spikelets that disarticulate below the glumes, often at the base of a stipe.

SELECTED REFERENCE Cope, T.A. 1982. Flora of Pakistan, No. 143: Poaceae (E. Nasir and S.I. Ali, eds.). Pakistan Agricultural Research Council and University of Karachi, Islamabad and Karachi, Pakistan. 678 pp.

1. Glumes with awns 3–12 mm long.
    2. Glumes deeply lobed, the lobes more than ⅙ the length of the glume body . . . . 7. *P. maritimus*
    2. Glumes not lobed or the lobes ¹/₁₀ or less the length of the glume body.
        3. Plants annual; glume apices rounded, lobed, the lobes 0.1–0.2 mm long; ligules 2.5–16 mm long . . . . 6. *P. monspeliensis*
        3. Plants perennial; glume apices acute to truncate, unlobed or the lobes shorter than 0.1 mm; ligules 1–6 mm long.
            4. Glume awns (3)4–6 mm long; longest blades 13–17 cm long . . . . 5. *P. australis*
            4. Glume awns 1.5–3.2 mm long; longest blades 5–9 cm long . . . . 2. *P. interruptus* (in part)
1. Glumes unawned or with awns to 3.2 mm long.
    5. Glumes unawned . . . . 1. *P. viridis*
    5. Glumes awned, the awns 0.2–3.2 mm long.
        6. Stipes 1.5–2.5 mm long; glumes tapering from about midlength to the acute, unlobed apices . . . . 3. *P. elongatus*

6. Stipes 0.2–1.5 mm long; glumes not tapering to the apices, the apices usually rounded
   to truncate, sometimes acute, often lobed.
   7. Lemmas 1–2 mm long; paleas about ¹/₂ as long as the lemmas; the lower glumes
      longer than the upper glumes . . . . 8. *P. imberbis*
   7. Lemmas 0.7–1.5 mm long; paleas from ³/₄ as long as to equaling the lemmas; glumes
      of each spikelet subequal to equal.
      8. Plants annual; glumes acute to rounded, lobed, the lobes 0.1–0.2 mm long . . . . 4. *P. fugax*
      8. Plants perennial, often flowering the first year; glumes acute to truncate, if lobed,
         the lobes to 0.1 mm long . . . . 2. *P. interruptus* (in part)

1. Polypogon viridis (Gouan) Breistr. [p. 664]
WATER BEARDGRASS

Plants perennial, often flowering the first year. **Culms** 10–90 cm, sometimes decumbent and rooting at the lower nodes. **Sheaths** glabrous, smooth; **ligules** to 5 mm; **blades** 2–13 cm long, 1–6 mm wide. **Panicles** 2–10 cm, ovate-oblong to pyramidal, dense but interrupted, pale green to purplish; **pedicels** not developed; **stipes** 0.1–0.6 mm. **Glumes** 1.5–2 mm, scabrous on the back and keel, apices obtuse or truncate, unawned; **lemmas** about 1 mm, erose, unawned; **paleas** subequal to the lemmas; **anthers** 0.3–0.5 mm. 2*n* = 28, 42.

Polypogon viridis grows in mesic habitats associated with rivers, streams, and irrigation ditches. It is native from southern Europe to Pakistan, but is now established in the *Flora* region, particularly the southwestern United States. Records from the Atlantic coast are based on plants found on ballast dumps; there have been no recent collections from these locations.

In Europe, *Polypogon viridis* hybridizes with *P. monspeliensis*, forming **P. ×adscendens** Guss. *ex* Bertol.; no such hybrids have been reported from the *Flora* region.

2. Polypogon interruptus Kunth [p. 664]
DITCH BEARDGRASS

Plants perennial, often flowering the first year. **Culms** 20–80 (90) cm, more or less decumbent. **Sheaths** smooth; **ligules** 2–6 mm, scabridulous-pubescent; **blades** 5–9 cm long, 3–6 mm wide. **Panicles** 3–15 cm long, 0.5–3 cm wide, usually interrupted or lobed; **pedicels** not developed; **stipes** 0.2–0.7 mm. **Glumes** 2–3 mm, subequal, scabrous, larger prickles extending up the keel beyond midlength, not tapering to the apices, apices acute to truncate, unlobed or the lobes to 0.1 mm, awned, awns 1.5–3.2 mm, those of the lower and upper glumes subequal; **lemmas** 0.8–1.5 mm, glabrous, smooth and shiny, apices obtuse, not emarginate, awned, awns 1–3.2 mm; **paleas** about ³/₄ as long as the lemmas; **anthers** 0.5–0.7 mm. 2*n* = 28, 42.

Polypogon interruptus grows in moist soil at lower elevations. It is native to the Western Hemisphere, extending south from the western United States into northern Mexico, and through the American tropics to Argentina and Bolivia. The more eastern records may indicate introductions; it is not known whether or not the species persists at these locations.

3. Polypogon elongatus Kunth [p. 664]
SOUTHERN BEARDGRASS

Plants perennial, often flowering the first year. **Culms** to 100 cm, erect or decumbent at the base. **Sheaths** smooth, glabrous; **ligules** 4–8 mm, scabridulous, lacerate; **blades** 10–30 cm long, 4–15 mm wide. **Panicles** 10–30 cm, erect or nodding, interrupted, dense; **pedicels** not developed; **stipes** 1.5–2.5 mm. **Glumes** 3–5 mm, hispidulous, tapering from about midlength to the acute apices, apices unlobed, awned, awns 1–3 mm; **lemmas** about 1.5 mm, awned, awns 1–2 mm, arising from above midlength; **paleas** ¹/₂–²/₃ as long as the lemmas; **anthers** 0.5–0.7 mm. 2*n* = 28, 56.

Polypogon elongatus is native from Mexico to Argentina. It now grows at scattered locations in the *Flora* region, primarily in California.

4. Polypogon fugax Nees *ex* Steud. [p. 666]
ASIAN BEARDGRASS

Plants annual. **Culms** (8.5) 15–60 cm, often decumbent at the base and rooting at the nodes. **Sheaths** smooth; **ligules** 2–8 mm; **blades** 2–16 cm long, 2–11 mm wide, scabrous. **Panicles** 3–15 cm long, 0.5–5 cm wide, narrowly ovoid, oblong, or cylindrical, dense, usually lobed, pale green or yellowish; **pedicels** absent or to 0.5 mm; **stipes** 0.2–1.3 mm. **Glumes** 1.8–2.4 mm,

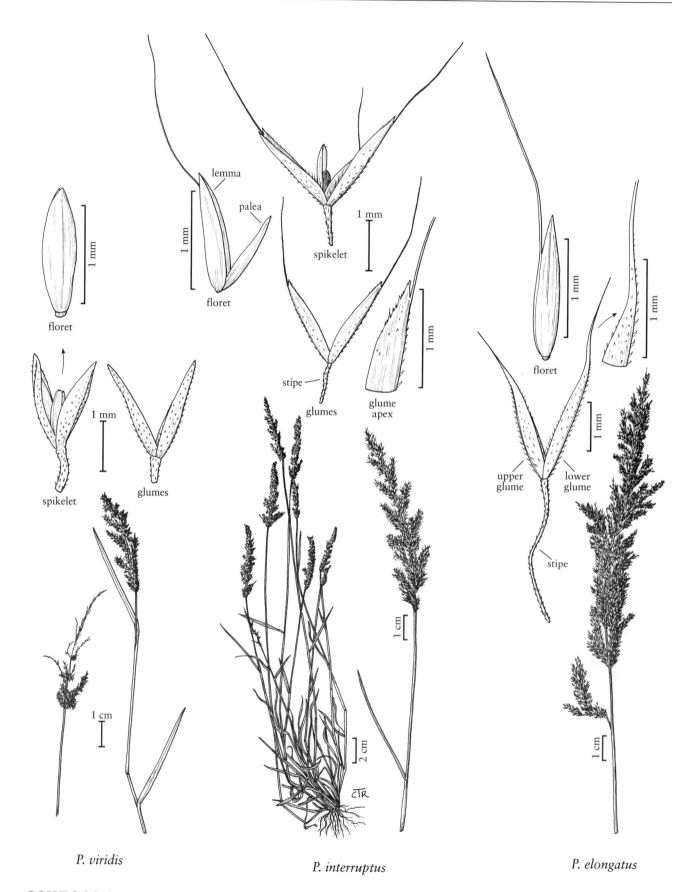

P. viridis   P. interruptus   P. elongatus

POLYPOGON

equal to subequal, scabridulous to echinate, not tapering to the apices, apices acute to rounded, lobed, lobes 0.1–0.2 mm, awned from the sinuses, awns 0.6–3 mm, those of the lower and upper glumes subequal to equal, flexuous; **lemmas** 0.9–1.2 mm, smooth, unawned or awned, awns to 2 mm, flexuous; **paleas** 0.7–1.2 mm, from ³/₄ as long as to equaling the lemmas; **anthers** 0.3–0.6 mm. 2*n* = 42.

Polypogon fugax is native from Iraq to Myanmar [Burma]. It was collected in Santa Barbara, California, and from salt marshes around Oakland, California, in the nineteenth century, and from Portland, Oregon, in the early twentieth century. There are no recent collections from the *Flora* region.

5. Polypogon australis Brongn. [p. 666]
CHILEAN BEARDGRASS

Plants perennial. **Culms** 20–100 cm. **Sheaths** smooth to scabridulous; **ligules** 1–3(4) mm, rounded to broadly acute, erose; **blades** 13–17 cm long, 5–7 mm wide, scabrous. **Panicles** 8–15 cm, lobed or interrupted, usually purplish; **pedicels** absent or vestigial; **stipes** 0.3–0.5 mm. **Glumes** 1.5–3 mm, smooth to echinate, margins ciliate, apices acute to truncate, unlobed or lobed, lobes to 0.1 mm, awned, awns (3)4–6 mm, flexuous; **lemmas** 1–1.3 mm, awned, awns 2–3.5 mm, flexuous; **paleas** from shorter than to subequal to the lemmas; **anthers** 0.3–0.5 mm. 2*n* = unknown.

Polypogon australis is native to South America. It has become established in western North America, where it grows alongside ditches and streams. The records from Washington and Oregon are from ballast dumps; it is not known from recent collections in those states.

6. Polypogon monspeliensis (L.) Desf. [p. 666]
RABBITSFOOT GRASS

Plants annual. **Culms** 5–65 (100) cm, erect to geniculately ascending. **Sheaths** glabrous, the uppermost sheaths sometimes inflated; **ligules** 2.5–16 mm; **blades** 1–20 cm long, 1–7 mm wide. **Panicles** 1–17 cm, narrowly ellipsoid, dense, sometimes lobed, greenish; **pedicels** absent or to 0.2 mm; **stipes** 0.1–0.2 mm. **Glumes** 1–2.7 mm, hispidulous throughout, largest prickles restricted to the lower ¹/₂, apices rounded, lobed, lobes 0.1–0.2 mm, ¹/₁₀ or less the length of the glume body, awned from the sinus, awns 4–10 mm, yellowish; **lemmas**

0.5–1.5 mm, glabrous, awned, awns 0.5–1(4.5) mm; **paleas** subequal to the lemmas; **anthers** 0.2–1 mm. 2*n* = 14, 28, 35, 42.

Polypogon monspeliensis is native to southern Europe and Turkey. It is now a common weed throughout the world, including much of the *Flora* region. It grows in damp to wet, often alkaline soils, particularly in disturbed areas. Vernon Harms (pers. comm., 2005) commented that the species' distribution in Saskatchewan appears to have increased greatly since the 1970s. The English-language name aptly describes the feel of the young panicles.

In Europe, *Polypogon monspeliensis* hybridizes with *Agrostis stolonifera*, producing the sterile ×*Agropogon lutosus* (p. 668); and with *P. viridis*, forming *P.* ×**adscendens** Guss. *ex* Bertol. Only ×*Agropogon lutosus* has been reported from the *Flora* region. It differs from *P. monspeliensis* in having more persistent spikelets, less blunt short-awned glumes, and lemmas with subterminal rather than terminal awns.

7. Polypogon maritimus Willd. [p. 667]
MEDITERRANEAN BEARDGRASS

Plants annual. **Culms** (5)20–40 (50) cm, geniculate. **Sheaths** glabrous, smooth, uppermost sheaths sometimes inflated; **ligules** to 7 mm; **blades** (1)3–9 (14) cm long, 0.5–5 mm wide. **Panicles** (1)2–8(15) cm, narrowly ellipsoid, dense, sometimes lobed, often purplish; **pedicels** to about 0.5 mm, capillary; **stipes** 0.1–1.2 mm. **Glumes** 1.8–3.2 mm, hispidulous basally, hairs sometimes strongly inflated and obtuse, apices lobed, lobes 0.3–1.2 mm, more than ¹/₆ the length of the glume body, awned from the sinus, awns (4)7–12 mm; **lemmas** 0.5–1.5 mm, unawned or awned, awns shorter than 1 mm; **paleas** subequal to the lemmas; **anthers** 0.4–0.5 mm. 2*n* = 14.

Polypogon maritimus grows in disturbed, moist places, from sea level to 700 m. It is a Mediterranean species that now occurs at scattered locations in North America, being particularly common in, or possibly just well-reported from, California. There are two varieties. Plants from the *Flora* region belong to **P. maritimus** Willd. var. **maritimus,** having stipes about as long as they are wide, glumes that never become strongly indurate at the base, and uninflated, acute hairs on the glume bases. Plants of **P. maritimus** var. **subspathaceus** (Req.) Bonnier & Layens have stipes that are 3–4 times as long as wide, glumes that become strongly indurate at maturity, and hairs on the glume bases that are strongly inflated and subobtuse.

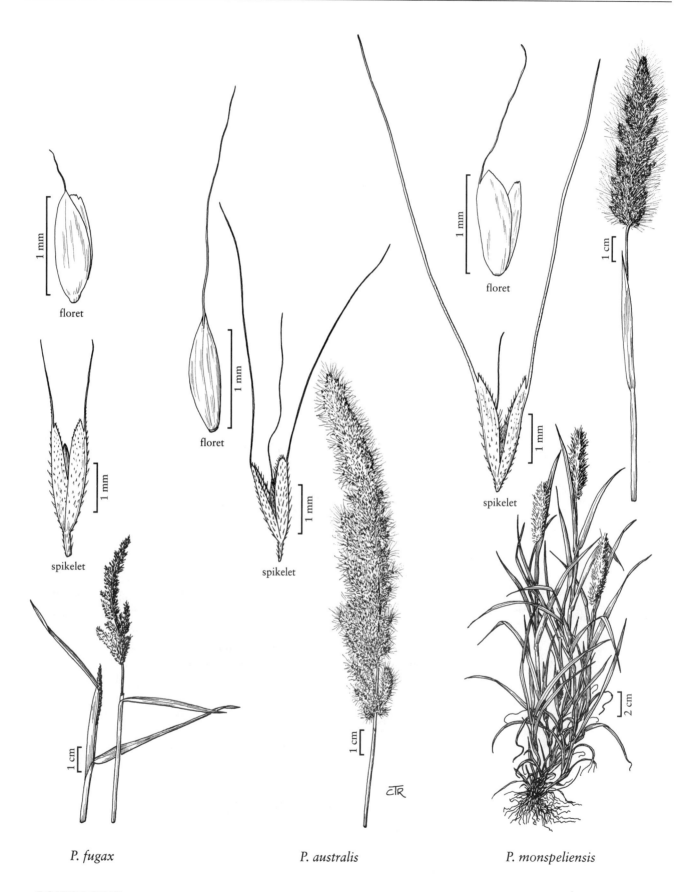

floret

spikelet

floret

spikelet

floret

spikelet

P. fugax

P. australis

P. monspeliensis

POLYPOGON

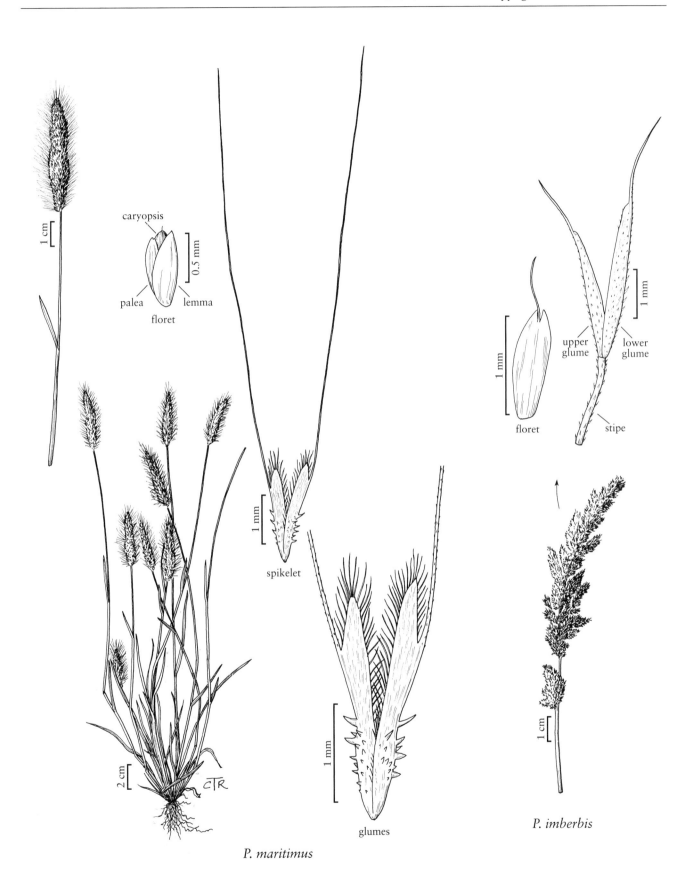

caryopsis

0.5 mm

palea   lemma
floret

1 mm
spikelet

1 mm
glumes

1 cm

2 cm

CTR

P. maritimus

1 mm

1 mm
upper   lower
glume   glume

floret   stipe

1 cm

P. imberbis

POLYPOGON

8. Polypogon imberbis (Phil.) Johow [p. 667]
SHORTHAIRED BEARDGRASS

Plants perennial. Culms 15–80 cm, ascending or geniculate, not branched. Sheaths smooth or scabrous; ligules 2–5 mm, membranous or hyaline, sometimes retrorsely scabridulous; blades 3–15 cm long, 1.5–8 mm wide, flat to convolute, scabrous, apices acute to sharp. Panicles 3–25 cm long, 1–8 cm wide, dense, glomerate, interrupted near the base; pedicels not developed; stipes 0.6–1.5 mm. Glumes 1.8–4 mm, scabridulous on the sides, keels echinate, not tapering to the apices, apices acute, unlobed, awned, awns 0.2–2.5 mm; lower glumes 1.8–4 mm; upper glumes 1.6–3.5 mm, usually shorter than the lower glumes; lemmas 1–2 mm, hyaline, unawned or awned, awns subterminal, to 1 mm; paleas 0.5–0.8 mm, about ¹/₂ as long as the lemmas; anthers 0.2–0.8 mm. Caryopses 1–1.5 mm long, 0.3–0.4 mm wide. $2n$ = unknown.

Polypogon imberbis is a South American species that has been collected at two locations in California, one from Oceano Beach, San Luis Obispo County, and the other near Martines, Contra Costa County. It does not appear to be established there, the last collections having been made before 1950. In South America, it grows in moist, sandy soils near streams, lagoons, and the coast.

14.29 ×AGROPOGON P. Fourn.

Mary E. Barkworth

Plants perennial; loosely cespitose to spreading, rhizomatous. Culms 8–60 cm, usually branched below, ascending from a decumbent base or the lower nodes geniculate. Sheaths smooth, open; auricles absent; ligules membranous, puberulent, acute to obtuse, often bifid or denticulate with age; blades usually glabrous, sometimes scabrous. Inflorescences panicles, moderately dense to very dense, often interrupted; disarticulation tardy, below the glumes. Spikelets laterally compressed, with 1 floret. Glumes similar, narrowly oblong to elliptic, 1-veined, apices notched, midveins extending into a short awn; calluses glabrous; lemmas shorter than the glumes, inconspicuously 5-veined, awned from just below the minutely toothed apices; paleas about ³/₄ as long as the lemmas; anthers usually indehiscent. x = 7.

×*Agropogon* comprises hybrids between *Agrostis* and *Polypogon*; one hybrid grows in the *Flora* region.

SELECTED REFERENCE Rúgolo de Agrasar, Z.E. and A.M. Molina. 1997. Presencia del híbrido ×*Agropogon littoralis* (Gramineae: Agrostideae) en Chile. Hickenia 11:209–214.

1. ×Agropogon lutosus (Poir.) P. Fourn. [p. 669]
PERENNIAL BEARDGRASS

Culms 8–60 cm. Sheaths sometimes inflated; ligules 3–7 mm; blades 3–20 cm long, 2–11 mm wide, usually glabrous, sometimes scabrous. Panicles 2–18 cm long, 0.6–7 cm wide, lanceolate to narrowly ovate, green or purplish; pedicels scabrous. Spikelets 2–3 mm. Glumes subequal, membranous, scabrous, acute, awns to about 2 mm, apical; calluses glabrous; lemmas 1–1.7 mm, awns to 3 mm, subterminal; paleas about ³/₄ as long as the lemmas; anthers about 1 mm. Caryopses not produced. $2n$ = 28.

×*Agropogon lutosus* is a sterile hybrid between *Agrostis stolonifera* and *Polypogon monspeliensis* that sometimes grows in locations where both parents occur, such as damp to wet, often alkaline soils on lakesides, marshes, ditches, and intermittently flooded fields. Some plants favor *A. stolonifera*, others *P. monspeliensis*. All differ from *Polypogon* in having more persistent spikelets, less blunt short-awned glumes, and lemmas with subterminal rather than terminal awns; and from *Agrostis* in having awned glumes and awned lemmas.

spikelet

2 mm

floret

1 mm

1 cm

×*A. lutosus*

×AGROPOGON

spikelet

2 mm

floret

2 cm

L. ovatus

LAGURUS

14.30 LAGURUS L.

Gordon C. Tucker

Plants annual; tufted. **Culms** 6–80 cm, erect or ascending, pilose or villous. **Leaves** pilose or villous; **sheaths** open nearly to the base; **auricles** absent; **ligules** truncate, erose, ciliate, abaxial surfaces densely pubescent; **blades** flat at maturity, convolute in the bud. **Inflorescences** terminal panicles, dense, ovoid. **Spikelets** laterally compressed, with 1 floret; **rachillas** prolonged beyond the base of the floret, pubescent; **disarticulation** above the glumes, beneath the floret. **Glumes** 2, equal, exceeding the florets, lanceolate, membranous, pilose, hairs 2–3 mm, 1-veined, veins acuminate and awned, awns pilose; **calluses** blunt, pubescent; **lemmas** lanceolate, pilose or scabridulous, 3-veined, 3-awned, apices bifid, lateral veins extending as slender awns, not twisted below, central awn arising from the distal $^1/_3$ of the lemma, twisted below; **paleas** nearly as long as the lemmas, veins scabridulous, extending as awnlike points; **lodicules** 2, narrowly elliptic, glabrous, apices minutely bilobed; **anthers** 3; **ovaries** glabrous; **styles** fused or separate. **Caryopses** stipitate, subterete, ellipsoid; **hila** $^1/_5$–$^1/_4$ as long as the caryopses, ovate. $x = 7$. Name from the Greek *lagos*, 'hare', and *oura*, 'tail', in reference to the densely pilose inflorescences.

Lagurus is a monotypic genus endemic to the Mediterranean region. *Lagurus ovatus* has been introduced in North America, as well as South America, southern Africa, and Australia. It is sometimes cultivated as an ornamental, and the dried flowering culms are used in floral arrangements.

SELECTED REFERENCES **Messeri, A.** 1942. Studio sistematico e fitogeografico di *Lagurus ovatus* L. Nuovo Giorn. Bot. Ital., n.s., 49:133–204; **Vergagno Gambi, O.** 1965. Osservazioni sullo sviluppo di *Lagurus ovatus* L. Giorn. Bot. Ital. 72:243–254.

1. Lagurus ovatus L. [p. 669]
HARETAIL GRASS

Culms (6)10–50(80) cm. **Sheaths** (0.5)2–8 cm, inflated, ribbed; **ligules** 1–1.7(3) mm; **blades** 2–6(10) cm long, 2.5–10 mm wide. **Panicles** 1.5–3 cm long, 1–2 cm wide; **branches** softly pilose, hairs white, about 0.4 mm, awns and glume veins sometimes purplish; **pedicels** 0.4–1(2) mm, pilose. **Glumes** (5.5)7–10 mm, awns 1.5–3 mm; **rachillas** prolonged 0.7–1.4 mm; **calluses** hairy, hairs to 0.8 mm; **lemmas** 3.5–4.5 mm, lateral awns 1–2(6) mm, central awns (8)12–22 mm; **paleas** 3–4 mm; anthers 1.4–2.2 mm. **Caryopses** 2–2.5 mm. $2n = 14$.

Lagurus ovatus grows in disturbed sites in full sun. It is a waif, or sparingly naturalized, in North Carolina and western Florida; it is established and plentiful in central California. It may not have persisted at the other sites shown. The earliest collection from the United States was made in California in 1903. It is an attractive species and is sometimes grown as an ornamental.

Natural populations of *Lagurus ovatus* in southern Europe contain both tall and short plants, flowering at about the same time. Protracted germination, followed by simultaneous floral induction at a 12–16 hour photoperiod, leads to the apparent "dimorphism". Plants are quite uniform in height when grown under controlled conditions indoors (Vergagno Gambi 1965).

14.31 PHLEUM L.

Mary E. Barkworth

Plants annual or perennial; cespitose, sometimes rhizomatous, occasionally stoloniferous. **Culms** 2–150 cm, erect or decumbent; **nodes** glabrous. **Sheaths** open; **auricles** absent or inconspicuous; **ligules** membranous, not ciliate; **blades** usually flat. **Inflorescences** dense, spikelike panicles, more than 1 spikelet associated with each node; **branches** often shorter than

2 mm, always shorter than 7 mm, stiff; **pedicels** shorter than 1 mm, sometimes fused; **disarticulation** above the glumes or, late in the season, beneath the glumes. **Spikelets** strongly laterally compressed, bases usually U-shaped, sometimes cuneate, with 1 floret; **rachillas** glabrous, sometimes prolonged beyond the base of the floret. **Glumes** equal, longer and firmer than the florets, stiff, bases not connate, strongly keeled, keels usually strongly ciliate, sometimes glabrous, sometimes scabrous, 3-veined, apices truncate to tapered, midveins often extending into short, stiff, awnlike apices; **calluses** blunt, glabrous; **lemmas** white, often translucent, not keeled, 5–7-veined, unawned, bases not connate, apices acute, entire, sometimes with a weak, subapical awn; **paleas** subequal to the lemmas, 2-veined; **lodicules** 2, free, glabrous, toothed; **anthers** 3; **ovaries** glabrous. **Caryopses** elongate-ovoid; **embryos** $^1/_6$–$^1/_4$ the length of the caryopses. $x = 7$. Name from the Greek *phleos*, the name of a reedy grass.

Phleum is a genus of approximately 15 species, most of which are native to Eurasia. One species, *P. alpinum*, is native to the *Flora* region and six are introduced. One of the introduced species, *P. pratense*, has been established in the region for a long time. It is also widely cultivated as a fodder grass, both in the *Flora* region and in other parts of the world. *Phleum phleoides* was first recognized as being present in the *Flora* region in 1990. *Phleum exaratum* has been reported from the United States. No specimens supporting the report have been seen. It resembles *P. arenarium* (p. 675), but has anthers 1.5–2 mm long rather than 0.3–1.2 mm, and an inflorescence that is rounded at the base.

Species of *Phleum* are sometimes mistaken for *Alopecurus*, but *Alopecurus* has obtuse to acute glumes that are unawned or taper into an awn, lemmas that are both awned and keeled, and paleas that are absent or greatly reduced. The species of *Phleum* that are most abundant in the *Flora* region are easily recognized by their strongly ciliate, abruptly truncate, awned glumes and adnate panicle branches. In addition, in *Phleum* the lemmas are not keeled, and the paleas are always subequal to the lemmas.

SELECTED REFERENCES Humphries, C.J. 1978. Notes on the genus *Phleum*. Bot. J. Linn. Soc. 76:337–340; Kula, A., B. Dudziak, E. Sliwinska, A. Grabowska-Joachimiak, A. Stewart, H. Golczyk, and A.J. Joachimiak. 2006. Cytomorphological studies on American and European *Phleum commutatum* Gaud. (Poaceae). Acta Biol. Cracov., Ser. Bot. 48:99–108.

1. Plants perennial.
  2. Panicles tapering distally; panicle branches free from the rachis; glumes scabrous to shortly ciliate on the keels; plants known, in the *Flora* region, only from British Columbia ... 3. *P. phleoides*
  2. Panicles not tapering distally; panicle branches adnate to the rachises; glumes conspicuously ciliate on the keels; plants widespread in the *Flora* region.
    3. Sheaths of the flag leaves not inflated; panicles 2–14(17) cm long, 5–20 times as long as wide; lower internodes of the culms frequently enlarged or bulbous; widespread in the *Flora* region .... 1. *P. pratense*
    3. Sheaths of the flag leaves inflated; panicles 1–6 cm long, usually 1.5–3 times as long as wide; lower internodes of the culms not enlarged or bulbous .... 2. *P. alpinum*
1. Plants annual.
  4. Glume apices abruptly truncate .... 4. *P. paniculatum*
  4. Glume apices gradually narrowed to tapered.
    5. Glumes semi-elliptical in outline, apices pointing towards each other; keels usually glabrous, sometimes scabrous .... 5. *P. subulatum*
    5. Glumes oblong-lanceolate, apices parallel or divergent; keels ciliate .... 6. *P. arenarium*

1. Phleum pratense L. [p. 673]

TIMOTHY, FLÉOLE DES PRÉS, PHLÉOLE DES PRÉS, MIL

Plants perennial; loosely to densely cespitose. **Culms** (20) 50–150 cm, usually erect, lower internodes frequently enlarged or bulbous. **Sheaths of the flag leaves** not inflated; **auricles** occasionally present, inconspicuous; **ligules** 2–4 mm, obtuse to acute; **blades** to 45 cm long, 4–8(10) mm wide, flat. **Panicles** (3)5–10(16) cm long, 5–7.5(10) mm wide, 5–20 times as long as wide, not tapering distally; **branches** adnate to the rachises. **Glumes** 3–4 mm, sides usually puberulent, keels pectinate-ciliate, apices awned, awns 1–1.5(2) mm; **lemmas** (1.2)1.7–2 mm, about $^{1}/_{2}$ as long as the glumes, usually puberulent; **anthers** 1.6–2.3 mm. $2n$ = 42 (21, 35, 36, 49, 56, 63, 70, 84).

Phleum pratense grows in pastures, rangelands, and disturbed sites throughout most of the mesic, cooler regions of North America. Originally introduced from Eurasia as a pasture grass, it is now well established in many parts of the world, including the *Flora* region. North American plants belong to the polyploid **Phleum pratense** L. subsp. **pratense**, which differs from the diploid *P. pratense* subsp. **bertolonii** (DC.) Bornm. in having obtuse ligules. Depauperate specimens of *P. pratense* are hard to distinguish from *P. alpinum*.

2. Phleum alpinum L. [p. 673]

ALPINE TIMOTHY, FLÉOLE ALPINE, PHLÉOLE ALPINE

Plants perennial; cespitose, sometimes shortly rhizomatous. **Culms** 15–50 cm, often decumbent, lower internodes not enlarged or bulbous. **Sheaths of the flag leaves** inflated; **auricles** not developed, leaf edges sometimes wrinkled at the junction of the sheath and blade; **ligules** 1–4 mm, truncate; **blades** to 17 cm long, 4–7 mm wide, flat. **Panicles** 1–6 cm long, 5–12 mm wide, usually 1.5–3 times as long as wide, subglobose to broadly cylindric, not tapering distally; **branches** adnate to the rachises. **Glumes** 2.5–4.5 mm, sides scabrous, keels hispid, apices awned, awns 0.8–2.5(3.2) mm; **lemmas** 1.7–2.5 mm, about $^{3}/_{4}$ as long as the glumes, mostly glabrous, keels hairy, hairs to 0.1 mm; **anthers** 1–1.5(2) mm. $2n$ = 14, 28.

Phleum alpinum grows along stream banks, on moist prairie hillsides, and in wet mountain meadows. It is a circumboreal species extending, in the *Flora* region, from northern North America southward through the mountains to Mexico and South America. It is also widespread in northern Eurasia. Isolated, depauperate plants of *P. pratense* may be difficult to distinguish from *P. alpinum*; there is never any difficulty in the field.

Kula et al. (2006) demonstrated that American and northern European plants of *Phleum alpinum* belong to the same taxon. They mistakenly identified the taxon as *P. commutatum* Gaudin. Because Humphries (1978) lectotypified *P. alpinum* on a plant from Lapland, it has priority over *P. commutatum*. North American plants belong to **P. alpinum** L. subsp. **alpinum** and are tetraploid. The count of $2n$ = 14 applies to **Phleum alpinum** subsp. **rhaeticum** Humphries, which grows in the mountains of central and southern Europe.

3. Phleum phleoides (L.) H. Karsten [p. 673]

PURPLE-STEM CAT'S TAIL

Plants perennial; densely cespitose, not rhizomatous. **Culms** 6–90 cm, erect or nearly so, lower internodes not enlarged or bulbous. **Sheaths of the flag leaves** inflated; **auricles** not developed; **ligules** 1–2 mm, truncate, rounded, or obtuse; **blades** to 12(26) cm long, 1–6 mm wide, flat or convolute. **Panicles** 2–14(17) cm long, 4–10 mm wide, narrowly cylindrical, tapering distally; **branches** not adnate to the rachises. **Glumes** 2–3.7 mm, oblong-lanceolate, keels usually scabrous or shortly ciliate, sometimes smooth, apices not abruptly narrowed, awned, awns 0.3–0.5 mm; **lemmas** $^{2}/_{3}$–$^{3}/_{4}$ as long as the glumes, glabrous or puberulent, apices acute; **anthers** about 1.5 mm. $2n$ = 14, 28.

Phleum phleoides is native to dry grasslands from Europe through central Asia. It was collected, in 1990, beside railroad tracks in Coquitlam, British Columbia.

4. Phleum paniculatum Huds. [p. 674]

BRITISH TIMOTHY

Plants annual. **Culms** 5–30(55) cm, erect. **Sheaths of the flag leaves** not inflated; **auricles** not developed; **ligules** 5–7 mm, acute; **blades** to 19 cm long, 5–7 mm wide, flat or folded. **Panicles** 1–12 cm long, 3.5–7 mm wide, narrowly cylindrical, tapering somewhat distally. **Glumes** 1.5–2.5 mm, scabrous, inflated, keels sometimes slightly ciliate, margins ciliate distally, apices abruptly truncate, mucronate to awned, awns to 0.6 mm, lower glumes pilose on the margins; **lemmas** about $^{2}/_{3}$ as long as the glumes, pubescent; **anthers** 0.3–0.5 mm. $2n$ = 28.

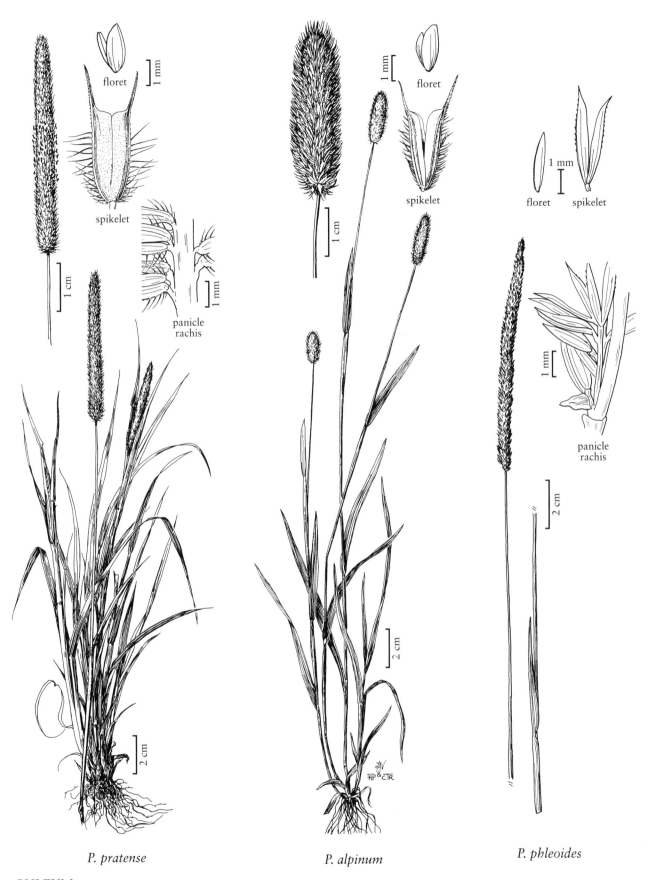

floret

spikelet

panicle
rachis

floret

spikelet

floret

spikelet

panicle
rachis

P. pratense

P. alpinum

P. phleoides

PHLEUM

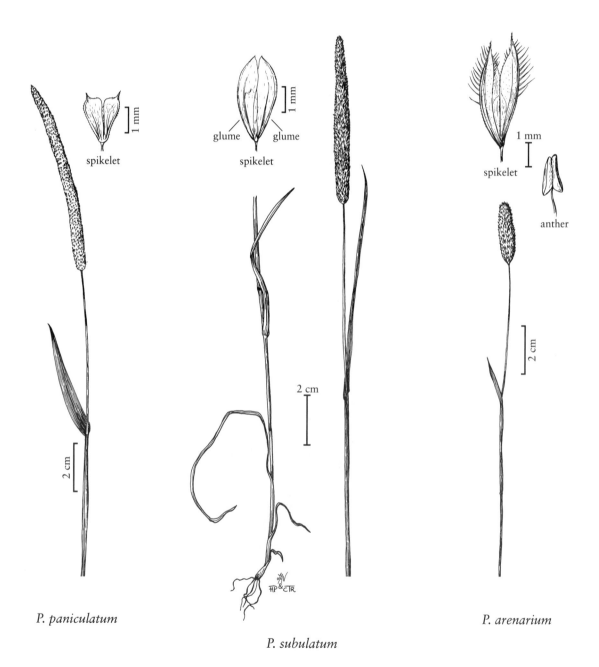

P. paniculatum

P. subulatum

P. arenarium

PHLEUM

Phleum paniculatum is native to dry habitats in southern and south-central Europe. In the *Flora* region, it is known only from old ballast dump records.

5. Phleum subulatum (Savi) Asch. & Graebn. [p. 674]
ITALIAN TIMOTHY

Plants annual. **Culms** 2–30(40) cm. **Sheaths** of the flag leaves not inflated; **auricles** not developed, leaf edges often wrinkled at the junction of the sheath and blade; **ligules** 4–7 mm, acute; **blades** 1.5–12 cm long, 1–4 mm wide, sometimes convolute. **Panicles** 0.8–11 cm long, 2–7 mm wide, narrowly cylindrical, parallel-sided. **Glumes** 2–4 mm, semi-elliptical in outline, keels usually glabrous, apices of the lower and upper glumes pointing towards each other; **lemmas** about $^2/_3$ as long as the glumes, glabrous, midveins scabridulous; **anthers** 1.5–2 mm. $2n = 14$.

Phleum subulatum is native to the grasslands of southern Europe. In the *Flora* region, it is known only from old ballast dump records.

6. Phleum arenarium L. [p. 674]
SAND TIMOTHY

Plants annual. **Culms** 2–35 cm. **Sheaths** of the flag leaves inflated; **auricles** not developed; **ligules** 1.5–3 mm, rounded to acute; **blades** 1–5(7) cm long, 1.5–5 mm wide, flat or folded. **Panicles** 0.5–5.5 cm long, 3–7 mm wide, ovoid to shortly cylindrical, cuneate at the base, widest at or above midlength. **Glumes** 2.2–4.4 mm, oblong-lanceolate, keels ciliate, tapered, apices parallel or divergent, awns 0.3–1 mm, usually parallel, sometimes divergent; **lemmas** about $^1/_3$ as long as the glumes, glabrous or pubescent; **anthers** 0.3–1.2 mm. $2n = 14$.

Phleum arenarium is native to maritime sands and shingles of southern and western Europe. It is known only from old ballast dump records in the *Flora* region.

14.32 GASTRIDIUM P. Beauv.

J.K. Wipff

Plants annual; tufted or the culms solitary. **Culms** 7–70 cm, erect to decumbent. **Leaves** cauline; **sheaths** open; **auricles** vestigial or absent; **ligules** membranous; **blades** flat. **Inflorescences** panicles, contracted or spikelike, more than 1 spikelet associated with each node; **branches** appressed to ascending. **Spikelets** pedicellate, laterally compressed, not U-shaped at the base, solitary, with 1 bisexual floret; **rachillas** prolonged as bristles beyond the base of the floret or absent; **disarticulation** above the glumes, beneath the floret. **Glumes** unequal, exceeding the florets, 1-veined, subcoriaceous and gibbous proximally, membranous distally, gradually narrowing to the apices, apices acuminate, unawned; **calluses** short, blunt, glabrous; **lemmas** membranous, pubescent or glabrous, 5-veined, lateral veins usually faint, not forming awns, apices more or less truncate and denticulate, unawned or awned, unawned and awned lemmas present in the same panicle, awns arising in the upper $^1/_3$, sometimes subterminal, shorter or longer than the lemmas, geniculate; **paleas** subequal to the lemmas, hyaline, 2-veined, bifid; **lodicules** 2, membranous, glabrous, entire, unlobed; **anthers** 3; **ovaries** glabrous. **Caryopses** shorter than the lemmas, concealed at maturity, slightly adhering to the lemmas and/or paleas. $x = 7$. Name from the Greek *gastridion*, 'small pouch', alluding to the gibbously swollen glumes.

Gastridium is a genus of two species that are native from Europe and North Africa east to Iran. They grow in grassy or disturbed sites. One species is established in the *Flora* region.

SELECTED REFERENCE Scholz, H. 1986. Bermerkungen zur Flora Griechenlands: *Gastridium phleoides* und *G. ventricosum* (Poaceae). Willdenowia 16:65–68.

1. Gastridium phleoides (Nees & Meyen) C.E. Hubb. [p. 677]

NIT GRASS

Culms 7–70 cm; nodes 3–5, glabrous, darkened. Sheaths 0.6–10.5 cm, smooth or scabridulous and minutely papillose; ligules 1–7 mm, veined, erose to lacerate, scabridulous and minutely papillose; blades (0.8) 1.5–20 cm long, (0.4)1.5–6 mm wide, antrorsely scabridulous, glabrous. Panicles (1.1)2–16.5 cm long, (1.5)3–37 mm wide; rachises glabrous; branches to 3.4 cm, appressed to ascending, antrorsely scabridulous, pedicels with enlarged apices. Rachilla prolongations 0.3–0.6 mm, densely pubescent. Glumes scabridulous and papillose proximally, scabriduous to scabrous distally, keels scabrous, margins hyaline; lower glumes 3–7 mm; upper glumes 2.7–5.5 mm; lemmas 1–1.5 mm, appressed-pubescent, unawned or awned, awns to 6 mm, twisted; paleas 1–1.3 mm, glabrous; anthers 0.5–0.9 mm, yellow-orange or purple. Caryopses 0.8–1 mm long, 0.4–0.5 mm wide, glabrous. $2n = 14$.

Native to southwest Asia and northeast Africa, *Gastridium phleoides* now grows in Australia, South Africa, North America, and South America, in dry, often disturbed areas. In the *Flora* region, it is established in Oregon and California; it has also been collected, but may not be established, in southwestern British Columbia, Arizona, western Texas (county unknown; US 843557, Texas, 1884, *Nealley s.n.*), Massachusetts, and South Carolina.

In North America, *Gastridium phleoides* has been mistakenly placed in *G. ventricosum* (Gouan) Schinz & Thell. It differs from that species in having densely pubescent lemmas and well-developed, densely pubescent rachilla prolongations.

14.33 ARCTAGROSTIS Griseb.

Susan G. Aiken

Plants perennial; rhizomatous. **Culms** 10–150 cm, erect, glabrous. **Sheaths** usually closed for up to $^1/_5$ their length; **auricles** absent; **ligules** membranous, obtuse, erose or lacerate; **blades** usually flat, glabrous, scabrous. **Inflorescences** terminal panicles, loose and open to narrow and contracted; **branches** 0.5–27 cm, scabridulous. **Spikelets** 3–6.5 mm, pedicellate, laterally compressed, usually with 1 floret, rarely with 2; **rachillas** hairy or scabrous, sometimes prolonged beyond the base of the single floret or terminating in a well-formed or vestigial floret; **disarticulation** above the glumes, beneath the floret(s). **Glumes** usually shorter than the lemmas, upper glumes sometimes equaling the lemma(s), ovate to lanceolate, glabrous, sometimes scabridulous, 1–3-veined, acute to obtuse, unawned; **calluses** weakly developed, glabrous; **lemmas** scabrous or puberulent, 3–5-veined, lateral veins sometimes obscure, not reaching the apices, apices obtuse to acute, sometimes mucronate, unawned; **paleas** about as long as the lemmas, folded along 1 prominent vein, the second vein faint; **lodicules** 2, glabrous; **anthers** 3; **ovaries** glabrous. **Caryopses** 1.7–3 mm, shorter than the lemmas, concealed at maturity; **hila** $^1/_4$–$^1/_3$ the length of the caryopses. $x = 7$. Name from the Greek *arktos*, 'north', a reference to the distribution of the genus, and *Agrostis*, from its similarity to that genus.

Arctagrostis is a circumpolar genus with one species and two subspecies.

SELECTED REFERENCES Aiken, S.G., L.L. Consaul, and M.J. Dallwitz. 1995 on. Grasses of the Canadian Arctic Archipelago: Descriptions, illustrations, identification and information retrieval. http://www.mun.ca/biology/delta/arcticf/poa/index.htm; Aiken, S.G. and L.P. Lefkovitch. 1990. *Arctagrostis* (Poaceae) in North America and Greenland. Canad. J. Bot. 68:2422–2432; Aiken, S.G., L.P. Lefkovitch, S.E. Gardiner, and W.W. Mitchell. 1994. Evidence against the existence of varieties in *Arctagostis latifolia* ssp. *arundinacea* (Poaceae). Canad. J. Bot. 72:1039–1050; Mitchell, W.W. 1992. Cytogeographic races of *Arctagrostis latifolia* (Poaceae) in Alaska. Canad. J. Bot. 70:80–83.

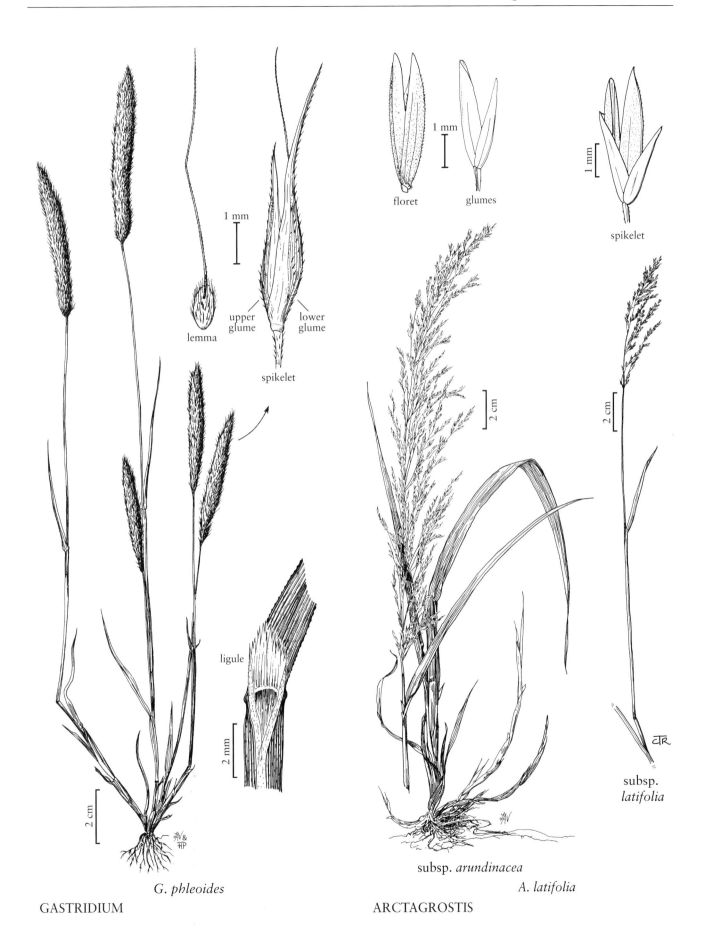

floret

glumes

1 mm

spikelet

1 mm

lemma

upper
glume

lower
glume

spikelet

1 mm

ligule

2 mm

2 cm

2 cm

2 cm

subsp.
latifolia

subsp. *arundinacea*

G. phleoides

A. latifolia

GASTRIDIUM

ARCTAGROSTIS

1. Arctagrostis latifolia (R. Br.) Griseb. [p. 677]
POLARGRASS

Rhizomes elongate or compact, 1.5–3 mm thick, with 8–35 mm scales, glabrous. **Culms** 10–150 cm tall, 1.2–7 mm thick. **Ligules** 2–7(15) mm; **blades** 1–36 cm long, 1.5–15 mm wide. **Panicles** 2.5–35(44) cm long, 0.7–10 (14+) cm wide, narrow and stiffly erect to spreading and diffuse; **branches** 0.5–14+ cm, with 3–140 spikelets. **Spikelets** 3–6.5 mm. **Lower glumes** 1.8–4.5 mm; **upper glumes** 2.3–5.5 mm; **lemmas** 2.4–6.5 mm; **anthers** 1.3–4 mm.

Arctagrostis latifolia grows in a wide range of habitats, from sea level to 2000 m, and exhibits considerable phenotypic plasticity. Infection by nematodes leads to the formation of enlarged dark purple ovaries. Such forms have been annotated as "forma *prolifera*" or "forma *vivipara*", but these names have no taxonomic validity.

Arctagrostis latifolia subsp. *arundinacea*, which grows in the low arctic of Alaska and northern Canada, is usually larger than subsp. *latifolia*, which grows on the islands of the arctic archipelago. The two subspecies are similar in their morphology in a narrow geographical zone, mainly because of phenotypic plasticity. Agronomic studies appear to have utilized only subsp. *arundinacea*.

1. Panicles (4)10–35(44) cm long; longest branches (1.4)3–27 cm long, with (5)50–140 spikelets per branch; secondary branches usually spreading; uppermost sheaths shorter than the blades; spikelets 3–5.2 mm long; glumes unequal, the upper glumes varying from shorter than to longer than the lemmas . . . . subsp. *arundinacea*
1. Panicles 2.5–10(17) cm long; longest branches 0.5–4(6.5) cm long, with 3–15(40) spikelets per branch; secondary branches usually appressed, spreading in warmer habitats; uppermost sheaths usually longer than the blades; spikelets (3)4–6.5 mm long; glumes subequal, about 1 mm shorter than the lemmas . . . . subsp. *latifolia*

Arctagrostis latifolia subsp. **arundinacea** (Trin.) Tzvelev [p. 677]
REED POLARGRASS

Culms (24)50–150 cm; **uppermost nodes** usually towards the middle of the culm. **Uppermost sheaths** shorter than the blades; **blades** 4–36 cm long, 2.4–15 mm wide. **Panicles** (4)10–35(44) cm long, 0.7–10(14+) cm wide; **longest branches** (1.4)3–27 cm, with (5)50–140 spikelets per branch; **secondary branches** usually spreading. **Spikelets** 3–5.2 mm. **Glumes** unequal; **lower glumes** 2.3–3.6 mm, 1(2)-veined; **upper glumes** 3–4.8 mm, shorter than to longer than the lemmas, 3-veined; **lemmas** 2.4–4.8 mm; **anthers** 1.3–3.5 mm. $2n = 28, 42$.

Arctagrostis latifolia subsp. *arundinacea* is the predominant subspecies from Alaska east to the western Northwest Territories, northern British Columbia, and Alberta. Cultivars for hay and revegetation mixes have been developed.

Arctagrostis latifolia subsp. **latifolia** [p. 677]
POLARGRASS, ARCTAGROSTIS À LARGES FEUILLES

Culms (10)15–50(95) cm; **uppermost nodes** usually in the lowest ⅓ of the culm. **Uppermost sheaths** usually longer than the blades; **blades** 1–16 cm long, 1.5–9 mm wide. **Panicles** 2.5–10(17) cm long, 0.7–3 cm wide; **longest branches** 0.5–4(6.5) cm, with 3–15(40) spikelets per branch; **secondary branches** usually appressed, spreading in warmer habitats. **Spikelets** (3)4–6.5 mm. **Glumes** subequal; **lower glumes** 1.8–4.5 mm, 1(3)-veined; **upper glumes** 2.3–5.5 mm, usually about 1 mm shorter than the lemmas, (1)3-veined; **lemmas** 3–6.5 mm; **anthers** 1.8–4 mm. $2n = 56$.

Arctagrostis latifolia subsp. *latifolia* is the common subspecies from the arctic coast of Alaska to Nunavut, northern Quebec, and northern Greenland.

14.34 SESLERIA Scop.

Mary E. Barkworth

Plants perennial; more or less cespitose, sometimes rhizomatous, rarely stoloniferous. **Culms** 3–80(100) cm. **Leaves** mostly basal; **sheaths** closed for most of their length, often shredding when mature; **auricles** absent; **ligules** 0.1–1 mm, membranous, truncate to obtuse, usually ciliolate; **blades** flat, plicate, or involute. **Inflorescences** single terminal panicles, cylindrical to globose, usually dense, sometimes subtended by ovate to round, scarious or hyaline, erose bracts, sometimes by 1–2 pubescent scales, sometimes without subtending scales or bracts, more than 1 spikelet associated with each node; **branches** shorter than 10 mm. **Spikelets** 3–9 mm, laterally compressed, with 2–5 florets, uppermost florets reduced; **rachillas** usually glabrous, rarely sparsely pilose, sometimes prolonged beyond the base of the distal floret, sometimes terminating in a reduced floret; **disarticulation** above the glumes, beneath the florets. **Glumes** unequal, usually shorter than the lowest lemmas, scarious to membranous, 1–3-veined, apices awned, awns glabrous; **calluses** very short, broadly rounded, glabrous or with scattered hairs; **lemmas** membranous, 5–7-veined, 3–5 veins usually extending into awnlike teeth, central teeth longer than the lateral teeth; **paleas** equaling or exceeding the lemmas; **lodicules** 2, free, glabrous, toothed; **anthers** 3; **ovaries** pubescent. **Caryopses** 1.5–3 mm; **embryos** $^1/_4$–$^1/_3$ the length of the caryopses. $x = 7$. Named for Lionardo Sesler (?–1785), a Venetian naturalist and director of a botanical garden at Santa Maria di Sala.

Sesleria has approximately 30 species. Most abundant in the Balkans, it extends from Iceland, Great Britain, and southern Sweden through central and southern Europe into northwest Asia. Four species are cultivated in North America.

SELECTED REFERENCES **Conert, H.J.** 1992. *Sesleria*. Pp. 473–480 *in* G. Hegi. Illustrierte Flora von Mitteleuropa, ed. 3. Band I, Teil 3, Lieferung 6 (pp. 401–480). Verlag Paul Parey, Berlin and Hamburg, Germany; **Conert, H.J.** 1994. *Sesleria* (continued). Pp. 481–486 *in* G. Hegi. Illustrierte Flora von Mitteleuropa, ed. 3. Band I, Teil 3, Lieferung 7 (pp. 481–560). Blackwell Wissenschafts-Verlag, Berlin, Germany; **Deyl, M.** 1980. *Sesleria* Scop. Pp. 173–177 *in* T.G. Tutin, V.H. Heywood, N.A. Burges, D.M. Moore, D.H. Valentine, S.M. Walters and D.A. Webb (eds.). Flora Europaea, vol. 5. Cambridge University Press, Cambridge, England. 439 pp.; **Pignatti, S.** 1982. *Sesleria*. Pp. 505–509 *in* S. Pignatti, Flora d'Italia, vol. 3. Edagricole, Bologna, Italy. 780 pp.

1. Panicles cylindrical, 4.5–10 cm long, 4–7 mm wide . . . . 1. *S. autumnalis*
1. Panicles ovoid, spherical, or cylindrical, 0.9–4(8) cm long, 5–15 mm wide.
  2. Glumes 5–6.5 mm long, lanceolate; anthers about 2.2 mm long; lemmas glabrous . . . . 4. *S. nitida*
  2. Glumes 3–5 mm long, ovate to ovate-lanceolate; anthers 2.3–4 mm long; lemmas pubescent.
    3. Glumes 3–4 mm long, ovate-lanceolate, coriaceous; anthers about 4 mm long . . . . 2. *S. heufleriana*
    3. Glumes 4–5 mm long, ovate, hyaline; anthers 2.3–3.2 mm long . . . . 3. *S. caerulea*

1. Sesleria autumnalis (Scop.) F.W. Schultz [p. 680]
AUTUMN MOORGRASS

Plants cespitose, rhizomatous. **Culms** 25–70 cm, erect, glabrous; **nodes** 4–5. **Basal leaves** distichous, keeled, folded; **sheaths** glabrous, lower sheaths scabrous, eventually disintegrating into wavy fibers, upper sheaths smooth; **ligules of lower leaves** to 1 mm, shortly ciliate; **blades** 25–35(40) cm long, 2–4 mm wide; **blades of cauline leaves** usually flat, rarely folded; **blades of flag leaves** 4–8(12) cm, diverging at about midculm. **Panicles** 4.5–10 cm long, 4–7 mm wide, cylindrical, lax, grayish to whitish, branches to 1 cm; **pedicels** with 1–2 hyaline, scalelike bracts at the base. **Spikelets** 5–7 mm long, wedge-shaped, with 2(3) florets. **Glumes** subequal, 4–5 mm, usually extending above the florets, narrowly lanceolate, hyaline, glabrous, 1-veined, acute, midveins extending into 1–2(4) mm awns; **lemmas** 3–5.5 mm, 3(7)-veined, broadly lanceolate to ovate, hyaline, bases shortly and finely pubescent on the veins and margins, glabrous between the veins, midveins

forming (0.5)1–1.5 mm teeth, marginal veins forming 0.5–1 mm teeth; **paleas** 4–5 mm, finely ciliate on the keels, apices rounded, notched, veins forming short awn tips; **anthers** 2.4–2.8 mm, white. **Caryopses** 2–3 mm, elliptical in outline, shortly and stiffly pubescent distally. $2n = 28$.

Sesleria autumnalis is native from northern and eastern Italy to central Albania, where it grows primarily on limestone-derived soils in deciduous, often littoral forests, and is frequently co-dominant with *Ostrya virginiana*. It sometimes forms intermediates with *S. nitida* (Deyl 1980). It is now popular in the *Flora* region as an ornamental, being an attractive, low-maintenance plant that provides excellent ground cover. It can be grown in open areas, and is drought- and shade-tolerant once established.

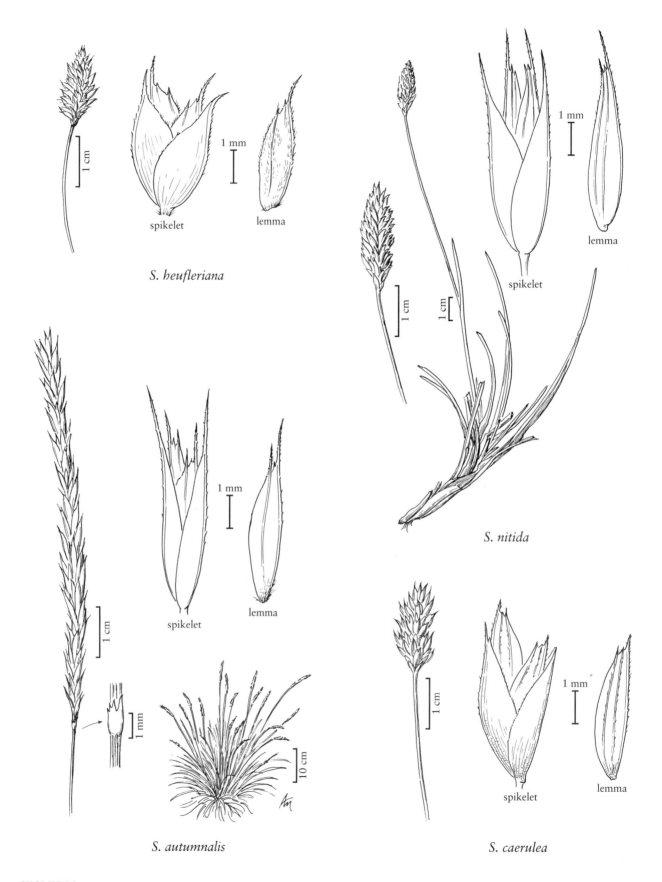

S. heufleriana

spikelet   lemma

S. nitida

spikelet   lemma

spikelet   lemma

S. autumnalis

S. caerulea

spikelet   lemma

SESLERIA

2. Sesleria heufleriana Schur [p. 680]
BLUE-GREEN MOORGRASS

Plants cespitose, not rhizomatous, rarely stoloniferous, green when young, becoming glaucous and blue in early summer. **Culms** 25–80(100) cm; **exposed nodes** 1–2. **Leaves** mostly basal; **sheaths** glabrous; **ligules** 0.2–0.5 mm, ciliate; **blades of cauline leaves** 0.8–5 cm long, 2–4(5.5) mm wide, acute or acuminate; **blades of flag leaves** 1–2.1 cm. **Panicles** (1)1.5–4(8) cm long, 7–11 mm wide, dense, ovoid to cylindrical. **Spikelets** 3–6 mm, with 2–4 florets, glaucous. **Glumes** 3–4 mm, ovate-lanceolate, coriaceous, sparsely ciliate on the veins and margins, 1-veined, narrowing to 0.5–2.5 mm awnlike apices; **lemmas** 3–4 mm, ovate to elliptic, pubescent, apices awned and toothed, central veins terminating in a 1–4 mm awn, lateral veins forming teeth shorter than 1 mm; **paleas** 3–5 mm, hairy, 2-awned, awns 0.2–1.5 mm; **anthers** about 4 mm. $2n$ = unknown.

Sesleria heufleriana is native to east-central Europe, where it grows in woods and on rocks, usually over calcareous substrates. It is grown as an ornamental in the *Flora* region.

3. Sesleria caerulea (L.) Ard. [p. 680]
BLUE MOORGRASS

Plants densely cespitose, shortly rhizomatous, glaucous. **Culms** to 40(60) cm, erect, stiff, strongly grooved, glabrous; **nodes** 4–6. **Leaves** mostly basal; **sheaths** grooved, glabrous, keeled distally, basal sheaths persistent, pubescent, disintegrating at maturity; **ligules** 0.3–0.5 mm, erose, often finely ciliate; **blades of innovations** and **lower cauline leaves** to 40 cm long, 2–4 mm wide, flat or slightly involute, divergent, apices rounded and slightly hooded, adaxial surfaces glaucous, particularly when young, midveins prominent, particularly basally, margins scabrous; **blades of flag leaves** 0.5–2.5 cm long, 2–4 mm wide. **Panicles** (0.9)1.2–1.4(2.4) cm long, 5–8(12) mm wide, dense, usually spherical or ovoid, usually purplish, branches

with 1–2 spikelets; **pedicels** 0.5–1 mm, scabrous; **bracts** 2, 2–4 mm long, equally broad, erose. **Spikelets** 4–6 mm, with (2)3 florets, usually purplish, rarely stramineous. **Glumes** 4–5 mm, ovate, hyaline, mostly glabrous, margins ciliate, particularly distally, midveins with stiff hairs, apices acute, awns to 1.5 mm; **lemmas** 3.5–5 mm, hyaline, ovate, midveins and marginal veins hairy at least distally, hairs about 0.1 mm, appressed hairy between the veins, midveins forming 0.2–2 mm teeth, lateral veins forming 0.1–0.4 mm teeth; **paleas** as long as the lemmas, margins and keels hairy distally, hairs about 0.1 mm, veins forming awnlike apices to 2.3 mm; **anthers** about 2.3–3.2 mm. **Caryopses** about 2 mm, obovoid, hairy distally. $2n$ = 28.

Sesleria caerulea is native to Europe, ranging from central Sweden to northwestern Russia and central Bulgaria. It usually grows in moist to wet, calcareous pastures and bogs. It is grown as an ornamental in the *Flora* region.

4. Sesleria nitida Ten. [p. 680]
GRAY MOORGRASS

Plants cespitose, not rhizomatous. **Culms** 20–70 cm, ascending. **Sheaths** glabrous or sparsely pubescent; **ligules** 0.6–2 mm, truncate to acute, ciliolate; **blades of cauline leaves** 3.5–7.5 cm long, 2–3 mm wide, glabrous, glaucous; **blades of flag leaves** 3.5–7.5 cm. **Panicles** 2–3.5 cm long, 9–15 mm wide, ovoid. **Spikelets** 5.6–6.5 mm, with 2–3 florets. **Glumes** 5–6.5 mm, lanceolate, veins hairy, apices 3–5-awned, central awns 1–2 mm, lateral awns about 0.5 mm; **lemmas** 5–6 mm, glabrous, 3–5-toothed, central teeth 1–1.5 mm, slightly divergent, lateral teeth 0.5–1 mm; **paleas** 4.5–5 mm, ciliate on the keels; **anthers** about 2.2 mm. $2n$ = unknown.

Sesleria nitida is native to the mountains of central Italy and Sicily, where it grows in broken, rocky, calcareous habitats, sometimes forming intermediates with *S. autumnalis* (Deyl 1980). It is grown as an ornamental in the *Flora* region.

14.35 DESMAZERIA Dumort.

Gordon C. Tucker

Plants annual. **Culms** to 60 cm, procumbent to erect, sparingly branched at the base. **Leaves** basal and cauline; **sheaths** open, glabrous; **auricles** absent; **ligules** longer than wide, acute; **blades** linear, usually flat, sometimes convolute when dry, glabrous. **Inflorescences** terminal, racemes or panicles, usually with 1 branch per node; **branches** stiff, not secund, pedicels 0.5–3 mm. **Spikelets** subsessile, tangential to the rachises, lanceolate to ovate, laterally compressed, with 4–25 florets, distal florets reduced; **disarticulation** above the glumes, beneath the florets; **rachillas** not prolonged beyond the base of the distal floret. **Glumes** unequal to subequal, shorter than or subequal to the adjacent lemmas, 1–5-veined, unawned; **calluses** blunt, rounded, glabrous; **lemmas** narrowly elliptic, coriaceous at maturity, inconspicuously 5-veined, glabrous, sometimes scabridulous towards the apices, apices acute to obtuse, sometimes bifid, often mucronate, unawned; **paleas** about as long as the lemmas, 2-veined; **lodicules** 2, free, lanceolate; **anthers** 3, only slightly exserted at anthesis; **ovaries** glabrous. **Caryopses** shorter than the lemmas, concealed at maturity, ellipsoid-oblong, dorsally flattened, falling with the paleas; **hila** about $^1/_{10}$ as long as the caryopses, ovate. $x = 7$. Named for Jean Baptiste Henri Joseph Desmazières (1786–1862), a French merchant, amateur botanist, and horticulturalist.

Desmazeria has six or seven species, all of which are native around the Mediterranean. There are two species in the *Flora* region, one established as a weed and one introduced as an ornamental.

One or two genera have sometimes been segregated from *Desmazeria*; current opinion favors the treatment presented here.

SELECTED REFERENCES **Hitchcock, A.S.** 1951. Manual of the Grasses of the United States, ed. 2, rev. A. Chase. U.S.D.A. Miscellaneous Publication No. 200. U.S. Government Printing Office, Washington, D.C., U.S.A. 1051 pp.; **Stace, C.A.** 1978. Notes on *Cutandia* and related genera. Bot. J. Linn. Soc. 76:350–352.

1. Lemmas 2–3 mm long, rounded on the back or weakly keeled distally; anthers 0.4–0.6 mm long; inflorescences usually panicles, sometimes racemes . . . . .1. *D. rigida*
1. Lemmas 3.5–4.5 mm long, strongly keeled; anthers 0.8–1.4 mm long; inflorescences racemes . . . . . 2. *D. sicula*

1. Desmazeria rigida (L.) Tutin [p. 683]
FERN GRASS

Culms to 60 cm, procumbent to erect, glabrous. **Sheaths** glabrous, upper margins membranous, continuous with the sides of the ligules; **ligules** 1.5–4 mm, lacerate; **blades** 2–8(12) cm long, 1–3(4) mm wide. **Inflorescences** usually panicles, sometimes racemes, 1–12(18) cm long, 12–30 mm wide; **branches** stiff, somewhat divaricate at maturity; **pedicels** 0.5–3 mm, appressed to divaricate. **Spikelets** 4–10 mm, narrowly ovate, with 5–12 florets; **rachillas** puberulent, hairs stiff. **Glumes** usually glabrous, more or less keeled, acute; **lower glumes** 1.3–2 mm, (1)3–veined; **upper glumes** 1.5–2.3 mm, 3-veined; **lemmas** 2–3 mm, rounded on the back or weakly keeled distally, glabrous, acute to obtuse, often shortly mucronate; **anthers** 0.4–0.6 mm. $2n = 14$.

Desmazeria rigida is native to Europe, and appears to have no distinctive habitat preferences. In the *Flora* region, it is now established as a weed in disturbed sites such as roadsides, ditches, and the edges of fields. It is probably more widespread than indicated on the map, because herbarium records of weed distributions are often poor.

2. Desmazeria sicula (Jacq.) Dumort. [p. 683]
SPIKE GRASS

Culms 8–25(35) cm, erect or geniculate, glabrous. **Sheaths** glabrous, upper margins membranous, continuous with the sides of the ligules; **ligules** 1.4–2.5 mm, erose; **blades** 2–13 cm long, 1–3 mm wide, blades of basal leaves usually folded, cauline blades flat. **Inflorescences** racemes, 1.5–6 cm, not stiff, spikelets borne in 2 rows; **pedicels** 0.5–1 mm, appressed. **Spikelets** 8–15(22) mm, ovate, crowded, with (5)12–25 florets; **rachillas** with capitate hairs. **Glumes** glabrous, usually smooth, keels sometimes scabridulous distally, apices obtuse to subacute; **lower glumes** 2.5–3 mm, 3-veined; **upper glumes** 3–4 mm, 3–5-veined; **lemmas** 3.5–4.5 mm, more or less cucullate, puberulent with capitate hairs, strongly keeled, keels sometimes scabridulous distally, apices subacute to obtuse, not or scarcely mucronate; **anthers** 0.8–1.4 mm. $2n = 14$.

Desmazeria sicula used to be cultivated as an ornamental in the *Flora* region (Hitchcock 1951), but it is not included in contemporary treatments of ornamental grasses.

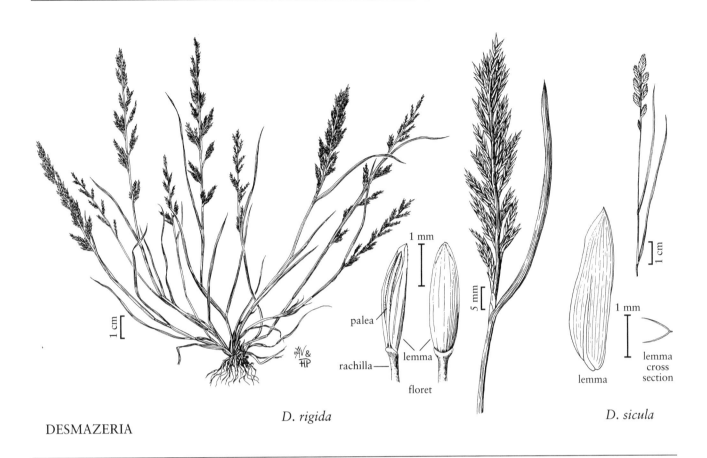

palea

rachilla

lemma

floret

1 mm

5 mm

D. rigida

DESMAZERIA

lemma

1 mm

lemma cross section

1 cm

D. sicula

14.36 VENTENATA Koeler

William J. Crins

Plants annual; tufted. **Culms** 10–75 cm, erect, puberulent below the nodes; **nodes** glabrous. **Leaves** mostly on the lower $^1/_2$ of the culm; **sheaths** open, glabrous or sparsely pubescent; **auricles** absent; **ligules** hyaline, acute or obtuse, usually lacerate; **blades** flat initially, involute with age. **Inflorescences** open or contracted panicles, with spikelets borne near the ends of the branches on clavate pedicels. **Spikelets** laterally compressed, with 2–10 florets; **disarticulation** above the first floret and between the distal florets; **rachillas** sometimes prolonged beyond the base of the distal floret, sometimes terminating in a reduced floret. **Glumes** unequal, lanceolate, hispidulous, similar in texture to the lemmas, margins scarious, apices acuminate; **lower glumes** 3–7-veined; **upper glumes** 3–9-veined; **calluses** of the lower florets shorter than those of the upper florets, sparsely hairy, calluses of the distal florets with a dense tuft of white hairs; **lemmas** lanceolate, chartaceous, 5-veined, margins scarious, apices entire or bifid, awned or unawned; **lowest lemma** within a spikelet awned or unawned, awns straight, terminal; **distal lemmas** within a spikelet awned, awns dorsal, geniculate; **paleas** shorter than the lemmas, membranous, keels ciliate distally; **lodicules** 2, membranous, glabrous, toothed or not toothed; **anthers** 3; **ovaries** glabrous. **Caryopses** shorter than the lemmas, concealed at maturity, glabrous. x = 7. Named for Etienne Pierre Ventenat (1757–1808), a French clergyman, librarian, and botanist.

Ventenata is native from central and southern Europe and north Africa to Iran. It has five species, all of which grow in dry, open habitats. Only one species is established in the *Flora* region.

SELECTED REFERENCES **Chambers, K.L.** 1985. Pitfalls in identifying *Ventenata dubia* (Poaceae). Madroño 32:120–121; **Old, R.R.** and **R.H. Callihan.** 1986. Distribution of *Ventenata dubia* in Idaho. Idaho Weed Control Rep. 1986:153.

1. **Ventenata dubia** (Leers) Coss. [p. 684]

<small>VENTENATA, NORTH-AFRICA GRASS</small>

Culms 15–75 cm, puberulent below the nodes; **nodes** 3–4, exposed, purple-black. **Ligules** 1–8 mm; **blades** 2–7(12) cm long, 0.8–2.5 mm wide. **Panicles** (7)15–20 cm, open, pyramidal, lower nodes with 2–5 branches; **branches** 1.5–7 cm, bearing 1–5 spikelets distally; **pedicels** 2–18 mm. **Spikelets** 9–15 mm, with 2–3 florets, the lowest usually staminate, the remainder bisexual; **rachillas** usually glabrous, sometimes pubescent abaxially, internodes mostly 1–1.5 mm, prolongation to 2 mm, empty or with a reduced floret. **Lower glumes** 4.5–6 mm; **upper glumes** 6–8 mm; **lemmas** 5–7.5 mm, awns of the lowest lemma within a spikelet to 4 mm, straight, distal lemmas within a spikelet bifid, teeth 1–2 mm, awns 10–16 mm, geniculate; **paleas** 4–5 mm; **anthers** 1–2 mm. **Caryopses** about 3 mm. $2n = 14$.

The first North American collection of *Ventenata dubia* was made in Washington in 1952. It is now established in crop and pasture lands of eastern Washington and western Idaho (Old and Callihan 1986), and has been found, but has not necessarily become established, at scattered locations elsewhere. Mature specimens can be confusing because the first, straight-awned floret remains after the distal, bisexual florets have disarticulated (Chambers 1985).

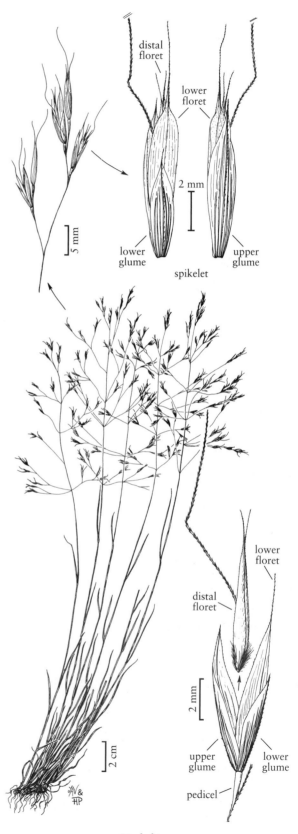

V. dubia

VENTENATA

14.37 CYNOSURUS L.

Sandy Long

Plants annual or perennial; sometimes rhizomatous. **Culms** 1.5–90 cm, erect. **Cauline leaves** 1–3; **sheaths** open to the base; **auricles** absent; **ligules** truncate, entire, erose, or ciliolate; **blades** flat. **Inflorescences** terminal panicles, condensed, often spikelike, linear to almost globose, more or less unilateral; **branches** and **pedicels** stiff, straight, smooth and glabrous or almost so, sometimes slightly scabridulous, scabrules/hairs to 0.1 mm. **Spikelets** dimorphic, usually paired, subsessile to shortly pedicellate, laterally compressed, proximal spikelet of each pair sterile, almost completely concealing the fertile spikelet, distal spikelet on each branch sometimes solitary; **disarticulation** above the glumes, beneath the florets. **Sterile spikelets** persistent, with 6–18 florets, florets reduced to narrow, linear-lanceolate lemmas, sometimes awned, awns terminal; **glumes** narrow, linear. **Fertile spikelets** adaxial to the sterile spikelets, with 1–5 florets; **rachillas** glabrous, prolonged beyond the base of the distal floret; **glumes** 2, subequal and shorter than the spikelets, thin, lanceolate, 1-veined, acute, sometimes awned; **calluses** short, blunt, glabrous; **lemmas** glabrous or pubescent, 5-veined, acute or bidentate, unawned to conspicuously awned, awns terminal; **paleas** about as long as the lemmas, bifid; **lodicules** 2, free, glabrous, ovate, bilobed; **anthers** 3; **ovaries** broadly ellipsoid, glabrous; **styles** separate. **Caryopses** oblong-ellipsoid, subterete, slightly dorsally compressed, sometimes adherent to the paleas; **hila** $^1/_5$–$^1/_2$ as long as the caryopses, oblong to linear. $x = 7$. Name from the Greek, *kynos*, 'dog', and *oura*, 'tail', referring to the shape of the panicle.

Cynosurus is a genus of eight species that grow in open, grassy, often weedy habitats. It is native around the Mediterranean and in western Asia. The affinities of the genus are obscure. Two species are established in the *Flora* region.

SELECTED REFERENCES Ennos, R.A. 1985. The mating system and genetic structure in a perennial grass, *Cynosurus cristatus* L. Heredity 55:121–126; Jirásek, V. and J. Chrtek. 1964. Zur frage des taxonomischen Wertes der Gattung *Cynosurus* L. Novit. Bot. Delect. Seminum Horti Bot. Univ. Carol. Prag. 20:23–27; Lodge, R.W. 1959. Biological flora of the British Isles: *Cynosurus cristatus* L. J. Ecol. 47:511–518.

1. Plants perennial; panicles linear; fertile lemmas unawned or with awns shorter than 3 mm . . . . . 1. *C. cristatus*
1. Plants annual; panicles ovoid to almost globose; fertile lemmas with awns 5–25 mm long . . . . . 2. *C. echinatus*

1. Cynosurus cristatus L. [p. 686]
CRESTED DOGTAIL, CYNOSURE ACCRÊTÉ, CRÉTELLE DES PRÉS

Plants perennial; cespitose, not rhizomatous. **Culms** (5)15–75 (90) cm. **Sheaths** smooth, glabrous; **ligules** 0.5–2.5 mm, truncate, erose or ciliolate; **blades** 3–15(19) cm long, 0.5–2(4.3) mm wide, glabrous or pubescent. **Panicles** (1) 3.5–14 cm long, 0.4–1 cm wide, linear, spikelike, unilateral. **Spikelets** 3–7 mm, subsessile or shortly pedicellate, pedicels to 1 mm. **Sterile spikelets** strongly laterally compressed, with 6–11(18) florets; **glumes** and **lemmas** similar, linear-lanceolate, keeled, keels ciliate, apices acuminate to awned, awns to 1 mm. **Fertile spikelets** with 2–5 florets, glumes and lemmas dissimilar; **glumes** 2.8–5.1 mm long, 0.6–0.9 mm wide,

1-veined, laterally compressed, hyaline, keeled, acute; **rachilla internodes** 0.4–0.6 mm; **lemmas** 3–4.5 mm long, 0.6–1.1 mm wide, dorsally compressed, membranous to subcoriaceous, not keeled, margins hyaline, ciliolate, apices obtuse to acute, unawned or awned, awns to 3 mm; **anthers** 1.8–3 mm. $2n = 14$.

Cynosurus cristatus is a European native that is now established in North America. It grows in a wide range of soils in dry or damp habitats. In Europe it is used for fodder and pasture, especially for sheep, but in North America it is regarded as a weedy species. It is self-incompatible.

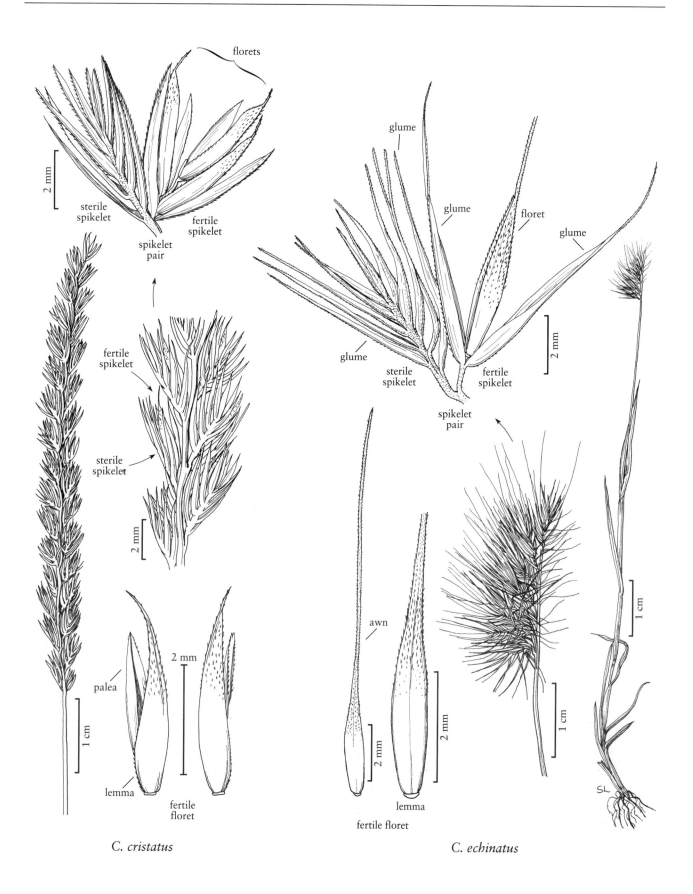

florets

sterile
spikelet

fertile
spikelet

spikelet
pair

2 mm

glume

glume

glume

floret

glume

sterile
spikelet

fertile
spikelet

spikelet
pair

2 mm

fertile
spikelet

sterile
spikelet

2 mm

palea

2 mm

lemma

fertile
floret

awn

lemma

fertile floret

2 mm

2 mm

1 cm

1 cm

1 cm

1 cm

C. cristatus

C. echinatus

SL

CYNOSURUS

2. Cynosurus echinatus L. [p. 686]

BRISTLY DOGTAIL

Plants annual; tufted. **Culms** 9–70 cm, clustered or solitary. **Sheaths** smooth, glabrous; **ligules** 2.5–5(10) mm, obtuse, entire; **blades** 3–13.5(23) cm long, 2.5–14 mm wide, scabrous. **Panicles** 1–4(8) cm long, 0.7–2 cm wide, ovoid to almost globose, unilateral. **Spikelets** 7–14 mm, subsessile or shortly pedicellate; **pedicels** to 1.6 mm. **Sterile spikelets** with 6–18 florets; **glumes** and **lemmas** similar, subulate to linear-lanceolate, glabrous, sometimes scabridulous, awned, awns to 8 mm. **Fertile spikelets** with 1–5 florets, glumes and lemmas dissimilar; **glumes** 5.5–12 mm long, 0.4–0.9 mm wide, narrowly lanceolate, laterally compressed, hyaline, glabrous, 1-veined, keeled, awned, awns 0.5–2.2 mm; **rachilla internodes** 0.9–1.3 mm; **lemmas** 4–7 mm, slightly dorsally compressed, chartaceous or coriaceous, not keeled, lower ¹/₂ smooth, glabrous, distal ¹/₂ villous or scabrous, apices hyaline, ciliolate, entire or bidentate, awned, awns 5–18(25) mm; **anthers** 1–4 mm. $2n = 14$.

Cynosurus echinatus is native to southern Europe. It is now established in dry, open habitats in North America, South America, and Australia.

14.38 PARAPHOLIS C.E. Hubb.

Thomas Worley

Plants annual; tufted. **Culms** 2–45 cm, erect to prostrate, sometimes branched above the base; **nodes** glabrous, usually purple; **internodes** hollow. **Leaves** not basally concentrated; **sheaths** not keeled; **auricles** sometimes present; **ligules** 0.3–2.3 mm, membranous, truncate; **blades** 0.3–3.5 mm wide, linear, flat to convolute. **Inflorescences** terminal and axillary spikes, often curved, with solitary spikelets sunk into the rachis; **disarticulation** in the rachis below each spikelet. **Spikelets** sessile, cylindrical, straight to strongly curved, tangential to the rachises, with 1 floret; **rachillas** sometimes prolonged beyond the base of the distal floret. **Glumes** 2, subequal, lying side by side, usually exceeding or sometimes slightly shorter than the floret and covering the rachis cavities, sometimes asymmetric and winged, coriaceous, stiff, veins conspicuous, margins translucent, apices unawned; **lemmas** 3.5–5.5 mm, membranous, translucent, glabrous, rounded on the back, 1–3-veined, unawned; **paleas** more or less equal to the lemmas, translucent; **lodicules** ovate-acute; **anthers** 3; **styles** 2, free to the base, white. **Caryopses** 3–3.6 mm, narrowly ovoid to ellipsoid. $x = 7, 9, 19$. Name from the Greek *para*, 'beside', and *pholis*, 'scale', referring to the two side-by-side glumes.

Parapholis is a genus of six species that grow in coastal habitats and salt marshes. Its native range extends from western Europe to India. Two species are established in the *Flora* region.

SELECTED REFERENCE **Runemark, H.** 1962. A revision of *Parapholis* and *Monerma* (Gramineae) in the Mediterranean. Bot. Not. 115:1–17.

1. Spikes curved and twisted; upper sheath margins expanded, enclosing the lowest spikelets; anthers 0.5–1.3 mm long . . . . 1. *P. incurva*
1. Spikes straight, not twisted; sheath margins usually all alike, the lowest spikelets generally exserted beyond the sheaths; anthers 1.5–4 mm long . . . . 2. *P. strigosa*

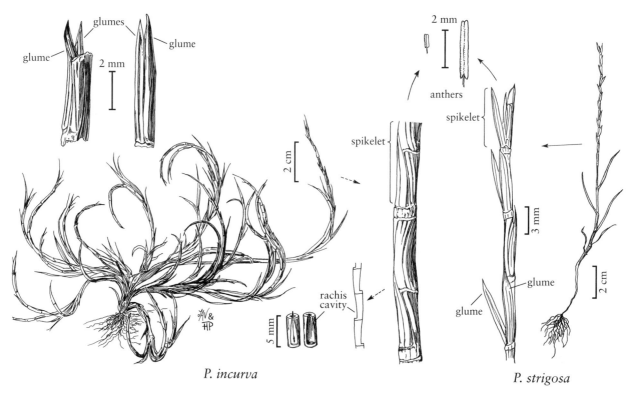

P. incurva   *P. strigosa*

PARAPHOLIS

1. Parapholis incurva (L.) C.E. Hubb. [p. 688]
CURVED SICKLEGRASS, SICKLEGRASS

Culms 2–35 cm, erect to decumbent, smooth, glabrous, branching at any node. **Sheaths of upper leaves** strongly inflated, margins expanded, enclosing the lowest spikelets; **ligules** to 1.5 mm; **blades** 1–3(10) cm long, 1–3 mm wide, adaxial surfaces scabrid. **Spikes** 1–15 cm, solitary, curved and twisted, rigid, with 2–20 spikelets. **Spikelets** 4.5–7.5 mm, usually slightly longer than the internodes, more or less cleistogamous. **Glumes** lanceolate, acuminate, keels rarely slightly winged; **anthers** 0.5–1.3 mm. $2n = 32, 36, 38, 42$.

Parapholis incurva is established at various locations on the coasts of the contiguous United States. It grows in both poorly drained and well-drained disturbed soils, at and above the high tide mark. It tends to grow in more saline soils, and at lower elevations with respect to the tide, than *P. strigosa*.

2. Parapholis strigosa (Dumort.) C.E. Hubb. [p. 688]
HAIRY SICKLEGRASS, STRIGOSE SICKLEGRASS

Culms 12–45 cm, erect to ascending, branching at the lower nodes. **Sheaths of upper leaves** usually with the margins all alike, not inflated, not enclosing the lowest spikelets; **ligules** to 2.3 mm; **blades** to 2(12) cm long, 1–3 mm wide, flat to inrolled, adaxial surfaces scabrid. **Spikes** 5–18 cm, straight, not twisted, with 5–25 spikelets. **Spikelets** 4.5–7 mm, usually slightly longer than the internodes, usually not cleistogamous. **Glumes** 4–6 mm, lanceolate, acuminate, keels obscure; **anthers** 1.5–4 mm. $2n = 14, 28$.

Parapholis strigosa has been found in disturbed areas of Humboldt Bay, California, and has also been reported from Del Norte, Mendicino, and Sonoma counties. It grows on moist soils above normal high tides, usually in well-compacted sandy loams. In general, it is found in less saline soils and at higher elevations than *P. incurva*.

14.39 SCRIBNERIA Hack.

James P. Smith, Jr.

Plants annual; tufted. **Culms** 3–35 cm, ascending to erect, often branched at the lowest nodes, glabrous; **nodes** purple. **Sheaths** open; **auricles** absent; **ligules** membranous; **blades** involute, nearly filiform. **Inflorescences** terminal distichous spikes, with 1 spikelet at all or most nodes, occasionally 2 at some nodes, very rarely 3 or 4 spikelets per node, lower spikelets sessile, upper spikelets pedicillate; **pedicels** shorter than 3 mm. **Spikelets** tangential to and partially embedded in the rachises, laterally compressed, with 1 floret; **rachillas** prolonged beyond the base of the floret; **disarticulation** above the glumes, beneath the florets. **Glumes** 2, exceeding the floret, glabrous, coriaceous, stiff, reddish- or purplish-tinged, 2-keeled, unawned; **lower glumes** longer and narrower than the upper glumes, 2–3-veined; **upper glumes** 3–4-veined; **calluses** pubescent; **lemmas** membranous, inconspicuously 5-veined, shortly bifid, awned from the sinus, awns 2–4 mm; **paleas** tightly clasped by the lemmas; **anthers** 1; **ovaries** glabrous. **Caryopses** about 2.5 mm, fusiform; **hila** punctiform; **embryos** about $^{1}/_{4}$ the length of the caryopses. $x = 13$. Named for Frank Lamson Scribner (1851–1938), an American agrostologist.

Scribneria is a monospecific genus native to North America.

SELECTED RERFERENCE **Crampton, B.** 1955. *Scribneria* in California. Leafl. W. Bot. 7:219–220.

1. **Scribneria bolanderi** (Thurb.) Hack. [p. 690]
   SCRIBNER GRASS

Culms (3)10–35 cm. **Ligules** 2–4 mm; **blades** 1–3 cm long, 0.8–1.6 mm wide, abaxial surfaces scabrous over the midveins. **Spikes** (2)4–11 cm long, 1–2.5 mm wide. **Spikelets** (3) 4–7 mm, slightly longer than the adjacent internodes. **Lemmas** glabrous or scabridulous distally and on the keels; **awns** 2–4 mm, inconspicuous; **paleas** generally smaller than the lemmas, apices notched. $2n = 26$.

Scribneria bolanderi grows between 500–3000 m, from Washington to Baja California, Mexico. It grows in diverse habitats, ranging from dry, sandy or rocky soils to seepages and vernal pools. It is often overlooked because it is relatively inconspicuous.

14.40 HAINARDIA Greuter

James P. Smith, Jr.

Plants annual. **Culms** 5–45 cm, branched above the base; **internodes** solid. **Uppermost sheath** open, often partially enclosing the inflorescences; **auricles** absent; **ligules** membranous, truncate; **blades** flat or convolute. **Inflorescences** single, terminal spikes, cylindrical, with solitary spikelets embedded in and radial to the rachises, the abaxial surface of the upper glumes exposed; **disarticulation** at the rachis nodes. **Spikelets** dorsiventrally compressed, with 1–2 florets, second floret reduced and sterile; **rachillas** sometimes prolonged beyond the base of the distal floret. **Lower glumes** absent from all but the terminal spikelets; **upper glumes** coriaceous, stiff, longer and firmer than the lemmas, rigid, 3–7(9)-veined, acute, unawned, sometimes mucronate; **lower lemmas** membranous, lanceolate, 3(5)-veined, unawned; **paleas** hyaline; **anthers** 1–3; **lodicules** 2, oblique, glabrous, fleshy basally. **Caryopses** shorter than the lemmas, oblong, somewhat dorsally compressed, with an apical appendage, concealed at maturity; **embryos** about $^{1}/_{5}$ the length of the caryopses; **hila** short, linear. $x = 13$. Named for Pierre Hainard (1936–), a Swiss phytogeographer.

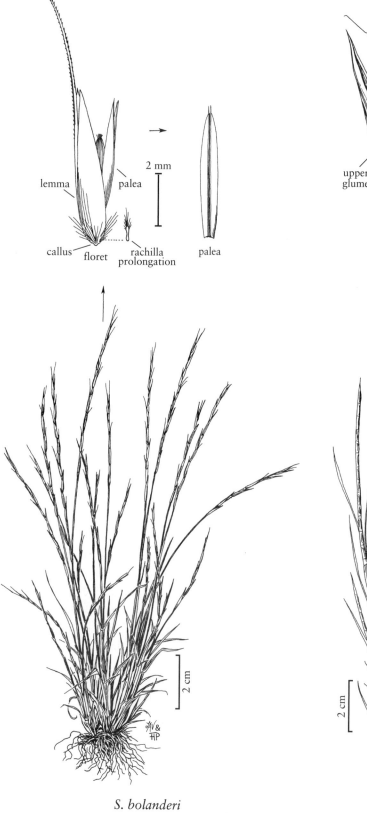

lemma

palea

2 mm

palea

callus   floret   rachilla
prolongation

spikelet

floret

upper
glume

floret

2 mm

upper
glume

rachis
internode

spikelet

2 cm

2 cm

S. bolanderi

H. cylindrica

SCRIBNERIA

HAINARDIA

Hainardia is a monospecific European genus that grows in saline and alkaline soils. It resembles *Parapholis*, which also occupies coastal salt marshes, but *Parapholis* differs in having spikelets with 2 glumes and culms with hollow internodes.

SELECTED REFERENCES **Greuter, W.** and **K.H. Rechinger.** 1967. Chloris Kythereia. Boissiera 13:22–196; **Gandhi, K.N.** 1996. Nomenclatural novelties for Western Hemisphere plants I. Harvard Pap. Bot. 8:63–66; **Runenark, H.** 1962. A revision of *Parapholis* and *Monerma* (Gramineae) in the Mediterranean. Bot. Not. 115:1–17.

1. Hainardia cylindrica (Willd.) Greuter [p. 690]

HARDGRASS, THINTAIL

Culms 5–40(45) cm, erect or ascending, smooth, glabrous. Sheaths 1–6 cm, uppermost sheath on each culm clearly expanded; **ligules** 0.2–1 mm, denticulate; **blades** to 7 cm long, 1.5–2.5 mm wide, adaxial surfaces scabrous. **Spikes** 8–25 cm, rigid, straight or somewhat curved. **Spikelets** 5–8 mm, slightly longer than the adjacent internodes, alternate, 2-ranked. **Glumes** ovate-lanceolate; **upper glumes** glabrous, midveins prominent, apices acute, sometimes mucronate; **lower lemmas** 4–6 mm; **paleas** 2-veined; **anthers** 2–3.5 mm. $2n = 14, 26, 52$.

Hainardia cylindrica is now established in California, along the Gulf coasts of Texas and Louisiana, in northern Baja California, Mexico, and in South Carolina. It grows in coastal salt marshes and alkaline soils below 300 m.

14.41 VAHLODEA Fr.

Jacques Cayouette

Stephen J. Darbyshire

Plants perennial; loosely cespitose. **Culms** 15–80 cm. **Sheaths** open nearly to the base; **auricles** absent; **ligules** membranous; **blades** rolled in the bud, flat. **Inflorescences** open or closed panicles; **branches** often flexuous, capillary, spikelets distal, some branches longer than 1 cm. **Spikelets** pedicellate, usually with 2 florets, sometimes with additional distal rudimentary florets; **rachillas** prolonged beyond the base of the distal florets about 0.5 mm or less, usually glabrous, sometimes with a few hairs; **first internodes** about 0.5 mm; **disarticulation** above the glumes, beneath the florets. **Glumes** subequal to equal, equaling or exceeding the florets, membranous, acute to acuminate, unawned; **lower glumes** 1(3)-veined; **upper glumes** 3-veined; **calluses** obtuse, pilose, hairs about $^1/_2$ as long as the lemmas; **lemmas** ovate, with 5(7) obscure veins, awned, awns attached near the middle of the lemmas, twisted, geniculate, visible between the glumes; **paleas** subequal to the lemmas; **lodicules** 2, membranous, toothed or not toothed; **anthers** 3; **ovaries** glabrous. **Caryopses** shorter than the lemmas, concealed at maturity, ellipsoid and irregularly triangular or ovate in cross section, deeply grooved, sometimes adhering to the lemmas and paleas; **hila** linear to oblong, $^1/_4$–$^1/_2$ the length of the caryopses. $x = 7$. Named for the Danish botanist Jens Laurentius Moestue Vahl (1796–1854), the son of botanist Martin Vahl.

Vahlodea is a monotypic genus with a discontinuous circumboreal distribution. It also grows in southern South America. The genus is sometimes included in *Deschampsia*, from which it differs in having loosely cespitose shoots and leaves that are usually mostly cauline, rather than mostly basal as in the perennial species of *Deschampsia*. The rachilla is prolonged beyond the upper floret for less than 0.5 mm in *Vahlodea* and is usually glabrous or has only a few hairs. In *Deschampsia*, the rachilla is usually prolonged more than 0.5 mm and is usually densely pubescent with long hairs. The caryopses of *Vahlodea* are ellipsoid, irregularly triangular or ovate in cross section, deeply grooved, and with a hilum $^1/_4$–$^1/_2$ the length of the caryopsis. In *Deschampsia*, the caryopses are narrowly ellipsoid to fusiform, elliptic or ovate in cross section, grooved or not, and with a hilum $^1/_{10}$–$^1/_3$ the length of the caryopsis.

SELECTED REFERENCE **Haraldsen, K.B.**, **M. Ødegaard**, and **I. Nordal.** 1991. Variation in the amphi-Atlantic plant *Vahlodea atropurpurea* (Poaceae). J. Biogeogr. 18:311–320.

1. **Vahlodea atropurpurea** (Wahlenb.) Fr. *ex* Hartm. [p. 692]

Mountain Hairgrass, Deschampsie Pourpre

Culms 15–80 cm, erect. **Leaves** glabrous or pilose; **lower sheaths** usually retrorsely hirsute, sometimes glabrous; **uppermost sheaths** smooth or scabridulous; **ligules** 0.8–3.5 mm, rounded to truncate, often lacerate and ciliate; **blades** flat, blades of the lower leaves to 30 cm long, 1–8.5 mm wide, blades of the flag leaves 1–10 cm long, 1–5 mm wide. **Panicles** 3–20 cm; **pedicels** smooth or scabrous-pubescent. **Spikelets** 4–7 mm. **Glumes** usually smooth or scabrous on the keels and marginal veins; **lower glumes** 4–5(6.5) mm; **upper glumes** 4–5.5(7) mm; **lemmas** 1.8–3 mm, apices scabrous, ciliate, awns 2–4 mm; **anthers** 0.5–1.2 mm. **Caryopses** 1–1.5 mm. $2n = 14$.

Vahlodea atropurpurea grows in moist to wet, open woods, forest edges, streamsides, snowbeds, and meadows, in montane to alpine and subarctic habitats. Plants from northwestern North America tend to have wider, more pubescent leaves and shorter lemma hairs than those elsewhere. They are sometimes treated as a distinct taxon, but the variation is continuous.

lemma

glumes

V. atropurpurea

VAHLODEA

14.42 PODAGROSTIS (Griseb.) Scribn. & Merr.

M.J. Harvey

Plants perennial; cespitose, sometimes rhizomatous. **Culms** 5–90 cm, erect or decumbent at the base. **Leaves** basally concentrated; **sheaths** open to the base, smooth, glabrous; **auricles** absent; **ligules** membranous, scabridulous dorsally, truncate to subacute, entire to lacerate; **blades** flat or involute. **Inflorescences** panicles, exserted at maturity, not disarticulating; **branches** ascending to erect. **Spikelets** pedicellate, weakly laterally compressed, with 1 floret; **rachillas** usually prolonged 0.1–1.9 mm beyond the base of the floret, sometimes absent, especially from the lower spikelets within a panicle, apices glabrous or with hairs, hairs to 0.3 mm; **disarticulation** above the glumes, beneath the floret. **Glumes** equal or the lower glumes longer than the upper glumes, flexible, acute to acuminate, sometimes apiculate, unawned; **calluses** glabrous or hairy, hairs to 0.5 mm; **lemmas** membranous, (3)5-veined, veins mostly obscure, sometimes prominent distally, apices truncate to rounded or acute, unawned or awned, awns to about 1.3 mm, usually subapical, occasionally attached near midlength; **paleas** more than $^1/_2$ as long as the lemmas, 2-veined, thinner than the lemmas; **anthers** 3. **Caryopses** shorter than the lemmas, concealed at maturity. $x = 7$. Name from the Greek *pous*, 'foot', and the genus *Agrostis*.

Podagrostis is a genus of six or more species that grow in cool, wet areas. In the past, its species have been included in *Agrostis*. Four or more species are native to Central and South America, and two to the *Flora* region. *Podagrostis* differs from *Agrostis* in its combination of a relatively long palea and, usually, the prolongation of the rachilla beyond the base of the floret. It differs from *Calamagrostis* in the poorly developed callus hairs and awns.

SELECTED REFERENCES Björkman, S.O. 1960. Studies in *Agrostis* and related genera. Symb. Bot. Upsal. 17:1–112; Carlbom, C.G. 1967. A biosystematic study of some North American species of *Agrostis* L. and *Podagrostis* (Griseb.) Scribn. & Merr. Ph.D. dissertation, Oregon State University, Corvallis, Oregon, U.S.A. 232 pp.

1. Glumes 2.3–4.3 mm long, generally equal; rachilla prolongations 0.5–1.9 mm long . . . . 1. *P. aequivalvis*
1. Glumes 1.6–2.3 mm long, lower glumes equal to or longer than the upper glumes; rachilla prolongations 0.1–0.6 mm long . . . . 2. *P. humilis*

1. Podagrostis aequivalvis (Trin.) Scribn. & Merr. [p. 695]

ARCTIC BENT

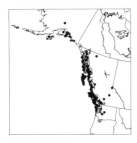

Plants rhizomatous. **Culms** 25–90 cm, erect; **nodes** 2–4(6). **Sheaths** smooth; **ligules** 0.4–4 mm, scabridulous, truncate to subacute, entire or lacerate; **blades** 4–18 cm long, 1–2.5 mm wide, flat. **Panicles** 5–15 cm long, 2–10 cm wide, lanceolate to ovate, often drooping, sparsely branched, lowest nodes with 1–4(5) branches; **branches** usually scabridulous, sometimes smooth, erect to ascending or spreading, spikelets usually restricted to the distal $^1/_2$; **lower branches** 3–6 cm; **pedicels** 2–10 mm. **Spikelets** narrowly ovate to lanceolate, usually purplish bronze, sometimes greenish purple; **rachilla prolongations** 0.5–1.9 mm, bristlelike, distal hairs shorter than 0.3 mm. **Glumes** 2.3–4.3 mm, veins sparsely scabridulous distally, apices acute to acuminate, sometimes apiculate; **lower glumes** usually equal to the upper glumes, usually 3-veined, lateral veins faint; **calluses** glabrous or with sparse hairs shorter than 0.1 mm; **lemmas** 2.5–3.5 mm, smooth, opaque, (3)5-veined, veins usually obscure, apices acute, entire or the veins minutely excurrent to about 0.3 mm, unawned; **paleas** 2–3 mm; **anthers** 3, 0.8–1.3 mm. **Caryopses** 1.2–1.5 mm; **endosperm** solid. $2n = 14$.

Podagrostis aequivalvis grows along lake, bog, and stream margins, and in forest fens. It is common in the coastal regions of Alaska and British Columbia, and occurs less frequently inland, as well as to about 1500 m in the Cascade Mountains south to Oregon.

2. Podagrostis humilis (Vasey) Björkman [p. 695]
ALPINE BENT

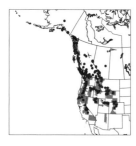

Plants sometimes rhizomatous. **Culms** 5–50 cm, erect to ascending, bases sometimes somewhat decumbent; **nodes** 2–4(7). **Sheaths** smooth; **ligules** (0.2)0.5–2(2.7) mm, scabridulous, truncate to obtuse, usually erose to lacerate, sometimes entire; **blades** 2–15 cm long, 1–4 mm wide, flat or involute; **flag blades** at about midculm. **Panicles** 1.5–14 cm long, 0.4–2.5 cm wide, narrowly oblong to ovate, somewhat lax, sometimes drooping, lowest nodes with 1–4(5) branches; **branches** scabridulous or smooth, loosely ascending to erect, branching in the distal ¹/₃–³/₄; **lower branches** 0.5–7 cm; **pedicels** 0.4–4 mm. **Spikelets** ovate to lanceolate, green to purple; **rachilla prolongations** 0.1–0.6 mm, glabrous or bristlelike, with a tuft of short hairs at the apex. **Glumes** 1.6–2.3 mm, 1-veined, veins smooth or slightly scabrous distally, apices acute; **lower glumes** equal to or longer than the upper glumes; **calluses** glabrous or with sparse hairs to 0.5 mm; **lemmas** 1.5–2.3 mm, relatively thick, smooth, opaque, 5-veined, veins obscure or prominent at the apices, apices truncate to rounded or acute, entire or erose, veins occasionally minutely excurrent to 0.4 mm, usually unawned, rarely awned, awns to about 1.3 mm, usually subapical, sometimes attached near midlength; **paleas** 0.9–1.6 mm; **anthers** 3, 0.4–0.8 mm. **Caryopses** 1–1.3 mm. 2*n* = 14.

Podagrostis humilis is a western North American species that grows in undisturbed alpine and subalpine meadows and screes at over 3500 m, down to meadows, fens, and open woodlands at less than 200 m.

As treated here, *Podagrostis humilis* includes *P. thurberiana* (Hitchc.) Hultén, a species that supposedly differs in having larger, more open panicles and wider leaves; there is complete intergradation between the two extremes. There is similar but less severe intergradation with *P. aequivalvis*. In the field, dwarf forms of *P. humilis* mimic *Agrostis variabilis*; they differ from that species in having paleas.

14.43 LACHNAGROSTIS Trin.

M.J. Harvey

Plants annual, or short-lived perennials; cespitose, sometimes rhizomatous. **Culms** 10–80 cm, erect or geniculately ascending. **Sheaths** open, rounded over the midvein; **auricles** absent; **ligules** membranous; **blades** flat or folded, margins sometimes involute. **Inflorescences** panicles, lax. **Spikelets** pedicellate, with 1(2) floret(s), laterally compressed; **rachillas** prolonged beyond the base of the floret, sometimes equaling the paleas and the apices hairy, or minute and glabrous; **disarticulation** above the glumes, beneath the floret, in perennial species the panicles detaching with a portion of the uppermost cauline nodes at maturity, in annual species the panicles persistent. **Glumes** equal to subequal, exceeding the florets, ovate-elliptic to lanceolate, membranous, 1(3)-veined, lateral veins much shorter than the midveins, keels scabridulous to scabrous, apices unawned; **calluses** minute, blunt, usually hairy, sometimes glabrous, hairs ¹/₅–²/₃ the length of the lemmas; **lemmas** usually shorter and more flexible than the glumes, rarely as long as and firmer than the glumes, usually hairy, sometimes glabrous, rarely scabrous, (3)5-veined, rounded over the midveins, apices often denticulate, marginal veins slightly excurrent, apices erose or toothed, usually awned, sometimes unawned, awns not or only slightly exceeding the glumes, attached near midlength or subapically, dorsal awns straight or geniculate, subapical awns straight; **paleas** from ¹/₂ as long as to equaling the lemmas, hyaline, weakly 2-keeled; **lodicules** 2, linear to lanceolate, glabrous; **anthers** 3, not penicillate; **ovaries** glabrous; **styles** 2. **Caryopses** shorter than the lemmas, concealed at maturity, fusiform, endosperm doughy or dry. Name from the Greek *lachnos*, 'wool', and the genus *Agrostis*.

Lachnagrostis includes about 20 species. It is native to the Southern Hemisphere, having its greatest concentration in Australasia. One species is established in the *Flora* region.

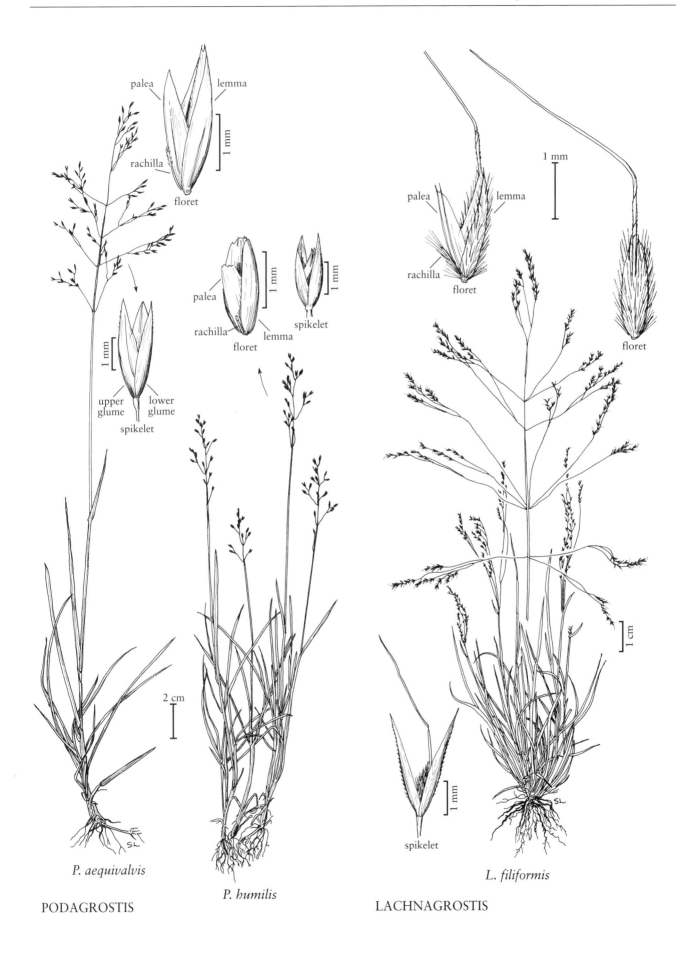

palea   lemma

rachilla

floret

1 mm

1 mm

palea

rachilla

lemma

floret

1 mm

spikelet

1 mm

upper glume   lower glume

spikelet

palea   lemma

rachilla

floret

1 mm

floret

2 cm

1 cm

1 mm

spikelet

P. aequivalis

P. humilis

L. filiformis

PODAGROSTIS

LACHNAGROSTIS

Lachnagrostis has usually been included in *Agrostis*. Its recognition as a segregate genus is supported by studies of both genera by Edgar (1995), Edgar and Connor (2000), Jacobs (2001), and Rúgolo de Agrasar and Molina (2002). *Lachnagrostis* differs from *Agrostis* in its combination of sometimes disarticulating panicles, paleas at least half as long as the lemmas, well-developed, sometimes hairy rachilla prolongations, and smooth lemma epidermes in which the walls of the long cells are wavy and more or less flush with the surface rather than raised. Both genera have individual species that resemble the other species in one of these features, but there is usually no difficulty in placing them in the correct genus.

SELECTED REFERENCES **Edgar, E.** 1995. New Zealand species of *Deyeuxia* P. Beauv. and *Lachnagrostis* Trin. (Gramineae: Aveneae). New Zealand J. Bot. 33:1–33; **Edgar, E. and H.E. Connor.** 2000. Flora of New Zealand, vol. 5. Manaaki Whenua Press, Lincoln, New Zealand. 650 pp.; **Jacobs, S.W.L.** 2001. The genus *Lachnagrostis* (Gramineae) in Australia. Telopea 9:439–448; **Rúgolo de Agrasar, Z.E. and A.M. Molina.** 2002. El género *Lachnagrostis* (Gramineae: Agrostideae) en América del Sur. Pp. 20–32 *in* A. Freire Fierro and D.A. Neill (eds.). 2002. La Botánica en el Nuevo Milenio: Memorias del Tercer Congreso Ecuatoriano de Botánica. FUNBOTANICA, Quito, Ecuador. 250 pp.

1. **Lachnagrostis filiformis** (G. Forst.) Trin. [p. 695]
PACIFIC BENT

Plants perennial; cespitose, sometimes rhizomatous. **Culms** 15–65 cm, erect, sometimes geniculate basally. **Sheaths** glabrous, sometimes scabridulous; **ligules** 2.2–7.8 mm, obtuse or acute, lacerate; **blades** 8–20 cm long, 1–3 mm wide, usually flat, sometimes involute, finely scabrous. **Panicles** 7–30 cm long, (2)5–25 cm wide, broadly ovate, loose and diffuse at maturity, bases sometimes not exserted at maturity; **branches** capillary, scabrous, often deflexed at maturity, branched above the middle, spikelets confined to the distal $^1/_3$; **lower branches** 5–15 cm; **pedicels** 0.5–7 mm; **disarticulation** above the glumes and below the panicles. **Spikelets** greenish to yellowish, often tinged with purple; **rachilla prolongations** about 1 mm, bristlelike, pilose distally, hairs about 0.6 mm. **Glumes** equal to subequal, 2.8–3.6 mm, scabrous on the midvein, at least distally, 1(3)-veined, apices long-acuminate; **callus hairs** to 0.5 mm, abundant; **lemmas** 1.8–2.3 mm, translucent to opaque, usually densely hairy at least on the lower $^1/_2$, hairs to 0.7 mm, 5-veined, apices truncate to obtuse, erose or 2–4-toothed, awned from the middle $^1/_3$ of the lemmas, awns 4–7.5 mm, geniculate; **paleas** (0.4)1.2–1.5 mm, usually more than $^2/_3$ the length of the lemmas, thin, veins usually not visible; **anthers** 3, about 0.5 mm. **Caryopses** about 1.2 mm; **endosperm** solid. $2n = 56$.

Lachnagrostis filiformis is native to New Guinea, Australia, New Zealand, and Easter Island. In North America, it grows in open, disturbed sites such as roadsides and burned areas, and has been spreading into vernal pools around San Diego, California. It was introduced to North America in the late nineteenth century, but is only known to be established in California. The most recent record located for Texas was collected in 1902. Records from South Carolina are from "waste areas around wool-combing mill; rare, perhaps only a waif." (Weakley, http: //www. herbarium.unc/edu/flora.htm)

14.44 **BROMIDIUM** Nees & Meyen

Zulma E. Rúgolo de Agrasar

Plants annual or perennial; cespitose, sometimes rhizomatous. **Culms** 5–50 cm, erect. **Sheaths** open almost to the base; **auricles** absent; **ligules** membranous; **blades** flat or subconvolute, smooth or scabrous. **Inflorescences** panicles, spikelike or subspikelike, dense, usually interrupted towards the base. **Spikelets** with 1 floret; **rachillas** not prolonged beyond the base of the floret; **disarticulation** above the glumes, beneath the floret. **Glumes** 2, exceeding the florets, equal or unequal, membranous, 1-veined, keeled, keels scabrous or scabrous-ciliate, apices unawned; **calluses** short, pilose on all sides or with 2 tufts of lateral hairs, hairs not reaching past the lower $^1/_3$ of the lemmas; **lemmas** membranous, glabrous or pilose, margins incurved, 5-veined, veins scabridulous on the distal $^1/_3$, lateral and marginal veins extending beyond the lemma margins as awns or 0.1–2 mm awnlike teeth, midvein forming an awn from the lower or upper ¹/₃ of the lemmas, awns exceeding the glumes, stout, geniculate, bases twisted, hygroscopic; **paleas** absent or much reduced; **lodicules** 2, membranous, acute; **anthers** 3, anther sacs separated in the distal ¹/₃ after dehiscence; **ovaries** oblong; **styles** 2; **stigmas** plumose. **Caryopses** shorter than the lemmas, concealed at maturity, fusiform, grooved, usually falling free of the lemmas and paleas; **hila** punctiform or oval; **embryos** small; **endosperm** lipid, liquid, doughy, or starchy. Name from the Greek *bromos*, a kind of oats, and the diminutive suffix *-idion*.

Bromidium includes five species, all of which are native to South America. One species, *B. tandilense*, is now established in California. None of the species is important for forage. Four of the five species are annual. The stout hygroscopic awn, apical awnlike teeth, and lemma pilosity all aid in its dispersal.

Bromidium is similar to *Agrostis*, and is sometimes included in it. It differs from *Agrostis* in its combination of dense, contracted, spikelike panicles with a relatively large number of spikelets, unawned glumes, and 4-awned or -toothed lemmas.

SELECTED REFERENCE **Rúgolo de Agrasar, Z.E.** 1982. Revalidación del género *Bromidium* Nees et Meyen emend. Pilger (Gramineae). Darwiniana 24:187–216.

1. Bromidium tandilense (Kuntze) Rúgolo [p. 697]

Plants annual, tufted. **Culms** 5–50 cm, slender, erect. **Sheaths** smooth; **ligules** 1.5–3 mm, membranous, rounded to acute, entire; **blades** 2–10 cm long, 0.5–4 mm wide, flat. **Panicles** 0.9–8 cm long, (0.2) 0.7–1.5 cm wide, cylindrical, generally straw-colored or brown, appearing bristly; **branches** to 1 cm, ascending to appressed, with 1 branch at the lowest node; **pedicels** 0.1–0.7 mm, scabrous. **Spikelets** 2.6–3.6 mm. **Glumes** lanceolate, subequal, keels scabrous, sides smooth, apices acute; **lower glumes** 2.6–3.6 mm; **upper glumes** (2)2.4–3.5 mm; **calluses** pilose on all sides, hairs to 0.5 mm, dense; **lemmas** (1)1.4–1.6 mm, pilose, 5-veined, lateral veins prolonged into awnlike teeth, outer teeth (0.8)1–2 mm, inner teeth 0.1–0.2(0.5) mm, awns 4.5–6 mm, arising from the lower ¹/₃ of the lemmas, exceeding the glumes by 2–2.5 mm; **paleas** absent; **anthers** 0.2–0.5 mm, sometimes differing in length within a floret, often remaining on the apices of the caryopses. **Caryopses** 1–1.3 mm; **hila** oval; **endosperm** liquid or doughy. 2*n* = unknown.

Bromidium tandilense is native to Argentina, Brazil, Uruguay, and Paraguay. It now grows around vernal pools in Solano, Monterey, and San Diego counties, California.

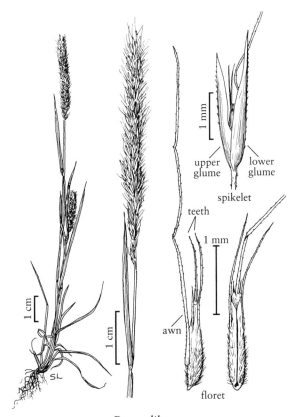

B. tandilense

BROMIDIUM

14.45 AVENULA (Dumort.) Dumort.

Gordon C. Tucker

Plants perennial; cespitose, sometimes stoloniferous. **Culms** 10–110 cm. **Sheaths** usually open, sometimes closed for most of their length; **auricles** absent; **ligules** membranous, acute to truncate; **blades** flat or folded, adaxial surfaces unribbed, with a furrow on either side of the midveins, margins sclerenchymatous. **Inflorescences** reduced panicles, many branches (all branches in depauperate specimens) with a single spikelet. **Spikelets** with 2–7 florets; **rachillas** glabrous on the side adjacent to the paleas, hairy elsewhere; **disarticulation** above the glumes, beneath the florets. **Glumes** as long as or longer than the adjacent lemmas, 1–3-veined, unawned; **calluses** acute; **lemmas** 5–7-veined, obtuse, bifid, awned from about midlength, awns geniculate, flattened or terete and twisted below the bend; **paleas** with lateral wings less than $^1\!/_2$ as wide as the intercostal region, apices shallowly bifid; **lodicules** 2, entire, unlobed; **anthers** 3. **Caryopses** more than twice as long as the hila, shorter than the lemmas, concealed at maturity; **endosperm** liquid or semi-liquid. $x = 7$. Name a diminutive of *Avena*.

Avenula is a genus of approximately 30 species, most of which are European. One species is native to the *Flora* region, and one has been introduced. The genus is frequently included in *Helictotrichon*, from which it differs in having acute cauline ligules, unribbed leaves, rachillas glabrous on one side, unlobed lodicules, short hila, liquid to semi-liquid endosperm, and no sclerenchyma ring in its roots.

SELECTED REFERENCES **Dixon, J.M.** 1991. Biological flora of the British Isles: *Avenula* (Dumort.) Dumort. J. Ecol. 79:829–865; **Gervais, C.** 1973. Contribution à l'étude cytologique et taxonomique des avoines vivaces. Denkschr. Schweiz. Naturf. Ges. 88:3–166; **Holub, J.** 1963. Ein Beitragzur Abgrenzung der Gattungen in der Tribus Aveneae: Die Gattung *Avenochloa* Holub. Acta Horti Bot. Prague 1962:75–86.

1. Sheaths closed for less than $^1\!/_3$ their length, sheaths and blades smooth to scabridulous; panicles 4–13 cm long, usually 0.8–2.5 cm wide; awns 10–17 mm long, flattened below the bend . . . . 1. *A. hookeri*
1. Sheaths closed to near the top, sheaths and blades usually pubescent; panicles 6–20 cm long, 2–6 cm wide; awns 12–26 mm long, terete below the bend . . . . 2. *A. pubescens*

1. Avenula hookeri (Scribn.) Holub [p. 699]
SPIKE OATGRASS

Plants cespitose, not stoloniferous. **Culms** 10–75 cm, erect. **Sheaths** closed for less than $^1\!/_3$ their length, smooth to scabridulous; **ligules** 3–7 mm, acute, usually lacerate; **blades** usually 4–20 cm long, 1–4.5 mm wide, smooth to scabridulous, margins cartilaginous and whitish. **Panicles** (4)6–11(13) cm long, usually 0.8–2.5 cm wide, erect or ascending; **branches** 10–25 mm, usually straight, stiff, usually with 1–2 spikelets. **Spikelets** 12–16 mm, with 3–6 florets; **rachilla internodes** usually 1.5–2.5 mm, hairs 0.2–1.5 mm. **Glumes** thin, acute; **lower glumes** 9–13 mm, 3-veined; **upper glumes** 9–14 mm, 3–5-veined; **calluses** bearded, hairs usually shorter than 1 mm; **lemmas** 10–12 mm, awned, awns 10–17 mm, flattened below the bend; **paleas** 6–8.75 mm; **anthers** 2.5–5 mm. $2n = 14$.

Avenula hookeri grows on mesic to dry, open prairie slopes, hillsides, forest openings, and meadows, in montane to subalpine zones, from the Yukon and Northwest Territories to northern New Mexico.

2. Avenula pubescens (Huds.) Dumort. [p. 699]
DOWNY ALPINE OATGRASS

Plants shortly stoloniferous. **Culms** 30–110 cm, erect or geniculate at the base. **Basal leaves: sheaths** closed to near the top; **ligules** 0.5–1 mm, truncate; **blades** 10–40 cm long, 2–6(8.5) mm wide, usually with hairs to 2 mm. **Cauline leaves: sheaths** closed for nearly their entire length, pubescent; **ligules** 5–8 mm, acute; **blades** 10–40 cm long, 2–6 mm wide, usually pubescent, margins very narrowly cartilaginous. **Panicles** 6–20 cm long, 2–6 cm wide, erect or nodding; **branches** 4–35 mm, flexuous or straight, with 1–4 spikelets. **Spikelets**

10–26 mm, with 2–4 florets; **rachilla internodes** about 2.5 mm, hairs 3–7 mm. **Glumes** scabrous on the veins. **Lower glumes** 7–20 mm, 1(3)-veined; **upper glumes** 10–26 mm, 3-veined; **calluses** bearded, hairs 2–5 mm; **lemmas** 8–16 mm, awned, awns 12–26 mm, terete below the bend; **paleas** 8–12 mm; **anthers** 5–7 mm. $2n$ = 14, 28.

Avenula pubescens is native to Eurasia, where it grows in meadows, pastures, and woodland clearings. The most widespread taxon is **Avenula pubescens** (Huds.) Dumort. subsp. **pubescens**, which differs from **Avenula pubescens** subsp. **laevigata** (Schur) Holub in having smaller spikelets (10–17 mm long with 2–3 florets versus 15–26 mm long with 3–4 florets). *Avenula pubescens* subsp. *pubescens* has been collected in southern Ontario, Anticosti Island in Quebec, and in New England, but is not known to be established in Canada.

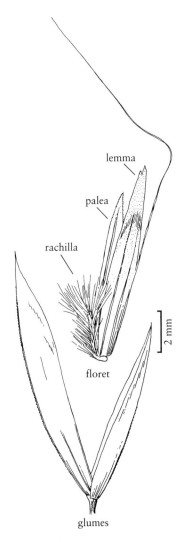

A. pubescens

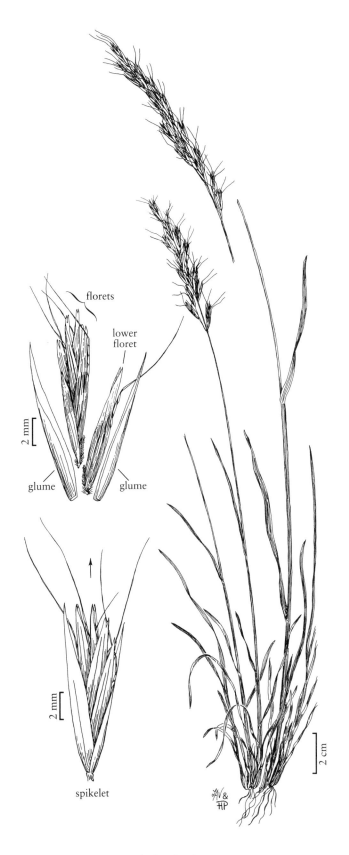

A. hookeri

AVENULA

14.46 DISSANTHELIUM Trin.

Nancy F. Refulio

Plants annual or perennial; cespitose, sometimes rhizomatous. **Culms** to 10(25) cm. **Sheaths** open, usually glabrous, lower sheaths shorter than the upper sheaths; **auricles** absent; **ligules** 2–6 mm, membranous. **Inflorescences** contracted panicles, some branches longer than 1 cm. **Spikelets** pedicellate, 2.5–5 mm, laterally compressed, with 2(3) florets, all florets bisexual or some florets bisexual and others pistillate; **disarticulation** above the glumes, beneath the florets. **Glumes** equal or subequal, usually exceeding all the florets, sometimes subequal to them, ovate or acuminate, margins scarious, apices unawned; **lower glumes** 1-veined; **upper glumes** 3-veined; **calluses** poorly developed, glabrous; **lemmas** oval or elliptic, 3(5)-veined, lateral veins near the margins, glabrous, scabrous, or pilose, apices acute or obtuse, sometimes denticulate, unawned; **paleas** slightly shorter than the lemmas; **lodicules** 2; **anthers** 3. **Caryopses** shorter than the lemmas, concealed at maturity. x = unknown. Name from the Greek *dissos*, 'double', and *anthelion*, 'a small flower', an allusion to the two small florets.

Dissanthelium contains about 20 species, and has an amphi-neotropical distribution. Most species grow in South America; two grow in North America. One of the two North American species, *D. mathewsii* (Ball) R.C. Foster & L.B. Sm., has a disjunct distribution, growing in both central Mexico and South America. The other, *D. californicum*, is discussed below.

SELECTED REFERENCES **Renvoize, S.A.** 1998. Gramíneas de Bolivia. Royal Botanic Gardens, Kew, England. 644 pp.; **Swallen, J.R.** and **O. Tovar.** 1965. The grass genus *Dissanthelium*. Phytologia 11:361–376; **Tovar, O.** 1993. Las gramíneas (Poaceae) del Perú. Ruizia 13:1–480.

1. **Dissanthelium californicum** (Nutt.) Benth. [p. 700]

CALIFORNIA DISSANTHELIUM

Plants annual. **Culms** to 25 cm, glabrous. **Blades** 2.5–15 cm long, 1–4 mm wide, flat, soft. **Panicles** 5–15 cm, contracted; **pedicels** scabrous. **Spikelets** with 2–3 florets, lower floret in each spikelet a little longer than the upper floret. **Glumes** subequal, 3–4 mm, exceeding the florets, lanceolate, keels scabrous, margins scarious, apices acuminate; **lemmas** 1.5–2 mm, pilose, obtuse to acute. $2n$ = unknown.

Dissanthelium californicum is known only from Santa Catalina and San Clemente islands, California, and Guadalupe Island, Baja California, Mexico. It grows in coastal sage-scrub. Until its rediscovery on Santa Catalina Island in 2005, it was thought to be extinct in California, not having been reported from the state since 1903.

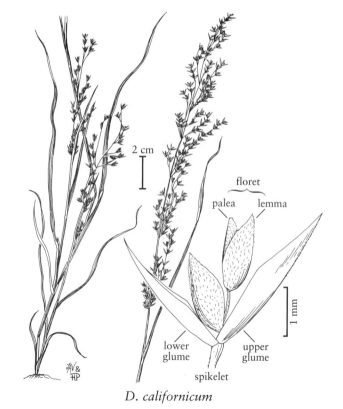

D. californicum

DISSANTHELIUM

14.47 HELICTOTRICHON Besser *ex* Schult. & Schult. f.

Gordon C. Tucker

Plants perennial; cespitose. **Culms** 5–150 cm, erect. **Sheaths** open nearly to the base; **auricles** absent; **ligules** about as long as wide, membranous, truncate to rounded, ciliate-erose; **blades** convolute or involute, adaxial surfaces ribbed over the veins. **Inflorescences** narrow panicles or racemes, some branches longer than 1 cm. **Spikelets** laterally compressed, with (1)2–8 florets; **rachillas** pilose on all sides, terminating in reduced florets; **disarticulation** above the glumes, beneath the florets. **Glumes** equaling or exceeding the adjacent lemmas, exceeded by the distal florets, 1–3(5)-veined; **calluses** acute, strigose; **lemmas** pilose or glabrous, 3–5-veined, apices acute, toothed, usually awned from about midlength, awns geniculate, twisted and terete below the bend, distal lemmas sometimes unawned; **paleas** shorter than the lemmas, wings more than $^{1}/_{2}$ as wide as the intercostal region; **lodicules** 2, lobed; **anthers** 3; **ovaries** pubescent distally. **Caryopses** shorter than the lemmas, concealed at maturity, with a solid endosperm, longitudinally grooved, with a terminal tuft of hairs; **hila** more than $^{1}/_{2}$ as long as the caryopses, linear. $x = 7$. Name from the Greek *helictos*, 'twisted', and *trichon*, 'awn', referring to the lemma awn.

Helictotrichon has about 15 species. Most are native to Europe; one is endemic to the *Flora* region, and one has been introduced as an ornamental. The genus is sometimes interpreted as including *Avenula*, from which it differs in having truncate to rounded ligules, ribbed leaves, rachillas that are pilose on all sides, lobed lodicules, long hila, solid endosperm, and sclerenchyma rings in its roots.

SELECTED REFERENCES Gervais, C. 1973. Contribution à l'étude cytologique et taxonomique des avoines vivaces. Denkschr. Schweiz. Naturf. Ges. 88:3–166; Hedberg, I. 1961. Chromosome studies in *Helictotrichon* Bess. Bot. Not. 114:389–396.

1. Culms 5–20 cm tall; panicles 2–8 cm long, most branches with 1 spikelet; plants native . . . 1. *H. mortonianum*
1. Culms 30–150 cm tall; panicles 8–20 cm long, most branches with 3–10 spikelets; plants
   cultivated as ornamentals . . . . 2. *H. sempervirens*

1. Helictotrichon mortonianum (Scribn.) Henrard [p. 702]
ALPINE OATGRASS

Culms 5–20 cm. **Ligules** 0.5–1 mm, truncate to rounded, ciliate; **blades** 3–6 cm long, 1–2 mm wide, involute or convolute, strigose, particularly on the adaxial surfaces. **Panicles** 2–5(8) cm; **branches** erect, usually with 1 spikelet each. **Spikelets** 8–12 mm, with (1)2(3) florets. **Glumes** equal or nearly so, 8–11 mm, acuminate or awn-tipped, awns 0.3–0.5 mm; **lowest lemmas** 7–10 mm, 3-veined, apices 4-toothed, awns 10–16 mm; **distal lemmas** unawned or awned; **anthers** 1.5–2 mm. $2n = 14, 28$.

Helictotrichon mortonianum grows in alpine and subalpine meadows and summits, at 3000–4200 m. It is restricted to the central and southern Rocky Mountains of the contiguous United States.

2. Helictotrichon sempervirens (Vill.) Pilg. [p. 702]
BLUE OATGRASS

Culms 30–100(150) cm. **Ligules** 0.5–1.5 mm, truncate, ciliate; **blades** 15–60 cm long, 2–4 mm wide, usually convolute and 0.9–1.5 mm in diameter, glaucous, scabridulous, basal blades deciduous when dead. **Panicles** 8–20 cm; **branches** ascending, with 3–7(10) spikelets. **Spikelets** 10–14 mm, with 3(5) florets. **Glumes** subequal to unequal; **lower glumes** 7–10 mm; **upper glumes** 10–12 mm; **lowest lemmas** 7–12 mm, awns about 15 mm; **distal lemmas** unawned; **anthers** about 5 mm. $2n = 28, 42$.

Helictotrichon sempervirens is a native of the southwestern Alps in Europe, where it grows on rocky soils and in stony pastures. In the *Flora* region, it is frequently grown as an ornamental species; it is not established in the region.

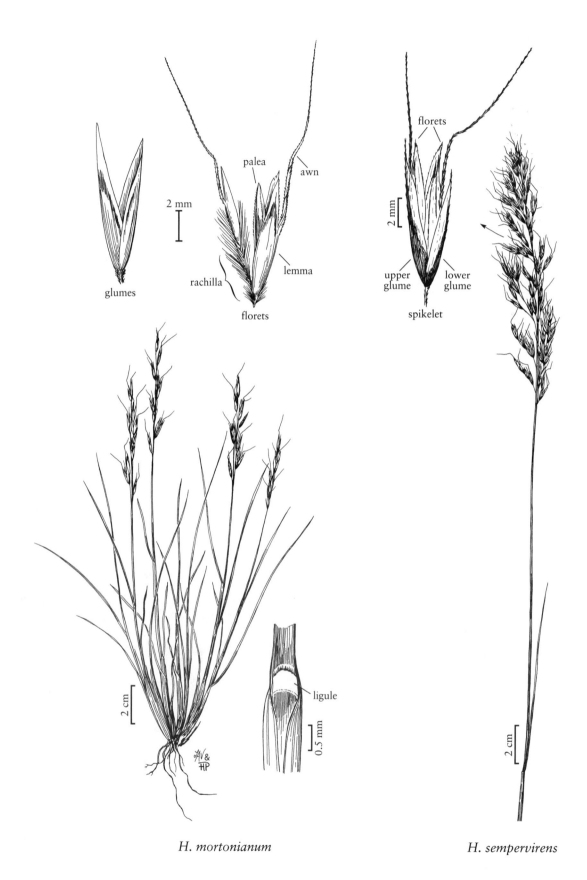

glumes

2 mm

palea

awn

rachilla

lemma

florets

florets

2 mm

upper
glume

lower
glume

spikelet

2 cm

ligule

0.5 mm

H. mortonianum

H. sempervirens

HELICTOTRICHON

14.48 AMPHIBROMUS Nees

Surrey W.L. Jacobs

Plants annual or perennial; cespitose, sometimes rhizomatous or stoloniferous. **Culms** to 180 cm, erect or geniculate. **Sheaths** open, lower sheaths often enclosing cleistogamous panicles with unawned spikelets having fewer florets and smaller anthers than the aerial spikelets; **auricles** absent; **ligules** elongate, membranous, becoming lacerate; **blades** flat or inrolled. **Terminal inflorescences** panicles, open to spikelike. **Spikelets** laterally compressed, with 2–10 florets, cleistogamous florets occasionally intermixed with the chasmogamous florets, distal florets often reduced, staminate; **rachillas** pubescent, prolonged beyond the base of the most distal pistillate floret, empty or terminating in a reduced floret; **disarticulation** above the glumes, beneath the florets. **Glumes** subequal to unequal, shorter than the adjacent lemmas, ovate to lanceolate, scarious, acute to obtuse, often erose, unawned; **lower glumes** shorter and narrower than the upper glumes, 1–5-veined; **upper glumes** 3–7-veined; **calluses** blunt, pubescent; **lemmas** chartaceous, smooth or scabrous, with 5–9 prominent veins, awned, apices with 2–4 teeth or lobes, outer lobes often smaller than the inner lobes, all lobes aristate to obtuse, awns arising from below midlength to near the apices, sometimes straight when young, geniculate at maturity, spreading or recurved; **paleas** subequal to or much shorter than the lemmas, bilobed; **lodicules** 2, free, glabrous, not lobed; **anthers** 3; **ovaries** glabrous. **Caryopses** shorter than the lemmas, concealed at maturity, terete, apices often with a few hairs; **hila** to $\frac{1}{2}$ the length of the caryopses. x = unknown. Name of uncertain origin, perhaps in part from the Greek *bromos*, a kind of oats. Nees (1843) wrote "Habitus potius *Avenæ* quam *Bromi*" [Appearance/habit rather/preferably of *Avena* than of *Bromus*], alluding to its similarity with *Avena*, not *Bromus*.

Amphibromus is a genus of 12 species, two native to South America, one to both New Zealand and Australia, and the remainder endemic to Australia; two have been introduced into the *Flora* region. Most species grow in open, damp habitats such as floodplains and other areas that are periodically flooded, and on the banks of, and sometimes in, inland and coastal rivers, marshes, lagoons, waterholes, and swamps.

Cleistogamy is common in *Amphibromus*. Cleistogamous spikelets in the terminal panicles resemble the chasmogamous spikelets, but have smaller anthers; those in the lower leaf sheaths are unawned. Plants usually flower in response to rain or flooding.

SELECTED REFERENCES **Calaway, M.L.** and **J.W. Thieret.** 1985. *Amphibromus scabrivalvis* (Gramineae) in Louisiana. Sida 11:207–214; **Jacobs, S.W.L.** and **L. Lapinpuro.** 1986. The Australian species of *Amphibromus* (Poaceae). Telopea 2:715–729; **Nees von Esenbeck, C.G.** 1843. Gramina Novæ Hollandiæ, præsertim Insulæ Van Diemen, collectionis Lindleyanæ, a v. cl. Drummond, Gunn, aliisque collecta. London J. Bot. 2:409–420; **Swallen, J.R.** 1931. The grass genus *Amphibromus*. Amer. J. Bot. 18:411–415.

1. Pedicels absent or to 10 mm long; lowest internodes usually swollen; awns 8–17 mm long . . . 1. *A. scabrivalvis*
1. Pedicels usually longer than 10 mm; lowest internodes not swollen; awns 12–26 mm long.
    2. Awns arising from the lower $^2/_5$–$^3/_5$ of the lemmas; lemma apices not appearing constricted
    . . . . 2. *A. nervosus*
    2. Awns arising from the upper $^2/_3$–$^3/_4$ of the lemmas; lemma apices appearing constricted . . . . 3. *A. neesii*

1. Amphibromus scabrivalvis (Trin.) Swallen [p. 705]
ROUGH AMPHIBROMUS

Plants perennial; rhizomatous. **Culms** 30–100+ cm, terete, erect or decumbent; **nodes** up to 10, lowest 3 or 4 nodes often underground, aerial nodes usually producing cleistogamous panicles; **lowest internodes** usually swollen. **Leaves** mostly cauline; **sheaths** usually longer than the internodes, smooth; **ligules** 6–15 mm; **lower blades** 5–25(40) cm long, 2–6 mm wide, flat, rather lax; **upper blades** reduced, sometimes to 1 cm. **Cleistogamous panicles** with 6–10 mm spikelets bearing 1–3 florets, lemmas unawned, sometimes mucronate, anthers about 0.7 mm. **Terminal panicles** 6–27 cm, often partially enclosed in the uppermost sheaths; **branches** 2–8 cm, ascending to drooping, often sinuous; **pedicels** absent or to 10 mm. **Spikelets** 12–25 mm, with 3–9 florets. **Glumes** $^1/_2$–$^2/_3$ the length of the adjacent lemmas; **lower glumes** 3.5–6.7 mm, 1–3-veined; **upper glumes** 5–8 mm, 3–5-veined; **lemmas** 5–11 mm, 7–9-veined, hispid or tuberculate, deeply bilobed, awned, awns 8–17 mm, arising near midlength; **paleas** 4–6 mm, chartaceous, margins scabrous to ciliolate distally; **anthers** 0.7–2 mm. $2n$ = unknown.

Amphibromus scabrivalvis is native to open grasslands of South America. It was discovered growing in strawberry patches in Tangipahoa Parish, Louisiana, in the late 1950s. Despite efforts to eradicate it, the species persists there; it is not known to have spread elsewhere. North American plants belong to **Amphibromus scabrivalvis** (Trin.) Swallen var. **scabrivalvis**.

2. Amphibromus nervosus (Hook. f.) Baill. [p. 705]
COMMON SWAMP WALLABYGRASS

Plants perennial; usually cespitose, occasionally rhizomatous. **Culms** 30–125 cm tall, 1–3 mm thick, erect, terete to flattened, glabrous; **nodes** 2–5; **lowest internodes** not swollen. **Sheaths** smooth or scabridulous, ribbed; **ligules** 10–20 mm, acute to acuminate; **blades** 10–30 cm long, 1.5–3.5 mm wide, flat or involute, smooth to scabrous, abaxial surfaces scabridulous, adaxial surfaces deeply ribbed, scabrous. **Terminal panicles** 15–40 cm, erect, sparse, the lower portion sometimes partially enclosed in the uppermost sheaths; **branches** usually 7–15 cm, ascending or appressed, often flexuous; **pedicels** usually 10–20 mm. **Spikelets** 10–16 mm, with 4–6 florets. **Glumes** unequal to subequal, green, sometimes purplish in the center, with

hyaline margins; **lower glumes** 2.5–5.5 mm, 3–5-veined; **upper glumes** 3–6.5 mm, 3–5-veined; **lemmas** 5–7.2 mm, 5–7-veined, scabrous, apices not appearing constricted, usually 4-toothed, 2 lateral teeth smaller than the 2 central teeth, awned from the lower $^2/_5$–$^3/_5$ of the lemmas, awns 12–22 mm, geniculate and twisted; **paleas** $^3/_4$ as long as to subequal to the lemmas, papillose; **anthers** of chasmogamous florets 2.2–3 mm, those of cleistogamous florets 0.3–1.4 mm. $2n$ = unknown.

Amphibromus nervosus is the most common species in the genus. It has frequently been misidentified as *A. neesii*, but has a lower awn insertion. Such misidentification is the basis of the report of *A. neesii* in North America; examination of the voucher specimens showed them to be *A. nervosus*. They were collected in 1990 from a vernal pool in Sacramento County, California. Its seeds had been found earlier as a contaminant in *Trifolium subterraneum* seed being imported from Australia. The discovery of living plants is of particular concern, because of their ability to invade and survive in vernal pools.

3. Amphibromus neesii Steud. [p. 705]
SOUTHERN SWAMP WALLABYGRASS

Plants perennial; usually cespitose, occasionally rhizomatous. **Culms** 30–150 cm tall, 1–2.5 mm thick, erect, terete to flattened, glabrous; **nodes** 2–4; **lowest internodes** not swollen. **Sheaths** smooth or scabridulous, ribbed; **ligules** 4.5–8 mm, acute to acuminate; **blades** 10–20(37) cm long, 2–4 mm wide, flat or involute, smooth to scabrous, abaxial surfaces scabridulous, adaxial surfaces deeply ribbed, scabrous. **Terminal panicles** 15–40 cm, erect, sparse, lower portion rarely partially enclosed in the uppermost sheaths; **branches** usually 7–14 cm, ascending or appressed, often flexuous; **pedicels** usually 10–20 mm. **Spikelets** 8–17 mm, with 2–6 florets. **Glumes** unequal, green, sometimes purplish in the center, with hyaline margins; **lower glumes** 3.5–6.5 mm, 3–5-veined; **upper glumes** 4.5–7.5 mm, 5–7-veined; **lemmas** 5–8.4 mm, 7-veined, papillose to scabrous, apices appearing constricted, 2–4-toothed, awned from the upper $^2/_3$–$^3/_4$ of the lemmas, awns 14–26 mm, geniculate and twisted; **paleas** $^8/_{10}$–$^9/_{10}$ the length of the lemmas, papillose; **anthers** of chasmogamous florets 1.3–2.3 mm. $2n$ = unknown.

Amphibromus neesii is an Australian species that grows on floodplains and river banks, and in marshes and lagoons. It was first reported as growing in North America in 1990; examination of the voucher specimens showed them to be *A. nervosus*, which differs from *A. neesii* in having a lower lemma awn insertion. Both species are included in this treatment to help prevent future misidentification of the two species in North America.

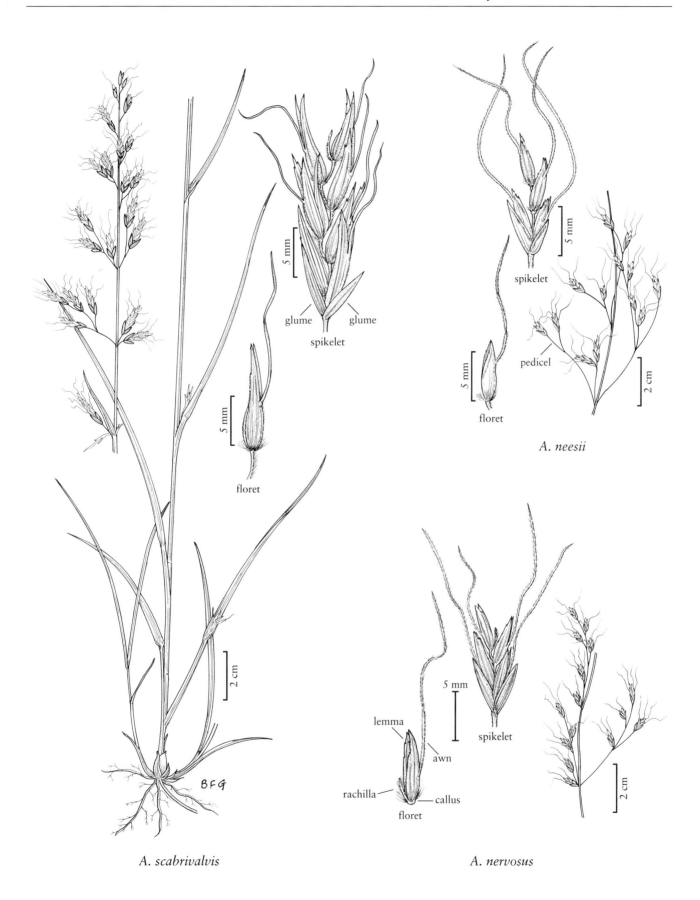

glume   glume

spikelet

floret

A. neesii

spikelet

floret

pedicel

A. scabrivalvis

BFG

lemma

awn

rachilla

callus

floret

spikelet

A. nervosus

AMPHIBROMUS

14.49 **CALAMAGROSTIS** Adans.

Kendrick L. Marr

Richard J. Hebda

Craig W. Greene†

Plants perennial; often cespitose, usually rhizomatous. **Culms** 10–210 cm, unbranched or branched, more or less smooth, **nodes** 1–8. **Sheaths** open, smooth or scabrous; **auricles** absent; **ligules** membranous, usually truncate to obtuse, sometimes acute, entire or lacerate, lacerations often obscuring the shapes; **blades** flat to involute, smooth or scabrous, rarely with hairs. **Inflorescences** panicles, open or contracted, sometimes spikelike; **branches** appressed to more or less drooping, some branches longer than 1 cm. **Spikelets** pedicellate, weakly laterally compressed, with 1(2) florets; **rachillas** prolonged beyond the base of the distal floret(s), usually hairy; **disarticulation** above the glumes. **Glumes** membranous, subequal, equal to, or longer than the lemmas, rounded or keeled, backs smooth or scabrous, rarely long-scabrous with bent projections, veins obscure to prominent, apices acute to acuminate, rarely awn-tipped or attenuate; **lower glumes** 1(3)-veined; **upper glumes** 3-veined; **calluses** hairy, hairs 0.2–6.5 mm, sparse to abundant; **lemmas** 3(5)-veined, smooth or scabrous, apices usually tapering into 4 teeth, awned; **awns** arising from near the base to near the apices, straight or bent, sometimes delicate and indistinct from the callus hairs, sometimes exserted beyond the lemma margins; **paleas** well developed, almost as long as to slightly longer than the lemmas, thin, 2-veined; **anthers** 3, sometimes sterile. **Caryopses** shorter than the lemmas, concealed at maturity, oblong, usually glabrous. *x* = 7. Name from the Greek *calamos*, 'reed', and *agrostis*, 'grass'.

Calamagrostis grows in cool-temperate regions and is especially diverse in mountainous regions. Its species grow in both moist and xeric habitats. There are about 100 species of *Calamagrostis*, if *Deyeuxia* Clarion *ex* P. Beauv. and *Lachnagrostis* are recognized as distinct from *Calamagrostis*. The latter two genera are often considered to be restricted to the Southern Hemisphere (Edgar 1995; Jacobs 2001). According to the criteria used by Phillips and Chen (2003) to distinguish *Calamagrostis* and *Deyeuxia*, most North American species of *Calamagrostis* fit within *Deyeuxia*. There has been insufficient time to evaluate the merits of their recommendation, adoption of which would require many new combinations.

Twenty-five species of *Calamagrostis* grow in the *Flora* region; one, *C. epigejos*, is introduced. Some species of *Calamagrostis* are rangeland forage grasses, but most occur too sparsely to be important for livestock. Agriculture Canada in Alberta experimented with cultivation of some western species during the 1960s and 1970s.

This treatment includes one cultivar, *Calamagrostis* ×*acutiflora* 'Karl Foerster', that is becoming increasingly popular in horticulture. A cultivar of *C. canadensis* has been registered for use in revegetation in arctic Alaska.

Interspecific hybridization is common; vivipary and agamospermy also occur in some species. Interspecific hybridization, polyploidy, and apomixis contribute to the taxonomic difficulty of the genus.

Some species of *Calamagrostis* are of interest because of their restricted distributions. These include: *C. howellii* (Columbia Gorge in Washington and Oregon); *C. tweedyi* (Washington, Oregon, and Montana); *C. tacomensis* (Washington and Oregon); *C. ophitidis*, *C. foliosa*, *C. muiriana*, and *C. bolanderi* (California); *C. breweri* (California and Oregon); and *C. cainii* (North Carolina and Tennessee).

An incomplete draft treatment of this genus was prepared by Craig W. Greene in 1993, with minor revisions made until 1999. After Greene's death in 2003, completion of the treatment was taken up by Marr and Hebda. The taxa recognized here essentially follow Greene's concepts, with the following exceptions: *Calamagrostis breweri sensu* Greene (1993) has been split into *C. muiriana* and *C. breweri sensu* Wilson and Gray (2002); *C. tacomensis* is recognized as a species distinct from *C. sesquiflora*; and *C. purpurascens* var. *laricina* Louis-Marie and *C. stricta* subsp. *borealis* (C. Laest.) Á. Löve & D. Löve are not recognized. Descriptions of eastern North American taxa are largely based on Greene's (1980) observations. Northwestern North American taxa are described on the basis of Marr and Hebda's data and field experience. Other western United States taxa were also examined as herbarium specimens; their descriptions include observations by Marr and Hebda. Greene's key was rewritten to conform with the new data.

There is a high degree of misidentification of taxa within this genus (30% for some species in some herbaria), and species distributions should be taken as a guide only. Much more field collecting is needed for several of the taxa in order to verify their distributions, especially near the limits of their ranges. *Calamagrostis* is sometimes confused with *Agrostis*; there is no single character that distinguishes all species of *Calamagrostis* from those of *Agrostis*. In general, *Calamagrostis* has larger plants with larger, more substantial lemmas and paleas than *Agrostis*, and tends to occupy wetter habitats.

Measurements of the rachilla and callus hairs reflect the longest hairs present. Panicle widths refer to pressed specimens. The following key will enable typical specimens to be identified readily, but atypical specimens are common. For this reason, most leads require observation of a combination of characters, notably awn length, length of callus hairs relative to the lemma, glume length and scabrosity, panicle size, and leaf width.

SELECTED REFERENCES **Edgar, E.** 1995. New Zealand species of *Deyeuxia* P. Beauv. and *Lachnagrostis* Trin. (Gramineae: Aveneae). New Zealand J. Bot. 33:1–33; **Greene, C.W.** 1980. The systematics of *Calamagrostis* (Gramineae) in eastern North America. Ph.D. dissertation. Department of Biology, Harvard University, Cambridge, Massachusetts, U.S.A. 238 pp.; **Greene, C.W.** 1984. Sexual and apomictic reproduction in *Calamagrostis* (Gramineae) from eastern North America. Amer. J. Bot. 71:285–293; **Greene, C.W.** 1993. *Calamagrostis*, Reed Grass. Pp. 1243–1246 *in* J.C. Hickman, ed. The Jepson Manual: Higher Plants of California. University of California Press, Berkeley, California, U.S.A. 1400 pp.; **Harmon, P.J.** 1981. The vascular flora of the ridge top of North Fork Mountain, Grant and Pendleton counties, West Virginia. Master's thesis, Southern Illinois University, Carbondale, Illinois, U.S.A. 434 pp.; **Hitchcock, C.L., A. Cronquist,** and **M. Ownbey.** 1969. Vascular Plants of the Pacific Northwest. Part 1: Vascular Cryptogams, Gymnosperms, and Monocotyledons. University of Washington Press, Seattle, Washington, U.S.A. 914 pp.; **Hultén, E.** 1968. Flora of Alaska and Neighboring Territories. Stanford University Press, Stanford, California, U.S.A. 1008 pp.; **Jacobs, S.W.L.** 2001. The genus *Lachnagrostis* (Gramineae) in Australia. Telopea 9:439–448; **Kawano, S.** 1965. *Calamagrostis purpurascens* R. Br. and its identity. Acta Phytotax. Geobot. 21:73–90; **Marr, K.L.** and **R.J. Hebda.** 2006. *Calamagrostis tacomensis* (Poaceae), a new species from Washington and Oregon. Madroño 53:290–300; **Phillips, S.M.** and **W.-L. Chen.** 2003. Notes on grasses (Poaceae) for the Flora of China, I: *Deyeuxia*. Novon 13:318–321. **Reznicek, A.A.** and **E.J. Judziewicz.** 1996. A new hybrid species, ×*Calammophila don-hensonii* (*Ammophila breviligulata* × *Calamagrostis canadensis*, Poaceae) from Grand Island, Michigan. Michigan Bot. 35:35–40; **Wilson, B.L.** and **S. Gray.** 2002. Resurrection of a century-old species distinction in *Calamagrostis*. Madroño 49:169–177.

1. Callus hairs more than 1.3 times as long as the lemmas; lemmas at least 2 mm shorter than the glumes, long-acuminate . . . . 1. *C. epigejos*
1. Callus hairs usually less than 1.2 times as long as the lemmas; if the callus hairs longer than the lemmas, then the lemmas less than 2 mm shorter than the glumes and not long-acuminate.
  2. Blades usually densely hairy on the adaxial surfaces; glumes keeled, scabrous; awns 4.5–9 mm long . . . . 2. *C. purpurascens*
  2. Blades glabrous or sparsely hairy on the adaxial surfaces; glumes keeled or rounded, scabrous or smooth; awns 0.5–17 mm long.
    3. Leaves with abundant white glands between the veins, visible at about 10×; awns 5–8 mm long; plants of California . . . . 4. *C. ophitidis*
    3. Leaves without abundant white glands between the veins; awns 0.5–17 mm long; plants from throughout the *Flora* region, including California.

4. Awns 5–17 mm long, always exserted and bent; if the awns 5–6 mm long, either some blades wider than 2 mm or the abaxial blade surfaces scabrous.
  5. Panicles open, (2)3.5–6.5(8) cm wide when pressed, branches spikelet-bearing only beyond midlength; awns 10–16 mm long . . . . 3. *C. howellii*
  5. Panicles usually contracted, (0.5)0.8–3 cm wide if open, the branches spikelet-bearing to below midlength, usually to the base; awns 5–17 mm long.
    6. Some leaf blades 6–13 mm wide; culms (47)60–120(150) cm tall . . . . 5. *C. tweedyi*
    6. All leaf blades (1.5)2–7 mm wide; culms (15)30–60(95) cm tall, if the culms taller than 60 cm, then the blades less than 4 mm wide.
      7. Awns 12–14(17) mm long; plants of California . . . . 6. *C. foliosa*
      7. Awns (5.4)7–11(13) mm long; plants of Alaska, British Columbia, Washington, and Oregon.
        8. Glume apices long-acuminate, usually twisted distally; glume keels usually scabrous for their whole length . . . . 7. *C. sesquiflora*
        8. Glume apices usually acute, if acuminate, not twisted distally; glume keels smooth or sparsely scabrous on the distal ¹/₂ . . . . 8. *C. tacomensis*
4. Awns 0.5–6 mm long, exserted or not, bent or straight; if the awns 5–6 mm long, then either all blades less than 2 mm wide or the abaxial blade surfaces smooth or nearly so.
  9. Awns attached on the distal ²/₅ of the lemmas, 0.5–2 mm long, straight; blades flat; panicles contracted, 0.7–2.5(3) cm wide.
    10. Lateral veins of the glumes prominent; rachillas hairy only distally . . . . 9. *C. cinnoides*
    10. Lateral veins of the glumes obscure; rachillas hairy throughout their length
    . . . . 10. *C. scopulorum*
  9. Awns attached on the lower ¹/₂(⁷/₁₀) of the lemmas, 0.9–6 mm long, straight or bent; blades flat or involute; panicles open or contracted, 0.4–5.5(9) cm wide.
    11. Blades 0.2–1.7 mm wide; panicles (1.5)1.9–8.5 cm long; callus hairs sparse.
      12. Blades involute, 0.2–0.4 mm in diameter; ligules 1–2.5 mm long . . . . 11. *C. muiriana*
      12. Blades flat, 0.9–1.7 mm wide, sometimes involute and 0.4–0.6 mm in diameter when dry; ligules 1.7–6 mm long . . . . 12. *C. breweri*
    11. Blades (1)1.5–20 mm wide, most wider than 2 mm; panicles (2)3–30(40) cm long; callus hairs sparse to abundant.
      13. Awns usually exserted, (2.8)3–6 mm long; callus hairs 0.1–0.7 times as long as the lemmas; leaf collars hairy or glabrous.
        14. Culms 10–55(60) cm tall; panicles open; blades (1)1.5–3(4) mm wide.
          15. Blades 2–8(15) cm long; panicles erect; plants of brackish arctic and subarctic coastal habitats . . . . 13. *C. deschampsioides*
          15. Blades (5)15–39 cm long; panicles often drooping; plants of rocky soils and disturbed sites in the mountains of Tennessee and North Carolina . . . . 14. *C. cainii*
        14. Culms (26)50–210 cm tall; panicles open or contracted; blades (1)1.5–8(12) mm wide, most blades wider than 3 mm.
          16. Panicles open, 2.5–6 cm wide; panicle branches with the spikelets confined to the distal ¹/₄–¹/₂; leaf collars glabrous . . . . 15. *C. bolanderi*
          16. Panicles usually contracted, 0.7–3(7) cm wide; panicle branches usually with the spikelets confined to the distal ¹/₂–²/₃, sometimes spikelet-bearing to the base; leaf collars hairy or glabrous.
            17. Culms 135–210 cm tall; plants cultivated ornamentals . . . . . 16. *C. ×acutiflora*
            17. Culms 26–120 cm tall; plants native.
              18. Awns 4–5.5 mm long; leaf collars glabrous; plants often densely cespitose; rhizomes usually 2–4 mm thick . . . . . 17. *C. koelerioides*
              18. Awns 2–4.5 mm long; leaf collars sometimes hairy; plants loosely cespitose; rhizomes 0.5–2 mm thick.

19. Blades (2)3–8(12) mm wide; panicle branches usually with the spikelets restricted to the distal ¹/₂–²/₃, sometimes spikelet-bearing to the base; plants from east of the 100th meridian . . . . 18. *C. porteri*

19. Blades (1)2–5(8) mm wide; panicle branches spikelet-bearing to the base; plants from west of the 100th meridian . . . . 19. *C. rubescens*

13. Awns usually not exserted, or if exserted, then barely so, 0.9–3.1(4) mm long; callus hairs (0.1)0.2–1.2(1.5) times as long as the lemmas; leaf collars glabrous or hairy, if hairy, then the callus hairs more than 0.7 times as long as the lemmas.

20. Callus hairs shorter than 1 mm, 0.2–0.3 times as long as the lemmas; awns bent . . . . 21. *C. pickeringii*

20. Callus hairs longer than 1 mm, (0.2)0.3–1.2(1.5) times as long as the lemmas; awns straight or bent.

21. Culms usually scabrous, rarely smooth; awns slightly bent; callus hairs 0.4–0.8 times as long as the lemmas; blades usually involute, 1–4 mm wide, the abaxial surfaces scabrous; nodes 1–2 . . . . 22. *C. montanensis*

21. Culms smooth to slightly scabrous; awns usually straight, rarely bent; callus hairs (0.2)0.5–1.2(1.5) times as long as the lemmas; blades flat or involute, 1–20 mm wide, the abaxial surfaces smooth or scabrous; nodes 1–8.

22. Lemmas (3)4–5 mm long; glumes keeled; blades flat . . . . 20. *C. nutkaensis*

22. Lemmas 2–4(5) mm long, if the lemmas longer than 4 mm, then the glumes rounded and the abaxial blade surfaces smooth; glumes keeled or rounded; blades flat or involute.

23. Panicle branches 2.7–6.5(12) cm long; ligules lacerate; glumes scabrous on the keels, often throughout; blades flat, the abaxial surfaces scabridulous or scabrous; nodes (2)3–7(8); panicles open.

24. Awns bent, stout, readily distinguished from the callus hairs; collars densely hairy; plants of New York State . . . . 23. *C. perplexa*

24. Awns usually straight, delicate, often difficult to distinguish from the callus hairs; collars rarely hairy; plants of northern and western North America . . . . 24. *C. canadensis*

23. Panicle branches (1)1.4–5(9.5) cm long; if the panicle branches longer than 3.7 cm, then the ligules usually entire; glumes smooth or scabrous only on the keels; blades flat or involute, the abaxial surfaces smooth or scabrous; nodes 1–3(4); panicles loosely contracted.

25. Glume lengths usually more than 3 times the widths, smooth, the keels rarely slightly scabrous, the lateral veins obscure; spikelets 3.5–5.5 mm long; awns usually slender and similar to the callus hairs . . . . 25. *C. lapponica*

25. Glume lengths usually less than 3 times the widths, usually smooth, rarely scabrous, the keels smooth or scabrous, the lateral veins prominent or obscure; spikelets 2–5 mm long; awns stout, usually readily distinguished from the callus hairs . . . . 26. *C. stricta*

1. Calamagrostis epigejos (L.) Roth [p. 711]

BUSHGRASS, CHEE REEDGRASS, FEATHERTOP,
CALAMAGROSTIDE ÉPIGÉIOS,
CALAMAGROSTIDE COMMUNE

Plants with sterile culms; cespitose, with numerous rhizomes 8+ cm long, 1.5–2 mm thick. **Culms** (50)100–150(160) cm, unbranched, slightly scabrous beneath the panicles; **nodes** (1)2–4(6). **Sheaths** and **collars** smooth or slightly scabrous; **ligules** (1.5)3–7(13) mm, truncate to obtuse, usually entire, infrequently lacerate; **blades** (6)25–40(55) cm long, (2.5)3.5–8(13) mm wide, flat, pale green, scabrous. **Panicles** (14)18–23(35) cm long, (2)2.5–4(6) cm wide, erect, contracted, greenish; **branches** (3.5)5–8(11) cm, smooth or slightly scabrous, spikelets usually confined to the distal $^3/_4$, infrequently confined to the distal $^1/_2$. **Spikelets** (4)4.5–5.5(8) mm; **rachilla prolongations** about 1 mm, hairs about 3 mm. **Glumes** slightly keeled, usually smooth, infrequently scabrous near the apices, lateral veins prominent, apices long-acuminate; **callus hairs** (2)3.5–5(6.5) mm, (1.3)1.5–2(2.5) times as long as the lemmas, abundant; **lemmas** 2–3.5(5) mm, (1.5)2–3(4.5) mm shorter than the glumes; **awns** (1.5)2–3(4) mm, attached to the lower $^1/_3$–$^2/_3$ of the lemmas, not exserted, delicate, not easily distinguished from the callus hairs, usually straight, infrequently bent; **anthers** about (1)1.5(2) mm. $2n = 28, 35, 42, 56, \pm70$.

Calamagrostis epigejos is an introduced Eurasian species that was first found in North America in the 1920s. It grows in waste places, along roadsides, in juniper swamps, sandy woods, and thickets, and on rehabilitated tailings and cinders of railway beds. It is known from scattered locations in southern Canada and the contiguous United States. It is probably more widespread than shown. In 2005, it was collected from the west coast, in southwestern British Columbia, for the first time.

In Ontario, *Calamagrostis epigejos* became established from impurities in seed mixtures used for highway roadcut revegetation. Rhizomes purchased from Manitoba have been used to stabilize gold mine tailings in Ontario. In Wisconsin, it was planted for erosion control at least as early as 1950. The Idaho record is from reseeded rangeland plots, and the Wyoming one from plants grown from purchased rhizomes.

Greene (1980) stated that almost all plants from the *Flora* region fit *Calamagrostis epigejos* var. *georgica* (K. Koch) Griseb. Because there is notable variability and overlap in the characteristics used to distinguish the varieties in Eurasia, no attempt has been made to provide a varietal treatment for the *Flora* region. Hybrids of *C. epigejos* with *C. arundinacea* (L.) Roth are called *C. ×acutiflora* (p. 721).

2. Calamagrostis purpurascens R. Br. [p. 711]

PURPLE REEDGRASS, CALAMAGROSTIDE POURPRE

Plants apparently without sterile culms; strongly cespitose, often with rhizomes 1–4 cm long, 1–2 mm thick. **Culms** (10)30–80 cm, usually unbranched, occasionally branched, usually slightly to strongly scabrous, sometimes puberulent beneath the panicles; **nodes** (1)2(3). **Sheaths** scabrous; **collars** usually scabrous or hairy, rarely smooth; **ligules** (1.5)2–4(9) mm, usually truncate and entire, sometimes lacerate. **Blades** (4)5–17(30) cm long, 2–5(6) mm wide, flat or involute, stiff, abaxial surfaces scabrous, adaxial surfaces usually densely long-hairy, rarely sparsely hairy. **Panicles** 4–13(15) cm long, 0.9–2(2.8) cm wide, erect, contracted, infrequently interrupted near the base, often red- or purple-tinged; **branches** 1.3–3.5 cm, scabrous, prickles long, almost hairlike, spikelet-bearing to the base. **Spikelets** (4.5)5.5–6.5(8) mm; **rachilla prolongations** about (1)2 mm, hairs about 2 mm. **Glumes** keeled, usually scabrous, rarely scabrous on the keels only, lateral veins obscure to prominent, apices acute; **callus hairs** (0.9)1.2–1.5(2.4) mm, 0.2–0.4(0.6) times as long as the lemmas, sparse; **lemmas** (3.5)4–4.5(5) mm, usually 1–2.5 mm shorter than, rarely equal to, the glumes; **awns** (4.5)6–7(9) mm, attached to the lower $^1/_{10}$–$^1/_3$ of the lemmas, usually exserted, stout, easily distinguished from the callus hairs, bent; **anthers** (1.3)1.7–2.5(2.9) mm. $2n = 42–58, 84$.

Calamagrostis purpurascens grows in alpine tundra, on subalpine slopes, in grasslands, sand dunes, meadows, coniferous and deciduous forests, and disturbed soils, usually on rocky ridgetops and slopes and, infrequently, on valley floors. It prefers well- to moderately-drained, medium- to coarse-textured substrates, including scree and talus, that are often calcareous, at elevations from 15–4000 m. Its range extends from Alaska through Canada to Greenland and Newfoundland, including the islands of the Canadian arctic, and south in the western mountains to California and northern New Mexico. It does not occur near the open coast except in the Aleutian Islands, the Arctic, and the Olympic Peninsula in Washington. In Asia, it ranges from eastern and central arctic Siberia south to the Kamchatka Peninsula and Sakhalin Island.

The hairy adaxial leaf surfaces are a reliable diagnostic characteristic for *Calamagrostis purpurascens*. Many specimens from Washington and Oregon

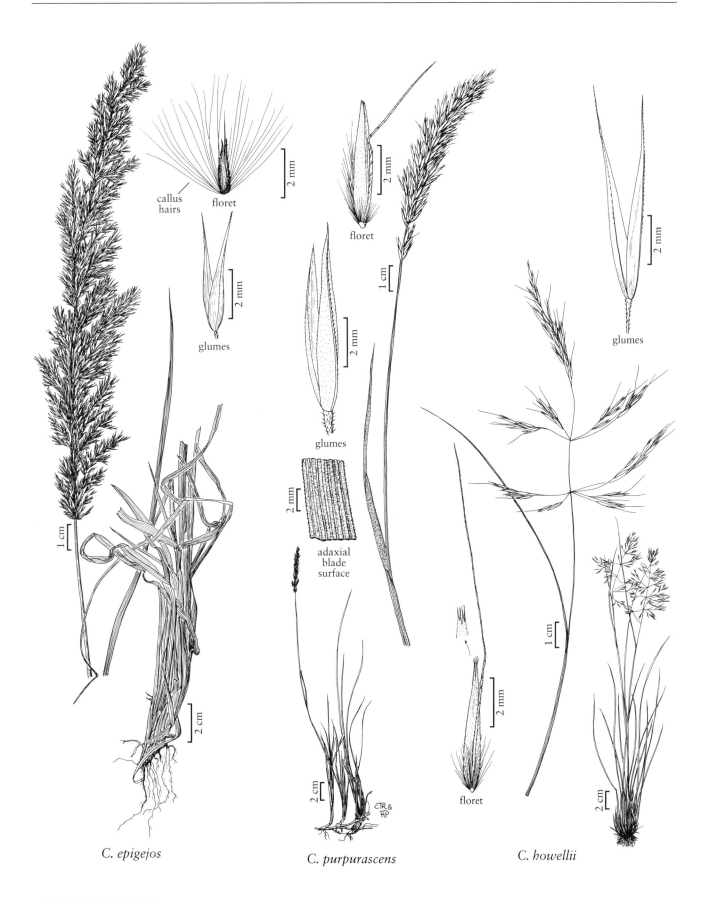

callus
hairs

floret

2 mm

glumes

2 mm

C. epigejos

1 cm

2 cm

floret

2 mm

glumes

2 mm

adaxial
blade
surface

2 mm

glumes

1 cm

C. purpurascens

2 cm

floret

2 mm

glumes

2 mm

1 cm

C. howellii

2 cm

CALAMAGROSTIS

currently identified as *C. purpurascens* belong to *C. tacomensis* (p. 716). In addition to differing in its leaf vestiture, *C. purpurascens* has shorter awns and panicle branches, and more scabrous glumes, than *C. tacomensis*. Plants of *C. purpurascens* that have short awns barely projecting beyond the lemma margins have been mistaken for *C. montanensis* (p. 724), but that species does not have hairy adaxial leaf surfaces.

Calamagrostis purpurascens var. *laricina* Louis-Marie supposedly has shorter glumes and awns, with the awns barely exserted, if at all. The variety is not recognized here because the range in variation in these two characters is continuous; plants that match the description of var. *laricina* are widely distributed throughout the range of the species. Some collections having both short and long awns are on the same sheet.

Calamagrostis lepageana Louis-Marie, collected only from Mont-Commis, Quebec, is here included within *C. purpurascens*. It differs from the typical form in that the panicle is more open; the branches are sparsely short-scabrous; and the glumes are at the shortest limit for the species, and are smooth or sparsely scabrous only on the keels. Other characteristics, including the densely hairy adaxial leaf surfaces, fit *C. purpurascens*.

According to Norwegian botanist Reidar Elven (pers. comm.), five specimens at ALA, collected from four sites on the Seward Peninsula in Alaska, are distinct from *Calamagrostis purpurascens* but are related to it, or are possibly a hybrid between *C. purpurascens* and *C. sesquiflora* (p. 714). They differ from typical *C. purpurascens* in having bristly (versus hairy) upper leaf surfaces, and acuminate, dark purple (versus acute, pale pink or lilac) glumes. They can be called *C. purpurascens* subsp. *arctica* (Vasey) Hultén; in eastern Russia this taxon is recognized at the species level as *C. arctica* Vasey.

3. Calamagrostis howellii Vasey [p. 711]
HOWELL'S REEDGRASS

Plants sometimes with sterile culms; usually densely cespitose, occasionally with rhizomes shorter than 1 cm. **Culms** (25)35–45(60) cm, unbranched, smooth or slightly scabrous beneath the panicles; **nodes** 1–2. **Sheaths** and **collars** smooth; **ligules** (2.5)3.5–6 mm, acute, lacerate; **blades** (9)12–20(25) cm long, 1–2.5(3) mm wide, flat to involute, abaxial surfaces smooth, adaxial surfaces finely scabrous, glabrous or sparsely hairy. **Panicles** (5)7–12(15) cm long, (2)3.5–6.5(8) cm wide, loose, open, straw-colored or green to purplish; **branches** (2)3.5–5(7) cm, smooth or sparsely scabrous, spikelets usually confined to the distal ¹/₂. **Spikelets** (5.5)6–8 mm; **rachilla prolongations** 1–1.5(2) mm, hairs

(1.5)2–2.5(3) mm. **Glumes** rounded to slightly keeled, smooth or scabrous distally, lateral veins usually prominent and raised, apices acuminate; **callus hairs** 2–3(4.5) mm, 0.4–0.6(0.7) times as long as the lemmas, abundant; **lemmas** 4.5–5 mm, about 2 mm shorter than the glumes; **awns** (10)13–16 mm, attached to the lower ¹/₅–²/₅ of the lemmas, exserted, stout, easily distinguished from the callus hairs, strongly bent; **anthers** (2)2.5–3(4) mm. $2n = 28$.

Calamagrostis howellii grows on dry rocky slopes, banks, ledges, and in cliff crevices, sometimes on basalt, from 100–500 m. It grows only in the Columbia River Gorge of Washington and Oregon.

4. Calamagrostis ophitidis (J.T. Howell) Nygren [p. 713]
SERPENTINE REEDGRASS

Plants usually with sterile culms; usually cespitose, often with rhizomes 2–15 cm long, 1–3 mm thick. **Culms** (30) 55–80(100) cm, unbranched, scabrous beneath the panicles; **nodes** (1)2(5). **Sheaths** and **collars** usually scabrous, rarely smooth; **ligules** (0.5)2–5.5(7) mm, usually truncate, entire to slightly lacerate; **blades** (8)10–20(27) cm long, (1.5)2–3(4) mm wide, usually involute, abaxial surfaces usually scabrous, rarely smooth, adaxial surfaces scabrous, glabrous or sparsely hairy, both surfaces with abundant white glands between the veins, visible only with magnification. **Panicles** (6)8–11(15) cm long, (1)1.2–1.5 cm wide, contracted, mostly erect, pale green to green; **branches** 2–4(4.5) cm, scabridulous, usually spikelet-bearing to the base. **Spikelets** (4.5)5–7(8) mm; **rachilla prolongations** about 1.5 mm, hairs 1–2 mm. **Glumes** keeled, usually scabrous over the entire surface, rarely only the keels scabrous distally, lateral veins prominent, apices acute to acuminate; **callus hairs** 1–1.5(2) mm, 0.2–0.4 times as long as the lemmas, sparse; **lemmas** 4.5–6.5 mm, 0–2 mm shorter than the glumes; **awns** 5–6(8) mm, attached to the lower ¹/₁₀–¹/₅ of the lemmas, exserted less than 2 mm, stout, distinguishable from the callus hairs, bent; **anthers** (2.5)3–3.5(4) mm. $2n = 28$.

Calamagrostis ophitidis grows in meadows, seeps, grasslands, and chaparral, as well as in coniferous forests, on serpentine outcrops and soils, at 50–1100 m. It is known only from Sonoma, Marin, Mendocino, Lake, and Napa counties in California.

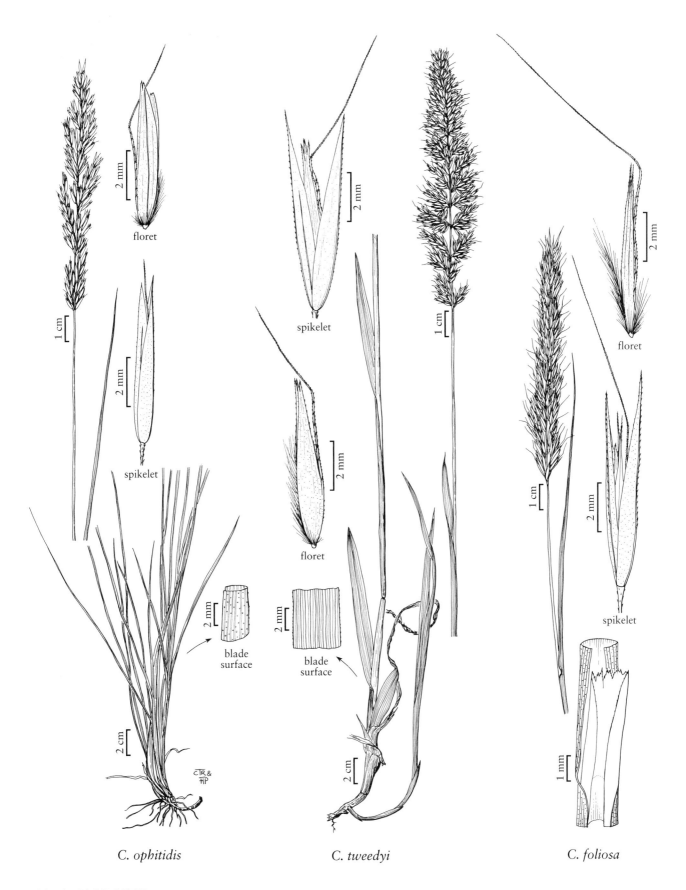

C. ophitidis

C. tweedyi

C. foliosa

CALAMAGROSTIS

5. Calamagrostis tweedyi (Scribn.) Scribn. [p. 713]
CASCADE REEDGRASS, TWEEDY'S REEDGRASS

Plants without sterile culms; loosely cespitose, with rhizomes 1–10 cm long, 2–4 mm thick. **Culms** (47)60–120(150) cm, unbranched, smooth, rarely slightly scabrous; **nodes** 2–3. **Sheaths** and **collars** smooth; **ligules** (1)3.5–6(8) mm, obtuse, lacerate; **blades** (3)4–20(38) cm long, (2)3–8(13) mm wide, culm blades wider than 6 mm, flat, abaxial surfaces smooth, adaxial surfaces smooth or slightly scabrous, glabrous or sparsely hairy. **Panicles** 7–16(19) cm long, (1)1.5–2 cm wide, erect, usually contracted, sometimes interrupted near the base, pale purple to purple; **branches** (0.2)2.4–6.7(7.7) cm, smooth, sometimes sparsely scabrous distally, spikelet-bearing to the base. **Spikelets** (4.5)5.5–8(9) mm; **rachilla prolongations** (0.5)1–2(4) mm, hairs 1.5–3 mm. **Glumes** keeled, smooth or the keels scabrous, lateral veins prominent, apices acute; **callus hairs** 0.8–1 mm, 0.2–0.3 times as long as the lemmas, sparse; **lemmas** (4)4.5–6.5(7.5) mm, 0–1.5 mm shorter than the glumes, scabridulous; **awns** 6–8 mm, attached to the lower $^1/_5$–$^3/_{10}$ of the lemmas, exserted more than 2 mm, stout, easily distinguished from the callus hairs, bent; **anthers** 2–3.5 mm. $2n$ = unknown.

Calamagrostis tweedyi grows in montane to subalpine moist meadows and coniferous forests, often in association with *Carex geyeri*, at 900–2000 m. Its range extends from Washington and Oregon to western Montana.

6. Calamagrostis foliosa Kearney [p. 713]
LEAFY REEDGRASS

Plants sometimes with sterile culms; cespitose, rhizomes occasionally present, shorter than 1 cm. **Culms** (25)30–60(70) cm, unbranched, mostly smooth, sparsely scabrous beneath the panicles; **nodes** 1–3. **Sheaths** and **collars** usually smooth; **ligules** (3)4–6(7) mm, usually truncate to obtuse, usually entire, sometimes lacerate; **blades** (10)11–21(27) cm long, (1.5)2–2.5(4) mm wide, mostly basal, flat or involute, abaxial surfaces smooth, adaxial surfaces slightly scabrous, glabrous or sparsely hairy. **Panicles** (9)10–12(19) cm long, 1–1.5(2.5) cm wide, erect to slightly nodding, contracted, branches sometimes slightly spreading at the base, usually pale green, rarely pale purple; **branches** (2)3–4(5) cm, sparsely scabrous, spikelet-bearing to the base. **Spikelets**

(7)8–11 mm; **rachilla prolongations** (1.5)2(3) mm, prominently bearded, hairs 2–3 mm. **Glumes** keeled, keels smooth or scabrous, lateral veins prominent, apices acuminate; **callus hairs** 2.5–3(4) mm, 0.4–0.6 times as long as the lemmas, abundant; **lemmas** (5)6–7(8) mm, (0.5)1–2(3) mm shorter than the glumes; **awns** 12–14(17) mm, attached to the lower $^1/_5$–$^2/_5$ of the lemmas, exserted more than 2 mm, easily distinguished from the callus hairs, bent; **anthers** 3–4.5 mm. $2n$ = 28.

Calamagrostis foliosa grows in coastal scrub and forest, and on rocks and crevices of bluffs and cliffs, from sea level to 1200 m. It is known only from Del Norte, Humboldt, Mendocino, and Sonoma counties in California.

7. Calamagrostis sesquiflora (Trin.) Tzvelev [p. 715]
ONE-AND-A-HALF-FLOWERED REEDGRASS

Plants rarely with sterile culms; strongly cespitose, usually without rhizomes, sometimes with rhizomes 1–2 cm long, 1–2 mm thick. **Culms** (15)30–46(50) cm, unbranched, usually smooth, rarely slightly scabrous beneath the panicles; **nodes** 1–2(3). **Sheaths** and **collars** smooth; **ligules** (0.5)2–5(6) mm, usually truncate, sometimes obtuse, usually entire, sometimes lacerate; **blades** (3)8–25(31) cm long, (2)3–7 mm wide, flat, abaxial surfaces usually scabrous, rarely smooth, adaxial surfaces smooth or slightly scabrous, glabrous or sparsely hairy. **Panicles** 4–11(12) cm long, 0.8–2.5(2.8) cm wide, erect, contracted to somewhat open, usually purple-tinged, sometimes brown or green; **branches** 1.5–3(4) cm, scabrous, prickles sometimes almost hairlike, usually spikelet-bearing to the base, lowest branches sometimes not so. **Spikelets** (5)5.5–8.5(9.5) mm; **rachilla prolongations** (1)1.5(2.2) mm, hairs 1–2.2 mm. **Glumes** keeled, keels usually scabrous for their whole length, sometimes the surfaces also scabrous, lateral veins prominent, apices long-acuminate, usually twisted distally; **callus hairs** (0.8)1.2–1.8(3) mm, 0.1–0.4 times as long as the lemmas, abundant; **lemmas** (3.5)4–4.5(6) mm, (0.5)1–2.5(4.5) mm shorter than the glumes; **awns** (5.4)7–11(13) mm, attached to the lower $^1/_{10}$–$^2/_5$ of the lemmas, exserted more than 2 mm, stout, easily distinguished from the callus hairs, bent; **anthers** (1.2)2.2–3(3.4) mm. $2n$ = 28.

Calamagrostis sesquiflora grows at 0–1000 m in open heath, meadows, and forest openings, on or at the base of open rocky cliffs and knolls, as well as in moist talus. It grows in strictly maritime habitats along the west coast of North America, from the Aleutian Islands in Alaska to the Queen Charlotte Islands and south to

lemma
apex

floret

2 mm

floret

glumes

spikelet

floret

spikelet

floret

rachilla

C. sesquiflora

C. tacomensis

C. cinnoides

CALAMAGROSTIS

Vancouver Island (Brooks Peninsula) in British Columbia. There is also a single collection from the coast of mainland British Columbia. In northeast Asia, it ranges into the Kamchatka Peninsula and Kuril Archipelago.

Some specimens from the northwestern United States are incorrectly identified, partly because an earlier name for *Trisetum spicatum* (L.) K. Richt. was *Trisetum sesquiflorum* Trin.

Calamagrostis sesquiflora has sometimes included *C. tacomensis* (see next) [as *C. vaseyi* Beal]. Several specimens that were previously identified as *C. sesquiflora* are actually *C. tacomensis*. *Calamagrostis sesquiflora* differs in preferring moister habitats, having wider leaves, callus hairs that are shorter relative to the lemmas, shorter panicle branches, and glumes that are often twisted at the apices.

8. Calamagrostis tacomensis K.L. Marr & Hebda [p. 715]
RAINIER REEDGRASS

Plants without sterile culms; cespitose, sometimes densely so, usually without rhizomes, sometimes with rhizomes about 2 cm long, 2–3 mm thick. **Culms** (20)30–55(95) cm, unbranched, smooth or slightly scabrous beneath the panicles; **nodes** (1)2(5). **Sheaths** and **collars** smooth or slightly scabrous; **ligules** (3)3.5–5.5(6) mm, usually truncate to obtuse, usually entire, sometimes lacerate; **blades** (6)7–14(30) cm long, (1.5)2–2.5(4) mm wide, flat, abaxial surfaces usually smooth, rarely slightly scabrous, adaxial surfaces usually slightly scabrous, rarely smooth, glabrous or sparsely hairy. **Panicles** (5)7–10(18) cm long, (0.5)1–2(3) cm wide, loosely contracted, sometimes open, erect to slightly nodding, shiny green and purple; **branches** (2)2.3–4(6) cm, scabrous, usually spikelet-bearing on the distal $^2/_3$, sometimes to the base. **Spikelets** (4)6–6.5(7) mm; **rachilla prolongations** 1.5–2(2.5) mm, hairs (1.5)2(3) mm. **Glumes** often green with a purple patch at the base, keeled, keels smooth or sparsely scabrous on the distal $^1/_2$, lateral veins usually prominent, apices usually acute, sometimes short-acuminate, not twisted; **callus hairs** (1.2)2(2.5) mm, (0.3)0.4–0.5(0.6) times as long as the lemmas, abundant; **lemmas** (3.5)4–5(5.5) mm, (0.5)1.5–2(3) mm shorter than the glumes; **awns** (5.5)7–8.5(10) mm, attached to the lower $^1/_{10}$–$^1/_3$ of the lemmas, exserted more than 2 mm, easily distinguished from the callus hairs, strongly bent; **anthers** (1)2–3(3.5) mm. 2n =unknown.

Calamagrostis tacomensis grows on montane to alpine slopes in dry or wet meadows, seeps, rocky talus

slopes, and cliff crevices, at 400–2200 m. It grows only in the mountains of western Washington and in the Steens Mountains of southeastern Oregon. It reaches its highest known elevations in the Steens Mountains.

This species has previously been identified as either *Calamagrostis purpurascens* (p. 710) (C.L. Hitchcock et al. 1969) or *C. sesquiflora* (p. 714) (Kawano 1965). It differs from *C. purpurascens* in having glabrous leaves, generally longer awns and inflorescence branches, and smoother glumes. It differs from *C. sesquiflora* in having narrower leaves, callus hairs that are longer relative to the lemmas, longer inflorescence branches, and glume apices that are not twisted, as well as in often preferring drier habitats.

9. Calamagrostis cinnoides (Muhl.) W.P.C. Barton [p. 715]
SMALL REEDGRASS

Plants without sterile culms; loosely cespitose, with rhizomes 2–3 cm long. **Culms** (60) 80–140(170) cm, often solitary, unbranched, scabrous; **nodes** 3–5(6). **Sheaths** smooth or slightly scabrous; **collars** smooth to densely scabrous; **ligules** (2)3–4(6) mm, usually truncate, rarely obtuse, usually entire, sometimes lacerate; **blades** (8)10–35(45) cm long, (3)3.5–7.5(10) mm wide, flat, pale green, smooth or slightly scabrous, adaxial surfaces glabrous or sparsely hairy. **Panicles** (8)12–20(25) cm long, (0.8)1.5–2.5(3) cm wide, erect, contracted, green to greenish purple; **branches** (2.5)4–7 cm, sparsely to densely scabrous, some prickles more than 10 times the length of the others, spikelets usually confined to the distal $^1/_2$–$^3/_4$. **Spikelets** 5–7(7.5) mm; **rachilla prolongations** about (0.5)1(1.5) mm, hairy only distally, hairs 2–4 mm. **Glumes** keeled, keels slightly scabrous, lateral veins prominent, apices acuminate, sometimes shortly awned; **callus hairs** 3–4 mm, 0.5–0.7 times as long as the lemmas, abundant; **lemmas** 4.5–5.5 mm, 1–2 mm shorter than the glumes; **awns** 1–2 mm, attached on the upper $^2/_5$ of the lemmas, not exserted, straight; **anthers** about 1.5 mm. 2n = unknown.

Calamagrostis cinnoides is found on roadsides, in ditches, pond edges, and boggy streamhead seepages, and along streams in oak or oak-pine woods on sandy to peaty soils, at 5–1100 m. Its range extends throughout eastern North America, from Nova Scotia and Maine to Georgia and Louisiana. It is adventive in Ohio.

Since *Arundo canadensis* Michx. was cited as a synonym for both *Arundo cinnoides* Muhl. and *Calamagrostis cinnoides*, both of these latter combinations are considered to be superfluous—and thus illegitimate—names, even though they differ taxonomically from *A. canadensis* Michx. A proposal is in preparation to conserve the name *C. cinnoides* over the less frequently used, but legitimate, combination *C. coarctata* Eaton.

10. Calamagrostis scopulorum M.E. Jones [p. 718]
JONES' REEDGRASS, DITCH REEDGRASS

Plants without sterile culms; loosely cespitose, with rhizomes to 2 cm long, 2–3 mm thick. **Culms** (40)50–92 cm, sometimes branched, sparsely to densely scabrous; **nodes** 2–3. **Sheaths** and **collars** smooth or scabrous; **ligules** (3)4–7(9) mm, obtuse, lacerate; **blades** 10–38 cm long, (2)3–4(7) mm wide, flat, scabridulous, adaxial surfaces glabrous or sparsely hairy. **Panicles** (4)7–16(18) cm long, (0.7)1.1–2(3) cm wide, nodding, contracted, pale green to purple-tinged; **branches** 1–5(6.5) cm, sparsely to densely scabrous, usually spikelet-bearing to the base. **Spikelets** (4)4.5–6 mm; **rachilla prolongations** 1–2 mm, hairy throughout, hairs 1.5–2.5 mm. **Glumes** keeled, mostly smooth, keels slightly scabrous distally, lateral veins obscure, apices acuminate; **callus hairs** 2–3 mm, 0.5–0.6 times as long as the lemmas, somewhat sparse; **lemmas** 3.5–5 mm, 0.5–1.5 mm shorter than the glumes; **awns** (0.5)1–1.5(2) mm, attached to the upper ²/₅ of the lemmas, not exserted, slender, straight, easily overlooked when short; **anthers** (1.8)2–2.7(3) mm. 2*n* = 28.

Calamagrostis scopulorum grows on canyon slopes and wash bottoms, and in dry to moist montane to alpine habitats, often on rocky, sandy to silty soil, at 1000–3550 m. It grows from western Montana and Wyoming south to Arizona and New Mexico.

11. Calamagrostis muiriana B.L. Wilson & Sami Gray [p. 718]
MUIR'S REEDGRASS

Plants sometimes with sterile culms; densely cespitose, often with rhizomes 1–3 cm long, 1–2 mm thick. **Culms** (10)12–35 cm, unbranched, smooth beneath the panicles; **nodes** 1–3. **Leaves** basally concentrated; **sheaths** and **collars** smooth or scabrous; **ligules** 1–2.5 mm, obtuse, entire to lacerate; **blades** (1)4–12 cm long, 0.2–0.4 mm in diameter, involute, abaxial surfaces scabrous, adaxial surfaces sparsely hairy. **Panicles** (1.5)1.9–5.7(7.5) cm long, 0.4–3 cm wide, contracted to open, usually dark purple, rarely straw-colored; **branches** (0.8)1.1–2(3.5) cm, smooth, spikelets usually confined to the ends of the branches. **Spikelets** (3)3.5–4.5(5) mm; **rachilla prolongations** about 2 mm, hairs 0.5–1 mm. **Glumes** rounded, midvein smooth or slightly scabrous, lateral veins obscure, apices acute to acuminate, rarely awn-tipped; **callus hairs** (0.2)0.3–0.6 mm, 0.1–0.2 times as long as the lemmas, sparse; **lemmas** (2.5)3–4 mm, 0.5–1 mm shorter than the glumes; **awns** 3.5–6 mm, attached to the lower ¹/₃ of the lemmas, exserted, bent, purple; **anthers** 0.9–2.5 mm. 2*n* = 28.

Calamagrostis muiriana grows in moist to dry, subalpine and alpine floodplain meadows, lake margins, and stream banks, at 2400–3900 m, in the Sierra Nevada south of Sonora Pass in central California. It differs from *C. bolanderi* (p. 719) in having basally concentrated leaves.

12. Calamagrostis breweri Thurb. [p. 718]
SHORTHAIR REEDGRASS

Plants sometimes with sterile culms; densely cespitose, often with rhizomes to 5 cm long, 1–2 mm thick. **Culms** (18)29–54 cm, unbranched, smooth beneath the panicles; **nodes** 1–2(3). **Leaves** basally concentrated; **sheaths** and **collars** smooth or slightly scabrous; **ligules** 1.7–4.1(6) mm, usually lacerate; **blades** (2)10–15 cm long, 0.9–1.7 mm wide when flat, 0.4–0.6 mm in diameter when dry and involute, abaxial surfaces scabrous, adaxial surfaces sparsely hairy. **Panicles** (4)5.7–8.5 cm long, 0.7–5.2 cm wide, usually open, sometimes contracted, mostly erect, dark purple; **branches** (1)2–3(3.5) cm, smooth, spikelets usually confined to the ends of the branches. **Spikelets** 3–5 mm; **rachilla prolongations** about 1.5 mm, hairs 1.5–2 mm. **Glumes** rounded, usually smooth, occasionally scabrous at the apices, lateral veins obscure, apices acute to attenuate; **callus hairs** 0.3–1.2 mm, 0.2–0.5 times as long as the lemmas, sparse; **lemmas** 2.5–4 mm, 0.5–1 mm shorter than the glumes; **awns** 3.5–5.5 mm, attached to the lower ¹/₁₀–³/₁₀ of the lemmas, exserted, bent; **anthers** 1.3–2.6 mm. 2*n* = 42.

Calamagrostis breweri grows in moist subalpine and alpine meadows, lake margins, and stream banks, at 1700–2600 m, from Mount Hood in Oregon south to north of the Carson Pass area in Alpine and Amador counties, California. It differs from *C. bolanderi* (p. 719) in having basally concentrated leaves.

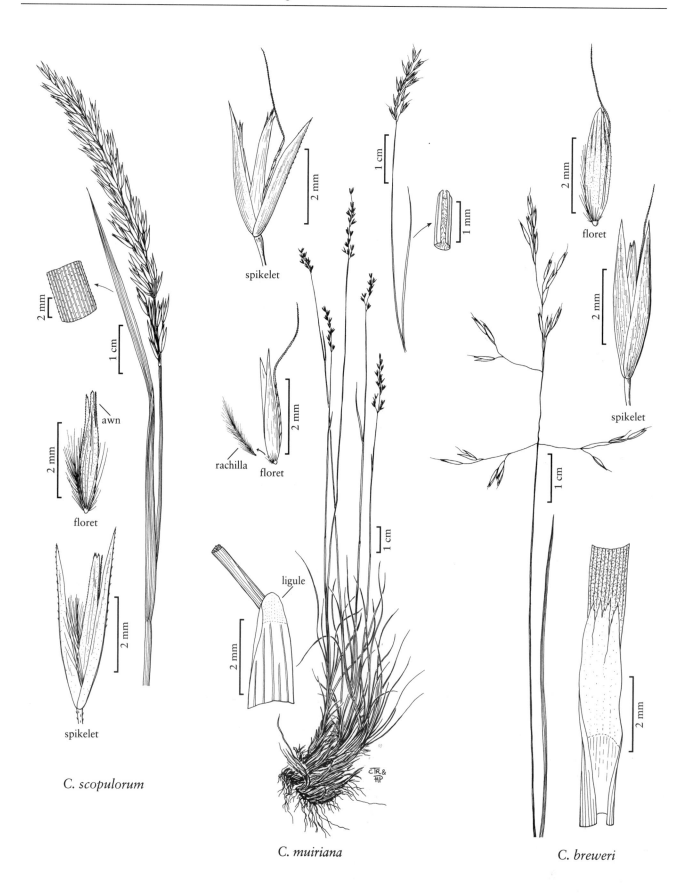

C. scopulorum

spikelet

spikelet

floret

awn

spikelet

C. muiriana

floret

rachilla

ligule

floret

spikelet

C. breweri

13. **Calamagrostis deschampsioides** Trin. [p. 720]
CIRCUMPOLAR REEDGRASS,
CALAMAGROSTIDE FAUSSE-DESCHAMPSIE

Plants sometimes with sterile culms; loosely cespitose, with rhizomes 7+ cm long, 1 mm thick. **Culms** (10)15–45(60) cm, unbranched, smooth beneath the panicles; **nodes** 1–2. **Sheaths** and **collars** smooth; **ligules** (0.5)1–2.5(3) mm, truncate to obtuse, usually entire, sometimes lacerate; **blades** (2)3–8(15) cm long, (1)1.5–2.5(3) mm wide, flat or somewhat involute, abaxial surfaces smooth, adaxial surfaces smooth or slightly scabrous, glabrous or sparsely hairy. **Panicles** 3–10(12) cm long, 1–4.5 cm wide, pyramidal, open, erect, green to dark purple; **branches** (2)2.5–4(5.5) cm, spreading, smooth or sparsely scabrous, spikelets usually confined to the distal $^1/_2$. **Spikelets** 4–5.5(7) mm; **rachilla prolongations** 1–2 mm, hairs (0.5)1–1.5 mm. **Glumes** rounded, usually smooth, sometimes scabrous along the midvein, lateral veins mostly obscure, apices acute to acuminate; **callus hairs** 2–3 mm, 0.4–0.7 times as long as the lemmas, abundant; **lemmas** 3.5–4.5(5.5) mm, 0.5–1(1.5) mm shorter than the glumes; **awns** 3–4.5(5.5) mm, attached to the lower $^1/_3$–$^1/_2$ of the lemmas, usually exserted, rarely included within the glumes, slender but distinguishable from the callus hairs, weakly to strongly bent; **anthers** (1.5)2–2.5 mm. 2*n* = 28.

Calamagrostis deschampsioides is a halophyte that grows, in the *Flora* region, on coastal dunes and beach ridges, gravel beaches, and in brackish coastal marshes, sometimes with *Carex lyngbyei*, at or near sea level. Its distribution is circumboreal, extending in North America from the islands of the Bering Sea and coastal Alaska, including the panhandle as far south as 56° N latitude, across the arctic coast to Hudson Bay and northern Labrador. It also extends from the arctic coast of Europe to Siberia and Japan. The alpine habitat reported for the Japanese plants suggests that they might belong to a different taxon.

14. **Calamagrostis cainii** Hitchc. [p. 720]
CAIN'S REEDGRASS

Plants sometimes with sterile culms; densely cespitose, without obvious rhizomes. **Culms** (30)45–55(60) cm, unbranched, smooth beneath the panicles; **nodes** 1–3. **Sheaths** and **collars** smooth; **ligules** 0.5–1.5(2) mm, usually truncate, sometimes obtuse, entire; **blades** (5)15–30(39)

cm long, (1)2–3(4) mm wide, flat, pale green, abaxial surfaces smooth, adaxial surfaces slightly scabrous, glabrous or sparsely hairy. **Panicles** (6)8–12(15) cm long, 1.5–3.5(4) cm wide, open, often drooping, pale green and purple; **branches** (2)2.5–4.5(5) cm, smooth or slightly scabrous, spikelets usually confined to the distal $^1/_2$–$^2/_3$. **Spikelets** 4–5(6) mm; **rachilla prolongations** 1–2 mm, hairs 1.5–2 mm. **Glumes** rounded to slightly keeled, usually smooth, keels rarely finely scabrous distally, lateral veins obscure, apices acute to acuminate; **callus hairs** 1.5–2 mm, 0.3–0.5 times as long as the lemmas, abundant; **lemmas** 3.5–4.5 mm, 0.5–1 mm shorter than the glumes; **awns** 4–5(6) mm, attached from near the base to the lower $^1/_3$ of the lemmas, exserted, stout and easily distinguished from the callus hairs, strongly bent; **anthers** (1)2–2.5 mm. 2*n* = 28.

Calamagrostis cainii grows on bouldery subtrates and in soil pockets, landslides, and disturbed sites, at 1200–2100 m. It has been found in only three locations: the slopes of Mount LeConte, Sevier County, Tennessee; and in North Carolina on Craggy Pinnacle, Buncombe County, and the summit of Mount Craig, Yancey County. The species is of conservation concern because of its limited distribution.

15. **Calamagrostis bolanderi** Thurb. [p. 720]
BOLANDER'S REEDGRASS

Plants sometimes with sterile culms; mostly cespitose, with rhizomes to 3 cm long, 1–2 mm thick. **Culms** 50–150 cm, unbranched, smooth or slightly scabrous beneath the panicles; **nodes** 2–4. **Leaves** distributed along the culms; **sheaths** and **collars** smooth or scabrous, glabrous; **ligules** (2)3(5) mm, more or less obtuse, entire to lacerate; **blades** (5)15–26(30) cm long, (2)3–7(10) mm wide, flat or involute, smooth or slightly scabrous, glabrous or sparsely hairy. **Panicles** (4)10–16(25) cm long, (2.5)4–5.5(6) cm wide, open, erect to nodding, pale green to bronze or purple; **branches** (4)5.5–8(9) cm, sparsely scabrous, spreading to ascending, spikelets confined to the distal $^1/_4$–$^1/_2$. **Spikelets** 3–4(5) mm; **rachilla prolongations** 0.5–1 mm, hairs 1–1.5 mm. **Glumes** rounded to slightly keeled, keels scabrous distally, lateral veins obscure, apices acute; **callus hairs** 0.5–1 mm, 0.2–0.4 times as long as the lemmas, appearing sparse primarily because of their shortness; **lemmas** 2.5–3 mm, 0–1(2) mm shorter than the glumes; **awns** 3.5–5(6) mm, attached to the lower $^1/_{10}$–$^1/_5$ of the lemmas, usually exserted, strongly bent; **anthers** (1.5)2–3 mm. 2*n* = 56.

C. *deschampsioides*   C. *cainii*   C. *bolanderi*

CALAMAGROSTIS

Calamagrostis bolanderi grows in marshes, swamps, bogs, fens, seeps, moist meadows, open and closed coniferous and broadleaf forests, prairies, and coastal scrub, from sea level to 500 m. It is known only from sites near the coast in Humboldt, Mendocino, and Sonoma counties, California. It differs from *C. breweri* and *C. muiriana* in having leaves evenly distributed along the culms.

16. Calamagrostis ×acutiflora (Schrad.) D.C. 'Karl Foerster' [p. 722]

FOERSTER'S REEDGRASS, FEATHER REEDGRASS

Plants often with robust sterile culms; densely cespitose, with rhizomes 1–3 cm long, about 2 mm thick. **Culms** 135–210 cm, smooth to slightly scabrous, usually unbranched; **nodes** about 3. **Sheaths** and **collars** smooth; **ligules** (1)2–7 mm, lacerate; **blades** (5)11–63(71) cm long, (1)1.5–7.5(8) mm wide, flat or involute, usually scabrous, rarely smooth, adaxial surfaces glabrous or sparsely hairy. **Panicles** 15–30 cm long, 0.75–1(3) cm wide, erect, contracted, pale green to purple; **branches** about 6 cm, slightly scabrous, longer branches with spikelets on the distal ²/₃, shorter branches with spikelets to the base. **Spikelets** 4–5 mm; **rachilla prolongations** about 0.5 mm, hairs about 1.5 mm. **Glumes** keeled, mostly smooth, slightly scabrous only on the keels, veins usually obscure, apices acute; **callus hairs** 2–2.5 mm, 0.4–0.6 times as long as the lemmas, abundant; **lemmas** 3–3.5 mm, 1–1.5 mm shorter than the glumes; **awns** about 3.5 mm, attached to the lower ¹/₁₀–¹/₅ of the lemmas, exserted, slender, usually distinguishable from the callus hairs, bent; **anthers** not observed.

Calamagrostis ×acutiflora is a hybrid of European origin that is now widely planted as an ornamental, especially in dry sites and gardens throughout northern North America. The parents are *C. arundinacea* (L.) Roth and *C. epigejos* (p. 710); the hybrids are seed-sterile.

17. Calamagrostis koelerioides Vasey [p. 722]

DENSE-PINE REEDGRASS

Plants without sterile culms; often densely cespitose, with rhizomes 2–6 cm long, 2–4 mm thick. **Culms** (26)60–85(120) cm, unbranched, slightly scabrous; **nodes** 2–3(5). **Sheaths** and **collars** usually scabrous, rarely smooth, glabrous; **ligules** (1.5) 2–4.5(7) mm, truncate to obtuse, entire or sometimes lacerate; **blades** (2)9–20(30) cm long, (2)2.5–4.5(8) mm wide, flat, slightly scabrous, adaxial surfaces glabrous or sparsely hairy. **Panicles** (4)10–13(16) cm long, about 1 cm wide, contracted, erect to slightly nodding, often slightly interrupted towards the base, straw-colored or pale green to pale

purple; **branches** (1.1)2.8–4(6) cm, scabrous, spikelet-bearing to the base. **Spikelets** (4)4.5–6(7) mm; **rachilla prolongations** 1.5–2.5(3) mm, hairs 1.5–2 mm. **Glumes** slightly keeled, keels smooth or slightly scabrous distally, lateral veins visible but not prominent, apices acute; **callus hairs** 1.5–2 mm, 0.3–0.4 times as long as the lemmas, sparse; **lemmas** (3.5)4–5(6) mm, 0.5–1.5 mm shorter than the glumes; **awns** 4–5.5 mm, attached to the lower ¹/₁₀–¹/₅ of the lemmas, exserted, sometimes barely so, stout, distinguishable from the callus hairs, bent; **anthers** 2–3.5 mm. 2*n* = 28.

Calamagrostis koelerioides grows in mountain meadows, chaparral, and Jeffrey pine and blue spruce forests, and on talus slopes, dry hills, and ridges, occasionally on serpentine soils, at 50–2100 m. It extends from Washington south to southern California and east to Montana and western Wyoming.

Calamagrostis koelerioides is similar to *C. rubescens* (p. 723). The two have traditionally been distinguished by the presence of hairs on the leaf collars in *C. rubescens*, and their absence in *C. koelerioides*; a more reliable differentiation is the longer lemmas, glumes, and awns of *C. koelerioides* compared to *C. rubescens*.

18. Calamagrostis porteri A. Gray [p. 722]

PORTER'S REEDGRASS

Plants with sterile culms; loosely cespitose, with rhizomes 5–7+ cm long, 0.5–1 mm thick. **Culms** (60)75–120 cm, unbranched, slightly scabrous; **nodes** 2–4(5). **Sheaths** smooth or slightly scabrous; **collars** smooth or hairy; **ligules** (1)2–5(6) mm, truncate to obtuse, entire or lacerate; **blades** 8–40 cm long, (2)3–8(12) mm wide, flat, abaxial surfaces smooth or scabrous, adaxial surfaces smooth or slightly scabrous, occasionally with hairs. **Panicles** (5)10–18(22) cm long, 0.8–3(7) cm wide, contracted to open, often slightly nodding, sometimes erect, green to pale purple; **branches** (1.5)2–7(7.5) cm, scabrous, usually spikelet-bearing on the distal ¹/₂–²/₃, sometimes to the base. **Spikelets** 4–5(6) mm; **rachilla prolongations** 0.5–1 mm, hairs 1.5–2(3.5) mm. **Glumes** rounded to slightly keeled, keels scabrous towards the apices, lateral veins obscure to slightly prominent, not raised, apices acute to acuminate; **callus hairs** 1.5–2(3) mm, 0.4–0.7 times as long as the lemmas, sparse; **lemmas** 3–4.5 mm, 0.5–1.5(2) mm shorter than the glumes; **awns** 3–4(4.5) mm, attached to the lower ¹/₁₀–³/₁₀ of the lemmas, exserted, stout, easily distinguished from the callus hairs, strongly bent; **anthers** 2–2.5 mm. 2*n* = 56, 84–±104.

Calamagrostis porteri grows in dry chestnut/oak forests, often on rocky ridgetops, piedmont bluffs, and slopes, at (100)600–1300 m. It is now restricted to the northeastern and central United States. Historically, its

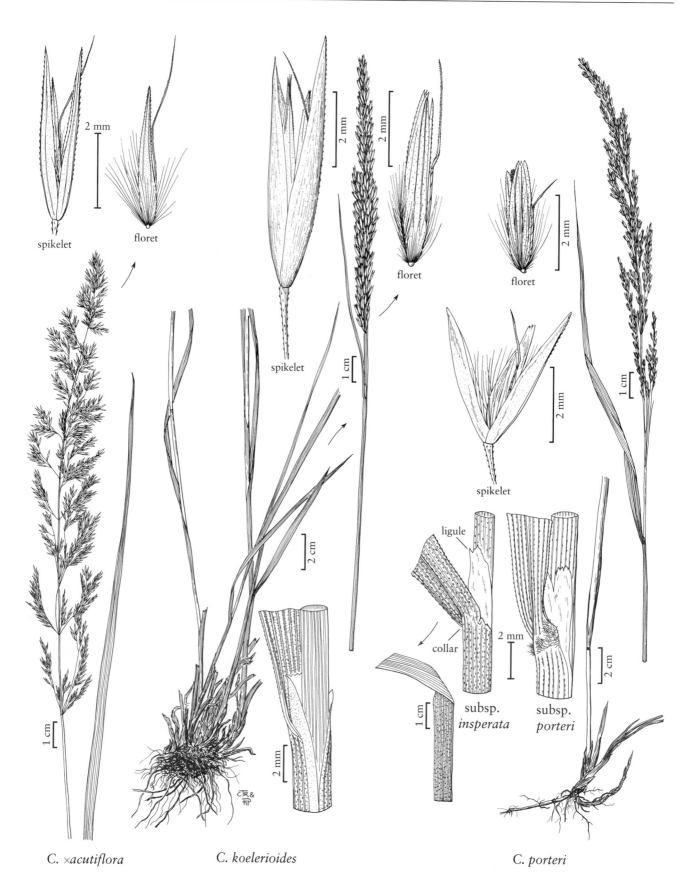

spikelet

floret

spikelet

floret

floret

spikelet

ligule

collar

subsp.
insperata

subsp.
porteri

C. ×*acutiflora*

C. *koelerioides*

C. *porteri*

CALAMAGROSTIS

range extended from Missouri and Arkansas east to New York and Alabama. Flowering appears to be a response to disturbance; plants in undisturbed habitats remain vegetative and may go unnoticed. Thus the species may be more widespread and abundant than reported.

Calamagrostis porteri and *C. rubescens* (see next) appear to be closely related. They may be part of the general phenomenon of eastern and western vicariants. The apparently sterile *C. perplexa* (p. 726) is intermediate between *C. porteri* and *C. canadensis* (Greene 1980).

1. Leaf blades glaucous on both surfaces; leaf collars glabrous . . . . subsp. *insperata*
1. Leaf blades light green and glaucous on the adaxial surfaces, darker green on the abaxial surfaces; leaf collars usually with prominent tufts of hair, rarely glabrous . . . . subsp. *porteri*

Calamagrostis porteri subsp. **insperata** (Swallen) C.W. Greene [p. 722]
BARTLEY'S REEDGRASS

Culms (85)90–120 cm. **Sheaths** smooth; **collars** glabrous; **ligules** (2)3–4(6) mm; **blades** (12)16–28(30) cm long, (2)6–8(12) mm wide, glaucous on both surfaces. **Panicles** (11)14–18(22) cm long, (1.5)2–3(3.5) cm wide; **branches** (2.5)5–7(7.5) cm. **Spikelets** (4)5(6) mm; **rachilla prolongations** with 1.5–2(3) mm hairs. **Callus hairs** 2(3) mm, 0.5–0.7 times as long as the lemmas; **lemmas** 3(4) mm, 1(2) mm shorter than the glumes; **awns** 3–3.5 mm, attached to the lower $^1/_{10}$–$^3/_{10}$ of the lemmas; **anthers** 2–2.5 mm. $2n = 56$.

Isolated populations of *Calamagrostis porteri* subsp. *insperata* grow from southern Missouri and Arkansas to Ohio and Kentucky. It is listed as possibly extirpated or of conservation concern in several states.

Calamagrostis porteri A. Gray subsp. **porteri** [p. 722]

Culms (60)75–100 cm. **Sheaths** smooth or slightly scabrous; **collars** usually with prominent tufts of hair, rarely glabrous; **ligules** (1)2–5(6) mm; **blades** 8–40 cm long, (2)3–6(8) mm wide, abaxial surfaces darker green than the adaxial surfaces, adaxial surfaces light green, glaucous. **Panicles** (5)10–12(15) cm long, 0.8–3(7) cm wide; **branches** (1.5)2–5 cm. **Spikelets** 4–5 mm; **rachilla prolongations** with 1.5–2(3.5) mm hairs. **Callus hairs** 1.5–2 mm, 0.4–0.6 times as long as the lemmas; **lemmas** 3.5–4.5 mm, 0.5–1.5 mm shorter than the glumes; **awns** 3–4(4.5) mm, attached to the lower $^1/_{10}$–$^1/_5$ of the lemmas; **anthers** about 2 mm. $2n = 84–\pm104$.

Calamagrostis porteri subsp. *porteri* grows from New York to Virginia, Tennessee, and Kentucky. One specimen from Claiborne County, Tennessee, lacks hairs around the collar, but the leaves are glaucous on both surfaces. There are other specimens from this county that have hairy collars.

Calamagrostis porteri subsp. *porteri* appears to form hybrids with the nearly sympatric *C. canadensis* (p. 726) in rocky wooded sites in central Virginia. These putative hybrids have hairy collars, relatively long callus hairs, and short awns.

19. **Calamagrostis rubescens** Buckley [p. 725]
PINEGRASS, PINE REEDGRASS

Plants sometimes with sterile culms; sometimes loosely cespitose, usually with rhizomes 15+ cm long, 1.5–2 mm thick. **Culms** (50)60–100(105) cm, unbranched, usually smooth, rarely slightly scabrous beneath the panicles; **nodes** (1)2–3(4). **Sheaths** smooth or slightly scabrous; **collars** often hairy, rarely glabrous; **ligules** (2)3–5(6) mm, truncate to obtuse, often lacerate; **blades** (6)8–40(42) cm long, (1)2–5(8) mm wide, usually flat, abaxial surfaces smooth or slightly scabrous, adaxial surfaces smooth or scabrous, glabrous or sparsely hairy. **Panicles** (5)6–15(25) cm long, (0.7)1.5–2(2.7) cm wide, contracted to somewhat open, erect, usually greenish, infrequently purplish; **branches** (1.2)2–4(10) cm, usually slightly scabrous, rarely densely long-scabrous, spikelet-bearing to the base. **Spikelets** (3)4–4.5(5.5) mm; **rachilla prolongations** 0.6–1.5(2) mm, hairs 1.2–2 mm. **Glumes** rounded to slightly keeled, mostly smooth, keels rarely slightly scabrous, lateral veins usually obscure, rarely prominent, apices acute; **callus hairs** (0.5)1–1.5(2.5) mm, 0.2–0.5(0.7) times as long as the lemmas, sparse; **lemmas** 2.5–3.5(4) mm, (0.5)1–2 mm shorter than the glumes; **awns** 2.8–3.5(4.5) mm, usually attached to the lower $^1/_5$ of the lemmas, rarely higher, exserted, stout and readily distinguished from the callus hairs, strongly bent; **anthers** (1)1.3–2(2.6) mm. $2n = 28, 42, 56$.

Calamagrostis rubescens grows at 50–2800 m, usually in open montane pine or aspen forests and parklands, infrequently in sagebrush steppes, chaparral, and meadows. It is primarily a species of interior western North America, although it reaches the Pacific coast in southern California. The distribution extends from central British Columbia and Alberta east to the Cypress Hills of eastern Alberta and the Pasquia and Cub hills of Saskatchewan, south to western California, Nevada, northeastern Utah, and central Colorado. It is considered threatened in Saskatchewan.

Calamagrostis rubescens is similar to *C. koelerioides* (p. 721). The two have traditionally been distinguished by the presence of hairs on the leaf collars of *C. rubescens*, and their absence from *C. koelerioides*; a more reliable differentiation is the shorter lemmas, glumes, and awns of *C. rubescens*.

Calamagrostis rubescens and *C. porteri* (p. 721) appear to be closely related. They may be part of the general phenomenon of eastern and western vicariants.

20. **Calamagrostis nutkaensis** (J. Presl) Steud. [p. 725]
PACIFIC REEDGRASS

Plants sometimes with sterile culms; densely cespitose, with rhizomes usually shorter than 3 cm, rarely to 6 cm long, 1.5–3 mm thick. **Culms** (42)55–105 (150) cm, stout, unbranched, smooth or slightly scabrous beneath the panicles; **nodes** 1–2(3). **Sheaths** and **collars** smooth; **ligules** (0.5)1–4(5.5) mm, usually truncate, entire, often hidden by the expanded collars; **blades** (4)10–41(56) cm long, (2)4–10(20) mm wide, flat, usually erect, abaxial surfaces smooth, adaxial surfaces smooth or slightly scabrous, glabrous or sparsely hairy. **Panicles** (8)12–23(33) cm long, (1.1)2–4.5(9) cm wide, contracted to somewhat loose, erect to slightly nodding, greenish yellow to purple-tinged; **branches** 2.7–7(10.5) cm long, sparsely scabrous, spikelets usually confined to the distal ¹/₂. **Spikelets** 4.5–6.5(8) mm; **rachilla prolongations** 0.5–1 mm, hairs 1–1.5 mm. **Glumes** keeled, smooth or infrequently scabrous on the keels, lateral veins somewhat prominent, apices acuminate; **callus hairs** (1)2–2.5(3) mm, (0.2)0.5–0.7 times as long as the lemmas, sparse; **lemmas** (3)4–4.5(5) mm, (0.4)0.8–1.2(1.9) mm shorter than the glumes; **awns** 1–3 mm, attached on the lower ¹/₃–¹/₂ of the lemmas, not exserted, easily distinguished from the callus hairs, straight or slightly bent; **anthers** (1)2.4–2.6(3.3) mm. 2n = 28.

Calamagrostis nutkaensis grows in wetlands and openings in coniferous forests, on marine and freshwater beaches and dunes, and, sometimes, on cliffs. It is usually found within a few kilometers of the marine shoreline at or near sea level, but it sometimes occurs as high as 800 m on the Brooks Peninsula of Vancouver Island, British Columbia; at 700 m and 1100 m in the Siskiyou Mountains, Oregon; and at 1100 m on Bald Mountain, California. It grows along the Pacific coast of North America from the Aleutian Islands in Alaska to San Luis Obispo County, California, and also in the Kamchatka Peninsula of Russia.

A hybrid between *Calamagrostis nutkaensis* and *Ammophila arenaria* (p. 777) was reported in the 1970s from the vicinity of Newport–Waldport in coastal Oregon, where both species grow (Kenton Chambers, pers. comm.). There is no voucher to support this report.

21. **Calamagrostis pickeringii** A. Gray [p. 725]
PICKERING'S REED BENTGRASS,
CALAMAGROSTIDE DE PICKERING

Plants without sterile culms; weakly cespitose, with rhizomes 2–8 cm long, about 1.5 mm thick. **Culms** (17)25–55(90) cm, solitary or in small clusters, unbranched, sparsely scabrous; **nodes** 1–3. **Sheaths** and **collars** smooth; **ligules** 2–3(5) mm, truncate to obtuse, entire, sometimes weakly lacerate; **blades** (3)6–17(38) cm long, (2)3–4(7) mm wide, flat, stiff, abaxial surfaces smooth or scabrous, adaxial surfaces slightly scabrous, glabrous or sparsely hairy. **Panicles** (3.5)5–12(15) cm long, (0.5)1–2(3.5) cm wide, contracted, erect, greenish to purplish; **branches** (1.5)2–3(4) cm, sparsely scabrous, usually spikelet-bearing on the distal ¹/₂–²/₃, sometimes to the base. **Spikelets** (2.5)3–4(4.5) mm; **rachilla prolongations** 0.2–1.3(1.5) mm, sparsely bearded, hairs about 1.5 mm. **Glumes** keeled, usually scabrous on the keel tips, lateral veins obscure, not raised, apices acute; **callus hairs** (0.3)0.5–1 mm, 0.2–0.3 times as long as the lemmas, sparse; **lemmas** 2.5–3(3.5) mm, (0)0.5–1 mm shorter than the glumes; **awns** 1.5–2(2.5) mm, attached to the lower ²/₅–³/₅ of the lemmas, sometimes slightly exserted, distinct from the callus hairs, bent; **anthers** 1.5–2 mm. 2n = 28.

Calamagrostis pickeringii grows in bogs, open white spruce scrub, wet meadows, coastal peatlands and lake shores, heaths, frost pockets (hollows), pitch pine barrens, and on sandy beaches, at 0–1600 m. It is found from Newfoundland and Nova Scotia south to the mountains of New Hampshire, New York, and New Jersey.

22. **Calamagrostis montanensis** (Scribn.) Vasey [p. 727]
PLAINS REEDGRASS

Plants with sterile culms; cespitose, with rhizomes 6+ cm long, 1–2 mm thick. **Culms** 15–50(54) cm, unbranched, usually scabrous, rarely smooth; **nodes** 1–2. **Sheaths** and **collars** smooth or slightly scabrous; **ligules** (1)2–3 mm, obtuse to acute, more or less lacerate; **blades** (5)8–19(23) cm long, (1)2–3(4) mm wide, usually involute, seldom reaching the panicles, abaxial surfaces scabrous, adaxial surfaces usually scabrous, rarely smooth, glabrous or sparsely hairy. **Panicles** 4–9(14) cm long, (0.7)1–2(3.5) cm wide, erect, contracted and not or only slightly interrupted, yellowish green with a light purple tinge; **branches** 1.3–3(3.7) cm, sparsely short-scabrous to densely long-

spikelet

floret

1 cm

C. rubescens

CALAMAGROSTIS

collar

2 cm

2 mm

spikelet

floret

1 cm

2 cm

spikelet

floret

2 mm

C. nutkaensis

2 mm

2 mm

spikelet

floret

1 cm

collar

2 mm

ligule

2 mm

2 cm

collar

2 mm

C. pickeringii

scabrous, spikelet-bearing to the base. **Spikelets** (3)3.5–4.5(7) mm; **rachilla prolongations** about 1 mm, densely bearded, hairs to 2 mm. **Glumes** keeled, smooth or scabrous throughout, lateral veins usually somewhat obscure, rarely prominent, apices acute to acuminate, rarely awn-tipped; **callus hairs** (1)1.5–2(2.5) mm, 0.4–0.8 times as long as the lemmas, abundant; **lemmas** (2.5)3–3.5(5.5) mm, 0.5–1(2) mm shorter than the glumes; **awns** (1)2–3(4) mm, usually attached to the lower $^1/_{10}$–$^2/_5$ of the lemmas, rarely above the middle, sometimes slightly exserted, stout, distinguishable from the callus hairs, slightly bent; **anthers** (1.1)1.8–2.4(3) mm. $2n = 28$.

Calamagrostis montanensis inhabits prairie grasslands and sagebrush flats, benchlands, valley bottoms, and occasionally woodlands, at 200–2600 m. It grows in the continental interior from eastern British Columbia (near Fort St. John and Invermere) and adjacent Alberta, south to southern Wyoming and east to Manitoba and western Minnesota. It has also been reported from Peace Point in Wood Buffalo National Park in northeastern Alberta, but this report has not been verified. *Calamagrostis montanensis* may be mistaken for *C. purpurascens* (p. 710), but the latter species has hairy adaxial leaf surfaces and longer awns.

23. Calamagrostis perplexa Scribn. [p. 727]
WOOD REEDGRASS

Plants with sterile culms; weakly cespitose, with rhizomes 8+ cm long, 1.5–2 mm thick. **Culms** (80)85–110(120) cm, sometimes branched, smooth or slightly scabrous beneath the panicles; **nodes** 4–6. **Sheaths** smooth; **collars** densely hairy; **ligules** (3)4–6(7) mm, lacerate; **blades** (10)15–30(35) cm long, (3)4.5–6.5(7) mm wide, flat, scabridulous, adaxial surfaces glabrous or sparsely hairy. **Panicles** 10–18(20) cm long, (1)2–3 cm wide, open, erect to nodding, green, purple-tinged; **branches** 5–6.5 cm, spikelets usually confined to the distal $^1/_4$–$^1/_2$. **Spikelets** (3)3.5–4 mm; **rachilla prolongations** about 1 mm, hairs about 2 mm. **Glumes** keeled, scabrous on the keels, lateral veins usually prominent, apices acuminate; **callus hairs** 2–3 mm, 0.7–1 times as long as the lemmas, somewhat sparse; **lemmas** 3–3.5 mm, 0.5–1 mm shorter than the glumes; **awns** 2–3 mm, attached to the lower $^1/_4$–$^1/_3$ of the lemmas, not exserted, stout, distinguishable from the callus hairs, bent; **anthers** 1–1.5 mm. $2n = 70$.

Calamagrostis perplexa grows on wet rocks and in dry woods at "Thatcher's Pinnacle" [Pinnacle Rock], Tompkins County, New York. There is also an unverified report of this species in Columbia County, New York. This apparently sterile species is intermediate between *C. porteri* (p. 721) and *C. canadensis* (see next) (Greene 1980). It is of conservation concern because of its limited distribution.

24. Calamagrostis canadensis (Michx.) P. Beauv. [p. 727]
BLUEJOINT, CALAMAGROSTIDE DU CANADA

Plants with sterile culms; cespitose, with rhizomes 2–15+ cm long, 1–3 mm thick. **Culms** (32)65–112(180) cm, often branching above the base, smooth or slightly scabrous beneath the panicles; **nodes** (2)3–7(8). **Sheaths** smooth or scabrous; **collars** usually scabrous, rarely smooth or hairy; **ligules** (1)3–8(12) mm, lacerate; **blades** (10)16–31(50) cm long, 2–8(11) mm wide, flat, lax, abaxial surfaces scabrous, adaxial surfaces usually strongly scabrous, rarely smooth or with scattered hairs, often glaucous. **Panicles** (6)9–17(25) cm long, (1)2–4(8) cm wide, often contracted when young, open at maturity, nodding, usually purplish, sometimes greenish to straw-colored; **branches** 2.7–6(12) cm, scabrous, spikelets sparsely to densely concentrated on the distal $^2/_3$. **Spikelets** 2–4.5(5.2) mm; **rachilla prolongations** 0.5–1 mm, hairs 1.5–3.2 mm. **Glumes** rounded or keeled, smooth or scabrous, keels often long-scabrous, lateral veins obscure to prominent, apices acute to acuminate; **callus hairs** (1.5)2–3.5(4.5) mm, (0.5)0.9–1.2(1.5) times as long as the lemmas, abundant; **lemmas** 2–3.1(4) mm, 0–2.1 mm shorter than the glumes; **awns** 0.9–3.1 mm, attached to the lower $(^1/_{10})^1/_5$–$^1/_2(^7/_{10})$ of the lemmas, usually not exserted, delicate, often difficult to distinguish from the callus hairs, usually straight; **anthers** (0.8)1.2–1.6(2.6) mm. $2n = 42$–66.

Calamagrostis canadensis is a species of moist meadows, thickets, bog edges, and forest openings. It grows from sea level to 3400 m. It occurs widely throughout the *Flora* region, except in Oklahoma, Texas, and the southeastern United States. Its range also extends from northern Asia to northeastern China and Japan, with additional scattered populations elsewhere in Asia.

Calamagrostis canadensis is closely related to, and possibly conspecific with, the European *C. purpurea* (Trin.) Trin. It hybridizes with *Ammophila breviligulata* (p. 777) in Grand Island, Michigan and on the adjacent mainland, forming ×**Calammophila don-hensonii** Reznicek & Judz. *Calamagrostis canadensis* also appears to form hybrids with the nearly sympatric *C. porteri* (p. 721) in rocky wooded sites in central Virginia. These putative hybrids have hairy collars, relatively long callus hairs, and short awns. The apparently sterile *C.*

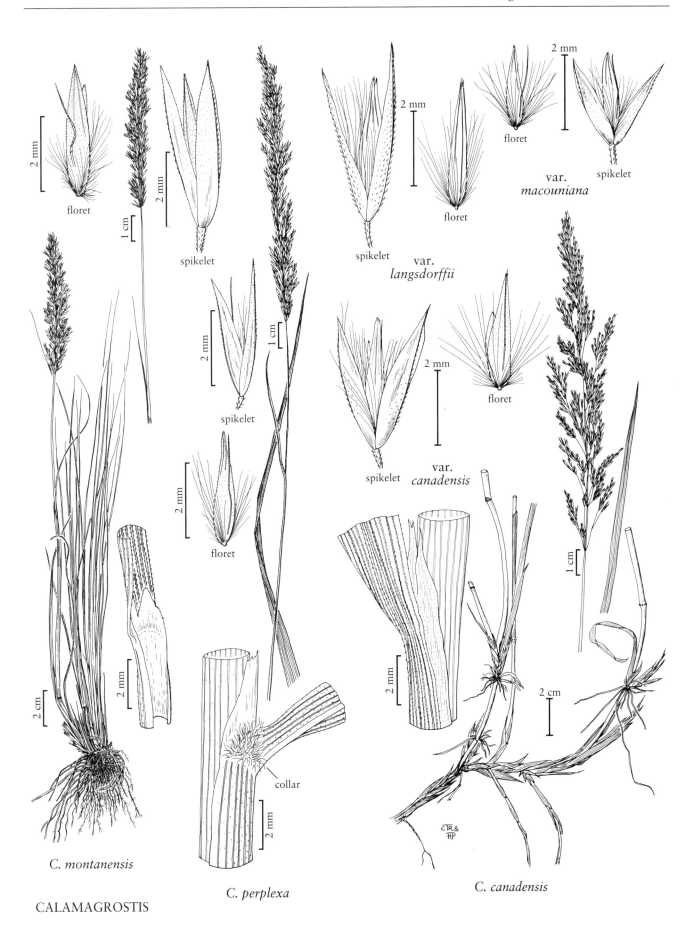

floret

2 mm

spikelet

1 cm

floret

2 mm

spikelet

var. *macouniana*

floret

spikelet

2 mm

var. *langsdorffii*

spikelet

floret

2 mm

spikelet

floret

2 mm

var. *canadensis*

floret

2 mm

1 cm

2 cm

2 mm

C. montanensis

collar

2 mm

C. perplexa

2 mm

2 cm

C. canadensis

CTR & HP

CALAMAGROSTIS

perplexa (see previous) is intermediate between *C. canadensis* and *C. porteri* (Greene 1980).

A high degree of pollen sterility has been documented in some populations, suggesting that seed formation via apomixis is common; sexual reproduction is also documented. The many forms, varieties, and subspecies that have been described for this species probably represent clones. The three varieties recognized here intergrade, and an argument could be made that their recognition is not warranted. Nevertheless, the extreme forms along the gradient of variation can be distinguished most readily by the glumes.

1. Spikelets (3.5)4–4.5(5.2) mm long; glumes usually scabrous over the entire surface, the prickles on the keels hairlike, often bent; glume apices distinctly acuminate ...... var. *langsdorffii*
1. Spikelets 2–4 mm long; glumes smooth or scabrous, often scabrous only on the keels, prickles straight; glume apices acute, rarely acuminate.
   2. Spikelets 2.5–4 mm long, lemmas usually shorter than the glumes; glumes rounded to broadly keeled, with raised midveins; glume apices usually acute, rarely acuminate .... var. *canadensis*
   2. Spikelets 2–3 mm long; lemmas usually about as long as the glumes; glumes rounded, midveins not raised; glume apices acute .... var. *macouniana*

Calamagrostis canadensis (Michx.) P. Beauv. var. **canadensis** [p. 727]

Culms (50)65–80(160) cm, often branching above the base, smooth or slightly scabrous beneath the panicles; **nodes** (2)3–4(6). **Sheaths** smooth or scabrous; **collars** usually scabrous, rarely smooth or hairy; **ligules** (1)4–6(12) mm; **blades** (11)16–31(41) cm long, (2)2.5–5(8) mm wide, adaxial surfaces usually strongly scabrous, rarely smooth, often glaucous. **Panicles** (9)11–14(19) cm long, (1)2–3(7) cm wide, often contracted when young, usually purplish, sometimes greenish to straw-colored; **branches** 2.9–4.5(5.7) cm. **Spikelets** 2.5–3.5(4) mm; **rachilla prolongations** 0.5–1 mm, hairs 2.5–3.2 mm. **Glumes** rounded to broadly keeled, smooth or scabrous, often only the keels scabrous, prickles straight, midveins raised, lateral veins obscure to prominent, apices usually acute, rarely acuminate; **callus hairs** (1.7)2.5–2.9(3.1) mm, (0.7)0.9–1.1(1.4) times as long as the lemmas; **lemmas** (2.2)2.5–3.1(4) mm, 0–1.6 mm shorter than the glumes; **awns** 0.9–2.6 mm, attached to the lower ($^1/_{10}$)$^1/_5$–$^2/_5$($^7/_{10}$) of the lemmas, usually not exserted, usually straight; **anthers** (0.8)1.2–1.3(2) mm. 2n = 42–66.

Calamagrostis canadensis var. *canadensis* is widespread throughout the range of the species, except on the arctic islands, Greenland, and the adjacent mainland.

Calamagrostis canadensis var. **langsdorffii** (Link) Inman [p. 727]

Culms (32)65–110(180) cm, often branching above the base, smooth or slightly scabrous beneath the panicles; **nodes** (2)3–5(8). **Sheaths** and **collars** smooth or scabrous; **ligules** (3)5–8(12) mm; **blades** (11)18–29(50) cm long, (2)3–7(11) mm wide, adaxial surfaces usually strongly scabrous, rarely smooth, often glaucous. **Panicles** (6)9–15(25) cm long, (1.5)2.5–4(8) cm wide, often contracted when young, usually purplish, sometimes greenish to straw-colored; **branches** (2.7)3.5–6 (12) cm. **Spikelets** (3.5)4–4.5(5.2) mm; **rachilla prolongations** 0.7–1 mm, hairs 2–3 mm. **Glumes** keeled, usually scabrous throughout, rarely smooth, prickles on the keels hairlike, often bent, lateral veins obscure to prominent, apices acuminate; **callus hairs** (1.5)3–3.3 (4.5) mm, (0.5)1–1.2(1.5) times as long as the lemmas; **lemmas** (2.3)2.5–3(4) mm, 0.4–2.1 mm shorter than the glumes; **awns** 1.2–3.1 mm, attached to the lower ($^1/_{10}$)$^1/_5$–$^2/_5$($^1/_2$) of the lemmas, not exserted, usually straight; **anthers** (0.9)1.2–1.6(2.6) mm. 2n = 28, 42, 56.

A circumboreal taxon, *Calamagrostis canadensis* var. *langsdorffii* grows from Alaska to Greenland and south to northern California, Colorado, Minnesota, New Hampshire, and New York.

Calamagrostis canadensis var. **macouniana** (Vasey) Stebbins [p. 727]
MACOUN'S REEDGRASS

Culms 100–112 cm, not branching above the base, smooth beneath the panicles; **nodes** 5–7. **Sheaths** and **collars** smooth or slightly scabrous; **ligules** (2)3–4(5) mm; **blades** (10)17–30(33) cm long, 2–6(8) mm wide, adaxial surfaces scabrous, sometimes with widely scattered long hairs, often glaucous. **Panicles** 14–17(20) cm long, (2.5)3(4) cm wide, usually purplish; **branches** 4.8–6(7) cm. **Spikelets** (2)2.5–3 mm; **rachilla prolongations** about 0.5 mm, hairs 1.5–2 mm. **Glumes** rounded, smooth, or sometimes slightly scabrous distally, often only on the midvein, prickles straight, midveins not raised, lateral veins obscure, apices acute; **callus hairs** 2–2.5 mm, nearly as long as the lemmas; **lemmas** 2–2.5 mm, 0–1 mm shorter than the glumes; **awns** 1.2–2 mm, attached to the lower $^1/_5$–$^1/_2$ of the lemmas, not exserted, straight; **anthers** 1–2 mm. 2n = unknown.

Calamagrostis canadensis var. *macouniana* is common in marshes and mud flats. It ranges eastward from Saskatchewan to Prince Edward Island and Nova Scotia, and from North Dakota and Missouri to New Jersey and Virginia. According to C.L. Hitchcock (Hitchcock et al. 1969), it may extend as far west as Idaho.

Vernon Harms (pers. comm.) reports that in the southern grassland region of Saskatchewan, intermediates between *Calamagrostis canadensis* var. *macouniana* and var. *canadensis* appear to be more common than plants that fit var. *macouniana* well. Some specimens that seem nearest to var. *macouniana* stretch across southern Saskatchewan south of 51° N latitude, from the Cypress Hills in the west to the Souris Valley in the east. Harms' observations suggest that it may not be practical to recognize this variety.

25. Calamagrostis lapponica (Wahlenb.) Hartm. [p. 731]
LAPLAND REEDGRASS, CALAMAGROSTIDE DE LAPPONIE

Plants rarely with sterile culms; loosely cespitose, with rhizomes 3–6+ cm long, 1–2 mm thick. **Culms** (12)35–50(90) cm, unbranched, smooth beneath the panicles; **nodes** 1–2(3). **Sheaths and collars** usually smooth, rarely with short hairs; **ligules** (0.5)2–4(5.5) mm, usually truncate, entire; **blades** (4)8–18(26) cm long, (1.5)2–3.5(4) mm wide, flat to involute, abaxial surfaces usually smooth, rarely slightly scabrous, adaxial surfaces usually smooth or scabrous, rarely sparsely hairy. **Panicles** (4)8–11(16) cm long, (0.7)1–2(2.8) cm wide, mostly erect, loosely contracted, purple; **branches** (2.1)2.5–3.5(5.4) cm, smooth or slightly scabrous, sometimes spikelet-bearing to the base, sometimes only on the distal ²/₃. **Spikelets** (3.5)4–5(5.5) mm; **rachilla prolongations** 0.4–1 mm, hairs 1.8–3 mm. **Glumes** usually more than 3 times as long as wide, rounded to slightly keeled, usually purple for most of their length and smooth, keels rarely slightly scabrous, lateral veins obscure, apices acute to acuminate; **callus hairs** (2)3–3.5(4.7) mm, (0.6)0.8–1(1.2) times as long as the lemmas, abundant; **lemmas** (2.5)3–4(5) mm, 0.3–1.5 (2.3) mm shorter than the glumes; **awns** 1.5–3 mm, attached to the lower ¹/₁₀–²/₅ of the lemmas, usually not exserted, usually slender and similar to the callus hairs, sometimes stouter, straight to somewhat bent; **anthers** (1.1)1.3–1.7(2) mm, usually poorly developed, sterile. 2*n* = 28, 42–112, 140.

Calamagrostis lapponica grows in northern and alpine tundra, particularly on ridgecrests and upper slopes, often with low shrubs including heathers, dwarf willows, and dwarf birch, usually on well-drained and coarse-textured (sand and gravel) soils, infrequently in meadows beside streams and lakeshores, very rarely in standing water, at 30–2300 m. It is circumboreal and circumpolar, ranging from Alaska to western Greeneland and Labrador, including the islands of the high arctic, south into the mountains of northern British

Columbia and the west-central Rocky Mountains of Alberta. In Europe it extends south to about 60° N latitude, and in Asia south to North Korea.

Calamagrostis lapponica is sometimes easily confused with *C. stricta* (see next), but the two grow in different habitats. In addition, the glumes of *C. lapponica* have a smoother, more glossy appearance than those of *C. stricta* and are typically purple for most of their length, including the apices; the glumes of *C. stricta* are generally brown at the apices. A specimen from Nakat Inlet, Alaska (ALA #V116195, *J. DeLapp* and *M. Duffy 93-339*) appears to be *C. lapponica*, although it is in a very different habitat and at an unusually low elevation for the species.

26. Calamagrostis stricta (Timm) Koeler [p. 731]
SLIMSTEM REEDGRASS

Plants rarely with sterile culms; cespitose, usually with rhizomes shorter than 5 cm, 1–1.5 mm thick. **Culms** (10)35–90(120) cm, usually unbranched, smooth to slightly scabrous; **nodes** 1–3(4). **Sheaths** usually smooth; **collars** usually smooth, sometimes scabrous, rarely pubescent; **ligules** (0.5)1–5.5(6) mm, truncate to obtuse, usually entire, sometimes lacerate; **blades** (5)11–25(34) cm long, (1)1.5–5(6) mm wide, flat or involute, usually scabrous, rarely smooth, sometimes puberulent. **Panicles** (2)4–18(29) cm long, (0.7)1–2(2.8) cm wide, erect, contracted, sometimes interrupted, pale green to purple; **branches** 1.4–5(9.5) mm, smooth or scabrous, usually spikelet-bearing to or near the base, sometimes only to midlength. **Spikelets** 2–4(5) mm; **rachilla prolongations** 0.5–1.5 mm, hairs 1.5–3 mm. **Glumes** usually less than 3 times as long as wide, rounded or keeled, usually smooth, rarely scabrous, keels smooth or scabrous, veins prominent to obscure, apices acute; **callus hairs** (1)1.5–3(4.5) mm, (0.5)0.7–0.9(1.3) times as long as the lemmas, abundant; **lemmas** 2–4(5) mm, 0.1–1.5 mm shorter than the glumes; **awns** 1.5–2.5 mm, usually attached to the lower ¹/₁₀–¹/₂ of the lemmas, rarely beyond the midpoint, equaling or exserted slightly beyond the margins of the glumes, usually stout, rarely slender, usually distinguishable from the callus hairs, straight or bent; **anthers** (0.9)1.2–1.8(2.4) mm, often sterile.

Calamagrostis stricta grows throughout northern North America; it also is found in Europe and northeastern Asia. It grows in habitats ranging from meadows and grassland to wetlands, sandy shorelines, and sand dunes, from sea level to 3400 m. Primarily a species of open settings, it is frequently found in association with shrubs. Both subspecies have a notable but not exclusive association with alkaline to saline substrates.

Calamagrostis stricta comprises both sexual and apomictic populations. Two subspecies, *C. stricta* subsp. *stricta* and subsp. *inexpansa*, intergrade but generally differ as described below. Greene (1984) treated subsp. *inexpansa* as consisting of the apomictic plants, probably derived from the sexual subsp. *stricta*. A number of apomictic variants were previously recognized at the species level; among these were *C. lacustris* (Kearney) Nash and *C. fernaldii* Louis-Marie, which are morphologically nearly indistinguishable from each other (Greene 1980, 1984).

Plants of short stature and short inflorescences, growing in the north, have been referred to as *Calamagrostis stricta* subsp. *borealis* (C. Laest.) Á. Löve & D. Löve or *C. stricta* var. *borealis* (C. Laest.) Hartm. These intergrade with taller plants; they are not recognized here as a distinct taxonomic entity.

Calamagrostis stricta is sometimes confused with *C. lapponica* (see previous). In addition to the differences noted in the descriptions and key, the glumes of *C. stricta* are not as smooth and glossy, and are generally brown at the tip; those of *C. lapponica* are typically purple.

1. Spikelets 3–4(5) mm long; callus hairs 2–4.5 mm long; rachilla prolongations 1–1.5 mm long; panicle branches 1.5–9.5 cm long; culms usually scabrous, sometimes smooth . . . subsp. *inexpansa*
1. Spikelets 2–2.5(3) mm long; callus hairs 1–3 mm long; rachilla prolongations 0.5–1 mm long; panicle branches 1.4–4 cm long; culms usually smooth, sometimes slightly scabrous
. . . . subsp. *stricta*

Calamagrostis stricta subsp. inexpansa (A. Gray) C.W. Greene [p. 731]

NORTHERN REEDGRASS, CALAMAGROSTIDE CONTRACTÉE

Plants apparently without sterile culms. **Culms** (29)35–75(120) cm, usually scabrous, sometimes smooth. **Sheaths** and **collars** usually smooth, collars sometimes scabrous, rarely pubescent; **ligules** (0.5)2–5.5(6) mm; **blades** (5)11–24(34) cm long, (1.5)2–5(6) mm wide, flat, usually stiff, sometimes puberulent. **Panicles** (6)8–11(29) cm long, (0.8)1–2(2.8) cm wide, pale green, sometimes purple-tinged; **branches** (1.5)1.6–5(9.5) cm, spikelet-bearing to the base. **Spikelets** 3–4(5) mm; **rachilla prolongations** 1–1.5 mm. **Glumes** broadly keeled or rounded; **callus hairs** (2)2.5–3(4.5) mm, (0.5)0.7–0.9(1.3) times as long as the lemmas; **lemmas** 2.5–3.5(5) mm, 0.1–1(1.4) mm shorter than the glumes; **awns** 2–2.5 mm, attached to the lower $^1/_{10}$–$^2/_5$ of the lemmas, rarely beyond the midpoint, straight or somewhat to strongly bent; **anthers** (0.9)1.5–1.8(2.4) mm, often poorly developed, sterile and indehiscent. $2n = 28, 56, 58, 70, 84–±120$.

Calamagrostis stricta subsp. *inexpansa* differs from subsp. *stricta* in its more robust growth and coarser habit. In North America, it extends from Alaska to Labrador and Newfoundland and south to California, Arizona, Minnesota, Iowa, Ohio, and New York. It also grows in northeastern Asia. This subspecies usually grows in moist meadows, sphagnum bogs, and grasslands associated with rivers and streams, and less frequently on grassy slopes, in open woods, and beside sand dunes. It is noted to grow at the edge of, rather than in, wetlands.

Plants of the Athabasca sand dunes in northern Saskatchewan, and similar habitats in northeast Alberta, have unusually long inflorescences and inflorescence branches.

A taxon with glumes that are thick and rounded at the base, rather than keeled, has been separated out as *Calamagrostis crassiglumis* Thurb., with a distribution along the Pacific coast from British Columbia to California. We have examined many such specimens, including the isotype, an unusually short specimen, at 15 cm tall, from a "sphagnum swamp". As far as the two character states described above are concerned, the specimens do not appear to be distinct from other individuals of *C. stricta* subsp. *inexpansa*. The specimens do differ in other characters from typical *C. stricta* subsp. *inexpansa* in that *C. crassiglumis* is generally shorter, and the culm leaves may diverge from the stem at a greater angle and be broader. Further investigation is warranted.

Calamagrostis stricta (Timm) Koeler subsp. stricta [p. 731]

CALAMAGROSTIDE DRESSÉE

Plants rarely with sterile culms. **Culms** (10)35–90(100) cm, usually smooth, sometimes slightly scabrous beneath the panicles. **Sheaths** and **collars** smooth; **ligules** (0.5)1–3.5(4) mm; **blades** (9)13–25 cm long, (1)1.5–2.5(3) mm wide, flat or often involute. **Panicles** (2)8–10(13) cm long, (0.7)1–2(2.5) cm wide, sometimes interrupted, purple-tinged; **branches** (1.4)2–2.5(4) cm, spikelet-bearing to below midlength, sometimes to the base. **Spikelets** 2–2.5(3) mm; **rachilla prolongations** 0.5–1 mm. **Glumes** keeled; **callus hairs** (1)1.5–2(3) mm, (0.5)0.7–0.8(1) times as long as the lemmas; **lemmas** 2–3 mm, 0.1–1.5 mm shorter than the glumes; **awns** 1.5–2.5 mm, attached to the lower $^1/_{10}$–$^1/_2$ of the lemmas, rarely slender, usually straight, sometimes bent; **anthers** (1.1)1.2–1.4(1.7) mm, usually fertile and well filled with pollen, dehiscent. $2n = 28, 42, 56, ±70$.

A circumboreal taxon, *Calamagrostis stricta* subsp. *stricta* favors moist meadows and fens, occurring less frequently in marshes and bogs, and sometimes grows near sand dunes. It is usually associated with fine-textured substrates. In the *Flora* region, its range

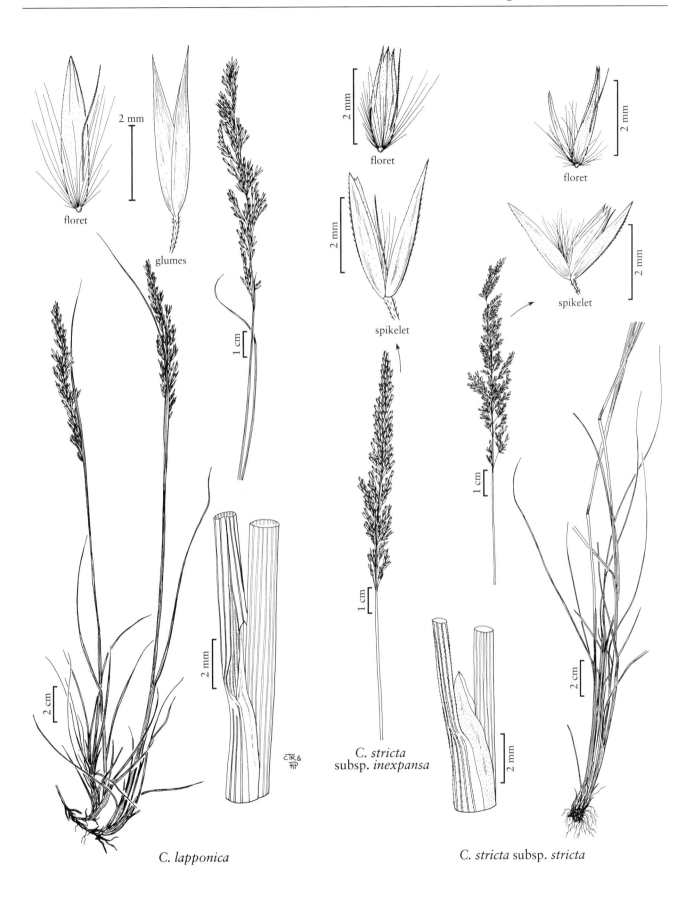

floret

glumes

2 mm

1 cm

floret

spikelet

floret

spikelet

2 mm

2 mm

2 mm

C. stricta
subsp. *inexpansa*

1 cm

2 mm

1 cm

C. lapponica

2 cm

2 mm

2 cm

C. stricta subsp. *stricta*

CALAMAGROSTIS

extends from Alaska to Newfoundland and Labrador and Greenland, and south to California, Utah, North Dakota, Minnesota, New Hampshire, and New York.

In Europe, it extends from 50° N latitude to Switzerland and the southern Caucasus.

14.50 GAUDINIA P. Beauv.

Thomas F. Daniel

Plants annual or perennial. **Culms** 15–120 cm. **Sheaths** open; **auricles** absent; **ligules** membranous; **blades** flat. **Inflorescences** solitary, distichous spikes; **disarticulation** in the rachis, immediately above the spikelets. **Spikelets** laterally compressed, sessile, tangential, more or less appressed to the rachis, with 3–11 florets; **rachillas** glabrous, prolonged beyond the base of the distal fertile floret, terminating in a reduced floret. **Glumes** 2, unequal, from shorter than to about as long as the spikelets, unawned; **calluses** blunt, glabrous; **lower glumes** 3(5)-veined; **upper glumes** 5–7(11)-veined; **lemmas** coriaceous, obscurely 7–9-veined, unawned or awned near the apices; **paleas** shorter than the lemmas; **lodicules** 2, free, membranous, glabrous, toothed; **anthers** 3; **ovaries** pubescent. **Caryopses** with a terminal tuft of hairs; **hila** round. $x = 7$. Named for Jean François Aimé Philippe Gaudin (1766–1833), a Swiss clergyman and botanist.

Gaudinia is a weedy genus of four species that are native to the Mediterranean, the Azores, and the Canary Islands. Its inflorescence is reminiscent of some *Triticeae*; it differs from members of that tribe in its manner of disarticulation and in having compound starch grains in its endosperm. One species has become established in North America.

SELECTED REFERENCE Daniel, T.F., C. Best, J. Guggolz, and B. Guggolz. 1992. Noteworthy collections: *Gaudinia fragilis*. Madroño 39:309–310.

1. Gaudinia fragilis (L.) P. Beauv. [p. 733]
FRAGILE OAT

Plants annual; usually tufted. **Culms** 15–80(120) cm, erect or ascending; **nodes** glabrous. **Sheaths** villous; **ligules** 0.5–0.7 mm, truncate; **blades** 1–6.5 cm long, 0.6–4 mm wide, villous. **Spikes** 6–15(35) cm. **Spikelets** 9–20 mm, with 3–6(8) florets; **rachillas** straight, proximal segments 2.6–4 mm, distal segments to 1.8 mm. **Glumes** scabrous over the veins, margins hyaline, apices unawned; **lower glumes** 3–5 mm; **upper glumes** 7–11 mm, about twice the length of the lower glumes; **calluses** glabrous; **lemmas** (3)5–8 mm, scabrous on the midveins, awned from above midlength, awns 4.5–15 mm, scabrous, twisted or geniculate; **anthers** 2–5 mm. **Caryopses** about 2.5 mm. $2n = 14$.

Gaudinia fragilis is the most widespread species of the genus in the Mediterranean region. The first record of its presence in the United States dates from a collection made by Karl Zimmer on ballast ground in Mobile, Alabama, in 1885. In 1991, it was discovered in Sonoma County, California, where it was found growing on an open, grassy hilltop in the thin, rocky soil of open oak woodlands, in a region that has long been used for agriculture.

14.51 SCOLOCHLOA Link

Mary E. Barkworth

Plants perennial; strongly rhizomatous, rhizomes succulent. **Culms** 70–200 cm, sometimes rooting at the lower nodes. **Sheaths** open; **auricles** absent; **ligules** membranous, truncate to rounded; **blades** 4–12 mm wide, flat. **Inflorescences** terminal, open panicles, some branches longer than 1 cm. **Spikelets** pedicellate, laterally compressed, with 3–4 florets; **rachillas**

floret

5 mm

terminal
spikelet

5 mm

palea

lemma

1 mm

floret

1 mm

spikelet

G. fragilis

2 cm

BFG

GAUDINIA

S. festucacea

2 cm

SCOLOCHLOA

prolonged beyond the base of the distal floret, internodes glabrous or with a few hairs to 0.2 mm long; **disarticulation** above the glumes, between the florets. **Glumes** unequal, acute to acuminate, unawned; **lower glumes** shorter than the adjacent lemma, 1–5-veined; **upper glumes** usually exceeding, sometimes equaling, the distal floret, 3–7-veined; **calluses** short, blunt, hairy; **lemmas** glabrous, rounded, 3–9-veined, veins excurrent, apices indistinctly 3-lobed or toothed, unawned; **paleas** about as long as the lemmas; **lodicules** free, glabrous; **anthers** 3; **ovaries** with pubescent apices. **Caryopses** shorter than the lemmas, concealed at maturity, about 2 mm, dorsiventrally compressed. $x = 7$. Name from the Greek *scolos*, 'cusp' or 'prickle', and *chloa*, 'grass', an allusion to the excurrent lemma veins.

Scolochloa is a monospecific genus that grows in shallow water and marshes of the temperate regions of North America and Eurasia. In the *Flora* region, it grows primarily in the northern Great Plains and prairie pothole region, from British Columbia to Manitoba and south through Nebraska and Iowa, with disjunct populations in Alaska, eastern Oregon, Montana, Idaho, and Wyoming. Other outlying populations probably reflect introductions.

SELECTED REFERENCES **Galatowitsch, S.M.** and **A. van der Valk**. 1994. Restoring Prairie Wetlands: An Ecological Approach. Iowa State University Press, Ames, Iowa, U.S.A. 246 pp.; **Neckles, H.A., J.W. Nelson**, and **R.L. Pederson**. 1985. Management of Whitetop (*Scolochloa festucacea*) Marshes for Livestock Forage and Wildlife. Technical Bulletin 1–1985. Delta Waterfowl and Wetlands Research Station, Portage la Prairie, Manitoba, Canada. 12 pp.

1. Scolochloa festucacea (Willd.) Link [p. 733]
COMMON RIVERGRASS, WHITETOP

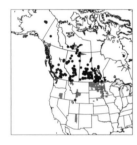

Culms 70–200 cm tall, 6–8 mm thick at the base. **Ligules** 3–7(9) mm; **blades** 20–45 cm long, 4–12 mm wide. **Panicles** 15–30 cm; **primary branches** ascending to divergent, with spikelets appressed to the branches. **Spikelets** 7–11 mm. **Lower glumes** 5–8 mm; **upper glumes** 6.5–10 mm; **calluses** about 0.5 mm; **lemmas** 4–9 mm; **paleas** about as long as the lemmas; **anthers** 2–4 mm. $2n = 28$.

Scolochloa festucacea grows in ponds, marshes, seasonally flooded basins, and the shallow margins of freshwater to moderately saline lakes and streams. It provides good nesting cover for some waterfowl and shorebirds, and can provide valuable forage for livestock and wildlife. It does not compete well with hybrid cattails (*Typha latifolia* × *T. angustifolia* or *T. domingensis*).

14.52 AVENA L.

Bernard R. Baum

Plants annual or perennial. **Culms** 8–200 cm, erect or decumbent. **Sheaths** open; **auricles** absent; **ligules** membranous; **blades** usually flat, sometimes involute, lax. **Inflorescences** panicles, diffuse, sometimes 1-sided, some branches longer than 1 cm. **Spikelets** 15–50 mm, pedicellate, laterally compressed, with 1–6(8) florets; **rachillas** not prolonged beyond the base of the distal floret; **disarticulation** above the glumes, usually also beneath the florets, cultivated forms not disarticulating. **Glumes** usually exceeding the florets, membranous, glabrous, 3–11-veined, acute, unawned; **calluses** rounded to pointed, with or without hairs; **lemmas** usually indurate and enclosing the caryopses at maturity, 5–9-veined, often with twisted, strigose hairs below midlength, apices dentate to bifid or biaristate, usually awned, sometimes unawned, awns dorsal, usually once-geniculate and strongly twisted in the basal portion; **paleas** bifid or entire, keels ciliate; **lodicules** 2, free, glabrous, toothed or not toothed; **anthers** 3; **ovaries** hairy.

Caryopses shorter than the lemmas, concealed at maturity, terete, ventrally grooved, pubescent; **hila** linear. $x = 7$. Name from the Latin *avena*, 'oats'.

Avena, a genus of 29 species, is native to temperate and cold regions of Europe, North Africa, and central Asia; it has become nearly cosmopolitan through the cultivation of cereal oats, and the inadvertent introduction of the weedy species. Six species have been introduced into the *Flora* region.

Reports of *Avena strigosa* Schreb. from California are based on misidentifications. The specimens involved belong to *A. barbata*.

SELECTED REFERENCES **Baum, B.R.** 1977. Oats: Wild and Cultivated; A Monograph of the Genus *Avena* L. (Poaceae). Biosystematics Research Institute Monograph No. 14. Supply and Services Canada, Ottawa, Ontario, Canada. 463 pp.; **Dore, W.G.** and **J. McNeill**. 1980. Grasses of Ontario. Research Branch, Agriculture Canada Monograph No. 26. Canadian Government Publishing Centre, Hull, Québec, Canada. 568 pp.

1. Florets not disarticulating from the glumes, remaining attached to the plant even at maturity; calluses glabrous . . . . 5. *A. sativa*
1. Florets disarticulating at maturity, only the glumes remaining attached; calluses bearded.
    2. Florets falling from the glumes as a unit . . . . 6. *A. sterilis*
    2. Florets falling separately.
        3. Lemma apices biaristate, 2 veins extending 2–4 mm beyond the apices . . . . 1. *A. barbata*
        3. Lemma apices erose to bifid, the veins not extending beyond the apices.
            4. Spikelets with 2(3) florets; disarticulation scar of the lower florets in a spikelet round to oval or triangular, that of the third floret, if present, similar . . . . 2. *A. fatua*
            4. Spikelets with 2–4(5) florets; disarticulation scar of the lower florets in a spikelet round to elliptic, those of the third and fourth florets (and sometimes the second) heart-shaped.
                5. Glumes 15–23 mm long . . . . 3. *A. hybrida*
                5. Glumes 28–40 mm long . . . . 4. *A. occidentalis*

1. **Avena barbata** Pott *ex* Link [p. 736]
SLENDER OATS, SLENDER WILD OATS

Plants annual. **Culms** 60–80 (150) cm, initially prostrate, usually becoming erect. **Sheaths** of the basal leaves pilose, upper sheaths usually glabrous; **ligules** 1–6 mm, obtuse; **blades** 6–30 cm long, 2–20 mm wide, glabrous or pilose. **Panicles** 15–35.5 (50) cm long, 6–12 cm wide, erect or nodding. **Spikelets** 21–30 mm, with 2–3 florets; **disarticulation** beneath each floret; **disarticulation scars** elliptic to triangular. **Glumes** subequal, 15–30 mm, 7–9-veined; **calluses** bearded, hairs 2–3 mm; **lemmas** 15–26 mm, densely strigose below midlength, apices acute, biaristate, 2 veins extending 2–4 mm beyond the apices, awns 30–45 mm, arising about midlength, geniculate; **lodicules** narrowly triangular, without lobes on the wings; **anthers** 2.5–4 mm. $2n = 28$.

Avena barbata is native to the Mediterranean region and central Asia. It has become naturalized in western North America, particularly California, displacing native grasses. It was collected once in Vancouver, British Columbia, but should be considered a waif there.

2. **Avena fatua** L. [p. 736]
WILD OATS, FOLLE AVOINE

Plants annual. **Culms** 8–160 cm, prostrate to erect when young, becoming erect at maturity. **Sheaths** of the basal leaves with scattered hairs, upper sheaths glabrous; **ligules** 4–6 mm, acute; **blades** 10–45 cm long, 3–15 mm wide, scabridulous. **Panicles** 7–40 cm long, 5–20 cm wide, nodding. **Spikelets** 18–32 mm, with 2(3) florets; **disarticulation** beneath each floret; **disarticulation scars** of all florets round to ovate or triangular. **Glumes** subequal, 18–32 mm, 9–11-veined; **calluses** bearded, hairs to $^1/_4$ the length of the lemmas; **lemmas** 14–22 mm, usually densely strigose below midlength, sometimes sparsely strigose or glabrous, veins not extending beyond the apices, apices usually bifid, teeth 0.3–1.5 mm, awns 23–42 mm, arising in the middle $^1/_3$ of the lemmas; **lodicules** without lobes on the wings; **anthers** about 3 mm. $2n = 42$.

Avena fatua is native to Europe and central Asia. It is known as a weed in most temperate regions of the world; it is considered a noxious weed in some parts of Canada and the United States.

lemma
apex

floret

disarticulation
scar

spikelet

floret

disarticulation
scar

glume   glume

florets

spikelet

ligule

A. barbata   *A. fatua*

AVENA

Avena fatua is sometimes confused with *A. occidentalis*, but differs in having shorter, wider spikelets, fewer florets, and a distal floret which does not have a heart-shaped disarticulation scar. Hybrids between *A. fatua* and *A. sativa* are common in plantings of cultivated oats. The hybrids resemble *A. sativa*, but differ in having the *fatua*-type lodicule; some also have a weak awn on the first lemma. They are easily confused with fatuoid forms of *A. sativa*.

3. Avena hybrida Peterm. [p. 738]

Plants annual. **Culms** erect; **nodes** often pubescent. **Sheaths** glabrous; **ligules** 4–5 mm, obtuse or acute; **blades** 12–25 cm long, 7–12 mm wide. **Panicles** 15–30 cm, equilateral, sometimes slightly secund. **Spikelets** 15–24 mm, with 2–4 florets; **disarticulation** beneath each floret; **disarticulation scars** of the lower floret(s) in a spikelet oval to round, those of the third and fourth (and sometimes the second) florets heart-shaped. **Glumes** equal, 15–23 mm, 7–9(11)-veined; **calluses** bearded; **lemmas** about 21 mm, usually glabrous, sometimes pubescent beneath the awn insertion, irregularly bidenticulate to bisubulate, veins not extending beyond the apices, awns about 30 mm, arising at midlength; **lodicules** with a small side lobe; **anthers** about 2 mm. 2*n* = 42.

Avena hybrida is native to western and central Asia; it grows as a weed in Europe. It has been reported from Essex County, Massachusetts, and Prince Edward Island.

4. Avena occidentalis Durieu [p. 738]
WESTERN OATS

Plants annual. **Culms** 50–80 cm, erect. **Sheaths** glabrous or sparsely pubescent, hairs 0.5–1 mm, sometimes confined to the margins; **ligules** of the lower leaves 3–5 mm, those of the upper leaves 1.5–2.5 mm, acute; **blades** 12–25 cm long, 5–11 mm wide. **Panicles** 15–26 cm. **Spikelets** 30–40 mm, with 3–4(5) florets; **disarticulation** beneath each floret; **disarticulation scars** of the lower florets in a spikelet round to elliptic, those of the third and fourth florets heart-shaped. **Glumes** subequal, 28–40 mm, 7–9-veined; **calluses** bearded, hairs 3–5 mm; **lemmas** 14–26 mm, usually densely strigose below midlength, sometimes sparsely strigose or glabrous, veins not extending beyond the apices, apices bifid, teeth sometimes shortly aristate, awns arising at midlength; **lodicules** without lobes on the wing; **anthers** 2–3.2 mm. 2*n* = 42.

Avena occidentalis is native to the Canary Islands, coastal North Africa, and Saudi Arabia; it is now established in western North America, from California to northern Mexico. It is often confused with *A. fatua*, but differs in its longer, narrower spikelets, greater number of florets, and the heart-shaped disarticulation scars of the distal florets.

5. Avena sativa L. [p. 738]
OATS, CULTIVATED OATS, NAKED OATS, AVOINE, AVOINE CULTIVÉE

Plants annual. **Culms** 35–180 cm, prostrate to erect when young, becoming erect at maturity. **Sheaths** smooth or scabridulous; **ligules** 2–8 mm, truncate to acute; **blades** 8–45 cm long, 3–14 (25) mm wide, scabridulous. **Panicles** (6)15–40 cm long, 5–15 cm wide, nodding. **Spikelets** (18)25–32 mm, to 50 mm in 'naked oats', with 1–2 florets (to 7 in 'naked oats'); **disarticulation** not occurring, the florets remaining attached even when mature. **Glumes** subequal, (18)20–32 mm, 9–11-veined; **calluses** glabrous; **lemmas** 14–18 mm, usually indurate, membranous in 'naked oats', usually glabrous, sometimes sparsely strigose, apices erose to dentate, longest teeth 0.2–0.5 mm, usually unawned, sometimes awned, awns 15–30 mm, arising in the middle ¹/₃, weakly twisted, not or only weakly geniculate; **lodicules** with a lobe or tooth on the wings, this sometimes very small; **anthers** (1.7)3–4.3 mm. 2*n* = 42.

Avena sativa, a native of Eurasia, is widely cultivated in cool, temperate regions of the world, including North America. Fall-sown oats are planted in the Pacific and southern states in the United States; spring-sown oats are more important elsewhere in North America. It is sometimes planted as a fast-growing soil stabilizer along roadsides. Several forms are grown, of which the most distinctive are 'naked oats'. These differ from typical forms as indicated in the description, and in having caryopses that fall from the florets. Escapes from cultivation are common but rarely persist.

Avena sativa hybridizes readily with *A. fatua*, forming hybrids with the *fatua*-type lodicule. The hybrids are easily confused with fatuoid forms of *A. sativa*, which differ in having the *sativa*-type lodicule.

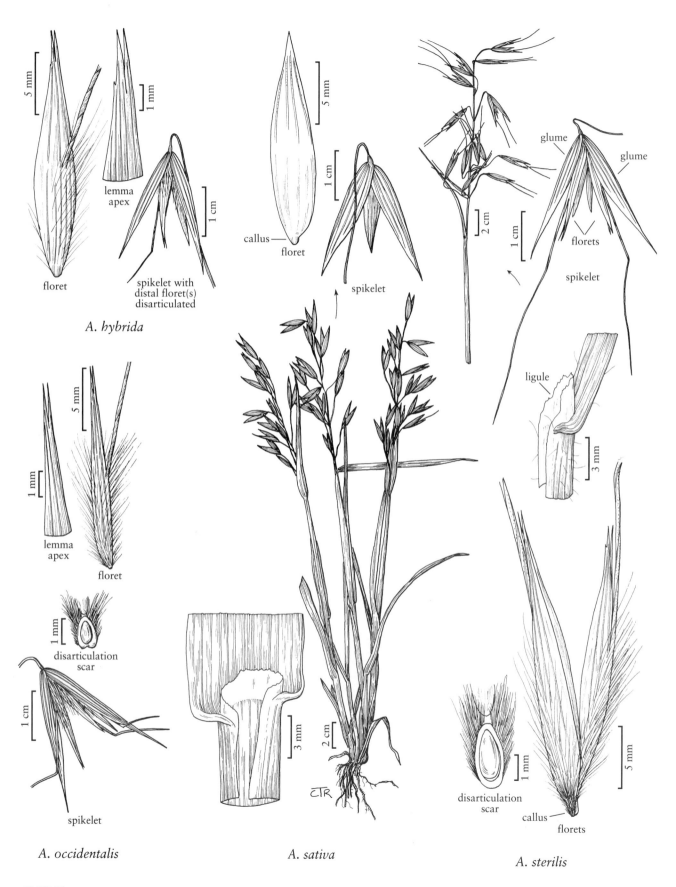

A. hybrida

A. occidentalis

A. sativa

A. sterilis

AVENA

6. Avena sterilis L. [p. 738]
ANIMATED OATS, AVOINE STÉRILE

Plants annual. **Culms** 30–120 cm, initially prostrate, becoming erect at maturity. **Sheaths** glabrous or hairy; **ligules** 3–8 mm, acute to truncate-mucronate; **blades** 8–60 cm long, 4–18 mm wide, scabridulous, often ciliolate on the margins. **Panicles** 10–45 cm long, 5–25 cm wide. **Spikelets** 24–50 mm, with 2–5 florets; **disarticulation** beneath the basal floret, the florets falling as a unit; **disarticulation scar** oval to round-elliptic. **Glumes** subequal, 20–50 mm, 9–11-veined; **calluses** bearded, hairs to $^1/_5$ the length of the lemmas; **lemmas** 17–40 mm, usually densely strigose below midlength, sometimes sparsely strigose, glabrous, or scabridulous, apices bidentate to bisubulate, teeth 1–1.5 mm, awns 30–90 mm, arising in the middle $^1/_3$; **lodicules** without a lobe on the wing; **anthers** 2.5–4 mm. $2n = 42$.

Avena sterilis is native from the Mediterranean region to Afghanistan; it now grows on all continents. It has become naturalized in California, where it can be found in fields, vineyards, orchards, and on hillsides. It has been reported from Oregon, but no specimens could be found to substantiate the report. Dore & McNeill (1980) also report it from Ottawa and Guelph, Ontario. It is listed as a noxious weed by the U. S. Department of Agriculture.

14.53 HOLCUS L.

Lisa A. Standley

Plants usually perennial, rarely annual; cespitose or rhizomatous, rarely both cespitose and rhizomatous. **Culms** (8)20–200 cm, glabrous or pubescent; **nodes** glabrous or retrorsely pubescent. **Sheaths** open; **auricles** absent; **ligules** 1–5 mm, membranous, entire or erose-ciliate, glabrous or puberulent; **blades** flat, pubescent. **Inflorescences** terminal panicles, contracted to open. **Spikelets** laterally compressed, with 2(3) florets, lower florets bisexual, upper floret(s) staminate or sterile; **rachillas** curved below the lowest florets, sometimes prolonged beyond the base of the distal florets; **disarticulation** below the glumes. **Glumes** equaling to exceeding the florets, strongly keeled, unawned; **lower glumes** 1-veined; **upper glumes** 3-veined; **calluses** glabrous or pubescent; **lemmas** firm, shiny, glabrous or pubescent, obscurely 3–5-veined, often bidentate; **lower lemmas** unawned; **upper lemmas** awned from below the apices, awns hooked or geniculate; **paleas** thin, subequal to the lemmas; **lodicules** 2, glabrous, toothed or not; **anthers** 3; **ovaries** glabrous. **Caryopses** shorter than the lemmas, concealed at maturity, glabrous. $x = 4$, 7. Name from the Greek *holkos*, a kind of grain, perhaps sorghum.

Holcus, a genus of eight species, is native to Europe, North Africa, and the Middle East. One species, *H. lanatus*, has become widely naturalized in the Americas, Japan, and Hawaii; a second, *H. mollis*, has become a troublesome weed in some areas of the *Flora* region.

SELECTED REFERENCES **Carroll, C.P.** and **K. Jones.** 1962. Cytotaxonomic studies in *Holcus*: III. A morphological study of the triploid F$_1$ hybrid between *H. lanatus* L. and *H. mollis* L. New Phytol. 61:72–84; **Jones, K.** 1958. Cytotaxonomic studies in *Holcus* L.: I. The chromosome complex in *Holcus mollis* L. New Phytol. 57:191–210; **Jones, K.** and **C.P. Carroll.** 1962. Cytotaxonomic studies in *Holcus* L.: II. Morphological relationships in *H. mollis* L. New Phytol. 61:63–71; **Zandee, M.** and **P.C.G. Glas.** 1982. Studies in the *Holcus lanatus–Holcus mollis* complex (Poaceae [= Gramineae]). Proc. Kon. Ned. Akad. Wetensch., C 85:413–437.

1. Awns 1–2 mm long, forming a curved hook at maturity; culms densely pilose adjacent to the lower nodes; plants cespitose, not rhizomatous . . . . .1. *H. lanatus*
1. Awns 3–5 mm long, straight or geniculate at maturity; culms glabrous or sparsely pubescent adjacent to the lower nodes; plants not cespitose, rhizomatous . . . . .2. *H. mollis*

1. Holcus lanatus L. [p. 741]

VELVETGRASS, YORKSHIRE FOG, HOULQUE LAINEUSE

Plants perennial; cespitose, not rhizomatous. **Culms** 20–100 cm, erect, sometimes decumbent; **lower internodes** densely pilose, hairs to 1 mm; **uppermost internode** often glabrous. **Sheaths** densely pubescent; **ligules** 1–4 mm, truncate, erose-ciliolate; **blades** 2–20 cm long, (3)5–10 mm wide, densely soft-pubescent. **Panicles** 3–15(20) cm long, 1–8 cm wide; **branches** hairy; **pedicels** 0.2–1.6(4) mm, pilose, hairs to 0.3 mm. **Spikelets** 3–6 mm; **rachillas** 0.4–0.5 mm, glabrous. **Glumes** exceeding and enclosing the florets, membranous, ciliate on the keels and veins, usually scabrous, puberulent, or villous between the veins, especially towards the apices, whitish green, often purple over the veins and towards the apices; **lower glumes** lanceolate, narrow, acute; **upper glumes** ovate, wider and longer than the lower glumes, midveins often prolonged as an awn to 1.5 mm, apices obtuse, somewhat bifid; **calluses** sparsely hirsute; **lemmas** 1.7–2.5 mm, acute, erose-ciliate; **upper lemmas** shallowly bifid, awns 1–2 mm, often purple-tipped, slightly twisted and forming a curved hook at maturity; **anthers** (1.2)2–2.5 mm. $2n = 14$.

Holcus lanatus grows in disturbed sites, moist waste places, lawns, and pastures, in a wide range of edaphic conditions and at elevations from 0–2300 m. A native of Europe, it was widely distributed in North America by 1800. It is an ancestor of the polyploid complex represented by *H. mollis*. In Europe, it hybridizes with tetraploids of *H. mollis* to form a sterile triploid that spreads vegetatively.

2. Holcus mollis L. [p. 741]

CREEPING VELVETGRASS, HOULQUE MOLLE

Plants perennial; not cespitose, rhizomatous, rhizomes to 40 cm. **Culms** 20–100(150) cm, usually decumbent at the base; **lower internodes** glabrous or sparsely pubescent. **Sheaths** glabrous or hairy; **ligules** 1–5 mm, obtuse, erose; **blades** 2–20 cm long, 3–10 mm wide, pubescent. **Panicles** 4–20(22) cm long, to 3 cm wide; **branches** puberulent or ciliate; **pedicels** to 5 mm long, pilose, hairs to 0.3 mm. **Spikelets** 4–6(7) mm; **rachillas** hairy. **Glumes** exceeding and enclosing the florets, subequal, nearly the same width, ovate, membranous, whitish green when young, straw-colored with age, veins ciliate, often purple, intercostal regions scabrous or glabrous, apices acuminate or acute, unawned; **calluses** densely to sparsely hairy; **lemmas** 2–2.5 mm, glabrous, acute; **upper lemmas** bifid, awned above midlength, awns 3–5 mm, scabrous, straight or geniculate at maturity; **anthers** about 2 mm. $2n = 28$ (35, 42, 49).

Holcus mollis grows in moist soil and disturbed sites, including lawns and damp pastures. It is a European introduction that has persisted in the *Flora* region, becoming a problematic weed in ungrazed pastures, prairie remnants, and oak savannahs in portions of the Pacific Northwest. It is also sold as an ornamental. There are two subspecies: **Holcus mollis** L. subsp. **mollis** (stems not thickened and tuberous at the base; panicles lax, brownish or purplish) and **H. mollis** subsp. **reuteri** (Boiss.) Malag. (stems thickened and tuberous at the base; panicles narrow, whitish). North American introductions belong to subsp. *mollis*.

14.54 ARRHENATHERUM P. Beauv.

Stephan L. Hatch

Plants perennial; cespitose, sometimes rhizomatous. **Culms** 30–200 cm, basal internodes occasionally globose. **Sheaths** open, not overlapping; **auricles** absent; **ligules** membranous, sometimes ciliate; **blades** flat or convolute. **Inflorescences** terminal, narrow panicles; **branches** spreading until after anthesis, then becoming loosely appressed to the rachises. **Spikelets** pedicellate, laterally compressed, with 2 florets, lower florets staminate, upper florets pistillate or bisexual, a rudimentary floret occasionally present distally; **rachillas** pubescent; **disarticulation** above the glumes, the florets usually falling together, rarely falling separately. **Glumes** unequal, hyaline, unawned; **lower glumes** less than ³/₄ the length of the upper glumes, 1- or 3-veined; **upper glumes** 3-veined; **calluses** short, blunt, pubescent; **lower lemmas** membranous, 3–7-veined, acute, awned below midlength, awns twisted and geniculate; **upper**

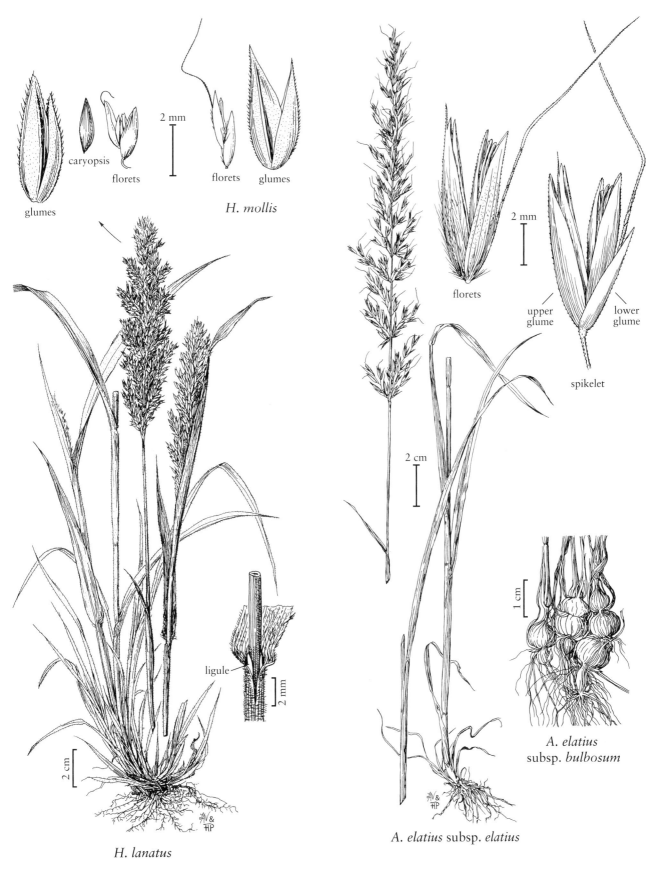

glumes   caryopsis   florets   florets   glumes

2 mm

H. mollis

florets

2 mm

upper glume   lower glume

spikelet

2 cm

ligule

2 mm

2 cm

A. elatius subsp. *bulbosum*

1 cm

H. lanatus

A. elatius subsp. *elatius*

HOLCUS   ARRHENATHERUM

lemmas membranous to subcoriacous, glabrous or hairy, 7-veined, acute, usually unawned, sometimes awned from near the apices, awns short, straight, rarely awned similarly to the lower lemmas; **paleas** subequal to the lemmas, 2-veined, 2-keeled, keels scabrous or hairy, apices notched; **lodicules** 2, free, linear, membranous, glabrous, entire; **anthers** 3, 3.4–6.5 mm; **ovaries** pubescent. **Caryopses** shorter than the lemmas, concealed at maturity, not grooved, dorsally compressed to terete, hairy; **hila** long-linear. $x = 7$. Name from the Greek *arren*, 'masculine', and *ather*, 'awn', referring to the awned staminate florets.

Arrhenatherum is a Mediterranean and eastern Asian genus of six species; one has become established in North America.

SELECTED REFERENCE Brandenburg, D.M. 1985. Systematic studies in the Poaceae and Cyperaceae. Ph.D. dissertation, University of Oklahoma, Norman, Oklahoma, U.S.A. 249 pp.

1. Arrhenatherum elatius (L.) P. Beauv. *ex* J. Presl & C. Presl [p. 741]

TALL OATGRASS, FENASSE, FROMENTAL

Plants loosely cespitose, sometimes rhizomatous, rhizomes to 3 mm thick. **Culms** 50–140 (180) cm, erect, glabrous, unbranched, basal internodes swollen or not; **nodes** usually glabrous, occasionally puberulent to densely hairy. **Sheaths** smooth; **ligules** 1–3 mm, obtuse to truncate, usually ciliate; **blades** 5–32 cm long, (1)3–8(10) mm wide, flat, usually glabrous, rarely shortly pilose, sometimes scabrous. **Panicles** 7–30(36) cm long, 1–6(10) cm wide, green, shiny, becoming stramineous, sometimes purple-tinged; **branches** 15–20 mm, ascending to divergent, verticillate, usually spikelet-bearing to the base; **pedicels** 1–10 mm. **Spikelets** 7–11 mm; **rachillas** stout, internodes to 0.7 mm, prolongations 1.2–2 mm, slender, apices often with a small, club-shaped rudiment. **Glumes** lanceolate to elliptic; **lower glumes** 4–7 mm; **upper glumes** 7–10 mm; **callus hairs** to 3.7 mm; **lemmas** (4)7–10 mm, apices bifid; **awns of lower lemmas** 10–20 mm, twisted below, often with alternating light and dark bands; **awns of upper lemmas** absent or to 5 mm and arising just below the apices, rarely to 15 mm and arising from above the

middle; **paleas** 0.5–1 mm shorter than the lemmas, acute; **anthers** 3.6–5(6) mm. **Caryopses** 4–5 mm long, about 1.2 mm wide, ellipsoid, densely hairy, yellowish. $2n = 14, 28, 42$.

Arrhenatherum elatius is grown as a forage grass, and yields palatable hay; it does not withstand heavy grazing. It readily escapes from cultivation, and can be found in mesic to dry meadows, the edges of woods, streamsides, rock outcrops, and disturbed areas such as fields, pastures, fence rows, and roadsides. Variegated forms, with the leaves striped green and white or yellow, are cultivated as ornamentals. There are two subspecies, both of which have been found in the *Flora* region. Plants in which both lemmas have long, geniculate awns have been called *A. elatius* var. *biaristatum* (Peterm.) Peterm., but do not merit formal taxonomic recognition.

Arrhenatherum elatius (L.) P. Beauv. *ex* J. Presl & C. Presl subsp. **elatius** has glabrous nodes and basal internodes 2–4 mm thick. It is more common than **Arrhenatherum elatius** subsp. **bulbosum** (Willd.) Schübl. & G. Martens, which has densely hairy nodes, and swollen basal internodes 5–10 mm thick. While both can be weedy, the latter subspecies is especially difficult to control in cultivated fields, as tilling the soil spreads the swollen internodes, which then propagate vegetatively.

14.55 CORYNEPHORUS P. Beauv.

John W. Thieret†

Plants annual or perennial. **Culms** to 60 cm, erect. **Sheaths** open; **auricles** absent; **ligules** membranous; **blades** flat or involute. **Inflorescences** terminal panicles. **Spikelets** pedicellate, laterally compressed, with 2 bisexual florets; **rachillas** prolonged beyond the base of the distal floret, pilose; **disarticulation** above the glumes, beneath the florets. **Glumes** subequal, exceeding the florets, lanceolate, membranous, acute, unawned; **lower glumes** 1-veined; **upper glumes** 3-veined at the base; **calluses** pilose; **lemmas** about $^1/_2$ as long as the glumes, ovate-lanceolate, membranous, 1-veined, acute, awned from just above the base, awns geniculate, articulated near the middle, with a ring of minute, conical protuberances near the joint, proximal segment yellow-brown to dark brown, smooth, thicker than the distal segment, distal segment pale green to whitish, clavate; **paleas** shorter than the lemmas, membranous; **lodicules** 2, free, glabrous, toothed; **anthers** 3; **ovaries** glabrous. **Caryopses** shorter than the lemmas, concealed at maturity, usually adhering to the lemmas and/or paleas, longitudinally grooved. $x = 7$. Name from the Greek *koryne*, 'club', and *phoreus*, 'bearer', an allusion to the club-shaped awns.

Corynephorus is a Eurasian and North African genus of five species, one of which has been introduced into the *Flora* region. Flowering specimens are easily recognized by their distinctive awns, but sterile plants, with their involute, glaucous leaves, resemble involute-leaved species of *Festuca*, such as *F. trachyphylla*.

SELECTED REFERENCES **Albers, F.** 1973. Cytosystematische Untersuchungen in der Subtribus *Deschampsiineae* Holub (Tribus Aveneae Nees): I. Zwei Arten der Gattung *Corynephorus* P.B. Preslia 45:11–18; **Douglas, G.W., G.B. Straley,** and **D.V. Meidinger** (eds). 1994. The Vascular Plants of British Columbia: Part 4–Monocotyledons. Research Branch, Ministry of Forests, Victoria, British Columbia, Canada. 257 pp.; **Jirásek, V.** and **J. Chrtek.** 1962. Systematische Studie über die Arten der Gattung *Corynephorus* Pal.-Beauv. (Poaceae). Preslia 34:374–386.

1. Corynephorus canescens (L.) P. Beauv. [p. 743]

GRAY HAIRGRASS

Plants perennial; cespitose, glaucous, with many sterile shoots. **Culms** 10–60 cm, glabrous, often geniculate; **nodes** dark brown to blackish. **Leaves** basally crowded, glabrous; **sheaths** to 8 cm, often purplish, scabridulous; **ligules** 2–4 mm, acute; **blades** to 10 cm long and 1 mm wide, rigid, involute, scabrous. **Panicles** 1.5–14 cm long, 0.4–3 cm wide, dense, purple or variegated with pale green; **branches** to 5 cm, spreading at anthesis, otherwise appressed; **pedicels** 1–3 mm. **Spikelets** 2.9–4.3 mm. **Glumes** often purplish, slightly to strongly scabridulous, particularly on the keels; **callus hairs** 0.2–0.4 mm; **lemmas** 1.4–2 mm, awns 2.3–2.7 mm, usually not exceeding the glumes, segments subequal, distal segments sometimes slightly exserted; **anthers** 1.4–1.6 mm, purple. $2n = 14$.

Corynephorus canescens is native to Europe. It grows on coastal sand dunes and inland on sandy soils, as well as in disturbed areas such as waste ground and ballast dumps. It has been recorded from scattered locations in North America, but its current status in these locations is not known. Douglas et al. (1994) reported that it no longer occurred in British Columbia, but it was later found near the original collection site (UBC 209521, *Lomer 94-256*).

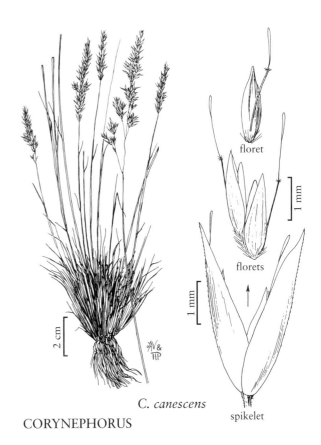

C. canescens

CORYNEPHORUS

14.56 TRISETUM Pers.

John H. Rumely

Plants annual or perennial; sometimes rhizomatous, sometimes cespitose. **Culms** 5–150 cm, glabrous or pubescent, basal branching extravaginal. **Sheaths** open the entire length or fused at the base; **auricles** absent; **ligules** membranous, often erose to lacerate, sometimes ciliolate; **blades** rolled in the bud. **Inflorescences** terminal panicles, open and diffuse to dense and spikelike; **branches** antrorsely scabrous. **Spikelets** 2.5–12 mm, usually subsessile to pedicellate, rarely sessile, laterally compressed, with 2–5 florets, reduced florets sometimes present distally; **rachillas** hairy, internodes evident, prolonged beyond the base of the distal bisexual florets; **disarticulation** usually initially above the glumes and beneath the florets, subsequently below the glumes, in some species initially below the glumes. **Glumes** subequal or unequal, keels scabrous, apices usually acute, unawned, often apiculate; **lower glumes** 1(3)-veined; **upper glumes** 3(5)-veined, lateral veins less than $^1/_2$ the glume length; **calluses** hairy; **lemmas** 3–7-veined, margins hyaline, unawned or awned from above the middle with a single awn, apices usually bifid, sometimes entire; **paleas** from subequal to longer than the lemmas, membranous, 2-veined, veins usually extending as bristlelike tips; **lodicules** 2, shallowly and usually slenderly lobed to fimbriate; **anthers** 3; **ovaries** glabrous or pubescent; **styles** 2. **Caryopses** shorter than the lemmas, concealed at maturity, elongate-fusiform, compressed, brown; **embryos** elliptic, to $^1/_3$ the length of the caryopses; **endosperm** milky. $x = 7$. Name from the Latin *tres*, 'three', and *seta*, 'bristle', alluding to the three-awned appearance of the lemmas of the type species, *Trisetum flavescens* (L.) P. Beauv.

Trisetum, a genus of approximately 75 species, occurs primarily in temperate, subarctic, and alpine regions. Eight species are native to the *Flora* region; two have been introduced, one of which is not known to have persisted. *Trisetum* usually differs from *Sphenopholis* in having longer awns that are inserted lower on the lemmas, and spikelets that disarticulate above the glumes. It differs from *Deschampsia* primarily in its more acute, bifid lemmas. In addition, all species of *Trisetum* have awns that are inserted at or above the midpoint of the lemmas; in *Deschampsia*, the awns are usually inserted at or below midlength, often near the base.

Trisetum spicatum is important as forage on native rangelands. Like other species of the genus, it is a significant component of natural food pyramids, especially in arctic and alpine regions and mountain parks. *Trisetum flavescens* was introduced from Europe as a pasture grass; *T. interruptum* is often weedy.

Spikelet measurements do not include the awns.

SELECTED REFERENCES **Dixon, J.M.** 1995. *Trisetum flavescens* (L.) Beauv. (*T. pratense* Pers., *Avena flavescens* L.). J. Ecol. 83:895–909; **Finot, V.L., P.M. Peterson, R.J. Soreng,** and **F.O. Zuloaga.** 2004. A revision of *Trisetum, Peyritschia,* and *Sphenopholis* (Poaceae: Poöideae: Aveninae) in Mexico and Central America. Ann. Missouri Bot. Gard. 91:1-30; **Hitchcock, A.S.** 1951. Manual of the Grasses of the United States, ed. 2, rev. A. Chase. U.S.D.A. Miscellaneous Publication No. 200. U.S. Government Printing Office, Washington, D.C., U.S.A. 1051 pp. **Hultén, E.** 1959. The *Trisetum spicatum* complex. Svensk. Bot. Tidskr. 53:203–228; **Louis-Marie, Father, O.C.** 1928. The genus *Trisetum* in America. Rhodora 30:209–228, 231–245; **Shelly, J.S.** 1987. Rediscovery and preliminary studies of *Trisetum orthochaetum*, Missoula County, Montana. Proc. Montana Acad. Sci. 47:3–4 [abstract].

1. Plants annual; without sterile shoots.
    2. Lower glumes 3-veined; spikelets 3–6 mm long; panicles 0.3–1.5 cm wide; plants native . . . 9. *T. interruptum*
    2. Lower glumes 1-veined; spikelets 2.5–3.5 mm long; panicles 0.5–3 cm wide; plants not native, not established . . . . 10. *T. aureum*
1. Plants perennial; usually producing both fertile and sterile shoots.
    3. Lemmas unawned or with inconspicuous straight awns up to 2 mm long that rarely exceed the lemma apices.

4. Panicles usually 2–4 cm wide, lax and nodding; callus and rachilla hairs 1.3–2 mm long; plants of eastern North America . . . . 1. *T. melicoides*

4. Panicles usually 1–1.5 cm wide, erect; callus hairs shorter than 0.5 mm; rachilla hairs up to 1 mm long; plants of western North America . . . . 2. *T. wolfii*

3. Lemmas with evident awns 3–14 mm long, these straight, curved, flexuous, or geniculate, exceeding the lemma apices.

   5. Plants rhizomatous; culms usually solitary.

      6. Lemma teeth usually 3–6 mm long; ligules 0.5–1(2) mm long . . . . 6. *T. flavescens* (in part)

      6. Lemma teeth usually shorter than 1 mm; ligules 1–5 mm long.

         7. Culms 15–65 cm tall; panicles 2–12(16) cm long; plants of Alaska and the Yukon Territory . . . . 8. *T. sibiricum*

         7. Culms 80–110 cm tall; panicles 13–20 cm long; known only from Montana . . . . 3. *T. orthochaetum*

   5. Plants not rhizomatous; culms clumped.

      8. Glumes usually subequal; both glumes lanceolate; upper glumes less than twice as wide as the lower glumes . . . . 7. *T. spicatum*

      8. Glumes usually unequal, sometimes subequal; lower glumes subulate to linear-lanceolate or lance-elliptic; upper glumes broadly lanceolate to ovate or obovate, at least twice as wide as the lower glumes.

         9. Upper glumes as long as or longer than the lowest florets; awns 3–9 mm long; rachilla hairs up to 1.5 mm long; ligules 0.5–2 mm long; panicles yellowish brown . . . . 6. *T. flavescens* (in part)

         9. Upper glumes shorter than the lowest florets; awns 7–14 mm long; rachilla hairs 0.7–2.5 mm long; ligules 1.5–6 mm long; panicles green or tan.

           10. Most panicle branches, except sometimes the lowermost, spikelet-bearing for their full length; panicles erect or nodding at the apices; branches ascending to somewhat divergent; upper glumes widest at or below the middle, tapering to the apices; lower glumes 3–5 mm long . . . . 4. *T. canescens*

           10. Most panicle branches, except sometimes the uppermost, spikelet-bearing only towards the apices; panicles nodding; branches of at least the lowest 1–3 whorls spreading or drooping; upper glumes widest at or above the middle, rounded to the apices; lower glumes 0.75–3 mm long . . . . 5. *T. cernuum*

1. Trisetum melicoides (Michx.) Scribn. [p. 747]

FALSE MELIC, TRISÈTE FAUSSE-MÉLIQUE

Plants perennial, with both fertile and sterile shoots; cespitose. **Culms** (20)40–80(100) cm, erect, smooth or scabridulous. **Leaves** concentrated below midlength on the culms; **sheaths** glabrous or pilose; **ligules** 1.5–3.5 mm, rounded or truncate; **blades** 10–20+ cm long, 2–9 mm wide, flat, lax. **Panicles** 8–20 cm long, usually 2–4 cm wide, lax, nodding, silvery-green or tan; **lower branches** to 5 cm, ascending, naked below, the spikelets imbricate distally. **Spikelets** 5–7(9) mm, pedicellate, lance-ovate, with 2(4) florets; **rachilla internodes** and **hairs** 1.3–2 mm. **Glumes** unequal, widest at or below the middle; **lower glumes** 4–5.5 mm; **upper glumes** 5–7 mm long, nearly as long as the florets, wider than the lower glumes; **callus hairs** 1.5–2 mm; **lemmas** 5–6 mm, smooth or scabridulous, apices usually minutely bifid, sometimes entire, unawned or awned, awns to 2 mm, arising just below and rarely

exceeding the apices; **paleas** shorter than the lemmas; **anthers** 0.6–0.8 mm. **Caryopses** about 3 mm, sparsely pubescent distally. $2n = 14$.

Trisetum melicoides is a native species that grows on moist, cool stream banks, gravelly shores, shaded rock ledges (especially calcareous ones), and in damp woods. It grows only in southeastern Canada and the northeastern United States. It is listed as endangered in Wisconsin, New York, and Maine. Plants with pilose sheaths have been called *T. melicoides* var. *majus* (A. Gray) Hitchc., but the trait varies within populations.

2. Trisetum wolfii Vasey [p. 747]

WOLF'S TRISETUM

Plants perennial, with both fertile and sterile shoots; shortly rhizomatous. **Culms** 20–80 (100) cm, erect, glabrous or retrorsely pubescent below the nodes. **Leaves** usually concentrated on the lower ⅓ of the culms; **sheaths** glabrous or sparsely retrorse-pilose, sometimes scabridulous; **ligules** (1.2)2.5–4(6) mm, truncate to rounded; **blades** to 15 cm long, 2–7 mm wide, flat, ascending, lax, smooth or scabrous, sometimes sparsely pilose, often involute near the sometimes prowlike apices. **Panicles** (10)20–40(50) cm long, usually 1–1.5 cm wide, stiffly erect, green, tan, or purple-tinged; **branches** appressed-ascending, the spikelets evenly distributed. **Spikelets** 4–7(8) mm, usually subsessile, rarely on pedicels to 4 mm, ovate, with 2(3) florets; **rachilla internodes** 1.5–2 mm; **rachilla hairs** to 1 mm. **Glumes** subequal, usually longer than the lowest florets; **lower glumes** 4–7 mm; **upper glumes** 4–6.5 mm, a little wider than the lower glumes; **callus hairs** shorter than 0.5 mm; **lemmas** 4–6.5 mm, lanceolate, firmer than the glumes, scabridulous-puberulent, obscurely bifid, unawned or awned, awns to 2 mm, arising just below and rarely exceeding the apices; **paleas** shorter than the lemmas; **anthers** (0.6)1(1.5) mm. **Caryopses** to 3 mm, pubescent. $2n = 14$.

Trisetum wolfii grows in moist meadows and marshes, and on stream banks in aspen groves and parks in the spruce-fir forest zone, at medium to high, but usually not alpine, elevations. It is restricted to southwestern Canada and the western United States.

3. Trisetum orthochaetum Hitchc. [p. 747]
BITTERROOT TRISETUM

Plants perennial, with both fertile and sterile shoots; shortly rhizomatous. **Culms** 80–110 cm, solitary, decumbent, often anthocyanic at the base, glabrous. **Leaves** evenly distributed; **sheaths** usually glabrous; **ligules** 3–5 mm, truncate or rounded, erose; **blades** 8–20 cm long, 3–7 mm wide, flat, lax, scabrous. **Panicles** 13–20 cm, narrow, moderately dense, nodding, pale green, slightly tinged with purple; **branches** loosely ascending, naked below for 1–2 cm, the spikelets closely and evenly distributed distally. **Spikelets** 7–9 mm, subsessile or on pedicels to 1 cm, oblong-ovate, with 2–3(4) florets; **rachilla internodes** to 2 mm; **rachilla hairs** about 1 mm. **Glumes** lanceolate or oblanceolate; **lower glumes** about 5.5 mm long, about 1 mm wide, widest near the base, slenderly acuminate; **upper glumes** to 6.3 mm long, about 2 mm wide at or just above the middle, acuminate; **callus hairs** about 0.5 mm; **lemmas** 5–6.5 mm long, about ⅓ as wide as long, apices bifid, teeth shorter than 1 mm, awned, awns 4–6 mm, arising about 1 mm below the teeth, not twisted basally, straight or flexuous, exceeding the lemma apices; **paleas** almost equaling the lemmas; **anthers** minute or to 1 mm, appearing non-functional; **ovaries** pubescent. **Caryopses** to 2.5 mm, malformed. $2n =$ unknown.

Trisetum orthochaetum is known only from Montana, in or near the edges of marshes, seeps, and creeksides, where it grows at about 1465 m. It may be a sterile hybrid between *T. canescens* and *T. wolfii* (Shelly 1987).

4. Trisetum canescens Buckley [p. 749]
TALL TRISETUM

Plants perennial, sometimes with both fertile and sterile shoots; cespitose, not rhizomatous. **Culms** 40–120 cm, clumped, erect, usually smooth. **Leaves** 3–4 per culm; **sheaths** crisped-pubescent to shaggy-pilose, scabrous or smooth; **ligules** (1.5)3–6 mm, rounded to truncate; **blades** 10–30 cm long, (3)7–10 mm wide, flat, erect, lax, margins and occasionally the surfaces with scattered 1–3 mm hairs. **Panicles** 10–25 cm long, (0.75)1–3(4) cm wide, erect or nodding at the apices, green or tan, occasionally purple-tinged; **branches** 1–5.5 cm, ascending to somewhat divergent, most spikelet-bearing for their full length, sometimes the lowermost branches naked below. **Spikelets** 7–9 mm, pedicellate, with 2–4 florets; **rachilla internodes** 1.5–3 mm; **rachilla hairs** 0.7–1 mm; **disarticulation** above the glumes, beneath the florets. **Glumes** unequal to subequal; **lower glumes** 3–5 mm, narrow, lanceolate to subulate, acute or long-tapered; **upper glumes** (3.5)5–7(9) mm long, shorter than the lowest florets, at least twice as wide as the lower glumes, broadly lanceolate to obovate, widest at or below the middle, tapering to the apices, acute; **callus hairs** about 0.5 mm; **lemmas** 5–7 mm, glabrous, apices bifid, teeth to 2.5(3.2) mm, setaceous, awned, awns 7–14 mm, usually arising on the upper ⅓ of the lemmas, exceeding the apices, geniculate; **paleas** as long as or slightly longer than the lemmas; **anthers** 1–3 mm. **Caryopses** usually to 3 mm, glabrous or finely hairy distally. $2n = 28, 42$.

Trisetum canescens grows on or near stream banks and in forest margins or interiors, in moist to dry areas in the western *Flora* region. It is especially abundant in ponderosa pine stands and spruce-fir forests. The vestiture of different parts varies throughout the range of the species. Plants from California with conspicuously interrupted panicles have been called *Trisetum cernuum* var. *projectum* (Louis-Marie) Beetle.

T. melicoides

T. wolfii

T. orthochaetum

TRISETUM

5. Trisetum cernuum Trin. [p. 749]
NODDING TRISETUM

Plants perennial, with both fertile and sterile shoots; cespitose, not rhizomatous. **Culms** (30)50–110 cm, clumped, erect, glabrous or pubescent. **Leaves** 2–3 per culm; **sheaths** scabridulous or pilose; **ligules** 1.5–3 mm, truncate, erose to lacerate; **blades** (8.5)15–20+ cm long, (3)7–12 mm wide, flat, ascending, lax at maturity, often scabridulous. **Panicles** 10–30 cm long, (1)2–9 cm wide, open, nodding, green or tan, occasionally purple-tinged; **branches** 2–12+ cm, most, except sometimes the uppermost, spikelet-bearing only towards the apices, with the basal ($^1/_5$)$^1/_3$–$^1/_2$ bare, filiform, flexuous, at least the lowest 1–3 whorls spreading or drooping. **Spikelets** 6–12 mm, subsessile to pedicellate, pedicels to 2 cm, usually with 2–3 functional florets below 1–2 reduced florets; **rachilla internodes** and **hairs** 1–2.5 mm; **disarticulation** above the glumes, beneath the florets. **Glumes** unequal; **lower glumes** 0.75–2(3) mm, subulate; **upper glumes** 3.5–5 mm long, shorter than the lowest florets, 2–3 times as wide as the lower glumes, widest at or above the middle, ovate or obovate, rounded to the acuminate apices; **callus hairs** to 1 mm; **lemmas** 5–6 mm, broadly lanceolate, glabrous, bifid, teeth to 1.3 mm, awned, awns (7)9–14 mm, arising from above midlength to just below the teeth, exceeding the lemma apices, arcuate to flexuous; **paleas** shorter than the lemmas; **anthers** about 1 mm. **Caryopses** 2.5–3.2 mm, densely to sparsely pubescent. $2n = 42$.

Trisetum cernuum grows in moist woods, on stream banks, lake and pond shores, and floodplains of the western *Flora* region. The hairiness of the leaf sheaths varies, often within a plant.

6. Trisetum flavescens (L.) P. Beauv. [p. 751]
YELLOW OATGRASS, AVOINE JAUNÂTRE

Plants perennial, sometimes with both fertile and sterile shoots; usually cespitose, sometimes rhizomatous, rhizomes usually short, to 7 cm in sandy soils. **Culms** (10)50–80(130) cm, solitary or clumped, erect or decumbent, glabrous, sometimes scabrous or pubescent near the upper nodes. **Leaves** usually evenly distributed; **sheaths** glabrous or pilose, throats often with 2+ mm hairs; **ligules** 0.5–1(2) mm, obtuse, lacerate, sometimes ciliolate, hairs to 0.5 mm; **blades** 5–15(18) cm long, 1.5–4(6) mm wide, flat or involute, lax, pubescent or pilose. **Panicles** 5–20 cm long, 1.5–7 cm wide, erect or nodding, glistening yellowish brown, sometimes purple-tinged or variegated; **branches** 2–4(6) cm, ascending to divergent, often flexuous, sometimes naked below. **Spikelets** 4–8 mm, subsessile or on pedicels to 5 mm, with (2)3(4) florets; **rachilla internodes** to 1+ mm; **rachilla hairs** to 1.5 mm; **disarticulation** above the glumes, beneath the florets. **Glumes** unequal, shiny; **lower glumes** 2.5–4.7 mm, narrowly lanceolate to subulate; **upper glumes** 4–7 mm long, as long as or longer than the lowest florets, twice as wide as the lower glumes, lance-elliptic, acute; **callus hairs** to 0.5 mm; **lemmas** 3.5–6.3 mm, ovate-lanceolate, minutely pubescent, bifid or bicuspidate, teeth conspicuous, usually 3–6 mm, awned, awns (3)5–9 mm, arising from the upper $^1/_3$ of the lemmas and exceeding the apices, geniculate, tightly twisted below; **paleas** 3–5.5 mm; **anthers** 1.3–2.8 mm. **Caryopses** 2.5–3 mm, glabrous. $2n = 28$.

Trisetum flavescens grows in seeded pastures, roadsides, and as a weed in croplands. Native to Europe, west Asia, and north Africa, it was introduced into the *Flora* region because of its drought resistance, wide soil tolerance, and high palatability to domestic livestock. It is one of the few range plants known to contain calcinogenic glycosides, which can lead to vitamin D toxicity in grazing animals (Dixon 1995). This species seems not to have persisted in southern Ontario (Michael Oldham, pers. comm.). Several infraspecific taxa have been recognized; no attempt has been made to determine which are present in the *Flora* region.

7. Trisetum spicatum (L.) K. Richt. [p. 751]
SPIKE TRISETUM, TRISÈTE À ÉPI

Plants perennial, with both fertile and sterile shoots; cespitose, not rhizomatous. **Culms** 10–120 cm, clumped, erect, usually glabrous, sometimes villous, sometimes scabridulous. **Leaves** mostly basal or evenly distributed; **sheaths** variously pubescent or glabrous; **ligules** 0.5–4 mm, truncate or rounded; **blades** (3)10–20(40) cm long, 1–5 mm wide, flat, folded, or involute, erect and stiff or ascending and lax. **Panicles** (5)20–30(50) cm long, (0.5)1–2.5(5) cm wide, spikelike to open, often interrupted basally, green, purplish, or tawny, usually silvery-shiny; **branches** with the spikelets evenly distributed. **Spikelets** 5–7.5 mm, sessile, subsessile, or on pedicels to 1.5(3.5) mm, with 2(3) florets; **rachilla internodes** 0.5–1.5 mm; **rachilla hairs** to 1 mm. **Glumes** subequal to unequal, lanceolate, usually smooth, sometimes sparsely scabrous, sometimes pilose, with

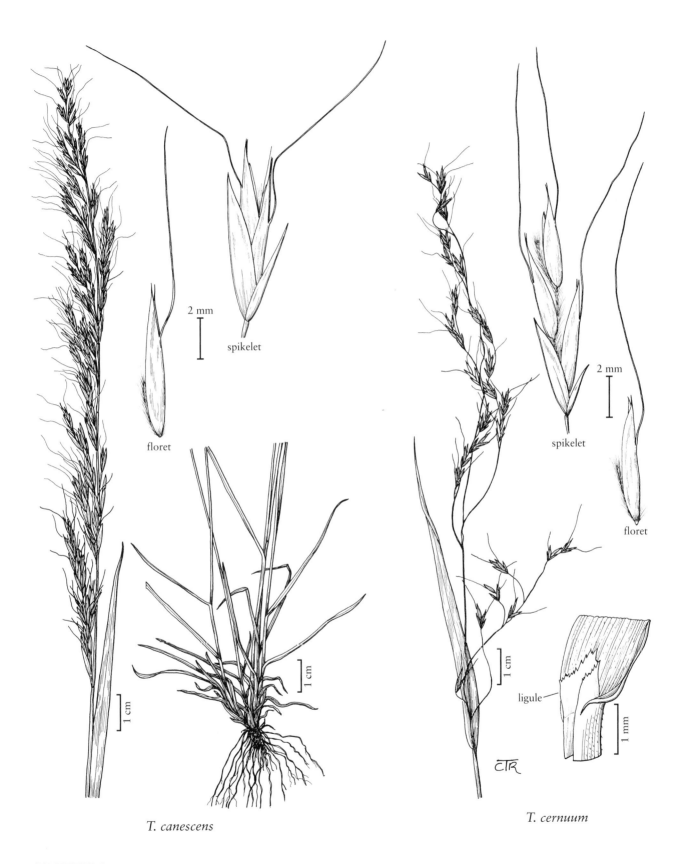

2 mm

spikelet

floret

1 cm

1 cm

2 mm

spikelet

floret

1 cm

ligule

1 mm

CTR

T. canescens

T. cernuum

TRISETUM

wide scarious margins, apices acute to acuminate, sometimes apiculate; **lower glumes** 3–4(5.5) mm; **upper glumes** 4–7 mm long, as long as or longer than the lowest florets, less than twice as wide as the lower glumes; **callus hairs** to 1 mm; **lemmas** 3–6(7) mm, narrowly to broadly lanceolate, glabrous or pilose, sometimes scabridulous, apices bifid, teeth usually shorter than 1 mm, awned, awns 3–8 mm, arising from the upper ⅓ of the lemmas and exceeding the apices, geniculate, twisted basally; **anthers** 0.7–1.4 mm. **Caryopses** 1.5–3(4) mm, glabrous. $2n = 14, 28, 42$.

Trisetum spicatum grows in moist meadows and forests, and on rock ledges, tundra slopes, and screes, at 0–4300 m. Its range includes both North and South America and Eurasia. Many infraspecific taxa have been based on the variation in vesture and openness of the panicle, but none appears to be justified (see Finot et al. 2004 for a different opinion). *Trisetum montanum* Vasey appears to represent no more than an extreme phase. *Trisetum spicatum* differs from *T. sibiricum* in its pubescent sheaths and denser, usually narrower panicles.

8. Trisetum sibiricum Rupr. [p. 752]
SIBERIAN TRISETUM

Plants perennial, sometimes with both fertile and sterile shoots; rhizomatous. **Culms** 15–40(65) cm, solitary, decumbent; **nodes** glabrous. **Sheaths** smooth; **ligules** 1–3.5 mm, truncate or slightly higher in the center, often lacerate; **blades** 8–15(24) cm long, 2.5–7 mm wide, flat, erect or ascending, stiff or lax, smooth. **Panicles** 2–12(16) cm long, (1)2–3(6) cm wide, ovate-spicate, sometimes basally interrupted, yellowish brown, often mottled, shiny; **branches** usually 0.1–2(4) cm, appressed-ascending, the spikelets distal; **disarticulation** above the glumes, beneath the florets. **Spikelets** 5–8 mm, subsessile or pedicellate, pedicels to 4 mm, with 2(3) florets; **rachilla internodes** usually 1–2 mm; **rachilla hairs** 0.5–1 mm, often curly and tangled. **Glumes** lanceolate, glabrous; **lower glumes** 3–4.5 mm; **upper glumes** 5–8 mm; **callus hairs** about 0.5 mm; **lemmas** 4.5–7 mm, glabrous, apices minutely bifid, teeth shorter than 1 mm, awned, awns 4–10+ mm, arising from the upper ⅓, exceeding the lemma apices, flexuous, curved, or bent and twisted basally; **paleas** subequal to or as long as the lemmas; **anthers** 2–3.5 mm. **Caryopses** 1–2 mm, ovate, smooth, brown. $2n = 14, 28$.

Trisetum sibiricum grows on coastal beaches and creek banks, and in moist meadows and open forests, from sea level to 300+ m. It is often abundant and has significant value as a pasture plant. Circumpolar in distribution, in the *Flora* region it grows in Alaska and the Yukon Territory. Most North American plants belong to **Trisetum sibiricum** subsp. **litorale** Rupr. *ex* Roshev., having culms 15–30 cm tall, leaf blades 2.5–4 mm wide, panicles 3–5 cm long and 2–3 cm wide, branches to 2 cm long, and lemma awns 5–8 mm long. **Trisetum sibiricum** Rupr. subsp. **sibiricum** occurs in the Yukon Territory and Eurasia. It differs from *T. spicatum* in its smooth culms and leaves, and its broad, less dense panicles.

9. Trisetum interruptum Buckley [p. 752]
PRAIRIE TRISETUM

Plants annual, without sterile shoots; tufted. **Culms** (5)10–40 (60) cm, erect or spreading, mostly glabrous, pilose below the nodes. **Leaves** basally concentrated; **sheaths** scabridulous or pilose; **ligules** 1–2.5 mm, truncate; **blades** 3–12 cm long, 1–4 mm wide, flat, folded, or involute distally when dry, ascending, glabrous or pubescent, margins frequently sparsely ciliate. **Panicles** 2–15 cm long, 0.3–1.5 cm wide, often interrupted, at least in the lower ⅓, green or tan; **branches** short, usually erect to appressed, the spikelets crowded. **Spikelets** 3–6 mm, often in pairs with 1 subsessile and 1 pedicellate, with 2–3 florets; **disarticulation** initially above the glumes, subsequently below; **rachilla internodes** usually 0.8–1 mm; **rachilla hairs** usually about 0.5 mm. **Glumes** subequal, 4–5 mm, about as long as the lowest lemmas, smooth or sparsely scabridulous; **lower glumes** 0.5–1 mm wide, lanceolate or elliptical, 3-veined, acuminate, sometimes apiculate; **upper glumes** about twice as wide as the lower glumes, elliptical or oblanceolate, acuminate; **callus hairs** 0.1–0.2(0.5) mm, sparse; **lemmas** 3–4.5 mm, usually glabrous, sometimes minutely pustulate-scabridulous, apices bifid, teeth to 1.7 mm, awned, awns usually 4–8 mm, arising from midlength to just below the teeth and exceeding the lemma apices, geniculate, twisted basally, rarely 2–4 mm, straight, arcuate, or flexuous; **paleas** usually ⅔ as long as the lemmas, hyaline; **anthers** about 0.2 mm. **Caryopses** 2–3 mm, longitudinally striate, sometimes with a few hairs distally. $2n = 14$.

Trisetum interruptum grows in open, dry or moist soil in deserts, plains, arid shrublands, and riparian woodlands, from the southern United States into Mexico. It is often weedy.

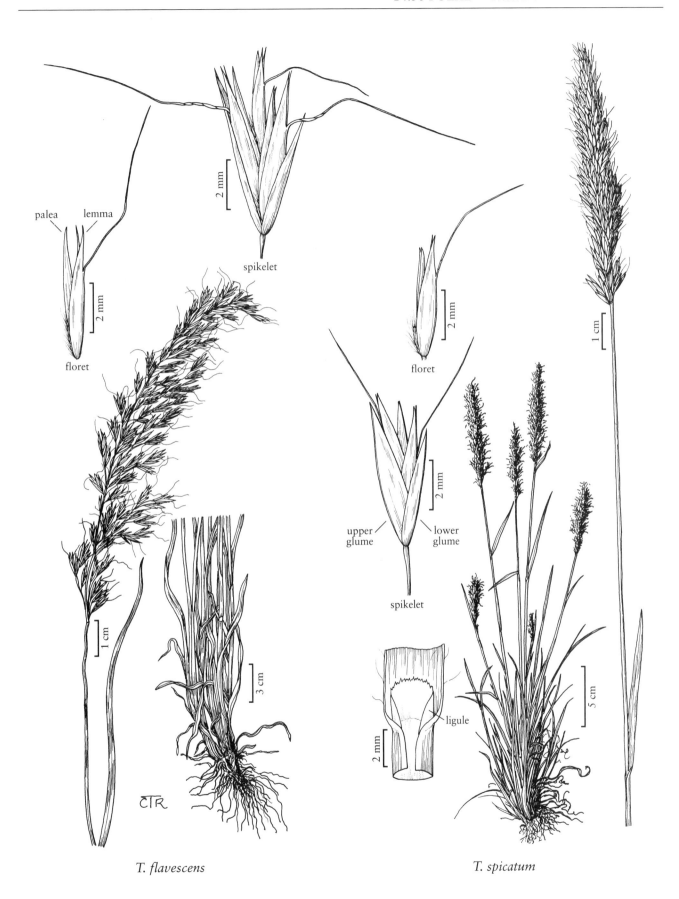

palea   lemma

floret

spikelet

2 mm

2 mm

floret

upper glume   lower glume

spikelet

ligule

2 mm

2 mm

2 mm

1 cm

1 cm

3 cm

5 cm

ĈTR

T. flavescens

T. spicatum

TRISETUM

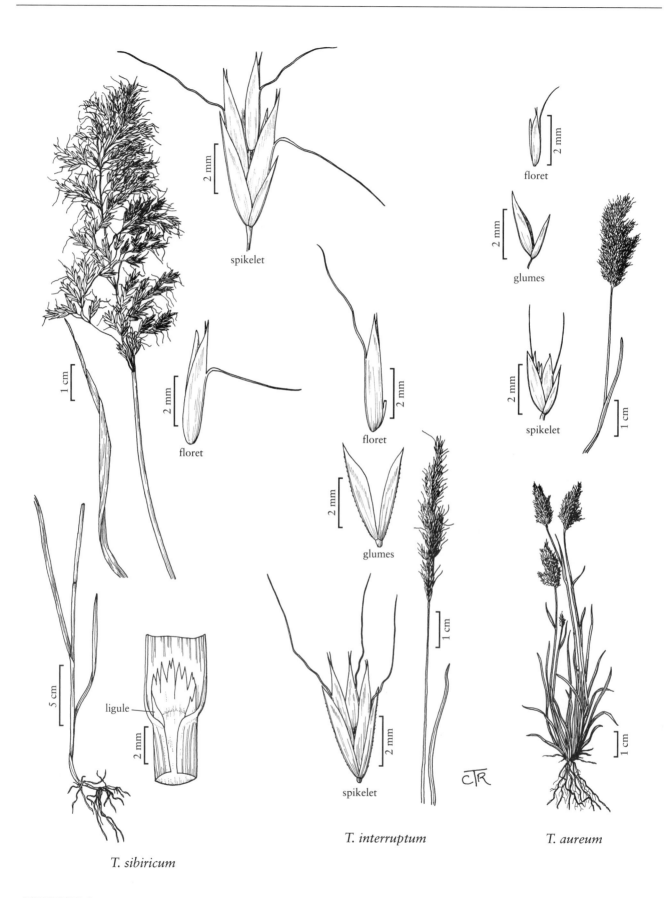

spikelet

floret

ligule

5 cm

1 cm

2 mm

T. sibiricum

floret

glumes

spikelet

1 cm

2 mm

T. interruptum

floret

glumes

spikelet

1 cm

2 mm

T. aureum

CTR

TRISETUM

10. Trisetum aureum (Ten.) Ten. [p. 752]

GOLDEN OATGRASS

Plants annual, without sterile shoots; tufted. **Culms** 7–30 cm, glabrous, erect, spreading, or geniculate. **Sheaths** somewhat inflated, glabrous or villous; **blades** to 10 cm long, to 3 mm wide, flat, subglabrous to villous. **Panicles** 1–5 cm long, 0.5–3 cm wide, pyramidal to ovoid, dense, shiny, yellowish to tan. **Spikelets** 2.5–3.5 mm, with 2–3 florets. **Glumes** unequal; **lower glumes** 2–2.5 mm long, narrower than the upper glumes, 1-veined; **upper glumes** 2.5–3 mm, 3-veined; **callus hairs** 0.3–0.4 mm; **lemmas** 1.6–2.7 mm, glabrous or hairy, with wide hyaline margins, apices bifid, teeth to about 0.5 mm, awned, awns 2–6 mm, arising from above midlength and exceeding the apices, slightly bent; **anthers** 1–1.5 mm. $2n$ = unknown.

Trisetum aureum is native to the Mediterranean region. It was collected from a ballast dump in Camden, New Jersey, in 1896 (Hitchcock 1951), and has not been reported since from the *Flora* region.

14.57 KOELERIA Pers.

Lisa A. Standley

Plants perennial; usually cespitose, sometimes weakly rhizomatous. **Culms** 5–130 cm, erect. **Sheaths** open; **auricles** absent; **ligules** membranous; **blades** flat to involute, pubescent or glabrous. **Inflorescences** panicles, erect, usually dense and spikelike, sometimes lax, stiffly and narrowly pyramidal at anthesis; **main rachis** and **branches** smooth, softly hairy. **Spikelets** laterally compressed, with 2–4 florets; **rachillas** to 1 mm, glabrous or pubescent, usually prolonged beyond the base of the distal floret, or with a vestigial floret; **disarticulation** above the glumes, beneath the florets. **Glumes** subequal to or slightly exceeding the lemmas, membranous, scabrid to tomentose, keels sometimes ciliate, unawned; **lower glumes** 1-veined, somewhat narrower and shorter than the upper glumes; **upper glumes** obscurely 3(5)-veined; **calluses** glabrous or hairy; **lemmas** thin, membranous, 5-veined, margins shiny, scarious, apices acute, sometimes mucronate or awned; **paleas** equaling or subequal to the lemmas, hyaline; **lodicules** 2, glabrous, toothed; **anthers** 3; **ovaries** glabrous. **Caryopses** glabrous. x = 7. Named for Georg Ludwig Koeler (1765–1807), a botanist at Mainz, Germany.

Koeleria is a cosmopolitan genus of about 35 species that grow in dry grasslands and rocky soils; two are native to the *Flora* region. *Koeleria pyramidata* is sometimes said to be in the *Flora* region; such reports seem to reflect a different interpretation of the species. In Europe, *Koeleria* forms a series of polyploid complexes in which cytotypes are morphologically and ecologically distinct, but species boundaries are not. *Koeleria* sometimes includes the genus *Rostraria*, which differs in its annual growth habit, and in having awned lemmas and paleas.

SELECTED REFERENCES Arnow, L.A. 1994. *Koeleria macrantha* and *K. pyramidata* (Poaceae): Nomenclatural problems and biological distinctions. Syst. Bot. 19:6–20; Greuter, W. 1968. Notulae nomenclaturales et bibliographicae, 1–4. Candollea 23:81–108; Ujhely, J. 1972. Evolutionary problems of the European Koelerias. Pp. 163–176 *in* G. Vida (ed.). Evolution in Plants. Symposia Biologica Hungarica No. 12. Akadémiai Kiadó, Budapest, Hungary. 231 pp.

1. Lemmas pubescent to densely tomentose, purple to almost black; panicles 1–5 cm long; culms 5–35 cm tall, finely pubescent throughout . . . . 1. *K. asiatica*
1. Lemmas usually glabrous, usually green when young, sometimes purple-tinged, stramineous at maturity; panicles 4–27 cm long; culms 20–130 cm tall, mostly glabrous, pubescent near the nodes and sometimes below the panicle.
  2. Spikelets 2.5–6.5 mm long; old sheaths usually breaking off with age or, if disintegrating, the fibers straight or nearly so; margins of the basal leaf blades glabrous or with hairs usually shorter than 1 mm near the base . . . . 2. *K. macrantha*

2. Spikelets 6–10 mm long; old sheaths weathering to wavy, curled, or arched fibers; margins
of the basal leaf blades frequently with hairs longer than 2 mm near the base .... 3. *K. pyramidata*

1. Koeleria asiatica Domin [p. 755]
EURASIAN JUNEGRASS

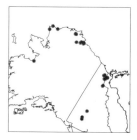

Plants cespitose, shortly rhizomatous. **Culms** 5–35 cm, densely and finely pubescent throughout. **Basal sheaths** remaining intact; **ligules** 0.5–2.5 mm; **basal blades** 2–20 cm long, 1–3 mm wide, involute, stiff, margins not ciliate, adaxial surfaces pubescent. **Panicles** 1–4(5) cm long, 0.7–1.5 cm wide, somewhat interrupted at the base; **branches** villous. **Spikelets** 4–6.5 mm, with 2–3 florets; **rachillas** pubescent. **Glumes** membranous, purple, scabrous, ciliate on the keels and upper margins; **lower glumes** 2.5–3.5 mm long, about 0.5 mm wide, lanceolate, narrow, apices acute or awn-tipped to about 0.6 mm; **upper glumes** 3–4.8 mm long, 0.8–0.9 mm wide, ovate; **calluses** pubescent; **lemmas** 4–5 mm, slightly longer than the glumes, pubescent to densely tomentose, purple to almost black, apices acute or awn-tipped to about 1.4 mm; **paleas** equaling or slightly shorter than the lemmas; **anthers** 1.2–2.5 mm. $2n = 28$.

Koeleria asiatica grows on the gravel bars of creeks, dry tundra, and scree slopes. Its range extends from the Ural Mountains through the Kamchatka Peninsula to northern Alaska and northwestern Canada.

2. Koeleria macrantha (Ledeb.) Schult. [p. 755]
JUNEGRASS, KOELERIE ACCRÊTÉ, KOELERIE À CRÊTES

Plants cespitose, sometimes loosely so. **Culms** 20–85(130) cm, mostly glabrous, pubescent below the panicles and near the nodes. **Leaves** primarily basal; **sheaths** pubescent or glabrous, usually breaking off with age, if disintegrating into fibers, then the fibers straight or nearly so; **ligules** 0.5–2 mm; **blades** 2–20 cm long, 0.5–3(4.5) mm wide, flat, involute when dry, minutely scabrous, occasionally glabrous or densely pubescent, margins of the basal blades glabrous or with hairs averaging shorter than 1 mm near the base. **Panicles** 4–27 cm long, 0.5–2 cm wide, interrupted at the base, otherwise dense; **branches** finely pubescent to villous. **Spikelets** 2.5–6.5 mm, obovate to obelliptic, with 2(3) florets; **rachillas** pubescent. **Glumes** 2.5–5 mm, ovate, membranous, green, scabrous except for the ciliate keels, apices acute; **calluses** pubescent; **lemmas** 2.5–6.5 mm, membranous, shiny, usually glabrous, sometimes scabrous, particularly on the keels, usually green when

young, sometimes purple-tinged, stramineous at maturity, acuminate, midveins prolonged into a 1 mm awn; **paleas** shorter than the lemmas; **anthers** 1–2.5(3) mm. $2n = 14, 28$.

Koeleria macrantha is widely distributed in temperate regions of North America and Eurasia. In North America, it grows in semi-arid to mesic conditions, on dry prairies or in grassy woods, generally in sandy soil, from sea level to 3900 m. It differs from *Sphenopholis intermedia*, with which it is frequently confused, in its less open panicles, and in having spikelets that disarticulate above the glumes.

The species is treated here as a polymorphic, polyploid complex. North American plants have sometimes been treated as a separate species, *Koeleria nitida* Nutt., but no morphological characters for distinguishing them from Eurasian members of the complex are known (Greuter 1968). Some plants from Oregon and Washington have densely pubescent culms, and high-elevation populations from western North America are often densely cespitose, with very short culms and purple leaves and inflorescences; both variants appear to intergrade with more typical plants.

3. Koeleria pyramidata (Lam.) P. Beauv. [p. 755]
CRESTED HAIRGRASS

Plants loosely cespitose, sometimes rhizomatous. **Culms** (30)40–90 cm, glabrous or puberulent above. **Sheaths** sparsely sericeous to densely pubescent, old sheaths persistent, disintegrating into wavy, curled, or arched fibers; **ligules** 0.5–1 mm; **blades** 5–24 cm long, 1–5 mm wide, flat, margins of the basal blades often ciliate below, with hairs usually longer than 2 mm. **Panicles** 5–22 cm long, 1–4(5) cm wide; **branches** villous. **Spikelets** 6–10 mm, with 2–4(5) florets; **rachillas** with scattered pubescence. **Glumes** acute, glabrous, smooth or scabrous; **lower glumes** 4–5 mm, 1-veined; **upper glumes** 5–6.5 mm, 3-veined; **calluses** broadly rounded, pubescent; **lemmas** 4–6 mm, glabrous or puberulent, rarely ciliate, usually green when young, stramineous at maturity, apices acuminate to shortly aristate; **anthers** 2–2.5 mm. **Caryopses** 2.5–3.8 mm. $2n = 14$.

Koeleria pyramidata, as interpreted here, is confined to Europe. Some North American records for *K. pyramidata* are based on robust specimens of *K. macrantha*; others reflect an interpretation of *K. pyramidata* that includes *K. macrantha*.

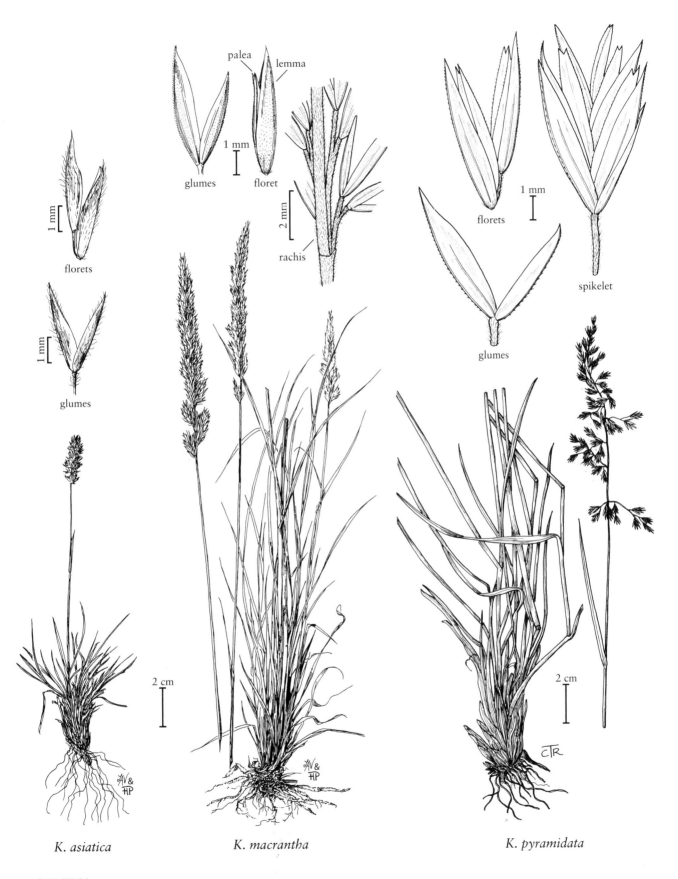

palea

lemma

glumes   floret

1 mm

rachis

2 mm

florets

1 mm

glumes

1 mm

florets

1 mm

spikelet

glumes

K. asiatica   *K. macrantha*   *K. pyramidata*

2 cm

2 cm

KOELERIA

14.58 ROSTRARIA Trin.

Lisa A. Standley

Plants annual; tufted or with solitary culms. **Culms** 3–60 cm, erect. **Leaves** mostly cauline; **sheaths** open, glabrous, pubescent, or pilose; **auricles** absent; **ligules** membranous; **blades** flat or involute, stiff. **Inflorescences** spikelike panicles, dense; **branches** scabrous to pubescent. **Spikelets** laterally compressed, with 2–7 florets; **rachillas** sparsely to moderately pubescent, prolonged or not beyond the base of the distal florets; **disarticulation** above the glumes, beneath the florets. **Glumes** unequal, membranous, glabrous or hirsute, keels ciliate, apices unawned; **lower glumes** shorter and narrower than the upper glumes, 1-veined; **upper glumes** subequal to the lowest lemmas, 3-veined; **calluses** usually glabrous, occasionally sparsely hairy; **lemmas** thin, membranous, 5-veined, glabrous or hirsute, margins shiny, scarious, apices acute, bifid, subterminally awned or unawned; **paleas** subequal to the lemmas, hyaline, veins sometimes extended into awnlike apices; **anthers** 3; **ovaries** glabrous. **Caryopses** glabrous. *x* = 7. Name from the Latin *rostrum*, 'beak', and *ari*, 'having the nature of', a reference to the beaked lemma of the type species.

Rostraria is a genus of approximately 10 species, all of which are native to the Mediterranean, southeastern Europe, and western Asia, where they grow in dry, disturbed sites. The genus is sometimes included in *Koeleria*; it differs in its annual growth habit and awned lemmas and paleas.

1. Rostraria cristata (L.) Trin. [p. 756]
MEDITERRANEAN HAIRGRASS

Culms 3–50 cm, glabrous. **Sheaths** glabrous or pilose; **ligules** 1–2 mm, erose-ciliolate; **blades** 2–15 cm long, 1–8 mm wide, pubescent. **Panicles** (0.5) 1.5–12 cm long, 0.5–1.6 cm wide, cylindrical, usually not interrupted at the base, green; **branches** and **pedicels** scabrous. **Spikelets** 2.8–4.5(7) mm, laterally flattened, with 3–6 florets; **rachillas** pubescent, not prolonged beyond the upper floret. **Glumes** scabrous to irregularly hirsute; **lower glumes** 1.8–3.5 mm; **upper glumes** 2–4.4 mm; **calluses** blunt, usually glabrous, occasionally with a few scattered hairs; **lemmas** 2.8–3.7 mm, ciliate over the keels; **lower lemmas** with papillose-based hairs or irregularly hirsute, awns 1–3 mm; **uppermost lemmas** merely scabrous and more shortly awned or unawned; **anthers** 0.2–0.5 mm; **ovaries** glabrous. 2*n* = 14, 26, 28.

Rostraria cristata is native to Europe. In the *Flora* region, it is found in California, around the Gulf of Mexico, and in New York State. Plants with hirsute glumes and lower lemmas belong to **Rostraria cristata** (L.) Trin. var. **cristata**; those with glumes and lemmas that are almost glabrous apart from their ciliate keels belong to **R. cristata** var. **glabriflora** (Trautv.) Doğan.

When included in *Koeleria*, *Rostraria cristata* is called *K. phleoides* (Vill.) Pers.

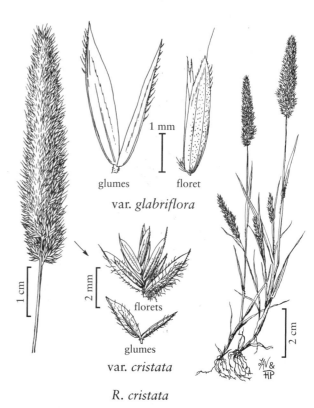

1 mm

glumes　　　　floret

var. *glabriflora*

florets

glumes

var. *cristata*

R. cristata

ROSTRARIA

14.59 MIBORA Adans.

Hans J. Conert

Plants annual; tufted. **Culms** 2–15 cm tall, to 0.3 mm wide, erect. **Leaves** primarily basal; **sheaths** closed almost to the top; **auricles** absent; **ligules** about 1 mm, hyaline, truncate; **blades** 0.3–1 mm wide, flat or involute. **Inflorescences** single, terminal, spikelike racemes; **rachises** smooth, glabrous, with 1 spikelet per node; **pedicels** 0.2–1 mm long, 0.1–0.2 mm thick. **Spikelets** imbricate, in 2 rows on 1 side of the rachis, slightly laterally compressed, with 1 floret; **rachillas** not prolonged beyond the base of the floret; **disarticulation** above the glumes, beneath the floret. **Glumes** subequal, exceeding the florets, membranous, glabrous, flexible, smooth, 1-veined, rounded on the back, unawned; **calluses** glabrous; **lemmas** about $^2/_3$ the length of the spikelets, elliptical in side view, thinner than the glumes, 5-veined, shortly and densely pubescent throughout, unawned, apices truncate, often denticulate; **paleas** about as long as the lemmas, 2-veined, shortly and densely pubescent between the inconspicuous keels; **lodicules** 2; **anthers** 3; **ovaries** glabrous; **styles** fused at the base, dividing into 2 feathery stigmas. **Caryopses** elliptical, terete, $^2/_3$ the length of the lemmas, enclosed by, but not fused to, the lemma and palea at maturity; **embryos** about $^1/_5$ the length of the caryopses, elliptic; **hila** punctate, basal. $x = 7$. Name of uncertain origin.

Mibora is a genus of two species, both of which grow in damp, sandy soils. *Mibora minima*, the species that has been introduced to the *Flora* region, grows throughout much of western Europe. The second species, *M. maroccana* (Maire) Maire, is restricted to northwest Africa.

SELECTED REFERENCE **Conert, H.J.** 1985. *Mibora*. Pp. 206–210 *in* G. Hegi. Illustrierte Flora von Mitteleuropa, ed. 3. Band I, Teil 3, Lieferung 3, Bg. 11–15 (pp. 161–240). Verlag Paul Parey, Berlin and Hamburg, Germany.

1. Mibora minima (L.) Desv. (p. 757)
EARLY SANDGRASS

Plants tufted. **Culms** 2–10(15) cm; **nodes** 2–3, glabrous. **Sheaths** delicate, rounded on the back, shallowly grooved; **ligules** 0.2–1 mm; **blades** 1–4(5) cm long, to 0.5 mm wide, flat or involute, obtuse. **Racemes** 0.5–2 (5) cm long, 1–2 mm wide, usually purplish; **pedicels** 0.5–1 mm. **Spikelets** 1.8–3 mm. **Glumes** 1.8–3 mm, obtuse to slightly emarginate; **lemmas** 1.4–2 mm, 5-veined, often denticulate at the apices; **paleas** 1.4–2 mm, narrowly elliptic, apices notched; **anthers** 1–1.7 mm. **Caryopses** 1–1.4 mm. $2n = 14$.

Mibora minima is native to western Europe, where it grows on sandy and other light, damp soils in places with mild winters. In central Europe, it usually flowers from February to April or May but, if the weather is warm, may flower in December.

Mibora minima was collected from ballast dumps in Plymouth, Massachusetts, in the first part of the twentieth century, and from an experimental farm in Sydney, British Columbia, in 1914. It has also been collected in Monroe County, New York. None of these populations appear to have led to the establishment of the species in North America. Recently, garden societies in both Canada and the United States have been offering the seed, recommending *M. minima* for rock gardens.

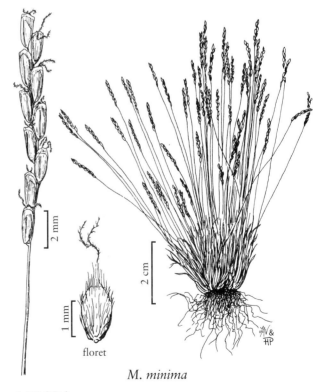

2 mm

1 mm

floret

2 cm

M. minima

MIBORA

14.60 ANTHOXANTHUM L.

Kelly W. Allred

Mary E. Barkworth

Plants perennial or annual; densely to loosely cespitose, sometimes rhizomatous; fragrant. **Culms** 4–100 cm, erect or geniculate, sometimes branched; **internodes** hollow. **Leaves** cauline or basally concentrated, glabrous or softly hairy; **sheaths** open; **auricles** absent or present; **ligules** membranous, sometimes shortly ciliate or somewhat erose; **blades** flat or rolled, glabrous or sparsely pilose. **Inflorescences** open or contracted panicles, sometimes spikelike. **Spikelets** pedicellate or sessile, 2.5–10 mm, laterally compressed, stramineous to brown at maturity, with 3 florets, lowest 2 florets staminate or reduced to dorsally awned lemmas subequal to or exceeding distal floret, distal floret bisexual, unawned; **rachillas** not prolonged beyond the base of the distal floret; **disarticulation** above the glumes, the florets falling together. **Glumes** unequal or subequal, equaling or exceeding the florets, lanceolate to ovate, glabrous or pilose, keeled; **calluses** blunt, glabrous or hairy. **Lowest 2 florets: lemmas** strongly compressed, 3-veined, strigose, hairs brown, apices bilobed, unawned or dorsally awned. **Distal florets: lemmas** somewhat indurate, glabrous or pubescent, shiny, inconspicuously 3–7-veined, unawned; **paleas** 1-veined, enclosed by the lemmas; **lodicules** 2 or absent; **anthers** 2 or 3. **Caryopses** shorter than the lemmas, concealed at maturity, tightly enclosed in the florets; **hila** less than $1/3$ the length of the caryopses, oval. $x = 5$. Name from the Greek *anthos*, 'flower', and *xanthos*, 'yellow', alluding to the golden color of the mature panicles.

Anthoxanthum is a cool-season genus of about 50 species that grow in temperate and arctic regions throughout the world. There are seven species in the *Flora* region, five of which are native. The fragrance emitted when fresh plants are crushed or burned is from coumarin. In addition to smelling pleasant, coumarin has anti-coagulant properties. It is the active ingredient in Coumadin, a prescription drug used to prevent blood clots in some patients after surgery. A disadvantage of coumarin is that it is metabolized by species of the fungal genus *Aspergillus* to dicoumarol, which induces vitamin K deficiency and a susceptibility to hemorrhaging in wounded animals. Because of this, using moldy hay containing *Anthoxanthum* as feed is dangerous.

This treatment follows the recommendation of Schouten and Veldkamp (1985) in merging what have traditionally be treated as two genera, *Anthoxanthum* and *Hierochloë*. In general, *Hierochloë* has less floral reduction, a less elaborate karyotype, and a higher basic chromosome number than *Anthoxanthum* (Weimarck 1971). The two genera appear distinct in North America but, when considered on a global level, Schouten and Veldkamp (1985) stated that the two genera overlap, with the placement of many species being arbitrary. *Phalaris* resembles *Anthoxanthum sensu lato* in its spikelet structure, differing only in the greater reduction of the lower florets. It also differs in lacking coumarin.

Anatomical studies (Pizzolato 1984) supported the close relationship of *Anthoxanthum* and *Phalaris*. Pizzolato also stated that although the bisexual florets of *Hierochloë* are described as terminal, a microscopic fourth floret is developed distal to the third (bisexual) floret.

Wherever they grow, the species that used to be treated as *Hierochloë* have been used by native peoples. Native Americans used them for incense, baskets, and decorations. In addition, they steeped them in water for a hair-, skin-, and eyewash, or for use as a cold medicine, analgesic, or insecticide. Early Europeans spread the species in churches at festivals. They can also be used to make ale (Stika 2003).

SELECTED REFERENCES Aiken, S.G., L.L. Consaul, and M.J. Dallwitz. 1995 on. Grasses of the Canadian Arctic Archipelago: Descriptions, illustrations, identification and information retrieval. http://www.mun.ca/biology/delta/arcticf/poa/index.htm; Belk, E. 1939. Studies in the anatomy and morphology of the spikelet and flower of the Gramineae. Ph.D. dissertation, Cornell University, Ithaca, New York, U.S.A. 183 pp.; Hedberg, I. 1990. Morphological, cytotaxonomic and evolutionary studies in *Anthoxanthum odoratum* L. s. lat.–A critical review. Sommerfeltia 11:97–107; Hitchcock, A.S. 1951. Manual of the Grasses of the United States, ed. 2, rev. A. Chase. U.S.D.A. Miscellaneous Publication No. 200. U.S. Government Printing Office, Washington, D.C., U.S.A. 1051 pp.; Norstog, K.J. 1960. Some observations on the spikelet of *Hierochloë odorata*. Bull. Torrey Bot. Club 87:95–98; Pizzolato, T.D. 1984. Vascular system of the fertile floret of *Anthoxanthum odoratum* L. Bot. Gaz. 145:358–371; Schouten, Y. and J.F. Veldkamp. 1985. A revision of *Anthoxanthum* including *Hierochloë* (Gramineae) in Malaysia and Thailand. Blumea 30:319–351; Stika, J. 2003. Sweetgrass ale. http://byo.com/feature/1067.html; Weimarck, G. 1971. Variation and taxonomy of *Hierochloë* (Gramineae) in the Northern Hemisphere. Bot. Not. 124:129–175; Weimarck, G. 1987. *Hierochloë hirta* subsp. *praetermissa*, subsp. *nova* (Gramineae), an Asiatic–E. European taxon extending to N and C Europe in the Northern Hemisphere. Symb. Bot. Upal. 2:175–181.

1. Glumes unequal, the lower glumes shorter than the upper glumes; lowest 2 florets sterile.
    2. Plants annual; ligules 1–3 mm long; blades 1–5 mm wide; panicles 1–4 cm long . . . . 1. *A. aristatum*
    2. Plants perennial; ligules 2–7 mm long; blades 3–10 mm wide; panicles 3–14 cm long . . . . . . 2. *A. odoratum*
1. Glumes subequal; lowest 2 florets staminate.
    3. Staminate lemmas awned, the awns of the upper staminate florets 4.5–10.5 mm long; plants densely to loosely tufted, with rhizomes rarely more than 2 cm long . . . . 3. *A. monticola*
    3. Staminate lemmas unawned or with an awn no more than 1 mm long; plants long-rhizomatous.
        4. Panicles spikelike, 0.3–0.5 cm wide, with 1–2 spikelets per branch; rhizomes 0.3–1 mm thick; plants of the high arctic . . . . 4. *A. arcticum*
        4. Panicles not spikelike, 1–10 cm wide, the longer branches usually with 3+ spikelets; rhizomes 0.7–3 mm thick; plants non-arctic or arctic.
            5. Lower staminate lemmas in each spikelet narrowly elliptic, lengths more than 5 times widths; glumes equaling or slightly exceeded by the apices of the bisexual florets; blades 3–15 mm wide . . . . 5. *A. occidentale*
            5. Lower staminate lemmas in each spikelet elliptic, lengths usually no more than 4 times widths; glumes exceeding the bisexual florets; blades 2–8 mm wide.
                6. Hairs on the distal portion of the bisexual florets mostly shorter than 0.5 mm, longer hairs, if present, concentrated near the midvein . . . . 6. *A. nitens*
                6. Hairs on the distal portion of the bisexual florets 0.5–1 mm long, evenly distributed around the apices . . . . 7. *A. hirtum*

1. **Anthoxanthum aristatum** Boiss. [p. 761]
VERNALGRASS

Plants annual. **Culms** 5–60 cm, often geniculate at the base, freely branched. **Auricles** to 0.5 mm, sometimes absent; **ligules** 1–2(3) mm, obtuse to acute; **blades** 0.8–6 cm long, 1–5 mm wide. **Panicles** 1–4 cm; **lowermost branches** 8–12(15) mm; **pedicels** 0.1–0.3 mm, pubescent. **Spikelets** (4)5–9 mm; **lower glumes** 3–5 mm; **upper glumes** 5–7 mm; **sterile florets** about 3 mm, awn of the first floret 3.5–5 mm, awn of the second floret 6–10 mm, exceeding the upper glumes by 2–3 mm; **fertile florets** about 2 mm; **anthers** 2, 2.8–4.1 mm. $2n = 10, 20$.

Anthoxanthum aristatum is native to Europe. It is now established but not common in the *Flora* region, being found in mesic to dry, open, disturbed habitats of western and eastern North America. North American plants belong to **Anthoxanthum aristatum** Boiss. subsp. **aristatum**, which differs from **Anthoxanthum aristatum**

subsp. **macranthum** Valdés in having well-exserted awns and deeply bifid, sterile lemmas.

Hitchcock (1951) stated that another annual species of *Anthoxanthum*, *A. gracile* Biv., is occasionally cultivated for dry bouquets, but it does not appear to be widely available at present. It differs from *A. aristatum* in having longer (10–12 mm) spikelets and simple or sparingly branched culms.

2. **Anthoxanthum odoratum** L. [p. 761]
SWEET VERNALGRASS, FLOUVE ODORANTE, FOIN D'ODEUR

Plants perennial. **Culms** (10) 25–60(100) cm, erect, simple or sparingly branched. **Auricles** 0.5–1 mm, pilose-ciliate, sometimes absent; **ligules** 2–7 mm, truncate; **blades** 1–31 cm long, 3–10 mm wide. **Panicles** (3) 4–14 cm, the spikelets congested; **lowermost branches** 10–25 mm; **pedicels** 0.5–1 mm, pubescent. **Spikelets** 6–10 mm; **lower glumes** 3–4 mm; **upper glumes** 8–10 mm; **sterile florets** 3–4 mm, awn of the first floret 2–4 mm, awn of the second floret 4–9 mm, equaling or only slightly exceeding the upper glumes; **bisexual florets** 1–2.5 mm; **anthers** 2, (2.9)3.5–4.8(5.5) mm. 2*n* = 10, 20.

Anthoxanthum odoratum is native to southern Europe. In the *Flora* region, it grows in meadows, pastures, grassy beaches, old hay fields, waste places, and openings in coniferous forests, occasionally in dense shade or as a weed in lawns. It is most abundant on the western and eastern sides of the continent, and is almost absent from the central region. In southern British Columbia, it is rapidly invading the moss-covered bedrock of coastal bluffs, and will soon exclude many native species. Diploids (2*n* = 10) have been referred to *A. odoratum* subsp. *alpinum* (Á. Löve & D. Löve) Hultén. Because the two ploidy levels can be distinguished only through cytological examination (Hedberg 1990), the two subspecies are not recognized here.

Anthoxanthum odoratum was often included in hay and pasture mixes to give fragrance to the hay, but this practice is waning. The aroma is released upon wilting or drying. By itself, the species is unpalatable because of the bitter-tasting coumarin.

3. **Anthoxanthum monticola** (Bigelow) Veldkamp [p. 761]
ALPINE SWEETGRASS, HIÉROCHLOÉ ALPINE

Plants perennial; densely to loosely cespitose, rhizomes to 2 cm long, rarely longer, about 2 mm thick. **Culms** 20–55(75) cm. **Basal sheaths** glabrous, brown to deep purple; **ligules** 0.2–1.5 mm, truncate, ciliate; **blades** 1–12 cm long, (0.7) 1–3(5) mm wide, flat or folded, abaxial surfaces glabrous and shiny, adaxial surfaces sparsely scabrous or pilose. **Panicles** 1–8.5 cm long, 1.2–2 cm wide, with (3)10–20(35) spikelets. **Spikelets** 5–8 mm, tawny; **rachilla internodes** about 0.1 mm, glabrous. **Glumes** subequal, 4.8–6.7 mm, about equal to the lemmas; **lowest 2 florets** staminate; **lemmas** 4–6.5(8) mm, moderately hairy, hairs to 1 mm, apices deeply bifid, first lemma awn 0.6–4(6.5) mm, second lemma awn 4.5–10.5 mm, usually geniculate, arising from near the base to near midlength; **bisexual lemmas** 3.5–5.2 mm, pubescent towards the bifid apices; **anthers** 1.5–2.7 mm. 2*n* = 56, 58, 63, 66, 72.

Anthoxanthum monticola is circumpolar, usually growing above or north of the tree line, occasionally in open forests. It occurs sporadically on well-drained, weakly acidic to neutral sand, gravel, and rocky barrens in most of arctic North America; it is not common to the south, even at high elevations. It is facultatively

apomictic, but slow to set seed. Revegetation is best accomplished vegetatively. It is listed as threatened or endangered in several parts of its range. There are two subspecies in the *Flora* region.

1. Awns of the upper staminate florets 5–10.5 mm long, attached from near the base to about midlength; awn usually strongly geniculate, the lower portion usually twisted, with 2–4 gyres . . . . subsp. *alpinum*
1. Awns of the upper staminate florets 4.5–7 mm long, attached at or above midlength, not or only weakly geniculate, the lower portion not twisted or twisted with 1–2 gyres . . . . . . subsp. *monticola*

Anthoxanthum monticola subsp. **alpinum** (Sw. *ex* Willd.) Soreng [p. 761]

Awns of upper staminate florets 5–10.5 mm, attached from near the base to midlength, usually strongly geniculate, the lower portion twisted, with (1)2–4 gyres. 2*n* = 56, 66, 72.

Anthoxanthum monticola subsp. *alpinum* is the common subspecies in the *Flora* region, extending from western Alaska to eastern Greenland. It extends south to the Canadian border in the Rocky Mountains but, east of these mountains, it is mostly north of 60° N latitude, except in Quebec and Labrador, where it extends south to about 53° N latitude. It usually grows above or north of the tree line, in places that are strongly exposed to the wind and have little snow cover during the winter.

Anthoxanthum monticola (Bigelow) Veldkamp subsp. **monticola** [p. 761]

Awns of upper staminate florets 4.5–7 mm, attached at or above midlength, straight or weakly geniculate, the lower portion not twisted or twisted with 1(2) gyres. 2*n* = 56, 58, 63.

Anthoxanthum monticola subsp. *monticola* grows from Greenland to the eastern side of Hudson Bay, through Labrador, and south to northern New England. It usually grows in similar, but wetter and more exposed, habitats than those occupied by subsp. *alpinum*.

4. **Anthoxanthum arcticum** Veldkamp [p. 763]
ARCTIC SWEETGRASS, HIÉROCHLOÉ PAUCIFLORE

Plants perennial; loosely cespitose or the culms solitary, rhizomes elongate, 0.3–1 mm thick. **Culms** 5–26(35) cm. **Sheaths** glabrous; **ligules** 0.4–1.3 mm, obtuse; **blades** 2–25 cm long, 0.7–2 mm in diameter when rolled, involute to convolute, abaxial surfaces glabrous,

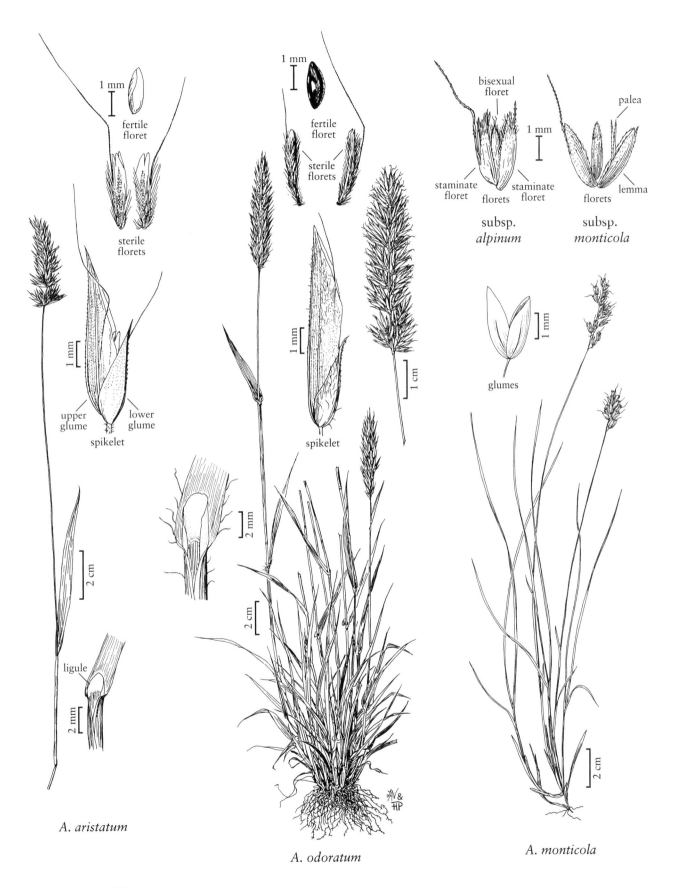

fertile floret

sterile florets

upper glume

lower glume

spikelet

ligule

A. aristatum

fertile floret

sterile florets

spikelet

A. odoratum

bisexual floret

staminate floret

staminate floret

florets

subsp. *alpinum*

palea

lemma

florets

subsp. *monticola*

glumes

A. monticola

ANTHOXANTHUM

adaxial surfaces pubescent. **Panicles** 1–3(4.5) cm long, 0.3–0.5 cm wide, spikelike, with 1–2 spikelets per branch. **Spikelets** 3.5–5 mm, green to purple; **rachilla internodes** about 0.1 mm, glabrous. **Glumes** subequal, 2.9–4.7 mm, shiny; **lowest 2 florets** staminate; **lemmas** sparsely hairy, acute or slightly notched, unawned or awned, awns to 1 mm; **bisexual lemmas** 2.9–4.4 mm, with sparse, spreading hairs towards the apices; **anthers** 1.5–3 mm. **Caryopses** about 2.5 mm. $2n = 28$.

Anthoxanthum arcticum is a coastal and lowland circumpolar species of the Alaskan, Canadian, and Russian arctic; it is absent from Greenland. It generally grows in wet tundra on acidic, peaty soils. In the warmest sectors of the western high arctic, it is rooted in mats of moss that are growing over carbonate substrates.

5. Anthoxanthum occidentale (Buckley) Veldkamp [p. 763]

CALIFORNIA SWEETGRASS

Plants perennial; loosely cespitose or the culms solitary, rhizomes elongate, 1–3 mm thick. **Culms** (4)60–90 cm. **Sheaths** scabrous to scabridulous; **ligules** 1.5–4(6) mm, rounded to truncate; **blades** 20–40 cm long, (3)5–15 mm wide, flat, rather stiffly erect, narrowing to the base, glabrous, often glaucous, veins widely spaced, cross venation evident on the abaxial surfaces; **flag leaf blades** 3.5–10 cm. **Panicles** 8–13 cm long, (1)2–6 cm wide, diffuse, with slender, often drooping branches and 3+ spikelets per branch. **Spikelets** 4.5–6 mm, tawny or green to olive-green, sometimes infused with purple; **rachilla internodes** 0.2–0.5 mm, glabrous. **Glumes** subequal, equaling or slightly exceeded by the apices of the bisexual florets; **lower glumes** 4.5–5 mm long, 0.7–1 mm wide; **upper glumes** 3.5–5.2 mm long, 1–1.8 mm wide; **lowest 2 florets** staminate; **lemmas** usually mostly glabrous on the body, sometimes scattered-pubescent, scabridulous near the apices, margins usually pilose, apices rounded and shallowly bilobed, unawned or awned, awns to 1 mm; **first lemma** 4–5 mm long, 0.75–1 mm wide, length more than 5 times width, narrowly elliptic; **bisexual lemmas** 3.5–4.5 mm, margins pilose, particularly distally; **anthers** 2–3.5 mm. $2n = 42$.

Anthoxanthum occidentale grows in moist to fairly dry forested areas, from Klickitat County, Washington, south to the coastal mountains of San Luis Obispo County, California. Its long flag leaf blades and more elongate spikelet parts make it easier to distinguish from *A. hirtum* than the key suggests.

6. Anthoxanthum nitens (Weber) Y. Schouten & Veldkamp [p. 763]

VANILLA SWEETGRASS, HIÉROCHLOÉ ODORANTE, FOIN D'ODEUR, HERBE SAINTE

Plants perennial; loosely cespitose or the culms solitary, rhizomes elongate, 0.7–2 mm thick. **Culms** (5)15–50(90) cm. **Sheaths** brownish or reddish, glabrous or puberulent; **ligules** 0.5–6.5(8) mm, truncate, obtuse, or acute; **blades** 10–30 cm long, 2–8 mm wide, usually flat, sometimes inrolled, abaxial surfaces glabrous or pubescent, without prominent cross venation, adaxial surfaces glabrous; **flag leaf blades** 0.3–1.5(4.5) cm long, 1.5–4.5(6) mm wide. **Panicles** (2)4–9(12.5) cm long, (1.5)2–5(7) cm wide, open, pyramidal, with 8–100 spikelets; **branches** with 3+ spikelets. **Spikelets** (2.5)3.5–7.5 mm, mostly tawny, sometimes tinged with green; **rachilla internodes** 0.15–0.3 mm, glabrous. **Glumes** subequal, exceeding the florets, glabrous; **lower glumes** with lengths 2–5 times widths, usually shorter and wider than the upper glumes; **lowest 2 florets** staminate; **lemmas** hairy, particularly distally, hairs brown, often papillose-based, to 0.8 mm on the margins and to 0.3 mm near the apices, margins with 11–26 hairs per mm, midvein usually terminating at the apices or extending beyond as an awn, apices acute to rounded, entire or bifid, unawned, mucronate, or with a thin awn, awns to 0.5 mm; **first lemma** 3.4–5 mm long, 1–1.5 mm wide, length usually less than 4 times width, elliptic; **bisexual lemmas** 2.5–4 mm, usually hairy distally, hairs 0.1–0.5 mm, appressed or almost so at maturity, longer, divergent hairs, if present, concentrated near the midvein; **anthers** of staminate florets 0.9–2.3 mm, those of bisexual florets 1.2–1.6 mm. $2n = 28, 42$.

Anthoxanthum nitens is primarily a European species. In the *Flora* region, it grows along the coast from northern Labrador to New England. It is not known from Greenland, although it grows in Iceland and northwestern Europe. It grows in wet meadows and at the edges of sloughs, marshes, roadsides, and fields. Only **A. nitens** (Weber) Y. Schouten & Veldkamp subsp. **nitens** is present in the region; it is also present in Europe. It differs from **A. nitens** subsp. **balticum** (G. Weim.) G.C. Tucker in being almost always awned and having $2n = 28$.

North American taxonomists have generally interpreted *Anthoxanthum nitens* as including *A. hirtum* (and treated both as *Hierochloë odorata* (L.) Wahlenb.). The two are distinct, although not easy to distinguish. Weimarck (1971) separated the two by the

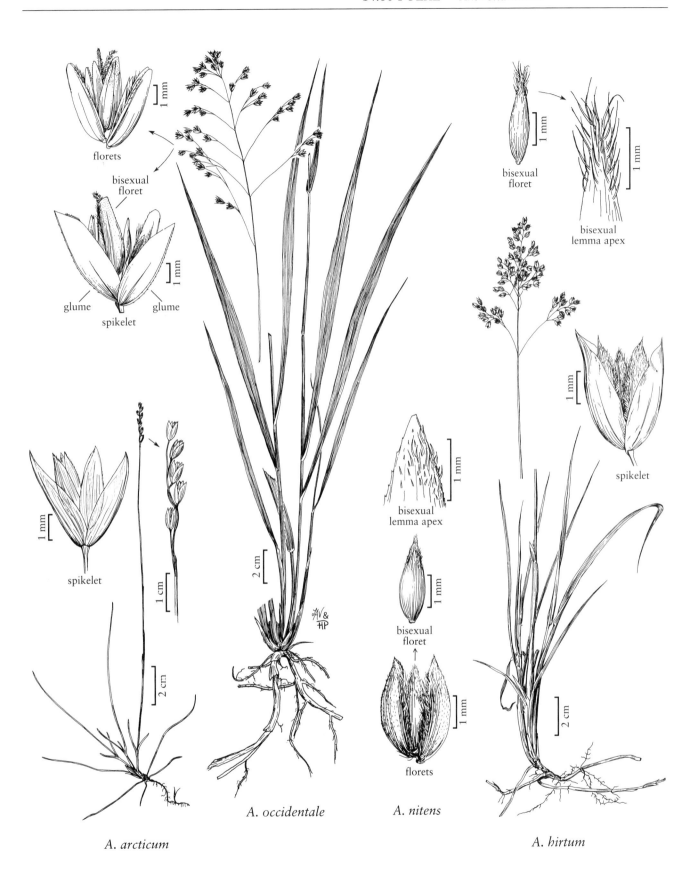

florets

bisexual floret

glume — glume

spikelet

bisexual floret

bisexual lemma apex

A. occidentale

bisexual lemma apex

bisexual floret

florets

A. nitens

spikelet

spikelet

A. arcticum

A. hirtum

ANTHOXANTHUM

density of the lateral hairs and development of the awns of the staminate florets. The difference in abundance and distribution of hairs more than 0.5 mm long on the apices of the bisexual florets is more reliable, *A. nitens* having few such long hairs concentrated near the midvein. M.J. Harvey (pers. comm.), who did not distinguish between the two species, found that plants from the Maritime Provinces collected near saltwater were uniformly $2n = 28$, whereas those from the interior of New Brunswick westward had $2n = 56$. This observation is consistent with Weimark's chromosome counts and distribution maps.

7. Anthoxanthum hirtum (Schrank) Y. Schouten & Veldkamp [p. 763]
HAIRY SWEETGRASS

Plants perennial; loosely cespitose or the culms solitary, rhizomes elongate, 0.7–2 mm thick. **Culms** 40–85(110) cm. **Sheaths** brownish or reddish; **ligules** 2.5–5.5 mm; **blades** 2.5–5.5 mm wide; abaxial surfaces glabrous, shiny, adaxial surfaces pilose; **flag leaf blades** 1–4.5(6) cm long, 3–4.5 mm wide. **Panicles** (5)7.5–15 cm long, 2–10 cm wide, open, pyramidal, with 20–100+ spikelets; **branches** with 3+ spikelets. **Spikelets** 4–6.3 mm, tawny at maturity; **rachilla internodes** 0.1–0.3 mm. **Glumes** subequal, exceeding the florets, glabrous, often somewhat purplish; **lowest 2 florets** staminate; **lemmas** 3–5 mm, with hairs to 0.5 mm towards the apices, margins with 16–30 hairs per mm, hairs 0.5–1 mm, apices acute, emarginate, or bifid; **first lemma** 3–5 mm long, 1.1–1.5 mm wide, length usually less than 4 times width, elliptic, awned, awns 0.1–1 mm; **bisexual lemmas** 2.9–3.5 mm, hairy distally, hairs 0.5–1 mm, evenly distributed around the apices, bases strongly divergent from the lemma surface; **anthers** of staminate florets 1.6–2.1 mm, those of bisexual florets 1.2–1.3 mm. $2n = 56$.

Anthoxanthum hirtum is the most widely distributed species of *Anthoxanthum* in the *Flora* region, extending from Alaska to northeastern Quebec and south to Washington and Colorado, South Dakota, Illinois, Ohio, and New York. It is not known from Newfoundland or Greenland. Outside the *Flora* region, it extends from Scandinavia south to Germany and east to Asiatic Russia. It grows in wet meadows and marshes with good water, not in salt- or brackish water. Because much of its native habitat has been drained, it is becoming less common. Its short flag leaf blades and more circular spikelets distinguish it from *A. occidentale*, and the relative abundance and even distribution of hairs longer than 0.5 mm distinguish it from *A. nitens*.

Weimark (1971, 1987) recognized three subspecies in *A. hirtum* (which he treated as *Hierochloë hirta*): subsp. *hirta*, subsp. *arctica* G. Weim., and subsp. *praetermissa* G. Weim. He stated that only *H. hirta* subsp. *arctica* grows in North America, but several North American specimens seem to fit within his circumscription of *H. hirta* subsp. *hirta*. Because the variation between the two appears continuous, no subspecies of *A. hirtum* are recognized here.

14.61 PHALARIS L.

Mary E. Barkworth

Plants annual or perennial; sometimes cespitose, sometimes rhizomatous. **Culms** 4–230 cm tall, erect or decumbent, sometimes swollen at the base, not branching above the base. **Leaves** more or less evenly distributed, glabrous; **sheaths** open for most of their length, uppermost sheaths often somewhat inflated; **auricles** absent; **ligules** hyaline, glabrous, truncate to acuminate, entire or lacerate; **blades** usually flat, sometimes revolute. **Inflorescences** terminal panicles, sometimes spikelike, ovoid to cylindrical, dense, sometimes interrupted, with 10–200 spikelets borne singly or in clusters, spikelets homogamous in species with single spikelets, heterogamous in species with spikelets in clusters, lower spikelets in the clusters usually staminate, rarely sterile, terminal spikelets bisexual or pistillate. **Spikelets** pedicellate, laterally compressed, with 1–3(4) florets, the terminal or only floret usually sexual, lower floret(s), if present, sterile; **disarticulation** above the glumes, beneath the sterile florets in species with solitary spikelets, in species with clustered spikelets usually at the base of the spikelet clusters, sometimes beneath the bisexual or pistillate spikelets. **Glumes** subequal, exceeding the florets, 1–5-veined, keeled, keels often conspicuously winged; **lower (sterile) florets** reduced, varying from knoblike projections on the calluses of the bisexual florets to linear or lanceolate lemmas less than $^3/_4$ as long as the bisexual florets; **terminal florets** usually bisexual, in the lower spikelets of a spikelet cluster the terminal florets pistillate or staminate, rarely sterile; **lemmas of terminal florets** coriaceous to indurate, shiny, glabrous or hairy, inconspicuously 5-veined, acute to acuminate or beaked, unawned; **paleas** similar to the lemmas in length and texture, enclosed by the lemmas at maturity, 1-veined, mostly glabrous, veins shortly hairy; **lodicules** absent or 2 and reduced; **anthers** 3; **ovaries** glabrous; **styles** 2, plumose. **Caryopses** shorter than the lemmas, concealed at maturity, with a reticulate pericarp, falling free of the lemma and palea; **hila** long-linear. $x = 6, 7$. The name of the genus is an old Greek name for a grass.

Phalaris has 22 species, most of which grow primarily in temperate regions. It is found in a wide range of habitats, although most species prefer somewhat mesic, disturbed areas. There are 11 species in the *Flora* region, 5 native and 6 introduced.

The sterile florets of *Phalaris* are frequently mistaken for tufts of hair at the base of a solitary functional floret. Close examination will reveal that the hairs are actually growing from linear to narrowly lanceolate pieces of tissue. Developmental studies have shown that these structures are reduced lemmas.

Many species of *Phalaris* are weedy. A few are cultivated for fodder, and one, *P. canariensis*, is grown for birdseed. In addition, the dense panicles of *P. paradoxa* are sometimes dyed green and used to simulate shrubs in landscape models.

SELECTED REFERENCES Anderson, D.E. 1961. Taxonomy and distribution of the genus *Phalaris*. Iowa State Coll. J. Sci. 36:1–96; **Baldini, R.M.** 1995. Revision of the genus *Phalaris* L. (Gramineae). Webbia 49:265–329; **Merigliano, M.F.** and **P. Lesica.** 1998. The native status of reed canarygrass (*Phalaris arundinacea* L.) in the inland northwest, USA. Nat. Areas J. 18:223–230; **Ross, E.M.** 1989. *Phalaris* L. Pp. 132–135 *in* T.D. Stanley and E.M. Ross. Flora of South-Eastern Queensland, vol. 3. Queensland Department of Primary Industries, Miscellaneous Publication QM88001. Queensland Department of Primary Industries, Brisbane, Queensland, Australia. 532 pp.; **Thellung, A.** 1911. La Flore Adventice de Montpellier. Imprimerie Émile Le Maout, Cherbourg, France. [Preprinted from Mém. Soc. Sci. Nat. Cherbourg 38:[57]–728 (1912)]; **Weiller, C.M., M.J. Henwood, J. Lenz,** and **L. Watson.** 1995 on. Poöideae (Poaceae) in Australia–Descriptions and Illustrations. http://delta-intkey.com/pooid/www/index.htm.

1. Spikelets in clusters, heterogamous, the lower 4–7 spikelets in each cluster with a staminate (rarely sterile) terminal floret, only the terminal spikelet in the clusters with a pistillate or bisexual terminal floret; disarticulation usually at the base of the spikelet clusters, sometimes beneath the bisexual or pistillate spikelets.
   2. Glumes of the bisexual or pistillate spikelets winged, the wings with 1 prominent tooth; plants annual; culms not swollen at the base . . . . 1. *P. paradoxa*
   2. Glumes of the bisexual or pistillate spikelets winged, the wings entire or irregularly dentate to crenate distally; plants perennial; culms with swollen bases . . . . 2. *P. coerulescens*
1. Spikelets borne singly, homogamous, all spikelets with a bisexual terminal floret; disarticulation above the glumes, beneath the sterile florets.
   3. Glume keels not winged or with wings no more than 0.2 mm wide.
      4. Plants perennial; bisexual florets with acute to somewhat acuminate apices.
         5. Panicles ovoid to cylindrical, 1.5–6 cm long, branches not evident; sterile florets usually more than $^1/_2$ as long as the bisexual florets . . . . 8. *P. californica*
         5. Panicles elongate, 5–40 cm long, evidently branched towards the base; sterile florets less than $^1/_2$ as long as the bisexual florets . . . . 9. *P. arundinacea*
      4. Plants annual; bisexual florets with beaked or strongly acuminate apices.
         6. Apices of the bisexual florets glabrous; glumes scabrous over the lateral veins and keels, and adjacent to the keels . . . . 7. *P. lemmonii*
         6. Apices of the bisexual florets hairy; glumes smooth or scabridulous over the lateral veins and keels, the wing surface smooth . . . . 10. *P. caroliniana* (in part)
   3. Glume keels broadly winged, the wings 0.2–1 mm wide.
      7. Sterile florets usually 1, if 2, the lower floret up to 0.7 mm long and the upper floret 1–3 mm long.

8. Plants annual; sterile florets 1, glabrous or almost so; wings of the glume keels irregularly dentate to crenate, varying within a panicle . . . . .3. *P. minor*

8. Plants perennial; sterile florets usually 1, sometimes 2, hairy; wings of the glume keels usually entire . . . . .4. *P. aquatica*

  7. Sterile florets 2, equal to subequal, 0.5–4.5 mm long.

    9. Panicles cylindrical, sometimes lobed; anthers 0.5–1.3 mm long . . . . 11. *P. angusta*

    9. Panicles usually ovoid to ellipsoid or oblong-ovoid, occasionally cylindrical, not lobed; anthers 1.5–4 mm long.

      10. Sterile florets 0.6–1.2 mm long, about ¹/₅ the length of the bisexual florets . . . . 5. *P. brachystachys*

      10. Sterile florets 1.5–4.5 mm long, ¹/₃ or more the length of the bisexual florets.

        11. Glumes 7–10 mm long, 2–2.5 mm wide; bisexual florets 4.5–6.8 mm long; anthers 2–4 mm long . . . . 6. *P. canariensis*

        11. Glumes 3.8–6(8) mm long, 0.8–1.5 mm wide; bisexual florets 2.9–4.7 mm long; anthers 1.5–2 mm long . . . . 10. *P. caroliniana* (in part)

1. Phalaris paradoxa L. [p. 767]
HOODED CANARYGRASS

Plants annual; tufted. **Culms** 20–100 cm, not swollen at the base. **Ligules** 3–5 mm, truncate to acute; **blades** 5–10(15) cm long, 2–5 mm wide. **Panicles** 3–9 cm long, about 2 cm wide, dense, obovoid to clavate, tapering at the base, rounded to truncate at the top; **branches** with groups of 5–6 usually staminate, rarely sterile spikelets clustered around a terminal pistillate or bisexual spikelet; **pedicels** hispid; **disarticulation** beneath the spikelet clusters. **Spikelets** heterogamous, with 3 florets, lower 2 florets sterile and highly reduced, terminal floret usually staminate, pistillate, or bisexual, rarely sterile. **Glumes of staminate** or **sterile spikelets** varying, those at the base of the panicle reduced to knobs of tissue terminating the pedicels, those higher up often clavate, those near the top of the panicle similar to the glumes of the sexual spikelets but somewhat narrower; **glumes of pistillate** or **bisexual spikelets** 4–8 mm long, about 1 mm wide, keeled, keels winged, wings 0.2–0.4 mm wide, terminating below the apices and forming a single, prominent tooth, lateral veins conspicuous, apices acuminate to awned, awns about 0.5 mm; **sterile florets** of all spikelets 0.2–0.4 mm, knoblike projections on the calluses of the terminal florets often with 1–2 hairs; **terminal florets** of all spikelets 2.5–3.5 mm long, 0.8–1.5 mm wide, indurate, shiny, glabrous or with a few short hairs near the tip; **anthers** 1.5–2.5 mm. 2*n* = 14.

Phalaris paradoxa is native to the Mediterranean region; it is now found throughout the world, primarily in harbor areas and near old ballast dumps. It is an established weed in parts of Arizona and California. Within an inflorescence, the most reduced sterile spikelets are located near the base, and the most nearly normal spikelets are near the top.

2. Phalaris coerulescens Desf. [p. 767]
SUNOLGRASS

Plants perennial; cespitose, not rhizomatous. **Culms** 70–200 cm, swollen at the base. **Ligules** 4–6 mm, rounded to narrowly acute; **blades** 4–20(25) cm long, 1–5(7) mm wide. **Panicles** 3–12 cm long, 1–2.3 cm wide, ovoid to cylindrical; **branches** with groups of 4–7 staminate (rarely sterile) spikelets clustered around a terminal pistillate or bisexual spikelet; **pedicels** glabrous or sparsely hispid; **disarticulation** below individual bisexual spikelets or below the spikelet clusters. **Spikelets** heterogamous, some staminate or sterile, others bisexual or pistillate, with 1–3 florets, if more than 1, lower floret(s) sterile and highly reduced, terminal (or only) floret staminate, pistillate, or bisexual. **Glumes** usually 5–9 mm long, to 3 mm long on spikelets near the base of the panicle, 1.1–2 mm wide, glabrous or hirsute, keels winged, wings 0.2–0.5 mm wide, entire or irregularly dentate to crenate distally, lateral veins conspicuous, scabrous, apices mucronate, mucros 0.3–0.7(1) mm; **sterile florets**, if present, to ¹/₁₀ as long the sexual florets, glabrous or almost so; **sexual florets** staminate, pistillate or bisexual, 2.5–4.5 mm long, 0.7–1.4 mm wide, glabrous or with a few short hairs at the base; **anthers** 2.5–3 mm. **Caryopses** 2.8–3.3 mm long, 1.2–1.4 mm wide. 2*n* = 14, 42.

Phalaris coerulescens is native around the Mediterranean; it is now established in northern Europe and South America. It was found in Contra Costa County, California, in 2000.

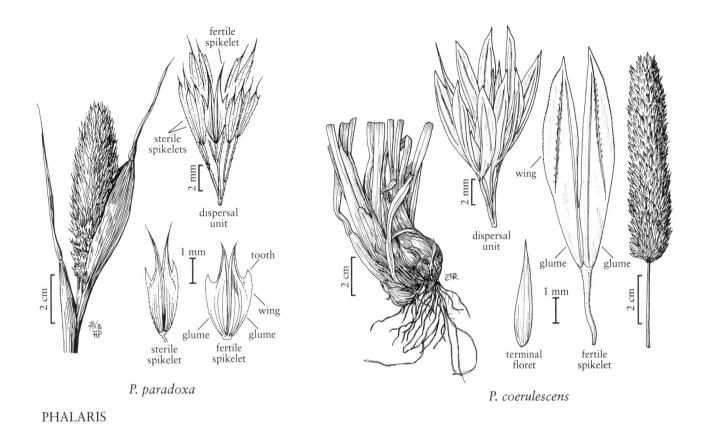

fertile
spikelet

sterile
spikelets

2 mm

dispersal
unit

1 mm

tooth

wing

glume

glume

sterile
spikelet

fertile
spikelet

P. paradoxa

wing

2 mm

dispersal
unit

glume

glume

1 mm

terminal
floret

fertile
spikelet

2 cm

P. coerulescens

PHALARIS

---

3. Phalaris minor Retz. [p. 769]
LESSER CANARYGRASS

Plants annual. **Culms** 10–100 cm, not swollen at the base. **Ligules** 5–12 mm, truncate to rounded, often lacerate; **blades** 3–15 cm long, 2–10 mm wide, smooth, shiny. **Panicles** 1–8 cm long, 1–2 cm wide, dense, ovoid-lanceoloid, truncate to rounded at the base, rounded apically, spikelets borne singly, not clustered. **Spikelets** homogamous, with 2 florets, 1 bisexual; **disarticulation** above the glumes, beneath the sterile florets. **Glumes** 3.5–6.5 mm long, 1.2–2 mm wide, keels winged distally, wings 0.3–0.5 mm wide, irregularly dentate or crenate, occasionally entire, varying within a panicle, lateral veins conspicuous, smooth; **sterile florets** 1, 0.7–1.8 mm, linear, glabrous or almost so; **bisexual florets** 2–4 mm long, 1–1.8 mm wide, hairy, dull yellow when immature, becoming shiny gray-brown at maturity, acute to somewhat acuminate; **anthers** 1–2 mm. $2n = 28, 29$.

Phalaris minor is native around the Mediterranean and in northwestern Asia, but is now found throughout the world. Even where it is native, it usually grows in disturbed ground, often around harbors and near refuse dumps. Although it has been found at numerous locations in the *Flora* region, it is only established in the southern portion of the region.

The compact panicle with its truncate to rounded base, and the rather variable edges of the glume wings, usually distinguish *Phalaris minor* from other species in the genus. It sometimes forms a polyploid hybrid with *P. aquatica*, **P. ×daviesii** S.T. Blake, which is cultivated for forage in Australia, Africa, and South America.

4. Phalaris aquatica L. [p. 769]
BULBOUS CANARYGRASS

Plants perennial; cespitose, shortly rhizomatous. **Culms** 60–200 cm, often swollen at the base, rooting at the lower nodes. **Ligules** 2–12 mm, truncate, lacerate; **blades** 5–15(20) cm long, 0.5–10 mm wide. **Panicles** 1.5–15 cm long, 1–2.5 cm wide, usually cylindric, sometimes ovoid, occasionally lobed at the base, spikelets borne singly, not clustered; **branches** not evident. **Spikelets** homogamous, with 2–3(5) florets, usually with 1 bisexual floret, occasionally with 2, occasionally the terminal floret viviparous; **disarticulation** above the glumes, beneath the sterile florets. **Glumes** 4.4–7.5 mm long, 1.2–1.5 mm wide, keels winged distally, wings 0.2–0.4 mm wide, usually entire, lateral veins conspicuous, smooth; **sterile florets** usually 1, hairy, if 2, lowest floret to 0.7 mm, upper or only sterile floret 1–3 mm; **bisexual florets** 3.1–4.6 mm long, 1.2–1.5 mm wide, hairy, stramineous, acute; **anthers** 3–3.6 mm. $2n = 28$.

A native of the Mediterranean region, *Phalaris aquatica* now grows in many parts of the world, frequently having been introduced because of its forage value. Even where it is native, it usually grows in disturbed areas, often those subject to seasonal flooding. It is now established in western North America, being most common along the coast, and as an invasive in disturbed wet prairies with clay soils.

Phalaris aquatica can hybridize with other species of *Phalaris*. The stabilized polyploid hybrid with *P. minor*, **P. ×daviesii** S.T. Blake, is cultivated as a forage grass in Australia, Africa, and South America. The hybrid with *P. arundinacea*, **P. ×monspeliensis** Daveau, is also a good forage grass. The name 'Toowoomba Canarygrass' has been applied to *P. ×monspeliensis* in North America, but Ross (1989) stated that it should be applied to *P. aquatica*. Using 'Bulbous Canarygrass' as the English-language name for *P. aquatica* avoids confusion, at least in the *Flora* region. In addition, it is descriptive, and is the name used by the U.S. Department of Agriculture for *P. aquatica*.

5. Phalaris brachystachys Link [p. 769]
SHORTSPIKE CANARYGRASS

Plants annual. **Culms** 80–100 cm. **Ligules** 4–6(7) mm, rounded, lacerate; **blades** 4–25 cm long, 3–8(10) mm wide. **Panicles** 1.5–5 cm long, 0.8–1.8 cm wide, usually ovoid to ellipsoid, occasionally cylindrical, continuous, not lobed, narrowly truncate at the base, rounded at the top; **branches** not evident, spikelets borne singly, not clustered. **Spikelets** homogamous, with 3 florets, terminal floret bisexual; **disarticulation** above the glumes, beneath the sterile florets. **Glumes** 6–8.5 mm long, 1.4–2.5 mm wide, glabrous or hairy, keels winged on the distal $^2/_3$, wings to 1 mm wide, entire, abruptly pointed; **sterile florets** 2, subequal to equal, 0.6–1.2 mm, about $^1/_5$ the length of the bisexual florets, with a tuft of hair at the base, otherwise glabrous; **bisexual florets** 4.4–5 mm long, 1.3–2 mm wide, hairy, shiny, brown to dark brown at maturity, acute; **anthers** 3–4 mm. $2n = 12$.

Phalaris brachystachys is native to the Mediterranean region and the Canary Islands, where it grows on waste ground, at the edges of cultivated fields, and on roadsides. It is adventive in northern Europe, Australia, and North America. It is known from a few locations in the *Flora* region, most of them in California.

6. Phalaris canariensis L. [p. 769]
ANNUAL CANARYGRASS, PHALARIS DES CANARIES, ALPISTE DES CANARIES

Plants annual. **Culms** 30–100 cm. **Ligules** 3–6 mm, rounded to obtuse, lacerate; **blades** 3–25 cm long, 2–10 mm wide. **Panicles** 1.5–5 cm long, 1.5–2 cm wide, ovoid to oblong-ovoid, continuous, not lobed, truncate at the base; **branches** not evident, spikelets borne singly, not clustered. **Spikelets** homogamous, with 3 florets, terminal floret bisexual; **disarticulation** above the glumes, beneath the sterile florets. **Glumes** 7–10 mm long, 2–2.5 mm wide, smooth, mostly glabrous, sometimes sparsely pilose between the veins, keels winged, wings to 0.6 mm, widening distally, lateral veins inconspicuous, smooth, apices rounded to acute, sometimes mucronate; **sterile florets** 2, equal or subequal, 2–4.5 mm, $^1/_3$ or more the length of the bisexual florets, lanceolate, sparsely hairy, acute; **bisexual florets** 4.5–6.8 mm, ovate, densely hairy, shiny, stramineous to gray-brown; **anthers** 2–4 mm. $2n = 12$.

Phalaris canariensis is native to southern Europe and the Canary Islands, but is now widespread in the rest of the world, frequently being grown for birdseed. The exposed ends of the glumes are almost semicircular in outline, making this one of our easier species of *Phalaris* to identify.

7. Phalaris lemmonii Vasey [p. 769]
LEMMON'S CANARYGRASS

Plants annual. **Culms** (7)25–150 cm. **Ligules** 1.5–8 mm, acute; **blades** to 14 cm long, 1–8 mm wide, smooth, shiny, sometimes revolute. **Panicles** (2)3–20 cm long, 0.6–1.5 cm wide, cylindrical, evidently branched below; **branches** to 2 cm, spikelets borne singly, not clustered. **Spikelets** homogamous, with (2)3 florets, terminal floret bisexual; **disarticulation** above the glumes, beneath the sterile florets. **Glumes** 4.5–6.7 mm long, 0.9–1.1 mm wide, acuminate, keels not or only slightly winged, wings to 0.2 mm wide, keels, lateral veins, and adjacent

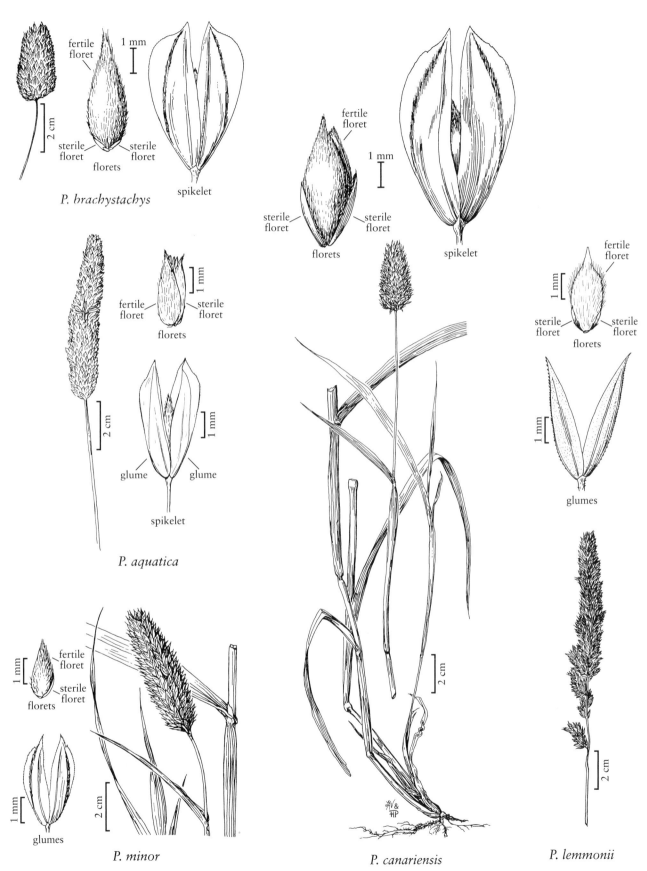

P. brachystachys

fertile floret

sterile floret

sterile floret

florets

spikelet

1 mm

2 cm

P. aquatica

fertile floret

sterile floret

florets

glume

glume

spikelet

2 cm

1 mm

fertile floret

sterile floret

sterile floret

florets

spikelet

1 mm

fertile floret

sterile floret

florets

glumes

1 mm

1 mm

P. minor

fertile floret

sterile floret

florets

glumes

1 mm

1 mm

2 cm

P. canariensis

2 cm

fertile floret

sterile floret

sterile floret

florets

glumes

1 mm

1 mm

P. lemmonii

2 cm

PHALARIS

surfaces scabrous; **sterile florets** (1)2, 1–1.6 mm, densely appressed-hairy; **bisexual florets** 2.7–5.1 mm long, 1.2–1.6 mm wide, shiny, stramineous to gray-brown, mostly hairy with spreading hairs, apices glabrous, strongly acuminate to beaked; **anthers** 0.7–2 mm. $2n = 14$.

Phalaris lemmonii is native to California, but it has also been found in Victoria, Australia. It grows in moist areas, and appears to hybridize with both *P. caroliniana* and *P. angusta* (Baldini 1995). The strongly beaked tips of the bisexual florets are a useful distinguishing feature.

Beecher Crampton noted on one unusually small specimen (UTC 230918) that it was the vernal pool ecotype of the species. He did not publish his observations.

Anderson (1961) and Baldini (1995) distinguished *Phalaris lemmonii* from **P. platensis** Henrard *ex* Wacht., a narrowly distributed South American taxon, arguing that it was slightly longer in the length of its ligules, glumes, florets, and anthers, but many California specimens fall within the range given for *P. platensis* rather than that for *P. lemmonii*. *Phalaris lemmonii* is the older name so, if further research shows that the two species should be combined, *P. lemmonii* will remain as the correct name for plants from the *Flora* region.

8. Phalaris californica Hook. & Arn. [p. 771]

CALIFORNIA CANARYGRASS

Plants perennial; cespitose, not rhizomatous. **Culms** 60–160 cm, swollen at the base. **Ligules** 3–5(8) mm, truncate to acute, irregularly erose; **blades** 5–35 (40) cm long, 3–12(18) mm wide, smooth. **Panicles** 1.5–6 cm long, 1–3 cm wide, ovoid to cylindrical, often purplish, often truncate at the base; **branches** not evident, spikelets borne singly, not clustered. **Spikelets** homogamous, with 3 florets, terminal floret bisexual; **disarticulation** above the glumes, beneath the sterile florets. **Glumes** 5–8 mm long, 0.9–1.6 mm wide, acute to acuminate, keels not or only narrowly winged distally, wings to 0.2 mm wide, scabrous, lateral veins conspicuous, smooth; **sterile florets** 2, equal or subequal, 1.8–3.5 mm, usually more than $^1/_2$ as long as the bisexual florets, densely hairy; **bisexual florets** 3.5–5 mm long, 1–1.5 mm wide, sparsely hairy, shiny, stramineous, becoming darker at maturity, apices acute to weakly acuminate; **anthers** 3–3.5 mm. $2n = 28$.

Phalaris californica is native to California and southwestern Oregon. It grows in ravines and on open, moist ground. Records from further north probably represent introductions. The relatively long, sterile

florets of *P. californica* distinguish it from other species of *Phalaris* in the *Flora* region.

9. Phalaris arundinacea L. [p. 771]

REED CANARYGRASS, ALPISTE ROSEAU, PHALARIS ROSEAU, ROSEAU

Plants perennial; not cespitose, rhizomatous, rhizomes scaly. **Culms** 40–230 cm. **Ligules** 4–10 (11) mm, truncate, lacerate; **blades** usually 10–30 cm long, 5–20 mm wide, flag leaf blades 4–15 cm, surfaces scabrous, margins serrate. **Panicles** 5–40 cm long, 1–4 cm wide, elongate, often dense, always evidently branched, at least near the base; **branches** to 5 cm, normally appressed but spreading during anthesis, spikelets borne singly, not clustered. **Spikelets** homogamous, with 3 florets, terminal floret bisexual; **disarticulation** above the glumes, beneath the sterile florets. **Glumes** subequal, 4–8.1 mm long, 0.8–1 mm wide, keels smoothly curved, usually scabrous, not or narrowly winged distally, wings to 0.2 mm wide, lateral veins conspicuous, apices acute; **sterile florets** 2, subequal to equal, 1.5–2 mm, less than $^1/_2$ as long as the bisexual florets, hairy; **bisexual florets** 2.5–4.2 mm, apices acute to somewhat acuminate; **lemmas** glabrate proximally, hairy distally and on the margins, dull yellow when immature, shiny gray-brown to brown at maturity, apices acute; **anthers** 2.5–3 mm. $2n = 27, 28, 29, 30, 31, 35$.

Phalaris arundinacea is a circumboreal species, native to north-temperate regions; it occurs, as an introduction, in the Southern Hemisphere. It grows in wet areas such as the edges of lakes, ponds, ditches, and creeks, often forming dense stands; in some areas it is a problematic weed. North American populations may be a mix of native strains, European strains, and agronomic cultivars (Merigliano and Lesica 1998).

The interpretation adopted here is that of Baldini (1995), who treated *Phalaris arundinacea sensu stricto* as the most widespread species in a complex of three species. The other two species are **P. rotgesii** (Husn.) Baldini, a diploid that is restricted to France and Italy, and **P. caesia** Nees, a hexaploid that grows in southern Europe, western Asia, and eastern to southern Africa. *Phalaris rotgesii* has glumes 2–3.8 mm long, sterile florets 1–1.5 mm long, bisexual florets 2–3 mm long, and anthers about 2 mm long. The corresponding measurements for *P. caesia* are 6–7 mm, about 2.5 mm, 4–5 mm, and 3.5–4 mm, respectively. Other taxonomists have included *P. rotgesii* and *P. caesia* in *P. arundinacea*. Only *P. arundinacea sensu stricto* has been found in North America.

fertile
floret

1 mm

sterile
floret

sterile
floret

florets

spikelet

fertile
floret

1 mm

sterile
floret

sterile
floret

florets

glumes

2 cm

1 mm

spikelet

2 cm

P. californica

P. arundinacea

PHALARIS

A sterile form of *Phalaris arundinacea* with striped leaves—*P. arundinacea* var. *picta* L., also referred to as *P. arundinacea* forma *variegata* (Parn.) Druce—is known as 'Ribbon Grass' or 'Gardener's Gaiters', and is sometimes grown as an ornamental. Baldini (1995) noted that it sometimes appears to escape, and is never found far from a cultivated stand.

Phalaris arundinacea hybridizes with other species of *Phalaris*. One hybrid, **P. ×monspeliensis** Daveau [= *P. arundinacea* × *P. aquatica*] is grown for forage.

10. Phalaris caroliniana Walter [p. 772]

CAROLINA CANARYGRASS

Plants annual. **Culms** to 150 cm. **Ligules** 1.5–7 mm, truncate to broadly acute; **blades** 1.5–15 cm long, 2–11 mm wide, smooth, shiny green, apices acuminate. **Panicles** 0.5–8(8.5) cm long, 0.8–2 cm wide, ovoid to subcylindrical, not lobed; **branches** not evident, spikelets borne singly, not clustered. **Spikelets** homogamous, with 3 florets, terminal floret bisexual; **disarticulation** above the glumes, beneath the sterile florets. **Glumes** 3.8–6(8) mm long, 0.8–1.5 mm wide, keels smooth or scabridulous, narrowly to broadly winged distally, wings 0.1–0.5 mm wide, entire, smooth, lateral veins prominent, usually smooth, sometimes scabridulous, apices acute or acuminate; **sterile florets** 2, equal to subequal, 1.5–2.5 mm, ½ or more the length of the bisexual florets, the basal 0.2–0.5 mm glabrous, the remainder hairy; **bisexual florets** 2.9–4.7 mm long, 0.9–1.8 mm wide, shiny, stramineous when immature, brown when mature, apices hairy, acuminate to beaked; **anthers** 1.5–2 mm. $2n = 14$.

Phalaris caroliniana grows in wet, marshy, and swampy ground. It is a common species in suitable habitats through much of the southern portion of the *Flora* region and in northern Mexico. It has also been found in Puerto Rico, where it may be an introduction, and in Europe and Australia, where it is undoubtedly an introduction. It appears to hybridize with *P. lemmonii* and *P. caroliniana* (Baldini 1995).

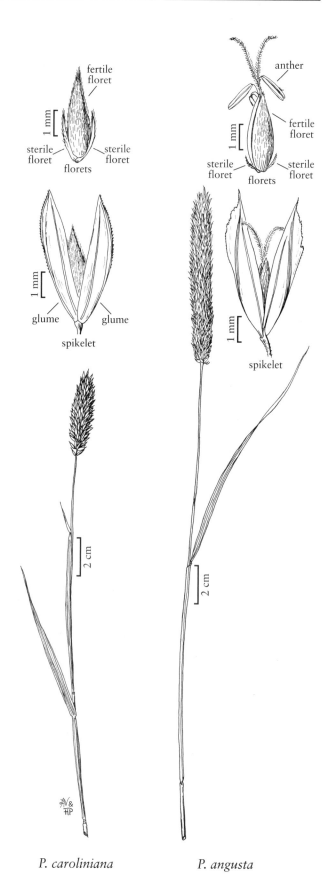

P. caroliniana   *P. angusta*

PHALARIS

11. Phalaris angusta Nees *ex* Trin. [p. 772]
NARROW CANARYGRASS

Plants annual. **Culms** 10–170 cm. **Ligules** 4–7 mm, truncate to rounded or obtuse, lacerate; **blades** 3–15 cm long, 2–12 mm wide. **Panicles** 2–20 cm long, 0.6–1.5 cm wide, cylindrical, discontinuous, sometimes lobed; **branches** sometimes evident, spikelets borne singly, not clustered. **Spikelets** homogamous, with 3 florets, terminal floret bisexual; **disarticulation** above the glumes, beneath the sterile florets. **Glumes** 2–6 mm long, 0.6–1.1 mm wide, rectangular, often purplish, keels winged, scabrous, wings about 0.4 mm wide, smooth, lateral veins conspicuous, scabrous, apices mucronate; **sterile florets** 2, equal, 0.5–1.5 mm, linear, sparsely and inconspicuously hairy; **bisexual florets** 2–3.8 mm long, 0.9–1.5 mm wide, laterally compressed, hairy, particularly distally, shiny, apices tapering; **anthers** 0.5–1.3 mm. $2n = 14$.

Phalaris angusta grows in the contiguous United States, primarily in the south. In South America, it is most abundant in a band from Chile to Argentina; it also grows in Ecuador, Peru, and Bolivia. Thellung (1911) considered it to be a South American species that is adventive in North America. Throughout its distribution, it tends to grow in open grasslands and prairies.

Baldini (1995) suggested that *Phalaris angusta*, *P. lemmonii*, and *P. caroliniana* are involved in reciprocal hybridization and introgression, particularly in California.

14.62 CINNA L.

David M. Brandenburg

Plants perennial; cespitose, sometimes rhizomatous. **Culms** 20–203 cm, solitary or clustered, often rooting at the lower nodes, usually glabrous. **Sheaths** open, glabrous; **auricles** absent; **ligules** scarious; **blades** flat, margins scabrous, surfaces scabrous or smooth. **Inflorescences** panicles; **branches** spreading to ascending, some branches longer than 1 cm; **pedicels** slightly flared, scabrous or smooth; **disarticulation** below the glumes. **Spikelets** laterally compressed, with 1 floret, rarely with a second rudimentary or fertile floret; **rachillas** usually prolonged beyond the base of the floret as a minute stub or bristle, smooth or scabridulous, sometimes not prolonged. **Glumes** from slightly shorter than to slightly longer than the floret, 1- or 3-veined, margins hyaline, keeled, keels scabrous, apices acute, sometimes minutely awn-tipped; **lower glumes** from somewhat shorter than to equaling the upper glumes, florets sessile or stipitate; **calluses** short, glabrous; **lemmas** 3- or 5-veined, sometimes obscurely so, apices acute, minutely bifid, usually awned, awns subterminal, sometimes unawned; **paleas** $^{3}/_{4}$ to nearly as long as the lemmas, 1-veined or with 2 closely spaced veins; **anthers** 1 or 2. **Caryopses** shorter than the lemmas, concealed at maturity, often beaked. $x = 7$. Name of uncertain origin.

Cinna is a genus of four species, all of which generally grow in damp woods, along streams, or in wet meadows. One species, *C. latifolia*, is northern temperate and circumboreal. The other three species are restricted to the Western Hemisphere. *Cinna poaeformis* (Kunth) Scribn. & Merr. extends from Mexico to Venezuela and Bolivia.

The reduction of *Limnodea* to synonymy under *Cinna* by Tucker (1996) introduced a markedly discordant element into *Cinna* (Brandenburg and Thieret 2000), and has not been followed here.

SELECTED REFERENCES **Brandenburg, D.M., W.H. Blackwell,** and **J.W. Thieret.** 1991. Revision of the genus *Cinna* (Poaceae). Sida 14:581–596; **Brandenburg, D.M., J.R. Estes, S.D. Russell,** and **J.W. Thieret.** 1991. One-nerved paleas in *Cinna arundinacea* L. (Poaceae). Trans. Kentucky Acad. Sci. 52:94–96; **Brandenburg, D.M. and J.W. Thieret.** 2000. *Cinna* and *Limnodea* (Poaceae): Not congeneric. Sida 19:195–200; **Tucker, G.C.** 1996. The genera of Poöideae (Gramineae) in the southeastern United States. Harvard Papers Bot. 9:11–90.

1. Anthers 2; lemmas 5-veined; florets more or less sessile . . . . 3. *C. bolanderi*
1. Anthers 1; lemmas 3(5)-veined; florets on a 0.1–0.65 mm stipe.
  2. Upper glumes prominently 3-veined; spikelets (3.5)4–6(7.5) mm long . . . . 1. *C. arundinacea*
  2. Upper glumes usually 1-veined, rarely 3-veined; spikelets (2)2.5–4(5) mm long . . . . 2. *C. latifolia*

1. Cinna arundinacea L. [p. 775]

STOUT WOODREED, SWEET WOODREED, CINNA ROSEAU

Culms 28–185 cm, somewhat bulbous at the base; **nodes** 5–13. **Ligules** 2–10 mm; **blades** to 34.5 cm long, 3–19 mm wide. **Panicles** 6.5–55 cm; **branches** ascending to spreading. **Spikelets** (3.5)4–6(7.5) mm; **rachilla prolongations** 0.1–0.4 mm, sometimes absent. **Lower glumes** (2.7)3.5–5(6.1) mm, somewhat shorter than the lemmas, 1-veined; **upper glumes** (3.5)4–6(7.5) mm, equal to or slightly longer than the lemmas, 3-veined; **stipes** 0.25–0.65 mm; **lemmas** (2.7)3.5–5(6.4) mm, 3(5)-veined, awns 0.2–1.5 mm, rarely absent; **paleas** 1-veined; **anthers** 1, 0.8–1.9 mm. **Caryopses** 2.1–2.8 mm. $2n = 28$.

Cinna arundinacea grows in southeastern Canada and throughout most of the eastern United States, at 0–900 m. It is most common in moist woodlands and swamps, depressions, along streams, and in floodplains and upland woods. It is less frequent in wet meadows, marshes, and disturbed sites. It flowers in late summer to fall. *Cinna arundinacea* is most easily distinguished from *C. latifolia* by its 3-veined upper glumes and larger spikelets.

2. Cinna latifolia (Trevir. *ex* Göpp.) Griseb. [p. 775]

DROOPING WOODREED, SLENDER WOODREED,
CINNA À LARGES FEUILLES

Culms 20–190 cm; **nodes** 4–9. **Ligules** 2–8 mm; **blades** to 28 cm long, 1–20 mm wide. **Panicles** 3–46 cm; **branches** usually spreading, sometimes ascending. **Spikelets** (2)2.5–4(5) mm; **rachilla prolongations** 0.1–1.3 mm, sometimes absent. **Lower glumes** (1.8)2.5–4(4.7) mm, 1-veined; **upper glumes** (1.9)2.5–4(5) mm, 1(3)-veined; **stipes** 0.1–0.45 mm; **lemmas** 1.8–3.8 mm, 3(5)-veined, awns 0.1–2.5 mm or absent; **paleas** 2-veined, with the veins very close together, or 1-veined; **anthers** 1, 0.4–1 mm. **Caryopses** 1.8–2.8 mm. $2n = 28$.

Cinna latifolia is a circumboreal species, extending from Alaska to Newfoundland in North America, and across Eurasia from Norway to the Kamchatka Peninsula, Russia. It grows in moist to wet soil in open coniferous or mixed forests, swamps, thickets, bogs, and streamsides, at 0–2600 m. It flowers in late summer and fall. *Cinna latifolia* differs from *C. arundinacea* in its 1(3)-veined upper glumes and its smaller spikelets; and from *C. bolanderi* in having 1 anther, shorter anthers and spikelets, and stipitate florets. A collection of *C. latifolia* from the Aleutian Islands had abnormally large (to 5.5 mm), often 2-flowered spikelets (Brandenburg et al. 1991). *Cinna latifolia* is a variable species for which varietal names have been proposed; because the variation is continuous, no varieties are recognized in this treatment.

3. Cinna bolanderi Scribn. [p. 775]

SIERRAN WOODREED, BOLANDER'S WOODREED

Culms 85–203 cm; **nodes** 4–8. **Ligules** 3.5–7 mm; **blades** to 40 cm long, 2–19 mm wide. **Panicles** 7.5–43 cm; **branches** spreading to ascending. **Spikelets** (3.6)4–5.5(6.3) mm; **rachilla prolongations** 0.4–0.9 mm, sometimes absent. **Lower glumes** (3.3)3.5–5.2(6) mm, 1-veined; **upper glumes** (3.6)4–5.5(6.3) mm, 1- or 3-veined; **stipes** essentially absent, florets more or less sessile; **lemmas** (2.7)3.2–4.6 mm, 5-veined, lateral veins underdeveloped and often faint, awns 0.2–1.5 mm or absent; **paleas** 2-veined, the veins approximate; **anthers** 2, 1.2–2.6 mm, rarely to 0.7 mm. **Caryopses** 2–2.9 mm. $2n = 28$.

Cinna bolanderi is endemic to meadows and streamsides at 1900–2400 m in Sequoia, Kings Canyon, and Yosemite national parks. It flowers from late summer to fall. It used to be included in *C. latifolia*, but it differs from that species in having 2 anthers, longer anthers and spikelets, and sessile florets. The two species do not overlap in distribution.

CINNA

C. latifolia

C. bolanderi

C. arundinacea

L. arkansana

LIMNODEA

14.63 LIMNODEA L.H. Dewey

Neil Snow

Plants annual; tufted. **Culms** to 60 cm, often prostrate and branching at the base, glabrous; **internodes** solid. **Sheaths** open or closed, rounded on the back, frequently hispid; **auricles** absent; **ligules** membranous, usually lacerate and minutely ciliate; **blades** flat, mostly ascending, abaxial and/or adaxial surfaces glabrous or hispid. **Inflorescences** panicles, loosely contracted; **branches** spikelet-bearing to the base or nearly so, some branches longer than 1 cm; **disarticulation** below the glumes. **Spikelets** pedicellate, subterete, with 1 floret; **rachillas** prolonged beyond the base of the floret as a slender bristle, glabrous. **Glumes** equal, rigid, coriaceous, hispid or scabrous throughout, rounded over the midvein, veinless or obscurely 3–5-veined, acute, unawned; **calluses** blunt, glabrous; **lemmas** equaling the glumes, chartaceous, smooth, veinless or inconspicuously 3-veined, minutely bifid or acute, awned, awns subterminal, exceeding the florets, geniculate near midlength, basal segment twisted; **paleas** shorter than the lemmas, veinless or the base 2-veined, hyaline; **lodicules** 2, glabrous, toothed or not; **anthers** 3; **ovaries** glabrous. **Caryopses** about 2.5 mm, shorter than the lemmas, concealed at maturity, linear. $x = 7$. Name from *Limnas*, a similar Eurasian genus, and the Greek *odes*, 'like'.

Limnodea is a monotypic genus of the southern United States and adjacent Mexico.

SELECTED REFERENCE **Brandenburg**, D.M. and **J.W. Thieret**. 2000. *Cinna* and *Limnodea* (Poaceae): Not congeneric. Sida 19:195–200.

1. Limnodea arkansana (Nutt.) L.H. Dewey [p. 775]
OZARKGRASS

Culms 15–60 cm. **Ligules** 1–2 mm; **blades** 3–12 cm long, 2–8 mm wide. **Panicles** 5–12 cm long, 1–5 cm wide, loose, usually erect. **Spikelets** 3–4 mm; **rachillas** prolonged 0.3–0.9 mm. **Glumes** 3–4 mm; **lemmas** 3–4 mm, awns (5)8–14.5 mm; **anthers** 0.2–0.4 mm. $2n = 14$.

Limnodea arkansana grows in dry, usually sandy soils of prairies, open woodlands, disturbed areas, and riverbanks. Along the Gulf coast, it grows on maritime shell mounds and middens, and upper beaches where shells accumulate. It has also been found, as an introduction, around a wool-combing mill in Jamestown, South Carolina.

14.64 AMMOPHILA Host

Mary E. Barkworth

Plants perennial; strongly rhizomatous. **Culms** 20–130 cm, erect, glabrous. **Leaves** mostly basal; **sheaths** open; **auricles** absent; **ligules** membranous, sometimes ciliolate; **blades** 0.5–8 mm wide, involute or convolute. **Inflorescences** terminal panicles, dense, cylindrical; **branches** strongly ascending and overlapping. **Spikelets** pedicellate, laterally compressed, with 1 floret; **rachillas** prolonged beyond the florets, glabrous or hairy; **disarticulation** above the glumes, beneath the florets. **Glumes** equaling or exceeding the florets, subequal, linear-lanceolate, papery, keeled, acute to acuminate; **lower glumes** 1-veined; **upper glumes** 3-veined; **calluses** short, pilose; **lemmas** chartaceous, linear-lanceolate, obscurely 3–5-veined, keeled, sometimes slightly rounded at the base, apices entire or minutely bifid, unawned or awned, awns 0.2–0.5 mm, subterminal; **paleas** equaling the lemmas, often appearing 1-keeled, 2- or 4-veined, central veins close together; **lodicules** 2, free, membranous, ciliate or glabrous, not toothed; **anthers** 3, 3–7 mm; **ovaries** glabrous. **Caryopses** enclosed by the hardened lemma and palea, ellipsoid, longitudinally grooved; **hila** about ²/₃ as long as the caryopses. $x = 7$. Name from the Greek *ammos*, 'sand', and *philos*, 'loving' a reference to its habitat.

Ammophila has two species, one native to the coast of Europe and northern Africa, and one to eastern North America. Both species are effective sand binders and dune stabilizers. They are sometimes mistaken for *Leymus arenarius* and *L. mollis*, which grow in the same habitats and have a similar habit, but species of *Leymus* have more than 1 floret per spikelet.

SELECTED REFERENCES **Cope, E.A.** 1994. Further notes on beachgrasses (*Ammophila*) in northeastern North America. Newsletter New York Fl. Assoc. 5(1):5–7; **Reznicek, A.A.** and **E.J. Judziewicz.** 1996. A new hybrid species, ×*Calammophila don-hensonii* (*Ammophila breviligulata* × *Calamagrostis canadensis*, Poaceae) from Grand Island, Michigan. Michigan Bot. 35:35–40; **Seabloom, E.W.** and **A.M. Wiedemann.** 1994. Distribution and effects of *Ammophila breviligulata* Fern. (American beachgrass) on the foredunes of the Washington coast. J. Coastal Res. 10:178–188; **Stern, R.J.** 1983. Morphometric and phenologic variability in *Ammophila breviligulata* Fernald. Master's thesis, University of Vermont, Burlington, Vermont, U.S.A. 29 pp.; **Walker, P.J., C.A. Paris,** and **D.S. Barrington.** 1998. Taxonomy and phylogeography of the North American beachgrasses. Amer. J. Bot. 85, Suppl.:87 [abstract].

1. Ligules 1–4.6 mm long, truncate to obtuse . . . . 1. *A. breviligulata*
1. Ligules 10–35 mm long, acute and bifid or lacerate . . . . 2. *A. arenaria*

1. **Ammophila breviligulata** Fernald [p. 779]

Culms 50–130 cm. **Ligules** 1–3 (4.6) mm, truncate to obtuse, ciliolate; **blades** 15–80 cm long, 4–8 mm wide when flat, usually involute, 0.5–2.5 mm in diameter when involute; **flag leaf blades** 5–39 cm. **Panicles** 9–40 cm long, 10–25 mm wide, stramineous. **Glumes** 8–15 mm, exceeding the florets; **callus hairs** 1–3 mm; **lemmas** 8–11.5(14) mm, somewhat thicker than the glumes, apices acute, entire, unawned, rarely mucronate, mucros to 0.2 mm, subterminal; **anthers** 3–7 mm.

The two subspecies overlap in their morphological characteristics and occupy similar habitats, the primary difference between them being their flowering time.

1. Flowering late July to September; panicles (9.5)13–40 cm long; anthers 3.5–7 mm long . . . . subsp. *breviligulata*
1. Flowering late June to early July; panicles 9–17 cm long; anthers 3–4.5 mm long . . . . subsp. *champlainensis*

Ammophila breviligulata Fernald subsp. **breviligulata** [p. 779]

AMERICAN BEACHGRASS, AMMOPHILE À LIGULE COURTE

Flag leaf blades 5–39 cm. **Panicles** (9.5)13–40 cm. **Glumes** 8–15 mm; **lemmas** 8–11.5(14) mm; **anthers** 3.5–7 mm. $2n = 28$.

Ammophila breviligulata subsp. *breviligulata* grows on sand dunes and dry, sandy shores from around the Great Lakes to the Atlantic coast from Newfoundland to South Carolina and, as an introduction, on the west coast. Anthesis is from late July to September. It hybridizes with *Calamagrostis canadensis* on Grand

Island, Michigan, and the adjacent mainland, to form ×**Calammophila don-hensonii** Reznicek & Judz. The hybrid is morphologically intermediate between its parents. Most florets appear to be sterile; some appear normal. Discovery of a plant that appears to be a backcross to *Calamagrostis canadensis* supports the possibility that some of the hybrid seed is functional.

Ammophila breviligulata subsp. **champlainensis** (F. Seym.) P.J. Walker, C.A. Paris & Barrington *ex* Barkworth [p. 779]

CHAMPLAIN BEACHGRASS

Flag leaf blades 8–22 cm. **Panicles** 9–17 cm. **Glumes** 8–11.5 mm; **lemmas** 8–10 mm; **anthers** 3–4.5 mm. $2n =$ unknown.

Ammophila breviligulata subsp. *champlainensis* differs from subsp. *breviligulata* primarily in its earlier anthesis, from June to early July rather than late July to September. It also tends to be smaller, but the two subspecies overlap in all their morphological characteristics.

2. **Ammophila arenaria** (L.) Link [p. 779]

EUROPEAN BEACHGRASS, MARRAMGRASS

Culms (50)60–120 cm. **Ligules** 10–35 mm, acute, bifid or lacerate; **blades** to 60(90) cm long and 6 mm wide when flat, usually tightly convolute and sharply pointed. **Panicles** 12–35 cm long, 15–20 mm wide, stramineous. **Spikelets** 12–14 mm. **Glumes** equaling or exceeding the florets; **callus hairs** 2–6 mm; **lemmas** 8–12 mm, apices shortly bifid, midveins slightly excurrent; **anthers** 4–7 mm. $2n = 14, 28, 56$.

Ammophila arenaria is a European species that has become naturalized in most temperate countries. It was introduced along the Pacific coast and in the interior of western North America as a sand binder. It is also reported to occur, as an introduction, on the east coast; the report has not been verified. It is known from a single 1941 collection on a sand dune in Erie County, Pennsylvania. *Ammophila arenaria* flowers from the end of June to August. It forms a strong association with the fungus *Psilocybe azurescens*, which grows, often prolifically, along the northern Oregon coast.

North American plants belong to **Ammophila arenaria** (L.) Link subsp. **arenaria**, in which the glumes exceed the lemma and the callus hairs are about 2–3 mm long. It is native from northern and western Europe to northwestern Spain. **Ammophila arenaria** subsp. **arundinacea** H. Lindb. has glumes that equal the lemma, and callus hairs 4–6 mm long. It is native around the Mediterranean.

In Europe, *Ammophila arenaria* hybridizes with *Calamagrostis*. The hybrids, which are sterile, have also been used as sand binders but not, so far as is known, in the *Flora* region. They differ from *A. arenaria* in having a less dense panicle, slender lemma awns up to 2 mm long, and callus hairs ¹/₂ as long as the lemmas or longer. In the *Flora* region, a hybrid with *C. nutkaensis* was reported in the 1970s from the vicinity of Newport–Waldport in coastal Oregon, where both species occur (K.L. Chambers, pers. comm.). There is no voucher to support this report.

14.65 MILIUM L.

William J. Crins

Plants annual or perennial; cespitose, sometimes rhizomatous. **Culms** 10–180 cm, glabrous or hispidulous; **nodes** 2–5, glabrous. **Sheaths** open, smooth or scabrous; **auricles** absent; **ligules** hyaline, glabrous, obtuse to acute; **blades** flat, smooth or scabrous over the veins. **Inflorescences** open panicles; **branches** drooping to ascending, smooth or scabrous, some branches longer than 1 cm. **Spikelets** pedicellate, dorsally compressed, with 1 floret; **rachillas** not prolonged beyond the base of the floret; **disarticulation** above the glumes, beneath the floret. **Glumes** equal, equaling or exceeding the lemmas, membranous, smooth or scabrous, unawned; **calluses** blunt, glabrous; **lemmas** dark, coriaceous, glabrous, lustrous, obscurely 5-veined, margins involute, apices unawned; **paleas** similar to the lemmas and partly enfolded by them; **lodicules** 2, free, glabrous, toothed or not; **anthers** 3; **ovaries** glabrous. **Caryopses** shorter than the lemmas, concealed at maturity, glabrous; **hila** ¹/₅ to nearly ¹/₂ the length of the caryopses. $x = 4, 5, 7, 9$. Name from an old Latin word for millet, an appellation which is associated with species in several different genera.

Milium is a circumtemperate genus of four species. All the species grow in mesic to dry mixed woods and dry open habitats. *Milium effusum* is native to the *Flora* region; *M. vernale* has become established.

SELECTED REFERENCES **Callihan, R.H.** and **D.S. Pavek.** 1988. *Milium vernale* survey in Idaho County, Idaho. Idaho Weed Control Rep. 1988:153; **Fernald, M.L.** 1950. The North American variety of *Milium effusum*. Rhodora 52:218–222.

1. Plants perennial; blades 8–17 mm wide; panicles 10–27 cm long; lemmas 2.3–3 mm long . . . . . . 1. *M. effusum*
1. Plants annual; blades 1.9–5 mm wide; panicles 4–11.5 cm long; lemmas 2–2.3 mm long . . . . 2. *M. vernale*

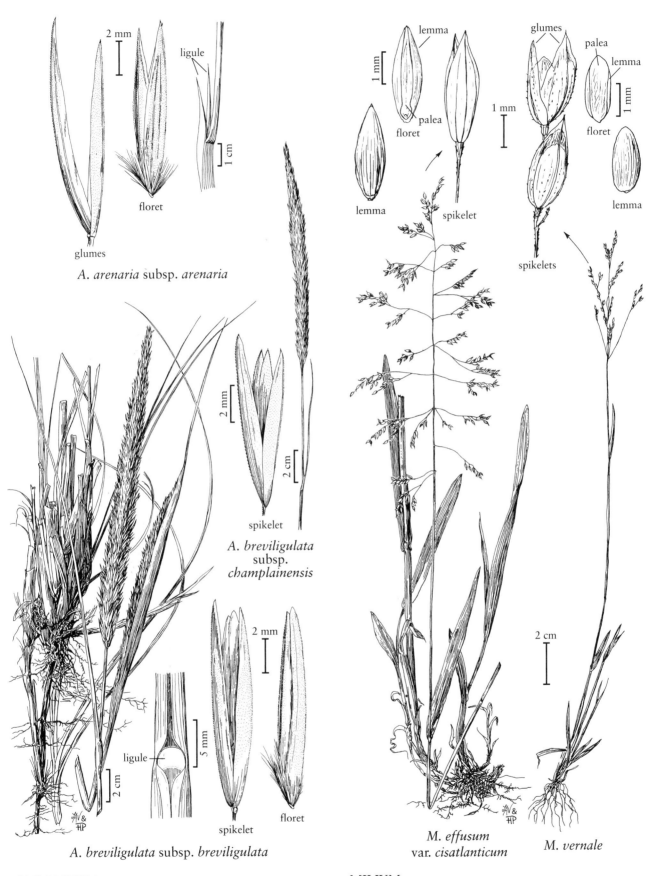

A. *arenaria* subsp. *arenaria*

A. *breviligulata*
subsp.
champlainensis

A. *breviligulata* subsp. *breviligulata*

AMMOPHILA

M. *effusum*
var. *cisatlanticum*

M. *vernale*

MILIUM

1. Milium effusum L. [p. 779]
WOOD MILLET, MILLET DIFFUS

Plants perennial; rhizomatous. **Culms** 55–140 cm, erect from decumbent bases, glabrous; **nodes** 3–5. **Sheaths** glabrous; **ligules** 3–9 mm, obtuse, erose; **blades** 5–26 cm long, 8–17 mm wide, flat, glabrous, equally distributed on the culms. **Panicles** 10–27 cm; **branches** 1–9 cm, in pairs or fascicles, flexuous, spreading or drooping, scabrous, with spikelets mainly near the distal ends. **Glumes** 2.5–5 mm, scabrous, 3-veined, acute to acuminate; **lemmas** 2.3–3 mm, acute; **anthers** 1.5–2 mm. 2*n* = 14, 28.

Milium effusum is widespread in temperate to subarctic regions in the Northern Hemisphere. North American plants belong to **M. effusum** var. **cisatlanticum** Fernald, an elegant native grass that grows in woodlands in eastern North America. It differs from **M. effusum** L. var. **effusum**, which grows from Europe to Asia and Japan, in having 2–3 panicle branches at most nodes and spikelets 2.5–5 mm long, rather than 4–5 panicle branches at most nodes and spikelets about 3 mm long. A cultivar of *M. effusum*, 'Aureum', is grown for its yellowish leaves.

2. Milium vernale M.-Bieb. [p. 779]
EARLY MILLET, SPRING MILLETGRASS

Plants annual. **Culms** 10–70 (90) cm, erect, solitary or clumped, retrorsely scabridulous; **nodes** 2–3, glabrous, green to purple. **Sheaths** retrorsely scabrous; **ligules** 2.5–4.5 mm, obtuse to acute, occasionally lacerate; **blades** 1.7–8.2 cm long, 1.9–5 mm wide, scabrous over the veins, concentrated on the lower portion of the culms. **Panicles** 4–11.5 cm; **branches** 0.5–6 cm, straight, ascending to erect, antrorsely scabrous, spikelets confined to the distal ½. **Glumes** 2.5–3.2 mm, scabridulous, 3-veined, acuminate; **lemmas** 2–2.3 mm; **anthers** about 1.5 mm. 2*n* = 18.

Native to Eurasia, *Milium vernale* was first detected in North America in 1987, when it was found infesting winter wheat and other crops in north-central Idaho (Callihan & Pavek 1988). The infested area has since increased.

14.66 ALOPECURUS L.

William J. Crins

Plants annual or perennial; sometimes cespitose, sometimes shortly rhizomatous. **Culms** 5–110 cm, clumped or solitary, erect or decumbent, occasionally cormlike at the base; **nodes** glabrous. **Leaves** inserted mostly on the lower ½ of the culms; **sheaths** open, upper sheaths sometimes inflated; **auricles** absent; **ligules** 0.6–6.5 mm, truncate to acute, membranous, puberulent or glabrous, entire to lacerate; **blades** 0.7–12 mm wide, flat or involute, glabrous or scabrous, blades of uppermost leaves sometimes short or absent. **Inflorescences** terminal panicles, spikelike, capitate to cylindrical; **branches** usually shorter than 5 mm, lower branches sometimes to 2 cm; **disarticulation** below the glumes. **Spikelets** 1.8–7 mm, pedicellate, strongly laterally compressed, oval in outline, with 1 floret; **rachillas** not prolonged beyond the base of the floret. **Glumes** equaling or exceeding the florets, membranous or coriaceous, free or connate in at least the lower ½, narrowing from above midlength, 3-veined, keeled, keels ciliate, at least basally, apices obtuse to acute or shortly awned; **calluses** blunt, glabrous; **lemmas** membranous, margins often connate in the lower ½, keeled, indistinctly 3–5-veined, apices truncate to acute, awned dorsally from just above the base to about midlength, geniculate or straight; **paleas** absent or greatly reduced; **lodicules** absent; **anthers** 3, 0.3–4.1 mm; **ovaries** glabrous; **styles** fused, with 2 branches. **Caryopses** shorter than the lemmas, concealed at maturity, glabrous; **hila** short. *x* = 7. Name from the Greek *alopex*, 'fox', and *oura*, 'tail', referring to the cylindrical panicles.

Alopecurus is a genus of 36 species that grow primarily in open, mesic habitats, and are native to the northern temperate zone and South America. Four species are native to the *Flora* region, four were introduced and have become established, and two were introduced and are not known to persist. Some species, including some native to the *Flora* region, have been introduced as pasture grasses outside of their native ranges. Of these, only *A. pratensis* has become widely naturalized.

Some species of *Alopecurus* can appear similar to *Phleum*, which has truncate glumes that are abruptly awned or mucronate, lemmas without awns or keels, and well-developed paleas; *Alopecurus* has glumes that are obtuse to acute and gradually awned or unawned, lemmas with both awns and keels, and paleas that are absent or greatly reduced.

SELECTED REFERENCES **Doğan, M**. 1999. A concise taxonomic revision of the genus *Alopecurus* L. (Gramineae). Turk. J. Bot. 23:245–262; **Moyer, J.R.** and **A.L. Boswall**. 2002. Tall fescue or creeping foxtail suppresses foxtail barley. Canad. J. Pl. Sci. 82:89–92; **Tsvelev, N.N.** 1995. *Alopecurus*. Pp. 106–114 *in* J.G. Packer (ed., English edition). Flora of the Russian Arctic, vol. 1, trans. G.C.D. Griffiths. University of Alberta Press, Edmonton, Alberta, Canada [English translation of A.I. Tolmachev (ed.). 1964. Arkticheskaya Flora SSSR, vol. 2. Nauka, Leningrad [St. Petersburg], Russia. 272 pp.].

1. Glume keels winged; glumes glabrous, pubescent over the veins.
    2. Glumes 4.5–7.5 mm long, connate in the lower $^1/_2$, the apices acute, convergent to parallel; lemma apices acute . . . . 9. *A. myosuroides*
    2. Glumes 3–4.5 mm long, connate in the lower $^1/_2$–$^4/_5$, the apices obtuse, mucronate, divergent; lemma apices truncate . . . . 10. *A. creticus*
1. Glume keels not winged; glumes usually sparsely to densely pubescent, sometimes glabrous.
    3. Plants annual, without rhizomes, not rooting at the lower nodes; blades 1–16 cm long, 0.9–4 mm wide; culms 5–50 cm tall.
        4. Glumes 5–6.4 mm long, coriaceous and dilated in the lower $^1/_2$; glume and lemma apices acute to acuminate; anthers about 3 mm long . . . . 6. *A. rendlei*
        4. Glumes 2.1–5 mm long, membranous, not dilated below; glume and lemma apices obtuse; anthers 0.3–1.8 mm long.
            5. Upper sheaths conspicuously inflated; glumes 3–5 mm long; lemmas 3–5 mm long, awns exceeding the lemmas by 3–6 mm; panicles 5.5–13 mm wide, excluding the awns . . . . 7. *A. saccatus*
            5. Upper sheaths not or only slightly inflated; glumes 2.1–3.1 mm long; lemmas 1.9–2.7 mm long, awns exceeding the lemmas by 1.6–4 mm; panicles 3–6 mm wide, excluding the awns . . . . 8. *A. carolinianus*
    3. Plants perennial, often rhizomatous, sometimes rooting at the lower nodes; blades 2–40 cm long, 1–12 mm wide; culms 5–110 cm tall.
        6. Glumes 1.8–3.7 mm long, the apices obtuse; anthers 0.5–2.2 mm long.
            7. Awns geniculate, exceeding the lemmas by 1.2–4 mm; anthers (0.9)1.4–2.2 mm long . . . . 4. *A. geniculatus*
            7. Awns straight, not exceeding the lemmas or exceeding them by less than 2.5 mm; anthers 0.5–1.2 mm long . . . . 5. *A. aequalis*
        6. Glumes 3–6 mm long, the apices acute; anthers 2–4 mm long.
            8. Glume margins connate in the lower $^1/_8$; glumes densely pilose throughout . . . . 3. *A. magellanicus*
            8. Glume margins connate in the lower $^1/_5$–$^1/_3$; glumes with long hairs mainly restricted to the veins.
                9. Lemma apices acute; glume apices parallel or convergent . . . . 1. *A. pratensis*
                9. Lemma apices obtuse to truncate; glume apices divergent . . . . 2. *A. arundinaceus*

1. Alopecurus pratensis L. [p. 783]
MEADOW FOXTAIL, VULPIN DES PRÉS

Plants perennial; shortly rhizomatous. **Culms** 30–110 cm, erect. **Ligules** 1.5–3 mm, obtuse to truncate; **blades** 6–40 cm long, 1.9–8 mm wide; **upper sheaths** not or scarcely inflated. **Panicles** 3.5–9 cm long, 6–10 mm wide. **Glumes** 4–6 mm, connate in the lower ¹/₅–¹/₄, membranous, sides pubescent, keels not winged, finely ciliate, apices acute, parallel or convergent; **lemmas** 4–6 mm, connate in the lower ¹/₃, usually glabrous, keels sometimes ciliate distally, apices acute, awns 5–10.5 mm, geniculate, exceeding the lemmas by (1)2.2–5.5 mm; **anthers** 2–4 mm, yellowish, orange, reddish, or purplish, sometimes varying within a population. **Caryopses** 1–1.2 mm. $2n = 28, 42$.

Alopecurus pratensis is native from temperate northern Eurasia south to North Africa. It is now widely naturalized in temperate regions throughout the world. It grows in poorly to somewhat drained soils in meadows, riverbanks, lakesides, ditches, roadsides, and fence rows. It has been widely introduced as a pasture grass; it may also have become established from ballast or imported hay. The earliest collections are from coastal New England; it is now established throughout much of the *Flora* region.

2. Alopecurus arundinaceus Poir. [p. 783]
CREEPING MEADOW FOXTAIL, VULPIN ROSEAU

Plants short-lived perennials; rhizomatous. **Culms** 30–110 cm, erect. **Ligules** 1.3–5 mm, truncate; **blades** 6–40 cm long, 3–12 mm wide; **upper sheaths** somewhat inflated. **Panicles** 3–10 cm long, 7–13 mm wide. **Glumes** 3.6–5 mm, connate in the lower ¹/₅–¹/₃, membranous, sparsely pubescent, keels not winged, ciliate, apices acute, divergent, pale green to lead-gray; **lemmas** 3.1–4.5 mm, connate in at least the lower ¹/₃, usually glabrous, sometimes with scattered hairs near the apices, apices truncate to obtuse, awns 1.5–7.5 mm, geniculate, exceeding the lemmas by 0–3 mm; **anthers** 2.2–3.5 mm. $2n = 26, 28, 30$.

Alopecurus arundinaceus is native to Eurasia, extending north of the Arctic Circle and south to the Mediterranean. It grows in wet, moderately acid to moderately alkaline soils, on flood plains, near vernal ponds, and along rivers, streams, bogs, potholes, and sloughs. It was introduced for pasture in North Dakota

and now occurs more widely, having been promoted as a forage species. It is sometimes used in seed mixtures for revegetation projects. It was evaluated for revegetation in Alberta, but there is no evidence that it was ever actually used in that province. *Alopecurus arundinaceus* suppresses *Hordeum jubatum*, a troublesome, unpalatable, weedy species, in irrigated pastures (Moyer and Boswall 2002).

3. Alopecurus magellanicus Lam. [p. 783]
ALPINE FOXTAIL, BOREAL FOXTAIL, VULPIN ALPIN, VULPIN BORÉALE

Plants perennial; shortly rhizomatous. **Culms** (6)10–80 cm, erect or decumbent. **Ligules** 1–2 mm, truncate; **blades** 4–22 cm long, 2.5–7 mm wide; **upper sheaths** inflated. **Panicles** 1–5 cm long, 8–14 mm wide. **Glumes** 3–5 mm, connate in the lower ¹/₈, membranous, densely pilose throughout, keels not winged, ciliate, apices acute and parallel; **lemmas** 2.5–4.5 mm, connate in the lower ¹/₂–²/₃, glabrous proximally, finely pubescent distally, apices usually obtuse, occasionally truncate, awns 2–6(8) mm, geniculate, exceeding the lemmas by 0–5 mm; **anthers** 2.3–3 mm, yellow. **Caryopses** 0.7–2 mm. $2n = 98, 100, 105, 112, 117, 119,$ ca. 120.

Alopecurus magellanicus has an arctic-alpine to subalpine circumpolar distribution, but it has not been found in Scandinavia or Iceland. It grows primarily in wet soils in tundra, meadows, along streams, shorelines, gravelbars, and floodplains, and occasionally in somewhat drier forest openings, in fine or silty to stony soils or moss. It is sometimes co-dominant with *Dupontia fisheri* in the arctic and subarctic portion of its range. The anthocyanic tint of the plant as a whole greatly increases to the north.

In the past, this species has been called *Alopecurus alpinus*, the name being attributed to Smith. Doğan (1999) pointed out that the name had first been used by Villars for a different species, which meant that this species had to have another name; Doğan used *A. borealis* Trin., listing *A. magellanicus* as a synonym but, because *A. magellanicus* was published first, it has priority and is therefore the correct name.

The morphological variability in *Alopecurus magellanicus* has prompted recognition of several segregate taxa, *A. stejnegeri* Vasey and *A. occidentalis* Scribn. & Tweedy being two of the more conspicuous extremes. The former are small plants occurring on enriched sites in the arctic, usually around seabird or seal colonies where high nutrient levels produce lush vegetative growth; the latter refers to tall-stemmed

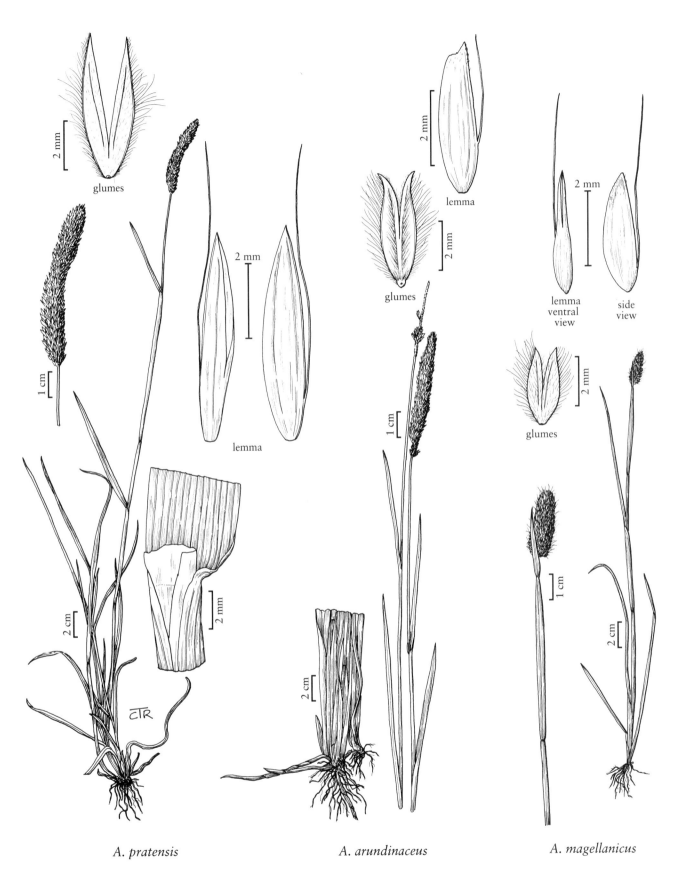

A. pratensis

A. arundinaceus

A. magellanicus

ALOPECURUS

plants found in the Rocky Mountains. Because such plants are simply extremes in a continuum of variation, they do not merit taxonomic recognition.

4. Alopecurus geniculatus L. [p. 785]
WATER FOXTAIL, VULPIN GÉNICULÉ

Plants perennial; cespitose. **Culms** (5)10–60 cm, erect or decumbent, rooting at the lower nodes. **Ligules** 2–5 mm, obtuse; **blades** 2–12 cm long, 1–4(7) mm wide; **upper sheaths** somewhat inflated. **Panicles** 1.5–7 cm long, 4–8 mm wide. **Glumes** 1.9–3.5 mm, connate at the base, membranous, pubescent, keels not winged, ciliate, apices obtuse, parallel, often purplish; **lemmas** 2.5–3 mm, connate in the lower $^1/_2$, glabrous or with a few scattered hairs at the apices, apices truncate to obtuse, awns 3–5(6) mm, geniculate, exceeding the lemmas by (1.2)2–4 mm; **anthers** (0.9)1.4–2.2 mm, yellow. **Caryopses** 1–1.5 mm. $2n = 28$.

Alopecurus geniculatus is native to Eurasia and parts of North America, growing in shallow water, ditches, open wet meadows, shores, and streambanks, from lowland to montane zones. It has been naturalized in eastern North America. The status of populations in the west, including the Queen Charlotte Islands in British Columbia, is less certain. Many occur in moist sites within native rangeland, but these areas have also been affected by European settlement, although less intensively and for a shorter period than those in eastern North America.

Alopecurus ×*haussknechtianus* Asch. & Graebn. is a hybrid between *A. geniculatus* and *A. aequalis*, which occurs fairly frequently in areas of sympatry, particularly in drier midcontinental areas from Alberta to Saskatchewan, south to Arizona and New Mexico. The hybrids are sterile and appear to have $2n = 14$.

5. Alopecurus aequalis Sobol. [p. 785]

Plants perennial; cespitose. **Culms** 9–75 cm, erect or decumbent. **Ligules** 2–6.5 mm, obtuse; **blades** 2–10 cm long, 1–5(8) mm wide; **upper sheaths** not inflated. **Panicles** 1–9 cm long, 3–9 mm wide. **Glumes** 1.8–3.7 mm, connate near the base, membranous, pubescent on the sides, keels not winged, ciliate, apices obtuse, sometimes erose, pale green, occasionally purplish; **lemmas** 1.5–2.5(3.5) mm, connate in the lower $^1/_3$–$^1/_2$, glabrous, apices obtuse, awns 0.7–3 mm, straight,

exceeding the lemmas by 0–2.5 mm; **anthers** 0.5–1.2 mm, usually pale to deep yellow or orange, rarely purple. **Caryopses** 1–1.8 mm. $2n = 14, 28$.

Alopecurus aequalis is native to temperate zones of the Northern Hemisphere. It generally grows in wet meadows, forest openings, shores, springs, and along streams, as well as in ditches, along roadsides, and in other disturbed sites, from sea level to subalpine elevations.

Alopecurus aequalis is the most widespread and variable species of *Alopecurus* in the *Flora* region. Despite its variability, the only phenotype meriting formal recognition is that found in the low marshes of Marin and Sonoma counties, California. Some high elevation plants of the Sierra Nevada have unusually long awns (exserted by up to 1.5 mm) and anthocyanic spikelets but, like the semi-aquatic ecotype *A. aequalis* var. *natans* (Wahlenb.) Fernald, they do not warrant taxonomic recognition.

Alopecurus ×*haussknechtianus* Asch. & Graebn. is a hybrid between *A. aequalis* and *A. geniculatus*, which occurs fairly frequently in areas of sympatry, particularly in drier midcontinental areas from Alberta to Saskatchewan, south to Arizona and New Mexico. The hybrids are sterile and apparently have $2n = 14$.

1　Panicles 3–6 mm wide; glumes 1.8–3 mm long; awns not exceeding the lemmas or exceeding them by less than 1 mm; anthers 0.5–0.9 mm long . . . . var. *aequalis*
1　Panicles 4–9 mm wide; glumes to 3.7 mm long; awns exceeding the lemmas by 1–2.5 mm; anthers 1–1.2 mm long . . . . var. *sonomensis*

Alopecurus aequalis Sobol. var. **aequalis** [p. 785]
SHORTAWN FOXTAIL, VULPIN À COURTES ARÊTES

Culms 9–75 cm. **Blades** 1–5(8) mm wide. **Panicles** 1–9 mm long, 3–6 mm wide. **Spikelets** usually not purplish-tinged; **glumes** 1.8–3 mm; **awns** not exceeding the lemmas or exceeding them by less than 1 mm; **anthers** 0.5–0.9 mm.

Alopecurus aequalis var. *aequalis* is the widespread variety in the *Flora* region. It is listed as threatened in the state of Connecticut.

Alopecurus aequalis var. **sonomensis** P. Rubtzov [p. 785]
SONOMA SHORTAWN FOXTAIL

Culms 30–75 cm. **Blades** to 7.5 mm wide. **Panicles** 2.5–9 cm long, 4–9 mm wide. **Spikelets** violet-gray to purplish-tinged, especially towards the apices; **glumes** to 3.7 mm; **awns** exceeding the lemmas by 1–2.5 mm; **anthers** 1–1.2 mm.

Alopecurus aequalis var. *sonomensis* is endemic to Marin and Sonoma counties, California, where it grows

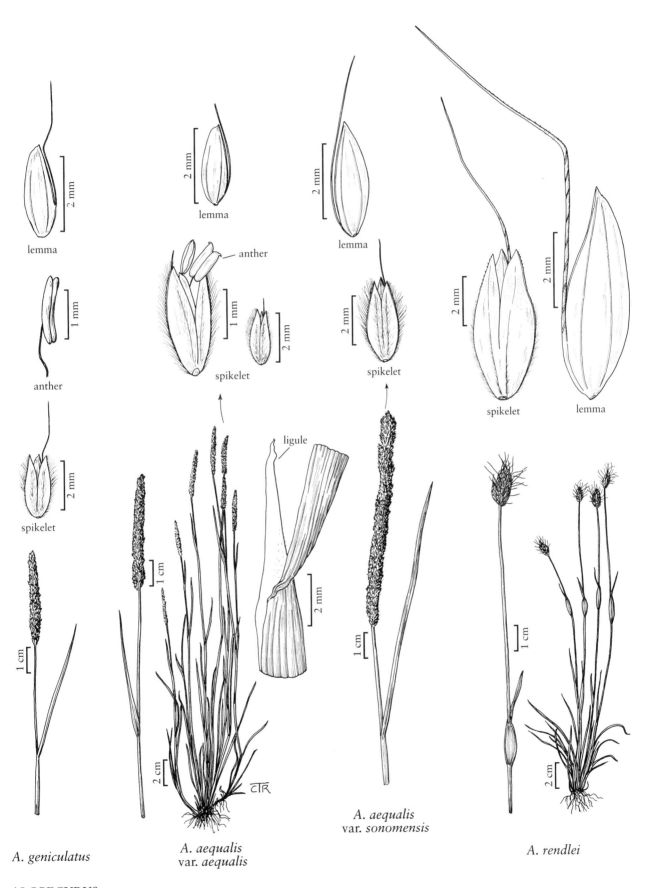

lemma

2 mm

lemma

anther

1 mm

anther

spikelet

2 mm

spikelet

lemma

2 mm

lemma

anther

spikelet

1 mm

2 mm

spikelet

lemma

2 mm

ligule

2 mm

spikelet

2 mm

spikelet

lemma

2 mm

1 cm

1 cm

1 cm

1 cm

2 cm

CTR

2 mm

1 cm

2 cm

A. aequalis
var. *sonomensis*

A. geniculatus

A. aequalis
var. *aequalis*

A. rendlei

ALOPECURUS

in shallow water and marshy or moist ground, usually in the open. It is listed as endangered by the U.S. Fish and Wildlife Service.

6. Alopecurus rendlei Eig [p. 785]
RENDLE'S MEADOW FOXTAIL

Plants annual; tufted. **Culms** 8–30(40) cm, erect or decumbent. **Ligules** 0.6–3 mm, obtuse; **blades** 1–16 cm long, 1–3 mm wide; **upper sheaths** inflated. **Panicles** 1–3.2 cm long, 5–11 mm wide. **Glumes** 5–6.4 mm, connate in the lower ¹/₄–¹/₃, dilated and coriaceous in the lower ¹/₂, pilose or glabrous in the lower ¹/₂, indurate and constricted above the middle, keels not winged, ciliate on all veins, apices acute, slightly divergent; **lemmas** 5.6–6.4 mm, connate basally, glabrous or puberulent distally, apices coriaceous, acuminate, awns 6.5–18 mm, geniculate, exceeding the lemmas by 0.8–7.2 mm; **anthers** about 3 mm. **Caryposes** 3.5–4.5 mm. $2n = 14$.

Alopecurus rendlei is native to wet meadows, and adventive in roadsides and waste places, in southern and western Europe. It was found growing on ballast in Philadelphia, Pennsylvania in 1880; it has not been collected in North America since then.

7. Alopecurus saccatus Vasey [p. 787]
PACIFIC MEADOW FOXTAIL

Plants annual; tufted. **Culms** 12–45 cm, erect or decumbent. **Ligules** 1.5–5.5 mm, obtuse; **blades** 4–12 cm long, 1.2–4 mm wide; **upper sheaths** conspicuously inflated. **Panicles** 1.5–6.5 cm long, 5.5–13 mm wide, often dense. **Glumes** 3–5 mm, connate at the base, not dilated, membranous, pubescent, keels not winged, veins ciliate, apices obtuse; **lemmas** 3–5 mm, connate in the lower ¹/₃–¹/₂, glabrous, apices obtuse, awns 6–10 mm, geniculate, exceeding the lemmas by 3–6 mm; **anthers** 0.7–1.8 mm, yellow to rusty brown. **Caryopses** 1.5–2 mm. $2n =$ unknown.

Alopecurus saccatus is a native annual that inhabits moist, open meadows, valley plains, and vernal pools, at elevations below 700 m, from Washington to California. Segregates have been treated as species in the past, but the variation between them appears to be continuous, and no habitat differentiation is evident.

8. Alopecurus carolinianus Walter [p. 787]
TUFTED FOXTAIL

Plants annual; tufted. **Culms** 5–50 cm, erect or decumbent. **Ligules** 2.8–4.5 mm, obtuse; **blades** 3–15 cm long, 0.9–3 mm wide; **upper sheaths** not or only slightly inflated. **Panicles** 1–7 cm long, 3–6 mm wide, always dense. **Glumes** 2.1–3.1 mm, connate at the base, membranous throughout, sparsely pubescent, not dilated below, keels not winged, ciliate, apices obtuse, pale green to pale yellow; **lemmas** 1.9–2.7 mm, connate in the lower ¹/₂, glabrous, apices obtuse, awns 3–6.5 mm, geniculate, exceeding the lemmas by 1.6–4 mm; **anthers** 0.3–1 mm, yellow or orange. **Caryopses** 1–1.5 mm. $2n = 14$.

Alopecurus carolinianus is native to the central plains, Mississippi valley, and southeastern United States, where it is common in wet meadows, ditches, wetland edges, and other moist, open habitats; it is occasionally a weed of rice fields. At the northern limit of its range it is clearly adventive, growing in gardens and nurseries. It also occurs in arid areas of the prairies and southwest, growing sporadically along sloughs and in ditches and vernal pools. Whether such populations are native or naturalized is not clear.

9. Alopecurus myosuroides Huds. [p. 787]
BLACKGRASS, SLENDER MEADOW FOXTAIL

Plants annual; tufted. **Culms** (10)40–85 cm, erect. **Ligules** 2–6 mm, obtuse; **blades** (2) 3.5–6 mm wide; **upper sheaths** somewhat inflated. **Panicles** 4–12 cm long, 3–7 mm wide. **Glumes** 4.5–7.5 mm, connate in the lower ¹/₂, coriaceous, sides glabrous, keels winged, ciliate, scabrous distally, lateral veins ciliate or glabrous proximally, apices acute, convergent to parallel; **lemmas** 4–7 mm, connate in the lower ¹/₃–¹/₂, glabrous, apices acute, awns to 12 mm, geniculate, exceeding the lemmas by 3–6 mm; **anthers** 2.4–4.1 mm, yellow. $2n = 14, 28$.

Alopecurus myosuroides is native to Eurasia, and grows in moist meadows, deciduous forests, and cultivated or disturbed ground. A significant weed species in temperate cereal crops, it is one of the most damaging weeds of winter cereals in England. It has been introduced repeatedly as a weed of cultivation into many parts of the *Flora* region, but apparently has not spread to a large degree outside of cultivation. *Alopecurus myosuroides* has been listed as a noxious weed in the state of Washington, one of the states where winter wheat is a major crop.

A. saccatus

A. carolinianus

A. creticus

A. myosuroides

ALOPECURUS

10. Alopecurus creticus Trin. [p. 787]

CRETAN MEADOW FOXTAIL

Plants annual; tufted. Culms 7–21(40) cm, erect. Ligules 1.4–2.3 mm, obtuse; blades 1–8 cm long, 1–3.5 mm wide; upper sheaths usually inflated. Panicles 2–4.3 cm long, 4–5.5 mm wide, dense. Glumes 3–4.5 mm, connate in the lower ¹/₂–⁴/₅, coriaceous, glabrous or pubescent over the veins, keels winged, ciliate, apices obtuse, mucronate, divergent; lemmas 2.5–3.5 mm, connate at the base, glabrous, apices truncate, awns 3–6 mm, geniculate, exceeding the lemmas by 0–4 mm; anthers 1.2–2.5 mm. $2n$ = unknown.

Alopecurus creticus is native to marshes and wet places in the southern part of the Balkan Peninsula. It was discovered on ballast dumps in Philadelphia, Pennsylvania in the nineteenth century; it has not persisted in the *Flora* region.

14.67 APERA Adans.

Kelly W. Allred

Plants annual; tufted or the culms solitary. Culms 5–120 cm, erect or geniculate, glabrous. Leaves mostly cauline; sheaths open, rounded to slightly keeled; collars glabrous, midveins continuous; auricles absent; ligules membranous, often lacerate to erose; blades flat or weakly involute, glabrous. Inflorescences terminal panicles; branches strongly ascending to divergent. Spikelets pedicellate, slightly laterally compressed, with 1 floret, rarely more, distal florets, if present, vestigial; rachillas prolonged beyond the base of the floret as a bristle, rarely terminating in a vestigial floret; disarticulation above the glumes, beneath the floret. Glumes unequal, lanceolate, scabrous on the distal ¹/₂, unawned; lower glumes 1-veined; upper glumes slightly shorter than to slightly longer than the florets, 3-veined; calluses blunt, glabrous or sparsely hairy; lemmas firmer than the glumes, folded to nearly terete, obscurely 5-veined, marginal veins not excurrent, apices entire, awned, awns subterminal; paleas about ³/₄ as long as to equaling the lemmas, hyaline, 2-veined; lodicules 2, free, glabrous, usually toothed; anthers 3; ovaries glabrous. Caryopses shorter than the lemmas, concealed at maturity, 1.2–2 mm, ellipsoidal, slightly sulcate; hila broadly ovate, ¹/₅ the length of the caryopses. $x = 7$. Name of uncertain origin, possibly from the Greek *a*, 'not', and *peros*, 'maimed'; Adanson provided no explanation.

Apera is genus of three species, native to Europe and western Asia. It is similar to *Agrostis*. It differs in its firm lemmas; paleas that are always present and equal to the lemma, or nearly so; and prolonged rachillas. In North America, two species have been introduced, growing as weeds in lawns and disturbed ground, and in grain fields.

SELECTED REFERENCE Warwick, S.I., B.K. Thompson, and L.D. Black. 1987. Genetic variation in Canadian and European populations of the colonizing weed species *Apera spica-venti*. New Phytol. 106:301–317.

1. Anthers 0.3–0.5 mm long; panicles contracted, 0.4–3 cm wide; most branches spikelet-bearing to within 2 mm of the base . . . . 1. *A. interrupta*
1. Anthers 1–2 mm long; panicles pyramidal, 2–15 cm wide; branches naked at the base for 5 mm or more . . . . 2. *A. spica-venti*

1. Apera interrupta (L.) P. Beauv. [p. 789]

INTERRUPTED WINDGRASS

Culms (5)10–50(75) cm, weak, slender, solitary or with several shoots, sometimes sparingly branched above the base; **internodes** usually longer than the sheaths. **Sheaths** often purplish; **ligules** 1.5–5 mm, acute to truncate, erose, margins decurrent; **blades** usually 4–12 cm long, 0.3–4 mm wide, flat or convolute when dry. **Panicles** 3–15(20) cm long, 0.4–1.5(3) cm wide, contracted, somewhat interrupted below; **branches** erect to ascending, most spikelet-bearing to within 2 mm of the base; **pedicels** 0.5–2 mm. **Spikelets** 2–2.8 mm, green or purplish; **rachillas** prolonged 0.2–0.6 mm. **Lower glumes** 1–2.2 mm; **upper glumes** 2–2.5(2.8) mm; **lemmas** 1.5–2.5 mm, slightly involute, awned, awns 4–10(16) mm; **anthers** 0.3–0.5 mm, often purplish brown. **Caryopses** 1–1.5 mm. $2n = 14, 28$.

Apera interrupta grows as a weed in lawns, grain fields (especially winter wheat), sandy open ground, and roadsides. Introduced from Europe, it now grows from British Columbia south to Arizona and New Mexico, as well as in Ontario and at a few scattered locations in the eastern part of the *Flora* region.

2. Apera spica-venti (L.) P. Beauv. [p. 789]

COMMON WINDGRASS, LOOSE SILKYBENT

Culms 20–80(120) cm, stout, usually with several shoots, sparingly branched; **internodes** shorter or longer than the sheaths. **Sheaths** often purplish; **ligules** 3–6(12) mm; **blades** usually 6–16(25) cm long, 2–5(10) mm wide, flat. **Panicles** (5)10–35 cm long, (2)3–15 cm wide, usually open, pyramidal; **branches** spreading, naked at the base for 5+ mm, with spikelets usually borne towards the distal ends; **pedicels** 1–3 mm. **Spikelets** 2.4–3.2 mm, often purplish; **rachillas** prolonged for about 0.5 mm. **Lower glumes** 1.5–2.5 mm; **upper glumes** 2.4–3.2 mm; **lemmas** 1.6–3 mm, folded, scabridulous above midlength, awned, awns 5–12 mm; **anthers** 1–2 mm, greenish to yellowish, often purple-tinged. **Caryopses** 1–1.5 mm. $2n = 14$.

Apera spica-venti grows as a weed in lawns, waste places, grain fields, sandy ground, and roadsides. Introduced from Europe, it is found in scattered locations in the *Flora* region.

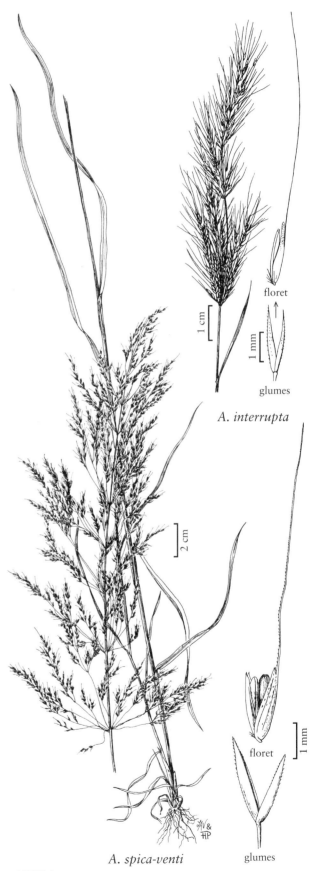

floret

glumes

A. interrupta

floret

glumes

A. spica-venti

APERA

Volume 25: Additions, Corrections, and Comments

The Editors

These pages present the substantive additions, corrections, and comments that we received after publication of volume 25 in 2003. The numbers given in parentheses refer to volume 25 and the page on which a particular item is found. Most of the changes affect the text, but two of the illustration plates (25: 223 and 25: 587) and three maps [*Muhlenbergia tenuifolia* (25: 160), *Setaria texana* (25: 546), and *Schizachyrium scoparium* (25: 669)] are incorrect. Files with the correct illustrations and maps can be downloaded from http://herbarium.usu.edu/grass_errors.htm. More current versions of the maps are available at http://herbarium.usu.edu/webmanual/. To save space, the additional references for volume 25 are listed alphabetically in the paragraph below. No illustrations have been prepared for the taxa added after publication of volume 25.

SELECTED REFERENCES **Aliscioni, S.S., L.M. Giussani, F.O. Zuloaga,** and **E.A. Kellogg.** 2003. A molecular phylogeny of *Panicum* (Poaceae: Paniceae): Tests of monophyly and phylogenetic placement within the Panicoideae. Amer. J. Bot. 90:746–821; **Anderson, H.M.** 1991. *Melinis* P. Beauv. Pp. 210–213 *in* G.E. Gibbs Russell, L. Watson, M. Koekemoer, L. Smook, N.P. Barker, H.M. Anderson, and M.J. Dallwitz. Grasses of Southern Africa (ed. O.A. Leistner). National Botanic Gardens, Botanical Research Institute, Pretoria, Republic of South Africa. 437 pp.; **Baird, J.R.** and **J.W. Thieret.** 1985. Notes on *Themeda quadrivalvis* (Poaceae) in Louisiana. Isleya 2:129–137; **Brummit, R.K.** 1998. Report of the Committee for Spermatophyta: 47. Taxon 47:869–870; **Catling, P.M., G. Mitrow, L. Black,** and **S. Caribyn.** 2004. Status of the alien race of common reed (*Phragmites australis*) in the Canadian maritime provinces. BEN (Botanical Electronic News) #324. http://www.ou.edu/cas/botany-micro/ben/; **Gerish, W.** 1979. Chromosomal analysis of a previously unidentified *Spartina* species. Master's thesis. Long Island University, Brookville, New York, U.S.A. [cited by Spicher and Josselyn]; **Giussani, L.M., H. Cota-Sanchez, F.O. Zuloaga,** and **E.A. Kellogg.** 2001. A molecular phylogeny of the subfamily Panicoideae (Poaceae) shows multiple origins of C4 photosynthesis. Amer. J. Bot. 88:1993-2012; **Gómez-Martínez, R.** and **A. Culham.** 2000. Phylogeny of the subfamily Panicoideae with emphasis on the tribe Paniceae: Evidence from the *trn*L-F cpDNA region. Pp. 136-140 *in* Grasses: Systematics and Evolution (S.W.L. Jacobs and J. Everett, eds.). CSIRO Publishing, Collingwood, Victoria, Australia. 408 pp.; **Gould, F.W.** 1958. Chromosome numbers in southwestern grasses. Amer. J. Bot. 45:757-767; **Marchant, C.J.** 1963. Corrected chromosome numbers for *Spartina* ×*townsendii* and its parent species. Nature 199(4896):929; **Marchant, C.J.** 1967. Evolution in *Spartina* (Gramineae): I. The history and morphology of the genus in Britain. J. Linn. Soc., Bot. 60:1–24; **Marchant, C.J.** 1968a. Evolution in *Spartina* (Gramineae): II. Chromosomes, basic relationships, and the problem of *S.* ×*townsendii* agg. J. Linn. Soc., Bot. 60:381–409; **Marchant, C.J.** 1968b. Evolution in *Spartina* (Gramineae): III. Species chromosome numbers and their taxonomic significance. J. Linn. Soc., Bot. 60:411–417; **Normile, D.** 2004. Expanding trade with China creates ecological backlash. Science 306:968–969; **McNamara, J.** and **J.A. Quinn.** 1977. Resource allocation and reproduction in populations of *Amphicarpum purshii* (Gramineae). Amer. J. Bot. 64:17–23; **Pfeiffer, L.G.** 1872–1873. Nomenclator Botanicus, vol. 1. Theodor Fisher, Kassel, Germany; **Quinn, J.A.** 1975. Variability among *Danthonia sericea* (Gramineae) populations in responses to substrate moisture levels. Amer. J. Bot. 62:884–891; **Quinn, J.A., J. Rotsettis,** and **D.E. Fairbrothers.** 1972. Inflorescence characters and reproductive proficiency in *Danthonia sericea* populations. Amer. J. Bot. 59:942–951; **Reeder, J.R.** 1977. Chromosome numbers in western grasses. Amer. J. Bot. 64:102–110; **Reeder, J.R.** 1984. *Poaceae.* Pp. 102–103 *in* Á. Löve (ed.). Chromosome number reports LXXXII. Taxon 33:126–134; **Reeder, J.R.** 1994. *Setaria villosissima* (Gramineae) in Arizona: Fact or fiction. Phytologia 77:452–455; **Reeder J.R.** and **D.N. Singh.** 1967. Validity of the tribe Spartineae (*Gramineae*) [Abstract]. Amer. J. Bot. 54:656; **Rotsettis, J., J.A. Quinn,** and **D.E. Fairbrothers.** 1972. Growth and flowering of *Danthonia sericea* populations. Ecology 53: 227–234; **Saltonstall, K.** 2002. Cryptic invasion by a non-native genotype of the common reed *Phragmites australis,* into North America. Proc. Natl. Acad. Sci. U.S.A. 99:2445-2449; **Saltonstall, K., P.M. Peterson,** and **R. J. Soreng.** 2004. Recognition of *Phragmites autralis* subsp. *americanus* (Poaceae: Arundinoideae) in North America: Evidence from morphological and genetic analyses. Sida 21:683–692; **Simon, B.K.** and **S.W.L. Jacobs.** 2003. *Megathyrsus,* a new generic name for *Panicum* subgenus *Megathyrsus*. Austrobaileya 6:571–574; **Spicher, D.** and **M. Josselyn.** 1985. *Spartina* (Gramineae) in northern California: Distribution and taxonomic notes. Madroño 32: 158–167; **Turner, B.L.** 2004. *Sporobolus coahuilensis* (Poaceae): A new record for the U.S.A. from Trans-Pecos, Texas. Sida 21:455–457; **Veldkamp, J.F.** 2004. Miscellaneous notes on mainly southeast Asian Gramineae. Reinwardtia 12:135–140; **Zuloaga, F.O., O. Morrone,** and **L.M. Giussani.** 2000. A cladistic analysis of the Paniceae: A preliminary approach. Pp. 123–135 *in* S.W.L. Jacobs and J. Everett (eds.). Grasses: Systematics and Evolution. CSIRO Publishing, Collingwood, Victoria, Australia. 408 pp.

Phragmites Adans. (25: 10)

The ligules of *Phragmites australis* are composed of hairs and are not membranous, and the chromosome numbers listed for the species should be modified to include $2n = 49$–54 and 120. Probably of greater interest is the discovery that *P. australis* is represented in the *Flora* region by at least three ecologically and genetically distinct lineages (Catling et al. 2004; Salstonstall 2002; Saltonstall et al. 2004). Saltonstall et al. (2004) formally recognized three taxa: **Phragmites australis** (Cav.) Trin. *ex* Steud., **Phragmites australis** subsp. **americanus** Saltonstall, P.M. Peterson & Soreng, and **Phragmites australis** var. **berlandieri** (E. Fourn.) C.F. Reed. They did not state how many subspecies they would recognize in *P. australis*, nor whether *P. australis* var. *berlandieri* belongs in subsp. *americanus*. The key below is based on their treatment.

1. Ligules 1–1.7 mm long; lower glumes 3–6.5 mm long; upper glumes 5.5–11 mm long; lemmas 8–13.5 mm long; leaf sheaths deciduous with age; culms exposed, smooth, shiny; plants rarely forming a monoculture . . . . subsp. *americanus* (native lineage)
1. Ligules 0.4–0.9 mm long; lower glumes 2.5–5 mm long; upper glumes 4.5–7.5 mm long; lemmas 7.5–12 mm long; leaf sheaths persistent; culms not exposed, smooth and shiny or ridged and dull; plants often forming a monoculture.
   2. Culms smooth and shiny; plants of southern California, Arizona, New Mexico, and Texas to Florida, extending through Mexico and Central America . . . . var. *berlandieri* (Gulf Coast lineage)
   2. Culms ridged, not shiny; from southern British Columbia and Quebec and throughout the contiguous United States . . . . *P. australis* (introduced lineage)

Monanthochloë Engelm. (25: 28)

The generic description should read "**Staminate spikelets** similar ... but smaller and the lemmas and paleas thinner."

Sporobolus R. Br. (25: 115)

An additional species of *Sporobolus* has been found in the region, *S. coahuilensis*. P.M. Peterson provided the following advice on how to modify the treatment in volume 25 to allow for its inclusion:

2. Lower panicle nodes with 7–20 branches.
   NEW. Pedicels 0.1–0.5(1) mm long, appressed . . . . 2. *S. pyramidatus*
   NEW. Pedicels (2)3–6(8) mm long, widely spreading . . . . *S. coahuilensis*
2. Lower panicle nodes with 1–3 branches.

Sporobolus coahuilensis Valdés-Reyna

Plants annual. **Culms** 15–60 cm, ascending, glabrous. **Sheaths** shorter than the internodes, glabrous; **ligules** 0.5–1 mm, ciliate; **blades** 4–12 cm long, 1.5–6 mm wide, flat, spreading, evenly distributed, adaxial surfaces sparsely ciliate-pustulate. **Panicles** 6–22 cm long, (1)5–13 cm wide, open, sometimes contracted; **branches** terminating in a spikelet, lowest branches whorled, in verticels of 7–20; **pedicels** (2)3–6(8) mm, widely spreading, capillary. **Spikelets** 1.1–1.5 mm. **Glumes** thin, acute; **lower glumes** about 0.5 mm; **upper glumes** 1.1–1.5 mm; **lemmas** 1.1–1.4 mm, acute; **paleas** 1–1.3 mm, hyaline. **Fruits** 0.6–0.9 mm, oblong, light brown; **embryos** 0.2–0.4 mm.

Sporobolus coahuilensis is primarily known from central Coahuila in Mexico. It has recently been found in Brewster and Hudspeth counties, Texas (Turner 2004). It is not clear whether it has been overlooked in the past or is a recent introduction. *Sporobolus coahuilensis* appears to be closely related to the widespread species *S. pyramidatus*, from which it differs in its long capillary pedicels and usually wider panicles.

Eustachys Desv. (25: 218)

The plate should be replaced with the one available at http://herbarium.usu.edu/grass_errors.htm.

Spartina Schreb. (25: 240)

J.R. Reeder (pers. com. 2003) noted that some of the information in Mobberly (1956) has not been supported by subsequent work. In particular, Reeder and Singh (1967) reported that lodicules are present in at least three species (*Spartina patens, S. pectinata,* and *S. spartinae*), and several taxonomists have obtained different chromosome counts. The generic description should be modified to read "**lodicules** sometimes present, truncate, vascularized". Most of the chromosome counts cited, which came from Mobberly (1956), were multiples of seven, but those obtained by other workers have all been multiples of 10. In the summary below, Mobberly's counts are listed in [square brackets]; those from other works are unbracketed. Sources of the other counts are included in the selected references.

Spartina alterniflora (25:244): 2*n* = 62 [56, 70]; *S. bakeri* (25:247): 2*n* = 40 [42]; *S.* ×*caespitosa* (25:249): 2*n* = 60, 60+2; *S. cynosuroides* (25:247): 2*n* = 40 [28, 42]; *S. densiflora* (25:247): 2*n* = 60 [this chromosome count was obtained by Gerish (1979), who reported it for *S. foliosa*, but Spicher and Josselyn (1985) demonstrated that the plants he worked with were almost certainly *S. densiflora*, a species that hitherto had been misidentified as the native *S. foliosa*]; *S. foliosa* (25:244): 2*n* = 60 [56]; *S. gracilis* (25:249): 2*n* = 40 [42]; *S. maritima* (25:246): 2*n* = 60 [56]; *S. patens* (25:249): 2*n* = [42]; *S. pectinata* (25:250): 2*n* = 40, 40+1, 80 [42, 70, 84]; *S. spartinae* (25:243): 2*n* = 40 [28, 42]; *S.* ×*townsendii* (25:246): 2*n* = 60, 60+2.

Spartina alterniflora is now a major weed problem in southeastern China (Normile 2004).

Cathestecum J. Presl (25: 272)

The treatment was prepared by Mary E. Barkworth.

Danthonia DC. (25: 301)

J.A. Quinn commented that the citation of one of his papers in arguing for the inclusion of *Danthonia epilis* in *D. sericea* was not appropriate. Additional papers are included in the references above. Further information can be found at http://herbarium.usu.edu/ webmanual/ (under the notes for *Danthonia*).

Rytidosperma Steud. (25: 309)

H.E. Connor identified two additional species of *Rytidosperma* among specimens that have been found as escapes in Alameda and San Mateo counties, California. They are **Rytidosperma caespitosa** (Gaudich.) Connor & Edgar, and **R. richardsonii** (Cashmore) Connor & Edgar. Both are native to Australia. *Rytidosperma caespitosa* differs from the three species included in volume 25 in having two rows of tufts of hair in which the upper row of hairs greatly exceeds the lemma body. Like *R. biannulare*, it has intravaginal branching. *Rytidosperma richardsonii* has lemma lobes that are shorter than the lemma body, and obovate paleas that are 2–2.5 mm wide.

PANICEAE R. Br. (25: 357)

In lead 25 in the key, change *Brachiaria* to **Moorochloa**. The reason is presented on p. 793.

Digitaria Haller (25: 358)

The description of *Digitaria filiformis* (25:364) should read "**primary branches** 20–25 cm, axes triquetrous, not winged, with spikelets in unequally pedicellate groups of 3(–5) on the basal ¹/₂ (J. Wipff, pers. comm.).

The illustration of *Digitaria longiflora* (25:371) was mistakenly based on a misidentified specimen of *D. fulvescens* (J. Presl) Henrard. The two species differ in little more than the pubescence of their lemmas, and might well be considered phases of a single species, in which case the name *D. longiflora* would have priority (J. Wipff, pers. comm.). The structure labeled "lower glume" in the illustration of *D. velutina* (25:379) is the lower lemma.

Amphicarpum Kunth (25: 387)

The generic comment should be modified to include the following (J.A. Quinn, pers. comm.): *Amphicarpum* differs from all other North American grass genera in its production of subterranean, cleistogamous spikelets. It has generally been reported that caryopses from the aerial spikelets fail to germinate, but McNamara and Quinn (1977) demonstrated that, at least in *A. amphicarpon*, this is not true. They found that the caryopses of aerial panicles and their seedlings were less robust than those of the subterranean spikelets. McNamara and Quinn concluded that the cleistogamous, subterranean spikelets usually contributed the largest number of plants to the populations, but that the chasmogamous aerial spikelets provided a potentially important source of genetic variability.

Echinochloa P. Beauv. (25: 398)

In volume 25, it was stated that the spelling of varietal epithet for one of the varieties of *Echinochloa crus-pavonis* should be "macra" rather than "macera". Kathryn Mauz, John and Charlotte Reeder, and Mark Dahmen have since argued convincingly against this decision. They noted that Wiegand spelled the epithet "macera" and described its spikelets as "awnless, soft-tipped" whereas "macra" would mean large or long. Nowhere does Wiegand suggest that the variety is longer or larger than *E. crus-pavonis* var. *pavonis*.

Dahmen, who is a Latin scholar, commented that, in classical Latin, the adjective macera means thin, lean, with a sense of having been attenuated. That leads to the English word macerate, one meaning of which is "chew up," that is, to soften as part of the digestive process. He continued, "So macer CAN imply soft. Why, however, other more obvious words for 'soft' were not used (mollis, elicata, tenera) is puzzling."

After reviewing the comments, K. Gandhi, nomenclatural editor, concluded that correct spelling of Wiegand's varietal epithet is "macera."

Dichanthelium (Hitchc. & Chase) Gould (25: 637)

The French vernacular name "Panic Laineux" should be associated with *Dichanthelium acuminatum* subsp. *fasciculatum* (25: 425), not subsp. *acuminatum* (also 25: 425).

Panicum L. (25: 450)

Add "Millet Commun" to the vernacular names for *Panicum miliaceum* (25: 456).

Brachiaria (Trin.) Griseb. (25: 488)

The name of this genus has to be changed from *Brachiaria* to **Moorochloa** Veldkamp, and the name of the species to **Moorochloa eruciformis** (Sm.) Veldkamp. Veldkamp (2004) pointed out that *B. eruciformis*, which was generally taken as the type of *Brachiaria*, was not included in Trinius' concept of *Panicum* sect. *Brachiaria* Trin., and so could not be taken as its type. Pfeiffer (1872–1873) lectotypified *Brachiaria* on *Panicum holosericum* R. Br., but this species is now included in *Urochloa*. Veldkamp recommended that *Brachiaria* be retained as the name of the few species left in the genus after most had been transferred to *Urochloa*, but the Nomenclatural Committee for Spermatophyta of the International Association for Plant Taxonomy rejected the suggestion, recommending that he publish a new name (Brummit 1998). He did so, reluctantly, in 2004. The name *Moorochloa* is derived from the Greek *mooros*, 'fool', and *chloa*, 'grass'; an expression of opinion concerning the decision not to permit conservation of the traditional, but invalid, typification which would have saved the name *Brachiaria* for these species.

Melinis P. Beauv. (25: 490)

A third species is now being sold in the *Flora* region as an ornamental, **Melinis nerviglumis** (Franch.) Zizka. The cultivar being marketed is 'Pink Crystal'. In the key on 25: 490, it will key out with *M. repens*. It differs from that species in having strongly overlapping leaf sheaths, rolled blades, and in usually being perennial. The description is based on Anderson (1991).

Melinis nerviglumis (Franch.) Zizka

Plants perennial; cespitose. **Culms** (25)40–120(150) cm. **Leaf sheaths** not strongly overlapping; **blades** (3)1–30(44) cm long, (1.3)2–3.5(4.5) mm wide, rolled. **Panicles** contracted. **Pedicels** scabrous, with hairs to 7 mm. **Spikelets** 3.2–5.7 mm long, 2 mm wide, often densely covered with hairs, hairs to 4 mm, white or pink to purple. **Glumes** separated by 0.3(0.6) mm; **lower glumes** about 0.5 mm, awns 1–2(3) mm. $2n = 36$.

In its native southern Africa, *Melinis nerviglumis* flowers from November to September [sic].

Urochloa P. Beauv. (25: 492)

Simon and Jacobs (2003) transferred *Urochloa maxima* (25: 507) to a new genus, **Megathyrsus**, as **Megathyrsus maximus** (Jacq.) B.K. Simon & S.W.L. Jacobs. They commented that papers by Gómez-Martínez and Culham (2000), Giussani et al. (2001), Zuloaga et al. (2000), and Aliscioni et al. (2003) all support inclusion of *U. maxima* in *Panicum* subg. *Megathyrsus* [= *Megathyrsus*] rather than *Urochloa*.

Pennisetum Rich. (25: 515)

A chromosome count of $2n = 36$ has been reported for *Pennisetum ciliare* (25: 525).

Setaria P. Beauv. (25: 539)

A.S. Hitchcock's (1951) report of *Setaria villosissima* (25: 548) from Arizona is based on misidentification of a specimen of *S. leucopila* (Reeder 1994).

Paspalum L. (25: 566)

A count of $2n = 36$ has been reported for *Paspalum almum* (25:575).

Reimarochloa Hitchc. (25: 599)

The treatment was prepared by Mary E. Barkworth.

Imperata Cirillo (25: 618)

The following addition should be made to the comment about *Imperata brevifolia* (25: 621): "In September 2003, G.F. Hrusa succeeded in persuading the California authorities that *Imperata brevifolia* should be taken off the state's noxious weed list. Endangered species status is now being sought."

Sorghum Moench (25: 626)

The authors of two of the subspecies of *Sorghum bicolor* were given incorrectly. The citations should be *S. bicolor* subsp. *arundinaceum* (Desv.) de Wet & J.R. Harlan *ex* Davidse (25: 628) and *S. bicolor* subsp. ×*drummondii* (Steud.) de Wet *ex* Davidse.

Themeda Forssk. (25: 682)

The comment under *Themeda quadrivalvis* (25: 684) should be modified to include "It is established in St. Landry and Iberia parishes, Louisiana, in addition to having escaped from cultivation in Florida (Baird and Thieret 1985)."

Geographic Bibliography

The Geographic Bibliography shows written sources of information and Web sites used in developing the geographic database from which the maps in volumes 24 and 25 were generated, as well as those placed on the Intermountain Herbarium's Grass Manual Web site (http://herbarium.usu.edu/webmanual/). Herbaria and individuals that provided files or information for the database are acknowledged individually in the Preface (see Data Contributors).

Book titles are listed in full; the names of journals are abbreviated according to Lawrence et al. (1968) and its supplement (Bridson and Smith 1991).

Aiken, S.G. and R.A. Buck. 2002. Aquatic leaves and regeneration of last year's straw in the arctic grass, *Arctophila fulva*. Canad. Field-Naturalist 116:81–86.

Aiken, S.G., L.L. Consaul, and M.J. Dallwitz. 1995 on. Grasses of the Canadian arctic archipelago: Descriptions, illustrations, identification, and information retrieval. http://www.mun.ca/biology/delta/arcticf/poa/index.htm.

Aiken, S.G., M.J. Dallwitz, C.L. McJannet, and L.L. Consaul. 1997. Fescue Grasses of North America: Interactive Identification and Information Retrieval. DELTA, CSIRO Division of Entomology, Canberra, Australia. CD-ROM. http://delta-intkey.com/festuca/.

Aiken, S.G., W.G. Dore, L.P. Lefkovitch, and K.C. Armstrong. 1989. *Calamagrostis epigejos* (Poaceae) in North America, especially Ontario. Canad. J. Bot. 67:3205–3218.

Alaska Exotic Plant Mapping Project [AKEPIC]. Viewed 2004. AKEPIC mapping project inventory field data. http://akweeds.uaa.alaska.edu/.

Albee, B.J., L.M. Shultz, and S. Goodrich. 1988. Atlas of the Vascular Plants of Utah. Utah Museum of Natural History Occasional Publication No. 7. Utah Museum of Natural History, Salt Lake City, Utah, U.S.A. 670 pp.

Alex, J.F. 1987. Quackgrass: Origin, distribution, description and taxonomy. Pp. 1–16 *in* Quackgrass Action Committee Workshop Technical Proceedings: March 10 and 11, Westin Hotel, Winnipeg, Manitoba (H.L. Glick, ed.). Monsanto Canada Ltd., Winnipeg, Manitoba, Canada.

Alford, M.H. 1999. The vascular flora of Amite County, Mississippi. Master's thesis, Duke University, Raleigh, North Carolina, U.S.A. 176 pp.

Alinot, S.F. 1973. The Vascular Flora of Glen Helen, Clifton Gorge, and John Bryan State Park. Ohio Biological Survey, Biological Notes No. 5. Ohio State University, Columbus, Ohio, U.S.A. 49 pp.

Allen, C.M., D.A. Newman, and H. Winters. 2004. Grasses of Louisiana, ed. 3. Allen's Nature Ventures, Pitkin, Louisiana, U.S.A. 374 pp.

Allen, C.M., C.S. Reid, and C.H. Doffitt. 1999. *Bouteloua rigidiseta* (Poaceae) new to Louisiana. Sida 18:1285.

Allen, L. and M.L. Curto. 1996. Noteworthy collections. Madroño 43:337–338.

Allred, K.W. 1997. A Field Guide to the Grasses of New Mexico, ed. 2. New Mexico Agricultural Experiment Station, Department of Agricultural Communications, New Mexico State University, Las Cruces, New Mexico, U.S.A. 258 pp.

Amoroso, J.L. and **W.S. Judd**. 1995. A floristic study of the Cedar Key Scrub State Reserve, Levy County, Florida. Castanea 60:210–232.

Anderson, D.E. 1961. Taxonomy and distribution of the genus *Phalaris*. Iowa State Coll. J. Sci. 36:1–96.

Anderson, L.C. 1995. Noteworthy plants from north Florida: VI. Sida 16:581–587.

Anderson, L.C. 2000. Noteworthy plants from north Florida: VII. Sida 19:211–216.

Anderson, L.C. and **D.W. Hall**. 1993. *Luziola bahiensis* (Poaceae): New to Florida. Sida 15:619–622.

Andreas, B.K. 1989. The Vascular Flora of the Glaciated Allegheny Plateau Region of Ohio. Bulletin of the Ohio Biological Survey, n.s., vol. 8, no. 1. College of Biological Sciences, Ohio State University, Columbus, Ohio, U.S.A. 191 pp.

Angelo, R. and **D.E. Boufford**. 1998. Atlas of the flora of New England: Poaceae. Rhodora 100:101–233. http://neatlas.huh.harvard.edu/.

Arizona Rare Plant Committee. 2001. Arizona Rare Plant Field Guide. Arizona Rare Plant Committee, Arizona, U.S.A. [Loose-leaf format, unpaginated.]

Associated Press. 2005. California botanists find rare grass species. [Associated Press wire article (31 May 2005) re. *Dissanthelium californicum*.]

Banks, D.L. and **S. Boyd**. 1998. Noteworthy collections. Madroño 45:85–86.

Barkley, T.M. (ed). 1977. Atlas of the Flora of the Great Plains. Iowa State University Press, Ames, Iowa, U.S.A. 600 pp.

Barrett, S.C.H. and **D.E. Seaman**. 1980. The weed flora of Californian rice fields. Aquatic Bot. 9:351–376.

Basinger, M.A. and **P. Rogertson**. 1996. Vascular flora and ecological survey of an old-growth forest remnant in the Ozark Hills of southern Illinois. Phytologia 80:352–367.

Bay, C. 1992. A phytogeographical study of the vascular plants of northern Greenland–north of 74° northern latitude. Meddel. Grønland, Biosci. 36:1–102.

Beatley, J.C. 1969. Vascular plants of the Nevada Test Site, Nellis Air Force Range, and Ash Meadows. University of California Laboratory of Nuclear Medicine and Radiation Biology, Los Angeles, California, U.S.A. 122 pp.

Beatley, J.C. 1970. Additions to vascular plants of the Nevada Test Site, Nellis Air Force Range, and Ash Meadows. University of California Laboratory of Nuclear Medicine and Radiation Biology, Los Angeles, California, U.S.A. 19 pp.

Beatley, J.C. 1971. Vascular plants of Ash Meadows, Nevada. University of California Laboratory of Nuclear Medicine and Radiation Biology, Los Angeles, California, U.S.A. 59 pp.

Beatley, J.C. 1973. Check list of vascular plants of the Nevada Test Site and central-southern Nevada. Department of Biological Sciences, University of Cincinnati, Cincinnati, Ohio, U.S.A. 42 pp.

Beck, K.A., **F.E. Caplow**, and **C.R. Bjork**. 2005. Noteworthy collections. Madroño 52:128–130.

Best, C., **J.T. Howell**, **W. Knight**, **I. Knight**, and **M. Wells**. 1996. A Flora of Sonoma County: Manual of the Flowering Plants and Ferns of Sonoma County, California. California Native Plant Society, Sacramento, California, U.S.A. 347 pp.

Best, K.F., **J.D. Banting**, and **G.G. Bowes**. 1978. The biology of Canadian weeds: 31. *Hordeum jubatum* L. Canad. J. Pl. Sci. 58:699–708.

Biek, D. 2002. Flora of Mount Rainier National Park. Oregon State University Press, Corvallis, Oregon, U.S.A. 506 pp.

Bittner, R.T. and **D.J. Gibson**. 1998. Microhabitat relations of the rare reed bent grass, *Calamagrostis porteri* subsp. *insperata* (Poaceae), with implications for its conservation. Ann. Missouri Bot. Gard. 85:69–80.

Blondeau, M. and **J. Cayouette**. 2002. La Flore Vasculaire de la Baie Wakeham et du Havre Douglas, Nord-du-Québec, Détroit d'Hudson. Provancheria No. 28. Herbier Louis-Marie, Québec City, Québec, Canada. 184 pp.

Böcher, T.W. 1938. Biological distributional types in the flora of Greenland: A study on the flora and plant-geography of South Greenland and East Greenland between Cape Farewell and Scoresby Sound [6. og 7. Thule-Expedition til sydøstgrønland 1931–1933. Leader: Knud Rasmussen]. Meddel. Grønland 106, nr. 2:1–339.

Böcher, T.W., **K. Holmen**, and **K. Jakobsen**. 1968. The Flora of Greenland, trans. T.T. Elkington and M.C. Lewis. P. Haase and Son, Copenhagen, Denmark. 312 pp.

Borowski, M. and **W.C. Holmes**. 1996. *Phyllostachys aurea* Riv. (Gramineae: Bambuseae) in Texas. Phytologia 80:30–34.

Bough, M., **J.C. Colosi**, and **P.B. Cavers**. 1985. The major weedy biotypes of proso millet (*Panicum miliaceum*) in Canada. Canad. J. Bot. 64:1188–1198.

Bowcutt, F. 1996. A floristic study of Delta Meadows River Park, Sacramento County, California. Madroño 43:417–431.

Boyd, S. 1998. Noteworthy collections. Madroño 45:326–328.

Boyle, W.S. 1945. A cyto taxonomic study of the North American species of *Melica*. Madroño 8:1–26.

Braun, E.L. 1967. The Monocotyledoneae: Cat-tails to Orchids. The Vascular Flora of Ohio, vol. 1. Ohio State University Press, Columbus, Ohio, U.S.A. 464 pp.

Brind'amour, M. and **V. Lavoie**. 1985. Addition à la flore vasculaire des marais intertidaux du Saint-Laurent (Québec): *Spartina ×caespitosa* A.A. Eaton. Naturaliste Canad. 112:431–432.

Brown, L.E. 1993. The deletion of *Sporobolus heterolepis* (Poaceae) from the Texas and Louisiana floras, and the addition of *Sporobolus silveanus* to the Oklahoma flora. Phytologia 74:371–381.

Brown, L.E. and **I.S. Elsik**. 2002. Notes on the flora of Texas with additions and other significant records: II. Sida 20:437–444.

Brown, L.E., B.R. MacRoberts, M.H. MacRoberts, P.A. Harcombe, W.W. Pruess, I.S. Elsik, and D. Johnson. 2005. Annotated checklist of the vascular flora of the Turkey Creek Unit of the Big Thicket National Preserve, Tyler and Hardin counties, Texas. Sida 21:1807–1827.

Brown, L.E., B.R. MacRoberts, M.H. MacRoberts, P.A. Harcombe, W.W. Pruess, I.S. Elsik, and S.D. Jones. 2006. Annotated checklist of the vascular flora of the Big Sandy Creek Unit, Big Thicket National Preserve, Texas. Sida 22:705-723.

Brown, L.E. and S.J. Marcus. 1998. Notes on the flora of Texas with additions and other significant records. Sida 18:315–324.

Brown, L.E. and J. Schultz. 1991. *Arthraxon hispidus* (Poaceae), new to Texas. Phytologia 71:379–381.

Bruner, J.L. 1987. Systematics of the *Schizachyrium scoparium* (Poaceae) complex in North America. Ph.D. dissertation, Ohio State University, Columbus, Ohio, U.S.A. 167 pp.

Bryson, C.T. 1993. *Sacciolepis indica* (Poaceae) new to Mississippi. Sida 15:555.

Calder, J.A. and R.L. Taylor. 1968. Flora of the Queen Charlotte Islands, part I: Systematics of the Vascular Plants. Canada Department of Agriculture Monograph No. 4, Part 1. Research Branch, Canada Department of Agriculture, Ottawa, Ontario, Canada. 659 pp.

CalFlora. Viewed 2000. Information on California plants for conservation, research, and education. http://www.calflora.org/.

Callihan, R.H. 1987. Eradication technology of matgrass (*Nardus strictus* L.) in the Clearwater National Forest. Department of Plant, Soil, and Entomological Sciences, University of Idaho, Moscow, Idaho, U.S.A. 51 pp.

Callihan, R.H. and T.W. Miller. 1998. Noxious weed identification guide. http://www.oneplan.org/Crop/noxWeeds/index.shtml.

Callihan, R.H. and D. Pavek. 1988. May, 1988 *Milium vernale* survey in Idaho County, Idaho. Department of Plant, Soil, and Entomological Sciences, University of Idaho, Moscow, Idaho, U.S.A. 6 pp.

Campbell, J.J.N. 1992. Atlas of the Kentucky Flora: First Release of Data on Ferns, Gymnosperms, and Monocots. Julian J.N. Campbell, Lexington, Kentucky, U.S.A. 296 pp.

Caplow, F.E. 2002. New species of *Spartina* in Washington. BEN [Botanical Electronic News] #286. http://www.ou.edu/cas/botany-micro/ben/.

Carlbom, C.G. 1967. A biosystematic study of some North American species of *Agrostis* L. and *Podagrostis* (Griseb.) Scribn. & Merr. Ph.D. dissertation, Oregon State University, Corvallis, Oregon, U.S.A. 223 pp.

Carter, J.L. 1962. The vascular flora of Cherokee County. Proc. Iowa Acad. Sci. 69:60–70.

Catling, P.M., D.S. Erskine, and R.B. MacLaren. 1985. The Plants of Prince Edward Island, with New Records, Nomenclatural Changes, and Corrections and Deletions. Research Branch, Agriculture Canada Publication No. 1798. Canadian Government Publishing Centre, Ottawa, Ontario, Canada. 272 pp.

Catling, P.M. and S.M. McKay. 1980. Halophytic plants in southern Ontario. Canad. Field-Naturalist 94:248–258.

Catling, P.M. and S.M. McKay. 1981. A review of the occurrence of halophytes in the eastern Great Lakes region. Michigan Bot. 20:167–179.

Catling, P.M., A.A. Reznicek, and J.L. Riley. 1977. Some new and interesting grass records from southern Ontario. Canad. Field-Naturalist 91:350–359.

Catling, P.M., A. Sinclair, and D. Cuddy. 2001. Vascular plants of a successional alvar burn 100 days after a severe fire and their mechanisms of re-establishment. Canad. Field-Naturalist 115:214–222.

Cayouette, J. 1987. La flore vasculaire de la région du Lac Chavigny (58° 12' N, 75° 08' O), Nouveau-Québec. Provancheria 20:1–51.

Cayouette, J. and S.J. Darbyshire. 1987. La répartition de *Danthonia intermedia* dans l'est du Canada. Naturaliste Canad. 114:217–220.

Cayouette, R. 1972. Additions à la flore adventice du Québec. Naturaliste Canad. 99:135–136.

Ceska, O. and A. Ceska. 2006. Lesser bluestem-- *Schizachyrium scoparium* (Poaceae)--in the Peace River area, British Columbia. [BEN (Botanical Electronic News) #366.] http://www.ou.edu/cas/botany-micro/ben/

Chambers, K.L. 1966. Notes on some grasses of the Pacific coast. Madroño 18:250–251.

Chester, E.W. 1996. Rare and noteworthy vascular plants from the Fort Campbell Military Reservation, Kentucky and Tennessee. Sida 17:269–274.

Chester, E.W., B.E. Wofford, H.R. DeSelm, and A.M. Evans. 1993. Atlas of Tennessee Vascular Plants. Pteridophytes, Gymnosperms, Angiosperms: Monocots, vol. 1. Austin Peay State University Miscellaneous Publication No. 9. The Center for Field Biology, Austin Peay State University, Clarksville, Tennessee, U.S.A. 118 pp. http://tenn.bio.utk.edu/vascular/vascular.html.

Chester, T.J. Viewed 2003. Flora of Torrey Pines State Reserve. http://tchester.org/plants/floras/coast/torrey_pines.html.

Clark, C. 1990. Vascular plants of the underdeveloped areas of California State Polytechnic University, Pomona. Crossosoma 16:1–7.

Clark, D.A. 1996. A Floristic Survey of the Mesa de Maya Region, Las Animas County, Colorado. Natural History Inventory of Colorado No. 17. University of Colorado Museum, Boulder, Colorado, U.S.A. 44 pp.

Clark, D.A. and T. Hogan. 2000. Noteworthy collections. Madroño 47:142–144.

Clay, K. 1995. Noteworthy collections. Castanea 60:84–85.

Clokey, I.W. 1951. Flora of the Charleston Mountains, Clark County, Nevada. Univ. Calif. Publ. Bot. 24:1–274.

Cochrane, T.S., M.M. Rice, and W.E. Rice. 1984. The flora of Rock County, Wisconsin: Supplement I. Michigan Bot. 23:121–133.

Cody, W.J. 1988. Plants of Riding Mountain National Park, Manitoba. Research Branch, Agriculture Canada Publication 1818/E. Canadian Government Publishing Centre, Ottawa, Ontario, Canada. 319 pp.

Cody, W.J. 1996. Flora of the Yukon Territory. NRC [National Resource Council of Canada], Ottawa, Ontario, Canada. 643 pp.

Cody, W.J., S.J. Darbyshire, and C.E. Kennedy. 1990. A bluegrass, *Poa pseudoabbreviata* Roshev., new to the flora of Canada, and some additional records from Alaska. Canad. Field-Naturalist 104:589–591.

Cody, W.J., C.E. Kennedy, and B. Bennett. 1998. New records of vascular plants in the Yukon Territory. Canad. Field-Naturalist 112:289–328.

Cody, W.J., C.E. Kennedy, and B. Bennett. 2000. New records of vascular plants in the Yukon Territory II. Canad. Field-Naturalist 114:417–443.

Cody, W.J., C.E. Kennedy, and B. Bennett. 2001. New records of vascular plants in the Yukon Territory III. Canad. Field-Naturalist 115:301–322.

Cody, W.J., K.L. MacInnes, J. Cayouette, and S.J. Darbyshire. 2000. Alien and invasive native vascular plants along the Norman Wells pipeline, District of Mackenzie, Northwest Territories. Canad. Field-Naturalist 114:126–137.

Consaul, L.L. and L.J. Gillespie. 2001. A re-evaluation of species limits in Canadian Arctic island *Puccinellia* (Poaceae): Resolving key characters. Canad. J. Bot. 79:927–956.

Consortium of California Herbaria. Viewed 2005. Database of specimens from eight California herbaria. http://ucjeps.berkeley.edu/chc_form.html.

Cope, E.A. 1994. Further notes on beachgrasses (*Ammophila*) in northeastern North America. Newslett. New York Fl. Assoc. 5:5–7.

Coxe, R., C.M. Morton, M.J. Haywood, B.L. Isaac, and J.A. Isaac. 2005. Checklist of the vascular plants of Greene County, Pennsylvania. Sida 21:1829–1859.

Crampton, B. 1955. *Scribneria* in California. Leafl. W. Bot. 7:219–220.

Crouch, V.E. and M.S. Golden. 1997. Floristics of a bottomland forest and adjacent uplands near the Tombigbee River, Choctaw County, Alabama. Castanea 62:219–238.

Curto, M.L. 1998. A new *Stipa* (Poaceae: Stipeae) from Idaho and Nevada. Madroño 45:57–63.

Curto, M.L. and L. Allen. 1992. California Native Plant Society grass walk: Cuyamaca Rancho State Park [16 pp. brochure].

Cusick, A.W., J.S. McCormac, and J. Beathard. 2001. Record number of rare plant species found in 2000. Nat. Ohio 23:1–2. [Spring 2001 newsletter of the Ohio Department of Natural Resources, Division of Natural Areas and Preserves.] http://www.dnr.ohio.gov/dnap/publications/default.htm.

Cusick, A.W., R.K. Rabeler, and M.J. Oldham. 2002. Hard or fairgrounds grass (*Sclerochloa dura*, Poaceae) in the Great Lakes region. Michigan Bot. 41:125–135.

Cusick, A.W. and G.M. Silberhorn. 1977. The Vascular Plants of Unglaciated Ohio. Bulletin of the Ohio Biological Survey, n.s., vol. 5, no. 4. College of Biological Sciences, Ohio State University, Columbus, Ohio, U.S.A. 157 pp.

Cusick, A.W. and M.A. Vincent. 2002. *Poa bulbosa* L. ssp. *bulbosa* (Poaceae) in North America. Michigan Bot. 41:19–23.

Cutler, H.C. and E. Anderson. 1941. A preliminary survey of the genus *Tripsacum*. Ann. Missouri Bot. Gard. 28:249–269.

Darbyshire, S.J. and J. Cayouette. 1989. The biology of Canadian weeds: 92. *Danthonia spicata* (L.) Beauv. in Roem. & Schult. Canad. J. Pl. Sci. 69:1217–1233.

Darbyshire, S.J. and J. Cayouette. 1995. Identification of the species in the *Panicum capillare* complex (Poaceae) from eastern Canada and adjacent New York State. Canad. J. Bot. 73:333–348.

Davis, J.I. 1990. *Puccinellia howellii* (Poaceae), a new species from California. Madroño 37:55–58.

Deam, C.C. 1940. Flora of Indiana. Department of Conservation, Division of Forestry, State of Indiana, Indianapolis, Indiana, U.S.A. 1236 pp.

Denny, G.C. 2003. *Muhlenbergia dubia* (Poaceae) in central Texas. Sida 20:1763–1764.

DeSelm, H.R. 1975. *Schizachyrium stoloniferum* Nash var. *wolfei* DeSelm. Sida 6:114–115.

Deshaye, J. and J. Cayouette. 1988. La flore vasculaire des îles et de la presqu'île de Manitounuk, Baie d'Hudson: Structure phytogéographique et interprétation bioclimatique. Provancheria 21:1–74.

Diamond, A.R., Jr. and J.D. Freeman. 1993. Vascular flora of Conecuh Co., Alabama. Sida 15:623–638.

Diamond, A.R., Jr., M. Woods, J.A. Hall, and B.H. Martin. 2002. The vascular flora of the Pike County Pocosin Nature Preserve. SouthE. Naturalist 1:45–54.

Diggs, G.M., Jr., B.L. Lipscomb, and R.H. O'Kennon. 1999. Shinners & Mahler's Illustrated Flora of North Central Texas. Sida Botanical Miscellany No. 16. Botanical Research Institute of Texas and Austin College, Fort Worth, Texas, U.S.A. 1626 pp.

Dignard, N. 2000. Additions récentes à la flore de l'île d'Anticosti. Ludoviciana 29:69–72.

Dix, W.L. 1945. Will the stowaway, *Molinia caerulea*, become naturalized? Bartonia 23:41–42.

Dore, W.G. and C.J. Marchant. 1968. Observations on the hybrid cord-grass, *Spartina* ×*caespitosa* in the maritime provinces. Canad. Field-Naturalist 82:181–184.

Dore, W.G. and J. McNeill. 1980. Grasses of Ontario. Research Branch, Agriculture Canada Monograph No. 26. Canadian Government Publishing Centre, Hull, Québec, Canada. 572 pp.

Dorn, R.D. 2001. Vascular Plants of Wyoming, ed. 3. Mountain West Publishing, Cheyenne, Wyoming, U.S.A. 412 pp.

Douglas, B.J., A.G. Thomas, I.N. Morrison, and M.G. Maw. 1985. The biology of Canadian weeds: 70. *Setaria viridis* (L.) Beauv. Canad. J. Pl. Sci. 65:669–690.

Douglas, G.W., D.V. Meidinger, and J.L. Penny. 2002. Rare Native Plants of British Columbia, ed. 2. Conservation Data Centre, Victoria, British Columbia, Canada. 358 pp.

Douglas, G.W., D.V. Meidinger, and J.J. Pojar. 2002. Illustrated Flora of British Columbia, vol. 8. Ministry of Sustainable Resource Management, Victoria, British Columbia, Canada. 457 pp.

Douglas, G.W., G.B. Straley, and D.V. Meidinger (eds). 1994. The Vascular Plants of British Columbia: Part 4—Monocotyledons. Research Branch, Ministry of Forests, Victoria, British Columbia, Canada. 257 pp.

Douglas, G.W. and R.J. Taylor. 1970. Contributions to the flora of Washington. Rhodora 72:496–501.

Doyon, D., C.J. Bouchard, and R. Néron. 1986. Répartition géographique et importance dans les cultures de quatre adventices du Québec: *Abutilon theophrasti, Amaranthus powellii, Acalypha rhomboïdea,* et *Panicum dichotomiflorum.* Naturaliste Canad. 113:115–123.

Doyon, D., C.J. Bouchard, and R. Néron. 1988. Extension de la répartition géographique de *Setaria faberii* au Québec. Naturaliste Canad. 115:125–129.

Doyon, D. and W.G. Dore. 1967. Notes on the distribution of two grasses, *Sporobolus neglectus* and *Leersia virginica,* in Québec. Canad. Field-Naturalist 81:30–32.

Drew, M.B., L.K. Kirkman, and A.K. Gholson, Jr. 1988. The vascular flora of Ichauway, Baker County, Georgia: A remnant longleaf pine/wiregrass ecosystem. Castanea 63:1–24.

Dubé, M. 1983. Addition de *Festuca gigantea* à la flore du Canada. Ludoviciana 14:213–215.

Dubé, M. 1986. La répartition de *Festuca pratensis* Hudson et de *F. arundinacea* Schreber (Poaceae) dans l'est du Canada. Naturaliste Canad. 113:325–330.

Easterly, N.W. 1951. The flora of Iowa County. Proc. Iowa Acad. Sci. 58:71–95.

Eddy, T.L. 1983. A vascular flora of Green Lake County, Wisconsin. Master's thesis, University of Wisconsin-Oshkosh, Oshkosh, Wisconsin, U.S.A. 130 pp.

Eddy, T.L. and N.A. Harriman. 1992. *Muhlenbergia richardsonis* in Wisconsin. Michigan Bot. 31:39–40.

Edgin, B., G.C. Tucker, and J.E. Ebinger. 2005. Vegetation and flora of American Beech Woods Nature Preserve, Clark County, Illinois. Sida 21:1861–1878.

Eilers, L.J. 1974. The flora of Brush Creek Canyon State Preserve. Proc. Iowa Acad. Sci. 81:150–157.

Eilers, L.J. and D.M. Roosa. 1994. The Vascular Plants of Iowa. University of Iowa Press, Iowa City, Iowa, U.S.A. 304 pp.

Elsley, J.E. and G. Halliday. 1971. Some plant records from southeast Greenland. Meddel. Grønland 178, nr. 8:1–15.

Enser, R.W. 2002. Rare native plants of Rhode Island. http://www.state.ri.us/dem/programs/bpoladm/plandev/heritage/pdf/plants.pdf/.

Erdman, K.S. 1965. Taxonomy of the genus *Sphenopholis.* Iowa State Coll. J. Sci. 39:259–336.

Ertter, B. 1997. Annotated Checklist of the East Bay Flora: Native and Naturalized Vascular Plants of Alameda and Contra Costa Counties, California. California Native Plant Society, East Bay Chapter, Berkeley, California, U.S.A. 114 pp.

Estes, D. 2005. The vascular flora of Giles County, Tennessee. Sida 21:2343–2388.

Estes, D. and J. Beck. 2005. *Sporobolus heterolepis* (Poaceae) new to Tennessee. Sida 21:1923–1926.

Estes, D. and J.L. Walck. 2005. The vascular flora of Rattlesnake Falls: A potential state natural area on the Western Highland Rim Escarpment in Tennessee. Sida 21:1753–1780.

False-Brome Working Group. 2006. False-Brome Working Group Home Page. [*Brachypodium sylvaticum.*] http://www.appliedeco.org/FBWG.htm

Fassett, N.C. 1951. Grasses of Wisconsin: The Taxonomy, Ecology, and Distribution of the Gramineae Growing in the State Without Cultivation. University of Wisconsin Press, Madison, Wisconsin, U.S.A. 173 pp.

Fay, M.J. 1951. The flora of Cedar County, Iowa. Proc. Iowa Acad. Sci. 58:107–131.

Fay, M.J. and R.F. Thorne. 1953. Additions to the flora of Cedar County, Iowa. Proc. Iowa Acad. Sci. 60:122–130.

Feilberg, J. 1984. A phytogeographical study of south Greenland: Vascular plants. Meddel. Grønland, Biosci. 15:1–70.

Felger, R., T.L. Burgess, S. Dorsi, J.R. Reeder, and T.R. Van Devender. 2005. *Dichanthium* (Poaceae) new to Arizona: Open door for a potentially invasive species. Sida 21:1905–1908.

Ferguson, E. and R.P. Wunderlin. 2006. A vascular plant inventory of Starkey Wilderness Preserve, Pasco County, Florida. Sida 22:635-659.

Ferlatte, W.J. 1974. A Flora of the Trinity Alps of Northern California. University of California Press, Berkeley, California, U.S.A. 206 pp.

Fernald, M.L. 1950. The North American variety of *Milium effusum.* Rhodora 52:218–222.

Fertig, W. 1999. Additions to the Flora of Wyoming-VI. Castilleja 18:3.

Fishbein, M. 1995. Noteworthy collections. Madroño 42:83.

Fleming, G.P. and C. Ludwig. 1996. Noteworthy collections. Castanea 61:89–95.

Fleming, K.M., J.R. Singhurst, and W.C. Holmes. 2002. Vascular flora of Big Lake Bottom Wildlife Management Area, Anderson County, Texas. Sida 20:355–372.

Fleming, P. and R. Kanal. 1995. Annotated checklist of vascular plants of Rock Creek Park, National Park Service, Washington, D.C. Castanea 60:283–316.

Fleurbec. 1985. Plantes Sauvages du Bord de la Mer: Guide d'Identification Fleurbec. Groupe Fleurbec, Saint-Augustin, Québec, Canada. 286 pp.

Flora of Texas Consortium. Viewed 2000. Herbarium Specimen Browser. http://www.csdl.tamu.edu/FLORA/tracy/main1.html.

Fox, W.E., III and S.L. Hatch. 1996. *Brachiaria eruciformis* and *Urochloa brizantha* (Poaceae: Paniceae) new to Texas. Sida 17:287–288.

Franklin, J.F. and C. Wiberg. 1979. Goat Marsh Research Natural Area. Federal Research Natural Areas in Oregon and Washington: A Guidebook for Scientists and Educators, Supplement No. 10. U.S. Department of Agriculture-Forest Service, Pacific Northwest Forest and Range Experiment Station, Corvallis, Oregon, U.S.A. 19 pp.

Freckmann, R.W. 1972. Grasses of Central Wisconsin. Reports on the Fauna and Flora of Wisconsin, Report No. 6. Museum of Natural History, Stevens Point, Wisconsin, U.S.A. 81 pp.

Fredskild, B. 1966. Contributions to the flora of Peary Land, north Greenland. Meddel. Grønland 178, nr. 2:1–23.

Fredskild, B. 1996. A phytogeographical study of the vascular plants of west Greenland (62° 20'–74° 00' N). Meddel. Grønland, Biosci. 45:1–157.

Freeman, C.C., R.L. McGregor, and C.A. Morse. 1998. Vascular plants new to Kansas. Sida 18:593–604.

Frelich, L. 1979. Vascular Plants of Newport State Park, Wisconsin. Department of Natural Resources Research Report No. 100. Department of Natural Resources, Madison, Wisconsin, U.S.A. 34 pp.

Friends of Mount Revelstoke and Glacier. 1996. Vascular Plant Checklist: Mount Revelstoke and Glacier National Parks. Natural History Handbook No. 1. Friends of Mount Revelstoke and Glacier, Revelstoke, British Columbia, Canada. 19 pp.

Gandhi, K.N. 1989. A biosystematic study of the *Schizachyrium scoparium* complex. Ph.D. dissertation, Texas A&M University, College Station, Texas, U.S.A. 188 pp.

Gardner, R.L., J.S. McCormac, and D. Minney. 2004. *Dichanthelium scoparium* and *Muhlenbergia glabrifloris*: New to the flora of Ohio. Sida 21:465–471.

Garlitz, R. 1998. Rare and interesting grass records from the northeastern Lower Peninsula. Michigan Bot. 28:67–71.

Garlitz, R. and D. Garlitz. 1986. *Bouteloua gracilis*, a new grass to Michigan. Michigan Bot. 25:123–124.

Gelting, P. 1934. Studies on the vascular plants of east Greenland between Franz Joseph Fjord and Dove Bay (lat. 73° 15'–76° 20' N.). Meddel. Grønland 101, nr. 2:1–336.

Gervias, C., R. Trahan, D. Moreno, and A.-M. Drolet. 1993. Le *Phragmites australis* au Québec: Distribution géographique, nombres chromosomiques et reproduction. Canad. J. Bot. 71:1386–1393.

Gillett, G.W., J.T. Howell, and H. Leschke. 1995. A Flora of Lassen Volcanic National Park, California (rev. Vernon H. Oswald, David W. Showers, and Mary Ann Showers), rev. ed. California Native Plant Society, Sacramento, California, U.S.A. 216 pp.

Gilman, A.V. 1993. Four recent additions to the vascular flora of Vermont. Maine Naturalist 1:31–32.

Gould, F.W. 1969. Taxonomy of the *Bouteloua repens* complex. Brittonia 21:261–274.

Gould, F.W. 1979. The genus *Bouteloua* (Poaceae). Ann. Missouri Bot. Gard. 66:348–416.

Gould, F.W., M.A. Ali, and D.E. Fairbrothers. 1972. A revision of *Echinochloa* in the United States. Amer. Midl. Naturalist 87:36–59.

Gould, F.W. and Z.J. Kapadia. 1964. Biosystematic studies in the *Bouteloua curtipendula* complex: II. Taxonomy. Brittonia 16:182–207.

Grant, M.L. 1950. Dickinson County flora: A preliminary check-list of the vascular plants of Dickinson County, Iowa, based largely on the herbarium of the Iowa Lakeside Laboratory. Proc. Iowa Acad. Sci. 57:91–129.

Grant, M.L. 1953. Additions to and notes on the flora of Dickinson County, Iowa. Proc. Iowa Acad. Sci. 60:131–140.

Gray, J.R. 1974. The genus *Tripsacum* L. (Gramineae): Taxonomy and chemosystematics. Ph.D. dissertation, University of Illinois at Urbana-Champaign, Urbana, Illinois, U.S.A. 191 pp.

Griffin, J.R. 1990. Flora of Hastings Reservation, Carmel Valley, California, ed. 3. Hastings Natural History Reservation, University of California-Berkeley, Berkeley, California, U.S.A. 98 pp.

Guala, G.F. 1988. *Poa bulbosa* L. (Poaceae) in Michigan. Michigan Bot. 27:13–14.

Guldner, L.F. 1960. The Vascular Plants of Scott and Muscatine Counties, with Some Reference to Adjoining Areas of Surrounding Counties in Iowa and to Rock Island and Whiteside Counties in Illinois. Davenport Public Museum Publications in Botany No. 1. Davenport Public Museum, Davenport, Iowa, U.S.A. 228 pp.

Haines, A. 2005. New combinations in *Poa*. Bot. Notes (Woodlot Alternatives) 10:1–5. http://www.woodlotalt.com/publications/BotNotesv1n10.pdf/.

Hall, D.W. 1982. *Sorghastrum* (Poaceae) in Florida. Sida 9:302–308.

Hall, H.H. and **S. Flowers**. 1961. Vascular plants found in the Navajo Reservoir basin, 1960: Colorado and New Mexico. Pp. 47–87 *in* Ecological Studies of the Flora and Fauna of Navajo Reservoir Basin, Colorado and New Mexico (A.M. Woodbury, ed.). University of Utah Anthropological Papers No. 55, Upper Colorado Series No. 5 (David M. Pendergast and Carol C. Stout, eds.). University of Utah Press, Salt Lake City, Utah, U.S.A. 203 pp.

Hall, H.M. 1902. A Botanical Survey of San Jacinto Mountain. University of California Publications in Botany, vol. 1, no. 1. University Press, Berkeley, California. 140 pp.

Halliday, G., L. Kliim-Nielsen, and **I.H.M. Smart**. 1974. Studies on the flora of the north Blosseville Kyst and on the hot springs of Greenland. Meddel. Grønland 199, nr. 2:1–47.

Hallsten, G.P., Q.D. Skinner, and **A.A. Beetle**. 1987. Grasses of Wyoming, ed. 3. Research Journal No. 202. Agricultural Experiment Station, University of Wyoming, Laramie, Wyoming, U.S.A. 432 pp.

Hämet-Ahti, L. 1965. Vascular plants of Wells Gray Provincial Park and its vicinity, in eastern British Columbia. Ann. Bot. Fenn. 2:138–164.

Hansen, B.F. and **R.P. Wunderlin**. 1992. Grasses of Florida: A Checklist of the Poaceae of Florida Along with Their Distribution by County. Institute for Systematic Botany, University of South Florida, Tampa, Florida, U.S.A. 133 pp. http://www.plantatlas.usf.edu/.

Harris, S.K. 1975. The Flora of Essex County, Massachusetts. Peabody Museum, Salem, Massachusetts, U.S.A. 81 pp.

Hartley, T.G. 1966. The Flora of the "Driftless Area". University of Iowa Studies in Natural History, vol. XXI, no. 1 (G.W. Martin, ed.). University of Iowa, Iowa City, Iowa, U.S.A. 174 pp.

Harvey, L.H. 1948. *Eragrostis* in North and Middle America. Ph.D. dissertation, University of Michigan, Ann Arbor, Michigan, U.S.A. 270 pp.

Harvill, A.M., Jr., T.R. Bradley, C.E. Stevens, T.F. Wieboldt, D.M.E. Ware, D.W. Ogle, G.W. Ramsey, and **G.P. Fleming**. 1992. Atlas of the Virginia Flora, ed. 3. Virginia Botanical Associates, Burkeville, Virginia, U.S.A. 144 pp.

Hatch, S.L. 1975. A biosystematic study of the *Schizachyrium cirratum—Schizachyrium sanguineum* complex (Poaceae). Ph.D. dissertation, Texas A&M University, College Station, Texas, U.S.A. 112 pp.

Hatch, S.L., W.E. Fox, III, and **J.E. Dawson, III**. 1998. *Triraphis mollis* (Poaceae: Arundineae), a species reported new to the United States. Sida 18:365–368.

Hatch, S.L., D.A. Kruse, and **J. Pluhar**. 2004. *Phalaris arundinacea* (Poaceae: Aveneae): A species new to Texas and a key to *Phalaris* in Texas. Sida 21:487–491.

Hatch, S.L., D.J. Rosen, J.A. Thomas, and **J.E. Dawson, III**. 1998. *Luziola peruviana* (Poaceae: Oryzeae) previously unreported from Texas and a key to Texas species. Sida 18:611–614.

Hawkins, T.K. and **E.L. Richards**. 1995. A floristic study of two bogs on Crowley's Ridge in Greene County, Arkansas. Castanea 60:233–244.

Hay, S.G. 1989. La migration récente de plantes halophytes dans la région Montréalaise. Quatre-Temps 13:7–11.

Hayden, A. 1943. A botanical survey in the Iowa lake region of Clay and Palo Alto counties. Iowa State Coll. J. Sci. 17:277–416.

Haynes, R.R. and **Z. Xu**. 1996. SERFIS [SouthEastern Regional Floristic Information System] database, University of Alabama herbarium (UNA).

Hays, J. 1995. A floristic survey of Falls Hollow sandstone glades, Pulaski County, Missouri. Phytologia 78:264–276.

Heidel, B. 1996a. Noteworthy collections. Madroño 43:436–440.

Heidel, B. 1996b. Questions and answers about sweetgrass. Kelseya 9:7.

Heil, K.D. and **S.L. O'Kane, Jr.** 2004. Catalog of the Four Corners flora: Vascular plants of the San Juan River drainage—Arizona, Colorado, New Mexico and Utah, ed. 8. [The FourCornersFlora.pdf file is part of the Bolack San Juan Basin Flora Project]. http://www.sjc.cc.nm.us/Herbarium/projects.htm.

Heise, K.L. and **A.M. Merenlender**. 1999. Flora of a vernal pool complex in the Mayacmas Mountains of southeastern Mendocino County, California. Madroño 46:38–45.

Hellquist, C.E. and **G.E. Crow**. 1997. The bryophyte and vascular flora of Little Dollar Lake peatland, Mackinac County, Michigan. Rhodora 99:195–222.

Henry, L.K. 1978. Vascular Flora of Bedford County, Pennsylvania: An Annotated Checklist. Carnegie Museum of Natural History, Pittsburgh, Pennsylvania, U.S.A. 29 pp.

Herrera-Arrieta, Y. 1998. A revision of the *Muhlenbergia montana* (Nutt.) Hitchc. complex (Poaceae: Chloridoideae). Brittonia 50:23–50.

Hinds, H. 2000. Flora of New Brunswick, ed. 2. Department of Biology, University of New Brunswick, Fredericton, New Brunswick, Canada. 695 pp.

Hitchcock, C.L., A. Cronquist, M. Ownbey, and **J.W. Thompson**. 1969. Vascular Plants of the Pacific Northwest, vol. 1: Vascular Cryptogams, Gymnosperms, and Monocotyledons. University of Washington Press, Seattle, Washington, U.S.A. and London, England. 914 pp.

Hoagland, B.W., P.H.C. Crawford, P.T. Crawford, and **F. Johnson**. 2004. Vascular flora of Hackberry Flat, Frederick Lake, and Suttle Creek, Tillman County, Oklahoma. Sida 21:429–445.

Hodgdon, A.R., G.E. Crow, and **F.L. Steele**. 1979. Grasses of New Hampshire: I. Tribes Poeae (Festuceae) and Triticeae (Hordeae). New Hampshire Agricultural Experiment Station Bulletin No. 512. New Hampshire Agricultural Experiment Station, University of New Hampshire, Durham, New Hampshire, U.S.A. 53 pp.

Holiday, S. 2000. A floristic study of Tsegi Canyon, Arizona. Madroño 47:29–42.

Holmen, K. 1957. The vascular plants of Peary Land, north Greenland. Meddel. Grønland 124, nr. 9:1–149.

Hough, M.Y. 1983. New Jersey Wild Plants. Harmony Press, Harmony, New Jersey. 414 pp.

Hounsell, R.W. and E.C. Smith. 1966. Contributions of the flora of Nova Scotia: VIII. Distribution of arctic-alpine and boreal disjuncts. Rhodora 68:409–419.

Howell, J.T. 1970. Marin Flora: Manual of the Flowering Plants and Ferns of Marin County, California, ed. 2. University of California Press, Berkeley, California, U.S.A. 366 pp.

Howell, J.T. 1979. A reconsideration of *Trisetum projectum* (Gramineae). Wasmann J. Biol. 37:21–23.

Howell, J.T., G.H. True, and C. Best. 1981. Notes on Marin County plants 1970–1980. Four Seasons 6:3–6.

Hrusa, G.F., B. Ertter, A.C. Sanders, G. Leppig, and E. Dean. 2002. Catalogue of non-native vascular plants occurring spontaneously in California beyond those addressed in *The Jepson Manual*—Part I. Madroño 49:61–98.

Hunsucker, R. and R.F. Mueller. 2003. Folly Mills calcareous wetland, Augusta County, Virginia. [Part of the Forests of the Central Appalachians Project by Virginians for Wilderness.] http://www.asecular.com/forests/folly.htm.

Hunt, D.M. and R.E. Zaremba. 1992. The northeastward spread of *Microstegium vimineum* (Poaceae) into New York and adjacent states. Rhodora 94:167–170.

Iverson, L.R., D. Ketzner, and J. Karnes. Viewed 2002. Illinois Plant Information Network (ILPIN) database. http://www.fs.fed.us/ne/delaware/ilpin/ilpin.html.

Jacobson, A.L., F.C. Weinmann, and P.F. Zika. 2001. Noteworthy collections. Madroño 48:213–214.

Johnson, A.E. and A. Blyth. 1988. Re-discovery of *Calamovilfa curtissii* (Gramineae) in the Florida panhandle. Sida 13:137–140.

Johnson-Groh, C.L. and D.R. Farrar. 1985. Flora and phytogeographical history of Ledges State Park, Boone County, Iowa. Proc. Iowa Acad. Sci. 92:137–143.

Johnson-Groh, C.L., D.Q. Lewis, and J.F. Shearer. 1987. Vegetation communities and flora of Dolliver State Park, Webster County, Iowa. Proc. Iowa Acad. Sci. 94:84–88.

Jones, S.B., Jr. and N.C. Coile. 1988. The Distribution of the Vascular Flora of Georgia. University of Georgia, Athens, Georgia, U.S.A. 230 pp.

Jones, S.D. and G. Jones. 1992. *Cynodon nlemfuënsis* (Poaceae: Chlorideae) previously unreported in Texas. Phytologia 72:93–95.

Jones, S.D. and J.K. Wipff. 1992. *Eustachys retusa* (Poaceae), the first report in Florida and a key to *Eustachys* in Florida. Phytologia 73:274–276.

Judziewicz, E.J. and R.G. Koch. 1993. Flora and vegetation of the Apostle Islands National Lakeshore and Madeline Island, Ashland and Bayfield Counties, Wisconsin. Michigan Bot. 32:43–193.

Jug Bay Wetlands Sanctuary. Viewed 2003. Wetlands plants of the Jug Bay Wetlands Sanctuary. http://www.jugbay.org/jugbay/wetplants.html.

Junak, S., T. Ayers, R. Scott, D. Wilken, and D. Young. 1995. A Flora of Santa Cruz Island. Santa Barbara Botanic Garden, Santa Barbara, California, U.S.A. 397 pp.

Kearney, T.H. and R.H. Peebles. 1951. Arizona Flora. University of California Press, Berkeley and Los Angeles, California, U.S.A. 1032 pp.

Kennedy, C.E., C.A.S. Smith, and D.A. Cooley. 2001. Observations of change in the cover of polargrass, *Arctagrostis latifolia*, and arctic lupine, *Lupinus arcticus*, in upland tundra on Herschel Island, Yukon Territory. Canad. Field-Naturalist 115:323–328.

Kershaw, L., J. Gould, D. Johnson, and J. Lancaster (eds). 2001. Rare Vascular Plants of Alberta. University of Alberta Press, Edmonton, Alberta, Canada. 486 pp.

Kiger, R.W. 1971. *Arthraxon hispidus* (Gramineae) in the United States: Taxonomic and floristic status. Rhodora 73:39–46.

Krings, A. 2002. Additions to the flora of Nags Head Woods (Dare County, North Carolina) and the Outer Banks of North Carolina. Sida 20:839–843.

Labrecque, J. and G. Lavoie. 2002. Les Plantes Vasculaires Menacées ou Vulnérables du Québec. Ministère de l'Environnement, Direction du Patrimoine Écologique et du Développement Durable, Québec City, Québec, Canada. 200 pp.

Lackschewitz, K. 1991. Vascular Plants of West-Central Montana—Identification Guidebook. General Technical Report INT-277. U.S. Department of Agriculture-Forest Service, Intermountain Research Station, Ogden, Utah, U.S.A. 648 pp.

Lammers, T.G. 1980. The vascular flora of Starr's Cave State Preserve. Proc. Iowa Acad. Sci. 87:148–158.

Lammers, T.G. 1983. The vascular flora of Des Moines County, Iowa. Proc. Iowa Acad. Sci. 90:55–71.

Landry, G.P. 1996. Noteworthy collections. Castanea 61:197.

Lange, K.I. 1998. Flora of Sauk County and Caledonia Township, Columbia County, South Central Wisconsin. Technical Bulletin No. 190. Department of Natural Resources, Madison, Wisconsin, U.S.A. 169 pp.

Lavin, M. and C. Seibert. Viewed 2003. Grasses of Montana. http://gemini.oscs.montana.edu/~mlavin/herb/mtgrass.pdf.

Lawrence, D.L. and J.T. Romo. 1995. Tree and shrub communities of wooded draws near the Matador Research Station in southern Saskatchewan. Canad. Field-Naturalist 108:397–409.

Layser, E.F., Jr. 1980. Flora of Pend Oreille County, Washington. Washington State University Cooperative Extension, Pullman, Washington, U.S.A. 146 pp.

LeBlond, R.J. 2001. Taxonomy of the *Dichotoma* group of *Dichanthelium* (Poaceae). Sida 19:821–837.

Leidolf, A., S. McDaniel, and **T. Nuttle.** 2002. The flora of Oktibbeha County, Mississippi. Sida 20:691–765.

Lelong, M.G. 1977. Annotated list of vascular plants in Mobile, Alabama. Sida 7:118–146.

Lelong, M.G. 1988. Noteworthy monocots of Mobile and Baldwin Counties, Alabama. Sida 13:101–113.

Lesica, P. 1985. Checklist of the Vascular Plants of Glacier National Park. Proceedings of the Montana Academy of Sciences Monograph No. 4. Montana Academy of Sciences, Missoula, Montana, U.S.A. 42 pp.

Lesica, P. 1994. Noteworthy collections. Madroño 41:231.

Lesica, P. 1998. Noteworthy collections. Madroño 45:328–330.

Lewis, M.E. 1971. Flora and major plant communities of the Ruby-East Humboldt Mountains, with special emphasis on Lamoille Canyon. Humboldt National Forest, Nevada Report to the U.S. Department of Agriculture-Forest Service, Region 4 [Intermountain Region], Ogden, Utah, U.S.A. 62 pp.

Lloyd, R.M. and **R.S. Mitchell.** 1965. Plants of the White Mountains, California and Nevada, rev. ed. University of California White Mountain Research Station, Berkeley, California, U.S.A. 61 pp.

Lomer, F. 1996a. Introduced bog plants, Vancouver, British Columbia. BEN [Botanical Electronic News] #128. http://www.ou.edu/cas/botany-micro/ben/.

Lomer, F. 1996b. Six new introduced species in British Columbia. BEN [Botanical Electronic News] #128. http://www.ou.edu/cas/botany-micro/ben/.

Lomer, F. 2001. Ephemeral introductions of vascular plants around Vancouver, British Columbia (Part 2). BEN [Botanical Electronic News] #270. http://www.ou.edu/cas/botany-micro/ben/.

Lonard, R.I. 1993. Guide to the Grasses of the Lower Rio Grande Valley, Texas. University of Texas-Pan American, Edinburg, Texas, U.S.A. 240 pp.

MacDonald, J. 1996. A survey of the flora of Monroe County, Mississippi. Master's thesis, Mississippi State University, Mississippi State, Mississippi, U.S.A. 162 pp.

MacRoberts, B.R. and **M.H. MacRoberts.** 1993. Vascular flora of sandstone outcrop communities in western Louisiana, with notes on rare and noteworthy species. Phytologia 75:463–480.

MacRoberts, B.R. and **M.H. MacRoberts.** 1995a. Vascular flora of two calcareous prairie remnants on the Kisatchie National Forest. Phytologia 78:18–27.

MacRoberts, B.R. and **M.H. MacRoberts.** 1995b. Noteworthy vascular plant collections on the Kisatchie National Forest, Louisiana. Phytologia 78:291–313.

MacRoberts, B.R. and **M.H. MacRoberts.** 1995c. Floristics of xeric sandhills in northwestern Louisiana. Phytologia 79:123–131.

MacRoberts, B.R. and **M.H. MacRoberts.** 1996a. Floristics of xeric sandhills in east Texas. Phytologia 80:1–7.

MacRoberts, B.R. and **M.H. MacRoberts.** 1996b. The floristics of calcareous prairies on the Kisatchie National Forest, Louisiana. Phytologia 81:35–43.

MacRoberts, B.R. and **M.H. MacRoberts.** 1997. Floristics of beech-hardwood forest in east Texas. Phytologia 82:20–29.

MacRoberts, B.R. and **M.H. MacRoberts.** 1998a. Noteworthy vascular plant collections on the Angelina and Sabine National Forests, Texas. Phytologia 84:1–27.

MacRoberts, B.R. and **M.H. MacRoberts.** 1998b. Floristics of wetland pine savannas in the Big Thicket National Preserve, southeast Texas. Phytologia 85:40–50.

MacRoberts, B.R. and **M.H. MacRoberts.** 1998c. Floristics of muck bogs in east central Texas. Phytologia 85:61–73.

MacRoberts, B.R., M.H. MacRoberts, and **L.E. Brown.** 2002. Annotated checklist of the vascular flora of the Hickory Creek unit of the Big Thicket National Preserve, Tyler County, Texas. Sida 20:781–795.

Madarish, D.M., J.L. Rodrigue, and **M.B. Adams.** 2002. Vascular flora and macroscopic fauna on the Fernow Experimental Forest. General Technical Report NE-291. U.S. Department of Agriculture-Forest Service, Northeastern Research Station, Newton Square, Pennsylvania, U.S.A. 37 pp.

Magee, D.M. 1993. Manual of the Vascular Flora of New England and Adjacent New York. Normandeau Associates, Inc., Bedford, New Hampshire, U.S.A. 49 pp.

Maley, A. 1994. A Floristic Survey of the Black Forest of the Colorado Front Range. Natural History Inventory of Colorado No. 14. University of Colorado Museum, Boulder, Colorado, U.S.A. 31 pp.

Mansfield, D.H. 1995. Vascular Flora of Steens Mountain, Oregon. Journal of the Idaho Academy of Science, vol. 31, no. 2. Idaho Academy of Science, Moscow, Idaho, U.S.A. 88 pp.

Marisco, T.D. 2005. The vascular flora of Montgomery County, Arkansas. Sida 21:2389–2423.

Marriott, H. 1985. Flora of the northwestern Black Hills, Crook and Weston counties, Wyoming. Master's thesis, University of Wyoming, Laramie, Wyoming, U.S.A. 93 pp.

Marsden, K.L. and **L.E. Hendrickson.** 2005. Noteworthy collections. Madroño 52:271.

Maryland Wildlife and Heritage Division-Department of Natural Resources. Viewed 2001. Rare, threatened, and endangered plants of Maryland. http://www.dnr.state.md.us/wildlife/rteplants.asp.

Matthews, M.A. 1997. An Illustrated Field Key to the Flowering Plants of Monterey County and Ferns, Fern Allies, and Conifers. California Native Plant Society, Sacramento, California, U.S.A. 401 pp.

Maxwell, C.L. 1991. Vascular flora of the Willapa Hills and lower Columbia River area of southwest Washington. Douglasia Occas. Pap. 4:28–76.

Maze, J. and K.A. Robson. 1996. A new species of *Achnatherum* (*Oryzopsis*) from Oregon. Madroño 43:393–403.

McClain, W.E. and J.E. Ebinger. 2002. A comparison of the vegetation of three limestone glades in Calhoun County, Illinois. SouthE. Naturalist 1:179–188.

McClintock, E., P. Reeberg, and W. Knight. 1990. A Flora of the San Bruno Mountains, San Mateo County, California. Special Publication No. 8. California Native Plant Society, Sacramento, California, U.S.A. 223 pp.

McLaughlin, S.P. 2006. Vascular floras of Sonoita Creek State Natural Area and San Rafael State Park: Arizona's first natural-area parks. Sida 22:661–704.

McNeill, J. 1981. *Apera*, silky-bent or windgrass, an important weed genus recently discovered in Ontario, Canada. Canad. J. Pl. Sci. 61:479–485.

Meeks, D.N. 1984. A floristic study of northern Tippah County, Mississippi. Master's thesis, Mississippi State University, Mississippi State, Mississippi, U.S.A. 69 pp.

Michener-Foote, J. and T. Hogan. 1999. The Flora and Vegetation of the Needle Mountains, San Juan Range, Southwestern Colorado. Natural History Inventory of Colorado No. 18. University of Colorado Museum, Boulder, Colorado, U.S.A. 39 pp.

Mitchell, R.S. and G.C. Tucker. 1997. Revised Checklist of New York State Plants. Contributions to a Flora of New York State, Checklist IV. New York State Museum Bulletin No. 490 (R.S. Mitchell, ed.). State Education Department, University of the State of New York, Albany, New York, U.S.A. 400 pp.

Mitchell, W.W. and H.J. Hodgson. 1965. The status of hybridization between *Agropyron sericeum* and *Elymus sibiricus* in Alaska. Canad. J. Bot. 43:855–859.

Mitchell, W.W. and A.C. Wilton. 1966. A new tetraploid brome, section *Bromposis*, of Alaska. Brittonia 18:162–166.

Moe, L.M. and E.C. Twisselmann. 1995. A Key to Vascular Plant Species of Kern County, California and A Flora of Kern County, California. California Native Plant Society, Sacramento, California, U.S.A. 620 pp.

Mohlenbrock, R.H. 2001. Grasses: Panicum to Danthonia, ed. 2. The Illustrated Flora of Illinois (R.H. Mohlenbrock, ed.). Southern Illinois University Press, Carbondale and Edwardsville, Illinois, U.S.A. 455 pp.

Mohlenbrock, R.H. 2002a. Grasses, Bromus to Paspalum, ed. 2. The Illustrated Flora of Illinois (R.H. Mohlenbrock, ed.). Southern Illinois University Press, Carbondale and Edwardsville, Illinois, U.S.A. 404 pp.

Mohlenbrock, R.H. 2002b. Vascular Flora of Illinois. Southern Illinois University Press, Carbondale and Edwardsville, Illinois, U.S.A. 490 pp.

Mohlenbrock, R.H. and D.M. Ladd. 1978. Distribution of Illinois Vascular Plants. Southern Illinois University Press, Carbondale, Illinois, U.S.A. 282 pp.

Molenaar, J.G.d. 1974. Vegetation of the Angmagssalik District, southeast Greenland: I. Littoral vegetation. Meddel. Grønland 198, nr. 1:1–79.

Molenaar, J.G.d. 1976. Vegetation of the Angmagssalik District, southeast Greenland: II. Herb and snow-bed vegetation. Meddel. Grønland 198, nr. 2:1–266.

Montana Natural History Program. Viewed 1996. Animal and plant species of Beaverhead County. http://nhp.nris.state.mt.us/plants/index.asp.

Montgomery County Department of Parks and Planning. 2003. Flora species for Montgomery County parks. The Maryland-National Capital Park & Planning Commission, Silver Spring, Maryland, U.S.A 15 pp.

Morris, M.W. 1987. The vascular flora of Grenada County, Mississippi. Master's thesis, Mississippi State University, Mississippi State, Mississippi, U.S.A. 123 pp.

Morris, M.W. 1997. Contributions to the flora and ecology of the northern longleaf pine belt in Rankin County, Mississippi. Sida 17:615–626.

Morrone, O., A.S. Vega, and F.O. Zuloaga. 1996. Revisión de las especies del género *Paspalum* L. (Poaceae: Panicoideae: Paniceae), grupo Dissecta (*s. str.*). Candollea 51:103–138.

Morrone, O. and F.O. Zuloaga. 1992. Revisión de las especies sudamericanas nativas e introducidas de los géneros *Brachiaria* y *Urochloa* (Poaceae: Panicoideae: Paniceae). Darwiniana 31:43–109.

Morrone, O. and F.O. Zuloaga. 1993. Synopsis del género *Urochloa* (Poaceae: Panicoideae: Paniceae) para México y América Central. Darwiniana 32:59–75.

Morton, C.M., J. Kartesz, B.L. Isaac, and R. Coxe. 2004. Additions to and noteworthy records for the vascular flora of West Virginia. Sida 21:481–485.

Morton, J.K. and J.M. Venn. 1984. The Flora of Manitoulin Island and the Adjacent Islands of Lake Huron, Georgian Bay and the North Channel, ed. 2, rev. Department of Biology, University of Waterloo, Waterloo, Ontario, Canada. 106 pp.

Moseley, R.K. 1996. Vascular flora of subalpine parks in the Coeur d'Alene River drainage, northern Idaho. Madroño 43:479–492.

Moss, E.H. 1983. Flora of Alberta: A Manual of Flowering Plants, Conifers, Ferns and Fern Allies Found Growing Without Cultivation in the Province of Alberta, Canada (rev. John G. Packer), rev. ed. University of Toronto Press, Toronto, Ontario, Canada. 687 pp.

Musselman, L.J., T.S. Cochrane, W.E. Rice, and M.M. Rice. 1971. The flora of Rock County, Wisconsin. Michigan Bot. 10:145–205.

Muzika, R.-M., R. Hunsucker, and T. DeMeo. 1996. Botanical reconnaissance of Big Run Bog Candidate Research Natural Area. General Technical Report NE-223. U.S. Department of Agriculture-Forest Service, Northeastern Forest Experiment Station, Radnor, Pennsylvania, U.S.A. 15 pp.

Naczi, R.F.C., R.L. Jones, F.J. Metzmeier, M.A. Gorton, and T.J. Weckman. 2002. Native flowering plant species new or otherwise significant in Kentucky. Sida 20:397–402.

Naumann, T. 1996. Plant List, Including Scientific and Common Names, Dinosaur National Monument. Dinosaur Nature Association, Vernal, Utah, U.S.A. 20 pp.

Negrete, I.G., A.D. Nelson, J.R. Goetze, L. Macke, T. Wilburn, and A. Day. 1999. A checklist for the vascular plants of Padre Island National Seashore. Sida 18:1227–1245.

Nelson, J.B. and K.B. Kelly. 1997. Noteworthy collections. Castanea 62:283–287.

Nesom, G.L. and L.E. Brown. 1998. Annotated checklist of the vascular plants of Walker, Montgomery, and San Jacinto counties, east Texas. Phytologia 84:98–106.

New Jersey Natural Heritage Program. Viewed 2003. Database of New Jersey rare species and natural community lists by county. http://www.state.nj.us/dep/parksandforests/natural/heritage/countylist.html.

New York Flora Association. 1990. Preliminary Vouchered Atlas of New York State Flora. New York State Museum Institute, Albany, New York, U.S.A. 496 pp. http://atlas.nyflora.org/.

New York Flora Association. 1990–1998. Newslett. New York Fl. Assoc. 1–9. [Miscellaneous articles and species lists containing additions to Preliminary Vouchered Atlas of New York State Flora.]

Neyland, R., B.J. Hoffman, M. Mayfield, and L.E. Urbatsch. 2000. A vascular flora survey of Calcasieu Parish, Louisiana. Sida 19:361–386.

Niemann, D.A. and R.Q. Landers, Jr. 1974. Forest communities in Woodman Hollow State Preserve, Iowa. Proc. Iowa Acad. Sci. 81:176–184.

North Carolina Natural Heritage Program. Viewed 2003. Database of grasses in North Carolina. http://207.4.179.38/nhp/county.html.

Northam, F.E. 1995. Range extension of southwestern cupgrass (Eriochloa acuminata) into Kansas. Trans. Kansas Acad. Sci. 98:68–71.

Northam, F.E. and R.H. Callihan. 1992. Morphology and phenology of interrupted windgrass in northern Idaho. J. Idaho Acad. Sci. 28:15–19.

Northam, F.E., R.H. Callihan, and R.R. Old. 1989. Sporobolus vaginiflorus (Torrey ex Gray) Wood: Biology and pest implications of an alien grass recorded in Idaho. J. Idaho Acad. Sci. 25:49–55.

Northam, F.E., R.H. Callihan, and R.R. Old. 1991. Range extensions of four introduced grasses in Idaho. J. Idaho Acad. Sci. 27:19–21.

Northam, F.E., R.R. Old, and R.H. Callihan. 1993. Little lovegrass (Eragrostis minor) distribution in Idaho and Washington. Weed Technol. 7:771–775.

Ohio Department of Natural Resources. Viewed 2002. Database of county distributions for Ohio's rare plants. http://www.dnr.state.oh.us/dnap/heritage/.

Old, R.R. and R.H. Callihan. 1986. Distribution of Ventenata dubia in Idaho. Idaho Weed Control Rep. 1986:153.

Oldham, M.J. and S.J. Darbyshire. 1993. The adventive grasses, Apera interrupta and Deschampsia danthonioides, new to Maine. Maine Naturalist 1:231–232.

Oldham, M.J., S.J. Darbyshire, D. McLeod, D.A. Sutherland, D. Tiedje, and J.M. Bowles. 1995. New and noteworthy Ontario grass (Poaceae) records. Michigan Bot. 34:105–132.

Ostenfeld, C.E.H. 1923. Two plant lists from Inglefield Gulf and Inglefield Land (77° 28' and 79° 10' N. lat.), N.W. Greenland. Meddel. Grønland 64, nr. 7:209–214.

Oswald, V. and L. Ahart. 1994. Flora of Butte County, California. California Native Plant Society, Sacramento, California, U.S.A. 348 pp.

Ownbey, G.B. and T. Morley. 1991. Vascular Plants of Minnesota: A Checklist and Atlas. University of Minnesota, Minneapolis, Minnesota, U.S.A. 307 pp.

Parker, C.L. 2001. Progress report, inventory and monitoring program, vascular plant inventory, summer 2001. Northwest Alaska Network (NWAN) [now Arctic Network], National Park Service, Fairbanks, Alaska, U.S.A. 6 pp.

Peck, J.H., L.J. Eilers, and D.M. Roosa. 1978. The vascular plants of Fremont County, Iowa. Iowa Bird Life 48:3–24.

Peck, J.H., B.W. Haglan, L.J. Eilers, D.M. Roosa, and D. vander Zee. 1984. Checklist of the vascular flora of Lyon and Sioux counties, Iowa. Proc. Iowa Acad. Sci. 91:92–97.

Peck, J.H., T.G. Lammers, B.W. Haglan, D.M. Roosa, and L.J. Eilers. 1981. A checklist of the vascular flora of Lee County, Iowa. Proc. Iowa Acad. Sci. 88:159–171.

Peck, J.H., D.M. Roosa, and L.J. Eilers. 1980. A checklist of the vascular flora of Allamakee County, Iowa. Proc. Iowa Acad. Sci. 87:62–75.

Pedersen, A. 1972. Adventitious plants and cultivated plants in Greenland. Meddel. Grønland 178, nr. 7:1–99.

Peet, R.K. 1993. A taxonomic study of Aristida stricta and A. beyrichiana. Rhodora 95:25–37.

Perkins, B.E. and T.S. Patrick. 1980. Status report on Tennessee populations of Calamovilfa arcuata. University of Tennessee, Knoxville, Tennessee, U.S.A. 13 pp.

Peterson, P.M. 1986. A flora of the Cottonwood Mountains, Death Valley National Monument, California. Wasmann J. Biol. 44:73–126.

Peterson, P.M., J. Cayouette, Y.S.N. Ferdinandez, B. Coulman, and R.E. Chapman. 2001. Recognition of Bromus richardsonii and B. ciliatus: Evidence from morphology, cytology, and DNA fingerprinting (Poaceae: Bromeae). Aliso 20:21–36.

Peterson, P.M., E.E. Terrell, E.C. Uebel, C.A. Davis, H. Scholz, and R.J. Soreng. 1999. *Oplismenus hirtellus* subspecies *undulatifolius*, a new record for North America. Castanea 64:201–202.

Plunkett, G.M. and G.W. Hall. 1995. The vascular flora and vegetation of western Isle of Wight County, Virginia. Castanea 60:30–59.

Pohl, R.W. 1959. Introduced weedy grasses in Iowa. Proc. Iowa Acad. Sci. 66:160–162.

Pohl, R.W. 1966. The grasses of Iowa. Iowa State Coll. J. Sci. 40:341–566.

Poole, J.P. 1978. An addition to the flora of the Gaspé Peninsula. Rhodora 80:154.

Popovich, S.J. and D. Henderson. 1994. Noteworthy collections. Madroño 41:149–150.

Popovich, S.J., W.D. Shepperd, D.W. Reichert, and M.A. Cone. 1993. Flora of the Fraser Experimental Forest, Colorado. General Technical Report RM-233. U.S. Department of Agriculture-Forest Service, Rocky Mountain Forest and Range Experiment Station, Fort Collins, Colorado, U.S.A. 62 pp.

Porsild, A.E. and W.J. Cody. 1980. Vascular Plants of the Continental Northwest Territories, Canada. National Museum of Natural Sciences, National Museums of Canada, Ottawa, Ontario, Canada. 667 pp.

Porsild, M.P. 1930. Stray contributions to the flora of Greenland I-V. Meddel. Grønland 77, nr. 1:1–44.

Powell, A.M. 1994. Grasses of the Trans-Pecos and Adjacent Areas. University of Texas Press, Austin, Texas, U.S.A. 377 pp.

Provance, M.C. and A.C. Sanders. 2000. Noteworthy collections. Madroño 47:139–141.

Radford, A.E., H.E. Ahles, and C.R. Bell. 1965. Atlas of the Vascular Flora of the Carolinas. North Carolina Agricultural Experiment Station Technical Bulletin No. 165. North Carolina Agricultural Experiment Station, University of North Carolina-Raleigh, Raleigh, North Carolina, U.S.A. 208 pp.

Rana Creek Habitat Restoration. 2002. San Mateo County Parks Vegetation Resources. Parks & Recreation Division, County of San Mateo Environmental Services Agency, Redwood City, California, U.S.A. 187 pp.

Raup, H.M. 1965. The flowering plants and ferns of the Mesters Vig District, northeast Greenland. Meddel. Grønland 166, nr. 2:1–119.

Raup, H.M. 1971. Miscellaneous contributions on the vegetation of the Mesters Vig District, northeast Greenland. Meddel. Grønland 194, nr. 2:1–97.

Raven, P.H., H.J. Thompson, and B.A. Prigge. 1986. Flora of the Santa Monica Mountains, California, ed. 2. Southern California Botanists, Special Publication No. 2. University of California-Los Angeles, Los Angeles, California, U.S.A. 181 pp.

Rawinski, T.J., M.N. Rasmussen, and S.C. Rooney. 1989. Discovery of *Sporobolous asper* (Poaceae) in Maine. Rhodora 91:220–221.

Read, J.C. and B.J. Simpson. 1992. Documented chromosome numbers 1992: 3. Documentation and notes on the distribution of *Melica montezumae*. Sida 15:151–152.

Redman, D.E. 1995a. Noteworthy collections. Castanea 60:82–84.

Redman, D.E. 1995b. Distribution and habitat types for Nepal microstegium [*Microstegium vimineum* (Trin.) Camus] in Maryland and the District of Columbia. Castanea 60:270–275.

Reed, C.F. 1964. A flora of the chrome and manganese ore piles at Canton, in the Port of Baltimore, Maryland and at Newport News, Virginia, with descriptions of genera and species new to the flora of the eastern United States. Phytologia 10:321–405.

Reeder, C.G. and J.R. Reeder. 1986. *Agrostis elliottiana* (Gramineae) new to Arizona and New Mexico. Phytologia 60:453–458.

Reeder, J.R. 1991. A new species of *Panicum* (Gramineae) from Arizona. Phytologia 71:300–303.

Reeder, J.R. 1994. *Stipa tenuissima* (Gramineae) in Arizona— A comedy of errors. Madroño 41:328–329.

Reeder, J.R. 2001. Noteworthy collections. Madroño 48:212–213.

Reeder, J.R. and C.G. Reeder. 1980. Systematics of *Bouteloua breviseta* and *B. ramosa* (Gramineae). Syst. Bot. 5:312–321.

Reeder, J.R. and C.G. Reeder. 1990. *Bouteloua eludens*: Elusive indeed, but not rare. Desert Pl. 10:19–22, 31.

Rhoads, A.F. and W.M. Klein, Jr. 1993. The Vascular Flora of Pennsylvania: Annotated Checklist and Atlas. American Philosophical Society, Philadelphia, Pennsylvania, U.S.A. 636 pp.

Rice, P. Viewed 2003. INVADERS database for early detection and tracking of invasive alien plants and weedy natives. http://invader.dbs.umt.edu/.

Richards, C.D., F. Hyland, and L.M. Eastman. 1983. Check-List of the Vascular Plants of Maine, ed. 2, rev. Bulletin of the Josselyn Botanical Society No. 11. Lincoln Press, Sanford, Maine, U.S.A. 73 pp.

Riefner, R.E., Jr. 2003. Noteworthy collections. Madroño 50:312–313.

Riefner, R.E., Jr. and D.R. Pryor. 1996. New locations and interpretations of vernal pools in southern California. Phytologia 80:296–327.

Riggins, R. 1977. A biosystematic study of the *Sporobolus asper* complex (Gramineae). Iowa State J. Res. 51:287–321.

Riley, J.L. 1979. Some new and interesting vascular plant records from northern Ontario. Canad. Field-Naturalist 93:355–362.

Riley, J.L. 2003. Flora of the Hudson Bay Lowland and Its Postglacial Origins. NRC [National Research Council of Canada] Research Press, Ottawa, Ontario, Canada. 236 pp.

Riley, J.L. and S.M. McKay. 1980. The Vegetation and Phytogeography of Coastal Southwestern James Bay. Life Sciences Contributions, Royal Ontario Museum No. 124. Royal Ontario Museum, Toronto, Ontario, Canada. 81 pp.

Rill, K.D. 1983. A vascular flora of Winnebago County, Wisconsin. Trans. Wisconsin Acad. Sci. 71:155–180.

Rivas-Martinez, S., D. Sánchez-Mata, and M. Costa. 1999. North American boreal and western temperate forest vegetation: Syntaxonomical synopsis of the potential natural plant communities of North America, II. Itinera Geobot. 12:3–311.

Roalson, E.H. and K.W. Allred. 1998. A floristic study in the Diamond Creek drainage area, Gila National Forest, New Mexico. Aliso 17:47–62.

Roberts, F.M., Jr. 1989. A Checklist of the Vascular Plants of Orange County, California. Museum of Systematic Biology Research Series No. 6. University of California-Irvine, Irvine, California, U.S.A. 58 pp.

Roché, C.T. and A. Rebischke. 2005. Noteworthy collections. Madroño 52:128.

Rogers, B.S. and A. Tiehm. 1979. Vascular Plants of the Sheldon National Wildlife Refuge, with Special Reference to Possible Threatened and Endangered Species. Department of the Interior, U.S. Fish and Wildlife Service, Region 1, Portland, Oregon, U.S.A. 87 pp.

Roosa, D.M., L.J. Eilers, and S. Zaber. 1991. An annotated checklist of the vascular plant flora of Guthrie County, Iowa. J. Iowa Acad. Sci 98:14–30.

Rosen, D.J., S.D. Jones, and J.K. Wipff. 2001. Phyllostachys bambusoides (Poaceae: Bambuseae) previously unreported from Louisiana. Sida 19:731–734.

Rosen, D.J. and J.K. Wipff. 2003. Andropogon glomeratus var. glaucopsis (Poaceae: Andropogoneae) documented in Louisiana. Sida 20:1723–1725.

Rouleau, E. and G. Lamoureux. 1992. Atlas of the Vascular Plants of the Island of Newfoundland and of the Islands of Saint-Pierre-et-Miquelon. Fleurbec, Québec City, Québec, Canada. 777 pp.

Rousseau, C. 1968. Histoire, habitat et distribution de 220 plantes introduites au Québec. Naturaliste Canad. 95:49–169.

Rousseau, C. 1974. Géographie Floristique du Québec-Labrador: Distribution des Principales Espèces Vasculaires. Les Presses de l'Université Laval, Québec City, Québec, Canada. 615 pp.

Rubtzoff, P. 1961. Notes on fresh-water marsh and aquatic plants in California. Leafl. W. Bot. 9:165–180.

Russell, N.H. 1956. A checklist of the vascular flora of Poweshiek County, Iowa. Proc. Iowa Acad. Sci. 63:161–176.

Saarela, J.M., P.M. Peterson, and J. Cayouette. 2005. Bromus hallii (Poaceae), a new combination for California, U.S.A., and taxonomic notes on Bromus orcuttianus and Bromus grandis. Sida 21:1997–2013.

Saarela, J.M., P.M. Peterson, R.J. Soreng, and R.E. Chapman. 2003. A taxonomic revision of the eastern North American and eastern Asian disjunct genus Brachyelytrum (Poaceae): Evidence from morphology, phytogeography and AFLPs. Syst. Bot. 28:674–692.

Saichuk, J.K., C.M. Allen, and W.D. Reese. 2000. Alopecurus myosuroides and Sclerochloa dura (Poaceae) new to Louisiana. Sida 19:411–412.

Saltonstall, K., P.M. Peterson, and R.J. Soreng. 2004. Recognition of Phragmites australis subsp. americanus (Poaceae: Arundinoideae) in North America: Evidence from morphological and genetic analysis. Sida 21:683–692.

Sanders, A.C. 1996. Noteworthy collections. Madroño 43:524–532.

Scheffer, T.H. 1945. The introduction of Spartina alternifolia to Washington with oyster culture. Leafl. W. Bot. 4:163–164.

Scott, R.W. 1995. The Alpine Flora of the Rocky Mountains, vol. 1: The Middle Rockies. University of Utah Press, Salt Lake City, Utah, U.S.A. 901 pp.

Seabloom, E.W. and A.M. Wiedemann. 1994. Distribution and effects of Ammophila breviligulata Fern. (American beachgrass) on the foredunes of the Washington coast. J. Coastal Res. 10:178–188.

Seagrist, R.V. and K.J. Taylor. 1998. Alpine vascular flora of Buffalo Peaks, Mosquito Range, Colorado, USA. Madroño 45:319–325.

Seagrist, R.V. and K.J. Taylor. 1999. Alpine vascular flora of Hasley Basin, Elk Mountains, Colorado, USA. Madroño 45:310–318.

Seidenfaden, G. 1933. The vascular plants of south-east Greenland 60° 04' to 64° 30' N. lat. Meddel. Grønland 106, nr. 3:1–129.

Sharma, M.P. and W.H. Vanden Born. 1978. The biology of Canadian weeds: 27. Avena fatua L. Canad. J. Pl. Sci. 58:141–157.

Simmons, M.P., D.M.E. Ware, and W.J. Hayden. 1995. The vascular flora of the Potomac River watershed of King George County, Virginia. Castanea 60:179–200.

Simmons, R. 2003. Flora of Araby Bog, Charles County, Maryland. http://www.mdflora.org/survey_data/arbybotsrvy_rod.html.

Simpson, M.G., S.C. McMillan, and B.L. Stone. 1995. Checklist of the Vascular Plants of San Diego County. San Diego State University Herbarium Press, San Diego, California, U.S.A. 80 pp. http://www.sdnhm.org/research/botany/sdplants/index.html.

Skojac, D., M.S. Devall, and B.R. Parresol. 2003. Additions to the flora of Cleveland County, Arkansas: Collections from Moro Bottoms Natural Area, a state-protected old-growth bottomland forest. Sida 20:1731–1736.

SMASCH Project. Viewed 2001. Specimen MAnagement System for California Herbaria database [University of California, Berkeley (UC) and Jepson (JEPS) herbaria]. http://www.mip.berkeley.edu/www_apps/smasch/.

Smith, C.F. 1976. A Flora of the Santa Barbara Region, California. Santa Barbara Museum of Natural History, Santa Barbara, California, U.S.A. 331 pp.

Smith, E.B. (ed). 1988. An Atlas and Annotated List of the Vascular Plants of Arkansas. Edwin B. Smith, Fayetteville, Arkansas, U.S.A. 489 pp.

Smith, G.L. and C.R. Wheeler. 1990–1991. A flora of the vascular plants of Mendocino County, California. Wasmann J. Biol. 48/49:63–81.

Snow, N. 1992–1994. The vascular flora of southeastern Yellowstone National Park and the headwaters region of the Yellowstone River. Wasmann J. Biol. 50:52–95.

Sørensen, T. 1933. The vascular plants of east Greenland from 71° 00' to 73° 30' N. lat. Meddel. Grønland 101, nr. 3:1–177.

Sørensen, T. 1943. The flora of Melville Bugt. Meddel. Grønland 124, nr. 5:1–70.

Sorrie, B.A. 1987. Notes on the rare flora of Massachusetts. Rhodora 89:113–196.

Sorrie, B.A. 1998. Noteworthy collections. Castanea 63:496–500.

Sorrie, B.A. and P.W. Dunwiddie. 1990. *Amphicarpum purshii* (Poaceae), a genus and species new to New England. Rhodora 92:105–107.

Sorrie, B.A. and S.W. Leonard. 1999. Noteworthy records of Mississippi vascular plants. Sida 18:889–908.

Sorrie, B.A. and P. Somers. 1999. The Vascular Plants of Massachusetts: A County Checklist. Natural Heritage and Endangered Species Program, Massachusetts Division of Fisheries and Wildlife, Westborough, Massachusetts, U.S.A. 189 pp.

Sorrie, B.A., B. Van Eerden, and M.J. Russo. 1997. Noteworthy plants from Fort Bragg and Camp MacKall, North Carolina. Castanea 62:239–259.

Soza, V. 2000. Noteworthy collections. Madroño 47:141–142.

Sparks, L.H., R. Del Moral, A.F. Watson, and A.R. Kruckeberg. 1976. The distribution of vascular plant species on Sergief Island, southeast Alaska. Syesis 10:5–9.

Spellenberg, R., D. Anderson, and R. Brozka. 1993. Noteworthy collections. Madroño 40:136–138.

Spicher, D. and M. Josselyn. 1985. *Spartina* (Gramineae) in northern California: Distribution and taxonomic notes. Madroño 32:158–167.

Spribille, T. 2002. Noteworthy collections. Madroño 49:55–58.

Stalter, R. and E. Lamont. 1996. Noteworthy collections. Castanea 61:396–397.

Stalter, R. and J. Tamory. 1999. The vascular flora of Biscayne National Park, Florida. Sida 18:1207–1226.

Steel, M.G., P.B. Cavers, and S.M. Lee. 1983. The biology of Canadian weeds: 59. *Setaria glauca* (L.) Beauv. and *S. verticillata* (L.) Beauv. Canad. J. Pl. Sci. 63:711–725.

Steury, B.W. 2000. Noteworthy collections. Castanea 65:168.

Stevens, O.A. 1963. Handbook of North Dakota Plants [third printing]. North Dakota Institute for Regional Studies, Fargo, North Dakota, U.S.A. 324 pp.

Stevenson, G.A. 1965. Notes on the more recently adventive flora of the Brandon area, Manitoba. Canad. Field-Naturalist 79:174–177.

Stewart, H. and R.J. Hebda. 2000. Grasses of the Columbia Basin of British Columbia. Research Branch, Ministry of Forests, British Columbia Working Paper No. 45. Ministry of Forests Research Program, Victoria, British Columbia, Canada. 228 pp. http://livinglandscapes.bc.ca/cbasin/cb_grasses/index_grasses.html.

Stickney, P.F. 1961. Range of rough fescue (*Festuca scabrella* Torr.) in Montana. Proc. Montana Acad. Sci. 20:12–17.

Stiles, B.J. and C.L. Howel. 1998. Floristic survey of Rabun County, Georgia, part II. Castanea 63:154–160.

Stuart, J.D., T. Worley, and A.C. Buell. 1996. Plant associations of Castle Crags State Park, Shasta County, California. Madroño 43:273–291.

Sundell, E., R.D. Thomas, C. Amason, and C.H. Doffitt. 2002. Noteworthy vascular plants from Arkansas: II. Sida 20:409–418.

Sundell, E., R.D. Thomas, C. Amason, R.L. Stuckey, and J. Logan. 1991. Noteworthy vascular plants from Arkansas. Sida 18:877–887.

Swink, F. and G. Wilhelm. 1994. Plants of the Chicago Region, ed. 4. The Morton Arboretum, Lisle, Illinois, U.S.A.

Talbot, S.S., B.A. Yurtsev, D.F. Murray, G.W. Argus, C. Bay, and A. Elvebakk. 1999. Atlas of Rare Endemic Vascular Plants of the Arctic. Conservation of Arctic Flora and Fauna (CAFF) Technical Report No. 3. U.S. Fish and Wildlife Service, Anchorage, Alaska, U.S.A. 73 pp.

Taylor, R.J. and C.E. Taylor. 1980. Status report on *Calamovilfa arcuata*. University of Tennessee, Knoxville, Tennessee, U.S.A. 27 pp.

Taylor, R.J. and C.E. Taylor. 1987. Additions to the vascular flora of Oklahoma—IV. Sida 12:233–237.

Terrell, E.E. and J.L. Reveal. 1996. Noteworthy collections. Castanea 61:95–96.

Terrell, E.E., J.L. Reveal, R.W. Spjut, R.F. Whitcomb, J.H. Kirkbride, and M.T. Cimino. 2000. Annotated List of the Flora of the Beltsville Agricultural Research Center, Beltsville, Maryland. U.S. Department of Agriculture-Agricultural Research Center, Beltsville, Maryland, U.S.A. 89 pp.

Texas A&M University. Viewed 1996. Bioinformatics Working Group herbarium specimen browser. http://www.csdl.tamu.edu/FLORA/.

Thomas, R.D. 2002. *Cynosurus echinatus* (Poaceae) new to Texas. Sida 20:837.

Thomasson, J.R. 1984. A new record of *Melica subulata* (Gramineae) from the northern Black Hills of South Dakota. Amer. Midl. Naturalist 112:208.

Thompson, R.L. 2001. Botanical Survey of Myrtle Island Research Natural Area, Oregon. General Technical Report PNW-GTR-507. U.S. Department of Agriculture-Forest Service, Pacific Northwest Research Station, Portland, Oregon, U.S.A. 27 pp.

Thompson, R.L., R.L. Jones, J.R. Abbott, and W.N. Denton. 2000. Botanical survey of Rock Creek Research Natural Area, Kentucky. General Technical Report NE-272. U.S. Department of Agriculture-Forest Service, Northeastern Research Station, Newton Square, Pennsylvania, U.S.A. 23 pp.

Thompson, R.L. and J.R. Skeese, III. 2005. Vascular flora of Golden and Silver Falls State Natural Area in the Oregon Coast Range, Coos County, Oregon. Madroño 52:215–221.

Thorne, R.F. 1955. The flora of Johnson County, Iowa. Proc. Iowa Acad. Sci. 62:155–196.

Thorne, R.F., B.A. Prigge, and J. Henrickson. 1981. A flora of the higher ranges and the Kelso Dunes of the eastern Mojave Desert in California. Aliso 10:71–186.

Titus, J.H., S. Moore, M. Arnot, and P.J. Titus. 1998. Inventory of the vascular flora of the blast zone, Mount St. Helens, Washington. Madroño 45:146–161.

Towne, E.G. 2002. Vascular plants of Konza Prairie Biological Station: An annotated checklist of species in a Kansas tallgrass prairie. Sida 20:269–294.

Towne, E.G. and I. Barnard. 2000. *Themeda quadrivalvis* (Poaceae: Andropogoneae) in Kansas: An exotic plant introduced from birdseed. Sida 19:201–203.

Tucker, G.C. 1996. The genera of Pooideae (Gramineae) in the southeastern United States. Harvard Pap. Bot. 9:11–90.

Turner, B.L. 2004. *Sporobolus coahuilensis* (Poaceae): A new record for the U.S.A. from Trans-Pecos, Texas. Sida 21:455–457.

Turner, B.L., H. Nichols, G.C. Denny, and O. Doron. 2003. Atlas of the Vascular Plants of Texas, vol. 2: Ferns, Gymnosperms, Monocots. Sida, Botanical Miscellany No. 24. BRIT [Botanical Research Institute of Texas] Press, Fort Worth, Texas, U.S.A. 888 pp.

U.S. Department of Agriculture-Animal and Plant Health Inspection Service. 1990. Weed alert! Be on the lookout for Serrated Tussock, a Federal noxious weed (*Nassella trichotoma*). U.S. Department of Agriculture-Animal and Plant Health Inspection Service [USDA-APHIS], in cooperation with the Illinois Department of Agriculture, Springfield, Illinois, U.S.A. 3 pp.

U.S. Department of Agriculture-Forest Service-Rocky Mountain Research Station. Viewed 1999. Manitou Experimental Forest web database: Species list of principal plants. [Website no longer available.]

University of Oklahoma. Viewed 2002. Atlas of the flora of Oklahoma. http://geo.ou.edu/botanical/.

University of South Carolina. Viewed 2001. South Carolina plant atlas. http://cricket.biol.sc.edu/herb/.

Urban, K.A. 1971. Common Plants of Craters of the Moon National Monument. Craters of the Moon Natural History Association, Arco, Idaho, U.S.A. 30 pp.

Vanderhorst, J.P. 1993. Flora of the Flat Tops, White River Plateau, and vicinity in northwestern Colorado. Master's thesis, University of Wyoming, Laramie, Wyoming, U.S.A. 129 pp.

Vega, A.S. 2000. Revisión taxonómica de las especies americanas del género *Bothriochloa* (Poaceae: Panicoideae: Andropogoneae). Darwiniana 38:127–186.

Vermont Botanical and Bird Club. 1973. Check List of Vermont Plants, Including All Vascular Plants Growing Without Cultivation. Vermont Botanical and Bird Club, Burlington, Vermont, U.S.A. 90 pp.

Villamil, C.B. 1969. El género *Monanthochloë* (Gramineae): Estudios morfológicos y taxonómicos con especial referencia a la especia Argentina. Kurtziana 5:369–391.

Voss, E.G. 1972. Michigan Flora: A Guide to the Identification and Occurrence of the Native and Naturalized Seed-Plants of the State: Part I, Gymnosperms and Monocots. Cranbrook Institute of Science, Bloomfield Hills, Michigan, U.S.A. 488 pp.

Wagenknecht, B.L. 1954. The flora of Washington County, Iowa. Proc. Iowa Acad. Sci. 61:184–204.

Wagnon, H.K. 1950. Three new species and one new form in *Bromus*. Leafl. W. Bot. 6:64–69.

Walker, S.A. 2001. Vascular plants of Vicksburg National Military Park. http://www.nps.gov/vick/visctr/flora.htm.

Ward, G.H. 1948. A flora of Chelan County, Washington. Master's thesis, State College of Washington [Washington State University], Pullman, Washington, U.S.A. 179 pp.

Warwick, S.I. 1979. The biology of Canadian weeds: 37. *Poa annua* L. Canad. J. Pl. Sci. 59:1053–1066.

Warwick, S.I. and S.G. Aiken. 1986. Electrophoretic evidence for the recognition of two species in annual wild rice (*Zizania*, Poaceae). Syst. Bot. 11:464–473.

Warwick, S.I. and L.D. Black. 1983. The biology of Canadian weeds: 61. *Sorghum halepense* (L.) Pers. Canad. J. Pl. Sci. 63:997–1014.

Warwick, S.I., L.D. Black, and B.F. Zilkey. 1985. The biology of Canadian weeds: 72. *Apera spica-venti* (L.) Beauv. Canad. J. Pl. Sci. 63:997–1014.

Watson, W.C. 1989. The vascular flora of Pilot Knob State Preserve. J. Iowa Acad. Sci 96:6–13.

Weber, W.A. 1984. A new genus of grasses from the western oil shales. Phytologia 55:1–2.

Weber, W.A. 1995. Checklist of Vascular Plants of Boulder County, Colorado. Natural History Inventory of Colorado No. 16. University of Colorado Museum, Boulder, Colorado, U.S.A. 68 pp.

Weimarck, G. 1971. Variation and taxonomy of *Hierochloë* (Gramineae) in the Northern Hemisphere. Bot. Not. 124:129–175.

Werner, P.A. and R. Rioux. 1977. The biology of Canadian weeds: 24. *Agropyron repens* (L.) Beauv. Canad. J. Pl. Sci. 57:905–919.

Whitney, K.D. 1996. Noteworthy collections. Madroño 43:336–337.

Wilder, G.J. and M.R. McCombs. 2006. New and significant records of vascular plants for Florida and for Collier County and Lee County, Florida. Sida 22:787–799.

Williams, A.H. 1997. Range expansion northward in Illinois and into Wisconsin of *Tridens flavus* (Poaceae). Rhodora 99:344–351.

Wilson, B.L. 1992. Checklist of the vascular flora of Page County, Iowa. J. Iowa Acad. Sci 99:22–33.

Wilson, B.L. and S. Gray. 2002. Resurrection of a century-old species distinction in *Calamagrostis*. Madroño 49:169–177.

Wilson, F.D. 1963. Revision of *Sitanion* (Triticeae, Gramineae). Brittonia 15:303–323.

Winstead, R. 1990. A taxonomic and ecological survey of the plant communities of Attala County, Mississippi. Master's thesis, Mississippi State University, Mississippi State, Mississippi, U.S.A. 347 pp.

Wipff, J.K. and S.L. Hatch. 1992. *Eustachys caribaea* (Poaceae: Chlorideae) in Texas. Sida 15:160–161.

Wipff, J.K. and S.D. Jones. 1994. *Melica subulata* (Poaceae: Meliceae): The first report for Colorado. Sida 16:210–211.

Wipff, J.K., S.D. Jones, and C.T. Bryson. 1994. *Eustachys glauca* and *E. caribaea* (Poaceae: Chlorideae): The first reports for Mississippi. Sida 16:211.

Wipff, J.K., R.I. Lonard, S.D. Jones, and S.L. Hatch. 1993. The genus *Urochloa* (Poaceae: Paniceae) in Texas, including one previously unreported species for the state. Sida 15:405–413.

Wolden, B.O. 1956. The flora of Emmet County, Iowa. Proc. Iowa Acad. Sci. 63:118–156.

Yatskievych, G. 2003. Steyermark's Flora of Missouri, vol. 1, rev. ed. Missouri Department of Convervation, Jefferson City, Missouri, U.S.A. 991 pp. http://biology.missouristate.edu/herbarium/.

Zebryk, T.M. 1998. Noteworthy collections. Castanea 63:78–79.

Zika, P.F. 1989. Noteworthy collections. Madroño 36:207.

Zika, P.F. 1990. Range expansion of some grasses in Vermont. Rhodora 92:80–89.

Zika, P.F. 2000a. Noteworthy collections. Madroño 47:214–216.

Zika, P.F. 2000b. Two more wccds in Maine. Rhodora 102:208–209.

Zika, P.F. and E.J. Marshall. 1991. Contributions to the flora of the Lake Champlain Valley, New York and Vermont, III. Bull. Torrey Bot. Club 118:58–61.

Zika, P.F., R.J. Stern, and H.E. Ahles. 1983. Contributions to the flora of the Lake Champlain Valley, New York and Vermont. Bull. Torrey Bot. Club 110:366–369.

Zika, P.F. and B.L. Wilson. 1998. Noteworthy collections. Madroño 45:86–87.

Zinck, M. 1998. Roland's Flora of Nova Scotia, vol. 2, ed. 3. Nova Scotia Museum and Nimbus Publishing, Province of Nova Scotia, Canada. 1297 pp.

Zobel, D.B. and C.R. Wasem. 1979. Pyramid Lake Research Natural Area. Federal Research Natural Areas in Oregon and Washington: A Guidebook for Scientists and Educators, Supplement No. 8. U.S. Department of Agriculture-Forest Service, Pacific Northwest Forest and Range Experiment Station, Corvallis, Oregon, U.S.A. 17 pp.

General Bibliography

The General Bibliography shows the complete citation for all of the "Selected References" in the present volume. The notes in square brackets refer to the treatment(s) in which a reference is cited, but many of them are of more general application. Book titles are listed in full; the names of journals are abbreviated according to Lawrence et al. (1968) and its supplement (Bridson and Smith 1991).

Aares, E., M. Nurminiemi, and C. Brochmann. 2000. Incongruent phylogeographies in spite of similar morphology, ecology, and distribution: *Phippsia algida* and *P. concinna* (Poaceae) in the North Atlantic region. Pl. Syst. Evol. 220:241–261. [**Phippsia**]

Aiken, S.G. and R.A. Buck. 2002. Aquatic leaves and regeneration of last year's straw in the arctic grass *Arctophila fulva*. Canad. Field-Naturalist 116:81–86. [**Arctophila**]

Aiken, S.G. and L.L. Consaul. 1995. Leaf cross sections and phytogeography: A potent combination for identifying members of *Festuca* subgg. *Festuca* and *Leucopoa* (Poaceae), occurring in North America. Amer. J. Bot. 82:1287–1299. [**Festuca**]

Aiken, S.G., L.L. Consaul, and M.J. Dallwitz. 1995 on. Grasses of the Canadian arctic archipelago: Descriptions, illustrations, identification, and information retrieval. http://www.mun.ca/biology/delta/arcticf/poa/index.htm. [**Anthoxanthum, Arctagrostis, Deschampsia, ×Dupoa, Phippsia**]

Aiken, S.G., L.L. Consaul, J.I. Davis and P.S. Manos. 1993. Systematic inferences from variation in isozyme profiles of arctic and alpine cespitose *Festuca* (Poaceae). Amer. J. Bot. 80:76–82. [**Festuca**]

Aiken, S.G., L.L. Consaul and L.P. Lefkovitch. 1995. *Festuca edlundiae* (Poaceae), a high arctic, new species compared enzymatically and morphologically with similar *Festuca* species. Syst. Bot. 20:374–392. [**Festuca**]

Aiken, S.G., M.J. Dallwitz, C.L. McJannet, and L.L. Consaul. 1997a. Biodiversity among *Festuca* (Poaceae) in North America: Diagnostic evidence from DELTA and clustering programs, and an INTKEY package for interactive, illustrated identification and information retrieval. Canad. J. Bot. 75:1527–1555. [**Festuca, Schedonorus**]

Aiken, S.G., M.J. Dallwitz, C.L. McJannet, and L.L. Consaul. 1997b. Fescue Grasses of North America: Interactive Identification and Information Retrieval. DELTA, CSIRO Division of Entomology, Canberra, Australia. CD-ROM. http://delta-intkey.com/festuca/index.htm. [**Lolium**]

Aiken, S.G. and S.J. Darbyshire. 1990. Fescue Grasses of Canada. Agriculture Canada Publ. 1844/E. Canadian Government Publishing Centre, Ottawa, Ontario, Canada. 113 pp. [**Festuca**]

Aiken, S.G., S.J. Darbyshire and L.P. Lefkovitch. 1985. Restricted taxonomic value of leaf sections in Canadian narrow-leaved *Festuca* (Poaceae). Canad. J. Bot. 63:995–1007. [**Festuca**]

Aiken, S.G. and G. Fedak. 1991. Cytotaxonomic observations on North American *Festuca* (Poaceae). Canad. J. Bot. 70:1940–1944. [**Festuca**]

Aiken, S.G. and S.E. Gardiner. 1991. SDS-PAGE of seed proteins in *Festuca* (Poaceae): Taxonomic implications. Canad. J. Bot. 69:1425–1432. [**Festuca**]

Aiken, S.G., S.E. Gardiner, and M.B. Forde. 1992. Taxonomic implications of SDS-PAGE analysis of seed proteins in North American taxa of *Festuca* subgenus *Festuca* (Poaceae). Biochem. Syst. & Ecol. 20:615–629. **[Festuca]**

Aiken, S.G., P.F. Lee, D. Punter, and J.M. Stewart. 1988. Wild Rice in Canada. Agriculture Canada Publication 1830. NC Press, Toronto, Ontario, Canada. 130 pp. **[Zizania]**

Aiken, S.G. and L.P. Lefkovitch. 1984. The taxonomic value of using epidermal characteristics in the Canadian rough fescue complex (*Festuca altaica*, *F. campestris*, *F. hallii*, "*F. scabrella*"). Canad. J. Bot. 62:1864–1870. **[Festuca]**

Aiken, S.G. and L.P. Lefkovitch. 1990. *Arctagrostis* (Poaceae) in North America and Greenland. Canad. J. Bot. 68:2422–2432. **[Arctagrostis]**

Aiken, S.G. and L.P. Lefkovitch. 1993. On the separation of two species within *Festuca* subg. *Obtusae* (Poaceae). Taxon 42:323–337. **[Festuca]**

Aiken, S.G., L.P. Lefkovitch, S.E. Gardiner, and W.W. Mitchell. 1994. Evidence against the existence of varieties in *Arctagostis latifolia* ssp. *arundinacea* (Poaceae). Canad. J. Bot. 72:1039–1050. **[Arctagrostis]**

Ainouche, M.L., R.J. Bayer, J.-P. Gourret, A. Defontaine, and M.-T. Misset. 1999. The allotetraploid invasive weed *Bromus hordeaceus* L. (Poaceae): Genetic diversity, origin and molecular evolution. Folia Geobot. 34:405–419. **[Bromus]**

Albers, F. 1973. Cytosystematische Untersuchungen in der Subtribus *Deschampsiineae* Holub (Tribus Aveneae Nees): I. Zwei Arten der Gattung *Corynephorus* P.B. Preslia 45:11–18. **[Corynephorus]**

Alexeev, E.B. 1977. To the systematics of Asian fescues (*Festuca*): I. Subgenera *Drymanthele*, *Subulatae*, *Schedonorus*, *Leucopoa*. Byull. Moskovsk. Obshch. Isp. Prir., Otd. Biol., n.s., 82(3):95–102. [In Russian]. **[Festuca]**

Alexeev, E.B. 1980. *Festuca* L.: Subgenera et sectiones novae ex America boreali et Mexico. Novosti Sist. Vyssh. Rast. 17:42–53. [In Russian]. **[Festuca]**

Alexeev, E.B. 1982. New and little known fescues (*Festuca* L.) of North America. Byull. Moskovsk. Obshch. Isp. Prir., Otd. Biol., n.s., 87(2):109–118. [In Russian]. **[Festuca]**

Alexeev, E.B. 1985. *Festuca* L. (Poaceae) in Alaska et Canada. Novosti Sist. Vyssh. Rast. 22:5–35. [In Russian]. **[Festuca]**

Aliscioni, S.S., L.M. Giussani, F.O. Zuloaga, and E.A. Kellogg. 2003. A molecular phylogeny of *Panicum* (Poaceae: Paniceae): Tests of monophyly and phylogenetic placement within the Panicoideae. Amer. J. Bot. 90:746–821. **[Panicum, Urochloa, Megathyrsus]**

Allred, K.W. 1993. *Bromus*, section *Pnigma*, in New Mexico, with a key to the bromegrasses of the state. Phytologia 74:319–345. **[Bromus]**

Anamthawat-Jónsson, K. 2005. The *Leymus* Ns-genome. Czech J. Genet. Pl. Breed. 41(Special Issue):13–20. **[Leymus, Psathyrostachys]**

Anderson, D.E. 1961. Taxonomy and distribution of the genus *Phalaris*. Iowa State Coll. J. Sci. 36:1–96. **[Phalaris]**

Anderson, H.M. 1991. *Melinis* P. Beauv. Pp. 210–213 *in* G.E. Gibbs Russell, L. Watson, M. Koekemoer, L. Smook, N.P. Barker, H.M. Anderson, and M.J. Dallwitz. Grasses of Southern Africa (ed. O.A. Leistner). National Botanic Gardens, Botanical Research Institute, Pretoria, Republic of South Africa. 437 pp. **[Melinis]**

Anderson, J.E. and A.A. Reznicek. 1994. *Glyceria maxima* (Poaceae) in New England. Rhodora 96:97–101. **[Glyceria]**

Anderson, L.C. and D.W. Hall. 1993. *Luziola bahiensis* (Poaceae): New to Florida. Sida 15:619–622. **[Luziola]**

Argus, G.W. and K.M. Pryer. 1990. Rare Vascular Plants in Canada: Our Natural Heritage. Canadian Museum of Nature, Ottawa, Ontario, Canada. 191 pp. **[Puccinellia]**

Arnow, L.A. 1994. *Koeleria macrantha* and *K. pyramidata* (Poaceae): Nomenclatural problems and biological distinctions. Syst. Bot. 19:6–20. **[Koeleria]**

Arriaga, M.O. 1983. Anatomía foliar de las especies de *Stipa* del subgénero *Pappostipa* (Stipeae-Poaceae) de Argentina. Revista Mus. Argent. Ci. Nat., Bernardino Rivadavia Inst. Nac. Invest. Ci. Nat., Bot. 6:89–141. **[Jarava]**

Arriaga, M.O. and M.E. Barkworth. 2006. *Amelichloa*: A new genus in the Stipeae (Poaceae). Sida 22:145–149. **[Achnatherum]**

Asay, K.H. and D.R. Dewey. 1979. Bridging ploidy differences in crested wheatgrass with hexaploid × diploid hybrids. Crop Science 19:519–523. **[Agropyron]**

Asay, K.H., K.B. Jensen, C. Hsiao, and D.R. Dewey. 1992. Probable origin of standard crested wheatgrass, *Agropyron desertorum* (Fisch. *ex* Link) Schultes. Canad. J. Pl. Sci. 72:763–772. **[Agropyron]**

Assadi, M. 1994. The genus *Elymus* L. (Poaceae) in Iran: Biosystematic studies and generic delimitation. Ph.D. dissertation, Lund University, Lund, Sweden. 104 pp. **[Thinopyrum]**

Assadi, M. and H. Runemark. 1995. Hybridization, genomic constitution and generic delimitation in *Elymus s.l.* (Poaceae, Triticeae). Pl. Syst. Evol. 194:189–205. **[Pseudoroegneria]**

Auquier, P. 1971. *Festuca rubra* subsp. *pruinosa* (Hack.) Piper: Morphologie, écologie, taxonomie. Lejeunia, n.s., 56:1–16. **[Festuca]**

Baden, C. 1991. A taxonomic revision of *Psathyrostachys* (Poaceae). Nordic J. Bot 11:3–26. **[Psathyrostachys]**

Baird, J.R. and J.W. Thieret. 1985. Notes on *Themeda quadrivalvis* (Poaceae) in Louisiana. Isleya 2:129–137. **[Themeda]**

Baldini, R.M. 1995. Revision of the genus *Phalaris* L. (Gramineae). Webbia 49:265–329. **[Phalaris]**

Barker, C.M. and C.A. Stace. 1982. Hybridization in the genera *Vulpia* and *Festuca*: The production of artificial F[1] plants. Nordic J. Bot. 2:435–444. **[Festuca]**

Barkworth, M.E. 1977. A taxonomic study of the large-glumed species of *Stipa* (Gramineae) in Canada. Canad. J. Bot. 56:606–625. [**Hesperostipa**]

Barkworth, M.E. 1981. Foliar epidermes and the taxonomy of North American Stipeae (Gramineae). Syst. Bot. 6:136–152. [**Achnatherum**]

Barkworth, M.E. 1982. Embryological characters and the taxonomy of the Stipeae (Gramineae). Taxon 31:233–243. [**Achnatherum**]

Barkworth, M.E. 1983. *Ptilagrostis* in North America and its relationship to other Stipeae (Gramineae). Syst. Bot. 8:395–419. [**Ptilagrostis**]

Barkworth, M.E. 1990. *Nassella* (Gramineae: Stipeae): Revised interpretation and nomenclatural changes. Taxon 39:597–614. [**Nassella**]

Barkworth, M.E. 1993a. *Nassella*. Pp. 1274–1276 *in* J.C. Hickman (ed.). The Jepson Manual: Higher Plants of California. University of California Press, Berkeley and Los Angeles, California, U.S.A. 1400 pp. [**Nassella**]

Barkworth, M.E. 1993b. North American Stipeae (Gramineae): Taxonomic changes and other comments. Phytologia 74:1–25. [**Achnatherum, Stipeae**]

Barkworth, M.E. 1997. Taxonomic and nomenclatural comments on the Triticeae in North America. Phytologia 83:302–311. [1997 on title page; printed in 1998]. [**Elymus, Thinopyrum**]

Barkworth, M.E. 2000. Changing perceptions in the Triticeae. Pp. 110–120 *in* S.W.L. Jacobs and J. Everett (eds.). Grasses: Systematics and Evolution. CSIRO Publishing, Collingwood, Victoria, Australia. 408 pp. [**Elymus**]

Barkworth, M.E. 2006. A new hybrid genus and 12 new combinations in North American grasses. Sida 22:495–501. [**×Elyleymus**]

Barkworth, M.E., L.K. Anderton, J. McGrew, and D.E. Giblin. 2006. Geography and morphology of the *Bromus carinatus* (Poaceae: Bromeae) complex. Madroño. 53:235–245. [**Bromus**]

Barkworth, M.E. and R.J. Atkins. 1984. *Leymus* Hochst. (Gramineae: Triticeae) in North America: Taxonomy and distribution. Amer. J. Bot. 71:609–625. [**Leymus**]

Barkworth, M.E. and R. von Bothmer. 2005. Twenty-one years later: Löve and Dewey's genomic classification proposal. Czech J. Genet. Pl. Breed. 41 (Special Issue):3–9. [**Elymus**]

Barkworth, M.E. and D.R. Dewey. 1985. Genomically based genera in the perennial Triticeae of North America: Identification and membership. Amer. J. Bot. 72:767–776. [**Triticeae**]

Barkworth, M.E. and J. Everett. 1987. Evolution in the Stipeae: Identification and relationships of its monophyletic taxa. Pp. 251–264 *in* T.R. Soderstrom, K.W. Hilu, C.S. Campbell, and M.E. Barkworth (eds.). Grass Systematics and Evolution. Smithsonian Institution Press, Washington, D.C., U.S.A. 473 pp. [**Hesperostipa, Stipa**]

Barkworth, M.E. and J. Linman. 1984. *Stipa lemmonii* (Vasey) Scribner (Poaceae): A taxonomic and distributional study. Madroño 31:48–56. [**Achnatherum**]

Barkworth, M.E. and M.A. Torres. 2001. Distribution and diagnostic characters of *Nassella* (Poaceae: Stipeae). Taxon 50:439–468. [**Nassella**]

Bartlett, E., S.J. Novak, and R.N. Mack. 2002. Genetic variation in *Bromus tectorum* (Poaceae): Differentiation in the eastern United States. Amer. J. Bot. 89:602–612. [**Bromus**]

Baum, B.R. 1977. Oats: Wild and Cultivated; A Monograph of the Genus *Avena* L. (Poaceae). Biosystematics Research Institute Monograph No. 14. Supply and Services Canada, Ottawa, Ontario, Canada. 463 pp. [**Avena**]

Baum, B.R. 1978. The status of *Hordeum brachyantherum* in eastern Canada, with related discussions. Canad. J. Bot. 56:107–109. [**Hordeum**]

Baum, B.R. and L.G. Bailey. 1990. Key and synopsis of North American *Hordeum* species. Canad. J. Bot. 68:2433–2442. [**Hordeum**]

Baum, B.R., C. Yen, and J.-L. Yang. 1991. *Roegneria*: Its generic limits and justification for its recognition. Canad. J. Bot. 69:282–294. [**Triticeae**]

Bay, C. 1992. A phytogeographical study of the vascular plants of northern Greenland—north of 74° northern latitude. Meddel. Grønland, Biosci. 36:1–102. [**Phippsia**]

Beal, W.J. 1896. Grasses of North America, vol. 2. Henry Holt & Company, New York, New York, U.S.A. 706 pp. [**Festuca**]

Belk, E. 1939. Studies in the anatomy and morphology of the spikelet and flower of the Gramineae. Ph.D. dissertation, Cornell University, Ithaca, New York, U.S.A. 183 pp. [**Anthoxanthum**]

Bennett, B.A. 2006. Siberian wild rye (*Elymus sibiricus* L., Poaceae) in western North America: Native or introduced? BEN [Botanical Electronic News] #366. http://www.ou.edu/cas/botany-micro/ben/. [**Elymus**]

Bernhardt, K.G., M. Koch, E. Ulbel, and J. Webhofer. 2004. The soil seed bank as a resource for in situ and ex situ conservation of extinct species. [Pagination unknown] *in* E. Robbrecht and A. Bogaerts (eds.). EuroGard III: Papers from the Third European Botanic Gardens Congress and the Second European Botanic Gardens Education Congress (BEDUCO II). Scripta Botanica Belgica Series, No. 29. National Botanic Garden, Meise, Belgium. 177 pp. [**Coleanthus**]

Björkman, S.O. 1960. Studies in *Agrostis* and related genera. Symb. Bot. Upsal. 17:1–112. [**Agrostis, Podagrostis**]

Blattner, F.R. 2006. Multiple intercontinental dispersals shaped the distribution area of *Hordeum* (Poaceae). New Phytol. 169:603–614. [**Hordeum**]

Bödvarsdóttir, S.K. and K. Anamthawat-Jónsson. 2003. Isolation, characterization, and analysis of *Leymus*-specific DNA sequences. Genome 46:673–682. [Elymus, Leymus, Psathyrostachys]

Boivin, B. 1967. Énumérations des plantes du Canada VI—Monopsides (2ème partie). Naturaliste Canad. 94: 471–528. [×Pascoleymus]

Bor, N.L. 1960. The Grasses of Burma, Ceylon, India and Pakistan (Excluding Bambuseae). International Series of Monographs on Pure and Applied Biology, Division: Botany, vol. 1. Pergamon Press, New York, Oxford, London, and Paris. 767 pp. [Hygroryza]

Borrill, M. 1955. Breeding systems and compatibility in *Glyceria*. Nature 175:561–563. [Glyceria]

Bothmer, R. von, N. Jacobsen, C. Baden, R.B. Jørgensen, and I. Linde-Laursen. 1995. An Ecogeographical Study of the Genus *Hordeum*, ed. 2. Systematic and Ecogeographic Studies on Crop Genepools No. 7. International Board of Plant Genetic Resources, Rome, Italy. 129 pp. [Hordeum]

Bothmer, R. von, N. Jacobsen, and R.O. Seberg. 1993. Variation and taxonomy in *Hordeum depressum* and in the *H. brachyantherum* complex (Poaceae). Nordic J. Bot. 13:3–17. [Hordeum]

Bowden, W.M. 1957. Cytotaxonomy of section *Psammelymus* of the genus *Elymus*. Canad. J. Bot. 35:951–993. [Leymus]

Bowden, W.M. 1958. Natural and artificial ×*Elymordeum* hybrids. Canad. J. Bot. 36:101–123. [×Elyhordeum, Elymus]

Bowden, W.M. 1959. Chromosome numbers and taxonomic notes on northern grasses: I. Tribe Triticeae. Canad. J. Bot. 37:1143–1151 [×Elyleymus, Leymus]

Bowden, W.M. 1960a. Chromosome numbers and taxonomic notes on northern grasses: III. Festuceae. Canad. J. Bot. 38:117–131. [Glyceria]

Bowden, W.M. 1960b. Typification of *Elymus macounii* Vasey. Bull. Torrey Bot. Club. 87:205–208. [×Elyhordeum]

Bowden, W.M. 1961. Chromosome numbers and taxonomic notes on northern grasses: IV. Tribe *Festuceae*; *Poa* and *Puccinellia*. Canad. J. Bot. 39:123–138. [Poa]

Bowden, W.M. 1964. Cytotaxonomy of the species and interspecific hybrids of the genus *Elymus* in Canada and neighboring areas. Canad. J. Bot. 42:547–601. [Elymus]

Bowden, W.M. 1967. Taxonomy of intergeneric hybrids of the tribe Triticeae from North America. Canad. J. Bot. 45:711–724. [Elymus, ×Leydeum]

Boyle, W.S. 1945. A cytotaxonomic study of the North American species of *Melica*. Madroño 8:1–26. [Melica]

Brandenburg, D.M. 1985. Systematic studies in the Poaceae and Cyperaceae. Ph.D. dissertation, University of Oklahoma, Norman, Oklahoma, U.S.A. 249 pp. [Arrhenatherum]

Brandenburg, D.M., W.H. Blackwell, and J.W. Thieret. 1991. Revision of the genus *Cinna* (Poaceae). Sida 14:581–596. [Cinna]

Brandenburg, D.M., J.R. Estes, and S.L. Collins. 1991. A revision of *Diarrhena* (Poaceae) in the United States. Bull. Torrey Bot. Club 118:128–136. [Diarrhena]

Brandenburg, D.M., J.R. Estes, S.D. Russell, and J.W. Thieret. 1991. One-nerved paleas in *Cinna arundinacea* L. (Poaceae). Trans. Kentucky Acad. Sci. 52:94–96. [Cinna]

Brandenburg, D.M., J.R. Estes, and J.W. Thieret. 1991. Hard grass (*Sclerochloa dura*, Poaceae) in the United States. Sida 14:369–376. [Sclerochloa]

Brandenburg, D.M. and J.W. Thieret. 1996. *Sclerochloa dura* (Poaceae) in Kentucky. Trans. Kentucky Acad. Sci. 57:47–48. [Sclerochloa]

Brandenburg, D.M. and J.W. Thieret. 2000. *Cinna* and *Limnodea* (Poaceae): Not congeneric. Sida 19:195–200. [Cinna, Limnodea]

Bremer, K. 2000. Early Cretaceous lineages of monocot flowering plants. Proc. Natl. Acad. Sci. U.S.A. [PNAS] 97:4707–4711. [Poaceae]

Bremer, K. 2002. Gondwanan evolution of the grass alliance of families (Poales). Evolution 56:1374–1387. [Poaceae]

Bridson, G.D.R. and E.R. Smith (eds.). 1991. B–P–H/S: Botanico–Periodicum–Huntianum/Supplementum. Hunt Institute for Botanical Documentation, Carnegie Mellon University, Pittsburgh, Pennsylvania, U.S.A. 1068 pp. [General Bibliography, Geographic Bibliography]

Briggs, B.G., A.D. Marchant, S. Gilmore and C.L. Porter. 2000. A molecular phylogeny of Restionaceae and allies. Pp. 661–671 *in* K.L. Wilson and D.A. Morrision (eds.). Monocots: Systematics and Evolution. CSIRO Publishing, Collingwood, Victoria, Australia. 738 pp. [Poaceae]

Brooks, R.E. 1974. Intraspecific variation in *Elymus virginicus* (Gramineae) in the central United States. Master's thesis, University of Kansas, Lawrence, Kansas, U.S.A. 112 pp. [Elymus]

Brown, W.V. 1952. The relation of soil moisture to cleistogamy in *Stipa leucotricha*. Bot. Gaz. 113:438–444. [Nassella]

Brown, W.V. and G.A. Pratt. 1960. Hybridization and introgression in the genus *Elymus*. Amer. J. Bot. 47:669–676. [Elymus]

Brummit, R.K. 1998. Report of the Committee for Spermatophyta: 47. Taxon 47:869–870. [Brachiaria]

Brummitt, R.K. and C.E. Powell (eds.). 1992. Authors of Plant Names: A list of authors of scientific names of plants, with recommended standard forms of their names, including abbreviations. Royal Botanic Gardens, Kew, England. 732 pp. [Introduction, Names and Synonyms]

Brysting, A.K., S.G. Aiken, L.P. Lefkovitch, and R.L. Boles. 2003. *Dupontia* (Poaceae) in North America. Canad. J. Bot. 81:769–779. [×Arctodupontia, Dupontia]

Brysting, A.K., M.F. Fay, I.J. Leitch, and S.G. Aiken. 2004. One or more species in the arctic grass genus *Dupontia*?—A contribution to the Panarctic Flora project. Taxon 53:365–382. [Arctophila, Dupontia]

Bush, B.F. 1926. The Missouri species of *Elymus*. Amer. Midl. Naturalist 10:49–88. [Elymus]

But, P.P.H. 1977. Systematics of *Pleuropogon* R. Br. (Poaceae). Ph.D. dissertation, University of California, Berkeley, California, U.S.A. 229 pp. [Pleuropogon]

But, P.P.-H., L.-C. Chia, H.-L.F. Chia, and S.-Y. Hu. 1985. Hong Kong Bamboos. Urban Council, Hong Kong. 85 pp. [Bambusa]

But, P.P.H., J. Kagan, V. Crosby, and J.S. Shelly. 1985. Rediscovery and reproductive biology of *Pleuropogon oregonus* (Poaceae). Madroño 32:189–190. [Pleuropogon]

Bystriakova, N., V. Kapos, C.M.A. Stapleton, and J. Lysenko. 2003. Bamboo Biodiversity: Information for Planning Conservation and Management in the Asia–Pacific Region. UNEP–WCMC Biodiversity Series No. 14. United Nations Environment Programme–World Conservation Monitoring Centre and International Network for Bamboo and Rattan, Cambridge, England. 71 pp. [Bambuseae]

Calaway, M.L. and J.W. Thieret. 1985. *Amphibromus scabrivalvis* (Gramineae) in Louisiana. Sida 11:207–214. [Amphibromus]

Calderón, C.E. and T.R. Soderstrom. 1980. The genera of Bambusoideae (Poaceae) of the American continent: Keys and comments. Smithsonian Contr. Bot. 44:1–27. [Bambuseae]

Callihan, R.H. and D.S. Pavek. 1988. *Milium vernale* survey in Idaho County, Idaho. Idaho Weed Control Rep. 1988:153. [Milium]

Campbell, C.S., P.E. Garwood and L.P. Specht. 1986. Bambusoid affinities of the north temperate genus *Brachyelytrum* (Gramineae). Bull. Torrey Bot. Club 113:135–141. [Brachyelytreae, Brachyelytrum]

Campbell, J.J.N. 2000. Notes on North American *Elymus* species (Poaceae) with paired spikelets: I. *E. macgregorii* sp. nov. and *E. glaucus* ssp. *mackenzii* comb. nov. J. Kentucky Acad. Sci. 61:88–98. [Elymus]

Campbell, J.J.N. 2002. Notes on North American *Elymus* species (Poaceae) with paired spikelets: II. The *interruptus* group. J. Kentucky Acad. Sci. 62:19–38. [Elymus]

Campbell, J.J.N. and A. Haines. 2002. Corrections and additions to: "Campbell, J.J.N. 2000. Notes on North American *Elymus* species (Poaceae) with paired spikelets: I. *E. macgregorii* sp. nov. and *E. glaucus* ssp. *mackenzii* comb. nov. J. Ky. Acad. Sci. 61:88–98." J. Kentucky Acad. Sci. 62:65. [Elymus]

Carlbom, C.G. 1967. A biosystematic study of some North American species of *Agrostis* L. and *Podagrostis* (Griseb.) Scribn. & Merr. Ph.D. dissertation, Oregon State University, Corvallis, Oregon, U.S.A. 232 pp. [Agrostis, Podagrostis]

Caro, J.A. and E. Sánchez. 1971. La identidad de *Stipa brachychaeta* Godron, *S. caudata* Trinius y *S. bertrandii* Philippi. Darwinia 16:637–653. [Amelichloa]

Caro, J.A. and E. Sánchez. 1973. Las especies de *Stipa* (Gramineae) del subgénero *Jarava*. Kurtziana 7:61–116. [Jarava]

Carroll, C.P. and K. Jones. 1962. Cytotaxonomic studies in *Holcus*: III. A morphological study of the triploid F_1 hybrid between *H. lanatus* L. and *H. mollis* L. New Phytol. 61:72–84. [Holcus]

Catalán, P., E.A. Kellogg, and R.G. Olmstead. 1997. Phylogeny of Poaceae subfamily Poöideae based on chloroplast *ndh*F gene sequences. Molec. Phylogenet. Evol. 8:150–166. [Meliceae, Poeae, Poöideae]

Catalán, P. and R.G. Olmstead. 2000. Phylogenetic reconstruction of the genus *Brachypodium* P. Beauv. (Poaceae) from combined sequences of chloroplast *ndh*F gene and nuclear ITS. Pl. Syst. Evol. 200:1–19. [Brachypodium]

Catalán, P., Y. Shi, L. Armstrong, and C.A. Stace. 1995. Molecular phylogeny of the grass genus *Brachypodium* P. Beauv. based on RFLP and RAPD analysis. Bot. J. Linn. Soc. 113:263–280. [Brachypodium]

Catalán, P., P. Torrecilla, J.Á. López Rodríguez, and R.G. Olmsted. 2004. Phylogeny of the festucoid grasses of subtribe Loliinae and allies (Poeae, Poöideae) inferred from ITS and *trn*L–F sequences. Molec. Phylogenet. Evol. 31:517–541. [Leucopoa, Poeae, Poöideae]

Catling, P.M., G. Mitrow, L. Black, and S. Caribyn. 2004. Status of the alien race of common reed (*Phragmites australis*) in the Canadian maritime provinces. BEN [Botanical Electronic News] #324. http://www.ou.edu/cas/botany-micro/ben/. [Phragmites]

Cayouette, J. and S.J. Darbyshire. 1993. The intergeneric hybrid grass "*Poa labradorica*". Nordic J. Bot. 13:615–629. [×Dupoa]

Chambers, K.L. 1985. Pitfalls in identifying *Ventenata dubia* (Poaceae). Madroño 32:120–121. [Ventenata]

Charmet, G., C. Ravel, and F. Balfourier. 1997. Phylogenetic analysis in the *Festuca–Lolium* complex using molecular markers and ITS rDNA. Theor. Appl. Genet. 94:1038–1046. [Schedonorus]

Cheeke, P.R. and L.R. Shull. 1985. Natural Toxicants in Feeds and Poisonous Plants. AVI Publishing Company, Westport, Connecticut, U.S.A. 492 pp. [Achnatherum]

Chen, S.-L., B. Sun, L. Liu, Z. Wu, S. Lu, D. Li, Z. Wang, Z. Zhu, N. Xia, L. Jia, G. Zhu, Z. Guo, G. Yang, W. Chen, X. Chen, S.M. Phillips, C. Stapleton, R.J. Soreng, S.G. Aiken, N.N. Tzvelev, P.M. Peterson, S.A. Renvoize, M.V. Olonova, and K.H. Ammann. 2006. Poaceae (Gramineae). Pp. 1–2 *in* Z.-Y. Wu, P.H. Raven, and D.-Y. Hong (eds.). Flora of China, vol. 22 (Poaceae). Science Press, Beijing, Peoples Republic of China and Missouri Botanical Garden Press, St. Louis, Missouri, U.S.A. 653 pp. http://flora.huh.harvard.edu/china/mss/volume22/index.htm. [Poaceae]

Chen, S.-L. and G.-H. Zhu. 2006a. *Agropyron*. Pp. 439–441 *in* Z.-Y. Wu, P.H. Raven, and D.-Y. Hong (eds.). Flora of China, vol. 22 (Poaceae). Science Press, Beijing, Peoples Republic of China and Missouri Botanical Garden Press, St. Louis, Missouri, U.S.A. 653 pp. http://flora.huh.harvard.edu/china/mss/volume22/index.htm. [Agropyron]

Chen, S.-L. and G.-H. Zhu. 2006b. *Leymus*. Pp. 387–394 *in* Z.-Y. Wu, P.H. Raven, and D.-Y. Hong (eds.). Flora of China, vol. 22 (Poaceae). Science Press, Beijing, Peoples Republic of China and Missouri Botanical Garden Press, St. Louis, Missouri, U.S.A. 653 pp. http://flora.huh.harvard.edu/china/mss/volume22/index.htm. [Leymus]

Chester, E.W., B.E. Wofford, H.R. DeSelm, and A.M. Evans. 1993. Atlas of Tennessee Vascular Plants, vol. 1. Austin Peay State University Miscellaneous Publication No. 9. The Center for Field Biology, Austin Peay State University, Clarksville, Tennessee, U.S.A. 118 pp. [Glyceria]

Chiapella, J. 2000. The *Deschampsia cespitosa* complex in central and northern Europe: A morphological analysis. Bot. J. Linn. Soc. 134:495–512. [Deschampsia]

Chiapella, J. and N.S. Probatova. 2003. The *Deschampsia cespitosa* complex (Poaceae: Aveneae) with special reference to Russia. Bot. J. Linn. Soc. 142:213–228. [Deschampsia]

Church, G.L. 1949. Cytotaxonomic study of *Glyceria* and *Puccinellia*. Amer. J. Bot. 36:155–165. [Glyceria]

Church, G.L. 1952. The genus *Torreyochloa*. Rhodora 54:197–200. [Torreyochloa]

Church, G.L. 1954. Interspecific hybridization in eastern *Elymus*. Rhodora 56:185–197. [Elymus]

Church, G.L. 1958. Artificial hybrids of *Elymus virginicus* with *E. canadensis, interruptus, riparius,* and *wiegandii*. Amer. J. Bot. 45:410–417. [Elymus]

Church, G.L. 1967a. Pine Hills *Elymus*. Rhodora 69:330–345. [Elymus]

Church, G.L. 1967b. Taxonomic and genetic relationships of eastern North American species of *Elymus* with setaceous glumes. Rhodora 69:121–162. [Elymus]

Cialdella, A.M. and M.O. Arriaga. 1998. Revisión de las especies sudamericanas del género *Piptochaetium* (Poaceae, Poöideae, Stipeae). Darwiniana 36:105–157. [Piptochaetium]

Cialdella, A.M. and L.M. Giussani. 2002. Phylogenetic relations of the genus *Piptochaetium* (Poaceae: Poöideae, Stipeae): Evidence from morphological data. Ann. Missouri Bot. Gard. 89:305–336. [Piptochaetium]

Clark, L.G. and E.J. Judziewicz. 1996. The grass subfamilies Anomochlooideae and Pharoideae (Poaceae). Taxon 45:641–645. [Phareae, Pharus]

Clark, L.G., W. Zhang, and J.F. Wendel. 1995. A phylogeny of the grass family (Poaceae) based on *ndh*F sequence data. Syst. Bot. 20:436–460. [Phareae]

Clarke, G.C.S. 1980. *Deschampsia* Beauv. Pp. 225–227 *in* T.G. Tutin, V.H. Heywood, N.A. Burges, D.M. Moore, D.H. Valentine, S.M. Walters. and D.A. Webb (eds.). Flora Europaea, vol. 5. Cambridge University Press, Cambridge, England. 452 pp. [Deschampsia]

Clayton, W.D. and S.A. Renvoize. 1986. Genera Graminum: Grasses of the World. Kew Bull., Addit. Ser. 13. Her Majesty's Stationery Office, London, England. 389 pp. [Oryzeae]

Conert, H.J. 1985. *Mibora*. Pp. 206–210 *in* G. Hegi. Illustrierte Flora von Mitteleuropa, ed. 3. Band I,Teil 3, Lieferung 3, Bg. 11–15 (pp. 161–240). Verlag Paul Parey, Berlin and Hamburg, Germany. [Mibora]

Conert, H.J. 1992a. *Coleanthus*. Pp. 434–437 *in* G. Hegi. Illustrierte Flora von Mitteleuropa, ed. 3. Band I, Teil 3, Lieferung 6 (pp. 401–480). Verlag Paul Parey, Berlin and Hamburg, Germany. [Coleanthus]

Conert, H.J. 1992b. *Glyceria*. Pp. 440–457 *in* G. Hegi. Illustrierte Flora von Mitteleuropa, ed. 3. Band I, Teil 3, Lieferung 6 (pp. 401–480). Verlag Paul Parey, Berlin and Hamburg, Germany. [Glyceria]

Conert, H.J. 1992c. *Sesleria*. Pp. 473–480 *in* G. Hegi. Illustrierte Flora von Mitteleuropa, ed. 3. Band I, Teil 3, Lieferung 6 (pp. 401–480). Verlag Paul Parey, Berlin and Hamburg, Germany. [Sesleria]

Conert, H.J. 1994. *Sesleria* (continued). Pp. 481–486 *in* G. Hegi. Illustrierte Flora von Mitteleuropa, ed. 3. Band I, Teil 3, Lieferung 7 (pp. 481–560). Blackwell Wissenschafts-Verlag, Berlin, Germany. [Sesleria]

Consaul, L.L. and S.G. Aiken. 1993. Limited taxonomic value of palea intercostal characteristics in North American *Festuca* (Poaceae). Canad. J. Bot. 71:1651–1659. [Festuca]

Consaul, L.L. and L.J. Gillespie. 2001. A re-evaluation of species limits in Canadian Arctic island *Puccinellia* (Poaceae): Resolving key characters. Canad. J. Bot. 79:927–956. [Puccinellia]

Consaul, L.L., L.J. Gillespie, and K.I. MacInnes. [in press]. Addition to the flora of Canada? A specimen from the Arctic Archipelago, Northwest Territories links two allopatric species of alkali grass. [Puccinellia]

Cope, E.A. 1994. Further notes on beachgrasses (*Ammophila*) in northeastern North America. Newsletter New York Fl. Assoc. 5(1):5–7. [Ammophila]

Cope, T.A. 1982. Flora of Pakistan, No. 143: Poaceae (E. Nasir and S.I. Ali, eds.). Pakistan Agricultural Research Council and University of Karachi, Islamabad and Karachi, Pakistan. 678 pp. [Bromus, Polypogon]

Cotton, R. and C.A. Stace. 1967. Taxonomy of the genus *Vulpia* (Gramineae): I. Chromosome numbers and geographical distribution of the Old World species. Genetica 46:235–255. [Vulpia]

Covas, G. 1949. Taxonomic observations on the North American species of *Hordeum*. Madroño 9:233–264. [×Leydeum]

Crampton, B. 1955. *Scribneria* in California. Leafl. W. Bot. 7:219–20. [Scribneria]

Curto, M.L. and D.H. Henderson. 1998. A new *Stipa* (Poaceae: Stipeae) from Idaho and Nevada. Madroño 45:57–63. [**Piptatherum**]

Cusick, A.W., R.K. Rabeler, and M.J. Oldham. 2002. Hard or fairgrounds grass (*Sclerochloa dura*, Poaceae) in the Great Lakes region. Michigan Bot. 41:125–135. [**Sclerochloa**]

Daniel, T.F., C. Best, J. Guggolz, and B. Guggolz. 1992. Noteworthy collections: *Gaudinia fragilis*. Madroño 39:309–310. [**Gaudinia**]

Dannhardt, G. and L. Steindl. 1985. Alkaloids of *Lolium temulentum*: Isolation, identification and pharmacological activity. Pl. Med. (Stuttgart) 1985:212–214. [**Lolium**]

Darbyshire, S.J. 1993. Realignment of *Festuca* subgenus *Schedonorus* with the genus *Lolium*. Novon 3:239–243. [**Schedonorus**]

Darbyshire, S.J. 1997. Tall Wheatgrass, *Elymus elongatus* subsp. *ponticus*, in Nova Scotia. Rhodora 99:161–165. [**Thinopyrum**]

Darbyshire, S.J., J. Cayouette, and S.I. Warwick. 1992. The intergeneric hybrid origin of *Poa labradorica* Steudel (Poaceae). Pl. Syst. Evol. 181:57–76. [**×Dupoa**]

Darbyshire, S.J. and S.I. Warwick. 1992. Phylogeny of North American *Festuca* (Poaceae) and related genera using chloroplast DNA restriction site variation. Canad. J. Bot. 70:2415–2429. [1992 on title page; printed in 1993]. [**Festuca, Leucopoa**]

Darke, R. 1999. The Color Encyclopedia of Ornamental Grasses: Sedges, Rushes, Restios, Cat-Tails, and Selected Bamboos. Timber Press, Portland, Oregon, U.S.A. 325 pp. [**Celtica, Macrochloa**]

Daubenmire, R.F. 1939. The taxonomy and ecology of *Agropyron spicatum* and *A. inerme*. Bull. Torrey Bot. Club 66:327–329. [**Pseudoroegneria**]

Daubenmire, R.F. 1960. An experimental study of variation in the *Agropyron spicatum–A. inerme* complex. Bot. Gaz. 122:104–108. [**Pseudoroegneria**]

Davies, R.S. 1980. Introgression between *Elymus canadensis* and *E. virginicus* L. (Triticeae, Poaceae) in south central United States. Ph.D. dissertation, Texas A&M University, College Station, Texas, U.S.A. 232 pp. [**Elymus**]

Davis, J.I. 1983. Phenotypic plasticity and the selection of taxonomic characters in *Puccinellia* (Poaceae). Syst. Bot. 8:341–353. [**Puccinellia**]

Davis, J.I. and R.J. Soreng. 1993. Phylogenetic structure in the grass family (Poaceae) as inferred from chloroplast DNA restriction site variation. Amer. J. Bot. 80:1444–1454. [**Nardeae**]

Davis, P.H. 1985. Flora of Turkey and the East Aegean Islands, vol. 9. Edinburgh University Press, Edinburgh, Scotland. 724 pp. [**Bromus**]

Decker, H.F. 1964. Affinities of the grass genus *Ampelodesmos*. Brittonia 16:76–79. [**Ampelodesmos, Stipeae**]

Desvaux, E. 1854. Gramineas. Pp. 233–469 *in* C. Gay. Flora Chilena [Historia Fisica y Politica de Chile], vol. 6. Museo Historia Natural, Santiago, Chile. 551 pp. [1853 on title page; printed in March 1854]. [**Nassella**]

Dewey, D.R. 1963. Natural hybrids of *Agropyron trachycaulum* and *Agropyron scribneri*. Bull. Torrey Bot. Club 90:111–120. [**Elymus**]

Dewey, D.R. 1964. Natural and synthetic hybrids of *Agropyron spicatum* × *Sitanion hystrix*. Bull. Torrey Bot. Club 91:396–405. [**×Pseudelymus**]

Dewey, D.R. 1965. Morphology, cytology, and fertility of synthetic hybrids of *Agropyron spicatum* × *Agropyron dasystachyum-riparium*. Bot. Gaz. 126:269–275. [**Elymus**]

Dewey, D.R. 1967a. Genome relations between *Agropyron scribneri* and *Sitanion hystrix*. Bull. Torrey Bot. Club 94:395–404. [**Elymus**]

Dewey, D.R. 1967b. Synthetic *Agropyron–Elymus* hybrids: II. *Elymus canadensis* × *Agropyron dasystachyum*. Amer. J. Bot. 54:1084–1089. [**Elymus**]

Dewey, D.R. 1968. Synthetic hybrids of *Agropyron dasystachyum* × *Elymus glaucus* and *Sitanion hystrix*. Bot. Gaz. 129:316–322. [**Elymus**]

Dewey, D.R. 1970. The origin of *Agropyron albicans*. Amer. J. Bot. 57:12–18. [**Elymus**]

Dewey, D.R. 1974. Cytogenetics of *Elymus sibiricus* and its hybrids with *Agropyron tauri*, *Elymus canadensis*, and *Agropyron caninum*. Bot. Gaz. 135:80–87. [**Elymus**]

Dewey, D.R. 1975a. Introgression between *Agropyron dasystachyum* and *A. trachycaulum*. Bot. Gaz. 136:122–128. [**Elymus**]

Dewey, D.R. 1975b. The origin of *Agropyron smithii*. Amer. J. Bot. 62:524–530. [**Pascopyrum**]

Dewey, D.R. 1976. Cytogenetics of *Agropyron pringlei* and its hybrids with *A. spicatum*, *A. scribneri*, *A. violaceum*, and *A. dasystachyum*. Bot. Gaz. 137:179–185. [**Elymus**]

Dewey, D.R. 1982. Genomic and phylogenetic relationships among North American perennial Triticeae. Pp. 51–58 *in* J.R. Estes, R.J. Tyrl, and J.N. Brunken (eds.). Grasses and Grasslands. University of Oklahoma Press, Norman, Oklahoma, U.S.A. 312 pp. [**Elymus**]

Dewey, D.R. 1984. The genomic system of classification as a guide to intergeneric hybridization in the perennial Triticeae. Pp. 209–279 *in* J.P. Gustafson (ed.). Gene Manipulation in Plant Improvement. Plenum Press, New York, New York, U.S.A. 668 pp. [**Pseudoroegneria, Triticeae**]

Dewey, D.R. 1986. Taxonomy of the crested wheatgrasses (*Agropyron*). Pp. 31–42 *in* K.L. Johnson (ed.). Crested Wheatgrass: Its Values, Problems and Myths; Symposium Proceedings. Utah State University, Logan, Utah, U.S.A. 348 pp. [**Agropyron**]

Dewey, D.R. and A.H. Holmgren. 1962. Natural hybrids of *Elymus cinereus* and *Sitanion hystrix*. Bull. Torrey Bot. Club 89:217–228. [**×Elyleymus**]

Deyl, M. 1980. *Sesleria* Scop. Pp. 173–177 *in* T.G. Tutin, V.H. Heywood, N.A. Burges, D.M. Moore, D.H. Valentine, S.M. Walters and D.A. Webb (eds.). Flora Europaea, vol. 5. Cambridge University Press, Cambridge, England. 439 pp. [**Sesleria**]

Dillman, A.C. 1946. The beginnings of crested wheatgrass in North America. J. Amer. Soc. Agron. 38:237–250. [**Agropyron**]

Dixon, J.M. 1991. Biological flora of the British Isles: *Avenula* (Dumort.) Dumort. J. Ecol. 79:829–865. [**Avenula**]

Dixon, J.M. 1995. *Trisetum flavescens* (L.) Beauv. (*T. pratense* Pers., *Avena flavescens* L.). J. Ecol. 83:895–909. [**Trisetum**]

Doğan, M. 1999. A concise taxonomic revision of the genus *Alopecurus* L. (Gramineae). Turk. J. Bot. 23:245–262. [**Alopecurus**]

Dore, W.G. 1950. Persian darnel in Canada. Sci. Agric. (Ottawa) 30:157–164. [**Lolium**]

Dore, W.G. 1969. Wild Rice. Canada Department of Agriculture Publication No. 1393. Information Canada, Ottawa, Ontario, Canada. 84 pp. [**Zizania**]

Dore, W.G. and J. McNeill. 1980. Grasses of Ontario. Research Branch, Agriculture Canada Monograph No. 26. Canadian Government Publishing Centre, Hull, Québec, Canada. 568 pp. [**Festuca, Glyceria, Oryzopsis**]

Dorofeev, V.F. and E.F. Migushova. 1979. Pshenitsa [Wheat]. (Vol. 1 *in* V.F. Dorofeev and O.N. Korovina (eds.). Kul'turnaia Flora SSSR [Flora of Cultivated Plants]). Kolos, Leningrad [St. Petersburg], Russia. 346 pp. [In Russian.]. [**Triticum**]

Douglas, G.W., G.B. Straley, and D.V. Meidinger (eds). 1994. The Vascular Plants of British Columbia: Part 4–Monocotyledons. Research Branch, Ministry of Forests, Victoria, British Columbia, Canada. 257 pp. [**Corynephorus**]

Dransfield, S. and E.A. Widjaja (eds.). 1995. Plant Resources of South-East Asia [PROSEA] No. 7: Bamboos. Backhuys, Leiden, The Netherlands. 189 pp. [**Bambusa**]

Dubé, M. 1983. Addition de *Festuca gigantea* (L.) Vill. (Poaceae) à la flore du Canada. Naturaliste Canad. 110:213–215. [**Festuca, Schedonorus**]

Dubé, M. and P. Morisset. 1987. Morphological and leaf anatomical variation in *Festuca rubra sensu lato* (Poaceae) from eastern Quebec. Canad. J. Bot. 65:1065–1077. [**Festuca**]

Dubé, M. and P. Morisset. 1995. La variation des caractères épidermiques foliaires chez le *Festuca rubra sensu lato* (Poaceae) dans l'est du Canada. Canad. J. Bot. 74:1425–1438. [**Festuca**]

Dubé, M. and P. Morisset. 1996. La plasticité phénotipique des caractères anatomiques foliaires chez le *Festuca rubra* L. (Poaceae). Canad. J. Bot. 74:1708–1718. [**Festuca**]

Dubé, M., P. Morisset and J. Murdock. 1985. Races chromosomiques chez le *Festuca rubra sensu lato* (Poaceae) dans l'est du Québec. Canad. J. Bot. 63:227–231. [**Festuca**]

Duckert-Henroid, M.M. and C. Favarger. 1987. Contribution à la Cytotaxonomie et à la Cytogéographie des *Poa* (Poaceae = Gramineae) de la Suisse. Mémoires de la Société Helvétique des Sciences Naturelles No. 100. Birkhäuser, Basel, Switzerland. 130 pp. [**Poa**]

Duistermaat, H. 1987. A revision of *Oryza* (Gramineae) in Malesia and Australia. Blumea 32:157–193. [**Oryza**]

Duvall, M.R. and D.D. Biesboer. 1988. Nonreciprocal hybridization failure in crosses between annual wild-rice species (*Zizania palustris* × *Z. aquatica*: Poaceae). Syst. Bot. 13:229–234. [**Zizania**]

Dvorak, J., Z.-C. Luo, Z.-L. Yang, and H.-B. Zhang. 1998. The structure of the *Aegilops tauschii* genepool and the evolution of hexaploid wheat. Theor. Appl. Genet. 97:657–670. [**Triticum**]

Dyksterhuis, E.J. 1949. Axillary cleistogenes in *Stipa leucotricha* and their role in nature. Ecology 26:195–199. [**Nassella**]

Edelman, D.K., T.R. Soderstrom, and G.F. Deitzer. 1985. Bamboo introduction and research in Puerto Rico. J. Amer. Bamboo Soc. 6: 43–57. [**Bambusa**]

Edgar, E. 1995. New Zealand species of *Deyeuxia* P. Beauv. and *Lachnagrostis* Trin. (Gramineae: Aveneae). New Zealand J. Bot. 33:1–33. [**Agrostis, Calamagrostis, Lachnagrostis**]

Edgar, E. and H.E. Connor. 2000. Flora of New Zealand, vol. 5. Manaaki Whenua Press, Lincoln, New Zealand. 650 pp. [**Agrostis, Ehrharteae, Lachnagrostis, Schedonorus**]

Elias, M.K. 1942. Tertiary Prairie Grasses and Other Herbs from the High Plains. Geological Society of America Special Paper No. 41. The [Geological] Society [of America, New York, New York]. 176 pp. [**Hesperostipa**]

Ennos, R.A. 1985. The mating system and genetic structure in a perennial grass, *Cynosurus cristatus* L. Heredity 55:121–126. [**Cynosurus**]

Environment Walkato. 2002–2007. Regional Pest Management Strategy. Walkato Regional Council, Hamilton East, New Zealand. http://www.ew.govt.nz/policyandplans/rpmsintro/rpms2002/operative5.2.7.htm. [**Zizania**]

Epstein, W., K. Gerber, and R. Karler. 1964. The hypnotic constituent of *Stipa vaseyi*, sleepy grass. Experientia (Basel) 20:390. [**Achnatherum**]

Erdman, K.S. 1965. Taxonomy of the genus *Sphenopholis* (Gramineae). Iowa State Coll. J. Sci. 39:289–336. [**Sphenopholis**]

False-Brome Working Group. Viewed 2006. Home page. http://www.appliedeco.org/FBWG.htm. [**Brachypodium**]

Farwell, O.A. 1919. *Bromelica* (Thurber): A new genus of grasses. Rhodora 21:76–78. [**Melica**]

Fernald, M.L. 1933. Recent discoveries in the Newfoundland flora. Rhodora 35:120–140. [**Festuca**]

Fernald, M.L. 1950. The North American variety of *Milium effusum*. Rhodora 52:218–222. [**Milium**]

Fernald, M.L. and G.A. Weatherby. 1916. The genus *Puccinellia* in eastern North America. Rhodora 18:1–23. [**Puccinellia**]

Finot, V.L., P.M. Peterson, R.J. Soreng, and F.O. Zuloaga. 2004. A revision of *Trisetum*, *Peyritschia*, and *Sphenopholis* (Poaceae: Poöideae: Aveninae) in Mexico and Central America. Ann. Missouri Bot. Gard. 91:1–30. [**Sphenopholis, Trisetum**]

Fox, A.M. and W.T. Haller. 2000. Production and survivorship of the functional stolons of giant cutgrass, *Zizaniopsis miliacea* (Poaceae). Amer. J. Bot. 87:811–818. [**Zizaniopsis**]

Frederiksen, S. 1978. *Festuca brevissima* Jurtz. in Alaska. Bot. Not. 131:409–410. [**Festuca**]

Frederiksen, S. 1979. *Festuca minutiflora* Rydb., a neglected species. Bot. Not. 132:315–318. [**Festuca**]

Frederiksen, S. 1981. *Festuca vivipara* (Poaceae) in the North Atlantic area. Nordic J. Bot. 1:277–292. [**Festuca**]

Frederiksen, S. 1982. *Festuca brachyphylla*, *F. saximontana* and related species in North America. Nordic J. Bot. 2:525–536. [**Festuca**]

Frederiksen, S. 1983. *Festuca auriculata* in North America. Nordic J. Bot. 3:629–632. [**Festuca**]

Frederiksen, S. 1986. Revision of *Taeniatherum* (Poaceae). Nordic J. Bot. 6:389–397. [**Taeniatherum**]

Frederiksen, S. 1991a. Taxonomic studies in *Dasypyrum* (Poaceae). Nordic J. Bot. 11:135–142. [**Dasypyrum**]

Frederiksen, S. 1991b. Taxonomic studies in *Eremopyrum* (Poaceae). Nordic J. Bot. 11:271–285. [**Eremopyrum**]

Frederiksen, S. and G. Petersen. 1998. A taxonomic revision of *Secale* L. (Triticeae, Poaceae). Nordic J. Bot. 18:399–420. [**Secale**]

Freitag, H. 1975. The genus *Piptatherum* (Gramineae) in southwest Asia. Notes Roy. Bot. Gard. Edinburgh 33:341–408. [**Oryzopsis, Piptatherum**]

Freitag, H. 1985. The genus *Stipa* in southwest and south Asia. Notes Roy. Bot. Gard. Edinburgh 42:355–487. [**Achnatherum, Stipa**]

Gabel, M.L. 1984. A biosystematic study of the genus *Elymus* (Gramineae: Triticeae). Proc. Iowa Acad. Sci. 91:140–146. [**Elymus**]

Galatowitsch, S.M. and A. van der Valk. 1994. Restoring Prairie Wetlands: An Ecological Approach. Iowa State University Press, Ames, Iowa, U.S.A. 246 pp. [**Scolochloa**]

Gandhi, K.N. 1996. Nomenclatural novelties for Western Hemisphere plants I. Harvard Pap. Bot. 8:63–66. [**Hainardia**]

Gerish, W. 1979. Chromosomal analysis of a previously unidentified *Spartina* species. Master's thesis. Long Island University, Brookville, New York, U.S.A. [**Spartina**]

Gervais, C. 1973. Contribution à l'étude cytologique et taxonomique des avoines vivaces. Denkschr. Schweiz. Naturf. Ges. 88:3–166. [**Avenula, Helictotrichon**]

Gibbs Russell, G.E. 1991. *Ehrharta* Thunb. Pp. 121–129 *in* G.E. Gibbs Russell, L. Watson, M. Koekemoer, L. Smook, N.P. Barker, H.M. Anderson, and M.J. Dallwitz. Grasses of Southern Africa (ed. O.A. Leistner). National Botanic Gardens, Botanical Research Institute, Pretoria, Republic of South Africa. 437 pp. [**Ehrharta**]

Gibbs Russell, G.E. and R.P. Ellis. 1987. Species groups in the genus *Ehrharta* (Poaceae) in southern Africa. Bothalia 17:51–65. [**Ehrharta**]

Gibbs Russell, G.E., L. Watson, M. Koekemoer, L. Smook, N.P. Barker, H.M. Anderson, and M.J. Dallwitz. 1991. Grasses of Southern Africa (ed. O.A. Leistner). National Botanic Gardens, Botanical Research Institute, Pretoria, Republic of South Africa. 437 pp. [**Ehrhartoideae**]

Gillespie, L.J., A. Archambault, and R.J. Soreng. 2006. Phylogeny of *Poa* (Poaceae) based on *trn*T-*trn*F sequence data: Major clades and basal relationships. Aliso 23:420–434. [**Poa**]

Gillespie, L.J. and R.J. Soreng. 2005. A phylogenetic analysis of the bluegrass genus *Poa* based on cpDNA restriction site data. Syst. Bot. 30:84–105. [**Poa**]

Gillett, J.M. and H.A. Senn. 1960. Cytotaxonomy and infraspecific variation of *Agropyron smithii* Rydb. Canad. J. Bot. 38:747–760. [**Elymus, Pascopyrum**]

Gillett, J.M. and H.A. Senn. 1961. A new species of *Agropyron* from the Great Lakes. Canad. J. Bot. 39:1169–1175. [**Elymus**]

Giorgi, D., R. D'Ovidio, O.A. Tanzarella, C. Ceoloni, and E. Porceddu. 2003. Isolation and characterization of S genome specific sequences from *Aegilops* sect. *Sitopsis* species. Genome 46:478–489. [**Triticum**]

Giussani, L.M., H. Cota-Sanchez, F.O. Zuloaga, and E.A. Kellogg. 2001. A molecular phylogeny of the subfamily Panicoideae (Poaceae) shows multiple origins of C4 photosynthesis. Amer. J. Bot. 88:1993-2012 [**Urochloa, Megathyrsus, Panicum**]

Godley, E.J. 1947. The variation and cytology of the British species of *Agropyron* and their natural hybrids. Master's thesis, Trinity College, Cambridge, England. 152 pp. [**Elymus**]

Gómez-Martínez, R. and A. Culham. 2000. Phylogeny of the subfamily Panicoideae with emphasis on the tribe Paniceae: Evidence from the *trn*L-F cpDNA region. Pp. 136-140 *in* Grasses: Systematics and Evolution (S.W.L. Jacobs and J. Everett, eds.). CSIRO Publishing, Collingwood, Victoria, Australia. 408 pp. [**Urochloa**]

Gould, F.W. 1958. Chromosome numbers in southwestern grasses. Amer. J. Bot. 45:757–767. [**Dichanthelium, Piptochaetium**]

Gould, F.W. 1965. Chromosome numbers in some Mexican grasses. Bol. Soc. Bot. México 29:49–62. [**Piptochaetium**]

Gould, F.W. 1975. The Grasses of Texas. Texas A&M University Press, College Station, Texas, U.S.A. 653 pp. [**Briza**]

Grass Phylogeny Working Group. 2000. A phylogeny of the grass family (Poaceae), as inferred from eight character sets. Pp. 3–7 *in* S.W.L. Jacobs and J. Everett (eds.). Grasses: Systematics and Evolution. CSIRO Publishing, Collingwood, Victoria, Australia. 408 pp. [**Diarrheneae**]

Grass Phylogeny Working Group. 2001. Phylogeny and subfamilial classification of the grasses (Poaceae). Ann. Missouri Bot. Gard. 88:373–457. [**Bambusoideae, Brachyelytreae, Diarrheneae, Meliceae, Nardeae, Phareae, Pharoideae, Poaceae, Poöideae, Stipeae**]

Greene, C.W. 1980. The systematics of *Calamagrostis* (Gramineae) in eastern North America. Ph.D. dissertation, Department of Biology, Harvard University, Cambridge, Massachusetts, U.S.A. 238 pp. [**Calamagrostis**]

Greene, C.W. 1984. Sexual and apomictic reproduction in *Calamagrostis* (Gramineae) from eastern North America. Amer. J. Bot. 71:285–293. [**Calamagrostis**]

Greene, C.W. 1993. *Calamagrostis*, Reed Grass. Pp. 1243–1246 *in* J.C. Hickman, ed. The Jepson Manual: Higher Plants of California. University of California Press, Berkeley, California, U.S.A. 1400 pp. [**Calamagrostis**]

Greuter, W. 1968. Notulae nomenclaturales et bibliographicae, 1–4. Candollea 23:81–108. [**Koeleria**]

Greuter, W. and K.H. Rechinger. 1967. Chloris Kythereia. Boissiera 13:22–196. [**Hainardia**]

Guo, Y.-L. and S. Ge. 2005. Molecular phylogeny of Oryzeae (Poaceae) based on DNA sequences from chloroplast, mitochondrial, and nuclear genomes. Amer. J. Bot. 92:1548–1558. [**Oryzeae**]

Hamilton, J.G. 1997. Changing perceptions of pre-European grasslands in California. Madroño 44:311–333. [**Nassella**]

Haraldsen, K.B., M. Ødegaard, and I. Nordal. 1991. Variation in the amphi-Atlantic plant *Vahlodea atropurpurea* (Poaceae). J. of Biogeogr. 18:311–320. [**Vahlodea**]

Harlan, J.R. 1945a. Cleistogamy and chasmogamy in *Bromus carinatus* Hook. & Arn. Amer. J. Bot. 32:66–72. [**Bromus**]

Harlan, J.R. 1945b. Natural breeding structure in the *Bromus carinatus* complex as determined by population analyses. Amer. J. Bot. 32:142–147. [**Bromus**]

Harmon, P.J. 1981. The vascular flora of the ridge top of North Fork Mountain, Grant and Pendleton counties, West Virginia. Master's thesis, Southern Illinois University, Carbondale, Illinois, U.S.A. 434 pp. [**Calamagrostis**]

Harms, V.L. 1985. A reconsideration of the nomenclature and taxonomy of the *Festuca altaica* complex (Poaceae) in North America. Madroño 32:1–10. [**Festuca**]

Hedberg, I. 1961. Chromosome studies in *Helictotrichon* Bess. Bot. Not. 114:389–396. [**Helictotrichon**]

Hedberg, I. 1990. Morphological, cytotaxonomic and evolutionary studies in *Anthoxanthum odoratum* L. *s. lat.*–A critical review. Sommerfeltia 11:97–107. [**Anthoxanthum**]

Hedberg, O. 1962. The genesis of *Puccinellia vacillans*. Bot. Tidsskr. 58:157–167. [**×Pucciphippsia**]

Hejný, S. 1969. *Coleanthus subtilis* (Tratt.) Seidl in der Tschechoslowakei. Folia Geobot. Phytotax. 4:345–399. [**Coleanthus**]

Hempel, W. 1970. Taxonomische und chorologische Untersuchungen an Arten von *Melica* L. subgen. *Melica*. Feddes Repert. 81:131–145. [**Melica**]

Hiesey, W.M. and M.A. Nobs. 1982. Interspecific Hybrid Derivatives Between Facultatively Apomictic Species of Bluegrass and Their Responses to Contrasting Environments. Experimental Studies on the Nature of Species 6, Publication No. 636. Carnegie Institution of Washington, Washington, D.C., U.S.A. 119 pp. [**Poa**]

Hitchcock, A.S. 1935. Manual of the Grasses of the United States. U.S. Government Printing Office, Washington, D.C., U.S.A. 1040 pp. [**Elymus**]

Hitchcock, A.S. 1951. Manual of the Grasses of the United States, ed. 2, rev. A. Chase. U.S.D.A. Miscellaneous Publication No. 200. U.S. Government Printing Office, Washington, D.C., U.S.A. 1051 pp. [1950 on title page; printed in 1951] [**Achnatherum, Agrostis, Anthoxanthum, Bromeae, Bromus, Desmazeria, Elymus, Macrochloa, Melica, Oryzopsis, Poa, Pharus, Piptochaetium, Secale, Setaria, Trisetum**]

Hitchcock, C.L. 1969. Gramineae. Pp. 384–725 *in* C.L. Hitchcock, A. Cronquist, and M. Ownbey. Vascular Plants of the Pacific Northwest, Part 1: Vascular Cryptogams, Gymnosperms, and Monocotyledons. University of Washington Press, Seattle, Washington, U.S.A. 914 pp. [**Bromus, Calamagrostis, Glyceria**]

Hodgson, H.J. and W.W. Mitchell. 1965. A new *Elymordeum* hybrid from Alaska. Canad. J. Bot 43: 1355–1358. [**×Leydeum**]

Hoge, P.S. 1992. Biosystematics of seven species of *Stipa* from the southwestern United States and northern Mexico. Master's thesis, Utah State University, Logan, Utah, U.S.A. 91 pp. [**Achnatherum**]

Hole, D.J., K.B. Jensen, R.R.-C. Wang, and S.M. Clawson. 1999. Molecular analysis of *Leymus flavescens* and chromosome pairing in *Leymus flavescens* hybrids (Poaceae: Triticeae). Int. J. Plant Sci. 160:371–376. [**Leymus**]

Holub, J. 1963. Ein Beitragzur Abgrenzung der Gattungen in der Tribus Aveneae: Die Gattung *Avenochloa* Holub. Acta Horti Bot. Prague 1962:75–86. [**Avenula**]

Holub, J. 1998. Reclassifications and new names in vascular plants 1. Preslia 70:97–122. [**Schedonorus**]

Hsiao, C., S.W.L. Jacobs, N.J. Chatterton, and K.H. Asay. 1999. A molecular phylogeny of the grass family (Poaceae) based on the sequences of nuclear ribosomal DNA (ITS). Austral. Syst. Bot. 11:667–688. [**Ampelodesmos, Stipeae**]

Hubbard, C.E. 1984. Grasses: A Guide to their Structure, Identification, Uses, and Distribution in the British Isles, ed. 3, rev. J.C.E. Hubbard. Penguin Books, Hammondsworth, Middlesex, England and New York, New York, U.S.A. 476 pp. [**Nardus**]

Hull, A.C., Jr. 1974. Species for seeding mountain rangelands in southeastern Idaho, northeastern Utah, and western Wyoming. J. Range Managem. 27:150–153. [**Brachypodium**]

Hultén, E. 1942. Flora of Alaska and Yukon [in part]. Acta Univ. Lund, n.s., 38:1–281. [**Poa**]

Hultén, E. 1959. The *Trisetum spicatum* complex. Svensk. Bot. Tidskr. 53:203–228. [**Trisetum**]

Hultén, E. 1960. Flora of the Aleutian Islands and Westernmost Alaska Peninsula with Notes on the Flora of Commander Islands. J. Cramer, Weinheim, Germany. 376 pp. [**Deschampsia**]

Hultén, E. 1968a. Comments on the flora of Alaska and Yukon. Ark. Bot., n.s., 7:1–147. [**×Leydeum**]

Hultén, E. 1968b. Flora of Alaska and Neighboring Territories. Stanford University Press, Stanford, California, U.S.A. 1008 pp. [**Calamagrostis, Elymus**]

Humphries, C.J. 1978a. Notes on the genus *Phleum*. Bot. J. Linn. Soc. 76:337–340. [**Phleum**]

Humphries, C.J. 1978b. Variation in *Taeniatherum caput-medusae* (L.) Nevski. Bot. J. Linn. Soc. 76:340–344. [**Taeniatherum**]

Jacobs, B.F., J.D. Kingston, and L.L. Jacobs. 1999. The origin of grass-dominated ecosystems. Ann. Missouri Bot. Gard. 86:590–643. [**Poaceae**]

Jacobs, S.W.L. 2001. The genus *Lachnagrostis* (Gramineae) in Australia. Telopea 9:439–448. [**Agrostis, Calamagrostis, Lachnagrostis**]

Jacobs, S.W.L., R. Bayer, J. Everett, M.O. Arriaga, M.E. Barkworth, A. Sabin-Badereau, M.A. Torres, F. Vázquez, and N. Bagnall. 2006. Systematics of the tribe Stipeae using molecular data. Aliso 23:349–361. [**Achnatherum, Nassella, Oryzopsis, Piptatherum, Stipa, Stipeae**]

Jacobs, S.W.L. and J. Everett. 1996. *Austrostipa*, a new genus, and new names for Australasian species formerly included in *Stipa* (Gramineae). Telopea 6:579–595. [**Austrostipa, Stipeae**]

Jacobs, S.W.L. and J. Everett. 1997. *Jarava plumosa* (Gramineae), a new combination for the species formerly known as *Stipa papposa*. Telopea 7:301–302. [**Jarava**]

Jacobs, S.W.L., J. Everett, and M.E. Barkworth. 1995. Clarification of morphological terms used in the Stipeae (Gramineae), and a reassessment of *Nassella* in Australia. Taxon 44:33–41. [**Nassella**]

Jacobs, S.W.L. and S.M. Hastings. 1993. *Ehrharta*. Pp. 652–654 *in* G.J. Harden (ed.). Flora of New South Wales, vol. 4. New South Wales University Press, Kensington, New South Wales, Australia. 775 pp. [**Ehrharta**]

Jacobs, S.W.L. and L. Lapinpuro. 1986. The Australian species of *Amphibromus* (Poaceae). Telopea 2:715–729. [**Amphibromus**]

Jakob, S.S. and F.R. Blattner. 2006. A chloroplast genealogy of *Hordeum* (Poaeae): Long-term persisting haplotypes, incomplete lineage sorting, regional extinction, and the consequences for phylogenetic inference. Mol. Biol. Evol. 23:1602–1612. [**Hordeum**]

Jarvie, J.K. 1992. Taxonomy of *Elytrigia* sect. *Caespitosae* and sect. *Junceae* (Gramineae: Triticeae). Nordic J. Bot. 12:155–169. [**Thinopyrum**]

Jarvie, J.K. and M.E. Barkworth. 1992a. Anatomical variation in some perennial Triticeae. Bot. J. Linn. Soc. 108:287–301. [**Thinopyrum**]

Jarvie, J.K. and M.E. Barkworth. 1992b. Morphological variation and genome constitution in some perennial Triticeae. Bot. J. Linn. Soc. 108:167–189. [**Thinopyrum**]

Jauhar, P.P. 1993. Cytogenetics of the *Festuca–Lolium* Complex. Monographs on Theoretical and Applied Genetics No. 18. Springer-Verlag, Berlin, Germany. 255 pp. [**Schedonorus**]

Jensen, K.B. 1993. *Elymus magellanicus* and its intra- and inter-generic hybrids with *Pseudoroegneria spicata*, *Hordeum violaceum*, *E. trachycaulus*, *E. lanceolatus*, and *E. glaucus* (Poaceae: Triticeae). Genome 36:72–76. [**Elymus**]

Jensen, K.B. and K.H. Asay. 1996. Cytology and morphology of *Elymus hoffmanni* (Poaceae: Triticeae): A new species from the Erzurum Province of Turkey. Int. J. Pl. Sci. 157:750–758. [**Elymus**]

Jensen, K.B., M. Curto, and K.H. Asay. 1995. Cytogenetics of Eurasian bluebunch wheatgrass and their relationship to North American bluebunch and thickspike wheatgrasses. Crop Sci. 35:1157–1162. [**Pseudoroegneria**]

Jensen, K.B, S.L. Hatch, and J. Wipff. 1992. Cytogenetics and morphology of *Pseudoroegneria deweyi* (Poaceae: Triticeae), a new species from the Soviet Union. Canad. J. Bot. 70:900–909. [**Pseudoroegneria**]

Jensen, K.B. and R.R.-C. Wang. 1997. Cytological and molecular evidence for transferring *Elymus coreanus* from the genus *Elymus* to *Leymus* and molecular evidence for *Elymus californicus* (Poaceae: Triticeae). Int. J. Pl. Sci. 158:872–877. [**Leymus**]

Jirásek, V. and J. Chrtek. 1962. Systematische Studie über die Arten der Gattung *Corynephorus* Pal.-Beauv. (Poaceae). Preslia 34:374–386. [**Corynephorus**]

Jirásek, V. and J. Chrtek. 1964. Zur frage des taxonomischen Wertes der Gattung *Cynosurus* L. Novit. Bot. Delect. Seminum Horti Bot. Univ. Carol. Prag. 20:23–27. [**Cynosurus**]

Johnson, B.L. 1945a. Cytotaxonomic studies in *Oryzopsis*. Bot. Gaz. 107:1–32. [**Oryzopsis, Piptatherum, Stipeae**]

Johnson, B.L. 1945b. Natural hybrids between *Oryzopsis hymenoides* and several species of *Stipa*. Amer. J. Bot. 32:599–608. [**Achnatherum**]

Johnson, B.L. 1960. Natural hybrids between *Oryzopsis* and *Stipa*: I. *Oryzopsis hymenoides* × *Stipa speciosa*. Amer. J. Bot. 47:736–742. [**Achnatherum**]

Johnson, B.L. 1962a. Amphiploidy and introgression in *Stipa*. Amer. J. Bot. 49:253–262. [**Achnatherum**]

Johnson, B.L. 1962b. Natural hybrids between *Oryzopsis* and *Stipa*: II. *Oryzopsis hymenoides* × *Stipa nevadensis*. Amer. J. Bot. 49:540–546. [**Achnatherum**]

Johnson, B.L. 1963. Natural hybrids between *Oryzopsis* and *Stipa*: III. *Oryzopsis hymenoides* × *Stipa pinetorum*. Amer. J. Bot. 50:228–234. [**Achnatherum**]

Johnson, B.L. 1972. Polyploidy as a factor in the evolution and distribution of grasses. Pp. 18–35 *in* V.B. Youngner and C.M. McKell (eds.). The Biology and Utilization of Grasses. Academic Press, New York, New York, U.S.A. 426 pp. [**Achnatherum**]

Johnson, B.L. and G.A. Rogler. 1943. A cytotaxonomic study of an intergeneric hybrid between *Oryzopsis hymenoides* and *Stipa viridula*. Amer. J. Bot. 30:49–56. [**×Achnella**]

Johnston, B.C. 2006. *Ptilagrostis porteri* (Rydberg) W.A. Weber (Porter's feathergrass): A technical conservation assessment. Species Conservation Project Report. U.S. Department of Agriculture-Forest Service, Rocky Mountain Region, Lakewood, Colorado, U.S.A. 62 pp. http://www.fs.fed.us/r2/projects/scp/assessments/ptilagrostisporteri.pdf. [**Ptilagrostis**]

Jones, K. 1958. Cytotaxonomic studies in *Holcus* L.: I. The chromosome complex in *Holcus mollis* L. New Phytol. 57:191–210. [**Holcus**]

Jones, K. and C.P. Carroll. 1962. Cytotaxonomic studies in *Holcus* L.: II. Morphological relationships in *H. mollis* L. New Phytol. 61:63–71. [**Holcus**]

Jones, S.B., Jr. and N.C. Coile. 1988. The Distribution of the Vascular Flora of Georgia. University of Georgia, Athens, Georgia, U.S.A. 230 pp. [**Elymus**]

Jordal, L.K. 1951. A floristic and phytogeographic survey of the southern slopes of the Brooks Range, Alaska. Ph.D. dissertation, University of Michigan, Ann Arbor, Michigan, U.S.A. 411 pp. [**×Elyhordeum**]

Jozwik, F.X. 1966. A biosystematic analysis of the slender wheatgrass complex. Ph.D. dissertation, University of Wyoming, Laramie, Wyoming, U.S.A. 112 pp. [**Elymus**]

Judziewicz, E.J. 1987. Taxonomy and morphology of the Tribe Phareae (Poaceae: Bambusoideae). Ph.D. dissertation, University of Wisconsin, Madison, Wisconsin, U.S.A. 557 pp. [**Pharus**]

Judziewicz, E.J. 1990. Flora of the Guianas: 187. Poaceae (series ed. A.R.A. Görts-van Rijn). Koeltz Scientific Books, Koenigstein, Germany. 727 pp. [**Lithachne**]

Judziewicz, E.J., L.G. Clark, X. Londoño, and M.J. Stern. 1999. American Bamboos. Smithsonian Institution Press, Washington, D.C., U.S.A. 392 pp. [**Arundinaria, Bambuseae, Bambusoideae, Lithachne, Olyreae**]

Jurtsev, B.A. 1995. *Dupontia*. Pp. 224–229 *in* J.G. Packer (ed., English edition). Flora of the Russian Arctic, vol. 1, trans. G.C.D. Griffiths. University of Alberta Press, Edmonton, Alberta, Canada. 330 pp. [English translation of A.I. Tolmachev (ed.). 1964. Arkticheskaya Flora SSSR, vol. 2. Nauka, Leningrad [St. Petersburg], Russia. 272 pp.]. [**Dupontia**]

Kam, Y.K. and J. Maze. 1974. Studies on the relationships and evolution of supraspecific taxa utilizing developmental data: II. Relationships and evolution of *Oryzopsis hymenoides*, *O. virescens*, *O. kingii*, *O. micrantha*, and *O. asperifolia*. Bot. Gaz. 135:227–247. [**Oryzopsis**]

Kawano, S. 1965. *Calamagrostis purpurascens* R. Br. and its identity. Acta Phytotax. Geobot. 21:73–90. [**Calamagrostis**]

Kawano, S. 1966. Biosystematic studies of the *Deschampsia caespitosa* complex with special reference to the karyology of Icelandic populations. Bot. Mag. (Tokyo) 79:293–307. [**Deschampsia**]

Kellogg, E.A. 1985. A biosystematic study of the *Poa secunda* complex. J. Arnold Arbor. 66:201–242. [**Poa**]

Kellogg, E.A. 1989. Comments on genomic genera in the Triticeae. Amer. J. Bot. 76:796–805. [**Triticeae**]

Kellogg, E.A. 1992. Tools for studying the chloroplast genome in the Triticeae (Gramineae): An *Eco*RI map, a diagnostic deletion, and support for *Bromus* as an outgroup. Amer. J. Bot. 79:186–197. [**Bromeae**]

Kerguélen, M. and F. Plonka. 1989. Les *Festuca* de la flore de France (Corse comprise). Bull. Soc. Bot. Centre-Ouest, Numéro Spécial 10:1–368. [**Festuca**]

Kerguélen, M., F. Plonka and É. Chas. 1993. Nouvelle contribution aux *Festuca* (Poaceae) de France. Lejeunia, n.s., 142:1–42. [**Festuca**]

Khan, M.A. and C.A. Stace. 1998. Breeding relationships in the genus *Brachypodium* (Poaceae: Poöideae). Nordic J. Bot. 19:257–269. [**Brachypodium**]

Kiang, A.-S., V. Connolly, D.J. McConnell, and T.A. Kavanagh. 1994. Paternal inheritance of mitochondria and chloroplasts in *Festuca pratensis–Lolium perenne* intergeneric hybrids. Theor. Appl. Genet. 87:681–688. [**Schedonorus**]

Kimber, G. and M. Feldman. 1987. Wild Wheat: An Introduction. Special Report No. 353. College of Agriculture, University of Missouri-Columbia, Columbia, Missouri, U.S.A. 142 pp. [**Aegilops**]

Kimber, G. and E.R. Sears. 1987. Evolution in the genus *Triticum* and the origin of cultivated wheat. Pp. 154–164 *in* E.G. Heyne (ed.). Wheat and Wheat Improvement, ed. 2. American Society of Agronomy, Madison, Wisconsin, U.S.A. 765 pp. [**Triticum**]

Komarov, V.L. 1963. Genus 176. *Glyceria* R. Br. Pp. 356–365 *in* R.Yu. Rozhevits [R.J. Roshevitz] and B.K. Shishkin [Schischkin] (eds.). Flora of the U.S.S.R., vol. 2, trans. N. Landau (series ed. V.L. Komarov). Published for the National Science Foundation and the Smithsonian Institution, Washington, D.C. by the Israel Program for Scientific Translations, Jerusalem, Israel. 622 pp. [English translation of Flora SSSR, vol. II. 1934. Botanicheskii Institut Im. V.L. Komarova, Akademiya Nauk, Leningrad [St. Petersburg], Russia. 778 pp.]. [Glyceria]

Komatsu, M., A. Chujo, Y. Nagato, K. Shimamoto, and J. Kyozuka. 2003. *FRIZZY PANICLE* is required to prevent the formation of axillary meristems and to establish floral meristem identity in rice spikelets. Development 130:3841–3850. [Oryza]

Kosarev, M.G. 1949. The variability of characters of crested wheatgrass. Selekts. & Semenov. 4:41–43. [Agropyron]

Koyama, T. 1987. Grasses of Japan and Its Neighboring Regions: An Identification Manual. Kodansha, Ltd., Tokyo, Japan. 370 pp. [Glyceria, Hygroryza]

Koyama, T. and S. Kawano. 1964. Critical taxa of grasses with North American and eastern Asiatic distribution. Canad. J. Bot. 42:859–864. [Brachyelytrum, Diarrhena, Schizachne, Torreyochloa]

Kula, A., B. Dudziak, E. Sliwinska, A. Grabowska-Joachimiak, A. Stewart, H. Golczyk, and A.J. Joachimiak. 2006. Cytomorpholgical studies on American and European *Phleum commutatum* Gaud. (Poaceae). Acta Biol. Cracov., Ser. Bot. 48:99–108. [Phleum]

Launert, E. 1971. *Oryza* L. Pp. 31–36 *in* A. Fernande, E. Launert, and H. Wild (eds.). Flora Zambesiaca, vol. 10¹. Crown Agents for Oversea Governments and Administrations, London, England. 152 pp. [Oryza]

Lawrence, G.H.M., A.F.G. Buchheim, G.S. Daniels, and H. Dolezal (eds.). 1968. B–P–H: Botanico–Periodicum–Huntianum. Hunt Botanical Library, Pittsburgh, Pennsylvania, U.S.A. 1063 pp. [General Bibliography, Geographic Bibliography]

Lawrence, W.E. 1945. Some ecotypic relations of *Deschampsia caespitosa*. Amer. J. Bot. 32:298–314. [Deschampsia]

Lepage, E. 1952. Études sur quelques plantes américaines: II. Hybrides intergénériques; *Agrohordeum* et *Agroelymus*. Naturaliste Canad. 79:241–266. [Elymus, ×Elyleymus]

Lepage, E. 1965. Revision généalogique de quelques ×*Agroelymus*. Naturaliste Canad. 92:205–216. [Elymus]

Lesica, P. and T.H. DeLuca. 1996. Long-term harmful effects of crested wheatgrass on Great Plains grassland ecosystems. J. Soil Water Conservation 51:408–409. [Agropyron]

Linnaeus, C. 1753. Species Plantarum. Impensis Laurentii Salvii, Stockholm, Sweden. 1200 pp. [Triticeae]

Liu, L. and S.M. Phillips. 2006a. *Oryzeae*. P. 183 *in* Z.-Y. Wu, P.H. Raven, and D.-Y. Hong (eds.). Flora of China, vol. 22 (Poaceae). Science Press, Beijing, Peoples Republic of China and Missouri Botanical Garden Press, St. Louis, Missouri, U.S.A. 653 pp. http://flora.huh.harvard.edu/china/mss/volume22/index.htm. [Oryzeae]

Liu, L. and S.M. Phillips. 2006b. *Zizania*. Pp. 187–188 *in* Z.-Y. Wu, P.H. Raven, and D.-Y. Hong (eds.). Flora of China, vol. 22 (Poaceae). Science Press, Beijing, Peoples Republic of China and Missouri Botanical Garden Press, St. Louis, Missouri, U.S.A. 653 pp. http://flora.huh.harvard.edu/china/mss/volume22/index.htm. [Zizania]

Liu, Z.W. and R.R.-C. Wang. 1993. Genome constitutions of *Thinopyrum curvifolium*, *T. scirpeum*, *T. distichum*, and *T. junceum* (Triticeae: Gramineae). Genome 36:641–651. [Thinopyrum]

Lodge, R.W. 1959. Biological flora of the British Isles: *Cynosurus cristatus* L. J. Ecol. 47:511–518. [Cynosurus]

Lonard, R.I. and F.W. Gould. 1974. The North American species of *Vulpia* (Gramineae). Madroño 22:217–230. [Vulpia]

Londo, J.P., Y.-C. Chiang K.-H. Hung, T.-Y. Chiang, and B.A. Schaal. 2006. Phylogeography of Asian wild rice, *Oryza rufipogon*, reveals multiple independent domestications of cultivated rice, *Oryza sativa*. Proc. Natl. Acad. Sci. U.S.A. [PNAS] 103:9578–9583. [Oryza]

Louis-Marie, Father, O.C. 1928. The genus *Trisetum* in America. Rhodora 30:209–228, 231–245. [Trisetum]

Löve, A. 1984. Conspectus of the Triticeae. Feddes Repert. 95:425–521. [Triticeae]

Love, R.M. 1946. Interspecific hybridization in *Stipa*: I. Natural hybrids. Amer. Naturalist 80:189–192. [Nassella]

Love, R.M. 1954. Interspecific hybridization in *Stipa*: II. Hybrids of *S. cernua*, *S. lepida*, and *S. pulchra*. Amer. J. Bot. 41:107–110. [Nassella]

Lu, B.-R., E.B. Naredo, A.B. Juliano, and M.T. Jackson. 2000. Preliminary studies on taxonomy and biosystematics of the AA genome *Oryza* species (Poaceae). Pp. 51–58 *in* S.W.L. Jacobs and J. Everett (eds.). Grasses: Systematics and Evolution. CSIRO Publishing, Collingwood, Victoria, Australia. 406 pp. [Oryza]

Lucchese, F. 1990. Revision and distribution of *Brachypodium phoenicoides* (L.) Roemer et Schultes in Italy. Ann. Bot. (Rome) 48:163–177. [Brachypodium]

Macfarlane, T.D. 1987. Poaceae subfamily Poöideae. Pp. 265–276 *in* T.R. Soderstrom, K.W. Hilu, C.S. Campbell and M.E. Barkworth (eds.). Grass Systematics and Evolution. Smithsonian Institution Press, Washington, D.C., U.S.A. 473 pp. [Poeae]

Marchant, C.J. 1963. Corrected chromosome numbers for *Spartina ×townsendii* and its parent species. Nature 199(4896):929. [Spartina]

Marchant, C.J. 1967. Evolution in *Spartina* (Gramineae): I. The history and morphology of the genus in Britain. J. Linn. Soc., Bot. 60:1–24. [**Spartina**]

Marchant, C.J. 1968a. Evolution in *Spartina* (Gramineae): II. Chromosomes, basic relationships, and the problem of *S. ×townsendii* agg. J. Linn. Soc., Bot. 60:381–409. [**Spartina**]

Marchant, C.J. 1968b. Evolution in *Spartina* (Gramineae): III. Species chromosome numbers and their taxonomic significance. J. Linn. Soc., Bot. 60:411–417. [**Spartina**]

Markgraf-Dannenberg, I. 1980. *Festuca* L. Pp. 125–153 *in* T.G. Tutin, V.H. Heywood, N.A. Burges, D.M. Moore, D.H. Valentine, S.M. Walters and D.A. Webb (eds.). Flora Europaea, vol. 5. Cambridge University Press, Cambridge, England. 439 pp. [**Festuca**]

Marr, K.L. and R.J. Hebda. 2006. *Calamagrostis tacomensis* (Poaceae), a new species from Washington and Oregon. Madroño 53:290–300. [**Calamagrostis**]

Marsh, V.L. 1952. A taxonomic revision of the genus *Poa* of the United States and southern Canada. Amer. Midl. Naturalist 47:202–250. [**Poa**]

Mason-Gamer, R.J. 2001. Origin of North American *Elymus* (Poaceae:Triticeae) allotetraploids based on granule-bound starch synthase gene sequences. Syst. Bot. 26:757–768. [**Elymus, Leymus**]

Matthei, O. 1965. Estudio crítico de las gramineas del género *Stipa* en Chile. Gayana, Bot. 13:1–137. [**Achnatherum, Jarava**]

Matthei, O. 1986. El género *Bromus* L. (Poaceae) en Chile. Gayana, Bot. 43:47–110. [**Bromus**]

Maze, J. 1962. A revision of the Stipas of the Pacific Northwest with special references to *S. occidentalis* Thurb. *ex* Wats. Master's thesis, University of Washington, Seattle, Washington, U.S.A. 95 pp. [**Achnatherum**]

Maze, J. 1965. Notes and key to some California species of *Stipa*. Leafl. W. Bot. 10:157–180. [**Achnatherum**]

Maze, J. 1981. A preliminary study on the root of *Oryzopsis hendersonii* (Gramineae). Syesis 14:151–153. [**Achnatherum, Piptatherum**]

McClure, F.A. 1955. *Bambusa* Schreb. Fieldiana, Bot. 24²: 52–60. [**Bambusa**]

McClure. F.A. 1957. Bamboos of the Genus *Phyllostachys* Under Cultivation in the United States. U.S. Department of Agriculture, Agricultural Research Service, Agriculture Handbook No. 114. U.S. Government Printing Office, Washington, D.C., U.S.A. 69 pp. [**Phyllostachys**]

McClure, F.A. 1966. The Bamboos: A Fresh Perspective. Harvard University Press, Cambridge, Massachusetts, U.S.A. 347 pp. [**Bambuseae**]

McClure, F.A. 1973. Genera of bamboos native to the New World (Gramineae: Bambusoideae). Smithsonian Contr. Bot. 9:1–148. [**Arundinaria, Bambuseae**]

McLachlan, K.I., S.G. Aiken, L.P. Lefkovitch, and S.A. Edlund. 1989. Grasses of the Queen Elizabeth Islands. Canad. J. Bot. 67:2088–2105. [**Deschampsia**]

McNamara, J. and J.A. Quinn. 1977. Resource allocation and reproduction in populations of *Amphicarpum purshii* (Gramineae). Amer. J. Bot. 64:17–23. [**Amphicarpum**]

McNeill, J. and W.G. Dore. 1976. Taxonomic and nomenclatural notes on Ontario grasses. Naturaliste Canad. 103:553–567. [**Schizachne**]

McVaugh, R. 1983. Flora Novo-Galiciana: A Descriptive Account of the Vascular Plants of Western Mexico, vol. 14; Gramineae (series ed. W.R. Anderson). University of Michigan Press, Ann Arbor, Michigan, U.S.A. 436 pp. [**Zizaniopsis**]

Mejia-Saulés, T. and F.A. Bisby. 2000. Preliminary views on the tribe Meliceae (Gramineae: Poöideae). Pp. 83–88 *in* S.W.L. Jacobs and J. Everett (eds.). Grasses: Systematics and Evolution. CSIRO Publishing, Collingwood, Victoria, Australia. 408 pp. [**Meliceae**]

Mejia-Saulés, T. and F.A. Bisby. 2003. Silica bodies and hooked papillae in lemmas of *Melica* species (Gramineae: Poöideae). Bot. J. Linn. Soc. 141:447–463. [**Melica**]

Melderis, A. 1980. *Elymus*. Pp. 192–198 *in* T.G. Tutin, V.H. Heywood, N.A. Burges, D.M. Moore, D.H. Valentine, S.M. Walters. and D.A. Webb (eds.). Flora Europaea, vol. 5. Cambridge University Press, Cambridge, England. 452 pp. [**Pseudoroegneria**]

Melderis, A. 1985. *Eremopyrum* (Ledeb.) Jaub. & Spach. Pp. 227–231 *in* P.H. Davis (ed.). Flora of Turkey and the East Aegean Islands, vol. 9. University Press, Edinburgh, Scotland. 724 pp. [**Eremopyrum**]

Merigliano, M.F. and P. Lesica. 1998. The native status of reed canarygrass (*Phalaris arundinacea* L.) in the inland northwest, USA. Nat. Areas J. 18:223–230. [**Phalaris**]

Messeri, A. 1942. Studio sistematico e fitogeografico di *Lagurus ovatus* L. Nuovo Giorn. Bot. Ital., n.s., 49:133–204. [**Lagurus**]

Michelangeli, F.A., J.I. Davis, and D.W. Stevenson. 2003. Phylogenetic relationships among Poaceae and related families as inferred from morphology, inversion in the plastid genome, and sequence data from the mitochondrial and plastid genomes. Amer. J. Bot. 90:93–106. [**Poaceae**]

Misra, K.C. 1961. Geography, morphology, and environmental relationships in certain *Stipa* species in the northern Great Plains. Ph.D. dissertation, University of Saskatchewan, Saskatoon, Saskatchewan, Canada. 141 pp. [**Hesperostipa**]

Misra, K.C. 1963. Phytogeography of the genus *Stipa* L. Trop. Ecol. 4:1–20 [reprint pagination]. [**Hesperostipa**]

Mitchell, W.W. 1992. Cytogeographic races of *Arctagrostis latifolia* (Poaceae) in Alaska. Canad. J. Bot. 70:80–83. [**Arctagrostis**]

Mitchell, W.W. and H.J. Hodgson. 1965. A new ×*Agrohordeum* from Alaska. Bull. Torrey Bot. Club 92:403–407. [**×Elyhordeum**]

Mitchell, W.W. and A.C. Wilton. 1966. A new tetraploid brome, section *Bromopsis*, of Alaska. Brittonia 18:162–166. [**Bromus**]

Morrison, L.A. 2001. The Percival Herbarium and wheat taxonomy: Yesterday, today, and tomorrow Pp. 65–80 *in* P.D.S. Caligari and P.E. Brandham (eds.) Wheat Taxonomy: The Legacy of John Percival. Linnean Special Issue No. 3. Academic Press, London, England. 190 pp. [**Triticum**]

Moyer, J.R. and **A.L. Boswall.** 2002. Tall fescue or creeping foxtail suppresses foxtail barley. Canad. J. Pl. Sci. 82:89–92. [**Alopecurus, Hordeum**]

Munz, P.A. 1959. A California Flora. University of California Press, Berkeley, California, U.S.A. 1681 pp. [**Poa**]

Murray, B.G. and **N.R.N. Barker.** 1988. Pollen/stigma interactions and hybridization in the genus *Briza* L. (Gramineae). Evol. Trends Pl. 2:107–110. [**Briza**]

Nečajev, A.P. and **A.A. Nečajev.** 1972. *Coleanthus subtilis* (Tratt.) Seidl in the Amur Basin. Folia Geobot. Phytotax. 7:339–347. [**Coleanthus**]

Neckles, H.A., J.W. Nelson, and **R.L. Pederson.** 1985. Management of Whitetop (*Scolochloa festucacea*) Marshes for Livestock Forage and Wildlife. Technical Bulletin 1–1985. Delta Waterfowl and Wetlands Research Station, Portage la Prairie, Manitoba, Canada. 12 pp. [**Scolochloa**]

Nees von Esenbeck, C.G. 1843. Gramina Novæ Hollandiæ, præsertim Insulæ Van Diemen, collectionis Lindleyanæ, a v. cl. Drummond, Gunn, aliisque collecta. London J. Bot. 2:409–420. [**Amphibromus**]

Nelson, E.N. and **R.J. Tyrl.** 1978. Hybridization and introgression between *Elymus canadensis* and *Elymus virginicus*. Proc. Oklahoma Acad. Sci. 58:32–34. [**Elymus**]

Nevski, S.A. 1963. Genus 193. *Brachypodium* P.B. Pp. 472–474 *in* R.Yu. Rozhevits. [R.J. Roshevitz] and B.K. Shishkin. [Schischkin] (eds.). Flora of the U.S.S.R., vol. 2, trans. N. Landau (series ed. V.L. Komarov). Published for the National Science Foundation and the Smithsonian Institution, Washington, D.C. by the Israel Program for Scientific Translations, Jerusalem, Israel. 622 pp. [English translation of Flora SSSR, vol. II. 1934. Botanicheskii Institut Im. V.L. Komarova, Akademiya Nauk, Leningrad [St. Petersburg], Russia. 778 pp.]. [**Brachypodium**]

Nihsen, M.E., E.L. Piper, C.P. West, R.J. Crawford, Jr., T.M. Denard, Z.B. Johnson, C.A. Roberts, D.A. Spiers, and **C.F. Rosenkrans, Jr.** 2004. Growth rate and physiology of steers grazing tall fescue inoculated with novel endophytes. J. Animal Sci. 82:878–883. [**Schedonorus**]

Normile, D. 2004. Expanding trade with China creates ecological backlash. Science 306:968–969. [**Spartina**]

Norstog, K.J. 1960. Some observations on the spikelet of *Hierochloë odorata*. Bull. Torrey Bot. Club 87:95–98. [**Anthoxanthum**]

Ogle, D. 2001. Intermediate wheatgrass, *Thinopyrum intermedium* (Host) Barkworth & D.R. Dewey. Plant Fact Sheet, U.S. Department of Agriculture, Natural Resource Conservation Service Plant Materials Program. http://plants.usda.gov/. [**Thinopyrum**]

Old, R.R. and **R.H. Callihan.** 1986. Distribution of *Ventenata dubia* in Idaho. Idaho Weed Control Rep. 1986:153. [**Ventenata**]

Paisooksantivatana, Y. and **R. Pohl.** 1992. Morphology, anatomy and cytology of the genus *Lithachne* (Poaceae: Bambusoideae). Revista Biol. Trop. 40:47–72. [**Lithachne**]

Papp, C. 1928. Monographie der Südamerikanischen Arten der Gattung *Melica* L. Repert. Spec. Nov. Regni Veg. 25:97–160. [**Melica**]

Parodi, L. 1944. Revisión de las gramíneas australes americanas del género *Piptochaetium*. Revista Mus. La Plata 6:213–310. [**Piptochaetium**]

Pavlick, L.E. 1983a. *Festuca viridula* Vasey (Poaceae): Re-establishment of its original lectotype. Taxon 32:117–120. [**Festuca**]

Pavlick, L.E. 1983b. Notes on the taxonomy and nomenclature of *Festuca occidentalis* and *F. idahoensis*. Canad. J. Bot. 61:337–344. [**Festuca**]

Pavlick, L.E. 1983c. The taxonomy and distribution of *Festuca idahoensis* in British Columbia and northwestern Washington. Canad. J. Bot. 61:345–353. [**Festuca**]

Pavlick, L.E. 1984. Studies of the *Festuca ovina* complex in the Canadian Cordillera. Canad. J. Bot. 62:2448–2462. [**Festuca**]

Pavlick, L.E. 1985. A new taxonomic survey of the *Festuca rubra* complex in northwestern North America, with emphasis on British Columbia. Phytologia 57:1–17. [**Festuca**]

Pavlick, L.E. 1995. *Bromus* L. of North America. Royal British Columbia Museum, Victoria, British Columbia, Canada. 160 pp. [**Bromus**]

Pavlick, L.E. and **J. Looman.** 1984. Taxonomy and nomenclature of rough fescues, *Festuca altaica*, *F. campestris* (*F. scabrella* var. *major*), and *F. hallii*, in Canada and the adjacent United States. Canad. J. Bot. 62:1739–1749. [**Festuca**]

Peñailillo, P. 2002. El género *Jarava* Ruiz et Pavón (Stipeae-Poaceae): Delimitacion y nuevas combinaciones. Gayana, Bot. 59:27–34. [**Achnatherum, Jarava**]

Percival, J. 1921. The Wheat Plant. Duckworth, London, England. 463 pp. [**Triticum**]

Petersen, G. and **O. Seberg.** 2003. Phylogenetic analyses of the diploid species of *Hordeum* (Poaceae) and a revised classification of the genus. Syst. Bot. 28:293–306. [**Hordeum**]

Peterson, P.M., J. Cayouette, Y.S.N. Ferdinandez, B. Coulman, and R.E. Chapman. 2001. Recognition of *Bromus richardsonii* and *B. ciliatus*: Evidence from morphology, cytology and DNA fingerprinting. Aliso 20:21–36. [Bromus]

Pfeiffer, L.G. 1872–1873. Nomenclator Botanicus, vol. 1. Theodor Fisher, Kassel, Germany. [Brachiaria, Moorochloa]

Phillips, S.M. and W.-L. Chen. 2003. Notes on grasses (Poaceae) for the Flora of China, I: *Deyeuxia*. Novon 13:318–321. [Calamagrostis]

Pignatti, S. 1982. *Sesleria*. Pp. 505–509 *in* S. Pignatti. Flora d'Italia, vol. 3. Edagricole, Bologna, Italy. 780 pp. [Sesleria]

Piper, C.V. 1906. North American species of *Festuca*. Contr. U.S. Natl. Herb. 10¹:1–48, vi–ix. [Festuca]

Pizzolato, T.D. 1984. Vascular system of the fertile floret of *Anthoxanthum odoratum* L. Bot. Gaz. 145:358–371. [Anthoxanthum]

Pohl, R.W. 1954. The allopolyploid *Stipa latiglumis*. Madroño 12:145–150. [Achnatherum]

Pohl, R.W. 1959. Morphology and cytology of some hybrids between *Elymus canadensis* and *E. virginicus*. Proc. Iowa Acad. Sci. 66:155–159. [Elymus]

Pohl, R.W. 1966. ×*Elyhordeum iowense*, a new intergeneric hybrid in the Triticeae. Brittonia 18:250–255. [Elymus]

Pohl, R.W. 1994a. *Bambusa* Schreber. Pp. 193–194 *in* G. Davidse, M. Sousa S, and A.O. Chater (eds.). Flora Mesoamericana, vol. 6: Alismataceae a Cyperaceae. Universidad Nacional Autónoma de México, Instituto de Biología, México, D.F., México. 543 pp. [Bambusa]

Pohl, R.W. 1994b. *Lithachne* P. Beauv. Pp. 215–216 *in* G. Davidse, M. Sousa S, and A.O. Chater (eds.). Flora Mesoamericana, vol. 6: Alismataceae a Cyperaceae. Universidad Nacional Autónoma de México, Instituto de Biología, México, D.F., México. 543 pp. [Lithachne]

Pohl, R.W. and G. Davidse. 2001. *Luziola*. Pp. 2072–2073 *in* W.D. Stevens, C.U. Ulloa, A. Pool, and M. Montiel (eds.). Flora de Nicaragua, vol. 3: Angiosperms (Pandanaceae–Zygophyllaceae). Missouri Botanical Garden Press, St. Louis, Missouri, U.S.A. 2666 pp. [for vols. 1–3]. [Luziola]

Porsild, A.E. 1964. Illustrated Flora of the Canadian Arctic Archipelago, ed. 2, rev. Bulletin of the National Museum of Canada No. 146 [Biological Series No. 50]. R. Duhamel, Queen's Printer, Ottawa, Ontario, Canada. 218 pp. [Puccinellia]

Porsild, A.E. and W.J. Cody. 1980. Vascular Plants of the Continental Northwest Territories, Canada. National Museum of Natural Sciences, National Museums of Canada, Ottawa, Ontario, Canada. 667 pp. [Elymus]

Prasad, V., C.A.E. Strömberg, H. Alimohammadian, and A. Sahni. 2005. Dinosaur coprolites and the early evolution of grasses and grazers. Science 310:1177–1180. [Poaceae]

Prokudin, Y.N., A.G. Vovk, O.A. Petrova, E.D. Ermolenko, and Y.V. Vernichenklo. 1977. Zlaki Ukrainy. Naukava Dumka, Kiev, Russia. 517 pp. [Stipeae]

Pyrah, G.L. 1969. Taxonomic and distributional studies in *Leersia* (Gramineae). Iowa State Coll. J. Sci. 44:215–270. [Leersia]

Pyrah, G.L. 1983. *Agropyron arizonicum* (Gramineae: Triticeae) and a natural hybrid from Arizona. Great Basin Naturalist 43: 131–135. [Elymus]

Quinn, J.A. 1975. Variability among *Danthonia sericea* (Gramineae) populations in responses to substrate moisture levels. Amer. J. Bot. 62:884–891. [Danthonia]

Quinn, J.A., J. Rotsettis, and D.E. Fairbrothers. 1972. Inflorescence characters and reproductive proficiency in *Danthonia sericea* populations. Amer. J. Bot. 59:942–951. [Danthonia]

Ramesar-Fortner, N.S., S.G. Aiken, and N.G. Dengler. 1995. Phenotypic plasticity in leaves of four species of arctic *Festuca* (Poaceae). Canad. J. Bot. 73:1810–1823. [Festuca]

Reeder, J.R. 1953. Affinities of the grass genus *Beckmannia* Host. Bull. Torrey Bot. Club 80:187–196. [Beckmannia]

Reeder, J.R. 1968. Notes on Mexican grasses VIII: Miscellaneous chromosome numbers–2. Bull. Torrey Bot. Club 95:69–86. [Piptochaetium]

Reeder, J.R. 1977. Chromosome numbers in western grasses. Amer J. Bot. 64:102–110. [Piptochaetium]

Reeder, J.R. 1984. *Poaceae*. Pp. 102–103 *in* Á. Löve (ed.). Chromosome number reports LXXXII. Taxon 33:126–134. [Spartina]

Reeder, J.R. 1994. *Setaria villosissima* (Gramineae) in Arizona: Fact or fiction. Phytologia 77:452–455. [Setaria]

Reeder J.R. and D.N. Singh. 1967. Validity of the tribe *Spartineae* (Gramineae) [Abstract]. Amer. J. Bot. 54:656. [Spartina]

Renvoize, S.A. 1998. Gramíneas de Bolivia. Royal Botanic Gardens, Kew, England. 644 pp. [Dissanthelium]

Reznicek, A.A. and E.J. Judziewicz. 1996. A new hybrid species, ×*Calammophila don-hensonii* (*Ammophila breviligulata* × *Calamagrostis canadensis*, Poaceae) from Grand Island, Michigan. Michigan Bot. 35:35–40. [Ammophila, Calamagrostis]

Rivas-Martinez, S., D. Sánchez-Mata, and M. Costa. 1999. North American boreal and western temperate forest vegetation. Itinera Geobot. 12:3–331. [Brachypodium]

Romero García, A.T., G. Blanca Lopez, and C. Morales Torres. 1988. Revisión del género *Agrostis* L. (Poaceae) en la Península Ibérica. Ruizia 7:1–160. [Agrostis]

Ross, E.M. 1989. *Phalaris* L. Pp. 132–135 *in* T.D. Stanley and E.M. Ross. Flora of South-Eastern Queensland, vol. 3. Queensland Department of Primary Industries, Miscellaneous Publication QM88001. Queensland Department of Primary Industries, Brisbane, Queensland, Australia. 532 pp. [Phalaris]

Rotsettis, J., J.A. Quinn, and D.E. Fairbrothers. 1972. Growth and flowering of *Danthonia sericea* populations. Ecology 53: 227–234. [Danthonia]

Rudall, P.J., W. Stuppy, J. Cunniff, E.A. Kellogg, and B.G. Briggs. 2005. Evolution of reproductive structures in grasses (Poaceae) inferred by sister-group comparison with their putative closest living relatives, Ecdeiocoleaceae. Amer. J. Bot. 92:1432–1443. [Poaceae]

Rúgolo de Agrasar, Z.E. 1982. Revalidación del género *Bromidium* Nees et Meyen emend. Pilger (Gramineae). Darwiniana 24:187–216. [Agrostis, Bromidium]

Rúgolo de Agrasar, Z.E. and A.M. Molina. 1997a. Las especies del género *Agrostis* L. (Gramineae: Agrostideae) de Chile. Gayana, Bot. 54:91–156. [Agrostis]

Rúgolo de Agrasar, Z.E. and A.M. Molina. 1997b. Presencia del híbrido ×*Agropogon littoralis* (Gramineae: Agrostideae) en Chile. Hickenia 11:209–214. [×Agropogon]

Rúgolo de Agrasar, Z.E. and A.M. Molina. 2002. El género *Lachnagrostis* (Gramineae: Agrostideae) en América del Sur. Pp. 20–32 *in* A. Freire Fierro and D.A. Neill (eds.). 2002. La Botánica en el Nuevo Milenio: Memorias del Tercer Congresso Ecuatoriano de Botánica. FUNBOTANICA, Quito, Ecuador. 250 pp. [Lachnagrostis]

Runemark, H. 1962. A revision of *Parapholis* and *Monerma* (Gramineae) in the Mediterranean. Bot. Not. 115:1–17. [Hainardia, Parapholis]

Saarela, J.M., P.M. Peterson, and J. Cayouette. 2005. *Bromus hallii* (Poaceae), a new combination for California, U.S.A., and taxonomic notes on *Bromus orcuttianus* and *Bromus grandis*. Sida 21:1997–2013. [Bromus]

Saarela, J.M., P.M. Peterson, R.J. Soreng, and R.E. Chapman. 2003. A taxonomic revision of the eastern North American and eastern Asian disjunct genus *Brachyelytrum* (Poaceae): Evidence from morphology, phytogeography and AFLPs. Syst. Bot. 28:674–692. [Brachyelytrum]

Saint-Yves, A. 1925. Contribution à l'étude des *Festuca* (subgen. *Eu-festuca*) de l'Amérique du Nord et du Mexique. Candollea 2:229–316. [Festuca]

Sakamoto, S. 1979. Genetic relationships among four species of the genus *Eremopyrum* in the tribe Triticeae, Gramineae. Mem. Coll. Agric. Kyoto Univ. 114:1–27. [Eremopyrum]

Sales, F. 1993. Taxonomy and nomenclature of *Bromus* sect. *Genea*. Edinburgh J. Bot. 50:1–31. [Bromus]

Salomon, B., R. von Bothmer, and N. Jacobsen. 1991. Intergeneric crosses between *Hordeum* and North American *Elymus* (Poaceae: Triticeae). Hereditas 114:117–122. [Elymus]

Saltonstall, K. 2002. Cryptic invasion by a non-native genotype of the common reed *Phragmites australis*, into North America. Proc. Natl. Acad. Sci. U.S.A. 99:2445–2449. [Phragmites]

Saltonstall, K., P.M. Peterson, and R.J. Soreng. 2004. Recognition of *Phragmites autralis* subsp. *americanus* (Poaceae: Arundinoideae) in North America: Evidence from morphological and genetic analyses. Sida 21:683–692. [Phragmites]

Sanders, T.S., J.L. Hamrick, and L.R. Holden. 1979. Allozyme variation in *Elymus canadensis* from the tallgrass prairie region: Geographic variation. Amer. Midl. Naturalist 101:1–12. [Elymus]

Scagel, R.K. and J. Maze. 1984. A morphological analysis of local variation in *Stipa nelsonii* and *S. richardsonii* (Gramineae). Canad. J. Bot. 62:763–770. [Achnatherum]

Schippmann, U. 1991. Revision der europäischen Arten der Gattung *Brachypodium* Palisot de Beauvois (Poaceae). Boissiera 45:1–249. [Brachypodium]

Scholander, P.F. 1934. Vascular plants from northern Svalbard with remarks on the vegetation in North-East Land. Skr. Svalbard Nordishavet 62:1–155. [Festuca]

Scholz, H. 1970. Zur Systematik der Gattung *Bromus* (Gramineae) mit einer Abbildung. Willdenowia 6:139–159. [Bromus]

Scholz, H. 1986. Bermerkungen zur Flora Griechenlands: *Gastridium phleoides* und *G. ventricosum* (Poaceae). Willdenowia 16:65–68. [Gastridium]

Scholz, H. 2003. Die Ackersippe der Verwechselten Trespe (*Bromus commutatus*). Botanik und Naturschutz in Hessen 16:17–22. [Bromus]

Schouten, Y. and J.F. Veldkamp. 1985. A revision of *Anthoxanthum* including *Hierochloë* (Gramineae) in Malesia and Thailand. Blumea 30:319–351. [Anthoxanthum]

Scoggan, H. 1978. Flora of Canada, part 2: Pteridophyta, Gymnosperms, Monocotyledoneae. National Museum of Natural Sciences Publications in Botany No. 7[2]. National Museums of Canada, Ottawa, Ontario, Canada. 545 pp. [Glyceria]

Scribner, F.L. and E.D. Merrill. 1910. The grasses of Alaska. Contr. U.S. Natl. Herb. 13[3]:47–92. [Puccinellia]

Seabloom, E.W. and A.M. Wiedemann. 1994. Distribution and effects of *Ammophila breviligulata* Fern. (American beachgrass) on the foredunes of the Washington coast. J. Coastal Res. 10:178–188. [Ammophila]

Shechter, Y. 1969. Electrophoretic investigation of the hybrid origin of *Oryzopsis contracta*. Pp. 19–25 *in* L. Chandra (ed.). Advancing Frontiers of Plant Sciences, vol. 23. Impex India, New Delhi, India. 201 pp. [Achnatherum]

Shechter, Y. and B.L. Johnson. 1968. The probable origin of *Oryzopsis contracta*. Amer. J. Bot. 55:611–618. [Achnatherum]

Shelly, J.S. 1987. Rediscovery and preliminary studies of *Trisetum orthochaetum*, Missoula County, Montana. Proc. Montana Acad. Sci. 47:3–4 [abstract]. [1987 on title page; printed in 1988]. [**Trisetum**]

Slageren, M.W. van. 1994. Wild Wheats: A Monograph of *Aegilops* L. and *Amblyopyrum* (Jaub. & Spach) Eig. Wageningen Agricultural University Papers 94–7. Wageningen Agricultural University and International Center for Agricultural Research in the Dry Areas (ICARDA), Wageningen, The Netherlands and Aleppo, Syria. 512 pp. [**Aegilops**]

Simon, B.K. and S.W.L. Jacobs. 2003. *Megathyrsus*, a new generic name for *Panicum* subgenus *Megathyrsus*. Austrobaileya 6:571–574. [**Panicum, Urochloa, Megathyrsus**]

Smith, E.B. (ed.). 1991. An Atlas and Annotated List of the Vascular Plants of Arkansas, ed. 2. Edwin B. Smith, Fayetteville, Arkansas, U.S.A. 489 pp. [**Elymus, Sphenopholis**]

Snyder, L.A. 1950. Morphological variability and hybrid development in *Elymus glaucus*. Amer. J. Bot. 37:628–636. [**Elymus**]

Snyder, L.A. 1951. Cytology of inter-strain hybrids and the probable origin of variability in *Elymus glaucus*. Amer. J. Bot. 38:195–202. [**Elymus**]

Soderstrom, T.R. 1980. A new species of *Lithachne* (Poaceae: Bambusoideae) and remarks on its sleep movements. Brittonia 32:495–501. [**Lithachne**]

Soderstrom, T.R. and R. Ellis. 1987. The position of bamboo genera and allies in a system of grass classification. Pp. 225–238 *in* T.R. Soderstrom, K.W. Hilu, C.S. Campbell, and M.E. Barkworth (eds.). Grass Systematics and Evolution. Smithsonian Institution Press, Washington, D.C., U.S.A. 473 pp. [**Bambuseae**]

Soderstrom, T.R. and R.P. Ellis. 1988. The woody bamboos (Poaceae: Bambuseae) of Sri Lanka: A morphological–anatomical study. Smithsonian Contr. Bot. 72:1–75. [**Bambusa**]

Soreng, R.J. 1985. *Poa* L. in New Mexico with a key to middle and southern Rocky Mountain species. Great Basin Naturalist 45:395–422. [**Poa**]

Soreng, R.J. 1991a. Notes on new infraspecific taxa and hybrids in North American *Poa* (Poaceae). Phytologia 71:340–413. [**Poa**]

Soreng, R.J. 1991b. Systematics of the "Epiles" group of *Poa* (Poaceae). Syst. Bot. 16:507–528. [**Poa**]

Soreng, R.J. 1993. *Poa* L. Pp. 1284, 1286–1291 *in* J.C. Hickman (ed.). The Jepson Manual of Higher Plants of California. University of California Press, Berkeley, California, U.S.A. 1400 pp. [**Poa**]

Soreng, R.J. 1998. An infrageneric classification for *Poa* in North America, and other notes on sections, species, and subspecies of *Poa*, *Puccinellia*, and *Dissanthelium* (Poaceae). Novon 8:187–202. [**Poa**]

Soreng, R.J. 2003. *Podagrostis*. Contr. U.S. Natl. Herb. 48:581. [**Agrostis**]

Soreng, R.J. 2005. Miscellaneous chromosome number reports for *Poa* L. (Poaceae) in North America. Sida 22:2195–2203. [**Poa**]

Soreng, R.J. and J.I. Davis. 1998. Phylogenetics and character evolution in the grass family (Poaceae): Simultaneous analysis of morphological and chloroplast DNA restriction site character sets. Bot. Rev. (Lancaster) 64:1–85. [**Ampelodesmos, Brachypodieae, Stipeae**]

Soreng, R.J. and J.I. Davis. 2000. Phylogenetic structure in Poaceae subfamily Poöideae as inferred from molecular and morphological characters: Misclassification versus reticulation. Pp. 61–74 *in* S.W.L. Jacobs and J. Everett (eds.). Grasses: Systematics and Evolution. CSIRO Publishing, Collingwood, Victoria, Australia. 408 pp. [**Leucopoa, Meliceae**]

Soreng, R.J., J.I. Davis, and J.J. Doyle. 1990. A phylogenetic analysis of chloroplast DNA restriction site variation in Poaceae subfam. Poöideae. Pl. Syst. Evol. 172:83–97. [**Catabrosa, Poeae, Torreyochloa**]

Soreng, R.J. and E.E. Terrell. 1997. Taxonomic notes on *Schedonorus*, a segregate genus from *Festuca* or *Lolium*, with a new nothogenus, ×*Schedololium*, and new combinations. Phytologia 83:85–88. [1997 on title page; printed in 1998]. [**Lolium, Schedonorus**]

Soreng, R.J., E.E. Terrell, J. Wiersema, and S.J. Darbyshire. 2001. Proposal to conserve the name *Schedonorus arundinaceus* (Schreb.) Dumort. against *Schedonorus arundinaceus* Roem. & Schult. (Poaceae: Poeae). Taxon 50:915–917. [**Schedonorus**]

Sørensen, T.J. 1953. A revision of the Greenland species of *Puccinellia* Parl. with contributions to our knowledge of the arctic *Puccinellia* flora in general. Meddel. Grønl. 136:1–169. [**Puccinellia**]

Sørensen, T.J. 1955. *Puccinellia agrostidea*, *Puccinellia bruggemannii*, *Puccinellia poacea*. Pp. 78–82 *in* A.E. Porsild. The Vascular Plants of the Western Canadian Arctic Archipelago. Bulletin of the National Museum of Canada No. 135 [Biological Series No. 45]. E. Cloutier, Queen's Printer, Ottawa, Ontario, Canada. 226 pp. [**Puccinellia**]

Spalton, L.M. 2001. Brome-grasses with small lemmas. B.S.B.I. [Botanical Society of the British Isles] News 87:21–23. [**Bromus**]

Spalton, L.M. 2002. An analysis of the characters of *Bromus racemosus* L., *B. commutatus* Schrad. and *B. secalinus* L. (Poaceae). Watsonia 24:193–202. [**Bromus**]

Stace, C.A. 1975. Wild hybrids in the British flora. Pp. 111–125 *in* S.M. Walters (ed.). European Floristic and Taxonomic Studies. E.W. Classey, Faringdon, England. 144 pp. [**Vulpia**]

Stace, C.A. 1978. Notes on *Cutandia* and related genera. Bot. J. Linn. Soc. 76:350–352. [**Cutandia, Desmazeria**]

Stace, C.A. 1987. Triticale: A case of nomenclatural mistreatment. Taxon 36:445–452. [×**Triticosecale**]

Stapleton, C.M.A. 1994. The bamboos of Nepal and Bhutan, Part I: *Bambusa, Dendrocalamus, Melocanna, Cephalostachyum, Teinostachyum,* and *Pseudostachyum* (Gramineae: Poaceae, Bambusoideae). Edinburgh J. Bot. 51:1–32. [**Bambusa**]

Stapleton, C.M.A. 1997. Morphology of woody bamboos. Pp. 251–267 *in* G.P. Chapman (ed.). The Bamboos. Academic Press, San Diego, California, U.S.A. 370 pp. [**Bambuseae**]

Stapleton, C.M.A. 1998. Form and function in the bamboo rhizome. J. Amer. Bamboo Soc.12:21–29. [**Bambuseae**]

Stapleton, C.M.A. 2002. *Bambusa ventricosa* versus *Bambusa tuldoides.* Bamboo 23:17–18. [**Bambusa**]

Stapleton, C.M.A., G.N. Chonghaile, and T.R. Hodkinson. 2004. *Sarocalamus,* a new Sino–Himalayan bamboo genus (Poaceae–Bambusoideae). Novon 14:345–349. [**Bambuseae**]

Stebbins, G.L., Jr. 1947. The origin of the complex of *Bromus carinatus* and its phytogeographic implications. Contr. Gray Herb. 165:42–55. [**Bromus**]

Stebbins, G.L., Jr. 1957. The hybrid origin of microspecies in the *Elymus glaucus* complex. Pp. 336–340 *in* International Union of Biological Sciences (ed.) Proceedings of the International Genetics Symposia, 1956: Tokyo & Kyoto, September 1956. Organizing Committee, International Genetics Symposia, Science Council of Japan, Tokyo, Japan. 680 pp. [**Elymus**]

Stebbins, G.L., Jr. and L.A. Snyder. 1956. Artificial and natural hybrids in the Gramineae, tribe Hordeae: IX. Hybrids between the western and eastern species. Amer. J. Bot. 43:305–312. [**Elymus**]

Stebbins, G.L., Jr. and H.A. Tobgy. 1944. The cytogenetics of hybrids in *Bromus:* 1. Hybrids within the section *Ceratochloa.* Amer. J. Bot. 31:1–11. [**Bromus**]

Stebbins, G.L., Jr. and D. Zohary. 1959. Cytogenetic and evolutionary studies in the genus *Dactylis.* Univ. Calif. Publ. Bot. 31:1–40. [**Dactylis**]

Steen, N.W., R. Elven, and I. Nordal. 2004. Hybrid origin of the arctic ×*Pucciphippsia vacillans* (Poaceae): Evidence from Svalbard plants. Pl. Syst. Evol. 245:215–238. [**Phippsia**]

Stephenson, S.N. 1971. The biosystematics and ecology of the genus *Brachyelytrum* (Gramineae) in Michigan. Michigan Bot. 10:19–33. [**Brachyelytreae, Brachyelytrum**]

Stern, R.J. 1983. Morphometric and phenologic variability in *Ammophila breviligulata* Fernald. Master's thesis, University of Vermont, Burlington, Vermont, U.S.A. 29 pp. [**Ammophila**]

Stevenson, G.A. 1965. Notes on the more recently adventive flora of the Brandon area, Manitoba. Canad. Field-Naturalist 79:174–177. [**Eremopoa**]

Steyermark, J.A. 1963. Flora of Missouri. Iowa State University Press, Ames, Iowa, U.S.A. 1725 pp. [**Elymus**]

Stika, J. 2003. Sweetgrass ale. http://byo.com/feature/1067.html. [**Anthoxanthum**]

Sun, G.-L. and B. Salomon. 2003. Microsatellite variability and heterozygote deficiency in the arctic-alpine Alaskan wheatgrass (*Elymus alaskanus*) complex. Genome 46:729–737. [**Elymus**]

Sun, G.-L., B. Salomon, and R. von Bothmer. 1998. Characterization of microsatellite loci from *Elymus alaskanus* and length polymorphism in several *Elymus* species (Triticeae: Poaceae). Genome 41:455–463. [**Elymus**]

Sun, G.L., H. Tang, and B. Salomon. 2006. Molecular diversity and relationships of North American *Elymus trachycaulus* and the Eurasian *E. caninus* species. Genetica 127:55–64. [**Elymus**]

Svitashev, S., T. Bryngelsson, X. Li, and R.R.-C. Wang. 1998. Genome-specific repetitive DNA and RAPD markers for genome identification in *Elymus* and *Hordelymus.* Genome 41:120–128. [**Elymus**]

Svitashev, S., B. Salomon, T. Bryngelsson, and R. von Bothmer. 1996. A study of 28 *Elymus* species using repetitive DNA sequences. Genome 39:1093–1101. [**Triticeae**]

Swallen, J.R. 1931. The grass genus *Amphibromus.* Amer. J. Bot. 18:411–415. [**Amphibromus**]

Swallen, J.R. 1941. New species, names, and combinations of grasses. Proc. Biol. Soc. Wash. 54:43–46. [**Leucopoa**]

Swallen, J.R. 1944. The Alaskan species of *Puccinellia.* J. Wash. Acad. Sci. 34:16–32. [**Puccinellia**]

Swallen, J.R. 1965. The grass genus *Luziola.* Ann. Missouri Bot. Gard. 52:472–475. [**Luziola**]

Swallen, J.R. and O. Tovar. 1965. The grass genus *Dissanthelium.* Phytologia 11:361–376. [**Dissanthelium**]

Tateoka, T. 1960. Cytology in grass systematics: A critical review. Nucleus (Calcutta) 3:81–110. [**Diarrhena, Diarrheneae**]

Tercek, M.T., D.P. Hauber, and S.P. Darwin. 2002. Genetic and historical relationships among thermally adapted *Agrostis* (Bentgrass) of North America and Kamchatka: Evidence for a previously unrecognized, thermally adapted taxon. Amer. J. Bot. 90:1306–1312. [**Agrostis**]

Terrell, E.E. 1968. A Taxonomic Revision of the Genus *Lolium.* Technical Bulletin, United States Department of Agriculture No. 1392. U.S. Government Printing Office, Washington, D.C., U.S.A. 65 pp. [**Lolium**]

Terrell, E.E. and L.R. Batra. 1982. *Zizania latifolia* and *Ustilago esculenta,* a grass-fungus association. Econ. Bot. 36:274–285. [**Zizania**]

Terrell, E.E., W.H.P. Emery, and H.E. Beaty. 1978. Observations on *Zizania texana* (Texas wildrice), an endangered species. Bull. Torrey Bot. Club 105:50–57. [**Zizania**]

Terrell, E.E., P.M. Peterson, J.L. Reveal, and M.R. Duvall. 1997. Taxonomy of North American species of *Zizania* (Poaceae). Sida 17:533–549. [**Zizania**]

Terrell, E.E., P.M. Peterson, and W.P. Wergin. 2001. Epidermal features and spikelet micromorphology in *Oryza* and related genera (Poaceae: Oryzeae). Smithsonian Contr. Bot. 91:1–50. [**Oryza**]

Terrell, E.E. and H. Robinson. 1974. Luziolinae, a new subtribe of oryzoid grasses. Bull. Torrey Bot. Club 101:235–245. [**Luziola, Zizaniopsis**]

Thellung, A. 1911. La Flore Adventice de Montpellier. Imprimerie Émile Le Maout, Cherbourg, France. [Preprinted from Mém. Soc. Sci. Nat. Cherbourg 38:[57]–728 (1912)]. [**Phalaris**]

Thomasson, J.R. 1978. Epidermal patterns of the lemma in some fossil and living grasses and their phylogenetic significance. Science 199:975–977. [**Achnatherum**]

Thomasson, J.R. 1979. Late Cenozoic Grasses and Other Angiosperms from Kansas, Nebraska, and Colorado: Biostratigraphy and Relationships to Living Taxa. Kansas Geological Survey, Bulletin No. 218. University of Kansas Publications, Lawrence, Kansas, U.S.A. 68 pp. [**Hesperostipa, Piptochaetium**]

Thurber, G. 1880. *Melica* Linn. Pp. 302–305 *in* S. Watson. Geological Survey of California: Botany, vol. 2. Little, Brown, Boston, Masssachusetts, U.S.A. 559 pp. [**Melica**]

Torres, M.A. 1993. Revisión del Género *Stipa* (Poaceae) en la Provincia de Buenos Aires. Monografia 12. Comisión de Investigaciones Científicas, Provincia de Buenos Aires, La Plata, Argentina. 62 pp. [**Achnatherum, Amelichloa**]

Tovar, O. 1993. Las gramíneas (Poaceae) del Perú. Ruizia 13:1–480. [**Dissanthelium**]

Tsvelev, N.N. 1975. [On the possibility of despecialization by hybridogenesis for explaining the evolution of the Triticeae (Poaceae)]. Zhurn. Obshchei Biol. 36:90–99. [In Russian; translation of article by K. Gonzales, available at the Intermountain Herbarium, Utah State University, Logan, Utah 84322–5305, U.S.A.] [**Triticeae**]

Tsvelev, N.N. 1976. Zlaki SSSR. Nauka, Leningrad [St. Petersburg], Russia. 788 pp. [**Achnatherum, Agropyron, Agrostis, Elymus, Leymus, Poa, Pseudoroegneria, Schizachne**].

Tsvelev, N.N. 1977. [On the origin and evolution of the feathergrasses (*Stipa* L.)]. Pp. 139–150 *in* Problemii Ekologii, Geobotaniki, and Botaniicheskoi Geografii i Floristickii. Nauka, Leningrad [St. Petersburg], Russia. 225 pp. [In Russian; translation of article by K. Gonzales, available at the Intermountain Herbarium, Utah State University, Logan, Utah 84322–5305, U.S.A.]. [**Achnatherum**]

Tsvelev, N.N. 1995. Gramineae. Pp. 87–306 *in* J.G. Packer (ed., English edition). Flora of the Russian Arctic, vol. 1, trans. G.C.D. Griffiths. University of Alberta Press, Edmonton, Alberta, Canada. 330 pp. [English translation of A.I. Tolmachev (ed.). 1964. Arkticheskaya Flora SSSR, vol. 2. Nauka, Leningrad [St. Petersburg], Russia. 272 pp.]. [**Alopecurus, Deschampsia, Leymus, Puccinellia**]

Tucker, G.C. 1996. The genera of Poöideae (Gramineae) in the southeastern United States. Harvard Papers Bot. 9:11–90. [**Cinna**]

Turner, B.L. 2004. *Sporobolus coahuilensis* (Poaceae): A new record for the U.S.A. from Trans-Pecos, Texas. Sida 21:455–457. [**Sporobolus**]

Tutin, T.G., 1952. Origin of *Poa annua*. Nature 169:160. [**Poa**]

Tutin, T.G. 1980a. *Nardus* L. P. 255 *in* T.G. Tutin, V.H. Heywood, N.A. Burges, D.M. Moore, D.H. Valentine, S.M. Walters, and D.A. Webb (eds.). Flora Europaea, vol. 5. Cambridge University Press, Cambridge, England. 452 pp. [**Nardus**]

Tutin, T.G. 1980b. *Piptatherum* Beauv. Pp. 246–247 *in* T.G. Tutin, V.H. Heywood, N.A. Burges, D.M. Moore, D.H. Valentine, S.M. Walters, and D.A. Webb (eds.). Flora Europaea, vol. 5. Cambridge University Press, Cambridge, England. 452 pp. [**Oryzopsis**]

Ujhely, J. 1972. Evolutionary problems of the European Koelerias. Pp. 163–176 *in* G. Vida (ed.). Evolution in Plants. Symposia Biologica Hungarica No. 12. Akadémiai Kiadó, Budapest, Hungary. 231 pp. [**Koeleria**]

Vaughan, D.A. 1989. The Genus *Oryza* L.: Current Status of Taxonomy. International Rice Research Institute Research Paper Series 138. International Rice Research Institute, Los Baños, Laguna, Philippines. 21 pp. [**Oryza**]

Vázquez, F.M. and M.E. Barkworth. 2004. Resurrection and emendation of *Macrochloa* (Gramineae: Stipeae). Bot. J. Linn. Soc. 144:483–495. [**Celtica, Macrochloa**]

Veldkamp, J.F. 1990. *Bromus luzonensis* is the correct name for *Bromus breviaristatus* Buckl. (Gramineae). Taxon 39:660. [**Bromus**]

Veldkamp, J.F. 2004. Miscellaneous notes on mainly southeast Asian Gramineae. Reinwardtia 12:135–140. [**Moorochloa, Brachiaria**]

Vergagno Gambi, O. 1965. Osservazioni sullo sviluppo di *Lagurus ovatus* L. Giorn. Bot. Ital. 72:243–254. [**Lagurus**]

Vilkomerson, H. 1950. The unusual meiotic behavior of *Elymus wiegandii*. Exp. Cell Res. 1:534–542. [**Elymus**]

Vogel, K.P., K.J. Moore, and L.E. Moser. 1996. Bromegrasses. Pp. 535–567 *in* L.E. Moser, D.R. Buxton, and M.D. Casier (eds.). Cool-Season Forage Grasses. Agronomy Monograph No. 34. American Society of Agronomy, Crop Science Society of America, and Soil Science Society of America, Madison, Wisconsin, U.S.A. 841 pp. [**Bromus**]

Voss, E.G. 1972. Michigan Flora: A Guide to the Identification and Occurrence of the Native and Naturalized Seed-Plants of the State, part 1. University of Michigan, Ann Arbor, Michigan, U.S.A. 488 pp. [**Glyceria, Pseudoroegneria**]

Wagnon, H.K. 1952. A revision of the genus *Bromus*, section *Bromopsis*, of North America. Brittonia 7:415–480. [**Bromus**]

Walker, P.J., C.A. Paris, and D.S. Barrington. 1998. Taxonomy and phylogeography of the North American beachgrasses. Amer. J. Bot. 85, Suppl.:87 [abstract]. [**Ammophila**]

Wang, C.-P., Z.-H. Yu, and G.-H. Ye. 1980. A taxonomical study of *Phyllostachys* in China [parts 1 & 2]. Acta Phytotax. Sin. 18:15–19, 168–193. [**Phyllostachys**]

Wang, R.R-C., D.R. Dewey, and C. Hsiao. 1986. Genome analysis of the tetraploid *Pseudoroegneria tauri*. Crop Sci. 26:723–727. [**Pseudoroegneria**]

Wang, R.R.-C. and K.B. Jensen. 1994. Absence of J genome in *Leymus* species (Poaceae: Triticeae): Evidence from DNA hybridization and meiotic pairing. Genome 37:231–235. [**Leymus**]

Wang, R.R.-C., J.-Y. Zhang, B.S. Lee, K.B. Jensen, M. Kishi, and H. Tsujimoto. 2006. Variations in abundance of 2 repetitive sequences in *Leymus* and *Psathyrostachys* species. Genome 49:511–519. [**Leymus**]

Warwick, S.I. and S.G. Aiken. 1986. Electrophoretic evidence for the recognition of two species in annual wild rice (*Zizania*, Poaceae). Syst. Bot. 11:464–473. [**Zizania**]

Warwick, S.I., B.K. Thompson, and L.D. Black. 1987. Genetic variation in Canadian and European populations of the colonizing weed species *Apera spica-venti*. New Phytol. 106:301–317. [**Apera**]

Watson, L. and M.J. Dallwitz. 1992. The Grass Genera of the World. C.A.B. International, Wallingford, England. 1038 pp. [**Hygroryza**]

Weiller, C.M., M.J. Henwood, J. Lenz, and L. Watson. 1995 on. Poöideae (Poaceae) in Australia–Descriptions and Illustrations. http://delta-intkey.com/pooid/www/index.htm. [**Phalaris**]

Weimarck, G. 1971. Variation and taxonomy of *Hierochloë* (Gramineae) in the Northern Hemisphere. Bot. Not. 124:129–175. [**Anthoxanthum**]

Weimarck, G. 1987. *Hierochloë hirta* subsp. *praetermissa*, subsp. *nova* (Gramineae), an Asiatic–E. European taxon extending to N and C Europe in the Northern Hemisphere. Symb. Bot. Upal. 2:175–181. [**Anthoxanthum**]

Wheeler, D.J.B., S.W.L. Jacobs, and R.D.B. Whalley. 2002. Grasses of New South Wales, ed. 3. University of New England, Armidale, New South Wales, Australia. 445 pp. [**Ehrharteae**]

Whipple, I.G., B.S. Bushman, and M.E. Barkworth. [in press]. *Glyceria* section *Glyceria* in North America. [**Glyceria**]

Widjaja, E.A. 1997. New taxa in Indonesian bamboos. Reinwardtia 11:57–152. [**Bambusa**]

Willemse, L.P.M. 1982. A discussion of the Ehrharteae (Gramineae) with special reference to the Malesian taxa formerly included in *Microlaena*. Blumea 28:181–194. [**Ehrharteae**]

Wilson, B.L. and S. Gray. 2002. Resurrection of a century-old species distinction in *Calamagrostis*. Madroño 49:169–177. [**Calamagrostis**]

Wilson, B.L., J. Kitzmiller, W. Rolle, and V.D. Hipkins. 2001. Isozyme variation and its environmental correlates in *Elymus glaucus* from the California floristic province. Canad. J. Bot. 79:139–153. [**Elymus**]

Wilson, F.D. 1963. Revision of *Sitanion* (Triticeae, Gramineae). Brittonia 15:303–323. [**Elymus**]

Woike, S. 1969. Beitrag zum Vorkommen von *Coleanthus subtilis* (Tratt.) Seidl (Feines Scheidenblütgras) in Europa. Folia Geobot. Phytotax. 4:401–413. [**Coleanthus**]

Wong, K.M. 1995. The Morphology, Anatomy, Biology, and Classification of Peninsular Malayan Bamboos. University of Malaya Botanical Monographs No. 1. University of Malaya, Kuala Lumpur, Malaysia. 189 pp. [**Bambusa**]

Xia, N.H. and C.M.A. Stapleton. 1997. Typification of *Bambusa bambos* (Gramineae, Bambusoideae). Kew Bull. 52:693–698. [**Bambusa**]

Yatskievych, G. 1999. Steyermark's Flora of Missouri, vol. 1, rev. ed. Missouri Department of Conservation, Jefferson City, Missouri, U.S.A. 991 pp. http://biology.missouristate.edu/herbarium/. [**Bromus, Festuca**]

Yen, C., J.-L. Yang, and B.R. Baum. 2005. *Douglasdeweya*: A new genus, with a new species and a new combination (Triticeae: Poaceae). Canad. J. Bot. 83:413–419. [**Pseudoroegneria**]

Yen, C., J.-L. Yang, and Y. Yen. 2005. Hitoshi Kihara, Áskell Löve and the modern genetic concept of the genera in the tribe Triticeae (Poaceae). Acta Phytotax. Sin. 43:82–93. [**Elymus**]

Zandee, M. and P.C.G. Glas. 1982. Studies in the *Holcus lanatus–Holcus mollis* complex (Poaceae [= Gramineae]). Proc. Kon. Ned. Akad. Wetensch., C 85:413–437. [**Holcus**]

Zhang, H.B. and J. Dvorák. 1991. The genome origin of tetraploid species of *Leymus* (Poaceae: Triticeae) inferred from variation in repeated nucleotide sequences. Amer. J. Bot. 78:871–884. [**Leymus**]

Zhang, X-Q., B. Salomon, and R. von Bothmer. 2002. Application of random amplified polymorphic DNA markers to evaluate intraspecific genetic variation in the *Elymus alaskanus* complex (Poaceae). Genet. Resources & Crop Evol. 49:397–407. [**Elymus**]

Zhu, Z., C-D. Chu, and C. Stapleton. 2006. *Arundinaria*. Pp. 113–115 *in* Z.-Y. Wu, P.H. Raven, and D.-Y. Hong (eds.). Flora of China, vol. 22 (Poaceae). Science Press, Beijing, Peoples Republic of China and Missouri Botanical Garden Press, St. Louis, Missouri, U.S.A. 653 pp. http://flora.huh.harvard.edu/china/mss/volume22/index.htm. [**Arundinaria**]

Zuloaga, F.O., O. Morrone, and L.M. Giussani. 2000. A cladistic analysis of the Paniceae: A preliminary approach. Pp. 123–135 *in* S.W.L. Jacobs and J. Everett (eds.). Grasses: Systematics and Evolution. CSIRO Publishing, Collingwood, Victoria, Australia. 408 pp. [**Paniceae**]

Names and Synonyms

The list of names that follows contains the scientific names of the taxa that are described in this volume, plus those already treated in volume 25.

The listing also shows synonyms of the taxa mentioned in both volumes. The listing of synonyms is not complete, containing only names which were encountered when creating the distribution maps from various current floras, herbarium databases, and miscellaneous other publications.

Accepted names are in **boldface**, followed by the name(s) of their author(s) and the volume and page number on which the taxon is described. The names of the authors are abbreviated according to Brummitt and Powell (1992), supplemented by information from IPNI, the International Plant Names Index (http://www.ipni.org/index.html/). Some of the accepted names shown, as well as some of the synonyms, have been inappropriately applied to taxa in the *Flora* region. These names are followed by the statement: "within the *Flora* region, misapplied to ...."

Synonyms are in *italics*, and are followed by their accepted names plus the volume and page number on which that taxon is treated. Names in standard typeface fall into two categories. Some are names that violate the International Code of Botanical Nomenclature (Greuter et al. 2000) in some way and therefore cannot be used; others are suspected by the author of the treatment concerned to apply to a hybrid for which no appropriate scientific name has been published.

Infraspecific names are listed alphabetically, based on the *infraspecific epithet*; the authors of tautonymic subspecies and varieties (subspecies and varieties in which the specific and infraspecific epithet are the same) are ignored as far as the alphabetic listing is concerned. In general, accepted names of forms are listed only if they are mentioned in the text.

ACHNATHERUM P. Beauv., 24: 114
Achnatherum aridum (M.E. Jones) Barkworth, 24: 131
Achnatherum arnowiae (S.L. Welsh & N.D. Atwood) Barkworth, 24: 141
Achnatherum ×bloomeri (Bol.) Barkworth, 24: 142
Achnatherum brachychaetum = **Amelichloa brachychaeta**, 24: 182
Achnatherum capense = **Stipa capensis**, 24: 155
Achnatherum caudatum = **Amelichloa caudata**, 24: 182
Achnatherum clandestinum = **Amelichloa clandestina**, 24: 184
Achnatherum contractum (B.L. Johnson) Barkworth, 24: 141
Achnatherum coronatum (Thurb.) Barkworth, 24: 127
Achnatherum curvifolium (Swallen) Barkworth, 24: 135
Achnatherum diegoense (Swallen) Barkworth, 24: 131
Achnatherum eminens (Cav.) Barkworth, 24: 133
Achnatherum hendersonii (Vasey) Barkworth, 24: 139
Achnatherum hymenoides (Roem. & Schult.) Barkworth, 24: 139
Achnatherum latiglume (Swallen) Barkworth, 24: 124
Achnatherum lemmonii (Vasey) Barkworth, 24: 125
  Achnatherum lemmonii (Vasey) Barkworth subsp. lemmonii, 24: 125

Achnatherum lemmonii subsp. **pubescens** (Crampton) Barkworth, 24: 125
Achnatherum lettermanii (Vasey) Barkworth, 24: 118
Achnatherum lobatum (Swallen) Barkworth, 24: 131
Achnatherum nelsonii (Scribn.) Barkworth, 24: 123
  Achnatherum nelsonii subsp. **dorei** (Barkworth & Maze) Barkworth, 24: 124
  Achnatherum nelsonii subsp. *longiaristatum* = **A. occidentale** subsp. **californicum**, 24: 121
  Achnatherum nelsonii (Scribn.) Barkworth subsp. **nelsonii**, 24: 124
Achnatherum nevadense (B.L. Johnson) Barkworth, 24: 120
Achnatherum occidentale (Thurb.) Barkworth, 24: 121
  Achnatherum occidentale subsp. **californicum** (Merr. & Burtt Davy) Barkworth, 24: 121
  Achnatherum occidentale (Thurb.) Barkworth subsp. **occidentale**, 24: 123
  Achnatherum occidentale subsp. **pubescens** (Vasey) Barkworth, 24: 123
Achnatherum papposum = **Jarava plumosa**, 24: 179
Achnatherum parishii (Vasey) Barkworth, 24: 127
  Achnatherum parishii subsp. **depauperatum** (M.E. Jones) Barkworth, 24: 129

Achnatherum parishii (Vasey) Barkworth subsp. **parishii**, 24: 129

Achnatherum perplexum Hoge & Barkworth, 24: 135

Achnatherum pinetorum (M.E. Jones) Barkworth, 24: 137

Achnatherum richardsonii (Link) Barkworth, 24: 133

Achnatherum robustum (Vasey) Barkworth, 24: 129

Achnatherum scribneri (Vasey) Barkworth, 24: 135

Achnatherum speciosum = **Jarava speciosa**, 24: 181

Achnatherum splendens (Trin.) Nevski, 24: 117

Achnatherum stillmanii (Bol.) Barkworth, 24: 118

Achnatherum swallenii (C.L. Hitchc. & Spellenb.) Barkworth, 24: 137

Achnatherum thurberianum (Piper) Barkworth, 24: 125

Achnatherum wallowaense J.R. Maze & K.A. Robson, 24: 138

Achnatherum webberi (Thurb.) Barkworth, 24: 137

×**ACHNELLA** Barkworth, 24: 169

×**Achnella caduca** (Beal) Barkworth, 24: 169

ACRACHNE Wight & Arn. *ex* Chiov., 25: 110

Acrachne racemosa (B. Heyne *ex* Roem. & Schult.) Ohwi, 25: 112

AEGILOPS L., 24: 261

Aegilops crassa Boiss., 24: 263

Aegilops cylindrica Host, 24: 263

Aegilops geniculata Roth, 24: 265

Aegilops neglecta Req. *ex* Bertol., 24: 265

Aegilops ovata = **A. geniculata**, 24: 265

Aegilops tauschii Coss., 24: 262

Aegilops triuncialis L., 24: 265

Aegilops ventricosa Tausch, 24: 263

×**AEGILOTRITICUM** P. Fourn., 24: 263

AEGOPOGON Humb. & Bonpl. *ex* Willd., 25: 273

Aegopogon tenellus (DC.) Trin., 25: 274

Aegopogon tenellus var. *abortivus* = **A. tenellus**, 25: 274

AELUROPUS Trin., 25: 25

Aeluropus littoralis (Gouan) Parl., 25: 27

Agrohordeum ×*macounii* = ×**Elyhordeum macounii**, 24: 284

Agrohordeum pilosilemma = ×**Elyhordeum pilosilemma**, 24: 284

×**AGROPOGON** P. Fourn., 379, **668**

×**Agropogon lutosus** (Poir.) P. Fourn., 643, 665, **668**, 669

AGROPYRON Gaertn., 24: 277

Agropyron alaskanum = **Elymus alaskanus**, 24: 326

Agropyron albicans = **Elymus albicans**, 24: 334

Agropyron albicans var. *griffithsii* = **Elymus albicans**, 24: 334

Agropyron arenicola = **Leymus pacificus**, 24: 358

Agropyron arizonicum = **Elymus arizonicus**, 24: 329

Agropyron bakeri = **Elymus bakeri**, 24: 330

Agropyron boreale = **Elymus alaskanus** subsp. **alaskanus**, 24: 326

Agropyron boreale subsp. *alaskanum* = **Elymus alaskanus** subsp. **alaskanus**, 24: 326

Agropyron caninum = **Elymus caninus**, 24: 322

Agropyron caninum var. andinum [applies to hybrids]

Agropyron caninum var. *hornemannii* = **Elymus trachycaulus** subsp. **trachycaulus**, 24: 322

Agropyron caninum var. *latiglume* = **Elymus violaceus**, 24: 324

Agropyron caninum subsp./var. *majus* = **Elymus trachycaulus** subsp. **trachycaulus**, 24: 322

Agropyron caninum var. *tenerum* = **Elymus trachycaulus** subsp. **trachycaulus**, 24: 322`

Agropyron caninum var. *unilaterale* = **Elymus trachycaulus** subsp. **subsecundus**, 24: 322

Agropyron cristatum (L.) Gaertn., 24: 279

Agropyron cristatum subsp. *desertorum* = **A. cristatum**, 24: 279

Agropyron cristatum subsp. *fragile* = **A. fragile**, 24: 279

Agropyron cristatum subsp. *pectinatum* = **A. cristatum**, 24: 279

Agropyron dasystachyum = **Elymus lanceolatus**, 24: 327

Agropyron dasystachyum var. *psammophilum* = **Elymus lanceolatus** subsp. **psammophilus**, 24: 327

Agropyron dasystachyum var. *riparium* = **Elymus lanceolatus**, 24: 327

Agropyron desertorum = **A. cristatum**, 24: 279

Agropyron elmeri = **Pascopyrum smithii**, 24: 351

Agropyron elongatum = **Thinopyrum ponticum**, 24: 376

Agropyron fragile (Roth) P. Candargy, 24: 279

Agropyron fragile var. *sibiricum* = **A. fragile**, 24: 279

Agropyrum griffithsii = **Elymus albicans**, 24: 334

Agropyron imbricatum = **A. cristatum**, 24: 279

Agropyron inerme = **Pseudoroegneria spicata**, 24: 280

Agropyron intermedium = **Thinopyrum intermedium**, 24: 376

Agropyron intermedium var. *trichophorum* = **Thinopyrum intermedium**, 24: 376

Agropyron junceum = **Thinopyrum junceum**, 24: 376

Agropyron latiglume = **Elymus violaceus**, 24: 324

Agropyron littorale = **Thinopyrum pycnanthum**, 24: 376

Agropyron macrourum = **Elymus macrourus**, 24: 324

Agropyron mongolicum = **A. fragile**, 24: 279

Agropyron parishii = **Elymus stebbinsii**, 24: 329

Agropyron parishii var. *laeve* = **Elymus trachycaulus** subsp. **subsecundus**, 24: 322

Agropyron pauciflorum = **Elymus trachycaulus** subsp. **trachycaulus**, 24: 322

Agropyron pectiniforme = **A. cristatum**, 24: 279

Agropyron pectinatum [within the *Flora* region, misapplied to A. cristatum], 24: 279

Agropyron pringlei = **Elymus sierrae**, 24: 332

Agropyron prostratum = **Eremopyrum triticeum**, 24: 254

Agropyron psammophilum = **Elymus lanceolatus** subsp. **psammophilus**, 24: 327

Agropyron ×*pseudorepens* = **Elymus** ×**pseudorepens**, 24: 340

Agropyron pungens = **Thinopyrum pycnanthum**, 24: 376

Agropyron repens = **Elymus repens**, 24: 334

Agropyron repens forma *aristatum* = **Elymus repens**, 24: 334

Agropyron repens forma *heberhachis* = **Elymus repens**, 24: 334

Agropyron repens var. *pilosum* = **Elymus repens**, 24: 334

Agropyron repens forma *setiferum* = **Elymus repens**, 24: 334

Agropyron repens var. *subulatum* = **Elymus repens,** 24: 334

Agropyron repens forma *trichorrhachis* = **Elymus repens,** 24: 334

Agropyron repens var. *typicum* = **Elymus repens,** 24: 334

Agropyron repens forma/var. *vaillantianum* = **Elymus repens,** 24: 334

Agropyron riparium = **Elymus lanceolatus** subsp. **lanceolatus,** 24: 327

Agropyron scribneri = **Elymus scribneri,** 24: 330

Agropyron semicostatum = **Elymus semicostatus,** 24: 338

Agropyron sericeum = **Elymus macrourus,** 24: 324

Agropyron sibiricum = **A. fragile,** 24: 279

Agropyron smithii = **Pascopyrum smithii,** 24: 351

   Agropyron smithii forma/var. *molle* = **Pascopyrum smithii,** 24: 351

   Agropyron smithii var. *palmeri* = **Pascopyrum smithii,** 24: 351

Agropyron spicatum = **Pseudoroegneria spicata,** 24: 280

   Agropyron spicatum "subsp."/var. *inerme* = **Pseudoroegneria spicata,** 24: 280

   Agropyron spicatum var. *molle* = **Pascopyrum smithii,** 24: 351

   Agropyron spicatum var. *pubescens* = **Pseudoroegneria spicata,** 24: 280

Agropyron squarrosum = **Eremopyrum bonaepartis,** 24: 254

Agropyron subsecundum = **Elymus trachycaulus** subsp. **subsecundus,** 24: 322

Agropyron subsecundum var. *andinum* = **Elymus trachycaulus** × **E. scribneri,** 24: 330

Agropyron tenerum = **Elymus trachycaulus** subsp. **trachycaulus,** 24: 322

Agropyron teslinense = **Elymus trachycaulus** subsp. **trachycaulus,** 24: 322

Agropyron trachycaulum = **Elymus trachycaulus,** 24: 322

   Agropyron trachycaulum var. *glaucum* = **Elymus trachycaulus** subsp. **subsecundus,** 24: 322

   Agropyron trachycaulum subsp. latiglume, *ined.* = **Elymus violaceus,** 24: 324

   Agropyron trachycaulum var. *majus* = **Elymus trachycaulus** subsp. **trachycaulus,** 24: 322

   Agropyron trachycaulum var. *novae-angliae* = **Elymus trachycaulus** subsp. **trachycaulus,** 24: 322

   Agropyron trachycaulum var. *pilosiglume* = **Elymus trachycaulus** subsp. **trachycaulus,** 24: 322

   Agropyron trachycaulum var. *typicum* = **Elymus trachycaulus** subsp. **trachycaulus,** 24: 322

   Agropyron trachycaulum var. *unilaterale* = **Elymus trachycaulus** subsp. **subsecundus,** 24: 322

Agropyron tricophorum = **Thinopyrum intermedium,** 24: 374

Agropyron triticeum = **Eremopyrum triticeum,** 24: 254

Agropyron vaseyi = **Pseudoroegneria spicata,** 24: 280

Agropyron violaceum = **Elymus violaceus,** 24: 324

   Agropyron violaceum var. *hyperarcticum* = **Elymus alaskanus** subsp. **hyperarcticus,** 24: 326

   Agropyron violaceum var. *latiglume* = **Elymus violaceus,** 24: 324

Agropyron vulpinum = **Elymus ×pseudorepens,** 24: 340

Agropyron yukonense auct. [within the *Flora* region, misapplied to **Elymus lanceolatus** subsp. **psammophilus**], 24: 327

Agropyron yukonense Scribn. & Merr. = **Elymus ×yukonensis,** 24: 340

×*Agrositanion saxicola* = **×Pseudelymus saxicola,** 24: 283

AGROSTIS L.

Agrostis aequivalvis = **Podagrostis aequivalvis,** 24: 693

Agrostis alaskana = **A. exarata,** 24: 651

Agrostis alba [within the *Flora* region, misapplied to **A. gigantea** and **A. stolonifera**], 24: 641

   Agrostis alba var. *aristata* = **A. capillaris,** 24: 639

   Agrostis alba forma *aristigera* = **A. stolonifera,** 24: 641

   Agrostis alba var. *major* = **A. gigantea,** 24: 641

   Agrostis alba var. *palustris* = **A. stolonifera,** 24: 641

   Agrostis alba var. *stolonifera* = **A. stolonifera,** 24: 641

   Agrostis alba var. *vulgaris* = **A. capillaris,** 24: 639

Agrostis altissima = **A. perennans,** 24: 647

Agrostis ampla = **A. exarata,** 24: 651

Agrostis anadyrensis Soczava, 24: 644

Agrostis aristiglumis = **A. hendersonii,** 24: 658

Agrostis avenacea = **Lachnagrostis filiformis,** 24: 696

Agrostis blasdalei Hitchc., 24: 656

   Agrostis blasdalei var. *marinensis* = **A. blasdalei,** 24: 656

Agrostis borealis = **A. mertensii,** 24: 644

Agrostis californica = **A. densiflora,** 24: 651

Agrostis canina L., 24: 643

Agrostis capillaris L., 24: 639

Agrostis castellana Boiss. & Reut., 24: 639

Agrostis clavata Trin., 24: 646

Agrostis clivicola var. *punta-reyensis* = **A. densiflora,** 24: 651

Agrostis densiflora Vasey, 24: 651

Agrostis diegoensis = **A. pallens,** 24: 651

Agrostis elliottiana Schult., 24: 661

Agrostis exarata Trin., 24: 651

   Agrostis exarata subsp./var. *minor* = **A. exarata,** 24: 651

   Agrostis exarata var. *monolepis* = **A. exarata,** 24: 651

   Agrostis exarata var. *pacifica* = **A. exarata,** 24: 651

   Agrostis exarata var. *purpurascens* = **A. exarata,** 24: 651

Agrostis filiculmis = **A. idahoensis,** 24: 649

Agrostis geminata = **A. scabra,** 24: 646

Agrostis gigantea Roth, 24: 641

   Agrostis gigantea var. *dispar* = **A. gigantea,** 24: 641

   Agrostis gigantea var. *ramosa* = **A. gigantea,** 24: 641

Agrostis glauca = **Calamagrostis cinnoides,** 24: 716

Agrostis hallii Vasey, 24: 653

Agrostis hendersonii Hitchc., 24: 658

Agrostis hiemalis = **A. hyemalis,** 24: 647

Agrostis hooveri Swallen, 24: 654

Agrostis howellii Scribn., 24: 654

Agrostis humilis = **Podagrostis humilis,** 24: 693

Agrostis hyemalis (Walter) Britton, Sterns & Poggenb.

   Agrostis hyemalis var. *geminata* = **A. scabra,** 24: 646

   Agrostis hyemalis var. *scabra* = **A. scabra,** 24: 646

   Agrostis hyemalis var. *tenuis* = **A. scabra,** 24: 646

Agrostis hyperborea = **A. vinealis,** 24: 643

Agrostis idahoensis Nash, 24: 649

Agrostis intermedia = **A. perennans**, 24: 647
Agrostis interrupta = **Apera interrupta**, 24: 789
Agrostis lepida = **A. pallens**, 24: 651
Agrostis longiligula = **A. exarata**, 24: 651
   Agrostis longiligula var. *australis* = **A. exarata**, 24: 651
Agrostis maritima = **A. stolonifera**, 24: 641
Agrostis mertensii Trin., 24: 644
   Agrostis mertensii subsp. *borealis* = **A. mertensii**, 24: 644
Agrostis microphylla Steud., 24: 661
Agrostis nebulosa Boiss. & Reut., 24: 661
Agrostis oregonensis Vasey, 24: 649
Agrostis pallens Trin., 24: 651
Agrostis palustris = **A. stolonifera**, 24: 641
Agrostis perennans (Walter) Tuck., 24: 647
   Agrostis perennans var. *aestivalis* = **A. perennans**, 24: 647
   Agrostis perennans var. *elata* = **A. perennans**, 24: 647
Agrostis retrofracta = **Lachnagrostis filiformis**, 24: 696
Agrostis rossae = **A. rossiae**, 24: 656
Agrostis rossiae Vasey, 24: 656
Agrostis rupestris = **A. mertensii**, 24: 644
Agrostis scabra Willd., 24: 646
   Agrostis scabra var. *geminata* = **A. scabra**, 24: 646
   Agrostis scabra var. *septentrionalis* = **A. scabra**, 24: 646
   Agrostis scabra forma *tuckermanii* = **A. scabra**, 24: 646
Agrostis schweinitzii = **A. perennans**, 24: 647
Agrostis semiverticillata = **Polypogon viridis**, 24: 663
Agrostis spica-venti = **Apera spica-venti**, 24: 789
Agrostis stolonifera L., 24: 641
   Agrostis stolonifera var. *compacta* = **A. stolonifera**, 24: 641
   Agrostis stolonifera var. *major* = **A. gigantea**, 24: 641
   Agrostis stolonifera var. *palustris* = **A. stolonifera**, 24: 641
Agrostis tandilensis = **Bromidium tandilense**, 24: 697
Agrostis tenuis Sibth. = **A. capillaris**, 24: 639
Agrostis tenuis Vasey = **A. idahoensis**, 24: 649
   Agrostis tenuis forma *aristata* = **A. capillaris**, 24: 639
Agrostis thurberiana = **Podagrostis humilis**, 24: 694
Agrostis tolucensis Kunth, 24: 656
Agrostis trinii = **A. vinealis**, 24: 643
Agrostis variabilis Rydb., 24: 654
Agrostis verticillata = **Polypogon viridis**, 24: 663
Agrostis vinealis Schreb., 24: 643
Agrostis viridis = **Polypogon viridis**, 24: 663
Agrostis vulgaris = **A. capillaris**, 24: 639

AIRA L.
Aira caryophyllea L., 24: 615
   Aira caryophyllea var. **capillaris** (Mert. & W.D.J. Koch) Bluff & Fingerh., 24: 616
   Aira caryophyllea L. var. **caryophyllea**, 24: 616
   Aira caryophyllea var. **cupaniana** (Guss.) Fiori, 24: 616
Aira elegans = **A. caryophyllea** var. **capillaris**, 24: 616
Aira elegantissima = **A. caryophyllea** var. **capillaris**, 24: 616
Aira flexuosa = **Deschampsia flexuosa**, 24: 631
Aira praecox L., 24: 616

ALLOLEPIS Soderstr. & H.F. Decker, 25: 27
Allolepis texana (Vasey) Soderstr. & H.F. Decker, 25: 28

ALLOTEROPSIS J. Presl, 25: 385
Alloteropsis cimicina (L.) Stapf, 25: 385

ALOPECURUS L., 24: 780
Alopecurus aequalis Sobol., 24: 784
   Alopecurus aequalis Sobol. var. **aequalis**, 24: 784
   Alopecurus aequalis var. *natans* = **A. aequalis** var. **aequalis**, 24: 784
   Alopecurus aequalis var. **sonomensis** P. Rubtzov, 24: 784
Alopecurus alpinus = **A. magellanicus**, 24: 782
   Alopecurus alpinus subsp. *glaucus* = **A. magellanicus**, 24: 782
Alopecurus arundinaceus Poir., 24: 782
Alopecurus borealis = **A. magellanicus**, 24: 782
Alopecurus carolinianus Walter, 24: 786
Alopecurus creticus Trin., 24: 788
Alopecurus geniculatus L., 24: 784
Alopecurus howellii = **A. saccatus**, 24: 786
Alopecurus magellanicus Lam., 24: 782
Alopecurus myosuroides Huds., 24: 786
Alopecurus occidentalis = **A. magellanicus**, 24: 782
Alopecurus pallescens = **A. geniculatus**, 24: 784
Alopecurus pratensis L., 24: 782
Alopecurus rendlei Eigf, 24: 786
Alopecurus saccatus Vasey, 24: 786
Alopecurus ventricosus = **A. arundinaceus**, 24: 782

AMELICHLOA Arriaga & Barkworth, 24: 181
Amelichloa brachychaeta (Godron) Arriaga & Barkworth, 24: 182
Amelichloa caudata (Trin.) Arriaga & Barkworth, 24: 182
Amelichloa clandestina (Trin.) Arriaga & Barkworth, 24: 184

AMMOPHILA Host, 24: 776
Ammophila arenaria (L.) Link, 24: 777
Ammophila breviligulata Fernald, 24: 777
   Ammophila breviligulata Fernald subsp. **breviligulata**, 24: 777
   Ammophila breviligulata subsp. **champlainensis** (F. Seym.) P.J. Walker, C.A. Paris & Barrington *ex* Barkworth, 24: 777
Ammophila champlainensis = **A. breviligulata** subsp. **champlainensis**, 24: 777
Ammophila longifolia = **Calamovilfa longifolia**, 25: 141

AMPELODESMOS Link, 24: 112
Ampelodesmos mauritanicus (Poir.) T. Durand & Schinz, 24: 112

AMPHIBROMUS Nees, 24: 703
Amphibromus neesii Steud., 24: 704
Amphibromus nervosa (Hook. f.) Baill., 24: 704
Amphibromus scabrivalvis (Trin.) Swallen, 24: 704

AMPHICARPUM Kunth, 25: 385
Amphicarpum amphicarpon (Pursh) Nash, 25: 387
Amphicarpum mühlenbergianum (Schult.) Hitchc., 25: 387
Amphicarpum purshii = **A. amphicarpon**, 25: 387

ANDROPOGON L., 25: 649
Andropogon L. sect. **Andropogon**, 25: 652
Andropogon sect. **Leptopogon** Stapf, 25: 653
Andropogon arctatus Chapm., 25: 655
Andropogon argenteus = **A. ternarius**, 25: 653

Andropogon barbinodis = **Bothriochloa barbinodis**, 25: 642

Andropogon bicornis L., 25: 655

Andropogon bladhii = **Bothriochloa bladhii**, 25: 646

Andropogon brachystachyus Chapm., 25: 659

Andropogon callipes = **A. virginicus** var. **glaucus**, 25: 661

Andropogon campyloracheus = **A. gyrans** var. **gyrans**, 25: 657

Andropogon cirratus = **Schizachyrium cirratum**, 25: 674

Andropogon contortus = **Heteropogon contortus**, 25: 680

Andropogon divergens = **Schizachyrium scoparium** var. **divergens**, 25 670

Andropogon elliottii = **A. gyrans**, 25: 657

Andropogon elliottii var. *projectus* = **A. gyrans**, 25: 657

Andropogon floridanus Scribn., 25: 655

Andropogon furcatus = **A. gerardii**, 25: 653

Andropogon gerardii Vitman, 25: 653

Andropogon gerardii var. *chrysocomus* = **A. hallii**, 25: 653

Andropogon gerardii var. *paucipilus* = **A. hallii**, 25: 653

Andropogon glaucopsis = **A. glomeratus** var. **glaucopsis**, 25: 664

Andropogon glomeratus (Walter) Britton, Sterns & Poggenb., 25: 661

Andropogon glomeratus var. **glaucopsis** (Elliott) C. Mohr, 25: 664

Andropogon glomeratus (Walter) Britton, Sterns & Poggenb. var. **glomeratus**, 25: 664

Andropogon glomeratus var. **hirsutior** (Hack.) C. Mohr, 25: 664

Andropogon glomeratus var. **pumilus** (Vasey) L.H. Dewey, 25: 664

Andropogon glomeratus var. **scabriglumis** C.S. Campb., 25: 664

Andropogon gracilis Spreng., 25: 653

Andropogon gyrans Ashe, 25: 657

Andropogon gyrans Ashe var. **gyrans**, 25: 657

Andropogon gyrans var. **stenophyllus** (Hack.) C.S. Campb., 25: 657

Andropogon hallii Hack., 25: 653

Andropogon hirtiflorus = **Schizachyrium sanguineum** var. **hirtiflorum**, 25: 674

Andropogon hirtiflorus var. *feensis* = **Schizachyrium sanguineum** var. **hirtiflorum**, 25: 674

Andropogon intermedius = **Bothriochloa bladhii**, 25: 646

Andropogon ischaemum = **Bothriochloa ischaemum**, 25: 646

Andropogon ischaemum var. *songaricus* = **Bothriochloa ischaemum**, 25: 646

Andropogon liebmannii, 25: 657

Andropogon liebmannii Hack. var. **liebmannii**, 25: 657

Andropogon liebmannii var. **pungensis** (Ashe) C.S. Campb., 25: 657

Andropogon longiberbis Hack., 25: 661

Andropogon macrourus var. *glaucopsis* = **A. glomeratus** var. **glaucopsis**, 25: 664

Andropogon maritimus = **Schizachyrium maritimum**, 25: 672

Andropogon mohrii = **A. liebmannii** var. **pungensis**, 25: 657

Andropogon nodosus = **Dichanthium annulatum**, 25: 638

Andropogon perangustatus = **A. gyrans** var. **stenophyllus**, 25: 657

Andropogon perforatus = **Bothriochola barbinodis**, 25: 642

Andropogon pertusus = **Bothriochloa pertusa**, 25: 646

Andropogon provincialis = **A. gerardii**, 25: 653

Andropogon rhizomatous = **Schizachyrium rhizomatum**, 25: 670

Andropogon saccharoides [Latin American species; within the *Flora* region, often misapplied to **Bothriochloa laguroides** subsp. **torreyana**, 25: 640]

Andropogon saccharoides var. *longipaniculata* = **Bothriochloa longipaniculata**, 25: 640

Andropogon saccharoides var. *torreyanus* = **Bothriochloa laguroides** subsp. **torreyana**, 25: 640

Andropogon scoparius = **Schizachyrium scoparium**, 25: 669

Andropogon scoparius var. *divergens* = **Schizachyrium scoparium** var. **divergens**, 25: 670

Andropogon scoparius var. *ducis* = **Schizachyrium littorale**, 25: 672

Andropogon scoparius var. *frequens* = **Schizachyrium scoparium** var. **scoparium**, 25: 670

Andropogon scoparius var. *littoralis* = **Schizachyrium littorale**, 25: 672

Andropogon scoparius var. *neomexicana* = **Schizachyrium scoparium** var. **scoparium**, 25: 670

Andropogon scoparius var. *septentrionalis* = **Schizachyrium scoparium** var. **scoparium**, 25: 670

Andropogon spadiceus = **Schizachyrium spadiceum**, 25: 669

Andropogon springfieldii = **Bothriochloa springfieldii**, 25: 644

Andropogon subtenuis = **A. gyrans**, 25: 657

Andropogon tener = **Schizachyrium tenerum**, 25: 672

Andropogon ternarius Michx., 25: 653

Andropogon ternarius var. **cabanisii** (Hack.) Fernald & Griscom, 25: 655

Andropogon ternarius L. var. **ternarius**, 25: 655

Andropogon tracyi Nash, 25: 659

Andropogon virginicus L., 25: 659

Andropogon virginicus var. *abbreviatus* = **A. glomeratus** var. **glomeratus**, 25: 664

Andropogon virginicus var. **decipiens** C.S. Campb., 25: 659

Andropogon virginicus var. *glaucopsis* = **A. glomeratus** var. **glaucopsis**, 25: 644

Andropogon virginicus var. **glaucus** Hack., 25: 661

Andropogon virginicus forma/var. *tenuispatheus* = **A. glomeratus** var. **pumilus**, 25: 664

Andropogon virginicus var. *tetrastachyus* = **A. virginicus** var. **virginicus**, 25: 661

Andropogon virginicus L. var. **virginicus**, 25: 661

ANDROPOGONEAE Dumort., 25: 602

ANISANTHA [included in **BROMUS**], 24: 193

Anisantha diandra = **Bromus diandrus**, 24: 224

Anisantha sterilis = **Bromus sterilis**, 24: 224

Anisantha tectorum = **Bromus tectorum**, 24: 226

ANTHENANTIA P. Beauv., 25: 384

Anthenantia rufa (Elliott) Schult., 25: 384

Anthenantia villosa (Michx.) P. Beauv., 25: 384

ANTHEPHORA Schreb., 25: 535

Anthephora hermaphrodita (L.) Kuntze, 25: 535

ANTHOXANTHUM L., 24: 758

Anthoxanthum aristatum Boiss., 24: 759

   Anthoxanthum aristatum Boiss. subsp. aristatum, 24: 759

   Anthoxanthum aristatum subsp. macranthum Valdés, 24: 759

Anthoxanthum monticola (Bigelow) Veldkamp, 24: 760

   Anthoxanthum monticola subsp. alpinum (Sw. *ex* Willd.) Soreng, 24: 760

   Anthoxanthum monticola (Bigelow) Veldkamp subsp. monticola, 24: 760

   Anthoxanthum monticola subsp. *orthanthum* = **A. monticola** subsp. **monticola**, 24: 760

Anthoxanthum nitens (Weber) Y. Schouten & Veldkamp, 24: 762

Anthoxanthum occidentale (Buckley) Veldkamp, 24: 762

Anthoxanthum odoratum L., 24: 759

   Anthoxanthum odoratum subsp. *alpinum* = **A. odoratum**, 24: 759

Anthoxanthum puelii = **A. aristatum**, 24: 759

APERA Adans., 24: 788

Apera interrupta (L.) P. Beauv., 24: 789

Apera spica-venti (L.) P. Beauv., 24: 789

APLUDA L., 25: 649

Apluda mutica L., 25: 649

ARCTAGROSTIS Griseb., 24: 676

Arctagrostis angustifolia = **A. latifolia** subsp. **arundinacea**, 24: 678

Arctagrostis arundinacea = **A. latifolia** subsp. **arundinacea**, 24: 678

Arctagrostis latifolia (R. Br.) Griseb., 24: 678

   Arctagrostis latifolia subsp. **arundinacea** (Trin.) Tzvelev, 24: 678

   Arctagrostis latifolia (R. Br.) Griseb. subsp. **latifolia**, 24: 678

Arctagrostis poaeoides = **A. latifolia** subsp. **arundinacea**, 24: 678

×**ARCTODUPONTIA** Tzvelev, 24: 604

×**Arctodupontia scleroclada** (Rupr.) Tzvelev, 24: 605

ARCTOPHILA (Rupr.) Andersson, 24: 605

Arctophila fulva (Trin.) Andersson, 24: 607

ARGILLOCHLOA [included in **FESTUCA**], 24: 389

Argillochloa dasyclada = **Festuca dasyclada**, 24: 443

ARISTIDA L., 25: 315

Aristida adscensionis L., 25: 330

   Aristida adscensionis var. *modesta* = **A. adscensionis**, 25: 330

Aristida affinis = **A. purpurascens** var. **purpurascens**, 25: 340

Aristida arizonica Vasey, 25: 335

Aristida barbata = **A. havardii**, 25: 324

Aristida basiramea Engelm. *ex* Vasey, 25: 326

   Aristida basiramea var. *curtissii* = **A. dichotoma** var. **curtissii**, 25: 328

Aristida beyrichiana = **A. stricta**, 25: 335

Aristida bromoides = **A. adscensionis**, 25: 330

Aristida brownii = **A. purpurea** forma **brownii**, 25: 335

Aristida californica Thurb., 25: 319

   Aristida californica Thurb. var. **californica**, 25: 321

   Aristida californica var. **glabrata** Vasey, 25: 321

Aristida condensata Chapm., 25: 340

Aristida curtissii = **A. dichotoma** var. **curtissii**, 25: 328

Aristida desmantha Trin. & Rupr., 25: 319

Aristida dichotoma Michx., 25: 328

   Aristida dichotoma var. **curtissii** A. Gray, 25: 328

   Aristida dichotoma Michx. var. **dichotoma**, 25: 328

Arstida divaricata Humb. & Bonpl. *ex* Willd., 25: 323

Aristida fendleriana = **A. purpurea** var. **fendleriana**, 25: 332

Aristida floridana (Chapm.) Vasey, 25: 321

Aristida glabrata = **A. californica** var. **glabrata**, 25: 321

Aristida glauca = **A. purpurea** var. **nealleyi**, 25: 333

Aristida gypsophila Beetle, 25: 326

Aristida gyrans Chapm., 25: 340

Aristida hamulosa = **A. ternipes** var. **gentilis**, 25: 323

Aristida havardii Vasey, 25: 324

Aristida intermedia = **A. longespica** var. **geniculata**, 25: 330

Aristida interrupta = **A. adscensionis**, 25: 330

Aristida lanosa Muhl. *ex* Elliott, 25: 338

Aristida longespica Poir., 25: 328

   Aristida longespica var. **geniculata** (Raf.) Fernald, 25: 330

   Aristida longespica Poir. var. **longespica**, 25: 330

Aristida longiseta = **A. purpurea** var. **longiseta**, 25: 332

   Aristida longiseta var. *robusta* = **A. purpurea** var. **longiseta**, 25: 332

Aristida mohrii Nash, 25: 337

Aristida necopina = **A. longespica** var. **geniculata**, 25: 330

Aristida oligantha Michx., 25: 326

Aristida orcuttiana = **A. schiedeana** var. **orcuttiana**, 25: 323

Aristida palustris (Chapm.) Vasey, 25: 338

Aristida pansa Wooton & Standl., 25: 324

   Aristida pansa var. *dissita* = **A. pansa**, 25: 324

Aristida parishii = **A. purpurea** var. **parishii**, 25: 333

Aristida patula Chapm. *ex* Nash, 25: 321

Aristida purpurascens Poir., 25: 338

   Aristida purpurascens var. *minor* = **A. purpurascens** var. **purpurascens**, 25: 340

   Aristida purpurascens Poir. var. **purpurascens**, 25: 340

   Aristida purpurascens var. **tenuispica** (Hitchc.) Allred, 25: 340

   Aristida purpurascens var. **virgata** (Trin.) Allred, 25: 340

Aristida purpurea Nutt., 25: 330

   Aristida purpurea forma **brownii** (Warnock) Allred & Valdés-Reyna, 25: 335

   Aristida purpurea var. **fendleriana** (Steud.) Vasey, 25: 332

   Aristida purpurea var. *glauca* = **A. purpurea** var. **nealleyi**, 25: 333

   Aristida purpurea var. *laxiflora* = **A. purpurea**, 25: 330

   Aristida purpurea var. **longiseta** (Steud.) Vasey, 25: 332

   Aristida purpurea var. **nealleyi** (Vasey) Allred, 25: 333

   Aristida purpurea var. **parishii** (Hitchc.) Allred, 25: 333

   Aristida purpurea var. **perplexa** Allred & Valdés-Reyna, 25: 333

   Aristida purpurea Nutt. var. **purpurea**, 25: 333

   Aristida purpurea var. rariflora, *ined.* = **A. purpurea** var. **longiseta**, 25: 332

   Aristida purpurea var. *robusta* = **A. purpurea** var. **longiseta**, 25: 332

   Aristida purpurea var. **wrightii** (Nash) Allred, 25: 333

Aristida ramosissima Engelm. *ex* A. Gray, 25: 326

Aristida ramosissima var. *chaseana* = **A. ramosissima**, 25: 326

Aristida rhizomophora Swallen, 25: 335

Aristida roemeriana = **A. purpurea**, 25: 330

Aristida schiedeana Trin. & Rupr., 25: 323

  Aristida schiedeana var. **orcuttiana** (Vasey) Allred & Valdés-Reyna, 25: 323

  Aristida schiedeana Trin. & Rupr. var. **schiedeana**, 25: 323

Aristida simpliciflora Chapm., 25: 337

Aristida spiciformis Elliott, 25: 330

Aristida stricta Michx., 25: 335

Aristida ternipes Cav., 25: 323

  Aristida ternipes var. **gentilis** (Henrard) Allred, 25: 323

  Aristida ternipes var. *hamulosa* = **A. ternipes** var. **gentilis**, 25: 323

  Aristida ternipes var. *minor* = **A. ternipes** var. **ternipes**, 25: 323

  Aristida ternipes Cav. var. **ternipes**, 25: 323

Aristida tuberculosa Nutt., 25: 319

Aristida virgata = **A. purpurascens** var. **virgata**, 25: 340

Aristida wrightii = **A. purpurea** var. **wrightii**, 25: 333

  Aristida wrightii var. *parishii* = **A. purpurea** var. **parishii**, 25: 333

ARISTIDEAE C.E. Hubb., 25: 314

ARISTIDOIDEAE Caro, 25: 314

ARRHENATHERUM P. Beauv., 24: 740

Arrhenatherum avenaceum = **A. elatius**, 24: 742

Arrhenatherum elatius (L.) P. Beauv. *ex* J. Presl & C. Presl, 24: 742

  Arrhenatherum elatius forma/var. *biaristatum* = **A. elatius**, 24: 742

  Arrhenatherum elatius subsp. **bulbosum** (Willd.) Schübl. & G. Martens, 24: 742

  Arrhenatherum elatius (L.) P. Beauv. *ex* J. Presl & C. Presl subsp. **elatius**, 24: 742

ARTHRAXON P. Beauv., 25: 677

Arthraxon hispidus (Thunb.) Makino, 25: 677

  Arthraxon hispidus var. *cryptatherus* = **A. hispidus**, 25: 677

ARUNDINARIA Michx., 24: 17

Arundinaria appalachiana Triplett, Weakley & L.G. Clark, 24: 18

Arundinaria gigantea (Walter) Muhl., 24: 18

  Arundinaria gigantea subsp. *macrosperma* = **A. gigantea**, 24: 18

  Arundinaria gigantea subsp. *tecta* = **A. tecta**, 24: 18

  Arundinaria japonica = **Pseudosasa japonica**, 24: 29

  Arundinaria macrosperma = **A. gigantea**, 24: 18

Arundinaria tecta (Walter) Muhl., 24: 18

  Arundinaria tecta var. *distachya* = **A. gigantea**, 24: 18

ARUNDINEAE Dumort., 25: 7

ARUNDINOIDEAE Burmeist., 25: 6

ARUNDO L., 25: 11

Arundo donax L., 25: 11

  Arundo donax var. *versicolor* = **A. donax**, 25: 11

AUSTROSTIPA S.W.L. Jacobs & J. Everett, 24: 184

Austrostipa elegantissima (Labill.) S.W.L. Jacobs & J. Everett, 24: 185

Austrostipa ramosissima (Trin.) S.W.L. Jacobs & J. Everett, 24: 185

AVENA L., 24: 734

Avena barbata Pott *ex* Link, 24: 735

Avena brevis [species not in the *Flora* region]

Avena fatua L., 24: 735

  Avena fatua var. *glabrata* = **A. fatua**, 24: 735

  Avena fatua var. *glabrescens* = **A. fatua**, 24: 735

  Avena fatua var. *sativa* = **A. sativa**, 24: 737

  Avena fatua var. *vilis* = **A. hybrida**, 24: 737

Avena hybrida Peterm., 24: 737

Avena mortoniana = **Helictotrichon mortonianum**, 24: 701

Avena nuda [species not in the *Flora* region]

Avena occidentalis Durieu, 24: 737

Avena pubescens = **Avenula pubescens**, 24: 699

Avena sativa L., 24: 737

Avena sterilis L., 24: 739

Avena striata = **Helictotrichon sempervirens**, 24: 701

Avena strigosa [within the *Flora* region, misapplied to **A. barbata**], 24: 735

AVENEAE [included in **POEAE**], 24: 378

AVENULA (Dumort.) Dumort., 24: 698

Avenula hookeri (Scribn.) Holub, 24: 698

Avenula pubescens (Huds.) Dumort., 24: 698

  Avenula pubescens subsp. **laevigata** (Schur) Holub, 24: 699

  Avenula pubescens (Huds.) Dumort. subsp. **pubescens**, 24: 699

AXONOPUS P. Beauv., 25: 565

Axonopus affinis = **A. fissifolius**, 25: 565

Axonopus compressus (Sw.) P. Beauv., 25: 566

Axonopus fissifolius (Raddi) Kuhlm., 25: 565

Axonopus furcatus (Flüggé) Hitchc., 25: 566

Axonopus scoparius (Humb. & Bonpl. *ex* Flüggé) Hitchc., 25: 566

BAMBUSA Schreb., 24: 21

Bambusa bambos (L.) Voss, 24: 22

Bambusa glaucescens = **B. multiplex**, 24: 22

Bambusa multiplex (Lour.) Raeusch. *ex* Schult. & Schult. f., 24: 22

Bambusa oldhamii Munro, 24: 25

Bambusa vulgaris Schrad. *ex* J.C. Wendl., 24: 22

  Bambusa vulgaris var. *aureovarigata* = **B. vulgaris**, 24: 22

BAMBUSEAE Nees, 24: 15

BAMBUSOIDEAE Luerss., 24: 14

BECKMANNIA Host, 24: 484

Beckmannia eruciformis subsp. **baicalensis** [an invalid name] = **B. syzigachne**, 24: 486

Beckmannia syzigachne (Steud.) Fernald, 24: 486

  Beckmannia syzigachne subsp. *baicalensis* = **B. syzigachne**, 24: 486

  Beckmannia syzigachne var. *uniflora* = **B. syzigachne**, 24: 486

BLEPHARIDACHNE Hack., 25: 48
Blepharidachne bigelovii (S. Watson) Hack., 25: 49
Blepharidachne kingii (S. Watson) Hack., 25: 49

BLEPHARONEURON Nash, 25: 47
Blepharoneuron tricholepis (Torr.) Nash, 25: 48

BOTHRIOCHLOA Kuntze, 25: 639
Bothriochloa alta (Hitchc.) Henrard, 25: 642
Bothriochloa barbinodis (Lag.) Herter, 25: 642
  Bothriochloa barbinodis var. *perforata* = **B. barbinodis**, 25: 642
Bothriochloa bladhii (Retz.) S.T. Blake, 25: 646
Bothriochloa edwardsiana (Gould) Parodi, 25: 644
Bothriochloa exaristata (Nash) Henrard, 25: 642
Bothriochloa hybrida (Gould) Gould, 25: 644
Bothriochloa intermedia = **B. bladhii**, 25: 646
Bothriochloa ischaemum (L.) Keng, 25: 646
  Bothriochloa ischaemum var. *songarica* = **B. ischaemum**, 25: 646
Bothriochloa laguroides (DC.) Herter, 25: 640
  Bothriochloa laguroides (DC.) Herter subsp. **laguroides**, 25: 640
  Bothriochloa laguroides subsp. **torreyana** (Steud.) Allred & Gould, 25: 640
Bothriochloa longipaniculata (Gould) Allred & Gould, 25: 640
Bothriochloa pertusa (L.) A. Camus, 25: 646
Bothriochloa saccharoides [Latin American species; within the *Flora* region, misapplied to **B. laguroides** subsp. **torreyana**], 25: 640
  Bothriochloa saccharoides var. *longipaniculata* = **B. longipaniculata**, 25: 640
  Bothriochloa saccharoides var. *torreyana* = **B. laguroides** var. **torreyana**, 25: 640
Bothriochloa springfieldii (Gould) Parodi, 25: 644
Bothriochloa wrightii (Hack.) Henrard, 25: 640

BOUTELOUA Lag., 25: 250
Bouteloua Lag. subg. Bouteloua, 25: 253
Bouteloua subg. Chondrosum (Desv.) A. Gray, 25: 261
Bouteloua aristidoides (Kunth) Griseb., 25: 255
  Bouteloua aristidoides (Kunth) Griseb. var. **aristidoides**, 25: 255
  Bouteloua aristidoides var. **arizonica** M.E. Jones, 25: 257
Bouteloua barbata Lag., 25: 265
  Bouteloua barbata Lag. var. **barbata**, 25: 265
  Bouteloua barbata var. **rothrockii** (Vasey) Gould, 25: 265
Bouteloua breviseta Vasey, 25: 267
Bouteloua chondrosioides (Kunth) Benth. *ex* S. Watson, 25: 257
Bouteloua curtipendula (Michx.) Torr., 25: 254
  Bouteloua curtipendula var. **caespitosa** Gould & Kapadia, 25: 254
  Bouteloua curtipendula (Michx.) Torr. var. **curtipendula**, 25: 254
Bouteloua eludens Griffiths, 25: 257
Bouteloua eriopoda (Torr.) Torr., 25: 262
Bouteloua filiformis = **B. repens**, 25: 259
Bouteloua glandulosa = **B. hirsuta** subsp. **hirsuta**, 25: 262
Bouteloua gracilis (Kunth) Lag. *ex* Griffiths, 25: 261

Bouteloua hirsuta Lag., 25: 261
  Bouteloua hirsuta var. *glandulosa* = **B. hirsuta** subsp. **hirsuta**, 25: 262
  Bouteloua hirsuta Lag. subsp. **hirsuta**, 25: 262
  Bouteloua hirsuta subsp. **pectinata** (Feath.) Wipff & S.D. Jones, 25: 262
Bouteloua kayi Warnock, 25: 264
Bouteloua oligostachya = **B. gracilis**, 25: 261
Bouteloua parryi (E. Fourn.) Griffiths, 25: 267
Bouteloua pectinata = **B. hirsuta** subsp. **pectinata**, 25: 262
Bouteloua procumbens = **B. simplex**, 25: 265
Bouteloua radicosa (E. Fourn.) Griffiths, 25: 259
Bouteloua ramosa Scribn. *ex* Vasey, 25: 267
Bouteloua repens (Kunth) Scribn. & Merr., 25: 259
Bouteloua rigidiseta (Steud.) Hitchc., 25: 259
Bouteloua rothrockii = **B. barbata** var. **rothrockii**, 25: 265
Bouteloua simplex Lag., 25: 265
Bouteloua trifida Thurb. *ex* S. Watson, 25: 264
  Bouteloua trifida var. **burkii** (Scribn. *ex* S. Watson) Vasey *ex* L.H. Dewey. 25: 264
  Bouteloua trifida Thurb. *ex* S. Watson var. **trifida**, 25: 264
Bouteloua uniflora Vasey, 25: 255
Bouteloua warnockii Gould & Kapadia, 25: 254

BRACHIARIA [*sensu* FNA vol. 25] = **MOOROCHLOA**, 25: 488
Brachiaria adsperca = **Urochloa adspersa**, 25: 497
Brachiaria arizonica = **Urochloa arizonica**, 25: 495
Brachiaria brizantha = **Urochloa brizantha**, 25: 499
Brachiaria ciliatissima = **Urochloa ciliatissima**, 25: 505
Brachiaria distichophylla = **Urochloa villosa**, 25: 501
Brachiaria eruciformis = **Moorochloa eruciformis**, 25: 488
Brachiaria extensa = **Urochloa platyphylla**, 25: 503
Brachiaria fasciculata = **Urochloa fusca**, 25: 495
Brachiaria mutica = **Urochloa mutica**, 25: 494
Brachiaria piligera = **Urochloa piligera**, 25: 499
Brachiaria plantaginea = **Urochloa plantaginea**, 25: 501
Brachiaria platyphylla = **Urochloa platyphylla**, 25: 503
Brachiaria platytaenia = **Urochloa oligobrachiata**, 25: 503
Brachiaria ramosa = **Urochloa ramosa**, 25: 497
Brachiaria reptans = **Urochloa reptans**, 25: 494
Brachiaria subquadripara = **Urochloa subquadripara**, 25: 501
Brachiaria texana = **Urochloa texana**, 25: 495

BRACHYELYTREAE Ohwi, 24: 59

BRACHYELYTRUM P. Beauv., 24: 59
Brachyelytrum aristatum = **B. aristosum**, 24: 60
Brachyelytrum aristosum (Michx.) P. Beauv. *ex* Branner & Coville, 24: 60
  Brachyelytrum aristosum var. *glabratum* = **B. erectum**, 24: 60
Brachyelytrum erectum (Schreb.) P. Beauv., 24: 60
  Brachyelytrum erectum var. *glabratum* [*sensu* T. Koyama & Kawano] [for the description] = **B. aristosum**, 24: 60
  Brachyelytrum erectum var. *glabratum* (Vasey) T. Koyama & Kawano [for the type specimen] = **B. erectum**, 24: 60
  Brachyelytrum erectum var. *septentrionale* = **B. aristosum**, 24: 60
Brachyelytrum septentrionale = **B. aristosum**, 24: 60

BRACHYPODIEAE Harz, 24: 187

BRACHYPODIUM P. Beauv., 24: 187
Brachypodium cespitosum = **B. rupestre**, 24: 192
Brachypodium distachyon (L.) P. Beauv., 24: 188
Brachypodium phoenicoides (L.) Roem. & Schult., 24: 190
Brachypodium pinnatum (L.) P. Beauv., 24: 190
Brachypodium rupestre (Host) Roem. & Schult., 24: 192
Brachypodium sylvaticum (Huds.) P. Beauv., 24: 190

BRIZA L., 24: 612
Briza maxima L., 24: 614
Briza media L., 24: 614
Briza minor L., 24: 614

BROMEAE Dumort., 24: 192

BROMELICA [included in **MELICA**], 24: 88
Bromelica bulbosa = **Melica bulbosa**, 24: 91
Bromelica spectabilis = **Melica spectabilis**, 24: 91
Bromelica subulata = **Melica subulata**, 24: 95

BROMIDIUM Nees & Meyen, 24: 696
Bromidium tandilense (Kuntze) Rúgolo, 24: 697

BROMOPSIS [included in **BROMUS**], 24: 193
Bromopsis canadensis = **Bromus ciliatus**, 24: 222
  Bromopsis canadensis subsp. *richardsonii* = **Bromus richardsonii**, 24: 222
Bromopsis ciliata = **Bromus ciliatus**, 24: 222
Bromopsis inermis = **Bromus inermis**, 24: 206
  Bromopsis inermis subsp. *pumpellianus* = **Bromus pumpellianus**, 24: 207
Bromopsis lanatipes = **Bromus lanatipes**, 24: 220
Bromopsis pacifica = **Bromus pacificus**, 24: 218
Bromopsis porteri = **Bromus porteri**, 24: 213
Bromopsis pubescens = **Bromus pubescens**, 24: 220
Bromopsis pumpelliana = **Bromus pumpellianus**, 24: 207
  Bromopsis pumpelliana var. *arctica* = **Bromus pumpellianus**, 24: 207
Bromposis richardsonii = **Bromus richardsonii**, 24: 222

BROMUS L., 24: 193
Bromus sect. **Bromopsis** Dumort., 24: 206
Bromus L. sect. **Bromus**, 24: 228
Bromus sect. **Ceratochloa** (P. Beauv.) Griseb., 24: 199
Bromus sect. **Genea** Dumort., 24: 224
Bromus sect. **Neobromus** (Shear) Hitchc., 24: 222
Bromus aleutensis Trin. *ex* Griseb., 24: 203
Bromus alopecuros = **B. caroli-henrici**, 24: 233
Bromus altissimus = **B. latiglumis**, 24: 209
Bromus anomalus Rupr. *ex* E. Fourn., 24: 213
  Bromus anomalus var. *lanatipes* = **B. lanatipes**, 24: 220
Bromus arenarius Labill., 24: 235
Bromus arizonicus (Shear) Stebbins, 24: 201
Bromus arvensis L., 24: 228
Bromus berteroanus Colla, 24: 224
  Bromus berteroanus var. *excelsus* = **B. arizonicus**, 24: 201
Bromus breviaristatus = **B. carinatus** var. **marginatus**, 24: 205
Bromus brizaeformis = **B. briziformis**, 24: 228
Bromus briziformis Fisch. & C.A. Mey., 24: 228
Bromus canadensis = **B. ciliatus**, 24: 222

Bromus carinatus Hook. & Arn., 24: 203
  Bromus carinatus var. *californicus* = **B. carinatus** var. **carinatus**, 24: 205
  Bromus carinatus Hook. & Arn. var. **carinatus**, 24: 205
  Bromus carinatus var. *hookerianus* = **B. carinatus** var. **carinatus**, 24: 205
  Bromus carinatus var. *linearis* = **B. carinatus** var. **carinatus**, 24: 205
  Bromus carinatus var. *maritimus* = **B. maritimus**, 24: 202
  Bromus carinatus var. **marginatus** Barkworth & Anderton, 24: 205
Bromus caroli-henrici Greuter, 24: 233
Bromus catharticus Vahl, 24: 199
  Bromus catharticus Vahl var. **catharticus**, 24: 199
  Bromus catharticus var. **elatus** (E. Desv.) Planchuelo, 24: 201
Bromus ciliatus L., 24: 222
  Bromus ciliatus var. *glaberrimus* = **B. vulgaris**, 24: 216
  Bromus ciliatus var. *intonsus* = **B. ciliatus**, 24: 222
  Bromus ciliatus forma/var. *laeviglumis* = **B. pubescens**, 24: 220
Bromus commutatus Schrad., 24: 230
Bromus danthoniae Trin. *ex* C.A. Mey., 24: 233
Bromus diandrus Roth, 24: 224
Bromus dudleyi = **B. ciliatus**, 24: 222
Bromus erectus Huds., 24: 218
Bromus frondosus (Shear) Wooten & Standl., 24: 213
Bromus grandis (Shear) Hitchc., 24: 213
Bromus grossus = **B. secalinus**, 24: 230
Bromus hallii (Hitchc.) Saarela & P.M. Peterson, 24: 211
Bromus hordeaceus L., 24: 232
  Bromus hordaceus subsp. *divaricatus* = **B. hordaceus** subsp. **molliformis**, 24: 232
  Bromus hordeaceus L. subsp. **hordeaceus**, 24: 232
  Bromus hordeaceus subsp. **molliformis** (J. Lloyd *ex* Billot) Maire & Weiller, 24: 232
  Bromus hordeaceus subsp. **pseudothominei** (P.M. Sm.) H. Scholz, 24: 232
  Bromus hordeaceus subsp. **thominei** (Hardouin) Braun-Blanq., 24: 232
Bromus inermis Leyss., 24: 206
  Bromus inermis forma *aristatus* = **B. inermis**, 24: 206
  Bromus inermis var. *purpurascens* = **B. pumpellianus**, 24: 207
  Bromus inermis subsp. *pumpellianus* = **B. pumpellianus**, 24: 207
  Bromus inermis forma *villosus* = **B. inermis**, 24: 206
Bromus japonicus Thunb., 24: 235
  Bromus japonicus var. *porrectus* = **B. japonicus**, 24: 235
Bromus kalmii A. Gray, 24: 209
Bromus laevipes Shear, 24: 209
Bromus lanatipes (Shear) Rydb., 24: 220
Bromus lanceolatus Roth, 24: 235
  Bromus lanceolatus var. lanuginosus [an illegitimate name] = **B. lanceolatus**, 24: 235
Bromus latiglumis (Scribn. *ex* Shear) Hitchc., 24: 209
  Bromus latiglumis forma *incanus* = **B. latiglumis**, 24: 209
Bromus lepidus Holmb., 24: 230
Bromus luzonensis = **B. carinatus** var. **marginatus**, 24: 205
Bromus macrostachys = **B. lanceolatus**, 24: 235

Bromus madritensis L., 24: 226
  Bromus madritensis subsp. *rubens* = **B. rubens**, 24: 226
Bromus marginatus = **B. carinatus** var. **marginatus**, 24: 205
  Bromus marginatus var. *latior* = **B. carinatus** var. **marginatus**, 24: 205
  Bromus marginatus var. *seminudus* = **B. carinatus** var. **marginatus**, 24: 205
Bromus maritimus (Piper) Hitchc., 24: 202
Bromus molliformis = **B. hordaceus** subsp. **molliformis**, 24: 232
Bromus mollis = **B. hordaceus**, 24: 232
  Bromus mollis forma *leiostachys* = **B. hordaceus** subsp. **pseudothominei**, 24: 232
Bromus mucroglumis Wagnon, 24: 218
Bromus nottowayanus Fernald, 24: 216
Bromus orcuttianus Vasey, 24: 211
  Bromus orcuttianus var. *grandis* = **B. grandis**, 24: 213
  Bromus orcuttianus var. *hallii* = **B. hallii**, 24: 211
Bromus pacificus Shear, 24: 218
Bromus polyanthus Scribn., 24: 205
Bromus porteri (J.M. Coult.) Nash, 24: 213
Bromus pseudolaevipes Wagnon, 24: 211
Bromus ×pseudothominei = **B. hordaceus** subsp. **pseudothominei**, 24: 232
Bromus pubescens Muhl. *ex* Willd., 24: 220
Bromus pumpellianus Scribn., 24: 207
  Bromus pumpellianus var. *arcticus* = **B. pumpellianus**, 24: 207
  Bromus pumpellianus subsp. dicksonii W.W. Mitch. & Wilton, 24: 207
  Bromus pumpellianus Scribn. **subsp. pumpellianus**, 24: 207
Bromus purgans [within the *Flora* region, misapplied to **B. pubescens**], 24: 220
  Bromus purgans forma *glabriflorus* = **B. pubescens**, 24: 220
  Bromus purgans forma *incanus* = **B. latiglumis**, 24: 209
  Bromus purgans var. *laeviglumis* = **B. pubescens**, 24: 220
  Bromus purgans forma *laevivaginatus* = **B. pubescens**, 24: 220
  Bromus purgans var. *latiglumis* = **B. latiglumis**, 24: 209
Bromus racemosus L., 24: 233
Bromus ramosus Huds., 24: 220
Bromus richardsonii Link, 24: 222
Bromus rigidus = **B. diandrus**, 24: 224
  Bromus rigidus var. *gussonei* = **B. diandrus**, 24: 224
Bromus riparius Rehmann, 24: 206
Bromus rubens L., 24: 226
Bromus scoparius L., 24: 235
Bromus secalinus L., 24: 230
  Bromus secalinus subsp. *decipiens* = **B. commutatus**, 24: 230
  Bromus secalinus var. hirsutus, *ined.* = **B. secalinus**, 24: 230
  Bromus secalinus var. *velutinus* = **B. secalinus**, 24: 230
Bromus sitchensis Trin., 24: 201
  Bromus sitchensis var. *aleutensis* = **B. aleutensis**, 24: 203
  Bromus sitchensis var. *marginatus* = **B. carinatus** var. **marginatus**, 24: 205
Bromus squarrosus L., 24: 237

Bromus stamineus = **B. catharticus** var. **elatus**, 24: 201
Bromus sterilis L., 24: 224
Bromus suksdorfii Vasey, 24: 218
Bromus tectorum L., 24: 226
  Bromus tectorum var. *glabratus* = **B. tectorum**, 24: 226
  Bromus tectorum var. *nudus* = **B. tectorum**, 24: 226
Bromus texensis (Shear) Hitchc., 24: 216
Bromus thominii = **B. hordaceus** subsp. **thominei**, 24: 232
Bromus trinii = **B. berteroanus**, 24: 224
  Bromus trinii var. *excelsus* = **B. berteroanus**, 24: 224
Bromus unioloides = **B. catharticus**, 24: 199
Bromus vulgaris (Hook.) Shear, 24: 216
Bromus willdenowii = **B. catharticus**, 24: 199

BUCHLOË Engelm., 25: 270
Buchloë dactyloides (Nutt.) Engelm., 25: 271

CALAMAGROSTIS Adans., 24: 706
Calamagrostis ×acutiflora (Schrad.) DC. 'Karl Foerster', 24: 721
Calamagrostis anomala = **C. canadensis**, 24: 726
Calamagrostis arctica = **C. sesquiflora**, 24: 714
Calamagrostis arundinacea var. *purpurascens* = **C. purpurascens**, 24: 710
Calamagrostis bolanderi Thurb., 24: 719
Calamagrostis brevipilis = **Calamovilfa brevipilis**, 25: 142
Calamagrostis breweri Thurb., 24: 717
Calamagrostis cainii Hitchc., 24: 719
Calamagrostis canadensis (Michx.) P. Beauv., 24: 726
  Calamagrostis canadensis var. *acuminata* = **C. stricta** subsp. **inexpansa**, 24: 730
  Calamagrostis canadensis (Michx.) P. Beauv. var. **canadensis**, 24: 728
  Calamagrostis canadensis var. *lactea* = **C. canadensis** var. **langsdorffii**, 24: 728
  Calamagrostis canadensis var. **langsdorffii** (Link) Inman, 24: 728
  Calamagrostis canadensis var. **macouniana** (Vasey) Stebbins, 24: 728
  Calamagrostis canadensis var. *pallida* = **C. canadensis** var. **canadensis**, 24: 728
  Calamagrostis canadensis var. *robusta* = **C. canadensis** var. **canadensis**, 24: 728
  Calamagrostis canadensis var. *scabra* = **C. canadensis** var. **langsdorffii**, 24: 728
  Calamagrostis canadensis var. *typica* = **C. canadensis** var. **canadensis**, 24: 728
Calamagrostis chordorrhiza = **C. stricta** subsp. **inexpansa**, 24: 730
Calamagrostis cinnoides (Muhl.) W.P.C. Barton, 24: 716
Calamagrostis coarctata = **C. cinnoides**, 24: 716
Calamagrostis confinis [name of uncertain application]
Calamagrostis crassiglumis = **C. stricta** subsp. **inexpansa**, 24: 730
Calamagrostis densa = **C. koeleroides**, 24: 720
Calamagrostis deschampsioides Trin., 24: 719
Calamagrostis epigejos (L.) Roth, 24: 710
  Calamagrostis epigejos var. **georgica** (K. Koch) Griseb., 24: 710

Calamagrostis fernaldii = **C. stricta** subsp. **inexpansa,** 24: 730

Calamagrostis foliosa Kearney, 24: 714

Calamagrostis howellii Vasey, 24: 712

Calamagrostis holmii = **C. stricta** subsp. **stricta,** 24: 730

Calamagrostis hyperborea [name of uncertain application, probably **C. stricta** subsp. **inexpansa**], 24: 730

Calamagrostis inexpansa = **C. stricta** subsp. **inexpansa,** 24: 730

  Calamagrostis inexpansa var. *barbulata* = **C. stricta** subsp. **inexpansa,** 24: 730

  Calamagrostis inexpansa var. *brevior* = **C. stricta** subsp. **inexpansa,** 24: 730

  Calamagrostis inexpansa var. *novae-angliae* = **C. stricta** subsp. **inexpansa,** 24: 730

  Calamagrostis inexpansa var. *robusta* = **C. stricta** subsp. **inexpansa,** 24: 730

Calamagrostis insperata = **C. porteri** subsp. **insperata,** 24: 723

Calamagrostis koelerioides Vasey, 24: 720

Calamagrostis lacustris = **C. stricta** subsp. **inexpansa,** 24: 730

Calamagrostis langsdorffii = **C. canadensis** var. **langsdorffii,** 24: 728

Calamagrostis lapponica (Wahlenb.) Hartm., 24: 729

  Calamagrostis lapponica var. *groenlandica* = **C. lapponica,** 24: 729

  Calamagrostis lapponica var. *nearctica* = **C. lapponica,** 24: 729

Calamagrostis laricina = **C. purpurascens,** 24: 710

Calamagrostis lepageana = **C. purpurascens,** 24: 710

Calamagrostis longifolia = **Calamovilfa longifolia** var. **longifolia,** 25: 142

Calamagrostis montanensis (Scribn.) Vasey, 24: 724

Calamagrostis muiriana B.L. Wilson & Sami Gray, 24: 717

Calamagrostis neglecta = **C. stricta,** 24: 729

  Calamagrostis neglecta var. *borealis* = **C. stricta,** 24: 729

  Calamagrostis neglecta var. *micrantha* = **C. stricta** subsp. **stricta,** 24: 730

Calamagrostis nutkaensis (J. Presl) Steud., 24: 724

Calamagrostis ophitidis (J.T. Howell) Nygren, 24: 712

Calamagrostis perplexa Scribn., 24: 726

Calamagrostis pickeringii A. Gray, 24: 724

  Calamagrostis pickeringii var. *debilis* = **C. pickeringii,** 24: 724

Calamagrostis poluninii = **C. purpurascens,** 24: 710

Calamagrostis porteri A. Gray, 24: 721

  Calamagrostis porteri subsp. **insperata** (Swallen) C.W. Greene, 24: 723

  Calamagrostis porteri subsp. *perplexa* = **C. perplexa,** 24: 726

  Calamagrostis porteri A. Gray subsp. **porteri,** 24: 723

Calamagrostis purpurascens R. Br., 24: 710

  Calamagrostis purpurascens subsp. *arctica* = **C. sesquiflora,** 24: 714

  Calamagrostis purpurascens var. *laricina* = **C. purpurascens,** 24: 710

  Calamagrostis purpurascens var. *ophitidis* = **C. ophitidis,** 24: 712

Calamagrostis purpurascens subsp. *tasuensis* = **C. sesquiflora,** 24: 714

Calamagrostis rubescens Buckley, 24: 723

Calamagrostis scopulorum M.E. Jones, 24: 717

Calamagrostis scribneri = **C. canadensis** var. **canadensis,** 24: 728

Calamagrostis sesquiflora (Trin.) Tzvelev, 24: 714

Calamagrostis sitchensis [name of uncertain application]

Calamagrostis stricta (Timm) Koeler, 24: 729

  Calamagrostis stricta var. *borealis* = **C. stricta,** 24: 729

  Calamagrostis stricta subsp. **inexpansa** (A. Gray) C.W. Greene, 24: 730

  Calamagrostis stricta (Timm) Koeler subsp. **stricta,** 24: 730

Calamagrostis tacomensis K.L. Marr & Hebda, 24: 716

Calamagrostis tweedyi (Scribn.) Scribn., 24: 714

Calamagrostis vaseyi [type belongs to **C. purpurascens**, name usually applied to **C. tacomaensis**], 24: 716

CALAMOVILFA (A. Gray) Hack., 25: 140

Calamovilfa (A. Gray) Hack. sect. **Calamovilfa,** 25: 141

Calamovilfa sect. **Interior** Thieret, 25: 141

Calamovilfa arcuata K.E. Rogers, 25: 142

Calamovilfa brevipilis (Torr.) Scribn., 25: 142

Calamovilfa curtissii (Vasey) Scribn., 25: 142

Calamovilfa gigantea (Nutt.) Scribn. & Merr., 25: 141

Calamovilfa longifolia (Hook.) Scribn., 25: 141

  Calamovilfa longifolia (Hook.) Scribn. var. **longifolia,** 25: 142

  Calamovilfa longifolia var. **magna** Scribn. & Merr., 25: 142

CAPRIOLA [included in **CYNODON**], 25: 235

Capriola dactylon = **Cynodon dactylon,** 25: 238

CATABROSA P. Beauv., 24: 610

Catabrosa aquatica (L.) P. Beauv., 24: 610

  Catabrosa aquatica var. *laurentiana* = **C. aquatica,** 24: 610

CATAPODIUM [included in **DESMAZERIA**], 24: 681

Catapodium rigidum = **Desmazeria rigida,** 24: 682

CATHESTECUM J. Presl, 25: 272

Cathestecum brevifolium Swallen, 25: 272

Cathestecum erectum Vasey & Hack., 25: 273

CELTICA F.M. Vázquez & Barkworth, 24: 152

Celtica gigantea (Link) F.M. Vázquez & Barkworth, 24: 154

CENCHRUS L., 25: 529

Cenchrus bambusoides = **C. spinifex,** 25: 533

Cenchrus biflorus Roxb., 25: 535

Cenchrus brownii Roem. & Schult., 25: 531

Cenchrus carolinianus = **C. longispinus,** 25: 534

Cenchrus ciliaris = **Pennisetum ciliare,** 25: 525

Cenchrus echinatus L., 25: 531

  Cenchrus echinatus var. *hillebrandianus* = **C. echinatus,** 25: 531

Cenchrus gracillimus Nash, 25: 533

Cenchrus incertus = **C. spinifex,** 25: 533

Cenchrus longispinus (Hack.) Fernald, 25: 534

Cenchrus myosuroides Kunth, 25: 534

Cenchrus pauciflorus = **C. spinifex,** 25: 533

Cenchrus setigerus = **Pennisetum setigerum**, 25: 525
Cenchrus spinifex Cav., 25: 533
Cenchrus tribuloides L., 25: 534

CENTOTHECEAE Ridley, 25: 344

CENTOTHECOIDEAE Soderstr., 25: 343

CERATOCHLOA [included in **BROMUS**], 24: 193
Ceratochloa carinata = **Bromus carinatus**, 24: 203
Ceratochloa marginata = **Bromus carinatus** var. **marginatus**, 24: 205
Ceratochloa polyantha = **Bromus polyanthus**, 24: 205
Ceratochloa unioloides = **Bromus catharticus**, 24: 199

CHAETOCHLOA [an invalid name, included in **SETARIA**], 25: 539
Chaetochloa verticillata = **Setaria verticillata**, 25: 554

CHASMANTHIUM Link, 25: 344
Chasmanthium latifolium (Michx.) H.O. Yates, 25: 345
Chasmanthium laxum (L.) H.O. Yates, 25: 346
  Chasmanthium laxum subsp. *sessiliflorum* = **C. sessiliflorum**, 25: 346
Chasmanthium nitidum (Baldwin) H.O. Yates, 25: 345
Chasmanthium ornithorhynchum Nees, 25: 346
Chasmanthium sessiliflorum (Poir.) H.O Yates, 25: 346

CHLORIDOIDEAE Kunth *ex* Beilschm., 25: 13

CHLORIS Sw., 25: 204
Chloris andropogonoides E. Fourn., 25: 216
Chloris argentina = **Eustachys retusa**, 25: 222
Chloris barbata (L.) Sw., 25: 208
Chloris berroi Arechav., 25: 207
Chloris canterae Arechav., 25: 208
  Chloris canterae Arechav. var. **canterae**, 25: 208
  Chloris canterae var. **grandiflora** (Roseng. & Izag.) D.E. Anderson, 25: 208
Chloris chloridea = **Enteropogon chlorideus**, 25: 225
Chloris ciliata Sw., 25: 207
Chloris crinita = **Trichloris crinita**, 25: 227
Chloris cucullata Bisch., 25: 214
Chloris dandyana = **C. elata**, 25: 208
Chloris distichophylla = **Eustachys distichophylla**, 25: 222
Chloris divaricata R. Br., 25: 212
Chloris elata Desv., 25: 208
Chloris floridana = **Eustachys floridana**, 25: 222
Chloris gayana Kunth, 25: 210
Chloris glauca = **Eustachys glauca**, 25: 220
Chloris inflata = **C. barbata**, 25: 208
Chloris latisquamae [applies to hybrids], 25: 216
Chloris pectinata Benth., 25: 214
Chloris petraea = **Eustachys petraea**, 25: 220
Chloris pilosa Schumach., 25: 210
Chloris pluriflora = **Trichloris pluriflora**, 25: 227
Chloris polydactyla = **C. elata**, 25: 208
Chloris radiata (L.) Sw., 25: 218
Chloris subdolichostachya [applies to hybrids], 25: 216
Chloris submutica Kunth, 25: 216
Chloris texensis Nash, 25: 216
Chloris truncata R. Br., 25: 212
Chloris ventricosa R. Br., 25: 212

Chloris verticillata Nutt., 25: 214
Chloris virgata Sw., 25: 210

CHONDROSUM Desv. [included in **BOUTELOUA**], 25: 250
Chondrosum barbatum = **Bouteloua barbata**, 25: 265
Chondrosum eriopodum = **Bouteloua eriopoda**, 25: 262
Chondrosum gracile = **Bouteloua gracilis**, 25: 261
Chondrosum hirsutum = **Bouteloua hirsuta**, 25: 261
Chondrosum prostratum = **Bouteloua simplex**, 25: 265

CHRYSOPOGON Trin., 25: 633
Chrysopogon aciculatus (Retz.) Trin., 25: 634
Chrysopogon fulvus (Spreng.) Chiov., 25: 634
Chrysopogon pauciflorus (Chapm.) Benth. *ex* Vasey, 25: 633
Chrysopogon zizanioides (L.) Roberty, 25: 634

CINNA L., 24: 773
Cinna arundinacea L., 24: 774
  Cinna arundinacea var. *inexpansa* = **C. arundinacea**, 24: 774
Cinna bolanderi Scribn., 24: 774
Cinna latifolia (Trevir. *ex* Göpp.) Griseb., 24: 774

CLADORAPHIS Franch., 25: 105
Cladoraphis cyperoides (Thunb.) S.M. Phillips, 25: 105

COELORACHIS Brongn., 25: 687
Coelorachis cylindrica (Michx.) Nash, 25: 688
Coelorachis rugosa (Nutt.) Nash, 25: 688
Coelorachis tessellata (Steud.) Nash, 25: 688
Coelorachis tuberculosa (Nash) Nash, 25: 688

COIX L., 25: 703
Coix lacryma-jobi L., 25: 704

COLEANTHUS Seidl, 24: 618
Coleanthus subtilis (Tratt.) Seidl, 24: 618

COLPODIUM [included in **PUCCINELLIA**, in part], 24: 459
Colpodium vahlianum = **Puccinellia vahliana**, 24: 465
Colpodium wrightii = **Puccinellia wrightii**, 24: 467

CORIDOCHLOA [included in **ALLOTEROPSIS**], 25: 385
Coridochloa cimicina = **Alloteropsis cimicina**, 25: 385

CORTADERIA Stapf, 25: 298
Cortaderia atacamensis = **C. jubata**, 25: 299
Cortaderia dioica = **C. selloana**, 25: 299
Cortaderia jubata (Lemoine) Stapf, 25: 299
Cortaderia selloana (Schult. & Schult. f.) Asch. & Graebn., 25: 299

CORYNEPHORUS P. Beauv., 24: 742
Corynephorus canescens (L.) P. Beauv., 24: 743

COTTEA Kunth, 25: 287
Cottea pappophoroides Kunth, 25: 289

CRITESION [included in **HORDEUM**], 24: 241
Critesion brachyantherum = **Hordeum brachyantherum**, 24: 243
Critesion glaucum = **Hordeum murinum** subsp. **glaucum**, 24: 250

Critesion jubatum = **Hordeum jubatum**, 24: 245
Critesion marinum subsp. *gussoneanum* = **Hordeum marinum** subsp. **gussoneanum**, 24: 248
Critesion murinum subsp. *glaucum* = **Hordeum murinum** subsp. **glaucum**, 24: 250
Critesion murinum subsp. *leporinum* = **Hordeum murinum** subsp. **leporinum**, 24: 250
Critesion murinum subsp. *murinum* = **Hordeum murinum** subsp. **murinum**, 24: 250
Critesion pusillum = **Hordeum pusillum**, 24: 243

CRYPSIS Aiton, 25: 139
Crypsis alopecuroides (Piller & Mitterp.) Schrad., 25: 139
Crypsis niliacea = **C. vaginiflora**, 25: 140
Crypsis schoenoides (L.) Lam., 25: 140
Crypsis vaginiflora (Forssk.) Opiz, 25: 140

CTENIUM Panz., 25: 232
Ctenium aromaticum (Walter) Alph. Wood, 25: 234
Ctenium floridanum (Hitchc.) Hitchc., 25: 234

CUTANDIA Willk., 24: 611
Cutandia memphitica (Spreng.) K. Richt., 24: 612

CYLINDROPYRUM [included in **AEGILOPS**], 24: 261
Cylindropyrum cylindricum = **Aegilops cylindrica**, 24: 263

CYMBOPOGON Spreng., 25: 664
Cymbopogon citratus (DC.) Stapf, 25: 666
Cymbopogon jwarancusa (Jones) Schult., 25: 665
Cymbopogon nardus (L.) Rendle, 25: 666

CYNODON Rich., 25: 235
Cynodon Rich. subg. **Cynodon**, 25: 237
Cynodon subg. **Pterolemma** Caro & E.A. Sánchez, 25: 237
Cynodon aethiopicus Clayton & J.R. Harlan, 25: 240
Cynodon dactylon (L.) Pers., 25: 238
  Cynodon dactylon var. **aridus** J.R. Harlan & de Wet, 25: 238
  Cynodon dactylon (L.) Pers. var. **dactylon**, 25: 238
Cynodon erectus = **C. dactylon**, 25: 238
Cynodon incompletus Nees, 25: 240
  Cynodon incompletus var. **hirsutus** (Stent) de Wet & J.R. Harlan, 25: 240
  Cynodon incompletus Nees var. **incompletus**, 25: 240
Cynodon ×magennisii Hurcombe, 25: 238
Cynodon nlemfuënsis Vanderyst, 25: 240
Cynodon plectostachyus (K. Schum.) Pilg., 25: 237
Cynodon transvaalensis Burtt Davy, 25: 237

CYNODONTEAE Dumort., 25: 14

CYNOSURUS L., 24: 685
Cynosurus cristatus L., 24: 685
Cynosurus dactylon = **Cynodon dactylon**, 25: 238
Cynosurus echinatus L., 24: 687

DACTYLIS L., 24: 482
Dactylis glomerata L., 24: 482
  Dactylis glomerata var. *ciliata* = **D. glomerata**, 24: 482
  Dactylis glomerata var. *detonsa* = **D. glomerata**, 24: 482
  Dactylis glomerata var. *vivipara* = **D. glomerata**, 24: 482

DACTYLOCTENIUM Willd., 25: 112

Dactyloctenium aegyptium (L.) Willd., 25: 113
Dactyloctenium geminatum Hack., 25: 113
Dactyloctenium radulans (R. Br.) P. Beauv., 25: 113

DANTHONIA DC., 25: 301
Danthonia alleni = **D. compressa**, 25: 303
Danthonia californica Bol., 25: 305
  Danthonia californica var. *americana* = **D. californica**, 25: 305
Danthonia canadensis = **D. intermedia**, 25: 303
Danthonia compressa Austin, 25: 303
Danthonia decumbens (L.) DC., 25: 302
Danthonia epilis = **D. sericea**, 25: 302
Danthonia intermedia Vasey, 25: 303
  Danthonia intermedia var. *cusickii* = **D. intermedia**, 25: 303
Danthonia macounii = **D. californica**, 25: 305
Danthonia parryi Scribn., 25: 305
Danthonia pilosa [within the *Flora* region, misapplied to **Rytidosperma penicillatum**], 25: 310
Danthonia sericea Nutt., 25: 302
Danthonia spicata (L.) P. Beauv. *ex* Roem. & Schult., 25: 303
  Danthonia spicata var. *longipila* = **D. spicata**, 25: 303
  Danthonia spicata var. *pinetorum* = **D. spicata**, 25: 303
Danthonia unispicata (Thurb.) Munro *ex* Vasey, 25: 305

DANTHONIEAE Zotov, 25: 298

DANTHONIOIDEAE N.P. Barker & H.P. Linder, 25: 297

DASYOCHLOA Willd. *ex* Rydb., 25: 45
Dasyochloa pulchella (Kunth) Willd. *ex* Rydb., 25: 47

DASYPYRUM (Coss. & Durieu) T. Durand, 24: 256
Dasypyrum villosum (L.) P. Candargy, 24: 256

DESCHAMPSIA P. Beauv, 24: 624
Deschampsia alpina (L.) Roem. & Schult., 24: 631
Deschampsia arctica = **D. brevifolia**, 24: 629
  Deschampsia arctica var. *pumila* = **D. sukatschewii**, 24: 629
Deschampsia atropurpurea = **Vahlodea atropurpurea**, 24: 692
  Deschampsia atropurpurea var. *latifolia* = **Vahlodea atropurpurea**, 24: 692
Deschampsia beringensis = **D. cespitosa** subsp. **beringensis**, 24: 626
Deschampsia borealis = **D. sukatschewii**, 24: 629
Deschampsia brevifolia R. Br., 24: 629
  Deschampsia brevifolia var. *pumila* = **D. sukatschewii**, 24: 629
Deschampsia caespitosa = **D. cespitosa**, 24: 626
  Deschampsia caespitosa subsp. *alpina* = **D. alpina**, 24: 631
  Deschampsia caespitosa var. *arctica* = **D. cespitosa**, 24: 626
  Deschampsia caespitosa var. beringensis, *ined.* = **D. cespitosa** subsp. **beringensis**, 24: 626
  Deschampsia caespitosa var. brevifolia, *ined.* = **D. brevifolia**, 24: 629
  Deschampsia caespitosa var. *genuina* = **D. cespitosa**, 24: 626

Deschampsia caespitosa var. *glauca* = D. **cespitosa** subsp. **cespitosa**, 24: 628

Deschampsia caespitosa subsp./"var." *littoralis* = D. **cespitosa** subsp. **cespitosa**, 24: 628

Deschampsia caespitosa var. *longiflora* = D. **cespitosa**, 24: 626

Deschampsia caespitosa var. *maritima* = D. **cespitosa**, 24: 626

Deschampsia caespitosa subsp. *orientalis* = D. **sukatschewii**, 24: 629

Deschampsia caespitosa subsp. *paramushirensis* = D. **sukatschewii**, 24: 629

Deschampsia caespitosa subsp./var. *parviflora* = D. **cespitosa**, 24: 626

Deschampsia cespitosa (L.) P. Beauv., 24: 626

　Deschampsia cespitosa subsp. **beringensis** (Hultén) W.E. Lawr., 24: 626

　Deschampsia cespitosa subsp. *brevifolia* = D. **brevifolia**, 24: 629

　Deschampsia cespitosa subsp. **holciformis** (J. Presl) W.E. Lawr., 24: 628

Deschampsia calycina = D. **danthonioides**, 24: 631

Deschampsia danthonioides (Trin.) Munro, 24: 631

Deschampsia elongata (Hook.) Munro, 24: 631

Deschampsia flexuosa (L.) Trin., 24: 631

　Deschampsia flexuosa forma *flavescens* = D. **flexuosa**, 24: 631

　Deschampsia flexuosa var. *montana* = D. **flexuosa**, 24: 631

Deschampsia glauca = D. **cespitosa** subsp. **cespitosa**, 24: 628

Deschampsia holciformis = D. **cespitosa** subsp. **holciformis**, 24: 628

Deschampsia mackenzieana Raup, 24: 628

Deschampsia paramushirensis = D. **sukatschewii**, 24: 629

Deschampsia pumila = D. **sukatschewii**, 24: 629

Deschampsia sukatschewii (Popl.) Roshev., 24: 629

　Deschampsia sukatschewii subsp. *borealis* = D. **sukatschewii**, 24: 629

　Deschampsia sukatschewii subsp. *orientalis* = D. **sukatschewii**, 24: 629

DESMAZERIA Dumort., 24: 681

Desmazeria rigida (L.) Tutin, 24: 682

Desmazeria sicula (Jacq.) Dumort., 24: 682

DIARRHENA P. Beauv., 24: 64

Diarrhena americana P. Beauv., 24: 65

　Diarrhenia americana var. *obovata* = D. **obovata**, 24: 65

Diarrhena obovata (Gleason) Brandenburg, 24: 65

DIARRHENEAE C.S. Campb., 24: 64

DICHANTHELIUM (Hitchc. & Chase) Gould, 25: 406

Dichanthelium sect. **Angustifolia** (Hitchc.) Freckmann & Lelong, 25: 442

Dichanthelium sect. **Clandestina** Freckmann & Lelong, 25: 418

Dichanthelium (Hitchc. & Chase) Gould sect. **Dichanthelium**, 25: 432

Dichanthelium sect. **Ensifolia** (Hitchc.) Freckmann & Lelong, 25: 436

Dichanthelium sect. **Lancearia** (Hitchc.) Freckmann & Lelong, 25: 441

Dichanthelium sect. **Lanuginosa** (Hitchc.) Freckmann & Lelong, 25: 422

Dichanthelium sect. **Linearifolia** Freckmann & Lelong, 25: 447

Dichanthelium sect. **Macrocarpa** Freckmann & Lelong, 25: 412

Dichanthelium sect. **Nudicaulia** Freckmann & Lelong, 25: 434

Dichanthelium sect. **Oligosantha** (Hitchc.) Freckmann & Lelong, 25: 419

Dichanthelium sect. **Pedicellata** (Hitchc. & Chase) Freckmann & Lelong, 25: 410

Dichanthelium sect. **Sphaerocarpa** (Hitchc. & Chase) Freckmann & Lelong, 25: 440

Dichanthelium sect. **Strigosa** Freckmann & Lelong, 25: 446

Dichanthelium aciculare (Desv. *ex* Poir.) Gould & C.A. Clark, 25: 442

　Dichanthelium aciculare (Desv. *ex* Poir.) Gould & C.A. Clark subsp. **aciculare**, 25: 444

　Dichanthelium aciculare subsp. **angustifolium** (Elliott) Freckmann & Lelong, 25: 444

　Dichanthelium aciculare subsp. **fusiforme** (Hitchc.) Freckmann & Lelong, 25: 444

　Dichanthelium aciculare subsp. **neuranthum** (Griseb.) Freckmann & Lelong, 25: 444

Dichanthelium acuminatum (Sw.) Gould & C.A. Clark, 25: 422

　Dichanthelium acuminatum (Sw.) Gould & C.A. Clark subsp. **acuminatum**, 25: 425

　Dichanthelium acuminatum subsp. **columbianum** (Scribn.) Freckmann & Lelong, 25: 425

　Dichanthelium acuminatum var. *consanguineum*, *ined.* = D. **consanguineum**, 25: 444

　Dichanthelium acuminatum var. *densiflorum* = D. **acuminatum** subsp. **spretum**, 25: 426

　Dichanthelium acuminatum subsp. **fasciculatum** (Torr.) Freckmann & Lelong, 25: 425

　Dichanthelium acuminatum subsp. **implicatum** (Scribn.) Freckmann & Lelong, 25: 425

　Dichanthelium acuminatum subsp. **leucothrix** (Nash) Freckmann & Lelong, 25: 426

　Dichanthelium acuminatum subsp. **lindheimeri** (Nash) Freckmann & Lelong, 25: 426

　Dichanthelium acuminatum subsp. **longiligulatum** (Nash) Freckmann & Lelong, 25: 426

　Dichanthelium acuminatum var. *septentrionale* = D. **acuminatum** subsp. **fasciculatum**, 25: 425

　Dichanthelium acuminatum subsp. **sericeum** (Schmoll) Freckmann & Lelong, 25: 426

　Dichanthelium acuminatum subsp. **spretum** (Schult.) Freckmann & Lelong, 25: 426

　Dichanthelium acuminatum subsp. **thermale** (Bol.) Freckmann & Lelong, 25: 426

　Dichanthelium acuminatum var. *thurowii* = D. **acuminatum** subsp. **acuminatum**, 25: 425

　Dichanthelium acuminatum var. *villosum* = D. **ovale** subsp. **villosissimum**, 25: 430

Dichanthelium acuminatum var. *wrightianum* = D. wrightianum, 25: 430

Dichanthelium angustifolium = D. aciculare subsp. angustifolium, 25: 444

Dichanthelium auburne = D. acuminatum subsp. acuminatum, 25: 425

Dichanthelium boreale (Nash) Freckmann, 25: 434

Dichanthelium boscii (Poir.) Gould & C.A. Clark, 25: 412

Dichanthelium chamaelonche (Trin.) Freckmann & Lelong, 25: 438

    Dichanthelium chamaelonche subsp. **breve** (Hitchc. & Chase) Freckmann & Lelong, 25: 440

    Dichanthelium chamaelonche (Trin.) Freckmann & Lelong subsp. **chamaelonche**, 25: 440

Dichanthelium clandestinum (L.) Gould, 25: 418

Dichanthelium commonsianum = D. ovale subsp. pseudopubescens, 25: 430

Dichanthelium commutatum (Schult.) Gould, 25: 414

    Dichanthelium commutatum subsp. **ashei** (T.G. Pearson *ex* Ashe) Freckmann & Lelong, 25: 414

    Dichanthelium commutatum (Schult.) Gould subsp. **commutatum**, 25: 414

    Dichanthelium commutatum subsp. **equilaterale** (Scribn.) Freckmann & Lelong, 25: 414

    Dichanthelium commutatum subsp. **joorii** (Vasey) Freckmann & Lelong, 25: 416

Dichanthelium consanguineum (Kunth) Gould & C.A. Clark, 25: 444

Dichanthelium depauperatum (Muhl.) Gould, 25: 450

Dichanthelium dichotomum (L.) Gould, 25: 432

    Dichanthelium dichotomum var. *barbulatum* = D. dichotomum subsp. dichotomum, 25: 433

    Dichanthelium dichotomum (L.) Gould subsp. **dichotomum**, 25: 433

    Dichanthelium dichotomum var. *ensifolium* = D. ensifolium, 25: 436

    Dichanthelium dichotomum subsp. **lucidum** (Ashe) Freckmann & Lelong, 25: 433

    Dichanthelium dichotomum subsp. **mattamuskeetense** (Ashe) Freckmann & Lelong, 25: 433

    Dichanthelium dichotomum subsp. **microcarpon** (Muhl. *ex* Elliott) Freckmann & Lelong, 25: 433

    Dichanthelium dichotomum subsp. **nitidum** (Lam.) Freckmann & Lelong, 25: 433

    Dichanthelium dichotomum var. *ramulosum* = D. dichotomum subsp. microcarpon, 25: 433

    Dichanthelium dichotomum subsp. **roanokense** (Ashe) Freckmann & Lelong, 25: 434

    Dichanthelium dichotomum var. *tenue* = D. tenue, 25: 438

    Dichanthelium dichotomum subsp. **yadkinense** (Ashe) Freckmann & Lelong, 25: 434

Dichanthelium ensifolium (Baldwin *ex* Elliott) Gould, 25: 436

    Dichanthelium ensifolium var. *breve* = D. chamaelonche subsp. breve, 25: 440

    Dichanthelium ensifolium subsp. **curtifolium** (Nash) Freckmann & Lelong, 25: 438

    Dichanthelium ensifolium (Baldwin *ex* Elliott) Gould subsp. **ensifolium**, 25: 438

Dichanthelium ensifolium var. *unciphyllum* = D. tenue, 25: 438

Dichanthelium erectifolium (Nash) Gould & C.A. Clark, 25: 440

Dichanthelium hirstii = D. dichotomum subsp. roanokense, 25: 434

Dichanthelium lanuginosum = D. acuminatum subsp. acuminatum, 25: 425

    Dichanthelium lanuginosum var. *sericeum* = D. acuminatum subsp. sericeum, 25: 426

Dichanthelium latifolium (L.) Harvill, 25: 412

Dichanthelium laxiflorum (Lam.) Gould, 25: 446

Dichanthelium leibergii (Vasey) Freckmann, 25: 416

Dichanthelium leucoblepharis = D. strigosum subsp. leucoblepharis, 25: 447

Dichanthelium leucothrix = D. acuminatum subsp. leucothrix, 25: 426

Dichanthelium lindheimeri = D. acuminatum subsp. lindheimeri, 25: 426

Dichanthelium linearifolium (Scribn.) Gould, 25: 449

    Dichanthelium linearifolium var. *werneri* = D. linearifolium, 25: 449

Dichanthelium longiligulatum = D. acuminatum subsp. longiligulatum, 25: 426

Dichanthelium malacophyllum (Nash) Gould, 25: 422

Dichanthelium mattamuskeetense = D. dichotomum subsp. mattamuskeetense, 25: 433

Dichanthelium meridionale var. *albemarlense* = D. acuminatum subsp. implicatum, 25: 425

Dichanthelium microcarpon = D. dichotomum subsp. microcarpon, 25: 433

Dichanthelium nitidum = D. dichotomum subsp. nitidum, 25: 433

Dichanthelium nodatum (Hitchc. & Chase) Gould, 25: 410

Dichanthelium nudicaule (Vasey) B.F. Hansen & Wunderlin, 25: 436

Dichanthelium oligosanthes (Schult.) Gould, 25: 419

    Dichanthelium oligosanthes var. *helleri* = D. oligosanthes subsp. scribnerianum, 25: 421

    Dichanthelium oligosanthes (Schult.) Gould subsp. **oligosanthes**, 25: 421

    Dichanthelium oligosanthes subsp. **scribnerianum** (Nash) Freckmann & Lelong, 25: 421

    Dichanthelium oligosanthes var. *wilcoxianum* = D. wilcoxianum, 25: 449

Dichanthelium ovale (Elliott) Gould & C.A. Clark, 25: 429

    Dichanthelium ovale var. *addisonii* = D. ovale subsp. pseudopubescens, 25: 430

    Dichanthelium ovale (Elliott) Gould & C.A. Clark subsp. **ovale**, 25: 429

    Dichanthelium ovale subsp. **praecocius** (Hitchc. & Chase) Freckmann & Lelong, 25: 429

    Dichanthelium ovale subsp. **pseudopubescens** (Nash) Freckmann & Lelong, 25: 430

    Dichanthelium ovale subsp. **villosissimum** (Nash) Freckmann & Lelong, 25: 430

Dichanthelium pedicellatum (Vasey) Gould, 25: 410

Dichanthelium perlongum (Nash) Freckmann, 25: 449

Dichanthelium polyanthes (Schult.) Mohlenbr., 25: 440

Dichanthelium portoricense (Desv. *ex* Ham.) B.F. Hansen & Wunderlin, 25: 441

    Dichanthelium portoricense subsp. **patulum** (Scribn. & Merr.) Freckmann & Lelong, 25: 442

    Dichanthelium portoricense (Desv. *ex* Ham.) B.F. Hansen & Wunderlin subsp. **portoricense**, 25: 442

Dichanthelium praecocius = **D. ovale** subsp. **praecocius**, 25: 429

Dichanthelium ravenelii (Scribn. & Merr.) Gould, 25: 421

Dichanthelium sabulorum var. *patulum* = **D. portoricense** subsp. **patulum**, 25: 442

Dichanthelium sabulorum var. *thinium* = **D. acuminatum** subsp. **implicatum**, 25: 425

Dichanthelium scabriusculum (Elliott) Gould & C.A. Clark, 25: 418

Dichanthelium scoparium (Lam.) Gould, 25: 419

Dichanthelium sphagnicola = **D. dichotomum** subsp. **lucidum**, 25: 433

Dichanthelium sphaerocarpon (Elliott) Gould, 25: 441

    Dichanthelium sphaerocarpon var. *isophyllum* = **D. polyanthes**, 25: 440

    Dichanthelium sphaerocarpon var. *polyanthes* = **D. polyanthes**, 25: 440

Dichanthelium spretum = **D. acuminatum** subsp. **spretum**, 25: 426

Dichanthelium strigosum (Muhl. *ex* Elliott) Freckmann, 25: 446

    Dichanthelium strigosum subsp. **glabrescens** (Griseb.) Freckmann & Lelong, 25: 447

    Dichanthelium strigosum subsp. **leucoblepharis** (Trin.) Freckmann & Lelong, 25: 447

    Dichanthelium strigosum (Muhl. *ex* Elliott) Freckmann subsp. **strigosum**, 25: 447

Dichanthelium subvillosum = **D. acuminatum** subsp. **fasciculatum**, 25: 425

Dichanthelium tenue (Muhl.) Freckmann & Lelong, 25: 438

Dichanthelium villosissimum = **D. ovale** subsp. **villosissimum**, 25: 430

    Dichanthelium villosissimum var. *praecocius* = **D. ovale** subsp. **praecocius**, 25: 429

    Dichanthelium villosissimum var. *pseudopubescens* = **D. ovale** subsp. **pseudopubescens**, 25: 430

Dichanthelium wilcoxianum (Vasey) Freckmann, 25: 449

Dichanthelium wrightianum (Scribn.) Freckmann, 25: 430

Dichanthelium xanthophysum (A. Gray) Freckmann, 25: 416

DICHANTHIUM Willemet, 25: 637

Dichanthium annulatum (Forssk.) Stapf, 25: 638

Dichanthium aristatum (Poir.) C.E. Hubb., 25: 638

Dichanthium sericeum (R. Br.) A. Camus, 25: 637

DIGITARIA Haller, 25: 358

Digitaria abyssinica (A. Rich.) Stapf, 25: 372

Digitaria adscendens = **D. ciliaris**, 25: 382

Digitaria arenicola (Swallen) Beetle, 25: 362

Digitaria bakeri (Nash) Fernald, 25: 364

Digitaria bicornis (Lam.) Roem. & Schult., 25: 380

Digitaria californica (Benth.) Henrard, 25: 368

Digitaria ciliaris (Retz.) Koeler, 25: 382

    Digitaria ciliaris var. **chrysoblephara** (Fig. & De Not.) R.R. Stewart, 25: 382

    Digitaria ciliaris (Retz.) Koeler var. **ciliaris**, 25: 382

Digitaria cognata (Schult.) Pilg., 25: 362

    Digitaria cognata subsp. *pubiflora* = **D. pubiflora**, 25: 362

Digitaria decumbens = **D. eriantha** subsp. **pentzii**, 25: 376

Digitaria didactyla Willd., 25: 376

Digitaria eriantha Steud., 25: 376

    Digitaria eriantha Steud. subsp. **eriantha**, 25: 376

    Digitaria eriantha subsp. **pentzii** (Stent) Kok, 25: 376

Digitaria filiformis (L.) Koeler, 25: 364

    Digitaria filiformis var. **dolichophylla** (Henrard) Wipff, 25: 366

    Digitaria filiformis (L.) Koeler var. **filiformis**, 25: 366

    Digitaria filiformis var. **laeviglumis** (Fernald) Wipff, 25: 366

    Digitaria filiformis var. **villosa** (Walter) Fernald, 25: 366

Digitaria floridana Hitchc., 25: 372

Digitaria gracillima (Scribn.) Fernald, 25: 364

Digitaria hitchcockii (Chase) Stuck., 25: 366

Digitaria horizontalis Willd., 25: 378

Digitaria insularis (L.) Mez *ex* Ekman, 25: 370

Digitaria ischaemum (Schreb.) Muhl., 25: 372

    Digitaria ischaemum var. *mississippiensis* = **D. ischaemum**, 25: 372

    Digitaria ischaemum var. *violascens* = **D. violascens**, 25: 372

Digitaria leucocoma (Nash) Urb., 25: 366

Digitaria longiflora (Retz.) Pers., 25: 370

Digitaria milanjiana (Rendle) Stapf, 25: 376

Digitaria nuda Schumach., 25: 378

Digitaria panicea [within the *Flora* region, misapplied to D. filiformis var. **dolichophylla**], 25: 366

Digitaria patens (Swallen) Henrard, 25: 368

Digitaria pauciflora Hitchc., 25: 374

Digitaria pentzii = **D. eriantha**, 25: 376

Digitaria pruriens = **D. setigera**, 25: 382

Digitaria pubiflora (Vasey) Wipff, 25: 362

Digitaria runyonii = **D. texana**, 25: 374

Digitaria sanguinalis (L.) Scop., 25: 380

    Digitaria sanguinalis var. *ciliaris* = **D. ciliaris**, 25: 382

Digitaria serotina (Walter) Michx., 25: 370

Digitaria setigera Roth, 25: 382

Digitaria simpsonii (Vasey) Fernald, 25: 374

Digitaria texana Hitchc., 25: 374

Digitaria tomentosa (J. König *ex* Rottler) Henrard, 25: 364

Digitaria velutina (Forssk.) P. Beauv., 25: 378

Digitaria villosa = **D. filiformis** var. **villosa**, 25: 366

Digitaria violascens Link, 25: 372

DINEBRA Jacq., 25: 63

Dinebra retroflexa (Vahl) Panz., 25: 64

DIPLACHNE [included in **LEPTOCHLOA**], 25: 51

Diplachne acuminata = **Leptochloa fusca** subsp. **fascicularis**, 25: 56

Diplachne dubia = **Leptochloa dubia**, 25: 54

Diplachne fascicularis = **Leptochloa fusca** subsp. **fascicularis**, 25: 56

Diplachne halei = **Leptochloa panicoides**, 25: 59

Diplachne maritima = **Leptochloa fusca** subsp. **fascicularis**, 25: 56

Diplachne uninerva = **Leptochloa fusca** subsp. **uninerva**, 25: 56

DISSANTHELIUM Trin., 24: 700
Dissanthelium californicum (Nutt.) Benth., 24: 700

DISTICHLIS Raf., 25: 24
Distichlis spicata (L.) Greene, 25: 25
  Distichlis spicata var. *borealis* = **D. spicata**, 25: 25
  Distichlis spicata subsp./var. *stricta* = **D. spicata**, 25: 25
  Distichlis stricta = **D. spicata**, 25: 25

×**DUPOA** J. Cay. & Darbysh., 24: 601
×**Dupoa labradorica** (Steud.) J. Cay. & Darbysh., 24: 602

DUPONTIA R. Br., 24: 602
Dupontia fisheri R. Br., 24: 604
  Dupontia fisheri subsp./var. *psilosantha* = **D. fisheri**, 24: 604
  Dupontia psilosantha = **D. fisheri**, 24: 604

EATONIA sensu Endl. [included in **SPHENOPHOLIS**], 24: 620
Eatonia annua = **Sphenopholis obtusata**, 24: 621

ECHINOCHLOA P. Beauv., 25: 390
Echinochloa colona (L.) Link, 25: 398
Echinochloa colonum = **E. colona**, 25: 398
Echinochloa crus-galli (L.) P. Beauv., 25: 400
  Echinochloa crus-galli subsp. *edulis* = **E. frumentacea**, 25: 400
  Echinochloa crus-galli var. *frumentacea* = **E. frumentacea**, 25: 400
  Echinochloa crus-galli forma *longiseta* = **E. muricata** var. **muricata**, 25: 396
  Echinochloa crus-galli var. *macera* = **E. crus-pavonis** var. **macera**, 25: 398
  Echinochloa crus-galli var. *mitis* = **E. crus-galli**, 25: 400
  Echinochloa crus-galli var. *oryzicola* = **E. oryzicola**, 25: 402
Echinochloa crus-pavonis (Kunth) Schult., 25: 398
  Echinochloa crus-pavonis (Kunth) Schult. var. **crus-pavonis**, 25: 398
  Echinochloa crus-pavonis var. **macera** (Wiegand) Gould, 25: 398
Echinochloa esculenta (A. Braun) H. Scholtz, 25: 402
Echinochloa frumentacea Link, 25: 400
Echinochloa microstachya = **E. muricata** var. **microstachya**, 25: 396
Echinochloa muricata (P. Beauv.) Fernald, 25: 396
  Echinochloa muricata var. *ludoviciana* = **E. muricata** var. **muricata**, 25: 396
  Echinochloa muricata var. **microstachya** Wiegand, 25: 396
  Echinochloa muricata (P. Beauv.) Fernald var. **muricata**, 25: 396
  Echinochloa muricata var. *occidentalis* = **E. crus-galli**, 25: 400
  Echinochloa muricata var. *wiegandii* = **E. muricata** var. **microstachya**, 25: 396
Echinochloa occidentalis = **E. crus-galli**, 25: 400

Echinochloa oplismenoides (Fourn.) Hitchc., 25: 398
Echinochloa oryzicola (Vasinger) Vasinger, 25: 402
Echinochloa oryzoides (Ard.) Fritsch, 25: 402
Echinochloa paludigena Wiegand, 25: 394
Echinochloa polystachya (Kunth) Hitchc., 25: 394
  Echinochloa polystachya (Kunth) Hitchc. var. **polystachya**, 25: 394
  Echinochloa polystachya var. **spectabilis** (Nees *ex* Trin.) Mart. Crov., 25: 394
Echinochloa pungens = **E. muricata** var. **muricata**, 25: 396
  Echinochloa pungens var. *microstachya* = **E. muricata** var. **microstachya**, 25: 396
  Echinochloa pungens var. *wiegandii* = **E. muricata** var. **microstachya**, 25: 396
Echinochloa pyramidalis (Lam.) Hitchc. & Chase, 25: 394
Echinochloa walteri (Pursh) A. Heller, 25: 396
  Echinochloa walteri forma *laevigata* = **E. walteri**, 25: 396
Echinochloa wiegandii = **E. muricata** var. **microstachya**, 25: 396
Echinochloa zelayensis = **E. crus-pavonis** var. **crus-pavonis**, 25: 398

ECTOSPERMA [included in **SWALLENIA**], 25: 24
Ectosperma alexandrae = **Swallenia alexandrae**, 25: 24

EHRHARTA Thunb., 24: 33
Ehrharta calycina Sm., 24: 34
Ehrharta erecta Lam., 24: 34
Ehrharta longiflora Sm., 24: 36

EHRHARTEAE Nevski, 24: 33

EHRHARTOIDEAE Link, 24: 32

ELEUSINE Gaertn., 25: 109
Eleusine coracana (L.) Gaertn., 25: 109
  Eleusine coracana subsp. **africana** (Kenn.-O'Byrne) Hilu & de Wet, 25: 110
  Eleusine coracana (L.) Gaertn. subsp. **coracana**, 25: 110
Eleusine indica (L.) Gaertn., 25: 109
Eleusine tristachya (Lam.) Lam., 25: 110

ELIONURUS Humb. & Bonpl. *ex* Willd., 25: 684
Elionurus barbiculmis Hack., 25: 685
  Elionurus barbiculmis var. *parviflorus* = **E. barbiculmis**, 25: 685
Elionurus tripsacoides Humb. & Bonpl. *ex* Willd., 25: 685

×**ELYHORDEUM** Mansf. *ex* Tsitsin & K.A. Petrova, 24: 283
×**Elyhordeum arcuatum** W.W. Mitch. & H.J. Hodgs., 24: 286
×**Elyhordeum californicum** (Bowden) Barkworth, 24: 286
×**Elyhordeum dakotense** (Bowden) Bowden, 24: 284
×**Elyhordeum iowense** R.W. Pohl, 24: 286
×**Elyhordeum jordalii** (Melderis) Tzvelev, 24: 284
×**Elyhordeum macounii** (Vasey) Barkworth & Dewey, 24: 284
×**Elyhordeum montanense** (Scribn. *ex* Beal) Bowden, 24: 286
×**Elyhordeum pilosilemma** (W.W. Mitch. & H.J. Hodgs.) Barkworth, 24: 284
×**Elyhordeum schaackianum** (Bowden) Bowden, 24: 284
×**Elyhordeum stebbinsianum** (Bowden) Bowden, 24: 286

×**ELYLEYMUS** B.R. Baum, 24: 343

×Elyleymus aleuticus (Hultén) B.R. Baum, 24: 346

×Elyleymus aristatus (Merr.) Barkworth & B.R. Dewey, 24: 344

×Elyleymus colvillensis (Lepage) Barkworth, 24: 344

×Elyleymus hirtiflorus (Hitch.) Barkworth, 24: 344

×Elyleymus hultenii (Melderis) Barkworth, 24: 348

×Elyleymus jamesensis (Lepage) Barkworth, 24: 348

×Elyleymus mossii (Lepage) Barkworth, 24: 346

×Elyleymus ontariensis (Lepage) Barkworth, 24: 346

×Elyleymus turneri (Lepage) Barkworth & D.R. Dewey, 24: 344

×Elyleymus uclueletensis (Bowden) B.R. Baum, 24: 346

×Elyleymus ungavensis (Louis-Marie) Barkworth, 24: 348

×*ELYMORDEUM* [included in ×**ELYHORDEUM**, in part], 24: 283

×Elymordeum macounii, *ined.* = ×Elyhordeum macounii, 24: 284

×*Elymordeum littorale* = ×Leydeum littorale, 24: 370

ELYMUS L., 24: 288

Elymus alaskanus (Scribn. & Merr.) Á. Löve, 24: 326

  Elymus alaskanus (Scribn. & Merr.) Á. Löve subsp. alaskanus, 24: 326

  Elymus alaskanus subsp. *borealis* = E. alaskanus, 24: 326

  Elymus alaskanus subsp. hyperarcticus (Polunin) Á. Löve & D. Löve, 24: 326

  Elymus alaskanus subsp. *latiglumis* = E. violaceus, 24: 324

Elymus albicans (Scribn. & J.G. Sm.) Á. Löve, 24: 334

  Elymus albicans var. *griffithsii* = E. albicans, 24: 334

Elymus ambiguus = Leymus ambiguus, 24: 366

  Elymus ambiguus var. *salina* = Leymus salina, 24: 364

  Elymus ambiguus var. *salmonis* = Leymus salina subsp. salmonis, 24: 364

Elymus angustus = Leymus angustus, 24: 360

Elymus arenarius = Leymus arenarius, 24: 356

  Elymus arenarius subsp./var. *mollis* = Leymus mollis, 24: 356

  Elymus arenarius var. *villosus* = Leymus mollis, 24: 356

Elymus arenicolus = Leymus flavescens, 24: 366

Elymus aristatus = ×Elyleymus aristatus, 24: 344

Elymus arizonicus (Scribn. & J.G. Sm.) Gould

Elymus arkansanus = E. villosus, 24: 302

Elymus australis = E. glabriflorus, 24: 296

Elymus bakeri (E.E. Nelson) Á. Löve, 24: 330

Elymus borealis = E. hirsutus, 24: 310

Elymus calderi = E. lanceolatus subsp. psammophilus, 24: 327

Elymus californicus = Leymus californicus, 24: 369

Elymus canadensis L., 24: 303

  Elymus canadensis var. brachystachys (Scribn. & C.R. Ball) Farw., 24: 305

  Elymus canadensis L. var. canadensis, 24: 305

  Elymus canadensis forma/var. *glaucifolius* = E. canadensis var. canadensis, 24: 305

  Elymus canadensis var. *hirsutus* = E. canadensis var. canadensis, 24: 305

  Elymus canadensis var. *interruptus* = E. interruptus, 24: 306

Elymus canadensis var. robustus (Scribn. & J.G. Sm.) Mack. & Bush, 24: 305

  Elymus canadensis var. *wiegandii* = E. wiegandii, 24: 305

Elymus caninus (L.) L., 24: 322

Elymus caput-medusae = Taeniatherum caput-medusae, 24: 258

Elymus ×cayouetteorum (B. Boivin) Barkworth, 24: 338

Elymus churchii J.J.N. Campb., 24: 314

Elymus ciliaris (Trin.) Tzvelev, 24: 336

Elymus cinereus = Leymus cinereus, 24: 364

Elymus condensatus = Leymus condensatus, 24: 362

Elymus curvatus Piper, 24: 300

Elymus dahuricus Turcz. *ex* Griseb., 24: 310

Elymus diversiglumis Scribn. & C.R. Ball, 24: 316

Elymus ×ebingeri G.C. Tucker, 24: 343

Elymus elongatus [within the *Flora* region, misapplied to Thinopyrum ponticum], 24: 376

  Elymus elongatus subsp./var. *ponticus* = Thinopyrum ponticum, 24: 376

Elymus elymoides (Raf.) Swezey, 24: 318

  Elymus elymoides subsp. brevifolius (J.G. Sm.) Barkworth, 24: 319

  Elymus elymoides subsp. californicus (J.G. Sm.) Barkworth, 24: 319

  Elymus elymoides (Raf.) Swezey subsp. elymoides, 24: 319

  Elymus elymoides subsp. hordeoides (Suskd.) Barkworth, 24: 319

  Elymus elymoides var. "*longifolius*" [error for "*brevifolius*"] = E. elymoides subsp. brevifolius, 24: 319

Elymus farctus = Thinopyrum junceum, 24: 376

Elymus flavescens = Leymus flavescens, 24: 366

Elymus giganteus = Leymus racemosus, 24: 356

Elymus glaber = E. elymoides, 24: 318

Elymus glabriflorus (Vasey *ex* L.H. Dewey) Scribn. & C.R. Ball, 24: 296

  Elymus glabriflorus var. *australis* = E. glabriflorus, 24: 296

Elymus glaucus Buckley, 24: 306

  Elymus glaucus var. *breviaristatus* = E. glaucus subsp. glaucus, 24: 308

  Elymus glaucus Buckley subsp. glaucus, 24: 308

  Elymus glaucus subsp. *jepsonii* = E. glaucus subsp. glaucus, 24: 308

  Elymus glaucus subsp. mackenziei (Bush) J.J.N. Campb., 24: 308

  Elymus glaucus var. *maximus* = E. glaucus, 24: 306

  Elymus glaucus subsp. virescens (Piper) Gould, 24: 308

Elymus ×hansenii Scribn., 24: 340

Elymus hirsutiglumis = E. virginicus var. intermedius, 24: 298

Elymus hirsutus J. Presl, 24: 310

Elymus hirtiflorus = ×Elyleymus hirtiflorus, 24: 344

Elymus hispidus = Thinopyrum intermedium, 24: 374

  Elymus hispidus subsp. *barbulatus* = Thinopyrum intermedium subsp. barbulatum, 24: 376

  Elymus hispidus var. *ruthenicus* = Thinopyrum intermedium subsp. barbulatum, 24: 376

Elymus hoffmannii K.B. Jensen & Asay, 24: 336

Elymus hyperarcticus = E. alaskanus subsp. hyperarcticus, 24: 326

Elymus hystrix L., 24: 316

Elymus hystrix var. *bigelovianus* = **E. hystrix**, 24: 316

Elymus innovatus = **Leymus innovatus**, 24: 366

  Elymus innovatus subsp. *velutinus* = **Leymus innovatus**, 24: 366

Elymus interruptus Buckley, 24: 306

Elymus junceus = **Psathyrostachys juncea**, 24: 372

Elymus lanceolatus (Scribn. & J.G. Sm.) Gould, 24: 327

  Elymus lanceolatus (Scribn. & J.G. Sm.) Gould subsp. **lanceolatus**, 24: 327

  Elymus lanceolatus subsp. **psammophilus** (J.M. Gillett & H. Senn) Á. Löve, 24: 327

  Elymus lanceolatus subsp. **riparius** (Scribn. & J.G. Sm.) Barkworth, 24: 329

  Elymus lanceolatus var. *riparius* = **E. lanceolatus** subsp. **riparius**, 24: 329

Elymus longifolius = **E. elymoides** subsp. **brevifolius**, 24: 319

Elymus macgregorii R.E. Brooks & J.J.N. Campb., 24: 295

Elymus macounii = ×**Elyhordeum macounii**, 24: 284

Elymus macrourus (Turcz. *ex* Steud.) Tzvelev, 24: 324

Elymus mollis = **Leymus mollis**, 24: 356

Elymus multicaulis = **Leymus multicaulis**, 24; 360

Elymus multisetus (J.G. Sm.) Burtt Davy, 24: 318

Elymus pacificus = **Leymus pacificus**, 24: 358

Elymus ×palmerensis (Lepage) Barkworth & D.R. Dewey, 24: 340

Elymus parishii = **E. glaucus** subsp. **glaucus**, 24: 308

Elymus ×pinalenoensis (Pyrah) Barkworth & D.R. Dewey, 24: 343

Elymus piperi = **Leymus cinereus**, 24: 364

Elymus pringlei Scribn. & Merr., 24: 312

Elymus ×pseudorepens (Scribn. & J.G. Sm.) Barkworth & D.R. Dewey, 24: 340

Elymus pubiflorus = **E. elymoides**, 24: 318

Elymys pycnanthus = **Thinopyrum pycnanthum**, 24: 376

Elymus racemosus = **Leymus racemosus**, 24: 356

Elymus repens (L.) Gould, 24: 334

Elymus riparius Wiegand, 24: 302

Elymus robustus = **E. canadensis** var. **robustus**, 24: 305

  Elymus sajanensis subsp. *hyperarcticum* = **E. alaskanus** subsp. **hyperarcticus**, 24: 326

Elymus salina = **Leymus salina**, 24: 364

Elymus salinus = **Leymus salina**, 24: 364

Elymus saundersii = **E. ×saundersii**, 24: 340

Elymus ×saundersii Vasey, 24: 340

Elymus saxicola = ×**Pseudelymus saxicola**, 24: 283

Elymus scribneri (Vasey) M.E. Jones, 24: 330

Elymus semicostatus (Nees *ex* Steud.) Melderis, 24: 338

Elymus sibiricus L., 24: 310

Elymus sierrae Gould, 24: 332

Elymus simplex = **Leymus simplex**, 24: 359

  Elymus simplex var. *luxurians* = **Leymus simplex**, 24: 359

Elymus sitanion = **E. elymoides** subsp. **elymoides**, 24: 319

Elymus smithii = **Pascopyrum smithii**, 24: 351

Elymus sp., aff. *pringlei* = **E. texensis**, 24: 312

Elymus sp., aff. *svensoniii* = **E. churchii**, 24: 314

Elymus spicatus = **Pseudoroegneria spicata**, 24: 280

  Elymus spicatus subsp. inermis, *ined.* = **Pseudoroegneria spicata**, 24: 280

Elymus stebbinsii Gould, 24: 329

Elymus stebbinsii subsp. **septentrionalis** Barkworth, 24: 329

  Elymus stebbinsii Gould subsp. **stebbinsii**, 24: 329

Elymus submuticus = **E. curvatus**, 24: 300

Elymus subsecundus = **E. trachycaulus** subsp. **subsecundus**, 24: 322

Elymus svensonii G.L. Church, 24: 314

Elymus texensis J.J.N. Campb., 24: 312

Elymus trachycaulus (Link) Gould *ex* Shinners, 24: 321

  Elymus trachycaulus subsp. *andinus* = **E. trachycaulus** × **E. scribneri**, 24: 330

  Elymus trachycaulus subsp. *bakeri* = **E. bakeri**, 24: 330

  Elymus trachycaulus subsp. *glaucus* = **E. trachycaulus** subsp. **subsecundus**, 24: 322

  Elymus trachycaulus var. *latiglume* = **E. violaceus**, 24: 324

  Elymus trachycaulus subsp. *major* = **E. trachycaulus** subsp. **trachycaulus**, 24: 322

  Elymus trachycaulus subsp. *novae-angliae* = **E. trachycaulus** subsp. **trachycaulus**, 24: 322

  Elymus trachycaulus subsp. **subsecundus** (Link) Á. Löve & D.Löve, 24: 322

  Elymus trachycaulus subsp. *teslinensis* = **E. trachycaulus** subsp. **trachycaulus**, 24: 322

  Elymus trachycaulus (Link) Gould *ex* Shinners subsp. **trachycaulus**, 24: 322

  Elymus trachycaulus var. *unilaterale* = **E. trachycaulus** subsp. **subsecundus**, 24: 322

  Elymus trachycaulus subsp. *violaceus* = **E. violaceus**, 24: 324

  Elymus trachycaulus subsp. **virescens** (Lange) Á. Löve & D. Löve, 24: 322

Elymus triticoides = **Leymus triticoides**, 24: 360

  Elymus triticoides subsp. *multiflorus* = **Leymus ×multiflorus**, 24: 362

  Elymus triticoides var. *pubescens* = **Leymus triticoides**, 24: 360

Elymus tsukushiensis Honda, 24: 336

Elymus ×vancouverensis = **Leymus ×vancouverensis**, 24: 358

Elymus villosus Muhl. *ex* Willd., 24: 302

  Elymus villosus var. *arkansanus* = **E. villosus**, 24: 302

Elymus violaceus (Hornem.) Feilberg, 24: 324

Elymus virescens = **E. glaucus** subsp. **virescens**, 24: 308

Elymus virginicus L., 24: 298

  Elymus virginicus forma/var. *australis* = **E. glabriflorus**, 24: 296

  Elymus virginicus var. *glabriflorus* = **E. glabriflorus**, 24: 296

  Elymus virginicus var. **halophilus** (E.P. Bicknell) Wiegand, 24: 298

  Elymus virginicus forma *hirsutiglumis* = **E. virginicus** var. **intermedius**, 24: 298

  Elymus virginicus var. **intermedius** (Vasey *ex* S. Watson & J.M. Coult.) Bush, 24: 298

  Elymus virginicus var. **jejunus** (Ramaley) Bush, 24: 300

  Elymus virginicus var. *jenkinsii* = **E. curvatus**, 24: 300

  Elymus virginicus var. *submuticus* = **E. curvatus**, 24: 300

  Elymus virginicus L. var. **virginicus**, 24: 300

Elymus virescens = **E. glaucus** subsp. **virescens**, 24: 308

Elymus wawawaiensis J.R. Carlson & Barkworth, 24: 332

Elymus wiegandii Fernald, 24: 305

Elymus ×yukonensis (Scribn. & Merr.) Á. Löve, 24: 340

ELYONURUS = **ELIONURUS**, 25: 684

×*ELYTESION* [included in ×**ELYHORDEUM**], 24: 283
×*Elytesion macounii* = ×**Elyhordeum macounii**, 24: 284

ELYTRIGIA Desv. [included in **THINOPYRUM**, in part], 24: 373
Elytrigia dasystachya = **Pascopyrum smithii**, 24: 351
Elytrigia elongata = **Thinopyrum ponticum**, 24: 376
Elytrigia intermedia = **Thinopyrum intermedium**, 24: 374
 Elytrigia intermedia subsp. *barbulata* = **Thinopyrum intermedium** subsp. **barbulatum**, 24: 376
Elytrigia juncea = **Thinopyrum junceum**, 24: 376
Elytrigia pontica = **Thinopyrum ponticum**, 24: 376
Elytrigia pungens, *ined.* = **Thinopyrum pycnanthum**, 24: 376
Elytrigia pycnantha = **Thinopyrum pycnanthum**, 24: 376
Elytrigia repens = **Elymus repens**, 24: 334
 Elytrigia repens forma *aristata* = **Elymus repens**, 24: 334
Elytrigia smithii = **Pascopyrum smithii**, 24: 351
Elytrigia spicata = **Pseudoroegneria spicata**, 24: 280

ENNEAPOGON Desv. *ex* P. Beauv., 25: 286
Enneapogon cenchroides (Licht.) C.E. Hubb., 25: 287
Enneapogon desvauxii P. Beauv., 25: 287

ENTEROPOGON Nees, 25: 224
Enteropogon chlorideus (J. Presl) Clayton, 25: 225
Enteropogon dolichostachyus (Lag.) Keng *ex* Lazarides, 25: 225
Enteropogon prieurii Kunth, 25: 225

EPICAMPES [included in **MUHLENBERGIA**], 25: 145
Epicampes rigens = **Muhlenbergia rigens**, 25: 194
Epicampes subpatens = **Muhlenbergia emersleyi**, 25: 185

ERAGROSTIS Wolf, 25: 65
Eragrostis abyssinica = **E. tef**, 25: 85
Eragrostis airoides Nees, 25: 103
Eragrostis amabilis (L.) Wight & Arn. *ex* Nees, 25: 72
Eragrostis arida = **E. pectinacea** var. **miserrima**, 25: 83
Eragrostis atherstonei = **E. trichophora**, 25: 76
Eragrostis atrovirens (Desf.) Trin. *ex* Steud., 25: 103
Eragrostis bahiensis (Schrad.) Schult., 25: 101
Eragrostis barrelieri Daveau, 25: 83
Eragrostis beyrichii = **E. secundiflora** subsp. **oxylepis**, 25: 99
Eragrostis brownii = **E. cumingii**, 25: 72
Eragrostis campestris = **E. refracta**, 25: 97
Eragrostis capillaris (L.) Nees, 25: 79
Eragrostis carolineana [within the *Flora* region, misapplied to E. pectinacea], 25: 81
Eragrostis chariis [within the *Flora* region, misapplied to E. atrovirens], 25: 103
Eragrostis chloromelas = **E. curvula**, 25: 76
Eragrostis cilianensis (All.) Vignolo *ex* Janch., 25: 83
Eragrostis ciliaris (L.) R. Br., 25: 71
 Eragrostis ciliaris (L.) R. Br. var. **ciliaris**, 25: 71
 Eragrostis ciliaris var. **laxa** Kuntze, 25: 71
Eragrostis cumingii Steud., 25: 72
Eragrostis curtipedicellata Buckley, 25: 89
Eragrostis curvula (Schrad.) Nees, 25: 76
 Eragrostis curvula var. *conferta* = **E. curvula**, 25: 76

Eragrostis cylindriflora Hochst., 25: 74
Eragrostis diandra = **E. elongata**, 25: 101
Eragrostis diffusa = **E. pectinacea** var. **pectinacea**, 25: 83
Eragrostis echinochloidea Stapf, 25: 87
Eragrostis elliottii S. Watson, 25: 99
Eragrostis elongata (Willd.) Jacq., 25: 101
Eragrostis erosa Scribn. *ex* Beal, 25: 97
Eragrostis flamignii = **E. gangetica**, 25: 87
Eragrostis frankii C.A. Mey. *ex* Steud., 25: 79
 Eragrostis frankii var. *brevipes* = **E. frankii**, 25: 79
Eragrostis gangetica (Roxb.) Steud., 25: 87
Eragrostis glomerata = **E. japonica**, 25: 74
Eragrostis hirsuta (Michx.) Nees, 25: 95
 Eragrostis hirsuta var. *laevivaginata* = **E. hirsuta**, 25: 95
Eragrostis horizontalis = **E. cylindriflora**, 25: 74
Eragrostis hypnoides (Lam.) Britton, 25: 72
Eragrostis intermedia Hitchc., 25: 97
Eragrostis japonica (Thunb.) Trin., 25: 74
Eragrostis lehmanniana Nees, 25: 76
Eragrostis lugens Nees, 25: 95
Eragrostis lutescens Scribn., 25: 79
Eragrostis major = **E. cilianensis**, 25: 83
Eragrostis megastachya = **E. cilianensis**, 25: 83
Eragrostis mexicana (Hornem.) Link, 25: 78
 Eragrostis mexicana (Hornem.) Link subsp. **mexicana**, 25: 78
 Eragrostis mexicana subsp. **virescens** (J. Presl) S.D. Koch & Sánchez Vega, 25: 78
Eragrostis minor Host, 25: 85
Eragrostis multicaulis = **E. pilosa**, 25: 81
Eragrostis neomexicana = **E. mexicana** subsp. **mexicana**, 25: 78
Eragrostis obtusiflora (E. Fourn.) Scribn., 25: 89
Eragrostis orcuttiana = **E. mexicana** subsp. **virescens**, 25: 78
Eragrostis oxylepis = **E. secundiflora** subsp. **oxylepis**, 25: 99
 Eragrostis oxylepis var. *beyrichii* = **E. secundiflora** subsp. **oxylepis**, 25: 99
Eragrostis palmeri S. Watson, 25: 93
Eragrostis pectinacea (Michx.) Nees, 25: 81
 Eragrostis pectinacea var. **miserrima** (E. Fourn.) Reeder, 25: 83
 Eragrostis pectinacea (Michx.) Nees var. **pectinacea**, 25: 83
 Eragrostis pectinacea var. **tracyi** (Hitchc.) P.M. Peterson, 25: 83
Eragrostis peregrina = **E. pilosa**, 25: 81
Eragrostis perplexa = **E. pilosa**, 25: 81
Eragrostis pilifera = **E. trichodes**, 25: 93
Eragrostis pilosa (L.) P. Beauv., 25: 81
 Eragrostis pilosa var. **perplexa** (L.H. Harv.) S.D. Koch, 25: 81
 Eragrostis pilosa (L.) P. Beauv. var. **pilosa**, 25: 81
Eragrostis plana Nees, 25: 91
Eragrostis poaeoides = **E. minor**, 25: 85
Eragrostis polytricha Nees, 25: 95
Eragrostis prolifera (Sw.) Steud., 25: 99
Eragrostis purshii = **E. pectinacea** var. **pectinacea**, 25: 83
Eragrostis refracta (Muhl.) Scribn., 25: 97
Eragrostis reptans (Michx.) Nees, 25: 74
Eragrostis scaligera Salzm. *ex* Steud., 25: 101
Eragrostis secundiflora J. Presl, 25: 99

Eragrostis secundiflora subsp. **oxylepis** (Torr.) S.D. Koch, 25: 99

Eragrostis secundiflora J. Presl subsp. **secundiflora**, 25: 99

Eragrostis **sessilispica** Buckley, 25: 103

Eragrostis **setifolia** Nees, 25: 78

Eragrostis **silveana** Swallen, 25: 91

Eragrostis **spectabilis** (Pursh) Steud., 25: 89

Eragrostis spectabilis var. *sparsihirsuta* = E. **spectabilis**, 25: 89

Eragrostis **spicata** Vasey, 25: 91

Eragrostis stenophylla = E. **gangetica**, 25: 87

Eragrostis **superba** Peyr., 25: 87

Eragrostis **swallenii** Hitchc., 25: 93

Eragrostis **tef** (Zucc.) Trotter, 25: 85

Eragrostis tenella = E. **amabilis**, 25: 72

Eragrostis tephrosanthos = E. **pectinacea** var. **miserrima**, 25: 83

Eragrostis tracyi = E. **pectinacea** var. **tracyi**, 25: 83

Eragrostis trichocolea [within the *Flora* region, misapplied to E. **polytricha**], 25: 95

Eragrostis **trichodes** (Nutt.) Alph. Wood, 25: 93

Eragrostis trichodes var. *pilifera* = E. **trichodes**, 25: 93

Eragrostis **trichophora** Coss. & Durieu, 25: 76

Eragrostis **unioloides** (Retz.) Nees *ex* Steud., 25: 85

Eragrostis virescens = E. **mexicana** subsp. **virescens**, 25: 78

Eragrostis virginica = E. **refracta**, 25: 97

Eragrostis weigeltiana = E. **hypnoides**, 25: 72

EREMOCHLOA Büse, 25: 688

Eremochloa **ciliaris** (L.) Merr., 25: 690

Eremochloa **ophiuroides** (Munro) Hack., 25: 690

EREMOPOA Roshev., 24: 618

Eremopoa **altaica** (Trin.) Roshev., 24: 619

EREMOPYRUM (Ledeb.) Jaub & Spach, 24: 252

Eremopyrum **bonaepartis** (Spreng.) Nevski, 24: 254

Eremopyrum **orientale** (L.) Jaub. & Spach, 24: 256

Eremopyrum **triticeum** (Gaertn.) Nevski, 24: 254

ERIANTHUS [included in **SACCHARUM**, in part], 25: 609

Erianthus alopecuroides = **Saccharum alopecuroides**, 25: 612

Erianthus brevibarbis = **Saccharum brevibarbe** var. **brevibarbe**, 25: 612

Erianthus coarctatus = **Saccharum coarctatum**, 25: 612

Erianthus contortus = **Saccharum brevibarbe** var. **contortum**, 25: 612

Erianthus giganteus = **Saccharum giganteum**, 25: 611

Erianthus ravennae = **Saccharum ravennae**, 25: 614

Erianthus strictus = **Saccharum baldwinii**, 25: 614

ERIOCHLOA Kunth, 25: 507

Eriochloa **acuminata** (J. Presl) Kunth, 25: 513

Eriochloa **acuminata** (J. Presl) Kunth var. **acuminata**, 25: 515

Eriochloa **acuminata** var. **minor** (Vasey) R.B. Shaw, 25: 515

Eriochloa **aristata** Vasey, 25: 513

Eriochloa **contracta** Hitchc., 25: 509

Eriochloa **fatmensis** (Hochst. & Steud.) Clayton, 25: 511

Eriochloa gracilis = E. **acuminata**, 25: 513

Eriochloa gracilis var. *minor* = E. **acuminata** var. **minor**, 25: 515

Eriochloa **lemmonii** Vasey & Scribn., 25: 511

Eriochloa lemmonii var. *gracilis* = E. **acuminata**, 25: 515

Eriochloa **michauxii** (Poir.) Hitchc., 25: 509

Eriochloa **michauxii** (Poir.) Hitchc. var. **michauxii**, 25: 509

Eriochloa **michauxii** var. **simpsonii** (Hitchc.) Hitchc., 25: 509

Eriochloa **polystachya** Kunth, 25: 515

Eriochloa procera [within the *Flora* region, misapplied to E. **fatmensis**], 25: 511

Eriochloa **pseudoacrotricha** (Stapf *ex* Thell.) J.M. Black, 25: 513

Eriochloa **punctata** (L.) Desv. *ex* Ham., 25: 511

Eriochloa **sericea** (Scheele) Munro *ex* Vasey, 25: 508

Eriochloa **villosa** (Thunb.) Kunth, 25: 509

ERIONEURON Nash, 25: 44

Erioneuron **avenaceum** (Kunth) Tateoka, 25: 45

Erioneuron avenaceum var. *nealleyi* = E. **nealleyi**, 25: 45

Erioneuron grandiflorum = E. **avenaceum**, 25: 45

Erioneuron **nealleyi** (Vasey) Tateoka, 25: 45

Erioneuron **pilosum** (Buckley) Nash, 25: 45

Erioneuron pulchellum = **Dasyochloa pulchella**, 25: 47

EUCHLAENA [included in **ZEA**], 25: 696

Euchlaena perennis = **Zea perennis**, 25: 699

Eulalia amaura = **Polytrias amaura**, 25: 623

Eulalia viminea = **Microstegium vimineum**, 25: 624

EUSTACHYS Desv., 25: 218

Eustachys **caribaea** (Spreng.) Herter, 25: 222

Eustachys **distichophylla** (Lag.) Nees, 25: 222

Eustachys **floridana** Chapm., 25: 222

Eustachys **glauca** Chapm., 25: 220

Eustachys **neglecta** (Nash) Nash, 25: 222

Eustachys **petraea** (Sw.) Desv., 25: 220

Eustachys **retusa** (Lag.) Kunth, 25: 222

FESTUCA L., 24: 389

Festuca sect. **Breviaristatae** Krivot., 24: 406

Festuca sect. **Elmera** E.B. Alexeev, 24: 404

Festuca L. sect. **Festuca**, 24: 412

Festuca sect. **Subulatae** Tzvelev, 24: 402

Festuca sect. **Subuliflorae** (E.B. Alexeev) Darbysh., 24: 406

Festuca sect. **Texanae** E.B. Alexeev, 24: 399

Festuca L. subg. **Festuca**, 24: 406

Festuca subg. **Montanae** (Hack.) Nyman, 24: 399

Festuca subg. **Obtusae** E.B. Alexeev, 24: 400

Festuca subg. **Subulatae** (Tzvelev) E.B. Alexeev, 24: 402

Festuca **altaica** Trin., 24: 407

Festuca altaica subsp. *hallii* = F. **hallii**, 24: 407

Festuca **amethystina** L., 24: 422

Festuca ammobia = F. **rubra** subsp. **pruinosa**, 24: 416

Festuca **arizonica** Vasey, 24: 438

Festuca arundinacea = **Schedonorus arundinaceus**, 24: 448

Festuca **auriculata** Drobow, 24: 424

Festuca **baffinensis** Polunin, 24: 432

Festuca **brachyphylla** Schult. & Schult. f., 24: 428

Festuca **brachyphylla** Schult. & Schult. f. subsp. **brachyphylla**, 24: 428

Festuca brachyphylla subsp. breviculmis Fred., 24: 428

Festuca brachyphylla subsp. coloradensis Fred., 24: 428

Festuca brevipila = **F. trachyphylla**, 24: 424

Festuca brevissima Jurtsev, 24: 426

Festuca bromoides = **Vulpia bromoidcs**, 24: 452

Festuca californica Vasey, 24: 410

Festuca californica Vasey subsp. **californica**, 24: 410

Festuca californica subsp. **hitchcockiana** (E.B. Alexeev) Darbysh., 24: 410

Festuca californica subsp. **parishii** (Piper) Darbysh., 24: 412

Festuca calligera (Piper) Rydb., 24: 437

Festuca campestris Rydb., 24: 408

Festuca capillata = **F. filiformis**, 24: 424

Festuca confinis = **Leucopoa kingii**, 24: 445

Festuca confusa = **Vulpia microstachys** var. **confusa**, 24: 452

Festuca dasyclada Hack. *ex* Beal, 24: 443

Festuca dertonensis = **Vulpia bromoides**, 24: 452

Festuca diffusa = **F. rubra** subsp. **fallax**, 24: 416

Festuca duriuscula = **F. trachyphylla**, 24: 424

Festuca earlei Rydb., 24: 420

Festuca eastwoodae = **Vulpia microstachys** var. **ciliata**, 24: 452

Festuca edlundiae S. Aiken, Consaul & Lefk., 24: 432

Festuca elatior = **Schedonorus arundinaceus**, 24: 448

Festuca elatior forma *aristata* = **Schedonorus arundinaceus**, 24: 448

Festuca elatior var. *arundinacea* = **Schedonorus arundinaceus**, 24: 448

Festuca elatior var. *pratensis* = **Schedonorus pratensis**, 24: 446

Festuca elmeri Scribn. & Merr., 24: 404

Festuca elmeri var. *conferta* = **F. elmeri**, 24: 404

Festuca filiformis Pourr., 24: 424

Festuca frederikseniae E.B. Alexeev, 24: 436

Festuca gigantea = **Schedonorus giganteus**, 24: 448

Festuca glauca Vill., 24: 422

Festuca grayi = **Vulpia microstachys** var. **ciliata**, 24: 452

Festuca groenlandica (Schol.) Fred., 24: 434

Festuca hallii (Vasey) Piper, 24: 407

Festuca heteromalla = **F. rubra** subsp. **fallax**, 24: 416

Festuca heterophylla Lam., 24: 420

Festuca howellii = **F. elmeri**, 24: 404

Festuca hyperborea Holmen *ex* Fred., 24: 432

Festuca idahoensis Elmer, 24: 438

Festuca idahoensis var. *oregona* = **F. idahoensis**, 24: 438

Festuca idahoensis var. *roemeri* = **F. roemeri**, 24: 440

Festuca kingii = **Leucopoa kingii**, 24: 445

Festuca lenensis Drobow, 24: 426

Festuca ligulata Swallen, 24: 408

Festuca longifolia var. *trachyphylla* = **F. trachyphylla**, 24: 424

Festuca megalura = **Vulpia myuros** forma **megalura**, 24: 449

Festuca microstachys = **Vulpia microstachys**, 24: 452

Festuca microstachys var. *ciliata* = **Vulpia microstachys** var. **ciliata**, 24: 452

Festuca microstachys var. *pauciflora* = **Vulpia microstachys** var. **pauciflora**, 24: 452

Festuca microstachys var. *simulans* = **Vulpia microstachys** var. **pauciflora**, 24: 452

Festuca microstachys var. *subappressa* = **Vulpia microstachys** var. **microstachys**, 24: 452

Festuca minutiflora Rydb., 24: 434

Festuca myuros = **Vulpia myuros**, 24: 449

Festuca myuros var. hirsuta, *ined.* = **Vulpia myuros** forma **megalura**, 24: 449

Festuca nigrescens = **F. rubra** subsp. **commutata**, 24: 416

Festuca nutans Biehler [an illegitimate name] = **F. paradoxa**, 24: 402

Festuca obtusa = **F. subverticillata**, 24: 400

Festuca obtusa forma *pilosifolia* = **F. subverticillata**, 24: 400

Festuca occidentalis Hook., 24: 437

Festuca octoflora = **Vulpia octoflora**, 24: 450

Festuca octoflora subsp. *hirtella* = **Vulpia octoflora** var. **hirtella**, 24: 450

Festuca octoflora var. *tenella* = **Vulpia octoflora** var. **glauca**, 24: 450

Festuca ovina L., 24: 422

Festuca ovina subsp. *alaskana* = **F. brevissima**, 24: 426

Festuca ovina var. *brachyphylla* = **F. brachyphylla**, 24: 428

Festuca ovina var. *brevifolia* = **F. brachyphylla**, 24: 428

Festuca ovina var. *duriuscula* = **F. trachyphylla**, 24: 424

Festuca ovina var. *glauca* [within the *Flora* region, misapplied to **F. trachyphylla**], 24: 424

Festuca ovina var. *rydbergii* = **F. saximontana**, 24: 430

Festuca ovina var. *saximontana* = **F. saximontana**, 24: 430

Festuca pacifica = **Vulpia microstachys** var. **pauciflora**, 24: 452

Festuca paradoxa Desv., 24: 402

Festuca pratensis = **Schedonorus pratensis**, 24: 446

Festuca prolifera (Piper) Fernald, 24: 419

Festuca prolifera var. *lasiolepis* = **F. prolifera**, 24: 419

Festuca pseudovivipara (Pavlick) Pavlick, 24: 419

Festuca reflexa = **Vulpia microstachys** var. **pauciflora**, 24: 452

Festuca richardsonii = **F. rubra** subsp. **arctica**, 24: 414

Festuca roemeri (Pavlick) E.B. Alexeev, 24: 440

Festuca roemeri var. klamathensis, *ined.* = **F. romeri**, 24: 440

Festuca rubra L., 24: 412

Festuca rubra var. *alaica* = **F. rubra** subsp. **arctica**, 24: 414

Festuca rubra subsp. **arctica** (Hack.) Govor., 24: 414

Festuca rubra subsp. **arenaria** (Osbeck) F. Aresch., 24: 414

Festuca rubra subsp. **aucta** (V.I. Krecz. & Bobrov) Hultén, 24: 414

Festuca rubra subsp. **commutata** Gaudin, 24: 415

Festuca rubra subsp. *cryophila* = **F. rubra** subsp. **arctica**, 24: 414

Festuca rubra subsp. **fallax** (Thuill.) Nyman, 24: 415

Festuca rubra var. *fraterculae* = **F. rubra** subsp. **fallax**, 24: 415

Festuca rubra var. *heterophylla* = **F. heterophylla**, 24: 420

Festuca rubra var. *juncea* [within the *Flora* region, misapplied to **F. rubra** subsp. **rubra**], 24: 418

Festuca rubra var. *lanuginosa* = **F. rubra** subsp. **arenaria**, 24: 414

Festuca rubra var. *littoralis* Vasey *ex* Beal [within the *Flora* region, misapplied to **F. rubra** subsp. **mediana**], 24: 415

Festuca rubra subsp. **mediana** (Pavlick) Pavlick, 24: 415

Festuca rubra subsp. *megastachys* [*sensu* Piper] [within the *Flora* region, misapplied to F. **rubra** subsp. **pruinosa**], 24: 416

Festuca rubra var. *mutica* = **F. rubra** subsp. **arctica**, 24: 414

Festuca rubra subsp./var. *multiflora* = **F. rubra** subsp. **fallax**, 24: 416

Festuca rubra subsp./var. *prolifera* = **F. prolifera**, 24: 419

Festuca rubra subsp. **pruinosa** (Hack.) Piper, 24: 415

Festuca rubra subsp. *richardsonii* = **F. rubra** subsp. **arctica**, 24: 414

Festuca rubra L. subsp. **rubra**, 24: 418

Festuca rubra subsp. **secunda** (J. Presl) Pavlick, 24: 418

Festuca rubra forma *squarrosa* [within the *Flora* region, misapplied to F. **rubra** subsp. **rubra**], 24: 418

Festuca rubra subsp. **vallicola** (Rydb.) Pavlick, 24: 418

Festuca saximontana Rydb., 24: 430

Festuca saximontana var. **purpusiana** (St.-Yves) Fred. & Pavlick, 24: 430

Festuca saximontana var. **robertsiana** Pavlick, 24: 430

Festuca saximontana Rydb. var. **saximontana**, 24: 432

Festuca scabrella = [within the *Flora* region, misapplied to **F. hallii**, **F. campestris**, and **F. altaica**], 24: 407, 408

Festuca sciurea = **Vulpia sciurea**, 24: 450

Festuca shortii = **F. paradoxa**, 24: 402

Festuca sororia Piper, 24: 404

Festuca subulata Trin., 24: 402

Festuca subuliflora Scribn., 24: 406

Festuca subverticillata (Pers.) E.B. Alexeev, 24: 400

Festuca suksdorfii = **Vulpia microstachys** var. **confusa**, 24: 452

Festuca tenuifolia = **F. filiformis**, 24: 424

Festuca thurberi Vasey, 24: 408

Festuca trachyphylla (Hack.) Krajina, 24: 424

Festuca tracyi = **Vulpia microstachys** var. **confusa**, 24: 452

Festuca valesiaca Schleich. *ex* Gaudin, 24: 420

Festuca versuta Beal, 24: 400

Festuca viridula Vasey, 24: 440

Festuca vivipara [within the *Flora* region, misapplied to F. **frederikseniae** and **F. viviparoidea**], 24: 436

Festuca vivipara subsp. *glabra* = **F. viviparoidea**, 24: 436

Festuca vivipara subsp. *hirsuta* = **F. frederikseniae**, 24: 436

Festuca viviparoidea Krajina *ex* Pavlick, 24: 436

Festuca viviparoidea subsp. **krajinae** Pavlick, 24: 437

Festuca viviparoidea Krajina *ex* Pavlick subsp. **viviparoidea**, 24: 437

Festuca washingtonica E.B. Alexeev, 24: 440

FINGERHUTHIA Nees, 25: 22
Fingerhuthia africana Lehm., 25: 22

FLUMINEA [included in **SCOLOCHLOA**], 24: 732
Fluminia festucacea = **Scolochloa festucacea**, 24: 734

GASTRIDIUM P. Beauv., 24: 675
Gastridium lendigerum [within the *Flora* region, misapplied to G. **phleoides**], 24: 676
Gastridium phleoides (Nees & Meyen) C.E. Hubb., 24: 676
Gastridium ventricosum [within the *Flora* region, misapplied to G. **phleoides**], 24: 676

GAUDINIA P. Beauv., 24: 732
Gaudinia fragilis (L.) P. Beauv., 24: 732

GLYCERIA R. Br., 24: 68
Glyceria R. Br. sect. **Glyceria**, 24: 81
Glyceria sect. **Hydropoa** (Dumort.) Dumort., 24: 71
Glyceria sect. **Striatae** G.L. Church, 24: 73
Glyceria acutiflora Torr., 24: 83
Glyceria alnasteretum Kom., 24: 71
Glyceria arkansana = **G. septentrionalis** var. **arkansana**, 24: 83
Glyceria borealis (Nash) Batch., 24: 81
Glyceria canadensis (Michx.) Trin., 24: 79
Glyceria canadensis (Michx.) Trin. var. **canadensis**, 24: 79
Glyceria canadensis var. **laxa** (Scribn.) Hitchc., 24: 80
Glyceria declinata Bréb., 24: 87
Glyceria elata (Nash) M.E. Jones, 24: 79
Glyceria erecta = **Torreyochloa erecta**, 24: 608
Glyceria fernaldii = **Torreyochloa pallida** var. **fernaldii**, 24: 608
Glyceria fluitans (L.) R. Br., 24: 85
Glyceria ×gatineauensis Bowden, 24: 77
Glyceria grandis S. Watson, 24: 71
Glyceria grandis S. Watson var. **grandis**, 24: 71
Glyceria grandis var. **komarovii** Kelso, 24: 71
Glyceria grandis forma *pallescens* = **G. grandis**, 24: 71
Glyceria laxa = **G. canadensis** var. **laxa**, 24: 80
Glyceria leptostachya Buckley, 24: 85
Glyceria maxima (Hartm.) Holmb., 24: 73
Glyceria maxima subsp. *grandis* = **G. grandis**, 24: 71
Glyceria melicaria (Michx.) F.T. Hubb., 24: 75
Glyceria nervata = **G. striata**, 24: 77
Glyceria notata Chevall., 24: 87
Glyceria nubigena W.A. Anderson, 24: 75
Glyceria obtusa (Muhl.) Trin., 24: 75
Glyceria ×occidentalis (Piper) J.C. Nelson, 24: 85
Glyceria otisii = **Torreyochloa pallida** var. **pauciflora**, 24: 608
Glyceria ×ottawensis Bowden, 24: 77
Glyceria pallida = **Torreyochloa pallida**, 24: 608
Glyceria pallida var. *fernaldii* = **Torreyochloa pallida** var. **fernaldii**, 24: 608
Glyceria pauciflora = **Torreyochloa pallida** var. **pauciflora**, 24: 608
Glyceria plicata = **G. notata**, 24: 87
Glyceria pulchella (Nash) K. Schum., 24: 77
Glyceria septentrionalis Hitchc., 24: 81
Glyceria septentrionalis var. **arkansana** (Fernald) Steyerm. & C. Kucera, 24: 83
Glyceria septentrionalis Hitchc. var. **septentrionalis**, 24: 83
Glyceria striata (Lam.) Hitchc., 24: 77
Glyceria striata subsp./var. *stricta* = **G. striata**, 24: 77

GYMNOPOGON P. Beauv., 25: 231
Gymnopogon ambiguus (Michx.) Britton, Sterns & Poggenb., 25: 231
Gymnopogon brevifolius Trin., 25: 231
Gymnopogon chapmanianus Hitchc., 25: 232
Gymnopogon floridanus = **G. chapmanianus**, 25: 232

GYNERIEAE Sánchez-Ken & L.G. Clark, 25: 352

GYNERIUM Willd. *ex* P. Beauv., 25: 353
Gynerium argenteum = **Cortaderia selloana**, 25: 299
Gynerium sagittatum (Aubl.) P. Beauv., 25: 353

HACKELOCHLOA Kuntze, 25: 691
Hackelochloa granularis (L.) Kuntze, 25: 693

HAINARDIEAE [included in **POEAE**], 24: 378

HAINARDIA Greuter, 24: 689
Hainardia cylindrica (Willd.) Greuter, 24: 691

HAKONECHLOA Makino *ex* Honda, 25: 8
Hakonechloa macra (Munro) Makino, 25: 8

HELEOCHLOA [included in **CRYPSIS**], 25: 139
Heleochloa alopecuroides = **Crypsis alopecuroides**, 25: 139
Heleochloa schoenoides = **Crypsis schoenoides**, 25: 140

HELICTOTRICHON Besser *ex* Schult. & Schult. f., 24: 701
Helictotrichon hookeri = **Avenula hookeri**, 24: 698
Helictotrichon mortonianum (Scribn.) Henrard, 24: 701
Helictotrichon pubsecens = **Avenula pubesens**, 24: 698
Helictotrichon sempervirens (Vill.) Pilg., 24: 701

HEMARTHRIA R. Br., 25: 685
Hemarthria altissima (Poir.) Stapf & C.E. Hubb., 25: 687

Hesperochloa kingii = **Leucopoa kingii**, 24: 445

HESPEROSTIPA (M.K. Elias) Barkworth, 42: 157
Hesperostipa comata (Trin. & Rupr.) Barkworth, 24: 158
  Hesperostipa comata (Trin. & Rupr.) Barkworth subsp. **comata**, 24: 158
  Hesperostipa comata subsp. **intermedia** (Scribn. & Tweedy) Barkworth, 24: 158
Hesperostipa curtiseta (Hitchc.) Barkworth, 24: 158
Hesperostipa neomexicana (Thurb.) Barkworth, 24: 158
Hesperostipa spartea (Trin.) Barkworth, 24: 161

HETEROPOGON Pers., 25: 680
Heteropogon contortus (L.) P. Beauv. *ex* Roem. & Schult., 25: 680
Heteropogon melanocarpus (Elliott) Benth., 25: 680

HIEROCHLOË [included in **ANTHOXANTHUM**], 24: 758
Hierochloë alpina = **Anthoxanthum monticola** subsp. **alpinum**, 24: 760
  Hierochloë alpina subsp. *orthantha* = **Anthoxanthum monticola** subsp. **monticola**, 24: 760
Hierochloë borealis = **Anthoxanthum nitens**, 24: 762
Hierochloë hirta = **Anthoxanthum hirtum**, 24: 764
  Hierochloë hirta subsp. *arctica* = **Anthoxanthum hirtum**, 24: 764
Hierochloë occidentalis = **Anthoxanthum occidentale**, 24: 762
Hierochloë odorata = **Anthoxanthum nitens**, 24: 762
  Hierochloë odorata subsp. *arctica* = **Anthoxanthum hirtum**, 24: 764
  Hierochloë odorata subsp. *hirta* = **Anthoxanthum hirtum**, 24: 764
Hierochloë orthantha = **Anthoxanthum monticola** subsp. **monticola**, 24: 760
Hierochloë pauciflora = **Anthoxanthum arcticum**, 24: 760

HILARIA Kunth, 25: 274
Hilaria belangeri (Steud.) Nash, 25: 278
  Hilaria belangeri (Steud.) Nash var. **belangeri**, 25: 278
  Hilaria belangeri var. **longifolia** (Vasey) Hitchc., 25: 278
Hilaria jamesii (Torr.) Benth., 25: 276
Hilaria mutica (Buckley) Benth., 25: 276
Hilaria rigida (Thurb.) Benth. *ex* Scribn., 25: 276
Hilaria swallenii Cory, 25: 278

HOLCUS L., 24: 739
Holcus lanatus L., 24: 740
Holcus mollis L., 24: 740

HORDEUM L., 24: 245
Hordeum adscendens = **H. jubatum** subsp. **intermedium**, 24: 245
Hordeum arizonicum Covas, 24: 248
Hordeum brachyantherum Nevski, 24: 243
  Hordeum brachyantherum Nevski subsp. **brachyantherum**, 24: 243
  Hordeum brachyantherum subsp. **californicum** (Covas & Stebbins) Bothmer, N. Jacobsen & Seberg, 24: 245
Hordeum bulbosum L., 24: 252
Hordeum caespitosum = **H. jubatum** subsp. **intermedium**, 24: 245
Hordeum californicum = **H. brachyantherum** subsp. **californicum**, 24: 245
Hordeum depressum (Scribn. & J.G. Sm.) Rydb., 24: 243
Hordeum distichon/"distichum" = **H. vulgare**, 24: 252
Hordeum geniculatum = **H. marinum** subsp. **gussoneanum**, 24: 248
Hordeum glaucum = **H. murinum** subsp. **glaucum**, 24: 250
Hordeum hystrix = **H. marinum** subsp. **gussoneanum**, 24: 248
Hordeum intercedens Nevski, 24: 242
Hordeum jubatum L., 24: 245
  Hordeum jubatum var. *caespitosum* = **H. jubatum** subsp. **intermedium**, 24: 245
  Hordeum jubatum subsp. **intermedium** Bowden, 24: 245
  Hordeum jubatum L. subsp. **jubatum**, 24: 245
Hordeum leporinum = **H. murinum** subsp. **leporinum**, 24: 250
Hordeum marinum Huds., 24: 248
  Hordeum marinum subsp. **gussoneanum** (Parl.) Thell., 24: 248
  Hordeum marinum Huds. subsp. **marinum**, 24: 248
Hordeum ×montanense = **×Elyhordeum montanense**, 24: 286
Hordeum murinum L., 24: 250
  Hordeum murinum subsp. **glaucum** (Steud.) Tzvelev, 24: 250
  Hordeum murinum subsp. **leporinum** (Link) Arcang., 24: 250
  Hordeum murinum L. subsp. **murinum**, 24: 250
Hordeum nodosum = **H. brachyantherum**, 24: 243
Hordeum pusillum Nutt., 24: 243
Hordeum secalinum Schreb., 24: 248
Hordeum stebbinsii = **H. murinum** subsp. **glaucum**, 24: 250
Hordeum vulgare L., 24: 252
  Hordeum vulgare subsp. *distichon* = **H. vulgare**, 24: 252
  Hordeum vulgare var. *trifurcatum* = **H. vulgare**, 24: 252

HYDROCHLOA [included in **LUZIOLA**], 24: 54
Hydrochloa caroliniensis = **Luziola fluitans**, 24: 54

HYGRORYZA Nees, 24: 46
Hygroryza aristata (Reta.) Nees *ex* Wright & Arn., 24: 46

HYMENACHNE P. Beauv., 25: 561
Hymenachne amplexicaulis (Rudge) Nees, 25: 563

HYPARRHENIA Andersson *ex* E. Fourn., 25: 678
Hyparrhenia hirta (L.) Stapf, 25: 678
Hyparrhenia rufa (Nees) Stapf, 25: 678

HYSTRIX [included in **ELYMUS**], 24: 288
Hystrix californica = **Leymus californicus**, 24: 369
Hystrix patula = **Elymus hystrix**, 24: 316
    Hystrix patula var. *bigeloviana* = **Elymus hystrix**, 24: 316

IMPERATA Cirillo, 25: 618
Imperata brasiliensis Trin., 25: 621
Imperata brevifolia Vasey, 25: 621
Imperata cylindrica (L.) Raeusch., 25: 621

ISCHAEMUM L., 25: 648
Ischaemum ciliare = **I. indicum**, 25: 648
Ischaemum indicum (Houtt.) Merr., 25: 648
Ischaemum rugosum Salisb., 25: 648

IXOPHORUS Schltdl., 25: 537
Ixophorus unisetus (J. Presl) Schltdl., 25: 537

JARAVA Ruiz & Pav., 24: 178
Jarava ichu Ruiz & Pav., 24: 179
Jarava plumosa (Nees) S.W.L. Jacobs & J. Everett, 24: 179
Jarava speciosa (Trin. & Rupr.) Peñailillo, 24: 181

KARROOCHLOA Conert & Türpe, 25: 308
Karroochloa purpurea (L. f.) Conert & Türpe, 25: 308

KOELERIA Pers., 24: 753
Koeleria asiatica Domin, 24: 754
Koeleria cairnesiana = **K. asiatica**, 24: 754
Koeleria cristata = **K. macrantha**, 24: 754
    Koeleria cristata var. *longifolia* = **K. macrantha**, 24: 754
    Koeleria cristata var. *major* = **K. macrantha**, 24: 754
    Koeleria cristata var. *pubescens* = **K. macrantha**, 24: 754
Koeleria gerardii = **Rostraria cristata**, 24: 756
Koeleria gracilis = **K. macrantha**, 24: 754
Koeleria macrantha (Ledeb.) Schult., 24: 754
Koeleria nitita = **K. macrantha**, 24: 754
Koeleria phleoides = **Rostraria cristata**, 24: 756
Koeleria pyramidata [within the *Flora* region, misapplied to **K. macrantha**], 24: 754
Koeleria pyramidata (Lam.) P. Beauv., 24: 754

LACHNAGROSTIS Trin., 24: 694
Lachnagrostis filiformis (G. Forst.) Trin., 24: 696

LAGURUS L., 24: 670
Lagurus ovatus L., 24: 670

LAMARCKIA Moench, 24: 484
Lamarckia aurea (L.) Moench, 24: 484

LASIACIS (Griseb.) Hitchc., 25: 387
Lasiacis divaricata (L.) Hitchc., 25: 389

Lasiacis ruscifolia (Kunth) Hitchc., 25: 389
    Lasiacis ruscifolia (Kunth) Hitchc. var. **ruscifolia**, 25: 389

LEERSIA Sw., 24: 42
Leersia hexandra Sw., 24: 44
Leersia lenticularis Michx., 24: 44
Leersia monandra Sw., 24: 42
Leersia oryzoides (L.) Sw., 24: 44
    Leersia oryzoides forma *glabra* = **L. oryzoides**, 24: 44
    Leersia oryzoides forma *inclusa* = **L. oryzoides**, 24: 44
Leersia virginica Willd., 24: 44
    Leersia virginica var. *ovata* = **L. virginica**, 42: 44

LEPTOCHLOA P. Beauv., 25: 51
Leptochloa acuminata = **L. fusca** subsp. **fascicularis**, 25: 56
Leptochloa attenuata = **L. panicea** subsp. **mucronata**, 25: 58
Leptochloa chinensis (L.) Nees, 25: 59
Leptochloa chloridiformis (Hack.) Parodi, 25: 54
Leptochloa dubia (Kunth) Nees, 25: 54
Leptochloa fascicularis = **L. fusca** subsp. **fascicularis**, 25: 56
    Leptochloa fascicularis var. *acuminata* = **L. fusca** subsp. **fascicularis**, 25: 56
    Leptochloa fascicularis var. *maritima* = **L. fusca** subsp. **fascicularis**, 25: 56
Leptochloa filiformis = **L. panicea** subsp. **brachiata**, 25: 58
    Leptochloa filiformis var. *attenuata* = **L. panicea** subsp. **mucronata**, 25: 58
Leptochloa fusca (L.) Kunth, 25: 54
    Leptochloa fusca subsp. **fascicularis** (Lam.) N. Snow, 25: 56
    Leptochloa fusca (L.) Kunth subsp. **fusca**, 25: 56
    Leptochloa fusca subsp. **uninervia** (J. Presl) N. Snow, 25: 56
Leptochloa maritima, *ined.* = **L. fusca** subsp. **fascicularis**, 25: 56
Leptochloa mucronata = **L. panicea** subsp. **mucronata**, 25: 58
Leptochloa nealleyi Vasey, 25: 58
Leptochloa panicea (Retz.) Ohwi, 25: 56
    Leptochloa panicea subsp. **brachiata** (Steud.) N. Snow, 25: 58
    Leptochloa panicea subsp. **mucronata** (Michx.) Nowack, 25: 58
Leptochloa panicoides (J. Presl) Hitchc., 25: 59
Leptochloa scabra Nees, 25: 58
Leptochloa uninervia = **L. fusca** subsp. **uninervia**, 25: 56
Leptochloa virgata (L.) P. Beauv., 25: 54
Leptochloa viscida (Scribn.) Beal, 25: 59

LEPTOLOMA [included in **DIGITARIA**, in part], 25: 358
Leptoloma arenicola = **Digitaria arenicola**, 25: 362
Leptoloma cognatum = **Digitaria cognata**, 25: 362

Lepturus cylindricus = **Hainardia cylindrica**, 24: 691

LERCHENFELDIA [included in **DESCHAMPSIA**], 24: 624
    Lerchenfeldia flexuosa subsp. *montana* = **Deschampsia flexuosa**, 24: 631

LEUCOPOA Griseb., 24: 443
Leucopoa kingii (S. Watson) W.A. Weber, 24: 445

×**LEYDEUM** Barkworth, 24: 369

×Leydeum dutillyanum (Lepage) Barkworth, 24: 370

×Leydeum littorale (H.J. Hodgs. & W.W. Mitch.) Barkworth, 24: 370

×Leydeum piperi (Bowden) Barkworth, 24: 370

LEYMUS Hochst., 24: 353

Leymus ambiguus (Vasey & Scribn.) D.R. Dewey, 24: 366

Leymus angustus (Trin.) Pilg., 24: 360

Leymus arenarius (L.) Hochst., 24: 356

Leymus californicus (Bol. *ex* Thurb.) Barkworth, 24: 369

Leymus cinereus (Scribn. & Merr.) Á. Löve, 24: 364

Leymus condensatus (J. Presl) Á. Löve, 24: 362

Leymus flavescens (Scribn. & J.G. Sm.) Pilg., 24: 366

Leymus innovatus (Beal) Pilg., 24: 366

Leymus innovatus (Beal) Pilg. subsp. **innovatus**, 24: 366

Leymus innovatus subsp. **velutinus** (Bowden) Tzvelev, 24: 366

Leymus mollis (Trin.) Pilg., 24: 356

Leymus mollis (Trin.) Pilg. subsp. **mollis**, 24: 358

Leymus mollis subsp. **villosissimus** (Scribn.) Á. Löve, 24: 358

Leymus multicaulis (Kar. & Kir.) Tzvelev, 24: 360

Leymus ×multiflorus (Gould) Barkworth & R.J. Atkins, 24: 362

Leymus pacificus (Gould) D.R. Dewey, 24: 358

Leymus racemosus (Lam.) Tzvelev, 24: 356

Leymus salina (M.E. Jones) Á. Löve, 24: 364

Leymus salina subsp. **mojavensis** Barkworth & R.J. Atkins, 24: 364

Leymus salina (M.E. Jones) Á. Löve subsp. **salina**, 24: 364

Leymus salina subsp. **salmonis** (C.L. Hitchc.) R.J. Atkins, 24: 364

Leymus simplex (Scribn. & T.A. Williams) D.R. Dewey, 24: 359

Leymus simplex var. **luxurians** (Scribn. & T.A. Williams) Beetle, 24: 359

Leymus simplex (Scribn. & T.A. Williams) D.R. Dewey var. **simplex**, 24: 359

Leymus triticoides (Buckley) Pilg., 24: 360

Leymus ×vancouverensis (Vasey) Pilg., 24: 358

Leymus villosissimus = **L. mollis** subsp. **villosissimus**, 24: 358

LIMNODEA L.H. Dewey, 24: 776

Limnodea arkansana (Nutt.) L.H. Dewey, 24: 776

LITHACHNE P. Beauv., 24: 30

Lithachne humilis Soderstr., 24: 30

Lithachne pauciflora (Sw.) P. Beauv., 24: 30

LOLIUM L., 24: 454

Lolium arundinaceum = **Schedonorus arundinaceus**, 24: 448

Lolium giganteum = **Schedonorus giganteus**, 24: 448

Lolium ×hubbardii Jansen & Wachter *ex* B.K. Simon, 24: 456

Lolium ×hybridum Haasskn., 24: 455

Lolium italicum = **L. multiflorum**, 24: 456

Lolium multiflorum Lam., 24: 456

Lolium multiflorum var. *diminutum* = **L. multiflorum**, 24: 456

Lolium multiflorum var. *ramosum* (Guss.) Parl. = **L. temulentum**, 24: 456

Lolium multiflorum forma *submuticum* = **L. multiflorum**, 24: 456

Lolium perenne L., 24: 455

Lolium perenne var. *aristatum* = **L. multiflorum**, 24: 456

Lolium perenne var. *cristatum* = **L. perenne**, 24: 455

Lolium perenne subsp./var. *italicum* = **L. multiflorum**, 24: 456

Lolium perenne subsp. *multiflorum* = **L. multiflorum**, 24: 456

Lolium perenne var. *perenne* = **L. perenne**, 24: 455

Lolium persicum Boiss. & Hohen. *ex* Boiss., 24: 459

Lolium pratense = **Schedonorus pratensis**, 24: 446

Lolium remotum = **L. temulentum** subsp. **remotum**, 24: 456

Lolium rigidum Gaudin, 24: 456

Lolium strictum = **L. rigidum**, 24: 456

Lolium temulentum L., 24: 456

Lolium temulentum var. *arvense* = **L. temulentum**, 24: 456

Lolium temulentum var. *leptochaeton* = **L. temulentum**, 24: 456

Lolium temulentum var. *macrochaeton* = **L. temulentum**, 24: 456

Lolium temulentum subsp. **remotum** (Schrank) Á. Löve & D. Löve, 24: 456

Lolium temulentum L. subsp. **temulentum**, 24: 459

LOPHOCHLOA [included in **ROSTRARIA**], 24: 756

Lophochloa cristata = **Rostraria cristata**, 24: 756

LOPHOPYRUM [included in **THINOPYRUM**], 24: 373

Lophopyrum elongatum = **Thinopyrum ponticum**, 24: 376

LUZIOLA Juss., 24: 54

Luziola bahiensis (Steud.) Hitchc., 24: 55

Luziola fluitans (Michx.) Terrell & H. Rob., 24: 54

Luziola peruviana J.F. Gmel., 24: 55

LYCURUS Kunth, 25: 200

Lycurus phleoides Kunth, 25: 203

Lycurus setosus (Nutt.) C. Reeder, 25: 202

MACROCHLOA Kunth, 24: 151

Macrochloa gigantea = **Celtica gigantea**, 24: 154

Macrochloa tenacissima (L.) Kunth, 24: 152

MANISURIS [a rejected name, included in **ROTTBOELLIA**], 25: 691

Manisuris altissima = **Hemarthria altissima**, 25: 687

Manisuris cylindrica = **Coelorachis cylindrica**, 25: 688

Manisuris exaltata = **Rottboellia cochinchinensis**, 25: 691

Manisuris rugosa = **Coelorachis rugosa**, 25: 688

Manisuris tessellata = **Coelorachis tessellata**, 25: 688

Manisuris tuberculosa = **Coelorachis tuberculosa**, 25: 688

MELICA L., 24: 88

Melica altissima L., 24: 100

Melica aristata Thurb. *ex* Bol., 24: 95

Melica bulbosa Geyer *ex* Porter & J.M. Coult., 24: 91

Melica bulbosa var. *inflata* = **M. bulbosa**, 24: 91

Melica bulbosa var. *intonsa* = **M. bulbosa**, 24: 91

Melica californica Scribn., 24: 93

Melica californica var. *nevadensis* = **M. californica**, 24: 93

Melica ciliata L., 24: 100

Melica frutescens Scribn., 24: 91

Melica fugax Bol., 24: 97
Melica fugax var. *inexpansa* = **M. fugax**, 24: 97
Melica geyeri Munro, 24: 93
Melica geyeri var. **aristulata** J.T. Howell, 24: 95
Melica geyeri Munro var. **geyeri**, 24: 95
Melica harfordii Bol., 24: 93
Melica harfordii var. *minor* = **M. harfordii**, 24: 93
Melica harfordii var. *tenuis* = **M. harfordii**, 24: 93
Melica harfordii var. *viridifolia* = **M. harfordii**, 24: 93
Melica imperfecta Trin., 24: 90
Melica imperfecta var. *flexuosa* = **M. imperfecta**, 24: 90
Melica imperfecta var. *minor* = **M. imperfecta**, 24: 90
Melica imperfecta var. *refracta* = **M. imperfecta**, 24: 90
Melica montezumae Piper, 24: 98
Melica mutica Walter, 24: 100
Melica nitens (Scribn.) Nutt. *ex* Piper, 24: 100
Melica porteri Scribn., 24: 98
Melica porteri var. **laxa** Boyle, 24: 98
Melica porteri Scribn. var. **porteri**, 24: 98
Melica smithii (Porter *ex* A. Gray) Vasey, 24: 95
Melica spectabilis Scribn., 24: 91
Melica stricta Bol., 24: 97
Melica stricta var. **albicaulis** Boyle, 24: 98
Melica stricta Bol. var. **stricta**, 24: 98
Melica subulata (Griseb.) Scribn., 24: 95
Melica subulata var. *pammelii* = **M. subulata**, 24: 95
Melica torreyana Scribn., 24: 90

MELICEAE (Link) Endl., 24: 67

MELINIS P. Beauv., 25: 490
Melinis minutiflora P. Beauv., 25: 490
Melinis nerviglumis (Franch.) Zizka, 25: added post-publication
Melinis repens (Willd.) Zizka, 25: 490

MIBORA Adans., 24: 757
Mibora minima (L.) Desv., 24: 757
Mibora verna = **M. minima**, 24: 757

MICROCHLOA R. Br., 25: 234
Microchloa kunthii Desv., 25: 235

MICROSTEGIUM Nees, 25: 623
Microstegium vimineum (Trin.) A. Camus, 25: 624

MILIUM L., 24: 778
Milium effusum [within the *Flora* region, misapplied to **M. effusum** var. **cisatlanticum**], 24: 780
Milium effusum var. **cisatlanticum** Fernald, 24: 780
Milium effusum L. var. **effusum**, 24: 780
Milium vernale M.-Bieb., 24: 780

MISCANTHUS Andersson, 25: 616
Miscanthus floridulus (Labill.) Warb. *ex* K. Schum. & Lauterb., 25: 617
Miscanthus nepalensis (Trin.) Hack., 25: 617
Miscanthus oligostachyus Stapf, 25: 618
Miscanthus sacchariflorus (Maxim.) Benth., 25: 618
Miscanthus sinensis Andersson, 25: 617

Mnesithea cylindrica = **Coelorachis cylindrica**, 25: 688
Mnesithea rugosa = **Coelorachis rugosa**, 25: 688

MOLINIA Schrank, 25: 7
Molinia caerulea (L.) Moench, 25: 8

MONANTHOCHLOË Engelm., 25: 28
Monanthochloë littoralis Engelm., 25: 28

MONERMA [an illegitimate name, included in HAINARDIA], 24: 689
Monerma cylindrica = **Hainardia cylindrica**, 24: 691

MONROA = **MUNROA**, 25: 51
Monroa squarrosa = **Munroa squarrosa**, 25: 51

MOOROCHLOA Veldkamp, 25: 488
Moorochloa eruciformis (Sm.) Veldkamp, 25: 28

MUHLENBERGIA Schreb., 25: 145
Muhlenbergia andina (Nutt.) Hitchc., 25: 156
Muhlenbergia appressa C.O. Goodd., 25: 164
Muhlenbergia arenacea (Buckley) Hitchc., 25: 181
Muhlenbergia arenicola Buckley, 25: 173
Muhlenbergia arizonica Scribn., 25: 171
Muhlenbergia arsenei Hitchc., 25: 169
Muhlenbergia asperifolia (Nees & Meyen *ex* Trin.) Parodi, 25: 179
Muhlenbergia brachyphylla = **M. bushii**, 25: 158
Muhlenbergia brachyphylla forma *aristata* = **M. bushii**, 25: 158
Muhlenbergia brevis C.O. Goodd., 25: 196
Muhlenbergia bushii R.W. Pohl, 25: 158
Muhlenbergia californica Vasey, 25: 154
Muhlenbergia capillaris (Lam.) Trin., 25: 188
Muhlenbergia capillaris var. *trichopodes* = **M. expansa**, 25: 188
Muhlenbergia crispiseta Hitchc., 25: 185
Muhlenbergia curtifolia Scribn., 25: 167
Muhlenbergia curtisetosa = **Muhlenbergia ×curtisetosa**, 25: 156
Muhlenbergia ×curtisetosa (Scribn.) Bush, 25: 156
Muhlenbergia cuspidata (Torr.) Rydb., 25: 171
Muhlenbergia depauperata Scribn., 25: 196
Muhlenbergia diffusa = **M. schreberi**, 25: 162
Muhlenbergia diversiglumis Trin., 25: 164
Muhlenbergia dubia E. Fourn., 25: 194
Muhlenbergia dubioides = **M. palmeri**, 25: 192
Muhlenbergia dumosa Scribn. *ex* Vasey, 25: 175
Muhlenbergia elongata Scribn. *ex* Beal, 25: 190
Muhlenbergia eludens C. Reeder, 25: 198
Muhlenbergia emersleyi Vasey, 25: 185
Muhlenbergia expansa (Poir.) Trin., 25: 188
Muhlenbergia filiculmis Vasey, 25: 181
Muhlenbergia filiformis (Thurb. *ex* S. Watson) Rydb., 25: 179
Muhlenbergia filipes = **M. sericea**, 25: 188
Muhlenbergia foliosa = **M. mexicana** var. **filiformis**, 25: 154
Muhlenbergia foliosa forma *ambigua* = **M. mexicana** var. **filiformis**, 25: 154
Muhlenbergia fragilis Swallen, 25: 200
Muhlenbergia frondosa (Poir.) Fernald, 25: 158
Muhlenbergia frondosa forma *commutata* = **M. frondosa**, 25: 158
Muhlenbergia glabrifloris Scribn., 25: 156

Muhlenbergia glauca (Nees) B.D. Jacks., 25: 165
Muhlenbergia glomerata (Willd.) Trin., 25: 154
   Muhlenbergia glomerata var. *cinnoides* = **M. glomerata**, 25: 154
Muhlenbergia ×involuta Swallen, 25: 187
Muhlenbergia jonesii (Vasey) Hitchc., 25: 183
Muhlenbergia lindheimeri Hitchc., 25: 192
Muhlenbergia longiligula Hitchc., 25: 187
Muhlenbergia marshii = **M. rigens**, 25: 194
Muhlenbergia metcalfei = **M. rigida**, 25: 190
Muhlenbergia mexicana (L.) Trin., 25: 154
   Muhlenbergia mexicana forma *ambigua* = **M. mexicana** var. *filiformis*, 25: 154
   Muhlenbergia mexicana subsp. *commutata* = **M. frondosa**, 25: 158
   Muhlenbergia mexicana var. **filiformis** (Torr.) Scribn., 25: 154
   Muhlenbergia mexicana (L.) Trin. var. **mexicana**, 25: 154
   Muhlenbergia mexicana forma *setiglumis* = **M. mexicana** var. *filiformis*, 25: 154
Muhlenbergia microsperma (DC.) Trin., 25: 162
Muhlenbergia minutissima (Steud.) Swallen, 25: 198
Muhlenbergia montana (Nutt.) Hitchc., 25: 183
Muhlenbergia monticola = **M. tenuifolia**, 25: 162
Muhlenbergia mundula = **M. rigens**, 25: 194
Muhlenbergia neomexicana = **M. pauciflora**, 25: 167
Muhlenbergia palmeri Vasey, 25: 192
Muhlenbergia palustris = **M. schreberi**, 25: 162
Muhlenbergia parviglumis = **M. spiciformis**, 25: 169
Muhlenbergia pauciflora Buckley, 25: 167
Muhlenbergia pectinata C.O. Goodd., 25: 164
Muhlenbergia peruviana (P. Beauv.) Steud., 25: 185
Muhlenbergia polycaulis Scribn., 25: 167
Muhlenbergia porteri Scribn. *ex* Beal, 25: 169
Muhlenbergia pulcherrima = **M. peruviana**, 25: 185
Muhlenbergia pungens Thurb. *ex* A. Gray, 25: 173
Muhlenbergia racemosa (Michx.) Britton, Sterns & Poggenb., 25: 153
   Muhlenbergia racemosa var. *cinnoides* = **M. glomerata**, 25: 154
Muhlenbergia ramulosa (Kunth) Swallen, 25: 200
Muhlenbergia repens (J. Presl) Hitchc., 25: 175
Muhlenbergia reverchonii Vasey & Scribn., 25: 190
Muhlenbergia richardsonis (Trin.) Rydb., 25: 177
Muhlenbergia rigens (Benth.) Hitchc., 25: 194
Muhlenbergia rigida (Kunth) Trin., 25: 190
Muhlenbergia schreberi J.F. Gmel., 25: 162
   Muhlenbergia schreberi var. *curtisetosa* = **M. ×curtisetosa**, 25: 156
Muhlenbergia sericea (Michx.) P.M. Peterson, 25: 188
Muhlenbergia setifolia Vasey, 25: 192
Muhlenbergia sinuosa Swallen, 25: 196
Muhlenbergia sobolifera (Muhl. *ex* Willd.) Trin., 25: 158
   Muhlenbergia sobolifera forma *setigera* = **M. sobolifera**, 25: 158
Muhlenbergia spiciformis Trin., 25: 169
Muhlenbergia straminea Hitchc., 25: 183
Muhlenbergia squarrosa = **M. richardsonis**, 25: 177
Muhlenbergia sylvatica (Torr.) Torr. *ex* A. Gray, 25: 160

Muhlenbergia sylvatica forma *attenuata* = **M. sylvatica**, 25: 160
Muhlenbergia sylvatica var. *robusta* = **M. sylvatica**, 25: 160
Muhlenbergia tenuiflora (Willd.) Britton, Sterns & Poggenb., 25: 160
   Muhlenbergia tenuiflora var. *variabilis* = **M. tenuiflora**, 25: 160
Muhlenbergia tenuifolia (Kunth) Trin., 25: 162
Muhlenbergia texana Buckley, 25: 198
Muhlenbergia thurberi (Scribn.) Rydb., 25: 165
Muhlenbergia torreyana (Schult.) Hitchc., 25: 179
Muhlenbergia torreyi (Kunth) Hitchc. *ex* Bush, 25: 173
Muhlenbergia uniflora (Muhl.) Fernald, 25: 181
Muhlenbergia utilis (Torr.) Hitchc., 25: 177
Muhlenbergia villiflora Hitchc., 25: 175
   Muhlenbergia villiflora Hitchc. var. **villiflora**, 25: 175
   Muhlenbergia villiflora var. **villosa** (Swallen) Morden, 25: 175
Muhlenbergia villosa = **M. villiflora** var. **villosa**, 25: 175
Muhlenbergia wolfii = **M. ramulosa**, 25: 200
Muhlenbergia wrightii Vasey *ex* J.M. Coult., 25: 171
Muhlenbergia xerophila = **M. elongata**, 25: 190

MUNROA Torr., 25: 51
Munroa squarrosa (Nutt.) Torr., 25: 51

NARDEAE W.D.J. Koch, 24: 62

NARDUS L., 24: 62
Nardus stricta L., 24: 63

NASSELLA E. Desv., 24: 170
Nassella cernua (Stebbins & Love) Barkworth, 24: 176
Nassella chilensis (Trin.) E. Desv., 24: 177
Nassella lepida (Hitchc.) Barkworth, 24: 174
Nassella leucotricha (Trin. & Rupr.) R.W. Pohl, 24: 172
Nassella manicata (E. Desv.) Barkworth, 24: 174
Nassella neesiana (Trin. & Rupr.) Barkworth, 24: 172
Nassella pulchra (Hitchc.) Barkworth, 24: 174
Nassella tenuissima (Trin.) Barkworth, 24: 176
Nassella trichotoma (Nees) Hack. *ex* Arechav., 24: 177
Nassella viridula (Trin.) Barkworth, 24: 177

NEERAGROSTIS [included in **ERAGROSTIS**], 25: 65
Neeragrostis reptans = **Eragrostis reptans**, 25: 74

NEOSTAPFIA Burtt Davy, 25: 294
Neostapfia colusana (Burtt Davy) Burtt Davy, 25: 295

NEYRAUDIA Hook. f., 25: 30
Neyraudia reynaudiana (Kunth) Keng *ex* Hitchc., 25: 31

OLYREAE Kunth, 24: 29

OPIZIA J. Presl, 25: 269
Opizia stolonifera J. Presl, 25: 269

OPLISMENUS P. Beauv., 25: 389
Oplismenus hirtellus (L.) P. Beauv., 25: 390
   Oplismenus hirtellus subsp. *fasciculatus* = **O. hirtellus**, 25: 390
   Oplismenus hirtellus subsp. *setarius* = **O. hirtellus**, 25: 390

Oplismenhus hirtellus subsp. *undulatifolius* = O. hirtellus, 25: 390

Oplismenus setarius = O. hirtellus, 25: 390

ORCUTTIA Vasey, 25: 290
Orcuttia californica Vasey, 25: 291
Orcuttia inaequalis Hoover, 25: 291
Orcuttia pilosa Hoover, 25: 292
Orcuttia tenuis Hitchc., 25: 292
Orcuttia viscida (Hoover) Reeder, 25: 291

ORCUTTIEAE Reeder, 25: 290

ORYZA L., 24: 37
Oryza longistaminata A. Chev. & Roehr., 24: 38
Oryza punctata Kotschy *ex* Steud., 24: 40
Oryza rufipogon Griff., 24: 38
Oryza sativa L., 24: 40

ORYZEAE Dumort., 24: 36

ORYZOPSIS Michx., 24: 167
Oryzopsis asperifolia Michx., 24: 168
Oryzopsis ×bloomeri = Achnatherum ×bloomeri, 24: 142
Oryzopsis canadensis = Piptatherum canadense, 24: 146
Oryzopsis contracta = Achnatherum contractum, 24: 141
Oryzopsis exigua = Piptatherum exiguum, 24: 146
Oryzopsis hendersonii = Achnatherum hendersonii, 24: 139
Oryzopsis hymenoides = Achnatherum hymenoides, 24: 139
Oryzopsis kingii = Ptilagrostis kingii, 24: 143
Oryzopsis micrantha = Piptatherum micranthum, 24: 148
Oryzopsis miliacea = Piptatherum miliaceum, 24: 151
Oryzopsis pungens = Piptatherum pungens, 24: 146
Oryzopsis racemosa = Piptatherum racemosum, 24: 148
Oryzopsis swallenii = Achnatherum swallenii, 24: 137
Oryzopsis webberi = Achnatherum webberi, 24: 137

PANICEAE R. Br., 25: 353

PANICOIDEAE Link, 25: 351

PANICULARIA [included in **POA**], 24: 486
Panicularia davyi = Glyceria leptostachya, 24: 85
Panicularia leptostachya = Glyceria leptostachya, 24: 85
Panicularia nervata = Glyceria striata, 24: 77
Panicularia pallida = Torreyochloa pallida, 24: 608
Panicularia septentrionalis = Glyceria septentrionalis, 24: 81

PANICUM L., 25: 450
Panicum sect. **Agrostoidea** (Nash) C.C. Hsu, 25: 475
Panicum sect. **Antidotalia** Freckmann & Lelong, 25: 482
Panicum sect. **Bulbosa** Zuloaga, 25: 481
Panicum sect. **Dichotomiflora** (Hitchc.) Honda, 25: 467
Panicum sect. **Hemitoma** (Hitchc.) Freckmann & Lelong, 25: 484
Panicum sect. **Monticola** Stapf, 25: 485
Panicum sect. **Obtusa** (Hitchc.) Pilg., 25: 480
Panicum L. sect. **Panicum**, 25: 456
Panicum sect. **Phanopyrum** Raf., 25: 484
Panicum sect. **Repentia** Stapf, 25: 470
Panicum sect. **Tenera** (Hitchc. & Chase) Pilg., 25: 480
Panicum sect. **Urvilleana** (Hitchc. & Chase) Pilg., 25: 475
Panicum sect. **Verrucosa** (Nash) C.C. Hsu, 25: 487
Panicum subg. **Agrostoidea** (Nash) Zuloaga, 25: 475

Panicum L. subg. **Panicum**, 25: 456
Panicum subg. **Phanopyrum** (Raf.) Pilg., 25: 482
Panicum abcissum = P. rigidulum subsp. abcissum, 25: 477
Panicum acroanthum = P. bisulcatum, 25: 485
Panicum aciculare = Dichanthelium aciculare, 25: 442
    Panicum aciculare var. *angustifolium*= Dichanthelium aciculare subsp. angustifolium, 25: 444
Panicum aculeatum [putative hybrid]
Panicum acuminatum = Dichanthelium acuminatum, 25: 422
    Panicum acuminatum var. *columbianum* = Dichanthelium acuminatum subsp. columbianum, 25: 425
    Panicum acuminatum var. *consanguineum* = Dichanthelium consanguineum, 25: 444
    Panicum acuminatum var. *densiflorum* = Dichanthelium acuminatum subsp. spretum, 25: 426
    Panicum acuminatum var. *fasciculatum* = Dichanthelium acuminatum subsp. fasciculatum, 25: 425
    Panicum acuminatum var. *implicatum* = Dichanthelium acuminatum subsp. implicatum, 25: 425
    Panicum acuminatum var. *leucothrix* = Dichanthelium acuminatum subsp. leucothrix, 25: 426
    Panicum acuminatum var. *lindeheimeri* = Dichanthelium acuminatum subsp. lindheimeri, 25: 426
    Panicum acuminatum var. *longiligulatum* = Dichanthelium acuminatum subsp. longiligulatum, 25: 426
    Panicum acuminatum var. *thurowii* = Dichanthelium acuminatum subsp. acuminatum, 25: 425
    Panicum acuminatum var. *unciphyllum* = Dichanthelium tenue, 25: 438
    Panicum acuminatum var. *villosum* = Dichanthelium ovale subsp. villosissimum, 25: 430
    Panicum acuminatum var. *wrightianum* = Dichanthelium wrightianum, 25: 430
Panicum addisonii = Dichanthelium ovale subsp. pseudopubescens, 25: 430
Panicum adspersum = Urochloa adspersa, 25: 497
Panicum agrostoides = P. rigidulum subsp. rigidulum, 25: 478
    Panicum agrostoides var. *condensum* = P. rigidulum subsp. rigidulum, 25: 478
    Panicum agrostoides var. *ramosius* = P. rigidulum subsp. rigidulum, 25: 478
Panicum albemarlense = Dichanthelium acuminatum subsp. implicatum, 25: 425
Panicum albomarginatum = Dichanthelium tenue, 25: 438
Panicum amarulum = P. amarum subsp. amarulum, 25: 472
Panicum amarum Elliott, 25: 472
    Panicum amarum subsp. **amarulum** (Hitchc. & Chase) Freckmann & Lelong, 25: 472
    Panicum amarum Elliott subsp. **amarum**, 25: 474
Panicum anceps Michx., 25: 478
    Panicum anceps Michx. subsp. **anceps**, 25: 478
    Panicum anceps subsp. **rhizomatum** (Hitchc. & Chase) Freckmann & Lelong, 25: 480
Panicum angustifolium = Dichanthelium aciculare subsp. angustifolium, 25: 444
Panicum annulum = Dichanthelium dichotomum subsp. mattamuskeetense, 25: 433
Panicum antidotale Retz., 25: 482

Panicum arenicoloides = **Dichanthelium aciculare** subsp. **angustifolium**, 25: 444

Panicum arizonicum = **Urochloa arizonica**, 25: 495

Panicum ashei = **Dichanthelium commutatum** subsp. **ashei**, 25: 414

Panicum auburne = **Dichanthelium acuminatum** subsp. **acuminatum**, 25: 425

Panicum barbulatum = **Dichanthelium dichotomum** subsp. **dichotomum**, 25: 433

Panicum benneri [putative hybrid]

Panicum bennettense [putative hybrid]

Panicum bergii Arechav., 25: 464

Panicum bicknellii [putative hybrid]

Panicum bisulcatum Thunb., 25: 485

Panicum boreale = **Dichanthelium boreale**, 25: 434

   Panicum boreale var. *michiganense* = **Dichanthelium boreale**, 25: 434

Panicum boscii = **Dichanthelium boscii**, 25: 412

   Panicum boscii var. *molle* = **Dichanthelium boscii**, 25: 412

Panicum brachyanthum Steud., 25: 487

Panicum bulbosum Kunth, 25: 481

   Panicum bulbosum var. *minor* = **P. bulbosum**, 25: 481

Panicum bushii [putative hybrid]

Panicum calliphyllum [putative hybrid]

Panicum capillare L., 25: 457

   Panicum capillare var. *barbipulvinatum* = **P. capillare** subsp. **capillare**, 25: 457

   Panicum capillare var. *brevifolium* = **P. capillare** subsp. **capillare**, 25: 457

   Panicum capillare var. *campestre* = **P. philadelphicum** subsp. **gattingeri**, 25: 459

   Panicum capillare L. subsp. **capillare**, 25: 457

   Panicum capillare subsp. **hillmanii** (Chase) Freckmann & Lelong, 25: 459

   Panicum capillare var. *occidentale* = **P. capillare** subsp. **capillare**, 25: 457

Panicum capillarioides Vasey, 25: 464

Panicum chamaelonche = **Dichanthelium chamaelonche**, 25: 438

Panicum ciliatissimum = **Urochloa ciliatissima**, 25: 505

Panicum ciliatum = **Dichanthelium strigosum** subsp. **leucoblepharis**, 25: 447

Panicum ciliosum = **Dichanthelium acuminatum** subsp. **acuminatum**, 25: 425

Panicum clandestinum = **Dichanthelium clandestinum**, 25: 418

Panicum clutei = **Dichanthelium dichotomum** subsp. **mattamuskeetense**, 25: 433

Panicum coloratum L., 25: 472

Panicum columbianum = **Dichanthelium acuminatum** subsp. **columbianum**, 25: 425

   Panicum columbianum var. *commonsianum* = **Dichanthelium ovale** subsp. **pseudopubescens**, 25: 430

   Panicum columbianum var. *oricola* = **Dichanthelium acuminatum** subsp. **implicatum**, 25: 425

   Panicum columbianum var. *siccanum* = **Dichanthelium acuminatum** subsp. **implicatum**, 25: 425

   Panicum columbianum var. *thinium* = **Dichanthelium acuminatum** subsp. **implicatum**, 25: 425

Panicum combsii = **P. rigidulum** subsp. **combsii**, 25: 477

Panicum commonsianum = **Dichanthelium ovale** subsp. **pseudopubescens**, 25: 430

   Panicum commonsianum var. *euchlamydeum* = **Dichanthelium ovale** subsp. **pseudopubescens**, 25: 430

Panicum commutatum = **Dichanthelium commutatum**, 25: 414

   Panicum commutatum var. *ashei* = **Dichanthelium commutatum** subsp. **ashei**, 25: 414

   Panicum commutatum var. *joorii* = **Dichanthelium commutatum** subsp. **commutatum**, 25: 414

Panicum condensum = **P. rigidulum** subsp. **rigidulum**, 25: 478

Panicum consanguineum = **Dichanthelium consanguineum**, 25: 444

Panicum cryptanthum = **Dichanthelium scabriusculum**, 25: 418

Panicum curtifolium = **Dichanthelium ensifolium** subsp. **curtifolium**, 25: 438

Panicum deamii [putative hybrid]

Panicum debile Elliott = **P. verrucosum**, 25: 487

Panicum depauperatum = **Dichanthelium depauperatum**, 25: 450

   Panicum depauperatum var. *involutum* = **Dichanthelium depauperatum**, 25: 450

   Panicum depauperatum var. *psilophyllum* = **Dichanthelium depauperatum**, 25: 450

Panicum dichotomiflorum Michx., 25: 469

   Panicum dichotomiflorum subsp. **bartowense** (Scribn. & Merr.) Freckmann & Lelong, 25: 469

   Panicum dichotomiflorum Michx. subsp. **dichotomiflorum**, 25: 469

   Panicum dichotomiflorum var. *geniculatum* = **P. dichotomiflorum** subsp. **dichotomiflorum**, 25: 469

   Panicum dichotomiflorum subsp. **puritanorum** (Svenson) Freckmann & Lelong, 25: 469

Panicum dichotomum = **Dichanthelium dichotomum**, 25: 433

   Panicum dichotomum var. *barbulatum* = **Dichanthelium dichotomum**, 25: 433

   Panicum dichotomum var. *clutei* = **Dichanthelium dichotomum** subsp. **mattamuskeetense**, 25: 433

   Panicum dichotomum var. *ensifolium* = **Dichanthelium ensifolium**, 25: 436

   Panicum dichotomum var. *ludicum* = **Dichanthelium dichotomum** subsp. **lucidum**, 25: 433

   Panicum dichotomum var. *mattamuskeetense* = **Dichanthelium dichotomum** subsp. **mattamuskeetense**, 25: 433

   Panicum dichotomum var. opacum, *ined.* = **Dichanthelium dichotomum** subsp. **ludicum**, 25: 433

   Panicum dichotomum var. *ramulosum* = **Dichanthelium dichotomum** subsp. **microcarpon**, 25: 433

   Panicum dichotomum var. *roanokense* = **Dichanthelium dichotomum** subsp. **roanokense**, 25: 434

Panicum diffusum Sw., 25: 466

Panicum divergens = **Dichanthelium commutatum**, 25: 414

Panicum ensifolium = **Dichanthelium ensifolium**, 25: 436

   Panicum ensifolium var. *breve* = **Dichanthelium chamaelonche** subsp. **breve**, 25: 440

Panicum ensifolium var. unciphyllum, *ined.* = Dichanthelium tenue, 25: 438

Panicum erectifolium = Dichanthelium erectifolium, 25: 440

Panicum fasciculatum = Urochloa fusca, 25: 495

   Panicum fasciculatum var. *reticulatum* = Urochloa fusca, 25: 495

Panicum filiforme = Digitaria filiformis var. filiformis, 25: 366

Panicum filipes = P. hallii subsp. filipes, 25: 467

Panicum firmulum = Setaria revershonii subsp. firmula, 25: 546

Panicum flexile (Gatt.) Scribn., 25: 460

Panicum fusiforme = Dichanthelium aciculare subsp. fusiforme, 25: 444

Panicum gattingeri = P. philadelphicum subsp. gattingeri, 25: 459

Panicum geminatum = Paspalidium geminatum, 25: 560

Panicum ghiesbreghtii E. Fourn., 25: 466

Panicum gymnocarpon Elliott, 25: 485

Panicum hallii Vasey, 25: 466

   Panicum hallii subsp. filipes (Scribn.) Freckman & Lelong, 25: 467

   Panicum hallii Vasey subsp. hallii, 25: 467

Panicum havardii = P. virgatum, 25: 474

Panicum helleri = Dichanthelium oligosanthes subsp. scribnerianum, 25: 421

Panicum hemitomon Schult., 25: 484

Panicum hians = Steinchisma hians, 25: 563

Panicum hillmanii = P. capillare subsp. hillmanii, 25: 459

Panicum hirstii = Dichanthelium dichotomum subsp. roanokense, 25: 434

Panicum hirsutum Sw., 25: 464

Panicum hirticaule J. Presl, 25: 460

   Panicum hirticaule J. Presl subsp. hirticaule, 25: 462

   Panicum hirticaule subsp. sonorum (Beal) Freckmann & Lelong, 25: 462

   Panicum hirticaule subsp. stramineum (Hitchc. & Chase) Freckmann & Lelong, 25: 462

Panicum huachucae = Dichanthelium acuminatum subsp. fasciculatum, 25: 425

   Panicum huachucae var. *fasciculatum* = Dichanthelium acuminatum subsp. fasciculatum, 25: 425

Panicum implicatum = Dichanthelium acuminatum subsp. implicatum, 25: 425

Panicum joorii = Dichanthelium commutatum subsp. joorii, 25: 416

Panicum lacustre Hitchc. & Ekman, 25: 467

Panicum lancearium = Dichanthelium portoricense subsp. patulum, 25: 422

Panicum lanuginosum Bosc *ex* Spreng. [an illegitimate name] = Dichanthelium scabriusculum, 25: 418

Panicum lanuginosum Elliott = Dichanthelium acuminatum, 25: 422

   Panicum lanuginosum var. *fasciculatum* = Dichanthelium acuminatum subsp. fasciculatum, 25: 425

   Panicum lanuginosum var. *huachucae* = Dichanthelium acuminatum subsp. fasciculatum, 25: 425

   Panicum lanuginosum var. *implicatum* = Dichanthelium acuminatum subsp. implicatum, 25: 425

Panicum lanuginosum var. *lindheimeri* = Dichanthelium acuminatum subsp. lindheimeri, 25: 426

Panicum lanuginosum var. *praecocium* = Dichanthelium ovale subsp. praecocius, 25: 429

Panicum lanuginosum var. *septentrionale* = Dichanthelium acuminatum subsp. fasciculatum, 25: 425

Panicum lanuginosum var. sericeum, *ined.* = Dichanthelium acuminatum subsp. sericeum, 25: 426

Panicum lanuginosum var. *tennesseense* = Dichanthelium acuminatum subsp. fasciculatum, 25: 425

Panicum latifolium = Dichanthelium latifolium, 25: 412

   Panicum latifolium var. *clandestinum* = Dichanthelium clandestinum, 25: 418

Panicum laxiflorum = Dichanthelium laxiflorum, 25: 446

Panicum leibergii = Dichanthelium leibergii, 25: 416

   Panicum leibergii var. *baldwinii* = Dichanthelium xanthophysum, 25: 416

Panicum lepidulum = P. hallii, 25: 466

Panicum leucothrix = Dichanthelium acuminatum subsp. leucothrix, 25: 426

Panicum lindheimeri = Dichanthelium acuminatum subsp. lindheimeri, 25: 426

Panicum linearifolium = Dichanthelium linearifolium, 25: 449

   Panicum linearifolium var. *werneri* = Dichanthelium linearifolium, 25: 449

Panicum lithophilum = P. philadelphicum subsp. lithophilum, 25: 459

Panicum longifolium = P. rigidulum subsp. pubescens, 25: 478

   Panicum longifolium var. *combsii* = P. rigidulum subsp. combsii, 25: 477

Panicum longiligulatum = Dichanthelium acuminatum subsp. longiligulatum, 25: 426

Panicum lucidum = Dichanthelium dichotomum subsp. lucidum, 25: 433

Panicum macrocarpon = Dichanthelium latifolium, 25: 412

Panicum malacon = Dichanthelium ovale subsp. ovale, 25: 429

Panicum malacophyllum = Dichanthelium malacophyllum, 25: 422

Panicum mattamuskeetense = Dichanthelium dichotomum subsp. mattamuskeetense, 25: 433

Panicum maximum = Urochloa maxima, 25: 505

Panicum meridionale = Dichanthelium acuminatum subsp. implicatum, 25: 425

   Panicum meridionale var. *albemarlense* = Dichanthelium acuminatum subsp. implicatum, 25: 425

Panicum microcarpon Muhl. [an illegitimate name] = Dichanthelium polyanthes, 25: 440

Panicum microcarpon Muhl. *ex* Elliott = Dichanthelium dichotomum subsp. microcarpon, 25: 433

Panicum miliaceum L., 25: 456

   Panicum miliaceum L. subsp. miliaceum, 25: 457

   Panicum miliaceum subsp. ruderale (Kitag.) Tzvelev, 25: 457

   Panicum miliaceum var. spontaneum, *ined.* = P. miliaceum subsp. ruderale, 25: 457

Panicum minimum = P. philadelphicum subsp. philadelphicum, 25: 460

Panicum mohavense Reeder, 25: 462

Panicum mundum [putative hybrid]

Panicum mutabile = **Dichanthelium commutatum** subsp. commutatum, 25: 414

Panicum nemopanthum [putative hybrid]

Panicum neuranthum = **Dichanthelium aciculare** subsp. neuranthum, 25: 444

Panicum niditum = **Dichanthelium dichotomum**, 25: 433

Panicum nodatum = **Dichanthelium nodatum**, 25: 410

Panicum nudicaule = **Dichanthelium nudicaule**, 25: 436

Panicum obtusum Kunth, 25: 481

Panicum occidentale = **Dichanthelium acuminatum** subsp. fasciculatum, 25: 425

Panicum oligosanthes = **Dichanthelium oligosanthes**, 25: 419

Panicum oligosanthes var. *helleri* = **Dichanthelium oligosanthes** subsp. scribnerianum, 25: 421

Panicum oligosanthes var. *scribnerianum* = **Dichanthelium oligosanthes** subsp. scribnerianum, 25: 421

Panicum oligosanthes var. wilcoxianum, *ined.* = **Dichanthelium wilcoxianum**, 25: 449

Panicum oricola = **Dichanthelium acuminatum** subsp. implicatum, 25: 425

Panicum ovale = **Dichanthelium ovale**, 25: 429

Panicum ovale var. *addisonii* = **Dichanthelium ovale** subsp. pseudopubescens, 25: 430

Panicum ovale var. *pseudopubescens* = **Dichanthelium ovale** subsp. pseudopubescens, 25: 430

Panicum ovale var. *villosum* = **Dichanthelium ovale** subsp. villosissimum, 25: 430

Panicum pacificum = **Dichanthelium acuminatum** subsp. fasciculatum, 25: 425

Panicum paludosum Roxb., 25: 470

Panicum pammelii [putative hybrid]

Panicum pampinosum = **P. hirticaule**, 25: 460

Panicum patentifolium = **Dichanthelium portoricense** subsp. patulum, 25: 442

Panicum pedicellatum = **Dichanthelium pedicellatum**, 25: 410

Panicum perlongum = **Dichanthelium perlongum**, 25: 449

Panicum philadelphicum Bernh. *ex* Trin., 25: 459

Panicum philadelphicum subsp. **gattingeri** (Nash) Freckmann & Lelong, 25: 459

Panicum philadelphicum subsp. **lithophilum** (Swallen) Freckmann & Lelong, 25: 459

Panicum philadelphicum Bernh. *ex* Trin. subsp. **philadelphicum**, 25: 460

Panicum philadelphicum var. *tuckermanii* = **P. philadelphicum** subsp. philadelphicum, 25: 460

Panicum pilocomayense = **P. bergii**, 25: 464

Panicum plenum Hitchc. & Chase, 25: 482

Panicum polyanthes = **Dichanthelium polyanthes**, 25: 440

Panicum portoricense = **Dichanthelium portoricense**, 25: 441

Panicum portoricense var. *nashianum* = **Dichanthelium portoricense** subsp. patulum, 25: 442

Panicum praecocius = **Dichanthelium ovale** subsp. praecocius, 25: 429

Panicum pseudopubescens = **Dichanthelium ovale** subsp. pseudopubescens, 25: 430

Panicum psilopodium Trin., 25: 462

Panicum purpurascens = **Urochloa mutica**, 25: 494

Panicum ramisetum = **Setaria reverchonii** subsp. **ramiseta**, 25: 546

Panicum ramosum = **Urochloa ramosa**, 25: 497

Panicum ravenelii = **Dichanthelium ravenelii**, 25: 421

Panicum recognitum [applies to hybrids]

Panicum repens L., 25: 470

Panicum reptans = **Urochloa reptans**, 25: 494

Panicum reverchonii = **Setaria reverchonii**, 25: 546

Panicum rhizomatum = **P. anceps** subsp. **rhizomatum**, 25: 480

Panicum rigidulum Bosc *ex* Nees, 25: 477

Panicum rigidulum subsp. **abscissum** (Swallen) Freckmann & Lelong, 25: 477

Panicum rigidulum subsp. **combsii** (Scribn. & C.R. Ball) Freckmann & Lelong, 25: 477

Panicum rigidulum subsp. **elongatum** (Pursh) Freckmann & Lelong, 25: 478

Panicum rigidulum subsp. **pubescens** (Vasey) Freckmann & Lelong, 25: 478

Panicum rigidulum Bosc *ex* Nees subsp. **rigidulum**, 25: 478

Panicum roanokense = **Dichanthelium dichotomum** subsp. roanokense, 25: 434

Panicum sabulorum var. *patulum* = **Dichanthelium portoricense** subsp. patulum, 25: 442

Panicum sabulorum var. *thinum* = **Dichanthelium acuminatum** subsp. implicatum, 25: 425

Panicum scabriusculum = **Dichanthelium scabriusculum**, 25: 418

Panicum scabriusculum var. *cryptanthum* = **Dichanthelium scabriusculum**, 25: 418

Panicum scoparium = **Dichanthelium scoparium**, 25: 419

Panicum scoparioides [applies to hybrids]

Panicum scribnerianum = **Dichanthelium oligosanthes** subsp. scribnerianum, 25: 421

Panicum shastense [applies to hybrids]

Panicum sonorum = **P. hirticaule** subsp. sonorum, 25: 462

Panicum sphaerocarpon = **Dichanthelium sphaerocarpon**, 25: 441

Panicum sphaerocarpon var. *inflatum* = **Dichanthelium sphaerocarpon**, 25: 441

Panicum sphaerocarpon var. *isophyllum* = **Dichanthelium polyanthes**, 25: 440

Panicum sphaerocarpon var. *polyanthes* = **Dichanthelium polyanthes**, 25: 440

Panicum sphagnicola = **Dichanthelium dichotomum** subsp. lucidum, 25: 433

Panicum spretum = **Dichanthelium acuminatum** subsp. spretum, 25: 426

Panicum stipitatum = **P. rigidulum** subsp. elongatum, 25: 478

Panicum stramineum = **P. hirticaule** subsp. stramineum, 25: 462

Panicum strigosum = **Dichanthelium strigosum**, 25: 446

Panicum strigosum var. *glabrescens* = **Dichanthelium strigosum** subsp. glabrescens, 25: 447

Panicum strigosum var. *leucoblepharis* = **Dichanthelium strigosum** subsp. leucoblepharis, 25: 447

Panicum subvillosum = **Dichanthelium acuminatum** subsp. fasciculatum, 25: 425

Panicum tenerum Beyr. *ex* Trin., 25: 480

Panicum tennesseense = Dichanthelium acuminatum subsp. fasciculatum, 25: 425

Panicum tenue = Dichanthelium tenue, 25: 438

Panicum texanum = Urochloa texana, 25: 495

Panicum thermale = Dichanthelium acuminatum subsp. thermale, 25: 426

Panicum thurowii = Dichanthelium acuminatum subsp. acuminatum, 25: 425

Panicum trichoides Sw., 25: 485

Panicum tsugetorum = Dichanthelium acuminatum subsp. columbianum, 25: 425

Panicum tuckermanii = P. philadelphicum, 25: 459

Panicum urvilleanum Kunth, 25: 475

Panicum verrucosum Muhl., 25: 487

Panicum villosissimum = Dichanthelium ovale subsp. villosissimum, 25: 430

Panicum villosissimum var. praecocius, *ined.* = Dichanthelium ovale subsp. praecocius, 25: 429

Panicum villosissimum var. *pseudopubescens* = Dichanthelium ovale subsp. pseudopubescens, 25: 430

Panicum virgatum L., 25: 474

Panicum virgatum var. *cubense* = P. virgatum, 25: 474

Panicum virgatum var. *spissum* = P. virgatum, 25: 474

Panicum webberianum = Dichanthelium portoricense subsp. patulum, 25: 442

Panicum werneri = Dichanthelium linearifolium, 25: 449

Panicum wilcoxianum = Dichanthelium wilcoxianum, 25: 449

Panicum wrightianum = Dichanthelium wrightianum, 25: 430

Panicum xalapense = Dichanthelium laxiflorum, 25: 446

Panicum xanthophysum = Dichanthelium xanthophysum, 25: 416

Panicum yadkinense = Dichanthelium dichotomum subsp. yadkinense, 25: 434

PAPPOPHOREAE Kunth, 25: 285

PAPPOPHORUM Schreb., 25: 285

Pappophorum bicolor E. Fourn., 25: 286

Pappophorum mucronulatum = P. vaginatum, 25: 286

Pappophorum vaginatum Buckley, 25: 286

PARAPHOLIS C.E. Hubb., 24: 687

Parapholis incurva (L.) C.E. Hubb., 24: 688

Parapholis strigosa (Dumort.) C.E. Hubb., 24: 688

×PASCOLEYMUS Barkworth, 24: 351

×Pascoleymus bowdenii (B. Boivin) Barkworth, 24: 353

PASCOPYRUM Á. Löve, 24: 348

Pascopyrum smithii (Rydb.) Barkworth & D.R. Dewey, 24: 351

PASPALIDIUM Stapf, 25: 558

Paspalidium chapmanii = Setaria chapmanii, 25: 545

Paspalidium geminatum (Forssk.) Stapf, 25: 560

Paspalidium geminatum var. *paludivagum* = P. geminatum, 25: 560

PASPALUM L., 25: 566

Paspalum acuminatum Raddi, 25: 572

Paspalum almum Chase, 25: 575

Paspalum bifidum (Bertol.) Nash, 25: 586

Paspalum blodgettii Chapm., 25: 577

Paspalum boscianum Flüggé, 25: 579

Paspalum bushii = P. setaceum var. stramineum, 25: 592

Paspalum caespitosum Flüggé, 25: 594

Paspalum ciliatifolium = P. setaceum var. ciliatifolium, 25: 590

Paspalum ciliatifolium var. *muhlenbergii* = P. setaceum var. muhlenbergii, 25: 590

Paspalum ciliatifolium var. *stramineum* = P. setaceum var. stramineum, 25: 592

Paspalum circulare = P. laeve, 25: 572

Paspalum conjugatum P.J. Bergius, 25: 572

Paspalum conspersum Schrad., 25: 581

Paspalum convexum Humb. & Bonpl. *ex* Flüggé, 25: 581

Paspalum coryphaeum Trin., 25: 586

Paspalum debile Michx. = P. setaceum, 25: 588

Paspalum denticulatum Trin., 25: 597

Paspalum difforme = P. floridanum, 25: 599

Paspalum dilatatum Poir., 25: 579

Paspalum dissectum (L.) L., 25: 572

Paspalum distichum L., 25: 575

Paspalum eggertii = P. setaceum var. ciliatifolium, 25: 590

Paspalum fimbriatum Kunth, 25: 577

Paspalum floridanum Michx., 25: 599

Paspalum floridanum var. *glabratum* = P. floridanum, 25: 599

Paspalum fluitans = P. repens, 25: 571

Paspalum geminum = P. pubiflorum, 25: 597

Paspalum hartwegianum E. Fourn., 25: 597

Paspalum hydrophilum = P. modestum, 25: 579

Paspalum intermedium Munro *ex* Morong & Britton, 25: 586

Paspalum laeve Michx., 25: 572

Paspalum laeve var. *australe* = P. laeve, 25: 572

Paspalum laeve var. *circulare* = P. laeve, 25: 572

Paspalum laeve var. *pilosum* = P. laeve, 25: 572

Paspalum langei (E. Fourn.) Nash, 25: 588

Paspalum laxum Lam., 25: 592

Paspalum lentiferum = P. praecox, 25: 597

Paspalum lividum Trin. *ex* Schltdl., 25: 597

Paspalum longepedunculatum = P. setaceum var. longepedunculatum, 25: 590

Paspalum longicilium = P. floridanum, 25: 599

Paspalum longipilum = P. laeve, 25: 572

Paspalum malacophyllum Trin., 25: 584

Paspalum minus E. Fourn., 25: 577

Paspalum modestum Mez, 25: 579

Paspalum monostachyum Vasey, 25: 594

Paspalum mucronatum = P. reptans, 25: 571

Paspalum nicorae Parodi, 25: 584

Paspalum notatum Flüggé, 25: 575

Paspalum notatum var. *latiflorum* Döll = P. notatum, 25: 575

Paspalum notatum var. *saurae* Parodi = P. notatum, 25: 575

Paspalum orbiculare [as "*orbiculatum*"] = P. scrobiculatum, 25: 571

Paspalum paniculatum L., 25: 577

Paspalum paspaloides = P. distichum, 25: 575

Paspalum pleostachyum Döll, 25: 594
Paspalum plicatulum Michx., 25: 581
Paspalum praecox Walter, 25: 597
   Paspalum praecox var. *curtisianum* = **P. praecox**, 25: 597
Paspalum psammophilum = **P. setaceum** var. **psammophilum**, 25: 590
Paspalum pubescens = **P. setaceum** var. **muhlenbergii**, 25: 590
Paspalum pubiflorum Rupr. *ex* E. Fourn., 25: 597
   Paspalum pubiflorum var. *glabrum* = **P. pubiflorum**, 25: 597
Paspalum quadrifarium Lam., 25: 586
Paspalum racemosum Lam., 25: 584
Paspalum repens P.J. Bergius, 25: 571
   Paspalum repens var. *fluitans* = **P. repens**, 25: 571
Paspalum saugetti = **P. cespitosum**, 25: 594
Paspalum scrobiculatum L., 25: 571
Paspalum setaceum Michx., 25: 588
   Paspalum setaceum var. **ciliatifolium** (Michx.) Vasey, 25: 590
   Paspalum setaceum var. **longepedunculatum** (Leconte) Alph. Wood, 25: 590
   Paspalum setaceum var. **muhlenbergii** (Nash) D.J. Banks, 25: 590
   Paspalum setaceum var. **psammophilum** (Nash) D.J. Banks, 25: 590
   Paspalum setaceum var. **rigidifolium** (Nash) D.J. Banks, 25: 590
   Paspalum setaceum Michx. var. **setaceum**, 25: 592
   Paspalum setaceum var. **stramineum** (Nash) D.J. Banks, 25: 592
   Paspalum setaceum var. **supinum** (Bosc *ex* Poir.) Trin., 25: 592
   Paspalum setaceum var. **villosissimum** (Nash) D.J. Banks, 25: 592
Paspalum stramineum = **P. setaceum** var. **stramineum**, 25: 592
Paspalum supinum = **P. setaceum** var. **supinum**, 25: 592
Paspalum texanum = **P. plicatulum**, 25: 581
Paspalum unispicatum (Scribn. & Merr.) Nash, 25: 599
Paspalum urvillei Steud., 25: 579
Paspalum vaginatum Sw., 25: 575
Paspalum villosum = **Eriochloa villosa**, 25: 509
Paspalum virgatum L., 25: 581
Paspalum virletii E. Fourn., 25: 594
Paspalum wrightii Hitchc. & Chase, 25: 584

PENNISETUM Rich., 25: 515
Pennisetum advena Wipff & Veldkamp, 25: 527
Pennisetum alopecuroides (L.) Spreng., 25: 521
Pennisetum americanum = **P. glaucum**, 25: 519
Pennisetum cenchroides = **P. ciliare**, 25: 525
Pennisetum ciliare (L.) Link, 25: 525
   Pennisetum ciliare var. *setigerum* = **P. setigerum**, 25: 525
Pennisetum clandestinum Hochst. *ex* Chiov., 25: 519
Pennisetum flaccidum Griseb., 25: 525
Pennisetum glaucum (L.) R. Br., 25: 519
Pennisetum latifolium Spreng., 25: 523
Pennisetum macrostachys (Brongn.) Trin., 25: 521
Pennisetum macrourum Trin., 25: 521

Pennisetum nervosum (Nees) Trin., 25: 521
Pennisetum orientale Willd. *ex* Rich., 25: 527
Pennisetum pedicellatum Trin., 25: 523
   Pennisetum pedicellatum subsp. *unispiculum* Brunken = **P. pedicellatum**, 25: 523
Pennisetum petiolare (Hochst.) Chiov., 25: 529
Pennisetum polystachion (L.) Schult., 25: 523
   Pennisetum polystachion (L.) Schult. subsp. **polystachion**, 25: 523
   Pennisetum polystachion subsp. **setosum** (Sw.) Brunken, 25: 523
Pennisetum purpureum Schumach., 25: 519
Pennisetum setaceum (Forssk.) Chiov., 25: 527
Pennisetum setigerum (Vahl) Wipff, 25: 525
Pennisetum setosum = **P. polystachion**, 25: 523
Pennisetum typhoideum = **P. glaucum**, 25: 519
Pennisetum villosum R. Br. *ex* Fresen., 25: 523

PHALARIS L., 24: 764
Phalaris angusta Nees *ex* Trin., 24: 773
Phalaris aquatica L., 24: 767
Phalaris arundinacea L., 24: 770
   Phalaris arundinacea var. *picta* = **P. arundinacea**, 24: 770
   Phalaris arundinacea forma *varigata* = **P. arundinacea**, 24: 770
Phalaris brachystachys Link, 24: 768
Phalaris californica Hook. & Arn., 24: 770
Phalaris canariensis L., 24: 768
Phalaris caroliniana Walter, 24: 772
Phalaris coerulescens Desf., 24: 766
Phalaris lemmonii Vasey, 24: 768
Phalaris minor Retz., 24: 767
Phalaris ×monspeliensis Daveau, 24: 768
Phalaris paradoxa L., 24: 766
   Phalaris paradoxa var. *praemorsa* = **P. paradoxa**, 24: 766
Phalaris stenoptera = **P. aquatica**, 24: 767
Phalaris tuberosa = **P. aquatica**, 24: 767
   Phalaris tuberosa var. *stenoptera* = **P. aquatica**, 24: 767

PHALAROIDES [included in **PHALARIS**], 24: 764
Phalaroides arundinacea = **Phalaris arundinacea**, 24: 770

PHANOPYRUM [included in **PANICUM**], 25: 450
Phanopyrum gymnocarpon = **Panicum gymnocarpon**, 25: 485

PHAREAE Stapf, 24: 11

PHAROIDEAE L.G. Clark & Judz., 24: 11

PHARUS P. Browne, 24: 12
Pharus glaber Kunth, 24: 13

PHIPPSIA (Trin.) R. Br., 24: 478
Phippsia algida (Sol.) R. Br., 24: 479
   Phippsia algida var. *algidiformis* = **P. concinna**, 24: 480
Phippsia concinna (Th. Fr.) Lindeb., 24: 480
Phippsia neoarctia = **Puccinellia phryganodes**, 24: 463

PHLEUM L., 24: 670
Phleum alpinum L., 24: 672
   Phleum alpinum subsp./var. *commutatum* = **P. alpinum**, 24: 672

Phleum arenarium L., 24: 675
Phleum commutatum = **P. alpinum**, 24: 672
  Phleum commutatum var. *americanum* = **P. alpinum**, 24: 672
Phleum exaratum Hochst. *ex* Griseb., 24: 671
Phleum graecum = **P. exaratum**
Phleum paniculatum Huds., 24: 672
Phleum phleoides (L.) H. Karsten, 24: 672
Phleum pratense L., 24: 672
  Phleum pratense subsp. **bertolonii** (DC.) Bornm., 24: 672
  Phleum pratense var. *nodosum* = **P. pratense**, 24: 672
  Phleum pratense L. subsp. **pratense**, 24: 672
Phleum subulatum (Savi) Asch. & Graebn., 24: 675

Pholiurus incurvus = **Parapholis incurva**, 24: 688

PHRAGMITES Adans., 25: 10
Phragmites australis (Cav.) Trin. *ex* Steud., 25: 10
Phragmites communis = **P. australis**, 25: 10
  Phragmites communis var. *berlandieri* = **P. australis**, 25: 10

PHYLLOSTACHYS Siebold & Zucc., 24: 25
Phyllostachys aurea Rivière & C. Rivière, 24: 27
Phyllostachys bambusoides Siebold & Zucc., 24: 27

PIPTATHERUM P. Beauv., 24: 144
Piptatherum canadense (Poir.) Dorn, 24: 146
Piptatherum exiguum (Thurb.) Dorn, 24: 146
Piptatherum micranthum (Trin. & Rupr.) Barkworth, 24: 148
Piptatherum miliaceum (L.) Coss., 24: 151
  Piptatherum miliaceum (L.) Coss. subsp. **miliaceum**, 24: 151
  Piptatherum miliaceum (L.) Coss. subsp. **thomasii** (Duby) Soják, 24: 151
Piptatherum pungens (Torr.) Dorn, 24: 146
Piptatherum racemosum (Sm.) Eaton, 24: 148
Piptatherum shoshoneanum (Curto & Douglass M. Hend.) P.M. Peterson & Soreng, 24: 148

PIPTOCHAETIUM J. Presl, 24: 161
Piptochaetium avenaceum (L.) Parodi, 24: 164
Piptochaetium avenacioides (Nash) Valencia & Costas, 24: 164
Piptochaetium fimbriatum (Kunth) Hitchc., 24: 164
Piptochaetium pringlei (Beal) Parodi, 24: 162
Piptochaetium setosum (Trin.) Arechav., 24: 166
Piptochaetium stipoides (Trin. & Rupr.) Hack., 24: 166
  Piptochaetium stipoides var. *purpurascens* = **P. stipoides**, 24: 166

PLEURAPHIS [included in **HILARIA**], 25: 274
Pleuraphis jamesii = **Hilaria jamesii**, 25: 276
Pleuraphis mutica = **Hilaria mutica**, 25: 276
Pleuraphis rigida = **Hilaria rigida**, 25: 276

PLEUROPOGON R. Br., 24: 103
Pleuropogon californicus (Nees) Benth. *ex* Vasey, 24: 105
  Pleuropogon californicus (Nees) Benth. *ex* Vasey var. **californicus**, 24: 107
  Pleuropogon californicus var. **davyi** (L.D. Benson) But, 24: 107
Pleuropogon davyi = **P. californicus** var. **davyi**, 24: 107

Pleuropogon hooverianus (L.D. Benson) J.T. Howell, 24: 107
Pleuropogon oregonus Chase, 24: 107
Pleuropogon refractus (A. Gray) Benth. *ex* Vasey, 24: 107
Pleuropogon sabinei R. Br., 24: 109

POA L., 24: 486
Poa sect. **Abbreviatae** Nannf. *ex* Tzvelev, 24: 579
Poa sect. **Alpinae** (Hegetschw. *ex* Nyman) Soreng, 24: 516
Poa sect. **Arctopoa** (Griseb.) Tzvelev, 24: 598
Poa sect. **Arenariae** (Hegetschw.) Stapf, 24: 515
Poa sect. *Bolbophorum* = **Poa** sect. **Arenariae**, 24: 515
Poa sect. **Dioicopoa** E. Desv., 24: 566
Poa sect. **Homalopoa** Dumort., 24: 533
Poa sect. **Madropoa** Soreng, 24: 542
Poa sect. **Micrantherae** Stapf, 24: 518
Poa sect. **Oreinos** Asch. & Graebn., 24: 568
Poa sect. **Pandemos** Asch. & Graebn., 24: 568
Poa L. sect. **Poa**, 24: 521
Poa sect. **Secundae** V.L. Marsh *ex* Soreng, 24: 585
Poa sect. **Stenopoa** Dumort., 24: 573
Poa sect. **Sylvestres** V.L. Marsh *ex* Soreng, 24: 510
Poa sect. **Tichopoa** Asch. & Graebn., 24: 579
Poa subg. **Arctopoa** (Griseb.) Prob., 24: 596
Poa L. subg. **Poa**, 24: 509
Poa subsect. **Epiles** Hitchc. *ex* Soreng, 24: 559
Poa subsect. **Halophytae** V.L. Marsh *ex* Soreng, 24: 592
Poa subsect. **Madropoa** Soreng, 24: 550
Poa subsect. **Secundae** Soreng, 24: 585
Poa abbreviata R. Br., 24: 582
  Poa abbreviata subsp. **abbreviata**, 24: 582
  Poa abbreviata subsp. *jordalii* = **P. abbreviata** subsp. **pattersonii**, 24: 582
  Poa abbreviata subsp. **marshii** Soreng, 24: 582
  Poa abbreviata subsp. **pattersonii** (Vasey) Á. Löve, D. Löve & B.M. Kapoor, 24: 582
Poa agassizensis = **P. pratensis** subsp. **agassizensis**, 24: 523
Poa alpigena = **P. pratensis** subsp. **alpigena**, 24: 525
  Poa alpigena var. *colpodea* = **P. pratensis** subsp. **colpodea**, 24: 526
  Poa alpigena var. prolifera, *ined.* = **P. pratensis** subsp. **colpodea**, 24: 526
Poa alpina L., 24: 518
  Poa alpina L. subsp. **alpina**, 24: 518
  Poa alpina subsp. **vivipara** (L.) Arcang., 24: 518
Poa alsodes A. Gray, 24: 512
Poa ammophila Porsild, 24: 592
Poa ampla = **P. secunda** subsp. **juncifolia**, 24: 586
Poa angustifolia = **P. pratensis** subsp. **angustifolia**, 24: 526
Poa annua L., 24: 519
  Poa annua var. *aquatica* = **P. annua**, 24: 519
  Poa annua var. *reptans* = **P. annua**, 24: 519
Poa aperta = **P. arctica** subsp. **aperta**, 24: 530
Poa arachnifera Torr., 24: 566
Poa arctica R. Br., 24: 529
  Poa arctica subsp. **aperta** (Scribn. & Merr.) Soreng, 24: 530
  Poa arctica R. Br. subsp. **arctica**, 24: 530
  Poa arctica subsp. **caespitans** Simmons *ex* Nannf., 24: 532
  Poa arctica subsp. **grayana** (Vasey) Á. Löve, D. Löve & B.M. Kapoor, 24: 532

Poa arctica subsp. **lanata** (Scribn. & Merr.) Soreng, 24: 532

Poa arctica subsp. *longiculmis* = P. arctica subsp. **arctica**, 24: 530

Poa arctica var. *vivipara* Hook. = P. arctica subsp. **caespitans**, 24: 532

Poa arctica subsp. *williamsii* = P. arctica subsp. **arctica**, 24: 532

Poa arida Vasey, 24: 599

Poa arnowiae Soreng, 24: 543

Poa atropurpurea Scribn., 24: 554

Poa autumnalis Muhl. *ex* Elliott, 24: 514

Poa bigelovii Vasey & Scribn., 24: 536

Poa bolanderi Vasey, 24: 533

Poa bolanderi var. *howellii* = P. **howellii**, 24: 534

Poa brachyanthera = P. **pseudoabbreviata**, 24: 580

Poa buckleyana = P. **secunda**, 24: 586

Poa bulbosa L., 24: 516

Poa bulbosa L. subsp. **bulbosa**, 24: 516

Poa bubosa subsp. **vivipara** (Koeler) Arcang., 24: 516

Poa canadensis = **Glyceria canadensis**, 24: 79

Poa canbyi = P. **secunda**, 24: 586

Poa chaixii Vill., 24: 540

Poa chambersii Soreng, 24: 548

Poa chapmaniana Scribn., 24: 534

Poa compressa L., 24: 579

Poa confinis Vasey, 24: 552

Poa curta auct. = P. **arnowiae**, 24: 543

Poa curta Rydb.= P. **wheeleri**, 24: 546

Poa curtifolia Scribn., 24: 589

Poa cusickii Vasey, 24: 559

Poa cusickii Vasey subsp. **cusickii**, 24: 560

Poa cusickii subsp. **epilis** (Scribn.) W.A. Weber, 24: 560

Poa cusickii subsp. **pallida** Soreng, 24: 560

Poa cusickii subsp. *pubens* = P. **×nematophylla**, 24: 562

Poa cusickii subsp. **purpurascens** (Vasey) Soreng, 24: 562

Poa cuspidata Nutt., 24: 545

Poa diaboli Soreng & D.J. Keil, 24: 552

Poa douglasii Nees, 24: 551

Poa douglasii subsp. *macrantha* = P. **macrantha**, 24: 551

Poa eminens J. Presl, 24: 598

Poa epilis = P. **cusickii** subsp. **epilis**, 24: 560

Poa epilis subsp. **paddensis**, *ined.* = P. **cusickii** subsp. **purpurascens**, 24: 562

Poa eyerdamii = P. **palustris**, 24: 574

Poa fendleriana (Steud.) Vasey, 24: 556

Poa fendleriana subsp. **albescens** (Hitchc.) Soreng, 24: 556

Poa fendleriana var. *arizonica* = P. **fendleriana** subsp. **longiligula**, 24: 558

Poa fendleriana (Steud.) Vasey subsp. **fendleriana**, 24: 556

Poa fendleriana subsp. **longiligula** (Scribn. & T.A. Williams) Soreng, 24: 558

Poa fernaldiana = P. **laxa** subsp. **fernaldiana**, 24: 572

Poa flava = **Tridens flavus**, 25: 39

Poa flexuosa Sm. = P. **laxa**, 24: 570

Poa flexuosa [*sensu* American authors] = P. **laxa × glauca**, 24: 572

Poa ×gaspensis Fernald, 24: 601

Poa glauca Vahl, 24: 576

Poa glauca var. **atroviolacea**, name of uncertain application

Poa glauca Vahl subsp. **glauca**, 24: 578

Poa glauca subsp. *glaucantha* = P. **glauca** subsp. **glauca**, 24: 578

Poa glauca var. **pekulnejensis** (Jurtzev & Tzvelev) Prob., 24: 578

Poa glauca subsp. **rupicola** (Nash) W.A. Weber, 24: 578

Poa glaucantha [*sensu* Hitchc.] = P. **glauca** subsp. **glauca**, 24: 578

Poa glaucifolia = P. **arida**, 24: 599

Poa gracillima = P. **secunda** subsp. **secunda**, 24: 588

Poa gracillima var. *multnomae* = P. **×multnomae**, 24: 545

Poa grayana = P. **arctica** subsp. **grayana**, 24: 532

Poa hansenii [*sensu* D.D. Keck] = P. **stebbinsii**, 24: 564

Poa hansenii Scribn. = P. **cusickii** subsp. **cusickii**, 24: 560

Poa hartzii Gand., 24: 589

Poa hartzii subsp. **alaskana** Soreng, 24; 591

Poa hartzii subsp. *ammophila* = P. **ammophila**, 24: 592

Poa hartzii forma *arenaria* = P. **hartzii** subsp. **hartzii**, 24: 591

Poa hartzii Gand. subsp. **hartzii**, 24: 591

Poa hartzii var. *vivipara* = P. **hartzii** subsp. **vrangelica**, 24: 591

Poa hartzii subsp. **vrangelica** (Tzvelev) Soreng & L.J. Gillespie, 24: 591

Poa hispidula = P. **macrocalyx**, 24: 527

Poa howellii Vasey & Scribn., 24: 534

Poa howellii var. *chandleri* = P. **bolanderi**, 24: 533

Poa incurva = P. **secunda** subsp. **secunda**, 24: 588

Poa infirma Kunth, 24: 519

Poa interior Rydb., 24: 576

Poa involuta = P. **strictiramea**, 24: 540

Poa jordalii = P. **abbreviata** subsp. **pattersonii**, 24: 582

Poa juncifolia = P. **secunda** subsp. **juncifolia**, 24: 586

Poa juncifolia var. *ampla* = P. **secunda** subsp. **juncifolia**, 24: 586

Poa keckii Soreng, 24: 584

Poa kelloggii Vasey, 24: 514

Poa labradorica = **×Dupoa labradorica**, 24: 602

Poa lanata = P. **arctica** subsp. **lanata**, 24: 532

Poa languida = P. **saltuensis**, 24: 510

Poa laxa Haenke, 24: 570

Poa laxa subsp. **banffiana** Soreng, 24: 570

Poa laxa subsp. **fernaldiana** (Nannf.) Hyl., 24: 572

Poa laxa Haenke subsp. **laxa**, 24: 570

Poa laxa × glauca, 24: 572

Poa laxiflora Buckley, 24: 538

Poa leibergii Scribn., 24: 563

Poa leptocoma Trin., 24: 573

Poa leptocoma subsp./var. *paucispicula* = P. **paucispicula**, 24: 538

Poa lettermanii Vasey, 24: 580

Poa ×limosa Scribn. & T.A. Williams, 24: 601

Poa longiligula = P. **fendleriana** subsp. **longiligula**, 24: 558

Poa longipila = P. **arctica** subsp. **arctica**, 24: 530

Poa macrantha Vasey, 24: 551

Poa macrocalyx Trautv. & C.A. Mey., 24: 527

Poa macroclada = P. **stenantha**, 24: 592

Poa malacantha = P. **arctica** subsp. **lanata**, 24: 532

Poa malacantha var. *vivipara* = P. **arctica** subsp. **lanata**, 24: 532

Poa marcida Hitchc., 24: 512
Poa montevansi = **P.** lettermanii, 24: 580
Poa ×multnomae, 24: 545
Poa napensis Beetle, 24: 594
Poa nascopieana [name of uncertain application]
Poa ×nematophylla Rydb., 24: 562
Poa nemoralis L., 24: 574
  Poa nemoralis subsp. *interior* = **P.** interior, 24: 576
Poa nervosa (Hook.) Vasey, 24: 545
  Poa nervosa var. *wheeleri* = **P.** wheeleri, 24: 546
Poa nevadensis = **P.** secunda subsp. juncifolia, 24: 586
  Poa nevadensis var. *juncifolia* = **P.** secunda subsp.
    juncifolia, 24: 586
Poa ×norbergii = **P.** macrocalyx, 24: 527
Poa occidentalis Vasey, 24: 536
Poa paludigena Fernald & Wiegand, 24: 572
Poa palustris L., 24: 574
Poa pattersonii = **P.** abbreviata subsp. pattersonii, 24: 582
Poa paucispicula Scribn. & Merr., 24: 538
Poa peckii = **P.** pratensis subsp. pratensis, 24: 526
Poa pekulnejensis = **P.** glauca var. pekulnejensis, 24: 578
Poa piperi Hitchc., 24: 554
Poa porsildii Gjaerev., 24: 563
Poa pratensis L., 24: 522
  Poa pratensis subsp. agassizensis (B. Boivin & D. Löve)
    Roy L. Taylor & MacBryde, 24: 523
  Poa pratensis subsp. alpigena (Lindm.) Hiitonen, 24: 525
  Poa pratensis subsp. angustifolia (L.) Lej., 24: 526
  Poa pratensis subsp. colpodea (Th. Fr.) Tzvelev, 24: 526
  Poa pratensis subsp. irrigata (Lindm.) H. Lindb., 24: 526
  Poa pratensis L. subsp. pratensis, 24: 526
Poa pringlei Scribn., 24: 564
Poa pseudoabbreviata Roshev., 24: 580
Poa reflexa Vasey & Scribn., 24: 538
Poa rhizomata Hitchc., 24: 546
Poa rigens Hartm. = **P.** pratensis subsp. alpigena, 24: 525
Poa rigens Trin. *ex* Scribn. & Merr. = **P.** eminens, 24: 598
Poa rupicola = **P.** glauca subsp. rupicola, 24: 578
Poa saltuensis Fernald & Wiegand, 24: 510
  Poa saltuensis subsp. languida (Hitchc.) A. Haines,
    24: 510
  Poa saltuensis var. *microlepis* = **P.** saltuensis, 24: 510
  Poa saltuensis Fernald & Wiegand subsp. saltuensis,
    24: 510
Poa sandbergii = **P.** secunda subsp. secunda, 24: 588
Poa scabrella = **P.** secunda subsp. secunda, 24: 588
Poa secunda J. Presl, 24: 586
  Poa secunda var. *elongata* = **P.** secunda subsp. secunda,
    24: 588
  Poa secunda var. *incurva* = **P.** secunda subsp. secunda,
    24: 588
  Poa secunda subsp. juncifolia (Scribn.) Soreng, 24: 586
  Poa secunda subsp. nevadensis, *ined.* = **P.** secunda subsp.
    juncifolia, 24: 586
  Poa secunda J. Presl subsp. secunda, 24: 588
  Poa secunda var. *stenophylla* = **P.** secunda subsp. secunda,
    24: 588
Poa serotina = **P.** palustris, 24: 574
Poa sierrae J.T. Howell, 24: 548
Poa stebbinsii Soreng, 24: 564

Poa stenantha Trin., 24: 592
  Poa stenantha Trin. var. stenantha, 24: 594
  Poa stenantha var. vivipara Trin., 24: 594
Poa strictiramea Hitchc., 24: 540
Poa subcaerulea = **P.** pratensis subsp. irrigata, 24: 526
Poa sublanata Reverd., 24: 527
Poa suksdorfii (Beal) Vasey *ex* Piper, 24: 584
Poa supina Schrad., 24: 521
Poa sylvestris A. Gray, 24: 512
Poa tenerrima Scribn., 24: 588
Poa ×tormentuosa = **P.** glauca subsp. glauca, 24: 578
Poa tracyi Vasey, 24: 543
Poa trivialis L., 24: 568
Poa turneri = **P.** macrocalyx, 24: 527
Poa unilateralis Scribn., 24: 594
  Poa unilateralis subsp. pachypholis (Piper) D.D. Keck *ex*
    Soreng, 24: 596
  Poa unilateralis Scribn. subsp. unilateralis, 24: 596
Poa vaseyochloa = **P.** leibergii, 24: 563
Poa wheeleri Vasey, 24: 546
Poa williamsii = **P.** arctica subsp. arctica, 24: 530
Poa wolfii Scribn., 24: 514

POACEAE Barnhart 24: 3

PODAGROSTIS (Griseb.) Scribn. & Merr., 24: 693
Podagrostis aequivalvis (Trin.) Scribn. & Merr., 24: 693
Podagrostis humilis (Vasey) Björkman, 24: 694
Podagrostis thurberiana = **Podagrostis humilis**, 24: 694

POEAE R. Br., 24: 378

POGONARTHRIA Stapf, 25: 105
Pogonarthria squarrosa (Licht.) Pilg., 25: 106

POLYPOGON Desf., 24: 662
Polypogon australis Brongn., 24: 665
Polypogon elongatus Kunth, 24: 663
Polypogon fugax Nees *ex* Steud., 24: 663
Polypogon imberbis (Phil.) Johow, 24: 668
Polypogon interruptus Kunth, 24: 663
Polypogon maritimus Willd., 24: 665
Polypogon monspeliensis (L.) Desf., 24: 665
Polypogon semiverticillatus = **P.** viridis, 24: 663
Polypogon viridis (Gouan) Breistr., 24: 663

POLYTRIAS Hack., 25: 623
Polytrias amaura (Büse) Kuntze, 25: 623

POÖIDEAE Benth., 24: 57

PSATHYROSTACHYS Nevski, 24: 372
Psathyrostachys juncea (Fisch.) Nevski, 24: 372

×PSEUDELYMUS Barkworth & D.R. Dewey, 24: 282
×Pseudelymus saxicola (Scribn. & J.G. Sm.) Barkworth &
  D.R. Dewey, 24: 283

PSEUDOROEGNERIA (Nevski) Á. Löve, 24: 279
Pseudoroegneria arizonica = **Elymus arizonicus**, 24: 329
Pseudoroegneria spicata (Pursh) Á. Löve, 24: 280
  Pseudoroegneria spicata forma inermis (Scribn. & J.G.
    Sm.) Barkworth, 24: 282

Pseudoroegneria spicata subsp. *inermis* = **P. spicata** forma **inermis**, 24: 282

Pseudoroegneria spicata forma **pubescens** (Elmer) Barkworth, 24: 282

Pseudoroegneria spicata subsp. pubescens, *ined.* = **P.** spicata forma **pubescens**, 24: 282

Pseudoroegneria spicata (Pursh) Á. Löve forma **spicata**, 24: 282

PSEUDOSASA Makino *ex* Nakai, 24: 27
Pseudosasa japonica (Siebold & Zucc. *ex* Steud.) Makino *ex* Nakai, 24: 29

PTILAGROSTIS Griseb., 24: 143
Ptilagrostis kingii (Bol.) Barkworth, 24: 143
Ptilagrostis porteri (Rydb.) W.A. Weber, 24: 144

PUCCINELLIA Parl., 24: 459
Puccinellia agrostidea = **P. arctica**, 24: 471
Puccinellia airoides = **P. nuttalliana**, 24: 475
Puccinellia ambigua = **P. pumila**, 24: 471
Puccinellia americana = **P. maritima**, 24: 463
Puccinellia andersonii Swallen, 24: 473
Puccinellia angustata (R. Br.) E.L. Rand & Redfield, 24: 467
   Puccinellia angustata var. *vaginata* = **P. vaginata**, 24: 477
Puccinellia arctica (Hook.) Fernald & Weath., 24: 471
Puccinellia borealis = **P. nuttalliana**, 24: 475
Puccinellia bruggemannii T.J. Sørensen, 24: 469
Puccinellia coarctata = **P. nutkaensis**, 24: 475
Puccinellia contracta = **P. angustata**, 24: 467
Puccinellia cusickii = **P. nuttalliana**, 24: 475
Puccinellia deschampsioides = **P. nuttalliana**, 24: 475
Puccinellia distans (Jacq.) Parl., 24: 473
   Puccinellia distans var. angustifolia [name of uncertain application]
   Puccinellia distans subsp. borealis [name of uncertain application]
   Puccinellia distans var. *tenuis* = **P. distans**, 24: 473
Puccinellia erecta = **Torreyochloa erecta**, 24: 608
Puccinellia fasciculata (Torr.) E.P. Bicknell, 24: 463
Puccinellia fernaldii = **Torreyochloa pallida** var. **fernaldii**, 24: 608
Puccinellia glabra = **P. nutkaensis**, 24: 475
Puccinellia grandis = **P. nutkaensis**, 24: 475
Puccinellia groenlandica T.J.Sørensen, 24: 467
Puccinellia hauptiana = **P. distans**, 24: 473
Puccinellia howellii J.I. Davis, 24: 475
Puccinellia hultenii = **P. nutkaensis**, 24: 475
Puccinellia interior = **P. nuttalliana**, 24: 475
Puccinellia kamtschatica var. *sublaevis* = **P. nutkaensis**, 24: 475
Puccinellia langeana = **P. tenella**, 24: 469
   Puccinellia langeana subsp. *alaskana* = **P. pumila**, 24: 471
Puccinellia laurentiana = **P. nutkaensis**, 24: 475
Puccinellia lemmonii (Vasey) Scribn., 24: 469
Puccinellia lucida = **P. nutkaensis**, 24: 475
Puccinellia macra = **P. nutkaensis**, 24: 475
Puccinellia maritima (Huds.) Parl., 24: 463
Puccinellia nutkaensis (J. Presl) Fernald & Weath., 24: 475
Puccinellia nuttalliana (Schult.) Hitchc., 24: 475

Puccinellia pallida = **Torreyochloa pallida** var. **pallida**, 24: 608
Puccinellia parishii Hitchc., 24: 465
Puccinellia phryganodes (Trin.) Scribn. & Merr., 24: 463
Puccinellia pauciflora = **Torreyochola pallida** var. **pauciflora**, 24: 608
   Puccinellia pauciflora var. *holmii* = **Torreyochloa pallida** var. **pauciflora**, 24: 608
   Puccinellia pauciflora var. *microtheca* = **Torreyochloa pallida** var. **pauciflora**, 24: 608
Puccinellia paupercula = **P. pumila**, 24: 471
   Puccinellia paupercula var. *alaskana* = **P. pumila**, 24: 471
Puccinellia ×phryganodes = **P. phryganodes**, 24: 463
Puccinellia poacea = **P. arctica**, 24: 471
Puccinellia porsildii = **P. nuttalliana**, 24: 475
Puccinellia pumila (Vasey) Hitchc., 24: 471
Puccinellia retroflexa = **P. distans**, 24: 473
Puccinellia rosenkrantzii = **P. nuttalliana**, 24: 475
Puccinellia rupestris (With.) Fernald & Weath., 24: 463
Puccinellia simplex Scribn., 24: 465
Puccinellia tenella (Lange) Holmb. *ex* Porsild, 24: 469
   Puccinellia tenella subsp. *alaskana* = **P. pumila**, 24: 471
   Puccinellia tenella subsp. *langeana* = **P. tenella**, 24: 469
Puccinellia triflora = **P. nutkaensis**, 24: 475
Puccinellia vaginata (Lange) Fernald & Weath., 24: 477
   Puccinellia vaginata var. *paradoxa* = **P. vaginata**, 24: 477
Puccinellia vahliana (Liebm.) Scribn. & Merr., 24: 465
Puccinellia wrightii (Scribn. & Merr.) Tzvelev, 24: 467

×**PUCCIPHIPPSIA** Tzvelev, 24: 477
×**Pucciphippsia vacillans** (Th. Fr.) Tzvelev, 24: 478

REDFIELDIA Vasey, 25: 41
Redfieldia flexuosa (Thurb. *ex* A. Gray) Vasey, 25: 41

REIMAROCHLOA Hitchc., 25: 599
Reimarochloa oligostachya (Munro *ex* Benth.) Hitchc., 25: 601

RHYNCHELYTRUM [included in **MELINIS**], 25: 490
Rhynchelytrum repens = **Melinis repens**, 25: 490
Rhynchelytrum roseum = **Melinis repens**, 25: 490

ROEGNERIA [included in **ELYMUS**], 24: 288
Roegneria borealis = **Elymus alaskanus**, 24: 326
   Roegneria borealis var. *hyperarctica* = **Elymus alaskanus** subsp. **hyperarcticus**, 24: 326
Roegneria hyperarctica = **Elymus alaskanus** subsp. **hyperarcticus**, 24: 326
Roegneria macroura = **Elymus macrourus**, 24: 324
Roegneria turuchanensis = **Elymus macrourus**, 24: 324
Roegneria villosa = **Elymus alaskanus**, 24: 326

ROSTRARIA Trin., 24: 756
Rostraria cristata (L.) Trin., 24: 756

ROTTBOELLIA L. f., 25: 691
Rottboellia cochinchinensis (Lour.) Clayton, 25: 691
Rottboellia exaltata = **R. cochinchinensis**, 25: 691

RYTIDOSPERMA Steud., 25: 309
Rytidosperma biannulare (Zotov) Connor & Edgar, 25: 311

Rytidosperma caespitosa (Gaudich.) Connor & Edgar,
25: added post-publication
Rytidosperma penicillatum (Zotov) Conner & Edgar,
25: 310
Rytidosperma racemosum (R. Br.) Connor & Edgar, 25: 312
Rytidosperma richardsonii (Cashmore) Connor & Edgar,
25: added post-publication

SACCHARUM L., 25: 609
Saccharum alopecuroides (L.) Nutt., 25: 612
Saccharum baldwinii Spreng., 25: 614
Saccharum bengalense Retz., 25: 616
Saccharum brevibarbe (Michx.) Pers., 25: 612
    Saccharum brevibarbe (Michx.) Pers. var. brevibarbe,
    25: 612
    Saccharum brevibarbe var. contortum (Baldwin) R.D.
    Webster, 25: 612
Saccharum ciliare = **S. bengalense**, 25: 616
Saccharum coarctatum (Fernald) R.D. Webster, 25: 612
Saccharum contortum = **S. brevibarbe** var. **contortum**,
25: 612
Saccharum giganteum (Walter) Pers., 25: 611
Saccharum officinarum L., 25: 614
Saccharum ravennae (L.) L., 25: 614
Saccharum spontaneum L., 25: 614

SACCIOLEPIS Nash, 25: 404
Sacciolepis indica (L.) Chase, 25: 404
Sacciolepis striata (L.) Nash, 25: 405

SCHEDONNARDUS Steud., 25: 228
Schedonnardus paniculatus (Nutt.) Trel., 25: 230

SCHEDONORUS P. Beauv., 24: 445
Schedonorus arundinaceus (Schreb.) Dumort., 24: 448
Schedonorus giganteus (L.) Holub, 24: 448
Schedonorus pratensis (Huds.) P. Beauv., 24: 446

SCHISMUS P. Beauv., 25: 307
Schismus arabicus Nees, 25: 307
Schismus barbatus (Loefl. *ex* L.) Thell., 25: 307

SCHIZACHNE Hack., 24: 103
Schizachne purpurascens (Torr.) Swallen, 24: 103
    Schizachne purpurascens var. *pubescens* = **S. purpurascens**,
    24: 103

SCHIZACHYRIUM Nees, 25: 666
Schizachyrium cirratum (Hack.) Wooton & Standl., 25: 674
Schizachyrium gracile = **Andropogon gracilis**, 25: 653
Schizachyrium littorale (Nash) E.P. Bicknell, 25: 672
Schizachyrium maritimum (Chapm.) Nash, 25: 672
Schizachyrium neomexicanum = **S. scoparium** var.
    scoparium, 25: 670
Schizachyrium niveum (Swallen) Gould, 25: 674
Schizachyrium rhizomatum (Swallen) Gould, 25: 670
Schizachyrium sanguineum (Retz.) Alston, 25: 674
    Schizachyrium sanguineum var. *brevipedicellatum* = **S.**
    sanguinem var. **hirtiflorum**, 25: 674
    Schizachyrium sanguineum var. domingense, *ined.* = **S.**
    sanguineum var. **hirtiflorum**, 25: 674
    Schizachyrium sanguineum var. hirtiflorum (Nees) S.L.
    Hatch, 25: 674

Schizachyriuym sanguineum (Retz.) Alston var.
    sanguineum, 25: 674
Schizachyrium scoparium (Michx.) Nash, 25: 669
    Schizachyrium scoparium var. divergens (Hack.) Gould,
    25: 670
    Schizachyrium scoparium var. *frequens* = **S. scoparium** var.
    scoparium, 25: 670
    Schizachyrium scoparium subsp./var. *littorale* = **S. littorale**,
    25: 672
    Schizachyrium scoparium var. *neomexicanum* = **S.**
    scoparium var. **scoparium**, 25: 670
    Schizachyrium scoparium var. *polycladus* = **S. scoparium**
    var. **scoparium**, 25: 670
    Schizachyrium scoparium var. rhizomatum, *ined.* = **S.**
    rhizomatum, 25: 674
    Schizachyrium scoparium (Michx.) Nash var. scoparium,
    25: 670
    Schizachyrium scoparium var. stoloniferum (Nash) Wipff,
    25: 670
    Schizachyrium scoparium var. *virile* = **S. scoparium** var.
    divergens, 25: 670
Schizachyrium sericatum = **Andropogon gracilis**, 25: 653
Schizachyrium spadiceum (Swallen) Wipff, 25: 669
Schizachyrium stoloniferum var. *wolfei* = **S. scoparium** var.
    scoparium, 25: 670
Schizachyrium tenerum Nees, 25: 672

SCLEROCHLOA P. Beauv., 24: 480
Sclerochloa dura (L.) P. Beauv., 24: 481

SCLEROPOA [included in **DESMAZERIA**], 24: 681
Scleropoa rigida = **Desmazeria rigida**, 24: 682

SCLEROPOGON Phil., 25: 42
Scleropogon brevifolius Phil., 25: 44

SCOLOCHLOA Link, 24: 732
Scolochloa festucacea (Willd.) Link, 24: 734

SCRIBNERIA Hack., 24: 689
Scribneria bolanderi (Thurb.) Hack., 24: 689

SECALE L., 24: 259
Secale cereale L., 24: 259
Secale strictum (C. Presl) C. Presl, 24: 259

SESLERIA Scop., 24: 678
Sesleria autumnalis (Scop.) F.W. Schultz, 24: 679
Sesleria caerulea (L.) Ard., 24: 681
Sesleria heufleriana Schur, 24: 681
Sesleria nitida Ten, 24: 681

SETARIA P. Beauv., 25: 539
Setaria subg. **Paurochaetium** (Hitchc. & Chase) Rominger,
    25: 545
Setaria subg. **Ptychophyllum** (A. Braun) Hitchc., 25: 543
Setaria subg. **Reverchoniae** W.E. Fox, 25: 545
Setaria P. Beauv. subg. **Setaria**, 25: 546
Setaria adhaerans (Forssk.) Chiov., 25: 554
Setaria arizonica Rominger, 25: 552
Setaria barbata (Lam.) Kunth, 25: 543
Setaria chapmanii (Vasey) Pilg., 25: 545
Setaria corrugata (Elliott) Schult., 25: 552

Setaria faberi R.A.W. Herrm., 25: 556

Setaria firmula = **S. reverchonii** subsp. **firmula**, 25: 546

Setaria geniculata = **S. parviflora**, 25: 556

Setaria glauca [within the *Flora* region, misapplied to **S. pumila** subsp. **pumila**], 25: 558

Setaria gracilis = **S. parviflora**, 25: 556

Setaria grisebachii E. Fourn., 25: 550

Setaria imberbis = **S. parviflora**, 25: 556

Setaria italica (L.) P. Beauv., 25: 556

    Setaria italica subvar. *metzgeri* = **S. italica**, 25: 556

    Setaria italica subsp. *stramineofructa* = **S. italica**, 25: 556

Setaria leucopila (Scribn. & Merr.) K. Schum., 25: 548

Setaria liebmannii E. Fourn., 25: 552

Setaria lutescens [within the *Flora* region, misapplied to **S. pumila**], 25: 558

Setaria macrosperma (Scribn. & Merr.) K. Schum., 25: 550

Setaria macrostachya Kunth, 25: 548

Setaria magna Griseb., 25: 552

Setaria megaphylla (Steud.) T. Durand & Schinz, 25: 543

Setaria pallidefusca = **S. pumila** subsp. **pallidefusca**, 25: 558

Setaria palmifolia (J. König) Stapf, 25: 543

Setaria parviflora (Poir.) Kerguélen, 25: 556

Setaria poiretiana [*sensu* Rominger] = **S. megaphylla**, 25: 543

Setaria pumila (Poir.) Roem. & Schult., 25: 558

    Setaria pumila subsp. **pallidefusca** (Schumach.) B.K. Simon, 25: 558

    Setaria pumila (Poir.) Roem. & Schult. subsp. **pumila**, 25: 558

Setaria ramiseta = **S. reverchonii** subsp. **ramiseta**, 25: 546

Setaria rariflora J.C. Mikan *ex* Trin., 25: 550

Setaria reverchonii (Vasey) Pilg., 25: 546

    Setaria reverchonii subsp. **firmula** (Hitchc. & Chase) Pilg., 25: 546

    Setaria reverchonii subsp. **ramiseta** (Scribn.) Pilg., 25: 546

    Setaria reverchonii (Vasey) Pilg. subsp. **reverchonii**, 25: 546

Setaria scheelei (Steud.) Hitchc., 25: 548

Setaria setosa (Sw.) P. Beauv., 25: 550

Setaria sphacelata (Schumach.) Stapf & C.E. Hubb., 25: 558

    Setaria sphacelata var. **aurea** (Hochst. *ex* A. Braun) Clayton, 25: 558

    Setaria sphacelata (Schumach.) Stapf & C.E. Hubb. var. **sphacelata**, 25: 558

Setaria texana Emery, 25: 546

Setaria verticillata (L.) P. Beauv., 25: 554

    Setaria verticillata var. *ambigua* = **S. verticilliformis**, 25: 554

Setaria verticilliformis Dumort., 25: 554

Setaria villosissima (Scribn. & Merr.) K. Schum., 25: 548

Setaria viridis (L.) P. Beauv., 25: 554

    Setaria viridis var. **major** (Gaudin) Peterm., 25: 556

    Setaria viridis var. *robustapurpurea* = **S. viridis** var. **major**, 25: 556

    Setaria viridis (L.) P. Beauv. var. **viridis**, 25: 556

    Setaria viridis var. *weinmannii* = **S. viridis** var. **viridis**, 25: 556

SETARIOPSIS Scribn., 25: 539

Setariopsis auriculata (E. Fourn.) Scribn., 25: 539

SIEGLINGIA [included in **DANTHONIA**], 25: 301

Sieglingia decumbens = **Danthonia decumbens**, 25: 302

SITANION [included in **ELYMUS**], 24: 288

Sitanion californicum = **Elymus elymoides** subsp. **californicus**, 24: 319

Sitanion hansenii = **Elymus ×hansenii**, 24: 340

Sitanion hystrix = **Elymus elymoides**, 24: 318

    Sitanion hystrix var. *brevifolium* = **Elymus elymoides** subsp. **brevifolius**, 24: 319

    Sitanion hystrix var. *californicum* = **Elymus elymoides** subsp. **californicus**, 24: 319

    Sitanion hystrix var. *hordeoides* = **Elymus elymoides** subsp. **hordeoides**, 24: 319

Sitanion jubatum = **Elymus multisetus**

Sitanion longifolium = **Elymus elymoides**

Sitanion pubiflorum = **Elymus elymoides**

SORGHASTRUM Nash, 25: 630

Sorghastrum apalachicolense = **S. elliottii**, 25: 631

Sorghastrum avenaceum = **S. nutans**, 25: 631

Sorghastrum elliottii (C. Mohr) Nash, 25: 631

Sorghastrum nutans (L.) Nash, 25: 631

Sorghastrum secundum (Elliott) Nash, 25: 631

SORGHUM Moench, 25: 626

Sorghum almum = **S. ×almum**, 25: 628

Sorghum ×almum Parodi, 25: 628

Sorghum arundinaceum = **S. bicolor** subsp. **arundinaceum**, 25: 628

Sorghum bicolor (L.) Moench, 25: 628

    Sorghum bicolor subsp. **arundinaceum** (Desv.) de Wet & J.R. Harlan, 25: 628

    Sorghum bicolor (L.) Moench subsp. **bicolor**, 25: 630

    Sorghum bicolor var. *caffrorum* = **S. bicolor** subsp. **bicolor**, 25: 630

    Sorghum bicolor var. *drummondii* = **S. bicolor** ×**drummondii**, 25: 630

    Sorghum bicolor subsp. **×drummondii** (Steud.) de Wet, 25: 630

    Sorghum bicolor subsp. halepense, *ined.* = **S. halepense**, 25: 628

    Sorghum bicolor var. *saccharatum* = **S. bicolor** subsp. **bicolor**, 25: 630

Sorghum caffrorum = **S. bicolor** subsp. **bicolor**, 25: 630

Sorghum caudatum var. *caudatum* = **S. bicolor** subsp. **bicolor**, 25: 630

Sorghum dochna var. *dochna* = **S. bicolor** subsp. **bicolor**, 25: 630

Sorghum dochna var. *technicum* = **S. bicolor** subsp. **bicolor**, 25: 630

Sorghum drummondii = **S. bicolor** subsp. **×drummondii**, 25: 630

Sorghum halepense (L.) Pers., 25: 628

    Sorghum halepense var. anatherum, *ined.* = **S. halepense**, 25: 628

Sorghum sudanense = **S. bicolor** subsp. **×drummondii**, 25: 630

Sorghum verticilliflorum = **S. bicolor** subsp. **arundinaceum**, 25: 628

Sorghum vulgare = **S. bicolor**, 25: 628

Sorghum vulgare var. *drummondii* = S. bicolor subsp. ×drummondii, 25: 630

Sorghum vulgare var. *sudanense* = S. bicolor subsp. ×drummondii, 25: 630

Sorghum vulgare var. *technicum* = S. bicolor subsp. bicolor, 25: 630

SPARTINA Schreb., 25: 240

Spartina alterniflora Loisel., 25: 244

Spartina alterniflora var. *glabra* = S. alterniflora, 25: 244

Spartina alterniflora var. *pilosa* = S. alterniflora, 25: 244

Spartina anglica C.E. Hubb., 25: 246

Spartina bakeri Merr., 25: 246

Spartina ×caespitosa A.A. Eaton, 25: 249

Spartina cynosuroides (L.) Roth, 25: 247

Spartina densiflora Brongn., 25: 247

Spartina foliosa Trin., 25: 244

Spartina gracilis Trin., 25: 247

Spartina juncea = S. patens, 25: 249

Spartina intermedia = S. maritima, 25: 246

Spartina maritima (Curtis) Fernald, 25: 246

Spartina patens (Aiton) Muhl., 25: 249

Spartina patens var. *monogyna* = S. patens, 25: 249

Spartina pectinata Link, 25: 250

Spartina pectinata var. *suttiei* = S. pectinata, 25: 250

Spartina spartinae (Trin.) Merr. *ex* Hitchc., 25: 243

Spartina ×townsendii H. Groves & J. Groves, 25: 246

SPHENOPHOLIS Scribn., 24: 620

Sphenopholis filiformis (Chapm.) Scribn., 24: 621

Sphenopholis intermedia (Rydb.) Rydb., 24: 623

Sphenopholis intermedia var. *pilosa* = S. intermedia, 24: 623

Sphenopholis longiflora (Vasey *ex* L.H. Dewey) Hitchc., 24: 624

Sphenopholis nitida (Biehler) Scribn., 24: 621

Sphenopholis obtusata (Michx.) Scribn., 24: 621

Sphenopholis obtusata var. *lobata* = S. obtusata, 24: 621

Sphenopholis obtusata var. *major* = S. intermedia, 24: 623

Sphenopholis obtusata var. *pubescens* = S. obtusata, 24: 621

Sphenopholis ×pallens (Biehler) Scribn., 24: 621

Sphenopholis pensylvanica (L.) Hitchc., 24: 621

SPODIOPOGON Trin., 25: 609

Spodiopogon sibiricus Trin., 25: 609

SPOROBOLUS R. Br., 25: 115

Sporobolus airoides (Torr.) Torr., 25: 126

Sporobolus argutus = S. pyramidatus, 25: 119

Sporobolus asper = S. compositus, 25: 121

Sporobolus asper var. *asper* = S. compositus var. compositus, 25: 122

Sporobolus asper var. *drummondii* = S. compositus var. drummondii, 25: 122

Sporobolus asper var. *hookeri* = S. compositus var. compositus, 25: 122

Sporobolus asper var. *macer* = S. compositus var. macer, 25: 122

Sporobolus asper var. *pilosus* = S. compositus var. drummondii, 25: 122

Sporobolus buckleyi Vasey, 25: 126

Sporobolus clandestinus (Biehler) Hitchc., 25: 122

Sporobolus clandestinus var. *canovirens* = S. clandestinus, 25: 122

Sporobolus compositus (Poir.) Merr., 25: 121

Sporobolus compositus var. *clandestinus* = S. clandestinus, 25: 122

Sporobolus compositus (Poir.) Merr. var. **compositus**, 25: 122

Sporobolus compositus var. **drummondii** (Trin.) Kartesz & Gandhi, 25: 122

Sporobolus compositus var. **macer** (Trin.) Kartesz & Gandhi, 25: 122

Sporobolus confusus = **Muhlenbergia minutissima**, 25: 198

Sporobolus contractus Hitchc., 25: 129

Sporobolus creber De Nardi, 25: 124

Sporobolus cryptandrus (Torr.) A. Gray, 25: 129

Sporobolus cryptandrus var. *fuscicolor* = S. cryptandrus, 25: 129

Sporobolus cryptandrus subsp. *fuscicolus* = S. cryptandrus, 25: 129

Sporobolus cryptandrus var. *strictus* = S. contractus, 25: 129

Sporobolus curtissii Small *ex* Kearney, 25: 135

Sporobolus depauperatus = **Muhlenbergia richardsonis**, 25: 177

Sporobolus diandrus (Retz.) P. Beauv., 25: 124

Sporobolus domingensis (Trin.) Kunth, 25: 126

Sporobolus fimbriatus (Trin.) Nees, 25: 124

Sporobolus flexuosus (Thurb. *ex* Vasey) Rydb., 25: 131

Sporobolus floridanus Chapm., 25: 137

Sporobolus giganteus Nash, 25: 131

Sporobolus gracillimus = **Muhlenbergia filiformis**, 25: 179

Sporobolus heterolepis (A. Gray) A. Gray, 25: 135

Sporobolus indicus (L.) R. Br., 25: 122

Sporobolus indicus var. *flaccidus* = S. diandrus, 25: 124

Sporobolus indicus var. *pyramidalis* = S. pyramidalis, 25: added post-publication

Sporobolus interruptus Vasey, 25: 133

Sporobolus jacquemontii Kunth, 25: 124

Sporobolus junceus (P. Beauv.) Kunth, 25: 133

Sporobolus microspermus = **Muhlenbergia minutissima**, 25: 198

Sporobolus nealleyi Vasey, 25: 131

Sporobolus neglectus Nash, 25: 121

Sporobolus neglectus var. *ozarkanus* = S. vaginiflorus var. ozarkanus, 25: 119

Sporobolus ozarkanus = S. vaginiflorus var. ozarkanus, 25: 119

Sporobolus patens = S. pyramidatus, 25: 119

Sporobolus pinetorum Weakley & P.M. Peterson, 25: 137

Sporobolus poiretii = S. indicus, 25: 122

Sporobolus pulvinatus = S. pyramidatus, 25: 119

Sporobolus purpurascens (Sw.) Ham., 25: 133

Sporobolus pyramidalus P. Beauv., 25: added post-publication

Sporobolus pyramidatus (Lam.) Hitchc., 25: 119

Sporobolus silveanus Swallen, 25: 137

Sporobolus tenuissimus (Mart. *ex* Schrank) Kuntze, 25: 118

Sporobolus teretifolius R.M. Harper, 25: 135

Sporobolus texanus Vasey, 25: 129
Sporobolus tharpii = S. airoides, 25: 126
Sporobolus vaginiflorus (Torr. *ex* A. Gray) Alph. Wood, 25: 119
    Sporobolus vaginiflorus var. *inaequalis* = S. vaginiflorus var. vaginiflorus, 25: 119
    Sporobolus vaginiflorus var. *neglectus* = S. neglectus, 25: 121
    Sporobolus vaginiflorus var. ozarkanus (Fernald) Shinners, 25: 119
    Sporobolus vaginiflorus (Torr. *ex* A. Gray) Alph. Wood var. vaginiflorus, 25: 119
Sporobolus virginicus (L.) Kunth, 25: 121
Sporobolus wrightii Munro *ex* Scribn., 25: 126

STEINCHISMA Raf., 25: 563
Steinchisma hians (Elliott) Nash, 25: 563

STENOTAPHRUM Trin., 25: 560
Stenotaphrum secundatum (Walter) Kuntze, 25: 561

STIPA L., 24: 154
Stipa arida = Achnatherum aridum, 24: 131
Stipa avenacea = Piptochaetium avenaceum
Stipa bloomeri = Achnatherum ×bloomeri, 24: 142
Stipa californica = Achnatherum occidentale subsp. californicum, 24: 121
Stipa capensis Thunb., 24: 155
Stipa cernua = Nassella cernua, 24: 176
Stipa clandestina = Amelichloa clandestina, 24: 184
Stipa columbiana [within the *Flora* region, misapplied to Achnatherum nelsonii and Achnatherum nelsonii subsp. dorei], 24: 123, 124
    Stipa columbiana var. *nelsonii* = Achnatherum nelsonii subsp. nelsonii, 24: 124
Stipa comata = Hesperostipa comata, 24: 158
    Stipa comata var. *intermedia* = Hesperostipa comata subsp. intermedia, 24: 158
    Stipa comata subsp. *intonsa* = Hesperostipa comata, 24: 158
Stipa coronata = Achnatherum coronatum, 24: 127
    Stipa coronata var. *depauperata* = Achnatherum parishii subsp. depauperatum, 24: 129
    Stipa coronata subsp. *parishii* = Achnatherum parishii subsp. parishii, 24: 129
Stipa curtiseta = Hesperostipa curtiseta, 24: 158
Stipa curvifolia = Achnatherum curvifolium, 24: 135
Stipa elegantissima = Austrostipa elegantissima, 24: 185
Stipa elmeri = Achnatherum occidentale subsp. pubescens, 24: 123
Stipa eminens = Achnatherum eminens, 24: 133
Stipa hendersonii = Achnatherum hendersonii, 24: 139
Stipa hymenoides = Achnatherum hymenoides, 24: 139
Stipa latiglumis = Achnatherum latiglume, 24: 124
Stipa lemmonii = Achnatherum lemmonii, 24: 125
    Stipa lemmonii var. *pubescens* = Achnatherum lemmonii subsp. pubescens, 24: 125
Stipa lepida = Nassella lepida, 24: 174
    Stipa lepida var. *andersonii* = Nassella lepida, 24: 174
Stipa lettermanii = Achnatherum lettermanii, 24: 118
Stipa leucotricha = Nassella leucotricha, 24: 172

Stipa litoralis = Amelichloa caudata, 24: 182
Stipa lobata = Achnatherum lobatum, 24: 131
Stipa neesiana = Nassella neesiana, 24: 172
Stipa nelsonii = Achnatherum nelsonii, 24: 123
    Stipa nelsonii subsp./var. *dorei* = Achnatherum nelsonii subsp. dorei, 24: 124
Stipa neomexicana = Hesperostipa neomexicana, 24: 158
Stipa nevadensis = Achnatherum nevadense, 24: 120
Stipa occidentalis = Achnatherum occidentale, 24: 121
    Stipa occidentalis var. *minor* [*sensu* C.L. Hitchc.] = Achnatherum nelsonii subsp. dorei, 24: 124
    Stipa occidentalis var. *montana* = Achnatherum occidentale, 24: 121
    Stipa occidentalis var. *nelsonii* = Achnatherum nelsonii, 24: 123
    Stipa occidentalis var. *pubescens* = Achnatherum occidentale subsp. pubescens, 24: 123
Stipa parishii = Achnatherum parishii, 24: 127
Stipa pinetorum = Achnatherum pinetorum, 24: 137
Stipa pringlei = Piptochaetium pringlei, 24: 162
Stipa pulcherrima K. Koch, 24: 155
Stipa pulchra = Nassella pulchra, 24: 174
Stipa richardsonii = Achnatherum richardsonii, 24: 133
Stipa robusta = Achnatherum robustum, 24: 129
Stipa scribneri = Achnatherum scribneri, 24: 135
Stipa spartea = Hesperostipa spartea, 24: 161
    Stipa spartea var. *curtiseta* = Hesperostipa curtiseta, 24: 158
Stipa speciosa = Jarava speciosa, 24: 181
Stipa stillmanii = Achnatherum stillmanii, 24: 118
Stipa tenuissima = Nassella tenuissima, 24: 176
Stipa thurberiana = Achnatherum thurberianum, 24: 125
Stipa vaseyi = Achnatherum robustum, 24: 129
Stipa viridula = Nassella viridula, 24: 177
Stipa webberi = Achnatherum webberi, 24: 137
Stipa williamsii = Achnatherum nelsonii, 24: 123

STIPEAE Dumort., 24: 109

×*Stiporyzopsis bloomeri* = Achnatherum ×bloomeri, 24: 142

SWALLENIA Soderstr. & H.F. Decker, 25: 24
Swallenia alexandrae (Swallen) Soderstr. & H.F. Decker, 25: 24

SYNTHERISMA [included in DIGITARIA], 25: 358
Syntherisma filiformis = Digitaria filiformis, 25: 364
Syntherisma ischaemum = Digitaria ischaemum, 25: 372
Syntherisma linearis = Digitaria ischaemum, 25: 372
Syntherisma sanguinalis = Digitaria sanguinalis, 25: 380

TAENIATHERUM Nevski, 24: 257
Taeniatherum asperum = T. caput-medusae, 24: 258
Taeniatherum caput-medusae (L.) Nevski, 24: 258

THEMEDA Forssk., 25: 682
Themeda arguens (L.) Hack., 25: 682
Themeda frondosa = T. arguens, 25: 682
Themeda quadrivalvis (L.) Kuntze, 25: 684
Themeda triandra Forssk., 25: 684

THINOPYRUM Á. Löve

Thinopyrum intermedium (Host) Barkworth & D.R. Dewey, 24: 374

   Thinopyrum intermedium subsp. barbulatum (Schur) Barkworth & D.R. Dewey, 24: 376

   Thinopyrum intermedium (Host) Barkworth & D.R. Dewey subsp. intermedium, 24: 376

Thinopyrum junceum (L.) Á. Löve, 24: 376

Thinopyrum ponticum (Podp.) Barkworth & D.R. Dewey, 24: 376

Thinopyrum pycnanthum (Godr.) Barkworth, 24: 376

THYSANOLAENA Nees, 25: 349

Thysanolaena latifolia (Roxb. *ex* Hornem.) Honda, 25: 350

THYSANOLAENEAE C.E. Hubb., 25: 349

TORREYOCHLOA G.L. Church, 24: 607

Torreyochloa erecta (Hitchc.) G.L. Church, 24: 608

Torreyochloa fernaldii = **T. pallida** var. **fernaldii**, 24: 608

Torreyochloa pallida (Torr.) G.L. Church, 24: 608

   Torreyochloa pallida var. fernaldii (Hitchc.) Dore, 24: 608

   Torreyochloa pallida (Torr.) G.L. Church var. pallida, 24: 608

   Torreyochloa pallida var. pauciflora (J. Presl) J.I. Davis, 24: 608

Torreyochloa pauciflora = **T. pallida** var. **pauciflora**, 24: 608

TRACHYPOGON Nees, 25: 624

Trachypogon montufari [within the *Flora* region, misapplied to **T. secundus**], 25: 626

Trachypogon secundus (J. Presl) Scribn., 25: 626

TRAGUS Haller, 25: 278

Tragus australianus S.T. Blake, 25: 280

Tragus berteronianus Schult., 25: 280

Tragus heptaneuron Clayton, 25: 281

Tragus racemosus (L.) All., 25: 281

TRIBOLIUM Desv., 25: 312

Tribolium obliterum (Hemsley) Renvoize, 25: 313

TRICHACHNE [included in **DIGITARIA**], 25: 358

Trichachne californica = **Digitaria californica**, 25: 368

Trichachne insularis = **Digitaria insularis**, 25: 370

Trichachne patens = **Digitaria patens**, 25: 368

TRICHLORIS E. Fourn. *ex* Benth., 25: 225

Trichloris crinita (Lag.) Parodi, 25: 227

Trichloris pluriflora E. Fourn, 25: 227

TRICHONEURA Andersson, 25: 61

Trichoneura elegans Swallen, 25: 63

TRIDENS Roem. & Schult., 25: 33

Tridens albescens (Vasey) Wooton & Standl., 25: 34

Tridens ambiguus (Elliott) Schult., 25: 36

Tridens buckleyanus (L.H. Dewey) Nash, 25: 36

Tridens carolinianus (Steud.) Henrard, 25: 34

Tridens chapmanii = **T. flavus** var. **chapmanii**, 25: 39

Tridens congestus (L.H. Dewey) Nash, 25: 36

Tridens elongatus = **T. muticus** var. **elongatus**, 25: 36

Tridens eragrostoides (Vasey & Scribn.) Nash, 25: 39

Tridens flavus (L.) Hitchc., 25: 39

   Tridens flavus var. chapmanii (Small) Shinners, 25: 39

Tridens flavus forma *cupreus* = **T. flavus**, 25: 39

   Tridens flavus (L.) Hitchc. var. **flavus**, 25: 39

Tridens grandiflorus = **Erioneuron avenaceum**, 25: 45

Tridens langloisii = **T. ambiguus**, 25: 36

Tridens muticus (Torr.) Nash, 25: 34

   Tridens muticus var. elongatus (Buckley) Shinners, 25: 36

   Tridens muticus (Torr.) Nash var. **muticus**, 25: 36

Tridens ×oklahomensis = **T. muticus** var. **elongatus**, 25: 36

Tridens pilosus = **Erioneuron pilosum**, 25: 45

Tridens pulchellus = **Dasyochloa pulchella**, 25: 47

Tridens strictus (Nutt.) Nash, 25: 34

Tridens texanus (S. Watson) Nash, 25: 39

Triodia albescens = **Tridens albescens**, 25: 34

Triodia elongata = **Tridens muticus** var. **elongatus**, 25: 36

Triodia flava = **Tridens flavus**, 25: 39

Triodia mutica = **Tridens muticus**, 25: 34

Triodia pilosa = **Erioneuron pilosum**, 25: 45

Triodia pulchella = **Dasyochloa pulchella**, 25: 47

Triodia sesleroides = **Tridens flavus**, 25: 39

Triodia stricta = **Tridens stricta**, 25: 34

TRIPLASIS P. Beauv., 25: 41

Triplasis americana P. Beauv., 25: 42

Triplasis purpurea (Walter) Chapm., 25: 42

TRIPOGON Roem. & Schult., 25: 61

Tripogon spicatus (Nees) Ekman, 25: 61

TRIPSACUM L., 25: 693

Tripsacum sect. Fasciculata Hitchc., 25: 695

Tripsacum L. sect. Tripsacum, 25: 695

Tripsacum dactyloides (L.) L., 25: 695

Tripsacum floridanum Porter *ex* Vasey, 25: 696

Tripsacum lanceolatum Rupr. *ex* E. Fourn., 25: 695

TRIRAPHIS R. Br., 25: 31

Triraphis mollis R. Br., 25: 31

TRISETUM Pers., 24: 744

Trisetum aureum (Ten.) Ten., 24: 753

Trisetum canescens Buckley, 24: 746

Trisetum cernuum Trin., 24: 748

   Trisetum cernuum subsp. *canescens* = **T. canescens**, 24: 746

Trisetum flavescens (L.) P. Beauv., 24: 748

Trisetum interruptum Buckley, 24: 750

Trisetum melicoides (Michx.) Scribn., 24: 745

   Trisetum melicoides var. *majus* = **T. melicoides**, 24: 745

Trisetum molle = **T. spicatum**, 24: 748

Trisetum montanum = **T. spicatum**, 24: 748

Trisetum nutkanense = **T. cernuum**, 24: 748

Trisetum orthochaetum Hitchc., 24: 746

Trisetum pensylvanicum (L.) P. Beauv. = **Sphenopholis pensylvanica**, 24: 621

Trisetum pensylvanicum (Spreng.) Trin. = **Sphenopholis nitida**, 24: 621

Trisetum sibiricum Rupr., 24: 750

   Trisetum sibiricum subsp. litorale (Rupr.) Roshev., 24: 750

   Trisetum sibiricum Rupr. subsp. **sibiricum**, 24: 750

Trisetum spicatum (L.) K. Richt., 24: 748

Trisetum spicatum subsp./var. *alaskanum* = **T. spicatum**, 24: 748

Trisetum spicatum subsp. *congdoni* = **T. spicatum**, 24: 748

Trisetum spicatum var. *maidenii* = **T. spicatum**, 24: 748

Trisetum spicatum subsp. *majus* = **T. spicatum**, 24: 748

Trisetum spicatum subsp./var. *molle* = **T. spicatum**, 24: 748

Trisetum spicatum subsp. *montanum* = **T. spicatum**, 24: 748

Trisetum spicatum var. *pilosiglume* = **T. spicatum**, 24: 748

Trisetum subspicatum = **T. spicatum**, 24: 748

Trisetum triflorum = **T. spicatum**, 24: 748

Trisetum triflorum subsp. *molle* = **T. spicatum**, 24: 748

Trisetum wolfii Vasey, 24: 745

TRITICEAE Dumort., 24: 238

×**TRITICOSECALE** Wittm. *ex* A. Camus, 24: 261

TRITICUM L., 24: 268

Triticum aestivum L., 24: 277

Triticum aestivum subsp. *vulgare* = **T. aestivum**, 24: 277

Triticum boeoticum Boiss., 24: 270

Triticum carthlicum Nevski, 24: 274

Triticum cylindricum = **Aegilops cylindrica**, 24: 263

Triticum dicoccoides (Körn.) Körn. *ex* Schweinf., 24: 270

Triticum dicoccum Schrank *ex* Schübl., 24: 272

Triticum durum Desf., 24: 272

Triticum monococcum L., 24: 272

Triticum monococcum subsp. *boeoticum* = **T. boeoticum**, 24: 270

Triticum polonicum L., 24: 274

Triticum spelta L., 24: 277

Triticum tauschii = **Aegilops tauschii**, 24: 262

Triticum vulgare = **T. aestivum**, 24: 277

Triticum timopheevii (Zhuk.) Zhuk., 24: 272

Triticum turgidum L., 24: 274

Triticum urartu Thumanjan *ex* Gandilyan, 24: 270

TUCTORIA Reeder, 25: 292

Tuctoria greenei (Vasey) Reeder, 25: 294

Tuctoria mucronata (Crampton) Reeder, 25: 294

UNIOLA L., 25: 22

Uniola latifolia = **Chasmanthium latifolium**, 25: 345

Uniola laxa = **Chasmanthium laxum**, 25: 346

Uniola nitida = **Chasmanthium nitidum**, 25: 345

Uniola paniculata L., 25: 24

Uniola sessiliflora = **Chasmanthium sessiliflorum**, 25: 346

UROCHLOA P. Beauv., 25: 492

Urochloa adspersa (Trin.) R.D. Webster, 25: 497

Urochloa arizonica (Scribn. & Merr.) Morrone & Zuloaga, 25: 495

Urochloa arrecta (Hack. *ex* T. Durand & Schinz) Morrone & Zuloaga, 25: 505

Urochloa brizantha (Hochst. *ex* A. Rich.) R.D. Webster, 25: 499

Urochloa ciliatissima (Buckley) R.D. Webster, 25: 505

Urochloa fasciculata = **U. fusca**, 25: 495

Urochloa fusca (Sw.) B.F. Hansen & Wunderlin, 25: 495

Urochloa maxima (Jacq.) R.D. Webster, 25: 505

Urochloa mosambicensis (Hack.) Dandy, 25: 497

Urochloa mutica (Forssk.) T.Q. Nguyen, 25: 494

Urochloa oligobrachiata (Pilg.) Kartesz, 25: 503

Urochloa panicoides P. Beauv., 25: 503

Urochloa piligera (F. Muell. *ex* Benth.) R.D. Webster, 25: 499

Urochloa plantaginea (Link) R.D. Webster, 25: 501

Urochloa platyphylla (Munro *ex* C. Wright) R.D. Webster, 25: 503

Urochloa ramosa (L.) T.Q. Nguyen, 25: 497

Urochloa reptans (L.) Stapf, 25: 494

Urochloa subquadripara (Trin.) R.D. Webster, 25: 501

Urochloa texana (Buckley) R.D. Webster, 25: 495

Urochloa villosa (Lam.) T.Q. Nguyen, 25: 501

VAHLODEA Fr., 24: 691

Vahlodea atropurpurea (Wahlenb.) Fr. *ex* Hartm., 24: 692

Vahlodea atropurpurea subsp. *latifolia* = **V. atropurpurea**, 24: 692

Vahlodea atropurpurea subsp. *paramushirensis* = **V. atropurpurea**, 24: 692

Vahlodea flexuosa = **V. atropurpurea**, 24: 692

VASEYOCHLOA Hitchc., 25: 106

Vaseyochloa multinervosa (Vasey) Hitchc., 25: 108

VENTENATA Koeler, 24: 683

Ventenata dubia (Leers) Coss., 24: 684

VETIVERIA [included in **CHYRSOPOGON**], 25: 633

Vetiveria zizanioides = **Chrysopogon zizanioides**, 25: 634

VULPIA C.C. Gmel., 24: 448

Vulpia bromoides (L.) Gray, 24: 452

Vulpia ciliata Dumort., 24: 454

Vulpia deteonensis = **V. bromoides**, 24: 452

Vulpia elliotea = **V. sciurea**, 24: 450

Vulpia megalura = **V. myuros** forma **megalura**, 24: 449

Vulpia microstachys (Nutt.) Munro, 24: 452

Vulpia microstachys var. **ciliata** (A. Gray *ex* Beal) Lonard & Gould, 24: 452

Vulpia microstachys var. **confusa** (Piper) Lonard & Gould, 24: 452

Vulpia microstachys (Nutt.) Munro var. **microstachys**, 24: 452

Vulpia microstachys var. **pauciflora** (Scribn. *ex* Beal) Lonard & Gould, 24: 452

Vulpia myuros (L.) C.C. Gmel., 24: 449

Vulpia myuros var. *hirsuta* [within the *Flora* region, misapplied to **V. myuros** forma **megalura**], 24: 449

Vulpia myuros forma **megalura** (Nutt.) Stace & R. Cotton, 24: 449

Vulpia myuros (L.) C.C. Gmel. forma **myuros**, 24: 449

Vulpia octoflora (Walter) Rydb., 24: 450

Vulpia octoflora var. **glauca** (Nutt.) Fernald, 24: 450

Vulpia octoflora var. **hirtella** (Piper) Henrard, 24: 450

Vulpia octoflora (Walter) Rydb. var. **octoflora**, 24: 450

Vulpia octoflora var. *tenella* = **V. octoflora** var. **glauca**, 24: 450

Vulpia pacifica = **V. microstachys** var. **pauciflora**, 24: 452

Vulpia reflexa = **V. microstachys** var. **pauciflora**, 24: 452

Vulpia sciurea (Nutt.) Henrard, 24: 450

WILLKOMMIA Hack., 25: 227
Willkommia texana Hitchc., 25: 228
   Willkommia texana Hitchc. var. **texana**, 25: 228

ZEA L., 25: 696
Zea sect. **Luxuriantes** Doebley & H.H. Iltis, 25: 698
Zea L. sect. **Zea**, 25: 698
Zea diploperennis H.H. Iltis, Doebley & R. Guzmán,
   25: 699
Zea luxurians (Durieu & Asch.) R.M. Bird, 25: 699
Zea mays L., 25: 701
   Zea mays subsp. **huehuetenangensis** (H.H. Iltis &
      Doebley) Doebley, 25: 701
   Zea mays L. subsp. **mays**, 25: 701
   Zea mays subsp. **mexicana** (Schrad.) H.H. Iltis, 25: 703
   Zea mays subsp. **parviglumis** H.H. Iltis & Doebley,
      25: 703
Zea mexicana = **Z. mays** subsp. **mexicana**, 25: 703
Zea perennis (Hitchc.) Reeves & Mangelsd., 25: 699

ZIZANIA L., 24: 47
Zizania aquatica L., 24: 48
   Zizania aquatica var. *angustifolia* = **Z. palustris** var.
      palustris, 24: 50
   Zizania aquatica L. var. **aquatica**, 24: 48
   Zizania aquatica var. **brevis** Fassett, 24: 48
   Zizania aquatica var. *interior* = **Z. palustris** var. **interior**,
      24: 50
   Zizania aquatica var. *subbrevis* = **Z. aquatica** var. **aquatica**,
      24: 48
Zizania interior = **Z. palustris** var. **interior**, 24: 50
Zizania latifolia (Griseb.) Turcz. *ex* Stapf, 24: 50
Zizania palustris L., 24: 48
   Zizania palustris var. **interior** (Fassett) Dore, 24: 50
   Zizania palustris L. var. **palustris**, 24: 50
Zizania texana Hitchc., 24: 50

ZIZANIOPSIS Döll & Asch., 24: 52
Zizaniopsis miliacea (Michx.) Döll & Asch., 24: 52

ZOYSIA Willd., 25: 281
Zoysia japonica Steud., 25: 283
Zoysia matrella (L.) Merr., 25: 283
Zoysia pacifica (Goudswaard) M. Hotta & Kuroi, 25: 283

Nomenclatural Index

Names of accepted taxa are listed in **boldface**; names of other taxa are in *italics*; vernacular names are in SMALL CAPS. Page numbers in **boldface** refer to the primary treatment; page numbers in *italics* refer to the illustration; other page references are in regular type.

Infraspecific names are listed alphabetically, based on the *infraspecific epithet*; the rank (subspecies, variety, or form) is ignored. The authors of tautonymic infraspecific taxa (subspecies, varieties, and forms in which the specific and infraspecific epithet are the same) are ignored as far as the alphabetic listing is concerned.

If the spellings are the same, vernacular names are listed after all the scientific names for that genus.

Achnatherum P. Beauv., 111, **114**, 167, 169, 177, 178

Achnatherum aridum (M.E. Jones) Barkworth, 117, **131**, *132*

Achnatherum arnowiae (S.L. Welsh & N.D Atwood) Barkworth, 116, *140*, **141**

Achnatherum ×bloomeri (Bol.) Barkworth, 141, **142**, *142*

Achnatherum contractum (B.L. Johnson) Barkworth, 116, *140*, **141**, 142, 148

Achnatherum coronatum (Thurb.) Barkworth, 116, **127**, *128*, 129

Achnatherum curvifolium (Swallen) Barkworth, 115, *134*, **135**

Achnatherum diegoense (Swallen) Barkworth, 117, 127, *130*, **131**

Achnatherum eminens (Cav.) Barkworth, 117, *132*, **133**, 176

Achnatherum hendersonii (Vasey) Barkworth, 116, *138*, **139**

Achnatherum hymenoides (Roem. & Schult.) Barkworth, 116, 137, **139**, *140*, 142, 148, 169, 177

Achnatherum latiglume (Swallen) Barkworth, 115, 116, 121, 124, *126*

Achnatherum lemmonii (Vasey) Barkworth, 117, 123, **125**, *126*

  Achnatherum lemmonii (Vasey) Barkworth subsp. **lemmonii**, **125**, *126*

  Achnatherum lemmonii subsp. **pubescens** (Crampton) Barkworth, **125**, *126*

Achnatherum lettermanii (Vasey) Barkworth, 117, 118, *119*, 120, 123, 137

Achnatherum lobatum (Swallen) Barkworth, 117, *130*, **131**, 135, 137

Achnatherum nelsonii (Scribn.) Barkworth, 117, 118, 121, **123**, *124*, 125, 137

  Achnatherum nelsonii subsp. **dorei** (Barkworth & J.R. Maze) Barkworth, 123, **124**, *124*, 129, 133

  Achnatherum nelsonii (Scribn.) Barkworth subsp. **nelsonii**, 121, **124**, *124*

Achnatherum nevadense (B.L. Johnson) Barkworth, 115, 120, *120*, 121, 125

Achnatherum occidentale (Thurb.) Barkworth, 115, 116, **121**, *122*, 123, 125, 142

  Achnatherum occidentale subsp. **californicum** (Merr. & Burtt Davy) Barkworth, 120, **121**, *122*, 123

  Achnatherum occidentale (Thurb.) Barkworth subsp. **occidentale**, 120, 121, *122*, **123**, 142

  Achnatherum occidentale subsp. **pubescens** (Vasey) Barkworth, 120, 121, *122*, **123**, 124

Achnatherum parishii (Vasey) Barkworth, 116, **127**, *128*, 135

  Achnatherum parishii subsp. **depauperatum** (M.E. Jones) Barkworth, *128*, **129**, 137

  Achnatherum parishii (Vasey) Barkworth subsp. **parishii**, *128*, **129**, 137

Achnatherum perplexum Hoge & Barkworth, 117, 118, 123, 124, 127, 131, *134*, **135**

Achnatherum pinetorum (M.E. Jones) Barkworth, 116, *136*, **137**

Achnatherum richardsonii (Link) Barkworth, 117, 124, *132*, **133**

Achnatherum robustum (Vasey) Barkworth, 117, **129**, *130*, 135, 142, 177

Achnatherum scribneri (Vasey) Barkworth, 116, 127, 131, *134*, **135**, 137, 142

Achnatherum splendens (Trin.) Nevski, 114, 116, **117**, *119*

Achnatherum stillmanii (Bol.) Barkworth, 117, **118**, *119*

Achnatherum swallenii (C.L. Hitchc. & Spellenb.) Barkworth, 116, *136*, **137**

Achnatherum thurberianum (Piper) Barkworth, 115, **125**, *126*, 142

Achnatherum wallowaense J.R. Maze & K.A. Robson, 116, **138**, *138*

Achnatherum webberi (Thurb.) Barkworth, 116, *136*, **137**

×Achnella Barkworth, 110, 116, **169**

×Achnella caduca (Beal) Barkworth, 142, **169**, *170*, 177

Aegilops L., 238, 240, **261**, 268
Aegilops crassa Boiss., 262, **263**, *264*
Aegilops cylindrica Host, 262, **263**, *266*, 277
Aegilops geniculata Roth, 262, **265**, *267*
Aegilops neglecta Req. *ex* Bertol., 262, **265**, *267*
Aegilops speltoides Tausch, 268
Aegilops tauschii Coss., **262**, *264*, 268
Aegilops triuncialis L., 262, **265**, *266*
    Aegilops triuncialis var. persica (Boiss.) Eig, 265
    Aegilops triuncialis L. var. triuncialis, 265, *266*
Aegilops ventricosa Tausch, 262, **263**, *264*
×Aegilotriticum sancti-andreae (Degen) Soó, 263, *266*
×*Agroelymus adamsii* Rousseau, 348
×*Agroelymus dorei* Bowden, 322
×*Agroelymus hodgsonii* Lepage, 340
×**Agropogon** P. Fourn., 379, **668**
×**Agropogon lutosus** (Poir.) P. Fourn., 643, 665, **668**, *669*
AGROPYRE À CHAUMES RUDES, 322
Agropyron Gaertn., 238, 239, 240, **277**
    Agropyron albicans var. *griffithsii* (Scribn. & J.G. Sm.) Beetle, 334
    Agropyron caninum (L.) P. Beauv., 336, 338
Agropyron cristatum (L.) Gaertn., 278, **279**, *280*
    Agropyron cristatum subsp. *pectinatum* (M.-Bieb.) Tzvelev, 278, 279
Agropyron dasystachyum (Hook.) Scribn., 344
Agropyron desertorum (Fischer *ex* Link) Schult., 278, 279
Agropyron fragile (Roth) P. Candargy, 278, **279**, *280*
Agropyron griffithsii Scribn. & J.G. Sm., 334
Agropyron latiglume (Scribn. & J.G. Sm.) Rydb., 324
Agropyron pectiniforme Roem. & Schult., 278
Agropyron pungens (Pers.) Roem. & Schult., 334
Agropyron sericeum Hitchc., 340
Agropyron sibiricum (Willd.) P. Beauv., 278
Agropyron smithii Rydb., 344, 353
Agropyron trachycaulum (Link) Malte, 344, 346
AGROPYRON À CRÊTE, 279
AGROPYRON ACCRÊTÉ, 279
AGROSTIDE BLANCHE, 641
AGROSTIDE D'HIVER, 647
AGROSTIDE DE MERTENS, 644
AGROSTIDE DES CHIENS, 643
AGROSTIDE FINE, 639
AGROSTIDE PERENNANT, 647
AGROSTIDE SCABRE, 646
AGROSTIDE STOLONIFÈRE, 641
AGROSTIDEAE Dumort., 379
Agrostis L., 379, 382, 383, 384, **633**, 662, 668, 693, 696, 697, 707, 788
Agrostis aequivalvis (Trin.) Trin., 634
Agrostis anadyrensis Soczava, 634, 636, **644**, *645*
Agrostis aristiglumis Swallen, 658
Agrostis avenacea J.F. Gmel., 633
Agrostis blasdalei Hitchc., 635, 637, 651, **656**, *657*
Agrostis canina L., 637, 638, 642, **643**, 644
    Agrostis canina var. *mutica* G. Sinclair, 643
Agrostis capillaris L., 635, **639**, *640*, 641, 644
Agrostis castellana Boiss. & Reut., 634, 635, **639**, *640*, 641, 643
Agrostis clavata Trin., 638, **646**, *646*, 647

Agrostis densiflora Vasey, 636, 637, **651**, *652*, 656
Agrostis elliottiana Schult., 636, 637, 660, **661**
Agrostis exarata Trin., 635, 636, 637, **651**, *652*, 661
Agrostis exigua Thurb., 661
Agrostis geminata Trin., 647
Agrostis gigantea Roth, 635, 639, **641**, *642*, 643
Agrostis hallii Vasey, 638, **653**, *653*
Agrostis hendersonii Hitchc., 635, **658**, *659*, 661
Agrostis hooveri Swallen, 636, **654**, *655*
Agrostis howellii Scribn., 636, **654**, *655*
Agrostis humilis Vasey, 634
Agrostis hyemalis (Walter) Britton, Sterns & Poggenb., 637, **647**, *648*, 661
Agrostis idahoensis Nash, 637, 638, 644, 647, **649**, *650*
Agrostis maritima Lam., 643
Agrostis mertensii Trin., 636, **644**, *645*, 647, 649
Agrostis microphylla Steud., 635, 658, *659*, **661**
    Agrostis microphylla var. *major* Vasey, 661
Agrostis nebulosa Boiss. & Reut., 635, 660, **661**
Agrostis oregonensis Vasey, 637, 638, **649**, *650*
Agrostis pallens Trin., 637, 638, **651**, *652*
Agrostis palustris Huds., 643
Agrostis pauzhetica Prob., 647, 656
Agrostis perennans (Walter) Tuck., 636, 638, **647**, *650*
Agrostis rossiae Vasey, 637, 647, **656**, *658*
Agrostis scabra Willd., 636, 637, 638, 644, **646**, 647, *648*, 649, 656, 661
    Agrostis scabra var. *geminata* (Trin.) Swallen, 647
Agrostis sesquiflora E. Desv., 634
Agrostis stolonifera L., 635, **641**, *642*, 644, 665, 668
Agrostis tandilensis (Kuntze) Parodi, 634
Agrostis tenuis 'Highland', 639
Agrostis tolucensis Kunth, 634, 635, **656**, *657*
Agrostis variabilis Rydb., 635, 637, **654**, *657*, 694
Agrostis vinealis Schreb., 637, 638, 639, **643**, *645*
Aira L., 381, 386, **615**
Aira caryophyllea L., **615**, *617*
    Aira caryophyllea var. **capillaris** (Mert. & W.D.J. Koch) Bluff & Fingerh., 615, **616**, *617*
    Aira caryophyllea L. var. **caryophyllea**, 615, **616**, *617*
    Aira caryophyllea var. **cupaniana** (Guss.) Fiori, 615, **616**, *617*
Aira cespitosa L., 626
Aira elegans Willd. *ex* Roem. & Schult., 616
Aira praecox L., 615, **616**, *617*
ALASKA ALKALI GRASS, 475
ALASKA BROME, 201
ALASKAN ONIONGRASS, 95
ALASKAN WHEATGRASS, 326
ALEUT BROME, 203
ALEUT FESCUE, 414
ALEUTIAN GLYCERIA, 71
ALKALI BLUEGRASS, 586, 588
ALKALI GRASS
    ALASKA, 475
    ANDERSON'S, 473
    ARCTIC, 471
    BRUGGEMANN'S, 469
    EUROPEAN, 473
    GREENLAND, 467

HOWELL'S, 475
LEMMON'S, 469
NUTTALL'S, 475
PACIFIC, 475
PARISH'S, 465
SHEATHED, 477
SMOOTH, 471
TALL, 467
TUNDRA, 469
VAHL'S, 465
WESTERN, 465
WRIGHT'S, 467
ALKALI WILDRYE, 359
Alopecurus L., 380, 383, 384, 671, **780**
Alopecurus aequalis Sobol., 781, **784**, *785*
   Alopecurus aequalis Sobol. var. aequalis, 784, *785*
   Alopecurus aequalis var. *natans* (Wahlenb.) Fernald, 784
   Alopecurus aequalis var. sonomensis P. Rubtzov, 784, *785*
Alopecurus alpinus Sm., 782
Alopecurus arundinaceus Poir., 781, **782**, *783*
Alopecurus borealis Trin., 782
Alopecurus carolinianus Walter, 781, **786**, *787*
Alopecurus creticus Trin., 781, **787**, **788**
Alopecurus geniculatus L., 781, **784**, *785*
Alopecurus ×haussknechtianus Asch. & Graebn., 784
Alopecurus magellanicus Lam., 781, **782**, *783*
Alopecurus myosuroides Huds., 781, **786**, *787*
Alopecurus occidentalis Scribn. & Tweedy, 782
Alopecurus pratensis L., 781, **782**, *783*
Alopecurus rendlei Eig, 781, **785**, **786**
Alopecurus saccatus Vasey, 781, **786**, *787*
Alopecurus stejnegeri Vasey, 782
ALPIGENE BLUEGRASS, 525
ALPINE BENT, 694
ALPINE BLUEGRASS, 518
ALPINE FESCUE, 428
ALPINE FOXTAIL, 782
ALPINE OATGRASS, 701
   DOWNY, 698
ALPINE SWEETGRASS, 760
ALPINE TIMOTHY, 672
ALPISTE DES CANARIES, 768
ALPISTE ROSEAU, 770
ALTAI FESCUE, 407
ALTAI GRASS, 619
ALTAI WILDRYE, 360
Amelichloa Arriaga & Barkworth, 111, 115, 178, **181**
Amelichloa brachychaeta (Godr.) Arriaga & Barkworth, **182**, *183*
Amelichloa caudata (Trin.) Arriaga & Barkworth, 181, **182**, *183*
Amelichloa clandestina (Hack.) Arriaga & Barkworth, 181, 182, *183*, **184**
AMERICAN BEACHGRASS, 777
AMERICAN BEAKGRAIN, 65
AMERICAN DUNEGRASS, 356
AMERICAN GLYCERIA, 71
AMERICAN MANNAGRASS, 71
AMERICAN SLOUGHGRASS, 486
Ammophila Host, 354, 381, 384, *552*, **776**

Ammophila arenaria (L.) Link, 724, **777**, *779*
   Ammophila arenaria (L.) Link subsp. **arenaria**, **777**, *779*
   Ammophila arenaria subsp. **arundinacea** H. Lindb., **777**, *779*
Ammophila breviligulata Fernald, 726, **777**, *779*
   Ammophila breviligulata Fernald subsp. **breviligulata**, **777**, *779*
   Ammophila breviligulata subsp. **champlainensis** (F. Seym.) P.J. Walker, C.A. Paris & Barrington *ex* Barkworth, **777**, *779*
AMOPHILE À LIGULE COURTE, 777
AMOUR DU VENT, 614
AMOURETTE COMMUNE, 614
Ampelodesmos Link, 110, **112**, 388
Ampelodesmos mauritanicus (Poir.) T. Durand & Schinz, **112**, *113*
AMPELOSDESMEAE (Conert) Tutin, 110
Amphicarpum Kunth, 792
Amphicarpum amphicarpon (Pursh) Nash , 792
Amphibromus Nees, 386, **703**
Amphibromus neesii Steud., 703, **704**, *705*
Amphibromus nervosus (Hook. f.) Baill., 703, **704**, *705*
Amphibromus scabrivalvis (Trin.) Swallen, 703, **704**, *705*
   Amphibromus scabrivalvis (Trin.) Swallen var. scabrivalvis, 704
AMPHIBROMUS
   ROUGH, 704
ANADYR BENT, 644
ANARTHRIACEAE D.F. Cutler & Airy Shaw, 6
ANDEAN TUSSOCKGRASS, 174
ANDERSON'S ALKALI GRASS, 473
ANDROPOGONEAE Dumort., 7
Anguillina agrostis, 634
ANIMATED OATS, 739
ANNUAL BLUEGRASS, 519
ANNUAL CANARYGRASS, 768
ANNUAL HAIRGRASS, 631
ANNUAL RYEGRASS, 456
ANNUAL SEMAPHOREGRASS, 107
ANNUAL VELDTGRASS, 36
ANNUAL WHEATGRASS, 254
ANOMOCHLOÖIDEAE Potztal, 6
Anthoxanthum L., 381, 383, **758**
Anthoxanthum arcticum Veldkamp, 759, **760**, *763*
Anthoxanthum aristatum Boiss., 759, *761*
   Anthoxanthum aristatum Boiss. subsp. **aristatum**, 759
   Anthoxanthum aristatum subsp. **macranthum** Valdés, 759
Anthoxanthum gracile Biv., 759
Anthoxanthum hirtum (Schrank) Y. Schouten & Veldkamp, 759, **762**, *763*, *764*
Anthoxanthum monticola (Bigelow) Veldkamp, 759, **760**, *761*
   Anthoxanthum monticola subsp. **alpinum** (Sw. *ex* Willd.) Soreng, **760**, *761*
   Anthoxanthum monticola (Bigelow) Veldkamp subsp. **monticola**, **760**, *761*
Anthoxanthum nitens (Weber) Y. Schouten & Veldkamp, 759, **762**, *763*, *764*
   Anthoxanthum nitens subsp. **balticum** (G. Weim.) G.C. Tucker, 762

Anthoxanthum nitens (Weber) Y. Schouten & Veldkamp subsp. nitens, 762

Anthoxanthum occidentale (Buckley) Veldkamp, 759, **762**, *763*, *764*

Anthoxanthum odoratum L., **759**, *761*

Anthoxanthum odoratum subsp. *alpinum* (Á. Löve & D. Löve) Hultén, 760

Apera Adans., 382, 384, **788**

Apera interrupta (L.) P. Beauv., 788, **789**, *789*

Apera spica-venti (L.) P. Beauv., 788, **789**, *789*

Arctagrostis Griseb., 384, **676**

Arctagrostis latifolia (R. Br.) Griseb., *677*, **678**

Arctagrostis latifolia subsp. **arundinacea** (Trin.) Tzvelev, *677*, **678**

Arctagrostis latifolia (R. Br.) Griseb. subsp. **latifolia**, *677*, **678**

ARCTAGROSTIS À LARGES FEUILLES, 678

ARCTIC ALKALI GRASS, 471

ARCTIC BENT, 693

ARCTIC BLUEGRASS, 529

ARCTIC BROME, 207

ARCTIC RED FESCUE, 414

ARCTIC SWEETGRASS, 760

ARCTIC WHEATGRASS, 324

×Arctodupontia Tzvelev, 379, **604**

×Arctodupontia scleroclada (Rupr.) Tzvelev, **605**, *606*, 607

Arctophila (Rupr.) Andersson, 379, 385, 387, 598, 604, **605**

Arctophila fulva (Trin.) Andersson, 604, 605, *606*, **607**

ARCTOPHILE FAUVE, 607

Argillochloa W.A. Weber, 443

ARISTIDEAE C.E. Hubb., 7

ARISTIDOIDEAE Caro, 6, 7

ARIZONA BARLEY, 248

ARIZONA BROME, 201

ARIZONA FESCUE, 438

ARIZONA WHEATGRASS, 329

ARNOW'S RICEGRASS, 141

Arrhenatherum P. Beauv., 383, **740**

Arrhenatherum elatius (L.) P. Beauv. *ex* J. Presl & C. Presl, *741*, **742**

Arrhenatherum elatius var. *biaristatum* (Peterm.) Peterm., 742

Arrhenatherum elatius subsp. **bulbosum** (Willd.) Schübl. & G. Martens, *741*, **742**

Arrhenatherum elatius P. Beauv. *ex* J. Presl & C. Presl subsp. **elatius**, *741*, **742**

ARROW BAMBOO
JAPANESE, 29

Arundinaria Michx., 15, 16, **17**

Arundinaria appalachiana Triplett, Weakley & L.G. Clark, **18**, *20*

Arundinaria gigantea (Walter) Muhl., 17, **18**, *19*

Arundinaria tecta (Walter) Muhl., **18**, *20*

ARUNDINEAE Dumort., 9, 10

ARUNDINOIDEAE Burmeist., 6, 9, 10

Arundo canadensis Michx., 716, 717

Arundo cinnoides Muhl., 716

ASIAN BEARDGRASS, 663

ASIAN WATERGRASS, 46

ASIAN WILDRICE, 50

Aspergillus, 758

ATLANTIC MANNAGRASS, 75

AUSTRALIAN BROME, 235

AUSTRALIAN FEATHERGRASS, 185

AUSTRALIAN PLUMEGRASS, 185

Austrostipa S.W.L. Jacobs & J. Everett, 111, **184**

Austrostipa elegantissima (Labill.) S.W.L. Jacobs & J. Everett, **185**, *186*

Austrostipa ramosissima (Trin.) S.W.L. Jacobs & J. Everett, **185**, *186*

AUTUMN BENT, 647

AUTUMN BLUEGRASS, 514

AUTUMN MOORGRASS, 679

Avena L., 4, 259, 383, 385, **734**

Avena barbata Pott *ex* Link, **735**, *736*

Avena fatua L., **735**, *736*, 737

Avena hybrida Peterm., **735**, *736*, 737

Avena occidentalis Durieu, 735, **737**, *738*

Avena sativa L., 735, **737**, *738*

Avena sterilis L., 735, *738*, **739**

Avena strigosa Schreb., 735

AVENEAE Dumort., 57, 379

Avenula (Dumort.) Dumort., 380, 385, **698**, 701

Avenula hookeri (Scribn.) Holub, **698**, *699*

Avenula pubescens (Huds.) Dumort., **698**, *699*

Avenula pubescens subsp. **laevigata** (Schur) Holub, 699

Avenula pubescens (Huds.) Dumort. subsp. **pubescens**, 699

AVOINE, 737

AVOINE CULTIVÉE, 737

AVOINE JAUNÂTRE, 748

AVOINE STÉRILE, 739

AWNED MELIC, 95

AWNLESS WILDRYE, 300

BAFFIN ISLAND FESCUE, 432

BAKER'S WHEATGRASS, 330

BAMBOO
COMMON, 22
FISHPOLE, 27
GIANT THORNY, 22
GIANT TIMBER, 27
GOLDEN, 27
HEDGE, 22
JAPANESE ARROW, 29
OLDHAM'S, 25

BAMBOOGRASS
STOUT, 185

Bambusa Schreb., 17, **21**

Bambusa arundinacea (Retz.) Willd., 22

Bambusa bambos (L.) Voss, 21, **22**, 23

Bambusa glaucescens (Willd.) Sieb. *ex* Munro, 22

Bambusa multiplex (Lour.) Raeusch. *ex* Schult. & Schult. f., 21, **22**, 23

Bambusa oldhamii Munro, 21, *24*, **25**

Bambusa tuldoides Munro, 22

Bambusa vulgaris Schrad. *ex* J.C. Wendl., 21, **22**

BAMBUSEAE Nees, 7, 14, **15**

BAMBUSOIDEAE Luerss. 6, 7, **14**, 59, 64, 110

BARBED GOATGRASS, 265

BARLEY, 252
  ARIZONA, 248
  BOBTAIL, 242
  BULBOUS, 252
  CALIFORNIA, 245
  FALSE-RYE, 248
  FOXTAIL, 245
  GENICULATE, 248
  INTERMEDIATE, 245
  LITTLE, 243
  LOW, 243
  MEADOW, 243, 248
  MEDITERRANEAN, 248
  MOUSE, 250
  NORTHERN, 243
  SEA, 248
  SMOOTH, 250
  SQUIRRELTAIL, 245
  WALL, 250
  WAY, 250
BARREN BROME, 224
BARTLEY'S REEDGRASS, 723
BAYOU WEDGEGRASS, 624
BEACHGRASS
  AMERICAN, 777
  CHAMPLAIN, 777
  EUROPEAN, 777
BEAKGRAIN
  AMERICAN, 65
  OBOVATE, 65
BEARDED FESCUE, 402
BEARDED WHEATGRASS, 322
BEARDGRASS
  ASIAN, 663
  CHILEAN, 665
  DITCH, 663
  MEDITERRANEAN, 665
  PERENNIAL, 668
  SHORTHAIRED, 668
  SOUTHERN, 663
  WATER, 663
BEARDLESS WILDRYE, 360
BEAUTIFUL FEATHERGRASS, 155
BEAUTIFUL GLYCERIA, 77
Beckmannia Host, 382, **484**, 598
Beckmannia syzigachne (Steud.) Fernald, *485*, **486**
BECKMANNIE À ÉCAILLES UNIES, 486
BELLY-SHAPED HARDGRASS, 263
BENT
  ALPINE, 694
  ANADYR, 644
  ARCTIC, 693
  AUTUMN, 647
  BLASDALE'S, 656
  BROWN, 643
  CALIFORNIA, 651
  CLAVATE, 646
  COLONIAL, 639
  CREEPING, 641
  DUNE, 651

  ELLIOTT'S, 661
  HALL'S, 653
  HENDERSON'S, 658
  HIGHLAND, 639
  HOOVER'S, 654
  HOWELL'S, 654
  MOUNTAIN, 654
  NORTHERN, 644
  PACIFIC, 696
  RHODE ISLAND, 639
  ROSS', 656
  ROUGH, 646
  SMALL-LEAF, 661
  SPIKE, 651
  UPLAND, 647
  VELVET, 643
  WINTER, 647
BENTGRASS
  PICKERING'S REED, 724
BEP clade, 6
BERINGIAN HAIRGRASS, 626
BIG BEND BLUEGRASS, 540
BIG BLUEGRASS, 586, 588
BIG QUAKINGGRASS, 614
BIG SQUIRRELTAIL, 318
BIGELOW'S BLUEGRASS, 536
BITTERROOT TRISETUM, 746
BLACKGRASS, 786
BLACKSEED SPEARGRASS, 164
BLASDALE'S BENT, 656
BLÉ COMMUN, 277
BLÉ CULTIVÉ, 277
BLÉ DUR, 272
BLÉ POULARD, 274
BLUE BUNCHGRASS, 438
BLUE FESCUE, 422
BLUE MOORGRASS, 681
BLUE OATGRASS, 701
BLUE WILDRYE, 306
BLUEBUNCH FESCUE, 438
BLUEBUNCH WHEATGRASS, 281
BLUEGRASS
  ALKALI, 586, 588
  ALPIGENE, 525
  ALPINE, 518
  ANNUAL, 519
  ARCTIC, 529
  AUTUMN, 514
  BIG, 586, 588
  BIG BEND, 540
  BIGELOW'S, 536
  BOLANDER'S, 533
  BULBOUS, 516
  CANADA, 579
  CANBY, 588
  CHAIX'S, 540
  CHAMBERS', 548
  CHAPMAN'S, 534
  COASTAL, 552
  COTTONBALL, 527

CUSICK'S, 559
DELICATE, 588
DIABLO, 552
DOUGLAS', 551
DUNE, 551
DWARF, 582
EARLY, 545
EASTERN BOG, 572
EMINENT, 598
FEW-FLOWER, 538
FOWL, 574
GLAUCOUS, 578
GROVE, 512
HARTZ'S, 589
HOWELL'S, 534
INTERIOR, 576
KECK'S, 584
KELLOGG'S, 514
KENTUCKY, 522
LABRADOR, 602
LARGE-GLUME, 527
LAX, 570
LAX-FLOWER, 538
LEIBERG'S, 563
LETTERMAN'S, 580
MARSH'S, 582
MOUNT WASHINGTON, 572
NAPA, 594
NARROW-FLOWER, 592
NEVADA, 586
NEW MEXICAN, 536
NODDING, 538
OLDPASTURE, 510
PACIFIC, 588
PATTERSON'S, 582
PEKULNEI, 578
PINE, 588
PIPER'S, 554
PLAINS, 599
PORSILD'S, 563
PRINGLE'S, 564
RHIZOME, 546
ROUGH, 568
SAN BERNARDINO, 554
SAND, 592
SANDBERG, 588
SEA-BLUFF, 594
SECUND, 586
SHORT-FLOWERED, 580
SIERRA, 548
SKYLINE, 560
STEBBINS', 564
SUKSDORF'S, 584
SUPINE, 521
TEXAS, 566
TIMBERLINE, 578
TRACY'S, 543
VEINY, 545
WASATCH, 543
WEAK, 519

WEEPING, 512
WENATCHEE, 589
WESTERN BOG, 573
WHEELER'S, 546
WOLF'S, 514
WOODLAND, 512, 574
BLUE-GREEN MOORGRASS, 681
BLUEJOINT, 726
BLUE OATGRASS, 701
BOBTAIL BARLEY, 242
BOG BLUEGRASS
EASTERN, 572
WESTERN, 573
BOLANDER'S BLUEGRASS, 533
BOLANDER'S REEDGRASS, 719
BOLANDER'S WOODREED, 774
BOREAL FOXTAIL, 782
BOREAL GLYCERIA, 81
BOREAL MANNAGRASS, 81
BOREAL WILDRYE, 366
BORRER'S SALTMARSH GRASS, 463
BOTTLEBRUSH
CALIFORNIA, 369
BOTTLEBRUSH GRASS, 316
Bouteloua curtipendula (Michx.) Torr., 98
Brachiaria (Trin.) Griseb., 793
Brachiaria eruciformis (Sm.) Griseb. 793
BRACHYELYTREAE Ohwi, 8, 58, **59**
Brachyelytrum P. Beauv., **59**
Brachyelytrum aristosum (Michx.) P. Beauv. *ex* Branner & Coville, 60, *61*
Brachyelytrum erectum (Schreb.) P. Beauv., 60, *61*, 61
Brachyelytrum erectum var. *glabratum* (Vasey) T. Koyama & Kawano, 60, 61
BRACHYELYTRUM DRESSÉ, 60
BRACHYPODIEAE Harz, 10, 58, **187**
Brachypodium P. Beauv., **187**, 193
Brachypodium cespitosum (Host) Roem. & Schult., 192
Brachypodium distachyon (L.) P. Beauv., **188**, *189*
Brachypodium phoenicoides (L.) Roem. & Schult., 188, **190**, *191*
Brachypodium pinnatum (L.) P. Beauv., 188, **190**, *191*
Brachypodium rupestre (Host) Roem. & Schult., 188, *191*, **192**
Brachypodium sylvaticum (Huds.) P. Beauv., 188, *189*, **190**
Brachypodium sylvaticum subsp. **glaucovirens** (St.-Yves) Murb., 190
Brachypodium sylvaticum (Huds.) P. Beauv. subsp. **sylvaticum**, 190
BRAZILIAN WATERGRASS, 55
BREAD WHEAT, 277
BRISTLY DOGTAIL, 687
BRISTLY RICEGRASS, 166
BRITISH TIMOTHY, 672
Briza L., 67, 385, 386, **612**
Briza maxima L., 613, *613*, 614
Briza media L., 613, *613*, 614
Briza minor L., 613, *613*, 614
BROME
ALASKA, 201

ALEUT, 203
ARCTIC, 207
ARIZONA, 201
AUSTRALIAN, 235
BARREN, 224
BROOM, 235
CALIFORNIA, 205
CANADA, 220
CHINOOK, 209
COLORADO, 205
COMMON, 216
COMPACT, 226
FIELD, 228
FLANGED, 209
FRINGED, 222
GREAT, 224
GREAT-BASIN, 205
HAIRY, 220
HALL'S, 211
HUNGARIAN, 206
JAPANESE, 235
KALM'S, 209
LANCEOLATE, 235
MARITIME, 203
MEADOW, 206, 218, 230
MEXICAN, 213
MOUNTAIN, 205
NODDING, 213
ORCUTT'S, 211
PACIFIC, 218
RATTLESNAKE, 228
RED, 226
RICHARDSON'S, 222
SCALY, 230
SHARPGLUME, 218
SITKA, 201
SMOOTH, 206, 233
SQUARROSE, 237
SUKSDORF'S, 218
TALL, 213
TEXAS, 216
THREE-AWNED, 233
UPRIGHT, 218
VIRGINIA, 216
WEEPING, 213
WOODLAND, 211
WOOLY, 220
BROME À GRAPPES, 233
BROME À LARGES GLUMES, 209
BROME CILIÉ, 222
BROME DE KALM, 209
BROME DES SEIGLES, 230
BROME DRESSÉ, 218
BROME FESCUE, 452
BROME INERME, 206
BROME MOU, 232
BROMEAE Dumort., 8, 58, 187, **192**
Bromidium Nees & Meyen, 381, 383, **696**
Bromidium tandilense (Kuntze) Rúgolo, 634, **697**, *697*
Bromus L., 192, **193**

Bromus sect. Bromopsis Dumort., 194, 195, **206**
Bromus L. sect. Bromus, 194, 197, **228**
Bromus sect. Ceratochloa (P. Beauv.) Griseb., 194, **199**
Bromus sect. Genea Dumort., 194, 197, **224**
Bromus sect. Neobromus (Shear) Hitchc., 194, 197, **222**
Bromus sect. *Pnigma* Dumort., 206
Bromus aleutensis Trin. *ex* Griseb., 195, 199, 201, *202*, 203
Bromus alopecuros Poir., 233
Bromus anomalus Rupr. *ex* E. Fourn., 196, **213**, *215*, 216
Bromus arenarius Labill., 198, **235**, *236*
Bromus arizonicus (Shear) Stebbins, 195, 199, **201**, *202*
Bromus arvensis L., 198, **228**, *229*
Bromus berteroanus Colla, 197, 222, *223*, **224**
Bromus biebersteinii Roem. & Schult., 206
Bromus briziformis Fisch. & C.A. Mey., 197, *227*, **228**
Bromus carinatus Hook. & Arn., 195, 199, 201, **203**, *204*
　　Bromus carinatus Hook. & Arn. var. **carinatus**, 201, *204*, **205**
　　Bromus carinatus var. **marginatus** (Nees) Barkworth & Anderton, 203, *204*, **205**
Bromus caroli-henrici Greuter, 198, 232, **233**, *234*
Bromus catharticus Vahl, 194, 195, **199**, *200*, 201
　　Bromus catharticus Vahl var. **catharticus**, **199**, *200*
　　Bromus catharticus var. **elatus** (E. Desv.) Planchuelo, 199, *200*, **201**
Bromus ciliatus L., 197, **222**, *223*
Bromus commutatus Schrad., 198, **229**, *230*
Bromus danthoniae Trin. *ex* C.A. Mey., 198, **233**, *234*
Bromus diandrus Roth, 197, **224**, *225*
Bromus erectus Huds., 196, **218**, *219*
Bromus frondosus (Shear) Wooton & Standl., 196, **213**, *214*
Bromus grandis (Shear) Hitchc., 195, 196, 197, **213**, *214*
Bromus hallii (Hitchc.) Saarela & P.M. Peterson, 196, **211**, *212*
Bromus hordeaceus L., 198, *231*, **232**
　　Bromus hordeaceus L. subsp. **hordeaceus**, *231*, **232**
　　Bromus hordeaceus subsp. **molliformis** (J. Lloyd) Maire & Weiller, *231*, **232**
　　Bromus hordeaceus subsp. **pseudothominei** (P.M. Sm.) H. Scholz, 230, *231*, **232**, 233
　　Bromus hordeaceus subsp. **thominei** (Hardouin) Braun-Blanq., *231*, **232**
Bromus inermis Leyss., 195, **206**, *208*
Bromus japonicus Thunb., 199, **235**, *237*
Bromus kalmii A. Gray, 195, **209**, *210*
Bromus laevipes Shear, 195, **209**, *210*
Bromus lanatipes (Shear) Rydb., 197, **220**, *221*
Bromus lanceolatus Roth, 198, **235**, *236*
Bromus latiglumis (Scribn. *ex* Shear) Hitchc., 195, **209**, *210*
　　Bromus latiglumis forma *incanus* (Shear) Fernald, 209
Bromus lepidus Holmb., 198, 230, **231**, 232
Bromus luzonensis J. Presl, 205
Bromus madritensis L., 197, **226**, *227*
Bromus maritimus (Piper) Hitchc., 195, 199, *202*, **203**
Bromus mucroglumis Wagnon, 196, **218**, *219*
Bromus nottowayanus Fernald, 196, *215*, **216**
Bromus orcuttianus Vasey, 196, **211**, *212*
Bromus pacificus Shear, 197, *217*, **218**
Bromus polyanthus Scribn., 195, 199, *204*, **205**
　　Bromus polyanthus var. *paniculatus* Shear, 205

Bromus polyanthus Scribn. var. *polyanthus*, 205
Bromus porteri (J.M. Coult.) Nash, 196, **213**, *215*
Bromus pseudolaevipes Wagnon, 195, **211**, *212*
Bromus pubescens Muhl. *ex* Willd., 197, **220**, *221*
Bromus pumpellianus Scribn., 195, 206, **207**, *208*
  Bromus pumpellianus subsp. **dicksonii** W.W. Mitch. & Wilton, **207**, *208*
  Bromus pumpellianus Scribn. subsp. **pumpellianus**, 206, **207**, *208*
Bromus racemosus L., 198, **233**, *234*
Bromus ramosus Huds., 197, **220**, *221*
Bromus richardsonii Link, 197, **222**, *223*
Bromus rigidus Roth, 224
Bromus riparius Rehmann, 195, **206**, 207, *208*
Bromus rubens L., 197, **226**, *227*
Bromus scoparius L., 198, **235**, *236*
Bromus secalinus L., 198, 229, **230**
  Bromus secalinus var. *velutinus* (Schrad.) W.D.J. Koch, 230
Bromus sitchensis Trin., 195, 199, **200**, **201**, 203
Bromus squarrosus L., 199, **237**, *237*
Bromus sterilis L., 197, **224**, *225*
Bromus suksdorfii Vasey, 196, **218**, *219*
Bromus tectorum L., 4, 197, **225**, **226**, 450
  Bromus tectorum forma *nudus* (Klett & Richt.) H. St. John, 226
Bromus texensis (Shear) Hitchc., 196, **216**, *217*
Bromus vulgaris (Hook.) Shear, 196, **216**, *217*
BROOKGRASS, 610
BROOM BROME, 235
BROWN BENT, 643
BROWNBEARD RICE, 38
BROWNTOP, 639
  DRYLAND, 639
BRUGGEMANN'S ALKALI GRASS, 469
'BUDDHA'S-BELLY', 22
BULBOUS BARLEY, 252
BULBOUS BLUEGRASS, 516
BULBOUS CANARYGRASS, 767, 768
BUNCH CUTGRASS, 42
BUNCHGRASS
  BLUE, 438
  MOUNTAIN, 440
BUSHGRASS, 710

CAIN'S REEDGRASS, 719
CALAMAGROSTIDE COMMUNE, 710
CALAMAGROSTIDE CONTRACTÉE, 730
CALAMAGROSTIDE DE LAPPONIE, 729
CALAMAGROSTIDE DE PICKERING, 724
CALAMAGROSTIDE DRESSÉE, 730
CALAMAGROSTIDE DU CANADA, 726
CALAMAGROSTIDE ÉPIGÉIOS, 710
CALAMAGROSTIDE FAUSSE-DESCHAMPSIE, 719
CALAMAGROSTIDE POURPRE, 710
Calamagrostis Adans., 384, 634, 693, **706**, 778
Calamagrostis ×acutiflora (Schrad.) DC. 'Karl Foerster', 706, 708, 710, **721**, 722
Calamagrostis arctica Vasey, 712
Calamagrostis arundinacea (L.) Roth, 710, 721

Calamagrostis bolanderi Thurb., 706, 708, 717, **719**, *720*
Calamagrostis breweri [*sensu* Greene], 707
Calamagrostis breweri Thurb., 706, 707, 708, **717**, *718*, 721
Calamagrostis cainii Hitchc., 706, 708, **719**, *720*
Calamagrostis canadensis (Michx.) P. Beauv., 706, 709, 723, **726**, *727*, 777
  Calamagrostis canadensis (Michx.) P. Beauv. var. **canadensis**, *727*, **728**
  Calamagrostis canadensis var. **langsdorffii** (Link) Inman, *727*, **728**
  Calamagrostis canadensis var. **macouniana** (Vasey) Stebbins, *727*, **728**
Calamagrostis cinnoides (Muhl.) W.P.C. Barton, 708, *715*, **716**
Calamagrostis coarctata Eaton, 717
Calamagrostis crassiglumis Thurb., 730
Calamagrostis deschampsioides Trin., 708, **719**, *720*
Calamagrostis epigejos (L.) Roth, 706, 707, **710**, *711*, 721
  Calamagrostis epigejos var. *georgica* (K. Koch) Griseb., 710
Calamagrostis fernaldii Louis-Marie, 730
Calamagrostis foliosa Kearney, 706, 708, *713*, **714**
Calamagrostis howellii Vasey, 706, 708, *711*, **712**
Calamagrostis koelerioides Vasey, 708, **721**, 722, 723
Calamagrostis lacustris (Kearney) Nash, 730
Calamagrostis lapponica (Wahlenb.) Hartm., 709, **729**, 730, *731*
Calamagrostis lepageana Louis-Marie, 712
Calamagrostis montanensis (Scribn.) Vasey, 709, 712, **724**, 727
Calamagrostis muiriana B.L. Wilson & Sami Gray, 706, 707, 708, **717**, *718*, 721
Calamagrostis nutkaensis (J. Presl) Steud., 709, **724**, *725*, 778
Calamagrostis ophitidis (J.T. Howell) Nygren, 706, 707, **712**, *713*
Calamagrostis perplexa Scribn., 709, 723, **726**, 727, 728
Calamagrostis pickeringii A. Gray, 709, **724**, *725*
Calamagrostis porteri A. Gray, 709, **721**, 722, 724, 726, 728
  Calamagrostis porteri subsp. **insperata** (Swallen) C.W. Greene, 722, **723**
  Calamagrostis porteri A. Gray subsp. **porteri**, 722, **723**
Calamagrostis purpurascens R. Br., 707, **710**, *711*, 716, 726
  Calamagrostis purpurascens subsp. *arctica* (Vasey) Hultén, 712
  Calamagrostis purpurascens var. *laricina* Louis-Marie, 707, 712
Calamagrostis purpurea (Trin.) Trin., 726
Calamagrostis rubescens Buckley, 709, 721, **723**, 725
Calamagrostis scopulorum M.E. Jones, 708, **717**, *718*
Calamagrostis sesquiflora (Trin.) Tzvelev, 707, 708, 712, **714**, *715*
Calamagrostis stricta (Timm) Koeler, 709, **729**, *731*
  Calamagrostis stricta subsp. *borealis* (C. Laest.) Á. Löve & D. Löve, 707, 730
  Calamagrostis stricta var. *borealis* (C. Laest.) Hartm., 730
  Calamagrostis stricta subsp. **inexpansa** (A. Gray) C.W. Greene, **730**, *731*
  Calamagrostis stricta (Timm) Koeler subsp. **stricta**, **730**, *731*

Calamagrostis tacomensis K.L. Marr & Hebda, 706, 707, 708, 712, *715*, **716**

Calamagrostis tweedyi (Scribn.) Scribn., 706, 708, *713*, **714**

Calamagrostis vaseyi Beal, 716

×Calammophila don-hensonii Reznicek & Judz., 726, 777

CALIFORNIA BARLEY, 245

CALIFORNIA BENT, 651

CALIFORNIA BOTTLEBRUSH, 369

CALIFORNIA BROME, 205

CALIFORNIA CANARYGRASS, 770

CALIFORNIA DISSANTHELIUM, 700

CALIFORNIA FESCUE, 410

CALIFORNIA MELIC, 93

   LITTLE, 90

CALIFORNIA NEEDLEGRASS, 121

CALIFORNIA SEMAPHOREGRASS, 105

CALIFORNIA SQUIRRELTAIL, 319

CALIFORNIA SWEETGRASS, 762

CALLUSED FESCUE, 437

Camassia, 563

CANADA BLUEGRASS, 579

CANADA BROME, 220

CANADIAN GLYCERIA, 79

CANADIAN MANNAGRASS, 79

CANADIAN PIPTATHERUM, 146

CANARYGRASS

   ANNUAL, 768

   BULBOUS, 767, 768

   CALIFORNIA, 770

   CAROLINA, 772

   HOODED, 766

   LEMMON'S, 768

   LESSER, 767

   NARROW, 773

   REED, 770

   SHORTSPIKE, 768

   TOOWOOMBA, 768

CANBY BLUEGRASS, 588

CANE

   GIANT, 18

   HILL, 18

   RIVER, 18

   SWITCH, 18

CANYONGRASS, 42

Carex geyeri, 714

Carex lyngbyei, 719

CAROLINA CANARYGRASS, 772

CASCADE REEDGRASS, 714

Cathestecum J. Presl, 792

CAT'S TAIL

   PURPLE-STEM, 672

Catabrosa P. Beauv., 67, 387, **610**

Catabrosa aquatica (L.) P. Beauv., **610**, *611*

CATABROSA AQUATIQUE, 610

CATCHFLY GRASS, 44

CEDAR WHITEGRASS, 42

Celtica F.M. Vázquez & Barkworth, 110, **152**

Celtica gigantea (Link) F.M. Vázquez & Barkworth, *153*, **154**

CENTOTHECEAE Ridl., 9

CENTOTHECOIDEAE Soderstr., 6, 9

CENTROLEPIDACEAE Endl., 6

CERNUOUS NASSELLA, 176

CHAIX'S BLUEGRASS, 540

CHAMBERS' BLUEGRASS, 548

CHAMPLAIN BEACHGRASS, 777

CHAPMAN'S BLUEGRASS, 534

CHEATGRASS, 226

CHEE REEDGRASS, 710

CHESS

   CHILEAN, 224

   DOWNY, 226

   FOXTAIL, 226

   HAIRY, 230

CHEWING'S FESCUE, 415

CHIENDENT, 334

CHIENDENT RAMPANT, 334

CHILEAN BEARDGRASS, 665

CHILEAN CHESS, 224

CHILEAN TUSSOCKGRASS, 177

CHINOOK BROME, 209

CHLORIDOIDEAE Kunth *ex* Beilschm., 6, 8, 9, 10

Chrysopogon zizanioides (L.) Roberty, 118

CHURCH'S WILDRYE, 314

CILIATE MELIC, 100

Cinna L., 383, 384, **773**

Cinna arundinacea L., **774**, *775*

Cinna bolanderi Scribn., **774**, *775*

Cinna latifolia (Trevir. *ex* Göpp.) Griseb., 773, **774**, *775*

Cinna poaeformis (Kunth) Scribn. & Merr., 773

CINNA À LARGES FEUILLES, 774

CINNA ROSEAU, 774

CIRCUMPOLAR REEDGRASS, 719

CLAVATE BENT, 646

CLOUD GRASS, 661

CLUSTER FESCUE, 402

COAST FESCUE, 404

COAST RANGE FESCUE, 406

COASTAL BLUEGRASS, 552

Coleanthus Seidl, 382, **618**

Coleanthus subtilis (Tratt.) Seidl, **618**, *619*

COLONIAL BENT, 639

COLORADO BROME, 205

COLORADO WILDRYE, 366

COMMON BAMBOO, 22

COMMON BROME, 216

COMMON RIVERGRASS, 734

COMMON SALTMARSH GRASS, 463

COMMON SQUIRRELTAIL, 319

COMMON SWAMP WALLABYGRASS, 704

COMMON WESTERN NEEDLEGRASS, 123

COMMON WESTERN WILDRYE, 306

COMMON WHEAT, 277

COMMON WINDGRASS, 789

COMPACT BROME, 226

CONE WHEAT, 274

CONFUSING FESCUE, 452

CONTRACTED RICEGRASS, 141

Cortaderia Stapf, 112

Corynephorus P. Beauv., 385, **742**

Corynephorus canescens (L.) P. Beauv., **743**, *743*

COTTONBALL BLUEGRASS, 527

COUCHGRASS, 334

CREEPING BENT, 641

CREEPING MANNAGRASS, 83

CREEPING MEADOW FOXTAIL, 782

CREEPING VELVETGRASS, 740

CRESTED DOGTAIL, 685

CRESTED HAIRGRASS, 754

CRESTED NEEDLEGRASS, 127

CRESTED WHEATGRASS, 279

CRETAN MEADOW FOXTAIL, 788

CRÉTELLE DES PRÉS, 685

CRINKLE-AWN FESCUE, 406

CRINKLED HAIRGRASS, 631

CULTIVATED OATS, 737

CURLYLEAF NEEDLEGRASS, 135

CURVED SICKLEGRASS, 688

CUSICK'S BLUEGRASS, 559

Cutandia Willk., 387, 610, **611**

Cutandia memphitica (Spreng.) K. Richt., *611*, **612**

CUTGRASS

  BUNCH, 42

  GIANT, 52

  RICE, 44

  SOUTHERN, 44

  WHITE, 44

Cynodon dactylon (L.) Pers., 4

CYNODONTEAE Dumort., 8, 9, 10

CYNOSURE ACCRÊTÉ, 685

Cynosurus L., 379, **685**

Cynosurus cristatus L., **685**, *686*

Cynosurus echinatus L., 685, **686**, 687

DACTYLE PELOTONNÉ, 482

Dactylis L., 382, **482**

Dactylis glomerata L., 379, **482**, *483*

Danthonia DC., 792

Danthonia epilis Scribn., 792

Danthonia sericea Nutt., 792

DANTHONIEAE Zotov, 9

DANTHONIOIDEAE N.P. Barker & H.P. Linder, 6, 9

DARNEL, 459

  FLAX, 456

  PERSIAN, 459

Dasypyrum (Coss. & Durieu) T. Durand, 240, **256**

Dasypyrum breviaristatum (H. Lindb.) Fred., 256

Dasypyrum villosum (L.) P. Candargy, **256**, *257*

DAVY'S SEMAPHOREGRASS, 107

DELICATE BLUEGRASS, 588

DELICATE HAIRGRASS, 616

DENSE-PINE REEDGRASS, 721

Deschampsia P. Beauv., 379, 385, 386, **624**, 691, 744

Deschampsia alpina (L.) Roem. & Schult., 625, *630*, **631**

Deschampsia brevifolia R. Br., 624, 625, 628, **629**, *630*

Deschampsia cespitosa (L.) P. Beauv., 144, 625, **626**, *627*, 629, 631

  Deschampsia cespitosa subsp. **beringensis** (Hultén) W.E. Lawr., **626**, 627, 628

Deschampsia cespitosa (L.) P. Beauv. subsp. **cespitosa**, 626, 627, **628**

Deschampsia cespitosa var. *glauca* (Hartm.) Lindm., 628

Deschampsia cespitosa subsp. **holciformis** (J. Presl) W.E. Lawr., 626, 627, **628**

Deschampsia cespitosa subsp. *parviflora* (Thuill.) K. Richt., 628

Deschampsia cespitosa var. *parviflora* (Thuill.) Coss. & Germ., 628

Deschampsia danthonioides (Trin.) Munro, 625, **631**, *632*

Deschampsia elongata (Hook.) Munro, 625, **631**, *632*

Deschampsia flexuosa (L.) Trin., 625, **631**, *632*

Deschampsia mackenzieana Raup, 625, **628**, *630*

Deschampsia sukatschewii (Popl.) Roshev., 624, 625, 628, **629**, *630*

  Deschampsia sukatschewii subsp. *borealis* (Trautv.) Tzvelev, 629

  Deschampsia sukatschewii subsp. *orientalis* (Hultén) Tzvelev, 629

  Deschampsia sukatschewii (Popl.) Roshev. subsp. *sukatschewii*, 629

DESCHAMPSIE ALPINE, 631

DESCHAMPSIE CESPITEUSE, 628

DESCHAMPSIE FLEXUEUSE, 631

DESCHAMPSIE NAINE, 629

DESCHAMPSIE POURPRE, 692

DESERT FESCUE, 452

DESERT NEEDLEGRASS, 181

Desmazeria Dumort., 380, 381, 387, 388, **681**

Desmazeria rigida (L.) Tutin, **682**, *683*

Desmazeria sicula (Jacq.) Dumort., **682**, *683*

Deyeuxia Clarion *ex* P. Beauv., 706

DIABLO BLUEGRASS, 552

Diarrhena P. Beauv., **64**

Diarrhena americana P. Beauv., **65**, *66*

Diarrhena obovata (Gleason) Brandenburg, **65**, *66*

DIARRHENEAE C.S. Campb., 10, 58, **64**

Dichanthelium (Hitchc. & Chase) Gould, 793

  Dichanthelium acuminatum (Sw.) Gould & C.A. Clark subsp. **acuminatum**, 793

  Dichanthelium acuminatum subsp. **fasciculatum** (Torr.) Freckmann & Lelong, 793

Digitaria Haller, 792

Digitaria filiformis (L.) Koeler, 792

Digitaria fulvescens (J. Presl) Henrard, 792

Digitaria longiflora (Retz.) Pers., 792

Digitaria velutina (Forssk.) P. Beauv., 792

DINKEL, 277

Dissanthelium Trin., 385, **700**

Dissanthelium californicum (Nutt.) Benth., **700**, *700*

Dissanthelium mathewsii (Ball) R.C. Foster & L.B. Sm., 700

DISSANTHELIUM

  CALIFORNIA, 700

DITCH BEARDGRASS, 663

DITCH REEDGRASS, 717

DOGTAIL

  BRISTLY, 687

  CRESTED, 685

DORE'S NEEDLEGRASS, 124

DOUGLAS' BLUEGRASS, 551

Douglasdeweya C. Yen, J.L. Yang & B.R. Baum, 281
DOWNY ALPINE OATGRASS, 698
DOWNY CHESS, 226
DOWNY RYEGRASS, 366
DOWNY WILDRYE, 302
DROOPING WOODREED, 774
DROPAWN, 169
DRYLAND BROWNTOP, 639
DUNE BENT, 651
DUNE BLUEGRASS, 551
DUNE RED FESCUE, 415
DUNEGRASS
   AMERICAN, 356
   EUROPEAN, 356
×Dupoa J. Cay. & Darbysh., 379, 598, **601**, 604
×Dupoa labradorica (Steud.) J. Cay. & Darbysh., **602**, *603*, 604
Dupontia R. Br., 379, 381, 384, 385, 598, **602**, 605
Dupontia fisheri R. Br., 602, *603*, **604**, 607
   Dupontia fisheri R. Br. subsp. *fisheri*, 604
   Dupontia fisheri subsp. *psilosantha* (Rupr.) Hultén, 604
DUPONTIE DE FISHER, 604
DURUM WHEAT, 272
DWARF BLUEGRASS, 582

EARLE'S FESCUE, 420
EARLY BLUEGRASS, 545
EARLY MILLET, 780
EARLY SANDGRASS, 757
EARLY WILDRYE, 295
EASTERN BOG BLUEGRASS, 572
EASTERN RIVERBANK WILDRYE, 302
EASTWOOD FESCUE, 452
ECDEIOCOLEACEAE D.F. Cutler & Airy Shaw, 6
ECDEIOCOLEA F. Muell., 6
Echinochloa P. Beauv., 792
   Echinochloa crus-pavonis var. macera (Wiegand) Gould, 792
EDLUND'S FESCUE, 432
Ehrharta Thunb., 33
Ehrharta calycina Sm., **34**, *35*
Ehrharta erecta Lam., **34**, *35*
Ehrharta longiflora Sm., **35**, 36
EHRHARTEAE Nevski, 10, 32, **33**
EHRHARTOIDEAE Link, 6, 8, 10, **32**
EINKORN, 272
   RED WILD, 270
   WILD, 270
ELLIOTT'S BENT, 661
ELMER'S FESCUE, 404
×Elyhordeum Mansf. *ex* Tsitsin & K.A. Petrova, 239, **283**, 288, 302
×Elyhordeum arcuatum W.W. Mitch. & H.J. Hodgs., **286**
×Elyhordeum californicum (Bowden) Barkworth, 286, *287*
×Elyhordeum dakotense (Bowden) Bowden, 284, *285*
×Elyhordeum iowense R.W. Pohl, 286, *287*
×Elyhordeum jordalii (Melderis) Tzvelev, 284, *285*
×Elyhordeum macounii (Vasey) Barkworth & D.R. Dewey, **284**, *285*, 322, 370

×Elyhordeum montanense (Scribn. *ex* Beal) Bowden, **286**, *287*
×Elyhordeum pilosilemma (W.W. Mitch. & H.J. Hodgs.) Barkworth, **284**, *285*, 326
×Elyhordeum schaackianum (Bowden) Bowden, **284**, *285*
×Elyhordeum stebbinsianum (Bowden) Bowden, **286**, *287*
×Elyleymus B.R. Baum, 239, 288, 308, **343**
×Elyleymus aleuticus (Hultén) B.R. Baum, **346**, *349*
×Elyleymus aristatus (Merr.) Barkworth & D.R. Dewey, **344**, *345*
×Elyleymus colvillensis (Lepage) Barkworth, **344**, *345*
×Elyleymus hirtiflorus (Hitchc.) Barkworth, **344**, *347*
×Elyleymus hultenii (Melderis) Barkworth, **348**, *349*
×Elyleymus jamesensis (Lepage) Barkworth, 322, **348**, *350*
×Elyleymus mossii (Lepage) Barkworth, **346**, *347*
×Elyleymus ontariensis (Lepage) Barkworth, 322, **346**, *347*
×Elyleymus turneri (Lepage) Barkworth & D.R. Dewey, **344**, *345*, 353
×Elyleymus uclueletensis (Bowden) B.R. Baum, **346**, *349*
×Elyleymus ungavensis (Louis-Marie) Barkworth, **348**, *350*
ÉLYME À CHAUMES RUDES, 322
ÉLYME DE VIRGINIE, 298
ÉLYME DE WIEGAND, 305
ÉLYME DES RIVAGES, 302
ÉLYME DES SABLES D'AMERIQUE, 356
ÉLYME DES SABLES D'EUROPE, 356
ÉLYME DU CANADA, 303
ÉLYME LATIGLUME, 324
Elymus L., 239, 240, 278, 281, 283, **288**, 351, 374
Elymus alaskanus (Scribn. & Merr.) Á. Löve, 291, 292, 321, 324, *325*, **326**, 340, 344
   Elymus alaskanus (Scribn. & Merr.) Á. Löve subsp. alaskanus, *325*, **326**, 348
   Elymus alaskanus subsp. hyperarcticus (Polunin) Á. Löve & D. Löve, *325*, **326**
   Elymus alaskanus subsp. *latiglumis* (Scribn. & J.G. Sm.) Á. Löve, 324
Elymus albicans (Scribn. & J.G. Sm.) Á. Löve, 282, 283, 291, 327, **334**, *335*
Elymus arenicolus Scribn. & J.G. Sm., 360, 369
Elymus arizonicus (Scribn. & J.G. Sm.) Gould, 282, 291, **329**, *331*, 343
Elymus bakeri (E.E. Nelson) Á. Löve, 292, **330**, *333*
Elymus canadensis L., 284, 295, 296, 302, **303**, *304*, 306, 314, 316, 318, 321, 340, 346
   Elymus canadensis var. brachystachys (Scribn. & C.R. Ball) Farw., *304*, **305**, 306
   Elymus canadensis L. var. canadensis, 303, *304*, **305**, 316
   Elymus canadensis var. *glaucifolius* (Muhl.) Torr., 305
   Elymus canadensis var. robustus (Scribn. & J.G. Sm.) Mack. & Bush, 296, *304*, **305**
   Elymus canadensis var. *villosus* Bates, 305
Elymus caninus (L.) L., 288, 292, 321, **322**, *323*, 336, 338
Elymus ×cayouetteorum (B. Boivin) Barkworth, 321, **338**, *341*
Elymus churchii J.J.N. Campb., 293, 306, **314**, *315*, 316, 318
Elymus ciliaris (Trin.) Tzvelev, 288, 289, 291, 324, **336**, *339*
Elymus curvatus Piper, 294, 300, *301*

Elymus dahuricus Turcz. *ex* Griseb., 288, 289, 295, 308, 310, *311*
  Elymus dahuricus Turcz. *ex* Griseb. var. **dahuricus**, 310
Elymus diversiglumis Scribn. & C.R. Ball, 293, 306, 314, *315*, **316**, 318
Elymus ×ebingeri G.C. Tucker, *342*, **343**
Elymus elymoides (Raf.) Swezey, 283, 286, 292, 308, **318**, *320*, 322, 330, 332, 340, 343, 344
  Elymus elymoides subsp. **brevifolius** (J.G. Sm.) Barkworth, **319**, *320*, 343
  Elymus elymoides subsp. **californicus** (J.G. Sm.) Barkworth, **319**, *320*
  Elymus elymoides (Raf.) Swezey subsp. **elymoides**, **319**, *320*
  Elymus elymoides subsp. **hordeoides** (Suksd.) Barkworth, **319**, *320*
Elymus glaberrimus, 281
Elymus glabriflorus (Vasey *ex* L.H. Dewey) Scribn. & C.R. Ball, 294, **296**, *297*, 298, 300, 302, 303, 305, 318
  Elymus glabriflorus var. *australis* (Scribn. & C.R. Ball) J.J.N. Campb., 296
Elymus glaucus Buckley, 286, 289, 291, 292, 295, 303, **306**, *309*, 310, 318, 319, 321, 329, 330, 340, 343, 346
  Elymus glaucus Buckley subsp. **glaucus**, 308, *309*
  Elymus glaucus subsp. *jepsonii* (Burtt Davy) Gould, 308
  Elymus glaucus subsp. **mackenziei** (Bush) J.J.N. Campb., **308**, *309*
  Elymus glaucus subsp. **virescens** (Piper) Gould, 308, *309*
Elymus ×hansenii Scribn., 319, **340**, *342*
Elymus hirsutus J. Presl, 284, 295, 308, **310**, *311*, 348
Elymus hoffmannii K.B. Jensen & Asay, 288, 289, 291, 334, **336**, *337*
Elymus hystrix L., 293, 296, 303, 306, 314, **316**, *317*, 343, 369
  Elymus hystrix var. *bigelovianus* (Fernald) Bowden, 316
Elymus innovatus Beal, 353
Elymus interruptus Buckley, 293, 294, **306**, *307*, 312, 314
Elymus lanceolatus (Scribn. & J.G. Sm.) Gould, 283, 291, 292, 322, **327**, *328*, 334, 340, 344, 346, 351
  Elymus lanceolatus (Scribn. & J.G. Sm.) Gould subsp. **lanceolatus**, **327**, *328*
  Elymus lanceolatus subsp. **psammophilus** (J.M. Gillett & H. Senn) Á. Löve, **327**, *328*, 340
  Elymus lanceolatus subsp. **riparius** (Scribn. & J.G. Sm.) Barkworth, 327, *328*, **329**
Elymus macgregorii R.E. Brooks & J.J.N. Campb., 294, **295**, *297*, 300
Elymus macrourus (Turcz. *ex* Steud.) Tzvelev, 284, 292, 321, **324**, *325*, 340
Elymus multisetus (J.G. Sm.) Burtt Davy, 283, 286, 292, 308, *317*, **318**, 319, 322, 340, 343
Elymus ×palmerensis (Lepage) Barkworth & D.R. Dewey, 321, 326, **340**, *341*
Elymus ×pinalenoensis (Pyrah) Barkworth & D.R. Dewey, 319, *342*, **343**
Elymus pringlei Scribn. & Merr., 293, **312**, *313*
Elymus ×pseudorepens (Scribn. & J.G. Sm.) Barkworth & D.R. Dewey, 321, **340**, *341*
Elymus repens (L.) Gould, 289, 291, **334**, 336, *337*
Elymus riparius Wiegand, 294, 300, *301*, **302**, 303

Elymus ×saundersii Vasey, 283, 319, 322, **340**, *342*
Elymus scribneri (Vasey) M.E. Jones, 291, 324, **330**, 332, *333*
Elymus semicostatus (Nees *ex* Steud.) Melderis, 288, 289, 291, **338**, *339*
Elymus sibiricus L., 286, 292, 295, **310**, *311*, 321, 340
Elymus sierrae Gould, 291, 319, **332**, *335*
Elymus stebbinsii Gould, 291, 308, 321, **329**, *331*
  Elymus stebbinsii subsp. **septentrionalis** Barkworth, **329**, *331*
  Elymus stebbinsii Gould subsp. **stebbinsii**, **329**, *331*
  Elymus striatus var. *ballii* Pammel, 302
Elymus submuticus (Hook.) Smyth, 286
Elymus svensonii G.L. Church, 293, 306, *313*, **314**, 316, 318
Elymus texensis J.J.N. Campb., 293, **312**, *313*
Elymus trachycaulus (Link) Gould, 284, 292, 303, 308, 319, **321**, *323*, 324, 326, 327, 329, 330, 340, 344, 346, 348
  Elymus trachycaulus subsp. **subsecundus** (Link) Á. Löve & D. Löve, 284, 321, **322**, *323*
  Elymus trachycaulus (Link) Gould subsp. **trachycaulus**, 284, **322**, *323*, 348
  Elymus trachycaulus subsp. **virescens** (Lange) Á. Löve & D. Löve, **322**, *323*
Elymus tsukushiensis Honda, 288, 289, 292, 310, 324, **336**, *337*
Elymus villosus Muhl. *ex* Willd., 286, 294, 296, *301*, **302**
  Elymus villosus var. *arkansanas* (Scribn. & C.R. Ball) J.J.N. Campb., 302
Elymus violaceus (Hornem.) Feilberg, 292, 321, 322, **324**, *325*, 330, 332, 348
Elymus virginicus L., 294, 296, **298**, *299*, 302, 303, 305, 314, 343
  Elymus virginicus var. **halophilus** (E.P. Bicknell) Wiegand, **298**, *299*
  Elymus virginicus var. **intermedius** (Vasey *ex* A. Gray) Bush, **298**, *299*, 302
  Elymus virginicus var. **jejunus** (Ramalay) Bush, 296, *299*, **300**, 302, 305
  Elymus virginicus forma *lasiolepis* Fernald, 298
  Elymus virginicus var. *micromeris* Schmoll, 300
  Elymus virginicus var. *submuticus* Hook., 302
  Elymus virginicus L. var. **virginicus**, 298, *299*, **300**, 303
Elymus wawawaiensis J.R. Carlson & Barkworth, 282, 291, *332*, *335*
Elymus wiegandii Fernald, 295, 305, *307*
  Elymus wiegandii forma *calvescens* Fernald, 306
Elymus ×yukonensis (Scribn. & Merr.) Á. Löve, **340**, *341*
Elytrigia Desv., 239, 281, 374
EMINENT BLUEGRASS, 598
EMMER, 272
  WILD, 270
ENGLISH RYEGRASS, 455
ENGLISH WATERGRASS, 73
ÉPEAUTRE, 277
Eragrostis Wolf, 4
Eremopoa Roshev., 388, **618**
Eremopoa altaica (Trin.) Roshev., **619**, *619*
Eremopoa persica (Trin.) Roshev., 620
Eremopyrum (Ledeb.) Jaub. & Spach, 240, **252**, 278
Eremopyrum bonaepartis (Spreng.) Nevski, **254**, *255*

Eremopyrum bonaepartis (Spreng.) Nevski subsp. *bonaepartis*, 254

Eremopyrum bonaepartis subsp. *hirsutum* (Bertol.) Melderis, 254

Eremopyrum bonaepartis subsp. *sublanuginosum* (Drobow) Á. Löve, 254

Eremopyrum bonaepartis subsp. *turkestanicum* (Gand.) Tzvelev, 254

Eremopyrum orientale (L.) Jaub. & Spach, **254**, *255*, **256**

Eremopyrum triticeum (Gaertn.) Nevski, **254**, *255*, **256**

ESPARTO GRASS, 152

ESTUARINE WILDRICE, 48

EURASIAN JUNEGRASS, 754

EUROPEAN ALKALI GRASS, 473

EUROPEAN BEACHGRASS, 777

EUROPEAN DUNEGRASS, 356

Eustachys Desv., 791

FAIRGROUND GRASS, 481

FALSE MANNAGRASS

    FERNALD'S, 608

    PALE, 608

    SPIKED, 608

FALSE MELIC, 103, 745

FALSE-RYE BARLEY, 248

FALSE SEMAPHOREGRASS, 109

FALSEBROME

    HEATH, 190

    PURPLE, 188

    SLENDER, 190

    THINLEAF, 190

    TUFTED, 192

FAR, 272

Fargesia Franch., 17

FARMER'S FOXTAIL, 250

FARRO, 272

FEATHER REEDGRASS, 721

FEATHERGRASS

    AUSTRALIAN, 185

    BEAUTIFUL, 155

    GIANT, 154

FEATHERTOP, 710

FENASSE, 742

FERN GRASS, 682

FERNALD'S FALSE MANNAGRASS, 608

FESCUE

    ALEUT, 414

    ALPINE, 428

    ALTAI, 407

    ARCTIC RED, 414

    ARIZONA, 438

    BAFFIN ISLAND, 432

    BEARDED, 402

    BLUE, 422

    BLUEBUNCH, 438

    BROME, 452

    CALIFORNIA, 410

    CALLUSED, 437

    CHEWING'S, 415

    CLUSTER, 402

    COAST, 404

    COAST RANGE, 406

    CONFUSING, 452

    CRINKLE-AWN, 406

    DESERT, 452

    DUNE RED, 415

    EARLE'S, 420

    EASTWOOD, 452

    EDLUND'S, 432

    ELMER'S, 404

    FINE-LEAVED SHEEP, 424

    FLATLEAF RED, 415

    FOXTAIL, 449

    FREDERIKSEN'S, 436

    FRINGED, 454

    GIANT, 448

    GRAY, 422

    GREEN, 440

    GREENLAND, 434

    GREENLEAF, 440

    GUADALUPE, 408

    HAIR, 424

    HARD, 424

    HOWELL'S, 440

    IDAHO, 438

    INTERMOUNTAIN, 443

    KING'S, 445

    LENA, 426

    LITTLE, 434

    LOBED, 424

    MEADOW, 446

    MOUNTAIN, 430

    MOUNTAIN RED, 418

    MOUNTAIN ROUGH, 408

    NODDING, 400

    NORTHERN, 432

    NORTHERN ROUGH, 407

    OPEN, 443

    OREGON, 440

    PACIFIC, 452

    PLAINS ROUGH, 407

    PROLIFEROUS, 419

    PSEUDOVIVIPAROUS, 419

    RATTAIL, 449

    RAVINE, 404

    RED, 412, 418

    ROCK, 415

    ROCKY MOUNTAIN, 430

    ROEMER'S, 440

    SECUND RED, 418

    SHEEP, 422, 424

    SHORT, 426

    SIXWEEKS, 450

    SMALL, 452

    SMALL-FLOWERED, 434

    SPIKE, 445

    SQUIRRELTAIL, 450

    TALL, 448

    TEXAS, 400

    THURBER'S, 408

TUFTED, 422
VALAIS, 420
VARIOUS-LEAVED, 420
VIVIPAROUS, 436
WASHINGTON, 440
WESTERN, 437
FESCUE-GRASS
    WATSON'S, 445
Festuca L., 379, 380, 386, 388, **389**, 443, 445, 446, 449, 743
Festuca sect. Breviaristatae Krivot., **406**
Festuca sect. Elmera E.B. Alexeev, **404**
Festuca sect. Fauria E.B. Alexeev, **400**
Festuca L. sect. Festuca, **412**
Festuca sect. Obtusae E.B. Alexeev, **400**
Festuca sect. Subulatae Tzvelev, **402**
Festuca sect. Subuliflorae (E.B. Alexeev) Darbysh., **406**
Festuca sect. Texanae E.B. Alexeev, **399**
Festuca L. subg. Festuca, **406**
Festuca subg. Montanae (Hack.) Nyman, **399**
Festuca subg. Obtusae E.B. Alexeev, **400**
Festuca subg. Subulatae (Tzvelev) E.B. Alexeev, **402**
Festuca altaica Trin., 394, 396, *405*, **407**
    Festuca altaica forma *pallida* Jordal, 407
    Festuca altaica forma *vivipara* Jordal, 407
Festuca amethystina L., 390, 396, **422**, *423*
Festuca ammobia Pavlick, 418
Festuca arizonica Vasey, 396, 422, **438**, *439*
Festuca auriculata Drobow, 398, 422, **424**, 426, *427*
Festuca baffinensis Polunin, 398, 399, 422, 428, **432**, *433*, 436
Festuca brachyphylla Schult. & Schult. f., 399, 422, **428**, *429*, 430, 432, 434, 436
    Festuca brachyphylla Schult. & Schult. f. subsp. brachyphylla, **428**, *429*
    Festuca brachyphylla subsp. breviculmis Fred., **428**, *429*
    Festuca brachyphylla subsp. coloradensis Fred., **428**, *429*
    Festuca brachyphylla forma *flavida* Polunin, 428
    Festuca brachyphylla subsp. *saximontana* (Rydb.) Hultén, 430
Festuca brevissima Jurtsev, 399, 422, **426**, *429*
Festuca californica Vasey, 393, 395, **410**, *411*
    Festuca californica Vasey subsp. **californica**, 410, *411*
    Festuca californica subsp. **hitchcockiana** (E.B. Alexeev) Darbysh., 410, *411*
    Festuca californica subsp. **parishii** (Piper) Darbysh., 410, *411*, **412**
Festuca calligera (Piper) Rydb., 397, 398, 422, **437**, 438, *439*
Festuca campestris Rydb., 394, 396, 407, **408**, *409*
Festuca canadensis E.B. Alexeev, 430
Festuca cinerea Vill., 390
Festuca dasyclada Hack. *ex* Beal, 393, 396, **442**, *443*
Festuca drymeia Mert. & W.D.J. Koch, 390
Festuca duriuscula L., 424
Festuca earlei Rydb., 395, 398, 413, **420**, *421*, 428
Festuca edlundiae S. Aiken, Consaul & Lefk., 399, 422, 428, **432**, *433*
Festuca elegans Boiss., 390
Festuca elmeri Scribn. & Merr., 394, **404**, *405*
    Festuca elmeri subsp. *luxurians* Piper, 406

Festuca filiformis Pourr., 390, 396, **424**, *425*
Festuca frederikseniae E.B. Alexeev, 395, 413, 419, 422, *435*, **436**
Festuca gauthieri (Hack.) K. Richt., 390
Festuca glauca Vill., 390, 397, **422**, 424, *425*
Festuca groenlandica (Schol.) Fred., 398, *433*, **434**
Festuca hallii (Vasey) Piper, 394, 395, **407**, 408, *409*
Festuca heterophylla Lam., 397, **420**, *423*
Festuca hyperborea Holmen *ex* Fred., 399, 422, 428, *431*, **432**
Festuca idahoensis Elmer, 397, 422, **438**, *441*
Festuca japonica Makino, 400
Festuca kasmiriana Stapf, 390
Festuca lenensis Drobow, 398, 422, **426**, *427*
Festuca ligulata Swallen, 394, 395, **408**, *409*
Festuca longifolia Thuill., 424
Festuca mairei St.-Yves, 390
Festuca minutiflora Rydb., 398, 399, 420, 422, 428, **434**, *435*
Festuca muelleri Vickery, 390
Festuca occidentalis Hook., 397, 398, **437**, *439*
Festuca ovina L., 397, 398, *422*, *425*, 426, 428, 432, 434, 436, 438
    Festuca ovina var. *duriuscula* (L.) W.D.J. Koch, 424
Festuca pallens Host, 390
Festuca paradoxa Desv., 394, 400, *401*, **402**
Festuca prolifera (Piper) Fernald, 395, **419**, *421*
    Festuca prolifera var. **lasiolepis** Fernald, 419
    Festuca prolifera (Piper) Fernald var. **prolifera**, 419
Festuca pseudoeskia Boiss., 390
Festuca pseudovina Hack. *ex* Wiesb., 420
Festuca pseudovivipara (Pavlick) Pavlick, 395, 415, **419**, *421*
Festuca richardsonii Hook., 414
Festuca roemeri (Pavlick) E.B. Alexeev, 398, **440**, *441*
Festuca rubra L., 389, 390, 392, 395, **412**, 413, *416*, *417*, 419, 420, 436
    Festuca rubra subsp. **arctica** (Hack.) Govor., 413, **414**, *416*
    Festuca rubra subsp. **arenaria** (Osbeck) F. Aresch., 413, **414**, *416*
    Festuca rubra subsp. *arenicola* E.B. Alexeev, 418
    Festuca rubra subsp. **aucta** (V.I. Krecz. & Bobrov) Hultén, **414**, *416*, 419
    Festuca rubra subsp. **commutata** Gaudin, 413, **415**, *416*
    Festuca rubra subsp. **fallax** (Thuill.) Nyman, 413, **415**, *416*
    Festuca rubra var. *fraterculae* Rasm., 415
    Festuca rubra subsp. *glaucodea* Piper, 413
    Festuca rubra var. *juncea* (Hack.) K. Richt., 413
    Festuca rubra var. *littoralis* Vasey *ex* Beal, 415
    Festuca rubra subsp. **mediana** (Pavlick) Pavlick, 414, **415**, *417*
    Festuca rubra var. *megastachys* (Gaudin) Hegi, 413
    Festuca rubra subsp. **pruinosa** (Hack.) Piper, 413, **415**, *417*
    Festuca rubra L. subsp. **rubra**, 390, 413, *417*, *418*
    Festuca rubra subsp. **secunda** (J. Presl) Pavlick, 414, *417*, **418**
    Festuca rubra forma *squarrosa* (Hartm.) Holmb., 413

Festuca rubra subsp. **vallicola** (Rydb.) Pavlick, 413, *417*, **418**

Festuca rupicaprina (Hack.) A. Kern., 390

Festuca saximontana Rydb., 392, 398, 422, 428, **430**, *431*

    Festuca saximontana var. **purpusiana** (St.-Yves) Fred. & Pavlick, **430**, *431*

    Festuca saximontana var. **robertsiana** Pavlick, **430**, *431*

    Festuca saximontana Rydb. var. **saximontana**, 430, *431*, **432**

Festuca sororia Piper, 394, *403*, **404**

Festuca spectabilis Jan, 390

    Festuca stricta subsp. *trachyphylla* (Hack.) Patzke, 424

Festuca subulata Trin., 393, **402**, *403*, 406

Festuca subuliflora Scribn., 393, 404, *405*, **406**

Festuca subverticillata (Pers.) E.B. Alexeev, 394, **400**, *401*

    Festuca subverticillata forma *pilosifolia* (Dore) Darbysh., 400

Festuca thurberi Vasey, 394, 395, **408**, *411*

Festuca trachyphylla (Hack.) Krajina, 390, 392, 397, 422, **424**, *427*, 743

Festuca valesiaca Schleich. *ex* Gaudin, 390, 398, **420**, *423*

Festuca varia Haenke, 390

Festuca versuta Beal, 394, **400**, *401*

Festuca villosa-vivipara (Rosenv.) E.B. Alexeev, 413, *419*, 436

Festuca viridula Vasey, 394, 397, **440**, *441*

Festuca vivipara (L.) Sm., 436

Festuca viviparoidea Krajina *ex* Pavlick, 395, 422, 428, 434, *435*, **436**

    Festuca viviparoidea subsp. **krajinae** Pavlick, *435*, **437**

    Festuca viviparoidea Krajina *ex* Pavlick subsp. **viviparoidea**, *435*, **436**, **437**

Festuca washingtonica E.B. Alexeev, 394, 395, 397, **440**, *442*

FESTUCEAE Dumort., 192

FÉTUQUE À COULEUR D'AMÉTHYSTE, 422

FÉTUQUE À FEUILLES COURTES, 428

FÉTUQUE CHEVELUE, 424

FÉTUQUE D'ALTAÏ, 407

FÉTUQUE DE FREDERIKSEN, 436

FÉTUQUE DE RICHARDSON, 414

FÉTUQUE DES MONTAGNES ROCHEUSES, 430

FÉTUQUE DES OVINS, 422

FÉTUQUE DES PRÉS, 446

FÉTUQUE DES ROCHEUSES, 430

FÉTUQUE DRESSÉE À FEUILLES SCABRES, 424

FÉTUQUE DU VALAIS, 420

FÉTUQUE ÉLEVÉE, 448

FÉTUQUE GLAUQUE, 422

FÉTUQUE HÉTÉROPHYLLE, 420

FÉTUQUE PRUINEUSE, 415

FÉTUQUE ROUGE, 412

FÉTUQUE ROUGE DES SABLES, 414

FÉTUQUE ROUGE TRAÇANTE, 418

FÉTUQUE TROMPEUSE, 415

FEW-FLOWER BLUEGRASS, 538

FIELD BROME, 228

FINE-LEAVED SHEEP FESCUE, 424

FINELEAVED NASSELLA, 176

FISHER'S TUNDRAGRASS, 604

FISHPOLE BAMBOO, 27

FLAGELLARIACEAE Dumort., 6

FLAGELLARIA L., 6

FLANGED BROME, 209

FLATLEAF RED FESCUE, 415

FLAX DARNEL, 456

FLÉOLE ALPINE, 672

FLÉOLE DES PRÉS, 672

FLORIDA SPEARGRASS, 164

FLOUVE ODORANTE, 759

FOERSTER'S REEDGRASS, 721

FOIN D'ODEUR, 759, 762

FOIN FOU, 646

FOLLE AVOINE, 50, 735

FOOTHILLS NASSELLA, 174

FOOTHILLS NEEDLEGRASS, 174

FOWL BLUEGRASS, 574

FOXTAIL

    ALPINE, 782

    BOREAL, 782

    CREEPING MEADOW, 782

    CRETAN MEADOW, 788

    FARMER'S, 250

    MEADOW, 782

    PACIFIC MEADOW, 786

    RENDLE'S MEADOW, 786

    SHORTAWN, 784

    SLENDER MEADOW, 786

    SONOMA SHORTAWN, 784

    TUFTED, 786

    WATER, 784

FOXTAIL BARLEY, 245

FOXTAIL CHESS, 226

FOXTAIL FESCUE, 449

FRAGILE OAT, 732

FREDERIKSEN'S FESCUE, 436

FRINGED BROME, 222

FRINGED FESCUE, 454

FROMENTAL, 742

'GAITERS'

    'GARDENER'S', 772

'GARDENER'S GAITERS', 772

Gastridium P. Beauv., 382, 383, **675**

Gastridium phleoides (Nees & Meyen) C.E. Hubb., **676**, *677*

Gastridium ventricosum (Gouan) Schinz & Thell., 676

Gaudinia P. Beauv., 380, **732**

Gaudinia fragilis (L.) P. Beauv., **732**, *733*

GENICULATE BARLEY, 248

GEYER'S ONIONGRASS, 93

GIANT CANE, 18

GIANT CUTGRASS, 52

GIANT FEATHERGRASS, 154

GIANT FESCUE, 448

GIANT GLYCERIA, 71

GIANT MANNAGRASS, 71

GIANT THORNY BAMBOO, 22

GIANT TIMBER BAMBOO, 27

GIANT WILDRYE, 362

GLAUCOUS BLUEGRASS, 578

GLUMELESS WILDRYE, 316

Glyceria R. Br., 67, **68**, 607
Glyceria R. Br. sect. **Glyceria**, 81, 83
Glyceria sect. **Hydropoa** (Dumort.) Dumort., 68, **71**
Glyceria sect. **Striatae** G.L. Church, 68, **73**
Glyceria acutiflora Torr., 70, *82*, **83**
Glyceria alnasteretum Kom., 69, **71**, 72
Glyceria borealis (Nash) Batch., 70, **81**, *82*, 87
Glyceria canadensis (Michx.) Trin., 70, 77, **79**, *80*
  Glyceria canadensis (Michx.) Trin. var. **canadensis**, 79, *80*
  Glyceria canadensis var. **laxa** (Scribn.) Hitchc., 77, 79, 80, *80*
Glyceria declinata Bréb., 70, 85, **86**, 87
Glyceria elata (Nash) M.E. Jones, 70, 71, 77, 78, **79**
Glyceria fluitans (L.) R. Br., 70, 84, **85**, 87
Glyceria ×gatineauensis Bowden, 69, 77
Glyceria grandis S. Watson, 69, **71**, 72, 73, 79
  Glyceria grandis S. Watson var. **grandis**, 71, 72
  Glyceria grandis var. **komarovii** Kelso, 71, 72
Glyceria ×laxa (Scribn.) Scribn., 77, 80
Glyceria leptostachya Buckley, 70, 84, **85**, 88
Glyceria maxima (Hartm.) Holmb., 68, 69, 71, **73**, 74
Glyceria melicaria (Michx.) F.T. Hubb., 69, **75**, 76, 77
Glyceria notata Chevall., 68, 70, 81, 83, 86, **87**
Glyceria nubigena W.A. Anderson, 70, **75**, 76
Glyceria obtusa (Muhl.) Trin., 69, **74**, 75
Glyceria ×occidentalis (Piper) J.C. Nelson, 69, 70, 84, **85**, 87
Glyceria ×ottawensis Bowden, 69, 77, 79, 80
Glyceria pulchella (Nash) K. Schum., 70, 76, **77**, 79
Glyceria septentrionalis Hitchc., 70, **81**, *82*, 87, 88
  Glyceria septentrionalis var. **arkansana** (Fernald) Steyerm.
    & Kučera, *82*, 83
  Glyceria septentrionalis Hitchc. var. **septentrionalis**, *82*, 83
Glyceria striata (Lam.) Hitchc., 70, **77**, 78, 79
  Glyceria striata (Lam.) Hitchc. var. *striata*, 77
  Glyceria striata var. *stricta* (Scribn.) Fernald, 77
GLYCERIA
  ALEUTIAN, 71
  AMERICAN, 71
  BEAUTIFUL, 77
  BOREAL, 81
  CANADIAN, 79
  GIANT, 71
  GREAT SMOKY MOUNTAIN, 75
  LOW, 87
  MARKED, 87
  NORTHERN, 81
  RIDGED, 77
  TALL, 73
GLYCÉRIE AQUATIQUE, 73
GLYCÉRIE BORÉALE, 81
GLYCÉRIE DE FERNALD, 608
GLYCÉRIE DU CANADA, 79
GLYCÉRIE FLOTTANTE, 85
GLYCÉRIE GÉANTE, 71
GLYCÉRIE MELICAIRE, 75
GLYCÉRIE SEPTENTRIONALE, 81
GLYCÉRIE STRIÉE, 77
GOATGRASS
  BARBED, 265
  JOINTED, 263

OVATE, 265
PERSIAN, 263
SWOLLEN, 263
TAUSCH'S, 262
THREE-AWNED, 265
GOLDEN BAMBOO, 27
GOLDEN OATGRASS, 753
GOLDENTOP, 484
GOOSE GRASS, 463
GRAMINEAE Adans., 3, 4
GRAND ÉPEAUTRE, 277
GRASS
  ALASKA ALKALI, 475
  ALTAI, 619
  ANDERSON'S ALKALI, 473
  ARCTIC ALKALI, 471
  BORRER'S SALTMARSH, 463
  BOTTLEBRUSH, 316
  BRUGGEMANN'S ALKALI, 469
  CATCHFLY, 44
  CLOUD, 661
  COMMON SALTMARSH, 463
  ESPARTO, 151
  EUROPEAN ALKALI, 473
  FAIRGROUND, 481
  FERN, 682
  GOOSE, 463
  GREENLAND ALKALI, 467
  HARETAIL, 670
  HOWELL'S ALKALI, 475
  JIJI, 117
  LEMMON'S ALKALI, 469
  MAURITANIAN, 112
  MEMPHIS, 612
  MOSS, 618
  MUD, 618
  NIT, 676
  NORTH-AFRICA, 684
  NUTTALL'S ALKALI, 475
  OATMEAL, 44
  PACIFIC ALKALI, 475
  PARISH'S ALKALI, 465
  PENDANT, 607
  RABBITSFOOT, 665
  RESCUE, 199
  'RIBBON', 772
  RIPGUT, 224
  SCRIBNER, 689
  SHEATHED ALKALI, 477
  SILVERLEAF, 54
  SMILO, 151
  SMOOTH ALKALI, 471
  SPIKE, 682
  SQUIRRELTAIL, 245
  STIFF SALTMARSH, 463
  TALL ALKALI, 467
  TOR, 190
  TUNDRA ALKALI, 469
  UPSIDEDOWN, 13
  VAHL'S ALKALI, 465

WESTERN ALKALI, 465
WINTER, 168
WRIGHT'S ALKALI, 467
GRAY FESCUE, 422
GRAY HAIRGRASS, 743
GRAY MOORGRASS, 681
GREAT-BASIN BROME, 205
GREAT BASIN WILDRYE, 364
GREAT BROME, 224
GREAT PLAINS WILDRYE, 303
GREAT SMOKY MOUNTAIN GLYCERIA, 75
GREAT SMOKY MOUNTAIN MANNAGRASS, 75
GREEN FESCUE, 440
GREEN NASSELLA, 177
GREEN NEEDLEGRASS, 177
GREENLAND ALKALI GRASS, 467
GREENLAND FESCUE, 434
GREENLEAF FESCUE, 440
GROVE BLUEGRASS, 512
GUADALUPE FESCUE, 408
GYNERIEAE Sánchez-Ken & L.G. Clark, 10

Hainardia Greuter, 380, **689**
Hainardia cylindrica (Willd.) Greuter, *690*, **691**
HAINARDIEAE Greut., 57, 379
HAIR FESCUE, 424
HAIRGRASS
   ALPINE, 631
   ANNUAL, 631
   BERINGIAN, 626
   CRESTED, 754
   CRINKLED, 631
   DELICATE, 616
   GRAY, 743
   MACKENZIE, 628
   MEDITERRANEAN, 756
   MOUNTAIN, 692
   SILVER, 616
   SILVERY, 616
   SLENDER, 631
   SPIKE, 616
   TUFTED, 628
   WAVY, 631
HAIRY BROME, 220
HAIRY CHESS, 230
HAIRY MELIC, 100
HAIRY SICKLEGRASS, 688
HAIRY SWEETGRASS, 764
HAIRY WHEATGRASS, 376
HAIRY WOODBROME, 209
HALL'S BENT, 653
HALL'S BROME, 211
HARD FESCUE, 424
HARD WHEAT, 272
HARDGRASS, 481, 691
   BELLY-SHAPED, 263
   ROUGH-SPIKED, 262
HARETAIL GRASS, 670
HARFORD MELIC, 93
HARTZ'S BLUEGRASS, 589

HEATH FALSEBROME, 190
HEDGE BAMBOO, 22
Helictotrichon Besser *ex* Schult. & Schult. f., 380, 385, 698, **701**
Helictotrichon mortonianum (Scribn.) Henrard, **701**, *702*
Helictotrichon sempervirens (Vill.) Pilg., **701**, *702*
HENDERSON'S BENT, 658
HENDERSON'S NEEDLEGRASS, 139
HERBE SAINTE, 762
Hesperostipa (M.K. Elias) Barkworth, 111, **157**
Hesperostipa comata (Trin. & Rupr.) Barkworth, 141, 157, **158**, *159*
   Hesperostipa comata (Trin. & Rupr.) Barkworth subsp. **comata**, 158, *159*
   Hesperostipa comata subsp. **intermedia** (Scribn. & Tweedy) Barkworth, **158**, *159*, 161
Hesperostipa curtiseta (Hitchc.) Barkworth, 157, **158**, *160*, 161
Hesperostipa neomexicana (Thurb.) Barkworth, 157, **158**, *159*
Hesperostipa spartea (Trin.) Barkworth, 157, *160*, **161**
Hierochloë R.Br., 758
Hierochloë hirta (Schrank) Borbás, 764
   Hierochloë hirta subsp. *arctica* G. Weim., 764
   Hierochloë hirta (Schrank) Borbás subsp. *hirta*, 764
   Hierochloë hirta subsp. *praetermissa* G. Weim., 764
Hierochloë odorata (L.)Wahlenb., 762
HIÉROCHLOÉ ALPINE, 760
HIÉROCHLOÉ ODORANTE, 762
HIÉROCHLOÉ PAUCIFLORE, 760
HIGH-ARCTIC WHEATGRASS, 326
HIGHLAND BENT, 639
HILL CANE, 18
HOFFMANN'S WHEATGRASS, 336
Holcus L., 385, 386, **739**
Holcus lanatus L., 739, **740**, *741*
Holcus mollis L., 739, **740**, *741*
   Holcus mollis L. subsp. **mollis**, 740
   Holcus mollis subsp. **reuteri** (Boiss.) Malag., 740
HOODED CANARYGRASS, 766
HOOVER'S BENT, 654
HOOVER'S SEMAPHOREGRASS, 107
Hordeum L., 4, 238, 239, **241**, 259, 283, 300, 308, 319, 369, 370
Hordeum sect. **Critesion** (Raf.) Nevski, 289
Hordeum arizonicum Covas, 242, **248**, *249*
Hordeum brachyantherum Nevski, 242, **243**, *246*, 248, 284, 286, 310, 370
   Hordeum brachyantherum Nevski subsp. **brachyantherum**, 243, *246*, 286
   Hordeum brachyantherum subsp. **californicum** (Covas & Stebbins) Bothmer, N. Jacobsen & Seberg, 243, **245**, *246*
Hordeum brevisubulatum (Trin.) Link, 252
Hordeum bulbosum L., 241, 242, **252**, *253*
Hordeum depressum (Scribn. & J.G. Sm.) Rydb., 242, **243**, *244*
Hordeum intercedens Nevski, **242**, *244*, 248
Hordeum intermedium Hausskn., 245
Hordeum ×intermedium Hausskn., 245

Hordeum jubatum L., 242, **245**, *247*, 248, 284, 286, 302, 322, 370, 782

    Hordeum jubatum subsp. intermedium Bowden, **245**, *247*

    Hordeum jubatum L. subsp. **jubatum**, **245**, *247*

    Hordeum leporinum var. *simulans* Bowden, 250

Hordeum marinum Huds., 241, 242, **248**, *249*

    Hordeum marinum subsp. **gussoneanum** (Parl.) Thell., **248**, *249*

    Hordeum marinum Huds. subsp. **marinum**, **248**, *249*

Hordeum murinum L., 241, 242, **250**, *251*

    Hordeum murinum subsp. **glaucum** (Steud.) Tzvelev, **250**, *251*

    Hordeum murinum subsp. **leporinum** (Link) Arcang., **250**, *251*

    Hordeum murinum L. subsp. **murinum**, **250**, *251*

Hordeum pusillum Nutt., 242, **243**, *244*, 248

Hordeum secalinum Schreb., 241, 243, **248**, *251*

Hordeum vulgare L., 241, 242, **252**, *253*

    Hordeum vulgare subsp. **spontaneum** (K. Koch) Thell., 252

    Hordeum vulgare L. subsp. **vulgare**, 252

HOULQUE LAINEUSE, 740

HOULQUE MOLLE, 740

HOWELL'S ALKALI GRASS, 475

HOWELL'S BENT, 654

HOWELL'S BLUEGRASS, 534

HOWELL'S FESCUE, 440

HOWELL'S REEDGRASS, 712

HUNGARIAN BROME, 206

Hygroryza Nees, 37, **46**

Hygroryza aristata (Retz.) Nees, **46**, *46*

Hystrix Moench, 239

ICEGRASS, 479

IDAHO FESCUE, 438

IDAHO REDTOP, 649

Imperata Cirillo, 793

Imperata brevifolia Vasey, 793

INDIAN RICEGRASS, 139

INTERIOR BLUEGRASS, 576

INTERIOR WILDRICE, 50

INTERMEDIATE BARLEY, 245

INTERMEDIATE WHEATGRASS, 376

INTERMOUNTAIN FESCUE, 443

INTERRUPTED WINDGRASS, 789

ITALIAN RYEGRASS, 456

ITALIAN TIMOTHY, 675

IVRAIE DU LIN, 456

IVRAIE ENIVRANTE, 459

IVRAIE MULTIFLORE, 456

IVRAIE VIVACE, 455

JAPANESE ARROW BAMBOO, 29

JAPANESE BROME, 235

Jarava Ruiz & Pav., 111, **178**

Jarava ichu Ruiz & Pav., 178, **179**, *180*

Jarava plumosa (Spreng.) S.W.L. Jacobs & J. Everett, 178, **179**, *180*

Jarava speciosa (Trin. & Rupr.) Peñail., 142, 178, 179, *180*, **181**

JIJI GRASS, 117

JOINTED GOATGRASS, 263

JOINVILLEACEAE Toml. & A.C. Sm., 6

JONES' REEDGRASS, 717

JUNEGRASS, 754

    EURASIAN, 754

KALM'S BROME, 209

KECK'S BLUEGRASS, 584

KELLOGG'S BLUEGRASS, 514

KENTUCKY BLUEGRASS, 522

Key to Tribes, 7

KING'S FESCUE, 445

KING'S PTILAGROSTIS, 143

Koeleria Pers., 385, 386, 388, **753**

Koeleria asiatica Domin, **753**, **754**, *755*

Koeleria macrantha (Ledeb.) Schult., 623, 753, **754**, *755*

Koeleria nitida Nutt., 754

Koeleria phleoides (Vill.) Pers., 756

Koeleria pyramidata (Lam.) P. Beauv., 753, **754**, *755*

KOLERIE ACCRÊTÉ, 754

KOLERIE À CRÊTES, 754

LABRADOR BLUEGRASS, 602

Lachnagrostis Trin., 384, 633, 634, **694**, 706

Lachnagrostis filiformis (G. Forst.) Trin., 633, *695*, **696**

Lagurus L., 379, **670**

Lagurus ovatus L., *669*, **670**

Lamarckia Moench, 379, **484**

Lamarckia aurea (L.) Moench, **484**, *485*

LANCEOLATE BROME, 235

LAPLAND REEDGRASS, 729

LARGE-GLUME BLUEGRASS, 527

LAX BLUEGRASS, 570

LAX-FLOWER BLUEGRASS, 538

LEAFY REEDGRASS, 714

Leersia Sw., 37, **42**

Leersia hexandra Sw., 42, **44**, *45*

Leersia lenticularis Michx., 42, **43**, **44**

Leersia monandra Sw., **42**, *43*

Leersia oryzoides (L.) Sw., 42, **44**, *45*

Leersia virginica Willd., 42, *43*, **44**

LÉERSIE DE VIRGINIE, 44

LÉERSIE FAUX-RIZ, 44

LEIBERG'S BLUEGRASS, 563

LEMMON'S ALKALI GRASS, 469

LEMMON'S CANARYGRASS, 768

LEMMON'S NEEDLEGRASS, 125

LENA FESCUE, 426

LESSER CANARYGRASS, 767

LETTERMAN'S BLUEGRASS, 580

LETTERMAN'S NEEDLEGRASS, 118

Leucopoa Griseb., 388, **443**

Leucopoa kingii (S. Watson) W.A. Weber, *444*, **445**

×**Leydeum** Barkworth, 239, 308, **369**

×**Leydeum dutillyanum** (Lepage) Barkworth, **370**, *371*

×**Leydeum littorale** (H.J. Hodgs. & W.W. Mitch.) Barkworth, **370**, *371*

×**Leydeum piperi** (Bowden) Barkworth, **370**, *371*

Leymus Hochst., 239, 240, 288, 308, 351, **353**, 369, 370, 372, 777

Leymus ajanensis (V.N. Vassil.) Tzvelev, 356

Leymus ambiguus (Vasey & Scribn.) D.R. Dewey, 355, **366**, *365*

Leymus angustus (Trin.) Pilg., 355, 360, *363*

Leymus arenarius (L.) Hochst., 354, 355, **356**, *357*, 777

Leymus californicus (Bol. *ex* Thurb.) Barkworth, 354, *368*, 369

Leymus cinereus (Scribn. & Merr.) Á. Löve, 344, 355, 360, 364, *365*, 372

Leymus condensatus (J. Presl) Á. Löve, 355, 360, **362**, *363*

Leymus flavescens (Scribn. & J.G. Sm.) Pilg., 355, 360, **366**, *367*

Leymus innovatus (Beal) Pilg., 322, 344, 346, 353, 355, **366**, *367*

    Leymus innovatus (Beal) Pilg. subsp. **innovatus**, 366, *367*

    Leymus innovatus subsp. **velutinus** (Bowden) Tzvelev, **366**, *367*

Leymus karelinii (Turcz.) Tzvelev, 362

Leymus mollis (Trin.) Pilg., 310, 322, 346, 348, 354, **356**, *357*, 360, 370, 777

    Leymus mollis (Trin.) Pilg. subsp. **mollis**, 348, *357*, **358**

    Leymus mollis subsp. **villosissimus** (Scribn.) Á. Löve, *357*, **358**

Leymus multicaulis (Kar. & Kir.) Tzvelev, 355, 360, *361*

Leymus ×multiflorus (Gould) Barkworth & R.J. Atkins, 355, **362**, *363*

Leymus pacificus (Gould) D.R. Dewey, 355, **358**, *359*

Leymus racemosus (Lam.) Tzvelev, 261, 355, **356**, *357*

Leymus salina (M.E. Jones) Á. Löve, 355, **364**, *365*

    Leymus salina subsp. **mojavensis** Barkworth & R.J. Atkins, **364**, *365*

    Leymus salina (M.E. Jones) Á. Löve subsp. **salina**, **364**, *365*

    Leymus salina subsp. **salmonis** (C.L. Hitchc.) R.J. Atkins, **364**, *365*

Leymus simplex (Scribn. & T.A. Williams) D.R. Dewey, 346, 355, **359**, 360, *361*

    Leymus simplex var. **luxurians** (Scribn. & T.A. Williams) Beetle, *359*, 360, *361*

    Leymus simplex (Scribn. & T.A. Williams) D.R. Dewey var. **simplex**, *359*, *361*

Leymus triticoides (Buckley) Pilg., 344, 351, 353, 355, **358**, 360, *361*, 362, 364, 369, 370

Leymus ×vancouverensis (Vasey) Pilg., 355, **358**, *359*, 360

Limnodea L.H. Dewey, 383, 773, **776**

Limnodea arkansana (Nutt.) L.H. Dewey, *775*, **776**

LIMP MANNAGRASS, 80

Lithachne P. Beauv., 30

Lithachne horizontalis Chase, 30

Lithachne humilis Soderstr., 30, *31*

Lithachne pauciflora (Sw.) P. Beauv., 30

Lithachne pineti (C. Wright *ex* Griseb.) Chase, 30

LITHACHNE
    SMALL, 30

LITTLE BARLEY, 243

LITTLE CALIFORNIA MELIC, 90

LITTLE FESCUE, 434

LITTLE MELIC, 97

LITTLE PIPTATHERUM, 146

LITTLE QUAKINGGRASS, 614

LOBED FESCUE, 424

LOBED NEEDLEGRASS, 131

Lolium L., 380, 445, 446, 449, **454**

Lolium ×**hubbardii** Jansen & Wacht. *ex* B.K. Simon, 456

Lolium ×**hybridum** Hausskn., **455**, 456

Lolium multiflorum Lam., **455**, **456**, *457*

Lolium perenne L., **455**, 456, *457*

Lolium persicum Boiss. & Hohen. *ex* Boiss., **455**, *458*, **459**

Lolium rigidum Gaudin, **455**, 456, *457*

Lolium temulentum L., **455**, **456**, *458*, 459

    Lolium temulentum subsp. **remotum** (Schrank) Á. Löve & D. Löve, **456**, *458*

    Lolium temulentum L. subsp. **temulentum**, 456, *458*, 459

LONGFLOWERED VELDTGRASS, 36

LONGLEAF SQUIRRELTAIL, 319

LONGSTAMEN RICE, 38

LONGTONGUE MUTTONGRASS, 558

LOOSE SILKYBENT, 789

LOPGRASS, 232

Lophochlaena Nees, 105

LOW BARLEY, 243

LOW GLYCERIA, 87

LOW NEEDLEGRASS, 129

Luziola Juss., 37, **54**

Luziola bahiensis (Steud.) Hitchc., 54, **55**, *56*

Luziola fluitans (Michx.) Terrell & H. Rob., **54**, *56*

    Luziola fluitans (Michx.) Terrell & H. Rob. var. **fluitans**, 55

    Luziola fluitans var. **oconnorii** (R. Guzmán) G. Tucker, 55

Luziola peruviana J.F. Gmel., 54, **55**, *56*

LYMEGRASS, 356
    SEA, 356

MACARONI WHEAT, 272

MACKENZIE HAIRGRASS, 628

MACKENZIE VALLEY MANNAGRASS, 77

MACOUN'S REEDGRASS, 728

Macrochloa Kunth, 110, **151**, 154

Macrochloa tenacissima (L.) Kunth, **152**, *153*, 155

MADAKE, 27

MAMMOTH WILDRYE, 356

MANNAGRASS
    AMERICAN, 71
    ATLANTIC, 75
    BOREAL, 81
    CANADIAN, 79
    CREEPING, 83
    FERNALD'S FALSE, 608
    GIANT, 71
    GREAT SMOKY MOUNTAIN, 75
    LIMP, 80
    MACKENZIE VALLEY, 77
    MELIC, 75
    NARROW, 85
    NORTHERN, 81
    PALE FALSE, 608
    RATTLESNAKE, 79
    SPIKED FALSE, 608

TALL, 79
WATER, 85
WEAK, 608
WESTERN, 85
MANY-FLOWERED WILDRYE, 362
MANY-STEM WILDRYE, 360
MARITIME BROME, 203
MARKED GLYCERIA, 87
MARRAMGRASS, 777
MARSH'S BLUEGRASS, 582
MATGRASS, 63
MAURITANIAN GRASS, 112
MEADOW BARLEY, 243, 248
MEADOW BROME, 206, 218, 230
MEADOW FESCUE, 446
MEADOW FOXTAIL, 782
    CREEPING, 782
    CRETAN, 788
    PACIFIC, 786
    RENDLE'S, 786
    SLENDER, 786
MEDITERRANEAN BARLEY, 248
MEDITERRANEAN BEARDGRASS, 665
MEDITERRANEAN HAIRGRASS, 756
MEDUSAHEAD, 258
Megathyrsus (Pilg.) B.K. Simon & S.W. L. Jacobs, 793
Megathyrsus maximus (Jacq.) B.K. Simon & S.W. L. Jacobs, 793
MELIC
    AWNED, 95
    CALIFORNIA, 93
    CILIATE, 100
    FALSE, 103, 745
    HAIRY, 100
    HARFORD, 93
    LITTLE, 97
    LITTLE CALIFORNIA, 90
    MONTEZUMA, 98
    PORTER'S, 98
    ROCK, 97
    SIBERIAN, 100
    SILKY-SPIKE, 100
    SMITH'S, 95
    TALL, 100
    THREE-FLOWER, 100
    TORREY'S, 90
    TWO-FLOWER, 100
    WOODY, 91
MELIC MANNAGRASS, 75
Melica L., 67, 68, **88**
Melica subg. **Bromelica** Thurb., 88
Melica L. subg. **Melica**, 88
Melica altissima L., 89, **100**, *102*
    Melica altissima var. *atropurpurea* Host, 100
Melica aristata Thurb. *ex* Bol., 89, **95**, *96*
Melica bulbosa Geyer *ex* Porter & J.M. Coult., 89, 90, **91**, *92*
Melica californica Scribn., 89, 90, 91, **93**, *94*
    Melica californica Scribn. var. *californica*, 93
    Melica californica var. *nevadensis* Boyle, 93

Melica ciliata L., 89, **100**, *102*
Melica frutescens Scribn., 90, **91**, *94*
Melica fugax Bol., 90, 91, **97**, *97*
Melica geyeri Munro, 89, 90, **93**, *94*
    Melica geyeri var. **aristulata** T.J. Howell, 93, *94*, 95
    Melica geyeri Munro var. **geyeri**, 93, *94*, 95
Melica harfordii Bol., 89, **93**, *94*
Melica imperfecta Trin., 89, 90, **92**, *93*
Melica montezumae Piper, 89, **98**, *101*
Melica mutica Walter, 89, **100**, *101*
Melica nitens (Scribn.) Nutt. *ex* Piper, 89, **100**, *101*
Melica porteri Scribn., 89, **98**, *99*
    Melica porteri var. **laxa** Boyle, **98**, *99*
    Melica porteri Scribn. var. **porteri**, **98**, *99*
Melica smithii (Porter *ex* A. Gray) Vasey, 89, **95**, *96*
Melica spectabilis Scribn., 89, 90, **91**, *92*
Melica stricta Bol., 89, **97**, *99*
    Melica stricta var. **albicaulis** Boyle, **98**, *99*
    Melica stricta Bol. var. **stricta**, **98**, *99*
Melica subulata (Griseb.) Scribn., 89, **95**, *96*, 97
Melica torreyana Scribn., 89, 90, **92**
MELICEAE Endl., 8, 58, 67, 110, 610
Melinis P. Beauv., **793**
Melinis nerviglumis (Franch.) Zizka, **793**
Melinis repens (Willd.) Zizka, **793**
MEMPHIS GRASS, 612
METAKE, 29
MEXICAN BROME, 213
MEXICAN NEEDLEGRASS, 184
MEXICAN WILDRYE, 312
Mibora Adans., 380, **757**
Mibora maroccana (Maire) Maire, **757**
Mibora minima (L.) Desv., **757**, *757*
MICRAIROIDEAE Pilg., 6
MIL, 672
Milium L., **778**
Milium effusum L., 778, 779, **780**
    Milium effusum L. var. **effusum**, 780
    Milium effusum var. **cisatlanticum** Fernald, 780
Milium vernale M.-Bieb., 778, 779, **780**
MILLET
    EARLY, 780
    WATER, 52
    WOOD, 780
MILLET COMMUN, 793
MILLET DIFFUS, 780
MILLETGRASS
    SPRING, 780
MOJAVE WILDRYE, 364
Monanthochloë Engelm., **791**
MONTANA WHEATGRASS, 334
MONTEZUMA MELIC, 98
MOORGRASS
    AUTUMN, 679
    BLUE, 681
    BLUE-GREEN, 681
    GRAY, 681
Moorochloa Veldkamp, **793**
Moorochloa eruciformis (Sm.) Veldkamp, **793**
MORMON NEEDLEGRASS, 131

MOSQUITOGRASS, 256
MOSS GRASS, 618
MOUNT WASHINGTON BLUEGRASS, 572
MOUNTAIN BENT, 654
MOUNTAIN BROME, 205
MOUNTAIN BUNCHGRASS, 440
MOUNTAIN FESCUE, 430
MOUNTAIN HAIRGRASS, 692
MOUNTAIN RED FESCUE, 418
MOUNTAIN RICEGRASS, 148
MOUNTAIN ROUGH FESCUE, 408
MOUSE BARLEY, 250
MUD GRASS, 618
Muhlenbergia tenuifolia (Willd.) Britton, Sterns & Poggenb.,
     790
MUIR'S REEDGRASS, 717
MUTTONGRASS
     LONGTONGUE, 558
     VASEY'S, 556

NAKED OATS, 737
NAPA BLUEGRASS, 594
NARDE RAIDE, 63
NARDEAE W.D.J. Koch, 8, 58, **62**
Nardus L., **62**
Nardus stricta L., **63**, *63*
NARROW CANARYGRASS, 773
NARROW-FLOWER BLUEGRASS, 592
NARROW MANNAGRASS, 85
Nassella (Trin.) E. Desv., 111, 157, 169, **170**, 184
Nassella cernua (Stebbins & Love) Barkworth, 133, 171,
     173, 174, **176**
Nassella chilensis (Trin.) E. Desv., 171, *175*, **177**
Nassella formicarum (Delile) Barkworth, 174
Nassella lepida (Hitchc.) Barkworth, 171, *173*, **174**
Nassella leucotricha (Trin. & Rupr.) R.W. Pohl, **172**, *173*,
     174
Nassella manicata (E. Desv.) Barkworth, 172, *173*, **174**, 176
Nassella neesiana (Trin. & Rupr.) Barkworth, **172**, *173*
Nassella pulchra (Hitchc.) Barkworth, 172, *173*, **174**, 176
Nassella tenuissima (Trin.) Barkworth, 171, *175*, **176**
Nassella trichotoma (Nees) Hack. *ex* Arechav., 171, *175*,
     177
Nassella viridula (Trin.) Barkworth, 129, 142, 169, 171,
     175, **177**
NASSELLA
     CERNUOUS, 176
     FINELEAVED, 176
     FOOTHILLS, 174
     GREEN, 177
     PURPLE, 174
     TEXAN, 172
NEEDLE-AND-THREAD, 158
NEEDLEGRASS
     CALIFORNIA, 121
     COMMON WESTERN, 123
     CRESTED, 127
     CURLYLEAF, 135
     DESERT, 181
     DORE'S, 124

     FOOTHILLS, 174
     GREEN, 177
     HENDERSON'S, 139
     LEMMON'S, 124
     LETTERMAN'S, 118
     LOBED, 131
     LOW, 129
     MEXICAN, 184
     MORMON, 131
     NELSON'S, 123
     NEVADA, 120
     NEW MEXICAN, 158
     NODDING, 176
     PARISH'S, 129
     PERPLEXING, 135
     PERUVIAN, 179
     PINEWOODS, 137
     PLUMOSE, 179
     PUNA, 182
     PURPLE, 174
     RICHARDSON'S, 133
     SAN DIEGO, 131
     SCRIBNER'S, 135
     SOUTH AMERICAN, 182
     SOUTHWESTERN, 133
     STILLMAN'S, 118
     SWALLEN'S, 137
     TEXAN, 172
     THURBER'S, 125
     WALLOWA, 138
     WEBBER'S, 137
     WESTERN, 123
     WIDE-GLUMED, 124
NELSON'S NEEDLEGRASS, 124
Neomolinia Honda, 65
Neotyphoidium coenophialum, 448
NEVADA BLUEGRASS, 586
NEVADA NEEDLEGRASS, 120
NEW MEXICAN BLUEGRASS, 536
NEW MEXICAN NEEDLEGRASS, 158
NIT GRASS, 676
NODDING BLUEGRASS, 538
NODDING BROME, 213
NODDING FESCUE, 400
NODDING NEEDLEGRASS, 176
NODDING SEMAPHOREGRASS, 107
NODDING TRISETUM, 748
NORTH-AFRICA GRASS, 684
NORTHERN BARLEY, 243
NORTHERN BENT, 644
NORTHERN FESCUE, 432
NORTHERN GLYCERIA, 81
NORTHERN MANNAGRASS, 81
NORTHERN REEDGRASS, 730
NORTHERN RIVERBANK WILDRYE, 305
NORTHERN ROUGH FESCUE, 407
NORTHERN SHORTHUSK, 60
NORTHERN STEBBINS' WHEATGRASS, 329
NORTHERN WHEATGRASS, 324
NORTHERN WILDRICE, 50

NORTHWESTERN WILDRYE, 310
NUTTALL'S ALKALI GRASS, 475

OAT
  FRAGILE, 732
OATGRASS
  ALPINE, 701
  BLUE, 701
  DOWNY ALPINE, 698
  GOLDEN, 753
  SPIKE, 698
  TALL, 742
  YELLOW, 748
OATMEAL GRASS, 44
OATS, 737
  ANIMATED, 739
  CULTIVATED, 737
  NAKED, 737
  SLENDER, 735
  SLENDER WILD, 735
  SWAMP, 621
  WESTERN, 737
  WILD, 735
OBOVATE BEAKGRAIN, 65
OLDHAM'S BAMBOO, 25
OLDPASTURE BLUEGRASS, 510
OLYREAE Kunth, 7, 14, **29**
ONE-AND-A-HALF-FLOWERED REEDGRASS, 714
ONE-SIDED WHEATGRASS, 322
ONIONGRASS, 91
  ALASKA, 95
  GEYER'S, 93
  PURPLE, 91
  TAPERED, 95
OPEN FESCUE, 443
ORCHARDGRASS, 482
ORCUTT'S BROME, 211
ORCUTTIEAE Reeder, 8
OREGON FESCUE, 440
OREGON REDTOP, 649
OREGON SEMAPHOREGRASS, 107
ORGE, 252
ORGE À ANTHÈRES COURTES, 243
ORGE AGRÉABLE, 245
ORGE DES PRÉS, 243
ORGE QUEUE D'ÉCUREUIL, 245
ORGE VULGAIRE, 252
Oryza L., 4, **37**
Oryza longistaminata A. Chev. & Roehr., 37, **38**, *39*
Oryza punctata Kotschy *ex* Steud., 37, 38, *39*, **40**
Oryza rufipogon Griff., 37, **38**, *39*
Oryza sativa L., 36, 37, 38, **40**, *41*
ORYZEAE Dumort., 8, 32, **36**
Oryzopsis Michx., 110, 114, 115, **167**
Oryzopsis asperifolia Michx., 167, **168**, *168*
ORYZOPSIS À FEUILLES RUDES, 168
ORYZOPSIS DU CANADA, 146
Ostrya virginiana, 679
Otatea (McClure & E.W. Sm.) C.E. Calderón & Soderstr., 16
OVATE GOATGRASS, 265

OZARKGRASS, 776

PACCAD clade, 6
PACIFIC ALKALI GRASS, 475
PACIFIC BENT, 696
PACIFIC BLUEGRASS, 588
PACIFIC BROME, 218
PACIFIC FESCUE, 452
PACIFIC MEADOW FOXTAIL, 786
PACIFIC REEDGRASS, 724
PACIFIC WILDRYE, 358
PACMCAD clade, 6
PALE FALSE MANNAGRASS, 608
PANIC LAINEUX, 793
PANICEAE R. Br., 7, 792
PANICOIDEAE Link, 6, 7, 10
PANICUM L., 793
Panicum sect. *Brachiaria* Trin., 793
Panicum subg. *Megathyrsus* Pilg., 793
Panicum holosericum R. Br., 793
Panicum miliaceum L., 793
PANIC VELDTGRASS, 34
PAPPOPHOREAE Kunth, 8
PARAPHOLIINAE Caro, 57
Parapholis C.E. Hubb., 380, **687**, 691
Parapholis incurva (L.) C.E. Hubb., 687, **688**, *688*
Parapholis strigosa (Dumort.) C.E. Hubb., 687, **688**, *688*
PARISH'S ALKALI GRASS, 465
PARISH'S NEEDLEGRASS, 129
×**Pascoleymus** (B. Boivin) Barkworth, 239, **351**
×**Pascoleymus bowdenii** (B. Boivin) Barkworth, 353, *353*
Pascopyrum Á. Löve, 239, 240, 278, 288, **348**, 351
Pascopyrum smithii (Rydb.) Á. Löve, 327, 344, **351**, *352*, 353
Paspalum L., 793
Paspalum almum Chase, 793
PATTERSON'S BLUEGRASS, 582
PÂTURIN DU LABRADOR, 602
PEKULNEI BLUEGRASS, 578
PENDANT GRASS, 607
Pennisetum Rich., 793
Pennisetum ciliare (L.) Link, 793
PERENNIAL BEARDGRASS, 668
PERENNIAL QUAKINGGRASS, 614
PERENNIAL RYEGRASS, 455
PERENNIAL VELDTGRASS, 34
PERPLEXING NEEDLEGRASS, 135
PERSIAN DARNEL, 459
PERSIAN GOATGRASS, 263
PERSIAN WHEAT, 274
PERUVIAN NEEDLEGRASS, 179
PERUVIAN WATERGRASS, 55
PETIT ÉPEAUTRE, 272
Phalarideae Dumort., 379
Phalaris L., 379, 381, 758, **764**
Phalaris angusta Nees *ex* Trin., 766, 770, 772, **773**
Phalaris aquatica L., 766, **767**, 769, 772
Phalaris arundinacea L., 765, 768, **770**, *771*
  Phalaris arundinacea var. *picta* L., 772
  Phalaris arundinacea forma *variegata* (Parn.) Druce, 772

Phalaris brachystachys Link, 766, **768**, *769*
Phalaris caesia Nees, 770
Phalaris californica Hook. & Arn., *765*, **770**, *771*
Phalaris canariensis L., *765*, 766, **768**, *769*
Phalaris caroliniana Walter, *765*, 766, 770, **772**, *772*
Phalaris coerulescens Desf., *765*, **766**, 767
Phalaris ×daviesii S.T. Blake, 767, 768
Phalaris lemmonii Vasey, *765*, **768**, *769*, 772
Phalaris minor Retz., 766, 767, **768**, *769*
Phalaris ×monspeliensis Daveau, 768, *772*
Phalaris paradoxa L., *765*, 766, **767**
Phalaris platensis Henrard *ex* Wacht., 770
Phalaris rotgesii (Husn.) Baldini, 770
PHALARIS DES CANARIES, 768
PHALARIS ROSEAU, 770
PHAREAE Stapf, 7, **11**
PHAROIDEAE L.G. Clark & Judz., 6, 7, **11**
Pharus P. Browne, **12**
Pharus glaber Kunth, 12, **13**, *13*
Pharus mezii Prod., 12
Pharus parvifolius Nash, 13
Phippsia (Trin.) R. Br., 379, 381, 382, 477, **478**
Phippsia algida (Sol.) R. Br., 467, 477, 478, **479**, *479*, 480
   Phippsia algida subsp. *algidiformis* (Harry Sm.) Á. Löve
     & D. Löve, 480
   Phippsia algida forma *vestita* Holmb., 479
Phippsia concinna (Th. Fr.) Lindeb., *479*, **480**
PHIPPSIE FROIDE, 479
PHLÉOLE ALPINE, 672
PHLÉOLE DES PRÉS, 672
Phleum L., 380, 381, **670**, 781
Phleum alpinum L., 671, **672**, *673*
   Phleum alpinum L. subsp. **alpinum**, 672
   Phleum alpinum subsp. **rhaeticum** Humphries, 672
Phleum arenarium L., 671, 674, **675**
Phleum commutatum Gaudin, 672
Phleum exaratum Hochst. *ex* Griseb., 671
Phleum paniculatum Huds., 671, **672**, *674*
Phleum phleoides (L.) H. Karsten, 671, **672**, *673*
Phleum pratense L., 379, 671, **672**, *673*, 781
   Phleum pratense subsp. **bertolonii** (DC.) Bornm., 672
   Phleum pratense L. subsp. **pratense**, *672*, *673*
Phleum subulatum (Savi) Asch. & Graebn., 671, *674*, **675**
Phragmites Adans., 791
Phragmites australis (Cav.) Trin. *ex* Steud., 791
   Phragmites australis subsp. **americanus** Saltonstall, P.M.
     Peterson & Soreng, 791
   Phragmites australis var. **berlandieri** (E. Fourn.) C.F. Reed,
     791
PHYLLORACHIDEAE C.E. Hubb., 32
Phyllostachys Siebold & Zucc., 17, **25**
Phyllostachys aurea Carrière *ex* Rivière & C. Rivière, **27**
Phyllostachys bambusoides Siebold & Zucc., 26, **27**
PICKERING'S REED BENTGRASS, 724
PILLAR-OF-SMOKE, 185
PINE BLUEGRASS, 588
PINE REEDGRASS, 723
PINEGRASS, 438, 723
PINEWOODS NEEDLEGRASS, 137
PINYON RICEGRASS, 164

PIPER'S BLUEGRASS, 554
PIPTATHERUM
   CANADIAN, 146
   LITTLE, 146
   SHARP, 146
   SHOSHONE, 148
   SMALL-FLOWERED, 148
Piptatherum P. Beauv., 111, 142, **144**, 167
Piptatherum canadense (Poir.) Dorn, 145, **146**, *147*
Piptatherum exiguum (Thurb.) Dorn, 145, **146**, *147*
Piptatherum micranthum (Trin. & Rupr.) Barkworth, 141,
   145, **148**, *149*
Piptatherum miliaceum (L.) Coss., 145, *150*, **151**
   Piptatherum miliaceum (L.) Coss. subsp. **miliaceum**, *150*,
     151
   Piptatherum miliaceum subsp. **thomasii** (Duby) Soják,
     150, **151**
Piptatherum pungens (Torr.) Dorn, 145, **146**, *147*
Piptatherum racemosum (Sm.) Eaton, 145, **148**, *150*
Piptatherum shoshoneanum (Curto & Douglass M. Hend.)
   P.M. Peterson & Soreng, 145, **148**, *149*
Piptochaetium J. Presl, 110, 157, **161**
Piptochaetium avenaceum (L.) Parodi, 162, *163*, **164**
Piptochaetium avenacioides (Nash) Valencia & Costas, 162,
   163, **164**
Piptochaetium fimbriatum (Kunth) Hitchc., 162, **164**, *165*
Piptochaetium leianthum (Hitchc.) Beetle, 164
Piptochaetium pringlei (Beal) Parodi, **162**, *163*, 166
Piptochaetium seleri (Pilg.) Henrard, 166
Piptochaetium setosum (Trin.) Arechav., 162, *165*, **166**
Piptochaetium stipoides (Trin. & Rupr.) Hack., 162, *165*,
   166
   Piptochaetium stipoides var. **echinulatum** Parodi, **166**
   Piptochaetium stipoides (Trin. & Rupr.) Hack. var.
     stipoides, **166**
PLAINS BLUEGRASS, 599
PLAINS REEDGRASS, 724
PLAINS ROUGH FESCUE, 407
Pleuropogon R. Br., 67, 68, **103**
Pleuropogon subg. **Lophochlaena** (Nees) But, 105
Pleuropogon R. Br. subg, **Pleuropogon**, 105
Pleuropogon californicus (Nees) Benth. *ex* Vasey, 105, *106*
   Pleuropogon californicus (Nees) Benth. *ex* Vasey var.
     californicus, 105, *106*, **107**
   Pleuropogon californicus var. **davyi** (L.D. Benson) But,
     105, *106*, **107**
Pleuropogon hooverianus (L.D. Benson) J.T. Howell, 105,
   106, **107**
Pleuropogon oregonus Chase, 105, **107**, *108*
Pleuropogon refractus (A. Gray) Benth. *ex* Vasey, 105, **107**,
   108
Pleuropogon sabinei R. Br., 105, *108*, **109**
PLEUROPOGON DE SABINE, 109
PLUMEGRASS
   AUSTRALIAN, 185
PLUMOSE NEEDLEGRASS, 179
Poa L., 4, 379, 388, **486**, 602
Poa sect. **Abbreviatae** Nannf. *ex* Tzvelev, 487, **579**
Poa sect. **Alpinae** (Hegetschw. *ex* Nyman) Stapf, **516**
Poa sect. **Aphydris** (Griseb.) Tzvelev, 598

Poa sect. Arctopoa (Griseb.) Tzvelev, **598**
Poa sect. Arenariae (Hegetschw.) Stapf, **515**
Poa sect. Dioicopoa E. Desv., **566**
Poa sect. Homalopoa Dumort., **533**, 540
Poa sect. Madropoa Soreng, **542**
Poa sect. Micrantherae Stapf, **518**, 568
Poa sect. Oreinos Asch. & Graebn., 538, **568**, 573
Poa sect. Pandemos Asch. & Graebn., **568**
Poa L. sect. Poa, 521, **585**, 601
Poa sect. Secundae V.L. Marsh *ex* Soreng, **568**, 585
Poa sect. Stenopoa Dumort., 568, **573**, 579
Poa sect. Sylvestres V.L. Marsh *ex* Soreng, **510**, 540, 598
Poa sect. Tichopoa Asch. & Graebn., **579**
Poa subg. Arctopoa (Griseb.) Prob., 509, **596**
Poa L. subg. Poa, **509**
Poa subsect. Epiles Hitchc. *ex* Soreng, 542, **559**
Poa subsect. Halophytae V.L. Marsh *ex* Soreng, 585, **592**
Poa subsect. Madropoa Soreng, **542**, 550
Poa subsect. Secundae Soreng, **585**, 592
Poa abbreviata R. Br., 469, 490, 491, 502, 580, **582**, *583*
    Poa abbreviata R. Br. subsp. **abbreviata**, **582**, *583*
    Poa abbreviata subsp. *jordalii* (A.E. Porsild) Hultén, 584
    Poa abbreviata subsp. **marshii** Soreng, **582**, *583*
    Poa abbreviata subsp. **pattersonii** (Vasey) Á. Löve, D. Löve & B.M. Kapoor, 578, **582**, *583*
Poa alpina L., 487, 489, 506, *517*, **518**, 522, 526, 530, 601
    Poa alpina L. subsp. **alpina**, *517*, **518**
    Poa alpina subsp. **vivipara** (L.) Arcang., *517*, **518**
Poa alsodes A. Gray, 490, **511**, *512*
Poa ammophila A.E. Porsild, 508, 591, **592**, *593*
Poa ampla Merr., 588
Poa annua L., 481, 487, 490, **519**, *520*, 521, 534
Poa arachnifera Torr., 494, **566**, *569*
Poa arctica R. Br., 487, 489, 497, 501, 522, 526, **529**, *531*, 578, 586, 594, 601
    Poa arctica subsp. **aperta** (Scribn. & Merr.) Soreng, **530**, *531*, 532, 599
    Poa arctica R. Br. subsp. **arctica**, **530**, *531*
    Poa arctica subsp. **caespitans** Simmons *ex* Nannf., **530**, *531*, 532
    Poa arctica subsp. **grayana** (Vasey) Á. Löve, D. Löve & B.M. Kapoor, 530, *531*, **532**, 601
    Poa arctica subsp. **lanata** (Scribn. & Merr.) Soreng, **529**, 530, *531*, 532
Poa arida Vasey, 488, 496, 530, 532, 586, **599**, *600*, 601
Poa arnowiae Soreng, 499, **543**, *544*, 545
Poa atropurpurea Scribn., 498, **554**, *557*
Poa autumnalis Muhl. *ex* Elliott, 492, 502, 505, **513**, *514*
Poa bigelovii Vasey & Scribn., 490, 534, **536**, *537*
Poa bolanderi Vasey, 490, 533, **534**, *535*
Poa bonariensis (Lam.) Kunth, 566
Poa bulbosa L., 487, 488, **516**, *517*
    Poa bulbosa L. subsp. **bulbosa**, **516**, *517*
    Poa bulbosa subsp. **vivipara** (Koel.) Arcang., **516**, *517*
Poa canbyi (Scribn.) Howell, 588
Poa chaixii Vill., 503, **540**, *541*
Poa chambersii Soreng, 494, 499, 546, **548**, *549*, 562
Poa chapmaniana Scribn., 490, 519, **534**, *535*, 536
Poa compressa L., 487, 494, **579**, *581*
Poa confinis Vasey, 494, 498, 523, 552, *554*, *555*

Poa curta Rydb., 543
Poa curtifolia Scribn., 504, 508, 564, 584, 586, **589**, *590*
Poa cusickii Vasey, 499, 502, 503, 505, 546, **559**, *561*, *562*, 564
    Poa cusickii Vasey subsp. **cusickii**, 560, *561*, *562*, 563, 564
    Poa cusickii subsp. **epilis** (Scribn.) W.A. Weber, 560, *561*, 562
    Poa cusickii subsp. **pallida** Soreng, 556, 560, *561*, 562, 564
    Poa cusickii subsp. **purpurascens** (Vasey) Soreng, 548, 560, *561*, *562*, 563
Poa cuspidata Nutt., 499, *544*, 545
Poa diaboli Soreng & D.J. Keil, 495, 552, *555*
Poa douglasii Nees, 498, **551**, 552, *553*
Poa eminens J. Presl, 493, 509, **598**, *599*, 602, 604
Poa fendleriana (Steud.) Vasey, 487, 495, 503, 506, **556**, *558*, 562, 601
    Poa fendleriana subsp. **albescens** (Hitchc.) Soreng, 556, *558*
    Poa fendleriana (Steud.) Vasey subsp. **fendleriana**, 556, *558*
    Poa fendleriana subsp. **longiligula** (Scribn. & T.A. Williams) Soreng, 556, 558, *558*
Poa flexuosa Sm., 572
Poa ×gaspensis Fernald, 488, 518, 526, 530, *600*, **601**
Poa glauca Vahl, 469, 487, 489, 491, 502, 505, 529, 532, 570, 572, *576*, **577**, 584, 586, 591
    Poa glauca Vahl subsp. **glauca**, *577*, **578**, 579
    Poa glauca var. **pekulnejensis** (Jurtzev & Tzvelev) Prob., *577*, 578
    Poa glauca subsp. **rupicola** (Nash) W.A. Weber, 576, *577*, 578
Poa glaucantha Gaud., 578
Poa glaucifolia Scribn. & T.A. Williams, 601
Poa gracillima Vasey, 588
Poa hansenii Scribn., 564
Poa hartzii Gand., 489, 506, 508, 578, **589**, 592, *593*
    Poa hartzii subsp. **alaskana** Soreng, **591**, *593*
    Poa hartzii Gand. subsp. **hartzii**, **591**, *593*
    Poa hartzii var. **vivipara** Polunin, **591**
    Poa hartzii subsp. **vrangelica** (Tzvelev) Soreng & L.J. Gillespie, 578, **591**, *593*
    Poa hartzii var. **vrangelica** (Tzvelev) Prob., **591**
Poa howellii Vasey & Scribn., 490, 514, **534**, *535*
Poa incurva Scribner & T.A. Williams, 588
Poa infirma Kunth, 490, **519**, *520*
Poa interior Rydb., 503, 505, 574, **576**, *577*, 578, 579
Poa involuta Hitchc., 540
Poa juncifolia Scribn., 588
Poa keckii Soreng, 491, 505, 507, 562, 564, *583*, **584**
Poa kelloggii Vasey, 494, *513*, **514**
Poa lanata var. *vivipara* Hultén, 533
Poa lanuginosa Poir., 566
Poa laxa Haenke, 492, 502, **570**, *571*, 572, 578
    Poa laxa subsp. **banffiana** Soreng, **570**, *571*, 578
    Poa laxa subsp. **fernaldiana** (Nannf.) Hyl., 570, *571*, **572**
    Poa laxa subsp. **flexuosa** (Sm.) Hyl., 570, *572*
    Poa laxa Haenke subsp. **laxa**, 570, *572*
Poa laxa × glauca, 491, 570, *571*, **572**, 578

Poa laxiflora Buckley, 489, 493, 514, **538**, *539*
Poa laxiuscula (Blytt) Lange, 572
Poa leibergii Scribn., 503, **563**, 564, *565*
Poa leptocoma Trin., 492, 493, 538, **573**, *575*
Poa lettermanii Vasey, 491, **580**, *581*, 582
Poa ×limosa Scribn. & T.A. Williams, 488, *600*, **601**
Poa longipila Nash, 532
Poa macrantha Vasey, 498, **551**, *553*
Poa macrocalyx Trautv. & C.A. Mey., 489, 500, 523, **527**, *528*, 532, 533
Poa macroclada Rydb., 594, *595*
Poa malacantha Kom., 532
Poa marcida Hitchc., 490, 510, **512**, *513*
Poa montevansii E.H. Kelso, 580
Poa ×multnomae Piper, 545
Poa napensis Beetle, 504, 509, **594**, *597*
Poa ×nematophylla Rydb., 506, 556, **562**, *565*
Poa nemoralis L., 491, 502, **574**, 576, *577*, 578
Poa nervosa (Hook.) Vasey, 497, 542, **545**, 546, *547*, 586, 589
Poa nevadensis Vasey *ex* Scribn., 588
Poa norbergii Hultén, 529
Poa occidentalis Vasey, 493, 533, **536**, *537*, 538, 543
Poa paludigena Fernald & Wiegand, 493, 512, 568, **572**, *575*
Poa palustris L., 499, 503, **574**, *575*, 576
Poa paucispicula Scribn. & Merr., 492, **538**, *539*, 573
Poa piperi Hitchc., 494, 548, **554**, *557*
Poa porsildii Gjaerev., 504, 562, **563**, *565*
Poa pratensis L., 379, 487, 489, 501, **522**, *524*, *525*, 529, 530, 546, 548, 552, 586, 601
    Poa pratensis subsp. **agassizensis** (B. Boivin & D. Löve) Roy L. Taylor & MacBryde, 522, **523**, *524*
    Poa pratensis subsp. **alpigena** (Lindm.) Hiitonen, 518, 522, 523, *524*, **525**, *526*, 530, 601
    Poa pratensis subsp. **angustifolia** (L.) Lej., 523, *524*, *525*, **526**, 527
    Poa pratensis subsp. **colpodea** (Th. Fr.) Tzvelev, 522, 523, *524*, **526**
    Poa pratensis subsp. **irrigata** (Lindm.) H. Lindb., 523, *524*, **526**, 527
    Poa pratensis L. subsp. **pratensis**, 523, *525*, **526**
Poa pringlei Scribn., 504, 560, 562, **564**, *567*, 584, 589
Poa pseudoabbreviata Roshev., 489, 490, **580**, *581*
Poa reflexa Vasey & Scribn., 492, 536, **538**, *539*, 573
Poa rhizomata Hitchc., 500, 523, 546, 548, *549*, 554
Poa saltuensis Fernald & Wiegand, 490, 502, 510, *511*, 512
    Poa saltuensis subsp. **languida** (Hitchc.) A. Haines, 510, *511*
    Poa saltuensis Fernald & Wiegand subsp. **saltuensis**, 510, *511*
Poa sandbergii Vasey, 588
Poa scabrella (Thurb.) Benth. *ex* Vasey, 588
Poa secunda J. Presl, 139, 319, 487, 496, 504, 507, 509, 522, 530, 545, 578, **586**, *587*, 589, 591, 594, 601
    Poa secunda subsp. **juncifolia** (Scribn.) Soreng, **586**, *587*, 588, 601
    Poa secunda J. Presl subsp. **secunda**, 586, *587*, **588**, 594
Poa sierrae J.T. Howell, 494, **548**, *550*
Poa stebbinsii Soreng, 504, 560, 563, **564**, *567*

Poa stenantha Trin., 489, 507, **592**, *595*
    Poa stenantha Trin. var. **stenantha**, 588, **594**, *595*
    Poa stenantha var. **vivipara** Trin., 530, **594**, *595*
Poa strictiramea Hitchc., 502, 503, 507, **540**, *541*
Poa sublanata Reverd., 501, **527**, *528*
Poa suksdorfii (Beal) Vasey *ex* Piper, 491, 505, 562, 564, **584**, *587*, 589
Poa supina Schrad., 495, 519, *520*, **521**
Poa sylvestris A. Gray, 492, 502, *511*, **512**, 514, 572
Poa tenerrima Scribn., 507, 545, 586, **588**, *590*
Poa tibetica Munro *ex* Stapf, 598
Poa tolmatchewii Roshev., 532
Poa tracyi Vasey, 500, 536, **543**, *544*, 545
Poa trichopoda Lange, 532
Poa trivialis L., 487, 495, 501, **568**, *569*, 572
    Poa trivialis L. subsp. **trivialis**, 568
Poa unilateralis Scribn., 504, 507, **594**, *597*
    Poa unilateralis subsp. **pachypholis** (Piper) D.D. Keck *ex* Soreng, **596**, *597*
    Poa unilateralis Scribn. subsp. **unilateralis**, **596**, *597*
Poa wheeleri Vasey, 487, 497, 522, 543, 545, **546**, *547*, 548
Poa wolfii Scribn., 493, 512, **514**, *515*
POACEAE Barnhart, 3, 4, 6, 7, 11, **57**, 188
POALES Small, 6
Podagrostis (Griseb.) Scribn. & Merr., 384, 634, **693**
Podagrostis aequivalvis (Trin.) Scribn. & Merr., **693**, 694, *695*
Podagrostis humilis (Vasey) Björkman, 656, 693, **694**, *695*
Podagrostis thurberiana (Hitchc.) Hultén, 694
POEAE R. Br., 8, 10, 57, 58, 192, **378**, 383, 385, 610
POLARGRASS, 678
    REED, 678
POLISH WHEAT, 274
Polypogon Desf., 379, 380, 383, 384, 634, **662**, 668
Polypogon ×adscendens Guss. *ex* Bertol., 663, 665
Polypogon australis Brongn., 662, **665**, 666
Polypogon elongatus Kunth, 662, *663*, 664
Polypogon fugax Nees *ex* Steud., *663*, 666
Polypogon imberbis (Phil.) Johow, 663, 667, **668**
Polypogon interruptus Kunth, 662, **663**, 664
Polypogon maritimus Willd., 662, 665, 667
    Polypogon maritimus Willd. var. **maritimus**, 665
    Polypogon maritimus var. **subspathaceus** (Req.) Bonnier & Layens, 665
Polypogon monspeliensis (L.) Desf., 643, 662, 663, **665**, 666, 668
Polypogon viridis (Gouan) Breistr., 662, *663*, 664, 665
POÖIDEAE Benth., 6, 8, 10, **57**, 59, 62, 64, 67, 110, 162, 187
PORCUPINEGRASS, 161
    SMALL, 158
PORSILD'S BLUEGRASS, 563
PORTER'S MELIC, 98
PORTER'S PTILAGROSTIS, 144
PORTER'S REEDGRASS, 721
PRAIRIE TRISETUM, 750
PRAIRIE WEDGEGRASS, 621
PRINGLE'S BLUEGRASS, 564
PRINGLE'S SPEARGRASS, 162
PROLIFEROUS FESCUE, 419

Psathyrostachys Nevski, 240, 354, **372**

Psathyrostachys juncea (Fisch.) Nevski, 364, **372**, *373*

×**Pseudelymus** Barkworth & D.R. Dewey, 239, **282**, 288, 334

×**Pseudelymus saxicola** (Scribn. & J.G. Sm.) Barkworth & D.R. Dewey, **283**, *283*

Pseudoroegneria (Nevski) Á. Löve, 239, 240, 278, **279**, 282, 288, 289

Pseudoroegneria elytrigioides (C. Yen & J.L. Yang) Bao-Rong Lu, 281

Pseudoroegneria geniculata (Trin.) Á. Love, 281

Pseudoroegneria spicata (Pursh) Á. Löve, **281**, *280*, 283, 303, 318, 327, 330, 332, 334, 336

    Pseudoroegneria spicata forma **inermis** (Scribn. & J.G. Sm.) Barkworth, 282

    Pseudoroegneria spicata forma **pubescens** (Elmer) Barkworth, 282

    Pseudoroegneria spicata (Pursh) Á. Löve forma **spicata**, 282

Pseudoroegneria strigosa (M.-Bieb.) Á. Löve, 281

Pseudosasa Makino *ex* Nakai, 17, **27**

Pseudosasa japonica (Siebold & Zucc. *ex* Steud.) Makino *ex* Nakai, *28*, *29*

Pseudoviviparous Fescue, 419

Psilocybe azuarescens Griseb., 778

Ptilagrostis Griseb., 111, **143**

Ptilagrostis kingii (Bol.) Barkworth, **143**, *144*

Ptilagrostis mongholica (Turcz. *ex* Trin.) Griseb., 144

Ptilagrostis porteri (Rydb.) W.A. Weber, 143, **144**, *144*

Ptilagrostis
    King's, 143
    Porter's, 144
    Rocky-Mountain, 144
    Sierra, 143

Puccinellia Parl., 68, 379, 385, 387, 388, **459**, 477, 607

Puccinellia agrostidea T.J. Sørensen, 473

Puccinellia alaskana Scribn. & Merr., 469, 471

Puccinellia andersonii Swallen, 460, 461, 462, **473**, *474*

Puccinellia angustata (R. Br.) E.L. Rand & Redfield, 461, **467**, *470*

Puccinellia arctica (Hook.) Fernald & Weath., 461, 462, **471**, *472*

Puccinellia borealis Swallen, 475

Puccinellia bruggemannii T.J. Sørensen, 461, **469**, *470*

Puccinellia distans (Jacq.) Parl., 460, 461, **473**, *474*

    Puccinellia distans subsp. **hauptiana** (Trin. *ex* V.I. Krecz.) W.E. Hughes, 473

Puccinellia fasciculata (Torr.) E.P. Bicknell, 460, **463**, *464*

Puccinellia grandis Swallen, 477

Puccinellia groenlandica T.J. Sørensen, 461, **467**, *468*

Puccinellia hauptiana (Trin. *ex* V.I. Krecz.) Kitag., 473

Puccinellia howellii J.I. Davis, 462, *474*, **475**

Puccinellia langeana (Berlin) T.J. Sørensen *ex* Hultén, 469, 471

Puccinellia lemmonii (Vasey) Scribn., 461, 462, **469**, *472*

Puccinellia lucida Fernald & Weath., 477

Puccinellia maritima (Huds.) Parl., 460, **463**, 466

Puccinellia nutkaensis (J. Presl) Fernald & Weath., 462, **475**, *476*

Puccinellia nuttalliana (Schult.) Hitchc., 462, 471, **475**, *476*, 477

Puccinellia parishii Hitchc., 461, **465**, *466*

    Puccinellia paupercula var. *longiglumis* Fernald & Weath., 471

Puccinellia phryganodes (Trin.) Scribn. & Merr., 460, **463**, *464*

Puccinellia poacea T.J. Sørensen, 473

Puccinellia pumila (Vasey) Hitchc., 460, 461, 469, **471**, *472*, 477

Puccinellia rupestris (With.) Fernald & Weath., 460, **463**, *464*

Puccinellia simplex Scribn., 461, **465**, *466*

Puccinellia tenella (Lange) Holmb. *ex* Porsild, 461, **469**, *470*, 471

    Puccinellia tenella subsp. *langeana* (Berlin) Tzvelev, 469

Puccinellia tenuiflora (Griseb.) Scribn. & Merr., 473

Puccinellia vaginata (Lange) Fernald & Weath., 462, *476*, **477**

Puccinellia vahliana (Liebm.) Scribn. & Merr., 460, 461, **465**, 467, 468, **478**, *479*

Puccinellia wrightii (Scribn. & Merr.) Tzvelev, 461, **467**, *468*, 477

Puccinellie à Fleurs Distantes, 473

Puccinellie Brillante, 475

Puccinellie de Vahl, 465

Puccinellie Engainée, 477

Puccinellie Étroite, 467

Puccinellie Maritime, 463

Puccinellie Naine, 471

Puccinellie Rampante, 463

Puccinellie Trompeuse, 463

×**Pucciphippsia** Tzvelev, 379, **477**

×**Pucciphippsia czukczorum** Tzvelev, 477

×**Pucciphippsia vacillans** (Th. Fr.) Tzvelev, 477, **478**, *479*

PUELIOIDEAE L.G. Clark, M. Kobay., S. Mathews, Spangler, & E.A. Kellogg, 6

Puna Needlegrass, 182

Purple Falsebrome, 188

Purple Nassella, 174

Purple Needlegrass, 174

Purple Oniongrass, 91

Purple Reedgrass, 710

Purple-Stem Cat's Tail, 672

Quackgrass, 334
    Tick, 376

Quakinggrass
    Big, 614
    Little, 614
    Perennial, 614

Queue d'Écureuil, 245

Rabbitsfoot Grass, 665

Rainier Reedgrass, 716

Rattail Fescue, 449

Rattlesnake Brome, 228

Rattlesnake Mannagrass, 79

Ravine Fescue, 404

Ray-Grass Anglais, 455

RAY-GRASS D'ITALIE, 456
RED BROME, 226
RED FESCUE, 412, 418
    ARCTIC, 414
    DUNE, 415
    FLATLEAF, 415
    MOUNTAIN, 418
    SECUND, 418
RED RICE, 38, 40
RED WILD EINKORN, 270
REDTOP, 641
    IDAHO, 649
    OREGON, 649
REED BENTGRASS
    PICKERING'S, 724
REED CANARYGRASS, 770
REED POLARGRASS, 678
REEDGRASS
    BARTLEY'S, 723
    BOLANDER'S, 719
    CAIN'S, 719
    CASCADE, 714
    CHEE, 710
    CIRCUMPOLAR, 719
    DENSE-PINE, 721
    DITCH, 717
    FEATHER, 721
    FOERSTER'S, 721
    HOWELL'S, 712
    JONES', 717
    LAPLAND, 729
    LEAFY, 714
    MACOUN'S, 728
    MUIR'S, 717
    NORTHERN, 730
    ONE-AND-A-HALF-FLOWERED, 714
    PACIFIC, 724
    PINE, 723
    PLAINS, 724
    PORTER'S, 721
    PURPLE, 710
    RAINIER, 716
    SERPENTINE, 712
    SHORTHAIR, 717
    SLIMSTEM, 729
    SMALL, 716
    TWEEDY'S, 714
    WOOD, 726
Reimarochloa Hitchc., 793
RENDLE'S MEADOW FOXTAIL, 786
RESCUE GRASS, 199
RESTIONACEAE R. Br., 6
RHIZOME BLUEGRASS, 546
RHODE ISLAND BENT, 639
'RIBBON GRASS', 772
RICE, 40
    BROWNBEARD, 38
    LONGSTAMEN, 38
    RED, 38, 40
RICE CUTGRASS, 44

RICEGRASS
    ARNOW'S, 141
    BRISTLY, 166
    CONTRACTED, 141
    INDIAN, 139
    MOUNTAIN, 148
    PINYON, 164
    ROUGHLEAF, 168
    STIPOID, 166
RICHARDSON'S BROME, 222
RICHARDSON'S NEEDLEGRASS, 133
RIDGED GLYCERIA, 77
RIPGUT GRASS, 224
RIVER CANE, 18
RIVERBANK WILDRYE
    EASTERN, 302
    NORTHERN, 305
RIVERGRASS
    COMMON, 734
RIVET WHEAT, 274
RIZ SAUVAGE, 50
ROCK FESCUE, 415
ROCK MELIC, 97
ROCKY MOUNTAIN FESCUE, 430
ROCKY-MOUNTAIN PTILAGROSTIS, 144
Roegneria K. Koch, 239, 288, 289
Roegneria glaberrima Keng & S.L. Chen, 281
ROEMER'S FESCUE, 440
ROSEAU, 770
ROSS' BENT, 656
Rostraria Trin., 381, 386, 753, **756**
Rostraria cristata (L.) Trin., **756**, 756
    Rostraria cristata (L.) Trin. var. **cristata**, 756
    Rostraria cristata var. **glabriflora** (Trautv.) Doğan, 756
ROUGH AMPHIBROMUS, 704
ROUGH BENT, 646
ROUGH BLUEGRASS, 568
ROUGH FESCUE
    MOUNTAIN, 408
    NORTHERN, 407
    PLAINS, 407
ROUGH-SPIKED HARDGRASS, 262
ROUGHLEAF RICEGRASS, 168
RUSH WHEATGRASS, 378
RUSSIAN WHEATGRASS, 376
RUSSIAN WILDRYE, 372
RYE, 259
RYEBROME, 230
RYEGRASS
    ANNUAL, 456
    DOWNY, 366
    ENGLISH, 455
    ITALIAN, 456
    PERENNIAL, 455
    STIFF, 456
Rytidosperma Steud., 792
Rytidosperma biannulare (Zotov) Connor & Edgar, 792
Rytidosperma caespitosa (Gaudich.) Connor & Edgar, 792
Rytidosperma richardsonii (Cashmore) Connor & Edgar, 792

SALINA WILDRYE, 364
Salix, 144
SALMON WILDRYE, 364
SALTMARSH GRASS
   BORRER'S, 463
   COMMON, 463
   STIFF, 463
SAN BERNARDINO BLUEGRASS, 554
SAN DIEGO NEEDLEGRASS, 131
SAND BLUEGRASS, 592
SAND-DUNE WHEATGRASS, 327
SAND TIMOTHY, 675
SANDBERG BLUEGRASS, 588
SANDGRASS
   EARLY, 757
Sasa Makino & Shibata, 17
SCALY BROME, 230
Schedonorus P. Beauv., 386, 388, **445**, 455
Schedonorus arundinaceus (Schreb.) Dumort., 446, *447*, *448*,
   455
Schedonorus giganteus (L.) Holub, 446, *447*, **448**
Schedonorus pratensis (Huds.) P. Beauv., **446**, *447*, 455
Schizachne Hack., 67, 68, **103**
Schizachne purpurascens (Torr.) Swallen, **103**, *104*
SCHIZACHNÉ POURPRÉ, 103
Schizachyrium scoparium (Michx.) Nash, 790
Sclerochloa P. Beauv., 380, 382, 387, **480**
Sclerochloa dura (L.) P. Beauv., **481**, *481*
Scolochloa Link, 385, 387, **732**
Scolochloa festucacea (Willd.) Link, *733*, **734**
SCRIBNER GRASS, 689
SCRIBNER'S NEEDLEGRASS, 135
SCRIBNER'S WHEATGRASS, 330
Scribneria Hack., 380, **689**
Scribneria bolanderi (Thurb.) Hack., **689**, *690*
SEA BARLEY, 248
SEA-BLUFF BLUEGRASS, 594
SEA LYMEGRASS, 356
Secale L., 238, 240, **259**
Secale cereale L., **259**, *260*, 261
   Secale cereale subsp. **ancestrale** Zhuk., 259
   Secale cereale L. subsp. **cereale**, 259
Secale strictum (C. Presl) C. Presl, **259**, *260*
Secale sylvestre Host, 259
SECUND BLUEGRASS, 586
SECUND RED FESCUE, 418
SEIGLE, 259
SEIGLE CULTIVÉ, 259
SEIGLE DE MER, 356
SEMAPHOREGRASS
   ANNUAL, 107
   CALIFORNIA, 105
   DAVY'S, 107
   FALSE, 109
   HOOVER'S, 107
   NODDING, 107
   OREGON, 107
SERPENTINE REEDGRASS, 712
SERRATED TUSSOCKGRASS, 177
Sesleria Scop., 381, **678**

Sesleria autumnalis (Scop.) F.W. Schulz, **679**, *680*, 681
Sesleria caerulea (L.) Ard., *679*, *680*, **681**
Sesleria heufleriana Schur, 679, *680*, **681**
Sesleria nitida Ten., *679*, *680*, **681**
Setaria P. Beauv., 793
Setaria leucopila (Scribn. & Merr.) K. Schum., 793
Setaria texana Emery, 790
Setaria villosissima (Scribn. & Merr.) K. Schum., 793
SHARP PIPTATHERUM, 146
SHARPGLUME BROME, 218
SHEATHED ALKALI GRASS, 477
SHEEP FESCUE, 422, 424
   FINE-LEAVED, 424
SHINY WEDGEGRASS, 621
SHORT FESCUE, 426
SHORT-FLOWERED BLUEGRASS, 580
SHORTAWN FOXTAIL, 784
   SONOMA, 784
SHORTHAIR REEDGRASS, 717
SHORTHAIRED BEARDGRASS, 668
SHORTHUSK
   NORTHERN, 60
   SOUTHERN, 60
SHORTSPIKE CANARYGRASS, 768
SHOSHONE PIPTATHERUM, 148
SIBERIAN MELIC, 100
SIBERIAN TRISETUM, 750
SIBERIAN WHEATGRASS, 279
SIBERIAN WILDRYE, 310
SICKLEGRASS, 688
   CURVED, 688
   HAIRY, 688
   STRIGOSE, 688
SIERRA BLUEGRASS, 548
SIERRA PTILAGROSTIS, 143
SIERRA WHEATGRASS, 332
SIERRAN WOODREED, 774
SILKYBENT
   LOOSE, 789
SILKY-SPIKE MELIC, 100
SILVER HAIRGRASS, 616
SILVERLEAF GRASS, 54
SILVERY HAIRGRASS, 616
Sitanion Raf., 288
SITKA BROME, 201
SIXWEEKS FESCUE, 450
SKYLINE BLUEGRASS, 560
SLEEPYGRASS, 129
SLENDER FALSEBROME, 190
SLENDER HAIRGRASS, 631
SLENDER MEADOW FOXTAIL, 786
SLENDER OATS, 735
SLENDER WEDGEGRASS, 623
SLENDER WHEATGRASS, 322
SLENDER WILD OATS, 735
SLENDER WOODREED, 774
SLIMSTEM REEDGRASS, 729
SLOUGHGRASS
   AMERICAN, 486
SMALL FESCUE, 452

SMALL-FLOWERED FESCUE, 434
SMALL-FLOWERED PIPTATHERUM, 148
SMALL-LEAF BENT, 661
SMALL LITHACHNE, 30
SMALL PORCUPINEGRASS, 158
SMALL REEDGRASS, 716
SMALL SPELT, 272
SMILO GRASS, 151
SMITH'S MELIC, 95
SMOOTH ALKALI GRASS, 471
SMOOTH BARLEY, 250
SMOOTH BROME, 206, 233
SNAKERIVER WHEATGRASS, 332
SNOWGRASS, 480
SOFT WHEAT, 277
SONOMA SHORTAWN FOXTAIL, 784
Sorghum Moench, 793
    Sorghum bicolor subsp. **arundinaceum** (Desv.) de Wet &
        J.R. Harlan *ex* Davidse, 793
    Sorghum bicolor subsp. **×drummondii** (Steud.) de Wet *ex*
        Davidse, 793
SOUTH AMERICAN NEEDLEGRASS, 182
SOUTHEASTERN WILDRYE, 296
SOUTHERN BEARDGRASS, 663
SOUTHERN CUTGRASS, 44
SOUTHERN SHORTHUSK, 60
SOUTHERN SWAMP WALLABYGRASS, 704
SOUTHERN WEDGEGRASS, 621
SOUTHERN WILDRICE, 48
SOUTHWESTERN NEEDLEGRASS, 133
SOUTHWESTERN WILDRYE, 306
SPEARGRASS
    BLACKSEED, 164
    FLORIDA, 164
    PRINGLE'S, 162
SPELT, 277
    SMALL, 272
Spartina Schreb., 791
Spartina alterniflora Loisel., 792
Spartina bakeri Merr., 792
Spartina ×caespitosa A.A. Eaton, 792
Spartina cynosuroides (L.) Roth, 792
Spartina densiflora Brongn., 792
Spartina foliosa Trin., 792
Spartina gracilis Trin., 792
Spartina maritima (Curtis) Fernald, 792
Spartina patens (Aiton) Muhl., 791, 792
Spartina pectinata Link, 791, 792
Spartina spartinae (Trin.) Merr. *ex* Hitchc., 791, 792
Spartina ×townsendii H. Groves & J. Groves, 792
Sphenopholis Scribn., 387, 388, **620**, 744
Sphenopholis filiformis (Chapm.) Scribn., 620, **621**, 622,
    623
Sphenopholis intermedia (Rydb.) Rydb., 621, **623**, 623, 624,
    754
Sphenopholis longiflora (Vasey *ex* L.H. Dewey) Hitchc., 621,
    623, **624**
Sphenopholis nitida (Biehler) Scribn., 620, **621**, 622, 623
Sphenopholis obtusata (Michx.) Scribn., 620, **621**, *623*
    Sphenopholis obtusata var. *major* (Torr.) Erdman, 624

Sphenopholis ×pallens (Biehler) Scribn., 621, 623
Sphenopholis pensylvanica (L.) Hitchc., 620, **621**, *622*, 623
SPHENOPHOLIS INTERMÉDIAIRE, 623
SPHENOPHOLIS OBTUS, 621
SPIKE BENT, 651
SPIKE FESCUE, 445
SPIKE GRASS, 682
SPIKE HAIRGRASS, 616
SPIKE OATGRASS, 698
SPIKE TRISETUM, 748
SPIKED FALSE MANNAGRASS, 608
SPIKEGRASS, 445
Sporobolus R. Br., 791
Sporobolus coahuilensis Valdés-Reyna, **791**
Sporobolus pyramidatus (Lam.) Hitchc., 791
SPRING MILLETGRASS, 780
SQUARROSE BROME, 237
SQUIRRELTAIL
    BIG, 318
    CALIFORNIA, 319
    COMMON, 319
    LONGLEAF, 319
SQUIRRELTAIL BARLEY, 245
SQUIRRELTAIL FESCUE, 450
SQUIRRELTAIL GRASS, 245
STARGRASS
    WATER, 46
STEBBINS' BLUEGRASS, 564
STEBBINS' WHEATGRASS, 329
    NORTHERN, 329
STIFF RYEGRAS, 456
STIFF SALTMARSH GRASS, 463
STILLMAN'S NEEDLEGRASS, 118
Stipa L., 111, 114, **154**, 157, 161, 171
Stipa subg. *Jarava* (Ruiz. & Pav.) Trin. & Rupr., 178
Stipa subg. *Pappostipa* Speg., 178
Stipa subg. *Ptilostipa* Speg., 178
Stipa capensis Thunb., **155**, *156*
Stipa columbiana Macoun, 124
Stipa litoralis Phil., 184
Stipa pennata L., 155
Stipa pulcherrima K. Koch, **155**, *156*
Stipa tenacissima L., 155
STIPEAE Dumort., 8, 10, 58, **109**, 112, 114, 143, 152, 159,
    171
STIPOID RICEGRASS, 166
STOUT BAMBOOGRASS, 185
STOUT WOODREED, 774
STREAMBANK WHEATGRASS, 329
STRIGOSE SICKLEGRASS, 688
SUKSDORF'S BLUEGRASS, 584
SUKSDORF'S BROME, 218
SUNOLGRASS, 766
SUPINE BLUEGRASS, 521
SVENSON'S WILDRYE, 314
SWALLEN'S NEEDLEGRASS, 137
SWAMP OATS, 621
SWAMP WALLABYGRASS, 704
    COMMON, 704
    SOUTHERN, 704

SWEET VERNALGRASS, 759
SWEET WOODREED, 774
SWEETGRASS
    ALPINE, 760
    ARCTIC, 760
    CALIFORNIA, 762
    HAIRY, 764
    VANILLA, 762
SWITCH CANE, 18
SWOLLEN GOATGRASS, 263

Taeniatherum Nevski, 240, **257**
Taeniatherum caput-medusae (L.) Nevski, 258, *258*
    Taeniatherum caput-medusae (L.) Nevski subsp. **caput-medusae**, 258
TALL ALKALI GRASS, 467
TALL BROME, 213
TALL FESCUE, 448
TALL GLYCERIA, 73
TALL MANNAGRASS, 79
TALL MELIC, 100
TALL OATGRASS, 742
TALL TRISETUM, 746
TALL WHEATGRASS, 378
TAPERED ONIONGRASS, 95
TAUSCH'S GOATGRASS, 262
TEXAN NASSELLA, 172
TEXAN NEEDLEGRASS, 172
TEXAN WILDRICE, 50
TEXAS BLUEGRASS, 566
TEXAS BROME, 216
TEXAS FESCUE, 400
TEXAS WILDRYE, 312
Themeda Forssk., 793
Themeda quadrivalvis (L.) Kuntze, 793
THICKSPIKE WHEATGRASS, 327
THINLEAF FALSEBROME, 190
Thinopyrum Á. Löve, 239, 240, 278, 288, **373**
Thinopyrum elongatum (Host) D.R. Dewey, 378
Thinopyrum intermedium (Host) Barkworth & D.R. Dewey, 374, *375*
    Thinopyrum intermedium subsp. **barbulatum** (Schur) Barkworth & D.R. Dewey, *375*, **376**
    Thinopyrum intermedium (Host) Barkworth & D.R. Dewey subsp. **intermedium**, *375*, **376**
Thinopyrum junceum (L.) Á. Löve, 374, *375*, **376**
Thinopyrum ponticum (Podp.) Barkworth & D.R. Dewey, 374, **378**, *377*
Thinopyrum pycnanthum (Godr.) Barkworth, 374, **376**, *377*
THINTAIL, 691
THORNY BAMBOO
    GIANT, 22
THREE-AWNED BROME, 233
THREE-AWNED GOATGRASS, 265
THREE-FLOWER MELIC, 100
THURBER'S FESCUE, 408
THURBER'S NEEDLEGRASS, 125
THYSANOLAENEAE Sánchez-Ken, 9
TICK QUACKGRASS, 376
TICKLEGRASS, 646

TIMBER BAMBOO
    GIANT, 27
TIMBERLINE BLUEGRASS, 578
TIMOPHEEV'S WHEAT, 272
TIMOTHY, 672
    ALPINE, 672
    BRITISH, 672
    ITALIAN, 675
    SAND, 675
TOOWOOMBA CANARYGRASS, 768
TOR GRASS, 190
TORREY'S MELIC, 90
Torreyochloa G.L. Church, 387, **607**
Torreyochloa erecta (Hitchc.) G.L. Church, 607, **608**, *609*
Torreyochloa pallida (Torr.) G.L. Church, 68, 71, 607, **608**, *609*
    Torreyochloa pallida var. **fernaldii** (Hitchc.) Dore, **608**, *609*
    Torreyochloa pallida (Torr.) G.L. Church var. **pallida**, **608**, *609*
    Torreyochloa pallida var. **pauciflora** (J. Presl) J.I. Davis, **608**, *609*
Trachnyia Link, 188
TRACY'S BLUEGRASS, 543
Tribes
    Key to, 7
Trichodium Michx., 634
Trifolium subterraneum L., 704
TRISÈTE À ÉPI, 748
TRISÈTE FAUSSE-MÉLIQUE, 745
Trisetum Pers., 386, 624, **744**
Trisetum aureum (Ten.) Ten., 744, 752, **753**
Trisetum canescens Buckley, 745, **746**, 749
Trisetum cernuum Trin., 745, **748**, 749
    Trisetum cernuum var. projectum (Louis-Marie) Beetle, 746
Trisetum flavescens (L.) P. Beauv., 744, 745, **748**, *751*
Trisetum interruptum Buckley, 744, **750**, 752
Trisetum melicoides (Michx.) Scribn., **745**, 747
    Trisetum melicoides var. majus (A. Gray) Hitchc., 745
Trisetum montanum Vasey, 750
Trisetum orthochaetum Hitchc., 745, **746**, 747
Trisetum sesquiflorum Trin., 716
Trisetum sibiricum Rupr., **750**, *745*
    Trisetum sibiricum subsp. litorale Rupr. *ex* Roshev., **750**
    Trisetum sibiricum Rupr. subsp. **sibiricum**, **750**
Trisetum spicatum (L.) K. Richt., 716, 744, 745, **748**, 750, *751*
Trisetum wolfii Vasey, 745, 747,
TRISETUM
    BITTERROOT, 746
    NODDING, 748
    PRAIRIE, 750
    SIBERIAN, 750
    SPIKE, 748
    TALL, 746
    WOLF'S, 745
TRITICALE, 261
TRITICEAE Dumort., 10, 58, 187, 192, **238**, 278, 289, 455
×**Triticosecale** Wittm. *ex* A. Camus, 239, 259, 260, **261**

Triticum L., 4, 238, 240, 241, 259, 262, **268**
Triticum aestivum L., 261, 263, 268, 269, 274, 276, **277**
   Triticum aestivum subsp. *tibetanum* J.Z. Shao, 277
Triticum araraticum Jakubz., 268, 272
Triticum baeoticum Boiss., 270
Triticum boeoticum Boiss., 268, **270**, 271, 272
Triticum carthlicum Nevski, 269, **274**, *276*
Triticum compactum Host, 277
Triticum dicoccoides (Körn.) Körn. *ex* Schweinf., **270**, *271*
Triticum dicoccum Schrank *ex* Schübl., 270, **272**, *273*
Triticum durum Desf., 263, 268, 269, **272** 274, *275*
Triticum militinae Zhuk. & Migush., 272
Triticum monococcum L., 269, 270, **272**, *273*
Triticum polonicum L., 269, **274**, *275*
Triticum sinskajae Filat. & Kurkiev, 272
Triticum spelta L., 268, 269, 276, **277**
Triticum timopheevii (Zhuk.) Zhuk., 269, 270, **272**, *273*
Triticum turanicum Jakubz., 274
Triticum turgidum L., 269, **274**, *275*
Triticum urartu Thumanjan *ex* Gandilyan, 268, 270
'TSUTSUMIANA', 29
TUFTED FALSEBROME, 192
TUFTED FESCUE, 422
TUFTED FOXTAIL, 786
TUFTED HAIRGRASS, 628
TUNDRA ALKALI GRASS, 469
TUNDRAGRASS
   FISHER'S, 604
TUSSOCKGRASS
   ANDEAN, 174
   CHILEAN, 177
   SERRATED, 177
   URUGUAYAN, 172
   YASS, 177
Typha angustifolia, 734
Typha domingensis, 734
Typha latifolia, 734
TWEEDY'S REEDGRASS, 714
TWO-FLOWER MELIC, 100

UNEQUAL-GLUMED WILDRYE, 316
UPLAND BENT, 647
UPRIGHT BROME, 218
UPSIDEDOWN GRASS, 13
Urochloa P. Beauv., 793
Urochloa maxima (Jacq.) R.D. Webster, 793
URUGUAYAN TUSSOCKGRASS, 172
Ustilago esculenta, 50

VAHL'S ALKALI GRASS, 465
Vahlodea Fr., 386, 624, **691**
Vahlodea atropurpurea (Wahlenb.) Fr. *ex* Hartm., **692**, *692*
VALAIS FESCUE, 420
VANCOUVER WILDRYE, 358
VANILLA SWEETGRASS, 762
VARIOUS-LEAVED FESCUE, 420
VASEY'S MUTTONGRASS, 556
VEINY BLUEGRASS, 545
VELDTGRASS
   ANNUAL, 36

   LONGFLOWERED, 36
   PANIC, 34
   PERENNIAL, 34
VELVET BENT, 643
VELVETGRASS, 740
   CREEPING, 740
Ventenata Koeler, 386, **683**
Ventenata dubia (Leers) Coss., **684**, *684*
VENTENATA, 684
VERNALGRASS, 759
   SWEET, 759
VIRGINIA BROME, 216
VIRGINIA WILDRYE, 298
VIVIPAROUS FESCUE, 436
Vulpia C.C. Gmel., 381, 382, 386, 388, **448**
Vulpia bromoides (L.) Gray, 449, **452**, *453*
Vulpia ciliata Dumort., 449, **454**, *454*
Vulpia microstachys (Nutt.) Munro, 449, 452, *453*
   Vulpia microstachys var. **ciliata** (A. Gray *ex* Beal) Lonard & Gould, **452**, *453*
   Vulpia microstachys var. **confusa** (Piper) Lonard & Gould, **452**, *453*
   Vulpia microstachys (Nutt.) Munro var. **microstachys**, **452**, *453*
   Vulpia microstachys var. **pauciflora** (Scribn. *ex* Beal) Lonard & Gould, **452**, *453*
Vulpia myuros (L.) C.C. Gmel., 449, *451*
   Vulpia myuros forma **megalura** (Nutt.) Stace & R. Cotton, 449
   Vulpia myuros (L.) C.C. Gmel. forma **myuros**, 449
Vulpia octoflora (Walter) Rydb., 449, 450, *451*
   Vulpia octoflora var. **glauca** (Nutt.) Fernald, 450, *451*
   Vulpia octoflora var. **hirtella** (Piper) Henrard, 450, *451*
   Vulpia octoflora (Walter) Rydb. var. **octoflora**, 450, *451*
Vulpia sciurea (Nutt.) Henrard, 449, **450**, *451*
VULPIN À COURTES ARÊTES, 784
VULPIN ALPIN, 782
VULPIN BORÉALE, 782
VULPIN DES PRÉS, 782
VULPIN GÉNICULÉ, 784
VULPIN ROSEAU, 782

WALL BARLEY, 250
WALLABYGRASS
   COMMON SWAMP, 704
   SOUTHERN SWAMP, 704
WALLOWA NEEDLEGRASS, 138
WASATCH BLUEGRASS, 543
WASHINGTON FESCUE, 440
WATER BEARDGRASS, 663
WATER FOXTAIL, 784
WATER MANNAGRASS, 85
WATER MILLET, 52
WATER STARGRASS, 46
WATER WHORLGRASS, 610
WATERGRASS, 54
   ASIAN, 46
   BRAZILIAN, *55*
   ENGLISH, 73
   PERUVIAN, *55*

WATSON'S FESCUE-GRASS, 445
WAVY HAIRGRASS, 631
WAY BARLEY, 250
WEAK BLUEGRASS, 519
WEAK MANNAGRASS, 608
WEBBER'S NEEDLEGRASS, 137
WEDGEGRASS
  BAYOU, 624
  PRAIRIE, 621
  SHINY, 621
  SLENDER, 623
  SOUTHERN, 621
WEEPING BLUEGRASS, 512
WEEPING BROME, 213
WENATCHEE BLUEGRASS, 589
WESTERN ALKALI GRASS, 465
WESTERN BOG BLUEGRASS, 573
WESTERN FESCUE, 437
WESTERN MANNAGRASS, 85
WESTERN NEEDLEGRASS, 123
  COMMON, 123
WESTERN OATS, 737
WESTERN WHEATGRASS, 351
WESTERN WILDRYE
  COMMON, 306
WHEAT, 277
  BREAD, 277
  COMMON, 277
  CONE, 274
  DURUM, 272
  HARD, 272
  MACARONI, 272
  PERSIAN, 274
  POLISH, 274
  RIVET, 274
  SOFT, 277
  TIMOPHEEV'S, 272
WHEATGRASS
  ALASKAN, 326
  ANNUAL, 254
  ARCTIC, 324
  ARIZONA, 329
  BAKER'S, 330
  BEARDED, 322
  BLUEBUNCH, 281
  CRESTED, 279
  HAIRY, 376
  HIGH-ARCTIC, 326
  HOFFMANN'S, 336
  INTERMEDIATE, 376
  MONTANA, 334
  NORTHERN, 324
  NORTHERN STEBBINS', 329
  ONE-SIDED, 322
  RUSH, 378
  RUSSIAN, 376
  SAND-DUNE, 327
  SCRIBNER'S, 330
  SIBERIAN, 279
  SIERRA, 332

  SLENDER, 322
  SNAKERIVER, 332
  STEBBINS', 329
  STREAMBANK, 329
  TALL, 378
  THICKSPIKE, 327
  WESTERN, 351
WHEELER'S BLUEGRASS, 546
WHITE CUTGRASS, 44
WHITEGRASS, 44
  CEDAR, 42
WHITETOP, 734
WHORLGRASS
  WATER, 610
WIDE-GLUMED NEEDLEGRASS, 124
WILD EINKORN, 270
  RED, 270
WILD EMMER, 270
WILD OATS, 735
  SLENDER, 735
WILDRICE
  ASIAN, 50
  ESTUARINE, 48
  INTERIOR, 50
  NORTHERN, 50
  SOUTHERN, 48
  TEXAS, 50
WILDRYE
  ALKALI, 359
  ALTAI, 360
  AWNLESS, 300
  BEARDLESS, 360
  BLUE, 306
  BOREAL, 366
  CHURCH'S, 314
  COLORADO, 366
  COMMON WESTERN, 306
  DOWNY, 302
  EARLY, 295
  EASTERN RIVERBANK, 302
  GIANT, 362
  GLUMELESS, 316
  GREAT BASIN, 364
  GREAT PLAINS, 303
  MAMMOTH, 356
  MANY-FLOWERED, 362
  MANY-STEM, 360
  MEXICAN, 312
  MOJAVE, 364
  NORTHERN RIVERBANK, 305
  NORTHWESTERN, 310
  PACIFIC, 358
  RUSSIAN, 372
  SALINA, 364
  SALMON, 364
  SIBERIAN, 310
  SOUTHEASTERN, 296
  SOUTHWESTERN, 306
  STEBBINS', 329
  SVENSON'S, 314

TEXAN, 312
UNEQUAL-GLUMED, 316
VANCOUVER, 358
VIRGINIA, 298
YELLOW, 366
WINDGRASS
    COMMON, 789
    INTERRUPTED, 789
WINTER BENT, 647
WINTER GRASS, 168
WOLF'S BLUEGRASS, 514
WOLF'S TRISETUM, 745
WOOD MILLET, 780
WOOD REEDGRASS, 726
WOODBROME
    HAIRY, 209
WOODLAND BLUEGRASS, 512, 574
WOODLAND BROME, 211
WOODREED
    BOLANDER'S, 774
    DROOPING, 774
    SIERRAN, 774
    SLENDER, 774
    STOUT, 774
    SWEET, 774

WOODY MELIC, 91
WOOLY BROME, 220
WRIGHT'S ALKALI GRASS, 467

YADAKE, 29
YASS TUSSOCKGRASS, 177
YELLOW OATGRASS, 748
YELLOW WILDRYE, 366
YORKSHIRE FOG, 740

Zea L., 4
Zizania L., 4, 37, 47
Zizania aquatica L., 47, 48, *49*
    Zizania aquatica L. var. aquatica, 48, *49*
    Zizania aquatica var. brevis Fassett, 48, *49*
Zizania latifolia (Griseb.) Turcz. *ex* Stapf, 47, 50, *51*
Zizania palustris L., 47, 48, *49*
    Zizania palustris var. interior (Fassett) Dore, *49*, 50
    Zizania palustris L. var. palustris, *49*, 50
Zizania texana Hitchc., 47, 50, *51*
ZIZANIE AQUATIQUE, 48
ZIZANIE DES MARAIS, 50
Zizaniopsis Döll & Asch., 37, 52
Zizaniopsis miliacea (Michx.) Döll & Asch., 52, *53*

Numerical Listing of the Family, Subfamilies, Tribes, and Genera, Volume 24

The list below shows the family, subfamilies, tribes, and genera of this volume in the order in which they are treated and the pages on which they are treated. For the tribes and genera, the numbers on the left show the relative position of the genera within the volume, and correspond to the numbers shown in the page headers.

POACEAE or GRAMINEAE, 3

PHAROIDEAE, 11
1.00 PHAREAE, 11
1.01 Pharus, 12

BAMBUSOIDEAE, 14
2.00 BAMBUSEAE, 15
2.01 Arundinaria, 17
2.02. Bambusa, 21
2.03 Phyllostachys, 25
2.04 Pseudosasa, 27
3.00 OLYREAE, 29
3.01 Lithachne, 30

EHRHARTOIDEAE, 32
4.00 EHRHARTEAE, 33
4.01 Ehrharta, 33
5.00 ORYZEAE, 36
5.01. Oryza, 37
5.02 Leersia, 42
5.03 Hygroryza, 46
5.04 Zizania, 47
5.05 Zizaniopsis, 52
5.06 Luziola, 54

POÖIDEAE, 57
6.00 BRACHYELYTREAE, 59
6.01 Brachyelytrum, 59
7.00 NARDEAE, 62
7.01 Nardus, 62
8.00 DIARRHENEAE, 64
8.01 Diarrhena, 64
9.00 MELICEAE, 67
9.01 Glyceria, 68
9.02 Melica, 88

9.03 Schizachne, 103
9.04 Pleuropogon, 103
10.00 STIPEAE, 109
10.01 Ampelodesmos, 112
10.02 Achnatherum, 114
10.03 Ptilagrostis, 143
10.04 Piptatherum, 144
10.05 Macrochloa, 151
10.06 Celtica, 152
10.07 Stipa, 154
10.08 Hesperostipa, 157
10.09 Piptochaetium, 161
10.10 Oryzopsis, 167
10.11 ×Achnella, 169
10.12 Nassella, 170
10.13 Jarava, 178
10.14 Amelichloa, 181
10.15 Austrostipa, 184
11.00 BRACHYPODIEAE, 187
11.01 Brachypodium, 187
12.00 BROMEAE, 192
12.01 Bromus, 193
13.00 TRITICEAE, 238
13.01 Hordeum, 241
13.02 Eremopyrum, 252
13.03 Dasypyrum, 256
13.04 Taeniatherum, 257
13.05 Secale, 259
13.06 ×Triticosecale, 261
13.07 Aegilops, 261
13.08 Triticum, 268
13.09 Agropyron, 277
13.10 Pseudoroegneria, 279
13.11 ×Pseudelymus, 282
13.12 ×Elyhordeum, 283
13.13 Elymus, 288
13.14 ×Elyleymus, 343
13.15 Pascopyrum, 348

13.16 ×Pascoleymus, 251
13.17 Leymus, 353
13.18 ×Leydeum, 369
13.19 Psathyrostachys, 372
13.20 Thinopyrum, 373
14.00 POEAE, 378
14.01 Festuca, 389
14.02 Leucopoa, 443
14.03 Schedonorus, 445
14.04 Vulpia, 448
14.05 Lolium, 454
14.06 Puccinellia, 459
14.07 ×Pucciphippsia, 477
14.08 Phippsia, 478
14.09 Sclerochloa, 480
14.10 Dactylis, 482
14.11 Lamarckia, 484
14.12 Beckmannia, 484
14.13 Poa, 486
14.14 ×Dupoa, 601
14.15 Dupontia, 602
14.16 ×Arctodupontia, 604
14.17 Arctophila, 605
14.18 Torreyochloa, 607
14.19 Catabrosa, 610
14.20 Cutandia, 611
14.21 Briza, 612
14.22 Aira, 615
14.23 Coleanthus, 618
14.24 Eremopoa, 618
14.25 Sphenopholis, 620
14.26 Deschampsia, 624
14.27 Agrostis, 633
14.28 Polypogon, 662
14.29 ×Agropogon, 668
14.30 Lagurus, 670
14.31 Phleum, 670
14.32 Gastridium, 675
14.33 Arctagrostis, 676
14.34 Sesleria, 678

14.35 Desmazeria, 681
14.36 Ventenata, 683
14.37 Cynosurus, 685
14.38 Parapholis, 687
14.39 Scribneria, 689
14.40 Hainardia, 689
14.41 Vahlodea, 691
14.42 Podagrostis, 693
14.43 Lachnagrostis, 694
14.44 Bromidium, 696
14.45 Avenula, 698
14.46 Dissanthelium, 700
14.47 Helictotrichon, 701
14.48 Amphibromus, 703
14.49 Calamagrostis, 706
14.50 Gaudinia, 732
14.51 Scolochloa, 732
14.52 Avena, 734
14.53 Holcus, 739
14.54 Arrhenatherum, 740
14.55 Corynephorus, 742
14.56 Trisetum, 744
14.57 Koeleria, 753
14.58 Rostraria, 756
14.59 Mibora, 757
14.60 Anthoxanthum, 758
14.61 Phalaris, 764
14.62 Cinna, 773
14.63 Limnodea, 776
14.64 Ammophila, 776
14.65 Milium, 778
14.66 Alopecurus, 780
14.67 Apera, 788

Numerical Listing of the Subfamilies, Tribes, and Genera, Volume 25

ARUNDINOIDEAE, 6
16.00 ARUNDINEAE, 7
16.01 Molinia, 7
16.02 Hakonechloa, 8
16.03 Phragmites, 10
16.04 Arundo, 11

CHLORIDOIDEAE, 13
17.00 CYNODONTEAE, 14
17.01 Fingerhuthia, 22
17.02 Uniola, 22
17.03 Swallenia, 24
17.04 Distichlis, 25
17.05 Aeluropus, 25
17.06 Allolepis, 27
17.07 Monanthochloë, 28
17.08 Neyraudia, 30
17.09 Triraphis, 31
17.10 Tridens, 33
17.11 Redfieldia, 41
17.12 Triplasis, 41
17.13 Scleropogon, 42
17.14 Erioneuron, 44
17.15 Dasyochloa, 45
17.16 Blepharoneuron, 47
17.17 Blepharidachne, 48
17.18 Munroa, 51
17.19 Leptochloa, 51
17.20 Tripogon, 61
17.21 Trichoneura, 61
17.22 Dinebra, 63
17.23 Eragrostis, 65
17.24 Cladoraphis, 105
17.25 Pogonarthria, 105
17.26 Vaseyochloa, 106
17.27 Eleusine, 109
17.28 Acrachne, 110
17.29 Dactyloctenium, 112
17.30 Sporobolus, 115

17.31 Crypsis, 139
17.32 Calamovilfa, 140
17.33 Muhlenbergia, 145
17.34 Lycurus, 200
17.35 Chloris, 204
17.36 Eustachys, 218
17.37 Enteropogon, 224
17.38 Trichloris, 225
17.39 Willkommia, 227
17.40 Schedonnardus, 228
17.41 Gymnopogon, 231
17.42 Ctenium, 232
17.43 Microchloa, 234
17.44 Cynodon, 235
17.45 Spartina, 240
17.46 Bouteloua, 250
17.47 Opizia, 269
17.48 Buchloë, 270
17.49 Cathestecum, 272
17.50 Aegopogon, 273
17.51 Hilaria, 274
17.52 Tragus, 278
17.53 Zoysia, 281
18.00 PAPPOPHOREAE, 285
18.01 Pappophorum, 285
18.02 Enneapogon, 286
18.03 Cottea, 287
19.00 ORCUTTIEAE, 290
19.01 Orcuttia, 290
19.02 Tuctoria, 292
19.03 Neostapfia, 294

DANTHONIOIDEAE, 297
20.00 DANTHONIEAE, 298
20.01 Cortaderia, 298
20.02 Danthonia, 301
20.03 Schismus, 307

20.04 Karroochloa, 308
20.05 Rytidosperma, 309
20.06 Tribolium, 312
ARISTIDOIDEAE, 314
21.00 ARISTIDEAE, 314
21.01 Aristida, 315

CENTOTHECOIDEAE, 343
22.00 CENTOTHECEAE, 344
22.01 Chasmanthium, 344
23.00 THYSANOLANEAE, 349
23.01 Thysanolaena, 349

PANICOIDEAE, 351
24.00 GYNERIEAE, 352
24.01 Gynerium, 353
25.00 PANICEAE, 353
25.01 Digitaria, 358
25.02 Anthenantia, 384
25.03 Alloteropsis, 385
25.04 Amphicarpum, 385
25.05 Lasiacis, 387
25.06 Oplismenus, 389
25.07 Echinochloa, 390
25.08 Sacciolepis, 404
25.09 Dichanthelium, 406
25.10 Panicum, 450
25.11 Brachiaria, 488
25.12 Melinis, 490
25.13 Urochloa, 492
25.14 Eriochloa, 507
25.15 Pennisetum, 515
25.16 Cenchrus, 529
25.17 Anthephora, 535
25.18 Ixophorus, 537
25.19 Setariopsis, 539
25.20 Setaria, 539

25.21 Paspalidium, 558
25.22 Stenotaphrum, 560
25.23 Hymenachne, 561
25.24 Steinchisma, 563
25.25 Axonopus, 565
25.26 Paspalum, 566
25.27 Reimarochloa, 599
26.00 ANDROPOGONEAE, 602
26.01 Spodiopogon, 609
26.02 Saccharum, 609
26.03 Miscanthus, 616
26.04 Imperata, 618
26.05 Polytrias, 623
26.06 Microstegium, 623
26.07 Trachypogon, 624
26.08 Sorghum, 626
26.09 Sorghastrum, 630
26.10 Chrysopogon, 633
26.11 Dichanthium, 637
26.12 Bothriochloa, 639
26.13 Ischaemum, 648
26.14 Apluda, 649
26.15 Andropogon, 649
26.16 Cymbopogon, 664
26.17 Schizachyrium, 666
26.18 Arthraxon, 677
26.19 Hyparrhenia, 678
26.20 Heteropogon, 680
26.21 Themeda, 682
26.22 Elionurus, 684
26.23 Hemarthria, 685
26.24 Coelorachis, 687
26.25 Eremochloa, 688
26.26 Rottboellia, 691
26.27 Hackelochloa, 691
26.28 Tripsacum, 693
26.29 Zea, 696
26.30 Coix, 703